2019 IEEE 69th Electronic Components and Technology Conference (ECTC 2019)

Las Vegas, Nevada, USA
28-31 May 2019

Pages 1575-2364

IEEE Catalog Number: CFP19ECT-POD
ISBN: 978-1-7281-1500-9

Copyright © 2019 by the Institute of Electrical and Electronics Engineers, Inc. All Rights Reserved

Copyright and Reprint Permissions: Abstracting is permitted with credit to the source. Libraries are permitted to photocopy beyond the limit of U.S. copyright law for private use of patrons those articles in this volume that carry a code at the bottom of the first page, provided the per-copy fee indicated in the code is paid through Copyright Clearance Center, 222 Rosewood Drive, Danvers, MA 01923.

For other copying, reprint or republication permission, write to IEEE Copyrights Manager, IEEE Service Center, 445 Hoes Lane, Piscataway, NJ 08854. All rights reserved.

*** This is a print representation of what appears in the IEEE Digital Library. Some format issues inherent in the e-media version may also appear in this print version.

IEEE Catalog Number: CFP19ECT-POD
ISBN (Print-On-Demand): 978-1-7281-1500-9
ISBN (Online):   978-1-7281-1499-6
ISSN: 0569-5503

Additional Copies of This Publication Are Available From:

Curran Associates, Inc
57 Morehouse Lane
Red Hook, NY 12571 USA
Phone: (845) 758-0400
Fax: (845) 758-2633
E-mail: curran@proceedings.com
Web: www.proceedings.com

Proceedings

The 2019 IEEE 69th Electronic Components and Technology Conference

Proceedings

IEEE 69th Electronic Components and Technology Conference

ECTC 2019

28 – 31 May 2019
Las Vegas, Nevada

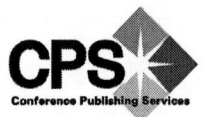

Los Alamitos, California

Washington • Tokyo

2019 IEEE 69th Electronic Components and Technology Conference (ECTC)
ECTC 2019

Table of Contents

Foreword lvii
Executive Committee lix
Program Committee lx

Session 1: Wafer-Level Fan-Out Process Integration

3D-MiM (MUST-in-MUST) Technology for Advanced System Integration 1
An-Jhih Su (Taiwan Semiconductor Manufacturing Company), Terry Ku (Taiwan Semiconductor Manufacturing Company), Chung-Hao Tsai (Taiwan Semiconductor Manufacturing Company), Kuo-Chung Yee (Taiwan Semiconductor Manufacturing Company), and Douglas Yu (Taiwan Semiconductor Manufacturing Company)

Construction of FO-MCM with C4 Bumps Built First Using Chip Last Assembly Technology 7
Chih-Hsun Hsu (Siliconware Precision Industries Co., Ltd), Wen-Yang Li (Siliconware Precision Industries Co., Ltd), Chi-Jen Chen (Siliconware Precision Industries Co., Ltd), Yih-Jenn Jiang (Siliconware Precision Industries Co., Ltd), Jui-Feng Tai (Siliconware Precision Industries Co., Ltd), Chang-Fu Lin (Siliconware Precision Industries Co., Ltd), and C. Key Chung (Siliconware Precision Industries Co., Ltd)

Feasibility Study of Fan-Out Panel-Level Packaging for Heterogeneous Integrations 14
Cheng-Ta Ko (Unimicron Technology Corporation), Henry Yang (Unimicron Technology Corporation), John H. Lau (ASM Pacific Technology Ltd), Ming Li (ASM Pacific Technology Ltd), Curry Lin (Unimicron Technology Corporation), Chieh-Lin Chang (Dow Chemical Company), Jhih-Yuan Pan (Dow Chemical Company), Hsing-Hui Wu (Dow Chemical Company), Iris Xu (Jiangyin Changdian Advanced Packaging Co., Ltd.), Tony Chen (Jiangyin Changdian Advanced Packaging Co., Ltd.), Zhang Li (Jiangyin Changdian Advanced Packaging Co., Ltd.), Kim Hwee Tan (Jiangyin Changdian Advanced Packaging Co., Ltd.), Penny Lo (ASM Pacific Technology Ltd), R. So (ASM Pacific Technology Ltd), Y. H. Chen (Unimicron Technology Corporation), Nelson Fan (ASM Pacific Technology Ltd), Eric Kuah (ASM Pacific Technology Ltd), Marc Lin (Dow Chemical Company), Y. M. Cheung (ASM Pacific Technology Ltd), Eric Ng (ASM Pacific Technology Ltd), Cao Xi (Huawei Technologies Co. Ltd.), Rozalia Beica (Dow Chemical Company), Sze Pei Lim (Indium Corporation), N. C. Lee (Indium Corporation), Mian Tao (Hong Kong University of Science and Technology), Jeffery Lo (Hong Kong University of Science and Technology), and Ricky Lee (Hong Kong University of Science and Technology)

Ultra-Thin FO Package-on-Package for Mobile Application 21
Hsiang-Yao Hsiao (Institute of Microelectronics, A*STAR), Soon Wee Ho (Institute of Microelectronics, A*STAR), Simon Siak Boon Lim (Institute of Microelectronics, A*STAR), Leong Ching Wai (Institute of Microelectronics, A*STAR), Ser Choong Chong (Institute of Microelectronics, A*STAR), Pei Siang Sharon Lim (Institute of Microelectronics, A*STAR), Yong Han (Institute of Microelectronics, A*STAR), and Tai Chong Chai (Institute of Microelectronics, A*STAR)

Development of Wafer Level Process for the Fabrication of Advanced Capacitive Fingerprint Sensors Using Embedded Silicon Fan-Out (eSiFO(R)) Technology 28
Shuying Ma (Huantian Technology (Kunshan) Electronics Co., Ltd., China), Chengqian Wang (Huantian Technology (Kunshan) Electronics Co., Ltd., China), Fengxia Zheng (Huantian Technology (Kunshan) Electronics Co., Ltd., China), Daquan Yu (Huantian Technology (Kunshan) Electronics Co., Ltd., China), Hong Xie (Filipchip International, USA), Xiaobing Yang (Huantian Technology (Kunshan) Electronics Co., Ltd., China), Li Ma (Huantian Technology (Kunshan) Electronics Co., Ltd., China), Ping Li (Huantian Technology (Xi'an) Electronics Co., Ltd., China), Weidong Liu (Huantian Technology (Xi'an) Electronics Co., Ltd., China), Jambo Yu (Synaptics, USA), and Jason Goodelle (Synaptics, USA)

Three-Dimensional Integrated Circuit (3D-IC) Package Using Fan-Out Technology 35
Jun Kyu Lee (NEPES Corporation), Sang Yong Park (NEPES Corporation), Young Ho Kim (NEPES Corporation), Jae Cheon Lee (NEPES Corporation), Sung Hyuk Lee (NEPES Corporation), Chul Hyo Lee (NEPES Corporation), Yong Tae Kwon (NEPES Corporation), Chang Woo Lee (NEPES Corporation), Jong Heon Kim (NEPES Corporation), Nam Chul Kim (NEPES Corporation), and Yun Hyun Sung (NEPES Corporation)

Ultra High Density IO Fan-Out Design Optimization with Signal Integrity and Power Integrity 41
  Keng Tuan Chang (Advanced Semiconductor Engineering, Inc.), Chih-Yi
  Huang (Advanced Semiconductor Engineering, Inc.), Hung-Chun Kuo
  (Advanced Semiconductor Engineering, Inc.), Ming-Fong Jhong (Advanced
  Semiconductor Engineering, Inc.), Tsun-Lung Hsieh (Advanced
  Semiconductor Engineering, Inc.), Mi-Chun Hung (Advanced Semiconductor
  Engineering, Inc.), and Chen-Chao Wang (Advanced Semiconductor
  Engineering, Inc.)

Session 2: Next-Generation Wirebonding and Die Attach

SB²-WB: A New Process Solution for Advanced Wire-Bonding 47
  Matthias Fettke (Pac Tech GmbH), Andrej Kolbasow (Pac Tech GmbH),
  Georg Friedrich (Pac Tech GmbH), Anna Palys (PacTech GmbH), Vinith
  Bejugam (Pac Tech GmbH), and Thorsten Teutsch (Pac Tech GmbH)

Smart Wire Bond Solutions for SiP and Memory Packages 55
  Basil Milton (Kulicke & Soffa, USA), Aashish Shah (Kulicke & Soffa,
  USA), Hui Xu (Kulicke & Soffa, USA), Odal Kwon (Kulicke & Soffa, USA),
  Gary Schulze (Kulicke & Soffa, USA), Ivy Qin (Kulicke & Soffa, USA),
  and Nelson Wong (Kulicke & Soffa, Singapore)

Preparation and Application of Cu-Ag Composite Preforms for Power Electronic Packaging 63
  Dongxiao Zhang (Wuhan University of Technology), Shengfa Liu (Wuhan
  University of Technology), Hui Xiang (Wuhan University of Technology),
  Li Liu (Wuhan University of Technology), Zhaoxia Zhou (Loughborough
  University), Stuart Robertson (Loughbrough University), Canyu Liu
  (Loughborough University), Zhiwen Chen (Wuhan University), and
  Changqing Liu (Loughborough University)

Au-Rich/Sn-Bi Interconnection in Chip-on-Module Package 69
  Jin Wang (Tsinghua University, China), Qian Wang (Tsinghua University,
  China), Xinnan Hou (GalaxyCore Inc., China), Ke Du (GalaxyCore Inc.,
  China), Lixin Zhao (GalaxyCore Inc., China), and Jian Cai (Tsinghua
  University, China)

The Properties of Cu Sinter Paste for Pressure Sintering at Low Temperature 76
  Jung-Lae Jo (Mitsui Engineered Materials Sector R&D Center, Mitsui
  Mining & Smelting Co., Ltd.), Kei Anai (Mitsui Engineered Materials
  Sector R&D Center, Mitsui Mining & Smelting Co., Ltd.), Sinichi
  Yamauchi (Mitsui Engineered Materials Sector R&D Center, Mitsui Mining
  & Smelting Co., Ltd.), and Takahiko Sakaue (Mitsui Engineered
  Materials Sector R&D Center, Mitsui Mining & Smelting Co., Ltd.)

Low Temperature Sintering of Dendritic Cu Based Pastes for Power Semiconductor Device Interconnection 81

Gang Li (Shenzhen Institutes of Advanced Technology, Chinese Academy of Sciences, Shenzhen College of Advanced Technology, University of Chinese Academy of Sciences), Jilei Fana (Shenzhen Institutes of Advanced Technology, Chinese Academy of Sciences, Shenzhen College of Advanced Technology, University of Chinese Academy of Sciences), Siyuan Liao (Shenzhen Institutes of Advanced Technology, Chinese Academy of Sciences, Shenzhen College of Advanced Technology, University of Chinese Academy of Sciences), Pengli Zhua (Shenzhen Institutes of Advanced Technology, Chinese Academy of Sciences), Baotan Zhang (Shenzhen Institutes of Advanced Technology, Chinese Academy of Sciences), Tao Zhao (Shenzhen Institutes of Advanced Technology, Chinese Academy of Sciences), Rong Sun (Shenzhen Institutes of Advanced Technology, Chinese Academy of Sciences), and Ching-Ping Wong (Shenzhen Institutes of Advanced Technology, Chinese Academy of Sciences)

A New Development of Direct Bonding to Aluminum and Nickel Surfaces by Silver Sintering in air Atmosphere 87

Ly May Chew (Heraeus Deutschland GmbH & Co. KG, Germany), Tamira Stegmann (Hochschule Aschaffenburg, University of Applied Sciences, Germany), Erika Schwenk (Hochschule Aschaffenburg, University of Applied Sciences, Germany), Monique Dubis (Hochschule Aschaffenburg, University of Applied Sciences, Germany), and Wolfgang Schmitt (Heraeus Deutschland GmbH & Co. KG, Germany)

Session 3: RDL and Additive Manufacturing

Submicron-Scale Cu RDL Pattering Based on Semi-Additive Process for Heterogeneous Integration 94
Takamasa Takano (DNP Co., Ltd.), Hiroshi Kudo (DNP Co., Ltd.), Masaya Tanaka (DNP Co., Ltd.), and Miyuki Akazawa (DNP Co., Ltd.)

Sub-Micron RDL Patterning for Advanced Packaging 101
Ken-Ichiro Mori (Canon Inc.), Yoshio Goto (Canon Inc.), Yasuo Hasegawa (Canon Inc.), Seiya Miura (Canon Inc.), and Douglas Shelton (Canon U.S.A Inc.)

Optimization of Electrolytic Plating Processes for Challenging Fan-Out Panel Level Package Designs 106
Ralph Zoberbier (Atotech Deutschland GmbH, Germany), Britta Scheller (Atotech Deutschland GmbH), and Christian Ohde (Atotech Deutschland GmbH)

3D Printed Substrates for the Design of Compact RF Systems 113
Mohd Ifwat Mohd Ghazali (Universiti Sains Islam Malaysia , Michigan State University), Saikat Mondal (Michigan State University), Saranraj Karuppuswami (Michigan State University), and Premjeet Chahal (Michigan State University)

Fully Additively Manufactured Tunable Active Frequency Selective Surfaces with Integrated On-package Solar Cells for Smart Packaging Applications 119
Syed Abdullah Nauroze (Georgia Institute of Technology), Xuanke He (Georgia Institute of Technology), and Manos M. Tentzeris (Georgia Institute of Technology)

First Demonstration of a Low Cost/Customizable Chip Level 3D Printed Microjet Hotspot-Targeted
Cooler for High Power Applications 126
    Tiwei Wei (IMEC & KU Leuven, Belgium), Herman Oprins (IMEC, Belgium),
    Vladimir Cherman (IMEC, Belgium), Ingrid De Wolf (IMEC & KU Leuven,
    Belgium), Eric Beyne (IMEC, Belgium), and Martine Baelmans (KU Leuven,
    Belgium)

Rapid Production of Customized 3D Electronics Via Hybrid Additive Manufacturing Technology 135
    Ji Li (Key Laboratory of MEMS of the Ministry of Education, Southeast
    University), Yang Wang (Key Laboratory of MEMS of the Ministry of
    Education, Southeast University), Peiren Wang (Key Laboratory of MEMS
    of the Ministry of Education, Southeast University), Jiangling He (Key
    Laboratory of MEMS of the Ministry of Education, Southeast
    University), Handa Liu (Key Laboratory of MEMS of the Ministry of
    Education, Southeast University), and Gengzhao Xiang (Key Laboratory
    of MEMS of the Ministry of Education, Southeast University)

Session 4: Advancements in Automotive and Power Devices

Solid-Liquid InterDiffusion (SLID) Bonding, for Thermally Challenging Applications 141
    Knut E Aasmundtveit (University of South-Eastern Norway, Norway),
    Thi-Thuy Luu (Zimmer and Peacock, Norway), Hoang-Vu Nguyen (University
    of South-Eastern Norway, Norway), Andreas Larsson (University of
    South-Eastern Norway, Norway), and Torleif A Tollefsen (TEGma, Norway)

Fluxless Bonding Technique of Diamond to Copper Using Silver-Indium Multilayer Structure 150
    Roozbeh Sheikhi (University of California, Irvine), Yongjun Huo
    (University of California, Irvine), and Chin C. Lee (University of
    California, Irvine)

Formulation and Processing of Conductive Polysulfide Sealants for Automotive and Aerospace
Applications 157
    Bo Song (Georgia Institute of Technology), Fan Wu (Georgia Institute
    of Technology), Kyoung-sik Moon (Georgia Institute of Technology), and
    CP Wong (Georgia Institute of Technology)

Challenges and Approaches to Developing Automotive Grade 1/0 FCBGA Package Capability 163
    Rajen Dias (Amkor Technology, USA), Mike Kelly (Amkor Technology,
    USA), Devarajan Balaraman (Amkor Technology, USA), Hideaki Shoji
    (J-Devices Corporation, Japan), Tomio Shiraiwa (J-Devices, Japan),
    KwangSeok Oh (Amkor Technology, Korea), and JoonYoung Park (Amkor
    Technology, Korea)

Advanced Substrates for GaN-Based Power Devices 168

Anthony Cibié (Univ. Grenoble Alpes, CEA, LETI, France), Julie Widiez (Univ. Grenoble Alpes, CEA, LETI, France), René Escoffier (Univ. Grenoble Alpes, CEA, LETI, France), Denis Blachier (Univ. Grenoble Alpes, CEA, LETI, France), Kremena Vladimirova (Univ. Grenoble Alpes, CEA, LETI, France), Jean-Philippe Colonna (Univ. Grenoble Alpes, CEA, LETI, France), Paul-Henri Haumesser (Univ. Grenoble Alpes, CEA, LETI, France), Stéphane Bécu (Univ. Grenoble Alpes, CEA, LETI, France), Perceval Coudrain (Univ. Grenoble Alpes, CEA, LETI, France), William Vandendaele (Univ. Grenoble Alpes, CEA, LETI, France), Jérôme Biscarrat (Univ. Grenoble Alpes, CEA, LETI, France), Charlotte Gillot (Univ. Grenoble Alpes, CEA, LETI, France), Matthew Charles (Univ. Grenoble Alpes, CEA, LETI, France), and Léa Di Cioccio (Univ. Grenoble Alpes, CEA, LETI, France)

A New Reliable, Corrosion Resistant Gold-Palladium Coated Copper Wire Material 175

Sandy Klengel (Fraunhofer Institute for Microstructure of Materials IMWS), Robert Klengel (Fraunhofer Institute for Microstructure of Materials IMWS), Jan Schischka (Fraunhofer Institute for Microstructure of Materials IMWS), Tino Stephan (Fraunhofer Institute for Microstructure of Materials IMWS), Matthias Petzold (Fraunhofer Institute for Microstructure of Materials IMWS), Motoki Eto (Nippon Micrometal Corporation), Noritoshi Araki (Nippon Micrometal Corporation), and Takashi Yamada (Nippon Micrometal Corporation)

Ultrasonic-Accelerated Intermetallic Joint Formation with Composite Solder for High-Temperature Power Device Packaging 183

Hongjun Ji (Harbin Institute of Technology at Shenzhen), Mingyu Li (Harbin Institute of Technology at Shenzhen), Weiwei Zhao (Harbin Institute of Technology at Shenzhen), and Wenwu Zhang (Harbin Institute of Technology at Shenzhen)

Session 5: Bonding Manufacturing Technologies

Comprehensive Study of Copper Nano-Paste for Cu-Cu Bonding 191

Ser Choong Chong (Institute of Microelectronics) and Pei Siang Lim Sharon (Insttitute of Microelectronics)

Enhanced Performance of Laser-Assisted Bonding with Compression (LABC) Compared with Thermal Compression Bonding (TCB) Technology 197

Kwang-Seong Choi (Electronics and Telecommunications Research Institute), Yong-Sung Eom (Electronics and Telecommunications Research Institute), Seok Hwan Moon (Electronics and Telecommunications Research Institute), Jiho Joo (Electronics and Telecommunications Research Institute), Ieeseul Jeong (Electronics and Telecommunications Research Institute), Kwangjoo Lee (LG Chem), Jung Hak Kim (LG Chem), Ju hyeon Kim (LG Chem), Gil-Sang Yoon (KITECH), Kwang-Hee Lee (Inha University), Chul-Hee Lee (Inha University), Geun-Sik Ahn (Protec), and Moo-Sup Shim (Protec)

A Study of 3D Packaging Interconnection Performance Affected by Thermal Diffusivity and Pressure Transmission 204

Jin-San Jung (Samsung Electronics Co., Ltd), Hyeong Gi Lee (Samsung Electronics Co., Ltd), Ji-Min Kim (Samsung Electronics Co., Ltd), Yong-Jin Park (Samsung Electronics Co., Ltd), Ji-In Yu (Samsung Electronics Co., Ltd), Yong Sung Park (Samsung Electronics Co., Ltd), Jun Su Lim (Samsung Electronics Co., Ltd), Hyun-Seok Choi (Samsung Electronics Co., Ltd), Sung-Il Cho (Samsung Electronics Co., Ltd), Dong wook Kim (Samsung Electronics Co., Ltd), and Sang-Ho An (Samsung Electronics Co., Ltd)

Vertical Laser Assisted Bonding for Advanced "3.5D" Chip Packaging 210

Andrej Kolbasow (Pac Tech GmbH), Timo Kubsch (Pac Tech), Matthias Fettke (Pac Tech GmbH), Georg Friedrich (Pac Tech GmbH), and Thorsten Teutsch (PacTech)

Optimization of a BEOL Aluminum Deposition Process Enabling Wafer Level Al-Al Thermo-Compression Bonding 218

Sebastian Schulze (IHP - Innovations for High Performance Microelectronics), Matthias Wietstruck (IHP - Innovations for High Performance Microelectronics), Mirko Fraschke (IHP - Innovations for High Performance Microelectronics), Peter Kerepesi (EV Group, Inc.), Helmut Kurz (EV Group, Inc.), Bernhard Rebhan (EV Group, Inc.), and Mehmet Kaynak (IHP - Innovations for High Performance Microelectronics)

Self-Assembly Process for 3D Die-to-Wafer using Direct Bonding: A Step Forward Toward Process Automatisation 225

Jouve Amandine (CEA, LETI), Loïc Sanchez (CEA, LETI), Clément Castan (CEA, LETI), Maxence Laugier (CEA, LETI), Emmanuel Rolland (CEA, LETI), Brigitte Montmayeul (CEA, LETI), Rémi Franiatte (CEA, LETI), Frank Fournel (CEA, LETI), and Severine Cheramy (CEA, LETI)

A Single Bonding Process for Diverse Organic-Inorganic Integration in IoT Devices 235

Tilo H. Yang (National Taiwan University), Yu-Shan Chiu (National Taiwan University), Hai-Yang Yu (National Taiwan University), Akitsu Shigetou (National Institute for Materials Science), and C. Robert Kao (National Taiwan University)

Session 6: Emerging Flexible Hybrid Electronics

Stretchable and Printable Medical Dry Electrode Arrays on Textile for Electrophysiological Monitoring 243

Yougen Hu (Shenzhen Institutes of Advanced Technology, Chinese Academy of Sciences), Hui Wang (Shenzhen Institutes of Advanced Technology, Chinese Academy of Sciences), Ommeaymen Sheikhnejad (AC2T Research GmbH), Yaoxu Xiong (Shenzhen Institutes of Advanced Technology, Chinese Academy of Sciences), Han Gu (Shenzhen Institutes of Advanced Technology, Chinese Academy of Sciences), Pengli Zhu (Shenzhen Institutes of Advanced Technology, Chinese Academy of Sciences), Guanglin Li (Shenzhen Institutes of Advanced Technology, Chinese Academy of Sciences), Rong Sun (Shenzhen Institutes of Advanced Technology, Chinese Academy of Sciences), and Ching-Ping Wong (Georgia Institute of Technology)

Screen-Printed Flexible Coplanar Waveguide Transmission Lines: Multi-physics Modeling and Measurement 249

Nahid Aslani Amoli (Georgia Institute of Technology), Sridhar Sivapurapu (Georgia Institute of Technology), Rui Chen (Georgia Institute of Technology), Yi Zhou (Georgia Institute of Technology), Mohamed L. F. Bellaredj (Georgia Institute of Technology), Paul A. Kohl (Georgia Institute of Technology), Suresh K. Sitaraman (Georgia Institute of Technology), and Madhavan Swaminathan (Georgia Institute of Technology)

Inkjet-Printed Filtering Antenna on a Textile for Wearable Applications 258

Hsuan-Ling Kao (Dept. of Electronic Engineering, Chang Gung University, Tao-Yuan, Taiwan), Chun-Hsiang Chuang (Dept. of Electronic Engineering, Chang Gung University, Tao-Yuan, Taiwan), and Cheng-Lin Cho (Dept. of Engineering and System Science, National Tsing Hua University, Hsin-Chu, Taiwan)

Mechanical and Electrical Characterization of FOWLP-Based Flexible Hybrid Electronics (FHE) for Biomedical Sensor Application 264

Yuki Susumago (Tohoku University), Qian Zhengyang (Tohoku University), Achille Jacquemond (Tohoku University), Noriyuki Takahashi (Tohoku University), Hisashi Kino (Tohoku University), Tetsu Tanaka (Tohoku University), and Takafumi Fukushima (Tohoku University)

A Wearable Fingernail Deformation Sensing System and Three-Dimensional Finite Element Model of Fingertip 270

Katsuyuki Sakuma (IBM Thomas J. Watson Research Center, U.S.A.), Bucknell Webb (IBM Thomas J. Watson Research Center, U.S.A.), Rajeev Narayanan (IBM Thomas J. Watson Research Center), Avner Abrami (IBM Thomas J. Watson Research Center), Jeff Rogers (IBM Thomas J. Watson Research Center), John Knickerbocker (IBM Thomas J. Watson Research Center), and Stephen J Heisig (IBM Thomas J. Watson Research Center)

Heterogeneous Integration of a Fan-Out Wafer-Level Packaging Based Foldable Display on Elastomeric Substrate 277

Arsalan Alam (University of California, Los Angeles), Amir Hanna (University of California, Los Angeles), Randall Irwin (University of California, Los Angeles), Goutham Ezhilarasu (University of California, Los Angeles), Hyunpil Boo (University of California, Los Angeles), Yuan Hu (University of California, Los Angeles), Chee Wei Wong (University of California, Los Angeles), Timothy S Fisher (University of California, Los Angeles), and S. S. Iyer (University of California)

A Study on the Flexible Chip-on-Fabric (COF) Assembly using Anisotropic Conductive Films (ACFs) Materials 283

Seung-Yoon Jung (KAIST) and Kyung-Wook Paik (KAIST)

Session 7: Advances in Flip Chip Packaging

7nm Chip-Package Interaction Study on a Fine Pitch Flip Chip Package with Laser Assisted Bonding and Mass Reflow Technology 289

Ian Hsu (MediaTek), Chi-Yuan Chen (MediaTek), Stanley Lin (MediaTek), Ta-Jen Yu (MediaTek), NamJu Cho (JCET), and Ming-Che Hsieh (JCET)

Ultra Large Area SIPs and Integrated mmW Antenna Array Module for 5G mmWave Outdoor Applications ... 294
Pouya Talebbeydokhti (Intel Corporation), Sidharth Dalmia (Intel Corporation), Trang Thai (Intel Corporation), Raanan Sover (Intel Corporation), and Sharon Tal (Intel Corporation)

Hybrid Approach for Large Size FC-BGA to Enhance Thermal and Electrical Performance Including Power Delivery 300
Heeseok Lee (Samsung Electronics, Co. Ltd.), Yunhyeok Im (Samsung Electronics, Co. Ltd.), Junghwa Kim (Samsung Electronics, Co. Ltd.), Jisoo Hwang (Samsung Electronics, Co. Ltd.), James Jeong (Samsung Electronics, Co. Ltd.), Youngsang Cho (Samsung Electronics, Co. Ltd.), Heejung Choi (Samsung Electronics, Co. Ltd.), and Youngmin Shin (Samsung Electronics, Co. Ltd.)

Package-on-Package Micro-BGA Microstructure Interaction with Bond and Assembly Parameter 306
Pascale Gagnon (IBM Canada Limited), Clément Fortin (IBM Canada Limited), and Thomas Weiss (IBM Systems)

Low Cost Flip-Chip Stack for Partitioning Processing and Memory 314
Fabian Hopsch (Fraunhofer Institute for Integrated Circuits IIS, Division Engineering of Adaptive Systems EAS) and Andy Heinig (Fraunhofer Institute for Integrated Circuits IIS, Division Engineering of Adaptive Systems EAS)

Impact of Low Temperature Solder on Electronic Package Dynamic Warpage Behavior and Requirement 318
Wei Keat Loh (Intel Technology Sdn Bhd), Ron W. Kulterman (Flex Ltd), Haley Fu (iNEMI), and Chih Chung Hsu (CoreTech System (Moldex3D))

High Density Ultra-Thin Organic Substrates for Advanced Flip Chip Packages 325
Nokibul Islam (JCET), KH Tan (JCET), Seung Wook Yoon (JCET), and Tony Chen (JCET)

Session 8: Material and Process Trends in FOWLP and PLP

Laser Releasable Temporary Bonding Film with High Thermal Stability 330
Yong-suk Yang (3M), Kyo-sung Hwang (3M), and Robin Gorrell (3M)

Design and Demonstration of 1µm Low Resistance RDL Using Panel Scale Processes for High Performance Computing Applications 334
Bartlet DeProspo (3D Systems Packaging Research Center), Aya Momozawa (Tokyo Ohka Kogyo Co. LTD.), Atsushi Kubo (Tokyo Ohka Kogyo Co. LTD.), Chandrasekharan Nair (3D Systems Packaging Research Center), Varun Rajagoapal (3D Systems Packaging Research Center), Jenefa Kannan (Georgia Institute of Technology), Emanuel Surillo (3D Systems Packaging Research Center), Fuhan Liu (3D Systems Packaging Research Center), Mohananlingam Kathaperumal (3D Systems Packaging Research Center), and Rao Tummala (3D Systems Packaging Research Center)

Advances in Temporary Carrier Technology for High-Density Fan-Out Device Build-up 340
Arnita Podpod (IMEC), Alain Phommahaxay (IMEC), Pieter Bex (IMEC), John Slabbekoorn (IMEC), Julien Bertheau (IMEC), Abdellah Salahoueldhadj (IMEC), Erik Sleeckx (IMEC), Andy Miller (IMEC), Gerald Beyer (IMEC), Eric Beyne (IMEC), Alice Guerrero (Brewer Science Inc), Kim Yess (Brewer Science Inc), and Kim Arnold (Brewer Science Inc)

xiii

Development of Novel Low-Temperature Curable Positive-Tone Photosensitive Dielectric Materials with High Reliability 346

Yutaro Koyama (Toray Industries, Inc.), Yu Shoji (Toray Industries, Inc.), Keika Hashimoto (Toray Industries, Inc.), Yuki Masuda (Toray Industries, Inc.), Hitoshi Araki (Toray Industries, Inc.), and Masao Tomikawa (Toray Industries, Inc.)

Highly Reliable Photosensitive Negative-Tone Polyimide with Low Cure Shrinkage 352

Daisaku Matsukawa (Hitachi Chemical DuPont MicroSystems, Ltd, Japan), Hiroko Yotsuyanagi (Hitachi Chemical DuPont MicroSystems, Ltd, Japan), Shiori Sakakibara (Hitachi Chemical DuPont MicroSystems, Ltd, Japan), Noriyuki Yamazaki (Hitachi Chemical DuPont MicroSystems, Ltd, Japan), Tetsuya Enomoto (Hitachi Chemical DuPont MicroSystems, Ltd, Japan), and Takeharu Motobe (Hitachi Chemical DuPont MicroSystems, Ltd, Japan)

High Rate and Low Damage Etching Method as Pre Treatment of Seed Layer Sputtering for Fan out Panel Level Packaging 358

Tetsushi Fujinaga (ULVAC, Inc.)

Investigation and Methods Using Various Release and Thermoplastic Bonding Materials to Reduce Die Shift and Wafer Warpage for eWLB Chip-First Processes 363

Michelle Fowler (Brewer Science, Inc.), John P. Massey (Brewer Science, Inc.), Tanja Braun (Fraunhofer Institute IZM), Steve Voges (Fraunhofer Institute IZM), Robert Gernhardt (Fraunhofer Institute IZM), and Markus Wohrmann (Fraunhofer institute IZM)

Session 9: Wearables and Thin-Package Reliability and Chip Package Interaction

Effect of Charging Cycle Elevated Temperature Storage and Thermal Cycling on Thin Flexible Batteries in Wearable Applications 370

Pradeep Lall (Auburn University), Amrit Abrol (Auburn University), Ben Leever (US AFRL), and Scott Miller (NextFlex)

Bladder Inflation Stretch Test Method for Reliability Characterization of Wearable Electronics 382

Benjamin G Stewart (Georgia Institute of Technology) and Suresh K Sitaraman (Georgia Institute of Technology)

Study of BEOL Failure Mode in Flip Chip Packages at High Temperature Conditions 392

Wei Wang (Qualcomm Technologies, Inc.), Yangyang Sun (Qualcomm Technologies, Inc.), Xuefeng Zhang (Qualcomm Technologies, Inc.), Lejun Wang (Qualcomm Technologies, Inc.), Lily Zhao (Qualcomm Technologies, Inc.), Mark Schwarz (Qualcomm Technologies, Inc.), Bill Stone (Qualcomm Technologies, Inc.), and Ahmer Syed (Qualcomm Technologies, Inc.)

A Novel Metal Scheme and Bump Array Design Configuration to Enhance Advanced Si Packages CPI Reliability Performance by Using Finite Element Modeling Technique 397

Kuo-Chin Chang (Taiwan Semiconductor Manufacturing Company Ltd.), Mirng-Ji Lii (Taiwan Semiconductor Manufacturing Company Ltd.), Steven Hsu (Taiwan Semiconductor Manufacturing Company Ltd.), Hao-Chun Liu (Taiwan Semiconductor Manufacturing Company Ltd.), Yen-Kun Lai (Taiwan Semiconductor Manufacturing Company Ltd.), Sheng-Han Tsai (Taiwan Semiconductor Manufacturing Company Ltd.), and Chieh-Hao Hsu (Taiwan Semiconductor Manufacturing Company Ltd.)

Assessment of CMP Fill Pattern Effect on the Thermal Performance of Interconnects in Integrated Circuits BEOL 405

Assaad Helou (Southern Methodist University), Peter Raad (Southern Methodist University), and Archana Venugopal (Texas Instruments, Inc.)

Three-Dimensional Simulation of the Thermo-Mechanical Interaction between the Micro-Bump Joints and Cu Protrusion in Cu-Filled TSVs of the High Bandwidth Memory (HBM) Structure 410

Jie-Ying Zhou (South China University of Technology), Shui-Bao Liang (South China University of Technology), Cheng Wei (South China University of Technology), Wen-Kai Le (South China University of Technology), Chang-Bo Ke (South China University of Technology), Min-Bo Zhou (South China University of Technology), Xiao Ma (South China University of Technology), and Xin-Ping Zhang (South China University of Technology)

Study of Design Optimization Method for Ultra-Low Power Micro Gas Sensor 417

Eiji Nakamura (IBM Research - Tokyo), Keiji Matsumoto (IBM Research - Tokyo), Andrea Fasoli (IBM Research - Almaden), Luisa Bozano (IBM Research - Almaden), and Hiroyuki Mori (IBM Research - Tokyo)

Session 10: Dicing and Encapsulation Technologies

A More Than Moore Enabling Wafer Dicing Technology 423

Jeroen van Borkulo (ASM Pacific Technologies Inc), Rogier Evertsen (ASM Pacific Technologies Inc), and Richard van der Stam (ASM Pacific Technologies Inc.)

Plasma Dicing Integration Schemes for Scribe Lane Layout and the Impact on Die Strength 428

David Parker (STMicroelectronics, France), Emmanuel Gourvest (STMicroelectronics, France), and Boris Bouillard (STMicroelectronics, France)

Advanced Dicing Technologies for Combination of Wafer to Wafer and Collective Die to Wafer Direct Bonding 437

Fumihiro Inoue (IMEC, Belgium), Alain Phommahaxay (IMEC, Belgium), Arnita Podpod (IMEC, Belgium), Samuel Suhard (IMEC, Belgium), Hitoshi Hoshino (Disco Hi-Tech Europe, Germany), Berthold Moeller (Disco Hi-Tech Europe, Germany), Erik Sleeckx (IMEC, Belgium), Kenneth June Rebibis (IMEC, Belgium), Andy Miller (IMEC, Belgium), and Eric Beyne (IMEC, Belgium)

Active Control of NCF Fillet Shape for 3D CoW by Multi Beam Laser Bonder 446

Keiko Ueno (Hitachi Chemical Co., Ltd.), Kazutaka Honda (Hitachi Chemical Co., Ltd.), Tsuyoshi Ogawa (Hitachi Chemical Co., Ltd.), and Toshihisa Nonaka (Hitachi Chemical Co., Ltd.)

Ultrafast Laser Scribe: An Improved Metal and ILD Ablation Process 453

Julia Chiu (Intel Corporation, USA), Aaron Gore (Intel Corporation, USA), Tyler Osborn (Intel Corporation, USA), Daragh Finn (Electro Scientific Industries, USA), Zhibin Lin (Electro Scientific Industries, USA), David Lord (Electro Scientific Industries, USA), and Jon Mellen (Electro Scientific Industries, USA)

Reliability and Benchmark of 2.5D Non-Molding and Molding Technologies 461

Yu-Hsiang Hsiao (Advanced Semiconductor Engineering, Group, Inc., Kaohsiung, Taiwan), Che-Ming Hsu (Advanced Semiconductor Engineering, Group, Inc., Kaohsiung, Taiwan), Yi-Sheng Lin (Advanced Semiconductor Engineering, Group, Inc., Kaohsiung, Taiwan), and Chien-Lin Chang Chien (Advanced Semiconductor Engineering, Group, Inc., Kaohsiung, Taiwan)

Laser-Induced Trench Design, Optimisation and Validation for Restricting Capillary Underfill Spread in Advanced Packaging Configurations 467

Gul Zeb (Université de Sherbrooke), David Danovitch (Université de Sherbrooke), and Eric Turcotte (IBM Canada)

Session 11: Automotive and Harsh-Environment Reliability

Effect of Substrate Preheating Treatment on Thermal Reliability and Micro-Structure of Ag Paste Sintering on Au Surface Finish 474

Zheng Zhang (Institute of Scientific and Industrial Research, Osaka University), Chuantong Chen (Institute of Scientific and Industrial Research, Osaka University), Katsuaki Suganuma (Institute of Scientific and Industrial Research, Osaka University), and Seigo Kurosaka (C. Uyemura & Co., Ltd.)

Package Material Selection Criteria for High Temperature Automotive Applications 479

Rene T.H. Rongen (NXP Semiconductors), A. Mavinkurve (NXP Semiconductors), G.M. O'Halloran (NXP Semiconductors), N. Owens (NXP Semiconductors), Y. Weber (NXP Semiconductors), P. Oberndorff (NXP Semiconductors), M-L Farrugia (NXP Semiconductors), E. van Olst (NXP Semiconductors), and M. van Soestbergen (NXP Semiconductors)

Solder Joint Reliability of Double-Side Mounted DDR Modules for Consumer and Automotive Applications... 486

Dongji Xie (NVIDIA), Joe Hai (Nvidia Corp.), Zhongming Wu (Nvidia Corp.), and Manthos Economou (Nvidia Corp.)

Reliability Investigation of Extremely Large Ratio Fan-Out Wafer-Level Package with Low Ball Density for Ultra-Short-Range Radar 493

P.S. Huang (MediaTek Inc.), C.K. Yu (MediaTek Inc.), W.S. Chiang (MediaTek Inc.), M.Z. Lin (MediaTek Inc.), Y.H. Fang (MediaTek Inc.), M.J. Lin (MediaTek Inc.), N.W. Liu (MediaTek Inc.), Benson Lin (MediaTek Inc.), and Ian Hsu (MediaTek Inc.)

Fatigue Behaviour of Lead-Free Solder Joints Under Combined Thermal and Vibration Loads 498

Meier Karsten (Technische Universität Dresden), Winkler Maria (Technische Universität Dresden), Leslie David (University of Maryland, Center for Advanced Life Cycle Engineering (CALCE)), Dasgupta Abhijit (University of Maryland, Center for Advanced Life Cycle Engineering (CALCE)), and Bock Karlheinz (Technische Universität Dresden)

Prognostication of Accrued Damage and Impending Failure Under Temperature-Vibration in Leadfree Electronics 505

Pradeep Lall (Auburn University), Tony Thomas (Auburn University), Jeff Suhling (Auburn University), and Ken Blecker (US Army ARDEC)

Electrochemical Impedance Spectroscopy (EIS) for Monitoring the Water Load on PCBAs Under Cycling Condensing Conditions to Predict Electrochemical Migration Under DC Loads 515

>Simone Lauser (Robert Bosch GmbH), Theresia Richter (Robert Bosch GmbH), Verdingovas Vadimas (Denmarks Technical University), and Rajan Ambat (Denmarks Technical University)

Session 12: Advanced Photonic Devices and Packaging

Micro-Fabricated SERF Atomic Magnetometer for Weak Gradient Magnetic Field Detection 522

>Xiang Yue (Southeast University), Jintang Shang (Southeast University), and Chen Ye (Southeast University)

Novel Solder Pads for Self-Aligned Flip-Chip Assembly 528

>Yves Martin (IBM T.J.Watson Research Center, Yorktown Heights USA), Swetha Kamlapurkar (IBM T.J.Watson Research Center, Yorktown Heights USA), Nathan Marchack (IBM T.J.Watson Research Center, Yorktown Height USA), Jae-Woong Nah (IBM T.J.Watson Research Center, Yorktown Height USA), and Tymon Barwicz (IBM T.J.Watson Research Center, Yorktown Height USA)

Collective Curved CMOS Sensor Process: Application for High-Resolution Optical Design and Assembly Challenges 535

>Bertrand Chambion (Univ. Grenoble Alpes, CEA, LETI), Christophe Gaschet (Univ. Grenoble Alpes, CEA, LETI), Marc Lombard (Univ. Grenoble Alpes, CEA, LETI), Maïlys Fernandez (Univ. Grenoble Alpes, CEA, LETI), Pierre Joly (Univ. Grenoble Alpes, CEA, LETI), Stéphane Caplet (Univ. Grenoble Alpes, CEA, LETI), Fabien Zuber (Univ. Grenoble Alpes, CEA, LETI), Aurélie Vandeneynde (Univ. Grenoble Alpes, CEA, LETI), Patrick Peray (Univ. Grenoble Alpes, CEA, LETI), Gilles Lasfargues (Univ. Grenoble Alpes, CEA), Marc Zussy (Univ. Grenoble Alpes, CEA, LETI), Jerôme Deschamps (Univ. Grenoble Alpes, CEA, LETI), Alexis Bedoin (Univ. Grenoble Alpes, CEA, LETI), and David Henry (Univ. Grenoble Alpes, CEA, LETI)

Integration and Characterization of InP Die on Silicon Interconnect Fabric 543

>Eric Sorensen (Center for Heterogeneous Integration and Performance Scaling (CHIPS) University of California, Los Angeles), Boris Vaisband (Center for Heterogeneous Integration and Performance Scaling (CHIPS) University of California, Los Angeles), SivaChandra Jangam (Center for Heterogeneous Integration and Performance Scaling (CHIPS) University of California, Los Angeles), Tim Shirley (Keysight Technologies), and Subramanian S. Iyer (Center for Heterogeneous Integration and Performance Scaling (CHIPS) University of California, Los Angeles)

Y-Branched Multimode/Single-Mode Polymer Optical Waveguides for Low-Loss WDM MUX Device: Fabrication and Characterization 550

>Takaaki Ishigure (Keio University), Tomoki Nakayama (Keio University), Fukino Nakazaki (Keio University), and Hiroki Hama (Keio University)

Vertically Stacked and Directionally Coupled Cavity-Resonator-Integrated Grating Couplers for Integrated-Optic Beam Steering 556

>Shogo Ura (Kyoto Institute of Technology, Japan), Junishi Inoue (Kyoto Institute of Technology, Japan), and Kenji Kintaka (AIST, Japan)

CiB(Chip in Board) Optical Engine Module Using Advanced Fan-Out Package Technology 563

Sang Yong Park (NEPES Corporation), Ju Hyun Nam (NEPES Corporation),
Ji Ni Shim (NEPES Corporation), Jun Kyu Lee (NEPES Corporation), Yong
Tae Kwon (NEPES Corporation), Chang Woo Lee (NEPES Corporation), Jong
Heon Kim (NEPES Corporation), and Nam Chul Kim (NEPES Corporation)

Session 13: Technologies Enabling 3D and Heterogeneous Integration

Active Interposer Technology for Chiplet-Based Advanced 3D System Architectures 569

Perceval Coudrain (Univ. Grenoble Alpes, CEA, LETI, France), Jean
Charbonnier (Univ. Grenoble Alpes, CEA, LETI, France), Arnaud Garnier
(Univ. Grenoble Alpes, CEA, LETI, France), Pascal Vivet (Univ.
Grenoble Alpes, CEA, LETI, France), Rémi Vélard (Univ. Grenoble Alpes,
CEA, LETI, France), Andrea Vinci (Univ. Grenoble Alpes, CEA, LETI,
France), Fabienne Ponthenier (Univ. Grenoble Alpes, CEA, LETI,
France), Alexis Farcy (STMicroelectronics, France), Roselyne Segaud
(Univ. Grenoble Alpes, CEA, LETI, France), Pascal Chausse (Univ.
Grenoble Alpes, CEA, LETI, France), Lucile Arnaud (Univ. Grenoble
Alpes, CEA, LETI, France), Didier Lattard (Univ. Grenoble Alpes, CEA,
LETI, France), Eric Guthmuller (Univ. Grenoble Alpes, CEA, LETI,
France), Giovanni Romano (Univ. Grenoble Alpes, CEA, LETI, France),
Alain Gueugnot (Univ. Grenoble Alpes, CEA, LETI, France), Frédéric
Berger (Univ. Grenoble Alpes, CEA, LETI, France), Jérôme Beltritti
(STMicroelectronics), Therry Mourier (Univ. Grenoble Alpes, CEA, LETI,
France), Mathilde Gottardi (Univ. Grenoble Alpes, CEA, LETI, France),
Stéphane Minoret (Univ. Grenoble Alpes, CEA, LETI, France), Céline
Ribière (Univ. Grenoble Alpes, CEA, LETI, France), Gilles Romero
(Univ. Grenoble Alpes, CEA, LETI, France), Pierre-Emile Philip (Univ.
Grenoble Alpes, CEA, LETI, France), Yorrick Exbrayat (Univ. Grenoble
Alpes, CEA, LETI, France), Daniel Scevola (Univ. Grenoble Alpes, CEA,
LETI, France), Didier Campos (STMicroelectronics), Maxime Argoud
(Univ. Grenoble Alpes, CEA, LETI, France), Nacima Allouti (Univ.
Grenoble Alpes, CEA, LETI, France), Raphaël Eleouet (Univ. Grenoble
Alpes, CEA, LETI, France), César Fuguet Tortolero (Univ. Grenoble
Alpes, CEA, LETI, France), Christophe Aumont (Univ. Grenoble Alpes,
CEA, LETI, France), Denis Dutoit (Univ. Grenoble Alpes, CEA, LETI,
France), Corinne Legalland (Univ. Grenoble Alpes, CEA, LETI, France),
Jean Michailos (STMicroelectronics, France), Séverine Chéramy (Univ.
Grenoble Alpes, CEA, LETI, France), and Gilles Simon (Univ. Grenoble
Alpes, CEA, LETI, France)

Process Development of Power Delivery Through Wafer Vias for Silicon Interconnect Fabric 579

Meng-Hsiang Liu (University of California, Los Angeles), Boris
Vaisband (University of California, Los Angeles), Amir Hanna
(University of California, Los Angeles), Yandong Luo (University of
California, Los Angeles), Zhe Wan (University of California, Los
Angeles), and Subramanian S. Iyer (University of California, Los
Angeles)

Active Through-Silicon Interposer Based 2.5D IC Design, Fabrication, Assembly and Test 587

Jayasanker Jayabalan (Institute of Microelectronics), Vivek
Chidambaram Nachiappan (Institute of Microelectronics), Sharon Lim Pei
Siang (Institute of Microelectronics), Wang Xiangyu (Institute of
Microelectronics), Jong Ming Chinq (Institute of Microelectronics),
and Surya Bhattacharya (Institute of Microelectronics)

xviii

System on Integrated Chips (SoIC(TM)) for 3D Heterogeneous Integration 594
    Ming-Fa Chen (Taiwan Semiconductor Manufacturing Company (TSMC)),
    Fang-Cheng Chen (Taiwan Semiconductor Manufacturing Company (TSMC)),
    Wen-Chih Chiou (Taiwan Semiconductor Manufacturing Company (TSMC)),
    and Doug C.H. Yu (Taiwan Semiconductor Manufacturing Company (TSMC))

Die-to-Wafer (D2W) Processing and Reliability for 3D Packaging of Advanced Node Logic 600
    Luke England (GLOBALFOUNDRIES), Daniel Fisher (GLOBALFOUNDRIES), Katie
    Rivera (GLOBALFOUNDRIES), Bill Guthrie (GLOBALFOUNDRIES), Ping-Jui Kuo
    (Advanced Semiconductor Engineering (ASE)), Chang-Chi Lee (Advanced
    Semiconductor Engineering), Che-Ming Hsu (Advanced Semiconductor
    Engineering), Fan-Yu Min (Advanced Semiconductor Engineering),
    Kuo-Chang Kang (Advanced Semiconductor Engineering), and Chen-Yuan
    Weng (Advanced Semiconductor Engineering)

Enabling Ultra-Thin Die to Wafer Hybrid Bonding for Future Heterogeneous Integrated Systems 607
    Alain Phommahaxay (IMEC), Samuel Suhard (IMEC), Pieter Bex (IMEC),
    Serena Iacovo (IMEC), John Slabbekoorn (IMEC), Fumihiro Inoue (IMEC),
    Lan Peng (IMEC), Koen Kennes (IMEC), Erik Sleeckx (IMEC), Gerald Beyer
    (IMEC), and Eric Beyne (IMEC)

The Thermal Dissipation Characteristics of The Novel System-In-Package Technology (ICE-SiP) for
Mobile and 3D High-end Packages 614
    Taejoo Hwang (Samsung Electronics Co., Ltd., Republic of Korea),
    Dan(Kyung Suk) Oh (Samsung Electronics Co., Ltd., Republic of Korea),
    Jaechoon Kim (Samsung Electronics Co., Ltd., Republic of Korea),
    Euseok Song (Samsung Electronics Co., Ltd., Republic of Korea), Taehun
    Kim (Samsung Electronics Co., Ltd., Republic of Korea), Kilsoo Kim
    (Samsung Electronics Co., Ltd., Republic of Korea), Joungphil Lee
    (Samsung Electronics Co., Ltd., Republic of Korea), and Taehwan Kim
    (Samsung Electronics Co., Ltd., Republic of Korea)

Session 14: Fine-Pitch Solderless Bonding

Fine-Pitch ($\leq$10 μm) Direct Cu-Cu Interconnects Using In-Situ Formic Acid Vapor Treatment 620
    SivaChandra Jangam (University of California Los Angeles), Adeel Ahmed
    Bajwa (Kulicke & Soffa Industries Inc), Umesh Mogera (University of
    California Los Angeles), Pranav Ambhore (University of California Los
    Angeles), Tom Colosimo (Kulicke & Soffa Industries Inc), Bob Chylak
    (Kulicke & Soffa Industries Inc), and Subramanian Iyer (University of
    California Los Angeles)

Low Temperature Cu Interconnect with Chip to Wafer Hybrid Bonding 628
    Guilian Gao (Xperi Corporation, USA), Laura Mirkarimi (Xperi
    Corporation, USA), Thomas Workman (Xperi Corporation, USA), Gill
    Fountain (Xperi Corporation, USA), Jeremy Theil (Xperi Corporation,
    USA), Gabe Guevara (Xperi Corporation, USA), Ping Liu (Xperi
    Corporation, USA), Bongsub Lee (Xperi Corporation, USA), Pawel Mrozek
    (Xperi Corporation, USA), Michael Huynh (Xperi Corporation, USA),
    Catharina Rudolph (Fraunhofer Institute for Reliability and
    Micro-Integration, IZM – ASSID, Germany), Thomas Werner (Fraunhofer
    Institute for Reliability and Micro-Integration, IZM – ASSID,
    Germany), and Anke Hanisch (Fraunhofer Institute for Reliability and
    Micro-Integration, IZM – ASSID, Germany)

Cu Microstructure of High Density Cu Hybrid Bonding Interconnection 636

Seokho Kim (Samsung Electronics Co., Ltd., Korea), Pilkyu Kang (Samsung Electronics Co., Ltd., Korea), Taeyeong Kim (Samsung Electronics Co., Ltd., Korea), Kyuha Lee (Samsung Electronics Co., Ltd., Korea), Joohee Jang (Samsung Electronics Co., Ltd., Korea), Kwangjin Moon (Samsung Electronics Co., Ltd., Korea), Hoonjoo Na (Samsung Electronics Co., Ltd., Korea), Sangjin Hyun (Samsung Electronics Co., Ltd., Korea), and Kihyun Hwang (Samsung Electronics Co., Ltd., Korea)

Low-Resistance and high-Strength Copper Direct Bonding in no-Vacuum Ambient Using Highly (111)-Oriented Nano-Twinned Copper 642

Jing Ye Juang (National Chiao Tung University, Taiwan), Kai Cheng Shie (National Chiao Tung University, Taiwan), Po-Ning Hsu (National Chiao Tung University, Taiwan), Yu Jin Li (National Chiao Tung University, Taiwan), K N Tu (University of California at Los Angeles), and Chih Chen (National Chiao Tung University, Taiwan)

Sub-10µm Pitch Hybrid Direct Bond Interconnect Development for Die-to-Die Hybridization 648

John P. Mudrick (Sandia National Laboratories, USA), Jonatan A. Sierra-Suarez (Sandia National Laboratories, USA), Matthew B. Jordan (Sandia National Laboratories, USA), T. A. Friedmann (Sandia National Laboratories, USA), Robert Jarecki (Sandia National Laboratories, USA), and M. David Henry (Sandia National Laboratories, USA)

Cu Pillar with Nanocopper Caps: The Next Interconnection Node Beyond Traditional Cu Pillar 655

Ramón A. Sosa (Georgia Institute of Technology - Packaging Research Center), Kashyap Mohan (Georgia Institute of Technology - Packaging Research Center), Luu Nguyen (Texas Instruments), Rao Tummala (Georgia Institute of Technology - Packaging Research Center), Antonia Antoniou (Georgia Institute of Technology), and Vanessa Smet (Georgia Institute of Technology - Packaging Research Center)

Cu-Cu Bonding by Low-Temperature Sintering of Self-Healable Cu Nanoparticles 661

Junjie Li (Huazhong University of Science and Technology, China), Qi Liang (Huazhong University of Science and Technology, China), Chen Chen (Huazhong University of Science and Technology, China), Tielin Shi (Huazhong University of Science and Technology, China), Guanglan Liao (Huazhong University of Science and Technology, China), and Zirong Tang (Huazhong University of Science and Technology, China)

Session 15: High-Bandwidth Packaging

Electrical Performance Limits of Fine Pitch Interconnects for Heterogeneous Integration 667

Ahmet C. Durgun (Assembly and Test Technology Development Intel Corporation), Zhiguo Qian (Assembly and Test Technology Development Intel Corporation), Kemal Aygun (Assembly and Test Technology Development Intel Corporation), Ravi Mahajan (Assembly and Test Technology Development Intel Corporation), Tim Tri Hoang (Programmable Solutions Group Intel Corporation), and Sergey Yuryevich Shumarayev (Programmable Solutions Group Intel Corporation)

A High-Bandwidth Fine-Pitch 2.57Tbps/mm In-package Communication Link Achieving 48fJ/bit/mm
Efficiency 674
  Nicolas Pantano (IMEC, KULeuven), Geert Van der Plas (IMEC), Pieter
  Bex (IMEC), Philip Nolmans (IMEC), Dimitrios Velenis (IMEC), Marian
  Verhelst (KU Leuven), and Eric Beyne (IMEC)

A New SI-PI co-Simulation Approach for Efficient Consideration of Coupling Between PDN and SDN 682
  Heesok Lee (Samsung Electronics, Co. Ltd.), Jisoo Hwang (Samsung
  Electronics, Co. Ltd.), Hoi-jin Lee (Samsung Electronics, Co. Ltd.),
  and Youngmin Shin (Samsung Electronics, Co. Ltd.)

Signal Integrity of Submicron InFO Heterogeneous Integration for High Performance Computing
Applications 688
  Chuei-Tang Wang (Taiwan Semiconductor Manufacturing Company Ltd.),
  Jeng-Shien Hsieh (Taiwan Semiconductor Manufacturing Company Ltd.),
  Victor C. Y. Chang (Taiwan Semiconductor Manufacturing Company Ltd.),
  Shih-Ya Huang (Taiwan Semiconductor Manufacturing Company Ltd.), T. Ko
  (Taiwan Semiconductor Manufacturing Company Ltd.), Han-Ping Pu (Taiwan
  Semiconductor Manufacturing Company Ltd.), and Douglas Yu (Taiwan
  Semiconductor Manufacturing Company Ltd.)

28GHz Through Glass Via (TGV) Based Band Pass Filter Using Through Fused Silica Via (TFV) Technology. 695
  Renuka Bowrothu (University of Florida), Seahee Hwangbo (University of
  Florida), Todd Schumann (University of Florida), and Yong-Kyu Yoon
  (University of Florida)

Innovative Packaging Solutions of 3D Double Side Molding with System in Package for IoT and 5G
Application 700
  Mike Tsai (Siliconware Precision Industries Co., Ltd., Taiwan), Ryan
  Chiu (Siliconware Precision Industries Co., Ltd., Taiwan), Dick Huang
  (Siliconware Precision Industries Co., Ltd., Taiwan), Feng Kao
  (Siliconware Precision Industries Co., Ltd., Taiwan), Eric He
  (Siliconware Precision Industries Co., Ltd., Taiwan), J. Y. Chen
  (Siliconware Precision Industries Co., Ltd., Taiwan), Simon Chen
  (Siliconware Precision Industries Co., Ltd., Taiwan), Jensen Tsai
  (Siliconware Precision Industries Co., Ltd., Taiwan), and Yu-Po Wang
  (Siliconware Precision Industries Co., Ltd., Taiwan)

Enhancing Efficiency of Antenna-in-Package (AiP) by Through-Silicon-Interposer (TSI) with Embedded
Air Cavity and Polyimide Dielectric Micro-Substrate 707
  Yunna Sun (Shanghai Jiao Tong University), Yunting Sun (Shanghai Jiao
  Tong University), Jiangbo Luo (Shanghai Jiao Tong University), Huiying
  Wang (Shanghai Jiao Tong University), Zhuoqing Yang (Shanghai Jiao
  Tong University), Yan Wang (Shanghai Jiao Tong University), Guifu Ding
  (Shanghai Jiao Tong University), and Kwangwoo Han (Samsung Electronics
  Co.)

Session 16: Advanced Materials for High-Speed Electronics

Low-Loss Glass Substrates Formulated with a Variety of Dielectric Characteristics for
Millimeter-Wave Applications 712
  Kazutaka Hayashi (AGC Inc.), Nobutaka Kidera (AGC Inc.), and Yoichiro
  Sato (AGC Inc.)

xxi

Evaluation of Fine-Pitch Routing Capabilities of Advanced Dielectric Materials for High Speed Panel-RDL in 2.5D Interposer and Fan-Out Packages 718

Shreya Dwarakanath (Georgia Institute of Technology), P. Markondeya Raj (Florida International University), Amit Agarwal (Microchips, USA), Daichi Okamoto (TAIYO INK MFG. CO., LTD. Japan), Atsushi Kubo (Tokyo Ohka Kogyo Co., Ltd., Japan), Fuhan Liu (Georgia Institute of Technology), Mohan Kathaperumal (Georgia Institute of Technology), and Rao R. Tummala (Georgia Institute of Technology)

Attenuation of high Frequency Signals in Structured Metallization on Glass: Comparing Different Metallization Techniques with 24 GHz, 77 GHz and 100 GHz Structures 726

Letz Martin (SCHOTT AG, Germany), Jost Matthias (TU Darmstadt), Brandon T. Gore (Samtec, Colorado Springs), William J. Kozlovsky (Samtec, Colorado Springs), Romeo Premerlani (Varioprint AG), Alex Bruderer (Varioprint AG), Manuel Martina (Schweizer Electronic AG), Thomas Gottwald (Schweizer Electronic AG), Tetsuya Onishi (Grand Joint Technology Ltd.), Shigeo Onitake (KOTO Electric Co.), Siddharth Ravichandran (Packaging Research Center), Holger Maune (Institute for microwave engineering and photonics), and Mathias Mydlak (SCHOTT AG)

The Highly Effective EMI Shielding Materials for Electric and Magnetic Fields Over the Wide Range of Frequency in Near-Field Region 733

Yoon-Hyun Kim (Ntrium Inc.), Kisu Joo (Ntrium Inc.), Kyu Jae Lee (Ntrium Inc.), Jung Woo Hwang (Ntrium Inc.), Seung Jae Lee (Ntrium Inc.), Se Young Jeong (Ntrium Inc.), and Hyun Ho Park (Ntrium Inc.)

Low Loss NCF Material for High Frequency Device 740

Kazutaka Honda (Hitachi Chemical Co., Ltd.), Keiko Ueno (Hitachi Chemical Co., Ltd.), Tsuyoshi Ogawa (Hitachi Chemical Co., Ltd.), and Toshihisa Nonaka (Hitachi Chemical Co., Ltd.)

In-Situ Redox Nanowelding of Copper Nanowires with Surficial Oxide Layer as Solder for Flexible Transparent Electromagnetic Interference Shielding 746

Xianwen Liang (Chinese Academy of Sciences), Jianwen Zhou (Chinese Academy of Sciences), Gang Li (Chinese Academy of Sciences), Tao Zhao (Chinese Academy of Sciences), Pengli Zhu (Chinese Academy of Sciences), Rong Sun (Chinese Academy of Sciences), and Ching-ping Wong (Georgia Institute of Technology)

Compartmental EMI Shielding with Jet-Dispensed Material Technology 753

Xuan Hong (Henkel Corporation), Qizhuo Zhuo Zhuo (Henkel Corporation), Xinpei Cao (Henkel Corporation), Dan Maslyk (Henkel Corporation), Noah Ekstrom (Henkel Corporation), Juliet Sanchez (Henkel Corporation), Selene Hernandez (Henkel Corporation), and Jinu Choi (Henkel Corporation)

Session 17: Materials and Design for Reliability of Next-Generation Packages:

Highly (111)-Oriented Nanotwinned Cu for High Fatigue Resistance in Fan-Out Wafer-Level Packaging 758

Yu-Jin Li (National Chiao Tung University), Chih-Han Theng (National Chiao Tung University), I-Hsin Tseng (National Chiao Tung University), Chih Chen (National Chiao Tung University), Benson Lin (MediaTek Inc), and Chia-Cheng Chang (MediaTek Inc)

WLCSP Package and PCB Design for Board Level Reliability 763

Jason Chiu (Taiwan Semiconductor Manufacturing Company, Ltd.), K.C. Chang (Taiwan Semiconductor Manufacturing Company, Ltd.), Steven Hsu (Taiwan Semiconductor Manufacturing Company, Ltd.), Pei-Haw Tsao (Taiwan Semiconductor Manufacturing Company, Ltd.), and M.J. Lii (Taiwan Semiconductor Manufacturing Company, Ltd.)

Assessing the Reliability of Highly Stretchable Interconnects for Flexible Hybrid Electronics 768

Rajesh Sharma Sivasubramony (Binghamton University), Ashwin Varkey Zachariah (Binghamton University), Mohammed Alhendi (Binghamton University), Manu Yadav (Binghamton University), Peter Borgesen (Binghamton University), Mark D. Poliks (Binghamton University), Nancy C. Stoffel (GE Global Research), David M. Shaddock (GE Global Research), and Liang Yin (GE Global Research)

The How and why of Biased Humidity Tests with Copper Wire 777

Amar Mavinkurve (NXP Semiconductors), R.T.H. Rongen (NXP Semiconductors), L. Goumans (NXP Semiconductors), M-L Farrugia (NXP Semiconductors), E. van Olst (NXP Semiconductors), Orla O'Halloran (NXP Semiconductors), and M. van Soestbergen (NXP Semiconductors)

Twist Testing for Flexible Electronics 785

Justin H. Chow (Georgia Institute of Technology), Jeffrey Meth (DuPont Electronics and Imaging), and Suresh K. Sitaraman (Georgia Institute of Technology)

Mechanical Properties and Microstructural Fatigue Damage Evolution in Cyclically Loaded Lead-Free Solder Joints 792

Sinan Su (Auburn University), Mohd Aminul Hoque (Auburn University), Md Mahmudur Chowdhury (Auburn University), Sa'd Hamasha (Auburn University), Jeffrey C. Suhling (Auburn University), John L. Evans (Auburn University), and Pradeep Lall (Auburn University)

Reliability Studies of Silicon Interconnect Fabric 800

Niloofar Shakoorzadeh (UCLA), Siva Chandra Jangam (UCLA), Kaysar Rahim (GlobalFoundries), Pranav Ambhore (UCLA), Han Chien (National Chiao Tung University), Amir Hanna (UCLA), and Subramanian S. Iyer (UCLA)

Session 18: Warpage and Material Performance

Improved Finite Element Modeling of Moisture Diffusion Considering Discontinuity at Material Interfaces in Electronic Packages 806

Lulu Ma (Lamar University), Rahul Joshi (AMD), Keith Keith Newman (AMD), and Xuejun Fan (Lamar University)

Study of Thermal Aging Behavior of Epoxy Molding Compound for Applications in Harsh Environments 811

Adwait Inamdar (Robert Bosch GmbH), Alexandru Prisacaru (Robert Bosch GmbH), Martin Fleischman (Robert Bosch GmbH), Erick Franieck (Robert Bosch GmbH), Przemyslaw Gromala (Robert Bosch GmbH), Agnes Veres (Robert Bosch Kft), Csaba Nemeth (Robert Bosch Kft), Yu-Hsiang Yang (University of Maryland), and Bongtae Han (University of Maryland)

Warpage Variation Analysis and Model Prediction for Molded Packages 819
Yuling Niu (Qualcomm Technologies, Inc., USA), Wei Wang (Qualcomm Technologies, Inc., USA), Zhijie Wang (Qualcomm Technologies, Inc., USA), Karthik Dhandapani (Qualcomm Technologies, Inc., USA), Mark Schwarz (Qualcomm Technologies, Inc., USA), and Ahmer Syed (Qualcomm Technologies, Inc., USA)

Peridynamics for Predicting Thermal Expansion Coefficient of Graphene 825
Erdogan Madenci (University of Arizona), Atila Barut (University of Arizona), and Mehmet Dorduncu (University of Arizona)

Machine Learning Approach to Improve Accuracy of Warpage Simulations 834
Cheryl Selvanayagam (Singapore University of Technology and Design), Pham Luu Trung Duong (Singapore University of Technology and Design), Rathin Mandal (Advanced Micro Devices Inc.), and Nagarajan Raghavan (Singapore University of Technology and Design)

Study on Warpage of Fan-Out Panel Level Packaging (FO-PLP) Using Gen-3 Panel 842
Fa Xing Che (Institute of Microelectronics A*STAR), Kazunori Yamamoto (Institute of Microelectronics A*STAR), Vempati Srinivasa Rao (Institute of Microelectronics A*STAR), and Vasarla Nagendra Sekhar (Institute of Microelectronics A*STAR)

Mechanical Properties of Intermetallic Compounds at Elevated Temperature by Nanoindentation 850
Fan Yang (The Institute of Technological Sciences, Wuhan University, Wuhan, China), Sheng Liu (Wuhan University), Zhaoxia Zhou (Loughborough University), Zhiwen Chen (Wuhan University), Li Liu (Wuhan University of Technology), Canyu Liu (Loughborough University), and Changqing Liu (Loughborough University)

Session 19: MEMS, Sensors, and IoT

A MEMS Microphone in a FOWLP 855
Horst Theuss (Infineon Technologies AG), Christian Geissler (Infineon Technologies AG), Franz-Xaver Muehlbauer (Infineon Technologies AG), Claus von Waechter (Infineon Technologies AG), Thomas Kilger (Infineon Technologies AG), Juergen Wagner (Infineon Technologies AG), Thomas Fischer (Infineon Technologies AG), Ulf Bartl (Infineon Technologies AG), Stephan Helbig (Infineon Technologies AG), Alfred Sigl (Infineon Technologies AG), Dominic Maier (Infineon Technologies AG), Bernd Goller (Infineon Technologies AG), Matthias Vobl (Infineon Technologies AG), Matthias Herrmann (Infineon Technologies AG), Johannes Lodermeyer (Infineon Technologies AG), Ulrich Krumbein (Infineon Technologies AG), and Alfons Dehe (Hahn-Schickard)

Fan-Out Wafer Level Packaging - A Platform for Advanced Sensor Packaging 861
Tanja Braun (Fraunhofer IZM), Karl-Friedrich Becker (Fraunhofer
Institute for Reliability and Microintegration), Ole Hoelck
(Fraunhofer Institute for Reliability and Microintegration), Steve
Voges (Fraunhofer Institute for Reliability and Microintegration),
Ruben Kahle (Fraunhofer Institute for Reliability and
Microintegration), Pascal Graap (Fraunhofer Institute for Reliability
and Microintegration), Markus Wöhrmann (Fraunhofer Institute for
Reliability and Microintegration), R. Aschenbrenner (Fraunhofer
Institute for Reliability and Microintegration), Tanja Braun
(Technical University Berlin), Marc Dreissigacker (Technical
University Berlin), Martin Schneider-Ramelow (Technical University
Berlin), and Klaus-Dieter Lang (Technical University Berlin)

3D-MID Evaluation and Validation for Space Applications 868
Etienne Hirt (Art of Technology AG), Klaus Ruzicka (Art of Technology
AG), Benedikt Wigger (Hahn – Schickard, Mikromontage), Maximilian
Barth (Hahn – Schickard), Rafat Saleh (Hahn – Schickard), Florian
Janek (Hahn – Schickard), and Ernst Müller (University Stuttgart)

High-Temperature Pressure Sensor Package and Characterization of Thermal Stress in the Assembly up
to 500 °C 878
Nilavazhagan Subbiah (University of Freiburg, Germany), Qingming Feng
(University of Freiburg, IMTEK), Juergen Wilde (University of
Freiburg, Germany), and Gudrun Bruckner (CTR AG, Austria)

Development of 3D WLCSP with Black Shielding for Optical Finger Print Sensor for the Application of
Full Screen Smart Phone 884
Daquan Yu (Huantian Technology (Kunshan) Electronics Co., Ltd), Yichao
Zou (Huantian Technology (Kunshan) Electronics Co., Ltd), Xirui Xu
(Huantian Technology (Kunshan) Electronics Co., Ltd), Aihua Shi
(Huantian Technology (Kunshan) Electronics Co., Ltd), Xiaobing Yang
(Huantian Technology (Kunshan) Electronics Co., Ltd), and Zhiyi Xiao
(Huantian Technology (Kunshan) Electronics Co., Ltd.)

Micro Fountain-Like Resonators 890
Jianfeng Zhang (Southeast University, China), Jintang Shang (Southeast
University, China), Bin Luo (Southeast University, China), and Zhaoxi
Su (Southeast University, China)

Novel Additively Manufactured Packaging Approaches for 5G/mm-Wave Wireless Modules 896
Tong-Hong Lin (Georgia Institute of Technology), Aline Eid (Georgia
Institute of Technology), Jimmy Hester (Georgia Institute of
Technology), Bijan Tehrani (Georgia Institute of Technology), Jo Bito
(Texas Instrument), and Manos M. Tentzeris (Georgia Institute of
Technology)

xxv

Session 20: Fanout and Heterogeneous Integration

Feasibility Study of Fan-Out Wafer-Level Packaging for Heterogeneous Integrations 903
John Lau (ASMPT), Ming Li (ASMPT), Iris Xu (Jiangyin Changdian Advanced Packaging Co., Ltd.), Tony Chen (Jiangyin Changdian Advanced Packaging Co., Ltd.), Kim Hwee Tan (Jiangyin Changdian Advanced Packaging Co., Ltd.), Zhang Li (Jiangyin Changdian Advanced Packaging Co., Ltd.), Nelson Fan (ASMPT), Eric Kuah (ASMPT), Raymond So (ASMPT), Penny Lo (ASMPT), Y. M. Cheung (ASMPT), Cao Xi (Huawei Technologies Co. Ltd.), Rozalia Beica (Dow Chemical Company), Sze Pei Lim (Indium Corporation), NC Lee (Indium Corporation), Cheng-Ta Ko (Unimicron Technology Corporation), Henry Yang (Unimicron Technology Corporation), YH Chen (Unimicron Technology Corporation), Mian Tao (Hong Kong University of Science and Technology), Jeffery Lo (Hong Kong University of Science and Technology), and Ricky Lee (Hong Kong University of Science and Technology)

Experiment of 22FDX(R) Chip Board Interaction (CBI) in Wafer Level Packaging Fan-Out (WLPFO) 910
Jae Kyu Cho (GLOBALFOUNDRIES, USA), Jens Paul (GLOBALFOUNDRIES, Germany), Simone Capecchi (GLOBALFOUNDRIES, Germany), Frank Kuechenmeister (GLOBALFOUNDRIES, Germany), and Ta-Chien Cheng (GLOBALFOUNDRIES, Singapore)

FOWLP Design for Digital and RF Circuits 917
Teck Guan Lim (Institute of Microelectronics, Singapore), David Soon Wee Ho (Institute of Microelectronics, Singapore), Eva Wai Leong Ching (Institute of Microelectronics, Singapore), Zihao Chen (Institute of Microelectronics, Singapore), and Surya Bhattacharya (Institute of Microelectronics, Singapore)

Next Generation of 2-7 Micron Ultra-Small Microvias for 2.5D Panel Redistribution Layer by Using
Laser and Photolithography Technologies 924
Fuhan Liu (Georgia Institute of Technology), Chandrasekharan Nair (Georgia Institute of Technology), Gaurav Khurana (Georgia Institute of Technology), Atom Watanabe (Georgia Institute of Technology), Bartlet H. DeProspo (Georgia Institute of Technology), Atsushi Kubo (Tokyo Ohka Kogyo Co. Ltd., Japan), Cheng Ping Lin (Panasonic Corporation, Japan), Toshiyuki Makita (Panasonic Corporation, Japan), Naoki Watanabe (Panasonic Industrial Devices Sales Company of America, USA), and Rao R. Tummala (Georgia Institute of Technology)

Multilayer RDL Interposer for Heterogeneous Device and Module Integration 931
Yi-Hang Lin (Taiwan Semiconductor Manufacturing Company Ltd.), M.C. Yew (Taiwan Semiconductor Manufacturing Company Ltd.), S.M. Chen (Taiwan Semiconductor Manufacturing Company Ltd.), M.S. Liu (Taiwan Semiconductor Manufacturing Company Ltd.), Pravin Kavle (Taiwan Semiconductor Manufacturing Company Ltd.), T.M. Lai (Taiwan Semiconductor Manufacturing Company Ltd.), C.T. Yu (Taiwan Semiconductor Manufacturing Company Ltd.), F.C. Hsu (Taiwan Semiconductor Manufacturing Company Ltd.), C.S. Chen (Taiwan Semiconductor Manufacturing Company Ltd.), T.J. Fang (Taiwan Semiconductor Manufacturing Company Ltd.), C.K. Hsu (Taiwan Semiconductor Manufacturing Company Ltd.), K.C. Lee (Taiwan Semiconductor Manufacturing Company Ltd.), C.H. Lin (Taiwan Semiconductor Manufacturing Company Ltd.), P.Y. Lin (Taiwan Semiconductor Manufacturing Company Ltd.), and Shin-Puu Jeng (Taiwan Semiconductor Manufacturing Company Ltd.)

Effects of Dielectric Curing Conditions on the Interfacial Adhesion of Cu RDL for Fan-Out Wafer Level Packaging 937

> Gahui Kim (Andong National University, Korea), Kirak Son (Andong National University, Korea), Dogeun Kim (Korea Institute of Materials Science, Korea), Seok-hyun Lee (SAMSUNG ELECTRONICS CO., LTD, Korea), and Young-Bae Park (Andong National University, Korea)

Al-Al Direct Bonding with Sub-µm Alignment Accuracy for Millimeter Wave SiGe BiCMOS Wafer Level Packaging and Heterogeneous Integration 942

> Matthias Wietstruck (IHP - Leibniz Institut für innovative Mikroelektronik), Sebastian Schulze (IHP - Leibniz Institut für innovative Mikroelektronik), Bernhard Rebhan (EV Group E. Thallner GmbH), Peter Kerepesi (EV Group E. Thallner GmbH), Helmut Kurz (EV Group E. Thallner GmbH), Gerald Silberer (EV Group E. Thallner GmbH), Josef Meiler (EV Group E. Thallner GmbH), Selin Tolunay Wipf (IHP - Leibniz Institut für innovative Mikroelektronik), Christian Wipf (IHP - Leibniz Institut für innovative Mikroelektronik), and Mehmet Kaynak (IHP - Leibniz Institut für innovative Mikroelektronik, Sabanci University)

Session 21: 5G, mm-Wave, and Antenna-in-Package

Vivaldi Antenna Array Fabricated Using a Hybrid Process 948

> Vincens Gjokaj (Michigan State University), Cameron Crump (Michigan State University), John Papapolymerou (Michigan State University), John Albrecht (Michigan State University), and Premjeet Chahal (Michigan State University)

Novel Multicore PCB and Substrate Solutions for Ultra Broadband Dual Polarized Antennas for 5G Millimeter Wave Covering 28GHz & 39GHz Range 954

> Trang Thai (Intel Corporation), Sidharth Dalmia (Intel Corporation), Josef Hagn (Intel Corporation), Pouya Talebbeydokhti (Intel Corporation), and Yossi Tsfati (Intel Corporation)

3D Glass Package-Integrated, High-Performance Power Dividing Networks for 5G Broadband Antennas 960

> Muhammad Ali (Georgia Institute of Technology), Atom Watanabe (Georgia Institute of Technology), Tong-Hong Lin (Georgia Institute of Technology), Markondeya Raj Pulugurtha (Florida International University), Manos M. Tentzeris (Georgia Institute of Technology), and Rao R. Tummala (Georgia Institute of Technology)

Advanced Wafer Level PKG Solutions for 60GHz WiGig (802.11ad) Telecom Infrastructure 968

> Dapeng Wu (Sivers IMA AB), Robin Dahlbäck (Sivers IMA AB), Erik Öjefors (Sivers IMA AB), Mats Carlsson (Sivers IMA AB), Francis Chee Peng Lim (STATS ChipPAC Pte. Ltd.), Yew Kheng Lim (STATS ChipPAC Pte. Ltd.), Aung Kyaw Oo (STATS ChipPAC Pte. Ltd.), Won Kyung Choi (STATS ChipPAC Pte. Ltd.), and Seung Wook Yoon (STATS ChipPAC Pte. Ltd.)

xxvii

Low-Loss Additively-Deposited Ultra-Short Copper-Paste Interconnections in 3D Antenna-Integrated
Packages for 5G and IoT Applications 972
*Atom O. Watanabe (Georgia Institute of Technology), Yiteng Wang
(Georgia Institute of Technology), Nobuo Ogura (Nagase & Co., LTD.,
Japan), P. Markondeya Raj (Florida International University), Vanessa
Smet (Georgia Institute of Technology), Manos M. Tentzeris (Georgia
Institute of Technology), and Rao R. Tummala (Georgia Institute of
Technology)*

Advanced Thin-Profile Fan-Out with Beamforming Verification for 5G Wideband Antenna 977
*Ricky Hsieh (Advanced Semiconductor Engineering (ASE), Inc., Taiwan),
Fu-Cheng Chu (Advanced Semiconductor Engineering (ASE), Inc., Taiwan),
Cheng-Yu Ho (Advanced Semiconductor Engineering (ASE), Inc., Taiwan),
and Chen-Chao Wang (Advanced Semiconductor Engineering (ASE), Inc.,
Taiwan)*

Integrated Compact Planar Inverted-F Antenna (PIFA) with a Shorting Via Wall for Millimeter-Wave
Wireless Chip-to-Chip (C2C) Communications in 3D-SiP 983
*Seahee Hwangbo (University of Florida), Renuka Bowrothu (University of
Florida), Hae-in Kim (University of Florida), and Yong-Kyu Yoon
(University of Florida)*

Session 22: Advanced Substrates and Interconnect Technology

Temporary SiC-SiC Wafer Bonding Compatible with High Temperature Annealing 989
*Fengwen Mu (The University of Tokyo), Tadatomo Suga (The University of
Tokyo), Miyuki Uomoto (Tohoku University), and Takehito Shimatsu
(Tohoku University)*

Ultrathin Glass to Ultrathin Glass Bonding Using Laser Sealing Approach 995
*Messaoud Bedjaoui (Univ. Grenoble Alpes, CEA-LETI, France), Johnny
Amiran (Univ. Grenoble Alpes, CEA-LETI, France), and Jean Brun (Univ.
Grenoble Alpes, CEA-LETI, France)*

Development of Resins for Bumpless Interconnects and Wafer-On-Wafer (WOW) Integration 1002
*Naoko Araki (Daicel Corporation), Shinji Maetani (Daicel Corporation),
Kim Young Suk (Disco Corporation), Shoichi Kodama (Disco Corporation),
and Takayuki Ohba (Tokyo Institute of Technology)*

Development of Novel Photosensitive Dielectric Material for Reliable 2.1D Package 1009
*Yune Kumazawa (Mitsubishi Gas Chemical Company, Inc.), Seiji Shika
(Mitsubishi Gas Chemical Company, Inc.), Shunsuke Katagiri (Mitsubishi
Gas Chemical Company, Inc.), Takuya Suzuki (Mitsubishi Gas Chemical
Company, Inc.), Tsuyoshi Kida (Mitsubishi Gas Chemical Company, Inc.),
and Shu Yoshida (Mitsubishi Gas Chemical Company, Inc.)*

High Reliability Solder Resist with Strong Adhesion and High Resolution for High Density Packaging 1015
*Sawako Shimada (TAIYO INK MFG. CO., LTD.), Kazuya Okada (TAIYO INK
MFG. CO., LTD.), Tomoya Kudo (TAIYO INK MFG. CO., LTD.), Chiho Ueta
(TAIYO INK MFG. CO., LTD.), and Yuya Suzuki (TAIYO INK MFG. CO., LTD.)*

Method for Mitigating the Warpage of Ultra-Thin FC-CSPs by Controlling of EMC Properties 1022
*Chika Arayama (Panasonic Corporation), Takahiro Akashi (Panasonic
Corporation), Yasunari Tomita (Panasonic Corporation), and Naoki
Kanagawa (Panasonic Corporation)*

Innovative Socketable and Surface-Mountable BGA Interconnections 1028

    Omkar Gupte (Georgia Institute of Technology - Packaging Research Center), Kristie Teoh (Georgia Institute of Technology - Packaging Research Center), Rao Tummala (Georgia Institute of Technology - Packaging Research Center), Gregorio Murtagian (Intel Corporation), and Vanessa Smet (Georgia Institute of Technology - Packaging Research Center)

Session 23: High-Bandwidth 3D and Photonic Integration

A Highly Reliable 1.4µm Pitch Via-Last TSV Module for Wafer-to-Wafer Hybrid Bonded 3D-SOC Systems . 1035

    Stefaan Van Huylenbroeck (IMEC), Joeri De Vos (IMEC), Zaid El-Mekki (IMEC), Geraldine Jamieson (IMEC), Nina Tutunjyan (IMEC), Karthik Muga (IMEC), Michele Stucchi (IMEC), Andy Miller (IMEC), Gerald Beyer (IMEC), and Eric Beyne (IMEC)

Nanoscale Topography Characterization for Direct Bond Interconnect 1041

    Bongsub Lee (Xperi Corporation), Pawel Mrozek (Xperi Corporation), Gill Fountain (Xperi Corporation), John Posthill (Xperi Corporation), Jeremy Theil (Xperi Corporation), Guilian Gao (Xperi Corporation), Rajesh Katkar (Xperi Corporation), and Laura Mirkarimi (Xperi Corporation, USA)

Fully-Filled, Highly-Reliable Fine-Pitch Interposers with TSV Aspect Ratio >10 for Future 3D-LSI/IC
Packaging 1047

    Murugesan Murugesan (Tohoku University, Japan), Takafumi Fukushima (Tohoku University, Japan), Kiyoharu Mori (T-Miro, Japan), Ai Nakamura (T-Micro, Japan), Yisang Lee (T-Micro, Japan), Makoto Motoyoshi (T-Micro, Japan), J.C Bea (Tohoku University), Shigeru Watariguchi (Meltex, Japan), and Mitsumasa Koyanagi (Tohoku University, Japan)

3D Silicon Photonics Interposer for Tb/s Optical Interconnects in Data Centers with Double-Side
Assembled Active Components and Integrated Optical and Electrical Through Silicon Via on SOI 1052

    Bogdan Sirbu (Fraunhofer Institute for Reliability and Microintegration IZM Berlin), Yann Eichhammer (Fraunhofer Institute for Reliability and Microintegration IZM Berlin), Hermann Oppermann (Fraunhofer Institute for Reliability and Microintegration IZM Berlin), Tolga Tekin (Fraunhofer Institute for Reliability and Microintegration IZM Berlin), Jochen Kraft (ams AG, Austria), Victor Sidorov (ams AG, Austria), Xin Yin (IMEC, Belgium), Johan Bauwelinck (IMEC, Belgium), Christian Neumeyr (Vertilas GmbH, Germany), and Francisco Soares (Fraunhofer Heinrich-Hertz-Institut HHI Berlin)

Flip-Chip III-V-to-Silicon Photonics Interfaces for Optical Sensor 1060

Yves Martin (IBM T. J. Watson Research Center, Yorktown Heights USA),
Jason S. Orcutt (IBM T. J. Watson Research Center, Yorktown Heights
USA), Chi Xiong (IBM T. J. Watson Research Center, Yorktown Heights
USA), Laurent Schares (IBM T. J. Watson Research Center, Yorktown
Heights USA), Tymon Barwicz (IBM T. J. Watson Research Center,
Yorktown Heights USA), Martin Glodde (IBM T. J. Watson Research
Center, Yorktown Heights USA), Swetha Kamlapurkar (IBM T. J. Watson
Research Center, Yorktown Heights USA), Eric J. Zhang (IBM T. J.
Watson Research Center, Yorktown Heights USA), William M.J. Green (IBM
T. J. Watson Research Center, Yorktown Heights USA), Victor
Dolores-Calzadilla (Fraunhofer Heinrich-Hertz Institute, Germany),
Ariane Sigmund (Fraunhofer Heinrich-Hertz Institute, Germany), and
Martin Moehrle (Fraunhofer Heinrich-Hertz Institute, Germany)

Extremely Low-Profile Single Mode Fiber Array Coupler Suitable for Silicon Photonics 1067

Mitsuharu Hirano (Sumitomo Electric Industries, Ltd., Japan), Akira
Furuya (Sumitomo Electric Industries, Ltd., Japan), Hideki Machida
(Sumitomo Electric Industries, Ltd., Japan), Koichi Koyama (Sumitomo
Electric Industries, Ltd., Japan), Yasunori Murakami (Sumitomo
Electric Industries, Ltd., Japan), and Kazunori Tanaka (Sumitomo
Electric Industries, Ltd., Japan)

Micro Lens Array Assembly for Optical Organic Substrate 1074

Patrick Jacques (IBM Bromont, Canada), Richard Langlois (IBM Bromont,
Canada), Élaine Cyr (IBM Bromont, Canada), Alexander Janta-Polczynski
(IBM Bromont, Canada), Paul Fortier (IBM Bromont, Canada), Koji Masuda
(IBM Tokyo Research, Japan), Masao Tokunari (IBM Tokyo Research), and
Hsiang-Han Hsu (IBM Tokyo Research)

Session 24: Advancements in Solder Joint Characterization and Reliability Evaluation

Effects of In and Zn Double Addition on Eutectic Sn-58Bi Alloy 1081
Shiqi Zhou (Osaka University, Japan), Yu-An Shen (Osaka University,
Japan), Tiffani Uresti (Texas A&M University at Qatar, Qatar), Vasanth
Shunmugasamy (Texas A&M University at Qatar, Qatar), Bilal Mansoor
(Texas A&M University at Qatar, Qatar), and Hiroshi Nishikawa (Osaka
University, Japan)

Microstructural Evolution in SAC+X Solders Subjected to Aging 1087
Jing Wu (Auburn University), Jeffrey C. Suhling (Auburn University),
and Pradeep Lall (Auburn University)

Microstructure Signature Evolution in Solder Joints, Solder Bumps, and Micro-Bumps Interconnection
in A Large 2.5D FCBGA Package During Thermo-Mechanical Cycling 1099
Arman Ahari (Portland State University), Andy Hsiao (Portland State
University), Greg Baty (Portland State University), Peng Su (Juniper
Networks, USA), and Tae-Kyu Lee (Portland State University)

Long-Term Reliability of Solder Joints in 3D ICs Under Near-Application Conditions 1106
Omar Ahmed (University of Central Florida), Golareh Jalilvand (University of Central Florida), Hector Fernandez (University of Central Florida), Peng Su (Juniper Networks), Tae-Kyu Lee (Portland State University), and Tengfei Jiang (University of Central Florida)

Experimental Investigation of the Correlation between a Load-Based Metric and Solder Joint Reliability of BGA Assemblies on System Level 1113
Fabian Schempp (Robert Bosch GmbH, University of Freiburg - IMTEK), Marc Dressler (Robert Bosch GmbH), Daniel Kraetschmer (Robert Bosch GmbH), Friederike Loerke (Robert Bosch GmbH), and Juergen Wilde (University of Freiburg - IMTEK)

Fatigue Life Prediction Model Development for Decoupling Capacitors 1121
Krishna Tunga (IBM Corporation, USA), Joseph Ross (IBM Corporation, USA), Kamal Sikka (IBM Corporation, USA), and Bakul Parikh (IBM Corporation, USA)

A Study of Substrate Models and Its Effect On Package Warpage Prediction 1130
Van-Lai Pham (Binghamton University), Huayan Wang (Binghamton University), Jiefeng Xu (Binghamton University), Jing Wang (Binghamton University), Chrandeep Singh (Corning Inc.), and Seungbae Park (Binghamton University)

Session 25: Wafer Level Packaging and Fan-In/Fan-Out Structures & Materials

3D Fan-Out Package Technology with Photosensitive Through Mold Interconnects 1140
Kentaro Mori (Toshiba Electronic Devices and Storage Corporation), Soichi Yamashita (Toshiba Electronic Devices and Storage Corporation), Takafumi Fukuda (Toshiba Development & Engineering Corporation), Masahiro Sekiguchi (Toshiba Electronic Devices and Storage Corporation), Hirokazu Ezawa (Toshiba Memory Corporation), and Shuzo Akejima (Toshiba Electronic Devices and Storage Corporation)

Effects of the Materials Properties of Epoxy Molding Films (EMFs) on Fan-Out Packages (FOPs) Characteristics 1146
Sangmyung Shin (KAIST), Hanmin Lee (KAIST), JunMo Kim (KAIST), Tae-Ik Lee (KAIST), Taek-Soo Kim (KAIST), Youjin Kyung (LG Chem), Minsu Jeong (LG Chem), Kwangjoo Lee (LG Chem), and Kyung-Wook Paik (KAIST)

Mechanism of Moldable Underfill (MUF) Process for RDL-1st Fan-Out Panel Level Packaging (FOPLP) ... 1152
Lin Bu (Institute of Microelectronics A*STAR), F. X. Che (Institute of Microelectronics A*STAR), Vempati Srinivasa Rao (Institute of Microelectronics A*STAR), and Xiaowu Zhang (Institute of Microelectronics A*STAR)

Study of the Board Level Reliability Performance of a Large 0.3 mm Pitch Wafer Level Package 1159
Bernd Waidhas (Intel Deutschland GmbH), Jan Proschwitz (Intel Deutschland GmbH), Christoph Pietryga (Intel Deutschland GmbH), Thomas Wagner (Intel Deutschland GmbH), and Beth Keser (Intel Deutschland GmbH)

Study of Board Level Reliability of eWLB (embedded Wafer Level BGA) for 0.35mm Ball Pitch 1165
  Kang Hai Lee (STATS ChipPAC Pte. Ltd.), Yeow Kheng Lim (STATS ChipPAC
  Pte. Ltd.), Seng Guan Chow (STATS ChipPAC Pte. Ltd.), Kang Chen (STATS
  ChipPAC Pte. Ltd.), Won Kyung Choi (STATS ChipPAC Pte. Ltd.), Seung
  Wook Yoon (STATS ChipPAC LTD PTE), NW Liu (Advanced Package
  Technology, Mediatek Inc.), Yenyao Chi (Advanced Package Technology,
  Mediatek Inc.), and Benson Lin (Advanced Package Technology, Mediatek
  Inc.)

Board Level Reliability Study of Fan-Out Single Die Package with 350um Bump Pitch 1170
  Chieh-Lung Lai (Siliconware Precision Industries Co., Ltd.), Gu-Yan
  Lin (Siliconware Precision Industries Co., Ltd.), Tz-Yuan Chao
  (Siliconware Precision Industries Co., Ltd.), Yih-Sin Chen
  (Siliconware Precision Industries Co., Ltd.), and Feng-Lung Chien
  (Siliconware Precision Industries Co., Ltd.)

The Analysis for Bump Resistance Improvement by Optimizing the Sputter Condition 1175
  Ming-Sin Su (Taiwan Semiconductor Manufacturing Company Ltd.),
  Chang-Ning Wang (Taiwan Semiconductor Manufacturing Company Ltd.),
  Clair Tsai (Taiwan Semiconductor Manufacturing Company Ltd.), T. L.
  Yang (Taiwan Semiconductor Manufacturing Company Ltd.), Rolance Yang
  (Taiwan Semiconductor Manufacturing Company Ltd.), W. C. Wu (Taiwan
  Semiconductor Manufacturing Company Ltd.), C. S. Liu (Taiwan
  Semiconductor Manufacturing Company Ltd.), J. M. Chiu (Taiwan
  Semiconductor Manufacturing Company Ltd.), Y. F. Chen (Taiwan
  Semiconductor Manufacturing Company Ltd.), Ponder Pang (Taiwan
  Semiconductor Manufacturing Company Ltd.), Harry Ku (Taiwan
  Semiconductor Manufacturing Company Ltd.), Kirin Wang (Taiwan
  Semiconductor Manufacturing Company Ltd.), C.H. Su (Taiwan
  Semiconductor Manufacturing Company Ltd.), Steven Hsu (Taiwan
  Semiconductor Manufacturing Company Ltd.), Calvin Lu (Taiwan
  Semiconductor Manufacturing Company Ltd.), K. C. Liu (Taiwan
  Semiconductor Manufacturing Company Ltd.), and Marvin Liao (Taiwan
  Semiconductor Manufacturing Company Ltd.)

Session 26: High-Speed Signaling for High-Performance Computing and Memory

Hybrid Prepreg Conventional Build-Up Laminate for 112Gbit/s SerDes 1179
  Kwang Won Choi (GLOBALFOUNDRIES US Inc.), Edmund Blackshear
  (GLOBALFOUNDRIES US Inc.), Eric Tremble (GLOBALFOUNDRIES US Inc.),
  David Stone (GLOBALFOUNDRIES US Inc.), Jean Audet (IBM Corporation,
  Canada), and Keiichi Hirabayashi (Shinko Electric Industries Co.,
  LTD., Japan)

PI/SI Analysis and Design Approach for HPC Platform Applications 1188
  Sungwook Moon (Samsung Electronics Co. Ltd.), Chanmin Jo (Samsung
  Electronics Co. Ltd.), and Seungki Nam (Samsung Electronics Co. Ltd.)

PoP LPDDR5 (6.4 Gbps) NTODT and 1-Tap DFE for Signal Integrity Enhancement 1194
  Sunil Gupta (Qualcomm Technologies, Inc.)

OpenCAPI Memory Interface Signal Integrity Study for High-Speed DDR5 Differential DIMM Channel with Standard Loss FR-4 Material and SNIA SFF-TA-1002 Connector 1200
> Biao Cai (IBM), Jose Hejase (IBM), Kyle Giesen (IBM), Junyan Tang
> (IBM), Brian Connolly (IBM), KyuHyoun Kim (IBM), Daniel Dreps (IBM),
> Zhineng Fan (Amphenol ICC), Rocky Huang (Amphenol ICC), Luyun Yi
> (Amphenol ICC), Qiaoli Chen (Amphenol ICC), Yifan Huang (Amphenol
> ICC), and Stephen Smith (Amphenol ICC)

Effectiveness of Equalization and Performance Potential in DDR5 Channels with RDIMM(s) 1208
> Nanju Na (Xilinx) and Hing "Thomas" To (Xilinx)

Inductive Links for 3D Stacked Chip-to-Chip Communication 1215
> Xiao Sun (IMEC, Belgium), Nicolas Pantano (IMEC, Belgium), Kim
> Soon-Wook (IMEC, Belgium), Geert Van der Plas (IMEC, Belgium), and
> Eric Beyne (IMEC, Belgium)

System Co-design of a 600V GaN FET Power Stage with Integrated Driver in a QFN System-in-Package (QFN-SiP) 1221
> Jie Chen (Texas Instruments, Inc), Yong Xie (Texas Instruments, Inc),
> Django Trombley (Texas Instruments Incorporated), and Rajen Murugan
> (Texas Instruments Incorporated)

Session 27: Advanced Biosensors and Bioelectronics

Flexible Probe for Electrical Neural Signal Recording 1227
> Sajay Bhuvanendran Nair Gourikutty (Institute of Microelectronics,
> A*STAR, Singapore) and Ruiqi Lim (Institute of Microelectronics,
> A*STAR, Singapore)

Stretchable, Implantable Nanomembrane Biosensor for Wireless, Real-Time Monitoring of Hemodynamics .. 1233
> Robert Herbert (Georgia Institute of Technology) and Woon-Hong Yeo
> (Georgia Institute of Technology)

A Wearable Passive pH Sensor for Health Monitoring 1240
> Saikat Mondal (Michigan State University), Saranraj Karuppuswami
> (Michigan State University), Rachel Steinhorst (Michigan State
> University), and Premjeet Chahal (Michigan State University)

Novel Packaging Structure and Processes for Micro-TFB (Thin Film Battery) to Enable Miniaturized Healthcare Internet-of-Things (IoT) Devices 1246
> Bing Dang (IBM Research), Qianwen Chen (IBM Research), Leanna Pancoast
> (IBM Research), Yu Luo (IBM Research), Hongqing Zhang (IBM Systems),
> Jae-woong Nah (IBM Research), John Knickerbocker (IBM Research), Andy
> Shih (Front Edge Technologies Inc.), Po Wen Cheng (Front Edge
> Technologies Inc.), Kai Liu (Front Edge Technologies Inc.), Mengnian
> Niu (Front Edge Technologies Inc.), and Simon Nieh (Front Edge
> Technologies Inc.)

Screen Printed Temporary Tattoos for Skin-Mounted Electronics 1252
> Samuli Tuominen (Tampere University) and Matti Mantysalo (Tampere
> University)

Thermoset Polymers for Bioelectronic Interfaces - Engineering of Thermomechanical Properties 1258
Adriana Carolina Duran-Martinez (The University of Texas at Dallas),
Seyedmahmoud Hosseini (The University of Texas at Dallas), Daniel Del
Nero (The University of Texas at Dallas), Alexandra Joshi-Imre (The
University of Texas at Dallas), Walter E. Voit (The University of
Texas at Dallas), and Melanie Ecker (The University of Texas at
Dallas)

Direct Heterogeneous Bonding of SiC to Si, SiO2, and Glass for High-Performance Power Electronics
and Bio-MEMS 1266
Jikai Xu (Harbin Institute of Technology), Chenxi Wang (Harbin
Institute of Technology), Qiushi Kang (Harbin Institute of
Technology), Shicheng Zhou (Harbin Institute of Technology), and
Yanhong Tian (Harbin Institute of Technology)

Session 28: Embedded and Integrated Technologies

Development of Flexible Hybrid Electronics Using Reflow Assembly with Stretchable Film 1272
Weifeng Liu (Flex), William Uy (Flex), Alex Chan (Flex), Dongkai
Shangguan (Flex), Andy Behr (Panasonic), Takatoshi Abe (Panasonic),
and Fukao Tomohiro (Panasonic)

Highly Compact RF Transceiver Module Using High Resistive Silicon Interposer with Embedded Inductors
and Heterogeneous Dies Integration 1279
G. Pares (Univ. Grenoble Alpes, CEA), Michel Jean-Philippe (CEA),
Deschaseaux Edouard (CEA), Ferris Pierre (CEA), Serhan Ayssar (CEA),
and Giry Alexandre (CEA)

Process Induced Wafer Warpage Optimization for Multi-chip Integration on Wafer Level Molded Wafer 1287
Chen-Yu Huang (Siliconware Precision Industries Co., Ltd, Taiwan),
Daniel Ng (Siliconware Precision Industries Co., Ltd, Taiwan), Hung-Ho
Lee (Siliconware Precision Industries Co., Ltd, Taiwan), Vito Lin
(Siliconware Precision Industries Co., Ltd, Taiwan), Chang-Fu Lin
(Siliconware Precision Industries Co., Ltd, Taiwan), and C. Key Chung
(Siliconware Precision Industries Co., Ltd, Taiwan)

Improved Structure for Package Substrates with Embedded Thin-Film Capacitor 1294
Tomoyuki Akahoshi (Fujitsu Laboratories Ltd.), Daisuke Mizutani
(Fujitsu Laboratories Ltd.), Kei Fukui (Fujitsu Interconnect
Technologies Ltd.), Seigo Yamawaki (Fujitsu Interconnect Technologies
Ltd.), Hidehiko Fujisaki (Fujitsu Interconnect Technologies Ltd.),
Manabu Watanabe (Fujitsu Advanced Technologies Ltd.), and Masateru
Koide (Fujitsu Advanced Technologies Ltd.)

3D Packaging with Embedded High-Power-Density Passives for Integrated Voltage Regulators 1300
Teng Sun (Georgia Institute of Technology), Robert G. Spurney (Georgia
Institute of Technology), Atom Watanabe (Georgia Institute of
Technology), P. Raj Pulugurtha (Florida International University),
Himani Sharma (Georgia Institute of Technology), Rao Tummala (Georgia
Institute of Technology), and Furukawa Yoshihiro (Nitto Denko
Corporation)

xxxiv

A Novel Panel Level Double Side Embedded Package for Small Size Power Devices 1306

Kunpeng Ding (Shenzhen Siptory Technologies Co., Ltd, China), Zhichao
Wu (Institute of Microelectronics, Tsinghua University, China), Mian
Huang (Shenzhen Siptory Technologies Co., Ltd, China), Bowei Zhang
(Wuxi Sky Chip Interconnection Technology Co., Ltd, China), and Jian
Cai (Institute of Microelectronics, Tsinghua University, China)

Chiplet Micro-Assembly Printer 1312

Bradley B. Rupp (PARC), Anne Plochowietz (PARC), Lara S. Crawford
(PARC), Matthew Shreve (PARC), Sourobh Raychaudhuri (PARC), Sergey
Butylkov (PARC), Yunda Wang (PARC), Ping Mei (PARC), Qian Wang (PARC),
Jamie Kalb (PARC), Yu Wang (PARC), Eugene M. Chow (PARC), and JengPing
Lu (PARC)

Session 29: Electromigration and Innovative Reliability Test Methods

Effect of Intermetallic Compound Growth on Electromigration Failure Mechanism in Low-Profile Solder
Joints 1316

Hossein Madanipour (University of Texas at Arlington), Yi-Ram Kim
(University of Texas at Arlington), Choong-Un Kim (University of Texas
at Arlington), Ninad Shahane (Texas Instruments, Inc.), Dibyajat
Mishra (Texas Instruments, Inc.), and Luu Nguyen (Texas Instruments,
Inc.)

Effect of Grain Orientation and Microstructure Evolution on Electromigration in Flip-Chip Solder
Joint 1324

Xing Fu (Science and technology on reliability physics and application
of electronic component laboratory, China), Bin Zhou (Science and
technology on reliability physics and application of electronic
component laboratory), Ruohe Yao (University of Technology), Yunfei En
(Science and technology on reliability physics and application of
electronic component laboratory), and Si Chen (Science and technology
on reliability physics and application of electronic component
laboratory)

High Electromigration Lifetimes of Nanotwinned Cu Redistribution Lines 1328

I-Hsin Tseng (National Chiao Tung University), Yu-Jin Li (National
Chiao Tung university), Benson Lin (MediaTek Inc), Chia-Cheng Chang
(MediaTek Inc), and Chih Chen (National Chiao Tung University)

Non-destructive Failure Analysis of Various Chip to Package Interaction Anomalies in FCBGA Packages
Subjected to Temperature Cycle Reliability Testing 1333

Vishnu V. B. Reddy (Georgia Institute of Technology), I. Charles Ume
(Georgia Institute of Technology), Jaimal Williamson (Texas
Instruments Inc., USA), and Luu Nguyen (Texas Instruments Inc., USA)

Assessment of Accelerometer Versus LASER for Board Level Vibration Measurements 1339

Varun Thukral (NXP Semiconductors), M. Cahu (NXP Semiconductors),
J.J.M. Zaal (NXP Semiconductors), J. Jalink (NXP Semiconductors), R.
Roucou (NXP Semiconductors), and R.T.H. Rongen (NXP Semiconductors)

Effect of Process Parameters on the Long-Run Print Consistency and Material Properties of Additively Printed Electronics 1347

Pradeep Lall (Auburn University), Nakul Kothari (Auburn University), Amrit Abrol (Auburn University), Jeff Suhling (Auburn University), Sudan Ahmed (Auburn University), Ben Leever (US AFRL), and Scott Miller (NextFlex)

A Viscoplastic-Based Fatigue Reliability Model for the Polyimide Dielectric Thin Film 1359

Yu-Chen Chang (National Cheng Kung University), Tz-Cheng Chiu (National Cheng Kung University), Yu-Ting Yang (Advanced Semiconductor Engineering Group, Inc.), Yi-Hsiu Tseng (Advanced Semiconductor Engineering Group, Inc.), and Xi-Hong Chen (Advanced Semiconductor Engineering Group, Inc.)

Session 30: Assembly and Process Modeling

Explicit FE Failure Prediction of Interfaces and Interconnect in Potted Electronics Assemblies Subject to High-g Acceleration Loads 1366

Pradeep Lall (Auburn University), Kalyan Dornala (Auburn University), Ryan Lowe (ARA Associates), and John Deep (US AFRL)

Numerical Simulation on the Formation Process of Metal Droplets by Pneumatic Diaphragm Drop-on Demand Technology 1377

Kun Ma (Wuhan University), Sheng Liu (Wuhan University), Zhiwen Chen (Wuhan University), Li Liu (Wuhan University of Technology), Hao Zheng (China Ship Development and Design Center), and Yao Zhang (China Ship Development and Design Center)

On Curing-Induced Residual Stresses After Molding Processes: Mold Shrinkage, Chemical Shrinkage or Both? 1382

Changsu Kim (University of Maryland), Sukrut Phansalkar (University of Maryland), Hyun-Seop Lee (University of Maryland), and Bongtae Han (University of Maryland)

Realistic Solder Joint Geometry Integration with Finite Element Analysis for Reliability Evaluation of Printed Circuit Board Assembly 1387

Chun Sean Lau (Western Digital Corporation), Ning Ye (Western Digital Corporation), and Hem Takiar (Western Digital Corporation)

Multi-physics Modelling and Experimental Investigation – An Original Approach for Laser-Dicing/Grooving Process Optimization 1396

Jeff Moussodji Moussodji (3IT-UdeS/C2MI/IBM), Oswaldo Chacon (IBM Canada Ltd), Francis Santerre (IBM Canada Ltd), and Dominique Drouin (3IT-Université de Sherbrooke)

Thermal Characteristics of Vertically-Integrated GaN/SiC-on-Si Assemblies: A Comparative Study 1405

Kimmo Rasilainen (Chalmers University of Technology, Sweden), Per Ingelhag (Ericsson AB, Sweden), Peter Melin (Ericsson AB, Sweden), Torbjörn M. J. Nilsson (Saab AB, Sweden), Mattias Thorsell (Chalmers University of Technology, Sweden and Saab AB, Sweden), and Christian Fager (Chalmers University of Technology, Sweden)

Comprehensive Investigation on Warpage Management of FOPLP with Multi Embedded Ring Designs 1413

Chang-Chun Lee (National Tsing Hua University), Yan-Yu Liou (National
Tsing Hua University), Pei-Chen Huang (National Tsing Hua University),
Fussen Hsu (Unimicron Technology Corporation), Puru Bruce Lin
(Unimicron Technology Corporation), Cheng-Ta Ko (Unimicron Technology
Corporation), and Yu-Hua Chen (Unimicron Technology Corporation)

Session 31: Automotive and Power Packaging

Development of High Power and High Junction Temperature SiC Based Power Packages 1419

Gongyue Tang (Institute of Microelectronics, A*STAR), Leong Ching Wai
(Institute of Microelectronics, A*STAR), Teck Guan Lim (Institute of
Microelectronics, A*STAR), Yong Liang Ye (Institute of
Microelectronics, A*STAR), Pal Singh Ravinder (Institute of
Microelectronics, A*STAR), Lin Bu (Institute of Microelectronics,
A*STAR), Boon Long Lau (Institute of Microelectronics, A*STAR), Tai
Chong Chai (Institute of Microelectronics, A*STAR), Kazunori Yamamoto
(Institute of Microelectronics, A*STAR), and Xiaowu Zhang (Institute
of Microelectronics, A*STAR)

New Developments of Copper Plating Technology for Embedded Power Chip Packages Challenges 1426

Yung-Da Chiu (Advanced Semiconductor Engineering (ASE) Inc.),
Shiu-Chih Wang (Advanced Semiconductor Engineering (ASE) Inc.), David
Tarng (Advanced Semiconductor Engineering (ASE) Inc.), An-Tai Wu
(Advanced Semiconductor Engineering (ASE) Inc.), Allenyl Chen
(Advanced Semiconductor Engineering (ASE) Inc.), Louis Chen (Advanced
Semiconductor Engineering (ASE) Inc.), and Chi-Tsung Chiu (Advanced
Semiconductor Engineering (ASE) Inc.)

Innovative Flip Chip Package Solutions for Automotive Applications 1432

Tom Tang (Siliconware Precision Industries Co., Ltd. Taiwan), Bo-Siang
Fang (Siliconware Precision Industries Co., Ltd. Taiwan), David Ho
(Siliconware Precision Industries Co., Ltd. Taiwan), B.H. Ma
(Siliconware Precision Industries Co., Ltd. Taiwan), Jensen Tsai
(Siliconware Precision Industries Co., Ltd. Taiwan), and Yu-Po Wang
(Siliconware Precision Industries Co., Ltd. Taiwan)

Reliability of Laminated Bond Structure Using (Cu, Ni)/Sn TLP Bonding with Al Interlayer for High
Temperature Power Electronics Packaging 1437

Yanghe Liu (Toyota Research Institute of North America), Shailesh N.
Joshi (Toyota Research Institute North America), and Ercan M. Dede
(Toyota Research Institute North America)

Silver Sintering on Organic Substrates for the Embedding of Power Semiconductor Devices 1443

Alexander Schiffmacher (IMTEK University of Freiburg, Germany),
Juergen Wilde (IMTEK University of Freiburg, Germany), Lorenz
Litzenberger (IMTEK University of Freiburg, Germany), Till Huesgen
(University of Applied Science Kempten, Germany), and Vladimir
Polezhaev (University of Applied Science Kempten, Germany)

High Temperature Resistant Packaging Technology for SiC Power Module by Using Ni Micro-Plating
Bonding 1451

    Kohei Tatsumi (Waseda University), Isamu Morisako (Waseda University),
    Keiko Wada (Waseda University), Minoru Fukuomori (Waseda University),
    Tomonori Iizuka (Waseda University), Nobuaki Sato (Mitsui High-tec
    Inc.), Koji Shimizu (Mitsui High-tec Inc.), Kazutoshi Ueda (Mitsui
    High-tec Inc.), Masayuki Hikita (Kyushu Institute of Technology),
    Rikiya Kamimura (Kitakyushu Foundation for the Advancement of
    Industry, Science and Technology), Naoki Kawanabe (WALTS Co., LTD.),
    Kazuhiko Sugiura (DENSO Corporation), Kazuhiro Tsuruta (DENSO
    Corporation), and Keiji Toda (TOYOTA Motor Corporation)

Pb-Free, High Thermal and Electrical Performance Driven Die Attach Material Development for Power
Packages 1457

    Kim Byong Jin (AMKOR), Dong Su Ryu (AMKOR), HyeongIl Jeon (AMKOR),
    Muhammad Hadhari Hazellah (AMKOR), Weng Tuck Chim (AMKOR), and Jin
    Young Khim (AMKOR)

Session 32: Power and Panel Assembly

An RDL-First Fan-Out Panel-Level Package for Heterogeneous Integration Applications 1463
    Yu-Min Lin (Industrial Technology Research Institute (ITRI), Taiwan),
    Sheng-Tsai Wu (Industrial Technology Research Institute (ITRI),
    Taiwan), Chun-Min Wang (Unimicron Technology Corporation, Taiwan),
    Chia-Hsin Lee (Brewer Science, Taiwan), Shin-Yi Huang (Industrial
    Technology Research Institute (ITRI), Taiwan), Ang-Ying Lin
    (Industrial Technology Research Institute (ITRI), Taiwan), Tao-Chih
    Chang (Industrial Technology Research Institute (ITRI), Taiwan), Puru
    Bruce Lin (Unimicron Technology Corporation, Taiwan), Cheng-Ta Ko
    (Unimicron Technology Corporation, Taiwan), Yu-Hua Chen (Unimicron
    Technology Corporation, Taiwan), Jay Su (Brewer Science, Taiwan), Xiao
    Liu (Brewer Science, USA), Luke Prenger (Brewer Science, USA), and
    Kuan-Neng Chen (National Chiao Tung University, Taiwan)

High Yield Precision Transfer and Assembly of GaN μLEDs Using Laser Assisted Micro Transfer Printing.... 1470
    Goutham Ezhilarasu (University of California Los Angeles), Amir Hanna
    (University of California Los Angeles), Ajit Paranjpe (Veeco
    Instruments Inc., USA), and Subramanian Iyer (University of California
    Los Angeles)

High-Density Flexible Substrate Technology with Thin Chip Embedding and Partial Carrier Release
Option for IoT and Sensor Applications 1475
    Kai Zoschke (Fraunhofer IZM), Piotr Mackowiak (Fraunhofer Institute
    for Reliability and Microintegration), Ha-Duong Ngo (Fraunhofer
    Institute for Reliability and Microintegration), Christian Tschoban
    (Fraunhofer Institute for Reliability and Microintegration), Carola
    Fritsche (Fraunhofer Institute for Reliability and Microintegration),
    Kevin Kröhnert (Fraunhofer Institute for Reliability and
    Microintegration), Thorsten Fischer (Fraunhofer Institute for
    Reliability and Microintegration), Ivan Ndip (Fraunhofer Institute for
    Reliability and Microintegration), and Klaus-Dieter Lang (Technical
    University of Berlin)

Advance Embedded Packaging for Power Discrete Device 1485
  Jia Ren Huo (Wuxi Sky Chip Interconnection Technology co., LTD), Song
  Guan Qiang (Wuxi Sky Chip Interconnection Technology co., LTD), Jing
  Jiang (Wuxi Sky Chip Interconnection Technology co., LTD), Wang Jun
  Tao (Wuxi Sky Chip Interconnection Technology co., LTD), and Ling Wen
  Kong (Wuxi Sky Chip Interconnection Technology co., LTD)

Large Panel Size Bonder with High Performance and High Accuracy 1492
  Hubert Selhofer (Besi Austria GmbH), Andreas Mayr (Besi Austria GmbH),
  and Hugo Pristauz (Besi Austria GmbH)

Advances in high Speed Plating for Vertical Glass Panel Fine-Line Plating 1498
  Christian Dunkel (Semsysco), Herbert Ötzlinger (Semsysco), Onishi
  Tetsuya (GJTech / Semsysco), and Raoul Schroeder (Semsysco)

Study of the Properties of AlN PMUT used as a Wireless Power Receiver 1503
  Dan Gong (Xiamen University), Shenglin Ma (Xiamen University),
  Yihsiang Chiu (Peking University), Hungping Lee (J-Metrics Technology,
  Shenzhen), and Yufeng Jin (Peking University)

Session 33: Fan-Out, Flip Chip, and WLCSP

A Sequential Finite Volume Method / Finite Element Analysis of a Power Electronic Semiconductor Chip..... 1509
  Mario Gschwandl (Polymer Competence Center Leoben GmbH, Austria),
  Peter Filipp Fuchs (Polymer Competence Center Leoben GmbH, Austria),
  Thomas Antretter (Montanuniversitaet Leoben, Austria), Martin Pfost
  (TU Dortmund University), Ivaylo Mitev (Polymer Competence Center
  Leoben GmbH, Austria), Tao Qi (Austria Technologie & Systemtechnik
  Aktiengesellschaft), Thomas Krivec (Austria Technologie &
  Systemtechnik Aktiengesellschaft), Angelika Schingale (CPT Group GmbH,
  Germany), and Michael Decker (CPT Group GmbH, Germany)

Failure Life Prediction of Wafer Level Packaging using DoS with AI Technology 1515
  P. H. Chou (National Tsing Hua University), H. Y. Hsiao (National
  Tsing Hua University), and K.N. Chiang (National Tsing Hua University)

Thermal Cycling Simulation and Sensitivity Analysis of Wafer Level Chip Scale Package with
Integration of Metal-Insulator-Metal Capacitors 1521
  Yi Zhou (Georgia Institute of Technology), Liangbiao Chen (ON
  Semiconductor), Yong Liu (ON Semiconductor), and Suresh Sitaraman
  (Georgia Institute of Technology)

Effect of Time-Dependent Bulk Modulus on Reliability Assessment of Automotive Electronic Control
Unit 1529
  Hyun Seop Lee (University of Maryland), Bongtae Han (University of
  Maryland), and Przemyslaw Gromala (Robert Bosch GmbH, Germany)

Thermal and Mechanical Simulations for Fan-Out Wafer-Level Packaging Technology: Introduction of a
"Solder Heatsink" 1535
  Jean-Philippe Colonna (CEA-Leti, Université Grenoble Alpes, France),
  Loic Marnat (Université Grenoble Alpes), Mathilde Cartier (Université
  Grenoble Alpes), Gabriel Pares (Université Grenoble Alpes), and
  Dominique Noguet (Université Grenoble Alpes)

xxxix

Wafer Level Warpage Modelling and Validation for FOWLP Considering Effects of Viscoelastic Material
Properties Under Process Loadings 1543

zhaohui chen (Institute of Microelectronics, A*STAR (Agency for
Science, Technology and Research)), Xiaowu Zhang (Institute of
Microelectronics, A*STAR (Agency for Science, Technology and
Research)), Sharon Pei Siang Lim (Institute of Microelectronics,
A*STAR (Agency for Science, Technology and Research)), Simon Siak Boon
Lim (Institute of Microelectronics, A*STAR (Agency for Science,
Technology and Research)), Boon Long Lau (Institute of
Microelectronics, A*STAR (Agency for Science, Technology and
Research)), Yong Han (Institute of Microelectronics, A*STAR (Agency
for Science, Technology and Research)), Ming Chinq Jong (Institute of
Microelectronics, A*STAR (Agency for Science, Technology and
Research)), Songlin Liu (A*STAR (Agency for Science, Technology and
Research)), Xiaobai Wang (A*STAR (Agency for Science, Technology and
Research)), and Yosephine Andriani (A*STAR (Agency for Science,
Technology and Research))

Ultra-Thin Package Board Level Drop Impact Modeling and Validation 1550

Shu-Shen Yeh (Taiwan Semiconductor Manufacturing Company (TSMC)), P.
Y. Lin (Taiwan Semiconductor Manufacturing Company (TSMC)), M. C. Yew
(Taiwan Semiconductor Manufacturing Company (TSMC)), W. Y. Lin (Taiwan
Semiconductor Manufacturing Company (TSMC)), K. C. Lee (Taiwan
Semiconductor Manufacturing Company (TSMC)), C. C. Yang (Taiwan
Semiconductor Manufacturing Company (TSMC)), J. H. Wang (Taiwan
Semiconductor Manufacturing Company (TSMC)), P. C. Lai (Taiwan
Semiconductor Manufacturing Company (TSMC)), C. K. Hsu (Taiwan
Semiconductor Manufacturing Company (TSMC)), and Shin-Puu Jeng (Taiwan
Semiconductor Manufacturing Company (TSMC))

Session 34: Emerging Materials and Processing

Flexible Graphene-Glass Fiber Composite Film with Ultrahigh Thermal Conductivity and Mechanical
Strength as Highly Efficient Thermal Spreader Materials 1556

Xiaoliang Zeng (Center for Advanced Material Research Shenzhen
Institutes of Advanced Technology, Chinese Academy of Sciences),
Linlin Ren (Center for Advanced Material Research Shenzhen Institutes
of Advanced Technology, Chinese Academy of Sciences), Rong Sun (Center
for Advanced Material Research Shenzhen Institutes of Advanced
Technology, Chinese Academy of Sciences), Jianbin Xu (Center for
Advanced Material Research Shenzhen Institutes of Advanced Technology,
Chinese Academy of Sciences), and Ching-Ping Wong (School of Materials
Science and Engineering Georgia Institute of Technology Atlanta, USA)

Highly Thermal Conductive and Electrically Insulated Graphene Based Thermal Interface Material with
Long-Term Reliability 1564

Nan Wang (SHT Smart High Tech AB), Ya Liu (Chalmers University of
Technology), Shujing Chen (Shanghai University), Lilei Ye (SHT Smart
High Tech AB), and Johan Liu (Chalmers University of Technology)

Further Enhancement of Thermal Conductivity through Optimal Uses of h-BN Fillers in Polymer-Based
Thermal Interface Material for Power Electronics 1569
    Han Jiang (Loughborough University), Han Zhou (Loughborough
    University), Stuart Robertson (Loughborough University), Zhaoxia Zhou
    (Loughborough University), Liguo Zhao (Loughborough University), and
    Changqing Liu (Loughborough University)

Wafer Level Integration of Thin Silicon Bare Dies Within Flexible Label 1575
    Jean-Charles Souriau (Univ. Grenoble Alpes, CEA, LETI), Ahmad Itawi
    (Univ. Grenoble Alpes, CEA, LETI), and Laetitia Castagné (Univ.
    Grenoble Alpes, CEA, LETI)

Laser Sintering of Aerosol Jet Printed Conductive Interconnects on Paper Substrate 1581
    Mohammed Alhendi (Binghamton University), Rajesh S. Sivasubramony
    (Binghamton University), Jack Lombardi (Binghamton University),
    Darshana L. Weerawarne (Binghamton University), Peter Borgesen
    (Binghamton University), Mark D. Poliks (Binghamton University), and
    Azar Alizadeh (General Electric Global Research)

In-Situ Investigation of Organic Additive Interactions in Copper Electroplating Solutions with
Surface Enhanced Raman Spectroscopy (SERS) 1588
    Nithin Nedumthakady (Georgia Institute of Technology), Bartlet
    DeProspo (Georgia Institute of Technology), Himani Sharma (Georgia
    Institute of Technology), Rahul Manepalli (Intel Corporation), Sashi
    Kandanur (Intel Corporation), Sajanlal Panikkanvalappil (Georgia
    Institute of Technology), Nasrin Hooshmand (Georgia Institute of
    Technology), and Rao Tummala (Georgia Institute of Technology)

C4 Compatible Ultra-Thick Cu On-chip Magnetic Inductor Architecture Integrated with Advanced
Polymer/Cu Planarization Process 1595
    C.H. Kuo (Taiwan Semiconductor Manufacturing Company, Ltd.), S.B. Yang
    (Taiwan Semiconductor Manufacturing Company, Ltd.), C.C. Kuo (Taiwan
    Semiconductor Manufacturing Company, Ltd.), Y.N. Chen (Taiwan
    Semiconductor Manufacturing Company, Ltd.), K.S. Yuan (Taiwan
    Semiconductor Manufacturing Company, Ltd.), G.C. Huang (Taiwan
    Semiconductor Manufacturing Company, Ltd.), C.N. Ke (Taiwan
    Semiconductor Manufacturing Company, Ltd.), Grace Chang (Taiwan
    Semiconductor Manufacturing Company, Ltd.), C.C. Hsu (Taiwan
    Semiconductor Manufacturing Company, Ltd.), H.L. Huang (Taiwan
    Semiconductor Manufacturing Company, Ltd.), Kirin Wang (Taiwan
    Semiconductor Manufacturing Company, Ltd.), Harry Ku (Taiwan
    Semiconductor Manufacturing Company, Ltd.), C.S. Chen (Taiwan
    Semiconductor Manufacturing Company, Ltd.), K.C. Liu (Taiwan
    Semiconductor Manufacturing Company, Ltd.), Alex Kalnitsky (Taiwan
    Semiconductor Manufacturing Company, Ltd.), and Marvin Liao (Taiwan
    Semiconductor Manufacturing Company, Ltd.)

Session 35: New Interconnects for Package Scaling

Development of 2.3D High Density Organic Package using Low Temperature Bonding Process with Sn-Bi Solder 1599
    Shota Miki (SHINKO ELECTRIC INDUSTRIES CO., LTD.), Hiroshi Taneda
    (SHINKO ELECTRIC INDUSTRIES CO., LTD.), Naoki Kobayashi (SHINKO
    ELECTRIC INDUSTRIES CO., LTD.), Kiyoshi Oi (SHINKO ELECTRIC INDUSTRIES
    CO., LTD.), Koji Nagai (SHINKO ELECTRIC INDUSTRIES CO., LTD.), and
    Toshinori Koyama (SHINKO ELECTRIC INDUSTRIES CO., LTD.)

PowerTherm Attach Process for Power Delivery and Heat Extraction in the Silicon-Interconnect Fabric Using Thermocompression Bonding 1605
    Pranav Ambhore (University of California, Los Angeles), Umesha Mogera
    (University of California, Los Angeles), Boris Vaisband (University of
    California, Los Angeles), Ujash Shah (University of California, Los
    Angeles), Timothy Fisher (University of California, Los Angeles), Mark
    Goorsky (University of California, Los Angeles), and Subramanian S.
    Iyer (University of California, Los Angeles)

Interconnect Scheme for Die-to-Die and Die-to-Wafer-Level Heterogeneous Integration for High-Performance Computing 1611
    Rabindra Das (MIT Lincoln Laboratory), Vladimir Bolkhovsky (MIT
    Lincoln Laboratory), Christopher Galbraith (MIT Lincoln Laboratory),
    Daniel Oates (MIT Lincoln Laboratory), Jason Plant (MIT Lincoln
    Laboratory), Renée Lambert (MIT Lincoln Laboratory), Scott Zarr (MIT
    Lincoln Laboratory), Ravi Rastogi (MIT Lincoln Laboratory), Dmitri
    Shapiro (MIT Lincoln laboratory), Manuel Docanto (MIT Lincoln
    Laboratory), Terence Weir (MIT Lincoln laboratory), and Leonard
    Johnson (MIT Lincoln Laboratory)

Ultra Wide Micro Bumps Interconnection Matrix for High Energy Particle Detection: Process and Assembly 1622
    Jean Charbonnier (CEA Leti), Myriam Assous (CEA-Leti), Thierry Mourier
    (CEA-Leti), Céline Ribière (CEA-Leti), Stéphane Minoret (CEA-Leti),
    Sophie Verrun (CEA-Leti), Pierre Tissier (CEA-Leti), Rémi Coquand
    (CEA-Leti), Mehmet Bicer (CEA-Leti), Fabienne Allain (CEA-Leti), Rémi
    Franiatte (CEA-Leti), and Gabriel Pares (CEA-Leti)

Growth Behavior and Orientation Evolution of Cu6Sn5 Grains in Micro Interconnect During Isothermal Reflow 1629
    S. Chen (Dalian University of Technology), N. Zhao (Dalian University
    of Technology), Y.Y. Qiao (Dalian University of Technology), Y.P. Wang
    (Dalian University of Technology), H.T. Ma (Dalian University of
    Technology), and C.M.L. Wu (City University of Hong Kong)

Development of a no Reflow Cu Pillar Bump to Improve Chip/Package Interactions (CPI) Process and Reliability Performance 1635
    Kuei Hsiao (Frank) Kuo (Siliconware Precision Industries Co., Ltd.
    (SPIL)), Jiunn Jie Wang (Siliconware Precision Industries Co., Ltd.
    (SPIL)), Yen Neng Wang (Siliconware Precision Industries Co., Ltd.
    (SPIL)), Feng Lung Chien (Siliconware Precision Industries Co., Ltd.
    (SPIL)), and Rick Lee (Siliconware Precision Industries Co., Ltd.
    (SPIL))

A Novel Interconnection Technology Using Ultra-Thin Under Barrier Metal for Multiple Chip-on-Chip Stacking Structure 1641

Takuya Nakamura (Sony Semiconductor Solutions), Kan Shimizu (Sony Semiconductor Solutions), Masataka Maehara (Sony Semiconductor Solutions), Toshihiko Hayashi (Sony Semiconductor Solutions), Kentaro Akiyama (Sony Semiconductor Solutions), Junichiro Fujimagari (Sony Semiconductor Solutions), Tomohiro Ohkubo (Sony Semiconductor Manufacturing), Atsushi Fujiwara (Sony Semiconductor Manufacturing), and Hayato Iwamoto (Sony Semiconductor Solutions)

Session 36: RF & Power Components and Modules

Multilayer Decoupling Capacitor using Stacked Layers of BST and LNO 1647
Todd Schumann (University of Florida), Sheng-Po Fang (University of Florida), Yong-Kyu Yoon (University of Florida), Jongmin Yook (Korea Electronics Technology Institute), and Dongsu Kim (Korea Electronics Technology Institute)

System Co-Design of a High Current (40A) Synchronous Step-Down Converter in an Innovative Multi-chip Module (MCM) LQFN-Type Packaging Technology 1653
Todd Harrison (Texas Instruments, Inc.), Jie Chen (Texas Instruments, Inc.), and Rajen Murugan (Texas Instruments, Inc.)

Integrating Solid State Protection with a RF-MEMS Switch for Achieving ESD Robustness 1660
Srivatsan Parthasarathy (Analog Devices, USA), Padraig Fitzgerald (Analog Devices, Ireland), Javier Salcedo (Analog Devices, USA), Ray Goggin (Analog Devices, Ireland), and Jean-Jacques Hajjar (Analog Devices, USA)

A Zero Height Small Size Low Cost RF Interconnect Substrate Technology For RF Front Ends For M.2 Modules And SiP 1666
Sidharth Dalmia (Intel Corporation), Kirthika Nahalingam (Intel Corporation), Swathi Vijayakumar (Intel Corporation), and Pouya Talebbeydokhti (Intel Corporation)

Open and Closed Loop Inductors for High-Efficiency System-on-Package Integrated Voltage Regulators 1672
Claudio Alvarez (Georgia Institute of Technology), Mohamed Bellaredj (Georgia Institute of Technology), and Madhavan Swaminathan (Georgia Institute of Technology)

RF Inductors Integrated in Organic Packaging 1680
Denis Mercier (CEA-Leti), Jean-Philippe Michel (CEA-Leti), Christine Raynaud (CEA-Leti), and Christophe Billard (CEA-Leti)

3D Printed Interposer Layer for High Density Packaging of IoT Devices 1687
Saikat Mondal (Michigan State University), Mohd. Ifwat Mohd. Ghazali (Universiti Sains Islam Malaysia), Kanishka Wijewardena (Michigan State University), Deepak Kumar (Michigan State University), and Premjeet Chahal (Michigan State University)

xliii

Session 37: Interactive Presentations 1

Comprehensive Solution for Micro Bump Coplanarity Control 1693
Chun-Chen Liu (Taiwan Semiconductor Manufacturing Company), J.H. Chen (Taiwan Semiconductor Manufacturing Company), Y.N. Hsu (Taiwan Semiconductor Manufacturing Company), Rung-De Wang (Taiwan Semiconductor Manufacturing Company), Yu-Cheng Wang (Taiwan Semiconductor Manufacturing Company), Bin-En Ho (Taiwan Semiconductor Manufacturing Company), Y.H. Wu (Taiwan Semiconductor Manufacturing Company), Ponder Pan (Taiwan Semiconductor Manufacturing Company), Harry Ku (Taiwan Semiconductor Manufacturing Company), Kirin Wang (Taiwan Semiconductor Manufacturing Company), Calvin Lu (Taiwan Semiconductor Manufacturing Company), K.C. Liu (Taiwan Semiconductor Manufacturing Company), and Marvin Liao (Taiwan Semiconductor Manufacturing Company)

Structural Enhancement for a CMOS-MEMS Microphone Under Thermal Loading by Taguchi Method 1697
Chun-Lin Lu (National Tsing Hua University) and Meng-Kao Yeh (National Tsing Hua University)

A Methodology to Correct in-Fixture Measurement of Impedance by a Machine Learning Model 1704
Bo-Siang Fang (Siliconware Precision Industries Co., Ltd. (SPIL), Taiwan), Chia-Chu Lai (Siliconware Precision Industries Co., Ltd. (SPIL), Taiwan), Ying-Wei Lu (Siliconware Precision Industries Co., Ltd. (SPIL), Taiwan), Kuan-Ta Chen (Siliconware Precision Industries Co., Ltd. (SPIL), Taiwan), Mike Tasi (Siliconware Precision Industries Co., Ltd. (SPIL), Taiwan), and Don-Son Jiang (Siliconware Precision Industries Co., Ltd. (SPIL), Taiwan)

Material and Structure Design Optimization for Panel-Level Fan-Out Packaging 1710
Dao-Long Chen (Advanced Semiconductor Engineering, Inc.), Ian Hu (Advanced Semiconductor Engineering, Inc.), KarenYU Chen (Advanced Semiconductor Engineering, Inc.), Meng-Kai Shih (Advanced Semiconductor Engineering, Inc.), David Tarng (Advanced Semiconductor Engineering, Inc.), Dinos Huang (Advanced Semiconductor Engineering, Inc.), and JY On (Advanced Semiconductor Engineering, Inc.)

The Microstructure and Mechanical Property of the High Entropy Alloy as a low Temperature Solder 1716
Li Pu (Beijing Institute of Technology), Quanfeng He (City University of Hong Kong), Yong Yang (City University of Hong Kong), Xiuchen Zhao (Beijing Institute of Technology), Zhuangzhuang Hou (Beijing Institute of Technology), K. N. Tu (University of California, USA), and Yingia Liu (Beijing Institute of Technology)

A Versatile Fan-Out Infrastructure Based on Die-Stencil Substrate Promoted by an Advanced Multifunctional Temporary Bonding Material 1722
Xiao Liu (Brewer Science, Inc.), Baron Huang (Brewer Science, Inc.), Hong Zhang (Brewer Science, Inc.), Lisa Kirchner (Brewer Science, Inc.), Arthur Southard (Brewer Science, Inc.), Rama Puligadda (Brewer Science, Inc.), and Tony Flaim (Brewer Science, Inc.)

Low Temperature and Pressureless Microfluidic Electroless Bonding Process for Vertical Interconnections 1729
Han-Tang Hung (National Taiwan University), Sean Yang (National Taiwan University), I-An Weng (National Taiwan University), Yan-Hao Chen (Unimicron Technology Corporation, Taiwan), and C. Robert Kao (National Taiwan University)

xliv

3D Integration of CMOS-Compatible Surface Electrode Ion Trap and Silicon Photonics for Scalable Quantum Computing 1735

Jing Tao (Nanyang Technological University), Yu Dian Lim (Nanyang Technological University), Hong Yu Li (Agency for Science, Technology and Research (A*STAR)), Nam Piau Chew (Nanyang Technological University), Anak Agung Alit Apriyana (Agency for Science, Technology and Research (A*STAR)), Lin Bu (Agency for Science, Technology and Research (A*STAR)), Peng Zhao (Nanyang Technological University), Luca Guidoni (Université Paris Diderot), and Chuan Seng Tan (Nanyang Technological University)

Integrated RTD Sensors for Maintaining Thermal Uniformity During TCB Process 1744

Salwa Ben Jemaa (Interdisciplinary Institute for Technological Innovation (3IT) Sherbrooke University), Julien Sylvestre (Interdisciplinary Institute for Technological Innovation (3IT) Sherbrooke University), and Pascale Gagnon (IBM Canada Bromont, QC, Canada)

Wireless Transfer of Power and Data Via a Single Resonant Inductive Link 1751

Shiang-Hwua Yu (National Sun Yat-sen University), Yi-Chen Hsieh (National Sun Yat-sen University), Chin-Wei Chan (National Sun Yat-sen University), I-Fang Lo (National Sun Yat-sen University), Heri Suryoatmojo (Institut Teknologi Sepuluh), and Lih-Tyng Hwang (National Sun Yat-sen University)

Adaptive Patterning of Optical and Electrical Fan-Out for Photonic Chip Packaging 1757

Ahmed Elmogi (Centre for Microsystems Technology, imec and Ghent University), Andres Desmet (Centre for Microsystems Technology, imec and Ghent University), Jeroen Missinne (Centre for Microsystems Technology, imec and Ghent University), Hannes Ramon (Ghent University-imec), Joris Lambrecht (Ghent University-imec), Peter De Heyn (imec), Marianna Pantouvaki (imec), Joris Van Campenhout (imec), Johan Bauwelinck (Ghent University-imec), and Geert Van Steenberge (imec and Ghent University)

Low Surface Reflectance Structure at Near Infrared Wavelength by Injection Molding 1764

Sho Yakabe (Sumitomo Electric Industries, Ltd.), Takuro Watanabe (Sumitomo Electric Industries, Ltd.), Takayuki Shimazu (Sumitomo Electric Industries, Ltd.), Ryohei Hokari (National Institute of Advanced Industrial Science and Technology), and Kazuma Kurihara (National Institute of Advanced Industrial Science and Technology)

A Novel Design of a Bandwidth Enhanced Dual-Band Impedance Matching Network with Coupled Line Wave Slowing 1770

Deepayan Banerjee (IIIT Delhi, India), Antra Saxena (IIIT Delhi, India), and Mohammad Hashmi (Nazarbayev University, Kazakhstan)

Effects of Electromigration on Microstructural Evolution and Mechanical Properties of Preferential Growth Intermetallic Compound Interconnects for 3D Packaging 1774

Mingliang L. Huang (Dalian University of Technology) and Lin Zou (Dalian University of Technology)

Telemetry for Implantable Biosensors 1782

Ryan B. Green (Virginia Commonwealth University) and Erdem Topsakal (Virginia Commonwealth University)

xlv

Ultra-Thin QFN-Like 3D Package with 3D Integrated Passive Devices 1789

Ayad Ghannam (3DiS Technologies S.A.S, France), Niek van Haare (Besi Netherlands, B.V., Netherlands), Julian Bravin (EV Group E.Thallner GmbH, Austria), Elisabeth Brandl (EV Group E.Thallner GmbH, Austria), Birgit Brandstätter (Besi Austria GmbH, Austria), Hannes Klingler (Besi Austria GmbH, Austria), Benedikt Auer (Besi Austria GmbH, Austria), Philippe Meunier (NXP Semiconductors, France), and Sebastiaan Kersjes (Besi Netherlands, B.V., Netherlands)

Low-Cost Non-TSV Based 3D Packaging Using Glass Panel Embedding (GPE) for Power-Efficient, High-Bandwidth Heterogeneous Integration 1796

Siddharth Ravichandran (Georgia Institute of Technology), Shuhei Yamada (Murata Manufacturing Co. Ltd, Kyoto, Japan), Fuhan Liu (Georgia Institute of Technology), Vanessa Smet (Georgia Institute of Technology), Mohanalingam Kathaperumal (Georgia Institute of Technology), and Rao Tummala (Georgia Institute of Technology)

Polylithic Integration of 2.5D and 3D Chiplets Using Interconnect Stitching 1803

Paul K. Jo (Georgia Institute of Technology), Ting Zheng (Georgia Institute of Technology), and Muhannad S. Bakir (Georgia Institute of Technology)

Characterization of the Current Mechanisms and Improved Leakage Current in Silver Doped Barium Strontium Titanate 1809

Todd Schumann (University of Florida), Kyoung-Tae Kim (University of Florida), Sheng-Po Fang (University of Florida), and Yong-Kyu Yoon (University of Florida)

High Temperature Aging Effects in SAC and SAC+X Lead Free Solders 1815

Mohammad S. Alam (Auburn University), KM Rafidh Hassan (Auburn University), Jeffrey C. Suhling (Auburn University), and Pradeep Lall (Auburn University)

Session 38: Interactive Presentations 2

Laundering Reliability of Electrically Conductive Fabrics for E-Textile Applications 1826
Jeffrey ChangBing Lee (iST-Integrated Service Technology Inc.), Weifeng Liu (FLEX Ltd.), ChangHo Lo (iST-Integrated Service Technology Inc.), and Cheng-Chih Chen (iST-Integrated Service Technology Inc.)

Preconditioning Technologies for Sputtered Seed Layers in FOPLP 1833
Johannes Weichart (Evatec AG), Jüergen Weichart (Evatec AG), Andreas Erhart (Evatec AG), and Kay Viehweger (Fraunhofer IZM ASSID)

Impact of Thermal Boundary Resistance on the Thermal Design of GaN-on-Diamond HEMTs 1842
Huaixin Guo (Nanjing Electronic Devices Institute), Yuechan Kong (Nanjing Electronic Devices Institute), and Tangsheng Chen (Nanjing Electronic Devices Institute)

Measuring the Electric Properties of Thin Film Shape Memory Polymers in Simulated Physiological Conditions 1848
Daniel Del Nero (The University of Texas at Dallas), Alexandra Joshi-Imre (The University of Texas at Dallas), and Walter Voit (The University of Texas at Dallas)

Evaluation of WLP Dielectrics for High Voltage Applications 1853
  Markus Wöhrmann (Fraunhofer IZM), Michael Toepper (Fraunhofer IZM),
  Marcus Paeck (Fraunhofer IZM), and Klaus-Dieter Lang (Technical
  University Berlin)

Mitigating the Effects of Microvortices in high-Re Deterministic Lateral Displacement by Using
Symmetric Airfoil-Shaped Pillars .... N/A
  Brian Dincau (Washington State University Vancouver), Kawkab Ahasan
  (Washington State University Vancouver), and Jong-Hoon Kim (Washington
  State University Vancouver)

Plasma Dry Process Technology Development of Glass-Epoxy Film on the Silicon Substrate to Fabricate
RDL for Future GPU/AI Application 1865
  Takahide Murayama (ULVAC, Inc.), Muneyuki Sato (ULVAC, Inc.), Akiyoshi
  Suzuki (ULVAC, Inc.), Atsuhito Ihori (ULVAC, Inc.), Tetsushi Fujinaga
  (ULVAC, Inc.), and Yasuhiro Morikawa (ULVAC, Inc.)

Fully Solid-State Integrated Capacitors Based on Carbon Nanofibers and Dielectrics with Specific
Capacitances Higher Than 200 nF/mm2 1870
  Amin Saleem (Smoltek AB, Sweden), Rickard Andersson (Smoltek AB,
  Sweden), Maria Bylund (Smoltek AB, Sweden), Charlotte Goemare (Smoltek
  AB, Sweden), Guilhem Pacot (Smoltek AB, Sweden), Mohammed Kabir
  (Smoltek AB, Sweden), and Vincent Desmaris (Smoltek AB, Sweden)

Application of Fan-Out Panel Level Packaging Techniques for Flexible Hybrid Electronics Systems 1877
  Wei-Yuan Cheng (ITRI), Shau-Fei Cheng (ITRI), Chen-Tsai Yang (ITRI),
  Shau-Fei Cheng (ITRI), Wei-Han Chen (ITRI), Hsin-Cheng Lai (ITRI),
  Tai-Jui Wang (ITRI), and Yuh-Zheng Lee (ITRI)

Structuring of Laser Activated Polymers for Sensor Applications 1883
  Sebastian Bengsch (University Hanover), Marc Christopher Wurz
  (University Hanover), Kevin Cromwell (University Hanover), and
  Maximilian Aue (University Hanover)

A Deep Learning Approach for Volterra Kernel Extraction for Time Domain Simulation of Weakly
Nonlinear Circuits 1889
  Thong Nguyen (University of Illinois at Urbana Champaign), Xinying
  Wang (University of Illinois at Urbana Champaign), Xu Chen (University
  of Illinois at Urbana Champaign), and Jose Schutt-Aine (University of
  Illinois at Urbana Champaign)

224G Package Interconnect Design Study - Based on Artificial Neural Network Modeling Approach 1897
  Hui Liu (Intel Corporation), Qian Ding (Intel Corporation), and
  Penglin Liu (Intel Corporation)

Enhanced Reliability of a RF-SiP with Mold Encapsulation and EMI Shielding 1902
  Chan-Yuan Liu (Advanced Semiconductor Engineering, Inc., Taiwan),
  Jason Chien (Advanced Semiconductor Engineering, Inc., Taiwan),
  Yu-Chou Tseng (Advanced Semiconductor Engineering, Inc., Taiwan),
  Kuo-Hsien Liao (Advanced Semiconductor Engineering, Inc., Taiwan),
  Alex Chan (Advanced Semiconductor Engineering, Inc., Taiwan), Dao-Long
  Chen (Advanced Semiconductor Engineering, Inc., Taiwan), Meng-Kai Shih
  (Advanced Semiconductor Engineering, Inc., Taiwan), and Mark Gerber
  (Advanced Semiconductor Engineering, Inc., U.S.)

Study of the Effect and Mechanism of a Cap Layer in Controlling the Statistical Variation of Via Extrusion 1909

Golareh Jalilvand (University of Central Florida) and Tengfei Jiang (University of Central Florida)

Three Dimensional Copper Foam-Filled Elastic Conductive Composites with Simultaneously Enhanced Mechanical, Electrical, Thermal and Electromagnetic Interference (EMI) Shielding Properties 1916

Tan Lu (Shenzhen Institutes of Advanced Technology, Chinese Academy of Sciences), Han Gu (Shenzhen Institutes of Advanced Technology, Chinese Academy of Sciences), Yougen Hu (Shenzhen Institutes of Advanced Technology, Chinese Academy of Sciences), Tao Zhao (Shenzhen Institutes of Advanced Technology, Chinese Academy of Sciences), Pengli Zhu (Shenzhen Institutes of Advanced Technology, Chinese Academy of Sciences), Rong Sun (Shenzhen Institutes of Advanced Technology, Chinese Academy of Sciences), and Ching-Ping Wong (Georgia Institute of Technology)

Vertical Interconnect Technology for Enlarging Capacity on Micro Solid Thin Film Rechargeable Battery 1921

Akihiro Horibe (IBM Research - Tokyo), Kuniaki Sueoka (IBM Research - Tokyo), Takahiro Mori (IBM Research - Tokyo), Risa Miyazawa (IBM Research - Tokyo), and Hiroyuki Mori (IBM Research - Tokyo)

Characterization of Fine Pitch Hybrid Bonding Pads using Electrical Misalignment Test Vehicle 1926

Imed Jani (CEA, LETI), Didier Lattard (CEA, LETI), Pascal Vivet (CEA, LETI), Lucile Arnaud (CEA, LETI), Severine Cheramy (CEA, LETI), Edith Beigné (CEA, LETI), Alexis Farcy (STMicroelectronics), Joris Jourdon (STMicroelectronics), Yann Henrion (STMicroelectronics), Emilie Deloffre (STMicroelectronics), and Halim Bilgen (STMicroelectronics)

Dynamic Characteristics Evaluation on NCF Under Challenging Conditions and Its Application 1933

Tomonori Nakamura (Shinkawa LTD, Japan), Hiromi Shibahara (Shinkawa LTD, Japan), Osamu Watanabe (Shinkawa LTD, Japan), Tetsuya Utano (Shinkawa LTD, Japan), Daisuke Tani (Shinkawa LTD, Japan), Sung Chenhsiu (Shinkawa LTD, Japan), Toru Maeda (Shinkawa LTD, Japan), Doug Day (Shinkawa LTD, Japan), Hidekazu Yagi (Dexerials Corporation, Japan), Ryoji Kojima (Dexerials Corporation, Japan), Daichi Mori (Dexerials Corporation, Japan), Tatsuo Nagamatsu (Dexerials Corporation, Japan), and Junichi Kaneko (Dexerials Corporation, Japan)

Study of Electrical and Mechanical Characteristics of Inkjet-Printed Patch Antenna Under Uniaxial and Biaxial Bending 1939

Yi Zhou (Georgia Institute of Technology), Sridhar Sivapurapu (Georgia Institute of Technology), Rui Chen (Georgia Institute of Technology), Nahid Aslani Amoli (Georgia Institute of Technology), Mohamed Bellaredj (Georgia Institute of Technology), Madhavan Swaminathan (Georgia Institute of Technology), and Suresh K. Sitaraman (Georgia Institute of Technology)

Effects of Oven and Laser Sintering Parameters on the Electrical Resistance of IJP Nano-Silver Traces on Mesoporous PET Before and During Fatigue Cycling 1946

G.S. Khinda (SUNY Binghamton), M.Z. Kokash (SUNY Binghamton), M. Alhendi (SUNY Binghamton), M. Yadav (SUNY Binghamton), J.P. Lombardi (SUNY Binghamton), D.L. Weerawarne (SUNY Binghamton), Mark D. Poliks (SUNY Binghamton), P. Borgesen (SUNY Binghamton), and Nancy C. Stoffel (General Electric Global Research Center)

Multilayer Glass Substrate with High Density Via Structure for All Inorganic Multi-chip Module 1952
    Toshiki Iwai (FUJITSU LABORATORIES LTD.), Taiji Sakai (FUJITSU
    LABORATORIES LTD.), Daisuke Mizutani (FUJITSU LABORATORIES LTD.),
    Seiki Sakuyama (FUJITSU LABORATORIES LTD.), Kenji Iida (FUJITSU
    INTERCONNECT TECHNOLOGIES LIMITED), Takayuki Inaba (FUJITSU
    INTERCONNECT TECHNOLOGIES LIMITED), Hidehiko Fujisaki (FUJITSU
    INTERCONNECT TECHNOLOGIES LIMITED), Akira Tamura (FUJITSU INTERCONNECT
    TECHNOLOGIES LIMITED), and Yoshinori Miyazawa (FUJITSU INTERCONNECT
    TECHNOLOGIES LIMITED)

The Poisson's Ratio of Lead Free Solder - The Often Forgotten But Important Material Property 1958
    KM Rafidh Hassan (Auburn University), Mohammad S. Alam (Auburn
    University), Jeffrey C. Suhling (Auburn University), and Pradeep Lall
    (Auburn University)

Additive Laser Metal Deposition Onto Silicon for Enhanced Microelectronics Cooling 1970
    Arad Azizi (Binghamton University (SUNY)), Matthias A. Daeumer
    (Binghamton University (SUNY)), Jacob C. Simmons (Binghamton
    University (SUNY)), Bahgat G. Sammakia (Binghamton University (SUNY)),
    Bruce T. Murray (Binghamton University (SUNY)), and Scott N. Schiffres
    (Binghamton University (SUNY))

Moisture Barrier, Mechanical, and Thermal Properties of PDMS-PIB Blends for Solar Photovoltaic (PV)
Module Encapsulant 1977
    Jinho Hah (Georgia Institute of Technology), Michael Sulkis (Georgia
    Institute of Technology), Chao Ren (Georgia Institute of Technology),
    Minsoo Kang (Georgia Institute of Technology), Kyoung-sik Moon
    (Georgia Institute of Technology), Samuel Graham (Georgia Institute of
    Technology), and C. P. Wong (Georgia Institute of Technology)

Session 39: Interactive Presentations 3

Modeling and Design of Power Distribution Network for a Heterogeneous Integrated Active Interposer
with Neuromorphic Computing Circuits 1983
    Min Miao (Beijing Information Science and Technology University),
    Tianfang Chen (Beijing Information Science and Technology University),
    Yang Yang (Peking University Shenzhen Graduate School), Jincan Zhang
    (Beijing Information Science and Technology University), Na Li
    (Beijing Information Science and Technology University), Kunkun Li
    (Beijing Information Science and Technology University), Liyuan Wang
    (Beijing Information Science and Technology University), Huan Liu
    (Peking University), Xiaole Cui (Peking University Shenzhen Graduate
    School), and Yufeng Jin (Peking University Shenzhen Graduate School)

PCB Microstrip Line Far-End Crosstalk Mitigation by Surface Mount Capacitors 1989
    Zhaoqing Chen (IBM Corporation)

New Cost-Effective Via-Last Approach by "One-Step TSV" After Wafer Stacking for 3D Memory
Applications 1996
    Masaya Kawano (Institute of Microelectronics, A*STAR), Xiangy-Yu Wang
    (Institute of Microelectronics, A*STAR), and Qin Ren (Institute of
    Microelectronics, A*STAR)

Microstructure and Property Changes in Cu/Sn-58Bi/Cu Solder Joints During Thermomigration 2003
Yu-An Shen (Joining and Welding Research Institute (JWRI), Osaka University), Shiqi Zhou (Osaka University), Jiahui Li (City University of Hong Kong), K. N. Tu (UCLA), and Hiroshi Nishikawa (Joining and Welding Research Institute (JWRI), Osaka University)

Simulation and Experimental Validations of EM/TM/SM Physical Reliability for Interconnects Utilized in Stretchable and Foldable Electronics 2009
Chang-Chun Lee (National Tsing Hua University), Oscar Chuang (National Tsing Hua University), Chia-Ping Hsieh (National Taiwan University), Wei-Yuan Cheng (Industrial Technology Research Institute), and Steve Chiu (Industrial Technology Research Institute)

A Complex Integrated Circuit Structure Transformation, Modeling and Simulation Method 2016
Daixing Wang (Peking University Shenzhen Graduate School), Wei Wang (Institute of Microelectronics Peking University), and Yufeng Jin (Peking University Shenzhen Graduate School)

A Study on the Oxygen Plasma Treatment on the Peel Adhesion Strength and Solder Wettability of SnBi58 Based Anisotropic Conductive Films 2022
Shuye Zhang (Harbin Institute of Technology), Mingliang Huang (Dalian University of Technology), Yang Wu (Dalian University of Technology), Ming Yang (Hisilicon Optoelectronics Co., Ltd), Tiesong Lin (Harbin Institute of Technology), Peng He (Harbin Institute of Technology), and Kyung-Wook Paik (Nano-Packaging and Interconnection Laboratory)

Numerical Analysis of the Influence of Polymeric Materials on a MEMS Package Performance Under Humidity and Temperature Loads 2029
Mahesh Yalagach (Polymer Competence Center Leoben GmbH, Leoben, Austria.), Peter Filipp Fuchs (Polymer Competence Center Leoben GmbH, Leoben, Austria.), Archim Wolfberger (Polymer Competence Center Leoben GmbH, Leoben, Austria.), Mario Gschwandl (Polymer Competence Center Leoben GmbH, Leoben, Austria.), Thomas Antretter (Montanuniversitaet Leoben, Institute of Mechanics, Leoben, Austria.), Michael Feuchter (Montanuniversitaet Leoben, Institute of Material Science and Testing of Polymers, Leoben, Austria.), Coen Tak (ams AG, Premstaetten, Austria), and Qi Tao (Austria Technologie & Systemtechnik Aktiengesellschaft, Leoben, Austria.)

Electromigration-Induced -Sn Grain Rotation in Lead-Free Flip Chip Solder Bumps 2036
Mingliang L. Huang (Dalian University of Technology), Jiameng M. Kuang (Dalian University of Technology), and Hongyu Y. Sun (Dalian University of Technology)

Low-Cost MT-Ferrule-Compatible Optical Connector for Co-packaged Optics Using Single-Mode Polymer Waveguide 2042
Akihiro Noriki (National Institute of Advanced Industrial Science and Technology (AIST)), Takeru Amano (National Institute of Advanced Industrial Science and Technology (AIST)), Masatoshi Tsunoda (Kyocera Corporation), and Toshiaki Michihiro (Kyocera Corporation)

Characterization of Coated Silver Wire Bond Interface Using TEM 2048
Murali Sarangapani (Heraeus Materials Singapore Pte. Ltd.,), Eric Tan Swee Seng (Heraeus Materials Singapore Pte. Ltd.,), and Jason Wong Chin Yeung (Heraeus Materials Singapore Pte. Ltd.,)

Research on Applied Reliability of BGA Solder Balls in Extreme Marine Environment 2054
  Liyuan Liu (China Electronic Product Reliability and Environmental
  Testing Research Institute), Tao Lu (China Electronic Product
  Reliability and Environmental Testing Research Institute), Daojun Luo
  (China Electronic Product Reliability and Environmental Testing
  Research Institute), and Hui Xiao (China Electronic Product
  Reliability and Environmental Testing Research Institute)

Influence of Single/Double Sweeping Mode and Sweeping Voltage Increment/Polarity on Measurement of
TSV Leakage Current 2061
  Qinghua Zeng (Peking University), Jing Chen (Peking University), and
  Yufeng Jin (Peking University)

Improving the Solder Wettability Via Atmospheric Plasma Technology 2067
  Sagung Dewi Kencana (National Taiwan University of Science and
  Technology), Yu-Lin Kuo (National Taiwan University of Science and
  Technology), Yee-Wen Yen (National Taiwan University of Science and
  Technology), Eckart Schellkes (Robert Bosch Taiwan Co., Ltd), and
  Wallace Chuang (Robert Bosch Taiwan Co., Ltd)

Orthogonal Quilt Packaging 3D Integration for High-Energy Particle Detectors 2072
  Jason Kulick (Indiana Integrated Circuits, LLC), Tian Lu (Indiana
  Integrated Circuits, LLC), Edit Varga (Indiana Integrated Circuits,
  LLC), Gary H. Bernstein (Indiana Integrated Circuits, LLC), Carlos
  Ortega (Indiana Integrated Circuits, LLC), Christopher Kenney (SLAC
  National Accelerator Laboratory), and Julie Segal (SLAC National
  Accelerator Laboratory)

Carbonized Electrodes for Electrochemical Sensing 2073
  Mohammad Aminul Haque (The University of Tennessee, Knoxville),
  Nickolay V. Lavrik (Oak Ridge National Laboratory), Dale Hensley (Oak
  Ridge National Laboratory), and Nicole McFarlane (The University of
  Tennessee, Knoxville)

Moldability Challenges Associated with the Assembly of Thicker IC Packages for High Voltage and
Power Applications 2079
  Sadia Naseem (Texas Instruments Inc.), Jack Chiang (Texas Instruments
  Inc.), Megan Chang (Texas Instruments Inc.), Bob Lee (Texas
  Instruments Inc.), and Jason Chien (Texas Instruments Inc.)

Highly Compact, Multiband Composite-Right/Left-Handed(CRLH) Transmission Line Based Stub for GPS
Applications 2085
  Hae-In Kim (University of Florida), Seahee Hwangbo (University of
  Florida), Renuka Bowrothu (University of Florida), and Yong-Kyu Yoon
  (University of Flordia)

Session 40: Interactive Presentations 4

Die Thickness Optimization for Preventing Electro-Thermal Fails Induced by Solder Voids in Power
Devices 2091
  Dario Vitello (STMicroelectronics), Andrea Albertinetti
  (STMicroelectronics), and Marco Rovitto (STMicroelectronics)

3-T (8-T) Decoupling Capacitors for Improved PDN in LPDDR4/4X/5 System 2097
  Sunil Gupta (Qualcomm Technologies, Inc.)

Improved Correlation Between Accelerated Board Level Reliability (BLR) Testing and Customer BLR Results Using a Hybrid Closed-Form/Finite Element Methodology 2103

Maxim Serebreni (DfR Solutions), Natalie Hernandez (DfR Solutions), Gil Sharon (DfR Solutions), Nathan Blattau (DfR Solutions), Craig Hillman (DfR Solutions), and Ken Symonds (Western Digital)

Fabrication and Reliability Demonstration of 3 μm Diameter Photo Vias at 15 μm Pitch in Thin Photosensitive Dielectric Dry Film for 2.5 D Glass Interposer Applications 2112

Daichi Okamoto (TAIYO INK MFG. CO., LTD.), Yoko Shibasaki (TAIYO INK MFG.CO.LTD), Daisuke Shibata (TAIYO INK MFG.CO.LTD), Tadahiko Hanada (TAIYO INK MFG.CO.LTD), Fuhan Liu (Georgia Institute of Technology), Mohanalingam Kathaperumal (Georgia Institute of Technology), and Rao R. Tummala (Georgia Institute of Technology)

Pre-Cure Modification of Electrically Conductive Adhesive for Low Temperature Interconnection 2117

Jinto George (University of Sherbrooke, Bromont, QC, Canada), David Danovitch (University of Sherbrooke, Bromont, QC, Canada), Alexandre Leblanc (IBM Canada Ltd, Bromont, QC, Canada), Eric Savage (IBM Canada Ltd, Bromont, QC, Canada), Michael Ayukawa (Redlen Technologies, Saanichton, BC, Canada), and Dexter Macaisa (Redlen Technologies, Saanichton, BC, Canada)

RDL-1st Fan-Out Panel Level Packaging (FOPLP) for Heterogeneous and Economical Packaging 2126

Nagendra Sekhar Vasarla (Institute of Microelectronics, A*STAR (Agency for Science, Technology and Research)), Vempati Srinivasa Rao (Institute of Microelectronics, A*STAR (Agency for Science, Technology and Research)), F. X. Che (Institute of Microelectronics, A*STAR (Agency for Science, Technology and Research)), Chong Ser Choong (Institute of Microelectronics, A*STAR (Agency for Science, Technology and Research)), and Kazunori Yamamoto (Institute of Microelectronics, A*STAR (Agency for Science, Technology and Research))

Epoxy Composites with Surface Modified Silicon Carbide Filler for High Temperature Molding Compounds. 2134

Fan Wu (Georgia Institute of Technology), Nicholas C Mitchell (Nicholas C), Bo Song (Georgia Institute of Technology), Kyoung-sik Moon (Georgia Institute of Technology), and CP Wong (Georgia Institute of Technology)

Ultra Low Resistivity and High Electrical Stability Silo-Ag ECAs Produced from Curing Chemistry Optimization for Flexible Electronics 2140

Xueqiao Wang (Georgia Institute of Technology), Bo Song (Georgia Institute of Technology), Kyoung-Sik Moon (Georgia Institute of Technology), and C. P. Wong (Georgia Institute of Technology)

Physics of Failure Based Simulation and Experimental Testing of Quad Flat No-Lead Package 2144

Jia-Shen Lan (National Sun Yat-sen University) and Mei-Ling Wu (National Sun Yat-sen University)

An Assessment of Electromigration in 2.5D Packaging 2150
  Jiefeng Xu (The State University of New York at Binghamton), Scott
  McCann (The State University of New York at Binghamton), Huayan Wang
  (The State University of New York at Binghamton), Jing Wang (The State
  University of New York at Binghamton), VanLai Pham (The State
  University of New York at Binghamton), Stephen R. Cain (The State
  University of New York at Binghamton), Gamal Refai-Ahmed (The State
  University of New York at Binghamton), and S.B. Park (The State
  University of New York at Binghamton)

Diffusion Enhanced Drive Sub 100 °C Wafer Level Fine-Pitch Cu-Cu Thermocompression Bonding for 3D IC
Integration 2156
  Asisa Kumar Panigrahy (Gokaraju Rangaraju Institute of Engineering &
  Technology, Hyderabad, India), Satish Bonam (Indian Institute of
  Technology Hyderabad, India), Tamal Ghosh (Indian Institute of
  Technology Hyderabad, India), Siva Rama Krishna Vanjari (Indian
  Institute of Technology Hyderabad, India), and Shiv Govind Singh
  (Indian Institute of Technology Hyderabad)

Development of Sheet Type Molding Compounds for Panel Level Package 2162
  Kenichi Ueno (Company), Kazuhiro Dohi (SANYU REC CO., LTD.), Yui
  Suzuki (SANYU REC CO., LTD.), and Masakazu Hirose (SANYU REC CO.,
  LTD.)

Defect Detection for the TSV Transmission Channel Using Machine Learning Approach 2168
  Huan Liu (Peking University), Runiu Fang (Peking University), Min Miao
  (Beijing Information Science and Technology University), Yang Yang
  (Shenzhen Graduate School, Peking University), and Yufeng Jin
  (Shenzhen Graduate School, Peking University)

Direct Printing of Heat Sinks, Cases and Power Connectors on Insulated Substrate Using Selective
Laser Melting Techniques 2173
  Rabih Khazaka (Safran SA), Donatien Martineau (Safran SA), Toni
  Youssef (Safran SA), Thanh Long Le (Safran SA), and Stephane Azzopardi
  (Safran SA)

Server CPU Package Design Using PoINT Architecture 2180
  Arun Chandrasekhar (Intel), Vijaya Boddu (Intel), Erich Chuh (Intel),
  Krishna Bharath (Intel), Farzaneh Yahyaei-Moayyed (Intel), Srikrishnan
  Venkataraman (Intel), Sriram Srinivasan (Intel), Ram Viswanath
  (Intel), Huthasana Kalyanam (Intel), and Ritesh Jain (Intel)

Highly Reliable Die-Attach Silver Joint with Pressure-Less Sintering Process 2186
  Sihai Chen (Indium Corporation, USA), William Shambach (Rochester
  Institute of Technology), Jordan Palmer (Rochester Institute of
  Technology), Christine Labarbera (Indium Corporation), Xuanyi Ding
  (Cornell University), and Ning-Cheng Lee (Indium Corporation)

3D Power Packaged Device Thermo-Mechanical Modeling and Stress Analysis After Reliability Trials 2194
  Lucrezia Guarino (STMicroelectronics, Italy), Lucia Zullino
  (STMicroelectronics, Italy), Luca Cecchetto (STMicroelectronics,
  Italy), Fiorella Pozzobon (STMicroelectronics, Italy), and Antonio
  Andreini (STMicroelectronics, Italy)

Millimeter Wave Dual Polarization Design Using Frequency Selective Surface (FSS) for 5G Base-Station Applications 2200

Chi-Hau Yang (National Sun Yat-Sen University), Chung-Yi Hsu (National Sun Yat-Sen University), and Lih-Tyng Hwang (National Sun Yat-Sen University)

Direct Bonding of low Temperature Heterogeneous Dielectrics 2206

Serena Iacovo (imec), Ian Peng (imec), Alain Phommahaxay (imec), Fumihiro Inoue (imec), Patrick Verdonck (imec), Soon-Wook Kim (imec), Erik Sleeckx (imec), Andy Miller (imec), Gerald Beyer (imec), and Eric Beyne (imec)

Session 41: Student Interactive Presentations

Low Temperature Transient Liquid Phase (TLP) Bonding using Eutectic Sn-In Solder Anisotropic Condctive Films (ACFs) for Flexible Ultrasound Transducer 2213

Jae-Hyeong Park (KAIST), Jongcheol Park (NanoFab Center), and Kyung-Wook Paik (KAIST)

Room-Temperature Bonding with Pd Coated Cu Wire on Al Pads: Ball Bond Optimization with 2-Stage Methodology 2219

Nicholas Kam (University of Waterloo), Michael David Hook (University of Waterloo), Celal Con (KA Imaging), Karim S. Karim (University of Waterloo), and Michael Mayer (University of Waterloo)

On-Chip ESD Monitor 2225

Kannan Kalappurakal Thankappan (University of California, Los Angeles), Boris Vaisband (University of California, Los Angeles), and Subramanian S. Iyer (University of California, Los Angeles)

Preparation and Characterization of Electroplated Cu/Graphene Composite 2234

Xin Wang (Tsinghua University), Qian Wang (Tsinghua University), Jian Cai (Tsinghua University), Changming Song (Tsinghua University), Yang Hu (Tsinghua University), Yang Zhao (University of Science and Technology of China), and Yu Pei (University of Science and Technology of China)

Quantifying the Impact of RF Probing Variability on TRL Calibration for LTCC Substrates 2240

Ömer Faruk Yildiz (Hamburg University of Technology, Germany), David Dahl (Hamburg University of Technology, Germany), and Christian Schuster (Hamburg University of Technology, Germany)

Effects of NCF and UBM Materials on Electromigration Reliabilities of Sn-Ag Microbumps for Advanced 3D Packaging 2246

Kirak Son (Andong National University), Gahui Kim (Andong National University), Hyodong Ryu (Andong National University), Gyu-Tae Park (Amkor Technology Korea Inc.), Ho-Young Son (SK hynix Inc.), Nam-Seog Kim (SK hynix Inc.), Cheol-Woong Yang (Sungkyunkwan University), Young-Cheon Kim (Andong National University), Jeong Sam Han (Andong National University), and Young-Bae Park (Andong National University)

Ag Diffusion Control Through Sn on a Sequential Plating-Based Bumping Process 2252

Abderrahim EL Amrani (Université de Sherbrooke), Etienne Paradis (Université de Sherbrooke), David Danovitch (Université de Sherbrooke), and Dominique Drouin (Université de Sherbrooke)

Mechanical Reliability Assessment of Cu_6Sn_5 Intermetallic Compound and Multilayer Structures in Cu/Sn Interconnects for 3D IC Applications 2258

Jui-Yang Wu (National Taiwan University), C. Robert Kao (National Taiwan University), and Jenn-Ming Yang (University of California, Los Angeles)

A Study on the Anchoring Polymer Layer (APL) Anisotropic Conductive Films (ACFs) with Self-Exposed Conductive Particles Surface for Ultra-Fine Pitch Chip-on-Glass (COG) Applications 2266

Dal-Jin Yoon (KAIST) and Kyung-Wook Paik (Korea Advanced Institute of Science and Technology)

Bending Properties of Fine Pitch Flexible CIF (Chip-in-Flex) Packages Using APL (Anchoring Polymer Later) ACFs (Anisotropic Conductive Films) 2272

Ji-Hye Kim (KAIST, Korea), Dal-Jin Yoon (KAIST, Korea), and Kyung-Wook Paik (KAIST, Korea)

Effects of the Curing Properties and Viscosities of Non-Conductive Films (NCFs) on the Sn-Ag Solder Bump Joint Morphology and Reliability 2278

HanMin Lee (KAIST, South Korea), SeYong Lee (KAIST, South Korea), SangMyung Shin (KAIST, South Korea), TaeJin Choi (Doosan Corporation Electro-Materials BG, South Korea), SooIn Park (Doosan Corporation Electro-Materials BG, South Korea), and Kyung-Wook Paik (KAIST, South Korea)

Experimental Investigations on Vertical Ultrasonic Assisted Low Temperature Sintering Process 2284

Henning Seefisch (Leibniz Universität Hannover) and Jens Twiefel (Leibniz Universität Hannover)

Pressureless Transient Liquid Phase Sintering Bonding of Sn-58Bi with Ni Particles for High-Temperature Packaging Applications 2290

Kyung Deuk Min (Sungkyunkwan University, Republic of Korea), Kwang-Ho Jung (Sungkyunkwan University, Republic of Korea), Choong-Jae Lee (Sungkyunkwan University, Republic of Korea), and Seung-Boo Jung (Sungkyunkwan University, Republic of Korea)

Epoxy/ Triazine Copolymer Resin System for High Temperature Encapsulant Applications 2296

Jiaxiong Li (Georgia Institute of Technology), Chao Ren (Georgia Institute of Technology), Kyoung-sik Moon (Georgia Institute of Technology), and Ching-ping Wong (Georgia Institute of Technology)

Low Temperature Ag-Ag Direct Bonding Technology for Advanced Chip-Package Interconnection 2302

Jiaqi Wu (University of California Irvine) and Chin C. Lee (University of California Irvine)

Reliability of Micro-Alloyed SnAgCu Based Solder Interconnections for Various Harsh Applications 2309

Sinan Su (Auburn University), Francy John Akkara (Auburn University), Anto Raj (Auburn University), Cong Zhao (Auburn University), Seth Gordon (Auburn University), Sharath Sridhar (Auburn University), Sivasubramanian Thirugnanasambandam (Auburn University), Sa'd Hamasha (Auburn University), Jeffery Suhling (Auburn University), and John Evans (Auburn University)

Wideband Low-Profile Ka-Band Microstrip Antenna with Low Cross Polarization Using Asymmetry AMC Structure 2318

Mei Xue (Institute of Microelectronics of the Chinese Academy of Sciences), Weikang Wan (Institute of Microelectronics of the Chinese Academy of Sciences), Qidong Wang (Institute of Microelectronics of the Chinese Academy of Sciences), and Liqiang Cao (Institute of Microelectronics of the Chinese Academy of Sciences)

Automatic Transient Thermal Impedance Tester for Quality Inspection of Soldered and Sintered Power Electronic Devices on Panel and Tile Level 2324

Maximilian Schmid (Technische Hochschule Ingolstadt), Bhogaraju Sri Krishna (Technische Hochschule Ingolstadt), and Gordon Elger (Technische Hochschule Ingolstadt)

Time 0 Void Evolution and Effect on Electromigration 2331

Jiefeng Xu (The State University of New York at Binghamton), Scott McCann (Xilinx, Inc.), Huayan Wang (The State University of New York at Binghamton), VanLai Pham (The State University of New York at Binghamton), Stephen R. Cain (The State University of New York at Binghamton), Gamal Refai-Ahmed (Xilinx, Inc.), and S.B. Park (The State University of New York at Binghamton)

Quintuple Band Lambda/4 Stub by using Unbalanced Bridged CRLH Transmission Lines 2337

Renuka Bowrothu (University of Florida), Seahee Hwangbo (University of Florida), Haein Kim (University of Florida), and Yong-Kyu Yoon (University of Florida)

Product Level Design Optimization for 2.5D Package Pad Cratering Reliability During Drop Impact 2343

Huayan Wang (State University of New York at Binghamton), Jing Wang (State University of New York at Binghamton), Jiefeng Xu (State University of New York at Binghamton), Vanlai Pham (State University of New York at Binghamton), Ke Pan (State University of New York at Binghamton), Seungbae Park (State University of New York at Binghamton), Hohyung Lee (Xilinx Inc, USA), and Gamal Refai-Ahmed (Xilinx Inc, USA)

Microstructures of Pb-Free Solder Joints by Reflow and Thermo-Compression Bonding (TCB) Processes 2349

Youngja Kim (Samsung Electronics), Jinho Hah (Georgia Institute of Technology), Patxi Fernandez-Zelaia (Georgia Institute of Technology), Sangil Lee (Georgia Institute of Technology), Leroy Christie (ASM Pacific Technology), Paul Houston (Engent Inc.), Shreyes Melkote (Georgia Institute of Technology), Kyoung-Sik Moon (Georgia Institute of Technology), and Ching-Ping Wong (Georgia Institute of Technology)

Reduction of Ag Corrosion Rate During Decapsulation of Ag Wire Bond Packages 2359

Young-Ja Kim (Samsung Electronics, Korea), Jinho Hah (Georgia Institute of Technology), Kyoung-Sik (Jack) Moon (Georgia Institute of Technology), and C. P. Wong (Georgia Institute of Technology)

Author Index

Foreword

On behalf of the Program Committee and Executive Committee, it is our pleasure to welcome you to the 69th Electronic Components and Technology Conference (ECTC) which will be held at The Cosmopolitan of Las Vegas in Las Vegas, Nevada from May 28-31, 2019. This premier international conference is sponsored by the IEEE Electronic Packaging Society (EPS). The ECTC Program Committee has selected over 350 papers which will be presented in 36 oral sessions and five interactive presentation session including one interactive presentation session exclusively featuring papers by student authors. The oral sessions will feature selected papers on key topics such as fan-out packaging, wafer-level packaging, flip-chip packaging, 3D/TSV technologies, design for RF performance and signal/power integrity, thermal and mechanical modeling, optoelectronics packaging, materials and reliability. Interactive presentation sessions will showcase papers in a format that encourages more in-depth discussion and interaction with authors about their work.

Authors from over twenty countries are expected to present their work at the 69th ECTC, covering ongoing technology development within established disciplines or emerging topics of interest for our industry such as additive manufacturing, heterogeneous integration, flexible and wearable electronics.

ECTC will also feature six special sessions with invited industry experts covering several important and emerging topic areas. On Tuesday, May 29 at 10 a.m., W. Hong Yeo and Mikel Miller will chair a special session covering "Transient Electronics: A Green Revolution for Packaging". On the same day at 2 p.m., Rena Huang and Soon Jang will chair a session focused on Photonics on the Cutting-Edge of Technology Evolution. Tuesday evening will also include the ECTC Panel Session at 7:30 p.m. chaired by IEEE EPS President Avi Bar-Cohen and Karlheinz Bock, where young researchers will share their visions of future packaging technologies and participate in discussions with experts in the field.

This conference will also feature a Women's Panel and Reception jointly organized by ECTC and ITherm on Wednesday, May 29 at 6:30pm. This year, panelists from around the globe will share their perspectives on efforts to enhance the participation of women in engineering, and the panel will be chaired by Kristina Young-Fisher and Cristina Amon. On the same day at 7:30 p.m., Tanja Braun will chair the ECTC Plenary Session titled "Sensors and Packaging for Autonomous Driving". In this plenary session, experts will address the challenges and demands for sensors and packages for autonomous driving along the value chain. On Thursday, May 31 at 8 p.m., the IEEE EPS Seminar titled "Roadmap of IC Packaging Materials to Meet Next-Generation Smartphone Performance Requirements" will be moderated by Yasumitsu Orii and Sheigenori Aoki from the High-Density Substrates & Boards Technical Committee of the IEEE EPS Society.

Supplementing the technical program, ECTC also offers Professional Development Courses (PDCs) and the Technology Corner exhibits. Co-located with the IEEE iTHERM Conference this year, the 69th ECTC will offer eighteen PDCs, organized by the PDC Committee chaired by Kitty Pearsall and Jeffrey Suhling. The PDCs will take place on Tuesday, May 28 and are taught by distinguished experts in their respective fields. The Technology Corner will showcase the latest technologies and products offered by leading companies in the electronic components, materials, packaging and services fields. More than one hundred Technology Corner exhibits will be open Wednesday and Thursday starting at 9 a.m. ECTC also offers attendees numerous opportunities for networking and discussion with colleagues during coffee breaks, daily luncheons and nightly receptions.

Whether you are an engineer, a manager, a student or an executive, ECTC offers something unique for everyone in the microelectronics packaging and components industry. I invite you to make your plans now to join us for the 69th ECTC and be a part of all the exciting technical and professional opportunities. I also take this opportunity to thank our sponsors, exhibitors, authors, speakers, PDC instructors, session chairs, and program committee members, as well as all the volunteers who help make the 69th ECTC a success. We look forward to meeting you in Las Vegas, Nevada May 28 –31, 2019.

Nancy Stoffel
69th ECTC Program Chair
General Electric Research
stoffel@ge.com

Mark Poliks
69th ECTC General Chair
Binghamton University
mpoliks@binghamton.edu

2019 Executive Committee

General Chair
Mark Poliks
Binghamton University
mpoliks@binghamton.edu

Vice-General Chair
Christopher Bower
X-Celeprint Inc.
cbower@x-celeprint.com

Program Chair
Nancy Stoffel
GE Research
nstoffel1194@gmail.com

Assistant Program Chair
Rozalia Beica
DuPont
rozalia.beica@dupont.com

Web Administrator
Ibrahim Guven
Virginia Commonwealth University
iguven@vcu.edu

Jr. Past General Chair
Sam Karikalan
Broadcom Inc.
sam.karikalan@broadcom.com

Sr. Past General Chair
Henning Braunisch
Intel Corporation
braunisch@ieee.org

Sponsorship Chair
Wolfgang Sauter
GLOBALFOUNDRIES
wolfgang.sauter@globalfoundries.com

Finance Chair
Patrick Thompson
Texas Instruments, Inc.
patrick.thompson@ti.com

Publications Chair
Steve Bezuk
sbezuk@gmail.com

Publicity Chair
Eric Perfecto
eric.perfecto.us@ieee.org

Treasurer
Tom Reynolds
T3 Group LLC
t.reynolds@ieee.org

Exhibits Chair
Joe Gisler
Vector Associates
gisler.h.dr@ieee.org

Exhibits Co-Chair
Alan Huffman
Micross Advanced Interconnect Technology
alan.huffman@micross.com

Arrangements Chair
Lisa Renzi Ragar
Renzi & Company, Inc.
lrenzi@renziandco.com

EPS Representative
C. P. Wong
Georgia Institute of Technology
cp.wong@mse.gatech.edu

2019 Program Committee

Applied Reliability

Chair
Deepak Goyal
Intel Corporation
deepak.goyal@intel.com

Assistant Chair
Darvin R. Edwards
Edwards Enterprise Consulting, LLC
darvin.edwards1@gmail.com

Tim Chaudhry
Amkor Technology, Inc.

Tz-Cheng Chiu
National Cheng Kung University

Vikas Gupta
Texas Instruments, Inc.

Sandy Klengel
Fraunhofer Institute for Microstructure of Materials and Systems

Pilin Liu
Intel Corporation

Varughese Mathew
NXP Semiconductors

Toni Mattila
Aalto University

Keith Newman
AMD

Donna M. Noctor
Nokia

S. B. Park
Binghamton University

Lakshmi N. Ramanathan
Microsoft Corporation

René Rongen
NXP Semiconductors

Scott Savage
Medtronic Microelectronics Center

Jeffrey Suhling
Auburn University

Pei-Haw Tsao
Taiwan Semiconductor Manufacturing Company, Ltd.

Dongji Xie
NVIDIA Corporation

Assembly & Manufacturing Technology

Chair
Mark Gerber
Advanced Semiconductor Engineering Inc.
mark.gerber@aseus.com

Assistant Chair
Jin Yang
Intel Corporation
jin1.yang@ieee.org

Sai Ankireddi
Soraa, Inc

Christo Bojkov
Qorvo

Garry Cunningham
NGC

Habib Hichri
Suss Microtech Photonic Systems Inc.

Paul Houston
Engent

Li Jiang
Texas Instruments

Chunho Kim
Medtronic Corporation

Wei Koh
Pacrim Technology

Ming Li
ASM Pacific Technology

Debendra Mallik
Intel Corporation

Jae-Woong Nah
IBM Corporation

Valerie Oberson
IBM Canada Ltd

Chandradip Patel
Schlumberger Technology Corporation

Shichun Qu
Intersil, a Renesas Company

Paul Tiner
Texas Instruments

Andy Tseng
JSR Micro

Jan Vardaman
Techsearch International

Yu Wang
Sensata Technologies

Shaw Fong Wong
Intel Corporation

Wei Xu
Huawei

Tonglong Zhang
Nantong Fujitsu Microelectronics Ltd.

Emerging Technologies

Chair
Florian Herrault
HRL Laboratories, LLC
fgherrault@hrl.com

Assistant Chair
Benson Chan
Binghamton University
chanb@binghamton.edu

Isaac Robin Abothu
Siemens Healthineers

Meriem Akin
Robert Bosch GmBH

Vasudeva P. Atluri
Renavitas Technologies

Karlheinz Bock
Technische Universitat Dresden

Vaidyanathan Chelakara
Acacia Communications

Rabindra N. Das
MIT Lincoln Labs

Dongming He
Qualcomm Technologies, Inc.

TengFei Jiang
University of Central Florida

Jong-Hoon Kim
Washington State University Vancouver

Ahyeon Koh
Binghamton University

Ramakrishna Kotlanka
Analog Devices

Santosh Kudtarkar
Analog Devices

Kevin J. Lee
Qorvo Corporation

Zhuo Li
Fudan University

Chukwudi Okoro
Corning

Bharat Penmecha
Intel Corporation

C. S. Premachandran
GLOBALFOUNDRIES

Jintang Shang
Southeast University

Rohit Sharma
IIT Ropar

Nancy Stoffel
GE Research

Liu Yang
IBM

Jimin Yao
Intel Corporation

W. Hong Yeo
Georgia Institute of Technology

Hongqing Zhang
IBM Corporation

High-Speed, Wireless & Components

Chair
Wendem Beyene
Intel Corporation
wendem.beyene@intel.com

Assistant Chair
Lianjun Liu
NXP Semiconductor, Inc.
lianjun.liu@NXP.com

Amit P. Agrawal
Microsemi Corporation

Kemal Aygun
Intel Corporation

Eric Beyne
IMEC

Prem Chahal
Michigan State University

Zhaoqing Chen
IBM Corporation

Charles Nan-Cheng Chen
HiSilicon Technologies

Craig Gaw
NXP Semiconductor

Abhilash Goyal
Velodyne LIDAR, Inc.

Xiaoxiong (Kevin) Gu
IBM Corporation

Rockwell Hsu
Cisco Systems, Inc.

Lih-Tyng Hwang
National Sun Yat-Sen University

Bruce Kim
City University of New York

Timothy G. Lenihan
TechSearch International

Rajen M Murugan
Texas Instruments

Nanju Na
Xilinx

Dan Oh
Samsung

P. Markondeya Raj
Florida International University

Hideki Sasaki
Renesas Electronics Corporation

Li-Cheng Shen
Wistron NeWeb Corporation

Jaemin Shin
Qualcomm Corporation

Manos M. Tentzeris
Georgia Institute of Technology

Maciej Wojnowski
Infineon Technologies AG

Yong-Kyu Yoon
University of Florida

Interconnections

Chair
Wei-Chung Lo
ITRI
lo@itri.org.tw

Assistant Chair
Dingyou Zhang
Broadcom Inc.
dingyouzhang.brcm@gmail.com

Thibault Buisson
Yole Développement

Jian Cai
Tsinghua University

William Chen
Advanced Semiconductor Engineering, Inc.

David Danovitch
University of Sherbrooke

Rajen Dias
Amkor Technology, Inc.

Bernd Ebersberger
Infineon Technologies

Takafumi Fukushima
Tohoku University

Tom Gregorich
Zeiss Semiconductor Manufacturing Technology

Kangwook Lee
Amkor Technology Korea

Steward Lee

Li Li
Cisco Systems, Inc.

Changqing Liu
Loughborough University

Nathan Lower
Rockwell Collins, Inc.

James Lu
Rensselaer Polytechnic Institute

Voya Markovich
Microelectronic Advanced Hardware Consulting, LLC

Lou Nicholls
Amkor Technology, Inc.

Peter Ramm
Fraunhofer EMFT

Katsuyuki Sakuma
IBM Corporation

Lei Shan
IBM Corporation

Ho-Young Son
SK Hynix

Jean-Charles Souriau
CEA Leti

Chuan Seng Tan
Nanyang Technological University

Matthew Yao
GE Energy Management

Materials & Processing

Chair
Mikel Miller
EMD Performance Materials
mikel.miller@emdgroup.com

Assistant Chair
Tanja Braun
Fraunhofer IZM
tanja.braun@izm.fraunhofer.de

Yu-Hua Chen
Unimicron

Qianwen Chen
IBM Research

Bing Dang
IBM Research

Yung-Yu Hsu
Apple Inc.

Lewis Huang
Senju Electronic

C. Robert Kao
National Taiwan University

Chin C. Lee
University of California, Irvine

Alvin Lee
Brewer Science

Yi (Grace) Li
Intel Corporation

Ziyin Lin
Intel Corporation

Yan Liu
Medtronic Inc. USA

Daniel D. Lu
Henkel Corporation

Joon-Seok Oh
Samsung Electro-Mechanics

Praveen Pandojirao-S
Johnson & Johnson

Mark Poliks
Binghamton University

Dwayne Shirley
Inphi

Ivan Shubin
Oracle

Bo Song
HP Inc.

Yoichi Taira
Keio University

Lejun Wang
Qualcomm Technologies, Inc.

Frank Wei
Disco Japan

Kimberly Yess
Brewer Science

Myung Jin Yim
Apple

Hongbin Yu
Arizona State University

Packaging Technologies

Chair
Dean Malta
Micross Advanced Interconnect Technology
Dean.Malta@micross.com

Assistant Chair
Luke England
GLOBALFOUNDRIES
luke.england@globalfoundries.com

Daniel Baldwin
H.B. Fuller Company

Bora Baloglu
Amkor Technology

Jie Fu
Apple

Mike Gallagher
DuPont

Ning Ge
Consultant

Allyson Hartzell
Veryst Engineering

Kuldip Johal
Atotech

Beth Keser
Intel Corporation

Young-Gon Kim
Integrated Device Technology, Inc.

Andrew Kim
Intel Corporation

John Knickerbocker
IBM Corporation

Albert Lan
Applied Materials

John H. Lau
ASM Pacific Technology

Jaesik Lee
Nvidia

Markus Leitgeb
AT&S

Luu Nguyen
Texas Instruments Inc.

Deborah S. Patterson
Harbor Electronics, Inc.

Raj Pendse
Facebook FRL (Facebook Reality Labs)

Subhash L. Shinde
Notre Dame University

Joseph W. Soucy
Draper Laboratory

Peng Su
Juniper Networks

Kuo-Chung Yee
Taiwan Semiconductor Manufacturing Corporation, Inc.

Christophe Zincke
Advanced Semiconductor Engineering, Inc.

Photonics

Chair
Ping Zhou
LDX Optronics, Inc.
pzhou@ldxoptronics.com

Assistant Chair
Z. Rena Huang
Rensselaer Polytechnic Institute
zrhuang@ecse.rpi.edu

Mark Beranek
Naval Air Systems Command

Stephane Bernabe
CEA Leti

Fuad Doany
IBM Research

Gordon Elger
Technische Hochschule Ingolstadt

Takaaki Ishigure
Keio University

Ajey Jacob
GLOBALFOUNDRIES

Soon Jang
ficonTEC USA

Harry G. Kellzi
Teledyne Microelectronic Technologies

Richard Pitwon
Resolute Photonics Ltd

Alex Rosiewicz
A2E Partners

Henning Schroeder
Fraunhofer IZM

Andrew Shapiro
JPL

Masato Shishikura
Oclaro Japan

Masao Tokunari
IBM Corporation

Shogo Ura
Kyoto Institute of Technology

Stefan Weiss
II-VI Laser Enterprise GmbH

Feng Yu
Huawei Technologies Japan

Thomas Zahner
OSRAM Opto Semiconductors GmbH

Thermal/Mechanical Simulation & Characterization

Chair

Przemyslaw Gromala
Robert Bosch GmbH
Przemyslawjakub.gromala@de.bosch.com

Assistant Chair

Ning Ye
Western Digital
ning.ye@wdc.com

Christopher J. Bailey
University of Greenwich

Kuo-Ning Chiang
National Tsinghua University

Xuejun Fan
Lamar University

Nancy Iwamoto
Honeywell Performance Materials and Technologies

Pradeep Lall
Auburn University

Chang-Chun Lee
National Tsing hua University (NTHU)

Yong Liu
ON Semiconductor

Sheng Liu
Wuhan University

Erdogan Madenci
University of Arizona

Tony Mak
Wentworth Institute of Technology

Karsten Meier
Technische Universität Dresden

Erkan Oterkus
University of Strathclyde

Sandeep Sane
Intel Corporation

Suresh K. Sitaraman
Georgia Institute of Technology

Wei Wang
Qualcomm Technologies, Inc.

G. Q. (Kouchi) Zhang
Delft University of Technology (TUD)

Tieyu Zheng
Microsoft Corporation

Jiantao Zheng
Hisilicon

Interactive Presentations

Chair
Michael Mayer
University of Waterloo
mmayer@uwaterloo.ca

Assistant Chair
Pavel Roy Paladhi
IBM Corporation
Pavel.Roy.Paladhi@ibm.com

Swapan Bhattacharya
Engent Inc.

Rao Bonda
Amkor Technology

Mark Eblen
Kyocera International SC

Ibrahim Guven
Virginia Commonwealth University

Alan Huffman
Micross Advanced Interconnect Technology

Jeffrey Lee
iST-Integrated Service Technology Inc.

Nam Pham
IBM Corporation

Mark Poliks
Binghamton University

Patrick Thompson
Texas Instruments, Inc.

Kristina Young-Fisher
GLOBALFOUNDRIES

Professional Development Courses

Chair
Kitty Pearsall
Boss Precision, Inc.
kitty.pearsall@gmail.com

Assistant Chair
Jeffrey Suhling
Auburn University
jsuhling@auburn.edu

Vijay Khanna
IBM Corporation

Eddie Kobeda
Nypro, A Jabil Company

Lakshmi N. Ramanathan
Microsoft Corporation

Wafer Level Integration of Thin Silicon Bare Dies Within Flexible Label

Jean-Charles Souriau, Ahmad Itawi, Laetitia Castagné
Univ. Grenoble Alpes, CEA, LETI
Grenoble, France
jcsouriau@cea.fr

Abstract— This paper presents new developments on the integration of ultra-thin silicon bare dies within a flexible label made on wafer carrier. It is proposed to interconnect silicon die on an electrical network by flip-chip and to perform collective thinning using conventional microelectronic equipment. After dies attach, they are collectively thinned down below 100µm and encapsulated in a polymer. Finally, the flex is take off from its wafer carrier. The total thickness of the label is approximately 150µm. Test flexible labels were made on 200 mm wafer and designed in order to test electrical interconnection between thin silicon die and electrical network on flex. The test label will be fully described in this paper. A technical focus will be done on the most important process steps for the silicon die integration and electrical results of four point Kelvin and daisy chain patterns will be presented and commented.

Keywords-Flexible electronics, Packaging, Flip-chip, Thin silicon die

I. INTRODUCTION

New developments on the integration of ultra-thin silicon bare dies within a flexible film open the way to a new paradigm. Indeed, thanks to the thinness and the flexibility of devices, it is conceivable to add a function around an object rather than to build an object around a function [1-5].

Mature solutions proposed by the manufacturers are generally to bond components, already packaged, on flexible substrate, which integrates an interconnection network. In this case, only conductive tracks are flexible because components are often several hundred microns thick. A very common product, in which bare die are bonded on a flexible substrate is the RFID TAG. However, silicon dies are not flexible there either. Their surface area so small (less than 1mm²) make them difficult to detect. In order to get fully flexible device, silicon dies have to be thin down below 100µm. Nevertheless, there is a challenge because of the manipulation of ultra-thin dies.

The goal of this article is to propose a generic process for manufacturing a flexible label, which integrate silicon components such as ASIC, sensor or actuator. The process is performed on wafer carrier in manufacturing microelectronic line. In this study, it was chosen to hybridize the components by flip-chip and to achieve the electrical interconnection thanks to gold stud-bumps. The main interest of stud-bumps is that they can be made on bare dies. In this study, test flexible labels were made on 200 mm wafer and designed in order to validate the electrical interconnection between thin silicon die and re-distribution layer on flex. The test label manufacturing is described in following chapter. A technical

focus is done on the most important process steps for the silicon die integration. Finally, electrical results of four-point Kelvin and daisy chain patterns are presented and commented.

II. MOTIVATIONS AND TECHNICALS CHOICES

Three types of format are possible to process flexible electronic: on roll to roll, on panel or on wafer. The first two of them are well adapted for large devices, low cost, high throughput. However patterning resolution is medium. Working at the wafer level allows taking benefic of high resolution of integration. It is adapted, for example to make Flexible Fan-Out Packaging. Our final target will be to build an heterogeneous flexible system which combine panel substrate including printed device and interconnection network with a silicon electronic die integrated within small flexible label (see Fig. 1).

This paper is essentially focus on the electronic label. It is proposed to interconnect silicon die in the label by flip-chip and to perform collective thinning on a wafer carrier using conventional microelectronic equipment. It was chosen to do the electrical interconnection thanks to gold stud-bumps and to perform the hybridization by thermocompression at low temperature (<150°C) in order to be compatible with polymer (Fig. 2). Indeed, the use of solder bump, such as SnAgCu, was not conceivable. Moreover, one interest of stud-bumps is that they can be made on bare dies [6].

The choice of the flexible material in which to integrate the silicon die is critical. In this study, we have tested a Siloxane polymer SINR commercialized by ShinEtsu. This material is photosensitive, available in spin on or dry film and has low stress and low cure temperature.

In this approach silicon, dies and re-distribution layer are encapsulated with an additional film in order to prevent stress on interconnection links (see Fig. 3).

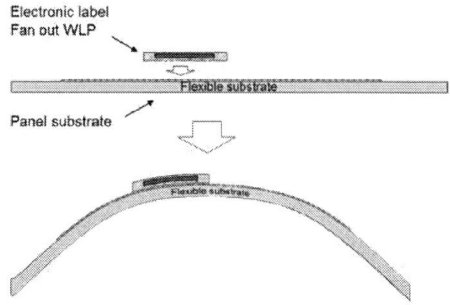

Figure 1. Heterogeneous flexible system.

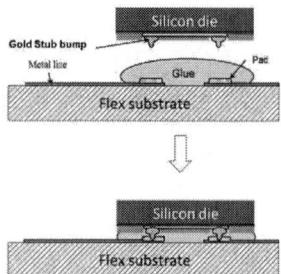

Figure 2. Principle of the interconnection of a silicon chip on a flexible substrate.

Figure 3. Silicon die encapsulated within a flexible label

III. PROCESS FLOW

The system integration was performed at the wafer level and the process flow is illustrated in the Fig. 4.

The carrier is a 200mm silicon wafer, which include a treatment in order to get a temporary adhesion layer. A SINR film 30 µm or 80µm thick was deposited by spin coating or by laminating. The electrical network was made of W50nm/WN50nm/Au200nm metallic. 50µm thick of silver glue was deposited on pads by serigraphy. Dies were aligned and attached using a DATACON 220APM+ (die on wafer) flip-chip tool. The equipment system enables dispense of dots of polymer glue and then aligns and mounts the components under a combination of heat and pressure. In this study the Epo-TekTM E505 glue was used because of its useful viscosity properties as a function of temperature. As it is shown in Fig. 5, the bonding was performed in two steps. All dies were attached with the flip-chip tool and then collectively bonded using a EVG 520IS thermo-compression Bonder.

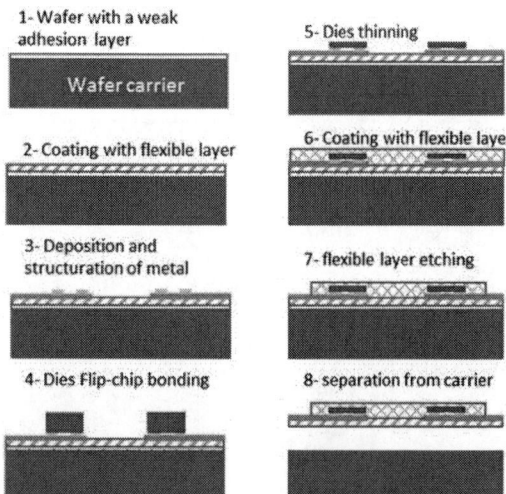

Figure 4. Wafer level process flow of silicon dies encapsulation.

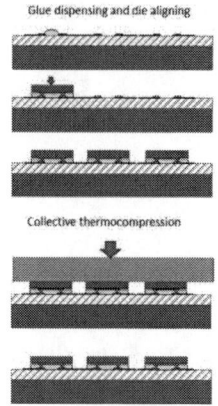

Figure 5. Bonding process flow.

Collective thinning was performed in order to reduce the die thickness below 100µm. An additional 80µm thick SINR layer was laminated under vacuum to encapsulate dies. Finally, a flexible label was diced by laser and took off from the wafer carrier.

In a first time, the process was performed without electrical interconnection. Silicon dies 5x5 mm² were attached on polymer thick 80µm laminated on wafer. Fig. 6 and Fig. 7 show results obtained on a label including a silicon die thinned down to 20µm.

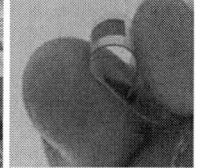

Figure 6. Thinned silicon chips in a flexible layer after take off from the wafer carrier.

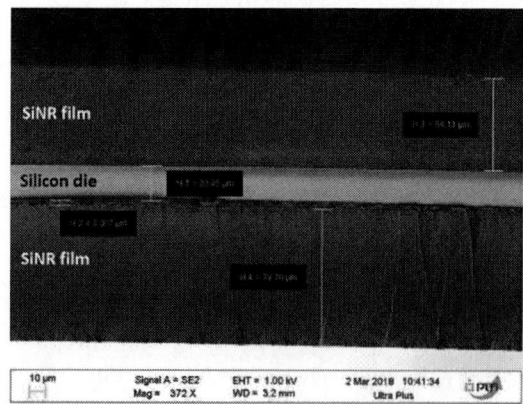

Figure 7. SEM observation of silicon chip thinned to 20µm in a flexible layer.

978-1-7281-1500-9/19 $31.00 © 2019 IEEE

IV. TEST VEHICLE

A silicon test vehicle was designed at CEA-LETI in order to mimic bare dies. Two sizes of chips have been designed, 5x5mm² and 10x10 mm². It includes a 0.6µm thick of AlSi lines, a passivation layers of SiO2 0.5 µm thick and SiN 0.6µm thick. Gold stud-bumps were formed on pads using a standard ball bumping equipment. Stud-bumps are approximately 70µm diameter and 30µm height (see Fig. 8).

The wafer substrate can be populated with 24 large and 24 small chips (Fig. 9). For the first made wafer of this study, the thickness of the bottom polymer layer was 30 µm. Dies were collectively thinned down to 100 µm using standard grinding equipment. A final polymer layer 80µm thick was deposited by lamination under vacuum in order fully cover dies. The Fig. 10 shows the result.

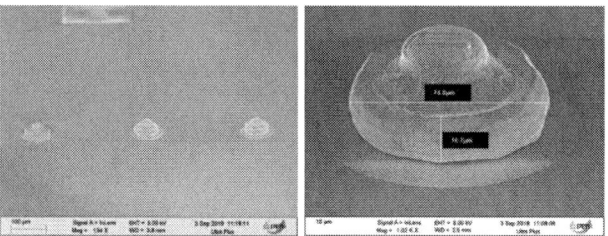

Figure 8. Gold stud-bumps on test vehicle.

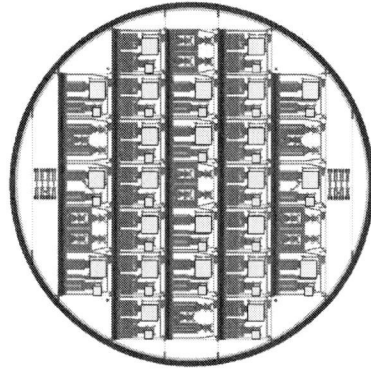

☐ Die: 10x10 mm²   ☐ Die: 5x5 mm²

Figure 9. 200mm test vehicle wafer with 24 sites including large and small dies.

Figure 10. Silicon dies flip-chip bonded and thinned down to 100µm in flexible layer.

The electrical continuity of silicon chips is testable from the wafer thanks to probe card at different stage of the process. The test vehicle was designed in order to test the following patterns:

- Peripheral four-point Kelvin pattern to measure the resistance of one contact between die and the substrate.
- Central four-point Kelvin pattern to measure the resistance of one contact between die and the substrate.
- Peripheral daisy chain pattern including 38 contacts in the large die and 22 contacts in the small die.
- Central daisy chain pattern including 22 contacts in the large die and 16 contacts in the small die.

One substrate wafer was fully populated and electrically characterized. Four-point Kelvin patterns resistance on wafers were measured after the main steps, flip-chip bonding, backside thinning and final encapsulation. Means values are presented in Fig.12.

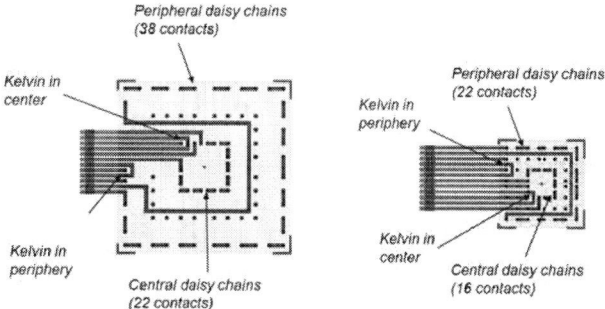

Figure 11. Layout overview of four-point Kelvin pattern and daisychains

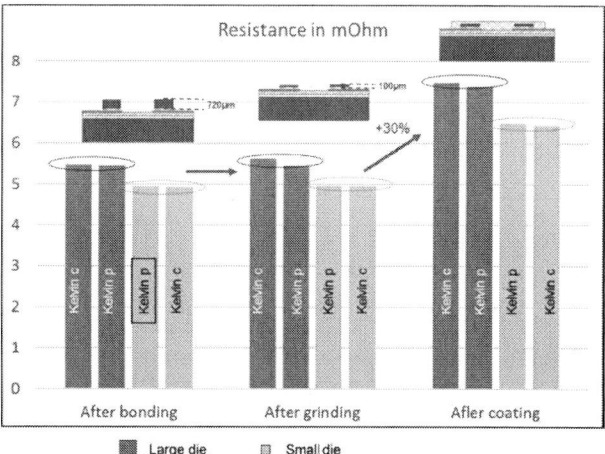

Figure 12. Resistance value of the kelvin pattern measured after main steps of the process.

The average resistance of one contact was found to be ~5.5 mOhms in large die and 5 mOhms in small die. It can be noted that the process of hybridization was also performed on wafer without polymer and the average resistance of one contact was found to be ~ 3.5 mOhms in small die. The presence of bottom polymer layer contributes to absorb the force on stud-bump during the thermocompression and so, probably reduce the quality of the contact. There is no difference between the center and the periphery of the die. It was observed an increase in the order of 30% of the resistance after final coating. An hypothesis of this increase is a softening of the glue during the polymer curing. However, more wafers have to be processed to confirm this phenomena.

Fig. 13 shows the mapping of central four-point kelvin pattern measured on large die. Values are higher on the right side of the wafer. This can be explain by a non-homogeneity of the pressure during the thermal-compression. This problem can be solved by adjusting the equipment parameters.

All daisy chains were tested. The functionality rates are presented in Table I. These results are very promising because 100% of daisy chain are functional after bonding and thinning. Yields are slightly reduced after coating. As it was seen on four point kelvin patterns, the contact between die and substrate throw stud-bump is deteriorated. Fig. 14 show the mapping of daisy chain resistance measured in a fix range of resistance from 30 to 50 Ohm for peripheral daisy chain and from 10 to 30 Ohm for central daisy chain.

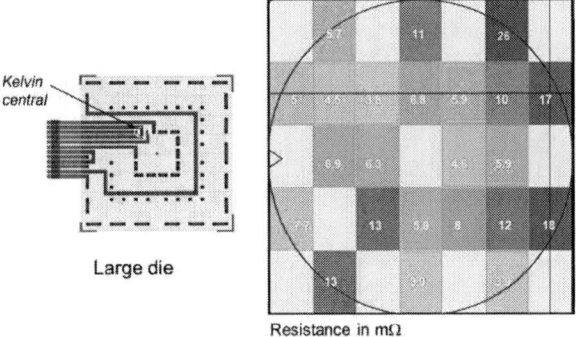

Figure 13. Resistance value distribution of the Kelvin pattern measured in the center of large die.

TABLE I.   FUNCTIONNALITY OF DAISYCHAINS PATTERNS

	Large dies		Small dies	
	Peripheral	Central	Peripheral	Central
After bonding	100%	100%	100%	100%
After thinning	100%	100%	100%	100%
After Coating	83%	75%	88%	83%

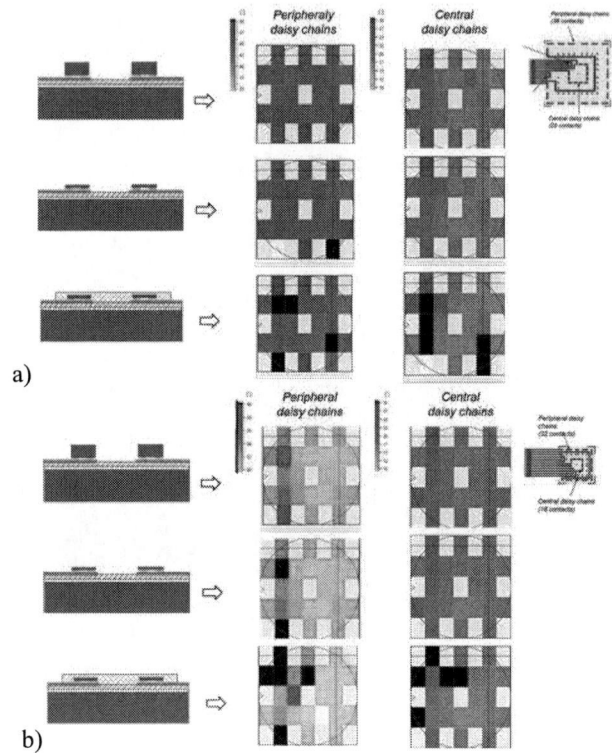

a)

b)

Figure 14. Resistance value distribution of the Daisychain pattern measured a) on large dies b) on small dies.

There is no major change after dies thinning. On the other hand, few daisy chains failed after the coating. Investigation into understanding this problem is already underway. New wafers are ongoing.

A flexible label 3x2.5 mm² was diced using laser and take off from the wafer carrier (Fig.15). To facilitate electrical characterization, a Printed Circuit Board (PCB) was designed and manufactured. Two options were considered to interconnect the label on PCB, a first one using a ZIF connector and a second one using an ACF (Fig. 16).

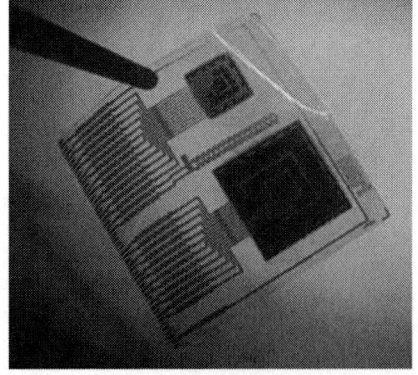

Figure 15. Electronic Flexible label (3x2.5 mm²).

978-1-7281-1500-9/19 $31.00 © 2019 IEEE   1578

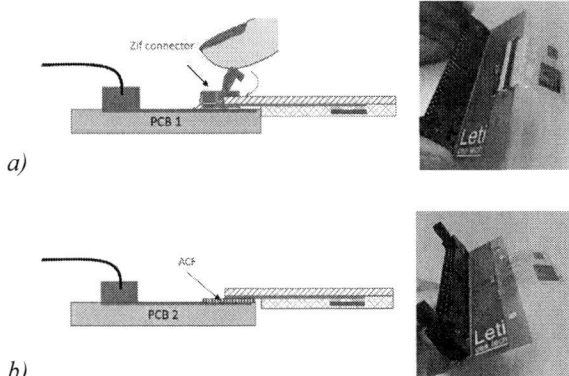

a)

b)

Figure 16. Printed Circuit Board (PCB) to interconnect the label a) using a ZIF connector and b) using an ACF.

6 test patterns were measured (see Fig. 17). The first two of them (straigth line and zigzag line) were just electrical lines without contact in silicon die. The goal was to make sure that metal lines were not damaged by the process of label take off from the carrier. Peripheral and central daisy chain patterns of large and small dies were measured. Electrical results are summarized in the Table II and compared with values measured with probe card at the wafer level and on label after take off from substrate with manual probes.

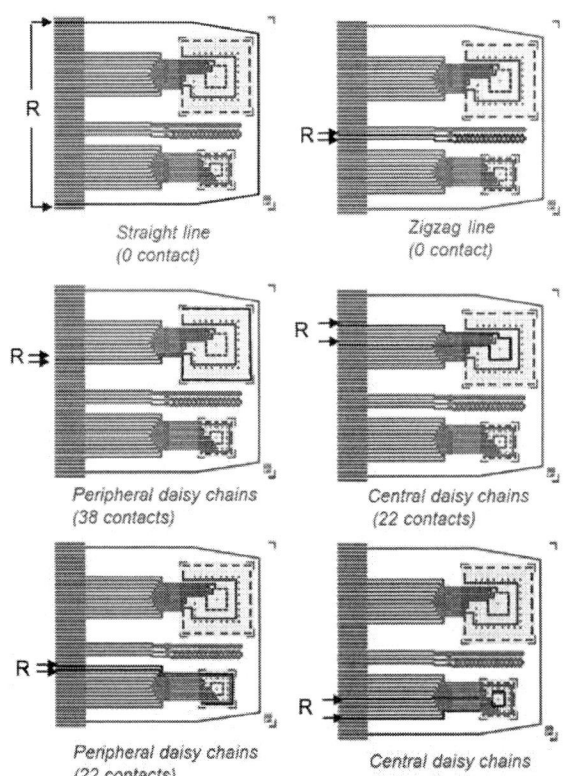

Straight line
(0 contact)

Zigzag line
(0 contact)

Peripheral daisy chains
(38 contacts)

Central daisy chains
(22 contacts)

Peripheral daisy chains
(22 contacts)

Central daisy chains
(16 contacts)

Figure 17. Overview of patterns tested on label

TABLE II.   RESISTANCE IN OHM MEASURED OF TEST PATTERNS

		Line		Large die		Small die	
		Straight	Zigzag	Periph. daisy chains	Central daisy chains	Periph. daisy chains	Central daisy chains
	Calculated value	56	51	67	39	59	43
1st label	On wafer	60	45	65	38	53	40
	After peeling	61	48	66	38	52	39
	On PCB using ZIF connector	62	50	65	40	53	41
2sd label	On wafer	58	45	66	38	52	38
	After peeling	58	45	66	38	52	39
	On PCB using ZIF connector	58	45	68	39	53	39

First, it has to be pointed out that 100% of test patterns are valid after take off of the label from the substrate. All central daisy chain are functional whatever the mode of interconnection on PCB. Moreover, measurements are in good agreement with calculated values. These results are very encouraging for a first trial.

V. CONCLUSION

In this paper it was presented a generic wafer level process for manufacturing a flexible label, which integrate silicon components. It is the first time that it was proposed flip-chip silicon dies interconnection within a flexible film and to perform collective thinning at the wafer level. The electrical interconnection was achieved with gold stud-bumps made on bare dies. Technological elementary bricks to manufacture the flexible label have been presented. The process was validated on a test vehicle with success. The electrical continuity of silicon chips within the label was tested thanks to specific test patterns. The resistance of one contact was found between 5 to 7 mOhm. It was observed an increase in the order of 30% of the resistance after final encapsulation with polymer. An hypothesis of this increase is a softening of the glue during the polymer curing. However, first electrical results are very promising. More electrical tests are ongoing on new labels and reliability tests will be considered. This study is the first step towards full electronic system in a flexible label.

ACKNOWLEDGMENT

This work was supported by the French National Research Agency (ANR) through Carnot funding and has been performed with the help of the "Plateforme Technologique Amont" from Grenoble, with the financial support of CNRS Renatech network.

REFERENCES

[1] M. Hassan, C. Schomburg, C. Harendt, E. Penteker, and J. N. Burghartz, "Assembly and Embedding of Ultra-Thin Chips in Polymers," Eur. Microelectron. Packag. Conf., 2013. pp. 1–6,

[2] T. Fukushima et al., "'FlexTrate ^TM' - Scaled Heterogeneous Integration on Flexible Biocompatible Substrates Using FOWLP,"

Proc. - Electron. Components Technol. Conf., 2017, pp. 649–654. 10.1109/ECTC.2017.226.

[3] M. Bedjaoui, S. Martin, and R. Salot, "Interconnection of Flexible Lithium Thin Film Batteries for Systems-in-Foil," Proc. - Electron. Components Technol. Conf., vol. 2016–August, 2016, pp. 2082–2088, doi: 10.1109/ECTC.2016.23.

[4] C. Van Hoof et al., "Design and integration technology for miniature medical microsystems," Tech. Dig. - Int. Electron Devices Meet. IEDM, 2008, doi:10.1109/IEDM.2008.4796683.

[5] Z. Huang et al., "Three-dimensional integrated stretchable electronics," Nat. Electron., vol. 1, no. 8, 2018, pp. 473–480, doi: 10.1038/s41928-018-0116-y.

[6] A. Itawi and J.-C. Souriau, "Development of a Flexible Label Integrating a Silicon Bare Die," 2018 7th Electron. Syst. Technol. Conf., 2018, pp. 18–21, doi: 10.1109/ESTC.2018.8546383.

Laser Sintering of Aerosol Jet Printed Conductive Interconnects on Paper Substrate

Mohammed Alhendi, Rajesh S. Sivasubramony, Jack Lombardi III,
Darshana L. Weerawarne, Peter Borgesen, Mark D. Poliks
Department of System Science and Industrial Engineering
Binghamton University, Binghamton, New York 13902
Email: mpoliks@binghamton.edu

Azar Alizadeh
General Electric Global Research
Niskayuna, New York 12309
Email: alizadeh@ge.com

Abstract—Growing demand for wearable and disposable electronics leads to a need for cost effective and compact sensor designs and fabrication. Most of the devices are multi-layered and require a carrier substrate to hold the sensors. Paper substrates have gained attention since they have the potential to act as both the sensor and the substrate itself. Paper-based printed sensors have been demonstrated and shown functional. However, device fabrication on paper is challenging because of the surface roughness, bleeding, and incompatibility with high temperature sintering processes needed to achieve high conductivity. The conductivity of the interconnects is therefore usually relatively low and imposes performance limitations. Here we report, for the first time, highly conductive silver nano-particle interconnects printed on a paper substrate and sintered with a continuous wave laser. The printing process was identified and the laser sintering parameters were optimized to achieve a conductivity of approximately 67% of the bulk material. As an example of application, interdigitated electrodes were printed and laser sintered. The leakage current was monitored while aging at 50°C /85% RH conditions and exposing to water and artificial sweat.

Keywords—Flexible Hybrid Electronics; Paper Substrate; Selective Laser Sintering; Silver Nano-particles

I. INTRODUCTION

Flexible Hybrid Electronics (FHE) have gained significant attention in many applications such as human performance wearable monitoring electronics, Internet of Things, and solar cells [1]. FHE enable the advantages of flexible printed electronics along with the conventional rigid electronics leading to a new generation of electronics industry. FHE advantages include flexibility, enabling new materials and hence a wide range of applications and functionalities, low power consumption, low cost and rapid manufacturing [2]. Among these applications, wearable electronics have attracted extensive attention such as vital signs monitoring platforms, tracking systems, gait analysis sensors, sweat sensors, microfluidic diagnosis platforms and other human healthcare related applications [3]. For all applications, substrate choice is very critical step in fabrication of flexible printed functional electronics. The most common substrates used for flexible hybrid electronics are poly-ethylene naphthalate (PEN) [4], polyimide (PI) [5], thermoplastic polyurethane (TPU) [6], polyethylene terephthalate (PET) [7], and glass [8]. For wearable and bio-sensing applications, paper is considered as a very promising substrate due to many advantages that are not offered by other plastic substrates. Paper is very cheap and available everywhere. Moreover, paper is made of environment friendly materials, which make it non-harmful for the environment and recyclable. In addition, paper substrates are thin, lightweight, flexible and can be modified for a wide range of applications by either surface modifications and/or structure functionalizing [9–12]. As the substrate choice matters, highly conductive interconnects are required for many FHE applications. Conductive interconnects are printed using either nano-particles [13] or flakes [14] based inks; however metal nano-particles are favorable because of the small size and the high conductivity that can be achieved [15]. Metal nanoparticle-based inks commonly used in printed electronics include dispersants, surfactants, solvents, and other additives. Even though some of the solvent evaporates during the printing process, these inks require thermal post processing to achieve conductivity closer to bulk. During this thermal post process, sometimes known as sintering, atomic diffusion takes place resulting in welding (necking) of nanoparticles, densification, recrystallization, and grain growth [16]. Commonly used sintering techniques such as convection oven and hotplates sometimes involve multiple hours of sintering time making them incompatible with high throughput Roll-to-Roll fabrication. Also, sintering nanoparticle-based inks requires high temperatures leading to incompatibility with many plastic polymeric substrates; for instance, TPU, PET, and specially coated papers. Different sintering techniques such as laser, intense pulsed light (IPL), microwaves, photonic, chemical, flash lamp, and electrical sintering have been used attempting to overcome these limitations [17–24]. Printing and laser sintering on paper has been reported in the literature [20–22] but the paper substrates used in those studies were surface coated to reduce the surface roughness. In addition, effects of laser sintering process parameters were not investigated in any detail. In this work, we report printing and laser sintering of thin and wide interconnects on an uncoated paper substrate. The effect of laser sintering process parameters on the conductivity of the printed interconnects was studied in detail.

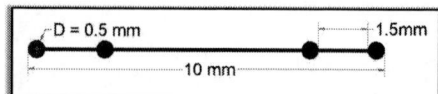

Fig. 1. Test structure design consists of four pads (0.5 mm diameter) and 5 mm sense line

II. METHODS

A. Printing Method

An Aerosol Jet Printer, AJP 300 from Optomec USA, was used in this work for fabricating interconnects on a paper substrate (GE Healthcare Fiberglass Grade D Paper). A commercially available silver nano-particles ink (PARU PG 007 AP, 60% solid weight) was used for printing with a pneumatic atomizer and 200 μm tip diameter. All printed interconnects were printed at room temperature. The printing process parameters consisted of 40, 680, and 640 sccm for the sheath, atomizer, and exhaust gas flow, respectively. To avoid/minimize oxidation and aging effects, printed samples were characterized and studied without significant delay after printing except as specified.

Fig. 1 shows the test structure used to assess the printability and the electrical conductivity of laser sintered traces on the paper substrate.

B. Laser Sintering Method

Laser sintering was performed using a CW laser system, installed on an Aerosol Jet Printer (AJP 5X, Optomec USA), which has 820 nm wavelength and spot size of approximately 80 μm. The same generated toolpath for printing was used for laser sintering, changing the tool from printing head to laser head.

C. Electrical Resistance Measurements

The electrical resistance of the interconnects was measured using a Keithley 2614B source meter at minimal source current to minimize any heat induced into the printed trace as this changes the resistance measurements. To eliminate the effect of the contact resistances and the source meter wire resistances, four-wire resistance measurement was employed. For each sample, five readings were recorded and averaged out to be statistically accurate.

III. RESULTS AND DISCUSSION

A. Substrate Characterization

Fig. 2 shows the surface texture of the paper substrate used. The high porosity, high surface roughness and 3D dimensional structure of this paper make printing quite challenging. Depending on the intended applications surface coating is commonly used to allow printing of continuous traces on such high roughness substrates. However, we demonstrate the feasibility of conformal printing on uncoated paper. The surface roughness was measured using a Keyence 3D laser confocal microscope (VK-X1050). The arithmetic mean height (S_a) and the root mean square height (S_q) were measured

Fig. 2. SEM image shows the paper substrate texture

Fig. 3. Line segments of 2 mm length printed at speed of 0.1 - 8 mm/s

to be 6.639 μm and 8.282 μm, respectively for 1.5 x 1.5 mm^2 scanned area. However, for the same scanned area, the measured highest (S_p) and lowest (S_v) peaks were 22.722 μm and 26.358 μm, respectively. The surface roughness was different for the same area size in a different location; 8.703 μm, 12.215 μm, 64.435 μm and 37.120 μm for S_a, S_q, S_p and S_v respectively. This difference indicates the high porosity and high surface roughness as well as the non-uniform surface texture distribution of this paper substrate. This affects the ink absorption on the paper surface leading to variations in the printing quality, width, thickness and eventually laser sintering results.

B. Printability of Conductive Traces

To assess the printability of conductive traces on such substrates, line segments of 2 mm length were printed at speeds of 0.1, 0.5, 1, 2, 3, 4, 5, 6, 7 and 8 mm/s. As shown in Fig. 3, we observed relatively higher bleeding and wetting of the paper substrate at low speed (0.1 and 0.5 mm/s) as the high absorption rate of the paper allows for a large amount of ink to diffuse into the substrate fiber network. On the other hand, high printing speed > 5 mm/s resulted in discontinuous lines with relatively less amount of ink. From this experiment,

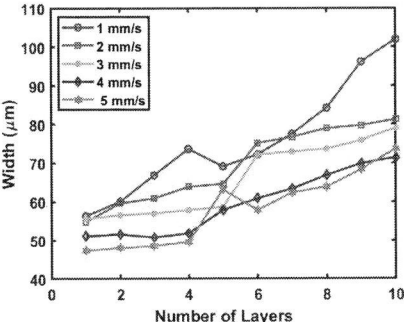

Fig. 4. Printed lines width versus number of layers at different speed

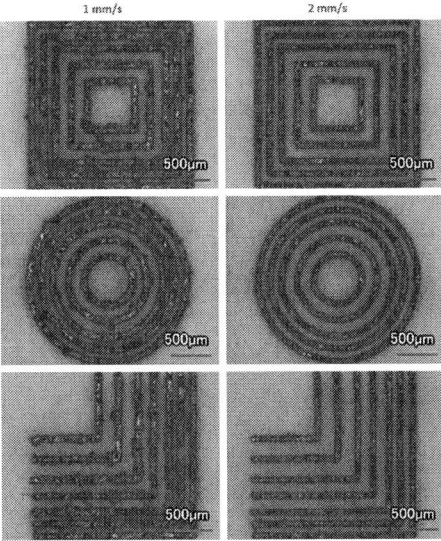

Fig. 5. Resolution patterns consist of different concentric geometries printed at 1 mm/s and 2 mm/s with five layers

we found that speeds of $1 - 5$ mm/s are promising for further experiments in term of amount of ink along the line length, but all the resulting printed lines were still non-conductive due to the high porosity surface. This high porosity allowed the ink solvent to be absorbed due to the capillary effect and left the metal nano-particles either on the surface or caused bleeding into the substrate pores. This resulted in silver nano-particles without a connecting path to achieve conductivity so achieving conductivity required printing of multiple layers.

Fig. 4 shows how the line width increases as the printing speed decreases and/or the number of printed layers increases. The relationship varies between different substrates such as PI and PET due to differences in the substrate surface texture. As shown in Fig. 4 a line width of less than 100 μm was achieved on paper which makes this printing technique attractive for printing high resolution patterns.

Printing speeds of 1 mm/s and 2 mm/s with five layers were considered further and a printing speed of 2 mm/s was seen to allow for relatively higher definition of trace edges, better

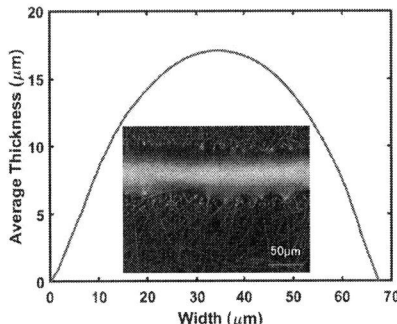

Fig. 6. Cross-section profile of a printed trace (shown inset) at 2 mm/s with five layers

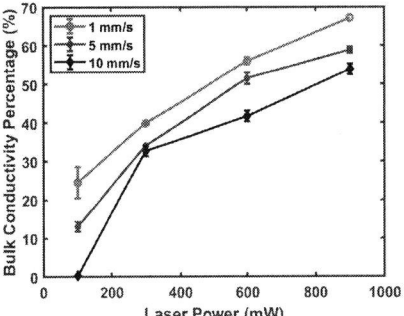

Fig. 7. Conductivity of the laser sintered traces as a function of laser parameters expressed as a percentage of the silver bulk conductivity

resolution and higher throughput as depicted in Fig. 5.

C. Laser Sintering of Printed Traces

The solvent in the silver nano-particle inks bleeds into the spaces within the fiber network of the paper resulting in interconnects with resistance on the order of MΩ. Therefore, welding (necking) of nanoparticles, densification, and grain growth of the silver nano-particles are required to obtain highly conductive traces. To realise that, traces of approximately 70 μm length and average thickness of 17 μm were printed at printing speed of 2 mm/s with five layers and laser sintered. Fig. 6 shows the cross-sectional profile of a trace printed at 2 mm/s with five layers. This profile was obtained by averaging a trace length of 270 μm with line scan interval of 300 nm. The laser sintering process was studied through a statistical design of experiment using a general full factorial design. The full factorial design consisted of three levels of laser scanning speed and four levels of laser power with three replicates for each combination. The laser power parameters consisted of 100 mW, 300 mW, 600 mW, and 900 mW and the speed consisted of 1 mm/s, 5 mm/s and 10 mm/s.

Fig. 7 shows the bulk conductivity percentage as a function of laser power and scanning speed levels. The electrical conductivity was calculated using the equation $\rho = L/(RA)$ where ρ is the resistivity which is the reciprocal of the conductivity, R is the four points resistance measurements (average of five readings) and A is the cross sectional area of the printed trace, i.e. length of the trace multiplied by the average thickness. The trace thickness was measured using a Keyence 3D laser confocal microscope and a Scanning Electron Microscope (SEM) to equal 20 μm approximately. However, we measured trace thickness printed on UPILEX PI 2 mil substrate to be around 15 μm. This difference is due to the substrate nature. In fact, identifying the actual thickness that contributes to the trace conductivity is quite challenging in the case of high surface roughness. From Fig. 7, the lowest bulk conductivity percentage, 0.20% of the silver bulk conductivity, was achieved at a speed of 10 mm/s and a laser power of 100 mW, indicating that a very small amount of the laser fluence was delivered into the printed trace. This energy was not high enough to melt the silver nano-particles and did not allow metal necking and grain growth. It could only enhance the evaporation of the ink organic solvent and other additives. On the other hand, the highest conductivity, 67.11% of the bulk conductivity, was achieved at 900 mW laser power and 1 mm/s scanning speed. However, the high delivered laser fluence into the trace caused thermal damage in a few samples, either by burning the substrate or by overheating the trace, leading to weak top layers. We observed a loss of conductivity at high power and slow speed due to delamination/cracking of the printed trace. This delamination could be due to the rapid evaporation of the solvent and subsequent thermal shock and volume shrinking. Therefore, such a high fluence was not always acceptable in this work. The conductivity of the laser sintered traces achieved with 600 mW and a speed of 5 mm/s was around 50% of the bulk conductivity. To the best of our knowledge this is the highest conductivity reported with a paper substrate [20, 21]. In contrast, sintering this ink, after printing on a PI substrate, in a convection oven required 1 hour and 20 min at 200 °C to yield only 7.0% of the bulk conductivity [13]. This illustrates the advantages of the laser sintering process for use with paper substrates.

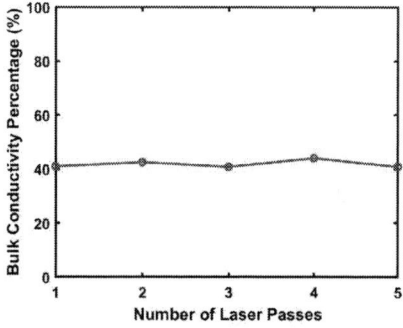

Fig. 8. Conductivity of the laser sintered thin traces as a function of multiple laser sweeps

Fig. 8 shows the effect of multiple sintering passes on the conductivity of thin printed silver traces. These printed traces were sintered at 600 mW and a scanning speed of 5 mm/s. Even though a single laser sweep resulted in higher

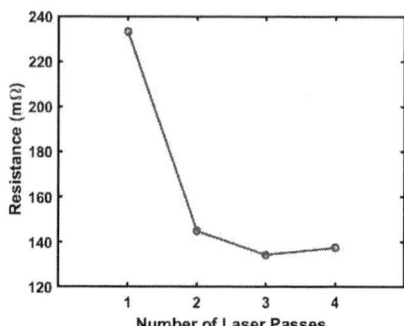

Fig. 9. Resistance of printed wide traces as a function of multiple laser sweeps

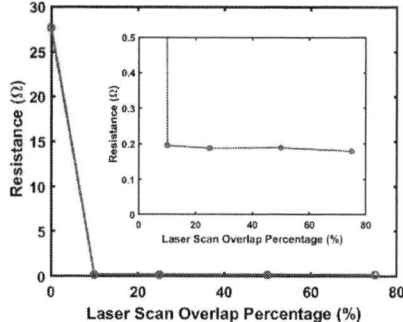

Fig. 10. Resistance of wide printed traces as a function of scanning overlap percentage. The inset figure shows smaller y-axis range for visualization

conductivity at 900 mW and 5 mm/s, we notice multiple passes had no significant effect on the conductivity at 600 mW and 5 mm/s. This may be due to the higher amount of surface reflection after the first laser sweep and resulting lower amount of transmitted/absorbed laser power in subsequent sweeps. However, this depends on the printed trace geometry; specifically, trace thickness. Lateral (on the substrate) and longitudinal (into the substrate) heat propagation also influence the sintering process via heat conduction.

Different traces, wider and thicker, were printed and laser sintered at a speed of 10 mm/s and a power of 600 mW with multiple passes. Fig. 9 shows the effect of multiple passes on these traces. The second laser scanning pass leads to an almost 60% drop in resistance compared to the first pass. However, more passes led to no further change in the resistance. We speculate that heat conduction during the second laser pass and resulting sintering may have played a role for these thicker traces. Multiple laser sweeps with an appropriate offset were required for wider traces. We studied the effect of the choice of laser scanning overlap on 350 μm wide traces. The overlap percentage was set to 0%, 10%, 25%, 50% and 75% of the laser spot size (80 μm). The laser energy transferred into the trace at 0% overlap was not high enough to make a strong connection between the silver nano- particles as indicated by the high resistance (27 Ω) in Fig. 10. The resistance decreased

978-1-7281-1500-9/19 $31.00 © 2019 IEEE

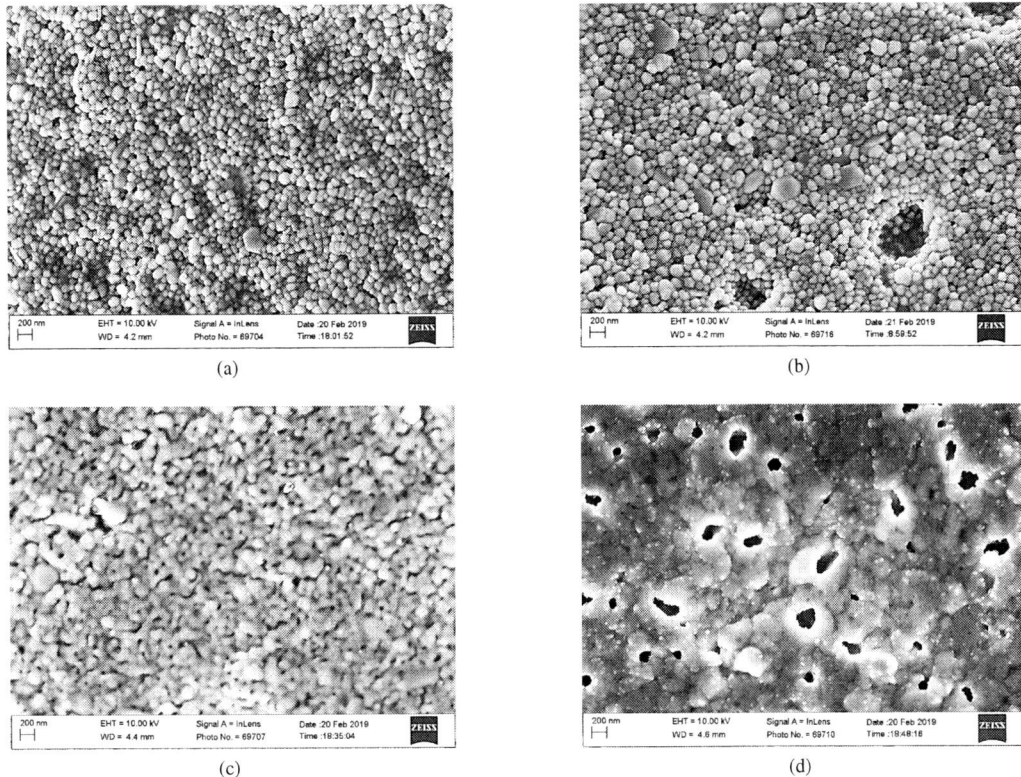

(a)   (b)

(c)   (d)

Fig. 11. SEM images show the microstructures morphology for printed traces a) without sintering (air dry) and sintering with b) 100 mW at 10 mm/s, c) 600 mW at 5 mm/s, and d) 900 mW at 1 mm/s

as the overlap percentage increased. However, the difference in the measured resistances was not significant above a 10% overlap; resistances dropping from 0.1957 Ω with 10% (72 μm offset) overlap to 0.1793 Ω with 75% (20 μm offset) as shown in the inset in Fig. 10. A minimum overlap is preferred to protect the paper from excessive heat generated by larger overlap percentages.

IV. MICRO STRUCTURE OF LASER SINTERED TRACES

Fig. 11 shows the micro-structure evolution of the printed traces as the laser sintering process parameters change. The unsintered trace (Fig. 11a) consists of individual silver nanoparticles protected by organic additives resulting in a high trace resistivity. Not surprisingly, the degree of sintering depends on the laser powers as well as the laser scanning speeds. The low laser power evaporates the organic additives and initiates the contact between the silver nanoparticles (Fig. 11b). At higher laser power, extensive necking and particles fusing were observed (Fig. 11c) and at even higher power (900 mW) and lower speed (1 mm/s), grain sizes increased (Fig. 11d) prompting very low resistivity. However, the high laser power also caused pores in the microstructure (Fig. 11d) which could be potential crack initiating sites.

V. APPLICATION EXAMPLE; INTERDIGITATED ELECTRODES FABRICATION

Interdigitated electrodes have been extensively used in many applications such as sweat, ion concentration, humidity, particle detection, and other chemical/biochemical sensors. As an example of an application we fabricated thin (100 μm wide) and wide (350 μm wide) interdigitated electrodes on the paper substrate with linear and diagonal direction patterns as shown in Fig. 12. The effect of the humidity and the temperature on printed patterns was studied. After aging the ID electrodes in an environmental chamber at 50 °C temperature and 85% relative humidity for 10 days, we found very consistent resistance between the ID electrodes terminals, i.e. there was only a negligible leakage current as a response to the humidity and temperature conditions. This could be as a reason of taping of the electrodes to a rigid surface, limiting the water vapor absorption to only one side; the printing side. Moreover, the hydrophobicity of this paper (fiberglass made) affected the water vapor absorption into the paper pore network.

Another experiment measured the changes in the resistance under two conditions; application of water and artificial sweat drops (Artificial Eccrine Perspiration from Pickering Laboratories) on the reverse side of the electrodes for one hour. The fluid drops diffused through the spaces between the glass

978-1-7281-1500-9/19 $31.00 © 2019 IEEE

Fig. 12. Interdigitated electrodes; linear and diagonal directions printed and laser sintered on the paper substrate

fibers, changing the resistance. As expected, the artificial sweat had more effect on the resistance than the water due to the chemical ion composition of the artificial sweat, Fig. 13.

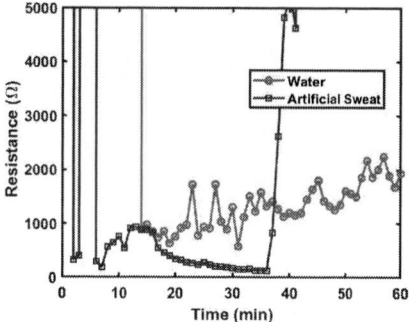

Fig. 13. Effect of water and artificial sweat drops on the resistance of interdigitated electrodes

VI. CONCLUSION

Interconnects were successfully aerosol jet printed on a very rough paper substrate. Lines with less than 100 μm width were achieved at different speeds up to 5 mm/s and with up to 10 layers. The laser sintering process parameters were optimized to achieve very high conductive interconnects; 67% of the silver bulk conductivity. The conductivity depended on the laser power, scanning speed, thickness of the printed traces, and the overlap percentage (wide traces). The conductivity increased with high laser power and slow scanning speed to a certain point where these parameters caused delamination and damage to the substrate. Multiple laser passes had no significant effect on the conductivity of thin traces. A spatial overlap between laser scans was required for sintering wide lines and the overlap was dependent on the laser spot size, the direction of the scanning, and the thickness of the printed trace.

ACKNOWLEDGMENT

This material is based on research sponsored by Air Force Research Laboratory under agreement number FA8650-15-2-5401. The U.S. Government is authorized to reproduce and distribute reprints for Governmental purposes notwithstanding any copyright notation thereon. The views and conclusions contained herein are those of the authors and should not

be interpreted as necessarily representing the official policies or endorsements, either expressed or implied, of Air Force Research Laboratory or the U.S. Government.

REFERENCES

[1] W. S. Wong and A. Salleo, *Flexible electronics: materials and applications.* Springer Science & Business Media, 2009, vol. 11.

[2] Y. Khan, M. Garg, Q. Gui, M. Schadt, A. Gaikwad, D. Han, N. A. Yamamoto, P. Hart, R. Welte, W. Wilson *et al.*, "Flexible hybrid electronics: direct interfacing of soft and hard electronics for wearable health monitoring," *Advanced Functional Materials*, vol. 26, no. 47, pp. 8764–8775, 2016.

[3] B. W. An, J. H. Shin, S.-Y. Kim, J. Kim, S. Ji, J. Park, Y. Lee, J. Jang, Y.-G. Park, E. Cho *et al.*, "Smart sensor systems for wearable electronic devices," *Polymers*, vol. 9, no. 8, p. 303, 2017.

[4] M. Berggren, D. Nilsson, and N. D. Robinson, "Organic materials for printed electronics," *Nature materials*, vol. 6, no. 1, p. 3, 2007.

[5] M. Poliks, J. Turner, K. Ghose, Z. Jin, M. Garg, Q. Gui, A. Arias, Y. Kahn, M. Schadt, and F. Egitto, "A wearable flexible hybrid electronics ecg monitor," in *2016 IEEE 66th Electronic Components and Technology Conference (ECTC)*. IEEE, 2016, pp. 1623–1631.

[6] B. Karaguzel, C. Merritt, T. Kang, J. Wilson, H. Nagle, E. Grant, and B. Pourdeyhimi, "Flexible, durable printed electrical circuits," *The Journal of The Textile Institute*, vol. 100, no. 1, pp. 1–9, 2009.

[7] D. Kim and J. Moon, "Highly conductive ink jet printed films of nanosilver particles for printable electronics," *Electrochemical and Solid-State Letters*, vol. 8, no. 11, pp. J30–J33, 2005.

[8] J. C. Switzer III and M. D. Poliks, *Roll-to-Roll Processing of Flexible Glass.* Wiley-Scrivener: Hoboken, NJ, 2017.

[9] D. Tobjörk and R. Österbacka, "Paper electronics," *Advanced Materials*, vol. 23, no. 17, pp. 1935–1961, 2011.

[10] A. C. Siegel, S. T. Phillips, M. D. Dickey, N. Lu, Z. Suo, and G. M. Whitesides, "Foldable printed circuit boards on paper substrates," *Advanced Functional Materials*, vol. 20, no. 1, pp. 28–35, 2010.

[11] D. D. Liana, B. Raguse, J. J. Gooding, and E. Chow, "Recent advances in paper-based sensors," *Sensors*, vol. 12, no. 9, pp. 11 505–11 526, 2012.

[12] J. Lombardi, M. D. Poliks, W. Zhao, S. Yan, N. Kang, J. Li, J. Luo, C.-J. Zhong, Z. Pan, M. Almihdhar *et al.*, "Nanoparticle based printed sensors on paper for detecting chemical species," in *2017 IEEE 67th Electronic Components and Technology Conference (ECTC)*. IEEE, 2017, pp. 764–771.

[13] R. S. Sivasubramony, N. Adams, M. Alhendi, G. S. Khinda, M. Z. Kokash, J. P. Lombardi, A. Raj, S. Thekkut, D. L. Weerawarne, M. Yadav *et al.*, "Isothermal fatigue of interconnections in flexible hybrid electronics based human performance monitors," in *2018 IEEE 68th Electronic Components and Technology Conference (ECTC)*. IEEE, 2018, pp. 896–903.

[14] M. Alhendi, J. P. Lombardi III, G. S. Khinda, M. Z. Kokash, D. L. Weerawarne, P. Borgesen, M. D. Poliks, N. C. Stoffel, and J. Iannotti, "Fatigue cycling of electrical interconnects dispensed on flexible substrate," in *International Symposium on Microelectronics*, vol. 2018, no. 1. International Microelectronics Assembly and Packaging Society, 2018, pp. 000 543–000 548.

[15] Y. Li, Y. Wu, and B. S. Ong, "Facile synthesis of silver nanoparticles useful for fabrication of high-conductivity elements for printed electronics," *Journal of the American Chemical Society*, vol. 127, no. 10, pp. 3266–3267, 2005.

[16] S.-J. L. Kang, *Sintering: densification, grain growth and microstructure.* Elsevier, 2004.

[17] W. Liu, R. An, C. Wang, Z. Zheng, Y. Tian, R. Xu, and Z. Wang, "Recent progress in rapid sintering of nanosilver for electronics applications," *Micromachines*, vol. 9, no. 7, p. 346, 2018.

[18] Z. Cao, N. Yoshikawa, and S. Taniguchi, "Microwave heating behavior of nanocrystalline au thin films in single-mode cavity," *Journal of Materials Research*, vol. 24, no. 1, pp. 268–273, 2009.

[19] K. Schroder, S. McCool, and W. Furlan, "Broadcast photonic curing of metallic nanoparticle films," *NSTI Nanotech May*, vol. 7, p. 11, 2006.

[20] D. Tobjörk, H. Aarnio, P. Pulkkinen, R. Bollström, A. Määttänen, P. Ihalainen, T. Mäkelä, J. Peltonen, M. Toivakka, H. Tenhu *et al.*, "Ir-sintering of ink-jet printed metal-nanoparticles on paper," *Thin Solid Films*, vol. 520, no. 7, pp. 2949–2955, 2012.

[21] E. Balliu, H. Andersson, M. Engholm, T. Öhlund, H.-E. Nilsson, and H. Olin, "Selective laser sintering of inkjet-printed silver nanoparticle inks on paper substrates to achieve highly conductive patterns," *Scientific reports*, vol. 8, no. 1, p. 10408, 2018.

[22] Q. Huang, W. Shen, Q. Xu, R. Tan, and W. Song, "Room-temperature sintering of conductive ag films on paper," *Materials Letters*, vol. 123, pp. 124–127, 2014.

[23] M. Zenou, O. Ermak, A. Saar, and Z. Kotler, "Laser sintering of copper nanoparticles," *Journal of physics D: Applied physics*, vol. 47, no. 2, p. 025501, 2013.

[24] M. L. Allen, M. Aronniemi, T. Mattila, A. Alastalo, K. Ojanperä, M. Suhonen, and H. Seppä, "Electrical sintering of nanoparticle structures," *Nanotechnology*, vol. 19, no. 17, p. 175201, 2008.

2019 IEEE 69th Electronic Components and Technology Conference (ECTC)

In-Situ Investigation of Organic Additive Interactions in Copper Electroplating Solutions with Surface Enhanced Raman Spectroscopy (SERS)

Nithin Nedumthakady, Bartlet DeProspo, Himani Sharma, Rahul Manepalli*,
Sashi Kandanur*, Sajanlal Panikkanvalappil, Nasrin Hooshmand, Rao Tummala

3D Microsystems Packaging Research Center
Georgia Institute of Technology, Atlanta, Georgia, USA
Email: nnedumthakady@gatech.edu

*Intel Corporation, Phoenix, AZ, USA

Abstract—To meet demand for increasingly higher-performance electronics in increasingly smaller form factors, achieving higher logic-memory bandwidth and higher I/O density is required. This necessitates finer copper lines, smaller copper microvias, and fine-pitch, copper-filled through-package vias (TPVs). However, low-quality copper filling in high-aspect ratio TPVs can lead to void formation and mechanical failures. This poses various material and process challenges for the achievement of high-quality copper metallization. These challenges can be addressed through a better understanding of copper deposition mechanisms and investigating the fundamental aspects of copper plating chemistry. In this study, in-situ interactions between different classes of organic additives in copper plating solutions were investigated with surface-enhanced Raman spectroscopy (SERS). A novel test set up is developed which allows for the direct observation of copper plating within a via. SERS was used to observe cases of competitive adsorption between suppressor, leveler, and accelerator additives within this 'via'. These observations are used to infer the impact of different classes of additives on via-filling performance and correlate the electrochemical behavior in the plating cell. A model is proposed to explain the relative tendencies of the differing classes of additives to adsorb to the copper surface within a via.

Keywords-copper electroplating; via fill; substrate plating; copper electrodeposition; plating additives;

I. INTRODUCTION

Increasing demands for high-performance electronics in smaller form factors are driving development of advanced packaging technologies. To meet this demand for improved system performance, achieving higher logic-memory bandwidth and higher I/O density is required. This necessitates finer copper lines, smaller copper microvias, and fine-pitch, copper-filled through-package vias (TPVs), as seen in recent 2.5D and 3D system packages [1], [2]. A cross-section of such as a system is given in Figure 1.

TPVs and blind-vias that are fully-filled with copper, also known as 'superfilled' are desired over conformally plated copper due to lower DC resistance, improved power

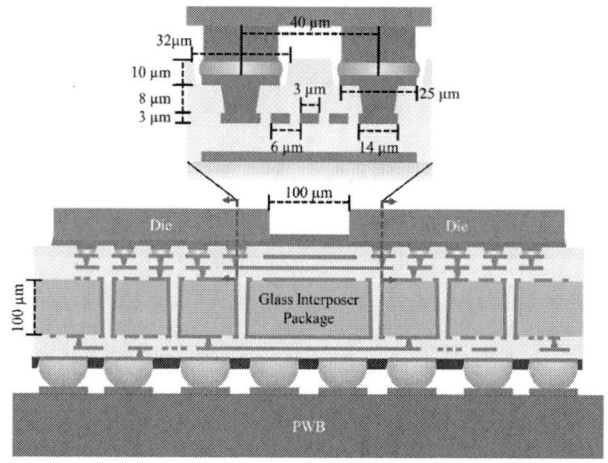

Figure 1. Schematic of 2.5D glass interposer package from Georgia Tech [1]

integrity, and superior thermal dissipation [2]. However, low-quality copper filling in high-aspect ratio TPVs can lead to void formation and mechanical failures. This poses various material and process challenges for the achievement of high-quality copper metallization. These challenges can be addressed through a better understanding of copper deposition mechanisms and investigating the fundamental aspects of copper plating chemistry.

Copper via-filling in electronic packaging is traditionally carried out with electrochemical deposition, or electroplating technique due to low cost, high scalability, reliability and processability. Fully-filled copper vias are typically fabricated by first depositing a nanoscale, conformal, copper seed layer and then electrolytic plating in an aqueous electrolyte that is made of an acid, copper salt, and organic additives. Challenges with electroplating fully-filled copper vias include: pinched vias, non-conformal deposition, voids, etc [3]. An example of voiding in copper deposition is shown

978-1-7281-1500-9/19 $31.00 © 2019 IEEE   1588

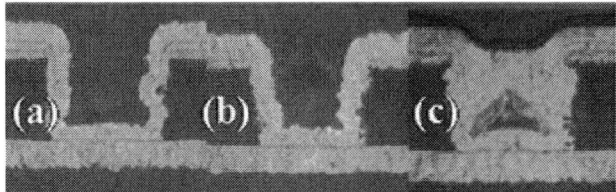

Figure 2.   Example of voiding within a copper via [4]

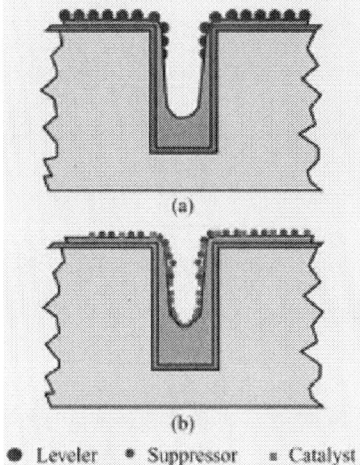

Figure 3.   Schematic of additive interactions promoting superfilling within a copper via [5]

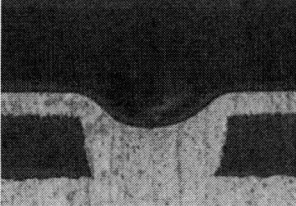

Figure 4.   Use of only suppressor (PEG) and accelerator (SPS) additivies resulting in conformal copper fill profile within the via (left). Addition of leveler additive helps promote superfilling of copper with the via (right). [6]

in Figure 2.

There is a need to refine the additives in plating to circumvent such challenges. Many additives have been studied in-depth and individually, and some combinations have been known to promote copper via superfilling. A schematic of additive interactions promoting superfilling of copper within vias is shown in Figure 3. However, there remains a lack of understanding of the relationship between additive structure, interactions with other additives, influence of additives on electrochemical behavior, and impact on filling performance.

Classical investigations into organic additive interactions are performed ex-situ and have limited insight into interactions that occur during the plating process and at the copper surface. In addition, limited insight is gained into the interactions of multi-additive systems. By developing a novel test set up, the organic additives interactions are observed in-situ using surface enhanced Raman spectroscopy (SERS) during the electrolytic copper plating process. This process helps in studying the changes in Raman signals from the copper surface while the additives are actively being adsorbed/ desorbed from it and allows for insight into interactions at the surface.

In these experiments, model organic additives were chosen based on industry standards. Depending on their function and influence on copper plating behavior and plating rate,

organic additives are generally categorized as suppressors, accelerators, or levelers. Accelerators form electroactive species responsible for an enhanced plating rate. Suppressors combine with chloride ions to inhibit plating on areas where a reduced plating rate is desired. Levelers polarize the areas with high current densities and balance current distribution, thus helping to control surface morphology. Combining these three types of additives in correct proportions results in healthy superfilling of copper vias; however, wrong proportions can result in defects [5]. An example of the effects of additives on electroplating profiles is given in Figure 4.

This study aims to explain the electrochemical behavior and adsorption/desorption characteristics of individual additives and their combinations. The adsorption and polarization characteristics of suppressor, accelerator, and leveler additives and their combinations were first studied through cyclic voltammetry and chronopotentiometry. Then, a novel set-up designed to allow the direct observation of copper plating in a via using SERS is fabricated and employed. Conclusions from the cyclic voltammetry and injection chronopotentiometry experiments were used to validate and explain observations and results generated from SERS. An empirical model is proposed that aims to explain the adsorption tendencies, electrochemical behavior, and competitive nature of these combined additives that promote void-free via-fill.

II. Experimental

A virgin make-up solution (VMS) was prepared as an additive-free base electrolyte. The VMS bath consisted of $CuSO_4$ (50 g/L), H_2SO_4 (100 g/L), and Cl^- ion (70 ppm).$CuSO_4$ was added in the form of $CuSO_4 \cdot 5(H_2O)$ (78 g/L) and Cl^- was added in the form of NaCl (115 ppm).

Polyethylene glycol (PEG) was selected as the suppressor, bis(sodium sulfopropyl) disulfide (SPS) was selected as the accelerator, and polyvinylpyrrolidone (PVP) was selected as the leveler to be studied. Within the plating bath, concentrations of these additives were varied, and different combinations of these additives were studied to understand the active polarization/depolarization of the copper surface as an effect of organic additive.

Electroanalytical characterization was performed using a platinum rotating disc electrode (RDE) with a diameter of

978-1-7281-1500-9/19 $31.00 © 2019 IEEE

0.4cm. The rotation speed was kept at 200RPM. A high-purity Cu sheet was used as the counter electrode. About 200 nm of copper was deposited by electrolytic plating in VMS prior to each measurement. All reported potentials are with reference to a $Hg/Hg_2SO_4/K_2SO_4$ (sat.) electrode.

Cyclic voltammetry was performed using a CH Instruments 600C Electrochemical Analyzer. The potential was swept from the open circuit potential to -1.0V and back at a rate of 10 mV/s. Cyclic voltammetry of solutions containing VMS, VMS+300ppm PEG, VMS+10ppm SPS, and VMS+300ppm PEG+ 10ppm SPS was performed. Please note that concentrations were based on industry standards.

Galvanostatic chronopotentiometry was used to monitor the potential of the copper working electrode while maintaining a plating current density of 20 mA/cm^2. For injection chronopotentiometry, additives were injected with a syringe containing the desired additives in VMS while plating on the Pt RDE at 200RPM with a current density of 20 mA/cm^2. All experiments began with plating the platinum RDE with fresh copper in pure VMS for 5 minutes. Galvanostatic chronopotentiometry was performed on solutions containing VMS, VMS+300ppm PEG, and VMS+10ppm SPS. Injection chronopotentiometry was performed on a solution containing VMS+300ppm PEG and 10ppm SPS was injected using a syringe. Once again, please note that concentrations were based on industry standards.

For SERS, the test vehicle was created by laminating photoresists of various thicknesses onto copper-clad boards. Trenches of varying lengths and widths were created in the photoresist lithographically. A copper seed layer is deposited on top of the photoresist and glass is bonded to the top of the copper-clad board. Adsorption kinetics can be best studied with surface enhanced Raman (SERS) where the copper seed layer was treated to enhance the Raman signal. By singulating these trenches, vias with a transparent fourth wall are created. This allows for the examination of all areas within a via over time and enables in-situ characterization with SERS. For the purposes of this study to provide initial proof of concept to determine that interactions can be visualized, trenches of dimensions 5mm x 25mm are fabricated. However, this concept can be translated to fabricate 'vias' on the microscale and understand interactions that occur within microvias. A process flow for test vehicle fabrication is given in Figure 5.

SERS spectra was measured using a Renishaw InVia Raman spectrometer. The 785nm laser was selected for superior organic ligand detection. The backscattered signals from the samples were collected by a CCD detector in the range of 400-1200 cm$_1^-$ with an integration time of 20s. Spectra was taken at 2 minutes, 5 minutes, and 10 minutes and changes in spectra were observed over time. In these experiments, please note that the concentration of the additives is raised significantly to produce a stronger signal in this proof of concept. SERS spectra was taken for solutions containing VMS+1000ppm PEG, VMS+1000ppm SPS. Two-additive systems modulating the ratios of PEG to SPS were also created and SERS spectra was taken. SERS spectra was taken for solutions containing VMS+1000ppm PEG+VMS+1000ppm SPS (1:1), VMS+1000ppm PEG+VMS+4000ppm SPS (1:4), VMS+4000ppm PEG+VMS+1000ppm SPS (4:1), and VMS+1000ppm PEG+VMS+100ppm SPS (100:1).

III. RESULTS

A. Cyclic Voltammetry

Cyclic voltammograms of VMS, VMS with 300ppm PEG, VMS with 10ppm SPS, and VMS with 300ppm PEG and 10ppm SPS are shown in Figure 6. Please note that concentrations were based on industry standards. The VMS CV curve gives a baseline by which the effects of the additives on the system can be observed.

In the VMS+PEG graph, there is an observable shift left as compared to just the VMS graph. The VMS+PEG solution shows strong suppressive effects, to about -0.6V, before beginning a steady, significant increase in current density. There is also minimal hysteresis observed, indicating that there is a singular effect on the system. In the VMS+SPS solution, there is an observable shift right as compared to just the VMS graph. The VMS+SPS solution shows strong accelerative effects, as the current density remains relatively high throughout the whole experiment. It is interesting to note that while the forward sweep is extremely similar to the forward sweep of the pure VMS system, the reverse sweep results in higher current densities throughout. Once again, there is minimal hysteresis observed, indicating only a singular effect on the system.

In the VMS+PEG+SPS solution, first a suppressive effect is observed in the forward sweep, followed by an accelerative effect in the reverse sweep. This hysteresis observed indicates two separate effects by the additives on the system as they adsorb and desorb from the copper surface. In addition, observing a final, accelerative effect in the reverse sweep indicates a dominance by the accelerator additive over the suppressor additive.

B. Injection Chronopotentiometry

Figure 7 shows the potential response of the copper working electrode when the RDE plating cell contained solutions of VMS, VMS+300ppm PEG, and VMS+10ppm SPS. Figure 7 also shows the potential response when the RDE plating cell contained a solution of VMS+300ppm PEG and 10ppm SPS was injected into the solution. Similarly to before, the VMS chronopotentiometry curve gives a baseline by which the effects of the additives on the system can be observed.

The potential of the solution containing VMS+PEG drops to about -0.7V. In other words, it takes more potential for plating to occur in the VMS+PEG system. This can be

Cu-clad FR4 Board

Laminate photoresist

Expose trenches, develop and cure
photoresist

Sputter Ti-Cu seed layer and singulate
trenches

Copper anneal and seal top with glass
slide

Figure 5.   Process flow for test vehicle fabrication for in-situ SERS

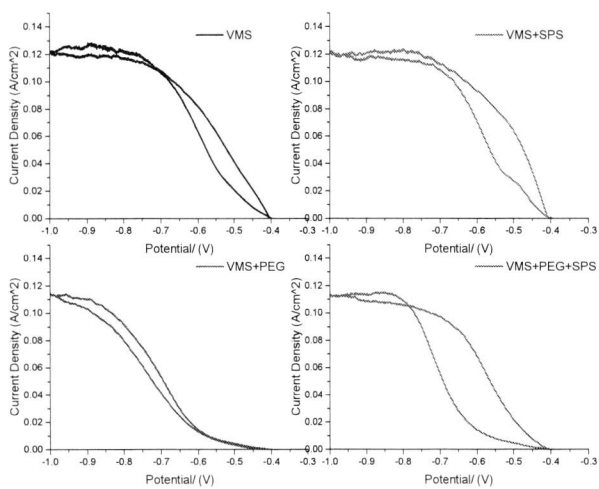

Figure 6.   Cyclic voltammetry curves for VMS, VMS+SPS, VMS+PEG, and VMS+PEG+SPS. Hysteresis in the 2-additive system indicates dominance of the SPS additive effects.

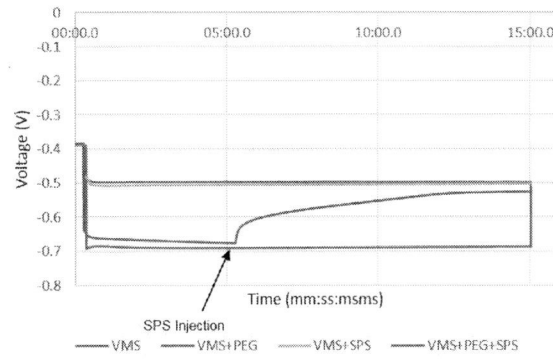

Figure 7.   Chronopotentiometry curves for VMS, VMS+PEG, VMS+SPS, and injection chronopotentiometry curves for VMS+PEG+SPS. Injecting SPS into the system causes the potential to rise indicating a dominance in the effect of SPS on the system.

attributed to the suppressive effects of PEG. The potential of the solution containing VMS+SPS increases very slightly. In other words, it takes slightly less potential for plating to occur in the VMS+SPS system. This can be attributed to the accelerative effects of SPS.

In the injection experiment, to start, PEG is the only additive in the RDE plating cell. The potential once again drops to approximately -0.7 and the same suppression of the system is observed. When SPS is injected into the system, the potential starts to rise until it reaches a stabilization point. This stabilization point is at a slightly lower potential than pure VMS or the VMS+SPS system. This can be attributed to SPS causing PEG to desorb from the copper surface and inducing an accelerative effect on copper plating; however, in this 2-additive system, copper plating is still slightly suppressed.

C. Surface Enhanced Raman Spectroscopy

Figure 8 shows the surface enhanced Raman spectra at 2, 5, and 10 minutes of the copper surface when plating in the VMS+PEG and VMS+SPS solutions. These spectra are used as baselines to look for specific peaks that are indicative of the presence of either PEG or SPS on the copper surface. When plating in VMS+PEG, specific peaks at 790, 985, and 1050 emerge. When plating in VMS+SPS, specific peaks at 622, 685, 731, 1034 emerge. These are the peaks that are used to determine the presence of SPS and PEG on the surface.

The 2-additive ratio modulation experiments were used to observe the relative dominance of the effects of SPS over PEG. Figures 8 and 9 show the surface enhanced Raman spectra of these ratio experiments. In the 1:1 PEG to SPS ratio case, spectra is repressed at the 2 minute time point, but by the 10 minute time point, the spectra is very reminiscent of that of just VMS+SPS. This indicates that as the plating process is carried out, SPS is the molecule that will preferentially adsorb to the copper surface. In the 1:4 PEG to SPS ratio case, the dominance of SPS is clear from the very beginning. In this experiment, the spectra is completely reminiscent of the VMS+SPS case at all time points. In the 4:1 PEG to SPS case, once again, peaks are repressed at the 2 minute time point, but a VMS+SPS-like spectra emerges by the 10 minute point. Even in the case of 100:1 PEG to SPS, the SPS peaks emerge by the 10 minute time point. It can be noted that these peaks are slightly repressed. These results correlate with the results gained from CV and chronopotentiometry. PEG, the suppressor additive, actively desorbs from the copper surface in the presence of SPS, the accelerator additive. PEG is shown to have a slight repression in the effects of SPS, but over time, SPS is the dominant additive within this 2-additive system. This adsorption/desorption process is what enables the accelerated plating effects that promote superfilling.

IV. FUTURE WORK

Future work involves the addition of the leveler additive in these experiments. Understanding the leveler additive's role in plating and how this molecule interacts with the suppressor and accelerator additives is crucial to understanding copper plating. By using SERS to monitor the copper surface during the plating process, the interactions between these three classes of additives can be made in real time and conclusions can be made. Work will also be completed to understand how the in-situ surface enhanced Raman spectroscopy peaks relate to the chemical bonds.

Further future work involves scaling down the 'via' size. By creating trenches on the microscale level, the interactions of these additives within the microvia can be explored.

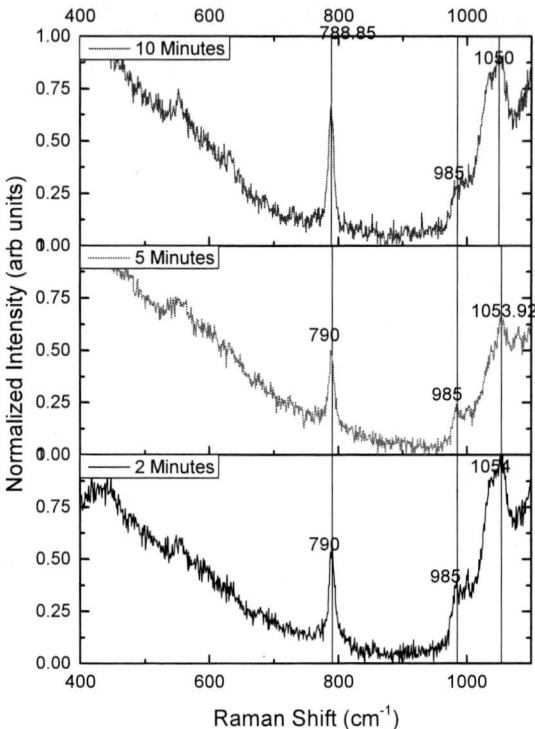

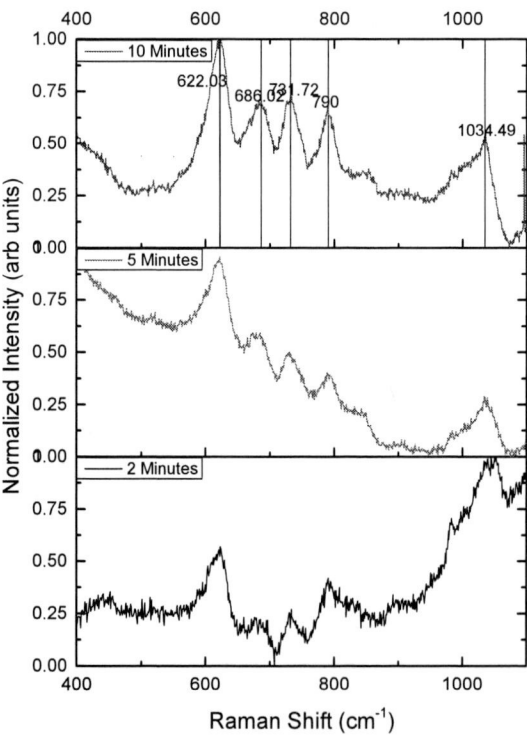

Figure 8.   SERS spectra for VMS+1000ppm PEG (top) and VMS+1000ppm SPS (bottom) taken at 2, 5, and 10 minutes.

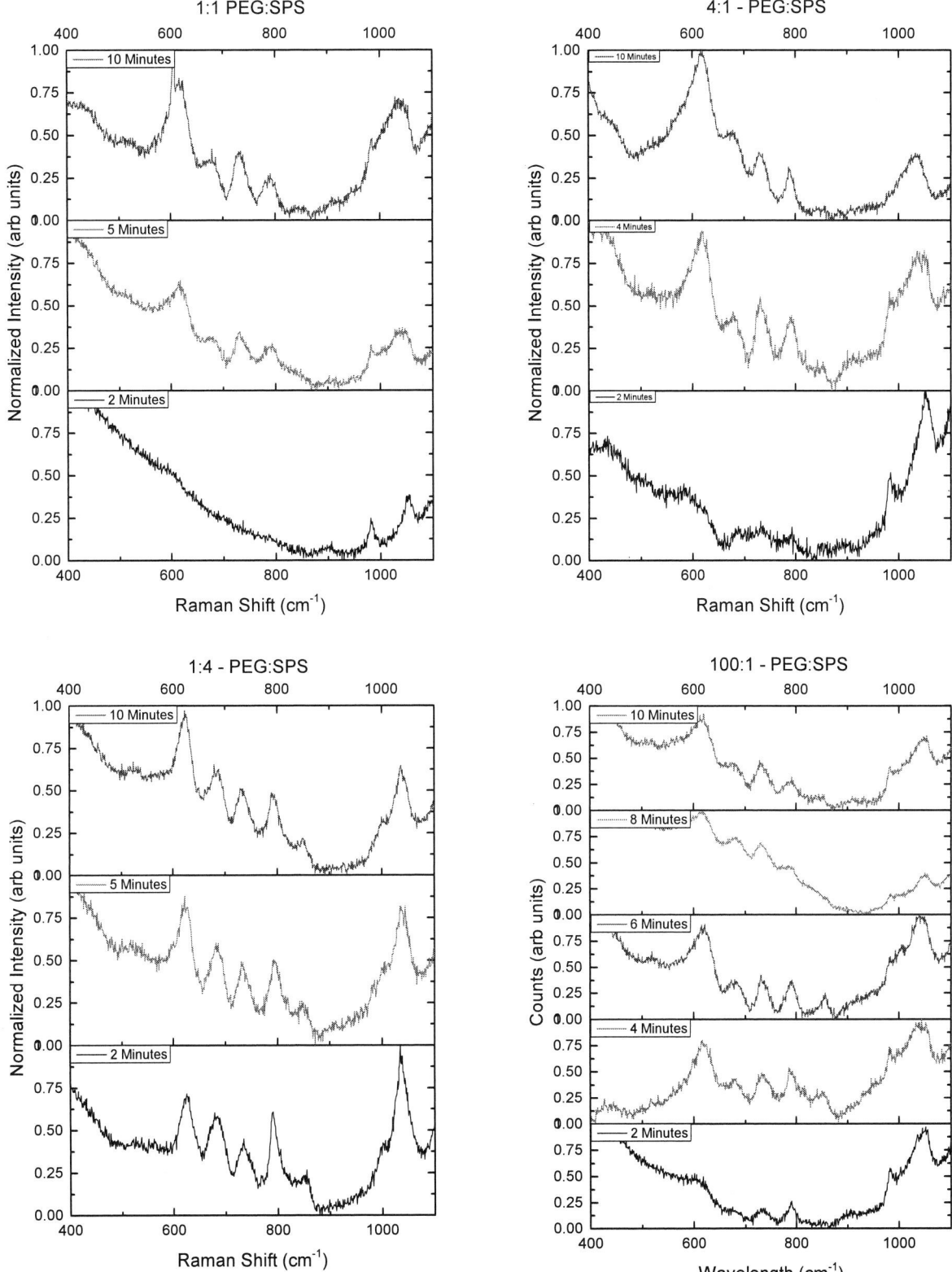

Figure 9. SERS spectra for 2-additive system ratio modulation experiments. In all cases, peaks that indicate the presence of SPS on the surface emerge over time. This indicates a preferential adsorption by SPS on the copper surface.

V. CONCLUSION

In summary, the in-situ interactions of different classes of additives within copper plating solutions is explored. Cyclic voltammetry and galvanostatic chronopotentiometry is used to make initial conclusions about the interactions between suppressor and accelerator types of additives. A novel test set up is developed which allows for the direct observation of copper plating within a via-like structure. Surface-enhanced Raman spectroscopy is used to observe the interactions that occur during plating within this via-like structure. Results obtained show the relative tendencies of the additives to adsorb and desorb on the copper surface over the plating process. The SERS spectra gave indications as to which additive molecule is present at the copper surface.

In conclusion, CV and chronopotentiometry of a system containing the suppressor (PEG) and accelerator PEG) show the dominance of the effect of SPS over PEG. SERS spectra of the two-additive system indicates the preferential presence of SPS on the copper surface over PEG over time. These results indicate the desorption of the suppressor additive and preferential adsorption of the accelerator molecule. Future work will seek to use SERS to observe in-situ interactions in a three-component system consisting of leveler, suppressor, and accelerator additives. Further work will involve scaling down the via sizes to observe interactions within microvias.

ACKNOWLEDGMENT

The authors would like to thank Intel Corporation for sponsoring this work. The authors would also like to thank Prof. El-Sayed and his research group for their assistance with in-situ Raman measurements. Finally, the authors would like to thank Dr. Timothy Huang who began this work.

REFERENCES

[1] B. M. Sawyer, Y. Suzuki, R. Furuya, C. Nair, T.-C. Huang, V. Smet, K. Panayappan, V. Sundaram, and R. Tummala, "Design and demonstration of a 2.5-d glass interposer bga package for high bandwidth and low cost," *IEEE Transactions on Components, Packaging and Manufacturing Technology*, vol. 7, no. 4, pp. 552–562, 2017.

[2] S. Cho, V. Sundaram, R. R. Tummala, and Y. K. Joshi, "Impact of copper through-package vias on thermal performance of glass interposers," *IEEE Transactions on Components, Packaging and Manufacturing Technology*, vol. 5, no. 8, pp. 1075–1084, 2015.

[3] U. Landau, "Copper metallization of semiconductor interconnects-issues and prospects," in *Proceedings–Electrochemical Society*, vol. 26, 2000, pp. 231–252.

[4] S. Ren, Z. Lei, and Z. Wang, "Investigation of nitrogen heterocyclic compounds as levelers for electroplating cu filling by electrochemical method and quantum chemical calculation," *Journal of The Electrochemical Society*, vol. 162, no. 10, pp. D509–D514, 2015.

[5] P. M. Vereecken, R. A. Binstead, H. Deligianni, and P. C. Andricacos, "The chemistry of additives in damascene copper plating," *IBM Journal of Research and Development*, vol. 49, no. 1, pp. 3–18, 2005.

[6] T. B. Huang, H. Sharma, R. Manepalli, S. Kandanur, V. Sundaram, and R. R. Tummala, "Electroanalytical study of organic additive interactions in copper plating and their correlation with via fill behavior," *Journal of Electronic Materials*, vol. 47, no. 12, pp. 7401–7408, 2018.

[7] P. C. Andricacos, C. Uzoh, J. O. Dukovic, J. Horkans, and H. Deligianni, "Damascene copper electroplating for chip interconnections," *IBM Journal of Research and Development*, vol. 42, no. 5, pp. 567–574, 1998.

[8] R. Akolkar and U. Landau, "A time-dependent transport-kinetics model for additive interactions in copper interconnect metallization," *Journal of The Electrochemical Society*, vol. 151, no. 11, pp. C702–C711, 2004.

C4 Compatible Ultra-Thick Cu On-chip Magnetic Inductor Architecture Integrated with Advanced Polymer/Cu Planarization Process

C.H. Kuo*, S.B. Yang, C.C. Kuo, Y.N. Chen, K.S. Yuan, G.C. Huang, C.N. Ke, T.C. Chang, C.C. Hsu, H.L. Huang, Kirin Wang, Harry Ku, C.S. Chen, K.C. Liu, Alex Kalnitsky, Marvin Liao

Taiwan Semiconductor Manufacturing Company, Ltd. Taiwan

Mail*:chkuozg@tsmc.com

Abstract—Inductor is an essential component in Integrated Voltage Regulator(IVR) which is strongly corresponding to the efficiency of power management. In 2018 ECTC, an on-chip solenoid inductor package integrated with high-permeability magnetic(MAG) film and two ultra-thick Cu trace was successfully demonstrated, which exhibits the benefits of higher inductance, lower resistance, and competitive packaging dimension [1]. Two ultra-thick Cu trace for coils of inductor device inevitably accompanies with higher topography that will induce some concerns on processes, one of them is like bubble defect from spin coating process of photo resist. On the other hand, uneven plan resulted from high topography will also cause poor co-planarity(COP) post bump formation, result in concerns like cold joint in assembly process. In order to resolve this hindrance, a polish process on polymer and Cu by chemical & mechanistic planarization (CMP) is introduced to improve the topography of the inductor package. In the paper, a key process issue, interface delamination, relating to CMP will be disclosed. And the key approach that is how to optimize Cu surface treatment and parameter of CMP process for delamination improvement will be focused. In short, we face a critical issue poor COP post bump formation that induced from high topography in ultra-thick Cu scheme. How to realize CMP process for planarization is a key and must be relied on optimization of key approaches for delamination. At last, compared to the air-core thick copper inductor package reported last year, the results show that CMP can improved COP around 3~5-fold, and have comparable performance of electric resistance before/post wafer level torture.

Keywords- Inductor, Magnetic, Thick Cu, CMP

I. INTRODUCTION

Typical high-end electronic products are, in general, composed of nearly 80% passive components, taking up to 40% of the total board area [2, 3]. Among the passives, inductor is undoubtedly important and has been widely used in the military, communication, automotive, and high-end portable electronic devices. There will be many eye-catching benefits such as size reduction, performance improvement, and cost reduction etc. for "inductors" if one imitates the concept of "MtM" to integrate inductor from the conventional planar substrate surface (2D) to a vertical direction (3D) through "embedded" or "on-chip" approaches [4-8].

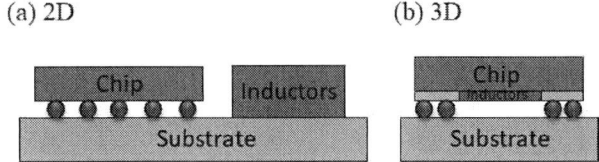

Figure 1. Integrate inductor from the (a) conventional planar substrate surface to a (b) vertical direction through "embedded" or "on-chip" approaches.

In this paper, an on-chip magnetic thick copper inductor integrated with proposed C4 packaging solution is introduced. The spotlight of this on-chip inductor package is twofold. The first is a MAG thin-film with high permeability is inserted into our solenoid structure. The second is the thickness of the Cu redistribution layer becomes 4~5 fold thicker than the Cu used in traditional WLCSP package. Figure 2 illustrates the overall inductor package infrastructure proposed in this study. The scheme starts with the bottom Cu(Metal-1) formation followed by the Metal-1 compatible polymer dielectric(Dielectric-1), Then the MAG thin-film is deposited and patterning. The second layer of dielectric(Dielectric-2) was coated followed by the contact opening for the connectivity in between Metal-1 and top Cu(Metal-2). Another dielectric(Dielectric-3) layer was coated on Metal-2, at last, bump formation for the I/O connection.

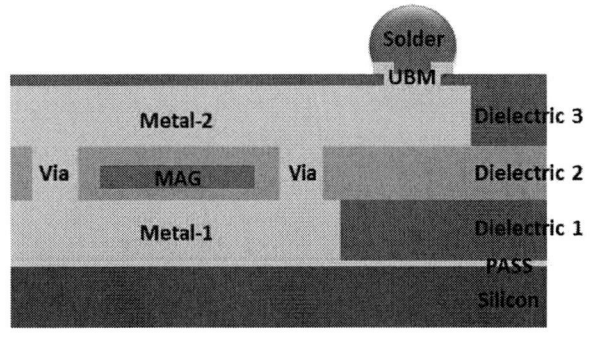

Figure 2. Schematic illustration of on-chip thick copper inductor package scheme.

In this scheme followed above process flow, even we use thicker polymer dielectric to fill the gap of thick Cu trace, the addressed thick Cu inductor device inevitably accompany with higher topography then results in 1. Poor COP post bump formation 2. Narrow process window of lithography and etching etc. 3. Cold joint in Flip-Chip process. Among these issues, COP is the primary concern in assembly process. In order to resolve this hindrance, CMP approach was introduced to improve the topography of the structure with inductor. Due to material CTE miss-match, an interfacial delamination from Cu sidewall to polymer dielectric is observed after CMP process, as can be seen in Fig. 3.

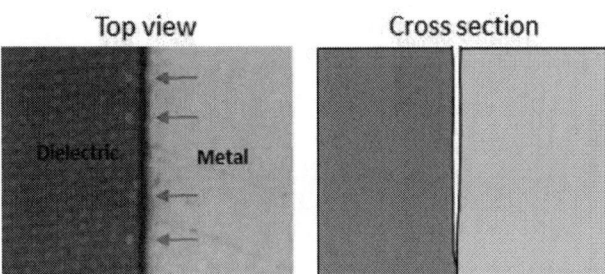

Figure 3. OM top view of post CMP process.

To enhance interfacial adhesion for delamination improvement, the common approaches are surface treatments for roughness enhancement or surface property modification. However, the surface enhancement is not enough to improve the delamination. One approach is proposed in this study including optimization of CMP process for improvement of the delamination, which will be discussed step-by-step in the next paragraph

II. EXPERIMENTAL

A 300mm wafer was designed and divided into 8.7 x 20 mm2 field block. As shown in Fig. 2, the scheme of the MAG inductor is consisted of two layers of Cu trace (labeled Metal-1 & Metal-2), three layers of polymer dielectric (labeled Dielectric-1, 2 and 3), one MAG thin film, and bump. The patterning of the Cu trace was sequentially defined by sputtering/ photo/ plating/ wet etching process, and the polymer dielectric layer was fabricated by photo/curing process, respectively. The fabrication process of Cu trace as well as dielectric layer are quite similar to common RDL process, which can be referred to other literatures [9-11] and just omitted here for shortly.

The MAG material was fabricated by thin film deposition and patterned by traditional Lithography process. Material and failure analysis tool such as Focused ion beam (FIB), Scanning electron microscopy (SEM), Energy dispersive X-ray spectrometry (EDX), Optical microscope (OM) and automated macro defect inspection tool were utilized to examine the microstructure and identify the delamination of the defect observed in this study.

III. RESULTS & DISCUSSIONS

A. Interfacial Delamination post CMP Process

In contrast to well-known RDL process, the thick dielectric layer used in this study, in order to have good step coverage on the peculiar thick Cu redistribution layer, and planarize dielectric and Metal layer by CMP process, Surprisingly, an interfacial delamination from Cu sidewall to dielectric observed after CMP process, as Fig. 3. In addition, it has been confirmed that the delamination only occurred post CMP process, and there is no delamination post dielectric layer formation, as can be seen in Fig. 4

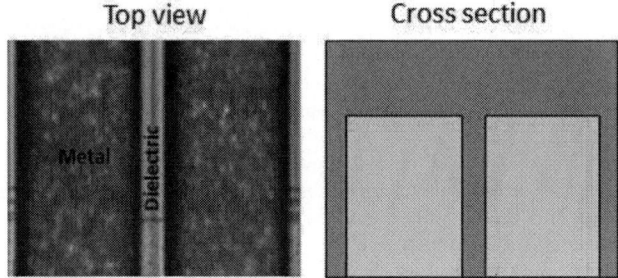

Figure 4. OM top view post dielectric formation.

The strategy for delamination improvement in our study as following: (a) How to enhance interface adhesion, (b) How to optimize CMP process.

The delamination failure for following experiment will be detected by optical scan tool with the same scanning parameter for comparison of different experiment conditions.

B. Enhance Interface Adhesion

The direction of interface adhesion improvement is to introducing chemical bonding by surface treatment. As well known, copper oxide is more favored for adhesion to general dielectric material. The surface treatment by O- radical can help CuOx formation. The result indicates CuOx is indeed helpful for delamination improvement and the best known method can be obtained by optimization of process time for each treatment step.

TABLE I. CONDITION SPLIT TABLE OF SURFACE TREATMENT FOR ROUGHNESS ENHANCEMENT OR SURFACE PROPERTY MODIFICATION.

Leg	Cu surface treatment					
	Step 1		Step 2		Step 3	
	Gas	Time	Gas	Time	Gas	Time
STD	A	1X	B	1X		
Leg 1	A	1X	B	1X	C	2X
Leg 2	C	2X	A	1X	C	0.5X
Leg 3	C	3X	A	3X	C	1.5X
Leg 4	C	3X	A	1X	C	1.5X
Leg 5	C	1X	A	1X	C	1.5X

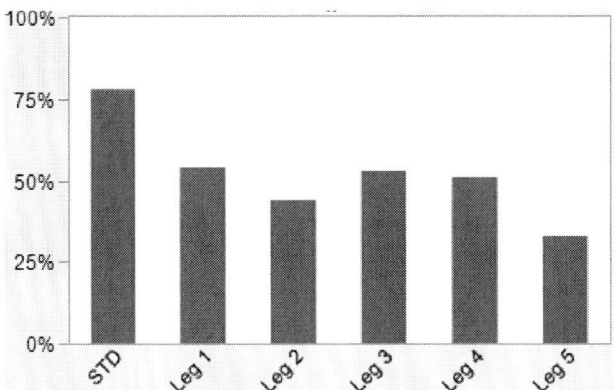

Figure 5. Delamination defect rate of Table 1

C. Optimization of CMP Process

Based on experiment results discussed in previous paragraph, even though surface treatment can mitigate delamination, but it does not resolve completely yet. The CMP process is another dominate factor for delamination and need to be optimized. To lower down force and spin speed is beneficial for delamination. Since lower down force and spin speed will decrease CMP removal amount then result in process time longer. The longer process time represents CMP slurry will have more time to attack the interface of Cu and polymer dielectric. Therefore, it is trade-off for lower down force and CMP process time, and needs to find appropriate sweet spot.

TABLE II. CONDITION SPLIT TABLE OF CMP PROCESS TUNING.

CMP process condition	Down force/ Spin speed	Parameter X
STD	STD	STD
Leg A	Lower	STD
Leg B	Lower+	STD
Leg C	Lower	Best

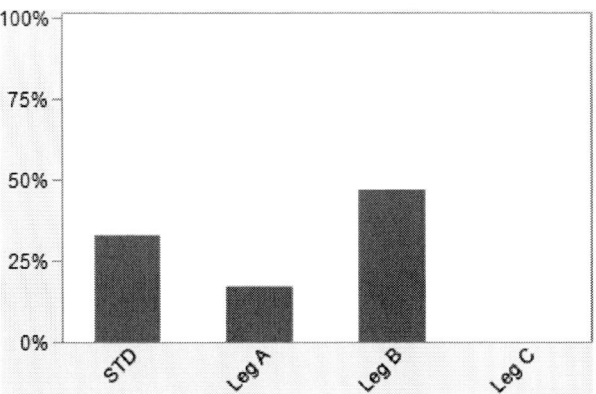

Figure 6. Delamination defect rate of Table 2

By using better condition of Cu surface treatment accompanied optimization of CMP process, the delamination can be improvement. As Fig. 7

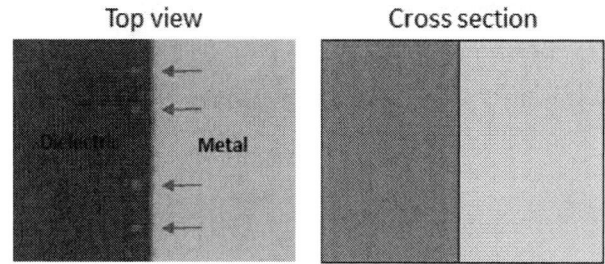

Figure 7. OM top view of post CMP process.

Compared to no planarization process, the results show that CMP can improve COP with 3~5-fold that can fulfill Assembly Joint requirement.

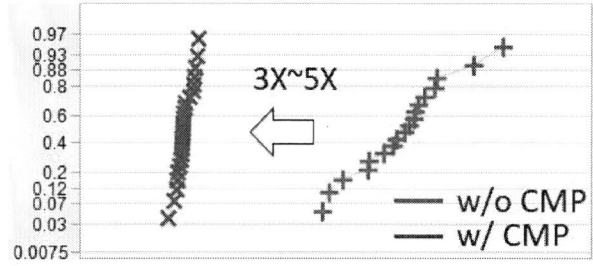

Figure 8. COP P-chart of w/o CMP(red) and w/ CMP(blue).

IV. CONCLUSIONS

In summary, the CMP plays an important role on realizing the proposed on-chip MAG thick Cu inductor package. In this paper, two novel concepts, "PPI surface treatment" and "CMP process tuning" are proposed. Without these two key approaches, the proposed on-chip MAG thick Cu inductor package with excellent COP performance will not be realized. In addition, electric resistance performance passed the wafer level torture (uHASTxx and TCC200) that also demonstrates C4 compatible on-chip package with advanced Polymer/Cu planarization can be innovated to integrate with passive device and collocate with thicker Cu to promote the electric performance. It is a brand new package solution for passive device.

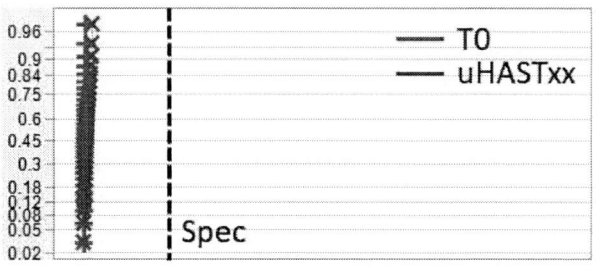

Figure 9. Electric resistance performance post uHASTxx.

978-1-7281-1500-9/19 $31.00 © 2019 IEEE

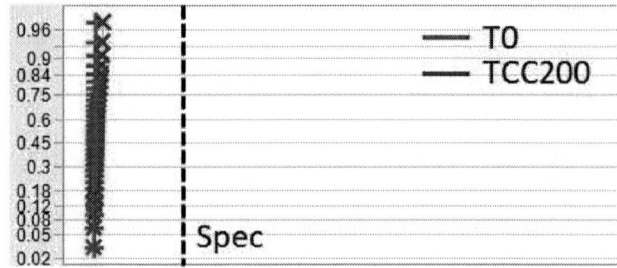

Figure 10. Electric resistance performance post TCC200.

REFERENCES

[1] T. L. Yang, S. B. Yang, W. L. Huang, C. C. Chen, C. C. Kuo, G. C. Huang, K. Y. Wu, T. C. Chang, C.C. Hsu, C. L. Chang, H. L. Huang, Edward Chen, K. C. Liu, Marvin Liao, Alex Kalnitsky, Harry Ku, "Realization of High Electrical Performance On-chip Thick Copper Inductor Package by Via Interface Process Improvement for Metal Contact", 68th Electronic Components and Technology Conference, 2018.

[2] R. K. Ulrich and L. W. Schaper , Integrated Passive Component Technology , Wiley-IEEE , New York , 2003

[3] J. Prymark , S. Bhattacharya , K. Paik , and R. R. Tummala , Fundamentals of Microsystems Packaging , McGraw-Hill , New York , 2001

[4] H. Wu, M. Lekas, R. Davies, K. L. Shepard, N. Sturcken", Integrated Transformations with Magnetic Thin Film s", IEEE Transations on Magnetics, Vol. 52, No. 7, 2016.

[5] N. Sturcken, R. Davies, H. Wu, M. Lekas, K. Shepard, K. W. Cheng, C. C. Chen, Y. S. Su, C. Y. Tsai, K. D. Wu, J. Y. Wu, Y. C. Wang, K. C. Liu, C. C. Hsu, C. L. Chang, W. C. Hua, Alex Kalnitsky, "Magnetic Thin-Film Inductors for Monolithic Integration with CMOS ", IEEE International Electron Devices Meeting (IEDM), 2015.

[6] N. Sturcken, E. J. O'Sullivan, N. Wang, P. Herget, B. C. Webb, L. T. Romankiw, M. Petracca, R. Davies, R. E. Fontana, J. G. M. Decad, I. J. Kymissis, A. V. Peterchev, L. P. Carloni, W. J. Gallagher, and K. L. Shepard, "A 2.5D Integrated Voltage Regulator Using Coupled-Magnetic Core Inductors on Silicon Interposer", IEEE Journal of Solid-State Circuits, Vol. 48, No.1, 2013.

[7] D. W. Lee, K. P. Hwang, and S. X. Wang, "Design and Fabrication of Integrated Solenoid Inductors with Magnetic Cores", 58th Electronic Components and Technology Conference, 2008.

[8] D. S. Gardner, G. Schrom, F. Paillet, B. Jamieson, T. Karnik, and S. Borkar, "Review of On-Chip Inductor Structures With Magnetic Films" IEEE Transactions on Magnetics, Vol. 45, No.10, 2009.

[9] S. B. Yang, C. C. Chen, W. L. Huang, T. L. Yang, G. C. Huang, T. Y. Chou, C. C. Hsu, C. L. Chang, H. L. Huang, C. C. Chou, C. Y. Ku, C. H. Chen, C. S. Chen, K. C. Liu, Alex Kalnitsky, Marvin Liao, "Implementation of Thick Copper Inductor Integrated into Chip Scaled Package", 67th Electronic Components and Technology Conference, 2017.

[10] J. H. Lau, "Critical issues of wafer level chip scale package (WLCSP) with emphasis on cost analysis and solder joint reliability", IEEE Transactions on Electronics Packaging Manufacturing , Vol. 25, Issue: 1, PP.42-50, 2002.

[11] T. Y. Tee, L. B. Tan, R. Anderson, H. S. Ng, J. H. Low, C. P. Khoo, R. Moody, B. Rogers, "Advanced Analysis of WLCSP Copper Interconnect Reliability under Board Level Drop Test", 10th Electronics Packaging Technology Conference, 2008.

Development of 2.3D High Density Organic Package using Low Temperature Bonding Process with Sn-Bi Solder

Shota Miki, Hiroshi Taneda, Naoki Kobayashi, Kiyoshi Oi, Koji Nagai, Toshinori Koyama

Process Development Dept. Research & Development Div.
SHINKO ELECTRIC INDUSTRIES CO., LTD.
Nagano-shi, Japan
sy.miki@shinko.co.jp

Abstract—Recently, 2.5D package structure such as using silicon interposer is used for high end sever and HPC (High Performance Computing) with high density connection between logic device and memory device. We are developing high density organic package called i-THOP® (integrated-Thin film High density Organic Package) 2.3D type combining organic interposer and build-up substrate. The organic interposer was mounted on build-up substrate. Interconnection between organic interposer and build-up substrate was consisted of copper post and Sn-Bi solder. The bonding process was applied by using the pre-applied NCF (Non-Conductive Film) method. Interconnection was achieved under 200 degrees C by using Sn-Bi solder. The good wettability of Sn-Bi solder was observed at joint portion with minimum 200 μm pitch and no NCF void was confirmed by SAT (Scanning Acoustic Tomography). After that, die assembly was evaluated. The good wettability of solder joints of assembled small die and large die was confirmed with minimum 40 μm pitch. Sn-Bi solder joints were also stable without re-melting after die assembly and stiffener attach process. The thermal warpages of i-THOP® 2.3D type was measured after organic interposer stacking, die assembly and stiffener attach. The warpage without die was small at approximately 20 μm at 30 degrees C and almost no change from 30 degrees C to 250 degrees C. However, the warpage with die was large at 300 μm at 30 degrees C. After stiffener attach process, the thick stiffener over 2.0 mm could reduce warpage below 200 μm that is the criteria of 1.0 mm pitch BGA (Ball Grid Array). The reliability tests of TC (Temperature Cycling), HTS (High Temperature Storage Life), u-HAST (unbiased-Highly Accelerated Temperature and Humidity Stress Test) and b-HAST (biased-Highly Accelerated Temperature and Humidity Stress Test) were carried out. The criteria of TC 1000cycles, HTS 1000 hours, u-HAST 192 hours and b-HAST 96 hours were passed.

Keywords-Organic interposer, Sn-Bi solder, Low temperature bonding, NCF

"i-THOP is registered trademark of SHINKO ELECTRIC INDUSTRIES CO., LTD."

I. INTRODUCTION

2.5D package structure such as using silicon interposer is used for high end sever and HPC (High Performance Computing) with high density connection between logic device and memory device [1]. In the future, it is expected that the package size will be expanded due to an increase of die size and the number of die mounted. Therefore, the package cost will be higher due to the necessity of larger size silicon interposer. Also, the package warpage will be increased due to CTE (Coefficient of Thermal Expansion) mismatch between silicon interposer and build-up substrate. It makes difficult to connect die to interposer. Minimizing CTE mismatch and low temperature bonding are the solutions to reduce warpage of the package. Currently, 2.1D type package with organic interposer have been developing to replace silicon interposer for low cost solution [2]. Shinko also develops organic interposer i-THOP® (integrated-Thin film High density Organic Package) as a 2.1D solution [3-4]. However, there are some issues of lower yield and longer lead time due to an increase of the number of layers in build-up substrate in the future. Shinko has been studing the low temperature bonding process using Sn-Bi solder for FC (Flip Chip) bonding [5-6].

We suggest high density organic package called i-THOP® 2.3D type. This package is consisted of the conventional build-up substrate and organic interposer prepared separately. The organic interposer is stacked on build-up substrate by pre-applied NCF (Non-Conductive Film) bonding method that we had experience [7-8]. Sn-Bi solder is used for organic interposer stacking to reduce package warpage. Therefore, i-THOP® 2.3D type enables the reduction of package warpage, the increase of quality yield and the shortening of lead time.

In this paper, we report on the result of the process, the package warpage and the reliability evaluations of i-THOP® 2.3D type.

II. PACKAGE STRUCTURE

Figure 1 shows the schematic cross-sectional image of i-THOP® 2.3D type. This package is consisted of the conventional build-up substrate and organic interposer.

Table I shows the specification of test substrates. The organic interposer with L/S (Line and Space) = 2/2 μm and φ10 μm via using the thin film technology was stacked on the build-up substrate. The build-up substrate was 65 x 65 mm size and 1820 μm in thickness. The organic interposer was 44 x 31 mm size and 80 μm in thickness. The bonding pitch between organic interposer and build-up substrate was minimum 200 μm and FC bonding pitch was minimum 40 μm.

For verification of assembly, the 4 small die and large die were bonded on the organic interposer and stiffener was attached on the build-up substrate. Table II shows the specification of 2 type die. The small die size was 8 x 11 mm with minimum 55 μm pitch like HBM (High Bandwidth Memory) die. The large die size was 25 x 28 mm with minimum 40 μm pitch. These die had copper pillar with Sn/Ag solder.

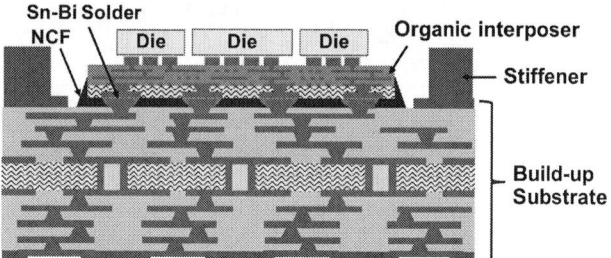

Figure 1. Schematic cross-sectional image of i-THOP® 2.3D type.

TABLE I.   SPECIFICATIONS OF 2.3D TYPE PACKAGE

	Build-up Substrate	*Organic interposer*
Size	65 x 65 mm	44 x 31 mm
Thickness	1,820 μm	80 μm
Design rule	Core thickness : 1,200 μm Through hole : φ250 μm Via diameter : φ60 μm	L/S = 2/2 μm (Min.) Via diameter = φ10 μm
Stack up	6 / 2 / 6	4 (thin film layer) + 1
Bump size	φ90 μm	φ150 μm
Bump material	Cu	Sn-Bi solder
Bump pitch	400, 200 μm	
Number of bumps	10,005	

TABLE II.   SPECIFICATIONS OF 2 TYPE DIE

	Small die	*Large die*
Die size	8 x 11 mm	25 x 28 mm
Thickness	775 μm	775 μm
Bump Pitch	55 μm (Min.)	40 μm (Min.)
Number of bumps	4,942	57,887
Bump size	φ23 μm	
Bump height	Cu : 25 μm , Sn/Ag : 13 μm	

III.   PROCESS AND EXPERIMENTAL METHOD

A.   *Organic interposer stacking process*

Figure 2 illustrates the organic interposer stacking process. First, the rigid layer was laminated on the carrier and the vias filled by Cu plating were formed. Then, the top side surface was smoothed by CMP (Chemical Mechanical Polishing) for the fine-wiring formation by semi-additive process. After removing carrier, Sn-Bi solder was formed on backside of organic interposer. The organic interposer was sawed after laminating NCF. Then, the organic interposer was mounted on the build-up substrate and was pushed flatly by TCB (Thermo-Compression Bonding) at 180 degrees C. After NCF cured, i-THOP® 2.3D type was completed.

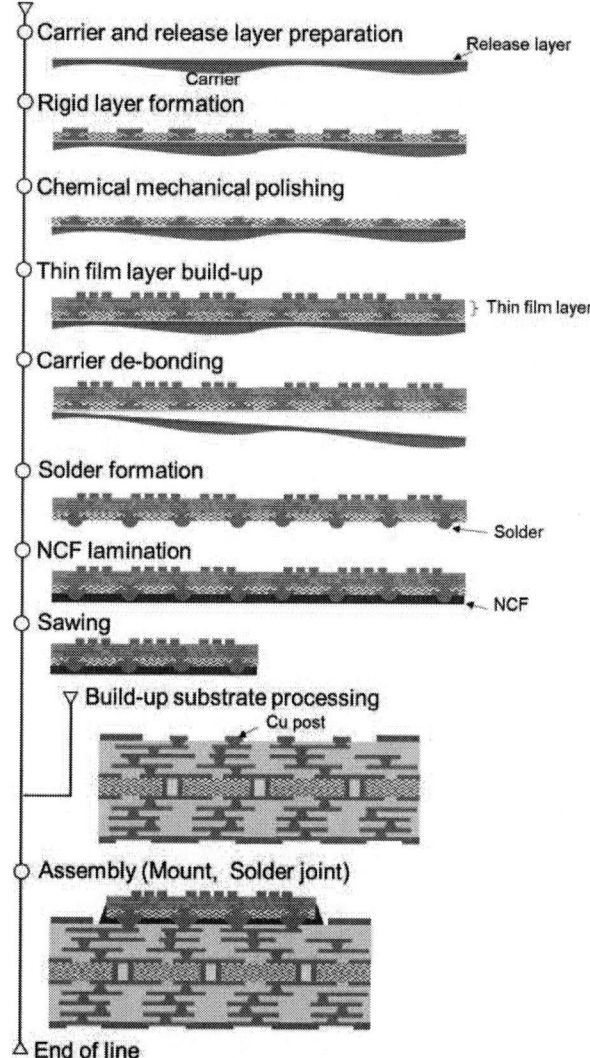

Figure 2. Process flow of organic interposer stacking on build-up substrate.

B. Die assembly process

Figure 3 illustrates the die assembly and the stiffener attach process. First, 4 small die were mounted to FC pad on the organic interposer and heated at 260 degrees C by reflow process. After that, the large die was bonded by TCB process. Then, UFR (Under Fill Resin) was applied and cured. Finally, the stiffener ring was attached on the build-up substrate to reduce the package warpage.

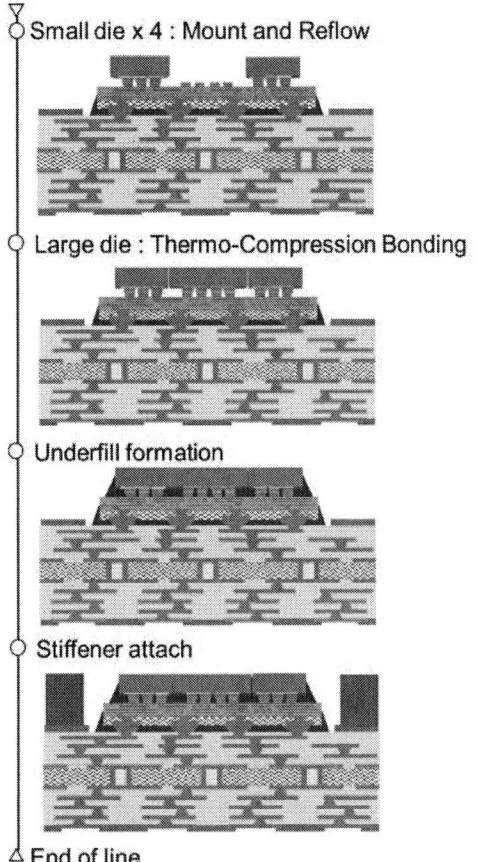

Figure 3.   Process flow of die assembly and stiffener attach.

C. Evaluation method

After thin film build-up process, the wiring, micro via and FC pad were checked by SEM (Scanning Electron Microscope). In organic interposer stacking process, the bonding condition was evaluated with different NCF viscosity. After stacking, NCF void and delamination was inspected by SAT (Scanning Acoustic Tomography) and the solder joints were checked by X-ray inspection machine. The cross-section of Sn-Bi solder joints between build-up substrate and organic substrate were observed by optical microscope and SEM. The IMCs (Inter-Metallic Compounds) of interconnection between Sn-Bi solder and Cu pad were analyzed by EPMA (Electron Probe Micro Analyzer). After the die assembly and stiffener attach process, the solder joints of small die and large die were

observed by SEM. Furthermore, the Sn-Bi solder joints were analyzed by EPMA. The package warpage was measured by laser measurement system to decide stiffener thickness.

Finally, reliability tests of TC (Temperature Cycling), HTS (High Temperature Storage Life), u-HAST (unbiased-Highly Accelerated Temperature and Humidity Stress Test) and b-HAST (biased-Highly Accelerated Temperature and Humidity Stress Test) shown in Table III were carried out to confirm joint of Sn-Bi solder and Cu pad.

TABLE III.   RELIABIRITY TEST CONDITION

Item	Condition	JEDEC Standard
TC	-55°C to 125°C	JESD22-A104E
HTS	Constant 150°C	JESD22-A103D
u-HAST	130°C / 85%RH / No bias	JESD22-A118
b-HAST	130°C / 85%RH / 3.5V	JESD22-A110D

IV.   RESULTS AND DISCUSSIONS

A. Organic interposer (Thin i-THOP ®)

Figure 4 shows the organic interposer after thin film build-up process. FC pads formed at the small die and large die mount area were successfully plated. The wiring with minimum L/S = 2/2 μm and φ10 μm micro via in the thin film layer were formed as designed.

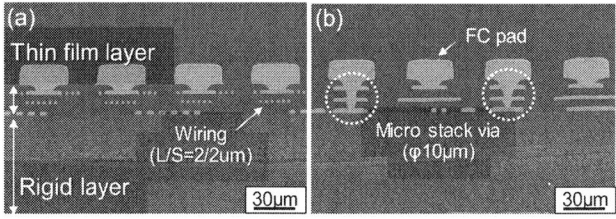

Figure 4.   Cross-sectional SEM image of organic interposer. (a) Wiring with minimum L/S = 2/2 μm, (b) φ10 μm micro vias.

B. Organic interposer stacking

Figure 5 shows the cross-section of Sn-Bi solder joints compared with different NCF viscosity. In the case of the high viscosity NCF, the center area of organic interposer could not be connected, although the corner area could be connected. On the other hand, all joints were achieved with low viscosity NCF. Figure 6 shows the appearance of i-THOP® 2.3D type after organic interposer stacking. The surface of organic interposer was flat. Figure 7 shows the inspection results of SAT and X-ray. There was no NCF void and delamination. Solder flow and solder bridge were not observed by X-ray. Figure 8 shows cross-sectional SEM images after the organic interposer stacking. The solder wettability was confirmed. Figure 9 illustrates the IMCs of the Sn-Bi solder joints. The joint portion consists of Cu_3Sn, Cu_6Sn_5, Sn and Bi. Bi and Sn were present together in part of joint portion. These were surrounded by Cu_6Sn_5 which the melting point is 435 degrees C. Also, each joint portion was surrounded by cured NCF.

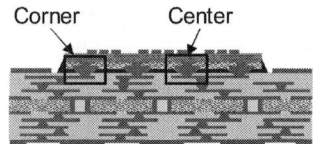

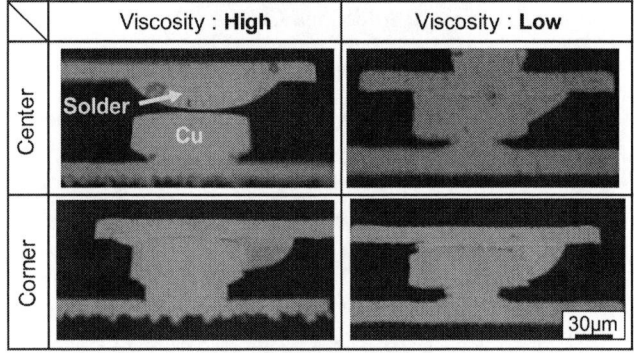

Figure 5. Cross-section of Sn-Bi solder jonts compared with different NCF viscosity.

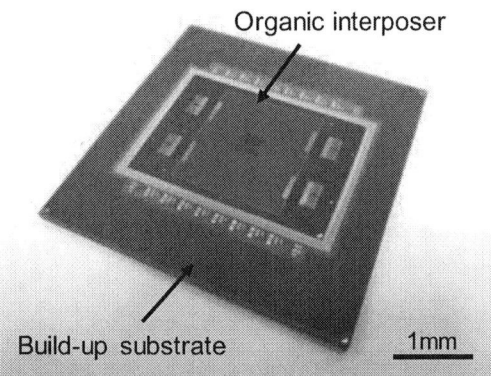

Figure 6. The appearance of i-THOP® 2.3D type.

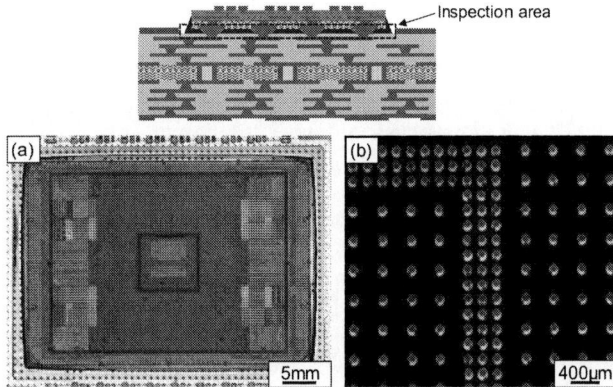

Figure 7. Inspection results after organic interposer stacking. (a) SAT image, (b) X-ray image of solder joints.

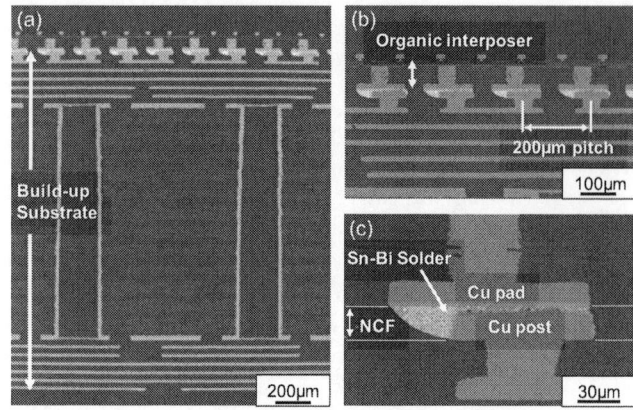

Figure 8. Cross-sectional SEM images of i-THOP® 2.3D type. (a) Overall view, (b) Sn-Bi soder joints between organic interposer and build-up substrate, (c) Enlarged view of Sn-Bi solder joint.

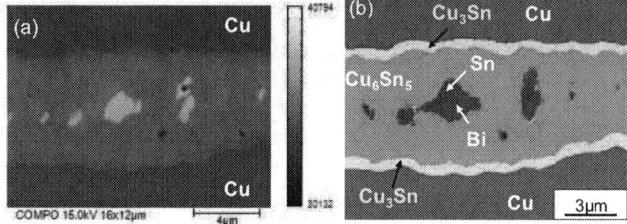

Figure 9. Result of EPMA analysis. (a) Enlarged image of Sn-Bi solder, (b) EPMA color map.

C.   Die assembly and Stiffener attach

The die assembly process was evaluated after organic interposer stacking. Figure 10 shows the appearance of i-THOP® 2.3D type attached die and stiffener. Figure 11 shows the cross-sectional SEM image of small die and large die. The solder wettability of small die with 55µm pitch bump and large die with 40 µm pitch bump was confirmed. Figure 12 shows the Sn-Bi solder joints between organic interposer and build-up substrate. In SEM image, there was no solder flow such as re-melting occurred by heat process of die assembly and stiffener attach. The joint portion was consisted of Cu_3Sn, Cu_6Sn_5, Bi and Sn as same as after organic interposer stacking. Therefore, the Sn-Bi solder joint was stable after die assembly and stiffener attach process.

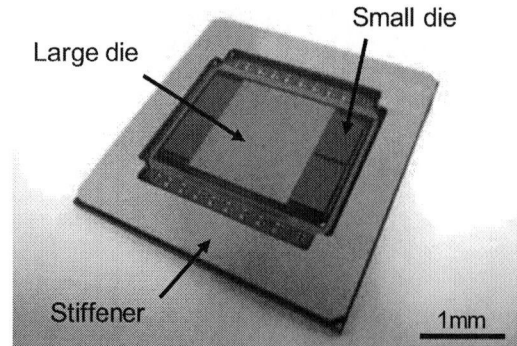

Figure 10. The appearance of i-THOP® 2.3D type attached die and stiffener.

978-1-7281-1500-9/19 $31.00 © 2019 IEEE

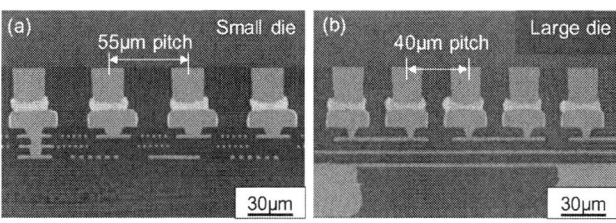

Figure 11. Cross-sectional SEM images after FC bonding. (a) Solder joints of small die, (b) Solder joints of large die.

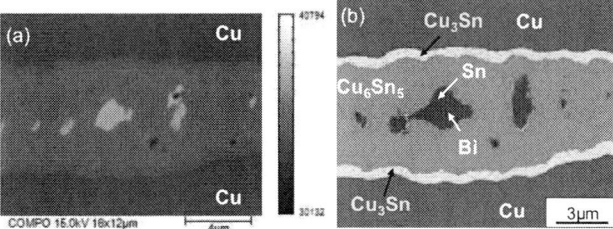

Figure 12. Sn-Bi solder joint after die assembly and stiffener attach process. (a) SEM image, (b) EPMA color map.

D. Package Warpage

The thermal warpage was measured in 63 x 63 mm area from BGA (Ball Grid Array) side. The criteria of warpage with 1.0 mm pitch BGA were under 200 µm. Figure 13 shows the warpage of the i-THOP® 2.3D type with die and without die. The warpage without die was small and nearly no change during heating condition. After die assembly, the thermal warpage was larger than the warpage without die and was shown from 330 µm in convex shape at room temperature to 200 µm in concave shape at 250 degrees C. The large warpage was caused by CTE miss match of the substrate and die. Figure 14 shows the thermal warpage of i-THOP® 2.3D type attached different kind of stiffener thickness. The stiffener was prepared for 3 type thickness in 1.1 mm, 2.0 mm and 3.0 mm. The warpage with each stiffener was smaller than without stiffener. The thick stiffener over 2.0 mm could reduce warpage below 200 µm. This result shows the stiffener thickness in 2.0 mm was suitable for this package because the warpage criteria were passed and the stress was lower than 3.0 mm type stiffener.

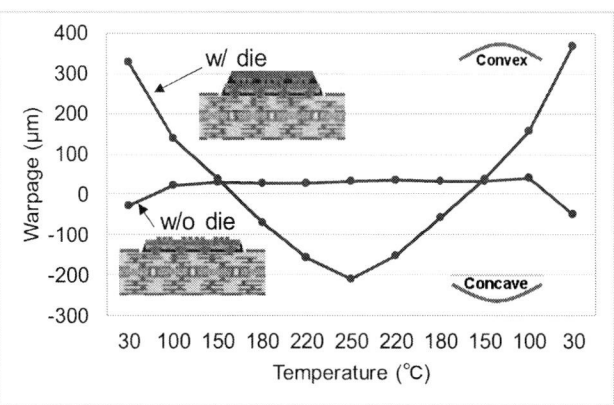

Figure 13. Thermal warpage of i-THOP® 2.3D type compared with die and without die.

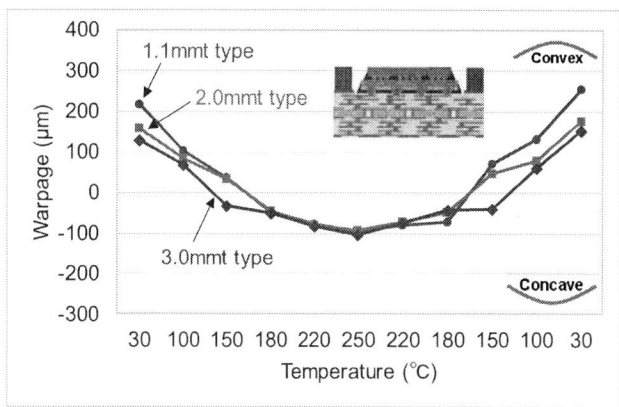

Figure 14. Thermal warpage of i-THOP® 2.3D type after stiffener attach with different type thickness.

E. Reliability Test

After stacking organic interposer, the reliability tests of Sn-Bi solder joints were evaluated by monitoring electrical resistance. Table IV shows the results of reliability tests of TC, HTS, u-HAST and b-HAST. The pass criteria of resistance change rate under 10 % were achieved up to TC 2000 cycles and 1500 cycles, HTS 1000 hours and u-HAST 192 hours. The pass criteria of insulation resistance over 10^6 Ω were achieved up to b-HAST 264 hours. Figure 15 shows the resistance change rate of TC, HTS and u-HAST. The resistance hardly changed at solder joints with 200 µm pitch. Figure 16 shows the cross-sectional SEM images of the Sn-Bi solder joints after the reliability tests. There was no solder crack. IMCs were grown after Pre-conditioning, especially HTS 1000 hours. Figure 17 shows the cross-sectional image of another solder joint after TC 1000 cycles. NCF entrapment was observed at SEM image. Although the pass criteria were achieved, it was necessary to reduce the NCF entrapment in joint portion for quality improvement.

TABLE IV. RESULTS OF RELIABILITY TESTS

Item	Requirement	Result	
TC	1000 cycles	2000 cycles	8/8 pass
		1500 cycles	26/26 pass
HTS	1000 hours	1000 hours	25/25 pass
u-HAST	192 hours	192 hours	26/26 pass
b-HAST	96 hours	264 hours	10/10 pass

*Criteria : Resistance change rate < 10 % : TC, HTS, u-HAST

*Criteria : Insulation resistance > 10^6 Ω : b-HAST

*Pre-Conditioning was carried out before each test.
  (125°C/24 hours + 60°C/60%RH/40 hours + 265°C Reflow x 3 cycles)

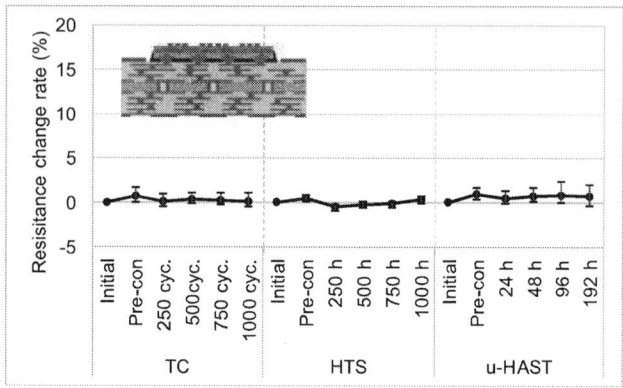

Figure 15. Resistance change rate of Sn-Bi solder joints with 200 μm pitch between organic interposer and build-up substrate.

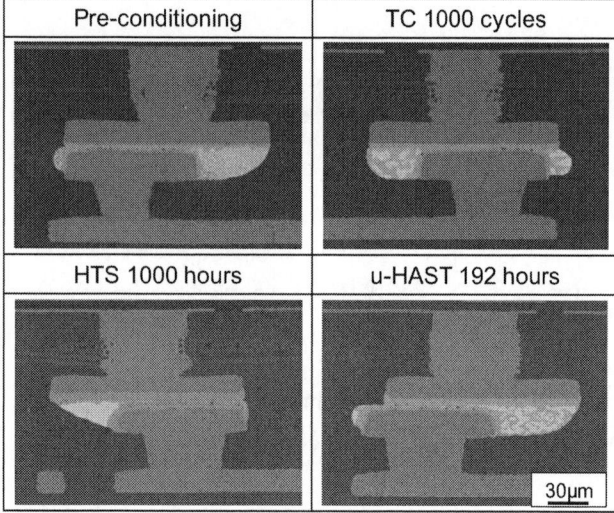

Figure 16. Cross-sectional SEM images of Sn-Bi solder joints after tests.

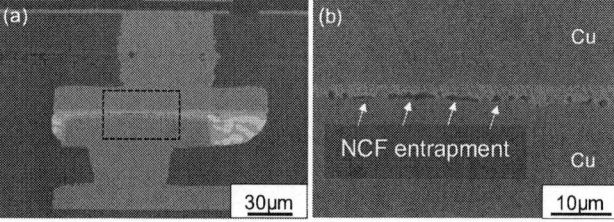

Figure 17. Cross-sectional SEM image of NCF entrapment of Sn-Bi solder joint after TC 1000 cycles. (a) Over view image, (b) Enlarged view image.

V. CONCLUSIONS

The i-THOP® 2.3D type was completed by low temperature bonding and pre-applied NCF method using Sn-Bi solder. The good wettability of Sn-Bi solder and no NCF void were confirmed.

After die assembly, the good wettability of solder joints between die and organic interposer of minimum 40 μm pitch was confirmed. The Sn-Bi solder joints between organic interposer and build-up substrate were stable without solder flow because of IMCs which has high melting point.

The thermal warpage of i-THOP® 2.3D type without die was almost no change in heating process. After die assembly and stiffener attach, the criteria of warpage below 200 μm were passed by attaching thick stiffener over 2.0 mm.

The pass criteria of TC 1000 cycles, HTS 1000 hours, u-HAST 192 hours and b-HAST 96 hours were achieved.

In the future work, it should be necessary to reduce the NCF entrapment in solder portion for the quality improvement. The reliability test will be evaluated in the structure with die assembly and stiffener attach.

ACKNOWLEDGMENT

The authors would like to thank Common Technology Development Division at Shinko for the analysis and reliability test support, and also co-workers who supported us on preparing the build-up substrate and the organic interposer.

REFERENCES

[1] Scott McCann, Ho Hyung Lee, Gamal Refai-Ahmed, Tom Lee and Suresh Ramalingam, "Warpage and Reliability Challenges for Stacked Silicon Interconnect Technology in Large Package," Proceedings of ECTC2018, San Diego, CA, USA(2018), pp.2339-2344.

[2] Wei-Chung Chen, Chiu-Wen Lee, Min-Hua Chung, Chaung-Chi Wang, Shang-Kun Huang, Yen-Sen Liao, Hung-Chun Kuo, Chen-Chao Wang and David Tarng, "Development of novel fine line 2.1D package with organic interposer using advanced substrate-based process," Proceedings of ECTC2018, San Diego, CA, USA(2018), pp.601-606.

[3] Noriyoshi Shimizu, Wataru Kaneda, Hiromu Arisaka, Naoyuki Koizumi, Satoshi Sunohara, Akio Rokugawa and Toshinori Koyama, "Development of Organic Multi Chip Package for High Performance Application," Proceedings of IMAPS2013, Orlando, FL, USA(2013), pp.414-419.

[4] Kiyoshi Oi, Satoshi Otake, Noriyoshi Shimizu, Shoji Watanabe, Yuji Kunimoto, Takashi Kurihara, Toshinori Koyama, Masato Tanaka, Lavanya Aryasomayajula and Zafer Kutlu, "Development of New 2.5D Package with Novel Integrated Organic Interposer Substrate with Ultra-fine Wiring and High Density Bumps," Proceedings of ECTC2014, Orlando, FL, USA(2013), pp.348-354.

[5] Kei Murayama, Mitsuhiro Aizawa, and Takashi Kurihara, "Low Stress Bonding for Large Size Die Application," Proceedings of ECTC2015, San Diego, CA, USA(2015), pp.1846-1853.

[6] Kei Murayama, Mitsuhiro Aizawa, and Takashi Kurihara, "Study of Crystal Orientation and Microstructure in Sn-Bi and Sn-Ag-Cu solder with Thermal Compression Bonding and Mass Reflow," Proceedings of ECTC2016, Las Vegas, NV USA, USA(2016), pp.909-916.

[7] Shota Miki, Takaharu Yamano, Sumihiro Ichikawa, Masaki Sanada and Masato Tanaka, "Development of 3D System-in-Package with TSV Technology," Proceedings of IMAPS2014, San Diego, CA, USA(2014) pp.612-617.

[8] Shota Miki and Takaharu Yamano "Fast Die Stacking Technology for 3D IC using the Collective Bonding Process," Proceedings of mate2017, Yokohama, Kanagawa, JAPAN(2017) pp.273-276.

PowerTherm Attach Process for Power Delivery and Heat Extraction in the Silicon-Interconnect Fabric using Thermocompression Bonding

Pranav Ambhore, Umesha Mogera, Boris Vaisband, Ujash Shah,
Timothy Fisher, Mark Goorsky and Subramanian S. Iyer

Center for Heterogeneous Integration and Performance Scaling (CHIPS),
Samueli School of Engineering and Applied Science,
University of California, Los Angeles, CA 90095
Email: pranavambhore@g.ucla.edu

Abstract – **High-density placement of dies on the Silicon Interconnect Fabric (Si-IF) demands high power delivery (1 W/mm^2) which in turn generates intense heat (~0.5-0.7 W/mm^2). To meet this power requirement and manage its thermal dissipation, we have introduced a novel architecture called PowerTherm which involves the attachment of electrically isolated copper terminal blocks to the back side of the Si-IF. The terminal blocks perform a dual function: they deliver high current at mission voltage and cool the Si-IF either passively or actively. This paper deals with PowerTherm bonding process and its characterization. The terminal blocks were attached to electrodeposited Cu on Si using thermocompression bonding in atmospheric conditions, however with confined air flow. Shear strength of ~ 17 MPa was achieved when the bonding was performed at the optimized conditions.**

Keywords- Silicon-Interconnect Fabric, Copper thermocompression bonding, power delivery, heat extraction

I. INTRODUCTION

Moore's law for packaging calls for several innovations in advanced packaging. One approach is the direct placement of dies, heterogeneously, on a single interconnected platform, termed 'Silicon Interconnect Fabric' (Si-IF) [1,2]. In this approach, interconnect pitch can be reduced by almost 100 times (<10 μm) [3,4], which is impossible to achieve by conventional packaging technologies and has the advantages of larger bandwidth, lower latency, and power [5]. However, such high placement density of dies requires enormous board-level power densities (~ 1 Wmm^{-2}) with a package-level current delivery of 10-100 kA. Delivering this amount of current through external connections is challenging given the dimensions of Si-IF (300 mm dia. wafer). At the same time, such a large but dense system will need efficient heat dissipation to avoid thermal degradation and to ensure reliability.

Typical power consumption density for high-performance chips is 1-2 W/mm^2[6]. For a 300 mm Si wafer with an area of ~7 x 10^4 mm^2 and with a 70% area coverage of chips, the total power consumption for a fully assembled Si-IF is estimated to be ~50 kW. To avoid the on-chip power conversion and subsequent power losses in Si-IF, the power is delivered directly at the mission voltages of chips using a platform called PowerBoard. If the mission voltage is assumed to be 1 V, for 50 kW power, 50 kA current needs to be delivered. On the other hand, a high-end chip, intended to be mounted on Si-IF has a typical heat flux density of 0.5-0.7 W/mm^2. Extracting such high heat flux is usually enabled by designing heat sinks with an area larger than the chip, typically two to five times the area of chip. However, putting large heat sinks for each chip on Si-IF is not feasible as inter-die space on Si-IF is on the order of ~100 μm.

These issues call for the construction of efficient unconventional architectures that not only delivers high power but also enables to manage the thermal loads generated in Si-IF. PowerTherm seeks to address these issues by bonding millimeter-sized Cu "terminal blocks" to Si-IF from the back side. A schematic of PowerTherm is shown in Figure 1. The voltage conversion will be done on a separate power board, and terminal blocks will deliver current. The resistance of thicker terminal blocks will be lower compared to wiring levels on Si-IF due to the use of thicker but shorter wires. Having numerous Cu blocks also support multiple voltage domains which will enable chips with different voltage requirements on the same Si-IF. This is a huge advantage as chips on Si-IF can receive power according to design voltage and no change in chip design is required to take advantage of Si-IF. Chips with different mission voltages can be mounted even adjacent to each other. The Si-IF will have lithographically defined pads to which Cu blocks are bonded via thermocompression bonding. The pads are fabricated on top of Through-Wafer Vias (TWVs) which deliver current to the die-

978-1-7281-1500-9/19 $31.00 © 2019 IEEE

side of Si-IF [7]. The terminal blocks may consist of pipes carrying coolant or may act as fins cooled by air or coolant vapors. We have investigated the heat extraction process by phase change cooling for PowerTherm [8]. Importantly, the backside focus of this integration leaves the Si-IF topside available for any supplemental cooling that might be required by specific components. Thus, the proposed PowerTherm architecture is general and agnostic to specific power and digital dies, but at the same time accommodates further customization of the power-thermal nexus in future systems.

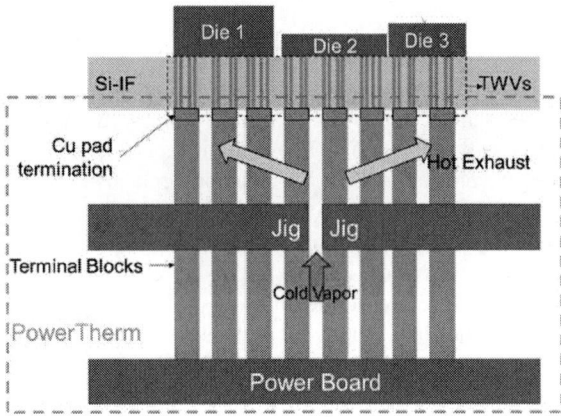

Figure 1: Schematic of PowerTherm cross-section. An array of terminal blocks connects the back side of Si-IF and to the Power Board which down-converts the as received voltage to mission voltages of individual dies. The terminal blocks also serve as fins or have coolant carrying pipes passing through them to extract heat.

The present paper focuses on the PowerTherm attachment (bonding) process. Using this procedure, we have bonded the Cu terminal blocks to a Silicon wafer in atmosphere and measured the shear strength. We opted to choose thermocompression bonding over other bonding techniques such as eutectic bonding because it provides excellent electrical resistivity as compared to solder alloy, which is usually Sn-based, which has ~10X higher resistance compared to copper. The shear strength of thermocompression bonded system is also expected to be higher compared to eutectic and other forms of bonding [9], [10]. Thermocompression bonding also helps in avoiding intermetallic compounds commonly encountered during solder bonding which affect the long-term reliability of bonded components, [11]. The bonding was performed at low pressure (2.5 MPa), in non-vacuum conditions without any reducing atmosphere. A lower pressure is desirable, to prevent buckling of terminal blocks.

II.   EXPERIMENTAL SECTION

a. Experimental Design

Higher temperatures are desirable in thermocompression bonding from a process perspective as this helps

- To breakdown and diffusion of contaminants away from the bonding interface
- To reduce the mechanical properties such as yield strength, Young's modulus. Reduced mechanical properties enable deformation of surface asperities at lower pressures
- Increased diffusion of metal atoms along and across the bond interface

Wafers coated with electroplated copper were bonded in air to determine the optimum bonding temperature. The optimum bonding temperature was decided on the basis of shear force values. The bonding between Cu terminal block to Cu coated wafer in atmosphere at optimum bonding temperature is demonstrated and shear strength is measured. Schematic showing the two bonding test structures is shown in Figure 2.

b. Sample Preparation

Samples were produced with 100 mm Si wafer with <100> orientation. A 3 μm thick thermal oxide layer was grown on wafers. The metal layers, 30 nm Ti as barrier and adhesion layer and 200 nm Cu as the seed layer, were sputter deposited on the oxide-coated wafers in CVC 601 metal deposition system. 2 μm Cu was electro-deposited on the seed layer using Technic electroplating setup. The wafers were then diced into 2×2 mm^2 (referred to as dies) and 10×10 mm^2 square samples. No polishing was performed on diced samples. Cu terminal blocks with dimensions $8 \times 8 \times 8$ mm^3, were lapped with a Logitech PM-5 polisher to prepare a flat and smooth surface.

c. Bonding Parameters and Procedure

Samples were cleaned ultrasonically in acetone for 5 min, water for 1 min and then dipped in acetic acid for 20 min to clean the surface and reduce the native oxide on the copper bonding surface. Bonding was performed on a hot plate from 200 to 350 °C with the ramp-up rate of 4 °C/min. These temperatures were chosen so that the PowerTherm bonding process is compatible with standard Si-IF die to wafer bonding. 2.5 MPa pressure was applied to all the samples. Samples bonded in air were wrapped in Al foil to diminish oxidation of wafer and block.

978-1-7281-1500-9/19 $31.00 © 2019 IEEE

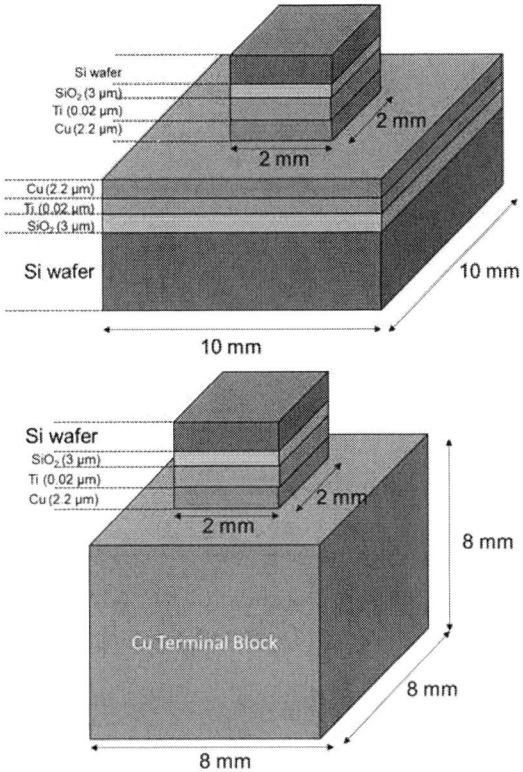

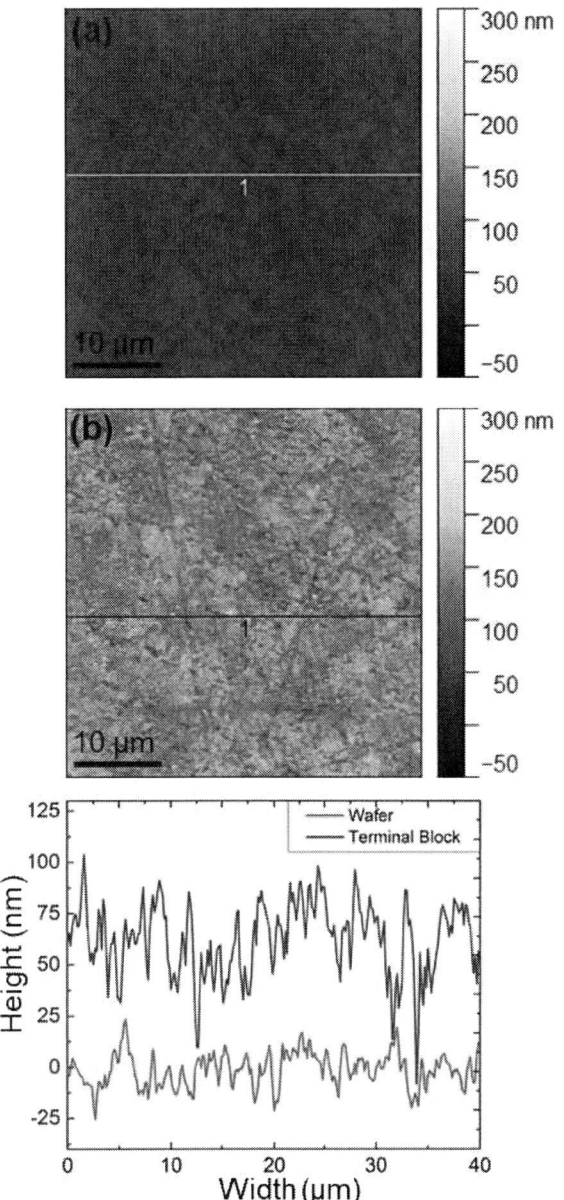

Figure 2: Schematic cross-section of bond test sites. Top: Die to wafer bonding. Bottom: Die to Terminal block bonding.

d. Characterization

Atomic force microscopy (AFM) was used to determine surface roughness of block and wafer. Quesant Q-Scope AFM system was used to collect roughness information, typically scanning 40 x 40 μm^2 randomly chosen area across the bonding surface. X-ray diffraction was used to determine phases present on the surface of block and wafer prior to bonding, to evaluate the grain size of electroplated Cu layer and the preferred orientation of bonding surfaces with the 2θ: ω scans. An ω offset of $3°$ was used to avoid the Si (004) peak. The scans were performed using a Jordan Valley D1 diffractometer, which had incident beam optics to produce a parallel beam of Cu Kα radiation. Shear strength was measured using Nordson Dage 4000 Plus shear tester. A test height of 5 μm above the bonding surface and a test speed of 10 $\mu m/s$ was applied. Three samples for each bonding condition were tested. Select fracture surfaces were investigated with Keyence VHX 6000 optical microscope.

Figure 3: AFM topology of (a) as-electrodeposited Cu on Si wafer surface with RMS roughness of 9 nm, (b) lapped terminal block with RMS surface roughness of 22 nm. Corresponding line profiles drawn in (a) and (b) are shown in (c).

III. RESULTS

The AFM scans shown in Figure 3a and 3b show surfaces of the plated wafer and terminal blocks post acetic acid treatment. The surface roughness of lapped Cu terminal block was 23 ± 2 nm and the roughness of electroplated wafers was 9 ± 0.1 nm. AFM scans along with line profiles can be seen in Figure 3c. X-ray diffraction of the unbonded wafer and terminal block are shown in Figure 4 and Figure 5. These scans reveal the presence of copper oxides and hydroxides at both wafer and block surfaces even after the acetic acid treatment. The preferred orientation of grains was determined using the following equation [12]:

$$C_i = \frac{\dfrac{I_i}{I_{Ri}}}{\dfrac{1}{N}\sum_1^N \dfrac{I_i}{I_{Ri}}} \ldots\ldots\ldots\ldots\ldots (1)$$

where i refers to plane orientation, C_i is texture coefficient, I_i is the integrated intensity of a peak, I_{Ri} is the intensity of randomly oriented sample, N is the number of reflections considered in the analysis. Peak intensities measured from wafer and terminal block are shown in Table 1. The electroplated wafer showed strong preferred orientation in (111) and (200) directions compared to (220) direction shown by the block. The grain sizes were also estimated from diffraction pattern and found to be 29 nm for wafer and 34 nm in terminal blocks.

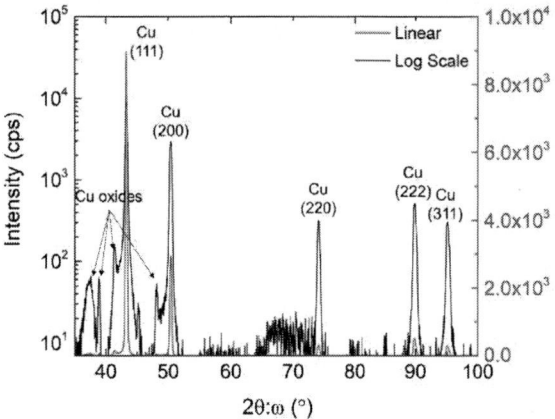

Figure 4: X-ray diffraction 2θ:ω scan of unbonded wafer with indexed peaks. Linear intensity scale (red) and its logarithmic intensity scale (black) are shown.

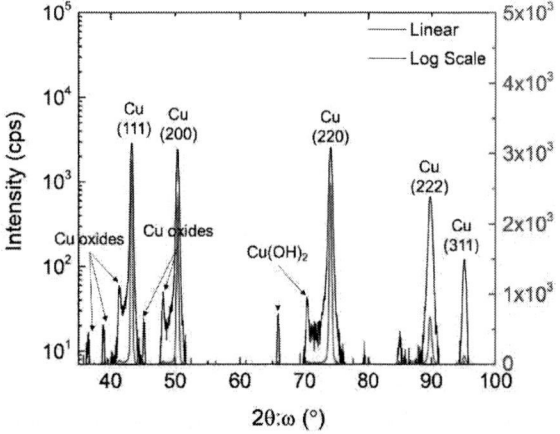

Figure 5: X-ray diffraction 2θ:ω scan of the unbonded terminal block with indexed peaks. Linear intensity scale (red) and its logarithmic intensity scale (black) are shown.

Plane	Normalized I_{Random}	I_{wafer} (cps)	$I_{Terminal\ Block}$ (cps)	Texture Coefficient	
				Wafer	Terminal Block
111	100	9012	2917	1.88	0.53
200	46	2971	2447	1.35	0.96
220	20	363.50	2605	0.38	2.36
311	17	324.80	137	0.4	0.15
Standard deviation				0.64	0.83

Table 1: Measured intensities of wafer and terminal block used for calculating texture coefficient.

Figure 6 shows an average die – wafer shear strength for different bonding temperatures. The shear strength almost tripled for samples bonded at 300 °C compared to those at 200 °C, whereas it decreased by a similar amount when bonded at 350 °C. The highest average shear strength among all bonding temperatures was 11.2 MPa at 300 °C. We used MIL 883 E criteria to estimate quality of our bonds. This standard requires 6 MPa or higher shear strength for bonding to be considered sufficient. Successful bonding was achieved when bonding temperature of 300 °C was used. At this temperature copper block to wafer bonding was performed which yielded bond with a shear strength of 16.9 MPa.

Selected fracture surfaces of Si and Cu were examined using optical microscopy and revealed a combination of cohesive fracture and copper indicating strong bonding and fracture interface differing from the bond interface. Figure 7 shows a selected fracture interface for bonding temperature of 300 °C. The fractures were classified as adhesive if the fracture occurred at Ti-Cu interface and cohesive – if the fracture occurred either in Si or Cu and mixed fracture which had a combination of adhesive and cohesive modes [13].

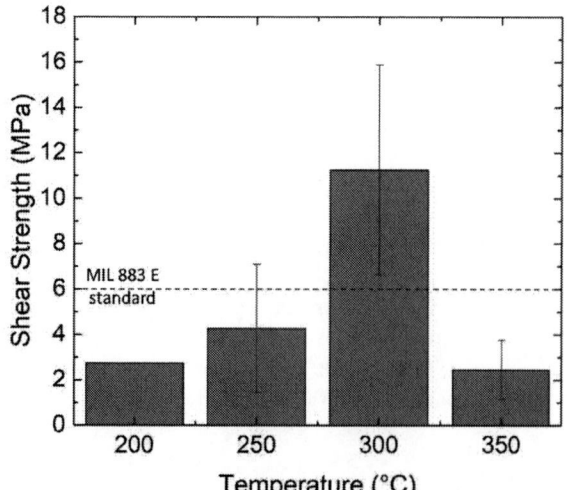

Figure 6: Shear strength of die-wafer bonding at 2.5 MPa, 200-350 °C for 60 min.

978-1-7281-1500-9/19 $31.00 © 2019 IEEE   1608

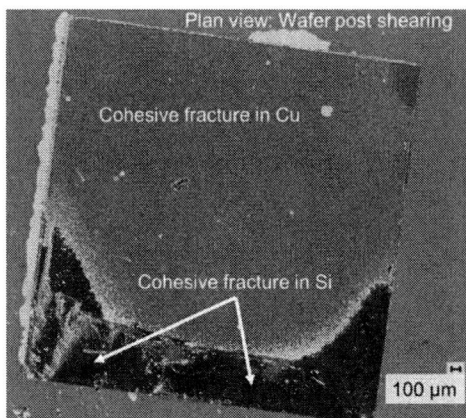

Figure 7: Optical microscope image of the wafer bonded at 300 °C, 1-hour post shear test. The image reveals a cohesive fracture in bulk Si and the copper layer.

IV. DISCUSSION

The roughness of the bonding surface, both at nano and macro-scale, plays a key role in bonding, especially at lower temperatures and pressures. Previous studies have studied the role of surface roughness on the threshold bonding temperature for Cu-Cu thermocompression bonding [14]. The temperature of bonding for acceptable strength (300 °C) as seen in Figure 6, agreed well with the previously reported study [14] for 9 nm RMS roughness despite differences in other factors such as surface impurities, the presence of oxides, grain size, and texture. These factors have been qualitatively characterized in this study. Wafer and terminal block both have small grain sizes (29 and 34 nm), which are expected to help in bonding due to higher grain boundary density at the interface which in turn promotes diffusion of copper atoms to the interface [15]. However, lack of predominant (111) texture, would slow down the bonding process as Cu has highest self-diffusion constant on (111) planes [16] compared to other planes such as (110) or (100).

Figure 8 shows a comparison of X-ray diffraction patterns of annealed wafer and unbonded wafer. The increased number and intensity of copper-oxide peaks and reduced intensity of elemental copper peaks in Figure 8 suggests a very thick surface oxide at the bond interface. Figure 9 shows oxide growth vs. time at different temperatures, plotted from empirically derived equation [3]:

$$T_{ox}(\text{Å}) = 0.0076 \times e^{0.022} \times T \times \log t$$

$$\cdots \cdots (2)$$

where T_{ox} is the copper oxide thickness, T is temperature in K and t is time in seconds.

This suggests that copper oxide thickness at 350 °C is similar in magnitude to deposited copper layer thickness. Copper oxides at the interface and other

contaminants need to be decomposed or dissolved before copper atoms can diffuse across the interface [17,18]. This breakdown is aided by increasing temperature and pressure. Therefore, the decrease in shear strength at 350 °C in Figure 6, might be a result of a thicker copper oxide layer at interface prior to initiation of bonding and before elemental copper surfaces can contact each other.

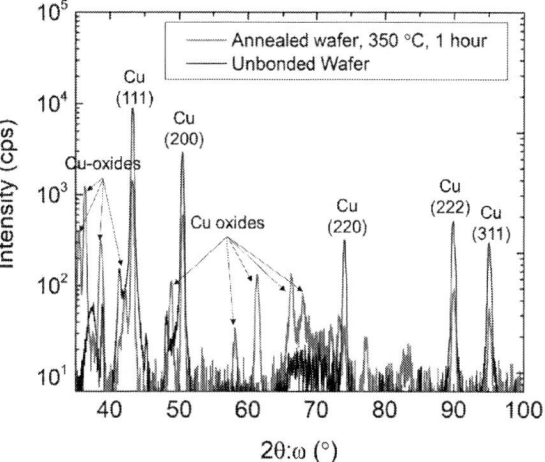

Figure 8: Comparison of X-ray diffraction scans for unbonded and annealed wafer (350 °C, 1 hour).

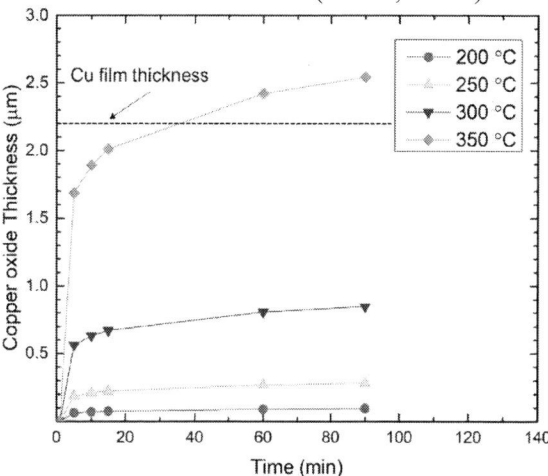

Figure 9: Copper oxide thickness growth at elevated temperatures plotted vs. time.

V. CONCLUSION

We have shown successful PowerTherm bonding process using thermocompression bonding performed in air. The electroplated Cu on Si was bonded to Cu terminal block at different temperature with a bonding pressure of 2.5 MPa. Highest shear strength of 16.9 MPa was achieved at the bonding temperature of 300°C above which shear strength decreased due to excessive oxidation. Future work involves thermocompression bonding of PowerTherm in vacuum and its characterization to meet power delivery requirements.

VI. ACKNOWLEDGMENT

This work was supported in part by ASCENT, one of six centers in JUMP, a Semiconductor Research Corporation (SRC) program sponsored by DARPA. Part of experimental work, especially fabrication of samples was performed in Nanoelectronics Research Facility (NRF), UCLA, and Integrated Systems Nanofabrication Cleanroom (ISNC) at California NanoSystems Institute (CNSI). The authors would take this opportunity to thank staff at NRF and ISNC.

VII. REFERENCES

1. S. S. Iyer, "Heterogeneous integration for performance and scaling," IEEE Trans. Compon., Packag. Manuf. Technol., vol. 6, no. 7, pp. 973–982, 2016.
2. S. Pal, S.S. Iyer, and P. Gupta, Advanced Packaging and Heterogeneous Integration to Reboot Computing. 2017 IEEE International Conference on Rebooting Computing (ICRC), 1-6, 2017.
3. A. Bajwa, S. Jangam, S. Pal, N. Marathe, T. Bai, T. Fukushima, M. Goorsky, and S. S. Iyer, "Heterogeneous Integration at Fine Pitch ($\leq$10μm) Using Thermal Compression Bonding," in IEEE 67th Electronic Components and Technology Conference (ECTC), pp. 1276–1284, May 2017.
4. A. Bajwa, S. Jangam, S. Pal, B. Vaisband, R. Irwin, M. Goorsky, and S. S. Iyer, "Demonstration of a Heterogeneously Integrated System-on-Wafer (SoW) assembly," IEEE 68th IEEE Electronic Components and Technology Conference (ECTC), May 29-June 1, 2018, San Diego, CA
5. S. Jangam, S. Pal, A. Bajwa, S. Pamarti, P. Gupta, and S. S. Iyer, "Latency, Bandwidth and Power Benefits of the SuperCHIPS Integration Scheme". 2017 IEEE 67th Electronic Components and Technology Conference (ECTC), 86-94, 2017.
6. S. C. Lin, and K. Banerjee, "Cool Chips: Opportunities and Implications for Power and Thermal Management", IEEE Trans. Electron Devices, Vol. 55, No. 1, January 2008
7. Meng-Hsiang Liu, B. Vaisband, A. Hanna, Y. Luo, Z. Wan and S. S. Iyer, "Process Development of Power Delivery Through Wafer Vias for Silicon Interconnect Fabric", IEEE 69th Electronic Components and Technology Conference (ECTC), Las Vegas, NV, 2019. (unpublished)
8. U. Shah, U. Mogera, P. Ambhore, B. Vaisband, S. S. Iyer, and T. S. Fisher, "Dynamic Thermal Management for Silicon Interconnect Fabric using Flash Cooling," Proceedings of the IEEE Intersociety Conference on Thermal and Thermomechanical Phenomena in Electronic Systems, May 2019 (unpublished).
9. M. Ohyama, M. Nimura, J. Mizuno, S. Shoji, M. Tamura, T. Enomoto, and A. Shigetou, "Hybrid bonding of Cu/Sn microbump and adhesive with silica filler for 3D interconnection of single micron pitch," in IEEE 65th Electronic Components and Technology Conference (ECTC), pp. 325–330, May 2015.
10. A. Munding, H. Hubner, A. Kaiser, S. Penka, P. Benkart, and E. Kohn, (2008). Cu/Sn Solid liquid Interdiffusion Bonding, Wafer Level 3-D ICs Process Technology, Springer
11. Yu, D. Q., Lee, C., Yan, L. L., Thew, M. L.&Lau, J. H.), "Characterization and Reliability Study of Low-Temperature Hermetic Wafer Level Bonding Using In/Sn Interlayer and Cu/Ni/Au Metallization", J. Alloy. Compd. 485(1-2): 444-450 (2009).
12. C. Barrett and T.B. Massalski, Structure of Metals (Pergamon, Oxford, 1980) p. 204
13. T. A. Tollefsen, A. Larsson, M. M. V. Taklo, E. Poppe and K. Schjølberg-Henriksen, "Die shear strength as a function of bond frame geometry — Au-Au thermo-compression bonding," 2012 4th Electronic System-Integration Technology Conference, Amsterdam, Netherlands, pp. 1-6, 2012.
14. B. Rebhan and K. Hingerl, "Physical mechanisms of copper-copper wafer bonding", J. Appl. Phys. 118, 135301 (2015)
15. Peng, L., Lim, D., Zhang, L. et al., "Effect of Prebonding Anneal on the Microstructure Evolution and Cu–Cu Diffusion Bonding Quality for Three-Dimensional Integration", J of Electron Mater. (2012) 41: 2567
16. Agrawal, P. M. et al. "Predicting trends in rate parameters for self-diffusion on FCC metal surfaces". Surf. Sci. 515, 21–35 (2003)
17. Dray AE. Ph.D. Thesis, University of Cambridge, Cambridge, 1985
18. A. A. Shirzadi, H. Assadi, and E. R. Wallach, "Interface evolution and bond strength when diffusion bonding materials with stable oxide films", Surf. Interface Anal., 31: 609-618, 2001

Interconnect Scheme for Die-to-Die and Die-to-Wafer-Level Heterogeneous Integration for High-Performance Computing

Rabindra N. Das, Vladimir Bolkhovsky, Christopher Galbraith, Daniel Oates, Jason J. Plant, Renée Lambert, Scott Zarr, Ravi Rastogi, Dmitri Shapiro, Manuel Docanto, Terence Weir and Leonard M. Johnson

Quantum Information and Integrated Nanosystems Group
MIT Lincoln Laboratory
E-mail: Rabindra.Das@ll.mit.edu

Abstract- Today, microbump-based flip-chip technology is a compelling option for heterogeneous integration in microelectronic packaging. Performance as well as density (smaller form factor) requirements continue to drive smaller, microbump-based, finer pitch interconnections. This paper presents heterogeneous integration based on microbump fabrication, characterization, and integration for fine pitch die-to-die and die-to-wafer flip-chip interconnection. A variety of microbump options including indium, indium-coated gold, gold and gold-on-gold were used for bump bonding of fine pitch flip-chip structures. A series of microbumps (2.5-15µm) were fabricated using contact and non-contact photolithography on superconducting multi-chip module (S-MCM) wafers. The integration process includes various microbump combinations to reduce microbump spreading / flow during the fine-pitch assembly process. In another study, gold-ball-based interconnects were developed to convert wire bondable chips to a flip-chip process. The use of microbumps to form 10µm-pitch flip-chip interconnects and their initial electrical performance are discussed. As a case study, we have developed indium microbump technology capable of interconnecting an array of 10-to-35-µm-pitch pads to a superconducting multi-chip module (S-MCM) fabricated on 200mm silicon wafers. The S-parameters of several back-to-back microstrip and grounded coplanar waveguide (GCPW) transitions were measured and simulated in a 2.5d (planar) full-wave electromagnetic software package (Sonnet *em*). The flip-chip configurations showed excellent impedance matching and low insertion loss to multi-gigahertz frequencies. For example, a microstrip-to-microstrip transition using 20µm-diameter bumps with 35µm pitch in a coaxial configuration (signal bump surrounded by ground bumps) had a simulated return loss of less than -20 dB from DC to 23 GHz. By adding some compensating inductance to the transitions, this simulated response was extended to 81 GHz. Smaller bump diameter and pitch yields a much better impedance match and less reflected energy, with the 10 µm pitch and 5 µm bump versions giving near-perfect performance to 100 GHz and beyond. Flip-chipped Josephson junctions and niobium lines interrupted by indium microbumps between the superconducting integrated circuits (ICs) and the S-MCM maintained their I-V characteristics. This paper also discusses large superconducting multi-chip module (S-MCM) fabrication and bonding approaches for future technology. Various fabrication options including jumper flip-chip, stitching and laser direct writing are considered for large S-MCM fabrication.

Keywords-Superconducting ICs, flip-chip, superconducting MCM, microbumps, stitching, laser direct writing.

I. INTRODUCTION

The demand for high-performance and low-energy-dissipation computing is driving the development of heterogeneous integration. Packaging multiple chips with different (heterogeneous) fabrication technologies has been a persistent challenge. Typically, individual packaged chips use a board-level assembly approach, and the associated parasitics become the limiting factor to system performance. Heterogeneous integration of multiple chips (CMOS, photonics, RF, superconducting), with fast and lossless connections between chips, is highly desirable for a hybrid computer architecture [1-3]. For cryogenic computing systems based on superconducting Josephson junction integrated circuit technology, an efficient way to interconnect the chips is through a passive Si-based superconducting multi-chip module (S-MCM) which distributes information between integrated circuits by low-loss transmission lines. Here, we show the implementation of such a Si-based S-MCM using lines with well-defined impedance to couple multiple ICs. We have developed multiple S-MCM technologies based on niobium and aluminum circuits for cryogenic and room-temperature operation respectively. The focus of this paper is to describe a heterogeneous integration approach for the chip-to-S-MCM-attachment process using microbump technology. For this integration, we are developing a mix-and-match interconnect scheme that performs three functions: (1) provides a versatile integration approach to any chip technology; (2) is able to convert any chip to flip-chip; and (3) provides an interconnect solution for multilevel assembly. This approach for system integration provides yield improvement and reworkability. A key advantage of fine pitch heterogeneous integration on the

978-1-7281-1500-9/19 $31.00 © 2019 IEEE

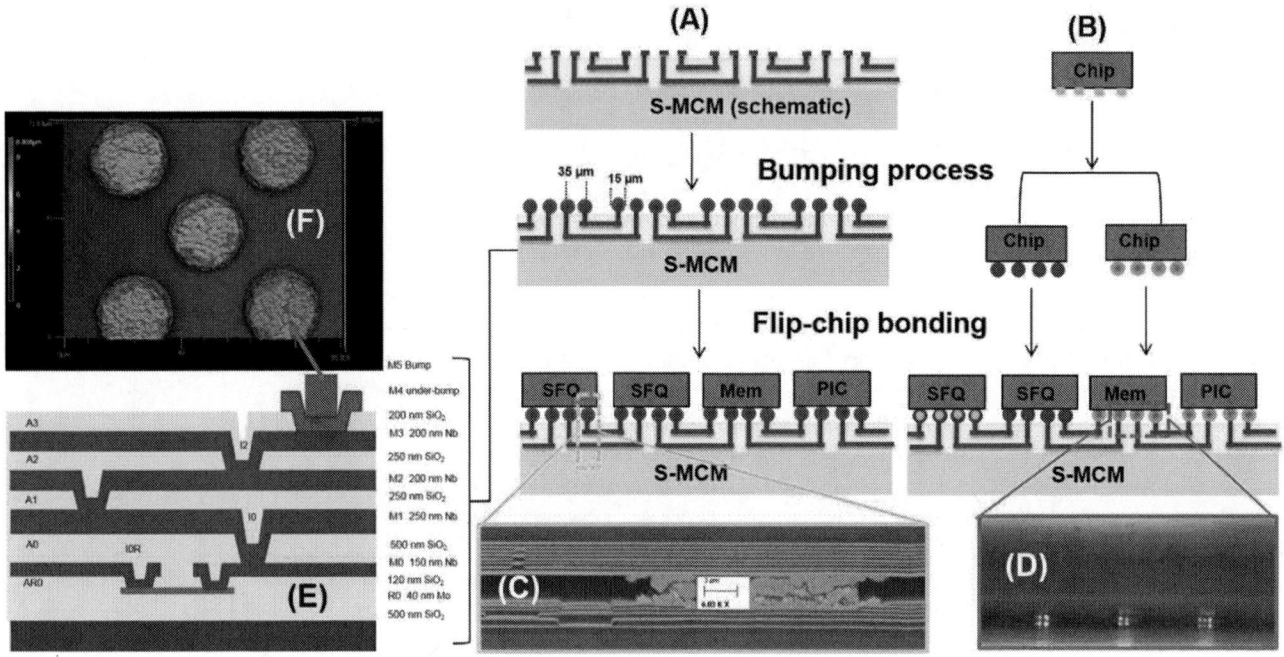

Figure 1: Flip-chip heterogeneous integration process for S-MCM using indium and gold bumps. Bumped multi-layer S-MCM flip-chip bonded to multiple ICs. (A) Schematic of indium bump-bonding process. (B) Schematic of gold and gold-to-solder bump-bonding process. (C) SEM FIB cross-section of SFQ (Single Flux Quantum) chip and S-MCM chip bonding wherein an indium bump is compressed to provide electrical connection between the S-MCM and SFQ chips. (D) Optical micrograph of flip-chip bonded chips using gold bumps. (E) Schematic cross-section view of S-MCM. (F) 3D confocal microscope image of Indium bump on S-MCM.

wafer is that the device may not need a redistribution layer (RDL) and thus existing die design with inline pitch can be directly applied for flip-chip applications.

The present S-MCM processes can be used to fabricate a wide range of die-to-die (2,3) and die-to-wafer level integrations using various microbump technologies. We present the design and fabrication of interconnect structures for flip-chip assembly to reduce interconnect spreading during flip-chip bonding that will allow us to integrate microbump interconnects suitable for smaller pitch (10-35μm) assembly. The topics that will be covered in the subsequent sections are the heterogeneous integration of (1) S-MCM with superconducting chips using 15μm diameter indium bump with 35μm bump-pitch, gold bump-bonding for (2) converting wire-bondable chip to flip-chip, and finer pitch (10-25μm) flip-chip daisy-chains with (3) a 5X5 mm² chip with a single-layer or multilayer superconducting module.

II.   HETEROGENEOUS INTEGRATION

In this section, we detail the integration processes for indium (35 μm bump-pitch) and gold bump-bonding. The 35μm bump-pitch fabrication processes had to be modified in order to fabricate finer pitch (10-25μm) flip-chip daisy-chains.

In order to efficiently build S-MCMs with different size dies, fabricated using different foundry processes on different size wafers, the heterogeneous integration process has been developed. The process uses a two-stack structure that consists of the bumped S-MCM and superconducting chips interconnected with each other. This heterogeneous integration is based on microbump design, fabrication, and integration of fine-pitch die-to-die and die-to-wafer flip-chip interconnection technology. A variety of microbump options including gold-to-gold, gold-to-solder (e.g., indium), solder-coated gold, solder, and low temperature solders (e.g., indium) were used for bump-bonding fine pitch flip-chip structures [4-6]. For test purposes, multiple S-MCM daisy chain structures with different (10-35μm) pitch were

978-1-7281-1500-9/19 $31.00 © 2019 IEEE   1612

designed and fabricated. A series of microbumps were deposited on 200mm S-MCM wafers using various contact and projection photolithography processes. Microbumps were characterized before and after bump bonding. Characterization includes alignment accuracy, micro-bump deformation during thermocompression bonding and reflow process, initial thickness of indium bump, bump to under bump metal (UBM) pad ratio, metal solder dam, pull and shear test [1,2]. Optimization of these parameters will help to provide design guidelines for finer pitch assembly.

Figure 1 shows the S-MCM circuits fabricated on 200mm silicon wafers. The S-MCM is based on four Nb metal layers and one resistor layer (optional), separated by PECVD silicon oxide dielectric and interconnected by superconducting Nb vias. It utilizes two signal and two ground layers. For example, the S-MCM uses 50-ohm and 20-ohm characteristic impedance lines for clock and data respectively. As a case study, heterogeneous integration based on 15μm indium bumps with 35μm pitch was developed (Figure 1A) for combining multiple superconducting chips on a large S-MCM. We also report the use of gold bumps to convert wire bondable chips to flip-chips. Gold bumps are deposited on the Al pads using ultrasonic ball bonding where the bonding breaks the native oxide, interdiffuses gold to aluminum and eliminates the need for UBM on Al pad. Gold-to-gold interconnects require high temperature thermal compression bonding and may not be suitable for superconducting electronics, which is easily degraded by high temperatures (above 170°C). In the present study, a variety of gold bonding options were evaluated for superconducting and cryogenic electronics. Figure 1 shows the process flow for heterogeneous integration which includes Indium, gold bump deposition (mechanically or lithographically), flip-chip bonding and cross-section views. For example, the schematic in Figure 1B shows gold mechanically deposited on an Al pad and flip-chip bonded to the S-MCM. Furthermore, Figure 1B shows a gold bump deposited on the S-MCM and flip-chip bonded to a solder-bumped chip to interconnect the chip and S-MCM. Gold based flip-chip bonding is a relatively high temperature (~280°C) process, but addition of a solder bump or pre-applied adhesive will reduce the flip-chip bonding temperature. Figure 1 shows a side-view SEM and optical micrograph of indium and gold interconnected chips bonded by using the higher-accuracy, force-controlled flip-chip bonder to produce a S-MCM flip-chip module. Although mechanically deposited gold bumps are important to convert wire bondable to flip-chip, these bump diameters are typically large and it is only a viable approach for ~100μm or larger pitch flip-chip.

In contrast to the gold bumps the use of Indium microbumps (15μm) and thermocompression bonding will provide full-wafer bonding, small bump pitch (35μm), low interconnect resistance, high adhesion strength and precise spacing between the chip and module. The indium microbumps are used for both electrical interconnection and mechanical support between the superconducting chip and the module.

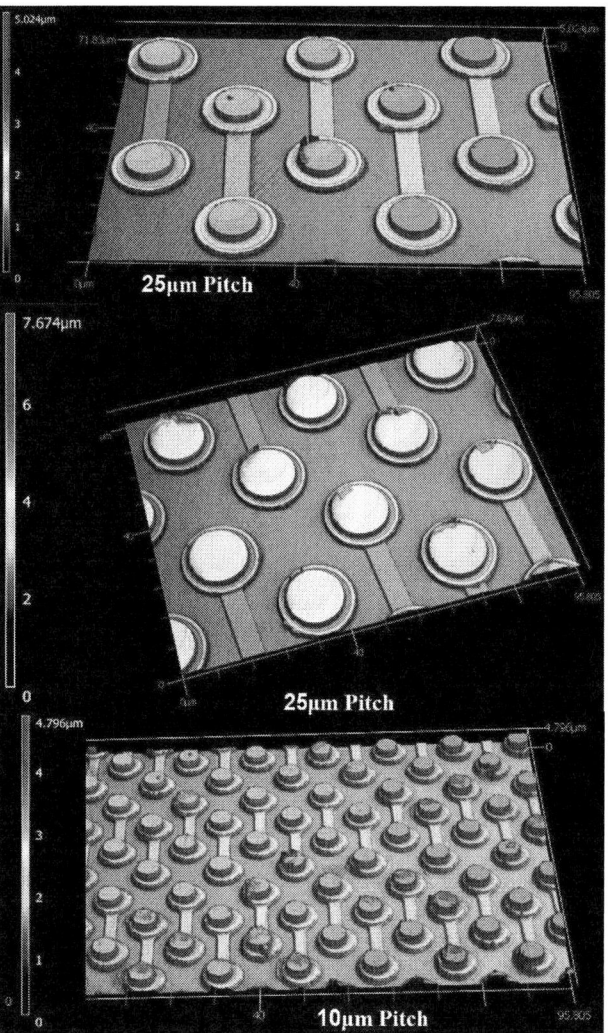

Figure 2: Confocal 3D image of patterned gold-on-gold bumps for 25μm and 10μm pitch flip-chip attachment.

III. FLIP-CHIP FINER PITCH DEVELOPMENT

Heterogeneous integration uses a variety of lithographic options to develop microbumps on S-MCMs to interconnect with superconducting or semiconducting chips. The baseline process uses 15μm diameter indium microbumps for 35μm pitch and extensions of this process have been developed to enable smaller bumps and pitch. The use of small-diameter bumps and small pitch has several advantages: For example, smaller bumps and pitch (≤10μm) eliminate the need for a redistribution layer; the chips will require less post processing compared to 3D integration. We have fabricated microbumps for 10-25μm pitch.

978-1-7281-1500-9/19 $31.00 © 2019 IEEE

Figure 3: Confocal 2D image of patterned-indium-coated gold bumps for 10μm pitch flip-chips

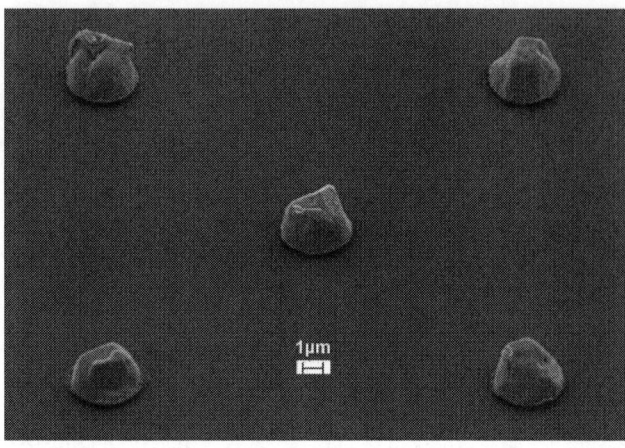

Figure 4: SEM of patterned indium for 10μm pitch flip-chips.

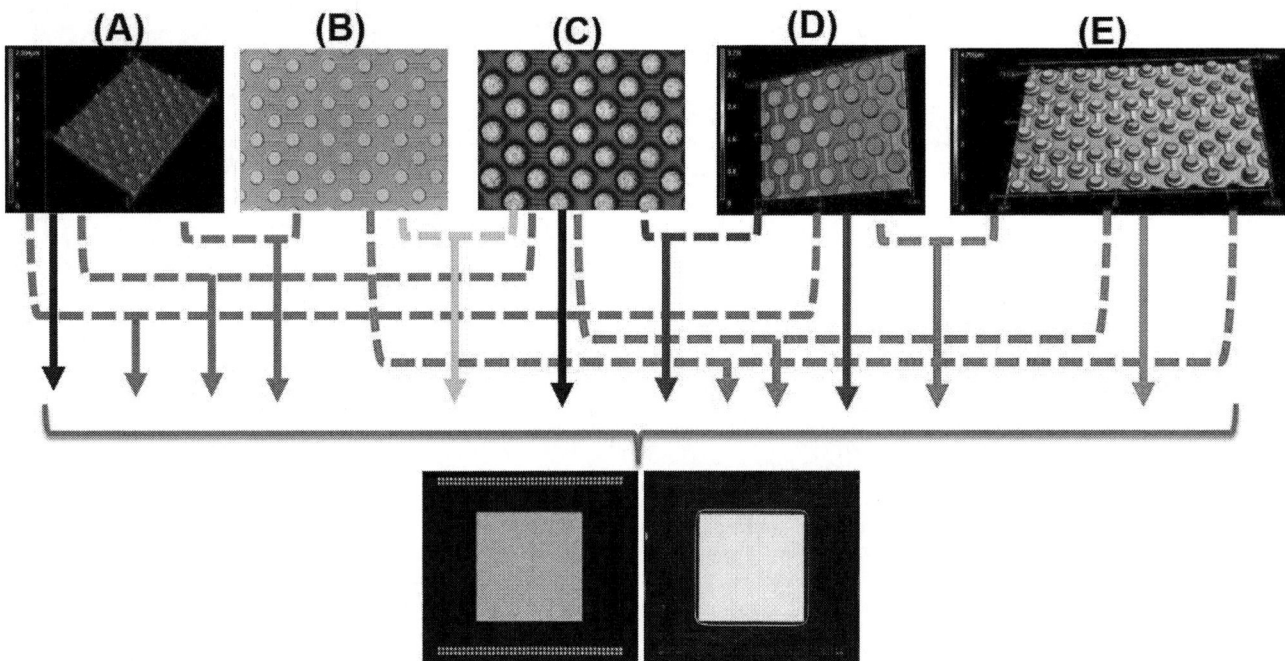

Figure 5: Various bump bonding schemes for producing 10μm pitch flip-chips. (A) Indium bump on UBM pad, (B) UBM pad only, (C) Indium-coated-gold bump on UBM pad, (D) Gold bump on UBM, and (E) gold-on-gold bump on UBM pad. Bottom two images represent 10μm pitch flip-chip attachements where a 10mmX10mm S-MCM was flip-chip bonded with a 5mmX5mm active circuit chip.

978-1-7281-1500-9/19 $31.00 © 2019 IEEE

Figures 2-4 show confocal and scanning electron micrographs (SEM) of gold, indium-coated-gold and indium micro-bumps for 10-25μm pitch flip-chip attachments as representative examples. Figure 2 shows confocal microscope images of gold-on-gold micro-bumps. Figure 3 is a confocal microscope image showing an area of indium-coated-gold microbumps with 10μm pitch. Indium coating can be controlled to minimize indium spreading during flip-chip bonding. Gold and indium-coated-gold utilizes I-line (365nm) photolithography to control bump-to-pad alignment for 10μm pitch with larger bump and smaller gap between the bumps. Figure 4 shows a SEM image of small (2. 5μm) indium microbumps and the image confirms larger bump separation for 10μm pitch. Small (2. 5μm) bump fabrication further supports the creation of smaller than 10μm pitch flip-chip capabilities.

Microbump diameters for both gold and indium-coated gold are relatively large (≥6 μm) in diameter and experience minimum deformation during bump-bonding that allows a larger bump approach for 10μm pitch. Indium microbumps with diameter 4μm and below are fabricated onto a three layer (20nm Ti - 50nm Pt - 150nm Au) under-bump metal (UBM) pad for 10μm pitch flip-chip applications. Altogether, there are four different micro-bumps fabricated to create a 10μm pitch flip-chip attachment. Four different micro-bumps can mix and match to produce 13 possible flip-chip combinations. Detailed bump-bonding schemes for gold, gold-on-gold, indium coated gold and indium microbumps are described in the bonding flowchart in Figure 5. For example, indium microbumps bonded in four different interconnect combinations (indium-to-indium, indium-to-UBM, indium-to-indium-coated-gold, indium-to-gold) are used to produce series connected flip-chip daisy-chain circuits.

IV. MEASUREMENT AND SIMULATION

High-performance computing based on superconducting integrated circuits is an attractive low-power technology with many potential advantages including fast switching (~1 ps), low dissipated energy per switch (<10^{-19} J), and low-loss interconnects in which information is transmitted using small current pulses through superconducting traces [1,7]. A first step towards large area flip-chip integration is to bump-bond flip-chip daisy chains between S-MCM and SFQ (Single Flux Quantum) chips. Daisy-chain and S-MCM-SFQ flip-chip modules were underfilled to measure I-V characteristics of niobium lines and Josephson junctions through the S-MCM at 4.2 K. Underfill will improve mechanical stability of flip-chip modules at 4.2K. The flip-chipped superconducting integrated circuit maintains a typical unshunted Josephson junction characteristic such as critical current, subgap voltage, and normal resistance (Rn) of Nb/Al-AlOx/Nb tunnel junctions [1]. Daisy chains showed resistance ranging from 0.25 milliohm/bump for 25-35 μm pitch and 3-5

milliohm/bump for 10-15μm pitch at 4.2K. Here the indium bump acts as a milliohm/microohm resistor attached to the superconducting niobium trace at 4.2K. Indium bumps with appropriate fabrication and bonding are able to maintain high niobium critical current in the flip-chip structure. Daisy chains with 10-15μm pitch showed higher bump resistance and smaller critical current than 25-35μm pitch samples due to smaller microbump contact area. The daisy-chain I-V curve is linear up to the niobium critical current, where only indium bumps contribute to the total resistance. Above the critical current, Nb becomes a normal conductor, and the I-V curve appears to be nonlinear. To assess critical current of 10μm pitch daisy-chains, flip-chips were underfilled and attached to a PCB and wire bonded for 4.2 K *I-V* measurements. Initial flip-chip daisy-chain results showed (Figure 6) *I-V* characteristics with critical current in the range of 10mA. Fabrication and bonding evaluation continue to optimize the niobium critical current in the flip-chip daisy-chains.

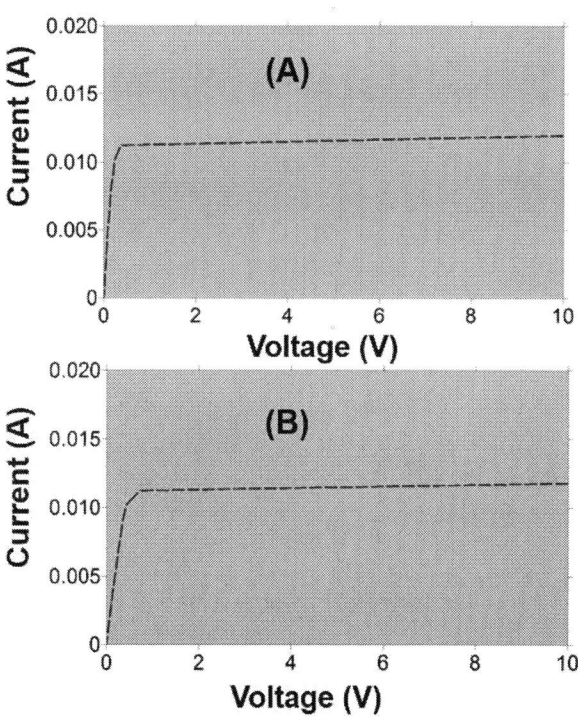

Figure 6: Representative I-V curves of 10μm-pitch flip-chip daisy chains with indium-to-indium (A) and indium-to-indium-coated-gold (B) bump bonding process.

The S-MCM bump-bonding process was verified by calibrated S-parameter [8] measurements at frequencies up to 4.5 GHz of a 50Ω superconducting niobium delay line that was bump bonded to an S-MCM. The bump-bonded line showed insertion loss within 0.1 dB of the same delay line conventionally packaged with wire bonds and comparable return loss, indicating that the impedance was well matched at 4.2 K.

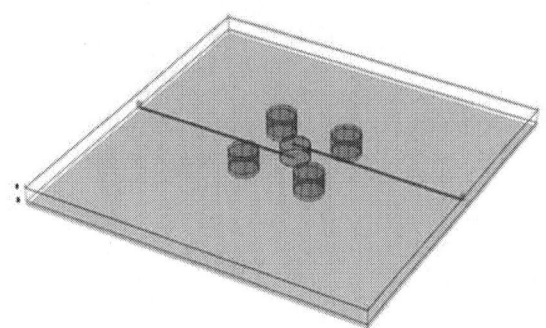

d = 20 μm, p = 35 μm, h = 2.5 μm

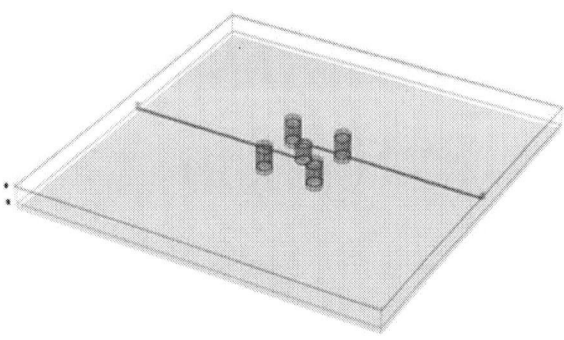

d = 10 μm, p = 25 μm, h = 2.5 μm

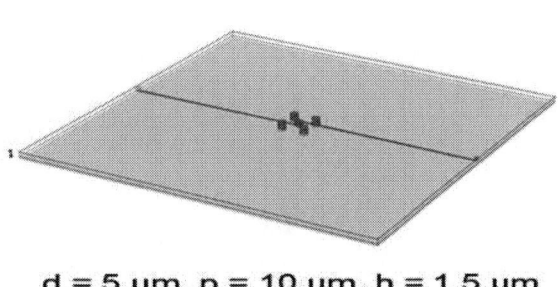

d = 5 μm, p = 10 μm, h = 1.5 μm

Figure 7: Sonnet *em* 2.5d models of three microstrip-to-microstrip flip-chip transitions. Microstrip lines (50 Ω) are 1.5 μm wide and 0.2 μm thick, over 0.7 μm SiO₂ dielectric above a ground plane. Microstrip lines and ground planes are modeled as Nb with a kinetic inductance of 0.1pH/square.

Several RF flip-chip transitions were designed using different geometries and design rules, and simulated to predict their performance. Microstrip transmission lines were designed for a characteristic impedance of 50 Ω using 1.5 μm wide, 200 nm thick Nb signal traces patterned over 0.7 μm of SiO_2 and a Nb ground plane (Figure 1E, M1:ground and M3:signal). This microstrip line (a "thru" line) was modeled and simulated in Sonnet *em* with a thick metal approximation composed of two sheets of Nb with 0.1pH/square kinetic inductance connected with perfectly electrical conductor (PEC) edge vias. The ground plane was simulated as zero thickness Nb (including kinetic inductance) also acting as the simulation box lid boundary. Next, a model was constructed with two microstrip lines, with one line "flipped" upside down, and connected by a set of PEC vias and bumps, representing a microstrip-to-microstrip flip-chip transition with short transmission line feeds. The flip-chip interface used a single pad/bump for the microstrip signal conductor with four surrounding pad/bumps that connect the upper and lower microstrip ground planes, approximating a coaxial structure.

Using variations of this modeling methodology, several flip-chip transitions were simulated with decreasing bump size, pitch, and height (Figure 7). The simulated input return loss (S_{11}) of each version is shown in Figure 8 along with the flip-chip pad geometry. Since the transmission lines are assumed superconducting and shielded, an input return loss of greater than 20 dB (i.e. $|S_{11}|_{dB} < $ -20 dB) is considered well matched and corresponds to negligible insertion loss.

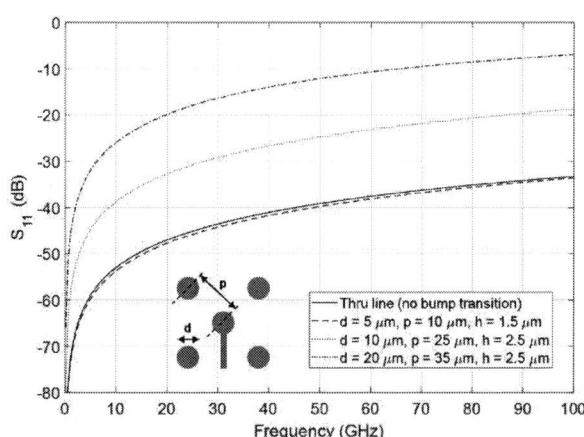

Figure 8: Simulated input return loss (S_{11}) of a microstrip ('thru') line, three microstrip lines and flip-chip transitions with different bump diameter, pitch, and height.

To improve the 20 μm diameter bump transition performance, an additional design and simulation was performed using inductive compensation. In this scheme (Figure 9), a small (100s of pH) inductance is placed in series with the microstrip lines immediately before and after the flip-chip transition to resonate with the transition's capacitance, extending the impedance matched bandwidth. In this case, adding meandered trace inductors improved the 20 dB return loss bandwidth from 23 GHz to 81 GHz.

While the flip-chip transition simulations shown here all involve microstrip lines, similar results were obtained for coplanar waveguide (CPW) and stripline structures. In the case of stripline, the stripline-to-microstrip transition was often the limiting factor in the overall structure's bandwidth.

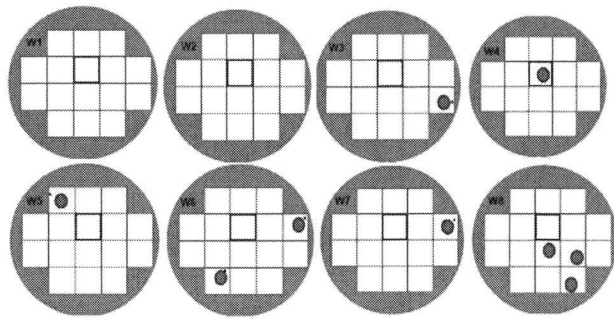

Figure 10: S-MCM wafer map. Each S-MCM run produces 8 wafers and each wafer consists of 16 individual 32mmX32mm S-MCMs. Red dots represent defect-based yield loss for the S-MCM.

V.   FUTURE   TRENDS   IN   CRYOGENIC PACKAGING

As defined above, we have used indium microbump technology, capable of large scale integration, to connect to a superconducting multi-chip module (S-MCM) fabricated on 200 mm silicon wafers. Figure 10 shows a representative example of wafer yield map for 32mmX32mm MCMs. The wafer map indicates that MCM yield is over 90%, which is high enough for fabricating large S-MCMs beyond the size possible with a single mask exposure field. The first step to fabricating large S-MCMs is to establish a bonding evaluation procedure to understand if we can handle large S-MCMs. Previously, a maximum of sixteen 5x5 mm²chips were bonded and underfilled to a 32 x 32 mm² S-MCM. A 200 mm wafer consists of 16 such S-MCMs, capable of each bonding 16 chips, that can be connected together to create a larger S-MCM. Figure 11 shows an example of full-wafer flip-chip bonding, where superconducting chips are attached using indium bumps to multiple S-MCM sites to create larger than existing S-MCMs (32mmX32mm). Each S-MCM has 16 SFQ chip bonding sites and there are pads at the edge of the S-MCM for each SFQ chip with different size JJs to

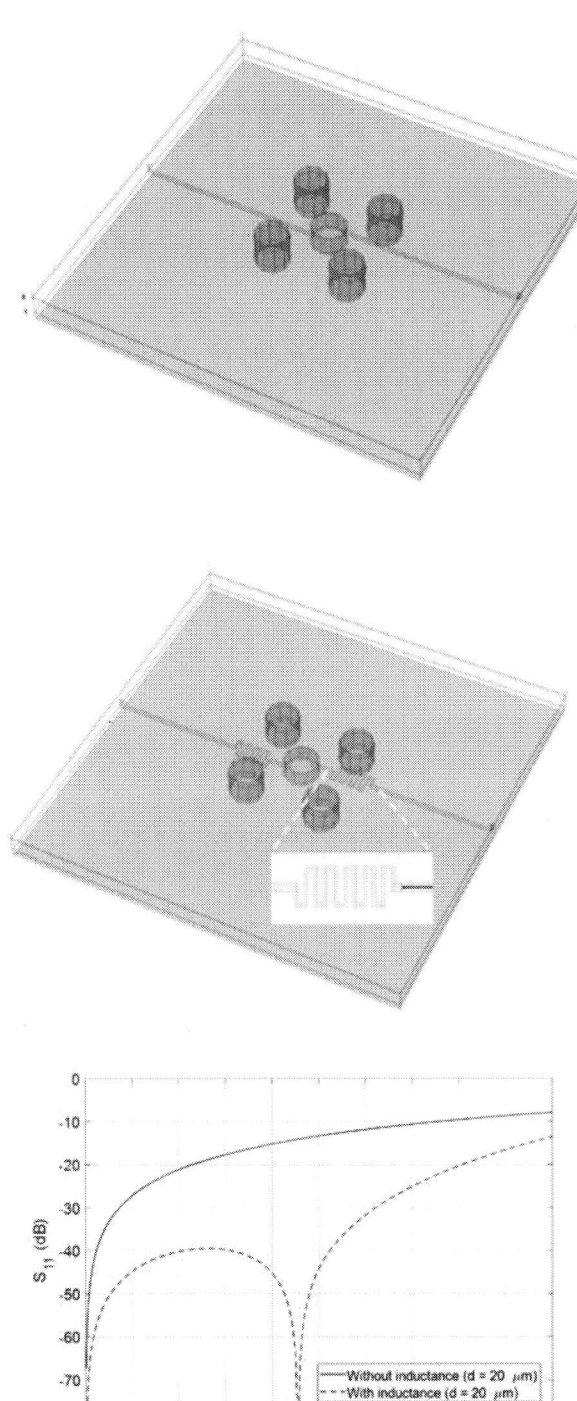

Figure 9: SEM Microstrip line and 20 μm bump flip-chip transition *without* (top) and *with* (center) series compensation inductance (inset, center). The addition of series inductors extends the 20-dB return loss frequency of the transition from 23 GHz to 81 GHz as seen in simulation (bottom).

978-1-7281-1500-9/19 $31.00 © 2019 IEEE   1617

individually measure and address test structures in order to characterize changes to the Josephson junction behavior. Full-wafer bonding capability for different size chips and different bump-pitch (10-35μm) supports the design of S-MCMs larger than a single mask exposure field.

We have used various lithographic and bonding-technology approaches that allow us to fabricate a large area "S-MCM" that can exceed the maximum imaging field of lithography tools. To realize a MCM larger than the reticle size, the following connection strategy can be applied:

- Jumper flip-chips to connect multiple reticles
- Lithographic stitching of multiple reticles
- Laser direct writing
- Combination of stitching and laser direct writing

The use of S-MCM jumper flip-chip technology using multilayer passive interconnects based on Niobium can be used for routing of the many signals from one S-MCM to the next S-MCM. Connections between the jumper chip and the S-MCM will be achieved using an indium bump-bonding process that is fully compatible with Josephson-junction-based electronics. Multiple passive wiring layers and 15um diameter indium bumps on the jumper chip will allow for large numbers of connections to be made while maintaining appropriate impedance for clock lines and data lines. The jumper flip-chip approach uses normal metal UBM and indium bump, which is not superconducting at 4.2K, and may require special design for superconducting electronics. Jumpers are used not only to connect multiple S-MCMs but also to add integration flexibility and re-workability. Furthermore, part of the large S-MCMs, if necessary, can be replaced or repaired, or even upgraded using a jumper approach. This approach may not be suitable for creating a superconducting path between the S-MCMs, but is useful for creating a low-resistance (micro-ohm) interconnect path, a cost effective and reworkable solution for creating large S-MCMs. Figure 12C shows that a part of the large S-MCM that is non-functional can be removed from the base S-MCM. Here jumper portions connect base S-MCM and jumper S-MCM portion used for SFQ flip-chip bonding (Figure 12 C) and thus possible to replace a non-functional S-MCM section.

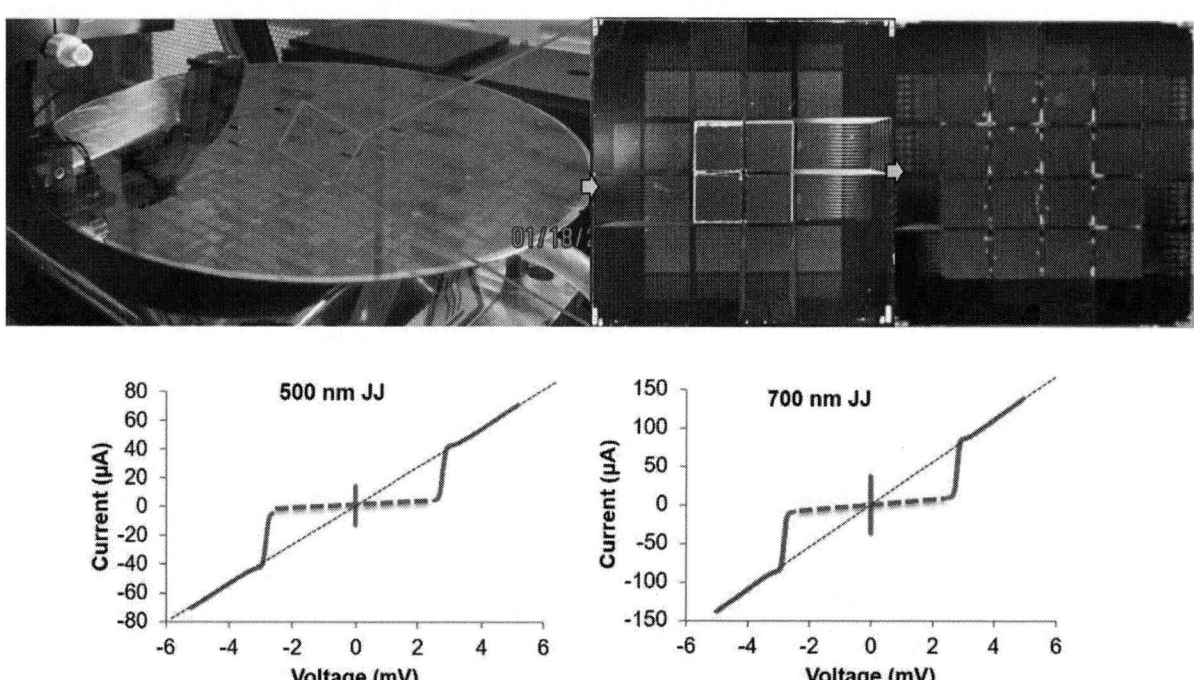

Figure 11: Full-wafer bonding demonstration. The 200mm diameter wafer has 16 MCMs and each MCM has up to 16 bonding sites for 5mmX5mm SFQ chips. Each SFQ chip can be measured individually. Plots on the bottom show the I-V curves of superconducting flip-chip bonded SFQ chips, to characterize the switching behavior of 500nm and 700nm Josephson junctions.

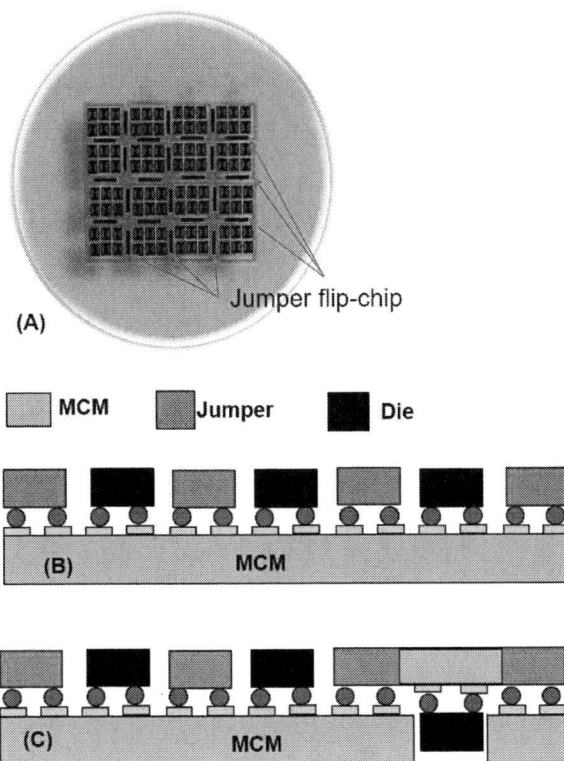

Figure 12: Schematic of full-wafer S-MCM using flip-chip Jumper approach. (A) Top view of Full-wafer S-MCM, (B) Cross-sectional View, and (C)Cross-sectional view of full-wafer S-MCM where part of the full-wafer S-MCM removed and replaced with jumper flip-chip S-MCM.

For stitching multiple reticles with single or multiple design blocks, reticles are lithographically stitch together to create larger S-MCM. Here each reticle (design blocks) use individual single exposure masks. Figure 13 shows four reticles stitched together to create large S-MCMs. We show various stitching options to create a large S-MCM. It can be single reticles repeated four times or two reticles repeated twice or four different reticles stitched together to create a S-MCM. For example, if the initial reticle size is 35mmX35mm, stitching will produce a 70mmX70mm S-MCM. The fabrication processes uses I-line (365 nm) photolithography to stitch patterns with a minimum line width of 0.8 micron [9]. At the stitch boundary, the photoresist will be double exposed which causes some line narrowing in the stitching area. By controlling appropriate overlap length, one can minimize the line width distortion. We have estimated the effects of line width distortion (stitching artifacts) at the stitch boundary. Even in the worst case, impedance discontinuities produced by the stitching artifacts are electrically small and have little effect below 1 THz. For

example, a 2μm-long stripline discontinuity is approximately 1.3% of a guided wavelength at 1 THz. At 100 GHz and below, the effects of the stitching discontinuities, which occur several mm apart at the reticle boundaries, are negligible on circuit performance. Simulation of 2μm-long stripline widened by 0.2μm produced a parasitic capacitance of 0.02 fF. Based on these findings, a stitching process can be considered for fabricating large S-MCM packages.

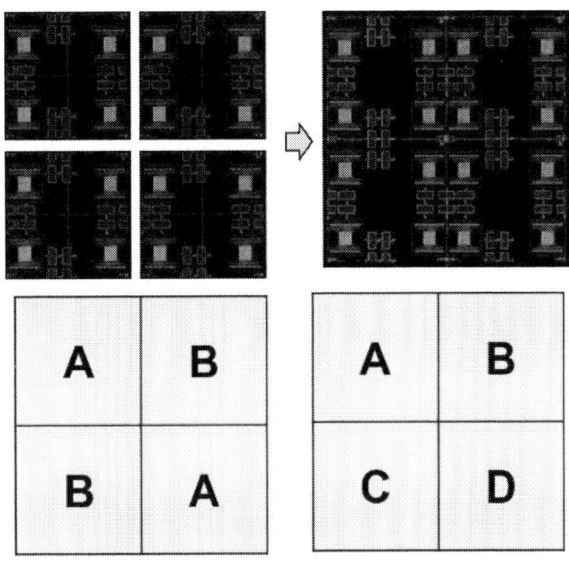

Figure 13: Schematic of large S-MCM fabrication using a reticle stitching approach. Four of the same reticles (top left) stitched together to create a large S-MCM (Top right). Bottom left: Schematic of two reticles stitched together. Bottom right: Schematic of four different reticles stitched together.

As an alternative approach, we have used laser direct writing, a maskless approach to create a large S-MCM. The laser writing study includes a variety of resist-patterning approaches. Figure 14 shows I-line (365nm) photolithography resist profiles using different processes, write and expose times. The write time depends on complexity and density of the features. The choice of resist materials can minimize write time. For example, laser writing on negative resist favors etching and laser writing on positive resist favors semi-additive plating or lift-off to create lines. Laser writing can be used to create small diameter (~2.5μm) bumps for 10μm or smaller pitch applications (Figure 4). We utilize I-line (365nm) photolithography resist for laser writing. A set of snake and comb structures were designed and fabricated to evaluate laser direct writing performance. Figure 14 show top view of I-line resist pattern of 0.8-1μm line structures.

We conducted multiple laser writing experiments to evaluate large S-MCM fabrication. In Figure 14, we show laser writing demonstrations for etching and lift-off processes and compared it with stitching. For this process, we patterned the I-line resist (Figure 14C) and used a lift-off process to create metal lines. In Figure 14B we show I-line resist patterning for etching to create the same metal lines by using the inverse mask. As we look further out to fabricating larger S-MCMs, we also consider customizable approaches to combine the laser writing with stitching for unique S-MCM fabrication and packaging solutions. As one option, large S-MCM fabrication could use the stitching process to pattern signal layers and use laser writing process to pattern ground layers. In another option, multiple reticles fabricated using single/multiple I-line masks can be stitched/fan-out using laser writing. In these schemes, S-MCM fabrication processes are combined to reduce the total number of masks as well as reduce laser write time for a given S-MCM size. The application of this combined approach will not only increase wiring density with controlled linewidth, but also provide a cost effective, re-workable solution for fabricating large S-MCMs.

VI. CONCLUSIONS

Superconducting computing system requirements for a large number of superconducting ICs with more functionality and higher wiring densities within a given cryogenic space are driving die-to-die and die-to-wafer level heterogeneous integration. Large or full-wafer S-MCMs with microbump-based interconnects can electrically couple many superconducting ICs and are capable of maintaining the I-V characteristics of JJs and superconducting traces interrupted by microbumps. This method provides benefits in design and fabrication enabling small-pitch and high-density interconnects. By designing and applying appropriate microbumps, scalable and high-wiring-density S-MCM based heterogeneous systems can be achieved. The S-MCM can be attached to various chips including, single flux quantum (SFQ), rapid SFQ (RSFQ), energy-efficient RSFQ (ERSFQ), reciprocal quantum logic (RQL), CMOS, and photonics to create a superconducting system in a package (SSiP). Furthermore, indium, indium-coated gold, gold and gold-on-gold microbump-based flip-chip interconnects have shown advantages for die-to-die and die-to-wafer packaging. Specially, a 10μm pitch flip-chip niobium based daisy chain with 10mA niobium critical current have been demonstrated. To drive broader adoption of finer pitch (10 μm) applications, non-contact lithography, and multiple interconnect materials need to be supported. Simulation of a microstrip-to-microstrip flip-chip transition with smaller bump in the range of 5-10 μm bump diameter compared with 20μm bump yields a much better impedance match and less reflected energy, with the near-perfect performance to 100 GHz and beyond. Overall, our results suggest that the heterogeneous integration approach comprising S-MCM and microbump technology may enable the design and realization of scalable computing.

ACKNOWLEDGEMENT

We greatly appreciate the support of Dr. M. Manheimer, Dr. D. S. Holmes and Dr. Deborah Van Vechten for this work. We gratefully acknowledge M.A. Gouker, E.A. Dauler, P. Gouker, S.K. Tolpygo, A. Wynn, R. D'Onofrio, C. Stark, D. Pulver, and P. Murphy for useful discussions and K. Magoon, P. Baldo, M. Townsend, M. Augeri, J. Liddell, B. Osadchy, M. Hellstrom, C. Thoummaraj, and J. Wilson for valuable technical assistance. This research was funded by the Intelligence Advanced Research Projects Activity (IARPA) and Office of Naval Research (ONR). Opinions, interpretations, conclusions and recommendations are those of the authors and are not necessarily endorsed by the United States Government.

REFERENCES

1. R. N. Das, V. Bolkhovsky, S. K. Tolpygo, P. Gouker, L. M. Johnson, E. A. Dauler, and M. A. Gouker, "Large-Scale Cryogenic Integration Approach for Superconducting High-Performance Computing" *IEEE Electronic Components & Technology Conference Proceedings*, 675-683 (2017).

2. R. N. Das, J. L. Yoder, D. Rosenberg, D. K. Kim, D. Yost, J. Mallek, D. Hover, V. Bolkhovsky, A. J. Kerman and W. D. Oliver, "Cryogenic Qubit Integration for Quantum Computing" *IEEE Electronic Components & Technology Conference Proceedings*, 504-514 (2018).

3. D. Rosenberg, D. Kim, R. Das, D. Yost, S. Gustavsson, D. Hover, P. Krantz , A. Melville, L. Racz, G. O. Samach, S. J. Weber, F. Yan, J. L. Yoder, A. J. Kerman and W. D. Oliver, "3D integrated superconducting qubits" *npj quantum information* 42 (2017) doi:10.1038/s41534-017-0044-0

4. R.N. Das, F.D. Egitto, S.G. Rosser, E. Kopp, B. Bonitz, R. Rai, "3D Integration of System-in-Package (SiP) Using Organic Interposer: Toward SiP-Interposer-SiP for High-End Electronics" *46th International Symposium on Microelectronics (IMAPS)*, 531-537(2013).

5. R.N. Das, F.D. Egitto, J. Lauffer, B. Bonitz, B. Wilson, M. D. Poliks, V. R. Markovich, "3D-Interconnect Approach for High End Electronics"

IEEE Electronic Components & Technology Conference Proceedings, 1333-1339 (2012).

6. F.D. Egitto, R.N. Das, G. E. Thomas, S. Bagen, "Miniaturization of Electronic Substrates for Medical Device Applications" *45th International Symposium on Microelectronics (IMAPS)*, 186-191(2012).

7. D. S. Holmes, A. L. Ripple, and M. M. Manheimer, "Energy efficient superconducting computing-power budges and requirements," *IEEE*

Transactions on Applied Superconductivity, 23(3), 2013, 1701610.

8. D. E. Oates, R. L. Slattery and D. J. Hover, "Cryogenic test fixture for two-port calibration at 4.2 K and above," *2017 89th ARFTG Microwave Measurement Conference (ARFTG)*, Honololu, HI, 2017, pp. 1-4.doi: 10.1109/ARFTG.2017.8000842.

9. R. N. Das, V. Bolkhovsky, R. D. Lambert, S. Zarr, R. Rastogi, D. Shapiro, M. Docanto, T. Weir, L. M. Johnson, and E. A. Dauler, To be submitted.

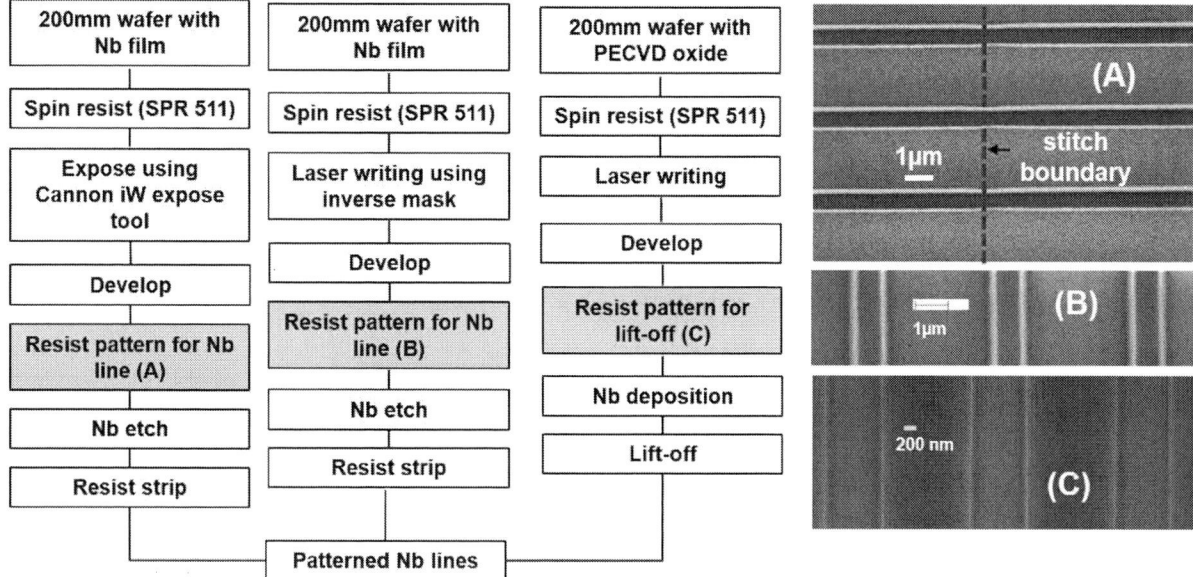

Figure 14: I-line (365nm) photolithography resist patterning using iW stepper and laser writing for creating Nb lines on 200mm wafer. (A-C) SEM of resist profile utilizing iW stepper and laser writing. (A) Stitching: 0.8μm line resist profile utilizing iW stepper. (B) Top view of resist profile utilizing laser writing. Here the inverse mask is used to create the resist pattern for etching Nb lines. (C) Top view of resist profile utilizing laser writing for lift-off process to create Nb lines.

Ultra Wide Micro bumps interconnection matrix for particle detection: process and assembly.

J. Charbonnier, M. Assous, T. Mourier, C. Ribière, S. Minoret, S. Verrun, P. Tissier, R. Coquand, B. Mehmet,
F. Allain, R. Franiatte, G. Pares

Univ. Grenoble Alpes, CEA, LETI, 38000 Grenoble, France

jean.charbonnier@cea.fr

Abstract—**Micro pillars and micro bumps interconnections are considered as mature technology for 3-D integration and chip stacking. However, in the framework of high-energy particles detection as ATLAS Large Hadron Collider new tracker project at CERN, very large area array of dense interconnections with an aggressive specification of total thickness variation (TTV) of +/-2μm are required to grant successful detectors assembly process. This paper will first describe the test vehicle that has been designed on purpose for this study. The studies undertaken including lithography mapping, seed layer and electrochemical deposition (ECD) process will then be detailed introducing a model of anode intensity contribution to the overall TTV. To conclude, daisy chain electrical test results and yields after stacking using die-to-wafer approaches will be discussed with respects to process parameters.**

Keywords - 3D interconnections for high energy particle detection; Copper pillars; Electrochemical deposition.

I. INTRODUCTION

Since 10 years now, 3D interconnects have made a breakthrough into global interconnects solutions for systems and package [1-3]. Concerning micro bumps interconnect, whereas some studies focus on pitch reduction [4-6] the industrial world seems to set standards between 40 μm and 50 μm for the pitch and 20 μm to 25 μm for the diameter, which is appropriate with most of technology nodes last rerouting metal design rules. In this framework, ATLAS Large Hadron Collider new tracker project at CERN [7-8] requires 2x2 cm^2 matrix of 25 μm diameter copper pillars type of interconnections at pitch 50 μm with a total thickness variations (TTV) of +/-2 μm over the full matrix. Considering such large area, electroplating process (ECD) is challenging since the load of current and the open surface may clearly influence the TTV of the pattern at die level but also at wafer level.

The first part of this article deals with design and layout considerations. It will detail the technological stack of the particular test vehicle developed. The second part of the article will describe the metrology set up for the study and will develop the layout and process parameters surveyed to establish the impact on micro bumps matrix TTV. Finally, electrical results of micro bumps/pillars chains after

stacking are discussed with respect to the optimized process and layout parameters.

II. TEST VEHICLE AND INTEGRATION

A silicon test vehicle has been designed for a bottom wafer and a top wafer and processed on 300 mm wafers process line including a standard damascene and passivation levels for metal lines.

A. Integration cross section

Fig. 1 shows a cross section of the stack with top and bottom wafers integrating a copper damascene level. The copper thickness of these lines is set to 1μm in order to be thick enough to estimate properly micro bumps resistance in the range of 4 mΩ.

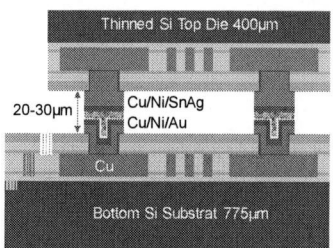

Figure 1. Schematic of the integration cross section including miror technlogies of copper damascene layers connected to micro bumps for the top wafer and micro pillars for the bottom substrat.

B. Matrix layout

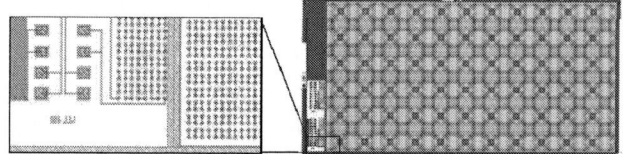

Figure 2. Detail and Global view of top die layout: 25μm diameter micro bumps matrix with 50μm pitch and 10x20mm^2 area.

The top wafer field consists in a single 1x2 cm^2 matrix of 25 μm diameter micro bumps, with 50 μm pitch and a stepping pitch of 22020 μm x 10710 μm, meaning that 85% of the wafer surface is occupied by micro bumps matrixes as introduced in Fig. 2. This micro bumps matrix is connected to the underneath damascene level through passivation

978-1-7281-1500-9/19 $31.00 © 2019 IEEE

openings of diameter 10 μm. Using the test pads row on the bottom left of the field (Fig. 2) it is possible to test several chains of micro bumps and damascene line to evaluate the resistivity of the contact through the passivation openings. When this top die is stacked on the bottom pair wafer, the interconnections matrix formed a chain of 80 000 micro bumps/pillars. This chain can be probed fully or partially thanks to dedicated test patterns.

C. Metrology of Micro bumps matrixes

A NSX Metrology system from Rudolph Technologies has been used for TTV measurements and deffectivity analysis. This tool hosts a monochrome camera for visual inspection and critical dimensions (CD), an infrared (IR) camera for thorough site location, and two interferometric point sensors for distance/thickness measurements using white-light bulb and infrared laser source. Thanks to these tool capabilities, a 35 chips cartography is defined at wafer level with 23 micro bumps measured per chip all along the matrix diagonal (Fig. 3).

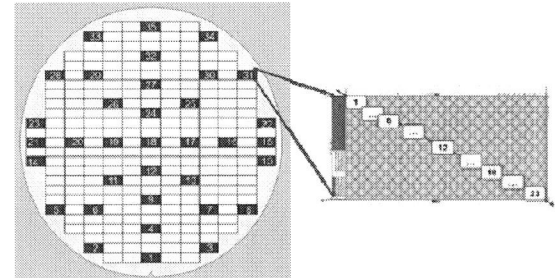

Figure 3.   Global view of the 35 chips cartography at wafer level and 23 measured micro bumps per matrix at chip level.

III.   MICRO BUMPS PROCESS AND LAYOUT SUTDY FOR ELECTRICAL AND TTV PROPERTIES

The engineering studies have been focusing on copper pillars, composed of copper, nickel and gold for the bottom wafers and copper bumps, composed of copper, nickel and tin silver alloy for the top wafer. Among all the process and layout parameters, the following have been selected based on Leti know-how: the type of process for Physical Vapor Deposition (PVD) of the seed layer, the edge exclusion of the insulation map, the growth rate of the electrochemical deposition (ECD) and the distribution of the current in the ECD tool's anodes.

A. Barrier and seed layer despostion process

In the micro bumps process flow, after the passivation opening, a titanium barrier and a copper seed layer are deposited to enable the further copper, nickel, tin silver ECD through the micro bumps resist lithography pattern. For this process, two types of deposition mode are evaluated: ionized physical vapor deposition mode of 100nm/400nm of titanium/copper or classical PVD (direct current) deposit 50nm/200nm of the same metals. In addition to this, several pre-clean treatments are also evaluated. A hot soft etch (HSE) which is a physical argon plasma etch equivalent to

the consumption of 20nm of thermal SiO_2 or a reactive hot soft etch (RHSE) which consists in a chemical mixed argon-hydrogen plasma etch. Additionally, a queue time of 4 weeks between passivation opening and the pre-clean is applied to produce a native copper oxide formation on the underneath damascene line as worst case condition. After completion of the micro bumps process, the wafers are observed and it appears that the process leg involving the seed layer DC deposit exhibits an erratic growth whatever the pre-clean (Fig. 4).

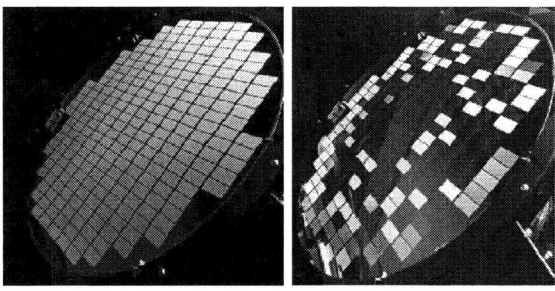

Figure 4.   Optical view at wafer level after Cu/Ni/SnAg ECD and stripping. Left: iPVD process and Right: DC process split .

This erratic behavior is evidenced at wafer level but also at die level for the DC split. The study then focuses on the iPVD process split, the DC split being not appropriated for the process.

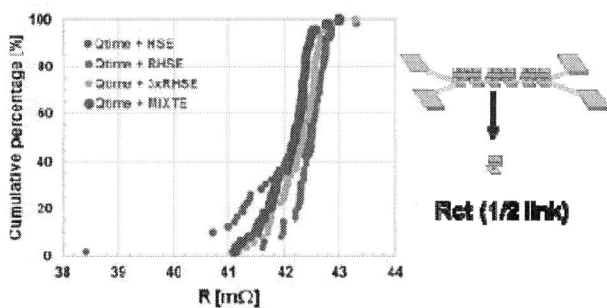

Figure 5.   Cumulative percentages of a unit resistance for a chain of 305 micro bumps and damascene level connections.

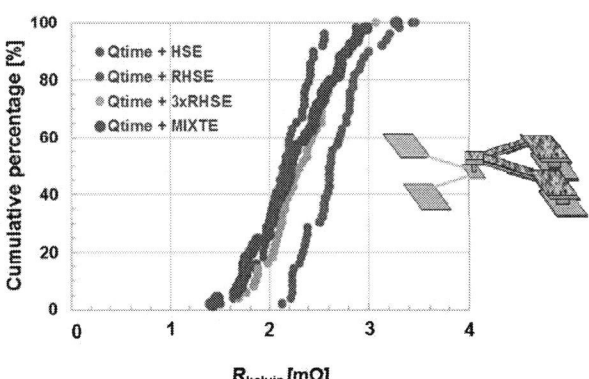

Figure 6.   Cumulative percentage of Kelvin resistance between micro bump and damascene level.

Chains of damascene line and micro bumps are tested with each pre-clean. Fig. 5 and 6 show resistance results of 305 connections chain and Kelvin pattern for a single bump and copper line, measured with 4 points probing approach.

It is possible to observe from Fig. 5 & 6 that all the pre-cleans are efficient even after a 4 weeks queue time between the passivation opening process and the barrier/seed layer deposit, the yield of each process split being close to 100%. On Fig. 6, the HSE pre-clean split exhibits a slightly lower contact resistance of 0.4 mΩ compare to the RHSE split (respectively brown and blue curve). Thus, the best process recipe considering the barrier and seed layer deposit is determined as iPVD process of 100nm of Ti and 400nm of copper using a Hot Soft Etch pre-clean.

B. ECD process: Copper Growth speed influence

After PMER 10µm thick resist coating and insulation using a Canon 300mm stepper tool with an overlay of 150 nm, the ECD process is performed using Enthone microfab bath for copper deposit, Dow Nikal chemistry for nickel deposit, and MMC Sula bath for tin silver in Semitool Raptor chambers (Fig. 7).

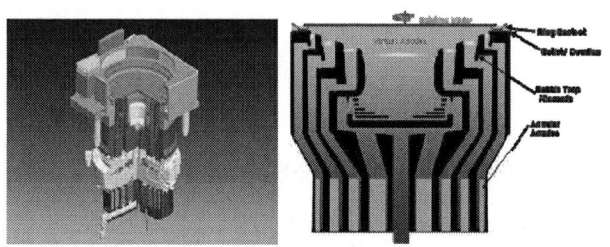

Figure 7.   Global view and schematic cross section of the ECD Raptor process chamber from Semitool (anodes in orange).

The micro bumps ECD stack is 7/2/9µm for respectively Cu/Ni/SnAg whereas the micro pillars ECD stack is 3/2/0.1µm for respectively Cu/Ni/Au metals stack. The first parameter evaluated for TTV optimization is the copper deposit rate during the ECD Process: 0.25 µm/min, 0.5 µm/min and 1 µm/min growth speed are tested on the bottom wafer layout. 15 chips are measured with 3 measurements per matrix: a copper bump at matrix center, one at the center of top side and one at the matrix corner. Copper height measurements results are presented in Fig.8.

1 µm/min is the standard copper deposit rate. The decrease of the copper deposit speed clearly improves the deposit uniformity from the center of the matrix to the matrix corner: from 1 µm range at 1 µm/min to 0.2 µm range at 0.25 µm/min deposit speed. Moreover, the standard deviation is also reduced from 0.7 µm to less than 0.3 µm with the decrease of the deposit speed. This means that reducing the copper deposit speed improves the chip level uniformity but also the wafer level uniformity.

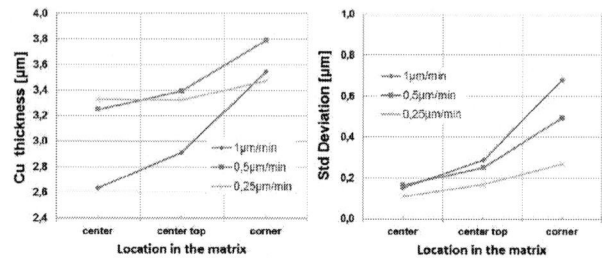

Figure 8.   Mean values (left) and standard deviation (right) of copper deposit thickness for 0.25, 0.5 and 1 µm/min growth rates.

For the rest of the study, 0.5 µm/min copper deposit rate is chosen being a tradeoff between uniformity and wafer process throughput. However, even with the deposit speed decrease, the micro bumps full stack uniformity at wafer level exhibits a dispersion of more than 10 µm from the targeted thickness due to edge overgrowth (Fig. 9).

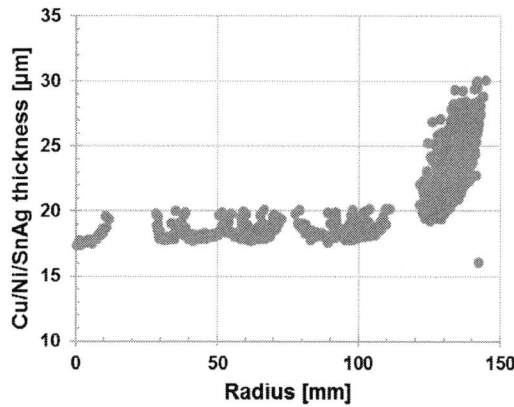

Figure 9.   Micro bumps (Cu/Ni/SnAg – 7/2/9 µm) thickness with the reference process at 0.5 µm/min in function of wafer radius.

Fig. 9 confirms that the micro bumps height profile is axi-symmetric, which is explained by the rotation of the wafer during the process. The micro bumps thickness profile through the wafers is uniform within 2 µm range until a radius of 120mm. The deviation in ECD mechanism occurring at 30 mm from wafer edge is evidenced thanks to the metrology developed for this study. This phenomenon could be due to a current density overshoot in this area.

C. Current distribution in Anodes for ECD process

To optimize the settings of the depositions to improve uniformity, without costly trial, which would waste wafers and machine time, a model of the ECD deposition steps has been developed. Micro bump height h_{bump} is assumed to be the sum of the contribution from the four anodes h_{1-4} (Fig. 7), taking into account their respective power repartition a_{1-4}. Overall, the micro bump height is written as:

$$h_{bump} = \sum_{i=1}^{i=4} h_i \cdot a_i$$

(1)

The modelling is done in two parts: calibration and prediction. Calibration consists in getting the "signature" of each of the four electrodes. Each anode power is set to 100% while the others are set to 0%, and the resulting profile is measured. This step is done for each compound: copper, nickel, and tin silver alloy. The four "signature" wafers bump heights are measured using the cartography described previously in section II. A polynomial fit is then applied to the data points with a six degrees polynomial equation which appears to be a good compromise between accuracy of the fit and low divergence at wafer center and edge.

Then, on the prediction step to obtain an optimized uniformity of the deposit, the coefficients applied to each electrode are defined in order to minimize the deviation from the bump targeted profile. This operation is repeated for each material of the stack, and the total height is taken as the sum of the heights of the three materials.

In order to validate the method, several wafers previously processed have been measured and compared to the simulated profile given by the model presented (Fig. 10).

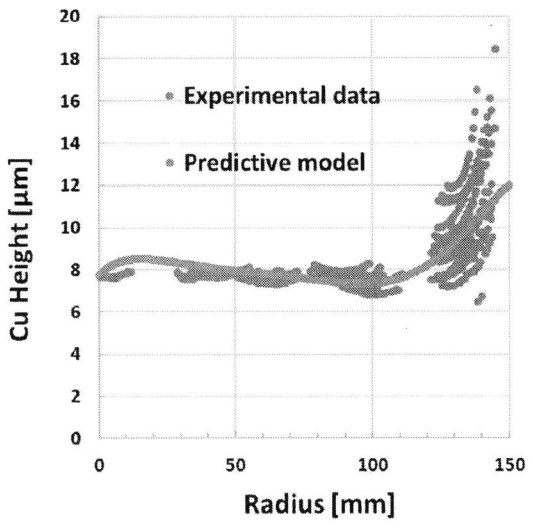

Figure 10. Comparison of experimental profil and predictive model for copper profile.

Fig. 10 shows the predicted copper bump height distribution (in orange), the measured copper bump height (in blue). The profiles are similar in the [25, 125 mm] range, and are close on the edges. This is attributed to the polynomial fitting causing divergence at the end of the fitting interval, especially at high orders. Finally, using several wafers with various process conditions, the agreement between the predictive model and the experimental points distribution has been validated.

D. ECD copper profile optimisation

Following this approach, the copper ECD process has been optimized. Fig. 11 introduces the copper bumps height for the copper contribution for each anodes separately. It is possible to identify easily each anode contribution to the copper height profile since their signature are link to the position of the annular anode above the wafer (Fig. 7). Thus, anode 1, which is close to wafer center, influences the copper profile at wafer center while anode 4 addresses wafer edge profile. Based on these profiles, each anode contribution is fitted to build the model. The same method is applied for nickel and tin silver contribution. It has to be mentioned that for these compounds a first 500 nm layer of copper ECD is preliminary deposited to ensure a proper deposit mechanism. Indeed a thin copper ECD layer helps the ECD growth of nickel or tin silver, which is delayed or uniform without. During this experiment, several heights measurement algorithms have been tested in order to fit properly the shape and roughness variations of each type of metal of micro bumps stack.

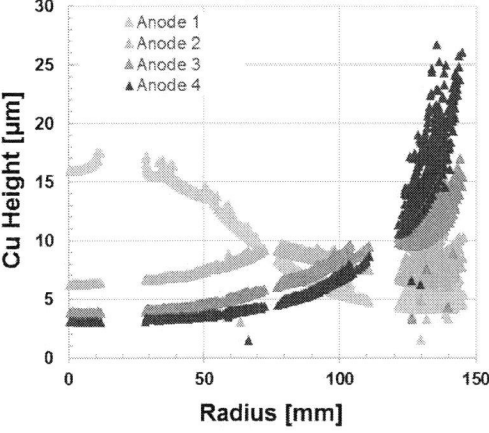

Figure 11. 4 Anodes signatures: micro bumps (target Cu 7μm) thickness in function of wafer radius.

Once all the data gathered, the different anode coefficients as well as the targeted heights after optimization are obtained (Tab. 1).

TABLE I.   ANODE COEFFICIENTS AND HEIGHT OF THE BUMP

Material	a_1	a_2	a_3	a_4	Height (μm)
Cu	0.20	0.71	0.09	0	7
Ni	0.22	0.50	0.28	0	2
SnAg	0.28	0.61	0.11	0	9

Even if there is some similitudes between heights profile for each anodes for each ECD bath, the resulting coefficients are different. This confirm the need for reference data for each bath or metal and to validate the global approach.

The comparison between the initial process and the optimized process using the coefficient from the model is presented in Fig. 12. In this figure, the predictive model profile (brown line) follows with a good agreement the experimental points (orange dots). Thanks to the model optimization, the edge divergence of height is not present anymore.

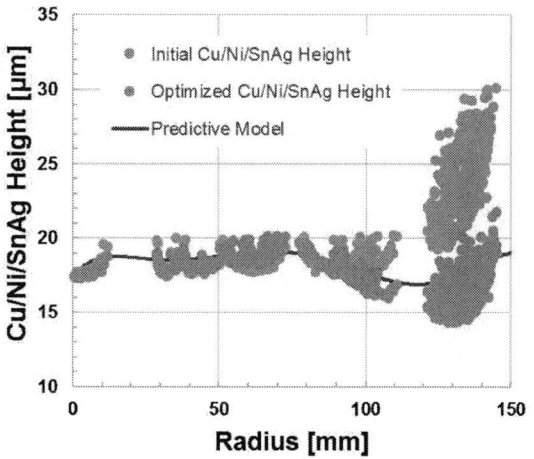

Figure 12. Micro bumps full stack (Cu/Ni/SnAg – 7/2/9μm) thickness in function of wafer radius for initial and optimised porcess.

E. Lithography Map influence

The lithography mapping influence has also been evaluated by increasing the edge exclusion area to potentially remove areas of current overflow. In that respect, a new job stepper map, named map B is elaborated with an edge exclusion of 10mm instead of the standard 3.6mm for map A. Fig. 13 introduces the 2 maps.

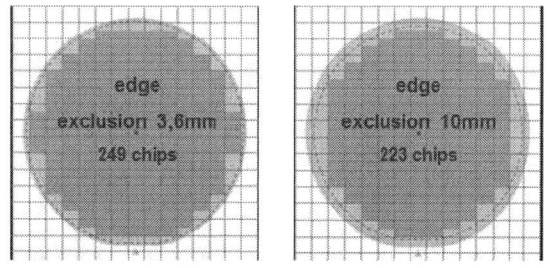

Figure 13. Reference map A and map B with an increased edge exclusion for micro bumps lithography stepper job

Wafers are processed with the two maps for micro bump lithography and the micro bumps heights are measured with a new cartography, called Carto B, of measured chips. Indeed using the original cartography, named Carto A (Fig. 3), is not possible since some of the chips measured at wafer edge disappear in map B. The results of copper bumps heights measurements are plotted in Fig. 14. The wafer 01 in Fig. 14 is processed with map A and measured with Carto A and B, whereas wafer 07 is processed with map B and measured with Carto B only.

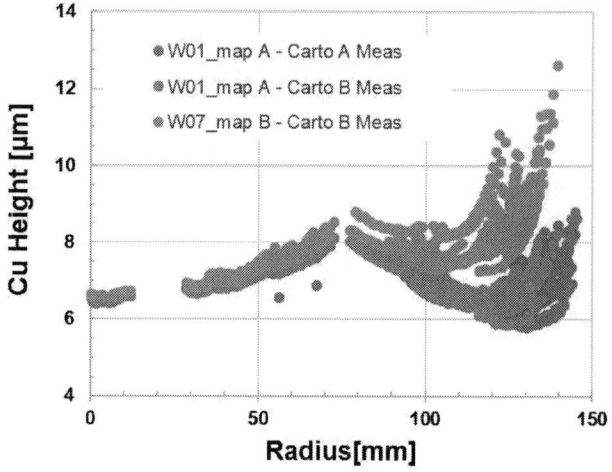

Figure 14. Comparison of cu thickness map A (Wafer 1) and and map B (Wafer 7). Copper thickness target: 7μm.

When using Carto B, the copper profile uniformity of wafer 07 is deviating from the target by 5 μm compared to the profile of wafer 01 which is deviating only by 1.5 μm. Moreover, if we compare the carto A and B measurements for W01 no major difference are evidenced. It is possible to conclude that the increase of the edge exclusion in lithography mapping of micro bumps is detrimental to copper height uniformity and also removes some active dices from the wafer. Then, this option has not been pursued.

IV. ELECTRICAL AND MORPHOLOGICAL RESULTS

After completion of the process flow for both top and bottom wafers, morphological characterizations are shown in Fig. 15.

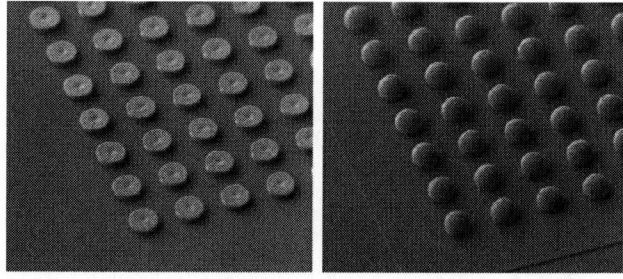

Figure 15. Microp pillars (Cu/Ni/Au) on left picture and Micro Bumps (Cu/Ni/SnAg) on right picture after full process and reflow..

Top wafers are then back grinded to 400μm and diced for stacking. The chip to wafer stacking process is operated on a Panasonic tool with liquid flux Kester TSF6592 and followed by a mass reflow on conveyor furnace. A typical example of result is shown in Fig. 16.

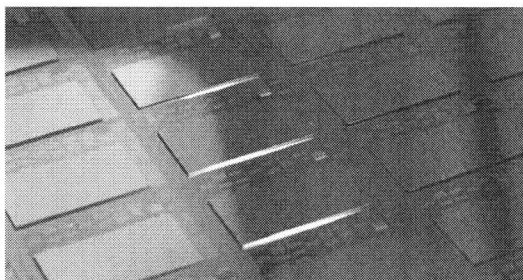

Figure 16. Stacked dice of 1x2cm² on the test vehicle for electrical 80 000 pillars chain yield evaluation

To check the alignment accuracy of the top dies on the bottom substrate infrared microcopy is performed. On Fig. 17, it is possible to confirm the quality of the alignment and to observe the top and bottom damascene lines forming the chain test pattern.

Two campaigns of stacking have been performed. The first campaign involves the top dies with the initial process and the second campaign the optimized process as described in section III to V. In between the two runs, the resist used for micro bumps ECD process has been changed from PMER to AZ 3DT which may also influence the results.

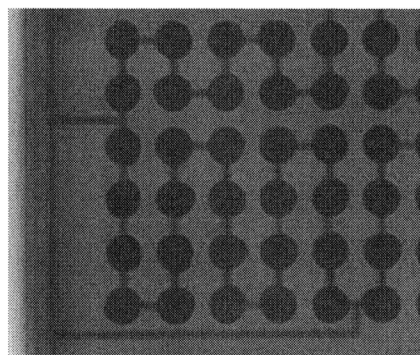

Figure 17. Infrared microscopy of the top die stacked on the bottom substrate: the 25 µm diameter micro bumps/pillars chain is visible.

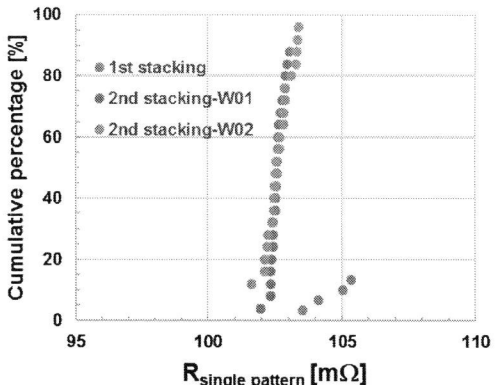

Figure 18. Cumulative percentages of 74880 micro bumps/pillars interconnections chain: before and after process optimisation (2ⁿᵈ stacking).

The electrical test results of the most aggressive chain for the 2 campaigns are plotted in Fig. 18. The electrical chain yield increases from 15% for the first campaign to more than 90% for the second campaign after optimization. The resistance per single pattern also decreases from 2 mΩ. These results validate the global optimization of the study.

V. CONCLUSION

In this study, thanks to the design of a dedicated test vehicle to micro bumps and micro pillars interconnections, impact of lithography map, barrier and seed layer deposit processes including several pre-clean treatments have been evaluated. For the ECD process, a model of anode intensity contribution to the overall TTV has been introduced and detailed. The ECD process of the full stack whether Cu/Ni/Au or Cu/Ni/SnAg has been optimized using this model resulting in a drastic reduction of micro bumps and micro-pillar TTV at chip level and at wafer level. Finally, daisy chain electrical test yields after stacking using die-to-wafer approach have benefited from this full integration study, increasing from 15% for the first campaign to more than 90% after optimization. To conclude, the optimized process flow and generic model developed in this study is now fully portable to any application involving very large interconnections matrixes and will be specially tuned for high-energy particles detection like ATLAS Large Hadron Collider new tracker project at CERN.

ACKNOWLEDGMENT

The project team would like to thanks Raphael Eleouët, Pierre Emile Philip, Bertrand Perrin and all our colleagues of the 3D300 process line of silicon platform department at Leti. A great thanks also to David Bouchu and Rémi Velard for their contributions to FIB/SEM morphological characterizations.

REFERENCES

[1] P. Coudrain, D. Henry, A. Berthelot, J. Charbonnier, S. Verrun, R. Franiatte, N. Bouzaida, G. Cibrario, F. Calmon, I. O'Connor, T. Lacrevaz, et al., "3D Integration of CMOS image sensor with coprocessor using TSV last and micro-bumps technologies" *Proc. - Electron. Components Technol. Conf.*, pp. 674–682, 2013.

[2] P. Coudrain, J.-P. Colonna, C. Aumont, G. Garnier, P. Chausse, R. Segaud, et al., "Towards efficient and reliable 300mm 3D technology for wide I/O interconnects", *Proc. - Electron. Packaging Technol. Conf.*, pp. 330–335, 2012.

[3] J. Charbonnier, M. Assous, J-P. Bally, K. Miyairi, M. Sunohara et al., "High density 3D silicon interposer technology development and electrical characterization for high end applications", *2012 4th Electron. Syst. Technol. Conf.*, pp. 01-07, 2012.

[4] A. Garnier, L. Arnaud, R. Franiatte, A. Toffoli, S. Moreau et al., "Electrical Performance of High Density 10 µm Diameter 20 µm Pitch Cu-Pillar with Chip to Wafer Assembly" *Proc. - Electron. Components Technol. Conf.*, pp. 999–1007, 2017.

[5] J. DeVos et al., " High density 20µm pitch CuSn Microbump Process for high End 3D Applications", *Proc. - Electron. Components Technol. Conf.*, pp. 27–31, 2011.

[6] M. Volpert, D. Henry, D. Taneja, T. Chaira, A. Gueugnot, F. Hodaj, "Cu pillar as interconnect for 10 µm pitch and below: Fabrication

issues and assembly results", *2018 7th Electron. Syst. Technol. Conf.*, pp. 01-07, 2018.

[7] R. L. Bates, C. Parkes, D. Pennicard, B. Rakotomiaramanana, et al., "Charge collection studies of heavily irradiated 3D double-sided sensors", *2009 IEEE Nuclear Science Symposium Conference Record (NSS/MIC)*, pp. 148-156, 2009.

[8] D. Henry, J. Alozy, A. Berthelot, R. Cuchet, C. Chantre, M. Campbel, "TSV last for hybrid pixel detectors: Application to particle physics and imaging experiments ", *Proc. - Electron. Components Technol. Conf.*, pp. 568–575, 2013.

Growth behavior and orientation evolution of Cu_6Sn_5 grains in micro interconnect during isothermal reflow

S. Chen, N. Zhao*, Y.Y. Qiao, Y.P. Wang, H.T. Ma

School of Materials Science and Engineering
Dalian University of Technology
Dalian 116024, China
*E-mail: zhaoning@dlut.edu.cn

C.M.L. Wu

Department of Materials Science and Engineering
City University of Hong Kong
Kowloon, Hong Kong SAR 999077, China

Abstract—At present, three-dimensional (3D) packaging technology is one of the effective methods to extend Moore's law. In 3D stacking, micro interconnects (below 30 μm) have been widely used. As for Cu/Sn/Cu micro interconnect, after repeated or long-term reflow, the micro interconnect will be mainly composed by Cu-Sn intermetallic compounds (IMCs) because of the small bonding gap and fast IMC growth rate under size effect. In the present research, a Cu/Sn(12.5 μm)/Cu micro interconnect was reflowed at 260 °C. Laminar scallop-like Cu_6Sn_5 IMC was the dominant produce at both interfaces. The total thickness of the two Cu_6Sn_5 layers and the consumption of the bottom Cu layer both showed a parabolic law with reflow time, and their logarithmic format followed a good linear relationship. It is also found that as the reaction time extended, the number of the Cu_6Sn_5 grains decreased and the grain orientation tended to gathering towards certain directions, resulting in texture structure. The angle between the <0001> direction of the Cu_6Sn_5 grains and the growth direction of IMC (perpendicular to the interface) was analyzed to characterize the grain orientation evolution. The anisotropic growth of the Cu_6Sn_5 grains along the <0001> direction was discussed to reveal the mechanism of orientation evolution.

Keywords-3D packaging; micro interconnect; IMC; Cu_6Sn_5; grain orientation

I. INTRODUCTION

In the trend of miniaturization of electronic products and packaged devices, the size of micro interconnects has continued to shrink, particularly for 3D stacking applications. In this case, the total growth rate of intermetallic compounds (IMCs) increases continuously due to the so called volume or size effect. Meanwhile, the proportion of IMCs in the whole micro interconnects is also increasing rapidly, even resulting in the formation of full IMC micro interconnects. At this point, the physical and mechanical properties of the micro interconnects will no longer be determined mainly by the Sn-based solders, but by the IMCs [1]. What's more, Cu-Sn IMCs, such as Cu_6Sn_5, have been found to possess better properties than Sn-based solders in terms of melting temperature, hardness, yield strength and Young's modulus [2-4]. However, since the Cu-Sn IMCs formed in micro interconnects have a non-cubic crystal structure, the IMC grains have a strong anisotropy along different crystal directions. Obviously, micro interconnects with desired preferred IMC grain orientation will show better physical

and mechanical properties. Therefore, it is significant to fully understand the growth and evolution of Cu-Sn IMC grains in micro interconnects during a reflow/bonding process.

In the present study, we investigated the growth behavior of Cu_6Sn_5 grains in a Cu/Sn/Cu micro interconnect during isothermal reflow until the Cu_6Sn_5 grains from the two interfaces merged together to form a full IMC interconnect. The orientation of the Cu_6Sn_5 IMC grains were characterized and the orientation evolution was discussed.

II. EXPERIMENTAL PROCEDURES

In order to fabricating Cu/Sn/Cu micro interconnects, we used an sequential electroplating method to deposit the above sandwich structure on a $15 \times 25 \times 1$ mm³ nickel substrate, as shown in Fig. 1. The thicknesses of the bottom Cu layer, the Sn layer and the top Cu layer are 34.6 μm, 12.5 μm and 35.0 μm, respectively. The Cu/Ni interface was used as the mark to trace the dissolution behavior of the bottom Cu layer during the reflow reaction. The plated sandwich structure was cut into several small pieces of $3 \times 5 \times 1$ mm³, and one of the pieces was selected as the micro interconnect for test. The micro interconnect was isothermal reflowed at 260 °C in an oil bath furnace for different durations. After each duration, the micro interconnect was taken out from the oil bath and cooled down to room temperature in water immediately. Then a fixed cross-section of the micro interconnect was carefully grinded and polished for microstructure and grain orientation characterization. After characterization, the micro interconnect was placed back in the oil bath furnace for the subsequent reflow. Electron probe microanalysis (EPMA, JXA-8530F PLUS) was used to analyze the microstructure and phase composition of the micro interconnect. All the EPMA images were done using backscatter electron (BSE) mode. Electron backscattered

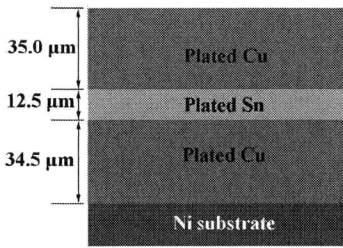

Figure 1. Schematic diagram of the Cu/Sn/Cu micro interconnect.

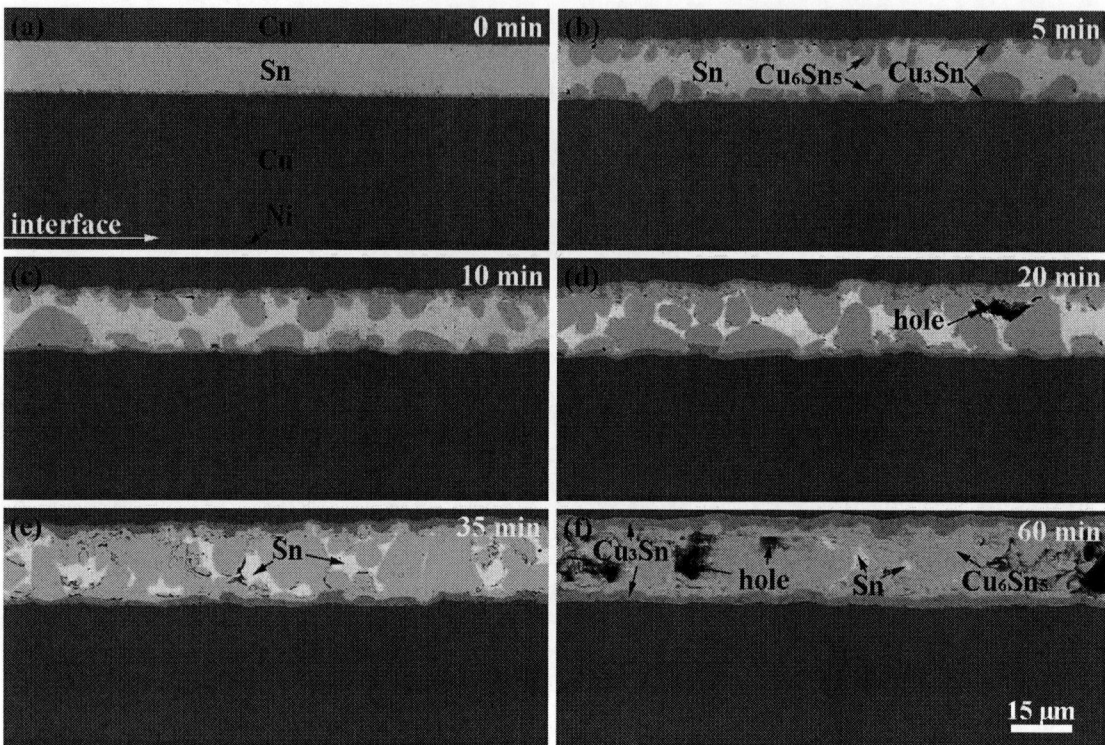

Figure 2.   Cross-sectional microstructure of the Cu/Sn/Cu micro interconnect after reflowed at 260 °C for: (a) 0 min (as-plated), (b) 5 min, (c) 10 min, (d) 20 min, (c) 35 min and (f) 60 min.

diffraction (EBSD, X-Max50) was used to analyze the IMC grain orientation. In the EBSD analysis, the growth direction of IMC grains was set parallel to the transverse direction (TD), and the Sn/Cu interface was parallel to the rolling direction (RD). All the EBSD data were processed using the HKL Channel5 software package.

III.   RESULTS AND DISCUSSION

A.   The growth behavior of interfacial Cu_6Sn_5 IMC during isothermal reflow

Fig. 2 shows the cross-sectional microstructure of the Cu/Sn/Cu micro interconnect after reflowed at 260 °C for different durations. As shown in Fig. 2(a), the two Sn/Cu interfaces are smooth and clear, indicating that the Cu/Sn/Cu sandwich structure has been well plated on the Ni substrate to form the micro interconnect. It is noted that a thin Cu_6Sn_5 layer has already formed at the bottom Sn/Cu interface in the as-plated state. After reflow for 5 min, many scallop-like Cu_6Sn_5 grains appear at both Sn/Cu interfaces with a total thickness of 5.96 μm. However, the Cu_6Sn_5 grains at the bottom interface are larger than those at the top interface, which may attribute to the preformed thin Cu_6Sn_5 layer in the as-plated state. Moreover, a thin inconsecutive Cu_3Sn layer forms at each Cu_6Sn_5/Cu interface, especially under larger Cu_6Sn_5 grains. After reflow for 10 min, the Cu_6Sn_5 grains grow larger with their number decreasing, indicating the occurrence of annexation between adjacent Cu_6Sn_5 grains. This annexation phenomenon has been in situ observed by using synchrotron radiation real-time imaging technique [5]. In Fig. 2(d), after reflow for 20 min, the Cu_6Sn_5 grains from different Sn/Cu interfaces continue to grow towards the opposite side and finally contact with each other at some areas. A continuous thin Cu_3Sn layer forms at the bottom interface. At this point, the total thickness of the Cu_6Sn_5 IMC increases to 10.96 μm and the consumption of the bottom Cu layer is 1.52 μm. With the extending of reaction time, most of the Cu_6Sn_5 grains are in contact with those from the opposite interface, leaving some Sn islands among the Cu_6Sn_5 grains. The total thickness of the Cu_6Sn_5 IMC becomes 14.09 μm and the consumption of the bottom Cu layer is 1.98 μm after reflow for 35 min. After 60 min reflow, as shown in Fig. 2(f), there are only a few residual Sn islands left in the micro interconnect, resulting in the formation of full IMC interconnect. The total thickness of the Cu_6Sn_5 IMC and the consumption of the bottom Cu layer reach to 17.21 μm and 2.5 μm, respectively. Herein, Sn is no longer the main component of the micro interconnect but Cu_6Sn_5 dominates. It should also be noted that a few holes form in the micro interconnect. Since $6Cu+5Sn \rightarrow Cu_6Sn_5$ and $Cu_6Sn_5+9Cu \rightarrow 5Cu_3Sn$, which are both volumetric contraction reactions, occur during the whole reflow process, holes could probably form as Sn is exhausted. The existence of such holes definitely degrade the reliability of the micro interconnect. Further studies are needed to investigate how to eliminate hole formation in the manufacturing of micro full Cu-Sn IMC interconnect. Applying bonding pressure during reflow seems a feasible method. Meanwhile, the two Cu_3Sn layers are slowly thickening throughout the reflow process.

Fig. 3(a) shows the total thickness, of the Cu_6Sn_5 IMC as a function of reflow time. It is obvious that the thickness of the Cu_6Sn_5 IMC increases parabolically with the reflow time. Generally, the growth kinetics of Cu_6Sn_5 layer at liquid-solid interface during reflow can be expressed by the following empirical relationship:

$$T = Kt^n \qquad (1)$$

where T is the thickness of the Cu_6Sn_5 layer, K is the Cu_6Sn_5 growth rate coefficient, t is the reflow time, and n is the time index of Cu_6Sn_5 growth. Available from (1):

$$\ln T = \ln K + n \ln t \qquad (2)$$

From (2), the item $\ln K$ can be regarded as a constant during the reflow. If $\ln T$ and $\ln t$ are plotted as the coordinates, the slope of the obtained straight line after linear fitting is the time index n of the growth of Cu_6Sn_5. As shown in Fig. 3(b), the fitting curve can be obtained as

$$\ln T = 0.43 \ln t + 1.10 \qquad (3)$$

The value of n is usually between 0.3 and 1. When $n = 1/3$, the growth kinetics of Cu_6Sn_5 obeys the parabolic law, and the growth mechanism is controlled by grain boundary diffusion; when $n = 1/2$, the growth kinetics of Cu_6Sn_5 also obeys parabolic law, but the growth mechanism is controlled by bulk diffusion; when $n = 1$, the growth kinetics of Cu_6Sn_5 shows the linear law, and the growth mechanism of Cu_6Sn_5 is controlled by reaction rate [6]. Thus, it can be concluded that the growth mechanism of Cu_6Sn_5 IMC in the micro interconnect is controlled by the combination of grain boundary diffusion and bulk diffusion.

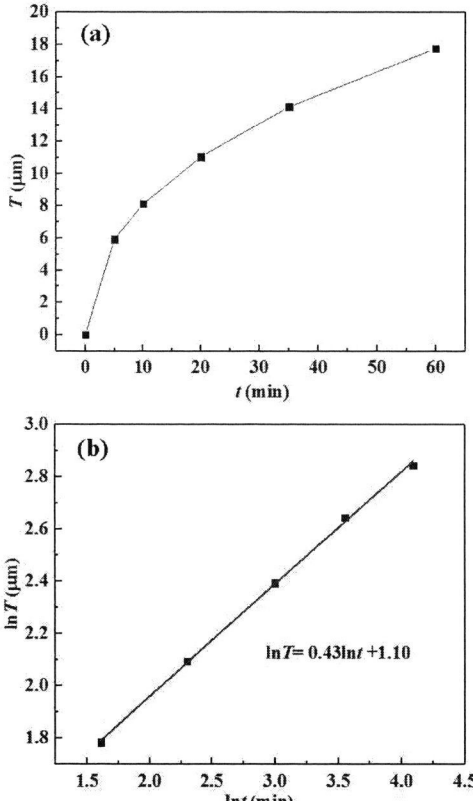

Figure 3. Total thickness of the Cu_6Sn_5 IMC as a function of reflow time: (a) T-t and (b) $\ln T$-$\ln t$.

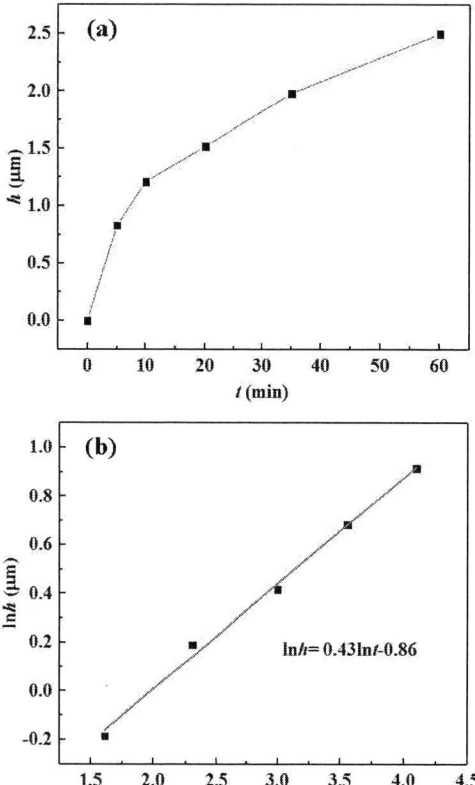

Figure 4. Consumption of the bottom Cu layer as a function of reflow time: (a) h-t and (b) $\ln h$-$\ln t$.

Fig. 4(a) shows the consumption, h, of the bottom Cu layer as a function of reflow time. Similarly, a parabolic relationship is obtained between the consumption of the bottom Cu layer and the reflow time, and a very good linear fitting is obtained in $\ln h$-$\ln t$ format. As shown in Fig. 4(b), the fitting curve can be given by

$$\ln h = 0.43 \ln t - 0.86 \qquad (4)$$

According to (3) and (4), the time index n of the Cu_6Sn_5 growth equals to that of the bottom Cu consumption.

B. The orientation evolution of interfacial Cu_6Sn_5 grains during isothermal reflow

Fig. 5 shows the EBSD maps of the micro interconnect after reflowed for 10 min. Fig. 5(a) is the phase map, clearly showing that the bottom Cu_6Sn_5 layer is thicker than the top one. Fig. 5(b) is the grain orientation map of the whole interconnect. The two Cu_3Sn layers both consist of fine Cu_3Sn grains. The Sn layer in the middle of the interconnect is composed of many β-Sn grains, which may be related to the fact that there are many grooves between the Cu_6Sn_5 grains providing heterogeneous nucleation sites for β-Sn and on the other hand the large Cu_6Sn_5 grains block the rapid growth of individual β-Sn grain. Fig. 5(c) is the grain orientation map of the Cu_6Sn_5 IMC. It is obvious that each scallop- like Cu_6Sn_5 is an independent single grain, and their orientation is relatively random. Table I lists the space angles of the marked Cu_6Sn_5 grains in Fig. 5(c). φ_1, Φ and

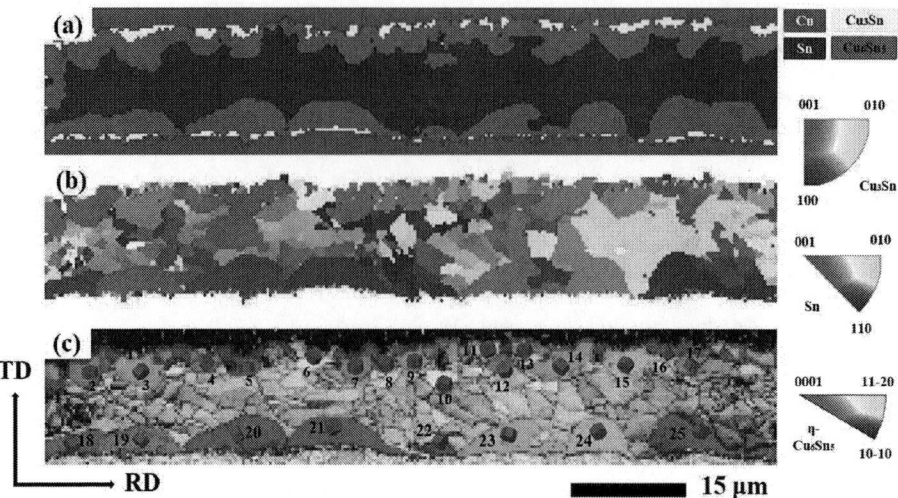

Figure 5. EBSD maps of the micro interconnect after reflowed for 10 min: (a) phase map, (b) the grain orientation map of the whole interconnect, and (c) the grain orientation map of the Cu_6Sn_5 IMC.

Figure 6. EBSD maps of the micro interconnect after reflowed for 35 min: (a) phase map, (b) the grain orientation map of the whole interconnect, and (c) the grain orientation map of the Cu_6Sn_5 IMC

φ_2 are the three Euler angles, and θ represents the angle between the <0001> direction of the Cu_6Sn_5 grains and TD. The value of θ can be calculated from φ_1 and Φ by

$$\beta = \arccos(-\cos\varphi_1 \times \sin\Phi)$$
$$\theta = \beta \qquad (\beta \leq 90°)$$
$$\theta = 180° - \beta \quad (90° < \beta) \qquad (5)$$

Fig. 6 shows the EBSD maps of the micro interconnect after reflowed for 35 min. Although the Cu_3Sn layers all become thicker, the Cu_3Sn grains are still quite fine. Most of the Cu_6Sn_5 grains from the top and bottom interfaces have already been contacted with each other, and some have been merged into large grains. As shown in Fig. 6(c), it is interesting that the Cu_6Sn_5 grains seem to have a tendency to form texture structure as they

have similar colors. The space angles of the marked Cu_6Sn_5 grains in Fig. 6(c) are shown in Table II.

To further analyze the grain orientation evolution of the Cu_6Sn_5 IMC, the θ values were converted into two histogram, as shown in Fig. 7. The horizontal axis represents each interval of the θ values and the vertical axis represents the ratio of the area of the Cu_6Sn_5 grains in each interval to the total area of all the Cu_6Sn_5 grains. Compared with the relatively stochastic distribution of θ in Fig. 7(a), the distribution of θ in Fig. 7(b) is mainly concentrated in the intervals of [10°, 40°), i.e., $10° \leq \theta < 40°$. In another words, during the later stage of the reflow reaction, the Cu_6Sn_5 grains have a tendency to epitaxially grow along the <0001> direction towards the liquid Sn.

978-1-7281-1500-9/19 $31.00 © 2019 IEEE

TABLE I. SPACE ANGLES OF THE MARKED CU₆SN₅ GRAINS IN FIG. 5.

Grain number	Euler angles (°)			θ (°)
	φ_1	Φ	φ_2	
1	29.1	85.4	53.3	29.4
2	4.2	109.1	20.8	19.4
3	40.7	66.0	37.7	46.2
4	44.4	103.3	51.2	45.9
5	7.5	91.6	37.0	7.7
6	149.3	41.5	27.9	55.3
7	176.4	60.0	51.9	30.2
8	15.9	120.9	38.9	34.4
9	23.1	130.5	11.9	45.6
10	160.8	132.2	35.1	45.6
11	107.3	121.3	43.3	75.3
12	24.9	143.8	54.9	57.6
13	19.0	118.6	39.8	33.9
14	135.3	113.1	29.2	49.2
15	141.9	124.6	46.4	49.6
16	124.7	84.2	52.3	55.5
17	150.2	81.4	59.3	30.9
18	118.5	109.5	12.9	63.3
19	138.8	102.6	11.9	42.8
20	117.5	86.1	53.4	62.6
21	131.0	43.4	29.0	63.2
22	68.0	59.6	12.7	71.1
23	161.4	110.2	24.6	27.2
24	34.8	49.5	52.0	51.4
25	63.2	47.2	21.7	70.7

TABLE II. SPACE ANGLES OF THE MARKED CU₆SN₅ GRAINS IN FIG. 6.

Grain number	Euler angles (°)			θ (°)
	φ_1	Φ	φ_2	
1	31.6	67.4	46.6	38.2
2	28.9	39.3	46.8	56.3
3	48.7	109.9	0.6	51.6
4	13.0	112.3	44.4	25.6
5	28.7	76.3	18.1	31.6
6	38.2	82.7	20.8	38.8
7	161.5	82.4	26.0	20.0
8	175.3	105.0	41.6	15.7
9	133.5	60.2	35.4	53.3
10	5.9	34.1	18.9	56.1
11	7.4	112.8	37.4	23.9
12	140.4	19.8	6.8	74.9
13	0.9	122.9	45.7	32.9
14	4.4	36.9	26.8	53.2
15	26.2	112.8	52.6	34.2
16	59.2	85.1	19.8	59.3
17	61.8	87.2	25.5	61.8
18	162.1	91.3	36.3	18.0
19	23.8	112.7	38.6	32.4
20	17.9	119.6	50.9	34.2
21	137.5	131.7	34.7	56.6
22	151.8	76.7	44.0	30.9

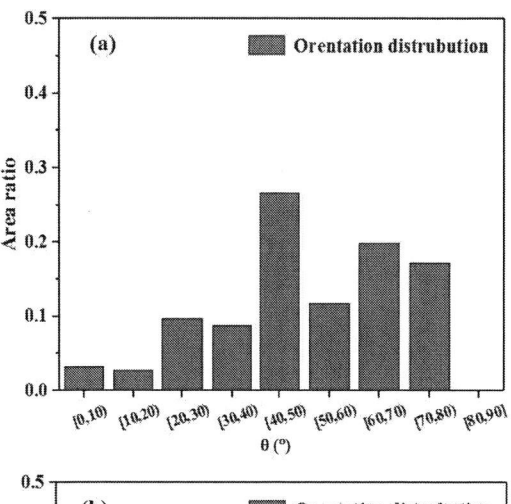

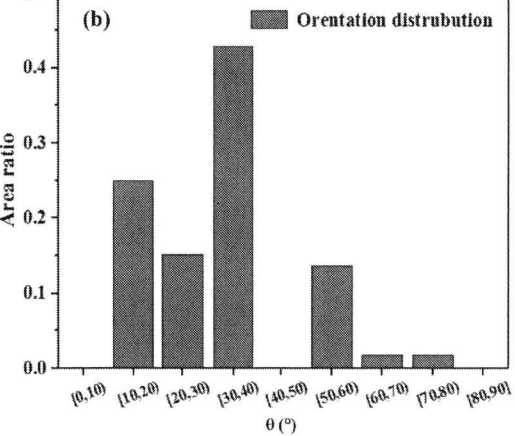

Figure 7. Distribution of angle θ between the <0001> direction of the Cu₆Sn₅ grains and TD after reflowed for: (a) 10 min and (b) 35 min.

As for solder joints, the dissolution of Cu atoms into liquid solder could not be negligible. During a soldering interfacial reaction, the Cu concentration in the liquid solder will continue to increase until reaching saturation. Hereafter, the interfacial reaction occurs actually between the Cu-rich solder and the Cu substrate [7]. For the current micro interconnect, the Cu layers dissolve fast and the liquid Sn is saturated with Cu quickly due to the size effect. It is reported that the interfacial Cu₆Sn₅ grain has an anisotropic growth mechanism of hexagonal prism under sufficient supply of Cu, and the center line of the hexagonal prism is parallel to the <0001> crystal orientation [8]. The micro liquid-solid interface structure at the top of the hexagonal prism is a rough interface, but the liquid-solid interface structure on the sides is a smooth interface. Therefore, in the top end of the Cu₆Sn₅ grain the growth mechanism is a continuous growth mechanism, while in the sides the growth mechanism is a two-dimensional nucleation growth or screw dislocation growth mechanism. So it is easy to exhibit a small plane feature on the grain sides of the hexagonal prism, and the growth rate on the top surface of the hexagonal prism is much larger than that on the side surface. Moreover, the grains will gain more growth space when the <0001> direction is toward the liquid solder. Therefore, when the angle between <0001> direction and TD, i.e., θ, is small, the Cu_6Sn_5 grains will have a faster growth rate and more growth space along the <0001> direction in the micro interconnect. As a result, more <0001> direction of the Cu_6Sn_5 grains has less misorientation with the TD as the reaction time increases.

IV. CONCLUSION

The Cu/Sn/Cu interconnect (with 12.5 μm Sn) was reflowed at 260 °C for different durations and its microstructure evolution was investigated after each duration. The results show that the scallop-like Cu_6Sn_5 grains from the two Sn/Cu interfaces grew gradually at the same time, and their grain orientations were random in the early stage. With extended reflow time, the Cu_6Sn_5 grains from different interfaces continued to grow towards the opposite side, contacted with each other at some areas, and finally merged into one grain. After the liquid Sn was nearly consumed away, the whole micro interconnect was composed of a thick Cu_6Sn_5 layer (middle) and two thin Cu_3Sn layers (side). During this process, the total thickness of the Cu_6Sn_5 IMC and the consumption of the bottom Cu layer both increased parabolically with reflow time, and the logarithm of thickness/consumption had a good linear relationship with the logarithm of time. Interestingly, the angle between the <0001> direction of the Cu_6Sn_5 grains and TD decreased undergoing the reflow reaction, which was attributed to faster growth rate and more growth space along the <0001> direction towards the liquid solder.

ACKNOWLEDGMENT

This work was supported by the National Natural Science Foundation of China (Grant Nos. 51675080 and U1732118) and the open project of Beijing Engineering Researching Center of Laser Technology (Grant No. BG0046-2018-08).

REFERENCES

[1] Y. Wang, D.T. Chu, and K.N. Tu, "Porous Cu_3Sn formation in Cu-Sn IMC-based micro-joints," Proc. IEEE Components Packaging & Mfg Technol Soc, IEEE Press, May. 2016, pp. 439-446, doi: 10.1109/ECTC.2016.359.

[2] X. Deng, N. Chawla, K.K. Chawla, and M. Koopman, "Deformation behavior of (Cu,Ag)-Sn intermetallics by nanoindentation," Acta Materialia, vol. 52, Aug. 2004, pp. 4291-4303, doi: 10.1016/j.actamat.2004.05.046.

[3] H. Frederikse, R. Fields, and A. Feldman, "Thermal and electrical properties of copper-tin and nickel-tin intermetallics," Journal of Applied Physics, vol. 72, Oct. 1992, pp. 2879-2882, doi: 10.1063/1.351487.

[4] Y. Zhong, R. An, and C. Wang, et al, "Low temperature sintering Cu_6Sn_5 nanoparticles for superplastic and superuniform high temperature circuit interconnections," Small, vol. 11, Sep. 2015, pp. 4097-4103, doi: 10.1002/smll.201500896.

[5] L. Qu, N. Zhao, H.J. Zhao, M.L. Huang, and H.T. Ma. "In situ study of the real-time growth behavior of Cu_6Sn_5 at the Sn/Cu interface during the soldering reaction," Scripta Materialia, vol. 72-73, Feb. 2014, pp. 43-46, doi: 10.1016/j.scriptamat.2013.10.013.

[6] J. Li., S. H. Mannan, and M. P. Clode, et al, "Interfacial reactions between molten Sn-Bi-X solders and Cu substrates for liquid solder interconnects," Acta Materialia, vol. 54, Jun. 2006, pp. 2907-2922, doi: 10.1016/j.actamat.2006.02.030.

[7] M.L. Huang, F. Yang, "Size effect model on kinetics of interfacial reaction between Sn-xAg-yCu solders and Cu substrate," Scientific Reports, vol. 4, Nov. 2014, doi: 10.1038/srep07117.

[8] J. Song, B.R. Huang, and C.Y. Liu, et al, "Nanomechanical responses of intermetallic phase at the solder joint interface–Crystal orientation and metallurgical effects," Materials Science and Engineering: A, vol. 534, Feb. 2012, pp. 53-59, doi: 10.1016/j.msea.2011.11.037.

Development of a no reflow Cu pillar bump to improve chip/package interactions (CPI) process and reliability performance

Kuei Hsiao Kuo (Frank), Jiunn Jie Wang, Yen Neng Wang, Feng Lung Chien and Rick Lee

Process Integration Engineering DEPT. 1
Siliconware Precision Industries Co., Ltd. (SPIL)
No. 19, Ke Ya rd., Daya, Taichung, Taiwan, R. O. C.
E-mail: frankkuo@spil.com.tw

Abstract—The no reflowed copper (Cu) pillar bump behavior during front-end process and performance of flat solder capped Cu pillar assembled with bump on trace (BOT) processed by mass reflow is measured and analyzed. The interconnections with two different solder cap structures were tested: Cu post capped with flat lead-free solder tip (without front-end reflow) and Cu post capped with round lead-free solder tip (with front-end reflow). Emphasis is placed on the risk of high solder tip height on small bump size in front-end process and assembly performance of the different solder capped Cu pillar bump structures. The package size of test vehicle (TV) is 116 mm^2 with a daisy-chain die size of 7 x 8 mm^2. The minimum pitch is 100μm with critical mixed bump size design in a die.

The bump stack-up with 4 different solder tip ratios (reflowed solder tip height divided by minimum UBM size); 0.45, 0.56, 0.67, 0.78 are investigated and the CPI performance based on bump stack-up with different solder tip ratios are collected and analyzed. All the different solder tip structures formed before and post front-end reflow have been evaluated and compared by employing package level reliability test following JEDEC include thermal cycling test, high temperature storage test and unbiased highly accelerated Test.

Keywords-component;Cu pillar bump; bump on trace (BOT); chip-package-interaction (CPI); under bump metal (UBM)

I. INTRODUCTION

The advancement of silicon technologies and complexities of device functionalities driven by wireless application such as application processor (AP), LTE Modem, baseband processor or RF application are continued to driving smaller flip chip bump pitch for Copper (Cu) pillar bump on trace (BOT) packages. The smaller Cu pillar bump size constrained by smaller pitch or increased escape trace counts between bumps in substrate design challenges the design of bump stack-up (Cu/Ni/solder tip) which will impact the front-end bumping process, chip-package-interaction (CPI) performance and package reliability. From the simulation input, the silicon ELK layer and PI peeling stress performance of a FCCSP package during die bump reflow process is reduced with higher solder tip height when the total Cu pillar stack-up height is fixed (Fig.1). Typically, the maximum solder tip height of a Cu pillar is proportional to the UBM (under bump metal) size. The smaller the UBM size is, the lower the solder tip height is to prevent the bump yield loss caused by solder flow down or bump bridge defects after front-end reflow process. Therefore, the solder tip height design of a Cu pillar bump is restricted and could not be optimized to improve assembly process window and package reliability performance when bump size is shrunk.

In this investigation, a novel package interconnection using no reflow Cu pillar bump (skip reflow step in bump front-end process) is studied. For the stack-up design of Cu pillar bump with high solder tip ratio (reflowed solder tip height divided by minimum UBM size), the no reflowed Cu pillar bump can not only improve potential bump yield loss after reflow but also reduce the process cycle time. The design of maximum solder tip height will not be decided by minimum UBM size but the thickness of photo-resist (PR) for Cu pillar bump without front-end reflow.

Leg	Cu pillar stack-up				ELK stress ratio	PI stress ratio
---	a. Cu height	b. Ni height	c. Solder tip height	Total height (a+ b+ c)	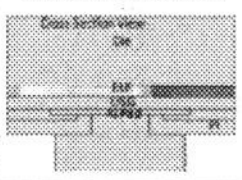	
1	High	Same	Low	Same	1	1
2	Low		High		0.98	0.98

Figure 1. Effect of solder tip height design on stress to ELK and PI layer.

II. TEST VEHICLE DESCRIPTION

The package size of test vehicle (TV) is 116 mm^2 with a daisy-chain die size of 7 x 8 mm^2. The minimum pitch is 100μm and mixed bump size in a die is designed to simulate the critical condition; the under bump metal (UBM) size of oblong bump located in die peripheral area is 45 x 75μm and 75μm UBM size of the round bump located in die center area (Fig. 2). The package samples were prepared following mainstream Cu pillar bump manufacturing technology with polyimide re-passivation applied as dielectric and stress buffer layer but the bump reflow & clean process were skipped post UBM etching process. The no reflow bumped wafer was back-grinded down to 150μm and went through typical assembly process to complete the packages. The daisy chain TV is capable of detecting the bump/substrate joint integrity with electrical open/short test. Fig. 3 shows the flow of CPI investigation and bump/assembly characterization.

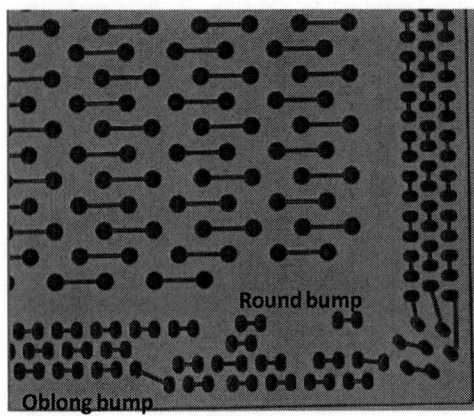

Figure 2. Test vehicle design with mixed bump size

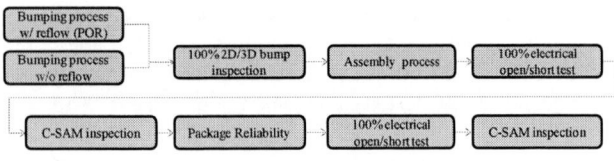

Figure 3. The experimental flow of bump/assembly characterization

III. EXPIRENMENTAL PLAN

The experimental matrix of no reflowed Cu pillar bump is summarized in table 1. The bumped wafers were reflowed in Leg 1 and Leg 2 to verify the impact of high solder tip ratio post reflow process. In contrast with no reflowed bumped wafers from Leg 3~6, the bumps were stacked up with four different solder tip ratios (predicted reflowed solder tip height divided by minimum UBM size); 0.44, 0.56, 0.67, 0.78 were plated to investigate the key features of solder surface profile, solder tip shear strength, bump yield, bump coplanarity and compared with reflowed legs in bumping process. The contact and effects of no reflowed Cu pillar bump on assembly process are investigated by electrical open/short test, X-ray, C-mode Scanning Acoustic Microscope (C-SAM) and Scanning Electron Microscope (SEM). All the study legs of different solder tip ratios with and without bump reflow process have been evaluated by employing package reliability test following JEDEC include thermal cycling test, high temperature storage test and unbiased highly accelerated test (uHAST)[1][2].

TABLE 1: THE EXPERIMENTAL MATRIX OF NO REFLOWED CU PILLAR BUMP

Leg	a. Cu post height (μm)	b. Ni height (μm)	c. Min. UBM size (μm)/short side	d. Plated solder height (μm)	e. bump reflow (Y/N)	f. Total bump height, μm (a+b+d)	g. Predicted solder tip height post reflow (μm)	h. Solder tip Ratio (g/c)
1 (control)	A	B	45	17	Y	63	25	0.56
2				27	Y	73	35	0.78
3				12	N	50	20	0.44
4				17	N	55	25	0.56
5				22	N	60	30	0.67
6				27	N	65	35	0.78

IV. RESULTS AND FAILURE ANALYSIS

A. Bump characterization results

The flat bump surface is formed post solder plated & etched which is different to the typical round solder tip post solder reflow (Fig. 4).

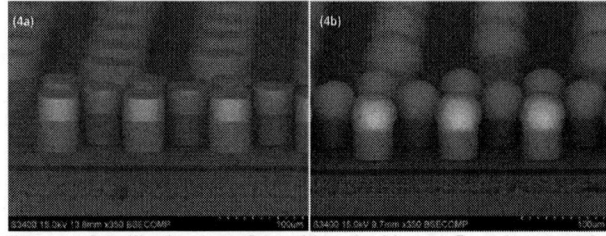

Figure 4. Cu pillar bump surface with and without reflow. (4a): Leg 6-no reflowed bumpe, (4b):Leg 2-reflowed bump.

The solder tip of different surface profiles, concave, flat, convex and round are evaluated to verify the contact qualities in subsequent assembly bonding process. The solder profile of no reflowed Cu pillar bump is formed and related to Cu/Ni profile plated under solder cap, plating parameters and type of leveler additive in plating solution. Fig. 5 compared the shape of solder tip in no reflowed legs. Step height profile meter was used to measure solder tip profile of Cu pillar bump (Fig. 5a, 5b); the concave tip was observed in leg 3, 4 (Fig. 5c, 5d) with small dimple, 5μm and 2μm, respectively in bump center area. The dimple turned into flat surface when more solder was plated in leg 5 (Fig. 5e) and solder tip protruded ~3 um found in leg 6 with highest solder tip (Fig. 5f) due to the total plated bump height (Cu post+ Ni layer+ solder tip) is close to the thickness of photo resist used.

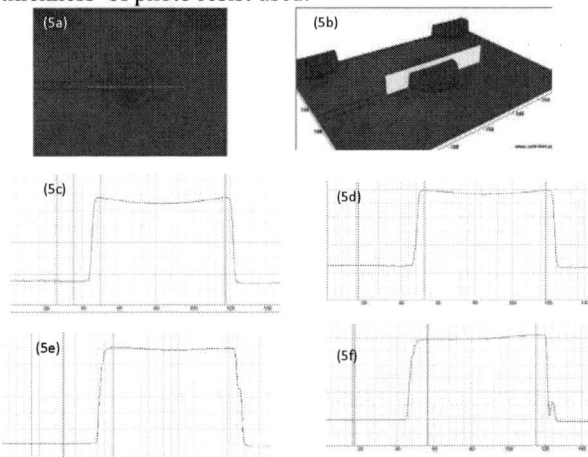

Figure 5. The surface profile of no reflowed bump. (5a, 5b): Movement path of step height profile meter on round pillar bump, (5c): surface profile of round bump in leg 3, (5d): surface profile of round bump in leg 4, (5e): surface profile of round bump in leg 5, (5f): surface profile of round bump in leg 6.

Fig. 6 showed the cross section of oblong bump (short UBM side in die edge) before and after reflow. The bump diameter after reflow (Fig. 6.3a) increases 17% as compared to no reflowed bump (Fig. 6.2a) when solder tip ratio is increased to 0.78. The larger bump diameter will increase

978-1-7281-1500-9/19 $31.00 © 2019 IEEE

the risk of solder joint bridge during assembly die bond process. The concave dimple was observed in leg 3 with low solder tip height (Fig. 6.1a) similar to the shape of round bump measured by step height profile meter (Fig. 5c); the solder tip height of leg 6 post plated was slightly over the top of photo-resist and caused tilt protruded solder tip (Fig. 6.2a). Almost no Ni_xSn_y inter-metallic compound (IMC) was observed from no reflowed bump (Fig. 6.1b, 6.2b) as compared to reflowed bump with IMC thickness ~1.1µm (Fig. 6.3b). The pure lead free solder deposited on Cu post after plating explained lower solder tip shear force and variation of no reflowed bump as showed in Fig. 7a, but the shear strength still can meet typical requirement ≥ 2.5g/mil^2. Similar shear failure mode with 100% solder residue (Fig. 7b, 7c) observed in both no reflowed and reflowed legs.

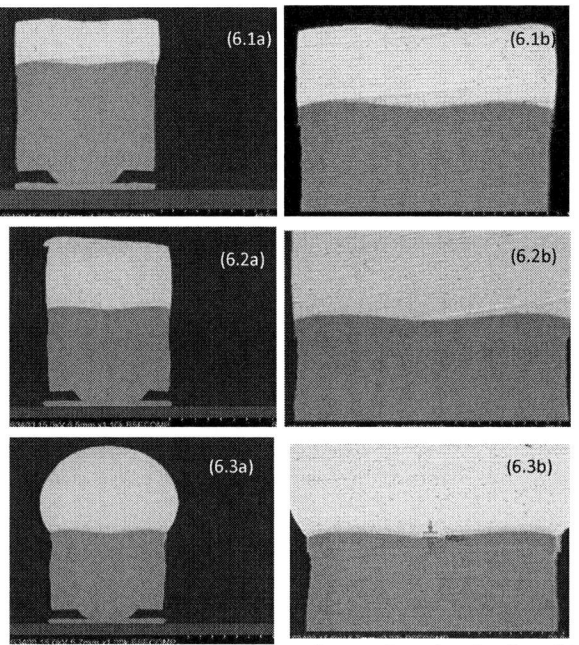

Figure 6. The bump cross-section of no reflowed and reflowed bump (round bump). (6.1a,6.1b): bump X-section of leg3, (6.2a, 6.2b): bump X-section of leg 6, (6.3a, 6.3b): bump X-section of leg 2.

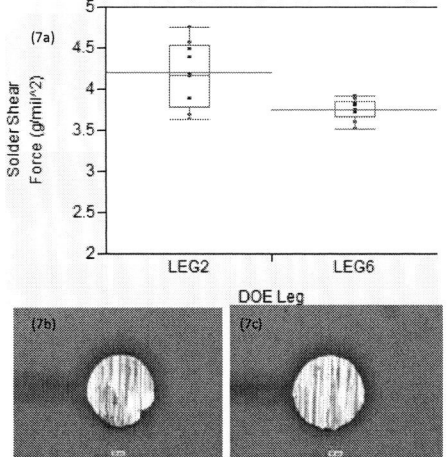

Figure 7. The solder shear test results. (7a): solder tip shear force of leg 2 and leg6, (7b): solder tip shear failure mode of leg 2 , (7c): solder tip shear failure mode of leg 6.

To verify killer yield loss and the impact of bump coplanarity related to the bump stack up with high solder tip ratio, all the bumped wafer in each leg were inspected and the results are summarized in table 2.

TABLE 2: Summary of killer defect and bump coplanarity

Leg	Cu height (µm)	Ni height (µm)	Min. UBM size (µm)/ short side	Plated solder height (µm)	Bump reflow (Y/N)	AOI Yield loss (%) Irregular bump defect	Bump Coplanarity Max.	Ave.	Min.
1 (control)	A	B	45	17	Y	0.00%	7.50	4.85	3.57
2				27	Y	1.67%	17.4	6.93	4.82
3				12	N	0.00%	5.74	2.55	1.06
4				17	N	0.00%	7.50	3.51	1.39
5				22	N	0.00%	8.90	2.97	1.25
6				27	N	0.00%	9.27	5.34	2.04

From the automatic optical inspection (AOI) results, it showed the major defect enhanced by higher solder tip ratio after reflow process in leg 2 was irregular bump. The 1.67% yield loss in leg 2 was caused by solder flow down to Cu post side wall as Fig. 8a, 8b showed. The surface tension during reflow cannot hold the increasing solder volume on top of Cu post area. All the irregular bump defect was found in oblong bump with smaller UBM size near die edge (high solder tip ratio), no such defect found in die center round bump (low solder tip ratio). As expected, no any irregular bump yield loss is found in all legs without reflow.

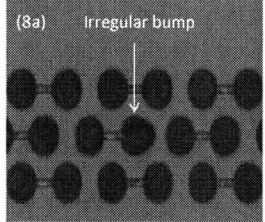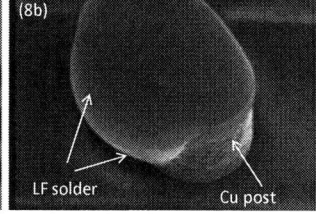

Figure 8. The irregular bump defect enhanced by high solder ratio after refow. (8a): irregular bump defect top view of leg 2 by AOI, (8b): irregular bump defect of leg 2 by SEM

The bump coplanarity within a die is critical to reduce bridge or non-wetting issues during assembly. The design of high solder tip ratio leads to worse bump coplanarity except irregular defect in reflowed bumps. As seen in table 2, the average bump coplanarity after reflowed is increased 42% when solder tip ratio increases from 0.56 to 0.78. The bump coplanarity performance of no reflowed is better than reflowed legs with >30% improvement (leg 1 vs leg 4, leg 2 vs leg 6). Compared with no reflowed legs, no significant coplanarity trend up when solder tip ratio increases from 0.44 to 0.67, but slightly increases from 2.97µm to 5.34µm when solder tip ratio increases from 0.67 to 0.78 (leg 5 vs leg 6. It is caused by non-uniform solder deposition near or over the top opening of photo-resist as showed in Fig. 5f,

6.2a. The no reflowed bump coplanarity with high solder tip design can be further improved when thickness of photo resist is increased and higher than the target plated height. Plot all the coplanarity data measured in each leg (Fig. 9), higher coplanarity results and deviation found in reflowed leg 2.

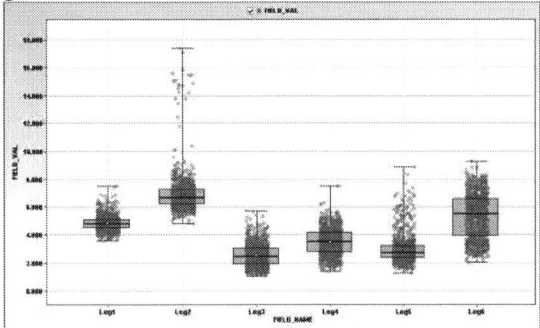

Figure 9. The bump coplanarity comparison for reflow/no reflow legs.

B. Assembly characterization results

As the Cu pillar with lead free solder tip is directly bonded onto the Cu trace of substrate, the volume (height) and diameter of the solder tip are critical to avoid either solder bridging or non-wetting during assembly process [3]. Typically, the bump diameter of no reflowed Cu pillar bump is close to UBM size no matter the solder tip ratio is. However, the bump diameter is increased for reflowed bump while the solder tip ratio is increased. As seen in previous Fig. 6.3a; the bump diameter after reflow is 17% larger than no reflow; that means the bump to trace space is smaller for reflowed bump before bonding. Fig. 10 illustrated the bump to trace space (B2T) and bump coplanarity between reflowed and no reflowed bumps before bonding

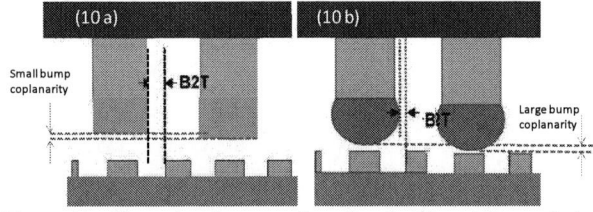

Figure 10. Illustration for no reflowed and reflowed bump during assembly die bond, (10a): no reflowed bump with larger B2T and smaller cop. before bonding, (10b): reflowed bump with smaller B2T and larger cop. before bonding.

The assembly results of study legs are summarized in table 3. No electrical open/short defect found in all reflowed and no reflowed legs. The bonding qualities between solder tip and Cu trace were not impacted by different tip shape found in no reflowed bumps. Besides, no any deformed bump (non-wetting, bridging related), white bump (bump crack related) and delamination/void found from X-ray (Fig. 11.1a, Fig.11.2a) and CSAM inspection (Fig. 11.1b, Fig.11.2b)

TABLE 3: ASSEMBLY O/S and CSAM RESULTS

Leg	Ass'y t0 results		
	O/S test (S.S.100%)		C-SAM (S.S. 80 pcs/leg)
	Open defect	Short defect	CSAM
1 (control)	0/300	0/300	0/80
2	0/300	0/300	0/80
3	0/300	0/300	0/80
4	0/300	0/300	0/80
5	0/300	0/300	0/80
6	0/300	0/300	0/80

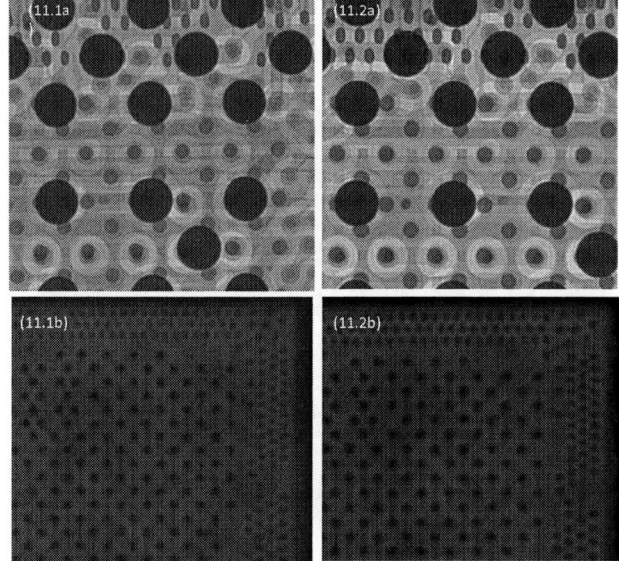

Figure 11. The X-ray and CSAM inspection results for leg 3 and leg 6. (11.1a, 11.2a): X-ray photos of leg 3, leg 6 respectively, (11.1b, 11.2b): CSAM photos of leg 3, leg 6 respectively.

The package cross-section post assembly is compared and analyzed in table 4. The stand-off height and solder joint size are increased while solder tip height is increased for both reflowed and no reflowed legs, but smaller solder joint size is observed for no reflowed bump (leg 2 vs leg 6). No any void trapped in-between Cu trace and solder tip. The bonding results of different solder tip profile; concave, convex and round demonstrated comparable joint qualities.

TABLE 4: SUMMARY OF PACKAGE X-SECTION RESULTS

Leg	Reflow	Package X-section results post assembly (3 bumps/leg, 1 bump/location)				
		Cu post height(μm)	Stand off height (μm)	Solder joint width (μm)	Bump to trace space (μm)	Void q'ty in joint interface
1 (control)	Yes	36.0	64.8	50.0	14.4	0
2	Yes	36.3	69.1	56.3	12.5	0
3	No	36.7	65.9	46.9	20.3	0
6	No	36.2	68.6	53.7	16.4	0

Fig.12 compares the solder joint profiles of oblong bumps located in die peripheral area. The solder joint size is close to UBM size when solder tip ratio is 0.44 (Leg 3, Fig. 12d, 12e, 12f) and obvious solder wetting Cu side wall was observed in reflowed bump when solder tip ratio is 0.78

(Leg2, Fig. 12a, 12b, 12c). The solder joint size is slightly increased when solder wetting Cu side wall, this explains the average joint size of reflowed bump is ~2.6μm larger than no reflowed bump (Fig. 12g, 12h, 12i) and the bump to trace space of no reflowed bump is expected increased which can further reduce bridging risk.

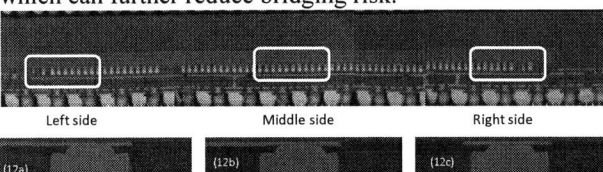

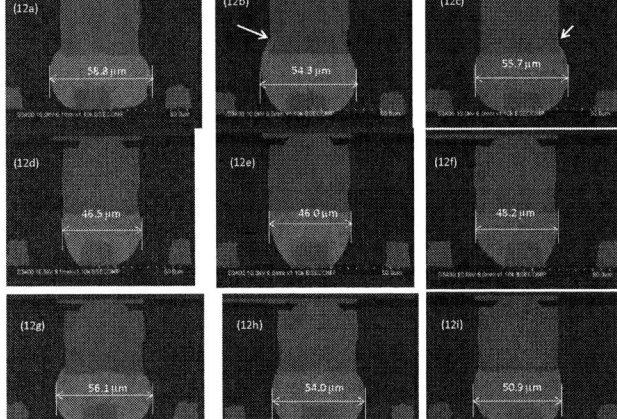

Figure 12. Package X-section of solder joints (oblong bump, short side) located in die left, middle and right side, (Fig. 12a, 12b, 12c): solder joint profile of Leg2, (Fig. 12d, 12e, 12f): solder joint profile of Leg3, (Fig. 12g, 12h, 12i): solder joint profile of Leg6.

C. Reliability results

The package samples of both reflowed and no reflowed bumps are subjected to JEDEC standard reliability tests, including Temperature Cycling (TCT, condition B), High Temperature Storage (HTS, 150℃) and Unbiased Highly Accelerated Test (uHAST, 130 ℃ & /85% RH). The temperature cycling test was extended from typical 1000 to 2000 cycles to verify the reliability capability. Table 5 summarizes the test results. All the legs passed uHAST 192 hrs, TCT 2000 cycles and HTS 1000 hrs inspected by 100% O/S electrical test and C-SAM, it showed comparable reliability performance between no reflowed and reflowed bump.

TABLE 5: PACKAGE RELIABILITY SUMMARY

Leg	Pre-condition MSL3		ubias HAST (130C/85%RH)				Temperature cycle test (-55~125C)				High temperature storage test (150C)	
			96 hrs		192 hrs		1000 cycles		2000 cycles		1000 hrs	
	O/S	SAT	O/S	SAT	O/S	SAT	O/S	SAT	O/S	SAT	O/S	SAT
1 (control)	0/154	0/44	0/77	0/22	0/77	0/22	0/77	0/22	0/77	0/22	0/77	0/22
2	0/154	0/44	0/77	0/22	0/77	0/22	0/77	0/22	0/77	0/22	0/77	0/22
3	0/154	0/44	0/77	0/22	0/77	0/22	0/77	0/22	0/77	0/22	0/77	0/22
4	0/154	0/44	0/77	0/22	0/77	0/22	0/77	0/22	0/77	0/22	0/77	0/22
5	0/154	0/44	0/77	0/22	0/77	0/22	0/77	0/22	0/77	0/22	0/77	0/22
6	0/154	0/44	0/77	0/22	0/77	0/22	0/77	0/22	0/77	0/22	0/77	0/22

From bump cross-sections post TCT 2000 cycles and HTS 1000 hrs to further proves the solder joint quality as shows in Fig.13, Fig 14; no any solder crack, delamination found between reflowed and no reflowed bump with low or high solder tip ratio. However, solder wetting Cu side wall observed in reflowed bump with higher solder tip ratio (Fig. 13b, Fig. 14c), this phenomenon aligned the finding in time zero post assembly and getting worse post HTS 1000 hrs (Fig. 14c). The effect of solder wetting Cu side wall needs to be further monitored especially for small Cu pillar bump (i.e. UBM size < 30μm) to reduce risk of package reliability issue.

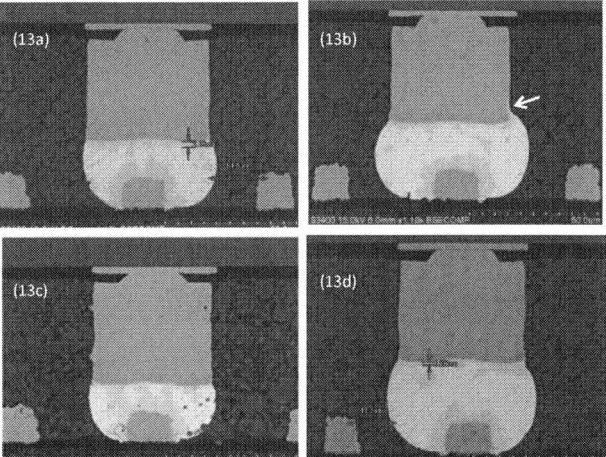

Figure 13. Package X-section of solder joints (oblong bump, short side) post TCT 2000 cycles test. (13a): solder joint of leg 1, (13b): solder joint of leg 2, (13c): solder joint of : leg 3, (13d): solder mjoint of leg 6

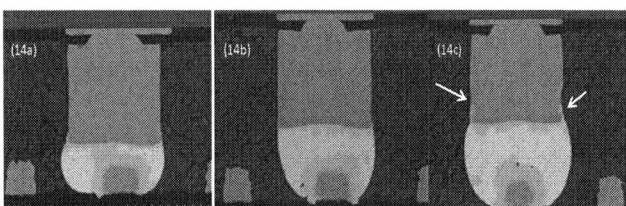

Figure 14. Package X-section of solder joints (oblong bump, short side) post HTS 1000 hrs test. (14a): solder joint of leg 3, (14b): solder joint of leg 6, (14c): solder joint of : leg 2.

V. CONCLUSIONS

A novel package interconnection verification using no reflowed Cu pillar bump is studied in this investigation. No reflowed Cu pillar bump can improve critical yield loss and design flexibility to meet high solder tip height request especially for shrunk bump size to reduce potential CPI and reliability risk. The interconnection risk of concave, flat and convex solder tip formed in no reflowed bump is verified and the bonding qualities are comparable to typical round solder tip post bump reflow during assembly process. The smaller bump diameter and lower bump coplanarity found in no reflowed bump can further reduce the risk of bridging or non-wetting in assembly die bond process. The comparable reliability performance between no reflowed and reflowed bumps is also demonstrated. Besides, shorter bump cycle time is achieved without reflow & clean process as compared with typical Cu pillar process.

ACKNOWLEDGMENT

The authors would like to thank SPIL bumping Yi Hong Lin, assembly Samuel teams for sample preparation and failure analysis lab for the reliability tests, cross sections/SEM check for this study.

[1] JEDEC Publication No. JESD22-A104E, "Temperature Cycling" October 2014.

[2] JEDEC Publication No. JESD22-A118B, "Accelerated Moisture Resistance-Unbiased HAST" July 2016.

[3] Yen-Liang Lin, Chung-ShiLiu, Douglas Yu, "Ultra Fine Pitch/ Low Cost FCCSP Package and Chip Package Interaction (CPI) for Advanced COMOS Nodes", pp. 595–599, 2016 Electronic Components Technology Conference (ECTC).

2019 IEEE 69th Electronic Components and Technology Conference (ECTC)

A Novel interconnection technology using ultra-thin under barrier metal for multiple chip-on-chip stacking structure

Takuya Nakamura[1], Kan Shimizu[1], Masataka Maehara[1] , Toshihiko Hayashi[1], Kentaro Akiyama[1]

Junichiro Fujimagari[1], Tomohiro Ohkubo[2], Atsushi Fujiwara[2] , Hayato Iwamoto[1]

1) Sony Semiconductor Solutions, Kanagawa Japan

2) Sony Semiconductor Manufacturing, Kumamoto Japan

e-mail: Takuya.Nakamura@sony.com

Abstract—In this report, connectivity and reliability between barrier metal and SnAg solder were evaluated from the viewpoint of intermetallic compound (IMC) growth and reliability behavior. On the basis of previous public research, Co, Ti and Ta were selected as the barrier metal. On the top chip side, 22,000 SnAg solder bumps with 20 umΦ/ 40 um pitch were formed. On the bottom chip side, Al pads with the selected barrier metal were fabricated as an electrode. Connectivity testing of the SnAg bump and barrier metal was carried out by Kelvin resistance, Daisy chain, and cross-sectional SEM. Reliability tests were carried out under high temperature and heat cycling conditions. The results of the connectivity verification revealed similar electric characteristics between SnAg solder and barrier in all barrier metal types. This successful demonstration of sub-micron thickness Co material as a barrier metal of SnAg solder bump connection will enable high reliability one-side soldering with reduced processing costs.

Keywords - barrier metal, Chip on Wafer, Co, Ti, Ta

1. Introduction

Next generation digital devices such as digital cameras, cellular phones, games, and high-definition TVs are required to be multifunctional and to have low power consumption. One potential solution is chip stacking technology for the integration of logic and memory ICs with sensors. We have previously reported on new Chip on Chip (CoC) interconnection technology and new interface technology [1][2][3][4]. For a chip stacked structure connected with solder bumps, an electroplating process was used in both top and bottom chips after fabricating the electrodes. However, solder bump formation by electroplating leads to yield loss. Moreover, a barrier metal over 1 um thick is generally needed for solder bumps to avoid diffusion of the barrier metal and the solder. It is difficult to form such barriers over films of this thickness by a wafer process. To resolve these issues and achieve high yield of the chip stacked structure, a one-side soldering structure has been studied (Fig. 1). This structure consists of one chip with a solder bump such as SnAg and another chip with an Al pad instead of a solder bump. Very thin metal with submicron thickness was introduced as the barrier metal layer of the Al pad (Fig. 2) [5][6][7], and the structure showed good connectivity and resistivity even after a 150°C storage test. In the present work, we evaluate connectivity and reliability between barrier metal and SnAg solder from the viewpoint of intermetallic compound (IMC) growth and reliability behavior.

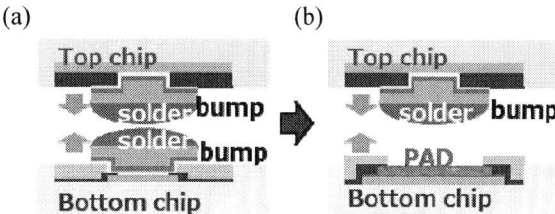

Figure 1. Schematic representation of bump structure: (a) conventional soldering structure and (b) one-side soldering structure.

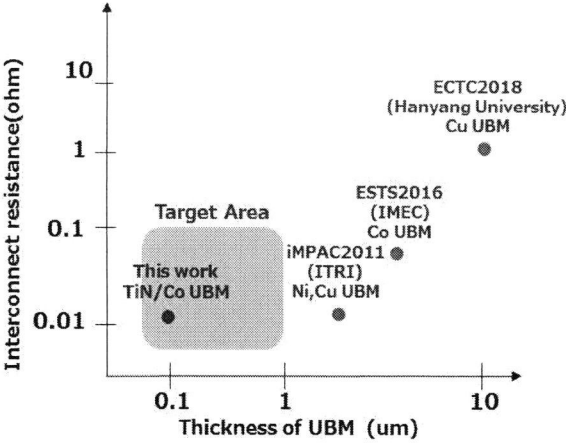

Figure 2. Relationship between UBM thickness and interconnect resistance in previous works.

978-1-7281-1500-9/19 $31.00 © 2019 IEEE   1641

2. Samples and Experimental Techniques

2-1 Test vehicle

A schematic of the test vehicle used in this study is shown in Fig. 3. Its specifications are listed in Table1.

Table 1 Test vehicle specifications.

Top chip size (mm)	**31 × 3**
Bottom chip size (mm)	**34 × 25**
Number of bumps	**21k × 2**
Bump dia./pitch	**20 μm/40 μm**
Four-terminal resistance patterns	**56 patterns ×2**
Daisy-chain resistance patterns	**21k bump ×2**

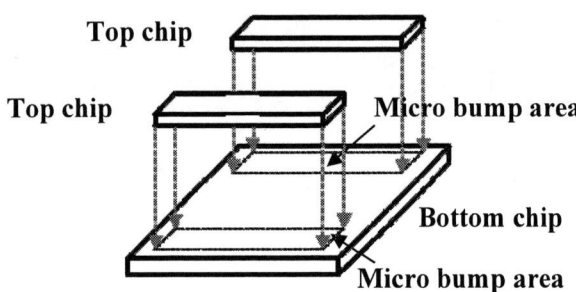

Figure 3. Schematic of CoC interconnection.

2-3 Experimental set up

The evaluation flow of this work is shown in Fig. 4. We formed 22,000 SnAg solder bumps on the top chip side and a barrier metal and electrode on the bottom chip side. After bonding with SnAg solder by the chip on wafer (CoW) process, evaluations on electric characteristics and reliability were carried out.

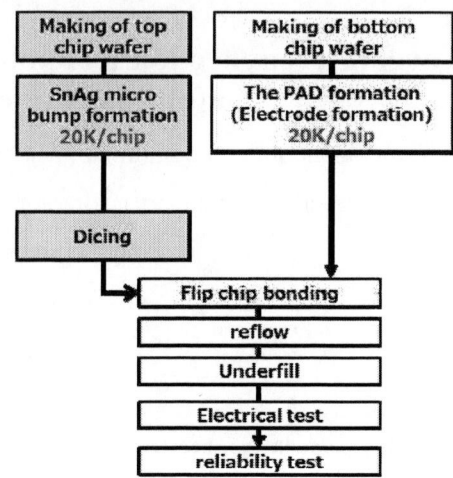

Figure 4. Evaluation flow.

2-4 Formation of barrier metal and electrodes for bottom chip

We selected which barrier metal material to evaluate on the basis of the following two criteria. In general, these are materials used in the semiconductor process.

1. Can be connected to SnAg solder.
2. No reports of barrier metal thinning.

On the basis of previous public research and phase diagrams [7][8][9], Co, Ti, and Ta were selected as the barrier metal and electrode (see Table 2). Al pads with the selected submicron thickness barrier metal were fabricated as an electrode by a wafer process. The thicknesses of each barrier metal and electrode were Ti: 200 nm, Ta: 20 nm, Co: 300 and 400 nm. The top view of the under barrier metal (UBM) is shown in Figure 5.

Table 2 Comparison of UBM materials.

	Criteria	Co	Ti	Ta	Cu	Ni	Pt	Au
Connect with SnAg solder	Yes	Yes	Yes	Yes	Yes	Yes	Yes	Yes
		O	O	O	O	O	O	O
Required barrier metal thickness	≦1 um	?	?	?	≧3 um	≧1 um	≧1 um	≧3 um
		?	?	?	x	x	x	x
PAD formation by a wafer process	Processable	Processable	Processable	Processable	Processable	Difficult	Difficult	Difficult
		O	O	O	O	x	x	x

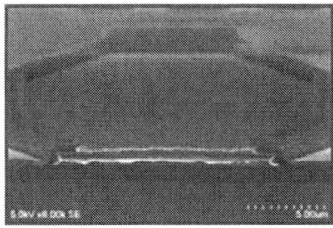

Figure 5. SEM image of UBM (top view).

2-5 Formation of SnAg micro bump

On the top chip side, 22,000 SnAg solder bumps with 20 umΦ/40 um pitch were formed by an elector plating method. SnAg solder height was 13 um and Ni UBM height was 3 um. The top view of the SnAg solder is shown in Fig. 6.

Figure 6. SEM image of SnAg solder.

2-6 Chip on wafer bonding process

We bonded between the SnAg solder bump and a barrier metal and electrode on all chips using a flux-less bonding process, as shown in Fig. 7. The SnAg solder and barrier metal were connected by reflow processing at 240°C. Underfill resin was applied after bonding the gap between the two connected chips to enhance reliability.

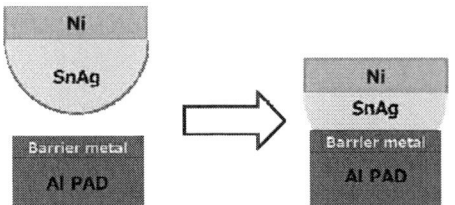

Figure 7. Bonding process.

2-7 Electrical test and analysis

We evaluated the four-terminal and daisy-chain resistance after the bonding process and after the reliability test.

The characteristics of each bonding interface were analyzed under various barrier metals by TEM/EDX.

3. Results and discussion

1) Initial bonding results

The connectivity test of the SnAg bump and various barrier metals was carried out by Kelvin resistance, Daisy chain, and cross-sectional SEM. Fig. 8 shows the results of measuring the bump resistance after bonding. The four-terminal resistances of all connecter bumps were 10–100 mohm and we confirmed the resistance of all connected bumps along a cumulative frequency distribution. Fig. 9 shows SEM images of a cross section of the connected micro bump. Good bondability without open or short failure regardless of barrier metal type was obtained.

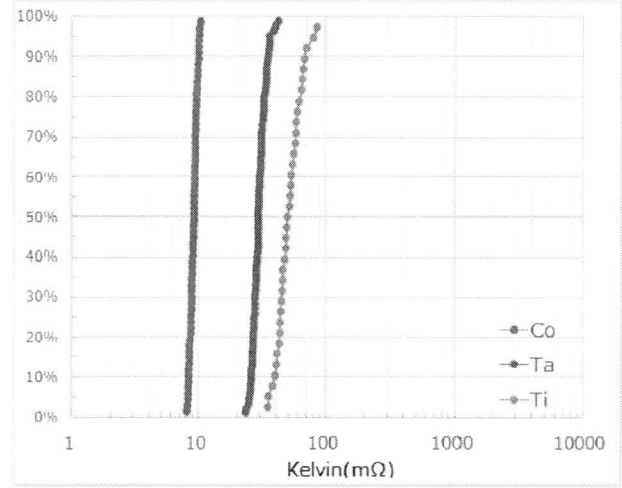

Figure 8. Kelvin resistance after bonding.

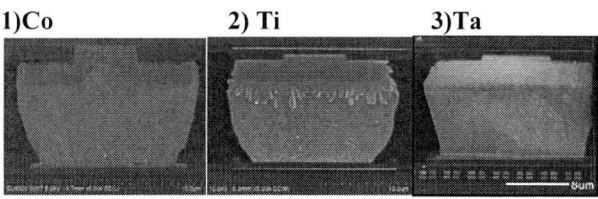

Figure 9. SEM images of connected micro bump.

2) Reliability test results

High temperature storage testing (150°C) was performed on CoW samples prepared using each barrier metal. Fig. 10 shows the measurement result of Kelvin resistance after reliability testing using various barrier metals. Ti and Ta experienced an open issue after 200 h and 300 h, respectively. SEM images of the cross section of these open bumps are shown in Fig. 11. We found that Ti had a high resistance layer on the surface, and Ta peeled off from the SnAg solder.

In contrast, using Co, we obtained a stable resistance until 1030 hours. Figure 12 shows cross section SEM images of the connection between the micro bump and Co barrier metal. Co reacted with the SnAg solder and even though it entirely changed into IMC (Sn_xCo_X), no delamination and no abnormal layer were observed, and good connectivity was secured regardless of Co thickness.

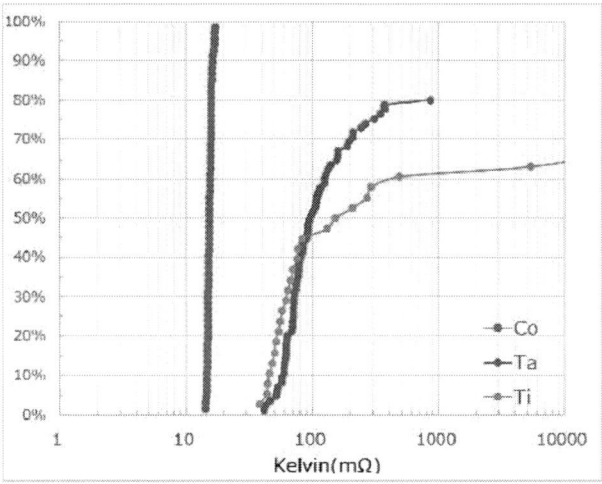

Figure 10. Kelvin resistance after reliability test.

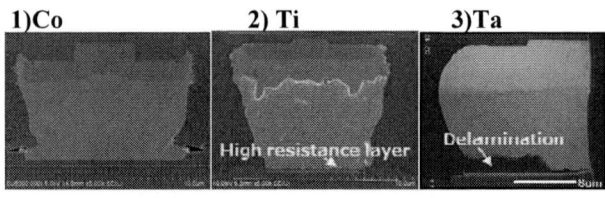

Figure 11. SEM images of cross section after reliability test. 1) Co had a good connection. 2) Ti had a high resistance layer. 3) Ta peeled off from the SnAg solder.

Figure 12 SEM images of cross section of Co film with changed thickness.

3) Analysis of bonding interface

Figure 13 shows the TEM/EDX analysis results of the connected micro bump in Ti barrier metal after CoW processing and the reliability test (150°C/200 h). As shown, an oxide layer was formed after the bump connection and reliability test. This indicates that growth occurs under the influence of the thermal history of the reliability test. For the barrier metal to be connected to the SnAg solder, it is necessary to select (by thermal history such as a reliability test) a metal that does not easily oxidize.

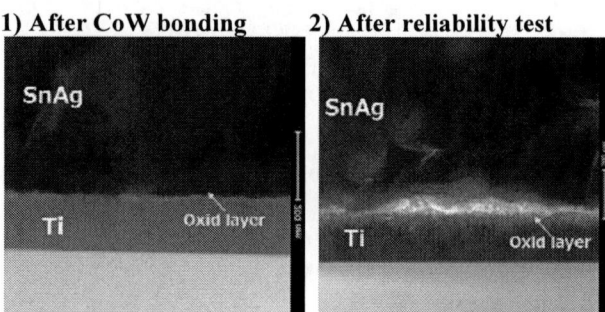

Figure 13. TEM/EDX analysis results of connected micro bump in Ti barrier metal.

Figure 14 shows the TEM/EDX analysis results of the connected micro bump in Ta barrier metal after the reliability test. As shown, the Ta and SnAg solder did not react with each other according to thermal history (e.g., the CoW process and reliability test). Therefore, Ta peeled off from the SnAg solder due to no interaction behavior between SnAg and Ta.

Figure 14. TEM/EDX analysis results of connected micro bump in Ta barrier metal.

In contrast, Co did react with the SnAg solder, and even though it entirely changed into IMC ($Sn_xNi_yCo_z$), no delamination and no abnormal layer were observed. To examine why this reliability could be secured, we conducted bonding interface analysis. Figure 15 shows the TEM image of the Co-Sn connection after the reliability test. We formed TiN as the second barrier metal under Co. As shown in the figure, the Co and SnAg solder reacted with each other, and then the pure Co layer was converted into IMC ($Sn_xNi_yCo_z$). In addition, according to the Fourier transform characteristics of each layer, which correspond to the electron diffraction pattern (Fig. 16), the crystal orientation of Co was affected along the TiN crystal orientation in the vicinity near the interface, and a sticky adhesion layer was formed.

After IMC($Sn_xNi_yCo_z$) completely reacted with the SnAg solder, it changed from the Co crystal structure after CoW bonding. However, we believe that this change is small, and there was no change in the adhesiveness near the Co-TiN interface. In future, a more detailed examination will be necessary.

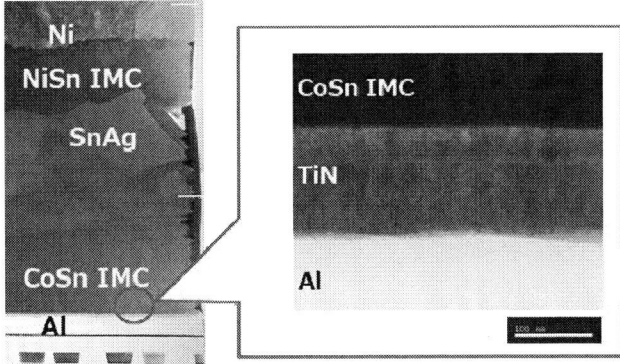

Figure 15. TEM image of Co-Sn connection after reliability test.

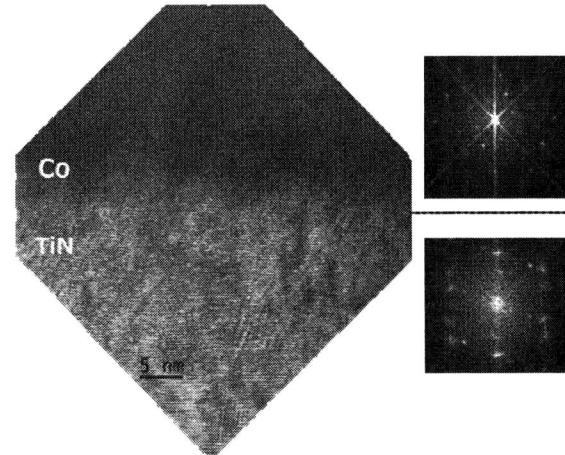

Figure 16. Fourier transform characteristics of each layer corresponding to electron diffraction pattern after reliability test.

4. Conclusion

In this work, we performed a reliability evaluation of the connectivity between barrier metal and SnAg solder from the viewpoint of IMC growth and reliability behavior. On the basis of previous public research, Co, Ti, and Ta were selected as the barrier metal and electrode. Al pads with specific submicron thicknesses were fabricated as an electrode by wafer process. We obtained good bondability without open or short failure regardless of the type of barrier metal. However, Ti and Ta experienced an open issue after 150°C/200 h and 300 h, respectively. We found that Ti had a high resistance layer on its surface and Ta peeled off from the SnAg solder. In contrast, with Co, we could obtain stable resistance until 150°C/1030 h. It also reacted with the SnAg solder, and even though it completely changed into IMC ($Sn_xNi_yCo_z$), no delamination and no abnormal layer were observed. In order to reduce the thickness of the barrier metal, we need to control the reactivity with the SnAg solder, maintain stability of the connection interface, and properly set the crystallinity parameters.

This successful demonstration of sub-micron thickness Co material as a barrier metal of the SnAg solder bump connection will contribute to high reliability one-side soldering with reduced processing costs. This technology shows promise in the creation of new chip stacking devices.

5. References

[1] T. Ezaki et al., "A 160Gb/s Interface Design Configuration for Multichip LSI", ISSCC Dig. Tech. Papers, pp. 140–141, Feb. 2004.

[2] S. Wakiyama et al., "Novel low-temperature CoC interconnection technology for Multichip LSI (MCL)", in 57th Electronic Components and Technology Conference, Proceedings, pp. 610–615, 2007.

[3] Y. Oike et al., "An 8.3M-pixel 480fps Global-Shutter CMOS Image Sensor with Gain-Adaptive Column ADCs and 2-on-1 Stacked Device Structure," in Symp. VLSI Circuits Dig. Tech. Papers, Jun. 2016.

[4] K. Akiyama et al., "A Front-illuminated Stacked Global-Shutter CMOS Image Sensor with Multiple Chip-on-Chip Integration", in 3D Systems Integration Conference (3DIC), 2016 IEEE International, pp. 1–3.

[5] J. Park et al., "The reliability and the effect of NCA trapping in thermo-compression flip-chip solder joints fabricated using Sn-Ag solder capped 40 μm pitch Cu pillar bumps and low temperature curable non-conductive adhesive (NCA)", in the 68th Electronic Components and Technology Conference, Proceedings, pp. 2377–5726, 2018.

[6] J. Juang et al., "The development of high through-put micro-bump-bonded process with Non-Conductive Paste (NCP)", in International Microsystems, Packaging, Assembly and Circuits Technology Conference 2011, Proceedings.

[7] I. Preter et al., "3D Stacking of Co-and Ni-based Microbumps", in Electronics System-Integration Technology Conference 2016, Proceedings.

[8] T. Studnizky et al., "Phase formation and reaction kinetics in M-Sn systems (M = Zr, Hf, Nb, Ta, Mo)", in International Journal of Materials Research, Volume 93, Issue 9, 2002.

[9] K. Kayukawa et al., "A role of Ti-Sn diffusion layer formed at the interface between Pb free solder and TiNiAu multi-layer", in 2005 International Symposium on Electronics Materials and Packaging, Proceedings.

Multilayer decoupling capacitor using stacked layers of BST and LNO

Todd Schumann, Sheng-Po Fang, Yong-Kyu Yoon
Electrical and Computer Engineering
University of Florida
Gainesville, FL, United States
toddschumann@gmail.com
ykyoon@ece.ufl.edu

Jongmin Yook, Dongsu Kim
Packaging Research Center
Korea Electronics Technology Institute
Bundang, Korea
kimds@keti.re.kr

Abstract—**Barium strontium titanate (BST) and lanthanum nickelate (LNO) are often used in conjunction to create high energy density capacitive devices. The LNO seeds and buffers the BST, reducing the effect of its dielectric "dead" layer. We show that this concept can be extended to a multilayer capacitor, where the second metal-insulator-metal stack is grown on top of, and buffered by, the first layer. Although this only doubles the effective area of the capacitor, the capacitance increases to 220%, 20% greater than the increase in surface area. Additional capacitive layers can be deposited to further take advantage of this effect.**

Keywords-barium strontium titanate (BST); lanthanum nickelate (LNO); perovskite materials; high dielectric constant; high energy density capacitor; chemical solution deposition; metal organic decomposition

I. INTRODUCTION

Barium strontium titanate (BST) is a perovskite ferroelectric/paraelectric material, noted for its extremely high permittivity, making it ideal for integrated capacitors. In bulk, its permittivity can reach as high as 20,000 [1]. In thin films, however, the permittivity is markedly lower, usually less than 1000. This is due to interfacial stress between the BST and the substrate associated with lattice constant mismatch, causing a non-crystalline "dead" layer directly at the interface, lowering the effective dielectric constant of the entire film [2]. This ultimately limits the capacitance of BST capacitors: as the BST layer thickness is reduced, which should increase capacitance in theory, the effect of the "dead" layer becomes more pronounced, reducing the observed capacitance.

This is often modelled as series capacitors consisting of two capacitors at the interfaces and a capacitor in the bulk [2]. The interface capacitances severely limit the capacitance behavior of the stressed films.

To reduce the effect of this "dead" layer, the BST layer is often grown on lattice matched substrates, such as sapphire [3] or lanthanum aluminate (LAO) [4,5]. When grown on silicon substrates, it must be buffered away from the substrate using one or more films of material, which attempt to bridge the difference in lattice constant [6] to achieve good performance.

In particular, the selection of a bottom electrode is particularly challenging. Because the BST films must be calcined at high temperatures in an oxygen environment, only noble elemental metals, which can withstand the calcination temperature, can be used as the bottom electrode [7]. While gold can be used, the calcination temperature is on the verge of gold's melting temperature and thus is unreliable [8]. This leaves platinum as the only realistic elemental electrode material. However, platinum is an

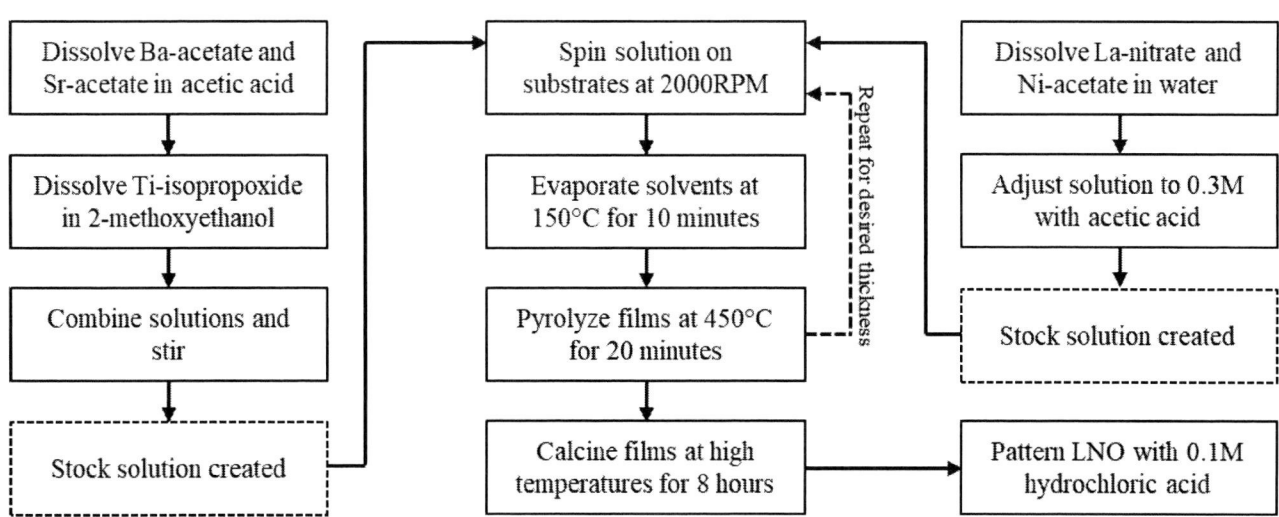

Figure 1.   Chemical solution deposition process flow for BST and LNO.

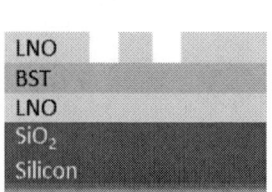

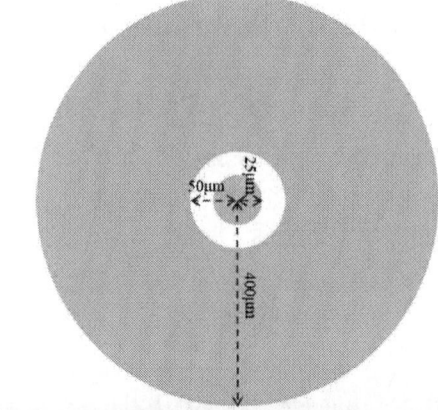

Figure 2. Two dimensional cross section of the two sample sets. Sample set 1 (left) consists of an LNO/BST/LNO MIM capacitor which represents the first layer of a double layer capacitor. Sample set 2 (right) has the same LNO/BST/LNO MIM capacitor except on top of additional LNO and BST layers to represent the second layer of a double layer capacitor.

entirely different crystal structure from BST and imparts stress at the interface itself [9].

Instead, conductive perovskite oxides, such as lanthanum nickelate (LNO), have been used to reduce the interfacial stress [10-12]. This has been shown to increase the dielectric properties of BST films grown on top. Not only does the LNO have the same perovskite crystal structure as BST with a lattice constant nearly matched, but the LNO is also buffering the BST layer away from the substrate such as Si or SiO_2. This further reduces the effect of the "dead" layer and acts as the lattice constant matched bottom electrode simultaneously.

In this paper, we report on the benefit of creating multilayer stacked capacitors. We use LNO as the bottom and top electrode to reduce the stress associated with lattice mismatch at the BST interface. Additionally, as the number of capacitor layers increases, the BST is further buffered from the Si substrate, leading to a further increased capacitance.

II. FABRICATION

The BST and LNO films were deposited using chemical solution deposition (CSD) [13], specifically metal organic decomposition (MOD), on <100> silicon with 100 nm of thermal oxide as a lattice buffer layer. The general process dissolves organic salts of the metals in the final crystal, and then, with various heating steps, gradually produces the films.

In the case of the BST, a 0.35M stock solution was created using an 80/20 molar mixture of barium acetate and strontium acetate. The acetates were first baked to reduce the hydrate component prior to dissolving in acetic acid. 2-methoxyethanol was added to the solution and allowed to stir. Finally, an equal molar amount of Ti-isopropoxide was added to the solution to create a stock. Ti-isopropoxide reacts with water, precipitating TiO_2, thus the acetates must be baked first to remove the water. The stock solution was stirred for 24 hours to allow molecular level mixing and to ensure that TiO_2 did not precipitate. The stock solution was used within 72 hours before disposal and remixed to ensure

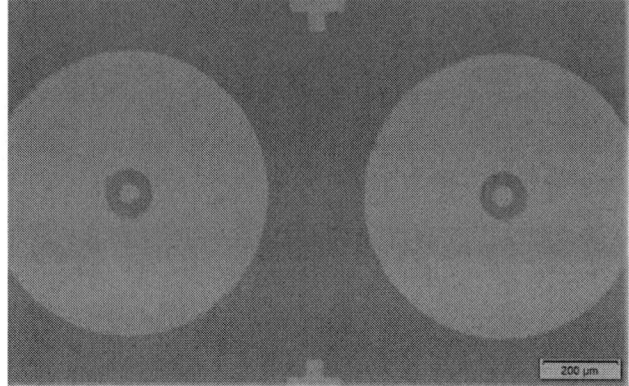

Figure 3. (top) Layout of the capacitor test structures etched into the topmost layer of LNO on each of the sample sets (not drawn to scale). (bottom) Optical image of the top LNO layer pattered with the capacitor test structure. The impedance between the inner circle and the outer ring is dominated by the inner circle, allowing a two terminal measurement without etching the BST dielectric.

that TiO_2 did not precipitate and alter the Ba/Sr:Ti molar balance.

Similarly, the LNO solution was created by dissolving equal molar amounts of La-nitrate and Ni-acetate in deionized water. The solution was also allowed to spin for 24 hours to ensure intermolecular mixing. The molarity of the solution was adjusted to 0.3M using acetic acid. It has been previously demonstrated that the LNO solution remains stable for extended periods, but to avoid changes in molarity due to evaporation, the solution was also remixed if set longer than 72 hours.

Both films were deposited from the stock solutions using the same process. First, the solution was spin coated onto samples at 2000 RPM for 40 seconds, with a 6 second 500 RPM spreading spin. The samples were baked at 150 °C for 10 minutes to evaporate the solvents. The samples were then baked at 450°C to pyrolyze the films and burn off the organic components. The resultant films are no longer dissolvable in the original solvents, so this process can be repeated to increase the film thickness or deposit subsequent films. Finally, the samples were annealed at 700 °C for 8 hours with 3 °C/min ramping up and down.

978-1-7281-1500-9/19 $31.00 © 2019 IEEE

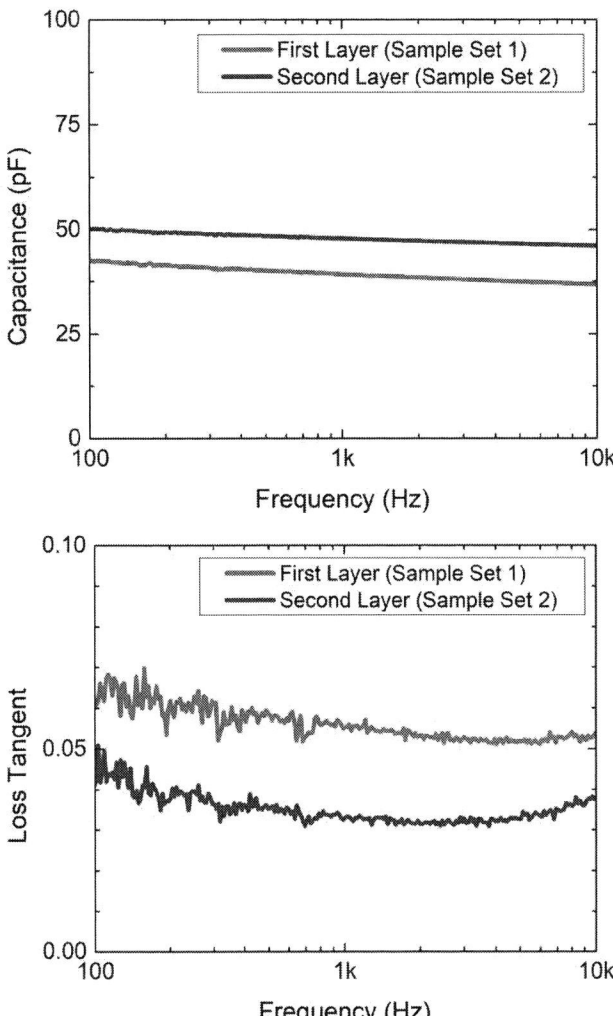

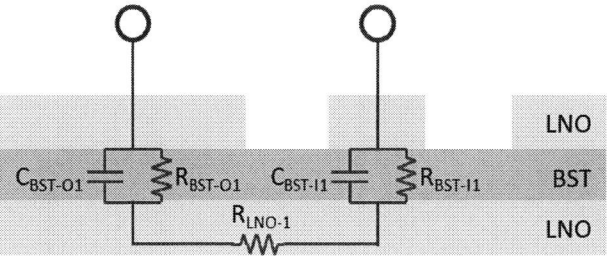

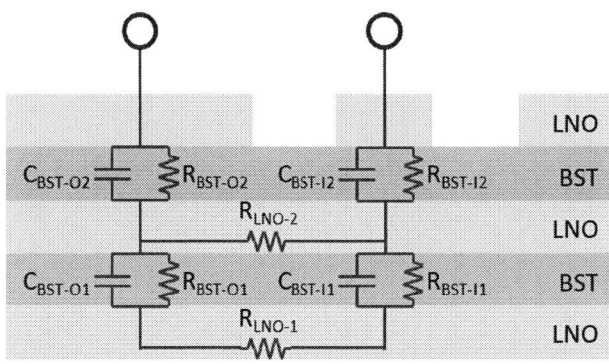

Figure 5. Equivalent low frequency circuit of the stacked LNO and BST layers for (top) sample set 1 and (bottom) sample set 2. The bottom LNO and BST layers of sample set two exhibit the same impedance network as sample set 1. As R_{LNO-2} is far less than the combined impedance of the sample set 1 network, as seen in the loss tangent data, the impedance of the bottom LNO and BST layers in sample set 2 are negligible.

Figure 4. (top) Capacitance of the first layer (red) and the second layer (blue). The second layer shows a 22% increase in capacitance over the first layer due to the extra buffering from the first layer. (bottom) Loss tangent of the first layer (red) and the second layer (blue). In addition to a higher capacitance, the second layer also shows a lower loss tangent.

The synthesis of LNO presents a particular challenge as the La-nitrate does not pyrolyze until the LNO crystal is formed [14]. Therefore, the pyrolysis step crystalizes the films into LNO. Similarly, the BST films are also crystalized during their respective pyrolysis steps as the same process was used.

Each layer was comprised of two spin coats of the respective solution. Separate samples were prepared containing only two spins of each material. The BST layer was etched with 5% HF and LNO was etched using 0.1 M HCl. The layer thicknesses were determined with a Dektak 150: the BST layers were 150 nm and the LNO layers were 50 nm while both films showed less than 10nm roughness.

Two sets of samples were prepared. The first set contained only a single LNO/BST/LNO stack, demonstrating the first layer of a capacitor. The second set had the same stack as the first sample set with an additional BST and LNO

layer, showing the resultant capacitance from a second capacitor layer. In addition, as the solution for LNO is water based, it does not wet the Si sample surface well, so an extra wetting layer of BST was spun prior to the first layer of LNO. This allowed the LNO solution to wet the samples and spin evenly. The subsequent spin coating of the bottom LNO layer was then able to spin evenly.

The second set of samples involved 11 sequential spin/ evaporate/pyrolyze steps, which repeatedly thermally shocks the films. As the films must be crystallized during the pyrolysis step, these thermal shocks progressively increase the chance of cracking in the films. This would cause a short circuit if in the BST layer or an open circuit if in the LNO layer. As a result, the thermal stress limited further spin coats (i.e. a third capacitive layer) from being tested.

Once the films were spin coated onto the samples, they were calcined at 700°C for 8 hours to allow the films to crystalize further and fill many of the oxygen vacancies. BST films typically crystalize at higher temperatures [11], but above 700°C LNO begins decomposing to nonconductive La_2NiO_4, severely reducing the conductivity.

After calcination, the top layer of LNO was patterned on both sample sets using lithography and 0.1M HCl. Due to plasmonic resonances in LNO, there is a high absorption in the ultraviolet range [15]. Because of this, a 3x higher dose was required to expose the photoresist. Additionally, once

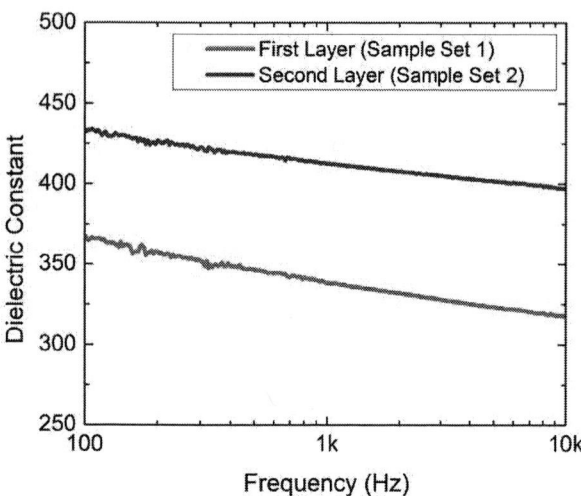

Figure 6. Extracted dielectric constant of the BST films for the first layer (red) and second layer (blue) of the capacitors. As the permittivity is directly proportional to the capacitance, the permittivity also increases 22% from the first to the second layer.

the films were calcined, the etch rate in the HCl reduced from 50 nm/min to 17 nm/min as the films are more crystallized.

III. ELECTRICAL PERFORMANCE

The top layer of LNO was patterned into concentric rings. The inner circle had a radius of 25 μm. The outer ring had an inner radius of 50 μm and an outer radius of 400 μm. This gives an area ratio of 1:250. The equivalent circuit model is shown in figure 5. The outer ring impedances are labelled with a subscript of "O" and the inner ring impedances are labelled with a subscript of "I".

With a surface area of 250 times that of the inner circle, the outer ring capacitance is 250 times the inner capacitance and the outer ring resistance is 250 times less than that of the inner circle. The resultant impedance of the outer ring is 250 times less than the impedance of the inner circle, allowing it to be ignored with little error. This allows probing of only the top LNO without the need to etch the BST to expose the bottom electrode.

Separate samples with the same LNO layer were measured using a four-point probe to determine the resistivity, which was 8.5×10^{-4} Ω·cm. Although this is higher than other common elemental electrodes, the series resistance added to the capacitor is far less than the capacitive impedance in the measured frequency range.

The impedance of both sets of samples were measured from 100 Hz to 100 kHz for their low frequency impedance using an HP 4294A Precision Impedance Analyzer. As seen in figures 4 and 5, the second sample set had a 22% higher capacitance as well as a reduced loss tangent.

The increased capacitance of the second sample set is not arising from the BST and LNO buffer layers beneath the measured capacitor. The underlying structure is identical to sample set 1. The full impedance of sample set 1 (dominated by C_{BST-I1} shown in figure 5) is far greater than the series

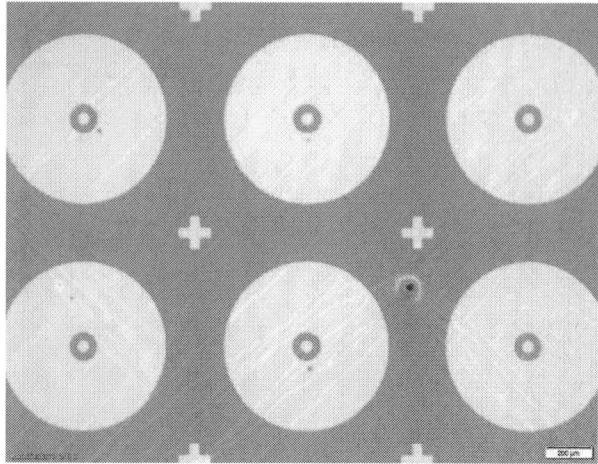

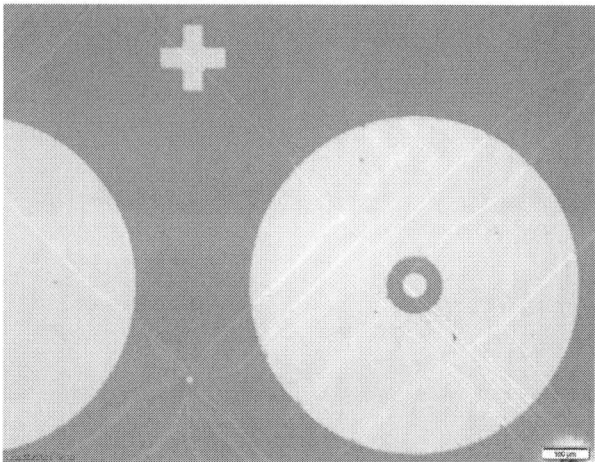

Figure 7. (top) Optical microscope image with a 5x objective showing cracks propogating from a spin defect following the BST/LNO crystal directions. (bottom) Optical microscope image with a 10x objective showing more closely showing the cracking running through the large ground pad as well as the capacitive electrode.

resistance of the middle LNO (R_{LNO-2}) in the measured frequency range as seen in the loss tangent of sample set 1. The resultant parallel combination of R_{LNO-2} and C_{BST-I1} does not differ significantly from R_{LNO-2}. Therefore, the impedance contribution from the first layers of LNO and BST is negligible, and the impedance measured in sample set two is dominated by C_{BST-I2}.

Instead, the increased capacitance is attributed to increased crystallinity in the BST layer and a reduction in the "dead" layer effect present at the interfaces. This effect has previously been shown with using LNO as the bottom electrode [14], however these results show that the effect will continue as the perovskite buffering layers are increased and more capacitive layers are added.

In addition to the reduction of the "dead" layer effect, as the BST and LNO layers are grown, they become progressively more <100> dominated through a process called Van der Drift orientation [16,17]. Since the (110) face of the crystals has a higher surface energy and grows faster, the <110> nucleations choke themselves off, leading to a

978-1-7281-1500-9/19 $31.00 © 2019 IEEE

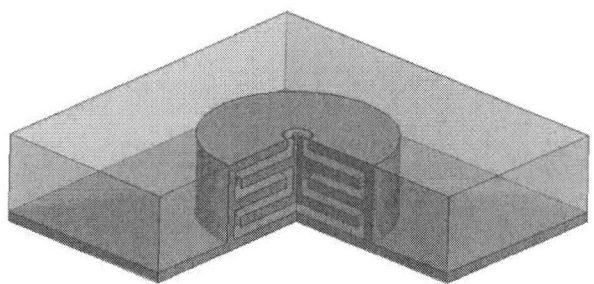

Figure 8. Example of a layered double layer capacitor structure with the BST shown in green and the LNO shown in grey. The structure could be fabricated using 4 lithography masks and need not stop at two layers as the patterned areas repeat.

larger percentage of <100> crystals as the films grow thicker [16] and a more uniformly crystallized film. XRD measurements of LNO grown on BST and BST grown on LNO confirm that whichever layer is grown on top has a higher percentage of <100> crystals. The Van der Drift orientation also adds to the increased electrical performance of the second capacitor layer.

The permittivity of the BST films from both sample sets was extracted and plotted in figure 6. It is known that the permittivity of BST will slightly decrease with increasing frequency due to ferroelectric relaxation [18]. As the permittivity is directly proportional to the capacitance, there was also a 22% increase in the BST permittivity.

A permittivity of 350-450 is marginal for BST films grown by CSD techniques. However, this report demonstrates the improvement possible by vertically stacking the capacitive layers. For example, if a multilayer capacitor was made using these exact layer stacks, a 2.2x improvement in the capacitance density would be achieved, instead of the expected 2x improvement if the BST permittivity did not improve. Additionally, the loss tangent would be reduced as the second layer exhibits less leakage through the dielectric.

The CSD approach to layering these materials is not the optimal fabrication method. On top of not being CMOS compatible, it also subjects the crystalized films to repeated thermal shocks, causing cracking on the micro/nano scale. This limits the number of achievable layers as the likelihood of measurable devices with no cracks becomes exceedingly small. In addition, the thermal shocks reduce the grain size of the films, limiting the performance of the individual layers themselves.

Figure 7 shows optical microscope images of one such area on a sample from sample set 2. Cracks tend to begin at spin defects and propagate along the BST/LNO crystal directions (roughly 45° off the <100> Si direction). As the number of layers increases, the density of these cracks increases dramatically. If a crack propagates through the capacitive electrode, the probed area no longer comprises the entire electrode, skewing the measured capacitance. Additionally, if cracks propagate through the large ground electrode, the impedance assumptions (i.e. that the impedance at the ground plane is sufficiently small to ignore), may not be valid. As the density of these cracks

increases, the yield of devices which do not contain cracks decreases dramatically.

Instead, a sputtering approach would give better performing films. The films can be *in situ* recrystallized during deposition with the addition of oxygen as a sputtering gas and substrate heating. The process is far less harsh on the films and gives better electrical performance [19]. The same improved BST quality has been demonstrated through the use of sputtered BST on sputtered LNO [20].

Additionally, the use of sputtered BST allows for a structured substrate to be used; instead of a flat substrate, it can be patterned with micro-scale pores to increase the surface area. A 1 mm^2 area filled with 100x 75 μm diameter 75 μm deep pores would have a 2.77x increased surface area over a flat substrate. These pores are easily fabricated in silicon using DRIE and a 1:1 step coverage on the micro scale is easily coated using sputtering. The crystalized BST and LNO are selectively etched using dilute HF and dilute HCl, respectively, allowing the final capacitor to be patterned.

IV. CONCLUSIONS

The buffering/templating of BST using perovskite electrodes has been shown in literature. We have shown that the effect will continue as more buffer layers are added. Using the bottom buffer layers as extra capacitive layers, the capacitance density per unit area can be increased. Additionally, the upper layers also exhibit better performance because of added buffering, so the overall increase in capacitance density is greater than the number of layers. For example, using a structure as in figure 8 with the BST films from this report, the capacitance density of the two-layer structure would be 122% greater than that of the single-layer structure.

This concept can be extrapolated and used in addition to other methods of increasing capacitance density, such as patterning the substrate with micropores to increase the surface area per unit area, to create extremely large capacitance density structures.

ACKNOWLEDGMENT

This work was supported in part by the Global R&D program of MOTIE/KIAT (N0000686). The devices have been fabricated using Nanoscale Research Facilities at the University of Florida.

REFERENCES

[1] T. M. Shaw, et al. "The effect of stress on the dielectric properties of barium strontium titanate thin films." Applied Physics Letters, vol. 75, no..14, pp. 2129-2131, 1999.

[2] M. Stengel and N. A. Spaldin, "Origin of the dielectric dead layer in nanoscale capacitors," Nature, vol. 443, no. 7112, pp. 679-682, 2006.

[3] G. Velu, J. C. Carru, E. Cattan, D. Remiens, X. Melique and D. Lippens. "Deposition of Ferroelectric BST Thin Films by Sol Gel Route in View of Electronic Applications," Ferroelectrics, vol. 288, no. 1, pp. 59-69, 2003.

[4] Y. L. Cheng, N. Chong, Y. Wang, J. Z. Liu, H. L. W. Chan and C. L. Choy, "Microwave Characterization of BST Thin Films on LAO Interdigital Capacitor," Integrated Ferroelectrics, vol. 55, no. 1, pp. 939-946, 2003.

[5] C. M. Carlson, T. V. Rivkin, P. A. Parilla, J. D. Perkins, D. S. Ginley, A. B. Kozyrev, V. N. Oshadchy, and A. S. Pavlov, "Large dielectric constant $(\varepsilon/\varepsilon0>6000)Ba_{0.4}Sr_{0.6}TiO_3$ thin films for high-performance microwave phase shifters," Applied Physics Letters, vol. 76, no. 14, pp. 1920-1922, 2000.

[6] K. T. Kim, C. Kim, D. E. Senior, D. Kim, Y. K. Yoon, "Microwave characteristics of sol-gel based Ag-doped $(Ba_{0.6}Sr_{0.4})TiO_3$ thin films," Thin Solid Films, vol. 565, pp. 172-178, August 2014.

[7] Emmanuel Arveux. "Surface and interface properties of BaTiO3 ferroelectric thin films studied by insitu photoemission spectroscopy." Material chemistry. Université Sciences et Technologies – Bordeaux I, 2009.

[8] S. Sheng, X.-Y. Zhang, P. Wang, C. K. Ong, "Effect of bottom electrodes on dielectric properties of high frequency Ba0.5Sr0.5TiO3 parallel plate varactor," Thin Film Solids, vol. 518, no. 10, pp. 2864-2866, March 2010.

[9] D. S. L. Pontes, R. A. Capeli, M. L. Garzim, F. M. Pontes, A. J. Chiquito, and E. Longo, "Structural, microstructural, optical and electrical properties of (Pb,Ba,Sr)TiO3 films growth on conductive $LaNiO_3$-coated $LaAO_3(100)$ and $Pt/Ti/SiO_2/Si$ substrates," Materials Letters, vol. 121, pp. 93-96, 2014.

[10] M.-S. Chen, T.-B. Wu, anf J.-M. Wu, "Effect of textured LaNiO3 electrode on the fatigue improvement of Pb(Zr0.53Ti0.47)O3 thin films," Applied Physics Letters, vol. 68, no. 10, 1996.

[11] A. Li, C. Ge, and P. Lu, "Preparation of perovskite conductive LaNiO3 films by metalorganic decomposition," Physical Review B, vol. 46, no. 10, pp. 6382, 1992.

[12] A. Li, et al, "Fabrication and electrical properties of sol-gel derived BaTiO3 films with metallic LaNiO3 electrode," Applied Physics Letters, vol. 70, no. 12, pp. 1616-1618, 1997.

[13] R. W. Schwartz, "Chemical Solution Deposition of Perovskite Thin Films," Chemistry of Materials, vol. 9, no. 11, pp. 2325-2340, 1997.

[14] H. Suzuki, T. Naoe, H. Miyazaki, T. Otac, "Deposition of highly oriented lanthanum nickel oxide thin film on silicon wafer by CSD," Journal of the European Ceramic Society, vol. 27, no. 13-15, pp. 3769-3773, 2007.

[15] T. Schumann, J. Neff, S. Breedlove, D. Look, K. Leedy, M. Allen, J. Allen, H. Zmuda, Y. K. Yoon, "Optical Transmittance and Reflectance of Lanthanum Nickelate at Telecommunication Frequencies," 2018 IEEE Research and Applications of Photonics In Defense Conference (RAPID), Miramar Beach, FL, USA, 2018.

[16] S. W. Boland, "Sol-gel synthesis of highly oriented lead barium titanate and lanthanum nickelate thin films for high strain sensor and actuator applications," Ph. D. Dissertation, California Institute of Technology, Pasadena, CA, USA, 2005.Bst 100

[17] A. Van der Drift, "Evolutionary selection, a principle governing growth orientation in vapour-deposited layers," Phillips Res. Rep., vol. 22, no. 3, pp. 267-288, 1967.

[18] B. Su, J. E. Holmes, C. Meggs, T. W. Button, "Dielectric and microwave properties of barium strontium titanate (BST) thick films on alumina substrates," Journal of the European Ceramic Society, vol. 23, no. 14, pp. 2699-2703, 2003.

[19] T. Horikawa, N. Mikami, T. Makita, J. Tanimura, M. Kataoka, K. Sato, and M. Nunoshita, "Dielectric Properties of (Ba, Sr)TiO_3 Thin Films Deposited by RF Sputtering," Japanese Journal of Applied Physics, vol. 32, no. 9S, pp. 4126, 1993.

[20] C. M. Chu and P. Lin, "Electrical properties and crystal structure of $(Ba,Sr)TiO_3$ films prepared at low temperatures on a $LaNiO_3$ electrode by radio-frequency magnetron sputtering," Applied Physics Letters, vol. 70, no. 2, pp. 249-251, 1997.

2019 IEEE 69th Electronic Components and Technology Conference (ECTC)

System Co-Design of a High Current (40A) Synchronous Step-Down Converter in an Innovative Multi-Chip Module (MCM) LQFN-type Packaging Technology

Todd Harrison[†], Jie Chen[‡], and Rajen Murugan[‡]

Texas Instruments Incorporated
[†]Cary, NC, USA
[‡]Dallas, TX, USA
r-murugan@ti.com

Abstract — The drive for multi-chip module (MCM) packaging technology essentially stems from the ever-increasing demand for miniaturization of power electronics. While promising, MCM packaging technologies present considerable design challenges (viz. electrical, thermal, reliability and manufacturing/assembly) if system co-design techniques are not adopted early in the design process. In this paper we present the electrical system co-design and measurement validation results of a high-efficiency, single channel, integrated FET, synchronous buck converter packaged in a 40-pin 7.00mm × 5.00mm MCM-in-LQFN-type innovative package. Due to the complex 3D level of integration of the monolithic control, drive circuitry, and the two discrete N-channel NexFET[TM] power MOSFETs, electromagnetic interactions, between die, package, and PCB, are exacerbated with potential impact to system-level performance. We detail here how optimization of the system, was achieved through a coupled circuit-to-electromagnetic co-design modeling and simulation methodology. Laboratory measurements on an integrated high current (40A) synchronous step-down converter are presented that validate the integrity of the co-design modeling and simulation methodology.

I. INTRODUCTION

Demand for the miniaturization of power electronics continues to drive the need for compact IC packaging without performance disruption. The innovative multi-chip module (MCM) in low-profile quad-flat no-lead (LQFN) package, i.e. PowerStack[TM], design supports multiple integrated circuits (ICs) with multiple benefits (viz. electrical, thermal, reliability, and lower cost) [1-2]. The PowerStack[TM] allows for stacking power MOSFETs in a "source-down" configuration which minimizes electrical parasitics and improves efficiency (Figure 1) [3-4]. The efficiency of high current, high frequency switching regulators is impacted by multiple power loss factors that are dependent on the system components. One critical factor, is the switching speeds of the MOSFETs in wide-V_{IN} source, synchronous step-down converters used to drive high current (e.g. 40A) load applications – e.g. in industrial, automotive, and communication infrastructure equipment. The fast-switching MOSFETs can experience significant voltage overshoots and ringing (i.e. spikes) on the switch or phase node [5-6]. The magnitude of the switch node ringing is a function of the MOSFET's switching speed, the magnitude of the inductor current I_L,

and the stray parasitic impedances in the MOSFET layout, package, and PCB system collectively (Figure 2). These high-voltage spikes, if not addressed early in the design process, will impact the system reliability and exacerbate EMI issues.

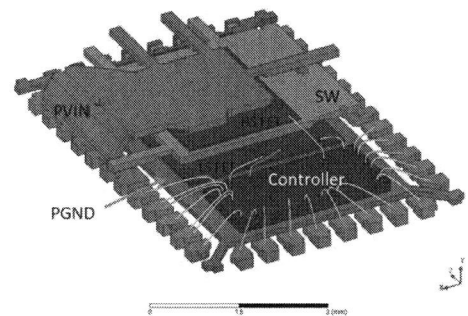

Figure 1. 3D view of the PowerStack MCM in LQFN package.

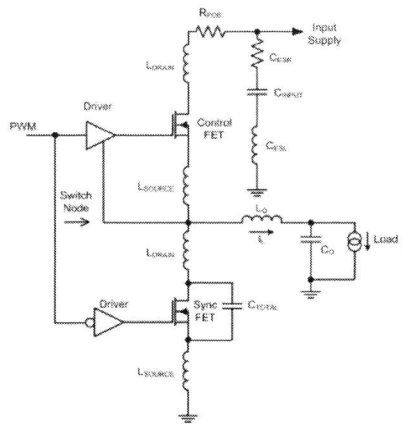

Figure 2. Schematic of device showing system parasitics contributions.

This paper focuses on the development and implementation of a co-design flow to quantify and optimize electrical parasitics of the system. After a brief overview of the device in Section II, the co-design methodology flow is detailed in Section III. System-level parasitics extraction and analysis of the power stage (drivers and FETs), package, and PCB are outlined in Section IV. Silicon measurements and system-level simulation correlation to measurement are covered in Section V and VI respectively.

978-1-7281-1500-9/19 $31.00 © 2019 IEEE   1653

II. DEVICE DESCRIPTIONS

The device is a high-efficiency, single channel, integrated FET, synchronous buck converter [7]. It is suitable for point-of-load applications with 40A or lower output current in storage, telecom, and similar digital applications. The device features a proprietary D-CAP3™ mode control combined with an adaptive, constant on-time architecture. This combination is ideal for building modern high/low duty ratio, ultra-fast load step response DC-DC converters. The device includes a power stage consisting of a half-bridge (control FET and sync FET) and their respective drivers (located on the controller IC in Figure 1), which involves the IC design portion of the co-design process.

One of the goals in the power stage design is to achieve the highest power supply efficiency while keeping the switch node ringing within component and system limitations. Power supply efficiency is defined as

$$\eta \equiv \frac{P_{OUT}}{P_{IN}} = \frac{(P_{IN}-P_{LOSS})}{P_{IN}}, \qquad (1)$$

where P_{IN} is the input power and P_{OUT} is the output power. One significant contributor to power loss (P_{LOSS}) in the converter IC is the switching loss. As shown in [8], switching losses are inversely proportional to the gate driver current.

$$P_{SW} \propto {1}/{I_g} \qquad (2)$$

The higher the gate current I_g, the faster the MOSFET will switch, thus leading to higher SW ringing. For example, the voltage across the parasitic inductance L_{DRAIN} in series with the drain of the control FET (Figure 2) is affected by the speed of the control FET turn-on.

$$V_{LDRAIN} = L_{DRAIN} \cdot \frac{dI_D}{dt}, \qquad (3)$$

where I_D is the drain current in the control FET. This voltage is added to the switch node spike and directly affects the maximum sync FET drain-source voltage V_{DS}, which must be limited to a voltage below its breakdown voltage. This component limitation must be carefully observed during the design.

III. EXTRACTION AND MODELING METHODOLOGY

To assess the electrical performance of the integrated solution, a coupled circuit-to-electromagnetic system-level extraction and analysis methodology was adopted [9-11]. Figure 3 below shows the high-level steps of the methodology. There are primarily two main components of the flow – physical and electrical co-designs respectively. The process is initiated by initially incorporate inputs from manufacturing/assembly and engineering/customer specifications and considerations. For the physical co-design component, layout and optimization of the power stage, package, and PCB designs are performed concurrently. As such the influence of the power stage, package, and PCB are comprehended early in the design cycle. Concurrent physical design is critical to bridging the gap between the different components of the system.

Once physical layout optimization and trade-offs have been performed for the power stage, package, and PCB, extraction algorithms and analyses are utilized for the electrical co-design. By extracting the appropriate parasitics of the components, system-level analysis can be performed to assess the performance to desired specifications.

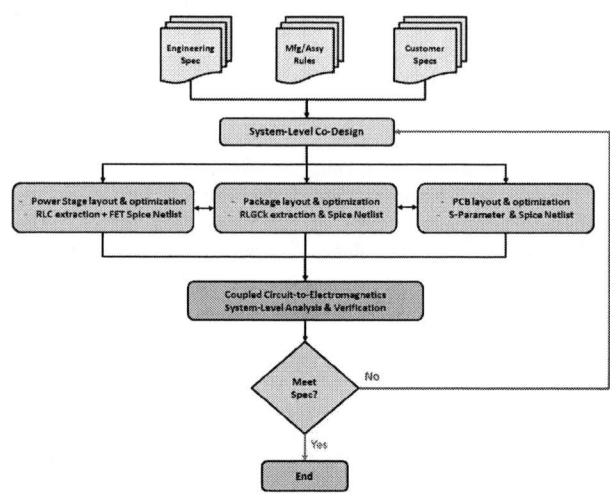

Figure 3. System-Level Co-Design and Modeling Methodology.

RLC/RLGCk extractions are performed for the MOSFET and package by using 3D quasi-static electromagnetic solvers respectively. Time-domain SPICE netlist are then generated. For the PCB, S-parameter extractions are performed using full-wave 3D frequency domain solvers. The S-parameter models are then converted, to time domain models, by using advanced broadband macromodeling algorithms for causality and passivity enforcements. Once these time domain models are generated, they are then assembled for coupled-circuit to electromagnetics system-level analysis. If the desired performance meets the desired specifications, then the flow is terminated. If not the process is re-iterated.

IV. SYSTEM-LEVEL MODELING & ANALYSIS

As mentioned in Section III, the power stage, package, and PCB are concurrently co-designed. Once the initial physical designs are available, the next step is the extraction, analysis, and SPICE netlist generation. In the next few sub-sections, the details of the extraction, analysis, and SPICE netlist generations are presented.

A. Power Stage Modeling and Analysis

The power stage is modeled in 3 basic partitions: controller IC, bond-wire interconnects, and power MOSFETs. The drivers which reside on the controller IC have SPICE models for each Silicon component. Post-layout IC parasitics (interconnect resistance and capacitance) are included in the netlist. The NexFET SPICE models also include additional modeling for accurate switching characteristics, such as the reverse recovery charge (Q_{RR}) [12], series gate resistance (R_G), and non-linear parasitic capacitances. Although it is possible to include the bond-wire parasitics in the extracted package model, it is sufficient to model these impedances as series resistor-inductor circuits.

The bond-wire parasitic impedance values were determined based on the length, diameter, and metallurgical composition of each of the bond-wires between the controller IC and the power MOSFETs. Both the control FET and the sync FET have similar connections with their respective drivers as shown in Figure 4 below.

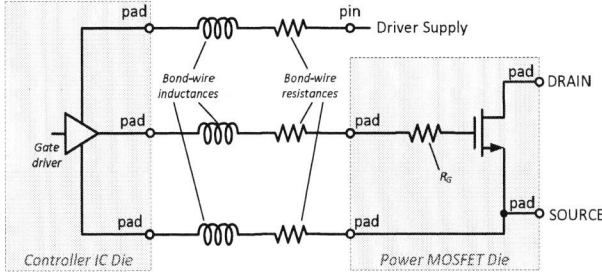

Figure 4. Half of the power stage model, showing bond-wire connections.

The bond-wire resistance was determined with a simplified calculation as:

$$R = \rho \cdot {}^{l}\!/_{A}, \qquad (4)$$

where ρ is the sheet resistance, l is the length, and A is the area of the bond-wire. The bond-wire self-inductance was approximated by

$$L_{bw} = 0.2 \cdot l \cdot \left[ln\left(4 \cdot {}^{l}\!/_{d}\right) - 1 \right], \qquad (5)$$

where d is the wire diameter and dimensions are in millimeters [13].

B. Package Modeling and Analysis

The device is packaged in a 7mm×5mm 40-pin MCM low-profile (LQFN-CLIP) as shown in Figure 5 below. Parasitics extraction is performed using a 3D quasi-static solver. The extraction process involved setting up ports at the junctures of the die-to-package and PCB-to-package. Once these ports are set-up with appropriate boundary conditions, extractions are done. The parasitics matrices for RLGCk (viz. resistance, inductance, conductance, capacitance, and mutual inductance and capacitances) are generated respectively. The model is then synthesized to generate a time-domain SPICE netlist representation of the model as shown in Figure 6. This time-domain model is then employed in the system-level circuit transient analysis.

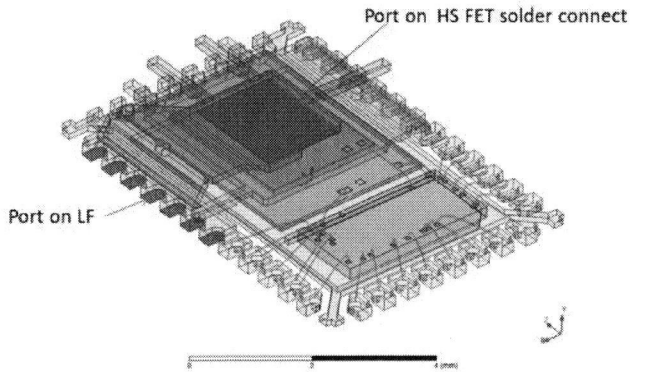

Figure 5. Ports set-up for package RLGCk extractions.

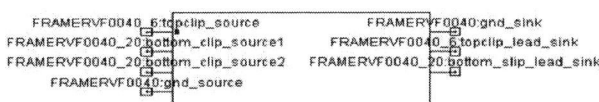

Figure 6. SPICE netlist representation of package RLGCK model.

In addition to the electrical parasitics extraction for transient analysis assessment of the switch (SW) node ringing, the impact of package and PCB direct current (DC) resistances on the efficiency performance of the DC-DC converters is critical to assess early in the design process. For the power MOSFET, $R_{DS(ON)}$ is the total resistance between the drain and the source during the ON state. $R_{DS(ON)}$ depends on the junction temperature and gate-source voltage (VGS). The power loss in any MOSFET is a combination of the conduction and switching losses. The conduction loss is shown in [8] as:

$$P_{cond} = R_{DS(ON)} \cdot I_{RMS}^2 \qquad (6)$$

As shown in equation (6), the practical approach to reduce conduction loss is to minimize the total resistance, $R_{DS(ON)}$. Typically, the primary components of $R_{DS(ON)}$ for a power MOSFET include the channel and electrical

978-1-7281-1500-9/19 $31.00 © 2019 IEEE

parasitics (viz. die metallization, package routing, and PCB interconnect) [14]. The DC resistance of the package is extracted by performing the volume integral of the three-dimensional 3D current density distribution. Figure 7 depicts the outcome of a current density analysis over the package. Assessing the current density distribution helps with identifying IR drop weakness in the package design.

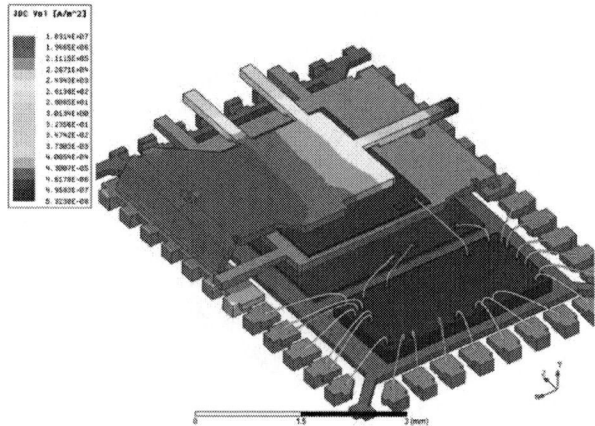

Figure 7. DC current density distribution through package interconnect.

C. PCB Modeling and Analysis

PCB layout design for a switching power supply IC is as important as the circuit design. Design guidelines for high current DC-DC converters are well covered in the literature [15-17]. In this work, the focus is on optimizing the power path loop inductance of the power distribution network (i.e. PDN) and associated passives, particularly the decoupling capacitors. It is critical to making sure that the total power path loop inductance is minimized to avoid high transient voltage ringing due to equation (7), where L is the total effective inductance of the power distribution network and di/dt is the rate of change of switching current of the DC-DC converter. It is evident from (7) that to minimize ringing, it is critical to minimize the total effective inductance.

$$V = L\ di/dt \qquad (7)$$

The concept of "loop inductance", equation (8), is a useful metric for quantifying the effectiveness of the decoupling capacitors of a power distribution network, PDN [18]. To calculate the "loop inductance" associated with the decoupling capacitor placement, equation (8) below was applied. To implement equation (8), the full PDN, which includes the power path routing and decoupling capacitors, was modeled using a full-wave 3D frequency domain solver. The solver generates frequency domain scattering parameter models of the PDN with accessible ports to integrate s-parameter models of the

decoupling capacitors. Figure 8 shows the modeling set-up for the PDN analysis with ports identified.

$$L_{eff} = \frac{imaginary\,(Z(power,gnd\ pads\ of\ decap))}{2*\pi*frequency} \qquad (8)$$

Where L_{eff} is the effective loop inductance, Z (power, gnd pads of decaps) represents the Z-parameters of the port defined across the power and ground pads of the corresponding decap.

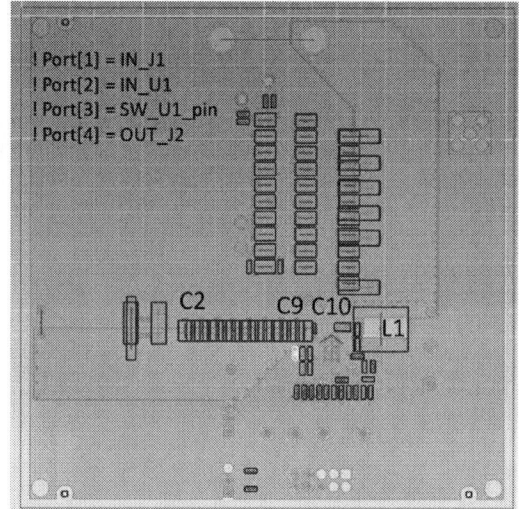

Figure 8. Ports set-up to extract S-parameter for PCB PDN.

Once the frequency domain s-parameter model is generated, they are converted over to time-domain model (Figure 9). The scattering parameter model is converted using advanced broadband macromodeling algorithms, while enforcing causality and passivity, to avoid oscillation/ringing in the final system-level transient analysis.

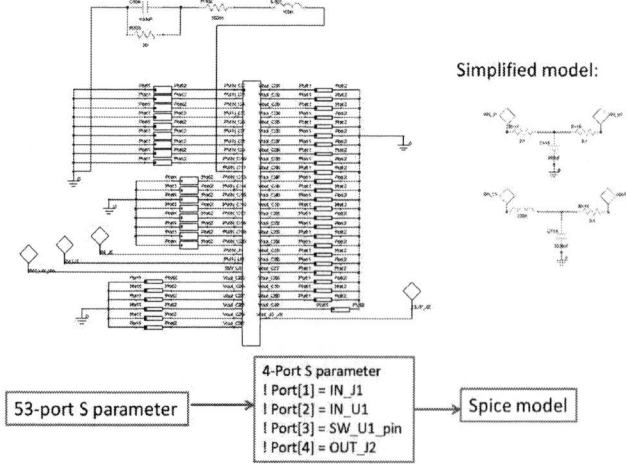

Figure 9. Conversion of S-parameter frequency model to time-domain.

In sub-sections IV.B and IV.C above, the package and PCB models were extracted separately, which were then later cascaded in the final system-level transient analysis. In previous work, the authors have demonstrated that the approach of separate extractions and cascading for circuit analysis of the package and PCB can lead to inaccuracies [19]. To verify if this phenomenon is inherent with this structure (package and PCB), the physical design of the package and PCB were merged and extracted as shown in Figure 10. The merged extracted model was compared with the cascaded models. The results were very similar for this structure. For this package and PCB combinations, both methods (viz. cascading/merging) seem to be viable options.

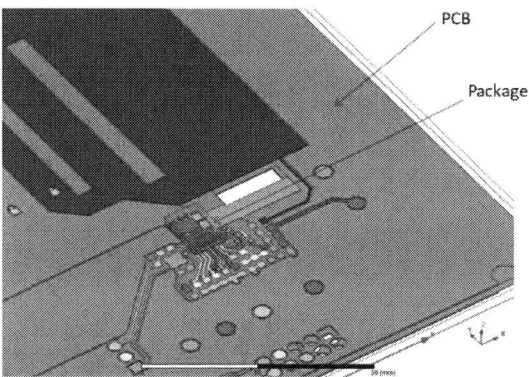

Figure 10. Package design merged to PCB design before extraction.

D. System (Power Stage + Package + PCB) Analysis

As per the extraction and analysis flow outlined in Section III, the final analysis step is the system-level transient analysis to assess ringing. With the package and PCB time domain models available, the system-level analysis involved importing and integrating these models in the system-level SPICE test bench. Figure 11 below shows the schematic of the system-level connectivity. The analysis includes the power stage SPICE model, package and PCB time domain models.

Due to the construction of this device having extremely low L_{SOURCE} and sync FET L_{DRAIN}, similarly described in [3] and [8], this section focuses on the control FET turn-on optimization (i.e. the switch node positive ringing spike with positive inductor current, I_L). Figure 12 shows the simulation result with $I_{OUT} = 40A$ and VIN=12V, where I_{OUT} is the converter output load current.

There are several key characteristics in the SW waveform of Figure 12. We are particularly interested in the magnitude of the positive spike because this is one of our design specifications to insure that the sync FET does not enter breakdown. If the models match the Silicon components and parasitic impedances well, then many parts of the SW waveform shape will match as well. The shape of the SW waveform before it starts rising can give

an indication of how well the driver timings and sync FET body diode is modeled. The dV/dt of SW while it is rising gives an indication of how well the control FET switching speed is modeled; the simulation predicts approximately 8 V/ns as seen in the waveform of Figure 12. The ringing frequency and decay provide a good indication of how well the parasitic impedances of the package and PCB are modeled. Figure 12 shows a ringing frequency of 133 MHz and a decay between the first 2 ringing peaks of 10.4V. The last component that we look carefully at is the DC resistance while the control FET is on and the ringing has stopped. This is to insure that the parasitic models are not skewing the DC behavior of the simulation.

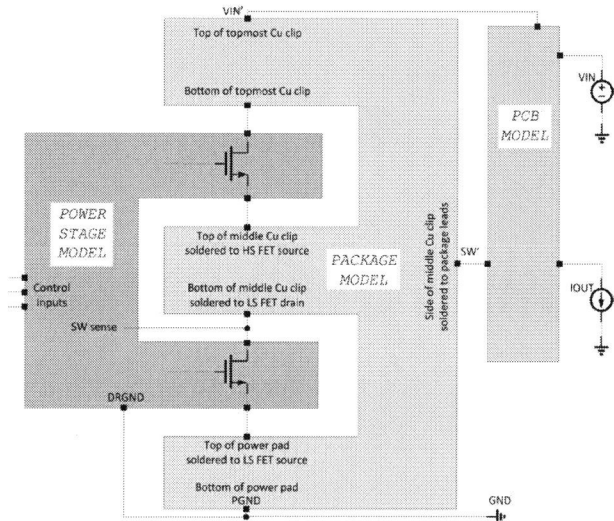

Figure 11. System-level set-up for transient analysis of SW ringing.

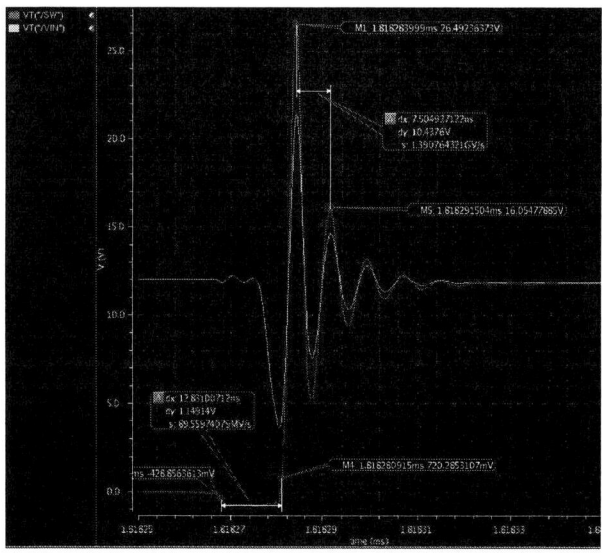

Figure 12. Transient simulation result for I_{OUT}=40A showing VIN and SW waveforms at the package pins.

V. SYSTEM MEASUREMENTS

Measurements were taken to verify the complete system model and to compare for accuracy. A Tektronix TDP1000 high-voltage differential probe was used on the evaluation module PCB as shown in Figure 13. Good measurement practices were followed as described in [20] to insure accurate readings. This insures that additional parasitic impedances and ground noise do not affect the reading. The measurement sense points for SW and PGND are the same as the ports placed at the package/PCB model interface.

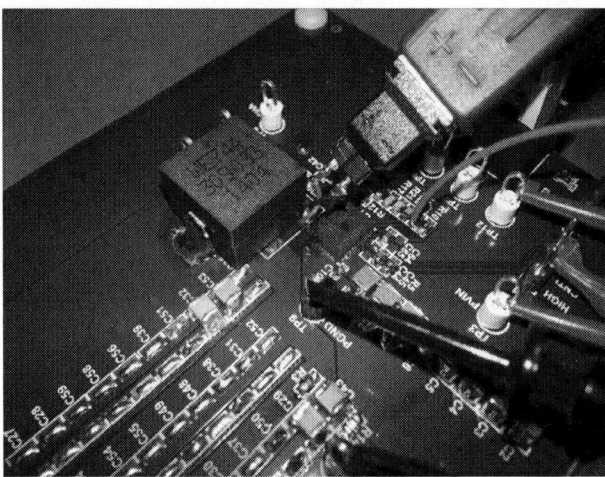

Figure 13. Measurement setup showing the differential probe, twisted wires, and sense points.

Figure 14 shows the oscilloscope capture for I_{OUT} = 40A and VIN=12V with persistence turned on. The observed SW rising edge has a positive SW spike that reaches 25.4V, or 13.4V above VIN. The ringing is characterized by a frequency of 100MHz and a decay, between the first and second pulse, of 9.8V.

Figure 14. Measurement of SW ringing.

VI. DISCUSSIONS

The simulation results are compared to the measurement results in TABLE I. The comparison shows reasonable errors that are good enough to allow design optimization before fabrication. The overall shape, amplitudes, and frequency measurements validate the use of this model in a sub-cell or complete DC-DC converter system simulation. However, further improvement would require a closer look at all partitions in the model. The ringing frequency is a good indication of the reactive impedances, which are affected by the control FET model, sync FET model, package model, and PCB model. While the decay is a good indication of the resistive and inductive portions of the control FET model, package model, and PCB model. Additionally, since the event is such high frequency, there is potential improvement in changing some ports from a single point to multiple distributed ports across the boundary of the model partitions.

TABLE I.   COMPARISON OF RESULTS

Characteristic	Simulation	Measurement
SW: +spike	Shape is very good 14.5V above VIN=12V	- 13.4V above VIN=12V
SW Shape: before rise	Looks like bench	-
SW dV/dt: rising	8 V/ns	9 V/ns
SW Shape: ringing frequency	133 MHz	100 MHz
SW Shape: decay between first 2 rings	10.4 V	9.8 V
SW Shape: after rise - $R_{DS(ON)}$	HS = 2.936 mΩ LS = 1.15 mΩ	HS = 2.9 mΩ LS = 1.2 mΩ

VII. CONCLUSIONS

The benefits of MCM package technology include the ability to achieve greater functionality and higher performance in a reduced time-to-market window that cannot be accomplished through direct silicon integration. Along with the many benefits of the MCM approach came new design challenges. As demonstrated in this work, it is critical to adopt system co-design modeling, analysis, and verification, early in the design process, to assess and minimize impact of the electromagnetic interactions that exist at system level. The successful implementation of the system co-design methodology was demonstrated for a high efficiency, single channel, integrated FET, synchronous buck converter which was packaged in a 40-pin 7.00mm × 5.00mm MCM-in-LQFN-type innovative package. Good correlation was observed between system-level transient SW node ringing simulations and laboratory characterization of the device. This methodology can be extended to assess and quantify current system-level mitigation techniques (viz. boot resistor, RC snubbers, ferrite beads, etc.) for switch node ringing impact on reliability and EMI.

978-1-7281-1500-9/19 $31.00 © 2019 IEEE

ACKNOWLEDGMENT

The authors would like to thank Gary Daum, John Duh, Tikno Harjono, and Vijay Krishnamurthy for their expertise in NexFET™ design and modeling.

REFERENCES

[1] R. R. Tummala, "Multichip packaging – a tutorial", Proceedings of the IEEE, Volume 80, Issue 12, pp. 1924-1941, December 1992.

[2] Sam Davis, "New breed of MCMs Optimize System Performance and Cost", Power Electronics Publications, October 1st, 2009, https://www.powerelectronics.com/power-management/new-breed-mcms-optimize-system-performance-and-cost.

[3] Matt Romig and Ozzie Lopez, "3D packaging advancements drive performance, power and density in power devices", July 2011, http://www.ti.com/lit/an/slit126/slit126.pdf.

[4] Sam Davis, "Innovative Packaging Yields 40A Synchronous Buck Converter IC", August 04/2016, Power Electronics Publications, https://www.powerelectronics.com/dc-dc-converters/innovative-packaging-yields-40a-synchronous-buck-converter-ic.

[5] Robert Taylor and Ryan Manack, "Controlling switch-node ringing in synchronous buck converters", Application Note – slyt465.pdf., http://www.ti.com/lit/an/slyt465/slyt465.pdf.

[6] Vijay Choudhary, "Controlling swtich-node ringing in DCDC converters", EDN, Power Management, February 21, 2016, https://www.edn.com/design/power-management/4441467/3/Controlling-switch-node-ringing-in-DC-DC-converters.

[7] TPS549D22 1.5-V to 16-V VIN, 4.5-V to 22-V VDD, 40-A SWIFT™ Synchronous Step-Down Converter with Full Differential Sense and PMBus™, Datasheet SLUSCI9A, Texas Instruments, Inc. 2016, http://www.ti.com/lit/ds/symlink/tps549d22.pdf

[8] David Jauregui, Bo Wang, and Rengang Chen, "Power Loss Calculation With Common Source Inductance Consideration for Synchronous Buck Converters", Texas Instruments, Inc., Application Report SLPA009A, June 2011, http://www.ti.com/lit/an/slpa009a/slpa009a.pdf.

[9] J. Xie et al., "Electrical-Thermal Co-analysis for Power Delivery Networks in 3D System Integration", 3DIC 2009. IEEE International Conference on, San Francisco, CA, 2009, pp. 1-4.

[10] Rajen Murugan, Nathan Ai, and C.T.Kao, "System-Level Electro-Thermal Analysis of RDS(ON) for Power MOSFET", IEEE SEMI-THERM33, March 13-17, 2017.

[11] Jie Chen, Rajen Murugan, Steven Kummerl, Usman Chaudhry, Edwin Lim, Tatsuhiro Shimizu, Thatcher Klumpp, and Jack Grantham (2018), "System ElectroThermal Co-Design of a Zero-Drift Current-Shunt Monitor with Precision Integrated Shunt Resistor", International Symposium on Microelectronics: Fall 2018, Vol. 2018, No. 1, pp. 000193-00019.

[12] K. J. Tseng and S. Pan, "Modified charge-control equation for more realistic simulation of power diode characteristics," Proceedings of Power Conversion Conference - PCC '97, Nagaoka, Japan, 1997, pp. 439-444 vol.1.

[13] Frederick W. Grover, Inductance Calculations : working formulas and tables, Mineola, New York, Dover Publications, Inc., 2009.

[14] J. Dodge, "Power MOSFET Tutorial", Application Note, APT-0403Rev. B, March 2, 2006. Available at: https://www.microsemi.com/document-portal/doc_download/14692-mosfet-tutorial

[15] Timothy Hegarty, "DC/DC converter PCB layout, Part 1-3", Power Management Design Center, How to Article, https://www.edn.com/design/power-management/4439695/DC-DC-converter-PCB-layout,June 15, 2015.

[16] Texas Instruments, Inc. Application note, SLVA773, Five Steps to a Good PCB Layout for a Boost converter", http://www.ti.com/lit/an/slva773/slva773.pdf, May 2015.

[17] Jim Perkins and Matthias Ulmann, "Reduced Size, Double-Sided Layout for High-Current DCDC Converters", Texas Instruments, Inc., Application Report SLVA963, http://www.ti.com/lit/an/slva963/slva963.pdf, April 2018.

[18] JAN 574: Printed Circuit Board (PCB) Power Delivery Network (PDN) Design Methodology, Altera, Application Note, AN-574-1.0, May 2009.

[19] Minhong Mi, Arlo Aude, Jie Chen, and Rajen Murugan, "SFF-8431 12.5Gbps Channel Return Loss (RL) Failure Debug: Simulation and Measurement Validation", 65th IEEE ECTC, May 2015.

[20] Minimizing Switching Ringing at TPS53355 and TPS53353 Family Devices, Application Report SLUA831A, Texas Instruments, Inc. 2017, http://www.ti.com/lit/an/slua831a/slua831a.pdf

Integrating Solid State Protection with a RF-MEMS Switch for Achieving ESD Robustness

Srivatsan Parthasarathy, Padraig Fitzgerald, Javier Salcedo, Ray Goggin and J-J Hajjar
Analog Devices, 804 Woburn Street, Wilmington, MA 01887, USA

Abstract— **The RF-MEMS (MicroElectroMechanical System) switch brings together the benefit of 0 Hz/DC precision and wideband RF performance in a small, surface-mountable form factor. The ESD (Electrostatic Discharge) sensitivity of this device can be a barrier for the adoption of RF-MEMS switches in certain applications, as sensitivity of this type of device to damage during assembly and end-application handling can be pervasive. This paper addresses this limitation in RF-MEMS applications, while preserving the benefits of the RF-MEMS performance, by integrating in a SIP (system-in-a-package) optimized RF solid-state ESD protection elements along with the RF switch. This is demonstrated with a device achieving insertion loss of -0.5 dB at 6 GHz with a -3 dB bandwidth of up to 13 GHz. Isolation performance of -20 dB is also demonstrated at 6 GHz. The integrated RF-MEMS SIP achieves industry leading ESD compliance levels of 3,000V HBM (Human Body Model) and 1,250V FICDM (Field-Induced Charged Device Model).**

I. INTRODUCTION

Microelectromechanical systems (MEMS) switches have long shown promise as a replacement for expensive and relatively large Electromechanical Relays or Solid-State Switches in radio frequency (RF) applications [1]. Fig.1 depicts the size difference between a MEMS switch and a solid-state relay. However, the key challenges for the adoption of the technology have been yield, reliability, manufacturability, and packaging. These design constraints are systematically addressed in this paper.

Figure 1: RFMEMS Switch vs Solid-State relay.

The MEMS switches in this work are assembled in standard lead frame plastic packages with wire bonds (Figure 2). This approach addresses limitations in the technology adoption and greatly reduces the development cost.

Figure 2: ASIC + MEMS switch die on leadframe with wirebonds.

Load boards for Automated Test Equipment (ATE) are key example applications in which using low power RF-MEMS switches as a replacement for bulky and power inefficient electromechanical relays is particularly beneficial. However, ESD robustness [2] is a barrier for the adoption of RF-MEMS switches in ATE, as electrical overstress induced by machine or human handling in such environments is common. To address this, the integration of optimized RF solid-state protection devices with RF-MEMS into a System-in-a-package (SIP) provides the necessary level of protection that the MEMS switch requires in these applications. In this paper, the design and manufacturability considerations in achieving this objective are presented. Key considerations in the development of this design solution included:

i) Maintaining the RF performance of the MEMS switch whilst achieving the required ESD robustness;

ii) Achieving high reliability and yield after integrating the protection as part of a system-in-a-package;

iii) Streamlining a low-cost manufacturing flow for high volume SIP production.

II. ARCHITECTURE AND DESIGN

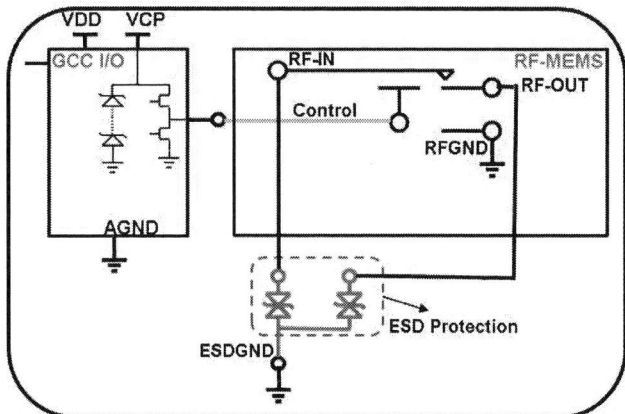

Figure 3: Block diagram showing a single input and output of the SP4T switch with the controller ASIC and ESD protection.

Fig. 3 depicts a high-level schematic of the RF-MEMS switch with the ESD protection architecture. The shunting element, inside the dotted box in the above figure, diverts the ESD charge away from the MEMS switch. The fundamental characteristics of this shunting element are: 1) non-conducting in the ±10 V range, 2) Minimal impact on the insertion loss of the switch at 6 GHz, and 3) robust to greater than 2,000V HBM. RFMEMS switches are typically built on very high resistivity substrates and the manufacturing process lacks the dopant diffusions necessary for developing effective RF-tolerant ESD protection networks. In this work an ESD protection device is built on a separate die. This approach is described in Fig. 4. The die is placed on the silicon insulating cap used for protecting the MEMS switches.

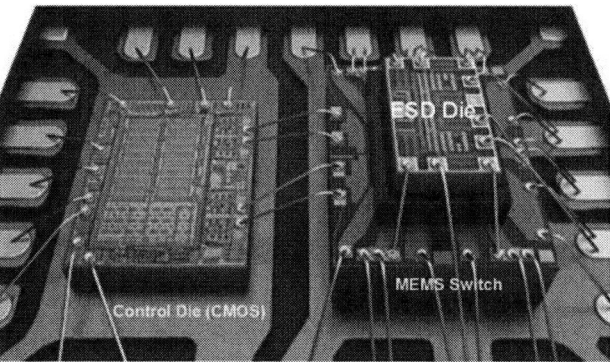

Figure 4: Fully integrated controller with RF-MEMS switch and optimized high voltage tolerant RF ESD protection.

The approach in Fig. 4 enables the integration of a separate "ESD Die", designed on a manufacturing process of choice, with a MEMS die. MEMS on CMOS can offer an even more integrated solution, [3] but there are other added manufacturing complexity constraints considerations in that option. Key among them is the low resistivity CMOS substrate that could severely degrade the RF performance. In this work, both the MEMS and ESD die are manufactured on

a high substrate resistivity process technology that minimizes RF signal loss to the substrate. In this approach the ESD die is wire-bonded to the RF ports that require ESD protection. It works efficiently to frequencies in the range of DC-10GHz.

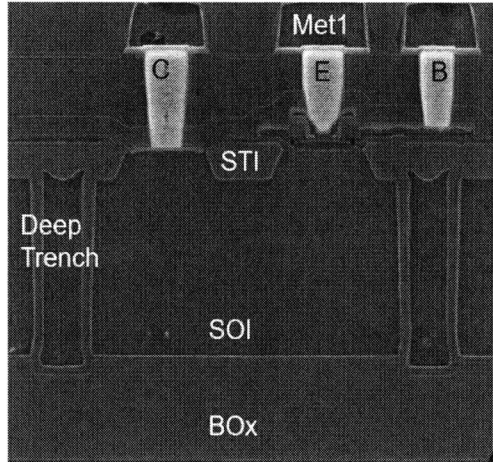

Figure 5: Silicon Process Technology Cross-section.

III. ESD DESIGN

The ESD protection cell is implemented on an Analog Devices proprietary 0.18 µm high-performance complementary SiGe BiCMOS process shown in Fig. 5. Deep oxide trenches and a buried SOI (silicon-on-insulator) substrate are used to isolate the devices from each other and from the substrate [4]. The ESD protection architecture is shown in the schematic in Fig 6. The RF signal swing specification utilized for this cell is ±20V. The ESD protection architecture chosen is a series combination of low capacitance diodes and SCR's (Silicon Controlled Rectifier).

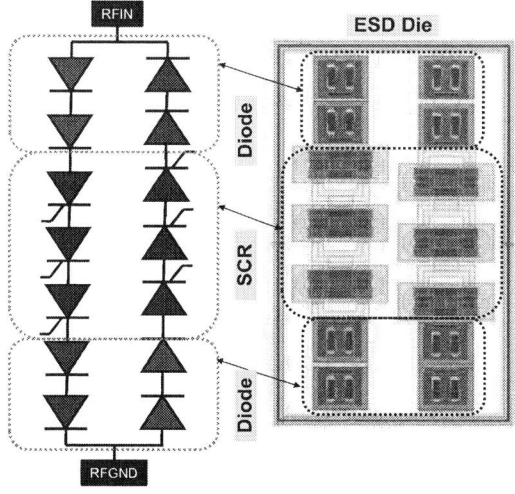

Figure 6: ESD protection schematic and top-level layout view.

The diode is fabricated using P+ and N+ diffusions into the epitaxial layer (EPI). Traditionally the focus of ESD diode development has been towards minimizing area/metallization. In this work, it was additionally observed that optimizing the junction can provide further benefits. Fig. 6 illustrates the complementary diode design. In a high resistivity RF BiCMOS process the intrinsic EPI has a very low doping concentration. A diode that uses such a layer will, in all practical purposes, behave like a PIN diode with very low capacitance. ESD diodes were traditionally not built in this manner due to the large series ON resistance. Additionally when subject to fast transient events initially there is very low conductivity in the EPI region. This results in a large forward voltage build-up (Overshoot). However, TCAD analysis indicated that during the high current ESD event the entire intrinsic EPI region becomes flooded with carriers that drastically reduces the resistance of the intrinsic region. This phenomenon is defined as conductivity modulation and enables the use of EPI layer in the diode. Apart from using the conductivity modulation phenomenon the diode overshoot is greatly minimized due to the use of N/P plug diffusions to contact the buried layers as shown in Fig. 7. The plug implants go much deeper and laterally out diffuse to a greater extent when compared to the NW/PW implants which are utilized in standard CMOS process. The lateral out diffusion of the plug implants beneath the STI enables the creation of a shorter path between the anode and cathode thereby reducing the off-state resistance which subsequently reduces the voltage overshoot.

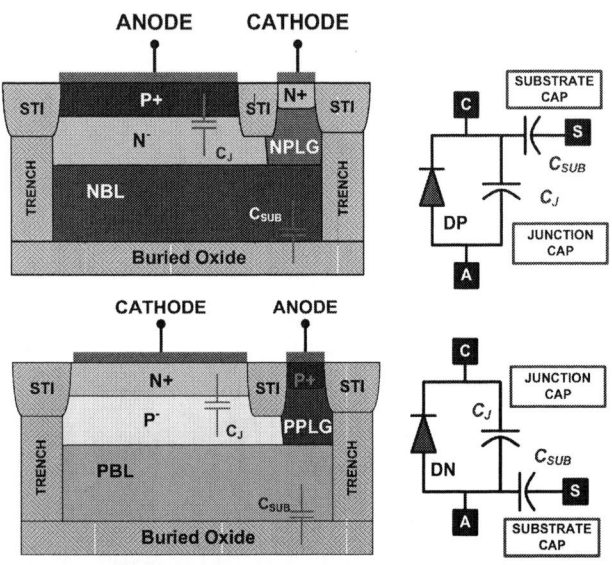

Figure 7: Diode architecture and schematic.

The second component of this ESD protection network is a poly-bounded Silicon Controlled Rectifier (SCR). The SCR allows for the large signal swing on the RF port while offering very low loading capacitance. Fig. 8 shows the cross-section associated with the SCR. An important feature of this SCR architecture is the introduction of a poly silicon layer in-lieu of STI between the anode P+ implant and N+ implant.

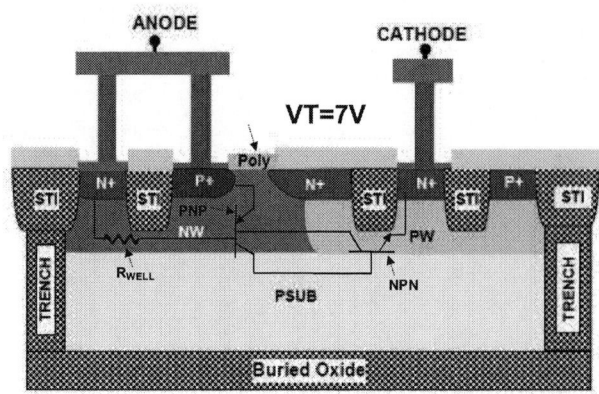

Figure 8: SCR cross-section with annotations [6].

The advantage of replacing the STI (Shallow Trench Isolation) region with a polysilicon layer is that it reduces the base width of the PNP (P+/NW/PW) and consequently enhances the SCR's turn-on time. The polysilicon layer additionally enhances the triggering action at the surface of the SCR which also helps to reduce the base transit time and improve the turn-on speed of the device. Notice that N+ and P+ implants have been alternatively utilized wherever the polysilicon gate is placed instead of the STI. This is to prevent the formation of MOS devices that could make the SCR leakage prone. Under this condition, the polysilicon gates are left floating to eliminate any kind of stress effects due to voltage over-shoots occurring on the supply lines without compromising the standing leakage. The presence of the STI results in a wider base width by forcing the trigger current to flow vertically into the SCR. The impact of poly gates is discussed in detail in reference [5].

IV. ESD DIE EXPERIMENTAL OBSERVATION

A. Quasi-Static Analysis

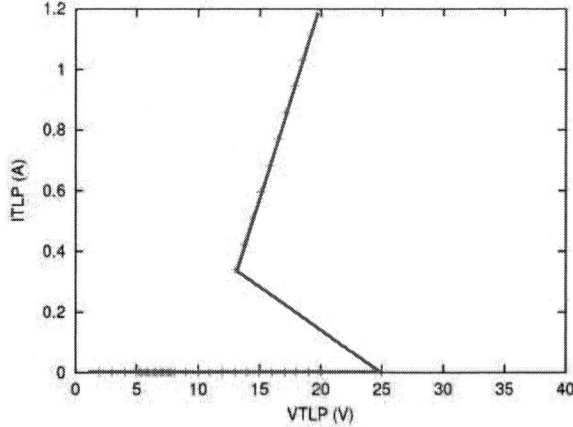

Figure 9: 100ns TLP characteristics.

To investigate the ESD robustness of the protection die under the standard HBM (Human Body Model) condition transmission line pulse (TLP) measurements are performed.

978-1-7281-1500-9/19 $31.00 © 2019 IEEE

The measurements are performed using a 100ns current pulse with a 0.2ns rise time. Post-stress leakage current is measured at 20V. Fig. 9 shows the 100ns Transmission Line Pulse (TLP) plot that depicts the clamping voltage of this cell. The clamping voltage across the ESD cell at 2,000V HBM stress levels which corresponds to ~1.2A TLP current is 20V. This is well below the MEMS structure breakdown voltage. The RFMEMS device utilized in this work has a 100V breakdown limit between the input and output ports.

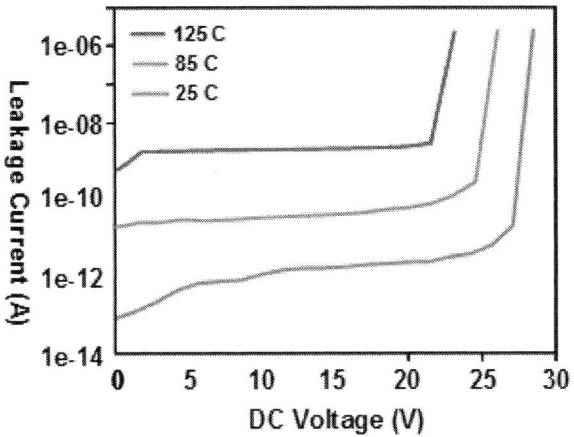

Figure 10: Measured DC leakage across temperature

B. DC Leakage Analysis

The DC current leakage performance is one of the additional benefits of this ESD device. Fig. 10 shows the DC leakage measured at various temperatures for a single ESD protection element as shown in figure 5. At 25C and 20V the leakage was measured to be less than 2pA and at 125C it was close to 1nA. It is observed additionally that with increases in temperature the breakdown voltage of the SCR decreases. The reason for this is attributed to the dominant breakdown mechanism, namely the collector to emitter breakdown (BVCEO) of the PNP. BJT BVCEO decreases with increasing temperature due to the enhanced current gain multiplication that causes breakdown. The DC leakage performance demonstrates the suitability of this protection die for low leakage analog and large signal swing RF I/O's.

C. Capacitance Analysis

The IO protection DC capacitance and linearity versus voltage is an important design parameter in large signal RF applications. Non-linearity is fundamentally introduced by having large reverse biased junctions formed between lightly doped implants. In order to analyze the overall capacitance of the ESD protection die in greater detail it is important to understand the capacitance contributions of the various junctions. In the SCR architecture the base of the NPN is left floating. This is illustrated in the schematic that is part of Fig. 11. This will introduce a capacitance C2 that will be in series

with the large capacitance C1 that is formed due to the NW/PW junction. Thus C-SCR is the series combination of C1 and C2, which is much smaller than C1.

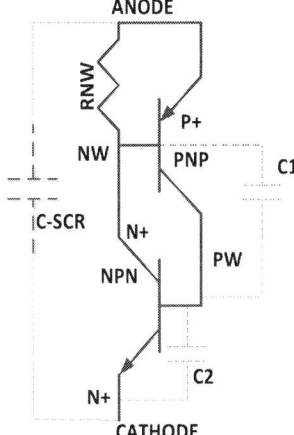

Figure 11: SCR schematic with capacitances.

Metallization layout is also a major contributor of ESD device's capacitance. It is also observed that the series combination of a forward biased and reverse biased junction ensures smaller variation in the capacitance of the device with voltage thus enabling higher linearity. This is explained by the fact that during a given DC voltage some portion of the potential appears across the forward biased junction which corresponds to a diffusion capacitance, and a major portion appears across the reverse biased junction which is depletion capacitance. Since the diffusion capacitance increases with voltage and depletion capacitance decreases with voltage a series combination of these two junctions will ensure a linear capacitance versus voltage characteristic. Fig. 12 illustrates the C-V characteristics of this network.

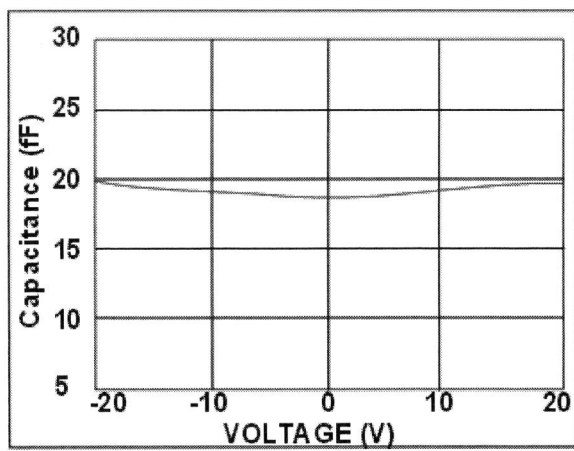

Figure 12: ESD network C-V characteristics.

978-1-7281-1500-9/19 $31.00 © 2019 IEEE

V. CHARACTERIZATION RESULTS

The maximum RF power handling capability of the MEMS switch after the addition of the ESD protection die is 32dBm, there is a 4dBm degradation when compared to the standalone MEMS switch whose maximum RF power handling is 36dBm. Detailed below is a summary of the RF characterization.

A. Insertion loss (IL)

IL is a key metric for the RF switch, is a measurement of the degree to which the switch introduces loss to the transmitted signal. Fig 13 is an IL performance comparison between a standalone MEMS switch and a MEMS switch with the integrated ESD protection devices. There is less than -0.3dB degradation in loss with the ESD protection integrated. Typically, the IL requirement for any such system is less than -1dB. This performance is enabled by the highly linear low capacitance ESD protection architecture.

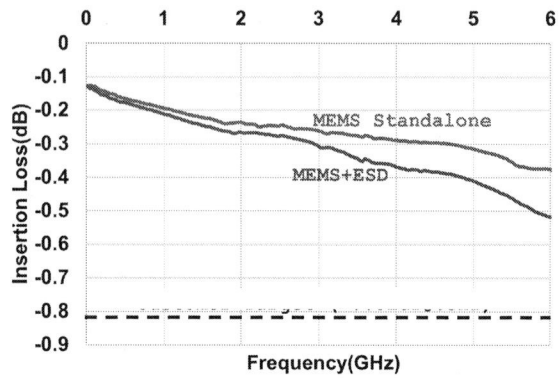

Figure 13: Insertion loss (S21) measurement.

B. Isolation (I_{ISO})

Multiple switches are used in any given system, requiring good isolation performance to 6 GHz. Isolation is the amount of signal transmitted between the input and output ports of the switch when the switch is in the off state. Fig 14 shows the isolation performance. The isolation performance shows very minimal degradation after the addition of the protection device.

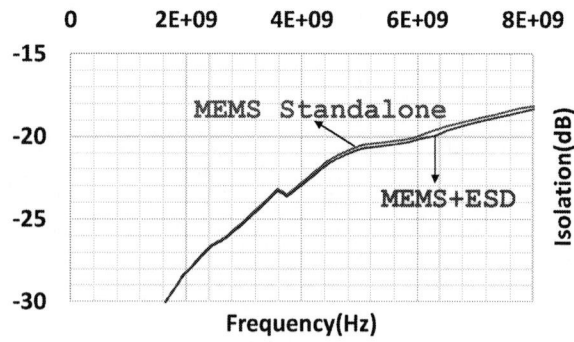

Figure 14: Isolation measurement.

C. Return Loss (RL)

Return loss (RL) is the amount of reflected signal relative to the incident signal. Fig. 15 illustrates the return loss performance showing -15dB return loss at 6GHz, with the protection device having minimal impact.

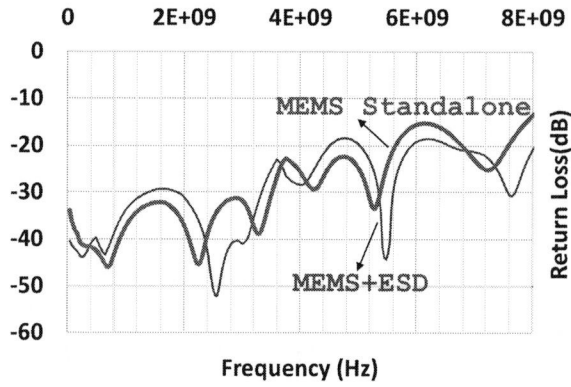

Figure 15: Return Loss measurement

D. Crosstalk (C_{TK})

Crosstalk (C_{TK}) is a measure of unwanted signal coupled through from one channel to another because of parasitic capacitance. This is an important metric since ESD protection is going to add additional capacitance between the RF lines. For the measurement, RF1 port is placed in the off position and RF2 port in the on position. All unused switches are in the off position and terminated with a load of 50 Ω. The crosstalk performance is ~-30dB at 6GHz and this is illustrated in Fig.16. There is no degradation after the addition of the protection device that is optimized to have a very low loading capacitance on the order of 20fF.

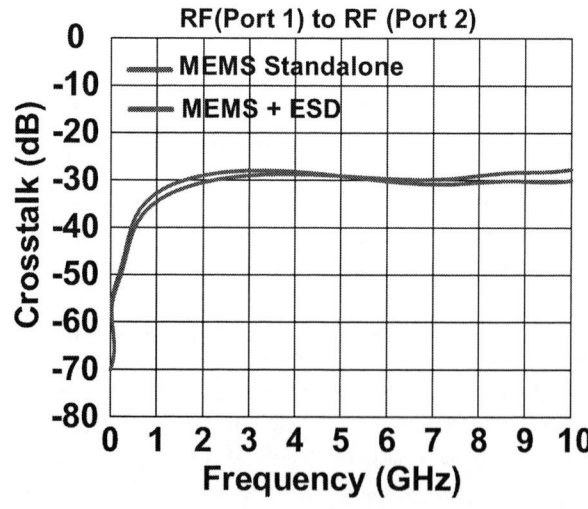

Figure 16: Crosstalk vs Frequency measurement

VI. CONCLUSION

A fully integrated high performance RF-MEMS switch with built-in control and solid-state ESD protection has been presented. This work has combined three proprietary lithographic technologies, with assembly and MEMS capping technology, to achieve breakthrough RF performance and ESD robustness. The MEMS switch enables simpler end application integration, and substantial reduction in solution footprint, while preserving excellent switch performance both in RF and DC applications. These included, achieving insertion loss of -0.5 dB at 6 GHz with a -3dB bandwidth of up to 13GHz, and isolation performance of -20dB at 6 GHz. The addition of the optimized solid-state protection die improved the ESD robustness of the MEMS switch from less than 100V sensitivity to the required >2,000V HBM compliance.

REFERENCES

[1] R. Goggin, P. Fitzgerald, J. Wong, B. Hecht and M. Schirmer, "Fully integrated, high yielding, high reliability DC contact MEMS switch technology & control IC in standard plastic packages," 2011 IEEE SENSORS Proceedings, 2011, pp. 958-961.

[2] A. Tazzoli, V. Peretti and G. Meneghesso, "Electrostatic Discharge and Cycling Effects on Ohmic and Capacitive RF-MEMS Switches," in IEEE Transactions on Device and Materials Reliability, pp. 429-437, Sept. 2007.

[3] S. Sangameswaran et al., "A SCR-based ESD protection for MEMS — Merits and challenges," EOS/ESD Symposium Proceedings, 2011, pp. 1-10.

[4] J.Steigerwald, P.Humphries, "TCAD assisted reflection on parameter extraction for compact modeling," IEEE Bipolar/BiCMOS Circuits and Technology Meeting (BCTM), pp.245, 2010.

[5] S. Parthasarathy, J. A. Salcedo and J. J. Hajjar, "A low leakage poly-gated SCR device for ESD protection in 65nm CMOS process," 2013 IEEE International Reliability Physics Symposium (IRPS), pp. EL.5.1-EL.5.5.

[6] J. A. Salcedo, S. Parthasarathy, Protection Devices for Precision mixed-Signal Electronic circuits and Methods of forming the Same, US Patent 9,362,265, June 7, 2016.

A Zero Height Small Size Low Cost RF Interconnect Substrate Technology For RF Front Ends For M.2 Modules And SiP

Sidharth Dalmia, Kirthika Nahalingam, Swathi Vijayakumar, Pouya Talebbeydokhti
Intel Communication and Devices Group (iCDG)
Intel Corporation
Santa Clara, CA USA
Sidharth.dalmia@intel.com, Kirthika.nahalingam@intel.com, Swathi.vijayakumar@intel.com, Pouya.talebbeydokhti.@intel.com

Abstract— a small size Radio Frequency Interconnect Substrate (RFIS) front end with 'zero height' and low cost has been designed for dual-band Wi-Fi filter solutions. In the conventional Wi-Fi front end solutions, discrete components are used, which are placed as other additional components externally to the System in Package (SiP) (or silicon) on the 1216 board. Conversely, in this new "zero height, low cost solution", the RFIS is placed in the SiP (System in Package) along with the Silicon. Since the RFIS can be integrated into the SiP, the device performance will be better in terms of linearity and power consumption. Also, it makes the manufacturing and assembly cost much lower, as it adds 'zero height' to the design and saves a lot of space on the board.

Keywords-component; passives, RF components, insertion loss, harmonics, front-end, M.2 Wi-Fi modules, SiP, substrate.

I. INTRODUCTION

In the past few years, the telecommunications industry has been experiencing accelerated developments in order to meet the growing demands for low cost, thinner and smaller form factor devices. Cellphone and personal computer manufacturers are striving to meet these demands to increase their market share and meet end user expectations.

One of the critical parts of consumer devices is the Radio-Frequency (RF) Front-end module (FEM). RF FEM components play an important role in the device performance, and hence its complexity has grown along with the growth of wireless technology. The need for smaller form factor devices has left very minimal room available on the physical device for FEM components. There have also been increasing requirements for bandwidth and higher throughput, as users would like to see higher internet upload and download speeds and faster data transfer, which are usually implemented with multi-band frequencies and M x N Multiple-Input Multiple-Output (MIMO) technology. But these features add components and circuitry to the module, thus increasing the size and cost, and also affects device power efficiency. This has always been a challenge to the design engineers since the evolution of wireless communication from 3G to 5G and 802.11a to 802.11ac. So,

the manufacturers have to forecast the future ahead and be proactive to be able to meet these market expectations.

Intel Corporation is one of the largest suppliers of 1216 and 2230 M.2 Wi-Fi cards for dual band 802.11ac and upcoming 802.11ax standards. 1216 measures about 12mm x 16mm in size and 2230 measures about 22mm x 33mm. They are connected to the M.2 interface in the personal computer's mother board for high speed internet. These modules support dual-band frequencies – 2.4GHz and 5GHz - and are lower profile, optimized for use with ultra-small form factor devices. More information on the 1216 and 2230 Wi-Fi modules from Intel can be found in [1].

As mentioned earlier, the RF front-end is a key component in providing powerful performance, while being required to be small in size and cost-effective. They comprise the functionality that exists between the antenna and baseband – Diplexers, Baluns, filters, and matching components. Conventional methodologies use discrete components for each of these functions, which leads to: (1) occupying more space on the board, (2) increased costs and (3) increases 'Z' height of the device. In this paper, the authors are presenting the RF Interconnect Substrate (RFIS) technology, a 'zero height' solution, that has been integrated for 2x2 WIFI 8021.11ac MIMO application, and demonstrating how it can be used in lieu of discrete components, thereby providing a solution to the aforementioned concerns.

Figure 1 shows a typical RF front end block diagram consisting of the diplexer, Baluns, filters and matching components required.

In simple terms, the RF front end consists of any circuitry that exists between the antenna and baseband portion. In the receiver chain, a diplexer splits the dual band signal from the antenna into 2.4GHz (low band) and 5GHz (high band), and passes them to the respective Band pass filters (BPF). The output signal from each BPF is converted to a double ended signal with a balun and then enters the silicon for baseband processing. This explains the basic architecture of an RF front-end Receiver chain and the indispensable purpose of each of these RF blocks.

978-1-7281-1500-9/19 $31.00 © 2019 IEEE

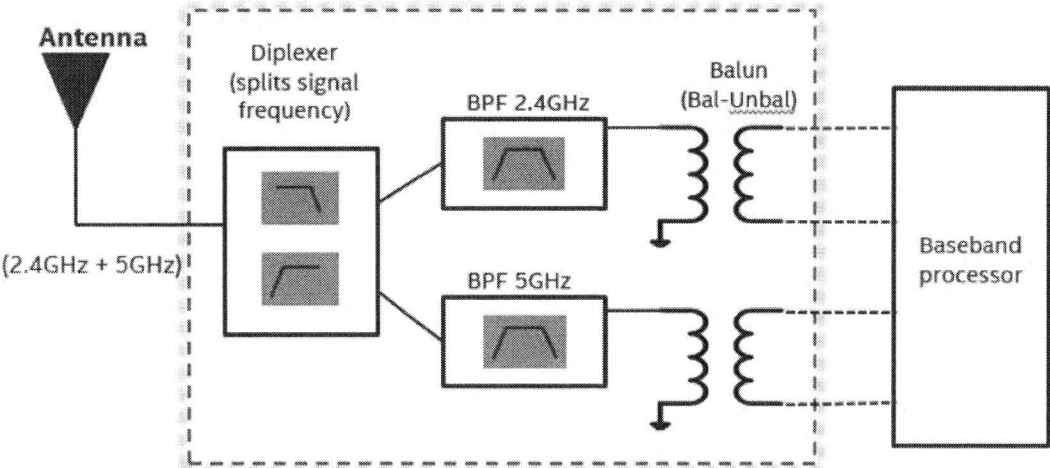

Figure 1. Block diagram of RF front-end Receiver

Figure 2 shows the Intel 1216 module with the RF front end comprised of the discrete components for two RF chains.

It is apparent that a 1216 module with two RF Wi-Fi chains or 2x2 MIMO needs at minimum six discrete components – 2 diplexers and 4 Baluns - and any other matching component would be additional. These components are placed on the 1216 module externally to the die package. It is obvious from Figure 2 that the components are occupying a lot of space on the 1216 board, and also add cost.

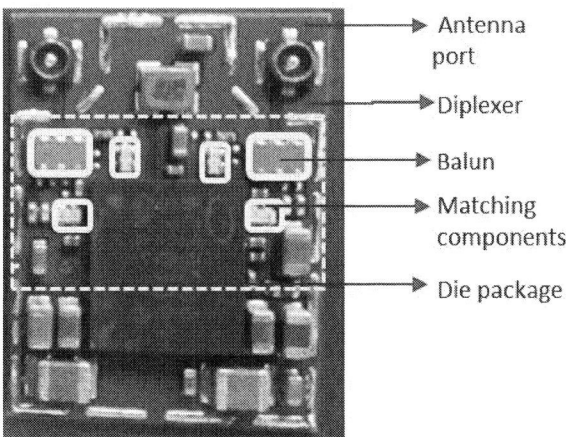

Figure 2. Intel 1216 module with discrete components

II. RF INTERCONNECT SUBSTRATE

RF Interconnect Substrate (RFIS) technology is a small size, low cost, 'zero height' solution designed for dual band Wi-Fi filter solutions. Figure 3 shows our fabricated RFIS design, which is a combination of inductors and capacitors printed directly on the substrate. The RFIS shown here measures about 6mm².

Figure 3. Fabricated RF Interconnect Substrate

RFIS, as discussed in this paper, is built using a 4 layer substrate. However, the design can also be done with 2 layer or 6 layer substrates, depending on the feasibility. A 2 layer package substrate would reduce cost and make the end unit much thinner. We have used 25um line spacing and 25um line width in certain areas of the design, and via sizes ranging from 60um to 70um, which helped make a highly dense passive design. At the frequency bands of interest, we have achieved Q's of about 50-100 and 100-300 for inductors (L) and capacitors (C) respectively.

RFIS was implemented using a low cost, low profile laminate substrate. Copper thicknesses for the substrate used were in the range of 8um to 24um. Figure 4 shows the substrate package stack-up with which the RFIS was designed. Material properties, and thickness of each dielectric and copper layer do play major roles in the design of RFIS.

A cross section of a portion of RFIS is shown in Figure 5. Cross sections of the substrates are done to verify the

978-1-7281-1500-9/19 $31.00 © 2019 IEEE   1667

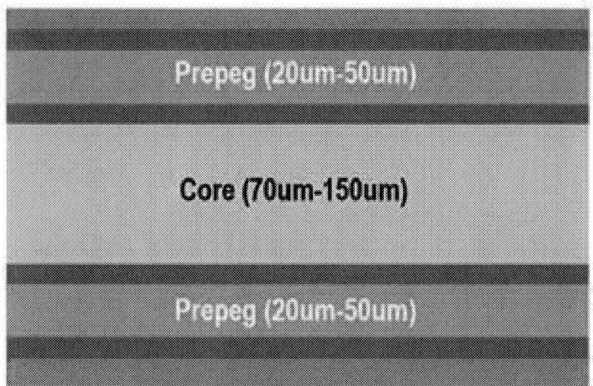

Figure 4.   4 layer Laminate stack-up

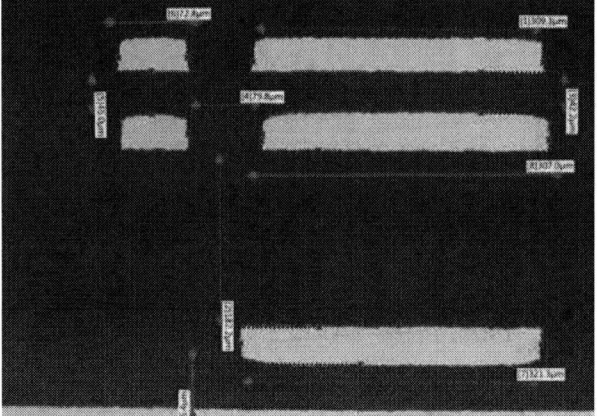

Figure 5.   Cross section of a portion of RFIS

targeted thickness of the Prepegs and core for which the RFIS was originally designed. A detailed study of correlation between cross section and measurement data was performed by including the cross section data into the stack-up.

The RF components described in Figures 1 and 2 are replaced by RFIS, and hence RFIS presents all of these functions cumulatively as a single device - thereby saving cost providing high system performance, low profile and light weight with no additional thickness. RFIS is a promising technology in terms of providing flexible, compact and cost-effective solutions for RF circuit design on System-In-Packages (SiP). In the past few years, a similar approach has been carried out using Liquid Crystal Polymer (LCP) [2] and Low Temperature Co-fired Ceramics (LTCC) [3] substrates.

This paper presents RFIS as the efficient alternative to the discrete components. RFIS is designed using the same stack-up as the substrate, integrated into the package, and therefore does not require a dedicated spot on the 1216. This can be clearly seen from the Figure 6, which shows a picture of SiP with RFIS integrated in the substrate for two RF chains. This makes the package size a bit bigger than nominal, but makes module size smaller. This is important, as we are driving toward a future where the end product size

is shrinking. The functions of the Baluns, filters and multiplexers are achieved using printed Ls and Cs. Consequently, this makes the device light-weight and low profile, and the cost of manufacturing will also become less. Intel has filed relevant patents, with another patent pending on a similar subject. Please refer to the patent applications US20190027432A1 [4] and WO201760280A1 [5] titled "Integrated Substrate Communication Frontend" for additional information.

The package size would be smaller still, if only one 1x1 RF chain is to be used. Conventional methods have been using substrates only for routing and die packaging. The fact that the package substrate can also be used to construct RF front ends makes it more valuable. The final package, including silicon die, printed front end balanced diplexers, and decoupling capacitors, allows further size reduction of M.2 1216 form factors. The arrows in Figure 7 show the space available after all the discrete components have been removed, and the RFIS performs the cumulative functions of the diplexers, Baluns and matching components.

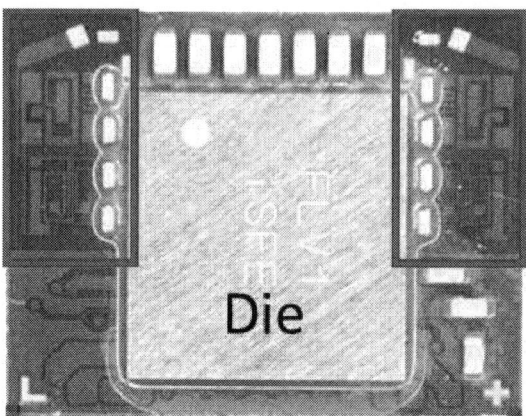

Figure 6.   Package with RFIS

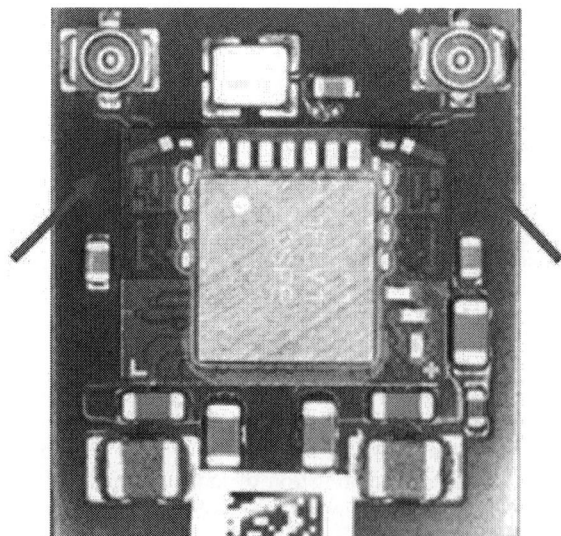

Figure 7.   1216 module with RFIS SiP

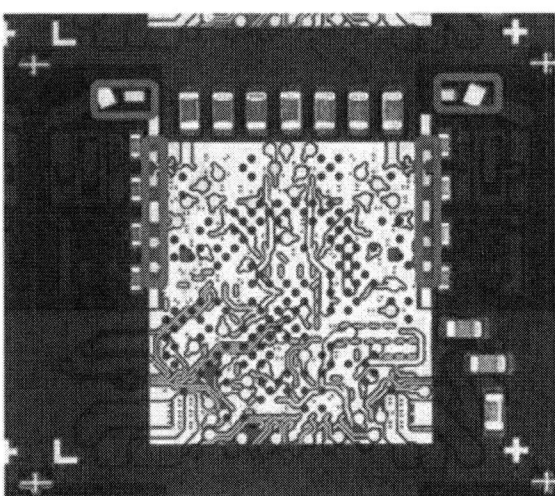

Figure 8. 1216 with RFIS showing the

The authors would like to point out another advantage of the RFIS. As seen in Figure 8, the ports are directly de-embedded to the silicon bump. De-embedding aids in removing any of the parasitics associated with connecting the front-end to the silicon bump. Since the RFIS output and input ports are directly on the bump, there is no inductance or capacitance effect added to the RF FEM.

III. ADVANTAGES OF RFIS OVER DISCRETE PASSIVES

The authors would like to point out the advantages of using RFIS compared to the discrete passives.

- Many conventional technologies use substrates and PCBs only for routing and interconnections. But RFIS is implemented on the substrate itself. So, a separate space or assembly for the front end is not required.
- They remove the need for diplexer, Baluns and harmonics filters as they have everything incorporated in the design. Due to this, several discrete RF components are removed and this results in assembly cost reduction.
- RFIS is printed during substrate assembly and hence there is no extra lead time required for manufacturing the front end.
- Any changes to front end can be dealt with at the package level without having to do expensive changes to the board.
- RFIS component adds zero height to the board. There is potential for 1216 M.2 module size reduction to a great extent in the future.

IV. RESULTS

S-parameter measurements on the RFIS were done using Cascade Microtech probe systems as shown in Figure 9.

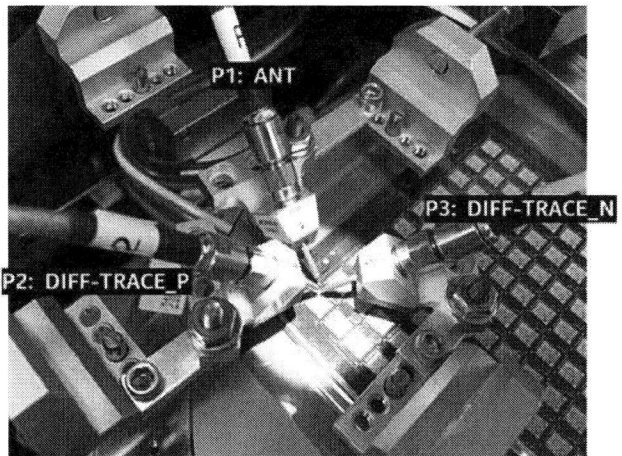

Figure 9. S-parameter measurement on Cascade Microtech

Figure 10 shows the measured results for RFIS on package substrate. Measurements were done over thousands of devices, and a typical result is shown here. Figures 10(a) and 10(b) show the low band 2.4GHz and high band 5GHz return loss and insertion loss plots. 2.4GHz has a -10dB bandwidth of about 80MHz. 5GHz return loss shows a bandwidth of about 750MHz from 5.1GHz to 5.85GHz. The insertion loss is about -1dB which is the most desirable compared to the discrete components. Achieving similar results would be unlikely with discrete components, as each of the diplexer, Baluns and filter components would present losses individually. RFIS presents -1dB insertion loss as the cumulative effect of all of these components together. We achieved harmonic rejections of approximately 25dB in the second and third harmonics of the 5GHz band, and the 2.4GHz frequency band has rejection of about -43dB in the second harmonic. Table 1 shows the low band and high band results summarized.

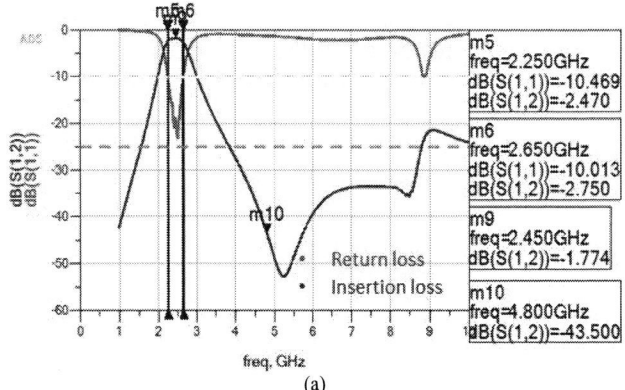

(a)

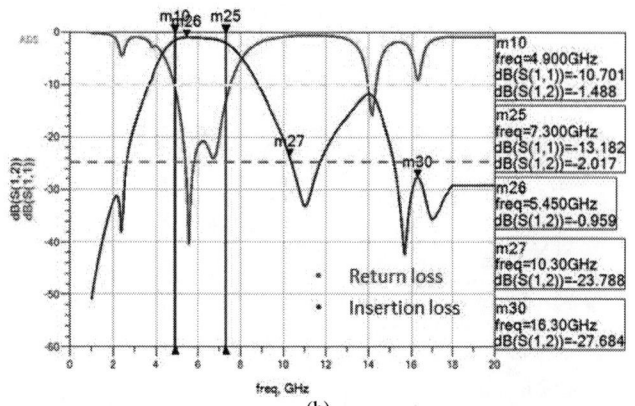

Figure 10. RFIS Measurement data on 1216 M.2
(a) 2.4 GHz   (b) 5 GHz

RFIS S-parameter Measured data		
Parameter	*Frequency (GHz)*	*Achieved*
Return Loss (-10dB) Bandwidth	2.4	2.4-2.48 GHz
	5	5.1-5.85 GHz
Insertion Loss	2.4	-1.9 dB
	5	-1 dB
Harmonic Rejection (2nd and 3rd harmonic)	2.4	-43 dB
	5	-31dB ,-28dB

Table 1. Summary of RFIS measured data in 2.4GHz & 5GHz

Figures 11 and 12 show the results for Manufacturing Validation Test (MVT) performed on the 1216 module. Transmit power (Tx) and Receiver (Rx) sensitivity tests are very important measurements, as they help understand the overall system performance. Transmit power measurement is done to give an idea of how much maximum power the device can offer when connected to the antenna. The greater the value, the better the user experience. The user can stay as far as away from the access points as desired, and still maintain a reliable communication. Figure 11 shows the comparison of Transmit power Error Vector Magnitude (EVM) measurements between RFIS (Figure 7) and discrete components (Figure 2) on a 1216 M.2 module. It is obvious that the RFIS measures 1-2dB better output than the discrete devices. This illustrates that the front-end has been matched well with the antenna. A 1dB betterment equals about 10% percent increased battery life for the device.

Receiver Sensitivity data is shown in Figure 12 as a comparison between RFIS and discrete on a 1216 module. Receiver Sensitivity is measured to determine the lowest possible signal the device can receive with lowering the Tx power, while maintaining an unimpeachable connection. This helps user download data as far as away from the access point as possible. It can be seen that the RFIS shows better packet error rate (PER) with lower power compared to the discretes, which means the RFIS has better receiver sensitivity results.

V. CONCLUSION

This paper outlines the implementation of front end integration of RFIS technology on laminate substrates for 1216 M.2 dual band Wi-Fi modules. The results have shown that the RFIS is the best possible candidate for dual band Wi-Fi RF front end when compared to the discrete components. RFIS has shown very low insertion loss, meets the required bandwidth specification with very good matching over the frequency range, better harmonic rejections, and desired Tx and Rx measurement data. Integrating the RF front end passives in the substrate shows cost saving, while maintaining its low-profile, and also has the potential to reduce 1216 M.2 module size by 20%-30%. This paves the way for availability of small size form factor devices in the future market. Although, the design has only been implemented for Wi-Fi frequencies, the design can be scaled for up to 65GHz on any application platform.

ACKNOWLEDGMENT

The authors would like to thank all the contributors from the Hardware team (Israel), especially Sharon, Eliav and Adi, and ATTD – Juan, Bill and Omkar - for their hard work and effort on successful execution of this project at Intel. We would like to express our appreciation to our managers Beth Keser and Paul Williams for their help and support.

REFERENCES

[1] https://www.intel.com/content/www/us/en/products/wireless/wireless-products.html

[2] Renbin Wu, Conrad Mmasi, Vinu Govind, Sidharth Dalmia, Camil Ghiu, and George White, "High Performance and Compact Balanced-Filter Design for WiMAX", IEEE/MTT-S International Microwave Symp., IEEE Press, June 2007, pp. 1619-1622.

[3] Li-Zheng Zhu, Xu-Bo Wei, Peng Wang, Song Ma, Zhi-Yi Zeng, and Bang-Chao Yang, "Compact LTCC Module for WLAN RF Front-End," 2011 International Conference on Computational Problem-Solving (ICCP), Oct. 2011, pp. 387-3892011 International Conference on Computational Problem-Solving (ICCP), Oct. 2011, pp. 387-389.

[4] Integrated Sustrate Communication Frontend, by S. Dalmia. (2016, Mar. 15). *US20190027432A1.* Accessed on: Aug. 13, 2018. [Online].Available: https://patentimages.storage.googleapis.com/e5/cc/f3/0af6fdbb6bc9ee/US20190027432A1.pdf

[5] Integrated Sustrate Communication Frontend with Balanced Diplexer, by S.Dalmia. (2016, Mar. 15). *WO201760280A1.* Accessed on: Sep. 21, 2017. [Online]. Available: https://patentimages.storage.googleapis.com/9a/50/2f/4af1bc9f851135/WO2017160280A1.pdf

978-1-7281-1500-9/19 $31.00 © 2019 IEEE

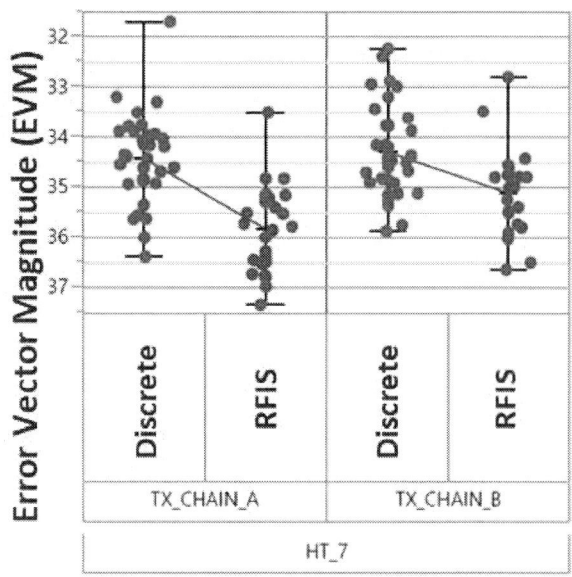

Figure 11. Comparison of Tx EVM between 1216 with discrete components and 1216 with RFIS

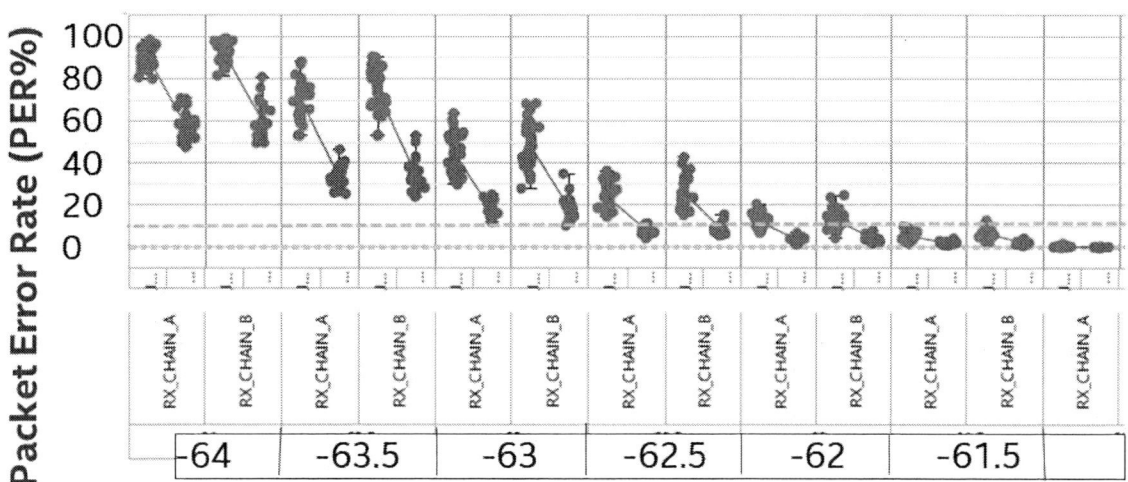

Figure 12. Comparison of Rx Sensitivity between 1216 with discrete components and 1216 with RFIS (blue – RFIS, red – discrete)

2019 IEEE 69th Electronic Components and Technology Conference (ECTC)

Open and Closed Loop Inductors for High-Efficiency System-on-Package Integrated Voltage Regulators

Claudio Alvarez *, Mohamed Bellaredj*, Madhavan Swaminathan*

*School of Electrical and Computer Engineering,

Georgia Institute of Technology, Atlanta, GA 30332, USA,

calvarez@gatech.edu

Abstract—Integrated Voltage Regulators (IVR) with integrated magnetics have been shown to provide better efficiency and performance as compared to conventional on-board solutions by reducing power delivery networks (PDN) parasitic, and DC and switching losses. However, IVR integration remains a challenging problem, with a major bottleneck being the integration of power inductors especially for high-efficiency and high-conversion ratio converters in applications such as personal computers and servers, where conversion ratios of 12:1V or 48:1V (with currents of 10A - 200A) are typical. In this paper, package-embedded open and closed loop inductors are designed for a 48:1V 2.5A single-phase converter (basic building block for multi-phase converters to support higher current levels) based on an analysis of the overall system efficiency, considering the simulated characteristics of the inductor dc resistance, ac resistance and inductance, and saturation current. We show that toroidal inductors outperform the solenoidal structures when they are embedded and surrounded by conducting planes, where the former has a higher inductance (32nH vs 16nH) and higher Q-factor (30 vs 21) at 10MHz, and similar DC resistance and saturation current between 5A - 8A.

I. INTRODUCTION

The increased demand in computational capacity and size miniaturization for emerging electronic devices are setting new levels of efficiency and thermal constraints on power delivery systems. Integrated Voltage Regulators (IVR) with integrated magnetics have been shown to provide better efficiency and performance as compared to conventional onboard multi-stage solutions by reducing power delivery networks (PDN) parasitic, and DC and switching losses [1]–[6]. However, IVR integration remains a challenging problem, with a major bottleneck being the integration of power inductors especially for high-efficiency and high-conversion ratio converters in applications such as personal computers and servers, where conversion ratios of 12:1V or 48:1V (with currents of 10A - 200A) are typical.

System on Package (SOP) solutions require converters with a switching frequency in the range of 10-100 MHz to reduce the inductance value and the inductor footprint to allow their integration closer to the switching devices [7]. Nevertheless, for input voltages around 48V the frequency has been kept below 2 MHz [8], [9] to obtain high-efficiency levels.

A better approach for inductor integration is by embedding it in the substrate, just underneath the active circuitry. This allows better usage of the available dielet volume as shown in figure 1. When the inductor is placed on-surface, its size needs to be as small as possible setting several constraints on the minimum value for DC and ac resistances, and inductance. However, when the inductor is embedded we just need to make it small enough so as to fit underneath the rest of the module [1], leaving more optimization space to lower the inductor losses.

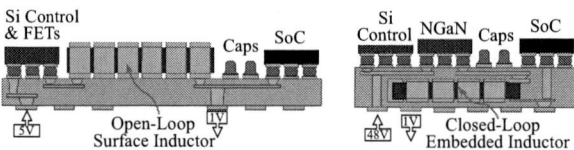

Figure 1: (left) Surface inductor versus (right) embedded inductors for high-density heterogeneous integration.

The inductor design depends on several parameters including, but not limited to, the IVR power stage topology. In this paper, substrate-embedded open and closed loop inductors are designed for a single-phase 48:1V 2.5A converter based on an analysis of the overall system efficiency, considering the simulated characteristics of the inductor dc resistance, ac resistance and inductance, and saturation current.

In [2], on surface solenoidal inductors with NiZn ferrite and CIP epoxy magnetic composite materials [10] have been designed with an inductance >25nH, low losses and high saturation current of 8A - 50A [11]. However, these inductors cannot be readily embedded due to their open loop structure, while embedded inductors are required to increase the components density in emerging 2.5D and 3D IC packaging platforms.

In this work, we show the impact of surrounding conducting surfaces, such as power planes, on package-embedded solenoidal inductors where the inductance can decrease by more than 35%. Then, we present a closed-loop toroidal design that provides higher inductance density with a minimal impact on inductance when embedded. We show that the DC resistance of both designs is similar, and the saturation

978-1-7281-1500-9/19 $31.00 © 2019 IEEE

current for the toroidal design is around 8A. In addition, we show that an air-gap typically used to increase the saturation current on toroidal inductors has minimal effect on low loss, high-frequency magnetic composite materials with moderate permeability, such as the ones used in [10].

II. IVR LOSS BREAKDOWN

The considered hard-switching buck converter topology is shown in figure 2. The duty cycle is calculated using

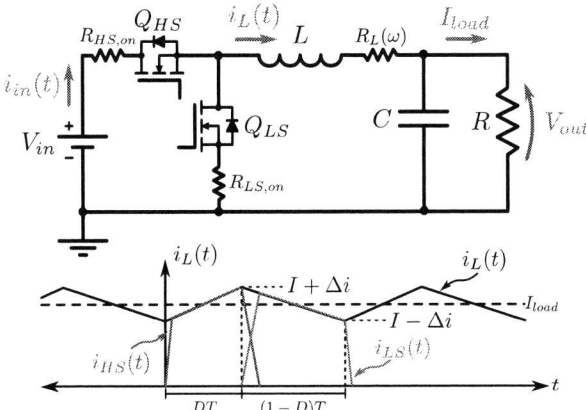

Figure 2: (*top*) Buck converter topology. (*bottom*) Current waveform.

equation (1), and the inductance required to operate the buck converter in Continuous Conduction Mode (CCM) is given by equation (2), where $R_{HS,on}$ and $R_{LS,on}$ are the high-side and low-side switch on-resistance, respectively, R_L is the inductor DC resistance, Δi is the inductor current ripple, D is the switching duty cycle, and f_s is the switching frequency.

$$D = \frac{V + I(R_L + R_{LS,on})}{V_g - I(R_{HS,on} - R_{LS,on})} \quad (1)$$

$$L = \frac{V_{in} - I_L(R_{HS,on} + R_L) - V}{2\,\Delta_i}\frac{D}{f_s} \quad (2)$$

By plotting the Inductance variation against f_s for three different levels of Δi, as shown in figure 3, it can be seen that at 10 MHz an inductance value less than 35nH is required if the current ripple is greater than 65% of I_{load}.

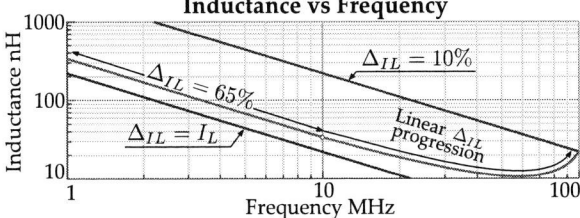

Figure 3: Buck converter required inductance vs frequency.

However, to determine the feasible working frequency the converter losses need to be evaluated over a large frequency range. The converter specifications that were used for this analysis are shown in Table I. With these parameters, the

Parameter	Value
V_{in}	48 V
V_{out}	1 V
I_{load}	2.5 A
f_s	0.5 - 5 MHz
L	338 nH @ 1.45 MHz
$R_{L,DC}$	15 mΩ
$R_{L,ac}$	186 mΩ @ 1.45 MHz
C	10 μF
R_C	10 mΩ
$C_{LS,oss}$	67.8pF
$R_{LS,on}$	34.3 mΩ
$C_{HS,oss}$	24.3pF
$R_{HS,on}$	85.8 mΩ

Table I: Buck converter parameters.

duty cycle D is 2.36% which allows the high-side switch to have a large $R_{DS,on}$ which is rendered to have small parasitic capacitance.

The main three types of losses are the *conduction losses*, *inductor ac losses*, and *FETs dynamic or switching losses*. The conduction losses P_{cond} are due to the DC component of the current and are produced mainly by the on-resistance of the low side FET $R_{LS,on}$ and the inductor's DC resistance $R_{L,DC}$. This loss component is calculated using equation (3).

$$P_{cond} = \left(I^2 + \frac{\Delta i_L^2}{3}\right)\left(D\,R_{HS,on} + (1-D)\,R_{LS,on}\right) \\ + I^2\,R_{L,DC} \quad (3)$$

The inductor *ac* losses P_{ac} are calculated using the Fourier expansion of the current ripple and result in the expression given in equation (4).

$$P_{ac} = \frac{2\Delta i^2}{D^2(1-D)^2}\sum_{n=1}^{N}\frac{sin^2(n\pi D)}{(n\pi)^4}R_{L,ac}(2n\pi f_s) \quad (4)$$

The Gallium-Nitride (GaN) FETs switching losses comprise [12]: gate charge losses P_G, turn-on and turn-off losses P_{sw}, output capacitance losses P_{oss}, and reverse conduction losses P_{SD}. For a GaN device in a buck converter operating with a input voltage $V_{in} = 48V$ the most significant losses are the conduction losses and the output capacitance losses, with the latter given by formula (5). All these loss sources are plotted as function of the switching frequency in figure 4.

$$P_{oss} = \int_0^{V_{in}} v_{ds}\,C_{ds}(v_{ds})\,dv_{ds} \cdot f_s \propto V_{in}^2 \cdot f_s \quad (5)$$

A SPICE simulation using LtSpice was used to validate the loss model developed in Matlab, where LtSpice gives

978-1-7281-1500-9/19 $31.00 © 2019 IEEE

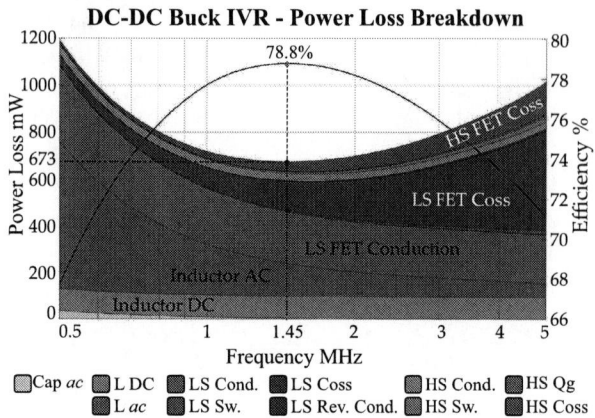

Figure 4: Buck converter power loss breakdown.

an efficiency of 78.89% and $V_{out} = 1.01V$ using a duty cycle $D = 2.45\%$. The common source inductance (CSI) was neglected but in [12] it is shown that its effect can reduce the overall efficiency by around 2% .

To design a low loss inductor we need to reduce their DC and ac resistance, which can be done by reducing the amount of inductance. To achieve this, the number of turns can be reduced and the tracks can be made wider. But for this work, the main requirements are to increase the inductance density and the switching frequency. However, we can see from figure 4 that the frequency is mainly limited by the FETs output capacitance losses P_{oss}, and for frequencies beyond 1.45Mhz the FETs losses start to dominate.

To operate at frequencies below 1.5 MHz, according to figure 3, we would need an inductance greater than 150nH, but from figure 4 it shows that we would need an inductance greater than 300nH to obtain some level of efficiency. However, such a high inductance is impractical for a package-embedded inductor, since it would lead to a large DC and ac resistance (due to the required high number of turns and magnetic material losses).

To operate at frequencies around 10MHz with an inductance below 35nH, soft-switching topologies need to be considered which also reduce the voltage stress and extends the duty cycle. This can be achieved by using a series-capacitor buck converter [13] and running the converter in triangular current mode (TCM) [14], where a negative current is allowed to flow in the inductor to enable Zero-Volt-Switching Clamped-Voltage (ZVS-CV). The design of a soft-switching topology will not be discussed in this paper.

III. INDUCTOR DESIGN

The inductor in this work was designed for an inductance around 35nH at 10 MHz, but with high DC and ac current capabilities using NiZn ferrite based magnetic composite material [10], [11]. In the next subsections, the inductance and saturation current modeling procedure is explained, with simulations results in the next sections.

A. Inductance

The magnetic material properties such as the relative permeability μ'_r and magnetic loss tangent δ_m as function of frequency are shown in figure 5, where the complex permeability is defined as follows,

$$\mu = \mu' - j\mu'' = \mu'\left(1 - j\frac{\mu''}{\mu'}\right) = \mu_0\mu'_r(1 - j\tan\delta_m) \quad (6)$$

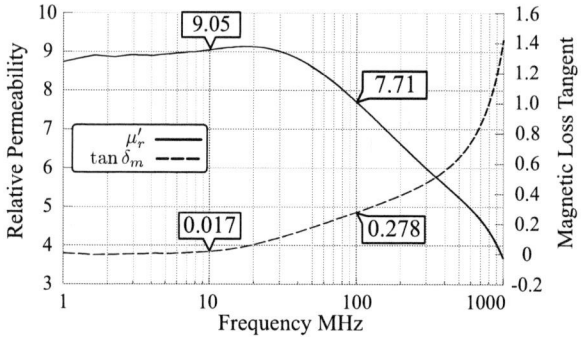

Figure 5: NiZn ferrite composite magnetic material properties [10].

Two inductor structures were designed: solenoids (open-loop) and toroids (closed-loop) as shown in figure 6. Also, each structure was simulated with a different number of turns.

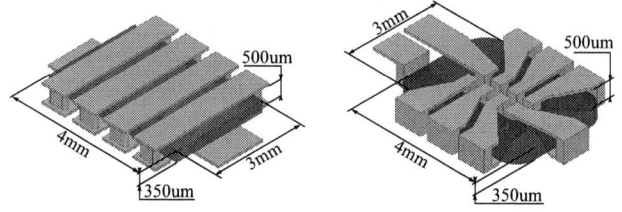

Figure 6: Solenoidal and toroidal inductor models.

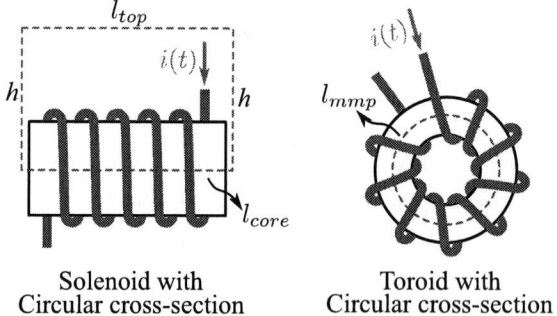

Figure 7: Ideal Solenoidal and toroidal inductors.

Assuming ideal solenoid and toroid structures, figure 7, the inductance can be found by solving for the flux $\phi(t)$ enclosed by the windings and can be approximated by,

$$\int_C \vec{H} \cdot dl = H\,l = N\,i(t)$$

$$\phi(t) = \mu_0 \mu_r' \int_S \vec{H} \cdot dA = \mu_0 \mu_r' \frac{N\, i(t)}{l} A$$

where N is the number of turns, l is the path for the contour integral which is l_{core} for the solenoid and l_{mmp} for the toroid, and the magnetic field intensity $\vec{H}$ was considered uniform over the integration path and surface. In the solenoid, we can always set the length h in the magnetic path arbitrarily large so that the contribution of $\vec{H} \cdot dl$ vanish, and only l_{core} contribute. The surface integral is over the cross-sectional area of each structure.

We can now express the developed voltage $v_L(t)$ across the inductor as function of the inductor current $i(t)$ as follows,

$$v_L(t) = N\frac{d\phi}{dt} = \mu_0 \mu_r' \frac{N^2 A}{l}\frac{di}{dt} \qquad (7)$$

to obtain the inductance given by equation (8).

$$L_{dsn} = \mu_0 \mu_e' \frac{N^2 A}{l} \kappa_L \qquad (8)$$

where μ_r' from equation (7) was replaced by μ_e' to adjust the permeability as the inductor is not completely filled with magnetic material. This parameter μ_e' can be used to estimate the usage factor of the magnetic material. The additional factor κ_L is a shape factor [15], [16] that corrects the inductance when the inductor structure is not close to the ideal.

The parameter μ_e' was approximated using,

$$\mu_e' = \frac{A_t - A_m}{A_t} + \frac{A_m}{A_t}\mu_r' \qquad (9)$$

where A_m is the crossectional area of the magnetic core and A_t is the total crossectional area of the inductor including track thickness and via width.

For the solenoid, the parameter κ_L was approximated by,

$$\kappa_L = \frac{length}{\sqrt{width^2 + length^2}} \qquad (10)$$

and for the toroid κ_L was approximated using,

$$\kappa_L = \frac{\pi\,\sqrt{width \cdot length}}{width + length} \qquad (11)$$

The approximated inductance values for both the solenoid and toroid are given in table II. The parameters μ_e' and κ_L are independent of the inductor number of turns.

	Solenoid			Toroid		
	3 Turns	4 Turns	5 Turns	6 Turns	8 Turns	10 Turns
L_{dsn}	16.8 nH	29.9 nH	46.8 nH	15.0 nH	26.6 nH	41.6 nH
L_{air}	3.8 nH	6.8 nH	10.7 nH	4.4 nH	7.8 nH	12.1 nH
μ_e'		4.38			3.43	
κ_L		0.80			1.55	

Table II: Inductor design parameters at $f_s = 10$MHz.

To extract the inductance from simulation (or measured data), the simple inductor model shown in figure 8 can

be used. The RLGC parameters are extracted from S-Parameters, which are converted to either Z or Y parameters, and are given by the formulae (12) to (13).

$$R = -Re\left\{\frac{1}{Y_{21}}\right\}, \quad G = Re\left\{\frac{1}{Z_{11} + Z_{21}}\right\} \qquad (12)$$

$$L = -Img\left\{\frac{1}{\omega Y_{21}}\right\}, C = Img\left\{\frac{1}{\omega(Z_{11} + Z_{21})}\right\} \qquad (13)$$

It was found that the calculated values for R,L,G,C are more accurate if R and L are computed using the Y-Parameters, and G and C computed from the Z-Parameters where Z and Y are obtained from S-parameters.

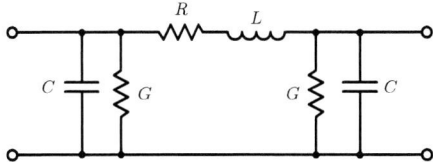

Figure 8: Inductor model for RLGC parameter extraction.

B. Saturation Current

For the saturation current, we used the BH characteristic curve for the NiZn ferrite composite material as shown in figure 9. This low-loss material has a very small hysteresis, and so only one magnetization curve is shown.

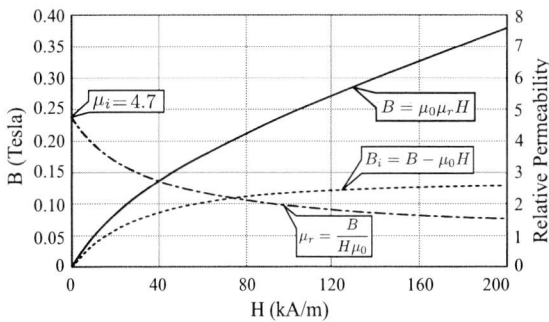

Figure 9: B-H and M-H curves for the NiZn ferrite based material.

The saturation current can be estimated by first determining the $\vec{H}$ field magnitude with a current of 1 A using equation (14),

$$H_{ref} = \left.\frac{N\, I}{l}\right|_{I=1A} \qquad (14)$$

At the point H_{ref} from the BH curve we obtain the reference permeability μ_{ref} from where we look for the permeability μ_{sat} that corresponds to the point where the permeability has dropped by a 10% from μ_{ref}. At this point in the curve we obtain the saturation flux density B_{sat}. Let define the effective permeability μ_e at saturation as follow,

$$\mu_e = \mu_{sat} \cdot \frac{\mu_e'}{\mu_r'} \qquad (15)$$

then, the saturation current can be approximated by the formula (16),

$$I_{sat} = H_{sat} \frac{l}{N \kappa_L} = \frac{B_{sat}}{\mu_0 \mu_e} \frac{l}{N \kappa_L} \quad (16)$$

To obtain a more accurate estimation of the saturation current, it was also calculated, by simulation, as the range where the inductance drops by 10% and 15% due to applied DC bias current. For this purpose, the inductance was calculated using equation (17) derived from the energy in an inductor and energy in a magnetic field.

$$L = \frac{1}{I^2} \iiint_V \vec{H} \cdot \vec{B} \, dv \quad (17)$$

The next correction to the permeability μ'_r was also considered. The permeability μ'_r vs frequency, for the magnetic material [10] considered in this work, is much higher than μ_r extracted from the BH curve. The next correction tries to combine both characteristics to obtain a more precise inductance in conditions where an ac and DC current is present.

The magnetic flux density B is related to H by the expression (18) (considering an isotropic material),

$$B = \mu_0 \mu'_r H = \mu_0 H + \mu_0 \chi_m(H) H \quad (18)$$

where $\chi_m(H)$ is a non-linear function of H. Let consider its Taylor expansion up to the second term as follow,

$$\chi_m(H_{DC} + \Delta h_{rms}) \approx \chi_m(H_{DC}) + \chi'_m(H_{DC})\Delta h \quad (19)$$

The term Δh is used to adjust the contribution of μ'_r, and is calculated using (20).

$$\Delta h = \frac{\mu'_r(\omega) - 1 - \chi_m(0)}{\chi'_m(0)} \quad (20)$$

The new corrected permeability μ_c is then given by,

$$\mu_c = 1 + \left(\chi_m(H_{DC}) + \chi'_m(H_{DC})\Delta h + k \right) \alpha \quad (21)$$

where the additional factors k and α are used to ensure that $\mu_c(H) \xrightarrow[H \to H_{sat}]{} \mu_r(H_{sat})$. With this new relation for μ_c a new range for the saturation current is obtained. The new BH curve is compared to the original one in figure 10, where we can see that the permeability is increased at low currents but maintains the saturation characteristics.

Finally, the approximated saturation current for both the solenoid and toroid are shown in table III.

	Isat$_{org}$	Isat$_{mod}$	Isat$_{org}$	Isat$_{mod}$	Isat$_{org}$	Isat$_{mod}$
	3 Turns		4 Turns		5 Turns	
Solenoid	13.6 A	12.1 A	10.9 A	9.7 A	9.3 A	8.3 A
	6 Turns		8 Turns		10 Turns	
Toroid	11.2 A	9.9 A	8.8 A	7.9 A	7.4 A	6.6 A

Table III: Saturation current approximation, using the original BH curve (Isat$_{org}$) and the modified BH curve from μ_c (Isat$_{mod}$).

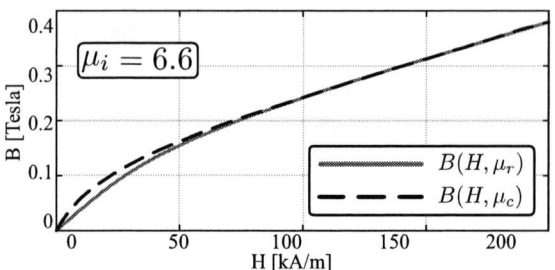

Figure 10: Original and Modified B-H curves for the NiZn ferrite composite material.

IV. SIMULATION RESULTS

A. Solenoid Inductance

Through simulation of the inductor structures, we obtained for the solenoid the inductance against frequency with 3, 4, and 5 turns as shown in figure 11. From the figure, we see that when the inductor is embedded (placed between two conducting planes) the inductance drops significantly. This also affects the quality factor Q, as shown in figure 12.

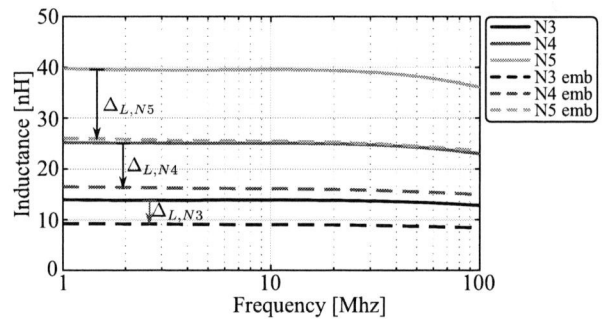

Figure 11: Solenoidal inductor inductance vs frequency.

Table IV summarizes the solenoid inductor characteristics, where L and L_{air} are the inductance with and without the magnetic material, respectively, and μ'_e and κ_L are the coefficients on equation (8).

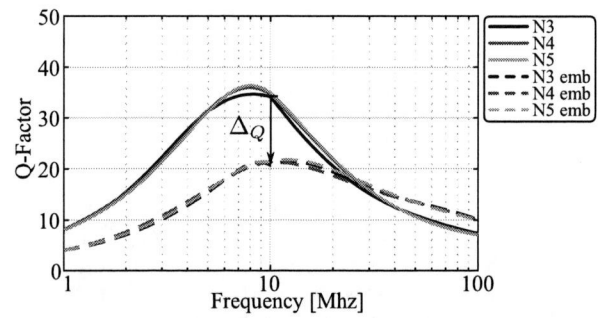

Figure 12: Solenoidal inductor Q-Factor vs frequency.

From the table, we see that both the inductance and Q-factor drop by more than 35% compared to the on-surface inductor values. This makes a solenoid structure

978-1-7281-1500-9/19 $31.00 © 2019 IEEE

unsuitable for high heterogeneous integration density, where the inductor must be placed embedded and underneath the active circuitry and power stage, as shown in figure 1. The inductance L_{dsn}, obtained with the relations in the previous section, compared to L was overestimated by around 17% mainly because the effective permeability, and the magnetic material usage, is less than expected.

| | 3 Turns | | 4 Turns | | 5 Turns | |
	Surf.	Emb.	Surf.	Emb.	Surf.	Emb.
L_{dsn}	16.8 nH		29.9 nH		46.8 nH	
L	13.9 nH	9.1 nH	25.1 nH	16.1 nH	39.6 nH	25.4 nH
L_{air}	4.5 nH	3.9 nH	7.9 nH	6.8 nH	12.3 nH	10.7 nH
μ_e'	3.12	2.34	3.18	2.37	3.22	2.38
κ_L	0.93	0.80	0.92	0.80	0.92	0.80
Δ_L	34.9%		35.8%		35.7%	
R_{DC}	5.3 mΩ		9.9 mΩ		16.8 mΩ	
R_{ac}	26 mΩ	27 mΩ	45 mΩ	47 mΩ	71 mΩ	75 mΩ
Q	34.2	20.7	34.7	21.3	35.0	21.4
Δ_Q	39.4%		38.6%		38.7%	

Table IV: Solenoidal inductor parameters at $f_s = 10$MHz.

In figure 13 both solenoid inductor scenarios – surface and embedded – are depicted, where with the plotted H-field it can be seen that the conduction planes reduce the field intensity, and decrease thus the inductance.

Surface
L = 25.1 nH @ 10Mhz
Q-Factor = 35 @ 10MHz

Embedded
L = 16.1 nH @ 10MHz
Q-Factor = 21 @ 10MHz

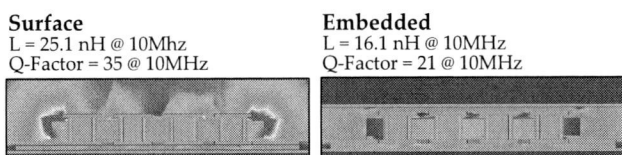

Figure 13: Surface and embedded solenoidal inductor H-Field.

B. Toroid Inductance

For the toroid, figure 14 shows the inductance against frequency for the inductor with 6, 8, and 10 turns. Note that the amount of copper is comparable to the solenoid structure, where each turn is cut by half. From the figure we see that when the inductor is embedded the inductance does not drop as significantly as compared to the solenoid, but also it is higher. The Q-factor is also not greatly affected, as is shown in figure 15.

Table V summarizes the toroidal inductor characteristics. The design inductance L_{dsn} was underestimated by 20%. From the parameter μ_e' we see that the magnetic material usage is higher than expected, and higher compared to the solenoid especially when it is embedded. In addition, from the table, we see that the DC characteristic is almost the same as the solenoid, but when it is embedded it outperforms the characteristics of the solenoidal structure.

In figure 16 both toroidal inductor scenarios – surface and embedded – are depicted, with the plotted H-field it can be seen that the conduction planes do not have a significant

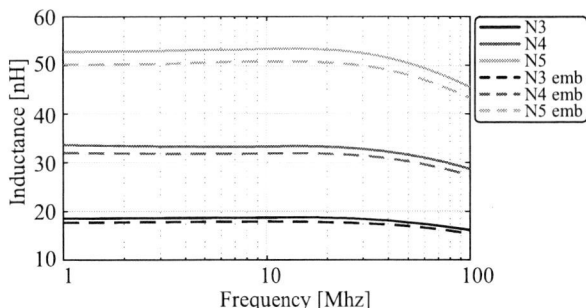

Figure 14: Toroidal inductor inductance vs frequency.

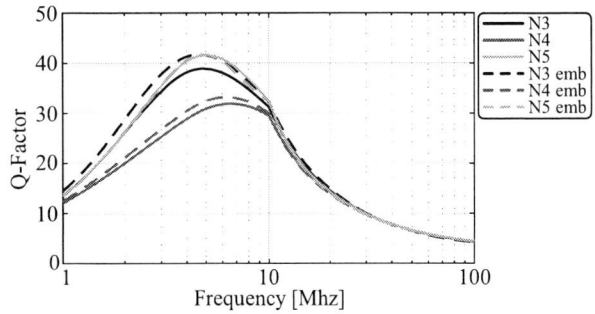

Figure 15: Toroidal inductor Q-factor vs frequency.

| | 6 Turns | | 8 Turns | | 10 Turns | |
	Surf.	Emb.	Surf.	Emb.	Surf.	Emb.
L_{dsn}	15.0 nH		26.6 nH		41.6 nH	
L	18.8 nH	17.9 nH	33.3 nH	31.9 nH	53.4 nH	50.8 nH
L_{air}	5.0 nH	4.4 nH	8.3 nH	7.6 nH	13.3 nH	12.3 nH
μ_e'	3.78	4.06	4.02	4.21	4.01	4.12
κ_L	1.76	1.57	1.66	1.52	1.70	1.58
Δ_L	4.393%		4.317%		4.821%	
R_{DC}	5.1 mΩ		9.7 mΩ		16.3 mΩ	
R_{ac}	38 mΩ	35 mΩ	71 mΩ	67 mΩ	103 mΩ	99 mΩ
Q	31.3	32.3	29.5	30.0	32.5	32.4
Δ_Q	3.229%		1.577%		0.234%	

Table V: Toroidal inductor parameters at $f_s = 10$MHz.

impact over the field intensity, as most of the energy is stored inside the toroid.

Surface
L = 33.3 nH @ 10MHz
Q-Factor = 30 @ 10MHz

Embedded
L = 31.9nH @ 10Mhz
Q-Factor = 30 @ 10MHz

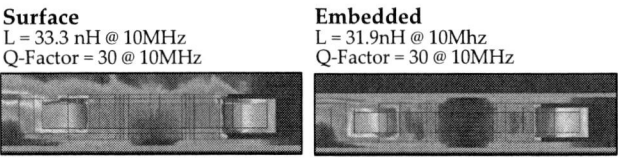

Figure 16: Surface and embedded toroidal inductor H-Field.

C. Saturation Current

With these two material definitions, with μ_r and μ_c, the simulated saturation current for the solenoidal and toroidal inductors are presented in tables VI and VII, respectively.

The Inductance vs DC bias current is shown in figure 17 and 18, respectively.

	3 Turns		4 Turns		5 Turns	
	μ_r	μ_{re}	μ_r	μ_{re}	μ_r	μ_{re}
L @ 1A	12.1 nH	13.3 nH	20.4 nH	22.7 nH	31.1 nH	34.7 nH
Isat 10%	12.9 A	16.5 A	10.4 A	12.0 A	9.1 A	9.7 A
Isat 15%	26.4 A	25.3 A	20.2 A	18.4 A	17.0 A	14.7 A

Table VI: Solenoidal inductor saturation current.

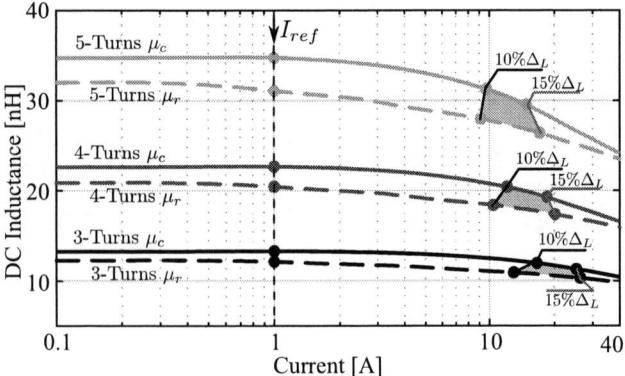

Figure 17: Solenoidal inductor saturation current vs DC bias current.

By comparison of table VI and table III we see that the predicted solenoid saturation current is very close to the simulation results for a 10% drop of inductance. Where for the toroid, by comparison of table VII and table III, the predicted saturation current lies in the range between 10% and 15% inductance drop region.

	6 Turns		8 Turns		10 Turns	
	μ_r	μ_{re}	μ_r	μ_{re}	μ_r	μ_{re}
L @ 1A	12.3 nH	14.5 nH	22.2 nH	26.3 nH	33.9 nH	40.5 nH
Isat 10%	5.7 A	6.5 A	5.2 A	5.2 A	4.6 A	4.3 A
Isat 15%	11.4 A	10.1 A	9.8 A	7.9 A	8.3 A	6.5 A

Table VII: Toroidal inductor saturation current.

Air-gaps are typically used in toroidal structures to increase the saturation current, as they tend to saturate at low currents. Figure 19 shows the results for three different types of magnetic cores (with and without air-gap) and 8 coil turns. For the NiZn ferrite compositve material used in this work, the air-gap has a minor effect on the saturation current, in contrast to other high-permeability bulk ferrite materials.

V. TOROIDAL INDUCTOR LOSSES

The inductor parameters and losses, when the embedded toroidal inductor with 8 turns is run at 10 MHz, are shown in table VIII. The parameter L_{ESR} is the power loss equivalent series resistance, $P_{L,DC}$ is the power loss due to inductor DC current, and $P_{L,ac}$ is the power loss due to inductor ac

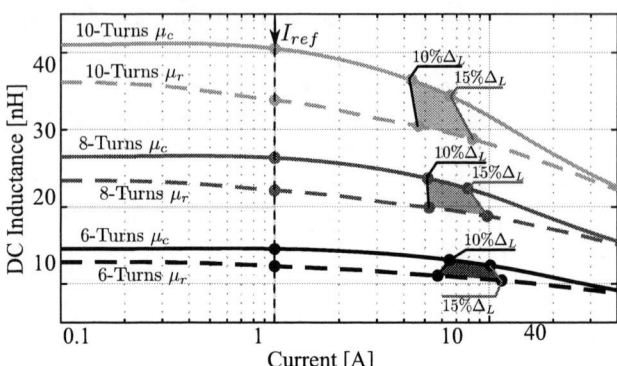

Figure 18: Toroidal inductor saturation current vs DC bias current.

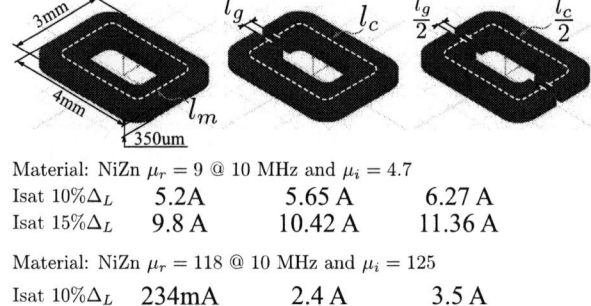

Material: NiZn $\mu_r = 9$ @ 10 MHz and $\mu_i = 4.7$
| Isat 10%Δ_L | 5.2A | 5.65 A | 6.27 A |
| Isat 15%Δ_L | 9.8 A | 10.42 A | 11.36 A |

Material: NiZn $\mu_r = 118$ @ 10 MHz and $\mu_i = 125$
| Isat 10%Δ_L | 234mA | 2.4 A | 3.5 A |

Figure 19: Toroidal inductor saturation current with air gaps.

Parameter	Value
Δi_L	1.708 A
$I_{L,RMS}$	2.687 A
L_{ESR}	75.9 mΩ
$P_{L,DC}$	61.9 mW
$P_{L,ac}$	486.6 mW

Table VIII: Toroidal inductor parameters and power loss at $f_s = 10$MHz and $I_{DC} = 2.5$A.

current. From the table we see that eventhough the inductor has very low ac resistance, the ac losses at 10 Mhz are too high due to the large current ripple (for a 80% efficiency, the maximum total converter losses is 500mW). Even with a inductor saturation current higher than 5A, the maximum current due to losses is limited by the inductor ac efficiency. This set a limit on the current ripple and thus the DC load current to operate in TCM to enable ZVS-CP transitions.

VI. CONCLUSION

In this work, a toroidal inductor design has been proposed with high Q, low DC and ac resistance, and saturation current between 5-8A, for the development of package-embedded inductors.

The effciency of a hard-switching topology was calculated to indentify the suitable working frequency, and what should be changed to operate at 10 MHz to allow the use of a low inductance inductor. This analysis identified the Output Capacitance Losses P_{oss} of the power FETs as the frequency limiting factor, and to evercome this a soft-switching topology using Triangular Current Mode (TCM) – where the current ripple is greater than the DC load current – was considered as an option to enable Zero Voltage Switching Clamped Voltage (ZVS-CV) transitions. In this context, two inductor structures were designed, with open (solenoid) and closed (toroidal) loop magnetic cores, considering a current ripple greater than 65% of I_{DC} running at 10 Mhz.

We showed that the solenoid inductance and Q-factor decrease by 35% when it is embedded, where the toroidal design is only affected by less than 5%. Also the toroid provides 50% more inductance, 33% higher Q-factor, with the same R_{DC}. This makes the toroidal design more suitable for heterogeneous package integration. However, eventhough the toroid inductor ac resistance is very low, only 67 mΩ, the power loss at 10 Mhz is 548 mW due to the high current ripple. In this application, the total loss budget for a 80% is 500mW and for a 90% efficiency is 250mW. This issue revert us back to rethink the power stage topology. More important, the inductor and power stage topology can't be designed as independent entities if high-efficiency at high-convertion-ratio wants to be achieved. The inductor design depends on several parameters including, but not limited to, the IVR power stage topology.

From the inductor structure side, there is more room for optimization to make the effective permeability (or magnetic material usage factor) $\mu'_e \approx 4$ be close to the magnetic permeability $\mu'_r \approx 9$. This would allow to increase the inductance density, reduce the number of copper turns, reduce the size, and decrease the DC and ac losses.

ACKNOWLEDGMENT

This work was supported by ASCENT, one of six centers in JUMP, a Semiconductor Research Corporation (SRC) program sponsored by DARPA.

REFERENCES

[1] Y. Su, "High frequency, high current 3d integrated point-of-load module," Ph.D. dissertation, Virginia Polytechnic Institute and State University, 2014.

[2] S. Mueller, K. Z. Ahmed, A. Singh, A. K. Davis, S. Mukhopadyay, M. Swaminathan, Y. Mano, Y. Wang, J. Wong, S. Bharathi, H. F. Moghadam, and D. Draper, "Design of high efficiency integrated voltage regulators with embedded magnetic core inductors," in *2016 IEEE 66th Electronic Components and Technology Conference (ECTC)*, May 2016, pp. 566–573.

[3] S. Mueller, A. K. Davis, M. L. F. Bellaredj, A. Singh, K. Z. Ahmed, M. Kar, S. Mukhopadhyay, P. A. Kohl, M. Swaminathan, Y. Wang, J. Wong, S. Bharathi, Y. Mano, A. Beece, B. Fasano, H. F. Moghadam, and D. Draper, "Modeling and design of system-in-package integrated voltage regulator with thermal effects," in *2016 IEEE 25th Conference on Electrical*

Performance Of Electronic Packaging And Systems (EPEPS), Oct 2016, pp. 65–68.

[4] E. A. Burton, G. Schrom, F. Paillet, J. Douglas, W. J. Lambert, K. Radhakrishnan, and M. J. Hill, "Fivr — fully integrated voltage regulators on 4th generation intel® core™ socs," in *2014 IEEE Applied Power Electronics Conference and Exposition - APEC 2014*, March 2014, pp. 432–439.

[5] W. J. Lambert, M. J. Hill, K. Radhakrishnan, L. Wojewoda, and A. E. Augustine, "Package inductors for intel fully integrated voltage regulators," *IEEE Transactions on Components, Packaging and Manufacturing Technology*, vol. 6, no. 1, pp. 3–11, Jan 2016.

[6] N. Sturcken, *Thin Film Inductors for Integrated Power Conversion*, March 2017 (accessed Dec 21, 2018), link.

[7] C. . Mathúna, N. Wang, S. Kulkarni, and S. Roy, "Review of integrated magnetics for power supply on chip (pwrsoc)," *IEEE Transactions on Power Electronics*, vol. 27, no. 11, pp. 4799–4816, Nov 2012.

[8] X. Zhao, C. Yeh, L. Zhang, J. Lai, and T. Labella, "A 2-mhz wide-input hybrid resonant converter with ultracompact planar coupled inductor for low-power integrated on-chip applications," *IEEE Transactions on Industry Applications*, vol. 54, no. 1, pp. 376–387, Jan 2018.

[9] EPC Co., *DrGaN plus Development Board EPC9201/3 Quick Start Guide*, August 2018 (accessed Dec. 21, 2018), link.

[10] M. L. F. Bellaredj, S. Mueller, A. K. Davis, P. Kohl, M. Swaminathan, and Y. Mano, "Fabrication, characterization and comparison of fr4-compatible composite magnetic materials for high efficiency integrated voltage regulators with embedded magnetic core micro-inductors," in *2017 IEEE 67th Electronic Components and Technology Conference (ECTC)*, May 2017, pp. 2008–2014.

[11] M. L. F. Bellaredj, S. Mueller, A. K. Davis, Y. Mano, P. A. Kohl, and M. Swaminathan, "Fabrication, characterization and comparison of composite magnetic materials for high efficiency integrated voltage regulators with embedded magnetic core micro-inductors," *Journal of Physics D: Applied Physics*, vol. 50, no. 45, p. 455001, 2017.

[12] A. Lidow, J. Strydom, M. de Rooij, and D. Reusch. John Wiley and Sons Ltd., 2015.

[13] P. S. Shenoy, *Introduction to the Series Capacitor Buck Converter*, April 2016 (accessed Jan., 2019), link.

[14] M. Ned, T. M. Undeland, and W. P. Robbins, *Power electronics Converters, Applications, and Design*, 3rd ed. John Wiley and Sons, Inc., 2002.

[15] H. A. Wheeler, "Inductance formulas for circular and square coils," *Proceedings of the IEEE*, vol. 70, no. 12, pp. 1449–1450, Dec 1982.

[16] H. Nagaoka, "The inductance coefficients of solenoids," *Journal of the College of Science, Imperial University, Tokyo, Japan*, vol. XXVII, article 6, 1909.

RF inductors integrated in organic packaging

Denis Mercier, Jean-Philippe Michel, Christine Raynaud, Christophe Billard
Univ. Grenoble Alpes, CEA, LETI,
38000 Grenoble, France
denis.mercier@cea.fr

Abstract— This paper presents RF inductors implemented in multilayer organic packaging substrates. The embedded inductors are designed using a 6 metal layer organic substrate based on a 2 build-up layers – 1 core layer – 2 build-up layers stack. The core and build-up layer materials have very low loss tangent at GHz frequencies and their thickness is chosen to assess two types of inductor structures: spiral and solenoid. The designed inductors present high quality factor with low foot print e.g. inductance of 7 nH with a 0.36 mm² footprint, a maximum quality factor of 28 at 1.5 GHz and a 5.8 GHz self-resonant frequency. Several inductors with different dimensions are presented and compared to both state of the art organic inductors and commercial inductors in term of electrical performances and footprint.

Keywords - Organic substrate; organic packaging; RF inductors; spiral inductors; solenoid; build-up layer; multilayer inductors;

I. INTRODUCTION

Any object connected to the cellular network contains several RF Front End Modules (RF FEM) which contain various functions (switches, power amplifier, filters, passive components) electrically connected by interconnections integrated into the organic packaging of the System in Package (SiP). The complexity and size of RF FEM in portable devices has increased with the number of frequency bandwidths and their miniaturization has become a strategic need. In RF FEM, a very large space of the organic packaging substrate remains dedicated to passive components (inductors, resistors, capacitors), mainly used for impedance matching. The passive components are most of the time Surface Mounted Devices (SMD) that are directly mounted on the surface of organic substrates next to silicon chips. To reduce size and cost of the RF FEM, an alternative solution to passive SMD consists in implementing some of the passive components directly into the organic substrate [1]-[16].

Organic packaging substrates used for SiP main advantages are: low manufacturing cost thanks to panel production, low RF loss of materials and thick copper metallization. Also, organic packaging allows interconnecting passive components with silicon chips which leads to high integration at low cost.

It should be remarked that lately, organic technologies have evolved a lot, not only in terms of materials but also in terms of design rules. It is now possible to fabricate conductors with line widths less than 10 microns and spacing also less than 10 microns. In addition, the use of thin dielectric layers (10 to 30 μm thick) leads to a diameter reduction of the vias (50 μm in diameter). All these improvements suggest that it is possible to replace some of the passive SMDs by components directly designed and fabricated thanks to the different layers of the organic packaging itself.

RF inductors are good candidates for integration in organic packaging as they require thick metallization and low loss dielectric materials to achieve high quality factors [9]-[14]. It is also worth mentioning that inductors in RF FEM have a large footprint and thickness and that the possibility to integrate them in the organic packaging of the RF FEM can potentially lead to important global size reduction of modules and to simplest and cheapest assembly processes

The present paper focuses on RF inductors embedded in organic packaging substrate. Inductors implemented on organic substrate using materials with very low loss tangent are presented. The substrate stack is detailed as well as the inductor structures. Several inductors with different dimensions have been designed using electromagnetic simulations which are compared to RF measurements. The inductors main parameters such as inductance, quality factor, self-resonant-frequency or footprint, are given and compared to state of the art organic inductors and also with commercial inductors. A comparison table is also provided to put in evidence the achievable electrical performances as a function of the footprint.

II. ORGANIC PACKAGING

The inductors have been developed using the 6 metal layer organic packaging substrate presented in Fig. 1. The stack is composed of a 400 μm thick core layer (E705G [17]) with 20 μm thick copper metallization on both sides. Two build-up layers (GL102 [18]) with a 30 μm thickness are stacked on each side of the core layer with their 15 μm thick copper metallization. Both sides of the organic substrate are protected using a 23 μm thick solder mask (PRS4000 [19]). The solder mask is opened to allow electrical contact with the copper layers and covered with Ni/Au metallization (≈3/0.1 μm thick). The core layer material has a 4.3 relative permittivity and a loss tangent of 0.008 up to 10 GHz. The build-up layer permittivity is 3.2 and its loss tangent is 0.0044. The permittivity and loss tangent of the solder mask

978-1-7281-1500-9/19 $31.00 © 2019 IEEE

are respectively 3.8 and 0.026. This organic stack has been chosen in order to assess on the same substrate both inductors based on multilayer spiral and solenoid through a core layer. It can also be noted that RF FEM are fabricated with organic substrates using the same materials properties even if the stack may vary depending on the applications and specifications targeted (e.g. coreless with N build-up layers, N core layers with N build-up layers, N core layers …).

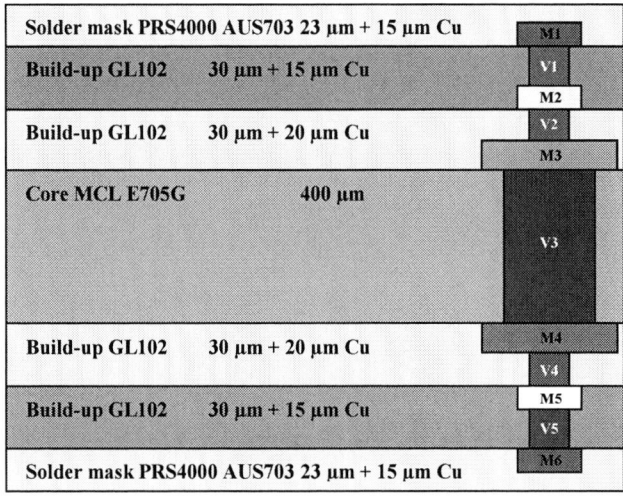

Figure 1. Substrate defintion with materials and thicknesses used (M for Metal and V for Via (and PTH).

III. INDUCTOR DESIGN

The basic design rules for the Fig. 1 substrate are defined both at the core layer level and at the build-up layer level. At the core layer level, the minimum diameter for a Plated Through Hole (PTH) is 100 µm with a 200 µm land pad diameter, the minimum width of the conductor is 40 µm and the minimum space between two conductors is also 40 µm. The minimum diameter of via at the build-up layer level is 50 µm with a 90 µm land pad diameter and the conductor width and the space between two conductors are 25 and 40 µm, respectively. Two types of inductor structures have been designed using these rules: multilayer spiral structures and solenoid structures.

A. Multilayer Spiral structures

Multilayers spiral inductors have been designed as presented in Fig. 2. In order to improve inductance and quality factor in a reduced size it has been decided to use only the build-up layers as they offer low loss tangent and lower conductor width. For a four layer spiral, it means that two spirals are drawn above the core layer and two other below. The size of the multilayer inductors is then defined by the internal diameter of the top spiral, the number of turns, the width of the conductors, the space between conductors and the metallized via. Several inductors were designed varying internal diameter (200, 300 and 400 µm), number of turns (3, 7, 11 and 15 total turns), conductor widths (25 or 50 µm) and number of layers (1, 2 and 4). The RF ports of all

the multilayer spiral inductors (green colored in Fig. 2) have been designed to be identical regardless of the shape and dimensions of the inductor. This allows an RF test to be performed automatically thanks to an automatic prober and GSG RF probes with a 250 µm pitch.

Figure 2. Example of a typical multilayer spiral inductor designed with the substrate presented in Fig. 1. The presented inductor has 7 turns distributed on 4 different metal layers.

B. Solenoid structures

Fig 3 presents the solenoid structure designed with the substrate given in Fig. 1. The metallization on each side of the core layer and the PTH through the core layer are used to form a coil wound. The solenoid structure is much larger than the multilayer spiral structures mainly because PTH diameter through the core layer is twice bigger than via diameter in the build-up layers. Several solenoids with different number of winding, width of winding (W in Fig 3) and width of conductors are designed. This solenoid structure is less interesting in term of footprint than the multilayer spiral structure but the better confinement of the electromagnetic field makes the quality factor better and less sensitive to electromagnetic interferences.

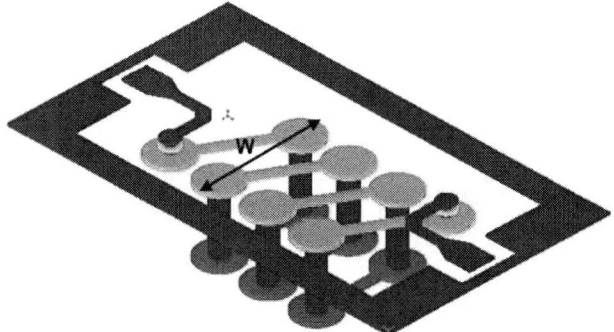

Figure 3. Example of a typical multilayer solenoid inductor designed with the substrate presented in Fig. 1. The presented inductor has 3.5 turns around the core layer.

C. Simulations

S parameters of the inductors are computed using electromagnetic simulations based on finite elements method (FEM). FEM (ANSYS HFSS) is used as it allows to compute electromagnetic fields inside the metal to accurately model the skin effect in copper. Knowing that the skin depth is 2.1 μm for copper at 1 GHz and 0.9 μm at 5 GHz, it affects significantly the quality factor of the inductors designed with copper thicknesses of 15 or 20 μm.

The simulation results for several multilayer spiral inductors are detailed in TABLE I. and in Fig. 4 to Fig. 7 where impedance, inductance and quality factor are calculated from S parameters with respect to (1) and (2). Multilayer spiral inductors with 4 layers, diameter ranging from 200 to 400 μm and with 3, 7, 11 or 15 turns have been simulated. The simulation results show that maximum quality factor from 24 to 34 are obtained with inductance values at 1 GHz between 2 and 49 nH depending on the inductor geometry.

$$L = \text{Im}(1/Y_{11})/(2\pi f) \qquad (1)$$

$$Q = \text{Im}(1/Y_{11})/\text{Re}(1/Y_{11}) \qquad (2)$$

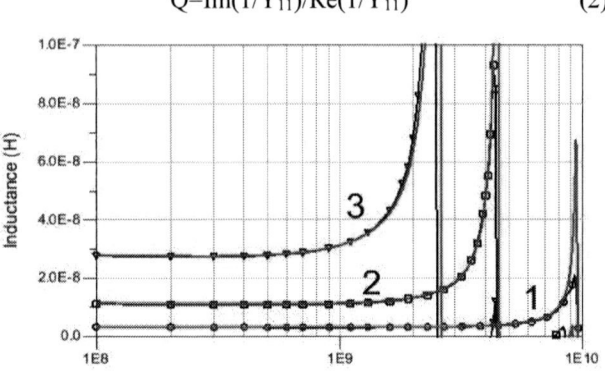

Figure 4. Measured and simulated inductance value for inductors with 1, 2 or 3 turns, with 4 metal layers and an internal diameter of 300 μm. The measurements are plotted in black and the simulations in red.

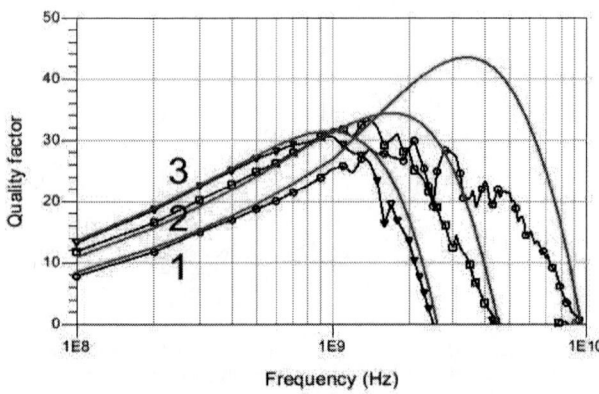

Figure 5. Measured and simulated quality factor for inductors with 1, 2 or 3 turns, 4 metal layers and an internal diameter of 300 μm. The measurements are plotted in black and the simulations in red.

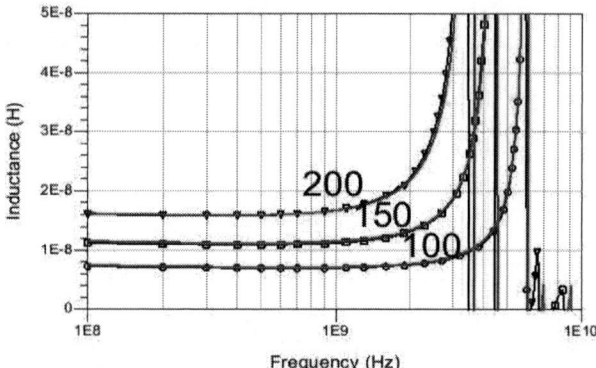

Figure 6. Measured and simulated inductance value for inductors with 200, 300 or 400 μm internal dimeter, with 4 metal layers and 2 turns. The measurements are plotted in black and the simulations in red.

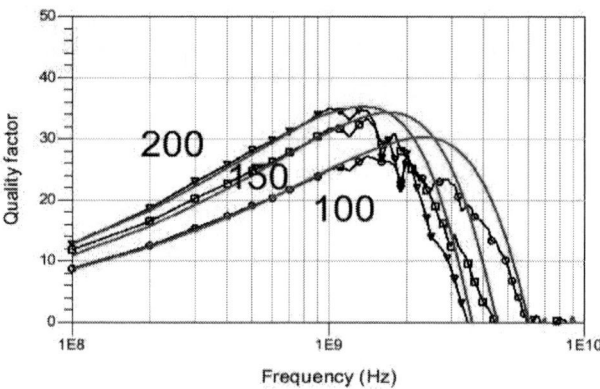

Figure 7. Measured and simulated quality factor for inductors with 200, 300 or 400 μm internal dimeter, with 4 metal layers and 2 turns. The measurements are plotted in black and the simulations in red.

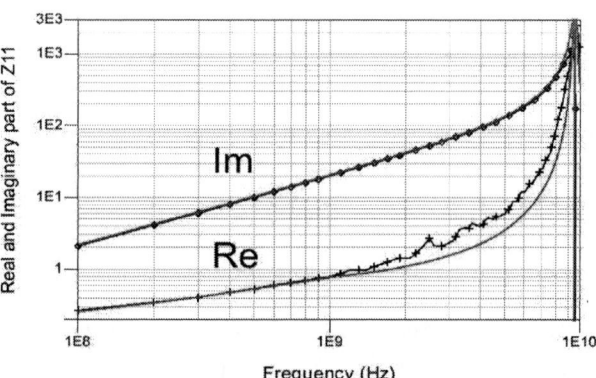

Figure 8. Measured and simulated real and imaginary part of Z_{11} for inductors with 300 μm internal dimeter, with 4 metal layers and 1 turn. The measurements are plotted in black and the simulations in red.

978-1-7281-1500-9/19 $31.00 © 2019 IEEE

TABLE I.   MEASUREMENT AND SIMULATION RESULTS FOR MULTILAYER SPIRAL INDUCTORS

3D Structure										
Turns	3	7	11	15	3	7	11	3	7	11
Metal levels	4	4	4	4	4	4	4	4	4	4
$\varnothing_{ind}$ (µm)	200	200	200	200	300	300	300	400	400	400
L_{1GHz} (nH)	2 (2)	7 (7)	19 (19)	46 (46)	3 (3)	11 (11)	31 (31)	4 (4)	17 (17)	49 (49)
Q_{max}	27 (41)	28 (30)	27 (30)	26 (26)	30 (44)	34 (34)	31 (32)	35 (52)	35 (35)	34 (33)
f_{Qmax} (GHz)	≈1.5 (4.5)	≈1.5 (2.3)	≈1.5 (2.3)	0.9 (0.8)	≈2.2 (3.4)	1.4 (1.8)	1 (1)	≈1.5 (3)	1.4 (1.3)	0.7 (0.8)
Q_{1GHz}	23 (24)	25 (25)	26 (25)	26 (25)	26 (27)	32 (31)	31 (32)	31 (29)	35 (34)	32 (32)
Q_{2GHz}	25 (33)	26 (30)	20 (30)	2 (5)	30 (39)	28 (34)	13 (17)	30 (45)	28 (32)	0 (4)
SRF (GHz)	>10(>10)	5.8 (5.8)	3.2 (3.4)	2.3 (2.1)	9.4 (9.4)	4.4 (4.5)	2.5 (2.6)	7.8 (7.9)	4.1 (4.1)	2 (7.9)
width (µm)	640	800	920	1050	750	900	1000	850	1000	1100
length (µm)	370	450	550	680	420	520	650	490	620	750
footprint (mm²)	0.24	0.36	0.51	0.71	0.31	0.47	0.65	0.42	0.62	0.82
density (nH/mm²)	8.3	19.4	37.3	57.3	9.7	23.4	43.8	9.5	27.4	47.6

The values in brackets are the simulated all others are measurements. The conductor width W is 25 µm and the space between two conductors S is 40 µm and the total height is 666 µm for all these inductors.

Density is calculated using the inductance value at a frequency far away from resonance where the value is constant with frequency (frequency can be lower than 1 GHz).

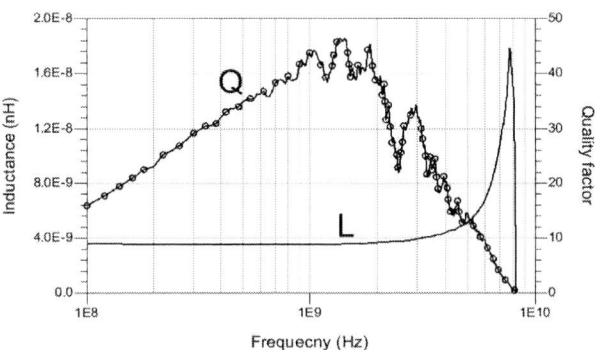

Figure 9.   Measured inductance value and quality factor for a 3 turn solenoid inductor.

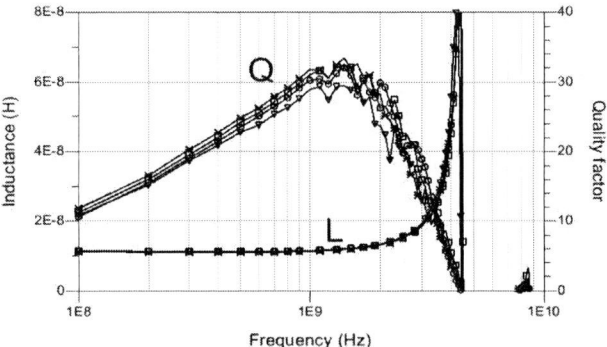

Figure 10. Measured repeatability of inductance value and quality factor of multilayer spiral inductors with 2 turns, 4 metal layers and an internal diameter of 300 µm.

IV.   CHARACTERIZATION

A.   RF Measurements

The RF measurements from 100 MHz up to 10 GHz are performed using a Keysight N5230A Vector Network Analyzer with GSG RF probes (250 µm pitch). A SOLT calibration is used and no deembedding of ports has been applied in order to ease comparison with simulation results.

The measurement results for many four layer inductors are presented in TABLE I. and Fig. 4 to Fig. 7. The plotted inductance value versus frequency presented in Fig. 4 and Fig. 6 show very good agreement with the simulated data. FEM can predict very accurately the inductance value but the quality factor is overestimated in the 1 turn case. Fig. 8 shows that it comes from the real part of Z_{11} which is underestimated by simulations. It might be due to dielectric loss that are underestimated at high frequencies, or due to roughness effects. Fig 9 presents the measurement results for a solenoid with 3 turns as shown on Fig. 3. The conductor width is 50 µm and the width (W in Fig 3) of the solenoid is 500 µm. Its total footprint is 1.09 mm², the maximum quality factor (Q_{max}) measured is 46 at 1.5 GHz and its Self-Resonant Frequency (SRF) is 8 GHz for an inductance of 3.6 nH. This type of inductor structure presents higher quality factor but a lower inductance density. A way to decrease the footprint consists in using coreless substrate where via size can be divided by two.

The repeatability of the fabrication process is assessed by measuring several identical multilayer spiral inductors both on the same substrate and on different substrates. Measurement results of the inductance value and quality factor for 2 turns, 4 layers and 300 µm internal diameter inductors are presented in Fig. 10. The measurements show a ±1% repeatability for the inductance value and ±5% for quality factor at 1 GHz.

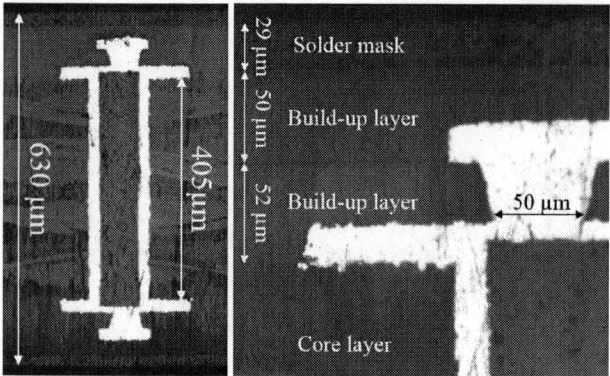

Figure 11. Left: cross section of a PTH with land pads. The PTH (through core layer) has 2 stacked via (through build-up layers) at each side. Right: cross section of a via through a build-up layer with land pads.

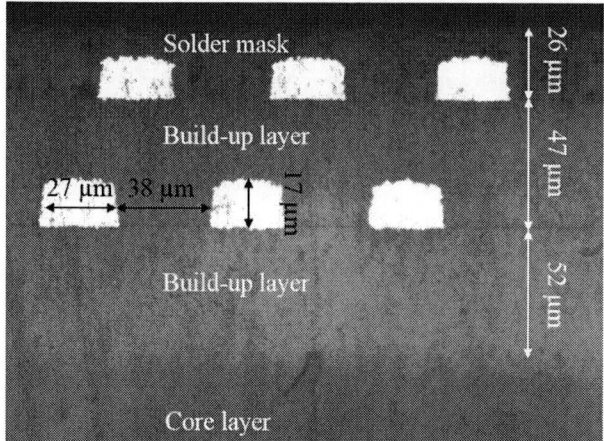

Figure 12. Cross section of a 2 layer spiral inductor with 3 turns at each metal layers.

B. Fabricated Susbtrate

Figure 11 presents cross sections with details of the stacked layer thicknesses and a zoom on PTH and on a via trough a build-up layer. Fig. 12 shows a cross section of a multilayer inductor. The cross section details the two top metallic layers of a 3 turns spiral inductor. The measured values are in good agreement with the values given in Fig. 1 considering measurement error of about 5% with optical microscope measurements. The relatively good agreement between measurements and simulation for the inductance value is directly due the good agreement between the layout and the fabricated substrate. In Fig. 11 it can be remarked that the roughness of the core layer is a few micron while the surface of the build-up layer is very smooth as seen in Fig. 12. The roughness of the metal layer in the build-up layers is close to 2 or 3 µm but is much higher in the core layer. Also, the roughness reduces thickness measurement accuracy as the interface between core and build-up layer is not clearly defined. The copper roughness can also explain an increase of loss (or a decrease of the quality factor) at high frequencies as explained in [11].

V. COMPARISON WITH STATE OF THE ART

TABLE II. gives a comparison of the designed inductors with both state of the art inductors implemented on organic packaging and commercial SMD inductors. First in can be noticed that footprint of the presented spiral inductors with equivalent inductance values is much lower than the footprint of previously published inductors on organic substrates. The footprint of the multilayer spiral inductors compared to these previous works is divided by 3 for inductor of about 2 nH, but it is twice bigger than the one of commercial non-organic SMD inductor. The quality factor of non-organic SMD inductors at equivalent footprint is also higher. Non-organic SMD inductors at equivalent footprint have better electrical performances but they require interconnections and a pick and place manufacturing step and soldering process that can lower the overall performances.

The quality factor of the multilayer spiral inductors cannot really be compared with the one of previously reported inductors as the footprints of the multilayer spiral inductors presented here are much smaller than those of previously published inductors integrated in organic packaging. As it can be seen on the grey zone in Fig. 13, all previously published inductors that are integrated in organic substrate have a density below 10 nH/mm². However, the quality factor can be compared to the quality factor of commercial SMD multilayer organic inductors with approximatively the same inductance and footprint [22]. The density of the multilayer inductors presented in this paper is just a little bit higher than the one of commercial organic inductors (as shown in TABLE II.) and their maximum quality factor is about 20% higher as presented in Fig. 13. Increasing the density is possible by using a coreless substrate and a higher number of build-up layers.

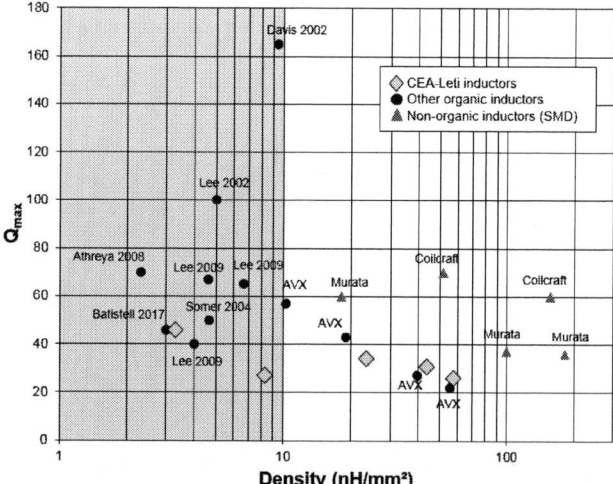

Figure 13. Maximum quality factor as a function of inductance density for inductors with working frequency of about 1 GHz.

978-1-7281-1500-9/19 $31.00 © 2019 IEEE

TABLE II. COMPARISON WITH STATE OF THE ART AND SMD INDUCTORS

Reference	W (µm)	S (µm)	Footprint (mm²)	Thickness (µm)	L (nH)	Q_{max}	Freq (Q_{max}) (GHz)	SRF (GHz	Density (nH/mm²)
[9]	100	80	1.86	≈650	8.6	67	0.8	4	4.6
[9]	100	80	1.29	≈650	8.6	65	0.8	4	6.7
[9]	100	80	≈4	≈650	16.0	40	0.8	3.1	≈4.0
[10]					2.4	90	2.0	>10	
[10]					10.0	80	2.0	4.9	
[11]	100		≈1.5		≈7.0	50	1.0	4.0	≈4.7
[12]	50	50	0.78	125	1.8	70	8.7	17.5	2.3
[13]			≈1.0	≈350	3.0	46	1.8	7.2	≈3
[13]			≈1.0	≈350	6.5	52	1.1	4	≈6.5
[14]	150		0.72	≈1050	3.6	100	1.8	10.6	5.0
[15]			0.36	1250	3.4	165		11.5	9.4
[20] LQP02HQ			0.11	320	2.0	>50		>10	18.2
[20] LQP02HQ			0.11	320	11.0	37	3.0	>4.0	100.0
[20] LQP02HQ			0.11	320	20.0	36	2.8	>3.0	181.8
[21] 0201DS			0.27	460	0.5-14.0	>50		17.9-5.1	51.8
[21] 0201DS			0.25	550	39.0	60	1.7	3.75	156.0
[22] HLQ02			0.58	350	6.0	57	2.3	5.5	10.3
[22] HLQ02			0.58	350	11.0	43	1.7	3.6	19
[22] HLQ02			0.58	350	23.0	27	1.2	2.1	39.6
[22] HLQ02			0.58	350	32.0	22	0.8	1.8	55.2
This work *	50	40	1.09	666	3.6	46	1.5	7.9	3.3
This work +	25	40	0.24	666	2.0	27	1.5	>10	8.33
This work +	25	40	0.47	666	11.0	34	1.4	4.4	23.4
This work +	25	40	0.65	666	28.5	31	1.0	2.5	43.8
This work +	5	40	0.71	666	40.7	26	0.9	2.3	57.3

"≈" is used when values are calculated from data available in the reference
+ Spiral structure, * solenoid structure

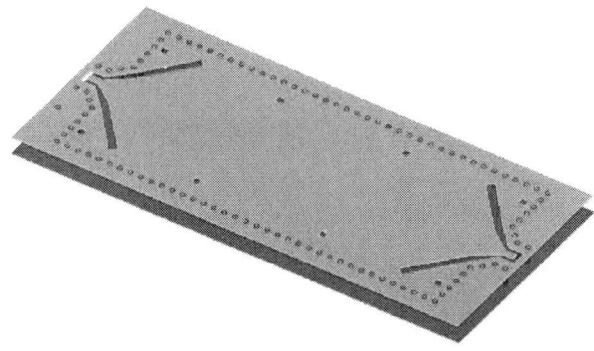

Figure 14. SIW designed with the substrate presented in Fig. 1.

VI. ORGANIC SUBSTRATE FOR MM WAVES

New opportunities for component design or integration at mm waves are emerging with 5G. Organic substrates have a role to play at these frequencies because they are low loss, low cost and with the recent improvements in fabrication process, it is now possible to design very small conductor lines (less than 10 µm in width) with high accuracy and reproducibility as it is required when frequency increases.

To assess the organic substrate presented in Fig. 1 for mm waves applications, a Substrate Integrated Waveguide (SIW) has been designed [22]. Fig 14 presents the SIW which is 4 mm wide and 8 mm long. The waveguide is formed with the core layer of the substrate. The top and bottom metallic walls of the waveguide are designed with the copper layers on the bottom and on the top of the core layer while the metallic side walls are made thanks to PTHs. Transitions from CPW to SIW have been designed in order to measure the SIW with RF GSG probes. RF measurements presented in Fig. 15 show that the losses of the SIW including the transitions from CPW to SIW are very low up to 40 GHz. S_{21} is higher than -2 dB and S_{11} is lower than -10 dB up to 40 GHz. It can also be noticed that simulations and measurements are in good agreement. It shows that the accuracy of the fabrication process is good enough to design SIW at mm waves. It should be mentioned that accuracy and roughness can be improved by using only build-up layers (coreless) as organic substrate.

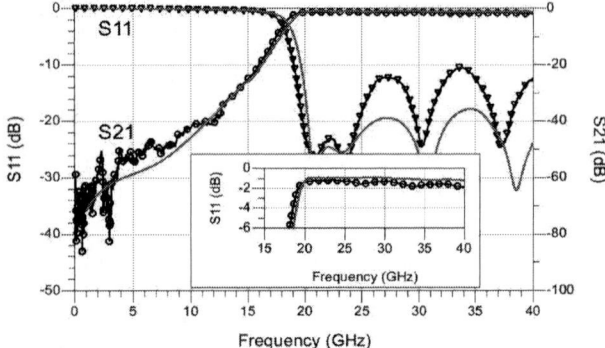

Figure 15. S_{21} and S_{11} measurements (black) and simulations (red) of the SIW designed with the substrate presented in Fig. 1.

VII. CONCLUSION

Inductors integrated in organic packaging have been designed, fabricated and characterized. The organic packaging is a 2 build-up layers – 1 core layer – 2 build-up layers stack composed of very low loss materials (tan $\delta \leq$ 0.008). Multilayer spiral inductors embedded in the organic packaging have been designed. The highest density measured is at the state of the art for inductors in organic packaging. It is $\approx$ 57.3 nH/mm² for a 39 nH inductor. The measured maximum quality factors are between 27 and 35 depending on the inductor geometry considered. Decreasing further the footprint is possible by using a coreless substrate and a higher number of build-up layers. A 3 turn solenoid inductor structure has been measured with a maximum quality factor of 46 at 1.5 GHz for an inductance value of 3.6 nH. A SIW has also been fabricated and characterized on the same substrate. It shows that recent improvements in the layout design rules of organic packaging make them suitable for designing at mm waves frequencies.

ACKNOWLEDGMENT

The authors acknowledge D. Bouchu for the cross section photographs and the thickness measurements.

REFERENCES

[1] J. S. Chang, A. F. Facchetti and R. Reuss, "A Circuits and Systems Perspective of Organic/Printed Electronics: Review, Challenges, and Contemporary and Emerging Design Approaches," in *IEEE Journal on Emerging and Selected Topics in Circuits and Systems*, vol. 7, no. 1, pp. 7-26, March 2017.

[2] M. Ali et al., "Miniaturized High-Performance Filters for 5G Small-Cell Applications," *2018 IEEE 68th Electronic Components and Technology Conference (ECTC)*, San Diego, CA, 2018, pp. 1068-1075.

[3] T. Kamgaing, E. Davies-Venn, et K. Radhakrishnan, "A compact 802.11 a/b/g/n WLAN front-end module using passives embedded in a flip-chip BGA organic package substrate", in *Microwave Symposium Digest, 2009. MTT'09. IEEE MTT-S International*, 2009, pp. 213–216.

[4] C. Romero, J. Lim, T. Kim, H. Kim, et K. Kim, "Design and characterization of fully embedded passive components on multilayer organic-based substrate for highly compact SOP applications", in *Electronic Components and Technology Conference, 2009. ECTC 2009. 59th*, 2009, pp. 1731–1736.

[5] E. Davies-Venn et T. Kamgaing, "Miniaturized rf transformer-based baluns for 802.11 a/b/g WLAN modules embedded in organic package substrate", in *Radio and Wireless Symposium, 2008 IEEE*, 2008, pp. 359–362

[6] J. Zhang *et al.*, "Fabrication and analysis of 2D embedded passive devices in PCB", in *Electronic Packaging Technology (ICEPT), 2013 14th International Conference on*, 2013, pp. 100–104

[7] J. Park, H. H. Lee, et J. Y. Park, "Q-Factor Improvement of FR-4 Embeeded RF Inductors", in *Intergated Ferroelectrics Journal*, Vol. 104, Issue 1, 2008, pp. 70-79.

[8] T. Lee, S. Cheon and J. Park, "Ultracompact UHF Tunable Filter Embedded Into Multilayered Organic Packaging Substrate," in *IEEE Transactions on Components, Packaging and Manufacturing Technology*, vol. 2, no. 1, pp. 46-52, Jan. 2012.

[9] H. Lee and J. Park, "Characterization of Fully Embedded RF Inductors in Organic SOP Technology," in *IEEE Transactions on Advanced Packaging*, vol. 32, no. 2, pp. 491-496, May 2009.

[10] Cheng-Hua Tsai et al., "Design and implementation of mobile communications SiP modules with embedded components in multilayer organic hybrid substrate," *2012 7th International Microsystems, Packaging, Assembly and Circuits Technology Conference (IMPACT)*, Taipei, 2012, pp. 419-422.

[11] G. Sommer, W. John, H. Reichl, "Influence of Technological Constrains of HDI Organic Substrates on RF Characteristics of Embedded Inductor Component," *EMC'04*, Sendai, 2004, pp. 257-260.

[12] D. Athreya, V. Sundaram, M. Iyer and R. Tummala, "Ultra high Q embedded inductors in highly miniaturized family of low loss organic substrates," *2008 58th Electronic Components and Technology Conference*, Lake Buena Vista, FL, 2008, pp. 2073-2080.

[13] G. Batistell, T. Holzmann, S. Leuschner, A. Wolter, A. Passamani, J. Sturm, "SiP Solutions for Wireless Transceiver Impedance Matching Networks," *2017 EuMW*, Nürnberg, 2017.

[14] S. H. Lee et al., "High performance spiral inductors embedded on organic substrates for SOP applications," *2002 IEEE MTT-S International Microwave Symposium Digest (Cat. No.02CH37278)*, Seattle, WA, USA, 2002, pp. 2229-2232 vol.3.

[15] M. F. Davis et al., "Integrated RF architectures in fully-organic SOP technology," in *IEEE Transactions on Advanced Packaging*, vol. 25, no. 2, pp. 136-142, May 2002.

[16] Y. S. Tsai, J. M. Wu, C. j. Fan and C. T. Chiu, "Wideband compact model of inductors for RF-SOP applications," *2012 Asia Pacific Microwave Conference Proceedings*, Kaohsiung, 2012, pp. 1262-1264.

[17] www.hitachi-chem.co.jp/

[18] https://www.ajinomoto.com/

[19] https://www.taiyo-hd.co.jp/

[20] https://www.murata.com/

[21] https://www.coilcraft.com/

[22] http://www.avx.com/

[23] H. Uchimura, T. Takenoshita and M. Fujii, "Development of a "laminated waveguide"," in *IEEE Transactions on Microwave Theory and Techniques*, vol. 46, no. 12, pp. 2438-2443, Dec. 1998.

3D Printed Interposer Layer for High Density Packaging of IoT Devices

Saikat Mondal[1], Mohd. Ifwat Mohd. Ghazali[2], Kanishka Wijewardena[1], Deepak Kumar[1] and Premjeet Chahal[1]

[1]Department of Electrical and Computer Engineering, Michigan Sate University, East Lansing, MI, USA.
[2]Faculty of Engineering and Built Environment, Universiti Sains Islam Malaysia (USIM), Nilai, Negeri Sembilan, Malaysia.
chahal@egr.msu.edu

Abstract—**3D printing has emerged as a potential solution to fabricate different RF components and embed prepackaged devices. In this work, 3D printed interposer layer is proposed to enable vertical stacking of different heterogeneous components and substrates. Different design parameters were first simulated using HFSS to optimize for maximum possible bandwidth. After optimization, the design parameters were used to fabricate a wireless battery less digital modulator as an example IoT device. Measurement results are provided to validate the design and simulation results.**

Index Terms—**interposer, energy harvesting, wireless, passive, 3D printing.**

I. INTRODUCTION

Continuous Internet of Things (IoT) has emerged as a potential solution to connect physical objects wirelessly and make them intelligent. Multi-module integration is an integral part of any IoT device, which usually consists of different modules such as 1) antennas for communication with RF signals, 2) a digital module for signal processing, 3) RF module to interface antenna with digital module, 4) sensors and actuators, and 4) power supply [1], [2]. Integration of all these different modules within a single package is a growing challenge to make the system compact. 3D printing has emerged as a potential solution to integrate thin film or packaged components on a package substrate for an efficient System-on-Package (SoP) solution [3]. 3D printing has revolutionized manufacturing due to the ability to custom print complex geometries with ease. Recently, 3D printing has been used to realize a number of planar and non-planar RF and microwave devices such as guided wave structures, probes, antennas, resonators and microstrip based circuits [4]–[7]. The freedom along z-axis allows customized vertical stacking for multi-layer structures and packaging solutions.

3D stacking of multiple layer ICs in SoP is very important as it reduces the device footprint. The implementation of vertical interconnect helps in reducing parasitic effects from long interconnects and thus helps in maintaining signal integrity. Interposer layer is an integral part of vertical IC stacking as it provides electrical connections in 3D multi-layer components. Interposer layer in silicon and glass such as through silicon via (TSV) and through glass via (TGV) are widely used as vertical interconnect layer in between stacked die [8], [9]. Although, modern fabrication technology has enabled very fine pitch TSV and TGV process, they are still complex and

costly and can be afforded for custom IC fabrication process. Additionally, fabrication of air substrate based RF components such as inductor or antenna is difficult in TSV or TGV based processes. As an emerging technology, it is not very common to obtain sub-micron resolution in 3D printing, which is prevalent in photo-lithography based fabrication technology. Hence, in recent days the practical application of 3D printing is limited as packaging substrate for modules rather than sub-micron size bare die ICs. Although having feature size limitation, 3D printing has a huge potential in vertical multi-module heterogeneous substrate integration. To carry signal from one layer to another layer, 3D printed interposer layer can be used overcoming the limitations of TGV and TSV technology.

To demonstrate the concept, a 3D printed interposer layer is implemented for low power and a wide range of RF spectrum (upto 3 GHz) transfer in between multi-functional module layers. The relative positions of different modules in 3D stacking is crucial as it can lead to unwanted parasitics leading to poor signal integrity. Hence, a systematic study was performed and a general guideline was proposed for optimum placement of different modules in a typical IoT device to reduce the signal latency and increase the bandwidth. As a test case, a passive 3D printed wireless digital modulator (IoT device) was designed and implemented using 3D printing for a small footprint.

II. DESIGN

The design methodology for a complete small feature IoT device with vertical stacking of different modules is shown in Fig. 1. As the device has to communicate wirelessly, antenna would be an integral part of the device. And for maximum radiation efficiency, the antenna should be at the top most layer. The high frequency RF processing layer should be closest to the antenna layer to minimize the loss due to long interconnects. The medium frequency digital layer can be beneath the RF layer as the conductor and dielectric loss decreases with frequency reduction. There can be more than single layer of digital layer, if the device requires more than one functionality. The sensor layer would be the lowest part of the IoT device touching the target for precise sensing. 3D printed interposer layer is proposed in between all the different functionality layers to ensure signal transmission among all the heterogeneous multi-functional modules. Apart from signal

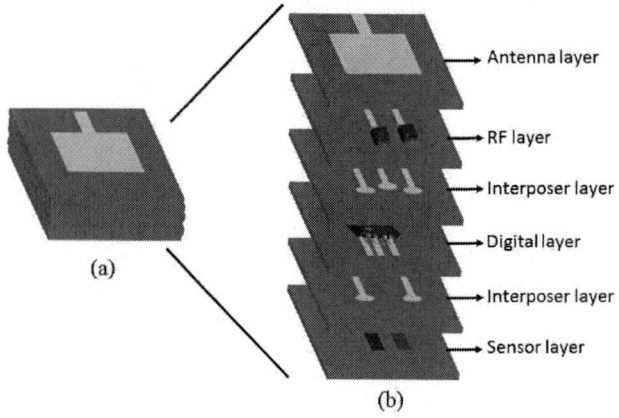

Fig. 1: Multi-layer (a) stacked and (b) different layer view of a complete IoT device.

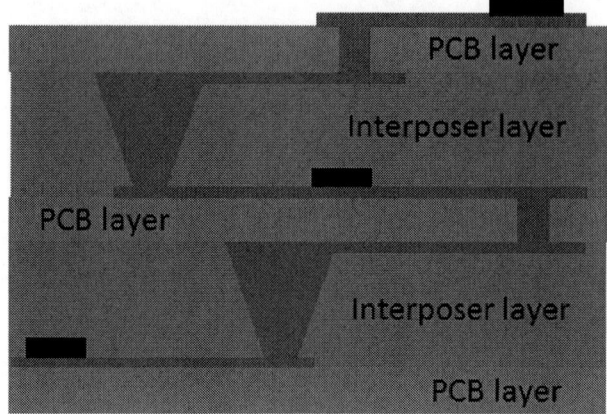

Fig. 2: Cross-section view of a 3D multi-layer device.

transmission, the interposer layers also provides mechanical stability among different functionality modules. With 3D-printing, packaged and bare die ICs can be stacked together within a single package along with thin film components such as inductors, capacitors and customizable different antenna structures. With reduction in feature size of 3D printing, a dramatic miniaturization can be achieved of the whole package. Earlier, embedding of active components using 3D printing was demonstrated in [3]. The complete IoT device can be prototyped and packaged rapidly with both packaged and bare die components.

The cross-sectional side view of a 3D multi-layer vertical device is shown in Fig. 2. In current fabrication or printed circuit board (PCB) technology, encapsulation of packaged IC is not possible. However, it is possible to customize the interposer layer at any desired shape using 3D printing. Hence, in addition to the through-substrate interconnect, it is possible to customize the shape of the 3D printed layer to encapsulate the packaged or bare die components. Furthermore, air substrate can be introduced in the 3D printed interposer layer for high frequency components such as inductors and antennas to minimize the substrate loss [10]. In TGV or TSV technology, the benefits of 3D printing can only be realized with slow etching process, which is time and material consuming process. In Fig. 2, three different module circuit board layers were integrated with two interposer layers sandwitched in between. The interposer layer can be printed to encapsulate the devices and thin planar signal layer. In place of vertical interconnects, the interposer layer is printed hollow, which is metalized later. To start the design of an IoT device, first the packaged components are mounted and connected on PCB substrate. Then the substrates are connected together through patterned metalized interposer layer in between.

The high frequency characteristics of a single 3D printed layer was shown earlier in [11]. It is important to estimate the characteristic loss during signal transition from one layer to

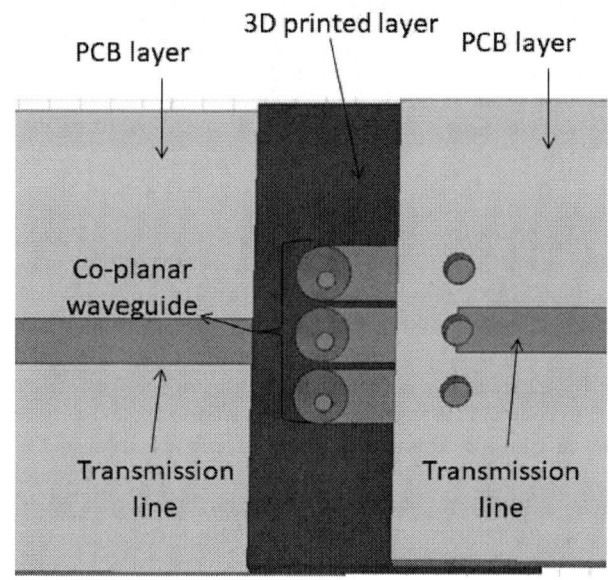

Fig. 3: Transition in between PCB transmission line and 3D printed CPW.

other layer. Hence, a test vehicle of multi-layer PCB substrate-3D printed layer - PCB substrate was created to find out the effect of different physical parameters on signal transmission loss and bandwidth. The test vehicle is shown in Fig. 3, where the transmission line in lower PCB is transposed into a co-planar waveguide (CPW) in the 3D printed layer and then again transposed back to transmission line in the upper PCB.

In the three layered test vehicle, the transmission line on PCB was designed upto 4 GHz. Two through vias were formed in the PCB layer to create signal transition from lower ground plane to top surface. Now, the 3D printed CPW layer was connected to the PCB through three vias, among which the center one is the signal line and the other two are ground lines. In the next layer, a through via PCB layer was connected to

978-1-7281-1500-9/19 $31.00 © 2019 IEEE   1688

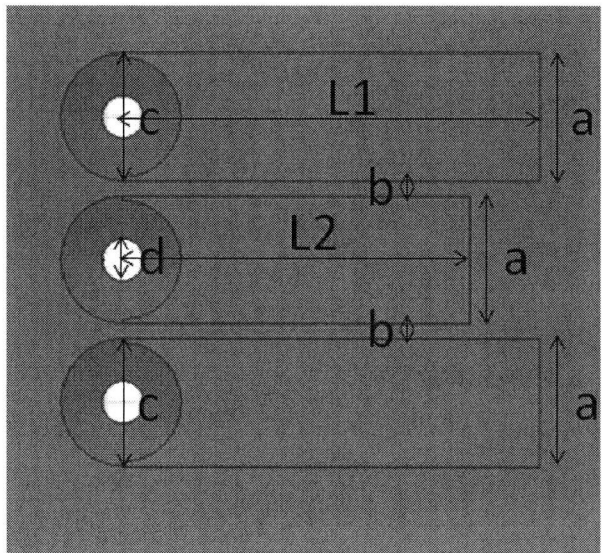

Fig. 4: Dimensions of different parameters in the interposer layer.

TABLE I: Description of different interposer layer parameters.

Parameter	Description
a	CPW conductor width
b	CPW ground and signal line separation
c	Top diameter of the interposer via
d	Bottom diameter of the interposer via
e	Interposer layer thickness
L1	CPW ground line length
L2	CPW signal line length

the 3D printed CPW layer in the similar manner. Thus, the three layer test vehicle was created which starts and ends with a transmission line and a CPW is sandwitched between them. The physical parameters of the interposer layer is summerized in Table I corresponding to Fig. 4.

Once the high frequency test vehicle was created, multiple simulations were carried our by varying the parameters as described in Table I in next section.

III. SIMULATION RESULTS

Before fabricating the designs, simulations were performed using full wave EM solver HFSS. The dielectric constant of the 3D printed interposer layer was chosen as 2.8 based on earlier literature [3]. FR4 material with dielectric constant of 4.4 was used for the PCB layer transmission line design. The three layer simulation test setup was used as described in earlier section.

In the first set of parametric sweep, the fixed parameters were a=3.6 mm, c=3.6 mm, d=1.6 mm, e=2 mm, L1=12 mm, L2=10 mm and b was varied from 0.4 mm to 2.4 mm with a step of 1 mm. For this sweep the number of layers were fixed at 3 and the simulation results are shown in Fig. 5. In the second set of parametric sweep, the fixed parameters were a=3.6 mm, b=0.4 mm, d=1.6 mm, e=2 mm, L1=12 mm, L2=10

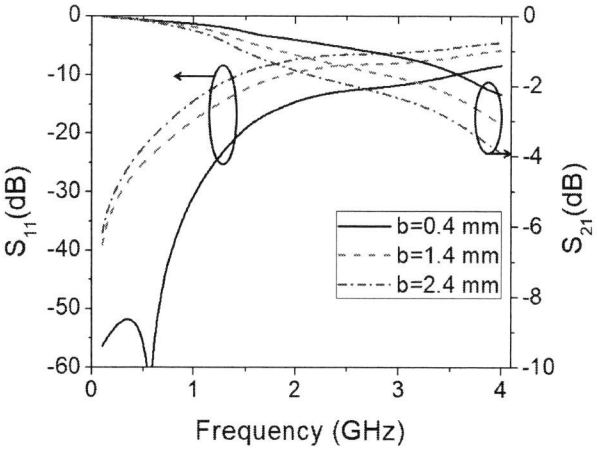

Fig. 5: S-parasmeters for different b values (Fig. 4) in three layers design.

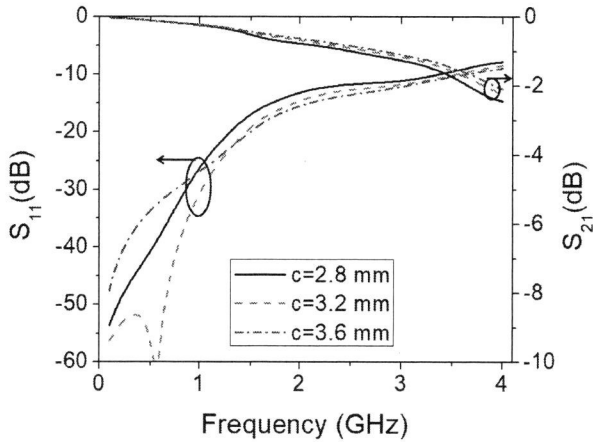

Fig. 6: S-parasmeters for different c values (Fig. 4) in three layers design.

mm and c was varied from 2.8 mm to 3.6 mm with a step of 0.4 mm. The results for this sweep is shown in Fig. 6. In the third set, the fixed parameters were a=3.6 mm, b=0.4 mm, c=3.6 mm, e=2 mm, L1=12 mm, L2=10 mm and d was varied from 1.6 mm to 2.4 mm with a step of 0.4 mm. The results for this sweep is shown in Fig. 7.

From the parametric sweep it was observed that the distance between signal and ground layers in the CPW impacts more strongly for a fixed conductor width. The parameter c effects moderately on the reflection coefficients and the parameter d has least effects on the S-parasmeters. After the parametric sweep, the dimensions were fixed as a=3.6 mm, b=0.4 mm, c=3.6 mm, d=1.6 mm, e=2 mm, L1=12 mm, and L2=10 mm. Next, the effect of multiple layers were shown in Fig. 8. For a three layers system, the layers were PCB-interposer-PCB and for the five layers system the layers were PCB-interposer-PCB-interposer-PCB. As expected, the signal transmission loss

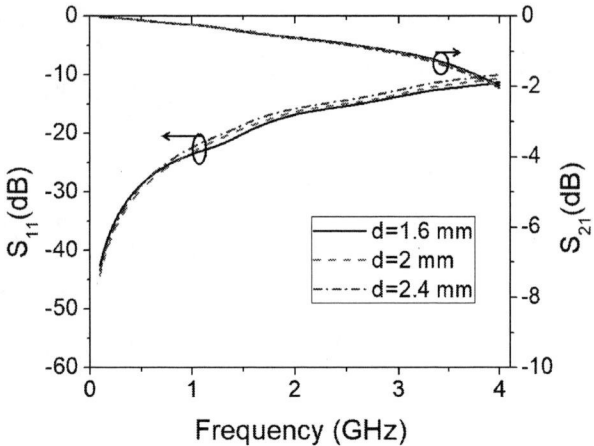

Fig. 7: S-parasmeters for different d values (Fig. 4) in three layers design.

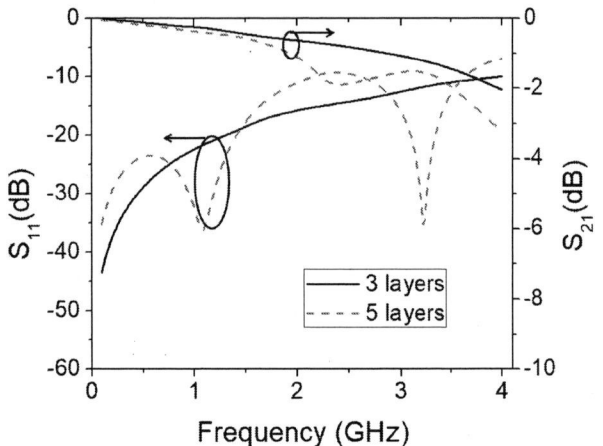

Fig. 8: S-parasmeters of three layers and five layers design.

increased with more number of layers.

IV. MEASUREMENT RESULTS

After the simulations, different parts were fabricated and integrated together to characterize the effect of 3D printed interposer layer on high frequency signal propagation. The fabricated different parts are shown in Fig. 9, which consists of top and bottom PCB layers and a 3D printed middle interposer layer. The PCB layers were fabricated on FR4 board with transmission lines working upto 4 GHz. The interposer layer parts are 3D printed using a commercial printer (Object Connex350) that uses a photopolymer resin VeraWhitePlus® with dielectric constant ($\epsilon_r \approx 2.8$) and loss tangent ($tan\delta \approx 0.04$) [5]. The 3D printed structures are metallized using a two step process, first, a seed layer of 500 nm thick Titanium (Ti) is sputtered for improving adhesion between the copper and the substrate followed by 5000 nm of Copper (Cu). Second, electroplating of Cu is performed to increase the metal

Fig. 9: Different pieces of PCB and 3d printed layers a) PCB bottom layer, b) 3d printed and metalized interposer middle layer and c) PCB top layer.

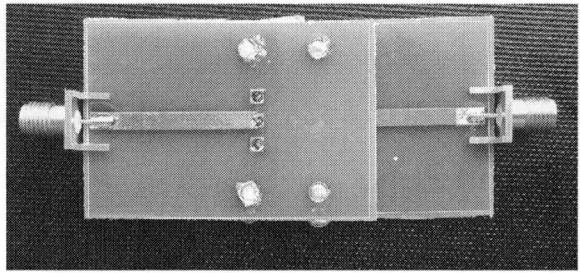

Fig. 10: Integrated three layers RF board with individual PCB and interposer layer parts.

thickness by 5-6 μm. Damascene like process was used to remove any redundant copper layer after metalization. Three pieces were electrically connected together using conductive silver paste. Four via holes are used as markers to ensure proper alignment of the three layers while vertical stacking. The combined view is shown in Fig. 10 with SMA connector mounted for high frequency measurement. The fabricated three layers RF board was characterized using a Network Analyzer and the S-parameters are plotted in Fig. 11. The transmission loss increased at 3 GHz compared to simulated result due to use of silver paste for electrical interconnect between multiple layers.

V. IoT DEVICE DESIGN

Based on the measurements results, the IoT device was fabricated with two PCB layers and 3D printed interposer layer stacked in the middle. The different parts are shown in Fig. 12(a). The right most PCB layer in Fig. 12(a) is the top RF layer and the left most PCB layer in Fig. 12(a) is the bottom digital layer. The two metalized 3D printed interposer layers are in the middle in Fig. 12(a). After stacking all the layers vertically, the side view is shown in Fig. 12(b). For a brief description, the top PCB layer consists a RF harvesting unit, which converts the RF energy into DC power to drive the digital circuit. The bottom PCB layer consists of a digital modulator to generate digital bit sequences. In Fig. 12(c) and Fig. 12(d), the device is shown with a meandered 930 MHz

978-1-7281-1500-9/19 $31.00 © 2019 IEEE   1690

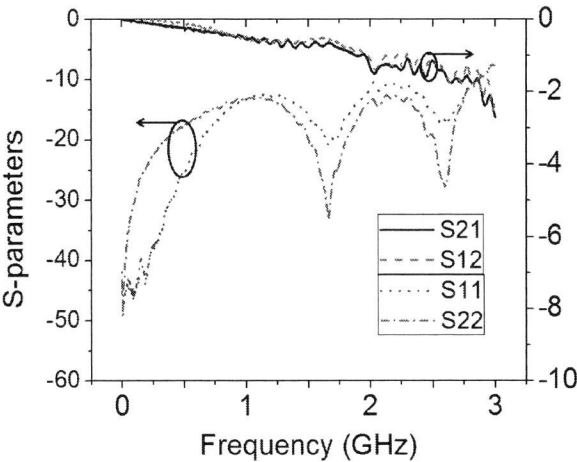

Fig. 11: Measured reflection coefficients of three layers RF board.

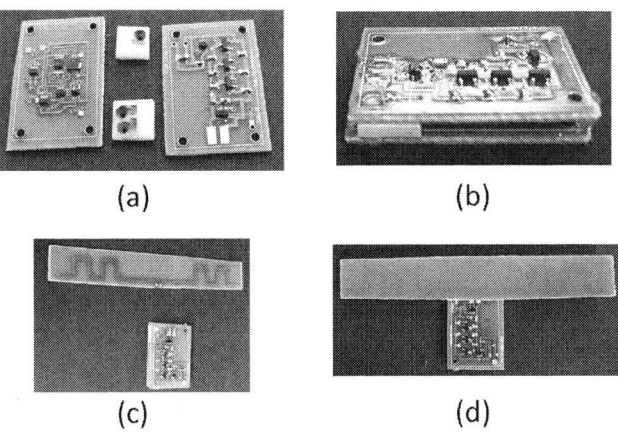

Fig. 12: Description of the IoT device (a) Two PCB layers with right most is the top RF layer and left most is bottom digital layer. Between the two 3D printed substrates, the bottom one is to transfer DC power with ground and the top one is to transfer digital signal. (b) The side view of the device after integration. (c) The device with separate antenna and (d) the device integrated with antenna.

meandered dipole antenna before and after integration. To minimize RF power loss, the RF layer was placed closest to the antenna. The DC power is transferred from the top PCB to the lower PCB layer through the interposer layer. Moderate to low frequency digital signal is produced by the lower PCB layer and transferred back to the top layer through the interposer layer for digital switching.

After fabrication, the antenna integrated wireless modulator was excited wirelessly with RF signal at 930 MHz and -4 dBm of transmitted power. The IoT device was kept at 6 cm from the transmitter antenna. The modulator perturbed the RF field, which is captured in Fig. 13. The digital bit sequences

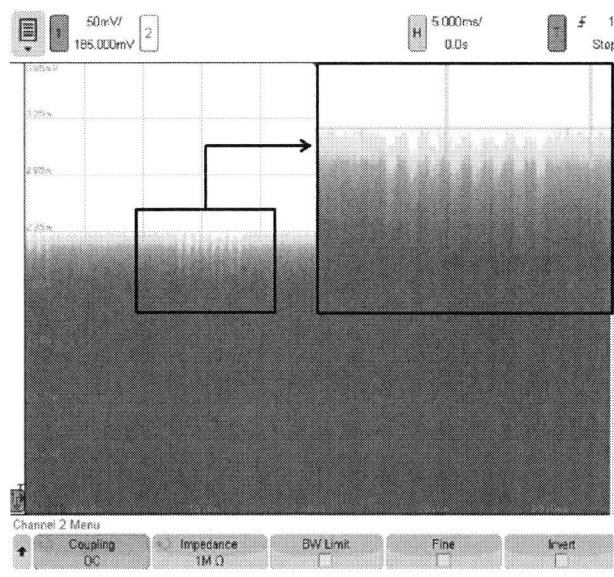

Fig. 13: RF field perturbation by the IoT device.

are shown in Fig. 13.

VI. DISCUSSION

The potential for 3D printing to fabricate interposer layer opens the opportunity to extend current 3D printing to complex electronic packagaing applications. The advantages of 3D printing is multi-fold ranging from relaizing complex light weight geometries to design flexibility and rapid customization on demand. Although, 3D printing has promising potential, it is challenging to realize fully 3D printed systems due to factors such as higher surface roughness, higher substrate dielectric loss, warpage due to heating, and lower conductive silver paste. The surface roughness can be minimized by adapting mechanical polishing before metallization. Using a high temperature polymer minimizes the effect of warpage and also allows direct soldering of the components restricting the use of lower conductive silver paste. Loss associated with substrate can be minimized by utilizing air based substrates as demonstrated in [10]. With further advances in developing high resolution printers with low loss print materials and multi-material printing will allow realizing a fully 3D printed SoP solution.

VII. CONCLUSION

In this paper, the application of 3D printed substrate is demonstrated as a low-cost potential interposer layer with the capability of pre-packaged IC encapsulation. A general design guideline is proposed to select different layers in multi-module device integration. High frequency RF signal should be restricted between a minimum number of vertical layers. However, low frequency digital signals can be allowed to travel through multiple vertical layers. Different parameters were varied in the design to maximize the RF bandwidth. Finally, a passive wireless digital modulator is designed, fabricated and measured to show the capability of 3D printing for potential interposer layer. With co-printing of dielectric and metal and reduction of feature size in the 3D printing technology, bare die ICs can be integrated within a single package along with different modules for heterogeneous integration.

ACKNOWLEDGMENT

This work was supported by the Axia Institute. The authors would like to thank Mr. Brian Wright from MSU ECEshop for his help with 3D printing and all members of MSU EMRG group for their helpful suggestions.

REFERENCES

[1] L. Lizzi, F. Ferrero, P. Monin, C. Danchesi, and S. Boudaud, "Design of miniature antennas for IoT applications," in *2016 IEEE Sixth International Conference on Communications and Electronics (ICCE)*. IEEE, 2016, pp. 234–237.

[2] N. Khalil, M. R. Abid, D. Benhaddou, and M. Gerndt, "Wireless sensors networks for Internet of Things," in *2014 IEEE ninth international conference on Intelligent sensors, sensor networks and information processing (ISSNIP)*. IEEE, 2014, pp. 1–6.

[3] M. I. M. Ghazali, S. Karuppuswami, S. Mondal, and P. Chahal, "Embedded Active Elements in 3D Printed Structures for the Design of RF Circuits," in *2018 IEEE 68th Electronic Components and Technology Conference (ECTC)*. IEEE, 2018, pp. 1062–1067.

[4] S. Karuppuswami, S. Mondal, M. I. M. Ghazali, and P. Chahal, "A multiuse fully 3D printed cavity sensor for liquid profiling," *IEEE Sensors Letters*, 2018.

[5] M. I. M. Ghazali, J. A. Byford, S. Karuppuswami, A. Kaur, J. Lennon, and P. Chahal, "3D printed out-of-plane antennas for use on high density boards," in *2017 IEEE 67th Electronic Components and Technology Conference (ECTC)*. IEEE, 2017, pp. 1835–1842.

[6] J. A. Byford, M. I. M. Ghazali, S. Karuppuswami, B. L. Wright, and P. Chahal, "Demonstration of RF and microwave passive circuits through 3-D printing and selective metalization," *IEEE Transactions on Components, Packaging and Manufacturing Technology*, vol. 7, no. 3, pp. 463–471, 2017.

[7] M. DAuria, W. J. Otter, J. Hazell, B. T. Gillatt, C. Long-Collins, N. M. Ridler, and S. Lucyszyn, "3-D printed metal-pipe rectangular waveguides," *IEEE Transactions on Components, Packaging and Manufacturing Technology*, vol. 5, no. 9, pp. 1339–1349, 2015.

[8] J. Kim, J. S. Pak, J. Cho, E. Song, J. Cho, H. Kim, T. Song, J. Lee, H. Lee, K. Park *et al.*, "High-frequency scalable electrical model and analysis of a through silicon via (TSV)," *IEEE Transactions on Components, Packaging and Manufacturing Technology*, vol. 1, no. 2, pp. 181–195, 2011.

[9] M. Töpper, I. Ndip, R. Erxleben, L. Brusberg, N. Nissen, H. Schröder, H. Yamamoto, G. Todt, and H. Reichl, "3-D Thin film interposer based on TGV (Through Glass Vias): An alternative to Si-interposer," in *2010 Proceedings 60th Electronic Components and Technology Conference (ECTC)*. IEEE, 2010, pp. 66–73.

[10] M. I. M. Ghazali, S. Karuppuswami, A. Kaur, and P. Chahal, "3-D printed air substrates for the design and fabrication of RF components," *IEEE Transactions on Components, Packaging and Manufacturing Technology*, vol. 7, no. 6, pp. 982–989, 2017.

[11] P. I. Deffenbaugh, T. M. Weller, and K. H. Church, "Fabrication and microwave characterization of 3-D printed transmission lines," *IEEE Microwave and wireless components letters*, vol. 25, no. 12, pp. 823–825, 2015.

Comprehensive Solution for Micro Bump Coplanarity Control

C.C. Liu, J.H. Chen, Y.N. Hsu, R.D. Wang*, Y.C. Wang, B.E. Ho, Y.H. Wu, Ponder Pan, Harry Ku, Kirin Wang, Calvin Lu, K. C. Liu, Marvin Liao

Taiwan Semiconductor Manufacturing Company, Ltd. Taiwan

Mail*: rdwanga@tsmc.com

Abstract

Chip-on-Wafer-on-Substrate (CoWoS®) technology is a promising candidate to enable most cutting edge applications. In CoWoS® architecture, enormous micro-bumps (uBumps) are adopted for interconnecting top-dies and interposer. Comparing to traditional flip chip bump, uBump encounters high risk for die-to-die joint failure (cold joint) due to limited solder volume. To improve die-to-die stacking quality, uBump coplanarity control is crucial. In this paper, a comprehensive solution is proposed for uBump coplanarity optimization. The solution includes "chip scale" and "uBump scale" approaches, which offers the design/ process flexibility for customized chip integration. Based on experiment results, more than 30% of uBump coplanarity reduction is determined. The comprehensive solution is proven as a successful path to mitigate cold joint risk and improve system integration quality.

Keywords- Chip-on-Wafer-on-Substrate (CoWoS®); micro-bump (uBump); coplanarity; cold joint.

I. INTRODUCTION

AI chip is the core of the next generation industrial revolution. CoWoS® technology emerges as the promising packaging solution to AI chips due to its advantages of multiple chip integration, small form factor, high I/O counts, and lower power consumption. uBump is the critical element to realize die-to-die stacking of CoWoS® technology.

Comparing to tradition flip chip bumps, uBump is characterized as less solder volume and fine pitch, which resulting in higher cold joint risk during assembly process. CoWoS® technology is attracting more attentions recently in ECTC [1-4]. However, few articles investigate the uBump formation optimization for die-to-die assembly risk mitigation. CoWoS® configuration consists of interposer and top-dies. To expand design flexibility and integration performance, different top-dies are adopted. Hence, interposer design shows a trend toward heterogeneous layout from homogeneous setup. The schematic CoWoS® is shown in Fig. 1.

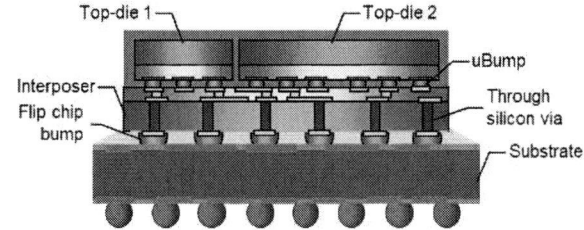

Figure 1.   Schematic CoWoS® architecture.

Typically, uBump coplanarity (similar to max. uBump height – min. uBump height) performance of top-die is ~20% less than which of interposer. The uBump coplanarity control of interposer is more challenging due to larger chip size, non-uniform uBump layout, and greater amount of uBump joints. According to these constraints, die-to-die stacking becomes more and more challenging. The control of uBump coplanarity is crucial to achieve high joint yield and realize heterogeneous integration. The comprehensive solution is proposed in this paper to overcome the above challenges and issues.

II. EXPERIMENT

In this paper, the uBump coplanarity improvement approaches will be divided into "chip scale" and "uBump scale" categories. Chip scale improvements are twofold, including "dummy uBump insertion" and "dynamic control of uBump width." Moreover, uBump scale approach is comprised of "plating current density optimization." The design of experiment will be described as following:

A.   Test Vehicle

In the experiment, a test vehicle with large die size (larger than 20mm x 20mm), small uBump width (less than 30um), multiple uBump metal stacking (including Cu & Solder layers), and heterogeneous uBump layout is designed. After uBump formation process, the uBump coplanarity will be measured. The schematic uBump formation flow is shown in Fig. 2.

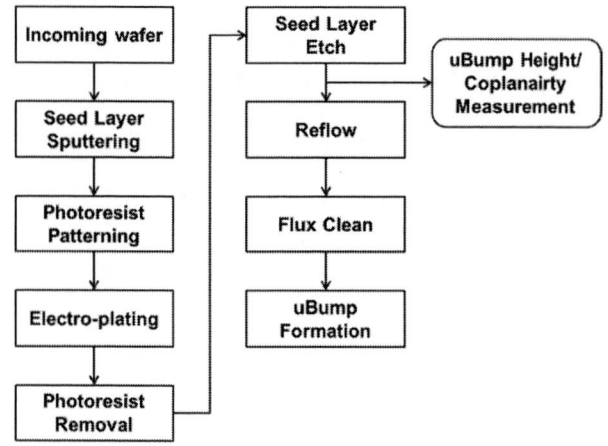

Figure 2.   Process flow of uBump.

B. Chip Scale Approach: Dummy uBump Insertion

uBump coplanarity simulation (MATLAB) is adopted for dummy uBump insertion effect study. Uniform layout of dummu uBumps are place in the peripheral area of original chip with different number of uBump strings. The schematic diagram is shown in Fig. 3.

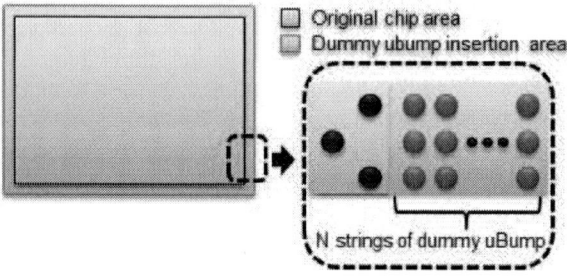

Figure 3.   Dummy uBump insertion in the chip peripheral area.

C. Chip Scale Approach: Dynamic Control of uBump Width

uBump height distribution is correlated to pattern density. For those uBumps located in isolated area, the uBump height tends to be higher. By contrast, uBumps positioned on dense area are characterized with lower height. In the experiment, the chip area is partitioned into several zones for local uBump density calculation, which is depicted as following [5]:

D = Local uBump density
A_i = Projecting area of individual uBump
L = Length of local area,
W = Width of local area

$$D = \sum_{i=1}^{n} A_i \Big/ (W \times L)$$

(1)

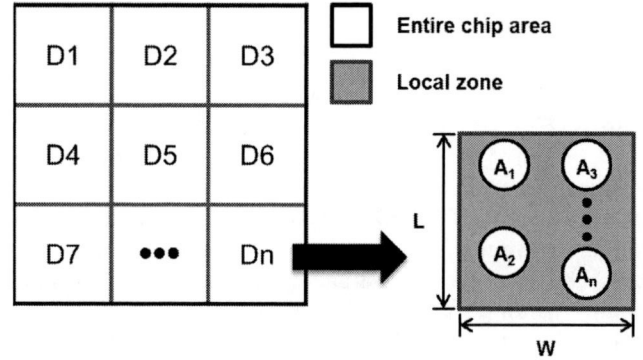

Local uBump density will be utilized for uBump width sizing design. The precision of local uBump density depends on the number of partitioned zones (array size). Proper array size for chip mapping is important to provide sufficient resolution for local uBump density calculation. Lithography mask with different pattern sizing is designed to achieve proactive control of uBump height. Dynamic sizing map of lithography mask is shown in Fig. 4.

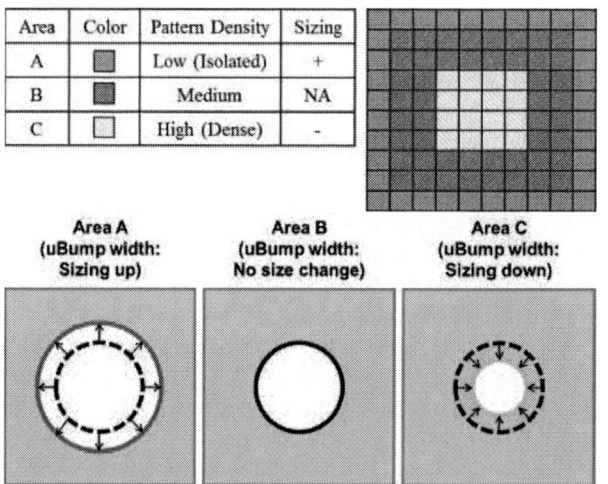

Figure 4.   Dynamic sizing map of lithography mask.

978-1-7281-1500-9/19 $31.00 © 2019 IEEE

D. uBump Scale Approach: Plating Current Density Optimization

Plating current density plays a key role for uBump height and coplanarity control. High plating current density facilitates fast metal grain growth with large grain size, and results in greater uBump height derivation (large uBump coplanarity). On the contrary, low plating current density generates slow metal growth speed with fine grain size. In the experiment, the plating current density is adjusted to validate the response for uBump coplanarity. Schematic diagram of plating density effect vs. metal grain size is shown in Fig. 5.

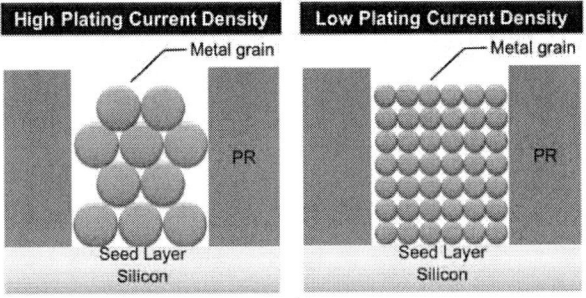

Figure 5. Plating density effect vs. metal grain size.

III. RESULTS & DISCUSSIONS

A. Chip Scale Approach: Dummy uBump Insertion

Based on the experiment, inserting dummy uBumps to the isolated area for local density compensation is beneficial for uBump coplanarity control. However, the isolated area is usually located at chip edge, and dummy uBump insertion will inevitably cause chip size expansion, which will lead to loss of gross chip count of each wafer. As shown in Fig. 6, ~35% coplanarity reduction is achieved with 10.7 % chip size expansion from Leg 4 (+ C strings at chip peripheral area.)

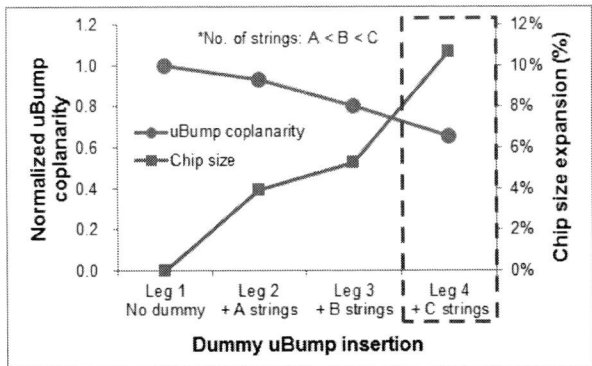

Figure 6. Simulation results of dummy uBump insertion.

B. Chip Scale Approach: Dynamic Control of uBump Width

This novel approach is proposed by delicately sizing up/ down uBump width of lithography mask under the specific range to meet process requirement. For isolated uBumps, uBump width will be sized up to slow down metal deposition speed during electroplating process. The uBump strength is improved after expanding uBump width, and uBump bridge concern is negligible due to isolated uBump pattern. For uBumps in dense area, uBump width is sized down to speed up metal stacking reaction. Even uBump width is smaller, no obvious strength impact of dense area due to loading effect. According to experiment results shown in Fig. 7, 20% uBump coplanarity reduction is determined. Moreover, this approach has successfully resolved the issue of chip size expansion issue.

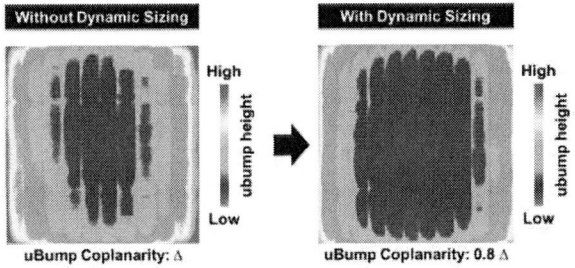

Figure 7. uBump height distribution across the chip.

C. uBump Scale Approach: Plating Current Density Optimization

The experiment results correspond to metal grain size effect during plating process. Lower plating current density generates smaller metal grain, and limits uBump height derivation. As shown in Fig. 8, approximately 38% uBump coplanarity reduction is determined after reducing 50% of plating current density. However, lower plating density will slow down the plating process, and cause the impact on tool efficiency.

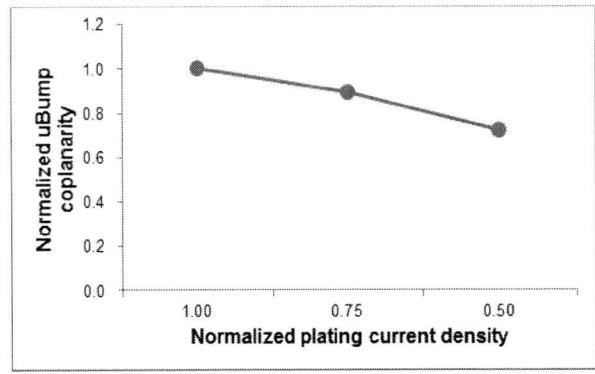

Figure 8. uBump coplanarity vs. plating current density.

D. Summary for uBump Coplanarity Reduction Actions

In this paper, a comprehensive solution is proposed to reduce uBump coplanarity, which contains chip scale and uBump scale approaches. The uBump coplanarity improvement summary is listed in TABLE I.

1) Chip scale approachs

a) Dummy uBump insertion: High effectiveness for coplanarity reduction. However, this approach will inevitably cause chip size expanion and loss of gross die.

b) Dynamic control of uBump width: Moderate ranking of effectiveness. By means of sizing up/ down uBump width, uBump coplanarity is improved. Moreover, the issue of chip size expansion that encountered in dummy uBump insertion has been fullly resolved.

E. uBump scale approach

Plating current density optimization: This approach also showed high effectivness. Reducing plating current density is favored for coplanarity reduction. The side effect is the loss of tool productivity due to prolonged plating process time for uBump formation.

TABLE I.   COPLANAIRTY IMPROVEMENT SUMMARY

Approaches		Ranking of uBump coplanairty improvement		Side effect
		Cu Layer	*Solder Layer*	
Chip scale	Dummy uBump insertion	⊙	⊙	Chip size expansion
	Dynamic control of uBump width	○	○	Mask pattern complexity
uBump scale	Plating current density optimization	⊙	⊙	Tool productivity impact

⊙: High, ○: Moderate

IV. CONCLUSIONS

The seeking of uBump coplanarity control is a trade-off between development cost and assembly risk, and the key to mitigate CoWoS® die-to-die stacking risk is to find a balance between chip design parameters and coplanarity improvement approaches. In this paper, a comprehensive solution that contains "chip scale" and "uBump scale" approaches is proposed to reduce uBump coplanarity. The adoption of different approaches provides the flexibility to adjust uBump coplanarity based on customized chip integration requirements. The experiment result indicates more than 30% reduction of uBump coplanarity is achieved from interposer, which is even better than top-die coplanarity performance. This comprehensive solution has significantly mitigated open/short risk during CoWoS® die-to-die stacking process, and provided a proven path to realize heterogeneous stacking for cutting edge applications.

REFERENCES

[1] Y. L. Chuang, C. S. Yuan, J. J. Chen, C. F. Chen, C. S. Yang, W. P. Changchien, C. C. Liu, and F. Lee, "Unified methodology for heterogeneous integration with CoWoS technology," 63rd Electronic Components and Technology Conference (ECTC), 2013, pp. 852–859.

[2] L. Lin, T. C. Yeh, J. L. Wu, G. Lu, T. F. Tsai, L. Chen, and A. T. Xu, "Reliability characterization of Chip-on-Wafer-on-Substrate (CoWoS) 3D IC integration technology," 63rd Electronic Components and Technology Conference (ECTC), 2013, pp. 366–371.

[3] G. Hariharan, R. Chaware, I. Singh, J. Lin, L. Yip, K. Ng, and S. Y. Pai, "A comprehensive reliability study on a CoWoS 3D IC package," 65th Electronic Components and Technology Conference (ECTC), 2015, pp. 573–577.

[4] R. Chaware, G. Hariharan, J. Lin, I. Singh, G. O'Rourke, K. Ng, and S. Y. Pai, C. C. Li, Z. Huang, S. K. Cheng, "Assembly challenges in developing 3D IC package with ultra high yield and high reliability," 65th Electronic Components and Technology Conference (ECTC), 2015, pp. 1447–1451.

[5] C. C. Liu, J. H. Chen, Y. N. Hsu, M. H. Tsai, C. C. Hung, R. D. Wang, C. S. Liu, T. H. Pan, C. S. Chen, K. C. Liu, and Harry Ku, "Micro bump height derivation control with dynamic sizing patterning," 29th Annual SEMI Advanced Semiconductor Manufacturing Conference (ASMC), 2018, pp. 103–105.

Structural Enhancement for a CMOS-MEMS Microphone under Thermal Loading by Taguchi Method

Chun-Lin Lu and Meng-Kao Yeh
Department of Power Mechanical Engineering
National Tsing Hua University
Hsinchu, Taiwan R.O.C.
cllu9733558@gmail.com; mkyeh@pme.nthu.edu.tw

Abstract—**Structural optimization is a necessary procedure to make progress toward mass production for a new device. Both of structural robustness and superior performance are targets for structural optimization. In this study the structural weakness of a complementary metal oxide semiconductor (CMOS) - microelectromechanical systems (MEMS) microphone chip with 4 by 3 microphone cells by TSMC 0.18 μm CMOS process during thermal loading was identified first by thermal cycling test and thermal stress analysis; then, the optimal structures of the microphone were discussed from viewpoints of thermal stress and sensitivity by Taguchi method. Therein, the finite element (FE) method was adopted for thermal stress analysis and capacitive sensitivity of the microphone was obtained from the equation of sensing capacitance. Moreover, the weakness spots at bottom of the diaphragm in the microphone chip from simulation were verified by the images of scanning electron microscope (SEM) for the chip after 500 cycles of thermal loading in experiment. The results of structural optimization by Taguchi method showed that the microphone with thicker metal and thinner SiO2, wider anchor, and larger diaphragm could reduce the thermal stress in the diaphragm up to 68% than that of the original design. However, for the capacitive sensitivity of microphone chip, the results indicated that the microphone with thicker metal and SiO2, narrower anchor, and larger diaphragm had 5.8 times increase of microphone capacitive sensitivity than that of the original design. This study could provide helpful suggestions for the design and structural robustness of MEMS microphone.**

Keywords- CMOS-MEMS microphone; thermal loading test; finite element analysis; optimal design; Taguchi method

I. INTRODUCTION

Nowadays, the microelectromechanical systems (MEMS) microphone is a common component in consuming electronic devices, such as smartphones, cameras, and laptops. More than two MEMS microphones are placed in a handset device for lower noise and higher quality of recording now [1]. Two major technologies for capacitive MEMS microphone were developed in both of commercial product and academic research. One is to package a surface micromachined MEMS microphone with a separated integrated circuit (IC) chip on a substrate for design flexibility. Another is to integrate active circuit and microphone in a single chip using complementary metal oxide semiconductor (CMOS) process for small form factor and to leverage current semiconductor manufacturing technology. Je et al. [2] focused on the design of surface-micromachined MEMS microphone in this technology and proposed a center-hole membrane for this type of microphone to improve the frequency response and sensitivity. The experimental results showed that the displacement of membrane with center-hole in microphone increased up to 2 times than that of the membrane without center-hole. They also presented that back-plate supporting beams in the surface-micromachined microphone could prevent deformation and increase resonant frequency by simulation [3]. Regarding the design with CMOS process, Huang et al. [4] implemented united microelectronics corporation (UMC) CMOS logic process to demonstrate a CMOS-MEMS condenser microphone with silicon substrate and corrugated diaphragm for enhancing the stiffness of back-plate and easing residual stress respectively. Tounsi et al. [5] investigated the electro-mechanical-acoustic behavior of a new CMOS-MEMS electrodynamic microphone. They built a theoretical model to evaluate the sensitivity of microphone and to control it by damping further. Mao et al. [6] proposed a design of CMOS-MEMS microphone without back-plate to lower deformation of the thin-film caused by residual stress during process. The measured results showed that the out-of-plane deformation of diaphragm, around 2 μm, had good stability under various temperatures.

However, unwanted thermal stress generated, either from fabricating process or operating condition, due to coefficient of thermal expansion (CTE) mismatch in multilayer is a critical factor for the commercialization of MEMS devices [7, 8]. Yew et al. [9] found that the buckling effect induced by thermal stress on membrane during process steps impacted the performance of CMOS-MEMS microphone. Wang et al. [10] developed a process modeling procedure by finite element analysis (FEA) and experiment for structure with multilayer design, such as MEMS microphone. Lu and Yeh [11] compared the thermal stress in diaphragm during thermal cycling test for a CMOS-MEMS microphone with various materials, including aluminum, copper, and polysilicon, by FEA. Optimization of design is considered in general for improving the device for structural robustness or performance of function. The Taguchi method is one of the optimal methods to employ in development of MEMS devices. Ya et al. [12] used Taguchi method to obtain optimal combination of geometric parameter for a capacitive RF-MEMS switch. Yeh et al. [13] demonstrated the optical inspection system used for human chorionic gonadotropin (hCG) concentration detection with optimal parameters by Taguchi method to increase the detection value from 6.25 to 50 mIU/mL. Bagherinia et al. [14] adopted Taguchi method to optimize the geometry of a microbeam in a Lorentz-force MEMS magnetometer for enhancing the performance, including sensitivity and power consumption. Lu et al. [15] lowered the stress generated by shock loading in the diaphragm of a CMOS-MEMS microphone around 35% than that of original design by FEA and Taguchi method.

However, rare literatures discuss the optimization of microphone chip by taking both of structural robustness under thermal loading and capacitive sensitivity performance into consideration. In this study, the robustness of a CMOS-MEMS microphone chip by Taiwan Semiconductor Manufacturing Company Limited (TSMC) 0.18 μm CMOS process under thermal loading was investigated by simulation and experiment. Furthermore, Taguchi method was employed not only to optimize the microphone design for structural enhancement, but to discuss the structural optimization from the viewpoint of performance.

II. MATERIALS AND METHODS

A. Finite Element Analysis

The thermal stress distribution of a CMOS-MEMS microphone chip by TSMC 0.18 μm CMOS process under thermal loading was analyzed by commercial finite element software ANSYS [16]. Fig. 1 shows the images of the microphone chip. A size of 2 x 1.5 x 0.7 mm³ MEMS microphone with 4 by 3 array microphone cell was mounted on print circuit board (PCB) by adhesive. Therein, electrical interconnection between the microphone and PCB was accomplished by wire bonding, as shown in Fig. 1(a). Each microphone cell consists of suspending diaphragm, supporting spring, and back-plate. Enlarged view of single microphone cell by optical microscope (OM) is shown in Fig. 1(b), after removing back-plate for clear vision. A movable diaphragm represents an acoustic plate and is connected to Si substrate by supporting spring and anchor. A total of thickness is 5μm for both of back-plate and diaphragm. Therein, three metal layers with the same thickness (0.6 μm) are in the diaphragm. The diameter of diaphragm is 0.202 mm. Table I shows the designed dimensions for the key structural elements in the microphone [11]. Actual thicknesses of the metal and SiO₂ in the microphone were confirmed by cross-sectional images of scanning electron microscope (SEM). Fig. 2 shows the cross-sectional images of the microphone chip. OM image of the microphone on a-a' cross section in Fig. 1(a) is shown in Fig. 2(a), displaying a total of three microphone cells with suspending diaphragm and an acoustic input hole at PCB. Enlarged SEM image at region b in diaphragm in Fig. 2(a) is shown in Fig. 2(b). The difference of metal thickness between the design and actual

dimensions is less than 5%. Therefore, the design dimensions are used for further analysis.

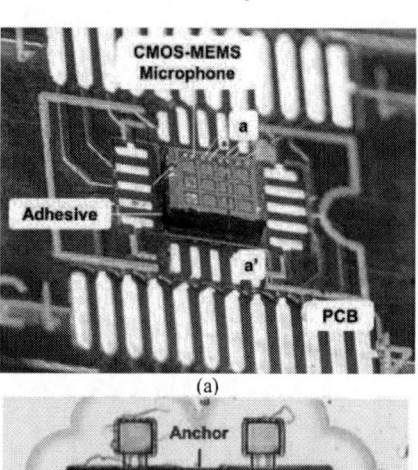

(a)

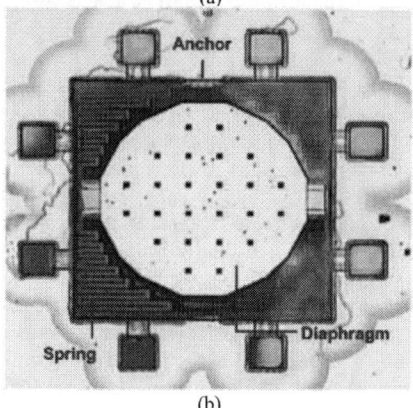

(b)

Figure 1. Images of a CMOS-MEMS microphone chip: (a) The microphone chip (b) Top view of single microphone cell without back-plate

TABLE I.   DESIGNED DIMENSIONS OF THE CMOS-MEMS MICROPHONE CHIP [11]

Items	Dimensions
CMOS-MEMS microphone	2 x 1.5 x 0.7 mm³
Back-plate thickness	0.005 mm (5 μm)
Diaphragm diameter	0.202 mm
Thickness of diaphragm and spring	0.005 mm (5 μm)
Spring width	0.005 mm (5 μm)
Thickness of metal layers in diaphragm	0.0006 mm (0.6 μm)

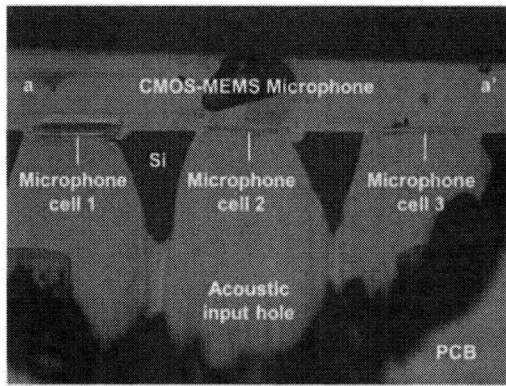

(a)

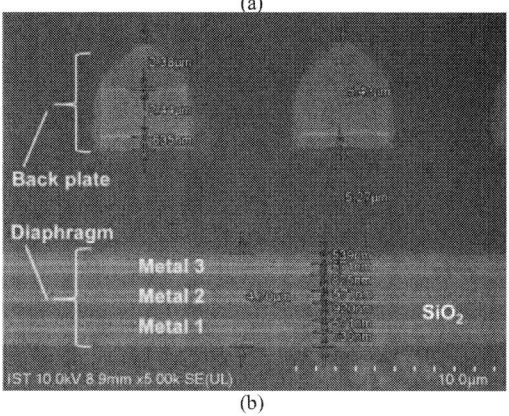

(b)

Figure 2. Cross-sectional images of the microphone chip: (a) OM image of chip on a-a' cross-section in Fig. 1; (b) Enlarged SEM image at region b in diaphragm in Fig. 2(a)

Structural weakness of the microphone chip under thermal loading was evaluated first by thermal stress analysis of finite element model for the original design. Then, to discuss the structural optimization of microphone chip for lowering the thermal stress, Taguchi method was used with finite element analysis. A quarter of finite element model of the microphone chip was built by ANSYS because of the symmetry of chip, in which three-dimensional element with eight nodes (Solid 45) was adopted for evaluation, as shown in Fig. 3(a). The back-plate, diaphragm, Si substrate, adhesive and PCB with an acoustic input hole were considered in thermal stress analysis, as shown in enlarged view in Fig. 3(a). A quarter of finite element model with 1,475,526 nodes (1,365,678 elements) was adopted for analysis after the mesh convergence test. The top view of diaphragm and supporting spring in a single microphone cell is shown in Fig. 3(b), in which the key structural parameters in the microphone chosen for structural optimization, including width of spring, width of anchor and diameter of diaphragm, are indicated.

Table II [11, 17, 18] summarizes the material properties of the microphone chip applied in finite element model, in which the temperature-dependent material property of adhesive and orthotropic property of FR4-PCB were considered in finite element analysis [17]. Referring to the condition G in Joint Electron Device Engineering Council (JEDEC) standard [19], a temperature ranging from -40 to 125 °C separated to five load steps was used as a thermal loading in finite element model and a stress-free state at room temperature was assumed. Moreover, the temperature was assumed to change uniformly during processing steps in analysis and all the residual stresses generated from the process were ignored. A symmetric condition was assumed for the boundary condition and two nodes at bottom of PCB were fixed in the quarter model to prevent rigid body motion. Then, the structural weakness of microphone chip was evaluated by analyzing the thermal stresses at -40 and 125 °C under thermal loading.

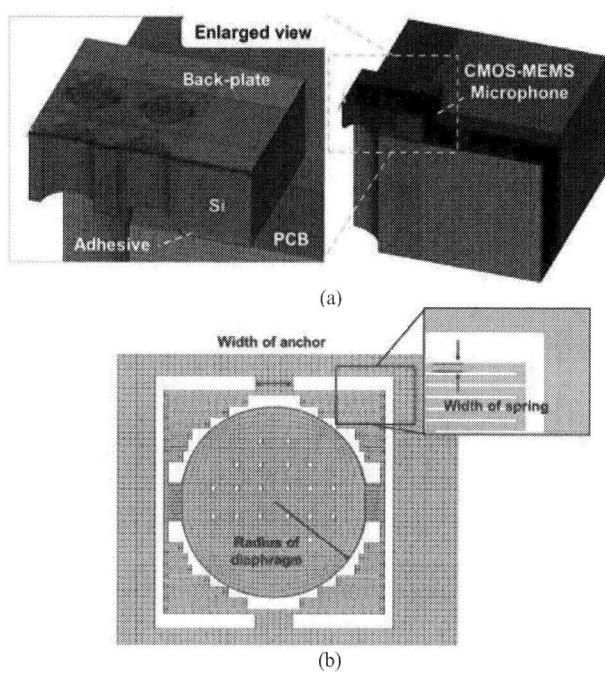

Figure 3. Finite element model of the CMOS-MEMS microphone chip: (a) A quarter of finite element model (b) Top view of single microphone cell model without back-plate

TABLE II. THE MATERIAL PROPERTIES OF FINITE ELEMENT MODELS [11, 17, 18]

Materials	Young's modulus (GPa)	CTE (ppm/°C)	Poisson's ratio
SiO2	75	0.55	0.17
Al	64	24	0.36
Silicon	165	2.5	0.22
Adhesive	3.56@-50 °C	74@-40°C	0.35
	2.76@25 °C	75@23 °C	
	0.15@150 °C	144@200 °C	
PCB	16.9 (XY)	16 (XY)	0.11 (XY)
	7.4 (Z)	84 (Z)	0.39 (YZ/XZ)

B. Experiments

To verify the finite element results, thermal cycling test was adopted to evaluate the structural weakness of the CMOS-MEMS microphone chip with 4 by 3 microphone cells. The prototype sample of the chip was subjected to a thermal loading according to condition G in JEDEC standard JESD22-A104C [19] with temperature ranging from -40 to 125°C in thermal cycling chamber. In order to enhance the thermal loading effect, the read point for the physical check on the test sample was set at 500 cycles directly. The cross-sectional samples before and after the thermal cycling test was prepared by the method of focused ion beam (FIB) cutting. Furthermore, the structural integrity of the sample was compared from the images of scanning electron microscope (SEM) before and after the test. According to the cross-sectional images, the structural weakness of microphone chip was identified and was verified with the stress concentration locations in the chip, obtained from finite element results.

C. Optimal Design

Both of optimization for structural robustness and electrical performance (capacitive sensitivity) of the CMOS-MEMS microphone were discussed by using Taguchi method and finite element analysis. Therein, the L9 orthogonal array in Taguchi method was chosen for the study. The effects of each key parameter were identified from the results of nine finite element models of microphone chip first and an optimal finite element model was obtained from viewpoint of structural robustness and electrical performance respectively.

1) Thermal stress analysis: Based on the finite element analysis results and experiment for the original design of microphone chip, four factors, including the thickness of metal in diaphragm (A), the thickness of SiO2 in diaphragm (B), the width of anchor (C), and the diameter of diaphragm (D), were selected for optimal analysis by Taguchi method. Each factor had three levels and the value increased in the order of level 1, level 2, and level 3. The combination of factor and levels with underline in Table III is the original design.

2) Sensitivity analysis: Besides the thermal stress analysis, the capacitive sensitivity of microphone chip was also considered. The capacitive sensitivity of the nine models was calculated from the difference between the initial capacitance C_0 and the capacitance under 1 Pa $C^{'}$ in the L9 orthogonal array. The equations of C_0 and $C^{'}$ are [20]

$$C_0 = \frac{\varepsilon_0 A}{(h_1 + h_2)\frac{\varepsilon_0}{\varepsilon_s} + (Z_0)} \qquad (1)$$

$$C^{'} = \frac{\varepsilon_0 A}{(h_1 + h_2)\frac{\varepsilon_0}{\varepsilon_s} + (Z_0 - d)} \qquad (2)$$

where h_1 and h_2 are the thicknesses of the single oxide layer in the diaphragm and back-plate, respectively; A represents the area of electrode; ε_0 and ε_s are the dielectric constants of air and SiO2, respectively; Z_0 is the initial gap between back-plate and diaphragm; and d is the displacement of diaphragm.

Better robustness under thermal loading and higher sensitivity based on the same factors were two target functions for the microphone chip. In this study, these two targets are evaluated by Taguchi method.

III. RESULTS AND DISCUSSION

The structural weakness in the CMOS-MEMS microphone chip with original design under thermal loading was identified by finite element analysis and thermal cycling test. Based on the validated finite element models, the optimal structures of microphone were discussed from viewpoints of thermal stress and capacitive sensitivity by implementing finite element analysis and Taguchi method.

A. Finite Element Results of the Original Design

The thermal stress of microphone chip with the original design was analyzed and the von Mises stress distribution of microphone under thermal loading was obtained. Since the maximum von Mises stress at the interface between metal and SiO2 in diaphragm at 125°C (159 MPa) was larger than that at -40°C (103 MPa), this study focused on the thermal stress at temperature 125°C. Fig. 4 shows the von Mises stress distribution of the original finite element model at 125°C. Therein, the stress distribution in diaphragm along a-a' cross-section was enlarged in Fig. 4(a). The results indicate that the stress of metal 1 is larger than those of metal 2 and metal 3. Fig. 4 (b) shows that the stress distribution of the metals in diaphragm and back-plate after finite element analysis. The higher stress occurs at the metal 1 located at the bottom of diaphragm.

B. Experimental Results

The samples of CMOS-MEMS microphone chip were tested under thermal cycling and their structures were examined from SEM image along a-a' cross-section in Fig. 1(a). Enlarged diaphragm SEM images of location c in Fig. 2(a) in CMOS-MEMS microphone chip before and after 500 cycles in thermal cycling test are shown in Fig. 5(a) and 5(b), respectively. No delamination occurred between layers in the suspending diaphragm before thermal cycling test, as shown in Fig. 5(a). However, serious delamination at the interface between metal 1 and SiO2 at the bottom of diaphragm is shown in Fig. 5(b). The experimental results are consistent with the results of finite element analysis for the original model in Fig. 4. Therefore, it concluded that the structural weakness of microphone chip under thermal loading is at the

TABLE III. MODELLING FACTORS AND LEVELS IN TAGUCHI METHOD

Factors (Unit: µm)		Level 1	Level 2	Level 3
A	Metal thickness	0.3	<u>0.6</u>	1.5
B	SiO₂ thickness	0.3	<u>0.9</u>	2
C	Anchor width	20	<u>40</u>	80
D	Diaphragm diameter	150	175	<u>200</u>

interface of different layers in diaphragm, especially at the bottom of diaphragm, according to simulation and experimental results. This study used this validated finite element model to discuss the optimization of microphone chip further.

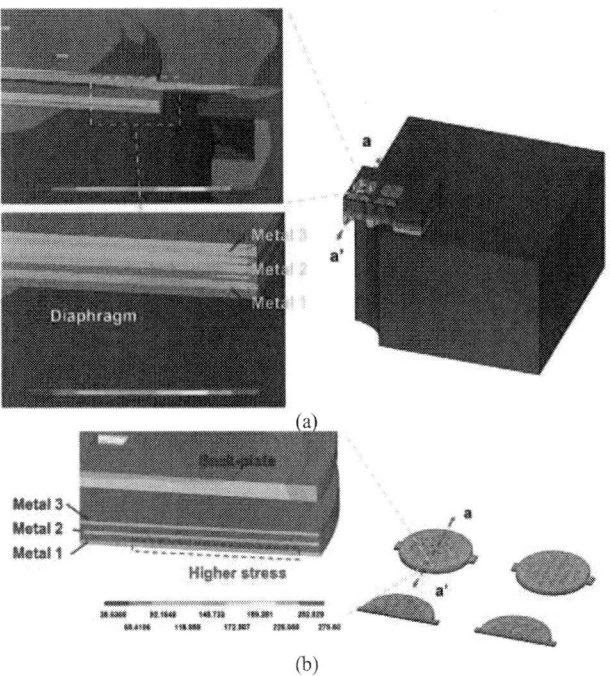

(a)

(b)

Figure 4. The von Mises stress distribution of the original finite element model at 125°C: (a) Enlarged diaphragm along a-a' cross-section (b) The von Mises stress distribution of metals in diaphragm and back-plate

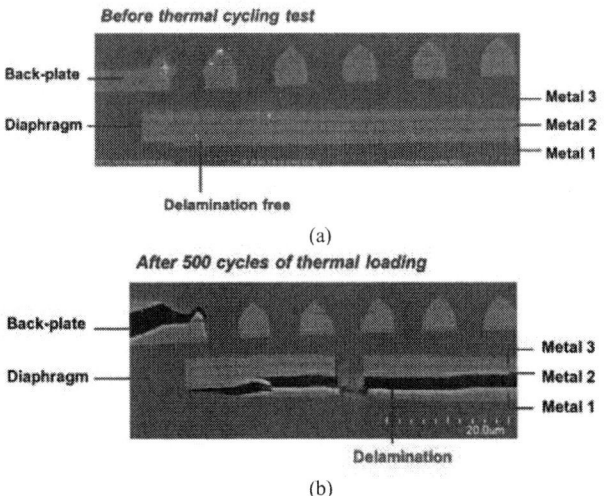

(a)

(b)

Figure 5. SEM images of diaphragm in CMOS-MEMS microphone chip before and after thermal loading: (a) Enlarged cross-sectional view at location c in Fig. 2(a) before thermal cycling test (b) Enlarged cross-sectional view at location c in Fig. 2(a) after 500 cycles thermal cycling test

C. Optimal Design by Taguchi Method

The optimal design of microphone chip under thermal loading based on the thermal stress and capacitive sensitivity are discussed by using Taguchi method with four geometric factors. The goal is to obtain a model with lower thermal stress and higher sensitivity. Fig. 6 shows the response graph of maximum von Mises stress in diaphragm to four factors of microphone models by Taguchi method. The order of the most sensitive factor for von Mises stress at the interface between metal 1 and SiO_2 in diaphragm is SiO_2 thickness (B), metal thickness (A), diaphragm diameter (D), and anchor width (C). And, the combination of optimal microphone model for minimizing the thermal stress is A3B1C3D3, which are highlighted in red in Fig. 6.

The von Mises stress distribution at the interfaces in diaphragm of the optimal model with optimal combination factors obtained from Fig. 6 at 125 °C is shown in Fig. 7. The von Mises stresses at the interface between metal 1 and SiO_2 along the b-b' cross section of diaphragm for both of original and optimal models are shown in Fig. 8. The maximum stress among the three metal layers is 51 MPa. It indicates that the optimal model reduces the maximum stress in the metal 1 of diaphragm up to 68%, when comparing to that of the original model (159 MPa).

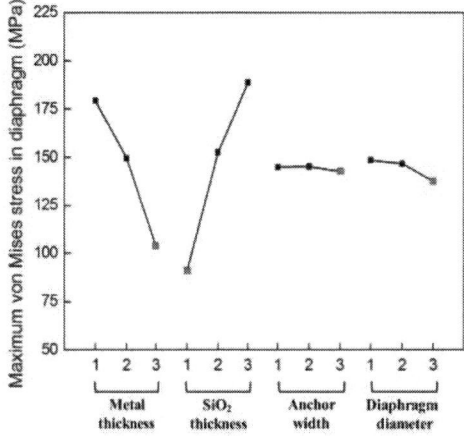

Figure 6. The response graph of maximum von Mises stress to factors of microphone models

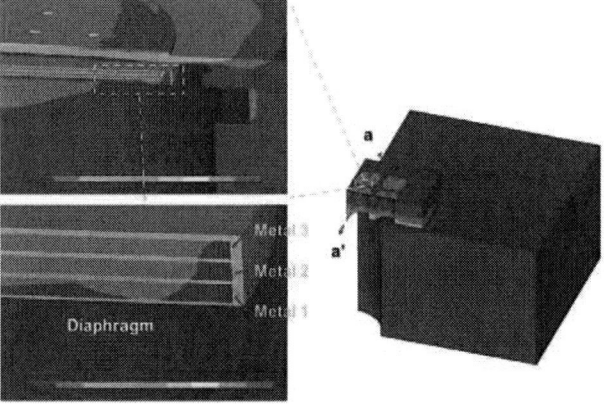

Figure 7. The von Mises stress distribution of optimal finite element model at 125°C

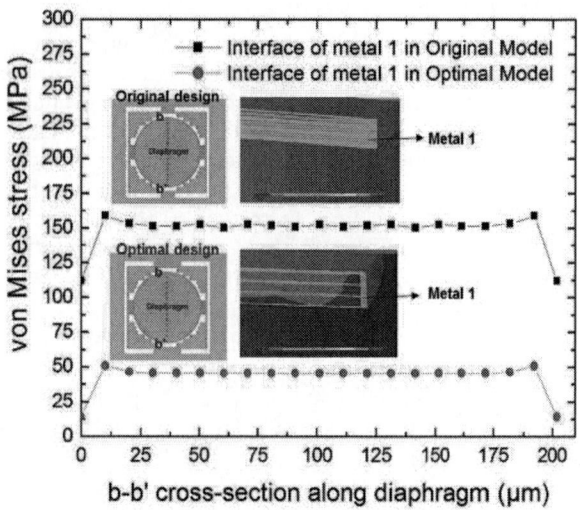

Figure 8. The von Mises stress distribution along the b-b' cross-section in diaphragm for original and optimal models at 125 °C

Nevertheless, the most sensitive factor for capacitive sensitivity of microphone model is anchor width (C) and followed by SiO_2 thickness (B), diaphragm diameter (D), and metal thickness (A). The order is different from that of the effect for thermal stress. Fig. 9 shows the response graph of the capacitive sensitivity to the factor of microphone models. The model with the combination A3B3C1D3, highlighted in red in Fig. 9, has the highest capacitive sensitivity 0.7 fF/Pa. When comparing to that of the original model (0.12 fF/Pa), it improves more than 5.8 times.

Fig. 10 shows the original and optimal microphone models based on the optimization of capacitive sensitivity. According to the results by Taguchi method, the trend of design for higher sensitivity is the model with narrower anchor between suspending diaphragm and Si substrate, and thicker metal and SiO_2. The design concept is different from the design for reducing thermal stress.

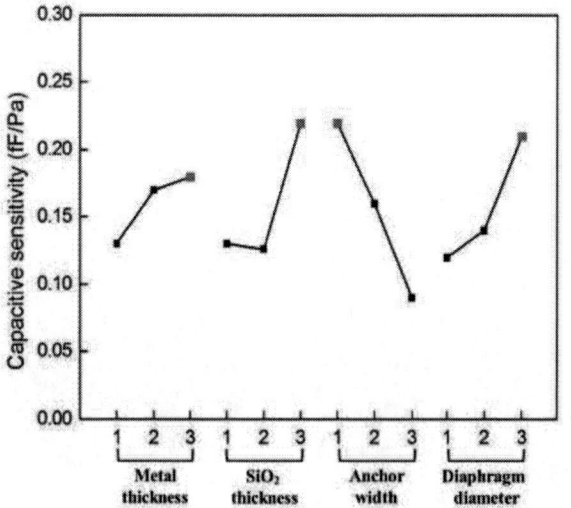

Figure 9. The response graph of capacitive sensitivity to factors of microphone models

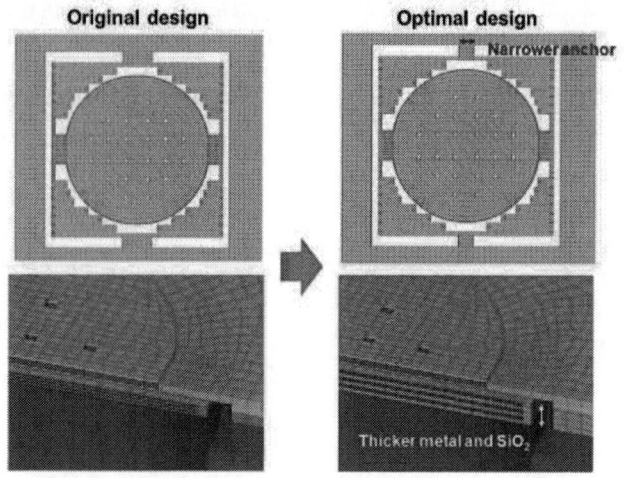

Figure 10. Original and optimal microphone models based on capacitive sensitivity optimization.

IV. CONCLUSIONS

The study identified the structural weakness of a CMOS-MEMS microphone chip by TSMC 0.18 μm CMOS process under thermal loading using finite element analysis and experiment. And optimal designs of the microphone chip from viewpoints of thermal stress and capacitive sensitivity by finite element analysis and Taguchi method were evaluated. The conclusions are summarized as follows:

Based on the thermal stress distribution in the original design, it indicates that the critical spot in microphone structure under thermal loading is the interface of metal at the bottom of diaphragm. And, the results of thermal cycling test were verified with the stress concentration locations in microphone chip, obtained from finite element results.

According to viewpoints of thermal stress, the optimal microphone model could lower the maximum von Mises stress at the interface of bottom metal and SiO_2 in diaphragm under thermal loading up to 68% when comparing to that of the original design. Therein, thicker metal and thinner SiO_2 layers are the trend to have lower stresses.

However, the optimal combination based on the capacitive sensitivity of microphone chip indicates that narrower anchor and thicker both of metal and SiO_2 is a favor design. And, the capacitive sensitivity obtained from the optimal model is 5.8 times than that of the original model.

The results show that the trend of optimal design based on the thermal stress and capacitive sensitivity is contrary from Taguchi method. This study provides useful suggestions for designing CMOS-MEMS microphone chip.

ACKNOWLEDGEMENTS

The authors would like to thank the support from Ministry of Science and Technology, Taiwan, R. O. C., through grant MOST-105-2221-E-007-031-MY3. The support is greatly acknowledged.

REFERENCES

[1] R. Bogue, "Recent developments in MEMS sensors:a review of applications, markets and technologies," Sensor Rev., vol. 33, No. 4, 2013, pp. 300-304, doi.org/10.1108/SR-05-2013-678.

[2] C. H. Je, J. Lee, W. S. Yang, and J. Kim, "The novel sensitivity improved surface micromachined MEMS microphone with the center-hole membrane," Procedia Engineer., vol. 25, 2011, pp. 583-586, doi.org/10.1016/j.proeng.2011.12.145.

[3] C. H. Je, J. Lee, W. S. Yang, J. Kim, and Y. H. Cho, "A surface-micromachined capacitive microphone with improved sensitivity," J. Micromech. Microeng., vol. 23, No. 5, 2013, 055018, doi:10.1088/0960-1317/23/5/055018.

[4] C. H. Huang, C. H. Lee, T. M. Hsieh, L. C. Tsao, S. Wu, J. C. Liou, M. Y. Wang, L. C. Chen, M. C. Yip, and W. Fang, "Implementation of the CMOS MEMS condenser microphone with corrugated metal diaphragm and silicon back-plate," Sensors, vol. 11, 2011, pp. 6257-62691, doi.org/10.3390/s110606257.

[5] F. Tounsi, B. Mezghani, L. Rufer, and M. Masmoudi, "Electroacoustic analysis of a controlled damping planar CMOS-MEMS electrodynamic microphone," Arch. Acoust., vol. 40, No. 4, 2015, pp. 527-537, doi: 10.1515/aoa-2015-0052

[6] W. J. Mao, C. L. Cheng, S. C. Lo, Y. S. Chen, and W. Fang, "Design and implementation of a CMOS-MEMS microphone without the back-plate," 2017 19th International Conference on Solid-State Sensors, Actuators and Microsystems (Transducers), Jun. 2017, pp. 1037-1040, doi.org/10.1109/TRANSDUCERS.2017.7994229.

[7] H. A. C. Tilmans, J. De Coster, P. Helin, V. Cherman, A. Jourdain, P. De Moor, B. Vandevelde, N. P. Pham, J. Zekry, A. Witvrouw, and I. De Wolf, "MEMS packaging and reliability: An undividable couple," Microelectron. Reliab., vol. 52, 2012, pp. 2228-2234, doi.org/10.1016/j.microrel.2012.06.029.

[8] J. Iannacci, "Reliability of MEMS: A perspective on failure mechanisms, improvement soultions and best practices at development level," Displays, vol. 37, 2015, pp. 62-71, doi.org/10.1016/j.displa.2014.08.003.

[9] M. C. Yew, C. W. Huang, W. J. Lin, C. H. Wang, and P. Chang, "A Study of Residual Stress Effects on CMOS-MEMS Microphone Technology," 2009 4th International Microsystems, Packaging, Assembly and Circuits Technology Conference (IMPACT), Oct. 2009, pp. 323-326, doi: 10.1109/IMPACT.2009.5382182.

[10] H. J. Wang, H. A. Deng, S. Y. Chiang, Y. F. Su, and K. N. Chiang, "Development of a process modeling for residual stress assessment of multilayer thin film structure," Thin Solid Films, vol. 584, 2015, pp. 146-153, doi.org/10.1016/j.tsf.2015.01.014.

[11] C. L. Lu and M. K. Yeh, "Thermal Stress and Failure Analysis for a CMOS-MEMS Microphone with Various Metallization and Materials," Proc. Materials for Advanced Metallization Conference (MAM2018), Milano, Italy, Mar. 2018.

[12] M. L. Ya, N. Soin, and A. N. Nordin, "Design and optimization of a low-voltage shunt capacitive RF-MEMS switch," Proc. Symposium on Design, Test, Integration and Packaging of MEMS/MOEMS (DTIP), France, Apr. 2014, doi: 10.1109/DTIP.2014.7056645.

[13] C. H. Yeh, Z. O. Zhao, P. L. Shen, and Y. C. Lin, "Optimization of an Optical Inspection System Based on the Taguchi Method for Quantitative Analysis of Point-of-Care Testing," Sensors, vol. 14, 2014, pp. 16149-16158, doi:10.3390/s140916148.

[14] M. Bagherinia, M. Bruggi, A. Corigliano, S. Mariani, and E. Lasalandra, "Geometry optimization of a Lorentz force, resonating MEMS magnetometer," Microelectron. Reliab., vol. 54, 2014, pp. 1192-1199, doi.org/10.1016/j.microrel.2014.02.020.

[15] C. L. Lu, P. R. Ni, and M. K. Yeh, "Stress analysis of CMOS-MEMS microphone under shock loading by Taguchi method," Microelectron. Reliab., vol. 88-90, Sep. 2018, pp. 824-828, doi.org/10.1016/j.microrel.2018.06.110.

[16] ANSYS, https://www.ansys.com/

[17] B.A. Zahn, "Finite element based solder joint fatigue life predictions for a same die size - stacked - chip scale - ball grid array Package," Proc. 27th Annual IEEE/SEMI International Electronics Manufacturing Technology Symposium, Jul. 2002, pp. 274-284, doi: 10.1109/IEMT.2002.1032767.

[18] M.K. Yeh and C.L. Lu, "Reliability analysis of 3D heterogeneous microsystem module by simplified finite element model," Microelectron. Reliab. vol. 63, 2016, pp. 111-119, doi.org/10.1016/j.microrel.2016.06.001.

[19] JESD22-A104C, Temperature Cycling. JEDEC Standard, 2005.

[20] C.T. Ko, S.H. Tseng, and S.C. Lu, "A CMOS micromachined capacitive tactile sensor with high-frequency output," J. Microelectromech. Syst., vol. 15, No. 6, Dec. 2006, pp. 1708-1714, doi: 10.1109/JMEMS.2006.883569.

2019 IEEE 69th Electronic Components and Technology Conference (ECTC)

A methodology to correct in-fixture measurement of impedance by a machine learning model

Bo-Siang Fang, Chia-Chu Lai, Ying-Wei Lu, Kuan-Ta Chen, Mike Tasi and Don-Son Jiang

Engineering Center
Siliconware Precision Industries Co., Ltd. (SPIL)
Taichung, Taiwan
boxiangfang@spil.com.tw

Abstract—it is usual to measure the characteristic impedance of the transmission line in the factory by the time-domain reflectometer (TDR) with the test fixture, but the insertion loss of the test fixture would raise the measurement error, especially when the transmission line designed with unconventional impedance that is deviated from the system impedance of the measurement instrument. The traditional calibration that requires the standard kits is not generally implemented in production line due to the complicated operation and high cost. Therefore, a method is proposed to correct the measurement by a generalized transformation that is a numeric model trained by the machine learning from the measurements corresponding to the well-chosen features of the test fixture. Based on the precise circuit model of test fixture presented in this paper, a large amount of data used for training can be produced by circuit simulation instead of real measurements. An experimental instance is given to demonstrate that the measurement error could be suppressed from around 20% to below 3%.

Keywords-Time domain reflectometry; Transmission line measurements; Calibration; Impedance measurement; Cables; Machine learning (key words)

I. INTRODUCTION

A. The impedance measurment

Engineers usually inspect the characteristic impedance of transmission lines to adjust the processing parameters when manufacturing package substrate or printed circuit board (PCB). In order to increase the production throughput by quickly loading and unloading device under test (DUT), the test fixture that connects the measurement instrument and the DUT is necessary. The key components of the test fixture are the probe interface and the coaxial cables. With the probe, the DUT can keep nondestructive after measurement; and with the cable, the instrument can be set flexibly at the arbitrary position assigned according to the production management. In Fig. 1, there is an example to display how the test system configured and a schematic drawing of the test system is presented in Fig. 2.

Usually, the measurement instrument of test system used in the production line is different from what is set in R&D site. The verification of impedance in R&D site is usually made by a vector network analyzer (VNA) that can be calibrated by the coplanar standard kits to fully remove the systematic errors in measurement [1]. The effect of the test fixture is included in systematic errors, so it is also removed from the measurement.

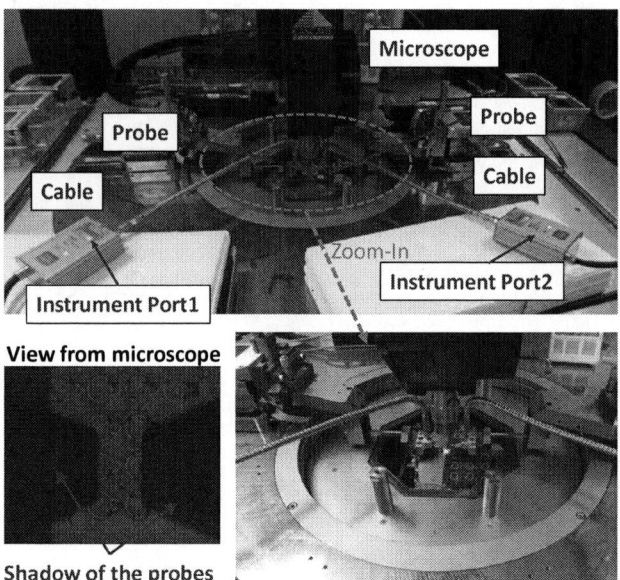

Figure 1. Example of a test system to measure impedance.

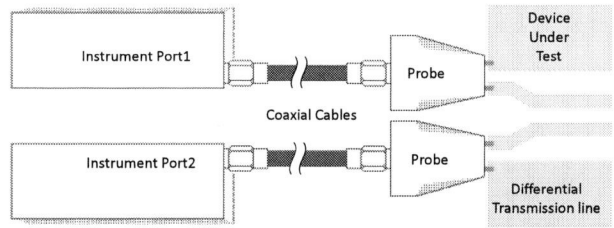

Figure 2. The schematic drawing of the test system presented in Fig. 1.

On the other hand, the TDR is preferred in production line because of operating convenience. The calibration of it is made by a coaxial terminator to zero the receiving signal reflected by the instrument itself, so the systematic errors caused by the cables and the probe are remained and degrade the measurement accuracy. Although a method had been developed for TDR in the past to fully remove the systematic errors [2], the complicated operation and the usage of coplanar standard kits make it expensive and unpopular in production line.

B. The measurement err

In the past, there were many experiences proofed that the difference of the measurement between R&D site and the

978-1-7281-1500-9/19 $31.00 © 2019 IEEE

production line was unobvious, so the impedance measurement in production line has been accurate enough even if no correction performed. With the growing of high-speed digital application, there are more and more novel transceivers presented with unconventional impedance. So the characteristic impedance of the transmission lines in package substrate or PCB should be also designed to match the input impedance of transceivers. However, we observed that the unconventional impedance raised the measurement error when we inspected impedance under the effect of the test fixture.

An experimental instance is given in Fig. 3 to demonstrate the error by measuring a 50 Ohm differential line both in R&D site and in production line. The measurement in the R&D site was made by a VNA (Anritsu MS4647B) that had been calibrated to shift the reference plane to the tip of the probe, so no effect of test fixture was residual. In the production line, the TDR (Tektronix DSA8200) was zeroed before it connected to the probes through the 1-meter long cables. The impedance measured at the R&D site was 49.6 Ohm that had only 0.8% deviation to reveal that it was made with precision. But the same device measured by TDR in the production line was 59.4 Ohm that was 18.8% deviation. The result in the production line had exceeded the specification which was normally defined as 10% or 15%. The error had led the process engineers to rework with the wrong treatments, then the reworked devices were still out of specification in the final inspection made by the R&D engineer. The situation not only delayed the lead-time of production, but also wasted the raw materials.

For the measurement error appeared in particular device, most engineers in production line would usually roughly add an offset value to compensate the error instead of doing the complete calibration. The offset value was usually determined by a golden sample that was measured in both R&D site and production line in advance. However, the compensation is not valid for the general usage. Once the device has changed, the correlation must be redone.

With the unconventional impedance become usual in novel high speed-digital circuit, a generalized correction on the measurement in the production line would be important. But the current solutions, neither the complicated calibration nor the particular compensation are not suitable for the production line. Therefore, we propose a method to correct the error accurately for generalized condition and the operation is simply to be exercised in production line.

II. THE ANALYSIS OF ERROR

The first assumption to reduce the complexity of the error analysis is to ignore the effect of the probes on a test fixture, because the insertion loss of the probe is significantly smaller than the longer cable. The assumption can be validated by comparing the actual TDR measurement with a transient simulation that only imported the Scattering parameters (S-parameters) of the cables and the DUT but no probe. If the equivalently measured impedance calculated from the simulation is very close to the real TDR measurement, the assumption should be valid.

The transient simulation was made by Keysight Advanced Design System (ADS) for the instance. The circuit diagram of the simulation is presented in Fig. 4 where we used a voltage sources with the step waveform and a 50 Ohm resistor that is the system impedance (Z_S) in every port to simulate the differential source of the measurement instrument. The elements of the coaxial cables and the DUT were set by importing the S-parameters which were made by the real measurements. After computing, the equivalently measured impedance was determined by a representative reading in the DUT section along the time-domain impedance (z_{TDR}) that was calculated by (1).

$$z_{TDR}(t) = \frac{Z_S \cdot V_{P1}(t)}{V_{P0}(t) - V_{P1}(t)} + \frac{Z_S \cdot V_{N1}(t)}{V_{N0}(t) - V_{N1}(t)}. \quad (1)$$

As a result, the difference between the simulation and the real TDR measurement was only 2.2% deviation. The deviation was accepted in the production line because it was quite smaller than the 18.8% error mentioned above. The simulation proofed that the main component of the measurement error should be caused by the effect of the cables, so the effect of the probe was ignored in the following analysis. Of course, if the characteristic of the probe can be taken into account, the accuracy of analysis can be improved. However, to characterize the probes is not easy, because it is required to take S-parameter measurements between the coaxial and coplanar interfaces at the same time.

The further analysis is to use a parameterized model of the differential lines to replace the really measured S-parameters

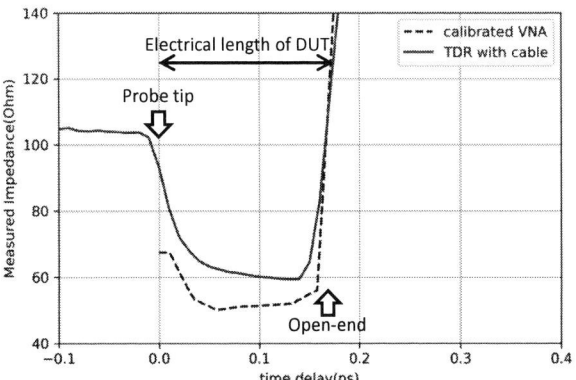

Figure 3. The inaccurate impedance measurements when using TDR.

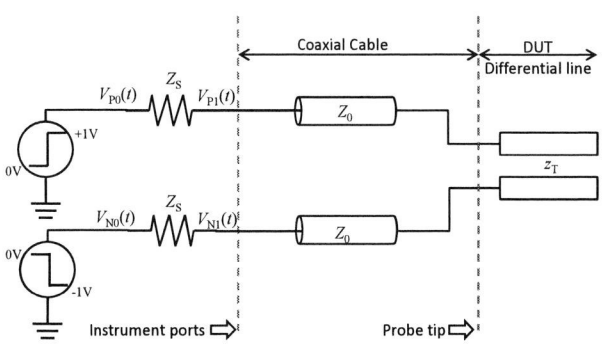

Figure 4. The circuit diagram for trainsient simualtion.

978-1-7281-1500-9/19 $31.00 © 2019 IEEE

of the DUT in ADS, so the equivalently measured impedance can be estimated versus the true impedance of the DUT. As the result shown in Fig. 5, the equivalently measured impedance is presented as the black dot line and the blue line is the true DUT impedance (z_T) that is set in the parameterized model as shown in Fig. 4. The deviation between the black dot line and the blue line is the measurement error. The minimum error of measurement is appeared at where the impedance of the DUT is close to the system impedance of the instrument, i.e. 100 Ohm for differential ports, but the error raises significantly when the impedance of DUT designed much higher or lower than the system impedance. The experiment can explain why the same test system worked well in the past, but it raised the error when the DUT was designed with unconventional impedance e.g. the differential 50 Ohm in the instance.

III. THE CORRECTION

The difference between the equivalently TDR measured impedance and the true DUT impedance presented in Fig. 5 is a conspicuously nonlinear curve. Instead of correcting it by a constant offset, we propose correcting the TDR measurement by a mathematical transformation that can be deployed to production line by a software tool, Microsoft Excel for example. Then it will be used to correct the TDR measurement of the DUT that designed with arbitrary impedance for every case. The suitable form of mathematical transformation implemented in Excel is a polynomial as shown in (2) where the variables of the polynomial is the TDR measurement (z). For each test system, the degree of the polynomial (N) and the coefficient of every term (c_N) should be determined by minimizing the difference between the predicted impedance of the DUT (z_P) and the true DUT impedance (z_T).

$$z_T \cong z_P = c_0 + c_1 z + c_2 z^2 + \cdots + c_N z^N. \quad (2)$$

Solving the polynomial equation presented in (2) by the linear regression is a well-used method, so a large amount of transmission lines that made with diverse impedances and the corresponding measurements may be required. Meanwhile, the large amount of DUT that were made into reality will induce high cost for an experiment, so we propose generating

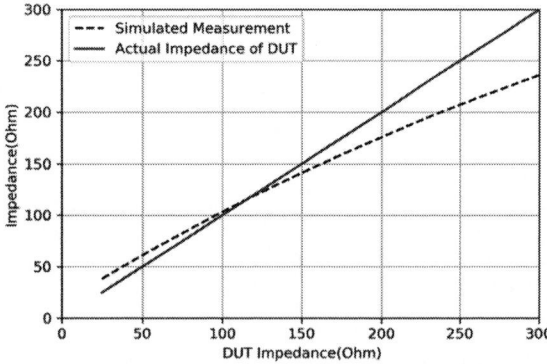

Figure 5. The equivalently TDR measured impedance estimated by the transient simulation.

the equivalently measurement by the circuit simulation in which the correlation between simulation and measurement had been proofed. The procedures of the work to determine the polynomial are presented as below steps:

1) Characterize the Test Fixture: To measure the S-parameter of the coaxial cables that are set in the test system.

2) Set a transient simulation: To configure a simulation as the schematic circuit presented in Fig. 4 where the characterisitics of cables are imported from the S-parameter meausred in 1st step and the DUT is set by a parameterized model with the specified impedance.

3) Execute simualtion with swept impedance of DUT: the swept range of the DUT impedance is usually determined according to the experience of the correlation done in reality, so, we set the swept range from 25 Ohm to 300 Ohm with 1 Ohm step for the instance.

4) Prepare the data set: To take the representative reading in the DUT portion of time-domain impedance from the result of the simulation. In the demonstration, we got 276 readings from simulation to be the equivalent TDR measurements that are corresponding to the 276 DUT impedances set in the model. In the following procedures, the equivalent TDR measurement would be treated as the feature data; and the actual impedances of the DUT would be treated as the label data. One dataset included one label data and the corresponding feature data, so there were 276 dataset used for machine learning.

5) Group the data set: To randomly select 20% of dataset as test data, and 80% of dataset as training data.

6) Normalize the data: To shift the feature data by the mean of the data, and to scale the feature data by the standard deviation of the data. Then a normalized feature data with zero mean and unit norm would be obtained.

7) Extend the features: To generate new feature data by executing inner product of the existing feature data for every dataset. After iterations, the interactive feature data would be obtained to compose a polynomial with specified degree. The degree of the polynomial would be set as a variable that will be determined by minimizing the prediction error within several trials in the following procedures. If the degree of polynomial is set too low, the error of correction would be high; if the degree of polynomial is set too high, there might be an extremely high error appeared when correcting the data that did not be taken for analysis i.e. the regression is overfitting.

8) Train and validate the regression model: To determine the coefficients of polynomial presented in (2) is now very easy by using diverse software toolkits developed for machine learning. For the instance, we selected the scikit-learn as the toolkit because of its open source, and the least absolute shrinkage and selection operator (LASSO) was selected as the regression model because of its good shrink of terms in the polynomial [3]. The LASSO model was fitted with the training data grouped in 5th step, and the result was

978-1-7281-1500-9/19 $31.00 © 2019 IEEE

validated by the test data to optimize the penalty parameter. Finally, the coefficients of the polynomial presented in (2) were determined after cross-validation [4].

9) Determine the degreee of the polynomial: To iterate the 7th and 8th steps with diverse values of degree to determine a reasonable number of degrees by minimizing the error calculated with the test data. For the instance, the history of the trials is shown in Fig. 6, then the degree of 4 is selected to guarantee that the maximum error after correction is below 2%.

10) Deploy to the production line: In the demonstration, the Microsoft Excel is the desired software tool to be used in the production line because of the convenience and popularity. The coefficients of the polynomial presented in (2) can be stored in the cells of an Excel worksheet. One cell defined as the input to represent the TDR measurement, and the polynomial equation is set in another cell to calculate the output which is the predicted impedance of the DUT. Of course, if the engineer write a custom program for operation in the production line, the model selected in 8th step can be more complex to gain more accuracy and wider application range.

IV. GENERALIZED CORRECTION

A. The Model of Test Cable

The polynomial determined by the above procedures is cable-bonded, so the management of correction will become a trouble when the products are made in different factories that might use different cables in test systems. If the cables of test systems were different, the engineer need to retrain their own polynomial for every site. Therefore, we would like to adjust several steps in the original procedures to make a generalized correction with the cable-independent polynomial to provide accurate correction among the factories.

The most important adjustment to achieve cable-independent correction is to replace the real measurement of the cable by a parameterized model at the 1st step in the original procedures. Because the insertion loss and the characteristic impedance of the cables are the key factors to raise the error in the TDR measurement of impedance, both

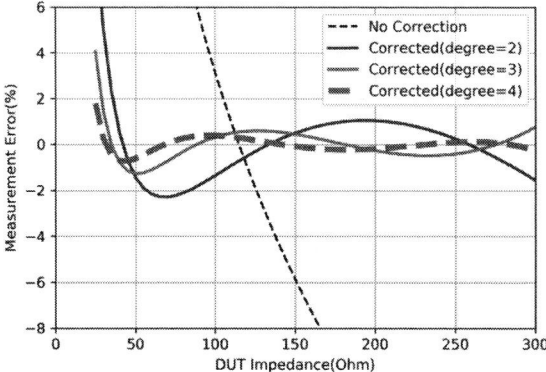

Figure 6. The prediction error versus the degree of polynomial.

factors should be set in the parameterized model for simulation and also will be taken as the new feature data that prepared at the 4th step in original procedures.

However, the insertion loss of cable changes with the frequency. If all the insertion loss sampled along the frequency axis were used for training, it would create large amounts of feature data in every dataset for training. The operation of the measurement on a cable usually takes more than 201 points sampled in frequency domain, so the number of feature data would be more than 203 for each dataset. If the number of the feature data in the dataset is too high, the possibility of overfitting will get higher. On the other hand, the large amount of data required high volume of memory for computing. Well-controlling the number of feature data in the dataset can make the result accurate and lower cost of computing.

The usual solution of machine learning to reduce the dimension of data is to use the principal component analysis (PCA) according to a statistical procedure [3]. Nevertheless, we preferred to reduce the dimension of data by a precise equation that is determined by the attenuation phenomenon of the coaxial cable. As the equation shown in (3), we used 2 features to represent the insertion loss of a coaxial cable along the frequency domain. The symbol, T denotes the insertion loss of the coaxial cable in dB; the symbol, f denotes the operating frequency in GHz; and the coefficients, c_1 and c_2 denote the factors of conductive loss and dielectric loss, respectively. The conductive loss of a coaxial cable is usually increased with the square root of frequency and the dielectric loss is increasing with the frequency increasing [5]. By using (3), we can use c_1 and c_2 as the features to represent a large amount of insertion loss sampling along the frequency domain, the efficiency and accuracy would be better than blindly using the PCA when reducing the number of features.

$$T = c_1\sqrt{f} + c_2 f. \quad (3)$$

The coefficients, c_1 and c_2 can be determined by solving (3) with 2 measurements of insertion loss that read from different frequencies. In the demonstration, we select 1GHz and 20GHz to solve the equation, then the coefficients were given as (4) where the symbol, T_{1GHz} and T_{20GHz} denoted the measurements of insertion loss in dB at 1GHz and 20GHz respectively. Directly using T_{1GHz} and T_{20GHz} to be the features instead of c1 and c2 would be much easier for the engineer in the production line.

$$c_1 = \frac{20}{20-\sqrt{20}} \cdot T_{1GHz} - \frac{1}{20-\sqrt{20}} \cdot T_{20GHz}$$
$$c_2 = -\frac{\sqrt{20}}{20-\sqrt{20}} \cdot T_{1GHz} + \frac{1}{20-\sqrt{20}} \cdot T_{20GHz} \quad (4)$$

According to the recommended practice of machine learning, the correlation between the features should be minimized to increase the accuracy of prediction after training. However, the insertion loss of a coaxial cable is always increasing with frequency, so the selected features T_{1GHz} and T_{20GHz} are correlated to each other. Therefore, we represented (4) by (5) where we use the symbol, L to present

978-1-7281-1500-9/19 $31.00 © 2019 IEEE

T_{20GHz} for short and the symbol, r to present the ratio of the insertion loss at 1GHz to that at 20GHz. Then the correlation between the new features, r and L would become lower.

$$c_1 = \left(\frac{20}{20-\sqrt{20}} \cdot r - \frac{1}{20-\sqrt{20}}\right) \cdot L$$
$$c_2 = \left(-\frac{\sqrt{20}}{20-\sqrt{20}} \cdot r + \frac{1}{20-\sqrt{20}}\right) \cdot L \qquad (5)$$

In order to verify the accuracy of (3) and (5), there were 3 different cables measured and modelled by the equations (3) and (5). The comparison between the real measurements and models are presented as shown in Fig. 7 where all the modeling responses almost fit to the measurements in wide frequency range.

B. The Adjustment of Traning Procedures

By the accurate equations (3) and (5) proofed in Fig. 7, the measurement of cable in 1st step can be replaced by a circuit

model in ADS. In the demonstration, we realized the parameterized model by an ideal transmission line with a specified characteristic impedance (Z_0) and the attenuation set according to (3) and (5). The cable model introduces not only 3 new parameters for simulation, but also 3 new feature data for machine learning. In the instance, these parameters were summarized in Table 1 where the particular swept range of simulation was also indicated. The swept ranges also defined the valid range of input parameters for correction. After computing for hours, a large amounts of equivalently measured impedance (z) corresponding to the input parameters were obtained.

As the 4th step in the original procedures, the actual impedance of the DUT (z_T) was set as the label data of training model, and the other parameters L, r, Z_0 and z are grouped as the feature data, i.e. the input parameters of the desired polynomial. According to the 6th step, the respective normalization of the feature data was taken and presented as the expressions in Table 1.

Thereupon the LASSO model was trained according to the original procedures from 7th step to 9th step. As a result, the desired polynomial was determined and shown in (6) where the predicted impedance of the DUT (z_P) was calculated from the TDR measurement and the properties of the test cable including the characteristic impedance and the insertion loss at 1GHz and 20GHz.

Finally, the worksheet presented in the 10th step should be modified with the additional inputs that are the properties of the cables. Thus, the engineer in production can type the information obtained from the vendor or the one-time measurement of the cables in reality. In order to make the

TABLE I.   INPUT AND OUTPUT PARAMETERS OF THE CORRECTION

Parameters	Unit	Swept range	Normalized features and the expressions
$L = T_{20GHz}$	dB	0.5-5.0	$L_n = (L-2.75)/1.4361$
$r = T_{1GHz}/T_{20GHz}$	-	0.1-0.3	$r_n = (r-0.2)/0.0632$
Z_0	Ohm	47-53	$Z_{0n} = (Z_0-50)/1.8708$
z	Ohm	-	$z_n = (z-146.7385)/63.9181$
z_T	Ohm	25-300	-

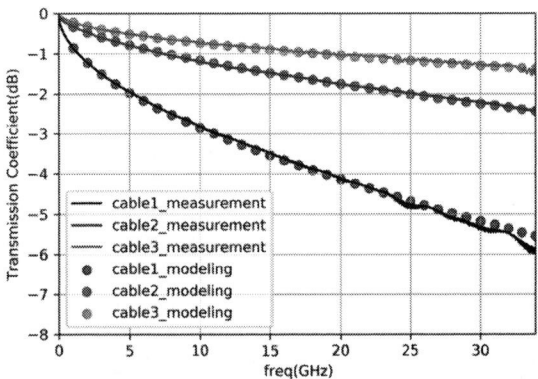

Figure 7.   The insertion loss of the cables and the circuit model.

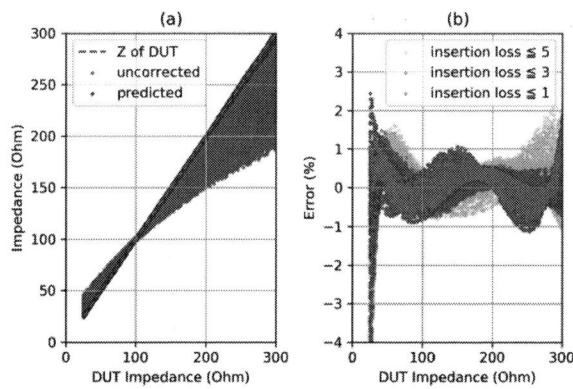

Figure 8.   The validation of prediction by (6) when using the test data.

$z_T \cong z_P(z, L, Z_0, r) = z_P(z_n, L_n, Z_{0n}, r_n) = z\,(0.0503\,L_n + 0.023\,r_n + 0.0919\,z_n - 0.0094\,Z_{0n} + 0.0019\,L_n^2 + 0.0159\,L_n r_n + 0.0657\,L_n z_n - 0.0055\,L_n Z_{0n} + 0.0329\,r_n z_n - 0.0024\,r_n Z_{0n} - 0.0059\,z_n^2 - 0.007\,z_n Z_{0n} + 0.0007\,L_n^3 + 0.0038\,L_n^2 r_n + 0.0095\,L_n^2 z_n - 0.0009\,L_n^2 Z_{0n} + 0.0013\,L_n r_n^2 + 0.0262\,L_n r_n z_n - 0.0022\,L_n r_n Z_{0n} - 0.0031\,L_n z_n^2 - 0.0033\,L_n z_n Z_{0n} + 0.0003\,L_n Z_{0n}^2 + 0.0015\,r_n^2 z_n - 0.0001\,r_n^2 Z_{0n} - 0.0006\,r_n z_n^2 - 0.0015\,r_n z_n Z_{0n} + 0.0001\,r_n Z_{0n}^2 + 0.0143\,z_n^3 - 0.0045\,z_n^2 Z_{0n} + 0.0008\,z_n Z_{0n}^2 - 0.0003\,Z_{0n}^3 + 0.0006\,L_n^4 + 0.0002\,L_n^3 r_n + 0.0007\,L_n^3 z_n - 0.0002\,L_n^3 Z_{0n} + 0.0004\,L_n^2 r_n^2 + 0.0033\,L_n^2 r_n z_n - 0.0002\,L_n^2 r_n Z_{0n} + 0.0041\,L_n^2 z_n^2 - 0.0007\,L_n^2 z_n Z_{0n} + 0.0001\,L_n^2 Z_{0n}^2 + 0.001\,L_n r_n^2 z_n + 0.0019\,L_n r_n z_n^2 - 0.001\,L_n r_n z_n Z_{0n} + 0.0001\,L_n r_n Z_{0n}^2 + 0.0061\,L_n z_n^3 - 0.0009\,L_n z_n^2 Z_{0n} + 0.0002\,L_n z_n Z_{0n}^2 - 0.0002\,L_n Z_{0n}^3 + 0.0002\,r_n^4 - 0.0001\,r_n^3 Z_{0n} + 0.0004\,r_n^2 z_n^2 + 0.0022\,r_n z_n^3 - 0.0002\,r_n z_n^2 Z_{0n} + 0.0001\,r_n z_n Z_{0n}^2 - 0.0002\,r_n Z_{0n}^3 - 0.0042\,z_n^4 + 0.0014\,z_n^3 Z_{0n} - 0.0002\,z_n Z_{0n}^3 + 0.0001\,Z_{0n}^4 + 1.0818)\ . \qquad (6)$

978-1-7281-1500-9/19 $31.00 © 2019 IEEE

TABLE II. THE VALIDATION BY A REAL EXPERIMENT.

DUT	Cable properties			Measurement			Measurement Error
	Loss @ 1GHz T_{1GHz}	Loss@ 20GHz T_{20GHz}	Z_0	TDR (z)	VNA (z_T)	This work (z_P)	TDR → improved by this work
#1	0.860 dB	4.138 dB	52 Ω	60.7 Ω	50.1 Ω	49.6 Ω	21.2%→-1.0%
#2				59.4 Ω	49.6 Ω	48.1 Ω	19.8%→-3.0%
#3				61.5 Ω	51.6 Ω	50.5 Ω	19.2% →-2.1%
#1	0.225 dB	1.046 dB	50 Ω	54.0 Ω	50.1 Ω	51.4 Ω	7.8%→2.5%
#2				53.0 Ω	49.6 Ω	50.3 Ω	6.9%→1.4%
#3				55.1 Ω	51.6 Ω	52.5 Ω	6.8%→1.8%

valid correction, the range of the input parameters should be in the valid ranges denoted in Table 1. Besides, the selected cables and probes should be qualified with the return loss larger than 20dB to meet the assumptions that simplified the modeling of the circuits.

C. The Validation

The predicted impedance (z_P) that is calculated by (6) is ideally equivalent to the impedance of the DUT (z_T). The accuracy of prediction can be validated by checking the test data that were not used for training, the result of validation is presented in Fig. 8(a). The green dots were presented as the uncorrected TDR measurement corresponding to the DUT impedance, the green dots were deviated from the red line that denoted the impedance of the DUT (z_T). So the TDR measurement is not accurate. The blue dots stand for the predicted impedance (z_P) that is very close to the red line, so the correction by (6) made significant improvement of accuracy on the raw TDR measurement.

The percentage of error after correction was shown in Fig. 8(b) where the test data were classified into 3 groups according to the loss of the cable at 20GHz. Generally, the equation (6) can improve the accuracy of measurement to below 3% error, except when the impedance of the tested DUT was small. In the instance, the smaller impedance of the DUT measured, the higher error obtained. But if the cables were selected with lower loss, the error can be suppressed. As the example presented in Fig. 8(b), if the loss of cable is 1dB at 20GHz, the maximum error would be below 2.5%.

On the other hand, the LASSO model that was selected in the 8th step in original procedures can be replaced by the other complex or nonlinear models. With a more complex model selected and higher degree of the polynomial, the accuracy of prediction can be improved further. But for the complex model, the limitation is the functionality of software tool for deployment. Besides, for the higher degree of the polynomial, the engineer should operate carefully to avoid overfitting.

V. EXPERIMENT

Because (6) was determined by the model trained with the data that generated from pure circuit simulation, an experiment was set to check the accuracy of correction by the real measurements on 3 DUT with 2 kinds of cables. The insertion loss of the cables and the impedance of the DUT had

been taken in advance and listed in Table 2. The actual impedance of each DUT was measured by a calibrated VNA to be the standard to estimate the error. As a result, the error can be improved from around 20% to below 3% when using the cable that is 4.1dB loss at 20 GHz; and from 7.8% to below 2.5% when using the high-end cable that is 1.04dB loss at 20 GHz.

However, the prediction errors in Table 2 are a little bit higher than what it expected in Fig. 8(b), the gap should be caused by the preciseness of the parameterized model of the test fixture in ADS. Because of simplification, we ignored the characteristic of the probe, and we only considered the characteristic impedance and the insertion loss of the cable instead of importing the full 2-port S-parameters. Once the parameterized model of test fixture can be estimated more precisely and accurately, the gap should be minimized. Even so, the 3% prediction error in the instance is acceptable for production line, because it is to trade the accuracy off against higher cost that would be used to collect and to characterize the diverse probes and cables.

VI. CONCLUSION

In this paper, a method was proposed to train a machine learning model with the data generated from the circuit simulation to correct TDR impedance measurement. An experiment demonstrated that the error of correction could be reduced to below 3% against the raw measurement of the TDR that has error around 20%. The uncertainty of measurement in production line could be reduced by this work to enhance production yield and efficiency without additional investment.

REFERENCES

[1] L. Hayden, "An enhanced line-reflect-reflect-match calibration," *67th ARFTG Microw. Meas. Conf., pp. 143-149, 2006.*

[2] W. Su and S. M. Riad, "Calibration of time domain network analyzers," *IEEE Transactions on Instrumentation and Measurement, Vol. 42, Issue: 2, 1993.*

[3] L. Buitink, et al, "API design for machine learning software: experiences from the scikit-learn project*", ECML PKDD Workshop, pp. 108-122, 2013.*

[4] M. Bowles, "*Machine learning in python: essential techniques for predictive analysis*", John Wiley & Sons, Inc., *pp.129-131, 2015.*

[5] HUBER+SUHNER, "*RF connector guide: understanding connector technology*", the 4th edition, pp. 32-33, 200

Material and Structure Design Optimization for Panel-Level Fan-Out Packaging

Dao-Long Chen, Ian Hu, KarenYU Chen, Meng-Kai Shih, David Tarng, Dinos Huang, JY On
Group R&D
Advanced Semiconductor Engineering, Inc.
Kaohsiung, Taiwan
e-mail: JimDL_Chen@aseglobal.com

Abstract—In recent years, the evolution of semiconductor packages is driven by the consumer demands of Internet of Things (IoT), mobile phones, wearable devices, and tablets. Manufacturing has evolved from wafer-level chip-scale packaging (WLCSP) to fan-out wafer-level processing (FOWLP) and 2.5/3D packaging. The panel-based fan-out processing was highly noticed. It is the best solution to have both high density and cost efficiency. In this study, the warpage behavior at mold and carrier release stages of the Panel-Level Fan-Out (PLFO) device was optimized by the theoretical and simulation tools. The panel-level (300×300 mm²) warpage of the PLFO test vehicle was measured by advanced metrology analyzer (aMA), which can not only do the global measure of warpage of the full panel, but also do the local measure of the unit package. The Finite Element Method (FEM) analysis was then conducted for the corresponding test vehicle with considering the temperature dependent material properties, including Young's modulus and Coefficient Thermal Expansion (CTE), and the simulated warpage was highly correlated to the measured result for modeling validation. Four-layer and three-layer curvature theoretical formulas were developed for mold and carrier release stages, respectively, and used to predict the trends versus different factors. These trends were verified by FEM simulations. The optimal curvature was then obtained by the optimized factors.

Keywords- panel-level; fan-out; warpage; advanced metrology analyzer; finite element method.

I. INTRODUCTION

In recent years, the evolution of semiconductor packages is driven by the consumer demands of Internet of Things (IoT), mobile phones, wearable devices, and tablets. Higher transmission date rates are required with rapid spread of portable devices and increase for video streaming, photo sharing, and other data-intensive applications. Advanced packaging has enabled integration of devices with new capabilities to meet the need for increased functionality that consumers demand. Manufacturing has evolved from wafer-level chip-scale packaging (WLCSP) [1] to fan-out wafer-level processing (FOWLP) [2] and 2.5/3D packaging [3].

The FOWLP has two branches of chip mount technologies. One is the so-called chip-first, and the other is the so-called chip-last [4]. The chip-first technology can further be classified as face-up and face-down [5]. Figure 1 shows the process flow of chip-first (face-down as an

example) and chip-last technologies. The advantage of chip-first is its lower cost than chip-last. However, it is sensitive to die shift problem, and if any processes fail, the known good die is lost. On the other hand, the chip-last technology fabricates the Re-Distribution Layers (RDL) first, and the known good dies are bonded on the known good RDL. Therefore, the chip-last (or the so-called RDL-first) is innately lower yield lost than the chip-first.

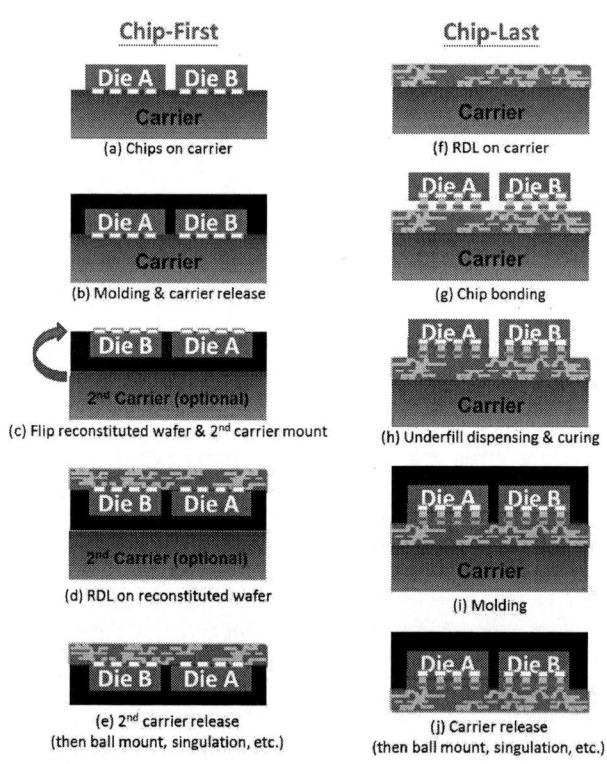

Figure 1. Process flow of chip-first and chip-last packaging.

Recently, the panel-based fan-out processing was highly noticed [6]. It is the best solution to have both high density and cost efficiency. As shown in Figure 2, (a) shows a 300 mm (12 inches) diameter reconstituted wafer with lesser chips, while (b) shows a square 300×300 mm² reconstituted panel with more chips. It is easily seen that the chip usage of a wafer is largely lower than that of a panel, especially for larger and rectangular chips. Migrating to panel-level fan-out packaging will increase the number of chips, and will result in significant productivity improvement over traditional wafer-level fan-out packaging.

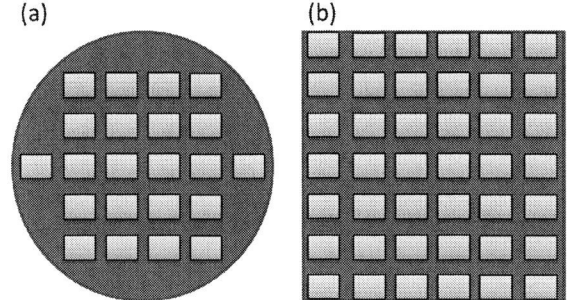

Figure 2. Fan-out packaging with (a) wafer form and (b) panel form.

However, the critical challenge of panel-level packaging is that its warpage during processing and final packaged product is larger than that of wafer-level packaging. The warpage of a wafer can be estimated by the formula

$$w_{\text{wafer}} = \kappa R^2, \qquad (1)$$

where κ is the curvature of the warpage, R is the radius of the wafer; while the warpage of a panel by

$$w_{\text{panel}} = 2\kappa \left(\frac{L}{2} \right)^2, \qquad (2)$$

where L is the length of the panel. If $R = L/2 = 150$ mm for example, and the curvatures are the same, then the warpage of a panel will be double of that of a wafer.

In this paper, the warpage behavior at key stages (mold and carrier release) of the Panel-Level Fan-Out (PLFO) device that developed by our research group are studied by optimal material and structure designs for panel-level processing with Finite Element Method (FEM) analysis. Figure 3 shows the typical PLFO device with top view and cross-section view. It can be applied to the high-level application processor (AP) with memory or graphics processing unit (GPU) products. It is noted that the line/space of RDL is 2/2 μm, which is advisable to the high density signal connection.

For the optimal material and structure design, the materials considered are the molding material (epoxy molding compound, EMC) and the glass carrier. The warpage is caused by the mismatch of Coefficient Thermal Expansion (CTE) between different components, and resisted by the modulus of each component. Therefore, the selection of suitable material properties is very important. On the other hand, the structure designs considered are the carrier thicknesses, chip thickness, EMC thickness, and chip to package area ratio (the inverse of the so-called fan-out ratio). The area ratio determines the contact area to cause the CTE mismatch, while the thicknesses determine the rigidities of

components. Theoretical estimation and trend analysis of warpage based on the effective model [7] and multilayer thin plate theory [8] will be provided to guide the improvement direction. In additional, a tool with deep depth of field to measure thermal dynamic panel warpage is crucial for providing sufficient data for modeling validation. Equipment based on three-dimensional (3D) digital image correlation method, named advanced metrology analyzer (aMA), is used to measure the panel level warpage at several interval statements during the process. Using the optimal design, the warpage at key process stages can be significantly reduced, and therefore minimize the yield loss of the fan-out devices.

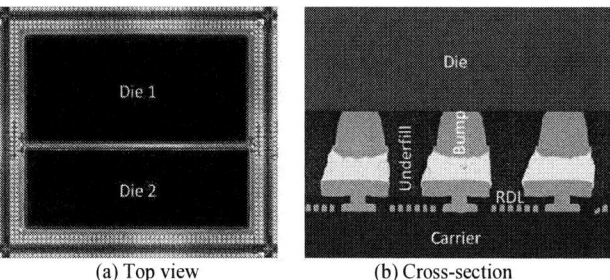

| (a) Top view | (b) Cross-section |

Figure 3. Package outline of PLFO device with (a) top view and (b) cross-section view.

II. TEST VEHICLE AND ANALYSIS METHOD

A PLFO package, which is composed of two chips, 2 μm line width / 2 μm space and 3 copper layers, is chosen to be the test vehicle for panel level chip-last fan-out process evaluation. The process flow of panel chip-last fan-out is shown in Figure 1. The RDL forms firstly on a glass carrier by a release layer, the RDL includes a deposited seed layer and multiple plated copper layers and coated dielectric layers. Then two chips are bonded on the known good RDL. After that, the tiny size bumps are protected by underfill. Compression molding process is conducted for package protection and handling. With the epoxy molding compound, the panel could be released from the carrier and then goes to seed layer etching to isolate under bump metallurgy (UBM). Finally, ball mount, optional grinding, and panel saw processes are conducted sequentially to result the fan-out package.

A measurement tool for large size panel and deep depth of field is required for measuring thermal dynamic panel warpage to provide validated data for modeling correlation. Stereoscope for 3D image reconstruction is a good method for measuring an instant warpage during dynamic thermal loading; and the measured digital images could be correlated to have the thermal deformation information.

Figure 4 shows the panel-level (300×300 mm^2) warpage of the PLFO test vehicle that measured by aMA. It can be seen that not only the global measure of warpage of the full panel, but also the local measure of the unit package. The local measure analysis can be used to confirm the uniformity of warpage behaviors at different locations.

978-1-7281-1500-9/19 $31.00 © 2019 IEEE

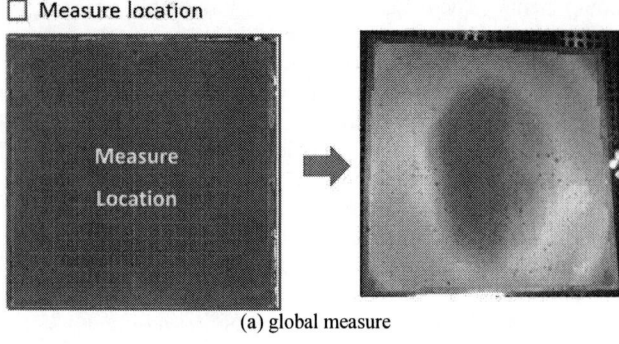

(a) global measure

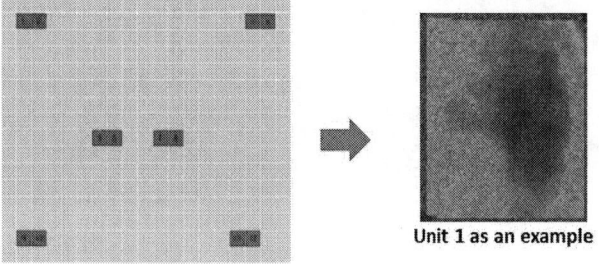

(b) local measure

Figure 4. Warpage measurements of (a) global measure and (b) local measure by aMA.

In order to use the FEM prediction for the optimal warpage, the verification between simulation and measurement should be implemented. Figure 5 (a) shows the panel-level (300×300 mm^2) warpage of the PLFO test vehicle at final stage that measured by aMA. The FEM is then conducted for the corresponding test vehicle with considering the temperature dependent material properties, including Young's modulus and CTE, and the simulated contour is shown in Figure 5 (b). As shown in these figures, the warpage of experiment and simulation has similar warpage contour; meanwhile, the simulated warpage is highly correlated to the measured result for modeling validation.

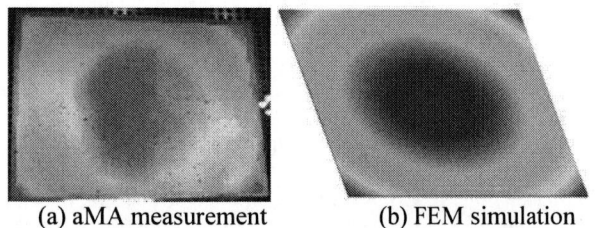

(a) aMA measurement   (b) FEM simulation

Figure 5. Panel-level warpage of measurement and simulation.

The fan-out structure can be simplified as a four-material structure as shown in Figure 6. The EMC has the material properties of modulus, E_1, CTE, α_1, and the structure thickness, h_1. The effective EMC layer with thickness h_2 is

mixed by the silicon dies (neglecting underfill) and EMC, and its material properties can be estimated by

$$E_2 = \frac{E_{\text{die}} A_{\text{die}} + E_1 A_{\text{EMC}}}{A_{\text{die}} + A_{\text{EMC}}} = \left(E_{\text{die}} - E_1\right)\phi + E_1,$$

$$\alpha_2 = \frac{\alpha_{\text{die}} A_{\text{die}} + \alpha_1 A_{\text{EMC}}}{A_{\text{die}} + A_{\text{EMC}}} = \left(\alpha_{\text{die}} - \alpha_1\right)\phi + \alpha_1, \quad (3)$$

where E_{die} and α_{die} are the modulus and CTE of die, $\phi = A_{\text{die}} / (A_{\text{die}} + A_{\text{EMC}})$ is the area ratio of die to package (i.e., the inverse of the so-called fan-out ratio). The effective RDL layer with thickness h_3 is mixed by the Cu and passivation layers, and its material properties can be approached by

$$E_3 = \frac{\sum_k E_k h_k}{\sum_k h_k}, \quad \alpha_3 = \frac{\sum_k \alpha_k h_k}{\sum_k h_k}, \quad (4)$$

where E_k, a_k, and h_k are the modulus, CTE, and thickness of the k-layer which may be the Cu layer or the passivation layer. And the carrier has the modulus, E_4, the CTE, α_4, and the thickness, h_4.

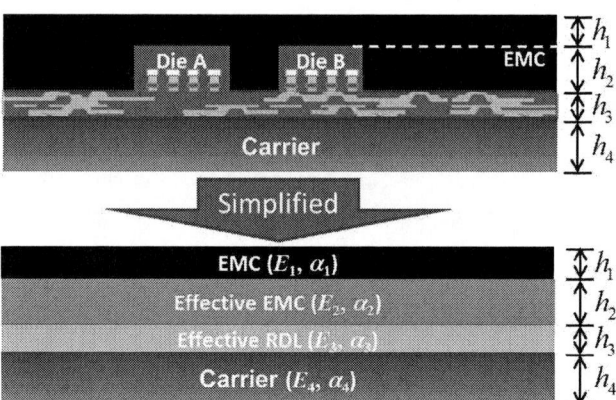

Figure 6. Simplification of a fan-out structure to a four-material structure.

For the multi-layer warpage estimation, Suhir [8] showed that the curvature, κ, of the warpage of a four-material structure with different thicknesses can be estimated by the formula

$$\kappa = -\frac{T - T_0}{D} \left[\alpha_{12}\lambda_3 \left(\beta_{12}\lambda_4 - \beta_{24}\lambda_1\right) + \alpha_{13}\lambda_2 \right. \quad (5)$$
$$\left. \times \left(\beta_{13}\lambda_4 - \beta_{34}\lambda_1\right) + \alpha_{23}\beta_{23}\lambda_1\lambda_4\right],$$

where

$$D = EI\lambda + \beta_{12}^2\lambda_3\lambda_4 + \beta_{13}^2\lambda_2\lambda_4 + \beta_{14}^2\lambda_2\lambda_3$$
$$+ \beta_{23}^2\lambda_1\lambda_4 + \beta_{24}^2\lambda_1\lambda_3 + \beta_{34}^2\lambda_1\lambda_2,$$
$$\lambda = \lambda_1\lambda_2\lambda_3 + \lambda_2\lambda_3\lambda_4 + \lambda_3\lambda_4\lambda_1 + \lambda_4\lambda_1\lambda_2, \quad (6)$$
$$\alpha_{12} = \alpha_1 - \alpha_2, \quad \alpha_{23} = \alpha_2 - \alpha_3, \quad \alpha_{13} = \alpha_1 - \alpha_3,$$
$$\beta_{12} = \frac{h_1 + h_2}{2}, \quad \beta_{23} = \frac{h_2 + h_3}{2}, \quad \beta_{34} = \frac{h_3 + h_4}{2},$$
$$\beta_{14} = \beta_{12} + \beta_{23} + \beta_{34}, \quad \beta_{24} = \beta_{23} + \beta_{34},$$
$$\lambda_1 = \frac{1}{E_1 h_1}, \lambda_2 = \frac{1}{E_2 h_2}, \lambda_3 = \frac{1}{E_3 h_3}, \lambda_4 = \frac{1}{E_4 h_4},$$

h_i, α_i, as well as E_i are the thickness, CTE, and modulus of the corresponding material i, respectively, EI is the flexural rigidity of the system; and T and T_0 are the current temperature and the reference temperature of the system, respectively. The panel warpage at molding stage can then be obtained by substituting the curvature into the equation (2). Furthermore, for the carrier release stage, the panel warpage can be estimated by letting $h_4 \rightarrow 0$. These two stages are the key of the warpage control, because for molding stage, the CTE mismatch is mainly caused by the mismatch of EMC and carrier; while for carrier release stage, the rigidity of whole system is significantly reduced so that the warpage increases correspondingly. The warpage of carrier release stage is also directly related to the warpage of final product.

III. RESULTS AND DISCUSSION

In order to evaluate the effects of EMC, the mold thickness, h_1, and its CTE, α_1, are studied at both mold and carrier release stages. In addition, the modulus, E_1, of EMC and CTE is assumed to have the relation, $E_1\alpha_1 = \text{constant}$, because the modulus of normal EMC usually decreases as CTE increases.

The curvature vs. the EMC thickness at mold and carrier release stages is analyzed and shown in Figure 7. For mold stage, the curvature is nearly unchanged because of the high rigidity of carrier. However, the influence of EMC thickness is significant for carrier release stage because if the thickness is thinner, the domination of EMC becomes lesser, and the influence of higher CTE passivation layers becomes important.

Figure 8 shows the curvature vs. CTE of EMC at mold and carrier release stages. For mold stage, the lower CTE means the higher modulus, so that the rigidity becomes higher; while for higher CTE, it induce not only the lower rigidity, but also the higher CTE mismatch to carrier. For carrier release stage, the main CTE mismatch is between EMC and passivation layers. Due to the higher CTE of passivation, the higher CTE of EMC is needed. From the calculation results, the range of α_1 from 8 to 20 ppm/°C can keep the curvature between ±0.1 ppm/μm^2 for mold stage, and the curvature from -0.8 to -0.5 ppm/μm^2 for carrier release stage.

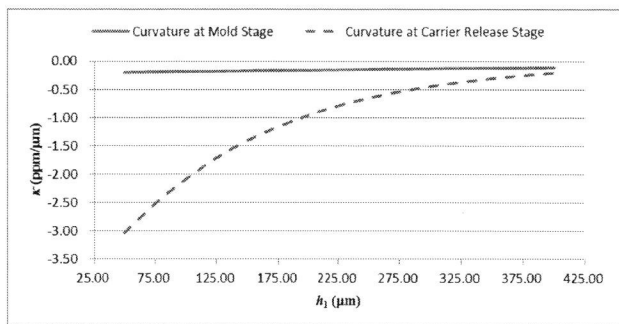

Figure 7. Curvature vs. EMC thickness at mold and carrier release stages

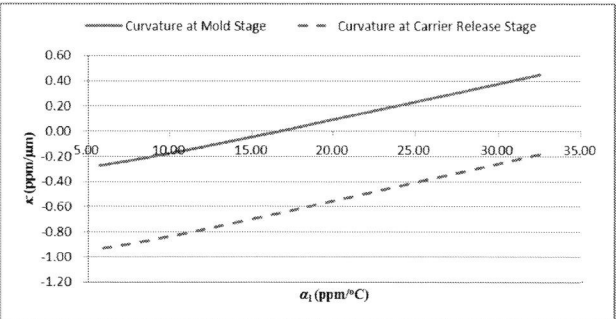

Figure 8. Curvature vs. CTE of EMC at mold and carrier release stages

For the die effects, Figure 9 shows the curvature vs. die thickness at mold and carrier release stages. For mold stage, the curvature is reversing when the die thickness is about 120 μm. However, the change of curvature is not significant for increasing the die thickness. On the other hand, the die thickness changes significantly the curvature for the carrier release stage. For thicker die thickness, the curvature, and thus the warpage, can be smaller. In summary, the thicker die thickness can improve the warpage significantly at carrier release stage, but only slightly worsens the warpage at mold stage.

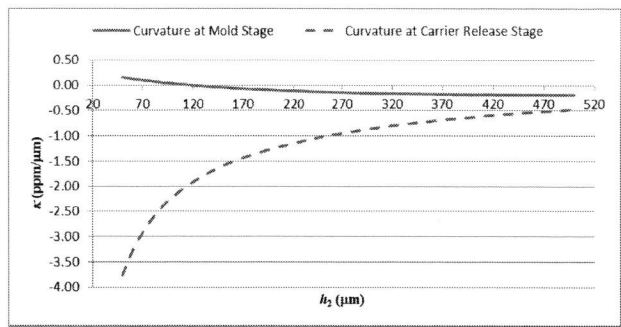

Figure 9. Curvature vs. die thickness at mold and carrier release stages

Figure 10 shows the curvature vs. die to package area ratio, ϕ, at mold and carrier release stages. For mold stage, the curvature is reversing when ϕ is about 0.3. However, the change of curvature is not significant for increasing the die to package area ratio. On the other hand, the die to package area ratio changes significantly the curvature for the carrier release stage. For larger ϕ, the curvature can be smaller. In summary, similar to the die thickness effect, the larger die to package area ratio can improve the warpage significantly at carrier release stage, but only slightly worsens the warpage at mold stage.

The thickness and modulus effects of carrier are shown in Figure 11. It is seen that the curvature is nonlinearly dependent on the carrier thickness, while it is nearly linearly dependent on the modulus of carrier. It is intuitive that the warpage is smaller as the flexural rigidity larger. The flexural rigidity can be defined mathematically as [9]

$$D_4 = \frac{E_4 h_4^3}{12(1-\nu_4^2)}, \qquad (7)$$

where ν_4 is the Poisson's ratio of the carrier. It can be seen that the rigidity is linearly dependent on the Young's modulus, while it is cubically dependent on the thickness. Thus, the thickness effect is significantly larger than modulus effect. In short, increasing carrier thickness and modulus can increase the flexural rigidity, and thus decrease the warpage.

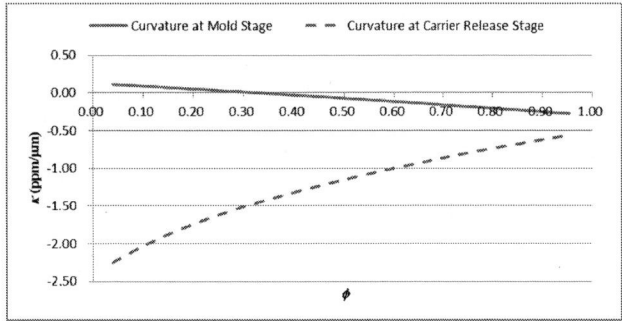

Figure 10. Curvature vs. die to package area ratio at mold and carrier release stages

Figure 12 shows the simulation verifications of the panel-level warpages at mold and carrier release stages. Each variable has three levels with symbol (-), (0), (+), and the values of three levels are (-) < (0) < (+). The simulation results are well matched with the trends of theoretical prediction. However, the predicted values may larger than the simulated values. It may be mainly caused by the mixed rules of material properties of die and EMC. For the situations of unit die and two dies that with the same area, the warpages are obviously different. However, for the mixed rules (3), only the total area is considered, no matter the die is unit or separated. Nevertheless, both analysis tools can be used complementarily; the simple theory tool used to predict the trends, while the simulation tool used to obtain the more accurate values.

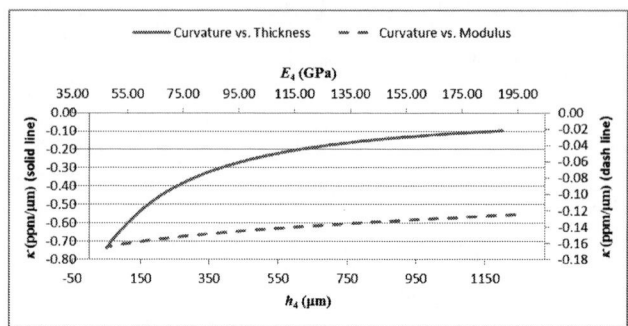

Figure 11. Curvature vs. carrier thickness (solid line) and modulus (dash line) at mold and carrier release stages

Once the trends are obtained, one can do the optimal design with the optimal factors by the simulation. The optimal curvature at mold stage is about -0.07 ppm/μm, while the one at carrier release stage is about -0.21 ppm/μm.

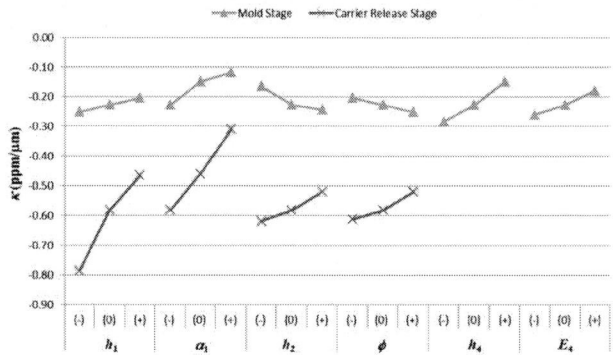

Figure 12. Simulation verifications for curvature vs. different variables with three levels ((-), (0), (+)) at mold and carrier release stages

IV. CONCLUSION

In this study, the warpage behavior at mold and carrier release stages of the PLFO device was optimized by the theoretical and simulation tools. The panel-level (300×300 mm^2) warpage of the PLFO test vehicle was measured by aMA, which can not only do the global measure of warpage of the full panel, but also do the local measure of the unit package. The FEM was then conducted for the corresponding test vehicle with considering the temperature dependent material properties, including Young's modulus and CTE, and the simulated warpage was highly correlated to the measured result for modeling validation. Four-layer and three-layer curvature theoretical formulas were developed for mold and carrier release stages, respectively, and used to predict the trends of warpages versus different factors including thickness and CTE of EMC, die thickness and die to package area ratio, thickness and modulus of carrier. Increasing the thickness and CTE of EMC can decrease the curvature, especially for carrier release stage. While decreasing the die thickness or increasing die to package area ratio is helpful for mold stage, but unhelpful for

978-1-7281-1500-9/19 $31.00 © 2019 IEEE

carrier release stage. The increasing of carrier modulus and thickness can reduce the curvature of mold stage. These trends were verified by FEM simulations. Based on the prediction of optimized trends, the optimal curvature/warpage can then be obtained.

REFERENCES

[1] M. K. Shih, H. C. Shih, Y. C. Lee, D. Tamg, and C. P. Hung, "Solder joint reliability analysis for large size WLCSP," *2017 International Conference on Electronics Packaging (ICEP)*, Yamagata, 2017, pp. 61-65.

[2] J. H. Lau *et al.*, "Warpage and Thermal Characterization of Fan-Out Wafer-Level Packaging," in *IEEE Transactions on Components, Packaging and Manufacturing Technology*, vol. 7, no. 10, pp. 1729-1738, Oct. 2017.

[3] C. C. Lee *et al.*, "An Overview of the Development of a GPU with Integrated HBM on Silicon Interposer," *2016 IEEE 66th Electronic Components and Technology Conference (ECTC)*, Las Vegas, NV, 2016, pp. 1439-1444.

[4] S. Chen, S. Wang, C. Lee, A. Hsieh, J. Hunt, and W. Chen, "Chip Last Fan Out as an Alternative to Chip First," International Symposium on Microelectronics: FALL 2015, Vol. 2015, No. 1, pp. 000245-000250.

[5] T. Braun *et al.*, "Trends in Fan-out wafer and panel level packaging," *2017 International Conference on Electronics Packaging (ICEP)*, Yamagata, 2017, pp. 325-327.

[6] T. Braun *et al.*, "Panel Level Packaging - A View Along the Process Chain," *2018 IEEE 68th Electronic Components and Technology Conference (ECTC)*, San Diego, CA, 2018, pp. 70-78.

[7] F. X. Che, D. Ho, M. Z. Ding and X. Zhang, "Modeling and design solutions to overcome warpage challenge for fan-out wafer level packaging (FO-WLP) technology," *2015 IEEE 17th Electronics Packaging and Technology Conference (EPTC)*, Singapore, 2015, pp. 1-8.

[8] E. Suhir, "Predicted Bow of Plastic Packages of Integrated Circuit (IC) Devices," J. Reinf. Plast. Comp., vol. 12(9), pp. 951–972. 1993.

[9] L. D. Landau and E. M. Lifshitz, Theory of Elasticity, 3rd ed., vol. 7. Butterworth-Heinemann, 1986, pp. 42.

The microstructure and mechanical property of the high entropy alloy as a low temperature solder

Li Pu[1], Quanfeng He[2], Yong Yang[2,3], Xiuchen Zhao[1], Zhuangzhuang Hou[1], K. N. Tu[4], and Yingxia Liu[1*]

1: Dept. of Materials Science and Engineering, Beijing Institute of Technology, Beijing, China
2: Dept. of Mechanical Engineering, City University of Hong Kong, Hong Kong, China
3: Dept. of Materials Science and Engineering, City University of Hong Kong, Hong Kong, China
4: Dept. of Materials Science and Engineering, University of California, Los Angeles, USA
*e-mail: yingxia.liu@bit.edu.cn

Abstract—In this paper, we studied the high entropy alloy (HEA) of SnBiInZn as a low temperature solder with a melting point about 80 °C. The microstructure of the HEA solder is very complicated. There are three phases at least in the solder, including Bi-rich phase, Sn-rich phase and InBi phase. The wetting reaction between the solder and Cu substrate was conducted at 100 °C and 120 °C. The wetting angle ranges from 35 to 40 degrees, indicating that the solder has a good wetting ability even at such low soldering temperature. The scallop and finger shape intermetallic compounds were determined to be Cu_6Sn_5. The thickness measurement of the IMC layer shows that the interfacial reaction rate is very slow. The shear stress ranges from 19 MPa to 28 MPa, which indicates that the solder joint is strong enough for electronic packaging technology.

Keywords-high entropy alloy; low temperature solder; microstructure; intermetallic compound; shear stress

I. INTRODUCTION

Nowadays, we are living in the information era. Smartphone, internet of things (IoT), artificial intelligence (AI) and big data have been rapidly developed. All of these technologies require the support of strong enough computational ability in microelectronic devices. However, the Moore's low has slowed down due to the miniaturization in transistors is approaching its physical limitation. 3-dimensional integration of circuits (3D IC) has been recognized as the most promising way to extend Moore's low [1-2]. In essence, 3D IC packaging is the stacking of multiple TSV (through-Si-Via) chips and multiple reflows are needed to achieve this structure. Thus, it's important to find solder materials with different ranges of melting point to avoid repeated melting and solidification during the multiple reflows. Soldering reaction needs to be conducted at low temperature (around 100 °C), middle temperature (around 200 °C) and high temperature (around 300 °C), respectively. According to the published research, high-Pb $Pb_{95}Sn_5$ solder and the eutectic SnAg solder have met the requirements of the high and middle reflow temperature in 3D IC technology [3-4]. However, there is no suitable and applicable solder material for the low temperature reflow.

The rate of the reaction between molten solder and Cu substrate has been a crucial issue in electronic packaging technology. It is a concern because the Cu layer thickness in the under bump metallization (UBM) is limited and will be reduced further as the size of packaging shrinking. Also, the IMC formation in bumps will affect its electrical and mechanical properties. The rate of Cu consumption must be controlled to avoid reliability problems. In most of the UBM, there is a Ni layer as a barrier to slow down the soldering reaction. However, to evaporate or sputter Ni films will increase the cost and also tend to have a high residue stress, which will crack the dielectric layer of SiO_2 on the surface of Si chip. It has also been reported that when Ni is adopted as the diffusion barrier, ENIG will be adopted as the surface finishing and lead to $(AuNi)Sn_4$ formation in microbumps, causing reliability issues [5]. Furthermore, in 3D IC, the diameter of microbumps shrinks 10 times from 100 μm to 10 μm, i.e. the volume of the solder shrinks 1000 times [6]. However, the thickness of IMC layer of microbumps is the same as it of C-4 solder for the same reflow time. The percentage of IMC volume will increase remarkably, which would induce reliability concerns. Thus, it is essential to slow down the rate of interfacial reaction between solder and Cu substrate for microbumps in the future advanced packages. In this study, we will show the application of low melting point HEA as solder with a very slow interfacial reaction rate.

Another reason for developing low temperature solder is that low temperature solder can effectively solve the warpage problems in large size chips. With the advent of the era of big data, setting up data centers has become crucial for major companies. The chips used in data centers and servers are different from which used in computers and mobile phones. They are characterized by high cost, high performance, high reliability, and no size limitation. Applying to servers, in order to improve chip performance and reduce power consumption, multiple chips are packaged in one integrated chip, which leads to huge chip size. The warpage of the chip and various internal stress increase as the chip size increases, and high warpage results in low yield and low reliability. Using low temperature solder can reduce the reflow temperature and thus reduce the warpage. It is the most direct and effective way to solve the problem of warpage in the surface mount technology (SMT) of large-size chips.

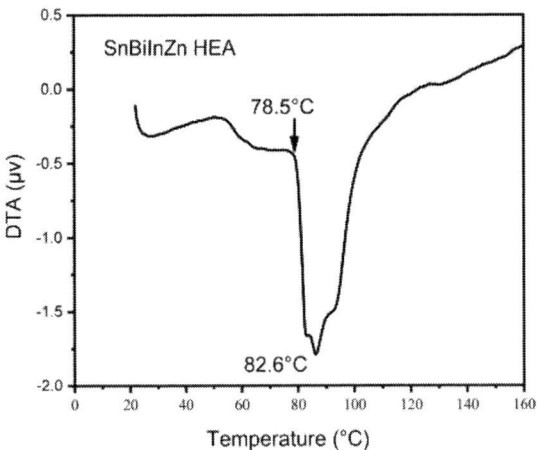

Figure.1 DTA results of the original HEA material

II. EXPERIMENTAL PROCEDURE

Some In and trace amounts of Zn were added into SnBi alloy to form a HEA of SnBiInZn. In order to identify the melting point of the HEA solder, about 5-10 mg pieces were cut from the bulk alloy. These pieces were analyzed by differential thermal analysis (DTA).

The solder bulk was cut into tiny pieces. These pieces were placed on polished and cleaned Cu foil. The system was placed on a hotplate and reflowed under flux at 100 °C and 120 °C for 1min, 5min, 10min and 20min, respectively. After soldering, the samples were removed from the hotplate and cooled to room temperature. Then pure alcohol was used to clean these samples.

The samples were mounted using epoxy resin and then polished. The cross-sectional interface of the polished samples were observed using scanning electron microscope (SEM) and focus ion beam (FIB). The elemental composition of IMC was analyzed using energy dispersive X-ray spectroscope (EDX). The software ImageJ was applied to measure the wetting angle and the thickness of the IMC layer. The thickness was obtained by dividing the area of the intermetallic layer to its length.

For the shear test, 5±0.5 mg of diced HEA solder pieces were reflowed on 1mm diameter circular Cu substrate. After soldering and cooling down to the room temperature, the samples were cleaned using alcohol. Subsequently the shear tests were performed using PTR-1100 shear test machine at room temperature with a shear strain race of 0.05 mm/s.

III. RESULTS AND DISCUSSION

A. Thermal analysis

As shown in Fig.1, the solder has a very low melting point about 80 °C. Comparing to traditional Sn based solders, the melting points of which are 183 °C (Sn38Pb), 138 °C (Sn58Bi) and 120 °C (Sn51In), it's obvious that the addition of In and Zn could highly reduce the melting point. According to the previous study, a trace amount addition of Zn element only has a slight effect on the melting point of SnBi solder. The melting point of Sn58Bi0.7Zn changes

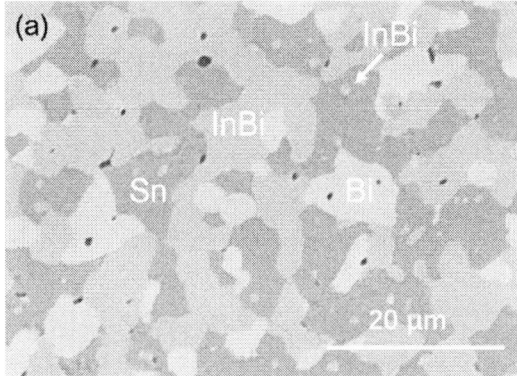

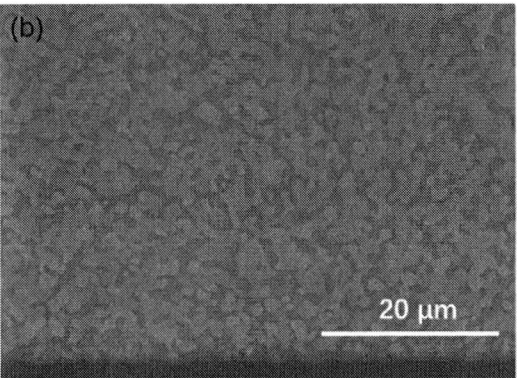

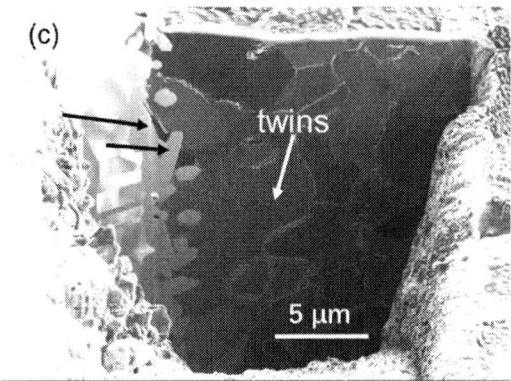

Figure.2 SEM images of (a) original HEA material, (b) the solder reflowed at 120 °C for 10 min and (c) FIB image of the HEA solder reflowed at 160 °C for 5 min.

from 138 °C to 136 °C [7]. However, the addition of In element can strongly reduce the melting point of SnBi solder. It has been reported that Sn38Bi12In solder has a low melting point of 101.5 -103.1 °C [8]. Nevertheless, the solder of SnBiIn has a poor mechanical property. It's reported the shear stress of Sn31.6Bi48.8In is between 2.4 - 11.3 MPa, especially only about 3 MPa when soldering at 100 °C [9].

B. Microstructure

The microstructure of the original HEA material is shown in Fig.2 (a). It's uniform and very complicated. Three phases and some tiny dots can be observed in the SEM micrograph. The three phases are Sn-rich phase, InBi phase

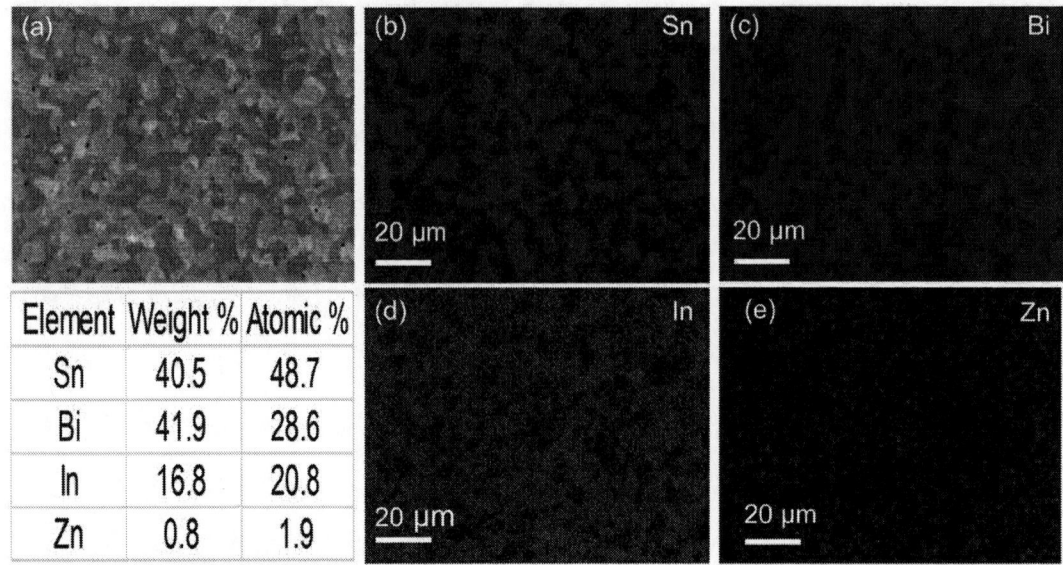

Figure 3. (a) The SEM image of the HEA solder with the corresponding EDS mapping analysis results for the elements: (b) Sn, (c) Bi, (d) In and (e) Zn.

Element	Weight %	Atomic %
Sn	40.5	48.7
Bi	41.9	28.6
In	16.8	20.8
Zn	0.8	1.9

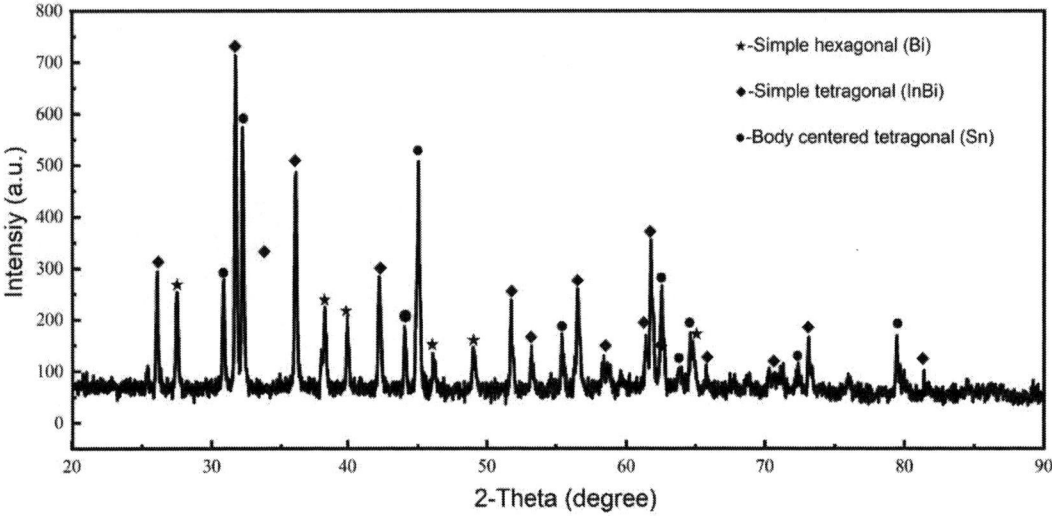

Figure 4. The XRD results of the original HEA material

and Bi-rich phase, according to the EDX results, marked in Fig.2 (a). The dots were also determined as InBi phase using EDX. Fig.3 shows the element mapping analysis results of the solder. Average element percentage is also shown in Fig.3. It's difficult to observe the distribution of Zn elements from the EDX mapping picture because the amount of Zn element is so tiny. The morphology of InBi phase is not clear in this picture and the reason is unknown. X-ray diffraction (XRD) was used to confirm the structure of the original material. It can be revealed from the XRD results, shown in Fig 4, there are three phases in the HEA, including simple hexagonal (Bi), simple tetragonal (InBi) and body centered tetragonal (Sn). It's consistent with the EDX results well.

Also, there are no Zn diffraction intensity peaks. Although no zinc was detected by XRD, it exists in the HEA solder in fact. Because the CuZn based IMC was observed on the fracture surface after shear test. More results will be discussed in section "E". For more detailed information, focus ion beam (FIB) was performed. The cross-sectional ion beam image of the HEA solder after reflow at 160 °C for 5min was shown in Fig.2 (c). There are more than three different phases we can observe in the FIB image. Nanotwin can be observed in one of the phases, as marked with white arrows. Comparing to the solder before and after soldering, as shown in Fig.2 (a) and Fig.2 (b), there is a morphology change after soldering and the microstructure becomes finer.

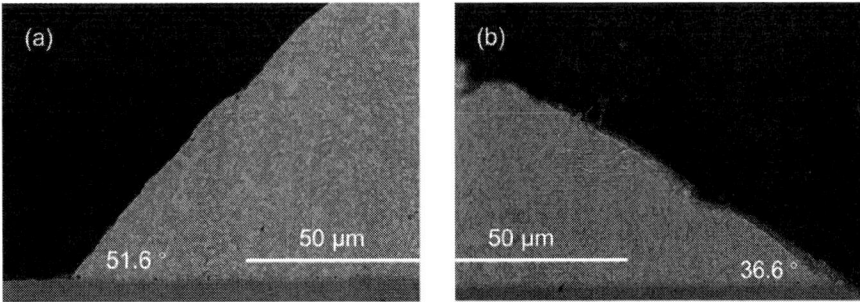

Figure 5. The wetting angles after being reflowed for 10 min at different temperatures, (a) 100 °C; (b) 120 °C.

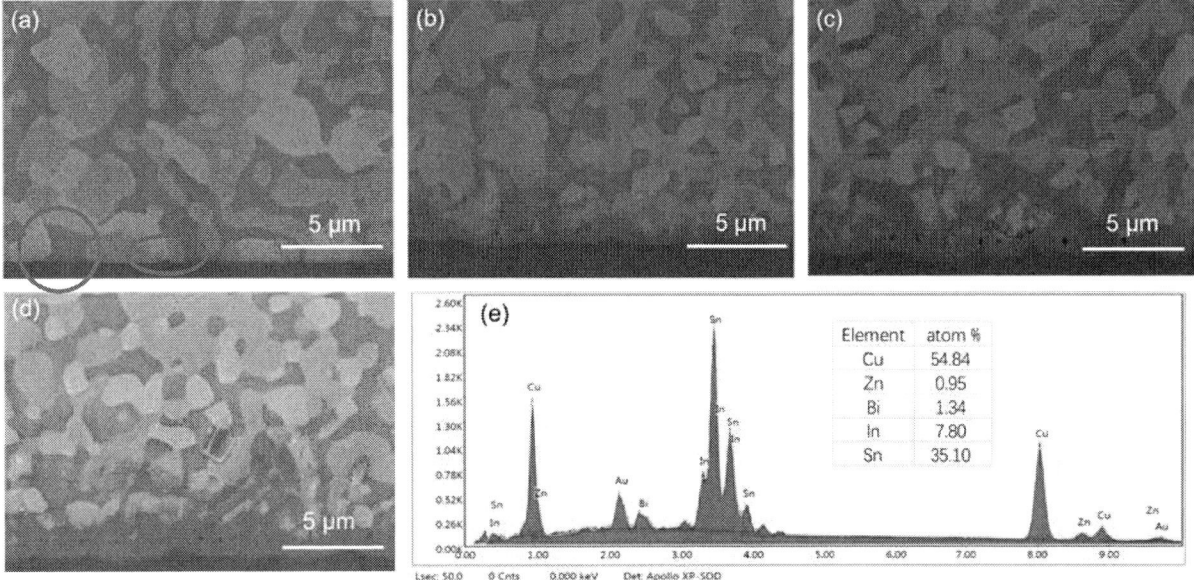

Figure 6. Cross-sectional SEM images of the HEA solder after reflowed at 120 °C for (a) 1 min, (b) 5 min, (c)10 min and (d) 20 min, respectively; (e) the EDX results of the IMC.

C. Wetting angle

After soldering, the wetting angle was measured from SEM micrograph, as shown in Fig.5. The wetting angle ranges from 35 to 40 degrees when soldering at 120 °C. But when we lowered the soldering temperature to 100 °C, the wetting angle changed to more than 50 degrees. It's worth noting that the flux cannot spread out well at such low temperature indicating that the changes of the wetting angle at different temperatures may be because of the incapability of the flux rather than the wetting ability of the HEA solder. In the future, we will try to find a new flux which can work well at low temperature.

D. Intermetallic compound

Fig.6 (a-d) are cross-sectional SEM images of the HEA solder after reflow at 120 °C for 1min, 5min, 10min and 20min. Intermetallic compound forms between the solder and Cu substrate, as shown in Fig. 6. According to the EDX results, the IMC layer is determined as Cu_6Sn_5 containing In and Bi atoms, as shown in Fig.6(e). As illustrated in Fig.6 (b, c and d), the IMC has a scallop shape or finger shape

protruding into the solder matrix. This interfacial reaction can be described by the following equation (1).

$$6Cu + 5Sn \rightarrow Cu_6Sn_5 \qquad (1)$$

When molten solder wets the Cu substrate, Cu atoms react directly with Sn atoms on the surface of Cu substrate. Cu substrate was consumed and subsequently formed a curved surface. It can be observed from the FIB micrograph that in the IMC layer, the layer next to the Cu and the layer next to the solder have different morphology, marked in Fig.2(c) using black arrow, indicating that there may be not only Cu_6Sn_5 based IMC in that area. In the future, we will collect more quantitative information to study the structure and composition of the IMC.

Regarding to the soldering for just 1 min, shown in Fig.6 (a), an extremely thin layer of IMC is observed. Although the intermetallic is mainly layer shape, some scallop shape IMC can be observed on the thin layer, as marked with red circle in Fig.6 (a). It's difficult to confirm the composition of this thin layer using EDX. In the further research, transmission electron microscope (TEM) will be performed to understand

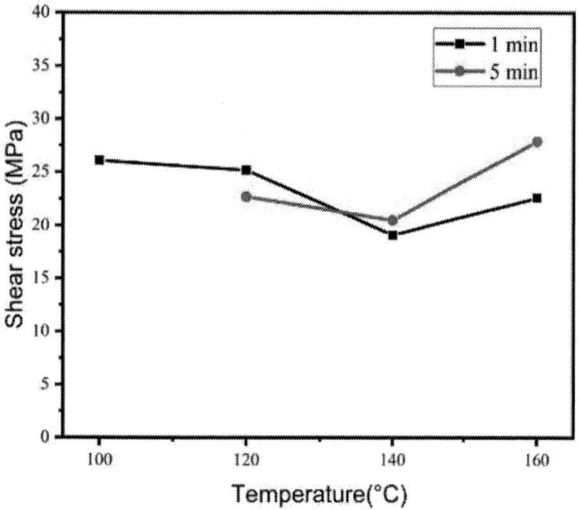

Figure.7 Shear sress of the HEA solder after reflowed on Cu substrate.

the composition. It's worth mentioning, shear test shows the shear stress of this solder joint is 25.2 MPa, indicating that the joint has a respectively good mechanical property though the unknown IMC layer is extremely thin.

The thickness of the IMC layer was measured using software of ImageJ. We can obtain average thicknesses of 1.22, 1.57 and 1.85 μm for 5, 10 and 20 min reflowing at 120 °C, respectively. We also tried to conduct wetting reaction

around 100 °C, however, the results cannot be repeated well. Only two sets of data we obtained from the soldering reaction, thickness of 1.32 and 1.49 μm for 10 and 20 min at 100 °C. What needs to be emphasized is that the thickness of the IMC layer is still only 1.49 μm after soldering for 20 min at 100 °C, indicating the rate of Cu substrate consumption in the soldering reaction is very slow. The rate of Cu consumption in the soldering reaction is highly concerned due to the limitation of the Cu thickness in the UBM. It can be speculated that the intermetallic layer will be thinner if we can conduct the soldering reaction at 100 °C for 1 min. As for why the formation of Cu_6Sn_5 is such slow, it may be related to the inhibition of Bi. It has been reported the IMC layer between Sn58Bi and Cu substrate is also thin, about only 1 μm after reflowing for several minutes [10]. On the other hand, the growth kinetics of IMC is also slow at such low temperature. It's unknown whether high entropy has an effect on the melting point of the solder. We will study more high entropy alloy as solders to clarify the influence in the future.

E. Shear test

Considering that portable and hand-held electronic devices exert a large external force from an impact or a shock, it's necessary to study the mechanical property of the low melting temperature solder. The shear stress of the HEA solder ranges from around 19 MPa to 28 MPa, as shown in Fig.7. Comparing to the values of 19.5 MPa, 32.5 MPa, 64 MPa, and 18.5-28.0 MPa previously reported for the solders

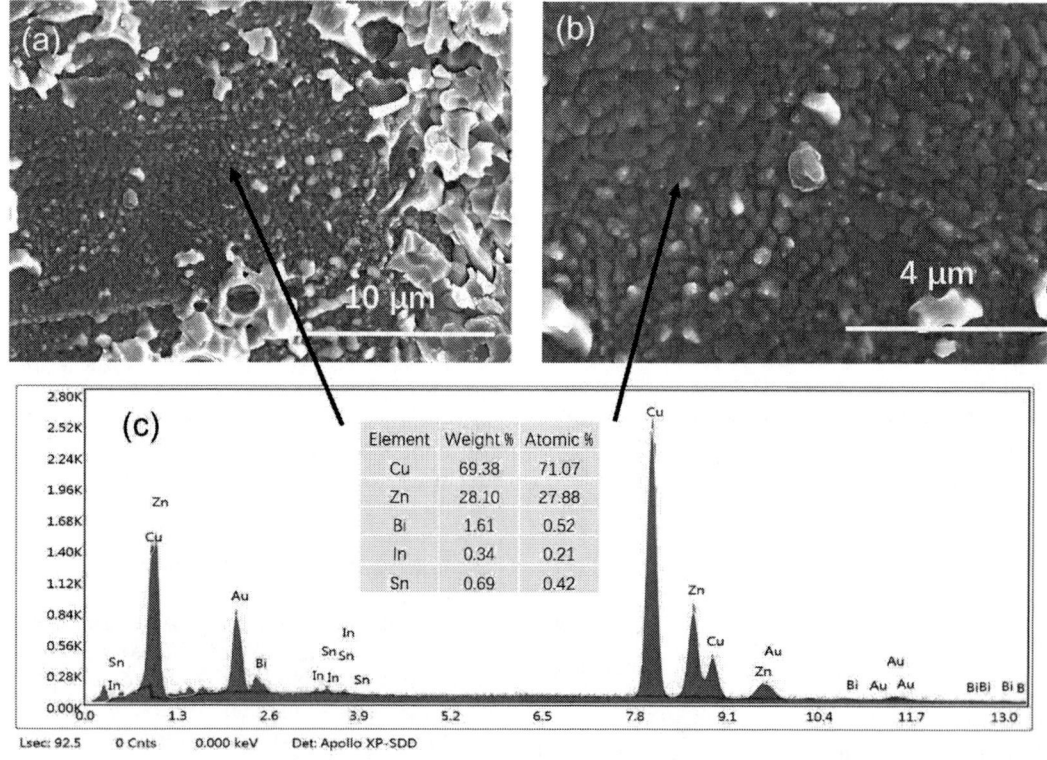

Element	Weight %	Atomic %
Cu	69.38	71.07
Zn	28.10	27.88
Bi	1.61	0.52
In	0.34	0.21
Sn	0.69	0.42

Figure. 8 (a) and (b) SEM images of the fracture surface after shear test reflowed at 120 °C for 1 min; (c) the EDX results of the IMC.

of different composition including Sn-0.4Cu, Sn-3Ag-0.4Cu, Sn-58Bi, and SnZnBi [11-13], the HEA solder joint has an eligible mechanical strength for electronic packaging technology.

The fracture surface after shear test was investigated by SEM and EDX. There is an extremely thin layer of CuZn IMC, as shown in Fig.8, indicating that a trace mount of Zn element exists in the original solder definitely. The IMC layer is too thin to accurately determine the percentage of each element, i.e. the percentage of Cu element in the IMC layer measured by EDX might bigger than it actually is. However, it can be determined qualitatively that the thin layer of IMC in Cu side is CuZn based IMC not CuSn.

IV. CONCLUSION

The microstructure and mechanical property of the HEA solder with a very low melting point have been investigated. Soldering reaction can occur at 100 °C and the solder joint is strong enough according to the shear test. The SnBiInZn solder meets the requirements of 3D IC as a low temperature solder in the multiple reflows. The formation of the intermetallic compound of Cu_6Sn_5 is very slow during the soldering reaction at 100 °C. After soldering at 120 °C for 1 min, an extremely thin layer of IMC was observed. As a result, this HEA solder has a potential application in controlling the percentage of IMCs in microbumps.

ACKNOWLEDGMENT

The authors at Beijing Institute of Technology would like to acknowledge the financial support of the "initiating funding" from Beijing Institute of Technology with the project number 3090011181815. The research of YY is supported by City University of Hong Kong with the project number 9610391.

REFERENCES

[1] Y. Liu, M. Li, D.W. Kim, S. Gu and K.N. Tu, "Synergistic effect of electromigration and Joule heating on system level weak-link failure in 2.5 D integrated circuits," Journal of Applied Physics, vol. 118, no. 13, 2015, pp. 135304.

[2] K.N. Tu, Y. Liu and M. Li, "Effect of Joule heating and current crowding on electromigration in mobile technology," Applied Physics Reviews, vol. 4, no. 1, 2017.

[3] A. Kroupa, N. Hoo, J. Pearce, A. Watson, A. Dinsdale and S. Mucklejohn, "Current Problems and Possible Solutions in High-Temperature Lead-Free Soldering," Journal of Materials Engineering and Performance, vol. 21, no. 5, 2012, pp. 629-637.

[4] R. Khazaka, L. Mendizabal, D. Henry and R. Hanna, "Survey of High-Temperature Reliability of Power Electronics Packaging Components," IEEE Transactions on Power Electronics, vol. 30, no. 5, 2015, pp. 2456-2464.

[5] Y. Liu, Y.T. Chen, S. Gu, D.W. Kim and K.N. Tu, "Fracture reliability concern of (Au, Ni)Sn 4 phase in 3D integrated circuit microbumps using Ni/Au surface finishing," Scripta Materialia, vol. 119, 2016, pp. 9-12.

[6] K.N. Tu and T. Tian, "Metallurgical challenges in microelectronic 3D IC packaging technology for future consumer electronic products," Science China Technological Sciences, vol. 56, no. 7, 2013, pp. 1740-1748.

[7] D. Ma and P. Wu, "Effects of Zn addition on mechanical properties of eutectic Sn–58Bi solder during liquid-state aging," Transactions of Nonferrous Metals Society of China, vol. 25, no. 4, 2015, pp. 1225-1233.

[8] Q. Li, Y. Lei, J. Lin and S. Yang, "Design and properties of Sn-Bi-In low-temperature solders," Proc. 2015 16th International Conference on Electronic Packaging Technology (ICEPT), IEEE, 2015, pp. 497-500.

[9] E.E.M. Noor, N.M. Sharif, C.K. Yew, T. Ariga, A.B. Ismail and Z. Hussain, "Wettability and strength of In–Bi–Sn lead-free solder alloy on copper substrate," Journal of Alloys Compounds, vol. 507, no. 1, 2011, pp. 290-296.

[10] J.F. Li, S.H. Mannan, M.P. Clode, D.C. Whalley and D.A. Hutt, "Interfacial reactions between molten Sn–Bi–X solders and Cu substrates for liquid solder interconnects," Acta Materialia, vol. 54, no. 11, 2006, pp. 2907-2922.

[11] J. Keller, D. Baither, U. Wilke and G. Schmitz, "Mechanical properties of Pb-free SnAg solder joints," Acta Materialia, vol. 59, no. 7, 2011, pp. 2731-2741.

[12] O. Mokhtari and H. Nishikawa, "Correlation between microstructure and mechanical properties of Sn–Bi–X solders," Materials Science Engineering: A, vol. 651, 2016, pp. 831-839.

[13] J. Zhou, Y. Sun and F. Xue, "Properties of low melting point Sn–Zn–Bi solders," Journal of alloys compounds, vol. 397, no. 1-2, 2005, pp. 260-264.

A Versatile Fan-Out Infrastructure Based on Die-Stencil Substrate Promoted by an Advanced Multifunctional Temporary Bonding Material

Xiao Liu*, Baron Huang, Hong Zhang, Lisa Kirchner, Arthur Southard, Rama Puligadda, and Tony Flaim

Wafer-Level Packaging Materials Business Unit
Brewer Science, Inc.
Rolla, Missouri, USA
e-mail: xliu@brewerscience.com

Abstract—The development of new advanced packaging technologies is always in high demand due to the continuous pursuit of smaller package sizes, higher performance, more integration of functionalities, lower power consumption, and reduced cost of ownership. As a result, various advanced packaging platforms have been established, such as flip chip, fan-in wafer-level packaging, fan-out wafer-level packaging, embedded die, and through-silicon via (TSV). Among them, fan-out wafer-level packaging (FOWLP) is quickly gaining prominence, especially in high-end use and for providing an efficient solution for system-in-package (SiP) and 3D integration. However, there are several well-known disadvantages of the traditional FOWLP process, such as die shift during molding, large warpage due to coefficient of thermal expansion (CTE) mismatch, time-consuming molding and curing processes, and outgassing and low thermal stability of epoxy resin. Therefore, people are seeking alternatives for existing FOWLP solutions, such as embedded silicon fan-out (eSiFO) and glass fan-out (GFO) packaging.

In this paper, we will introduce a new versatile fan-out packaging structure as an alternative solution for the traditional FOWLP. This new emerging technology is based on a polymeric die-stencil substrate, in which cavities are created by laser ablation or by mechanical means to accommodate different dies. There are several advantages of this new FO packaging platform: 1) more flexibility on choice of polymeric die-stencil substrates in order to achieve optimized performance; 2) ease of configuration of stencil to accommodate different die sizes, especially for fabrication of SiP; and 3) improved throughput by eliminating the time-consuming molding and curing process in traditional epoxy resins. In addition, we will introduce new multifunctional materials that are under development specifically to facilitate the demonstration of this new stencil-based FO technology.

Keywords: fan-out; laser release; temporary bonding material; adhesive; stencil

I. INTRODUCTION

The entire semiconductor industry has been following Moore's Law as a golden law since 1965. The law, proposed by Gordon Moore, claims that the number of transistors on a chip doubles every two years while the costs are halved. However, Moore's law is a prediction based on an observation of the semiconductor industry and is not a natural law. This prediction made by Moore has proved accurate for the past

five decades and has been used as a long-term guideline for the semiconductor industry to chase smaller, faster, and more efficient chips until recent years. Since Moore's Law is an observation, and not a natural law, it is unlikely to continue indefinitely which has been noticed by many experts and companies; Moore's Law is hitting its limitation both physically and economically.

With the increase in investment to develop the next-generation node, there are fewer companies and foundries that can afford to follow Moore's Law, especially after the 28-nm process node. The cost to build a leading-edge fab has risen 11 percent per year in the past 25 years, ending up with about $7 billion to construct a new fab in 2017. [1] Fig. 1 shows the change in number of manufacturers with leading-edge production capabilities with node technology since 2002. With the announcement that both GLOBALFOUNDRIES and UMC are stopping the development of 7 nm technology, there are only three foundry/integrated device manufacturers (IDMs) still investing on the next-generation nodes. Taiwan Semiconductor Manufacturing Company (TSMC), Samsung, and Intel, which has delayed their 10-nm technology for several years, are still significantly active in advanced packaging marketing recently.

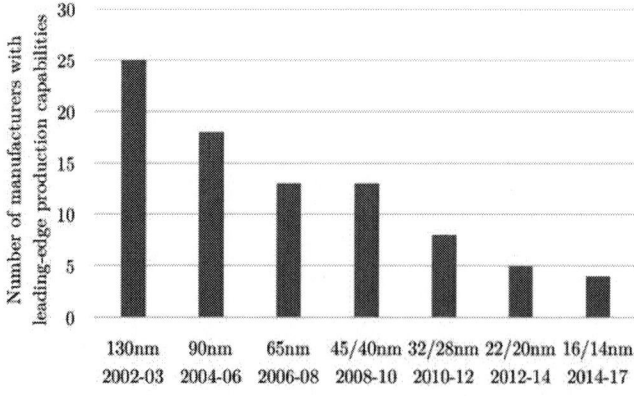

Figure 1. The change of number of manufactures with leading-edge production capabilities with process node technology. [1]

Intel announced their new 3D advanced packaging technology named "FOVEROS" for the first time in

December 2018 on their Architecture Day. This advanced packaging technology utilizes the concept of chiplets with an active interposer that can contain active parts of the system. Therefore, the system can be optimized by using the best processor components with an optimum (performance/cost) production node. [2] In addition, TSMC has been trying to build up their own advanced packaging platforms since 2008, called Wafer Level System Integration, including Chip on Wafer on Substrate (CoWoS), Integrated Fan-Out (InFO), Under-Bump-Metallurgy Free Integration (UFI), System-on-Integrated Chips (SoIC), and Wafer-on-Wafer (WoW). Furthermore, Samsung also developed their package solutions in order to keep pace with Moore's Law, such as, FOPLP-PoP, I-Cube™ (2.5D), and 3D SiP. With the investment into packaging by the big players in the front end, it is a very clear sign they are looking for all the possible ways to improve performance of devices, not only from development of next-generation process nodes, but also from innovation of advanced packaging technology.

Among various advanced packaging platforms, fan-out packaging technology has recently received intensive attention, especially after TSMC applied their InFO technology on the Apple A10 application processor which is the first product using FO technology in a high-performance chip. [3] In addition, FO technology also provides an efficient solution for SiP and 3D integration. However, in spite of the encouraging performance of FO, there are several disadvantages associated with the traditional FOWLP process, such as die shift during molding, large warpage due to CTE mismatch, time-consuming molding and curing processes, and outgassing from low thermal stability of epoxy resin. [4] Therefore, people are seeking alternatives for existing FOWLP solutions.

In 2015, Huatian Technology introduced a new FO packaging technology using silicon instead of epoxy mold compound (EMC) wafers as embedding material, called embedded silicon fan-out (eSiFO) packaging. [5] One of the major advantages of eSiFO over traditional EMC-based FO is that the wafer warpage is significantly improved due to good CTE match of die with the silicon wafer. In addition, no molding or temporary bonding and debonding processes are needed in the eSiFO process. [6] However, it is hard to scale up to panel size using eSiFO for the further cost reduction because of the limited availability of large Si panels beyond 300 mm. Furthermore, other types of FO package technologies were also developed, such as the glass fan-out (GFO) package. It has been demonstrated that GFO is well-suited to the panel level with higher I/O and component density, lower interconnect loss, and higher board-level reliability. [7] The low-loss tangent of glass by a factor of ~2-3x compared to mold compound, makes GFO an ideal candidate for RF and mm-wave modules. [8] However, there are still some challenges for cavity formation on glass and handling, especially with ultrathin glass panels, due to their fragility. Therefore, there is still a continuous demand for new and innovative FO packaging technologies with better design flexibility, simpler processing, and reduced cost of ownership. All of these aforementioned requirements and problems can be addressed using a temporary bond die-stencil process.

II. DESIGN OF DIE-STENCIL FO

Fig. 2 demonstrates the process flow for a die–stencil based FO packaging technology. A release layer is first applied to a carrier, followed by a curable bonding adhesive as backing layer, to which the die stencil film is laminated or bonded. The dies are then placed in the stencil openings and attached to the adhesive backing layer during thermal curing. The gaps between the dies and stencil are then filled with a flexible yet curable polymeric material, yielding a stable reconstituted substrate. This is followed by construction of the RDLs while still supported on the carrier. Finally, the reconstituted substrate is released from the carrier. Fig. 3 gives an overview of the input parameters of materials for a die-stencil FO process. [9]

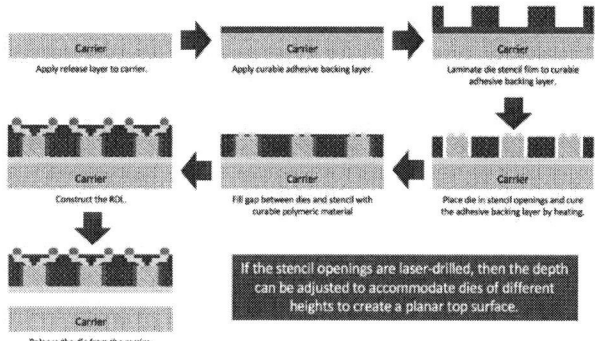

Figure 2.   Process flow for a Die-Stencil FO concept.

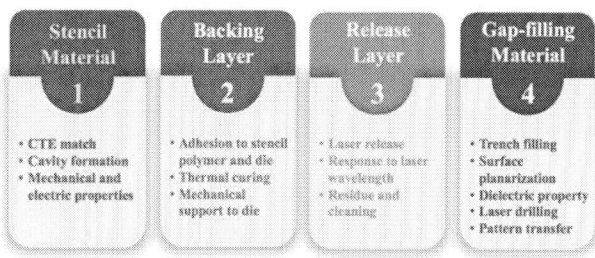

Figure 3.   Input parameters of materials for the Die-Stencil FO process.

There are several advantages of the die-stencil FO packaging platform: 1) more flexibility on the choice of polymeric die-stencil substrates in order to achieve optimized performance; 2) ease of configuration of stencil to accommodate different die sizes, especially for fabrication of SiP; and 3) improved throughput by eliminating time-consuming molding and curing process in traditional epoxy resins.

III. RESULTS AND DISCUSSION

A. Characterization of Material A for Stencil Fabriation

One of the advantages of the die-stencil FO is the flexibility of material choice for die packaging, besides epoxy mold compound resin, silicon, and glass, the three most commonly used materials in FO platforms. Material A, a polymer-based material, is a good example for this goal. As shown in Fig. 4, Material A possesses extremely high thermal stability with initial thermal degradation temperature of 520°C

978-1-7281-1500-9/19 $31.00 © 2019 IEEE

in nitrogen, as compared to 298°C for the commonly used EMC resin. This high thermal stability will minimize material degradation, reduce outgassing, and ensure chemical stability of packaging material during redistribution layer (RDL) build up, which usually involves high temperature and lengthy processing conditions. In addition, Material A also exhibits good mechanical stability up to 300°C as indicated from its rheology in Fig. 5, to guarantee material integrity for die protection. The property summary and comparison of Material A and EMC resin are listed in Table I.

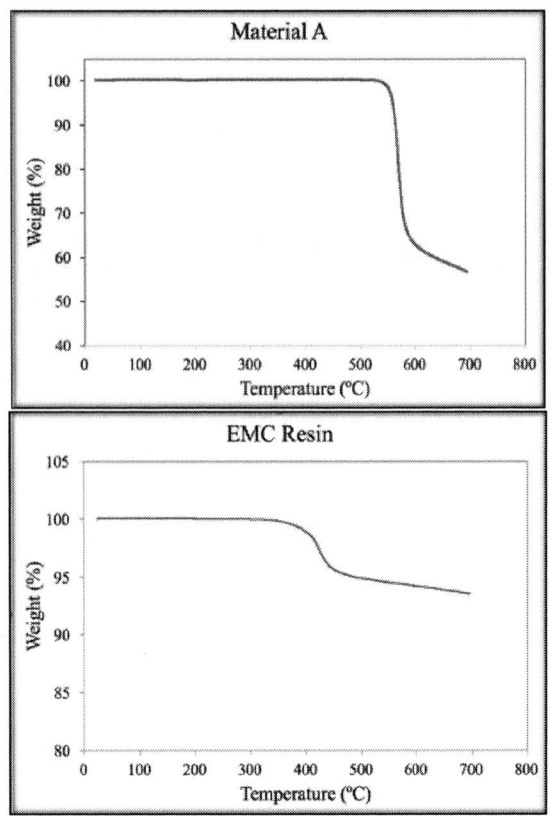

Figure 4.   TGA of Material A (top) and EMC resin (bottom) in nitrogen with a ramping rate of 10°C/min.

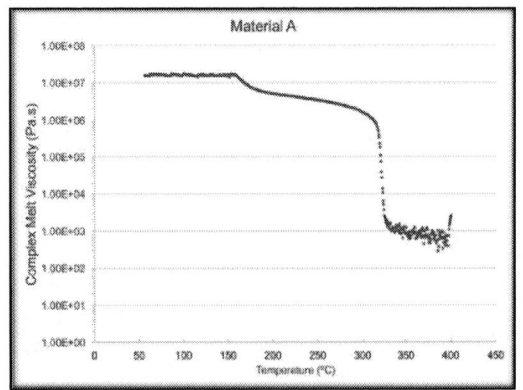

Figure 5.   Rheology of Material A.

One crucial challenge in FOWLP packaging technology is EMC wafer warpage handling during processing. Studies and research have been conducted in order to resolve this issue. The previous studies showed that the major factors that contribute to wafer warpage include Young's modulus, CTE, and glass transition temperature (T_g). Low warpage will be the result of the combination of low Young's modulus, low CTE, and high T_g. [10] However, it is not easy to achieve all of these three requirements for low wafer warpage when using epoxy-based resin that contains a high amount of inorganic fillers. In contrast, it is possible to obtain this type of material with desired properties as a packaging substrate for better warpage control as in die-stencil FO technology which is one of its advantages over EMC-based platforms. For example, it has been reported that a polyimide-based material with a specific structure shows very low CTE of 3 ppm/°C, which matches well with the CTE of a silicon wafer. In addition, low Young's modulus (6 GPa) and high T_g (>300°C) can also ensure good warpage control. [11]

TABLE I.   SUMMARY OF MATERIAL PROPERTIES

Properties	EMC	Material A
$T_{d,ini}$	298°C	520°C
T_g	175°C	150°C
T_m	N/A	340°C
CTE1	7 ppm/°C	56 ppm/°C
CTE2	28 ppm/°C	190 ppm/°C
Young's Modulus	16.7 GPa	1.95 GPa

B. Cavity Formation on Material A

Cavity formation on Material A with well-controlled size and shape is required for accommodation of single die or even multiple die, as in SiP, to achieve FO packaging structure. There are several methods for cavity formation on polymeric film. One of the most straightforward ways is mechanical punching, which lacks the flexibility for different cavity sizes. It also makes it difficult to achieve a more complicated cavity shape. Another way for cavity formation is hot embossing, but it usually requires a time-consuming process consisting of heating, embossing, cooling, and demolding steps, to form cavities within a material. In addition, it is very costly to customize a mold for specific applications. Therefore, laser cutting is a very attractive method for cavity formation on Material A due to its speed, precision, and ease of design for complicated shapes with precise size control.

A CO_2-laser cutter from Trotec® was used to form cavities on Material A with two different sizes, 6 mm × 6 mm, and 18 mm × 18 mm, using a pre-designed feature on Auto-CAD. A total of 49 cavities with a size of 6 mm x 6 mm can be easily created on a 70 mm x 70 mm Material A sheet within 30 seconds. The photo image of this stencil is shown in Fig. 6. The formed cavities were investigated under microscope and a fine cut was formed on both edges and corners as in Fig 7. An O_2 plasma surface treatment was conducted for both debris cleaning and surface activation for bonding to a glass wafer.

Fig. 8 showed efficient cleaning of carbon debris by O_2 plasma on Material A surface.

Figure 6. Photo images of 6 mm × 6 mm cavities formde on a 70 mm × 70 mm Material A sheet.

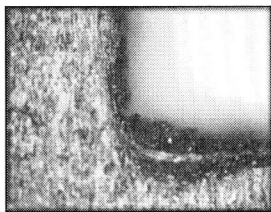

Figure 7. Microscope images of corner and edge of cavities formed on Material A.

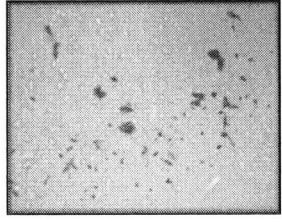

Figure 8. Microscope images of Material A surface before (left) and after (right) O_2 plasma treatment.

C. Bonding of Die-Stencil Material A on Carrier

The die stencil made from Material A was bonded on to a glass carrier wafer for better handling, warpage control, and die placement in the following steps.

Figure 9. Die stencil made from Material A bonded on glass carrier wafer.

The bonding process is illustrated in Fig. 2. A laser release material was coated on glass carrier wafer, on top of which a curable layer was coated for stencil bonding and die placement, followed by curing to secure bonding and reduce die shift. The bonding of the die stencil with pre-cut cavities to a 4" glass wafer with the bonding and laser release layer was demonstrated successfully as shown in Fig. 9. Additionally, a lamination process can also be applicable for the bonding process to mount die stencil on carrier wafer as another bonding option. As a result, it is also possible to scale up this process to panel size for further cost savings.

D. Die Placement

A die placement test was conducted using the die stencil on carrier wafer prepared in Section C. The thickness and size parameters of die stencil, die, bonding and release materials are summarized in Table II. The cavities formed by laser cutting can accommodate dummy silicon dies very well.

TABLE II. PARAMTERS OF DIE STENCIL, DIE, AND MATERIALS

	Die Stencil	Cavity	Dummy Die	Bonding Material	Laser Release
Thickness	200 μm	200 μm	200 μm	5 μm	600 nm
Size	70 mm × 70 mm	6 mm × 6 mm	5 mm × 5 mm	N/A	N/A

The die placement with two different sizes of dummy dies were tested. Fig. 10 (left) demonstrated the die placement with the size of 5 mm × 5 mm in a cavity of 6 mm × 6 mm on a 7 mm × 7 mm stencil sheet. The placement was conducted using the conditions of 6 N bonding force at 50ºC. A good bond line between the Si die and bonding material was achieved without any defect. Additionally, the die placement test was also conducted on the large stencil size with 18 mm × 18 mm to accommodate a 17 mm × 17 mm Si dummy as shown in Fig. 10 (right). Furthermore, the placed die was also investigated under a 3D optical profiler. From the measurement, the average distance between die and stencil material is 602 μm. The 3D image was presented in Fig. 11 (right). This test effectively demonstrated the diversity of die stencil FO technology for packaging dies with various shape, thickness, and size. With the help of laser cutting, irregular stencil shapes can be easily generated as required in the application of SiP in order to accommodate multiple dies with different dimensions.

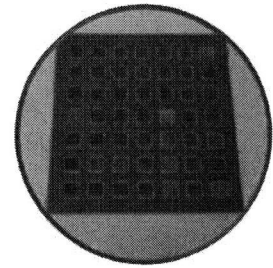

Figure 10. Photo images of die stencil after die placment in cavities: 6 mm × 6 mm (left) and 18 mm × 18 mm (right).

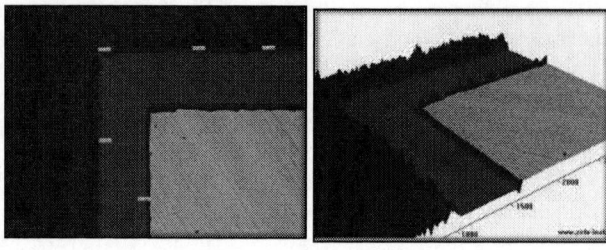

Figure 11. Investigation of the stencil with die under 3D optical profiler.

E. Embossing on Dielectric Material

The prepared carrier wafer with die stencil will be subjected to gap filling and redistribution layer (RDL) build up on top. The most commonly used method for pattern formation in RDL building up in fan-out process is photolithography. The low-density fan-out applications require 8 μm/8 μm line/space (L/S) or greater. In contrast, for high-density fan-out packages used in servers and smartphones, the L/S needs to be smaller than 8 μm/8 μm, even 1 μm/1 μm or below, with more demands for devices with smaller form factor and better performance. [12]. However, as the RDL dimensions shrink, more issues will arise when using the photolithography method, with the largest disadvantage being the cost. As a result, people are looking for an innovative alternative method for RDL build up. For example, a novel process of RDL formation by excimer laser ablation was invented by SUSS MicroTech. [13,14] This technology offers a shorter, faster and more cost-effective process which overcomes the challenges of conventional semi-additive processes as used in photolithography, with a 30% saving on cost and 59% saving on cycle time. With the demand of cost-effective processes for RDL fabrication, a new RDL build up method is demonstrated in the following section.

This new idea for the RDL formation process is based on embossing technology. The process flow of an embossing process is illustrated in Figure 12. First, the desired patterns are formed on template material by laser ablation, followed by pattern transfer to form a stamp material. Later, the stamp is used for final pattern imprinting on a target material. In the end, vertical interconnect access (vias) are opened by laser ablation for connection.

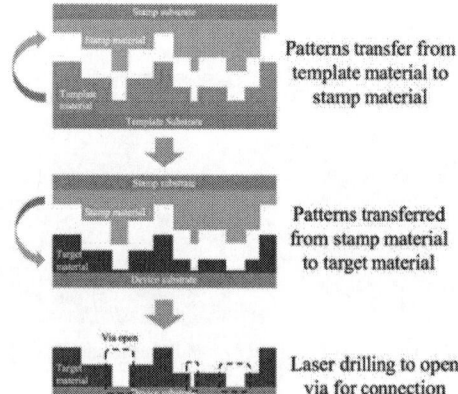

Patterns transfer from template material to stamp material

Patterns transferred from stamp material to target material

Laser drilling to open via for connection

Figure 12. Process flow of embossing for RDL pattern formation.

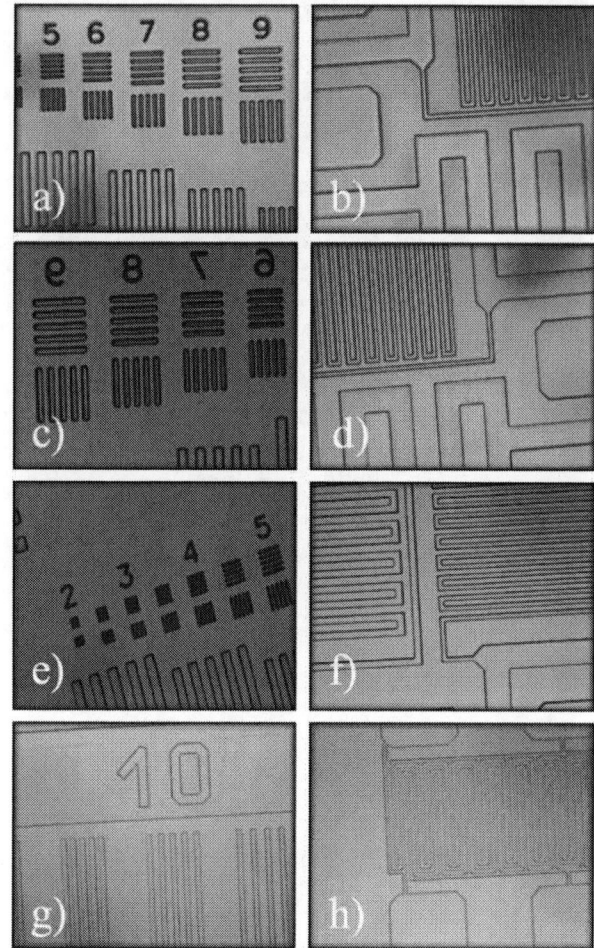

Figure 13. Microscope images of embossing process: a) & b) patterns formed on template material; c) & d) patterns stamped on stamp material; e) & f) patterns transferred on target Material A; g) & h) patterns transferred on target Material B.

In order to achieve good embossing and pattern transfer, there are some requirements for the properties of materials used in process. As for the template material, high response to laser ablation with no or little generation of residual debris is needed for good pattern formation and minimum effort in surface cleaning. Secondly, the stamp material should possess good releasing capability to various surfaces to ensure precise feature transfer from template material to target material without cross contamination. Finally, the target material must be able to be embossed under relatively low temperature (≤ 100°C) to avoid high temperature exposure of the device substrate, be curable by either UV light or heat to retain transfer pattern during following processes, and be responsive to laser ablation for via opening.

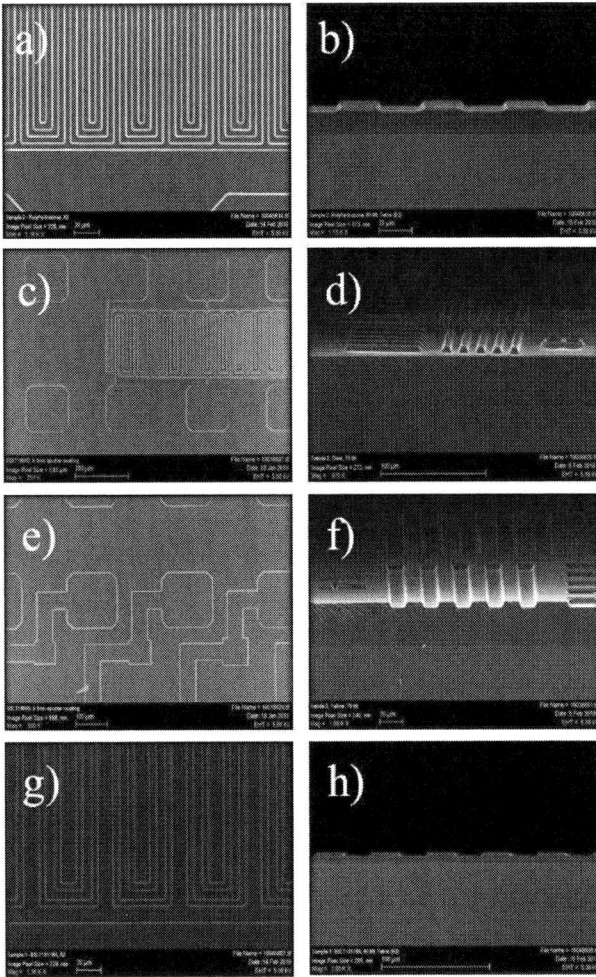

Figure 14. SEM images of embossing process: a) & b) patterns formed on template material; c) & d) patterns stamped on stamp material; e) & f) patterns transferred on target material A; g) & h) patterns transferred on target material B.

fan-out applications, but also for high-density routings with a smaller-than-5-µm L/S for high-end use.

IV. CONCLUSION

In the present work, a new, versatile fan-out packaging structure is proposed as an alternative solution for the traditional FOWLP. This new emerging technology is based on a polymeric die-stencil substrate, in which various cavities can be created by using time-saving and efficient CO_2 laser cutting method to accommodate different dies. A good die placement in die stencil openings was successfully demonstrated with the assist of bonding and release materials on the top of carrier wafer. In combination the the new and cost-effective embossing technology for RDL layer build up, the newly developed die-stencil FO technology could be a promising technical replacement for the current FOWLP for both low- and high-density fan-out applications. A further study is currently undergoing for continuous development of this technology which will be reported in the future.

ACKNOWLEDGMENT

We would like to thank SUSS MicroTech and Georgia Institute of Technology - PRC for their support for this work.

REFERENCES

[1] M. Feldman, "The Era of General Purpose Computers is Ending", The Next Platform, 2019.

[2] K. Morris, "Intel FOVEROS 3D Packaging", Electronic Engineering Journal, 2019.

[3] F. Hsu, J. Lin, S. Chen, P. Lin, J. Fang, and J. Wang, S. Jeng, "3D Heterogeneous Integration with Multiple Stacking Fan-Out Package" Electronic Components and Technology Conference (ECTC 2018), IEEE Press, Aug. 2018, pp.337-342, doi: 10.1109/ECTC.2018.00058.

[4] D. Yu, Z. Huang, Z. Xiao, L. Yang, and M. Xiang, "Embedded Si Fan Out: A Low-Cost Wafer Level Packaging Technology Without Molding and De-bonding Processes" Electronic Components and Technology Conference (ECTC 2018), IEEE Press, Aug. 2018, pp. 28-34, doi: 10.1109/ECTC.2017.166.

[5] D. Yu, "Embedded silicon fan-out package and the method of forming the same," Chinese Patent 201510486674.1, filed on Aug. 11, 2015.

[6] C. Chen, T. Wang, D. Yu, S. Ma, K. Zhu, Z. Xiao, and L. Wan, "Reliability of Ultra-thin Embedded Silicon Fan-out (eSiFO) Package Directly Assembled on PCB for Mobile Applications" Electronic Components and Technology Conference (ECTC 2018), IEEE Press, Aug. 2018, pp. 1600-1606, doi: 10.1109/ECTC.2018.00242.

[7] T. Shi, C. Buch,V.a Smet, Y. Sato, L. Parthier, F. Wei, C. Lee, V. Sundaram, and R. Tummala, "First Demonstration of Panel Glass Fan-out (GFO) Packages for High I/O Density and High Frequency Multi-Chip Integration", Electronic Components and Technology Conference (ECTC 2018), IEEE Press, Aug. 2018, pp. 41-46, doi: 10.1109/ECTC.2017.287.

[8] L. Y. Ying, D. H. S. Wee, K. H. Joon, and P. Damaruganath, "Low cost characterization of the electrical properties of thin film and mold compound for embedded wafer level packaging (EMWLP)" Electronics Packaging Technology Conference (EPTC 2011), IEEE press, Apr. 2011, pp. 401-405, doi: 10.1109/EPTC.2011.6184454.

[9] K. Yess, "Material innovations for advancements in fan-out packaging" Solid State Technology, 2018.

[10] K. Kwon, Y. Lee, J. Kim, J. Y. Chung, K. Jung, . Park, D. Lee, and S. K. Kim, "Compression molding encapsulants for wafer-level embedded active devices Wafer warpage control by epoxy molding

Fig. 13 shows microscope images of the patterns on different materials in specific stages. No residual debris was observed on the template material after laser ablation for pattern generation (Fig. 13 a) and b)). The formed patterns were well copied by the stamp material to form mirror images on top (Fig. 13 c) and d)), which successfully transferred to the target Materials A and B (Fig. 13 e) to h)) later. In addition, the properties of the target material A and B have been presented in the previously published literature to show their capabilities as new advanced multifunctional materials to facilitate the new die-stencil FO technology. [15]

Furthermore, the patterns were also characterized by scanning electron microscopy (SEM) for different stages during process flow as shown in Fig. 14. Well-defined line/space and via structures were generated with 5 µm depth without obvious defects. The results showed that a good resolution was obtained for transferring a 5 µm L/S or even lower. Therefore, by using the embossing technology, it is possible to achieve a finer L/S feature not only for low-density compounds" Electronic Components and Technology Conference (ECTC 2017), IEEE Press, Aug. 2018, pp. 319-323, doi: 10.1109/ECTC.2017.266.

[11] M. Hasegawa, and S. Horii, "Low-CTE Polyimides Derived from 2,3,6,7-Naphthalenetetracarboxylic Dianhydride", Polymer Journal, vol. 39, Apr. 2007, pp. 610–621, doi: 10.1295/polymj.PJ2006234.

[12] H. Pu, H. J. Kuo, C. S. Liu, and D. C. H. Yu, "A Novel Submicron Polymer Re-distribution Layer Technology for Advanced InFO Packaging", Electronic Components and Technology Conference (ECTC 2018), IEEE Press, Aug. 2018, pp. 45-51, doi: 10.1109/ECTC.2018.00015.

[13] H. Hichri, M. Arendt, and M. Gingerella, "Novel Process of RDL formation for Advanced Packaging by Excime Laser Ablation"

Electronic Components and Technology Conference (ECTC 2016), IEEE Press, Aug. 2016, pp. 1733-1739, doi: 10.1109/ECTC.2016.225.

[14] R. Hollman, O. Dimov, Sanjay Malik, Habib Hichri, and M. Arendt, "Ultra fine RDL structure fabrication using alternative patterning and bottom-up plating processes" Electronic Components and Technology Conference (ECTC 2018), IEEE Press, Aug. 2018, pp. 58-63, doi: 10.1109/ECTC.2018.00017.

[15] H. Zhang, X. Liu, S. Rickard, R. Puligadda, and T. Flaim, "Novel Temporary Adhesive Materials for RDL-First Fan-Out Wafer-Level Packaging" Electronic Components and Technology Conference (ECTC 2018), IEEE Press, Aug. 2018, pp. 1931-1936, doi: 10.1109/ECTC.2018.00289.

Low Temperature and Pressureless Microfluidic Electroless Bonding Process for Vertical Interconnections

H. T. Hung[1*], S. Yang[1], I A. Weng[1], Y. H. Chen[2], C. R. Kao[1]

[1]Department of Materials Science and Engineering, National Taiwan University, Taipei, TAIWAN
[2]Unimicron Corp., Taoyuan, TAIWAN
*d05527004@ntu.edu.tw

Abstract—Thermocompression bonding (TCB) process is now being adopted for high density interconnections but the necessity of applying force and heat causes a lot of problems, such as warpage-induced defects, cracking of delicate chips and thermal drift. To address the above issues, we proposed a novel bonding technique called microfluidic electroless interconnection (MELI) process to directly fabricate interconnection between Cu pillars at the temperature below 80°C and without applying any pressure on the chips. It has been shown in our previous researches that the MELI process using electroless Ni and electroless Au could bond vertical interconnection under controlled flow. In order to extend the application range of the MELI to fine pitch, in this study we analyze the feasibility of selective electroless deposition in microchannel by adding stabilizers into electroless Ni and electroless Au solution. Electroless plating can provide a uniform conformal coating on all parts of the surface that have been catalytically activated. However, the extended coating from the sides of the Cu pillar bump shortens the interconnect pitch, which may cause the risk of bridging. In this paper, we successfully achieve selective electroless Ni plating in microchannel by adding 1.5 ppm of lead acetate into the plating bath. As for electroless Au plating, selective deposition in the microchannel can be accomplished by narrowing the gap. In summary, the innovative MELI process provides a low-temperature and pressureless fine pitch bonding technique.

Keywords-component; Electroless plating; Low temperature bonding; Pressureless bonding;

I. INTRODUCTION

In the trend of high-performance and continuous miniaturization of microelectronics, going beyond the Moore's law as well as elevating the performance have become the first priority for the semiconductor industry. [1, 2] Among all approaches, the development of 3D ICs integration has received a great deal of interest. The basic concept of 3D IC integration is accommodating functional chips through vertical stacking using micro joints and through-silicon vias. There are many advantages of 3D IC integration, e.g., small form factor, low power consumption, less delay-related problem, and heterogenous integration of different kinds of chips. [3, 4, 5] However, many challenges behind 3D IC integration arises from the shrinkage of the pitch size and the increasing of interconnect density. When pitch size decreases to only 10 μm, which is the average size between micro bumps, traditional solder-based technologies would face many issues, such as low standoff height, current crowding effect, and high risk of bridging. [6]

To overcome these challenges and for fine pitch application, copper pillar TCB has arisen as the key assembling technology. Although it can provide high die assembly yield, as the name suggests, utilizing heat and force to bond two metals together would also leave many issues. High temperature bonding process would cause warpage concern due to large coefficient of thermal expansion (CTE) mismatch between die and board. [7] This warpage effect could lead to open joints and bridging and non-wetting of solder bumps. [8] In addition, delamination of low-k dielectric layer is another reliability challenge. [9] Thus, in the past few decades, various bonding methods have been studied extensively for application of low temperature and low bonding pressure assembly technology. These processes could be divided into surface-activated bonding (SAB), TCB with passivation mechanisms, and texturing Cu crystal orientation bonding. However, these technologies either relied on an ultra-vacuum environment to prevent the surface oxidation or a strict bonding surface cleanness and flatness condition to attain high bonding quality. In summary, a novel bonding technology, which could be conducted at low temperature and low pressure, is needed in next generation high-performance microelectronics.

In our previous studies [10-13], we provided an innovative vertical interconnection bonding method called microfluidic electroless interconnection (MELI), and have successfully fabricated interconnection at temperature below 80°C without applying any pressure on chips. Our proposed method is a combination of microfluidic device and electroless plating, which integrates the merits of both processes. It has been shown that MELI process using electroless Ni and Au could fabricate interconnection under controlled flow in the microchannel. However, the conformal electroless deposition of MELI process is a key issue that limits the process capability for fine pitch interconnect. Reducing the process time by downsizing the gap of the joint to below 1 μm could be one of the solutions to overcome the bridging issue but is not a perfect solution. In such case, the uniformity of entire chip will become a major concern. Further, the urgent requirement for very fine pitch interconnects in microelectronic packaging is a major development trend in industry. Even if the gap can be reduced to below 1 μm, a higher susceptibility to cause bridging issue in pitches below 5-10 μm cannot meet the requirement for very fine pitch interconnections.

Selective electroless deposition that only targets the gap of joint could be an ideal solution to enable fine pitch interconnections. Organic and inorganic additives are often added to electroless solution to modify electroless deposition mechanism. Yang et al. found that when the concentration of EPE-8000 additive reached 1.0 mg/L, a strong inhibition for electroless copper deposition would occur. [14] They are able to achieve void-free bottom up filling for submicron trench by addition of EPE-8000, which results in a concentration gradient of EPE-8000 in the submicron trench that the deposition rate in the trench is faster than that of the surface of substrate. Similarly, Osborn et al. found that excess concentration of PEG would lead to a significant decrease in deposition rate of electroless copper and that the diffusion of PEG in the narrow trenches was limited, which would result in a higher deposition rate in restricted spaces. [15] Consequently, they are able to achieve seamless deposition in copper pillar joints.

In this report, we aim to investigate the feasibility of selective electroless Ni and Au deposition in the microchannel, so as to provide a route of fine pitch application for MELI process. The results demonstrate that selective electroless plating in microchannel could be achieved by adding optimal stabilizers into electroless solution and controlling the gap between pillars. To summarize, stabilizers extend the application range of MELI process to fine pitch by preventing the risk of bridging during bonding process.

II. Experiment

A. Dome-shaped copper pillar fabrication

The fabrication flow of copper pillar bumps is shown in Fig. 1. A seed layer composed of Cr (30 nm) and Cu (300 nm) were sputtered on to a 4-inch p-type glass wafer. Before dry film coating, the glass wafer was dipped in 5% sulfuric acid solution for 1 min and isopropyl alcohol for 5 min respectively to remove the native oxides and to clean the surface. Afterward, the wafer was put on a 120℃ hotplate for 10 min to dehydrate the Cu surface and to ensure a better adhesion between the wafer and dry film. A 25 μm dry film photoresist (RY-5125EE) was coated on the wafer using hot-roll laminator at 120℃ and with 0.5 MPa pressure. After coating, the wafer was hold at 25℃ for 1 h to cool down. Subsequently, the laminated wafer was lithographically patterned using 80 mj/cm^2 ultraviolet light. Prior to the development process, a holding time of 1 h under 25°C was applied to keep the adhesion between the resist and the wafer. Subsequently, the wafer was developed by 1 % sodium carbonate aqueous solution for 5 min. The circle holes on the wafer was therefore defined for the further fabrication of copper pillar bumps. Dome-shaped copper pillar bumps were applied in this study, because it has been shown in previous study that it could provide void-free and seamless interconnection. [10] The solution for dome-shaped pillars was supplied by JCU Taiwan Corporation. Before electrolytic copper plating, the wafer was dipped in 5% sulfuric acid solution for 1 min, and deionized water (DI) for

1 min to remove copper oxides. Copper pillars of 50 μm in diameter and 23 μm in height were then plated on the wafer using 2 ASD current density as shown in Fig. 2. After copper electroplating, the dry film was removed using an aqueous amine type stripper for 5 min. Thereafter, the Cu seed layer was etched by sodium persulfate and sulfuric acid solution. The Cr seed layer was etched by hydrochloric acid solution. At last, the glass wafer was diced into 88 dies.

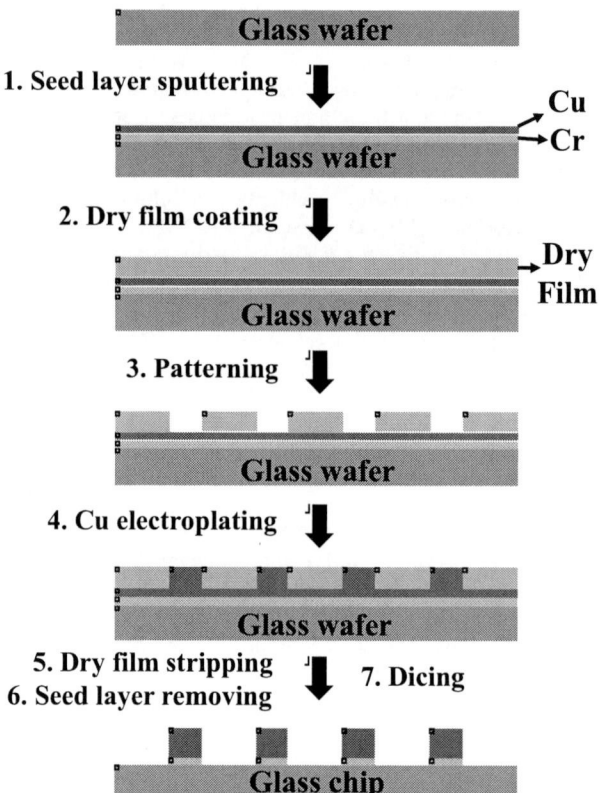

Figure 1.   Fabrication flow of copper pillars.

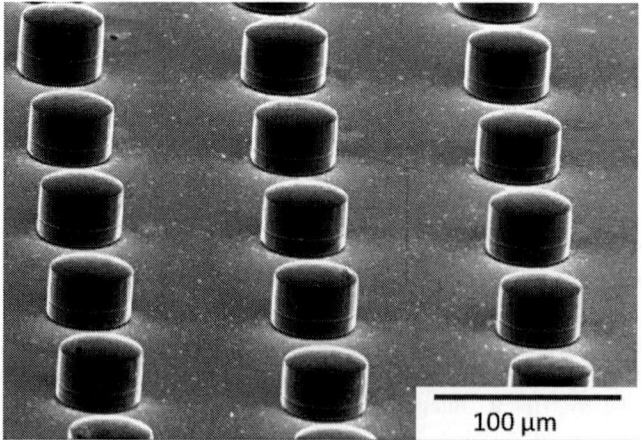

Figure 2.   The SEM micrograph of copper pillar bumps.

B. Test vehicle fabrication

After dicing the wafer into individual dies, 40-50 μm metal wires, which served as spacers, were put parallelly on two sides of the die. To ensure a good rectangle microchannel so as to precisely control the flow and the flow rate between dies, these two wires were straightened. Then, UV adhesive was applied near the wires. Before curving, the other die with the same copper pillar layout was flipped, aligned, and placed on the prepared die in T-3002-FC2 die bonder as shown in Fig. 3. After stacking, UV adhesive was solidified by UV light. The aligned dies were bonded together and a microchannel with a length of 8 mm, a width of 5.5 mm, and a height of 40-50 μm was formed. Since copper pillars from both the top and the bottom dies were with the same height, the gap between pillars was therefore determined by the diameter of the metal wire. The MELI process was thereafter used to fabricate the interconnection in the gap.

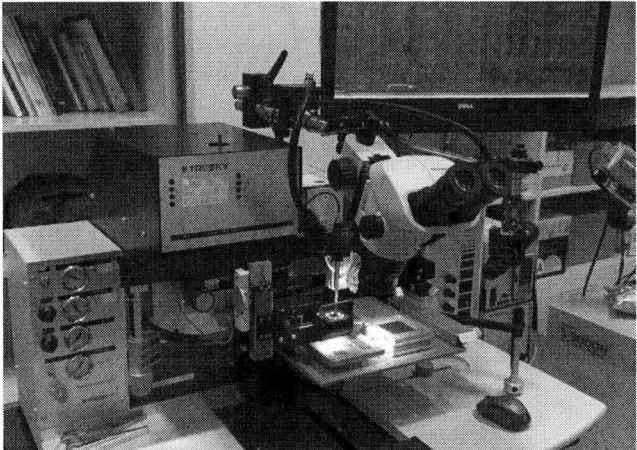

Figure 3. T-3002-FC2 die bonder.

C. MELI process

After the test vehicle preparation, PDMS was utilized to make a fixture for the MELI process. First, liquid PDMS was poured into a metal mold comprising a structure of a rectangle channel and a test vehicle. After baking at 60℃ for 90 min, it was solidified into a patterned PDMS mold. The test vehicle was then put between the patterned PDMS mold and a piece of glass substrate. Next, we used O_2 plasma to modify the surface of the patterned PDMS mold and the glass substrate, so as to bond them together rigidly. The fixture, which could prevent leaking when the fluids passing through the microchannel, was completed. Before conducting electroless plating process, a pretreatment process was adopted to clean the surface of copper pillars, remove native copper oxides, and deposit a thin catalyst layer. Table 1 shows the pretreatment process for electroless Ni plating on copper pillars. All these pretreatment process solutions were injected into the microchannel using syringe pump (KDS100, Kd scientific, MA). A thin layer of immersion palladium was deposited at the last step to activate the plating process. However, for electroless Au plating, another pretreatment process was applied. In this study, we choose electroless nickel and immersion gold (ENIG) as the catalytic layer to activate the electroless Au plating process. Thereafter, the microfluidic fixture was immersed in a thermostatic bath at 80℃ for electroless Ni plating and at 50℃ for electroless Au plating. A mildly acidic electroless Ni plating bath (NPR-4, Uyemura) was then carrying out electroless Ni plating process, as shown in Fig. 4, using intermittent flow. Just like the previous work, the plating solution was injected into the microchannel for 6 s at a flow rate of 60 ml/h with a delay of 6 s [11]. A non-cyanide electroless Au solution (TMX-23, Uyemura) was chosen in this research. Both continuous and intermittent flow were adopted in this research. To attain the goal of selective electroless deposition, some stabilizers were added into the electroless plating solution. The electroless Ni plating solution we used in this study already contains a small amount of Pb^+. Because it has been reported that such stabilizer can poise sites on the edges, which could potentially achieve a concentration gradient in the restricted spaces. Therefore, the present study aims to achieve selective electroless Ni deposition that only plate in the gap region of the joint by increasing the amount of Pb^+ concentration from 0 to 2 ppm. As for electroless Au plating, the solution we used in this study already contains a small amount of stabilizer, which we don't know what it is. Therefore, we aim to complete selective deposition through controlling the flow pattern inside the microchannel. After the MELI process, the bonded dies were mounted in epoxy. Before using scanning electron microscopy (SEM) to observe the cross-section, samples were undergone the polishing processes. Focused ion beam (FIB) ion image was also used in this study for observation.

TABLE I.   THE PRETREATMENT PROCESS OF ELECTROLESS NI PLATING ON COPPER

Pretreatment Process	Composition	Temp. (℃)	Time (min)	Flow rate (ml/h)
Cleaner	Soapy water	25	5	10
	DI	50	1	10
Rinse	DI	25	2	10
Soft Etching	Sodium persulfate and sulfuric solution	25	1	10
Rinse	DI	25	2	10
Acid Dipping	10% sulfuric acid solution	25	1	10
Rinse	DI	25	2	10
Pre-dip	5% sulfuric acid solution	25	1	10
Activator	$PdCl_2$ solution	25	3	Still
Rinse	DI	25	2	10

Figure 4. MELI process using electroless Ni plating.

III. RESULTS AND DISCUSSION

A. Selective electroless Ni plating in microchannel

Previously, Putten and Bakker found that anisotropic deposition of electroless Ni can be achieved by addition of appropriate amount of Pb^{2+}. [16, 17] The function of Pb^{2+} in electroless Ni bath is used as a stabilizer to prevent the bath from decomposing or improve the deposit morphology. They tend to adsorb on the metal surface that would poison the metal surface and inhibit deposition. Therefore, this type of stabilizer usually presents in a very low concentration to avoid this problem. Putten and Bakker also discovered that, with addition of some Pb^{2+} stabilizer in the electroless Ni bath, bevel plating could be achieved and the adsorption of Pb^{2+} stabilizer would be shown as a function of time during electroless Ni plating. The adsorbed Pb^{2+} stabilizer on edges is far more than that in the center region because of the enhanced mass transport caused by non-linear diffusion, leading to poisoning of those high Pb^{2+} concentration areas. Consequently, a pyramid-shape of electroless Ni(P) was formed. Besides, different Pb^{2+} concentration would introduce different inhibition effect. The angle of bevel increases with increasing concentration of Pb^{2+}. The deposit morphology indeed can be changed by addition of additives. However, there is limited research regarding the effect of such additives on the deposition behavior of electroless Ni plating in restricted spaces.

The present study aims to achieve selective electroless Ni deposition that only plate in the gap region of the joint by increasing the amount of Pb^{2+} concentration. We added 0.5 ppm, 1 ppm, and 1.5 ppm and 2 ppm of lead acetate into electroless Ni bath respectively. For samples conducted MELI process using electroless Ni plating with addition of 0.5 and 1 ppm lead acetate for 1 h, there would be no selective deposition property. Electroless Ni would deposit on the side wall of copper pillars. However, when the addition of lead acetate increased to 1.5 ppm, there were no Ni deposition on the side wall. Figure 5 shows pillar-to-pillar joints that have been plated by electroless Ni plating with addition of 1.5 ppm lead acetate for 1 h. The gap of the joint was reduced to about 1.5 μm with an intention to achieve a

concentration gradient of Pb^{2+} in the small gap. After 1 h of plating time, it can be seen that the sides of copper pillar were poised against deposition and that the electroless Ni would only deposit in the gap region between pillars. Figure 6 (a) and (b) show ion images of the joints. No void was observed in the deposit. Because the gap of the joint was really small, the diffusion of Pb^{2+} into the narrow gap region was very limited. Therefore, it is reasonable to assume that there was a concentration gradient in the gap region. The concentration of Pb^{2+} inside the gap is low, whereas the concentration of Pb^{2+} on the sides of copper pillar is increased, causing poisoning in those areas. In this way, selective deposition can be achieved in order to enable fine pitch process capability.

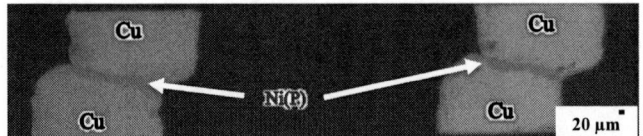

Figure 5. Selective Ni(P) deposition by addition of 1.5 ppm of lead accetate (1 h of plating time).

B. Selective electroless Au plating in microchannel

Figure 7 shows the plating results that underwent electroless Au plating with 0.02 mm/s, 0.6 mm/s, and 1.2 mm/s flow velocity. It was disclosed that Au deposition tended to fill the bottom gaps of copper pillars. This skipping phenomenon was also discovered on the electroless process on micropads. [18] When the pad size decrease, the skipping problem becomes more critical due to the higher concentration near the pad, which causes the nonlinear diffusion of stabilizers. They would adsorb firmly on metal surfaces and impede the deposition of reactants. [17]

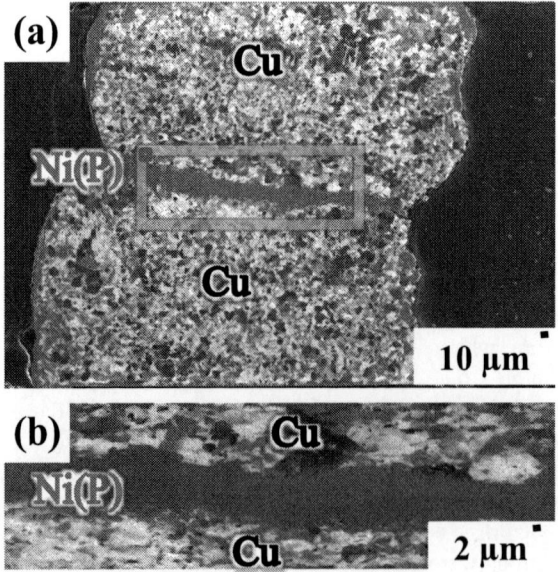

Figure 6. (a) FIB ion image of the joint after selective Ni(P) deposition by addition of 1.5 ppm of lead accetate for1 h, (b) an enlarged view of (a).

978-1-7281-1500-9/19 $31.00 © 2019 IEEE

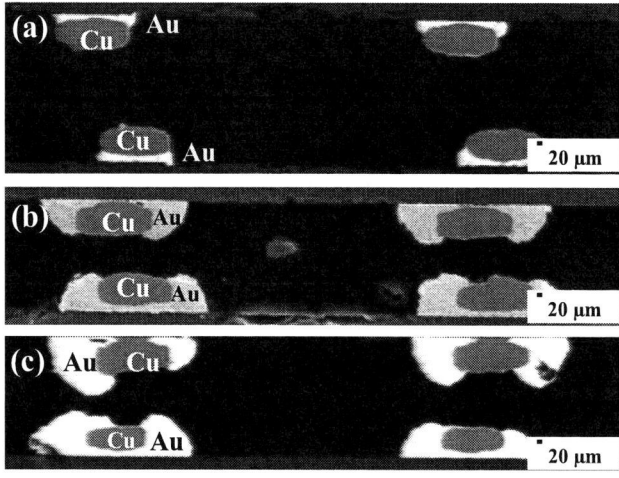

Figure 7. SEM micrographs of joints after selective Au deposition. (a) with a flow velocity of 0.02 mm/s for 7 h, (b) with a flow velocity of 0.6 mm/s for 7 h, (c) with a flow velocity of 1.2 mm/s for 4 h.

When the dimension of the surface reaches the thickness of the diffusion layer, nonlinear diffusion will arise, which leads to a higher surface density of adsorbed stabilizers. In addition, if the size of pad continuously decreases, the stabilizers adsorbed on the surface would prohibit the plating process. In this research, the skipping problem of electroless Au plating in the microchannel was also attributed to the nonlinear diffusion of stabilizers. It seems that electroless Au tends to deposit only at the small gap. Therefore, to address this problem, the gap between copper pillars was reduced to about 2 μm. Figure 8 shows the SEM micrograph of joints after selective Au deposition with a flow rate of 0.3 ml/h for 12 h. The flow direction was from left to the right. It is noticed that electroless Au only deposited on the gaps, which was corresponding to the results of electroless Ni. Thus, selective electroless Au deposition can be achieved for fine pitch application.

Figure 8. SEM micrograph of joints after selective Au deposition with a flow rate of 0.3 ml/h for 12 h.

IV. CONCLUSION

The present study aimed at achieving selective electroless Ni and Au deposition in the microchannel. It was found that the key to preventing the deposition on the sides of copper pillar includes the addition of stabilizer into the electroless Ni and Au bath and the decrease in gap between Cu pillars. With the addition of more than 1.5 ppm lead acetate, electroless Ni presents high selectivity during deposition in microchannel as the high concentration of stabilizer on sidewall of pillars inhibits the deposition. When the gap

between copper pillars is reduced to about 2 μm, electroless Au only deposit on the Cu surface with narrow gap. This research developed a selective electroless plating approach, providing a new pathway for fine pitch application.

ACKNOWLEDGMENT

This research was financially supported by the Ministry of Science and Technology, Taiwan and Unimicron Corporation under grant No. 107-2622-E-002 -004 -CC2. The authors would like to thank and acknowledge the technical support from Taiwan Uyemura Co., Ltd, and JCU Taiwan Corporation. The authors would also like to express their gratitude to Hitachi chemical and Thirty Party Technology Corporation for their help on dry film coating.

REFERENCES

[1] J. Wu, Y. L. Shen, K. Reinhardt, H. Szu, and B. Dong, "A nanotechnology enhancement to Moore's Law," Applied computational Intelligence and Soft computing, vol. 2013, pp. 1–13, 2012.

[2] T. N. Theis, and H. S. P. Wong, "The end of Moore's Law: A New Beginning for Information Technology," Computing in Science & Engineering, vol. 19, pp. 41–50, 2017.

[3] J. Lau, "TSV manufacturing yield and hidden costs for 3D IC integration," The IEEE 60th Electronic Components and Technology Conference (ECTC), pp. 1031-1042, 2010.

[4] J. Lau, "Evolution, challenge, and outlook of TSV, 3D IC integration" 2011 International Symposium on Advanced Packaging Materials, pp. 462-488, 2011.

[5] S. W. Yoon, J. H. Ku, N. Suthiwongsunthorn, P. C. Marimuthu, and F, Carson, "Fabrication and packaging of microbump interconnections for 3D TSV," 2009 IEEE International Conference on 3D System Integration, 2009.

[6] E. C. C. Yeh, W. J. Choi, K. N. Tu, P. Elenius, and H. Balkan "Current-crowding-induced electromigration failure in flip chip solder joints," Applied Physics Letter, vol. 80, pp. 580-582, 2002.

[7] V. Jayaram, S. McCann, T. C. Huang, R. Pulugurtha, V. Smet, R. Tummala, and S. Kawamoto "Thermocompression bonding process design and optimization for warpage mitigation of ultra-thin low-cte package assemblies," The IEEE 66th Electronic Components and Technology Conference (ECTC), pp. 101-107. 2016.

[8] K. Sakuma, K. Tunga, B. Webb, K. Ramachandran, M. Interrante, H. Liu, M. Angyal, D. Berger, J. Knickerbocker, and S. Iyer, "An enhanced thermo-compression bonding process to address warpage in 3D integration of large die on organic substrates," The IEEE 65th Electronic Components and Technology Conference (ECTC), pp. 318-324. 2015.

[9] K. N. Tu, H. Y. Hsiao, and C. Chen, "Transition from flip chip solder joint to 3D IC microbump: Its effect on microstructure anisotropy," Microelectronics Reliability, vol. 53, pp. 53-57, 2013.

[10] H. T. Hung, S. Yang, Y. B. Chen, and C. R. Kao, "Chip-to-chip direct interconnections by using controlled flow electroless Ni plating," Journal of Electronic Materials, vol. 46, pp. 4321-4325, 2017.

[11] S. Yang, H. T. Hung, P. Y. Wu, Y. W. Wang, H. Nishikawa, and C. R. Kao, "Materials merging mechanism of microfluidic electroless interconnection process," Journal of The Electrochemical Society, 165, pp. D273-D281, 2018

[12] I A. Weng, H. T. Hung, S. Yang, Y. H. Chen, and C. R. Kao, "Self-assembly of reduced Au atoms fro vertical interconnections in three dimensional integrated circuits," Scripta Materialia, vol. 159, pp. 119-122, 2019

[13] S. Yang, H. T. Hung, Y. B. Chen and C. R. Kao, "Low-Temperature, Pressureless Cu-to-Cu Bonding By Electroless Ni Plating," The IEEE

66th Electronic Components and Technology Conference (ECTC), pp. 111-114, 2016

[14] Z. Yang, N. Li, X. Wang, Z. Wang, and Z. Wang, "Bottom-Up Filling in Electroless Plating with an Addition of PEG–PPG Triblock Copolymers," Electrochemical and Solid-State Letters, vol. 13, pp. D47, 2010.

[15] T. Osborn, N. Galiba, and P. Kohl, "Electroless copeer deposition with PEG supression for all-copper flip-chip connections,"Journal of The Electrochemical Society, vol.156, pp. D226–D230, 2009.

[16] A. M. T. van der Putten, and J. W. G. de Bakker, "Geometrical effects in the electroless metallization of fine metal patterns," Journal of Electrochemical Society, vol. 140, pp. 2221-2228, 1993.

[17] A. M. T. van der Putten, and J. W. G. de Bakker, "Anisotropic deposition of electroless nickel bevel plating," Journal of Electrochemical Society, vol. 140, pp. 2229-2235, 1993.

[18] Zhamg, J. D. Baets, M. Vereeken, A. Vervaet, and A. V. Calster, "Stabilizer coneentration and local environment: Their effects on electroless nickel plating of PCB micropads," Journal of Electrochemical Society, vol. 146, pp. 2870-2875, 1999.

2019 IEEE 69th Electronic Components and Technology Conference (ECTC)

3D Integration of CMOS-compatible Surface Electrode Ion Trap and Silicon Photonics for Scalable Quantum Computing

Jing Tao[1], Yu Dian Lim[1], Hong Yu Li[2], Nam Piau Chew[1], Anak Agung Alit Apriyana[2], Lin Bu[2], Peng Zhao[1], Luca Guidoni[3] and Chuan Seng Tan[1*]

[1]Nanyang Technological University, Singapore 639798
[2]Institute of Microelectronics, Agency for Science, Technology and Research (A*STAR), Singapore 117685
[3]Laboratoire Matériaux et Phénomènes Quantiques, Université Paris Diderot, France, 75205
*E-mail: tancs@ntu.edu.sg

Abstract—In this work, we report ion trap fabrication using prevailing foundry copper back-end-of-line process on a 300mm Si platform. Surface electrodes comprising of ~3.7 μm thick Cu and ~0.2 μm thick of Au surface finish are electroplated above a ~3 μm thick of SiO_2 layer on a high-resistivity Si substrate. The innovative process, which is fully compatible with CMOS back-end, enables a fine gap trench structure between the electrodes, such that the exposed dielectric area to the trapped ions is reduced. By optimizing the electroplating process, a relatively flat Cu surface is created with a thin Au layer deposited as an effective protective layer to prevent surface oxidation. The fabricated trap is wire-bonded in a CPGA package for DC and RF testing. Small size Si traps show a good RF dissipation property which is a prerequisite for ion trapping. The further integration of TSV and Si photonics shows a promising prospect in terms of electrical and optical performance enhancement of ion trap.

Keywords-surface electrode ion trap; electroplating; leakage current; RF dissipation; TSV integration; photonics integration

I. INTRODUCTION

Quantum computers have the potential to perform complex computational tasks that are challenging or impossible using conventional von Neumann computers [1]. Among various qubit gates proposed for quantum computing, trapped ion quantum computer received much attention due to its exquisite coherence properties and high efficiency [2]. In ion trap quantum computing, the qubits, i.e. the trapped ions, are held in free space with electromagnetic forces and controlled by strong laser field [3]. The apparatus to generate the trapping electrical field is so-called radio-frequency ion trap, which is evolved from conventional macroscopic linear radio-frequency (RF) Paul trap [4] to modern microfabricated surface electrode ion traps [5][6].

The standard linear RF Paul trap consists of four parallel rod electrodes equidistance from the trap axis, where RF potential is applied to two opposite electrodes, while the other two electrodes are held at RF ground. The RF field creates a nearly harmonic ponderomotive pseudopotential to confine the ion in the radial direction (perpendicular to the trap axis). The confinement in the trap axis is provided by a static potential applied to two end electrodes placed along the trap axis. Despite the stability and deep trap depth of Paul trap [6], the macroscopic trap may not provide a path toward

a scalable and complicated trap design that is necessary for large-scale quantum computing application with trapped ions.

Surface electrode trap is a planar version of Paul trap [7], where all trap electrodes reside on a single surface and the ions are trapped above this surface. The trap is constructed with standard microfabrication processes with the potential of chip integration and scaling to make traps with multiple trapping zones with specialized functions such as detection, storage and logic gates to perform a particular quantum computing algorithm [8]. The principle of surface electrode trap to trap ions is analogy to that of linear Paul trap, where two surface RF electrodes provide a pseudopotential to confine the ions in the radial direction and the segmented DC electrodes to confine the ion the axial direction.

Over the past decades, there has been continuous effort in the miniaturization of ion traps from mechanically-assembled metallic traps to planar electrodes traps, using insulator or semiconductor substrates. The examples of insulator substrate based surface traps include gold-on-quartz surface trap [8], sliver-on-quartz surface trap [9], gold-on-sapphire surface trap [10] and gold-on-silica surface trap [11]. The insulator substrates have the advantage of low RF loss but it is not readily for chemical etching process to make complex trap structures. On the other hand, the ion traps can be fabricated in a semiconductor chip. Both gallium arsenide [12] and low-resistivity silicon [13] substrate have been utilized as the electrode/supporting material to fabricate the three-dimensional ion traps. However, this technology involves heavy substrate etching processes and MEMS fabrication techniques which may not be viable for further electronics and photonics integration in the same chip.

In order for ion traps to be viable for large-scale quantum computing applications, its fabrication should be compatible with the well-established CMOS processes for process repeatability and to enable hybrid integration. Ion trap fabricated on a Si substrate using CMOS-compatible processes have been demonstrated by several research groups [14][15][16]. MIT demonstrated an aluminum-on-Si surface trap with active and passive layers beneath the trap-electrode layer. Georgia Tech used patterned layers of aluminum and silicon dioxide (SiO_2) to form surface traps on Si. Sandia National Laboratories fabricated the surface traps on Si substrate with multilayer metal structures. All reported Si traps shows a good trap performance with a relatively long ion lifetime (on the order of several hours with ion cooling)

978-1-7281-1500-9/19 $31.00 © 2019 IEEE
1735

and low heating rate (as low as 81 quanta/s), which paves the way for next-generation Si-based surface trap for scalable quantum computing.

In this paper, we demonstrate a CMOS-compatible surface trap fabrication with prevailing foundry copper back-end-of-line process on a 300 mm Si platform. Copper instead of aluminum is used to fabricate the trap electrodes to lower the trap resistance. To reduce the RF loss and the substrate induced trap capacitance, high-resistivity Si substrate is employed for trap fabrication and a thick layer structure of SiO_2 and metal is designed to form trap electrode. The main challenges of fabricating surface trap with CMOS-compatible backend process are discussed. RF resonator test is conducted to evaluate the power dissipation loss of the Si traps. Finally, the future perspectives and the ongoing work of integrating TSV and Si photonics with surface traps are given.

II. SURFACE ELECTRODE ION TRAP

A. Trap Design and Pseudopotential Simulation

The trap design is based on a 5-wire symmetric geometry as reported in [17], consisting of two RF electrodes and three pairs of DC electrodes for multi-directional ion confinement. A variation of RF width (120, 80, 40, 20 μm) is designed to fabricate the traps in different size and with different trap height and depth. With the miniaturization of the trap size, the electrodes induced capacitance is reduced due to smaller capacitive coupling area [17]. The minimum gap sizes between the electrodes in the central trapping area is 5 μm for trap variations. The small gap distance is preferred due to the reduced dielectric area exposed to the trapped ions and therefore less stray field induced by ion accumulation in the dielectric surface.

The RF pseudopotential is simulated to obtain the trap position and depth of $^{88}Sr^+$ ion by applying a RF voltage of 200 V, 56 MHz to RF electrodes and ground the DC electrodes. Fig. 1 illustrate the 3D pseudopotential color contour above the trap surface, where the pseudopotential forms a confining tube (blue color) along the trap axis. For a typical trap geometry of 40 μm RF line width, the trapped ion radial position is simulated to be 40 μm above the trap

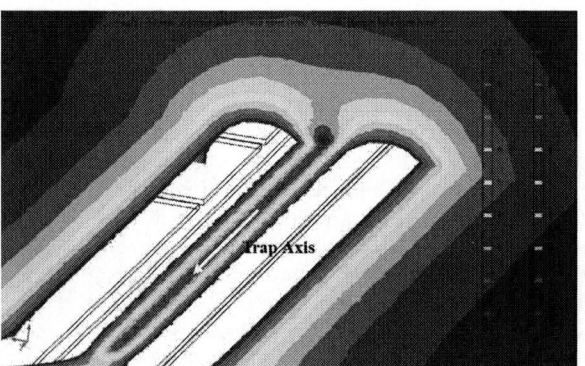

Figure 1. Simulated RF pseudopotential color contour above the trap surface.

surface. A set of DC voltages needs to be applied to DC electrodes to confine ions along the axis of the tube, which is not shown here.

B. Trap Fabrication

The fabrication begins with a 300 mm high-resistivity (> 750 Ω·cm) Si wafer. An insulation layer of 3 μm thick low stress SiO_2 is deposited by PECVD before electrode fabrication. To reduce the exposed dielectric surface in the trap surface, the SiO_2 in the gap area is patterned and dry etched. Then, Cu electrode is patterned and fabricated on top of oxide layer using Cu electroplating (ECP) process with Ti/Cu as the seed and adhesion layer. To prevent Cu surface oxidation, a thin layer of Au is sequentially electroplated followed by Cu ECP to form a final electrode. Fig. 2 (a) and (b) show the SEM top view of the fabricated trap with an overview and a close-up view of the central RF-DC-RF lines of 20 μm line width and 5 μm gap distance.

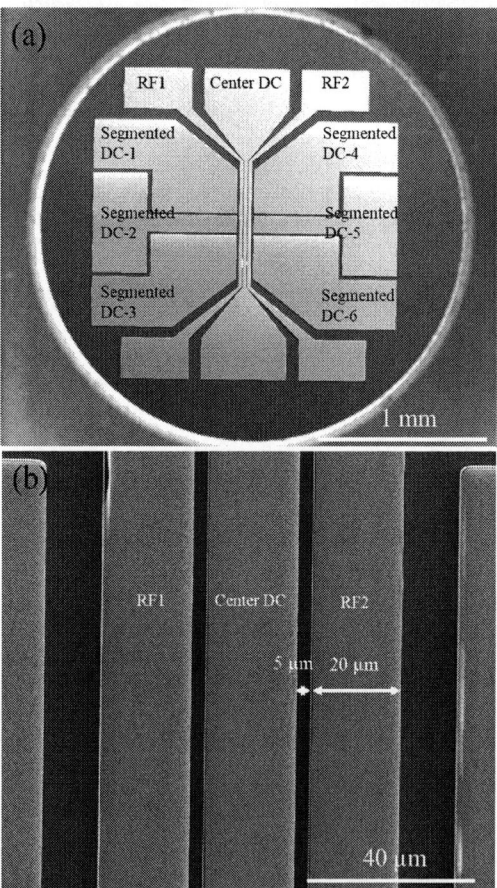

Figure 2. SEM image showing (a) the top view of the whole trap geometry and (b) magnified view of central RF-DC-RF lines.

Fig. 3 (a) shows the cross-sectional SEM image of the fabricated trap electrode. The actual dimensions of the fabricated layers comprising a trap electrode is shown in the figure. The metal layer consists of a 3.7μm ECP Cu with a

978-1-7281-1500-9/19 $31.00 © 2019 IEEE

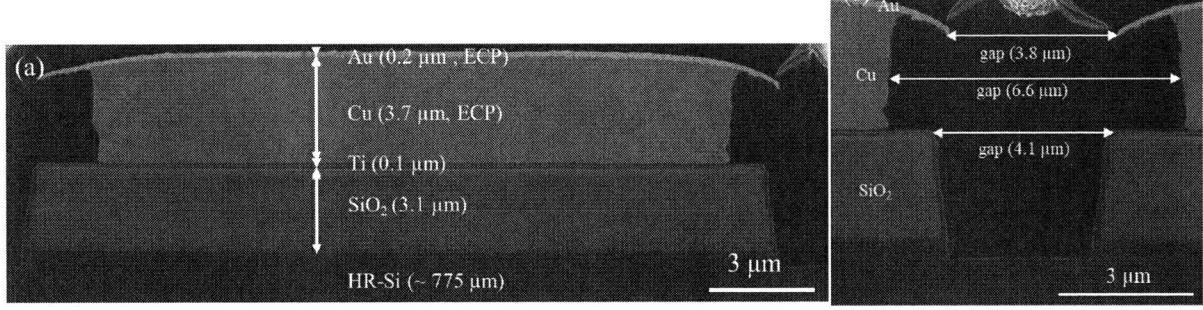

Figure 3. SEM image showing (a) the cross-section of trap electrode and (b) the trench gap.

top 0.2 µm thick ECP Au layer. Ti layer of 0.1 µm thickness is used as the adhesion layer between Cu and SiO_2. The insulation layer is 3.1 µm SiO_2 between the metal and Si substrate. Fig. 3 (b) shows the fabricated trench gap area between two electrodes. The minimum gap distance is formed by top Au layer and it is 3.8 µm. An undercut of Cu to Au layer is formed due to seed layer etching process and results in a lateral Cu undercut distance of ~1.4 µm. This brings an Au overhang structure which is suspected to induce the unwanted leakage current between the fine-gap electrodes. This will be discussed in the next section.

C. Trap Testing

One challenge to fabricate thick trap electrodes with ECP process is with the seed etching process, especially with a high length-to-width RF line structure of surface traps. Table I gives the specifications of different trap variations with designed RF line dimensions. It is noticeable that a relatively narrow but long line structure is designed with a fine line gap. This poses the challenge for ECP process to remove the seed layer between the long lines. A high leakage current may be induced due to the metal residue which is not fully removed during wet etching.

TABLE I. TRAP SPEC AND SHORT SCREENING TEST RESULTS

Trap Type	RF Line Width	RF Line Length	RF-DC Line Gap	No. of Dies Pass Short Screening
Trap_20-5	20	730	5	4/6
Trap_40-5	40	1460	5	3/7
Trap_80-5	80	2920	5	1/3

An IV measurement is conducted with a DC bias of 0-150 V at atmospheric conditions to screen every adjacent metal pads on the traps. With the 9 separate metal pads as shown in Fig. 2 (a), 8 measurements are conducted for a whole trap electrodes set. To define the pass/fail criteria, the leakage current above 10^{-8} between the electrodes is considered to maintain sufficient insulation. Table I gives the

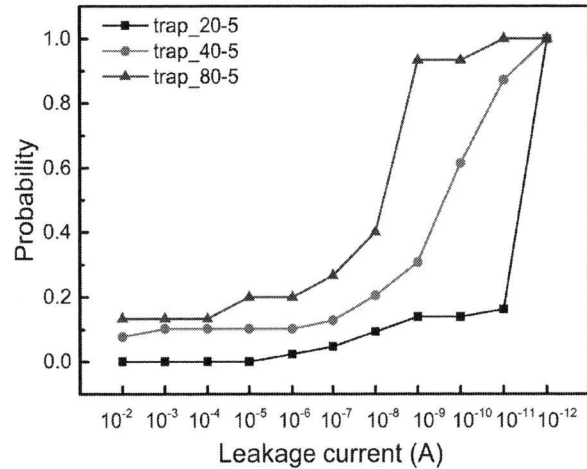

Figure 4. Probability plot of the leakage current for each trap type. The points in the curves indicates the actual percentage of measured data falls into the correponding leakage current range.

number of dies which pass the short screening test. It is noted that the big traps (trap_80-5 and 40-5) with longer RF lines are prone to fail the test. To understand the probability of the leakage current induced between the electrodes in each trap type, Fig. 4 shows the probability plot of the leakage current based on measured data. We found that more short occurred for big traps and a higher leakage current is resulted for big traps compared to small traps (trap_20-5). Except for the seed residual which may cause the short, the other possible reason is that the thin overhang Au layer may easily peel off due to mechanical instability and create the short bridge in the fine gap.

A failure analysis into the shorted electrodes found the peeled Au layer as shown in Fig. 5 (a). One possible solution to solve the problem is to use ultrasonic (US) clean method to remove the peeled Au and the electrodes after US clean is shown in Fig. 5 (b). We found that the short test yield is significantly improved after US clean is applied to the trap dies. The optimization in fabrication process is still needed to minimize the Au overhang to further improve the short test pass yield.

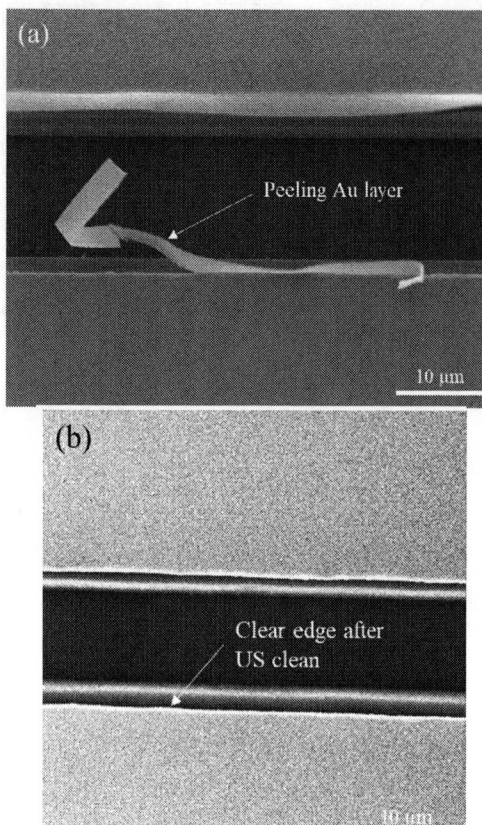

Figure 5. SEM images of (a) a peeling Au layer to possibly cause the short between the electrodes and (b) a clear electrode edge (without Au overhang layer) after ultrasonic clean.

D. Trap Packaging

For planar trap packaging, wire-bonding is used to connect the surface electrodes to the ceramic pin grid array (CPGA) package. The surface trap is die attached with a vacuum-compatible non-conductive paste to the metal ground surface of CPGA with a glass spacer of a ~0.5mm thick to raise the height of the trap chip above the CPGA surface. Standard 25 μm Au wire is used to wire-bond the trap electrodes to CPGA pads. A discrete thin-film capacitor (820 pF) is used as a bypass capacitor to filter the RF noise for each DC electrode. Each capacitor is die attached with silver conductive paste to the CPGA ground, which is RF grounded. A wire-bonding configuration of a surface trap in CPGA package is shown in Fig. 6.

III. ION TRAP RF PERFORMANCE

A. RF Dissipation of Si Trap

One possible drawback of using Si substrate for ion trap is the possible high RF dissipation resulted from the lossy

Figure 6. Planar surface trap assembly in a CPGA package.

substrate. The equation of dissipated power in the trap is given in equation (1) by [6]:

$$P = I^2_{RMS}(R_{Lead} + R_{ESR}) = (C\Omega V_{RMS})^2 [R_{Lead} + \tan\delta/(C\Omega)] \quad (1)$$

where R_{Lead} is the total resistance in the trap system, C is the capacitance between the RF and RF grounded electrodes and $\tan\delta$ is loss tangent of the substrate. To have a low power dissipation, we select high-resistivity Si with a low loss tangent and design the trap structure to minimize the R and C in the trap.

To understand RF transmission property of the metal lines in a Si trap structure, we conduct S-parameter measurement in GSG configuration by probing the central RF-DC-RF lines. The measurement frequency is corresponding to the RF drive frequency range of 10 to 100 MHz for ion trapping. Fig. 7 shows the measured insertion loss, S21 for different trap types. All trap types show a relatively low insertion loss below < 3 dB up to 100 MHz. In

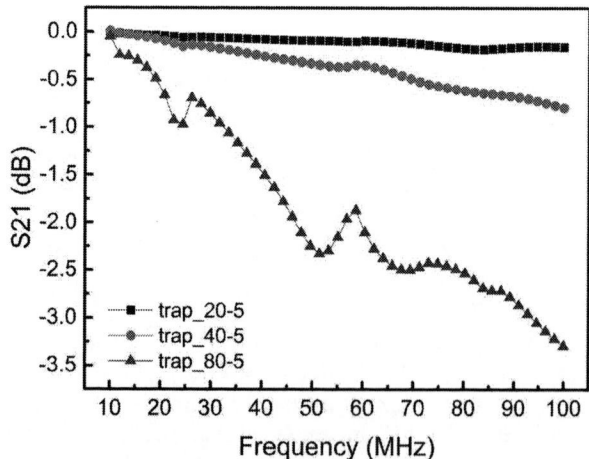

Figure 7. On-chip S-parameter measurement for three trap types.

a typical RF driver frequency of 56 MHz, the insertion loss for trap type 20-5, 40-5 and 80-5 are -0.1, -0.4 and -2 dB, respectively. We found the insertion loss becomes smaller with smaller trap size. This is explained by the short line length and smaller surface area in the small traps which results in a lower R and C.

B. RF Resonator Circuit

In the ion trapping experiment, the trap is connected as a capacitor to an external inductor to form a *LCR* resonator circuit in order to obtain the required trapping voltage. A Cu coil is used as the external inductor to provide the inductive component L, while capacitance between the RF and RF-grounded electrodes in the ion trap provides the capacitive component C. The resonance frequency is determined by L and C as given in equation (2):

$$f_0 = 1/[2\pi \times \sqrt{(LC)}] \qquad (2)$$

The quality factor Q is an important property of the resonator which is proportional to the voltage gain for resonator circuit. Q is determined by L, C and R in the trap circuit as given in equation (3):

$$Q = 1/R \times \sqrt{(L/C)} \qquad (3)$$

where R is the total resistance in the trap circuit including trap resistance, wire resistance and coil resistance. A simplified *LCR* circuit diagram for ion trap is shown in Fig. 8, where L is mainly the coil inductance, C_{trap} is trap capacitance between RF and RF ground and R is the total resistance in the circuit. To measure the output power from the resonator circuit, C_1 and C_2 is connected as capacitive voltage divider next to the trap, which is further connected to a RF signal analyzer for read out. In the actual measurement, the trap is connected in series to a Cu coil shielded in a steel box with the RF signal fed from signal generator into the coil box via a coaxial cable and the divided output voltage is connected to signal analyzer via coaxial cable for read out.

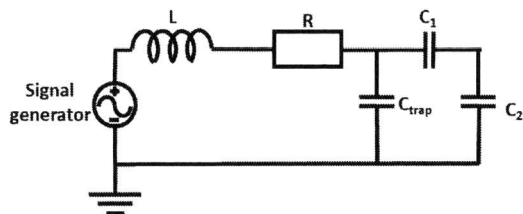

Figure 8. Simplified circuit diagram of ion trap *LCR* resonator circuit.

C. RF Resonator Test

RF resonator test is conducted by connecting the packaged ion trap in CPGA to a resonator box comprising of a Cu coil and capacitor dividers. An input RF power of -10 dBm is supplied to the circuit with a frequency sweep of 10

to 100 MHz. Fig. 9 shows the resonant curves which is the output power as function of frequencies for different trap types. We obtain the resonance frequency from the curves, which are listed in Table II. For trap_80-5, no obvious resonance peak is obtained, which may be due to the shorting existing between RF and ground. Hence, the data is not analyzed further. For trap_20-5 and 40_5, the resonance frequency is 54.8 and 46.8 MHz, respectively. From the resonance curve, we can also obtain the quality factor by equation (4):

$$Q = f_0 / \text{FWHM} \qquad (4)$$

where f_0 is the resonance frequency and FWHM is the Full width at half maximum of the normalized power. The calculated quality factor is given in Table II, which is 16 and 10 for the respective trap_20-5 and 40-5 connected in the circuit. The dissipated power in dB is also given in table II and the power loss of a small trap_20-5 is comparable to a reference 3.3 pF capacitor if connected in the circuit. In contrast, the trap_40-5 shows a relatively large dissipated power.

In order to find out *LCR* in the circuit, we measure the trap capacitance and calculate the other two components. *CV* measurement is conducted to measure the trap capacitance

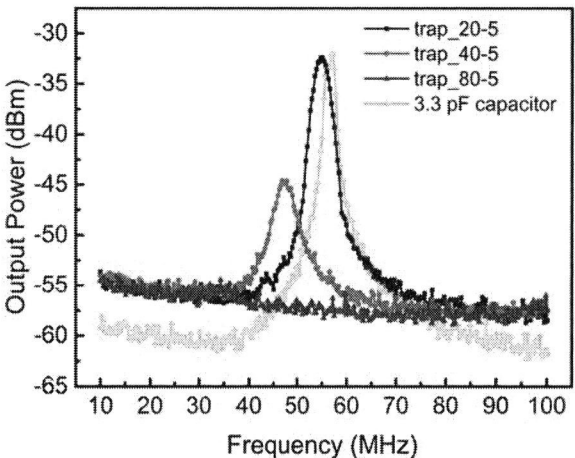

Figure 9. Resonant curves of three trap types. The resonator curve of 3.3 pF capacitor is used as reference.

with a voltage sweep from -30 to 30 V, at a measuring frequency of 10 MHz. The measured capacitance, C for trap_20-5 and 40-5 is 23 and 28 pF, respectively. We consider this capacitance as the major capacitance in the circuit. Using equation (2), we can obtain an estimated

TABLE II. RF RESONATOR TEST RESULTS AND LCR ANALYSIS

Trap Type	Resonance frequency f_0 (MHz)	Quality factor Q	Dissipated power P (dB)	Measured trap capacitance C (pF, @ 10 MHz)	Estimated circuit inductance L (µH)	Estimated circuit resistance R (Ω)
Trap_20-5	54.8	16	-22.41	23	0.37	7.9
Trap_40-5	46.8	10	-34.64	28	0.41	12.1
3.3 pF capacitor	56.9	24	-22.16	-	-	-

inductance, L in the circuit, which is ~ 0.4 µH. Then using equation (3), we can obtain an estimated resistance, R, which is 8 and 12 Ω for the respective trap_20-5 and 40-5. Generally, the small trap (20-5) results in a better power dissipation compared to the big trap (40-5), due to a smaller trap capacitance and resistance resulted from the corresponding geometry. To further minimize the power dissipation, especially for big traps 40-5 and 80-5, a ground layer structure can be inserted between the electrodes and Si substrate to shield the lossy effect of the substrate.

IV. TOWARDS 3D HYBRID INTGERATION

A 3D hybrid integration architecture for ion trap chip is proposed in this work to mitigate scalability concerns related to conventional wire bond interconnects and also to integrate the scalable optical interface on the same Si platform. The schematics illustrate the hybrid integration is shown in Fig.

10 (a)-(e), where through silicon via (TSV) is used for electrical integration to efficiently deliver the DC/RF signals to trap electrodes and the passive photonics components (waveguide and grating coupler) are used to guide the light to address the ions with high precision. The integration process is fully compatible with the current planar trap fabrication process.

A. TSV Integration

The integration of 3D through-Si-via (TSV) in the ion trap is considered in this work. Instead of doped silicon filled TSV and AuSn bump reported in [18], CMOS-compatible Cu filled TSV and CuSn micro-bump are integrated with surface electrodes of ion trap for better interconnect RC performance. The other advantage of replacing wire-bonding on the trap surface with TSVs is to minimize the wire obstruction to laser access. Also by removing the bond pads for wire-bonding, the designed TSV trap geometry result in a

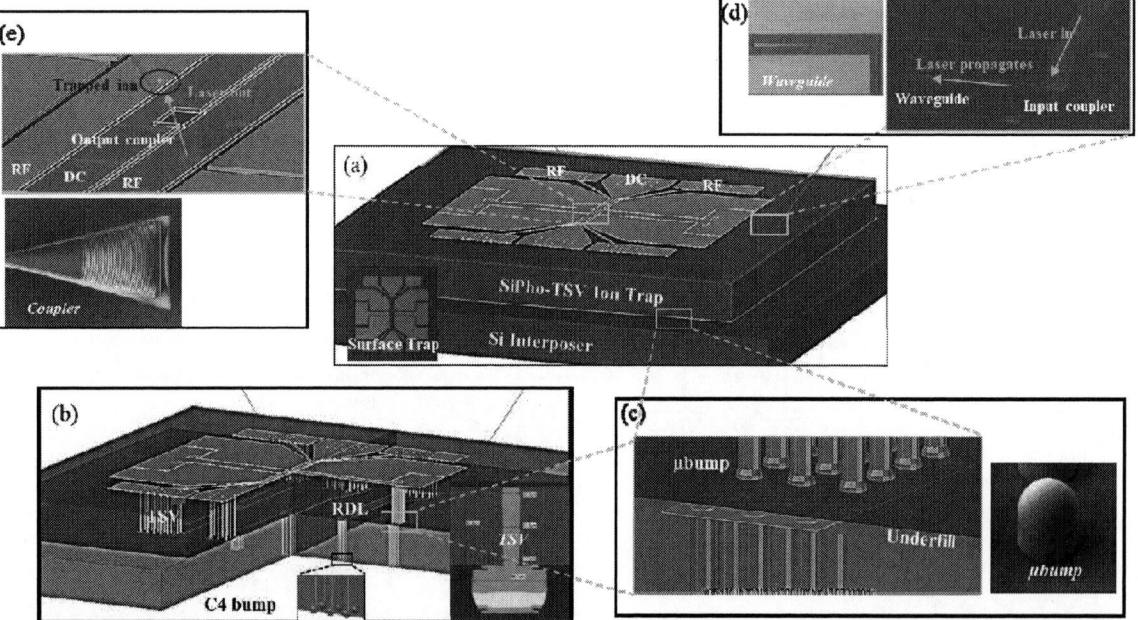

Figure 10. The schematics illustrate the 3D hybrid integration of TSV and Si photonics with surface ion trap for scalable quantum computing: (a) SiPhotonics-TSV ion trap on a Si interposer, (b) TSV connecting surface electrodes and RDL, (c) micro-bump for chip-on-chip bonding, (d) Si photonics for input laser and (e) Si photonics for output laser.

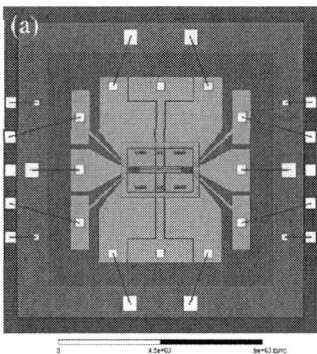 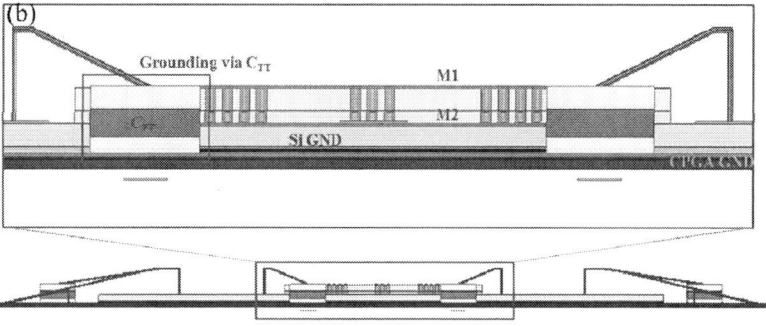

Figure 11. HFSS simulation of TSV ion trap: (a) TSV trap in package model and (b) a cross-sectional view showing the TSV in trap and wire-bonding connection in interposer.

form factor reduction of 3 ~ 8 times as compared to the planar traps.

TSV is designed to connect the surface electrodes to a Si interposer through micro-bumps as illustrated in Fig. 10 (b) and (c). TSV with a dimension of 20 μm diameter and 100 μm deep is fabricated in a standard foundry line. RF property of the designed TSV trap structure is simulated using HFSS to obtain the insertion and reflection loss. One of the TSV trap-in-package model is shown in Fig. 11 (a) and (b) with the top view and cross-sectional view showing the trap layout and the TSV connection between the trap and interposer. Note that in this design the wire-bonding is still employed for electrical connection from interposer to CPGA. The simulated S-parameter indicates the improvement of insertion loss of 0.5-1.5 dB for TSV traps compared to planar traps. For future integration, Chip on Wafer on Substrate (CoWoS) or Fan Out Wafer Level Packaging (FOWLP) are potential candidates to eventually eliminate wire-bonding in the package.

B. Photonics Integration

To realize the computing operation of the trapped qubit, such as Doppler cooling/readout, state transition, repumping, laser sources of specified wavelengths is needed [19]. However, it becomes a challenge to accurately "aim" the laser onto the trapped ion qubit in on-chip ion trap which possesses complicated wire-bonding. Furthermore, in complicated ion trap systems involving multiple ion qubits, it exhibits greater challenge to address each ion qubits individually with laser sources with several wavelengths (typically 3 ~ 4 wavelengths). Therefore, a scalable architecture to integrate the surface electrodes for ion trapping and photonics for addressing ions has to be considered. From the state-of-the-art, there has been some preliminary efforts in integrating the photonics components to the ion trap on the same substrate [20]. However, the reported ion trap-photonics integration work by Mehta et al. is carried out on a quartz substrate, which is not compatible with the conventional CMOS fabrication technologies. In this work, we design the integrated photonics with a laser-coupling system comprises of grating coupler and bent

waveguide in combination of surface electrodes to route laser beam on-chip to the trapped ion position. The schematic diagram of the photonics system integrated with ion trap is illustrated in Fig. 10 (d) and (e).

For the trapped $^{88}Sr^+$ ion system, laser sources of 422, 674, and 1,092 nm are needed for Doppler cooling/readout, state transition, and repumping, respectively. Among the required laser sources, we begin with 1092 nm wavelength design, which is close to the standard wavelength design for Si photonics of 1,550 nm. A commercial 220 nm layer Silicon on Insulator (SOI) wafer is utilized for photonics integration. The photonics system for quantum computing operation comprises of an input grating coupler, a straight waveguide, and an output grating coupler. The coupling efficiencies of both input and output couplers are simulated using Lumerical FDTD solution. Fig. 12 (a) and (b) show the input coupler model and the simulation results of laser propagation. With an optimized coupler geometry design, the simulated overall coupling efficiency for input coupler can reach 20% at 1,092 nm wavelength with an input angle, θ_{in}, of ~ 16°.

On the other hand, the overall coupling efficiency of the optimized output coupler can reach 80% in the simulation, Fig. 12 (c) and (d). On top of efficiency, another important figure-of-merit for the output coupler for ion trapping is the output angle, θ_{out}. As it is crucial for the output laser to "hit" the trapped ion for repumping operation, θ_{out} helps to estimate the output coupling direction of laser and serves as a clear guidance in positioning the output coupler. From the simulation, θ_{out} for output coupler design is estimated to be ~16°. Meanwhile, laser refraction also happened on the top SiO_2 layer, where the final output laser is calculated to be ~22° from the vertical by using Snell's Law. Since Sr^+ ion is expected to be trapped at ~40 μm height for the trap type 40-5 as an example, to accurately "hit" the ion with laser, the coupler is to be placed at a distance of ~16 μm from the center of the electrode. To further ensure that the output laser "hits" the ion qubit, the length of the output coupler can be increased. The optimization of the proposed photonics system is still on-going, and will be reported in our future studies.

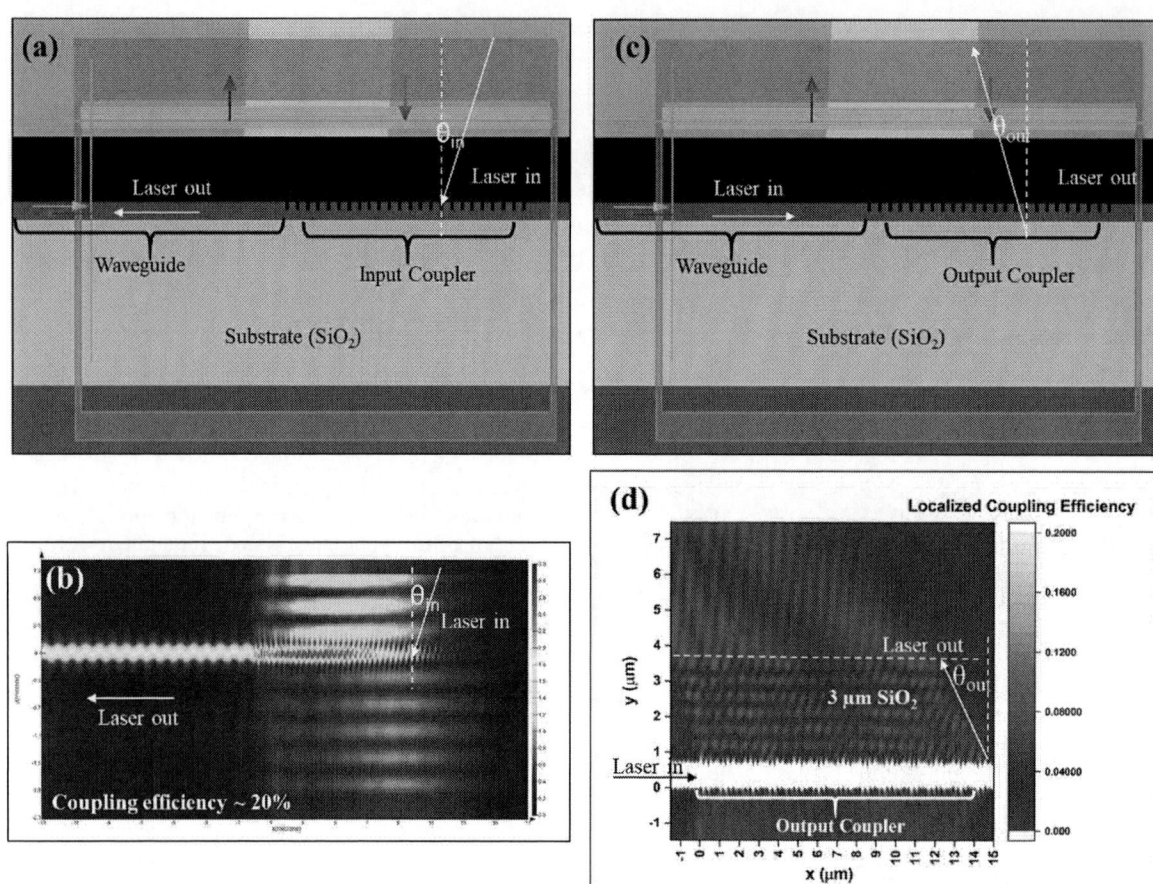

Figure 12. FDTD simulation of (a) input coupler on SOI substrate model and (b) the simulated incident laser coupling and propagation mode; (c) output coupler on SOI substrate model and (d) simulated outgoing laser coupling and propagation mode.

V. CONCLUSIONS

We demonstrate a CMOS-compatible surface trap fabrication with prevailing foundry copper back-end-of-line process on a 300 mm Si platform. Trap electrodes are fabricated with electroplating of Cu and Au on a SiO₂ layer. Fabrication difficulty lies in short occurrence due to the very long line and narrow gap geometry for the surface electrodes and process optimization is needed to improve the shorting test yield especially for the big traps. The fabricated traps show low insertion losses. Small trap (20-5) shows a lower power dissipation loss in RF resonator test as compared to big traps which is the prerequisite for ion trapping. The preliminary simulation results of TSV and Si photonics integration with ion trap show promising prospects for future works.

ACKNOWLEDGMENT

We greatly thank WW Seit for his technical supports on wire-bonding and packaging. The work was supported by A*STAR Quantum Technology for Engineering (A1685b0005).

REFERENCES

[1] K. R. Brown, J. Kim, and C. Monroe, "Co-Designing a Scalable Quantum Computer with Trapped Atomic Ions," pp. 1–11, 2016.

[2] T. P. Harty et al., "High-fidelity preparation, gates, memory, and readout of a trapped-ion quantum bit," Phys. Rev. Lett., vol. 113, no. 22, pp. 2–6, 2014.

[3] W. Lange, "Quantum computing with trapped ions," Comput. Complex. Theory, Tech. Appl., vol. 9781461418, no. September, pp. 2406–2436, 2013.

[4] W. Paul, "Electromagnetic traps for charged and neutral particles," Rev. Mod. Phys., vol. 62, no. 3, pp. 531–540, 1990.

[5] D. Kielpinski, C. Monroe, and D. J. Wineland, "Architecture for a large-scale ion-trap quantum computer," Nature, vol. 417, p. 709, 2002.

[6] J. Chiaverini et al., "Surface-Electrode Architecture for Ion-Trap Quantum Information Processing," vol. 5, no. 6, pp. 419–439, 2005.

[7] S. Seidelin et al., "Microfabricated surface-electrode ion trap for scalable quantum information processing," Phys. Rev. Lett., vol. 96, no. 25, pp. 1–4, 2006.

[8] J. M. Amini et al., "Toward scalable ion traps for quantum information processing," New J. Phys., vol. 12, 2010.

[9] J. Labaziewicz, Y. Ge, P. Antohi, D. Leibrandt, K. R. Brown, and I. L. Chuang, "Suppression of heating rates in cryogenic surface-electrode ion traps," Phys. Rev. Lett., vol. 100, no. 1, pp. 1–4, 2008.

[10] N. Daniilidis et al., "Erratum: Fabrication and heating rate study of microscopic surface electrode ion traps (New Journal of Physics (2011) 13 (013032))," New J. Phys., vol. 14, 2012.

[11] D. T. C. Allcock et al., "Implementation of a symmetric surface-electrode ion trap with field compensation using a modulated Raman effect," New J. Phys., vol. 12, 2010.

[12] D. Stick, W. K. Hensinger, S. Olmschenk, M. J. Madsen, K. Schwab, and C. Monroe, "Ion trap in a semiconductor chip," Nat. Phys., vol. 2, no. 1, pp. 36–39, 2006.

[13] G. Wilpers, P. See, P. Gill, and A. G. Sinclair, "A monolithic array of three-dimensional ion traps fabricated with conventional semiconductor technology," Nat. Nanotechnol., vol. 7, no. 9, pp. 572–576, 2012.

[14] K. K. Mehta et al., "Ion traps fabricated in a CMOS foundry," vol. 1749011749, no. 2015, 2016.

[15] S. C. Doret et al., "Controlling trapping potentials and stray electric fields in a microfabricated ion trap through design and compensation," New J. Phys., vol. 14, 2012.

[16] D. Stick et al., "Demonstration of a microfabricated surface electrode ion trap," pp. 1–4, 2010.

[17] J. Tao, N. P. Chew, L. Guidoni, Y. D. Lim, P. Zhao, and and C. S. Tan, "Fabrication and Characterization of Surface Electrode Ion Trap for Quantum Computing," in 20th Electronics Packaging Technology Conference (EPTC), 2018.

[18] N. D. Guise et al., "Ball-grid array architecture for microfabricated ion traps," J. Appl. Phys., vol. 117, no. 17, 2015.

[19] C. Monroe and J. Kim, "Scaling the ion trap quantum processor," Science (80-.)., vol. 339, no. 6124, pp. 1164–1169, 2013.

[20] K. K. Mehta, C. D. Bruzewicz, R. Mcconnell, R. J. Ram, J. M. Sage, and J. Chiaverini, "Integrated optical addressing of an ion qubit," Nat. Nanotechnol., vol. 11, no. 12, pp. 1066–1070, 2016.

Integrated RTD Sensors for maintaining thermal uniformity during TCB process

S. Ben Jemaa, J. Sylvestre
Interdisciplinary Institute for Technological Innovation
(3IT)
Sherbrooke University
Sherbrooke, QC, Canada
salwa.ben.jemaa@usherbrooke.ca

P.Gagnon
IBM Canada
Bromont, QC, Canada
pgagnon@ca.ibm.com

Abstract—Thermocompression bonding (TCB) is an advantageous method adapted for system miniaturization, for instance to assemble microelectronic devices in 3D stacks. Low throughput and yield are the main drawbacks of TCB, which consequently can lead to higher production costs. Throughput can be improved by reducing the process cycle time using faster heating rates, generally at the cost of reducing the temperature uniformity of the assembly and therefore inducing more assembly defects and incurring higher yield losses. A rapid heating rate allows solder joints to achieve their melting point faster, but can cause large temperature variations between the edge and the center of the die. This significantly affects the quality of bonds and leads to the formation of non-wet (open) and bridge (short) defects. These defects are a barrier to achieving a high-quality product at low cost. We report a novel methodology to quantify the thermal limits of the TCB process in terms of achievable temperature uniformity across the chip area, as a function of heating rate. We investigate the effect of the heating rate on the surface temperature uniformity using a specially designed temperature sensor with high spatial and temporal resolutions, and which further minimally perturbs the TCB process when it is inserted into industrial TCB equipment for testing.

Keywords-component; Thermocompression bonding (TCB); temperature uniformity; heating rate; temperature sensor

I. INTRODUCTION

One of the biggest challenges for the industry is to optimize the whole life cycle of electronic product from design to commissioning with production cost reduction as a main objective. The assembly phase of microelectronic components is one of the key steps to achieve this goal. The most common chip assembly processes are traditional mass reflow (MR) and Thermocompression bonding (TCB). The conventional mass reflow technique has several problems such as warpage caused by the coefficient of thermal expansion (CTE) mismatch between chip and substrate [1-2]. These problems become more apparent when the application calls for larger devices, higher interconnection densities, and thinner package substrates. This is why new TCB processes are developed to face those difficulties. There is no doubt that TCB can provide a high level of reliability and more energy-efficient devices for the next generation of packaging [3]. However, the big issue hindering the adoption of TCB in the industry is throughput limitations. This is attributed to the long

TCB process, which includes several steps: alignment, flux dipping, heating, joint formation and cooling. High-throughput TCB bonding processes attracted the attention of several researchers through the investigation of various approaches such as mixed-techniques bonding, gang bonding, and 3D integrated packages [4-7]. However, the success of such approaches relies on the optimization of bonding parameters. The most crucial parameter to optimize is temperature. Good control of the temperature profile ensures a uniform thermal distribution, which typically results in more robust and reliable joints [8]. The throughput of the TCB process can be significantly increased when higher transfer temperatures or shorter dwell times are applied. Indeed, the dwell time parameter does not affect the quality or reliability of bonds as much as the heating rate does [9]. A rapid heating rate allows solder joints to achieve their melting point faster, but can cause large temperature variations between the edge and the center of the die. This significantly affects the quality of bonds and leads to the formation of non-wet (open) and bridge (short) defects. Some researchers focus on the development of methods aimed at ensuring a better control of the temperature profile, especially for the pre-applied underfill process [10-12]. However, these methods are not very accurate to predict temperature distributions near interconnections and do not provide a clear overview of the temperature across the chip surface. It is impossible to attach a large number of thermocouples to precise locations on the chip surface, given that a commercial thermocouple has a thickness of ~100μm. Two alternative solutions were proposed recently by Athia et al. [13, 14] and Bex et al.[8] to measure the interface temperature and characterize the heat transfer for fine pitch interconnect TCB processes. Athia et al. tried to integrate RTDs and force sensors in custom chip with an 8×8 array of Au stud bumps pads. Bex et al. have used only three RTD'S temperature sensors integrated in both chips and substrate. Certainly, these methods enable in situ temperature measurements without thermocouple insertion, but they show that developing a sensor-embedded chip for each optimisation of the TCB process is not straightforward.

In this paper, we present a novel methodology to quantify the thermal limits of the TCB process in terms of achievable temperature uniformity across the chip area, as a function of heating rate. We present the feasibility, fabrication and measurement of temperature by a micro-fabricated resistance thermal detector (RTD). The originality of our RTD sensor

978-1-7281-1500-9/19 $31.00 © 2019 IEEE

resides in its ability to measure temperature over an area of 30x30 mm² with high accuracy (±1°C), short response time (<200ms) in the range 25°C-260°C, and high spatial resolution (144 RTDs on a 2.4mm pitch in a (12x12) grid. The RTD sensor can be placed at various positions in the TCB stack, including between the chip mounted to a bonder head and the laminated substrate on the stage of the TCB machine (Fig. 1). As the RTD is an independent and accurate temperature sensor, it offers the possibility of reliable in-situ temperature measurements during the TCB process. A correlation study between the RTD thermal measurements and bond reliability is carried out to provide an excellent approach to optimize the bonding processes. Pre-process thermal responses measured via the sensor have allowed the identification of the steady-state temperature. The rate of heat flow was also analyzed with respect to the lateral temperature distributions. First, the static measurements were carried out in order to obtain the peak temperature that must be set for the tool head to reach the equilibrium melting temperature of the solder joints. Second, dynamic heat transfer experiments were used to investigate the effect of heating rate as well as thermal uniformity on the solder joint quality. The effects of the critical bonding temperature and heating rate on joint quality was performed through experiments on SnAg solder bumps. The assembly of the test vehicle was conducted by varying only the heating rate while keeping all other parameters constant (bonding dwell time and applied force). To find the higher transfer temperature providing good solder joint, an evaluation of the joints quality was performed. Tensile pull tests and dye-and-pry processes were used to identify the non-wet defects and bridges locations. The cross-section of the solder joints in the best and the worst-case conditions were observed using an optical microscope.

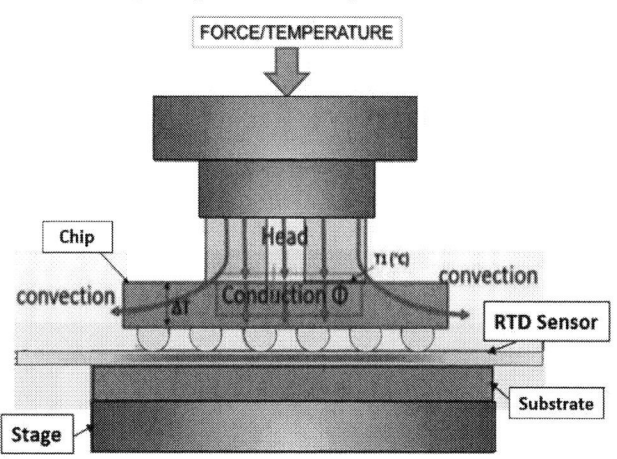

Figure 1.   Schematic of temperature measurements with the sensor.

The testing results show that there is a significant difference between the peak temperature achieved in the solder bumps and the bonding head peak temperature. More importantly, the transient temperature measurements allow the precise measurement of temperature non-uniformity at faster heating rates. It was quantitatively observed that the bond quality was reduced with faster heating rates. The reduction in temperature uniformity could be correlated to the decrease in bond quality at larger heating rates, thus enabling the precise optimization of the TCB process for both throughput and quality.

II.   EXPERIMENTAL PROCEDURE

A. Design of the RTD Sensor Chip

Our micro-fabricated sensor consists of serial RTDs (144 RTD) made of electrically conductive materials on a 4-inch silicon wafer. Each RTD sensor has a number of associated connection wires and each wire must be connected to a contact pad in order to connect the test chip to the measurement devices. Electrical industrial connectors, selected to be able to withstand high temperatures (+150°C), connect the acquisition system to the sensors. These connectors are spread on the peripherals of the wafer, outside the bonding area so their temperature does not exceed 150°C during normal use of the sensor. The layout of the sensor is shown in Fig. 2a. The description of the stack of layers is presented in Fig. 2b. First, a thin layer of SiO2 (150 nm thick) was used as the bottom insulation layer. An e-beam evaporator was then used to deposit Cr (50 nm) as an adhesive layer. Then, the copper (0,5μm-thick) was deposited as the sensing layer, for its good linearity and sensitivity. To protect the connection pads from oxidation a thin layer of gold (40 nm) was deposited on the copper. A thin layer of Cr (5nm), serves as an adhesion layer, was also deposited on the gold layer. Finally, the active zone of the sensor was coated with SiO2 (300 nm thick) to protect the fabricated thin-films.

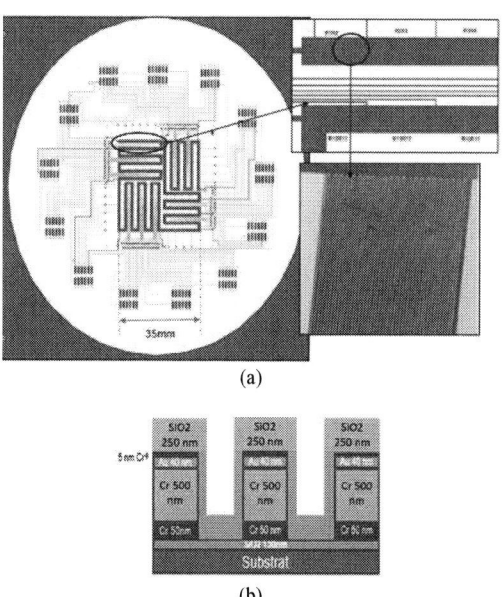

(a)

(b)

Figure 2.   Schematic image of: (a) layout of RTD temperature sensor array (b) sensor layer stack.

The sensor was then annealed at 260°C under a N2 atmosphere to improve the Cu resistivity and prevent drift in the RTD response.

B. Sensor Calibration

RTD sensors simple to manufacture and integrate, providing a high-precision temperature measurement. The accuracy depends on the operating temperature; it is the largest error expected between the measured temperature values and the ideal temperature values. To minimize measurement errors and ensure good accuracy, it is important to calibrate the RTDs correctly. The relationship between resistance and temperature can be written as follows:

$$R(T) = R_0 \times (1 + TCR \times (T - T_0)) \quad (1)$$

$$T = \frac{(R - R_0)}{R_0} \times \frac{1}{TCR} + T_0, \quad (2)$$

where R_0 is the resistance at the reference temperature, TCR is the Temperature Coefficient of Resistance, and T_0 is a reference temperature.

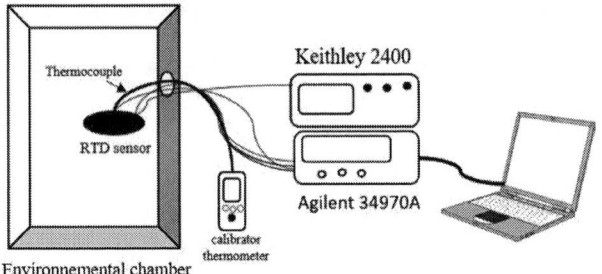

Figure 3.   Experimental set-up for the sensor calibration.

To obtain the temperature-resistance relationship and determine the value of the slope coefficient ($\Delta R/\Delta T$), a calibration was performed in a thermal chamber. The imposed temperature was controlled by a thermocouple (type-T) fixed on the sensor. A constant current ($10\mu A$) derived from a Keithley 2400 series source meter was applied. The voltage drop across each resistor was monitored using the four-point measurement method through an Agilent acquisition system (34970A model). The experimental setup is shown in Fig. 3.

Temperature calibration was performed for 9 RTDs over a temperature range from 0°C to 120°C. This range was chosen in order to limit exposure of the connectors to temperatures above their rating. The measurement current set at $10\mu A$ to prevent the effect of self-heating. The fabricated sensor showed a linear behaviour for all RTDs. This linear behaviour was confirmed with calibration points at high temperature (around 230°C) using a Finetech flip-chip bonder. Fig. 4 shows an example of a single RTD component calibration curve. The temperature value can be extracted from the voltage value obtained using the calibration curve for each RTD.

C. TCB: Test Vehicle and Thermal Characterization Setup

In this study, the test vehicle a flip-chip Ball Grid Array. The size of the chip was approximately 17x17 mm² with SnAg solder bumps (Fig. 5). The bumps on the chip were $85\mu m$ in diameter and were placed on a $150\mu m$ pitch. The organic

substrates used in this study were 50×50 mm² in size, 0,85mm in thickness, and they had SnAgCu solder pads.

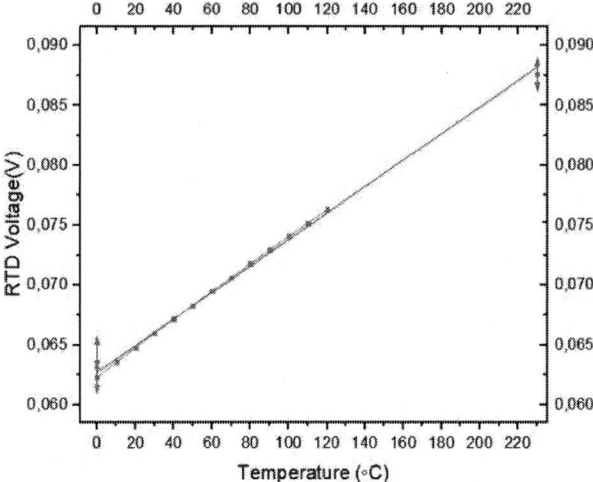

Figure 4.   Calibration curves for RTD sensor; the pink line correspond to the linear fit of data up to 230°C, purple line correspond to fit line through the points at 0°C and 120°C.

The bonding characterization experiments were conducted on a high-accuracy TORAY-FC3000WS flip-chip bonder. A Gap-controlled thermo-bonding method was used to minimize the solder squeezing phenomenon during the TCB process [15]. Before the start of the bonding process, the bonding head was maintained below 80°C to minimize the oxide formation on the SnAg solder surface. The substrate was held steady at 100°C (Fig. 1). During the bonding, the bond head picked up the chip and placed it on the active zone of the sensor. The sensor was powered and the real time in-situ temperature profile was recorded. Afterward, the temperature of the bonding head was increased with a given heating rate up to the equilibrium melting temperature of SnAg alloys (221°C). Once this temperature was reached, the bonder head dwells at peak bonding temperature was maintained for 5 sec. The choice of the 5 sec dwell time was strictly linked to the need of high throughput. Finally, the bonding head was cooled down to the initial temperature in 1 sec, with some of the cooling occurring after the tool lifted off the die. After extraction of the temperature profiles, TCB bonding experiment under the same conditions was conducted to verify the experimental thermal responses.

Figure 5.   TCB test vehicle.

978-1-7281-1500-9/19 $31.00 © 2019 IEEE

III. EXPERIMENTAL RESULTS AND DISCUSSION

A. Steady State Temperature Measurements

Due to symmetry, only a quarter of the die was considered in the experimental characterization measurements (Fig. 6). The temperature sensor map contained nine RTDs along the ¼ of the die. As shown in Fig. 7, the steady-state temperature achieved in the solder bumps was collected for different bond head temperature conditions.

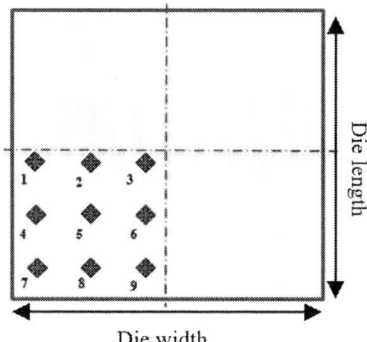

Figure 6.   Number and location of RTDs sensors in the test chip

The evaluation of the steady state temperature in the solder as function of the setting tool temperature was performed for the edge (black line) and the center of the chip (red line), to measure the thermal gradient effect. It was found that to achieve the melting point of SnAg alloys (221°C), a peak tool temperature setting of ~390°C was required. Accordingly, a temperature (in bump) between 235°C and 250°C was chosen to be slightly above the melting point of the solder.

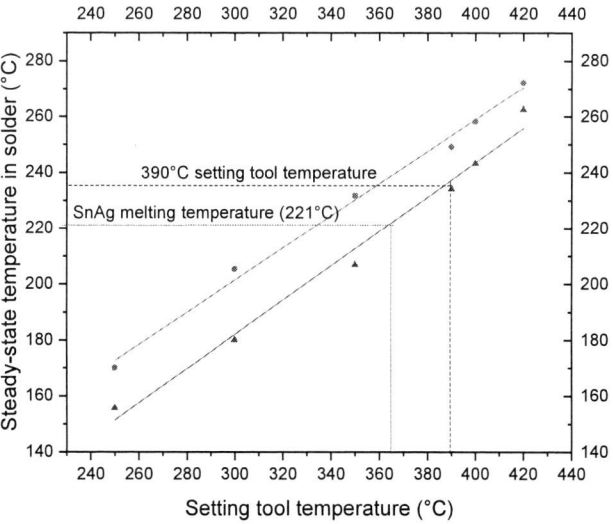

Figure 7.   Predictions of the steady- state temperature in solder as a function of the setting tool temperature; red line correspond to the linear fit of data at the center of chip, black line correspond to the linear fit of data at the edge.

B. Dynamic Heat Transfer

Once the required steady-state temperature was reached, a transient temperature measurement was carried out. The heating capability of the bonding tool is up of 200°C/s. The heating rates evaluated in this study were: 60°C/s, 100°C/s, 150°C/s, and 200°C/s, the latest corresponding to production rates in manufacturing processes. The process conditions were ramped from 60°C to the steady state temperature (390°C) and back to 60°C with a 5s dwell time at 390°C. Fig. 8 shows the bond head thermal profile measured for all RTDs sensors for each heating rate. As expected, the measured data show non-uniform temperatures distributions and significant thermal gradients between the center and the corners of the chip. For the 60°C/s heating rate (Fig. 8a), the surface temperature reached between 230.7°C and 248.9°C after 9.2s. The heating duration for 100°C/s decreased to 8.3s to reach a temperature between 224.3°C and 246.7°C (Fig. 8b). With higher heating rate of 150°C/sec and 200°C/sec, the temperature at the edges did not reach the melting point of the solder (221°C); this led to incomplete or partial bonds during TCB process. A surface temperature between 216.2°C - 247.8°C and 204.6°C - 251.3°C was found under 150°C/s, and 200°C/s, after 7.6s and 6.6s, respectively (Fig. 8c-d). A heating temperature of 60°C/s and 100°C/s should be chosen to reach the target temperature (~220°C) on all the surface of the chip, but the process time remains long for high throughput requirements.

The temperature-heating rate as function of the surface temperature non-uniformity is plotted in Fig. 9. The non-uniformity is founded by calculating the difference between the maximum and minimum temperatures (ΔT) within the chip surface. The surface temperature non-uniformity shows strong correlations with the heating rate. It can be found that ΔT increases with higher heating rates.

C. Bond Quality Characterization

To validate the thermal measurements, a dye-and-pry destructive analyses was conducted on four samples bonded at 60°C/s, 100°C/s, 150°C/s and 200°C/s. The samples were fully immersed in a red dye under high vacuum and baked at 90°C for 3 hours. The chips were then pulled off and observed under an optical-microscopy to identify the non-wet and bridge defects.

Table I gives the bonding inspection results for each heating rate condition, as well as corresponding percentage defects. Fig. 10 shows optical images of joints results for the investigated heating rates. At the slow heating rate (60°C/s), some solders bridging located in the center of the chip was observed (Fig. 10a). No solder non-wetting and bridging was found near the corners of the chip (Fig. 10b). Fig. 10c shows an acceptable soldering at 100°C/s heating rate, with 92% full solder joints and 8% partially bonded joints (located in the edge region). The samples bonded at faster heating rates (150°C/s and 200°C/s) gave a higher amount of non-wets at the edge of the chip: 43% of solder joints exhibited non-wetting at 200°C/s.

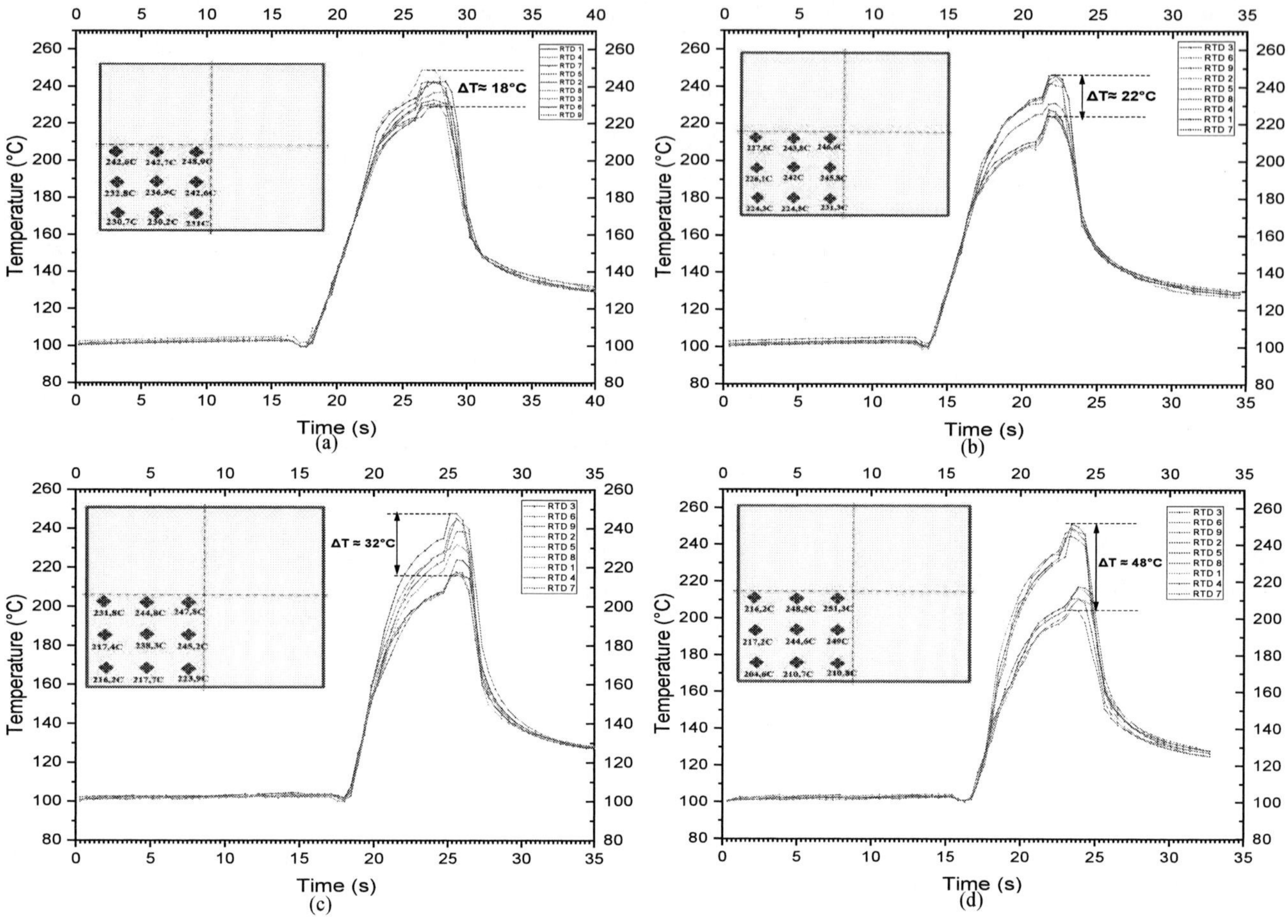

Figure 8. RTDs thermal profiles responses with different heating rates: a) 60 °C/s, b) 100°C/s, c) 150°C/s, d) 200°C/s

In the 60°C/s and 100°C/s heating rates, the chips was lifted from the NiAu-Under- bump-metallization (UBM) (Fig. 10 b-c), suggesting the formation of strong joints at the interfaces.

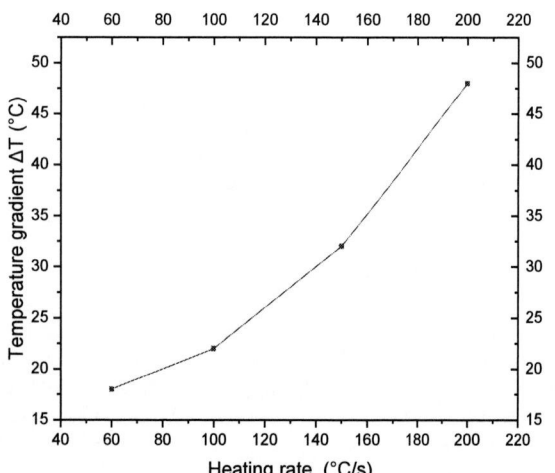

Figure 9. Surface temperature non-uniformity as function of temperature-heating rate.

TABLE I. INSPECTION TEST RESULT

Heating rate (°C/s)	Bonding results	Frequency of occurrence %	Observation results
60	Solder bridges	0-0.1	Fig. 10a-b
100	Partial-wet	8	Fig. 10c
150	Non-wet	32	Fig. 10d
200	Non-wet	43	Fig. 10e

Furthermore, a measure of the pull forces for each heating rate confirms the previous investigations. Fig. 11 plots the relationship between non-wet unit percentage and tensile strength at all heating rates. It becomes clear that increasing the heating rate shows a role in affecting the non-wet defects probability of occurrence and therefore the tensile strength. These results show good agreements with the RTDs thermal response measurements. Indeed, the thermal response measurements predict and explain the trends seen in the TCB bonding experiments results. The variations between the edge and the center of the chip have led to significant effects on the bonds quality, causing the formation of non-wet defects, which decrease joints strength.

978-1-7281-1500-9/19 $31.00 © 2019 IEEE

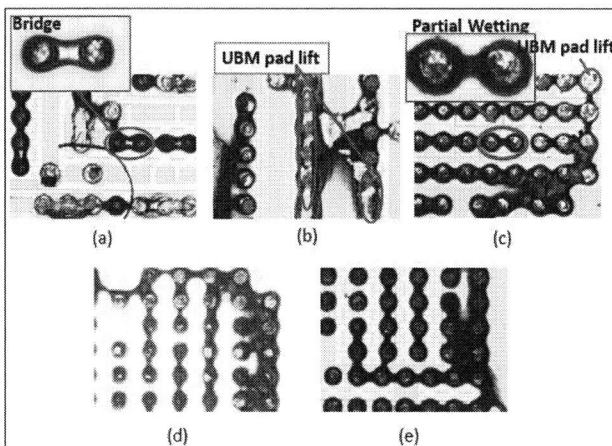

Figure 10. Dye and pry results: a) chip bonded at 60°C/s at the center joint; b) chip bonded at 60°C/s at the corner joint, c) chip bonded at 100°C/s at the corner, d) chip bonded at 150°C/s at the corner, e) chip bonded at 200°C/s at the corner.

Cross-section samples were obtained for both 60°C/s and 200°C/s heating rates to verify the results of the dye-and-pry analyses. Fig. 12 shows typical cross-section images of the corner joints taken under optical microscope. It is clear that the sample bonded at a 200°C/s heating rate was more susceptible to non-wet defects, but none was found in the sample bonded at 60°C/s.

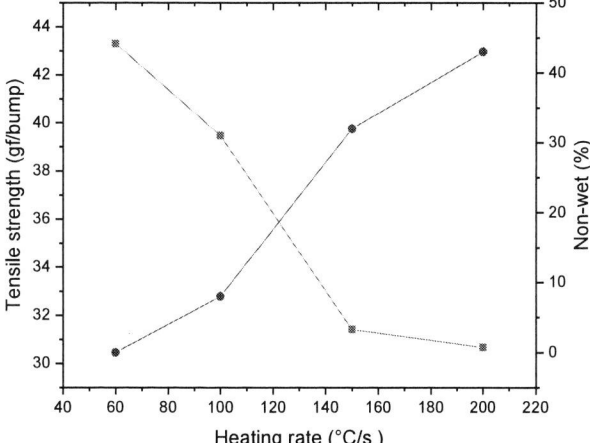

Figure 11. Tensile strength and non-wets percentage as function of temperature-heating rate; red line correspond to the non-wets percentage, blue line correspond to tensile strength.

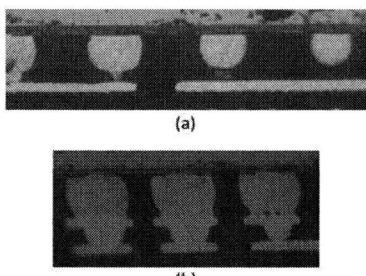

Figure 12. Cross Section of TCB bond: a) at 200°C/s, b) at 60°C/s.

IV. CONCLUSION

This paper presents a new methodology to investigate the thermal aspects of the TCB process. A RTD sensor has been developed, enabling in situ temperature measurements. The measurement responses extracted from the RTD sensor provided viable information about temperature variations. The interfacial measurements with the RTD showed significant temperature differences between the center and the edge of chip. It was detected that slow heating rates helped to reduce the temperature gradient between the center and periphery. Four heating rate profiles were investigated: 60°C/s, 100°C/s, 150°C and 200°C/s. Only 100°C/s provided acceptable results without any non-wet and bridge defects. The robust solder joints can therefore be obtained with an appropriate heating rate profile (between 60°C/s and 100°C/s). If the process-heating rate exceeds 100°C/s, large amount of non-wets defects appear. These defects seems degrade the chip pull strength. All these measurements were carried out to provide guidelines on heating profiles for the optimization of the TCB process.

ACKNOWLEDGMENT

The authors thank the IBM team for their support, in particular Steve Whitehead , Barbara Perkins and Edgar Tremblay. This research was supported by the NSERC-IBM Industrial Research Chair in Smarter Microelectronics Packaging for Performance Scaling, as well as by Prompt Quebec and Mitacs.

REFERENCES

[1] J. Sylvestre, A. Blander, V. Oberson, E. Perfecto, and K. Srivastava, "The impact of process parameters on the fracture of device structures during chip joining on organic laminates," *Proc. - Electron. Components Technol. Conf.*, pp. 82–88, 2008.

[2] V. D. Khanna and S. M. Sri-Jayantha, "Novel mass reflow method for organic substrates," *Proc. - Electron. Components Technol. Conf.*, vol. 2015–July, pp. 200–207, 2015.

[3] M. Gerber *et al.*, "Next generation fine pitch Cu Pillar technology - Enabling next generation silicon nodes," *Proc. - Electron. Components Technol. Conf.*, pp. 612–618, 2011.

[4] J. Y. Juang *et al.*, "Development of Micro-Bump-Bonded Processes for 3DIC Stacking with High Throughput," *Proc. Tech. Pap. - Int. Microsystems, Packag. Assem. Circuits Technol. Conf. IMPACT*, pp. 366–369, 2011.

[5] H. Clauberg *et al.*, "High Productivity Thermocompression Flip Chip Bonding," *Proc. - Electron. Components Technol. Conf.*, vol. 2015–July, pp. 22–29, 2015.

[6] N. Asahi, Y. Miyamoto, M. Nimura, Y. Mizutani, and Y. Arai, "High productivity thermal compression bonding for 3D-IC," *2015 Int. 3D Syst. Integr. Conf. 3DIC 2015*, p. TS7.3.1-TS7.3.5, 2015.

[7] J. H. Lau, "Recent Advances and New Trends in Flip Chip Technology," *J. Electron. Packag.*, vol. 138, no. 3, p. 030802, 2016.

[8] P. Bex *et al.*, "Thermal compression bonding : understanding heat transfer by in situ measurements and modeling," 2017.

[9] Xuejun Fan, G. Rasier, and V. S. Vasudevan, "Effects of Dwell Time and Ramp Rate on Lead-Free Solder Joints in FCBGA Packages," *Proc. Electron. Components Technol. 2005. ECTC '05.* vol. 2, pp. 901–906, 2005.

[10] Y. Jeong, J. Choi, Y. Choi, N. Islam, and E. Ouyang, "Optimization of Compression Bonding processing temperature for fine pitch Cu-column flip chip devices," *Proc. - Electron. Components Technol. Conf.*, pp. 836–840, 2014.

[11] H. Clauberg *et al.*, "Thermocompression flip chip bonding

optimization for pre-applied underfill," *Proc. Electron. Packag. Technol. Conf. EPTC*, vol. 2016–Febru, pp. 1–4, 2016.

[12] H. Clauberg, A. Rezvani, V. Venkatesan, G. Frick, B. Chylak, and T. Strothmann, "Chip-To-Chip and Chip-To-Wafer Thermocompression Flip Chip Bonding," *Proc. - Electron. Components Technol. Conf.*, vol. 2016–Augus, pp. 600–605, 2016.

[13] D. Athia, A. Rezvani, H. Clauberg, I. Qin, and M. Mayer, "Numerical Simulations of Joint-to-Joint Temperature Variation during Thermo-Compression Bonding," *Proc. - Electron. Components Technol. Conf.*, pp. 1906–1915, 2017.

[14] A. Laor, D. Athia, A. Rezvani, H. Clauberg, and M. Mayer, "Monitoring of thermo-mechanical stress via CMOS sensor array: Effects of warpage and tilt in flip chip thermo-compression bonding," *Microelectron. Reliab.*, vol. 73, pp. 60–68, 2017.

[15] C. J. Zhan, C. C. Chuang, J. Y. Juang, S. T. Lu, and T. C. Chang, "Assembly and reliability characterization of 3D chip stacking with 30um pitch lead-free solder micro bump interconnection," *Proc. - Electron. Components Technol. Conf.*, pp. 1043–1049, 2010.

2019 IEEE 69th Electronic Components and Technology Conference (ECTC)

Wireless Transfer of Power and Data via a Single Resonant Inductive Link

Shiang-Hwua Yu[1], Yi-Chen Hsieh[1], Chin-Wei Chan[1], I-Fang Lo[1], Heri Suryoatmojo[2], and Lih-Tyng Hwang[1],[*]

[1]Department of Electrical Engineering, National Sun Yat-sen University, Kaohsiung, Taiwan
[2]Department of Electrical Engineering, Institut Teknologi Sepuluh, Surabaya, Indonesia
[*]The correspondence author's email: fiftyohm@mail.nsysu.edu.tw

Abstract— A novel circuit design for wireless power and signal transfer using a single resonant inductive link is presented. On the transmitter side, a self-oscillating relay feedback circuit automatically generates a carrier signal near the resonance and drives the transmitter coil. A tunable phase shifter made of two cascaded first-order allpass filters is inserted in the relay feedback loop to modulate the carrier amplitude in accordance with the input signal. On the receiver side, a bridge rectifier and a voltage regulator are used for AC-to-DC power conversion, and also a bandpass filter is added to perform amplitude demodulation for extracting the signal from the carrier. The proposed 1.1MHz inductive link has been successfully implemented on a 2.5W wireless speaker. It allows music being played on a speaker without any connection to a power cable and an audio line. The measurement shows that the total harmonic distortion plus noise for a 1kHz sinusoidal signal being transferred can be as low as 1%.

Keywords— *resonant inductive link, wireless power and signal transfer, relay feedback circuit, amplitude modulation*

I. INTRODUCTION

Wireless power transfer have found uses in a variety of commercial products, such as water-resistant devices, biomedical implants, electric vehicles, internet of things (loT), and wearable and portable devices [1-7]. The earliest concept of wireless power transfer (WPT) can be traced back to the turn of the 20th century when Nikola Tesla pursued ideas of wireless lighting and worldwide wireless electric power distribution. A century later in the 1980s, the technique of wireless power transfer started to take off in the biomedical field [8-10]. In 2007, a team from MIT demonstrated power transfer in a longer distance, with a light bulb connected to a receiving coil and lit by another transmitting coil 2 meters away [11, 12]. Since then, WPT has attract a lot of attentions. There are near field WPT, which includes inductive coupling [13, 14] and magnetic resonant coupling [15], and far-field WPT, which includes electromagnetic radiation and laser power beaming.

In some applications, the devices require not only power but also data to be transferred wirelessly. A straightforward and common design is to adopt multiple separate wireless systems, one for power transfer and the others for data transfer, at the cost of system complexity and larger device size [7, 14, 16, 17]. The other way of achieving this aim is to transfer power and data using the same inductive link but different carriers, thereby reducing the system complexity while still allowing power and data to be transmitted independently without sacrificing the power efficiency or

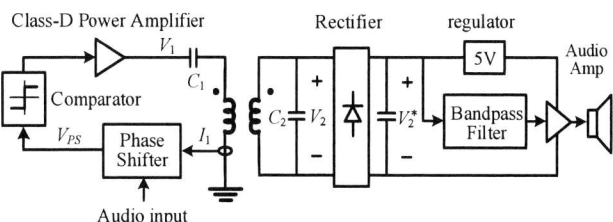

Figure 1. Schematic diagram of a resonant inductive link for a wireless loudspeaker

the date rate [18, 19]. The third option is to use the single inductive link with the only one carrier to transfer power and data; Yilmaz *et al.* [20] proposed a method for the bidirectional data transfer by directly modulating the power carrier in the frequency shifting keying (FSK) and the load-shift keying (LSK), and the drawback of their design is a slight decrease in the power efficiency.

This paper proses a novel circuit for simultaneously transmitting power and a signal through the same inductive link with a single carrier, with an aim to reduce the circuit complexity while still attaining good performance. The novelty of the proposed design is a self-oscillating relay feedback circuit with phase control, which is responsible for generating a carrier, modulating the amplitude and frequency of the carrier, and then driving the transmission coil. A conceptual design of the power and data link has been successfully demonstrated on a wireless speaker. It allows music being played on a speaker without any connection to a power cable and an audio source line, and nor is a battery used to power the power amplifier.

II. CIRCUIT DESIGN

Figure 1 displays the proposed resonant inductive link intended for transferring power and an audio signal to a wireless speaker. Two coupled coils, together with capacitors C_1 and C_2, make a resonant inductive link. The capacitance values are chosen to tune both the transmitter and receiver coils to almost equal resonance frequencies of about 1.1 MHz.

A. Design Concept

The core design is a self-oscillating relay feedback circuit on the transmitter side, which is responsible for generating a carrier signal near the link resonance frequency, performing the carrier amplitude modulation, and also

978-1-7281-1500-9/19 $31.00 © 2019 IEEE

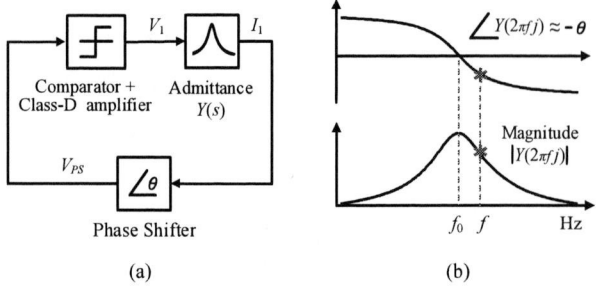

(a)   (b)

Figure 2. Equivalent relay feedback system on the transmitter side: (a) Block diagram; (b) Bode plot of the admittance function Y. The desired operating range of frequencies should be higher than the resonance frequency f_0=1.1 MHz, like that shown as the asterisk on the graph, where the power amplifier sees the link as an inductive load.

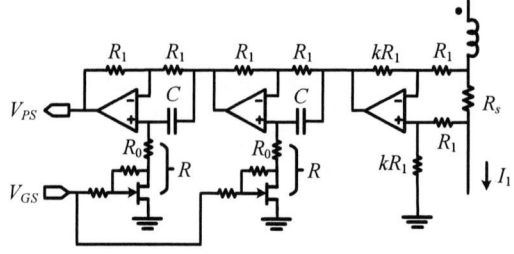

Figure 3. Tunable phase shifter.

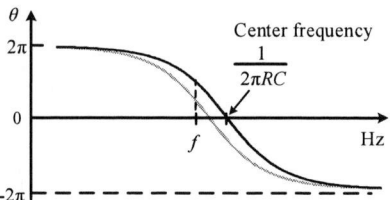

Figure 4. Phase response of the tunable phase shifter.

powering the transmitter coil. The relay feedback circuit consists of a current detector, a phase shifter, a comparator, and a class-D power amplifier. The design idea is illustrated in Fig. 2. The equivalent relay feedback system comprises a nonlinear part (a relay) that performs a signum function and a linear part that includes the phase shifter and an admittance function $Y=I_1/V_1$ seen by the class-D power amplifier. To prevent possible damage of the power transistors, the resonant inductive link is driven in its inductive region where the class-D power amplifier sees the link as an inductive load, by setting the desired oscillation frequency f slightly higher than the resonance frequency f_0 of the link [5]. Figure 2(b) shows the frequency response of the admittance Y seen by the power amplifier. The response is very similar to that of a high-Q second-order bandpass filter. The admittance has a very sharp response with a narrow bandwidth; in this case, the relay feedback system will oscillate roughly at the frequency where the total phase shift of its linear part is zero [21, 22]. For example, we may set the nominal phase of the phase shifter to, say $\theta=\pi/4$, and thus at the oscillation frequency $\angle Y(2\pi f) \approx -\theta$, where a negative phase angle for the admittance Y means that at the driving frequency Y behaves like an inductive load. During operation, θ varies around its nominal value according to the audio input signal, which results in the modulation of the frequency and the amplitude of the coil driving current I_1. One can see in Fig. 2(b) that an increase in θ will cause an increase in the operating frequency f, which results in a decrease in the current amplitude. In this way, the amplitude modulation of the driving current I_1 can be done by modulating θ.

On the receiver side, an amplitude-modulated AC voltage is inductively coupled to the receiver coil. A Schottky bridge rectifier together with a 5V voltage regulator converts inductive AC voltage V_2 into a 5V DC voltage, for use as a supply voltage for an audio class-D power amplifier. The audio power amplifier is used to drive a loudspeaker, and its audio input comes from an amplitude

demodulator, which is realized by the same bridge rectifier for power conversion followed by a bandpass filter.

B. Phase Modulation in the Feedback Loop

The tunable phase shifter in the relay feedback system has two important functions: first, setting a nominal phase shift θ for the relay feedback loop to oscillate at a proper frequency at which the resonant inductive link behaves like an inductive load to the coil driver; second, modulating the phase shift around its nominal value in accordance with the audio input signal so as to modulate the amplitude of the coil driving current.

Figure 3 shows the detailed circuit design for the current sensing circuit and the tunable phase shifter. R_s is a current sensing resistor. A differential amplifier is to measure the voltage drop over R_s. Following the differential amplifier is two cascaded first-order allpass filters, in which two matched JFETs are used as voltage-controlled resistors. The transfer function $H=V_{PS}/I_1$ for the circuit in Fig. 3 is

$$H(s) = -kR_s \left(\frac{RCs-1}{RCs+1} \right)^2, \qquad (1)$$

where the resistance R is affected by the gate voltage V_{GS} of the JFET [23]:

$$R = R_0 + \frac{V_p^2}{I_{DSS}(V_{GS} - 2V_p)}, \qquad (2)$$

978-1-7281-1500-9/19 $31.00 © 2019 IEEE   1752

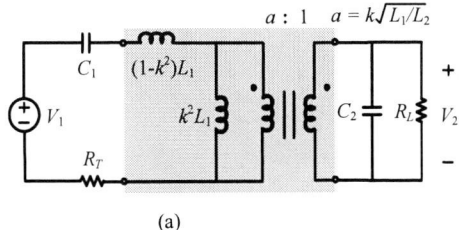

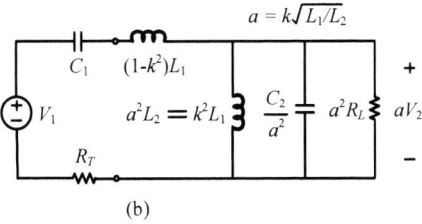

Figure 5. (a) Circuit model for the resonant inductive link, where the shaded circuit block is to model coupled coils. (b) Equivalent circuit referred to the primary side of the transformer.

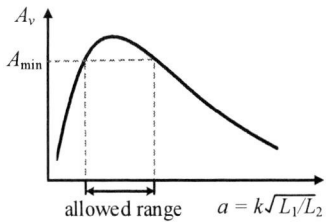

Figure 6. The range of parameter a for a minimum required link gain $A_{\min}$.

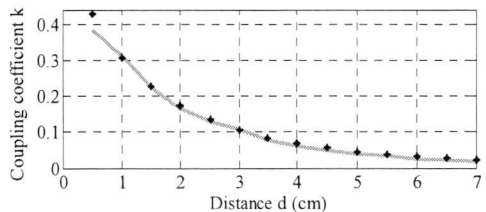

Figure 7. Coupling coefficient estimation (grey) and measurement (asterisk) for the two coils separated by d cm.

I_{DSS} is the JFET drain saturation current at $V_{GS}=0$, and V_p is the pinch off voltage.

The phase shifter has a constant gain $|H| = kR_s$ over the entire frequency band, and its phase $\angle H = \theta$ is plotted against frequency in Fig. 4. The zero phase occurs at the frequency $1/(2\pi RC)$ Hz, which is called the center frequency. Varying V_{GS} causes a change in the center frequency, thereby resulting in a phase shift at the operating frequency f (We can assume that f is a fixed value here, because during operation f changes only very slightly due to the high Q-factor of the admittance Y). For example, as V_{GS} increases, R decreases and thus the center frequency increases, which in turn raises the phase θ at f. It leads to a decrease in the amplitude of the coil driving current I_1.

III. COIL DESIGN

Figure 5 displays a circuit model for the resonant inductive link. The shaded circuit block, which consists of two inductors and an ideal transformer, is to model two coupled coils [24, 25]. L_1 and L_2 are the self-inductances of the transmitter and receiver coils, respectively, and k is the coupling coefficient of the two coils. R_T is the sum of the transmitter coil resistance, the turn-on resistances of power transistors, and the current sensing resistance. R_L accounts for the conduction loss of the load resistance and the receiver coil.

There are three major considerations for designing the coils: shapes and sizes, link gain, and power efficiency. Here we focus more on the former two: choosing the circular coil with sizes that can be accommodated by the main circuit cabinet and speaker cabinet, and also with a high enough link gain so that the inductive AC voltage

appears the receiver coil can be easily rectified and regulated to a desired DC supply voltage. First, the link gain $A_v=V_2/V_1$ is derived to facilitate the coil design. Under the assumption of weak coupling $k\ll1$, the two LCRs in Fig. 5(b) have almost equal resonance frequencies. Also, it is the usual case that the quality factor of the parallel resonant LCR is much smaller than that of the series resonant LCR in Fig. 5(b). So, at the operating frequencies near the resonance, the parallel resonant LCR behaves more like a single resistor a^2R_L, and thus the admittance Y seen by the voltage source V_1 is very much like that made up of C_1, $(1-k^2)L_1$, and $R_T+a^2R_L$ in series. From the analysis in Sec. II, we know that the driving frequency f will roughly satisfy $\angle Y(2\pi f) \approx -\theta$, and so the link gain can be approximated by

$$ A_v = \frac{V_2}{V_1} \approx \frac{aR_L}{(R_T + a^2 R_L)\sqrt{1+\tan^2\theta}}, \qquad (3) $$

Obviously, to do the phase modulation, we have to have the nominal operating point θ deviate from zero (exact resonance point), at a cost of a reduced link gain. Given a lower bound $A_{\min}$ on the link gain, one can calculate the allowable range for parameter a from (3):

$$ \frac{1}{b} - \sqrt{\frac{1}{b^2} - \frac{R_T}{R_L}} \le a \le \frac{1}{b} + \sqrt{\frac{1}{b^2} - \frac{R_T}{R_L}}, \qquad (4) $$

where

$$ a = k\sqrt{L_1 / L_2} , \qquad (5) $$

TABLE I
PARAMETERS OF RESONANT INDUCTIVE LINK

Transmitter side		Receiver side	
Turns	8.25	Turns	4
Diameter	6.8 cm	Diameter	6 cm
L_1	10.5 μH	L_2	2.4 μH
C_1	2 nF	C_2	8.8 nF
R_T	1.75 Ω	R_L	20 Ω
Link gain			>0.5
Resonance frequency			1.1 MHz

$$b = 2A_{min}\sqrt{1+\tan^2\theta}, \qquad (6)$$

The design procedure is to first assign the sizes and the numbers of turns for the two coils, then estimate their self-inductances and mutual inductance for different separation distances, and finally see if the resulting parameter, turns ratio a of the transformer, lies in the range specified in (4) for a required lower bound A_{min} of the link gain. If not, modify the sizes and/or the turn numbers of the coils.

Example 1. A tentative design is as follows: the transmitter coil consists of 8.25 turns, each of diameter 6.8 cm, and the receiver coil consisting of 4 turns, each of diameter 6 cm. The resulting inductances are 10.5 μH and 2.4 μH, respectively. Figure 7 displays the estimated and measured coupling coefficients for the two coils separated by different distances. The coupling coefficients are estimated by the Neumann's formula for mutual inductance [26]. The estimations are in excellent agreement with the measured results. The total series resistance seen on the transmitter side is R_T=1.75 Ω. The parallel resistance seen on the receiver side is assumed to be R_L=20 Ω. Assume the required minimum link gain is A_{min}=0.5 and $\theta=\pi/4$, then by (4), the allowed range for parameter a is

$$0.0648 \le a \le 1.3494.$$

It leads to the following required range for the coupling coefficient k

$$0.0325 \le k \le 0.6758,$$

We can tell from Fig. 7 that the designed inductive link is expected to have a voltage gain higher than 0.5 within the coil separation of 5.6 cm.   □

The coupling coefficient measurement in the above example is made by 1) (without coupling to the receiver coil) measuring the self-inductance L_1; 2) (with coupling to the receiver coil) shorting the receiver coil (V_2=0) and measuring the effective self-inductance value $(1-k^2)L_1$ seen from the transmitter coil; 3) comparing the above two inductance values and calculating coupling coefficient k.

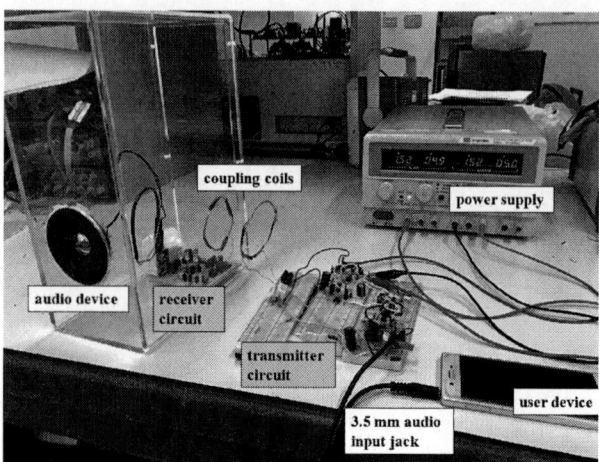

Figure 8. Prototype of a wireless speaker with the proposed resonant inductive link for power and signal transfer. Note that the speaker was powered (green light indicating presence of musical sound) by the same power supply that drove the musical device.

IV. EXPERIMENTAL RESULTS

Figure 8 shows a photo of the designed resonant inductive link for a 2.5W wireless speaker. The detailed coil and circuit design and experimental results are described in the following.

A. Design Description

We use the same coil design as in Example 1. The capacitors in resonance with the coils are chosen as C_1=2 nF and C_2=8.8 nF (Fig. 1), setting the resonance frequency to about 1.1 MHz. Table I summarizes the design parameters of the resulting resonant inductive link.

The self-oscillating relay feedback driving circuit runs from supply voltages ±5 V. A current sensing resistor of 0.05 Ω is connected in series with the transmitter coil for measuring its current I_1. With reference to Fig. 3, the differential amplifier for measuring the voltage across the shunt resistor has the gain k=0.4, and the phase shifter has the component values: R_0=20 Ω, C= 180 pF, JFET 2N5952 (V_p= -2.4 V, I_{DSS}=6.77 mA). The gain k should be carefully chosen to ensure the operation of the JFETs in the linear region ($V_{DS}<V_{GS}-V_p$). Overly large V_{DS} will saturate the JFETs, and overly small will pinch off them. The DC bias voltage for V_{GS} is set to -0.4 V to render a phase lead of about 124 degrees, for the compensation of the inevitable total circuit lag of -73 degrees in the entire feedback loop. That is, taking into the unavoidable circuit lag into account, we have the effective nominal total phase shift of 51 degrees, which ascertains the driving of the resonant inductive link in its inductive region. A comparator AD8561, following the phase shifter, is to convert an analog signal into a binary switching command for the class-D power amplifier. The class-D power amplifier, made of a full bridge of IRF9540 and RF512 MOSFET transistors, is driven by Gate driver

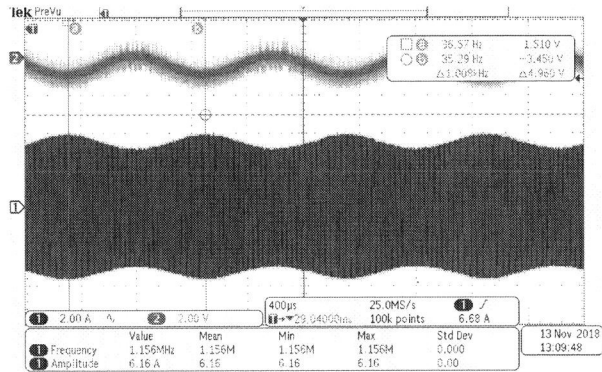

Figure 9. 1kHz sinusoidal test input signal (red) and the amplitude-modulated coil driving current I_1 (blue).

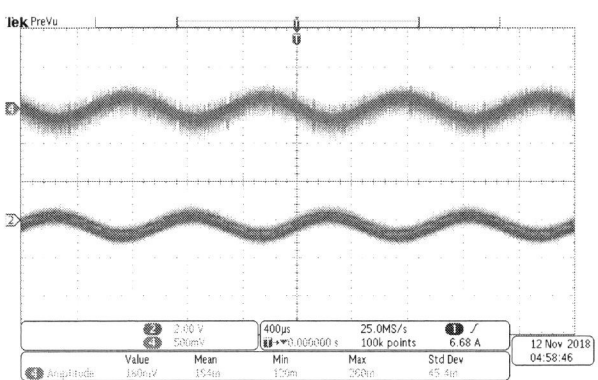

Figure 10. 1-kHz sinusoidal test input signal (red) on the transmitter side and its recovered signal (pink) on the receiver side. The measurement is carried out with the two coils placed 2 cm apart.

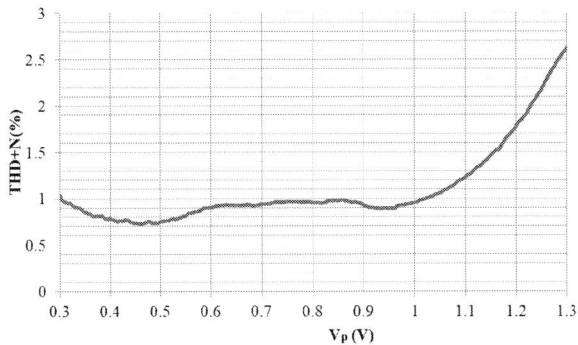

Figure 11. THD+N versus input amplitude for 1kHz sinusoidal test input signals. The experiments are conducted with the two coils placed 2 cm apart and a 20 Ω resistive load.

ICs TC 4424 with a deadtime of about 60 ns. Figure 9 shows the scope plot of a 1 kHz sinusoidal test input v_{in} and the corresponding coil driving current I_1. As expected, the driving current I_1 has a frequency of 1.156 MHz (higher than the resonance frequency 1.1 MHz) and an amplitude of 6.16 A. The driving current I_1 has an envelope following the inverse waveform of the test input signal v_{in}.

On the receiver side, a full-wave Schottky rectifier followed by a parallel 2.2μF capacitor is used to convert the inductive AC voltage V_2 into a fluctuated DC voltage V_2^*. Then, a low-dropout regulator LM2937 regulates V_2^* into a steady 5V voltage, intended for use as a supply to a class-D audio power amplifier. Also, a bandpass filter, with lower cutoff 20 Hz and upper cutoff 20 kHz, is used to extract the audio signal from V_2^*. The class-D audio power amplifier IC that we use to drive the speaker is SSM2377, with a maximum output power of 2.5 W. Experiments show that the allowable distance of the two coils indeed can be extended up to 5 cm with an effective load R_L=20 Ω as predicted. Figure 10 plots the extracted signal associated with a 1kHz test input signal.

B. Evaluation of Transmitted Signal quality

The signal transfer via the inductive link is evaluated by measuring the total harmonic distortion plus noise (THD+N) of the demodulated output signal within a bandwidth of 20 kHz by an audio analyzer AP2700. Figure 11 displays the measurement result in the audio band given a 1 kHz test input signal with different amplitudes. The THD+N is less than 1% for the input amplitude of less than 1 V, beyond which the THD+N increases dramatically, mainly owing to the non-linearity of the JFET resistances under large signal swing.

V. CONCLUSION

A novel tunable-phase relay feedback circuit is proposed to drive a resonant inductive link for wireless power and signal transfer. The proposed design has the following two attractive features:

1) (*Circuit simplicity*) Only a single inductive link is required for wireless power and signal transfer, and the carrier generation, amplitude modulation, and coil driving are all performed by a simple relay feedback circuit.
2) (*Resonance tracking*) The circuit is capable of compensating for unavoidable phase lag caused by circuit non-ideality and automatically adjusting the carrier frequency according to the resonance of the inductive link.

Apparently the coil design here is not fully optimized with respect to the power efficiency and the link gain. However, it demonstrated the validity of a power and data single link approach. A systematic coil optimization method and more rigorous circuit analysis will be reported in the future.

ACKNOWLEDGMENT

The authors would like to acknowledge financial supports provided by Taiwan Ministry of Science and Technology (grants MOST 107-2221-E-110-019 and MOST 107-2221-E-110-029).

REFERENCES

[1] E. Waffenschmidt and T. Staring, "Limitation of inductive power transfer for consumer application," *Eur. Conf. Power Electron. Appl.*, pp. 1- 10, Sep. 2009.

[2] R. F. Xue, K. W. Cheng, and M. Je, "High-efficiency wireless power transfer for Biomedical implants by optimal resonant load transformation," *IEEE Trans. Circuits Syst. I*, vol. 60, no. 4, pp. 867-874, 2013.

[3] B. L. Cannon, 1. F. Hoburg, D. D. Stancil, and S. C. Goldstein, "Magnetic resonant coupling as a potential means for wireless power transfer to multiple small receivers," *IEEE Trans. Power Electron.*, vol. 24, no.7, pp. 1819-1825, Jul. 2009.

[4] C.-J. Chen, T.-H. Chu, C.-L. Lin, and Z.-C. Jou, " A study of loosely coupled coils for wireless power transfer," *IEEE Trans. Circuits Syst. II*, vol. 57, no. 7, pp. 536-540, Jul. 2010.

[5] D.M. Beams and S.G. Annam, "Failure mechanisms in MOSFET square-wave drivers for wireless power transfer applications," *IEEE Symposium on System Theory* pp.1-5, 2013.

[6] M. Rana, W. Xiang, E. Wang, X. Li, and B. J. Choi, "Internet of things infrastructure for wireless power transfer systems," *IEEE Access*, vol. 6, pp. 19295-19303, 2018.

[7] W. P. Choi, W. C. Ho, X. Liu, and S. Y. R. Hui, "Bidirectional communication techniques for wireless battery charging systems & portable consumer electronics," *Proc. IEEE Conf. Applied Power Electron.*, pp. 2241-2257, CA, 2010.

[8] N. de N. Donaldson, "Voltage regulators for implants powered by coupled coils," *Med. Biol. Eng. Comput.*, vol. 21, pp. 756-761, 1983.

[9] P.E.K. Donaldson, "Three separation-insensitive radio frequency inductive links," *J. Med. Eng. Technol.*, vol. 11, pp. 23-29, 1987.

[10] D.C. Galbraith, M. Soma, and R.L. White, "A wide-band efficient inductive transdermal power and data link with coupling insensitive gain," *IEEE Trans. Biomed. Eng.*, vol. 34, pp. 265-275, 1987.

[11] A. Karalis, J. Joannopoulos, and M. Soljacic, "Efficient wireless non-radiative mid-range energy transfer," *Ann. Phys.*, 323, pp. 34-48, 2008.

[12] A. Kurs, A. Karalis, R.Moffatt, J. Joannopoulos, P. Fisher, and M. Soljacic, "Wireless power transfer via strongly coupled magnetic resonance," *Science*, vol. 317, no. 5834, pp. 83-86, 2007.

[13] G. A. Covic and J. T. Boys, "Modern trends in inductive power transfer for transportation applications," *IEEE J. Emerg. Sel. Topics Power Electron.*, vol. 1, no. 1, pp. 28–41, Mar. 2013.

[14] Y. Zeng, B. Clerckx, and R. Zhang, "Communications and signals design for wireless power transmission," *IEEE Trans. Communications*, vol. 65, no. 5, pp. 2264-2290, May 2017.

[15] M. Kiani and M. Ghovanloo, "The circuit theory behind coupled-mode magnetic resonance-based wireless power transmission," *IEEE Trans. Circuits Syst. I*, vol. 59, no. 9, pp. 2065 – 2074, 2012.

[16] M. Ghovanloo and S. Atluri, "A wideband power-efficient inductive wireless link for implantable microelectronic devices using multiple carriers," *IEEE Trans. Circuits Syst. I*, vol. 54, no. 10, pp. 2211–2221, Oct. 2007.

[17] R. Sarpeshkar, W. Wattanapanitch, S. K. Arfin, B. I. Rapoport, S. Mandal, M. W. Baker, M. S. Fee, S. Musallam, and R. A. Andersen, "Low-power circuits for brain-machine interfaces," *IEEE Trans. Biomed. Circuits Syst.*, vol. 2, Issue 3, pp. 173-183, Sept. 2008.

[18] J. Wu, C. Zhao, Z. Lin, J. Du, Y. Hu, and X. He, "Wireless power and data transfer via a common inductive link using frequency division multiplexing," *IEEE Trans. Ind. Electron.*, vol. 62, Issue 12, pp. 7810-7820, 2015.

[19] J. Hirai, T. W. Kim, and A. Kawamura, "Study on intelligent battery charging using inductive transmission of power and information," *IEEE Trans. Power Electron.*, vol. 15, no. 2, pp. 335-345, 2000.

[20] G. Yilmaz, O. Atasoy, and C. Dehollain, "Wireless energy and data transfer for in-vivo epileptic focus localization," *IEEE Sensors Journal*, vol. 13, no. 11, pp. 4172-4179, 2013.

[21] A. Gelb and W. V. Velde, *Multiple-Input Describing Functions and Nonlinear System Design*, McGraw-Hill, 1968.

[22] S. H. Yu and P. H. Wu, "Two kinds of self-oscillating circuits mechanically demonstrated," *Int. J. Electronics and Communication Engineering*, vol. 8, no. 7, pp. 1334-1338, 2014.

[23] D. Kleinfeld, "The field effect transistor as a voltage controlled resistor," https://neurophysics.ucsd.edu/courses/physics_120/ The%20 Field%20Effect%20Transistor%20as%20a%20Voltage%20Controlled %20Resistor.pdf

[24] R. A. DeCarlo and P. M. Lin, "Coupled inductors modeled with an ideal transformer", *Linear Circuit Analysis*, Ch. 18, Sec. 7, 1995.

[25] R. R. Harrision, "Designing efficient inductive power links for implantable devices," *Proc. IEEE Int. Symposium on Circuits and Systems*, pp. 2080-2083, 2007.

[26] M. Soma, D. C. Galbraith, and R. L. White, "Radio-frequency coils in implantable devices: misalignment analysis and design procedure," *IEEE Trans. Biomed. Eng.*, vol. BME-34, no. 4, pp. 276-282, 1987.

Adaptive Patterning of Optical and Electrical Fan-out for photonic chip packaging

Ahmed Elmogi[1], Andres Desmet[1], Jeroen Missinne[1], Hannes Ramon[2], Joris Lambrecht[2], Peter De Heyn[3], Marianna Pantouvaki[3], Joris Van Campenhout[3], Johan Bauwelinck[2], and Geert Van Steenberge[1]

[1]Centre for Microsystems Technology, imec and Ghent University, Technologiepark-Zwijnaarde 126, 9052 Ghent, Belgium
[2]IDLab Department of Information Technology, Ghent University-imec, 9052 Ghent, Belgium
[3]imec, 3001 Leuven, Belgium
E-mail: AhmedG.Elmogi@ugent.be

Abstract—Packaging and assembly challenges for photonic chips still need to be addressed in order to enable rapid deployment in mass-market production. Integration and assembly solutions that not only enable ease of packaging but also allow a dense co-integration of the electronic and photonic ICs are essential. In that context, we demonstrate an adaptive patterning of both optical and electrical fan-out for face-up electronic-photonic integration. For the optical fan-out, we developed an approach based on adiabatic optical coupling between single-mode polymer waveguides and silicon waveguides on a silicon photonic chip. The polymer waveguides were directly patterned on the silicon photonic chip by direct-write lithography (DWL). The electrical interconnects between a photonic chip and electronic IC are realized by employing high-speed silver interconnects using aerosol-jet printing (AJP), as a promising alternative for the traditional bond-wires. Furthermore, a direct comparison between the AJP interconnects and the conventional bondwires is established. Finally, an NRZ optical transmitter has been successfully demonstrated based on the AJP interconnection and clear open eye diagrams were obtained at 56 Gb/s.

Keywords: electronic-photonic integration, packaging, photonic integrated circuits, polymer waveguides, direct-write lithography, adiabatic coupling, electrical interconnects, aerosol-jet printing (AJP)

I. INTRODUCTION

The increasing demand for high-performance optoelectronic devices largely rises from the need for high data rates across communication and transceiver systems for future data centers. Packaging of photonic integrated circuits (PICs) is one of the most critical obstacles that still need to be addressed in order to enable true mass-market production. The performance and the overall cost of the electro-optical devices highly depend on how efficiently the electronic and photonic ICs are co-packaged. Photonics packaging" refers to all technical aspects related to the optical, electrical, thermal, and mechanical interfaces between the PICs and the outside world [1-4]. Monolithic integration of electronic and photonic functions onto a single chip is considered as the ultimate solution to provide a very tight electronic-photonic integration [5]. However, hybrid integration of photonic and electronic functionalities on two separate chips is still the most economical approach because it enables using different technologies for both PICs and EICs [6]. There has been a lot of research in this area in order to operate the photonic devices at high-speed and with very high bandwidth to meet the data rate requirements for interconnect applications [7-9]. Fiber-to-chip coupling and electronic-photonic integration are considered the most prominent challenges for photonics packaging.

For fiber-to-chip coupling, the main challenge is to couple the light from single-mode fibers (SMF) and silicon photonic waveguides due to the large mode mismatch. There are two main approaches for coupling the light from SMF to PIC; surface grating couplers [10] and edge couplers [11]. Grating couplers enable wafer-level optical testing and they can be placed at any location on the wafer without requiring any additional post-processing such as dicing and polishing. They are also more alignment-tolerant than edge couplers. However, they are sensitive to polarization and wavelength-dependent. The alternative approach to grating couplers is using edge couplers. Typically, edge couplers offer lower insertion loss than grating couplers. Unlike grating couplers, edge couplers are typically broadband and polarization-tolerant. However, they should be located at the PIC edge and hence they need post processing to create the coupling facet using dicing and polishing if necessary. Both edge and grating couplers approaches have their own advantages and drawbacks. Therefore, the application and the cost will determine which approach is chosen to meet the specific requirements defined by the application. In this work, we developed a process to demonstrate adiabatic optical coupling by laser direct writing of single-mode (SM) polymer waveguides on silicon-on-insulator (SOI) waveguides. The SM polymer waveguides in this approach act as an optical fan-out and a spot size converter (SSC) between SMF and SOI waveguides and they are adaptively patterned by direct-write lithography (DWL), directly on the PIC.

As for electronic-photonic integration, most of the photonic ICs are interfaced with the electronic IC using traditional wire-bonding technology owing to their flexibility and reliability. However, at high frequency, the bondwire can be a major problem due to the parasitic inductance induced by the wire itself and the longer the bondwire the higher the parasitic inductance. This parasitic inductance at high frequency will cause high reflections and losses and eventually will limit the bandwidth. To reduce parasitic induction effects, bondwires should be as short and straight as possible, which can be very difficult to achieve. Flip-chip technology is proven to be a good alternative particularly at high frequencies [12] as it replaces the long wire (100-500 μm) with very short vertical microbump interconnects (10 μm), which minimize the effect of the parasitic inductances. However, the flip-chip approach is still less flexible compared to the wire-bonding technology. Therefore, new integration and automated packaging techniques which not only sustain higher operating frequencies, but also maintain the flexibility (in assembly) and reliability are essential. In this context, we present a flexible face-up 2.5D packaging method to interconnect electronic and photonic IC dies using aerosol-jet printing (AJP). AJP technology enables printing very short and straight electrical interconnects between PIC and EIC (100-300 μm) with a pitch of 50 μm, thus minimizing the parasitic inductance for high-speed applications. It also offers tuning the characteristic impedance of the printed transmission lines (TLs) by controlling their dimensions and spacing in order to design impedance-mated TLs, thus reducing signal reflections and attenuation.

II. ADIABATIC COUPLING BETWEEN POLYMER AND SOI WAVEGUIDES

In edge couplers, the light from the optical fiber is first coupled to an intermediate waveguide with a mode size matching that of a SMF. This requires the intermediate waveguide to possess a lower index contrast (e.g. polymer waveguide) in order to match the mode of the optical fiber. In adiabatic coupling conditions, the geometries of tapered Si waveguide are gradually changed so that the fundamental propagating mode can not transfer the optical power to any higher order or radiation modes. Fig. 1(a) shows a schematic of the adiabatic coupling approach using polymer waveguides as a spot-size converter between SMF and Si waveguides. For coupling from an optical fiber to Si waveguides, the mode within the optical fiber is almost matched with the mode of the polymer waveguide and hence the light propagates in the polymer core until it reaches the Si taper. In the taper, the width is horizontally widened in order to convert the effective index gradually from that of the polymer mode to that of the Si waveguide, resulting in a confined mode in the Si waveguide. The opposite approach occurs during the coupling from Si waveguide to an optical fiber. The inverse taper is gradually narrowed so that the waveguide effective index is reduced, which results in expanding the mode until it couples to the polymer

waveguide and eventually matching the mode of optical fiber. Fig. 1(b) shows the cross section view for the adiabatic coupling approach between polymer waveguides and Si waveguides.

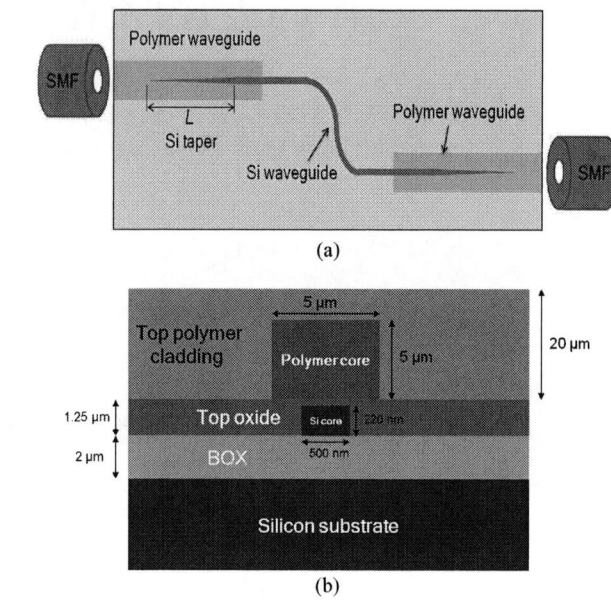

Figure 1: Schematics of the adiabatic coupling approach (a) top view (b) cross section view

A. Fabrication process flow

In order to experimentally realize the adiabatic coupling approach, we process the polymer waveguides after embedding SOI chip in an epoxy layer. The fabrication process flow is shown in Fig. 2. The goal is to embed the SOI chip in a flat epoxy surface so that polymer waveguides can be defined by direct-write lithography (DWL). First, the top surface of the SOI chip is attached to a PDMS stamp (50 μm PDMS layer residing on a borosilicate glass). Afterwards, an epoxy polymer (Epotek OG142-112) is dispensed on the backside of the SOI chip. The epoxy will flow around the sides of SOI chip but will not reach the top surface since the top side is protected by the PDMS layer. Another borosilicate glass substrate is pressed onto the dispensed polymer to create a flat uniform surface. Next, the polymer is cured using a UV lamp at 30 mW/cm^2 for 2 minutes by illuminating from the backside of the glass substrate. The PDMS stamp is then removed after UV exposure since the epoxy does not adhere to PDMS. In this case, the SOI chip is embedded in an epoxy layer with a flat surface which has the same thickness as the SOI chip. Finally, single-mode polymer waveguides are patterned using direct-write lithography (DWL) in a similar way as introduced in [13]. An EpoCore layer with a thickness of about 5 μm was spinned on the sample after treating the surface with oxygen plasma to ensure good wettability and adhesion properties and hence create a uniform polymer core layer. The EpoCore layer is then directly exposed by DWL. The core layer in this case is required to be aligned with respect to the Si waveguide, which can be accurately done using the DWL alignment system. Next an EpoClad layer with a thickness of 20 μm is applied on top of the core layer. The requirement of creating a flat epoxy surface is essential particularly at the edges of the SOI chip so that polymer waveguides can cross through the edge of the SOI chip without having any cuts or interruptions in the waveguide structures. The epoxy also acts as an under-cladding for the EpoCore waveguides in this case. The polymer waveguides were directly patterned by DWL at both sides of the Si waveguides so that the light can be coupled in and out of the Si waveguides. Fig. 3 shows the laser direct written polymer waveguides at both sides of the S-bend Si waveguides. The fabricated polymer waveguides have a width of 6 μm and a thickness of 5.6 μm. An offset of 2 μm was observed in the lateral direction of the DWL polymer waveguides with respect to the SOI waveguides, which can be corrected in future realizations. However, it has been recently reported that the additional optical loss due to an offset of 2 μm is limited to about 1 dB [14]. On the same SOI chip there were empty straight regions where there were no Si waveguides. Those regions were utilized to pattern straight waveguides in order to provide a reference for the adiabatically coupled waveguides as shown in Fig. 3.

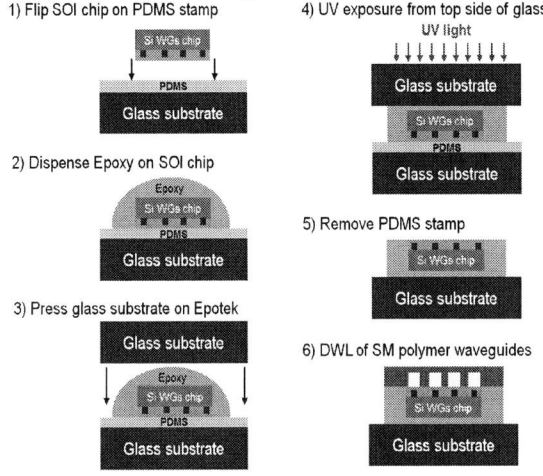

Figure 2: Fabrication process flow for the adiabatic coupling approach

B. Characterization results

The facets of the fabricated samples were created by mechanical dicing using a blade dedicated for glass substrates. The samples were diced at a feed speed of 1 mm/s and blade rotation speed of 10000 rpm. A sample with a length of 4 mm was prepared for the loss measurements. The sample is mounted the vacuum chuck. Laser diode sources (QDFBLD-1300-10 & QDFBLD-1550-5 from QPhotonics) operating at 1310 nm and 1550 nm respectively were used as the light sources. The light is coupled from a standard single-mode fiber (SMF-28) at 1310 nm and 1550 nm to the polymer waveguides from one side and the light is captured by a Xenics IR camera at the out-coupling side. Fig. 4 shows the IR camera image of the coupled light in the reference

polymer waveguide (PWG), while fig. 5 shows the IR image of the adiabatically coupled light from the PWG to SiWG then to PWG again. We can see that the IR profile for the reference PWG has a very clear light spot with a minimal scattered light around the waveguide core. For the IR image of the adiabatically coupled light from the input PWG to the SiWG then to output PWG, there is a scattered light next to the detected light spot due to the uncoupled light in the S-bend and it propagates in the cladding layer instead. The region where light scatters is the 100 μm-spacing between the two branches of the S-bend. Once the in-coupling fiber is aligned, the IR camera is replaced with the out-coupling fiber. An index matching liquid was used at the waveguide end facets in order to minimize the coupling loss. The insertion loss of the reference waveguides was set as a reference. For the 4 mm-long reference polymer waveguides, an insertion loss of 2.7 dB and 3 dB was measured at 1310 nm and 1550 nm respectively. We measured the total insertion loss (SMF-PWG-SiWG-PWG-SMF) both at 1310 nm and 1550 nm. The total insertion loss was found to be 9.2 dB and 9.6 dB at 1310 nm and 1550 nm respectively. Hence, the insertion loss per facet corresponds to 4.6 dB and 4.8 dB at 1310 nm and 1550 nm respectively. This indicates that the adiabatic coupling loss per facet (PWG-SiWG or SiWG-to-PWG) corresponds to 3.25 dB and 3.3 dB at 1310 nm 1550 nm respectively.

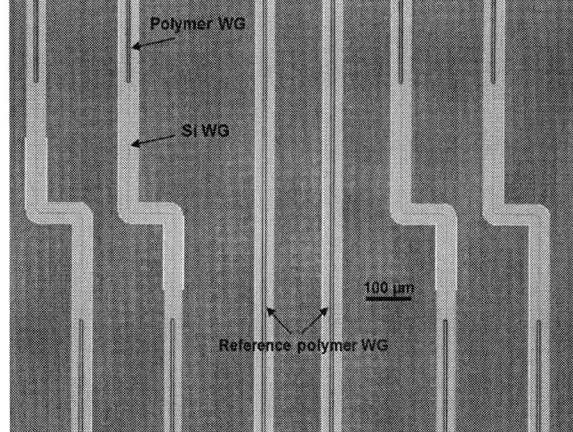

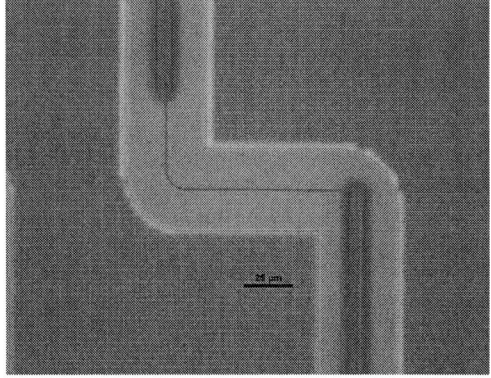

Figure 3: Laser direct written polymer waveguides on the SOI chip

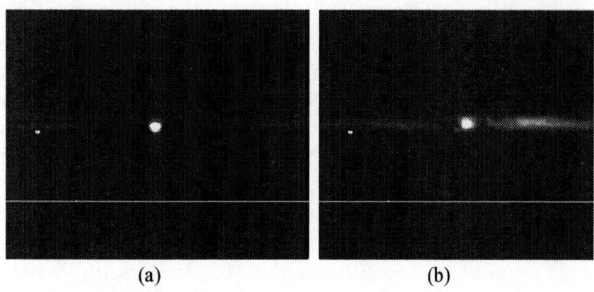

(a)   (b)

Figure 4: IR camera images (a) the reference polymer waveguide (b) the adiabatic coupling (PWG to Si WG to PWG)

The resulting adiabatic coupling loss can be further reduced by tuning the dimensions of the polymer waveguide to achieve higher optical coupling. Additionally, the alignment accuracy of the DWL polymer waveguides can still be improved by compensating the offset in the lateral direction. The taper design can also be further optimized to achieve higher optical coupling.

III. ELECTRONIC-PHOTONIC INTEGRATION USING AEROSOL-JET PRINTING

Recently, we developed an interconnection process to print high-speed electrical interconnects between the electronic and photonic ICs using AJP [15, 16]. First, we will describe the experimental process flow. Afterwards, a comparison between the AJP interconnects and the bonding wires is established. At last, the assembly of the high-speed CMOS driver and the microring modulator is presented.

A. Fabrication process flow for electronic-photonic AJP interconnection

The fabrication process flow for electronic and photonic integration is shown in Fig. 5. In order to interconnect the photonic and electronic chips by AJP, a mechanical polymer support was created to bridge the gap between the chips and subsequently the electrical interconnects were printed on top. The process flow can be described as follows. First, the chips were die-bonded to the PCB followed by an epoxy polymer dispensing (Epotek OG 142-112). The epoxy was locally dispensed on the chips by a fine needle. Next, a flat PDMS stamp was gently pressed onto the dispensed epoxy. Then, the epoxy was cured using a UV lamp at 30 mW/cm^2 for 2 minutes. The stamp can be easily removed after UV exposure since the epoxy does not adhere to PDMS. Since, the polymer shrinks after UV exposure, the polymer imprinting step could be repeated few times until uniform and sufficient coverage for the chips could be achieved. The thickness of the epoxy layer on top of the chips should be less than 10 μm, by applying a sufficient manual pressure during UV exposure. Next, vias were opened on the contact pads by excimer laser ablation. At last the electrical interconnects were precisely printed between the chips (pad-to-pad) using aerosol-jet printing. In order to ensure that the

vias were completely filled, at least two printing passes (5 μm per pass) were applied to connect between the chips. The epoxy covering the electronic and photonic chips is transparent so that it would have no negative effect on the light coupling efficiency in or out of the photonic chips. However, in this work, the printed electrical interconnects are realized only between EIC & PIC. In principal, the electrical interconnects can also be extended till the traces on the printed circuit board (PCB) to act as an electrical fan-out. In this case a ramp needs to be created at the chip edges. This ramp can be realized using laser ablation by tilting the laser beam under an angle. Hence, this will enable a complete printed assembly.

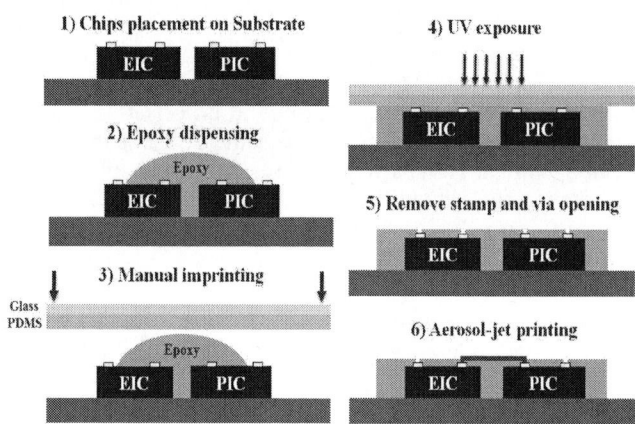

Figure 5: Fabrication process flow for electronic-photonic AJP interconnection

B. Comparison between bonding wires and aerosol-jet printed interconnects

To compare the aerosol-jet interconnects with bonding wires, we developed two assembly schemes; one is interconnected by aerosol-jet printing and the other is wire-bonded. In this comparison, we test the interconnection to a PIC (microring modulator) [17] separately. The assembly consists of a ring modulator interconnected to a silicon interposer (daisy-chain), which was realized using the two techniques. Fig. 6 shows the realized assemblies by AJP and wire bonding. The interposer has contact pads with a pitch of 150 μm. Hence, this required making the interconnection in both cases under an angle to enable probing the assembly with a 150 μm-pitch RF GS probe. The aerosol-jet interconnects were realized using the same procedure as discussed above. For the aerosol-jet interconnection, it was possible to place the modulator and the interposer chips very close together. For the wire bonding interconnection, it was required to leave an additional gap of about 100 μm to enable creating the wire loop.

978-1-7281-1500-9/19 $31.00 © 2019 IEEE

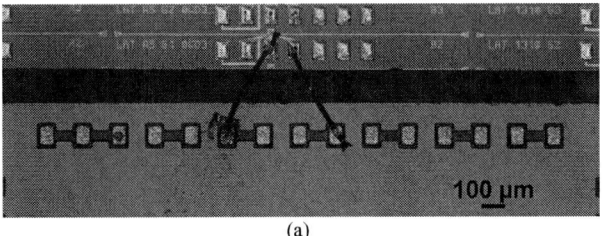

(a)

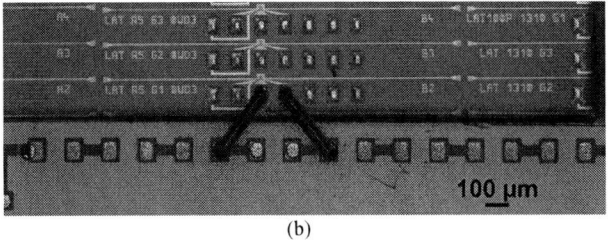

(b)

Figure 6: The microring modulator interconnected by (a) wire bonding (b) aerosol-jet printing

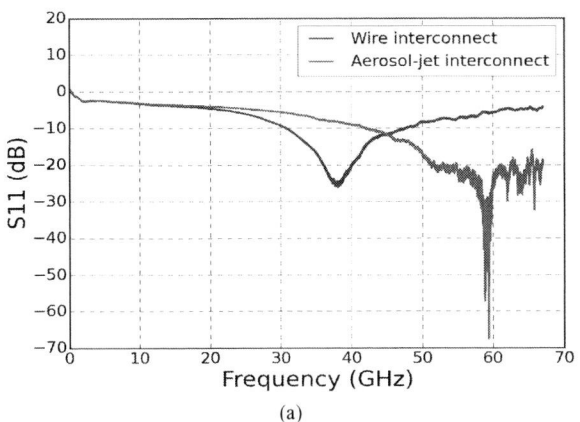

(a)

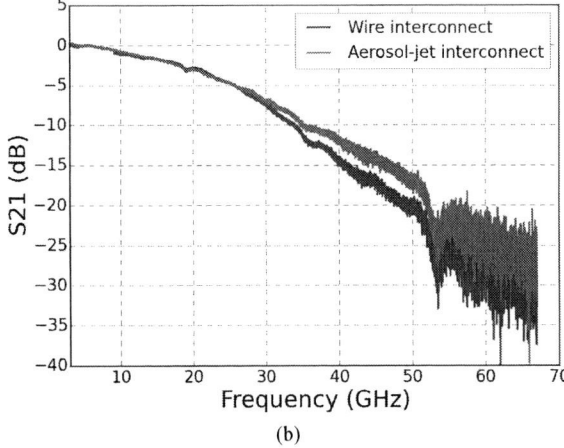

(b)

Figure 7. (a) measured S_{11} for the ring modulator with aerosol-jet and bondwire interconnects (b) measured S_{21} for the ring modulator with aerosol-jet and bondwire interconnects

We measured the reflection coefficient S_{11} for the aerosol-jet and the bond wire interconnection as illustrated in Fig. 7(a). The reflection coefficient S_{11} shows the same behavior up to 20 GHz for both the aerosol-jet and wire interconnection. However, at higher frequencies, the wire interconnection started to suffer from a resonance at 38 GHz and afterwards the S_{11} performance starts to degrade till it reaches -5 dB at 67 GHz. On the other hand, the aerosol-jet interconnection shows much better S_{11} performance as the frequency increases. The resonance frequency of the AJP interconnect is situated at a much higher frequency of 59 GHz and the reflections remain below -20 dB until 67 GHz. The S_{11} measurements clearly reflect that aerosol-jet printed interconnects have less parasitic inductance than the wire interconnection. Additionally, the optical transmission coefficient S_{21} was measured for both the AJP interconnect and the wire interconnect as illustrated in Fig. 7(b). Although the 3-dB BW for the ring modulator interconnected by AJP and wire-bonding is approximately the same around 20 GHz, the AJP interconnect shows 4 dB higher transmission than the wire interconnect at 50 GHz.

D. CMOS driver & electro-absorption modulator (EAM) assembly

Afterwards, we demonstrated our interconnection technology by developing an optical transmitter assembly. The optical transmitter consists of an electro-absorption modulator (EAM) and a CMOS driver. The EAM was selected from imec's silicon photonics platform which includes high-speed EAMs among many other devices [17]. The driver chip was fabricated in a 28 nm fully depleted silicon-on-insulator (FDSOI) CMOS process [18]. Since the driver and the modulator chips have Al contact pads, they were electro-plated with NiAu bumps to avoid the interfacial corrosion and high-contact resistance between the silver interconnects and the Al contact pads [19]. The modulator chip was thinned-down to 250 µm to achieve the same height as the driver chip. The distance between the CMOS driver and the modulator was kept as short as possible (almost edge-to-edge) to minimize the parasitics. Fig. 8 shows the resulting assembly of the EAM modulator and the driver. The light is coupled into and out of the EAM modulator by two grating couplers at 1560 nm. Fig. 9 shows the measured back to back optical eye-diagrams at 40 Gb/s, 50 Gb/s and 56 Gb/s. An extinction ratio (ER) of 2.83 dB was achieved while applying a drive voltage of 1 Vpp. The eye diagrams were clearly open even after transmission distances of 1 km and 2 km of SSMF as illustrated in Fig. 10 and Fig. 11.

978-1-7281-1500-9/19 $31.00 © 2019 IEEE   1761

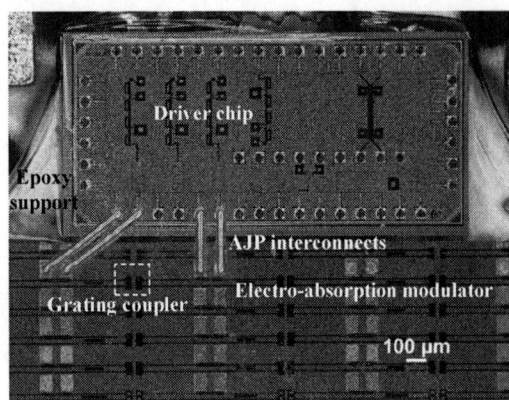

Figure 8: The assembly of the EAM and CMOS driver interconnected by aerosol-jet printing

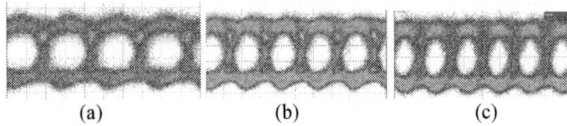

Figure 9: The measured back to back eye diagrams (a) 40 Gb/s, (b) 50 Gb/s, and (c) 56 Gb/s

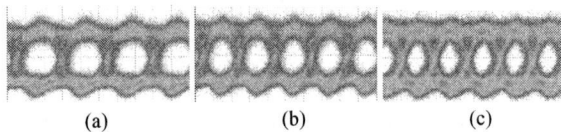

Figure 10. The measured eye diagrams after 1 km of fiber (a) 40 Gb/s, (b) 50 Gb/s, and (c) 56 Gb/s

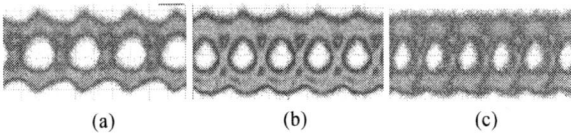

Figure 11. The measured eye diagrams after 2 km of fiber (a) 40 Gb/s, (b) 50 Gb/s, and (c) 56 Gb/s

IV. CONCLUSIONS

In this paper, we presented an adaptive patterning process to develop optical and electrical interconnects for photonics packaging. The optical interconnects were used as a fanout for PIC based on adiabatic optical coupling between SM polymer waveguides and SOI waveguides. A fiber-to-chip coupling losses of 4.6 and 4.8 dB were achieved at 1310 nm and 1550 nm respectively. The high-speed electrical interconnects between EIC and PIC were realized using AJP, providing a higher degree of freedom in tuning the interconnect geometries and hence the characteristic impedance. The AJP interconnects showed lower inductive parasitics than the traditional bondwires while interconnecting a microring modulator using the two methods. Finally, an optical transmitter assembly

based on an EAM and a CMOS driver was successfully demonstrated at 56 Gb/s.

ACKNOWLEDGMENT

The authors would like to acknowledge the financial support from the Special Research Fund of Ghent University (BOF14/GOA/034 project), from the European Commission through the H2020 project Streams (Contr. No. 688172), and from the Hercules Foundations Flanders (ZW11-15: Helicom). The microring and EAM modulators were developed as part of imec's industry affiliation R&D program on Optical I/O.

REFERENCES

[1] Tolga Tekin, Review of Packaging of Optoelectronic , Photonic, and MEMS Components. *IEEE Journal of Selected Topics in Quantum Electronics*, 17(3):704–719,

[2] Lee Carroll, Jun-su Lee, Carmelo Scarcella, Kamil Gradkowski, Matthieu Duperron, Huihui Lu, Yan Zhao, Cormac Eason, Padraic Morrissey, Marc Rensing, Sean Collins, How Yuan Hwang, and Peter O Brien. Photonic Packaging Transforming Silicon Photonic Integrated Circuits into Photonic Devices simple. *MDPI applied sciences*, 6 (426):1–21, 2016..

[3] Tymon Barwicz, Ted W Lichoulas, Yoichi Taira, Yves Martin, Shotaro Takenobu, Alexander Janta-polczynski, Hidetoshi Numata, Eddie L Kimbrell, Jae-woong Nah, Bo Peng, Darrell Childers, Robert Leidy, Marwan Khater, Swetha Kamlapurkar, Elaine Cyr, Sebastian Engelmann, Paul Fortier, and Nicolas Boyer. Automated, high-throughput photonic packaging. *Optical Fiber Technology Elsevier*, 44, 24–35, 2018.

[4] Bradley Snyder, Nivesh Mangal, Guy Lepage, Sadhishkumar Balakrishnan, Xiao Sun, Nicolas Pantano, Michal Rakowski, Lieve Bogaerts, Peter De Heyn, Peter Verheyen, Andy Miller, Marianna Pantouvaki, Philippe Absil, and Joris Van Campenhout. Packaging and Assembly Challenges for 50G Silicon Photonics Interposers. *Optical Fiber Communications Conference and Exposition (OFC)*, pages 1–3, 2018.

[5] Amir H Atabaki, Sajjad Moazeni, Fabio Pavanello, Hayk Gevorgyan, Jelena Notaros, Luca Alloatti, Mark T Wade, Chen Sun, Seth A Kruger, Huaiyu Meng, Kenaish Al Qubaisi, Imbert Wang, Bohan Zhang, Miloš A Popović, Vladimir M Stojanović, Rajeev J Ram, Anatol Khilo, and V Christopher, Integrating photonics with silicon nanoelectronics for the next generation of systems on a chip, *Nature*, 556:349–354, 2018.

[6] Kaushik Sengupta, Tadao Nagatsuma, and Daniel M Mittleman, Terahertz integrated electronic and hybrid electronic photonic systems, *Nature Electronics*, 2018.

[7] Jun Su Lee, Lee Carroll, Carmelo Scarcella, Nicola Pavarelli, Sylvie Menezo, Enrico Temporiti, and Peter O Brien. Meeting the Electrical , Optical , and Thermal Design Challenges of Photonic-Packaging. *IEEE Journal of Selected Topics in Quantum Electronics*, 22(6), 2016

[8] Bradley Snyder, Brian Corbett, and Peter O Brien. Hybrid Integration of the Wavelength-Tunable Laser With a Silicon Photonic Integrated Circuit. *Journal of Lightwave Technology*, 31(24):3934–3942, 2013.

[9] C. Li, T. Li, G. Guelbenzu, B. Smalbrugge, R. Stabile and O. Raz "Chip Scale 12-Channel 10 Gb/s Optical Transmitter and Receiver Subassemblies Based on Wet Etched Silicon Interposer," *Journal of lightwave technology*, vol. 35, pp. 3229-3236, 2017.

[10] G Roelkens, D Vermeulen, D Van Thourhout, R Baets, S Brision, P Lyan, P Gautier, J Fedeli, High efficiency diffractive grating couplers for interfacing a single mode optical fiber with a nanophotonic silicon-on-insulator waveguide circuit, *Applied Physics Letters*, 92, pp.1–4, 2008.

[11] J. Shu, C. Qiu, X. Zhang, and Q. Xu, "Efficient coupler between chip-level and board-level optical waveguides," *Optics Letters*, 36, 3614-3616, 2011.

[12] S. Kanazawa, T. Fujisawa, K. Takahata, Y. Ueda, H. Ishii, R. Iga, W. Kobayashi, and H. Sanjoh., "Flip-chip interconnection technique for beyond 100- Gb/s (4 x 25.8-Gb/s) EADFB laser array transmitter," *J. Lightw. Technol.*, vol. 34, no. 2, pp. 296–302,.2016.

[13] Ahmed Elmogi, Erwin Bosman, Jeroen Missinne and Geert Van Steenberge "Comparison of epoxy- and siloxane-based single-mode optical waveguides defined by direct-write lithography" *OPTICAL MATERIALS Elsevier*. Vol. 52. 26-31, 2016.

[14] Roger Dangel, Antonio La Porta, Daniel Jubin, Folkert Horst, Norbert Meier, Marc Seifried, Bert J Offrein, and Senior Member. Polymer Waveguides Enabling Scalable Low-Loss Adiabatic Optical Coupling for Silicon Photonics. *IEEE Journal of Selected Topics in Quantum Electronics*, 24(4), 2018.

[15] Ahmed Elmogi, Wouter Soenen, Hannes Ramon, Xin Yin, Jeroen Missinne, Silvia Spiga, Markus-ChristianAmann, Ashwyn Srinivasan, Peter De Heyn, Joris Van Campenhout, Johan Bauwelinck and Geert Van Steenberge, Aerosol-Jet Printed Interconnects for 2.5D Electronic and Photonic Integration, *Journal of Lightwave Technology*, vol. 36, no. 16, pp. 3528-3533, 2018.

[16] Ahmed Elmogi, Hannes Ramon, Joris Lambrecht, Peter Ossieur, Guy Torfs, Jeroen Missinne, Peter De Heyn, Yoojin Ban, Marianna Pantouvaki, Joris Van Campenhout, and Geert Van Steenberge, Aerosol-Jet Printed Interconnects for 60 Gb/s CMOS Driver and Microring Modulator Transmitter Assembly, *IEEE Photonics Technology Letters,* vol. 30, no. 22, pp. 1944-1947, 2018.

[17] M. Pantouvaki, S. A. Srinivasan, Y. Ban, P. De Heyn, P. Verheyen, G. Lepage, H. Chen, J. De Coster, N. Golshani, S. Balakrishnan, P. Absil, and J. Van Campenhout, "Active Components for 50 Gb/s NRZ-OOK optical interconnects in a silicon photonics platform," *J. Lightw. Technol.*, vol. 35, no. 4, pp. 631–638, Feb. 2017.

[18] H. Ramon, M. Vanhoecke, J. Verbist, W. Soenen, P. De Heyn, Y. Ban, M. Pantouvaki, J. Van Campenhout, P. Ossieur, X. Yin , and J. Bauwelinck., "Low-Power 56Gb/s NRZ Microring Modulator Driver in 28nm FDSOI CMOS," in *IEEE Photonics Technology Letters*, vol. 30, no. 5, pp. 467-470, March1, 1 2018.

[19] JS. Cho, KA. Yoo, JT. Moon, SB. Son, SH. Lee, and KH. Oh, "Pd effect on reliability of Ag bonding wires in microelectronic devices in high-humidity environments," *Met Mater Int*, Vol. 18(5), pp. 881–885. 2012

Low surface reflectance structure at near infrared wavelength by injection molding

Sho Yakabe, Takuro Watanabe, Takayuki Shimazu
Sumitomo Electric Industries, Ltd.
Yokohama, Japan
yakabe-syou@sei.co.jp

Ryohei Hokari, Kazuma Kurihara
National Institute of Advanced Industrial Science and Technology
Tsukuba, Japan
k.kurihara@aist.go.jp

Abstract— **Plastic lenses are utilized for optical communication devices such as AOC (Active Optical Cable), transceivers and connectors in order to improve coupling efficiency. In order to reduce the optical loss of the communication device one must also, reduce the Fresnel loss due to the refractive index of the transparent thermoplastic resin, the typical solution is, use of an Anti-reflection (AR) coating. Although it is possible to greatly lower the reflectance at the end face of the lens by this AR coating, the cost of the device is increased. To solve this problem we propose, reduction of reflectance in the near infrared wavelength band without AR coating by transferring the nanoscale concavo-convex structure through injection molding alone. Furthermore, to achieve low reflectance in various transparent thermoplastic resin, we propose a new method to drastically improve the transferability of the surface shape during injection molding. Utilizing this both methods, we successfully developed surface structure with a concavo-convex of 0.3 μm or more and a 2% or less reflectance at over 1 μm wavelength.**

Keywords-component; Nanostructure , Low surface reflectance structure，Injection Molding，

I. INTRODUCTION

Optical fiber communication technologies have been developed to help realize higher density bandwidth while maintaining low cost for high performance computers and general-purpose server applications in a large data center. Especially in recent years the development of on-boat optics technology has advanced drastically. An example of this technology is the optical engine consisting of a transmitter (semiconductor laser), receiver (photodiode), driver IC, TIA and optical fiber within a compact integrated package. Such an optical engine, connector and other optical devices utilized plastic lens by injection molding [1-4]. Plastic lens is one of the solutions that is cost effective and allows for a flexible design in positioning of the lens to the fiber connecting portion. However, Fresnel reflection occurs at the interface between transparent thermoplastic resin and air, and the surface reflectance due to Fresnel reflection depends on the refractive index difference between transparent thermoplastic resin and air. most of transparent thermoplastic resins have a high refractive index of 1.50 to 1.65. Fresnel reflection loss of 0.18 dB (4%) to 0.27 dB (6%) per surface. As the transmission speed increases over 10 Gbps, it is essential to reduce the Fresnel loss which depends on the high refractive index of the plastic lens. Fig.1 shows the cross-section of the lens module mounted on the PCB. For example, in such an optical system coupled to a Multi-Mode fiber with such two lenses, the maximum Fresnel loss is 1 dB in the entire optical link of optical engine. Therefore, Anti-reflection (AR) coating on the lens is used to suppress the Fresnel loss, unfortunately the tradeoff is higher cost.

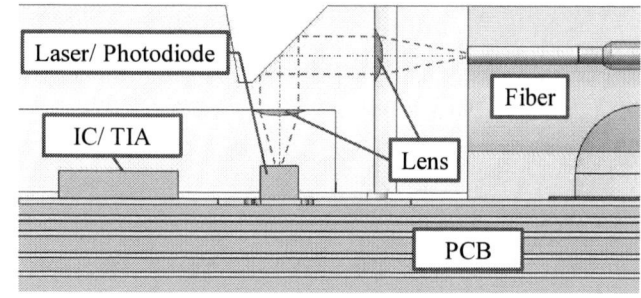

Figure 1. Cross-section of lens module

To solve traditional lens performance gaps, we have proposed a manufacturing method that realizes a low reflectance lens by injection molding [5], but the reduction of the reflectance was not sufficient at the 1 μm band wavelength used for single mode fiber communication. Further, in order to realize a low reflectance structure, it was limited to a thermoplastic resin with excellent Melt Flow Rate (MFR). Therefore, in this study, in order to realize a low cost lens with low reflectance <1% at over 1 μm wavelength, we combined the previously proposed method of forming a nanoscale concave-convex structure in an injection molding die and the special molding die which can improve transferability.

II. INJECTION MOLDING DIE WITH NANOSCALE CONCAVO-CONVEX STRUCTURE

One of the anti-reflection structures recently attracting attention is a method of adding an anti-reflection effect by nanoscale structures. Since the structure size is optical wavelength or less, the nanoscale structure forms a refractive index distribution in which the refractive index varies more gradually. Therefore, Fresnel reflection caused by the difference in refractive index is suppressed, so that the reflectance can be reduced. As one of the methods for creating this nanoscale structure, and in order to realize a

978-1-7281-1500-9/19 $31.00 © 2019 IEEE

anti-reflection structures on the lens surface, using self-organized metal nano particles, the National Institute of Advanced Industrial Science and Technology have developed a mold fabrication technology that can produce an anti-reflection structure to a lens surface or complicated shape [6] [7]. Fig.2 shows an injection molding die manufacturing process with nanoscale structure using metal nano particles. At first, metal nano particles are deposited on the surface of an injection molding die having a lens shape, then the surface of the molding die is etched using a reactive ion etching apparatus. Here the metal nano particles function as a mask for etching the nanoscale structure, so that it is possible to create the nanoscale structure on the surface of the molding die having the lens shape. Finally, by injection molding using this molding die with nanoscale structures, it is possible to make a plastic lens having an anti-reflection effect by only the injection molding process.

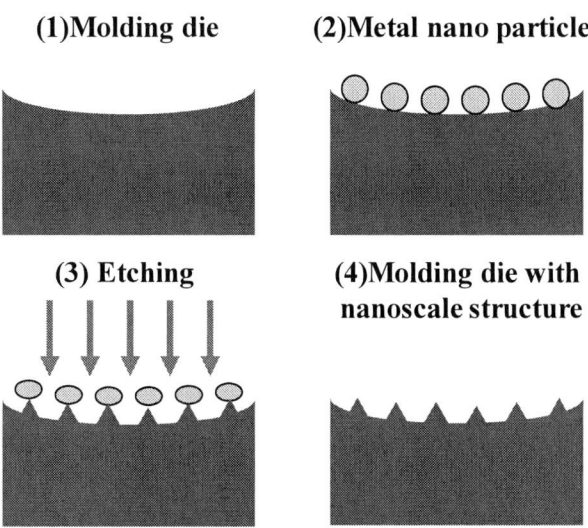

Figure 2. Fabrication process of injection molding die with anti-reflection nanostructure

By using this manufacturing process, it is possible to manufacture a plastic lens having an anti-reflection structure at a lower cost only versus traditional AR coating processes.

In previous work, we demonstrated low surface reflectance on micro lens array that was fabricated by this technology. Fig.3 shows micro lens with low surface reflectance. In order to evaluate this nanoscale structure on the lens surface, the surface topography of the lens vertex and the lens cross section was observed using a scanning electron microscope (SEM). It was confirmed that the nanoscale structure was clearly formed versus the lens surface having no nanoscale structure.

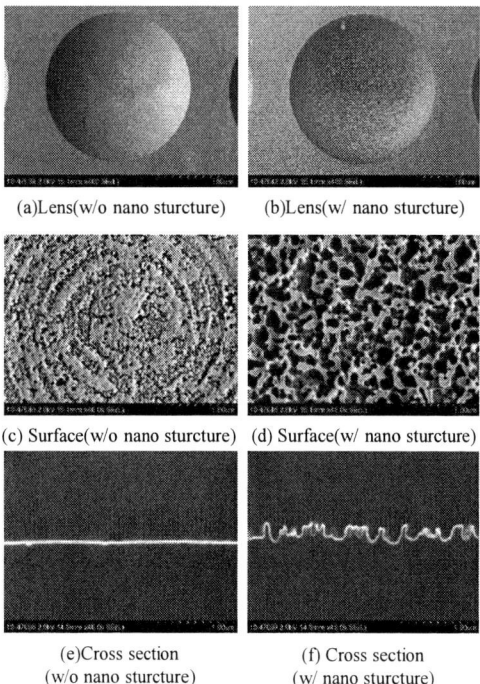

Figure 3. Nanoscale concavo-convex structure on lens surface

Furthermore, Fig.4 shows surface roughness index of each condition measured by atomic force microscope (AFM). The arithmetic mean height (Sa), which is one of the index for evaluating the surface roughness, indicates that the nanoscale structure is well transferred along with an increase in molding die temperature. Likewise, since the correlation with the molding die temperature is observed also in the maximum height (Sz). From these results, it is indicated that the molding die temperature and resin temperature are extremely important parameters for improving the transferability of the anti-reflection nanostructure.

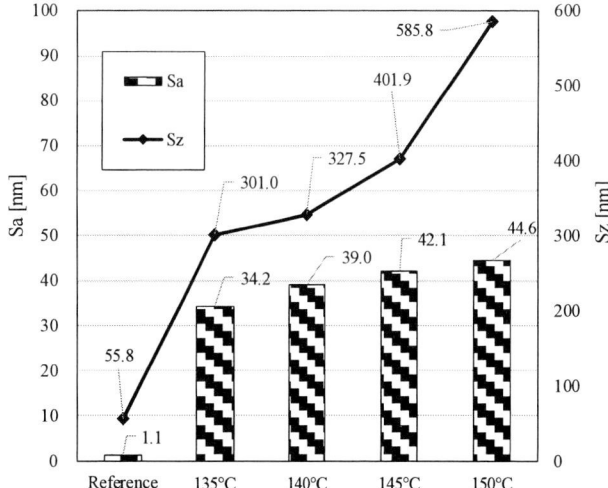

Figure 4. Surface roughness of nanoscale concavo-convex structure

978-1-7281-1500-9/19 $31.00 © 2019 IEEE

On the other hand, in the previous work, only the multi-mode communication wavelength (0.85 μm) is examined for the plastic lens having the anti-reflection effect by injection molding, and the reflectance is not enough in the optical communication wavelength at over 1 μm. Furthermore, in the previous work, A lens having an anti-reflection nanostructure could not be fabricated unless a resin excellent in transferability.

In order to fabricate low reflectance lens by various transparent thermoplastic resin, we combined the previously proposed method of forming a nanoscale concave-convex structure in an injection molding die and the special molding die which can improve transferability. Therefore, even in a thermoplastic resin having a low MFR, it is possible to maintain a high temperature state with good fluidity, so that transfer of the nanostructure becomes easy. By combined both technologies, we aimed for a lens with a concavo-convex structure of 2% or less reflectance at over 1 μm wavelength.

III. OPTICAL ANALYSIS OF ANTI-REFLECTION NANOSTRUCTURE

At first, optimum anti-reflection structure was investigated by optical analysis of anti-reflection nanostructure in the 1 μm wavelength band. Fig. 5 shows a model of the anti-reflection nanostructure used for optical analysis. In this optical analysis, reflection on the back side of the substrate is ignored because near-infrared micro spectrometer separates back side reflection. For this reason, only reflection on the anti-reflection structure surface was analyzed. The anti-reflection nanostructures were analyzed using an analytical model assuming conical shape that is considered to be the most approximate shape from the shape measured in the past and assumed conical shape as close‐packed structure as shown in Fig. 5 (b). In this model, reflectance on surface was affected by height (h), diameter (φ) and wavelength (λ). Fig. 6 shows these parameter and optical analysis model. In order to compare the actual reflectance measurement value with the optical analysis result, the incident light was plane wave in the direction perpendicular to the substrate.

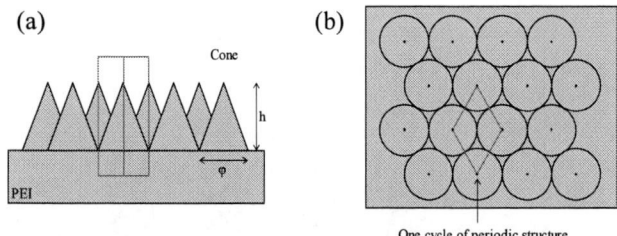

Figure 5. Anti-reflection structure model of optical analysis
(a)Side view (b)Top view

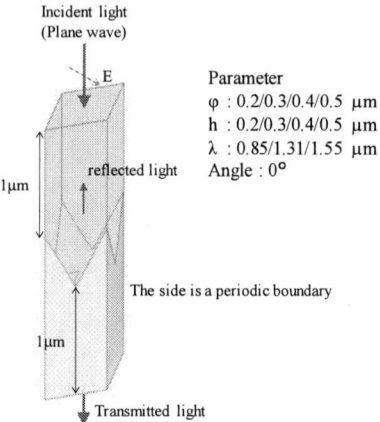

Figure 6. optical analysis model and parameters

The optical analysis result of the surface reflectance is shown in Fig. 7. These results show the difference in surface reflectance when height and diameter are used as parameters in each wavelength. From this result, it was found that the height of the nanostructure has the greatest influence on the surface reflectance, and it was confirmed that a low reflectance of -30 dB (0.1 %) or more can be obtained at the wavelength of 0.85 μm. On the other hand, at the wavelengths of 1.31 μm and 1.55 μm the surface reflectance is around -20 dB (1 %), indicating that the effect of reducing the reflectance sufficiently by the nanostructures is not obtained. If the height of the nanostructure is half of the wavelength, the phase of the reflected light becomes 0 to 360 degrees on the same plane, indicating that the cancellation effect of the reflected light becomes strong.

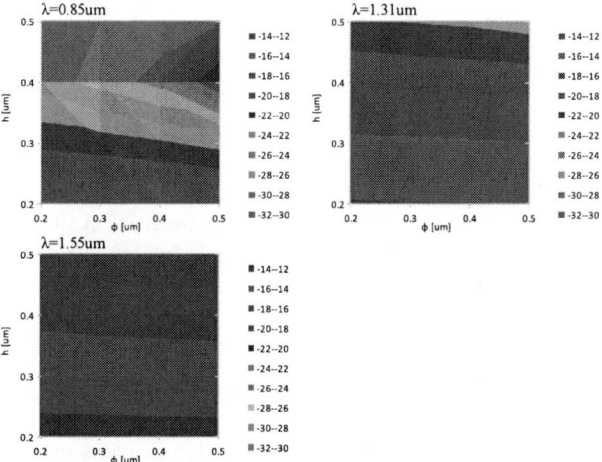

Figure 7. Surface reflectance of each wavelength

Furthermore, Fig. 8 shows relationship between wavelength and surface reflectance in nanostructures. From this result, the minimum value of the surface reflectance is confirmed at the wavelength 0.8 μm and 1 μm corresponding to twice the height. On the other hand, If the diameter of the nanostructure is too small, a sufficient effect of reducing the surface reflectance cannot be obtained. If it is too large, the surface reflectance decreases due to scattering, but the transmittance decreases.

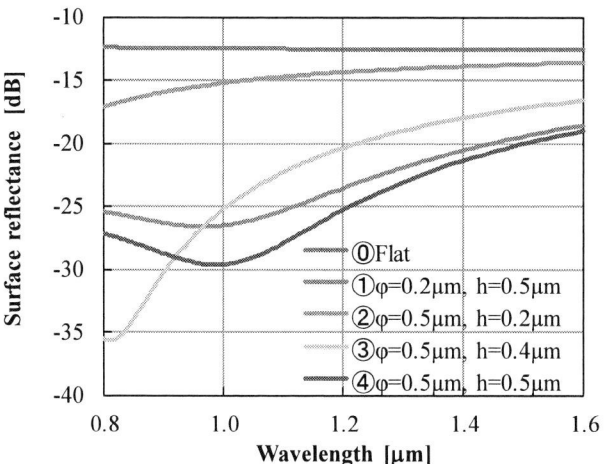

Figure 8. Relationship between wavelength and surface reflectance

Finally, the surface reflectance due to the shape of the anti-reflection nanostructure by optical analysis was investigated. Regarding the shape of the nanostructure, a model shown in Fig. 9 was prepared for a cone and a parabolic surface, and surface reflectance was compared. In this comparison, the height of the nanostructure is 0.5 µm, the diameter is 0.5 µm, and the only difference in the refractive index gradient depending on the shape influences the surface reflectance. Fig. 10 shows the surface reflectance of a cone and paraboloid. In the conical shape, the minimum value of the surface reflectance is confirmed at the wavelength 1 µm corresponding to twice the height, as discussed above. On the other hand, in the parabolic shape, the minimum value of the surface reflectance was confirmed between the wavelengths of 1.4 and 1.5 µm. From this optical analysis result, it becomes possible to realize a very low surface reflectance if it is possible to freely shape the shape of the anti-reflection nanostructure.

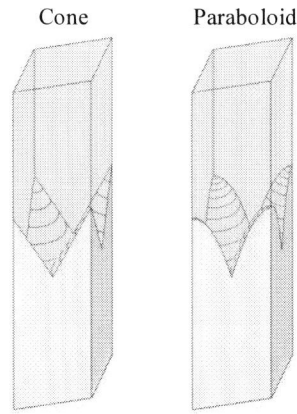

Figure 9. Shape comparison of nanostructures

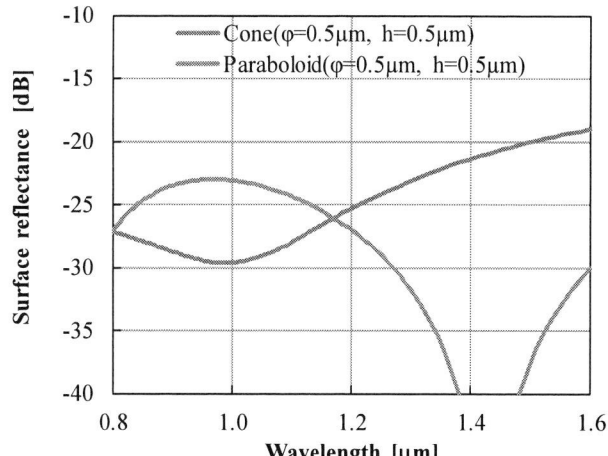

Figure 10. Surface reflectance of a cone and paraboloid

IV. TRANSFER OF NANO-CONCAVE-CONVEX STRUCTURE USING MOLDED ARTICLE OF HIGH HEAT-RESISTANT PLASTIC

In order to fabricate a high heat resistant and low reflectance micro lens array applicable to optical communication wavelength, transferability by injection molding using polyetherimide (PEI) which is a high heat resistant material having a high Glass-transition temperature (Tg) was investigated. For this study, a plate-shaped molded sample (40 mm × 40 mm, thickness 1.5 mm) was used. Evaluation was carried out by transferring nanostructures of size less than the wavelength on the front and back surfaces of the sample in the region of diameter 5 mm of the plate-shaped molded sample. Fig. 11 shows a picture of the sample.

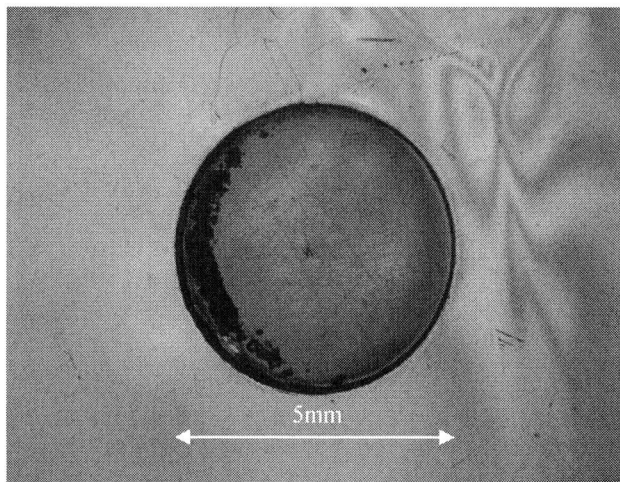

Figure 11. Picture of molded products

The PEI used as a molding material has a high Tg of 217 °C. and is excellent in heat resistance and has a high refractive index of 1.63, so it is widely utilized as a micro lens for optical communication. On the other hand, transfer of microstructure is difficult because melt flow rate (337 °C / 66 N) is low (9 g / 10 min).

978-1-7281-1500-9/19 $31.00 © 2019 IEEE

Even in previous work, injection molding experiments were carried out using plate molds with anti-reflective nanostructures on the surface and PEI material. A plate sample was fabricated by changing the mold temperature in the range of 140 °C to 160 °C and the reflectance of the surface was measured. Fig. 12 shows the measurement results of reflectance. Even under the highest mold temperature of 160 °C molding conditions, the decrease in the reflectance of the surface was low and it was difficult to transfer the anti-reflective nanostructure to the surface.

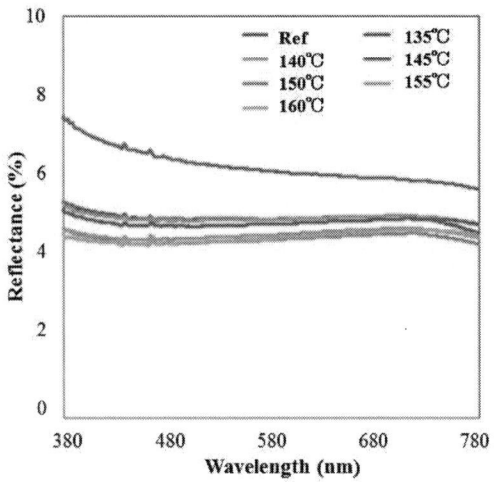

Figure 12. Previous experiments using PEI molded products

In this study, samples were fabricated by changing the temperature of molding die at 160 to 165 °C. and the melting temperature of PEI within the range of 350 to 370 °C. First, the measurement results of reflectance at each mold temperature when the melting temperature of PEI is fixed at 360 °C is shown in FIG 13.

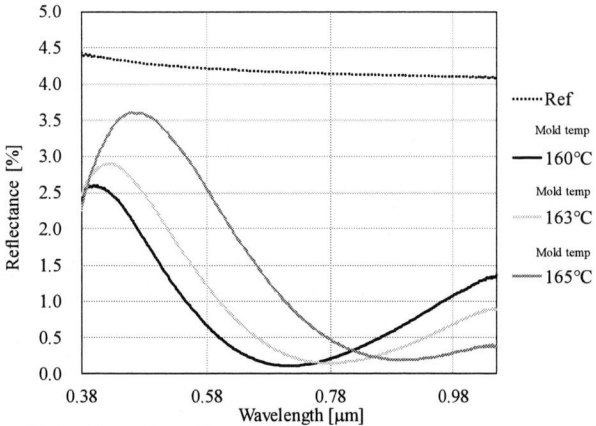

Figure 13. Surface reflectance of nanoscale concavo-convex structure

As the mold temperature increased, the wavelength of the minimum reflectance shifted in the longer wavelength direction. This is probably because the transfer rate of the anti-reflection nanostructure was improved by the rise of the mold temperature.

Next, samples were fabricated by fixing the mold temperature to 165 °C and changing the melting temperature of PEI. Fig. 14 shows the measurement results of reflectance.

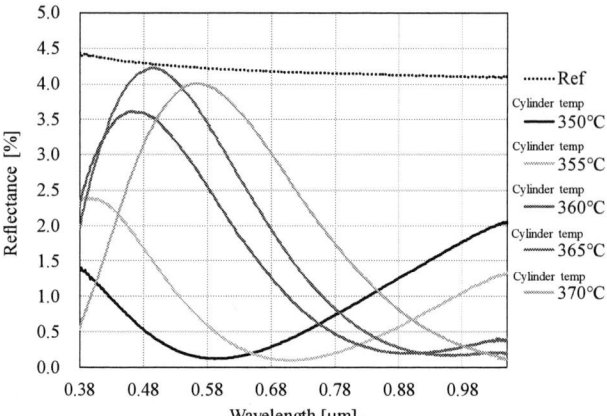

Figure 14. Surface reflectance of nanoscale concavo-convex structure

As the melting temperature increased, the transfer rate of the anti-reflective nanostructure improved. As a result, at the cylinder temperature of 370 °C, the reflectance decreased to 0.5 % or less at a wavelength of 1 μm or more.

By using the special molding die which can improve transferability, it became possible to transfer the anti-reflection nanostructure even in the case of using the PEI material which was difficult to transfer the anti-reflection nanostructure in the conventional mold. It is thought that this can be expected for controlling the reflectance of the optical communication wavelength used for long distance transmission.

Finally, the surface shape of the sample fabricated under the conditions of a mold temperature of 165 °C and a cylinder temperature of 370 °C was observed. Fig. 15 shows measurement results using an AFM.

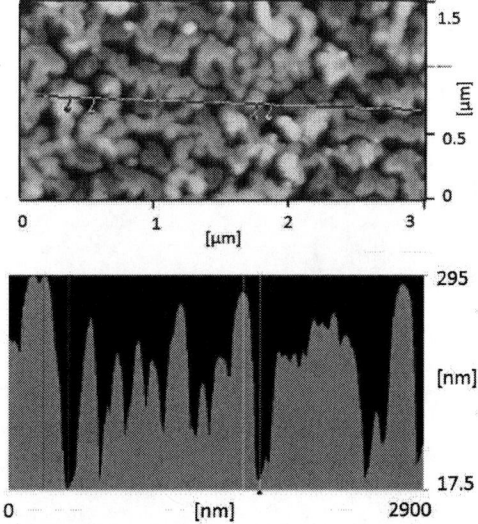

Figure 15. AFM measurement of fine asperity structure

978-1-7281-1500-9/19 $31.00 © 2019 IEEE

From this result, it was confirmed that an uneven structure having a height of about 0.3 μm and a pitch of about 0.1 to 0.2 μm was transferred onto the surface.

Comparison of the reflectance measurement result obtained in this experiment with the calculated value of reflectance by optical analysis revealed a difference in reflectance at a wavelength of 0.85 μm. In the analysis result, the reflectance at the wavelength of 0.85 μm was 0.88 %, whereas the result of the experiment was 1.05%, which was slightly difference. Although it was assumed that the anti-reflection nanostructure was an ideal conical shape in the analysis, it is actually thought that the shape is different, and this difference in shape is considered to be a factor in the difference in the reflectance.

CONCLUSION

This molding die fabrication technology for forming microstructure on metal mold surface using nano metal particles realizes transfer of antireflection structure on molding die surface and can be applied to existing standard injection molding processes. Furthermore, it is possible to transfer the antireflection structure with low reflectance (< 2 %) at 1 μm or more, and even in a thermoplastic resin having a low MFR. This is because antireflective structures can be transferred with various resins according to the specifications of the product, and it is expected to be applied to various injection molded parts.

In addition, from the optical analysis of the antireflection structure, it was confirmed that the reflectance mainly depends on the height of the structure, and in the conical shape, it is confirmed that the wavelength which is twice the height becomes the minimum reflectance. It can be expected that design of an appropriate antireflection structure becomes possible by using this analysis.

By applying this technology to optical components of optical devices like AOC, it is possible to further improve the coupling efficiency. As a result, this technology can contribute to reduce the requirement of mounting accuracy and lower power consumption.

REFERENCES

[1] D. Childers, E. Childers, J. Graham, M. Hughes, D. Schoellner and A. Ugolini, "Miniature Detachable Photonic Turn Connector for Optical Module Interface," Electronic Components and Technology Conference (2011), pp1922-1927

[2] T Shimazu, M Harumoto, T Sano, "Reflowable Thermoplastic Optical Lens Module for 10-Gbit/s Transmission with 850-nm VCSEL" Optical Fiber Communication Conference (OFC), March 2015

[3] D. Childers et al., "Next-generation, high-density, low-cost, multimode optical backplane interconnect," Proc. SPIE, vol. 8267, 826700, 2012

[4] A. Nakama et al., "High Density Optical Connector with Unibody Lensed Resin Ferrule," Proc. IWCS2015, 8-3.

[5] S. Yakabe, "Low Reflectance and Reflowable Thermoplastic Optical Lens without AR Coating" in Proceedings of 68th Electronic Components and Technology Conference (2018).

[6] K Kurihara, Japanese Journal of Optics KOGAKU, vol.1, pp.24−29, 2011.

[7] K Kurihara, Seikei-Kakou, Vol.25, 4, pp.171−174, 2013.

A Novel Design of a Bandwidth Enhanced Dual-Band Impedance Matching Network With Coupled Line Wave Slowing

Deepayan Banerjee[+], Antra Saxena[+] and Mohammad Hashmi[+$]

[+] *Circuit Design Research Lab., IIIT Delhi, New Delhi 110020, India*
[$] *School of Engineering, Nazarbayev University, Astana 010000, Kazakhstan*

Abstract—**This paper presents a novel design of a bandwidth enhanced dual-band impedance matching network utilizing the principle of wave slowing. Coupled-line sections have been used in their all-pass configuration to incorporate the same. The proposed design is generalized for real as well as complex loads with simple closed form design equations. The design is compact and robust and solves the hurdle of bandwidth crunch at GSM and near-GSM frequencies. To validate the proposed concept, prototypes have been fabricated on RO5880 ($\epsilon_r = 2.2$), which demonstrate wideband performance for real loads at 900MHz and 2.4GHz.**

Index Terms—**Dual-band, Impedance Transformer, Slow Wave Structure, Wideband.**

I. INTRODUCTION

The last decade has seen a huge surge of wireless device usage in all aspects of modern life. With a huge host of application areas that include, but are not limited to communication systems, military, medical, mining, agriculture etc, wireless systems prove to be the backbone of the modern "connected" life. With the recent emergence of Internet of Things (IoT) multi-band wireless systems have come to central focus. Connecting almost anything to the internet requires effective and efficient use of the available bandwidth and this forms the foundation behind the craze for enhanced bandwidth circuits. Much of research is being carried out on multi-band and multi-functional devices globally with a target to either increase the overall bandwidth of the device or have multiple devices operating within a fixed bandwidth. A multi-frequency impedance matching network forms one of the most important building blocks of any multi-band system. They form basic structures of power dividers, couplers, baluns, crossovers, power amplifiers, low noise amplifiers and antennae. Thus, designing an efficient multi-frequency impedance matching network is very crucial to maintain the performance of any system.

Dual-band designs have come up in literature and are extensively used in many applications. These networks came up in 2002 with the Chow transformer [1] which proposed matching schemes at a frequency and its harmonics. The design limitations include non-availability of closed form design equations and the harmonic matching

relation. The same was addressed by Monzon [2] in 2003, which proposed an architecture that could match at any two arbitrary frequencies of interest for real loads.The other popular dual-band matching architectures, to name a few, include the T- [3], Pi-type [4], Virtual Impedance [5], Common Reference Frequency [6], Impedance Bridge [7], unequal susceptance stubs [8], all-pass coupled line transformers [9] and the ones transforming Frequency Dependent Complex Loads (FDCL) to real ones.

It is often observed that the architectures available in literature operating particularly at lower frequencies (GSM and WiFi as example), have limited bandwidth for the operating band, which indeed limits the effective data rate over the channel. With increasing demands for high data rates and for low-latency applications like satellite communication and wireless non-adhoc medical surgeries, this is a huge limitation. The desire is always for high bandwidth such that sophisticated data multiplexing schemes can be incorporated to effectively enhance the system efficiency. As for the best of the authors' knowledge, the existing architectures have not addressed this issue. This paper, for the first time, proposes a dual-band impedance matching network that provides a substantial bandwidth increment over the GSM and WiFi bands. The network, being generalized, applies for both real as well as FDCL and the same has been verified and tested with fabricated prototypes (tested real load).

Section-II describes the theoretical concept behind the proposed design, section-III- the mathematical formulation and the design approach, section-IV provides case studies, simulation and design examples with measurement results of a fabricated prototype and section-VI concludes the paper.

II. THE PROPOSED IMPEDANCE TRANSFORMER

The schematic representation of the proposed dual-band impedance matching network is depicted in Fig. 1.

The load Z_L can be of real or FDCL type and is matched to the conventional 50Ω real source Z_S using three Slow Wave Structures (SWSn). Z_{L1} and Z_{L2} are two intermediate impedance values. The matching is obtained

978-1-7281-1500-9/19 $31.00 © 2019 IEEE

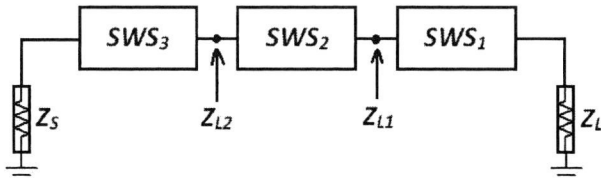

Fig. 1: Schematic of the Proposed Impedance Transformer

in a stepwise fashion using symmetrical $\lambda/4$ meanders. Firstly Z_L is matched to Z_{L1} using SWS_1, followed by Z_{L1} with Z_{L2} using SWS_2, and finally, Z_{L2} with the source Z_S using SWS_3. This essentially slows down the group velocity of the travelling wave in the bands of interest, thereby enhancing the bandwidth. For implementing the SWS, any topology can be used, considering the electrical length relationship for dual-band matching. In this work, coupled-lines have been used in their all-pass configuration (quarter wave meandered lines) as they are compact and provide easy closed form design equations for robust prototyping.

III. Design and Mathematical Formulation

The design is based on recursive blocks of wave slowing structures- in this case being quarter wave meandered lines. The circuit representation of the network is presented in Fig. 2. Sections- A, B and C are the SWSs, implemented using $\lambda/4$ meandered lines with even/odd impedance values and electrical lengths as depicted in Fig. 2.

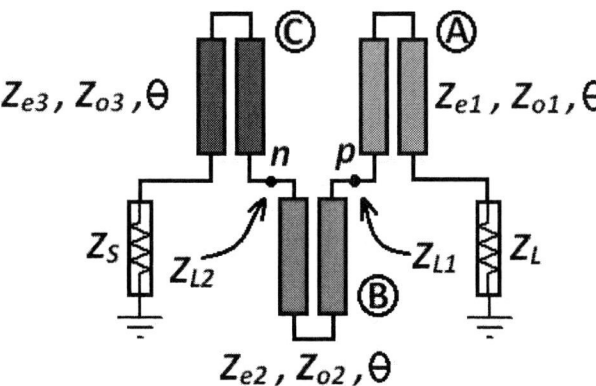

Fig. 2: Circuit Representation of the Proposed Design

The design is based on correct determination of the values of Z_{L1} and Z_{L2} to facilitate the step-matching criteria. The choice can be made such that the impedances lie equidistant to one another linearly, as provided in (1).

$$Z_{L1} = Z_L - \frac{Z_L - Z_S}{3} \tag{1a}$$

$$Z_{L2} = Z_S + \frac{Z_L - Z_S}{3} \tag{1b}$$

This essentially means that the impedance difference $(Z_L - Z_{L1})$ is equal to the impedance difference $(Z_{L1} - Z_{L2})$ which indeed equals the impedance difference $(Z_{L2} - Z_S)$ Following that, the determination of the meandered line parameter values is quite straightforward and are provided in (2) below [12].

$$Z_{e1} = \sqrt{Z_L . Z_{L1}} \tan \theta \tag{2a}$$

$$Z_{o1} = \frac{\sqrt{Z_L . Z_{L1}}}{\tan \theta} \tag{2b}$$

where θ is given as:

$$\theta = \frac{m\pi}{1+r} \tag{3}$$

and r is the frequency ratio, denoted by $r = f_2/f_1$ and $m \in [1, 2, 3, ...]$.

The design of sections- B and C are similar as the SWSs are symmetric. Thus, the design parameters Z_{e2}, Z_{o2}, Z_{e3}, and Z_{o3} can be determined as:

$$Z_{e2} = \sqrt{Z_{L1} . Z_{L2}} \tan \theta \tag{4a}$$

$$Z_{o2} = \frac{\sqrt{Z_{L1} . Z_{L2}}}{\tan \theta} \tag{4b}$$

and

$$Z_{e3} = \sqrt{Z_{L2} . Z_S} \tan \theta \tag{5a}$$

$$Z_{o3} = \frac{\sqrt{Z_{L2} . Z_S}}{\tan \theta} \tag{5b}$$

The proposed scheme is applicable for frequency dependent complex loads as well. The only addition to the existing architecture would be a complex to real impedance transformation network at two frequencies [5]. The same can be a pi-, L- or T- network and are available quite easily in literature. The conversion network would be placed between the load and Section-A (from Fig. 2), and would transform the complex load impedance $Z_{complex}$ into real load Z_L. While chosing the complex to real impedance transformer it is recommended to select a network that offers high bandwidth. As this would be the first stage in the matching network, a slight bandwidth crunch at this stage would decrease the overall bandwidth of the network. The rest circuit operation would follow as discussed. The design flow for the above discussed theory is illustrated in Fig. 3.

978-1-7281-1500-9/19 $31.00 © 2019 IEEE

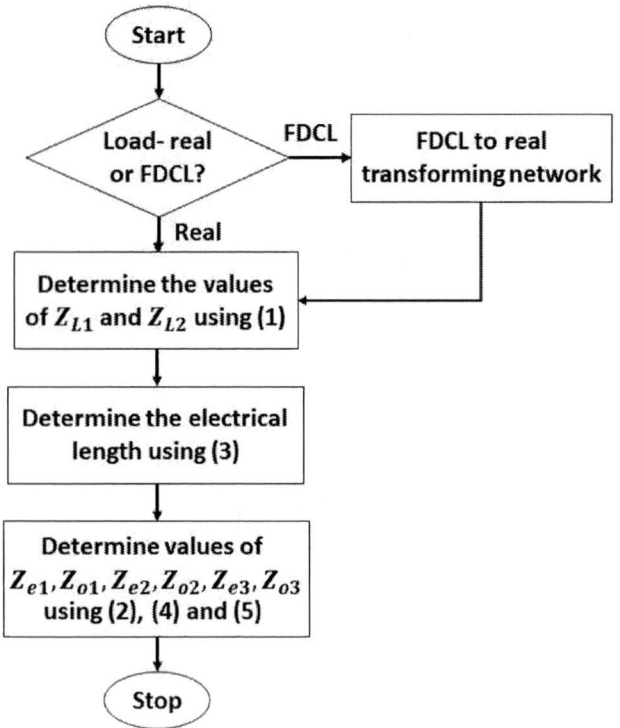

Fig. 3: Design Flow of the Proposed Matching Network

IV. SIMULATION AND DESIGN EXAMPLES

Based on the above discussed theory, CAD simulations have been performed and illustrated in Table-I.

TABLE I: Case Studies For Different Real Loads at Different Frequencies

	f_1(GHz)	f_2(GHz)	$Z_L(\Omega)$
Case-I	0.9	1.8	200
Case-II	0.9	2.4	100
Case-III	0.9	3.5	80

Three cases have been considered for three different real loads at three different sets of frequencies. The design parameters are listed in Table-II and the corresponding simulation results are provided in Fig. 4. Fractional Bandwidth is defined as the bandwidth achieved at a particular band of interest, i.e. $(\Delta f/f_{desired})X100$. Table-III illustrates the same for all the cases considered here. It is evident that as the band separation $(\Delta f = f_2 - f_1)$ increases, the fractional bandwidth increases, which is obvious. The interesting point to note is that even when the bands are closely spaced (as in Case-I), the fractional bandwidth is still above 30%, considering a -15dB reference level. In general, bandwidth calculation uses a -10dB reference level but in this work, -15dB has been chosen intentionally.

TABLE II: Design Parameters for Table-I

	Section-A	Section-B	Section-C
Case-I	$Z_{e1} = 300\Omega$ $Z_{o1} = 100\Omega$ $\theta = 60°$	$Z_{e2} = 212.13\Omega$ $Z_{o2} = 70.71\Omega$ $\theta = 60°$	$Z_{e3} = 122.47\Omega$ $Z_{o3} = 40.82\Omega$ $\theta = 60°$
Case-II	$Z_{e1} = 96.55\Omega$ $Z_{o1} = 72.49\Omega$ $\theta = 49.09°$	$Z_{e2} = 74.78\Omega$ $Z_{o2} = 56.16\Omega$ $\theta = 49.09°$	$Z_{e3} = 63.21\Omega$ $Z_{o3} = 47.46\Omega$ $\theta = 49.09°$
Case-III	$Z_{e1} = 56.02\Omega$ $Z_{o1} = 99.96\Omega$ $\theta = 36.82°$	$Z_{e2} = 48.52\Omega$ $Z_{o2} = 86.57\Omega$ $\theta = 36.82°$	$Z_{e3} = 41.00\Omega$ $Z_{o3} = 73.16\Omega$ $\theta = 36.82°$

The fact that lower the S_{11}, better is the match, is well known. For a high bandwidth, the S_{11} should be lower for a larger band of frequencies. If the reference level is brought down to -15dB, this would mark a worst case consideration for bandwidth. A better performance at -15dB reference would assure even higher bandwidth at -10dB.

TABLE III: Fractional Bandwidth at Two Frequencies for the Case Study in Table-I (Ideal Lines)

	$(\Delta f/f_1)X100$	$(\Delta f/f_2)X100$
Case-I	62%	31%
Case-II	122%	46%
Case-III	153%	48%

Based on the above discussed theory, a prototype (Fig. 5) operating on 900MHz and 2.4GHz (Case-II) has been fabricated on RO5880 substrate. The simulated and the measured results are provided in Fig. 6 and Table-IV.

TABLE IV: EM-Simulated & Measured Results

Results	Fractional Bandwidth	
Simulated	85.56% @ f_1	23.43% @ f_2
Measured	50% @ f_1	25% @ f_2

V. CONCLUSION

A novel design for bandwidth increment in dual-band networks has been proposed. The described theory addresses the limitations of present multi-band impedance matching networks and introduces closed form design equations for bandwidth enhanced robust designs. To illustrate the relevance of the research, a comparison of the design with the current state-of-the-art has been provided in Table-V. Measured results from the fabricated prototype go well in hand with the simulations, thus proving the validity of the proposed theory. The future of this research

978-1-7281-1500-9/19 $31.00 © 2019 IEEE

TABLE V: Comparison With the State-of-The-Art

Ref., Year	Venue	Method used	Design Procedure	*$(\Delta f/f_1)X100$	+$(\Delta f/f_2)X100$
[3], (2011)	IET-EL	T-model	simple	4%	3.6%
[4], (2015)	MWCL	Pi-model	tedious	6%	4%
[11], (2009)	MWCL	Three-Section TL	tedious	8%	6%
This work	———	**Coupled-Line Wave Slowing**	**simple**	**50%**	**25%**

*fractional BW at f_1; +fractional BW at f_2

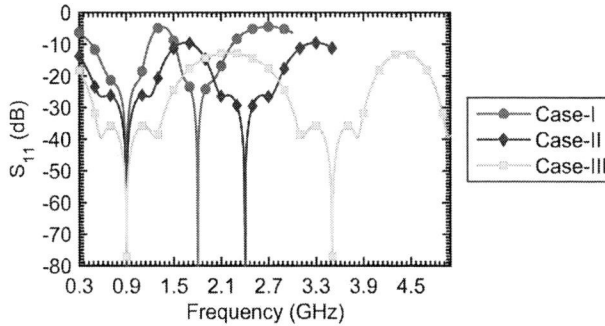

Fig. 4: Cases for Different Impedances and Frequencies

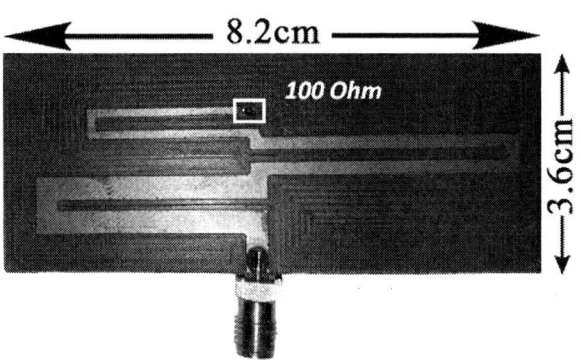

Fig. 5: The Fabricated Prototype

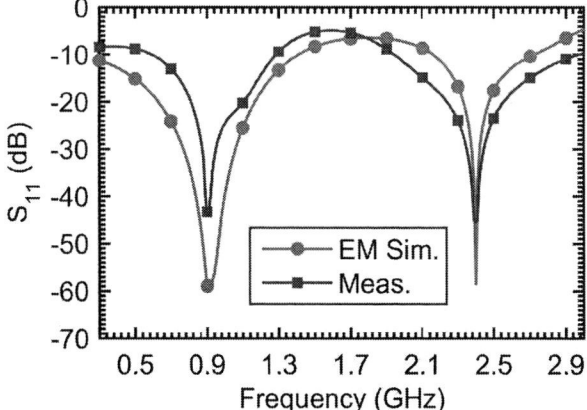

Fig. 6: EM Simulation and Measurement Results

points well into high bandwidth dual-band Power Amplifiers for communication and space applications.

VI. ACKNOWLEDGEMENT

The authors would like to thank Mr. Rahul Gupta, Senior Lab. Engineer at Circuit Design Research Lab. (CDRL), IIIT Delhi for fabricating the prototype and assisting with the measurements.

REFERENCES

[1] Y.L. Chow and K.L Wan, "A Transformer of One-Third Wavelength in Two Sections For a Frequency and its First Harmonic" *IEEE Micro. and Wireless Comp. Lett.*, Vol. 12, No. 1, p.p. 22 – 23, 2002

[2] C. Monzon, "A Small Dual-Frequency Transformer in Two Sections", *IEEE Trans. on Micro. Theory and Tech.*, Vol. 51, No. 4, p.p. 1157 – 1161, 2003

[3] M.A Nikravan, Z. Atlasbaf, "T-Section Dual-Band Impedance Transformer for Frequency Dependent Complex Impedance Loads", *Electronics Letters*, Vol. 47, Issue 9, Apr. 2011

[4] O. Manoochehri, A. Asoodeh, and K. Forooraghi, "Pi-Model Dual-Band Impedance Transformer for Unequal Complex Impedance Loads", *IEEE Micro. Wireless Comp. Lett.*, Vol. 25, No. 4, p.p. 238 – 240, 2015

[5] D. Banerjee, A. Saxena, and M.S. Hashmi, "A Novel Concept of Virtual Impedance for High Frequency Tri-Band Impedance Matching Networks". *IEEE Transactions on Circuits and Systems II: Express Briefs*, Vol. 65 Iss. 9, pp.1184-1188.

[6] D. Banerjee, A. Saxena and M.S. Hashmi, "A Novel Compact Triband Matching Network Utilizing two Dual-Band Transformers at a Common Reference Frequency", *2017 IEEE Asia Pacific Microwave Conference (APMC)* (pp. 1080-1083)

[7] D. Banerjee, M.S. Hashmi and F.M. Ghannouchi, "A Novel Design of a Tri-Band Impedance Matching Network Based on the Concept of an Impedance Bridge", *2018 Asia-Pacific Microwave Conference (APMC)* (pp. 318-320)

[8] A. Saxena, D. Banerjee, M.S. Hashmi and F.M. Ghannouchi, "Design of Compact Dual-Band Matching Network with Single Unequal Susceptance Cancellation Stub", *2018 Asia-Pacific Microwave Conference (APMC)* (pp. 300-302)

[9] D. Banerjee, A. Saxena, M.S. Hashmi and F.M. Ghannouchi, "A Compact Dual-Band Impedance Matching Network Based on All-Pass Coupled Lines", *2018 IEEE 61st International Midwest Symposium on Circuits and Systems (MWSCAS)* (pp. 937-939)

[10] R. Gupta and M.S. Hashmi, "High impedance transforming simplified Balun architecture in microstrip technology", *Microw Opt Technol Lett.* 2018; 60: 3019– 3023. https://doi.org/10.1002/mop.31450

[11] X. Liu, Y. Liu, S. Li, F. Wu, and Y. Wu, "A Three-Section Dual-Band Transformer For Frequency-Dependent Complex Load Impedances", *IEEE Micro. & Wire. Comp. Lett.*, Vol. 19, No. 10, p.p. 611 – 613, 2009

[12] D.M Pozar, "Microwave Engineering", *2nd Ed.*, Wiley, New York, 1998

Effects of Electromigration on Microstructural Evolution and Mechanical Properties of Preferential Growth Intermetallic Compound Interconnects for 3D Packaging

M.L. Huang*, L. Zou

Electronic Packaging Materials Laboratory
School of Materials Science & Engineering
Dalian University of Technology
Dalian 116024, China
E-mail: huang@dlut.edu.cn

Abstract—The full preferential growth intermetallic compound (IMC) interconnects are fabricated on a (111) Cu single crystal substrate by the method named current driven bonding (CDB), and the morphology, orientation, electromigration resistance and mechanical properties of the full preferential growth Cu_6Sn_5 grains in the (111) Cu/IMC (30 μm Cu_6Sn_5)/Cu interconnects are investigated. The CDB method successfully controls the crystal orientation and maintains the preferential growth of Cu_6Sn_5 grains on (111) Cu single crystal substrate. The prism-type Cu_6Sn_5 grains show a texture feature and the continuous preferential epitaxial growth of Cu_6Sn_5 form the full IMC interconnect with $<11\bar{2}0>_{Cu_6Sn_5}$ directions paralleling to the current flowing direction. The fabrication of full preferential growth IMC interconnects provides an approach to unify the orientations of the IMC interconnects, which effectively eliminates the random distribution of grain orientations and thus the anisotropy of interconnects. The full (111) Cu/Cu_6Sn_5/Cu IMC interconnects exhibit an excellent electromigration resistance and high mechanical reliability even after having experienced high temperature aging and high current stressing. There is no obvious damage after aging and current stressing (2.0×10^4 A/cm²) at 150 °C and 180 °C even for 500 h. The average tensile strength of full preferential growth IMC interconnects remaines unchanged, i.e., 111.1 MPa and 108.1 MPa, even after aging at 150 °C for 500 h and current stressing (2.0×10^4 A/cm²) at 150 °C for 500 h, respectively, which are similar to that of the as-soldered state (118.8 MPa). This work is expected to provide theory support and guidance for the application of full preferential growth and high strength IMC interconnects in 3D IC packaging.

Keywords-current driven bonding (CDB); electromigration; preferential growth; IMC interconnect; mechanical property; 3D packaging

I. INTRODUCTION

To achieve the demands for high performance and multi-functions in electronic devices, the three-dimensional (3D) integrated circuits (IC) packaging technology is undergoing extensive research and development in the microelectronics industry [1]. In 3D packaging, micro bumps and through silicon vias (TSVs) are used for the interconnections of the stacked Si chips. With the downscaling of the electric devices, the size of the solder joints continues to decrease to microscale (about 1~100 μm). A full intermetallic compound (IMC) interconnect will be formed during the soldering reaction and the following service process. Since IMCs have higher melting points, better electrical and mechanical properties, etc., compared with traditional Sn-based solder joints, many kinds of researches are focused on the physical properties and reliabilities of the IMCs as an interconnect. Methods like transient liquid phase (TLP) bonding [2, 3] or solid-liquid-inter-diffusion (SLID) bonding [4] have been provided to form high-melting full IMC interconnects.

However, the grain orientations of the IMC interconnects prepared by the above bonding methods are usually random. The achievement of a uniform grain orientation IMC interconnects is expected to contribute a significant advantage to future 3D IC manufacturing. Moreover, with the advantages of no grain boundary defects, no impurities and low atomic diffusion rate, etc., single crystal Cu is introduced to be used as the substrate to form the preferential growth Cu_6Sn_5 IMC interconnects.

In our previous work, a novel method to form the preferred orientation IMC interconnect using single crystal Cu was proposed and reported at 2016 ECTC conference [5], which was named current driven bonding (CDB). The CDB method achieved the rapid growth of preferred orientation IMC on (001) Cu single crystal substrate to fabricate the full IMC interconnects with no voids or other defects at the interface, and the whole preferred orientation IMC interconnects exhibited an excellent electromigration resistance and mechanical properties even at high temperature and under high current density.

Similar to the (001) Cu substrate [6], the prism-type Cu_6Sn_5 grains also form on the (111) Cu single crystal substrate but with an intersecting angle of 60° [7]. Whether there were preferred orientation IMC grains formed on (111) Cu during current stressing was still unknown and the microstructural evolution during the current driven bonding (CDB) process was not clear yet, which needed further studies.

In the present work, the microstructural evolution of (111) Cu/Sn/Cu interconnects under a current density of 1.0×10^4 A/cm² at 300 °C was investigated. In order to compare with

978-1-7281-1500-9/19 $31.00 © 2019 IEEE

the electromigration effect, the reference (111) Cu/Sn/Cu interconnects were aged at the same temperature for the same time. The electromigration study focused on the growth kinetics, the morphological evolution and orientation relationship of interfacial Cu_6Sn_5 IMC and (111) Cu single crystal substrate. The electromigration resistance and the mechanical properties of the full preferential growth IMC interconnects under a current density of 2.0×10^4 A/cm^2 at 150 and 180 °C were also evaluated. These properties of the full preferential growth IMC interconnects were expected to provide a guidance for the promising applications in 3D IC packaging.

II. EXPERIMENTAL

The linear single crystal (111) Cu/Sn/polycrystalline Cu interconnects were prepared by immersion soldering. The thickness of solder (Sn) in the interconnects was 30 μm, which was controlled by spacers. The fabrication of the full preferential growth IMC interconnects using CDB method was described in the previous study [5, 8].

To reveal the growth progress of the whole IMC interconnect, the synchrotron radiation in situ real-time imaging technology was used, which was carried out at the Beamline BL13W1 of the Shanghai Synchrotron Radiation Facility. The resolution ratio and exposure time of charged couple device (CCD) camera were set as 0.325 μm/pixel and 3 s/frame, respectively, to catch and collect the images.

In order to evaluate the electromigration reliability of the full preferential growth IMC interconnects, the (111) Cu/IMC(Cu_6Sn_5)/Cu interconnects were current stressed under a current density of 2.0×10^4 A/cm^2 at 150 °C and 180 °C, respectively. In order to compare with the electromigration effect, the reference (111) Cu/IMC(Cu_6Sn_5)/Cu interconnects were aged at the same temperatures for the same time. The electromigration samples were immersed in silicone oil to keep a constant temperature and minimize oxidation of the sample surface.

The tensile strength of the full preferential growth IMC interconnects was measured using an Instron 5948 high precision materials testing system at a strain rate of 10^{-3} s^{-1} at 25 °C. Scanning electron microscope (SEM), energy dispersive X-ray (EDX) spectrometer and electron backscatter diffraction (EBSD) were used to examine the microstructural evolution, the growth behavior of interfacial IMCs and the orientation relationship between the (111) Cu single crystal substrate and the IMCs in the full preferential growth IMC interconnects.

III. RESULTS AND DISCUSSION

Fig. 1 shows the top-view of Cu_6Sn_5 grains on (111) Cu single crystal substrates in an as-soldered (111) Cu/Sn/polycrystalline Cu interconnect. Regular prism-type Cu_6Sn_5 grains on (111) Cu single crystal substrate elongating along three directions with an intersecting angle of 60° were observed, which exhibited a typical texture feature [7].

Fig. 2 shows the synchrotron radiation image of a 30 μm thick (111) Cu/Sn/polycrystalline Cu interconnect undergoing current stressing with a current density of 1.0×10^4 A/cm^2 at 300 °C for different times. The electrons flowed from the

polycrystalline Cu to the (111) Cu single crystal substrate. The white dashed lines showed the initial interfaces in the interconnect. After current stressing for 2 min, a layered IMC formed at the anode (111) Cu/Sn interface, while at the cathode interface, the IMC layer was too thin to be observed. With the increasing current stressing time, the interfacial IMCs at the anode continuously coarsened into large prism-type Cu_6Sn_5 grains. Compared with the high consumption of the polycrystalline Cu substrate, almost no consumption of the anode (111) Cu single crystal substrate was observed. Finally, the (111) Cu/Sn/polycrystalline Cu interconnect formed a full IMC interconnect after current stressing for 14 min. The preferential growth rate of Cu_6Sn_5 grains was 2.3 μm/min.

Fig. 3 shows the top-view morphology of prism-type Cu_6Sn_5 grains on the anode (111) Cu single crystal substrate and the cross-sectional EBSD maps of the (111) Cu/Sn/polycrystalline Cu interconnect after current stressing under 1.0×10^4 A/cm^2 at 300 °C for 5 min and 14 min, respectively. Fig. 3(a) shows that large faceted prism-type Cu_6Sn_5 grains on the anode (111) Cu single crystal substrate had a texture feature with an intersecting angle of 60°. Fig. 3(b) shows the EBSD maps of the (111) Cu/Sn/polycrystalline Cu interconnect in the rolling direction (RD), transverse direction (TD) and normal direction (ND) after current stressing for 5 min. Each color represents the orientation of one single crystal Cu_6Sn_5 grain in the direction. There were 6 Cu_6Sn_5 IMC grains in the (111) Cu/Sn/polycrystalline Cu interconnect after current stressing for 5 min, and the misorientations of these Cu_6Sn_5 IMC grains were around 60°. Fig. 3(c) shows the EBSD maps of the (111) Cu/Cu_6Sn_5/polycrystalline Cu full IMC interconnect after current stressing for 14 min. Similarly, the orientations of the 6 Cu_6Sn_5 grains exhibited a strong texture feature with the misorientations of 60°, which corresponds to the EBSD maps in Fig. 3(b). The Cu_6Sn_5 grains with different orientations formed on the (111) Cu single crystal substrate did not merge into one single crystal grain during the formation of the full IMC interconnects undergoing current stressing. However, the Cu_6Sn_5 grain orientations maintained during the continuous epitaxial growth towards the cathode. Fig. 3(d) shows the inverse pole figure of the Cu_6Sn_5 IMC grains in RD, indicating that the $<11\bar{2}0>_{Cu_6Sn_5}$ directions of the Cu_6Sn_5 IMC grains were parallel to the current flowing direction.

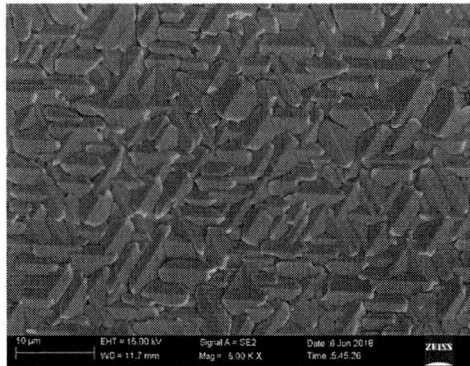

Figure 1. Top-view SEM image of the prism-type preferential growth of Cu_6Sn_5 grains on the as-soldered (111) Cu substrate.

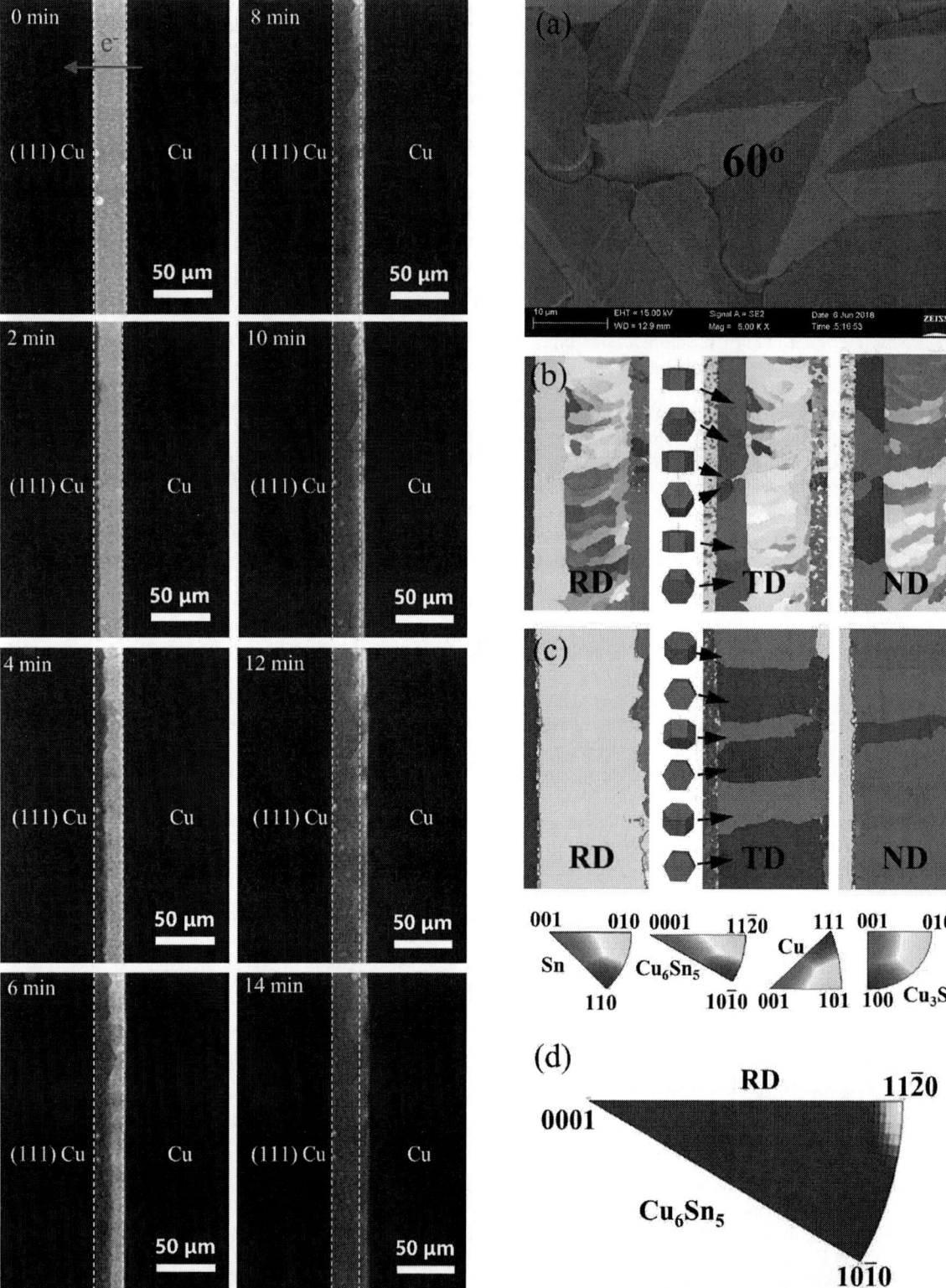

Figure 2. Synchrotron radiation images of preferential growth of Cu₆Sn₅ grains to form full (111) Cu/Cu₆Sn₅ IMC/Cu interconnect by CDB method.

Figure 3. (a) Top-view SEM image of prism-type Cu₆Sn₅ grains on the anode (111) Cu single crystal substrate after current stressing for 5 min; (b) and (c) cross-sectional EBSD maps of the (111) Cu/Sn/polycrystalline Cu interconnect after current stressing for 5 min and 14 min, respectively; (d) corresponding inverse pole figure of Cu₆Sn₅ in RD.

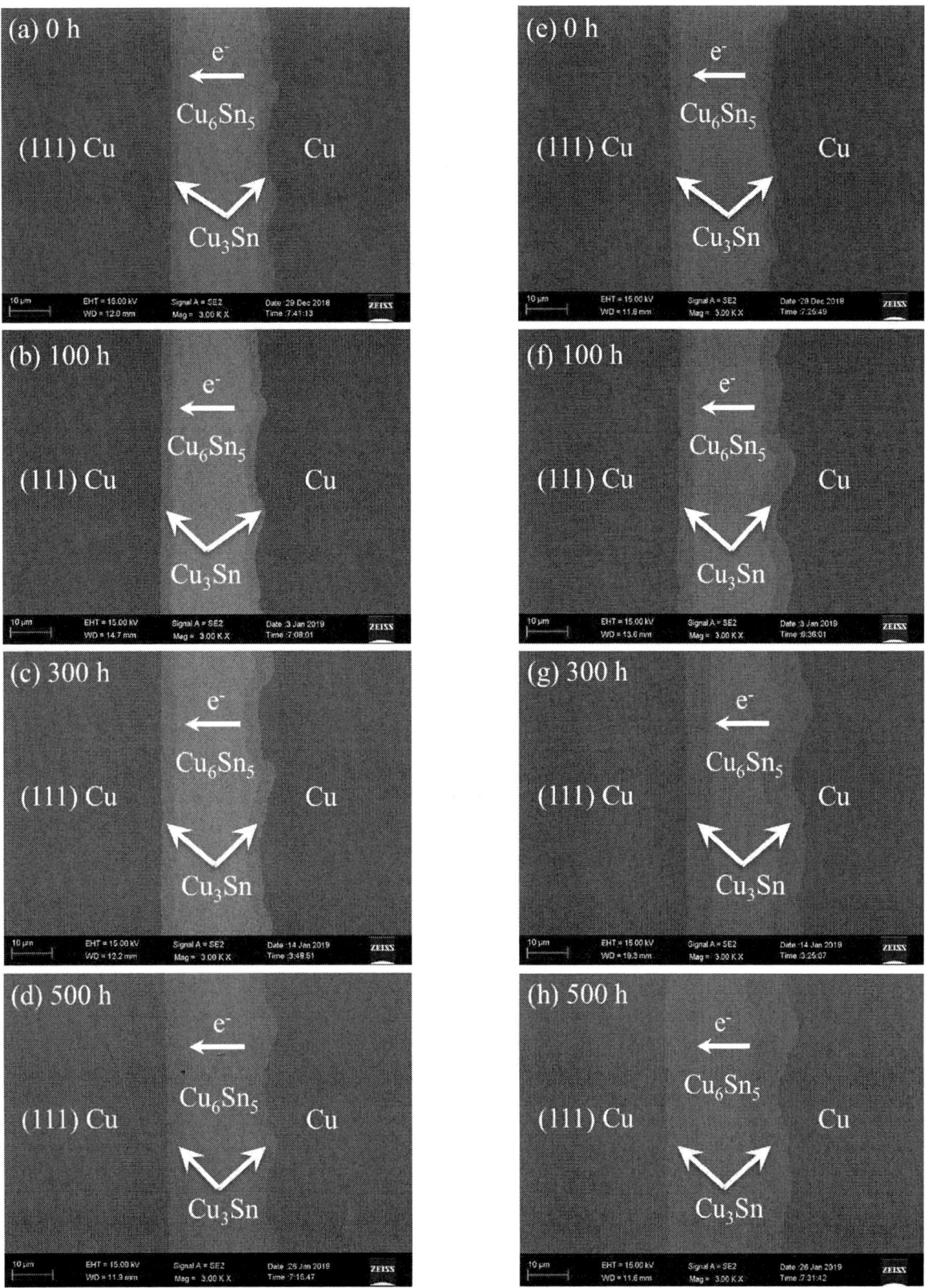

Figure 4. Cross-sectional SEM images of full preferential growth (111) Cu/Cu$_6$Sn$_5$/Cu IMC interconnects after aging at 150 ((a)-(d)) and 180 °C ((e)-(h)) for different times.

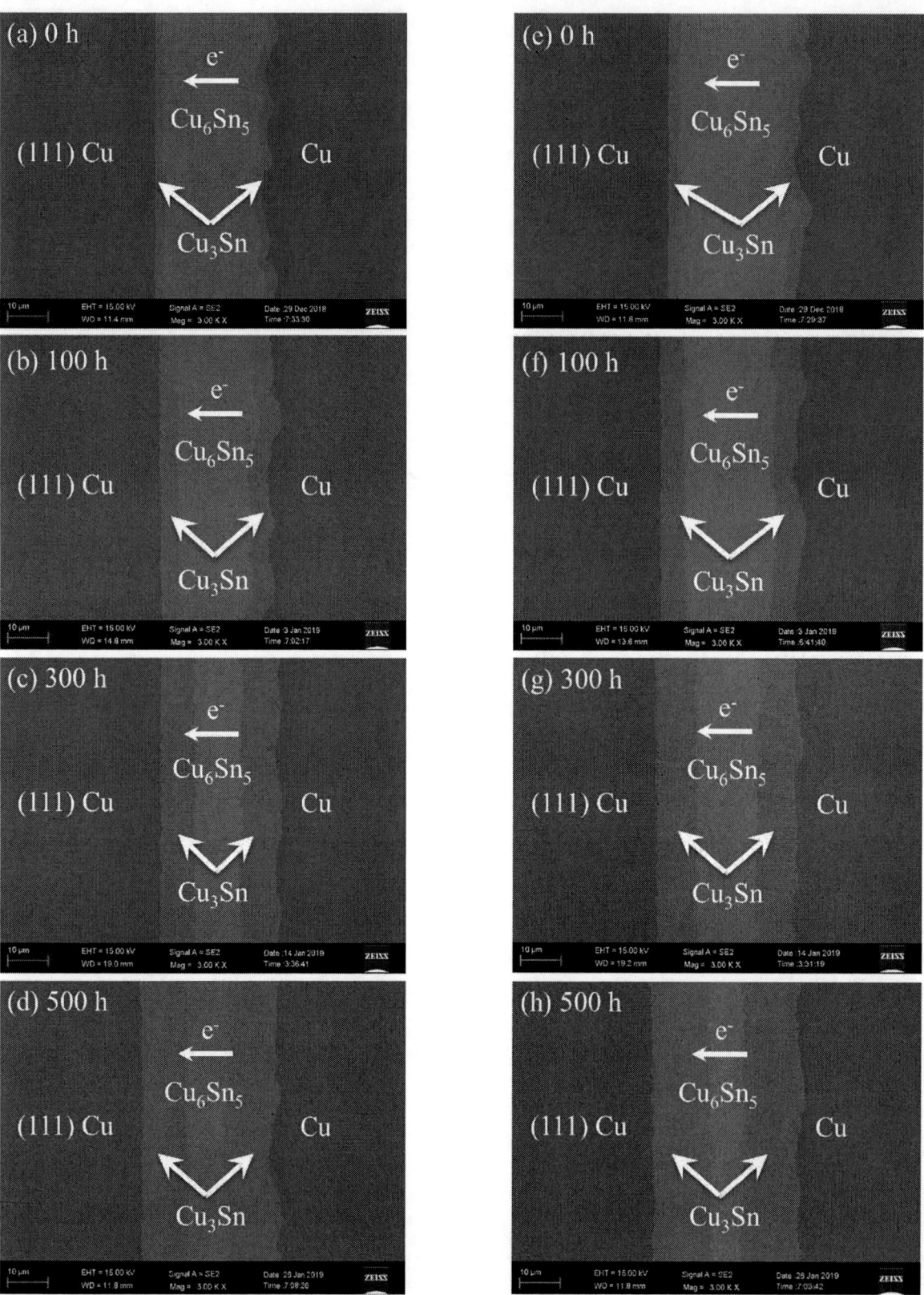

Figure 5. Cross-sectional SEM images of full preferential growth (111) $Cu/Cu_6Sn_5/Cu$ IMC interconnects after current stressing at 150 ((a)-(d)) and 180 °C ((e)-(h)) under 2.0×10^4 A/cm^2 for different times.

It is reported that the texture of Cu_6Sn_5 IMC grains would disappear and the prism-type Cu_6Sn_5 IMC grains would change into scallop-type IMC grains with increasing aging time [9]. However, in the present work, the CDB method successfully controlled the crystal orientation and maintained the preferential growth Cu_6Sn_5 grains on (111) Cu single crystal substrate undergoing current stressing. The preferential growth of prism-type Cu_6Sn_5 grains on (111) Cu single crystal substrate undergoing current stressing was maintained and thus the strong texture feature in the interconnect was kept.

Fig. 4 shows the cross-sectional SEM images of the full preferential growth (111) $Cu/Cu_6Sn_5/Cu$ IMC interconnects after aging for different times at 150 and 180 °C, respectively. As shown in Fig. 4(a) and Fig. 4(e), the as-soldered interconnects consisted of a Cu_6Sn_5 IMC and a thin Cu_3Sn IMC layer of 2 to 3 μm in thickness at both interfaces of polycrystalline Cu and (111) Cu single crystal substrates. The interfacial Cu_3Sn IMC layer aged at 180 °C was obviously thicker than that aged at 150 °C for the same aging time. Except for some tiny Kirkendall voids, no crack or damage was observed in the full preferential growth IMC interconnects even after aging at 180 °C for 500 h.

Fig. 5 shows the cross-sectional SEM images of the full preferential growth (111) $Cu/Cu_6Sn_5/Cu$ IMC interconnects after current stressing under 2.0×10^4 A/cm^2 for different times at 150 and 180 °C, respectively. Under current stressing, Cu_3Sn layers at both interfaces grew much faster than that under aging at the same temperature for the same time. The thickness of Cu_3Sn layers after current stressing at 180 °C increased obviously compared with those at 150 °C. Similarly, except for some tiny Kirkendall voids, no crack or electromigration-induced damage was observed in the IMC interconnect even after current stressing at 180 °C for 500 h.

Fig. 6 shows the evolution of interfacial Cu_3Sn IMC layer at both the anode and cathode interfaces in the full preferential

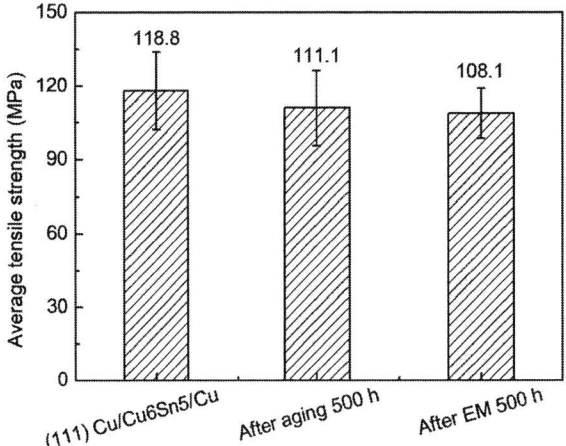

Figure 7. Average tensile strength of as-soldered full preferential growth (111) $Cu/Cu_6Sn_5/Cu$ IMC interconnects and that after aging and current stressing.

growth (111) $Cu/Cu_6Sn_5/Cu$ IMC interconnects during aging (150 and 180 °C) and current stressing (150 and 180 °C, 2.0×10^4 A/cm^2). After aging at 150 °C for 500 h, the average thicknesses of interfacial Cu_3Sn IMCs at the anode and cathode increased by 1.76 ± 0.4 μm and 2.05 ± 0.3 μm, respectively. The average increasing thicknesses of the interfacial Cu_3Sn IMCs at the anode and cathode undergoing current stressing were 8.55 ± 0.5 μm and 8.78 ± 0.3 μm, respectively. The interfacial Cu_3Sn IMC growth was accelerated undergoing current stressing compared with that undergoing aging.

After aging at 180 °C for 500 h, the average thicknesses of interfacial Cu_3Sn IMCs at the anode and cathode increased by 6.11 ± 0.3 μm and 6.61 ± 0.3 μm, respectively. After current stressing at 180 °C for 500 h, the average increasing thicknesses of the interfacial Cu_3Sn IMCs at the anode and cathode were 11.96 ± 0.5 μm and 11.71 ± 0.5 μm, respectively. Similarly, current stressing accelerated the growth of interfacial Cu_3Sn IMCs.

According to our previous study, the high temperature and high current density promoted the growth of interfacial Cu_3Sn IMCs in the full (001) $Cu/Cu_6Sn_5/Cu$ IMC interconnects, and the phenomenon was also found in the full preferential growth (111) $Cu/Cu_6Sn_5/Cu$ IMC interconnects. Moreover, no "polarity effect" was observed in the growth of the Cu_3Sn IMCs undergoing current stressing.

Fig. 7 shows the average tensile strength of the as-soldered full preferential growth (111) $Cu/Cu_6Sn_5/Cu$ IMC interconnect and that after aging (150 °C) and current stressing (150 °C, 2.0×10^4 A/cm^2) for 500 h. The average tensile strength of the as-soldered full preferential growth (111) $Cu/Cu_6Sn_5/Cu$ IMC solder interconnects was 118.8 MPa. The average tensile strength of the full preferential growth (111) $Cu/Cu_6Sn_5/Cu$ IMC interconnects after aging and current stressing at 150 °C for 500 h were 111.1 MPa and 108.1 MPa, respectively, which almost had no reduction compared with that of the as-soldered (111) $Cu/Cu_6Sn_5/Cu$ IMC interconnects,

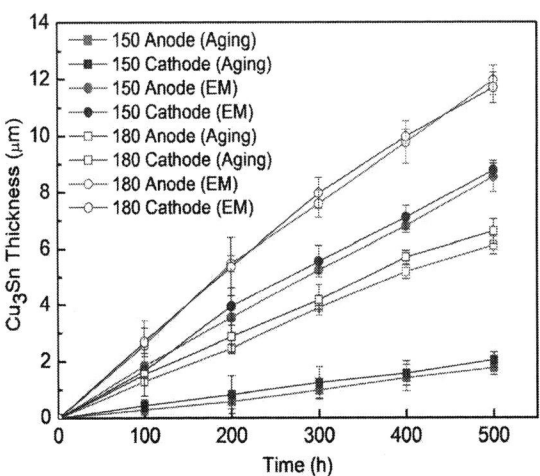

Figure 6. Growth kinetics of interfacial Cu_3Sn at both cathode and anode in the full preferential growth (111) $Cu/Cu_6Sn_5/Cu$ IMC interconnects during aging (150 and 180 °C) and current stressing (150 and 180 °C, 2.0×10^4 A/cm^2).

978-1-7281-1500-9/19 $31.00 © 2019 IEEE

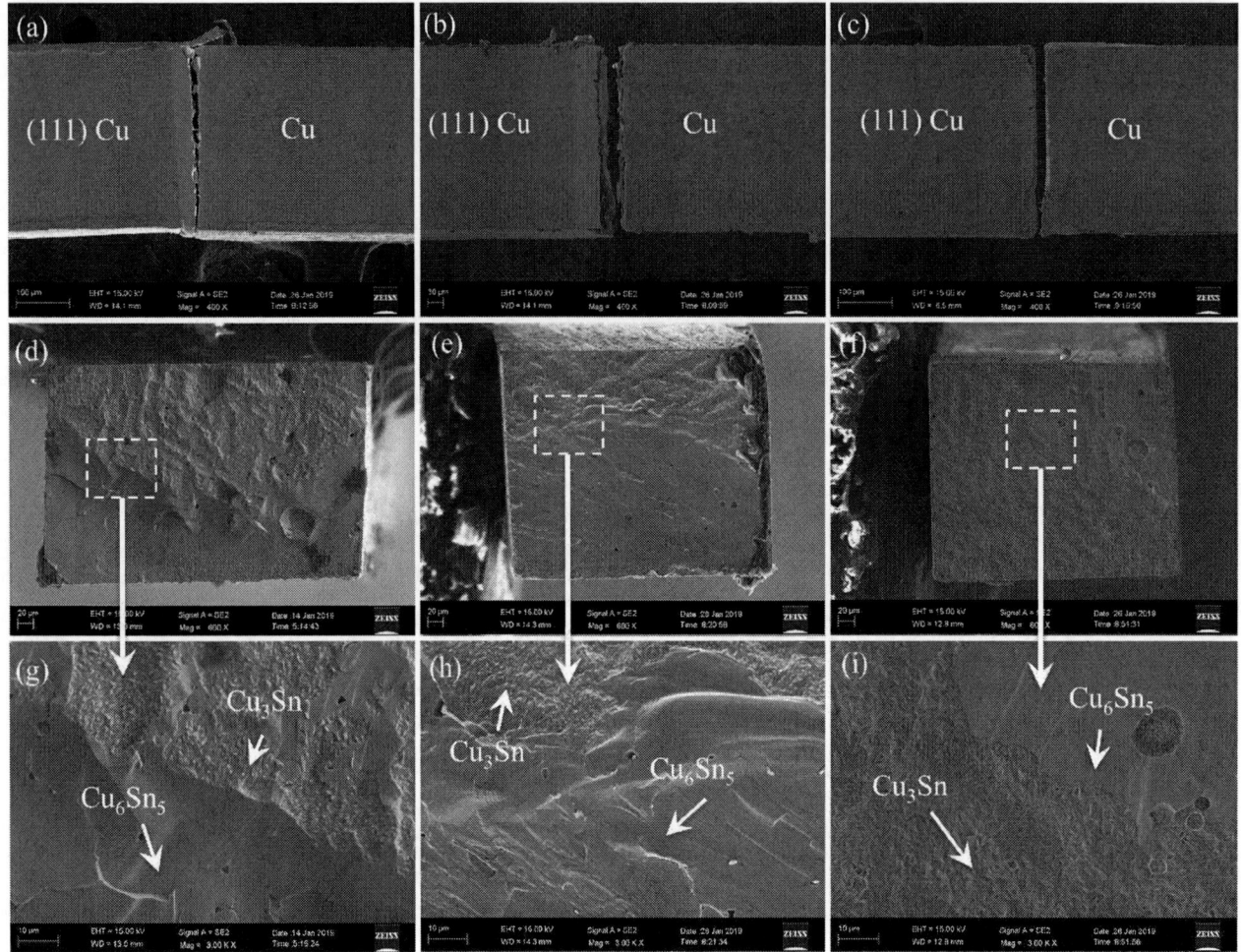

Figure 8. Fractographs of full preferential growth (111) Cu/Cu$_6$Sn$_5$/Cu interconnects: (a) and (d) as-soldered, (b) and (e) after aging for 500 h, (c) and (f) after current stressing for 500 h, (g)-(i) higher magnification.

showing an excellent electromigration resistance at high temperature and under high current density.

Fig. 8 shows the fractographs of the as-soldered (111) Cu/Cu$_6$Sn$_5$/Cu IMC interconnect and that after aging (150 °C) and current stressing (150 °C, 2.0×10^4 A/cm^2) for 500 h. As shown in Figs. 8(a)-8(c), the failure mainly occurred at the Cu$_6$Sn$_5$/Cu$_3$Sn interface of polycrystalline Cu side. The failure modes of the (111) Cu/Cu$_6$Sn$_5$/Cu IMC interconnects under the above three conditions were all brittle fracture.

The interface of two phases is the weak position where stress concentration and strain discontinuity usually occurred during the tensile process, due to the mismatch in material elastic modulus. Since the difference in elastic modulus between Cu substrate (106-115 GPa [10]) and Cu-Sn IMCs (Cu$_3$Sn: 132 GPa and Cu$_6$Sn$_5$: 119 GPa [11]) is large and the interfacial Cu$_3$Sn IMC is relatively thin, stress concentrates near the interfacial Cu$_3$Sn IMCs. As to the (111) Cu/IMC/Cu interconnect, both the Cu$_6$Sn$_5$ IMCs and the cathode Cu substrate consisted of several grains with different orientations. Since Cu$_6$Sn$_5$ and Cu have anisotropic properties, the elastic modulus varies in the grains with different

orientations. When loading tensile stress, initial fracture locations will form at the larger mismatch locations. Moreover, under current stressing, Kirkendall voids moved from the anode (111) Cu single crystal substrate to the cathode Cu substrate through the grain boundaries of the Cu$_6$Sn$_5$ IMCs and accumulated at the cathode Cu$_6$Sn$_5$/Cu$_3$Sn interface. So, the failure occurred at the interface of cathode Cu$_6$Sn$_5$/Cu$_3$Sn IMCs.

IV. CONCLUSION

The full preferential growth (111) Cu/Cu$_6$Sn$_5$ (30 μm Cu$_6$Sn$_5$)/Cu IMC interconnects were successfully fabricated on a (111) Cu single crystal substrate by the CDB method, which maintained the preferential epitaxial growth of Cu$_6$Sn$_5$ grains on (111) Cu single crystal substrate to form the full IMC interconnects.

The prism-type Cu$_6$Sn$_5$ grains exhibited a texture feature with an intersecting angle of 60° and the epitaxial growth of Cu$_6$Sn$_5$ IMC grains followed a relationship with the current flowing direction, i.e., <11$\bar{2}$0>$_{Cu6Sn5}$ directions were parallel to the current flowing direction.

978-1-7281-1500-9/19 $31.00 © 2019 IEEE

The fabrication of full preferential growth IMC interconnects provided an approach to unify the orientations of the solder (IMC) interconnects, which effectively eliminated the random distribution of grain orientations and thus the anisotropy of interconnects.

The full preferential growth (111) $Cu/Cu_6Sn_5/Cu$ IMC interconnects exhibited an excellent electromigration resistance and high mechanical reliability even after having experienced high temperature aging and high current stressing. There was no obvious damage after aging and current stressing (2.0×10^4 A/cm^2) at 150 °C and 180 °C even for 500 h. The average tensile strength of full preferential growth IMC interconnects remained unchanged, i.e., 111.1 MPa and 108.1 MPa, even after aging at 150 °C for 500 h and current stressing (2.0×10^4 A/cm^2) at 150 °C for 500 h, respectively, which were similar to that of the as-soldered state (118.8 MPa).

Finally, the theoretical analyses and experimental results demonstrated a novel way to produce the full preferential growth and high strength IMC interconnects, which is expected to guide the practical application in industry.

ACKNOWLEDGMENT

This work was supported by the National Natural Science Foundation of China (Grant Nos. 51671046 and U1837208), the Fundamental Research Funds for the Central Universities (Grant No. DUT17ZD202), and the BL13W1 beam line of Shanghai Synchrotron Radiation Facility (SSRF).

REFERENCES

[1] H. Y. Hsiao, C. M. Liu, H. W. Lin, C. L. Lu, Y. S. Huang, C. Chen, and K. N. Tu, "Unidirectional growth of microbumps on (111)-oriented and nanotwinned copper," Science, vol. 336, May. 2012, pp. 1007-1010, doi:10.1126/science.1216511.

[2] K. Chu, Y. Sohn, and C. Moon, "A comparative study of Cn/Sn/Cu and Ni/Sn/Ni solder joints for low temperature stable transient liquid phase bonding," Scripta Materialia, vol. 109, Dec. 2015, pp. 113-117, doi:10.1016/j.scriptamat.2015.07.032.

[3] W. L. Chiu, C. M. Liu, Y. S. Haung, and C. Chen, "Formation of plate-like channels in Cu_6Sn_5 and Cu_3Sn intermetallic compounds during transient liquid reaction of Cu/Sn/Cu structures," Materials Letters, vol. 164, Feb. 2016, pp. 5-8, doi:10.1016/j.matlet.2015.10.056.

[4] H. Huebner, S. Penka, B. Barchmann, M. Eigner, W. Gruber, M. Nobis, S. Janka, G. Kristen, and M. Schneegans, "Microcontacts with sub-30 μm pitch for 3D chip-on-chip integration," Microelectronic Engineering, vol. 83, Nov. 2006, pp. 2155-2162, doi:10.1016/j.mee.2006.09.026.

[5] M. L. Huang, Z. J. Zhang, F. Yang, and N. Zhao, "Novel growth of whole preferred orientation intermetallic compound interconnects for 3D IC packaging," Proc. Electronic Components and Technology Conference (ECTC), May 2016, pp. 1216-1221, doi:10.1109/ECTC.2016.8.

[6] J. O. Suh, K. N. Tu, and N. Tamura, "Dramatic morphological change of scallop-type Cu_6Sn_5 formed on (001) single crystal copper in reaction between molten SnPb solder and Cu," Applied Physics Letters, vol. 91, Jul. 2007, pp. 051907, doi:org/10.1063/1.2761840.

[7] H. F. Zou, H. J. Yang, and Z. F. Zhang, "Morphologies, orientation relationships and evolution of Cu_6Sn_5 grains formed between molten Sn and Cu single crystals," Acta Materialia, vol. 56, Jun. 2008, pp. 2649-2662, doi:org/10.1016/j.actamat.2008.01.055.

[8] M. L. Huang, L. Zou, S. Q. Yin, "Electromigration behavior and mechanical properties of the whole preferred orientation intermetallic compound interconnects for 3D packaging," Proc. Electronic Components and Technology Conference (ECTC), California, USA, May 2018, pp. 2035-2042, doi:10.1109/ECTC.2018.00306.

[9] H. W. Lin, C. L. Lu, C. M. Liu, C. Chen, D. Chen, J. C. Kuo, and K. N. Tu, "Microstructure control of unidirectional growth of η-Cu_6Sn_5 in microbumps on<111> oriented and nanotwinned Cu," Acta Materialia, Aug. 2013, vol. 61, pp. 4910-4919, doi:org/10.1016/j.actamat.2013.04.056.

[10] S. N. Dub, Y. Y. Lim, and M. M. Chaudhri, "Nanohardness of high purity Cu (111) single crystals: the effect of indenter load and prior plastic sample strain," Journal of Applied Physics, vol. 107, Feb. 2010, doi:10.1063/1.3290970.

[11] P. F. Yang, Y. S. Lai, S. R. Jian, J. Chen, and R. S. Chen, "Nanoindentation identifications of mechanical properties of Cu_6Sn_5, Cu_3Sn, and Ni_3Sn_4 intermetallic compounds derived by diffusion couples," Materials Science and Engineering: A, vol. 485, Jun. 2008, pp. 305-310, doi:10.1016/j.msea.2007.07.093.

Telemetry for Implantable Biosensors

Ryan B. Green
Department of Electrical and Computer Engineering
Virginia Commonwealth University
Richmond, VA, USA
greenrb@vcu.edu

Erdem Topsakal
Department of Electrical and Computer Engineering
Virginia Commonwealth University
Richmond, VA, USA
etopsakal@vcu.edu

Abstract— This paper presents the design, fabrication, and test of antennas intended to be used for implantable biosensors. With the rising number of Americans managing chronic diseases, implantable biosensors offer better access to vital information for the management of these diseases. Diabetes, particularly, is managed by many through the means of a finger prick monitor, offering only a snap shot of blood glucose at a particular time without showing trends of a rise or fall in the glucose levels. Implantable sensors not only offer these trends via a continuous monitor, a fully implantable sensor offers these trends without the inconvenience of a monitoring system extending outside of the human body. Such appendages hinder the normal activity of patients monitoring their chronic diseases. Due to their full implantation, the sensor monitor needs to broadcast the analyte date from the human body outward to an external data collector. This broadcast requires an antenna made of a biocompatible antenna that operates on frequency bands allocated for medical telemetry as well as frequency bands allocated for unlicensed communication (e.g. WiFi and Bluetooth). This paper presents antennas designed using a biocompatible material Titanium Nitrite (TiN) for subcutaneous implant. The antennas presented operate on the Wireless Medical Telemetry System (WMTS) frequency band and the 2.4 GHz Industrial, Scientific, and Medical (ISM) frequency band.

Keywords- Titanium Nitrite, Implantable, Biosensor, Antenna, WMTS

I. INTRODUCTION

Nearly 70% of deaths in the United States are due to chronic diseases such as cancer, heart disease, and diabetes [1]. Particularly, it is estimated that 30.3 million adults within the United States have diabetes, a quarter of whom do not know they have the disease [3]. Each patient that is diagnosed manages prescriptions and monitoring regimens to keep blood glucose within healthy levels. The most common method to monitor glucose levels is through a finger prick monitor. While the industry standard for glucose monitoring, the finger prick monitors only offer a snap shot of what the glucose level is doing over time. With the finger prick method, it is hard, if not impossible, to show trends over hours without an inordinate amount of money to spend on test strips and an inordinate amount of time spent to gather this data. Such trends are necessary in developing new

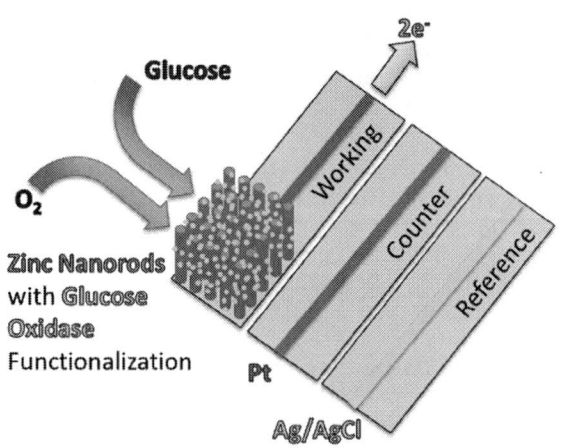

Figure 1. Glucose sensor diagram

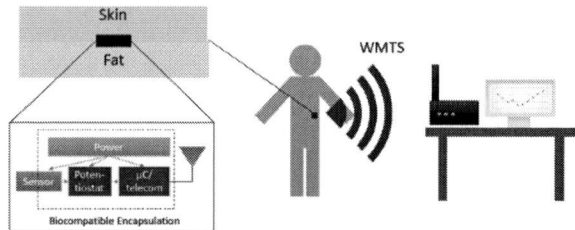

Figure 2. Implantable biosensor system diagram

medication, diet, and exercise regimens for diabetic patients. As a result, there is a need for a continuous monitoring system that can show trends over months, days, and even hours and minutes. Implantable biosensors offer this capability through an electrochemical process where when an analyte (e.g. glucose) is exposed to a corresponding oxidase, electrons are released (Figure 1). The more electrons are released in a corresponding fashion as the analyte concentration increases. Such a sensor can be monitored via an amperometric potentiostat and connected to a monitoring system. Additionally, if this system is fully implanted, this continuous monitor can collect data with little impact to the patient's normal activities. While fully implanted, this continuous data will need to be collected and transmitted from inside the body to a station nearby for use by doctors and the patient. A corresponding implantable antenna is necessary to link the internal sensor to an external monitoring and data acquisition center. Such a monitoring system flow chart can be seen in Figure 2.

When an implantable antenna is considered for a subcutaneous implant several factors need also be considered. One factor is the operation frequency band. The antennas used in implantable monitoring systems operate on the MedRadio band (401 MHz to 406 MHz) [7]. While the MedRadio band is excellent for deep tissue telemetry, this frequency band offers a challenge in antenna miniaturization for subcutaneous implant. As a result a new frequency band, operating on a smaller wavelength, can be explored as a potential alternative to MedRadio for such circumstances to aid antenna miniaturization. One such alternative is to use either existing ISM band, or the newly opened Wireless Medical Telemetry System band (608 MHz to 614 MHz, 1.395 GHz to 1.432 GHz). Another factor to consider is the materials used to fabricate such antennas. Traditionally, implantable antennas are fabricated using biologically incompatible metals (such as copper) and coated in a medically compatible material to isolate the antenna from the body. Unfortunately, many of these coatings have low dielectric properties compared to that of human tissue. Using the high permittivities of human tissues can help with antenna miniaturization. In order to accomplish this, biocompatible conductive materials need to be explored. One such material is Titanium Nitrite (TiN), is a highly conductive ($\sigma = 2 \times 10^5$ S/m to 2×10^7 S/m [4] [5]) and highly durable [6] biocompatible material used in medical implants. TiN is a promising candidate for an alternative material for implantable devices.

This study presents the design, simulation, and measurement of implantable antennas using TiN as a radiating conductor. Two antennas were explored, one operating on the WMTS band, and the other operating on the 2.4 GHz ISM band for Bluetooth telemetry. The WMTS antenna design was validated *in vitro* (via tissue mimicking gels). The ISM band antenna was validated *in vivo* (via porcine animal model).

II. ANTENNA DESIGNS

There are several conditions to take into account when designing a TiN antenna for medical implant. One condition is the antenna size. Because the antenna is being implanted into a dielectric material (human tissue), the operational wavelength changes. Therefore, in order for an antenna to continue to operate at the proper frequency needed, the antenna must be changed in size to compensate. The design wavelength of the implanted antenna is presented below in equation (1) [8]. In equation (1), λ is the design wavelength, λ_0 is the free space wavelength, μ_r is the relative permeability, and ε_r is the relative permittivity. In the context of human tissue, μ_r is assumed to always be 1.

$$\lambda = \lambda_0/(\mu_r \varepsilon_r)^{0.5} \tag{1}$$

Another consideration for the design of an implantable antenna made of TiN is the film thickness. Because TiN is grown on the order of 100 nm to 400 nm, the film is well below the skin depth at the WMTS band (approximately 10μm) and the skin depth at the 2.4 GHz ISM band (approximately 7 μm), the antenna has added resistance. The TiN film has a sheet resistance (R_S) that is calculated via equation (2). In equation (2), σ is the conductivity of the film, and t is the thickness of the film.

$$R_S = 1/\sigma t \tag{2}$$

This sheet resistance is used to determine the extra resistance added in thin film antennas. The added resistance is dependent on the shape of the antenna (see Figure 3). This resistance is calculated via equation (3), where R is the total resistance of the shape, l is the differential length, and $W(l)$ is the width of the shape at the length l. If the antenna takes the shape of a rectangle, the resistance reduces to R = R_S*L/W. The resistance is used in the simulation, Ansys HFSS, by dividing the total topology into various shapes, and applying the resistance of each shape via a "lumped RLC" boundary.

$$R = R_S/[dl/W(l)] \tag{3}$$

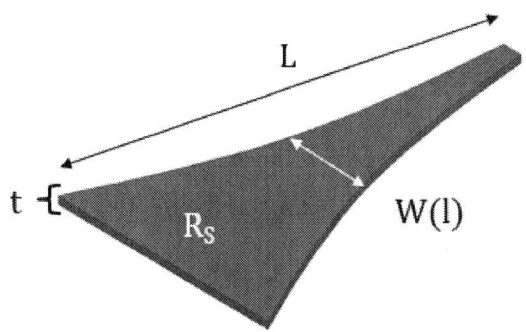

Figure 3. Antenna design on sapphire wafer

The antennas designed for this study are coplanar monopoles on a sapphire substrate (ε_r = 10, thickness of 1mm for ISM band antenna and 375 μm for WMTS antenna). One antenna is designed to operate on the WMTS band while the other is designed to operate on the 2.4 GHz ISM band. The antenna design template is shown below in Figure 4 and the antenna design attributes are detailed in Table 1. The TiN thickness used in the simulation is 400 nm and the conductivity is 1.4×10^7 S/m. The antenna is simulated to replicate subcutaneous placement with a 3mm thick skin(at 1 GHz, ε_r = 45.711 and σ = 88.181x10^{-2} S/m, at 3 GHz ε_r = 42.112 and σ = 1.9474 S/m [7]) layer surrounding the antenna. Additionally, the antenna was simulated to have a small layer of medical grade silicone (ε_r = 2.5) on the ground and port due to feeding the antenna with a brass U.FL connector. The brass component is not biocompatible and must be biologically isolated from the

978-1-7281-1500-9/19 $31.00 © 2019 IEEE   1783

environment it is implanted. The return loss resulting from these simulations are shown below in Figure 6.

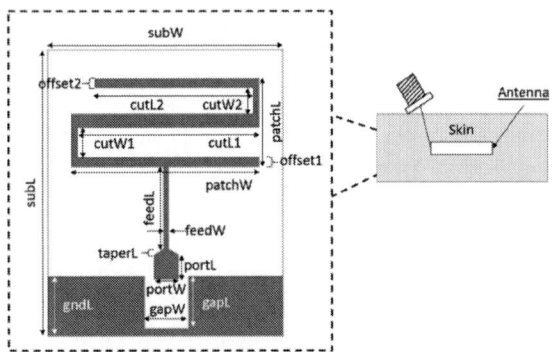

Figure 4. Antenna design on sapphire wafer

Table I. Antenna design attributes for WMTS and ISM designs

Attribute	Dual Band Design (mm)	ISM Band Design (mm)
cutL1	7.75	4.725
cutL2	7	4.725
cutW1	1.26	1
cutW2	1.12	1
feedL	3.45	1.5
feedW	0.2	0.5
gapL	2.2	2.2
gapW	1.9	1.9
gndL	2.5	2.5
offset1	0.4	0.5
offset2	0.35	0.5
patchL	3.68	3.3
patchW	8	5
portL	1	1
portW	1	1
subL	12	10
subW	10	10
taperL	0.25	0.25

Figure 5. Fabricated Antennas (Top dual band Design, Bottom 2.4 GHz ISM Design)

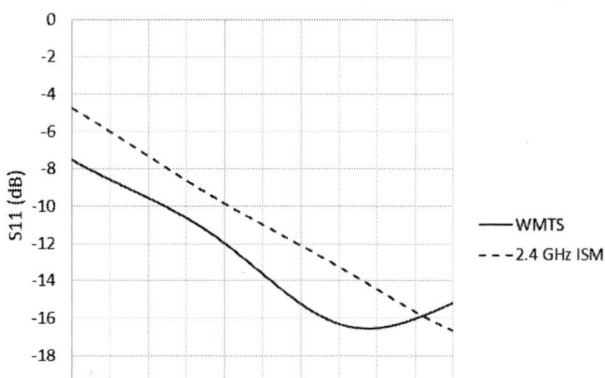

Figure 6. Simulated for WMTS and 2.4 GHz ISM Band antennas

The antennas were etched on a sapphire wafer with a 400 nm thick TiN thin film grown on the surface. After each antenna was etched, a UMCX surface mount connector (Pasternack PE44594) was connected to the antenna feed using a silver conductive epoxy, baked for 4 hours for curing, and connected to a 3 inch long coaxial cable (Pasternack PE3CA1002). Medical grade silicone (Silastic MDX4-4210) was applied to the connector (as described above) and to the coaxial cable and SMA connector intended to extent from the surgical suture in *in vivo* tests. In addition to the biocompatibility reasoning, the medical grade silicone adds extra structural support to SMA to coaxial joint and the UMCX to TiN joint. The fabricated antennas are shown in Figure 5.

III. IN VITRO TESTS

After fabrication, the dual antenna was tested in vitro in a skin mimicking gel. The gel is made from a mixture of triton-x, soap, water, gelatin-a, table salt, vegetable oil, and food coloring. The dielectric properties are shown in Figure 7. The relative permittivity of human skin ranges between 48.6 at 500 MHz to 33.5 at 10 GHz [7]. The conductivity of human skin ranges between 0.7 S/m at 500 MHz and 8.98 S/m at 10 GHz [7]. The relative permittivity of the gel ranges between 48.2 at 500 MHz to 28.9 at 10 GHz. The conductivity of the gel ranges between 0.46 S/m at 500 MHz to 10.7 S/m at 10 GHz. At 1.4 GHz, the relative permittivity of human tissue is 44.6 and the conductivity is 1.04 S/m [7]. At 1.4 GHz, the gel has a relative permittivity of 46.4 and a conductivity of 0.89 S/m. At 2.45 GHz, the relative permittivity of human skin is 42.8 and the conductivity is 1.61 S/m [7]. At 2.45 GHz, the relative permittivity of the gel is 44.3 and the conductivity is 1.60 S/m. In both cases for the WMTS band and 2.4 GHz ISM band, the dielectric properties of the human tissue and the tissue mimicking gel.

978-1-7281-1500-9/19 $31.00 © 2019 IEEE

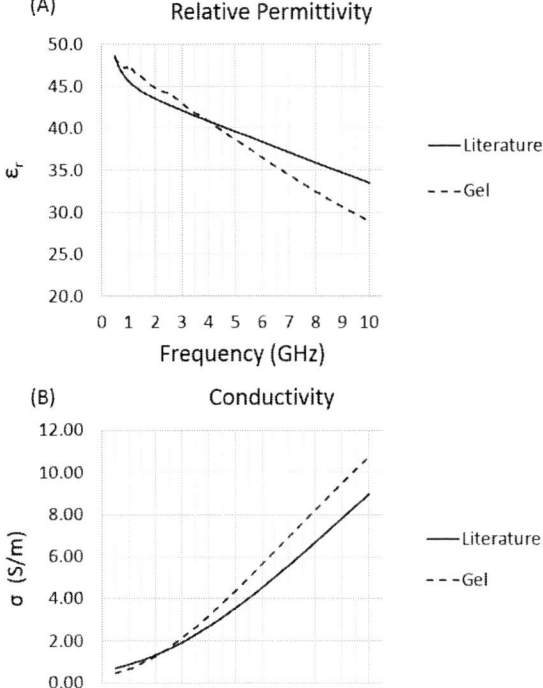

Figure 7. Dielectric properties of human skin compared to tissue mimicking gel (A) relative permittivity (B) conductivity

The antenna was inserted into the tissue mimicking gel and tested for both return loss and range testing (Figure 10). The implanted dual band antenna is shown in Figure 8 while the return loss is shown in Figure 9 and the link budget analysis is shown in Figure 11.

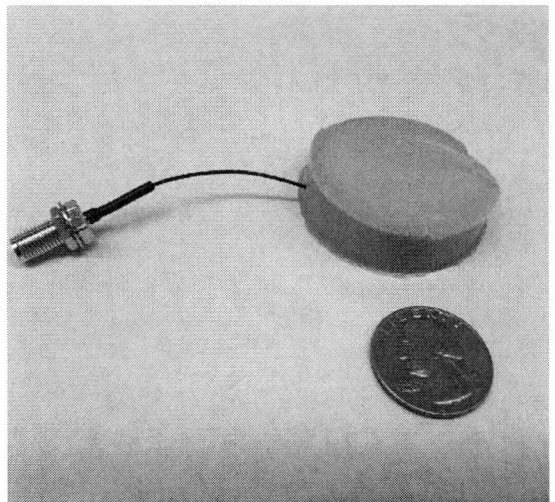

Figure 8. WMTS antenna implanted in tissue mimicking gel.

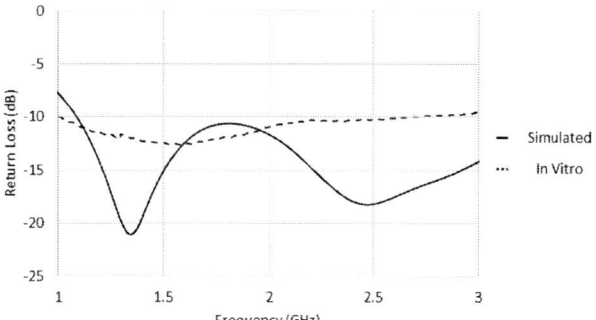

Figure 9. Return loss of dual band antenna tested *in vitro*.

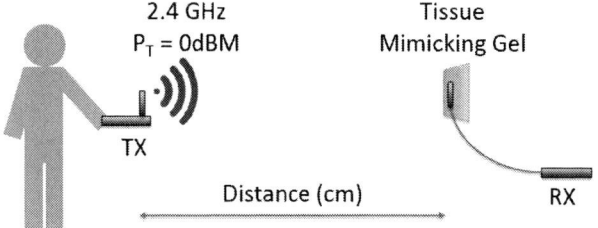

Figure 10. Link budget analysis test setup

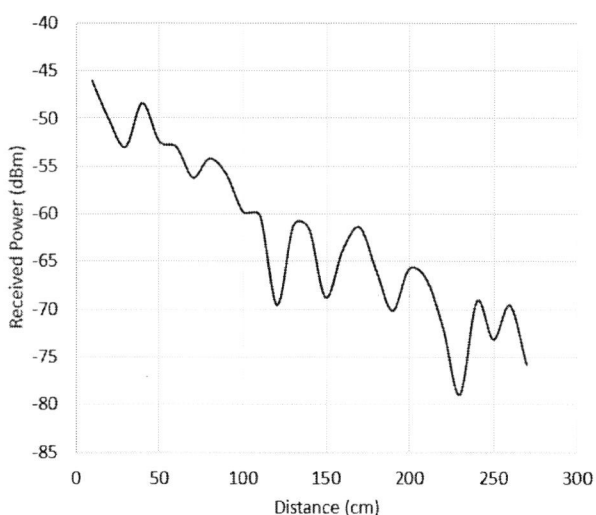

Figure 11. Link budget analysis of dual band antenna at 2.4 GHz antenna.

In Figure 9, the measured and simulated return loss results show some correlation with some variation between the two. The differences can be due to several factors including the small dielectric differences, feeding mismatches, and even some inconsistencies in film thickness. Despite these differences, the antenna operates sufficiently in both the 1.4 GHz WMTS band and 2.4 GHz ISM band. The budget link analysis of the antenna was measured through the Texas Instruments' CC2538EM evaluation boards. The implanted antenna was connected to one module set as a receiver. A monopole antenna was connected to the other module set as a transmitter, to broadcast at 2.405 GHz with a 0 dBm transmit power. The

978-1-7281-1500-9/19 $31.00 © 2019 IEEE

received power on the implanted antenna was measured over various distances between the transmit antenna and receive antenna. The results are shown in Figure 11. At a distance of 10 cm, the received power is -46 dBm. At a distance of 270 cm, the received power is -75.8 dBm. This design, tested *in vitro*, shows that this antenna has a sufficient transmission within the 2.7 meter test range.

Next, the 2.4 GHz ISM antenna was tested in vitro for its return loss. The antenna was inserted into the tissue mimicking gel and tested via a network analyzer. The results are shown in Figure 12. The measured results show a clear resonance on the 2.4 GHz ISM band frequencies (2.4 GHz to 2.5 GHz). At 2.4 GHz the return loss of the fabricated antenna is -16.5 dB and at 2.5 GHz the return loss of the antenna is -15.0 dB.

other module set as a transmitter, to broadcast at 2.405 GHz with a 0 dBm transmit power. The test setup for the link budget analysis in vivo is shown in Figure 15. The antenna was tested for its broadcast distance between 1 and 20 meters. The received power on the implanted antenna was measured over various distances between the transmit antenna and receive antenna. The measured power is seen below in Figure 16. At a distance of 1 meter, the received power was -66 dBm. At the distance of 8 meters, the receiver power dropped to -85 dBm. After 8 meters, the receiver power displays primarily an ambient signal (noise floor of the device is approximately -90 dBm) at 20 meters. While not adequate for long range communications, the antenna would performs adequately for shorter distances for potential body area networks (< 2 m).

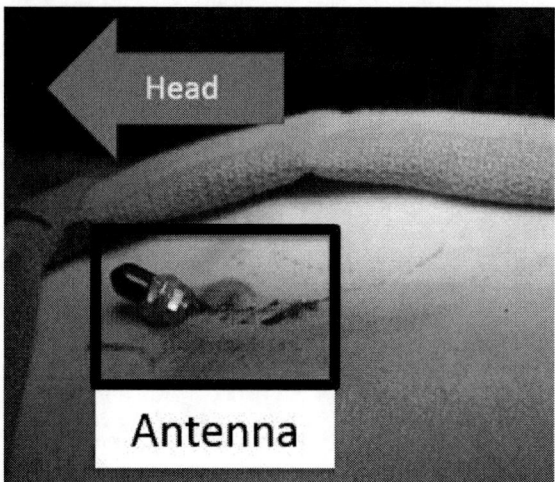

Figure 13. Antenna implanted in porcine animal model

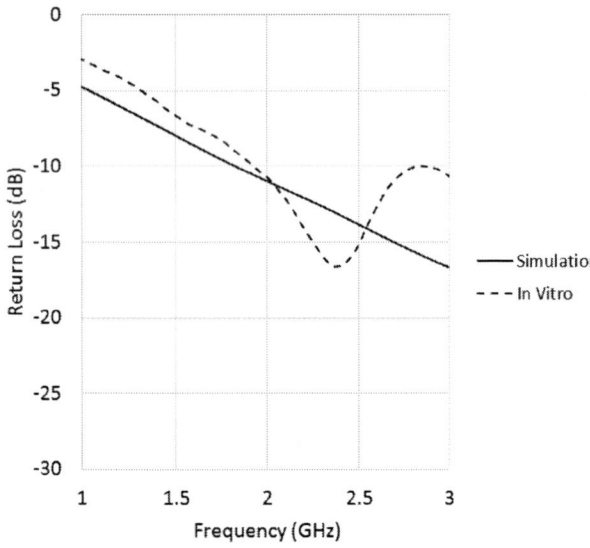

Figure 12. Return loss of 2.4 GHz ISM band antenna tested *in vitro*.

IV. IN VIVO TESTS

After fabrication, the 2.4 GHz antenna was sterilized in an autoclave chamber. The dual band antenna was not selected to move forward with the in vivo tests as the antenna substrate was deemed too thin to test with a porcine animal model without fracturing. After being sterilized, the antenna was surgically placed beneath the skin in a porcine animal model. The incision was sutured such SMA coaxial connector was exposed outside of the body for testing (Figure 13). The in vivo antenna return loss is shown in Figure 14 (dashed line). The measured and simulated return loss show a correlation with some variation between the two. This variation can be due to many factors including dielectric mismatch due to temperature, water content, or even due to the amount of blood in the incision area.

The budget link analysis of the antenna was measured through the Texas Instruments' CC2538EM evaluation boards. The implanted antenna was connected to one module set as a receiver. A monopole antenna was connected to the

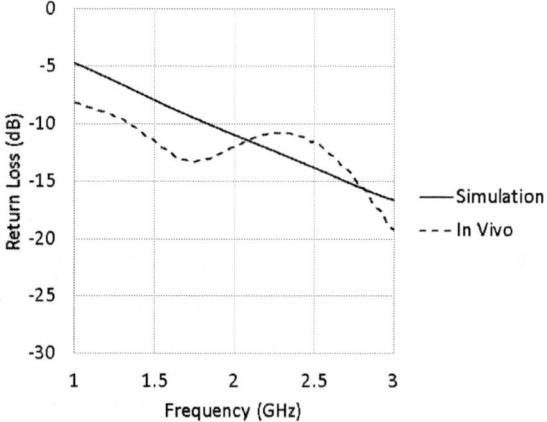

Figure 14. Return loss of ISM band antenna implanted tested *in vivo*.

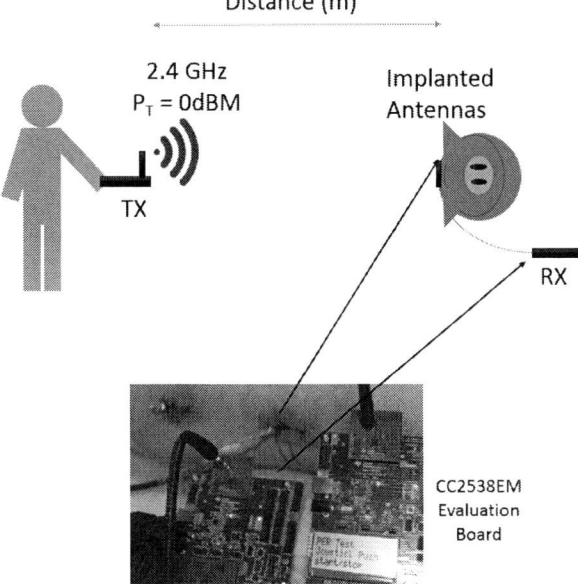

Figure 15. Meaured link budget analysis of impleanted antenna

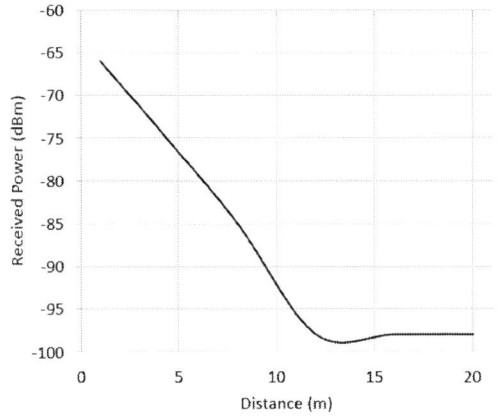

Figure 16. Meaured link budget analysis of impleanted antenna

V. CONCLUSION

In this paper, we present the design and fabrication of two biocompatible antennas for subcutaneous implant. In particular we present antennas made from a highly conductive, biocompatible material called Titanium Nitrite (TiN). The antennas were designed to operate on the 2.4 GHz ISM band and the newly proposed Wireless Medical Telemetry System (WMTS) band. The return loss and link budget shows that the fabricated antennas operate adequately within the 2.4 GHz band. While the broadcast distance is low, the antennas are adequate for local body area network distances (< 2 meters) for wearable medical sensors.

Future research aims for implantable Titanium Nitrite antennas include different antenna designs, alternative substrate materials, and other various analytes than glucose. One of the bottlenecks of using thin films for implantable antennas is the extra losses due to the thickness of the films. Extra resistances added due to conductor losses add to the low gain already present in implantable antennas. In the same vain, alternative fabrication methods need to be explored. The methods used to grow and etch the antenna do not yield the ability to produce multi-layer designs (e.g. Patch antennas) for implant. Alternative topologies, other than coplanar designs, could yield a higher efficiency and allow for longer broadcast distances. In the same vain, new substrate materials and fabrication methods for TiN antennas could yield smaller antennas for medical implant. Additionally, further research is needed in the context of long term medical implant of the antennas and how animal models respond to the long term implant of such antennas. Such research questions including encapsulation effect on antenna performance and local toxicity of TiN antennas needs to be explored as the technology advances Additionally, SAR and cellular health (stress or death) need to be investigated for long term implantable antennas at various operation frequencies and implantation sites. Lastly, TiN antennas need to be explored with different frequency bands for deep tissue telemetry as well as commercial applications ranging from consumer electronics to military and space applications. While more research needs to be done to better understand the antenna, the implantable wireless analyte sensor method can extend further than the application of glucose to the monitoring of L-lactate, blood alcohol, and various other vital signs.

REFERENCES

[1] Center for Disease Control and Prevention, "About Chronic Diseases," accessed Nov 7 2018, available: https://www.cdc.gov/chronicdisease/about/index.htm

[2] T. Karacolak, A. Z. Hood and E. Topsakal, "Design of a Dual-Band Implantable Antenna and Development of Skin Mimicking Gels for Continuous Glucose Monitoring," in IEEE Transactions on Microwave Theory and Techniques, vol. 56, no. 4, pp. 1001-1008, April 2008

[3] Center for Disease Control and Prevention, " About Diabetes," accessed Nov 7 2018, https://www.cdc.gov/diabetes/basics/diabetes.html

[4] W. M. Mohammed, A.I. Gumarov, I.R. Vakhitov, I.V. Yanilkin, A.G. Kiiamov, S.I. Nikitin, R.V. Yusupov, "Electrical properties of titanium nitride films synthesized by reactive magnetron sputtering," IOP Conference Seris: Journal of Physics: Conference Series 927, 2017.

[5] J. P. Carmo, N. S. Dias, H. R. Silva, P. M. Mendes, C. Couto and J. H. Correia, "A 2.4-GHz Low-Power/Low-Voltage Wireless Plug-and-Play Module for EEG Applications," in IEEE Sensors Journal, vol. 7, no. 11, pp. 1524-1531, Nov. 2007.

[6] Ruud P. van Hove, Inger N. Sierevelt, Barend J. van Royen, and Peter A. Nolte, "Titanium-Nitride Coating of Orthopaedic Implants: A Review of the Literature," BioMed Research International, vol. 2015, Article ID 485975, 9 pages, 2015.

[7] N Carrara, "Dielectric Properties of Body Tissues in the Frequency Range 10Hz – 100 GHz", Italian National Research Council, Institute for Applied Physics. (2015)

[8] F.T. Ulaby, E. Michielssen, U. Ravaioli, *Fundamentals of Applied Electromagnetics*, 6th ed., USA, Prentice Hall, 2010.

2019 IEEE 69th Electronic Components and Technology Conference (ECTC)

Ultra-thin QFN-Like 3D Package with 3D Integrated Passive Devices

Ayad Ghannam[1*], Niek van Haare[2], Julian Bravin[3], Elisabeth Brandl[3], Birgit Brandstätter[4], Hannes Klingler[4], Benedikt Auer[4], Philippe Meunier[5], Sebastiaan Kersjes[2]

[1] 3DiS Technologies S.A.S, 478 rue de la Découverte – Mini Parc 3 - CS 67624, F-31676 Labège, France
[2] Besi Netherlands, B.V., 6921 RW Duiven, The Netherlands
[3] EV Group E.Thallner GmbH, DI Erich Thallner Str. 1, 4782 St. Florian am Inn, Austria
[4] Besi Austria GmbH, Innstrasse 16, 6241 Radfeld, Austria
[5] NXP Semiconductors France, 2 esplanade Anton Philips - BP 20000, F-14906 CAEN, France
*Email: ghannam@3dis-tech.com

Abstract— In this work, a new wafer-level 3D packaging technology is developed to enable integration of an ultra-thin QFN-like (quad-flat no-leads) 3D package that targets both effective electrical and thermal properties and a thickness smaller than 200 µm. The proposed architecture allows 3D interconnection of stacked staggered dies and integration of compact, high-performance 3D integrated passive devices inside the package for added functionality and electrical performance. The developed technology consists of using debonding from a temporary carrier, Cu 2D-RDL (Redistribution Layer), accurate thin die pick & place, 3D-RDL and overmolding processes to integrate a QFN-like 3D package. Interconnection between die and package I/O is achieved using conformal 3D-RDL, thus without wire-bond, flip-chip or TSV.

Keywords- 3D-WLSiP; 3D-WLP; 3D-RDL; 3D-IPD; High-Q; QFN-like; Ultra-thin package; WLP.

I. Introduction

Today, electronics is everywhere, in computers, tablets, mobile phones, cars, home appliances, IoT objects, etc. To communicate, all these devices need to have an RF communication module. The main challenges for RF modules which integrate switches, a low noise amplifier, and a power amplifier to be used for the high-volume consumer market, are small size, small thickness, low cost, outstanding (or good) RF performance, and good thermal performance.

The latter two challenges are mainly dependent on packaging technology and a trade-off between losses up to 6 GHz and thermal dissipation. Heatsink very-thin quad-flat no-leads (HVQFN) solutions on one hand are known to be really effective in terms of thermal properties. High frequency applications, however, suffer from high insertion losses and poor matching due to wire-bonds. On the other hand, flip-chip based solutions are really effective from an RF point of view but have bad thermal properties. Excellent RF grounding is key for the successful implementation of many RF functions, but the best solution in terms of grounding is generally the worst one from a thermal point of view. We present an innovative 3D wafer-level package (3D-WLP) technology combining a low-cost multilayer 3D-RDL

[1], ultra-compact, high-performance integrated passives [2] and a die in order to overcome this trade-off. This package makes it also possible to integrate a 3D system with planar or stacked (staggered) dies inside using 3D-RDL.

So far, 3D system packaging and 3D integrated passive device (3D-IPD) solutions have included a substrate (silicon, glass, flex, PCB…)[3], [4]. To further enhance electrical and thermal performances of system packages and to reduce their thickness, this work is looking at developing novel ultra-thin, substrate-less packages. Eliminating the substrate and integrating ultra-thin 3D stacked dies and compact 3D inductive devices inside will result in a package having a thickness that it is less than 200 µm.

This paper focuses on reporting a new package concept, material selection, process building blocks development and first results of successful demonstration of a 200 µm thick package with a 100 µm thick die inside as well as 3D integrated passive devices.

II. Package Concept & Process Steps

The QFN-like ultra-thin 3D package with 3D-IPD inside is formed using two thick copper layers, three dielectric layers and an epoxy molding compound (EMC) as shown in Fig. 1. Material choice and thicknesses are explained in section III.

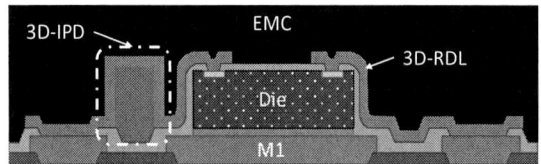

Figure 1. Cross-section view of QFN-like thin 3D package with 3D-IPD inside

In order to fabricate the package, a redistribution layer first (RDL-first) approach is employed. Thus, the process requires working on a temporary carrier with a coated layer of temporary bonding/debonding material (TBDB). The debonding method is selected to be UV laser-based, given its advantages for fast debonding of thin devices without introducing mechanical or thermal stress. After coating of the

978-1-7281-1500-9/19 $31.00 © 2019 IEEE
1789

TBDB material, a first interlayer dielectric (ILD1) is processed on top of the carrier to create openings for package I/O and thermal pad (Fig. 2.a). Afterwards, a planar thick copper layer is electroplated to form these pads (Fig. 2.b). Then, the dies are placed on top of the thermal pad (Fig. 2.c). A conformal dielectric layer (ILD2) is coated and processed to create openings over the die's and the package's I/O pads (Fig. 2.d). To integrate 3D-IPD devices (3D inductors), an additional thick dielectric layer (ILD3) is formed (Fig. 2.e). The connections between die and package and the 3D-IPD are formed using a conformal 3D-RDL layer (Fig. 2.f). Finally, the wafer is molded using transfer molding technique and laser debonded and cleaned (Fig. 2.g).

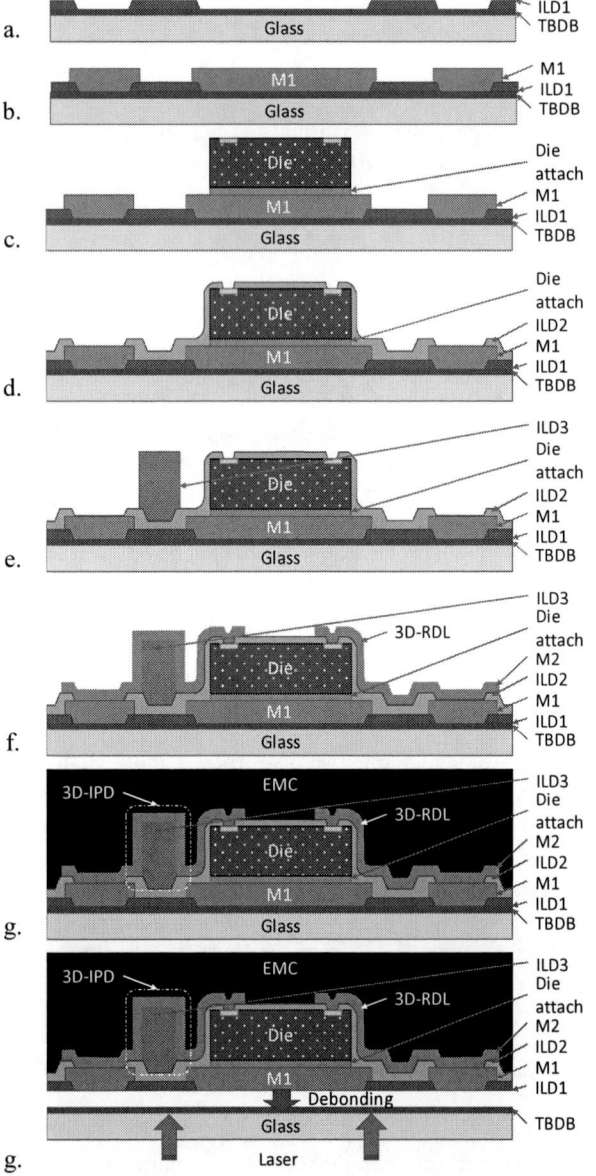

Figure 2. QFN-like 3D package with 3D-IPD process steps

Cleaning of the debonded wafer is dependent on the TBDB material. This step is explained in following section.

III. DEVELOPMENT OF TECHNOLOGY BRICKS

Development of technology bricks is performed first on prototype #0 which has a 100-μm-thick dummy die inside. This prototype #0 allowed to fix the overall process and layer thicknesses for prototype #1 which has a 150-μm-thick silicon die. For this prototype #1, a target thickness of 230 μm is set due to die thickness.

A. Planar RDL layer

The first RDL layer is used to form the I/O pads of the package as well as the heat spreader underneath the die. This layer is comprised of a 5-μm thick polymer dielectric layer (ILD1) and a 15-μm thick copper layer (M1). Although ILD1 can be omitted to reduce packaging cost, it was decided to be implemented to allow routing of unexposed RDL lines, such as the ones used to form the first (planar) half of 3D-IPD devices. The choice of M1 layer thickness was made to enhance both, electrical and mechanical performance of the package, as well as to remove the need of a UBM/ ENIG layer. For the initial version of the package, it was decided to use Direct Immersion Gold (DIG) as a wetting layer for package I/O as it yields good result according to Hashimoto et al. [5].

B. Pick & Place Process

Die placement accuracy is very critical for this wafer-level packaging (WLP) process as it impacts landing of 3D-RDL vias inside die's bond pads. The smaller the size of bond pad opening, the tighter the accuracy. Additionally, it is very important to have very low bleeding distance of die attach in order to improve 3D-RDL fidelity and RF electromagnetic (EM) simulation model as well as to reduce minimum distance between die edge and package I/O. A target distance smaller than 10 μm for bleeding was set for this step. Thus, using a dicing die attach film is more compatible with this requirement compared to liquid dispensed die attach.

The pick and place machine was set up for pick and place assembly. The dies back sides are laminated with conductive die attach foil. The process flow for direct attach pick and place is the following: Dies are ejected from the wafer with an ejector comprised of one plastic needle with large tip diameter to avoid damaging the die attach foil. Dies are directly picked by the bond head without flipping. Subsequently, the exact chip position with respect to the tool center point is determined by an up looking camera. Before the die placement, the exact bonding position on the wafer substrate is determined with the substrate camera. Finally, the die is placed and its position with respect to the substrate position is measured.

Process tests were done to determine the ideal process parameters. In peel tests, the wetting of the DAF is evaluated, and microscopy evaluation shows if bleeding is homogeneous around the die, see Fig. 3. The tests showed that bond head tilt and bond tool selection are critical for yielding good wetting results. Additionally, a substrate temperature of 80°C, a placement force of 1.5 N, and a placement time of 200 ms were determined to yield the best process results. Finally, a bleeding distance smaller than 6 μm is obtained.

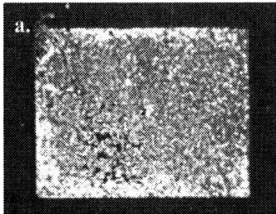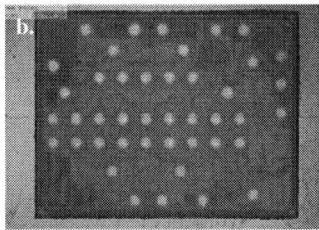

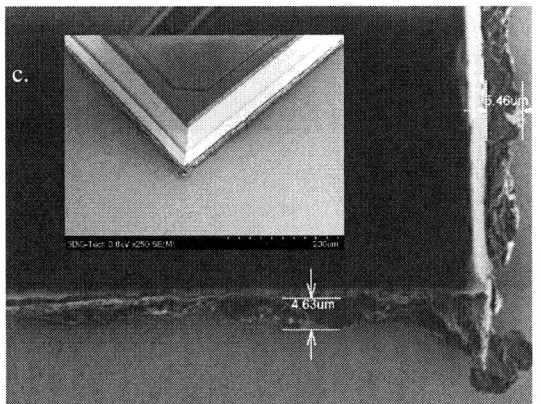

Figure 3. Wetting and bleeding results. a. Good wetting confirmed by die peel off, b. The bleeding around the die is symmetric, c. SEM image showing bleeding distance less than 6 μm

C. 3D-RDL for 3D interconnects & 3D-IPD process

The connection between the die and package I/O as well as integration of 3D part of 3D-IPD devices are done using conformal 3D-RDL technology [1]. This RDL layer is comprised of a 7.5 μm thick conformal 3D dielectric layer (ILD2) and a 15 μm thick 3D copper interconnect layer (M2).

The main challenge for ILD2 layer is to: 1/ achieve a uniform thickness on horizontal and vertical faces of the die with little resist accumulation at the interface between die and thermal pad while having a thickness at the top edges of the die 40% bigger than that of the layer; 2/ resolve small via openings, both, on package bottom and on die top, simultaneously. A spray-coating technique was used to apply the ILD2 layer. After both, deposition and lithography process optimization, a near vertical profile with good conformity was obtained with via opening sizes of 17.5 μm, as shown on Fig. 4. On the top edge of die, the thickness of ILD2 layer is 3.5 μm, thus, achieving the target objective.

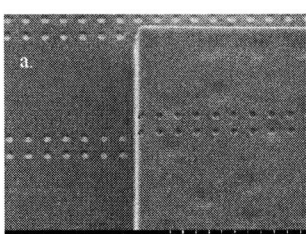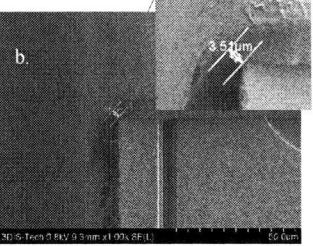

Figure 4. SEM micrographs of conformably coated ILD2 layer with 17.5 μm resolved vias (a) and its 3D thickness profile (b)

After successful development of the conformal ILD2 layer, the copper 3D interconnects process was optimized to achieve a line/space resolution of 15/15 μm over a vertical step of 160 μm (150μm thick die + 10 μm die attach) as shown on Fig. 5. The figure image shows good definition of lines as well as a small rounding effect at the interface between die and thermal pad, indicating again that ILD2 had little resist accumulation at this area. This characteristic is very important as it allows to ease EM modeling of these 3D interconnects while increasing accuracy with real world result.

Figure 5. SEM micrograph of 15-μm line/space 3D-RDL over 160 μm vertical step

D. Overmolding process

1) Introduction of overmolding

Traditionally, wafer level transfer molding is done with mold caps larger than 200 μm thickness. The Epoxy Molding Compounds (EMC) used for these caps cure fast inside the mold and have relatively large filler content to prevent warpage. The latter results in high viscosity of the EMC.

For the ultra-thin package, discussed here, one has to make a mold cap of 100 μm or less above the dies. One also has to be able to fill the small features inside the 3D-RDL structure containing 15 μm thin slits between the features. To keep warpage under control and fill the thin cap and 3D-RDL features, a low warpage EMC with low viscosity is needed. This combination is not common for EMC's used for wafer level molding. Therefore, EMC selection is essential for this thin cap mold process.

Further, the pre-molding processes have been set-up as 4" test wafers are used, where common wafer level molding takes place at 8" or 12' wafer sizes. Hence, a test mold environment has been set-up for molding a 4" wafer in a standard strip based molding system. This test environment requires a carrier for the wafer and an adapted mold filling strategy for the selected EMC. This will be discussed in the molding strategy paragraph.

2) Material selection

In [6], it is demonstrated that EMC material is the key parameter for warpage control. The α_1 value is the Coefficient of Thermal Expansion (CTE) of the EMC below the glass temperature. The closer the α_1 value to the CTE of the carrier's material, glass, the less warpage one will get. Secondly, filler size and distribution in the compound have influence on the viscosity of the EMC and the resulting thin slit filling capability. Measurements of the viscosity of the EMC's under actual molding conditions have therefore been done with a Fico viscosity meter as described in [7]. Based on evaluations on dummy samples a viscosity < 10 Pa·s during 40 s is needed for filling of ~100 μm thin cap above the dies for prototype #0 and ~47.5 μm above prototype #1. From a pre-selection four EMC's have been characterized, see Fig. 6.

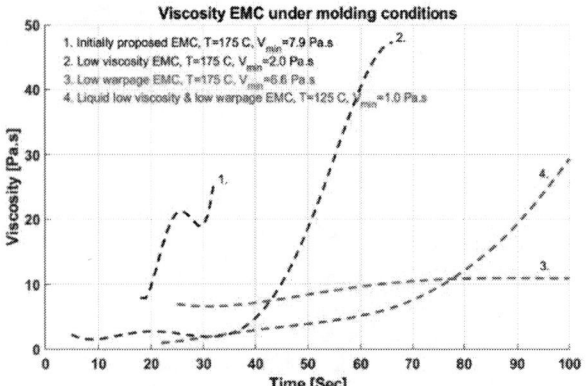

Figure 6. Viscosity in time of initial proposed EMC (1. black), low viscosity EMC (2. blue), low warpage EMC (3. magenta) and low viscosity & warpage liquid EMC (4.red)

In Fig. 6, one can see that the EMC No.1 cures within 13 s and is quickly above 10 Pa·s viscosity. EMC No.1 does not meet the viscosity criteria, $\mu < 10$ Pa·s for 40 s, to transfer the EMC over the wafer.

The low viscosity EMC No.2 cures slow enough, but has as disadvantage a high α_1 resulting in high warpage as has been demonstrated on a prototype sample.

The low warpage EMC No.3 has higher initial viscosity as the low viscosity EMC. The viscosity does stay long enough below 10 Pa·s to fill the cavity and small 3D-RDL features.

Finally, an EMC liquid at room temperature (#4) has been characterized. This liquid EMC combines low warpage and low viscosity. Liquid EMC does however need a new approach of dispensing the EMC into the molding machine and therefore will be part of future work.

The choice made is to use the low warpage EMC No.3 for prototype #1 on the 4" wafer, as theoretically its viscosity profile allows filling the cavity with <2 MPa pressure.

3) Molding strategy

Fig. 7 displays the molding sequence and discussed pre-processing. It starts with sample preparation where the 4" wafer is placed inside a rectangular steel carrier containing center pin holes for alignment (1.). To prevent the EMC from getting between the glass carrier wafer and the bottom mold, a Kapton foil is laminated to the bottom side of the wafer and carrier (2.). The sample is pre-baked for one hour at 125 °C to increase adhesion of the bottom foil (3.). Next, the sample is plasma cleaned with argon hydrogen to increase EMC adhesion and so reliability (4.). After this, the prepared sample is pre-heated 2 minutes inside the mold tool to thermally stabilize it (5.).

After sample preparation and loading, the molding process starts. First, the EMC is loaded in the transfer section

of the machine. The mold closes and plungers push the now molten EMC into the cavity. The now liquid EMC is pushed through the gate into the cavity. At this point the EMC has to flow over the wafer with the dies and 3D-RDL structure.

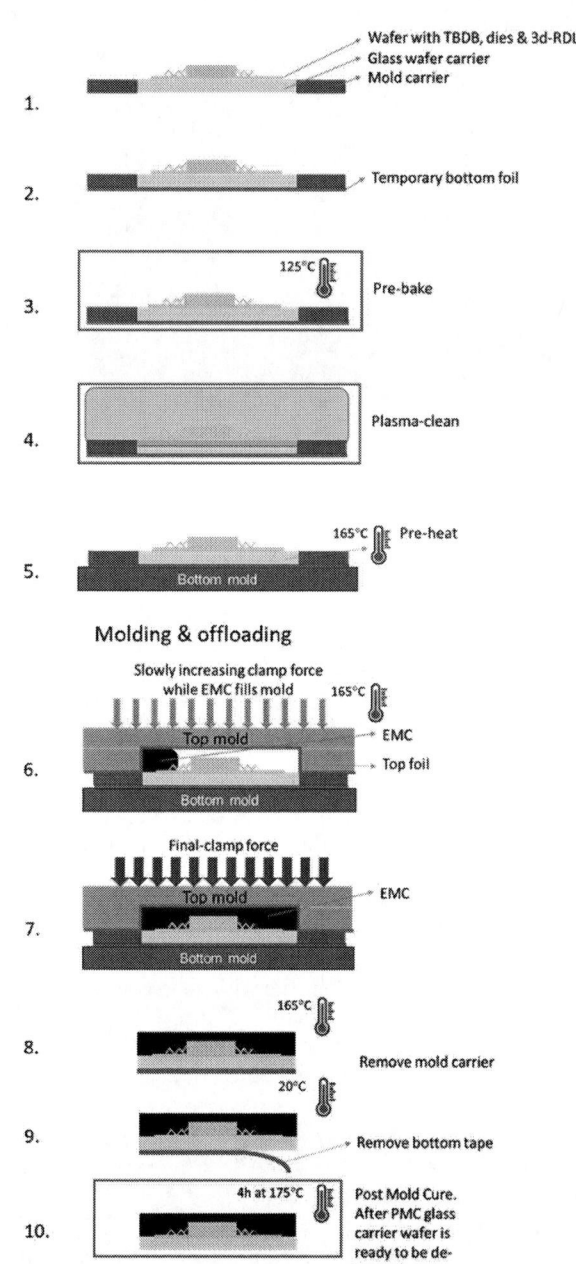

Figure 7. Schematic process flow

To keep the mold closed, a dynamic clamp force controller (DCC) is used. The DCC uses the transfer pressure as input. If the transfer pressure is still low, one will clamp with limited force (pre-clamp). This limited force allows to gently clamp the sample, keeping the mold at the intended cap height, despite having 2 soft foil layers on top and bottom of the sample. During the filling process EMC pressure is build up in the cavity. Normally this would open the mold, but the transfer pressure also increases and so the DCC reacts by increasing the clamp force (6.). An optimization study has been done to find the optimal filling profile for thin cap with selected compound. Fig. 8 gives an example of a transfer logging obtained in this optimization study. By reducing the speed, the filling pressure is controlled while at larger pressures the DCC increases the clamp force.

The plunger keeps pushing until the cavity is full and a final cavity pressure of 8 MPa is measured by the machine. At that point DCC ensures that the final clamp force is reached to keep the mold closed (7.). In mold curing for 600 s starts for proper release of the mold cap from the top foil during mold opening. The sample is removed from the mold. To prevent carrier wafer breakage the mold carrier is removed from the sample at mold temperature (8.). The bottom foil is peeled off from the carrier wafer at room temperature (9.). Finally, a post mold cure process is used to fully cure and thermally stabilize the sample (10.). TABLE I summarizes thin-cap molding process parameters for EMC No.3.

TABLE I. MOLDING PARAMETER FOR THIN CAP

Parameter	Value
EMC	No.3 Low warpage compound
Temperature	165 °C
Cure time	600 sec
Pre-clamping	75 kN
Final Clamping	230 kN
Final packing pressure	8 MPa
Cavity vacuum	< 4 hPa
Post Mold Cure	4 hours at 175 °C

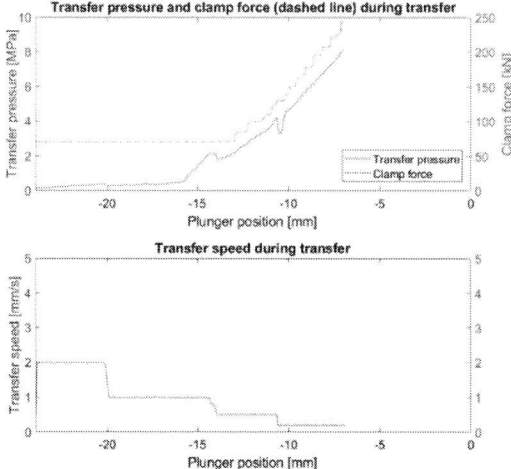

Figure 8. Example of transfer logging observed in the optimization study. Top: transfer pressure and clamp force related to the plunger position. Bottom: the transfer speed related to plunger position

The molding set-up and process have been demonstrated on a wafer for both prototypes. Fig. 9 shows molding result for prototype #1 with ~47.5 μm thick mold cap above the dies.

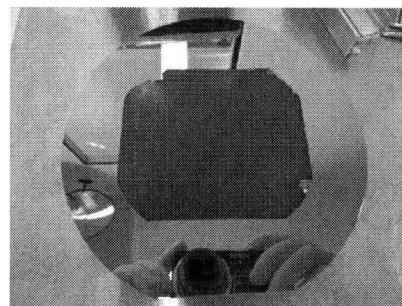

Figure 9. Photo of a molded wafer with ~47.5 μm thick mold cap

E. Laser debonding & cleaning

The laser debonding is performed with a UV solid state laser by scanning the laser beam with high repetition rate across the wafer. The UV laser treatment results in a dramatical drop in the adhesion force because of material decomposition of the TBDB material and a force free separation is possible. As the debond typically takes place close to the carrier wafer the remaining TBDB material on the device side has to be removed. For TBDB #1, neither plasma cleaning with O2 nor O2/CF4 were successful because it either had no effect or it etched ILD1 layer. Wet etching using NMP-based solution was successful after process optimization as shown on Fig. 10.

Figure 10. Photos of debonded wafer with TBDB #1 before (a) and after cleaning (b) – no laser marks on ILD1

For TBDB #2, cleaning with O2 plasma was successful as shown on Fig. 11. However, it revealed that laser slightly attacked ILD1 layer (Fig. 11.b). This is mainly due to TBDB #2 transmitting some of laser's energy. Increasing TBDB #2 thickness in the future could help overcome this artefact.

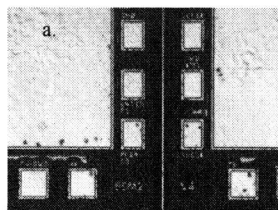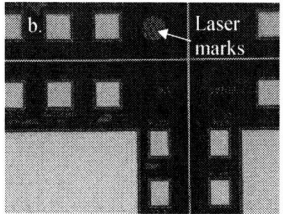

Figure 11. Photos of debonded wafer with TBDB #2 before (a) and after cleaning (b) revealing laser marks on ILD1

For both materials, successful debonding was achieved, allowing to attain objectives for this work.

IV. PACKAGE DEMONSTRATION

A. Fabrication of prototypes

After development of technology bricks, process steps were successfully applied to fabricate prototype #0 and prototype #1. As explained previously, prototype #0 used dummy dies in order to demonstrate the ultra-thin package feasibility as well as substrate-less, molded 3D-IPD devices. Fig. 12 shows a cross-section of debonded prototype #0 with a total thickness of 210 µm.

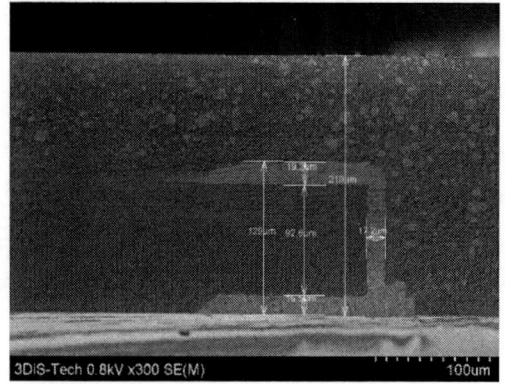

Figure 12. SEM micrograph showing cross-section of prototype #0

Prototype #1 with the 150-µm-thick die was produced on a temporary glass carrier and successfully debonded as shown on Fig. 13.a. The debonded wafer exhibited a slight warpage (Fig. 13.b). However, since production finished at this stage, such warpage does not incur any processing difficulties to clean the wafer and dice the packages. Fig. 13.c shows that a total package thickness of ~233 µm was successfully obtained, demonstrating thus the capability of producing an ultra-thin package with a very thin cap (< 50 µm thickness). Finally, the packaged device is presented in Fig. 14.

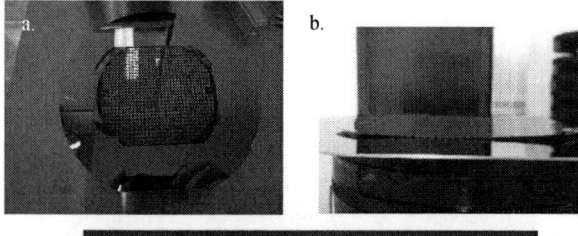

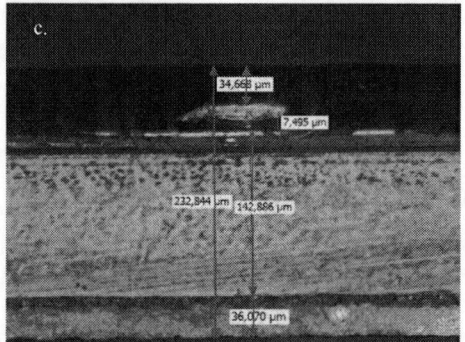

Figure 13. Photos of molded and debonded prototype #1 wafer (a), its warpage (b) and its cross-section showing thin mold cap (c)

Figure 14. Photo of prototype #1 ultra-thin package

B. Electrical performance of 3D-IPD

To evaluate the performance of molded, substrate-less 3D-IPD integrated inside the ultra-thin package (Fig. 15), RF measurements were performed on 3D inductors using GSG probe pads in order to extract their electrical performance.

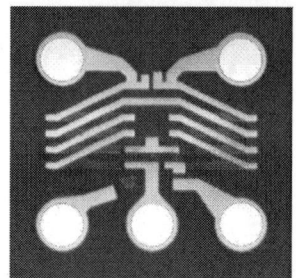

Figure 15. Molded, substrate-less 3D-IPD devices

Fig. 16 shows that for a 3-truns inductor, a Q_{max} of 51 at 4.7 GHz and an inductance of 2.8 nH are achieved with a resonance frequency of 12 GHz. The inductor occupies an area of 300x430 µm². For 6-turns inductor, a Q_{max} of 43 at 3.3 GHz and an inductance of 5.5 nH are achieved with a resonance frequency of 8.3 GHz. The inductor occupies a surface area of 450x475 µm². This result demonstrates excellent RF performance of these very small devices inside the ultra-thin package, contributing thus to enhancing RF electrical performance while reducing package area and cost.

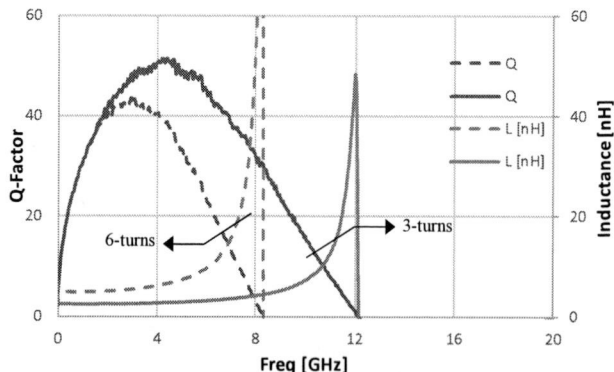

Figure 16. Measure quality factor and inductance of 3-turns and 6-turns inductors inside ultra-thin package

V. CONCLUSION

In this work, a novel ultra-thin, QFN-like 3D package with integrated 3D-IPD devices was developed and presented. The process relied on laser temporary debonding, accurate pick & place, high-density 3D-RDL over vertical sidewalls, and very thin cap transfer-molding to produce the package. Target total thickness of 200 µm or less for this package (tolerance $< \pm 10\%$) and a cap thickness smaller than 50 µm were successfully demonstrated. Additionally, high-performance 3D-IPD devices were successfully integrated inside the package and exhibited Q values as high as 51. This result show heterogeneous 3D integration capability of this concept, making it highly suitable for RF 3D wafer-level system-in-package (3D-WLSiP).

This package concept was used to produce a prototype with an RF device inside. The objective was to improve electrical performance compared to wire-bond-based solutions through short 3D interconnects and thermal dissipation compared to flip-chip-based ones by spreading heat through the PCB. This is ongoing work which will be published soon.

ACKNOWLEDGMENT

This work was performed within the project EuroPAT-MASiP, which has received funding from the ECSEL JU under grant agreement No 737497. The ECSEL JU is supported by the European Union's Horizon 2020 research and innovation programme and by national programmes of Austria, Finland, France, Germany, Hungary, Ireland, Netherlands, Portugal, and Sweden.

This work was partly supported by the French RENATECH network.

REFERENCES

[1] A. Ghannam, D. Bourrier, L. Ourak, C. Viallon, and T. Parra, "3-D Multilayer Copper Interconnects for High-Performance Monolithic Devices and Passives," *Components, Packaging and Manufacturing Technology, IEEE Transactions on*, vol. 3, pp. 935–942, 2013.

[2] A. Ghannam, A. Magnani, D. Bourrier, and T. Parra, "Ultra-Compact, High-Performance, 3D-IPD Integrated Using Conformal 3D Interconnects," in *2018 IEEE 68th Electronic Components and Technology Conference (ECTC)*, 2018, pp. 1082–1088.

[3] S. Kuramochi and Y. Hobie, "3D IPD on thru Glass via substrate using panel Manufacturing Technology for RF applications," presented at the 2017 Pan Pacific Microelectronics Symposium (Pan Pacific), 2017, pp. 1–6.

[4] T. C. Lee *et al.*, "Glass Based 3D-IPD Integrated RF ASIC in WLCSP," presented at the 2017 IEEE 67th Electronic Components and Technology Conference (ECTC), 2017, pp. 631–636.

[5] S. Hashimoto, M. Kiso, Y. Oda, H. Otake, G. Milad, and D. Gudaczauskas, *Direct immersion gold (DIG) as a final finish*, vol. 32. 2006.

[6] S. H. M. Kersjes, J. L. J. Zijl, N. de Jong, and H. Wensink, *Exposed Die Fan-Out Wafer Level Packaging by Transfer Molding*. 2018.

[7] S. H. M. Kersjes, J. L. J. Zijl, W. G. J. Gal, H. A. M. Fierkens, and H. Wensink, "Exposed die wafer level encapsulation by transfer molding," in *2016 6th Electronic System-Integration Technology Conference (ESTC)*, 2016, pp. 1–4.

Low-Cost Non-TSV based 3D Packaging using Glass Panel Embedding (GPE) for Power-efficient, High-Bandwidth Heterogeneous Integration

Siddharth Ravichandran*, Shuhei Yamada†, Fuhan Liu*, Vanessa Smet*, Mohanalingam Kathaperumal*, and Rao Tummala*

*School of Electrical and Computer Engineering
Georgia Institute of Technology, Atlanta, Georgia 30332
Email: siddharth.ravichandran@gatech.edu
†Murata Manufacturing Co. Ltd, Kyoto, Japan

Abstract—High density Logic-HBM integration, today, is built predominantly using 2.5D interposers which are fundamentally limited by long interconnect lengths, and they also are expensive as the package sizes increase. Although 3D ICs enable lowest possible latencies and power-efficiencies, they are challenged by power-delivery and thermal management issues. This paper presents, for the first time, a non-TSV based 3D face-to-face Logic-memory integration using Glass Panel Embedding (GPE) technology for high-density large package applications achieving excellent bandwidth and power-efficiency that are not possible in current approaches. The proposed architecture also achieves small form-factors, and reliability at low cost for high-bandwidth applications. Direct-board attach of such packages enabled by tailorable CTE of glass also provides radical benefits to power delivery from reduced loop-inductances. By studying the fundamental issues of panel-warpage in fan-out packages, and through process improvements achieve <80 μm across 100x100 mm panel enabling HBM assembly at 40 μm pitch.

Keywords-High Density Fan-out; Panel Fan-out; 3D; Glass Panel Embedding; Heterogeneous Integration

I. INTRODUCTION

The advent of multi-chip heterogeneously integrated modules in today's electronics are driven primarily by the increasing need to move large chunks of data between the processor and memory. Starting from integration on the board and multi-chip modules to today's state-of-the-art Silicon interposer, the trend has always been to increase bandwidth while consuming lesser energy [1]. Figure 1 shows the historic progression of multi-chip integration along the key metrics of bandwidth, I/O pitch, power efficiency and thermal flux. As CMOS nodes become smaller, the die sizes grow smaller too, while the number of I/Os tend to increase in order to meet bandwidth requirements, thus, establishing a need for packaging technologies supporting and interconnecting at finer I/O pitches.

2.5D Silicon interposers (shown in Figure 2a) changed the path of the semiconductor industry, interconnecting logic and high bandwidth memory (HBM) with advanced packaging rather than transistor scaling enabling scaling of power-efficient bandwidth at system level for the first

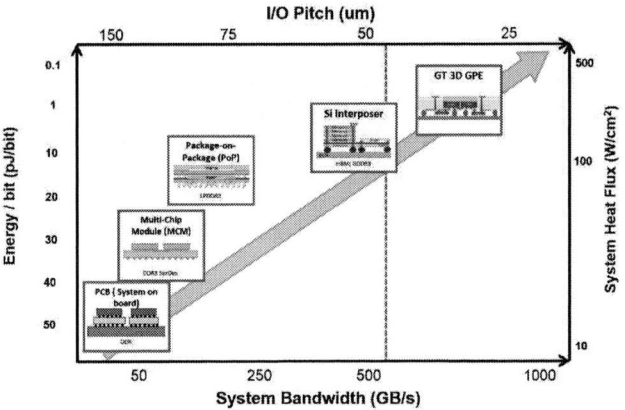

Figure 1: Comparison of proposed approach with current packaging technologies

time [2]. Silicon interposers, however, are limited in cost-efficient package size to approximately 15x15 mm, since they come from 300 mm wafers [3]. Wafer-based chip-first embedding such as Integrated Fan-out (InFO) by TSMC and RDL-first high density fan-out packages are also being developed as lower cost alternatives to silicon interposers [4], [5]. However, both these 2.5D packaging technologies are fundamentally limited by long interconnect lengths that tend to become lossy given the line/space requirements. Tremendous efforts have gone towards the integration of the Application Processor (AP) and memory by means of vertical 3D stacking using expensive Through Silicon Vias (TSVs)(shown in Figure 2b) [6]. Although this enables excellent bandwidth and power efficiencies, this technology is fundamentally challenged by inefficient power delivery and poor thermal performance. TSVs are not only expensive to make, but they also cost a lot of IC real-estate which becomes critical moving to smaller and smaller nodes [7]. Today, 3D die-stacking remains limited to homogeneous integration of stacking DRAM dies in HBM. Face-to-face chip assemblies have also been explored previously, but such a package is still not robust and is challenged by fine-line

978-1-7281-1500-9/19 $31.00 © 2019 IEEE

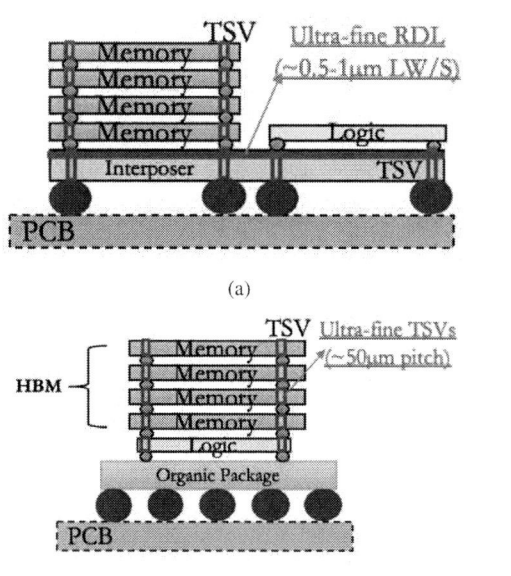

(a)

(b)

Figure 2: Schematic of (a) 2.5D Silicon Interposer and (b) 3D IC stacks

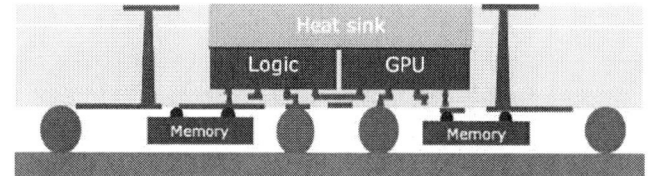

Figure 3: Proposed non-TSV 3D Glass Panel Embedded (GPE) package

RDL integration [8], [9].

WLFO packages have disrupted the entire semiconductor industry due to its benefits in size, cost, electrical and thermal performance, reliability and potential for heterogeneous integration when compared to traditional flip-chip and wire bond packages. WLFO packages by embedding dies in epoxy mold compounds (EMC) which also forms the substrate [4], [10]. 3D Logic-Memory integration in WLFO, today, is primarily a package-on-package (PoP) architecture realized through TEVs or through-encapsulant vias which is a vertical interconnect through the mold compound in the fan-out area. Although this technology is well suited for today's smartphone needs, such a PoP architecture cannot scale to I/O, power and electrical requirements of high-performance computing. The use of mold compounds, also, limits the scope of current WLFO packages to be used in applications requiring superior electrical performance, reliability and also panel-scale manufacturing [11]. This presents a need for a 3D packaging architecture that offers ultra-high I/O densities with very low latencies without being limited by TSVs, power-delivery and thermal management issues.

Figure 3 shows the proposed Glass Panel Embedding (GPE) based 3D architecture for power-efficient high-bandwidth computing. Since the memory dies are assembled at fine-pitch on embedded Glass substrates, the interconnects between the Logic/GPU and the memory is shortened multi-fold. Such an architecture does not need TSVs in the logic die to establish such interconnect lengths and this adds real-estate on expensive dies also reducing costs. The proposed architecture also has potential to address the thermal issues that is created in 3D IC stacks. Georgia Tech

has been developing the key technologies for Glass Panel Embedding (GPE) addressing the limitations of EMC-based WLFO packages such as reduced die-shift for high I/O densities, surface planarity for fine-line RDL and near-zero thermal resistance back-side heat slug for improved thermal-management in embedded fan-out packages [12]–[14]. GPE can hence be explored for non-TSV based high-performance 3D face-to-face logic-memory integration. Such a architecture is not possible in EMC WLFO packages due the high panel warpage which prevents fine-pitch assembly. The CTE mismatch between the die and the EMC also creates reliability challenges for such an assembly process. This paper will go beyond glass panel embedding demonstrated earlier and study the fundamental issue of panel-warpage in embedded substrates and demonstrate assembly at 40 μm achieving 3D vertical interconnections between embedded AP and the assembly of HBM on top.

This paper is organized as follows: Section 2 describes the design analyses for bandwidth and power efficiency, Section 3 describes the process flow to fabricate embedded glass panels and assemble dies at 40 μm pitch. Section 4 lays out the panel-warpage results and presents a discussion. Section 5, in conclusion, summarizes the proposed approach and results on this work.

II. ELECTRICAL DESIGN ANALYSES

In order to compare the bandwidth and power-efficiency, 4 architectures are considered: 2.5D Silicon Interposer, 2.5D Glass Panel Embedded Package, 3D Logic-Memory IC stack and the proposed 3D GPE package. 3D-stacked High Bandwidth Memory (HBM) has emerged as the widely accepted standard for memory in high-performance computing [15], and hence has been used in this architectural comparison. Also, in order to measure the performance benefits coming from the architecture alone, the bump pitch (55 μm) and RDL Line/Space (2 μm/2 μm) are fixed across all four cases. All the results are also normalized to 2.5D Silicon Interposer which is the most widely used architecture in production today. Figure 4a and 4b show the schematic of interconnects in 2.5D GPE and 3D GPE. ANSYS Q3D was used to extract the interconnect parasitics of each of these architectures.

A. Bandwidth

From the extracted parasitics, the τ or the time constant of each interconnect is calculated in order to obtain the

978-1-7281-1500-9/19 $31.00 © 2019 IEEE

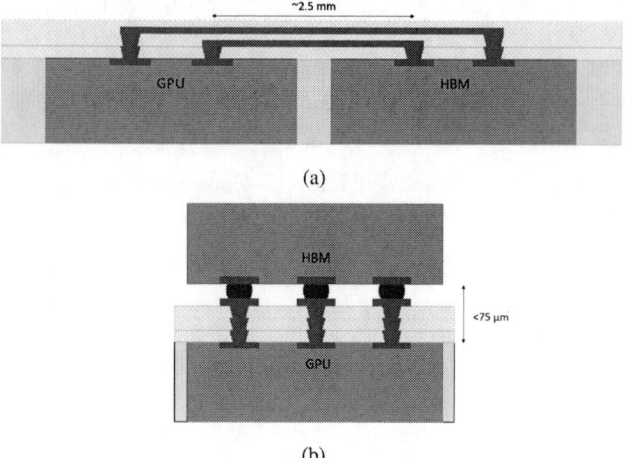

(a)

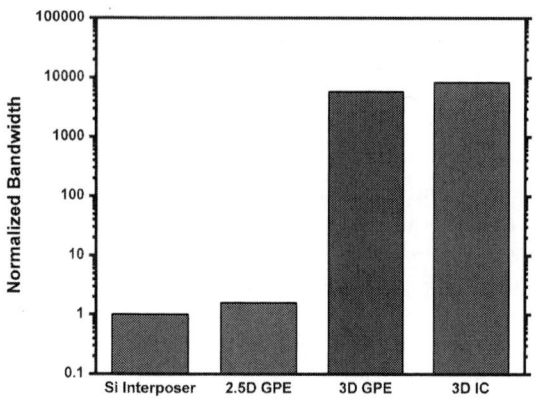

(b)

Figure 4: Schematic for design analyses of (a) 2.5D GPE and (b) 3D GPE

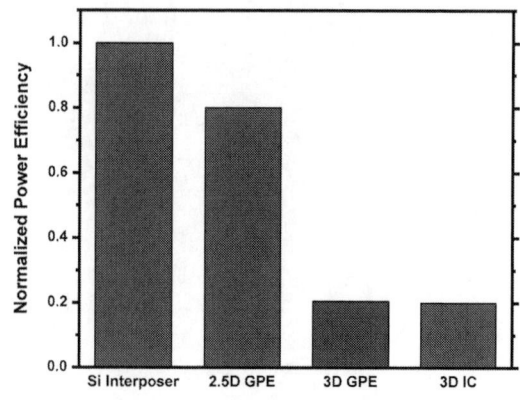

Figure 6: Comparison of power efficiency of different architectures normalized to Silicon interposer

B. Power Efficiency

The energy consumed to move one bit of data from logic to memory or vice-versa is depends on the capacitance on that trace, the voltage on the trace and the frequency of signaling (as shown in Equation 2).

$$P = C \times V^2 \times f \qquad (2)$$

Reduction in the interconnect length, reduces the capacitance on the trace along with the potential to reduce the voltage of operation given the greater signal-to-noise ratio. This is evident in Figure 6 where 3D architectures show over 5x better power efficiency. Again, 3D GPE offers performance similar to 3D ICs without being limited by power delivery and thermal management [16].

III. DEMONSTRATION OF GLASS PANEL EMBEDDING WITH 3D INTEGRATION

Figure 7 shows the fabrication process of the embedded Glass substrate along with the assembly to form 3D vertical interconnections between the logic and memory die. Each of the processing steps are detailed in this section.

A. Test Vehicle Design

To emulate embedded ICs, daisy chain test dies, with a mean size of 7.2 mm x 7.2 mm and mean thickness of 100 μm, provided by Global Foundries were used (shown in Figure 8a). The dies are un-bumped with a pad pitch of 50 μm in one direction and 40 μm on the other. Based on this, blind-cavities were designed at 7.5 mm x 7.5 mm and 200 μm glass panels were structured by Asahi Glass Company (AGC) with cavities depths of 110 μm. As shown in Figure 8b, TGVs with μm vias are drilled with respect to the position of the cavity. RDL layers utilize daisy chains to will evaluate the yield of the module (shown in Figure 8c).

Figure 5: Comparison of bandwidth of different architectures normalized to Silicon interposer

maximum allowed frequency on a trace given by:

$$f_{max} = \frac{0.885}{\tau} \qquad (1)$$

f_{max} is then used to plot of the potential bandwidth allowed by the architecture (Figure 5). It can be seen that there is a dramatic increase in the bandwidth achievable in 3D architectures and this comes from the ultra-short interconnects not possible in 2.5D architectures. As the interconnect densities are scaled, the traces in 2.5D interposers become finer and more resistive which affects the allowable bandwidth on each trace. It can also be noted that the proposed 3D GPE achieves bandwidths comparable to 3D IC stacks while eliminating the aforementioned limitations associated with 3D ICs.

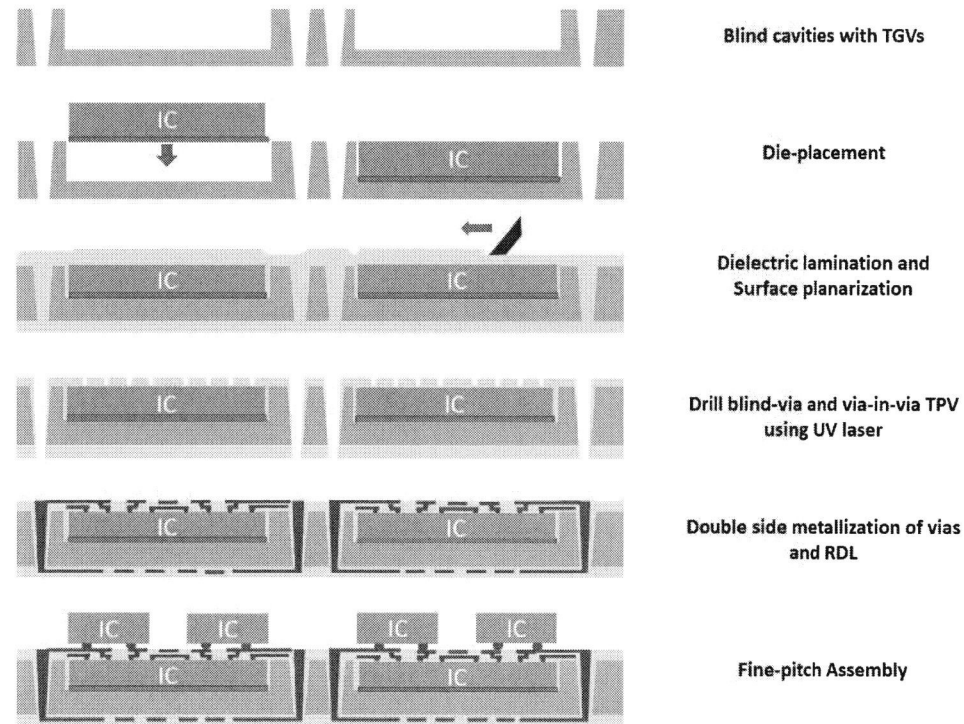

Figure 7: Process flow of Glass Panel Embedding for 3D Logic-Memory Integration

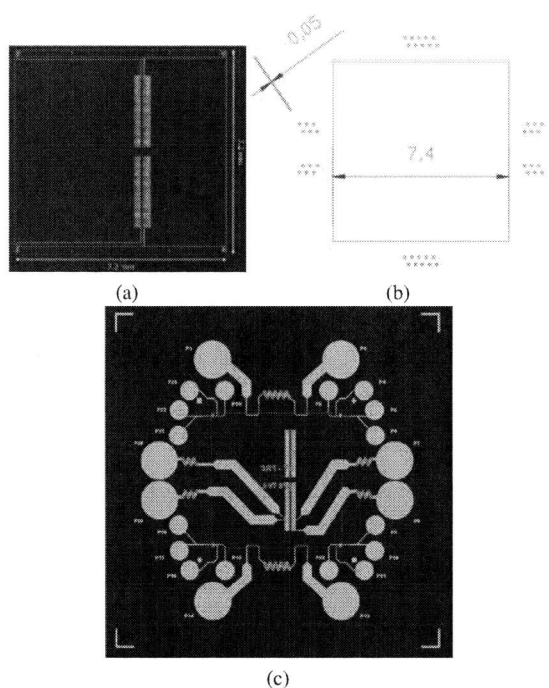

Figure 8: Design schematics of the (a) emulator die (b) cavity with TGV design and (c) test vehicle mask

B. Embedded Substrate Fabrication

Cavities in glass panels can be formed using two techniques: 1) Laminated-cavities and 2) Blind-cavities. Laminated-cavities are formed by laminating glass cavity panels on-to bare glass panels using polymer films [12]. Blind-cavities are, basically like wells, formed on bare glass by wet-etching methods [13]. Following this, TGVs are drilled with positions respective to the cavities to enable double-side RDL. Dies are adhesively placed in these cavities using a pick-and-place tool and polymer dry-films are laminated on both sides. Blind-vias and Via-in-vias are drilled using a UV laser tool befoe starndard SAP process is used to form RDL on both sides. Figure 9 shows the cross-section of the embedded glass substrate before assembly.

C. Assembly and Integration

After the completion of the substrate fabrication, bumped emulator dies with the same pad pattern are used. The dies are assembled using Thermocompression Bonding (TCB) on the Finetech Semi-automatic bonder with pre-applied underfill. Figure 11 and Figure 12 show the top view and the cross section of the final 3D embedded glass package respectively. A 100 μm thick die embedded in glass cavity is connected to other die assembled on top through RDL and blind micro-vias plated directly onto the pads of the embedded IC. Such an architecture achieves ultra-short interconnects by eliminating the long interconnects in 2.5D

978-1-7281-1500-9/19 $31.00 © 2019 IEEE

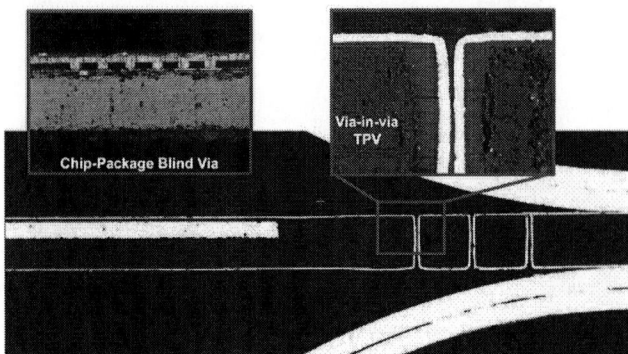

Figure 9: Cross-section of the die-embedded substrate after 1ML RDL formation

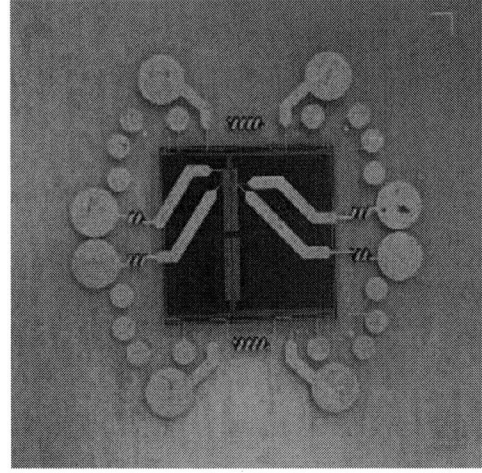

Figure 10: Top view showing the die embedding in glass cavity with Cu RDL on top

Figure 11: Top view of the 3D GPE package with emulator die assembled on top of embedded glass panel

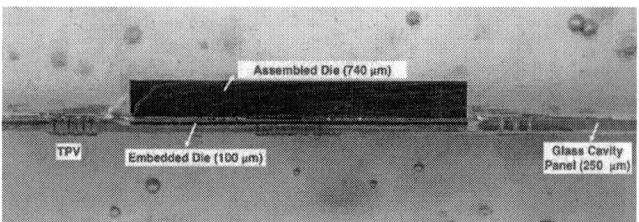

Figure 12: Cross-section showing the 3D GPE with die embedding in glass cavity and another die assembled on top

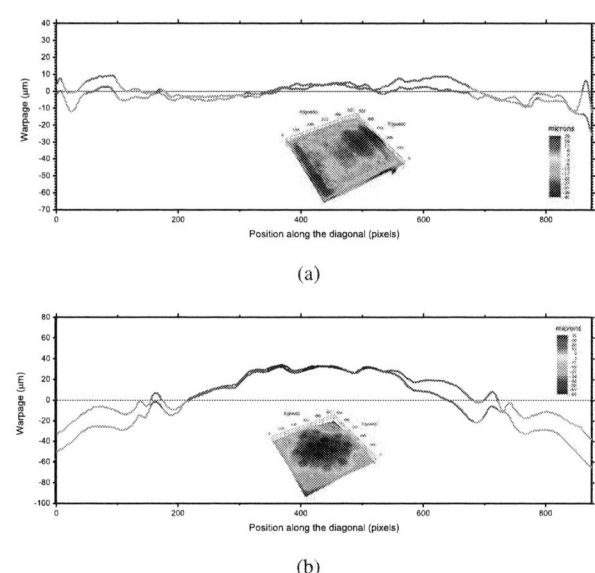

(a)

(b)

Figure 13: Panel warpage measured on Blind-cavities using Shadow Moiré (a) after die-placement and (b) after dielectric lamination and curing

IV. RESULTS AND DISCUSSION

In order to ensure high-yielding fine-pitch assembly on embedded packages, it is critical to study the warpage during the entire processing of the panel. The structured glass panels present an imbalance in distribution of material through the panel and hence are inherently susceptible to warpage during processing. Polymer dielectrics shrink while curing and hence double-side processing is always preferred to mitigate warpage. However, in the architecture proposed in this paper, the polymer dielectrics fill the cavities around the embedded dies, increasing the chances of warpage. Hence, it is important to study the progression of warpage through the processing steps. Also, warpage on the panel determines the number of coupons allowed on a panel, which ultimately determines the throughput of the process.

Figure 13 shows the RT warpage across a 100x100 mm glass blind-cavity panel measured using the Akrometrix Shadow Moiré tool. The 2 lines in the plot correspond to

architectures.

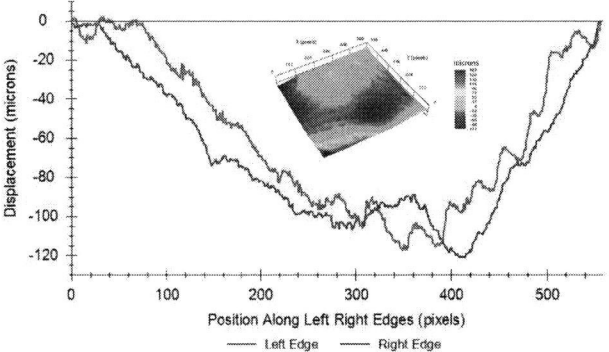

Figure 14: Panel warpage measured on laminated-cavities formed using low-stress bonding film

the 2 diagonals along the panel where typically warpage tends to be the maximum. The full field of the panel was captured at 621x621 pixels which results in a diagonal of 875 pixels shown in the x-axes. Figure 13a captures the warpage after die-placement and DAF cure. The loading of Silicon on the panel did not affect the warpage as the results were similar to starting glass warpage. Figure 13b shows the warpage after polymer dielectric lamination and cure. Each panel has 16 cavities, but dies were not placed in the corner 4 to help mounting of the panel for plating. It can be seen that the warpage is largely in control in the coupons with dies in them ($<$20 μm eliminating the effect of the corner coupons). This is because the polymer dielectrics fill up the empty cavities completely in the corner coupons resulting in an imbalance of dielectric thickness at the corners. This can, however, be mitigated by using a different mounting frame for plating or by designing the cavity locations appropriately. The total warpage observed for such a thin package is still very much lower than warpages observed in mold-compound based WLFO packages [17]. Warpage for assembly at 40 μm pitch is also within tolerable limits [18]. Figure 14 shows the RT warpage across a 100x100 mm laminated-cavity panel formed using a low-stress bonding film. Since both the glass are bonded, there is also flexibility of CTE on the 2 panels which can be tailored to improve reliability based on the specific package size. The use of low-stress bonding film as opposed to standard RDL epoxy polymers improved the RT warpage across the panel by 4.7 times.

V. CONCLUSION

This paper demonstrated for the first time a radical new concept in next generation of high-bandwidth packaging using 3D face-to-face assembly at 40 μm through glass panel embedding (GPE) with better I/O density, performance, cost and reliability than current 2.5D architectures and 3D IC stacks. Through design analyses, this paper demonstrates that 3D GPE achieves the performance of 3D IC stacks without being limited by power-delivery and thermal man-

agement. TSVs in 3D ICs are not only expensive but also occupy a lot of critical real-estate on the processor. This paper presented a new novel architecture to overcome TSVs while also achieving ultra-short interconnects for power-efficient ultra-high-bandwidth. A key enabler of such an architecture is the reduced warpage in embedded panel substrates which enables high-yields for HBM assembly and also fine-line RDL lithography. Panel warpage was shown to be considerably lower than EMC-based WLFO technology. Such a TSV-less 3D architecture has tremendous potential to simultaneously scale bandwidth and power-efficiency at a lower cost.

ACKNOWLEDGMENT

The authors would like to thank Dr. Venky Sundaram for his valuable inputs in this work. The authors would also like to acknowledge the tool, material and process support of Asahi Glass Company, ESI, Finetech, Ajinomoto, Panasonic materials and Dow Chemicals, Global Foundries, Schott AG, and all other members of the Packaging Research Center consortium. The authors would also like to thank Omkar Gupte, Chris White and Lila Dahal in assisting with the fabrication processes.

REFERENCES

[1] B. Sabi, *Advanced Packaging in the New World of Data*, ECTC Luncheon Keynote, 2017.

[2] S. Hou *et al.*, "Wafer-level integration of an advanced logic-memory system through the second-generation CoWoS technology," *IEEE Transactions on Electron Devices*, vol. 64, no. 10, pp. 4071–4077, 2017.

[3] R. Mahajan, R. Sankman, N. Patel, D.-W. Kim, K. Aygun *et al.*, "Embedded Multi-die Interconnect Bridge (EMIB)–A High Density, High Bandwidth Packaging Interconnect," in *Electronic Components and Technology Conference (ECTC), 2016 IEEE 66th*. IEEE, 2016, pp. 557–565.

[4] C.-F. Tseng, C.-S. Liu, C.-H. Wu, and D. Yu, "InFO (wafer level integrated fan-out) technology," in *Electronic Components and Technology Conference (ECTC), 2016 IEEE 66th*. IEEE, 2016, pp. 1–6.

[5] V. S. Rao *et al.*, "Development of high density fan out wafer level package (HD FOWLP) with multi-layer fine pitch RDL for mobile applications," in *Electronic Components and Technology Conference (ECTC), 2016 IEEE 66th*. IEEE, 2016, pp. 1522–1529.

[6] S. Das, A. Fan, K.-N. Chen, C. S. Tan, N. Checka, and R. Reif, "Technology, performance, and computer-aided design of three-dimensional integrated circuits," in *Proceedings of the 2004 international symposium on Physical design*. ACM, 2004, pp. 108–115.

[7] A.-C. Hsieh and T. Hwang, "TSV redundancy: Architecture and design issues in 3-D IC," *IEEE Transactions on Very Large Scale Integration (VLSI) Systems*, vol. 20, no. 4, pp. 711–722, 2012.

978-1-7281-1500-9/19 $31.00 © 2019 IEEE

[8] J. Sutanto, "POSSUMTM Die Design as a Low Cost 3D Packaging Alternative," *3D Packaging*, vol. 25, pp. 16–18, 2012.

[9] J. Sutanto, D. Kang, S. Ma, J. Yoon, K. Oh, M. Oh, K. R. Park, R. Lanzone, and R. Huemoeller, "Development of chip-on-chip with face to face technology as a low cost alternative for 3D packaging," in *2013 IEEE 63rd Electronic Components and Technology Conference.* IEEE, 2013, pp. 955–965.

[10] C. C. Liu *et al.*, "High-performance integrated fan-out wafer level packaging (InFO-WLP): Technology and system integration," in *2012 International Electron Devices Meeting.* IEEE, 2012, pp. 14–1.

[11] R. Tummala, V. Sundaram, P. M. Raj, and V. Smet, "Future of embedding and fan-out technologies," in *2017 Pan Pacific Microelectronics Symposium (Pan Pacific).* IEEE, 2017, pp. 1–9.

[12] S. Ravichandran *et al.*, "2.5 D Glass Panel Embedded (GPE) Packages with Better I/O Density, Performance, Cost and Reliability than Current Silicon Interposers and High-Density Fan-Out Packages," in *2018 IEEE 68th Electronic Components and Technology Conference (ECTC).* IEEE, 2018, pp. 625–630.

[13] Ravichandran, Siddharth and others, "Design and demonstration of Glass Panel Embedding for 3D System Packages for heterogeneous integration applications," in *International Symposium on Microelectronics*, vol. 2018, no. 1. International Microelectronics Assembly and Packaging Society, 2018, pp. 000 331–000 336.

[14] N. Nedumthakady *et al.*, "Integrated copper heat spreaders in glass panel embedded packages with near-zero thermal interface resistance," in *2018 IEEE 68th Electronic Components and Technology Conference (ECTC).* IEEE, 2018, pp. 2013–2018.

[15] H. Jun, J. Cho, K. Lee, H.-Y. Son, K. Kim, H. Jin, and K. Kim, "HBM (High Bandwidth Memory) DRAM technology and architecture," in *2017 IEEE International Memory Workshop (IMW).* IEEE, 2017, pp. 1–4.

[16] P. Jain, P. Zhou, C. H. Kim, and S. S. Sapatnekar, "Thermal and power delivery challenges in 3d ics," in *Three Dimensional Integrated Circuit Design.* Springer, 2010, pp. 33–61.

[17] F. Che, D. Ho, M. Z. Ding, and D. R. MinWoo, "Study on process induced wafer level warpage of fan-out wafer level packaging," in *2016 IEEE 66th Electronic Components and Technology Conference (ECTC).* IEEE, 2016, pp. 1879–1885.

[18] K. Oi *et al.*, "Development of new 2.5 D package with novel integrated organic interposer substrate with ultra-fine wiring and high density bumps," in *2014 IEEE 64th Electronic components and technology conference (ECTC).* IEEE, 2014, pp. 348–353.

2019 IEEE 69th Electronic Components and Technology Conference (ECTC)

Polylithic Integration of 2.5D and 3D Chiplets Using Interconnect Stitching

Paul K. Jo, Ting Zheng, and Muhannad S. Bakir
School of Electrical and Computer Engineering
Georgia Institute of Technology
Atlanta, GA 30318, USA
mbakir@ece.gatech.edu

Abstract— This paper explores polylithic integration of heterogeneous dice (chiplets) for high-density electronic systems. In this approach, stitch-chips are used to enable 2.5D integration by providing dense signal pathways between assembled 'anchor chips,' while surface-embedded chips provide 3D face-to-face electrical interconnection with corresponding anchor chips. Multi-height Compressible MicroInterconnects (CMIs) are used to enable low-loss and mechanically robust interfaces between the anchor chips and the stitch-chips as well as the surface-embedded chips. Fabrication and assembly of a testbed is reported and demonstrates robust interconnection. In an effort to characterize the CMIs and stitch-chip channels at high-frequency, electromagnetic simulations are carried out and demonstrate less than 0.6 dB insertion loss for 90 μm tall CMIs and 500 μm long channels on a fused silica stitch-chip.

Keywords-Compliant interconnects, 2.5D and 3D ICs, heterogeneous integration

I. INTRODUCTION

In the era of ubiquitous computing, Internet-of-Things (IoT), Artificial Intelligence (AI), 5G, self-driving vehicles, and big data, the number of connected electronic devices and volume of data generated are sharply increasing [1], [2]. These have pushed the semiconductor industry towards ever more complex and sophisticated chip designs for better computing capabilities. Over the last several decades, this has been enabled by system-on-chip (SoC) designs in which various functionalities have been integrated into a single die using monolithic processes along with technology scaling [3]. However, Moore's Law is slowing down and approaching its limits while SoC design complexity and fabrication costs continue to increase [4], [5]. These emerging challenges have introduced significant research in high-density multi-die integration technologies by virtue of their integration flexibility, potential reduced fabrication costs, and ability to mix-and-match across different technologies (i.e., chiplet design approach) that include silicon interposer, Embedded Multi-chip Interconnect Bridge (EMIB), and Foveros 3D integration technologies [6], [7]. The benefits of these heterogeneous integration approaches have been show recently. For example, AMD and GLOBALFOUNDRIES have reported approximately 40% and 63% reduction in silicon fabrication cost along with improved system-level performance through 2.5D and 3D integration, respectively [8], [9].

In this paper, we extend our prior work [10] to demonstrate a polylithic integration technology of heterogeneous dice to

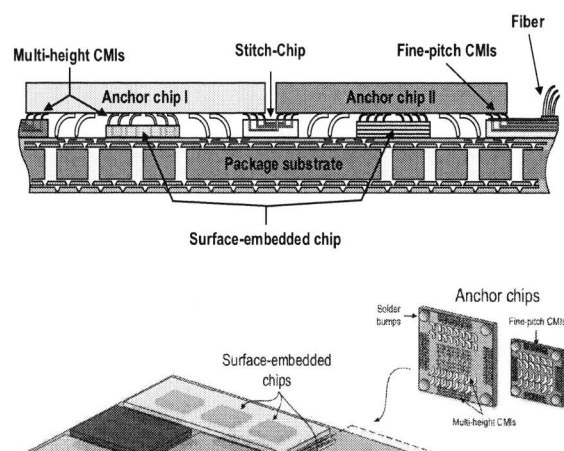

Fig. 1. Schematic of polylithic integration for heterogeneous dice using multi-height CMIs

enable *the ultimate flexibility* in 2.5D/3D multi-die chiplet integration yet providing monolithic-like performance and low-loss dense signal interconnections. Fig. 1 illustrates a schematic of the proposed integration technology. Firstly, multiple 'anchor chips' are concatenated by 'stitch-chips' with dense interconnects, which are placed between the anchor chips and the package substrate. If needed, 'surface-embedded chips,' which may be passive or active dice, can be integrated underneath the anchor chips. Vertically flexible off-chip interconnects, which we call Compressible MicroInterconnects (CMIs) [11], are used to provide signal interconnections between the assembled dice. CMIs on the edge of the anchor chips are fine-pitch to provide high bandwidth interconnection between the anchor chips through the stitch-chips [10]. Multi-height CMIs are used to enable interconnections between the anchor chips and the surface-embedded chips. Power delivery and signal interconnections from the package substrate can also be enabled by the multi-height CMIs. Mechanical bonding between the anchor chips and the package substrate is enabled by using large solder bumps on each of the four die corners to enable reworkability of the assembled die. The stitch-chips, in the simplest form, provide high-bandwidth density signal pathways between the anchor chips, yet they can include high-quality passives

978-1-7281-1500-9/19 $31.00 © 2019 IEEE   1803

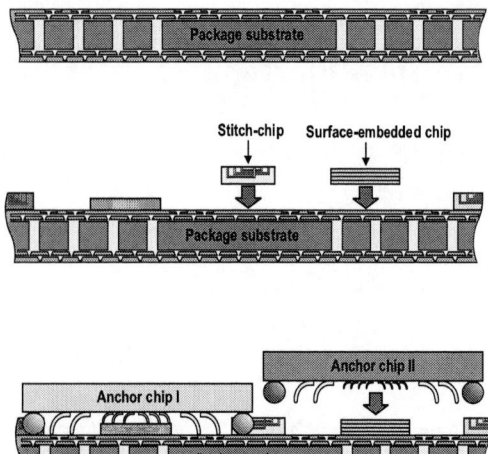

Fig. 2. The assembly process flow

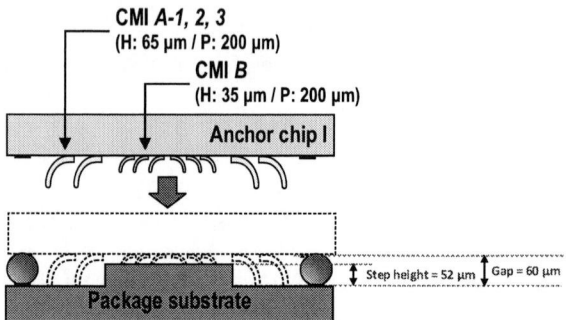

Fig. 3. Schematic of the fabricated testbed

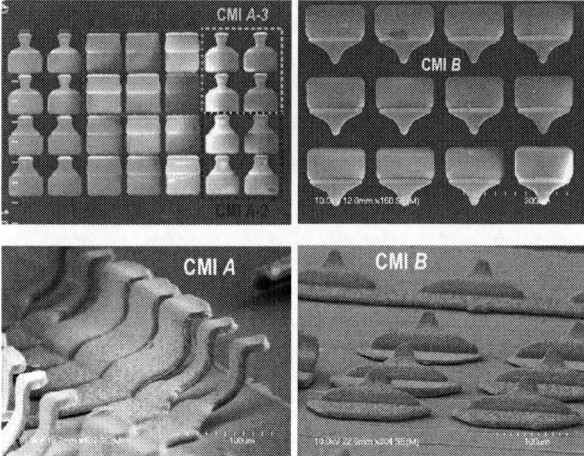

Fig. 4. SEM images of the fabricated multi-height CMIs

and/or active circuits as well. This approach can also enable direct interconnection between the anchor chip and a silicon photonic integrated circuit (PIC) with direct fiber assembly, as illustrated in Fig. 1. The anchor chips may be an ASIC, CPU, GPU, FPGA, MMIC, or photonic die, and the surface-embedded chips may be memory dice, power conversion chips, or Integrated Passive Device (IPD) dice, for example.

Since the CMIs are elastically compressible unlike conventional solder bumps, the proposed polylithic integration technology can compensate for any possible off-chip interconnection distance differences resulting from chip thickness differences; this enables both 2.5D and 3D face-to-face interconnection in one platform, as shown in Fig. 1. In addition, CMIs can provide temporary interconnection, which can improve package and system yield as CMIs facilitate die replacement/rework. The mechanical compliance of the CMIs can also improve the thermomechanical reliability of the assembled system [12] as well as enabling flexibility in dice and substrates to be stitched together irrespective of CTE mismatch (e.g., silicon, glass, organic, and GaN).

This paper is organized as follows: Section II describes the fabrication and assembly process of the proposed integration technology. In Section III, HFSS simulations of stitch-chip links are demonstrated. Finally, in Section IV, concluding remarks are stated.

II. FABRICATION AND ASSEMBLY OF STITCH-CHIP BASED POLYLITHIC INTEGRATION

Fig. 2 shows the overall assembly process flow of the proposed polylithic integration technology. The integration process begins with forming traces and pads on the package substrate. Next, the stitch-chips and/or surface-embedded chips are attached onto the package substrate. The anchor chips with multi-height CMIs and relatively large solder bumps are flip-chip bonded onto the package substrate, as shown in Fig. 2 (note, CMIs can be on the stitch-chips instead). Finally, the assembly is completed by reflowing the solder bumps. NiW is used to form the core of the CMIs due

to its high yield strength of 1.93 GPa [11]. This high yield strength enables the CMI to tolerate larger stress before experiencing plastic deformation during compression. The NiW CMIs are electroless gold plated as the final fabrication step to prevent oxidation.

The testbed is fabricated and assembled in order to demonstrate the key features of the proposed technology: assembly of an anchor chip with multi-height CMIs onto a substrate with a surface-embedded chip and mechanical bonding using solder bumps. In the testbed, the surface-embedded chip on the package substrate is emulated by forming a tall step on the silicon substrate. Four solder bumps are also fabricated on the silicon substrate to mechanically secure the assembled testbed. For the anchor chip with multi-height CMIs, two CMI designs with different heights, which we refer to as CMI A and CMI B in this paper, are fabricated on a silicon substrate. Both CMI A and B designs are formed on a 200 μm pitch while the heights are different. Specifically, for CMI A, three different CMI designs (A-1, A-2, and A-3) of the same height are designed to demonstrate the simplicity of CMI compliance engineering. Fig. 3 shows a schematic of the fabricated testbed. Note, in prior work, we demonstrated the fabrication of CMIs with 20 μm of in-line pitch [10].

SEM images of the fabricated anchor chip are shown in Fig. 4. An approximately tapered design is used for the

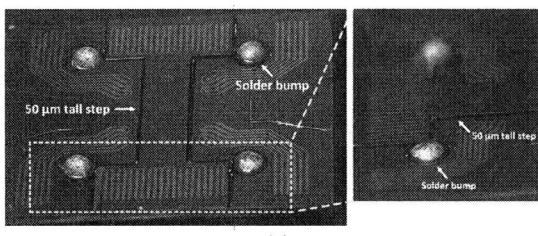

(a)

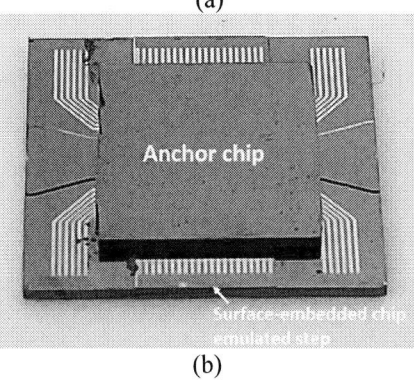

(b)

Fig. 5. Optical images of the testbed: (a) the silicon substrate with the step and solder bumps and (b) the assembled testbed

TABLE I
DIMENSIONS AND MECHANICAL AND ELECTRICAL CHARACTERIZATION
RESULTS OF THE FABRICATED MULTI-HEIGHT CMIs

Interconnect	Pitch (μm)	Height (μm)	Compliance (out-of-plane) (mm /N)	Four-point resistance (mΩ)
CMI *A-1*			2.42	190.8
CMI *A-2*	200	65	3.86	-
CMI *A-3*			6.1	-
CMI-*B*	200	35	5.4	64.4

CMIs and the pads contributes to this difference in the average resistances; as shown in Fig. 3, the 60 μm gap (between the silicon substrate and the anchor chip) deforms the 65 μm tall CMI *A* by approximately 5 μm while the 8 μm gap (between the step and the anchor chip) deforms the 35 μm tall CMI *B* by approximately 27 μm; this will affect the contact force and hence contact resistance of the CMIs. Table I summarizes the compliance and four-point resistance characterization results and the dimensions of the fabricated multi-height CMIs.

III. STITCH-CHIP LINK SIMULATION

In order to characterize the high-frequency properties of the multi-height CMIs and the stitch-chip links both in aggregate form and individually, a carefully constructed testbed must be designed, fabricated, and experimentally tested to validate the models. In this paper, we present the initial design and simulation results of such a testbed. Since the L-2L de-embedding method has been widely utilized for the characterization of transmission lines with through silicon via (TSV) in the RF range [13], [14], our stitch-chip based high-frequency testbed is designed to be compatible with L-2L de-embedding.

Fig. 6 shows the details of our ANSYS HFSS testbed model. The model contains a 2.5D signal link, whose ABCD-matrix is denoted as [2.5D-Link] in Fig. 6 (a). As shown in Fig. 6 (a), this 2.5D channel can be partitioned into a single CPW intermediate channel, two G-S-G pairs of CMIs whose ABCD-matrices are denoted as [CMI] and two extended CPW T-line whose ABCD-matrices are denoted as [TL]. The two extended CPW T-lines can be de-embedded and the remaining structure, as shown in Fig. 6 (b), is called the stitch-chip link whose ABCD-matrix is [Link1']. In practice, the tip of the CMIs would touch the CPW T-lines and deform elastically.

different CMI designs in order to distribute the stress along their length during deformation. The upward-curved cross-sectional design ensures that the tip of the CMI maintains contact with the corresponding pad during assembly. Both fabricated CMI *A* and *B* have 200 μm pitch while the heights are approximately 65 μm and 35 μm, respectively.

Fig. 5 shows optical images of the fabricated silicon substrate with emulated surface-embedded chips (and solder bumps) and the assembled testbed. In this testbed, the step height is approximately 52 μm. Spherical solder balls with a diameter of 500 μm are manually attached and reflowed on the metal pads before assembly (though electroplating can be used to fabricate the solder bumps as well). Thermocompression bonding is used to assemble the dice. Once the anchor chip is aligned to the substrate, the solder balls are reflowed again to provide mechanical interconnection between the anchor chip and the substrate while maintaining approximately 60 μm of gap.

Mechanical compliance, a key property of CMIs, was measured using a Hysistron Triboindenter. The measured average compliance was 2.42 mm/N, 3.86 mm/N, 6.1 mm/N, and 5.4 mm/N, respectively for CMI *A-1*, CMI *A-2*, CMI *A-3*, and CMI *B* designs. These results illustrate the simplicity of CMI mechanical compliance engineering through only photomask geometry re-design (though there are additional parameters that impact compliance including thickness, material, and three-dimensional curved geometry). The four-point resistances of the interconnections after assembly were measured using a Karl-Suss probe station. The four-point resistance values of CMI *A-1* and CMI *B* were measured, and their average four-point resistance, including contact resistance with the gold pads, is 190.8 mΩ and 64.4 mΩ, respectively. We believe the contact resistance between the

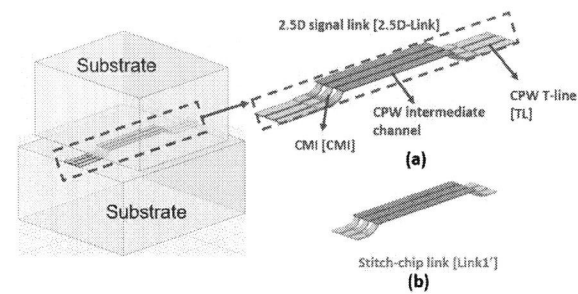

Fig. 6. HFSS model of stitch-chip channel

978-1-7281-1500-9/19 $31.00 © 2019 IEEE

However, this deformation is not accounted for in this work. L-2L de-embedding requires two 2.5D signal links, one of which has an intermediate channel length twice as long as the other (ABCD-matrices [2.5D-Link1] and [2.5D-Link2]). [2.5D-Link1] and [2.5D-Link2] can be derived from the S-matrices measured from a vector network analyzer. After de-embedding the two extended CPW T-lines, the stitch-chip link's ABCD-matrix can be obtained. The procedure of this L-2L de-embedding can be summarized as follows [13], [14]:

$$[Link1'] = [TL]^{-1}[2.5D - Link1][TL]^{-1} \quad (1)$$

$$[Link2'] = [TL]^{-1}[2.5D - Link2][TL]^{-1} \quad (2)$$

$$[CMI] = \left(\sqrt{[Link1']^{-1}[Link2'][Link1']^{-1}}\right)^{-1} \quad (3)$$

where [Link1'] and [Link2'] represent [2.5D-Link1] and [2.5D-Link2] after de-embedding the CPW T-lines at both ends. It should be noted that [Link1'] represents an ABCD-matrix of a stitch-chip link that consists of a CPW with CMIs on both ends. Thus, based on L-2L de-embedding, our designed testbed can potentially extract the electrical properties of both CMIs and stitch-chip links with high accuracy.

In the simulated testbed, the substrate material used for the stitch-chip is fused silica, which provides relatively low dielectric constant (~3.9) and loss tangent (~0.0002) within the RF range. The low dielectric constant and loss tangent enable low-loss transmission line design. Second, all CPWs are made using copper and their characteristic impedances are optimized to 50 Ω. In Fig. 6, the CPW intermediate channel will have two lengths, one of which is twice the other. The extended CPW T-line is fixed at 250 μm length while the total stitch-chip link has 500 μm-long CPW, as shown in Fig. 6 (b). Lastly, several design versions for the CMIs have been included. Fig. 7 illustrates the cross-sectional view of one nickel-core CMI (thickness varies from 5 μm to 7 μm) coated with 500 nm-thick gold to avoid oxidation and to decrease resistance (DC and AC) [15]. Table II summarizes the key dimensions for different CMI designs. In Table II, AR refers to the aspect ratio of the CMI and is defined as the ratio of the CMI's horizontal length (L) and vertical height (H) (see Fig. 7). Arc length (AL) represents aggregate physical length of the CMI.

Simulations from DC to 30 GHz are conducted utilizing above models. Following conversion of S-matrices to ABCD-matrices and using model (1), an ABCD-matrix and S-matrix of the stitch-chip link (CMI+CPW+CMI) can be extracted. In order to validate the de-embedding method, a standalone case where a stitch-chip link has the same dimensions as the de-embedded stitch-chip link is set up as a reference (REF link). S_{21} and S_{11} for different stitch-chip link designs are shown in Fig. 8. The magnitudes of the insertion loss (S_{21}) and the return loss (S_{11}) of the reference link and the testbed link after de-embedding are compared and show agreement. The stitch-chip links exhibit an insertion loss of less than 0.6 dB within 30 GHz and maintains good impedance matching (return loss better than -10 dB) even with largest CMIs (90 μm-high).

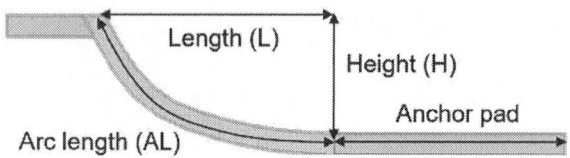

Fig. 7. Cross-section view of CMI with design parameters

TABLE II
DESIGN POINTS FOR DIFFERENT CMI VERSIONS

CMI design	AR = L / H	H (μm)	L (μm)	AL (μm)
60 μm-high CMI	2	60	120	145.3267
90 μm-high CMI	2	90	180	217.9901
90 μm-high CMI	1	90	90	141.3717

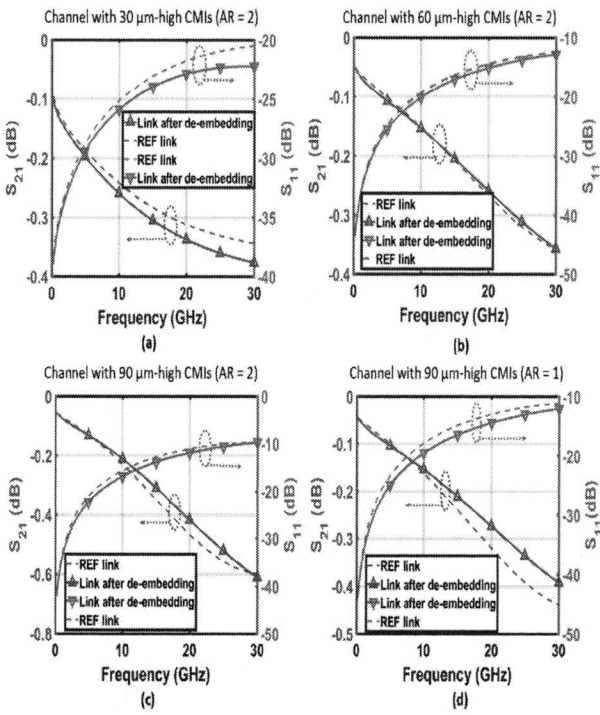

Fig. 8. S-parameters results compared with standalone link (CMI+Tline+CMI). De-embedded link with (a) 30 μm-high CMIs (AR = 2); (b) 60 μm-high CMIs (AR = 2); (c) 90 μm-high CMIs (AR = 2); (d) 90 μm-high CMIs (AR = 1)

Another objective of this testbed is to extract the electrical parasitics of the CMIs. After converting S-matrices to ABCD-matrices and following the models in (1) to (3), an ABCD-matrix and S-matrix of CMIs can be easily extracted. Fig. 9 shows the S_{21} and S_{11} for different CMI designs. As for S_{21}, CMIs provide low-loss 3D interconnect solution within the RF range (insertion loss better than -0.11 dB). For S_{11}, CMIs still maintain acceptable impedance matching thanks to their electrically short structure. Note that by only reducing AR, the

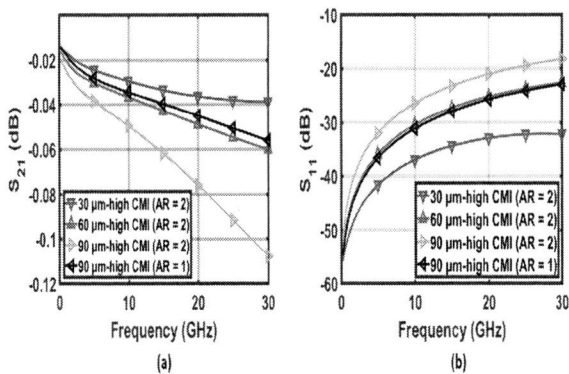

Fig. 9. S-parameters comparison for CMIs with different heights: (a) Insertion loss; (b) Return loss

TABLE III
PARASTICS EXTRACTION @ 30 GHZ FOR DIFFERENT CMIS

CMI design	R (mΩ/μm)	L (pH/μm)	G (μS/μm)	C (fF/μm)
60 μm-high CMI (AR = 2)	2.75	0.42	0.07	0.06
90 μm-high CMI (AR = 2)	2.2	0.45	0.03	0.06
90 μm-high CMI (AR = 1)	2.69	0.43	0.04	0.07

90 μm-high CMI's loss is improved. The reason is that shrinking AR down from 2 to 1 results in approximately 35% reduction in physical length. Because of their short length relative to wavelength, a lumped model [16] can be used to extract RLGC parasitics. The RLGC parasitics can be extracted using [16]:

$$R = real\left(\frac{-2}{Y_{12}+Y_{21}}\right) \qquad (4)$$

$$L = \frac{imag\left(\frac{-2}{Y_{12}+Y_{21}}\right)}{2\pi f} \qquad (5)$$

$$G = real\big(2Y_{11} + (Y_{12} + Y_{21})\big) \qquad (6)$$

$$C = \frac{imag\big(2Y_{11}+(Y_{12}+Y_{21})\big)}{2\pi f} \qquad (7)$$

where f is frequency. Table III summarizes the RLGC parasitics for some of the CMI designs at 30 GHz. Most of the CMI interconnect is surrounded by air, which results in low parasitic capacitance and conductance.

IV. CONCLUSION

This paper explores the fabrication and assembly process of a stitch-chip based polylithic integration enabled by multi-height CMIs. Experimental characterization of the proposed approach was performed by assembling the testbed using multi-height CMIs and surface-embedded chips emulated using a silicon step. The experimental results show that the multi-height CMIs can enable robust electrical interconnection irrespective of off-chip interconnection distance differences. These results demonstrate a new degree of freedom in system-level integration when compared to conventional solder bumps. In this paper, we also present HFFSS-based simulations of the CMIs and stitch-chips to gain initial insight into their high-frequency response.

ACKNOWLEDGMENT

This work was performed in part at the Georgia Tech Institute for Electronics and Nanotechnology, a member of the National Nanotechnology Coordinated Infrastructure, which is supported by the National Science Foundation (Grant ECCS-1542174). This work was support by the National Science Foundation Grant 1810081.

REFERENCES

[1] A. Steegen, "Technology innovation in an IoT Era," in *Proc. IEEE VLSI Technol. Symp.*, Jun. 2015, pp. C170-C172.

[2] M. Bohr, "The evolution of scaling from the homogeneous era to the heterogeneous era," *IEDM Tech. Dig.*, Dec. 2011, pp. 1.1.1-1.1.6.

[3] G. Yeric, "Moore's law at 50: Are we planning for retirement?" *IEDM Tech. Dig.*, Dec. 2015, pp. 1.1.1-1.1.8.

[4] L. T. Su, S. Naffziger, and M. Papermaster, "Multi-Chip Technologies to Unleash Computing Performance Gains over the Next Decade" *IEDM Tech. Dig.*, Dec. 2017, pp. 1.1.1-1.1.8.

[5] L. England and I. Arsovski, "Advanced Packaging Saves the Day! - How TSV Technology Will Enable Continued Scaling" *IEDM Tech. Dig.*, Dec. 2017, pp. 3.5.1-3.5.4.

[6] G. Hellings, *et al.*, "Active-lite interposer for 2.5 & 3D integration," in *Proc. IEEE VLSI Technol. Symp.*, Jun. 2015, pp. T222-T223.

[7] R. Mahajan, *et al.*, "Embedded multi-die interconnect bridge (EMIB) - - a high density, high bandwidth packaging interconnect," *IEEE 66th Electronic Components and Technology Conference (ECTC)*, May 2016, pp. 557-565.

[8] L. T. Su, S. Naffziger, and M. Papermaster, "Multi-Chip Technologies to Unleash Computing Performance Gains over the Next Decade" *IEDM Tech. Dig.*, Dec. 2017, pp. 1.1.1-1.1.8.

[9] L. England and I. Arsovski, "Advanced Packaging Saves the Day! - How TSV Technology Will Enable Continued Scaling" *IEDM Tech. Dig.*, Dec. 2017, pp. 3.5.1-3.5.4.

[10] P. K. Jo, X. Zhang, J. L. Gonzalez, G. S. May, and M. S. Bakir, "Heterogeneous Multi-Die Stitching Enabled by Fine-Pitch and Multi-Height Compressible Microinterconnects (CMIs)," *IEEE Trans. Electron Devices*, vol. 65, no. 7, pp. 2957-2963, July. 2018.

[11] P. K. Jo, M. Zia, J. L. Gonzalez, H. Oh, and M. S. Bakir, "Design, fabrication, and characterization of dense compressible microinterconnects," *IEEE Trans. Compon., Packag., Manuf. Technol.*, vol. 7, no. 7, pp. 1003-1010, May 2017.

[12] M. O. Hossen, J. L. Gonzalez, and M. S. Bakir, "Thermomechanical Analysis and Package Level Optimization of Mechanically Flexible Interconnects (MFIs) for Interposer-on-Motherboard Assembly," *IEEE Trans. Compon., Packag., Manuf. Technol.*, vol. 8, no. 12, pp. 2081-2089, Dec. 2018.

[13] Y. Li *et al.*, "Electromagnetic Characteristics of Multiport TSVs Using L-2L De-Embedding Method and Shielding TSVs," *IEEE Trans. Electromagn. Compat.*, vol. 59, no. 5, pp. 1541-1548, Oct. 2017.

[14] H. Yen *et al.*, "TSV RF de-embedding method and modeling for 3DIC," *SEMI Advanced Semiconductor Manufacturing Conference*, May. 2012, pp. 394-397.

978-1-7281-1500-9/19 $31.00 © 2019 IEEE

[15] C. Zhang, H. S. Yang and M. S. Bakir, "Highly Elastic Gold Passivated Mechanically Flexible Interconnects," *IEEE Trans. Compon., Packag., Manuf. Technol.*, vol. 3, no. 10, pp. 1632-1639, Oct. 2013.

[16] I. Ndip *et al.*, "Analytical, Numerical-, and Measurement–Based Methods for Extracting the Electrical Parameters of Through Silicon Vias (TSVs)," *IEEE Trans. Compon., Packag., Manuf. Technol.*, vol. 4, no. 3, pp. 504-515, Mar. 2014.

Characterization of the current mechanisms and improved leakage current in silver doped barium strontium titanate

Todd Schumann, Kyoung-Tae Kim, Sheng-Po Fang, and Yong-Kyu Yoon
Electrical and Computer Engineering
University of Florida
Gainesville, FL, United States
toddschumann@gmail.com
ykyoon@ece.ufl.edu

Abstract—Barium strontium titanate (BST) has gained attention as a high dielectric constant material for high capacitance density. The films are doped with silver, which can reduce the leakage current by an order of magnitude. In this report, we characterize the current mechanisms of doped and undoped BST films as space charge limited (SCL) transport and Schottky emission, respectively. Additionally, by extracting the trap density from the SCL flow, we show that the optimal doping concentration of silver for leakage reduction is calculated, which matches well with experimental data.

Keywords- barium strontium titanate (BST); perovskite materials; high dielectric constant; Schottky emmission, space charge limited (SCL) current

I. INTRODUCTION

Barium strontium titanate is a perovskite dielectric material known for its extremely high dielectric constant, even when deposited as a thin film [1]. Additionally, as a ferroelectric material, the permittivity can be tuned by an electric field [2]. This allows for creating tunable radio frequency (RF) devices such as transmission lines [3], antennas [4], and matching networks [5].

Of particular note in RF devices created with BST is the loss, which is directly related to leakage through the BST films. To minimize leakage through the BST, it is critically important to understand the current mechanisms exhibited. Meantime, it has been shown that by doping the BST with metals such as silver [6] and gold [7], the leakage through the films can be reduced by over an order of magnitude.

In this report, we show that as BST films are doped with silver, the leakage switches from Schottky emission to space charge limited (SCL) current. In addition, the trap density resulting from oxygen vacancies in the films is extracted and used to determine the optimal doping concentration, which matches well with the experimental data in [6].

II. FABRICATION PROCESS

Perovskite materials are described in an oxide compound chemical form of ABO_3. They are often deposited using a chemical solution deposition technique [8]. Although a sputtering process generally yields better performing films [9] and is within the CMOS thermal budget, the molar ratio between the A-site atoms and the B-site atoms in the

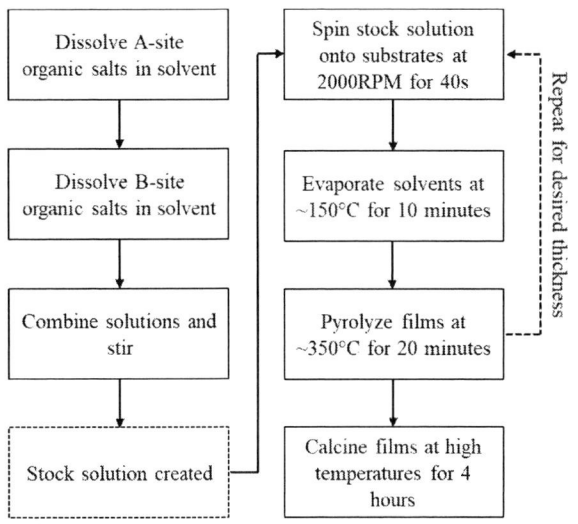

Figure 1. Generalized process flow for chemical solution deposition (CSD) of perovskite materials. Oxygen vacancies are introduced because of non-uniform molecular level mixing being unable to be compensated with high temperature calcinations.

perovskite crystal is not necessarily preserved [10]. This effect is complicated further when the final perovskite film consists of a combination of two perovskites with different A or B site atoms, such as the case with barium strontium titanate: $Ba_xSr_{1-x}TiO_3$ (BST). In this case, not only must the (Ba + Sr):Ti molar ratio be maintained, but also the Ba:Sr molar ratio, which can vary significantly from the target material based on the sputtering parameters [10].

In the case of CSD techniques, the molar ratio used in the preceding solution is preserved in the final film, making studies of the effect of the Ba:Sr ratio much more convenient. This also allows adjustment of the Ba:Sr ratio to be done rapidly without investment in an entirely different target.

The generalized process flow for CSD techniques is shown in figure 1. It involves creating a solution containing ions of the metals used in the final film, then through a series of heat treatments, removing all organic elements before finally crystalizing the film into the perovskite crystal [8]. In the case of BST, a common recipe is to dissolve the desired

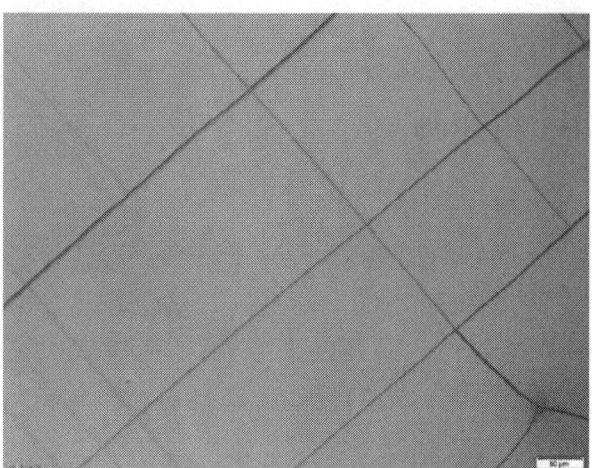

Figure 2. Example of film cracking due to thermal stress that can occur if the calcination temperatures used are too high. The difference in coefficients of thermal expansion leads to high interfacial stress during cooling after the calcination leading to cracking along the crystal directions.

molar ratio of Ba-acetate and Sr-acetate in acetic acid and combine them with a solution of an equal molar amount of Ti-isopropoxide dissolved in 2-methoxyethanol [6]. This solution is then spin coated on samples to yield a controllable thickness with low surface roughness.

The first heating step is to evaporate the solvents (acetic acid and 2-methoxyethanol in the case of BST). At this point, the organic metal salts are precipitated on the surface of the sample. However, there is a large discrepancy between the physical size of the organic metal salts and the final lattice spacing of the BST crystal. Attempting to immediately crystallize the BST at this point would yield a porous film.

Instead a second heating step is used to pyrolyze the films. This burns off the organic elements of the salts, leaving only the oxides of the metals, which are much more comparable to the size of the final crystal. Additionally, the metal oxides are not dissolvable in the acetic acid or 2-methoxyethanol, so the process can be repeated. By keeping the stock solution molarity relatively low (approximately 0.3M) and performing repeating steps of spin/evaporate/pyrolyze, thicker films can be grown with sufficient density to calcine into good quality film [8].

The final heating stage is the calcination, where the BST crystal is formed from the metal oxides. Although the chemical equation shows that the precursor metal oxides have sufficient oxygen to form the final crystal as seen in equation 1, it assumes truly molecular level mixing with no randomness in the order (i.e. every BaO is surrounded by TiO_2 and vice versa). In reality, this is not the case, which leads to the formation of oxygen vacancies in the films. To minimize the number of oxygen vacancies, the calcination is performed in an oxygen environment at high temperatures [8]. However, the temperatures are limited due to the thermal mismatch between the films and the substrate. If the films are calcined at too high a temperature, they will crack when cooled to room temperature. Figure 2 shows an example of a film which cracked due to the thermal stress during cooling after calcining at 800 °C.

$$x(BaO) + (1-x)(SrO) + TiO_2 \rightarrow Ba_xSr_{1-x}TiO_3 \quad (1)$$

The oxygen vacancies act as deep level traps within the BST films, contributing a small, but observable number of free carriers (electrons in the case of BST [11]). These carriers allow conduction through the film, ultimately leading to losses.

To counteract the effects of the oxygen vacancies, the BST can be doped with additional metals during the synthesis of the stock solution. Using a metal with a lower oxidation state than either of the cations will substitute one of the cations with the lower oxidation state metal [11]. For example, it has been shown that silver will substitute either the barium or strontium A-site location in BST films [6]. Since the silver has a lower oxidation state (1+) than the barium or strontium (2+), an oxygen vacancy with two corresponding silver substitutions would nullify the oxygen vacancy.

Understanding the effect this has on the leakage current is critical to optimize the doping levels. For example, doping up to a certain concentration will reduce the leakage current, but surpassing this limit will cause the leakage to increase higher than the intrinsic films [6].

III. CHARACTERIZATION OF LEAKAGE CURRENT

Kim, et al. showed that doping BST with a higher mol% of silver resulted in drastically reduced leakage current up to 5 mol% where the leakage current increased to higher than the original leakage [6]. In order to determine the optimal dose of silver, it is important to understand the current mechanism.

A. Without silver doping (0 mol% Ag)

The current flow without doping to reduce the leakage is characteristic of emission over a Schottky barrier. As seen in the top graph of figure 3, the JV trend shows a very linear region between 1-5 V when plotted on a semilogarithmic scale. This matches the exponential trend of thermionic emission at the reverse biased interface [12]:

$$J = A^* T^2 \exp\left(-\phi_b / k_B T\right) \exp\left(\beta \sqrt{E_0}\right). \quad (2)$$

$$A^* = \frac{4\pi q m^* k_B^2}{h^3}. \quad (3)$$

$$\beta = \left(\frac{q}{k_B T}\right) \sqrt{\frac{q}{4\pi \epsilon_{opt} \epsilon_0}}. \quad (4)$$

where J is the current density, T is temperature, φ_b is the barrier height, k_B is the Boltzmann constant, E_0 is the electric field at the metal-BST interface, q is the electron charge, m* is the carrier effective mass, h is Planck's constant, ϵ_{opt} is the optical dielectric constant of BST, and ϵ_0 is the vacuum permittivity.

978-1-7281-1500-9/19 $31.00 © 2019 IEEE

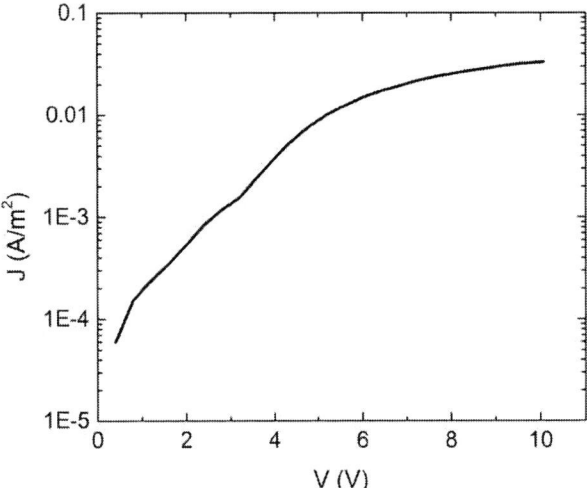

Figure 4. JV data for 0.5 mol% Ag doping from [6] plotted on a semilogarithmic scale. Although the trend looks exponential, it is better modelled by SCL currents given by regions of power law with different exponents as seen in figure 5.

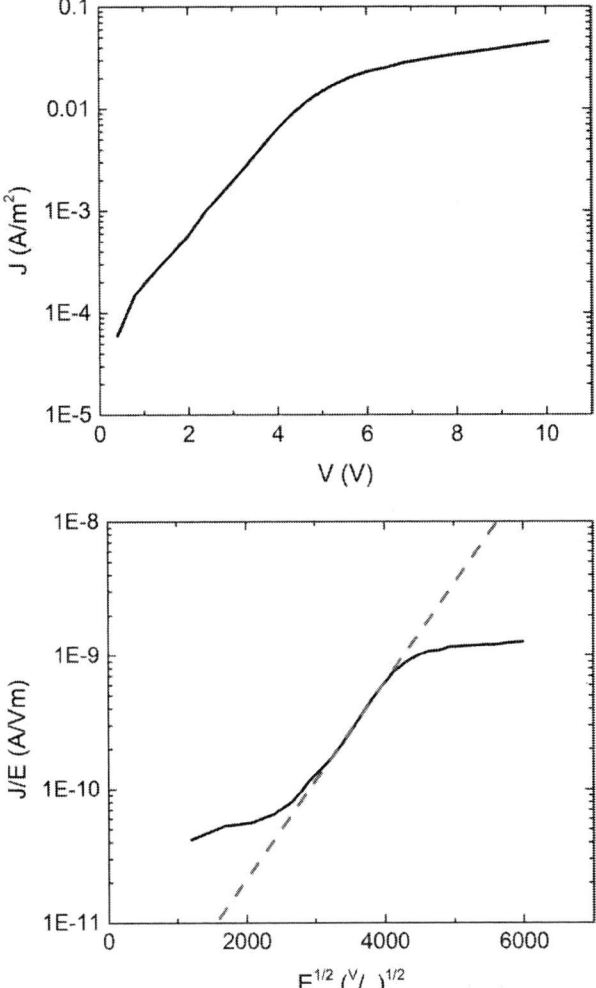

Figure 3. (top) JV data from [6] plotted on a semilogarithmic scale. The pronounced linear trend between 1-5 V indicates Schottky emmision. (bottom) The same data plotted on axes that allow fitting to equation 5. The corresponding fit line is shown as the dashed red line, which was used to extract a barrier height of 0.74 eV.

Traditionally, this equation is fit to the linear region in the semilogarithmic plot and used to extract the barrier height and Richardson constant, A^*. However, in the case of BST, the extracted Richardson constant is unrealistically low, indicating an unrealistic electron effective mass of approximately 1×10^{-4} m_0 [13].

Equations 2-4 were originally applied to carriers escaping into vacuum from metals and it has been shown that it only applies to dielectrics if the mean free path of the carriers is greater than the dielectric's thickness [14]. As the BST is oxygen depleted, this is not a safe assumption, thus a modified form of the Schottky equation must be used [13]:

$$J = 2q \left(\frac{2\pi m^* k_B T}{h^2} \right)^{3/2} \mu E_0 \exp\left(\frac{-\phi_b}{k_B T} \right) \exp\left(\beta \sqrt{E_0} \right). \quad (5)$$

where μ is the carrier mobility and the other symbols have the same meaning as in equations 2-4. As in [13], the BST films are assumed to be fully depleted as has been shown in other reports [15]. Thus, E_0 is taken to be V/t where t is the thickness of the BST, 280 nm.

The bottom graph in figure 3 shows the same data plotted on axes to reflect equation 5. The linear portion of the plot was fit and used to extract the barrier height, making the assumptions of m^*/m_0 and μ being equal to 5 and 1×10^{-3} cm²/Vs, respectively, the bulk values for the effective mass and mobility [13]. This results in a barrier height of 0.74 eV.

B. With silver doping

With the introduction of the silver dopant, Kim, et al. showed that the leakage current density decreases, even with relatively low concentrations. Figure 4 shows the leakage current data when the BST films were doped with 0.5 mol% Ag on semilogarithmic axes.

Initially, the data appears to follow the same Schottky emission trend. However, the JV trend follows a space charge limited (SCL) trend as seen in figure 5.

Prior to the onset of SCL flow, ohmic behavior is dominant, indicated by a slope of 1 on a log-log plot of JV (red dashed line in figure 5). At the onset of SCL flow, the slope abruptly increases to 2 (blue dashed line in figure 5), following the SCL trend [16,17]:

$$J = \frac{9}{8} \epsilon \mu \frac{V^2}{L^3}. \quad (6)$$

where V is the applied voltage and L is the thickness of the film, 280 nm.

As the voltage approaches the trap level of the oxygen vacancies, there is a large increase in the current, exhibited by a slope greater than 2 at approximately 3 V [18,19] corresponding to the trap levels being completely filled and

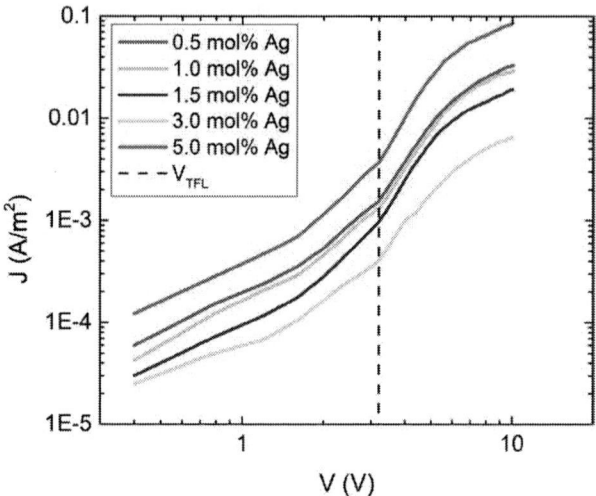

Figure 6. JV data plotted on a log-log scale for the silver doping in [6]. All measurements of the silver doped samples yielded SCL current trends. In addition, the V_{TFL}, where the trend changes from quadratic to superquadratic remains constant in all doping levels. This indicates that the silver doping is changing the trap cross-section area instead of the number of traps itself.

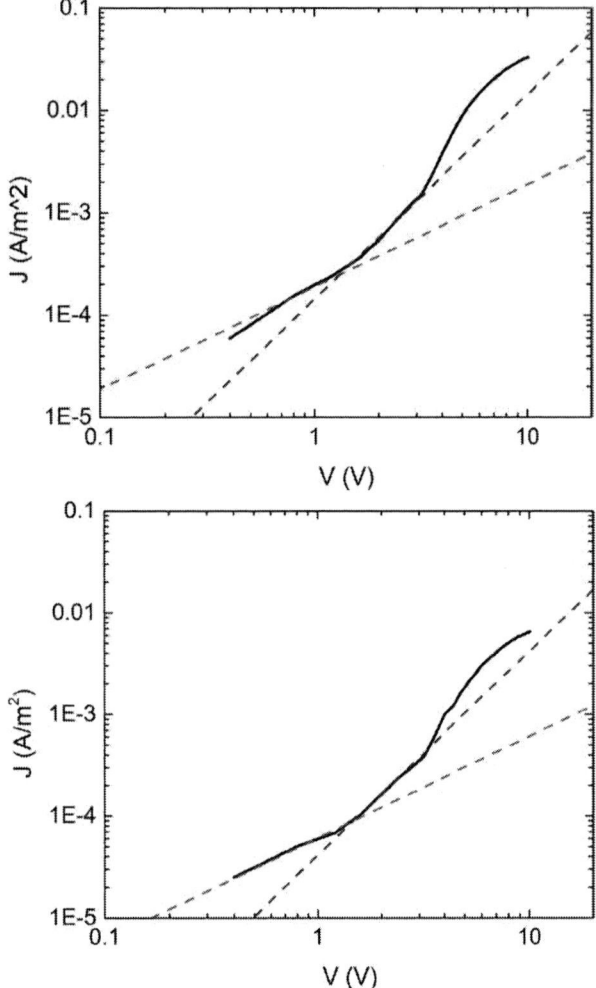

Figure 5. Data for 0.5 mol% Ag (top) and 3 mol% Ag (bottom) silver doping from [6] plotted with both axes on a logarithmic scale. The individual power law regions are clearly visible. The ohmic region is fitted with the red dashed line and the quadratic region is fitted with the blue dashed line in both plots.

additional carriers being injected directly into the conduction band. Finally, the series resistance becomes the limiting factor and the slope relaxes back to unity.

It should be noted that the current trend changes from V^1 to V^2 to $V^{>2}$. As the exponent in this voltage range is increasing, it also appears to approximately fit an exponential trend. However, as these are distinct sections of power law behavior, the trend would jump above and below the best fit line with each change to the power law, as can by the distinctive "humps" seen in figure 4.

On the other hand, the distinct sections of the power law are clearly visible in figure 5 for 0.5 mol% Ag doping and even more so for 3 mol% Ag. These distinct sections are absent in the intrinsic BST JV trend. In fact, all silver doping levels except the intrinsically 0 mol% Ag show SCL currents as the limiting current mechanism.

The transition from Schottky emission to SCL currents can be explained by the silver compensation of the oxygen vacancies. As the silver compensates the oxygen vacancies, the free carriers in the BST film are reduced, reducing the bulk conductivity. This allows SCL currents to become the limiting current mechanism over the interfacial Schottky emission.

When the JV curves defined by SCL currents are plotted together in figure 6, it can be seen that the voltage where the transition between quadratic and superquadratic occurs remains the same. This voltage is termed as the trap filled limit voltage, V_{TFL}, in the SCL with deep traps case and is directly related to the density of traps in the bulk [18,19]. As V_{TFL} remains the same for differing silver doping concentrations, the trap density, N_t is also remaining constant. Thus, the silver doping is not adding or removing traps as it is incorporated into the lattice. Instead, it is reducing the effective leakage cross-section by compensating the traps, resulting in reduction of leakage current [18]. As higher percentages of the total traps are compensated by silver, the leakage current reduces.

IV. DETERMINING THE OPTIMAL SILVER DOPING LEVEL

As the silver continues to compensate the oxygen vacancy traps, the leakage current continues to reduce up through 3 mol% Ag. Once the doping is increased to 5 mol%, the leakage current increases past the intrinsic level. This shift indicates that there is an optimal doping level of silver to minimize the leakage current (i.e. the silver doping is not creating a different perovskite crystal which simply has lower leakage).

To determine this doping level, the density of the oxygen vacancies is extracted. With the assumption that $p_{t,0} \gg n_{t,0}$, where $n_{t,0}$ is determined from the voltage at the onset of SCL flow and $p_{t,0}$ is determined from the voltage at the onset of

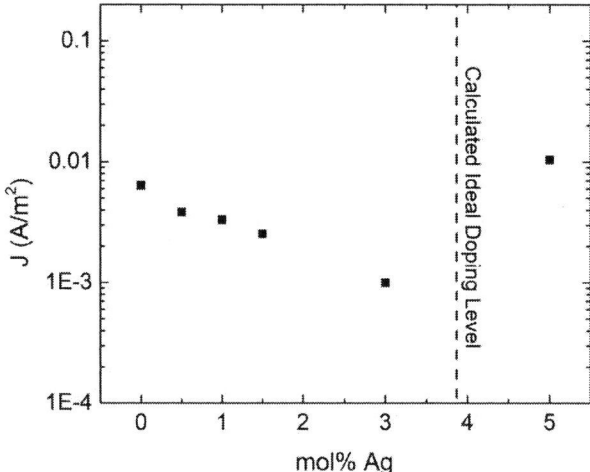

Figure 7. Plot of the leakage current taken at 4V for the different doping concentrations in [6] (the current trends are roughly parallel, so the voltage at which they are taken is arbitrary). The calculated ideal doping concentration from equation 8 is shown as the dashed line. It can be seen that the leakage current is reduced as the doping level approaches the ideal level, then increases beyond as the doping overcompensates the oxygen vacancies.

superquadratic JV trend, the trap density, N_t, can be determined by:

$$V_{TFL} = \frac{qN_tL^2}{2\epsilon}. \qquad (7)$$

where the symbols retain the same meaning as in previous equations [19].

In this case, it is also prudent to use the bulk value of the BST dielectric constant, ϵ. The significant difference in the dielectric constant between thin films and bulk BST arises from surface effects, which are often modeled as a set of series capacitors [20]. As the SCL current is flowing through the bulk, well beyond the shallow surface effects, the bulk value for the dielectric constant, approximately 20,000 [1], is used instead of the thin film effective dielectric constant. Solving equation 7 for the trap density yields $N_t = 9.01 \times 10^{19}$ cm^{-3} for the films.

In an ideal sense, each oxygen vacancy would be compensated with two surrounding silver impurities. Thus, to calculate the amount of silver required to compensate the oxygen vacancies, we can derive equation 8:

$$mol\% \, Ag_{ideal} = \frac{6N_tM}{\rho N_A}. \qquad (8)$$

where N_t is the trap density derived from equation 7, M is the molar mass of the BST crystal, ρ is the density of BST, and N_A is Avogadro's constant. In addition, the prefactor 6 comes from the fact that there are three oxygen in the BST crystal and two silver are required to compensate each oxygen vacancy.

Using equation 8, an ideal silver doping concentration of 3.87 mol% Ag can be calculated for $Ba_{0.6}Sr_{0.4}TiO_3$. Figure 7

shows a plot of the leakage current taken at 4V for each of the different doping concentrations along with the extracted ideal doping level. It can be seen that the calculated ideal doping concentration falls between 3 mol% and 5 mol%, which empirical results confirm bound the ideal doping concentration.

V. CONCLUSIONS

In this work we modelled the current flow in intrinsic and silver doped BST. The current model switched from Schottky emission to space charge limited current as silver was incorporated into the lattice. This was attributed to a reduction in the free carriers, allowing SCL bulk conduction to be the limiting current mechanism.

From the critical V_{TFL} in the silver doped films, the trap density, resulting from the oxygen vacancies, was extracted. This trap density was used to calculate the optimal silver doping concentration. The calculated value is in good agreement with the empirical results from [6] as it is bound by the lowest leakage current and where the leakage current increases past the intrinsic level.

This analysis allows the calculation of the ideal doping level for BST for an individual process. The number of oxygen vacancies will vary from process to process, even depending on fabrication non-idealities (e.g. cold spots in an oven used for calcination). Such an analysis can be used to tune an individual process to minimize the leakage current of the perovskite film.

Additionally, neither silver doping nor this analysis is limited to CSD deposition of BST or other perovskite films. The same analysis can be used for sputtered films, allowing the leakage current to be reduced in films fabricated using CMOS compatible techniques.

REFERENCES

[1] T. M. Shaw, et al. "The effect of stress on the dielectric properties of barium strontium titanate thin films," Applied Physics Letters, vol. 75, no..14, pp. 2129-2131, 1999.

[2] B. Su, J. E. Holmes, C. Meggs, and T. W. Button, "Dielectric and microwave properties of barium strontium titanate (BST) thick films on alumina substrates," Journal of the European Ceramic Society, vol. 23, no. 14, 2003

[3] J. Nath, D. Ghosh, J.-P. Maria, A. I. Kingon, W. Fathelbab, P. D. Franzon, and M. B. Steer, "An electronically tunable microstrip bandpass filter using thin-film Barium-Strontium-Titanate (BST) varactors," IEEE Transactions on Microwave Theory and Techniques, vol. 53, no. 9, 2005

[4] G. Wang, T. Polley, A. Hunt, and J. Papapolymerou, "A high performance tunable RF MEMS switch using barium strontium titanate (BST) dielectrics for reconfigurable antennas and phased arrays," IEEE Antennas and Wireless Propagation Letters, vol 4, pp. 217-220, 2005

[5] L.-Y. Vicki Chen, R. Forse, D. Chase, and R. A. York, "Analog tunable matching network using integrated thin-film BST capacitors," 2004 IEEE MTT-S International Microwave Symposium Digest, June 2004

[6] K.-T. Kim, C. Kim, D. Senior, D. Kim, Y. K. Yoon, "Microwave characteristics of sol-gel based Ag-doped $(Ba_{0.6}Sr_{0.4})TiO_3$ thin films," This Solid Films, vol. 565, pp. 172-178, 2014

[7] H.-W. Wang, S.-W Nien, and K.-C. Lee, "Enhanced tunability and electrical properties of barium strontium titanate thin films by gold doping in grains," Applied Physics Letters, vol. 84, no. 15, 2004

[8] R. W. Schwartz, "Chemical Solution Deposition of Perovskite Thin Films," Chemistry of Materials, vol. 9, no. 11, pp. 2325-2340, 1997.

[9] T. Horikawa, N. Mikami, T. Makita, J. Tanimura, M. Kataoka, K. Sato, and M. Nunoshita, "Dielectric Properties of (Ba, Sr)TiO$_3$ Thin Films Deposited by RF Sputtering," Japanese Journal of Applied Physics, vol. 32, no. 9S, pp. 4126, 1993.

[10] M. S. Tsai, S. C. Sun, and T. Y. Tseng, "Effect of oxygen to argon ratio on properties of (Ba,Sr)TiO$_3$ thin films prepared by radio-frequency magnetron sputtering," Journal of Applied Physics, vol. 82, no. 7, 1997

[11] M. W. Cole, C. Hubbard, E. Ngo, M. Ervin, and M. Wood, "Structure–property relationships in pure and acceptor-doped Ba$_{1-x}$Sr$_x$TiO$_3$ thin films for tunable microwave device applications," Journal of Applied Physics, vol. 92, no. 1, 2002

[12] A. J. Dekker, Solid State Physics (Prentice-Hall, Inc. , Englewood Cliffs, New Jersey, 1957), pp. 220-226.

[13] S. Zafar, R. E. Jones, B. Jiang, B. White, V. Kaushik, S. Gillespie, "The electronic conduction mechanism in barium strontium titanate thin films," Applied Physics Letters, vol. 73, no. 24, 1998

[14] J. G. Simmons, "Richardson-Schottky Effect in Solids," Physics Review Letters, vol. 15, no. 25, 1965

[15] G. W. Dietz, M. Schumacher, and R. Waser, "Leakage currents in Ba$_{0.7}$Sr$_{0.3}$TiO$_3$ thin films for ultrahigh-density dynamic random access memories," Journal of Applied Physics, vol, 82, no. 5, pp. 2359-2364, 1997

[16] P. Mark, "Space-Charge-Limited Currents in Organic Crystals," Journal of Applied Physics, vol. 33, no. 1, pp. 205-215, 1962

[17] M. Lampert, "Simplified Theory of Space-Charge-Limited Currents in an Insulator with Traps," Physical Review, vol. 103, no. 6, 1956

[18] W. Xu, H. Rao, and G. Bosman, "Evidence of space charge limited flow in the gate current of AlGaN/GaN high electron mobility transistors," Applied Physics Letters, vol. 100, no. 22, 2012

[19] M. Lampert, "Volume-controlled current injection in insulators," Reports on Progress in Physics, vol. 27, no. 1, 1964

[20] M. Stengel and N. A. Spaldin, "Origin of the dielectric dead layer in nanoscale capacitors," Nature, vol. 443, no. 7112, pp. 679-682, 2006.

High Temperature Aging Effects in SAC and SAC+X Lead Free Solders

Mohammad S. Alam, KM Rafidh Hassan, Jeffrey C. Suhling, Pradeep Lall
Department of Mechanical Engineering, and
Center for Advanced Vehicle and Extreme Environment Electronics (CAVE[3])
Auburn University
Auburn, AL 36849
E-mail: jsuhling@auburn.edu

Abstract— Lead free solders are common as interconnects in electronic packaging due to their relatively high melting point, attractive mechanical properties, good thermal cycling reliability, and environmentally friendly chemical properties. The mechanical behavior and reliability of a lead free solder is highly dependent on the operating temperature. Previous investigations on mechanical characterization of lead free solders have mainly emphasized stress-strain and creep testing at temperatures up to T = 125 °C. However, electronic devices sometimes experience harsh environment applications including well drilling, geothermal energy, automotive power electronics, and aerospace engines, where solders are exposed to very high temperatures from T = 125-200 °C. Knowledge on the mechanical properties of lead free solders at elevated temperatures is limited.

In our prior work presented at ECTC 2018, we investigated the mechanical behavior of several SAC and SAC+X lead free solder alloys including SAC305 (96.5Sn-3.0Ag-0.5Cu), SAC_Q (SAC+Bi), and Innolot (SAC+Bi+Ni+Sb) at extreme high temperatures up to 200 °C. In the current study, we have extended our prior work to consider extreme high temperature aging effects in lead free solders (SAC305 and SAC_Q). Before testing, the solder uniaxial specimens were aged (preconditioned) at the extreme high temperature of either T = 125 °C or T = 200 °C. At each of these aging temperatures, several durations of aging were considered including 0, 1, 5, and 20 days. Stress-strain and creep tests were then performed on the aged specimens. Using the measured data, the evolutions of the stress-strain and creep behaviors were determined as a function of aging temperature and aging time, and models describing the evolution of the mechanical properties with extreme aging were established. Microstructural evolution of the solder alloys during extreme high temperature aging has also been explored. In particular, aging induced coarsening of the IMCs has been studied using Scanning Electron Microscopy (SEM), and correlated to our material property evolution findings.

Our experimental measurements show that a 40-50% drop in strength and modulus occurred for both alloys between T = 125 °C and T = 200 °C. Comparisons between SAC305 and SAC_Q indicate that SAC_Q had significantly better mechanical properties at all test temperatures and prior aging conditions. Substantial degradations of the mechanical properties (initial modulus, ultimate tensile strength, and secondary creep strain rate) occur in SAC305 during extreme high temperature aging, whereas, SAC_Q exhibited relatively small variations in its properties during aging. Quantitative analysis of the SAC305 microstructural evolution during aging has shown that the number of IMC particles decreases during aging, while the average diameter of the particles increases significantly. The changes in particle size were rapid during the first 5 days of aging, and then became slower. This matches the aging induced degradations in the mechanical behavior of lead free solders.

Keywords: Microelectronics Reliability, Lead-Free Solder, High Temperature Aging, Mechanical Testing, Stress-Strain Curve, Creep, Modulus, Ultimate Tensile Strength, Secondary Creep Strain Rate, Microstructure, SEM Analysis.

I. INTRODUCTION

The reliabilities of electronic products depend strongly on the environmental conditions experienced during field use. Consumer electronics typically experience maximum operating temperatures of 100 °C. However, there are several harsh environment applications such as oil and gas exploration, avionics, automotive, and defense applications where electronics are exposed to much higher temperatures than those experienced by consumer electronics [1-2]. Insuring high reliability of such electronic products is very challenging. In under-the-hood automotive applications, electronic modules are operated at temperatures over 150 °C [3-4]. In oil and gas exploration applications, electronic systems experience ambient temperatures above 150 °C, and often up to 200 °C [5]. The electronics used in the logging tools are lowered into the wellbore during wireline logging applications and experience extremely low or high temperatures based on the location. In addition, this process typically lasts 2 to 6 hours at a time. Finally, electronic systems used in ground military vehicles, high-speed civil transport, supersonic aircraft, commercial and defense aircraft can be exposed to temperatures up to 200 °C [2].

The melting points of the widely used Sn-Ag-Cu (SAC) lead free solder alloys are typically in the range of T = 210 to T = 225 °C. Thus, the mechanical properties of lead free solders exposed to harsh environments can be degraded significantly, as the alloys are being used at temperatures near their melting points. In addition, longer exposures at such extreme high temperatures lead to aging-induced degradations to their properties [6]. Exposure to isothermal aging conditions results in microstructural evolution including coarsening and coalescing of IMC particles and sub grains [7-9], and breakdown of dendrite structures, as well as potential recrystallization at Sn grain boundaries. This leads to significant degradations in several areas including ball shear strength [10], elastic modulus [11], drop reliability [12], fracture behavior [13], creep behavior [14-17], thermal cycling reliability [17-23], Anand model parameters [21-22, 24-25], nanoindentation joint modulus and hardness [26-30], Poisson's ratios [31], high strain rate mechanical properties [32], and uniaxial and shear cyclic stress-strain curves and fatigue life [33-36].

SAC305 (96.5Sn-3.0Ag-0.5Cu) is commonly used in industry as solder interconnects during surface mount assembly because of its attractive mechanical properties, relatively high melting point, and thermal cycling reliability. However, addition of some new elements such as Bi, Sb, Ni, Mn, In, Co, Mg, Zn, La, Ce, and Ti to the traditional SAC alloys has been reported to facilitate improved wettability, shock/drop reliability, melting temperature, creep properties, and microstructure [16, 28, 37-40]. For example, addition of Bismuth (Bi) as a dopant can improve strength by means of precipitation hardening, reduce the solidification temperature, and also reduce of IMC layer thicknesses in lead free solder alloys. Matahir and coworkers [31] reported that an increase of Bi up to 2 wt% can improve the shear strength of SAC (Sn3.5Ag0.9Cu) alloy. Beyond that point, the shear strength reduced with increasing Bi% (wt), since higher Bi content led to the evolution of Bi-rich phase and fragmentation of the IMC.

In a studies conducted by Cai, et al. [16, 39], it was reported that addition of 0.1% Bi in SAC0307 solder considerably decreased the effects of isothermal aging. Witkin [40] reported on the mechanical and microstructural properties of SAC305 and some Bi-doped alloys, and found enhanced properties with Bi addition. Bismuth exists as a single phase in the microstructure, and it goes into solid solution within the Sn matrix when exposed to isothermal aging. Also, Bi doesn't form any intermetallic compounds with other constituents, and reduces IMC layer thicknesses [41-42].

Microalloy additions can also affect the microstructure of a SAC solder alloy significantly. Zhao, et al. [43] reported that the addition of 0.02% Ni to SAC105 increased the NiCuSn IMC and reduced the localized grain size at SAC/NiAu pad interfaces. Also, Sousa, et al. [44] conducted a detail study on SAC305 BGA solder joints with various doping elements (i.e. Co, Ni, Fe, In, Zn, and Cu) and concluded that addition of low levels of Zn improves the properties significantly. Lee, et al. [45] found that additions of low concentrations of Ni and Bi to SAC alloys improve thermal fatigue life and reduces impact resistance. Yeung, et al. [46] investigated the properties of SAC_Q (SAC+Bi) and reported a superior solder joint reliability during thermal cycling.

Most prior researches on lead free solder have been restricted to testing temperatures at or below T = 125 °C. Extreme high temperatures properties up to T = 200 °C are relatively unexplored. As discussed above, investigations on the mechanical and microstructural behaviors of these alloys at such extreme high temperatures are necessary to support several harsh environment electronics applications. Previously, we have investigated the mechanical behavior of several SAC and SAC+X lead free solder alloys including SAC305 (96.5Sn-3.0Ag-0.5Cu), SAC_Q (SAC+Bi), Innolot (SAC+Bi+Ni+Sb) at extreme high temperatures up to 200 C [47-50]. In our most recent work [51], we have reported on the temperature dependent stress-strain behavior of SAC305 and SAC_Q solders subjected to high temperature aging at T= 125 °C. In the current study, we have extended that work to also explore extreme high aging at T = 200 °C. Before testing, the solder uniaxial specimens were aged (preconditioned) at the extreme high temperature of either T = 125 °C or T = 200 °C. At each of these aging temperatures, several durations of aging were considered including 0, 1, 5, and 20 days. Stress-strain and creep tests were then performed on the aged specimens. Using the measured data, the evolutions of the stress-strain and creep behaviors were determined as a function of aging temperature and aging time, and models describing the evolution of the mechanical properties with extreme aging were established. Microstructural evolution of the solder alloys during extreme high temperature aging has also been explored. In particular, aging induced coarsening of the IMCs has been studied using Scanning Electron Microscopy (SEM), and correlated to our material property evolution findings.

II. EXPERIMENTAL PROCEDURE

A. High Temperature Test Matrix

Figures 1 and 2 show the experimental test matrices for the uniaxial tensile and creep tests, respectively. The solder alloys tested were SAC305 (96.5Sn-3.0Ag-0.5Cu), and SAC_Q (92.8Sn-3.4Ag-0.5Cu-3.3Bi). The chemical compositions were verified using Energy Dispersive X-Ray Spectroscopy (EDX). Test specimens were preconditioned with aging at T = 125 or T = 200 °C in a box oven for three different aging times (1 Day, 5 Days, and 20 Days). Tests have been performed on the preconditioned/aged samples along with the no aging samples at four testing temperatures including T = 125, 150, 175, and 200 °C.

Alloy	Strain Rate (sec⁻¹)	Test Temperature (°C)	Aging Condition (Aging T = 125 or T = 200 °C)			
			No Aging	1 Day Aging	5 Days Aging	20 Days Aging
SAC305	10⁻³	125	√	√	√	√
		150	√	√	√	√
		175	√	√	√	√
		200	√	√	√	√
SAC_Q		125	√	√	√	√
		150	√	√	√	√
		175	√	√	√	√
		200	√	√	√	√

Figure 1. Aging Test Matix for High Temperature Tensile Tests.

B. Uniaxial Test Sample Preparation

Uniaxial tension specimens of dimensions 80 x 3 x 0.5 mm were created using rectangular cross section glass tubes and a vacuum suction system as described in references [47-51]. In this study, the samples were initially quenched in a water bath, and then they were subsequently reflowed. The reflow profile was chosen to closely mimic profiles used for BGA assemblies, so that the obtained microstructures in the solder samples were similar to those found in typical solder joints. The test samples were stored in a low temperature freezer at T = -40 °C after solidification to avoid any unintentional aging effects on their properties.

Alloy	Stress Level (MPa)	Test Temperature (°C)	Aging Condition (Aging T = 125 °C)			
			No Aging	1 Day Aging	5 Days Aging	20 Days Aging
SAC305	4	125	√	√	√	√
		150	√	√	√	√
		175	√	√	√	√
		200	√	√	√	√
SAC_Q		125	√	√	√	√
		150	√	√	√	√
		175	√	√	√	√
		200	√	√	√	√

Figure 2.   Aging Test Matix for High Temperature Creep Tests.

C. Mechanical Testing System and Data Processing

The testing machine and thermal chamber shown in Fig. 3 were used to perform the mechanical testing in this work. Using thermocouples attached to trial specimens, we determined the times required for the specimen to sit in the heating chamber and come to the desired test temperatures. The applied forces were measured using a precision six-axis robotic load cell, while the average specimen strains (engineering strains) were measured using the monitored cross-head displacements. A gage length of 60 mm was utilized for all tests.

Figure 3.   Mechanical Test System and High Temperature Chamber.

A total of 8-10 uniaxial stress-strain tests were performed for each solder alloy and leg of the test matrix in Fig. 1. The raw data from all of the tests at a particular set of conditions were fitted using a hyperbolic tangent empirical model

$$\sigma = C_1 \tanh(C_2\varepsilon) + C_3 \tanh(C_4\varepsilon) \qquad (1)$$

and an "average" stress-strain curve was determined.

For the creep testing, 5 constant load/stress tests were performed for each leg of the test matrix. The four parameter Burger's (spring-dashpot) model given by

$$\varepsilon = C_0 + C_1 t + C_2(1 - e^{-C_3 t}) \qquad (2)$$

was used to fit the raw experimental data, where C_0, C_1, C_2 and C_3 are fitting constants. Constant C_1 represents the "steady state" creep strain rate.

III. Experimental Results

A. Uniaxial Tensile Test Results

The recorded experimental tensile test results for SAC305 and SAC_Q solder alloys with different aging durations at T = 200 °C are plotted in Fig. 4. Each of the graphs shows a comparative analysis of the uniaxial stress-strain behavior of both alloys at a particular test temperature, with the various colored curves representing the different aging durations (no aging, and 1, 5 , and 20 Days of aging at T = 200 °C). As mentioned before, the curves in the plots are the "average" stress-strain curves found by fitting the raw experimental data with the empirical model in eq. (1). The results are grouped in two sets of curves in each graph. The first set (top four curves) shows the aging dependent stress-strain behavior of SAC_Q, and the second set (bottom four curves) represents the same for SAC305. For each material, the top (blue) curve represents the average stress-strain curve for no aging, and the bottom (red) curve shows the average stress-strain curve at 20 days of aging.

At each test temperature, SAC_Q exhibited the best mechanical properties, and typically both the ultimate tensile strength (UTS) and effective elastic modulus (E) of SAC_Q were twice as large of the corresponding values for SAC305. Furthermore, the SAC305 alloy demonstrated large aging induced degradations with significant drop in mechanical properties after only 1 day of high temperature aging, while the SAC_Q alloy appeared to have negligible aging effects for all aging durations.

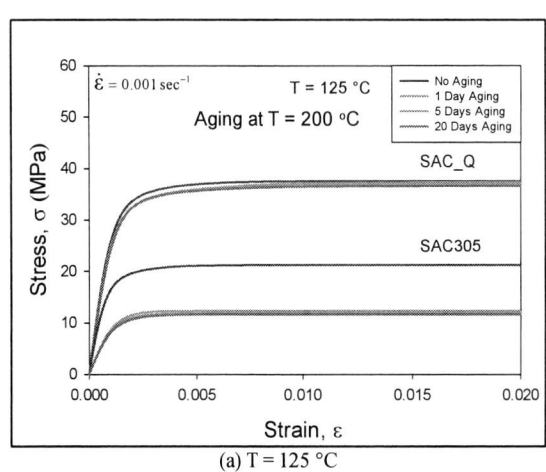

(a) T = 125 °C

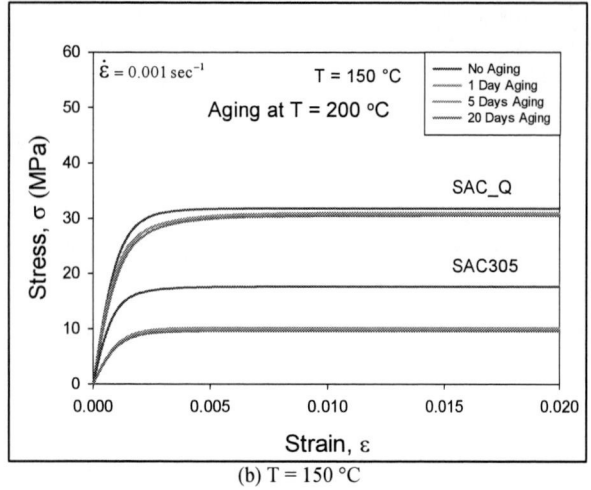

(b) T = 150 °C

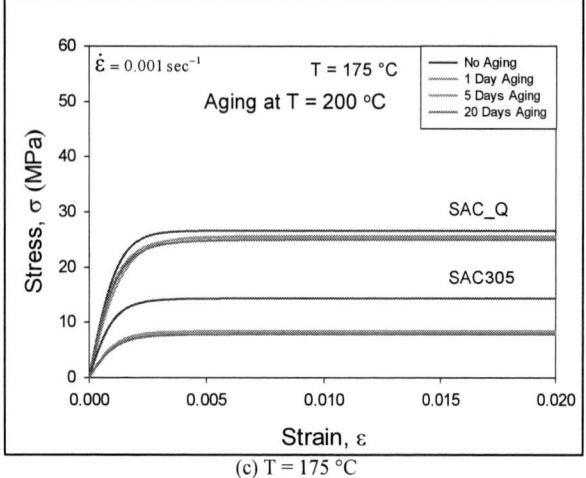

(c) T = 175 °C

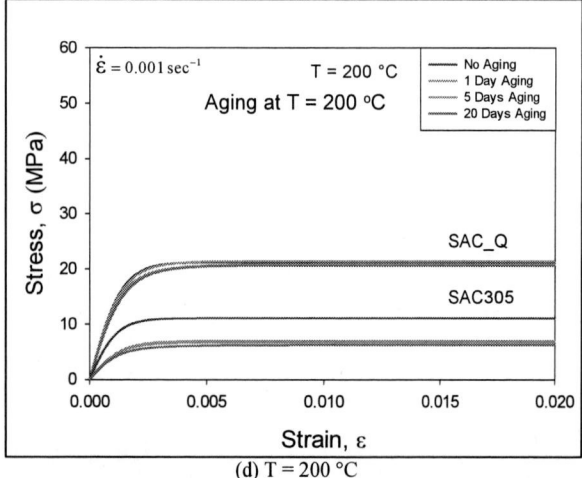

(d) T = 200 °C

Figure 4.   Comparison of Stress-Strain Curves for SAC305 and SAC_Q and Aging at T = 200 °C

Fig. 5 show tables of the aging dependent mechanical properties of SAC 305 and SAC_Q solder alloys for aging at 200 °C. These properties include effective elastic modulus and ultimate tensile strength extracted from the average stress-strain data. The mechanical properties of these same solder alloys aged at 125 °C from our previously published results [51] have been included as well for reference in Fig. 6. These properties are plotted as a function of aging time in Fig. 7. Large aging induced degradations are evident for SAC305 with a significant drop in properties after 5 days of aging at 200 °C. After that point, the properties degrade at a much slower rate. Conversely, the properties of SAC_Q were found to degrade very slowly or remain nearly constant at all aging conditions.

Alloy	Testing Temp (°C)	Effective Modulus, E (GPa)				Ultimate Strength, UTS (MPa)			
		No Aging	1 Day Aging	5 Days Aging	20 Days Aging	No Aging	1 Day Aging	5 Days Aging	20 Days Aging
SAC305	125	25.1	11.2	10.8	10.5	21.2	12.3	11.9	11.6
	150	18.4	9.0	8.8	8.4	17.9	10.2	10.0	9.6
	175	14.0	6.9	6.7	6.4	14.3	8.4	8.1	7.8
	200	8.6	5.2	5.0	4.7	10.9	6.9	6.7	6.2
SAC_Q	125	34.0	32.2	32.4	30.5	37.6	37.3	36.9	36.6
	150	27.5	26.1	25.2	23.4	31.8	31.0	30.9	30.5
	175	21.5	20.0	19.4	18.3	26.5	25.4	25.6	24.9
	200	15.6	15.5	14.7	14.0	21.2	21.4	20.9	20.6

Figure 5.   Aging Dependent Mechanical Properties of SAC305 and SAC_Q (Aging T = 200 °C)

Alloy	Testing Temp (°C)	Effective Modulus, E (GPa)				Ultimate Strength, UTS (MPa)			
		No Aging	1 Day Aging	5 Days Aging	20 Days Aging	No Aging	1 Day Aging	5 Days Aging	20 Days Aging
SAC305	125	25.1	9.4	9.0	8.6	21.2	12.8	12.1	11.6
	150	18.4	8.1	7.6	7.6	17.9	11.1	10.4	9.7
	175	14.0	6.6	6.4	6.2	14.3	9.5	8.8	8.4
	200	8.6	5.1	4.7	4.4	10.9	7.7	7.6	7.4
SAC_Q	125	34.0	32.9	32.0	30.0	37.6	36.7	36.1	35.7
	150	27.5	26.0	24.7	23.7	31.8	30.6	29.9	29.3
	175	21.5	19.7	18.8	17.6	26.5	25.4	24.7	24.2
	200	15.6	14.9	14.4	13.7	21.2	20.5	20.0	19.7

Figure 6.   Aging Dependent Mechanical Properties of SAC305 and SAC_Q (Aging T = 125 °C) [51]

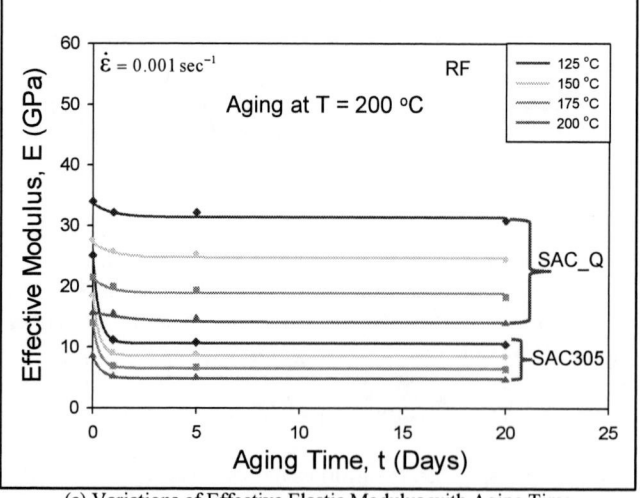

(a) Variations of Effective Elastic Modulus with Aging Time

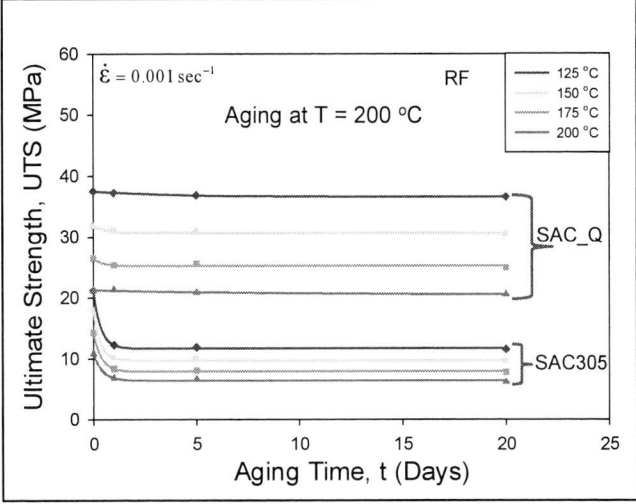

(a) Variations of Ultimate Tensile Strength with Aging Time

Figure 7. Comparison of Mechanical Properties of SAC305 and SAC_Q

B. Creep Test Results

The creep study in this work involved specimens with prior aging at T = 125 °C. Creep tests with aging at T = 200 °C are currently being explored and will be published at a later date. Fig. 8 illustrates the aging dependent creep curves for SAC305 and SAC_Q under a fixed stress level of 4 MPa. Each of the graphs shows a comparative analysis of the creep behavior of both alloys at a particular test temperature, with the various colored curves representing the different aging durations (no aging, and 1, 5, and 20 Days of aging at T = 125 °C). The solid curves represent average creep curves for SAC305 and the dashed curves correspond to SAC_Q. In all cases, the creep rates for SAC_Q are lower than those for SAC305 under comparable test conditions. Comparing the different graphs, we can see significant increases in the secondary creep strain rate for both alloys as the temperature increases. For testing at T = 200 °C, the samples experienced tertiary creep when the load was applied for about 2 hours. Increased dislocation movements occur for the higher creep rates at higher temperatures.

It is observed that SAC305 shows a significant degradation in its creep properties (sharp increase of the secondary creep strain rate) as the aging time increases. However, the creep rate for SAC_Q varied only slightly with aging, similar to the small changes observed in the tensile test properties with aging time. In addition, the creep behaviors of SAC_Q for all aging conditions were comparable to the creep behavior of SAC305 with no aging (the solid blue curve is within all the dashed curves at each test temperature). The average secondary creep strain rates extracted from the data are tabulated in Fig. 9. Also, the variations of secondary creep strain rate with testing temperature and aging time are plotted in Fig. 10.

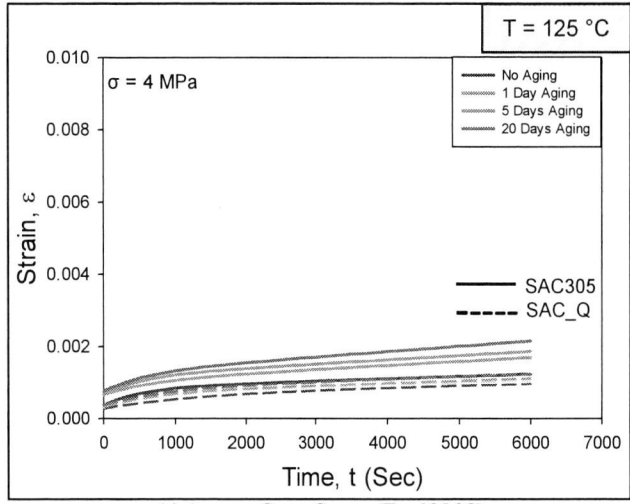

(a) Average Creep Curves (T = 125 °C)

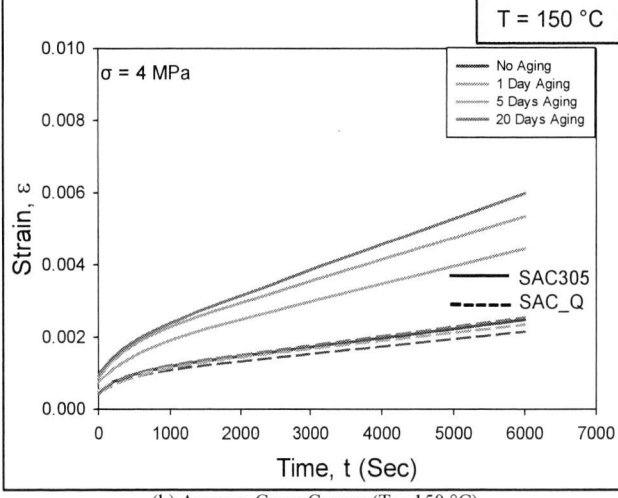

(b) Average Creep Curves (T = 150 °C)

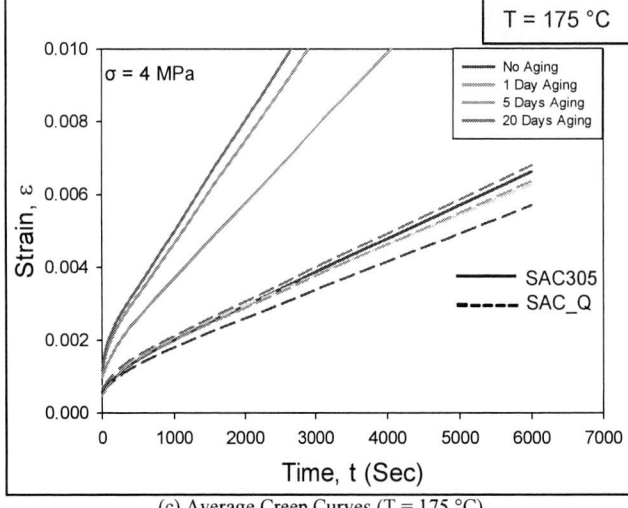

(c) Average Creep Curves (T = 175 °C)

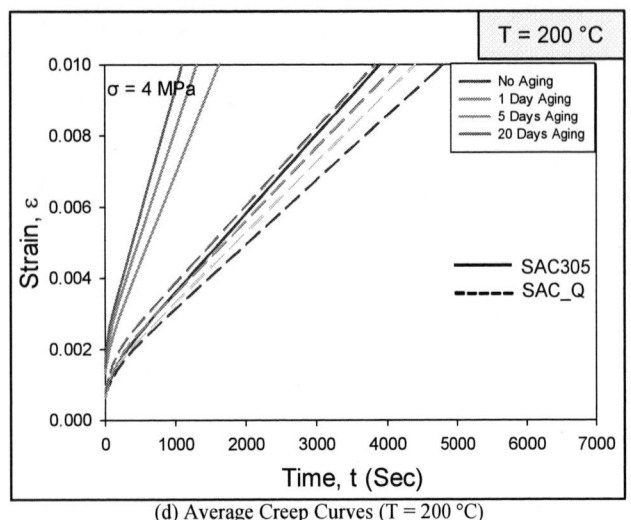

(d) Average Creep Curves (T = 200 °C)

Figure 8. Comparison of Creep Behavior of SAC305 and SAC_Q
[Aging at T = 125 °C]

Alloy	Stress Levels (MPa)	Test Temperature (°C)	Strain Rate (x 10^{-8} sec^{-1})			
			No Aging	1 Day Aging	5 Days Aging	20 Days Aging
SAC305	4	125	6.5	9.1	11.7	14.8
		150	25.2	48.2	60.0	71.2
		175	92.3	206.7	281.3	304.3
		200	220.4	496.4	597.5	705.1
SAC_Q	4	125	5.5	6.2	6.9	7.7
		150	20.5	22.0	22.4	25.8
		175	77.8	83.1	87.3	93.6
		200	181.2	197.1	205.1	217.3

Figure 9.   Secondary Creep Strain Rate (Aging at T = 125 °C)

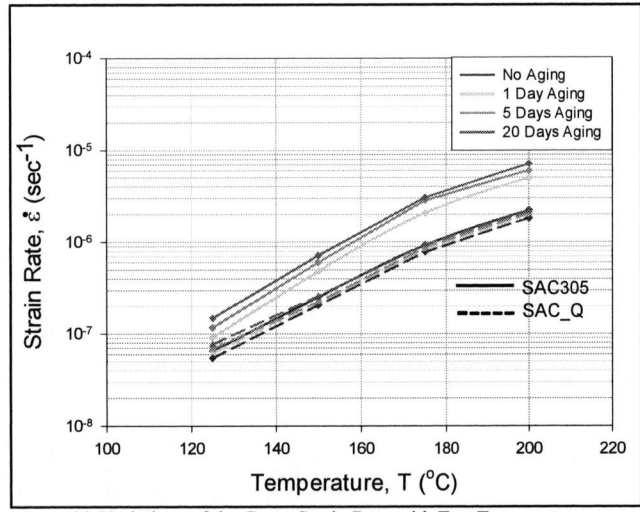

(a) Variations of the Creep Strain Rate with Test Temperature

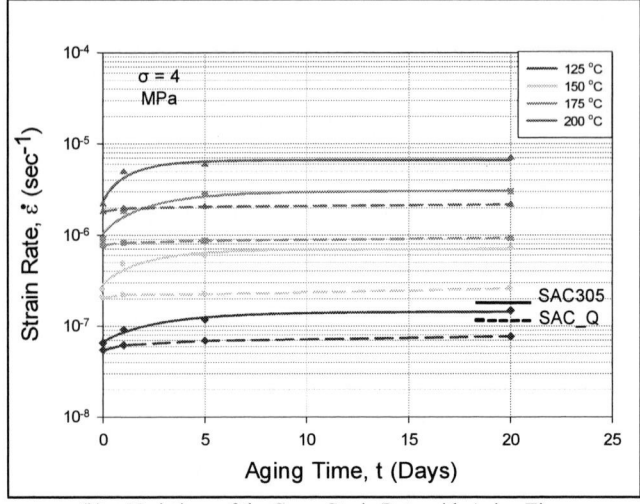

(b)   Variations of the Creep Strain Rate with Aging Time

Figure 10. Creep Properties of SAC305 and SAC_Q

C. Microstructural Evolution Due to Aging

In this study, the evolution of microstructure in SAC305 lead free solder due to high temperature aging has been examined.   Solder joint samples were prepared using a technique similar to that mentioned in the sample preparation section. Solder cross-sectional specimens were then encapsulated in an epoxy mold and polished to get a mirror-like surface. Several regions of interest consisting of β-Sn dendrites surrounded by eutectic regions (Ag_3Sn and Cu_6Sn_5 IMC particles) were identified.   Nanoindentation indents were used to mark the outside of these regions so they could be located again in the future [7-8].   SEM analysis of the selected regions was then performed on the non-aged as-polished samples. The same samples were then aged at three different extreme high temperatures (T= 125, 150, and 175 °C) for four different durations of aging (0, 1, 5, and 20 Days). The microstructures of the selected regions were captured after each aging interval. One SAC305 solder joint was prepared for each of the three aging temperatures, and three regions were studied in each joint. Example SEM images from one region at each aging temperature are shown in Fig. 11.

As expected, coarsening of intermetallic particles occurred due to aging. Some IMC particles grew in size, some split into several smaller particles, and some decreased in size and disappeared in the end. In other words, the number of IMC particles decreased, the average particle size increased, and the average particle separation distance increased.   Also, the particles shifted to more spherical shapes. The shape and size of the dendrites remained nearly unchanged.

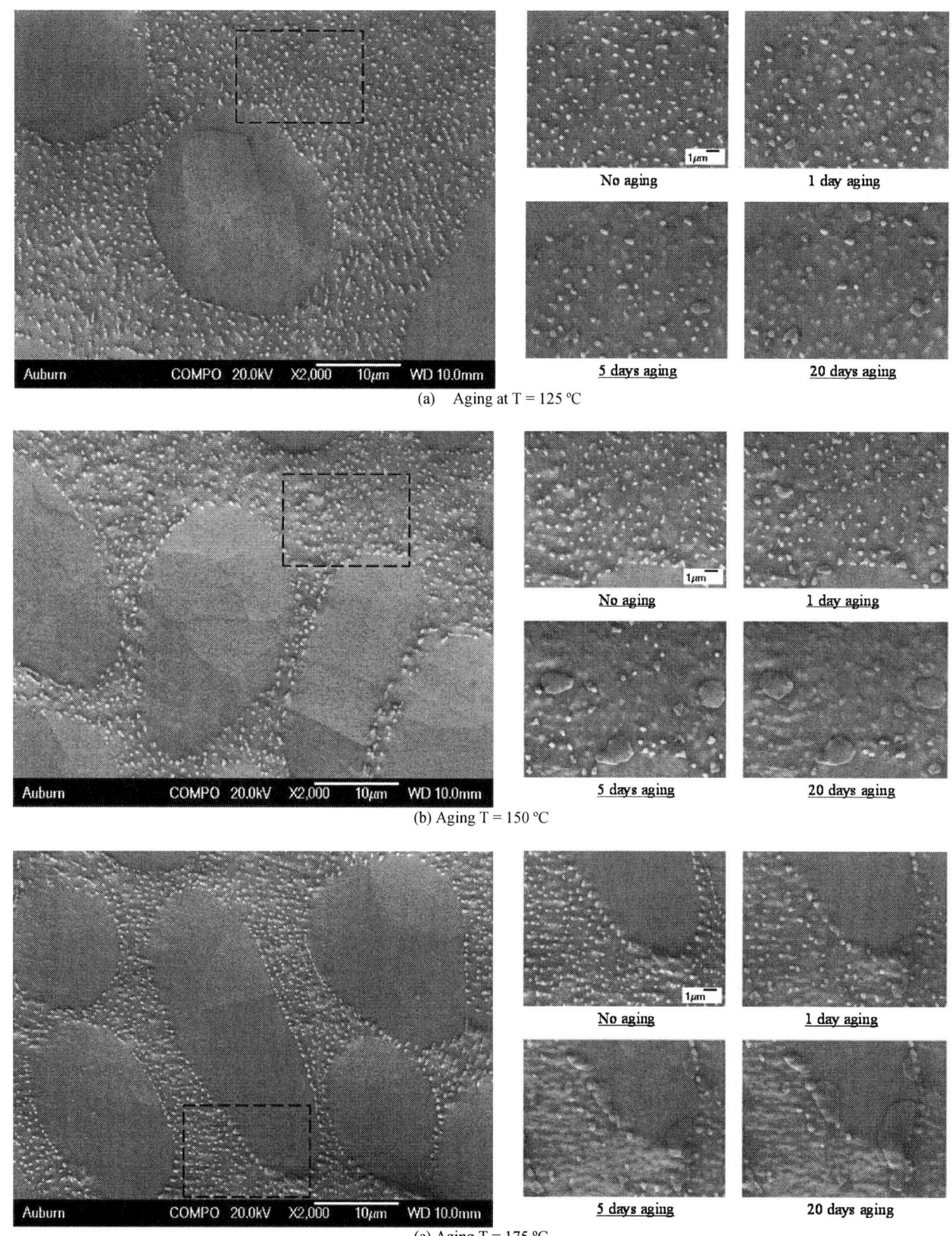

(a) Aging at T = 125 ºC

(b) Aging T = 150 ºC

(c) Aging T = 175 ºC

Figure 11. Microstructural Evolution in SAC305 Subjected to Extreme High Temperature Aging

Quantitative analyses of the size metrics of the IMC particles were performed with all of the SEM images at different aging times and aging temperatures. Image analysis software (ImageJ) and Matlab were used to determine the area of each particle, the total area of all of the particles, as well as the total number of particles in each selected region. These values were used to estimate the variation of the average particle diameter with aging. For an ideal spherical IMC particle, the amount of the particle that is visible on the polished cross-sectional surface is actually unknown. Using the methods discussed in ref. [7], the measured (apparent) average particle diameter has been adjusted to calculate the actual average particle diameter.

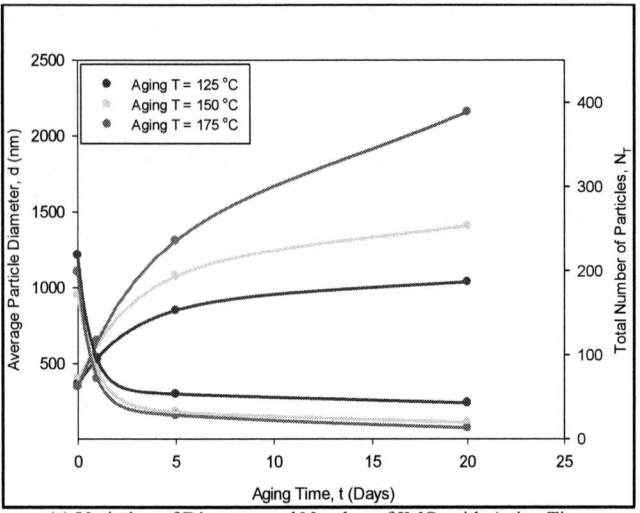

(a) Variation of Diameter and Number of IMCs with Aging Time

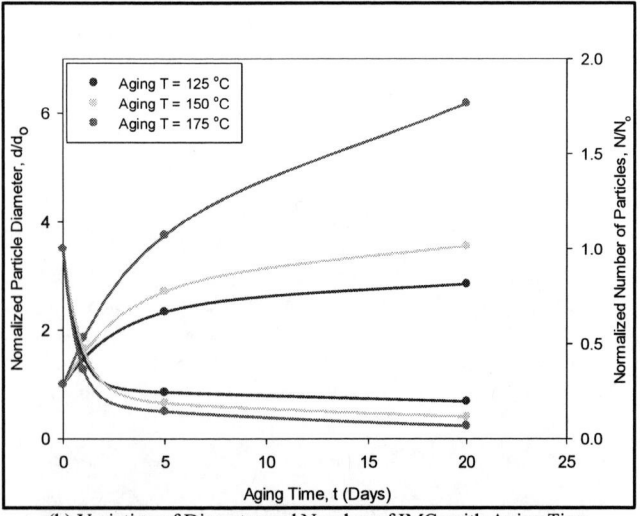

(b) Variation of Diameter and Number of IMCs with Aging Time (Normalized)

Figure 12. Quantitative Analysis of IMC Particles During High Temperature Aging of SAC305

The variations of the average IMC particle diameter and the number of IMC particles with aging time are shown in Fig 12. The colors of the data points and associated fitting curves indicate the aging temperature. The data were fitted using a two term exponential relation

$$\frac{d}{d_o} = k_1 e^{k_2 t} + k_3 e^{k_4 t} \qquad (4)$$

where d is the average diameter of all particles at time t, d_o is the average diameter of all particles at time t = 0, and K_1, K_2, K_3, and K_4 are fitting constants.

As with the mechanical properties, the major changes in IMC particle diameter occurred with the first 5 days of high temperature aging. The coarsening of IMC particles in the SAC305 solder alloy due to aging plays an important role in the degradations of its mechanical properties. IMC particles pin and block the dislocations movement. However, aging leads to both a smaller number of larger IMC particles and increased spacing between the IMC particles and thus facilitates movement of dislocation and degradation of strength of the SAC alloys.

Additional work to explore the evolution of SAC_Q microstructure during high temperature aging is currently underway. Results for aging at T = 125 °C have shown that the microstructure of SAC_Q is relative insensitive to aging [52]. After aging, the Bi phase in SAC_Q dissolves into the Sn matrix. Hence, Bi contributes to some enhancement in strength of the doped alloy by the solid solution strengthening mechanism. Also, the presence of Bi mitigates the coarsening of the Ag_3Sn IMCs during aging.

IV. SUMMARY AND CONCLUSIONS

In this paper, we have presented a comparative study of the aging dependent mechanical behavior of SAC305 and SAC_Q solder alloys at extreme high temperatures (T= 125, 150, 175, and 200 °C). Before testing, the solder uniaxial specimens were aged (preconditioned) at the extreme high temperature of either T = 125 °C or T = 200 °C, and several different durations of aging were considered (no aging, and 1 day, 5 days, and 20 days of aging). Stress-strain and creep tests were then performed on the aged specimens. Using the measured data, the evolutions of the stress-strain and creep behaviors were determined as a function of aging temperature and aging time, and models describing the evolution of the mechanical properties (effective elastic modulus and ultimate tensile strength, secondary creep strain rate) with extreme aging were established. Microstructural evolution of the solder alloys during extreme high temperature aging has also been explored. In particular, aging induced coarsening of the IMCs in SAC305 has been studied for aging at T = 125, 150, and 175 °C using Scanning Electron Microscopy (SEM), and correlated to our material property evolution findings.

Our experimental measurements show that a 40-50% drop in strength and modulus occurred for both alloys between T = 125 °C and T = 200 °C. Comparisons between SAC305 and SAC_Q indicate that SAC_Q had significantly better mechanical properties at all test temperatures and prior aging conditions. Substantial degradations of the mechanical properties (initial modulus, ultimate tensile strength, and secondary creep strain rate) occur in SAC305 during extreme high temperature aging, whereas, SAC_Q exhibited relatively small variations in its properties during aging. Quantitative analysis of the SAC305 microstructural evolution during aging has shown that the number of IMC particles decreases during aging, while the average diameter of the particles increases significantly. The changes in particle size were rapid during the first 5 days of aging, and then became slower. This matches the aging induced degradations in the mechanical behavior of lead free solders.

ACKNOWLEDGMENTS

This work was supported by the NSF Center for Advanced Vehicle and Extreme Environment Electronics (CAVE[3]).

REFERENCES

[1] McCluskey, P., Grzybowski, R., Podlesak, T., *High Temperature Electronics*, CRC Press, 1997.

[2] Ganesan, S., Pecht, M., *Lead-Free Electronics*, John Wiley and Sons, 2006.

[3] Hattori, M., "Needs and Applications of High-Temperature LSIs for Automotive Electronic Systems," *Proceedings of the Third European Conference on High Temperature Electronics (HITEN)*, pp. 37-43, Berlin, Germany, 1999.

[4] Johnson, R., Evans, J., Jacobsen, P., Thompson, J., Christopher, M., "The Changing Automotive Environment: High-Temperature Electronics," *IEEE Transactions on Electronics Packaging Manufacturing*, Vol. 27(3), pp. 164-176, 2004.

[5] Parmentier, B., Vermesan, O., Beneteau, L., "Design of High Temperature Electronics for Well Logging Applications," *Proceedings of the International Conference on High Temperature Electronics (HITEN)*, Oxford, UK, July 8-11, 2003.

[6] Ma, H., Suhling, J. C., "A Review of Mechanical Properties of Lead-Free Solders for Electronic Packaging," *Journal of Materials Science*, Vol. 44, pp. 1141-1158, 2009.

[7] Fu, N., Ahmed, S., Suhling, J. C., Lall, P., "Visualization of Microstructural Evolution in Lead Free Solders During Isothermal Aging Using Time-Lapse Imagery" *Proceedings of the 67th Electronic Components and Technology Conference*, pp. 429-440, Orlando, FL, May 30 - June 2, 2017.

[8] Ahmed, S., Wu, J., Fu, N., Suhling J. C., and Lall, P.; "Quantification and Modeling of Microstructural Evolution in Lead Free Solders During Long Term Isothermal Aging," *Proceedings of the 68th Electronic Components and Technology Conference (ECTC)*, pp. 162- 171, San Diego, CA, May 29 – June 1, 2018.

[9] Ahmed, S., Suhling, J. C., Lall, P., "Evaluation of Aging Induced Microstructural Evolution in Lead Free Solders Using Scanning Probe Microscopy," *Proceedings of ITherm 2018*, pp. 1062-1070, San Diego, CA, May 29 - June 1, 2018.

[10] Lee, C. B., Jung, S. B., Shin, Y. E., and Chang, C. C., "Effect of Isothermal Aging on Ball Shear Strength in BGA Joints with Sn-3.5Ag-0.75Cu Solder," *Materials Transactions*, Vol 43(8), pp. 1858-1863, 2002.

[11] Hasegawa, K., Noudou, T., Takahashi, A., Nakaso, A., "Thermal Aging Reliability of Solder Ball Joint for Semiconductor Package Substrate," *Proceedings of the 2001 SMTA International*, pp.1-8, 2001.

[12] Chiu, T. C., Zeng, K., Stierman, R., Edwards, D., Ano, K., "Effect of Thermal Aging on Board Level Drop Reliability for Pb-Free BGA Packages," *Proceedings of the 54th Electronic Components and Technology Conference (ECTC)*, pp. 1256-1262, 2004.

[13] Ding, Y., Wang, C., Li, M., Bang, H. S., "Aging Effects on Fracture Behavior of 63Sn37Pb Eutectic Solder during Tensile Tests under the SEM," *Materials Science and Engineering*, Vol. A384, 314-323, 2004.

[14] Ma, H., Suhling, J. C., Zhang, Y., Lall, P., and Bozack, M. J., "The Influence of Elevated Temperature Aging on Reliability of Lead Free Solder Joints," *Proceedings of the 57th IEEE Electronic Components and Technology Conference*, pp. 653-668, Reno, NV, May 29-June 1, 2007.

[15] Zhang, Y., Cai, Z., Suhling, J. C., Lall, P., Bozack, M. J., "The Effects of Aging Temperature on SAC Solder Joint Material Behavior and Reliability," *Proceedings of the 58th IEEE Electronic Components and Technology Conference*, pp. 99-112, Orlando, FL, May 27-30, 2008.

[16] Cai, Z., Zhang, Y., Suhling, J. C., Lall, P., Johnson, R. W., Bozack, M. J., "Reduction of Lead Free Solder Aging Effects Using Doped SAC Alloys," *Proceedings of the 60th Electronic Components and Technology Conference*, pp. 1493-1511, 2010.

[17] Zhang, J., Hai, Z., Thirugnanasambandam, S., Evans, J. L., Bozack, M. J., Sesek, R., Zhang, Y., Suhling, J. C., "Correlation of Aging Effects on Creep Rate and Reliability in Lead Free Solder Joints," *SMTA Journal*, Vol. 25(3), pp. 19-28, 2012.

[18] Zhang, J., Hai, Z., Thirugnanasambandam, S., Evans, J. L., Bozack, M. J., Zhang, Y., Suhling, J. C., "Thermal Aging Effects on Thermal Cycling Reliability of Lead-Free Fine Pitch Packages," *IEEE Transactions on Components and Packaging Technologies*, Vol. 3(8), pp. 1348-1357, 2013.

[19] Hai, Z., Zhang, J., Shen, C., Snipes, E. K., Suhling, J. C., Bozack, M. J., and Evans, J. L., "Reliability Degradation of SAC105 and SAC305 BGA Packages Under Long-Term, High Temperature Aging," *SMTA Journal*, Vol. 27(2), pp. 11-18, 2014.

[20] Hai, Z., Zhang, J., Shen, C., Evans, J. L., Bozack, M. J., Basit, M. M., Suhling, J. C., "Reliability Comparison of Aged SAC Fine-Pitch Ball Grid Array Packages Versus Surface Finishes," *IEEE Transaction on Components, Packaging, and Manufacturing Technology*, Vol. 5(6), pp. 828-837, 2015.

[21] Motalab, M., Cai, Z., Suhling, J. C., Zhang, J., Evans, J. L., Bozack, M. J., Lall, P., "Improved Predictions of Lead Free Solder Joint Reliability That Include Aging Effects," *Proceedings of the 62nd IEEE Electronic Components and Technology Conference*, pp. 513-531, San Diego, CA, May 30 - June 1, 2012.

[22] Motalab, M., Cai, Z., Suhling, J. C., Zhang, J., Evans, J. L., Bozack, M. J., Lall, P., "Correlation of Reliability Models Including Aging Effects with Thermal Cycling Reliability Data," *Proceedings of the 63rd IEEE Electronic Components and Technology Conference*, pp. 986-1004, Las Vegas, NV, May 28-31, 2013.

[23] Basit, M., Motalab, M., Suhling, J. C., Hai, Z., Evans, J. L., Bozack, M. J., and Lall, P., "Thermal Cycling Reliability of Aged PBGA Assemblies - Comparison of Weibull Failure Data and Finite Element Model Predictions," *Proceedings of the 65th IEEE Electronic Components and Technology Conference*, pp. 106-117, San Diego, CA, May 27-29, 2015.

[24] Basit, M. M., Motalab, M., Suhling, J. C., and Lall, P., "The Effects of Aging on the Anand Viscoplastic Constitutive Model for SAC305 Solder," *Proceedings of ITherm 2014*, pp. 112-126, Orlando, FL, May 28-30, 2014.

[25] Basit, M. M., Ahmed, S., Motalab, M., Roberts, J. C., Suhling, J. C., and Lall, P., "The Anand Parameters for SAC Solders after Extreme Aging," *Proceedings of ITherm 2016*, pp. 440-447, Las Vegas, NV, June 1-3, 2016.

[26] Hasnine, M., Mustafa, M., Suhling, J. C., Prorok, B. C., Bozack, M. J., Lall, P., "Characterization of Aging Effects in Lead Free Solder Joints Using Nanoindentation," *Proceedings of the 63rd IEEE Electronic Components and Technology Conference*, pp. 166-178, Las Vegas, NV, May 28-31, 2013.

[27] Hasnine, M., Suhling, J. C., Prorok, B. C., Bozack, M. J., Lall, P., "Exploration of Aging Induced Evolution of Solder Joints Using Nanoindentation and Microdiffraction," *Proceedings of the 64th IEEE Electronic Components and Technology Conference*, pp. 379-394, Orlando, FL, May 28-30, 2014.

[28] Hasnine, M., Suhling, J. C., Prorok, B. C., Bozack, M. J., Lall, P., "Nanomechanical Characterization of SAC Solder Joints - Reduction of Aging Effects Using Microalloy Additions," *Proceedings of the 65th IEEE Electronic Components and Technology Conference*, pp. 1574-1585, San Diego, CA, May 27-29, 2015.

[29] Hasnine, M., Suhling, J. C., Prorok, B. C., Bozack, M. J., and Lall, P., "Anisotropic Mechanical Properties of SAC Solder Joints in Microelectronic Packaging and Prediction of Uniaxial Creep Using Nanoindentation Creep," *Experimental Mechanics*, Vol 57(4), pp. 603-614, 2017.

[30] Ahmed, S., Hasnine, M., Suhling, J. C., and Lall, P., "Mechanical Characterization of SAC Solder Joints at High Temperature Using Nanoindentation," *Proceedings of the 67th IEEE Electronic Components and Technology Conference*, pp. 1128-1135, Orlando, FL, May 30 - June 2, 2017.

[31] Hassan, KM. R., Alam, M. S., Basit, M. M., Suhling, J. C., Lall, P., "The Effects of Temperature, Strain Rate, and Aging on the Poisson's Ratio of SAC Lead Free Solders," *Proceedings of InterPACK 2018*, Paper InterPACK2018-8410, pp. 1-11, San Francisco, CA, August 28-30, 2018.

[32] Lall, P., Shantaram, S., Suhling, J., Locker, D., "Effect of Aging on the High Strain Rate Mechanical Properties of SAC105 and SAC305 Leadfree Alloys," *Proceedings of the 63rd IEEE Electronic Components and Technology Conference*, pp. 1277-1293, Las Vegas, NV, May 28-31, 2013.

[33] Mustafa, M., Cai, Z., Suhling, J. C., and Lall, P., "The Effects of Aging on the Cyclic Stress-Strain Behavior and Hysteresis Loop Evolution of Lead Free Solders," *Proceedings of the 61st IEEE Electronic Components and Technology Conference*, pp. 927-939, Orlando, FL, June 1-3, 2011.

[34] Mustafa, M., Roberts, J. C, Suhling, J. C, Lall, P., "The Effects of Aging on The Fatigue Life of Lead Free Solders," *Proceedings of the 64th IEEE Electronic Components and Technology Conference*, pp. 666-683, Orlando, FL, May 28-30, 2014.

[35] Mustafa, M., Cai, Z., Roberts, J. R., Suhling, J. C., Lall, P., "Evolution of the Tension/Compression and Shear Cyclic Stress-Strain Behavior of Lead-Free Solder Subjected to Isothermal Aging," *Proceedings of ITherm 2012*, pp. 765-780, San Diego, CA, May 30 - June 1, 2012.

[36] Mustafa M., Suhling J. C., Lall P., "Experimental Determination of Fatigue Behavior of Lead Free Solder Joints in Microelectronic Packaging Subjected to Isothermal Aging," *Microelectronics Reliability*, Vol. 56, pp. 136-147, 2016.

[37] Huang, M. L. and Wang, L., "Effects of Cu, Bi and In on Microstructure and Tensile Properties of SnAg (Cu, Bi, In) Solders," *Metallurgical and Materials Transactions A*, Vol. 36(6), pp. 1439-1446, 2005.

[38] Matahir, M., Chin, L.T., Tan, K. S., Olofinjana, A. O., "Mechanical Strength and its Variability in Bi-modified Sn-Ag-Cu Solder Alloy," *Journal of Achievement in Materials and Manufacturing Engineering*, Vol. 46, pp. 50-56, 2011.

[39] Cai, Z., Suhling, J. C., Lall, P., Bozack, M. J., "Mitigation of Lead Free Solder Aging Effects using Doped SAC-X Alloys," *Proceedings of ITherm 2012*, pp. 896-909, San Diego, CA, May 30 - June 1, 2012.

[40] Witkin, D. B., "Influence of Microstructure on Mechanical Behavior of Bi-Containing Pb-Free Solders," *Proceedings of IPC APEX EXPO Conference and Exhibition*, pp. 540-547, San Diego, CA, Feb 19-21, 2013.

[41] Li, G., Shi, X., "Effects of Bismuth on Growth of Intermetallic Compounds in Sn-Ag-Cu Pb-Free Solder Joints," *Transactions of Nonferrous Metals Society of China*, Vol. 16, pp. 739-743, 2006.

[42] Hodulova, E., Palcut, M., Lechovic, E., Simekova, B., Ulrich, K., "Kinetics of Intermetallic Phase Formation at the Interface of Sn–Ag–Cu–X (X = Bi, In) Solders with Cu Substrate," *Journal of Alloys and Compounds*, Vol. 509, pp. 7052-7059, 2011.

[43] Zhao, Z., Wang, L., Xie, X., Wang, Q., Lee, J., "The Influence of Low Level Doping of Ni on the Microstructure and Reliability of SAC Solder Joint," *Proceedings of 9th International Conference on Electronic Packaging Technology & High Density Packaging*, pp. 1-5, Shanghai, China, 2008.

[44] Sousa, I., Henderson, D. W., Patry, L., Kang, S. K., Shih, D. Y., "The Influence of Low Level Doping on The Thermal Evolution of SAC Alloy Solder Joints with Cu Pad Structures," *Proceeding of the 56th Electronic Components and Technology Conference*, San Diego, CA, pp. 1454-1461, 2006.

[45] Lee, J. H., Kumar, S., Kim, H. J., Lee, Y. W. and Moon, J. T., "High Thermo-Mechanical Fatigue and Drop Impact Resistant Ni-Bi Doped Lead Free Solder," *Proceedings of the 64th Electronic Components and Technology Conference*, pp. 712-716, Orlando, FL, May 27-30, 2014.

[46] Yeung, T-S., Sze, H., Tan, K., Sandhu, J., Neo, C-W., Law, E., "Material Characterization of a Novel Lead-Free Solder Material-SACQ," *Proceedings of the 64th Electronic Components and Technology Conference*, pp. 518-522, Orlando, FL, May 27-30, 2014.

[47] Alam, M. S., Basit, M. M., Suhling, J. C., Lall, P., "Mechanical Characterization of SAC305 Lead Free Solder at High Temperatures," *Proceedings of ITherm 2016*, pp. 755-760, Las Vegas, NV, May 31- June 3, 2016.

[48] Alam, M. S., Suhling, J. C., Lall, P., "High Temperature Tensile and Creep Behavior of Lead Free Solders," *Proceedings of ITherm 2017*, pp. 1229-1237, Orlando, FL, May 30 - June 2, 2017.

[49] Alam, M. S., Hassan, KM. R., Suhling, J. C., Lall, P., "High Temperature Mechanical Behavior of SAC and SAC+X Lead Free Solders," *Proceedings of the 68th IEEE Electronic Components and Technology Conference*, pp. 1781- 1789, San Diego, CA, May 29 - June 1, 2018.

[50] Alam, M. S., Hassan, KM. R., Suhling, J. C., Lall, P., "A Comparative Study of the High Temperature Mechanical Behavior of Lead Free Solders," *Proceedings of ITherm 2018*, pp. 1314- 1323, San Diego, CA, May 29 – June 1, 2018.

[51] Alam, M. S., Hassan, KM. R., Suhling, J. C., Lall, P., "Investigation of the Effects of High Temperature Aging on the Mechanical Behavior of Lead Free Solders," *Proceedings of InterPACK* 2018, Paper InterPACK2018-8396, pp. 1-9, San Francisco, CA, August 27-30, 2018.

[52] Wu, J., Suhling, J. C., Lall, P., "Microstructural Evolution in SAC+X Solders Subjected to Aging," *Proceedings of the 69th IEEE Electronic Components and Technology Conference*, Las Vegas, NV, May 28-31, 2019.

Laundering Reliability of Electrically Conductive Fabrics for E-Textile Applications

Jeffrey ChangBing Lee[1], ChangHo Lo[1], Cheng-Chih Chen[1], Weifeng Liu[2]

[1]iST-Integrated Service Technology Inc.,Taiwan
[2]FLEX Ltd., US

E-mail: jeffrey_lee@istgroup.com

Abstract—This paper presents studies on the launderability of 4 types of conductive fabrics made by weaving polyester and nylon yarns with metal coatings (Cu, Ag, Ni/Cu, Ni/Cu/Co). They are laminated with thermoplastic urethane (TPU) film under hot compression on the 3 common fabrics (Spandex, Nylon, Denim) with different elongation and flexibility. Electrical resistance as a function of laundry cycles is used to characterize the performance of the conductive materials. The laundry procedure is to follow AATCC M6 test standard by using the AATCC compliant laundry machine and dryer with the factor control such as detergent, water temperature, agitation speed and spin speed and so on.

After intended wash and dry cycles, test samples are measured in electrical resistance with 4 point probe ohm meter to detect the resistance stability. The rise of resistance of conductive material over the wash/dry cycle can be compared among 4 metal coating on the 3 common fabrics. In general, the unstable resistance and electrical open can be reflected with the microstructure fracture observation under 3D optical microscope and SEM to provide further insight on the performance of these conductive materials withstanding laundering process. Furthermore, Non-destructive analysis of low angle XRD and XRF are also adopted to analyze the metal crystalline lattice structure and thickness change after laundering process for the understanding of degradation mechanisms. The elongation effect of base fabric of Spandex, Polyester and Nylon under certain washing shear force during agitation on the conductive fabric launderability can be concluded.

Keywords-E-textile, Launderability, Conductive Fabric, AATCC, Reliability

INTRODUCTION

Technology innovation has created the combination of electronics and textile together in order to satisfy the requirement monitoring accurately physical and vital signals for human being such as athletes, patients, soldiers, and other consumers interacting with their clothes daily.

The smart fabrics, also called "E-textile", is capable of interacting with their user or the environment, including tracking and communicating data to other devices through embedded sensors and conductive yarns, film or fabrics, which has been used currently in the fields of defense, sports/fitness, health, and public safety [1]. There are many technical challenges to overcome in order to identify and address fundamental quality and reliability issues when incorporating electronics into fabrics due to the distinctly different industries.

The launderability of E-textiles is one of the biggest reliability challenges for wearable application [2-5]. Many

critical components of E-textiles are the conductive materials to connect different sensors, modules, MCU and power supplies to form a body area network (BAN). In their lifetime, E-textiles will definitely experience many laundering and drying cycles, so that the electronic component and conductive materials may fail to deliver stable performance after multiple washing.

Limited data is available on the performance of conductive materials going through the washing and drying cycles. IPC released a white paper in Aug 2018 to provide insight for the industry brainstorming on reliability and washability of E-textile structures – Readiness for the Market [6], which highlighted the washable reliability issue for the E-textile product. In terms of reliability, E-textile structures should be well functioning over a period of several years according to the product guidelines. However, the additional issues of BoM integration including textile material, connector, actuators, passive, conductive trace material, embedded sensor, antenna and cross industry assembly are all critical for its success. IPC white paper emphasizes all the problems inherent in creating effective E-textiles, encompassing efforts that industry and research laboratories must cooperate to make E-textile structures more robust. Home laundering and commercial cleaning is an everyday reality for millions of textile wearable products. Explosive growth in E-textiles is just beginning to break through performance market segments making standardization urgently necessary. More and more e-textile laundering research identifies with the procedure and requirements gap of these merging manufacturing technologies [6].

In this paper, some understanding of the challenges and issues on the launderability reliability with the conductive fabrics applied in the E-textile can be outlined and improved accordingly. Further harmonized laundering test standard need to be developed by way of the industry alliance collaboration between the electronics and textile industries, such as IPC, AATCC, ASTM, IEC and so on, so that the related components and conductive trace such as conductive thread, yarn, fabric and conductive ink can be validated in order to pass the acceptable launderability test.

EXPERIMENTAL PROCEDURE

Test Vehicle Description

4 types of conductive fabrics with certain thickness metal coated (Ag, Cu, Ni/Cu, Ni/Cu/Co) on the polyester and nylon yarns are selected for launderability test, as listed in Table 1. Photos of the samples are shown in Figure 1. Ag is expected to have highest electrical resistance among them due to lowest

thickness. Figure 2 is the schematic cross section of NiCuCo coating on the polyester yarn with proprietary process technology.

Table1. The description of the conductive fabric

Sample	Cu	Ag	NiCu	NiCuCo
Fabric	Polyester	Nylon	Polyester	Polyester
Metal Coating	Copper	Nylon	Nickel/Copper	Nickel/Copper/Cobalt
Resistivity (Ohm/sq)	0.05	<0.25	0.03	<0.1
Thickness (um)	3.44	0.49	2.43	2.64

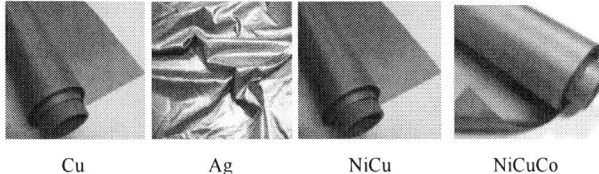

| Cu | Ag | NiCu | NiCuCo |

Figure 1 The appearance of conductive fabric

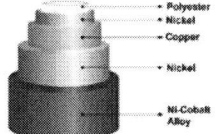

Figure 2 The NiCuCo layer structure coated on the polyester.

The 4 conductive fabrics are cut into 3cmX10cm dimension then subjected to sandwich lamination with thermoplastic urethane (TPU, Bemis 3206) film under hot compression at 120⁰C on the 3 common base fabrics with 5cmX12cm dimension (Spandex, Nylon, Denim) respectively, as shown the test specimens outline in Figure 3. The base fabrics are designed with different elongation and flexibility in order to achieve certain severity of shear and torsion strain with the conductive fabrics due to mechanical mismatch during laundering process. Figure 4 shows the fabric stress-strain behavior under tensile test of ASTM D5034 at 100mm/min. Denim fabric presents stiffer property but lower elongation than Nylon and Spandex. Nylon containing certain percentage Spandex exhibits similar elongation behavior but lower strength than Spandex. It is expected the mechanical strain to the conductive fabric caused by Denim based fabric will be constrained less than that by Nylon and Spandex if the agitation force is gentle enough during laundering process. But once the agitation force during washing is strong enough to make Denim fabric deformed, the conductive fabric will still suffer the mechanical strain caused by deformed Denim.

Test Procedure

Electrical resistance as a function of laundering cycles is used to characterize the performance of the conductive materials. Electrical resistance of every specimen is measured

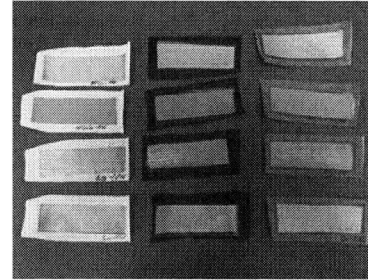

Figure3. The test specimens for the laundering test

before and after a standard laundering procedure by using 4 point probing digital Ohm meter with 1 μΩ resolution. The resistance is measured along the diagonal of specimens with distance between marked points 10.44 cm, as shown in figure5.

The laundering procedure is to follow AATCC M6 test standard with the factors control such as detergent, water temperature, agitation speed and spin speed and so on, and the AATCC compliant washing machine and dryer are used here[7,8,9]. All test specimens are loaded in the washing tank with enough ballast pieces to make a total load weight of 1.8 ± 0.1 kg (4.0 ± 0.2 lb), and distribute them evenly around center agitator, as shown in figure 6. After the tank is filled to specified water level, add 66 ± 1 g of AATCC 1993 Standard Reference Detergent into washing tank and agitate briefly to dissolve completely.

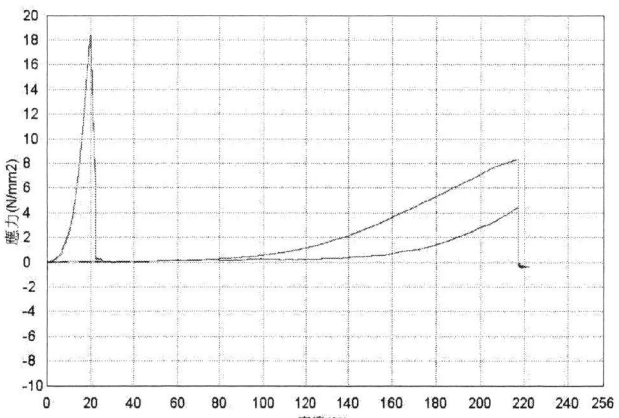

Base Fabric	Max stress (N/mm2)	Max Strain (%)
Denim	18.5	19.7
Nylon	4.4	217.3
Spandex	8.3	217.3

Figure4 The S-S curve of 3 base fabrics

Table 2 is the standard washing machine parameters with normal procedure defined in AATCC Monograph M6,

Traditional Top Loading Washing, which is used in the study. After each cycle laundering, the specimens are taken out for drying in the tumble dryer, as shown in table3. The electrical resistance of every specimen is measured after washing and drying at 1,3,5,7,10 cycles to monitor its change in comparison with initial resistance prior to laundering.

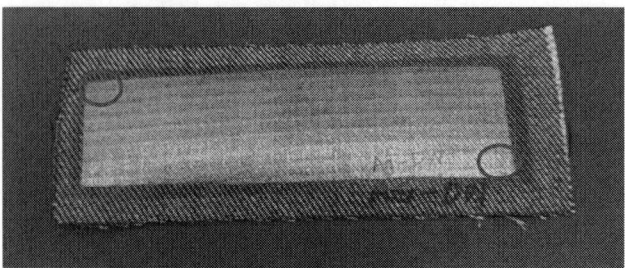

Figure5. Marking the probing distance for resistance measurement

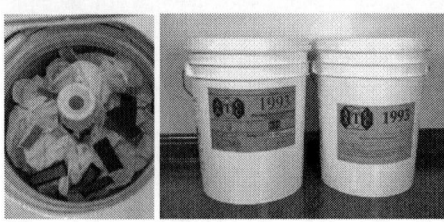

Figure6. All specimens are loaded into the washing tank with ballast pieces (Left). AATCC 1993 Standard Reference Detergent (Right)

Table2 Standard Washing Machine Parameters

Washing Cycle	Normal
Water Level, L (gal)	$72 \pm 4 (19 \pm 1)$
Agitation Speed, strokes/min.	86 ± 2
Washing Time, min.	16 ± 1
Final Spin Speed, rpm	660 ± 15
Final Spin Time, min.	5 ± 1
Wash Temp, °C (°F)[1]	**Cold:** $27 \pm 3 (80 \pm 5)$

Table3. Standard Tumble Dryer Parameters

Cycle	Normal
Max. Exhaust Temp, °C (°F)	$68 \pm 6 (155 \pm 10)$
Duration, min	30
Cool Down Time, min.	≤ 10

Material Characterization and Failure Mode Observation

The XRF and low angle XRD are employed to identify the change of metal coating thickness and lattice constant respectively prior and post 10 times laundering. Cross-Section followed by SEM observation is also performed to understand the metal coating quality with fiber or fabric among the 4 conductive materials. Furthermore, the failure mode related to electrical resistance increase is inspected by 3D OM and SEM as well.

Moreover, of the conductive metal coatings are suspected to be potentially dissolved by the alkaline detergent so that the coating thickness will be decreased to result in the resistance increase. In order to investigate the anti-detergent performance of various conductive fabrics, the 3cm by 10cm dimension specimens are prepared then soaked into solution with 66g

AATCC compliant detergent in the washing tank with same water level. The specimens are taken out for resistance measurement at 1,3,5,7 days after clean water rinsing.

RESULTS AND DISCUSSION

The electrical resistance is measured in all specimens after periodical laundering cycles, as shown in Table 4. Apparently, resistance increase with the laundering cycle number, which illustrates the metal conductive structure do encounter certain destruction caused by the mechanical force or detergent attacking and so on during the laundering process. The resistance in Ag seems to be quite stable in comparison with the other metal material although its electrical conductivity is the lowest.

Table4. Conductive resistance change over the laundering cycle
(R0: Prior to laundering, R10: Resistance after 10 times laundering)

Specimens	Conductive Fabric	Base Fabric	R0	R1	R3	R5	R7	R10
1	Ni+Cu	Spandex	170 mΩ	221mΩ	360mΩ	1.24Ω	1.08~1.1Ω	4~4.5Ω
2	Ni+Cu	Nylon	175 mΩ	208mΩ	280mΩ	421mΩ	840~880 mΩ	7.1Ω
3	Ni+Cu	Denim	180 mΩ	220mΩ	1.38Ω	23~28Ω	42~47Ω	820Ω~1Ω
4	Ni+Cu+Co	Spandex	252 mΩ	359mΩ	557mΩ	678mΩ	1~1.2 Ω	1.07Ω
5	Ni+Cu+Co	Nylon	229 mΩ	404mΩ	561mΩ	1.41Ω	1.92Ω	4.6Ω
6	Ni+Cu+Co	Denim	250 mΩ	500mΩ	1.08Ω	16.3Ω	28~30Ω	57~60Ω
7	Ag	Spandex	1.69Ω	1.67Ω	2.097Ω	2.11Ω	2.35Ω	2.54Ω
8	Ag	Nylon	1.65 Ω	1.76Ω	2.17Ω	2.24Ω	2.5Ω	2.6Ω
9	Ag	Denim	1.59 Ω	1.69Ω	2.48Ω	3.18Ω	4.05Ω	3.11Ω
10	Cu	Spandex	213 mΩ	635mΩ	316Ω	1.55 KΩ	900 mΩ~1Ω	960~9800Ω
11	Cu	Nylon	209 mΩ	322mΩ	149Ω	680~730Ω	620~720Ω	1.2~1.23KΩ
12	Cu	Denim	232 mΩ	452mΩ	1.16kΩ	2.4~2.7KΩ	4.2KΩ	6.3KΩ

(*After the probes are contacted to the sample surface, wait for 10 seconds before reading the resistance; if the resistance is not stable, record the range of resistance in the first 10 seconds)

The percentage of resistance change is calculated after laundering using following Equation.

$$RC = (Rf - Ri)/Ri \times 100\%$$

Where:
RC = % resistance change after laundering,
Rf = Mean final resistance, after laundering (Ω),
Ri = Mean initial resistance (Ω).

Figure 7 show the trend of the percentage of resistance change over laundering cycle among different fabric bases. Denim based fabric presents significant resistance change over the laundering cycle no matter what any type of metal material. Nylon and Spandex based fabric with close stress-strain mechanical behavior present similar resistance change over the laundering cycle. The resistance change caused by Denim based fabric is more significant than by Nylon and Spandex. This finding is against the previous assumption that stiffer base fabric can keep conductive material more stable during mechanical force operation from laundering unless the mechanical force is strong enough to make all fabric deform significantly. Moreover, the resistance change of Cu is most severe than other metal materials, which means the Cu should not be good candidate in the E-textile application for the signal or power transmission if the cleaning process is required. There is certain improvement for NiCu and NiCuCo and NiCuCo even outperforms NiCu quite a lot due to additional NiCo layer as protective out layer. Ag is still most stable material on the resistance change over laundering with its excellent stretchability and flexibility for the stress release.

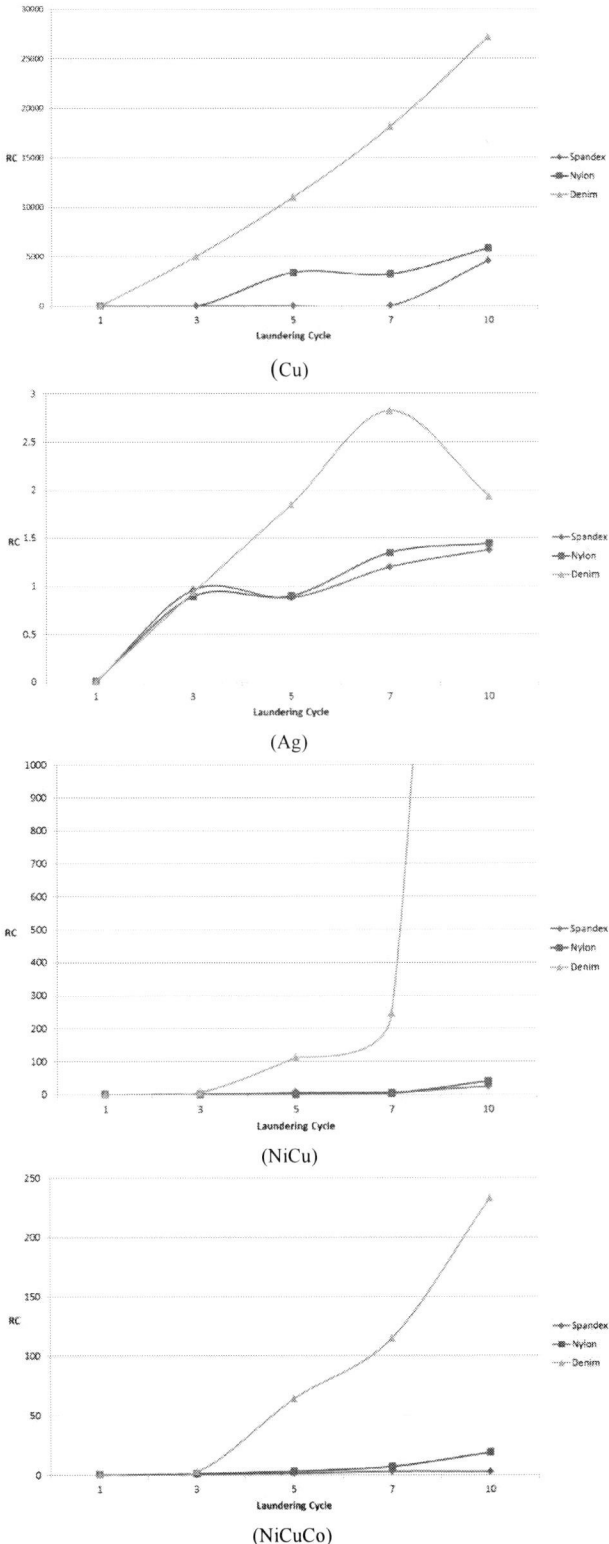

(Cu)

(Ag)

(NiCu)

(NiCuCo)

Figure7. Percentage of resistance change over laundering cycle among different fabric bases

Figure 8 is the 3D OM top view on the Spandex based fabric prior to laundering. The fabric seems to be well coated by the

metal material with different color appearance, which is also confirmed by SEM view at higher magnification in Figure 9. The metal coating process on the polyester and nylon fabric is proprietary for the product vendor, which will not be discussed here. From the SEM view of cross section of conductive fabric in Figure 10, the NiCuCo is well coated around the polyester fiber which is corresponding to the structure design as shown in figure 2, the other 3 metals seem to be coated on the surface of fabric by certain process. Moreover, the Ag thickness is found to be very thin in the view of cross section.

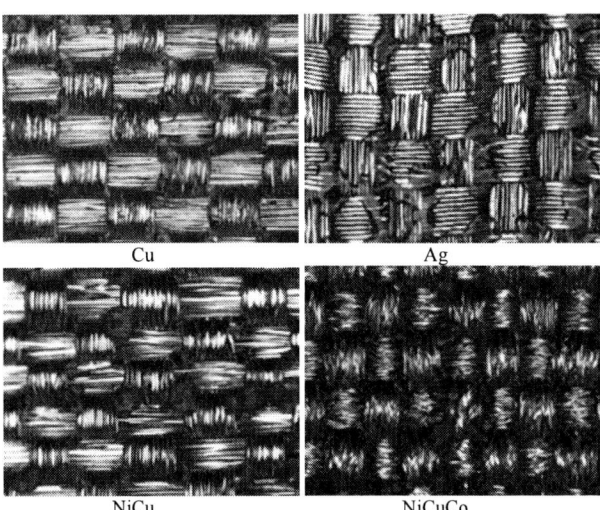

Figure8. The 3D OM surface view (X100) for the 4 metal coatings on Spandex based fabric prior to laundering

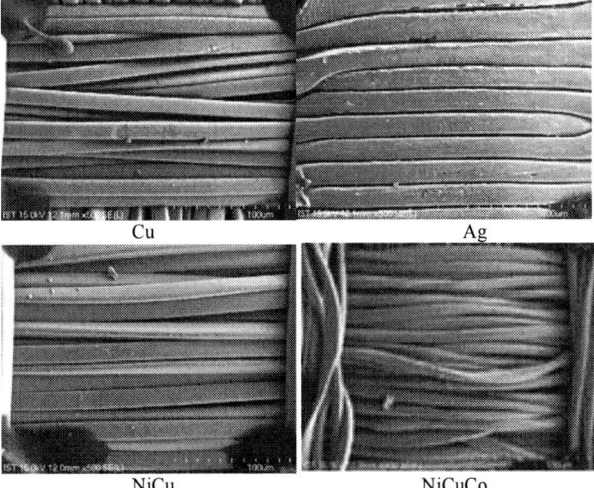

Figure 9. The SEM surface view (X500) for the 4 metal coatings on Spandex base fabric prior to laundering

Denim based fabric demonstrates the most significant resistance change over the laundering cycle. From the top SEM view at X100 in figure10, partial metal surface damage is observed and Cu seems to present more damage than other metals in Denim based fabric after 10 times laundering. When observing at higher magnification X500 for localized area view in Figure 10, the metal peeling and scratch with the fiber is found obviously due to mechanical force during laundering agitation, which illustrates the adhesion between the metal and fiber is not strong enough to sustain the 10 times laundering process. In terms of Ag, the severity level outperforms others. When observing at X10000 in Figure 11, the failure mode of Ag seems different from others to show more metal scratch rather than metal peeling. This is attributed possibly to its lower thickness and better adhesion with Nylon, as well as superior stretchability.

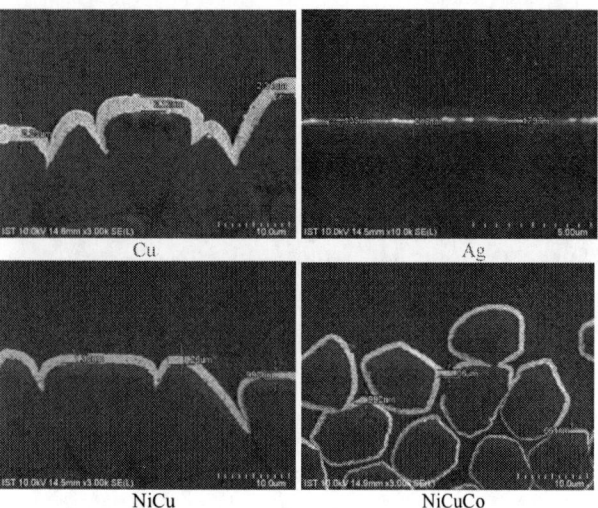

Figure10. The SEM XS view (X3000) for the 4 metal coating on Spandex based fabric prior to laundering

In terms of NiCuCo observed in Figure 12, the metal layer around the polyester fiber is peeled and there is no good adhesion seen between them. When subjected to mechanical force caused from agitation, the stiffer and thicker metal layer is cracked due to poor stretchability. The NiCo alloy out layer seems to be crack initiation.

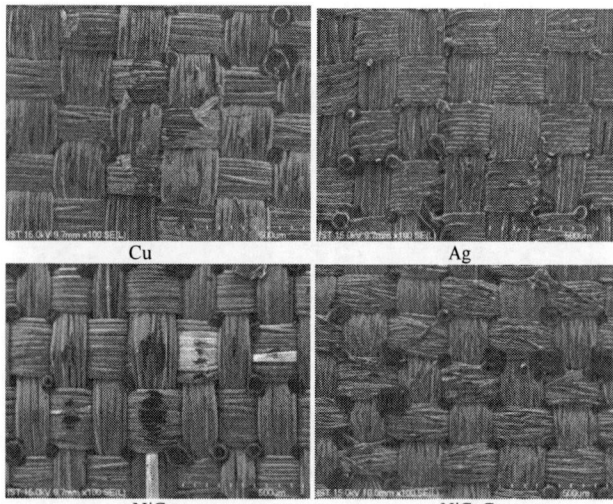

Figure10. The SEM surface view (X100) for the 4 metal coatings on Denim base fabric post 10 times laundering

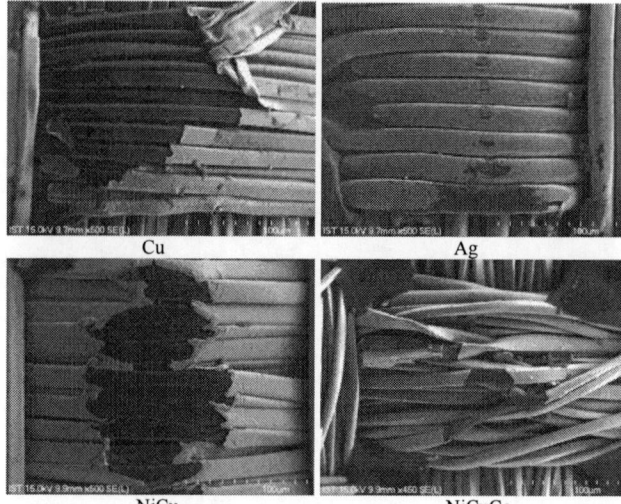

Figure10. The SEM surface view (X500) for the 4 metal coatings on Denim based fabric post 10 times laundering.

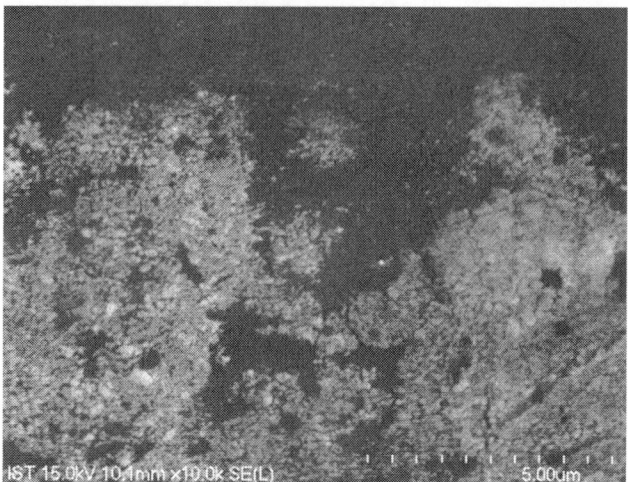

Figure11. The SEM surface view (X10000) for the Ag metal coating on Denim base fabric post 10 times laundering.

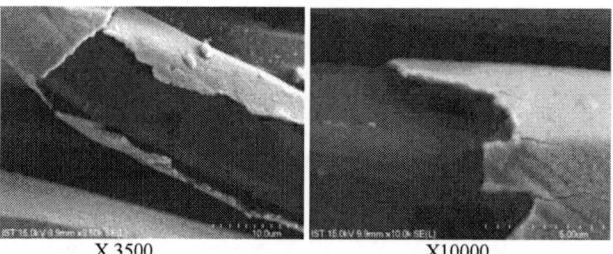

Figure12. The SEM surface view for the NiCuCo metal coating on Denim base fabric post 10 times laundering.

Figure 13 shows the SEM surface view at X50 for the 4 metal coatings on Spandex based fabric after 10 times laundering. There is some black TPU dot seen between the fiber thread caused from resin extrusion from thermal compression. Not sure the TPU dot will have any negative effect to constrain the fabric deformation, but it is worthy to do further investigation to make TPU not to penetrate into fabric by optimizing thermal compression process.

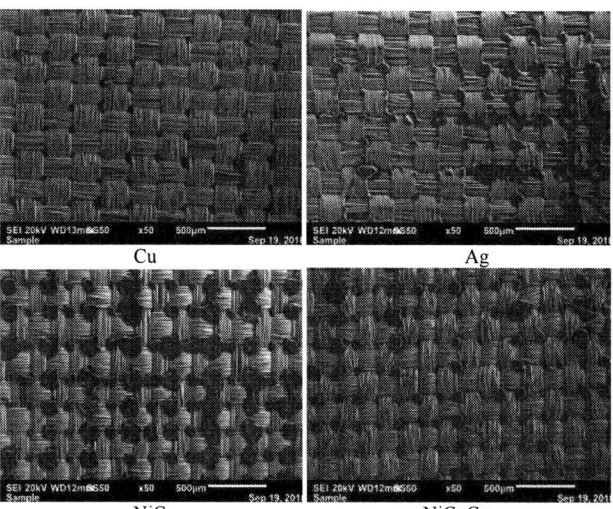

Figure13. The SEM surface view (X50) for the 4 metal coatings on Spandex based fabric after 10 times laundering

Figure 14 illustrates the less severity level of metal peeling on Spandex based fabric after 10 times laundering than the Denim based fabrics, which is consistent with the result in Figure 7.

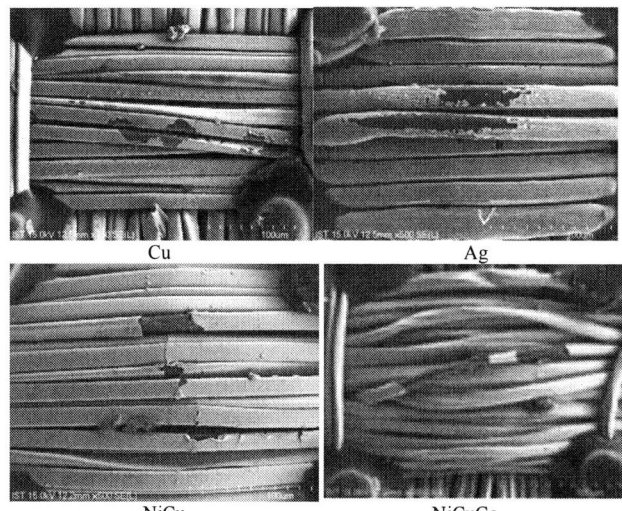

Figure14. The SEM surface view (X500) for the 4 metal coatings on Spandex based fabric after 10 times laundering

In addition to above discussion about the metal peeling or scratch for the resistance raise, the metal anti-detergent performance is also hypothesized to be one of reasons for the resistance increase. Usually, the metal will be potentially dissolved by the certain acids and alkalis solvent to generate porous structure or thickness reduction so that the conductive resistance will be increased. Figure 15 shows the resistance trend over the soaking days among various metal coating fabrics. There is no significant resistance change over the soaking days until 7 days, which illustrates the all metals are

quite stable under AATCC detergent soaking in terms of resistance measured.

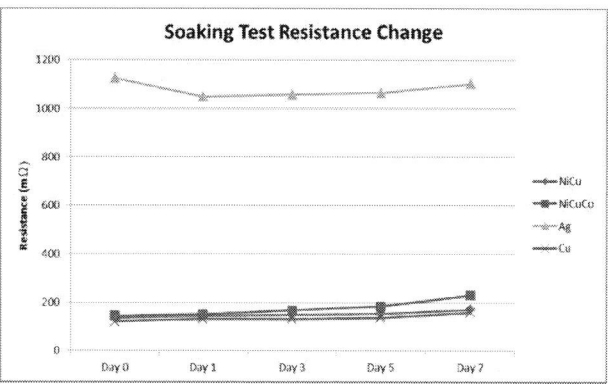

Figure15. Resistance change over soaking days in terms of conductive fabric

XRF is also applied to verify the change of metal coating thickness through non-destructive analysis at High voltage 50 kV, Apertures (Collimators) ø0.2 and 30sec duration prior and post 10 times laundering on Denim and Nylon based fabrics, as shown the result in the Table 5 and Table 6. 5 points with 4 at corners and 1 at center in every specimen are measured for average metal thickness confirmation.

Table5. Metal thickness change on Denim based fabric prior and post 10 times laundering (Unit: µm)

	Cu	Ag	NiCu	NiCuCo
prior	2.236	0.469	2.468	2.974
Post	2.655	0.477	2.688	3.032

Table6. Metal thickness change on Nylon based fabric prior and post 10 times laundering(Unit: µm).

	Cu	Ag	NiCu	NiCuCo
prior	2.159	0.468	2.536	2.955
Post	2.407	0.518	2.740	2.964

There is no change in the metal thickness concluded after 10 times laundering, which means the washing process will not affect the metal coating thickness through the quantitative non-destructive analysis.

Combining the XRF result and the study on detergent soaking test, the metal coating thickness is quite stable during washing. The resistance raise over the laundering cycle is primarily due to metal peeling or scratch. Therefore, enhance the metal adhesion with fabric and reduce metal stiffness may promote laundering reliability of E-textile.

Moreover, additional test using low angle XRD is adopted to investigate the metal lattice change prior and post laundering due to mechanical force from agitation. In terms of NiCu on Spandex based fabric, Figure 15 shows comparison of the XRD diagrams prior and post 10 times laundering. The diffraction curve broadens with lower peaks after 10 times laundering, which can be attributed to the fact that deformation from mechanical agitation causes increased dislocation density and formation of smaller grains, thus the intensity of a set of diffracted beams decreases. Grain refinement from smaller grains formation inside the NiCu layer can cause an increase in electron scattering by grain boundaries for the latent loss of electrical conductivity.

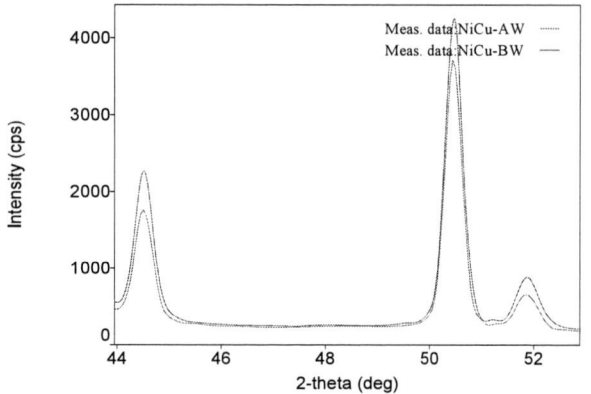

Figure 15. The XRD diagram of NiCu on Spandex based fabric prior and post 10 times laundering (BW: Before washing, AW: After washing)

Table 7 also explains the identified phases prior and post 10 times laundering in Spandex based fabric. It is found that the change of lattice constant prior and post laundering is not significant in all metals, confirming the most resistance increase is still attributed to metal peeling or scratch.

Table7 The identified phases in low angle XRD prior and post 10 times laundering in Spandex based fabric

Sample description		Phase Identified	measured Lattice constant	Lattice constant based on card info	Card No. used
Ag	Before laundry	Ag	4.087774	4.0855	01-071-3762
	After laundry	Ag	4.089885		
Cu	Before laundry	Cu	3.616797	3.6150	00-004-0836
	After laundry	Cu	3.621808		
NiCu	Before laundry	Cu	3.616143	3.6150	00-004-0836
		Ni*	3.524996	3.5238	00-004-0850
	After laundry	Cu	3.614672	3.6150	00-004-0836
		Ni	3.523856	3.5238	00-004-0850
NiCuCo**	Before laundry	Cu	3.617158	3.6150	00-004-0836
	After laundry	Cu	3.621500		

*Pure Ni was the best fit for this graph however a compound with chemical formula Cu 0.32 Ni 3.68 was also close
**For these samples no peaks corresponding to either Ni or Co were detected.

CONCLUSION

The E-textile has been emerging for many field applications including defense, sports/fitness, health, and public safety. The conductive material on the textile responsible for signal and power transmission unlike traditional and mature Cu trace on PCB, will encountermore challenges in quality and reliability, such as flexibility, conformability, stretchability and washability in addition to deliver power and signal. The conductive material developed for it, no matter yarns, thread, film, fabrics, paste, metal wire, will all have to meet above requirements at reasonable cost. At the moment, the test standard developed for E-textiles is still in earlier stage, especially for the laundering reliability. In the study, the existing AATCC laundering test standard is followed to understand conductive fabric with 4 types of metal coatings for the laundering reliability. It is concluded Ag outperforms other 3 metals on the resistance change, but the resistance of the silver coated fabrics is the highest so as to limit its application. How to improve the other 3 metal candidates or identify a new alternative to achieve steady resistance over laundering will be necessary.

Moreover, as e-textile technology keep evolving, more specialized methods may be developed to make e-textile structures more reliable to be washed similarly to daily textile products such as underwear, clothing, home textiles and technical textiles. The entire BoM will be facing more and more uncertain challenges and need more collaboration in the supply chain of electronic and textile industry.

ACKNOWLEDGEMENTS

The authors would like to thank Prof. Chin Lee and Mr. Sheikhi shooshtari of UC Irvine for low angle XRD analysis, iST FA lab for SEM inspection, Bemis for TPU samples and providing guidance on thermal compression process, and Textile and Material Industrial Research Center, Feng Chia University Taiwan for their providing different fabric samples.

REFERENCES

[1] McCann, J., & Bryson, D. (Eds.). Smart Clothes and Wearable Technology. Oxford: Woodhead Publishing. 2009.
[2] Weifeng Liu and Jeffrey Lee etc,"Laundering Electrically Conductive Fabrics for E-Textile Application", IFAI EXPO Conference, Dallas, 2018.
[3]A.Lee, M Seo, S. Yang, J. Koh, and H. Kim. 2008 . The Effects of Mechanical Actions on Washing Efficiency. Fibers and Polymers (2008) Vol. 9, No. 1, pages 101-106.
[4]L. Buechley, and M. Eisenberg Fabric PCBs, Electronic Sequins, and Socket Buttons: Techniques for E-textile Craft. Journal of Personal and Ubiquitous Computing, 13(2), pp 133-150, 2009
[5] L. Buechley. A construction kit for electronic textiles. In International Symposium on Wearable Computers ISWC, Montreux, Switzerland, October 2006.
[6] IPC-WP-024, IPC White Paper on Reliability and Washability of Smart Textile Structures – Readiness for the Market, August 2018.
[7]AATCC Laboratory Procedure (LP) 1, Home Laundering: Machine Washing
[8] ISO 6330, Textiles -- Domestic washing and drying procedures for textile testing
[9]AATCC EP13-2018, Evaluation Procedure for Electrical Resistance of Electronically-Integrated Textiles

Preconditioning Technologies for Sputtered Seed Layers in FOPLP

Johannes Weichart, Jürgen Weichart, Andreas Erhart
Evatec AG
9477 Truebbach, Switzerland
johannes.weichart@evatecnet.com

Kay Viehweger
Fraunhofer IZM ASSID
01468 Moritzburg, Germany

Abstract—The shift from Fan-Out Wafer Level Packaging (FOWLP) towards Fan-Out Panel Level Packaging (FOPLP) is mainly driven by the expected cost reduction. While the same quality as in FOWLP is requested however, larger, more flexible substrates based on mold compound causing heavy outgassing need to be handled successfully. The main quality criteria for a sputtered seed layer in packaging are adhesion, contact resistance (R_C) and yield (particles and other substrate damages). Reliable substrate handling and good process uniformities are however the key enablers a hardware and process supplier needs to guarantee to enter the FOPLP market. In this paper we focus on the influence of preconditioning technologies, namely degassing and physical etching, on contact resistance. Degassing technologies are compared and a newly developed dual frequency capacitive coupled plasma (DF CCP) etch source with a dual electrode is tested. This DF CCP source combines the good characteristics of different etching technologies, offering an optimized solution for panel preconditioning. Experiments were performed on a fully automated panel level production tool with cluster architecture. R_C devices with Aluminum and Copper contact pad materials were tested, having pad sizes of 5 - 30 μm. The influence of process variations on the R_C was tested and the data compared with residual gas analysis measurements to explain process performance results. The results enabled the definition of design criteria for preconditioning technologies. Experiments were run in parallel on a state of the art FOWLP tool confirming no relative reduction in performance for panel processing and therefore paving the way for FOPLP.

Keywords-FOPLP, Panel, Seed layer, Preconditioning, Degassing, Dry Etch, Sputtering, Contact resistance

I. INTRODUCTION

Fan-Out Panel Level Packaging (FOPLP) is expected to be the next trend after Fan-Out Wafer Level Packaging (FOWLP) allowing parallel processing of more devices due to larger substrate sizes, resulting in cost reduction. [1] While FOWLP brought the challenges of heavily outgassing polymers and high substrate warpages for seed layer deposition systems, additional challenges arise in FOPLP which are summarized in Table I. The same quality as in FOWLP is requested. The main quality criteria for a sputtered seed layer in packaging are adhesion, contact resistance (R_C) and yield (particles and other substrate damages), while reliable substrate handling and good process uniformities are the key enablers a hardware and process supplier needs to guarantee to enter the FOPLP market.

Table I
OVERVIEW OF CHALLENGES IN FOPLP

Driver	Main challenges
Substrate size	Process uniformity tuning
	Higher outgassing levels
Large & thin substrates	Substrate handling (flexible and large warpage)
Panel Value	Higher yield requested (high MTBF & low particle count)
Various substrate sizes and thickness	Flexible and easily tunable systems

A. About Preconditioning Technologies

For sputtered seed layers these quality criteria are mainly related to the preconditioning technologies used, namely degassing of substrates and surface etching/activation. This section describes and compares state of the art technologies in relation to their applicability to FOPLP. For degassing, the primary technologies are vacuum degassing [2, 3] and the recently introduced atmospheric degassing in N_2 atmosphere [4]. Table II summarizes the pros and cons of these technologies, Figure 1 shows the principles of both technologies in a batch degassing process. Parallel degassing in a batch degasser is mandatory to achieve high throughputs, as degassing times for Fan-Out substrates are long. Bulk degassing of the thick polymer substrates plays an important role in minimizing vacuum contamination by water or organic volatiles.

Table II
COMPARISON OF DEGASSING TECHNOLOGIES FOR LARGE AREA PROCESSING (5 = BEST RATING, 1 = WORST RATING)

Criteria	Vacuum Degassing	Atmospheric (N2) degassing
Complexity	2	4
Scaling simplicity	3	4
Temperature control (overshoot & uniformity)	2	5
Influence of substrate emissivity	2	5
Vacuum cross contamination	2	5
Degassing temperature level	5	4

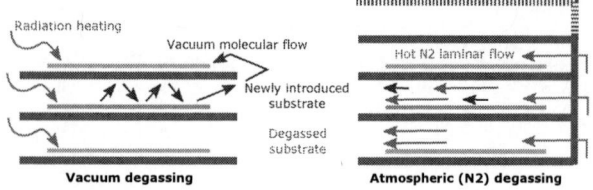

Figure 1. Basic comparison of batch degassing in vacuum and atmosphere. Desorped molecules are symbolized as blue vectors. Orange substrates are hot and degassed, blue ones are cold and not degassed.

In the degassing step, elevated temperature results in desorption of volatile components and diffusion of volatiles in the bulk material towards the surface, avoiding uncontrolled desorption in later vacuum processing steps. In vacuum degassing, the substrates are transferred to a vacuum chamber, where they are heated mainly by radiation, as vacuum does not allow other heating mechanisms like conduction or convection. Moving the degassing into a N_2 atmosphere at atmospheric pressure greatly simplifies the process, as no vacuum system is needed. A laminar flow of N_2 over the surface additionally helps to achieve uniform heating independent of substrate type while avoiding cross contamination between substrates in a batch process. An additional advantage is the minimized contamination of the vacuum system with volatile components, as contaminants are efficiently removed in the atmospheric part. Hot substrates from the degassing process can also be efficiently cooled during transfer into the vacuum system, whereas cooling in a vacuum system after vacuum degassing is slow. The use of atmospheric degassing decreases the initial thermal budget for the following processes, allowing higher throughputs.

Surface treatment can be done with a variety of etch sources such as capacitive or inductively coupled plasma (CCP & ICP) sources and ion milling sources. [5] Relative advantages and disadvantages of these technologies for large area etching are displayed in Table III for an overview.

Table III
COMPARISON OF SURFACE ETCH TECHNOLOGIES FOR LARGE AREA
PROCESSING (5 = BEST RATING, 1 = WORST RATING)

Criteria	CCP	ICP	Scanning source	DF CCP
Complexity	5	1	3	4
Scaling simplicity	4	1	4	4
Uniformity tuning	3	1	5	3
Plasma Density (Rate)	2	5	2	4
Particle control	4	4	1	4
Recontamination (Simultaneous etching)	4	4	1	4

CCP sources are comparatively simple in their design and scalability, but plasma densities are low and there are few parameters for uniformity tuning. CCP sources have, with reasonable etch rates, the highest bias voltage which may

cause damage on devices. [6] ICP sources with RF bias are state of the art for plasma etching on wafer size due to their high plasma densities and low bias voltage, but scaling to large area reactors is difficult as large inductors and dielectric windows are needed. [7] Uniformity tuning with inductor modification is also challenging. Scanning sources are common in large surface area etching, e.g. the display industry. Different types are available but in the packaging industry ion milling or ICP sources are reported. [6, 8] While tuning of uniformities is much easier with a moveable, smaller or one dimensional source, two major disadvantages arise. The surface is not etched simultaneously and areas previously etched are recontaminated during the ongoing etch process. Additionally, particle control is very challenging due to mechanical movement of the source or substrate and undefined zones of redeposition of etched material.

Using a dual frequency CCP source, a good tradeoff between the advantages of CCP & ICP sources can be found, namely an etch source with reduced complexity but increased plasma density. The drawback of difficult uniformity tuning can be improved with an innovative dual electrode concept. Details of this technology are discussed further in section II-A.

B. About Contact Resistance Measurements

The reduction of contact resistance (R_C) between metallization layers in electronic manufacturing is getting more important due to an increasing number of redistribution layers (RDL) and shrinking of the contact size. This is made more challenging by trends to introduce more polymer material as insulation layers, like in FOWLP/FOPLP technology. The R_C is usually denoted in Ω for a certain via size connecting the RDL layers and can be measured by several methods. Daisy chain structures can be used, where several contacts of the same kind are connected in series. They do however have the drawback that absolute numbers of the R_C are difficult to calculate due to the contribution of the transmission (connecting) lines. Additional transmission line measurements TLM [9] can be used for R_C evaluation, which is usually applied for metal-semiconductor contacts with a higher resistance range. Recent works [10] demonstrate that TLM may become even applicable for very low R_C.

The most popular method however is the cross-bridge Kelvin resistor (CBKR) method based on a 4-wire measurement of a single contact, since it allows direct access to the contact and can, due to negligible space requirements, even be implemented easily in production substrates. Nevertheless, since resistances in the range of 1 $m\Omega$ or less have to be measured, the electrical setup as well as the contacting of the needles to the pad require care. In addition it is also apparent that the evaluated contact resistances strongly depend on the design of the CBKR structure [11], in particular the overlap and the width of the connecting lines to the contact, which is explained by the current crowding effect. This means that a geometric correction term has to be added to the R_C to get the measured resistance, which has been proposed as:

$$R_k = R_C + R_{geom} \tag{1}$$
$$= \frac{\rho_C}{A} + \frac{4R_{s,0}\delta^2}{3W_xW_y}\left[1 + \frac{\delta}{2(W_x - \delta)}\right] \tag{2}$$

with A being the contact area, $R_{s,0}$ the sheet resistance of the underlying layer, δ the overlap of the pad area and $W_{x,y}$ the width of the connecting lines. As a consequence the comparison of CBKR structures with different designs has to take these effects into account. It was also found [12], that the current crowding effect becomes more severe when the interface specific contact resistivity ρ_C decreases, as R_{geom} gets more weight in the proposed equation.

II. EXPERIMENTAL

A. Hardware

An Evatec PNL tool was used to precondition the substrates and deposit the seed layer. It is a fully automated seed layer deposition system with a cluster architecture. Panel sizes up to 620 x 620 mm are possible in high volume manufacturing. Figure 2 shows the configuration of the lab system used to generate the results discussed in this report. Table IV gives a short overview on its components and their role in the processing.

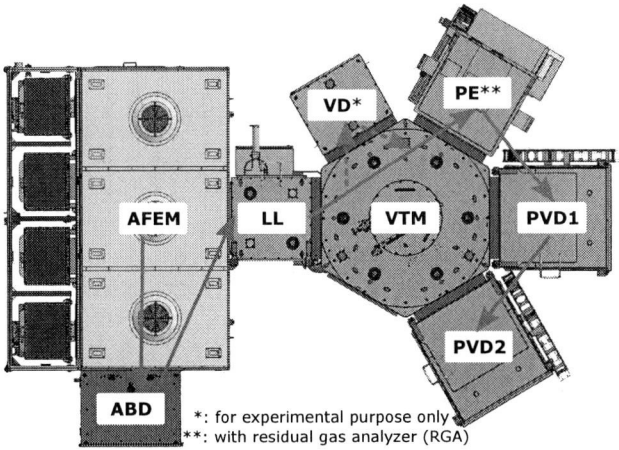

Figure 2. Layout of the PNL lab tool. Red arrows symbolize the process flow.

Three modules are of major interest in this paper. On one side there are the two versions of degassing modules, the ABD and VD. The ABD is a scaled version of the atmospheric degassing technology presented in section I-A with 22 slots. The VD is a single slot chamber for vacuum degassing using a radiation heater. The PE module is a newly developed CCP source. Figure 3 gives a schematic

Table IV
COMPONENTS OF THE LAB TOOL

Module	Description
AFEM	Atmospheric front end module
ABD	Atmospheric batch degasser. Multi-slot degassing chamber for high throughputs
LL	Load lock
VTM	Vacuum transfer module
VD	Vacuum degasser. Separate chamber with radiation heater for experimental comparison
PE	Pre-etch. Dual frequency, dual electrode CCP source
PVD 1&2	Physical vapor deposition (Ti & Cu). Array of rotary cathodes for sputtering of seed layers with low cost of ownership and a static substrate for minimized particle load

overview on the source design. A dual frequency approach for independent control of ion energy and ion density is used, giving a flexibility equivalent to ICP sources and not achievable with a standard CCP. [5, 13] A high frequency (HF) is used to control the ion energy, while a very high frequency (VHF) basically controls the ion density. The substrate bias can be varied between 0 – 600V, depending on the required process conditions, while typical physical surface etching processes are done between 300 – 400V. Increasing the VHF power can compensate the etch rate while substrate damage from high substrate bias can be avoided.

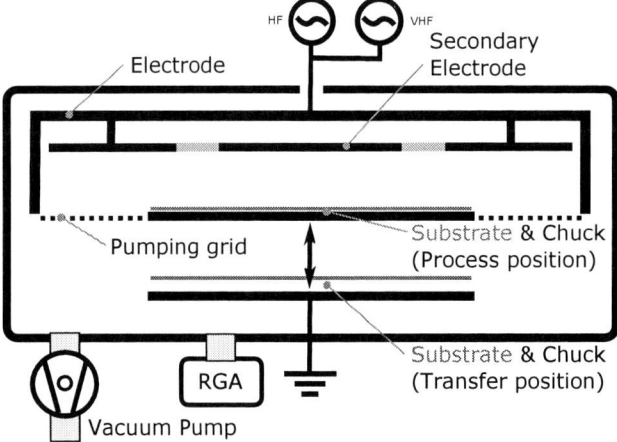

Figure 3. Schematic of the DF CCP etch source with a dual electrode

A dual electrode design is used to shape the plasma density and act as the required control parameter for uniformity tuning in a large area CCP reactor. [14] The secondary electrode is a structured two dimensional plate which presents a simple method for uniformity tuning. Finally, as pumping capacity is critical for high outgassing materials and there is a large organic load generated in the etching process, an innovative pumping grid is used to maximize the pumping capacity from the reactor. [15]

B. R_C device and Experiment Description

For contact resistance measurements a dedicated R_C test vehicle has been developed, which consists of 784 test dies on a 300mm wafer. Figure 4 shows the architecture of the single contacts between a subjacent redistribution layer (RDL_0) and a top redistribution layer (RDL_1) layer separated by a passivation layer. RDL_0 is a $1\mu m$ layer of either Cu or Al, two different experiment series were run to validate the effect of preconditioning. RDL_1 is a $3\mu m$ layer of Cu, which is electroplated on the sputtered seed layer. The $6\mu m$ polybenzoxazole (PBO) passivation layer was applied to investigate the effect on contact resistance during the seed layer application, either due to volatile products outgassing or by recontamination of contacts through etch residues. Two dimensions are of special importance in the design of the via, first the via diameter d_0 and second the pad size $d_1 = d_2$. Both dimensions were varied on the test vehicle:

1) $d_0 = 5 - 30\mu m$
2) $d_1 - d_0 = \delta$. The overlap size was additionally varied between $\delta = 5 - 25\mu m$ to study the effect of current crowding in the connecting structures. This investigation is not further discussed in this paper, all results shown have a constant overlap size of $5\mu m$.

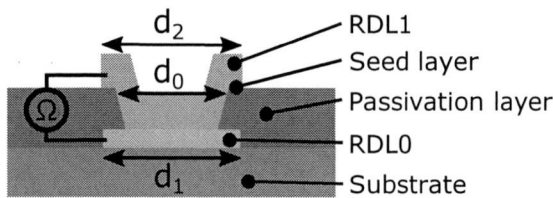

Figure 4.   Sketch of the R_C device structure and its major components.

The process steps run on the R_C test vehicles are discussed in more detail in [16], but a short overview is given here:

1) Via preparation: Vias are opened on the RDL_0 contact pads with photolithography.
2) Preconditioning for seed layer sputtering process: First the substrates are preconditioned. A degassing process removes moisture and organic volatiles, a surface etch remaining contaminations on the contact pads. Both processes are investigated in more detail in this paper and run with varying process conditions.
3) Seed layer sputtering process: 100nm Ti and 200nm Cu layers are sputtered on the cleaned contacts without breaking the vacuum in the same tool.
4) Patterning and electroplating: RDL_1 is reinforced by Cu plating.

5) The contact pads are electroplated with Ni/Au for reliable R_C measurements.
6) The R_C test wafers are measured with an automatic prober using a probe card.

To characterize the preconditioning in the panel size deposition tool, the R_C wafers were placed in the center of a 510 x 515 mm carrier. The material of the carrier was varied. While the wafer material itself results in a basic organic load contaminating the cleaned contacts in the seed layer fabrication, the material of the carrier was also varied for additional organic load. Three carrier materials were tested:

- Glass: Float glass (1 mm). No outgassing.
- ABF: Ajinomoto build-up film GY50 laminated on a copper clad laminate (CCL, 1 mm). Polymers with low outgassing levels designated for vacuum processes.
- FR4: Low quality FR4 plates (1 mm). High outgassing to simulate a worst case.

RGA analysis in the PE chamber was done to show the relation between vacuum condition and the contact resistance. Hereby criteria can be defined for in-situ process control and quality assurance. Mass 18 (H_2O), mass 28 (organic volatiles, C_2H_4) and mass 32 (O_2) were found to be characteristic and monitored and the partial pressures compared with the contact resistance results. Mass 18 can be seen as a good indicator for moisture removal in the previous degassing step. Mass 28 is a good indicator for the organic load present in the vacuum system. These can be desorbing volatiles or etch residues. Mass 32 was logged to track additional sources of O_2 which might result in contact oxidation.

III. RESULTS

Two series of experiments were run to evaluate the achieved contact resistance. Selected results of vias with Copper or Aluminum as a RDL_0 layer are presented. All graphs show the average of 30 measurements per via diameter. Standard deviation is not displayed for the measurement points but measurement results showed good consistency.

A. Cu as RDL_0 layer

Figure 5 shows R_C data of wafers run with and without preconditioning on a glass carrier. The organic load is comparably small, with only the outgassing wafer being a source of contamination. The glass carrier can be neglected as an outgassing source. It can be seen that both preconditioning technologies are needed to achieve a good contact resistance. Skipping the etching step, meaning the contact cleaning, results in an increase of R_C by about one order of magnitude. If the vacuum system is additionally contaminated with outgassing from the polymer of the wafer by abandoning the degassing, R_C levels increase by 2 - 3 orders of magnitude. This is even the case for well cured PBO wafers used in this experiment.

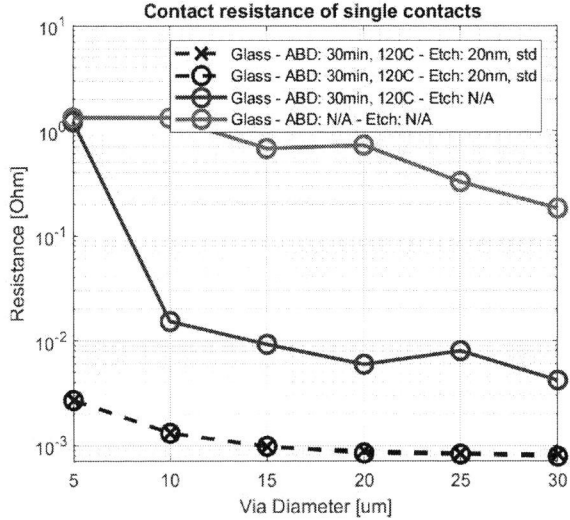

Figure 5. R_C data – with and without preconditioning technologies. Legend entries are structured as follows: 'carrier material' - 'degassing process conditions' - 'etch process conditions, etch removal in SiO_2 equivalents'

If wafers are processed with the standard process defined in this experiment (30 min degassing at 120°C and 20nm etch removal in SiO_2 equivalents), results are very consistent throughout the test. A total of 5 R_C wafers were run in a marathon test of 250 polymer (FR4) substrates etched and no shift in the R_C data was observable. No pasting or conditioning was done during this experiment, a loading effect from accumulating polymer debris was not observable. For better comparison two series of these baseline data are displayed in all graphs. They show the very small deviation of wafers processed with the same parameters, allowing a better differentation of other process conditions runs.

Two series of experiments on Cu-Ti-Cu contacts are discussed in this paper. The first series presents the influence of the etch process on the R_C value. Figure 6 shows the measurement data. It can be observed that a decrease in etch amount (green) did not modify the R_C. The amount of contamination on the contact must have been small and removal fast. If the etch rate is decreased (red), R_C values can be improved. This directly affects the throughput of the tool however. Another approach investigated was the reduction of the DC bias and increase of plasma density, which is achievable in the DF CCP reactor by changing the frequency ratio towards the higher frequency. It is observable that a lower DC bias, while compensating the etch rate with a higher plasma density, improves the R_C (blue). Figure 7 shows the RGA data logged during the etch process. Mass 18 (H2O) shows almost no increase during the etch process. The wafers are well degassed in the previous ABD process. Mass 28 (organic load) shows differences between the etch processes. The wafers processed with more gentle etch steps

such as a slow process or a lower DC bias show less organic load in the etch process and a fast decline after the process. Mass 32 (O2) shows low signals for all cases.

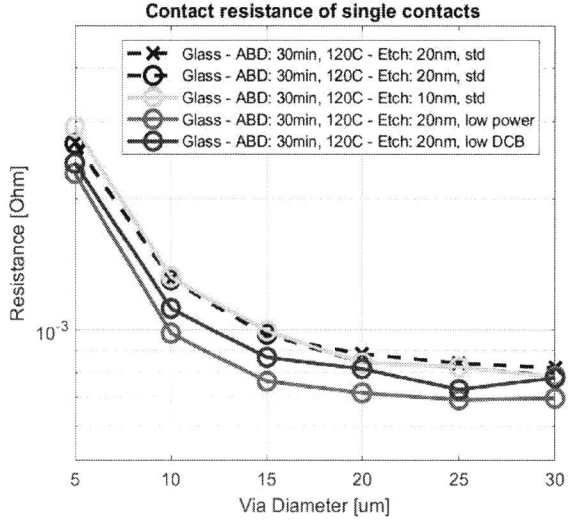

Figure 6. R_C data - different etch process on Cu-Ti-Cu contacts.

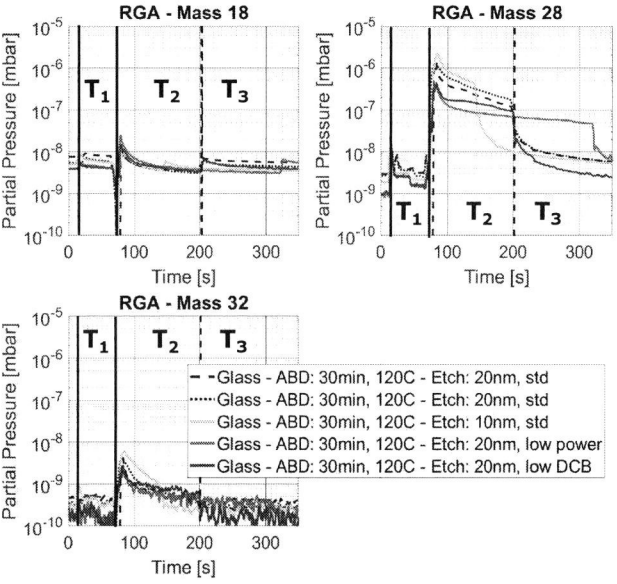

Figure 7. RGA data - different etch processes. The different process steps are marked between the vertical lines: T_1=Transfer in, T_2=Etch process (varying length in this experiment), T_3=Transfer out.

A second series investigates the influence of the degassing process on the R_C data. Atmospheric and vacuum degassing are compared, as well as short and long degassing times. The carrier material is varied as well to investigate the robustness of the degassing process in relation to varying intensities of outgassing. R_C results are presented in Figure 8.

978-1-7281-1500-9/19 $31.00 © 2019 IEEE

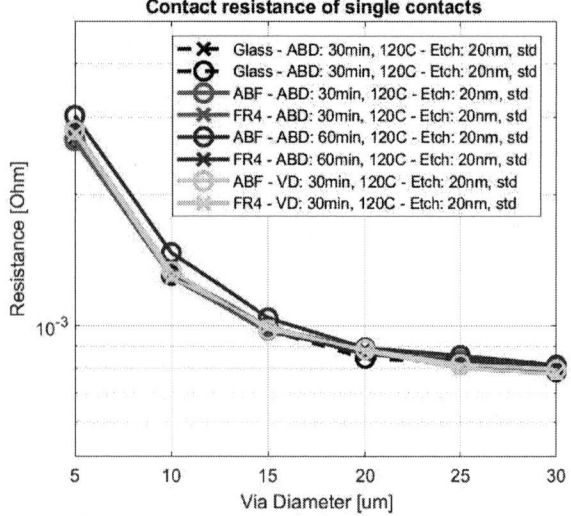

Figure 8. R_C data - different carriers and degas processes on Cu-Ti-Cu contacts

No influence of the outgassing process on the R_C is observable in this case. The additional organic load and outgassing levels do not increase the R_C level in comparison to the glass carriers. RGA data was logged and is very similar to the data presented in Figure 10 for the Al-Ti-Cu contacts.

B. Al as RDL_0 layer

The experiment with varying carriers and degas processes was repeated with the RDL_0 layer being changed to an Aluminum layer. The R_C data is displayed in Figure 9.

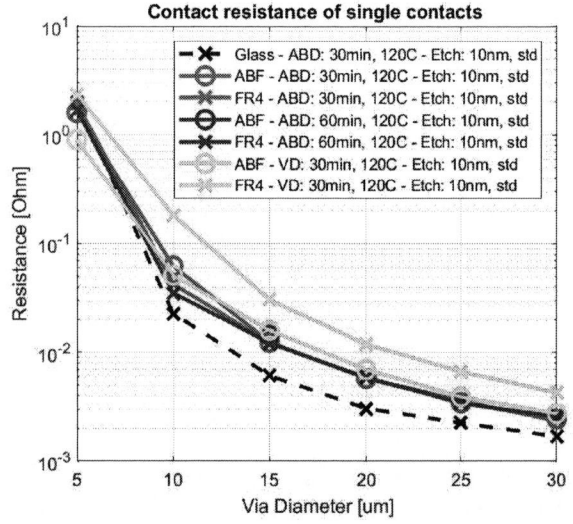

Figure 9. R_C data - different carriers and degas processes on Al-Ti-Cu contacts

It is observable that all R_C values are higher than for the Cu-Ti-Cu contacts, especially if going towards smaller via sizes. Results for $10\mu m$ vias are about one order of magnitude higher. Resolution problems in the fabrication process influenced the $5\mu m$ vias, making the results in this case not completely trustworthy. R_C values vary depending on the carrier and degassing processes. While a glass substrate results in the lowest R_C data, a FR4 panel degassed in the VD module shows higher R_C data. Figure 10 shows the RGA data logged in this experiment series, also representative for the similar experiments run on the Cu-Ti-Cu wafers. For mass 18 it can be observed that glass carriers as well as ABF carriers, which show lower outgassing, behave in a similar way in the etch process. FR4 panels with high outgassing show a higher signal after the etching process. This behavior is attributable to the architecture of the etch chamber. During the etch process the outgassing panel stays in the reactor compartment and the flow of outgassing material is partly obstructed by the pumping grid which separates the reactor area from the lower chamber compartment (figure 3). In the transfer period the substrate is lowered to the transfer compartment which also includes the pump and RGA, resulting in a higher signal. Mass 18 is especially high for the FR4 panel being degassed in the VD module. It can be seen that the level is very high already during transfer-in period. The reason is the transfer of contamination from the VD module to the PE module. The experimental VD module was not capable of pumping the large load of outgassing volatiles from the FR4 panel, resulting in contamination of the VTM and PE module in the transfer time. This is in direct correlation with the statement of vacuum system contamination when doing vacuum degassing, discussed in section I-A. The organic load (mass 28) is comparable for the polymer substrates etched, while being smaller for the glass substrate. Oxygen levels (mass 32) are higher for polymer substrates than for glass carriers.

C. Summary

Table V gives an overview on the quantitative data collected, only selected data is shown. The Table is completed with substrates run in the same R_C test on a wafer level processing tool, where the Evatec HEXAGON tool was used. This tool proved to have very good R_C data. [16] Additionally, results from two customer experiments are displayed for Al-Ti-Cu data, where a different device design and metrology has been used. The R_C data are average values of several samples run, and are lower than for the R_C vehicle developed for internal analysis.

IV. DISCUSSION

The R_C data collected includes two major sources of information. On one side it shows the dependence of the R_C value on the via diameter. It is obvious that smaller vias will result in higher R_C, and better preconditioning

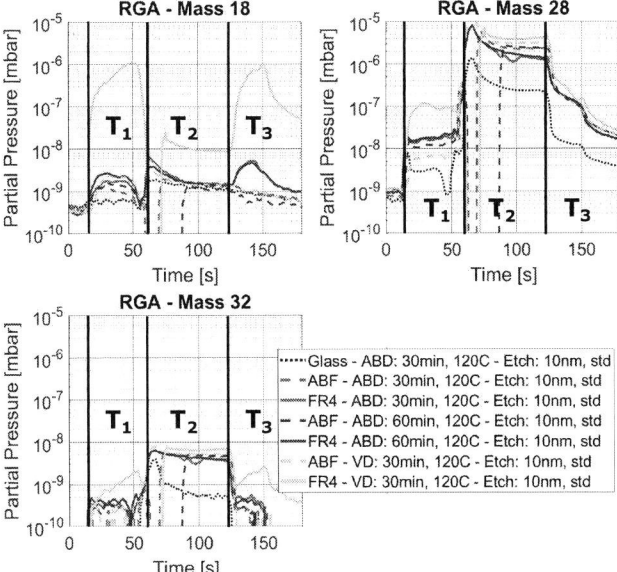

Figure 10. RGA data - different carriers and degassing processes. The different process steps are marked between the vertical lines: T_1=Transfer in, T_2=Etch process, T_3=Transfer out.

Table V
SUMMARY OF R_C RESULTS, EXPANDED BY RESULTS ON A WAFER PROCESSING TOOL AND ADDITIONAL CUSTOMER SAMPLES RUN

RDL0	Carrier / Tool	Degassing process	Etch process	RC 10um [mOhm]	RC 30 um [mOhm]
Cu	N/A wafer tool	ABD 30 min	20 nm, Std	1.35	0.83
	Glass	ABD 30 min	20 nm, Std	1.31	0.80
	Glass	ABD 30 min	20 nm, low power	0.98	0.70
	Glass	ABD 30 min	20 nm, low DCB	1.12	0.78
	FR4	ABD 30 min	20 nm, Std	1.29	0.78
	FR4	VD 30 min	20 nm, Std	1.22	0.79
Al	N/A wafer tool	ABD 30 min	20 nm, Std	10.70	3.38
	Glass	ABD 30 min	20 nm, Std	22.70	1.70
	FR4	ABD 30 min	20 nm, Std	42.60	2.35
	FR4	VD 30 min	20 nm, Std	182.00	4.26
Al Customer 1	Glass	ABD 30 min	20 nm, Std		0.59
Al Customer 2	Glass	ABD 30 min	20 nm, Std		0.52

technologies will allow moving towards smaller via sizes while still achieving the R_C required for the device design. In this paper we focus however on the second source of information, the interpretation of varying levels of the R_C depending on the preconditioning technology. While it is well known that preconditioning technologies are needed for good R_C as shown in Figure 5, further optimization might not be straightforward. Figure 6 shows that the R_C could be improved for "softer" etch processes. Slower etching or etching with a lower DC bias (resulting in weaker surface

bombardment) resulted in better R_C. Comparing the RGA data in Figure 7 one can see that mass 18 & 32 levels are very similar for all processes, while mass 28 is lower for the softer processes. Differences of the R_C in this case cannot be explained with varying re-oxidation levels of the contacts, but probably more by recontamination from the organic load remaining in the vacuum chamber after the etch process. The amount of monolayers deposited on the substrate is inversely proportional to the vacuum level, resulting in a contact barrier building up after the pre-clean process before the Ti layer is sputtered. Slower etch processes or split etch processes do reduce the amount of organic load per time in the vacuum system, improving the vacuum level. The longer processing time however directly affects the throughput. The modification of the DC bias for improved R_C might be promising, as the rate could be compensated by increasing the plasma density. The influence on other process quality criteria however needs to be further investigated. An independent factor for the R_C improvement is the improvement of the reactor pumping capacity. The large area pumping grid discussed in section II-A helps to improve the removal of organic volatiles from the substrate area.

When analyzing the second series of experiments which compared the influence of substrates and degassing, result interpretation is not straightforward. While it is obvious that mass 28 signals in Figure 10 are higher for the runs with polymer carriers etched, this explains the better R_C on the glass carrier in the Al-Ti-Cu contacts (figure 9). It is however not clear why the R_C in the Cu-Ti-Cu contact (figure 8) on the glass carrier is invariant to the polymer substrates etched. The mass 18 signals show that both degassing technologies are efficiently removing volatiles from the substrate. Compared to not degassed substrates in Figure 5, R_C values are by orders of magnitude lower and the RGA shows negligible H_2O levels in the etch process. This statement excludes highly outgassing materials (FR4) in the vacuum degasser. The high H_2O signal results in increased re-oxidation of the cleaned contacts, adding an additional contact barrier. This effect is expected to be stronger for Al contacts due to its higher affinity to oxygen and the higher resistivity of AlO_x in comparison to CuO_x. While the resistivity of Al_2O_3 is in the range of 10^{14} Ωcm [17], the resistivity of CuO_x is in the range of $3 \cdot 10^3$ Ωcm [18]. Even though the exact stochiometry of the forming oxide layer is not known, a resistivity difference of several orders of magnitude can be expected. While the Cu contacts were prone mainly to recontamination of the contacts by organic residues, the Al contacts show additional sensitivity to oxygen concentrations, which result in AlO_x layers with higher resistance. Good degassing technologies are therefore more relevant for Al contacts.

Results imply that higher pumping capacity would have been required for the designed VD module to efficiently remove the high outgassing levels, while the ABD efficiently removed this contamination outside of the vacuum system, avoiding any vacuum contamination in the first instance. Additional pumping capacity would have improved the results for the VD, the general problem of vacuum contamination however still arises as stated in section I-A. The efficient removal of contaminants in both degassing technologies with consistently low R_C values therefore leaves no obvious reason why a VD should be chosen over the ABD, as the atmospheric degassing temperature of 120°C allows the removal of all problematic volatiles.

By carrying out the experiments in parallel on a tool for FOWLP and FOPLP we could show that the new panel level tool performs in the same league as state of the art wafer level tools. The Evatec HEXAGON tool taken for comparison is discussed further in [16] and showed best in class R_C values on 300mm wafers. R_C values run on the panel tool show even better performance, a reason could be the improvement of reactor pumping efficiency explained in [15]. The new DF CCP reactor achieves the same performance as state of the art ICP etch reactors in FOWLP platforms. Degassing technology on both platforms is the same. The statements and findings about the atmospheric degassing technology presented are therefore applicable for both platforms.

As stated in section I-B, the R_C value depends strongly on the design of the single contact structures. The test structure developed for these test series show higher R_C than similar structures in other devices. As the same process was run on all samples, the differences must originate from the device design or other steps in the processing chain, which are not within the scope of this paper. Variations in the width of the connecting lines $W_{x,y}$ might be a plausible explanation.

V. CONCLUSION

Preconditioning in sputtering consists of two processes, degassing and pre-etching. In section I-A possible solutions for panel level processing were discussed. For degassing, atmospheric and vacuum degassing were compared. Both technologies were implemented in the fully automated Evatec PNL production tool to compare their performance. R_C and RGA results showed that both technologies are capable of degassing substrates, however the limits of the vacuum degassing technology could be observed for highly outgassing FR4 substrates. High outgassing contaminated the vacuum system, resulting in lower R_C values on Al-Ti-Cu contacts for vacuum degassing. Al contact pads proved to be especially sensible to variations in degassing performance, while Cu contact pads are more robust.

R_C levels on the investigated Cu contact pads were lower. Variations of degassing showed no influence in the investigated process regime. The results however could be influenced by the etching technology. It could be observed that slower etching processes or processes with a lower

DC bias but higher plasma density could improve the R_C. This is explained by less recontamination of the contacts after processing due to lower organic load in the reactor. The versatility of the DF CCP reactor can be used for additional process tuning, namely by tuning plasma density and ion intensity similar to ICP reactors on the wafer level. The measurements additionally support the importance of parallel processing of the full substrate surface. Scanning sources, which are often found for surface etching of large area substrates, show higher levels of recontamination after processing, as already etched surfaces are recontaminated while other areas of the substrate are etched.

By comparing the results with substrates run on a state of the art FOWLP tool it could be shown that same or slightly better performance could be reached on both Al and Cu contacts. Panel level technology is therefore able to deliver the same performance for seed layer deposition as on the wafer level. The investigation of other process quality criteria like adhesion and yield is beyond the scope of this paper, but experiments to date indicate that very good results could be achieved. Adhesion strengths of 11 - 21 N/cm could be observed on different polyimide films and the consequent processing of the substrates without moving parts in the reactors showed low particle levels. This proves that the presented FOPLP equipment can perform state the art Fan-Out processes, allowing for future trends towards smaller line-spaces in packaging.

ACKNOWLEDGMENT

The authors would like to thank the PNL tool team for the great collaboration in developing the new technology, especially Raphael Hidber for the preparation and execution of the experiments. Special thanks also go to the advanced packaging team for their support in technological topics.

REFERENCES

[1] T. Braun, K.-F. Becker, S. Voges, J. Bauer, R. Kahle, V. Bader, T. Thomas, R. Aschenbrenner, and K.-D. Lang, "24× 18 fan-out panel level packing," in *Electronic Components and Technology Conference (ECTC), 2014 IEEE 64th*. IEEE, 2014, pp. 940–946.

[2] K. K. Barber, M. Fissel, S. Y. Joh, M. Khosla, K. B. Levy, R. Martinson, M. Meyers, and D. Shrivastava, "Apparatus and method for enhanced degassing of semiconductor wafers for increased throughput," Dec. 24 2002, US Patent 6,497,734.

[3] J. Zhao, D. Quach, T. Weidman, R. J. Roberts, F. Moghadam, and D. Maydan, "Apparatus and method for heating substrates," Jun. 3 2008, US Patent 7,381,052.

[4] J. Weichart, "Chamber for degassing substrates," 2018, US Patent 9,934,992.

[5] B. N. Chapman, *Glow discharge processes: sputtering and plasma etching.* Wiley, 1980, p. 139.

[6] T. Fujinaga, "High productivity sputtering system for seed layer of printed circuit board," in *International Conference on Electronics Packaging (ICEP), 2014*. IEEE, 2014, pp. 26–29.

[7] M. Barnes, N. Benjamin, J. Holland, R. Beer, and R. Veltrop, "Plasma processor for large workpieces," Dec. 31 1996, US Patent 5,589,737.

[8] R. Mullapudi, D. Smith, E. Strepka, and S. Dasaradhi, "Multistation sputtering and cleaning system," 2012, US Patent 8,092,659.

[9] G. Tuttle, "Contact resistance and TLM measurements," 2014.

[10] H. Yu, M. Schaekers, T. Schram, N. Collaert, K. De Meyer, N. Horiguchi, A. Thean, and K. Barla, "A simplified method for (circular) transmission line model simulation and ultralow contact resistivity extraction," *IEEE Electron Device Lett*, vol. 35, no. 9, pp. 957–959, 2014.

[11] N. Stavitski, J. H. Klootwijk, H. W. van Zeijl, A. Y. Kovalgin, and R. A. Wolters, "Cross-bridge kelvin resistor structures for reliable measurement of low contact resistances and contact interface characterization," *IEEE transactions on semiconductor manufacturing*, vol. 22, no. 1, pp. 146–152, 2009.

[12] P. Zhang, Y. Lau, and R. Gilgenbach, "Analysis of current crowding in thin film contacts from exact field solution," *Journal of Physics D: Applied Physics*, vol. 48, no. 47, p. 475501, 2015.

[13] J. Weichart and J. Weichart, "Rf capacitive coupled dual frequency etch reactor," Patent WO2018/121 896, 2017.

[14] ——, "Rf capacitive coupled etch reactor," Patent WO2018/121 898, 2017.

[15] ——, "Vacuum plasma workpiece treatment apparatus," Patent WO2018/121 897, 2017.

[16] P. Carazzetti, F. Balon, M. Hoffmann, J. Weichart, A. Erhart, E. Strolz, and K. Viehweger, "Impact of process control on UBM/RDL contact resistance for next-generation fan-out devices," in *2017 IEEE 67th Electronic Components and Technology Conference (ECTC)*. IEEE, 2017, pp. 909–916.

[17] M. Barsoum and M. Barsoum, *Fundamentals of ceramics*. CRC press, 2002, p. 371.

[18] C. Guillén and J. Herrero, "Single-phase Cu2O and CuO thin films obtained by low-temperature oxidation processes," *Journal of Alloys and Compounds*, vol. 737, pp. 718–724, 2018.

Impact of thermal boundary resistance on the thermal design of GaN-on-Diamond HEMTs

Huaixin Guo
Science and Technology on
Monolithic Integrated Circuits and
Modules Laboratory
Nanjing Electronic Devices Institute
Nanjing, China
guohuaixin@gmai.com

Yuechan Kong
Science and Technology on
Monolithic Integrated Circuits and
Modules Laboratory
Nanjing Electronic Devices Institute
Nanjing, China
kycfly@163.com

Tangsheng Chen
Science and Technology on
Monolithic Integrated Circuits and
Modules Laboratory
Nanjing Electronic Devices Institute
Nanjing, China
chentsh@vip.sina.com

Abstract—The thermal boundary resistance is the key feature of innovative approach of high-thermal-conductivity diamond substrate for high-power GaN HEMTS, and has significant impact on the thermal design of GaN chip. A three-dimensional thermal simulation for analysis of the heat dissipation capability of different thermal designs is presented by finite element method. The model accounts for the nonlinear thermal conductivity of GaN and diamond materials by employing Kirchhoff's transformation with the aims to improve calculation accuracy. We investigate impact rules of thermal boundary resistance on the thermal design of the GaN chip, including the GaN buffer, diamond substrate, gate-gate pitch spacing, and chip size. Those results indicate that the impacts of diamond thickness and chip size on junction temperature are have a rule, the thermal boundary resistance has no influence upon the rules, but it influences the variation of the junction temperature. In addition, the rule for impact of gate-gate pitch spacing on junction temperature is limited by the thermal boundary resistance, and the impact is proportional to the value of thermal boundary resistance. What is worth our special attention is the impact of GaN thickness on junction temperature, the optimal GaN thickness should exist for the junction temperature, and is also affected by the thermal boundary resistance. Overall, we provide thermal structure design guidelines for GaN-based diamond substrate.

Keywords- thermal simulation; thermal boundary resistance; GaN-on-Diamond HEMTs; junction temperature

I. INTRODUCTION

GaN high electron mobility transistors (HEMTs) have demonstrated over 40 W/mm RF output power representing the potential for high output power densities, but are typically de-rated to operation at 8-10 W/mm, or less, in functional systems largely because of the accompanying high heat fluxes that cause unacceptably high device junction temperatures [1-3]. Managing high heat fluxes at such high RF output power densities is currently one of the key challenges for further developing GaN-based RF devices [1-5]. Several approaches of thermal management have been proposed to address this issue. GaN-on-Diamond technology has been developed, in which the diamond substrate with high thermal conductivities is placed immediately adjacent to

the hot spot of the chip to minimize temperature rise and the total thermal resistance of near junction region of GaN HEMTs is largely reduced [3-8]. However, the approach of GaN-on-Diamond HEMTs introduces a thermal boundary resistance (TBR) of GaN/diamond interfacial layer, and this TBR is the contribution to the total conduction thermal resistance of GaN chip which effect the heat dissipation capability. Successful thermal design of the GaN-on-Diamond HEMTs therefore requires a comprehensive understanding of the impact of the TBR on the channel temperature rise which has not been thoroughly previously investigated thus far, this underscores the need for a dependable temperature estimation method [9-13]. Here, we want to characterize the impacts of the TBR of GaN/diamond interfacial layer on the thermal design of GaN HEMTs through a theoretical model based on finite element analysis.

In this letter, to understand of contributions of the TBR of GaN/diamond interfacial layer to overall thermal management of GaN HEMTs and deliver insight into how to exploit the thermal design of high power transistor applications, the relationships between junction temperature and geometric parameters of GaN-on-diamond chip, which effected by the TBR of GaN/diamond interfacial layer, are analyzed by the finite element method which implemented in a commercial simulation software (COMSOL). In Section 2, the simulation details are described, such as the geometry of typical multi-fingers GaN and analytical equations for the temperature-dependent thermal conductivities of the materials. Numerical results and discussion shown in section 3 focuses on illustrating the impact of the TBR on the thermal design. Finally, some conclusions are presented in section 4.

II. SIMULAION DETAILS

The simulated GaN device geometries includes AlGaN barrier layer, GaN buffer layer, interface layer of GaN/diamond, and diamond substrate layer, and the GaN chip was soldered to a CuMo heatsink with a AuSn joint layer shown in Fig.1 (a). The sizes of the GaN chip and the device are design fixed structure as Table.1, the length and width of chip are the L and W, and the transistor is made up of 10 gate fingers. To avoid the effect of overall heat dissipation capacity affected by the macro size of the device, the length and width of the heatsink are more than twice that of the chip, and the thickness is also placed by means of the infinite element domain model because of the millimeter level size. The heat sources represent the constant heat flux generated by dissipated power directly under the gate fingers as shown in Fig. 1 (b), and gate/drain/source metallization is omitted due to the small structural complexity effect [4, 13-18]. Moreover, boundary conditions are assumed that the bottom of the substrates is an isothermal surface plane with a constant temperature of 293.15 K, and other external surfaces are applied to natural convection [15-22]. And the TBR of GaN/diamond interfacial layer is defined as a list of values, including 3, 5, 10, 15, 20, 30, 40, 60, 80, 120, and 140 m²K/GW, in order to understand of the effects of the TBR of GaN/diamond interfacial layer on overall thermal management of GaN HEMTs [9-13]. The calculations are carried out by the three-dimensional finite element method with COMSOL multiphysics under conditions identical those in a previous paper.

AlGaN	$25*(T/300)^{-1.44}$	490	6070
GaN	$150*(T/300)^{-1.42}$	490	6150
Diamond	$1185*(T/300)^{-0.55}$ (in plane) $1480*(T/300)^{-0.55}$ (vertical plane)	520	3515
AuSn	57	128	14700
CuMo	167	134	9900

III. NUMERICAL RESULTS AND DISCUSSION

A. Comparison of Junction Temperature Depending on TBR for Diamond Thicknesses

Here, we investigate the effects of junction temperature on the TBR for different diamond substrate thicknesses, The junction temperature was calculated by varying the TBR from 3 to 140 m² K/GW at diamond thicknesses of 50 μm and 110 μm (step:15μm), the power density of 6 W/mm, and the other structure parameters are shown Table 1. The results are analyzed as Fig.2.

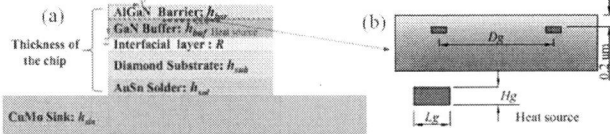

Figure 1. (a) Schematic of cross section, (b) the heat source of a multi-finger AlGaN/GaN HEMTs.

TABLE I. GEOMETRICAL PARAMETERS OF THE ALGAN/GAN HEMT

	Definition	Value		Definition	Value
L	Length of chip	920 μm	h_{bar}	AlGaN barrier thickness	22 nm
W	Width of chip	950 μm	h_{buf}	GaN buffer thickness	1.5 μm
Lg	Heat source length	120 μm	h_{sub}	Diamond thickness	80 μm
Wg	Heat source width	0.5 μm	h_{sol}	AuSn solder thickness	20 μm
Hg	Heat source thickness	0.2 μm	h_{sin}	CuMo heat sink thickness	1.5 mm
Dg	Gate-gate pitch spacing	20 μm	R	TBR of GaN/diamond	3-140 m²K/GW
Ld	Width of sink	2.2 mm	Wd	Length of sink	2.2 mm

For accurate simulations, the temperature-dependent thermal conductivities of chip materials are introduced in to the model. AlGaN, GaN and diamond materials have been considered with nonlinear thermal conductivities by employing Kirchhoff's transformation, and model parameters for given temperature ranges are provided in Table 2 [4,16,18,20].

TABLE II. THERMAL PARAMETERS USED IN SIMULATION

Material	Thermal conductivity k (W/m·K)	Heat capacitance c (J/kg·K)	Mass desity ρ (kg/m³)

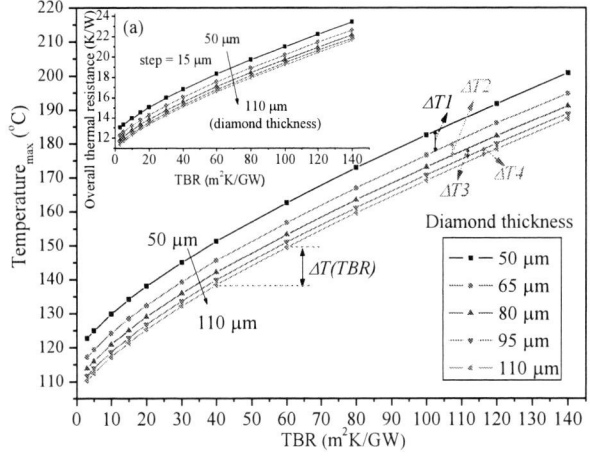

Figure 2. Dependence of junction temperature on the diamond thicknesses at different values of TBR, (a) the change rule of the overall thermal resistances.

The junction temperature is reduced by increasing the diamond thickness but with a decreasing trend at a constant TBR, and it is embodied in the temperature variation which the value decreases in the order $\Delta T1$, $\Delta T2$, $\Delta T3$, $\Delta T4$ at the same TBR, for instance, at the TBR of 3 m²K/GW, the corresponding values are 5.5 K, 3.3 K, 2.2 K and 1.5 K, respectively. Meanwhile, the value of $\Delta T4$ under different TBR is almost constant, and ranges from 1.5 K to 1.6K within the investigated TBR range of 3-140 m²K/GW. This trend is reflected in the values of $\Delta T1$, $\Delta T2$ and $\Delta T3$ under different TBR, this demonstrates that the rule for impact of diamond thickness on junction temperature is not bound by the TBR. The overall thermal resistances R_{TH} are calculated by $R_{TH} = (T_j-T_0)/P_d$ (where T_j is the junction temperature, T_0 is the reference ambient temperature, and P_d is power dissipation), it is noteworthy that the rules of overall thermal resistance and junction temperature depending on TBR for diamond thicknesses are perfectly aligned as illustrated in the inset Fig. 2 (a).

The sensitivity of junction temperature to TBR at different diamond thickness is additional analyzed as Fig.3, and represented by the average ratio (δ) of junction temperature to TBR. The results show that the average ratio is the most value as 1.1 K per m^2K/GW when the value of TBR increases from 3 m^2K/GW to 5 m^2K/GW, meaning as TBR rises by 1 m2K/GW at the range of 3-5 m^2K/GW, the corresponding junction temperature rise will be 1.1 K. In the following, the average ratio (δ) reduces faster to 0.56 K per m^2K/GW while the range of TBR is up to 40-60 m^2K/GW, and then is slightly slower at 0.45 per m^2K/GW with the TBR range of 120-140 m^2K/GW. The results demonstrates that there is a critical TBR value, when the TBR is less than the critical value, the effects of junction temperature to TBR at different diamond thickness become sharp stronger as the TBR decreased; but when the TBR is more than the critical value, the effects are a saturation behavior. Meanwhile, the average ratio (δ) is almost equal with the diamond thicknesses of 50-110 μm, this indicates that the sensitivity is irrelevant to the diamond thickness

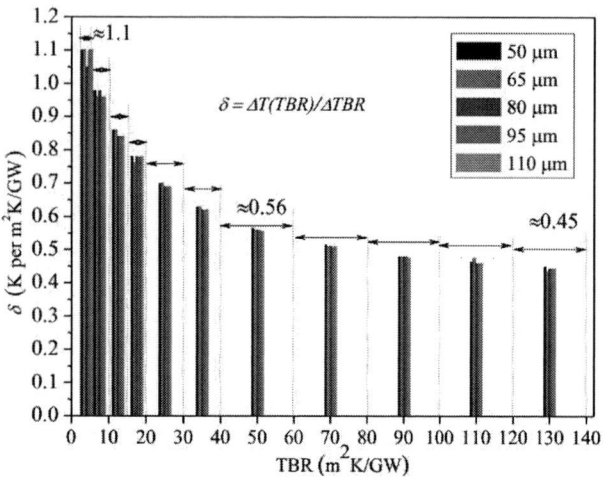

Figure 3. The sensitivity of junction temperature to TBR at different diamond thicknesse.

B. Comparison of Junction Temperature Depending on TBR for GaN Thicknesses

The effects of junction temperature on GaN thicknesses were investigated by varying the GaN thicknesses from 0.9 μm to 2.1 μm. The junction temperature was calculated by varying the TBR from 3 to 140 m^2 K/GW at the power density of 6 W/mm, and the other structure parameters are shown Table 1.The results was analyzed as Fig. 4, there is a cross value of the TBR, meaning that the junction temperature of two different GaN thicknesses is the same, and the cross value of the TBR decreases with increasing GaN thickness. When the TBR is less the cross value and at a constant value, the junction temperature decreases with the rise of GaN thickness, but the trend of change is reversed

while the TBR is more the cross value. The temperature variation of two different GaN thicknesses is as follows $\Delta T1$, $\Delta T2$, $\Delta T3$, and $\Delta T4$, the value of $\Delta T1$ is influenced by TBR, and the absolute value is small close to the cross value of TBR. This trend is reflected in the values of $\Delta T2$, $\Delta T3$, and $\Delta T4$ under different TBR, and becomes more significant when the GaN thickness is smaller. This demonstrates that the rule for impact of GaN thickness on junction temperature is limited by the TBR. Additionally, the rules of overall thermal resistance and junction temperature depending on TBR for GaN thicknesses are perfectly same as illustrated in the inset Fig. 4 (a).

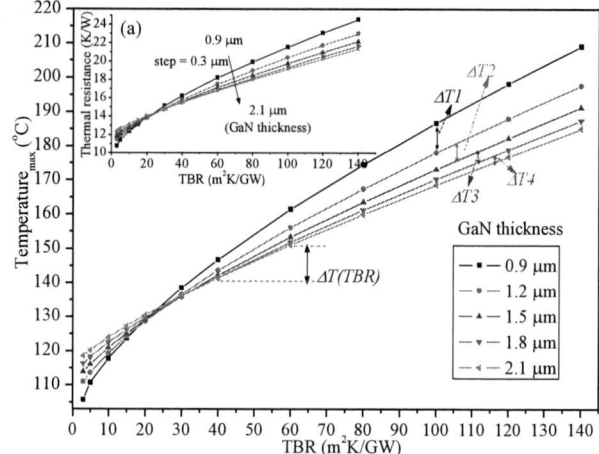

Figure 4. Dependence of junction temperature on the GaN thicknesses at different values of TBR, (a) the change rule of the overall thermal resistances.

Meanwhile, the sensitivity of junction temperature to TBR at different GaN thicknesses is also analyzed and represented by the average ratio (δ) of junction temperature to TBR in the Fig. 5.

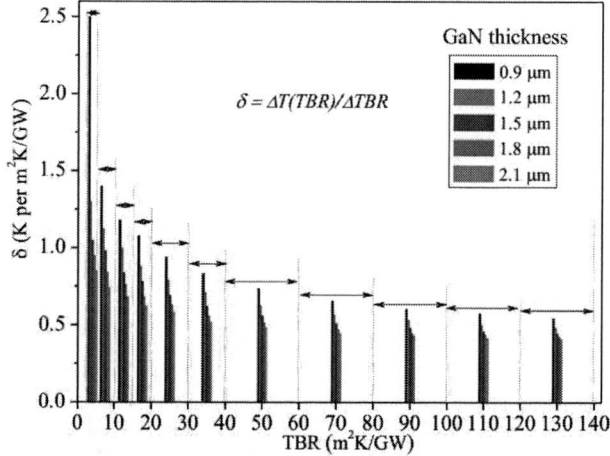

Figure 5. The sensitivity of junction temperature to TBR at different GaN thicknesses

As Fig. 5 shows, when the GaN thickness is 0.9μm, the average ratio is the most value as 2.5 K per m^2K/GW when the value of TBR increases from 3 m²K/GW to 5 m²K/GW, and then reduces faster to 0.74 K per m²K/GW as the range of TBR increasing of 40-60 m²K/GW, but finally becomes a slow decline as 0.55 per m²K/GW at the TBR range of 120-140 m²K/GW. However, when the GaN thickness is up to 2.1 μm, that trend reduces from the original 0.85 K per m²K/GW to 0.48 K per m²K/GW, and finally 0.41 K per m²K/GW. The results indicate that the sensitivity of junction temperature to TBR is reduced by increasing the TBR but with a decreasing trend, and presents nearly a linear relation in later stage. Meanwhile, this sensitivity is influenced not only by the value of TBR itself, but also by the GaN thickness, the average ratio (δ) decreases with the increasing GaN thickness, and this trend will become sharp stronger as the TBR decreased.

C. Comparison of Junction Temperature Depending on TBR for Gate-gate Pitch Spacing

The effects of junction temperature on the TBR for gate-gate pitch spacing are investigated and analyzed as Fig.6, the junction temperature was calculated by varying the TBR from 3 to 140 m² K/GW at the power density of 6 W/mm, the gate-gate pitch spacing is from 10 μm to 30 μm with the step of 5 μm, and the other structure parameters are shown Table 1.

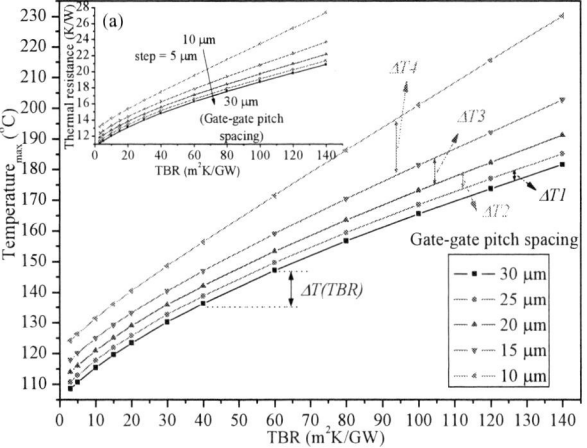

Figure 6. Dependence of junction temperature on the gate-gate pitch spacing at different values of TBR, (a) the change rule of the overall thermal resistances

The results show that the junction temperature increases while the gate-gate pitch spacing decreases with a tendency of increase at a constant TBR value, and it is embodied in the temperature variation which the value increases in the order ΔT1, ΔT2, ΔT3, ΔT4 under the same TBR. Meanwhile the values are affected by TBR, in specific, the ΔT1 increases slightly from 2.3 K to 2.7 K with the TBR increasing from 3 to 80 m²K/GW, and then increases to 3.6 K when the TBR is 140 m²K/GW; but the ΔT4 increases slightly from 6.2 K to 6.7 K with the TBR increasing from 3 to 15 m²K/GW, and then increases to 27.4 K when the TBR is 140 m²K/GW. The results indicate that the values will rise sharply when the

TBR goes up to some extent, and the TBR extent relate to the gate-gate pitch spacing, and this trend becomes more significant when gate-gate pitch spacing is smaller. This demonstrates that the rule for impact of gate-gate pitch spacing on junction temperature is limited by the TBR. Besides, the rules of overall thermal resistance and junction temperature depending on TBR for gate-gate pitch spacing are perfectly aligned as illustrated in the inset Fig. 6 (a).

The sensitivity of junction temperature to TBR at different gate-gate pitch spacing was also analyzed and shown Fig.7. When the gate-gate pitch spacing is 30 μm, the average ratio (δ) of junction temperature to TBR reduces faster from 1.05 to 0.54 K per m²K/GW with the TBR range increasing from 3-5 m²K/GW to 40-60 m²K/GW, and then drops to 0.40 per m²K/GW at the TBR range of 120-140 m²K/GW, this result shows that there is a saturation behavior when the TBR is more than 60 m²K/GW. Not only that, the sensitivity of junction temperature to TBR is also affected by gate-gate pitch spacing, when the gate-gate pitch spacing shrinks down to 10 μm, the δ value reduces faster from 1.1 to 0.76 K per m²K/GW with the TBR range increasing from 3-5 m²K/GW to 40-60 m²K/GW, and then drops to 0.73 per m²K/GW at the TBR range of 120-140 m²K/GW. This results demonstrate the sensitivity junction temperature to TBR is reduced by increasing the TBR but with a decreasing trend, and presents nearly a linear relation in later stage. Meanwhile, this sensitivity is also influenced by the gate-gate pitch spacing, the average ratio (δ) at the same TBR increases with the decreasing gate-gate pitch spacing, and this trend will become sharp stronger as the TBR increases.

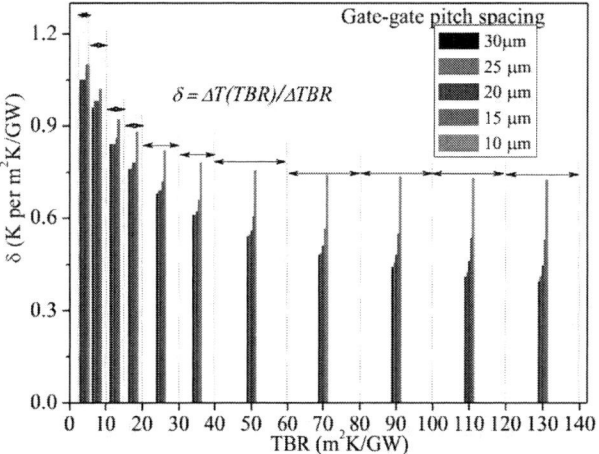

Figure 7. The sensitivity of junction temperature to TBR at different gate-gate pitch spacing

D. Comparison of Junction Temperature Depending on TBR for Chip Sizes

The most heat of GaN chip is spread by the GaN buffer and diamond substrates, and transferred to heatsink. So, the chip size is also an important factor that influences the amount of thermal sperading. The percentage (η) of the length (L) and width (W) of the chip is designed as 0.8, 0.9, 1.0, 1.1, and 1.2, respectively, the power density is 6 W/mm, and and the other structure parameters are shown Table 1.

The results, as shown in Fig.8. present that the junction temperature is reduced by increasing the chip size but with a decreasing trend at a constant TBR value, and it is embodied in the temperature variation which the value decreases in the order $\Delta T1$, $\Delta T2$, $\Delta T3$, and $\Delta T4$ under the same TBR. Meanwhile, the values of $\Delta T1$ under different TBR are almost constant, and this trend is reflected in the values of $\Delta T2$, $\Delta T3$, and $\Delta T4$ under different TBR, this demonstrates that the rule for impact of chip size on junction temperature is not bound by the TBR. Furthermore, the rules of overall thermal resistance and junction temperature depending on TBR for chip sizes are perfectly same as illustrated in the inset Fig. 8 (a).

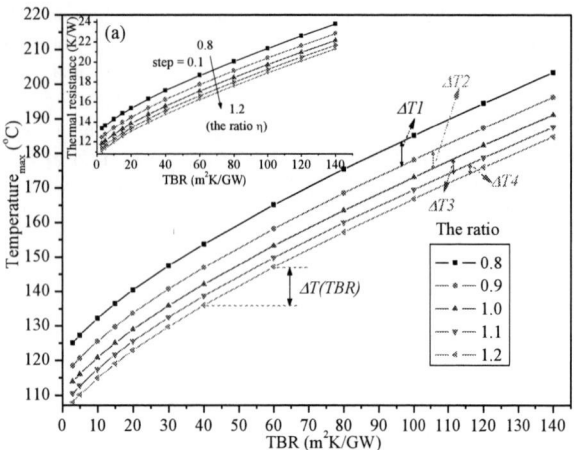

Figure 8. Dependence of junction temperature on the chip size at different values of TBR, (a) the change rule of the overall thermal resistances.

At last the sensitivity of junction temperature to TBR at different chip size is analyzed as Fig.9.

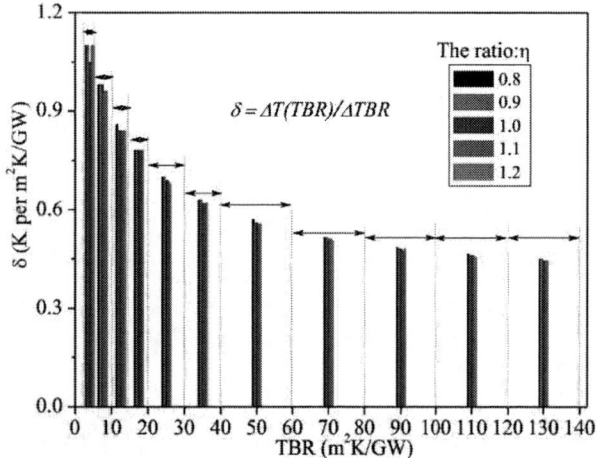

Figure 9. The sensitivity of junction temperature to TBR at different chip sizes.

The results show that the average ratio of junction temperature to TBR is the most value as 1.1 K per m²K/GW when the value of TBR increases from 3 m²K/GW to 5 m²K/GW, and then reduces faster to 0.57 K per m²K/GW while the range of TBR is up to 40-60 m²K/GW, ultimately,

to 0.45 per m²K/GW at the TBR range of 120-140 m²K/GW, meaning that the trend is saturation behavior when the TBR is more 60 m²K/GW. Meanwhile, the δ value are almost constant at the same TBR, this demonstrates that the sensitivity of junction temperature to TBR is not associated with the chip size. The trend of junction temperature depending on TBR for diamond thicknesses and that of chip size are similar.

IV. CONCLUSUINS

In this paper, a theoretical thermal model based on finite element analysis is presented for study the impact of the TBR on the thermal design of GaN-on-Diamond HEMTs, including the GaN buffer, the diamond substrates, gate-gate pitch spacing and as well as the chip size. The calculation results indicate that the junction temperature reduces by increasing the diamond thickness but with a decreasing trend, and this trend is not bound by the TBR, but the increment of the junction temperature reduces from 1.1 to 0.45 K per m²K/GW with the TBR increasing from 3 to 140 m²K/GW. The thermal impact trend of the TBR on the chip sizes is the same with that of the diamond thickness. For the gate-gate pitch spacing, the junction temperature increases while the gate-gate pitch spacing decreases with a tendency of increase, this trend becomes more significant when gate-gate pitch spacing is smaller, and the sensitivity of junction temperature is also affected by TBR. Meanwhile, the rule for impact of GaN thickness on junction temperature is limited by the TBR, there is a cross value of TBR that the junction temperature of two different GaN thicknesses is the same, and the cross value of TBR decreases with increasing GaN thickness, and the sensitivity of junction temperature to is affected by both TBR and GaN thickness, Simply put, the simulation highlights the expected great potential of GaN-on-diamond with significant impact on the junction temperature and the design rules.

REFERENCES

[1] .G. Felbinger, M.V.S. Chandra, Y. Sun, L.F. Eastman, J. Wasserbauer, F. Faili, D. Babic, D. Francis, F. Ejeckam, "Comparison of GaN HEMTs on diamond and SiC substratess," IEEE Electron. Dev. Lett, vol. 28, pp. 948-950, Nov 2007.

[2] J. Kuzmík, M. Tapajna, L. Válik, M. Molnár, D. Donoval, C. Fleury, D. Pogany, G. Strasser, O. Hilt, F. Brunner, J. Würfl, "Self-heating in GaN transistors designed for high-power operation," IEEE Trans. Electron. Device. vol.61, pp. 3429-3434, Oct 2014.

[3] J.T. Asubar, Z. Yatabe,T. Hashizume, "Reduced thermal resistance in AlGaN/GaN multi-mesa-channel high electron mobility transistors," Appl. Phys. Lett, vol. 105, pp. 053510, 2014.

[4] H. Guo, Y. Kong, T Chen, "Thermal simulation of high power GaN-on-diamond substrates for HEMT applications," Diam. Relat. Mater, vol. 73, pp. 260-266, 2017.

[5] G. Pavlidis, G. Pavlidis, E.R. Heller, E.A. Moore, R. Vetury, S. Graham, "Characterization of AlGaN/GaN HEMTs using gate resistance thermometry," IEEE Trans. Electron Devices, vol. 64, pp. 78-83, 2017.

[6] G. Agarwal, T. Kazior, T. Kenny, D. Weinstein, "Modeling and analysis for thermal management in gallium nitride HEMTs using microfluidic cooling," J. Electron. Packag, vol. 139, pp. 1–11, 2017.

[7] A. Nigam, T.N. Bhat, S. Rajamani, S.B. Dolmanan, S. Tripathy, M. Kumar, "Effect of self-heating on electrical characteristics of AlGaN/

GaN HEMT on Si (111) substrate," AIP Adv. vol. 7, pp. 085015, 2017.

[8] J.Z. Wu, J.M. Min, W.Lu, P.K.L. Yu, "Thermal resistance extraction of AlGaN/GaN depletion-mode HEMTs on diamond," Journal of Electronic Materials, vol. 44, pp. 1275-1280, 2015.

[9] B.K.Schwitter, A.E. Parker, S.J. Mahon, A.P. Fattorini, M.C. Heimlich, "Impact of bias and device structure on gate junction temperature in AlGaN/GaN-on-Si HEMTs," IEEE Trans. Electron. Device, vol. 61, pp. 1327-1334, May 2014.

[10] P.C. Chao, Kanin Chu and Carlton Creamer, "A New high power GaN-on-Diamond HEMT with Low-Temperature bonded substrates technology," CS MANTECH Conference, pp. 179-182, May 2013.

[11] J.W. Pomeroy, M. Bernardoni, D.C. Dumka, D.M. Fanning, M. Kuball, "Low thermal resistance GaN-on-diamond transistors characterized by three-dimensional Raman thermography mapping," Appl. Phys. Lett., vol. 104, pp. 083513, 2014.

[12] J. Cho, Z. Li, E.B. Grayeli, T. Kodama, D. Francis, F. Ejeckam,F. Faili, M. Asheghi, K.E. Goodson, "Improved thermal interfaces of GaN–Diamond composite substrates for HEMT applications," IEEE Trans. Compd. Packag.Technol, vol. 3 pp. 79-85, Jan 2013.

[13] Y. Won, J. Cho, D. Agonafer, M. Asheghi, K.E. Goodson, "Fundamental cooling limits for high power density gallium nitride electronics." IEEE T. Comp. Pack. Man., vol. 5, pp. 737-744, Jun 2015

[14] X.Y Guan, H.K. Lee, S.H. Lee, J.S. Yu, "Device characteristics and thermal analysis of GaN-based vertical light-emitting diodes with different types of packages." Solid-State Electronics, vol. 127, pp. 51–56, 2017.

[15] K. Park, C. Bayram, "Thermal resistance optimization of GaN/substrate stacks considering thermal boundary resistance and temperature-dependent thermal conductivity", Appl. Phys. Lett., vol. 109, pp. 151904, 2016.

[16] A. Darwish, A.J. Bayba, H.A. Hung, "Channel temperature analysis of GaN HEMTs with nonlinear thermal conductivity," IEEE Trans. Electron. Device, vol. 62 pp. 840-846, Mar 2015.

[17] M. Garven, J.P. Calame, "Simulations and optimization of gate temperatures in GaN-on-SiC monolithic microwave integrated circuits, IEEE Trans." Compd. Packag. Technol. vol. 32, pp. 63-72, Mar 2009.

[18] R. Zhang, W.S. Zhao, W.Y. Yin, Z.G. Zhao, H.J. Zhou, "Impacts of diamond heat spreader on the thermo-mechanical characteristics of high-power AlGaN/GaN HEMTs," Diamond & Related Materials. vol. 52, pp. 25-31, 2015.

[19] T. Ishizaki, M. Yanase, A. Kuno, T. Satoh, M. Usui, F. Osawa, Y. Yamada, "Thermal simulation of joints with high thermal conductivities for power electronic devices," Microelectron. Reliab., vol. 55, pp. 1060-1066, 2015.

[20] A. Wang, M.J. Tadjer, T.J. Anderson, R. Baranyai, J.W. Pomeroy, T.I. Feygelson, K.D. Hobart, B.B. Pate, F. Calle, "Impact of intrinsic stress in diamond capping layers on the electrical behavior of AlGaN/GaN HEMTs," IEEE Trans. Electron. Device., vol. 60, pp. 3149-3156, Oct 2013.

[21] K. Lee, J.S. Moon, T. Oh, S. Kim, P. Asbeck, "Analysis of heat dissipation of epitaxial graphene devices on SiC," Solid State Electronics, vol. 101, pp. 4-9, Jul 2014.

[22] S.M. Horcajo, A. Wang, M. F. Romero, M.J. Tadjer, F. Calle, "Simple and accurate method to estimate channel temperature and thermal resistance in AlGaN/GaN HEMTs," IEEE Trans. Electron. Device, vol. 60, pp. 4105-4111, Dec 2013.

Measuring the electric properties of thin film shape memory polymers in simulated physiological conditions

Daniel Del Nero
Electrical Engineering
The University of Texas at Dallas
Richardson, United States
daniel.delnero@utdallas.edu

Alexandra Joshi-Imre
Engineering Innovation
The University of Texas at Dallas
Richardson, United States
alexandra.joshi-imre@utdallas.edu

Walter E. Voit
Material Science and Engineering
The University of Texas at Dallas
Richardson, United States
walter.voit@utdallas.edu

Abstract—Shape memory polymers (SMP) exhibit unique physical properties attractive for potential application in medical devices. These materials can assume a rigid solid state but soften after exposure to saline solution and heat, a feature which may aid the implantation process. Exposure to the biological environment leads to fluid absorption by the polymer coating resulting in the desired change in mechanical properties. The fluid uptake, however, also results in a change of electric properties that need to be taken into account when designing electrical components integrated with SMP. We developed procedures based on electrochemical impedance spectroscopy (EIS) to study the electric properties of saline-soaked SMP coatings. With these, we can observe the softening process as well as long-term electrical behavior. Simple circuit models are used to model the coating properties and calculate relative permittivity and electrical resistivity of the SMP material.

Keywords— Electrochemical impedance spectroscopy, organic coatings, shape memory polymers, relative permittivity, resistivity

I. Introduction

Shape memory polymers (SMP) have shown potential in use as substrate material for electronics that interface with neural tissue [1]. Neural interfaces provide recording or modulating of neural activity, allowing doctors to better diagnose or treat certain illnesses. The role of SMPs in these electronic interfaces is to allow for an easy implantation in their resting rigid solid state but to provide for a soft, more mechanically compliant implant once inside the body, preventing irritation and damage to the neural tissue. This state transition from being stiff to flexible is caused by the absorption of extracellular fluids within the body by the polymers coating the electrode in combination with the increase in temperature. The uptake of electrolyte also affects the electric properties of the SMPs. In this work, we investigate the electric properties of SMPs in the soft state and explore the process of transitioning. Our results have implications to the electrical performance of the whole implant.

Here, we present results from thiol-ene/acrylate SMP showing a decrease in electric impedance over three days of exposure to saline solution. We use potentiostatic electrochemical impedance spectroscopy (EIS) to determine the electric properties of SMP coatings in a flat plate, clamp-on, O-ring seal test cell previously described by J. N. Murray [2]. Briefly, in this method, the SMP material is prepared as the thin film dielectric of a capacitor, with one side contacted by a metal plate, and the other side exposed to electrolyte.

Phosphate buffered saline (PBS) is used as an electrolyte to represent extracellular fluids, and the samples are measured in an oven at 37° C to simulate the environment of the human body. We perform EIS in a two-electrode setup and we calculate SMP coating capacitance and leakage resistance by fitting the EIS data with an equivalent circuit model. From the coating capacitance, we estimate the dielectric constant, and from the leakage resistance, we estimate resistivity of the polymer. We measure changes in EIS data over time to learn about the transitioning of the SMP. Our experimental setup was validated by measuring solid-state capacitors and resistors of known value.

II. Materials and Methods

A. Sample preparation

Four samples were prepared on large glass slide substrates that are 1.2 mm thick and 75 x 51 mm area. The slides were cleaned using Alconox detergent followed by an acetone rinse and a deionized water rinse. After drying, the slides were deposited with thin film metals consisting of titanium (30 nm) and gold (300 nm). The SMP monomer solution is prepared as described as "FS-SMP composition" in [3][4]. Briefly, acrylate, ene, and thiol monomers were mixed with a photoinitiator, spin-coated, and UV-polymerized using a sequence of a 3 minute exposure to 254 nm UV followed by a 60 minute exposure to 365 nm UV with a post-cure in a vacuum oven at 120 °C. A small portion of the metal-coated glass slide was masked off from the polymer coating and later used as a site for wire attachment. Polymer thickness was

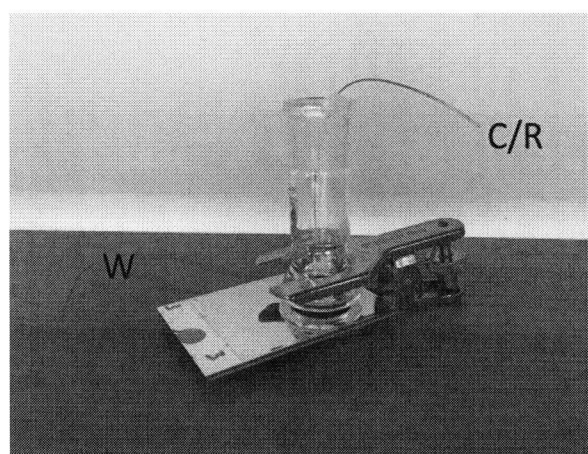

Fig. 1. Photograph of a polymer sample assembled for measurement.

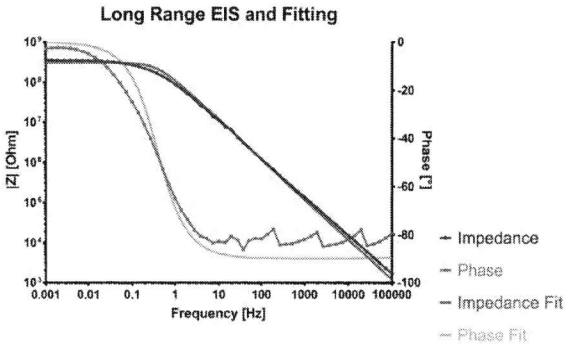

Fig. 2. EIS from 1mHz to 100kHz of sample B alongside a paint model fitting. The paint model equivalent circuit is shown in the top right.

measured on each slide using contact profilometry and was found to be approximately 18 μm.

After the polymer sample is assembled with the clamp-on glass adapter, as seen in Fig.1, PBS is poured into the adapter. The polymer separates the conductive PBS from the conductive metal coating, making the system similar to a capacitor. The behavior of this capacitor is labeled as the coating capacitance and is the focus of our work on characterizing the electric properties of our polymer coating, as this value changes over time while it is exposed to PBS. The coating capacitance is also responsible for the permittivity of the polymer coating.

B. Modeling

To model the behavior of these polymers during and after soaking we use an equivalent circuit developed to model organic paint coatings, as described in [5][6][7]. This equivalent circuit, described as the "paint model," consists of a collection of real components that best match the response of the electrochemical cell, as seen in Fig. 2. This figure also shows a representative EIS Bode plot from a sample. In the model, R_s describes the solution resistance of the electrolyte volume between the surface of the polymer and the electrode placed inside the solution. R_s does not dominate in our Bode plots, being only a few hundred Ohm. R_{po} describes the resistance across pores in the polymer coating that allow ions to conduct through. The ions may exchange electrons with the gold film underneath the polymer coating, which process is represented by the R_{ct}, charge transfer resistance and the C_{dl} double layer capacitance (that assumes the formation of a liquid pocket at the interface). Most importantly, C_c describes the coating capacitance.

C. Electrochemical Impedance Spectroscopy

To identify the absorption period's effects on the electric properties, we use Electrochemical Impedance Spectroscopy (EIS) performed by a potentiostat. EIS involves measuring the impedance of the system for a range of frequencies at a given number of points per decade, as described in [8][9][10].

The potentiostat is connected to an electrochemical cell placed inside a Faraday cage inside an oven at 37° C to simulate physiological conditions. This electrochemical cell is made by clamping a glass tube onto the surface of the polymer-coated sample and filling the tube with PBS. A platinum wire rests within this tube and acts as a counter. These EIS are run in a two-electrode configuration, which consists of a working electrode connected to the gold layer underneath the SMP and the platinum electrode connected to the potentiostat as the counter and the reference [11]. The ground is connected to the Faraday cage. EIS measurements in the frequency range of 1 mHz to 100 kHz took approximately two hours to perform. The short range EIS is performed with an initial frequency range of 25 kHz to 50 kHz to capture the effects of water absorption into the polymer coating. This range is measured constantly for the first ten minutes of soaking and with a 20 second delay for the next two hours. This is to better capture the absorption that occurs immediately after the PBS is introduced to the surface of the polymer, the effects if which will be discussed later in this paper. After two hours, a second EIS is performed from 1mHz to 100kHz along with the short range EIS to capture the effects of leakage pathways forming on the polymer surface.

D. Equipment Validation

To identify the EIS measurement limits of our potentiostat and wiring, we loaded a high value (100 GOhm) resistor in place of the sample. We found that the setup was able to measure the resistor value correctly below 100 mHz, while it was limited by its input impedance at frequencies above. By isolating the limited portion of the impedance curve and fitting a capacitor model to it, we were able to calculate the impedance from the input. The equivalent capacitance of this input impedance was approximately 8.1 pF.

978-1-7281-1500-9/19 $31.00 © 2019 IEEE

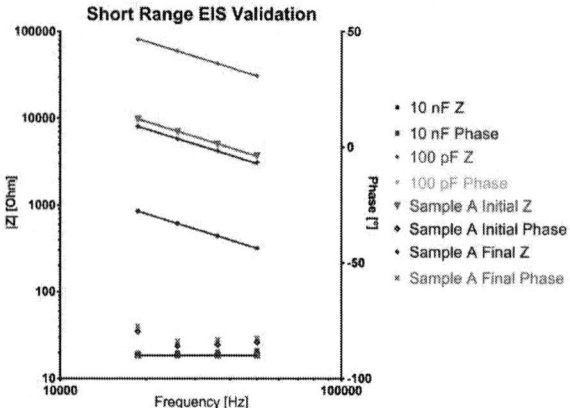

Fig. 3. Bode plot of the first and last EIS tests run on two solid state capacitors (100pF and 10nF) and sample A with capacitor fittings shown by lines.

To check if our model fittings are correct, we loaded two solid-state capacitors in place of our sample, one of 100 pF and one of 10 nF. Fitting capacitor models to these impedance results yielded capacitance values close to the listed solid-state value, validating the accuracy of our capacitor model. We selected the values of our solid-state capacitors by choosing one with a capacitance above the polymer coating capacitance and one below. We then performed a sequence of short EIS measurements over time on our solid-state capacitors and found that our potentiostat experienced no electronic drift as there was no difference between the first and last capacitor tests. A comparison between solid-state capacitors and sample A is show in Fig. 3. In this figure, we also observe a decrease in impedance between the initial and final EIS performed on the sample over two hours, which change we could track by performing a sequence of short range EIS measurements.

III. RESULTS AND DISCUSSION

A. SMP Impedance and Permittivity

The decrease in SMP impedance is related to an increase in capacitance due to water permeating the surface of the polymer. Because water has a higher permittivity than the polymer coating, we expect this increase in capacitance as the polar water molecules find their way into and the polymer coating. In fact, our obtained impedance data show a large initial decrease within the first 10 minutes of exposure to PBS that slows to a gradual decrease over two hours. Fig. 4a illustrates this rapid initial decrease in impedance, with differences between samples stemming from varying thicknesses and surface defects allowing for faster absorption of water. This short period of time corresponds with the softening of the material after it is exposed to saline as mentioned previously. While the impedance does continue to decrease after the initial 10 minutes of soak time, it is at a slow rate. Fig. 4b shows that most of the change in permittivity occurs in the first 10 minutes.

We use data from this initial period of water intake to calculate the permittivity of the material. This is done by fitting a capacitor to the Bode plot, which yields a value close to the coating capacitance. This capacitance is then used to calculate permittivity using the equation for the capacitance of a parallel plate capacitor rearranged to find permittivity:

$$E_r = (C_c\, d) / (E_o\, A) \qquad (1)$$

where d represents the thickness of the polymer coating, E_o is the vacuum permittivity and A is the surface area of the polymer in contact with the PBS solution.

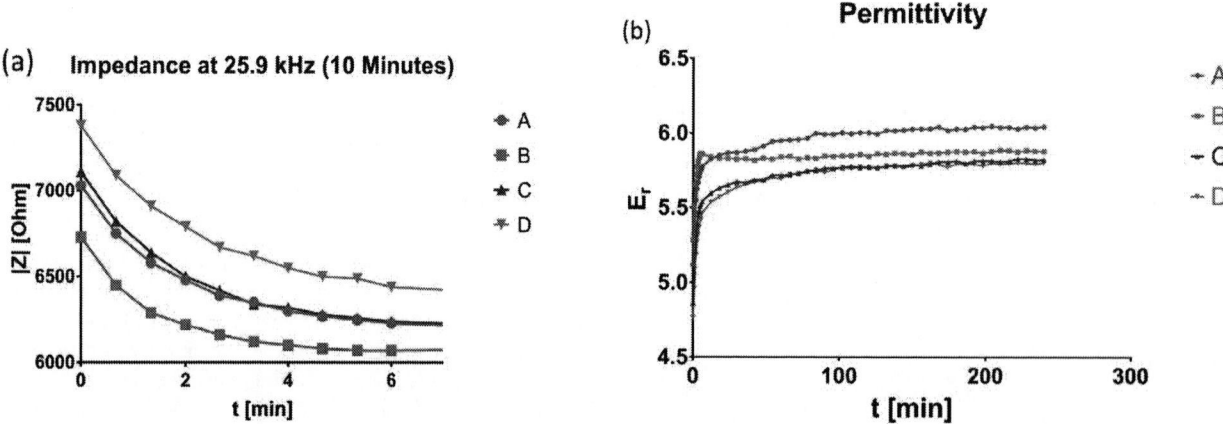

Fig. 4a. Impedance data over 10 minutes showing a large initial decrease after exposure to PBS
Fig. 4b. Permittivity data over two hours showing a large initial increase.

Table I compares permittivity values calculated from coating capacitances that were obtained by 1) fitting the paint model using long range EIS data and 2) calculating equivalent capacitance of the impedance measured at 25.9kHz, as described below.

Table I.

Permittivity at 240 minutes		
Sample:	Paint Model	Capacitor
A	6.092092	6.045069
B	5.90963653	5.880177
C	5.915398	5.823625
D	5.901634	5.804648

While the values are not exactly the same, they still fall within a close range and the capacitor calculation is a much quicker method of characterizing the sample's permittivity, especially as it evolves over time. The calculation for capacitance involves using the impedance data from each sample and calculating capacitance using the given impedance at that point and a selected frequency. In our case, we use the impedance measured at 25.9 kHz to calculate the coating capacitance using the equation:

$$|Z| = 1 / (2\pi f C) \qquad (2)$$

In this equation, $|Z|$ represents the impedance, C represents the capacitance, and f represents the frequency. Calculated relative permittivity appears to stabilize after the initial increase for all samples, as seen in Fig. 4b.

We estimate the resistivity of our polymer to be approximately 5.9E9 Ωm based on the values of R_{po} taken from paint model fittings on the curves in Fig. 5. Note, that large defect sites may introduce significant current pathways and can affect the value obtained for R_{po}, for example in the case of sample A in Fig. 5. Sample B provides a good example of a non-defect-dominated sample, where the leakage pathway is clearly visible and R_{po} is easy to determine. For more information about the evaluation of Bode plots, please refer to Murray in [7].

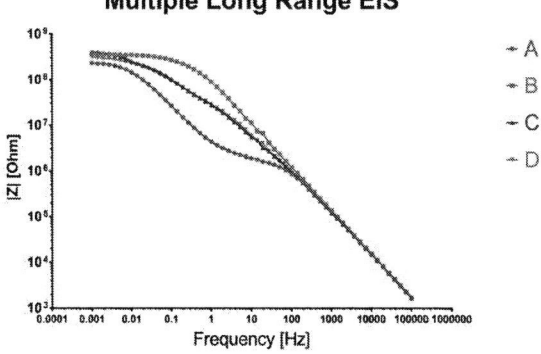

Fig. 5. An EIS performed over a range of 1 mHz to 100 kHz after two hours of soaking. The formation of leakage pathways is visible in the flattening of the curve at lower frequencies.

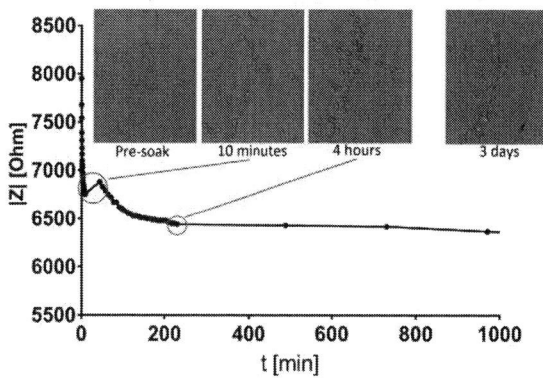

Fig. 6. Part of a three-day EIS with microscope images taken before soak and after 10 minutes, 4 hours, and 3 days. Images taken mid-soak correspond with jump in impedance due to water evaporation while sample was under microscope. Real size of the sample images is 400 µm by 600 µm

B. Long-term Soak

In general, we have observed a slight but steady decrease of impedance of our samples over longer time scales. For example, Sample C exhibited a decrease of impedance at a slope of -0.05 between two hours to three days of soaking. To better understand this long-term trend, we performed optical microscopy at various stages of the soak test on an additional sample. The same sequence of short range and long range EIS measurements was employed, but it was interrupted twice in order to make observations of the coating quality. The optical microscope images allowed us to track the development of water pockets and compare them with the impedance data taken immediately before and after the image, as shown in Fig. 6. We observe a development of pockets corresponding directly with a decrease in impedance over time, confirming that the continuing decrease in impedance is due to the development of pockets of water in or under the SMP coating.

IV. CONCLUSION

EIS is a powerful method in characterizing the electric properties of organic coatings and polymer encapsulation, and many have developed their own procedures for measurement [12][13][14][15][16]. While these reports may be similar in their methods using EIS to characterize the impedance behavior materials, they may differ in purpose. For example, one may estimate water absorption in a material [14][16], or study corrosion protection properties of a coating [2]. In evaluating EIS results, some articles discuss the usage of a constant phase element or a faradaic impedance in their equivalent circuit to account for the phase angle at higher frequencies being slightly above -90°, indicating that part of the curve cannot be represented using a solid-state capacitor [14][15]. Some have also proposed distributed models [16]. In this work, we opted to use simple circuit models that were sufficient in describing coating behavior.

Here, we studied the electric properties of samples from a single thiol-ene/acrylate SMP composition. Our purpose is to develop procedures for EIS-based measurements that are capable of tracking changes in the electric properties of the polymer. To summarize the method we developed, EIS is applied to an electrochemical cell composed of a thin metal film with a polymer coating that is exposed to PBS. EIS is taken over a short range of frequencies for the first two hours of soaking, and it is taken over a longer range of frequencies for the remaining 68 hours. The short range EIS allows us to capture the rapid change in impedance that occurs as water absorbs into the polymer coating and the polymer softens. The long EIS allows us to identify the formation of leakage pathways that can only be seen at lower frequencies. Using the data from EIS, we calculate the permittivity and resistivity of our polymer coating using an equivalent circuit model.

ACKNOWLEDGMENT

The University of Texas at Dallas is acknowledged for supporting the Center of Engineering Innovation, where this work was performed. The authors would like to thank Vindhya Reddy Danda for her help with sample preparation.

REFERENCES

[1] Ecker, M., Joshi-Imre, A., Modi, R., Frewin, C. L., Garcia-Sandoval, A., Maeng, J. Gutierrez-Heredia, G., Pancrazio, J. J., Voit, W. E. (2018). "From softening polymers to multimaterial based bioelectronic devices," Multifunctional Materials, vol. 2, no. 1, December 2018.

[2] Murray, J. N., "Electrochemical test methods for evaluating organic coatings on metals: An update. Part I. Introduction and generalities regarding electrochemical testing of organic coatings," Progress in Organic Coatings, vol. 30, pp. 225–233, 1997 https://doi.org/10.1016/S0300-9440(96)00677-7

[3] Black, B. J., Ecker, M., Stiller, A., Rihani, R., Danda, V. R., Reed, I., ... Pancrazio, J. J. "In vitro compatibility testing of thiol-ene/acrylate-based shape memory polymers for use in implantable neural interfaces," Journal of Biomedical Materials Research - Part A, vol. 106, pp. 2891–2898. October 2018. https://doi.org/10.1002/jbm.a.36478

[4] Ecker, M., Danda, V., Shoffstall, A. J., Mahmood, S. F., Joshi-Imre, A., Frewin, C. L., ... Voit, W. E. "Sterilization of thiol-ene/acrylate based shape memory polymers for biomedical applications," Macromolecular

Materials and Engineering, vol. 302, pp. 1–10, 2017. https://doi.org/10.1002/mame.201600331

[5] "EIS of organic coatings and paints," Gamry Instruments Application Note, 2017.

[6] Loveday, D., Peterson, P., & Rodgers, B., "Evaluation of organic coatings with electrochemical impedance spectroscopy. Part 2: application of EIS to coatings," Journal of Coatings Technology, vol. 1, no. 10, pp. 88–93, October 2004.

[7] Murray, J. N., "Electrochemical test methods for evaluating organic coatings on metals: an update. Part III: Multiple test parameter measurements," Progress in Organic Coatings, vol. 31, pp. 375–391, 1997. https://doi.org/10.1016/S0300-9440(97)00099-4

[8] "Basics of Electrochemical Impedance Spectroscopy," Gamry Instruments Application Note, 2010.

[9] Loveday, D., Peterson, P., & Rodgers, B., "Evaluation of organic coatings with electrochemical impedance spectroscopy Part 1: fundamentals of electrochemical impedance spectroscopy," Journal of Coatings Technology, vol. 1, no. 8, pp. 88–93, August 2004.

[10] O'Donoghue, M., Garrett, R., Datta, V., Roberts, P., & Aben, T., "Electrochemical impedance spectroscopy- testing coatings for rapid immersion service," Material Performance, vol. 28, pp. 36–41, September 2003.

[11] "Two-, three-, and four-electrode experiments," Gamry Instruments Application Note, 2015.

[12] Bouvet, G., Nguyen, D. D., Mallarino, S., & Touzain, S., "Analysis of the non-ideal capacitive behaviour for high impedance organic coatings," Progress in Organic Coatings, vol. 77, no. 12, pp. 2045–2053, 2014. https://doi.org/10.1016/j.porgcoat.2014.02.008

[13] Kittel, J., Celati, N., Keddam, M., & Takenouti, H., " New methods for the study of organic coatings by EIS: New insights into attached and free films," Progress in Organic Coatings, vol. 41, no. 1-3, pp. 93–98, 2001. https://doi.org/10.1016/S0300-9440(00)00155-7

[14] Miszczyk, A., & Darowicki, K., "Water uptake in protective organic coatings and its reflection in measured coating impedance," Progress in Organic Coatings, vol. 124, pp. 296–302, August 2017. https://doi.org/10.1016/j.porgcoat.2018.03.002

[15] Macedo, M. C. S. S., Margarit-Mattos, I. C. P., Fragata, F. L., Jorcin, J. B., Pébère, N., & Mattos, O. R., "Contribution to a better understanding of different behaviour patterns observed with organic coatings evaluated by electrochemical impedance spectroscopy," Corrosion Science, vol. 51, no. 6, pp. 1322–1327, 2009. https://doi.org/10.1016/j.corsci.2009.03.016

[16] Castela, A. S., & Simoes, A. M., "An impedance model for the estimation of water absorption in organic coatings. Part II: A complex equation of mixture," Corrosion Science, vol. 45, no. 8, pp. 1647–1660, 2003. https://doi.org/10.1016/S0010-938X(03)00015-5

Evaluation of WLP Dielectrics for High-Voltage Applications

Marcus Paeck, Markus Woehrmann, Michael Teopper
Fraunhofer Institute for Reliability and Microintegration
Berlin, Germany
e-mail: Markus.Woehrmann@izm.fraunhofer.de

Klaus-Dieter Lang
Technical University Berlin,
Berlin, Germany

Abstract—Due to the excellent mechanical and electrical properties together with a cost efficient processing, the thin film polymers are used in various applications as an interlayer dielectric (ILD) material where the device is generally working at CMOS like low voltages. The integration of high voltages in the redistribution layer becomes more attractive in future to realize higher system miniaturization and integration. Regarding to this, the power transformations could be integrated on wafer level for direct support of the CMOS device without an additional power transformation unit. Also high voltage III-V semiconductor dies will be integrated in Fan-Out packages. Thin film polymers have the potential to meet the demands for high voltage applications like high break down voltage and low leakage current. Well established thin film polymers like CYCLOTENE (Benzocyclotbutene – BCB) show a break down voltage in the range of 530 V/um. Compared with inorganic passivation such as thermal SiO_2 with a breakdown voltages from 400 up to 1000 V/μm the BCB is quiet compatible regarding to the processing windows of layer thickness from BCB in the range from 5 to 20 um. BCB has been widely adopted in electronic applications, due to its low dielectric constant, excellent chemical resistance, and high thermal stability (glass transition temperature, T g > 350 °C) after hard-curing. The BCB has several advantages, especially the good process capabilities such as a photo structuring capability, the low temperature and the vacuum-less processing, might be interesting to use it in high voltage applications. This paper contains the profound investigation of the behavior of BCB at high voltages. Regarding the usability analyzing for the BCB as high voltage dielectric the break down voltage behavior was estimated and a time depending breakdown voltage was observed regarding the duration of the voltage application. This behavior is not yet published in the literature for BCB. MIM structures with 1 mm and 5 mm radius were manufactured on wafer level for this investigation. The leakage current of a few pA is measured and the time-dependent dielectric breakdown (TDDB) is extracted from Current-Time characteristics and shown in Weibull-plots. The measurement shows a spectrum from 350 V/μm to 450 V/μm, depending of the duration of the voltage and the level of the applied voltage. It is found that there is an exponential linkage between field strength and the time till the breakthrough occurs. The break through at 450 V/μm occurs after about 280 seconds, while the sample at 350 V/μm withstands longer by the factor of 13. These results will a very important milestone for future high power applications. *(Abstract)*

Keywords: WLP; BCB; break down voltage; TDDB; Fan-In; Fan-Out

I. INTRODUCTION

The continuous development of microelectronic components, along with continuous miniaturization and simultaneous increase in performance, relates not only the low voltage range, but it is also for power electronics of increasing interest. The driving industries including wind energy, solar energy, electric mobility and railway technology are using power electronics, where the systems and vehicles have to be withstand high voltages. Insofar the used materials must have a high dielectric strength or the dielectric layers have to be relative thick.. Furthermore, in particular in vehicle technology, there is a desire for even more compact and efficient components.

Most of the electronic devices in high power applications are separated in logic und controlling units. Logic units are often low power chips and control the high voltages which are leads by the controlling units. But there are a lot of applications where the free space is strictly limited [5]. To meet this requirement, the logic and controlling unit have to build in one single device. This mean the redistribution layer on chip should withstand high voltages. A pure inorganic passivation like silicon dioxide is limited in reaching crack-free, low stress thick insulation layers for this application [6] [7].

Unfilled polymers are state of the art for realizing inter dielectric layer in Fan-In or Fan-Out redistribution layer, which have the potential to withstand also higher applied voltage due to a realization of a thicker layer in the range of 5 to 20 μm. Related to chip embedding application, now power-MOS and CMOS could be combined together in fan-out packages with degree of miniaturization. For this reason, unfilled thin film polymer as a dielectric insulator for high voltage electronics is in the focus of this work.

For applications that are supposed to work in the higher power range, attributes such as the breakdown voltage and the leakage current of the polymer must be considered. The breakdown voltage is usually sparely specified by the manufacturer. For application at high voltage a denser analyzing of the material behavior is demanded like the time characteristic or influence of the voltage increasing rate. In addition to the leakage currents at higher voltages may degrades the polymer over time. This may results in a cleavage of existing polymer chains or in dissociation of foreign molecules such as water molecules. The effects

978-1-7281-1500-9/19 $31.00 © 2019 IEEE

would be reflected in an increased leakage current, up to the age-related breakdown.

In this work capacitor structures are built with the BCB polymer as dielectric and the electrical behavior in relation to applied high voltage is analyzed which is close to the dielectric strength of material.

II. THE CHOICE OF THE INTER DIELECTRIC POLYMER MATERIAL

In these work the BCB is used as dielectric material for high voltage application. The material is usable for fan-in and due the low cure temperature also for fan-out application. The BCB has a low water uptake of 0.1 %. The water uptake decrease the break down toughness, where other dielectric materials like polyimide, PBO and Epoxy have a significant uptake in the range of 1.5 to 3 %. Table 1 shows the volume resistivity of a polyimide layer at ambient condition and after drying in an Oven for 2h at 120°C. After the oven bake the material has nearly no water in the polymer matrix which leads to a higher resistivity and also a lower leakage current. This effect could not observed for the BCB material.

Table 1: Volume resistivity measurement of a polyimide layer

	Volume resistivity at r.Hd. of 35 % [Ω cm]	Volume resistivity at r.Hd. of 0 % [Ω cm]
Polyimide 5 μm	1,13E+10	2,00E+14

The polymer Benzocyclobutene (BCB) is a thermoset resin which is sold by DUPOND. It based on the monomer divinyl siloxane bisbenzocyclobutene (DVS-bis-BCB).

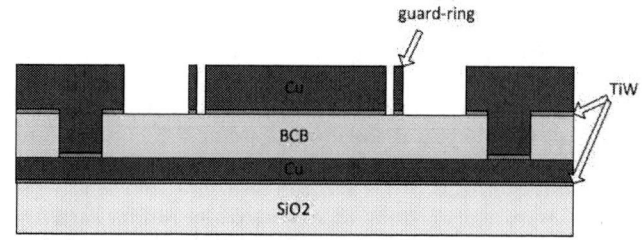

Figure 1: Chemical structure of the monomer DVS-bis-BCB [4]

The functional group that holds the both benzocyclobutene groups together is siloxane tetramethyldivinylsilane. These groups are highly covalent which leads to a very low dielectric constant. The pre-polymer is a b-staged material to provide a proper viscosity for spin-coating.

The BCB polymerization mechanism starts at 250°C and doesn't provide any byproducts. At least a highly cross-linked thermoset accrues with good electric, thermic and mechanic properties.

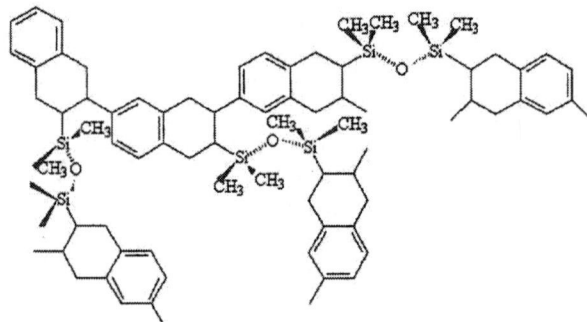

Figure 2: Detail of the polymerized BCB [4]

BCB has a lower Young's modulus of 2.1 GPa compared to the inorganic insulators. This results in lower stresses in the thin films which could be benefit for the reliability (also the higher elongation allows to use the layer as stress buffer). Planarization is an important property, to be sure that the thickness of the layer is nearly constant. If the thickness would change over the capacitance area, the determination of the capacity and also of the leakage current and the breakdown voltage are not precise. Studies of planarization of BCB used in packaging applications show excellent results (>90%). Furthermore BCB films absorb no significant amounts of moisture. This is also a crucial characteristic and would have an impact of the parameters which are under investigation [4]. These properties are important reasons for using BCB as an inter-dielectric in integrated circuits.

III. MEASUREMENT SETUP

The electrical test structures are integrated capacitors (MIM) consisting of the BCB layer as the dielectric material sandwiched by copper electrodes.

Figure 3: Schematic structure of a thin film capacitor

There are MIM structures with a radius of 1 mm and 5 mm to investigate the influence of the size. The lower electrodes are copper films with a thicknesses of 1 um and were produced by magnetron sputter deposition. The lower layer was not enforced by a galvanic copper layer to realize a smooth and defect free surface for the MIM stack. The lower copper layer was therefore structured by a subtractive etching step. After that, the BCB is applied via spin coating with a thickness of 4 um. At least the upper electrode is built by an initial sputter layer followed by galvanic copper deposition. The galvanic step is necessary to create a 5 µm thick layer to prevent the MIM structure against mechanical damaging by the prober tips.

Additionally, the upper electrode is surrounded by a guard-ring to realize a homogeneous electric field in the dilectric.

IV. CHARACTERIZATION OF THE MIM

In advanced of the electrical characterization the measurement of the capacitance of MIM is done in order to verify the dielectric constant. The measurements of the two different sizes of the MIMs are shown in Figure 4.

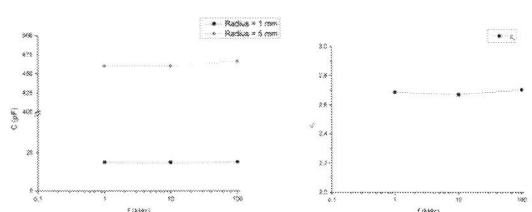

Figure 4: left - Capacitance measurement in the frequency range of 1 kHz to 100 kHz; right - Relative permittivity of BCB in the frequency range from 1 kHz to 100 kHz

The dielectric constant is calculated by

$$\epsilon_r = \frac{C * d}{\epsilon_0 * A}$$

where C is the capacitance, d the thickness of the dielectric layer, A the area of the capacitor and ϵ_0 the vacuum permittivity. Furthermore the measurement of the capacity is an indicator for the quality of the MIM structures. Possible defects could be detected. The relative dielectric constant was in the frequency range 1 kHz to 100 kHz 2.69 which is in correlation to the datasheet of the BCB.

V. LEAKAGE CURRENT AT HIGH VOLTAGE

The leakage current is measured over time. Therefor a voltage is applied to the electrodes and the current answer is recorded. The high voltage source is from the company Heizinger and can generate up to 10 kV. The voltage was pre-defined and instantaneously set to the sample. The current detection was done with the Agilent 4339B.

In order to protect the measurement against external electrical influences, the specimen chamber was electrical shielded. Another challenge when working with high voltages is the flash over of a spark between the contact electrodes. Under standard conditions it can be assumed that about 3 kV / mm are required in air until it comes to a flash over at the surface. The structures used in this work had partial distances in the micrometer range, so flash overs were observed, which made the measurement impossible.

So the specimen was covered with silicon oil having a dielectric strength up to 23 kV / mm to prevent electrical flash overs.

The measurement of the current response to a voltage step leads to a time dependent current whose curve is shown in Figure 5. It is chosen a field strength at which even after long exposure no breakthrough is expected.

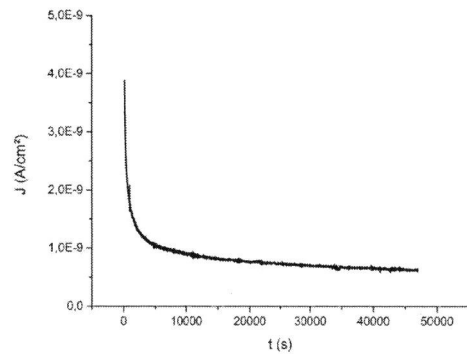

Figure 5: Current response at field strength of 2,75MV/cm. Measurement was performed on a sample with a radius of 1 mm

The curve has a hyperbolic characteristic with a decrease of the leakage current even over longer time frames where no clear saturation could be observed.

To compare the curves at certain voltage steps the leakage current after 1000 sec was recorded. This delay time was chosen because this is a good compromise between the gradient of the current profile and the duration of the measurement process.

The current response at different field strengths is shown in Figure 6

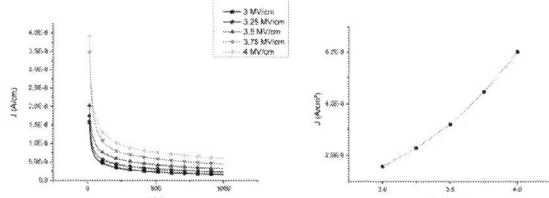

Figure 6: left - Current response at different field strengths. Measurements were performed on samples with a radius of 1 mm; right - Leakage current after 1000 seconds to different field strengths. Measurements were performed on samples with a radius of 1 mm

The stress condition are different electrical fields in the range from 3 to 4 MV/cm at room temperature. It could be shown that there is an exponential linkage between the applied field and the leakage current. That correspond to the electron conduction calculations with the Poole–Frenkel-

Effect, Schottky-Effect and Fowler-Nordheim-Effect [3] [8] [9].

Figure 7 compares the leakage current of the 1 mm radius and 5 mm radius MIM structures.

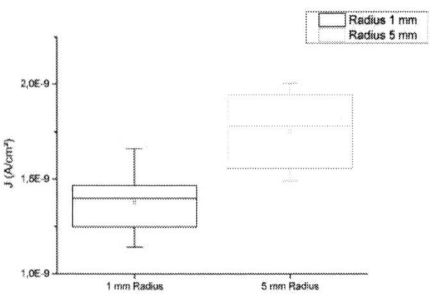

Figure 7: Leakage currents of different samples at a field strength of 3.75 MV/cm

The determined values obtained for the capacitors with a radius of 5 mm have nearly 27% larger leakage current. The reason is that the leakage current is defined after 1000 sec. Due to the larger area there is a higher capacitance and so the charge time of the capacitor is quit larger. Consequently, the charging currents of different sized capacitors do not match.

VI. TIME DEPENDED BREAK DOWN VOLTAGE ANALYZING

the literature, four breakdown mechanisms are described, which can also take place superimposed in their effect:

- Electrical breakdown
- Thermal breakdown
- Breakdown due to partial discharges
- Electromechanical breakdown

Because of the required conditions for a thermal breakdown, this breakthrough is very unlikely [2]. Furthermore, no evidence of an electromechanical breakdown could be found by any investigations [1].

There are several approaches to describe a dielectric breakthrough. O'Dwyer gives a good concept of breakthrough [3]:

- Is a homogeneous field applied across a dielectric, there is ionization due to electrons collisions.
- The ionization creates more mobile electrons and relative immobile holes. The holes drift to the cathode and create a space charge region, which partially distort the electrical field.
- This distortion enhances the electrical field, which leads to a higher amount of emitted electrons from the electrode.

- At least, the dielectric will be destroyed due to a massive emission of electrons close to the cathode.

Figure 8 will illustrate this behavior.

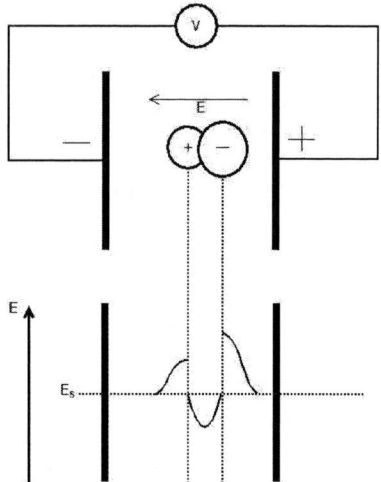

Figure 8: Schematic representation of the space charge field when voltage is applied

So, if the structure is exposed to electrical stress, the space charge region will grow over the time or with increasing voltage.

The test setup correspond to the leakage current analysis, wherein the measurement of the breakdown voltage, the applied voltage is sustained until a breakdown occurs. The results are displayed as Weibull distributions and shown in Figure 9.

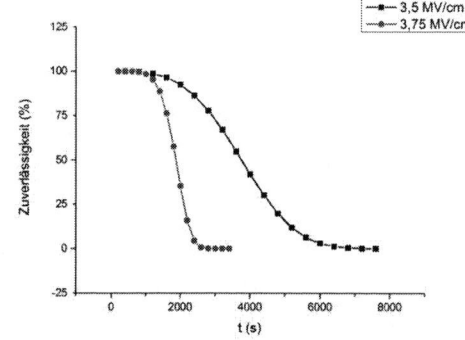

Figure 9: Weibull distribution for the field strengths 3.5 MV/cm and 3.75 MV/cm. Measurements were performed on samples with 1 mm radius

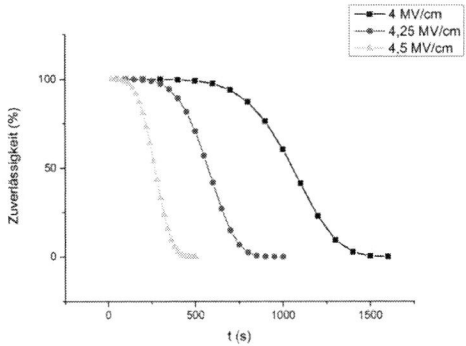

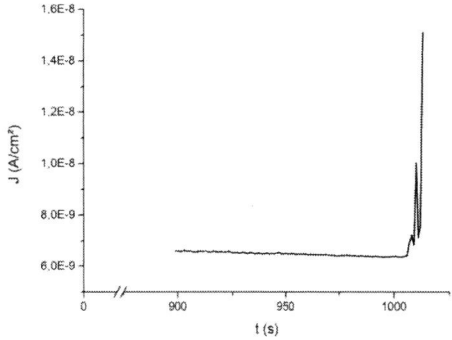

Figure 10: Weibull distribution for the field strengths 4 MV/cm, 4.25 MV/cm and 4.5 MV/cm. Measurements were performed on samples with 1 mm radius

These plots give an exemplary overview of the behavior of BCB at high voltages. The Figure 11 shows the time dependence at different field strength, if the reliability of 63,2% is set.

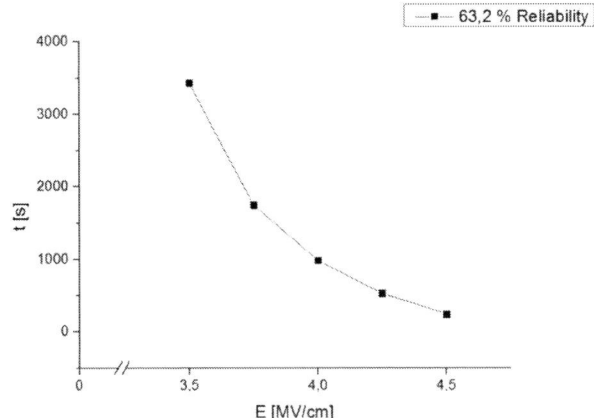

Figure 11: Weibull data for different field strengths, when a reliability of 63.2% is required

Furthermore, the breakdown phenomenon should be clarified. Of interest is to consider the current profiles close to a breakthrough. At all measurements there are volatile curves of the current answers briefly before the breakdown occurs as shown in Figure 12.

Figure 12: Current response at a field strength of 4 MV/cm. Presentation of the last 100 seconds before the breakthrough. Measurements were performed on samples with 1 mm radius.

The most likely reason for the ever decreasing current and the surge at the end of the measurement could be described by the theory of space-charge-limited current. There are a lot of unoccupied states in the dielectric material which could be occupied by the injected. So the material is no longer neutral and a space charge is created. Also the ionization caused by accelerated electrons creates more electrons and relative immobile holes which lead to a drift of the holes to the cathode. Consequently, with sustained load the space charge region will grow and the current flow will restricted. The curve of the current from Figure 12 suggests that there have been at the time of the first jump a pre-breakthrough in polymer. Thus, a current path has formed over a section of the dielectric thickness. That means the electrons will accelerate over these pathway and gains sufficient kinetic energy to overcome a potential barrier necessary to break an interface state (hot-electrons). Consequently, the rapidly increasing conduction electron density leads to the collective breakdown of the total polymer.

However, the capacitors with a radius of 5 mm reveal no meaningful breakthrough behavior if a voltage step is applied. There are instant breakthroughs if field strength of 2 MV/cm is direct applied. Whereas continuously increasing field strength could be reach field strength up to 7 MV/cm. A possible reason is that instant applied field strength injects more electrons from the electrode than it would be with continuous increasing field strength. So the possibility is higher to get a hot-electron which leads to an ionization and creates more electrons. At least the breakthrough occurs due to a massive electron emission.

Another phenomenon is the location of the breakthrough. It could be shown that 92% of the breakthroughs occur near the edge of the capacitor. This behavior could have two reasons. First, the edge of the capacitor is a discontinuity and prevents homogenous field lines. The impact of the guard-ring gives over the area of the capacitor homogenous field lines, but at the edge are still distorted field lines. Second, BCB absorb very low amounts of moisture, but it still does. Especially at the edges is storage of moisture highly probable that leads to a lower breakdown voltage.

VII. FTIR SPECTROSCOPY

The FTIR-Spectroscopy should exposure how the degradation of the polymer occurs. Either the polymer degrades due to the oxidation of the BCB or there is a formation of an electrical tree. The comparison of an oxidized BCB sample with an after breakdown BCB sample will give information about the behavior of BCB at high field strengths. Figure 13 shows both samples.

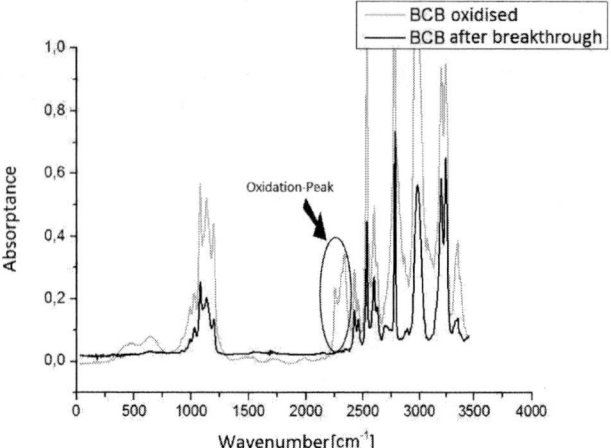

Figure 13: Measurement of an oxidized sample and a sample with breakthrough

The characteristic wavenumber for oxidation is between 2250 and 2400 cm-1. There is no oxidation peak in the sample with an electrical breakthrough. This mean there is no evidence for electric field driven degradation of the BCB. The measurement was carried out right next to the breakthrough by etching away the top electrode. Thus, it is excluded that the capacitor is influenced over the entire surface. It can therefore only have come to degradation in the form of an "electrical tree" at the location of the breakthrough. However, this evidence turns out to be problematic, since no external alterations would indicate a breakthrough. For this reason the measurement immediately before the breakthrough could not performed.

VIII. CONCLUSION

The behavior of polymer dielectric layers is analyzed to understand the behavior and the limits of high voltage application in a redistribution layer. The dielectric polymer BCB is chosen for the measurement regarding there well known behavior of low humidity uptake. A fundamental understanding of the conducting and the break

down process is started based on the analyzed behavior of BCB at high voltages. The measurement of the leakage current indicates the generation of space-charge zone in the polymer. It is measured a leakage value of $3.4 \pm 1 \times 10^{-10} A/cm^2$ at 1 MV/cm after 1000 seconds for the BCB. The data sheet of BCB indicates a leakage current of $4.70 \pm 1.6 \times 10^{-10}$ A/cm² without any information after which time it was measured. The size of the MIM structure and there different capacity could affect the leakage current value in the relation to the recorded time period. The generation of space-charge-regions in the polymer could lead to change of the electrical field in the MIM structure. The theory of electric field driven degradation of polymer could be neglected for the case of BCB. After electrical stressing no change of the polymer matrix could be observed by FTIR spectroscopy.

The TDDB for BCB is measured, where also the exponential relation between the maximum field strength and the time to break down was shown. For a field strength of 3.5 MV/cm the break down occurs in a time frame from 2000 to 7000 seconds, whereas for the higher field strength of 4.5 MV/cm a break down will be observed after 100 to 300 seconds. This behavior is important to take into account for the applications like a voltage transformer, where a high peak voltage could occur in the first few micro seconds after switch on. The TDDB of the polymer is highly depending how the high voltage is applied. For the larger 5 mm MIM structure a strong decreased breakthrough toughness is observed for the case of suddenly applied high voltage in contrast to a slow continues increase. These works set the basics for the realization of reliable high voltage wafer-level packages.

ACKNOWLEDGMENT

The authors would like to thank all personnel of Fraunhofer IZM and TU Berlin involved in this work but not explicitly mentioned as co-authors.

REFERENCES

1. J. Artbauer, *Elektrische Festigkeit von Polymeren*, Kolloid-Zeitschrift und Zeitschrift für Polymere, März 1965, Volume 202, Issue 1, pp 15-25
2. M. Kahle, *Elektrische Isoliertechnik*, VEB Verlag Technik Berlin, 1988, pp. 111
3. J.J. O'Dwyer, *The Theory of Electrical Conduction and Breakdown in Solid Dielectric*s, Clarendon Press, Oxford, 1973
4. M. Töpper, *Entwicklung einer auf Photo-BCB basierenden Technologie für das Waferlevel Packaging*, Shaker Verlag, 2004
5. L. Boettcher, S. Karaszkiewicz, D. Manessis, E. Hoene und A. Ostmann, „Next Generation High Power Electronic Modules Based on Embedded Power Semiconductors," in *IMAPS 10th*

International Conference on Device Packaging, Fountain Hills, AZ USA, 2014.

6. R. Ghodssi, L. G. Frechette, S. F. Nagle, X. Zhang, A. A. Ayón, S. D. Senturia, and M. A. Schmidt: Thick buried oxide in silicon (TBOS): " An integrated fabrication technology for multi-stack wafer-bonded MEMS processes," *Dig. Tech. Papers 10th Int. Conf. Solid-State Sensors and Actuators (Tansducers'99)*, Sendai, Japan, 1999, pp. 1456–1459

7. X. Zhang, R. Ghodssi, K.-S. Chen, A. A. Ayón, and S. M. Spearing: "Residual stress characterization of thick PECVD TEOS film for power MEMS applications," *Tech. Dig. Solid-State Sensor and Actuator Workshop*, 2000, pp. 316–319

8. L. A. Dissado, J. C. Fothergill: *Electrical degradation and Breakdown in Polymers*, The Redwood Press, Wiltshire, UK, 1992

9. W. Vollmann: *Poole-Frenkel Conduction in Insulators of Large Impurity Densities*, phys. stat. sol. (a) 22, 1974, pp. 195

Gap in pagination due to withheld paper.

Pages 1860-1864

Plasma Dry Process Technology Development of Glass-epoxy film on the Silicon substrate to fabricate RDL for Future GPU/AI Application.

Takahide Murayama[1], Muneyuki Sato[1], Akiyoshi Suzuki[1], Atsuhito Ihori[1], Tetsushi Fujinaga[1]
and Yasuhiro Morikawa[2]

ULVAC, Inc.
1: Institute of Semiconductor and Electronics Technologies
1220-1 Suyama, Susono, Shizuoka, 410-1231, Japan
2: Global Market & Technology Strategy Division
2500 Hagisono, Chigasaki, Kanagawa, 253-8543, Japan
takahide_murayama@ulvac.com

Abstract—In recent years, discussion on power consumption and latency of GPU used for AI application has started. In order to realize further high-speed processing and low power consumption of the GPU processing a huge amount of data, it is necessary to consider the packaging structure of the GPU [1]. The current GPU package structure is based on the package substrate using flip chip PoP (Package on Package) technology and Si interposer. In this structure applied, the wiring distance is increased due to the structural restriction of signal transmission through the Si interposer on the package substrate, which is the cause of the increase in power consumption and latency. Therefore, the packaging structure around the Si interposer has been focused, and expected structures that does not use the Si interposer have been proposed [2]. A method of directly forming fine wiring layers which plays a role of RDL (Redistributed Layer) by using a photosensitive insulation material on a build-up substrate without using a Si interposer has been reported [3]. Furthermore, in view of the high frequency trend of the signal frequency, the development of glass-epoxy materials having low D_f (dielectric loss constant) and low D_k (dielectric constant) material properties as a build-up film is proceeding [4]. It is expected that it will be a more effective method to effectively utilize the characteristics of low D_f and low D_k and to form fine wiring on the build-up layer using semiconductor fine wiring technology. For future high density packaging, plasma dry etching technology aiming fabrication of multilayer wiring on build-up film has been developed [5].

In this paper, the results of microfabrication of build-up thickness of 5 μm are reported for the purpose of fabricating fine wiring on build-up film using dry process. This technology has been developed as one of new SiP (System in Package) technologies for realizing

future heterogeneous integration. The process results of dry etching and Cu electroplating are described. In order to adapt to chip mounting, the size of the wiring formed in the build-up layer is targeted at line / space = 2 μm / 2 μm. The reason for using Si substrate instead of mold panel is because it is suitable for use of expensive NGD (known good die). In Si semiconductor packaging, very stable technology corresponding to Si substrate of 300 ㎜ size has been established up to today. And, for Cu fine wiring formation on a build-up film using a dry process, it is also necessary to ensure sufficient adhesion between the Cu seed layer and the build-up film. In order to manufacture highly reliable fine Cu wiring, it is necessary to evaluate the controllability of good adhesion of the seed Cu layer / glass epoxy film interface. Fluorine compound gas is used for dry etching of build-up film. There are residues containing fluorine on the surface to be etched. These residual fluorine compounds reduce the adhesion between the build-up film and the seed layer for Cu plating. Therefore, it is necessary to construct a method of dry process to improve the adhesion to the seed layer by eliminating the effect of residual fluorine compound. The change in the surface free energy before the seed sputtering process is compared with the peel test result of the Cu seed layer. Basic investigation results on the surface condition of the build-up film and the adhesion of the seed film are reported.

Keywords-component; High-density package, Plasma etching, Build-up, Seed layer adhesion, Surface modification

I. INTRODUCTION

In present GPU module, Si interposer is mounted on the package substrate, GPU chip and stacked DRAM chip are mounted on RDL (Fig.1). Considering the long signal wiring distance and the signal loss in this package form, it can be attractive that GPU chip and stacked DRAM chip are directly mounted on the build-up layer of the package

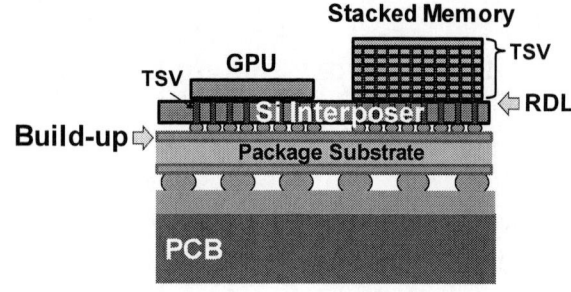

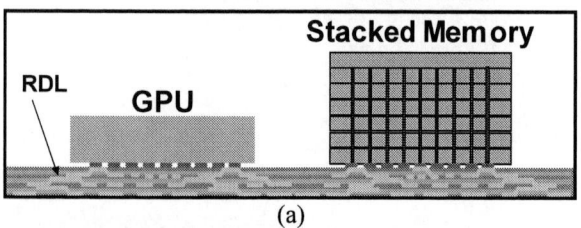

(a)

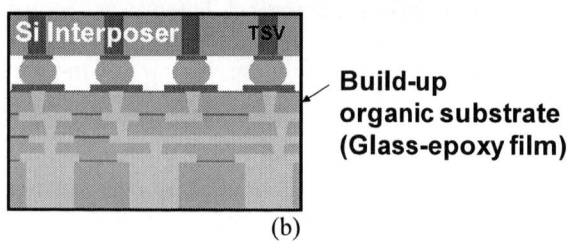

(b)

Figure 1. GPU Module Package.
(a) GPU chip and stacked DRAM chip on RDL
(b) Build-up multilayer wiring

substrate. In order to make this possible, it is necessary to form fine wiring in accordance with the mounted chip in the build-up layer. However, there will be three main technological challenges in the case of forming trenches and vias for wiring formation on build-up film using conventional technology (Fig.2 (a)). Currently, laser drilling processes are commonly used to form vias in build-up films [6]. Firstly, there is an issue of controlling the roughness of the laser drilled surface of the build-up film. It will be difficult to form a smooth surface on the side walls of the fabricated trenches and vias for wiring. Roughness of the sidewall influences the signal transmission and the reliability of the thin seed Ti / Cu film for Cu plating. Secondly, in the wet desmear process for smear removal of build-up film after laser drilling, there is issues of swelling and silica residue. And thirdly, it can be difficult to fabricate fine vias and line / space due to the limitation of the laser wavelength. The above issues can be avoided and microfabrication of build-up film can be realized by applying the dry process technologies developed in the semiconductor microfabrication field (Fig.2 (b)). In order to fabricate fine Cu wiring having sufficient adhesion with build-up film, process optimization of the dry surface treatment for seed Ti

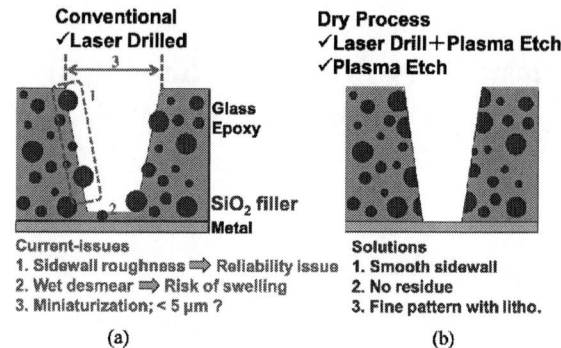

Figure.2 (a) Issues in microfabrication of build-up film and (b) dry process feasibility.

/ Cu layer sputtering is also important to control interface condition between build-up film and seed Ti / Cu layer.

II. EXPERIMENTAL

The process flow for fabricating fine Cu wiring pattern in build-up film is shown in figure 3. In this process, 200 ㎜ Si wafer is used, build-up film is laminated on this Si wafer. Ajinomoto build-up film GY series is selected from the point of relatively lower D_k and D_f properties. The thickness of GY series build-up film is 5 μm. To fabricate 2 μm thickness Cu mask for plasma etching of build-up film, lithography and Cu electroplating are carried out, then Cu is etched by wet process after photoresist removed. Thus, Cu mask is fabricated. The line / space is 2 / 2 μm. By using this Cu mask, dry etching for 5 um thickness build-up film through is tried using fluorine compound gas plasma. After build-up film etched, pre-treatment processes of surface modification / degas heating / pre-etching are applied, seed Ti / Cu layer is deposited by using sputtering process. Ti / Cu thickness is 50 / 500 nm. Finally, Cu electroplating is conducted and top Cu layer is removed. Based on above process, one layer of Cu wiring in build-up film is fabricated. Multilayer Cu wiring layer fabrication is expected by repetition of this process.

Next, the evaluation of adhesion property between build-up film and seed Ti / Cu layer is carried out. The evaluation flow is shown in figure 4. In this evaluation, the purpose is to investigate the effect of surface modification process. Surface modification process is expected to be applied for pretreatment before sputtering of seed Ti / Cu layer formation. Surface modification process using oxygen radical is used to after plasma etched build-up film surface in actual process flow to remove residual fluorine just after etched build-up film surface. In this evaluation, non-etched initial build-up film surface is tested as early basic investigation. The build-up film surface condition with surface modification and without surface modification are compared by the results of surface free energy change and peel strength of seed Ti / Cu layer. Surface free energy is calculated using the measurement value of liquid contact angle [7] on the build-up film surface. Contact angle is measured in KYOWA DMS-401 system, surface free energy

978-1-7281-1500-9/19 $31.00 © 2019 IEEE   1866

is calculated by using KYOWA FAMAS (interface Measurement and Analysis System) software. Seed layer peel strength on the build-up film surface is measured by using Orientec STA-1150 system. Seed Ti (50 nm) / Cu (300 nm) deposition to the build-up film surface by sputtering and 20 μm thickness Cu electroplating is carried out, then peel strength is measured in 90 degree angle pulling test.

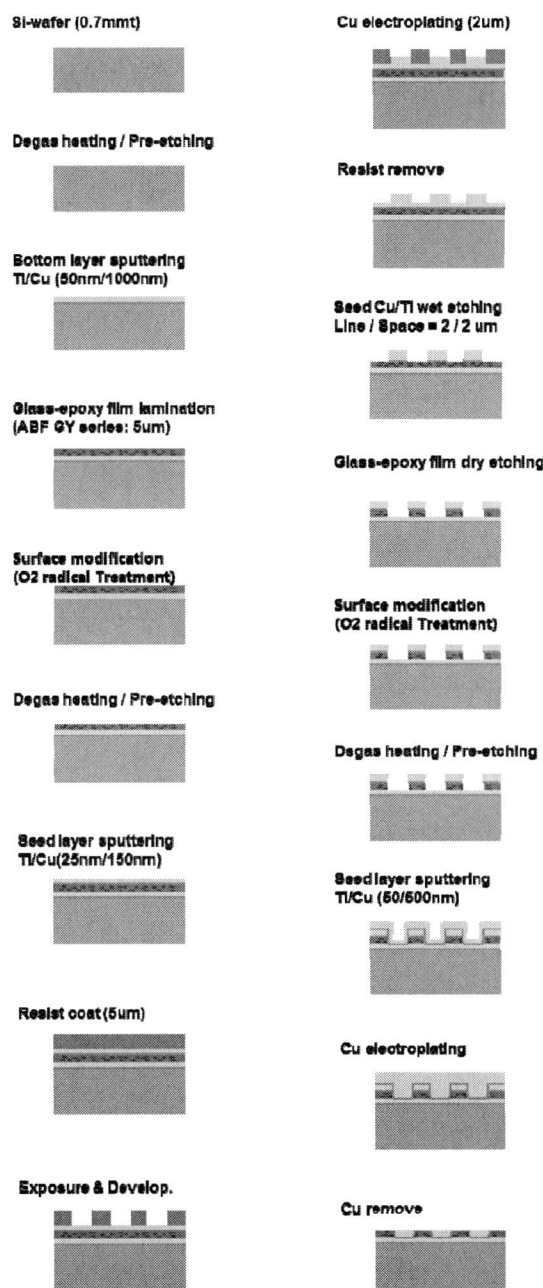

Figure 3. Process flow for Cu wiring microfabrication to build-up film by dry process

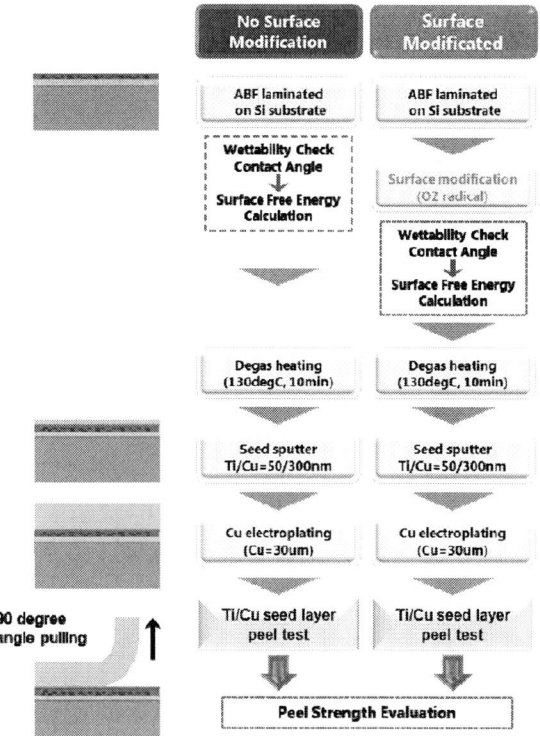

Figure 4. Evaluation flow of surface modification process effect

III. RESULT AND DISCUSSION

A. Cu Wiring Pattern Fabrication in Build-up Film by using Plasma Dry Process

Overview of 200 ㎜ wafer which Cu mask pattern fabricated according to flow of figure 3 is shown in figure 5 (a). Optical microscope observation image of Cu mask line is shown figure 5 (b). Cross-section SEM image of build-up film with Cu mask is shown in figure 5 (c). Actual fabricated Cu mask width is 1.4 μm, opened space (build-up film exposure) of pattern is 2.6 μm width. Figure 6 shows the result of etched through 5μm thickness build-up film pattern by plasma etching. It can be also observed that smooth etched surface is obtained on the etched sidewall by plasma etching process. And the result of Cu electroplating is shown in figure 7. The capability of dry process technology for microfabrication to build-up film can be confirmed from above results.

B. Investigation of Adhesion Property between Build-up Film and Seed Ti / Cu layer

The result of wettability / contact angle on the build-up film is shown in figure 8. It is shown that hydrophilic control of build-up film surface can be available for initial surface state and even after plasma etched surface by surface

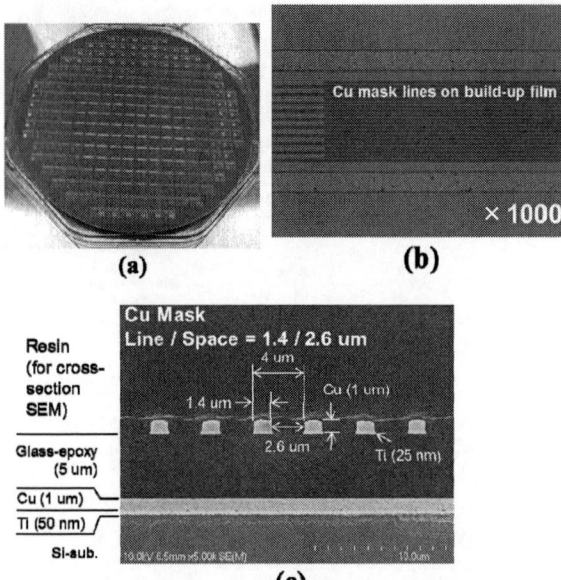

(a)　　　　　**(b)**

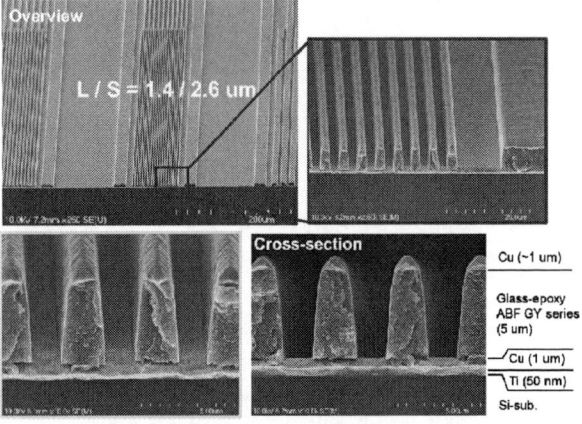

(c)

Figure 5. Fabricated Cu pattern for etch mask on the build-up film (GY series) laminated on 200 ㎜ Si wafer. (a) Wafer overview, (b) Cu mask pattern overview, (c) Cu mask cross-section

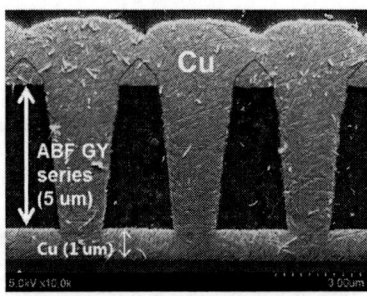

Figure 6. Plasma etched build-up film with smooth sidewall.

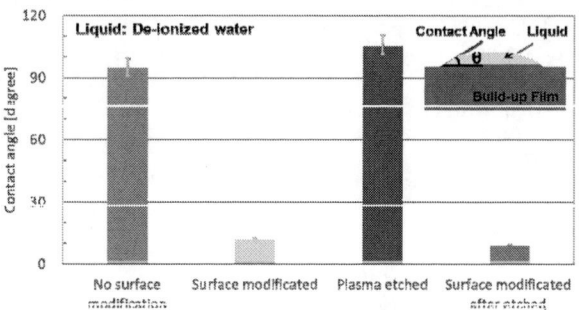

Figure 8. Hydrophilic controllability of surface modification process to build-up film (GY series)

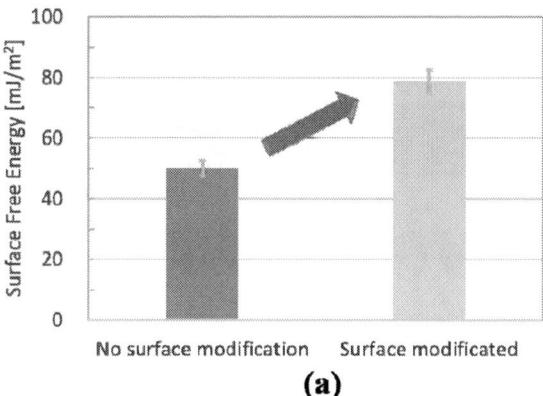

(a)

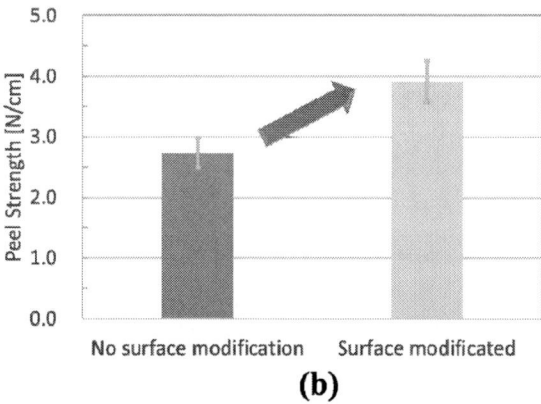

(b)

Figure 9. The effect of surface modification process. (a) Surface energy of build-up surface (b) Seed Ti / Cu layer peel strength

modification process using oxygen radical. It can be considered that surface modification by oxygen radical has hydrophilic controllability with removing residual fluorine on the surface. Figure 9 (a) shows build-up film surface free energy calculation results of initial state (no surface modification condition) and surface modificated condition. The increase of surface free energy is confirmed after surface modification. The Ti / Cu peel strength measurement results is shown in figure 9 (b). Compared with initial state, the peel strength after surface modification shows increase.

Figure 7. Cu electroplating result.

978-1-7281-1500-9/19 $31.00 © 2019 IEEE　　　　1868

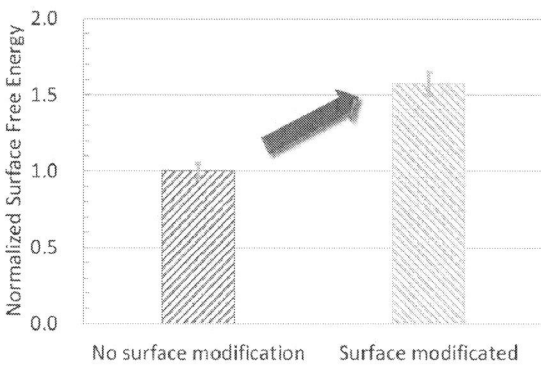

(a)

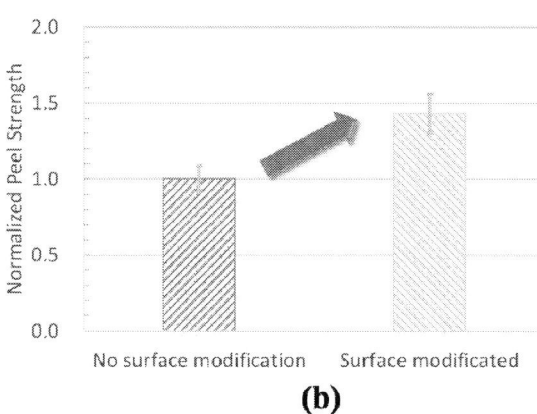

(b)

Figure 10. Normalized plot for
(a) surface energy of build-up surface
(b) seed Ti / Cu layer peel strength

Normalized plot is shown in figure 10 for comparison simply in the view point of surface free energy increasing ratio. It can be confirmed that surface free energy after surface modification is about 1.5 times higher than the initial state (Fig.10 (a)). And also, there is an increase in seed Ti / Cu layer peel strength at the similar degree (Fig.10 (b)). It can be considered that activated build-up film surface can be improved in the adhesion property for seed Ti / Cu layer by sputtering.

IV. CONCLUSION

In this study, it was shown that plasma etching can fabricate Cu wiring pattern in line / space = 2.6 / 1.4 μm size with depth of 5 μm in build-up film. It was also confirmed that the surface free energy of the build-up film increased by a simple surface modification treatment by the dry process, and accordingly the adhesion of the Ti / Cu seed layer increased. It is expected that introducing dry process technology for semiconductor process to fabricate fine wiring in multilayer wiring fabrication process for build-up film is a feasible method.

ACKNOWLEDGMENT

This research was partly supported by Ajinomoto Fine-Techno Co., Inc. (ABF material supply) and Research Center for Three-Dimensional Semiconductors (Cu mask formation for plasma etching).

REFERENCES

[1] C. -T. Wang, T. -C. Tang, C. -W. Lin, C. -W. Hsu, J. -S. Hsieh, C. -H. Tsai, K. -C. Wu, H. -P. Pu, and Douglas C. -H. Yu "InFO_AiP Technology for High Performance and Compact 5G Millimeter Wave System Integration," Proc. Electronic Components and Technology Conference (ECTC), pp. 202 - 297, 2018

[2] Christian Romero, Jeongho Lee, Kyungseob Oh, Kyoungmoo Harr and Toungdo Kweon "A Small Feature-Sized Organic interposer for 2.1D Packaging Solutions," Proc. International Microelectronics Assembly and Packaging Society (IMAPS), 2014

[3] Kiyoshi Oi, Satoshi Otake, Noriyoshi Shimizu, Shoji Watanabe, Yuji Kunimoto, Takashi Kurihara, Toshinori Koyama, Masato Tanaka, Lavanya Aryasomayajula and Zafer Kutlu "Development of New 2.5D Package with Novel Integrated Organic Interposer Substrate with Ultra-fine Wiring and High Density Bumps," Proc. Electronic Components and Technology Conference (ECTC), pp. 348 - 353, 2014

[4] Hirohisa Narahashi "Low Df Build-up Material for High Frequency Signal Transmission of Substrates," Electronic Components and Technology Conference (ECTC), 2013

[5] Yasuhiro Morikawa, Muneyuki Sato, Takahide Murayama1 and Tetsushi Fujinaga "Fabrication of Fine Via and Line / Space in Low CTE Film for Panel Fan-out Using A Dry Etching Technology," Proc. Electronic Components and Technology Conference (ECTC), pp. 1000 - 1004, 2018

[6] Yyua Suzuki, Venky Sundaram, Rao Tummala, Ye Chen, Kwon Sang Lee, Frank Wei, Habib Hichiri, Lee Seongkuk, Markus Arendt, Ognian Dimov, Deepak Arora and Sanjay Malik "Embedded Trench Redistribution Layers (RDL) by Eximer Laser Ablation and Surface Planer Process," Proc. Electronic Components and Technology Conference (ECTC), 2017

[7] M. Zenkiewicz "Methods for the calculation of surface free energy of solids," Journal of Achievements in Materials and Manufacturing Engineering, Volume 24 Issue 1 September 2007, p. 137 - 145

Fully solid-state integrated capacitors based on carbon nanofibers and dielectrics with specific capacitances higher than 200 nF/mm²

A. M. Saleem, R. Andersson, M. Bylund, C. Goemare, G. Pacot, M. S. Kabir and V. Desmaris

Smoltek AB
Gothenburg, Sweden
vincent@smoltek.com

Abstract— Complete on-chip fully solid-state 3D integrated capacitors using vertically aligned carbon nanofibers as electrodes to provide a large 3D surface in a MIM configuration have been manufactured and characterized in terms of capacitance per device footprint area, equivalent series resistance (ESR), breakdown voltage and leakage current. The entire manufacturing process of the capacitors is completely CMOS compatible, which along with the low device profile of about 4 µm makes the devices readily available for integration on a CMOS-chip, in 3D stacking, or redistribution layers in a 2.5D interposer technology. Capacitances of 200 nF/mm², ESR of about 100 mΩ, breakdown voltages of 25 V and leakage current of the order of 0.004 A/F have been measured.

Keywords- CNFs; CMOS; MIM; capacitor; integrated; solid-state; on-chip

I. INTRODUCTION

The constant demand for miniaturization, added functionality, and increased performance of electronic devices drives the need for more devices on a single chip. This is even more true with the uprising of the Internet of Things (IoT) pushing the system integration into unprecedented form factors. Furthermore, the fact that the standard downscaling of the components according to Moore's law [1] does not result in a better performance to costs ratio anymore results in a growing interest and innovation in component and system integration technologies in the form of 3D and 2.5D [2] packaging technologies.

Regardless of which novel system integration technology is implemented, more chips would always be needed to be fitted into the same physical space [3] and as a consequence, there is a rising need for on-chip, or in-package, capacitors to be used not only in traditional integrated circuits, but also for integrated components on interposers.

For this purpose, high energy density capacitors have been developed to serve diverse functions in ICs and integrated systems, such as decoupling or RF filtering. Decoupling capacitors act as a charge reservoir and are used to discharge energy at times of high power demand and recharge during periods of lower power demand, to account for spikes in power demand due to simultaneous switching of

many devices [4]. However, in order to operate optimally, decoupling capacitors must be placed within a critical distance to both the current load and the power supply, to be effective [5]. As a result, decoupling capacitors tend to strive for being integrated directly on a CMOS chip for best operation. Further, with the surge of the 2.5D interposer technology featuring embedded components, RF filtering is becoming a very important function requiring capacitors.

Electric double layer capacitors (EDLCs), commonly referred to as supercapacitors, are interesting due to their high capacitance per footprint area. For such devices carbon nanostructures are widely investigated as electrode materials due to their high surface area to volume ratio, their chemical stability, and their electrical properties [6] – [9]. However, the electrolytes conventionally required for the EDLCs are often toxic and corrosive, and being in a liquid state complicate their packaging and definitely hinders direct integration in an IC. Yet few EDLCs employing sol-gel electrolytes that solidify after application combined with carbon nanofibers (CNFs) have been demonstrated but at the cost of limited breakdown performances and limited capacitance per footprint area [10], [11].

A more flexible and suitable option is to use a fully solid state device, with the parallel plate capacitor being the simplest example. In order to avoid consuming precious space on the chip surface, clever methods need to be implemented to increase the surface area of the electrode without increasing the footprint area of the device. By etching deep trenches into the Si substrate and using atomic layer deposition (ALD) or low pressure chemical vapor deposition (LPCVD) to deposit the materials a 3D structure with increased surface area can be achieved. IPDiA, now Murata, has previously used this method to get capacitance values in excess of 500 nF/mm² by etching 100 µm deep trenches [12]. The disadvantage is that this method is time consuming and expensive, and it also weakens the substrate, making it more susceptible to defects. In a similar way, some attempts to include capacitors using via walls in interposers have been suggested with large capacitance per unit area, but since such devices need a via to be created, the actual usage of chip area is in fact rather large [13]

Similar to EDLCs, carbon nanostructures can be used to enhance the surface area of the electrode of a solid state device. Carbon nanotubes (CNTs) have been investigated as an electrode material for fully solid-state capacitor, yet the

978-1-7281-1500-9/19 $31.00 © 2019 IEEE

problems with uniform coating of the CNT walls limited the reliability of the demonstrated devices and the high temperature required to synthetize the CNTs hinders further integration into ICs. In fact, Pint et al. achieved capacitance values of ca 120 nF/mm^2, however the CNTs were grown at 700 – 750 °C and then transferred to the active site [14]. Fiorentino et al. were able to grow CNTs at a lower temperature of 500 °C and got a capacitance of 42 nF/mm^2, however that temperature is still too high to be CMOS compatible and the test device used was impractically small (0.0025 mm^2, giving an absolute capacitance of 106 pF) [15].

Previous studies have investigated the feasibility of using CNFs in combination with solid dielectrics by growing vertically aligned carbon nanofibers (VACNFs) on a substrate and then depositing Al_2O_3, HfO_x and ZnO by means of ALD. It was shown that the CNFs were conformally coated [16] and even operating devices with improved capacitance densities up to 50 nF/mm^2 have been demonstrated [17], [18].

In the present work, we present the fabrication and characterization of the performance of fully CMOS compatible capacitors based on VACNFs and ALD deposited dielectrics. Practically, two specific standard dielectric materials, Al_2O_3 and HfO_2, of different thicknesses have been investigated, as well as different CNF-based electrode configurations. As a result, truly solid-state, 3D integrable, on-chip capacitors based on CNFs as electrode material [18] have been manufactured and characterized. The capacitors exhibit specific capacitance in excess of 200 nF/mm^2 (per device footprint area), low equivalent series resistance of about 100 mΩ with overall profile height lower than 4 microns, and breakdown voltages up to 25 V. Further, leakage currents of the order of 0.004 A/F have been measured which supersedes the leakage current performance of standard polymer-based capacitors used for decoupling purposes.

II. EXPERIMENT

The fabrication process of the solid-state capacitors is shown schematically in Fig. 1, leading up to the complete design shown in Fig. 2. A substrate consisting of high resistivity float zone Si is used, and the top surface is first covered with SiO_2 deposited via plasma enhanced chemical vapour deposition (PECVD). The bottom electrode of the capacitor is then fabricated through evaporation of a Ti/Cu stack followed by a lift-off process, Fig. 1a. Each chip contained several different capacitors of different dimensions ranging from 210 μm x 420 μm up to 840 μm x 840 μm. Electron beam lithography is then used to pattern a matrix of openings, 100 nm in diameter, with two different pitches, 250 nm and 300 nm, which are then used to deposit patterned catalyst dots via evaporation and lift-off. Vertically aligned carbon nanofibers (VACNFs) are then grown directly on the bottom electrode selectively on the catalyst in a DC-PECVD system at 390 °C using ammonia and acetylene as process gasses, Fig. 1b. Further details regarding the growth can be found in [19]. The resulting CNF length is 2-3 μm.

Using atomic layer deposition (ALD), the devices are then coated with a dielectric layer, conformally coating the CNFs as well as the rest of the bottom electrode, Fig. 1c. Different combinations of HfO_2 and Al_2O_3 are used as the dielectric materials in the resulting capacitors, both as pure materials and different stacks of $Al_2O_3/HfO_2/Al_2O_3$. Thermal ALD at 250 °C is used for the dielectric depositions.

After the dielectric deposition the top electrode and probing structure is formed via metal deposition followed by lithography defined etching, Fig. 1d. Plasma enhanced ALD at 250 °C is first used to deposit a layer of TiN to ensure fully contacting the sidewalls of the CNFs. A deeper discussion of the effect of TiN deposited via ALD is available in [18]. A stack of Ti/Al is then sputtered to form the bulk of the top electrode. After patterning, the Al is chemically etched away, and the TiN/Ti is etched using reactive etching using a plasma containing fluoride ions.

Finally, the dielectric material is opened to allow for probing, Fig. 1 d. This is again done via lithography defined reactive ion etching.

The physical properties of the CNFs after growth are characterized using scanning electron microscopy (SEM), and for the electrical characterization of the capacitors an HP HP4285A LCR meter is used, as well as a low frequency vector network analyzer (VNA).

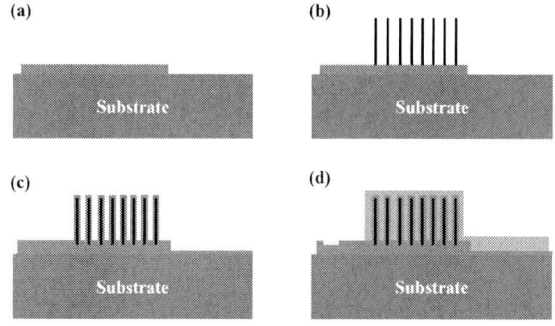

Figure 1. Schematic overview of the capacitor manufacturing process: (a) Bottom electrode formation. (b) CNF growth directly on the bottom electrode. (c) Conformal coating of dielectric material. (d) Top electrode formation and opening of dielectric for probing of bottom electrode.

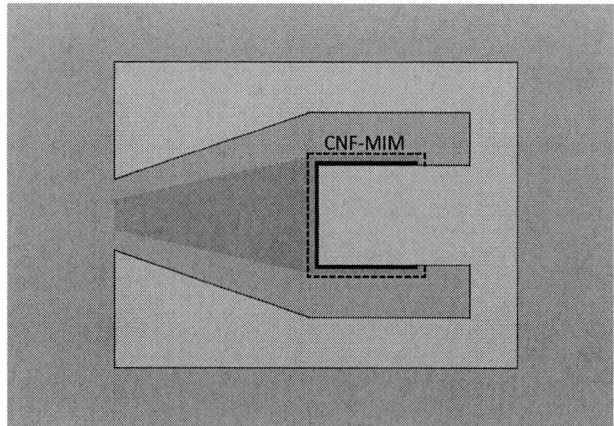

Figure 2. Architecture of the CNF-MIM

The sheet resistance of different metal stacks before and after subjecting them to the growth conditions is systematically studied to ensure low electrical series resistance in the capacitors. For this purpose, samples are prepared by depositing different metal stacks, making two identical samples for each metal combination. One sample out of each pair is then subjected to the conditions of the CNF growth, and afterwards the sheet resistance of both samples are measured using an automated CMT-SR2000N 4 point probe station with a probe tip radius of 200 μm and a probe spacing of 1 mm.

III. RESULTS AND DISCUSSION

A. Physical properties

As seen in Fig. 3 the CNFs grow according to the patterned matrix of catalyst, and they are vertically aligned. Specifically, the CNFs are individually grown on each of the catalyst dots in the matrix, unlike [18], where the CNFs were grown on an uninterrupted uniform film of catalyst, resulting in more closely packed fibers. The extra spacing created when using a patterned catalyst layer and selective growth results in a larger effective surface area for the top electrode, despite reducing the amount of vertical area along the sidewalls of the CNFs (fewer CNFs per footprint area leading to less area generated along the fiber side walls).

The fiber length is only $2 - 3$ μm, meaning that the total height profile of the complete device is in the range of 4 - 5 μm. This makes the capacitors readily available for integration onto a CMOS chip or in 3D stacking.

Fig. 4 shows the CNFs after the dielectric coating via ALD, the image showing a coating of $Al_2O_3/HfO_2/Al_2O_3$ (5/3/5 nm). It is clear that the dielectric layer is uniformly covering the individual CNFs. An SEM image of the complete devices is shown in Fig. 5 and Fig. 6 depicts a zoomed in view of a corner of a capacitor, showing the layered structure of first the bottom electrode (area 1), then CNFs coated with a dielectric material (area 2), and finally the CNFs that are also coated with the top electrode metal (area 3).

Figure 4. SEM image of the CNFs after being coated with $Al_2O_3/HfO_2/Al_2O_3$ (5/3/5 nm) via ALD.

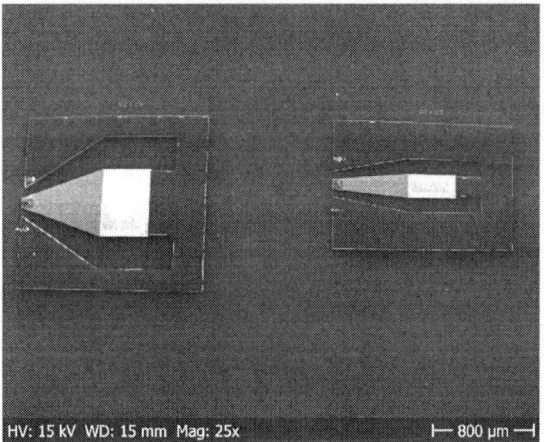

Figure 5. SEM image of two complete CNF-MIM devices of different dimensions.

Figure 6. SEM image of the corner of a CNF-MIM taken at 40 tilt showing the bottom electrode (area 1), dielectric covered CNFs (area 2), and the top electrode covering the CNFs (area 3).

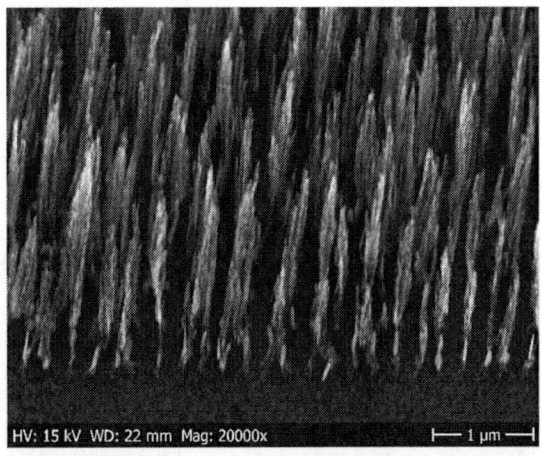

Figure 3. SEM image of the vertically aligned CNF growth taken at 40° tilt.

B. Capacitance density

For reference, a set of samples was produced using an unpatterned catalyst layer, where the CNFs would randomly grow as in [18]. The reference samples were manufactured using 20 nm HfO_2 as dielectric material, and TiN deposited via ALD to form the first layer of the top electrode. The capacitance for those devices was measured via an LCR meter and was found to be 81.3 ± 1.6 nF/mm^2 (per device footprint area) at 1 kHz.

Different patterning of the catalyst and hence CNFs densities were later investigated: 100 nm catalyst dots were used instead of a continuous film for the catalyst layer with two different values of pitches between two consecutive dots (250 nm and 300 nm). The lengths of the grown CNFs is given in Table I. Two different thicknesses of HfO_2 were implemented: 15 nm and 20 nm. The capacitance measured at 1 kHz frequency is plotted versus the footprint area of the active devices, Fig 7. The capacitances is found to increase almost linearly with footprint area for all types of capacitors. Moreover, the capacitance is lower for the higher pitch because of the presence of fewer CNFs per unit area and thus a lower surface area provided by the CNFs. Similarly, the capacitance is lower for thicker dielectric because of the inverse relation of the thickness with capacitance.

By extrapolating the data, the capacitance density (per footprint area) is calculated. A capacitance density value of the order of 200 nF/mm^2 is realized from the devices containing 250 nm pitch, and 15 nm dielectric thickness, Table I. High capacitance value of 152 nF is realized from a single device consisting of 0.7056 mm^2 footprint area which is 14 times higher than the theoretical value of a parallel plate capacitor of the same configuration but without CNFs if the dielectric constant of HfO_2 is considered to be 25 which is highest value of the range presented [20].

The predicted capacitance densities are calculated as well, using the measured lengths of the grown CNFs for specific configurations and the thickness of the dielectric are used, Table I.

C. Dielectric properties

The energy safely stored in a capacitor is limited by the breakdown voltage, which is determined by the dielectric strength of the material. This work investigates two different dielectric materials, Al_2O_3 and HfO_2, with different properties regarding dielectric constants, breakdown voltage, etc. Where Al_2O_3 provides higher breakdown voltage and HfO_2 contributes higher capacitance since the dielectric constant is much higher. Our previous investigations [18] of the dielectric properties of the ALD deposited dielectrics show that the dielectric constants of HfO_2 and Al_2O_3 were 19.3 and 8.2 respectively.

Starting with devices having pure HfO_2 or Al_2O_3 as dielectric and using two different thicknesses for each dielectric material, two sets of devices were produced for each thickness one with CNFs and one without, i.e. regular parallel plate devices. Then the breakdown voltage was measured for all devices, as summarized in, Table II. The measurements were performed using a Keithley 4200SCS Parameter Analyzer, using a voltage sweep from 0 to 38 V with a compliance of 50 mA.

As expected, the devices with HfO_2 have a lower breakdown voltage than the devices with Al_2O_3. It is also clear that the breakdown voltage is slightly lower for the CNF enhanced capacitors than it is for the parallel plate devices. The most probable explanation being that the sharp tips of the fibers will lead to a focusing of the electric field around the tip. Thus causing a locally concentrated field, which will lead to a local breakdown of the material. The parallel plate configuration would not experience the same local concentrations given the same applied voltage and would therefore have a higher breakdown voltage than devices containing CNFs.

Bearing in mind the inherent breakdown–capacitance trade-off originating from the dielectric properties of the oxides, several oxide stacks have been investigated to produce capacitors with capacitance densities and breakdown voltages suitable for decoupling purposes in digital ICs. The measured and modelled breakdown voltages based on linear extrapolations from the performance of pure HfO_2 and Al_2O_3 are summarized in Table III. The modelled and measured data are in good agreement and show that both oxides actually conformally cover the edges of the CNFs.

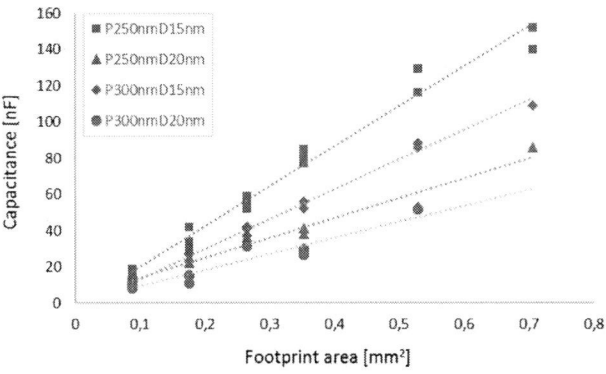

Figure 7. Measured capacitance, and fit to the data, for devices of different footprint areas using HfO₂ as dielectric material

TABLE I.   DEVICE DATA AND SPECIFIC CAPACITANCE FOR DIFFERENT CAPACITORS USING HfO₂ AS DIELECTRIC MATERIAL

CNF Pitch [nm]	CNF Length [μm]	Dielectric thickness [nm]	Capacitance [nF/mm²]	
			Measured	*Estimation*
250	3	15	221 ± 6	207
	2	20	109 ± 12	110
300	2.5	15	164 ± 10	124
	2.2	20	88	86

TABLE II.   BREAKDOWN VOLTAGE OF DIFFERENT DIELECTRIC THICKNESSES

Device configuration	Breakdown voltage [V]			
	HfO₂, 20 nm	*HfO₂, 51.6 nm*	*Al₂O₃, 23.5 nm*	*Al₂O₃, 49.8 nm*
Parallel plate	7.7 ± 0.4	19.8 ± 1.3	14.3 ± 3.5	33.3 ± 1.6
With CNFs	7.2 ± 0.6	16.0 ± 2.9	9.7 ± 0.5	22.3 ± 3.1

To further investigate the usefulness of the capacitors, measurements of their leakage current were performed. The devices were biased at 1 and 2 V for one minute before the leakage current was recorded to avoid transients. The measurements are shown in Table IV and with a leakage current to capacitance ratio of 0.004 A/F, the CNF-based capacitors supersede the performance of polymer capacitors, proving their usefulness.

D. ESR

Typically, integrated large capacitance density devices often present relatively large equivalent series resistance (ESR), compared to their discrete counterparts. In fact, for the CNF-based capacitors, using the proper metal stack is crucial to achieve the lowest possible ESR. It is actually not enough to consider the usual conductivity of the different metals, it also has to remain low after the CNF growth. Samples with films of different metal stacks were produced to test the sheet resistance before and after growth. For each combination, there was one sample that was subjected to the growth conditions and one that was not. The results are seen in Table V.

It is clear that most metal stacks did not perform well after the growth. The majority more than doubled their sheet resistance by undergoing growth, whereas the Al based electrodes increased the resistance by a factor of 11 and 48 respectively. The metal combination that performed the best, only increasing its sheet resistance by ca 20 %, was Ti/Cu, which is why that particular combination was chosen for the bottom electrodes of the capacitors. The top electrode, however, is not deposited until after the CNF growth is already done, and thus Al is suitable since it is a readily available material that is easily removed via chemical etching when patterning the top electrode.

To characterize the ESR of the devices, the VNA was used to measure the impedance of the devices for different frequencies ranging up to 3 GHz. The resulting Smith chart of the reflection factor for a 420 μm x 840 μm device with

$Al_2O_3/HfO_2/Al_2O_3$ (3/12/3 nm) as dielectric material is seen in Fig. 8. A simple equivalent circuit of a resistor, a capacitor, and an inductor in series is used as a model to fit to the measurement data. The behavior is close to that of an ideal capacitor. The architecture of the device seen in Fig. 2 is needed to enable measurements using a ground-signal-ground (GSG) probe for the VNA. However, the current leads, and the shape of the top electrode, creates parasitic impedances that are not inherent to the capacitor itself, but to the metal structure related to probing. To remove this part of the measured data, de-embedding devices were manufactured without any CNFs or dielectric material, i.e. the top and bottom electrodes form a short circuit. These devices are then measured the same way as the CNF-MIM capacitor, resulting in the Smith chart seen in Fig. 9. It is evident that the design of the current leads generates an inductive behavior at higher frequencies, seen as the part of the curve with a positive imaginary part. The measured data of the short circuit devices can then be used for de-embedding, to isolate the impedance for the capacitor itself. The resulting de-embedded impedance for the CNF-MIM capacitor mentioned above is plotted in Fig. 10. The ESR of the device is extracted as 115 mΩ.

TABLE V.   SHEET RESISTANCE OF DIFFERENT METAL STACKS BEFORE AND AFTER CNF GROWTH.

Metal stack	Sheet resistance [Ohm/sq]	
	Before growth	*After growth*
Ti/Cu	0.3417	0.4166
Ti/Al	0.7181	7.9202
Cr/Cu	0.3578	0.7849
Cr/Al	0.6953	33.7524
Ti/TiN	5.7607	5.7872
Ti/TiN/Cu	0.2980	0.6387
Ti/Pd	1.3647	3.4840

TABLE III.   MEASURED AND ESTIMATED BREAKDOWN VOLTAGES FOR DIFFERENT DIELECTRIC STACKS.

Thickness of layers in $Al_2O_3/HfO_2/Al_2O_3$ stack [nm]	Breakdown voltage [V]	
	Measured	*Estimation*
0/15/0 (Pure HfO_2)	5.2 ± 1.8	5.0
0/15/0 (Pure HfO_2)	6.3 ± 1.3	6.0
6/3/6	5.5 ± 1.3	6.3
3/9/3	5.9 ± 1.3	5.7
3/12/3	7.6 ± 0.9	6.7

TABLE IV.   MEASURED DC LEAKAGE CURRENT (DCL) AT DIFFERENT VOLTAGES

Measured Cap. [nF]	DCL @1V [nA]	DCL @2V [nA]	DCL/Cap. @1V [A/F]	DCL/Cap. @2V [A/F]
27.1	0.15	0.47	0.0055	0.017
13.2	0.05	0.18	0.0038	0.014
41.6	0.12	0.35	0.0029	0.008
6.2	0.02	0.16	0.0032	0.026
56.8	0.13	0.90	0.0023	0.016
12.9	0.06	0.33	0.0047	0.026

Impedance Smith chart

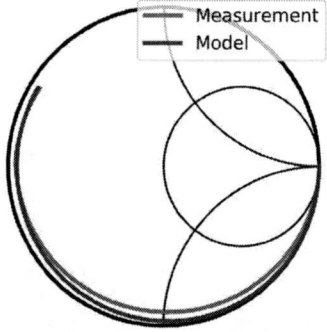

Figure 8.   Smith chart of the reflection factor for a CNF-MIM capacitor before de-embedding

978-1-7281-1500-9/19 $31.00 © 2019 IEEE

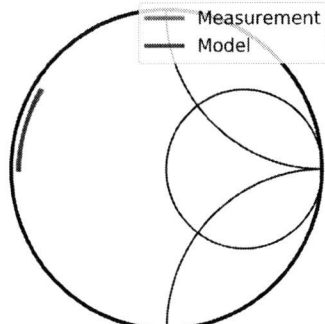

Figure 9. Smith chart of the reflection factor for the de-embedding device where top and bottom electrode form a short circuit

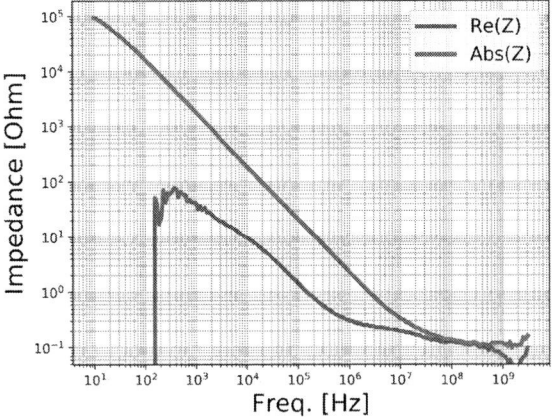

Figure 10. Plot of the impedance for the CNF-MIM capacitor after de-embedding.

IV. CONCLUSIONS

Complete on-chip fully solid-state 3D integrated capacitors using VACNFs as electrodes to provide a large 3D surface in a MIM configuration have been manufactured and characterized in terms of capacitance per device footprint area, equivalent series resistance (ESR), breakdown voltage and leakage current. A capacitance density in the order of 200 nF/mm^2 has been realized from devices with a 250 nm pitch, and 15 nm of HfO$_2$ as dielectric material. A high capacitance value of 152 nF was realized from a single device with a footprint area of 0.7056 mm^2, which is 14 times higher than that of a parallel plate device with the same configuration. Ti/Cu was found to be the best metal combination for the bottom electrodes of the capacitors, and by de-embedding the ESR of the devices was found to be about 100 mΩ. The measured breakdown voltage of up to 25 V followed the expected trends when varying the dielectric thickness and combinations of dielectric materials. DC

leakage current was measured, with a leakage current to capacitance ratio of 0.004 A/F, superseding the performance of current polymer capacitors.

The entire manufacturing process of the capacitors is completely CMOS compatible, which along with the low device profile of about 4 µm makes the devices readily available for integration on a CMOS-chip, in 3D stacking, or redistribution layers in a 2.5D interposer technology.

REFERENCES

[1] G. E. Moore, " Cramming More Components onto Integrated Circuits," Electronics, vol. 38, no. 8, Apr. 1965,

[2] W. Chen, W.R. Bottoms, "Heterogeneous integration Roadmap", International Conference on Electronics Packaging (ICEP), 2017, pp. 302-305, doi: 10.23919/ICEP.2017.7939380

[3] M. M. Waldrop, "The chips are down for Moore's law," Nature, vol. 530, no. 7589, Feb. 2016, p. 144, doi:10.1038/530144a.

[4] H. Hu, S. Harb, N. Kutkut, I. Batarseh, and Z. J. Shen, "A Review of Power Decoupling Techniques for Microinverters With Three Different Decoupling Capacitor Locations in PV Systems," IEEE Transactions on Power Electronics, vol. 28, no. 6, Jun. 20 [1]13, pp. 2711–2726, doi:10.1109/TPEL.2012.2221482.

[5] M. Popovich, M. Sotman, A. Kolodny, and E. G. Friedman, "Effective Radii of On-Chip Decoupling Capacitors," IEEE Transactions on Very Large Scale Integration (VLSI) Systems, vol. 16, no. 7, Jul. 2008, pp. 894–907, doi:10.1109/TVLSI.2008.2000454.

[6] V. Desmaris, M. A. Saleem, and S. Shafiee, "Examining Carbon Nanofibers: Properties, growth, and applications.," IEEE Nanotechnology Magazine, vol. 9, no. 2, Jun. 2015, pp. 33–38, doi:10.1109/MNANO.2015.2409394

[7] M. Beidaghi and Y. Gogotsi, "Capacitive energy storage in micro-scale devices: recent advances in design and fabrication of micro-supercapacitors," Energy Environ. Sci., vol. 7, no. 3, Feb. 2014, pp. 867–884, doi:10.1039/C3EE43526A

[8] A. M. Saleem, A. Boschin, D.-H. Lim, V. Desmaris, P. Johansson, and P. Enoksson, "Coin-cell Supercapacitors Based on CVD Grown and Vertically Aligned Carbon Nanofibers (VACNFs)," Int. J. Electrochem. Sci., vol. 12, Jun. 2017, pp. 6653–6661, doi:10.20964/2017.07.46

[9] A. M. Saleem, V. Desmaris, and P. Enoksson, "Performance Enhancement of Carbon Nanomaterials for Supercapacitors," Journal of Nanomaterials, 2016, doi:10.1155/2016/1537269

[10] A. M. Saleem, R. Andersson, V. Desmaris, B. Song, and C. P. Wong, "On-Chip Integrated Solid-State Micro-Supercapacitor," Proc. 2017 IEEE 67th Electronic Components and Technology Conference (ECTC), 2017, pp. 173–178, doi:10.1109/ECTC.2017.135.

[11] R. Andersson, A. M. Saleem, V. Desmaris, B. Song, and C. P. Wong, "On-Chip Solid-State CMOS Compatible Micro-Supercapacitors," Proc. 2018 IEEE 68th Electronic Components and Technology Conference (ECTC), 2018, pp. 1382–1388, doi: 10.1109/ECTC.2018.00211

[12] F. Nodet, "Ultra-low ESL and Ultra-low Profile Silicon Capacitor interest in decoupling high-speed IC," IPDiA.com, 19-Oct-2016. [Online]. Available: http://www.ipdia.com/index.php?page=news&item_id=97. [Accessed: 09-Feb-2018].

[13] Y. Lin, C. S. Tan, "Dielectric Quality of 3D Capacitor Embedded in Through-Silicon Via (TSV)" Proc. 2018 IEEE 68th Electronic Components and Technology (ECTC), 2018, pp. 1153-1163, doi: 10.1109/ECTC.2018.00178

[14] C. L. Pint et al., "Three dimensional solid-state supercapacitors from aligned single-walled carbon nanotube array templates," Carbon, vol. 49, no. 14, Nov. 2011, pp. 4890–4897, doi:10.1016/j.carbon.2011.07.011.

[15] G. Fiorentino, S. Vollebregt, F. D. Tichelaar, R. Ishihara, and P. M. Sarro, "Impact of the atomic layer deposition precursors diffusion on solid-state carbon nanotube based supercapacitors performances," Nanotechnology, vol. 26, no. 6, 2015, p. 064002, doi:10.1088/0957-4484/26/6/064002.

[16] G. A. Malek *et al.*, "Atomic Layer Deposition of Al-Doped ZnO/Al2O3 Double Layers on Vertically Aligned Carbon Nanofiber Arrays," ACS Appl. Mater. Interfaces, vol. 6, no. 9, May 2014, pp. 6865–6871, doi:10.1021/am5006805.

[17] A. M. Saleem, R. Andersson, V. Desmaris, and P. Enoksson, "Integrated on-chip solid state capacitor based on vertically aligned carbon nanofibers, grown using a CMOS temperature compatible process," Solid-State Electronics, vol. 139, Jan. 2018, pp. 75–79, doi:10.1016/j.sse.2017.10.037.

[18] R. Andersson, A. M. Saleem and V. Desmaris, "Integrated fully solid-state capacitor based on carbon nanofibers and dielectrics," Proc. 2018 IEEE 68th Electronic Components and Technology (ECTC), 2018, pp. 2307-2312, doi: 10.1109/ECTC.2018.00348.

[19] A. M. Saleem *et al.*, "Low temperature and cost-effective growth of vertically aligned carbon nanofibers using spin-coated polymer-stabilized palladium nanocatalysts," Sci. Technol. Adv. Mater., vol. 16, no. 1, 2015, p. 015007, doi:10.1088/1468-6996/16/1/015007.

[20] E. P. Gusev, C. Cabral, M. Copel, C. D'Emic, and M. Gribelyuk, "Ultrathin HfO2 films grown on silicon by atomic layer deposition for advanced gate dielectrics applications," Microelectronic Engineering, vol. 69, no. 2, Sep. 2003 pp. 145–151, doi:10.1016/S0167-9317(03)00291-0.

Application of Fan-Out Panel Level Packaging Techniques for Flexible Hybrid Electronics Systems

Shau-Fei Cheng, Chen-Tsai Yang, Wei-Yuan Cheng*, Shau-Fei Cheng, Wei-Han Chen, Hsin-Cheng Lai, Tai-Jui Wang, Yuh-Zheng Lee

Electronic and Optoelectronic System Research Laboratories, Industrial Technology Research Institute, Hsinchu, Taiwan, 30010, R.O.C

*e-mail: weiyuan.cheng@itri.org.tw

Abstract—*Flexible Hybrid Electronics (FHE) is in the limelight for next generation electronics, it has good potential for further development as the priority areas for focused study, not only can it create new products, but also meet the needs of flexible, bendable, and stretchable characteristics. Furthermore, FHE brings new business opportunities to the semiconductor industry, and it has niche growth potential in the fields of healthcare, aviation, soft robotics and Internet of Things (IoT).*

Fan-out panel-level packaging (FOPLP) technology have the advantages of thinner packaging, good heat dissipation and high electrical performance. In this paper, the RDL-first FOPLP techniques is adopted to develop the FHE systems with ITRI's core technologies, i.e., FlexUP^{TM} and mechanical de-bonding. With ITRI's FlexUP^{TM} and flexible display technology, the related problems can be overcame such as broken circuit and poor electrical performance, which were caused by bending and stretching the circuit of the RDL substrate. The results showed that ITRI's core technology could meet the demand of FHE systems. It also provide a solution to solve the contradiction between the characteristics of the rigid structure and flexible requirements.

Keywords- FHE; FlexUP^{TM}; RDL-first

I. INTRODUCTION

FOPLP with thinning, high I/O pin counts, high efficiency signal transmission and other advantages. Because it's a large area package, so it has a significant benefit in reducing process costs[1-4]. Therefore, it is an ideal packaging technology choice for consumer electronics and IoT devices.

According to the ITRI's FlexUP^{TM} and flexible display technology, we had integrated development of 4-layer redistribution layer (RDL) for FOPLP, In addition to supporting high I/O pin counts for this technology. Also has the feasibility of mass production, thinner packaging and power saving, especially suitable for High-Performance Computing (HPC) such as smart handheld devices and the IoT devices. Therefore, FOPLP provides an IC packaging strategy for wearable, portable mobile device with high accuracy of calculation, so it is solve the bottleneck of high pin count that is difficult to improve in previous technologies, and it can be used in wearable products that are freely stretchable and bendable for users curve.

FHE is based on flexible materials to form a flexible electronic circuit and system. Therefore, this study focuses on the flexible RDL manufacturing and the stress analysis with heterogeneous structure, and use the FOPLP packaging technology for the FHE system, Integration of FOPLP and FHE for future applications in automotive electronics, communications, health care and other fields.

II. RDL FIRST FOPLP PROCESS

Now ITRI's FOPLP platform can complete the process of RDL first FOPLP, in addition, the molding process with SMC (Sheet Molding Compound) and mechanical de-bonding that can finished the panel-level package structure successfully[5], The following are single-layer RDL with daisy chain die structure and related processes.

A. 1 layer RDL FOPLP with daisy chain die and SMC molding

The purpose of this experiment is to complete the package structure of FOPLP, as shown in Fig. 1, and the rate of change of resistance is ≤5% after mechanical de-bond the glass carrier

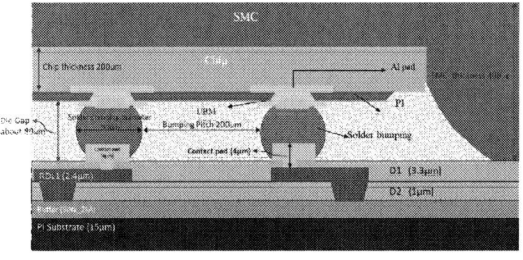

Fig.1. single layer RDLwtih daisy chain die package structure

Base on ITRI's G2.5 LCD facilities [6], we develop complete packaging process for single-layer RDL with daisy chain die is used as the test vehicle, that include front end process of RDL substrate and back end process of thermal-compression flip-chip (TCFC) technology with molding processe, as shown in Fig. 2(a).

For the front end process, first, apply a 15μm polyimide (PI) layer to the glass carrier, then coated 2kÅ buffer layer, next finished the dielectric layer coated with 1μm, and 4 μm

RDL is completed through whole procedure of photoresist coating, baking, exposure, development, seed layer sputtering, metal plating and etching, subsequently coated with a dielectric layer of 3.3 μm. Finally, through the photoresist coating, baking, exposure, development, etching and other processes, complete Ti / Cu (1kÅ /2kÅ) UBM layer and 4μm Cu pad, as shown in Fig. 2 (b) and (c).

For the back end process, after the RDL substrate is completed, the daisy chain die is bonded to the RDL substrate by a TCFC technology, as shown in Fig. 2 (d). And underfill encapsulate material is applied in the gap between the daisy chain die and RDL substrate, then attach the SMC over the whole structure to complete the FOPLP manufacturing, finally, removed the glass carrier by mechanical debond method, after that, test key holes were generated from the back side of FOPLP structure by laser drill for double checking the resistance change of RDL film after it was de-bonded.

(a)

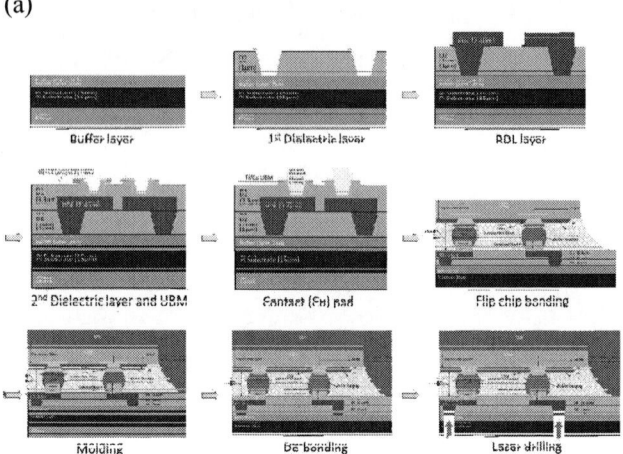

(b)

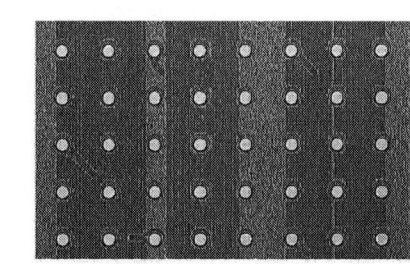

(c)

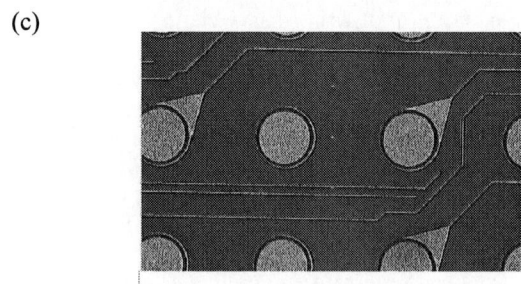

(d)

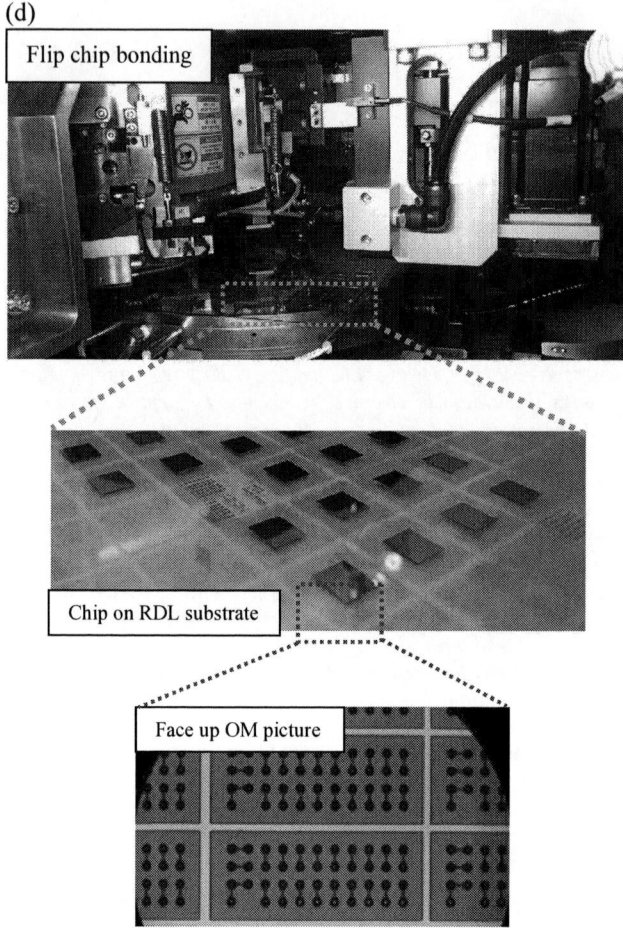

Fig. 2. RDL first FOPLP (a) front-end single-layer RDL and back-end packaging process (b) RDL Etching (c) RDL Cu Pad (d) TCFC process

Summary of the results of this study is to complete the FOPLP for heterogeneous integration daisy chain die packaging and de-bonded the glass carrier successfully, as shown in Fig. 3.

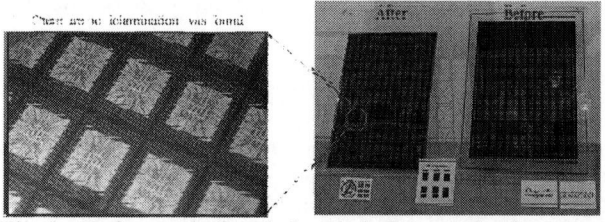

Fig. 3. Before and after de-bonded glass carrier for 370mm by 470mm FOPLP, and there are no delamination was found

For the measurement of the resistance change, five areas are selected on the back side of the package structure, and the position of 12 die is selected for each area, a total of 60 positions, after that, test key holes were generated by laser drilling method, finally, taking the measured test key for resistance change, as shown in Fig. 4. Average the resistance change of the test key part of the 12 die position in 5 areas. Summarize the average resistance change rate to meet the <5% requirement after de-bonded glass carrier, as shown in Table 1.

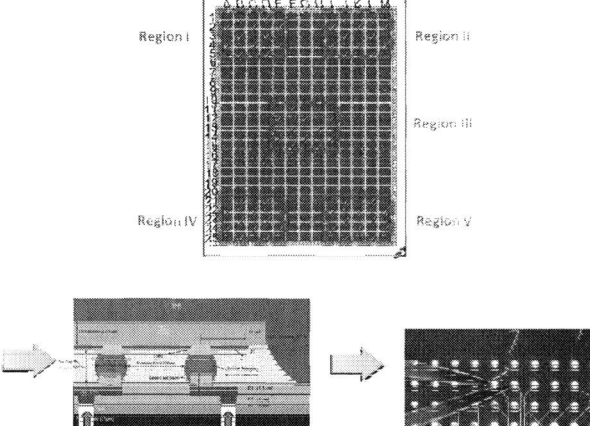

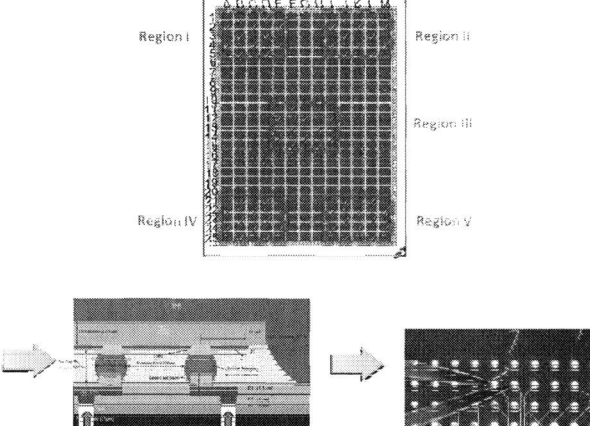

test key opening by laser drilling   measuring resistance

Fig.4 Select five areas on the back side of the package structure, with 12 die positions selected for each area, then test key holes were generated by laser drilling, finally, taking the measured test key for resistance change

Table 1. The results in average resistance change rates that are before and after de-bonding

Region	Average Resistance		Resistance Change Rate
	Before de-bonding	After de-bonding	
I	5.98	6.09	2%
II	6.53	6.61	1%
III	6.12	6.23	2%
IV	5.55	5.77	4%
V	5.64	5.42	4%

This experiment was successfully verified 1-layer RDL FOPLP with daisy chain die and SMC molding process, In addition, the result of the resistance change rate <5% after the glass carrier was de-bonded was also successfully completed.

B. 1 layer RDL FOPLP with daisy chain die and flexible molding compound

The flexible FOPLP packaging can be further completed by the ITRI's FOPLP technology, the purpose of this experiment was to introduce PDMS (Polydimethylsiloxane) as a molding material and apply it to FOPLP process, that can enables flexible package structure, as shown in Fig. 5.

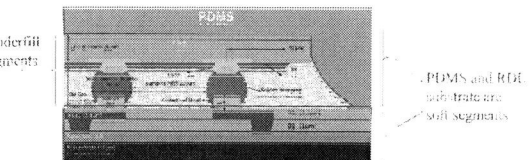

Fig. 5 1-layer RDL with daisy chain die and PDMS is used as the molding material

This verification is based on the FOPLP process of the previous section, that replacing the molding material SMC with PDMS to achieve flexible requirements, in terms of process, the PDMS with a thickness of 800μm is uniformly coated on the surface of the die and the RDL substrate by blade coater, it can form the outermost protective layer in the package structure, and can also make the whole structure flexible, as shown in Fig.6.

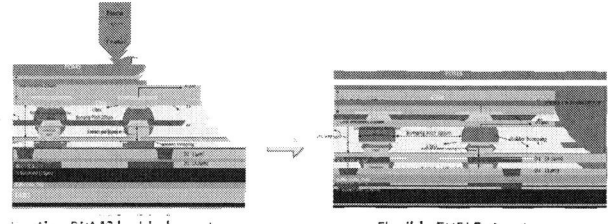

Fig. 6 PDMS coated on die and RDL substrate surface

This experiment successfully completed the G2.5 panel-level flexible packaging structure based on the use of PDMS as the molding material and ITRI's FlexUPTM and flexible display technology, simultaneous realization of rigid and flexible heterogeneous bonding technology and transparent effect.

On the application and market sides, 30mm bending radius of flex circuit construction, it has applications in a wide range of consumer products including wearables and mobile devices in the future. As a result, 1 layer RDL FOPLP with daisy chain die and flexible molding process were successfully validated in this experiment.

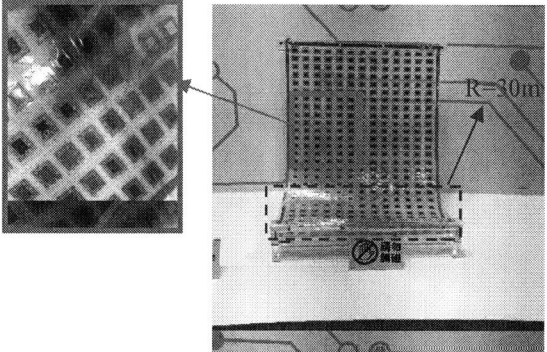

Fig.7 Introducing PDMS molding material instead of SMC to realize the integration of rigid and flexible heterogeneous bonding technologies

C. 3 layer RDL with 1 layer UBM

Considering the demand for flexible electronics in the future, this research will integrate multi-die with flexural multi-layer RDL, therefore, the behavior of resistance change of multilayer RDL under bending conditions is further discussed.

In this experiment, the results of 3-layer RDL and 1-layer UBM affected by bending stress were verified, and the check method is the resistance change rate after bending.

The process of the 4-layer RDL is the same as that of the single-layer RDL described above, thus, RDL2 and RDL1 are repeating step 1 to step 8 procedures to complete the 3-layer RDL and 1-layer UBM process. as shown in Fig. 8, where RDL3 is 3.3μm, RDL2 and RDL1 are all 2.4μm, on the other hand, the dielectric layer D4 is 1μm, and D3, D2 and D1 are all 5μm, the whole 3-layer RDL and 1-layer UBM as shown in step 9 of Fig.8.

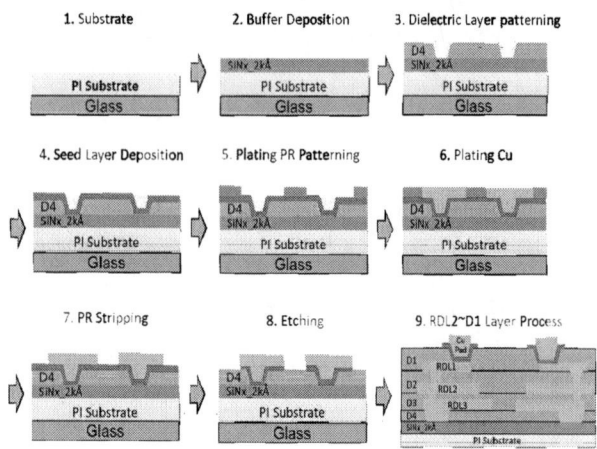

Fig. 8 3-layer RDL and 1-layer UBM structure process flow and schematic diagram

The strength and electrical properties for different thickness structures in a fixed bending radius condition are verified, Fig. 9 shows the relationship between the strain and the thickness of structure and the bending radius.

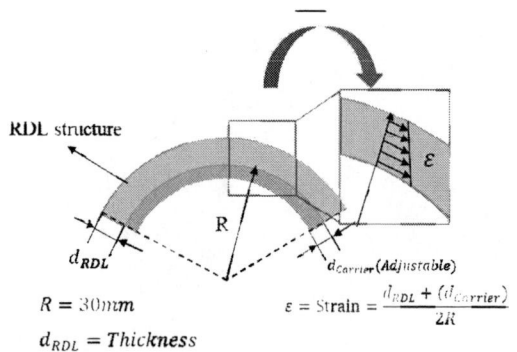

$$R = 30mm$$
$$\varepsilon = Strain = \frac{d_{RDL} + (d_{Carrier})}{2R}$$
$$d_{RDL} = Thickness$$

Fig. 9 The strain is proportional to the bending radius and structural thickness

This study prepared two test samples with different thickness structures, one is the original RDL structure with a thickness of about 25μm as an experimental group, the other is the 100μm PET film used by ITRI in the development of flexible AMOLED [7], then attached to the back size of the original RDL structure as a control group, as shown in Fig. 10.

(a)
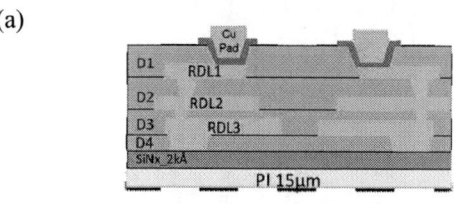

(b)
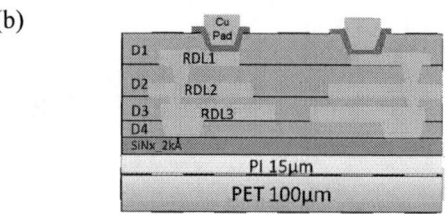

Fig. 10 (a) The original structure of the experimental group with a thickness of about 25μm (b) The structure of the control group (RDL with PET) with a thickness of about 125μm

For the control group with 100 μm PET film attached to the original structure, after 10,000 times of repeated bending test, a significant crack was generated on the surface of the RDL substrate, as shown in Fig. 11, and using SEM to confirm the significantly increased average crack size from 5μm to 11μm, as shown in Fig. 12.

For the experimental group with a structure thickness of about 25μm during bending test was 100,000 times by the bending radius is 30 mm condition, as shown in Fig. 13. Besides check the each heterogeneous interface has no delamination, also check resistance change rate, the measurement method of resistance is to design the test key on the RDL substrate and distribute it in 6 regions, as shown in Fig. 14, when measuring, the bias voltages of 0.5V, 1V, and 1.5V are input respectively, and after the calculation by Ohm's law, the corresponding relationship between the average resistance for before and after bending can be obtained, as shown in table 2, the difference in resistance before and after bending is slightly increased by 2.8Ω, 3.05Ω, and 4.1Ω with applying voltage of 0.5V, 1.0V, and 1.5V, respectively. However, the resistance change rate can meet the demand for ≤5%, as shown in Table 3.

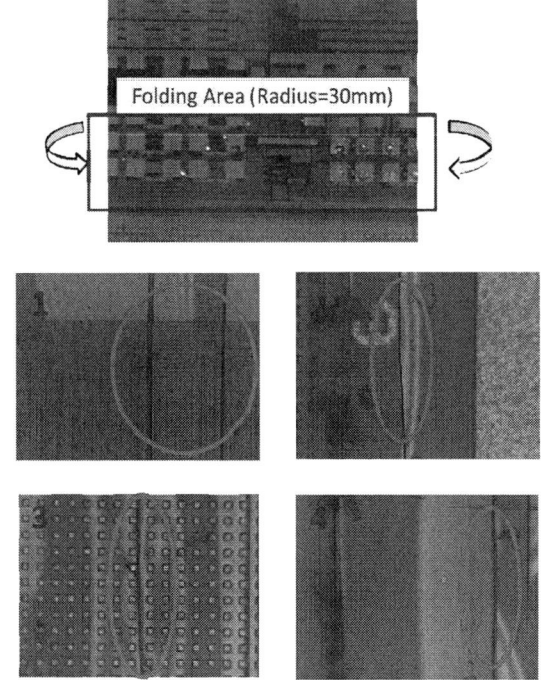

Fig. 11 Fail sites detected by optical microscopic image from the control group (RDL with PET)

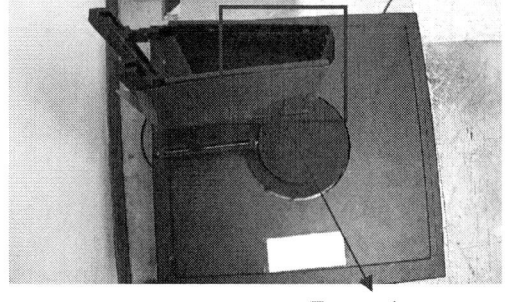

Fig. 12 The crack size of about 5~11μm can be observed by SEM images

Front view

Fig. 13 The experiment platform for the bending test

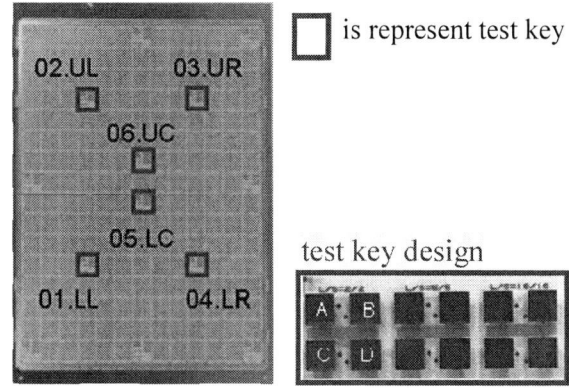

□ is represent test key

test key design

Fig. 14 The test key design and its positions

Table 2. The resistance varation before and after bending

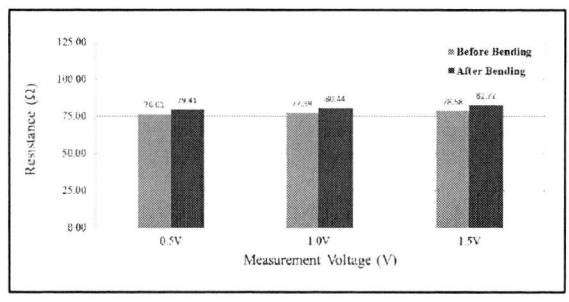

Table 3. After 100,000 times bending test, in different bias conditions, the average resistance change rate can meet the demand for ≤ 5%

Measurement Voltage	Resistance		Resistance Change Rate
	Before Bending	After Bending	
0.5V	76.61	79.41	4%
1.0V	77.39	80.44	4%
1.5V	78.58	82.72	5%

III. CONCLUSIONS

With ITRI's core technology and FOPLP facility platform, RDL-first FOPLP can be successfully fabricated.

The FHE is based on a flexible material and rigid IC to form a flexible hybrid electronic circuit system, therefore, this research also focuses on the process integration and structure evaluation.

Through FlexUP™ technology and introduction of PDMS molding material, flexible FOPLP package can be successfully completed. Finally, considering the demand for flexible electronics in the future, bending test of multi-layer RDL had also been evaluated. During 100,000 times bending, the resistance variation of RDL layers is ≤5%.

978-1-7281-1500-9/19 $31.00 © 2019 IEEE

With ITRI's technology, the related problems can be overcame such as broken circuit and poor electrical performance, which were caused by bending and stretching the circuit of the RDL substrate. The results showed that ITRI's core technology could meet the demand of FHE systems. It also provide a solution to solve the contradiction between the characteristics of the rigid structure and flexible requirements in the near future.

REFERENCES

[1] C. F. Tseng, C. S. Liu, C. H. Wu, and D. Yu, "InFO (Wafer Level Integrated Fan-Out) Technology", p.1 (ECTC 2016).

[2] R. McCleary, P. Cochet, T. Swarbrick, C. T. Sim, Gurvinder Singh, Yong Chang Bum, and Andy Kyawoo Aung, "Panel Level Advanced Packaging", p.25 (ECTC 2016).

[3] H. D. Chang, D. Chang, K. Liu, H. S Hsu, R. F. Tai, H. C. Huang, Y. C. Lai, C. L. Lu, C. T. Lin, S. Chiu, "Development and Characterization of New Generation Panel Fan-out (P-FO) Packaging Technology", p.947 (ECTC 2014).

[4] T. Braun, S. Raatz, S. Voges, R. Kahle, V. Bader, J. Bauer, K. F. Becker, T. Thomas, R. Aschenbrenner, K. D. Lang, "Large Area Compression Molding for Fan-out Panel Level Packing", p.1077 (ECTC 2015).

[5] H. W. Liu, Y. W. Liu, J. Ji, J. Liao, A. Chen, Y. H. Chen, N. Kao, and Y. C. Lai, "Warpage Characterization of Panel Fan-out (P-FO) Package", p.1750 (ECTC 2014).

[6] W.-Y. Cheng, C.-T. Yang, J.-M. Lin, W.-H. Chen, T.-J. Wang, Y.-Z. Lee, "Evaluation of Chip-last Fan-out Panel Level Packaging with G2.5 LCD Facility using FlexUP™ and Mechanical De-bonding Technologies", p.386 (ECTC 2017).

[7] J.-L. Chen, and C.-T. Liu, "Technology Advances in Flexible Displays and Substrates," vol. 1, no. 10, pp. 2169 (IEEE Access. 2013).

Structuring of Laser Activated Polymers for Sensor Applications

Bengsch, Sebastian
Institute of Micro Production Technology
Leibniz Universitaet Hanover
Hannover, Germany
bengsch@impt.uni-hannover.de

Aue, Maximilian
Institute of Micro Production Technology
Leibniz Universitaet Hanover
Hannover, Germany

Cromwell, Kevin
Institute of Micro Production Technology
Leibniz Universitaet Hanover
Hannover, Germany

Dr.-Ing. Wurz, Marc Christopher
Institute of Micro Production Technology
Leibniz Universitaet Hanover
Hannover, Germany
wurz@impt.uni-hannover.de

Abstract— Previous publications have shown that polymer-based materials hold a great potential for the use as substrates for sensors, for example for AMR sensors. Polymers can substitute expensive substrates such as silicon, and pre-structured substrates can eliminate substantial cleanroom and micro-technological processes, especially photolithography. Ultimately, process optimization can yield manufacturing processes without expensive procedures (through-silicon vias), guaranteeing a complete abandonment of processes such as photolithography, CMP and the like.

At this point, injection molding with laser direct structuring (LDS) polymers offers distinct advantages, such as the electroless and selective deposition of metals, which can be used to implement through-vias. In this context, the LDS-capable polymer polyetheretherketone (PEEK) is employed. The thermoplastic polymer features a high glass transition temperature and chemical resistances to many solvents. As a result, the substrate material can potentially cover a wide range of applications, especially due to the ease of integration into micro-technological processes.

The presented sensor structures were produced by micro-technological processes and contacted using LDS-realized vias. The cost-effective polymer-based module, or rather the substrate, can be integrated directly into other processes and modules, such as a system on a chip system, without the need for costly processes for mounting the devices. To verify this, temperature and magnetic field sensors based on the AMR effect were manufactured and evaluated. This article demonstrates that approach using a process developed at the Institute of Micro Production Technology (IMPT).

Keywords—injection molding; manufacturing technology; polymer substrate; AMR-sensor; sensors; PEEK; polyetheretherketone; thermoplastic

I. INTRODUCTION

To open up the opportunity for low cost sensor manufacturing and with regards to the market of a trillion sensors [1], the Institute of Micro Production Technology developed a manufacturing method previously presented at the IEEE ECTC 2018 [2]. The basic principle of the manufacturing process consists of a polymer substrate structuring process by injection molding. Lithography can be completely neglected when creating micro structured systems such as e.g. AMR sensors. However, to create micro systems using injection molding, additional steps are necessary which must not be performed in a clean room environment. For example, a sputter deposition step is employed for the sensor thin film layer (e.g. nickel-iron (NiFe) for an AMR element) inside the cavities on the polymer substrates surface. A final polishing step excavates the desired structures. The comparison of the common micro technological process chain vs. the newly developed concept can be seen in fig. 1. The common manufacturing consists of at least seven process steps in comparison to the three-step process using the injection molding strategy. An additional feature is the use of laser activatable plastics like PEEK (Ensinger®). The technology of the laser direct structuring allows creating laser punch holes that are activated for electroless copper deposition. Fig. 3 shows the detailed process chain of laser direct structuring. The copper-filled holes can serve as via-structures, therefore supplying backside contacting for the sensor system. The IMPT has already gained experience in the field of LDS technology [4]. Resulting from these process steps, a fully packaged sensor can be mounted by die bonding using e.g. reflow soldering or anisotropic gluing combined with common pick and place technology (fig. 4). This handling opportunity is not only a result of the manufacturing process but also due to the properties of the thermoplastic material of choice. PEEK withstands temperatures up to 260°C permanently and up to 300°C for short times. Additionally, it has excellent chemical resistance against solvents commonly used in micro technological environments and biocompatibility for the use in implants is given. Another interesting property of PEEK is the thermal expansion coefficient, which is closer to NiFe than the one of Kapton, for example (PEEK= 18 x $10^{-6}K^{-1}$, Kapton® = 20 x $10^{-6}K^{-1}$; NiFe = 13.5 x $10^{-6}K^{-1}$). The first test setup consists of an AMR sensor built on a conventional polycarbonate substrate. It serves to proof that the process developed herein is fit to produce a magnetoresistiv sensor system on plastic substrates. This process merely expands the institute's portfolio of magnetoresitive manufacturing processes on different substrates, including flexible polymer substrate materials such as Kapton® [5]-[10]. In the next step, the AMR sensor system is built on a PEEK substrate. To do so, the newly developed material TECACOMP® PEEK LDS black 1047045 by Ensinger® was investigated regarding its process parameters using injection molding. Following these parameter studies, the LDS structuring was performed and a sensor demonstrator tested. In the end, all process steps were combined and evaluated. A conceptual drawing of the envisioned demonstrator is shown in fig. 2.

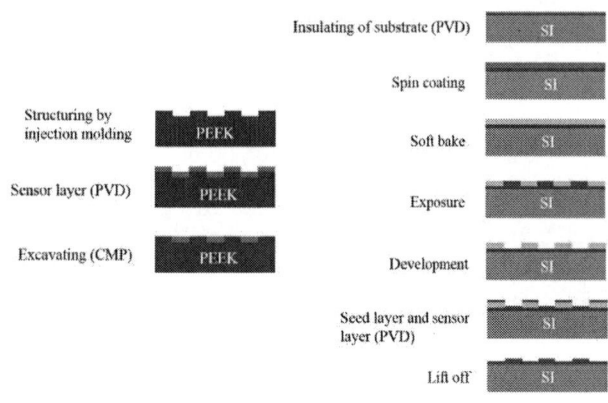

Figure 1. Comparison of new process flow vs. lithography based manufacturing of sensors

Figure 2. Concept drawing of sensor incl. via structures

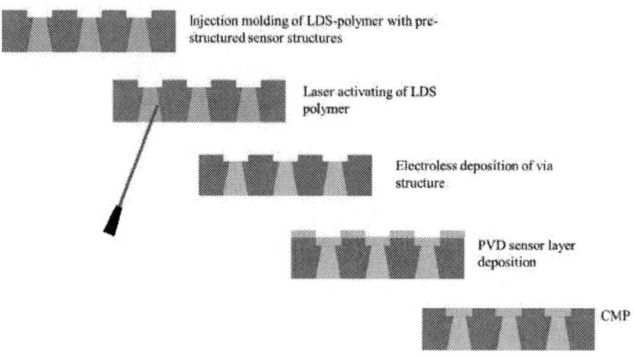

Figure 3. Combination of injection molded manufacturing and LDS structuring

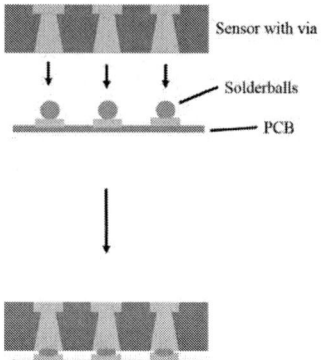

Figure 4. Contacting of sensor on PCB (example using soldering)

II. PROCESS DEVELOPMENT

A. Injection Molding

Concerning the process of injection molding, it is crucial to perform a design of experiment study for every new polymer substrate with respect to the injection molding parameters to create a sufficient substrate quality. The goal is to achieve a polymer substrate surface roughness as close as possible to commonly used silicon substrates. As shown before [2], the surface roughness greatly affects the final sensor quality. Values down to $R_a \sim 20$ nm were reached using Polycarbonate (Apec 1897 by Bayer®). In comparison, R_a values of silicon substrates center around 0.1 nm. Following the previous works [2], the IMPT investigated the process parameters of PEEK-LDS by Ensinger® (TECACOMP® PEEK LDS black 1047045 by Ensinger®), which is a state of the art material in terms of laser structurable PEEK. The parameter studies, performed using design of experiment methods, show, that the screw temperature, injection speed, mold temperature and cooling time are the most critical values for the desired substrate quality. It was shown in recent investigations [2] that a temperature field of 140°C on the top and bottom of the mold creates a sufficient surface planarity but does not satisfy expectations regarding roughness and structural quality. Therefore, the form has to be cooled down, after the plastic injection, to temperatures below the glass transition temperature of PEEK (PEEK T_g: 143°C). An additional cooling ramp from 140°C to 100°C in 2 min. and 30 sec.

creates a sufficient surface quality to perform the thin film deposition. The minimum roughness realized on a PEEK substrate in this manner is approximately $R_a = 30$ nm. The maximum mold temperature is limited to 140°C (BOY 55EVV) due to a water cooling system in the injection molding machine. Fig. 5-8 display the increased planarity and surface quality. The injection molding process-parameters are shown in Table 1.

TABLE I.

Parameter	Injection Molding Parameters of Polyetheretherketone (DR BOY 55EVV)
	Value
Closing Force	550 kN
Mold Temperature Top	140°C (Cooling Profile 140°C - 100°C)
Mold Temperature Buttom	140°C (Cooling Profile 140°C - 100°C)
Screw Temperature	400;390;380;370;360
Cooling Time	150 sec
Changeover Point Dwell Pressure	0 mm
Injection Pressure	180-170 bar
Way of Screw	26 mm
After Pressure	70 bar
After Pressure Time	1 sec

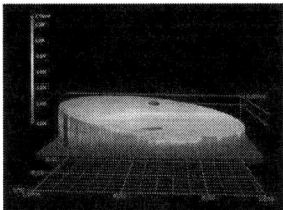

Figure 5. Planarity before DoE

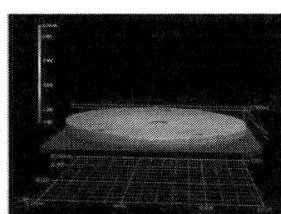

Figure 6. Planarity after DoE

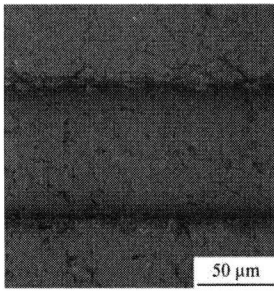

Figure 7. Surface of PEEK before DoE

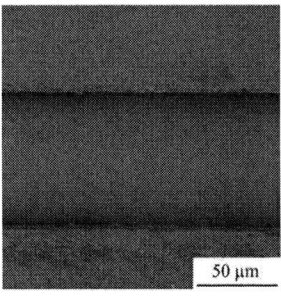

Figure 8. Surface of PEEK after DoE

B. Sensor Manufacturing on Test Substrate (Polycarbonate)

For the investigation of an AMR sensor system produced on an injection molded polymer substrate, polycarbonate was chosen initially, because of its simple processibility. Firstly, an injection mold inlay was produced on a polished steel substrate (C45, 1.1191) using lithography and electroplating to yield 13 μm Ni structures. The surface quality of the electroplated structures was improved performing a CMP step on the Ni structures, creating R_a values below 20 nm (resolution limits of the tactile and confocal measurement). The results of the surface measurements are summarized in Table 2. The process flow for the form inlay manufacturing is shown in fig. 9, while fig. 10 displays the resulting injection molding form inlay. After manufacturing the form inlay by lithography and electroplating, the injection molding process was performed using adapted process parameters (Table 3) [2].

TABLE II.

Substrates	Ra in nm	Rz in nm	Rq in nm	Rv in nm	Rp in nm
Mold Inlay	8	75	11	37	39
PC Substrate	14	56	16	27	30
PEEK Substrate	30	238	32	135	102

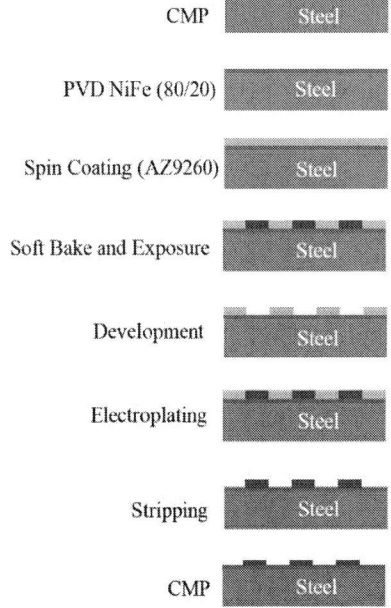

Figure 9. Process flow for form inlay manufacturing

TABLE III.

Parameter	Injection Molding Parameters of Polycarbonate (DR BOY 55EVV)
	Value
Closing Force	550 kN
Mold Temperature Top	120°C
Mold Temperature Buttom	130°C
Screw Temperature	330;320;310;300;290
Cooling Time	120 sec
Changeover Point Dwell Pressure	2 mm
Injection Pressure	180-170 bar
Way of Screw	26 mm
After Pressure	70 bar
After Pressure Time	1 sec

Following the injection molding, 150 nm, 100 nm and 50 nm NiFe (80/20) were sputter deposited (von Ardenne® Cluster System CS 730 S®) to create different sensor layer set-ups. CMP processes were employed to excavate the structures. 150 nm NiFe layers did not withstand the final CMP process. Fig. 11 shows the substrate after the final CMP step. A polished sensor structure can be found in fig. 12, while fig. 13 displays cross-section of a manufactured sensor structure.

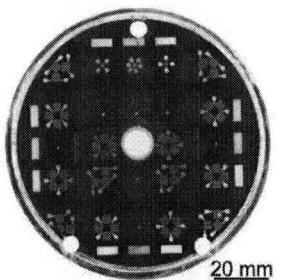 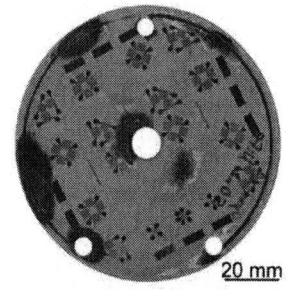

Figure 10. Manufactured injection molding inlay

Figure 11. Polished sputtered PC substrate

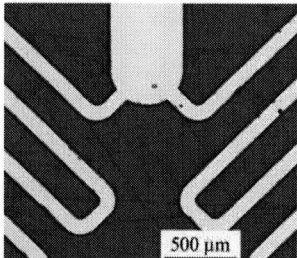 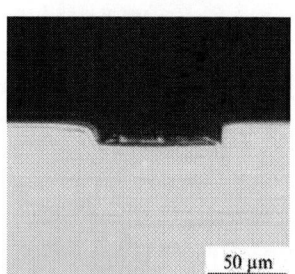

Figure 12. Polished sputtered sensor structure

Figure 13. Cavity of sensor structure (cross-section)

After sensor manufacturing, the systems were tested using a Helmholtz Coil. The most promising results concerning the average change in resistance applying a magnetic field could be achieved by using sensor systems build up with 100 nm NiFe thin film layers. The results of the magnetic behavior testing can be seen in fig. 14. The average change in resistance $\Delta R/R_0$ can reach up to 1.2 percent.

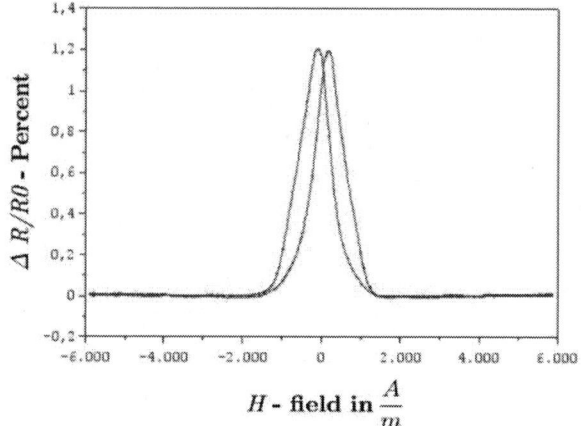

Figure 14. Behavior of a 100 nm NiFe AMR sensor on PC Substrate

III. System Manufacturing on LDS Material

After the proof of concept of an AMR sensor on injection molded polycarbonate substrates, an AMR sensor supposed to be manufactured on TECACOMP® PEEK LDS black 1047045 by Ensinger® substrate to combine backside contacting of the substrate, enabled by electroless deposited via holes, punched by laser. Therefore, the system manufacturing starts with the production of an AMR sensor and later combined with laser structured vias and backside contacting.

A. AMR Sensor on PEEK

Following the manufacturing steps of the AMR sensors processed on polycarbonate substrate, the AMR sensors were manufactured accordingly only using 100 nm structures of NiFe (80/20) reasoning in enabling the best performance on plastic substrates. The used sputtering device is a von Ardenne® Clustersystem CS 730 S®. The sputter deposition parameters for 100 nm NiFe (80/20) can be seen in table 4. The designed structures and deposited thin film realize a resistance of each sensor between 900 Ω and 1100 Ω. The differences of the resistivities occur due to the thin film deposition height-difference depending on the position on the substrate. A rotating substrate holder or target can control this. In addition, the structuring method is dealing with sidewall deposition of metallic thin films, inside the cavities, due to the sputter deposition process, which lowers the electrical resistance due to the increase in conductive diameter of the sensor structure. Therefore, is the calculated resistivity for a 100 nm NiFe AMR sensor is 1650 Ω in comparison to about 1000 Ω for a manufactured structure.

TABLE IV.

Parameter	Sputter Deposition Values (Ardenne® Cluster System CS 730 S®)
	100 nm NiFe (80/20)
Pressure	1×10^{-7} bar
Power	520 Watts
Time	113 sec
Gas flow	50 sccm Ar
Substrate Temperature	RT

Following the sensor manufacturing process, the sensors were tested in a Helmholtz Coil set-up. The tested specimen could reach average changes in resistance $\Delta R/R_0$ up to 1.4 percent. The results of the magnetic behavior test of a 100 nm NiFe (80/20) AMR sensor on PEEK substrate is shown in fig. 15. In comparison, fig. 16 shows the AMR effect with the same parameter settings on a silicon dioxide substrate. The graph shows an average change in resistance $\Delta R/R_0$ up to 2.3 percent.

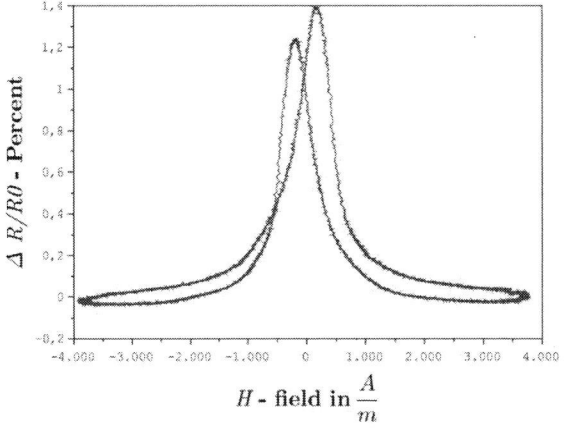

Figure 15. Behavior of a 100 nm NiFe AMR sensor on PEEK substrate

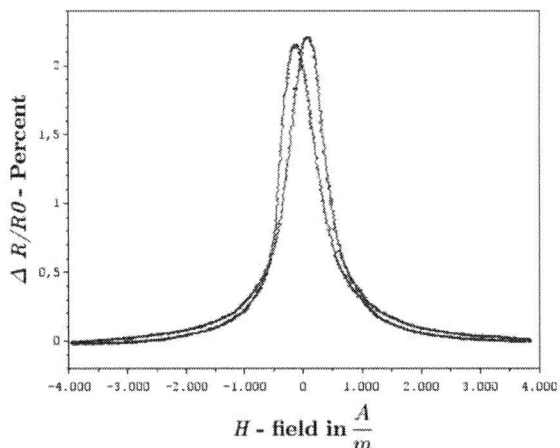

Figure 16. Behavior of a 100 nm NiFe AMR sensor on a silicon dioxide substrate

B. LDS of vias on PEEK

To verify the concept of LDS via structuring and metallization, the contact pad areas of the PEEK substrates were laser treated (fig. 17) and metallized afterwards (fig. 18). A through-hole contacting was realized and verified by backside contacting of the sensors using isotropic conductive epoxy glue. The resistance of the via structure is less than 1 Ω and the epoxy contact resistance is about 3 Ω. The used laser system (ML160iF) is a fiber-laser offering a wavelength of 1054 nm at a spot diameter of 80 μm. The performed power output was 4.2 W at a frequency of 30 kHz. The speed of the laser was set to 2 m/s and 6000 repetitions. The substrate was cleaned afterwards in an ultra-sonic bath using a cleaning solution of 7% MacDermid XB Strike Bath at 50°C for a duration of 90 seconds. The subsequent metallization of the substrate was performed at 56°C in a MacDermid XB Strike bath for 60 minutes resulting in a copper layer inside the punch hole of about 2 – 3 μm.

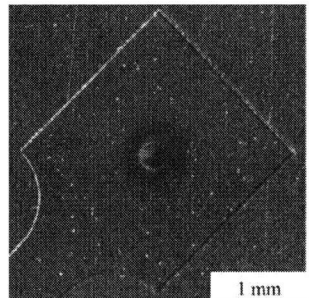 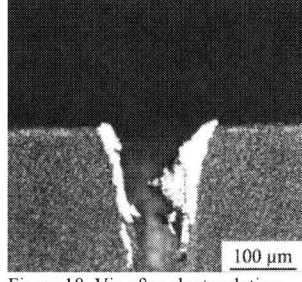

Figure 17. Laser punched hole before electroplating

Figure 18. Via after electroplating (cross-section)

C. Combination of AMR-Sensor and Via-Structure

The final step for the demonstration of the concept of injection molded based sensor structures and laser direct structured vias, is the combination of the sensor manufacturing method and the backside contacting. Fig. 19 shows the manufactured sensor system with through hole via contacting combined.

After combining the AMR Sensor with the metallized via substrates, the set up will be bonded onto a PCB board using conductive epoxy glue (CW2400© epoxy with conductive silver particles by Chemtronics©) This was tested beforehand using a dummy LDS Structure as mentioned in chapter B.

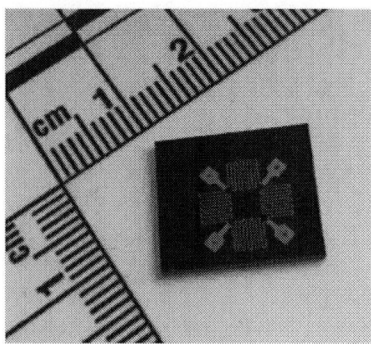

Figure 19. Final demonstrator set up

IV. CONCLUSION AND OUTLOOK

The concept of pre-structured polymer substrates manufactured by injection molding was verified in this work. The IMPT realized a pre-structured AMR-sensor on polyetheretherketone (PEEK) substrate. The polymer of choice shows high resistivity against solvents and long-term temperature stability of up to 260°C as well as maximum temperatures of up to 300°C for short times. The produced substrates were optimized in design of experiment injection molding parameter studies resulting in surface qualities crucial for the deposition of sensor thin film layers on the PEEK substrate. It was shown in prior experiments, that polycarbonate as a substrate is capable of being processed to be suitable for AMR sensors. The newly developed LDS capable PEEK material is now processable at the IMPT in terms of injection molding and laser direct structuring as well. Pre-structured PEEK substrates were manufactured using the same AMR layout as used for polycarbonate specimen. Injection mold inlays were fabricated at IMPT using common lithography technology and electroplating. Fabricated pre-structured PEEK-LDS substrates were further processed using LDS at LPKF. A demonstrator set-up showed that PEEK holds great potential as substrate material for AMR sensor manufacturing. The magnetoresistiv effect in the sensor structures was as high as $\Delta R/R_0 = 1.4\,\%$ on the PEEK substrate in comparison to $\Delta R/R_0 = 1.2\,\%$ using previously studied PC substrates. In summary, the results support the use of plastics as substrate material for thin film based AMR sensor manufacturing. A pre-structuring using injection molding can offer a lithography and clean room free manufacturing process, which decreases manufacturing and infrastructural costs. A contacting method using LDS structuring to create via structures enables conventional mounting onto PCB boards, while state of the art anisotropic gluing or soldering facilitates the direct packaging on PCB substrates without developing new handling technologies. Future work will focus on the investigation of temperature behavior when using different temperature cycles, as well as mechanical load and bending tests, to investigate the sensor behavior for different application areas, such as automotive, consumer or medical devices. Additional research will center on developing suitable soldering alloys to enable the use of copper or gold contact vias resulting from the LDS process. For example, in terms of reflow soldering, a viable solution will result in the negation of one of the last arguments against using thermoplastics like PEEK as sensor substrates.

ACKNOWLEDGMENT

I would like to thank the Institute of Micro Production Technology as well as the Leibniz University Hanover for the opportunity and the equipment to develop the structuring method of laser activated polymers for sensor applications. Further thanks goes out to my co-author and supervisor Dr.-Ing. M.C. Wurz for the excellent supervision regarding this project as well as the support from Maximilian Aue, Sascha de Wall, and especially Dörthe Leifheit for the support regarding sputter deposition processes. The IMPT would also like to thank Bernd Rösener from LPKF Laser & Electronics AG for his support with the laser direct structuring and electroless deposition technology and equipment.

REFERENCES

[1] Robert Bogue, "Towards the trillion sensors market", Sensor Review, Vol. 34 Issue: 2, pp.137-142 (2014)

[2] S. Bengsch; M. Aue, S. DeWall, M.C. Wurz, Structuring Methods of Polymers for low Cost Sensor Manufacturing, IEEE ECTC (2018).

[3] S. Bengsch; M. Rechel; E. Asadi; M.C. Wurz, Structuring Methods of Plastic Substrates for Electroplating Applications ECS (2016)

[4] M.C. Wurz; S. Bengsch; S.Beringer, Concept for using MID Technology for Advanced Packaging IEEE ECTC (2018)

[5] A. Belski, M. Wurz, J. Rittinger, L. Rissing: Development, Micro Fabrication and Test of Flexible Magnetic Write Head for Gentelligent Applications, Microelectronic Engineering Journal, (spec. issue), Proc. 38th Int. Conf. on Micro & Nanoengineering, Toulouse, France (2013)

[6] P. Taptimthong, J. Rittinger, M. C. Wurz, L. Rissing: Flexible Magnetic Writing / Reading System: Polyimide Film as Flexible Substrate, Procedia Technology, Vol. 15, 2014, pp. 230–237 (2014)

[7] J. Rittinger, P. Taptimthong, L. Jogschies, M. C. Wurz, L. Rissing: Impact of different polyimide-based substrates on the soft magnetic properties of NiFe thin films, SPIE Microtechnologies (pp. 95171R-95171R). International Society for Optics and Photonics (2015)

[8] Johannes Rittinger, Piriya Taptimthong, Lisa Jogschies, Marc C. Wurz, Lutz Rissing: Magnetic microstructure analysis of sputter deposited permalloy thin films on a spin-on polyimide substrate, Microsystem Technologies: 1-6. (2016)

[9] Taptimthong P, Rissing L, Wurz MC: Flexible Magnetic Reading/Writing System: Characterization of a Read/Write Head, 18. GMS/ITG-Fachtagung Sensoren und Messsysteme, Nürnberg, pp. 769-775 (2016)

[10] Taptimthong P, Düsing JF, Rissing L, Wurz MC: Flexible Magnetic Reading/Writing System: Heat-Assisted Magnetic Recording, 3rd International Conference on System-integrated Intelligence: New Challenges for Product and Production Engineering, SysInt 2016, Procedia Technology 26, pp. 72-78 (2016)

A Deep Learning Approach for Volterra Kernel Extraction for Time Domain Simulation of Weakly Nonlinear Circuits

Thong Nguyen, Xinying Wang, Xu Chen and Jose Schutt-Aine
Department of Electrical and Computer Engineering.
University of Illinois Urbana-Champaign
Urbana, IL 61801, USA
tnnguye3, xinying, xuchen1, jesa@illinois.edu

Abstract—Volterra kernels are well known to be the multi-dimensional extension of the impulse response of a linear time invariant (LTI) system. It can be used to accurately model weakly nonlinear, specifically, polynomial nonlinearity systems. It has been used in the past for white-box model order reduction (MOR) to model frequency-domain performance metric quantities such as distortion in power amplifiers (PA). In this paper, we train a neural network from time-domain response of high-speed link buffers to extract multiple high-order kernels at once. Once the kernels are extracted, they can fully characterize the dynamics of the buffers of interest. Using the kernels, we demonstrate that time-domain response is straight-forward to obtain using super-, or multi-dimensional convolution. Previous work has used a shallow feed-forward neural network to train the system by using Gaussian noise as the identification signal. This is not convenient for the method to be compatible with existing computer-aided design tools. In this work, we directly use a pseudo random bit sequence (PRBS) to train the network. The proposed technique is more challenging because the PRBS has flat regions which have highly rich frequency spectrum and requires longer memory length, but allows the method to be compatible with existing simulation programs. We investigate different topologies including feed-forward neural network and recurrent neural network. Comparisons between training phase, inference phase, convergence are presented using different neural network topologies. The paper presents a numerical example using a 28Gbps data rate PAM4 transceiver to validate the proposed method against traditional simulation methods such as IBIS or SPICE level simulation for comparison in speed and accuracy. Using Volterra kernels promises a novel way to perform accurate nonlinear circuit simulation in the LTI system framework which is already well known and well developed. It can be conveniently incorporated into existing EDA frameworks.

Keywords-Volterra Series, Volterra Kernels, Weakly Nonlinear Time-invariant, Neural Network, Behavioral Modeling, Signal Integrity, PAM4, High-speed channel.

I. Introduction

As input/output (I/O) buffer circuits grows in complexity and speed, design verification under intellectual property (IP) protection becomes more challenging because of the accuracy requirement that is put on macromodels of I/O buffers. IBIS [1] has been used as a standard in the industry to exchange I/O buffer designs while protecting IP for many years. It is popular because it is the exchanging data

agreement between vendors and customers, however, there are quite some improvements needed in the accuracy of IBIS modeling compared to transistor-level simulation [2]. The need for a more accurate method is imperative.

Volterra theory of nonlinear systems [3], [4] is the extension of the well-known LTI system theory and has been successfully used to characterize nonlinear circuits such as communication channels [5], PA [6], low noise amplifier (LNA) [7], audio filters [8] or power electronics converters [9]. It provides a novel way to simulate a large class of nonlinear circuits using input/output relationship. The robustness of Volterra framework has been proven for both commensurate and non-commensurate excitation data [10], [11]. Works done in the past have been focusing on creating white-box Volterra macromodels, where transistor-level circuits are available, to avoid Newton-Ralphson alike iterative solution in SPICE solvers for time-domain simulation [6], [8], [9], [11]–[16], or distortion analysis in frequency-domain [5], [14], [16]–[24]. There were fewer efforts put on black-box, i.e. only input/output signals are accessible, Volterra macromodeling [25].

Neural network has been used to model input/output mapping for a long while. In the past, there were attempts to use neural network to extract Volterra kernels using input/output data [26]–[29]. However, due to limitation of computational resources and the fact that training a neural network was not efficient, the method did not gain much popularity. With recent advancement in deep learning, training a neural network with back-propagation [30] is more efficient and faster compared to the old method such as the Levenberg-Marquardt (LM) algorithm [36]. In this paper, we propose to leverage modern development of deep learning to extract Volterra kernels for I/O buffers. Once the Volterra kernels are extracted, time domain simulation of high-speed channels can be performed accurately.

The paper is structured as follows: Section II reviews the Volterra theory for weakly nonlinear system, Section III presents the connection between Volterra kernels and neural network, consequently, formulates the process to extract Volterra kernels from trained neural network weights. We discuss the comparison between two different choice of activation function and, consequently, the difference in kernel extraction formulation. Section IV demonstrates the proposed method with two examples, including a PAM4 buffer high-speed link circuit. Section V concludes the paper, discusses about possible extension of this work.

II. VOLTERRA THEORY OF WEAKLY NONLINEAR SYSTEM

A non-linear time-invariant (NLTI) system is considered weakly non-linear if it possesses *fading memory* property. That is, the present output does not depend on the infinitely long past [31] information. For a weakly non-linear time-invariant (NLTI) system, memory effects can be well approximated by an N-term truncated Volterra series [32], the output response $y(t)$ of the system under the input excitation $x(t)$ is

$$y(t) = \sum_{n=1}^{N} y_n(t) \quad (1)$$

$$y_n(t) = \int_{[-\infty,t]^n} h_n(\tau_1, \tau_2, ..., \tau_n) \prod_{i=1}^{n} x(t - \tau_i) \, d\tau_i \quad (2)$$

where $h_n(\tau_1, \tau_2, ..., \tau_n)$ is the nth order Volterra kernel (VK). A frequency-domain VK, or generalized frequency response function (GFRF) [33], can be obtained by multi-dimensional Fourier transform.

$$H_n(\omega_1, \omega_2, ..., \omega_n) = \int_{\mathbb{R}^n} h_n(\tau_1, \tau_2, ..., \tau_n) \times \\ \exp\left(-j \sum_{k=1}^{n} \omega_k \tau_k\right) \prod_{i=1}^{n} d\tau_i \quad (3)$$

Many previous works have used GFRF for distortion and inter-modulation analysis [7], [34].

Inversely, time-domain kernels can be recovered by multi-dimensional inverse Fourier transform

$$h_n(t_1, t_2, ..., t_n) = \int_{\mathbb{R}^n} H_n(\omega_1, \omega_2, ..., \omega_n) \times \\ \exp\left(j \sum_{k=1}^{n} \omega_k t_k\right) \prod_{i=1}^{n} d\omega_i \quad (4)$$

It is worth noted that Volterra kernels representing a system are, in general, not unique [3], [32]. However, it can be seen in Equation (2) that any permutation of τ's will leave the output $y(t)$ unchanged. Thus, symmetrical kernels are unique, mathematically, it is found by taking the average over all $n!$ possible permutations π of τ's.

$$h_{n_{sym}}(\tau_1, \tau_2, ..., \tau_n) = \frac{1}{n!} \sum_{\pi} h_{n_{asym}}(\tau_{\pi(1)}, \tau_{\pi(2)}, ..., \tau_{\pi(n)}) \quad (5)$$

For simplicity, all VK notations used in this paper refer to the symmetric kernel though the subscript "sym" is omitted.

In discrete time, the contribution from n^{th} order Volterra kernel response is given by the nth-dimensional discrete convolution

$$y_n(t) = \sum_{\tau_n=-\infty}^{t} ... \sum_{\tau_1=-\infty}^{t} h_n(\tau_1, \tau_2, ..., \tau_n) \prod_{i=1}^{n} x(t - \tau_i) \, d\tau_i \quad (6)$$

III. VOLTERRA KERNEL EXTRACTION USING NEURAL NETWORK

In this section, connections to neural network are discussed. Training schemes for different neural network topologies including feed-forward and recurrent ones will be presented.

A. Feed-forward Neural Network (FNN)

We now present the process for VK extraction. The prediction $\widetilde{y}$ of a 1-hidden layer feed-forward neural network (FNN) of width M, i.e. there are M neurons, at input and hidden layer is represented by

$$\widetilde{y} = \boldsymbol{C} f(\boldsymbol{W} \boldsymbol{x} + \boldsymbol{b}) \quad (7)$$

where $x \in \mathbb{R}^{M \times 1}$ is the input to FNN, $W \in \mathbb{R}^{M \times M}$ is the input weight matrix, $C \in \mathbb{R}^{1 \times M}$ is the output weight vector and f is a chosen nonlinear activation function. Such an FNN is shown in Figure 1. To match the formulation in Equation (7) to Equation (6), $\boldsymbol{x}$ must be a segment of input signal of length M, the memory length of the system of interests. The FNN, hence, represents the mapping from $\begin{bmatrix} x(t-M) & x(t-(M-1)) & \cdots & x(t) \end{bmatrix}^T$ to $y(t)$.

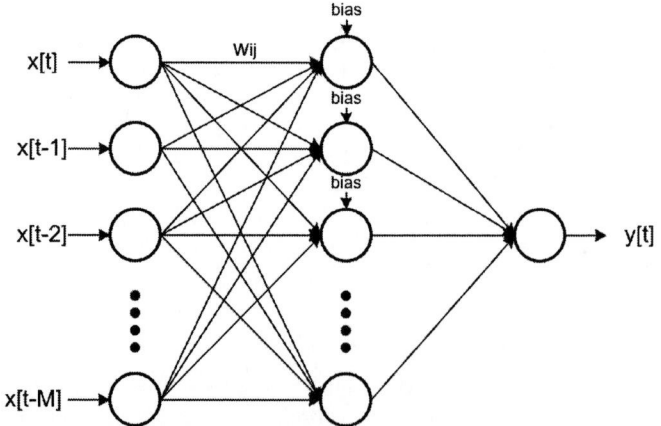

Figure 1: 1-hidden layer FNN

In previous works, f is chosen to be tanh. The kernels are then can be extracted from [26]–[29]:

$$
\begin{aligned}
h_0 &= \sum_{i=1}^{M} c_i a_{0i} \\
h_1(\tau) &= \sum_{i=1}^{M} c_i a_{1i} w_{\tau i} \\
h_2(\tau_1, \tau_2) &= \sum_{i=1}^{M} c_i a_{2i} w_{\tau_1 i} w_{\tau_2 i} \\
&\;\;\vdots \\
h_n(\tau_1, \tau_2, ..., \tau_n) &= \sum_{i=1}^{M} c_i a_{ni} w_{\tau_1 i} w_{\tau_2 i} ... w_{\tau_n i}
\end{aligned}
\tag{8}
$$

where $a_{ni} = \dfrac{1}{n}\tanh^{(n)}(b_i)$, the nth-order derivative of tanh. However, deriving the kernels this way has two serious disadvantages. First, the kernel information is scattered in M kernels the nth of which is an n-dimensional hypercube of length M on each dimension. This means for applications that have long memory effects, the number of kernels needed will be large (i.e, $n = M$) and this approach becomes impractical. Second, when the value of b_i's drives tanh to enter the saturation region, extracting the kernels becomes unstable due to inaccurate calculations of a_{ni}'s which asymptotically approach 0. For these reasons, the earlier works are prone to numerical instability and costly numerical effort. Thus, power series activation function is introduced [35], i.e. for any input signal z, the activation function f is chosen as

$$
f(z) = z^n
\tag{9}
$$

For the nth-order kernel in Equation (8), the coefficient a_{ni} can be obtained from the nth power term in the multi-nominal expansion, which is given by [35]

$$
a_{ni} = b_i^{k_b} \begin{pmatrix} n \\ k_0, k_1, k_2, ..., k_{n-1} \end{pmatrix}_{k_b + k_0 + ... + k_{n-1} = n}
\tag{10}
$$

where k_b, $k_j \in \{0, 1, 2..., n\}$, k_j is the power of the jth delayed time response, k_b is the power of the bias term.

The proposed method employs the same FNN structure in which only one hidden layer is used. This method allows multiple hidden layers to achieve higher order Volterra kernel identification.

Training a FNN, in the past, was done using the LM algorithm to iteratively optimize the nonlinear error function between the NN prediction and the training output $\epsilon(t) = |\widetilde{y}(t) - y(t)|^2$. However, LM algorithm is a second-order gradient-based algorithm which requires the expensive computation and storage for the Hessian matrix. In this paper, we use the now popular back-propagation algorithm [30] to train the NN. It is much cheaper and have been shown to work on many modern problems.

B. Recurrent Neural Network (RNN)

This section discusses about the performance of RNN in the same task of modeling the high-speed channel. Detailed discussions about RNN, comparisons between different RNN topologies and training schemes can be found in [37], [38] and the references therein. Specifically, among two topologies, namely, output feed-back RNN and Elman RNN [39], the latter is favorable due to its acknowledgment of a hidden state that connects between the input and output. ERNN is presented as

$$
\begin{cases}
h_t &= \phi_h(W_{ih} x_t + W_{hh} h_{t-1}) \\
y_t &= f_o(h_t)
\end{cases}
\tag{11}
$$

where ϕ_h is the nonlinear activation function for the recurrent connection, it can be either Tanh, ReLU or any other nonlinear functions. W_{ih} and W_{hh} are the recurrent weight matrices from current input and from the past state to the current state respectively. f_o is weighted activating operator acts on the hidden state to generate the output of an RNN. t is discrete time index. Training an RNN is not straightforward as it is the case for FNN. Because of the recurrence, the weights are shared across time steps, which easily leads to vanishing or exploding gradient problems, prevents the network from learning [40]. RNN is shown to work best when trained with truncated back-propagation through time. Long-short term memory (LSTM) unit [41] is an improved topology of the RNN that enables fast and stable learning by introducing *gates* to alleviate the gradient flow problem. In this work, adapted from [38], we use ERNN structure built on LSTM cells for a stable training process.

IV. NUMERICAL EXAMPLE

Two examples are shown in this section, one synthetic and one practical, to show the effectiveness of the proposed method.

A. Analytical Wiener system

The proposed method is now testified using an analytical system describes by

$$
y(t) = \left(\int_{-\infty}^{t} h(t-\tau) x(\tau) \, d\tau \right)^2
\tag{12}
$$

where $h(t) = h_0 e^{-kt} \sin(\omega t)$. The response for $h_0 = 4$, $\omega = 0.5$ and $k = 1$ is shown in Figure 2. Notice that this system is a 2nd order Wiener system. Schematically, it is shown in Figure 3. It can be shown that for a Wiener system, the nth-order VK kernel can be derived from the LTI response. In particular,

$$
h_n(\tau_1, \tau_2, ..., \tau_n) = \prod_{i=1}^{n} h(\tau_i)
\tag{13}
$$

The FNN is then trained on a white-noise signal input and its corresponding output as shown in Figure 4.

The order of FNN is chosen to be 3, which is higher than the highest nonlinear order of the system. It is, thus, expected that the first and third order kernel is ideally zero. The extracted kernels are shown in Figure 5. As shown, the method is able to capture the exact nonlinear order of the system. Only the second order kernel is significant and matches the analytical kernel.

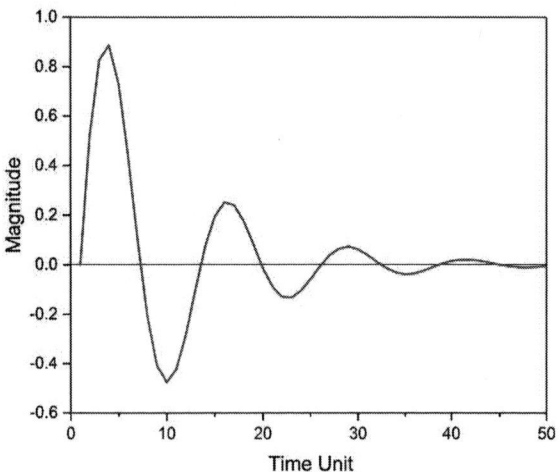

Figure 2: Impulse response used in Wiener system example

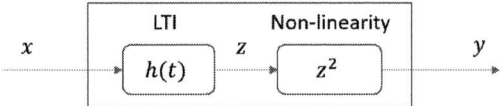

Figure 3: Second order Wiener system

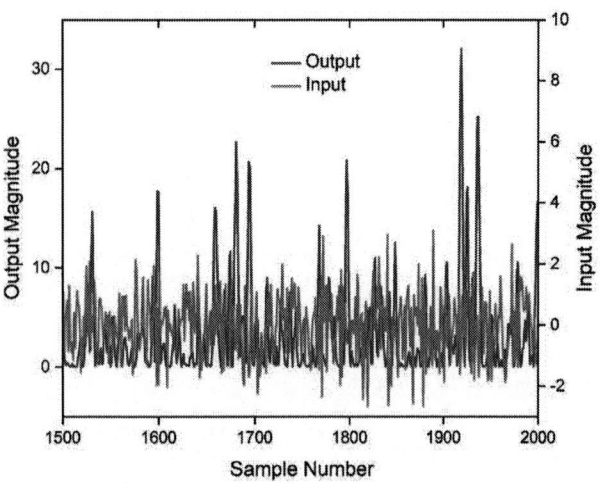

Figure 4: Training signals for the second order Wiener system example

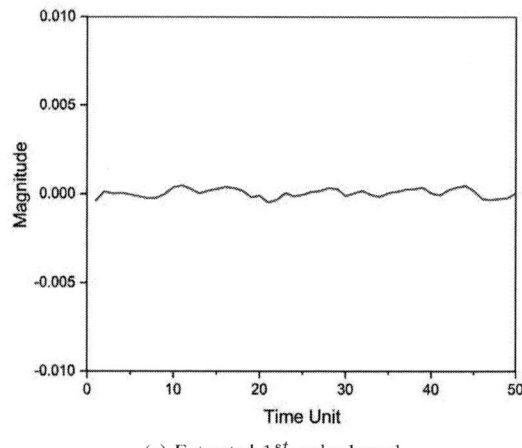

(a) Extracted 1st order kernel

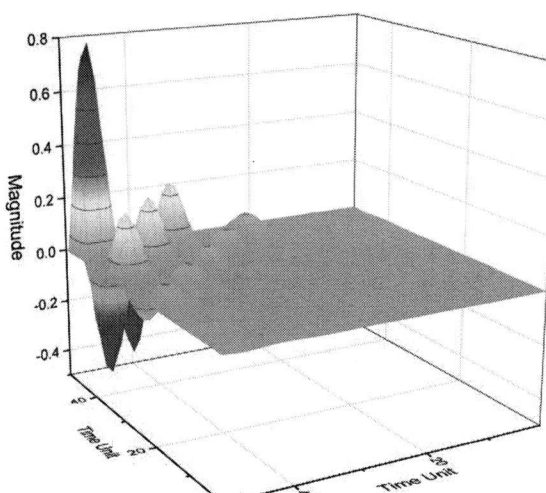

(b) Extracted 2nd order kernel

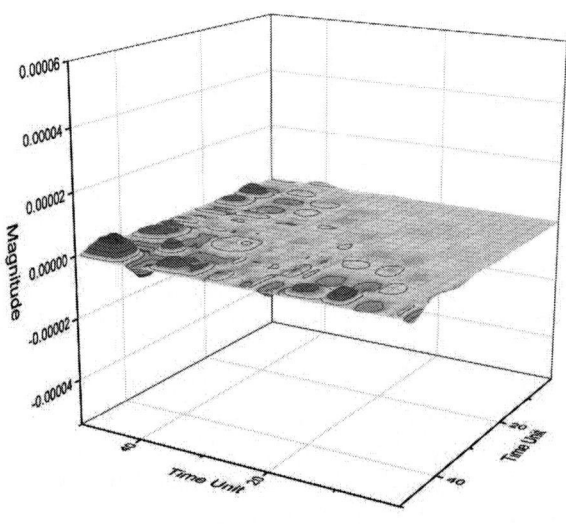

(c) Extracted 3rd order kernel (averaged over the 3rd time dimension)

Figure 5: Extracted kernels for 2nd order Wiener system example

B. PAM4 I/O buffer

The proposed method is now validated on a high-speed channel PAM4 buffer circuit transmitting data at 28Gbps rate. A typical schematic of a high-speed channel is shown in Figure 6. The input voltage to the transmitter TX is denoted as V_{T0}, input and output voltage to the channel is denoted as V_{TX} and V_{RX} respectively. V_{RO} is the output the receiver RX.

Figure 6: A high-speed channel with I/O buffer

Differentiating from the previous example, as mentioned, the training signal for this buffer circuit is a PRBS similar to that exists in the working condition of the channel. A portion of the training signals is shown in Figure 7. Due to many flat regions in the signals, especially the input signal, the memory length M has to be increased up to 150 in order for the training to converge. Thanks to the proposed method, the number of kernels remains 3, the order of chosen activation function, instead of M as in the existing method in the literature [26]–[29]. In addition, Adam [42] optimizer is found to outperform other stochastic optimization methods and gives the best convergence for the training in this example.

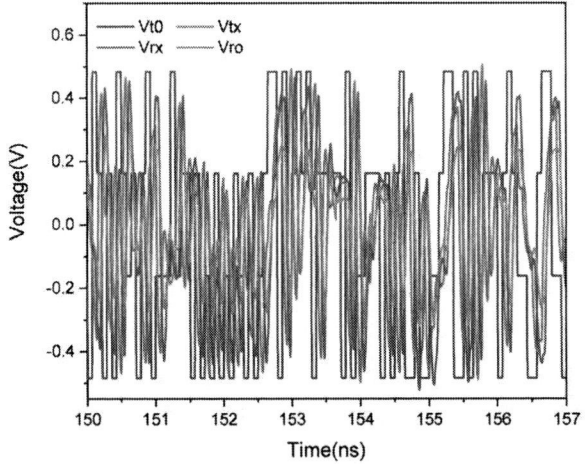

Figure 7: Training signals in PAM4 buffer example.

Extracted kernels are shown in Figure 8. Once the kernels are extracted, an unseen PRBS is used to obtain the output response of the PAM4 channel and verify it against simulation result from SPICE-based simulators. Up to 3-dimensional direct convolution was used to produce the response.

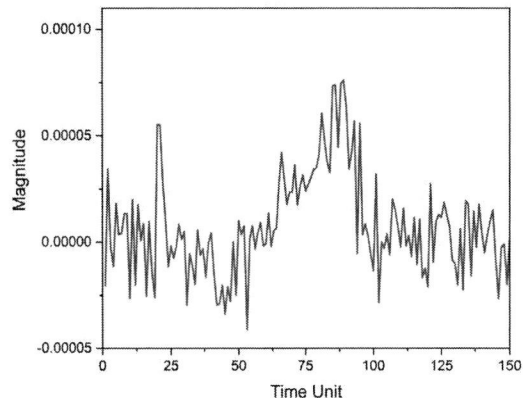

(a) Extracted 1st order kernel

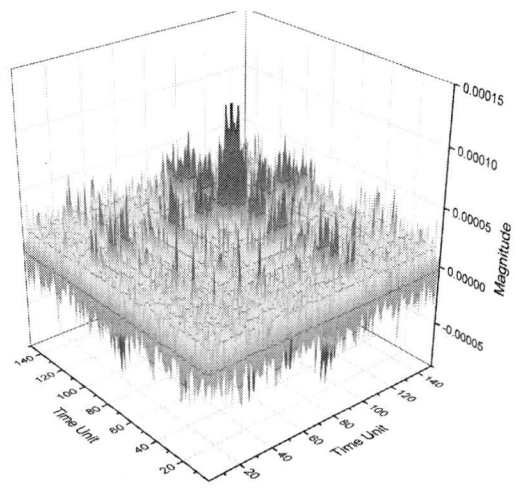

(b) Extracted 2nd order kernel

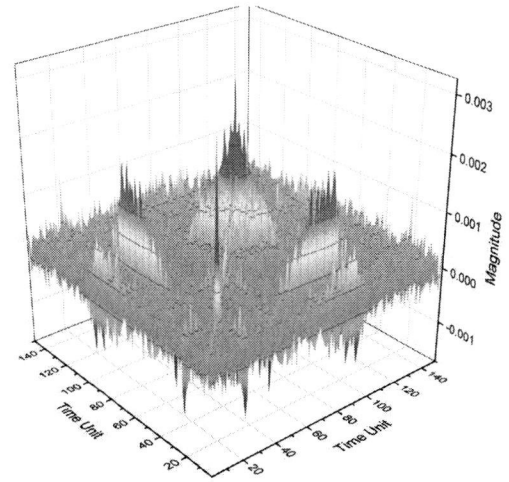

(c) Extracted 3rd order kernel (averaged over the 3rd time dimension)

Figure 8: Extracted kernels for PAM4 buffer example

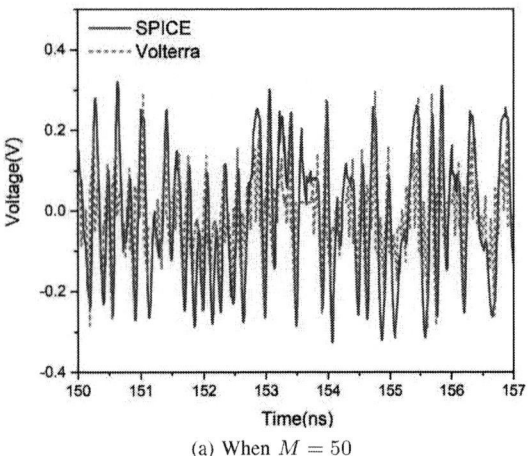

(a) When $M = 50$

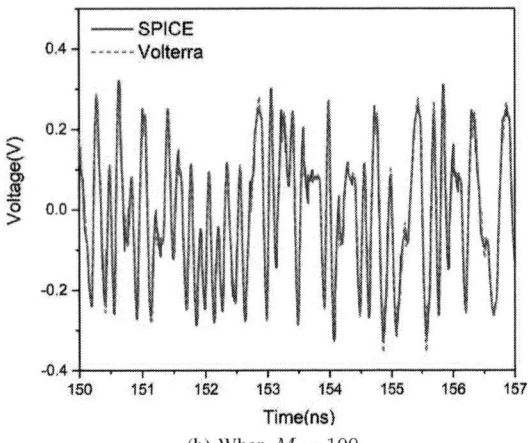

(b) When $M = 100$

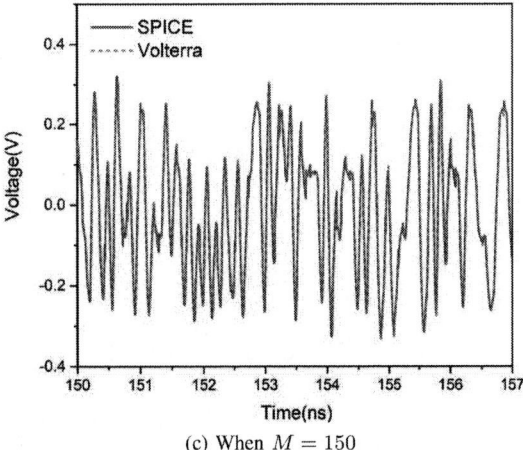

(c) When $M = 150$

Figure 9: Output response of PAM4 channel under unseen PRBS excitation using extracted VK kernels.

Different results obtained from different sets of VKs trained using various memory lengths are reported in Figure 9 to compare the effect of memory length M to the accuracy of the output response. As shown, when $M = 150$, the information contained in 3 extracted kernels is sufficient

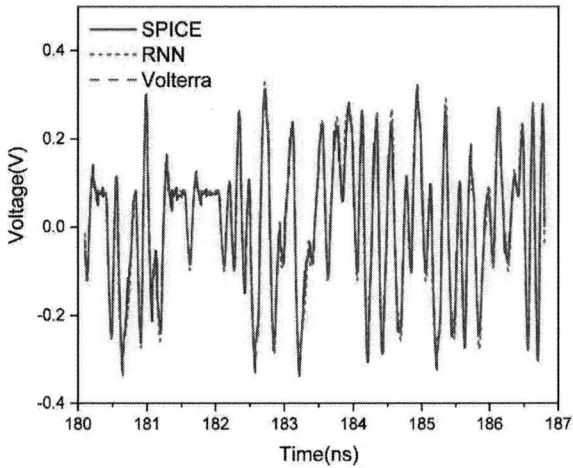

Figure 10: Output response comparison between RNN model and VK model for PAM4 example

so that the output obtained from them matches the SPICE level simulation very well.

Figure 10 shows a comparison of the output voltage after equalization in PAM4 channel obtained from an RNN model and the above VK model. They were both trained with memory length $M = 150$, the RNN has 6 layers, the hidden state size is 30. Adam with initial learning rate of 0.001 and dropout regularization of 0.3 were used to train the RNN [38].

V. CONCLUSION AND FUTURE WORK

The connection between the weights of a trained neural network and VK is established in this paper. A closed form expression is available to extract the VK from the feed-forward neural network. It has been pointed out that using tanh activation function and training the neural network with random signal as prior works would lead to unstable extraction process and high numerical cost when using the kernels for transient simulation afterward. The robustness of the proposed method was demonstrated with two different examples, one of which is a 28 Gbps PAM4 high-speed link circuit. Due to the long delay of the channel, a memory length of at least $M = 100$ was needed to obtain an acceptable accuracy using VK compared with SPICE simulation. In contrast with previous works, which will require up to the 100th-order kernel, the kernel extraction process with power series neural network could be limited the number of needed kernels to 3 only as shown above. Comparison with RNN approach was also presented. Vanilla structure of RNN is ill-conditioned, very difficult to train and cannot handle long memory effect. Modern structures such as LSTM are able to capture the long-term memory but has no closed form relationship to VK. In addition, the proposed method works well when the training signal is a PRBS that is typically used to characterize and validate high-speed channel circuits. This promises a simple integration of this method into existing design verification framework in contrast to prior works which use Gaussian noise signals.

In the future work, a different representation of VK, namely the Volterra–Laguerre model, and the extraction process of them will be studied to further reduce the number of parameters needed to represent the VK. This would reduce a substantial amount of numerical effort for use of VK models in simulation, hence speed up the simulation process.

ACKNOWLEDGMENT

This material is based upon work supported by the National Science Foundation under Grant No. CNS 16-24811 for Center for Advanced Electronics through Machine Learning (CAEML), the U.S Army Small Business Innovation Research (SBIR) Program office and the U.S. Army Research Office under Contract No.W911NF-16-C-0125 and by Zhejiang University under grant ZJU Research 083650.

REFERENCES

[1] "IBIS (I/O Buffer Information Specification)." [Online]. Available: https://ibis.org/about/

[2] Y. Ji, K. Mouthaan, and N. V. Venkatarayalu, "Evaluation of ibis modelling techniques for signal integrity simulations without and with package parasitics," in *2010 IEEE Electrical Design of Advanced Package Systems Symposium*, Dec 2010, pp. 1–4.

[3] W. J. Rugh, *Nonlinear system theory: the Volterra/Wiener approach*. Johns Hopkins University Press, 1981.

[4] S. Boyd, L. O. Chua, and C. A. Desoer, "Analytical foundations of volterra series," *IMA J Math Control Info*, vol. 1, no. 3, pp. 243–282, jan.

[5] V. Pavlenko, O. Fomin, P. Sergey, and G. Yuriy, "Identification accuracy of nonlinear system based on volterra model in frequency domain," vol. 4, 04 2013.

[6] P. Jantunen, "Modelling of nonlinear power amplifiers for wireless communications," Master's thesis, Helsinki University of Technology, 2004.

[7] J. Yang and S. X.-D. Tan, "Nonlinear transient and distortion analysis via frequency domain volterra series," *Circuits, Systems and Signal Processing*, vol. 25, no. 3, pp. 295–314, Jun 2006. [Online]. Available: https://doi.org/10.1007/s00034-004-0819-3

[8] T. Hélie, "On the use of volterra series for real-time simulations of weakly nonlinear analog audio devices: application to the moog ladder filter," in *in Proceedings of the 9th International Conference on Digital Audio Effects (DAFx '06)*, 2006, pp. 7–12.

[9] J. Leonard and C. S. Edrington, "Modeling and simulation of shipboard nonlinear dynamic loads using volterra kernels," in *2013 IEEE Electric Ship Technologies Symposium (ESTS)*, April 2013, pp. 249–255.

[10] X. Y. Z. Xiong, L. J. Jiang, J. E. Schutt-Aine, and W. C. Chew, "Volterra series-based time-domain macromodeling of nonlinear circuits," *IEEE Transactions on Components, Packaging and Manufacturing Technology*, vol. 7, no. 1, pp. 39–49, Jan 2017.

[11] P. M. Lavrador, J. C. Pedro, and N. B. Carvalho, "A new volterra series based orthogonal behavioral model for power amplifiers," in *2005 Asia-Pacific Microwave Conference Proceedings*, vol. 1, Dec 2005, pp. 4 pp.–.

[12] J. Dooley, B. O. Brien, and T. J. Brazil, "Behavioural modelling of rf power amplifiers using modified volterra series in the time domain," in *High Frequency Postgraduate Student Colloquium*, Sep. 2004, pp. 169–174.

[13] C. Crespo-Cadenas, J. Reina-Tosina, and M. J. Madero-Ayora, "Volterra behavioral model for wideband rf amplifiers," in *IEEE Transactions on Microwave Theory and Techniques*, vol. 5, no. 3, 2007, pp. 449–457.

[14] W. A. Gardner and T. L. Archer, "Simplified methods for identifying the volterra kernels of nonlinear systems," in *[1991] Proceedings of the 34th Midwest Symposium on Circuits and Systems*, May 1991, pp. 98–101 vol.1.

[15] T. Nguyen, J. E. Schutt-Aine, and Y. Chen, "Volterra kernels extraction from frequency-domain data for weakly non-linear circuit time-domain simulation," in *2017 IEEE Radio and Antenna Days of the Indian Ocean (RADIO)*, Sept 2017, pp. 1–2.

[16] C. Xin, "Radio frequency circuits for wireless receiver frontends," Ph.D. dissertation, Texas A&M University, 2004.

[17] P. Wambacq, G. Gielen, P. Kinget, and W. Sansena, "High-frequency distortion analysis of analog integrated circuits," in *IEEE Trans. Circuits Syst. II: Analog Digit.Signal Process*, vol. 46, no. 3, 1999, pp. 335–345.

[18] Y. Hu and K. Muyaram, "A modified-volterra-seriestechnique for improving the accuracy of quasistatic harmonic balance analysis in coupled device and circuit simulation," in *IEEE 2004 CUSTOM INTEGRATED CIRCUITS CONFERENCE*, Apr. 2004, pp. 125–128.

[19] L. Liu, L. Li, Y. Huang, K. Cui, Q. Xiong, F. N. Hauske, C. Xie, and Y. Cai, "Intrachannel nonlinearity compensation inverse volterra series transfer function," *Journal of Lightwave Technology*, vol. 30, no. 3, pp. 310–316, Feb. 2012.

[20] G. Bicken, G. F. Carey, and R. O. Stearman, "Frequency domain kernel estimation for 2nd-order volterra models using random multitone excitation," *Vlsi Design - VLSI DES*, vol. 15, pp. 701–713, 12 2002.

[21] L. Li and S. Billings, "Volterra series truncation and reduction in the frequency domain for weakly nonlinear system," University of Sheffield, Tech. Rep., 2006.

[22] X. Li, P. Li, Y. Xu, R. Dimaggio, and L. Pileggi, "A frequency separation macromodel for system-level simulation of rf circuits," in *Proceedings of the ASP-DAC Asia and South Pacific Design Automation Conference, 2003.*, Jan 2003, pp. 891–896.

[23] A. Soury and E. Ngoya, "A two-kernel nonlinear impulse response model for handling long term memory effects in rf and microwave solid state circuits." in *2006 IEEE MTT-S International Microwave Symposium Digest*, June 2006, pp. 1105–1108.

[24] M. Youssef, E. Chong, and K. Phang, "Distortion analysis using signal flow graphs and volterra series," in *2003 46th Midwest Symposium on Circuits and Systems*, vol. 1, Dec 2003, pp. 84–89 Vol. 1.

[25] P. J. Lawrence, "Estimation of the volterra functional series of a nonlinear system using frequency-response data," *IEE Proceedings D - Control Theory and Applications*, vol. 128, no. 5, pp. 206–210, Sep. 1981.

[26] G. Stegmayer, "Volterra series and neural networks to model an electronic device nonlinear behavior," in *IEEE International Joint Conference on Neural Networks*, Jul. 2004, pp. 2907–2910.

[27] ——, "Neural networks and volterra series for modeling new wireless communication devices," in *International Joint Conference on Neural Networks*, Aug. 2007.

[28] J. Wray and G. G. R. Green, "Calculation of the volterra kernels of non-linear dynamic systems using an artificial neural network," *Biological Cybernetics*, vol. 71, no. 3, pp. 187–195, Jul. 1994.

[29] J. Misic, V. Markovic, and Z. Marinkovic, "Volterra kernels extraction from neural networks for amplifier behavioral modeling," in *International Symposium on Telecommunications*, Oct. 2014.

[30] Y. LeCun, Y. Bengio, and G. Hinton, "Deep learning," *Nature*, vol. 521, pp. 436–444, May 2015, doi:10.1038/nature14539.

[31] N. Wiener, *Nonlinear problems in random theory.* MIT Press, Cambridge, MA, 1958.

[32] M. Schetzen, *The Volterra and Wiener Theories of Nonlinear Systems.* Melbourne, FL, USA: Krieger Publishing Co., Inc., 2006.

[33] X. Jing and Z. Lang, *The Generalized Frequency Response Functions and Output Spectrum of Nonlinear Systems*, 12 2015, vol. 119, ch. 2, pp. 9–30.

[34] L. Li and S. Billings, "Estimation of generalized frequency response functions for quadratically and cubically nonlinear systems," *Journal of Sound and Vibration*, vol. 330, no. 3, pp. 461 – 470, 2011.

[35] X. Wang, T. Nguyen, and J. E. Schutt-Aine, "Volterra Kernel Extraction Through Power Series Feed Forward Neural Network for Behavior Modeling of High Speed I/O Buffer," 2019, (Accepted).

[36] D. Marquardt, "An algorithm for least-squares estimation of nonlinear parameters," *Journal of the Society for Industrial and Applied Mathematics*, vol. 11, no. 2, pp. 431–441, 1963. [Online]. Available: https://doi.org/10.1137/0111030

[37] T. Nguyen, T. Lu, J. Sun, Q. Le, K. We, and J. Schut-Aine, "Transient Simulation for High-Speed Channels with Recurrent Neural Network," in *2018 IEEE 27th Conference on Electrical Performance of Electronic Packaging and Systems (EPEPS)*, Oct 2018, pp. 303–305.

[38] T. Nguyen, T. Lu, K. Wu, and J. Schutt-Aine, "Fast Transient simulation of High-Speed Channels using Recurrent Neural Network," 2019. [Online]. Available: https://arxiv.org/abs/1902.02627

[39] J. L. Elman, "Finding structure in time," *COGNITIVE SCIENCE*, vol. 14, no. 2, pp. 179–211, 1990.

[40] R. Pascanu, T. Mikolov, and Y. Bengio, "On the Difficulty of Training Recurrent Neural Networks," in *Proceedings of the 30th International Conference on International Conference on Machine Learning - Volume 28*, ser. ICML'13. JMLR.org, 2013, pp. III–1310–III–1318. [Online]. Available: https://arxiv.org/abs/1211.5063

[41] S. Hochreiter and J. Schmidhuber, "Long Short-Term Memory," *Neural Comput.*, vol. 9, no. 8, pp. 1735–1780, Nov. 1997. [Online]. Available: http://dx.doi.org/10.1162/neco.1997.9.8.1735

[42] D. P. Kingma and J. Ba, "Adam: A method for stochastic optimization," *CoRR*, vol. abs/1412.6980, 2014. [Online]. Available: http://arxiv.org/abs/1412.6980

224G Package Interconnect Design Study

Based on Artificial Neural Network Modeling Approach

Hui Liu, Qian Ding, Penglin Liu
PSG of the Data Center Group
Intel Coporation
San Jose, CA 95134, USA
hui1.liu@intel.com

Abstract — a new modeling methodology for package interconnect study based on abstraction from physical model to circuit model to Artificial Neural Network (ANN) is proposed here. The mapping between circuit model and physical model is done through S-parameter correlation and the training data for ANN is generated through sweeping variables of the circuit model. The trained network is used to explore design space of package interconnect and to predict its electrical performance. The methodology is applied to 224G PAM4 package interconnect design study. 15 variables are defined for the circuit model and 74K data points are generated for training the neural network. The trained network gives good predictions for the differential insertion loss of package interconnect with a testing Mean Squared Error (MSE) of 0.004 at 56GHz and prediction results can be generated instantly for any variable combinations defined in the circuit model. The proposed modeling methodology has demonstrated effectiveness and good accuracy.

Keywords— signal integrity, interconnect, Artificial Neural Network (ANN), S-parameter, 224G PAM4

I. INTRODUCTION

Artificial Neural Network (ANN) has been used in signal integrity and power integrity (SIPI) modeling and analysis in recent years and is gaining popularity [1] [2] [3] [4]. The major advantages of ANN modeling approach are the wide design space coverage and almost instant result generation. The main challenge is to generate a big enough training data set so that the behavior model of the trained neural network is meaningful with acceptable accuracy. For traditional 3D full-wave electro-magnetic simulation-based data generation in high-speed interconnect design, it is not practical to generate hundreds of thousands of data points needed for neural network training. Furthermore, due to the ever-increasing interconnect speed from 32G, to 56G, to future 112G/224G [5], 3D full wave simulation at extreme high bandwidth becomes even more challenging. Due to these constraints, most ANN based interconnect studies can only focus on micro structures, such as a pair of differential VIA, a few connector pins, with a small set of variables in the network modelling.

To address the previous challenges, a new modeling methodology based on circuit model to physical interconnect mapping and circuit model to artificial neural network abstraction is proposed here. The methodology is applied to a 224G PAM4 package interconnect design study covering 15 variables of multiple values to show its effectiveness and accuracy.

Section II of this paper gives an overview of the proposed modeling methodology. Sections III to VI covers the application of the methodology in a 224 Gbps package interconnect design study. Section VII summarizes the methodology development engineering effort. Section VIII is the conclusion.

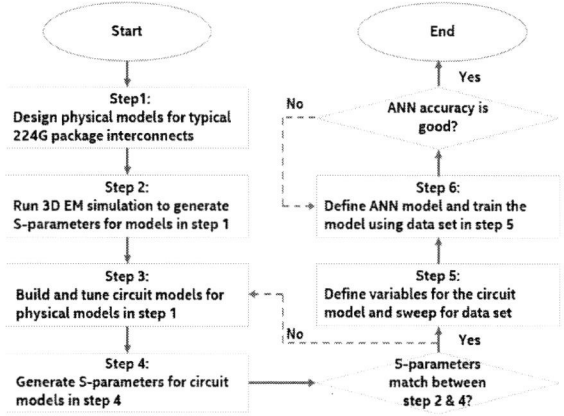

Fig. 1. Modeling methodology flow

II. MODELING METHODOLOGY

Figure 1 shows the modeling methodology flow. The first step is to design a physical interconnect in package. The design should be representative for its typical application. In this study, three physical interconnects of different structures are designed for better model correlation between steps 2 and 4. The second step is to run 3D full-wave EM simulation for the whole interconnect from die bump to package ball to generate S-parameters. This step is time consuming in general but is only one-time effort. The third step is to build a circuit model based on the physical nature of the interconnect followed by S-parameter generation in a circuit simulator in the step 4 of the flow. Then the circuit model is tuned to match its S-parameters to those in step 2. This is an iterative process and but is only one-time effort as well. Once the circuit model is correlated to the physical interconnect, physical parameters, material properties, and electrical parameters, defined as variables, can be swept (step 5) to generate S-parameters at frequencies of interest (for example at 56GHz for PAM4 application) in reasonable amount of time.

Step 6 defines a neural network, including the number of network layers and the number of neurons in each layer. Then the data generated in step 5 is used to train the network. Different training algorithms can be used for searching the best modeling accuracy. This is an iterative process but is one-time effort as well. Once the network model is trained to acceptable accuracy it can be used to instantly predict the performance of an interconnect for arbitrary combinations of the variables defined in step 5.

Implementation details of the methodology in a 224 Gbps package interconnect design study will be discussed in the following sections.

978-1-7281-1500-9/19 $31.00 © 2019 IEEE

III. PHYSICAL MODEL

To ensure the neural network model is representative and accurate, the physical baseline interconnect must be carefully designed for its application. Three trial interconnects of different structures are designed and figure 2 shows their insertion loss performance. The red and purple interconnects have good performance up to 60GHz. The blue interconnect, a less ideal design, has good performance up to 50GHz but with very significant loss drop above 50GHz due to some impedance mismatch in a micro structure. This design cannot be used as the baseline physical model.

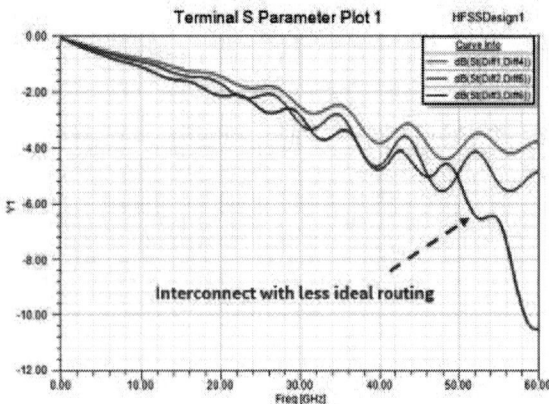

Fig. 2.   Insertion losses of trial interconnects

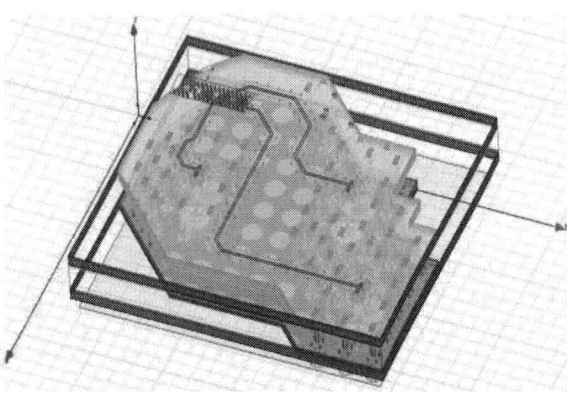

Fig. 3.   Physical design of the three interconnects on package

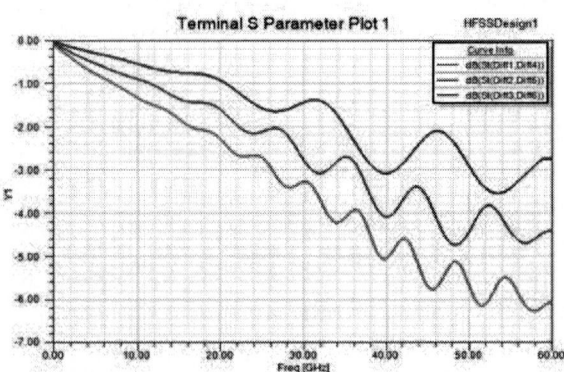

Fig. 4.   Insertion losses of the phycial models for interconnects

Three new well-defined interconnects of routing length about 5mm, 9mm, and 13mm respectively are designed for the study. Figure 3 shows the 3D physical design and figure 4 shows insertion losses of the interconnects simulated in HFSS. This step of building physical models and running 3D full-wave EM simulation for them takes a few weeks. It is slow but is one-time effort in the process.

One thing learned in building the physical interconnect is that every single detail of the whole interconnect needs to be carefully designed for such a high-speed application. That means there are many variables to tune for acceptable performance. The traditional design trial and simulation iteration method is not effective for exploring large solution space in early pathfinding for large and complex designs. A more effective modeling methodology such as the one proposed in this paper is needed. Final design verification still needs to be done through 3D full-wave EM simulation based on physical artwork.

IV. CIRCUITE MODEL

The next step in the process is to build a circuit model representing the physical model discussed in section III. A sample circuit model is shown in figure 5. Vertical structures (C4 bump, package solder ball, uVIA, plated through hole (PTH), etc.) are modelled as short transmission lines (T-lines) and horizontal traces are modelled as long T-lines. The circuit model can be relatively simple or can be very complicated depending on accuracy need and the number of variables for physical and electrical parameters that will be chosen for the neural network in step 6.

Then the circuit model is simulated and turned so that its S-parameters match those of the physical model in section III. Theoretically only one physical model and one circuit model is needed for the process. However, to ensure the methodology have the best accuracy three physical models of different length and structure, and three circuit models are designed for correlation. Figure 6 shows the insertion losses of the circuit models generated in ANSYS circuit simulator. Comparing the two plots in figure 5 and figure 6 one can see correlations between the circuit models and physical models are fairly good for all three interconnects in terms to trend and value up to 60GHz. One noticeable difference is the oscillation amplitude that deserves more investigation. One possible source of this discrepancy is anti-pad modeling.

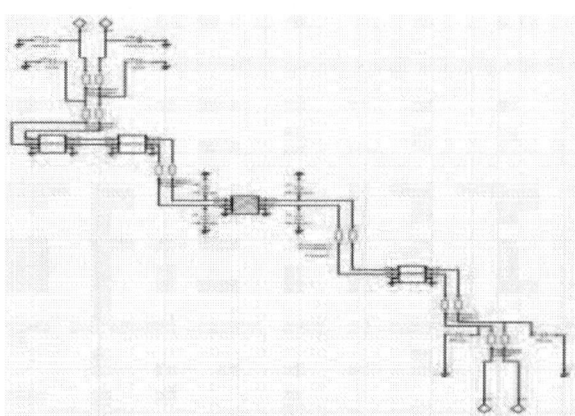

Fig. 5.   Circuite model representing the phycial model

978-1-7281-1500-9/19 $31.00 © 2019 IEEE

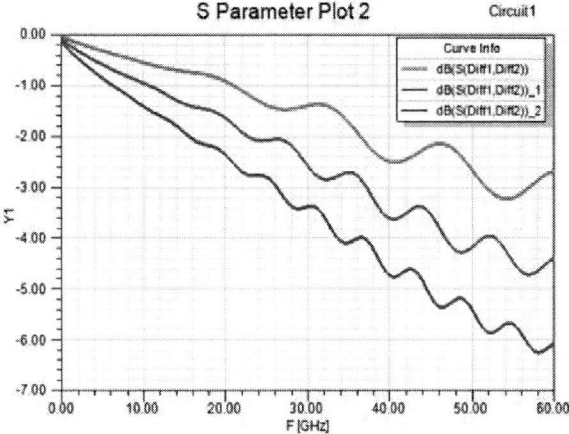

Fig. 6. Insertion losses of the circuit models for interconnects

one output node in the output layer, defined as the differential insertion loss of the interconnect. The numbers of hidden layers and hidden neurons are set and tuned through the model training process.

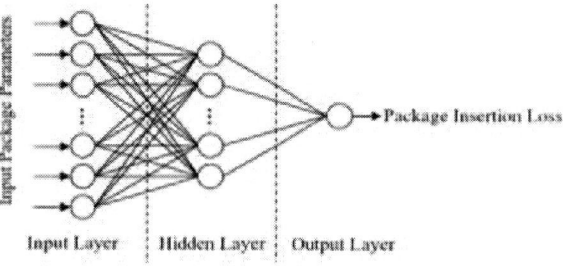

Fig. 8. Three-layer neural network

V. MODELING VARIABLE

A baseline circuit model with medium routing length is defined after the good correlation with the physical model in the step 3 and 4 iteration. One major challenge now is to define the number of variables that need to be swept for training date for the neural network. This depends on what interconnect parameters, including both physical parameters and electrical parameters one likes to investigate from design space coverage point of view, and on modeling time and accuracy. In this study, 15 variables are defined as shown in figure 7. Some variable has two values (min and max) and some variables have three values (min, nominal, and max), such as for package solder ball. The combinations of these 15 variables give about 74K data points for differential insertion loss at a single frequency point, the Nyquist frequency of 56GHz for 224G PAM4 for this study.

Figure 9 shows more details of the hidden neuron and the output node. Each neuron has a weight factor (w) for each of its input and associated bias factor (b). Each neuron has multiple inputs. In this study, one hidden layer is chosen based on trials. The number of neurons in the hidden layer is tuned based on performance monitoring. The performance function is Mean Squared Error (MSE) that is the average squared difference between outputs and target.

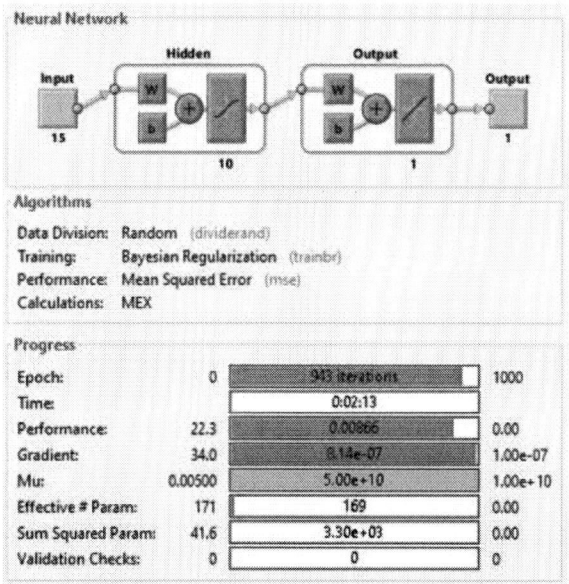

Fig. 9. Bayesian Regularization method for 10 neurons

Structure	Die Pad	C4 Bump	uVIA1	
Parameter	Capacitance	Impedance	Impedance	Length
Values	min, max	min, max	min, max	min, max

Structure	Diff T-line1			uVIA2
Parameter	Length	Width	Gap btwn P&N	Impedance
Values	L1, L2, L3	W1, W2	min, max	min, max

Structure	uVIA2	PTH		uVIA3
Parameter	Length	Radius	Length	Impedance
Values	min, max	min, max	min, max	min, max

Structure	uVIA4	Package Solder Ball		
Parameter	Impedance	Impedance	Height	
Values	min, max	min, max	H1, H2, H3	

Fig. 7. 15 variables of the circuit model

VI. ANN MODEL

The next step of the process is to define the structure of the neural network model, including the number of network layers and the number of neurons in each layer. Figure 8 shows the network used for this study. It has three layers, one input layer, one hidden layer, and one output layer. There are 15 input nodes in the input layer, corresponding to the 15 variables defined for the circuit model in section V. There is

	🍥 Samples	🖼 MSE	☑ R
● Training:	51610	8.66040e-3	9.97351e-1
● Validation:	11059	0.00000e-0	0.00000e-0
● Testing:	11059	8.57795e-3	9.97390e-1

Fig. 10. Training results based on Bayesian Regularization method

978-1-7281-1500-9/19 $31.00 © 2019 IEEE

The data set generated in the circuit model simulation are used as samples for training the network. The samples are randomly divided into three groups (training, validation, and testing groups) for different model development purpose as shown in figure 11.

Sample type	Sample count	Sample usage
Training	51610	Used to train the network (tuning the w and b factors of neuron in figure 9)
Validation	11059	Used to validate the trained network
Testing	11059	Used to test the trained network

Fig. 11. Training samples and their usage

Different training algorithms can be used for searching for the best modeling accuracy in reasonable amount of time. Readers can refer to these articles [6] to [11] for more details of different training algorithms. Figures 9 and 10 show the training results based on Bayesian Regularization method for ten-neuron case. Figures 12 and 13 show the training results based on Levenberg-Marquardt method. There is no significant performance difference between these two training methods. The first one has MSE of 0.00866 while the second one has MSE of 0.00863. Figures 10 and 12 also show the testing results and again, there is no significant difference between the two methods. However, the first on has 943 iterations vs. the 288 iterations of the second one.

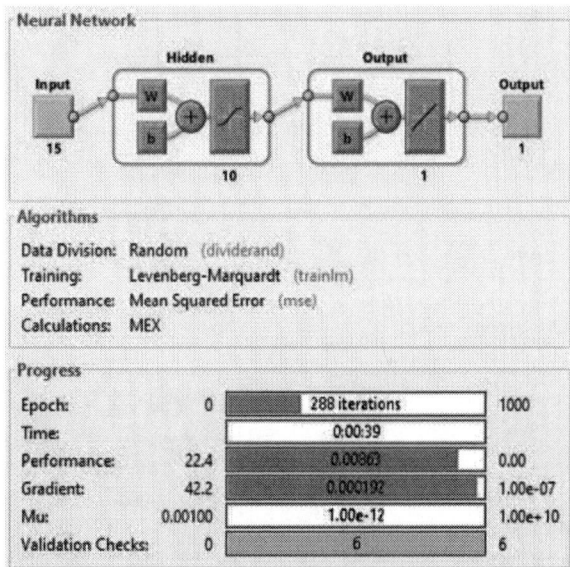

Fig. 12. Levenberg-Marquardt method for Performance of 10 neurons

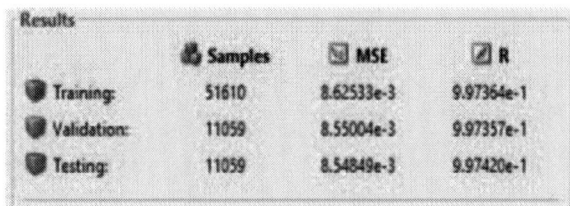

Fig. 13. Training results based on Levenberg-Marquardt method

Figures 14 and 15 show the training results for 15 hidden neurons based on Levenberg-Marquardt method. The MSE is very good at 0.004 and regression results are all better than 0.99 for training, validation, and testing. Figure 16 shows the error histogram for training, validation, and testing results using 20 bins. The statistical data seem to follow a Gaussian distribution. Network with 12 hidden neurons has been trained as well. The performance result is between 10-neuron and 15-neuron cases. It seems for such a large data set more neurons give better results. The correlation between the number of neurons and the size of the training data set is worth more investigation.

For the final network model of one hidden layer with 15 neurons, MSE of less than 0.004 and tight error distribution has demonstrated the good accuracy of this neural network based modeling methodology.

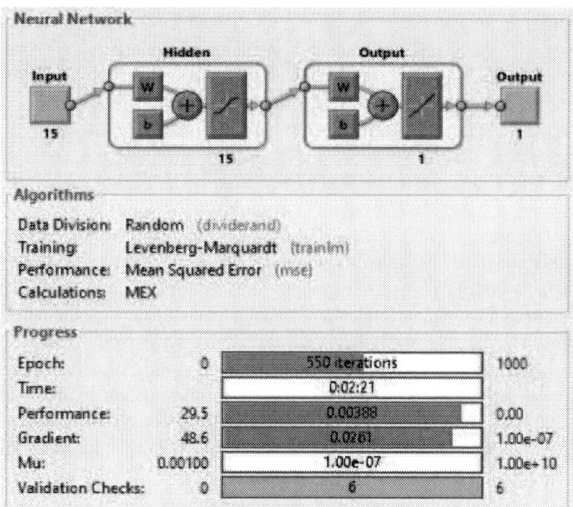

Fig. 14. Levenberg-Marquardt method for Performance of 15 neurons

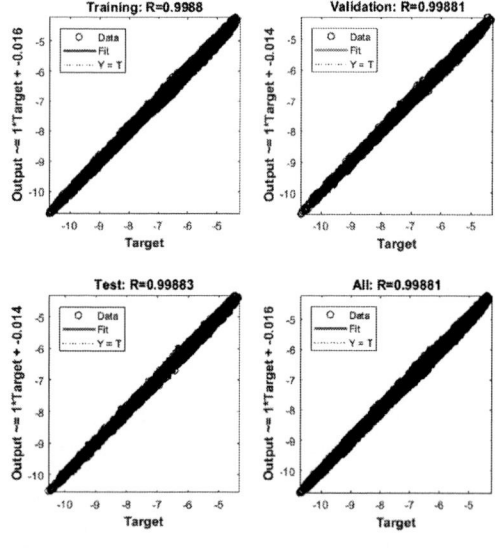

Fig. 15. Regression results based on Levenberg-Marquardt method

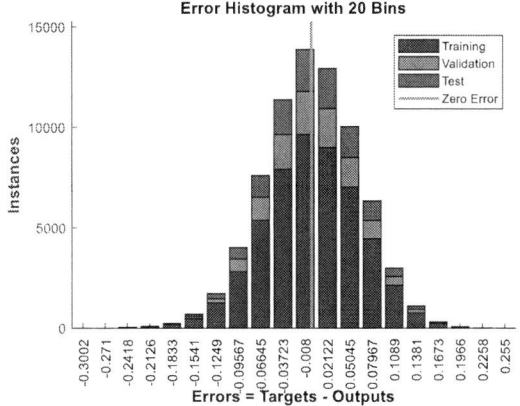

Fig. 16. Error histogram of the 15-neuron model

VII. MODELING EFFICIENCY

The table in figure 17 shows the rough engineering effort (including computing time) for each step in the modeling flow in figure 1. In the real development of this modeling methodology, there are iterations between steps 1 and 2 and the total effort is about 5 weeks. Steps 3 and 4 have multiple iterations as well. Using the trained neural network to predict outcome for different input parameter combination is instant. This table shows the importance of using behavior model (ANN in this case) in exploring large design space with many variables. It is not practical to trial many combinations of many variables with multiple values (such as the case in this study for 224G package interconnect design) through physical modeling and 3D full-wave EM simulation even though this approach is more accurate and should be used for final design verification.

Step	Task	Effort
1	Design physical model	3 weeks
2	Run 3D full-wave EM simulation	2 weeks
3	Build and tune circuit model	1 weeks
4	Generate S-parameters for circuit model	5 minutes
5	Sweep variables to generate data set for ANN	2 day
6	Train ANN (including sample preparation)	1 day
7	Predict performance	Instantly

Fig. 17. Engineering effort for each step in the methodology flow

VIII. CONCLUSING

New and emerging applications is pushing transceiver signaling speed into the range of 56G to 224G, with much higher aggregated bandwidth. For on package interconnect design at these speeds, it is necessary to have an effective modeling method for exploring target solution space at the early path finding stage. An artificial neural network (ANN) based modeling methodology is presented in this paper for that purpose. The methodology includes a few major steps: first, map the physical model of an interconnect into circuit model; second, sweep selected physical and electrical parameters of the circuit model to generate S-parameter data

set; third, use those parameters as inputs and the S-parameter data set as target to train the ANN to acceptable accuracy level; then, then the trained network can be used to predict interconnect performance in S-parameter for arbitrary input parameter combinations.

The methodology is applied to a package 224G PAM4 interconnect. 15 parameters are defined as inputs, including die pad capacitance, uVIA and package ball impedance and length, transmission line and PTH geometries. Differential insertion loss of the interconnect is defined as the output of the neural network. The network, trained through 74K data points, has a testing MSE of 0.004 for interconnect differential insertion loss prediction at 56GHz (the Nyquist frequency of 224G PAM4 application). The proposed modeling methodology has demonstrated effectiveness and good accuracy. The same methodology can be used for other similar package design study as well.

For future work, some modeling discrepancies found in this study deserve more investigation. Better circuit model to physical model correlation and verifying neural network prediction based on independent physical package models are in consideration.

ACKNOWLEDGMENT

The authors would like to thank our colleagues for their helpful discussion.

REFERENCES

[1] N. Ambasana, G. Anand, B. Mutnury, and D. Gope, "Eye Height/Width Prediction From S-Parameters Using Learning-Based Models," IEEE Transactions on Components, Packaging and Manufacturing Technology, vol. 6, pp. 873-885, 2016

[2] W. T. Beyene, "Application of Artificial Neural Networks to Statistical Analysis and Nonlinear Modeling of High-Speed Interconnect Systems", IEEE Transactions on Computer-Aided Design of Integrated Circuits and Systems, vol. 26, pp. 166-176, 2007

[3] T. Lu, J. Sun, K. Wu, and Z. Yang, "High-Speed Channel Modeling With Machine Learning Methods for Signal Integrity Analysis", IEEE Transactions on Electromagnetic Compatibility, vol. PP, pp. 1-8, 2018

[4] Jun Xu, Ling Zhang, Mike Sapozhnikov, Jun Fan, "Application of Deep Learning for High-speed Differential Via TDR Impedance Fast Prediction", IEEE EMC-SIPI conference, Long Beach, 2018

[5] http://www.ieee802.org/3/ad_hoc/ngrates/public/17_05/goergen_nea_01_0517.pdf

[6] A. J. Smola and B. Scholkopf, "A tutorial on support vector regression," "Statist. Comput., vol. 14, no. 3, pp. 199–222, 2004.

[7] V. Vapnik, The Nature of Statistical Learning Theory. New York, NY, USA, Springer, 2013.

[8] H. Borchani, G. Varando, C. Bielza, and P. Larranaga, "A survey on multi-output regression," Wiley Interdiscip. Rev., Data Mining Knowl. Discov., vol. 5, no. 5, pp. 216–233, 2015.

[9] J. Schmidhuber, "Deep learning in neural networks: An overview," Neural Netw., vol. 61, pp. 85–117, 2015.

[10] Zao Liu, "Introduction to Machine Learning and its Applications in High Speed Serial Link Design and Analysis", DesignCon 2018

Enhanced Reliability of a RF-SiP with Mold Encapsulation and EMI Shielding

Chan-Yuan Liu[1], Jason Chien[1], Yu-Chou Tseng[1], Kuo-Hsien Liao[1], Alex Chan[1], Dao-Long Chen[1], Meng-Kai Shih[1], and Mark Gerber[2]

[1]Advanced Semiconductor Engineering, Inc., Taiwan
[2] Advanced Semiconductor Engineering, Inc., U.S.
ChanYuan_Liu@aseglobal.com

Abstract

This paper will discuss the details on how the molding of a SiP (System In Package), with a coated shielding layer, were shown to have an improved reliability on a functional SiP module. The EMI shielding structure was created by using a conformal metal film over the molded area and is subsequently contacted to the substrate ground inner layers on the package edge. The study includes process integration details, reliability test data, as well as the RF performance of the SiP module with EMI package shielding; with integration of high density surface-mount technology (SMT), Cu wire bond, mold encapsulation, conformal shielding and bottom side pre-solder processes. It was shown by molding this type of package structure in combination with an EMI shielding structure, a reduced moisture absorption rate can be achieved when comparing a non-EMI shielding structure. Temperature humidity test, with 85°C/ 85% RH environment, was carried out to benchmark the effect of moisture sensitivity of a module. In addition, mechanical testes such as shock, vibration, and drop tests were performed to compare the molded module as to a metal lid type module. Thermal performance differences between modules with and without EMI shielding was also investigated to understand performance capabilities at higher operating temp. The thermal simulation of the molded type SiP module with EMI shielding result was equivalent to thermal measurement result using a baseline at 85°C ambient temperature. Based on the study, it was found molded SiP module, with EMI shielding, showed an enhanced thermal rate that was 6.63% higher than the metal lid type module and 1.23% higher than the molded only encapsulated SiP package individually.

Keywords - SiP module, mold encapsulation, EMI shielding, enhanced reliability, stress test, thermal enhancment.

I. INTRODUCTION

SiP technologies have been expanding applications among the consumer, industrial, and automotive applications. Smaller form factor modules that can handle power and thermal challenges have been an important requirement for product integration and development.

Based on earlier fabrication results, the molded type SiP module with coated EMI shielding is easier to achieve size miniaturization, enhanced reliability, and thermal enhancement than the metal lid type SiP module. Metal lid type package requires underfill underneath the IC and this drives a certain enlarged dispensing space inside the module area. Also considering the design rules that are required for a a solder mounted metal lid, the metal lid module package size will be 25 ~ 50% larger than that of a molded type package. It was also found that molded package could sustain a higher level mechanical reliability tests (i.e.: shock, vibration, and free fall.), waterproof and dustproof test criteria. Meanwhile, the molded package with coated metal film could improve the EMI shielding performance and improve the thermal management as well. The JEDEC thermal measurement spec. was followed to design the printed circuit board (PCB), and the thermal test condition is shown in Table I.

TABLE I.   THERMAL MEASUREMENT AND SIMULATION CONDITION

PCB Substrate	Ta (℃)	Power Condition
JEDEC 2S2P PCB	25	5G@50% (0.73W)
		5G@95% (1.0W)

As per the fabrication results, molded SiP with EMI shielding is a good candidate for an inexpensive module which could integrates perfect volume utilization, and no need metal lid, nor underfill and related expensive processes.

TABLE II.   ENHANCED RELIABILITY TEST CONDITION

#	Item	Reliability test		SS	Function test		
		Condition			-40℃	25℃	85℃
A1-1	Pre-con.	MSL3: 30℃/60%RH 192 hrs & 2 times reflow				V	
A1-2		MSL3: 30℃/60%RH 192 hrs & 3 times reflow					
A2	THB	After A1-1 85℃/85%RH	500, 1000 hrs	25		V(1)	V(2)
A3	TH	After A1-2 85℃/85%RH	500, 1000 hrs	25		V	
A4	TC	After A1-2 -55℃ to 125℃	500, 1000 cycles	25			V
A6	HTSL	150℃	500, 1000 hrs	25		V(1)	V(2)
B1	HTOL	85℃	500, 1000 hrs	25	V(2)	V(1)	V(3)
G1	Shock	3 drops/face , 6 faces , total 18 drops ; duration : 0.25 msec ; 10000G peak acceleration.		15		V	
G2	Vibration	20 Hz to 2 KHz to 20 Hz in > 4 minutes , 4X in each orientation, 50G peak acceleration.		15		V	
G5	Free Fall	Drop part on each of 6 axes once from a height of 1.2m onto a concrete surface.		5		V	

The structure of a RF-SiP was shown in Figure 1. It integrated the SMT, Cu wire bonding, molding, conformal shielding coating and bottom side pre-solder processes were used to complete the module, and were used to verify the enhanced reliability criteria which was based on AEC [1] & industrial JEDEC [2] specs. in Table II.

In addition to room temperature functional test, high and low temperature functional tests were also carried out when at each reliability test read points to data comparison, in which max. operation temperature 85℃ was used.

The SiP module was mounted on a ref. board, as shown in Figure 2, and powered to carry out the temperature and humidity bias test (THB) verification. In additional, the module was mounted on board to do the high temperature operating life test (HTOL). The test boards were compatible for both the THB and the HTOL tests. Before THB and HTOL tests, 2 times reflow pre-cond. (shown as A1-1 in table II) was applied to the HTOL and HTB samples, considering there will be additional 1 more reflow will be when module mount on board.

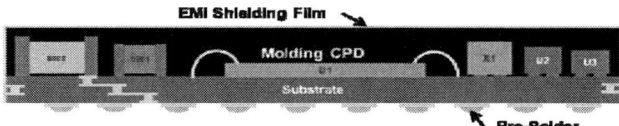

Figure 1. Molding type SiP module with EMI shielding.

Figure 2. Molding type SiP module with EMI shielding mount on function test board

Figure 3. Molded SiP modules with EMI shielding mount on JEDEC board

Figure 3 showed the molded modules were mounted onto the JEDEC board for the shock [2] and vibration [7] tests.

II. FUNCTIONAL TEST AT HIGH/ ROOM/ LOW TEMP.

At time zero functional test, power test result show that power at high temperature is pretty consistent to that in room

temperature. While the power test results in lower temperature is around 1~2dBm higher than that at room temperature, the results were shown in Figure 4.

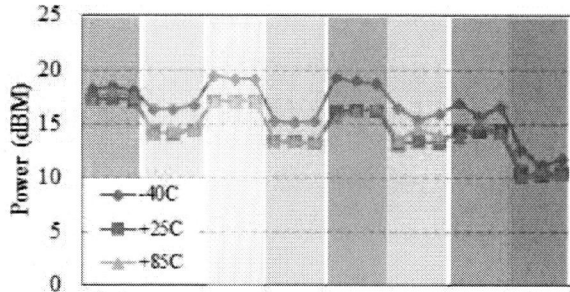

Figure 4. Power test result at 8 type transmitter modes at 3 temp. levels

The EVM test results were concluded in Figure 5. It was found the EVM does not change between different temperature levels.

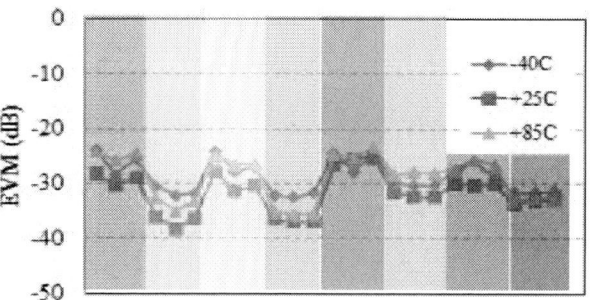

Figure 5. EVM test result at 8 type transmitter modes at 3 temp. levels

Furthermore, receiver transferability test results shown in Figure 6, also conclude the module performs consistent receiver performance at different temperature levels.

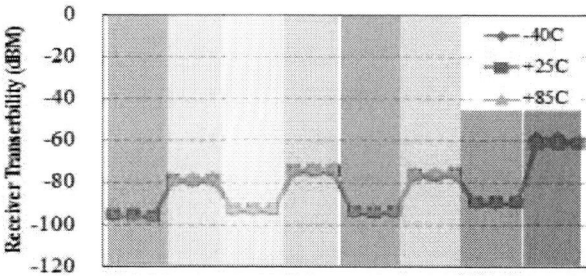

Figure 6. Test result at 8 type receiver transerbility modes at 3 temp. levels

III. STRESS SIMULATION AND MEASUREMENT

A. Shock test and simulation

For this test, we opted to utilize a more stringent test requirement based on customer feedback- 10kg shock test condition (shown as G1-1 in table III) was used in the test to molding type SiP module. Shock test setup was shown in Figure 7. And the shock parameter at 10000G/ 0.25msec was recorded in Figure 7. It was verified 15 unit modules all pass the open/ short test after shock test, the results shown in Table IV.

Figure 7.   Shock test image at –Z direction and accelerometer contact at PCB center.

TABLE III.   SHOCK TEST CONDITION

Item	Shock Test Condition	Spec
G1-1	3 drops/face , 6 faces , total 18 drops ; Duration: 0.25 msec; 10000G peak acceleration.	End Customer
G1-2	3 drops/face , 6 faces , total 18 drops ; Duration : 0.5 msec ; 1500G peak acceleration.	AEC Q-100

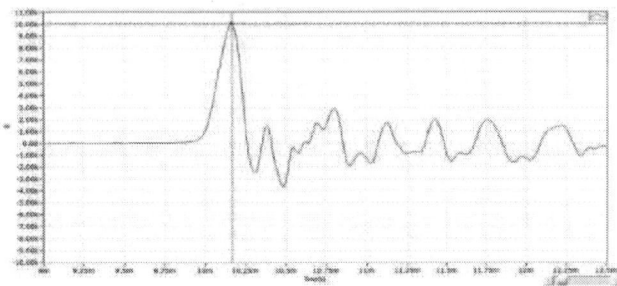

Figure 8.   Shock accelerometer measure: 10080.97 G's / 0.25 msec

TABLE IV.   SHOCK TEST RESULT BY G1-1 TEST CONDITION

Position	Test order 3 times -Z status	Test order 3 times +Z status	Test order 3 times -X status	Test order 3 times +X status	Test order 3 times -Y status	Test order 3 times +Y status
Sample 1	Pass	Pass	Pass	Pass	Pass	Pass
Sample 2	Pass	Pass	Pass	Pass	Pass	Pass
Sample 3	Pass	Pass	Pass	Pass	Pass	Pass
Sample 4	Pass	Pass	Pass	Pass	Pass	Pass
Sample 5	Pass	Pass	Pass	Pass	Pass	Pass
Sample 6	Pass	Pass	Pass	Pass	Pass	Pass
Sample 7	Pass	Pass	Pass	Pass	Pass	Pass
Sample 8	Pass	Pass	Pass	Pass	Pass	Pass
Sample 9	Pass	Pass	Pass	Pass	Pass	Pass
Sample 10	Pass	Pass	Pass	Pass	Pass	Pass
Sample 11	Pass	Pass	Pass	Pass	Pass	Pass
Sample 12	Pass	Pass	Pass	Pass	Pass	Pass
Sample 13	Pass	Pass	Pass	Pass	Pass	Pass
Sample 14	Pass	Pass	Pass	Pass	Pass	Pass
Sample 15	Pass	Pass	Pass	Pass	Pass	Pass

Shock stress simulation was also carried out, based on the given shock parameter shown in table III, to compare the main chip join stress level in between the molded package and to the metal lid type one. The data was focused on the corner module, and the main chip in it, as shown in figure 9. From the shock stress simulation data shown in figure 10, main chip joint stress of a metal lid type package was higher than that of the molded package.

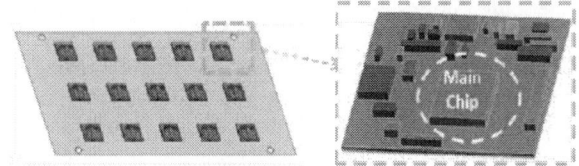

Figure 9.   Shock simulation site focus on main chip bump in the SiP module at PCB corner.

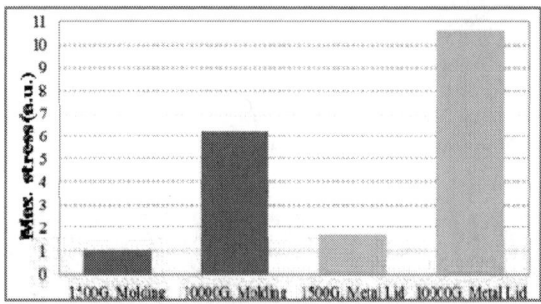

Figure 10.   Normalized stress of 2 shock test conditions between molding type SiP module and metal lid type SiP module.

B.   Vibration test and simulation

Figure 11 and 12 showed the molded module vibration test setup in X-Y and Z direction respectively. Test conditions were shown in table II. During the vibration test (G2), parameters of 20 Hz to 2 KHz to 20 Hz in > 4 minutes, 50G peak acceleration were recorded and shown in Figure 13 and Figure 14. Resonance from the tooling and machine was found, but it was still within the JEDEC spec.

Open/ short test results after the vibration test was shown in Table V. It's verified all 15 unit modules all pass and no damage found.

Figure 11.   Vibration test setup, X and Y directions.

Figure 12.   Vibration test setup, Z direction.

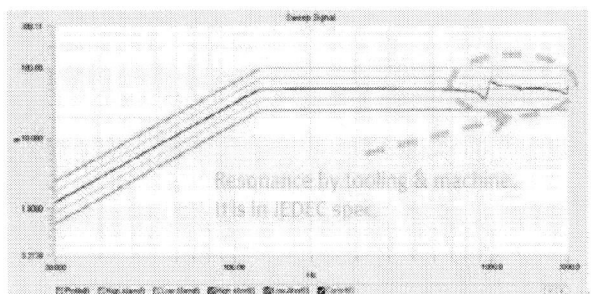

Figure 13. Graph of X and Y axises for vibration test. (Resonance by tooling and machine is within JEDEC spec.)

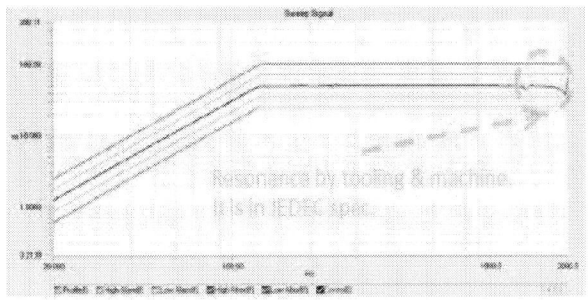

Figure 14. Graph of Z axises for vibration test. (Resonance by tooling and machine is within JEDEC spec.)

TABLE V.   VIBRATION TEST RESULT BY G2 TEST CONDITION

Position	Test order X status	Test order Y status	Test order Z status
Sample 1	Pass	Pass	Pass
Sample 2	Pass	Pass	Pass
Sample 3	Pass	Pass	Pass
Sample 4	Pass	Pass	Pass
Sample 5	Pass	Pass	Pass
Sample 6	Pass	Pass	Pass
Sample 7	Pass	Pass	Pass
Sample 8	Pass	Pass	Pass
Sample 9	Pass	Pass	Pass
Sample 10	Pass	Pass	Pass
Sample 11	Pass	Pass	Pass
Sample 12	Pass	Pass	Pass
Sample 13	Pass	Pass	Pass
Sample 14	Pass	Pass	Pass
Sample 15	Pass	Pass	Pass

Vibration stress simulation was carried out to compare the molded package to that with metal lid type module. The parameters were shown in G2 of Table II. There are 8 vibration modes vibration that were simulated from between 20 Hz to 2K Hz in Figure 15. The simulation data was concluded in Table VI, It was found molding type SiP module had one pick of discrepancy (> 800%) at 1050 (Mode 5) Hz, while the metal lid type SiP module had two picks of discrepancy (> 800%) at 745 (Mode 3) & 1380 (Mode 6) Hz individually. This again concluded the vibration stress distribution of a molded module is better than a metal lid type module.

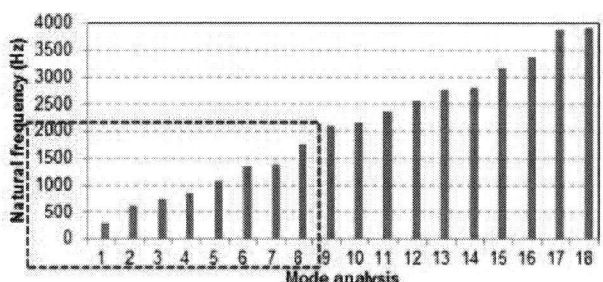

Figure 15. Vibration simulation modes

TABLE VI.   VIBRATIOB TEST RESULT

		Mode 1	Mode 2	Mode 3	Mode 4
Stress Dynamic Analysis					
Frequency (Hz)		300	600	745	840
Discrepancy (%)	Molding	248.3	692.1	445.9	548.3
	Metal Lid	219.7	670.5	828.2	589.2
		Mode 5	Mode 6	Mode 7	Mode 8
Stress Dynamic Analysis					
Frequency (Hz)		1050	1380	1450	1750
Discrepancy (%)	Molding	845.9	569.2	602.3	483.6
	Metal Lid	578.1	801.4	677.8	578.8

C. Free fall test

Figure 16 showed the free fall test setup onto a concrete surface, with the parameter given in table II. Function test results were concluded in table VII, in which showed all 5 units after free fall all passed.

TABLE VII.   FREE FALL TEST RESULT BY G3 TEST CONDITION

Position	Test order Bottom	Test order Top	Test order Right	Test order Left	Test order Front	Test order Rear
Sample 1	Pass	Pass	Pass	Pass	Pass	Pass
Sample 2	Pass	Pass	Pass	Pass	Pass	Pass
Sample 3	Pass	Pass	Pass	Pass	Pass	Pass
Sample 4	Pass	Pass	Pass	Pass	Pass	Pass
Sample 5	Pass	Pass	Pass	Pass	Pass	Pass

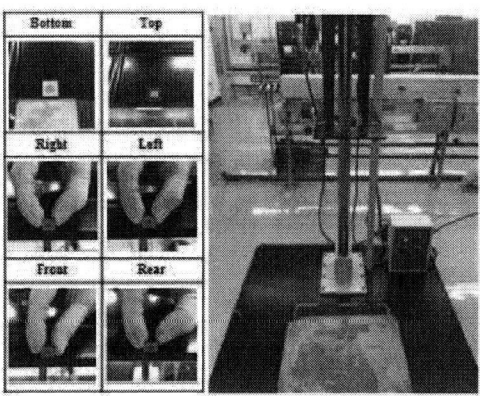

Figure 16. Free fall test setup onto a concrete surface.

IV. THERMAL ENHANCEMENT SIMULATION AND MEASUREMENT

A. Thermal simulation align with measurement result

The molded SiP module with EMI shielding had one heat source at main chip as shown in Figure 17. A 4 layer thermal board was used following JEDEC spec., also shown in Figure 18. [9]

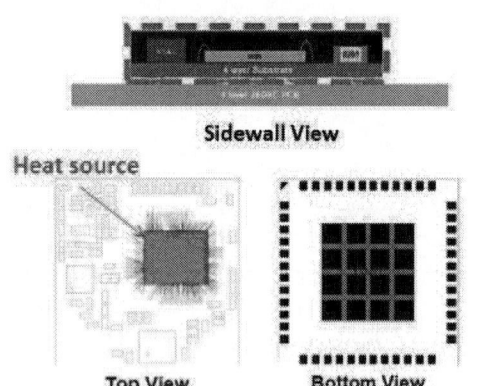

Figure 17. Molding type SiP module with EMI shielding had one heat source at main chip.

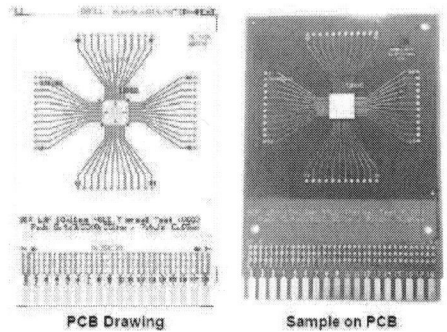

Figure 18. Thermal board, followed JEDEC spec. design.

Thermal measurement environment is still air chamber with heater and thermostat in Figure 19. Parameters shown in Table I was used in the thermal test.

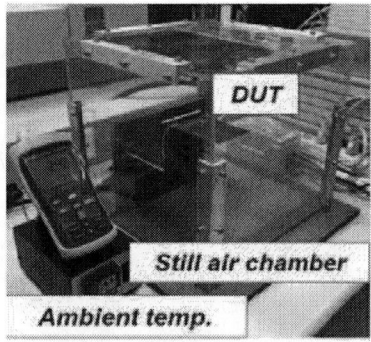

Figure 19. Thermal measurement setup.

Equation (1) was used to calculate θ_{JA} (Junction-to-ambient thermal resistance) for thermal measurement and simulation. It was found the simulation results are close fitting to the empirical thermal measurement ones, as shown in Table VIII and Figure 20.

TABLE VIII. THERMAL MEASUREMENT AND SIMULATION RESULTS

PCB Structure	Test Condition	Power Dissipation (W)	Still Air Condition			Simulation Result	θ_{JA} Gap (%)
			T_a (℃)	T_J (℃)	θ_{JA} (℃/W)	θ_{JA} (℃/W)	
JEDEC 2s2p PCB	5G @ 50%	0.735	23.4	48.12	33.61	35.76	6.38%
	5G @ 95%	1.005	23.6	56.52	32.77	35.46	8.23%

Junction-to-ambient Thermal Resistance:
$$\theta_{JA} = (T_J - T_a) / P_H. \tag{1}$$
T_J: Max. junction temperature
T_a: Ambient temperature
P_H: Total power dissipation

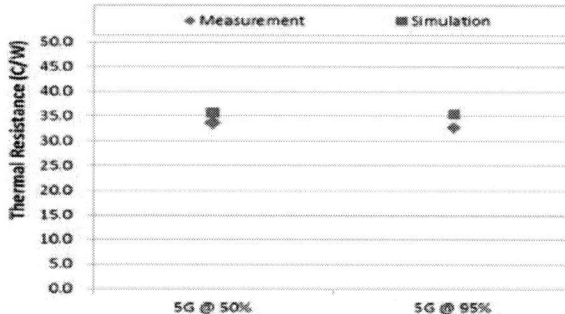

Figure 20. Thermal thata-JA simulation & measurement results for molding type SiP module.

Thermal simulation show that temp. contour of molding type SiP module and metal lid type SiP module which are in 5G mode at 95% power at ambient temperature 25℃ in Figure 21.

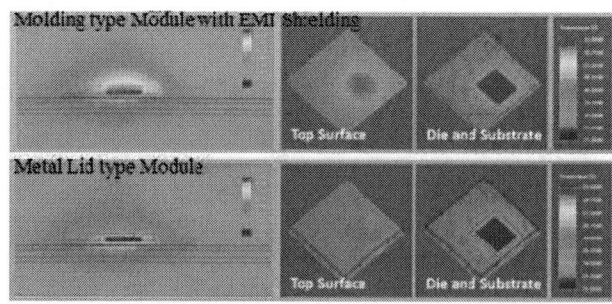

Figure 21. Temperature contour (still air, T_{amb} = 25C, 5G@95%)

B. Thermal simulation by different structures

There are 4 structures that we performed thermal simulations on. Sample A structure is molded type SiP module with EMI shielding. Sample B is metal lid type SiP module. Sample C is molded type SiP module reducing 50% thermal via with EMI shielding due to confirm thermal effectiveness by different thermal via distribution below the main chip. Sample D is molded type SiP module without EMI shielding structure. Thermal simulation DoE plan and results are in Table IX and Figure 22.

TABLE IX. THERMAL SIMULATION RESULTS FOR 4 STRUCTURES

PCB Structure	Test Condition	Power Dissipation (W)	Sample A	Sample B	Sample C	Sample D
			Molding type SiP module with EMI shielding	Metal lid type SiP module	Molding type SiP module reducing 50% thermal via with EMI shielding	Molding type SiP module without EMI shielding
			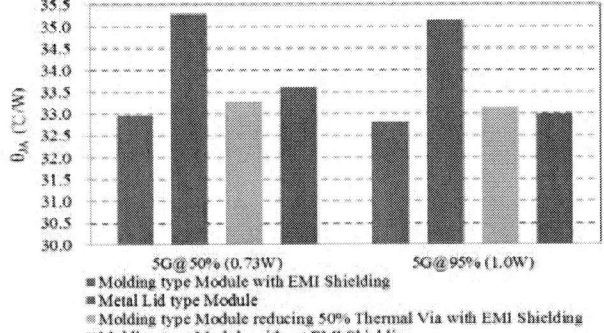			
			θ_{JA} (°C/W)	θ_{JA} (°C/W)	θ_{JA} (°C/W)	θ_{JA} (°C/W)
JEDEC 2s2p PCB	5G @ 50%	0.735	32.96	35.29	33.29	33.60
	5G @ 95%	1.005	32.80	35.14	33.14	32.98
Thermal enhanced rate base on sample A			Standard	6.63%	1.01%	1.23%

Figure 22. Thermal performace comparision by 4 structures

Figure 22 showed that molded SiP module with EMI shielding (Sample A) performed the best thermal enhanced effectiveness. It was found molded SiP with EMI shielding (sample A) showed 6.63% thermal enhancement when compared to that of a metal lid type module (Sample B). While also found the type A showed 1.01% thermal enhancement to type C, which is 50% reduced thermal via below main chip, and 1.23% better than that of type D, which is molded but without EMI shielding.

V. RELIABILITY TEST

THT by 85°C temperature & 85%RH humidity was carried out to demonstrate the moisture resistance capability of a molded and shielded SiP. After 1000hrs THT test, scanning acoustic tomography (SAT) was used and found all 25 units pass, as shown in Figure 23 and Figure 24. Special notice was remarked in Figure23, where it is was easy to mis read the SAT results.

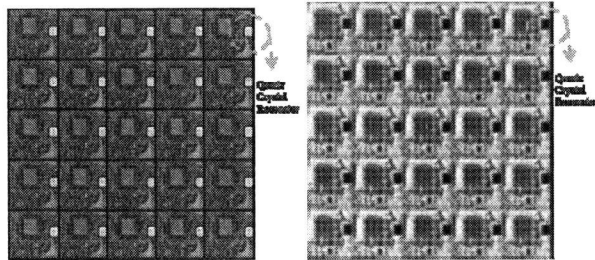

Figure 23. SAT C-scan & T-scan result after THT 1000hrs.

Functional tests were carried out to units after THT 1000hrs test. The results were concluded in Figure 24-26 and found all PASS

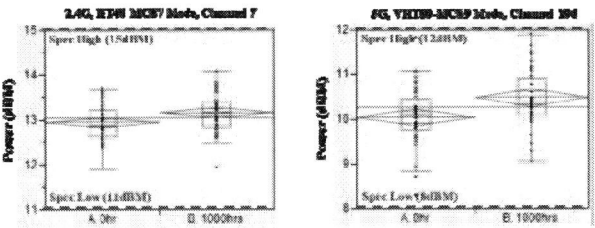

Figure 24. Power test result at time zero & THT 1000hrs.

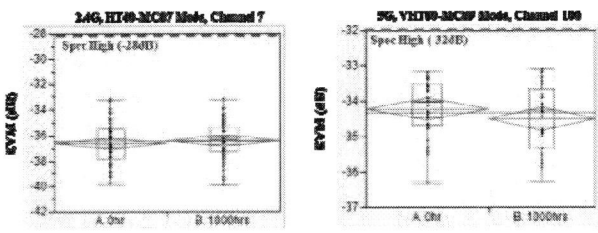

Figure 25. EVM test result at time zero & THT 1000hrs.

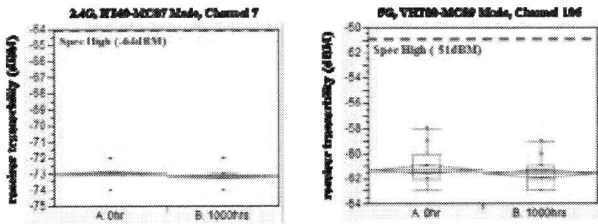

Figure 26. Test result of receiver transerbility at time zero & THT 1000hrs.

VI. RESULTS AND SUMMARY

From the study, it was found the Molded and EMI shielded SiP module revealed enhanced waterproof, dustproof, and qualify harsh mechanical stress test conditions including shock, vibration, and free fall. Meanwhile, the molded and shielded SiP could realize better miniaturization, enhanced reliability and thermal enhancement when compared to that of a metal lid SiP module.

For shock test simulation, stress of molded and shielded SiP module is around 70% compared to that of a metal lid SiP module at G1-2 test condition (Shown in Figure 10).

For vibration test simulation, molded and shielded SiP module is with only one peak with high discrepancy (> 800%). While 2 peaks were found in the metal lid SiP, which will show higher vibration failure risk.

For thermal simulation, molded and shielded SiP module was found 6.63% thermal enhancement compared to that of a metal lid module (in Figure 22).

Based on the functional test at different temperatures, low temp. power measurement is 1~2dBm higher than that at room temperature (in Figure 4); While the other testes showed no big difference among low, room, and high temperature level, as concluded in Figure 4, 5 and 6.

ACKNOWLEDGMENT

The authors would like to say thank you to all members who participate in this molding type SiP module with EMI shielding development for their continuous support and technical insights from ASE SiP BU, and ASE RD Lab groups.

REFERENCES

[1] AEC, "Failure Mechanism Based Stress Test Qualification For Integrated Circuits (base document)," AEC - Q100 - Rev-H, Sep. 11th, 2014

[2] JEDEC, "Board level drop test method of components for handheld electronic products," JES22-B111A, Nov.,2016

[3] J.E. Luan and T.Y. Tee, "Novel Board Level Drop Test Simulation using Implicit Transient Analysis with Input-G Method," Electronics Packaging Technology Conference, pp. 671-677, 2004.

[4] X. Qu, Z. Chen, B. Qi., T. Lee and J. Wang, "Board level drop test and simulation of leaded and lead-free BGA-PCB assembly," Microelectronics Reliability, Vol. 47, pp. 2197–2204, 2007.

[5] W. Wang, C. Glancey, D. Robbins, "Simulation Model to Predict Failure Cycles in Board Level Drop Test," 66th Electronic Components and Technology Conference, pp. 1886–1891 2016.

[6] Long Zhang, Chun-yue Huang, Wei Huang, Tian-ming Li and Jianwei Hua, "Study of Package-on-Package solder joints under random vibration load based on Patran," 17th International Conference on Electronic Packaging Technology (ICEPT), pp. 443–447, 2016.

[7] Liu Yang, Sun Fenglian, Zhang Hongwu, Zhou Zhen and Qin Yong, "Harmonic vibration test for accelerated reliability assessment of board level packaging," 7th International Forum on Strategic Technology (IFOST), pp. 1–5, 2012.

[8] I. Hu, M. Shih, G. Kao, "Performance and Reliability of TIM in High Power HFCBGA," International microsystems, in: Packaging, Assembly and Circuits Technology Conference, IMPACT'15, Taipei, Taiwan, October 21-23, 2015.

[9] JEDEC, "Integrated circuits thermal test method environmental conditions – natural convection (still air)," JESD51-2A, January, 2008.

[10] M. Springborn, B. Wunderle, D. May, R. Mrossko, C.A. Manier, M.A. Ras et al., "Transient thermal management by using double-sided assembling thermo-electric cooling and phase-change based thermal buffer structures," Design technology and application, pp. 1-7, 2015.

[11] Bo-Syun Chen, Tang-Yuan Chen, Jin-Feng Yang, Chin-Li-Kao, Yu-Chang Chen, Chan-Lin Yeh and Meng-Kai Shih, "Thermal characterization for dual side SiP module technology," IEEE 19th Electronics Packaging Technology Conference (EPTC), pp. 1-4, 2017.

Study of the Effect and Mechanism of a Cap Layer in Controlling the Statistical Variation of Via Extrusion

Golareh Jalilvand and Tengfei Jiang
Department of Materials Science and Engineering
Advanced Materials Processing and Analysis Center
University of Central Florida
Orlando, USA
g_jalilvand@knights.ucf.edu

Abstract—This work examines the effect of a metallic cap layer in controlling via extrusion and explores the underlying mechanisms. Ta was deposited as the cap material, which was very effective in reducing the statistical spread of via extrusion. The correlation between extrusion and microstructure of the vias was investigated and compared for the reference uncoated vias and the Ta-capped vias. Thermo-mechanical characterization as well as TEM characterization of the Ta cap/via interface were also carried out. Void formation at grain junctures were observed. The results suggest that mass transport through grain boundaries plays an important role in causing the statistical variation of extrusion, which can be effectively suppressed by the Ta cap. Void formation was also reduced by the cap layer. Additional factors affecting extrusion, including interfacial diffusion and dislocation glide, were also discussed.

Keywords-Through Silicon Via; Extrusion; Mass transport; Microstructure; Cap layer

I. Introduction

Cu through-silicon via (TSV) is an essential structural and functional element in three-dimensional integration circuits (3DICs), which brings substantial improvement in integration density, form factor, device performance, and power efficiency [1-3]. In the commonly used via-middle fabrication process, Cu-filled TSVs are exposed to multiple thermal cycles, which lead to the development of considerable thermal stress in and around the vias since the difference in the coefficient of thermal expansion (CTE) between Cu (Cu=17ppm/$^\circ$C) and Si (Si=2.3ppm/$^\circ$C) is very large. The stress subsequently raises reliability concerns during the fabrication, testing and operation of the TSV structure [4-6], among which, via extrusion is an important one. Via extrusion, also known as via protrusion or Cu pumping, refers to the irreversible residual deformation of Cu near the top of the via after thermal processing. Via extrusion appears as the outward protrusion of Cu vias in the axial direction, which can deform and crack the back-end-of-line (BEOL) interconnect layers to degrade the 3DIC component [7,8]. Recent studies also revealed that for a large population of TSVs, there is a wide spread in the values of via extrusion [8-10]. Although a few approaches such as post-plating annealing and via dimension scaling have been used to reduce the average values of via extrusion, these methods are not effective in controlling the statistical spread of via extrusion [7,11,12]. Controlling via extrusion, especially its statistical spread, remains an important issue, as the reliability of a 3DIC containing many TSVs will be determined by the weakest link, i.e., the 0.1% of TSVs with the highest extrusion heights.

In our previous work, we proposed and demonstrated a promising approach that can reduce both the average extrusion height and the statistical spread of extrusion by applying a metallic cap layer on the top surface of the vias [13,14]. In the present study, the effect of the cap layer and its mechanism in controlling via extrusion were further studied. Ta was chosen as the cap material and was deposited on a TSV sample (Ta-capped sample), which was then annealed to induce extrusion. A sample cut from the same TSV wafer but without any surface coating (uncoated sample) was also annealed for comparison. For both the uncoated and the Ta-capped samples, the extrusion behaviors were studied for over 300 vias using white light interferometry (WLI). For selected vias on each sample, focused ion beam (FIB) was used to reveal the microstructure on the top surface of the via. The microstructure at the via cross-section was also examined by FIB. Substrate curvature measurement was used to study the isothermal relaxation behaviors of each sample during annealing. High-resolution transmission electron microscopy (HRTEM) was used to characterize the interface between the Ta cap and the surface of the via. Based on the experimental observations, the role of the cap layer in suppressing Cu diffusion and reducing the statistical spread of extrusion was discussed.

II. Experiments

A. TSV test vehicle

The TSVs used in this study were blind Cu vias embedded in a 780 μm thick (001) Si wafer. The vias were fabricated using the standard etching and electroplating processes,

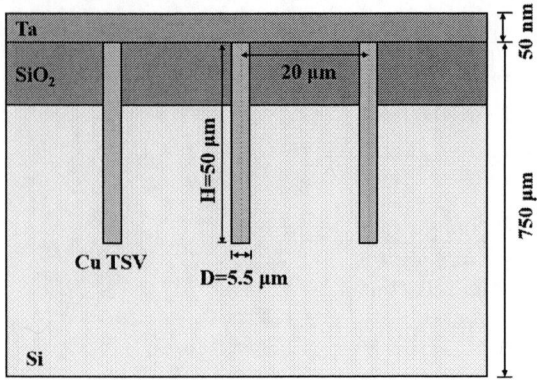

Figure 1. Illustration of the via structure with the Ta cap layer.

which yielded vias with a diameter of 5.5 μm, depth of 50 μm, and the pitch distance of 20 μm. The thickness of the TaN barrier was 0.1 μm and that of the oxide liner was about 0.4 μm. No post-plating annealing was carried out after via filling. Two pieces of samples were cut from the same TSV wafer. For one piece, a thin layer of Ta was deposited on the surface of the sample by sputtering to serve as the metallic cap. The structure of the Ta-capped vias is illustrated in Fig. 1. Prior to Ta deposition, the sample was dipped in acetic acid, followed by ultrasonic cleaning with Acetone, IPA and DI water, to remove the natural Cu oxide formed at the via top surface. The thickness of the Ta cap was determined by a quartz crystal thickness monitor and confirmed by atomic force microscopy (AFM) to be 50 nm.

B. Via extrusion measurement

Both the uncoated and the Ta-capped samples were annealed at 400°C for one hour in a forming gas atmosphere ($Ar - 4\%H_2$) with a heating rate of 6°C/min to induce extrusion. Afterwards, a 150 nm Al layer was deposited on both samples to enhance the surface reflectivity for WLI measurement, which was carried out for over 300 vias on each sample, obtaining the surface profiles of the vias. Based on the measured surface profiles, the average height (h_{avg}) was determined for vias in both the uncoated and the Ta-capped samples. Separately, using an un-annealed and uncoated sample cut from the same TSV wafer, the height of the vias before annealing was measured and determined to be h_{avg}= 3.7 nm. The average extrusion of the vias was then determined as $\Delta_{avg} = h_{avg} - h_0$. The statistics of via extrusion was analyzed for each sample and will be further discussed in section III.

C. Microstructure characterization

To examine the microstructure at the top surface of the vias, the surface coatings (Al reflection layer on the uncoated vias and Al + Ta on the Ta-capped vias) and the extruded Cu were removed by FIB milling. The FIB milling was carried out with the via axis tilted at 90° with respect to the ion beam. Care was taken with low current ion beam to remove less than 100 nm of Cu below the original surface of the via. 48 uncoated vias and 30 Ta-capped vias, which have extrusion values that correspond to the low tail and middle of the extrusion distribution curve, were accessible by FIB milling. After milling, low current FIB secondary electron images were taken for these vias, where grains with different orientations showed different contrasts due to the ion channeling effect [15]. Cross-sectional FIB milling and imaging was carried out for a few vias in both the uncoated and the Ta-capped samples to examine the microstructure inside the via. TEM samples were prepared by FIB milling and lift-off and the interface between the Ta cap and the Cu via was examined.

D. Substrate curvature measurement

To study the effect of the cap layer on the thermo-mechanical behavior of the vias, substrate curvature measurements were performed on both the uncoated and the Ta-capped samples. The measurements obtained the curvature change of the samples as a function of temperature during temperature excursion using an optical lever setup [16]. The same thermal profile as described in section B was applied and the isothermal curvature relaxation behaviors at 400°C were obtained.

III. RESULTS

A. Via extrusion

Based on the extrusion profiles obtained by WLI, the inverse standard normal cumulative distribution function (CDF), in the form of rankit Z_i, were plotted for the average extrusion values, Δ_{avg}, for 358 vias in the uncoated sample and 318 vias in the Ta-capped sample (Fig. 2). The

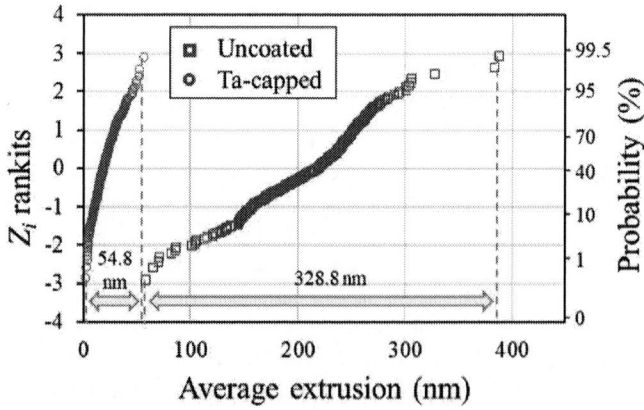

Figure 2. Inverse normal cumulative distribution function (Z_i) and normal probability plot for average extrusion of 358 uncoated vias and 318 Ta-capped vias.

978-1-7281-1500-9/19 $31.00 © 2019 IEEE

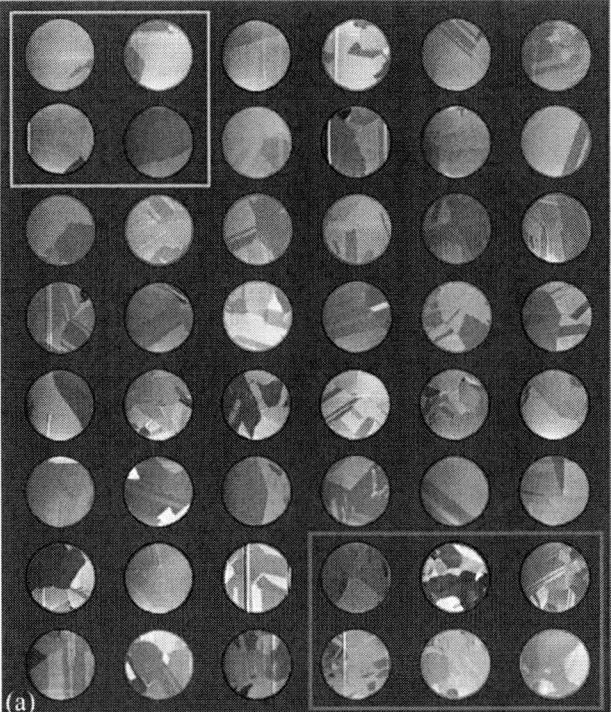

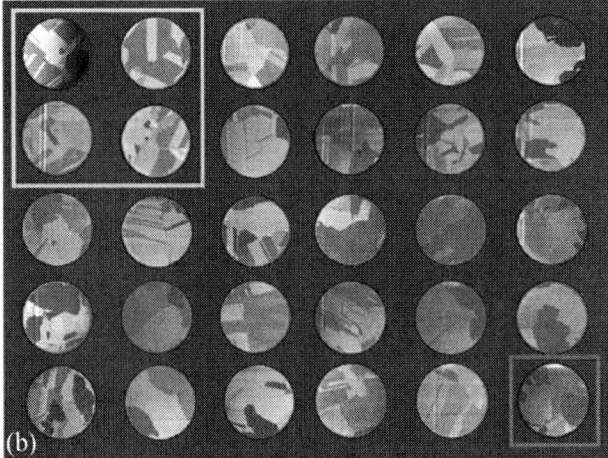

Figure 3. (a) FIB images for 48 uncoated vias. (b) FIB images for 30 Ta-capped vias. Vias with small extrusion values (low tail of the distribution plot) are circled in green, and vias with large extrusion values (middle to high tail of the distribution plot) are circled in red.

Table I
The probability for a via with n grains to have extrusion values of Δ for the 48 uncoated vias.

Average extrusion, nm	Number of Grains			
	$n < 5$	$5 \leq n < 10$	$10 \leq n < 15$	$n \geq 15$
$\Delta < 100$	1	0	0	0
$100 \leq \Delta < 150$	0	0.46	0.26	0.14
$\Delta \geq 150$	0	0.54	0.74	0.86

Table II
The probability for a via with n grains to have extrusion values of Δ for the 30 Ta-capped vias.

Average extrusion, nm	Number of Grains			
	$n < 5$	$5 \leq n < 10$	$10 \leq n < 15$	$n \geq 15$
$\Delta < 15$	0	0	0.15	0.4
$15 \leq \Delta < 30$	1	0.83	0.7	0.5
$\Delta \geq 30$	0	0.17	0.15	0.1

compared, the spread in the Ta-capped sample (54.8 nm) was notably smaller than the spread in the uncoated sample (328.8 nm), as marked by the double arrows in Fig. 2.

B. Microstructure-Extrusion correlation

The FIB images of the 48 uncoated vias and the 30 Ta-capped vias were shown in Fig. 3a and 3b, respectively. In the figures, the vias enclosed by the green boxes have low extrusion values that correspond to the low tails of their respective distribution curves, and vias enclosed by the red boxes have large extrusion values that belong to the middle-to high tail of the distribution curves. The rest of the vias have extrusion values that fall in between.

Although qualitative, the grain contrast in the FIB images can be used to examine the size and distribution of the grains on the surface of the vias. In the uncoated sample (Fig. 3a), the four vias with small extrusion values (enclosed in the green box) have large grains, or a small number of grains. In comparison, the six uncoated vias that have large extrusion values (circled by the red box) have small grains, or a large number of grains. Overall, in the uncoated sample, vias with a larger number of grains tend to have higher extrusion values, suggesting a correlation between the number of grains at the surface of the via and the extrusion values. On the other hand, in the Ta-capped sample (Fig. 3b), no clear difference in terms of the number of grains can be seen between the four vias with small extrusion values and the one via with the large extrusion value. Unlike the uncoated vias, the Ta-capped vias did not seem to show a clear correlation between the number of grains on the via surface and the magnitude of extrusion.

Using an image analysis tool, the number of grains on the surface of each via was determined from the FIB images for a more quantitative comparison with the extrusion values. For both samples, the vias were divided into four categories based on the number of grains, n, on the via surface, which are: $n < 5$, $5 \leq n < 10$, $10 \leq n < 15$, and $n \geq 15$. The

equivalent normal probability values were shown on the secondary vertical axis in the figure.

Comparing the two distribution curves, a significant reduction of Δ_{avg} in the Ta-capped vias can clearly be seen. The average extrusion at the 99th percentile of the distribution was 387.5 nm for the uncoated sample and 56.9 nm for the Ta-capped sample, which was a seven-fold reduction. Individually, the uncoated and the Ta-capped samples both showed a spread in their extrusion values. However, when extrusion values for each sample were also grouped into three bins, which are $\Delta < 100nm$, $100nm \leq \Delta < 150nm$, and $\Delta \geq 150nm$ for the uncoated sample, and $\Delta < 15nm$, $15nm \leq \Delta < 30nm$, $\Delta \geq 30nm$ for the Ta-capped sample. Next, the probability for a via in each category of n grains to have extrusion values in each bin of Δ was calculated. The results are shown in Table. 1 and Table. 2 for the uncoated and the Ta-capped samples, respectively. In the uncoated sample (Table 1), a via with fewer grains is more likely to have smaller extrusion values, and the probability for a via to have larger extrusion increases when the via has more grains. In the Ta-capped sample, the probability for a via to have extrusion values belong to each bin does not seem to be directly related to the number of grains in the via. This calculation corroborates the observation in Fig. 3b, where no direct correlation could be seen between the number of grains and extrusion in the Ta-capped sample.

C. Curvature-Temperature behaviors

In Fig. 4, the isothermal curvature relaxation behaviors at $400°C$ were plotted for both the uncoated and the Ta-capped sample. Both samples had negative curvatures at the beginning of the isothermal annealing, suggesting that both vias had reached the compressive stress states. Comparing to the uncoated sample, the Ta-capped sample was subjected to a larger compression, as seen by the relative positions of the two curves in Fig. 4. During isothermal annealing, the curvature of the uncoated sample began to relax to become less negative, while the curvature of the Ta-capped sample continued to decrease to become more negative at the end of annealing.

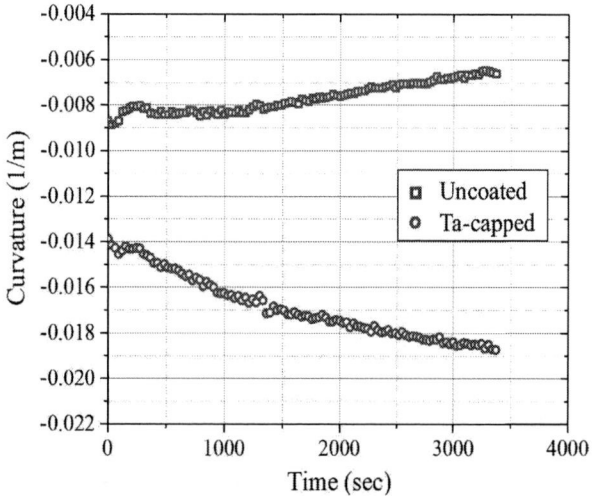

Figure 4. Comparison of the isothermal curvature behaviors for uncoated and Ta-capped sample during 1 hr of annealing at $400°C$.

IV. DISCUSSION

A. Grain boundary diffusion and via extrusion

At elevated temperatures, the expansion of the Cu vias is constraint by the surrounding Si, which has a lower CTE than Cu. This CTE mismatch subjects the Cu vias to large compressive stress, which at elevated temperatures can drive diffusive mass flow to relax the stress [17]. Several diffusion paths are available in a TSV structure, including the Cu lattice, the Cu grain boundaries, and the Cu via/TaN liner interface. Among them, diffusion along Cu grain boundary has the lowest reported activation energies ($\sim 0.7 - 0.9eV$) [18,19] comparing to the Cu lattice diffusion ($\sim 2.2eV$) [20] and the Cu/TaN interfacial diffusion ($\sim 1.1 - 1.8eV$) [21-24]. Diffusive flow of Cu atoms along the grain boundaries towards the surface of the via would serve as a preferred stress relaxation mechanism for a via during heating. When the surface of the vias is exposed, as is the case for the uncoated vias, the free surface provides vacancy sinks for the diffusion of Cu atoms along the grain boundary towards the surface, which is manifested as via extrusion. Using the number of grains obtained in the FIB images as an indication of the grain boundary area at the top of the via, qualitatively, vias with more grains can be assumed to contain a larger area of grain boundaries. The observed correlation between the number of grains and extrusion in the uncoated vias is then consistent with the proposed role of grain boundary diffusion on extrusion. Similar observation has also been reported in literature [8,25-28].

Within one via, the extrusion height profile is not uniform but instead shows a random undulation, as shown in Fig. 5. It is reasonable to assume that the grain boundaries at the surface of the via have a range of diffusivity depending on the grain misorientation angles. The surface undulation can be understood as a natural consequence of the grain boundary diffusion. When a one-to-one comparison was made between the height profiles and the FIB images, peaks in the height profile almost always coincide with locations that have high grain contrasts. Qualitatively, these locations are likely random high angle grain boundaries with large diffusivity values. Combined with the stochastic nature of grains in TSVs, the role of grain boundary diffusion on extrusion also suggests that the statistical spread of via

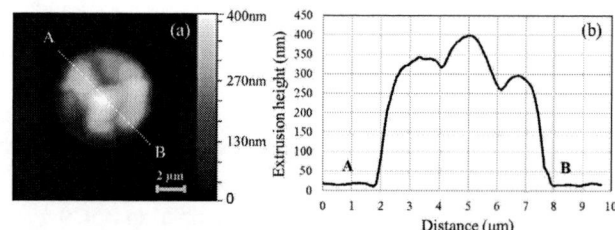

Figure 5. (a) AFM extrusion profile for an uncoated via. (b) Height profile along A-B in (a).

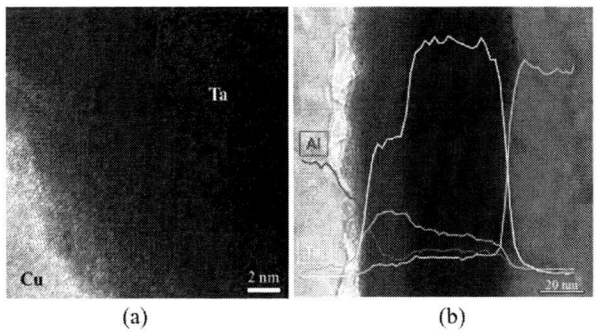

Figure 6. (a) Bright field high resolution TEM image of Cu/Ta interface after the Ta-capped via was annealed at 400C for 1 hour. b) EDS line scan of the interface in (a).

extrusion is an intrinsic problem that will not be eliminated by existing methods developed to control via extrusion, such as scaling of the via diameters and post-plating annealing [7,11,12].

B. The effect of the cap layer in controlling extrusion

The results in Fig. 2 showed that the thin Ta cap layer was very effective in reducing the magnitude of via extrusion. Ta is known to form good adhesion to Cu and has been used to reduce the mass transport at the Ta/Cu interface during electromigration [29]. The HRTEM image in Fig. 6 shows the formation of a dense and continuous interface between the Ta cap and the Cu via. A TEM energy dispersive x-ray spectroscopy (EDS) line scan was taken across the interface from the cap layer to the via, which further revealed a sharp interface with no intermixing (Fig. 6b). The detection of oxygen in the EDS indicated that oxidation of the Ta film had occurred after the deposition, however the Ta-O did not reach the Ta/Cu interface and was expected to have negligible influence on the integrity of the Ta/Cu interface.

The applied Ta cap will be very effectively in reducing the vacancy sources at the top of the via, which in turn suppressed the mass transport along grain boundaries, leading to the observed reduction of extrusion in the Ta-capped vias. This effect is illustrated in Fig. 7. Furthermore, by limiting mass transport along grain boundaries, the cap layer was

also able to reduce the spread of extrusion, as observed in Fig. 2. This underlying mechanism of the cap layer in suppressing diffusion also explains why no correlation was observed between the number of grains and extrusion for the Ta-capped vias.

The observed curvature behavior was consistent with the proposed role of the Ta cap in suppressing Cu diffusion. While the uncoated sample experienced a decrease in compressive stress by diffusional creep, the obstruction of grain boundary diffusion effectively led to increased compressive stress in the capped via, an effect that has been seen in passivated thin films and confined lines [27,28,30]. The CTE mismatch in Ta and Cu could also contribute to local compressive stress near the top the via.

C. Void formation

The cross-sectional FIB images for an uncoated via and a Ta-capped via, which have extrusion values corresponding to the middle of their individual distribution plots, are shown in Fig. 8. Formation of voids was observed in the uncoated via but not in the Ta-capped via. It should be pointed out that the contrast in the middle of the via in Fig. 8b was an artifact due to FIB cutting, not any physical features in the via. Several voids in the uncoated via are also highlighted in Fig. 8a. The voids seem to have formed at the junctures where multiple grains meet. Voids located at the junctures can also be clearly seen in the TEM image in Fig. 9.

Grain growth promoted by annealing is a possible cause for void formation [31]. However, comparing the microstruc-

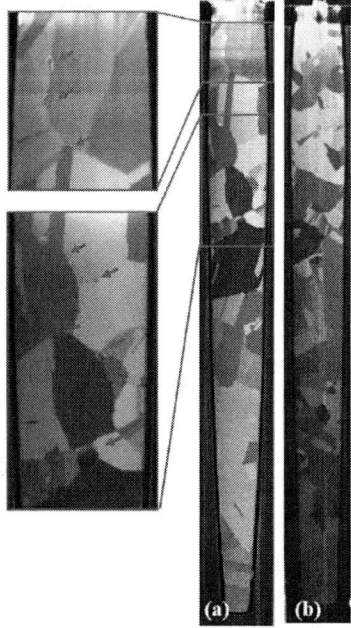

Figure 8. Cross-sectional FIB image of a) an annealed uncoated via with normal extrusion b) an annealed Ta-capped via with normal extrusion.

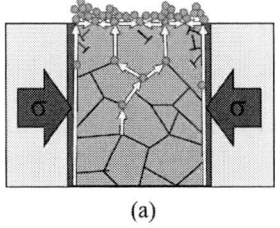 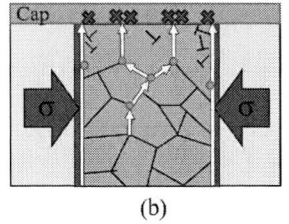

Figure 7. Illustration of the effect of the cap layer in suppressing grain boundary diffusion in the Cu vias. (a) An uncoated via. (b) A Ta-capped via.

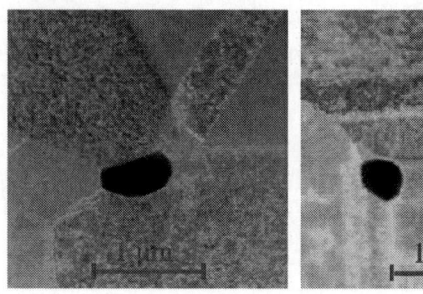 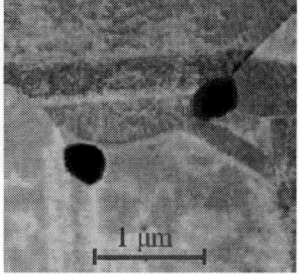

Figure 9. TEM image from the void formation at grain multiple-points after annealing.

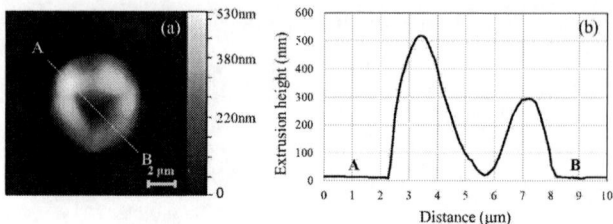

Figure 10. (a) AFM extrusion profile for an uncoated via showing the donut extrusion shape. (b) Height profile along A-B in (a).

ture between the uncoated and the Ta-capped vias, no qualitative difference in the grain size can be seen. The extensive void formation in the uncoated vias, which was not seen in the Ta-capped vias, suggests that grain growth was unlikely the main cause for void formation. Instead, void can be formed by the diffusive mass flow discussed in previous sections. As a simple estimation, the characteristic diffusion length of grain boundary diffusion can be calculated as $x = \sqrt{Dt}$, where D is the diffusivity and t is the annealing time. Using an activation energy of Q=0.9eV [19] and a prefactor of $D_0 = 0.3 cm^2/sec$ [32] for the annealing condition of T=400°C and t=1hr, the diffusion length is estimated to be 140 μm, which is larger than the dimension of the via. It is therefore possible for the free surface to supply vacancies through grain boundary diffusion, which coalesce at grain boundaries as voids. In the Ta-capped vias, void formation due to diffusion was minimized due to the grain boundary diffusion being suppressed.

During cooling, both the uncoated and the Ta-capped samples have linear curvatures [33] with values crossed over from negative to positive at different temperatures. The cross-over temperature of the uncoated sample was higher than that of the Ta-capped sample, suggesting that a larger tensile stress will be induced in the uncoated vias during cooling. The tensile stress in the via can act as the driving force for stress induced voiding in both the uncoated and the Ta-capped vias [31,32] and the effect will be stronger for the uncoated vias. The driving force for stress induced voiding will be affected by local stress gradient and the anisotropic elastic properties of the Cu grains. Further analysis of the voiding problem will require more detailed stress analysis and more quantitative microstructure characterization.

D. Additional factors for via extrusion

In many of the vias studied, whether with or without the Ta coating, relatively large extrusion at the peripherals of the vias was observed. A representative profile of a via having this donut shaped extrusion is shown in Fig. 10. Observation of the extrusion behavior in Fig. 10 suggests there were other factors besides grain boundary diffusion that had contributed to via extrusion. One possible factor is the diffusion along

the TaN/Cu via sidewall interface, which, despite of having a larger activation energy than grain boundary diffusion, can still be a fast path for mass flow. Interfacial diffusion may also contribute to the variation of extrusion as the interfacial energy between the TaN/Cu will vary depending on the orientation of Cu grains along the sidewall. Dislocation glide based mechanism is another important factor to consider. FEA has shown that the von Mises stress is non-uniform in the Cu vias and reaches the yield strength only in a small region between the sidewall and the via top surface [16,17,34]. This localized stress concentration can activate dislocation glide based mechanisms to accumulate permanent deformation, i.e. extrusion, near the perimeter of the vias. The magnitude of extrusion by dislocation mechanisms will also be affected by the microstructure of Cu. More detailed microstructure analysis will be needed to investigate these additional mechanisms for via extrusion.

V. CONCLUSION

In this work, we demonstrated that the application of a Ta cap layer can effectively reduce the extrusion magnitude and statistical variation in Cu TSVs. The correlation between the extrusion and grain structure was examined for both the uncoated and Ta-capped vias. The important role of grain boundary diffusion on the statistical variation of via extrusion is discussed. Void formation at grain multi-junction points was observed and the possible mechanisms due to diffusion and stress-induced voiding were discussed. Different extrusion shapes were observed, suggesting additional factors such as interfacial diffusion and dislocation glide based plasticity were also important for via extrusion. Further studies involving detailed microstructure characterization and stress analysis would be required to elucidate the via extrusion mechanisms.

ACKNOWLEDGMENT

The financial support by UCF Startup fund is gratefully acknowledged.

REFERENCES

[1] J.U. Knickerbocker, C.S. Patel, P.S. Andry, C.K. Tsang, L.P. Buchwalter, E.J. Sprogis, H. Gan, R.R. Horton, R.J. Polastre, S.L. Wright, J.M. Cotte, "3-D silicon integration and silicon

packaging technology using silicon through-vias," *IEEE J Solid-St Circ*, 41(8), pp.1718-1725 (2006).

[2] K. Banerjee, S.J. Souri, P. Kapur, K.C. Saraswat, "3-D ICs: a novel chip design for improving deep-submicrometer interconnect performance and systems-on-chip integration," *Proc. IEEE*, 89(5), pp.602-633 (2001).

[3] J.U. Knickerbocker, P.S. Andry, B. Dang, R.R. Horton, M.J. Interrante, C.S. Patel, R.J. Polastre, K. Sakuma, R. Sirdeshmukh, E.J. Sprogis, S.M. Sri-Jayantha, A.M. Stephens, A.W. Topol, C.K. Tsang, B.C. Webb, S.L. Wright, "Three-dimensional silicon integration," *IBM J. Res. Dev.*, 52(6), pp. 553-569 (2008).

[4] S. Ryu, K. Lu, X. Zhang, J. Im, P. S. Ho, R. Huang, "Impact of near-surface thermal stresses on interfacial reliability of through-silicon vias for 3-D interconnects," *IEEE Trans.Dev. Mater. Reliab.*, 11(1), pp 3543 (2011).

[5] N. Ranganathan, K. Prasad, N. Balasubramanian, K. L. Pey, "A study of thermo-mechanical stress and its impact on through-silicon vias," *J. Micromech. Microeng.*, 18(7), pp 1-13 (2008).

[6] C. S. Selvanayagam, J. H. Lau, X. Zhang, S. K. W. Seah, K. Vaidyanathan, T. C. Chai, "Nonlinear thermal stress/strain analyses of copper filled TSV (Through Silicon Via) and their flip-chip microbumps," *IEEE Trans. Adv. Packag.*, 32(4), pp 720-728 (2009).

[7] J. De Messemaeker, O. Varela Pedreira, B. Vandevelde, H. Philipsen, I. De Wolf, E. Beyne, K. Croes, "Impact of post-plating anneal and through-silicon via dimensions on Cu pumping," *Proc. IEEE ECTC*, pp 586-591 (2013).

[8] J. De Messemaeker, O. Varela Pedreira, H. Philipsen, E. Beyne, I. De Wolf, T. Van der Donck, K. Croes, "Correlation between Cu microstructure and TSV Cu pumping," *Proc. IEEE ECTC*, pp 613-619 (2014).

[9] D. Smith, S. Singh, Y. Ramnath, M. Rabie, D. Zhang, L. England, "TSV residual Cu step height analysis by white light interferometry for 3D Integration," *textitProc. IEEE ECTC*, pp 578-584 (2015).

[10] L. Spinella, M. Park, J. Im, P.S. Ho, N. Tamura, T. Jiang, "Effect of scaling copper through-silicon vias on stress and reliability for 3D interconnects," *IEEE Int. Interconnect Tech. Conf.*, pp 80-82 (2016).

[11] I. De Wolf, K. Croes, O. Varela Pedreira, R. Labie, A. Redolfi, M. Van De Peer, K. Vanstreels, C. Okoro, B. Vandevelde, E. Beyne, "Cu pumping in TSVs: effect of pre-CMP thermal budget," *Microelectron. Reliab.*, 51(9-11), pp 1856-1859 (2011).

[12] A. Heryanto, W.N. Putra, A. Trigg, S. Gao, W.S. Kwon, F.X. Che, X.F. Ang, J. Wei, R. I Made, C.L. Gan, K.L. Pey, "Effect of copper TSV annealing on via protrusion for TSV wafer fabrication," *J. Electron. Mater.*, 41(9), pp 2533-2542 (2012).

[13] G. Jalilvand, O. Ahmed, K. Bosworth, C. Fitzgerald, Z. Pei, T. Jaing, "Application of a metallic cap layer to control Cu TSV extrusion," *Proc. IEEE ECTC*, pp 61-66 (2017).

[14] G. Jalilvand, O. Ahmed, L. Spinella, L. Zhou, T. Jiang, "The effective control of Cu through-silicon via extrusion for three-dimensional integrated circuits by a metallic cap layer," *Scrip. Mater.*, 164, pp 101-104 (2019).

[15] B. W. Kempshall, S. M. Schwarz, B. I. Prenitzer, L. A. Giannuzzi, R. B. Irwin, F. A. Stevie, "Ion channeling effects on the focused ion beam milling of Cu," *J. Vac. Sci. Technol. B*, 11(2), pp 749-754 (2001).

[16] T. Jiang, S.K. Ryu, Q. Zhao, J. Im, R. Huang, P.S. Ho, "Measurement and analysis of thermal stresses in 3D integrated structures containing through-silicon-vias," *Microelectron. Reliab.*, 53(1), pp. 53-62 (2013).

[17] T. Jiang, J. Im, R. Huang, P. S. Ho, "Through-silicon via stress characterization and reliability impact on 3D integration circuits," *MRS Bulletin*, 40(3), pp 248256 (2015).

[18] T. Surholt, C. Herzig, "Grain boundary self-diffusion in Cu polycrystals of different purity," *Acta Mater.*, 45(9), pp 3817-3823 (1997).

[19] T. Surholt, Y. Mishin, C. Herzig, "Grain-boundary diffusion and segregation of gold in copper: investigation in the Type-B and Type-C kinetic regimes," *Phys. Rev. B*, 50(6), pp 3577-3587 (1994).

[20] S. Rothman, N. Peterson, "Isotope effect and divacancies for self-diffusion in copper," *Physica Status Solidi*, 35(1), pp 305-312 (1969).

[21] J. Lin, S. Park, K. Pfeifer, R. Augur, V. Blaschke, S. Shue, H. Yu, M. Liang, "Electromigration reliability study of self-ionized plasma barriers for dual damascene Cu metallization," *Mater. Resear. Soc. Proc.*, pp 233-237 (2003).

[22] C. Hu, D. Canaperi, S. Chen, L. Cignac, B. Herbst, S. Kaldor, M. Krish- nan, E. Liniger, D. Rath, D. Restaino, R. Rosenberg, J. Rubino, S.-C. Seo, A. Simon, S. Smith, W. Tseng, "Effects of overlayers on electromigration reliability improvement for Cu/low K interconnects," *IEEE Inter. Reliab. Phys. Symp. Proc.*, pp 222-228 (2004).

[23] S. Demuynck, Z. Tokei, C. Bruynseraede, J. Michelon, K. Maex, "Alpha- Ta formation and its impact on electromigration," *AMC*, pp 355-359 (2004).

[24] C. Hu, L. Gignac, E. Liniger, B. Herbst, D. Rath, S. Chen, S. Kaldor, A. Simon, W. Tseng, "Comparison of Cu electromigration lifetime in Cu interconnects coated with various caps," *Appl. Phys. Lett.*, 83(5), pp 869-871 (2003).

[25] R-M. Keller, S. P. Baker, E. Arzt, "Quantitative analysis of strengthening mechanisms in thin Cu films: Effects of film thickness, grain size, and passivation," *J. Mater. Res.*, 13(5), pp 1307-1317 (1998).

[26] D. Gan, "Thermal stress and stress relaxation in copper metallization for ULSI interconnects," *Ph.D. thesis*, University of Texas at Austin (2005).

[27] D. Weiss, "Deformation mechanisms in pure and alloyed copper films," *Ph.D. thesis*, Universitt Stuttgart (2000).

[28] S.P. Baker, R.-M. Keller, A. Kretschmann, E. Arzt, "Deformation mechanisms in thin Cu films," *MRS Proc.*, 516, pp 287-298 (1998).

[29] M. W. Lane, E. G. Liniger, J. R. Lloyd, "Relationship between interfacial adhesion and electromigration in Cu metallization," *J. Appl. Phys.*, 93(3), pp 14171421 (2003).

[30] M. A. Korhonen, P. Bo/rgesen, K. N. Tu, C. Li, "Stress evolution due to electromigration in confined metal lines," J. Appl. Phys., (73)8, pp 37903799 (1993).

[31] J. A. Nucci, "Effects of linewidth, microstructure, and grain growth on voiding in passivated copper lines," *Appl. Phys. Lett.*, 66(26), pp 3585-3587 (1995).

[32] E. Ogawa, J. McPherson, J. Rosal, K. Dickerson, T. Chiu, L. Tsung, M. Jain, T. Bonifield, J. Ondrusek, W. McKee, "Stress-induced voiding under vias connected to wide Cu metal leads," *Reliab. Phys. Symp. Proc.*, pp 312-321 (2002).

[33] S.-K. Ryu, T. Jiang, K. H. Lu, J. Im, H.-Y. Son, K.-Y. Byun, R. Huang, P. S. Ho, "Characterization of thermal stresses in through-silicon vias for three-dimensional interconnects by bending beam technique," *Appl. Phys. Lett.*, 100(4), pp 041901 (2012).

[34] T. Jiang, C. Wu, L. Spinella, J. Im, N. Tamura, M. Kunz, H-Y. Son, B. G. Kim, R. Huang, and P.S. Ho, "Plasticity mechanism for copper extrusion in through-silicon vias for three-dimensional interconnects," *Appl. Phys. Lett.*, 103(21), 211906 (2013).

Three Dimensional Copper Foam-Filled Elastic Conductive Composites with Simultaneously Enhanced Mechanical, Electrical, Thermal and Electromagnetic Interference (EMI) Shielding Properties

Tan Lu[1,2], Han Gu[1,3], Yougen Hu[1*], Tao Zhao[1], Pengli Zhu[1], Rong Sun[1*], Ching-Ping Wong[4]

[1] Shenzhen Institutes of Advanced Technology, Chinese Academy of Sciences, Shenzhen 518055, China
[2] College of Materials Science and Engineering, Shenzhen University, Shenzhen 518055, China
[3] Nano Science and Technology Institute, University of Science and Technology of China, Suzhou 215123, China
[4] School of Mechanical Engineering, Georgia Institute of Technology, 771 Ferst Drive, Atlanta, Georgia 30332, USA
e-mail: yg.hu@siat.ac.cn; rong.sun@siat.ac.cn

Abstract—**With the rapid growth of modern electronic devices towards higher power, higher integration, thinner, lighter, and smaller, electrical and thermal conductive as well as electromagnetic interference (EMI) shielding issues are attracted more and more concerns. Thermal interface materials (TIMs) with high thermal conductivity and excellent EMI shielding efficiency are desired to solve heat emission and EMI problems of the electronic devices. So far, most of studies were independently focused on TIMs or EMI shielding materials, which have many limits for some practical applications. In this work, to address the challenges, a unique material with above dual functions was developed. The material composed of Cu foam skeleton and filled thermoplastic polyurethane/silver (TPU/Ag) elastic conductive composite, which shows better mechanical flexibility, higher thermal conductivity and higher EMI shielding effectiveness compared with sole Cu foam or TPU/Ag composite. The outstanding performance of the Cu foam/TPU/Ag composite will see a promising application in the EMI shielding and heat management of electronic devices.**

Keywords-electromagnetic interference (EMI), copper foam; thermoplastic polyurethane; conductive polymer composites; thermal conductivity

I. INTRODUCTION

With the demand of high bandwidth, high stacked density and high power of modern electronic devices, electrical and thermal conductive as well as electromagnetic interference (EMI) shielding become a big problem due to higher operation speeds and increased chip count in the devices [1-3]. Increasing operating temperature of the high-performance devices or chips provides significant challenges associated with thermal management, EMI pollution and device reliability, etc [4]. In order to solve these problems, thermal interface materials (TIMs) with high thermal conductivity and excellent EMI shielding performance are urgently required and even simultaneously needed for some special cases [5, 6].

Foamed composites are good candidates for thermal management and EMI shielding due to they possess both high thermal conductivity and electrical conductivity [7-10]. Among the common foam-metal materials, three dimensional copper (3D Cu) foam is much suitable for EMI

shielding because Cu posses good magnetic, electrical, thermal properties, and also low-cost. Although Cu foam has been extensively studied in recent years, there are still many obstacles to be solved. For example: 1) Cu foam usually has a poor and small interfacial contact area owing to its rough surface, generating a high interfacial thermal resistance between heat sinks and heat sources. 2) Cu foam is easy permanently deformed under mechanical compressing due to its brittle skeleton structure and lack of elasticity. 3) The 3D pores in the Cu foam reduce its EMI shielding ability because electromagnetic wave can penetrate the large open pores.

Many efforts were made with the help of improving thermal conductive and EMI performance of the metal foam. For instance, Ji *et al.* reported open-cell foam of Cu-Ni alloy integrated with carbon nanotubes (CNTs) which was successively prepared by electroless copper plating, nickel (Ni) electroplating and electrophoretic deposition of CNTs to enhance its electromagnetic interference shielding performance [11]. Zhao *et al.* embedded phase change materials (PCMs) into Cu foam to enhance heat transfer and thermal energy storage performance [12]. Ye *et al.* presented a strategy for preparing thermal heat packs through impregnating organic PCMs within carbon-coated copper foams (CCFs) [13]. Huang *et al.* fabricated partial reduced graphene oxide (P-rGO) sheets-wrapped nickel foams (NF@P-rGO) by a drop-wetting process, which can effectively improve the compatibility between Ni foam and epoxy resin and thus enhancing its thermal conductivity from 0.226 W/(m·K) of neat epoxy to 0.535 W/(m·K) [14]. However, the thermal conductivity is still too low to rapidly spread concentrated heat.

In this work, we present a facile and scalable fabrication process of 3D Cu foam-filled elastic conductive composite with simultaneously enhanced mechanical, electrical, thermal and EMI shielding properties to overcome the above challenges. The elastic conductive composite was composed of thermoplastic polyurethane (TPU) and micro-sized silver (Ag) flakes. The TPU/Ag composite is intrinsic elasticity and electrical conductive due to TPU matrix providing excellent mechanical flexibility and dense Ag fillers offering abundant electrical conductive paths. The elastic and robust TPU matrix protects the Cu foam skeletons from fracture during mechanical deformations of the hybrids. Moreover, the thermal conductivity and EMI shielding effectiveness (SE) values of 1.15 W/(m·K) and 58.1 dB for the pure Cu foam are significantly increase to 4.42 W/(m·K) and 83.9 dB in the frequency range of 8.2-12.5 GHz for the Cu foam/TPU/Ag hybrids with the similar thickness, respectively. The outstanding performance of the Cu foam/TPU/Ag composite will see a promising application in the EMI shielding and heat management of electronic devices.

II. EXPERIMENTAL SECTION

A. Fabrication of the 3D Cu foam/TPU/Ag hybrids

Firstly, thermoplastic polyurethane (TPU) resins were dissolved into N, N-dimethylformamide (DMF) solvent at room temperature under mechanical stirring to obtain TPU/DMF solution with desired concentration of 25 wt%. Then the slurry composites of TPU/Ag-DMF were formed by mixing a certain amount of silver micro-flakes with TPU/DMF solution (mass ratio of TPU to Ag is 15: 85, i.e. 85wt% of Ag content). The TPU/Ag paste was subsequently pressed into Cu foam (porosity = 98%, density = 0.4 g cm^{-3}) sheets cut into suit size and placed in a mould with a groove, followed by heat solidifying at 80 ℃ for 4 h to achieve the free-standing 3D Cu foam/TPU/Ag hybrids.

B. Characterization of the 3D Cu foam/TPU/Ag hybrids

The morphologies, microstructures and elements mapping analysis of the printed TPU/Ag electrodes were characterized by a field emission scanning electron microscope (FE-SEM) with an energy dispersive detector (EDS). The mechanical performance of the samples was measured by an electronic universal testing machine. The electrical resistance of the electrode was measured by a digital multimeter and the corresponding electrical conductivity is calculated by the following equation:

$$\sigma = \frac{1}{R} \times \frac{l}{t \times w} \qquad (1)$$

where σ is the electrical conductivity, R is the electrical resistance, l is the length, w is the width and t is the thickness of the measured samples, respectively. The thermal conductivity was tested using a TIM thermal resistance and conductivity measurement apparatus designed according to ASTM D 5470 standard and is calculated by

$$K = \frac{1}{A} \times \frac{t}{R} = \frac{1}{A} \times \frac{Q}{T_h - T_c} \qquad (2)$$

where K is the thermal conductivity (W m^{-1} K^{-1}), A is effective heat transfer area (m^2), t is thickness (m), R is thermal resistance (K W^{-1}), Q is the heat flux (W), T_h is the temperature of hot meter bar (℃), and T_c is the temperature of cold meter bar (℃), respectively. Before measurement, the samples were cut into square specimens with size of 25.4 mm × 25.4 mm (length × width). The temperature of hot meter bar was set to 80 ℃ and the contact force was set to 40 psi during measurement of the thermal conductivity.

The EMI shielding performance measurements of the hybrids were carried out using a two-point vector network analyzer (VNA) combining with two waveguide-to-coaxial adaptors connected face to face at room temperature. X-band frequency range of 8.2-12.5 GHz was selected as tested range due to this wave range is widely used in commercial applications. Before measurement, a standard calibration was performed. Samples were cut into rectangular with 25 mm × 12 mm (length × width) and well placed to the holder which was tightened up with proper screw. EMI shielding effectiveness (SE) is the ability to shield devices from the electromagnetic wave source and was used to measure the ability of the material to attenuate the electromagnetic wave strength, which is defined as the logarithm of incoming power (P_i) to transmitted power (P_t) of an electromagnetic wave in decibels (dB) as equation 3 [15, 16]:

$$SE(dB) = 10\log(\frac{P_i}{P_t}) \qquad (3)$$

There are three typical processes when an electromagnetic wave propagates to a shielding material: reflected on the surface of the material, absorbed and dissipated in the form of a leaking current or heat, and multi-reflected inside the shielding material, and the other electromagnetic waves penetrate through the material. The total EMI shielding effectiveness (SE_T) is dominated by two functions: reflection (SE_R) and absorption (SE_A). The S scattering parameters (i.e., S_{11} and S_{21}) of each sample were recorded, and the reflection coefficient (R), transmission coefficient (T), and absorption coefficient (A) were calculated by the S parameters (S11 and S21 or S12 and S22) according to the following equations:

$$R = |S_{11}|^2 = |S_{22}|^2 \qquad (4)$$

$$T = |S_{12}|^2 = |S_{21}|^2 \qquad (5)$$

$$A + R + T = 1 \qquad (6)$$

Finally, SE_R, SE_A and SE_T are obtained:

$$SE_R = 10\log\left(\frac{1}{1-R}\right) = 10\log\left(\frac{1}{1-|S_{11}|^2}\right) \qquad (7)$$

$$SE_A = 10\log\left(\frac{1}{1-A}\right) = 10\log\left(\frac{1-|S_{11}|^2}{|S_{21}|^2}\right) \qquad (8)$$

$$SE_T = SE_{ref} + SE_{abs} = 10\log\left(\frac{1}{|S_{21}|^2}\right) \qquad (9)$$

III. THE PERFORMANCE OF THE FLEXIBLE INTERLOCKED PRESSURE SENSORS

Figure 1 shows the SEM images of the Cu foam and the Cu foam/TPU/Ag hybrid composites. It can be seen from a low resolution SEM image of Figure 1a, the Cu foam exhibits a porous and interconnected 3D framework structure similar to the typical sponge with a pore size of 550 ± 50 μm. And the magnified view of Figure 1b reveals that the Cu foam skeleton with width of about 50~65 μm has a smooth surface and classical trifurcated structure. After TPU/Ag paste was pressed into the Cu foam pores and heat solidified, the Cu foam/TPU/Ag hybrid composite was obtained, and the cross-sectional SEM image was shown in Figure 1c. It can be clearly seen that the TPU/Ag composites were full filled the Cu foam pores and closely surrounded the Cu skeleton, leaving a relative smooth cross-section. Figure 1d presents a partial magnified SEM image of the TPU/Ag, which shows the micro-sized silver flakes were densely and uniformly distributed in the TPU matrix, forming abundant electrical conductive paths and high electrical conductivity of the finally hybrids.

Figure 1. SEM images of the Cu foam (a, b) and Cu foam/TPU/Ag hybrid composite (c, d) with different magnifications, respectively.

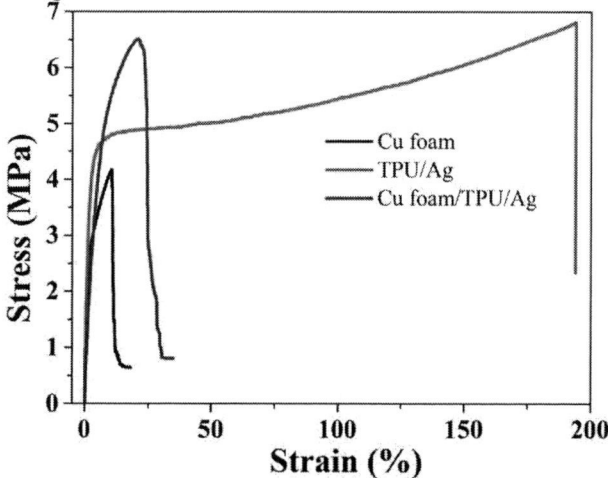

Figure 2. Cross-sectional SEM images of the printed TPU/Ag conductive composites on a TPU film.

In order to further confirm the microstructure and the chemical components of the Cu foam/TPU/Ag hybrids, elements mapping analysis was carried out and the result was exhibited in Figure 2. From this element map, we can clearly distinguish the difference between the Cu skeleton area and TPU/Ag region. There are three elements of Ag, Cu and C in the hybrids, which originated from silver flakes, Cu skeleton and TPU polymer, respectively. Moreover, the tightly bonding between TPU/Ag composites and Cu skeleton was confirmed again from their interfacial area without obvious cracks and delaminations.

The mechanical performance of the hybrids is important for their practical applications. To evaluate the mechanical performance of the Cu foam/TPU/Ag hybrids, stress-strain curves of the Cu foam/TPU/Ag samples were measured. Cu foam and TPU/Ag samples were also tested as control. All the results were collected and shown in Figure 3. It can be seen that, for the pure Cu foam, there is a low fracture strength value of about 4.17 MPa at strain of 10.5% and a small tensile breaking elongation of about 15%. For the TPU/Ag sample without Cu foam, the fracture strength and breaking elongation are about 6.8 MPa at strain of 20.5% and 194%, respectively, showing remarkable difference to the Cu foam. After filled TPU/Ag composites into Cu foam, the Cu foam/TPU/Ag hybrids present a fracture strength value of about 6.5 MPa, which is slightly lower than that of the TPU/Ag but much higher than that of the pure Cu foam. And the maximum elongation of the Cu foam/TPU/Ag hybrids is about 30%, which is twice that of pure Cu foam. In short, the Cu foam/TPU/Ag hybrids posses higher fracture strength and larger breaking elongation compared to Cu foam, indicating a noteworthy improvement of mechanical performance of the hybrids. The excellent mechanical property of the Cu foam/TPU/Ag hybrids is mainly attributed to the surrounding of the Cu skeletons by elastic TPU/Ag composites, which will protect Cu skeletons from fracture under a low tensile strain and enhance the fracture strength.

Figure 3. The stress-strain curves of the Cu foam, TPU/Ag and Cu foam/TPU/Ag hybrids samples.

The electrical performance of the Cu foam, TPU/Ag and Cu foam/TPU/Ag samples was simultaneously measured during mechanical performance measurement, and the results were shown in Figure 4 (insert is a partial magnified figure). It can be seen that there are all low resistance values for these three samples, indicating excellent electrical performance of them. The initial electrical conductivity of the Cu foam, TPU/Ag and Cu foam/TPU/Ag samples without tensile strain is calculated by equation (1) to be high values of about 2.21×10^6, 2.53×10^6 and 2.71×10^6 S m^{-1}, respectively. The above results prove that, after filling TPU/Ag composites into Cu foam, not only the mechanical performance but also the electrical performance of the hybrids are improved. Moreover, they exhibit different mechanical-electrical behaviors during tensile test. For the pure Cu foam, the resistance almost keeps constant in breaking strain, and rapidly increases to overload with somewhat vibration of resistance nearly the maximal breaking strain of 20.5%. This abnormal change of resistance may be originated from the partial cracks of Cu skeletons during tensile process especially near the final breaking, which will damage the initial and whole electrical conductive paths, and finally completely break at maximal strain. For the TPU/Ag sample, the resistance gradually increases to overload with increasing of the tensile strain until breaking of the sample. It is a classic mechanical-electrical behavior for conductive polymer composites (CPCs) composed by polymer matrix and conductive fillers. When the CPCs are suffered external tensile strain, the distances among conductive fillers will be increased and the conductive paths will be gradually damaged, resulting in resistance increase of the CPCs. However, for the Cu foam/TPU/Ag hybrids sample, its resistance changes behavior more like Cu foam rather than TPU/Ag. From Figure 4 and its insert, we can seen that the resistance of the Cu foam/TPU/Ag also keeps constant in the whole strain range before breaking, and then abruptly and steeply increases to overload. It is interesting that there is no resistance vibration for the Cu foam/TPU/Ag sample near the breaking strain.

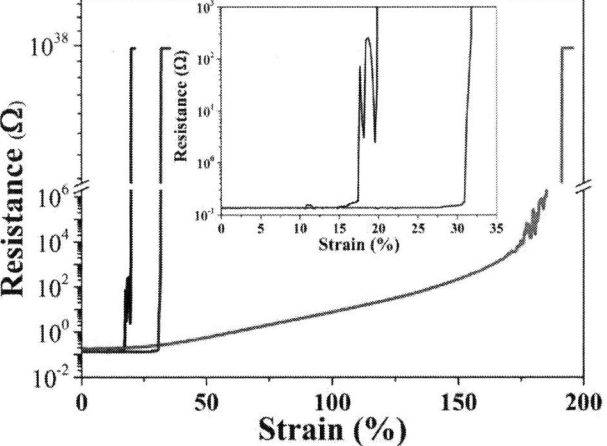

Figure 4. Mechanical-electrical performance of the Cu foam, TPU/Ag and Cu foam/TPU/Ag hybrids samples.

The thermal performance is crucial for thermal interface materials. Out-of-plane thermal conductivities of the Cu foam, TPU/Ag, Cu foam/TPU/Ag as well as TPU samples were measured by following ASTM D 5470 standard, as shown in Figure 5. It can be seen that there is a low thermal conductivity of 0.23 W/(m·K) for TPU sample. It is consistent with many previous reposts for polymer materials.

The thermal conductivity of Cu foam is 1.15 W/(m·K), which is observably higher than that of TPU, but much less than that of theoretical value of pure Cu (about 380 W/(m·K)). It can be explained as follows: 1) the purity and crystal structure integrity of Cu foam are lower than that of pure Cu; 2) there are a large number of open pores in the Cu foam, significantly decrease its thermal conductivity; 3) there is a relatively high surface roughness of the Cu foam and low contact area between surface Cu skeleton and hot/cold meter bar, resulting into large interface thermal resistance. The thermal conductivity of TPU/Ag sample with 85wt% silver loading is 3.85 W/(m·K), which is 16.74 and 3.35 times that of TPU and Cu foam, respectively. After TPU/Ag paste was filled into Cu foam and solidified, the Cu foam/TPU/Ag sample shows a high thermal conductivity of 4.42 W/(m·K), which is 3.84 times that of pure Cu foam, revealing a remarkable enhancement of thermal conductance ability of Cu foam. The improvement of thermal conductivity of Cu foam can be mainly attributed to: 1) the TPU/Ag composites with a relative high thermal conductivity were fully filled the open pores of Cu foam and eliminated the air in Cu foam; and 2) the surface roughness of the Cu foam decreases after TPU/Ag composites were filled and the effective contact area increase. The Cu/foam/TPU/Ag with high thermal conductivity provides promise potential as thermal interface materials.

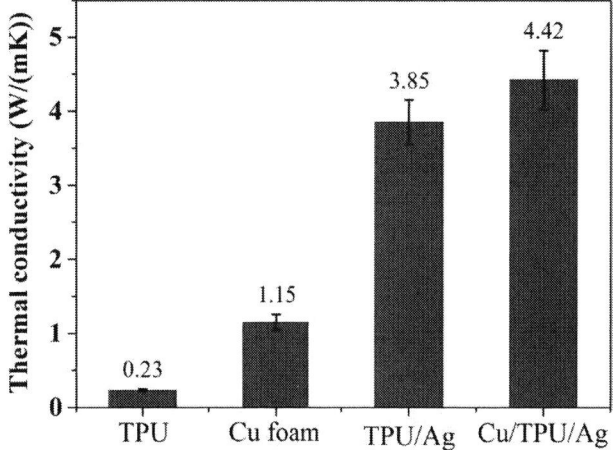

Figure 5. Out-of-plane thermal conductivity of the Cu foam, TPU/Ag composites and Cu foam/TPU/Ag hybrids samples.

The electromagnetic shielding effectiveness of the Cu foam (thickness of 1.18 mm) and Cu foam/TPU/Ag (thickness of 1.25 mm) samples was measured as Figure 6. The average total EMI shielding effectiveness (SE$_T$) of Cu foam and Cu foam/TPU/Ag in frequency range of 8.2-12.5 GHz is about 58.1 dB and 83.9 dB, respectively, exhibiting a remarkable enhancement of SE value for Cu foam by filling TPU/Ag composites into with high electrical conductivity its open pores. The high shielding effectiveness values of Cu foam and Cu foam/TPU/Ag samples are originated from absorption effectiveness.

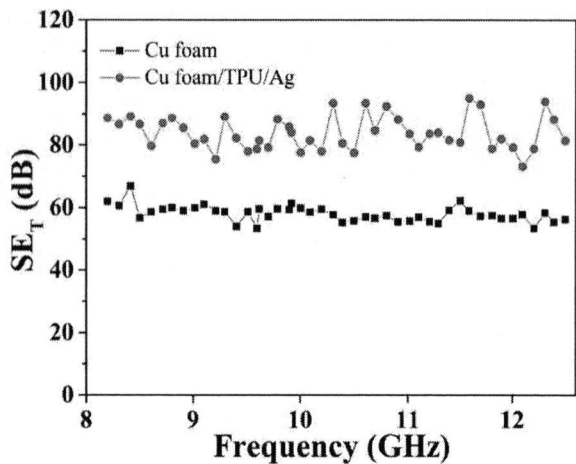

Figure 6. Total EMI shielding effectiveness of the Cu foam and Cu foam/TPU/Ag hybrids samples.

IV. CONCLUSIONS

In summary, we present a facile and scalable fabrication process of 3D Cu foam-filled elastic conductive composite with simultaneously enhanced mechanical, electrical, thermal and EMI shielding properties. The elastic conductive composite was composed of TPU and Ag flakes. The TPU/Ag composite with 85wt% Ag loading is intrinsic elasticity and electrical conductive with a high electrical conductivity of 2.53×10^6 S m^{-1}. The Cu foam/TPU/Ag hybrids obtained higher electrical conductivity of 2.71×10^6 S m^{-1}, larger breaking strain of about 30% and higher fracture strength of 6.5 MPa compared with Cu foam with electrical conductivity of 2.21×10^6 S m^{-1}, breaking strain of about 20.5% and fracture strength of 4.17 MPa. Moreover, the thermal conductivity and EMI shielding effectiveness values of 1.15 W/(m·K) and 58.1 dB for the pure Cu foam are significantly increase to 4.42 W/(m·K) and 83.9 dB in the frequency range of 8.2-12.5 GHz for the Cu foam/TPU/Ag hybrids with the similar thickness, respectively. The outstanding performance of the Cu foam/TPU/Ag composite will see a promising application in the EMI shielding and heat management of electronic devices.

ACKNOWLEDGMENT

The authors acknowledge the financial support from the National Natural Science Foundation of China (61701488), Shenzhen Basic Research Plan (JCYJ20170818162548196), and the National key R&D project from minister of science and technology of China (2016YFA0202702).

REFERENCES

[1] Y. L. Yang, and M. C. Gupta, "Novel Carbon Nanotube-Polystyrene Foam Composites for Electromagnetic Interference Shielding," Nano Letter, vol. 5, no. 11, pp. 2131-2134, Sept. 2005, doi:10.1021/nl051375r.

[2] N. Li, Y. Huang, X B. He, X. Lin, H. J. Gao, Y. F. Ma, F. F. Li, Y. S. Chen, and P. C. Eklund, "Electromagnetic Interference (EMI) shielding of Single-Walled Carbon Nanotube Epoxy Composites,"

Nano Letter, vol. 6, no. 6, pp. 1141-1145, May 2006, Doi: 10.1021/nl0602589.

[3] P. S. Liu, H. B. Qing, H. L. Hou, Y. Q. Wang, and Y. L. Zhang, "EMI Shielding and Thermal Conductivity of A High Porosity Reticular Titanium Foam," Materials & Design, vol. 92, pp. 823-828, Feb. 2016, doi: 10.1016/j.matdes.2015.12.105.

[4] N. Burger, A. Laachachi, M. Ferriol, M. Lutz, V. Toniazzo, and D. Ruch, "Review of Thermal Conductivity in Composites: Mechanisms, Parameters and Theory," Progress in Polymer Science, vol. 61, pp. 1-28, Oct. 2016, doi: 10.1016/j.progpolymsci.2016.05.001

[5] B. Shen, W. T. Zhai, and W. G. Zheng, "Ultrathin Flexible Graphene Dilm: An Excellent Thermal Conducting Materials with Efficient EMI Shielding," Advanced Functional Materials, vol. 24, no. 28, pp. 4542-4548, Jul. 2014, doi: 10.1002/adfm.201400079.

[6] X. Z. Ye, J. Hu, B. Li, M. Hong, and Y. F. Zhang, "Graphene loaded with nano-Cu as a highly efficient foam interface material with excellent properties of thermal-electronic conduction, anti-permeation and electromagnetic interference shielding," Chemical Engineering Journal, vol. 361, pp. 1110-1120, Apr. 2019, doi: 10.1016/j.cej.2018.12.047.

[7] J. He, R. Zhao, N. Zhang, X. J. Jin, X. F. Lu, and C, Wang, "Lightweight and flexible electrospun polymer nanofiber/metal nanoparticle hybrid membrane for high-performance electromagnetic interference shielding," NPG Asia Materials, vol. 10, pp. 749-760, Agu. 2018, doi: 10.1038/s41427-018-0070-1.

[8] S. T., Wu, M. C. Zou, Z. C. Li, D. Q. Chen, H. Zhang, Y. J. Yuan, Y. M. Pei, and A. Y. Cao, "Robust and Stable Cu Nanowire@Graphene Core-Shell Aerogels for Ultraeffective Electromagnetic Interference Shielding," Small, vol. 14, no. 23, pp. 1800634, Jun. 2018, doi:10.1002/smll.201800634.

[9] S. H. Lee, S. Yu, F. Shahzad, W. N. Kim, C. Park, S. M. Hong, and C. M. Koo, "Density-tunable lightweight polymer composites with dual-functional ability of efficient EMIshielding and heat dissipation," Nanoscale, vol. 9, no. 36, pp. 13432-13440, Sep. 2017, doi: 10.1039/c7nr02618h.

[10] J. Lee, Y. N. Liu, Y. Liu, S. J. Park, M. Park, and H. Y. Kim, "Ultrahigh electromagnetic interference shielding performance of lightweight, flexible, and highly conductive copper-clad carbon fiber nonwoven fabrics," Journal of Materials Chemistry C, vol. 5, no. 31, pp. 7853-7861, Aug. 2017, doi: 10.1039/c7tc02074k.

[11] K. J. Ji, H. H. Zhao, J. Zhang, J. Chen, and Z. D. Dai, "Fabrication and electromagnetic interference shielding performance of open-cell foam of a Cu-Ni alloy integrated with CNTs," Applied Surface Science, vol. 311, pp. 351-356, Aug. 2014, doi: 10.1016/j.apsusc.2014.05.067.

[12] C.Y. Zhao, W. Lu, and Y. Tian, "Heat transfer enhancement for thermal energy storage using metal foams embedded within phase change materials (PCMs)," Solar Energy, vol. 84, no. 8, pp. 1402-1412, Aug. 2010, doi: 10.1016/j.solener.2010.04.022.

[13] Q. X. Ye, P. Tao, C. Chang, L. Y. Zhou, X. L. Zeng, C. Y. Song, W. Shang, J. B. Wu, and T. Deng, "Form-stable solar thermal heat packs prepared by impregnating phase-changing materials within carbon-coated copper foams," ACS Applied Materials & Interface, vol. 11, no.3, pp. 3417-3427, Dec. 2018, doi: 10.1021/acsami.8b17492.

[14] L. Huang, P. L. Zhu, G. Li, and R. Sun, "Improved wetting behavior and thermal conductivity of the three-dimensional nickel foam/epoxy composites with graphene oxide as interfacial modifier," Applied Physics A, vol. 122, pp. 515, May 2016, doi. 10.1007/s00339-016-0048-1.

[15] Y. Bhattacharjee, I. Arief, and S. Bose, "Recent trends in multi-layered architectures towards screening electromagnetic radiation: challenges and perspectives," Journal of Materials Chemistry C, vol. 5, no. 30, pp. 7390-7403, Aug. 2017, doi: 10.1039/c7tc02172k.

[16] S. Sankaran, K. Deshmukh, M. B. Ahamed, S. K. K. pasha,"Recent Advances in Electromagnetic Interference Shielding Properties of Metal and Carbon Filler Reinforced Flexible Polymer," Composites Part A: Applied Science and Manufacturing, vol. 114, pp. 49-71, Nov. 2018, doi: 10.1016/j.compositesa.2018.08.006.

978-1-7281-1500-9/19 $31.00 © 2019 IEEE

Vertical interconnect technology for enlarging capacity on micro solid thin film rechargeable battery

Akihiro Horibe, Kuniaki Sueoka, Takahiro Mori, Risa Miyazawa, Hiroyuki Mori
IBM Research – Tokyo
Kawasaki, Japan
hory@jp.ibm.com

Abstract—A vertical interconnect technology for stacking micro solid thin film batteries (STFB) in IoT devices is proposed. This technology consists of stacking glass substrates with a layer of solid thin film battery, drilling by laser machining, filling the holes with solder to connect the stacked batteries, and dicing the stacked wafer to 1×1 mm micro stacked batteries. In this work, various metals are evaluated to identify the appropriate electrode drilled with glass and resin. The effects of the laser wavelength are also investigated.

Keywords - vertical interconnect, wafer stacking, through glass via, laser machining, solid thin film battery, micro rechargeable battery, non-alkaline glass, molten solder injection

I. INTRODUCTION

In the near future, high capacity micro batteries will be in high demand for use in intelligent IoT infrastructures. Recent trends in IoT devices include smaller form factors and lower power consumption. This may prove difficult since such devices require a combination of processors, memory, sensors, wireless communication, and power sources including batteries and / or energy harvesters. The solid thin film lithium ion rechargeable battery [1] is a potential power source that could achieve a smaller footprint and thinner thickness than commercially available coin cell batteries. In addition, small devices like this need to be rechargeable because the energy capacity is limited by the number of active ions in the reduced volume of the device. The vertical stacking concept shows promise as a definitive solution to enlarging the capacity and power density while reducing the inherent internal and interface resistances of the solid battery components in the limited footprint of the micro battery. The main advantages of the stacked STFB are that it is non-flammable, has no liquid leakage, is rechargeable, and has high power density, high energy capacity, and low internal voltage depression. These features are a good fit for the batteries of self-powered IoT devices.

In this paper, we demonstrate the simple vertical interconnecting process for stacked STFB. We have previously proposed a low-cost vertical interconnection method called vertical integration after stacking (ViaS) [2–4], in which vias are filled with solder using IBM's proprietary solder injection tool that can directly inject molten solder into holes on a wafer with high throughput. This ViaS technology does not require complex process steps for each substrate (unlike the commercial through-silicon-via (TSV) process) or an excess connection area like wire bonding does. Thus, this technology is well suited for stacked STFB. This novel vertical interconnect technology for the micro battery is newly developed on a laser drilling method [5–10] for stacked glass substrates with electrodes based on the ViaS technology. The future trends of the micro battery and its potential applications in future IoT infrastructures are also discussed.

II. SOLID THIN FILM RECHARGEABLE BATTERY

The solid thin film battery (Fig. 1) consists of only solid-state materials, including solid electrolytes. All materials can be deposited by a physical deposition process. The stacked battery prototyped in this paper adopts $LiCoO_2$ as a cathode, which requires high temperature annealing (600–700°C) for crystallization, LiPON as an electrolyte, and a non-lithium anode to enable not only high temperature operation but also high safety, which is especially important in medical and healthcare applications. The planar dimensions of this type of battery can be arbitrary, and the thickness is quite thin (< ~15 μm), not including the substrate. The substrate cost and sputtering rate dominate the cost structure of the battery. Thus, a low-cost glass substrate like non-alkaline glass, which is widely used for display panels, is suitable for affordable IoT applications.

The capacity of the battery is limited by the volume of the battery components (e.g., the cathode). However, the internal stress of cathode film such as $LiCoO_2$ limits the maximum thickness [11]. This has a significant effect on the capacity limitation of solid thin film batteries fabricated by physical deposition. To overcome this issue, we have developed a new stacking technology for the STFB.

III. VERTICAL INTERCONNECT TECHNOLOGY FOR STACKED BATTERY

In our previous work [2–4], we proposed vertical interconnection technology for stacked silicon wafers. In this paper, we present a new vertical interconnect technology for stacked glass substrates on which battery layers are formed. First, the solid thin film battery is formed on a glass substrate by sputtering with metal masks for each layer. These glass substrates containing the thin film battery layers are adhered to another battery substrate by alignment through the glass substrates. A laser is then used to drill holes in the center of the stacked electrodes, penetrating the glasses and adhesive layers. Finally, the holes are filled by solder. As a result, the stacked cathode and anode electrodes are electrically connected by the filled solder in the vertical direction (Fig. 2).

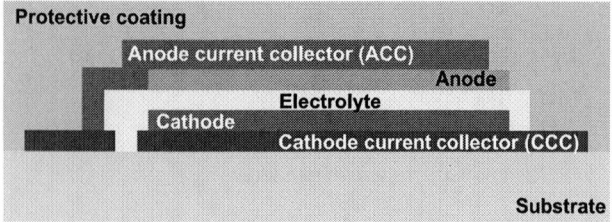

Figure 1. Structure of solid thin film battery.

This process is performed as a wafer-level or panel-level process. The glass thinning process is one of the key steps to maximize the energy capacity per volume of the battery module. Through this assembly process, the thinned substrate to be stacked is mechanically supported by the thick first battery substrate, and the risk of defect by ion diffusion (as occurs with silicon devices) does not exist. Thus, the only influence on the glass thinning process is the precision of the machine.

A. Metal mask designs for solid battery layers

Figure 3 shows an example of a metal mask design for the base metal layer (anode electrode and cathode current collector with cathode electrode), which has cell sizes of 10×10, 5×5, and 1×1 mm. The square pads located at the corner of each battery are anode electrodes. The clearance between the electrodes and the other cathode area is 10% of the cell size of each battery. Thus, the clearance for the 1×1 mm battery is 100 μm. Iron-nickel alloy is used as the metal for the mask so as to decrease the difference of thermal expansion between glass substrate and mask. Each metal mask is 50 μm thick, which minimizes the shadow effect caused by using thicker masks. As a result, we achieved an 86% active area ratio in the 1×1 mm battery footprint and could theoretically keep such small clearance even in wafers of the 300-mm size due to the small thermal expansion mismatch.

B. Solid-battery-layers deposition and substrate stacking

The deposition process flow of the STFB is shown in Fig. 4. In this process, high temperature annealing is required for crystalizing the $LiCoO_2$ layer. This anneal temperature limits the selection of the substrates. Non-alkaline glass, which has a high strain point around 670°C, is used as the substrate. Figure 5 shows a battery sample just after the high temperature annealing step. The two electrodes are located at the diagonal corners of each battery. Both pads are made of the same metal and are covered by no other layers except the protection layer, as this makes the following drilling condition the same for both pads. In the case of conventional mono-layer solid batteries, the protection layer is coated to avoid exposing the water-sensitive electrolyte layer to humidity. In contrast, with our approach, this protection layer is used as an adhesion layer to bond between the substrates. The thickness should be designed appropriately to absorb the uneven height of the deposited battery layers. In this case, we set it to around 10 μm.

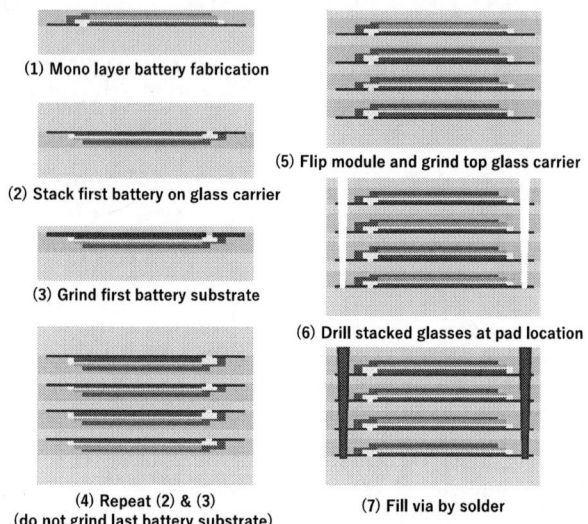

Figure 2. Assembly process for stacked STFB.

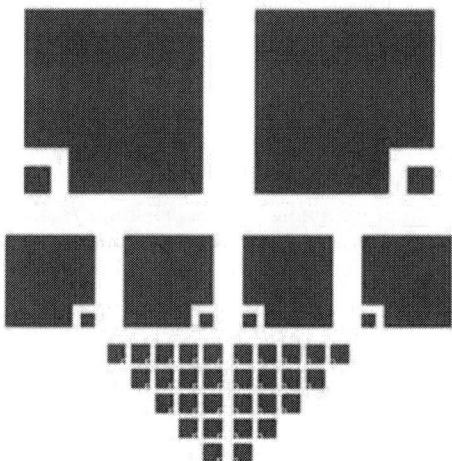

Figure 3. Metal mask design for anode electrode and cathode current corrector with cathode electrode.

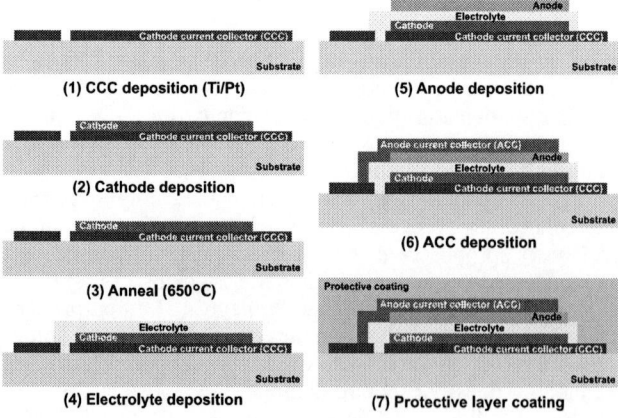

Figure 4. Deposition process flow for STFB.

978-1-7281-1500-9/19 $31.00 © 2019 IEEE

Figure 5. Solid battery sample just after annealing process shown in Fig. 4 (3).

C. Laser drilling

To drill non-alkaline glass, CO_2 (wavelength: 9.3 μm) or excimer (wavelength: 248 nm (KrF)) lasers are candidates because non-alkaline glass has high absorbance in UV regions shorter than 250 nm or IR regions over 6.5 μm. The target via diameter is 100 μm, and the depth is 100–300 μm depending on the number of batteries to be stacked. Each stacked glass substrate is thinned by a grinding process just after the bonding step for each substrate. Followed by substrate stacking, vias are formed by laser at the center of the stacked electrodes through the adhesives and the glass substrates, then filled with solder.

The performance of the laser drilling is affected by the structure to be drilled and the target materials, which are glass substrate, adhesive resin, and metal pad. In this paper, we evaluate two types of structure (Fig. 6) and various metals as the electrodes (Table 2). Figure 6 (1) shows a test sample structure with the resin coated on the metal, and (2) shows another structure with the resin located under the metal. The former assumes the case of a face-up battery and the latter assumes the opposite. Table 1 lists the physical properties of the metals evaluated. Each metal has a different melting point and thermal conductivity and optical absorption for the excimer (KrF) laser. During the laser drilling process, a higher etching resistance of the metal pad compared with the resistance of the glass and resin is preferable due to the protruding metal pad from the side wall of the via so as to ensure reliable electrical connection between the solder via and the electrodes. In terms of material properties, a high melting point, high thermal conductivity, and low absorption are preferable, but such a material encompassing the advantages of all properties does not exist among the materials listed in Table 1.

Figure 7 is an optical microscope image of the cross section of the laser-drilled via on an aluminum pad performed by excimer laser. The diameter of the via is 97 μm at the top and 69 μm at the bottom, and the depth is 289 μm. The aluminum pad is depicted in the figure as a thin white layer at which three arrows are pointed. The edge of the aluminum pad is completely aligned to the side wall of the via. This result is not preferable when the goal is to achieve reliable electrical contact. This demonstrates that, although aluminum has a relatively high thermal conductivity and low optical absorption, these advantages are not much use under these drilling conditions.

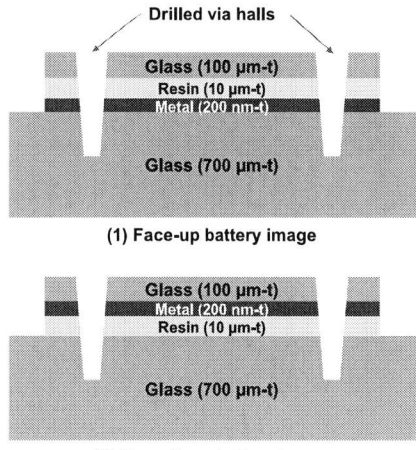

Figure 6. Test sample structures for laser drilling evaluation.

TABLE I. PHYSICAL PROPERTIES OF METALS.

	M.P. (°C)	T.C. (W m^{-1} K^{-1})	Absorption (%@248nm)
Al	660	236	10–15
Ag	961	428	~40
Cu	1083	403	75–80
Ni	1455	94	55–60
Ti	1727	22	75–80
Pt	1774	72	70

(M.P.: Melting point, T.C.: Thermal conductivity)

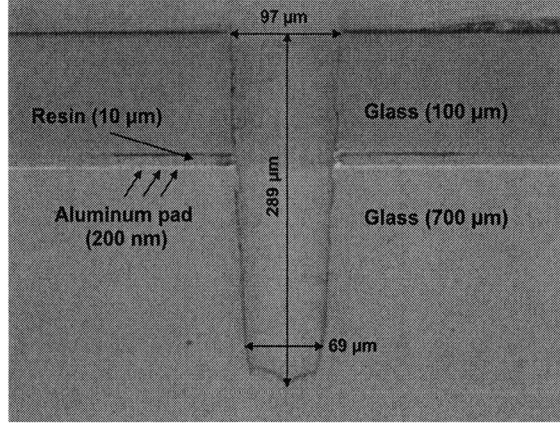

Figure 7. Cross section of laser drilled via (eximer laser, aluminum pad under resin layer).

In contrast, platinum has the highest melting point in Table 1. Figure 8 shows the cross-section images of platinum electrode samples. (1) and (3) in the figure show the via cross section drilled by the CO_2 laser. The via wall angle was around 82°. The platinum pads remained clearly visible near the via hole and were shaped like a bead, possibly due to the high energy treatment by the laser and high heat resistance of the platinum metal. (2) and (4) show the samples drilled by excimer laser. The wall angle was around 86.5°. The platinum pad shown in (4) did not remain around the via, possibly because it had been removed by the laser mechanically.

During the drilling process, the resin around the via was heated and dissipated by laser, and a dimple shape with a depth of around 10 μm was formed in the side wall of the via. After that, the electrode adhered to the upper glass could be removed. In the case of (3), the pad remained and the via diameter of the underside of the top glass was expanded. Figure 9 shows a schematic drawing of the via structure of the platinum-under-resin system drilled by excimer laser. In general, laser processing generates fume or debris that are then deposited on the surface of the drilled sample. The metal pad is also covered by at least some of them. Thus, if the pad is outcropped like (3) from the top view of the via, the fume or debris on the pad can be removed by reactive ion etching.

D. Solder injection into via holes

The drilled via holes are filled with metal to connect all pads through the via. The conventional metal filling process is electrical plating. As an alternative process, injection-molded-solder (IMS) technology (shown in Fig. 10) can be used to fill molten solder into vias directly. This technology has various advantages compared with electrical plating. Perhaps most importantly, a dry process like IMS does not use any chemicals, which makes it suitable for battery module fabrication where we want to avoid dipping the battery in a hot plating bath.

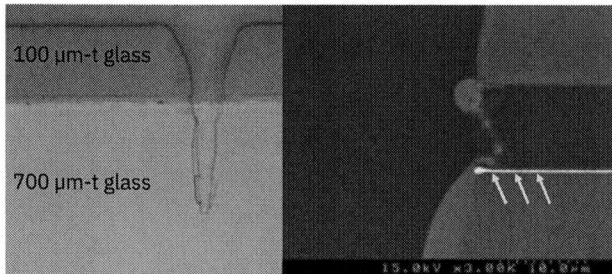

(1) CO₂ laser, platinum under resin

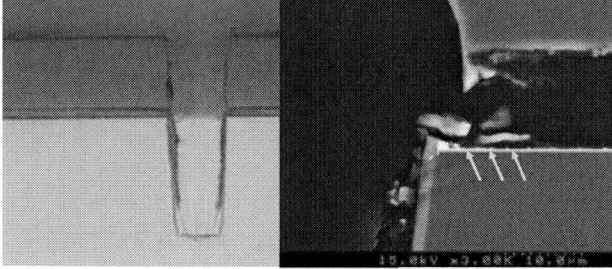

(2) Excimer laser, platinum under resin

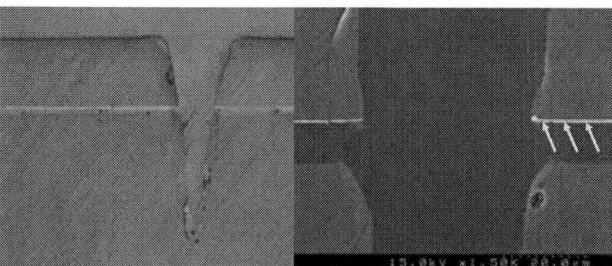

(3) CO₂ laser, platinum on resin

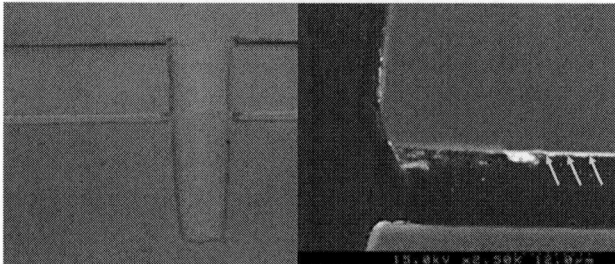

(4) Excimer laser, platinum on resin

Figure 8. Cross sections of laser drilled via (CO₂/excimer lasers, platinum pad under/on resin layer).

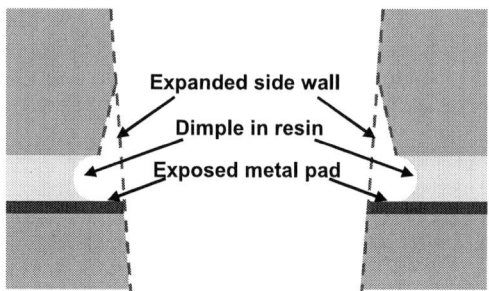

Figure 9. Schematic of side wall structure of platinum-under-resin system drilled by excimer laser.

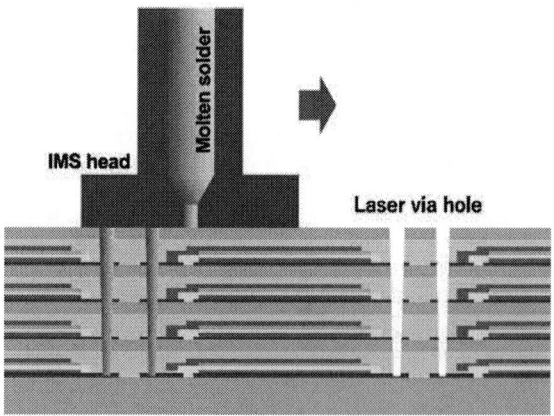

Figure 10. Schematic of via filling process by injection-molded-solder (IMS) system.

IV. DISCUSSION

Vertical interconnection technology for micro solid thin film battery was developed. We adopted laser machining for via drilling and filled the vias by solder to connect the stacked electrodes. We evaluated two kinds of laser, two structures, and six metals. Results showed that the excimer laser could make side walls that were more perpendicular than the CO_2 laser could. The perpendicular shape enables us to fabricate small diameter vias that maximize the active area in the battery with a small footprint. The two structures we examined assumed face-up and face-down batteries, respectively. In terms of forming electrodes protruding from the via side wall, the face-up battery structure (metal-under-resin) was better than the face-down one (metal-on-resin), at least in the case of a platinum electrode. Six metals were also evaluated. The platinum pad located under the resin was not damaged by excimer laser, and it seemed that platinum also contributed to expanding the side wall of nearby glass just on the electrode.

This vertical interconnection technology can contribute to enlarging the capacity of the STFB. The energy capacity of the current solid thin film lithium ion battery with 3-μm-thick $LiCoO_2$ is about 200 μAh/cm^2 [12]. Thus, a 4-layer micro battery with 1-mm footprint could achieve 3.4 μAh.

V. CONCLUSION

We demonstrated a novel vertical interconnection technology for a stacked micro solid thin film lithium ion rechargeable battery. In the technology demonstration, the process flow, the layer structure around the via, the electrode metal, and the laser conditions were identified. This technology enables low cost three dimensional (3D) STFB that can overcome the limited capacity of a mono-layer STFB while ensuring a limited footprint.

A high performance micro battery will be in strong demand for various IoT applications in the near future. In particular, medical and healthcare devices such as human activity monitoring devices attached to the skin or implanted in the body require a highly safe power source. Also, for industrial applications such as vibration or temperature logging in manufacturing tools or chemical plant reactors, high temperature operation which is one of the key advantages of STFB is required. The high heat resistance also enables the soldering process which is necessary for IoT module assembly.

In the future, small and intelligent IoT devices consisting of sensors, data processing units, transceivers, and a power source will help enable higher efficiency and higher quality manufacturing processes and reveal unexplored and complex medical and healthcare mechanisms.

ACKNOWLEDGMENTS

We would like to thank Four Universities [Keio, Waseda, TIT, and U. Tokyo] Nano/Micro Fabrication Consortium in NANOBIC for helpful support and technical advice on usage of their various process equipment.

REFERENCES

[1] N.J. Dudney, "Solid-state thin-film rechargeable batteries," pp. 245-249, 2005.

[2] A. Horibe, K. Sueoka, et al., "Through silicon via process for effective multi-wafer integration," Proc. IEEE 66th Electronic Components and Technology Conference (ECTC), May 2015, pp. 1808-1812, doi: 10.1109/ECTC.2015.7159844.

[3] K.Sueoka, A. Horibe, et al., "Vertical integration after stacking (ViaS) process for low-cost and low-stress 3D silicon integration," Proc. International 3D Systems Integration Conference (3DIC), Aug. 2015, pp. 147-151, doi: 10.1109/3DIC.2015.7334608.

[4] A. Horibe ; K. Sueoka ; R. Miyazawa; et al., "Solder injected through via for multi stacked wafers," Proc. IEEE 66th Electronic Components and Technology Conference (ECTC), May 2016, pp. 2176-2181, doi: 10.1109/ECTC.2016.232.

[5] Wolfgang shultz, Urs Eppet, and Reinhart Poprawe, "Review on laser drilling I. Fundamentals, modeling, and simulation," Journal of Laser Applications 25(1), 2013.

[6] Lucas A. Hof and Jana Abou Ziki, "Micro-hole drilling on glass substrates—A Review," *Micromachines 2017, 8, 53.*

[7] Lars Brusberg, Marco Queisser, Clemens Gentsch, Henning Schröder, Klaus-Dieter Lang, "Advances in CO_2-laser drilling of glass substrates," Physics Procedia, 39 2012, pp. 548-555.

[8] W. Lin, *et al.*, "Optimization of parameters CO_2 laser for drilling different types of glass," *New Journal of Glass and Ceramics*, 2015, 5, pp. 75-83.

[9] Deepa Bhatt, Karen Williams, David A. Hutt, and Paul P. Conway, "Process Optimisation and Characterization of Excimer Laser Drilling of Microvias in Glass," 9th Electronics Packaging Technology Conference, 2015, pp. 196-201.

[10] Mikael Broas, Kaya Demir, Yoichiro Sato, Venkatesh Sundaram, Rao Tummala, "A comparative reliability study of copper-plated glass vias, drilled with CO_2 and ArF excimer lasers," Proceedings of the 5th Electronics System-integration Technology Conference (ESTC), 2014, pp. 1-5.

[11] Y. Matsuda, N. Kuwata, J. Kawamura, "Thin-film lithium batteries with 0.3–30μm thick LiCoO2 films fabricated by high-rate pulsed laser deposition," J of Solid State Ionics 320 (2018), pp.38-44.

[12] A. Suzuki, S. Sasaki, I. Kimura, T. Jimbo, "Improvement of power characteristics of all-solid-state thin-film rechargeable lithium batteries," Abstract#499, Proceeding of 226th Electrochemical Society meeting (2015).

2019 IEEE 69th Electronic Components and Technology Conference (ECTC)

Characterization of fine pitch Hybrid Bonding pads using electrical misalignment test vehicle

Imed Jani, Didier Lattard, Pascal Vivet,
Lucile Arnaud, Severine Cheramy, Edith Beigné
CEA, LETI, MINATEC Campus
38054 Grenoble, France
imed.jani@cea.fr

Alexis Farcy, Joris Jourdon, Yann Henrion,
Emilie Deloffre, Halim Bilgen
STMicroelectronics
38926 Crolles, France
alexis.farcy@st.com

Abstract— Cu/oxide Hybrid Bonding (HB) technology is currently the ultimate fine pitch 3D interconnect solution to reach submicron pitches. It's an attractive technique to address the needs of several applications such as smart imagers, high-performance computing and memory-on-logic folding. But test and characterization of such fine-grained 3D interconnect is still an open issue; Cu-Cu interconnects are prone to many structural defects due to fabrication process, such as misalignment, which needs to be thoroughly tested to ensure the performance of 3D-ICs. In this work, we focus on testing and characterizing, on-wafer, misalignment defect induced at the bonding step. A misalignment test structure was fabricated in a Wafer-to-Wafer (W2W) assembly configuration with a pitch of 3.42µm and 1.44µm using a very small measurement step for an accurate misalignment measurement (respectively 45nm and 22nm). Electrical tests have been performed using five multi-pitch wafers with 71 measurements points per wafer. The experimental results show that the results of the proposed test structure are aligned with conventional overlay measurements. Finally, the impact of misalignment defect on resistance and capacitance parameters was demonstrated.

Keywords: 3D-IC, high density interconnects, Cu-Cu hybrid bonding, misalignment, RC parameters.

I. INTRODUCTION

The semiconductor industry is continuously demanding products with higher integration density, higher performance, lower power consumption and reduced cost. In recent years, 3D stacking integration has been providing an alternative way for higher performance and lower cost by vertically integrating multiple dice to reduce signal delays and power consumption. The 3D chips can be stacked using two main techniques: Wafer-to-Wafer (W2W) to address the needs of several applications such as smart imagers [1] and Die-to-Wafer (D2W) for high-performance computing applications [2]. The functional units are vertically stacked using inter-die interconnects (µ-bumps or Cu/oxide Hybrid Bonding (HB) interconnects) and intra-die interconnects Through Silicon Vias (TSV).

Compared to µ-bumps, Cu/oxide hybrid bonding is currently the ultimate fine pitch 3D interconnect solution with target a pitch of 1µm [3] and improved physical properties [4]. High interconnection density is an attractive technique for reliably connecting fine grained architectures

(processor, memory, digital and analog IC or MEMS...). However, Cu-Cu hybrid bonding interconnects are prone to many structural defects due to fabrication process. They thus need to be thoroughly tested to ensure the performance of 3D-ICs. With pitches in the range of 1µm, misalignment becomes a clear issue. Despite the rapid development of these new stacking technologies providing more and more interconnection density and even if stacking tools accuracy has been improved, providing accuracy down to ~200nm for Wafer-to-Wafer bonding and down to ~1µm for Die-to-Wafer bonding, some bonding defects between chips will still remain especially when targeting smaller pitches [5-7]. Therefore it is necessary to measure the alignment after bonding and electrically characterize the vertical interconnections in order to estimate the potential impact on the global performances.

Figure 1 shows misalignment and micro-voids defects in a Cu-Cu 3D-IC. Translation, rotation and the magnification effect (run-out) are the causes of local misalignment; the global translation and rotation value depend on bonding equipment accuracy and are steadily minimized by continued developments of commercial bond alignment tools. But the run-out effect, which relate to the wafer expansion due to the thermal stress, is still one of the most challenging issues in 3D-IC bonding process [8]. Misalignment affects electrical characteristics and can cause leakage current overheads, resulting in an undesirable conductive path between two adjacent Cu-Cu interconnects which decrease the life time of the 3D-IC [9]. µ-voids defects are in the form of cavity caused by practices at the contact surface.

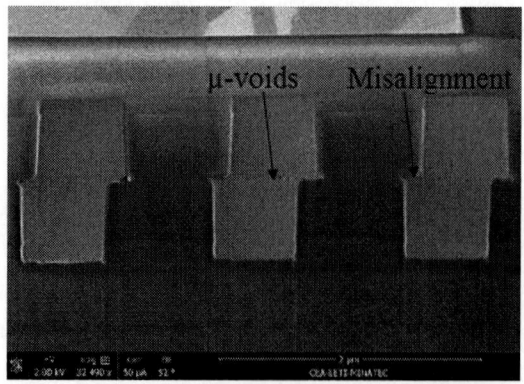

Figure 1.   Misalignment and µ-voids defects in Cu-Cu 3D-IC

978-1-7281-1500-9/19 $31.00 © 2019 IEEE

Many test methods have been proposed in the literature to control and evaluate bonding quality; the most used method is the Infra-Red (IR) metrology tool (IR light system and alignment marks) [10]. This module allows to control the fine movement of the wafers before bonding process [11]. Although both translational and rotational errors could be controlled and minimized during the alignment step, the run-out error remains a key contributor to the entire overlay error. Therefore, to handle a wafer without damaging it and to achieve higher alignment and overlay accuracy, the wafer should be affixed to flat using an electrostatic chuck [12]. After bonding process, the misalignment measurements for all wafer reticles require a long testing time. Other morphological investigations using and expensive tools such SEM (Scanning Electron Microscope) help visualizing defects at the bonding interface [13], but this method is destructive. Another alternative to detect misalignment is based on electrical measurements (resistance and capacitance) with two classical test structures (Kelvin and Daisy chain) [14], [15] but this method cannot measure accurately the bonding defects. These techniques are focusing on process characterization and cannot be combined in the 3D-IC test architecture to measure bonding misalignment.

This paper details the implementation, measurements and validation of a specific electrical structure firstly proposed in [16]. We obtain misalignment information such as direction and magnitude as well as an estimation of resistance values. The proposed structure implemented for technology development, before fabricating the 300 mm wafers with all Front-End-Of-Line (FEOL) and Back-End-Of-Line (BEOL) layers, using only the top metal layer (short-loops) to enable incremental test and characterization. Several test structures are necessary to validate the technological process and perform electrical characterization of Cu-Cu interconnects: yield, reliability and electro-migration [17]. For the purpose of our study, we deliberately generate a misalignment between the Cu-pads, for which we need to know the direction and the misalignment value after bonding. This test structure, used to target process development and characterization, has to be used in conjunction with more usual 3D DFT architecture such as [18] [19] to perform functional and structural test of 3D functional interfaces and to assess performance of 3D-ICs. In this work, we illustrate for the first time, the experimental test results using five wafers with 71 measurements points per wafer. Misalignment vector maps and statistical analyses are used to determine the cause of global misalignment. In a second part, in order to validate the misalignment test structure, we compare the test results with the overlay measurements managed by Infra-Red (IR) imaging using specific marks. In the third part, we analyze the impact of misalignment on resistance and capacitance parameters.

The paper is organized as follows: in part II we present the misalignment test structure; in part III, test results are detailed to validate the proposed test structure; in part IV we present the impact of misalignment defect on R & C parameters.

II. PROPOSED MISALIGNMENT TEST STRUCTURE

A. Overview

First of all, we developed a specific tool using MATLAB to exhibit defects on a wafer map. Figure 2 shows an example of simulation, the misalignment vectors shown in the wafer map are the results of the superposition of translation effect (T_X=150nm and T_Y=-250nm), rotation of 1µRad and a run-out of 4.3ppm (part per million). We show that the maximum local misalignment value is equal to 750nm. Equation (1) describes the local misalignment value. In complement of misalignment prediction, the developed tool will also be used later to analyze the fabricated wafers.

$$\text{Misalignment} = \sqrt{(X^2 + Y^2)} \qquad (1)$$

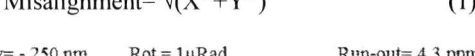

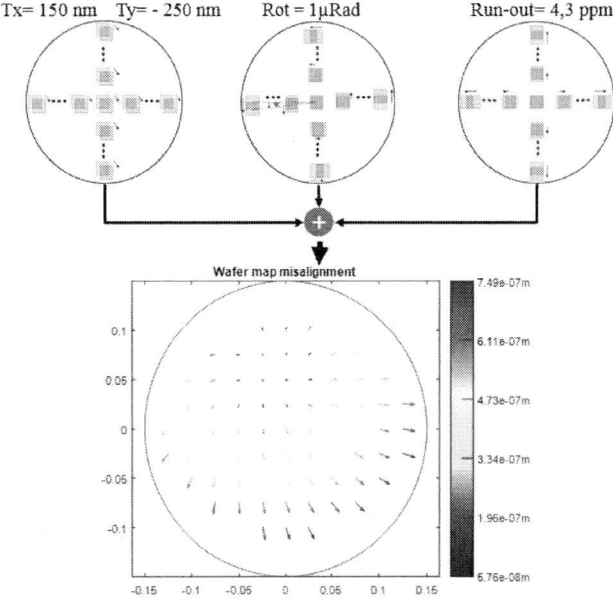

Figure 2. Misalignment wafer map

B. Principle

In order to measure the misalignment as shown on the wafer map of Figure 2, we described in [16] a specific structure based on alignment reference patters. A reference pattern (Figure 3), is composed by a single top Cu-pad and four bottom Cu-pads and Hybrid Bonding Vias (HBV).

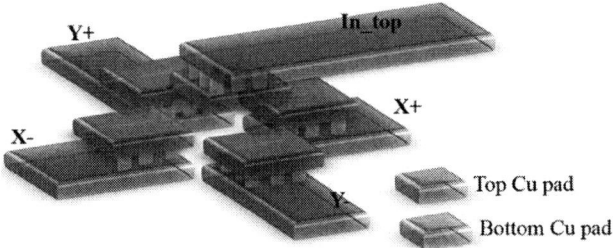

Figure 3. The reference Pattern of misalignment test structure (3D view).

To test the bonding misalignment and its direction, we inject a signal at the input (in_TOP) and we observe the binary digital outputs: X+, Y+, X-, and Y-. In case of misalignment, the top Cu-pad is in contact with one or more bottom Cu-pads. In case there is no contact between top and bottoms pads, we can conclude that the alignment is perfect.

C. Test structure architecture

The duplication of the reference pattern, using different spacing values (Offset_N) between bottom Cu-pads, offers the possibility to estimate the misalignment value. Figure 4 shows the architecture of the misalignment test structure. It contains 18 patterns (PA_0...PA_17) and a probing test row. To measure misalignment values, electrical test consists in current injection through X+, X-, Y+ and Y- and sensing outputs. The misalignment value in direction (X+,X-,Y+ or Y-) is equal to the offset value of the last pattern which has top/bottom connections (contact between top pad and the tested direction bottom pad).

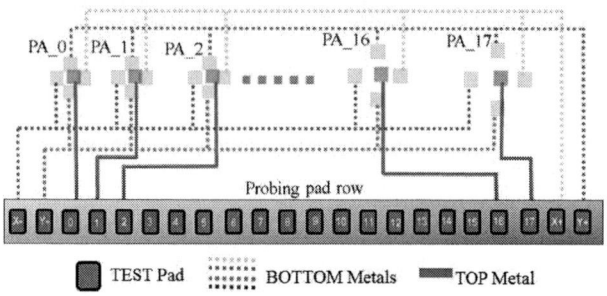

Figure 4. Misalignment test structure architecture

TABLE I. OFFSET VALUES (NM)

	Bloc X only 3.42μm Offset	Bloc Y only 3.42μm Offset	Step 3.42 (nm)	Bloc 1.44μm Offset (nm)	Step 1.44 (nm)	Z
PA_0	22.5			22.5		
PA_1		45		45		
PA_2	67.5			67.5		
PA_3		90		90		
PA_4	112.5			112.5		
PA_5		135	45	135		
PA_6	157.5			157.5	22.5	
PA_7		180		180		Z1
PA_8	202.5			202.5		
PA_9		225		225		
PA_10	247.5			247.5		
PA_11		270		270		
PA_12	292.5			292.5		
PA_13		315		315		
PA_14	360		X:67	360	45	
PA_15		540	Y:225	405		Z2
PA_16	810		X:450	495	90	Z3
PA_17		1260	Y:720	585		

The test vehicle was fabricated in a Wafer-to-Wafer (W2W) assembly configuration with either a pitch of 3.42μm and 1.44μm using a very small measurement step for an accurate misalignment test structure. TABLE I. shows the accuracy (step) and the offset values of the proposed misalignment test structure. We distinguish four zones; Z1, Z2 and Z3 corresponding to the coverage of the test structure while Z4 is out of the structure coverage. Due to test chip constraints, for the pitch of 3.42μm we have implemented X direction only and Y direction only with a double step of 45nm in Z1. While for the finest pitch (1.44μm), we have implemented X and Y with a step of 22.5nm to provide the best accuracy measurement.

III. TEST STRUCTURE VALIDATION

A. Test results: pitch 3.42μm and 1.44μm

1) Pitch 3.42μm

After the bonding process using W2W assembly, an electrical test was performed for five multi-pitch wafers. Figure 5 shows the misalignment distributions of the five wafers using misalignment test structure (pitch =3.42μm) with 71 measurement points for each wafer (superposition for some measurements). We distinguish four zones; Z1, Z2 and Z3 corresponding to the coverage of the test structure (see TABLE I) while Z4 is out of the structure coverage. The misalignment values of W2 and W14 are located in Z1 and those of W3, W4 and W13 are in Z2 and Z3 area. For instance, in real wafers misalignment values are all included in the capability of our test structure. In addition, our design provides high resolution (45nm) for a misalignment values between 0 and 300nm (Zone 1) but this resolution decreases for higher misalignment values. The test coverage of the misalignment test structure is equal to 0.8μm for X direction and 1.2μm for Y direction.

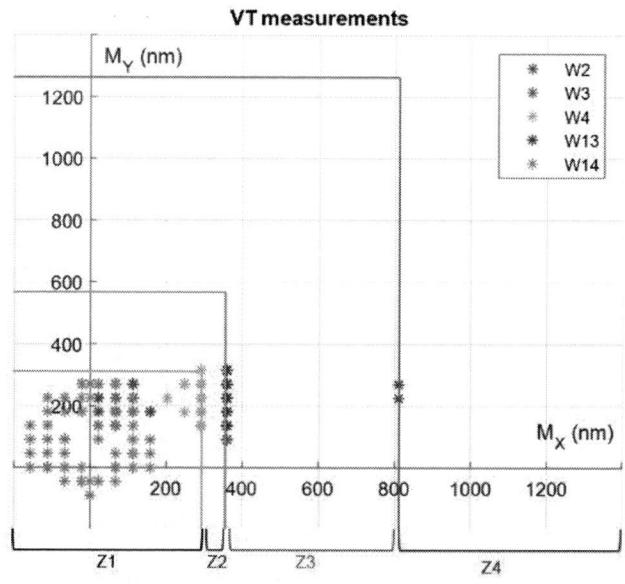

Figure 5. Misalignment test vehicle results (Pitch=3.42μm)

TABLE II shows minimum, maximum, mean and 3sigma values for wafers. We observe that the five wafers exhibit a different range of misalignment values; the minimum misalignment values are observed for W2 and W14 and the maximum misalignment values are observed for W13. The 3-sigma value represents the dispersion of misalignment values and explain the distorted wafers (run-out effects). For W2, W3, W4 and W13 it happens that translation is the predominant effect whereas run-out is predominant for W14.

TABLE II. STATICAL DATA OF MISALIGMENT TEST RESULTS (PITCH=3.42μM)

	W2		W3		W4		W13		W14	
Axis	X	Y	X	Y	X	Y	X	Y	X	Y
Max	158	270	360	315	360	320	810	320	158	270
Min	23	140	293	90	203	140	360	140	-160	-90
Mean	81	200	350	161	309	240	370	270	0	90
3 σ	102	90	72	147	117	120	210	150	267	330
Effect	Translation		Translation		Translation		Translation		Run-out	

2) Pitch 1.44μm

Figure 6 shows the misalignment distribution of the five wafers using the test structure (pitch=1.44μm) with 71 measurement points for each wafer (superposition for some measurements). We show that the misalignment values of W2 and W14 are in Z1 area, those of W3 are in Z2 and Z3 and that and those of W4 are in Z2 (including some results in Z1). For W13 the misalignments values are in Z4 (off cover zone). Moreover, for W13 (in X direction), there are some misalignment measurements that exceed the pad size (PS) and induce an artifact on the results of Y direction.

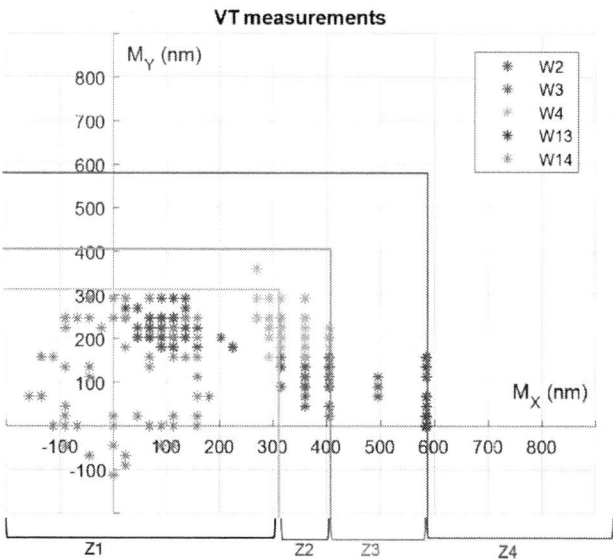

Figure 6. Misalignment test vehicle results (Pitch=1.44μm)

TABLE III shows the minimum, maximum, mean and 3sigma values for each wafer. W2, W3, W4 and W14 the measurements are very close to the 3.42μm pitch results, due to the small distance between the test structures as well as the offset values. W13 misalignment values (in X direction) are greater than our actual structure's coverage. For the rest

of the paper we will not take the W13 measurements into consideration. Figure 7 shows the wafer maps for W3 (Figure 7.a translation) and W14 (Figure 7.b run-out).

TABLE III. STATICAL DATA OF MISALIGNMENT TEST RESULTS (PITCH=1.44μM)

	W2		W3		W4		W13		W14	
Axis	X	Y	X	Y	X	Y	X	Y	X	Y
Max	225	293	495	292	405	360	585	157	180	293
Min	22	180	315	22	270	158	585	0	-158	-113
Mean	98	227	377	139	325	232	585	14	16	123
3 σ	118	91	122	176	103	113	0	117	292	364
Effect	Translation		Translation		Translation		Translation		Run-out	

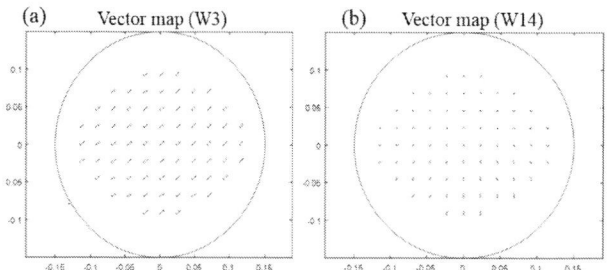

Figure 7. Misalignment wafer map:
(a) Wafer 3 and (b) Wafer 14

B. Comparison of misalignment test structure results with overlay measurements

1) Overlay measurements

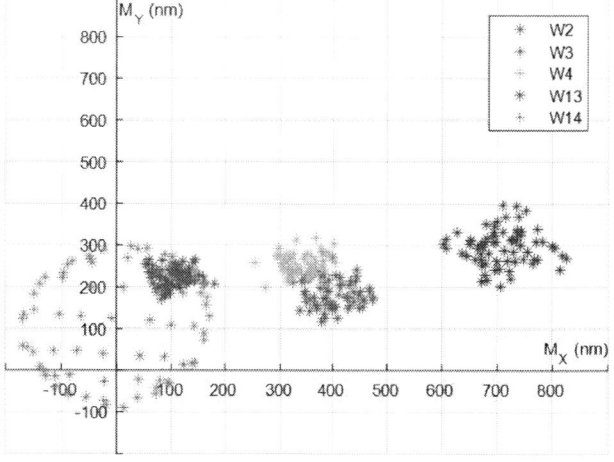

Figure 8. Overlay measurements destribution

The test results of the five wafers for the pitches of 3.42 and 1.44 are similar and offer a different range of misalignment values. In the next part, to validate the proposed test structure, we compare the test results with the overlay measurements obtained using IR imaging tool [20].

Figure 8 shows the overlay measurements distribution of the five wafers with 71 measurement points for each wafer.

2) Pitch 3.42μm

The comparison between overlay measurements and test structures results is done using "boxplot" plots. On each box, the central mark indicates the median, and the bottom and top edges of the box indicate the 25th and 75th percentiles, respectively. The whiskers extend to the most extreme data points not considered outliers, and the outliers are plotted individually using the '+' symbol. We differentiate between overlay measurements and misalignment test vehicle results for each wafer and for 71 measurement points. Figure 9 shows the comparison results for the five wafers in X and Y direction. For W2 and W14, we show that the most of difference values are smaller than the resolution in Z1 (45 nm), for W3 and W4 the difference is smaller than the resolution of Z2 (90nm) and Z3 (X/Y: 450/720nm) and for W13 the difference values are very high since the measurements are in Z3.

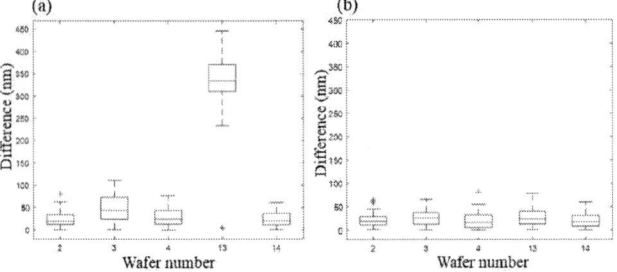

Figure 9. Comparison of test structure results (3.42μm pitch) and overlay measurements in X and Y direction

3) Pitch 1.44μm

Figure 10shows the comparison for the five wafers in X and Y direction; concerning W2 and W14 we show that the majority of difference values are less than the resolution in Z1 (22.5 nm), for W4 and W3 the differences is respectively less than the resolution of Z2 (45nm) and Z3 (90nm). Then for W13 the differences values are very high. Because for X direction the misalignment values are in Z4 (off cover zone) and some misalignment values exceeds the pad size, it is not possible to measure misalignment in Y direction using this test structure. Therefore, we do not include the results of W13 in the comparison. In conclusion for the wafers W2, W3, W4 and W14, we can affirm that the test structure results are aligned with conventional overlay measurements.

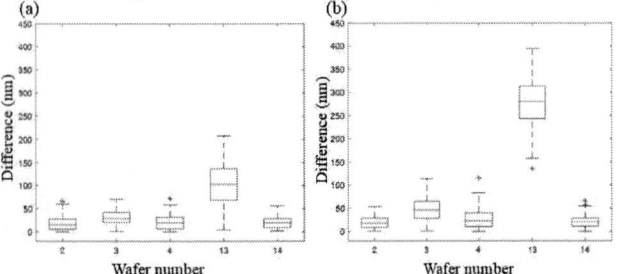

Figure 10. Comparison of test structure results (1.44μm pitch) and overlay measurements in X and Y direction

After comparison, we can claim that the test results of misalignment test structure with 3.42μm and 1.44μm pitch are aligned with conventional overlay measurements. The accuracy of misalignment measurements, using the proposed test structure, depends on offset values.

IV. IMPACT OF MISALIGNMENT ON R&C PARAMETERS

In this section, we theoretically investigate the impact of misalignment defect on electrical parameters (R and C) of Cu-Cu interconnect. Figure 11.a illustrates the equivalent electrical model of Cu-Cu interconnect used in simulation to define our test structures. The pad resistance (R_pad) is a function of the effective contact surface. The resistance of Hybrid Bonding Vias (R_HBV) is a function of the number of vias connecting the last metal layer to the Cu-pads. Cp is the parasitic capacitance due to the distance between two adjacent Cu pads.

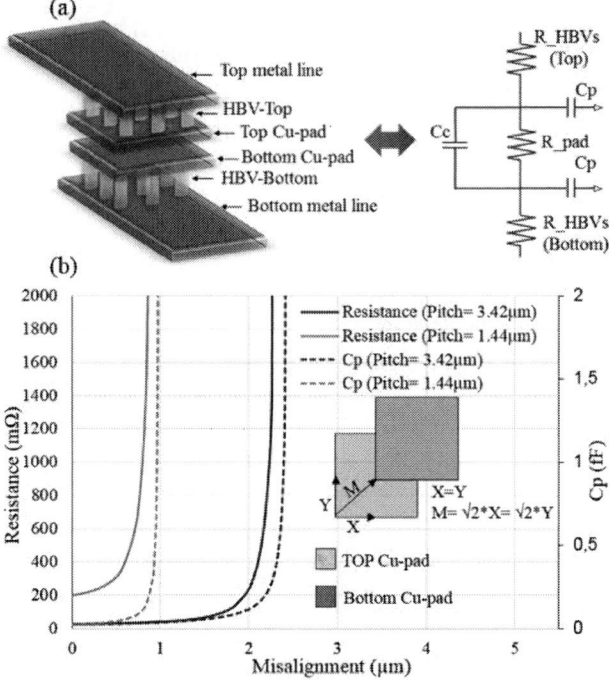

Figure 11. Impact of misalignment defect:

Based on the electrical model of Cu-Cu interconnects [19], we studied the impact of misalignment and μ-voids on the capacitance and resistance values. Figure 11.b shows the variation of pad resistance (R_pad) and parasitic capacitance (Cp) as a function of the misalignment value for the pitches of 3.42 and 1.44 obtained using (1) (X=Y). The highest values of R_pad and Cp are obtained at the limit of misalignment for a given pitch; i.e. for a pitch of 1.4 μm (Pad-size=Pitch/2), the limit of misalignment $\approx$ (0.7² + 0.7²)1/2 $\approx$ 0.99μm. The optimal pad resistance values (corresponding to the perfect alignment M $\approx$ 0) vary from one pitch to another: they depend on the pad size and the number of HBVs (equal respectively to 9 and 1 for the pitches of 3.42 and 1.44μm). For a given misalignment value, small pitch pads are clearly more sensitive to the increase of R and C values.

In second step, we study the impact of misalignment defect on R and C parameters for the wafers used in III example W14 with a pitch of 1.44μm (because of the variation of misalignment values in the case of the run-out effect) using MATLAB tool. Figure 12.a and Figure 12.b show respectively the resistance (of Cu-Cu pads and HBVs) and capacitance wafer map distribution (the run-out is the predominant effect). The maximum misalignment values in X and Y direction are equal to 180nm and 293nm respectively and the pad size (PS) is equal to 720nm. While the effective contact surface remains large, there is not a great variation in the resistance values; the minimum value is equal to 202mΩ and the maximum is equal to 237mΩ. Concerning capacitance values, the distance between two adjacent Cu pads is large, so we do not have a big variation (min=0.042fF, max=0.047fF). On the contrary, the simulation of the electrical model of Cu-Cu pad (with a pitch of 1.44μm) shows that a 700nm misalignment value in X and Y direction increases significantly the resistance from 197 mΩ to 50Ω and the capacitance from 0.02fF to 0.74fF.

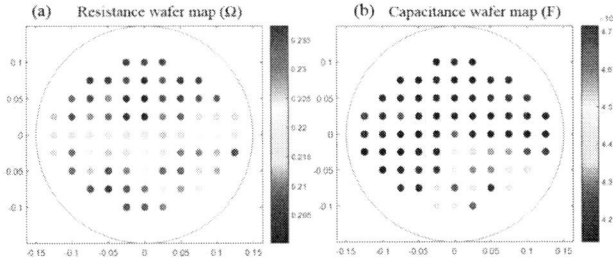

Figure 12. Resistance and capacitance values destribution (W14 with a pitch=1.44μm)

V. CONCLUSION

High density 3D-IC poses new challenges for test and characterization. In this work, we presented a test structure dedicated to bonding misalignment measurements with high accuracy for process development and characterization. The test structure was fabricated in a W2W assembly configuration with pitches of 3.42μm and 1.44μm. Silicon measurements of misalignments values using the proposed test structure are consistent with conventional overlay measurements. The impact of misalignment defect on electrical parameters of Cu-Cu interconnects has been studied, the simulations show that high density 3D-IC interconnects are very sensitive to misalignment defect. So it is very interesting to use the same low cost test structure, with several test patterns to increase the accuracy of misalignment measurements, in conjunction with more usual 3D Design-For-Test (3D-DFT) architecture to perform functional and structural test of 3D interfaces in 3D-IC.

ACKNOWLEDGMENT

This work was funded thanks to the French national program "Programme d'Investissements d'Avenir, IRT Nanoelec" ANR-10-AIRT-05 and IPCEI DEMO3S project.

REFERENCES

[1] T. Haruta et al., "4.6 A 1/2.3inch 20Mpixel 3-layer stacked CMOS Image Sensor with DRAM," 2017 IEEE International Solid-State Circuits Conference (ISSCC), San Francisco, CA, 2017, pp. 76-77.

[2] P. D. Franzon et al., "Computing in 3D," 2015 International 3D Systems Integration Conference (3DIC), Sendai, 2015, pp. TS6.1.1-TS6.1.2.

[3] J. Jourdon et al., " Hybrid bonding for 3D stacked image sensors: impact of pitch shrinkage on interconnect robustness," IEDM, 2018

[4] C.Sart, R. Estevez, V. Fiori, S. Lhostis, G. Parry, R. Gonella, "Numerical and experimental investigations on the hybrid bonding of Cu/SiO2 patterned surfaces using a cohesive model," in 2016 17th International Conference on Thermal, Mechanical and Multi-Physics Simulation and Experiments in Microelectronics and Microsystems (EuroSimE), 2016, pp. 1–8.

[5] B. Rebhan et al., "< 200 nm Wafer-to-wafer overlay accuracy in wafer level Cu/SiO2 hybrid bonding for BSI CIS," in 2015 IEEE 17th Electronics Packaging and Technology Conference (EPTC) , pp.1-4.

[6] I. Sugaya, H. Mitsuishi, H. Maeda, and K. Okamoto, "New precision wafer bonding technologies for 3DIC," in 3D Systems Integration Conference (3DIC), 2015 International, 2015, pp. TS7–1.

[7] W. H. Teh et al., "Post-bond sub-500 nm alignment in 300 mm integrated face-to-face wafer-to-wafer Cu-Cu thermocompression, Si-Si fusion and oxideoxide fusion bonding," in 3D Systems Integration Conference (3DIC), 2010 IEEE International, 2010, pp. 1–6.

[8] J. De Vos et al., "Importance of alignment control during permanent bonding and its impact on via-last alignment for high density 3D interconnects," in 3D Systems Integration Conference (3DIC), 2016 IEEE International, 2016, pp. 1–5.

[9] F. Kurz et al., "High Precision Low Temperature Direct Wafer Bonding Technology for Wafer-Level 3D ICs Manufacturing," ECS Transactions, Sep. 2016, vol. 75, no. 9, pp. 345–353.

[10] T. Fukushima et al., "Multichip Self-Assembly Technology for Advanced Die-to-Wafer 3-D Integration to Precisely Align Known Good Dies in Batch Processing," IEEE Transactions on Components, Packaging and Manufacturing Technology, vol. 1, no. 12, pp. 1873–1884, Dec. 2011.

[11] L. Xie, S. Wickramanayaka, S. C. Chong, V. N. Sekhar and D. I. Cereno, "High-Throughput Thermal Compression Bonding of 20 um Pitch Cu Pillar with Gas Pressure Bonder for 3D IC Stacking," 2016 IEEE 66th Electronic Components and Technology Conference (ECTC), Las Vegas, NV, 2016, pp. 108-114.

[12] I. Sugaya et al., "High-Precision Wafer-Level Cu–Cu Bonding for 3-DICs," in IEEE Transactions on Electron Devices, vol. 62, no. 12, pp. 4154-4160, Dec. 2015.

[13] A. K. Panigrahi, S. Bonam, T. Ghosh, S. R. K. Vanjari and S. G. Singh, "High Quality Fine-Pitch Cu-Cu Wafer-on-Wafer Bonding with Optimized Ti Passivation at 160°C," 2016 IEEE 66th Electronic Components and Technology Conference (ECTC), Las Vegas, NV, 2016, pp. 1791-1796.

[14] Lan Peng et al., "3D-SoC integration utilizing high accuracy wafer level bonding," 2016 IEEE Electronics Packaging Technology Conference (EPTC) ECS Transactions, 30 November 2016, Singapore.

[15] K. Shibin, V. Chickermane, B. Keller, C. Papameletis, and E. J. Marinissen, "At-Speed Testing of Inter-Die Connections of 3D-SICs in the Presence of Shore Logic," in Asian Test Symposium (ATS), 2015, pp. 79–84.

[16] I. Jani, D. Lattard, P. Vivet, and E. Beigné, "Innovative structures to test bonding alignment and characterize high density interconnects in 3D-IC," in New Circuits and Systems Conference (NEWCAS), 2017 15th IEEE International, 2017, pp. 153–156.

[17] L. Arnaud et al., "Fine pitch 3D interconnections with hybrid bonding technology: From process robustness to reliability," 2018 IEEE International Reliability Physics Symposium (IRPS), Burlingame, CA, 2018, pp. 4D.4-1-4D.4-7.

[18] J. Durupt, P. Vivet, and J. Schloeffel, "IJTAG supported 3D DFT using chiplet-footprints for testing multi-chips active interposer system," in Test Symposium (ETS), 2016 21th IEEE European,2016,pp. 1-6

[19] I. Jani, D. Lattard, P. Vivet, L. Arnaud and E. Beigné, "BISTs for post- bond test and electrical analysis of high density 3D interconnect defects," 2018 IEEE 23rd European Test Symposium (ETS), Bremen, 2018, pp.1-6.

[20] A. Jouve et al., "1μm Pitch direct hybrid bonding with< 300nm wafer-to-wafer overlay accuracy," in SOI-3D-Subthreshold Microelectronics Technology Unified Conference (S3S), 2017 IEEE, 2017, pp. 1–2.

Dynamic Characteristics Evaluation on NCF under Challenging Conditions and Its Application

Tomonori Nakamura, Hiromi Shibahara,
Osamu Watanabe, Tetsuya Utano, Daisuke Tani,
Sung Chenhsiu, Toru Maeda, Doug Day
Die Attach Business Unit
Shinkawa LTD
Tokyo, Japan
tomo-nakamura@shinkawa.com

Hidekazu Yagi, Ryoji Kojima, Daichi Mori,
Tatsuo Nagamatsu, Junichi Kaneko
Product Development Section 7,
Product Development Department
Dexerials Corporation
Tokyo, Japan
Daichi.Mori@dexerials.com

Abstract—**We demonstrate a dynamic characteristics evaluation system using a flip chip bonder for non-conductive film (NCF) and its applications for analysis and design of the processes. Under thermally transient conditions, quasi-static viscosity of NCF which has minimum viscosity of 2900Pas at 139°C, as measured by a rheometer under a ramp rate of 0.167°C/s, drastically decreases to 52.6 Pas at 296.4°C when accompanied by increased ramp rate up to 200°C/s. These dynamic conditions are similar to the actual thermal compression bonding (TCB) process, allowing the conclusion that analysis and simulation of the TCB process should be based on the dynamic characteristics. We also show analytical results for solder flow-out, which is problematic with the TCB process, and propose a new TCB process simulation sequence based on dynamic characteristics evaluation.**

Keywords-component; 3D IC; Thermal compression bonding; NCF; Reliability; Analysis and optimization of process;

I. INTRODUCTION

For semiconductor packages, numerous resins are used for bonding dice to base materials such as lead frames, build-up substrates, and wafers including interposers. All of the resins have multiple functions including providing physical strength for devices, protection from contamination, and suppression of stress.

For 3D stacked ICs typical of high-end memory like High Bandwidth Memory (HBM), non-conductive film (NCF) is widely used because of its feasibility for finer dice pitches. Thermal compression bonding using NCF (TC-NCF) is typically applied for HBMs to interconnect fine solder bumps through processes that melt and harden solder while simultaneously curing the NCF [1-6]. Evaluation of the dynamic properties of the NCF at a temperature near the melting point of the solder, specifically the temperature transitional properties, is very important to ensure the optimum solder shape and the gaps between dice. With the TC-NCF process, NCF is exposed to challenging conditions including a snap thermal ramp up rate (RR) of about 300°C/s, and a cooldown rate of 100°C/s. Because there is usually only a referential rheometric database available for NCF, which is measured under a slow ramp rate below 0.1°C/s, the dynamic NCF characteristics exhibited during actual bonding is only assumption. This is one of the most difficult challenges to optimizing the NCF bonding process.

In this study, we propose an evaluation method for the dynamic characteristics of NCF based on flip chip bonding. We discuss the basic dynamic characteristics of NCF during the TC-NCF process, also propose a new process optimization method to determine optimum process parameters based on the NCF dynamic characteristics.

II. DYNAMIC CHARACTERISTICS EVALUATION SYSTEM

A key parameter of NCF is viscosity, which is a factor affecting the gap between a die and a substrate, as well as solder shape. To measure viscosity, there are 3 major systems: indentation method, parallel plate method, and rotation plate method, as shown in Fig. 1 [7]. When considering these measurement systems compared to actual bonding, the parallel plate method is the most similar to the bonding system and thus most suitable for measuring the dynamic characteristics. Therefore, we selected this method to evaluate the materials' dynamic characteristics and developed an NCF compression method using a flip chip bonder.

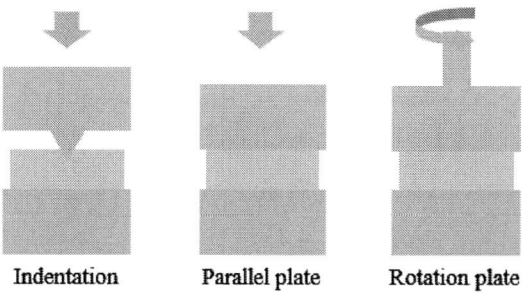

Figure 1. Schematic illustration of viscosity evaluation methods.

Fig. 2 shows a schematic illustration of the evaluation system. The flip chip bonder closely controls compression force, as well as temperatures of the stage and tool, and the ramp rate of the tool. Under a constant force, the bond head position reflects the NCF viscosity. The equation is written as:

$$\eta = 2\pi F H^5 / 3V(-dH/dt)(2\pi H^3 + V)\ldots(1)$$

Where, η is viscosity, F is force, H is thickness of NCF, and V is volume of NCF, respectively. Based on this fundamental system and equation, the evaluation experiment is constructed. Equation (1) does not have a temperature component. To evaluate dynamic characteristics for transient temperature, instantaneous values should be obtained by controlling temperatures of stage and tool, and the ramp rate of the tool during the evaluation.

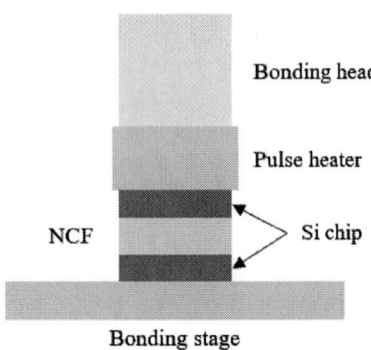

Figure 2. Schematic illustration of evaluation system using flip chip bonder.

III. DYNAMIC CHARACTERISTICS EVALUATION OF NCF

Firstly, we will clarify the differences between the quasi-static and dynamic characteristics of NCF. We measured NCF samples with a thickness of 20um as supplied by Dexerials Corporation, having minimum viscosity of 2900Pas at 139°C measured by a rheometer as illustrated in Fig. 3. The RR was 0.167°C/s. We define the quasi-static characteristics as the values obtained by the rheometer. To evaluate the dynamic characteristics, we utilized an LFB-1102 Super, which is a flip chip bonder produced by Shinkawa Ltd., having high precision control units for force and head position, in addition to a high-speed pulse heater with an RR over 200°C/s. With these controlled process systems, we can readily manage the experimental conditions. The NCF is attached between bare Si dice to simulate bonding, as shown in Fig. 2. The sample was placed onto the LFB-1102 Super bonding stage and compressed with a constant force of 30N. The stage temperature was set to 80°C, which is close to the actual bonding condition, while pulse heater temperatures were varied up to 300°C, depending on experimental conditions.

The dynamic characteristics of NCF under the highest RR condition of 200°C/s are shown in Fig. 4, which is close to the RR of an actual TCB process. The minimum viscosity of

52.6Pas at 296.4°C is obtained from Fig. 4. To investigate the details, we measured the RR dependency of the dynamic characteristics.

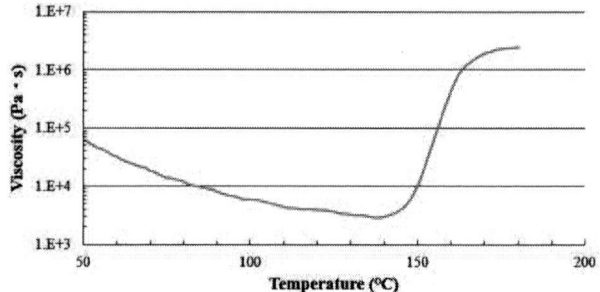

Figure 3. Viscosity of NCF measured by rheometer.

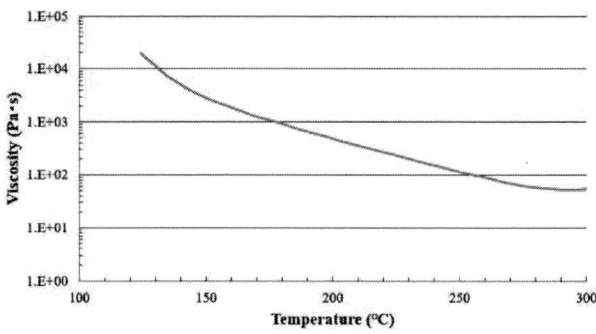

Figure 4. Dynamic characteristics of NCF at ramp rate of 200°C/s.

We varied the pulse heater RR as a function of NCF viscosity. The RRs were 3°C/s, 30°C/s, and 100°C/s. Fig. 5 depicts these data and obtained results from Fig. 4. The minimum viscosity and the temperatures were strongly dependent upon the ramp rates. These measured minimum viscosities and the related temperatures are 947.2Pas and 179.2°C at 3°C/s, 199.2Pas and 190.1°C at 30°C/s, and 92.5Pas and 245.9°C at 100°C/s. The NCF cure rate becomes slower than its softening rate under a high ramp rate. In case of RR of 3°C/s, a difference in viscosity is still observed although the values become closer to those shown in Fig. 3, implying that the chemical reaction mechanism differs from that of the quasi-static case. Incidentally, a peak before hardening was observed. We speculate that these peaks relate to the glass transition point. More detailed studies are needed to determine the cause.

To clarify the variation of minimum viscosity, those data are plotted as a function of heating time. Fig. 6 shows the heating time effect on viscosity at the RR levels described above. Increase of NCF viscosity depends on heating time. It is suggested that NCF curing is affected by total heat application, namely the integral of temperature and heating time.

Summarizing dynamic characteristics evaluation, behavior of NCF viscosity is drastically changed by RR. Although quasi-static characteristics evaluation using a rheometer is a good method to identify the physical properties of the material, dynamic characteristics evaluation should be utilized to understand the total TC-NCF process. We propose this evaluation in the following sections.

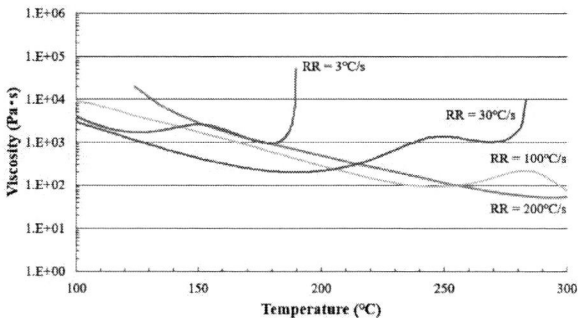

Figure 5. RR dependence on NCF dynamic characteristics.

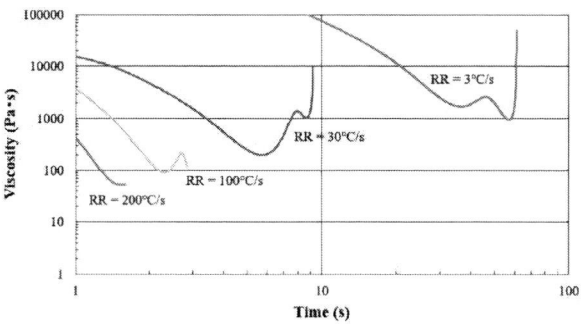

Figure 6. Viscosity dependence on heating time at various RR.

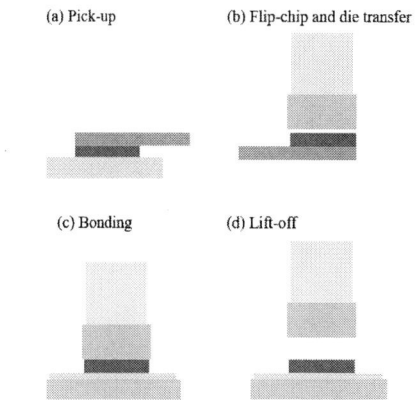

Figure 7. Schematic illustration of flip-chip bonding.

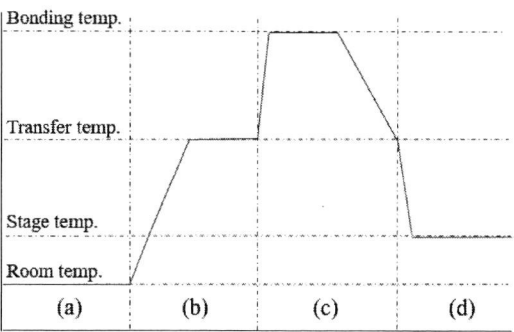

Figure 8. Die temperature excursion for TCB process sequence along fig. 7.

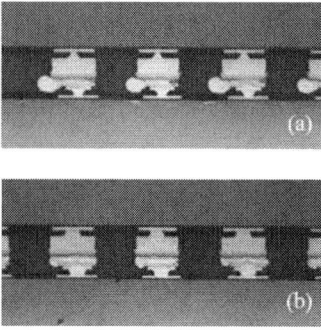

Figure 9. (a) Solder flow-out and (b) without solder flow-out.

IV. TCB PROCESS ANALYSIS USING DYNAMIC CHARACTERISTICS EVALUATION OF NCF

Dynamic characteristics evaluation is necessary for analysis of an actual TCB process. Fig. 7 depicts the TCB process sequence using a flip chip bonder while the bonded die temperature excursion is shown in Fig. 8. After pick up, the die is transferred to the bonding head. During production, the temperature at die transfer from the pick-up tool to the bond head (TT) is one of the key parameters because cooling time, along with its impact on overall productivity, depends on the temperature difference between bonding and transfer temperatures. After the bond head receives the die, it is bonded onto a substrate or a wafer and then the bond head is removed from the die. Bonding includes both the heating up and cooling down processes.

To analyze NCF characteristics, we have to trace the bonding sequence. The proposed evaluation system has good reproducibility of the process sequence due to usage of a flip chip bonder, which closely controls each process step. This section describes the acquisition method of actual NCF characteristics during execution of the bonding process.

One well-known problem of TC-NCF is solder flow-out. In Fig. 9(a) solder has flown out of the joints and the joint shapes are asymmetric. The asymmetry affects signal transmission between dice. In the worst case, electrical shorting is caused between pillars. To address the problem, we discover that TT plays an important role in suppressing the solder flow-out, as shown in Fig. 9(b).

A simple model is proposed for the origin of the flow-out, as shown in Fig. 10. NCF is compressed before the solder caps touch down on the bumps, then solder melting and NCF curing occur at the same time. In case of high flowability of NCF depending on the viscosity, the NCF flows from the die center to the outside easily due to compression displacement and the NCF carries melted solder with it. Controlling NCF hardness is important for suppression of flow-out. From experimental results, TT affects this phenomenon.

Low force bonding below 100N/cm^2 has emerged as a recent TC-NCF process requirement since the latest devices are constructed with fragile parts. In this case, the solder flow-out problem becomes an even more significant cause of potential defects. In our case study of the phenomenon, the NCF was tuned by Dexerials Corporation to facilitate low force bonding. Therefore the viscosity during bonding is lower than other cases and may have further exacerbated this problem. In addition, through experimentation, we found there was TT dependence for the flow-out problem. The TCB sequence, when solder shape in Fig. 9(b) is achieved, is shown in Fig.11.

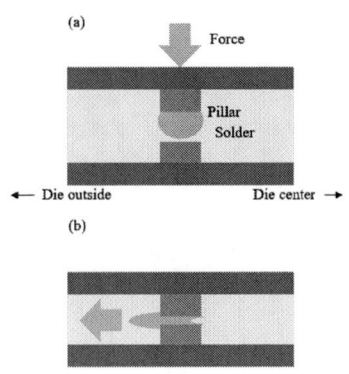

Figure 10. Model of solder flow-out.

Fig. 11 shows the bonding force and temperature profiles without solder flow-out after bond head touchdown. Where bonding temperature was 250°C, TT was 160°C, stage temperature was 120°C and bonding force was 70N. The bonding was done using force control mode. After a die is transferred to the bond head, it takes 200ms to transport it from the transfer position until touch down on the substrate. At around 0.7s, the force is decreased drastically, with the expectation that the solder melts at this timing. The force is increased again to 70N for 0.5s, suggesting that NCF resistance is generated, namely, NCF hardening may have occurred.

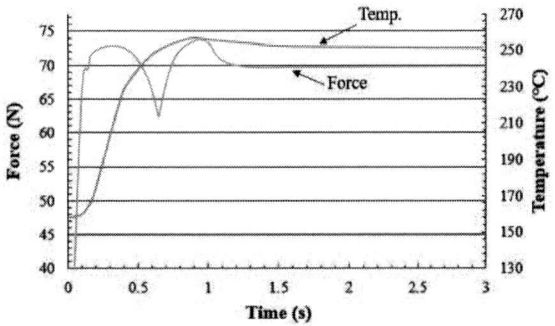

Figure 11. Bonding profile without solder flow-out.

During the sequence, we performed dynamic characteristics evaluation for the NCF. The obtained result is shown in Fig. 12. Minimum NCF viscosity was estimated to be 83.2Pas at 0.7s and 236.9°C. Under that condition, solder was melted when following the profile in Fig.11.

Another problem with the TCB process is that entrapped NCF filler is subject to being trapped between solder and bumps [4]. However, in Fig. 9(b), we did not observe these defects, therefore, minimization of the NCF viscosity at solder melting plays a key role in elimination of encapsulated filler. At 1.5s, the viscosity reached around 1000Pas. The increase of NCF viscosity is important in maintaining solder shape.

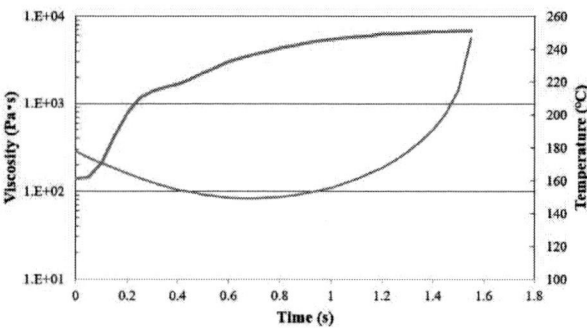

Figure 12. Dynamic characteristics measurement of sequence of bonding profile shown in Fig.8.

Clearly, dynamic characteristics evaluation provides data on the behavior of NCF under actual bonding conditions, while no such information is obtained in quasi-static rheometric measurement. The obtained results lead to the conclusion that dynamic characteristics evaluation is a useful method to confirm operation of the actual bonding system. As further application of the method, we propose TCB process fabrication using the approach described in the next section.

V. PROCESS FABRICATION AS APPLICATION OF DYNAMIC CHARACTERISTICS EVALUATION

In the previous section, we demonstrated that dynamic characteristics evaluation is useful for analysis of the TCB process. The reverse is also true: by first understanding the NCF characteristics, we can more easily optimize the TCB process.

As described in section IV, TT is a key parameter for suppression of solder flow-out. TT dependence on NCF viscosity can be acquired by evaluation with a flip chip bonder. A simplified experimental profile for temperature and force is shown in Fig. 13. First, the bonding head, which is heated up to TT, touches down on a substrate with low force of 5N for 2s to simulate the die transfer process. Fig. 13 depicts TT of 150°C. After annealing, the NCF is compressed at 30N and the bonding tool is ramped up to 260°C at a ramp rate of 300°C/s.

The experimental results are shown in Fig.14. After the bond head was placed onto the NCF-attached samples at a bond force of 5N, the bond head was kept at 150°C, 175°C, 200°C, and 225°C, each corresponding to its respective TT, for 2s. The dynamic characteristics shown in Fig. 14 appear during compression at 30N. The time required to reach a viscosity of 1000 (T1000) is strongly dependent upon TT, allowing the conclusion that NCF curing time depends on TT. In actual bonding, this information is important for selecting process parameters. To determine the viscosity under other TT

978-1-7281-1500-9/19 $31.00 © 2019 IEEE   1936

conditions, the obtained data is plotted according to the Arrhenius equation.

The Arrhenius plot for T1000 is shown in Fig.15. From that plot, the activation energy of T1000s was calculated to be 0.085eV. By using the activation energy, optimized TT for a particular device can be obtained without any further experiments. The obtained results indicate that dynamic characteristics evaluation is effective in determining the NCF process window for reliable process optimization.

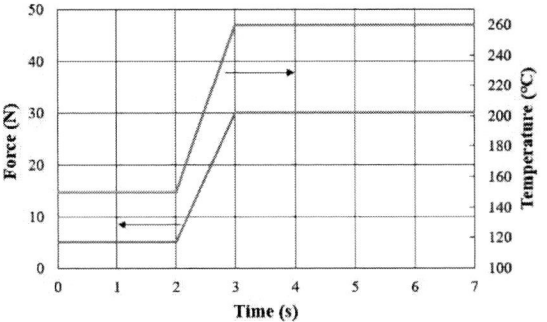

Figure 13. Schematic illustration of the bonding profile.

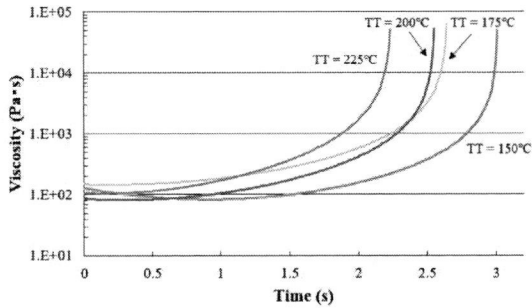

Figure 14. TT dependence of NCF viscosity during bonding sequence.

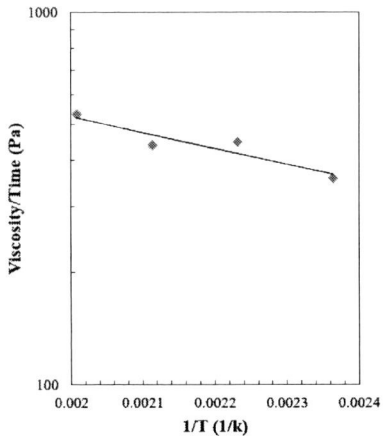

Figure 15. Arrhenius plot for Viscosity/T1000 on Fig.14.

Process optimization sequences could be drastically improved by using dynamic characteristics evaluation. As described above, we can measure the characteristics modeled on an actual bonding sequence. For example, we can set the process parameters using the flowchart shown in Fig. 16.

When process optimization is started, the bonding tool temperature should be measured, since it relates to the solder melting point on real devices and also because it depends on machine and device structures. After that, TT dependence on viscosity is measured and the Arrhenius plot is made. Incidentally, under this measurement, some data collection will be needed depending on the available measurement system in the machine. The Arrhenius plot can indicate suitable TT for an actual device. Final process establishment is determined by checking solder shape and other defects on actual device using the TT.

As NCF dynamic characteristics become observable, the process optimization sequence should be simplified. The described measurement system is also applicable for other resins, such as Non-Conductive Paste (NCP), Die Attach Film (DAF), Ag paste, and Anisotropic Conductive Film (ACF) and Paste (ACP).

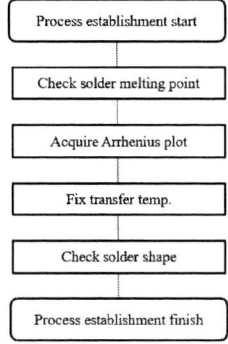

Figure 16. Process establishment flowchart using dynamic characteristics evaluation.

VI. CONCLUSION

We demonstrated that a dynamic characteristics evaluation system using a flip chip bonder for NCF is applicable for analysis and optimization of a TCB process through basic experimental results. Dynamic characteristics evaluation clarifies the viscosity behavior under high RR over 200°C/s, similar to actual bonding conditions. For problems with TCB such as solder flow-out, the evaluation leads to an analytical solution. The TCB process optimization sequence is also dynamically changed because ideal NCF conditions matching the desired device structure should be determined by the evaluation.

Expanding the range of possible applications of the method, dynamic characteristics evaluation can support development of new NCF materials. Moreover, for other resins including NCP, DAF, Ag paste, ACF, and ACP, the evaluation method can also be applicable. The proposed system is relevant for the design of packaging processes and related materials.

REFERENCES

[1] M. Koyanagi, H. Kurino, K.W. Lee, K. Sakuma, N. Miyakawa, and H. Itani, "Future System-on-Silicon LSI chips," IEEE Micro, 18 (4), 1998, pp.17-22.

[2] S. Joblot, A. Farcy, N. Hotellier, A. Jouve, F. de Crecy, A. Carnier, C. Ferrandon, J. P. Colonna, R. Franiatte, C. Laviron, and S. Cheramy,"Wafer Level Encapsulated Materials Evaluation for Chip on Wafer (CoW) Approach in 2.5D Si Interposer Integration," in Proc. IEEE 3D Systems Integration Conf. (3DIC), 2013, pp.1-7.

[3] M. Lee, M. Yoo, J. Cho, S. Lee, J. Kim, C. Lee, D. Kang, C. Zwenger, and R. Lanzone, "Study of Interconnection Process for Fine Pitch Flip Chip," in Proc. IEEE Electronic Components and Technol. Conf. (ECTC), 2009, pp.720-723

[4] T. Nagamatsu, K. Honjo, K. Ebisawa, T. Ishimatsu, T. Saito, D. Mori, D. Motomura, and H. Yagi, "Use of Non-Conductive Film (NCF) with Nano-Sized Filler Particles for Solder Interconnect: Research and Development on NCF Material and Process Characterization," in Proc. IEEE Electronic Components and Technol. Conf. (ECTC), 2016, pp.923-928

[5] V. D. Khanna and S. M. Sri-Jayantha, "Novel Mass Reflow Method for Organic Substrate," in Proc. IEEE Electronic Components and Technol. Conf. (ECTC), 2015, pp.200-207

[6] H. Clauberg, A. Rezvani, V. Venkatesan, G. Frick, B. Chylak and T. Strothmann, "Chip-toChip and Chip-to-Wafer Thermocompression Flip Chip Bonding," in Proc. IEEE Electronic Components and Technol. Conf. (ECTC), 2016, pp.601-605.

[7] Y. Shiraishi, S. Nagasaki and M. Yamashiro, "Ultra wide Range Viscometer Based on Indentation Parallel Plate Creep Rotational Methods and Its Characteristics," in Journal of Japan Institution of Metals, Vol. 60, No. 2 (1996), pp.184-191.

Study of Electrical and Mechanical Characteristics of Inkjet-Printed Patch Antenna under Uniaxial and Biaxial Bending

Yi Zhou[1], Sridhar Sivapurapu[2], Rui Chen[1], Nahid Aslani Amoli[2],

Mohamed Bellaredj[2], Madhavan Swaminathan[2], and Suresh K. Sitaraman[1]

[1]George W. Woodruff School of Mechanical Eng.
Georgia Institute of Technology
Atlanta, GA, USA
suresh.sitaraman@me.gatech.edu

[2]School of Electrical and Computer Eng.
Georgia Institute of Technology
Atlanta, GA USA

Abstract— **This paper discusses the electrical and mechanical characterization of an inkjet-printed patch antenna under uniaxial and biaxial bending. A 30 mm × 40 mm patch antenna design with a truncated copper ground plane was designed and fabricated by inkjet printing on a polyethylene terephthalate (PET) substrate. The fabricated antenna samples were then fixtured on cylindrical polycarbonate mandrels as well as on 3D printed saddle-like structures using polylactic acid (PLA). S_{11} was measured in both bent and flat configurations, and a shift in resonant frequency was observed. A maximum decrease of about 12.6% in resonant frequency under cylindrical bending was observed, and this decrease can be attributed to the decrease in conductance in the printed components due to small radius of some of the mandrels and due to sequential repetitive bending. The change in resonant frequency under biaxial saddle-like bending was small due to larger radii of the saddle-like surfaces and due to one-time bending. Electrical simulations in ANSYS^{TM} HFSS were carried out to determine S_{11} of the patch antenna under flat configuration and was compared favorably against experimental data. Mechanical finite-element models in ANSYS^{TM} Workbench were carried out to determine the strain distribution under both uniaxial and biaxial bending. A maximum strain of 0.0169 under uniaxial bending and 0.0029 under biaxial bending was observed.**

Keywords-flexible electronics; patch antenna; inkjet printing; finite-element model; bending

I. INTRODUCTION

Flexible hybrid electronics (FHE) has wide range of applications including medical devices, wearable technology, communication devices, automotive and aerospace sensors, and various consumer Internet of Things (IoT). FHE takes advantage of printing technologies, such as inkjet printing, screen printing, aerosol jet printing, gravure printing and flexographic printing with benefits in cost-efficiency, fast prototyping, and customization [1- 6]. This work focuses on inkjet-printed elements, and inkjet printing is a maskless, material-saving and fully additive technique which allows a variety conductive inks to be deposited on a wide range of flexible substrates. Several electronic components, such as strain sensors [7], inductors [8], capacitors [9] and antennas [10] can be fabricated using commercially available inkjet printers. During usage, the FHE components are often stretched, bent, folded, and/or twisted to conform to underlying structure. Therefore, the electrical and mechanical

characteristics of flexible printed electronic components should be studied under such deformation in operation.

In this study, patch antenna samples were fabricated by inkjet printing on a polyethylene terephthalate (PET) substrate. The response of the antenna (S_{11}) was measured by a vector network analyzer (VNA). An electrical model of the patch antenna design was built in ANSYS^{TM} HFSS, and key parameters were examined to correlate the measurement and the simulation at the flat configuration.

Antenna's response was characterized in a uniaxial mandrel bending test and a biaxial saddle-like surface bending test. Uniaxial mandrel bending test was performed using polycarbonate mandrels of different sizes. Special fixtures were developed by 3D printing with polylactic acid (PLA) material to sandwich the antenna sample in the biaxial saddle-like surface bending test. S_{11} of the antenna was recorded at the bent configurations and was compared to S_{11} in the flat or undeformed configuration.

II. SAMPLE FABRICATION

A. Antenna Fabrication

Fig. 1 shows the geometry details of the proposed design of the patch antenna. The antenna had a 30 mm × 40 mm rectangular patch with a feedline length of 22.6 mm. The patch antenna and the feedline were printed on the silicone-coated front side of a 135-µm thick polyethylene terephthalate (PET) substrate. An Epson^{TM} C88+ inkjet printer was used to print the antenna using a silver nanoparticle ink (JS-B25P) from Novacentrix^{TM}. To achieve a full ink coverage of the patch, four passes were printed. As indicated in Fig. 1, the direction of printing was alternated between directions along the feedline and perpendicular to the feedline. In between each pass, the printed structure was air-cured in ambient temperature for 30 minutes. As illustrated in Fig. 1, the other side of the substrate had a copper-foil ground plane that fully overlapped the feedline and partially overlapped the patch antenna through a length of 5 mm across the width. The copper foil was attached to the backside of the PET substrate with a 3M^{TM} 966 adhesive tape. A vertical launch VLF40-002 SMA was connected by screws to excite the printed antenna.

978-1-7281-1500-9/19 $31.00 © 2019 IEEE

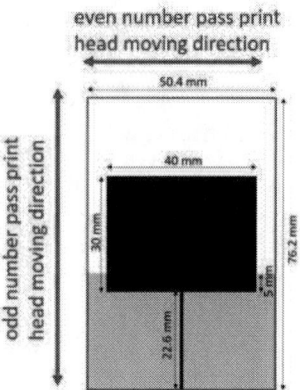

Figure 1. Geometry design details and inkjet printing methods of the patch antenna sample.

B. Fabrication Deviation

Fig. 2a shows two fabricated antenna samples. As seen, the top side of the PET substrate has the patch antenna with a feedline, while the bottom side of the substrate has the ground plane. The copper ground plane is seen through the transparent PET substrate. Fig. 2b shows the optical image of the top view of the edge of the antenna where the darker region on the right is the PET substrate, and the brighter region on the left is the silver ink. As shown, the printed structure does not have a well-defined edge, with a maximum serration width of 281 µm. Optical inspection and measurement of the antennas indicate that the patch's dimensions are approximately 30 × 40 mm, while the measured average feedline width is about 0.8 mm for sample 1 and 0.6 mm for sample 2. Cross-sectioning was done to determine the thickness of each layer, and as shown in Fig. 2c, the bottom copper layer has a thickness of 107.54 µm, the adhesive layer on top has a thickness of 59.66 µm, the PET substrate with the silicone coating has a thickness of 134.79 µm, and the silver ink has an average thickness of 1.75 µm.

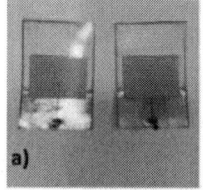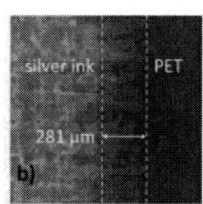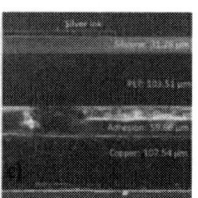

Figure 2. Fabrication results of the proposed antenna design: a) two fabricated samples of the patch antenna; b) top view of the optical inspection of the right edge of the patch; c) cross-section view of the of the optical inspection and measurement of antenna stacking.

III. ANTENNA ELECTRICAL SIMULATION AND CORRELATION WITH THE MEASUREMENT

The antenna electrical simulation was done using the design geometry and dimensions given in Figs.1 and 2c. The dielectric material properties were taken from literature and are shown in Table I. The key parameters to correlate the measured S_{11} with simulation values are the ink's conductivity, feedline width, and ink layer thickness. These parameters are most impacted by fabrication non-uniformity. With the measured width ranging from about 0.6 to about 1.0 mm at various locations of the feedline, simulations were carried out at 0.6, 0.8, and 1.0 mm. On the other hand, the patch antennas were simulated using 30 × 40 mm dimensions. This is because the serration of the patch was 281 µm which is less than 1% of the antenna length and width implying that it has a minimal impact on the patch electrical characteristics. The ink layer thickness is also impacted by the non-uniformity seen in the fabrication process as the thickness ranged from 0.46 µm to 2.46 µm. It should be pointed out that the printer used in this work is inexpensive, and thus, tighter control of dimensions was not possible. The purpose of this work is to demonstrate the simulation and measurement methodology, as opposed to creating a high-quality printed sample. However, in the ongoing work, other antenna geometries are being printed with tighter fabrication tolerances with significantly less variations. Therefore, in this current work, as a consequence of process variations, the conductivity, width, and thickness will have a wide range of values. Parameter sweeps for three variables were completed in ANSYS™ HFSS. Fig. 3a shows S_{11} through a frequency range of 10 MHz to 6 GHz with the conductivity set at 1.75×10^6 S/m, while varying the ink layer thickness and feedline width. For the various cases simulated, the dominant resonant frequency is in the range of 1 to 2 GHz, and the resonant frequency of the antenna increases, when the ink layer thickness or the feedline width increases. Similarly, when the ink layer thickness is fixed at 1.5 µm, the resonant frequency increases when the conductivity of the ink or the feedline width increases, as shown in Fig. 3b. Furthermore, as shown in Fig. 3c, when the feedline width is fixed at 0.8 mm, the resonant frequency increases when the conductivity or the thickness of the ink increases.

TABLE I.   MATERIAL ELCTRICAL PROPERTIES

Material	Relative Permittivity	Dielectric Loss Tangent
PET [11]	4	0.01
Adhesive [12]	2.92	0.025

After these sweeps, the best match between the measured and the simulated S_{11} occurred for a conductivity of 2.00×10^6 S/m, a feedline width of 0.8 mm, and a thickness of 0.5 µm for the antenna sample 1; and for a conductivity of 1.75×10^6 S/m, a feedline width of 0.6 mm, and a thickness of 0.5 µm for the antenna sample 2 as seen in Fig. 3d. In Fig. 3d, a similar match was also found when the conductivity of the ink was halved, while the thickness of the ink was doubled. This indicates that for a given feedline, as long as the effective conductance per square is 1 S/sq. for antenna sample 1 and 0.875 S/sq. for antenna sample 2, the antenna will have a similar S_{11}. This will hold true as long as the conductor thickness is below the skin depth, which is determined using $\delta = \frac{1}{\sqrt{\sigma f \pi \mu_0 \mu_r}}$, where δ is the skin depth, σ is the conductivity, f is the frequency of operation, and $\mu_0 * \mu_r$ is the magnetic permeability of the material (and the conductor is not magnetic). Details of the experimental measurements of S_{11} are presented in the next section.

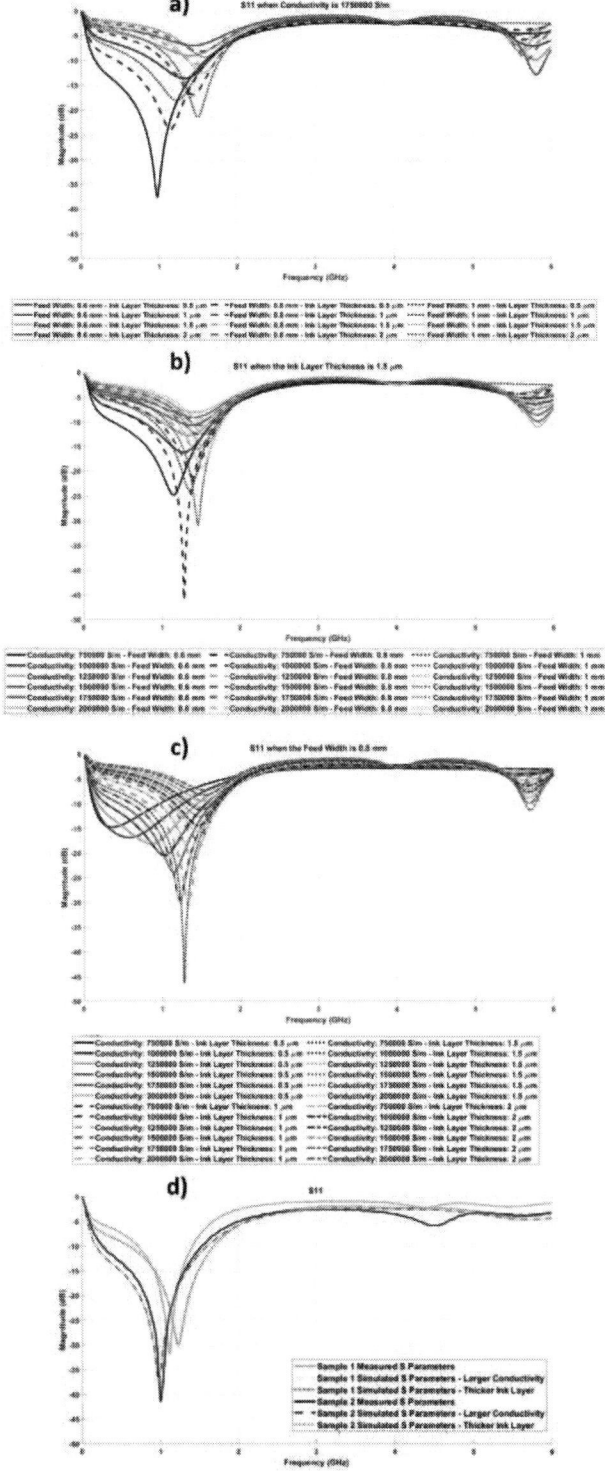

Figure 3. S_{11} simulation results by ANSYS™ HFSS model with different parameters: a) constant conductivity value of 1.75×10^6 S/m with values variation of ink layer thickness and feedline width; b) constant ink layer thickness of 1.5μm with variation of conductivity and feedline width; c) constant feedline width of 0.8 mm with variation of conductivity and ink layer thickness; d) simulated S_{11} compared against experimental data for the two antenna samples.

IV. UNIAXIAL MANDREL BENDING

A. Experimental Setup and Results

In this paper, for all of the high frequency measurements, an Agilent N9923A RF Vector Network Analyzer (VNA) was used and calibrated with an HP 85052D Calibration Kit, using the SOLT (Short-Open-Load-Thru) standards [13]. At the start of the bending experiments, the VNA was used to measure S_{11} of the antenna in the flat configuration. The sample was placed on a flat polycarbonate sheet as shown in Fig. 4a. The polycarbonate sheet was used in the flat configuration to account for the potential effect on the electrical characteristics of the antenna in the proximity of another material and then to compare the electrical characteristics of the same antenna under bent configuration over a cylindrical mandrel of the same material. After the flat-configuration measurements, the antenna was sequentially wrapped around various polycarbonate cylindrical mandrels with an outer radius of 4, 3, 2.5, 2, 1.5, 1, and 0.625 in. In all of the experiments, the substrate was wrapped such that the ink layer would face outward as demonstrated in Fig. 4b, and therefore, circumferential tensile strain was applied on the silver ink layer. The experiment was started from the largest radius to minimize potential damage to the printed structure, and the measurements were then sequentially conducted at smaller radii. After measuring S_{11} while wrapped around a cylinder of a particular radius, S_{11} was measured again in the flat configuration before the antenna was wrapped around the next smaller cylinder. This sequence of bent and flat configuration measurements was repeated through all cylinders of decreasing radii. Once S_{11} was measured around the smallest cylinder of radius 0.625 in. and once it was observed that S_{11} had minimal change, the experiment was then repeated wrapping the antenna around the cylinders of increasing radii. This completes Cycle 1 of S_{11} measurements, and the results are shown in Fig. 5a. To avoid overcrowding of curves, the flat measurements are shown at the beginning and the end of Cycle 1. This sequence of measurements was repeated five cycles, and the measured data are shown in Figs. 5a through 5e. Fig. 5f shows measured S_{11} for flat configurations at the beginning and end of all cycles.

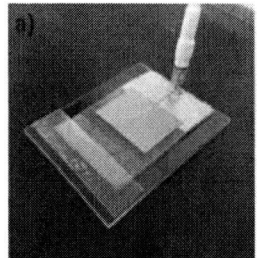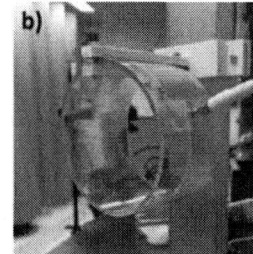

Figure 4. Experimental setup of the uniaxial bending test: a) the antenna was seated on a sheet of polycarbonate for the flat measurements; b) the antenna was wrapped around the polycarbonate mandrel for the bent measurements.

In Fig. 5, the initial flat measurements of each cycle are denoted as "F0," and the final flat measurements are denoted as "FF." The in-situ bending S_{11} measurements are denoted as "R" followed by bending radius dimension which is followed by a – sign or a + sign. The – sign indicates that the radius is decreasing from one measurement to the next, while the + sign indicates that the radius is increasing from one measurement to the next. As shown in Fig. 5, the initial S_{11} measurement shows a resonant frequency of 1.11 GHz. Throughout the experiments, the overall shape of S_{11} is similar, but the resonant frequency and the fractional bandwidth show minor changes. In general, with smaller radius bending, the resonant frequency decreases in a given cycle, and as the cycling progresses, the resonant frequency continues to decrease for a given radius bending. This resonant frequency shift is possibly due to increasing damage and thus decreasing conductance in the printed elements at smaller radii and with repeated cycling. The maximum decrease in resonant frequency was observed to be 140 MHz at Cycle 5 (last cycle) at a bending radius of 0.625 in. It was also seen that there was practically no change in the resonant frequency in the flat configuration regardless of the number of times the antenna has been bent around the cylinders. This means that as long as the number of cycles is fairly small (in this case five cycles), the potential cracking in the printed ink probably closes in the flat configuration leading to no change in the conductance. All of the experiments were conducted using antenna sample 1.

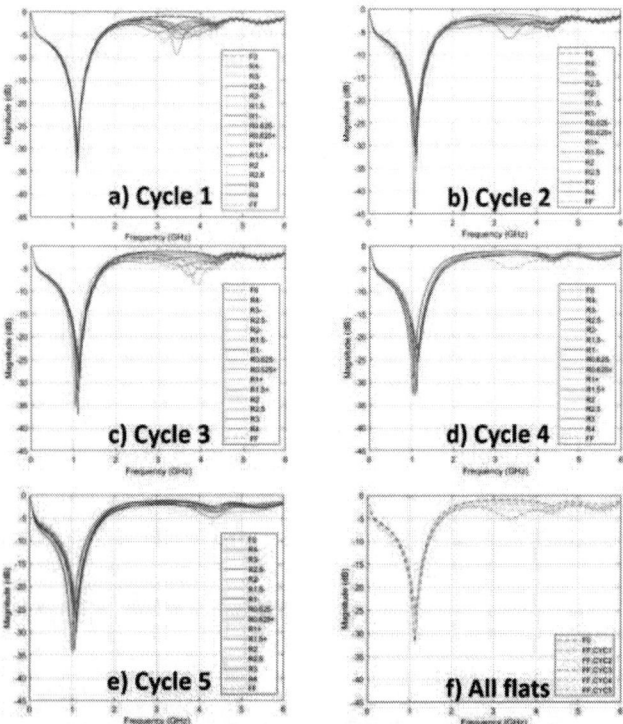

Figure 5. Five cycles of S_{11} experimental measurement of antenna sample 1 during mandrel bending. The initial flat measurements of each cycle are denoted as "F0," and the final flat measurements are denoted as "FF." The in-situ bending S11 measurements are denoted as "R" followed by bending radius dimension which is followed by a – sign or a + sign. The – sign indicates that the radius is decreasing from one measurement to the next, while the + sign indicates that the radius is increasing from one measurement to the next.

B. Finite-Element Model

A mechanical 3D finite-element model was built in ANSYS™ Workbench to determine the strain distribution, when the antenna is bent over a cylindrical mandrel. The antenna model was built using the geometry and dimensions shown in Figs. 1 and 2c. The material mechanical properties for the simulation are shown in Table II. The silver ink modulus was measured in-house using nanoindentation, while the other properties were obtained from literature. The adhesive layer was assumed to have a modulus of 2 GPa. The cylindrical mandrel was assumed to have a very high modulus to indicate its rigid nature in the simulations.

TABLE II.   MATERIAL MECHANICAL PROPERTIES

Material	Young's Modulus (GPa)	Poisson's Ratio
Silver ink [14]	4.61*	0.37
PET [15]	2	0.4
Copper [16][17]	110	0.34

* measured in-house using nanoindentation

Fig. 6 shows the model setup and the loading details. Simulation was carried out for a mandrel of radius 0.625 in. which was the smallest mandrel size used in the experiment that induced the largest strain. Fig. 6a shows the isometric view of the initial position of the model. To mimic the mandrel bending test, the bottom edge of the substrate without the ground plane was simulated to be bonded to the mandrel. To make sure that the sample could fully conform to the surface of the mandrel, a rotational displacement was applied to the mandrel. A 1° increment per step was applied through 288° when the entire sample was fully wrapped around the mandrel. A small force of 3N was applied to the end surface of the substrate to keep the sample flat during rolling. The bottom surfaces of the substrate and the ground plane were assigned frictionless contact with the surface of the mandrel. Figs. 6b and 6c demonstrate two isometric views of the final position of the simulation as the mandrel is rotated to 288°.

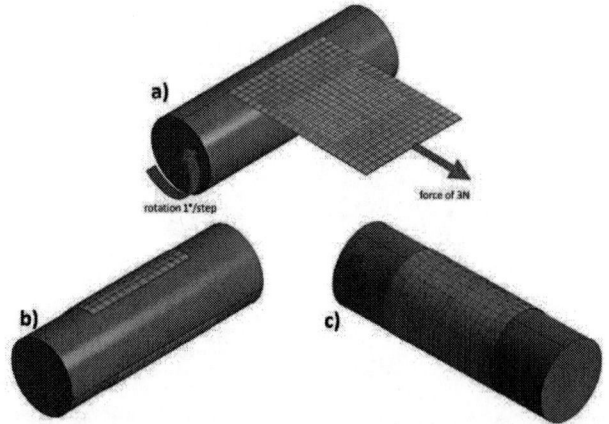

Figure 6.   Uniaxial mandrel bending mechanical simulation model: a) isometric view of the initial position of the model; b) and c) isometric views of the final position of the model.

Bending strain at any location can be approximately computed by $\varepsilon = y/\rho$, where ε is the strain, y is the distance from the neutral axis of the multilayer structure and ρ is the bending radius at neutral axis. The calculated bending strain is 0.0154 for the ink layer with a copper ground plane underneath. For the sake of expediency, first principal strain contours of the silver ink layer, obtained from numerical simulations, are shown in Fig. 7 and are compared against the hand-calculated values. As discussed earlier, it is seen that the strains are maximum where there is an underlying copper ground plane moving the neutral plane closer to the cylindrical outer surface, and thus making the distance to the outer surface larger. This magnitude 0.0169 is close to the hand-calculated value of 0.0154. The portion of the ink with no copper ground plane underneath has strain values of 0.0053 to 0.0072.

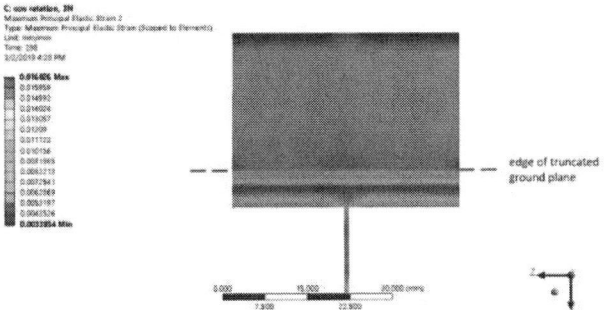

Figure 7. Simulation result of the maximum principal strain distribution of the ink layer in the uniaxial bending model.

The strain sensitive silver nanoparticle film printed by a desktop printer on the silicone coated PET has been reported to show an increase of 16% in resistance when the applied bending strain is 0.08, and the change in resistance has a linear relationship with the strain at the strain range of 0 to 0.08 [18]. Therefore, in the current work the change in electrical resistance for the smallest cylinder under the first cycle is expected to be a maximum of 3.4%, and is expected to increase in subsequent cycling for the same radius bending. The electrical simulations presented in Fig. 3 are for flat configuration; simulations will be carried out for bent configuration with the change in electrical resistance, and the results from such simulations will be presented in a future publication.

V. BIAXIAL SADDLE-LIKE SURFACE BENDING

A. Experimental Setup and Results

Special experimental fixtures were designed and fabricated by 3D printing in polylactic acid (PLA) material to conduct the biaxial bending test. Fig. 8a show the 3D model of the fixture design. The fixtures were designed as a two-part system that the electronics can be sandwiched to the desired biaxial curved surface as presented in Fig. 7a. A saddle-like mating surfaces were designed such that a 4 in. radius convex curve was swept along a 40 in. radius concave curve in the transverse direction.

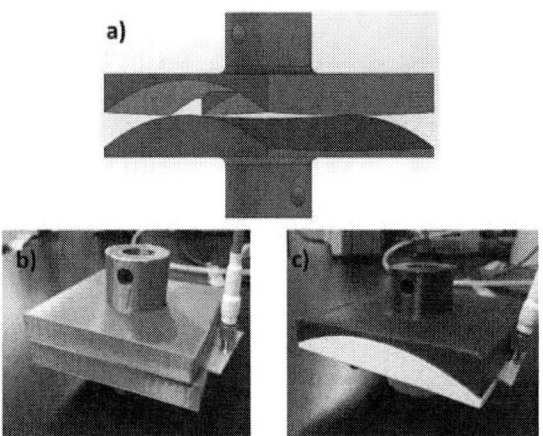

Figure 8. Experimental setup of the biaxial bending test: a) biaxial saddle-like surface bending test fixture model; b) the antenna was seated between two PLA printed blocks for the flat measurement; c) the antenna was seated in between the PLA printed fixture with the ink layer facing up when taking the bent measurement.

At the start of the biaxial bending experiment, S_{11} of the antenna was measured in the flat configuration. The sample was placed between the two blocks with flat mating surfaces as shown in Fig. 8b. The two blocks were 3D-printed using PLA with a thickness of 15 mm, and this thickness roughly corresponds to the average thickness of each of the saddle fixtures to account for the potential effect on the electrical characteristics of the antenna being sandwiched by another material. After the measurements in flat fixtures, the antenna sample was placed in between the saddle-like fixtures with the silver ink layer facing up. The two fixtures were clamped together as demonstrated in Fig. 8c. In such scenario, the antenna experienced tensile bending along the feedline line direction with a bending radius of 4 in. and a compressive bending in the direction perpendicular to the feedline with a bending radius of 40 in. After measuring S_{11} while the antenna being sandwiched by the fixtures, S_{11} was measured again in the flat configuration. This biaxial bending experiment was conducted using another pristine antenna sample (antenna sample 2).

The measured S_{11} results are presented in Fig. 9. The initial flat measurement is denoted as "F0," the bent measurement is denoted as "B," and the after-bending flat measurement is denoted as "FF." As shown, the initial S_{11} measurement shows a resonant frequency of 0.99 GHz. A small increase of 37 MHz in resonant frequency and no change in fractional bandwidth were observed, when the antenna was sandwiched in the saddle-like fixtures. This was due to the large radii used in the saddle-like fixtures, and thus, minimal change in the electrical resistance. All of the S_{11} measured in this experiment almost overlapped against one another. This indicates that no damage of the ink was initiated during the single biaxial bending test.

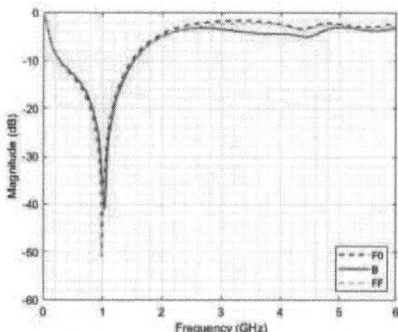

Figure 9. S_{11} response measurements of antenna sample 2 in the biaxial saddle-like surface bending test. The initial flat measurement is denoted as "F0," the bent measurements are denoted as "B," and the after bending flat measurement is denoted as "FF."

B. Finite-Element Model

A mechanical 3D finite-element model was built to determine the strain distribution, when the antenna is sandwiched between the saddle-like surfaces. The antenna model in the biaxial bending model was the same as the one built in the uniaxial bending model. Similar to the mandrel bending model, the fixtures were assigned a very large elastic modulus to indicate their rigid nature in the simulation. Fig. 10a shows the front view of the initial position of the simulation, where the antenna sample was seated between the simplified saddle fixtures. At the initial position, the antenna sample was flat, and the top and bottom surfaces of the antenna were just in contact with the fixtures. Upward displacement of 0.05 mm increment per step was applied to the bottom fixture, while the top fixture was held rigid. In addition to that, the node where the antenna sample was in contact with the top fixture, and the three adjacent nodes in substrate layer were fixed to prevent rigid body movement of the antenna. The contact surfaces between the antenna and the fixtures were assigned to be frictionless. A total of 6.8 mm displacement was needed to be applied to the bottom fixture for the sample to fully conform into the curved surface as shown in Fig. 10b. Fig. 10c shows the shape of the antenna at the final position.

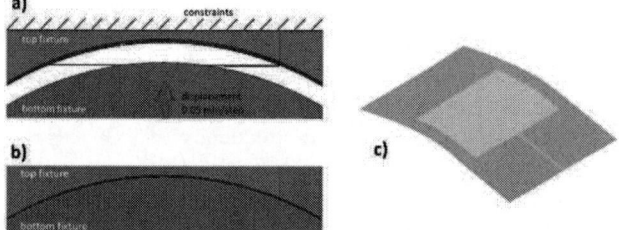

Figure 10. Biaxial saddle-like surface bending mechanical simulation model: a) front view of the initial position of the model; b) front view of the final position of the model; c) isometric view of the antenna sample at the final position.

First principal strain contours of the silver ink layer are shown in Fig. 11. It is seen that the maximum strain value of 0.0029 occurs near the edge of the copper ground plane. Other portions of the ink layer have the strain values ranging from 0.0005 to 0.0014. Thus, the strain values in the whole ink layer are fairly small due to the large bending curvatures.

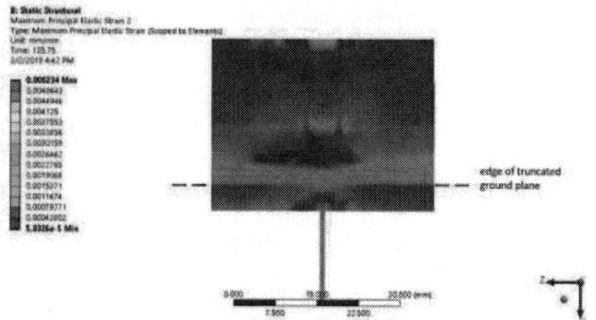

Figure 11. Simulation result of the maximum principal strain distribution of the ink layer in the biaxial bending model.

VI. CONCLUSION

In this paper, the electrical and mechanical behavior of an inkjet-printed patch antenna under uniaxial and biaxial bending was studied. Two inkjet-printed antenna samples were fabricated, with a measured resonant frequency of 1.11 GHz and 0.99 GHz. In flat configuration, the electrical models of S_{11} agree with the experimental measurements, and for a given feedline width, a good match between simulations and experiments was obtained when the conductance per square of 1 S/sq. was used for antenna sample 1 and 0.875 S/sq. was used for antenna sample 2. In uniaxial cyclic bending test, when the antenna sample was wrapped around the cylindrical mandrels, a decrease in resonant frequency was observed at the smaller mandrel size, and the resonant frequency continued to decrease with repeated cycling, especially at smaller radii. The shift in resonant frequency is possibly due to increasing damage and thus decreasing conductance in the printed structure. Not much resonant frequency shift was observed in biaxial bending test due to the larger radii used in the fixtures and due to the fact that the experiment was conducted through only one cycle. In this work, the radiation pattern of the antenna was not studied and may be explored in the future. In addition to electrical characterization, mechanical finite-element analysis was conducted to determine the strain distribution over the ink layer. A maximum strain of 0.0169 was seen in uniaxial bending model, and a maximum strain of 0.0029 was seen in biaxial bending model. Future work will be conducted to study the relationship between the strain, conductance change and thus, the resonant frequency shift.

VII. ACKNOWLEDGEMENT

This material is based, in part, on research sponsored by Air Force Research Laboratory under agreement number FA8650-15-2-5401, as conducted through the flexible hybrid electronics manufacturing innovation institute, NextFlex. The U.S. Government is authorized to reproduce and distribute reprints for Governmental purposes notwithstanding any copyright notation thereon. The views and conclusions contained herein are those of the authors and should not be interpreted as necessarily representing the

978-1-7281-1500-9/19 $31.00 © 2019 IEEE

official policies or endorsements, either expressed or implied, of Air Force Research Laboratory or the U.S. Government.

REFERENCES

[1] M. Stoppa and A. Chiolerio, "Wearable electronics and smart textiles: A critical review," Sensors, vol. 14, no. 7, pp. 11957–11992, 2014.

[2] J. van den Brand et al., "Flexible and stretchable electronics for wearable health devices", Solid-State Electronics, vol. 113, pp. 116-120, 2015.

[3] G. Yang et al., "A Health-IoT Platform Based on the Integration of Intelligent Packaging, Unobtrusive Bio-Sensor, and Intelligent Medicine Box," in IEEE Transactions on Industrial Informatics, vol. 10, no. 4, pp. 2180-2191, Nov. 2014.

[4] A. Eid et al., "Nanotechnology-Empowered Flexible Printed Wireless Electronics: A Review of Various Applications of Printed Materials," in IEEE Nanotechnology Magazine, vol. 13, no. 1, pp. 18-29, Feb. 2019.

[5] S. Khan, L. Lorenzelli and R. S. Dahiya, "Technologies for Printing Sensors and Electronics Over Large Flexible Substrates: A Review," in IEEE Sensors Journal, vol. 15, no. 6, pp. 3164-3185, June 2015.

[6] V. Misra et al., "Flexible Technologies for Self-Powered Wearable Health and Environmental Sensing," in Proceedings of the IEEE, vol. 103, no. 4, pp. 665-681, April 2015.J. Clerk Maxwell, A Treatise on Electricity and Magnetism, 3rd ed., vol. 2. Oxford: Clarendon, 1892, pp.68–73.

[7] B. Andò and S. Baglio, "All-Inkjet Printed Strain Sensors," in IEEE Sensors Journal, vol. 13, no. 12, pp. 4874-4879, Dec. 2013.

[8] A. B. Menicanin, N. P. Ivanisevic, L. D. Zivanov, M. S. Damnjanovic, A. M. Maric and D. V. Randjelovic, "Improved Performance of Multilayer CPW Inductors on Flexible Substrate," in IEEE Transactions on Magnetics, vol. 50, no. 11, pp. 1-4, Nov. 2014, Art no. 8401204.

[9] B. S. Cook, J. R. Cooper and M. M. Tentzeris, "Multi-Layer RF Capacitors on Flexible Substrates Utilizing Inkjet Printed Dielectric Polymers," in IEEE Microwave and Wireless Components Letters, vol. 23, no. 7, pp. 353-355, July 2013.

[10] S. Ahmed, F. A. Tahir, A. Shamim and H. M. Cheema, "A Compact Kapton-Based Inkjet-Printed Multiband Antenna for Flexible Wireless Devices," in IEEE Antennas and Wireless Propagation Letters, vol. 14, pp. 1802-1805, 2015.

[11] D. Lane, "Conductive Inkjet Printed Ultra-Wideband (UWB) Planar Monopole Antenna On Low Cost Flexible Polyethelyne Terephthalate (Pet) Substrate Material", Master of Science, San Diego State University, 2015.

[12] "3M™ Technical Data", 3m.citrination.com, 2019. [Online]. Available: https://3m.citrination.com/pif/000314?locale=en-US. [Accessed: 26- Feb- 2019].

[13] "Understanding VNA Calibration", 2012. [Online]. Available: http://anlage.umd.edu/Anritsu_understanding-vna-calibration.pdf [Accessed 18 Jan. 2019].

[14] E. Lam, "Fabrication and Material Characterization of Silver Cantilevers via Direct Surface Micromachining", Undergraduate, Massachusetts Institute of Technology, 2005.

[15] "Polyethylene terephthalate - online catalogue source - supplier of research materials in small quantities – Goodfellow", Goodfellow.com, 2019. [Online]. Available: http://www.goodfellow.com/E/Polyethylene-terephthalate.html. [Accessed: 27- Feb- 2019].

[16] D. Yu and F. Spaepen, "The yield strength of thin copper films on Kapton", Journal of Applied Physics, vol. 95, no. 6, pp. 2991-2997, 2004.

[17] J. Zhu, J. Feng and Z. Guo, "Mechanical properties of commercial copper current-collector foils", RSC Adv., vol. 4, no. 101, pp. 57671-57678, 2014.

[18] P. Joshi and V. Santhanam, "Strain-Sensitive Inkjet-Printed Nanoparticle Films on Flexible Substrates," in IEEE Sensors Letters, vol. 2, no. 1, pp. 1-4, March 2018, Art no. 2500304.

*** Formatting Issue - Best Available Paper/Graphic ***

2019 IEEE 69th Electronic Components and Technology Conference (ECTC)

Effects of Oven and Laser Sintering Parameters on the Electrical Resistance of IJP Nano-Silver Traces on Mesoporous PET Before and During Fatigue Cycling

G.S. Khinda, M.Z. Kokash, M. Alhendi, M. Yadav,
J.P. Lombardi, D.L. Weerawarne,
M.D. Poliks, P. Borgesen
Systems Science and Industrial Engineering
SUNY Binghamton, NY 13902, USA
mpoliks@binghamton.edu

Nancy C. Stoffel
General Electric Global Research Center, Niskayuna,
NY 12309, USA
stoffel@ge.com

Abstract—- Inkjet printing of conducting traces offers well established advantages and disadvantages as an alternative to electroplating of interconnects in flexible electronics. Assessment and optimization of their reliability is, however, often more complicated than commonly recognized. This is the case for an approach based on the deposition of silver nano-particle inks onto mesoporous PET substrates. In this case heating leads the trace resistance to drop not only because of the shrinkage and cure of the organic matrix holding the particles together, but also because some of that matrix 'disappears' into the substrate pores. The substrates can however only sustain relatively brief excursions above their glass transition, nominally 75°C, so it is not always practical to sinter the traces completely by conventional means. That has consequences such as ongoing reductions in resistance over time or under cyclic loading. Laser sintering does however offer the opportunity for much better fusing of the particles without excessive heating of the PET. The present work addresses effects of sintering parameters such as time/temperature and power/speed in oven and laser sintering, respectively, on the initial resistance and its evolution in subsequent low cycle fatigue testing. Interconnects of an average width of 80 μm and thickness of 550 nm were printed and post processed by one of two different sintering techniques: a) Convection oven sintering, and (b) Laser sintering. The resulting resistances were quantified, and samples finally subjected to tensile cycling with amplitudes of 1-2% and in-situ monitoring of the resulting resistance changes using a four-point probe.

As expected, the resistance increased in each cycle as the substrate was stretched and it decreased again during unloading. However unlike for other kinds of traces, even though a remaining viscoelastic strain on the substrate prevented the complete elimination of the strain on the trace, the resistance of oven sintered traces usually ended up slightly *lower* after each cycle than before it. This effect was stronger for higher strain amplitudes, but it could be reduced or eliminated by longer preceding sintering of the traces. While a reduction in resistance may seem preferable to an increase, an even better solution would be a lower initial resistance that remained insensitive to subsequent fatigue cycling. This could be achieved by laser sintering, but careful optimization was required as too low a power did not prevent further resistance drops in cycling while too

high ones led to significant degradations in fatigue resistance

Keywords- Inkjet printing, mesoporous PET, Laser sintering, Convection oven sintering, nano-silver ink, Fatigue Cycling

Introduction

Flexible electronics are getting more popular in many IOT applications. Electronic circuits are printed using different techniques such as screen, gravure, flexography, aerosol jet and inkjet printing. The inkjet printing technique offers well known advantages for the deposition of interconnects on flexible and rigid substrates, but the properties are very different from what is achieved by electroplating. A lot of literature reports on the mechanical and electrical behavior of printed conductive interconnects on non-porous flexible substrates like PET, polyimide etc. In the present study, we address the robustness and reliability of printed nano-silver ink interconnects on mesoporous PET based substrate.

Inkjet printing is a non-contact drop on demand printing technology. The final quality of nano-silver ink deposits on polymeric substrates is optimized by selection of surface properties favoring high adhesion and resolution while controlling the deposition process [1,2]. The roughness of the porous substrate dictates the final deposition of colloidal drops by combination of particle motion, solvent evaporation, along with solvent infiltration into nanometer sized pores, resulting in an elimination of the coffee-ring effect because of faster solvent infiltration into pores than evaporation [3]. A mesoporous surface with pore sizes less than that of an ink particle results in conductivity of traces at room temperature through a combination of ink's solvent infiltration, silver particles stacking, and coffee ring effect elimination [4]. The long-term reliability of printed Ag traces, on the other hand, changes with sintering conditions. A sintering temperature more than decomposition temperature of organics (binders, fillers etc.) shows good conductivity due to the neck formation among the nanoparticles and further increase in temperature makes nano-porous microstructure denser and coarse. The long post heating at higher than decomposition temperature results in high conductivity and improves microstructure connectivity but reduces the plastic deformation resulting in

978-1-7281-1500-9/19 $31.00 © 2019 IEEE

formation of fatigue damages [5]. The adhesion and fracture energy between the substrate and particles depend on the thermal sintering time-period and temperature. The fracture energy can be increased by optimizing the sintering conditions between time and temperature because organic material present during the moment of fatigue testing determines the facture rate. The movement of organics to the bottom of the substrate at particular temperatures is the reason of increase in fracture energy of organic residual bridges [6]. The general fatigue cycling trend for printed silver nanoparticle is a damaging trend during the first cycle which keep on accumulating in consecutive cycles. This behavior is widely dependent on the material microstructure (particle size, metal weight percentage, solvent content, etc.), substrate properties, printing technique, pre-& post processing, and mechanical history. The whole printed system fatigue behavior interpretation is more complicated due to time-dependent mechanical properties like viscoelasticity and modulus of substrate [7,8].

I. EXPERIMENT

A. Printing

The test vehicle (Fig. 1) was printed using FUJIFLIM Dimatix-2831 inkjet printer on mesoporous PET substrate from Novacentrix Novele™ JS-220 having an average thickness of 140 μm. A 10pL 16 nozzles cartridge was used to print Novacentrix Metalon® JS-B25HV (25% by weight) nano silver (particle size ranging 60-80 nm) aqueous dispersion conductive ink which is filtered through 0.45 μm nylon filter.

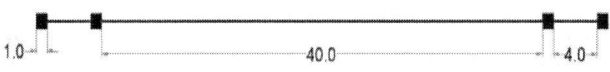

Figure 1. Test Vehicle

The traces (interconnects) were printed using single nozzle at a platen temperature of 40°C, which at printing height of 750 μm gives a drop size of ~ 40 μm. Adjacent drops were overlapped at 50% for printing of traces. Single layer traces had an average thickness of 550 nm. TGA of the ink shows that a temperature 110°C or more is required to decompose organics and solvent material in the ink.

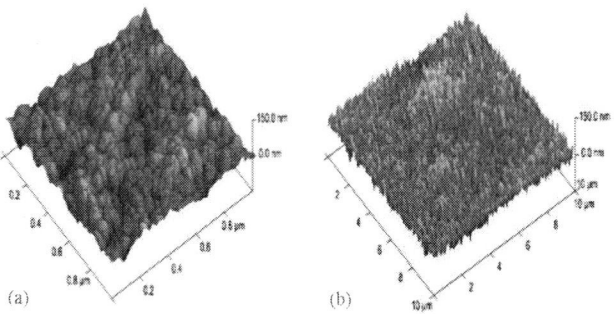

Figure 2. AFM of substrate a) 1μm×1μm b) 10μm×10μm area with the z range of 118 nm

The substrate has nanometer scale pores at the surface which tunnel to depths of ~20 μm (Fig. 3). The substrate has an average roughness (Ra) of 10.7 nm and root mean square roughness (Rq) of 13.8 nm (Fig. 2). The combination of particle size, surface roughness and pore dimension, limits the flow of ink over the substrate by acting as barrier/anchor and flush out the solvent into the pores. This helps the stacking of the silver particles over each other into close contact, making the trace conductive just after printing.

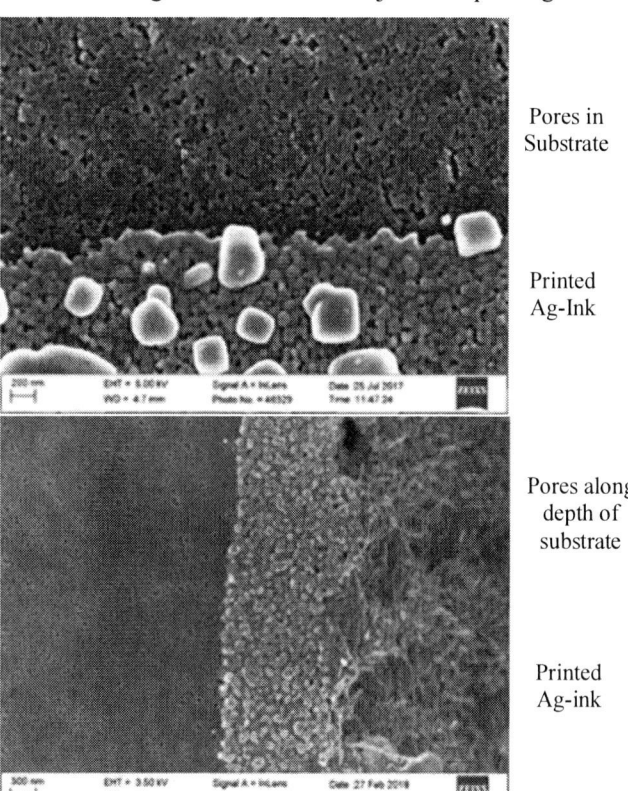

Figure 3. SEM image of PET substrate a) top-view of surface b) cross-section showing Nano scale pores

B. Sintering

Silver ink is consisting of solvent, organic materials (binders, fillers etc.) and silver nanoparticles. To increase conductivity of the printed traces, the solvent of ink and organic shells has to be evaporated and decomposed respectively, to provide the proper necking among silver nanoparticles, which needs curing and sintering of the printed trace. The degree of the necking and amount of organic material around the silver nanoparticles is determined by the sintering tempearture which also defines the electrical and mechanical properties of the trace. Sintering temperature can be achieved by different methods. In this study we are focusing on Convection oven sintering in air environment and laser sintering.

In convection oven sintering, the forced circulating hot air is used to achieve the required sintering temperature. The rate of solvent evaporation and organics decomposition is

*** Formatting Issue - Best Available Paper/Graphic ***

determined by the temperature and time-period. In laser sintering, a highly focused intensity continuous/pulsed laser beam of defined wavelength and spot size is illuminated on the printed trace to achieve an instantaneous high sintering temperature. The sintering temperature achieved on the trace can be controlled by the laser power and the scanning speed of the laser. Laser sintering is a selective technique, used for sintering trace without effecting the substrate as compare to oven sintering where full printed system is in contact to high temperature atmosphere.

C. Experimental setup

The fatigue cycling of printed traces were done using an Instron-3344 Tensile and Compression tester and *In-situ* 4-wire measurement of resistance using a Keithley® Multimeter (Fig. 4). The printed samples were cut manually using a paper cutter along printed alignment marks and then connecting wires for electrical measurement were attached using conductive silver epoxy and left overnight to cure. The samples are followed up with two types of sintering process a) Convection Oven sintering b) Laser sintering

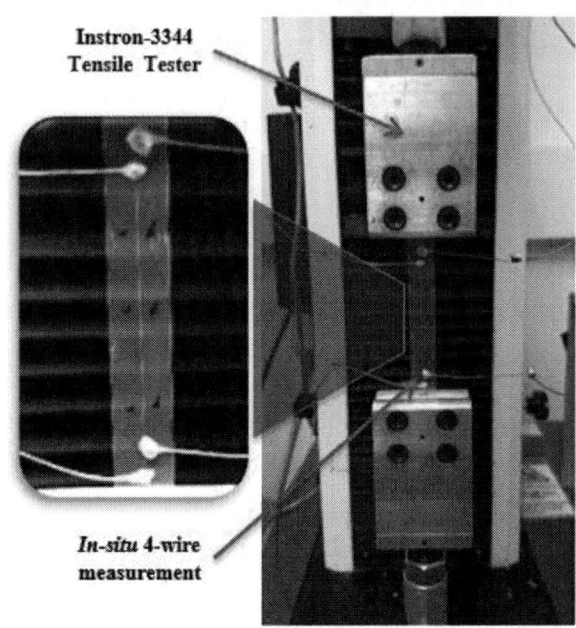

Figure.4 Experimental Setup

II. EXPERIMENTS AND RESULTS

A. Convection Oven Sintering

The printed interconnects are cured for 15 minutes: with 5 minutes' ramp to 100°C from 30°C and 10 minutes at constant 100 °C in a convection Binder® oven in air to remove excess of organic material and increase the conductivity between the nano particles. Fatigue cycling at 2.00% strain amplitudes for 100 cycles were performed. Samples with similar resistance were selected for reproducibility with assumption that equal probability of defects over the trace due to printing. After fatigue cycling, the relative resistance vs number of cycles plots were

plotted at 0.5% strain amplitude during loading cycle. The measurement is done at this point to eliminate the effect of permanent elongation/damage of the substrate during initial fatigue cycles as viscoelastic strain of the substrate increases with cycles (Fig. 6).

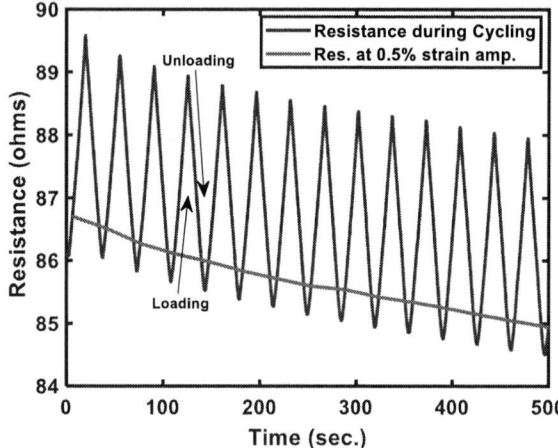

Figure 5. Resistance behavior during loading and unloading cycle representing the damage in each and consecutive cycles

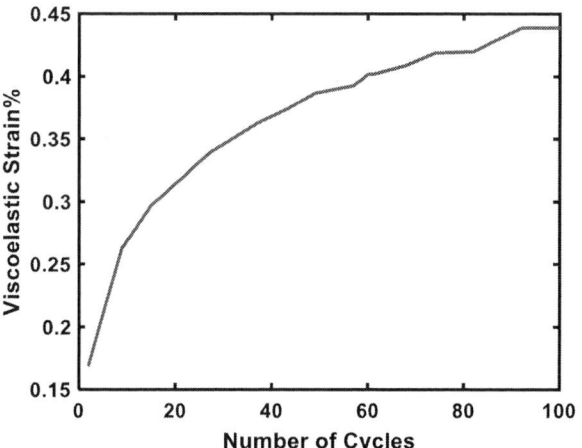

Figure 6. Viscoelastic strain at 2.00% strain amplitude

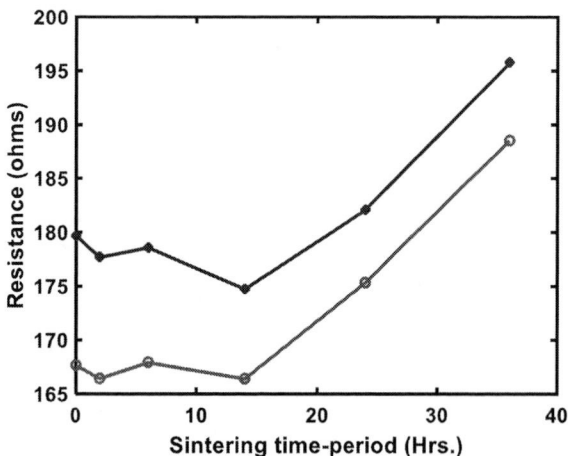

Figure 7. Resistance vs Aging Time for 1-Layer traces at 75°C

978-1-7281-1500-9/19 $31.00 © 2019 IEEE   1948

*** Formatting Issue - Best Available Paper/Graphic ***

Within each cycle, it is noted that resistance increases initially during loading and then decreases during unloading in each cycle (Fig. 5). But the final resistance after each cycle is less as compared to after the previous cycle. This trend continues with consecutive cycles. The drop-in resistance is faster as the strain amplitude of the cycling increases.

After seeing the effect of fatigue cycling on as-cured samples, tests were then done to assess the effect of prolonged exposure to a moderately elevated temperature on the fatigue performance of the samples. Sintering in the convection oven at 75°C for 2hrs, 10hrs, 18hrs, 24hrs and 36hrs time periods, respectively, was followed by fatigue cycling for 100 cycles at 2.00% strain amplitude. The temperature of 75°C was chosen to minimize effects on the properties of the PET substrate which had a glass transition temperature near that.

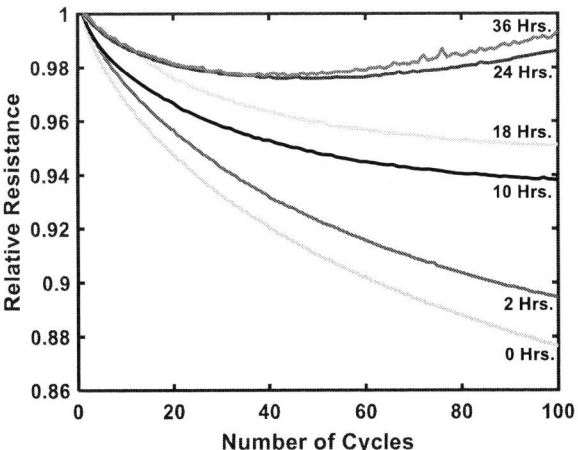

Figure.8 SEM images at (a) 0 Hrs. (b) 10 Hrs. (c) 24 Hrs. and (d) 36 Hrs. of sintering at 75 °C (51KX mag.)

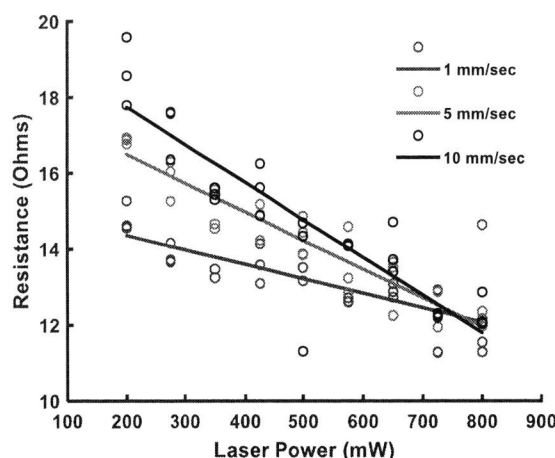

Figure 9. Relative Resistance vs. Number of cycles of 1-layer traces fatigue cycled at 2.0% strain amplitude after annealing at 75°C

Fig. 7 shows a small decrease in resistance with time up to 14 hours. After 14 hours, traces show increase in resistance. The top view SEM images of the traces show the particles are aggregating into clusters on the surface of trace with time, albeit at a slow rate but there are still nanoscale silver particles are visible in-depth of trace (Fig. 8).

Sintering also affected the subsequent sensitivity to a limited number of fatigue cycles. Fig. 10 reflects the relatively small effect of sintering on the initial resistance and normalizing to this resistance. Fig. 9 shows a systematic reduction in the relative variation with number of tensile cycles with a 2% strain amplitude

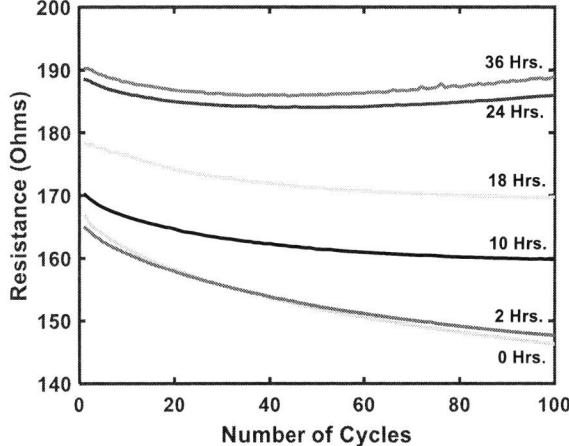

Figure 10. Absolute Resistance vs. Number of cycles of 1-layer traces fatigue cycled at 2.0% strain amplitude after annealing at 75°C

B. Laser Sintering

The sheet resistance obtained by convection oven sintering was 9-12 times that of bulk silver. Higher sintering temperatures are expected to damage the PET substrate. An alternative approach to obtain low sheet resistance without affecting the PET substrate is selective laser sintering.

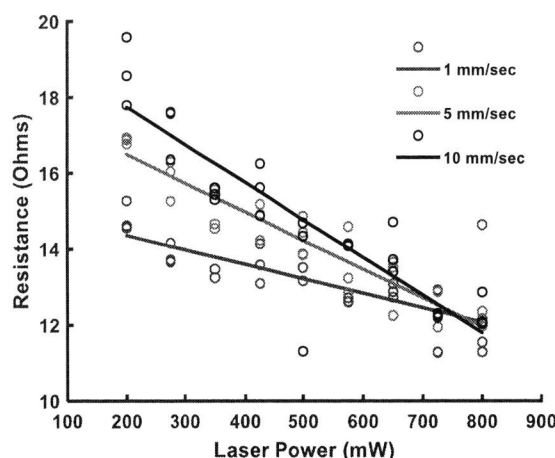

Figure 11. Resistance vs Laser power for different scanning speeds (20mm effective length, 80μm width, 1-layer)

Printed samples were sintered with a continuous laser beam of wavelength 830nm, a spot size of ~ 90 μm, at

*** Formatting Issue - Best Available Paper/Graphic ***

powers between 50mW to 700mW with scanning speed of 1-10 mm/sec. Depending on the specific parameters this led to sheet resistances of 2.5-4.0 times that of bulk silver. As expected, low speed and high power led to the lowest sheet resistance Fig. 11.

Top view SEM images of air dried, oven cured and laser sintering with different powers at 10 mm/sec scanning speed show the latter leading to growth of silver particles into big clusters and broad necking among adjacent particles, explaining the low sheet resistance (Fig.11). However, high power also seems to decompose the organics around the nano-particles as high power generates high localized temperature resulting in decomposition of organics.

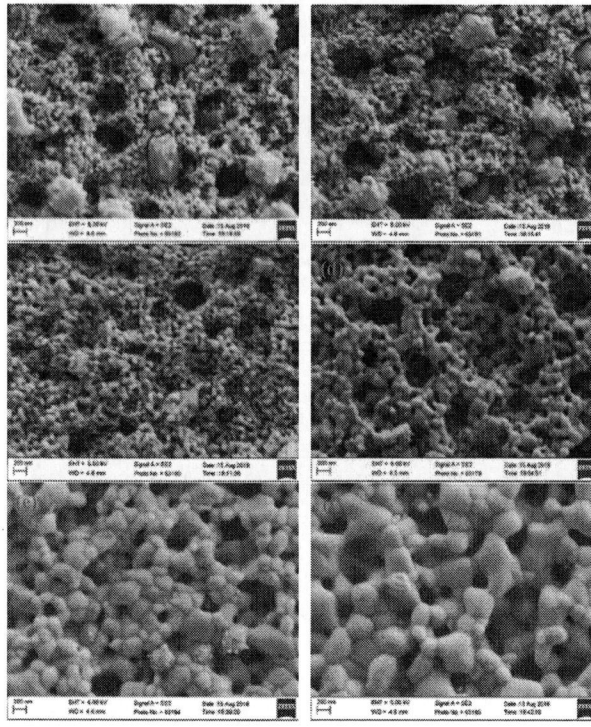

Figure 12. SEM images at WD of 4.6mm with 65KX magnification a) After printing b) Oven Curing for 15 minutes c) 100mW d) 250mW e) 550mW f) 700mW at scanning speed 10mm/sec. With power increment, necking increases among adjacent particles and big clusters formation.

Like oven sintering, the laser sintering also had a systematic effect on the behavior in tensile cycling to a 2% strain amplitude. Even the lowest power and the highest speed (50mW and 10mm/s) virtually eliminated the reduction in resistance in 100 tensile cycles to a strain amplitude of 2%. In fact, raising the power to 400mW or more led to major reductions in initial resistance (Fig. 11), but ever more rapid increases in resistance in cycling (Fig. 13). Considering the absolute resistance (Fig. 14) thus points to a 'sweet spot', a power of 50-150mW leading to the lowest resistance after a very limited number of cycles. In fact, while a low resistance may seem attractive a *stable* resistance is likely to be more so. 50-150mW leads to less

than ± 2% change in resistance in cycling (Fig. 15) with 100mW being the most stable.

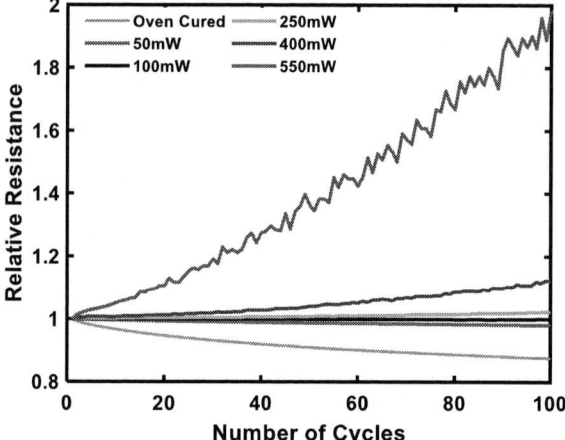

Figure 13. Relative Resistance Vs No. of Cycles Fatigue behavior of Laser sintered traces at different powers for 2.0% strain amplitude

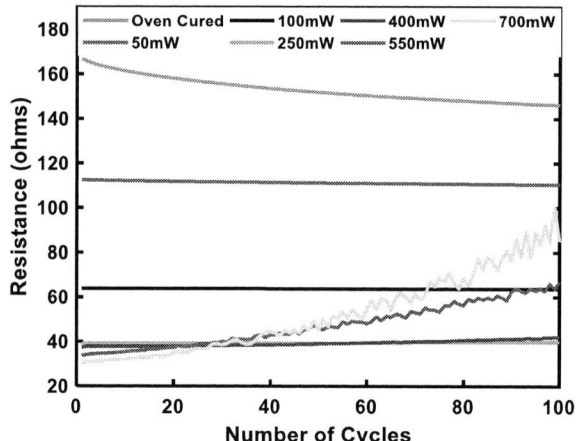

Figure 14. Absolute Resistance Vs No. of Cycles Fatigue behavior of Laser sintered traces at different powers for 2.0% strain amplitude

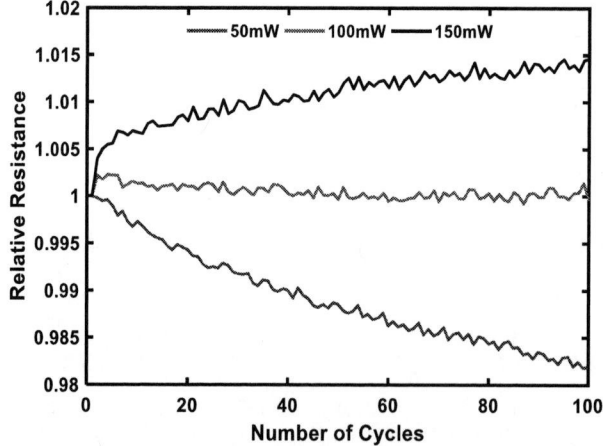

Figure 15. Relative Resistance Vs No. of cycles fatigue behavior of Laser sintered traces at close range power of 50mW to 150mW at 2.0% strain amplitude.

III. CONCLUSION

Mesoporous substrates offer advantages in terms of high-resolution inkjet printing of conductive traces, but the subsequent stability of the resistance of oven cured traces under relatively mild cyclic loading poses a challenge. This is at least partly because PET substrates do not allow for very high curing temperatures. We show that lower and more stable resistances can be achieved by laser sintering with properly optimized parameters.

ACKNOWLEDGMENT

This material is based on research sponsored by Air Force Research Laboratory under agreement number FA8650-15-2-5401. The U.S. Government is authorized to reproduce and distribute reprints for Governmental purposes notwithstanding any copyright notation thereon. The views and conclusions contained herein are those of the authors and should not be interpreted as necessarily representing the official policies or endorsements, either expressed or implied, of Air Force Research Laboratory or the U.S. Government.

REFERENCES

[1] Grzegorz Tomaszewski and Jerzy Potencki, "Drops forming in inkjet printing of flexible electronic circuits" Circuit World, Vol. 43, Number-1, pp. 13–18, 2017.

[2] Alena Pietrikova, Peter Lukacs, Dagmar Jakubeczyova, Beata Ballokova, Jerzy Potencki, Grzegorz Tomaszewski, Jan Pekarek, Katerina Prikrylova and Martin Fides, "Surface analysis of polymeric substrates used for inkjet printing technology" Circuit World, Vol. 42, Number-1, pp. 9–16, 2016.

[3] Min Pack, Han Hu, Dong-Ook Kim, Xin Yang, and Ying Sun, "Colloidal Drop Deposition on Porous Substrates: Competition among Particle Motion, Evaporation, and Infiltration", American Chemical Society *Langmuir*, Vol. 31 (29), pp. 7953–7961, 2015.

[4] Anna Schuppert, Moritz Theilen, Ingo Reinhold," Inkjet printing of conductive silver tracks from nanoparticles inks on mesoporous substates" Society of imaging and Technology.

[5] Byoung-Joon Kim, Thomas Haas, Andreas Friederich, Ji-Hoon Lee, Dae-Hyun Nam, Joachim R Binder, Werner Bauer, In-Suk Choi, Young-Chang Joo, Patric A Gruber and Oliver Kraft, "Improving mechanical fatigue resistance by optimizing the nanoporous structure of inkjet printed Ag electrodes for flexible devices" Nanotechnology, Vol. 25, Number-12, 125706, 2014.

[6] Inhwa Lee, Sanghyeok Kim, Jeonghoon Yum, "Interfacial toughening of solutions processed Ag nanoparticles thin films by organic residuals", Nanotechnology, Vol. 23, 485704, 2012.

[7] R. S. Sivasubramony, N. Adams, M. Alhendi, G. S. Khinda, M. Z. Kokash, J. P. Lombardi, A. Raj, S. Thekkut, D. L. Weerawarne, M. Yadav, A. V. Zachariah, N. C. Stoffel, D. M. Shaddock, L. Yin, M. D. Poliks, and P. Borgesen, "Isothermal fatigue of interconnections in flexible hybrid electronics based human performance monitors," in 2018 IEEE 68th Electronic Components and Technology Conference (ECTC), pp. 896–903, IEEE, 2018.

[8] Mohammed Alhendi, Jack P. Lombardi III, Guvinder S. Khinda, Maan Z. Kokash, Darshana L. Weerawarne, Peter Borgesen, Mark D. Poliks, Nancy C. Stoffel, and Joe Iannotti (2018) "Fatigue Cycling of Electrical Interconnects Dispensed on Flexible Substrate. International Symposium on Microelectronics" No. 1, pp.543-548. Fall 2018, Vol. 2018.

Multilayer Glass Substrate with High Density Via Structure for All Inorganic Multi-chip Module

Toshiki Iwai, Taiji Sakai,
Daisuke Mizutani, and Seiki Sakuyama
FUJITSU LABORATORIES LTD.
10-1, Morinosato-Wakamiya, Atsugi, Kanagawa, Japan
Email: iwai.toshiki@fujitsu.com

Kenji Iida, Takayuki Inaba, Hidehiko Fujisaki,
Akira Tamura, and Yoshinori Miyazawa
FUJITSU INTERCONNECT TECHNOLOGIES LIMITED
36, Oaza Kitaowaribe, Nagano-shi, Nagano, Japan
Email: krem@fujitsu.com

Abstract—Silicon interposer (Si-IP) technology has been used in accelerated processing units such as graphic processing units in high-performance computing because it can package a system-on-chip and high bandwidth memories. However, the conventional Si-IP has difficulty developing larger packages because of the mismatch in the coefficient thermal expansions (CTE) of the Si-IP and the organic substrate. Therefore, the Si-IP has limited capacity for improving computing performance by the application which requires more chips. We developed a multilayer glass substrate (Glass-ST) that features a stacked glass core and propose to apply this Glass-ST to a computer board. The proposed structure has no CTE mismatch and can use high density wiring. Thus, the Glass-ST enables the assembly of more large chips than is possible using the conventional Si-IP. In this study, we prepared a 100×100 mm Glass-ST with a 5/5 μm line/space and 20 μmΦ vias. We mounted nine 21×21 mm chips with 40 μm pitch micro bumps. The results revealed that conformal plated through glass vias and a fine wiring pattern had been fabricated in the Glass-ST, and that the nine chips and Glass-ST were connected by micro bumps. The maximum warpage of the nine chips was 23 μm between temperatures of 30 $^\circ$C and 250 $^\circ$C. This means that the Glass-ST can mount chips with micro bumps due to the very slight resulting warpage. In addition, we performed thermo-mechanical simulation to investigate the stress experienced by the micro bumps. The results show that the maximum stresses of micro bumps with pitches ranging between 10 μm and 55 μm are very similar to that of 40 μm pitch micro bumps with which the real sample was packaged. We believe the improvements in the computing performance are significant by the Glass-ST technology compared to that of the conventional Si-IP technology.

Keywords-Glass Package; Stacked Glass Substrate; High Bandwidth Memory;

I. INTRODUCTION

In recent years, there has been wide recognition of the need for improved performances by computing technologies such as artificial intelligence and deep learning with respect to their high density packaging. For example, improvement in the computing performance of high-end servers was realized by the graphic processing unit (GPU). Using 2.5D silicon interposer technology, the GPU is composed of a system on a chip and high bandwidth memories (HBMs). Therefore,

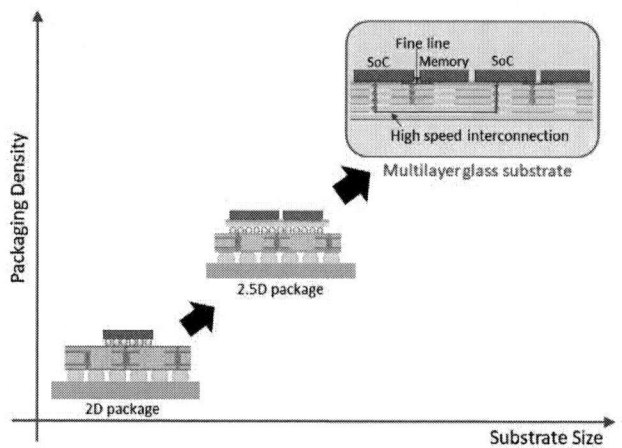

Figure 1.   Proposing high density packaging

researchers investigated the possibility of developing an Si-IP technology that features fine wiring, fine vias, and micro bumps [1], [2]. However, the Si-IP has limited capability for improving computing performance by the application which requires more chips. This is because the Si-IP cannot be enlarged due to the coefficient thermal expansion (CTE) mismatch between the silicon substrate and the organic package substrate. Thus, in previous work, we proposed and verified a structure comprising a glass multilayer substrate (Glass-ST) that has no CTE mismatch. The Glass-ST has a stacked glass core and uses conductive paste to connect the interlayer vias. We reported the results of reliability tests, the warpage characteristics during the reflow process, and the CTE and electrical characteristics [3]–[5]. We also proposed an all-inorganic packaging structure(Figure 1). In this manner, we were able to develop a large multi-chip module with no CTE mismatch between substrates.

In this study, we investigated the implementation of the Glass-ST using micro bumps and a fine line pattern to achieve high density packaging of the HBMs and processors. We prepared a Glass-ST of 100×100 mm with 20 μmΦ

Table I
PACKAGE SPECIFICATIONS

	Item	Specification (μm)
Substrate	Line and space	5 / 5
	Copper thickness	5
	TGV diameter	20
	Glass thickness	100
	Insulating film thickness	10
	Solder resist	10
	Number of layer	10
Chip	size	21×21 mm
	Bump pitch	40
	Copper pillar	20
	Solder	10

and a 5/5 μm line/space. We mounted nine 21×21 mm test chips with 40 μm pitch micro bumps on a Glass-ST. We then evaluated the resistance of the daisy chain between the chips and Glass-ST, the cross section of micro bumps after assembly, and the warpage characteristics of the chips assembled on the Glass-ST. Moreover, to investigate the assembly of micro bumps on the Glass-ST, we used a thermo-mechanical simulator to simulate the stresses on micro bumps with pitches from 10 μm to 55 μm. Based on the result of these investigations, we can now realize a large multi-chip packaging structure to replace that of the Si-IP. We can conclude, therefore, that the Glass-ST technology can improve the computing performance achieved the use of Si-IP technology.

II. MANUFACTURING PROCESS AND SAMPLE DESCRIPTION

Figure 2 and Table I show the manufacturing process and specifications, respectively. First, we fabricated the through glass vias (TGVs) by laser induced deep etching (LIDE) [6]. Then we sputtered the Ti/Cu seed layer. We fabricated the wiring pattern using a semi-additive process. We laminated a dry film of resin and drilled a hole in this film by laser. To print the conductive paste, we used a screen printing process. Single-layer substrates were stacked and then pressed by a vacuum hot press to melt the conductive paste and dry film. The solder resist was fabricated on the substrate. Next, using a flip-chip bonder, we aligned the nine chips (21×21 mm). Then, we mounted the chips using a formic-acid reflow oven at 250 °C. Finally, we applied the underfill and cured it at 165 °C for two hours.

Table II shows the material properties. The glass is alkali-free, and the conductive paste is composed of solder grain and resin. The resin dry film is epoxy-based and the underfill is also an epoxy-based material. The micro bump consists of a 20 μm copper pillar and a 10 μm AgSn solder.

A. Test Samples

A cross section of the TGVs and the wiring pattern surface are shown in Figures 3 and 4, respectively. We confirmed the 40 μm pitch of the conformal-plated TGVs and the

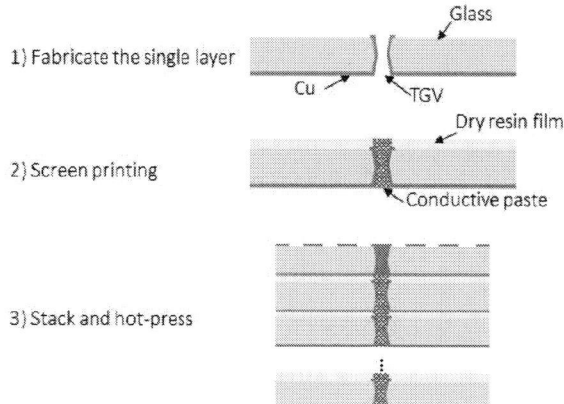

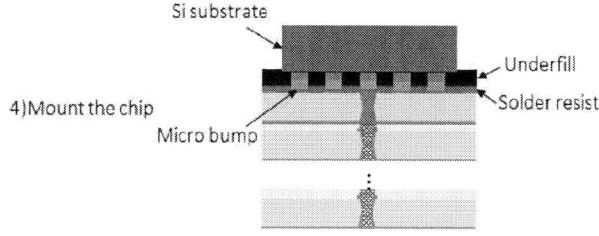

Figure 2. Manufacturing Process

Table II
MATERIAL PROPERTIES OF PACKAGE COMPOUNDS AT ROOM TEMPERATURE

Material	CTE (ppm/°C)	Elastic Modulus (GPa)
Glass	4	77
Copper	16.8	117
Insulating film	37	7
Conductive Paste	37	3.6
Si	3.6	157
Underfill	36	4.7
Solder resist	11	3.8
solder	20.8	47

wiring pattern of the 40 μm pitch pad. Figures 5 and 6 show a scanning microscope image of a test chip with micro bumps and a photograph of a sample, respectively. We confirmed that the nine test chips (21×21 mm size with 40 μm pitch micro bump grid array) were mounted on the Glass-ST. Figure 7 shows a cross section of the micro bumps after the chips had been mounted. We measured 20 points along the daisy chain of 24 bumps for which the resistance averaged 2.2 Ω and the standard deviation was 0.10 Ω. We confirmed that the micro bumps had been properly connected because the standard deviation was less than 5%. Thus, we demonstrated the mounting of large chips on the Glass-ST. This size would be difficult to achieve using the conventional Si-IP.

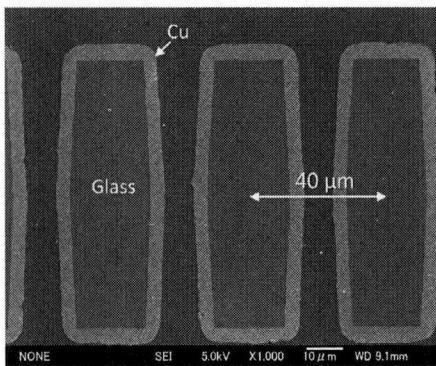

Figure 3. Cross section of TGVs

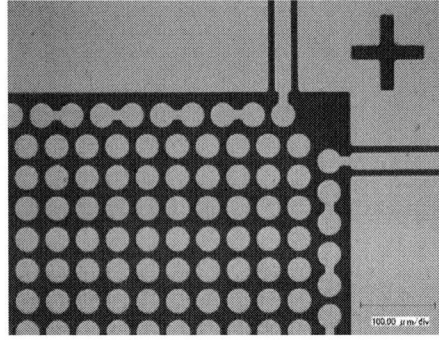

Figure 4. Image of pads on Glass-ST

III. WARPAGE UNDER THE REFLOW PROCESS

To determine the influence of the use phase and packaging process, we investigated the warpage characteristics occurring between 30 °C and 250 °C. To do so, we used the Shadow-Moiré AXP (Akrometrix). In the measurement preparation, we sprayed the sample with a white coating and left it to dry in a heat oven for 12 hours at 125 °C. Figure 6 shows a photograph of the sample with the chip numbers identified. Figures 8 and 9 show contour plots of warpage at 30 °C and that with respect to different temperatures, respectively. The maximum warpage was 23 μm in the

Figure 5. SEM image of test chip

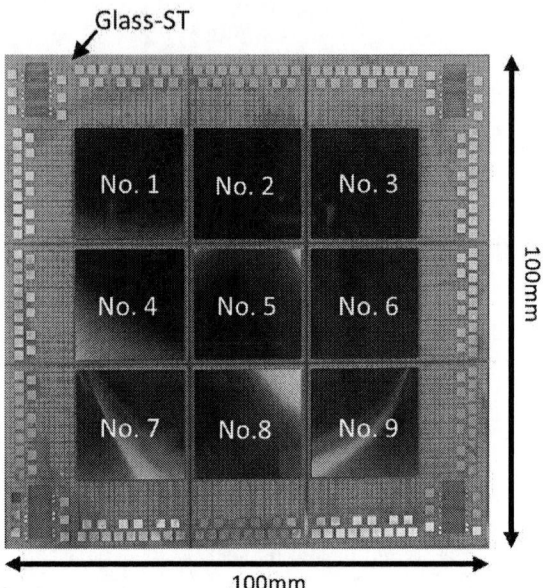

Figure 6. Photograph of sample with 21×21 mm test chips

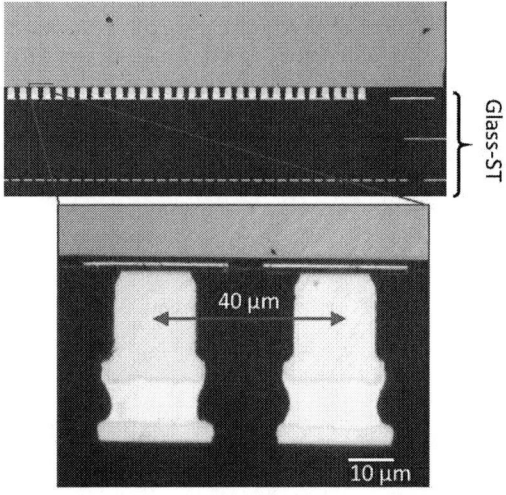

Figure 7. Cross section of 40 μm pitch micro bumps

measured temperature range. We consider the warpage of the chips to be less than the measured results due to the micro unevenness of the spray in the contour plots. The warpage of all the chips decreased as temperature increased until approximately 150 °C. At temperatures higher than 150 °C, the warpage characteristics remained stable. We consider the warpage characteristics at 30 °C and 150 °C to be due to the CTE mismatch of the underfill and substrate, because the cure temperature of the underfill was 165 °C.

978-1-7281-1500-9/19 $31.00 © 2019 IEEE

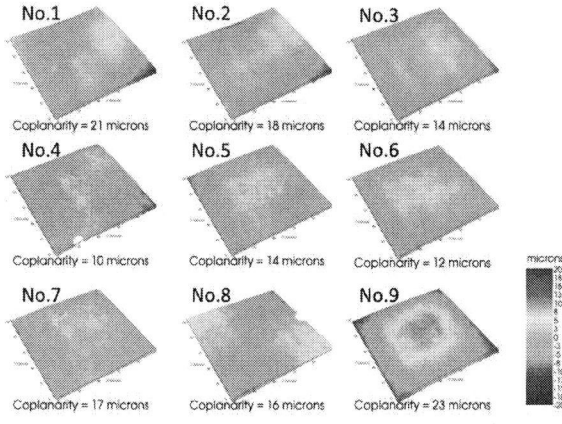

Figure 8. Warpage of chips

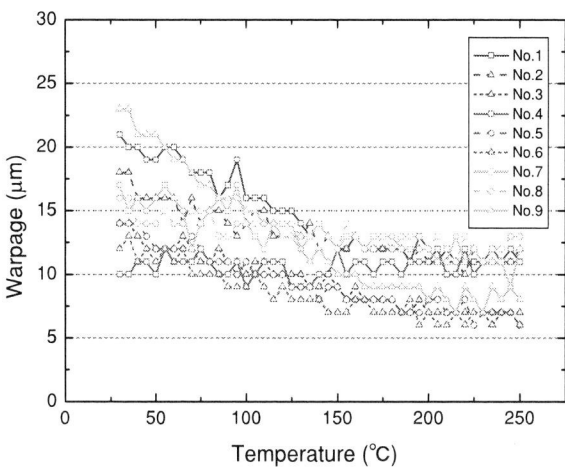

Figure 9. Warpage of chips between 25°C and 250°C

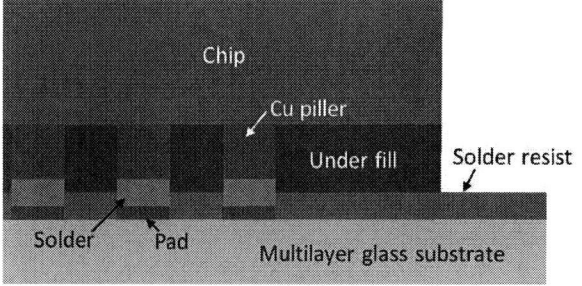

Figure 10. Simulation model

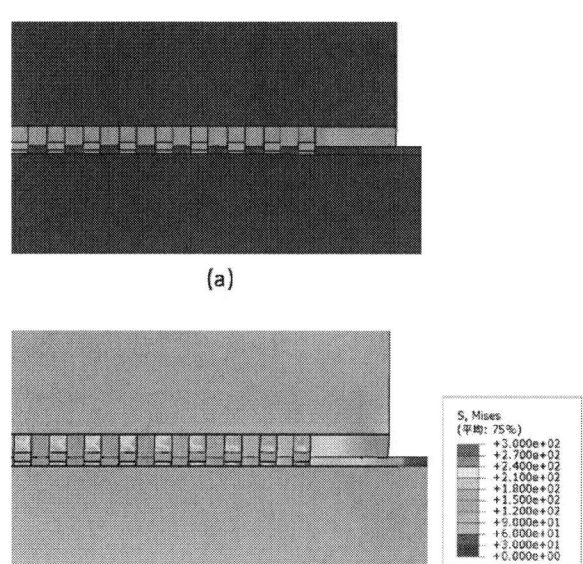

Figure 11. Comparison of stress of organic substrate and Glass-ST; (a) stress contour plots of Glass-ST, (b) stress contour plots of organic substrate

IV. THERMO-MECHANICAL SIMULATION

Several studies have reported that edge stress on micro bumps occurs during the mounting process [7]–[9]. We investigated stress on 40 μm pitch micro bumps under thermal-cycle test conditions (125 °C to -55 °C) We also investigated the stress on micro bumps with pitches ranging between 10 μm and 55 μm because we would like to confirm the capability for finer pitch assembly of micro bumps with pitches other than 40 μm. The CTE of the Glass-ST was the same as that of glass since the glass is relatively thicker than the wiring pattern and resin [4]. Therefore, for the CTE of the Glass-ST, we used 3.6 ppm/°C and for that of the organic substrate, we used 16 ppm/°C at room temperature. In the simulation, for the material properties we used the temperature-dependent data of Young's modulus and the CTE. Table II shows the material properties at room temperature. Figure 10 shows a model of micro bumps with a 40 μm pitch. In the chip model, we used a quarter-symmetry representation of the full 21×21 mm model, and a grid of 10×10 micro bumps were arranged at the edge of the chip. In the models of micro bump pitches between 10 μm and 55 μm, we used the same aspect ratio as that in the 40 μm pitch micro bump model with respect to the copper pillar, solder, solder resist, underfill, and pad. We investigated the stresses at 125 °C, 20 °C, and -55 °C as compared to the stress-free 165 °C, which is the underfill cure temperature. We simulated the models by elastic analysis. Figure 11 shows stress contour plots of the 40 μm bump pitch at -55 °C.

The stress contour plots of the Glass-ST reveal that the stress was concentrated on the micro bumps, whereas the stress of the organic substrate was concentrated at the edge of the silicon chip. The reason for this is that the stress on the Glass-ST is caused by the CTE mismatch between the underfill and micro bumps, and that of the organic substrate is due to the CTE mismatch between the chip and organic substrate. Figure 12 shows the stress contours of the bumps.

978-1-7281-1500-9/19 $31.00 © 2019 IEEE   1955

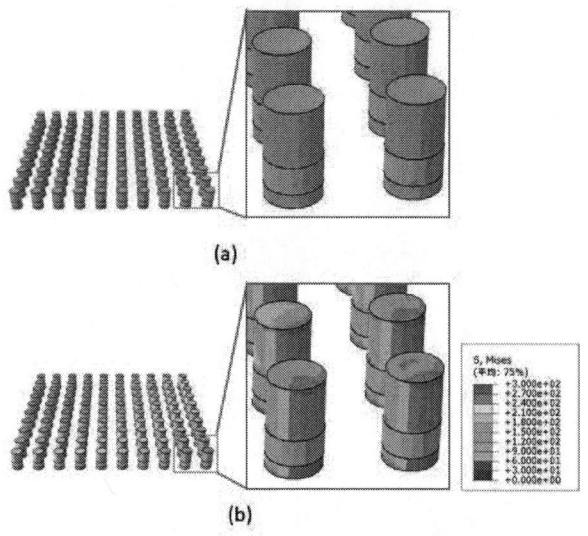

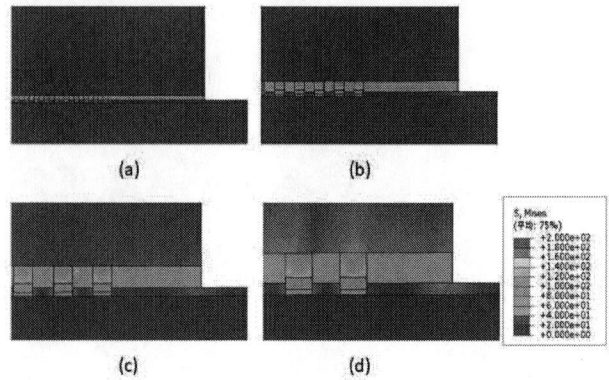

Figure 14. Stress contour plots of micro bumps with pitches between 10 μm and 55 μm; (a) stress on 10 μm pitch micro bumps, (b) stress on 20 μm pitch micro bumps, (c) stress on 40 μm pitch micro bumps, (d) stress on 55 μm pitch micro bumps

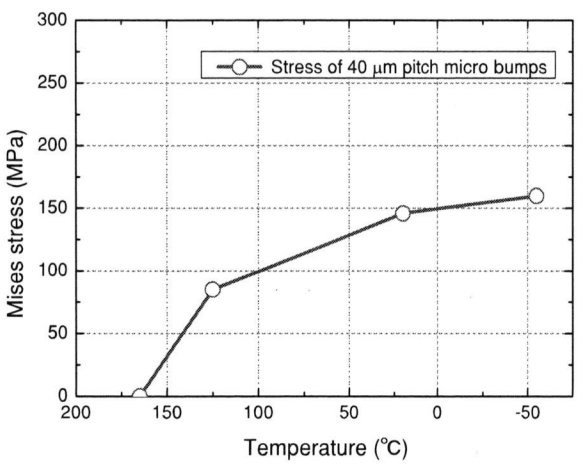

Figure 12. Contour plots of micro bump stress; (a) micro bumps on Glass-ST, (b) micro bumps on organic substrate

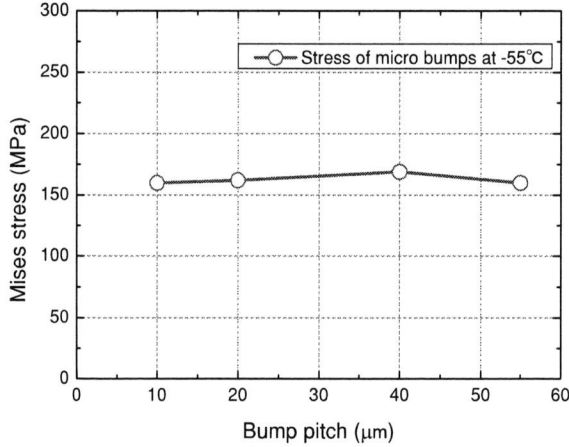

Figure 15. Stress characteristics with respect to bump pitch

Figure 13. Stress characteristics from 165 °C to -55 °C

The micro bump stress in the Glass-ST was concentric, whereas the bump stress on the organic substrate showed a gradient in the direction of the chip edge. We consider that stress occurred due to the CTE mismatch between the chip and organic substrate. Therefore, we suggest that bumps in chips mounted on an organic substrate exhibit failure modes such as a crack due to the shear stress at the interface, whereas chips mounted on a Glass-ST experience no shear stress. Figure 13 shows the stress characteristics, in which we can see that the maximum stresses occur at 125 °C, 20 °C, and -55 °C. The difference in the stresses at 165 °C and 125 °C was about 90 MPa, but that at 125 °C and 20 °C was about 50 MPa. The difference in the stress at 20 °C and -55 °C was only several MPa. Thus, the results of the

thermal-cycle test (125 °C to -55 °C) indicate that the stress on the bumps is small. We consider that the actual stress will be even smaller than that in the simulation results. Further research should be done to perform viscoelastic analysis to simulate actual stress.

Figure 14 shows the contour plots of micro bumps with pitches ranging from 10 μm to 55 μm. Figure 15 shows the maximum stress on micro bumps with respect to pitch, in which we can see that the maximum stresses for all bump pitches were almost the same. In other words, the stress on the mounted Glass-ST did not change between 10 μm and 55 μm when the aspect ratio of the materials was held constant. Thus, we conclude that the Glass-ST can improve the bandwidth between chips because the large number of connections between micro bumps on the Glass-ST are ensured.

978-1-7281-1500-9/19 $31.00 © 2019 IEEE

V. CONCLUSION

In this paper, we investigated the connection of micro bumps on a glass multilayer substrate instead of a silicon interposer. The results confirm the 40 μm pad pitch pattern and 20 μm conformal-plated through glass vias. We mounted nine 21×21 mm chips with 40 μm pitch micro bumps on a 100×100 mm glass multilayer substrate. Our results indicate that this multilayer glass substrate is more than five times larger than the conventional silicon interposer. The resistance of the daisy chain of 24 bumps between the chip and substrate averaged 2.2 Ω with a standard deviation of 0.1 Ω. This resistance was stable within a 5% standard deviation. The maximum warpage of the chips was measured to be 23 μm between 25 °C and 250 °C temperature range and was stable with respect to temperature change. Moreover, using thermo-mechanical simulation, we investigated the stress on the micro bumps at temperatures between 125 °C and -55 °C. The results reveal that the glass multilayer substrate experiences less stress than the organic substrate. This is because the stress on the glass multilayer is concentric, but the organic substrate experiences a stress gradient in the direction of the chip edge. The stresses on micro bumps with pitches between 10 μm and 55 μm were the same as that of a micro bump with a 40 μm pitch mounted on the real sample. The results show that the glass multilayer substrate was resistant to thermal stress. As such, we predict the multilayer glass substrate will be the next high-end packaging structure.

ACKNOWLEDGMENT

The authors would like to thank the Vitrion team from LPKF for fabricating the TGVs.

REFERENCES

[1] M. Matsuo, et al., "Silicon interposer technology for high-density package", Electronic Components and Technology Conference (ECTC), IEEE 50th, pp.1445 - 1449, 2002

[2] V. Sukumaran, et al., "Low cost, high performance, and high reliability 2.5D silicon interposer", Electronic Components and Technology Conference (ECTC), IEEE 63rd, pp.342 - 347, 2013

[3] T. Iwai, et al., "A Novel Inorganic Substrate by Three Dimensionally Stacked Glass Core Technology", Electronic Components and Technology Conference (ECTC), IEEE 68th, pp.1981 - 1986, 2018

[4] T. Iwai, et al., "Glass Multilayer Package Substrate using Conductive Paste Via Connection", IEEE CPMT Symposium Japan 2018 (ICSJ2018), pp.105 - 108, 2018

[5] T. Iwai, et al., "Glass Multilayer Substrate Using Conductive Paste for Interstitial Via Hole(Japanese)", Microjoining and Assembly Technology in Electronics (MATE), 25th, pp.153 - 158, 2019

[6] R. Ostholt, et al., "High speed through glass via manufacturing technology for interposer", Electronics System-integration Technology Conference (ESTC), 5th, 2014

[7] J. Lee, et al., "Micro Bump System for 2nd Generation Silicon Interposer with GPU and High Bandwidth Memory (HBM) Concurrent Integration", Electronic Components and Technology Conference (ECTC), 68th, pp.607 - 612, 2018

[8] C. Lee, et al., "An Overview of the Development of a GPU with integrated HBM on Silicon Interposer", Electronic Components and Technology Conference (ECTC), 66th, pp.1439 - 1444, 2016

[9] B. Banijamali, et al., "Advanced reliability study of TSV interposers and interconnects for the 28nm technology FPGA ", Electronic Components and Technology Conference (ECTC), IEEE 61st, pp. 285 - 290, 2011

The Poisson's Ratio of Lead Free Solder - The Often Forgotten but Important Material Property

KM Rafidh Hassan, Mohammad S. Alam, Jeffrey C. Suhling, Pradeep Lall
Department of Mechanical Engineering, and
Center for Advanced Vehicle and Extreme Environment Electronics (CAVE³)
Auburn University
Auburn, AL 36849
E-mail: jsuhling@auburn.edu

Abstract—The Poisson's ratio (PR = ν) is a basic mechanical property of a material that relates the transverse contraction strain to the axial extension strain in a uniaxial tension stress-strain test. It is a required input parameter in almost all material constitutive models (e.g. elastic, elastic-plastic-creep, Anand model, etc.) used for solder in finite element simulations. Since it is more difficult to measure relative to standard stress-strain and creep curves, it is often guessed or assumed to be a fixed value (e.g. ν = 0.3), independent of temperature, strain rate, and material composition. Using finite element simulations of several PBGA configurations subjected to thermal cycling, we have shown that the plastic strain and energy dissipation results predicted by the FEA calculations are strongly dependent on the specified value of the solder Poisson's ratio. Several sizes (5, 10, 15 mm) of PBGA components with SAC305 solder joints with 0.4 and 0.8 mm spacing were modeled, and the package assemblies were subjected to a time dependent cyclic temperature distribution from -40 to 125 °C. Our simulation results have demonstrated that for specified values of 0.15 < ν < 0.40, the solder plastic work varied over 20% and the predicted reliability varied over 50% when using a Morrow-Darveaux energy-based fatigue model. Thus, the FEA results are highly sensitive to the specified value of the solder Poisson's ratio, so that it is important to carefully characterize the Poisson's effect in lead free solders.

In this work, we have conducted an experimental investigation to characterize the Poisson's ratio of several SAC and SAC+X lead free solders. Uniaxial tensile stress-strain tests were carried out on SAC305 (96.5Sn3.0Ag0.5Cu), SAC405 (95.5Sn4.0Ag0.5Cu), and SAC_Q (SAC+3.3Bi) specimens using a micro tension/torsion testing machine with two strain rates (0.0001, and 0.00001 (sec⁻¹)) and four testing temperatures (T = 25, 50, 75, 100 °C). Deformations and strains in the axial and transverse directions were measured using miniature strain gages with automatic data acquisition from LabVIEW software. The recorded transverse strain vs. axial strain data for each test were then fit with a linear regression analysis to determine the Poisson's ratio value. A large test matrix of experiments was developed to study the effects of temperature, strain rate, alloy composition, solidification cooling profile, and isothermal aging exposure on the value of solder Poisson's ratio; and to create a material property database for finite element simulations. The Poisson's ratio was found to increase with increasing temperature, and decrease with increasing strain rate, alloy silver content, and cooling rate after solidification. Finally, the microstructural coarsening that

occurs during isothermal aging led to an increase in the Poisson's ratio (up to 15% increase with one day of aging). Considering all of the possible branches of the test matrix (temperatures, strain rates, prior aging conditions, and microstructure), the value of the Poisson's ratio of SAC305 ranged from 0.31 < ν < 0.44. Analogous results were found for the other solder alloys.

Keywords: Poisson's Ratio, Lead Free Solder, SAC, SAC+X, Isothermal Aging, Reliability

I. INTRODUCTION

With the emergence of the modern electronic packaging technology over the last few decades, solder alloys have been the prime interconnect material used in electronic packaging. Among various alloy systems that are considered as lead-free solder candidates, Sn-Ag-Cu (SAC) alloys have been recognized as the most promising because of their relatively low melting temperature (compared with the Sn-Ag binary eutectic lead free solder), superior mechanical properties, and good compatibility with other components [1-3]. As the electronic industries transition to lead free soldering by the motivation of environmental concerns, legislative mandates, and market differentiation, great efforts have been undertaken to develop desirable lead free solders and establish a corresponding database of key material properties.

Finite element analysis (FEA) is a key modeling technology used for predicting the reliability of electronic assemblies and packaging. The accuracy of the results obtained by FEA simulations are strongly dependent on the correctness of the input material properties used in the selected constitutive and failure models. For example, there is an abundance of literature where researchers have explored the influence of the specified material properties on the FEA results (e.g. [4-10]).

The Poisson's ratio (PR = ν) is a basic mechanical property of a material that relates the transverse contraction strain to the axial extension strain in a uniaxial tension stress-strain test. It is a required input parameter in almost all material constitutive models (e.g. elastic, elastic-plastic-creep, Anand model, etc.) used for solder in finite element simulations. Since it is more difficult to measure relative to standard stress-strain and creep curves, it is often guessed or

978-1-7281-1500-9/19 $31.00 © 2019 IEEE

assumed to be a fixed value (e.g. ν = 0.3), independent of temperature, strain rate, and material composition. The importance of the Poisson's ratio is typically ignored or not understood, so that it often becomes forgotten in the analysis.

Using finite element simulations of several PBGA configurations subjected to thermal cycling, we have shown that the plastic strain and energy dissipation results predicted by the FEA calculations are strongly dependent on the specified value of the solder Poisson's ratio [10]. Several sizes (5, 10, 15 mm) of PBGA components with SAC305 solder joints with 0.4 and 0.8 mm spacing were modeled, and the package assemblies were subjected to a time dependent cyclic temperature distribution from -40 to 125 °C. Our simulation results have demonstrated that for specified values of 0.15 < ν < 0.40, the solder Plastic Work varied over 20% and the predicted reliability varied over 50% when using a Morrow-Darveaux energy-based fatigue model. Thus, the FEA results are highly sensitive to the specified value of the solder Poisson's ratio, so that it is important to carefully characterize the Poisson's effect in lead free solders.

The constitutive and failure behaviors of lead free solder joints are highly dependent on the solder microstructure. In addition, the microstructures of popular SAC lead free solders are inherently unstable at relatively low temperatures, and continually evolve when utilized in electronic packaging assemblies. These microstructural changes lead to degradations of the constitutive and failure behaviors of lead free solder joints and evolutions of their material properties [11], and these combined effects are typically referred to as solder aging.

The recent literature contains a large selection of research on lead free solder aging. For example, aging leads to large reductions in bulk mechanical properties such as modulus and strength [12-15]. Other mechanical responses affected by aging include creep behavior [12-16], viscoplastic Anand model parameters [7-9, 17-20], high strain rate mechanical behavior and drop reliability [21-22], ball shear strength [23], fracture behavior [24], high temperature behavior [25], uniaxial cyclic stress-strain curves and fatigue life [26-30], shear cyclic stress-strain curves and fatigue life [31-32], and thermal cycling reliability [8-9, 33-36]. In addition, tests on joints using nanoindentation have revealed aging induced degradations in modulus, hardness, and creep rate [37-40]. Several investigations [15, 20, 29-30, 39, 42] have also reported that the use of dopants (SAC+X), particularly bismuth (X = Bi), can help mitigate aging induced degradations in lead free SAC alloys. Recently, our group has explored the evolution of solder microstructure during aging quantitatively using processing of digital image data from SEM and Scanning Probe Microscopy, and correlated the results to the observed material property changes [43-45]. While it is a key material property, the effects of aging on the Poisson's ratio of solder materials have not been explored in the literature.

In this work, we have conducted an experimental investigation to characterize the Poisson's ratio of several SAC and SAC+X lead free solders. Uniaxial tensile stress-strain tests were carried out on SAC305 (96.5Sn3.0Ag0.5Cu), SAC405 (95.5Sn4.0Ag0.5Cu), and SAC_Q (SAC+3.3Bi) specimens using a micro tension/torsion testing machine with two strain rates (0.0001, and 0.00001 (sec^{-1})) and four testing temperatures (T = 25, 50, 75, 100 °C). Deformations and strains in the axial and transverse directions were measured using miniature strain gages with automatic data acquisition from LabVIEW software. The recorded transverse strain vs. axial strain data for each test were then fit with a linear regression analysis to determine the Poisson's ratio value. A large test matrix of experiments was implemented to study the effects of temperature, strain rate, alloy composition, solidification cooling profile, and isothermal aging exposure on the value of solder Poisson's ratio; and to create a material property database for finite element simulations.

II. FINITE ELEMENT ANALYSIS

A. Overview of Models

Using finite element simulations of several PBGA configurations subjected to thermal cycling, we have shown that the plastic strain and energy dissipation results predicted by the FEA calculations are strongly dependent on the specified value of the solder Poisson's ratio. The details of these analyses are presented in references [9-10]. In particular, several sizes (5, 10, 15 mm) of PBGA components with SAC305 solder joints with 0.4 and 0.8 mm spacing were modeled, and the package assemblies were subjected to a time dependent cyclic temperature distribution from -40 to 125 °C. Figures 1 and 2 show a typical BGA component construction and the corresponding finite element mesh, respectively; while Fig. 3 illustrates the thermal cycling profile. The Anand model [46] was used for the post yielding viscoplastic behavior of the SAC305 solder joints, and crack initiation and growth models proposed by Darveaux [47] were used as failure criteria.

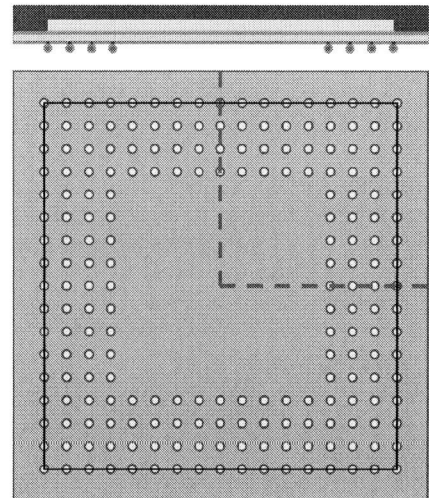

Figure 1.   BGA (15 mm) Component Construction

978-1-7281-1500-9/19 $31.00 © 2019 IEEE

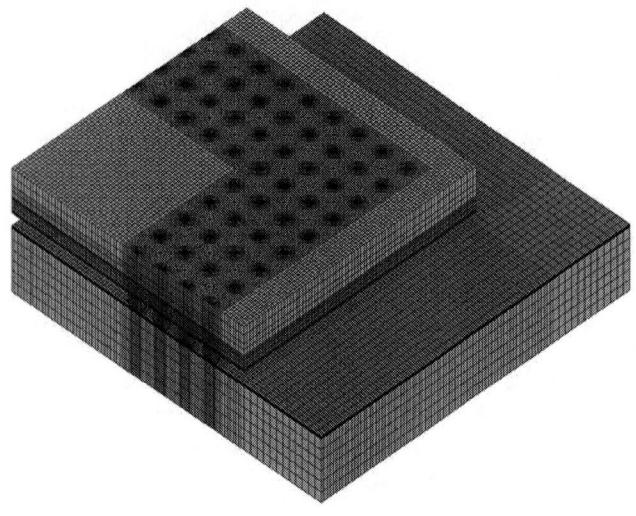

Figure 2. Finite Element Mesh (15 mm BGA)

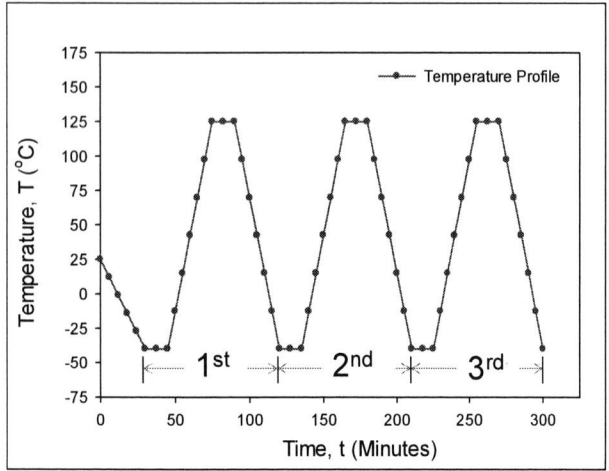

Figure 3. Thermal Cycling Profile Used in FEA

the solder Poisson's ratio were determined for the three types of assembled PBGA components. The results for the 15, 10 mm, and 5 mm components are tabulated in Fig. 4, and plotted as function of Poisson's ratio in Fig. 5.

Poisson's Ratio (ν) (Input to the FEA Model)	15 mm BGA		10 mm BGA		5 mm BGA	
	Plastic Work Per Cycle ΔW (MPa)	No. of Cycles to Crack Initiation (N_f)	Plastic Work Per Cycle ΔW (MPa)	No. of Cycles to Crack Initiation (N_f)	Plastic Work Per Cycle ΔW (MPa)	No. of Cycles to Crack Initiation (N_f)
0.15	0.171	5297	0.167	5670	0.139	9470
0.20	0.190	3980	0.186	4232	0.158	6688
0.25	0.191	3914	0.187	4160	0.159	6554
0.30	0.195	3715	0.190	3944	0.162	6157
0.35	0.197	3583	0.193	3801	0.165	5897
0.40	0.209	3032	0.205	3206	0.177	4836

Figure 4. Predicted Plastic Work and Cycles to Crack Initiation for Various Values of the Specified Solder Joint Poisson's Ratio

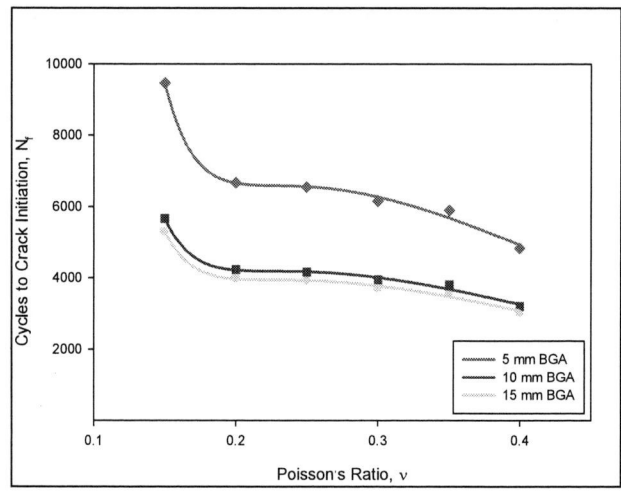

Figure 5. Predicted Dependence of the Cycles to Crack Initiation on the Solder Joint Poisson's Ratio

Values of the utilized Anand parameters for SAC305 solder and of the elastic constants of the other packaging materials are listed in Figures 12 and 13 of reference [9]. In that work, ν = 0.3 was used for the Poisson's ratio of the SAC305 solder. In our more recent study [10], several different values of Poisson's ratio were chosen including ν = 0.15, 0.20, 0.25, 0.30, 0.35, and 0.40. For each value of Poisson's ratio, FEA analyses were performed for the three types of PBGAs. The plastic energy dissipated per cycle in the critical solder ball was calculated for different values of the solder Poisson's ratio, and then the reliability was estimated from these values using the energy based fatigue criterion proposed by Darveaux [17]

B. *Reliability Calculations from FEA Results*

Using the FEA results for different values of the input solder Poisson's ratio, the predicted dependencies of the plastic work per cycle and the cycles to crack initiation on

As seen in Figs. 4 and 5, the value of the input Poisson's ratio specified in the FEA simulation affected the results for both the plastic work per cycle and the cycles to crack initiation considerably. The variations in the plastic work per cycle for different Poisson's ratio values were ~22% for the 15 mm BGA, ~23% for the 10 mm BGA, and ~27% for the 5 mm BGA). The effect was more significant in terms of the reliability. For the predicted cycles to crack initiation, the variations for different Poisson's ratio values were ~43% for the 15 mm BGA, ~44% for the 10 mm BGA, and ~49% for the 5 mm BGA. In summary, our simulation results have demonstrated that for specified values of 0.15 < ν < 0.40, the solder Plastic Work varied over 20% and the predicted reliability varied over 50% when using a Morrow-Darveaux energy-based fatigue model. Thus, the FEA results are highly sensitive to the specified value of the solder Poisson's ratio, so that it is important to carefully characterize the Poisson's effect in lead free solders.

III. EXPERIMENTAL PROCEDURE

A. Uniaxial Test Sample Preparation

Uniaxial stress-strain testing has been performed to measure and characterize the Poisson's ratio experimentally. The uniaxial specimens had nominal dimensions of 80 x 3 x 0.5 mm, and they were produced by casting within rectangular cross-section glass tubes [12-15]. Molten solder was drawn into the tubes by a vacuum suction process, and then the tubes were cooled and the solder was solidified. Solidification was initially performed using water quenching. This yielded a first set of test samples with extremely fine water quenched (WQ) microstructure and high (peak) mechanical properties. Some of the quenched samples were also subsequently reflowed with a controlled temperature profile in a SMT reflow oven. This yielded a second set of reflowed (RF) samples with a coarser and more realistic microstructure similar to that found in BGA solder joints. After solidification, the samples were carefully extracted from the glass tubes, and Fig. 6 shows a typical final specimen. Fig. 7 illustrates the water quenched (WQ) and reflowed (RF) cooling profiles used to solidify the samples.

Figure 6. Solder Uniaxial Test Specimen

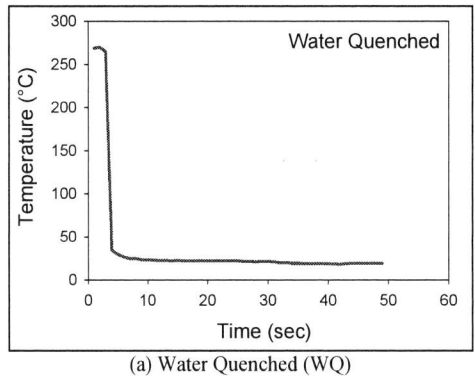

(a) Water Quenched (WQ)

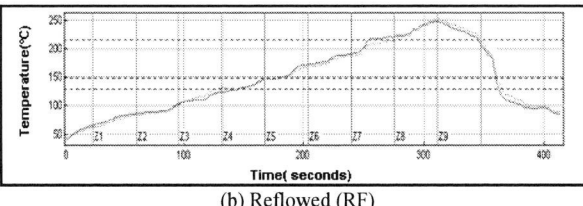

(b) Reflowed (RF)

Figure 7. Solder Specimen Cooling Profiles

To measure the Poisson's ratio, both axial and transverse strains must be monitored during uniaxial testing. In this work, we used bonded strain gages to make the strain measurements. Therefore, two strain gages were mounted on each sample, one on the top surface in the axial direction along the length of the specimen, and one on the bottom surface in the transverse direction that is perpendicular to the length of the specimen. The selected strain gages (Micro Measurements C2A-13-015LW-120) were very small (0.38 mm active gage length) so that they would easily fit within the 3 mm wide test specimens. Prior to mounting, careful preparation on the surfaces of the specimens was done using standard stain gage methodology, and then a very thin layer of adhesive was utilized to bond the gages to the solder specimens. Figure 8 shows a completely prepared solder specimen with mounted strain gages.

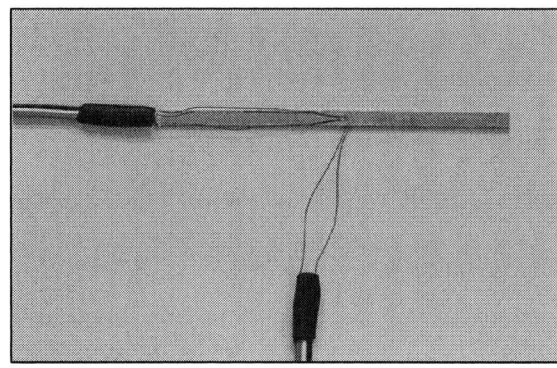

Figure 8. Uniaxial Solder Specimen with Mounted Strain Gages

B. Mechanical Testing and Data Processing

In this study, the solder specimens were loaded uniaxially using a micro tension/torsion thermo-mechanical test system (MT-200 from Wisdom Technology, Inc), as shown in Figs. 9 and 10. This system facilitates testing at elevated temperatures up to +300 °C by using an integral environmental chamber. In early testing, thermocouples were attached on trial samples to determine the required heating time necessary to achieve the desired test temperature. In subsequent tests, the heating time was kept at a minimum to reduce unwanted aging effects during the experiments. The exerted forces during uniaxial testing were measured using a precision universal six-axis automated load cell. Samples with an active gage length of 60 mm were used for all tests.

Figure 9. Mechanical Test System and High Temperature Chamber.

Figure 10. Strain Gage Mounted Sample Gripped within the Heating Chamber (Enlarged View)

Data acquisition was performed using a National Instruments NI-9237 strain gage bridge module along with NI-9944 connector hardware. The experiments were controlled using LabVIEW software, and the strain gage data from each test were stored for subsequent analysis. Post processing consisted of performing a linear regression fit to the recorded transverse vs. axial strain data from each test. From the definition of a linear Poisson's ratio, the relationship between the two strains is:

$$\varepsilon_y = -\nu\varepsilon_x \qquad (1)$$

or

$$|\varepsilon_y| = +\nu\varepsilon_x \qquad (2)$$

where ε_y is the transverse strain, ε_x is the axial strain, and ν is the Poisson's ratio. Typical transverse strain vs. axial strain raw data for an example test for SAC305 is shown in Figure 11 along with the corresponding linear regression fit. The variation of the raw data was quite linear, and the slope is the desired Poisson's ratio. Similar results (good linearity) were found for all tests performed in this study. All of the subsequent plots in this paper will feature the linear regression fits to the transverse vs. axial strain data.

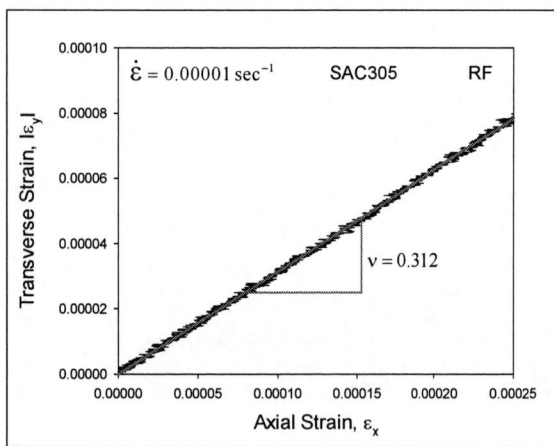

Figure 11. Example Poisson's Ratio Raw Data and Regression Fit

IV. POISSON'S RATIO MEASUREMENT RESULTS

A. Results for Non-Aged Solder Samples

A large test matrix of experiments was developed to study the effects of temperature, strain rate, alloy composition, and solidification cooling profile on the value of the solder Poisson's ratio. These results were used to create a material property database for FEA simulations. In particular, uniaxial tensile stress-strain tests were carried out on three different lead free solder alloys including SAC305 (96.5Sn3.0Ag0.5Cu), SAC405 (95.5Sn4.0Ag0.5Cu), and SAC_Q (SAC+3.3Bi). The SAC305 and SAC405 solders were tested with both water quenched (WQ) and reflowed (RF) microstructures, while experiments for SAC_Q were preformed using only the RF microstructure. No aging was considered for these experiments. For each solder alloy and the non-aged RF microstructure, testing was performed using four testing temperatures (T = 25, 50, 75, 100 °C) and two strain rates (0.0001 and 0.00001 (sec^{-1})). For the non-aged WQ microstructure and SACN05 samples, testing was performed using the same four testing temperatures (T = 25, 50, 75, 100 °C) and a strain rate of 0.00001 (sec^{-1}). The test matrix for the non-aged specimens is shown in Fig. 12.

Temperature (°C)	Strain Rate	
	10^{-4} sec^{-1}	10^{-5} sec^{-1}
25	√	√
50	√	√
75	√	√
100	√	√

(a) SAC305 (RF), SAC405 (RF), SAC_Q (RF)
RF = Reflowed Microstructure (No Aging)

Temperature (°C)	Strain Rate
	10^{-5} sec^{-1}
25	√
50	√
75	√
100	√

(b) SAC305 (WQ) and SAC405 (WQ)
WQ = Water Quenched Microstructure (No Aging)

Figure 12. Poisson's Ratio Test Matrix for Non-Aged Specimens

The measured results for the various solder alloys, temperatures, and strain rates are tabulated in Figs. 13 and 14 for the water quenched and reflowed microstructures, respectively. In all cases, the Poisson's ratio was seen to increase significantly with temperature. For example, temperature dependent Poisson's ratio data are shown in Fig. 15 for the reflowed microstructure and SR = 0.00001 sec^{-1}. For each solder alloy, the slope of the Poisson's ratio curves is seen to increase steadily with temperature. The extracted

978-1-7281-1500-9/19 $31.00 © 2019 IEEE

Poisson's ratio values change in the range of 15-25% between T = 25 °C to T = 100 °C.

When comparing solder alloys with the same solidification profile and strain rate, the Poisson's ratio for SAC305 is the greatest, followed by that for SAC405, and then finally that for SAC_Q. This can be easily visualized from the plots in Fig. 16, which show the dependence of the Poisson's ratio on temperature. In each graph, the solidification profile and strain rate are the same, and the three different alloys are represented by the different colors (red = SAC305, green = SAC405, and blue = SAC_Q). It is clear that the Poisson's ratio of a SAC solder is reduced when adding additional silver. The observed increases in the Poisson's ratio values for SAC305 relative to SAC405 are typically in the range of 5-8%. More silver (Ag) in the composition leads to more Ag_3Sn IMC particles, which stiffens to the material increases its resistance to dislocation movement. The smallest Poisson's ratio values were found for SAC_Q, which has more silver than SAC305, but also includes bismuth (Bi). The bismuth ion SAC_Q has been observed to go into solution at elevated temperatures [42], which provides additional solid solution strengthening of the material relative to SAC305 and SAC405, which have no bismuth.

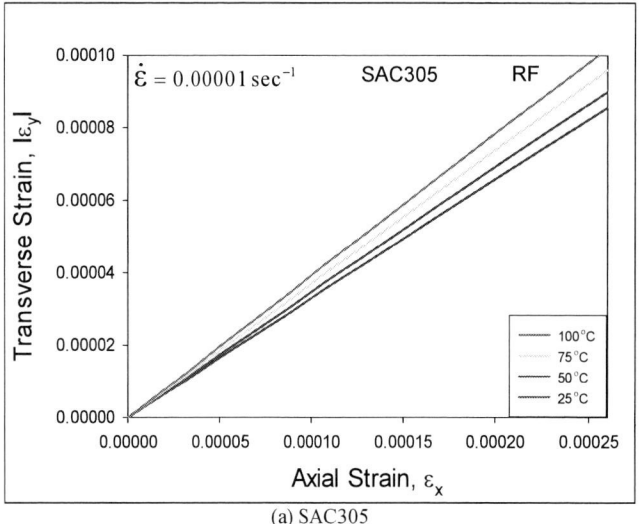

(a) SAC305

Alloy	Strain Rate	Poisson's Ratio, v			
		25 °C	50 °C	75 °C	100 °C
SAC305	10^{-5} sec^{-1}	0.307 (0.002)	0.319 (0.002)	0.342 (0.002)	0.361 (0.003)
SAC405	10^{-5} sec^{-1}	0.284 (0.002)	0.302 (0.002)	0.327 (0.003)	0.348 (0.002)
SAC_Q	10^{-5} sec^{-1}	0.261 (0.002)	0.289 (0.001)	0.311 (0.003)	0.331 (0.001)

Figure 13. Poisson's Ratio Values for Lead Free Solders with WQ Microstructures (No Aging)

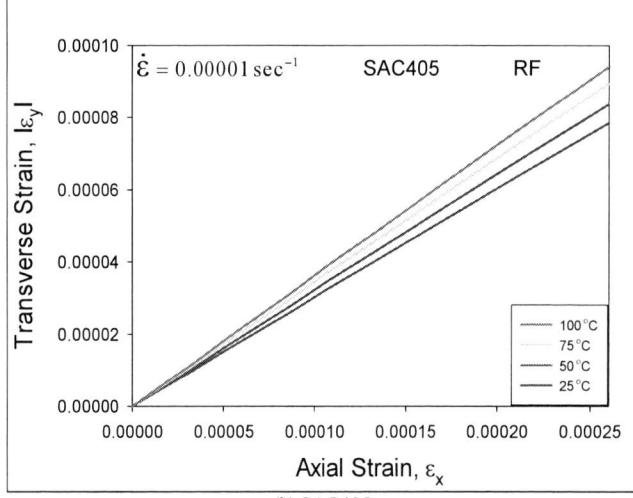

(b) SAC405

Alloy	Strain Rate	Poisson's Ratio, v			
		25 °C	50 °C	75 °C	100 °C
SAC305	10^{-4} sec^{-1}	0.304 (0.001)	0.325 (0.001)	0.349 (0.001)	0.371 (0.001)
	10^{-5} sec^{-1}	0.329 (0.002)	0.346 (0.001)	0.369 (0.001)	0.383 (0.001)
SAC405	10^{-4} sec^{-1}	0.292 (0.001)	0.308 (0.002)	0.332 (0.001)	0.351 (0.002)
	10^{-5} sec^{-1}	0.305 (0.001)	0.322 (0.001)	0.344 (0.003)	0.364 (0.001)
SAC_Q	10^{-4} sec^{-1}	0.274 (0.001)	0.298 (0.001)	0.318 (0.003)	0.335 (0.002)
	10^{-5} sec^{-1}	0.289 (0.002)	0.309 (0.003)	0.326 (0.002)	0.346 (0.002)

Figure 14. Poisson's Ratio Values for Lead Free Solders with RF Microstructures (No Aging)

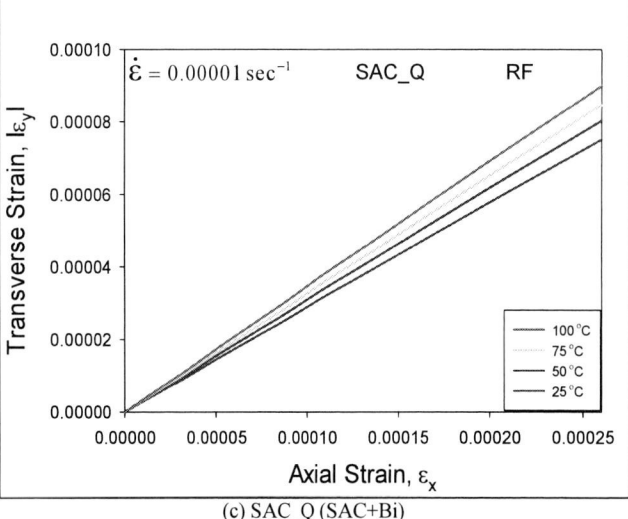

(c) SAC_Q (SAC+Bi)

Figure 15. Example Poisson's Ratio Test Data for Different Temperatures

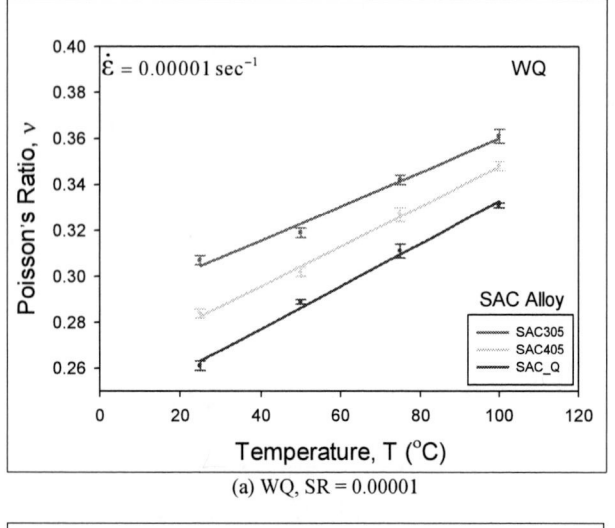

(a) WQ, SR = 0.00001

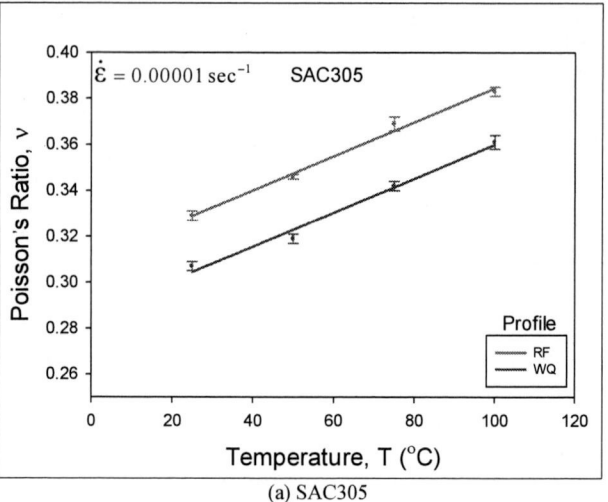

(a) SAC305

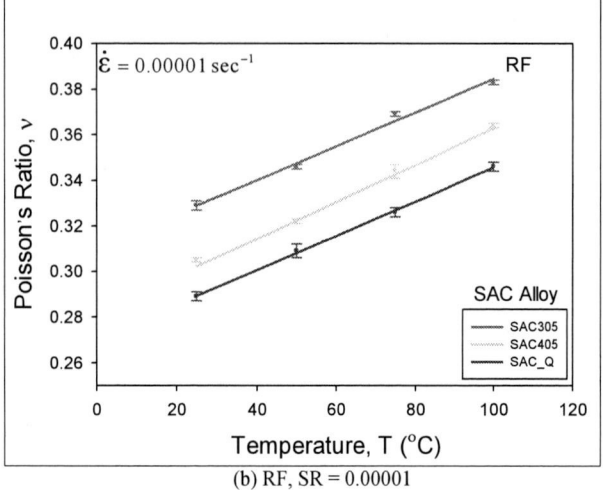

(b) RF, SR = 0.00001

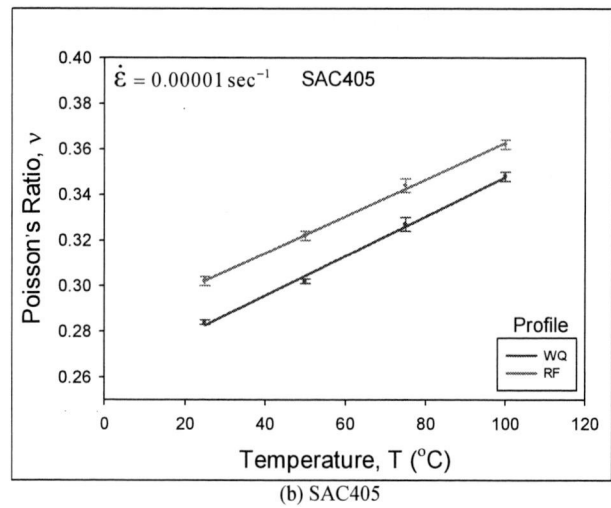

(b) SAC405

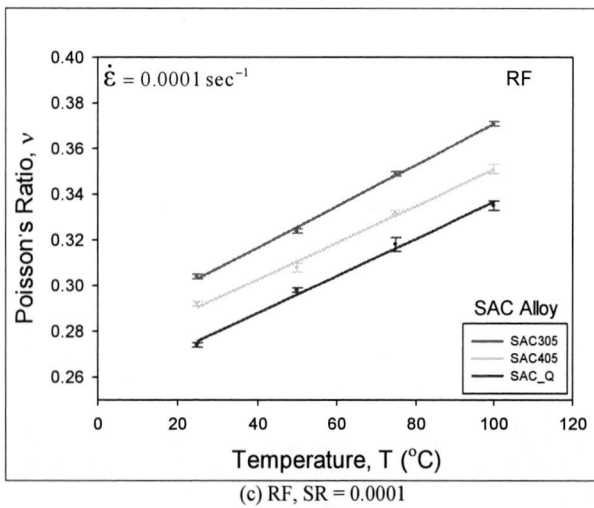

(c) RF, SR = 0.0001

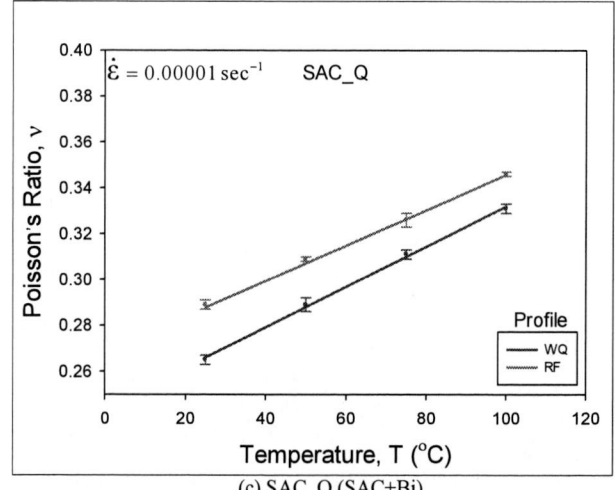

(c) SAC_Q (SAC+Bi)

Figure 16. Temperature Dependence of the Poisson's Ratio for Different Solder Alloys (SAC305, SAC405, SAC_Q)

Figure 17. Temperature Dependence of the Poisson's Ratio for Different Microstructures (WQ and RF)

978-1-7281-1500-9/19 $31.00 © 2019 IEEE

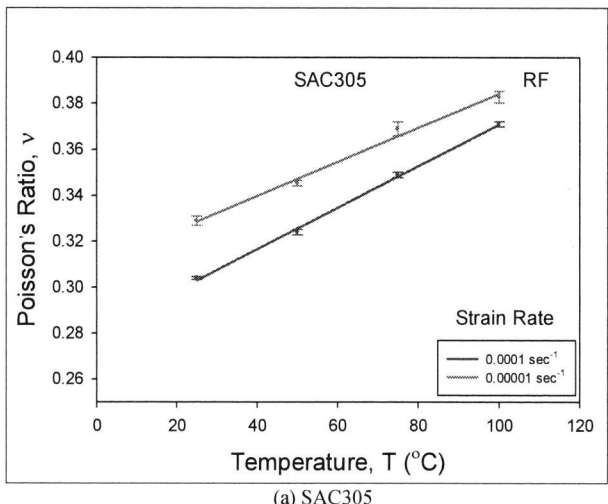

(a) SAC305

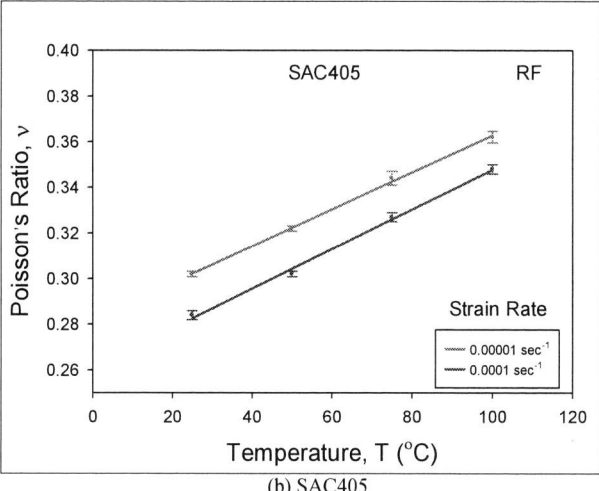

(b) SAC405

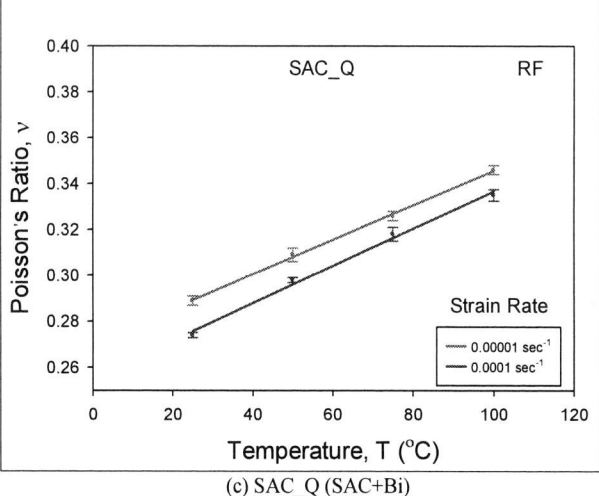

(c) SAC_Q (SAC+Bi)

Figure 18. Temperature Dependence of the Poisson's Ratio for Different Strain Rates (0.0001 and 0.00001 sec^{-1})

The dependence of the Poisson's ratio on solidification profile can be visualized for each alloy using the data in Figs. 14-15 and the plots in Fig. 17. In general, the Poisson's ratio value decreases when increasing the rate of cooling in the solder solidification profile. In other words, the Poisson's ratio values are lower for the very fine microstructures obtained with the WQ profile relative to the values obtained with the coarser (more realistic) microstructures obtained with the RF profile.

The dependence of the Poisson's ratio on strain rate can be visualized for each alloy using the data in Fig. 14 and the plots in Fig. 18. In general, the Poisson's ratio decreases when increasing the strain rate. For the two strain rates considered (10^{-4} and 10^{-5} sec^{-1}), the Poisson's ratio value for the slower strain rate was typically 5-10% higher than that for the higher strain rate.

B. Effects of Aging

A set of additional experiments was performed to determine the effects of aging (microstructural coarsening) on the Poisson's ratio values. For the aging experiments, the starting point was the reflowed (RF) microstructure for each of the three solder alloys. The samples were aged at T = 125 ^{o}C for various durations including 0 (no aging), 1, 2, 3, 4, 5, 6, 12, and 24 hours. For each aging duration, testing was performed using the same four testing temperatures (T = 25, 50, 75, 100 ^{o}C) and a strain rate of 0.00001 (sec^{-1})). The test matrix is shown in Fig. 19.

Test Temperature (^{o}C)	Aging Time at T = 125 ^{o}C (Hours)								
	0	1	2	3	4	5	6	12	24
25	√	√	√	√	√	√	√	√	√
50	√	√	√	√	√	√	√	√	√
75	√	√	√	√	√	√	√	√	√
100	√	√	√	√	√	√	√	√	√

Figure 19. Poisson's Ratio Aging Test Matrix
(SR = 0.00001 sec^{-1}; SAC305, SAC405, SAC_Q)

Typical variations in the extracted Poisson's ratio values with aging are illustrated by the tablulated values in Figs. 20, 21, and 22 for SAC305, SAC405, and SAC_Q, respectively. In these tables, the measured Poisson's ratio values for 0, 3, 6, 12, and 24 hours of aging are included. With one day of aging, the PR value increases by 10-15% for SAC305 and SAC405, while only 6-8% for SAC_Q. As with other material properties [15, 20, 29-30, 39, 42], the presence of Bi in the SAC_Q alloy appears to mitigate aging induced degradations (increases) in the value of Poisson's ratio. Overall, the SAC_Q alloy has the lowest PR values for all temperatures, strain rates, solidification profiles, and durations of aging.

Temperature (°C)	Aging Time at T = 125 °C				
	No Aging	3 Hours	6 Hours	12 Hours	24 Hours
25	0.329 (0.002)	0.341 (0.002)	0.359 (0.001)	0.361 (0.002)	0.362 (0.001)
50	0.346 (0.001)	0.361 (0.001)	0.380 (0.002)	0.382 (0.002)	0.385 (0.002)
75	0.369 (0.001)	0.382 (0.001)	0.412 (0.001)	0.414 (0.001)	0.417 (0.003)
100	0.383 (0.001)	0.402 (0.003)	0.431 (0.002)	0.433 (0.002)	0.437 (0.002)

Figure 20. Poisson's Ratio Values for SAC305
with Different Aging Conditions

Temperature (°C)	Aging Time at T = 125 °C				
	No Aging	3 Hours	6 Hours	12 Hours	24 Hours
25	0.305 (0.001)	0.326 (0.003)	0.348 (0.001)	0.351 (0.002)	0.355 (0.002)
50	0.322 (0.001)	0.349 (0.001)	0.371 (0.001)	0.376 (0.001)	0.379 (0.003)
75	0.344 (0.003)	0.366 (0.002)	0.384 (0.002)	0.388 (0.002)	0.392 (0.002)
100	0.364 (0.001)	0.388 (0.001)	0.411 (0.002)	0.414 (0.003)	0.416 (0.002)

Figure 21. Poisson's Ratio Values for SAC405
with Different Aging Conditions

Temperature (°C)	Aging Time at T = 125 °C				
	No Aging	3 Hours	6 Hours	12 Hours	24 Hours
25	0.289 (0.001)	0.298 (0.001)	0.307 (0.001)	0.308 (0.003)	0.310 (0.001)
50	0.309 (0.003)	0.319 (0.003)	0.328 (0.001)	0.330 (0.002)	0.331 (0.003)
75	0.326 (0.002)	0.337 (0.002)	0.345 (0.002)	0.348 (0.001)	0.351 (0.003)
100	0.346 (0.002)	0.355 (0.002)	0.364 (0.003)	0.366 (0.002)	0.368 (0.002)

Figure 22. Poisson's Ratio Values for SAC_Q
with Different Aging Conditions

Figure 23 illustrates the evolution of the Poisson's ratio value with aging time for each of the three solder alloys. The various colored curves in each graph represent the different testing temperatures. It is observed that the most dramatic changes occur in the first 5-6 hours of aging, and then the degradation rates becomes smaller. In addition, the more gradual nature of the aging induced changes for SAC_Q relative to SAC305/SAC405 are clearly evident. We are currently exploring longer aging times and other alloy systems.

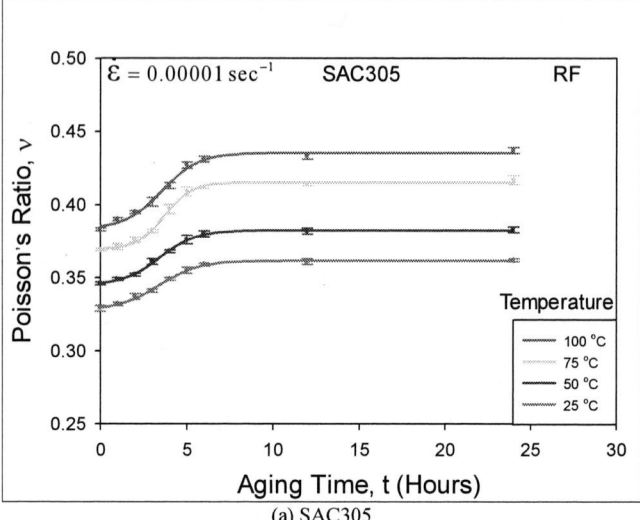

(a) SAC305

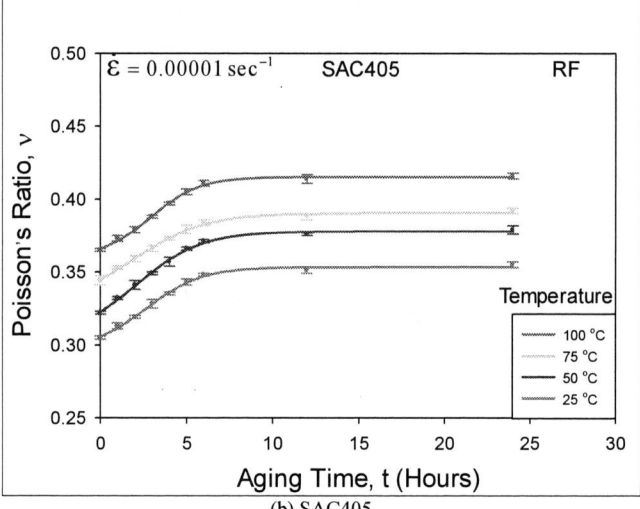

(b) SAC405

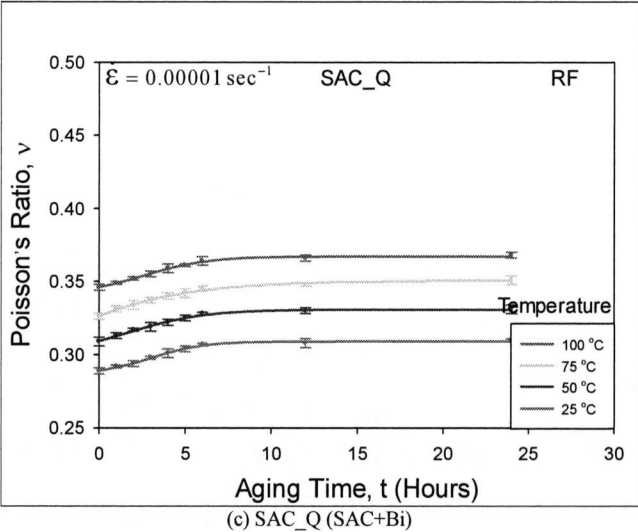

(c) SAC_Q (SAC+Bi)

Figure 23. Example Poisson's Ratio Test Data for Different Temperatures

C. SUMMARY AND CONCLUSIONS

In this work, we have conducted an experimental investigation to characterize the Poisson's ratio of several SAC and SAC+X lead free solders. Uniaxial tensile stress-strain tests were carried out on SAC305 (96.5Sn3.0Ag0.5Cu), SAC405 (95.5Sn4.0Ag0.5Cu), and SAC_Q (SAC+3.3Bi) specimens using a micro tension/torsion testing machine with two strain rates (0.0001, and 0.00001 (sec^{-1})) and four testing temperatures (T = 25, 50, 75, 100 °C). Deformations and strains in the axial and transverse directions were measured using miniature strain gages with automatic data acquisition from LabVIEW software. The recorded transverse strain vs. axial strain data for each test were then fit with a linear regression analysis to determine the Poisson's ratio value.

A large test matrix of experiments was developed to study the effects of temperature, strain rate, alloy composition, solidification cooling profile, and isothermal aging exposure on the value of solder Poisson's ratio; and to create a material property database for finite element simulations. The Poisson's ratio was found to increase significantly (15-25%) with increasing temperature. It was also found to decrease with increasing strain rate, alloy silver content, and cooling rate after solidification. Finally, the microstructural coarsening that occurs during isothermal aging led to an increase in the Poisson's ratio (up to 15% increase with one day of aging). Considering all of the possible branches of the test matrix (temperatures, strain rates, prior aging conditions, and microstructure), the value of the Poisson's ratio of SAC305 ranged from $0.31 < \nu < 0.44$. Similarly, the Poisson's ratio of SAC405 ranged from $0.28 < \nu < 0.42$, and the Poisson's ratio of SAC_Q ranged from $0.26 < \nu < 0.37$. Overall, the SAC_Q alloy had the lowest PR values for all temperatures, strain rates, solidification profiles, and durations of aging.

ACKNOWLEDGMENTS

This work was supported by the NSF Center for Advanced Vehicle and Extreme Environment Electronics (CAVE^3).

REFERENCES

[1] Kim, K. S., Huh, S. H., Suganuma, K., "Effects of Intermetallic Compounds on Properties of Sn–Ag–Cu Lead-Free Soldered Joints," *Journal of Alloys and Compounds,* Vol. 352, pp. 226-236, 2003.

[2] Yoon, J.-W., Kim, S.-W., Jung, S.-B., "IMC Morphology, Interfacial Reaction and Joint Reliability of Pb-Free Sn–Ag–Cu Solder on Electrolytic Ni BGA Substrate," *Journal of Alloys and Compounds,* Vol. 392, pp. 247-252, 2005.

[3] Wu, C. M. L., Yu, D. Q., Law, C. M. T., Wang, L., "Properties of Lead-Free Solder Alloys with Rare Earth Element Additions," *Materials Science and Engineering: R: Reports,* Vol. 44, pp. 1-44, 2004.

[4] Schubert, A., Dudek, R., Auerswald, E., Gollhardt, A., Michel, B., Reichel, H., "Fatigue Life Models for SnAgCu

and SnPb Solder Joints Evaluated by Experiments and Simulations," *Proceedings of the 53rd IEEE Electronic Components and Technology Conference*, pp. 603-610, New Orleans, LA, 2003.

[5] Syed, A., "Updated Life Prediction Models for Solder Joints with Removal of Modeling Assumptions and Effect of Constitutive Equations," *Proceedings of the 7th EuroSimE Conference,* pp. 1-9, 2006.

[6] Dudek, R., Faust, W. Ratchev, R., Roellig, M., Albrecht, H-J., Michel, B., "Thermal Test and Field Cycling Induced Degradation and its FE-Based Prediction for Different SAC Solders," *Proceedings of ITherm 2008*, pp.668 – 675, 2008.

[7] Motalab, M., Cai, Z., Suhling, J. C., Zhang, J., Evans, J. L., Bozack, M. J., Lall, P., "Improved Predictions of Lead Free Solder Joint Reliability That Include Aging Effects," *Proceedings of 62nd IEEE Electronic Components and Technology Conference*, pp. 513-531, San Diego, CA, 2012.

[8] Motalab, M., Mustafa, M., Suhling, J. C., Zhang, J., Evans, J. L., Bozack, M. J., Lall, P., "Correlation of Reliability Models Including Aging Effects with Thermal Cycling Reliability Data," *Proceedings of the 63rd IEEE Electronic Components and Technology Conference*, pp. 986-1004, Las Vegas, NV, May 28-31, 2013.

[9] Basit, M., Motalab, M., Suhling, J. C., Hai, Z., Evans, J. L., Bozack, M. J., and Lall, P., "Thermal Cycling Reliability of Aged PBGA Assemblies - Comparison of Weibull Failure Data and Finite Element Model Predictions," *Proceedings of the 65th IEEE Electronic Components and Technology Conference*, pp. 106-117, San Diego, CA, May 27-29, 2015.

[10] Hassan, KM. R., Alam, M. S., Basit, M. M., Suhling, J. C., Lall, P., "The Influence of Poisson's Ratio on the Reliability of SAC Lead Free Solder Joints," *Proceedings of ITherm 2018*, pp. 1207-1216, San Diego, CA, May 29-June 1, 2018.

[11] Ma, H., and Suhling, J. C., "A Review of Mechanical Properties of Lead-Free Solders for Electronic Packaging," *Journal of Materials Science*, Vol. 44, pp. 1141-1158, 2009.

[12] Ma, H., Suhling, J. C., Lall P., Bozack, M. J., "Reliability of the Aging Lead-Free Solder Joint," *Proceeding of the 56th IEEE Electronic Components and Technology Conference*, pp. 849-864, San Diego, California, 2006

[13] Ma, H., Suhling, J. C., Zhang, Y., Lall, P., and Bozack, M. J., "The Influence of Elevated Temperature Aging on Reliability of Lead Free Solder Joints," *Proceedings of the 57th IEEE Electronic Components and Technology Conference*, pp. 653-668, Reno, NV, May 29-June 1, 2007.

[14] Zhang, Y., Kurumaddali, K., Suhling, J. C., Lall, P., and Bozack, M. J., "Analysis of the Mechanical Behavior, Microstructure, and Reliability of Mixed Formulation Solder Joints," *Proceedings of the 59th IEEE Electronic Components and Technology Conference*, pp. 759-770, San Diego, CA, May 27-29, 2009.

[15] Cai, Z., Zhang, Y., Suhling, J. C., Lall, P., Johnson, R. W., Bozack, M. J., "Reduction of Lead Free Solder Aging Effects Using Doped SAC Alloys," *Proceedings of the 60th IEEE Electronic Components and Technology Conference*, pp. 1493-1511, Las Vegas, NV, June 2-4, 2010.

[16] Zhang, J., Hai, Z., Thirugnanasambandam, S., Evans, J. L., Bozack, M. J., Sesek, R., Zhang, Y., Suhling, J. C., "Correlation of Aging Effects on Creep Rate and Reliability in Lead Free Solder Joints," *SMTA Journal*, Volume 25(3), pp. 19-28, 2012.

[17] Motalab, M., Cai, Z., Suhling, J. C., Lall, P., "Determination of Anand constants for SAC Solders using Stress-Strain or Creep Data," *Proceedings of ITherm 2012*, pp. 910-922, San Diego, CA, May 30 - June 1, 2012.

[18] Basit, M. M., Motalab, M., Suhling, J. C., and Lall, P., "The Effects of Aging on the Anand Viscoplastic Constitutive Model for SAC305 Solder," *Proceedings of ITherm 2014*, pp. 112-126, Orlando, FL, May 28-30, 2014.

[19] Basit, M. M., Ahmed, S., Motalab, M., Roberts, J. C., Suhling, J. C., and Lall, P., "The Anand Parameters for SAC Solders after Extreme Aging," *Proceedings of ITherm 2016*, pp. 440-447, Las Vegas, NV, June 1-3, 2016.

[20] Ahmed, S., Suhling, J. C., Lall, P., "The Anand Parameters of Aging Resistant Doped Solder Alloys" *Proceedings of ITherm 2017*, pp. 1416-1424, Orlando, FL, May 30 - June 2, 2017.

[21] Lall, P., Shantaram, S., Suhling, J., and Locker, D., "Effect of Aging on the High Strain Rate Mechanical Properties of SAC105 and SAC305 Leadfree Alloys," *Proceedings of the 63rd IEEE Electronic Components and Technology Conference*, pp. 1277-1293, Las Vegas, NV, May 28-31, 2013.

[22] Chiu, T. C., Zeng, K., Stierman, R., Edwards, D., and Ano, K., "Effect of Thermal Aging on Board Level Drop Reliability for Pb-Free BGA Packages," *Proceedings of the 54th IEEE Electronic Components and Technology Conference*, pp. 1256-1262, 2004.

[23] Lee, C. B., Jung, S. B., Shin, Y. E., and Chang, C. C., "Effect of Isothermal Aging on Ball Shear Strength in BGA Joints with Sn-3.5Ag-0.75Cu Solder," *Materials Transactions*, Vol 43(8), pp. 1858-1863, 2002.

[24] Deng, X., Sidhu, R. S., Johnson, P., and Chawla, N., "Influence of Reflow and Thermal Aging on the Shear Strength and Fracture Behavior of Sn-3.5Ag Solder/Cu Joints," *Metallurgical and Materials Transactions A*, Vol. 36A, pp. 55-64, 2005.

[25] Alam, M. S., Hassan, KM. R., Suhling, J. C., Lall, P., "High Temperature Mechanical Behavior of SAC and SAC+X Lead Free Solders," *Proceedings of the 68th IEEE Electronic Components and Technology Conference*, pp. 1781-1789, San Diego, CA, May 29-June 1, 2018.

[26] Mustafa, M., Cai, Z., Suhling, J. C., and Lall, P., "The Effects of Aging on the Cyclic Stress-Strain Behavior and Hysteresis Loop Evolution of Lead Free Solders," *Proceedings of the 61st IEEE Electronic Components and Technology Conference*, pp. 927-939, Orlando, FL, June 1-3, 2011.

[27] Mustafa, M., Roberts, J. C, Suhling, J. C., and Lall, P., "The Effects of Aging on the Fatigue Life of Lead Free Solders," *Proceedings of the 64th IEEE Electronic Components and Technology Conference*, pp. 666-683, Orlando, FL, May 28-30, 2014.

[28] Fu, N., Suhling, J. C., Lall, P., "Cyclic Stress-Strain Behavior of SAC305 Lead Free Solder: Effects of Aging, Temperature, Strain Rate, and Plastic Strain Range," *Proceedings of the 66th IEEE Electronic Components and Technology Conference*, pp. 1119-1127, Las Vegas, NV, May 31 - June 3, 2016.

[29] Chowdhury, M. M. R., Fu, N.; Suhling, J. C., Lall, P., "Evolution of the Cyclic Stress-Strain Behavior of Doped SAC Solder Materials Subjected to Isothermal Aging" *Proceedings of ITherm 2017*, pp. 1369-1379, Orlando, FL, May 30 - June 2, 2017.

[30] Chowdhury, M. M. R., Hoque, M. A., Fu, N., Suhling, J. C., Hamasha, S., and Lall, P., "Characterization of Material Damage and Microstructural Evolution Occurring in Lead Free Solders Subjected to Cyclic Loading," *Proceedings of the 68th IEEE Electronic Components and Technology Conference*, pp. 865-874, San Diego, CA, May 29 - June 1, 2018.

[31] Mustafa, M., Cai, Z., Roberts, J. R., Suhling, J. C., Lall, P., "Evolution of the Tension/Compression and Shear Cyclic Stress-Strain Behavior of Lead-Free Solder Subjected to Isothermal Aging," *Proceedings of ITherm 2012*, pp. 765-780, San Diego, CA, May 30 - June 1, 2012.

[32] Mustafa, M., Suhling, J. C., Lall, P., "Experimental Determination of Fatigue Behavior of Lead Free Solder Joints in Microelectronic Packaging Subjected to Isothermal Aging," *Microelectronics Reliability*, Vol. 56, pp. 136-147, 2016.

[33] Zhang, J., Hai, Z., Thirugnanasambandam, S., Evans, J. L., Bozack, M. J., Zhang, Y., Suhling, J. C., "Thermal Aging Effects on Thermal Cycling Reliability of Lead-Free Fine Pitch Packages," *IEEE Transactions on Components, Packaging, and Manufacturing Technology*, Vol. 3(8), pp. 1348-1357, 2013.

[34] Hai, Z., Zhang, J., Shen, C., Snipes, E. K., Suhling, J. C., Bozack, M. J., and Evans, J. L., "Reliability Degradation of SAC105 and SAC305 BGA Packages Under Long-Term, High Temperature Aging," *SMTA Journal*, Vol. 27(2), pp. 11-18, 2014.

[35] Hai, Z., Zhang, J., Shen, C., Evans, J. L., Bozack, M. J., Basit, M. M., and Suhling, J. C., "Reliability Comparison of Aged SAC Fine-Pitch Ball Grid Array Packages Versus Surface Finishes," *IEEE Transactions on Components, Packaging, and Manufacturing Technology*, Vol. 5(6), pp. 828-837, 2015.

[36] Zhao, C., Shen, C., Hai, Z., Basit, M. M., Zhang, J., Bozack, M. J., Evans, J. L., and Suhling, J. C., "Long Term Aging Effects on the Reliability of Lead Free Solder Joints in Ball Grid Array Packages with Various Pitch Sizes and Ball Arrangements," *SMTA Journal*, Vol. 29(2), pp. 37-46, 2016.

[37] Hasnine, M., Mustafa, M., Suhling, J. C., Prorok, B. C., Bozack, M. J., Lall, P., "Characterization of Aging Effects in Lead Free Solder Joints Using Nanoindentation," *Proceedings of the 63rd IEEE Electronic Components and Technology Conference*, pp. 166-178, Las Vegas, NV, May 28-31, 2013.

[38] Hasnine, M., Suhling, J. C., Prorok, B. C., Bozack, M. J., and Lall, P., "Exploration of Aging Induced Evolution of Solder Joints Using Nanoindentation and Microdiffraction," *Proceedings of the 64th IEEE Electronic Components and Technology Conference*, pp. 379-394, Orlando, FL, May 28-30, 2014.

[39] Hasnine, M., Suhling, J. C., Prorok, B. C., Bozack, M. J., and Lall, P., "Nanomechanical Characterization of SAC Solder Joints - Reduction of Aging Effects Using Microalloy Additions," *Proceedings of the 65th IEEE Electronic Components and Technology Conference*, pp. 1574-1585, San Diego, CA, May 27-29, 2015.

[40] Hasnine, M., Suhling, J. C., Prorok, B. C., Bozack, M. J., and Lall, P., "Anisotropic Mechanical Properties of SAC Solder Joints in Microelectronic Packaging and Prediction of

Uniaxial Creep Using Nanoindentation Creep," *Experimental Mechanics*, Vol 57(4), pp. 603-614, 2017.

[41] Ahmed, S., Hasnine, M., Suhling, J. C., and Lall, P., "Mechanical Characterization of SAC Solder Joints at High Temperature Using Nanoindentation," *Proceedings of the 67th IEEE Electronic Components and Technology Conference*, pp. 1128-1135, Orlando, FL, May 30 - June 2, 2017.

[42] Ahmed, S., Basit, M., Suhling, J. C., Lall, P., "Effects of Aging on SAC-Bi Solder Materials" *Proceedings of ITherm 2016*, pp. 746-754, Las Vegas, NV, May 30 - June 3, 2016.

[43] Fu, N., Ahmed, S., Suhling, J. C., Lall, P., "Visualization of Microstructural Evolution in Lead Free Solders During Isothermal Aging Using Time-Lapse Imagery" *Proceedings of the 67th IEEE Electronic Components and Technology Conference*, pp. 429-440, Orlando, FL, May 30 - June 2, 2017.

[44] Ahmed, S., Wu, J., Fu, N., Suhling, J. C., Lall, P., "Quantification and Modeling of Microstructural Evolution in Lead Free Solders During Long Term Isothermal Aging" *Proceedings of the 68th IEEE Electronic Components and Technology Conference*, pp. 162-171, San Diego, CA, May 29 - June 1, 2018.

[45] Ahmed, S., Suhling, J. C., Lall, P., "Evaluation of Aging Induced Microstructural Evolution in Lead Free Solders Using Scanning Probe Microscopy," *Proceedings of ITherm 2018*, pp. 1062-1070, San Diego, CA, May 29 - June 1, 2018.

[46] Anand, L., "Constitutive Equations for the rate dependent Deformation of Metal at Elevated Temperatures,' *Journal of Engineering Materials and Technology,* Vol. 104(1), pp. 12-17, 1982.

[47] Darveaux, R., "Effect of Simulation Methodology on Solder Joint Crack Growth Correlation," *Proceedings of the 50th IEEE Electronic Components and Technology Conference*, pp. 1048-1058, 2000.

Additive Laser Metal Deposition onto Silicon for Enhanced Microelectronics Cooling

Arad Azizi [1]
Department of Mechanical Engineering & Materials
Science and Engineering Program
Binghamton University (SUNY)
Binghamton, NY, 13902, United States

Bahgat G. Sammakia
Department of Mechanical Engineering & Materials
Science and Engineering Program
Binghamton University (SUNY)
Binghamton, NY, 13902, United States

Matthias A. Daeumer
Department of Mechanical Engineering & Materials
Science and Engineering Program
Binghamton University (SUNY)
Binghamton, NY, 13902, United States

Bruce T. Murray
Department of Mechanical Engineering & Materials
Science and Engineering Program
Binghamton University (SUNY)
Binghamton, NY, 13902, United States

Jacob C. Simmons
Department of Mechanical Engineering & Materials
Science and Engineering Program
Binghamton University (SUNY)
Binghamton, NY, 13902, United States

Scott N.Schiffres *
Department of Mechanical Engineering & Materials
Science and Engineering Program
Binghamton University (SUNY)
Binghamton, NY, 13902, United States
Sschiffr@binghamton.edu

[1] First author
* Corresponding author

Abstract—**We previously demonstrated how the Sn3Ag4Ti alloy can robustly bond onto silicon via selective laser melting (SLM). By employing this technology, thermal management devices (e.g., micro-channels, vapor chamber evaporators, heat pipes) can be directly printed onto the electronic package (silicon die) without using thermal interface materials. Under immersion two-phase cooling (pool boiling), we compare the performance of three chip cooling methods (conventional heat sink, bare silicon die and additively manufactured metal micro-fins) under high heat flux conditions (100 W/cm^2). Heat transfer simulations show a significant reduction in the chip temperature for the silicon micro-fins. Reduction of the chip operating temperature or increase in clock speed are some of the advantages of this technology, which results from the elimination of thermal interface materials in the electronic package. Performance and reliability aspects of this technology are discussed through experiments and computational models.**

Keywords- Electronic Cooling; Thermal Management; Additive Manufacturing; Laser Metal Deposition; Performance and Reliability.

I. INTRODUCTION

Based on Moore's Law, the trend towards higher computational power in microprocessors, heat fluxes double every ~3.5 years [1]. Current conventional chip level cooling methods that rely on thermal interface materials may simply fail to meet this demand, even with optimization in cooling design [2]–[4]. Moreover, current electricity reduction can also be reduced by better cooling. Every year ~73 billion kWh of electricity is consumed in data centers in the US alone. Reducing the core temperature of microprocessors by 10 °C can improve computational efficiency by over 5%. Furthermore, this temperature reduction would save $450 million per year in electricity, and greatly reduce CO_2 emissions and e-waste.

In conventional microprocessor packages, the silicon die is attached to the lid/heat spreader using a thermal interface material (TIM 1), and the heat spreader is attached to the heat sink via a second TIM (TIM 2) (Figure 1). The TIMs in applications with low heat fluxes are usually thermal pastes, greases, or pads with conductivities up to 10 W/m-K [5]. Even in high-performance applications that utilize high conductivity materials such as indium foils, appreciable temperature drops are observed (thermal resistance of ~0.13 cm^2-°C/W at ~70 PSI for solid Indalloy®19 with thickness of 76.2 μm) [6]. The thermal resistances of TIMs in the cooling paradigm has become a key issue as the chip heat fluxes move beyond 100 W/cm^2.

978-1-7281-1500-9/19 $31.00 © 2019 IEEE

In high-end packages, at least one of the TIMs are replaced with indium foil with thermal conductivity of up to 86 W/m-K [6], [7]. However, indium films have certain limitations such as cost and the die metallizations are deposited by expensive and time-consuming methods such as chemical or physical vapor deposition or sputtering.

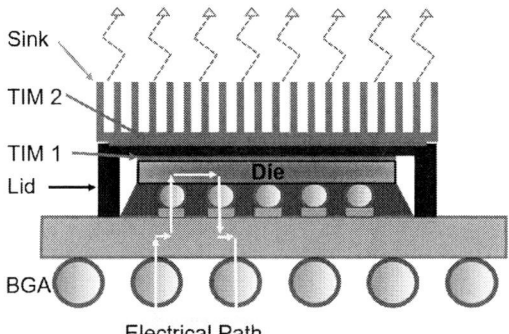

Figure 1 Conventional packaging strategy in microprocessors.

We have recently demonstrated how additive laser metal deposition can be used to bond Sn3Ag4Ti brazing alloy onto a silicon substrate [8], [9]. In this process, a thin layer of metal powder (~20 μm) consisting of Sn3Ag4Ti is deposited onto the backside of a silicon die or wafer and bonded by a laser beam in a protective N_2 gas environment. The temperature of the metal powder being exposed to the laser will reach thousands of degrees Celsius in microseconds and melt the powder. The localized heating will enable the reactive element of the alloy, which in this case is titanium, to overcome the energy barrier of reaction and react with silicon substrate to create a strong silicide bond [8].

It is also possible to print another metal, such as copper or aluminum, onto the Sn3Ag4Ti alloy to take advantage of their higher thermal conductivity [9]. In this case, the alloy acts as an interlayer which bonds both to the silicon substrate via silicide bonding and to the copper via intermetallics. The resolution of this technique is limited by the optics and powder size. Our current printer has an x-y minimum feature of 100 um, but different optics and powders can reduce this further. Various structures such as fins, pillars and lattice structures can be built on top of the interlayer alloy to optimize the fluid flow and heat transfer characteristics. Modulated porosity can be produced by changing the laser power, which is of interest for wick design. These fine features can be used to increase the surface area of a chip for single phase or two phase cooling approaches (e.g. pool boiling or impinging jets) [10]–[17].

II. PERFORMANCE CHARACTRIZATION

In order to better understand how this printing technique may yield enhanced thermal performance, three configurations subject to immersion cooling were computationally modeled. The first case is a conventional heat sink with thermal interface materials and heat spreaders (Figure 2). The second case has micro-fins directly

manufactured onto the silicon via selective laser melting (Figure 2). The third case is a fully immersed bare chip (Figure 2). ANSYS Fluent 19.2 was used for this study. Mesh independency studies were performed for each case. The boundary conditions are prescribed to mimic realistic conditions. Pool boiling with phase change was represented by using a constant heat transfer coefficient of 15,000 W/m²-K. The working fluid was assumed to have free stream temperature of 40 °C for all cases. All simulations were performed under steady state conditions. The thermal conductivities used in the simulations are provided in table 1.

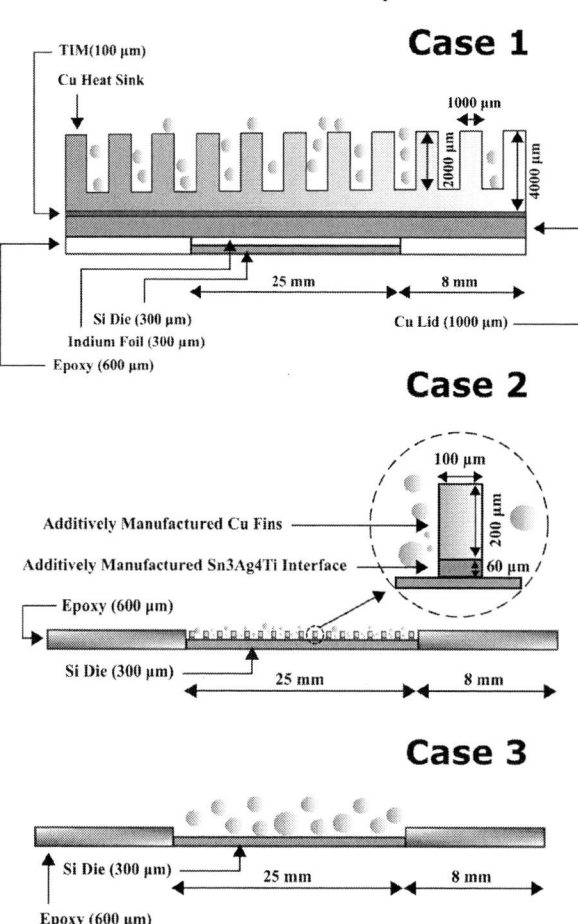

Figure 2 Configuration of case 1, 2 and 3.

Table 1 Thermal conductivity of materials.

	Material	Thermal conductivity [W/m-K]	Reference
1	Indium (pure)	86	[6]
2	Thermal paste	8.4	[5]
3	Copper (C10400)	384	[18]
4	Sn3Ag4Ti alloy	39.4 (*)	[8]
5	Silicon	148	[19], [20]
6	High conductivity epoxy	0.49	[21]
7	AISI 1018	51.9	[22]

(*) With process improvement, should be able to increase up to 70 W/m-K.

A. Case 1: Si + Indium + Cu Lid + TIM + Cu Heat Sink

In this case, a silicon die with a heat flux of 100 W/cm^2 is considered with an area of 25 × 25 mm and a thickness of 300 μm. On top of the die, an indium foil is placed with the same thickness as the silicon die. On top of the indium foil, a copper lid with thickness of 1mm is stacked and afterwards a conventional TIM (100 μm) in the form of thermal paste and finally on top a copper heat sink consisting of fins with 2 mm height and 1 mm spacing is installed (Figure 2) and immersed on the working fluid. The near-junction temperature reached 85.5°C (Figure 3, 4).

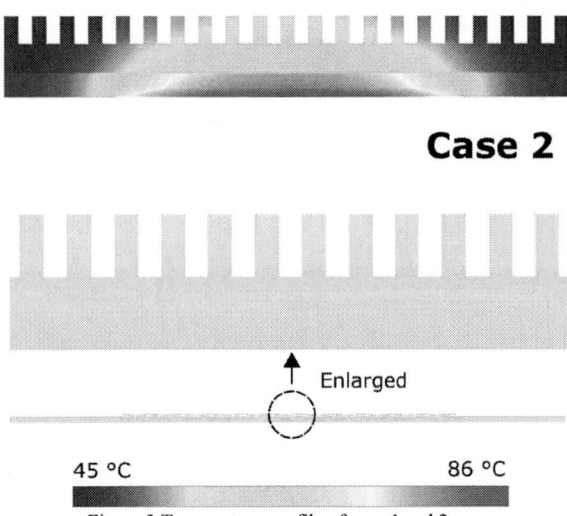

Figure 3 Temperature profile of case 1 and 2.

B. Case 2: Si + Sn3Ag4Ti Interface + Cu micron sized Fins by Additive Manufacturing (AM)

The motivation of this research is direct printing of micro-fins on top of the silicon chip. Copper fins with height of 200 μm and width of 100 μm and 100 μm spacing is additively manufactured by laser deposition onto silicon die with heat flux of 100 W/cm^2 with 60 μm of Sn3Ag4Ti as an interfacial alloy (Figure 2) and then immersed for pool boiling, as were the previous scenarios. Near-junction temperature in this case reached 62.4 °C (Figure 3, 4). The predicted decrease in near-junction temperature in the simulation with same boundary conditions as the other cases is due to minimization of interfacial thermal resistances and the large increase of the active surface area for heat transfer by printing micron sized fins.

C. Case 3: Bare Si

Two-phase immersion cooling of the bare chip (no heat sink) has been available in the industry for some time [23], [24]. In this case study, a bare silicon die (300 μm) is immersed in working fluid with the same heat flux and heat transfer coefficient as case 1. This configuration takes advantage of minimal conduction thermal resistance as no

TIMs are involved. However, the near-junction temperature was estimated as 108.5 °C due to low surface area negating any benefit of no TIM (Figure 4). This number is not physically true as in reality critical heat flux is reached and the assumed heat transfer coefficient is no longer valid.

Based on the simulations we have observed that the additively manufactured fins have higher performance compared to conventional methods of chip thermal management (Figure 4). This technology can be used to either increase the microprocessor performance or save energy by the chip operating at lower junction temperatures.

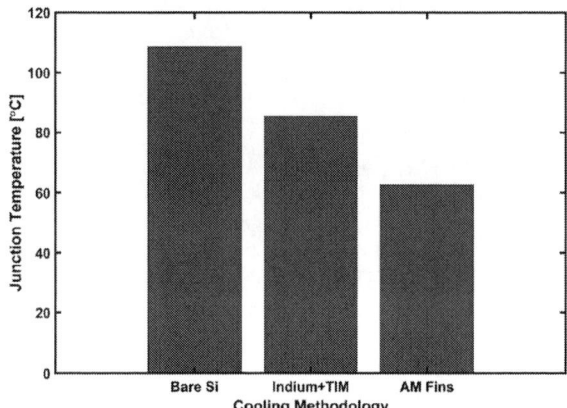

Figure 4 Comparison of near-junction temperature of three configurations with same boundary conditions.

Furthermore, by employing AM fins, the size of the package can significantly decrease due to the removal of heat spreaders and heat sinks, which is important for applications that tightly pack electronics (*eg* future high-density computing servers). For instance, the AM fins printed directly on silicon die (case 2) are 5 mm shorter than the conventional heat sink with TIMs and a heat spreader (case 1) (Figure 5).

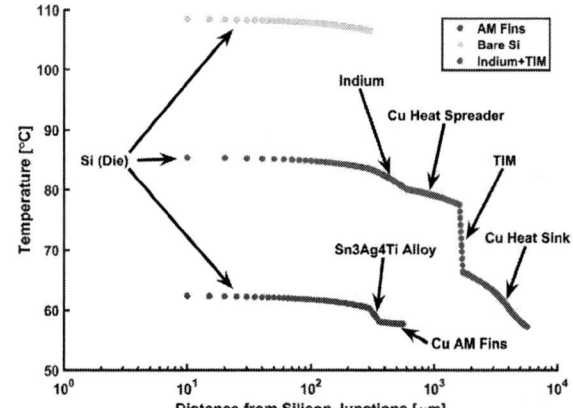

Figure 5 Comparison of different cases in terms of cooling device height.

III. RELIABILITY AND TESTING

The process of selective laser melting heat removal devices onto silicon must not damage the underlying transistors. One reliability concern is how much the transistors will heat up during laser processing. In order to answer this question, two approaches are considered. When the metal powder on the back of the die is exposed to the laser beam, an extremely fast (~μs) and localized (~μm) temperature rise will occur [25]. The amount of heat being absorbed, depends on the optical absorption coefficient of the metal powder at laser wavelength, laser power and scanning rate [26], [27]. A melt pool is created during this process which cools down at the rate of millions of degrees per second. As the heating and cooling process occurs extremely fast, it is not straightforward to measure instantaneous temperature rise accurately. As a result, transient numerical simulation for fast temperature changes and experimental temperature sensing for slower variations of the thermal wave is used.

A. Experimental thermal measurement of die heating during additve laser metal deposition

An experiment was designed in order to evaluate the temperature rise at the transistor side of the Si substrate during the laser deposition of Sn3Ag4Ti alloy. On one side of a 4-inch silicon wafer (553 μm thick) two E Type thermocouples (gauge 36 and PFA insulation from Omega™) were attached using high conductivity thermal grease to ensure fastest and most accurate response during laser processing within the accuracy of the temperature measurement technique (Figure 6, 7) [28]. LabJack U6-Pro with sampling rate of 3.58 ms and effective resolution of 0.6 μV was used for datalogging [29]. The response rate of the thermocouple is slower than 10 ms, as a result, all signals are picked up real-time by the datalogger. Furthermore, the Seebeck voltage of the E type thermocouple in temperatures between 20 °C to 500 °C is 1 mV to 37 mV respectively which provide a strong signal for the datalogger to pick up [30].

The sample is installed on an AISI 1018 steel wafer holder with the thermocouple located at the interface between the wafer holder and Si wafer (Figure 7). A thin layer of powder with ~20 μm thickness was deposited on the silicon wafer and was then exposed twice by the laser beam at a power of 120 W and a scan velocity of 1700 mm/s in skin exposure setting (Figure 8).

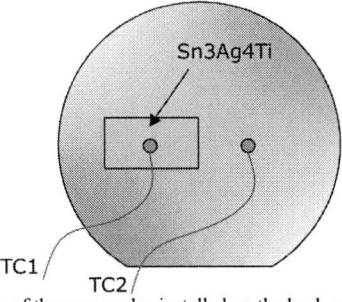

Figure 6 Schematic of thermocouples installed on the back of Si wafer and the footprint of Sn3Ag4Ti surface bring scanned on the other side.

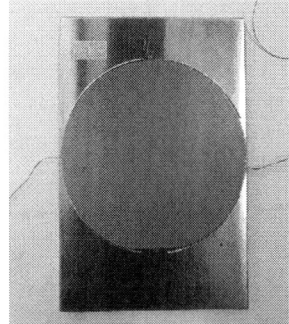
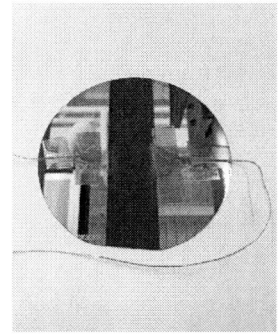

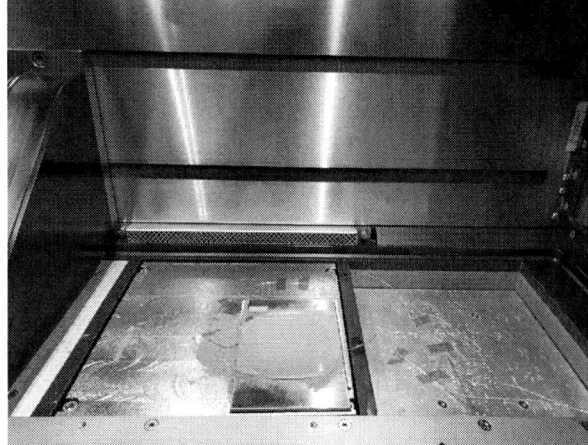

Figure 7 Experimental setup installed on the EOS M290 machine for temperature measurement on the back of silicon wafer.

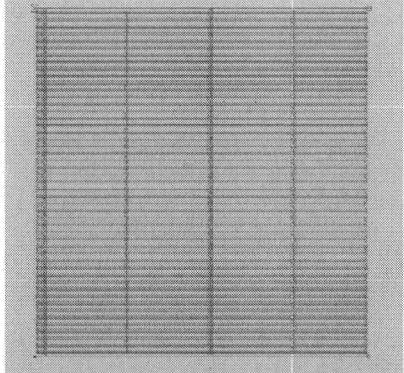

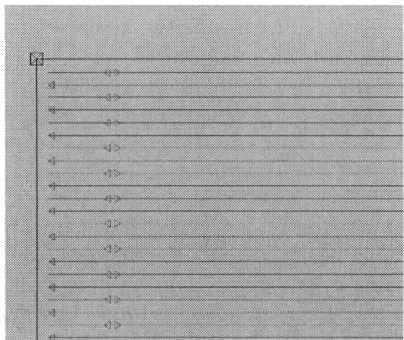

Figure 8 Snapshot of the laser scan map in EOSPRINT© V1.3 software. The laser scan lines are shown as green.

978-1-7281-1500-9/19 $31.00 © 2019 IEEE

Thermocouple 1 is placed directly under the laser exposure area and thermocouple 2 is installed away from the exposure area (Figure 6). The first and second exposure peaks were detected by thermocouple 1. Substrate heating by conduction were also sensed by thermocouple 2. The maximum temperature rise of 231 °C was recorded on the back of the silicon wafer (Figure 9). This is a worst-case heating, as we printed a large area at one time. By changing the geometry printed from a cube to fins or pillars, the temperature rise will decrease. We can also refine the print path to avoid heat buildup.

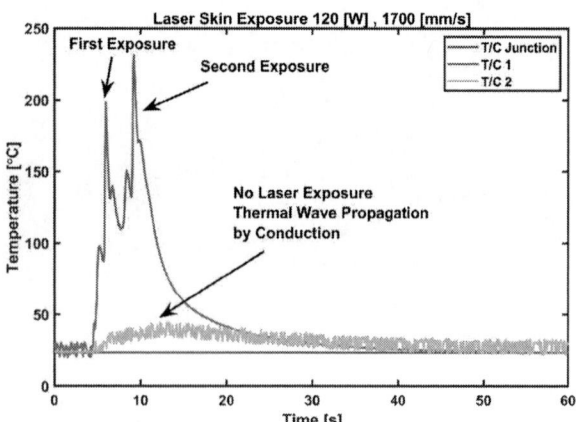

Figure 9 Temperature measurement by thermocouples on the back of silicon wafer during laser processing.

B. Numerical simulation of die heating during additve laser metal deposition

A set of numerical simulations to model the effects of the laser melting on temperature rise through the Si substrate was conducted using ANSYS Fluent 19.2. The computational (solid) zones were set to move at the constant laser velocity of 1700 mm/s through the domain. Constant heat flux at the substrate surface is applied on a fixed area to model the gaussian beam with power of 120 W and diameter of 80 μm at focal plane. Melting in general is a transient problem due to the time dependent physics involved in phase transformation. However, by employing the above methodology, it is possible to treat this problem as steady state and decrease computational requirements without a loss of accuracy.

Three separate zones (solids) are created for this simulation. The first zone is a 20 μm thick Sn3Ag4Ti alloy which represents a single layer of powder deposited on the die surface. The next zone is silicon with 500 μm thickness which is created from the bottom of the previous zone. Finally, the third zone is steel substrate with 500 μm thickness that represents the steel wafer holder which the Si substrate is attached to during the print. The size of the computational domain is 0.5 (x-direction) × 2 (y-direction) ×1.5 (z-direction) mm³. The mesh is created with a gradient

which increases in cell size moving away from the laser spot. After mesh independency study, the size of cells in final mesh at the location of laser spot were 1×1×1 μm³. Total number of 11,594,880 computational cells are used in this simulation.

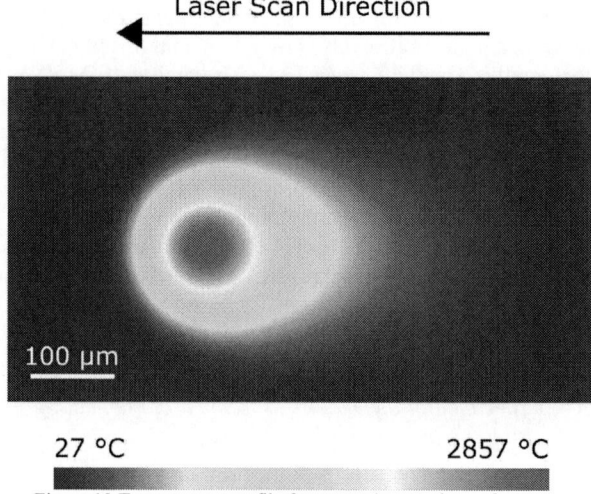

Figure 10 Temperature profile from top view on the surface.

Radiative losses at the surface are also accounted for by Stefan-Boltzmann law for black body radiation. Optical absorptivity and emissivity of the Sn3Ag4Ti alloy is approximated by pure Sn metal which is 0.18 and 0.05, respectively. The maximum temperatures were recorded at depths of 10 μm to 500 μm below the Sn3Ag4Ti-Si interface (Figure 12).

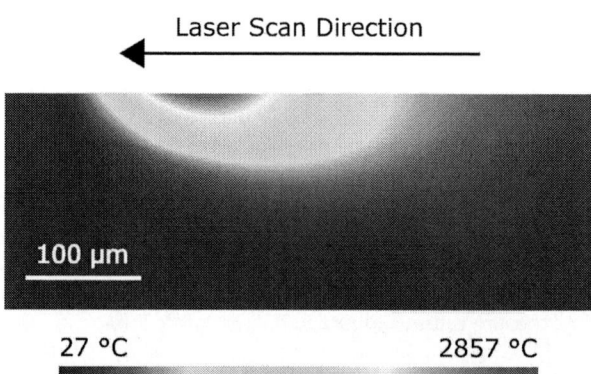

Figure 11 Temperature profile cross-section view. Penetration of thermal wave into the sample can be seen.

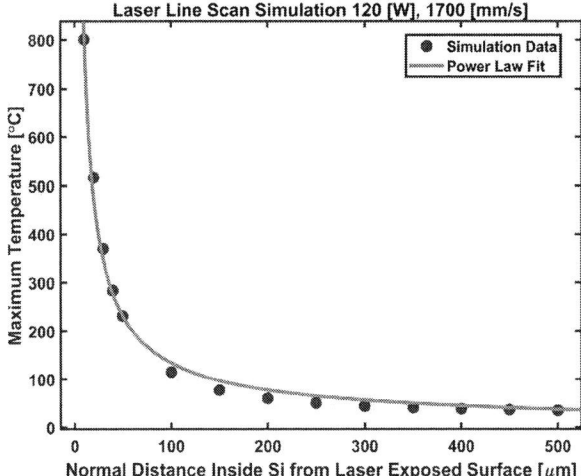

Figure 12 Maximum temperature rise inside Si substrate during additive laser deposition of a single line scan.

Another concern is whether the structure being deposited on the die surface can handle the thermal stresses of laser processing and the temperature fluctuations during computational cycles. Thermal stresses occur due to the large difference in coefficient of thermal expansion (CTE) in materials. Silicon has low CTE as oppose to metals such as copper or aluminum which have at least 5 times bigger CTEs. By using a low melting point interlayer alloy which in this study is Sn3Ag4Ti, which melts at ~250 °C, less thermal stress is imposed to the Sn3Ag4Ti-Si interface during manufacturing and usage. Our previous work thermal cycled a 316L structure built onto silicon with Sn3Ag4Ti as an interlayer. This part was thermal cycled from -40 °C to 130 °C 100 times. Afterwards, the part was visually inspected for signs of delamination or fracture, which was not observed [8].

In the case of additive laser deposition of copper on silicon, at the interface between the Sn3Ag4Ti interlayer alloy and Cu heat sink, intermetallic phases in Sn-Cu couples will form. Intermetallic compounds are phases of precise constituent stoichiometry and are common in lead-free Cu-Sn solders. Cu_6Sn_5 and Cu_3Sn are the primary phases found in isothermal and transient liquid phase bonding. However, in selective laser melting, both the interlayer alloy and copper will melt. Amongst the Cu-Sn intermetallic phases, Cu_3Sn has the most exothermic enthalpy of formation and will thus form first and constitute the main IMC phase at the interface [31]. The exact morphology and properties of Cu_3Sn formed during laser melting has not yet been thoroughly studied. However, Frederikse et al. reported the thermal conductivity of Cu_3Sn to be 70.4 W/m-K.

AUTHOR CONTRIBUTIONS

S.N.S. conceived the initial idea of this research. A.A. developed the processing technique and performed all experimental and numerical characterizations and testing.

M.A.D. carried out the FDTR thermal measurement and contributed towards intermetallics section. J.C.S performed laser melting simulations. S.N.S. guided the work. All authors contributed to the writing of this paper.

ACKNOWLEDGMENT

We gratefully acknowledge the support of SUNY Binghamton through NYS startup funds, the ADL's Small Grant (ADLG173), the NSF Grant (1846157).

REFERENCES

[1]   S. Krishnan, S. V. Garimella, G. M. Chrysler, and R. V. Mahajan, "Towards a Thermal Moore's Law," *IEEE Transactions on Advanced Packaging*, vol. 30, no. 3, pp. 462–474, Aug. 2007.
[2]   Z. Li and S. G. Kandlikar, "Current Status and Future Trends in Data-Center Cooling Technologies," *Heat Transfer Engineering*, vol. 36, no. 6, pp. 523–538, Oct. 2014.
[3]   X. C. Tong, *Advanced Materials for Thermal Management of Electronic Packaging*, vol. 30. New York, NY: Springer New York, 2011.
[4]   D. Lu and C. P. Wong, Eds., *Materials for Advanced Packaging*. Springer US, 2009.
[5]   T. A. Howe, C.-K. Leong, and D. D. L. Chung, "Comparative evaluation of thermal interface materials for improving the thermal contact between an operating computer microprocessor and its heat sink," *Journal of Electronic Materials*, vol. 35, no. 8, pp. 1628–1635, Aug. 2006.
[6]   D. Saums, B. Jarrett, A. C. Mackie, and J. Ross, "Thermal Management Materials Choices," Indium Corporation, Indium Corporation Tech Paper.
[7]   M. P. Renavikar *et al.*, "Materials Technology for Environmentally Green Micro-electronic Packaging," *Intel Technology Journal*, vol. 12, no. 1, 2008.
[8]   A. Azizi, M. A. Daeumer, and S. N. Schiffres, "Additive laser metal deposition onto silicon," *Additive Manufacturing*, vol. 25, pp. 390–398, Jan. 2019.
[9]   S. N. Schiffres and A. Azizi, "ADDITIVE LASER OR ELECTRON BEAM METAL DEPOSITION ONTO DISSIMILAR SUBSTRATES," Provisional Patent Filed to USPTO, 62717444.
[10]   L. Qiu, S. Dubey, F. H. Choo, and F. Duan, "Confined jet impingement boiling in a chamber with staggered pillars," *Applied Thermal Engineering*, vol. 131, pp. 724–733, Feb. 2018.
[11]   L. Qiu, S. Dubey, F. H. Choo, and F. Duan, "Recent developments of jet impingement nucleate boiling," *International Journal of Heat and Mass Transfer*, vol. 89, pp. 42–58, Oct. 2015.
[12]   A. S. Rattner, "General Characterization of Jet Impingement Array Heat Sinks With Interspersed Fluid Extraction Ports for Uniform High-Flux Cooling," *Journal of Heat Transfer*, vol. 139, no. 8, p. 082201, Apr. 2017.
[13]   M. J. Rau and S. V. Garimella, "Confined Jet Impingement With Boiling on a Variety of Enhanced Surfaces," *Journal of Heat Transfer*, vol. 136, no. 10, p. 101503, Jul. 2014.
[14]   Y. Hadad *et al.*, "Three-objective shape optimization and parametric study of a micro-channel heat sink with discrete non-uniform heat flux boundary conditions," *Applied Thermal Engineering*, vol. 150, pp. 720–737, Mar. 2019.
[15]   M. Iyengar *et al.*, "Server liquid cooling with chiller-less data center design to enable significant energy savings," in *2012 28th Annual IEEE Semiconductor Thermal Measurement and Management Symposium (SEMI-THERM)*, San Jose, CA, USA, 2012, pp. 212–223.
[16]   M. Iyengar and R. Schmidt, "Analytical Modeling for Thermodynamic Characterization of Data Center Cooling Systems," *Journal of Electronic Packaging*, vol. 131, no. 2, p. 021009, 2009.
[17]   M. J. Cannell, R. Cooley, R. W. Garman, G. Green, P. N. Harrison, and J. D. Walters, "Fluid-cooled heat sink with turbulence-enhancing support pins," US6729383B1, 04-May-2004.
[18]   J. R. Davis, *Copper and Copper Alloys*. ASM International, 2001.

[19]   C. J. Glassbrenner and G. A. Slack, "Thermal Conductivity of Silicon and Germanium from 3°K to the Melting Point," *Physical Review*, vol. 134, no. 4A, pp. A1058–A1069, May 1964.

[20]   H. R. Shanks, P. D. Maycock, P. H. Sidles, and G. C. Danielson, "Thermal Conductivity of Silicon from 300 to 1400°K," *Physical Review*, vol. 130, no. 5, pp. 1743–1748, Jun. 1963.

[21]   Y. Takezawa, "Novel High Thermal Conductive Epoxy Resins," *EINA (Electrical Insulation News in Asia)*, vol. 12, pp. 43–44, 2005.

[22]   S. A. David, *Trends in Welding Research: Proceedings of the 8th International Conference, June 1-6, 2008, Callaway Gardens Resort, Pine Mountain, Georgia, USA*. ASM International, 2009.

[23]   R. E. Simons, "The evolution of IBM high performance cooling technology," in *Proceedings of 1995 IEEE/CPMT 11th Semiconductor Thermal Measurement and Management Symposium (SEMI-THERM)*, San Jose, CA, USA, 1995, pp. 102–114.

[24]   I. Mudawar, "Direct-immersion cooling for high power electronic chips," in *[1992 Proceedings] Intersociety Conference on Thermal Phenomena in Electronic Systems*, Austin, TX, USA, 1992, pp. 74–84.

[25]   R. Dayal, "Numerical Modelling of Processes Governing Selective Laser Sintering," Ph.D. Thesis, TU Darmstadt, Darmstadt, Germany, 2014.

[26]   S. A. Khairallah, A. T. Anderson, A. Rubenchik, and W. E. King, "Laser powder-bed fusion additive manufacturing: Physics of complex melt flow and formation mechanisms of pores, spatter, and denudation zones," *Acta Materialia*, vol. 108, pp. 36–45, Apr. 2016.

[27]   C. D. Boley, S. A. Khairallah, and A. M. Rubenchik, "Calculation of laser absorption by metal powders in additive manufacturing," *Applied Optics*, vol. 54, no. 9, p. 2477, Mar. 2015.

[28]   "Thermocouple Response Time." [Online]. Available: https://www.omega.com/techref/ThermocoupleResponseTime.html. [Accessed: 21-Feb-2019].

[29]   "Appendix B - Noise and Resolution Tables [U6 Datasheet] | LabJack." [Online]. Available: https://labjack.com/support/datasheets/u6/appendix-b. [Accessed: 21-Feb-2019].

[30]   "Type E Thermocouple - Type E Thermocouples - E Type Thermocouples." [Online]. Available: https://www.thermocoupleinfo.com/type-e-thermocouple.htm. [Accessed: 21-Feb-2019].

[31]   H. Flandorfer, U. Saeed, C. Luef, A. Sabbar, and H. Ipser, "Interfaces in lead-free solder alloys: Enthalpy of formation of binary Ag–Sn, Cu–Sn and Ni–Sn intermetallic compounds," *Thermochimica Acta*, vol. 459, no. 1–2, pp. 34–39, Jul. 2007.

Moisture Barrier, Mechanical, and Thermal Properties of PDMS-PIB Blends for Solar Photovoltaic (PV) Module Encapsulant

Jinho Hah[1], Michael Sulkis[2], Chao Ren[1], Minsoo Kang[1], Kyoung-sik Moon[1], Samuel Graham[2], and C. P. Wong[1*]

[1]School of Materials Science and Engineering, Georgia Institute of Technology, 771 Ferst Drive, Atlanta, GA 30332
[2]School of Materials Science and Engineering, Georgia Institute of Technology, 771 Ferst Drive, Atlanta, GA 30332
*e-mail: cp.wong@mse.gatech.edu

Abstract—**In this study, we have investigated and screened the performance of our adhesive material that does not require an edge seal. This paper primarily focuses on the synergistic properties of PDMS and PIB via physical blending. The UV blocking nature of PDMS and the high adhesion and superior barrier property of PIB were considered. Our polymer blend also maintains the transparency in the visible range. This polymer blends were performed with several kinds of material characterizations such as morphology, hardness, thermal profiles, moisture ingress properties, and adhesion strength. This paper serves to provide discussions on preliminary evaluations for transparent PV module encapsulant.**

Keywords-PV Encapsulant; Moisture Ingress; PDMS-PIB Blend; Adhesion Strength

I. INTRODUCTION

Perovskite Solar Cells (PSCs) have demonstrated outstanding performance, achieving power conversion efficiencies (PCE) as high as 23.7% and theoretical efficiencies up to 31% [1, 2]. In addition, an even higher PCE can be obtained by combining the two silicon and PSC devices together [2]. However, the instability of this organic-inorganic hybrid device remains a significant hurdle to commercialization. Currently, PSCs only lasts up to 6 months at an outdoor environment, because PSCs are susceptible to harsh environmental conditions such as extreme amount of absorbed UV rays, temperature, and high humidity. Therefore, encapsulation of the device is one of the top priorities to improve its stability [2-5].

Traditional silicon PV module packaging materials include protective frontsheets and backsheets, adhesive encapsulant materials and edge seals, which are required to ensure the reliability of solar photovoltaic (PV) modules [6-8]. Currently a glass-to-glass encapsulation method has been used for the 1st and 2nd generation encapsulation of PSCs, where glass slides serve as both frontsheet and backsheet, and getter-filled PIB and poly(dimethyl-siloxane) (PDMS) are used as the edge seal and encapsulant respectively [9]. This type of structure, utilizing PIB as an edge seal, is required to protect the PSCs from adverse environmental conditions, especially from moisture. PIB is known for having one of the lowest water vapor transmission rates (WVTR) reportedly 0.01 to 0.001 gm^{-2} day^{-1} [10], whereas PDMS elastomer (Sylgard 184), is ~900 gm^{-2} day^{-1} [11]. It is also reported that a 1.25 cm wide getter-filled PIB edge seal as a moisture barrier can pass IEC 61646 (1000 h at 85 °C/85% R.H.), which is equivalent to a 25-year lifetime at an outdoor environment [12].

However, for the next generation encapsulation method it would be valuable to minimize or eliminate the need for an opaque edge sealant. This will increase the available area for light-harvesting regions of PSCs and will simplify the encapsulation process, presumably aiding high-volume manufacturing as it removes an additional lamination step. It is also crucial that frontsheets and backsheets be replaced by transparent and flexible polymer barrier films for the next generation encapsulation method to be compatible with roll-to-roll processing. This would provide for fast, efficient, and large-scale processing of flexible thin film solar cells [13]. In this work, we have designed a novel transparent polymer blend of PIB and PDMS as an encapsulant (adhesive) material for PV module packaging. Fig. 1 illustrates the difference between the conventionally used encapsulation method for PSC package and our proposed method. We report here enhanced moisture transport properties and interfacial adhesion strength of our transparent polymer blends compared to commercially available PDMS-based adhesives for PV module packaging applications.

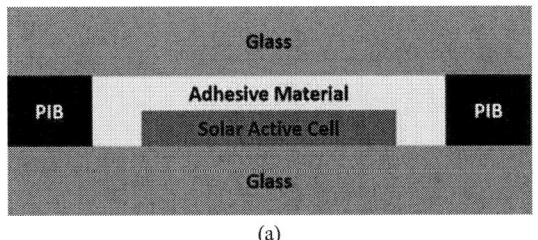
(a)

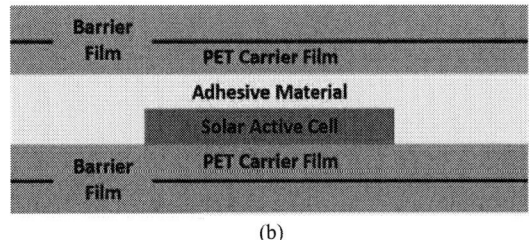
(b)

Figure 1. Representative schematic design of PSC package (a) of 1st and 2nd generation encapsulation technique which utilizes opaque PIB and (b) towards next generation encapsulation technique.

978-1-7281-1500-9/19 $31.00 © 2019 IEEE

The designed transparent adhesive in this study was a mixture of thermoplastic transparent pure butyl rubber (PIB) without any fillers and thermally-cured thermoset PDMS-based adhesive with UV absorber and UV stabilizer fillers for PV applications. In this study, we characterized the moisture ingress and performed interfacial adhesion strength through designed transparent polymer blends of different mixing ratios via optical calcium measurement tests and T-Peel tests, respectively. The moisture ingress results were compared to the commercially available PIB edge seal and PDMS-based adhesive to benchmark our data. This study serves to provide an improvement in packaging materials and in encapsulation processes with the introduction of a new type of adhesive composite.

II. EXPERIMENTAL SECTION

A. Materials

A transparent, two part-addition cure PDMS adhesive and a transparent thermoplastic PIB melt (45,000 gmol^{-1}) were used as encapsulants. Multilayer barrier films with a PET carrier were used as a transparent and flexible polymer frontsheet and backsheet for T-Peel test coupons. Optical calcium samples were deposited on a borosilicate glass and a cover glass was used as a frontsheet for optical calcium test coupons.

B. Polymer Blend Preparation

PDMS and PIB stock solutions were prepared by mixing part A of PDMS adhesive and PIB melt, respectively, with toluene solvent into 1:2 ratio by weight. The individual stock solutions were mixed thoroughly using a magnetic stirrer for overnight to fully dissolve the polymer into the solvent at 80 °C and at 1600 rpm. PDMS and PIB blends were mixed into separate glass vials in 3 ratios by weight as shown in Table I. Note that these ratios in Table I are respective to PDMS:PIB. These polymer blends were also mixed thoroughly using the same stirring conditions. Note that part B (cross-linking agent with catalyst) of PDMS adhesive was added to the polymer blend solution according to the manufacturer's specification before utilizing the solution for fabricating either optical calcium test coupons or T-Peel test coupons.

TABLE I. PDMS AND PIB BLEND RATIOS

PDMS Stock Solution	Blend A	Blend B	Blend C	PIB Stock Solution
1:0	3:1	1:1	1:3	0:1

C. Optical Calcium Sample Preparation and Screening Test Conditions

The polymer blend was spin-coated onto the cover glass at room temperature at 500 rpm for 60 s to ensure full coverage of polymer blend solution on the cover glass and 1500 rpm for another 60 seconds to ensure uniform thickness of the polymer blend and to remove residual toluene solvent. During each spin-coating step, 1 mL of polymer solution was dropped onto the cover glass before initiation.

The cover glass was bonded with the calcium coated borosilicate glass using a press with a force of 65 psi, and the bonded test coupon was cured at 100 °C for 1 h (manufacture's specification for PDMS). Fig. 2 illustrates a representative schematic of as-prepared test coupon for optical calcium screening test. Note from Fig. 2b that our formulated barrier adhesive (encapsulant) is optically transparent as both the gray calcium on glass and Texwipe are visible.

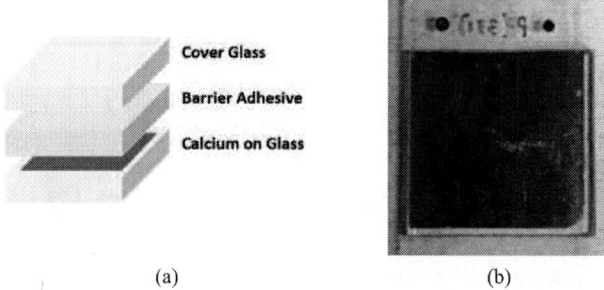

Figure 2. (a) Schematic layer-by-layer structure of an optical calcium test coupon and (b) Top-view of the fabricated optical calcium test coupon.

Calcium test coupons were then put into the environmental chamber set at 85 °C/85% R.H., and these samples were scanned periodically to quantitatively calculate the effective moisture ingress. How optical calcium sample screening test could be utilized to deduce moisture ingress through the sample can be explained using (1). In this equation, calcium before reacting with water is optically opaque at visible light; however, upon reacting with moisture, calcium corrodes into calcium (II) hydroxide, which is optically transparent at visible light. As the samples were periodically scanned, calculating the effective area loss would be used to extrinsically calculate for moisture ingress values.

$$Ca_{(s)} + 2H_2O \rightarrow Ca(OH)_{2\,(aq)} + H_{2\,(g)} \qquad (1)$$

D. T-Peel Test Sample Preparation and Test Conditions

A Universal Testing Machine (Test Resources, Shakopee, MN) was used to perform displacement-controlled T-peel experiments at 100 mm/min. Mean peel-off forces were calculated using MATLAB software. All peel test samples were fabricated with 1 cm in width. Also, it was ensured that uniform thickness of adhesive was deposited as 50 μm-thick polyimide Kapton tape was used as a dam. A schematic of T-peel test coupons and test set-up is shown in Fig. 3.

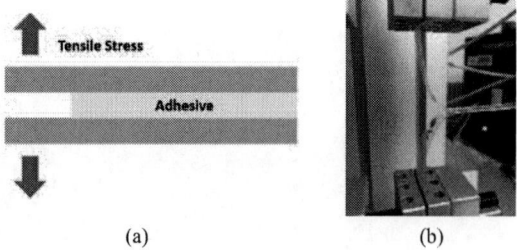

Figure 3. Representative schematic design of (a) T-Peel test coupons and (b) T-Peel test set up.

E. Thermal Characterization of the Polymer Blend

Thermogravimetric analysis (TGA) and differential scanning calorimetry (DSC) characterizations were performed. A TGA Q5000 (TA Instruments, New Castle, DE) was used to perform TGA characterization tests for all polymer samples. Temperature was reached from 25 °C to 800 °C with a ramping rate of 20 °C/min. Weight change was calculated using the software. A DSC Q2000 (TA Instruments, New Castle, DE) equipped with a cooling system was used to perform DSC characterization test for all polymer samples. Temperature was reached from -80 °C to 0 °C with a ramping rate of 10 °C/min. Glass transition temperature (T_g) was obtained using the software.

III. RESULTS AND DISCUSSION

A. Morphology of Barrier Adhesives (PDMS-PIB Blend)

The morphology of polymer blends of PDMS and PIB were characterized using optical microscope. Fig. 4 compares different morphologies across different polymer blends used in this study. In Fig. 4a, a typical sea-island type second phase of PIB is observed in PDMS matrix (Blend A). This morphology is also observed from the study Peng and his co-workers [14]. When added more PIB into the PDMS matrix (Blend B with 1:1 ratio), the second phase of PIB coalesced into a larger size (Fig. 4b). Finally, this coalescence reached the saturation point and the phase inversion was occurred (Fig. 4c). Therefore, when more PIB concentration was increased, the aspect ratio of PIB in the PDMS matrix increased and lost the droplet structure. Also note that the scale bar for Fig. 4a is much smaller than Fig. 4b and Fig. 4c. It is hypothesized from this morphology that Blend A would have the lowest moisture ingress as the smaller and more uniformly confined droplet structure of PIB is well dispersed within the PDMS matrix that will suppress facile-ingress of moisture through this complex.

B. Moisture Ingress through Barrier Adhesives

The moisture ingress values (K values) were obtained after processing the calcium corrosion test. In Fig. 5a, ingress distance of the three polymer blends is plotted against exposure time under the damp heat environment (85 °C/85% R.H.). Also, they were compared to a commercially available encapsulants for PV module packaging applications. The K values are plotted in Fig. 5b using the information from Fig. 5a. Results indicate that our polymer blends have K values in between the two commercially available products. Note that the previously mentioned getter-filled PIB that can sustain a 25-year lifetime at an outdoor environment [12] have K value of 0.013, a magnitude lower than the polymer blend that has the lowest K value of 0.23 (Blend A). It was previously hypothesized that Blend A would have the lowest moisture ingress through the calcium coupons from the morphology analysis. Although the blends with higher concentration of moisture-blocking PIB were added into the PDMS matrix, due to the phase inversion and high-aspect ratio PIB morphology, higher K values were obtained for Blend B (0.50) and Blend C (0.26).

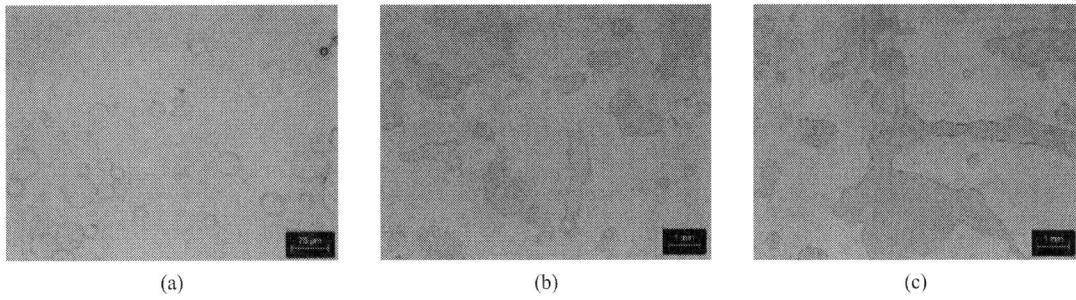

Figure 4. Optical microscope images showing the morphologies of (a) Blend A, (b) Blend B, and (c) Blend C.

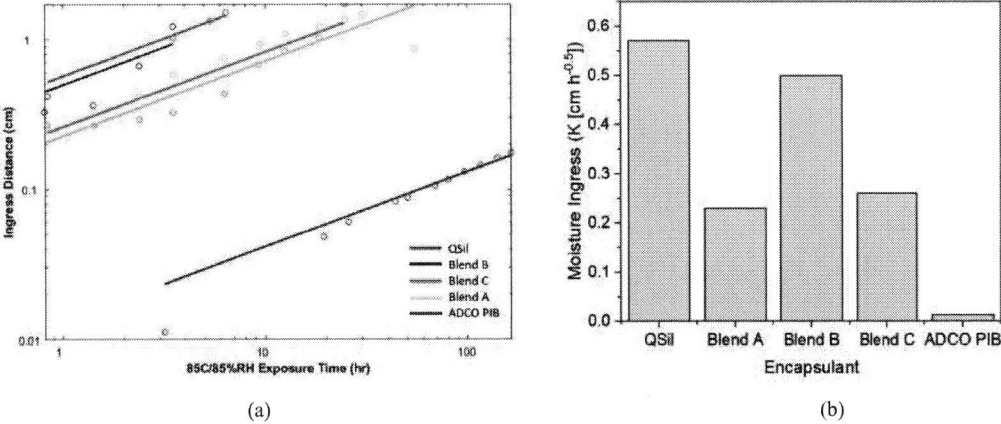

Figure 5. (a) Moisture permeation through polymer blends via calcium corrosion testing at 85 °C/85% R.H. and (b) their calculated K values.

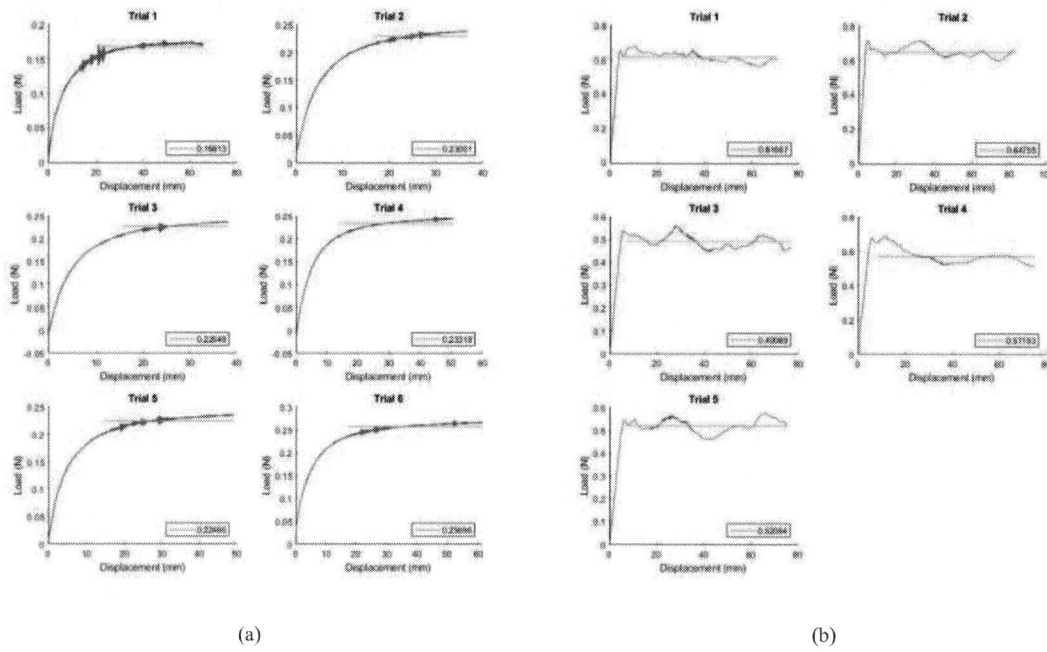

(a)   (b)

Figure 6. T-Peel test data of (a) PDMS samples and (b) PIB samples.

C. Adhesion Strength between Barrier Adhesives and PET Carrier Films

Adhesion strength between the PET (lamination side of the barrier film) and the barrier adhesive is one of the crucial factors that affects moisture ingress through the interfaces [15, 16]. Stronger the adhesion, more reliable bonds are formed at the interface due to the stronger network generated between the functional groups present from each layer. In Fig. 6 and Fig. 7, adhesion strength between two polymeric materials used in this study (PDMS and PIB) and polymer surface (PET) were evaluated using T-Peel test method. Note that PIB-PET has a much stronger adhesion strength (0.57 N/cm) than that of PDMS-PET (0.22 N/cm). Delamination profile (Fig. 6) shows that PDMS delaminates much more uniformly and cleanly while PIB delaminates in random fashion, but for sure, more energy is required to de-bond PIB from a PET carrier. Although Fig. 6 does not define a specific delamination pattern, it can provide an information that there are relatively weaker regions at the interface (Fig. 6b). This explanation could also suggest why moisture ingress was not uniform throughout as some regions with stronger interface would require more time for the moisture to ingress through the blends that contained PIB. Furthermore, lower adhesion strength values have higher susceptibility for delamination, which can reduce the moisture barrier efficiency and corrosion of metals [17, 18]. Therefore, the stronger interface formed for PIB-PET samples could explain for slower moisture ingress (delaying calcium corrosion) observed in Fig. 5 for the polymer blends and pure PIB polymer.

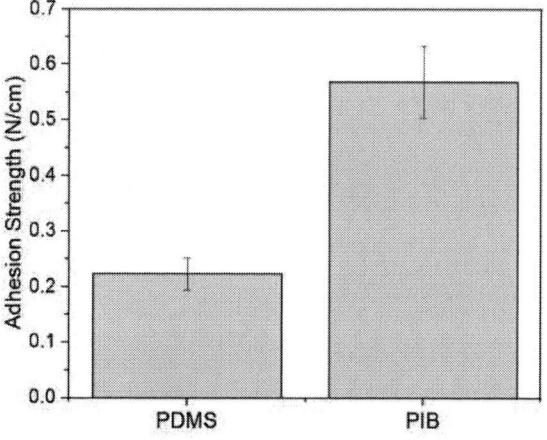

Figure 7. Calculated adhesion strength of PDMS-PET and PIB-PET.

D. Barrier Adhesive Characterizations

Polymer blends synthesized in this study were also characterized using TGA and DSC analysis to understand their thermal profiles and thermal stabilities for PV encapsulant applications. From Fig. 8a, the pyrolysis temperature of PIB is much lower than that of PDMS. The TGA plot of the PDMS-PIB blend has two on-set decomposition temperatures: the earlier being that of PIB and the latter being that of the PDMS. Therefore, it is supported from this TGA plot that polymer blend was not mixed at the molecular level and was rathe mixed physically. In another words, the synthesized PDMS-PIB blend has a phase separation, which was observed earlier from the sea-island structure in Fig. 4. The remaining weight from PDMS-PIB blend and pure PDMS is the O-Si-O backbone from the PDMS.

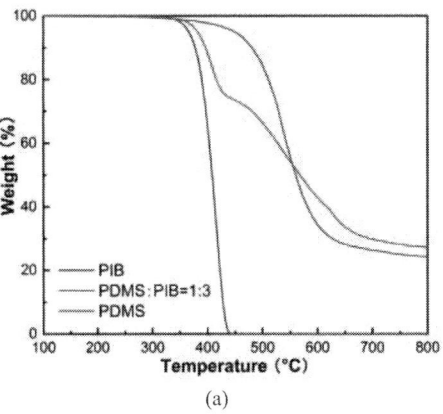 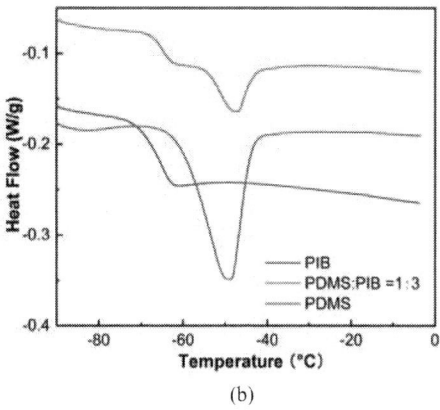

(a) (b)

Figure 8. Characterization plots of polymer blends using (a) TGA and (b) DSC.

From Fig. 8b, PDMS-PIB blend is shown with an endothermic peak known as the melting temperature (T_m), as well as the glass transition temperature (T_g). The endothermic peak of PDMS is attributed to the melts of the solid phase that was formed from devitrification and cold crystallization, which is typically observed in rubber materials; however, the mechanism has not been clearly understood yet. It was obtained from thermal analysis that T_g for the pure PIB is -65.29 °C, and that of this blend (Blend C) is slightly higher -66.07 °C. Because a typical T_g of PDMS is around -125 °C, it was not observed from our DSC analysis. In addition, T_m increases when PIB was added to PDMS as the endothermic peaks are shifted to the right. The DSC curve also suggests that the polymer blend was not mixed at the molecular level. It is suggested that in order to mix them at the molecular level, homogenizing agents must be used. In short, the polymer blends are well-suited as the encapsulants for thin-film PV applications because they are stable at temperature greater than 200 °C, and they are also flexible at room temp as the T_g for all samples were much below sub-zero temperature (less than -60 °C).

E. Hardness of the Barrier Adhesives

Mechanical testing of the polymer blends was conducted using the hardness testing method. From Fig. 9, hardness of the polymer decreases with increased PIB content. Note that hardness value of pure PIB was unmeasurable due to a highly viscous character at room temperature.

Because the thermoset PDMS polymer is crosslinked, it is expected to have higher hardness values compared to a thermoplastic PIB, which is visco-elastic at room temperature. This can be supported by TGA thermal analysis (Fig. 8a), where PDMS has a higher thermal stability than PIB as the on-set decomposition temperature is much higher.

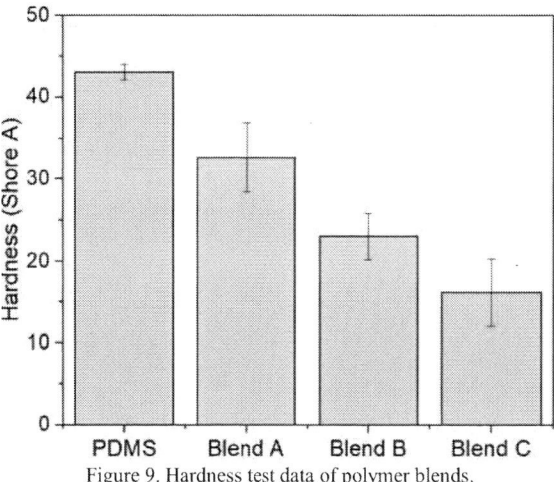

Figure 9. Hardness test data of polymer blends.

IV. CONCLUSION

Currently, poly(isobutylene) (PIB) based thermoplastic polymers, known for having the lowest moisture vapor transmission rate (MVTR) among various sealant materials, is used as an edge seal to protect the PV modules from moisture ingress. However, due to its nature of the vulnerability to UV exposure, a plethora of carbon-black fillers is incorporated as UV absorbers and blockers, which in turn deprives the PIB polymer of its transparency. In this study, we have developed a PDMS-PIB mixture for PV module encapsulants. Noticeable results indicated that adding 33% of PIB into the PDMS mixture increases the moisture stability, maintains the thermal stability, and showing the least amount of hardness lost among the PDMS-PIB blends. Also, the formulated transparent adhesives do not require an edge sealant such as the conventional non-transparent PIB with a high filler content. However, it is important to understand when even lower PIB content could even achieve better moisture blocking property and have better UV stability as PDMS content would increase. Our future work would later focus on the UV stability of the formulated adhesives using the accelerating test under the UV lamp and observe their UV-degradation profiles.

978-1-7281-1500-9/19 $31.00 © 2019 IEEE   1981

ACKNOWLEDGMENT

The authors would like to thank DuraMAT for their financial support on this research.

REFERENCES

[1] M. A. Green et al., "Solar cell efficiency tables (Version 53)," Progress in Photovoltaics: Research and Applications, vol. 27, no. 1, pp. 3-12, 2019.

[2] Y. Y. a. J. You, "Make Perovskite Solar Cells Stable," Nature, vol. 544, no. 7649, 2017.

[3] B. Li, Y. Li, C. Zheng, D. Gao, and W. Huang, "Advancements in the stability of perovskite solar cells: degradation mechanisms and improvement approaches," RSC Advances, 10.1039/C5RA27424A vol. 6, no. 44, pp. 38079-38091, 2016.

[4] Z. Wang, Z. Shi, T. Li, Y. Chen, and W. Huang, "Stability of Perovskite Solar Cells: A Prospective on the Substitution of the A Cation and X Anion," Angewandte Chemie International Edition, vol. 56, no. 5, pp. 1190-1212, 2017.

[5] T. A. Berhe et al., "Organometal halide perovskite solar cells: degradation and stability," Energy & Environmental Science, 10.1039/C5EE02733K vol. 9, no. 2, pp. 323-356, 2016.

[6] D. L. King, M. A. Quintana, J. A. Kratochvil, D. E. Ellibee, and B. R. Hansen, "Photovoltaic module performance and durability following long-term field exposure," Progress in Photovoltaics: Research and Applications, vol. 8, no. 2, pp. 241-256, 2000.

[7] C. Tuan, F. Wu, K. Moon, and C. Wong, "Epoxy/Cyanate Ester Copolymer Material for Molding Compounds in High-Temperature Operations," in 2017 IEEE 67th Electronic Components and Technology Conference (ECTC), 2017, pp. 1328-1333.

[8] B. Song, K. Moon, and C. Wong, "Stretchable and Electrically Conductive Composites Fabricated from Polyurethane and Silver Nano/Microstructures," in 2017 IEEE 67th Electronic Components and Technology Conference (ECTC), 2017, pp. 2181-2186.

[9] R. Cheacharoen et al., "Encapsulating perovskite solar cells to withstand damp heat and thermal cycling," Sustainable Energy & Fuels, 10.1039/C8SE00250A vol. 2, no. 11, pp. 2398-2406, 2018.

[10] L. Shi et al., "Accelerated Lifetime Testing of Organic–Inorganic Perovskite Solar Cells Encapsulated by Polyisobutylene," ACS Applied Materials & Interfaces, vol. 9, no. 30, pp. 25073-25081, 2017.

[11] S. Kirsten, M. Schubert, M. Braunschweig, G. Woldt, T. Voitsekhivska, and K. Wolter, "Biocompatible packaging for implantable miniaturized pressure sensor device used for stent grafts: Concept and choice of materials," in 2014 IEEE 16th Electronics Packaging Technology Conference (EPTC), 2014, pp. 719-724.

[12] M. D. Kempe, D. Panchagade, M. O. Reese, and A. A. Dameron, "Modeling moisture ingress through polyisobutylene-based edge-seals," Progress in Photovoltaics: Research and Applications, vol. 23, no. 5, pp. 570-581, 2015.

[13] K. Hwang et al., "Toward Large Scale Roll-to-Roll Production of Fully Printed Perovskite Solar Cells," Advanced Materials, vol. 27, no. 7, pp. 1241-1247, 2015.

[14] X. Peng, Y. Huang, T. Xia, M. Kong, and G. Li, "Shapes of dispersed phase in confined PIB/PDMS blends with different compositions during shear flow," European Polymer Journal, vol. 47, no. 10, pp. 1956-1963, 2011.

[15] J. A. del Cueto and T. J. McMahon, Analysis of leakage currents in photovoltaic modules under high - voltage bias in the field. 2002, pp. 15-28.

[16] J. Hah, B. Song, K. Moon, S. Graham, and C. P. Wong, "Design and Surface Modification of PET Substrates Using UV/Ozone Treatment for Roll-to-Roll Processed Solar Photovoltaic (PV) Module Packaging," in 2018 IEEE 68th Electronic Components and Technology Conference (ECTC), 2018, pp. 2397-2403.

[17] V. Šály, M. Ružinský, and P. Redi, Testing of Photovoltaic Modules and Encapsulations at Elevated Voltage, Temperature and Humidity. 2000, pp. 2053-2056.

[18] K. R. McIntosh, N. E. Powell, A. W. Norris, J. N. Cotsell, and B. M. Ketola, "The effect of damp-heat and UV aging tests on the optical properties of silicone and EVA encapsulants," Progress in Photovoltaics: Research and Applications, vol. 19, no. 3, pp. 294-300, 2011.

Modeling and Design of Power Distribution Network for a Heterogeneous Integrated Active Interposer with Neuromorphic Computing Circuits

Min MIAO, Tianfang CHEN，Jincan ZHANG,
Na LI, Kunkun LI, Liyuan WANG
Information Microsystem Institute
Beijing Information Science & Technology University
Beijing, 100101, China
e-mail: miaomin@bistu.edu.cn;
Tel: +86-10-64884695

Yang YANG, Xiaole CUI, Yufeng JIN
Shenzhen Graduate School of Peking University
Shenzhen, 518055, China

Huan LIU
Peking University, Beijing, 100871, China

Abstract—**Ultra-high density 2.5/3D heterogeneous integration has been considered an essential solution for rebooting computation applications like high performance computing, machine learning, and brain-mimicking neuromorphic computing, in both cloud and edge modes. Interposers with active auxiliary circuitry such as tunable power distribution network (PDN), phase locked loops, active signaling equalizers/ buffers, and even neuromorphic units, are emerging as much more attractive and flexible platforms than passive interposers for these applications. In this paper, an active interposer conception acting as a platform for heterogeneous integration of logic, memory and neuromorphic computing circuits, is proposed for rebooting computation purpose, featuring functioning units such as active PDN, data switching and signal conditioning circuits, and neuromorphic units based on so-called neural TSVs. New features such as pulse operation of neuron circuit and high speed data switching may induce new power integrity issues. Fortunately, the introduction of neural TSVs with large capacitance may also acting as the in-situ decoupling capacitors whose value can be modulated by the switching on/off of the MOSFETs associated with the neural TSVs in the same neuron cell, and thus a compact and tunable PDN can be constructed, whose impedance and anti-resonance characteristics may be reconfigured flexibly for an optimal power distribution and minimal simultaneous switching noise. Principles are disclosed and compact circuit modeling is set up for the PDN. Circuit and full-wave simulation results are demonstrated, confirming the conception and its effectiveness.**

Keywords-2.5/3D heterogeneous integration; active interposers; tunable power distribution network (PDN); simultaneous switching noise; power integrity (PI); neuromorphic circuit; through silicon via

I. INTRODUCTION

Ultra-high density 2.5/3D heterogeneous integration of logic computational units and memory units has been considered an essential solution for rebooting computing applications like high performance computation, artificial intelligence (AI)/machine learning in both cloud and edge computing modes. Interposers with active auxiliary (AUX) circuitry such as tunable power distribution network (PDN), phase locked loops (PLLs), active signaling equalizers and buffers, known as active interposers, are emerging as more attractive and flexible platforms than passive interposers[1-3] for the aforementioned integration architecture. In addition, neuromorphic computing mimicking information processing of brains on the fundamental physical structure level with high energy efficiency and processing capability for non-structural data, has recently found its significance in rebooting computation domain, and thus neuromorphic units are now becoming integral parts of 2.5/3D heterogeneously integrated modules, both as a discrete chip or as units monolithically integrated on the active interposer.

A 3D circuit cell design mimicking a brain one was demonstrated in ECTC2018 by the authors, featuring TSVs (through semiconductor/substrate vias) with unconventional lateral configuration. Each TSV consists of coaxial Cu filling, a liner and a heavily N+ doped Si sidewall from the inside out, which is equivalent to a coaxial capacitance, as shown in Fig.1 [4]. These TSVs, each acting as an input capacitor, are combined with ordinary on-chip MOSFET to form a vertical threshold summator whose output shall switch its state or issue a pulse when weighted summation of all the input voltages goes over a threshold set by the TSV and MOSFET parameters. The TSVs-MOSFET combination may therefore be seen as a neuron FET as a whole, and the TSVs themselves are thus called neural TSVs. Furthermore, it can be naturally conceived that the neural TSVs may also act as decoupling capacitors (decaps) for the power integrity (PI) of a heterogeneous integrated 3D module by suppressing power noise in the active interposer PDN, while leaving precious surface areas for active devices and signal interconnects.

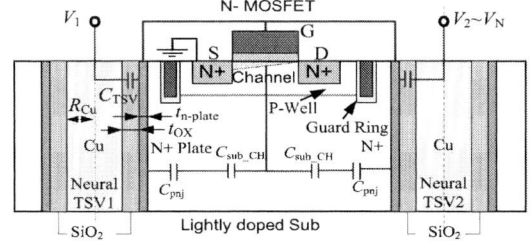

Figure 1. Schematic of a neuron FET (cell) as a basic neuromorphic circuit element, with neural TSVs designed as input coupling capacitors of an associated N-channel MOSFET [4]. The TSVs may also act as decaps for tunable (active) on-interposer PDN.

Therefore, in this paper, an active interposer integrated with the neuron circuit cell as mentioned above, is proposed, featuring on-interposer AUX circuits and neuromorphic circuits with neuron TSVs, as shown in Fig.2. Combined with circuit cells mimicking neurons, highly efficient 3D interconnects, active power distribution network, high speed data switching and signal conditioning circuit, this active interposer may be used as an ideal platform for the module-level 2.5D/3D heterogeneous integration of logic, memory and neuromorphic computing devices for rebooting computation. However, new features such as random pulse operation of neuron circuit may induce new power integrity issues and even challenges. Fortunately, the introduction of neuron TSVs with large capacitance may also provide the in-situ decaps whose value can be modulated by the switching on/off of the MOSFETs associated with the neural TSVs in the same neuron cell, and thus a compact and tunable PDN can be constructed, whose characteristics may thus be reconfigured for an optimal power delivery and minimal simultaneous switching noise (SSN). Principles are disclosed and compact circuit modeling is set up for the PDN. Circuit and full-wave simulation results are demonstrated, confirming the conception and its effectiveness.

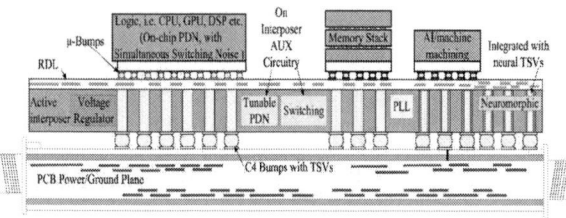

Figure 2. Schematic of a 2.5D heterogeneous integration scheme based on an active interposer featuring neuromorphic circuit and integrated tunable (active) on-interposer PDN .

II. WORKING PINCIPLES AND AN INTUITIVE MODELING OF THE TUNABLE PDN BASED ON THE NEURAL TSVS

Simultaneous switching current (SSC) induced by the synchronous switching operation of digital transistors, is one of the major source of PDN noise, i.e. SSN. Eventually, SSN causes ground/power bounce, leading to signal distortion, noise margin reduction, and power integrity (PI) degradation. The magnitude of SSN is determined both by transient SSC and PDN impedance. In particular, when major SSC spectral components coincide with the anti-resonance frequencies of PDN impedance, a critical SSN can be generated.

A tunable PDN has been demonstrated most recently, based on the switching between decap arrays and the anti-resonance peak shift thus induced [5]. Therefore, SSN can be suppressed as the SSC spectral fundamental frequency component deviates from the major anti-resonance peak. A similar tunable PDN conception is proposed in this paper. The tuning of the PDN is based on the variation of PDN decap loading, realized by switching the NMOS FET inside a neuron FET cell and associated neural TSV capacitors in the same cell, with the equivalent circuit shown in Fig.3(a). That is, the total loading effect of TSV capacitors as decaps

and equivalent PDN impedance are modulated by sending enabling signals to switch various numbers of NMOS FETs (on/off) inside neuron FETs (referring to Fig.1) according to the monitoring of SSN. The principles can be intuitively explained by a simple lumped circuit model also shown in Fig.3 (b) and quick circuit simulation.

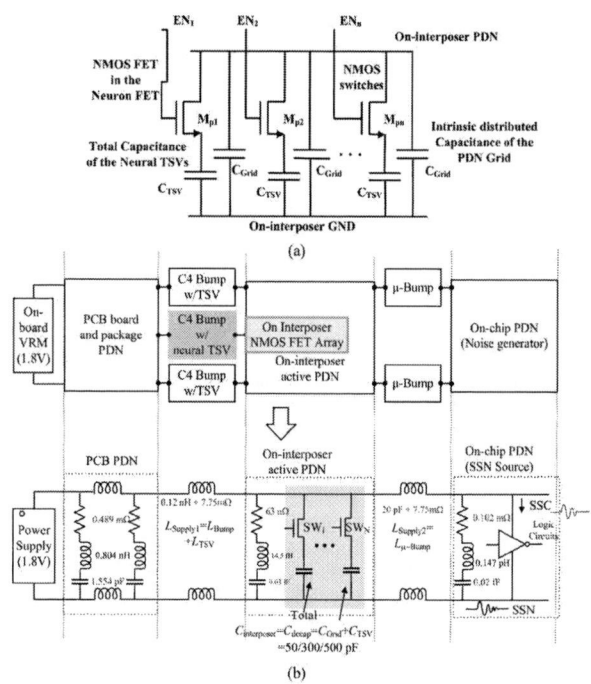

Figure 3. (a) Equivalent circuit of switchable decap array realized by combining TSV capacitors and associated NMOS transistors acting as the switches for on-interposer PDN tuning inside the same neuron FET; (b) the PCB-interposer-chip hierarchy of a PDN for a typical 2.5D heterogeneous integration scheme and the intuitive lumped circuit model, with the tunable PDN part implemented on the interposer level (the part in the yellow shadow). Parameters for simlation are also shown.

In this model, decap capacitance C_{decap} plus the intrinsic PDN capacitance C_{grid}, i.e. $C_{interposer}$, is set to 50 pF, 300 pF and 500 pF, dubbed as Case A1, B1 and C1 respectively. The simulation results of the frequency dependence of PDN impedance taken at the interposer-chip port are displayed in Fig.4. It can be seen that, anti-resonances frequencies are 202.1 MHz, 164.7 MHz, and 151.9 MHz in case A1, B1 and C1, respectively. That is, with the increase of decap value, the anti-resonance peak decreases and moves to lower frequency. On this basis, a current source is added at the chip load port as SSC source to reveal the SSN behavior of on-interposer PDN. The current source is set as triangular pulse wave with a period of 5 ns, peak current of 20mA, rise/fall time of 1 ns, and periodic quiescence of 3ns, and its spectrum is also shown Fig.4. As seen in Fig.5, the time-domain peak-to-peak voltage of SSN is 69.4, 85.7 and 14.2 mV for Case A1, B1 and C1 respectively. Comparing Fig.4 with Fig.5, it is found that, at the SSC spectrum peak frequency, impedance in case C1 is the minimum while that in case B1 is the maximum, which is consistent with the simulated SSN peak-to-peak voltage.

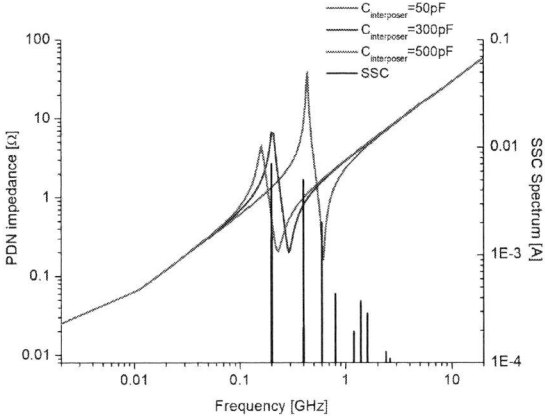

Figure 4. PDN impedance vs frequency curves for Case A1~C1, with $C_{\text{interposer}}$ =50, 300, and 500pF respectively, and the SSC spectrum, for the analysis of the intuitive model in Fig. 3

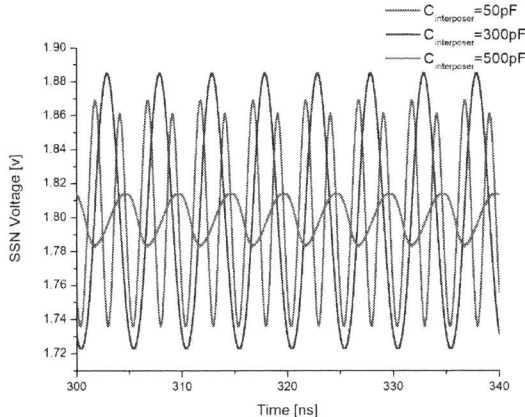

Figure 5. SSN voltage waveforms for Case A1~C1.

III. MODELING OF THE NEURAL TSVS AS DECAPS AND THE TUNABLE ACTIVE INTERPOSER PDN ANALYSIS SETUP

A. Modeling of the Neural TSV as Decaps

Table I lists the typical values taken for the neural TSV and neuron FET, based on the practical manufacturing process of choice.

TABLE I. BASIC PARAMETER SETUP FOR MODELING AND ANALYSIS OF THE NEURAL TSV AND THE 3D NERON CIRCUIT CELL

Parameters	Value for the configuration
Process	TSV middle; 180nm node CMOS
TSV height: h_{TSV}	100 μm
TSV Cu radius: R_{Cu}	5~15μm
Oxide liner thickness: t_{OX}	0.1μm
Oxide liner permissivity: ε_{OX}	4
Outer plate thickness and radius: $t_{\text{n-plate}}$, $r_{\text{n-plate}}$	0.2μm, R_{Cu}+ t_{OX}+ $t_{\text{n-plate}}$
TSV pitch: p_{TSV}	>15μm

Parameters	Value for the configuration
Substrate doping: N_{A} (p type well)	$1.2 \times 10^{15} \text{cm}^{-3}$
Outer plate doping: $N_{\text{n-plate}}$	$1.0 \times 10^{20} \text{cm}^{-3}$
Power Supply: V_{DD}	1.8V

It can be seen from Fig. 1, that the neural TSV is a standard co-axial capacitor formed by the copper filling, oxide and the N+ outer plate, and the value can be precisely expressed as

$$C_0 = C_{\text{TSV}\_i} = \frac{2\pi\varepsilon_{\text{ox}}}{\ln(1 + t_{\text{ox}} / R_{Cu})} h_{\text{TSV}}. \tag{1}$$

B. PDN Power Grid Design

The process taken for the proposed active interposer is a typical 180nm node CMOS line which is a cost-effective one for active interposer implementation, with the M5 and M6 metal layers that hold the thick metallization dedicated to the crossbar mesh grid of power and ground layer, referred to as PWR layer and GND layer respectively.

The planar layout, inspired by [5], is shown in Fig.6, with a total size of 3mm × 3mm. As the PDN mesh grid is design with perfect mirror symmetry, only the lower-left quadrant is displayed. As one unit cell is 300μm×300μm large, totally 10×10=100 unit cells exist in the interposer PDN.

Under each unit cell, TSVs are densely placed into decap arrays, and the number of TSVs connected to the on-interposer PDN is controlled by a neuron circuit, so as to adjust the magnitude of decap array loading on the PDN according to the on-line SSN monitoring. The capacitance of each TSV is designed to be C_0=1pF by carefully choosing its geometry. According to the spectral characteristics of the noise, the frequency of the impedance anti-resonance peak of on-interposer PDN is changed to prevent critical SSN at the anti-resonance peak.

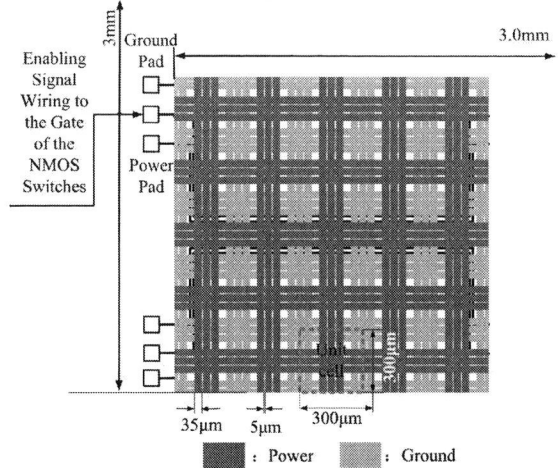

Figure 6. Planar layout of the power Grid for on-interposer PDN (only a quadrant is shown) with neural TSVs embedded under the M1 layer (TSV middle).

C. Circuit Modeling of the PDN Unit Cell

Each unit cell of the on-interposer PDN is modeled with the transmission-line modeling method, as shown in Fig.7, and combined to the complete on-interposer PDN structure based on the segmentation method [5, 6, 7]. The parasitic parameters for the neural TSVs are extracted on the basis of a combination of analytical and 3D full wave analysis. The parameters used are listed in Table II.

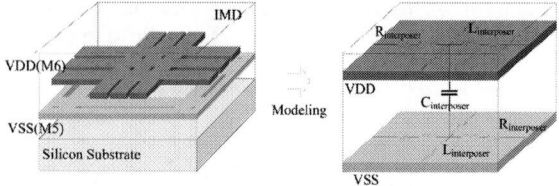

Figure 7.   Circuit Equivalents for a Unit Cell.

TABLE II.   Parameters for Circuit Modeling of the Unit Cell and PDN

No.	Parameters for Circuit Modeling		
	Object	*Parameter*	*Value*
1	Power and ground plane of PCB	R_{PCB}	0.489 mΩ
2		L_{PCB}	0.804 nH
3		C_{PCB}	1.554 pF
4	BGA (C4 Bump)	R_{BGA}	7.75 mΩ
5		L_{BGA}	0.12 nH
6	Unit cell of the on-interposer PDN (intrinsic)	$R_{interposer}$	63 μΩ
7		$L_{interposer}$	14.5 fH
8		$C_{interposer}$	0.03 fF
9	TSV	R_{TSV}	7.75 mΩ
10		L_{TSV}	20pH
11	Unit cell of the on-chip PDN	R_{Chip}	0.102 mΩ
12		L_{Chip}	0.147 pH
13		C_{Chip}	0.02 fF

D. Set up of the PCB-Interposer-Chip Hierarchy for Simulation

Three main chip modules that can be integrated on the interposer are considered, including a surface-mounted GPU (graphic processing unit, GPU) and a memory stack (shortly as M hereafter), together with a neuron computing unit consisting of a flip-chip mounted neuron/synapse chip and neuron FETs integrated in the active interposer for artificial intelligence purpose (shortly as AI hereafter). The GPU is assumed to takes a planar footprint of 5×5 unit cells, the memory 2×3 unit cells and the AI 3×2 unit cells of the interposer, as shown in Fig.8. The PDN of the unit cells under the center part of a module is connected to the module, providing multiple power supplies (with each unit cell providing one set of power/ground connection) to satisfy the power need of the GPU/M/AI modules. Therefore an

interposer-chip port exists for each unit cell. Under the central area of the three modules, each unit cell is densely packed with 25 TSVs, while a set of 4 TSVs are placed evenly in each of the rest unit cells of the entire interposer. The 4 TSV-sets distributed globally are always connected to the PDN mesh grid. An on-line SSN monitor decides in real time how many TSVs out of the 25 TSVs in each unit cell shall be connected to the PDN for proper decap loading by enabling the switching of the NMOS transistors in the neuron FETs, according to the spectrum characteristics of SSC. In this way, the impedance regulation of the entire interposer PDN is realized. Four power supplies are provided at the four corners of the active interposer from the PCB, with the on-PCB PDN taken into account in the form of lumped circuit elements.

The setup shown here is intended both for SPICE model-based circuit simulation and full-wave simulation.

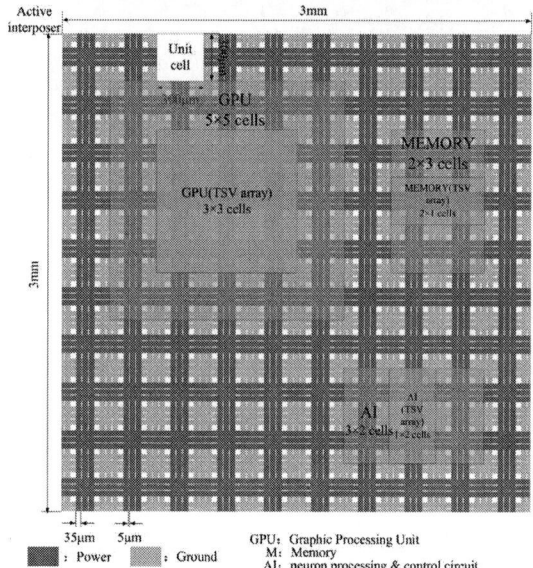

Figure 8.   Set up of the PCB-Interposer-Chip Hierarchy for Simulation.

IV. CIRCUIT SIMULATION RESULTS FOR THE ACTIVE INTERPOSER PDN

A. Impedance Characteristics Analysis Results

The impedance is simulated for three cases. In Case A2, 5 TSVs are connected to the PDN in each unit cells at central area of each chip module; that is, TSV decap are nearly well-distributed in the entire on-interposer PDN (referred to as $C_{T\text{-}TSV}$=5pF). In Case B2, 15 TSVs are connected ($C_{T\text{-}TSV}$=15pF). In Case C2, all of the 25 TSVs are connected ($C_{T\text{-}TSV}$=25pF). The scheme is designed to suppress SSN at critical locations on the interposer. The frequency dependence of the on-interposer PDN impedance taken at the interposer-chip port below GPU and the triangle-wave SSC spectrum (taken according to the clocking) are both shown in Fig.9. The anti-resonances frequencies are 351.1MHz, 311.7MHz, and 281.4MHz in case A2, B2 and C2, respectively. With the increase of decap value, the anti-resonance peak decreases and moves to lower frequency.

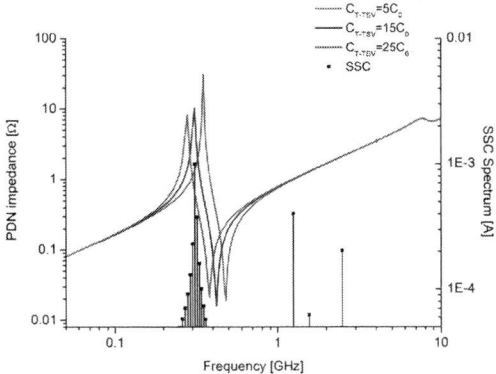

Figure 9. PDN impedance vs frequency curves for the port under GPU, which is typical for the interposer PDN as in Case A2~C2, obtained with circuit simulation, and the triangle wave SSC spectrum.

B. SSN Analysis Results

For the analysis of SSN, frequency-dependent current sources are added to the ports between interposer PDN and GPU, M (memory) and AI modules in the central areas. SSC with triangle time-domain waveforms are taken as the noise stimuli for GPU and M module. There are many power-to-signal-noise issues at the I/O interfaces with random signals transmitted and received, while the switching of neuron computing unit can be considered a random process, with the switching currents coming in the form of sharp electrical pulse. So it is more reasonable to use PRBS (pseudo random binary sequence) current source to reflect SSC at chip I/O and AI module. The spectrum of PRBS current source is shown in Fig. 10.

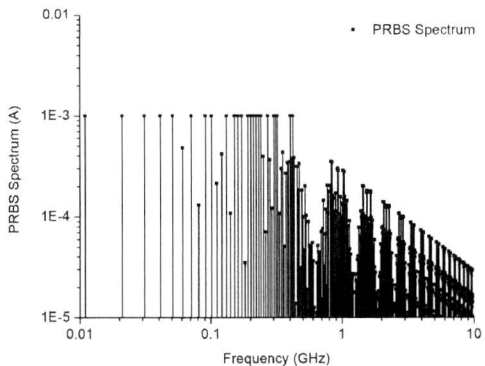

Figure 10. spectrum of PRBS current source

The SSN at the center of the GPU which is typical for the interposer PDN SSN distribution, for Case A2~C3, are compared in Fig. 11. It can be seen, that the peak-to-peak voltage of time-domain SSN is 84.6, 278.0 and 70.1 mV in Case A2, B2 and C2. As seen from Figs.9, at the frequency of SSC peak, impedance in case C is the minimum while that in case B is the maximum, which is consistent with the simulation results of SSN voltage. It can be concluding that, by adjusting the number of connected TSVs to set appropriate decap value, SSN can be effectively suppressed.

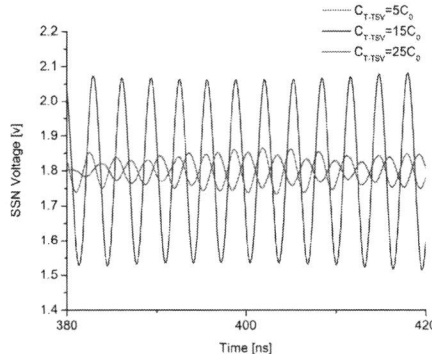

Figure 11. SSN voltage waveforms for Case A2~C3 by circuit simulate.

V. FULL WAVE ANALYSIS OF THE TUNABLE PDN

A. Impedance Analysis Results

Full wave analysis based on the solid modeling is made to further verify the conception proposed. The tool taken is SIWave[TM] from Ansys, Inc [8]. As in the Case A2~C2 in the circuit simulation-based analysis, Case A3~C3 are defined, corresponding to C_{T-TSV}=5pF, 15pF and 25pF. The results are shown in Fig.12. The first anti-resonance frequencies obtained through circuit simulation and full-wave simulations respectively are compared in Table III. It can be seen that the two set of results match well with each other.

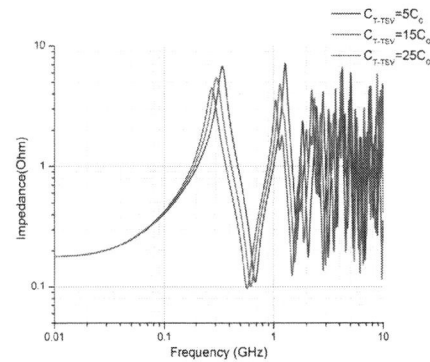

Figure 12. PDN impedance vs frequency curves for the port under GPU, which is typical for the interposer PDN as in Case A3~C3, obtained with full-wave simulation.

TABLE III. COMPARISON BETWEEN THE ANALYSED ANTI-RESONANCE FREQUENCIES THROUGH CIRCUIT AND FULL-WAVE SIMULATION

Main Antiresonances Frequencies	Case A3	Case B3	Case C3
ADS Simulation results (MHz)	344.3	305.4	277.3
SIWave Simulation results (MHz)	351.1	311.7	281.4
Error	2.0%	2.1%	1.5%

B. DC Analysis

DC analysis results are shown below in Fig.13 and Fig.14. It can be seen that the max IR voltage drop and current density are less than 0.1V and 1.3×10^{-3}A/μm^2 respectively, which is satisfying.

978-1-7281-1500-9/19 $31.00 © 2019 IEEE   1987

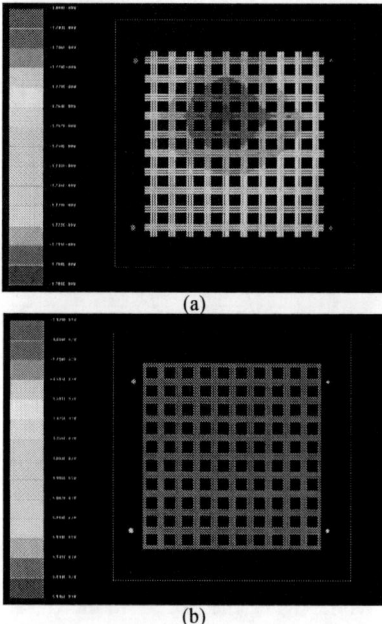

Figure 13. DC voltage IR drop on the (a) power level (PWR) and (b) ground level (GND) of the PDN power grid.

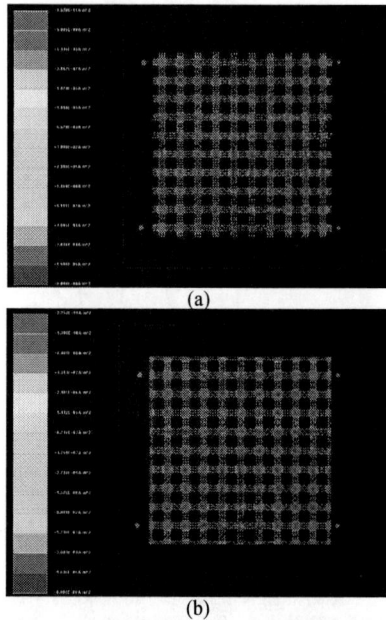

Figure 14. Current Density on the (a) power level (PWR) and (b) ground level (GND) of the PDN power grid.

VI. DISCUSSIONS AND CONCLUSIONS

In this paper, an active interposer conception acting as a platform for heterogeneous integration of logic, memory and neuromorphic computing devices, is proposed for rebooting computation purpose, which may feature functioning units such as active PDN, data switching and signal conditioning, neurons circuit cells based on so-called neural TSVs. New features such as random pulse operation of neuron circuit and fast switching of data exchange may induce new power integrity issues. Fortunately, the introduction of neural TSVs with large capacitance may also acting as the in-situ decaps whose value can be modulated by the switching of the associated MOSFETs in the same neuron cell, and thus a compact and tunable PDN can be constructed, whose impedance and anti-resonance characteristics may be reconfigured flexibly for optimal power delivery and minimal SSN. Principles are explained and compact circuit modeling is set up for the PDN. Circuit and full-wave simulation results are demonstrated, confirming the conception and its effectiveness. The tuning range of on-interposer PDN depends highly on the TSVs integrated on the interposer, which can be extended further by optimal layout of TSVs. As the PDN impedance characteristics vary with the operation modes and conditions of the whole microsystem, a real-time monitoring unit is demanded for the effective optimization of the PDN impedance on line, which is the focus under further research.

ACKNOWLEDGMENT

The work was supported in part by National Natural Science Foundation of China (Grant No. 61674016), in part by the State Key Development Program for Basic Research of China (973) under Grant 2015CB057201, in part by the Importation and Development of High-Caliber Talents Project of Beijing Municipal Institutions (Great Wall Scholar, No. CIT&TCD20150320) and Beijing Nova Program Interdisciplinary Studies Cooperative projects (No. Z161100004916036).

REFERENCES

[1] J. Kim, "Active Si interposer for 3D IC integrations", 2015 IEEE International 3D Systems Integration Conference (3DIC), Aug.-Sept., 2015, pp. TS11. 1.1-TS11. 1.3, 10.1109/3DIC.2015.7334619

[2] Pascal Vivet, "Aspects Consommation & Thermique dans les Circuits 3D", ECOFAC' 2014, 2014, invited speech.

[3] Pascal Viaud, "3DIC & 2.5D Interposer Market trends and technological evolutions", SEMICON CHINA, 2013, invited speech.

[4] M. Miao, L. Wang, T. Chen, X. Duan, J. Zhang, N. Li, L. Sun, R. Fang, X. Sun, H. Liu, and Y. Jin. "Modeling and Design of A 3D Interconnect Based Circuit Cell Formed with 3D SiP Techniques Mimicking Brain Neurons for Neuromorphic Computing Applications", 2018 IEEE 68th Electronic Components and Technology Conference (ECTC 2018), Sheraton San Diego Hotel and Marina, San Diego, USA, May 29-June 1, 2018, pp. 490-497.

[5] S. Kim, Y. Kim, K. Cho, J. Song, and J. Kim, "Design and measurement of a novel on-interposer active power distribution network for efficient simultaneous switching noise suppression in 2.5-D/3-D IC" IEEE Transactions on Components, Packaging and Manufacturing Technology, vol.9, no.2, Feb. 2019, pp. 317-328.

[6] B. Bae, Y. Shim, K. Koo, J. Cho, J. S. Pak, and J. Kim, "Modeling and measurement of power supply noise effects on an analog-to digital converter based on a chip-PCB hierarchical power distribution network analysis," IEEE Trans. Electromagn. Compat., vol. 55, no. 6, Dec. 2013, pp. 1260–1270.

[7] M. Swaminathan and A. Engin, Power Integrity Modeling and Design for Semiconductors and Systems, Pearson Education, 2008.

[8] https://www.ansys.com/products/electronics/ansys-siwave.

PCB Microstrip Line Far-End Crosstalk Mitigation by Surface Mount Capacitors

Zhaoqing Chen
IBM Corporation
Poughkeepsie NY 12601
zhaoqing@us.ibm.com

Abstract—In this paper, a technique by adding SMT capacitors at the spacing between two adjacent microstrip lines is applied to mitigate far-end crosstalk. By adding SMT capacitors with optimized capacitance value taking into account the SMT parasitic parameters and frequency-dependent solder-mask-coated microstrip line *RLGC* per-unit length parameters, the far-end crosstalk can be minimized. That will make microstrip line structure a realistic channel physical layer option for high-speed system card and board design. This technique can also be applied to differential net applications by adding the SMT capacitors only at the off-pair spacing. Potential application examples are shown for 32Gb/s cases. Vendor SMT capacitor SPICE model including parasitic parameters is used for linear network analysis and transient circuit simulations to optimize and verify the FEXT mitigation.

Keywords- signal integrity/ power integrity/ EMI; electric modeling/ analysis/ design/ characterization of interconnects; high speed and wireless electronics indluding RF, millimeter wave to THz.

I. INTRODUCTION

In high-speed server system card and board design, microstrip line[1][2] is applied more and more widely in addition to stripline. Microstrip line has significant advantages such as lower loss than stripline and eliminating vias and via stubs for on-PCB surface connection. However, due to a slightly non-TEM nature caused by mixed air and dielectric regions, microstrip line has stronger than stripline far-end crosstalk (FEXT) which harms signal integrity of the system interconnect. Microstrip line FEXT mitigation is critical for high-speed system card and board SI design. Microstrip line per-unit-length inductance and capacitance parameters are "unbalanced" namely the capacitive coupling is much smaller than the inductive coupling which causes FEXT. To make it more balance, additional mutual capacitance was introduced in single ended net application by using tabbed routing [3][4] which usually provides limited range of additional mutual capacitance and restricted by routing space and PCB manufacture technology limitation such as minimum wire spacing. Application of lumped capacitors in single ended net application was discussed briefly in [5]. In this paper, a technique by adding vendor surface mount (SMT) capacitors at the spacing between two adjacent microstrip lines is applied. By adding SMT capacitors with optimized capacitance value taking into account the SMT parasitic parameters and frequency-dependent *RLGC* per-unit length parameters of the solder-mask-coated microstrip line, the far-end crosstalk can be minimized that will make microstrip line structure a realistic channel physical layer option for high-speed system card and board design. A special treatment of differential net is also proposed by adding SMT capacitors only at the off-pair spacing so that the method can be used for both single-ended net and differential net applications.

In this paper, an *LC* balanced cell of single-ended net coupled microstrip lines made by adding an SMT capacitor is proposed. An equation for estimate the capacitance initial value of the SMT capacitor as a starting point value for capacitance optimization is proposed. A parameter sweep or optimization for mitigating FEXT on microstrip line cell cascading and arraying in linear network and transient simulations is proposed for practical single-ended and differential net applications in on-PCB signal channel design. Application examples are shown for 32Gb/s cases. Vendor SMT capacitor SPICE model including parasitic parameters will be used for linear network analysis and transient circuit simulations to optimize and verify the FEXT mitigation.

II. FEXT MITIGATION BY SMT CAPACITORS

In uniform microstrip lines, far-end crosstalk (FEXT) can be expressed by Equation 1 from [3].

$$V_{fext}(t)=0.5t_f(C_m/C_s-L_m/L_s)dV_i(t-t_f)/dt \qquad (1)$$

Where t_f is the time of flight, V_i is the aggressor input voltage, C_m, C_s, L_m and L_s are the per-unit-length mutual capacitance, self- capacitance, mutual inductance, and self-inductance, respectively.

Fig. 1 shows a two-coupled microstrip line section cell with length of l_{cell}. An SMT capacitor is placed at the center of the cell in between the two microstrip lines. Although we do not show here in Fig.1 some practical application details such as a solder-mask layer on top of the dielectric substrate and trapezoidal wire metal cross section shape, we do include these in the 2D per-unit-length *RLGC* modeling by using commercial electromagnetic simulation tools.

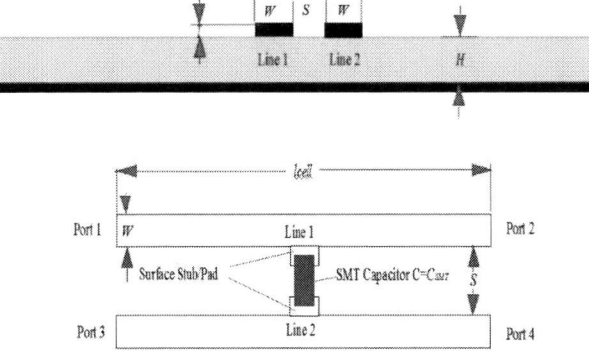

Fig.1 Two-coupled micristrip line section cell for signle-ended net application

978-1-7281-1500-9/19 $31.00 © 2019 IEEE

Based on Equation 1, Equation 2 can be derived for the optimal value of C_{SMT} of minimum FEXT. In practical case, the per-unit-length *RLGC* are frequency-dependent and the SMT capacitor has some parasitic parameters such as R_s, L_s, and R_p as shown in Fig.2 which come from a vendor Single Layer Capacitor [6]. We need to optimize the value of C_1 in Fig.2 by either parameter sweep or optimization in linear network or transient circuit simulations [7] based on the initial value by Equation 2.

$$C_{SMT} = (L_m C_s / L_s - C_m) \; l_{cell} \qquad (2)$$

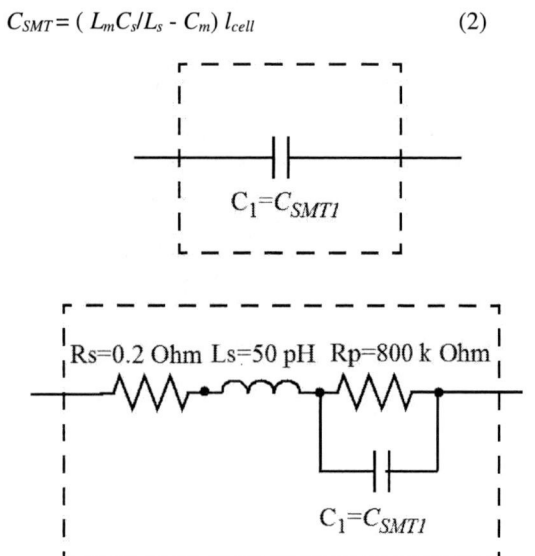

Fig.2 Ideal C_{SMT} (upper) and practical C_{SMT} (lower) with parasitic parameters

In practical single-ended net application, usually we use the two coupled microstrip line cell as a building block for cascading and arraying. The parameter sweep or optimization can also be performed on a larger array as show in Figs. 3 and 4 for more accurate modeling. The linear network simulation circuit schematic for a single-section length array which is a section of 5 cascaded sections in Fig.3 is shown by Fig.5.

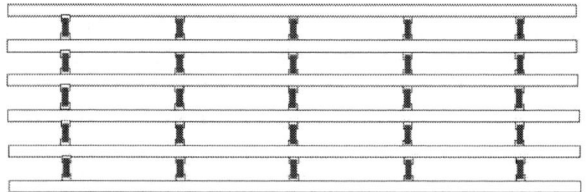

Fig.3 Array (lower) of coupled microstrip cells for single-ended net applications with grey rectangles representing SMT capacitors

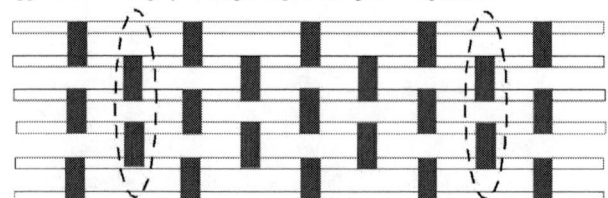

Fig.4 Offset array of two-coupled microstrip cells for single-ended net applications with SMT capacitors offset for larger physical sized capacitor (grey rectangles)

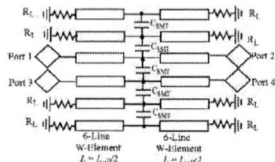

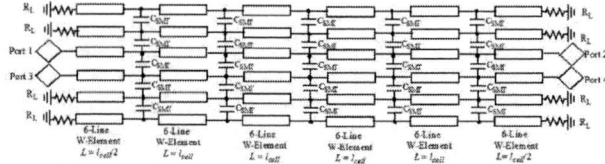

Fig.5 Linear network simulation circuit schematic for one-section six-line array and five-section six-line array based on 6-line W-Element model

For differential net application, we don't need to mitigate the FEXT between the two legs of a differential pair since the in-pair FEXT is a part of the differential signal transmission. We need to care about only the off-pair FEXT. The SMT capacitor is only needed at the off-pair spacing to mitigate the FEXT between differential pairs. Since differential net has intrinsic FEXT mitigation mechanism, Equation 2 does not apply directly to differential FEXT application. We have to use parameter sweep or optimization by linear network or transient simulation on a differential two-pair section cell as shown in Fig.6 or an array as sown in Fig.7 to get an optimal value of C_1 in C_{SMT}.

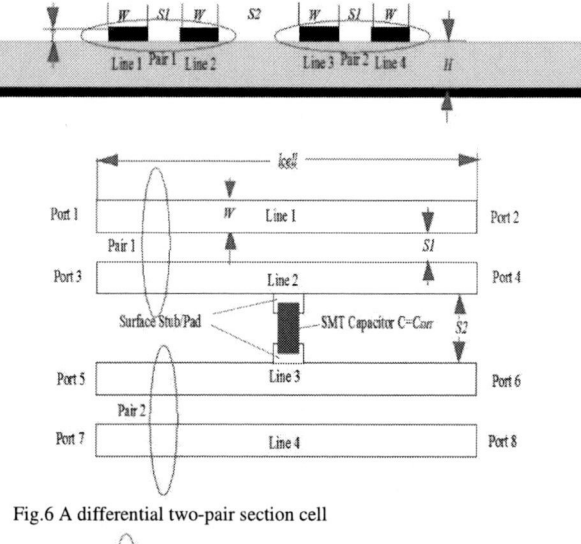

Fig.6 A differential two-pair section cell

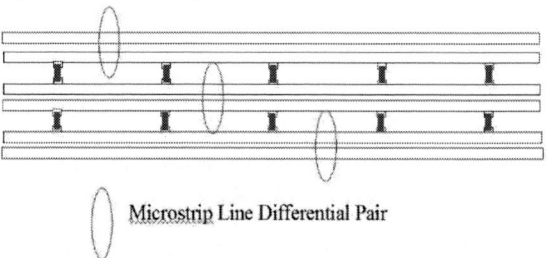

Fig.7 A differential net 3-pair 5-section cell array

978-1-7281-1500-9/19 $31.00 © 2019 IEEE

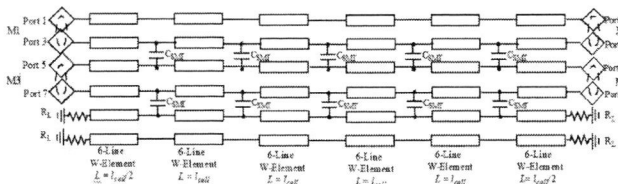

Fig.8 Circuit schematic of a differential two-pair section cell based on 6-line W-Element model

For some applications, we may need to use 3D electromagnetic simulation tool [7] [8] to deal with real 3D structures and some features implemented in 3D tools only such as materials with anisotropic dielectric constant[9]. We can use small length unit cells or the whole structure with individual SMT capacitors mounting ports such as Ports 5-9 in Fig.9 for interfacing to the SMT capacitors in parameter sweep or optimization for optimal FEXT. In Fig.9, the solder mask layer was set invisible for better view of the microstrip structure.

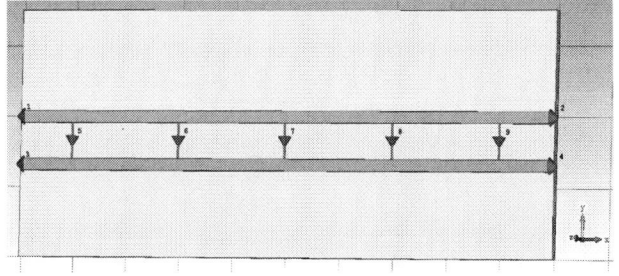

Fig.9 Modeling using 3D electromagnetic simulation tool with SMT capacitor mounting ports added

III.   APPLICATION SIMULATION EXAMPES

Two microstrip line application examples by using 2D electromagnetic simulation tool are given first, one for single-ended net application, the other differential net application. Then another example by using 3D electromagnetic simulation tool will be given.

A quasi-static 2D EM simulation tool is used to extract *RLGC* per-unit-length parameters based on frequency-dependent dielectric constant and loss tangent. The substrate thickness is 3.9mil with dielectric constant 3.94 at f=1MHz, loss tangent 0.016 at 20GHz. The solder mask thickness is 1.7mil with dielectric constant 4.3 at f=1MHz and loss tangent 0.037 at f=20GHz. The microstrip line width is 6.6mil with metal thickness of 2.0mil and spacing S_1=20.0mil for single-ended case, and S_1=5.4mil (in pair spacing) and S_2=20.0mil (off-pair spacing) for differential net case.

The 2D simulated per-unit-length self and mutual *LC* of Lines 3 and 4 of the 6-Line microstrip of the single-ended net case at f=10GHz are L_s=2.933nH/cm, L_m=0.117nH/cm, C_s=1.204pF/cm, and C_m=0.0135pF/cm. By Equation 2, the initial value for optimal C_{SMT} is 6.9fF for l_{cell}=2mm.

The parameter sweep results on C_{SMT1} in Fig.5 by linear network analysis of FEXT is shown in Fig.10. The ideal capacitors in Fig.5 are replaced by the vendor SMT capacitor models with parasitic parameters R_s, L_s and R_p as shown in Fig.2. The cell length is set to l_{cell}=2mm. The FEXT without the

SMT capacitor is also shown in Fig.10 as reference. It can be detected that the optimal value of C_{SMT1} is 7.0fF from FEXT point of view which is almost same as calculated by Equation 2 based on per-unit-length L and C parameters from quasi-static 2D electromagnetic simulation tool. However, the corresponding Near End Crosstalk (NEXT) is worse than the case without the SMT capacitors as shown in Fig.11. If the NEXT is more critical than FEXT to the signal integrity property in a design, we should not use the SMT capacitor to mitigate the FEXT.

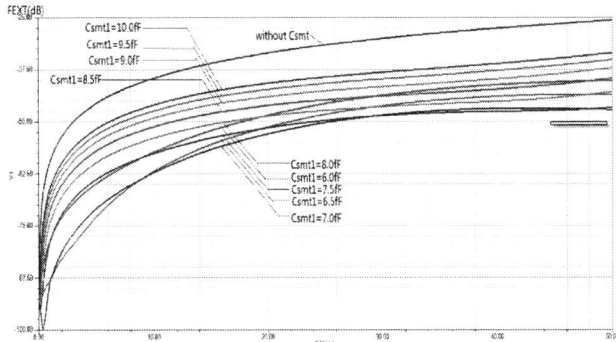

Fig.10 Adjacent wire frequency-domain FEXT of a 6-line cell by parameter sweep of C_{SMT1}

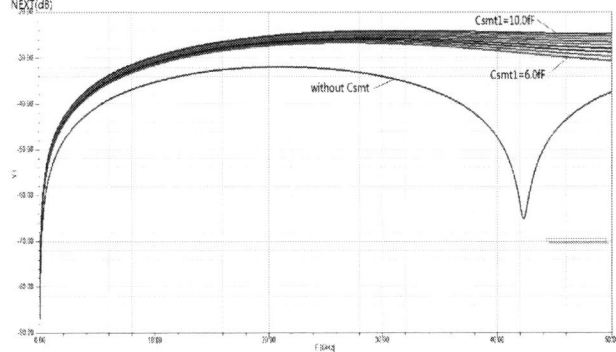

Fig.11 Adjacent wire frequency-domain NEXT of a 6-line cell by parameter sweep on C_{SMT1}

The parameter sweep on C_{SMT1} can also be performed on a multi-line multi-section microstrip cell array as shown in Figs. 3 and 4 instead of a single cell. Using multi-line multi-section micristrip array in parameter sweep will get more accurate results because of taking into account some complex interactions between multi lines and sections. Fig 12 shows the C_{SMT1} sweep to get optimum FEXT resulting in a value of 7fF which is the same as by the single cell parameter sweep in Fig.10. The corresponding NEXTs become worse than the case without the SMT capacitors as shown in Fig.13.

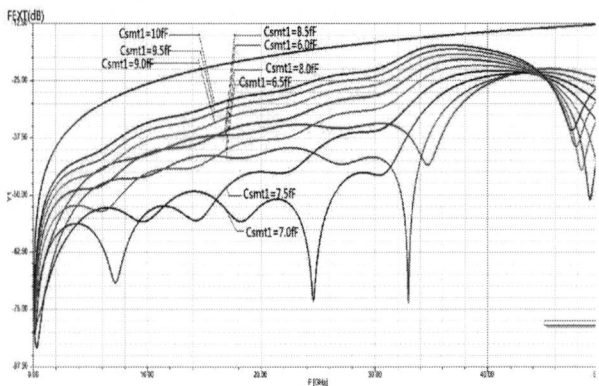

Fig.12 Adjacent wire frequency-domain FEXT of a 6-line 5-section array by parameter sweep of C_{SMT1}

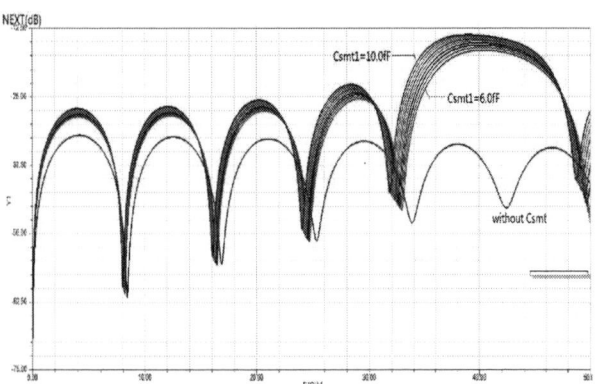

Fig.13 Adjacent wire frequency-domain NEXT of a 6-line 5-section array by parameter sweep of C_{SMT1}

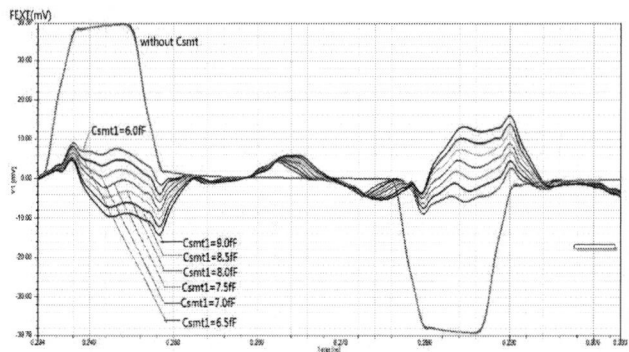

Fig.14 Transient adjacent wire FEXT of a 6-line 5-section array by parameter sweep of C_{SMT1}

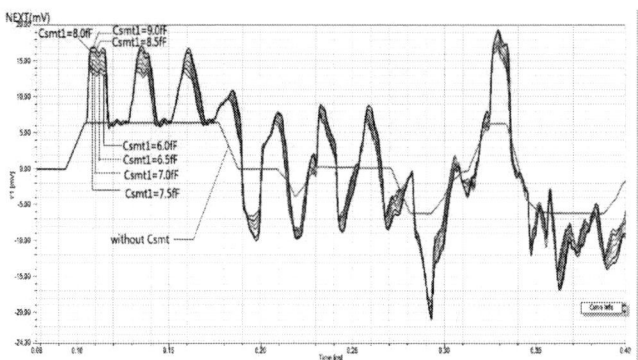

Fig.15 Transient adjacent wire NEXT of a 6-line 5-section array by parameter sweep of C_{SMT1}

For more direct estimation, the other option is to use transient simulation to observe the FEXT and NEXT in time domain by using a real time-domain signal with application bit rate, bit pattern, and signal rise/fall time. A lot of vendor signal integrity tools support the parameter sweep in transient simulations. We can make use of this kind of powerful feature. The FEXT at the receiver input location of a 32Gb/s application is displayed in Fig.14 with C_{SMT1} parameter sweep. We can easily see the optimal C_{SMT1} value is 7.0fF corresponding to minimum FEXT. In the time domain, we can also see NEXT become worse after adding the SMT capacitors (Fig.15). However, in most unidirectional signal channel applications, the NEXT is not sensitive to signal integrity properties.

In the above linear network and transient simulations with parameter sweep, the SMT capacitor vendor models are used including RL parasitic parameters. Using a "good" Single Layer Capacitor, the parasitic parameters Rs, Rp, and Ls do not significantly affect the FEXT and NEXT according the vendors models as shown in Fig.16.

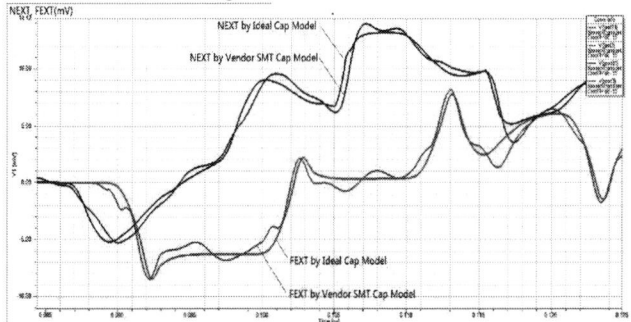

Fig.16 Transient adjacent wire FEXT and NEXT of a 6-line 5-section array by parameter sweep of C_{SMT1} based on ideal SMT capacitor model and vendor SMT capacitor model with parasitic RL.

As mentioned earlier in this paper, the multi-line multi-section array is preferred for higher accuracy. In Fig.17, the parameter sweep results are compared between using regular SMT capacitor array as shown in Fig.3 and the offset SMT capacitor array as shown in Fig.4. In the offset SMT capacitor array as shown in Fig.4, the circled capacitor groups have a larger capacitance value as $1.5C_{SMT1}$ because of longer section length $1.5l_{cell}$ they have. As we can see in Fig.17, the optimum capacitor parameter C_{SMT1} of the offset SMT array 6.0fF is different from the regular SMT array 7.0fF with respecting to minimum FEXT.

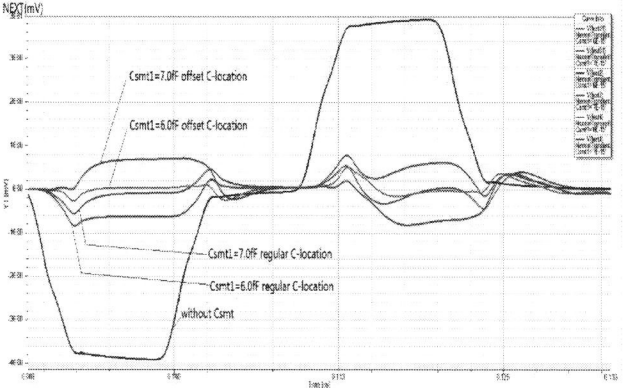

Fig.17 Transient adjacent wire FEXT of a 6-line 5-section array by parameter sweep of C_{SMT1} based on regular SMT capacitor array as in Fig.3 and offset SMT capacitor array as in Fig.4

For differential net applications, a similar procedure can be used by using SMT capacitors at off-pair spacing only. We don't provide an equation for estimating the initial capacitance value of the SMT capacitor. By parameter sweeping, the optimum value of C_{SMT1} can be found to mitigate the differential FEXT in time-domain or differential-to-differential FEXT in the frequency domain. Since the differential net has smaller FEXT by intrinsic crosstalk cancellation mechanism, the FEXT mitigation effect is smaller than single ended case. The parameter sweep linear network simulation results are shown in Fig.18. To make a clearer display, only optimum curve and reference curve of FEXT are shown in Fig.19. The corresponding NEXT curves are shown in Fig.20. Again, they become worse than the case without the SMT capacitors.

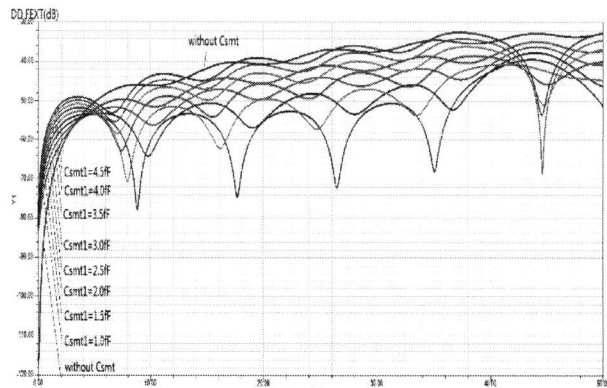

Fig.18 Adjacent pair frequency-domain differential-differential FEXT of a 6-line 5-section array by parameter sweep of C_{SMT1}

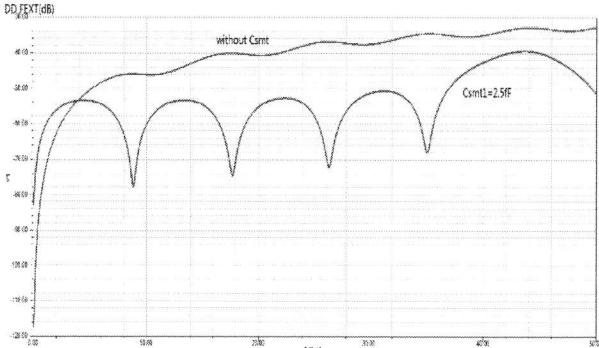

Fig.19 Adjacent pair frequency-domain differential-differential FEXT of a 6-line 5-section array by optimum parameter C_{SMT1}=2.5fF

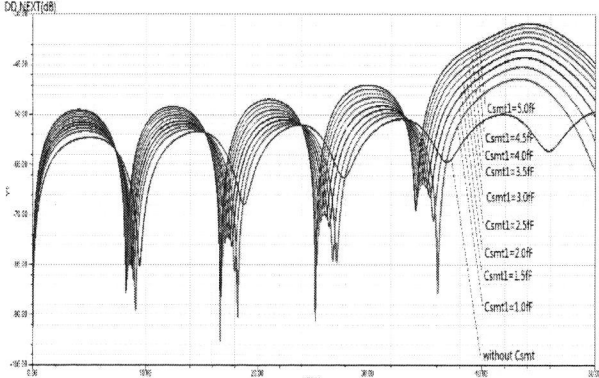

Fig.20 Adjacent pair frequency-domain differential-differential NEXT of a 6-line 5-section array by parameter sweep of C_{SMT1}

The parameter sweep in transient simulations can also be performed to detect the optimum capacitance value for the SMT capacitor placed in the off-pair spacing. Figs. 21-22 show the FEXT and corresponding NEXT in time domain, respectively for a 32Gb/s application. The optimum value of C_{SMT1} is 2.5fF with respect to minimum differential FEXT. Equation 2 cannot be used for estimating the initial value of C_{SMT1} because of the more complex crosstalk mechanism in the differential FEXT case. For example, if we consider Lines 2 and 3, the spacing is off-pair, the 2D simulated per-unit-length self and mutual LC at f=10GHz are L_s=2.90nH/cm, L_m=0.114nH/cm, C_s=1.233pF/cm, and C_m=0.0125pF/cm. By Equation 2, the estimated initial value for optimal C_{SMT} is 7.2fF for l_{cell}=2mm which has large difference from the actual optimal value of 2.5fF.

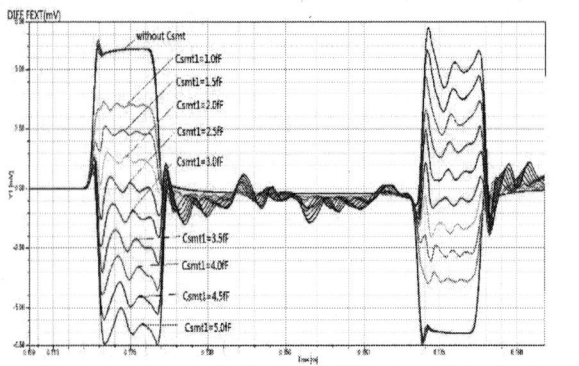

Fig.21 Transient adjacent pair differential FEXT of a 6-line 5-section array by parameter sweep of C_{SMT1}

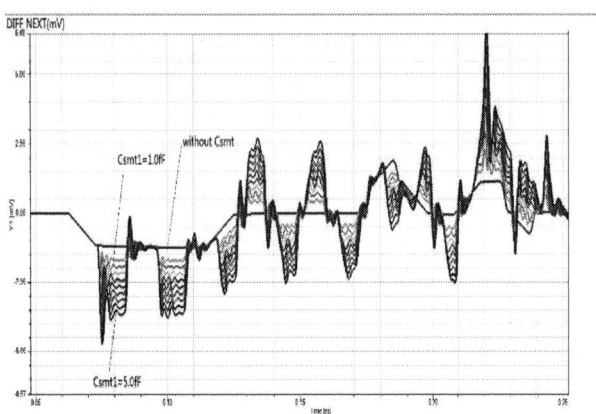

Fig.22 Transient adjacent pair differential NEXT of a 6-line 5-section array by parameter sweep of C_{SMT1}

The last application example takes into account an anisotropic substrate using a 3D electromagnetic simulation tool to derive the S-parameter model first. The 3D electromagnetic tool supports anisotropic dielectric material even with frequency dependent dielectric constant Dk and loss tangent Df. The microstrip structure is similar to the single-ended one used above except only two lines are included for simplicity as shown in Fig.9. By CST MWS we can make use of the anisotropic dielectric property with different Dk and Df for xy and z directions as discussed in [9]. In this example, $Dkxy$=3.91 and Dkz=3.69 at f=10GHz. By using 3D electromagnetic simulation tool, we are able to take more 3D features such as details of capacitor pad and nearby vias into account.

After the 9-port S-parameter model is derived by CST MWS, the five SMT capacitors are connected to Ports 5-9. By linear network analysis, the final 4-port S-parameter model can be calculated (Fig.24). For comparison, another structure with an isotropic dielectric substrate is also simulated for reference. The single-ended FEXT can be derived by S_{41} of the final 4-port S-parameter model.

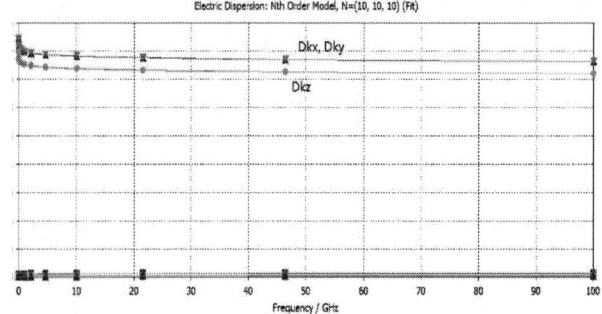

Fig.23 Frequency dependent property curve fitting of an anisotropic dielectric material by CST MWS

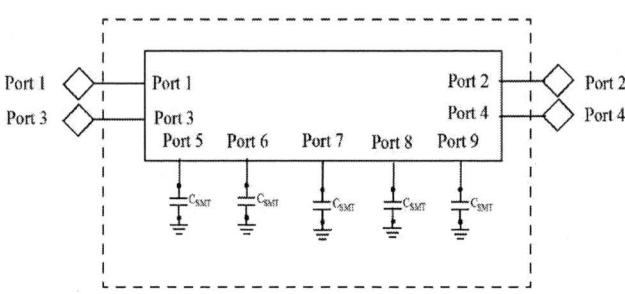

Fig.24 Linear Network Analysis setup for anisotropic and isotropic substrates by using 9-port S-parameter models with SMT capacitors connected to 5 of the 9 ports of each schematic

Four of the curves by parameter sweep results for minimizing FEXT are shown in Fig.27. We can see at f=16GHz, about 20dB FEXT mitigation can be realized by adding five 5fF SMT capacitors. We also see that using anisotropic dielectric material model does not have significant impact on the microstrip line FEXT modeling. This is consistent with [9].

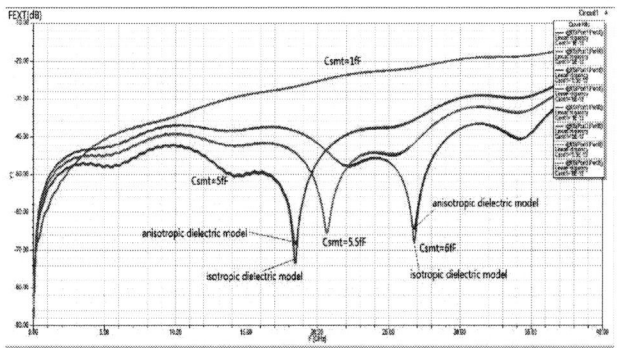

Fig.25 Parameter sweep of C_{smt} on FEXT of the microstrip line structure as shown in Fig.9

CONCLUSIONS

A technique by adding vendor SMT capacitors at the spacing between two adjacent microstrip lines is proposed to mitigate far-end crosstalk. By adding SMT capacitors with optimized capacitance value taking into account the SMT parasitic parameters and frequency-dependent solder-mask-coated microstrip line $RLGC$ per-unit length parameters, the far-end crosstalk can be minimized. This technique can also be applied to differential net applications by adding the SMT capacitors only at the off-pair spacing. The 3D electromagnetic simulation tool can also be used for some features the 2D tools do not have.

Potential application examples are shown for 32Gb/s cases. Vendor SMT capacitor SPICE model including parasitic parameters is used for linear network analysis and transient circuit simulations to optimize and verify the FEXT mitigation.

REFERENCES

[1] Gunter Kompa, *Practical Microstrip Design and Applications*, Artech House, 2005.

[2] Ramesh Garg, Inder Bahl, and Maurizio Bozzi, *Microstrip Lines and Slotlines* (3rd Ed.), Artech House, 2013.

[3] San K. Chhay, Richard K. Kunze, and Yunhui Chu, "Crosstalk Mitigation and Impedance Management Using Tabbed Lines," in *Proc. 2013 IEEE 22nd Conference on Electrical Performance of Electronic Packaging and Systems*, pp.231-234, 2013.

[4] Nevin Altunyurt, Kemal Aygun, Kevin J.Dorran, and Yidnekachew S.Mekonnen, *Mitigation of Far-End Crosstalk Induced by Routing and Out-of-Plane Interconnects*, US Patent Application US 2014/0203417 A1.

[5] Bao-Ren Huang, Kuan-Chung Chen, and Chun-Long Wang, "Far-end crosstalk noise reduction using decoupling capacitor, " *IEEE Trans. on Electromagnetic Compatibility*, vol. 58, no.3, 2016, pp.836-848.

[6] Knowles, www.knowlescapacitors.com

[7] ANSYS, *Electronics Desktop*, https://www.ansys.com

[8] CST, *CST Microwave Studio*, https://www.cst.com/Products/CSTMWS

[9] Biyao Zhao, Zhaoqing Chen, and Dale Becker, "Impacts of anisotropic permittivity on PCB trace and via modeling," in *Proc. 2018 IEEE 27th Conference on Electrical Performance of Electronic Packaging and Systems*, pp.39-41, 2018.

2019 IEEE 69th Electronic Components and Technology Conference (ECTC)

New Cost-effective Via-last Approach by "One-step TSV" after Wafer Stacking for 3D Memory Applications

Masaya Kawano*, Xiang-Yu Wang, and Qin Ren

Institute of Microelectronics

A*STAR (Agency for Science, Technology and Research)

2 Fusionopolis Way, #08-02, Innovis Tower, Singapore 138634

*e-mail: kawanom@ime.a-star.edu.sg

Abstract—A new cost-effective via-last through silicon via (TSV) process after multiple wafer stacking named "one-step TSV" is proposed and developed for 3D stacked memory module applications. According to cost modeling, one-step TSV could reduce more than 50% process cost compared to other stacking technology using hybrid bonding with conventional TSV. 4-wafer-stack fusion bonding combining face-to-face and back-to-back bonding was successfully demonstrated without any interfacial voids. Good TSV connections at the sidewall of the Al pads were also confirmed by electrical test. This cost-effective 3D integration approach has huge potential to expand application area not only for stacked DRAM but also stacked NAND flash, Memory/Logic 3D-SiP, 3D-FPGA, and image sensor devices etc.

Keywords- One-step TSV; Wafer-to-Wafer bonding; 3D integration; 3D stacked memory; Cost modeling

I. INTRODUCTION

For graphic processing systems as well as other high performance computing systems, it is strongly desired to improve performance without increasing its power consumption or operating frequency, because high frequency operation needs high power consumption and generates cooling problems. To improve memory access bandwidth with power consumption efficiency, wide IO memory bus is necessary instead of high frequency interface. In addition, larger memory capacity is required along with system performance improvement. Because of that, several types of high bandwidth memories have been proposed and introduced to the market.

In order to achieve large memory capacity and high bandwidth, through silicon vias (TSVs) are used to enable 3D stack of DRAM chips and inter-chip wide bus connections [1, 2]. For example, High Bandwidth Memory (HBM) has been introduced to the market in particular graphic accelerator market. However, this kind of memory product is implemented to only limited area such as high-end computing, due to its expensive process cost necessary to fabricate TSVs and multi-chip stacking using micro-bump bonding.

From process cost point of view, wafer to wafer stacking approach is more attractive than chip to chip stacking, since multiple chips can be 3D stacked simultaneously at a wafer level. In addition, fabrication cost of micro-bumps and TSVs occupy large portion of assembly cost [3, 4]. A cost model result shows fabrication costs of front/back side micro-bumps and TSV per 300 mm wafer are around US$230 and

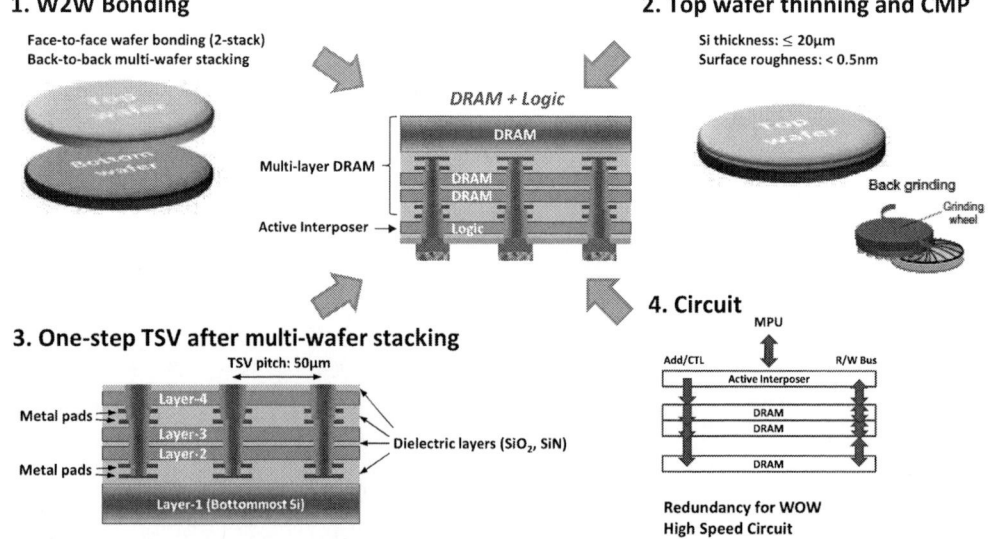

Figure 1. Cost-effective 3D integration technology for high speed 3D-DRAM proposed in this paper. Wafer-level 3D integration is done combining with one-step TSV fabrication process after multiple wafer stacking.

US$430, respectively. To reduce the assembly cost without scarifying device performance, bump-less wafer stacking process [5] is more attractive for low-cost solutions. Even bump-less wafer bonding technologies, however, still require TSVs for each layer. In addition, two or more TSVs might be required to connect different layers each other in case that TSVs connect landing pads in different layers after wafer stacking. Furthermore, the metal patterns of landing pads should be located at different location. Each device wafer should have different metal design for this purpose. These design limitations consume effective device area resulting higher cost.

To solve aforementioned problems, a via-last (VL) TSV process after wafer stacking has been proposed recently [6]. The new process combines face-to-face wafer-to-wafer (W2W) bonding and VL-TSV after top wafer thinning. By making a hole in an upper layer metal pad in advance, one TSV is able to connect two different layers simultaneously. This TSV concept, however, has low tolerance for wafer bonding misalignment, because VL-TSV etching needs to hit exact location of the hole, i.e. "hole-in-one." In addition, the TSV hole diameter will be significantly reduced at the hole position depending on misalignment, which limits allowable via depth below the upper pad.

In this paper, a new cost-effective TSV concept which we call "one-step TSV" is proposed and demonstrated. The new process fabricates TSV at a time after multiple wafer stacking. During TSV dry etching, multiple layers will be punched through including metal pads. After dry etching and metallization, the one TSV electrically connects multiple metal layers. The target application is stacked DRAM for this study. The one-step TSV has potential to expand application area not only for high-end 3D-DRAM but also middle range to low-end DRAM because of its cost-effectiveness.

II. Concept of New Cost-Effective 3D Integration Technology

A schematic illustration is shown in Fig. 1 to explain new 3D integration concept. It consists of multiple W2W stacking which is combination of face-to-face (F2F) wafer bonding for 2-stack and back-to-back (B2B) wafer bonding for further multiple stack, thinning and polishing of bonded wafers, one-step TSV fabrication after multi-wafer stacking, and topmost metallization such as micro-bumps or Al pads. Most important aspect of this concept is that TSV fabrication is not required for each layer. The larger the number of layers is, the more cost reduction is realized. Of course TSV hole dry etching will be challenging in this case, because multiple layers including Si, intermetallic dielectric layers, and metal layers are needed to be etched away. From etching point of view, thinner layer thickness is desired. We are targeting 20 μm or even thinner Si thickness. Another important aspect is this fabrication process does not require thin wafer handling technique such as temporary bonding and debonding, which is commonly used for TSV process but increases the fabrication cost much. Even after top wafer thinning, the thick bottom wafer supports whole structure

throughout the process. The wafer stack can be handled almost same as a standard Si wafer.

Yield is commonly known issue for all wafer-level 3D integration technologies. Redundancy circuits are necessary to avoid this yield issue. Certain minor DRAM failures such as single bit/word and single column/row/bank failures could be repaired. The redundancy circuits replace the failure with the associated redundant one. In addition, active interposer layer would be necessary especially for high-speed DRAM application. The active interposer, actually CMOS logic device layer, could be placed topmost layer (layer-4) which is nearest position to external connection. The layer acts as interface, e.g. between main processing unit (MPU) and DRAM core.

All device layers must have Al top layer in this concept, because the metal pads will be dry etched during TSV fabrication process. Because Al top pad is most commonly used for any types of CMOS devices, this will not be a problem. In addition, the Al pads which are used for 3D interlayer connection, need to be designed in a mirror-image layout between odd number and even number layers.

III. Process Flow of One-step TSV

Fig. 2 shows proposed 3D wafer-level integration flow in this study. TSV is fabricated after multiple wafer stacking, which we call "one-step TSV." This enables significant cost reduction for TSV fabrication as mentioned. There are 4 types of wafers from layer-1 to layer-4. Top Al pads which will be used for 3D interlayer connection need to be aligned after F2F and B2B wafer bonding. This could be possible even if same wafers are stacked as far as the top Al pad layout is mirror image. Before bonding, the Al pads should be covered by SiO$_2$ layer and planarized by CMP to enable fusion bonding afterwards. The top and bottom wafers are directly bonded at a wafer level in face to face manner. After bonding and annealing, top Si wafer is thinned by backgrinding. Then ring trench via isolation is fabricated. Because TSV isolation is done before via hole etching, there is no metal contamination risk into Si substrate during one-step TSV etching process. In addition, parasitic capacitance of TSV can be reduced compared to conventional isolation. After multiple wafer staking process by B2B wafer bonding and wafer thinning, via holes are formed by dry etching.

More detail 4-layer 3D integration flow step-by-step is explained as follows:

1) Depositing SiO$_2$ on all of incoming wafers. The film thickness is depending on initial topography of the wafers. Oxide CMP follows after deposition to planarize the top surface suitable for fusion bonding;

2) F2F fusion bonding of the top and bottom wafers. Plasma activation and DIW cleaning are done as pre-bond treatments to obtain enough bond strength even at low temperature post-bond annealing less than 400 °C. After bonding and annealing, thinning backside of Si substrate is done for the top wafer by backgrinding and Si CMP;

3) Depositing SiO$_2$ as passivation on the backside of Si, Si trench fabrication for TSV isolation. Si trench is filled by SiO$_2$, and oxide CMP follows to make flat surface for further fusion bonding;

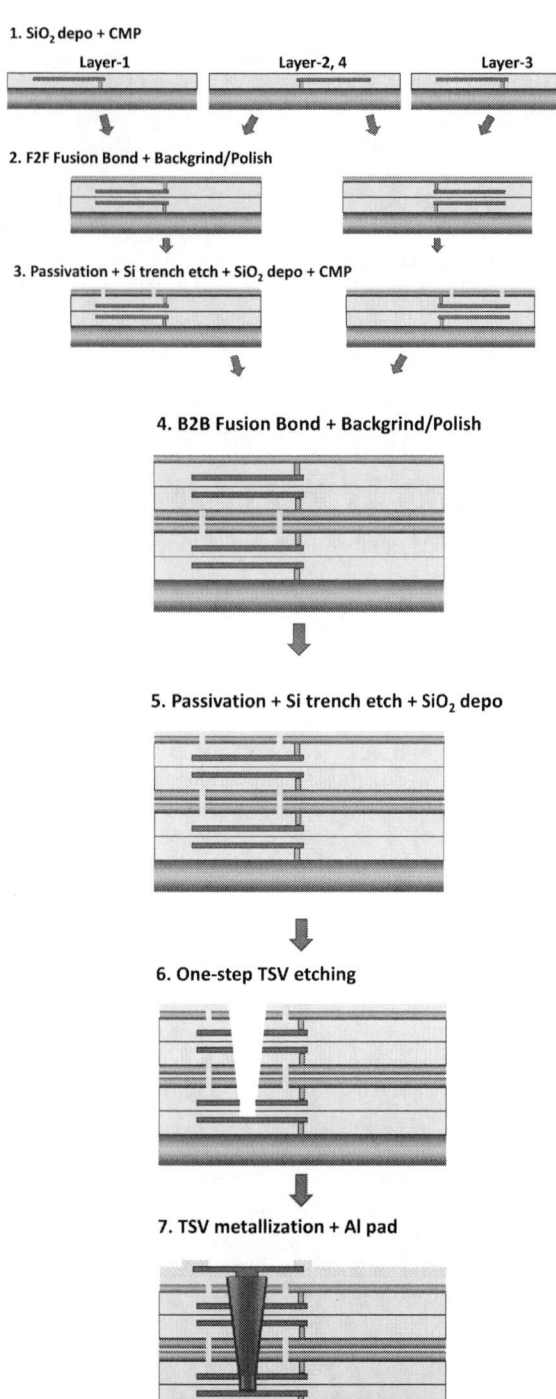

1. SiO$_2$ depo + CMP

Layer-1   Layer-2, 4   Layer-3

2. F2F Fusion Bond + Backgrind/Polish

3. Passivation + Si trench etch + SiO$_2$ depo + CMP

4. B2B Fusion Bond + Backgrind/Polish

5. Passivation + Si trench etch + SiO$_2$ depo

6. One-step TSV etching

7. TSV metallization + Al pad

Figure 2. W2W stacking and one-step TSV process integration flow. Si trench is fabricated from the backside of each layer for the purposes of metal contamination protection, reduction of paracitic capacitance and B2B bond alginment.

4) B2B fusion bonding to make 4-layer stack. After bonding and annealing, top wafer thinning is done by backgrinding and Si CMP same as step-2;

5) Passivation and Si trench fabrication are done as step-3. Oxide CMP will not be necessary if there is no further stacking;

6) One-step TSV etching. This etching process consists of multiple etching steps of Si, dielectric layers, and metal layers;

7) TSV metallization follows such as barrier/seed PVD, Cu filling by electroplating, and Cu CMP. Finally top electrodes are fabricated for testing and packaging.

IV. COST COMPARISON OF WAFER-LEVEL 3D INTEGRATION

There are several types of wafer-level 3D integration which have been proposed. In this section, process cost comparison is done for three types of 3D integration flows: a) One-step TSV combining with 4-layer wafer stacking by fusion bonding (this study); b) VL-TSV for each layer combining with hybrid bonding; c) Via-middle (VM)-TSV for each layer combining with hybrid bonding as shown in Fig. 3.

Top Al pads are fabricated for all types of 3D integration process. Top metallization will not affect cost comparison results regardless of final metallization such as Al pads, UBM or solder bumps etc. Production volume of 30k wafers/months is assumed for the calculations. Other cost components such as labor cost, tool depreciation, machine time and operation time are taken into account in our standard cost model.

Cost comparison for above 3 different 3D integration flows were summarized in Fig. 4. The results showed that one-step TSV could achieve less than half of the manufacturing cost compared to other W2W 3D integration methods. Bonding cost alone is same for fusion bonding and

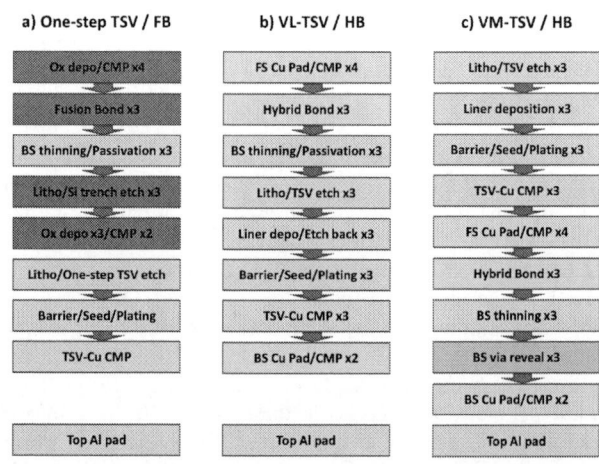

Figure 3. Four layer W2W 3D stacking flows used for cost modeling: a) One-step TSV combining with fusion bonding (this study), b) VL-TSV combining with hybrid bonding, c) VM-TSV combining with hybrid bonding. This chart does not represent exact process sequence, because some of process modules are repeated to make multple stack.

978-1-7281-1500-9/19 $31.00 © 2019 IEEE   1998

hybrid bonding. However, process cost for front side (FS) and backside (BS) Cu pads significantly increases total 3D integration cost for hybrid bonding.

We also extracted only TSV process cost portion and compared it for 3 types of 4-layer 3D stacking process (Fig. 5). Large portion of process cost for hybrid bonding is occupied by FS and BS Cu pad fabrication as mentioned before. Although one-step TSV cost is higher than that of

VL-TSV and VM-TSV per layer and Si trench fabrication is necessary for each layer on top of that, total TSV fabrication cost is still less than 40% of other 2 technologies.

Cost components of 4-layer stacking and one-step TSV process were also analyzed as shown in Fig. 6. Largest portion is for dry etching process which is come from one-step TSV etching. Other large cost components are lithography, CMP, PVD/Electroplating, wet cleaning in descending order.

Fig. 7 shows cost analysis of one-step TSV fabrication only, which includes passivation CVD, lithography, TSV etching, barrier/seed sputtering, TSV Cu filling, and CMP. The results clearly show that dry etching step is most expensive part. This costs more than half of one-step TSV process cost. This is mainly due to long time etching for thick inter-metal dielectric layers. Oxide dry etching takes time in general, and high power is required compared to Si dry etching. If high speed dielectric dry etching is realized, we can expect drastic process cost reduction for one-step TSV.

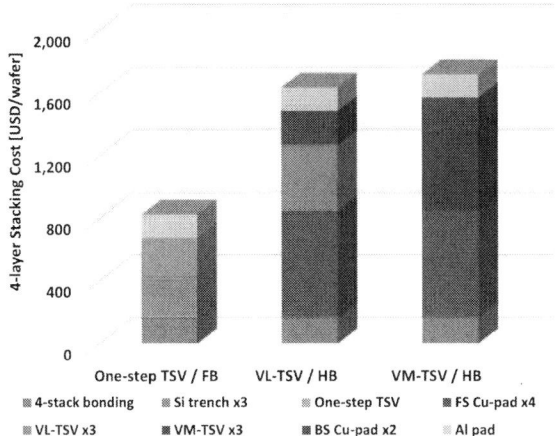

Figure 4. 3D integration cost comparison for 3 types of 4-layer stacking technologies. One-step TSV cost shows 50% cost reduction compared to other W2W 3D integration meothods.

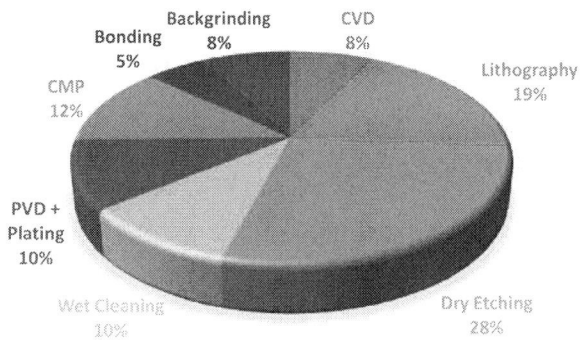

Figure 6. 3D integration cost components for 4-layer stacking and one-step TSV.

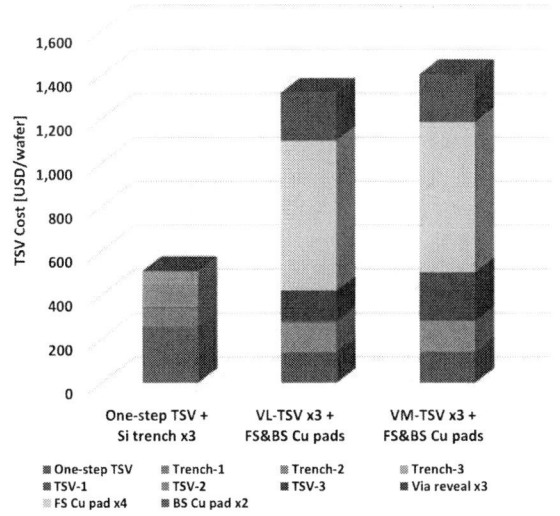

Figure 5. TSV process cost comparison for 3 types of 4-layer 3D stacking technologies. Large portion of cost for hybrid bonding is occupied by FS and BS Cu pad fabrication. Although one-step TSV cost is higher than that of VL-TSV and VM-TSV per layer and Si trench fabrication is necessary for each layer, total TSV fabrication cost is less than 40% of other 2 technologies.

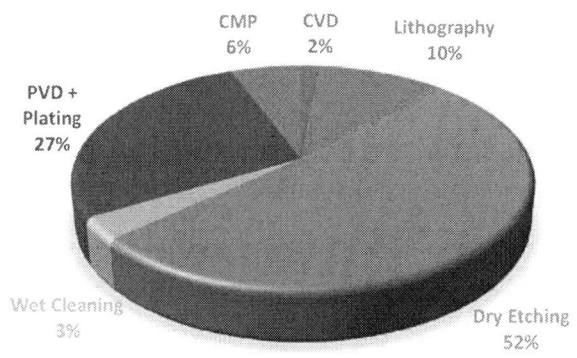

Figure 7. Cost components of one-step TSV only, which includes passivation CVD, lithography, TSV etching, barrier/seed sputtering, TSV Cu filling, and CMP.

978-1-7281-1500-9/19 $31.00 © 2019 IEEE

V. TEST VEHICLE

A test vehicle was designed and fabricated for 4-layer stacking evaluation as well as one-step TSV process development. The test vehicle design was aligned to stacked DRAM requirements. Minimum TSV pitch and diameter range are 51.2 μm and 10 ~ 20 μm, respectively. The connection pads are designed at the same location regardless of layer number in the stack. One Al layer with 1 μm

thickness was fabricated, where the metal layer was used for one-step TSV connection pads and electrical test structure such as daisy chains. The thickness of inter-metal dielectric layer was 8 μm, which was set as same thickness of DRAM devices which we are targeting. The Si thickness was targeted 20 μm after bonding and thinning in this study.

VI. FOUR LAYER WAFER STACKING

The front surface of incoming wafers need to be carefully planarized by CMP for subsequent fusion bonding. This

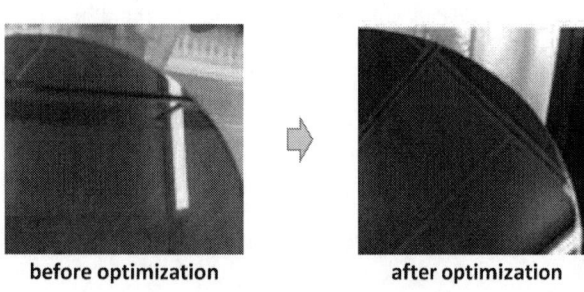

before optimization   **after optimization**

Figure 8. Wafer inspection images after backgrindinng/polishing and passivation. Standard backgrinding and polishing process generated waveness so called "grinding marks." Optimization for backgrinding/polishing is necessary for B2B fusion bonding.

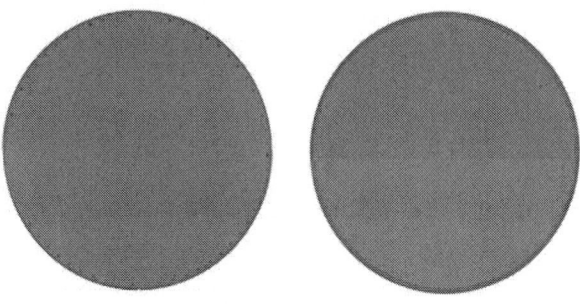

Figure 9. Through-scan SAM images after 4-layer wafer stacking. In the case of insufficient backgrinding/polishing optimization, periodical voids appeared at wafer edge (left).

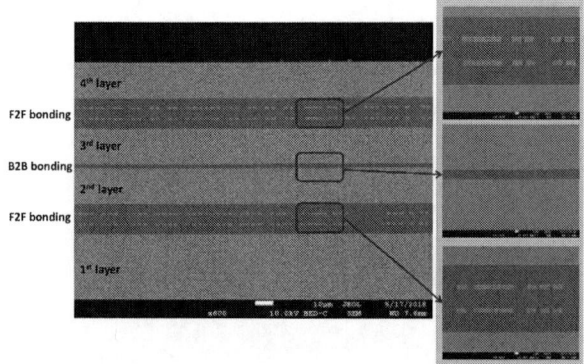

Figure 10. Four layer wafer stacking cross section SEM images. There are 3 bond interfaces in total, which are two F2F bond interfaces and one B2B bond interface. According to cross sectional images, no interfacial voids were observed for all bond interfaces.

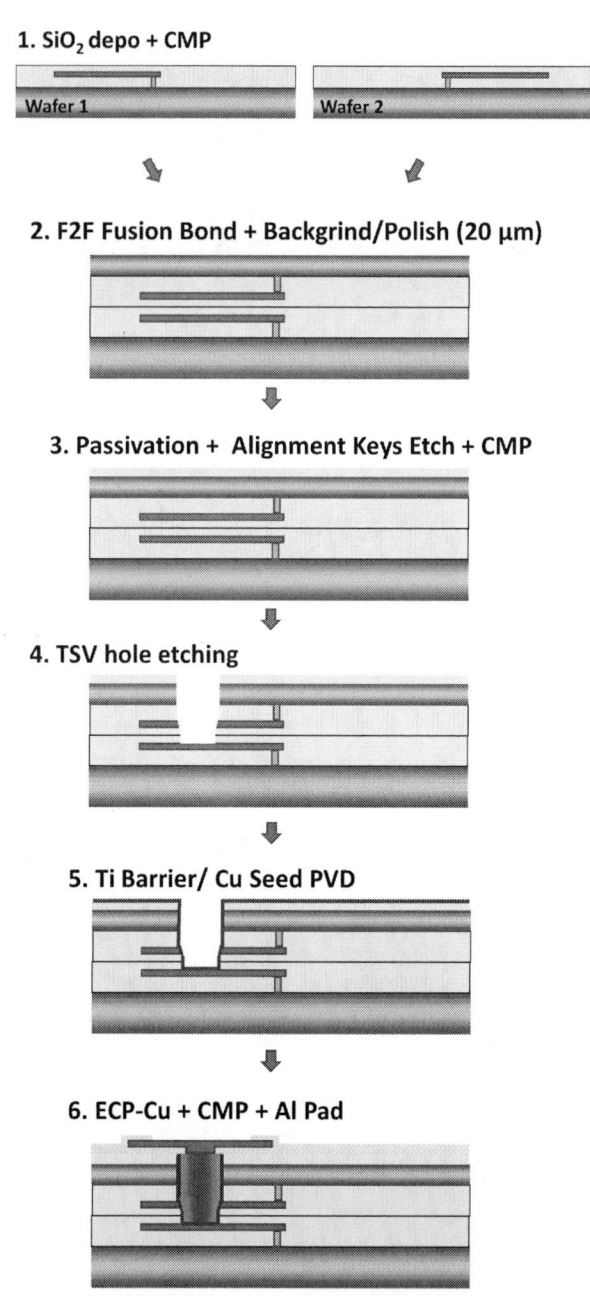

Figure 11. Process flow for one-step TSV process evaluation with 2-layer stack structure.

planarization technique for F2F fusion bonding has been intensively studied for long time in this field. A new challenge for 4-layer stacking is B2B bonding. In general, a backgrinder is not designed for B2B fusion bonding, and creates grinding marks which are with periodical and radial macro scale waviness on background Si surface. Fig. 8 shows images after backgrinding/polishing and passivation. With standard backgrinding and Si polishing processes, the grinding marks were clearly generated. The fusion bonding of these wafers were not successful because of grinding marks. We optimized both of backgrinding and Si polishing processes, and obtained grinding mark-less results as shown in Fig. 8. After this optimization, B2B fusion bonding was done successfully. Good bond quality was obtained by through-scan result of scanning acoustic microscopy (SAM) except for minor particle related voids (Fig. 9). In the case of insufficient backgrinding/polishing optimization, periodical voids appeared at wafer edge as shown in left image of Fig. 9, which is attributed to remaining grinding marks. The through-scan images contain all bond interface information.

Using optimized backgrinding and polishing conditions, 4-layer wafer stacking of test vehicle wafers by fusion bonding was demonstrated. Fig. 10 shows cross sectional SEM images of 4-layer wafer stacked wafers. There are 3 bond interfaces in total, which are two F2F bond interfaces and one B2B bond interface. According to cross sectional images, no interfacial voids were observed for all bond interfaces.

VII. ONE-STEP TSV FABRICATION FOR 2-LAYER STACK

To confirm process feasibility of the new 3D integration scheme, we developed one-step TSV process for 2-layer stack structure first. Process flow of this evaluation is shown in Fig. 11. Si trench was not included in the evaluation to focus on TSV fabrication development.

By the nature of one-step TSV structure, the contact area at the connection pad is small except for bottom-most pad. Sidewall angle of Al pads after dry etching is required to be carefully controlled so that Al oxide can be properly removed by pre-cleaning process of barrier PVD. The sidewall angle is preferably positive angle, and overhang angle is not accepted. The cross sectional SEM images are shown in Fig. 12. We could achieve sidewall angle of 70 ~ 80 deg. by optimizing Al etching.

Electrical test was conducted by using Kelvin test structures for 15 μm and 20 μm diameter one-step TSVs. The obtained resistance values were consistent with theoretical values which were calculated based on Cu and Al resistivity (Fig. 13). The barrier layer resistance was not considered for resistance calculation, because its resistance is negligible small. This clearly indicates that one-step TSV process is robust from electrical connection point of view. Electrical connection yield was also evaluated for double stack layer by using 100 TSV daisy chains. Although open failures were not observed for both of 15 μm and 20 μm diameter one-step TSVs, higher resistance was observed at wafer edge. This indicates there is still room for improvement especially for dry etching process.

VIII. SUMMARY AND CONCLUSIONS

We proposed new cost-effective via-last TSV process "one-step TSV" in which TSVs are fabricated at a time after multiple wafer stacking. Cost modeling results showed the one-step TSV could reduce more than 50% process cost from stacking technology using hybrid bonding with VL or VM TSV. There is large room for further cost reduction, if high speed dielectric etching is enabled.

A test vehicle was designed and fabricated for 4-layer stacking evaluation and 4-wafer-stack fusion bonding was successfully demonstrated without any interfacial voids. Optimizing backgrinding and polishing process after F2F stacking is one of keys to success. Regarding TSV, side wall angle of intermediate Al layer was confirmed within range of 70 ~ 80 deg with positive angle, which enables stable electric contact by Ar pre-cleaning and subsequent PVD. The TSV resistance showed almost same as theoretically estimated values. 100% electrical connection was confirmed for double stack layer by 100 TSV daisy chains. We are working on 4-stack 3D process development with one-step TSV by using DRAM device wafers.

The proposed new TSV concept can be applied not only to stacked DRAM but also stacked NAND flash, Memory/Logic 3D-SiP, 3D-FPGA, and image sensor devices etc.

REFERENCES

[1] Masaya Kawano, Nobuaki Takahashi, Yoichiro Kurita, Koji Soejima, Masahiro Komuro, and Satoshi Matsui, "Three-Dimensional Packaging Technology for Stacked DRAM With 3-Gb/s Data Transfer," IEEE Trans. Electron Devices, Vol. 55, No. 7, July 2008, pp. 1614–1620.

[2] Uksong Kang, Hoe-Ju Chung, Seongmoo Heo, Soon-Hong Ahn, Hoon Lee, Soo-Ho Cha, Jaesung Ahn, DukMin Kwon, Jin Ho Kim, Jae-Wook Lee, Han-Sung Joo, Woo-Seop Kim, Hyun-Kyung Kim, Eun-Mi Lee, So-Ra Kim, Keum-Hee Ma, Dong-Hyun Jang, Nam-Seog Kim, Man-Sik Choi, Sae-Jang Oh, Jung-Bae Lee, Tae-Kyung Jung, Jei-Hwan Yoo, Changhyun Kim, "8Gb 3D DDR3 DRAM Using Through-Silicon-Via Technology," in Proc. International Solid-State Circuits Conference (ISSCC), Feb 2009, pp. 130–131.

[3] K.-J. Chui, H.-Y. Li, Ka-Fai Chang, Surya Bhattacharya, Mingbin Yu, "A Cost Model Analysis Comparing Via-Middle and Via-Last TSV Processes," in Proc. 17th Electronics Packaging Technology Conference (EPTC), Dec 2015, pp. 1–4.

[4] King-Jien Chui, Woon Leng Loh, Chunmei Wang, Ka-Fai Chang, Qin Ren, Gilho Hwang, Hung-Ming Chua, Mingbin Yu, "A Cost-Effective, CMP-Less, Via-Last TSV Process for High Density RDL Applications," in Proc. 66th Electronic Components and Technology Conference (ECTC), May 2016, pp. 277–282.

[5] Takayuki Ohba, Youngsuk Kim, Yoriko Mizushima, Nobuhide Maeda, Koji Fujimoto, and Shoichi Kodama, "Review of wafer-level three-dimensional integration (3DI) using bumpless interconnects for tera-scale generation," IEICE Electronics Express, Vol. 12, No. 7, April 2015, pp. 1–14.

[6] Joeri De Vos, Stefaan Van Huylenbroeck, Anne Jourdain, Nancy Heylen, Lan Peng, Geraldine Jamieson, Nina Tutunjyan, Stefano Sardo, Andy Miller, Eric Beyne, " "Hole-in-one TSV", a new via last concept for high density 3D-SOC interconnects," in Proc. 68th Electronic Components and Technology Conference (ECTC), May 2018, pp. 1499–1504.

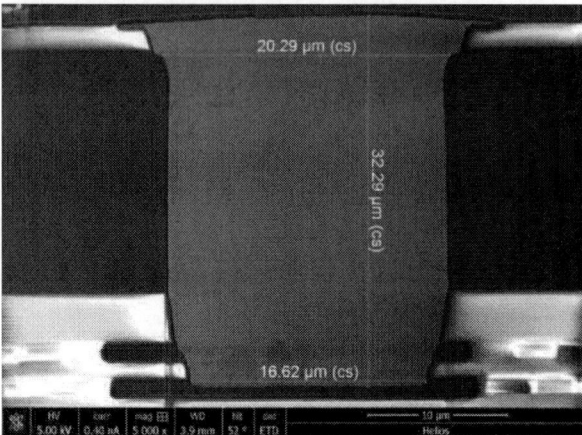

Figure 12. SEM cross sectional images for one-step TSV for 15 μm (top) and 20 μm (bottom) diameters.

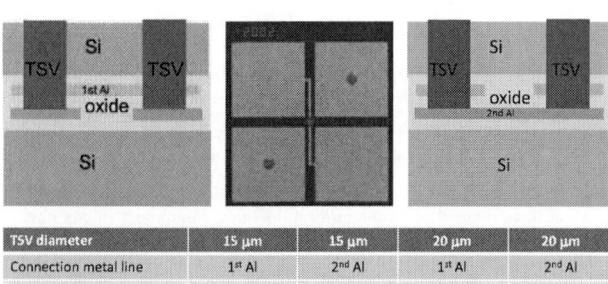

TSV diameter	15 μm	15 μm	20 μm	20 μm
Connection metal line	1st Al	2nd Al	1st Al	2nd Al
Resistance [mΩ]	5.5	4.8	3.2	3.7
Theoretical value [mΩ]	3.4		1.9	

Figure 13. Resistance of one-step TSV measured by Kelvin test structures for 15 μm and 20 μm (bottom) diameters.

(a)   Daisy chains by one-step TSV and Al layer.

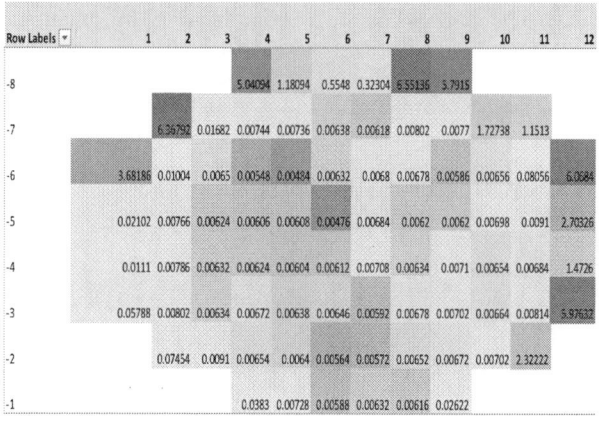

Row Labels	1	2	3	4	5	6	7	8	9	10	11	12
-8				5.04094	1.18094	0.5548	0.32304	6.55136	5.7915			
-7		6.36792	0.01682	0.00744	0.00736	0.00638	0.00618	0.00802	0.0077	1.72738	1.1513	
-6	3.68186	0.01004	0.0065	0.00548	0.00484	0.00632	0.0068	0.00678	0.00586	0.00656	0.08056	6.0684
-5	0.02102	0.00766	0.00624	0.00606	0.00608	0.00476	0.00684	0.0062	0.0062	0.00698	0.0091	2.70326
-4	0.0111	0.00786	0.00632	0.00624	0.00604	0.00612	0.00708	0.00634	0.0071	0.00654	0.00684	1.4726
-3	0.05788	0.00802	0.00634	0.00672	0.00638	0.00646	0.00592	0.00678	0.00702	0.00664	0.00814	5.97632
-2		0.07454	0.0091	0.00654	0.0064	0.00564	0.00572	0.00652	0.00672	0.00702	2.32222	
-1			0.0383	0.00728	0.00588	0.00632	0.00616	0.02622				

(b)   Chain resistance map for 15 μm diameter TSV

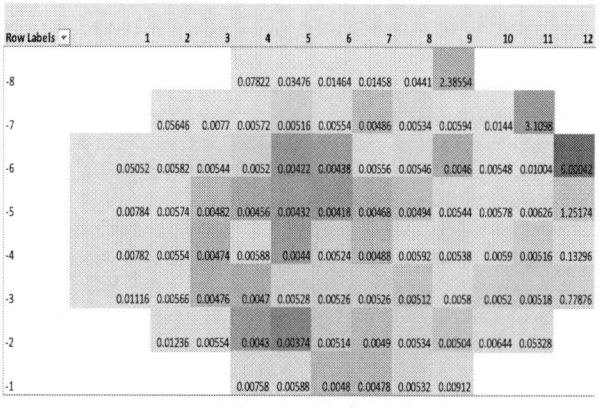

Row Labels	1	2	3	4	5	6	7	8	9	10	11	12
-8				0.07822	0.03476	0.01464	0.01458	0.0441	2.38554			
-7		0.05646	0.0077	0.00572	0.00516	0.00554	0.00486	0.00534	0.00594	0.0144	3.1098	
-6	0.05052	0.00582	0.00544	0.0052	0.00422	0.00438	0.00556	0.00546	0.0046	0.00548	0.01004	6.00042
-5	0.00784	0.00574	0.00482	0.00456	0.00432	0.00418	0.00468	0.00494	0.00544	0.00578	0.00626	1.25174
-4	0.00782	0.00554	0.00474	0.00588	0.0044	0.00524	0.00488	0.00592	0.00538	0.0059	0.00516	0.13296
-3	0.01116	0.00566	0.00476	0.0047	0.00528	0.00526	0.00526	0.00512	0.0058	0.0052	0.00518	0.77876
-2		0.01236	0.00554	0.0043	0.00374	0.00514	0.0049	0.00534	0.00504	0.00644	0.05328	
-1			0.00758	0.00588	0.0048	0.00478	0.00532	0.00912				

(c)   Chain resistance map for 20 μm diameter TSV

Figure 14. Daisy chain structure and resistance wafer maps of one-step TSV for 15 μm and 20 μm diameters.

Microstructure and Property Changes in Cu/Sn-58Bi/Cu Solder Joints during Thermomigration

Yu-An Shen[1,*], Shiqi Zhou[1,2], Jiahui Li[3], K. N. Tu[4], and Hiroshi Nishikawa[1]

1. Joining and Welding Research Institute (JWRI), Osaka University, Osaka, 5600047, Japan
2. Graduate School of Engineering, Osaka University, Osaka 565-0871, Japan
3. Department of Electronic Engineering, City University of Hong Kong, Hong Kong, China
4. Department of Materials Science and Engineering, UCLA, Los Angeles, CA, USA
*yashen@jwri.osaka-u.ac.jp

Abstract—**Thermomigration is very important for electronic packaging. To conduct heat away, it requires a temperature gradient. Just 1°C difference across a microbump of 10 μm in diameter induces a considerable thermal gradient in solder bumps, known as thermomigration, especially in eutectic Sn58Bi solder joints with a low melting point. However, the study of thermomigration in eutectic Sn-58Bi solder joints is rare, not to mention the effect on thermal-stress distribution in Sn-58Bi solder. In this study, Cu/Sn-58Bi/Cu solder joints are examined between different thermogradients of 1284 °C/cm and 890 °C/cm. We found the effect of thermogradients on the microstructures in the Sn-58Bi solders. The considerable Bi accumulations at the cold end and smaller grains formation near the hot end are caused by the higher thermogradient. Conversely, the fewer Bi accumulations at the cold end and grain growth at the hot end are induced by the lower thermogradient. Additionally, phase segregation would lead the crack propagation along the boundary due to the thermal-stress crowding on the phase boundaries. The findings provide a new sight of reliability issues for electronic packaging.**

Keywords- Thermomigration; Sn-Bi solder joint; Phase segregation; Thermal stress; Finite element method (FEM)

I. INTRODUCTION

Thermomigration (TM) is a kind of atomic migration by a temperature gradient. It is critical for Pb-free solders joints [1-4]. When TM occurred in the Sn-rich solder joints, a numerous formation of IMC at the cold end and a severe dissolution of under bump metallization (UBM) lead the reliability issue in the solder joints [5-7]. This is because the brittle IMC could affect the mechanical property, and the heavy UBM consumption could affect the electrical resistance in Pb-free solder joints.

Recently, three-dimensional integrated circuit (3D IC) with the benefits of high performance, small form factor, and heterogeneous integration is the promising direction to extend Moore's law [8]. However, because the cooling design leads the considerable heat-flux out from the chipset to the interposer, the thermal gradient is avoidless in solder microbumps which are the joints for chips with through-

silicon-vias (TSVs). That is the reason why TM has drawn the attention in electronic packaging. Moreover, because of the miniaturization of devices, the microbump will have a thermogradient of ~7000 °C/cm between a slight temperature difference (ΔT) of 3.8°C [9]. Thus, the TM reliability in microbump is urgent to investigate [10,11].

On the other hand, because of the step soldering process after multiple reflows, low-temperature solder is very important. And eutectic Sn-58Bi solder, with good mechanical properties and wettability on Cu substrate [12-14], is promising to be the low-temperature solder in electronic packaging. However, because a Sn58Bi solder has a lower melting point (Tm = 139 °C) than the Sn-rich solder (Tm = ~220 °C), the atomic migration between thermogradient is serious and rapid. Therefore, the thermomigration reliability of Sn-58Bi solder joint is a critical concern. Although Gu et al. have investigated the Sn-Bi phase segregation during electromigration in eutectic Sn58Bi solder joint [15], the phenomenon by thermomigration is seldom. Additionally, although phase segregation by thermomigration in the eutectic Sn58Bi solders is observed in the reference article [16], the effect of various thermogradient on microstructure changes has not been investigated. And the changes of thermal-stress distribution by thermomigration have not yet been considered deeply because of the substantial phase boundaries causing severe mismatches of the coefficient of thermal expansion (CTE), accreting the crack propagation.

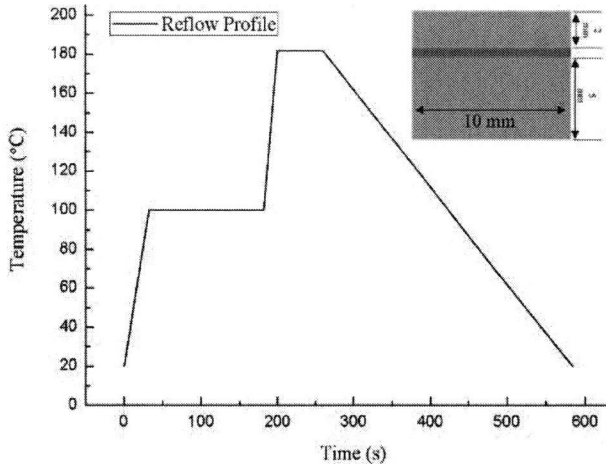

Figure 1 Reflow profile for the fabrication of solder joints with the dimension

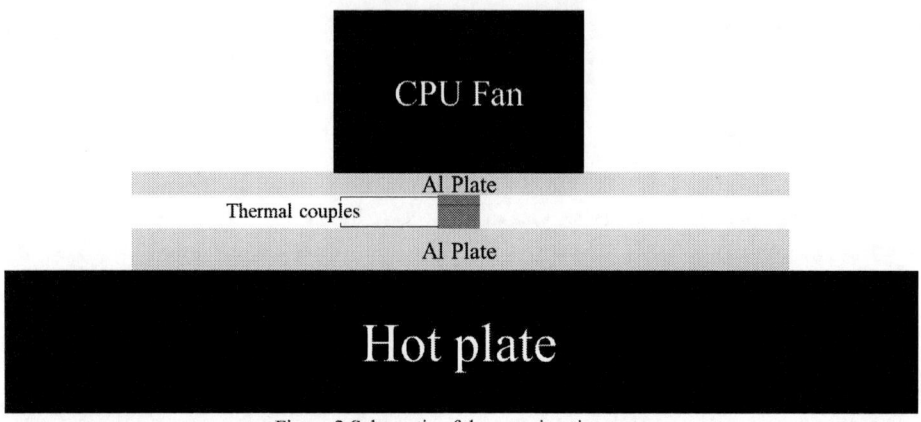

Figure 2 Schematic of thermomigration test.

In this study, Cu/Sn-58Bi/Cu solder joints were examined for thermomigration with various thermal gradients. Through the TM tests, the different levels of Sn-Bi phase segregation and Bi accumulation at the cold end were demonstrated clearly. The dissolution rate of Cu UBM and the IMC growth at the hot end is equal to that at the cold end.

Then the thermal-stress distribution in a Sn-58Bi solder, referred to the TM-tested microstructure, was evaluated and analyzed by a two-dimensional (2D) simulation of FEM. The effect of thermomigration caused the thermal-stress crowding along the boundary of Sn-Bi phase segregation. The findings provide not only a much clearer TM mechanism in Sn-58Bi solder joint, but also the related issues of the effect of TM on the thermal-stress distribution in the Sn-58Bi solder joint.

II. EXPERIMENTAL

A solder joint of 2-mm-Cu/eutectic Sn-58Bi /5-mm-Cu with 10 mm diameter was assembled by reflow at 182°C. The reflow profile and dimension of solder joints are shown in Fig. 1. For thermomigration studies with various thermal gradients, solder joints, placed between an Al plate on a hot plate (the hot end,) and an Al plate with a heat sink of a cooling fan (the cold end), were maintained by the thermogradients for 300 h, as shown in Fig 2. When we put tested specimens far away the center of the CPU fan, the various temperatures at the cold end were obtained. K-type thermocouples were used to measure the temperatures of the interfaces between the Cu substrates and Al plates. Simulations by finite element method (Ansys Workbench) were used for evaluating the temperature distributions in the eutectic Sn-58Bi solder joints during the thermomigration tests. Microstructures of each specimen were observed by a scanning electron microscope (SEM, JOEL-7100, Japan) with backscattered electron images (BEIs). The distributions of elements were identified by Energy dispersive X-ray spectrometer (EDS, JOEL, Japan). In 2D simulation, a 2D model of Sn-58Bi solder with the referred microstructure of TM-tested solder was fabricated to understand the effect of phase segregation on the thermal-stress redistribution after thermomigration. The thermal stress was caused by a uniform surrounding at 100°C in the 2D simulation. The properties of materials in the simulations are summarized in Table 1.

Table 1: Materials properties of simulations.

Materials	Thermal conductivity, W/(m·K)	CTE, μm/ (m·k)	Young's modulus, GPa	Poisson's Ratio
Sn-58Bi Solder	19	-	-	-
Bi	7.97	13.4	32	0.33
Sn	66.8	22	45	0.335
Cu	401	-	-	-

III. RESULTS AND DISCUSSION

A. Thermomigration

Fig. 3 shows a BEI image of an as-reflowed solder joint with 32 μm height. We observed the lamellae structure of

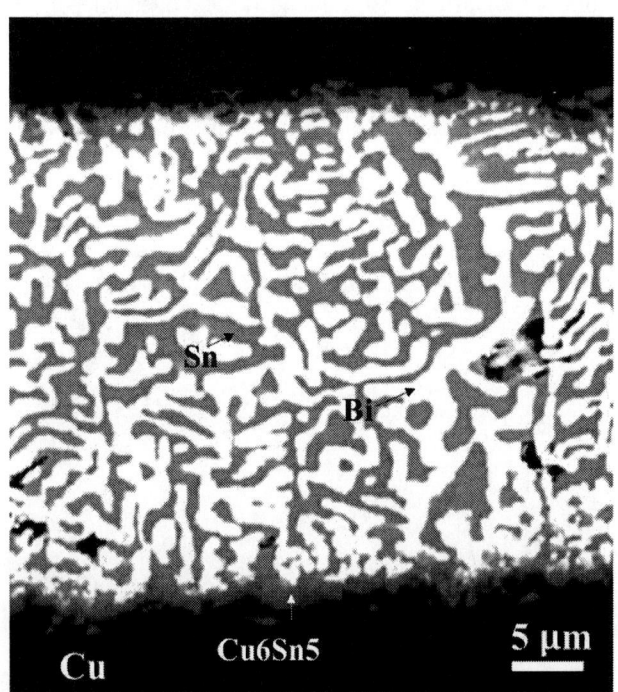

Figure 3 Cross-sectional microstructure in an as-reflow Sn-58Bi solder joint.

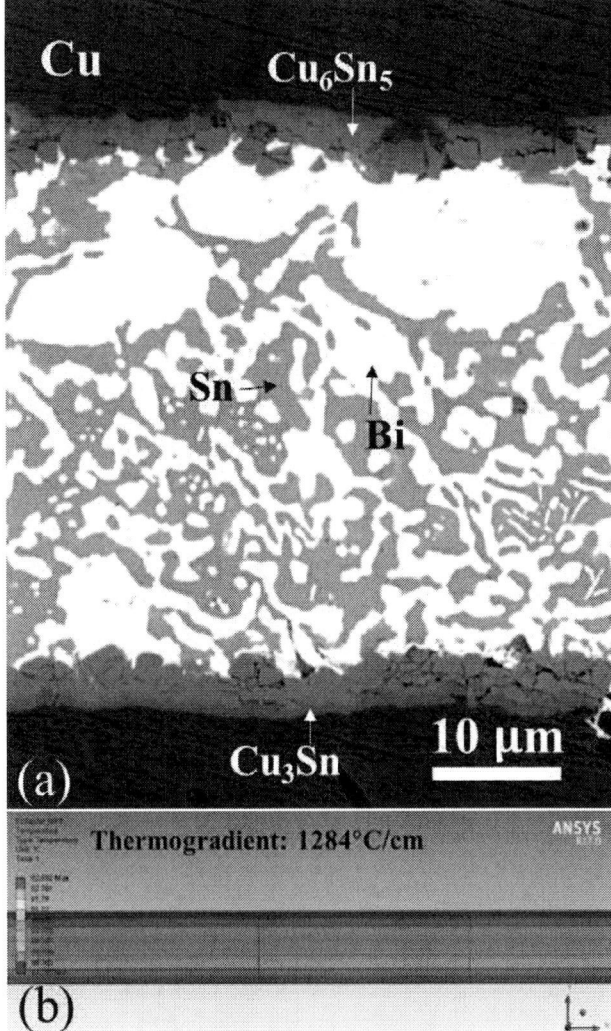

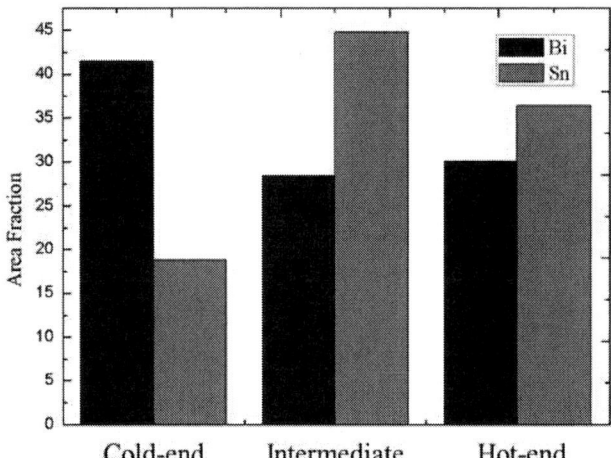

Figure 5 Distributions of Sn and Bi in the cold-end, intermediate and hot-end regions after thermomigration under the thermogradient of 1284 °C/cm.

Figure 4 (a) Cross-sectional microstructures of a TM-tested solder joint. (b) Its thermogradient of 1284 °C/cm.

Sn and Bi phases in the solder. And the orientations of Bi were mostly parallel or the Cu surface. This microstructure is very similar to that of the reference article [17]. Fig. 4a shows a BEI image of a TM-tested solder joint with a solder thickness of 40 μm. The joint was examined by a thermogradient of 1284°C/cm between a hot end temperature of the solder at 93°C and a cold end temperature at 88°C, as shown in Fig. 4b. We could observe the accumulation of Bi at the cold end. And the orientations of Bi are mostly parallel to the direction of thermogradient. The effect of thermomigration on microstructure changes of the eutectic Sn58Bi solder is significant.

If we divided the solders of Fig. 3 and Fig. 4a by three parts, near cold-end, intermediate and hot-end, the accumulation of Bi is so significant. The area of Bi at the cold end is 43%, and that of Sn is 22%, as shown in Fig. 5. This is the critical evidence to prove that Bi is the dominant diffusing species during thermomigration. Because of the interdiffusion, the passive migration (Sn) would migrate in a direction opposite to the direction of the dominant species of diffusion (Bi). And the migrated Bi would also squeeze the Sn, inducing Sn migration from the cold end to the hot

end. However, the driving force for Sn migration was not enough; the uniform distribution of in the intermediate and the hot end was observed. Moreover, the smaller Bi predicated out at the cold-end and intermediate regions. That results are similar to those in the reference article [15].

Furthermore, we put the solder joints between a thermogradient of 890 °C/cm for 300 h thermomigration. The solder thickness is 37 μm, and the temperature distribution of the solder is shown in Fig. 6b. The accumulation of Bi at the cold end and the Bi orientation parallel to the direction of thermogradient were also observed in Fig. 6a. The lower driving force of TM in Fig. 6 induced a lower level of Bi accumulation at the cold end than that in Fig. 4. This phenomenon is similar to the results with various current densities during electromigration. However, interestingly, the Bi grains at the hot-end and

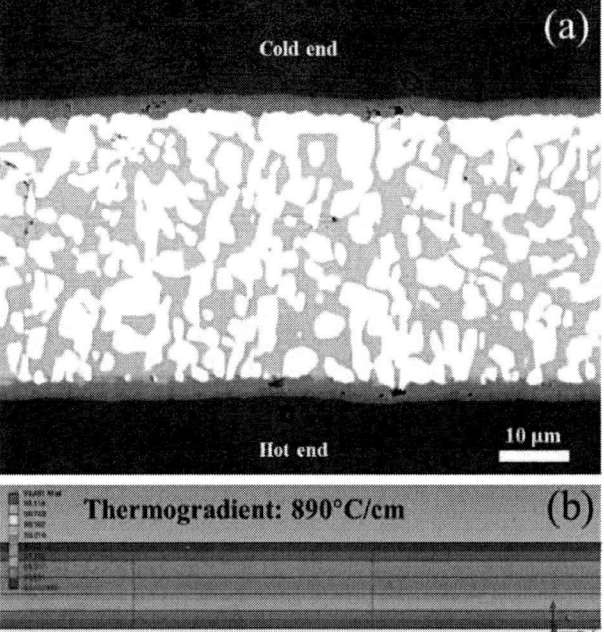

Figure 6 (a) Cross-sectional microstructures of a TM-test solder between (b) a thermogradient of 890 °C/cm.

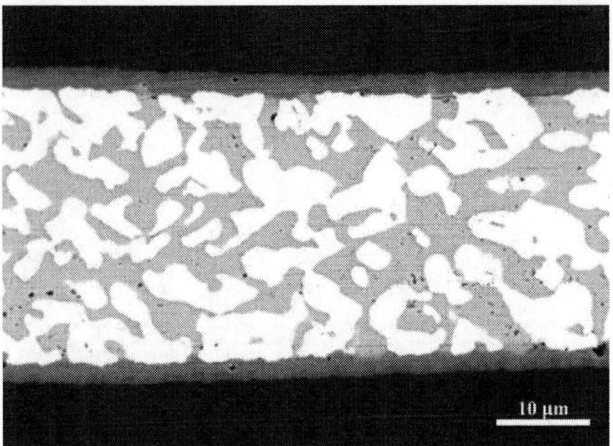

Figure 7 Cross-sectional microstructures of a Sn-58Bi solder joint after thermal aging at 100 °C for 300 h.

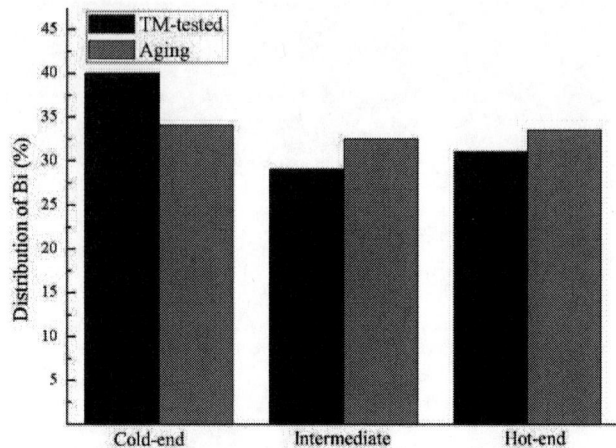

Figure 8 Bi distribution in the TM-tested solder of Fig. 6(a) and the solder after aging test of Fig. 7.

intermediate regions are much larger than those in Fig. 4. The effect of different thermogradients is clear in the above results. In the eutectic Sn-58Bi solder with a lower thermogradient (Fig. 6a), the Bi migrations pushed by the driving force of are relatively limited. Therefore, there were considerable Bi atoms left at the hot end. According to the evaluation of the simulation in Fig. 6b, the temperature at the hot was about 99.5°C during the thermomigration test, inducing a kind of thermal aging at 100°C happening. When a eutectic Sn-58 Bi solder in a surrounding at a constant temperature much higher than the room temperature, the gathering and growth of Bi would occur simultaneously. A piece of clear evidence is shown in Fig. 7 that is a BEI image of a Sn-58Bi solder after a thermal-aging test at 100°C for 300 h. There were 40% Bi in the cold-end region and the uniform distributions of Bi in the intermediate and hot-end regions after thermomigration, as shown in Fig 6a. They are different from the uniform distributions of Bi in the three regions in the Sn-58Bi solder after thermal aging at 100°C in Fig. 7. Their Bi distributions are shown in Fig. 8. Obviously, the Bi grain growth and gathering occurred anywhere in the solder, and there was not the Bi accumulation at one side because of the thermogradient free. Conversely, if your thermogradient was as high as that in Fig. 4, the number of Bi atoms was not enough to grow up as large as that in Fig. 6a in the hot-end and intermediate regions due to the considerable migrations of Bi to the cold

end. The effect of thermogradients on the microstructure changes is clearly demonstrated in the results.

However, no matter the solder joints were between high or low thermogradient, their IMC growths are symmetrical, because the numerous phase boundaries stopped the Cu migrations from the hot end to cold end. This phenomenon is completely different from that in Sn-rich solder. In the latter, UBM dissolutions at the hot end and the growth of brittle IMC at the cold end were critical [7]. But that should not be concerned in Sn-58Bi solder joints as the results in this study. Following the above results, the effect of phase segregation on the current crowding in the solder would be presented.

B. FEM analysis of current-density change in Sn-58Bi solder by thermomigration

A 2D model with a microstructure of the serious phase segregation in TM-tested Sn-58Bi solder is used for simulating the distribution of thermal stress. Another 2D model with an as-reflowed Sn-58Bi solder is served as a benchmark. Their dimension is of 6 μm×12 μm. We set the top and bottom side for fix support. The models were in a uniform temperature surrounding at 100°C. Because of the thermal expansions, there are thermal stresses distributed in the 2D models. As can be seen, in Fig. 9, thermal stresses are crowding at the phase boundaries in the TM-tested and as-reflow Sn-58Bi solders. The crowding is caused by the

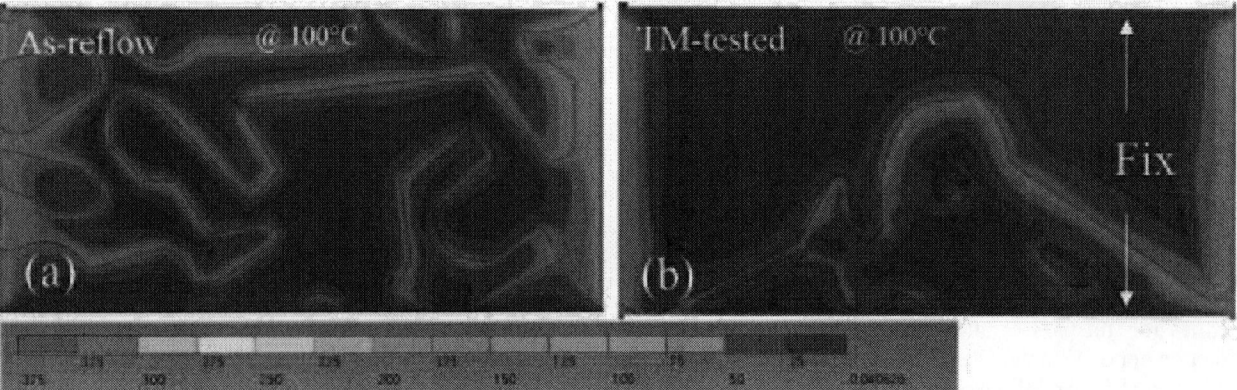

Figure 9 Distributions of the thermal stresses in (a) as-reflow solder and (b) TM-tested solder of SnBi solder.

CTE mismatch between Sn and Bi phases. In Table 1, the CTE of Bi is 13 μm/(m·k) much lower than that of Sn (22 μm/(m·k)). When the thermal expansion occurred during a long-term thermal aging or thermal cycling tests, the thermal stresses would cause crack propagation. In Sn-rich solder, the crack propagation on Sn grain boundaries are commonly observed and induces the severe increase in the resistivity of Sn-rich solder joints [18,19]. Although the levels of thermal stresses along the boundaries of TM-tested and as-reflow soldered were identical, the thermal stresses along the long phase boundaries between the phase segregation of Sn-Bi caused by thermomigration are critical because of the extension of crack propagation. In the as-reflow Sn-58Bi solder, crack propagation is sporadic, so the impact of cracks induced by the thermal stresses is limited. On the other hand, when the crack propagation occurs along the long phase boundaries of phase segregation, the critical increase in the resistivity of TM-tested Sn-58Bi solder can be expected reasonably. Another issue by the long crack propagation is the mechanical reliability of Sn-58Bi solder joints during thermomigration. Recently, smart phone is a kind of most popular consumer electronic devices. The crisis of smart phone drop-down always exists to every user. If there is a long crack between the phase segregation when the smart phone drop to the ground, the destruction in the solders will easily induce the failure of the smart phone. Therefore, our findings in the 2D simulation of thermal-stress distribution provide a critical concern for the reliability of electronic packaging as the effect of thermomigration on the thermal-stress crowding.

IV. CONCLUSION

Cu/Sn-58Bi/Cu solder joints are examined by various thermomigration gradients of 1284 °C/cm and 890 °C/cm for 300 h, respectively. In this study, the different levels of phase segregation caused by different levels of thermogradient are demonstrated. In higher thermogradient, considerable Bi accumulations occurred at the cold end, and smaller grains of Bi formed in the intermediated and hot-end regions. Conversely, fewer accumulations of Bi occurred in the Sn-58Bi solder between a lower thermogradient. Alternatively, grain growth of Bi could be observed at the hot end because of enough atoms at the temperature of 99.5°C, similar to the aging test at 100°C. The IMC growths between the solders and Cu substrates were not affected by thermogradient because of the Cu diffusions blocked by the phase boundaries of Sn and Bi. In 2D simulation, the thermal-stress crowding on the phase boundaries. However, in the TM-tested solder with serious phase segregation, the crack propagation along the long phase boundary could be expected to affect the resistivity and mechanical reliability of Sn-58Bi solder joints during thermomigration.

V. ACKNOWLEDGEMENTS

The software support of Ansys Workbench from the group of N.T. Tsou in Department of Materials Science and Engineering, National Chiao Tung University, Taiwan, is acknowledged.

VI. REFERENCES

[1] R.A.Oriani, "Thermomigration in solid metals," J. Phys. Chem. Solids, 30 (1969) 339-351.

[2] H. Ye, C. Basaran, D. Hopkins, "Thermomigration in Pb–Sn solder joints under joule heating during electric current stressing," Appl. Phys. Lett. 82 (2003) 1045.

[3] C. Chen, H.M. Tong, K.N. Tu, "Electromigration and Thermomigration in Pb-Free Flip-Chip Solder Joints," Annu. Rev. Mater. Res. 40 (2011) 531-555.

[4] C. Chen, H.Y. Hsiao, Y.W Chan, F.Y. Ouyang, K.N. Tu, "Thermomigration in solder joints," Mater. Sci. Eng. R-Rep. 73 (2012) 85-100.

[5] W.N. Hsu, F.Y. Ouyang, "Effects of anisotropic β-Sn alloys on Cu diffusion under a temperature gradient," Acta. Mater. 81 (2014) 141-150.

[6] T.L. Yang, T. Aoki, K. Matsumoto, K. Toriyama, A. Horibe, H. Mori., Y. Orii, J.Y. Wu, C.R. Kao, "Full intermetallic joints for chip stacking by using thermal gradient bonding," Acta. Mater. 113 (2016) 90-97.

[7] Y.A. Shen, F.Y. Ouyang, C. Chen, "Effect of Sn grain orientation on growth of Cu-Sn intermetallic compounds during thermomigration in Cu-Sn2.3Ag-Ni microbumps," Mater. Lett. 236, (2019) 190-193.

[8] K.N. Tu, "Reliability challenges in 3D IC packaging technology," Microelectron. Rel. 51 (2011) 517-523.

[9] Y.P. Su, C.S. Wu, F.Y. Ouyang, "Asymmetrical Precipitation of Ag3Sn Intermetallic Compounds Induced by Thermomigration of Ag in Pb-Free Microbumps During Solid-State Aging," J. Electron. Mater., 45 (2016) 30-37.

[10] M.Y. Guo, C.K. Lin, C. Chen, K.N. Tu, "Asymmetrical growth of Cu6Sn5 intermetallic compounds due to rapid thermomigration of Cu in molten SnAg solder joints," Intermetallics 29 (2012) 155-158.

[11] M.L. Huang, F. Yang, N. Zhao, "Thermomigration-induced asymmetrical precipitation of Ag3Sn plates in micro-scale Cu/Sn–3.5Ag/Cu interconnects," Mater. Des., 89 (2016), 116-120.

[12] S. Zhou, C.H. Yang, S.K. Lin, A.N. Aihazaa, O. Mokhtari, X. Liu, H. Nishikawa, "Effects of Ti addition on the microstructure, mechanical properties and electrical resistivity of eutectic Sn58Bi alloy," Mater. Sci. Eng. A 744 (2019) 560-569.

[13] S. Zhou, Y.A Shen, T. Uresti, V.C. Shunmugasamy, B. Mansoor, H. Nishikawa, "Improved mechanical properties induced by In and In & Zn double additions to eutectic Sn58Bi alloy," J. Mater. Sci. Mater. Electron., (2019), https://doi.org/10.1007/s10854-019-01056-y.

[14] S. Zhou, C.H. Yang, Y.A Shen, S.K. Lin, H. Nishikawa, "The newly developed Sn–Bi–Zn alloy with a low melting point, improved ductility, and high ultimate tensile strength," Materialia, 6 (2019), pp. 599-607.

[15] C.M. Chen, L.T. Chen, Y.S. Lin, "Electromigration-induced Bi segregation in eutectic SnBi Solder joint," J. Electron. Mater., 36 (2007) 168-172.

[16] Y.A. Shen, S. Zhou, J. Li, K.N. Tu, H. Nishikawa, "Thermomigration induced microstructure and property changes in Sn-58Bi solders," Mater. Des., 166 (2019), 107619.

[17] S. Zhou, O. Mokhtari, M.G. Rafique, V.C. Shumugasamy, B. Mansoor, H. Nishikawa, J. Alloy Compd. 765 (2018) 1243-1252.

[18] D.A. Shnawah, M.F.M. Sabri, I.A. Badruddin, "A review on thermal cycling and drop impact reliability of SAC solder joint in portable electronic products," Microelectron. Reliab., 52, (2006) 90-99.

[19] Y.C. Chu, C. Chen, N. Kao, D.S. Jiang, "Effect of Sn grain orientation and strain distribution in 20-μm-diameter microbumps

on crack formation under thermal cycling tests," Electron. Mater. Lett., 13-6, (2017) 457-462.

Simulation and Experimental Validations of EM/TM/SM Physical Reliability for Interconnects Utilized in Stretchable and Foldable Electronics

Chang-Chun Lee[1], Oscar Chuang[1,2], Chia-Ping Hsieh[3], Wei-Yuan Cheng[2], Steve Chiu[2]

[1] Microsystems Mechanical Design & Reliability Analysis Laboratory,
Dept. of Power Mechanical Engineering, National Tsing Hua University, Hsinchu 30013, Taiwan, R.O.C.

[2] Electronic and Optoelectronic System Research Laboratories (EOSL), Industrial Technology Research Institute, Hsinchu, Taiwan 30013, R.O.C.

[3] Dept. of Mechanical Engineering, National Taiwan University, Taipei, Taiwan 10617, R.O.C.

Phone: 886-3-5162410. Fax: 886-3-5722840, and E-mal: cclee@pme.nthu.edu.tw

Abstract—**There are many portable electronic products made by fan-out wafer/panel level package (FOWLP/FOPLP). The redistribution layer(RDL) design with fine pitch to connect inter each application is used. However, the above packaging products usually face the issue of electromigration (EM). Hence, a finite-element-based simulation method to predict the EM/ thermomigration (TM)/ stressmigration (SM) physical reliability and experimental validation are proposed. In this research, the EM simulation method also considers other physical behaviors including stress and temperature effects for future applications such as stretchable and foldable electric products.**

Keywords- Interconnects; Flexible electrics; Electromigration; Redistribution layers

I. INTRODUCTION

Portable electronic products face the urgent requirements including the miniaturization and high input/output (I/O) counts. Therefore, high-density fan-out packaging technology is widely used in fan-out wafer or panel level package based on these advantages consisting higher I/Os, thinner total structure thickness, faster electrical performance, and so on. However, fan-out package following of tinier line space and smaller line width would expect to suffer electromigration (EM) failure situation, accompanies the occurrences of thermomigration (TM) and stressmigration (SM). It would affect gradually electrical performance and mechanical behaviors while the Joule-heating effect combined with the accelerated loading tests, such as elevated temperature or precipitous operated environments, external mechanical tensile, high power voltage, are taken into account. In order to address the abovementioned issue, this research proposes a powerful and reliable simulated methodology to implement the failure prediction of the comprehensive reliability phenomenon induced from EM for the next generation flexible electronic via SiP package technology. The key simulated procedure of the present EM estimation is to introduce atomic flux divergence (AFD) theory, which has the capability of dealing with the resultant effects of TM and SM. In the meanwhile, these coupling problems for structural designs when flexible electronic suffers both mechanical bending

and electric effect even temperature influence can be resolved. With regard to experimental validation, the panel-form plating equipment is adopted to fabricate multi-layers of redistribution layer (RDL) based on ITRI FlexUP TM substrate with specifications of 2μm line space and 2μm line width. Meanwhile, different bending loads and power effects are adopted to demonstrate the samples related to the interconnects of RDLs. From the compared results between experimental data and simulated estimations, the proposed simulated methodology is validated to be suitable for the structural designs of advanced flexible electronics. Through the foregoing validations, the combination of the tinier line space and smaller line width for instance multi-layers of RDLs with 0.8μm line space and 0.8μm line width could be examined and realized. The related process flow of flexible electric components is shown in Fig. 1.

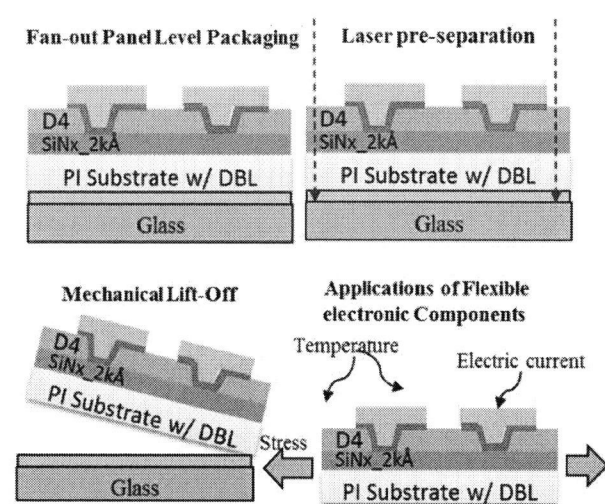

Figure 1. Process flow of flexible electric components.

From previous literatures related to EM, it has shown that local atomic flux accelerates the generation of adjacent pores when void induced behavior of EM is caused in the surface of interconnector [1]. In addition, the type of geometry about interconnector also affects the occurrence of EM [2-3]. Furthermore, the corners of interconnectors are

978-1-7281-1500-9/19 $31.00 © 2019 IEEE

more prone to induce current concentration. It has been proved by finite element simulation and experiments [4-5]. The study provides a beneficial solution for future layout designs and material selections of flexible electrics, including high-sensitivity precision intelligent rehabilitation robots, automated cars, and smart wearable devices.

II. THEORY

For EM research, the mean-time-to-failure (MTTF) is usually to estimate the functional time for flexible electric component. Moreover, the concerned components may suffer stretchable or flexible stress. Therefore, the diffusion flux considering different thermal stress, mechanical stress, and electric flux are revised.

A. ElectroMigration (EM)

After the metal atom collides the electron under high current density, momentum is transferred causing the EM phenomenon. The diffusion theory is usually used to describe the foregoing behavior. For the next generation flexible electronics, it is expected that flexible components need to face complex physical effects including not only EM but the influences of temperature as well as mechanical stress. Consequently, the diffusion theory has been revised including additional applies stress and applied temperature, respectively. The concerned flex can be expressed as shown in Eq. (1).

$$J = J_{chem} + J_{elec} + J_{me} = -D\frac{dC}{dx} + C\frac{D}{k_B T}Z^* eE - C\frac{D}{k_B T}\frac{\Omega d\sigma}{dx} \quad (1)$$

where symbol D means diffusion coefficient, C is electron concentration, k_B is Boltzmann constant, T indicates local temperature, Ω is the atomic volume, Z^* is the effective valence electrons with regard to electric field E, and e is charge of electron.

B. Mean-Time-to-Failure(MTTF)

Starting with Black's formula with regard to (MTTF) [6], it has been demonstrated to estimate EM. By MTTF, the functional time of electric components has been predicted before arriving the failure. The formula of MTTF is wrote as shown in Eq. (2)

$$MTTF = A\, j^{-n} \exp(\frac{Q}{k_B T}) \quad (2)$$

where A is material property related to conductor geometry and structure, j is the electric density (A/cm^2), n is the index of electric density from the experiment, Q is active energy obtained from the experiment, k_B is Boltzmann constant (8.617e-5 eV/K), and T is absolute temperature. Through the use of different absolute temperature, the EM reaction is able to be accelerated to acquire material characteristics of MTTF needed via related experimental designs.

C. Atomic fluxes divergence

Because metal wire has been set up in the environment with high temperature in order to increase the EM reaction. During high temperature, the atomic threshold has been broken by metal atom. Moreover, it causes the activation energy of atom to decrease. This behavior would like to lead the atom movement of atom. The concerned driving force has been named atomic fluxes divergence (AFD). The main reasons which cause AFD include electronic wind, thermal energy, and mechanical stress, separately. These physical phenomenon are separately represented by the following atomic flux equations listed in Eq. 3.

$$\vec{J_A} = \frac{N}{k_B T}eZ^*\vec{j}\rho D_0 exp\left(-\frac{E_A}{k_B T}\right) \quad (3)$$

$$\vec{J_{th}} = -\frac{NQ^* D_0}{k_B T^2}exp\left(-\frac{E_A}{k_B T}\right)\nabla T \quad (4)$$

$$\vec{J_S} = \frac{N\Omega D_0}{k_B T}exp\left(-\frac{E_A}{k_B T}\right)\nabla\sigma_H \quad (5)$$

In the above-mentioned equations, T is temperature, D_0 means diffusion coefficient, k_B means Boltzmann constant, N is the density of atom, e means valence electron number, Z^* means effective charge number, j is the density of current, E_A is activation energy, Q^* is heat transfer coefficient, ∇T is temperature gradient, Ω is the volume of atom, $\nabla\sigma_H$ is the stress gradient. In order to judge the physical phenomenon of atom including movement or gather, the above three equations have been modified by differential types.

$$div\left(\vec{J_A}\right) = \left(\frac{E_A}{k_B T^2} - \frac{1}{T} + \alpha\frac{\rho_0}{\rho}\right)\vec{J_A} \cdot \nabla T \quad (6)$$

$$div\left(\vec{J_{th}}\right) = \left(\frac{E_A}{k_B T^2} - \frac{3}{T} + \alpha\frac{\rho_0}{\rho}\right)\vec{J_{th}} \cdot \nabla T + \frac{NQ^* D_0}{3k_B{}^3 T^3}j^2\rho^2 e^2 exp\left(-\frac{E_A}{k_B T}\right) \quad (7)$$

$$div\left(\vec{J_s}\right) = \left(\frac{E_A}{k_B T^2} - \frac{1}{T}\right)\vec{J_s} \cdot \nabla T + \frac{2EN\Omega D_0\alpha_1}{3(1-v)k_B T}exp\left(-\frac{E_A}{k_B T}\right)\left(\frac{1}{T} - \alpha\frac{\rho_0}{\rho}\right)\nabla^2 T$$
$$+ \frac{2EN\Omega D_0\alpha_1}{3(1-v)k_B T}exp\left(-\frac{E_A}{k_B T}\right)\frac{j^2\rho^2 e^2}{3k_B{}^2 T} \quad (8)$$

where α is the electric resistance related to temperature, E is Young's modulus. v is the passion's ratio. ρ_0 is the electric resistance in room temperature, the relationship between ρ_0 and ρ are listed in Eq. (9).

$$\rho = \rho_0\,[\,1 + \alpha\,(\,T - T_0\,)\,] \quad (9)$$

III. SIMULATION METHODOLOGY AND VALIDATION

A novel modeling simulation of EM prediction for complex conditions including electric force, thermal stress, and applied stress is proposed in this research. To validate the reliability of simulated approach, two types of structure designs including interconnects and SAC solder joint are considered to compare the simulated results between other research groups and the presented analysis of this study.

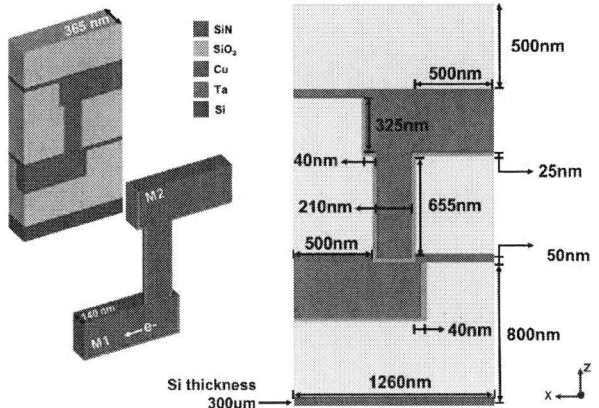

Figure 2. Geometries of RDL trace line.

TABLE 1 SIMULATION REALITY VALIDATION

Width of trace lines	NTHU Max current density (A/mm²)	Reference (Tan, 2005) Max current density (A/mm²)	Simulation result difference(%)
0.28 μm	24690	23964	3.00%
0.7 μm	72532	72152	0.53%

A. EM simulation validation of RDL trace line

In fan out panel/wafer level package, EM failure mode prediction for structural design of Cu RDL interconnect has been validated first [7-9]. To validate the present EM simulated methodology is reliable, the Cu RDL trace line shown in Fig. 2 is considered. With regard to the given boundary condition in this simulation, that the electric flux flows from M2 to M1 trace is assumed. Moreover, a 0.8 MA/cm² current density is applied to the cross section of Cu trace line. As compared the predicted results with the Tan's research group [10] shown in Fig. 3, both contours of current density are almost identical. In addition, a difference of maximum current density lower than 3.1% is achieved as listed in Table 1. Consequently, the proposed simulated methodology is validated to be suitable for such structural designs of RDL trace lines.

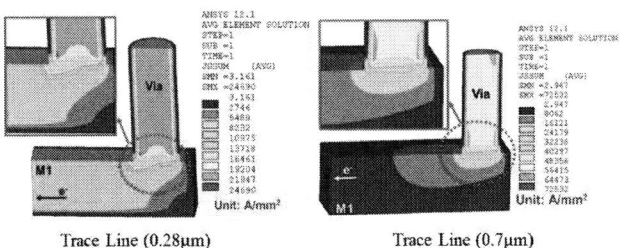

Figure 3. EM contours of Cu trace line predicted by proposed simulation.

B. EM simulation validation of SAC solder joint

The EM situation in fan-out packaging usually happens around Cu RDL interconnects and SAC solder joint [11-12]. Moreover, the structural design of solder joints is also considered to demonstrate the proposed simulated approach. As shown in Fig 4, the geometries of solder joints are labeled in detail. From the simulated result shown in Fig. 5, a difference lower than 3.1% is obtained between the present simulated approach and Lee's group [13]. Therefore, the proposed simulation methodology in this research is also workable for the EM prediction of solder joints at packaging level.

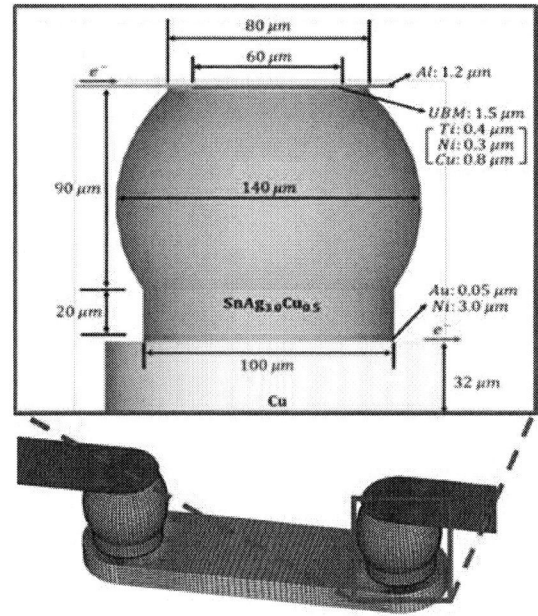

Figure 4. Geometries of SAC solder joints for simulation validation.

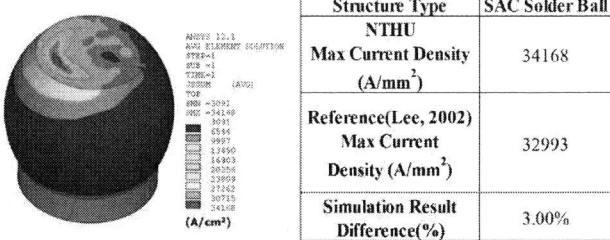

Structure Type	SAC Solder Ball
NTHU Max Current Density (A/mm²)	34168
Reference(Lee, 2002) Max Current Density (A/mm²)	32993
Simulation Result Difference(%)	3.00%

Figure 5. Compared simulated results of EM simulation for solder joint.

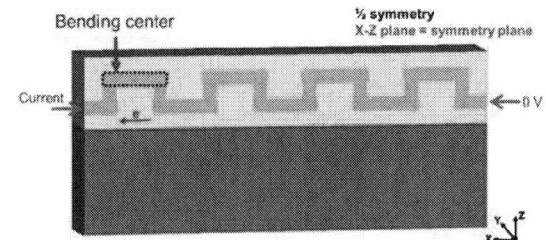

Figure 6. Structural modeling of simulation validation (EM/TM/SM).

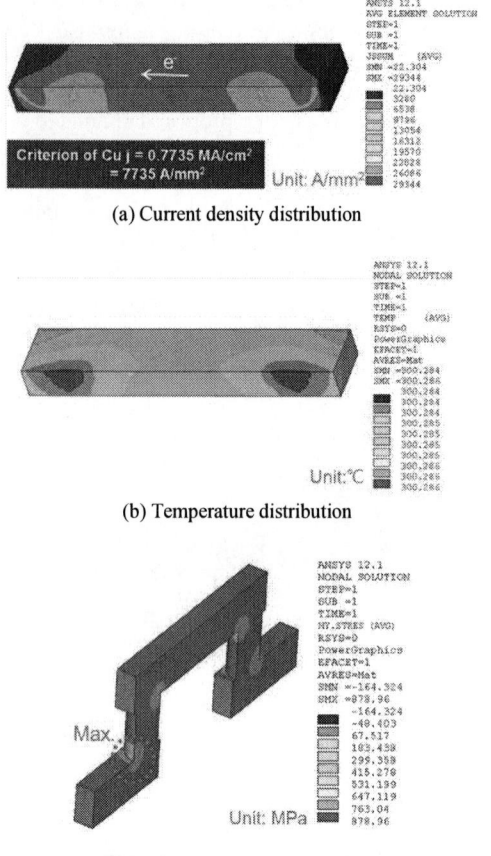

(a) Current density distribution

(b) Temperature distribution

(c) Hydrostatic stress distribution

Figure 7. Each simulated contour for EM/TM/SM.

C. Coupling EM simulation (EM/TM/SM)

A new estimated methodology has been established to predict EM complex condition with thermal and applied mechanical stresses. Flexible electronic components of soft RDL are expected to work in such complex environment. Therefore, the revised traditional EM simulation method is proposed to meet the foregoing requirements. During folding or stretchable situations, the thermal and applied external stresses should be considered in the revised EM methodology. When implementing this simulation, double layers of RDL structure, whose line space/line width of 2/2μm, total thickness of 36.57μm, are considered. As the modeling geometries shown in Fig. 6, the length and depth of RDL are 66.7μm and 12.181μm, respectively. While a 258mA current and a 1mm of bending radius are stressed at 300℃, the corresponding distributions of double-layered RDL regarding current density, temperature, and hydrostatic stress are separately obtained and shown the Fig. 7. In addition, through the adoption of AFD, simulation result has transferred into atom driving force and divergence of atom driving. The simulated results shown in Fig. 8 point out the different range of current from 5mA to 25mA under the condition of applying a 1mm of bending radius. From the

analytic results, this current structure design of RDL has the risk of inducing void when applied current exceeds ~6.5mA.

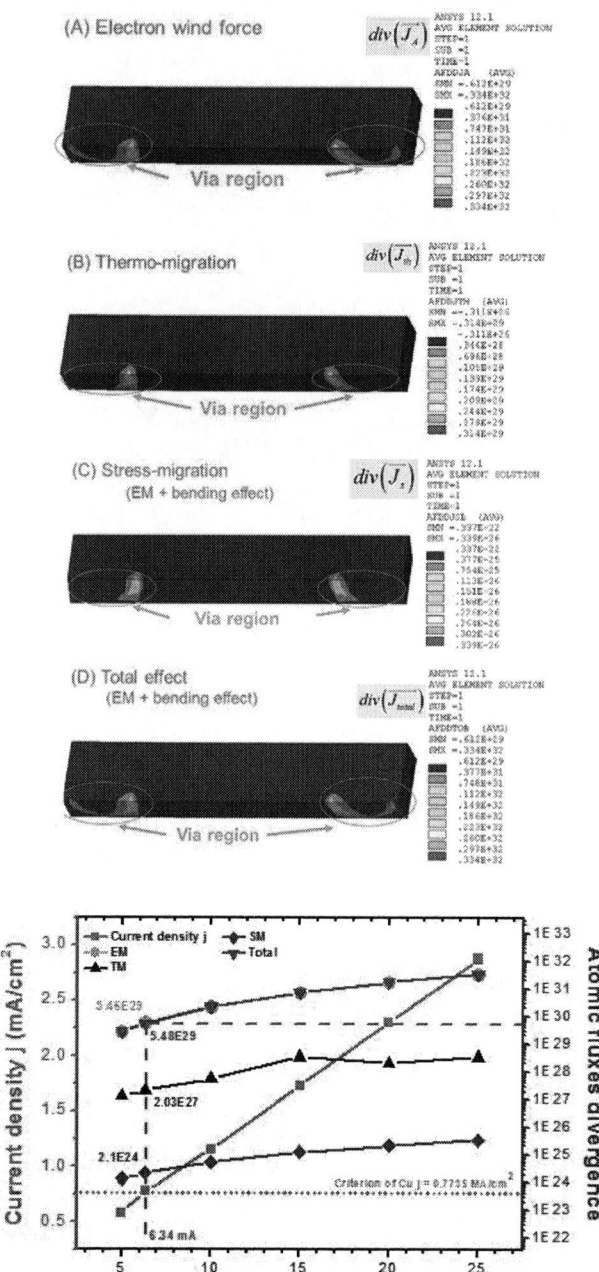

Figure 8. Estimated results of combined effect of EM/TM/SM.

IV. EM/TM/SM SIMULATION PREDICTS FOR FIN-PITCH INTERCONNECTS OF RDLS OF FAN-OUT PACKAGES

After confirming the estimated accuracy of combined effect of EM/TM/SM, fin-pitch interconnects of RDLs adopted in fan-out packages as shown in Fig. 9 are also taken into account. a 3D finite element analysis with a quarter model based on the biaxial symmetry is therefore performed. It should be noted that several widths of RDLs, such as 2, 5, 10μm, are separately considered.

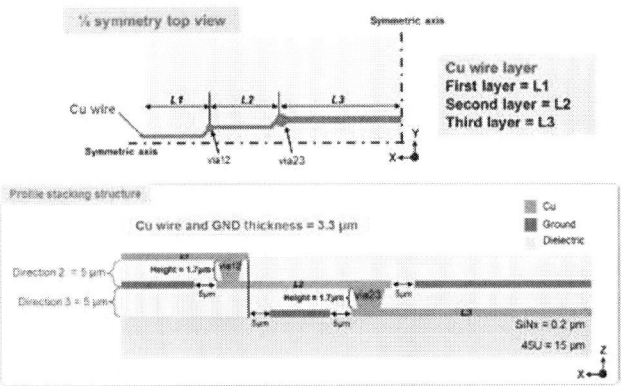

Figure 9. Geometrical scheme of three layers RDL FOPLP structure.

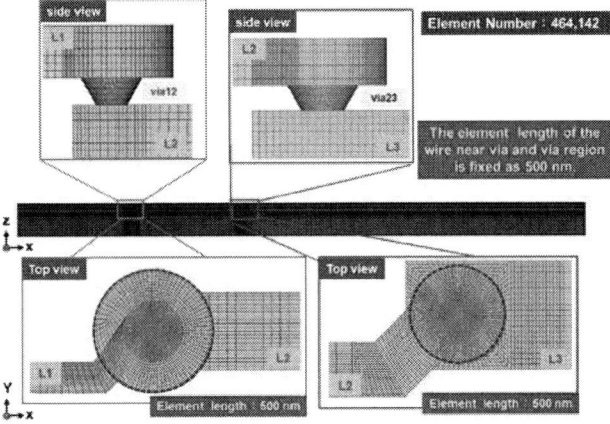

Figure 10. The detailed modeling conditions appeared in the three-layered RDLs of fan-out packaging.

As shown in Fig. 10, the structural design of vias is the focus location during simulation. Hence, an increase in the mech number adjacent to via structures, including via12 and via23, and adjacent to interconnects, are needed to make sure the numerical convergence of analysis. In this case, the total element numbers are 464,142, and the minimum size of mesh is 500nm.

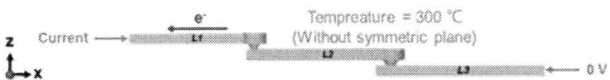

Figure 11. Applied boundary conditions of RDLs within FOPLP structure.

The boundary conditions used in simulation are pointed out in Fig. 11. The current from the left hand side of L1 interconnect and 0 volt in the right hand side of L3 interconnect are separately given. Consequently, a electrical potential difference is generated. Beside the symmetry plane,

a 300 ℃ is exerted on the surface of concerned structure. The temperature distribution obtained from electrothermal simulation is regarded as the boundary conditions for subsequent mechanical analysis. In addition, bending radius is applied from the symmetric center at this moment. Table 2 lists the analytic results. When stressing two types of currents, 5mA and 25mA, combined with identical bending radius of 30mm, the AFD distributions in the location via 12 and via 23 are found. The AFD distributions of Via12 and Via23 applied in 25mA current are higher than in 5mA. A comparison of these results are also shown in Fig. 12.

TABLE 2 AFD SIMULATION RESULTS OF VIA12 AND VIA23 UNDER DIFFERENT TYPES OF CURRENTS

Condition	5mA	25mA
	R=30mm	R=30mm
Location	AFD	AFD
Via12	0.221×10^{33}	0.343×10^{35}
Via23	0.298×10^{32}	0.488×10^{34}

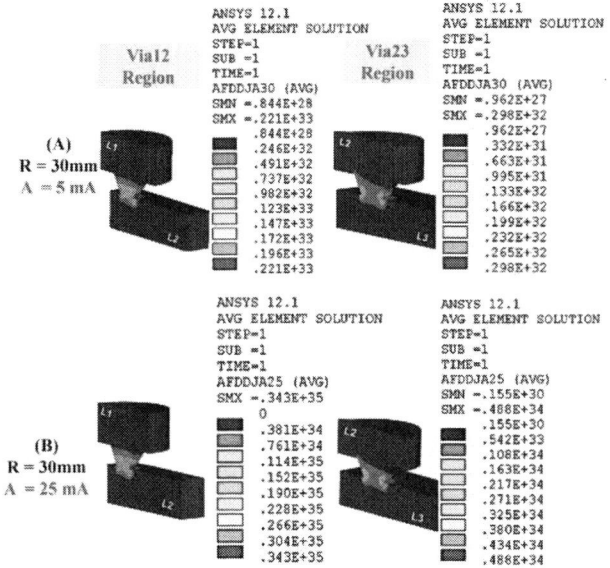

Figure 12. AFD distributions under different currents combined with an identical bending radius of 30mm: (a) 5mA; (b) 25mA.

In addition to currents, the effects of different bending radii and PI substrate thicknesses are also analyzed. From the results listed in Table 3, the maximum AFD occurs at Via 12. In addition, AFD distribution of RDL is influenced by the thickness of PI substrate. From the contour results in Fig. 13, the maximum AFD is adjacent to the bottom of via design including Via 12 and Via 23. Hence, the structural design between interconnects adjacent to vias needs to check carefully and to validate about EM failure mode. In addition, various conditions include different radius R from 0mm to 50mm and different thicknesses of PI substrate are performed. From the compared results shown in Fig. 14, this structural design of 15μm thick PI substrate has the risk of void generation under the condition of that applied current is large than 10mA and the bending radius is 30mm. In Fig. 15, our research team uses this simulation approach to validate the structural design of FOPLP related to foldable RDLs of flexible devices displayed in 2018 SEMICON Taiwan.

TABLE 3 AFD ESTIMATED RESULTS OF VIA12 AND VIA23 UNDER DIFFERENT TYPES OF SUBSTRATE THICKNESS AND BENDING RADIUS

Condition	5mA	5mA	5mA
	15μm PI substrate	15μm PI substrate	30μm PI substrate
	R=0mm	R=10mm	R=10mm
Location	AFD	AFD	AFD
Via12	0.650×10^{24}	0.602×10^{24}	0.636×10^{24}
Via23	0.245×10^{24}	0.235×10^{24}	0.243×10^{24}

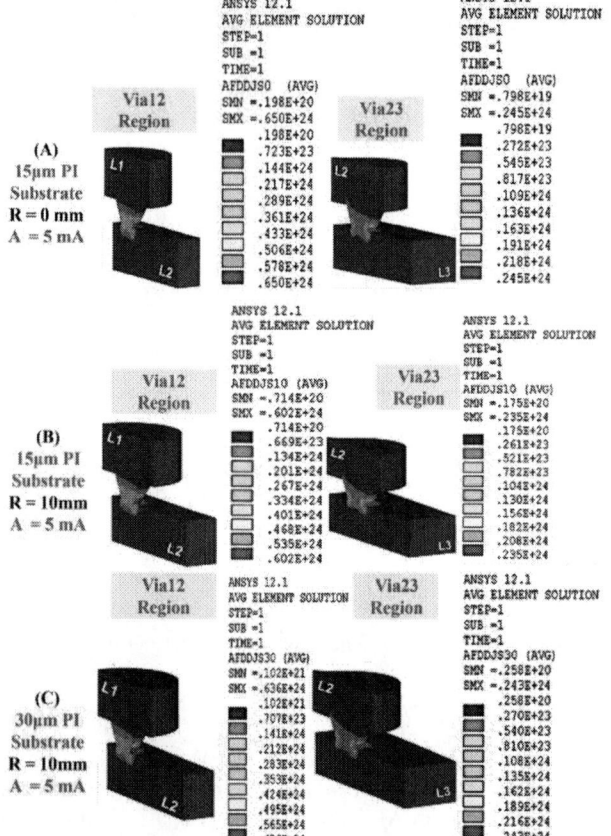

Figure 13. The influences of different bending radius for AFD distribution in same current 5mA: (a) 15μm PI substrate/ 0 mm bending radius; (b) 15μm PI substrate/ 10 mm bending radius; (c) 30μm PI substrate/ 10 mm bending radius.

V. CONCLUSIONS

A study of EM/TM/SM combined behavior has performed for flexible electronic components. By revised simulation methodology, it has the excellent capability to predict structural design of flexible electronic components under complex environment factors in the future when more compound failure modes are suffered [14-17]. In common, for high power packaging component applications or fine pitch wire package designs, key factor only considers EM. However, with customer demand including more application in SiP or flexible hybrid electronics (FHE) and material development such as flexible substrate, it has been expected that physical phenomena would faces not only EM situation and also stress effect and temperature influence. Therefore, a novel simulation prediction method for future flexible electric components is proposed. In addition, the present simulated approach is also suitable for the reliability estimation of multi-physical phenomena of EM/TM/SM.

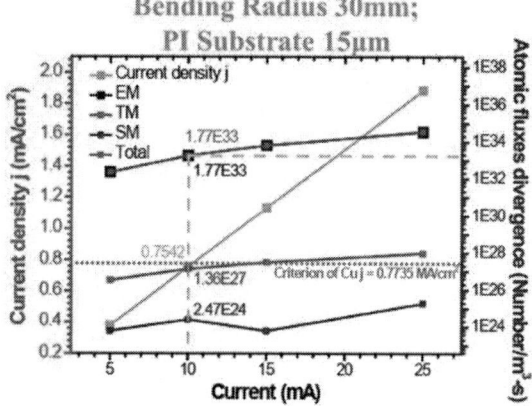

Figure 14. Estimated effect of EM/SM/TM under the applications of different currents from 5mA to 25mA.

Figure 15. Fine-pitch interconnects of foldable RDLs in flexible devices.

ACKNOWLEDGMENT

The authors would like to thank the Ministry of Economic Affairs (MOEA) and the Ministry of Science and Technology (MOST), Taiwan, R.O.C., for providing financial support under contract numbers MOEA 107-EC-17-A-24-1442, MOST 105-2628-E-007-015-MY3, and MOST 107-2622-E-007-010-CC3.

REFERENCES

[1] H. S. Hsu et al., "Electromigration induced spontaneous Ag whisker growth in fine Ag-alloy bonding interconnects: Novel polarity effect," Materials Letters, vol. 182, pp. 55-58, 2016.

[2] Y. C. Lee et al., "Fan-out chip on substrate device interconnection reliability analysis," Electronic Components and Technology Conference (ECTC), pp. 22-27, 2017.

[3] F. Liu et al., "Next generation panel-scale RDL with ultra small photo vias and ultra-fine embedded trenches for low cost 2.5D interposers and high density Fan-Out WLPs," Electronic Components and Technology Conference (ECTC), pp. 1515-1521, 2016.

[4] C. M. Tan et al., "Revisit to the finite element modeling of electromigration for narrow interconnects," Journal of Applied Physics, pp. 1-7, 2007.

[5] C. M. Tan et al., "Investigation of the effect of temperature and stress gradients on accelerated EM test for Cu narrow interconnects," Thin Solid Films, vol. 504, no. 1-2, pp. 288-293, 2006.

[6] J. R. Black, "Electromigration - A brief survey and some recent results," IEEE Transactions on Electron Devices, pp. 338-347, 1969.

[7] C. C. Lee et al., "Electromigration characteristic of SnAg3.0Cu0.5 Flip Chip interconnection," IEEE Transactions on Device and Materials Reliability, vol. 5, no. 2, pp. 198-205, 2006.

[8] M. A. Rui et al., "Electromigration simulation of flip chip CSP LED," International Conference on Electronic Packaging Technology (ICEPT), pp. 1133-1137, 2017.

[9] S. Yokogawa et al., "Electromigration induced incubation, drift and threshold in single-damascene copper interconnects," IEEE Interconnect Technology Conference, 2002.

[10] C. M. Tan et al., "Current crowding effect on copper dual damascene via bottom failure for ULSI applications." IEEE Transactions on Device and Materials Reliability, pp. 198-205, 2005.

[11] G. A. Rinne, "Electromigration in SnPb and Pb-free solder bumps," Electronic Components and Technology Conference (ECTC), 2004.

[12] K. N. Chiang et al., "Current crowding-induced electromigration in Sn Ag3.0Cu0.5 microbumps," Applied Physics Letters, 88, 072102, 2006.

[13] T. Y. Lee et al., "Electromigration of eutectic SnPb solder interconnects for flip chip technology," Journal of Applied Physics, pp. 3189-3194, 2001.

[14] Y. Chin et al., "Effects of underlayer dielectric on the thermal characteristics and electromigration resistance of copper interconnect," Japanese Journal of Applied Physics, vol. 42, no. 1, pp. 7502-7509, 2003.

[15] Y. Liu et al., "Joule heating enhanced electromigration failure in redistribution layer in 2.5D IC," Electronic Components and Technology Conference (ECTC), pp. 1359-1363, 2016.

[16] Ebersberger et al., "Reliability of lead-free SnAg solder bumps: influence of electromigration and temperature," Electronic Components and Technology Conference (ECTC), 2005.

[17] E. C. C. Yeh et al., "Current-crowding-induced electromigration failure in flip chip solder joints," Applied Physics Letters, pp. 580-582, 2002.

A complex integrated circuit structure transformation, modeling and simulation method

Daixing Wang, Yufeng Jin
School of Electronic and Computer Engineering
Peking University Shenzhen Graduate School
Shenzhen, Guangdong, China

Wei Wang*
Institute of Microelectronics
Peking University
Beijing, China
e-mail: w.wang@pku.edu.cn

Abstract—The performance degradation and thermal-mechanical problems caused by the huge heat in integrated circuits, especially high power integrated circuits, such as RF circuits, are gradually attracting attention. Therefore, thermal simulation is particularly important in the circuit design process. However, the current integrated circuit design software and the finite element thermal simulation software are not related to each other, and the complexity of the circuit files can't be modeled in the thermal simulation software in detail. As a result, the simulation method only focus on the total power density and the main circuit substrate materials, but don't pay attention to the high thermal conductivity path from hot spot to heat sink composed of wiring, through silicon via (TSV) and Bump. The absence of this electro-thermal collaborative design will lead to great simulation errors, which will affect the design process and performance of the circuit. In order to simulate the structure which has a great effect on the heat dissipation of integrated circuits, this paper provides a conversion method from circuit file to three dimensional(3D) geometry structure. In this method, we can extract the detailed multilayer circuit structure from circuit files in popular circuit design software, and model the circuit structure in finite element simulation software, so as to achieve accurate simulation of high thermal conductivity structure of integrated circuits. To verify the conversion effect and simulation accuracy of the method, this paper takes an actual Low Temperature Co-fired Ceramic(LTCC) radio frequency(RF) circuit as an example to compare the thermal test and simulation results. The results show that thermal simulation based on circuit reconstruction is more accurate than rough simulation ignoring through holes and wiring layers, and the error is less than 4.84% compared with the actual thermal test.

Keywords-circuit file; transformation; 3D geometric structure; thermal; simulation;

I. INTRODUCTION

With the development of Moore's Law, the integration and power density of integrated circuits(IC) are gradually increasing, which leads to more and more serious thermal problems [1-3]. In high power density integrated circuits, especially in RF circuits, thermal issues have a more significant impact on device performance. For example, in dedicated circuits, the power density has reached 1 kW/cm² [4,5]. The thermal problem is not only caused by high power density integration, but also by the low thermal conductivity substrate in the circuit.

The LTCC substrates is widely used in RF circuits. LTCC has excellent high frequency, high speed transmission and wide passband characteristics, which is beneficial to improve the quality factor of the circuit system. In addition, LTCC has high current and high temperature characteristics which can manufacture three-dimensional circuit substrates, and increases circuit integration density. However, the thermal conductivity of the LTCC substrate is 2-6 W/(m·K), which depends on the composition of the material and is much lower than the thermal conductivity of the Si substrate (130 W/(m·K)), gold wiring and copper wiring (400 W/(m·K)). Therefore, in high integration, multi-layer, high-power-density circuits, the LTCC substrate with the low thermal conductivity will cause the substrate temperature to rise sharply, which will affect the circuit performance [6-7].

In current simulation software or engineering cases, due to the complexity of circuit wiring layer and the low efficiency of fluid-solid coupling thermal simulation, researchers generally only consider the uniform heat source and the uniform substrate, but ignore the metal wiring layer in the substrate that plays a significant role in heat transfer. This leads to a large deviation in the thermal simulation results of the circuit. In addition, the accurate microfluidic layout can't be carried out, which reduces the effect of microfluidic cooling [8].

In this paper, we explored the thermal problems of actual LTCC RF substrates. Firstly, we developed the software for multi-layer layout of integrated circuits to three-dimensional modeling and simulation, and conduct thermal simulation of substrate structure. Next, the microfluidic design is carried out to further reduce the temperature. Finally, the thermal distribution of LTCC substrate is studied by thermal testing and the simulation results are compared with the actual test results.

II. LTCC SUBSTRATE INFORMATION

In this paper, our research team explores the heat transfer performance of the substrate and the three-dimensional modeling of the substrate. Figure 1(a) shows the LTCC RF substrate to be researched. The LTCC RF substrate consists of 15 layers of circuit. Since the structure of this substrate is repeatable, we only study a quarter of the substrate as shown in Figure 1(b).

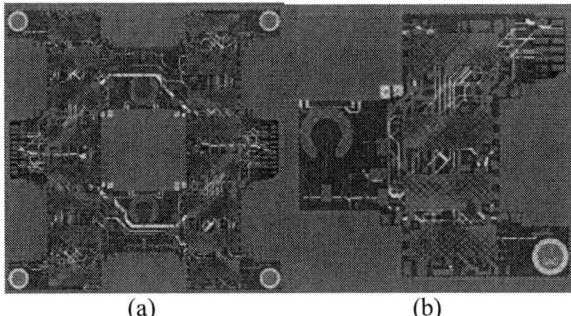

| (a) | (b) |

Figure 1. The LTCC RF substrate circuit: (a) the whole circuit structure, (b) the 1/4 circuit structure.

Table 1 shows the detailed structure size and material distribution of the LTCC RF substrate. The RF substrate material is LTCC with a size of 4 cm×4 cm×0.21 mm. There are 16 GaN power chips on the surface of the substrate. The chip area is 1.6×0.8 mm and the power density is 30 W/cm². LTCC is a layered structure formed by sintering ceramic powder with a thickness of about 96 μm. The thermal conductivity of LTCC substrate is only 2 W/(m·K) (FerroA6M). The working environment temperature of LTCC RF substrate is 60°C which will lead to serious thermal problems and affect the normal operation of the device.

TABLE I.   THE LTCC RF SUBSTRATE MATERIAL DISTRIBUTION.

Component	Material	Size (μ m)	Thermal conductivity [W/(m · K)]
LTCC	ceramic	Height: 2100	2
Wire	Au	Width: 127~300 Height: 10	310
Through-hole	Au	Radius: 0.1~0.75 Height:96	310
Heat source chip	GaN	Length: 1600 Width: 800 Height: 400	230

III.   SUBSTRATE THERMAL MANAGEMENT

In order to ensure the normal work of LTCC substrate, we must carry out thermal management for it. As hot spots, 16 power chips conduct heat through natural convection heat transfer with air, and the other part conducts heat through the substrate to the heat sink at the bottom. Because of the high working environment temperature and the low thermal conductivity of LTCC substrate limitation, the effect of these two heat transfer methods is not enough to meet the requirements. In order to carry out the accurate substrate heat transfer analysis and efficient heat dissipation design, 3D geometric reconstruction and thermal simulation of the detailed structure of the substrate is required. At the same time, the design of microchannel can further increase the heat dissipation effect.

A.   The Software for 3-D Structural Reconstruction

The biggest difficulty in developing general software for electrical and thermal designers is to completely reproduce the 3D electrical structure, so that detailed structural modeling and thermal simulation can be established. Figure 1(a) is the circuit structure of the LTCC RF substrate. In order to analyze the thermal problem of LTCC RF substrate, we have to reproduce the structure in finite element thermal simulation software. However, it is almost impossible for us to model the 15-layer circuit board manually, which is the necessity for us to develop software.

Taking a quarter circuit as an example, different colors represent different circuit layers. This structure consists of 15 layers. Each layer contains different sizes and quantities of through holes, wiring and irregular RF grounded. The structure of the RF substrate processing file is shown in Figure 2. Detailed structural types such as position, size and shape (through hole, rectangle, polygon) are indicated in the file. In order to build a software bridge between electrical and thermal designers, we must analyze the files by computer.

G04 Layer_Physical_Order=1*	G36* //Build Polygon	D10* //Build Via
G04 Layer_Color=255*	X828Y703D02*	X1950Y1991D02*
%FSLAX42Y42*%	X828Y703D01*	Y1781D02*
%MOMM*%	Y703D01*	D03*
%TF.Part,Single*%	Y683D01*	D11*
G75*	X828Y683D01*	X1920Y1921D02*
%TA.AperFunction,NonConductor*	X808D01*	D03*
%	Y673D01*	D12* //Build wire
%ADD10C,0.14*%	X808Y673D01*	X2048Y1921D02*
%TA.AperFunction,Conductor*%	X778D01*	D03*
%ADD11C,0.14*%	X778Y673D01*	D13*
%TA.AperFunction,SMDPad,CuDef*	Y673D01*	X2072Y1781D02*
%ADD12R,1.00X0.40*%	X778Y703D01*	D03*
%ADD13R,0.60X0.40*%	X828Y703D01*	D14*
%TA.AperFunction,Via*%	D02*	X1511Y538D02*
%ADD14C,3.00*% a	G37* b	D03* c

Figure 2. The circuit file structure diagram: (a) circuit structure statement, (b) polygonal structure, (c) establishment of via and wiring.

For the circuit file, the software we developed is mainly divided into six modules. Figure 3 shows the flow chart of the software system module. The first module is used to read the circuit file information and save the circuit information in the computer to prepare for the follow-up operation. The second module is the reconstruction system of 3D geometry module, which reconstructs the three-dimensional structure of the metal holes and wiring layers according to the circuit file information. The third module is material distribution, which gives different material attributes according to the detailed structure of the circuit to prepare for the later thermal simulation. The fourth module establishes the finite element mesh. The fifth module is to build physical field, which can be added according to different simulation types, such as solid heat transfer or fluid-solid coupling heat transfer. The sixth module is the data export module. According to the requirements, the temperature distribution of points, lines and surfaces can be derived as the next design guidance.

978-1-7281-1500-9/19 $31.00 © 2019 IEEE   2017

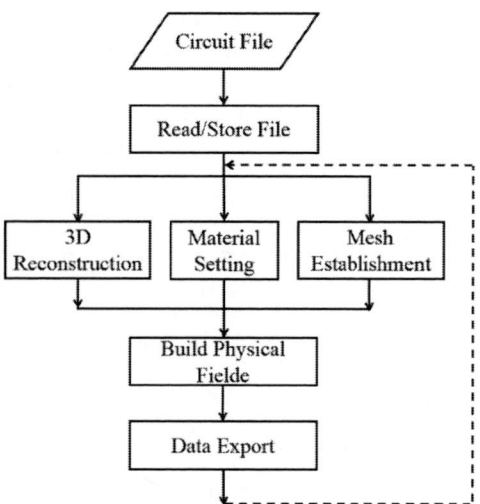

Figure 3. Flow chart of software system module.

Figure. 4 is the 3D image of the LTCC RF substrate which is reconstructed by the conversion software. In the picture, different colors represent different materials. Compared with the circuit structure of Figure. 1 (b), the through holes, wiring and the polygon area in the circuit structure can be fully modeled. After obtaining detailed geometry and material distribution, we can perform thermal simulation by adding physical field.

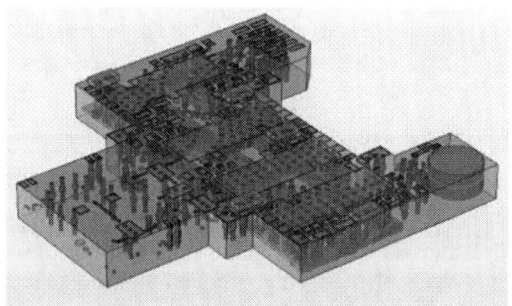

Figure 4. 3D Reconstruction of LTCC Substrate(Golden: wiring, Gray: LTCC, Blue: water, Green: GaN)

B. The Microchannel Structure

After 3D modeling of LTCC substrate, we can obtain he detailed heat transfer structure of the substrate. However, there isn't an effective method for heat dissipation. Therefore, for the irregular shape of the LTCC RF substrate and the distribution of 16 hotspots, we built a microfluidic structure for powerful heat dissipation.

The chip distribution of the LTCC RF substrate is special, and the assembly position of the chip is at the edge of the substrate. In order to increase the heat dissipation effect of the microchannel, we hope to distribute the microchannel directly below the hotspots as far as possible so as to increase the heat dissipation effect of the microchannel. However, due to the limitation of machining accuracy and material characteristics of LTCC, the edge of LTCC will be damaged in the process of laser cutting, resulting in the leakage phenomenon. LTCC substrate consists of 15 layers of circuit. In order to ensure the integrity of electrical signals, 5 layers can be added to the bottom of the substrate for microchannel layout, as shown in Figure 5. In the actual design of microchannel, the microchannel is 0.5mm away from the LTCC edge and directly below the metal through holes, which can ensure the integrity of the substrate and heat dissipation effect.

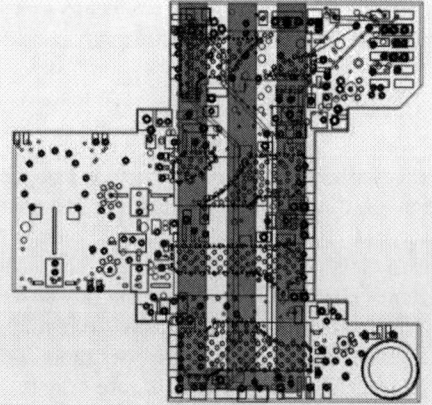

Figure 5. The structure of microchannel

At present, the pressure drop of portable microfluidic pump is about 50kPa, so the heat dissipation effect and equipment limitation should be taken into account in the design of microchannel. Figure 6 shows the relationship between the width of microchannel, the maximum temperature and pressure drop. It can be seen from the figure that the pressure drop decreases gradually with the increase of the width of the microchannel at the flow rate of 180 ml/min. In addition to the pressure drop limitation, LTCC substrate involves co-firing and hot pressing processes in the manufacturing process, and excessive width of microchannel may lead to collapse. So considering comprehensively, the width of micro-channel is set to 1.5 mm. Limited by the thickness of the substrate, the microchannel consists of two layers of LTCC with a thickness of about 0.2 mm.

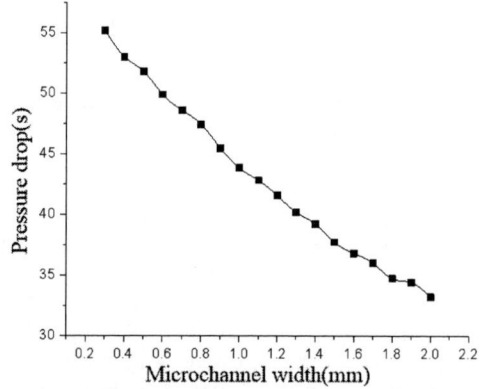

Figure 6. The relationship between pressure drop and microchannel width.

IV. SIMULATION AND TEST RESULT

After developing the software, we can reproduce the 3D model of the circuit structure. After obtaining the specific structure and material distribution of LTCC substrate, we can carry out the finite element thermal simulation to obtain the thermal distribution of the whole substrate by meshing and applying physical field. The simulation results provide guidance for electrical design. At the same time, LTCC substrate is processed, and the actual thermal test results are used to verify the accuracy of the simulation. Figure.7 (a) is a diagram of substrate on which power chips had been placed. Figure 7(b) is the reverse of substrate. There are four inlet and outlet channels in the figure. Figure 7 (c) shows the layout of substrate processing. The green rectangle in the figure is the microchannel. Figure. 7(d) is a cross-sectional view of a 1500×200 μm microchannel in LTCC substrate. It can be seen that the substrate contains a large number of gold wiring layers.

Figure 7. LTCC RF Substrate: (a) Substrate surface, (b) Reverse side of substrate, (c) LTCC Layout, (d) Substrate cross-sectional view.

A. Solid Heat Transfer Results

After obtaining the three-dimensional structure of LTCC RF substrate, the heat transfer performance of the substrate is studied by thermal simulation. A quarter of the substrate in Figure 4 contains five power chips (30 W/cm²). The working temperature of LTCC RF substrate is 60°C, and the boundary condition is natural convection heat transfer (25 W/(m·K)). Figure 8 shows the thermal simulation results of the substrate.

Figure 8 (a) is the thermal simulation result of the 3D reconstruction of the LTCC RF substrate. Figure 8(b) shows the rough simulation result ignoring through holes and wiring layers considering only the hotspot and the substrate

material without considering the metal wiring layer and via hole in the substrate.

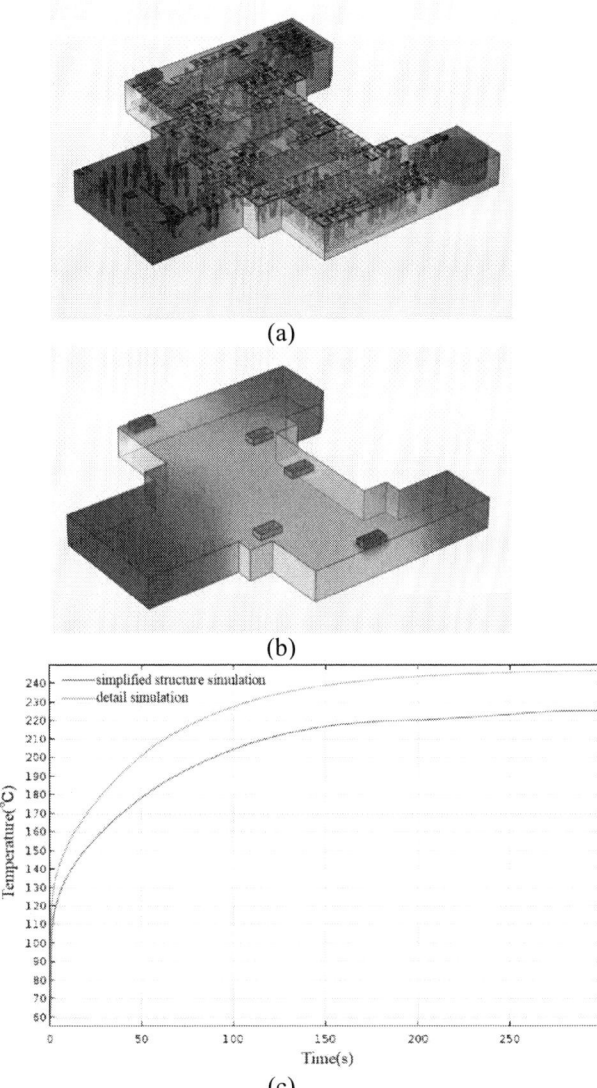

Figure 8. Solid heat transfer results: (a) Detailed thermal simulation of substrate structure, (b) Rough thermal simulation ignoring through holes and wiring layers of substrate, (d) Maximum temperature curve of two simulation results.

It can be seen that the highest temperature of the substrate appears at the power chip. In the detailed simulation of Figure 8(a), the maximum temperature can reach 225.53°C, while in the rough simulation of Figure 8(b), the maximum temperature of the substrate is 248.26°C. The simulation error has reached 10%. In Figure 8(a), the through hole under the hot spot can transfer heat from the hotspots to the bottom of the substrate. In the rough simulation ignoring through holes and wiring layers, the temperature will be higher than the detailed simulation because the substrate ignores the existence of high thermal conductivity path. The temperature change proves that the metal wiring layers and the through holes in the circuit structure play a great role in heat dissipation and it is necessary for 3D modeling reconstruction of circuit structure.

LTCC RF substrate is tested by infrared thermal imager. Figure 9 shows the test results of the RF substrate. Figure 9(a) is a thermal infrared image of the substrate in stable state. The uneven temperature distribution on the surface of the substrate shows that the wiring layer and the through hole effectively disperse the heat generated by power chips. The low temperature areas on the substrate surface is metal material areas, which is better heat dissipation effect with higher thermal conductivity (310 W/(m·K)) than LTCC (2 W/(m·K)). It also shows that the detailed structure of the substrate is indispensable in the thermal simulation process.

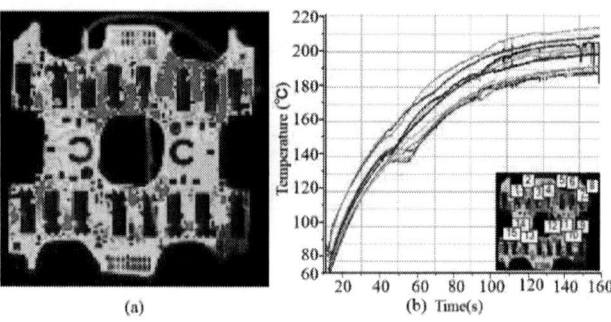

(a)

(b) Time(s)

Figure 9.   The result Infrared thermal test: (a) substrate heat distribution, (b) hotspots temperature curve.

Figure 9(b) shows the temperature rise of hotspots in the substrate. With the increase of time, the hot spot temperature gradually increases, reaching a stable state after 160s, and the highest temperature is 214.62°C. Compared with the test result, there is a 4.84% error in the detailed thermal simulation result(225.53°C), and a 15.67% error in the rough simulation result(248.26°C).

B.   Microchannel Heat Dissipation Results

The maximum temperature of the LTCC RF substrate under solid heat transfer simulation and test has reached 210°C. The maximum temperature exceeds the maximum operating temperature of the device. At this high temperature, it has seriously affected the electrical and mechanical properties of itself, the surrounding devices and the substrate. In order to reduce the temperature of substrate and power chip, microfluidic thermal simulation and test are carried out. The boundary conditions of microfluidic heat dissipation are: flow rate 180 ml/min, water temperature 20°C, environment temperature 60°C, heat transfer coefficient 25 W/(m·K).

Figure 10 is the heat distribution diagram obtained by microfluidic finite element thermal simulation. Figure 10(a) is the microfluidic simulation of the reconstructed circuit structure of the substrate. The chips with the highest temperature are located at the outlet of microfluids, and the heat dissipation effect is the worst. The highest temperature here is 91.24°C, and the temperature drop is 123.38°C. Microfluids not only reduce the chip temperature, but also reduce the LTCC substrate temperature from 120°C to about 45°C. Figure 10(b) is the microfluidic heat dissipation simulation result for the rough substrate structure ignoring

through holes and wiring layers. It can be clearly seen from the figure that the overall temperature of the substrate is higher than that of the detailed simulation. The reason is that the metal wiring layers and holes with high thermal conductivity disappear in the LTCC RF substrate after simplification, which reduces the thermal conductivity from the surface to the inside of the substrate. Therefore, the rough circuit structure has large errors in the thermal design process, which is not enough to provide guidance for thermal design.

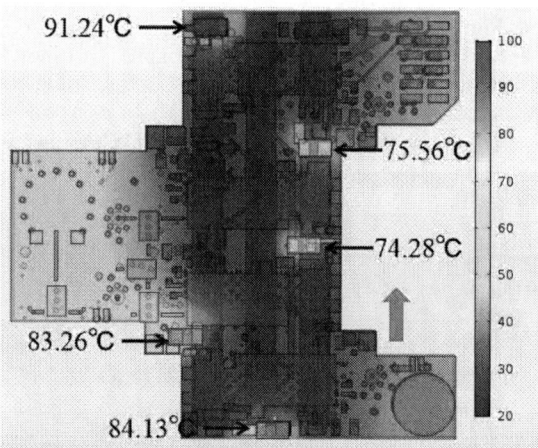

(a) Detail structure simulation

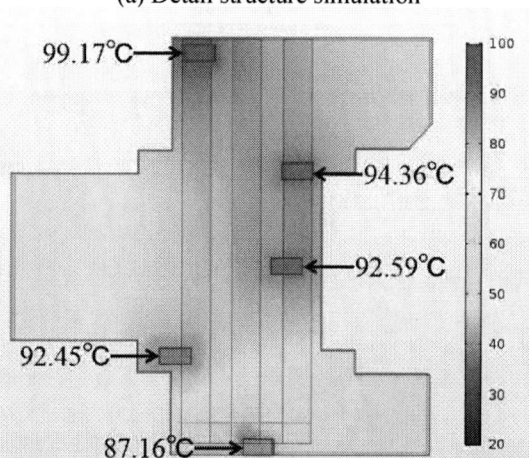

(b) Rough structure simulation

Figure 10. Thermal simulation results of microfluids: (a) detail structure simulation, (b) rough structure simulation.

Figure 11 is the result of microfluidic heat dissipation of LTCC RF substrate under infrared thermal imager. It can be seen from the temperature curve of Figure 11(b) that the microfluid heat dissipation is performed when the power chip temperature is no longer increased at the steady state, and the maximum temperature of the chip is reduced from 214.62°C to about 80°C after 20 s. At this time, the maximum temperature of the whole substrate has been lowered to below 100°C, which ensures the normal operation of the LTCC RF substrate. The highest temperature on the substrate after microfluidic cooling is at the microfluidic

978-1-7281-1500-9/19 $31.00 © 2019 IEEE   2020

outlet. From Figure 11(b), the maximum temperature is 95.48°C. Compared with the test results, there is only 4.25% error in the finite element thermal simulation of the detailed structure of the substrate (91.42°C). Therefore, the detailed high thermal conductivity structure of the substrate must be considered in the thermal design process.

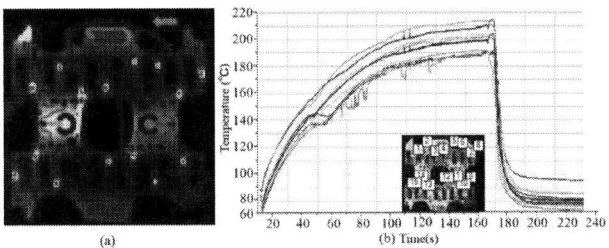

Figure 11. The Infrared microfluidic thermal test: (a) substrate heat distribution, (b) hotspots temperature curve.

V. CONCLUSION

In this paper, a software for modeling the circuit three dimensional geometric structure is developed. The software can reconstruct the large-scale complex circuit structures in three-dimensional geometry. And material properties are also set according to the detailed circuit structure. Then physical fields are added according to different simulation types. Finally, the temperature distribution of the circuit is obtained by the finite element thermal simulation.

This paper also manufactures LTCC RF substrate to verify the accuracy of the software. Through the solid heat transfer test and the microchannel heat transfer test, the thermal simulation error of three-dimensional reconstruction of circuit structure is controlled at 4.84% and 4.25% respectively. In the face of shrinking circuits and increasingly serious thermal problems, this circuit structure reconstruction software has great application prospects.

ACKNOWLEDGMENT

This work was supported by grants from the National Natural Science Foundation of China (Grant No. U1613215 and Grant No. U1537208).

This work was supported from the TSV 3D integrated micro/Nanosystem Lab.

REFERENCES

[1] Banerjee, Kaustav, M. Pedram, and A. H. Ajami. "Analysis and optimization of thermal issues in high-performance VLSI." International Symposium on Physical Design ACM, 2001, pp. 230-237, doi:10.1145/369691.369779.)

[2] Im S, Banerjee K. "Full chip thermal analysis of planar (2-D) and vertically integrated (3-D) high performance ICs." In: International Electron Devices Meeting (IEDM), San Francisco, 2000, pp. 727-730, doi:10.1109/IEDM.2000.904421.

[3] Wang T Y, Tsai J L, Chen C P. "Thermal and power integrity based power/ground networks optimization." In: Design, Automation and Test in Europe (DATE), France, 2004, pp. 1530-1591, doi: 10.1109/DATE.2004.1268986.

[4] Fodor, A., et al. "Guidelines on thermal management solutions for modern packaging technologies-a review." Design and Technology in Electronic Packaging (SIITME), 2015 IEEE 21st International Symposium for. IEEE, 2015.

[5] J. Q. Lu, K. Rose, S. Vitkavage, "3D Integration: Why, what, who, when? ", Future Fab Int, 2007, 23, pp. 25-26.

[40] J. Jang, H. S. Kim, W. Cho, et al., "Vertical cell array using TCAT (Terabit Cell Array Transistor) technology for ultra-high density NAND flash memory", Symposium on VLSI Technology, 2009, pp. 192-193.

[6] Golonka L J, Technology and Applications of Low Temperature Cofired Ceramic(LTCC) Based Sensors and Microsystems, Faculty of Microsystem Electronics and Photonics, Wroclaw University of Technology, Bulletin of the Polish Academy of Sciences Technical Sciences Vol. 54, No. 2, 2006

[7] Dominik Jurków, Thomas Maeder, Arkadiusz Dąbrowski, Marina Santo Zarnik, Darko Belavič, Heike Bartsch, Jens Müller. "Overview on low temperature co-fired ceramic sensors." Sensors and Actuators A-physical 2015 : Volume 233, pp. 125-146.

[8] Cong, J., Luo, G., & Shi, Y. "Thermal-aware cell and through-silicon-via co-placement for 3D ICs". In Proceedings of the 48th Design Automation Conference , 2011, June. pp. 670-675.

A Study on the Oxygen Plasma Treatment on the Peel Adhesion Strength and Solder Wettability of SnBi58 based Anisotropic Conductive Films

Shuye Zhang[1], Mingliang Huang[2], Yang Wu[2], Ming Yang[3], Tiesong Lin[1], Peng He[1] and Kyung-Wook Paik[4]

[1]State Key Laboratory of Advanced Welding and Joining, Harbin Institute of Technology, Harbin 150001, China
[2]Department of Materials Science and engineering, Dalian University of Technology, Dalian, China
[3]Hisilicon Optoelectronics Co., Ltd, Wuhan 430073, China
[4]Nano-Packaging and Interconnection Laboratory, KAIST, Daejeon 305-701, South Korea
syzhang@hit.edu.cn

Abstract— O₂ plasma treatment is a useful way to increase adhesion strength between polymer resins and adhered substrates in electronic packaging areas. However, a low adhesion strength of polymer adhesives on the adhered surfaces is a critical issue for anisotropic conductive film (ACF) joint reliability, especially when it comes to moisture-induced effects. This paper discusses the effects of oxygen plasma treatment (100W 20mTorr 3min) on the wettability and reliability of solder ACFs joints on Au/Ni metal electrodes. We carried out the surface analysis using surface energy, AFM, XPS, FTIR and joint resistances, peel adhesion strength, and reliability. By using the contact angle and surface energy, the un-wetted solder joining was explained from a spontaneous wettability to a hindered wettability, as a reason for poor electrical performance and reliability after oxygen treatment. Although the resin and electrode adhesion was increased, it was proven that solder part played a more important role in determining joint reliability and mechanical property.

Keywords-O₂ plasma; adhesion strength; solder wettability; anisotropic conductive films

I. INTRODUCTION

Anisotropic conductive films (ACFs) are conventionally made up of micro-sized conductive fillers and polymer matrixes.[1] Conductive fillers can be applied by Ag or Ni based metal balls[2-3], metal coated polymer balls,[4] and Ag or Cu nanowires[5-6]. Polymer matrixes are usually divided into 3 types, such as epoxy resin, polyurethane and silicone resin.[7-9] Polymer resins are always as a thermosetting base, providing a mechanical protect when fully cured at high temperature.[10]

Recently, solder particles have been added into ACFs to support a solder metallurgical joint between Sn-base alloy and Cu or Ni metal substrates for a higher reliability. [11] Although metallic Au, Ag or Cu conductive fillers exhibit excellent comprehensive properties, such as electrical conductivity, thermal conductivity and chemical stability, on the contrary, the high cost of the products is not able to be solved. [12] What's more, Cu conductive fillers are easily to

be oxidized[13] and Ag nanowires have a promising future due to unique electrical, optical, and thermal properties and their potential applications in microelectronic, optoelectronic devices and sensors.[14-15] However, the poor reliability of degradation in terms of ultraviolet radiation and unwanted migration of Ag atoms is a crucial issue to Ag nanowire reliability.[16]

In the history of solder ACFs developments, industrial engineers started to use a low melting point solder into anisotropic conductive adhesives to perform a flip-chip bonding process on 2006.[17,18] As solder oxide layer is a big issue for solder joining, Dr. Kiwon Lee invented an ultrasonic method to wipe off solder oxide layer and enhance solder wettability during solder ACFs joints bonding in 2011,[19] Afterwards, Dr. Yoo-Sun Kim optimized the ultrasonic bonding parameters on ACFs joint properties and investigated an ultrasonic-assisted thermo-compression bonding method of solder ACF joints for reliable camera module packaging on 2013. [20-21]

More recently, Dr. Sang-Hoon Lee cooperated solder joints with nanofibers to control solder ball movement during ACF bonding process, obtaining fine-pitch capability of ACFs under 50 um.[22] Dr. Tae-Wan Kim utilized nanofiber/solder ACFs joints into flex-on-flex applications and found out the bending performance reliability.[23-24] In addition, Dr. Yoo-Sun Kim investigated the bonding pressures effects on solder joint morphology[25] and Dr. Zhang optimized the best solder ball size and content for flex-on-board assembly of solder ACFs joints [26]. Hydro-swelling is a critical reliability issue for ACFs joints, Dr. Zhang addressed the failure mechanism of SnBi58 solder ACFs joints in case of hydro-swelling.[27] It was found out the same trend of Sn-3Ag-0.5Cu solder alloy in epoxy ACFs joints.[28]

Most researchers are optimizing the solder joints, however, the adhesion strength of ACFs is still low. Especially, adhesion strength is very important for ACF bending reliability.[29] Kim Ji-Hye explored the ACFs adhesion on the bending reliability of chip-in-flex packages

978-1-7281-1500-9/19 $31.00 © 2019 IEEE

and Lin Chao-Ming investigated the fracture and conductivity of flex-on-flim flexible bonding using repeated bending.[30] In the old days, Chan Y. C. and Noh Bo-In had investigated the adhesive joint behavior under 3-point-bneding and 4-point bending test, respectively.[31-32] Although the ACF modulus and Au electrode gaps were discussed, there were few studies about the epoxy/electrode interfaces to enhance the adhesion performances.

In order to enhance the adhesion strength, 2 ways are usually used. One is to use a mechanical interlock in rough interfaces to increase forcing areas and interface areas [33], another is to use a chemical way to increase the Van der Waal forces and Hydrogen bonds[34-35]. Although some researchers managed to use a high modulus of polymer resin to increase the peel adhesion strength, that is not a common method.[36] In addition, Table I summarized the adhesive forces and bond energy in various types from previous research.[37-38] It indicated that intermolecular interactions played a more important role in adhesive forces. So it is a wise way to increase Van der Waal forces and hydrogen bonds for adhesives.[39]

Table I Adhesive forces in various types

Types		Bond length in nm	Bond energy in kJ/mol
Chemical bonds	Covalent	0.1-0.2	150-950
	Metallic	0.3-0.5	100-400
	Ionic	0.2-0.3	400-800
Intermolecular interactions	Van der Waal forces	0.4-0.5	2-15
	Hydrogen bonds	0.2	20-30

On the other hands, surface cleaning by using plasma to remove small molecules and oils at flexible printed circuits (FPCs) [40] substrate and printed circuit board (PCB)[41] have been investigated to increase the adhesive adhesion. According to the previous results, the adhesion energy was increased by plasma treatment from 55 to 86J/m^2, which was by over 50%. However, what if the residual O_2 meets with the solder alloy. The morphology of the micron sized solder ACFs joint would be changed and the reliability results could be undesirable, though the mechanical adhesion was increased.

In this study, we investigated the influences of oxygen plasma on the wettability of solder ACFs joint and 90° peel adhesion strength. The oxygen plasma was set as 100W 20mTorr 3min and solder joint shapes were evaluated by a scanning electron microscope (SEM), and the conductivity of solder ACFs joint was examined by a contact resistance method as well. The failure mechanism of solder ACFs joints after 90° peel adhesion test and pressure cooker test (121°C 100% humidity and 2atm) by oxygen plasma treatment was illustrated and found out in the last.

II. EXPERIMENT

A. Materials and test vehicles

In this study, we used a 500-um-pitch PCB and FPC substrates and a thermo-compression bonding method same as the previous study.[42] Sn-58Bi solder alloy was used as solder particles with 32-45um size and cationic epoxy was selected to provide the sufficient thermomechanical property[43], in order to prevent polymer rebound after the release of bonding pressure and causing solder joint cracks[44]. The weight percent of Sn-58Bi alloy was 30% among the ACFs weight. And a novel flux material was added to remove solder oxide layer during thermocompresion bonding. The bonding condition was set as 200°C 10s 2MPa on Sn-58Bi solder ACFs joint. Cu electrodes coated by an electroless nickel/immersion gold (ENIG) layer were patterned on PCB and FPC. The Au is around 50nm thickness and the Ni is 1.5mm thickness.

B. Plasma treatment

Plasma treatment was carried out by using an oxygen reactive-ion etcher (RIE). FR-4 PCB and FPC substrates were put in the RIE chamber and treated by a 100-W power for 3 min. The working pressure of the oxygen plasma was 20 mTorr with a constant oxygen gas flow of 100 sccm. Both the PCB and FPC were treated on the side of Au/Ni metal electrode, the schematic was shown in Fig. 1. In order to find out the mechanism of solder wettability and the changed morphologies, three types of specimens were prepared: (a) no treatment, (b) plasma-treated FPC substrate, (c) plasma-treated PCB substrate, (d) both plasma-treated FPC and PCB substrate. After plasma treatment, ACFs were directly and immediately bonded by the treated substrates.

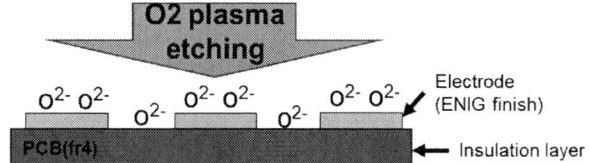

Fig. 1. Schematic of oxygen plasma treatment on electrode side

C. Solder Joint Morphology

We used a scanning electron microscope (SEM) in a back scattered electron (BSE) imaging mode to characterize solder ACFs joint morphology, as well as solder wettability, solder joint shape, intermetallic compounds between Sn-58Bi solder and Au/Ni interfaces. Since there is a clear boundary between Sn-rich and Bi-rich phase in BSE scanning, an obvious contrast of Sn-rich and Bi-rich phase can be observed in this BSE mode.[45] In addition, we compared 4 types of oxygen plasma treatment on substrates and found out the mechanism of solder morphology changes, such as no-treatment, oxygen plasma treatment on FPC, oxygen plasma treatment on PCB, and oxygen plasma treatment both on FPC and PCB.

D. Peel Adhesion Strength

A 90° peel adhesion strength test was used to evaluate FOB solder ACFs joints before and after oxygen plasma treatment. The displacement speed in the horizontal dimension is 10 mm/min. The width of FOB sample was 2-cm-wide and the peeling force was recorded as the unit of kfg. So the final peel adhesion strength should be divided by 2cm into kfg/cm. The data in each condition was tested by 3 samples, in addition, the maximum, the moderate and the minimum peeling force curves were illustrated. After peeling test, fractured morphologies and failure mode was analysis by an OLYMPUS optical scope.

E. Surface Energy

In order to precisely analyze the surface energy of the substrates, contact angle should be measured. A drop of water and a drop of glycerol was placed on the top surface of Cu plate coated with Au/Ni surface, polyimide film and FR-4 PCB. Interfacial energy between a solder and a resin γ_{SR} (249.17mJ/m^2) from previous result, interfacial energy between an electrode and a resin γ_{ER}, and interfacial energy between a solder and an electrode γ_{SE} can be obtained. The schematic of interfacial energy at solder ACFs joints was illustrated in Fig. 2.

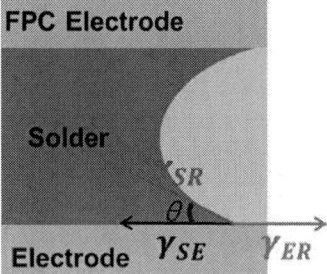

Fig. 2. Schematic of interfacial energy at solder ACFs joints

In addition, an atomic force microscope (AFM) (Nanoman, VEECO) characterized the morphology and the roughness of Au/Ni electrode surface treated by the oxygen plasma. In AFM analysis, an area of $10\times10\ \mu\text{m}^2$ was scanned using a tapping mode and the root-mean-square surface roughness (Ra) was calculated from the roughness profile measured. A fourier transform infrared spectroscopy (FTIR) was used to characterize the chemical molecule at Au/Ni metal surface after the oxygen plasma treatment.

F. Reliability

In order to analyze the solder ACFs joints reliability after the oxygen plasma treatment, we used a severe reliability test at a harsh environment as 121°C 100% humidity 2atm for 5 days to evaluate the moisture-introduced reliability for solder ACFs joints influenced by solder joint morphologies and the oxygen remaining at the interface. Contact resistance for Sn-58Bi solder ACFs joint was measured before and after the PCT reliability. During the reliability, contact resistances were measured every 24 hours. The joint morphologies and interfacial reaction were carried out.

III. RESULTS AND DISCUSSION

A. Sureface Charactierization

Fig. 3 shows the morphologies and roughness results of Au at PCB side before and after oxygen plasma treatment. It was clear that small molecule at gold surface was removed by oxygen plasma etching for 100W 20mTorr 3min and the morphology was cleaned. In details, before the oxygen plasma etching, surface roughness Ra was 97.506 nm. However, after the oxygen plasma etching, surface roughness Ra was still 97.448 nm. As a result, it was only

around 0.15 nm difference in case of surface roughness. Therefore, we regarded that surface roughness was not changed by this oxygen plasma treatment, despite the small molecules at surface was perfectly removed.

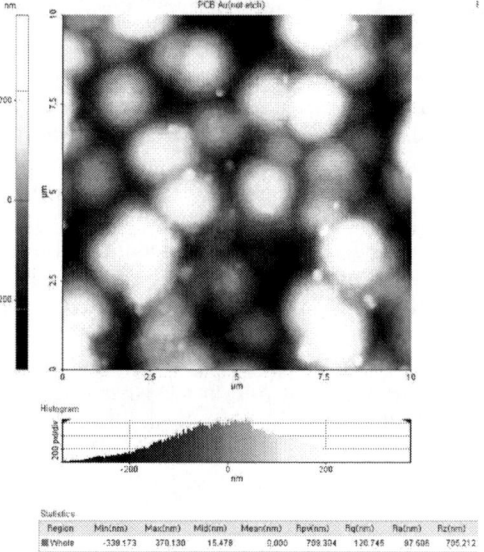

(a) PCB Au (not etch)

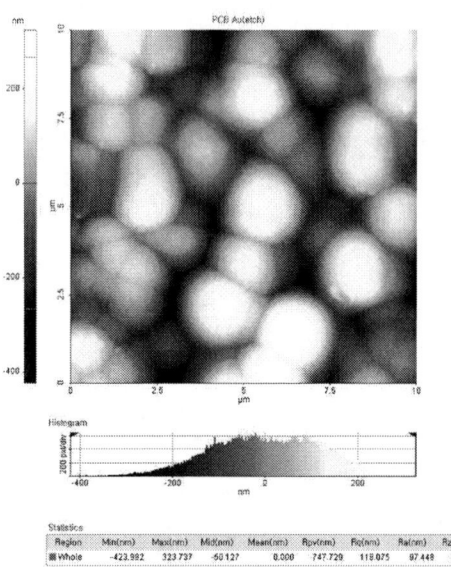

(b) PCB Au (etch)

Fig. 3. Morphologies and roughness of PCB Au by 100W 20mTorr 3min oxygen plasma etching

Fig. 4 gives the contact angle measurement suing water and glycerol on the top surface of resin and Au surface. Depending on the different dispersion and polar energy of water and glycerol, substrate surface energy can be measured using a following equation [46]:

$$\gamma_i\,(1+cos\theta_i)= 2(\gamma^d_i\gamma^d_R)^{1/2}+2(\gamma^p_i\gamma^p_R)^{1/2}$$

Where θ_i: contact angle of testing drop i; γ_i: surface energy of testing drop i; γ^d_i γ^p_i: dispersion and polar energy of testing drop i; γ^d_R γ^p_R: dispersion and polar energy of testing surface. Surface energy is the sum of dispersion and polar energy. From Fig. 4, the surface energy of epoxy resin was 34.12mJ/m^2 and that of Au surface was 80.69 mJ/m^2.

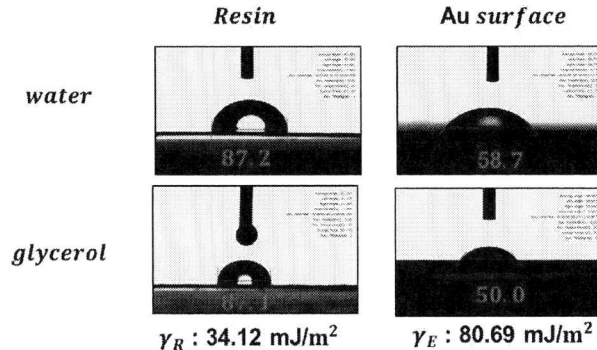

Fig. 4. Contact angle and surface energy of resin and Au surface

In addition, we used the following equation (1) to calculate the interface adhesion energy between any two interfaces.[47] After oxygen plasma treatment, the adhesion energy increased to 48.5mJ/m^2, which was the double amount of the previous result. On the contrast, solder wetting energy on oxygen plasma treated Au surface was reduced to 249.9 mJ/m^2 from the previously untreated 384.1 mJ/m^2. Fig.5 gives the wetted angle of solder on O$_2$ plasma untreated and treated surfaces. 18.6° wetting angle was achieved at non-O$_2$ plasma treated Au surface, and 66.5° wetting angle was obtained at O$_2$ plasma treated Au surface. It indicated the residual O$_2$ blocked the solder wetting. Fig. 6 gives the evidence of O$_2$ on ENIG metal surface after oxygen plasma surface. The wavenumber 2300-2400 cm^{-1} was believed as the oxygen from previous result.[48]

Finally, the γ_{SE} was reduced by 40% and γ_{ER} was increased to 200%. From the simple calculation in Table II, 110.9 mJ/m^2 interfacial energy indicated a spontaneous wettability before plasma etching. While -47.7 mJ/m^2 interfacial energy was regarded as a hindered wettability by oxygen plasma. Fig. 7 illustrated the schematic of un-wetted solder joint shapes by O$_2$ plasma treatment. Due to the residual oxygen on ENIG metal surface, solder wettability was hampered into a convex solder joint shape.

$$\gamma_{ER} = \gamma_E + \gamma_R - 4\left(\frac{\gamma^P_E \gamma^P_R}{\gamma^P_E + \gamma^P_R}\right) - 4\left(\frac{\gamma^d_E \gamma^d_R}{\gamma^d_E + \gamma^d_R}\right) \quad (1)$$

Table II Interfacial energy between solder, electrode and resin

	Not etching	O$_2$ plasma
γ_{ER}	24.1 mJ/m^2	48.5 mJ/m^2
γ_{SE}	384.1 mJ/m^2	249.9 mJ/m^2
γ_{SR}	249.1 mJ/m^2	249.1 mJ/m^2
γ_{SE}- γ_{ER}- γ_{SR}	110.9 mJ/m^2	-47.7 mJ/m^2

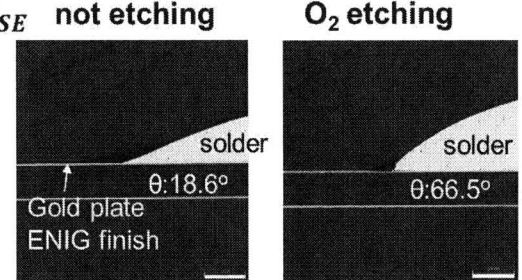

Fig. 5. Wetted solder angle on O$_2$ plasma untreated and treated Au surfaces

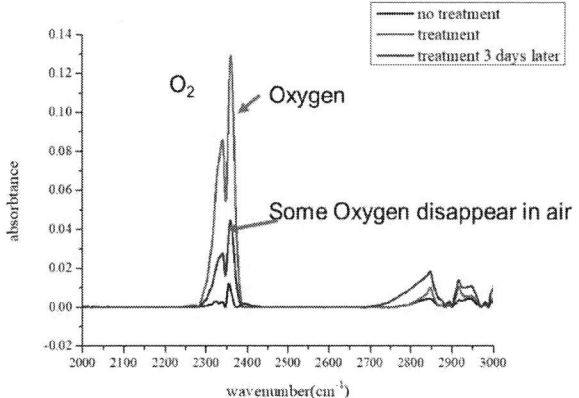

Fig. 6. FTIR measurement of O$_2$ on ENIG metal surface

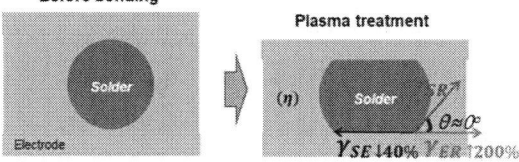

Fig. 7. Schematic of un-wetted solder joint shapes by O$_2$ plasma treatment

B. Solder Joint Morphology

Fig. 8 shows the Sn-58Bi solder ACFs joint morphologies at various conditions. It was interesting that solder joint shape changed from a concave shape to a convex shape by oxygen plasma treatment. And the solder wettability was resisted on the oxygen plasma treated side. This was probably due to the solder wettability decrease in Table II analysis. Except the un-wetted solder part, the other part would not be affected by the bonding process. Therefore, oxygen plasma treatment is the key issue for solder ACFs joints shapes.

In addition, the contact resistance of Sn-58Bi solder ACFs joints was increased. In no-treatment case, 8.78 mΩ was the initial data, but 10.84 mΩ and 23.6 mΩ was achieved by one side plasma treatment. O$_2$ treatment at PCB side played more important role in increasing contact resistances. 41.3 mΩ was formed by both PCB and FPC side oxygen plasma etching, which was obviously un-wetted in Fig. 8.

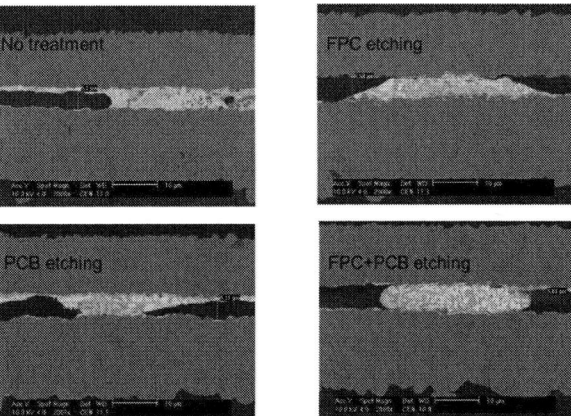

Fig. 8. Sn-58Bi solder ACFs joint morphologies by oxygen plasma etching

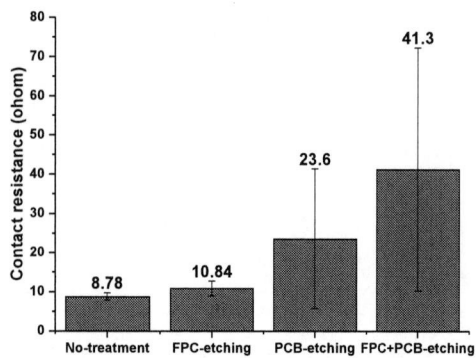

Fig. 9. Contact resistance of Sn-58Bi solder ACFs joints
by oxygen treatment

C. Peel Adhesion Strength

Fig. 10 gives the peel adhesion strength of Sn-58Bi solder ACFs joints as a function of displacement. It was showing that solder joints played a more important role in the joint adhesion strength. Although the adhesion energy between resin and the oxygen plasma treated electrode increased double amount, it was found the peel adhesion strength was reduced by oxygen plasma etching. Especially for both FPC and PCB plasma treatment, adhesion strength was decreased from 650kfg/cm to 400kfg/cm.

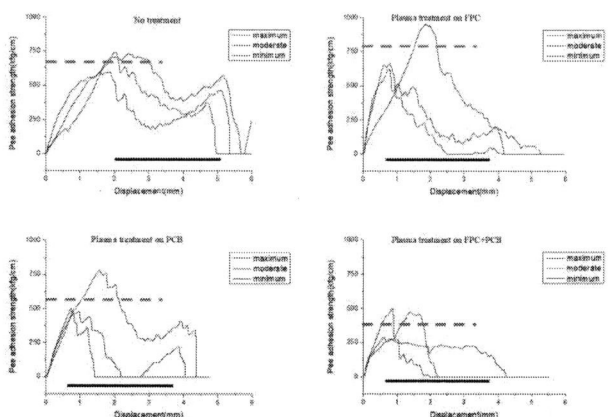

Fig. 10. Peel adhesion strength of Sn-58Bi solder ACFs joints by oxygen treatment

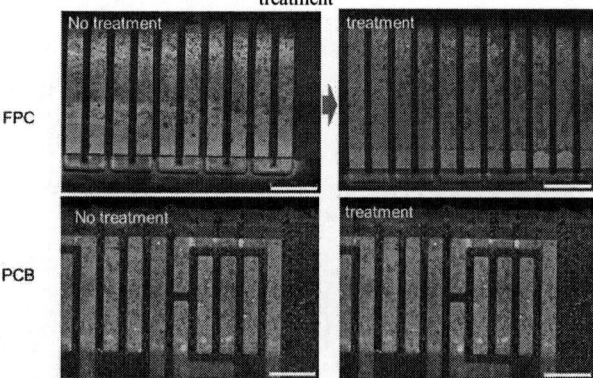

Fig. 11. Peeled surface of solder ACFs Flex-on-board test vehicles after O_2 plasma treatment at FPC side

Fig.11 shows the peeled surfaces of FOB samples using solder ACFs after oxygen plasma treatment on FPC side. It was clear that the solder wetting area was reduced at FPC side after oxygen plasma treatment, compared with no treatment samples. This matched the contact resistance result and indicated solder part played a more important role in determining joint 90° peel adhesion strength.

D. Reliability

Fig. 12 shows the joint morphologies of oxygen treated Sn-58Bi solder ACFs joints during 5 days PCT test. It was found that Sn was quickly consumed in intermetallic compound (IMC) layer and Bi-rich phase was remained in the middle of Sn-58Bi solder joint after reliability. However, at the oxygen treated electrode sides, solder joint hampered the wettability and did not have a complete metallurgical joining with the oxygen plasma treated Au surfaces, but also SnO layer was detected by SEM EDS. We gave the possible mechanism to form SnO in the humidity environment by the residual oxygen at the treated interface. Compared with the non-treated samples, poor solder joint morphologies and bad reliability was obtained in the pictures of Fig. 12. Fig. 13 shows the contact resistances changes of oxygen plasma etched Sn-58Bi solder ACFs joints during 5 days PCT. Due to the residual O_2 at interfaces, contact resistances increased largely and oxygen treated solder joints showed a more severe resistance increase than non-treated one.

$$O_2 + H_2O^+ = H_3O^+$$
$$H_3O^+ + Sn = SnO + H_2O^+$$

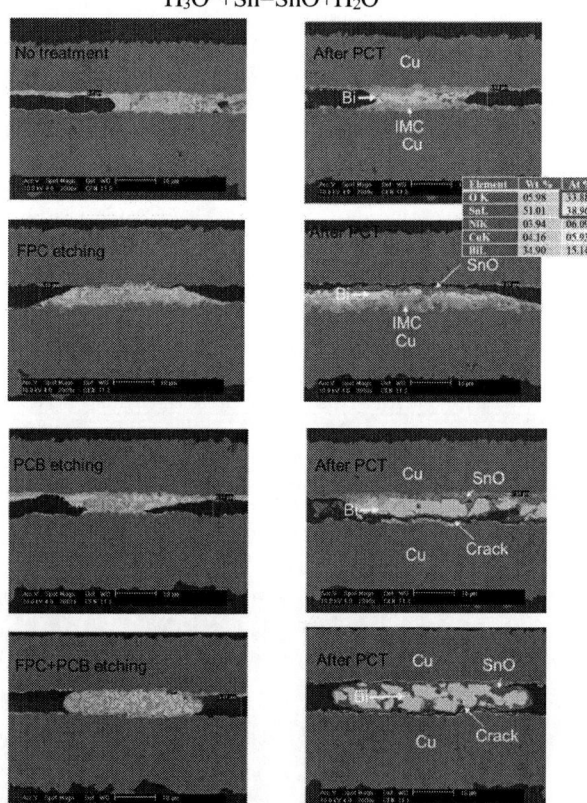

Fig. 12. Joint morphologies of oxygen plasma treated Sn-58Bi solder ACFs joints during 5 days PCT test

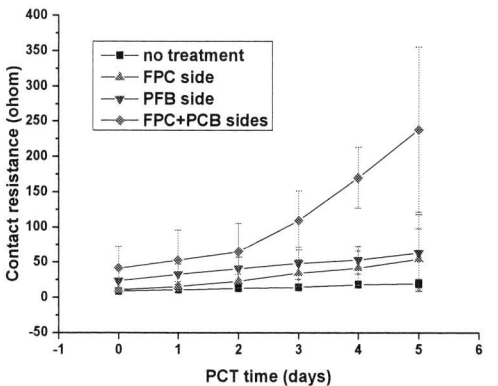

Fig. 13. Contact resistances of oxygen plasma treated Sn-58Bi solder ACFs joints during 5 days PCT test

I. CONCLUSION

In this study, we investigated the oxygen plasma treatment on the Sn-58Bi solder wettability, reliability and $90°$ peel adhesion strength. It was found that the residual oxygen would remain at the plasma treated interface and hamper the solder wetting on the surface. By oxygen plasma treatment, the interfacial adhesion energy between resin and electrode was increased to the double amount by 48.5 mJ/m^2, however, the solder and metal surface interfacial energy was decreased around 40% to 249.9 mJ/m^2. As a result, a negative solder wetting energy was achieved as -47.7 mJ/m^2, resulting into a convex solder joint shape and un-wetted metallurgical joining at the oxygen plasma treated interfaces. In addition, at the oxygen treated surfaces, the joint electrical resistance and reliability would be worse than non-treated samples. Although the resin and electrode adhesion energy was enhanced, the $90°$ peel adhesion strength was decreased as oxygen treated. Therefore, solder part played an more important role in determining ACF joint reliability, mechanical property and contact resistance.

ACKNOWLEDGMENT

The authors would like to thank National Natural Science Foundation of China (Grant 51805115) for research funding support.

REFERENCES

[1] Li, Yi Grace, Daniel Lu, and C. P. Wong. Electrical conductive adhesives with nanotechnologies. Springer Science & Business Media, 2009.

[2] Lee H H, Chou K S, Shih Z W. Effect of nano-sized silver particles on the resistivity of polymeric conductive adhesives[J]. International Journal of Adhesion and Adhesives, 2005, 25(5): 437-441.

[3] Goh C F, Yu H, Yong S S, et al. The effect of annealing on the morphologies and conductivities of sub-micrometer sized nickel particles used for electrically conductive adhesive[J]. Thin Solid Films, 2006, 504(1-2): 416-420.

[4] Pan Y, Song L, Zhang S, et al. Effects of Polymer Conductive Particle Contents on the Electrical Performance and Reliability of 50-

um-Pitch Flex-on-Flex Assemblies Using Anisotropic Conductive Films[J]. IEEE Transactions on Components, Packaging and Manufacturing Technology, 2017, 7(11): 1759-1764.

[5] Ji Y H, Liu Y, Huang G W, et al. Ternary Ag/epoxy adhesive with excellent overall performance[J]. ACS applied materials & interfaces, 2015, 7(15): 8041-8052.

[6] Razeeb K M, Tao J, Stam F. Nanowire ACF for Ultrafine-Pitch Flip-Chip Interconnection[M]//Nanopackaging. Springer, Cham, 2018: 701-723.

[7] Shin Y J, Shin M J, Shin J S. Encapsulation of imidazole with synthesized copolymers for latent curing of epoxy resin[J]. Polymer International, 2017, 66(6): 795-802.

[8] Gupta K K, Abbas S M, Abhyankar A C. Carbon black/polyurethane nanocomposite-coated fabric for microwave attenuation in X & Ku-band (8–18 GHz) frequency range[J]. Journal of Industrial Textiles, 2016, 46(2): 510-529.

[9] Ardebili H, Zhang J, Pecht M. Encapsulation technologies for electronic applications[M]. William Andrew, 2018.

[10] Wu X, Yang X, Yu R, et al. Highly crosslinked and uniform thermoset epoxy microspheres: Preparation and toughening study[J]. Polymer, 2018, 143: 145-154.

[11] Zhang L, Liu Z, Chen S W, et al. Materials, processing and reliability of low temperature bonding in 3D chip stacking[J]. Journal of Alloys and Compounds, 2018.

[12] Wiley, Benjamin, Yugang Sun, and Younan Xia. "Synthesis of silver nanostructures with controlled shapes and properties." Accounts of Chemical Research 40.10 (2007): 1067-1076.

[13] Park, Bong Kyun, et al. "Direct writing of copper conductive patterns by ink-jet printing." Thin Solid Films 515.19 (2007): 7706-7711.

[14] Miller M S, O'Kane J C, Niec A, et al. Silver nanowire/optical adhesive coatings as transparent electrodes for flexible electronics[J]. ACS applied materials & interfaces, 2013, 5(20): 10165-10172.

[15] Yang M, Kim S W, Zhang S, et al. Facile and highly efficient fabrication of robust Ag nanowire–elastomer composite electrodes with tailored electrical properties[J]. Journal of Materials Chemistry C, 2018, 6(27): 7207-7218.

[16] Wang J, Jiu J, Zhang S, et al. The comprehensive effects of visible light irradiation on silver nanowire transparent electrode[J]. Nanotechnology, 2018, 29(43): 435701.

[17] Eom Y S, Baek J W, Moon J T, et al. Characterization of polymer matrix and low melting point solder for anisotropic conductive film[J]. Microelectronic Engineering, 2008, 85(2): 327-331.

[18] Li Y, Wong C P. Recent advances of conductive adhesives as a lead-free alternative in electronic packaging: Materials, processing, reliability and applications[J]. Materials Science and Engineering: R: Reports, 2006, 51(1-3): 1-35.

[19] Lee K, Saarinen I J, Pykari L, et al. High power and high reliability flex-on-board assembly using solder anisotropic conductive films combined with ultrasonic bonding technique[J]. IEEE Transactions on Components, Packaging and Manufacturing Technology, 2011, 1(12): 1901-1907.

[20] Kim Y S, Lee K, Paik K W. Effects of ACF bonding parameters on ACF joint characteristics for high-speed bonding using ultrasonic bonding method[J]. IEEE Transactions on Components, Packaging and Manufacturing Technology, 2013, 3(1): 177-182.

[21] Kim Y S, Kim S H, Lee K, et al. Ultrasonic-Assisted Thermocompression Bonding Method of Solder Anisotropic Conductive Film Joints for Reliable Camera Module Packaging[J]. IEEE Transactions on Components, Packaging and Manufacturing Technology, 2013, 3(12): 2156-2163.

[22] Lee S H, Suk K L, Lee K, et al. Study on fine pitch flex-on-flex assembly using nanofiber/solder anisotropic conductive film and ultrasonic bonding method[J]. IEEE Transactions on Components, Packaging and Manufacturing Technology, 2012, 2(12): 2108-2114.

[23] Kim T W, Suk K L, Lee S H, et al. Low temperature flex-on-flex assembly using polyvinylidene fluoride nanofiber incorporated

Sn58Bi solder anisotropic conductive films and vertical ultrasonic bonding[J]. Journal of Nanomaterials, 2013, 2013: 7.

[24] Kim T W, Lee T I, Pan Y, et al. Effect of nanofiber orientation on nanofiber solder anisotropic conductive films joint properties and bending reliability of flex-on-flex assembly[J]. IEEE Transactions on Components, Packaging and Manufacturing Technology, 2016, 6(9): 1317-1329.

[25] Kim Y S, Zhang S, Paik K W. Highly reliable solder ACFs FOB (flex-on-board) interconnection using ultrasonic bonding[J]. J. Microelectron. Packag. Soc., 2015, 22(1): 35-41.

[26] Zhang S, Kim S H, Kim T W, et al. A study on the solder ball size and content effects of solder ACFs for flex-on-board assembly applications using ultrasonic bonding[J]. IEEE Transactions on Components, Packaging and Manufacturing Technology, 2015, 5(1): 9-14.

[27] Zhang S, Paik K W. A study on the failure mechanism and enhanced reliability of Sn58Bi solder anisotropic conductive film joints in a pressure cooker test due to polymer viscoelastic properties and hydroswelling[J]. IEEE Transactions on Components, Packaging and Manufacturing Technology, 2016, 6(2): 216-223.

[28] Zhang, Shuye, et al. "A study on the optimization of anisotropic conductive films for Sn-3Ag-0.5 Cu-Based flex-on-board application at a 250 C bonding temperature." IEEE Transactions on Components, Packaging and Manufacturing Technology 8.3 (2018): 383-391.

[29] Kim J H, Lee T I, Kim T S, et al. The effect of anisotropic conductive films adhesion on the bending reliability of chip-in-flex packages for wearable electronics applications[J]. IEEE Transactions on Components, Packaging and Manufacturing Technology, 2017, 7(10): 1583-1591.

[30] Lin C M, Chen D C, Liu Y C. Investigation on fracture and conductivity of flex-on-film flexible bonding using anisotropic conductive film considering repeated bending[J]. Microsystem Technologies, 2018: 1-10.

[31] Rizvi M J, Chan Y C, Bailey C, et al. Study of anisotropic conductive adhesive joint behavior under 3-point bending[J]. Microelectronics Reliability, 2005, 45(3-4): 589-596.

[32] Noh B I, Yoon J W, Kim J W, et al. Reliability of Au bump flip chip packages with adhesive materials using four-point bending test[J]. International Journal of Adhesion and Adhesives, 2009, 29(6): 650-655.

[33] Hirsch F, Kästner M. Microscale simulation of adhesive and cohesive failure in rough interfaces[J]. Engineering Fracture Mechanics, 2017, 178: 416-432.

[34] Uddin M A, Alam M O, Chan Y C, et al. Plasma cleaning of the flex substrate for flip-chip bonding with anisotropic conductive adhesive film[J]. Journal of electronic materials, 2003, 32(10): 1117-1124.

[35] Khanna V K. Adhesion–delamination phenomena at the surfaces and interfaces in microelectronics and MEMS structures and packaged devices[J]. Journal of Physics D: Applied Physics, 2010, 44(3): 034004.

[36] Kim J H, Lee T I, Kim T S, et al. Effects of ACFs Modulus and Adhesion Strength on the Bending Reliability of CIF (Chip-in-Flex) Packages at Humid Environment[C]//2018 IEEE 68th Electronic Components and Technology Conference (ECTC). IEEE, 2018: 2319-2325.

[37] Owens D K, Wendt R C. Estimation of the surface free energy of polymers[J]. Journal of applied polymer science, 1969, 13(8): 1741-1747.

[38] Evans E, Ritchie K. Dynamic strength of molecular adhesion bonds[J]. Biophysical journal, 1997, 72(4): 1541-1555.

[39] Parhizkar N, Shahrabi T, Ramezanzadeh B. A new approach for enhancement of the corrosion protection properties and interfacial adhesion bonds between the epoxy coating and steel substrate through surface treatment by covalently modified amino functionalized graphene oxide film[J]. Corrosion Science, 2017, 123: 55-75.

[40] Mandolfino C, Lertora E, Genna S, et al. Effect of laser and plasma surface cleaning on mechanical properties of adhesive bonded joints[J]. Procedia Cirp, 2015, 33: 458-463.

[41] Shin D K, Song Y H, Im J. Effect of PCB surface modifications on the EMC-to-PCB adhesion in electronic packages[J]. IEEE Transactions on Components and Packaging Technologies, 2010, 33(2): 498-508.

[42] Zhang, Shuye, et al. "A Study on the Bonding Conditions and Nonconductive Filler Contents on Cationic Epoxy-Based Sn–58Bi Solder ACFs Joints for Reliable Flex-on-Board Applications." IEEE Transactions on Components, Packaging and Manufacturing Technology 7.12 (2017): 2087-2094.

[43] Zhang S, Lin T, He P, et al. Effects of acrylic adhesives property and optimized bonding parameters on Sn58Bi solder joint morphology for flex-on-board assembly[J]. Microelectronics Reliability, 2017, 78: 181-189.

[44] Zhang S, Lin T, He P, et al. A Review: Solder Joint Cracks at Sn-Bi58 Solder ACFs Joints[M]//Lead Free Solders. IntechOpen, 2019.

[45] Zhang S, Park J H, Paik K W. Joint morphologies and failure mechanisms of anisotropic conductive films (ACFs) during a power handling capability test for flex-on-board applications[J]. IEEE Transactions on Components, Packaging and Manufacturing Technology, 2016, 6(12): 1820-1826.

[46] Janczuk B, Zdziennicka A. A study on the components of surface free energy of quartz from contact angle measurements[J]. Journal of materials science, 1994, 29(13): 3559-3564.

[47] Tyson W R, Miller W A. Surface free energies of solid metals: Estimation from liquid surface tension measurements[J]. Surface Science, 1977, 62(1): 267-276.

[48] Wang Y X, Pan Z Y, Ho Y K, et al. Nuclear instruments and methods in physics research section B: beam interactions with materials and atoms[J]. Nuclear Instruments and Methods in Physics Research B, 2001, 180(1-4): 251-256.

Numerical analysis of the influence of polymeric materials on a MEMS package performance under humidity and temperature loads

Mahesh Yalagach*, Peter Filipp Fuchs*, Archim Wolfberger*, Mario Gschwandl*, Thomas Antretter†, Michael Feuchter‡, Coen Tak§, Tao Qi¶

*Polymer Competence Center Leoben GmbH
Leoben, Austria
mahesh.yalagach@pccl.at

† Montanuniversitaet Leoben, Institute of Mechanics
Leoben, Austria

‡ Montanuniversitaet Leoben, Institute of Material Science and Testing of Polymers
Leoben, Austria

§ams AG
Premstaetten, Austria

¶Austria Technologie & Systemtechnik Aktiengesellschaft
Leoben, Austria

Abstract—The rapid expansion of the Internet of Things (IoT) and consumer electronics is driving the demand for microelectromechanical systems (MEMS) in the area of wearables, smartphones, and home and building applications. MEMS sensor packages feature a variety of polymeric materials which can significantly affect their behavior under environmental loads as humidity or temperature. A broad range of different polymeric materials can be applied in the packages, but to get a good MEMS sensor performance, an optimized application tailored material combination should be applied. To analyze the effect of the applied material types an advanced simulation approach has to be considered. A hygro-thermo-mechanical simulation based on measured temperature and moisture dependent material properties is presented. As MEMS sensor packages are complex multimaterial systems, the main challenge is the modeling of the moisture discontinuities in the interfaces. The discontinuity of moisture concentration has been solved by using the solubility approach.

Keywords-hygro-thermal, hygro-mechanical characterization; hygroscopic swelling; solubility; hygro-thermomechanical simulation; deformation;

I. INTRODUCTION

The advancement in MEMS sensor packages towards their applications has gained interests in developing new polymeric materials. In operation, these MEMS sensor packages are often exposed to environmental and mechanical loads like temperature, humidity or bending. Under these loads, the thermo-hygro-mechanical material behavior of the applied polymeric materials strongly affects the MEMS sensor precision due to influences from materials, interfaces and package geometries [1]. The significant difference between the properties of the applied non-polymeric and polymeric

materials leads to a mismatch in hygroscopic swelling, thermal expansion, and bending strain. To optimize the material choice and design, simulation models accurately describing the resulting stress and strain field need to be established. The models can then be used as a basis for the development of tailored polymer materials for MEMS applications by studying the sensitivity on changed properties in the virtual simulation environment.

The effect of material combinations on the MEMS sensor package signal is studied in this work. Doing so, an extensive material characterization was performed, and a thermo-hygro-mechanical simulation strategy for a multimaterial system was implemented. To solve the discontinuity in the moisture concentration, the solubility approach [9] is applied. The detailed local die deformation behavior was analyzed using a sub-modeling approach.

II. EXPERIMENTAL

To understand the behavior of polymeric materials under different temperature and moisture loads, the materials were subjected to thermo-mechanical, hygro-thermal and hygromechanical material characterization. The following section explains briefly the different measurement techniques used:

A. Thermo-mechanical and Hygro-thermal characterization

The temperature dependent Elastic modulus (E) for the critical materials were measured using the uni-axial tensile tests as well the dynamic mechanical analysis (DMA). The thermal conductivity (k) was measured using *Netzsch LFA 467 HyperFlash*® (Netzsch GmbH, Selb, Germany) and the heat capacity (C_p) was measured using differential scanning calorimetry (DSC). The test conditions and dimensions of the material samples for all the thermo-mechanical analysis were considered according to [2][3][4]. To measure the in-plane coefficient of thermal expansion (CTE) the digital image correlation (DIC) method was employed using the system *Q400 TCT* (Dantec Dynamics, Ulm, Germany) [5]. The hygro-thermal material properties like the moisture diffusion coefficient (D) and saturated concentration (M_∞) were determined based on gravimetric humidity conditioning methods according to JEDEC-JESD22-A120B standards [6]. The obtained conditioning curves were fitted using single and dual Fickian diffusion models [2].

B. Coefficient of moisture expansion (CME)

The hygro-mechanical material parameter CME was measured using an approach based on a DIC system monitoring the hygroscopic strains during the drying process. The hygroscopic swelling strains ϵ^β have a linear relation with change in moisture concentration ΔC [2][7] as shown in (1).

$$\epsilon^\beta = \beta \Delta C, \qquad (1)$$

where ϵ^β is the hygroscopic strain, β is the coefficient of moisture expansion (CME) and ΔC is the concentration change.

The following procedure was followed: *a*) humidity conditioning of two samples of the same material and dimension (sample A and sample B) by immersing them in demineralized water, *b*) desorption of moisture at constant temperature using *Q400 TCT* (Dantec Dynamics, Ulm, Germany) using sample A and a conventional temperature oven for sample B. Doing so the swelling strains ϵ^β and concentration change ΔC can be derived from sample A and sample B respectively. The desorption of the saturated material samples is carried out at $90°C$, $120°C$ and $160°C$ for six hours. A linear curve fit on the ϵ^β over the ΔC is performed to compute the CME β.

III. HYGRO-THERMO-MECHANICAL SIMULATION

The hygro-thermo-mechanical simulations were performed using Abaqus (Abaqus 6.17, Dassault Systemes Simulia Corp., Providence, USA). The user-defined subroutines USDFLD and UEXPAN are used in this model to compute the total moisture concentration and total volumetric strains respectively. The overall simulation approach except for the consideration of the moisture discontinuity is described in detail in a preceding publication [2]. The following section explains briefly an approach to solve the problem of moisture discontinuity in the interface region in multi-material systems using solubility S approach.

A. Solubility approach for multimaterial interface

To solve the discontinuities in the multi-materials system, several approaches have been listed in the literature considering the thermal and moisture analogy: wetness theory,

normalization concentration analogy [8], "Direct" analogy [9] and advanced normalized concentration analogy [10]. The normalization concentration analogy was applied in this work because it is best suited to describe our load cases. However, a transient hygro-mechanical load is not possible to be simulated. Therefore, the advanced normalized concentration analogy would have to be used.

The mass diffusion analysis in Abaqus is described by the Fick's second Law as shown in (2)

$$\dot{C} = \nabla \cdot (D\nabla C), \qquad (2)$$

where C is the moisture concentration and D is the diffusivity. In heterogeneous materials, the normalization concentration method was used to solve the discontinuity. In this approach, the moisture concentration C is normalized by the solubility S. The normalized concentration φ [9] is defined as,

$$\varphi = \frac{C}{S} \qquad (3)$$

Using Eq.(3) in Eq.(2), and assuming that the solubility S is both uniform ($\nabla S = 0$) and time independent ($\dot{S} = 0$), the Fick's second law yields,

$$S\dot{\varphi} = \nabla \cdot (DS\nabla\varphi), \qquad (4)$$

In the solubility approach, the Henry's Law [11] is used to determine the S.

$$M_\infty = M_{sat} = SP_{VP}, \qquad (5)$$

where P_{VP} is the ambient vapor pressure, as the saturated concentration M_∞ at $23°C$, $60°C$ and $90°C$ is known from the hygro-thermal material characterization, and the ambient vapor pressure can be adopted from the literature [12] at this temperature levels the solubility S can be calculated using 4. Using the above definition, the solubility S was computed. With D and M_∞ derived from the experimental data, all parameters needed for the mass diffusion analysis are defined.

B. Global and local simulation approach

1) Global simulation approach: In a global simulation approach, the effect of temperature, moisture and mechanical loads were evaluated considering the printed circuit board (PCB). Since PCB's have a complex structure and detailed FE model of an entire PCB would be computationally very expensive. Hence, a homogenization approach is applied. Doing so, an in-house development for an automated PCB FE-model generation is used [13]. The automated model generation uses segmentation and clustering algorithms to automatically generate FEA-models of arbitrarily complex build-ups using their design data. A conventional PCB (Figure 1) designed according to IPC/ JEDEC-9702 standards [14] with dimensions, $135 \times 77 \times 0.83\ mm$ was considered for this modeling approach. The used element type is a linear

978-1-7281-1500-9/19 $31.00 © 2019 IEEE

hexahedron continuum element of type C3D8R, and the total number of elements is 69692.

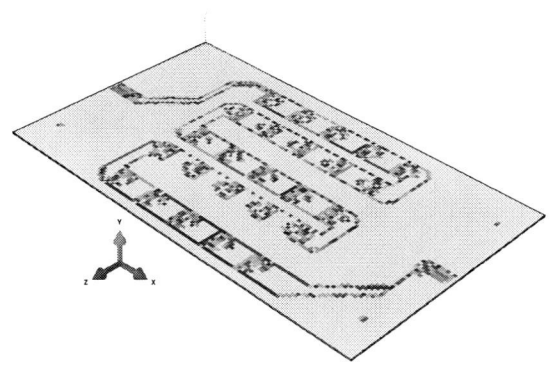

Figure 1. The conventional PCB used for global simulation modeling.

2) Local simulation approach: In this approach, the MEMS sensor soldered on to the PCB was considered, and the influence of temperature, moisture and mechanical loads on the MEMS sensor was evaluated. A commonly used MEMS sensor package with dimensions $2 \times 2 \times 0.7\ mm$ was considered and the dimensions of the submodel PCB is $9 \times 9 \times 0.83\ mm$. The complete local model setup including the MEMS sensor and PCB with solder lands is depicted in Figure 2. The used element type for MEMS sensor package is a linear tetrahedron continuum element of type C3D4, and the total number of elements are 829724. For the local PCB, 580608 linear hexahedral elements of type C3D8

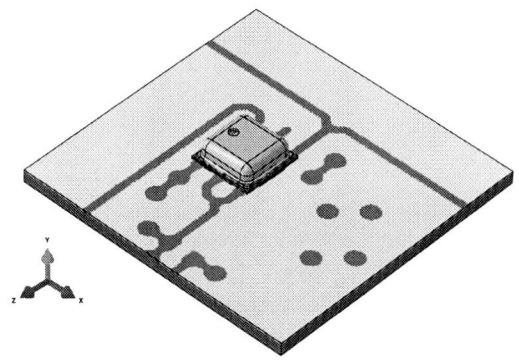

Figure 2. MEMS sensor package soldered on a PCB for local simulation modeling.

C. Materials

The MEMS sensor is made up of a variety of materials as shown in Figure 3. The material properties for non-polymeric materials like metal-lid, Au-wire, silicon-die and copper were taken from literature [15][16]. The polymeric

materials in this MEMS sensor are the conductive adhesive, insulating adhesive, solder mask, and a prepreg.

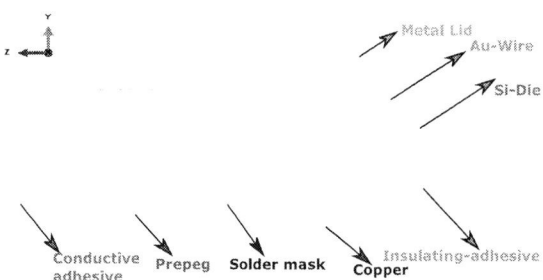

Figure 3. Materials in the considered MEMS sensor package.

The Table I, shows the two different polymeric material combination's (MAT-1 and MAT-2) considered for the MEMS sensor. For prepreg and insulating adhesive two different materials were considered. The prepreg's, CCL-1078 and PP1-1037, are glass fiber fabric woven reinforced epoxy resins. They differ concerning to the matrix resin and the fabric type. PP1-1037 features a higher fiber volume content than CCL-1078.

Table I
CONSIDERED MATERIAL COMBINATION'S USED IN MEMS SENSOR PACKAGE.

Materials	MAT-1	MAT-2
Prepreg	CCL-1078	PP1-1037
Insulating adhesive	ADH-X-01	ADH-Y-01
Conductive adhesive	ADH-Z-02	
Solder mask	SLM-01	

The insulating adhesives ADH-X-01 and ADH-Y-01 are both made of silicone as a base material, but the curing process for both the materials differs. The insulating adhesive is used in the package to fix the silicon die on to the substrate. The influence of prepreg and the insulating adhesive was chosen to be evaluated as their material properties are expected to have the most significant effect on the MEMS sensor performance.

The conductive adhesive (ADH-Z-02) and the solder mask (SLM-01) were kept the same for both material combinations. The ADH-Z-02 is a silver-filled epoxy resin and is used to fix the metal lid to the substrate. The moisture uptake in these materials is significantly higher compared to other polymeric materials during the humidity conditioning. The PCB used for the global and local sub-modeling approach is made up of multiple layers of commonly used FR4 prepreg (PP2-1037), solder masks (SLM-01) and copper layers.

D. Evaluation line

The MEMS sensor precision depends on the ability to minimize as much as possible the influence of external factors as temperature, moisture and mechanical loads on the die deformation. Thus, the criteria for a suitable material combination and package design is the local deformation at the upper die surface. The defined evaluation line in this contribution is shown in Figure 4.

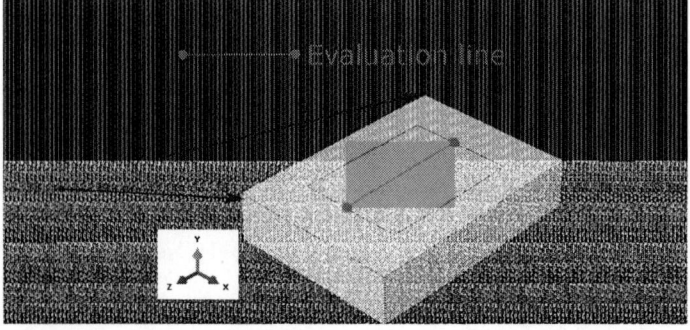

Figure 4.   Evaluation line in the silicon die of the MEMS sensor.

E. Boundary conditions

The hygro-thermo-mechanical simulation involves a heat transfer simulation, two mass diffusion simulations and a thermo-mechanical step to evaluate the total volumetric strains. All loads were applied in a linear way over 3600 seconds. For the heat transfer simulation, the initial condition is $23°C$, and the temperature of $85°C$ is applied to the surfaces exposed to the environment. For the dual Fickian mass diffusion simulation, the normalized saturated concentrations ($M_{1\infty}$ & $M_{2\infty}$) are applied to the polymeric materials exposed to the environment. The initial moisture concentration of 0% was considered for all the materials. For the non-polymeric materials like the metal lid, Au-wire, Si-die, and copper, the moisture dependent material properties like D, M_∞ and S are considered to be negligible. In total four different simulations were carried out. For all the simulations, the defined moisture ($100\% RH$) and temperature ($85°C$) loads were considered. In CASE-1 and CASE-2, the simulations are based on a local model only, and the influence of the material combinations (MAT-1 and MAT-2) is analyzed. In CASE-3 the effect of considering the global model is evaluated. In CASE-4 additionally, a bending load in the global model according to the IPC/JEDEC-9702 standards [14] of 3mm is taken into account. The four simulations are summarized in Table II.

IV. RESULTS AND DISCUSSION

A. Material Characterization

The results obtained from thermo-mechanical and hygro-thermal material characterization techniques have been discussed in [2]. The additional results obtained for in-plane CTE (α) and CME (β) are discussed in this contribution.

Table II
LOAD CASES CONSIDERED IN LOCAL SUBMODELING APPROACH.

	Material	**Loads**	**Simulation approach**
CASE-1	MAT-1	$85°C$, $100\% RH$	Local model
CASE-2	MAT-2		
CASE-3	MAT-1	$85°C$, $100\% RH$	Global model + Local submodel
CASE-4		$85°C$, $100\% RH$ + $3mm$ bending load	

The Figure 5 shows the CTE results obtained for the insulating adhesives ADH-X-01 and ADH-Y-01, and the conductive adhesive ADH-Z-02. Within the measurement range of $-50°C$ to $+180°C$ there is a substantial difference between the adhesives behavior. ADH-X-01 shows the highest average CTE and ADH-Y-01 shows the lowest CTE. The CTE of ADH-Z-02 is in between. Maximum differences of about 50% emphasize the importance of the material choice.

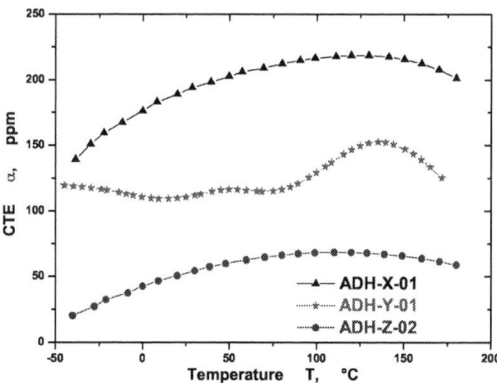

Figure 5.   The in-plane CTE measured for different adhesive materials.

For the coefficient of moisture expansion (CME) measurements, the measured change in concentration ΔC is depicted in the Figure 6 and 7. It is observed from Figure 6, that the prepreg PP2-1037 shows the highest concentration change of about 1.2% compared to CCL-1078 (0.7%) and PP1-1037 (0.9%).

The moisture adsorption in the adhesives depicted in Figure 7 shows even more pronounced differences between the analyzed materials. Maximum moisture uptakes varied between 1.7% (ADH-Z-02), 1.1% (ADH-X-01) and 0.55% (ADH-Y-01). As the operating temperature for ADH-Y-01 is limited to $130°C$, no measurement was performed at $160°C$ for this material.

The measured dimensional change or the hygroscopic swelling strains ϵ^β for prepregs and adhesives are depicted in Figure 8 and 9. In Figure 8, the prepreg hygroscopic

978-1-7281-1500-9/19 $31.00 © 2019 IEEE

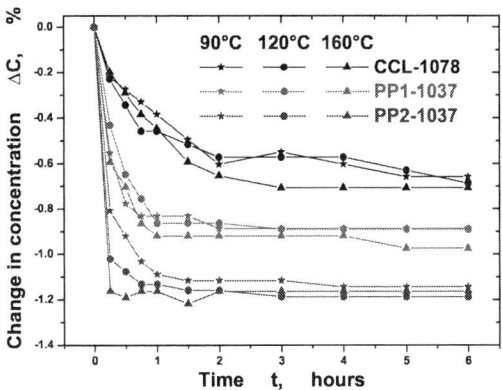

Figure 6. Change in the moisture concentration ΔC for prepregs from desorption process at $90°C$, $120°C$ and $160°C$ for six hours.

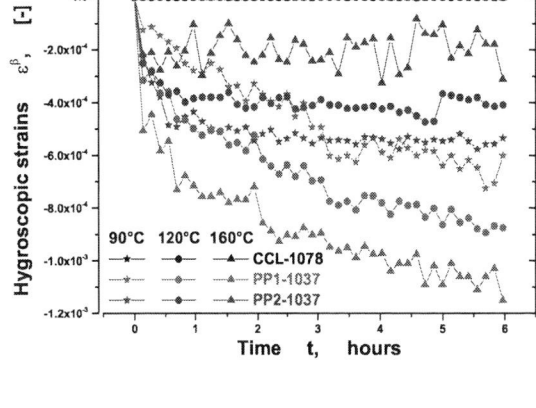

Figure 8. Change in the hygroscpic swelling strains ϵ^β for prepregs at $90°C$, $120°C$ and $160°C$ for six hours.

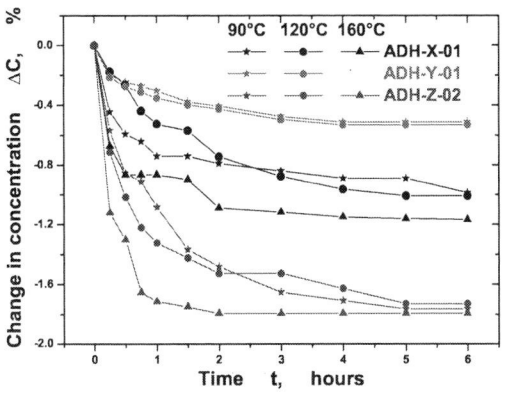

Figure 7. Change in the moisture concentration ΔC for adhesives from desorption process at $90°C$, $120°C$ and $160°C$ for six hours.

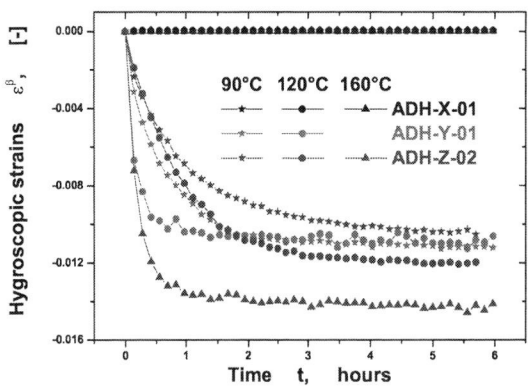

Figure 9. Change in the hygroscpic swelling strains ϵ^β for adhesives at $90°C$, $120°C$ and $160°C$ for six hours.

swelling is very low for all materials. However, a maximum hygroscopic strain of 0.12% can be observed for prepreg PP1-1037 while no hygroscopic strain was measured for prepreg CCL-1078 at all. It is interesting to note that the temperature trend for PP1-1037 and PP2-1037 is vice-versa. That is assumed to be caused by the different glass transition (T_g) regions. The T_g values were determined in DMA measurements: T_g of PP1-1037 is $270°C$ and the T_g of PP2-1037 is $170°C$. While the tested temperatures for CME are well below the T_g level for PP1-1037, for PP2-1037 they are within the onset of the transition region. It is known from the literature that the temperature trend changed in this region [8].

The ϵ^β in adhesives are significantly higher (Fig. 9) compared with prepreg materials. For the conductive adhesive ADH-Z-02, the ϵ^β is greatest (1.4%) compared with insulating adhesives ADH-X-01 and ADH-Y-01. The ADH-

Y-01 material showed a ϵ^β of about 1% while the ADH-X-01 did not show any strains. Hence, in this contribution, based on the measurement results, only the moisture dependence from PP1-1037, PP2-1037, ADH-Y-01, and ADH-Z-02 materials has been considered for the hygro-thermo-mechanical simulation.

The CME (β) was computed from a linear curve fit between the hygroscopic swelling strains ϵ^β and the change in concentrations ΔC.

B. Simulations

In this section, the results from the hygro-thermo-mechanical simulations are discussed. The results obtained from CASE-1 and CASE-2, using different material combinations (MAT-1 and MAT-2) are depicted in the Figure 10. The strain levels in CASE-2 are significantly higher compared to CASE-1. The materials in MAT-2 experience hygro-thermal expansion during the measurements, while the materials in MAT-1 were observed to show no hygro-thermal expansion in the experiments. Additionally, the materials applied in MAT-1 also show a lower CTE. The simulation results give the expected trend and confirm that a lower thermal and hygroscopic expansion in the polymeric materials is preferable.

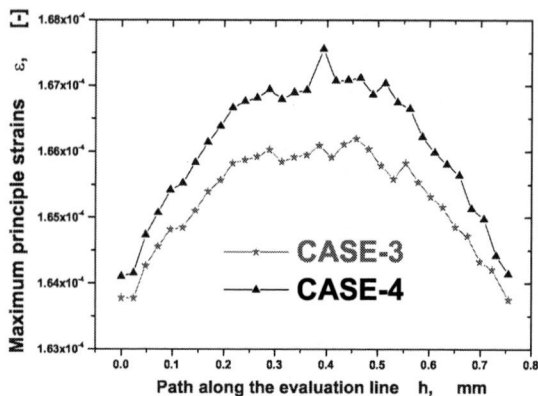

Figure 12. Strain levels along the evaluation line using CASE-3 and CASE-4 in local submodeling approach.

From the Figure 12, the strain levels in the evaluation line are significantly higher for CASE-4 compared with CASE-3. It is shown that already small global mechanical loads affect the local silicon die deformation and have to be considered in detail in local submodeling approach.

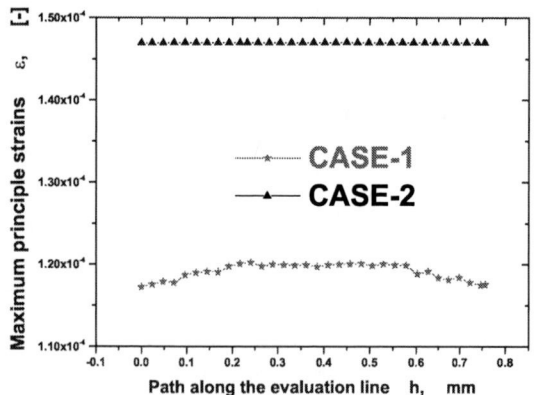

Figure 10. Strain levels along the evaluation line using the CASE-1 (MAT-1) and CASE-2 (MAT-2) material combinations in local modeling approach.

In the Figure 11, a comparison between the CASE-1 and CASE-3, showing the difference in strain levels is given. This difference is due to the submodel boundary conditions in CASE-3. The influence of the deformations in the global model (CASE-3) is significant. Thus, to correctly describe the local deformations of the silicon die, it is essential to consider the global model results.

V. CONCLUSION

In this research work, the results from the thermo-mechanical and hygro-mechanical material characterization for different polymeric materials used in MEMS sensor packages are presented. Based on the experimental results, using global and local simulation modeling approaches, hygro-thermo-mechanical simulations were performed. The local deformation of the silicon die under $85°C$ and 100% RH was analyzed in a specified evaluation region. To this end, the influence of different material combinations was evaluated. Obtained results emphasize the importance of the material combination choice. Furthermore, the significance of considering the total system including the whole PCB could be shown. A submodeling approach featuring the results of a global model gives different results than a local model only. The suitability to include additional mechanical loads on a global level and to derive its influence is also shown by considering a bending load.

The following open points will be worked on in the future. Absorption at different humidity levels is planned to be tested. Additionally to the yet considered 100% RH (immersion in water) conditioning in a climate chamber ($85°C$ and 85% RH) will be performed. Furthermore, validation experiments for the simulation models are planned. The die deformation under defined loads (humidity, temperature, and bending) will be derived from a measured sensor signal and compared to the simulation results.

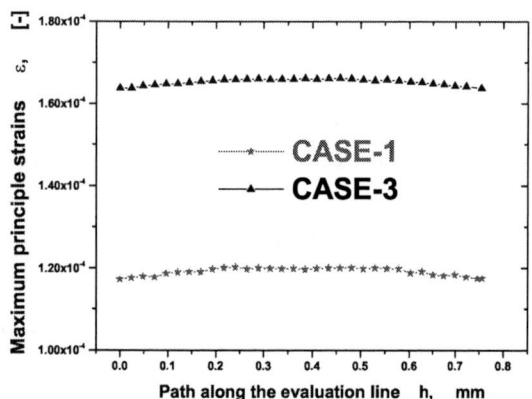

Figure 11. Strain levels along the evaluation line using the MAT-1 in a local (CASE-1) and to local submodeling (CASE-3) approach.

The results obtained from CASE-3 and CASE-4 in a hygro-thermo-mechanical step of the local sub modeling approach using the MAT-1 material combination is depicted in the Figure 12.

ACKNOWLEDGMENT

The research work was performed within the K-Project "PolyTherm" at the Polymer Competence Center Leoben GmbH (PCCL, Austria) within the framework of the COMET-program of the Federal Ministry for Transport, Innovation and Technology and the Federal Ministry for Digital and Economic Affairs with contributions by University of Leoben, ams AG and by AT&S Austria Technologie & Systemtechnik Aktiengesellschaft. Funding is provided by the Austrian Government and the State Government of Styria.

REFERENCES

[1] Zarnik, M. Santo; Belavic, D. (2012): An Experimental and Numerical Study of the Humidity Effect on the Stability of a Capacitive Ceramic Pressure Sensor.

[2] M. Yalagach et al., "Influence of environmental factors like temperature and humidity on MEMS packaging materials.," 2018 7th Electronic System-Integration Technology Conference (ESTC), Dresden, 2018, pp. 1-6. doi: 10.1109/ESTC.2018.8546484.

[3] M. Gschwandl et al., "Modeling of manufacturing induced residual stresses of viscoelastic epoxy mold compound encapsulations," 2017 IEEE 19th Electronics Packaging Technology Conference (EPTC), Singapore, 2017, pp. 1-8. doi: 10.1109/EPTC.2017.8277557.

[4] M. Morak, P. Marx, M. Gschwandl, P. Fuchs, M. Pfost, and F. Wiesbrock, "Heat dissipation in epoxy/amine-based gradient composites with alumina particles: A critical evaluation of thermal conductivity measurements," Polymers, vol. 10, no. 10, p. 1131, 2018.

[5] M. Gschwandl et al., "Evaluation of Digital Image Correlation Techniques for the Determination of Coefficients of Thermal Expansion for Thin Reinforced Polymers.", Proceedings of 20th International Conference on Electronic Materials and Packaging, 2018.

[6] JESD22-A120A, Test method for the measurement of moisture diffusivity and water solubility in organic materials used in electronic devices, JEDEC Standards 2001.

[7] Stellrecht, E., Han, B., Pecht, M.G., "Characterization of the hygroscopic swelling behavior of mold compounds and plastic packages", IEEE Transactions on Components and Packaging Technologies, 27, 499506, 2004.

[8] E.H. Wong, Y.C. Teo, T.B. Lim, Moisture diffusion and vapor pressure modeling of IC packaging Proc. of Electronic Components and Technology Conference 1998, pp. 1372-1378.

[9] Yoon, S., Han, B., and Wang, Z., 2007, "On Moisture Diffusion Modeling Using Thermal Diffusion Analogy," J. Electron. Packag., 129, pp. 421-426.

[10] Jang C, Park S, Han B, Yoon S. Advanced Thermal-Moisture Analogy Scheme for Anisothermal Moisture Diffusion Problem. ASME. J. Electron. Packag. 2008;130(1):011004-011004-8. doi:10.1115/1.2837521.

[11] Osswald, T., and Menges, G., 1996, Materials Science of Polymers for Engineers, Hanser-Gardener, Munich.

[12] http://www.wiredchemist.com/chemistry/data/vapor-pressure

[13] M. Gschwandl et al., "Finite Element Analysis of Arbitrarily Complex Electronic Devices," In: IEEE 18th Electronics Packaging Technology Conference (EPTC) 2016, S. 497-500.

[14] IPC/JEDEC-9702, Monotonic Bend Charactrization of Board-Level Interconnects, JEDEC Standards 2004.

[15] Dwight E. Gray (Ed.) American Institute of Physics Handbook McGraw-Hill Book Company Inc. 1957.

[16] Handbook of Stainless Steels, Donald Peckner and I. M. Bernstein, McGraw-Hill Book Company, New York, NY, (1977)

Electromigration-induced β-Sn grain rotation in lead-free flip chip solder bumps

M.L. Huang*, J.M. Kuang, H.Y. Sun
Electronic Packaging Materials Laboratory
School of Materials Science & Engineering
Dalian University of Technology
Dalian 116024, China
E-mail: huang@dlut.edu.cn

Abstract—The electromigration-induced β-Sn grain rotation behavior in Ni/ Sn-3.0Ag-0.5Cu/ENEPIG(OSP) lead-free flip chip solder joints under a current density of 1×10^4 A/cm² at 150 ºC were investigated. The occurrence of β-Sn grain rotation resulted in an obvious morphological evolution on the surface of solder joints with both ENEPIG and OSP finishes. The mechanism on β-Sn grain rotation was proposed to explain the experimental results. The β-Sn grains with different orientations in the solder bumps led to different vacancy flux through different grains and the saturation situation of vacancies at the grain boundaries. No grinding-polishing process on the surface of the solder bump during in-situ experiment makes it stress-free and a good sink/source of vacancy, where there is an equilibrium vacancy concentration. Redundant Sn atoms diffused inward or outward the surface, resulting in the floating or sinking of the β-Sn grains. For real flip chip solder joints, both the cracks and the size of Sn grains have a significant effect on the β-Sn grain rotation. The present work is expected to provide a theoretical guidance for the electromigration (EM) life prediction of solder joints.

Keywords-electromigration; β-Sn grain; diffusion; anisotropy; grain rotation; vacancy flux

I. INTRODUCTION

With the increasing demands of high density, high performance and high reliability in electronic packaging, the solder bumps that interconnect Si chips and substrates are downsizing to micron scale (micro bumps). The cross-sectional area of solder bumps will be reduced by a factor of four as the diameter is reduced by a factor of two [1]. As a result, the current density through each bump increases up to 10^4-10^6 A/cm², inducing a significant EM failure [2].

Each micro bump consists of a few or even one single β-Sn grain with the continuous miniaturization. For instance, Sn-Ag-Cu solder bumps form a few large β-Sn grains after reflow soldering [3]. Since β-Sn has a lattice structure of body-centered tetragonal (a=b=5.83 Å and c=3.18 Å), it exhibits a highly anisotropic behavior in mechanical, thermal and electrical properties [4], as well as in diffusivity [5-7], as listed in Table I. For instance, the diffusion coefficient of Ni atoms along the *c*-axis of β-Sn grain at 150 ºC is 3×10^4 times larger than that along the *a*-axis of β-Sn grain, showing a remarkable anisotropy in diffusivity.

Electromigration (EM) is the atomic motion driven by the collisions of electrons under the influence of electric current. Since the diffusivity of solute atoms in β-Sn is dominated by the grain orientation, EM behavior in micro scale β-Sn bumps is closely related to the grain orientations. Lu et al. [8] reported that different failure modes occurred at the same surface in a solder joint, due to the different β-Sn grain orientation in the neighboring area of the cathode. Huang et al. [9] demonstrated the effect of β-Sn grain orientation on the IMC precipitation and the failure mode in Cu/Sn-Ag-Cu/Cu line type solder joints. β-Sn grain orientation was a dominant factor in determining the dissolution of cathode Cu. Excessive dissolution of cathode Cu occurred in β-Sn grains with small angle θ (between the *c*-axis of Sn grain and electron flowing direction), while for large angle θ β-Sn grains, the consumption was retarded. On grain rotation, the grains would align themselves along the low resistance direction with an obvious morphological evolution in the surface of the solder joints under the stress of EM.

Furthermore, during EM, Sn grains rotations occur undergoing EM, which was reported in several studies. Harris et al. [10] reported the Sn grain orientation and related rotation rate with the grain size. The rotation might cause a grain emergence in polycrystalline materials [11]. Chen et al. [12] researched on the evolution of grain orientation and electric resistivity of Sn grain under EM in flip chip interconnects by synchrotron radiation. During EM at 75 ºC, Sn grain growth was not obvious with little decrease of the electric resistivity, while the grains at current crowding region rotated at a constant speed. Lloyd et al. [13] noticed a voltage dropping (about 12.8 % at most) in Sn thin film deposited on Si wafer under direct current associated with the reorientation of the microstructure. However, Wu et al. [14] reported that no similar effect appeared when applying equivalent alternating current or long-time storage at the same temperature, which attributed the voltage drop to the rearrangement of the microstructure of β-Sn grains, instead of merely heat activation. During EM, β-Sn grains rearranged to decrease the resistance, showing the grain rotation morphologically, whereas, not all the grains would rotate during EM. The grains that rotated depended not only on its own grain orientation, but also on the orientation of its neighboring grains.

978-1-7281-1500-9/19 $31.00 © 2019 IEEE

Table I. Anisotropic properties of β-Sn

axis	σ [μΩ·cm]	γ [ppm/°C]	E [Gpa]	D (150°C) [cm²/s]			DT (150°C) [cm²/s]
				Ag	Cu	Ni	
a-axis	13.25	15.45	22.90	5.60×10^{-11}	1.99×10^{-7}	3.85×10^{-9}	8.70×10^{-13}
c-axis	20.27	30.50	68.90	3.13×10^{-9}	8.57×10^{-6}	1.17×10^{-4}	4.71×10^{-13}

Up to now, limit studies focus on the characteristics of the grain rotation and its effect on the property of interconnects, and none or less on the mechanism of the grain rotation. In the present work, in situ observation and a developed finite element model were carried out to investigate the β-Sn grain orientation behavior in solder bumps undergoing EM.

II. EXPERIMENTAL

The test vehicles were flip chip packages, whose surface finishes on printed circuit board (PCB) side, including an electroless nickel electroless palladium immersing gold (ENEPIG) and a Cu pad covered by a thin organic solderability preservative (OSP), which would disappear during reflow soldering.

Fig. 1 shows the schematic configuration of the Ni/Sn-3.0Ag-0.5Cu/ENEPIG(OSP) lead-free flip chip solder joints, in which Sn-3.0Ag-0.5Cu solder balls of 300 μm in diameter were used.

The outermost solder bumps were ground and polished to their central plane and stressed under a current density of 1.0×10^4 A/cm² in silicone oil at 150 °C for various times, as shown in Fig. 1 (a). The specimens were taken out from the silicone oil for microstructural and compositional examinations using a scanning electron microscope (SEM) and an electron probe microanalyzer (EPMA).

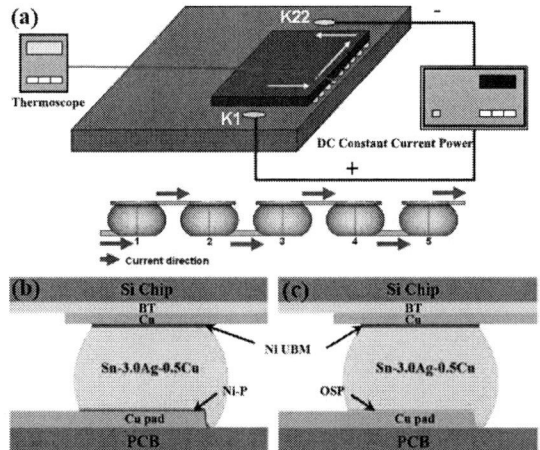

Figure 1. (a) configuration of the experiments; (b) schematic of the Ni/Sn-3.0Ag-0.5Cu/ENEPIG solder joint; (c) schematic of the Ni/Sn-3.0Ag-0.5Cu/OSP.

The electron backscattered diffraction (EBSD) analyses were also carried out to characterize the crystal orientation of β-Sn grains.

The angle θ between the *c*-axis of β-Sn grain and the electron flowing direction is calculated using a mathematical relationship between the Euler angle of the crystal orientation and the Cartesian coordinate system transformation matrix.

III. RESULTS AND DISCUSSION

A. Occurrence of grain rotations in solder bumps

After long-period EM, obvious β-Sn grain rotations occurred in several solder bumps no matter the finish types are ENEPIG or OSP. To facilitate discussion, four typical solder bumps were taken into consideration. Fig. 2 shows the cross-sectional microstructure of the Nos. 1, 2, 3 and 4 solder bumps after EM for 500 h and the corresponding EBSD maps in RD. Nos. 1 and 4 solder bumps were taken from Ni/Sn-3.0Ag-0.5Cu/ENEPIG solder joints, and the other two were taken from Ni/Sn-3.0Ag-0.5Cu/OSP solder joints. These solder bumps were all consisted of two or three β-Sn grains with different orientations.

B. Microstructural evolution of solder bumps with occurance of grain rotations

Fig. 3 shows the cross-sectional microstructure of No. 1 solder bump after EM for the various time as well as the corresponding EBSD map in RD. No. 1 solder bump consisted of two β-Sn grains: grain 1 (79.6°) and grain 2 (74.5°). Because the *c*-axis of the grain was nearly perpendicular to electron flowing direction, the self-diffusion of Sn atoms was enhanced undergoing EM with a weaken Ni diffusion. As a result, micro voids formation and crack propagation occurred at the cathode interface with increasing EM time. The β-Sn grain rotation occurred at the grain boundary of two grains, with grain 2 moving inward and grain 1 out ward.

Fig. 4 shows the cross-sectional microstructure of No. 2 solder bump after EM for the various time as well as the corresponding EBSD map in RD. No. 2 solder bump consisted of two β-Sn grains: grain 1 (79.1°) and grain 2 (72.6°). The *c*-axis of Sn grain trended to be perpendicular to electron flowing direction for grain 2. Cu atoms diffused slowly so that even Cu pad in current crowding region wound not dissolve with extended EM time. Grain rotation occurred at the grain boundary between grain 1 and 2, with grain 2 moving inward and grain 1 out ward.

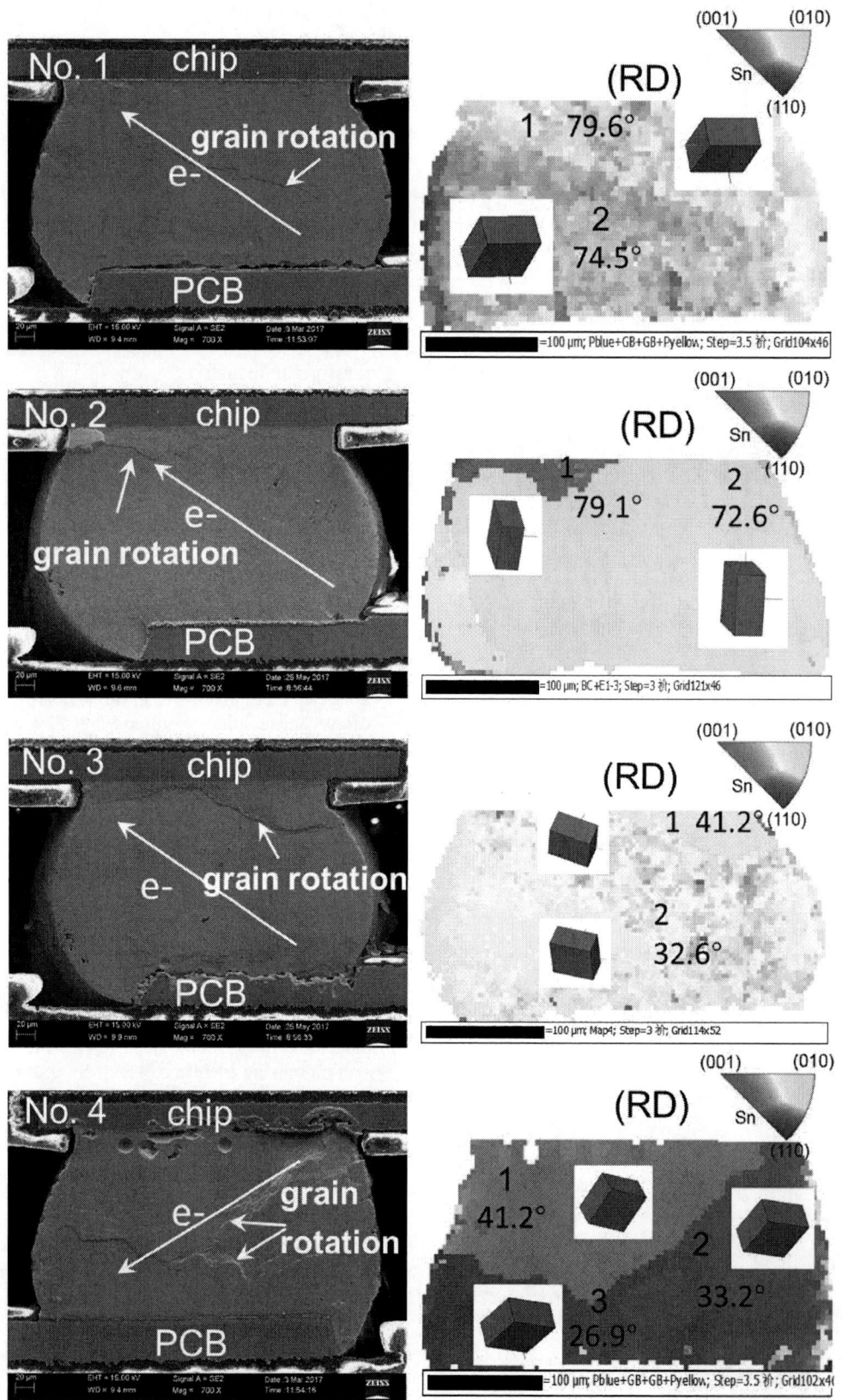

Figure 2. Cross-sectional microstructure of the Nos. 1, 2, 3 and 4 solder bumps and (b) their corresponding EBSD maps in RD.

978-1-7281-1500-9/19 $31.00 © 2019 IEEE

Fig. 5 shows the cross-sectional microstructure of No. 3 solder bump after EM for the various time as well as the corresponding EBSD map in RD. No. 3 solder bump consisted of two β-Sn grains: grain 1 (41.2°) and grain 2 (32.6°). The c-axes of Sn grain were nearly parallel to electron flowing direction for grain 2.

When EM time was 150 h, dissolution happened to Cu pad at the cathode corresponding to grain 2. After 500 h, Cu pad dissolved seriously. Grain rotation occurred at the grain boundary between grain 1 and 2, with grain 2 moving inward and grain 1 outward.

Fig. 6 shows the cross-sectional microstructure of No. 4 solder bump after EM for the various time as well as the corresponding EBSD map in RD. No. 4 solder bump consisted of three β-Sn grains: grain 1 (41.2°), grain 2 (33.2°), and grain 3 (26.9°) in No. 4 solder bump.

After EM for 150 h, the Ni UBM corresponding to grain 2 dissolved rapidly, while for grain 1, no Ni UBM dissolution was observed. Then after 250 h, the Ni UBM corresponding to grain 1 began to dissolve. Ni UBM dissolved quicker as EM time went on, with a similar solder collapse and grain rotation to Nos. 1 and 2 bumps.

C. Mechanism on EM-induced β-Sn grain rotation along the grain boundaries

For all the above solder bumps, the major diffusion elements are Cu, Ni, Sn. As seen from Table I, Sn atoms are the most difficult to diffuse through the grain boundaries during EM, even as the major element of the inner solder. Vacancies left by the diffusion of Sn atoms are more likely to be retarded by the grain boundaries.

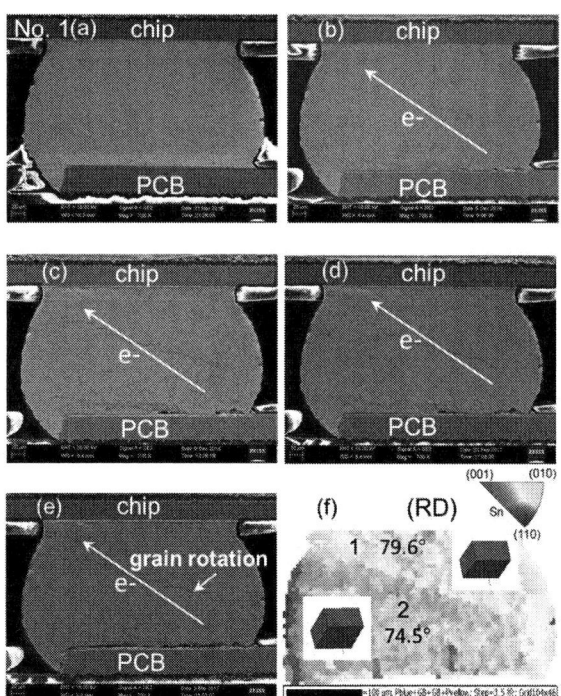

Figure 3. Cross-sectional microstructure of No. 1 flip chip lead-free solder bump after EM at 150 °C under a current density of 1.0×10^4 A/cm² for: (a) 0 h, (b) 150 h, (c) 250 h, (d) 350 h, (e) 500 h, and (f) the corresponding EBSD map in RD.

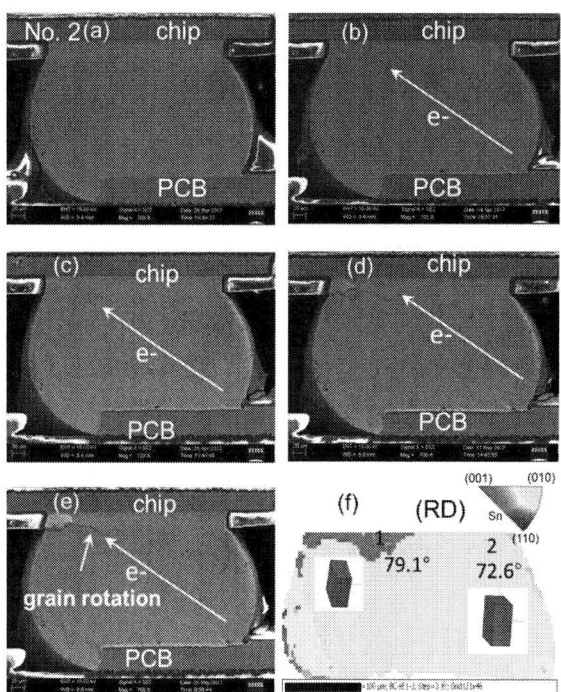

Figure 4. Cross-sectional microstructure of No. 2 flip chip lead-free solder bump after EM at 150 °C under a current density of 1.0×10^4 A/cm² for: (a) 0 h, (b) 150 h, (c) 250 h, (d) 350 h, (e) 500 h, and (f) the corresponding EBSD map in RD.

Due to the self-diffusion coefficient of Sn atoms differs along the c-axis and a-axis of Sn grain, the corresponding vacancy fluxes will be different:

$$J_v^c = \frac{C_v D_v^c}{kT} eZ^* \rho_{Sn}^c j \tag{1}$$

$$J_v^a = \frac{C_v D_v^a}{kT} eZ^* \rho_{Sn}^a j \tag{2}$$

Where J_V^c and J_V^a are vacancy fluxes along c-axis and a-axis (atoms/m²s). D_V^c and D_V^a are diffusion coefficient of Sn atoms along c-axis and a-axis (m²/s). Z* works as the measure of the magnitude of the driving force under EM and is considered the same in both directions [15].

Fig. 7 shows the schematic diagrams of EM-induced grain rotation in Sn-3.0Ag-0.5Cu flip chip solder bumps. As shown in Fig. 7 (a), abstracted from No. 1 solder bump, the model I includes grain 1(up) and grain 2(down) in the bump, with a-axis and c-axis parallel to the current direction respectively. The electrons flow up from PCB side to chip side. During EM, Vacancies diffuse downward through grain 1, the grain boundary and lastly grain 2. Vacancies diffuse hard through grain 2, so they accumulate at the grain boundary to reach supersaturation. No grinding-polishing process to the surface of the bump during in-situ experiment makes it stress-free and a good sink/source of vacancy, where there is an equilibrium vacancy concentration.

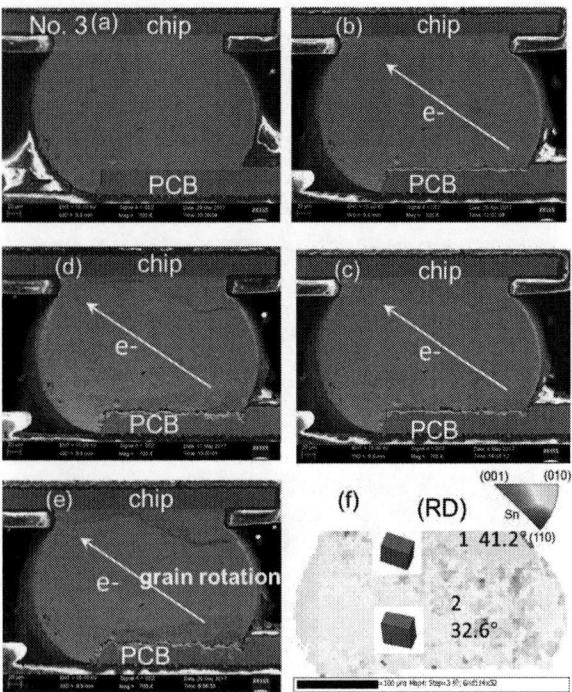

Figure 5. Cross-sectional microstructure of No. 3 flip chip lead-free solder bump after EM at 150 °C under a current density of 1.0×10^4 A/cm^2 for: (a) 0 h, (b) 150 h, (c) 250 h, (d) 350 h, (e) 500 h, and (f) the corresponding EBSD map in RD.

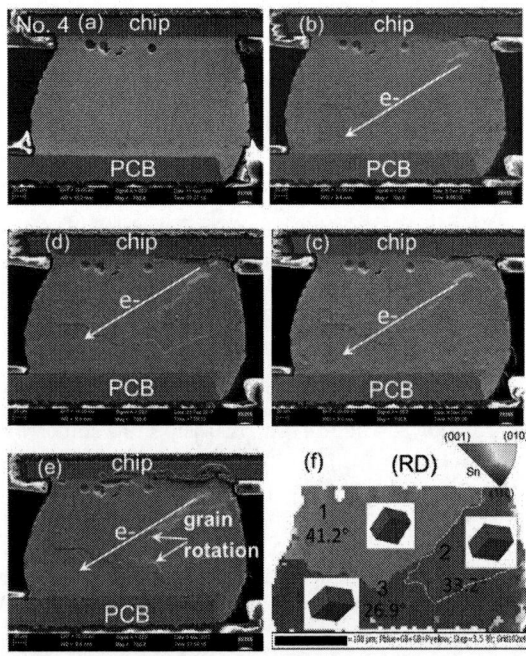

Figure 6. Cross-sectional microstructure of No. 4 flip chip lead-free solder bump after EM at 150 °C under a current density of 1.0×10^4 A/cm^2 for: (a) 0 h, (b) 150 h, (c) 250 h, (d) 350 h, (e) 500 h, and (f) the corresponding EBSD map in RD.

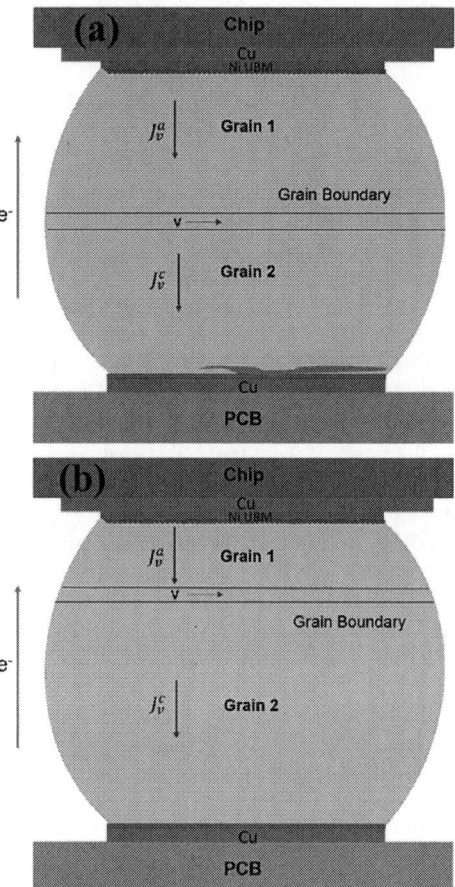

Figure 7. Schematic diagrams of the grain rotation induced by EM in a Sn-3.0Ag-0.5Cu flip chip lead-free solder bump: (a) crack-controlled type, (b) size-controlled type.

As a result, the redundant vacancies will diffuse outward along the grain boundary to the surface. Correspondingly, Sn atoms diffuse inward, and hence two grains are forced inward. Due to the crack near grain 2, the restriction to the grain 2 is weaken that grain 2 is easier to be rotate inward, resulting in floating and sinking of grain 1 and 2 along the boundary, respectively.

Considering the grain size, model II is proposed based on the results of No. 2 solder bump. Under the circumstance in Fig. 7 (b), small grain 1 (up) and big grain 2 (down) exist in the solder bump, also with different orientations. The same derivation gives that two grains are forced inward. To grain 2, the big size makes it longer distance from its point to the interface and hence easier to rotate, resulting in floating and sinking of grain 1 and 2 along the boundary, respectively.

For No.3 solder bump, it consists two grains, 1: small, 2: big with a crack. According to model I and II, grain 2 is more likely to rotate inward compared to the situation of Nos. 1 and 2 solder bumps. For No. 4 solder bump, a serious crack exists near grain 1 at the chip side, and the diffusion coefficients of Sn atoms in grain 1, 2 and 3 are decreasing.

Vacancies diffuse from the PCB side to chip side, making undersaturation at three grain boundaries. The three grains are all forced outward, but the restrictions are totally different. Grain 2 situated in the middle of the solder with no restriction, and the interface remains intact. As a result, to grain 1, the side of grain 1 near grain 3 will float and the side of grain 2 near grain 1 will float more obviously.

In summary, the proposed mechanism in model I and II can illustrate well the existing grain rotation occurred in Ni/Sn-3.0Ag-0.5Cu/ENEPIG(OSP) lead-free flip chip solder joints. Both the diffusion related to solder property and the restriction related to interfaces and grain position jointly affect the grain rotation.

IV. CONCLUSIONS

This study presents the results of electromigration behavior of Ni/Sn-3.0Ag-0.5Cu/ENEPIG(OSP) lead-free flip chip solder joints with respect to EM-induced β-Sn grain rotation. The mechanism on EM-induced β-Sn grain rotation was proposed to be related to diffusion and restriction. The main conclusions are:

(1) The occurrence of EM-induced β-Sn grain rotation is independent of the finish types (ENEPIG and OSP).

(2) Different grain orientations result in different saturation degree of vacancies at the grain boundaries, resulting in Sn atoms flux inward or outward the surface, and hence the floating or sinking of the β-Sn grains.

(3) Cracks at the interface lead to a weaken restriction to the β-Sn grains. The β-Sn grain near the crack is easier to rotate.

(4) The position of the β-Sn grains also has an effect on the EM-induced β-Sn grain rotation, floating or sinking.

ACKNOWLEDGMENT

This work is supported by the National Natural Science Foundation of China (Grant Nos. 51475072, 51511140289, 51671046 and U1837208) and the Fundamental Research Funds for the Central Universities (grant No. DUT17ZD202).

REFERENCES

[1] M.L. Huang, F. Yang, "Size effect model on kinetics of interfacial reaction between Sn-xAg-yCu solders and Cu substrate," Sci. Rep., vol. 4, Nov. 2014, pp. 7117, doi:10.1038/srep07117.

[2] M.L. Huang, S.M. Zhou, L.D. Chen, "Electromigration-Induced Interfacial Reactions in Cu/Sn/Electroless Ni-P Solder Interconnects," J. Electron. Mater., vol. 41, Apr. 2012, pp. 730-740, doi:10.1007/s11664-012-1952-6.

[3] K.S. Kima, S.H. Huha, K. Suganumab, "Effects of intermetallic compounds on properties of Sn-Ag-Cu lead-free soldered bumps," J. Alloys. Compd., vol. 352, Mar. 2003, pp. 226-236, doi:10.1016/s0925-8388(02)01166-0.

[4] S.K. Seo, S.K. Kang, M.G. Cho, D.Y. Shih, H.M. Lee, "The crystal orientation of β-Sn grains in Sn-Ag and Sn-Cu solders affected by their interfacial reactions with Cu and Ni(P) under bump metallurgy," J. Electro. Mater., vol. 38, Dec. 2009, pp. 2461-2469, doi:10.1007/s11664-009-0902-4.

[5] B.F. Dyson, T.R. Anthony, D. Turnbull, "Interstitial diffusion of copper in tin," J. Appl. Phys., vol. 37, Apr. 1967, pp. 3408-3409, doi:10.1063/1.1710127.

[6] D.C. Yeh, H.B. Huntington, "Extreme fast-diffusion system: nickel in single-crystal tin," Phys. Rev. Lett., vol. 53, Oct. 1984, pp. 1469-1472, doi:10.1103/PhysRevLett.53.2185.

[7] F.H. Huang, H.B. Huntington, "Diffusion of Sb124, Cd109, Sn113, and Zn65 in tin," Phys. Rev. B., vol. 9, Feb. 1974, pp. 1479-1488, doi:10.1103/PhysRevB.9.1479.

[8] M.H. Lu, D.Y. Shih, P. Lauro, C. Goldsmith, D.W. Henderson, "Effect of Sn Orientation on electromigration degradation mechanism in high Sn-based Pb-free solders," Appl. Phys. Lett., vol. 92, May. 2008, pp. 211909, doi:10.1063/1.2936996.

[9] M.L. Huang, J.F. Zhao, Z.J. Zhang, N. Zhao, "Dominant effect of high anisotropy in β-Sn grain on electromigration-induced failure mechanism in Sn3.0Ag0.5Cu interconnect," J. Alloys Compd., vol. 678, Sep. 2016, pp. 370-374, doi:10.1016/j.jallcom.2016.04.024.

[10] K.E. Harris, V.V. Singh, A.H. King, "Grain rotation in thin films of gold," Acta Mater., vol. 46, May. 1998, pp. 2623-2633, doi:10.1557/PROC-403-15.

[11] D. Moldovan, V. Yamakov, D. Wolf, S.R. Phillpot, "Scaling Behavior of Grain-Rotation-Induced Grain Growth," Phys. Rev. Lett., vol. 89, Oct. 2002, pp. 35-55, doi:10.1103/PhysRevLett.89.206101.

[12] K. Chen, N. Tamura, K.N. Tu, "In-situ study of electromigration-induced grain rotation in Pb-free solder joint by synchrotron microdiffraction," MRS Proceeding., vol. 1116, 2008, pp. 19-24, doi:10.1557/PROC-1116-I05-06.

[13] J.R. Lloyd, "Electromigration induced resistance decrease in Sn conductors," J. Appl. Phys., vol. 94, Oct. 2003, pp. 6483-6486, doi:10.1063/1.1623632.

[14] A.T. Wu, A.M. Gusak, K.N. Tu, "Electromigration-induced grain rotation in anisotropic conducting beta tin," Appl. Phys. Lett., vol. 86, Jun. 2005, pp. 1-3, doi:10.1063/1.1941456.

[15] M.L. Huang, J.F. Zhao, Z.J. Zhang, N. Zhao, "Role of diffusion anisotropy in β-Sn in microstructural evolution of Sn-3.0Ag-0.5Cu flip chip bumps undergoing electromigration," Acta Mater., vol. 100, Nov. 2015, pp. 98-106, doi:10.1016/j.actamat.2015.08.037.

Low-cost MT-ferrule-compatible Optical Connector for Co-packaged Optics Using Single-mode Polymer Waveguide

Akihiro Noriki and Takeru Amano
Electronics and Photonics Research Institute
AIST
1-1-1 Umezono, Tsukuba, Ibaraki, Japan
e-mail: a-noriki@aist.go.jp

Masatoshi Tsunoda and Toshiaki Michihiro
Connector Products Development Department
Kyocera Corporation
2-4-1 Nakayama, Midori-ku, Yokohama, Kanagawa,
Japan

Abstract—In high performance computer systems and large-scale data centers, data movement becomes a critical problem. To overcome this problem, we have studied an optical connector and an optical/electrical hybrid package substrate for a single-mode co-packaged optics of LSI packages. To realize low-cost self-aligned assembling of the optical connector on the hybrid package substrate, we proposed unique alignment structures which can be fabricated precisely with low-cost mass-productive processes. Using the proposed technique, the alignment structures were fabricated in precise position without any long-tact-time high-cost alignment processes. We fabricated the optical connector and the hybrid package substrate and demonstrated the self-aligned assembling process. We successfully assembled the optical connector to the waveguide-integrated package substrate with an error of 1.57-micrometers. Considering the obtained alignment error, low loss optical coupling with 0.5-dB misalignment loss was expected.

Keywords-MT-connector-compatible; optical connector; co-packaged optics; single-mode polymer waveguide; injection molding; self-alignment; assembling; optical/electrical hybrid package; low-cost mass-productive

I. INTRODUCTION

In high performance computer systems and large-scale data centers, data movement becomes a critical problem [1]. However, increasing data rates of conventional electrical links causes low latency tolerance, high power consumption, and poor signal integrity, especially for long distance interconnects. To extremely reduce the length of high-data-rate electrical links, co-packaging technologies of optics chips (e.g. silicon (Si) photonics) and high-performance large-scale integration (LSI) chips are attracted much attention. For example, Rockley Photonics showed a switch application specific integration circuits (ASICs) where Si photonics chips were co-packaged at OFC 2018, and dozen ribbon optical fiber cables were connected to it [2]. Such massively parallel optical input/outputs (I/Os) will be necessary for high performance LSIs like upcoming high-capacity switch ASICs for 25.6 and/or 51.2 Tbps. For the massively parallel co-packaged optics, we have worked on an optical and electrical hybrid package substrate which includes polymer waveguides as shown in Fig. 1. [3-5]. Using the polymer waveguides, the photonics chips can be placed closer to LSI chips, and thus, high-data-rate electrical links can be extremely shortened. The polymer waveguides also enable pitch conversion of optical I/O interfaces from low-density fiber ribbon I/Os to high-density Si photonics I/Os. In previous works, we reported a low-loss multi-mode hybrid package substrate and demonstrated its 25-Gbps operation. Low-loss optical connectors were also placed at the edge of the package substrate and the 24-channel multi-mode polymer waveguides was connected to ribbon fiber cables [3,4].

However, upcoming high-performance LSIs demand more and more data-rate capacity. To catch up such demands by using the multi-mode optical links, physical optical channels must be scale up with the data-rate capacity. Therefore, photonics-chip size, fiber count, and optical-I/O footprint for the co-packaged optics will be increased and its total cost will continue to raise up with the data-rate capacity. To overcome this problem, co-packaged optics for single-mode optical link is very attractive solution. By using wavelength division multiplexing (WDM) of the single-mode links, physical channel count can be reduced, and the data-rate capacity can be scaled up at low-cost. Total capacity of more than 10 Tbps will be realized in an optical connector by using the WDM.

To realize a hybrid package substrate for single-mode co-packaged optics, we have studied a polymer waveguide and an optical connector. We developed a low-loss single-mode polymer waveguide on a standard glass epoxy substrate [6]. Several types of single-mode optical connectors to connect waveguides and optical fibers were studied. Y. Taira, et al. demonstrated an optical connector fabricated by injection molding to connect a polymer waveguide film to a standard mechanical transfer (MT) fiber connector [7]. R. Krähenbühl, et al. demonstrated an optical connector based on a silicon V-groove to connect a polymer waveguide on a printed circuit board (PCB) to a standard LC-connector [8]. All of these connectors adopted self-alignment structures to align the optical connectors to the polymer waveguides. However, for efficient optical connections, alignment structures to be used for the self-alignment should be fabricated at precise positions within an error of 1–2 μm. Such precise fabrication would be less compatible with low-cost mass production of the optical connectors and polymer waveguide integrated package substrates.

In this work, we developed an optical connector and a package substrate to realize the single-mode co-packaged optics as shown in Fig. 1. For the connector and the package substrate, we proposed precise alignment structures which can be fabricated with low-cost mass-productive processes and demonstrated fine self-alignment of them.

978-1-7281-1500-9/19 $31.00 © 2019 IEEE

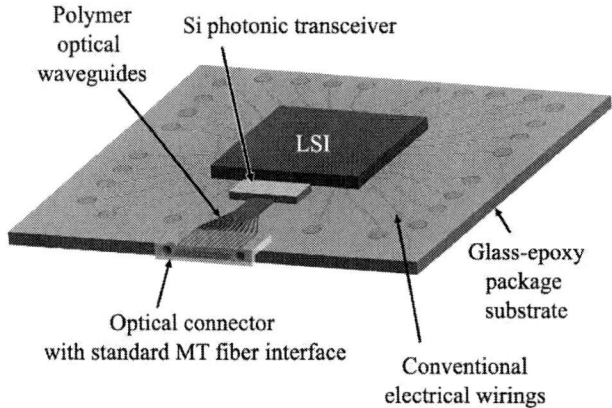

Figure 1. Concept of co-packaged optics for LSI packages.

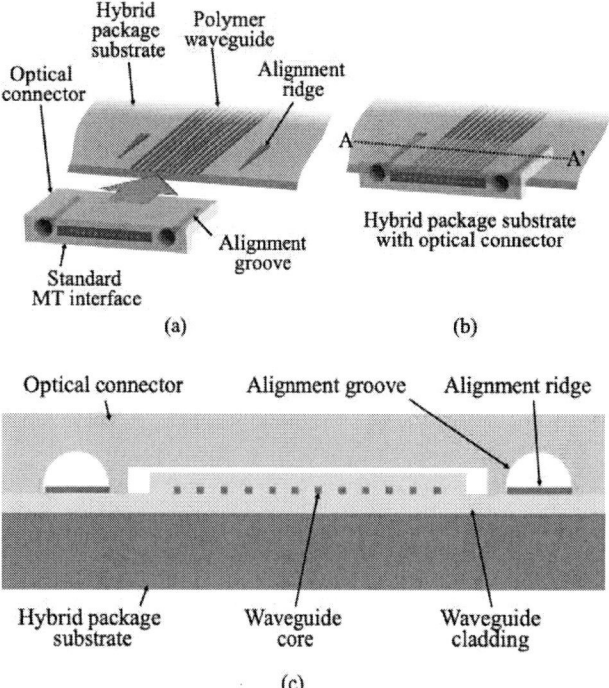

Figure 2. Concept of optical connector for hybrid package substrate. (a) Optical connector is slid on package substrate and (b) assembled with self-alignment. (c) Cross sectional image of assembled optical connector.

II. CONCEPT OF OPTICAL CONNECTOR

To develop the single-mode co-packaged optics as shown in Fig. 1, the optical connector for the polymer waveguide integrated hybrid package substrate was developed. Cost-efficiency, scalability, and manufacturing volume of photonic packaging are important [9]. The same can be said for the optical connector. To have the scalability, the optical connector compatible with standard MT fiber connector was designed. To increase cost-efficiency and manufacturing volume, the optical connector was also designed to be fabricated with injection molding.

Concept of the optical connector was shown in Fig. 2. Like the other optical connectors in ref. 6, we also use self-alignment method to assemble the optical connector on the hybrid package substrate at low cost. Groove structures of the optical connector and ridge structures on the package substrate were used for the self-alignment. As shown in Fig. 2(a), the optical connector was slid on the package substrate, and its alignment groove was guided by the alignment ridge on the package substrate. To detect the end point of the self-alignment, the applied force to slide the optical connector was monitored, and the slide motion was stopped when the monitored force was over a target value. Thus, the lateral self-alignment was achieved by the sidewalls of the alignment ridges and grooves. On the other hand, the vertical alignment was achieved by the contact area between the connector and the package substrate. The contact area of the connector was its bottom surface and that of the package substrate was the top surface of the bottom cladding layer.

As a result, the optical connector was assembled at the edge of the package substrate as shown in Fig. 2(b). Fig. 2(c) shows the cross-sectional image on A-A' line depicted in Fig. 2(b). The alignment ridge of the package substrate was fabricated in a polymer waveguide core layer. Only the ridge structure was exposed, and the other waveguide cores were covered with a top cladding layer. Detail of the alignment structures are described in following section.

III. DESIGNS AND FABRICATIONS

A. Optical connector

To achieve single-mode efficient optical connection from the polymer waveguide to the MT fiber connector (including propagation in the opposite direction), these components must be precisely aligned with an alignment error of 1–2 μm through the optical connector. For the optical connector, the MT fiber connector and the polymer waveguides are aligned using guide-pin holes and the alignment grooves, respectively. Thus, the alignment grooves must be positioned precisely against the guide-pin holes.

To satisfy this requirement, we considered an optical connector where guide-pin holes is also used as alignment grooves. Specifically, we used the same part of the mold to form the guide-pin hole and the alignment groove (part of the mold serves as the core part). Fig. 3 shows a simplified schematic of the mold structure and shows the core part and the optical connector. We fabricated the core part by grinding, which offers high precision and concentricity. Using this technique, the optical connector's guide-pin holes and alignment grooves can be fabricated with submicron precision.

Fig. 4 shows the schematic image of optical connector's surface (interface to the MT fiber connector). The guide-pin hole and alignment groove on the left and right sides are concentric. Generally, the MT ferrule measures accuracy based on the deviation from the axis, which is defined by the guide-pin hole and alignment groove on the left and right sides. Thus, the position of the optical connector may be

978-1-7281-1500-9/19 $31.00 © 2019 IEEE

measured using the MT ferrule method and its measuring instrument.

Furthermore, to facilitate mass production, we fabricated the optical connector via injection molding. Additionally, we designed the optical connector's interface to be compatible with the interface of the MT fiber connector, which is a popular commercial fiber cable. Thus, the proposed system, using the optical connector, can be fabricated at a low cost.

Fig. 5 shows the final molded optical connector. Fig. 6 shows a histogram of the frequency of groove-diameter errors, which is the deviation from the design value. These results indicated a groove error in the optical connector of ~0.5 μm.

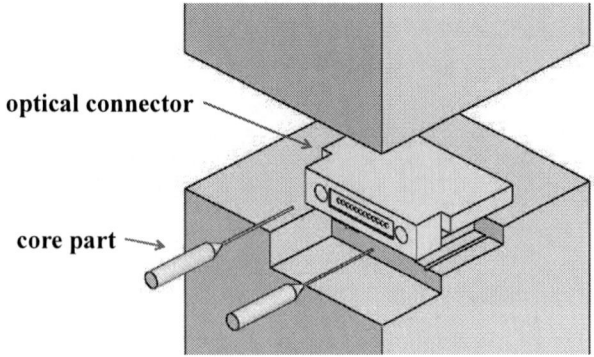

Figure 3.   Simplified schematic image of mold structure.

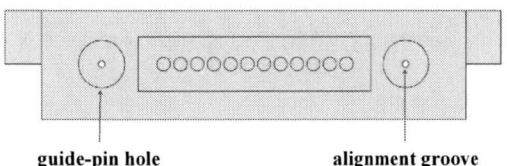

guide-pin hole   alignment groove

Figure 4.   The schematic image of optical connector's surface (interface to the MT fiber connector).

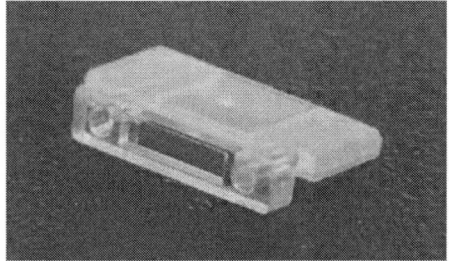

Figure 5.   Photograph of final molded optical connector.

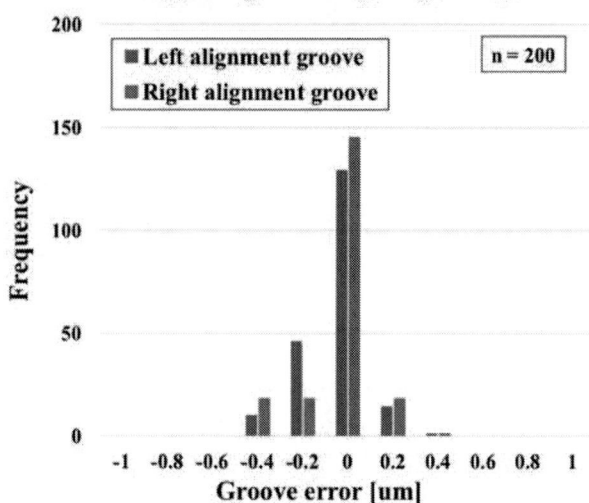

Figure 6.   Frequency histogram of groove errors.

B.   Hybrid package substrate

On the package substrate, the alignment structures should be also placed very accurately against the polymer waveguide cores. However, it is difficult to fabricate the alignment structures within 1-μm position error. To realize such fabrication, long-tact-time and high-cost alignment process would be required. Therefore, the alignment ridge was fabricated with the same photomask for the waveguide core. That is, the ridges were fabricated in the core layer. As the relative position error of commercially available photomasks is in the order of 0.1 μm, the ridges were fabricated in the accurate position.

The height of the alignment ridge was very small since it was fabricated in the core layer. Thus, the self-alignment like ref. 6 requires very accurate pre-alignment. To overcome this problem, the self-aligned assembling was carried out by sliding the optical connector on the package substrate as shown in Fig. 2(a). Lateral self-alignment was achieved by the sidewalls of the alignment ridges and grooves. By using this style, large taper ridge patterns were available for self-alignment, and thus, required pre-alignment accuracy was moderated to be over 50 μm. For the vertical alignment, the top surface of the bottom cladding layer was used as the contact area. Because of the waveguide cores were fabricated on the bottom cladding surface, it was good reference plane to the waveguide cores. Only a core thickness must be controlled accurately. As the waveguide was single-mode, the thickness of the core layer is 5–10 μm, and thus, its thickness can be easily controlled within 1-μm error.

A fabrication process of the hybrid package substrate is shown in Fig. 7. As a package substrate, we used a popular glass-epoxy (FR4) substrate. After bottom cladding formation, waveguide cores and the alignment ridges were fabricated by spin-coating and photo lithography. Then top cladding was also formed by spin-coating and photo lithography to expose the alignment ridges.

The polymer waveguide fabricated with the above process was shown in Fig. 8. There was large roughness in the surface of the FR4 substrate. However, a smooth surface was obtained just after the bottom-cladding formation. The insertion losses of the fabricated polymer waveguides were measured as shown in Fig. 9. The low propagation loss of ~0.4 dB/cm was obtained for 1.31-μm wavelength. Therefore, the high-quality single-mode polymer waveguide was successfully fabricated on the FR4 substrate.

1. Spin coating of bottom cladding layer

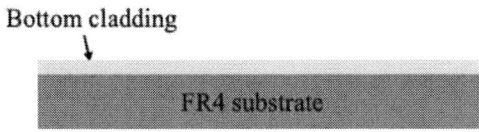

2. Spin coating and photolithography of core layer

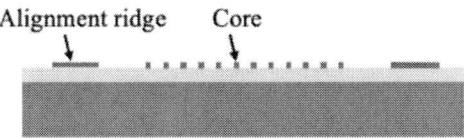

3. Spin coating of top cladding layer

4. Photolithography of top cladding layer
(Only expose alignment ridges)

Figure 7.   Fabrication process of hybrid package substrate.

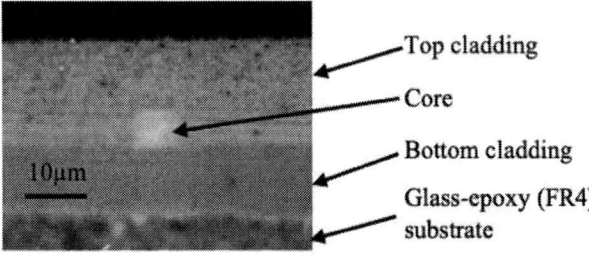

Figure 8.   Photograph of fabricated polymer waveguide.

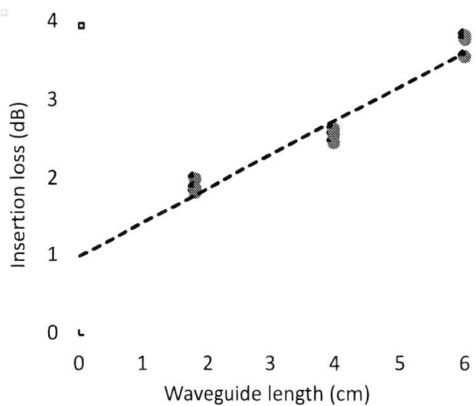

Figure 9.   Insertion loss of fabricated polymer waveguides.

IV.   EVALUATION

To evaluate the self-alignment accuracy, we carried out the assembling process and evaluated the alignment error between the alignment groove and ridge as shown in Fig. 10. The optical connector was roughly put on the package substrate and slid by a loadcell attached tool (Fig. 10(a)). During the assembling process, applied force to slide the optical connector was monitored by the loadcell. The slide motion was stopped when the monitored force was over a target value (Fig. 10(b)). Finally, the alignment groove and ridge were observed through the guide-pin hole by optical microscope and the alignment error was measured as shown in Fig. 10(c).

The monitored applied force was shown in Fig. 11. As shown in this profile, times to start the slide motion and the self-alignment were clearly observed. The slide motion was stopped when the applied force was achieved to a target force. A maximum force which can be applied without any damage to the alignment ridge was used for the target force.

The assembled optical connector was shown in Fig. 12. The connector was successfully assembled at the edge of the package substrate and evaluated its alignment error by optical-microscope observation. The optical microscopic image of the alignment groove and ridge, which was observed through the guide-pin hole, was shown in Fig. 13. Because of the edge of the groove and ridge structure were not in the same plane, the groove structure was not focused. Therefore, the groove edge was observed before the alignment ridge observation, and it was shown as an inserted circle in the Fig. 13. The center of the circle was shown as a cross line. The taper tip of the alignment ridge was clearly observed. Alignment accuracy of one assembled pair was directly evaluated by measuring the alignment error between the groove and ridge structures. That is, the relative position between the center of the alignment ridge and the cross line was measured as the alignment error. The lateral and vertical alignment error was 0.22 and 1.55 μm, respectively. Thus, we successfully detected the end point of the self-alignment by monitoring the applied force, and the optical connector was passively assembled to the waveguide-integrated package substrate within 1.57-μm error. Considering the obtained alignment error, an optical coupling with 0.5-dB misalignment loss was expected. By optimizing several assembling parameters such as the target force to stop the assembling and taper pattern of the alignment ridge, further accurate alignment would be realized.

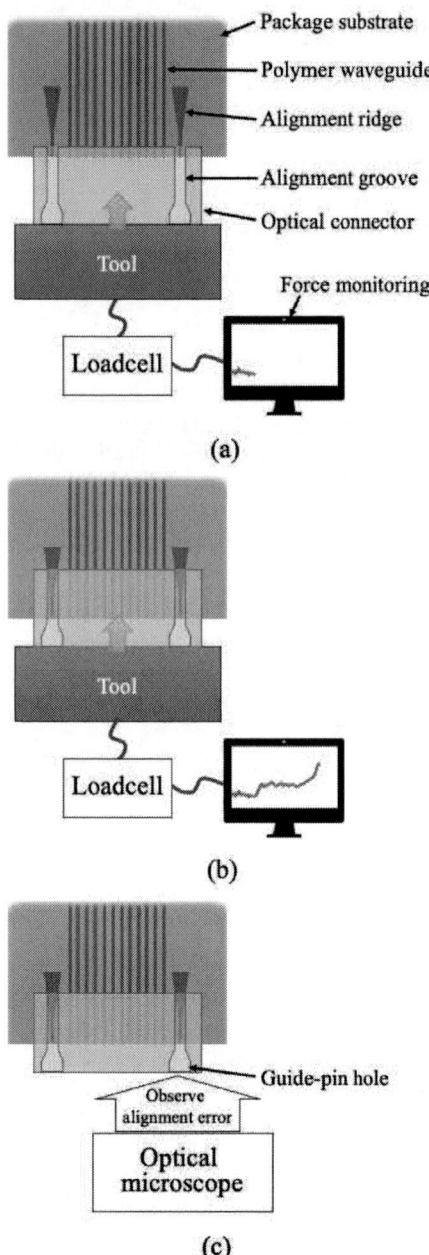

(a)

(b)

(c)

Figure 10. Evaluation method. (a) shows start of assembling, (b) shows end of assembling, and (c) shows measuring method of alignment error.

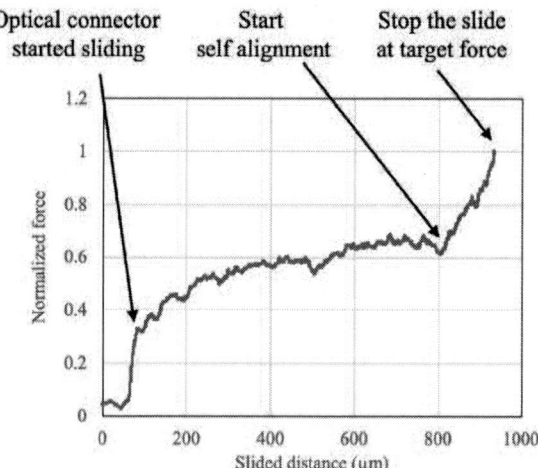

Figure 11. Profile of monitored force.

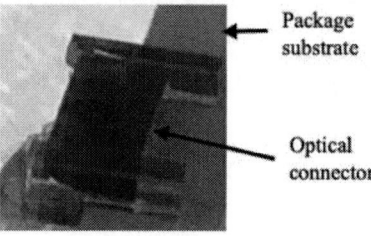

Figure 12. Assembled optical connector on package substrate.

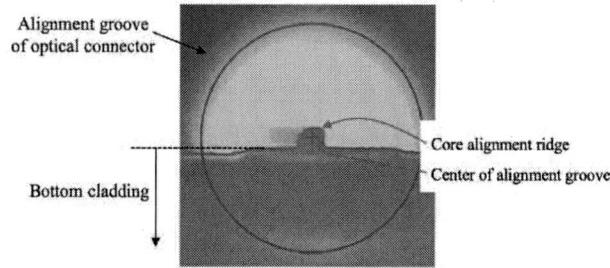

Figure 13. Alignment groove and ridge structures after self-alined assembling process.

V. CONCLUSION

We have studied the optical connector and the optical/electrical hybrid package substrate to realize the single-mode co-packaged optics for the LSI packages. For low-cost self-assembling of the optical connector on the hybrid package substrate, we proposed precise alignment structures which can be fabricated with low-cost mass-productive processes.

In the optical connector, the alignment grooves for the self-alignment and the guide-pin holes for the MT fiber connectors were fabricated with the same molding part. Thus,

978-1-7281-1500-9/19 $31.00 © 2019 IEEE 2046

these structures can be self-aligned with low-cost mass-productive injection molding with submicron accuracy. To evaluate the accuracy of the molding process, we measured the frequency of the groove diameter errors. We obtained small errors of ~0.5 μm. On the package substrate, the alignment ridge for the self-alignment was fabricated with same photomask for the waveguide core. Thus, these structures can be fabricated with precise position without the long-tact-time high-cost alignment process.

We demonstrated the optical connector assembling with the self-alignment. The optical connector was passively assembled to the waveguide-integrated package substrate within 1.57-μm error. Considering the obtained alignment error, low loss optical coupling with 0.5-dB misalignment loss was expected.

REFERENCES

[1] R. Dangel, J. Hofrichter, F. Horst, D. Jubin, A. L. Porta, N. Meier, I. M. Soganci, J. Weiss, and B. J. Offrein, "Polymer waveguides for electro-optical integration in data centers and high-performance computers," Opt. Express, vol. 23, pp. 4736-4750, 2015.

[2] "Rockley Photonics showcases its in-packaged design at OFC," Gazettabyte, 2018 [Online], Available: http://www.gazettabyte.com/home/2018/3/15/rockley-photonics-showcases-its-in-packaged-design-at-ofc.html. [Accessed: 11-Mar-2018]

[3] T. Amano, S. Ukita, Y. Egashira, M. Sasaki, A. Noriki, M. Mori, K. Kurata, and Y. Sakakibara, "Low-Loss Characteristics of a Multimode Polymer Optical Waveguide at 1.3 um Wavelength on an Electrical Hybrid LSI Package Substrate," 2016 IEEE 66th ECTC, 2016, pp. 477-482.

[4] T. Amano, S. Ukita, Y. Egashira, M. Sasaki, A. Noriki, M. Mori, K. Kurata, and Y. Sakakibara, "25-Gb/s Operation of a Polymer Optical Waveguide on an Electrical Hybrid LSI Package Substrate With Optical Card Edge Connector," J. Lightwave Technol., vol. 34, pp. 3006-3011, 2016.

[5] A. Noriki, T. Amano, M. Mori, Y. Sakakibara, K. Kurata, A. Ukita, M. Kurihara, and K. Takemura, "Evaluation of optical coupling characteristics for optoelectronic hybrid LSI package," 2016 IEEE CPMT Symposium Japan (ICSJ), 2016, pp. 227-230.

[6] D. Hashimoto, T. Amano, T. Kubota, Y. Okano, and A. Noriki, "Low propagation loss of a single mode polymer optical waveguide on a glass epoxy substrate," 2018 IEEE CPMT Symposium Japan (ICSJ), 2018, pp. 131-132.

[7] Y. Taira, H. Numata, S. Takenobu and T. Barwicz, "Improved connectorization of compliant polymer waveguide ribbon for silicon nanophotonics chip interfacing to optical fibers," 2015 IEEE 65th ECTC, 2015, pp. 1640-1645.

[8] R. Krähenbühl, T. Lamprecht, E. Zgraggen, F. Betschon, and A. Peterhans, "High-Precision, Self-Aligned, Optical Fiber Connectivity Solution for Single-Mode Waveguides Embedded in Optical PCBs," J. Lightwave Technol., vol. 33, pp. 865-871, 2015.

[9] T. Barwicz, T. W. Lichoulas, Y. Taira, Y. Martin, S. Takenobu, A. Janta-Polczynski, H. Numata, E. L. Kimbrell, J. Nah, B. Peng, R. Leidy, M. Khater, S. Kamlapurkar, S. Engelmann, P. Fortier, and N. Boyer, "High-Throughput Photonic Packaging," Optical Fiber Communication Conference, 2017, p. Tu3K.4.

Characterization of Coated Silver Wire Bond Interface Using TEM

Sarangapani Murali, Tan Swee Seng *Eric and Wong Chin Yeung Jason,*

Innovation, Heraeus Materials Singapore Pte. Ltd., Blk-5002#04-07,

Ang Mo Kio Ave 5, TECHplaceII, Singapore – 569871.

Corresponding author email: murali.sarangapani@heraeus.com

Abstract— **Coated silver (Ag) wire is designed and developed to replace gold (Au) wire of soft nature. Ion-milling the reliability tested coated-Ag bonded device revealed intact Ag ball bonds to Al-0.5wt%Cu pad exhibiting good compatibility with green epoxy mold compound. The paper primarily reports the characterization of bond interfaces of the reliability tested coated-Ag ball bonds. Where, thin samples parted from the bond interface using focus-ion beam (FIB) and analyzed using selected area electron diffraction (SAED) phase analysis, attached to transmission electron microscope (TEM). During thermal cycling ((TC) -40 to 125°C, 2000cycles)) and biased highly accelerated stress test (bHAST - 130°C, 85%RH, +3.3V, 192h) bond interfaces revealed predominant Ag_2Al phase formation along with growth of $AlAu_4$ phase. Moreover, thermal ageing (high temperature storage ((HTS), 175°C, 3000h)) revealed predominant Ag_3Al phase growth along with $AuAl/Au_2Al$ phases and pure Au/Ag regions. Relatively, Ag diffuses rapidly compared to Au and Al at the interface, thus consuming entire Al pad. Aging at high temperature for longer periods transform Ag_3Al to Au_2Al phases, perhaps stable than other gold/silver-aluminide phases. This indicates that the phases formed at the bond interface Ag_3Al, Ag_2Al, $AuAl$, Au_4Al, Au_2Al are harmless and do not contribute to the reliability failures.**

Keywords- Coated-Ag bonding wire, reliability of Ag wire bonds, silver-aluminide, gold-aluminide, intermetallics

I. Introduction

Novel coated-Ag wire is invented and introduced to support memory application, chain bonding on stacked devices. Coated-Ag wire revealed formation of axi-symmetrical spherical FAB under atmosphere similar to gold wire [1-5]. FAB surface is clean and bondability of coated-Ag wire is comparable to Au wire. One of the principal property of the wire is its soft nature inherently deforming easily, especially in forming soft Ag FAB similar to Au FAB. The Ag wire is coated with Au and its core is alloyed with noble metals. Stitch bond bondability of the coated-Ag wire is equivalent to Au wire satisfying the

requirements of stacked devices. The bondability and reliability test results are published in detail [1-5]. The scope of the present paper is to characterize the ball bond interface using FIB sectioning and TEM-SEAD phase analysis, especially to identify the crystal structures of intermetallics which are formed at the bond interface when exposed to varying growth kinetics during reliability evaluations (TC, bHAST, HTS).

II. Experimental Details

FAB were processed using coated-Ag wire and Kulicke & Soffa iConn bonder under atmosphere without purging protective or inert gases, firing 50 to 70mA EFO current, targeting 1.8-2.3 ball to wire ratio (BSR), and maintaining 760µm (30mil) wand gap. Coated-Ag wires were bonded on BGA2x2, ball pull/shear and stitch pull were tested using DAGE4000-plus. FABs and bonded balls were observed in LEO-1450VP scanning electron microscope (SEM) for axi-symmetrical spherical shape and off-centered ball morphology, respectively.

Leica Tic3x model was used for ion-milling the bonded ball to center from outer epoxy molded device. FEI Helios 600i FIB model was used to part the sample by applying 5kV. SAED pattern phase analysis was conducted using high resolution TEM (HRTEM) Talos F200X model. The samples were also observed and imaged using STEM at 200kV accelerated voltage attached with Oxford X-max 80-mm2 EDX unit.

III. Results & Discussion

A. Reliability of Coated-Ag

Green epoxy molding compound (EMC) with chlorine (Cl) less than 15ppm, sulfur (S) less than 10ppm, formulated with ion-catcher and pH between 5 and 7 (manufacturer's non-standard specification), free of bromine and antimony, revealed reliable Ag ball bonds without failures:

> ➤ until 3000h of HTS test at 175°C. Silver-aluminide at the interface grows to a thickness of 3µm for 1000h of storage at 175°C and it remains constant for further storage (Fig.1).

978-1-7281-1500-9/19 $31.00 © 2019 IEEE

➤ until 192h on biased HAST at 130°C, 85%RH, +3.3V

➤ until 2000cycles on thermal cycling at -40°C to +125°C

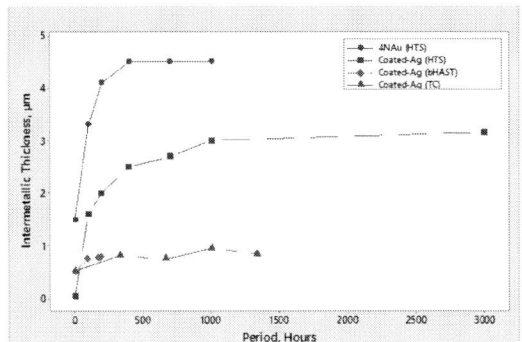

Fig.1 Rate of growth of intermetallic phase at the bond interface for varying test temperature.

Coated-Ag wire of 20μm diameter is bonded to Al-0.5wt%Cu 1μm pad thickness for bHAST and TC tests, while the same wire is bonded to 0.7μm pad for HTS study. The trend of growth rate of silver-aluminide is similar to Au ball bond following square-root power law [2] possessing an exponential increase until 1000h and remains constant thereafter for further storage of up to 3000h at 175°C. Statistical best curve fitting for the four plots of Fig.1 to square-root power law (Equ.1), and predicting the constant value "a" in the equation is shown in the Table.1

$$Y = a\sqrt{t} \quad ....Equ.1$$

Where, Y and t stands for intermetallic thickness and period reported in Fig.1, respectively. The constant value differed for coated-Ag bond in second decimal compared to Au bond. Also, the constant value strongly depends on the test temperature of 130°C/150°C, thus significantly behaves differently than that of 175°C ageing.

Table.1 Calculated value of the constant "a" from Fig.1 plots

4NAu (HTS)	Coated-Ag		
	HTS	bHAST	TC
0.184	0.080	0.062	0.028

Growth of the thickness of intermetallic (aluminide) phase is 0.65μm for lower temperature exposure (130°C to 150°C) during TC and bHAST testing. Initial curing of the

epoxy molding compound (175°C for 4h) may create aluminide and further growth is negligible during reliability testing.

FIB parting of thin samples from the four quadrants of FAB, and STEM-EDX analysis revealed Au diffuses to a depth of ~4μm from the surface of FAB along the periphery. Excellent miscibility of Au into Ag forms into a solid-solution with a composition gradient ranging from 10 to 58wt%. Higher content of Au in Ag FAB is observed near the neck of FAB (Table.2, Fig.2) [2]. Therefore, any reactions at the bond interface are expected/related to the diffusion of the three atoms present at the interface Ag, Au and Al.

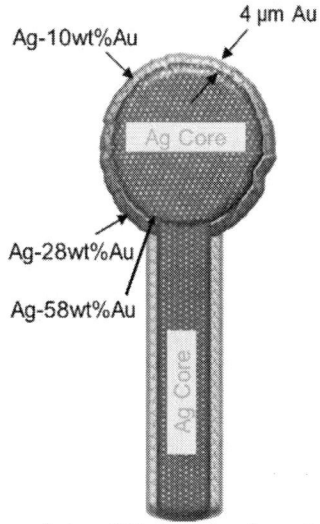

Fig.2 Illustration of Au diffusion depth and existence of AgAu solid-solution along the FAB periphery [2].

Table.2 Line scan EDX revealing Au diffusion depth in Ag FAB [2].

Au diffusion length, μm	Au level, wt%	Au diffusion length, μm	Au level, wt%
Quadrant1 (Top)		Quadrant2 (Top)	
1.5 to 4.6	10	1.5 to 4.8	10 to 28
Quadrant3 (Neck)		Quadrant4 (Neck)	
1.0 to 2.0	20 to 58	2 to 4	15 to 25

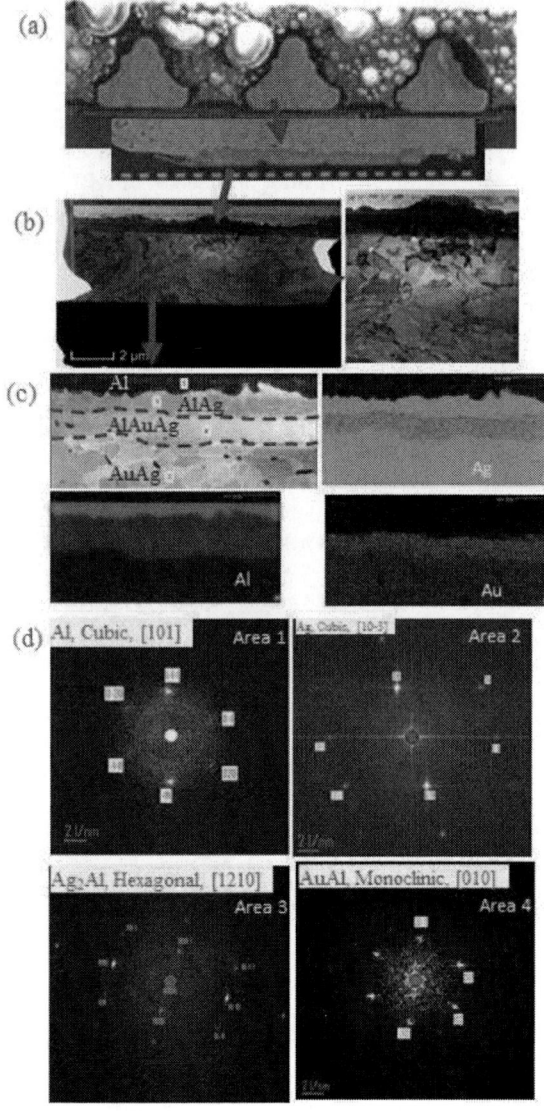

Fig.3 Observations of time zero bonded ball, epoxy molded and cured samples for the conditions: (a) ion-milled section of bonded ball, (b) TEM image of FIB parted thin sample, (c) elemental dot mapping using EDX attached to STEM and (d) SAED pattern phase analysis of four spots.

B. FIB Sectioning and TEM/HRTEM Phase Analysis of the Reliability Tested Ball Bond Interface

Chiefly, the bond interface is characterized using energy dispersive X-ray analysis (EDX) and selected area electron diffraction (SAED) pattern phase analysis to identify the intermetallic compounds formed at time zero bonding and reliability tested samples. All the three samples, time zero, bHAST and TC evaluated devices were cured for 4h at 175°C. Hence, expected to observe intermetallic phase formation in the interface for time zero bonded sample without reliability testing. Fig.3a shows the ion-milled

cross-section of time zero bond with intermetallic phase formation at the interface. Parting a thin sample with FIB and observing in TEM revealed dark and light contrast images, perhaps related to dislocation densities and stress levels (Fig.3b). EDX elemental dot mapping showed bright contrast of Al layer which is about 200nm thickness. Moreover, inter-diffusion depth of Al and Ag is approximately 400nm. Thus, silver-aluminide exists to a thickness of around 400nm. Dot mapping showed a layer of bright Au of 200nm close to silver-aluminide layer, likely showing the existence of gold-silver solid solution layer (Fig.3c). Four spots are selected for phase analysis using TEM/HRTEM as shown in Fig.3c. As expected, regions (also referred as spots or areas) 1 and 2 confirmed face centered cubic (FCC) of Al and Ag corresponding to [101] and [10-3] zone axis, respectively. It is interesting to note, regions 3 and 4 revealed hexagonal Ag_2Al and monoclinic AuAl phases, respectively (Fig.3d). Perhaps, AuAl intermetallic co-exists with Au-Ag mixture. The bonded ball Ag matrix is alloyed with low fraction of palladium (Pd), but finds difficult to trace the presence of it using TEM.

Similarly, thermal aged (HTS, 175°C, 3000h) bond interface is observed in TEM/HRTEM after ion-milling and FIB parting of thin samples (Fig.4). Where, two bonded balls are observed for six regions. Monoclinic AuAl phase is formed in Al region indicating Au and Al inter-diffuse and forms into gold-aluminide when aged for long time (Fig.4a). There is no other second phases formed in the observed Al bond pad and Ag bonded ball regions (Fig.4a (c) & (d)).

Fig.4b shows interface of neighbor bonded ball to Fig.4a, where abundantly available Ag deeply diffuse into Al, transforming the entire Al into stable Ag_3Al intermetallic phase with cubic structure identified as [111] zone axis (Fig.4b (c)). A localized region showed formation of Au_2Al phase along with Ag_3Al phase area. Interestingly, low symmetry orthorhombic Au_2Al is formed in a high symmetry cubic Ag_3Al phases. This indicates that bonded ball interface when exposed to an elevated temperature, phase transforms into two types consisting of gold-aluminide and silver-aluminide. Perhaps, phase transformation to Ag_3Al and Au_2Al are relatively stable compounds. Due to dynamic e-beam scattering, it is a challenge to analyze spot10.

STEM images of thermal cycled sample showed (Fig.5) fine pores in Ag bonded ball regions. However, zero pores are noticed in Al bond pad regions with two distinct aluminide layers: silver-aluminide of ~500nm thickness and ~800nm of gold-aluminide (Fig.5b). SAED phase analysis of these two layers revealed hexagonal Ag_2Al and monoclinic AuAl. Spot 11 and 12 indicates that the entire Al is transformed to silver aluminide Ag_2Al, where Ag, Al and Au inter-diffuse, rearrange and form (transform) compounds of Ag_2Al and $AuAl$ layers. All the three atoms have same atomic diameter of 1.44Å, hence should have no hindrance in the mobility to form stoichiometric compounds (Ex. Ag_2Al and $AuAl$). Remarkably, pure Au spots are observed in Ag bonded ball near to gold-aluminide layer (Fig.5d). Thus, further supply of thermal energy these phases could be transformed into low energy stoichiometric compounds such as Ag_3Al, Au_2Al, etc.

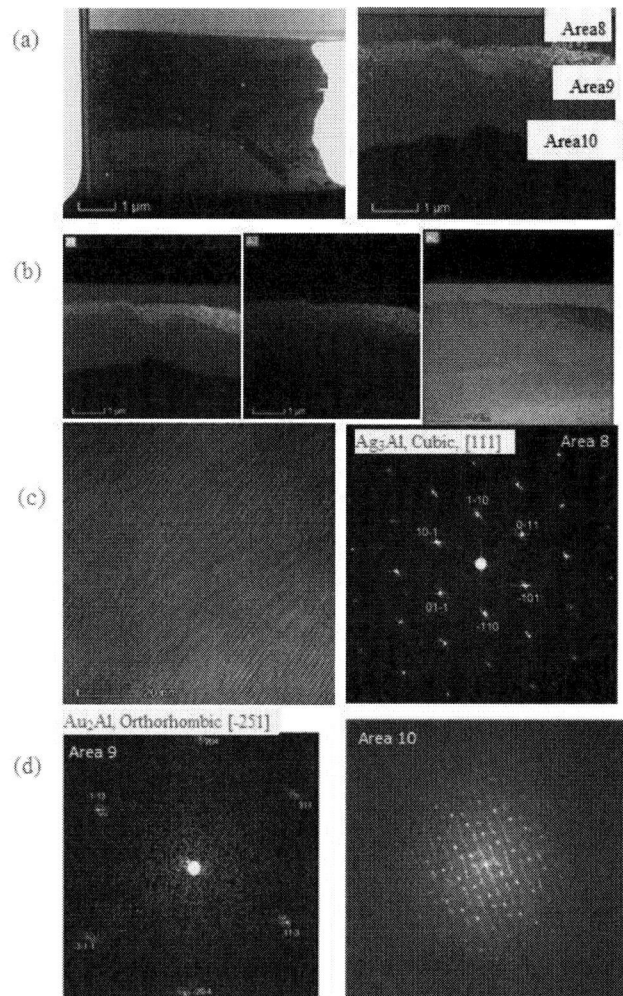

Fig.4b Observations of thermal aged bonded ball at 175°C for 3000h sample for the conditions: (a) TEM image of FIB parted thin sample from another neighbor bonded ball to Fig.4a, (b) elemental EDX dot mapping revealed the formation and co-existence of silver-aluminide and gold-aluminide, (c) spot8 showed Ag_3Al cubic structure identified with [111] zone axis, and (d) spot9 confirmed the formation of Au_2Al phase in and around Ag_3Al phase region.

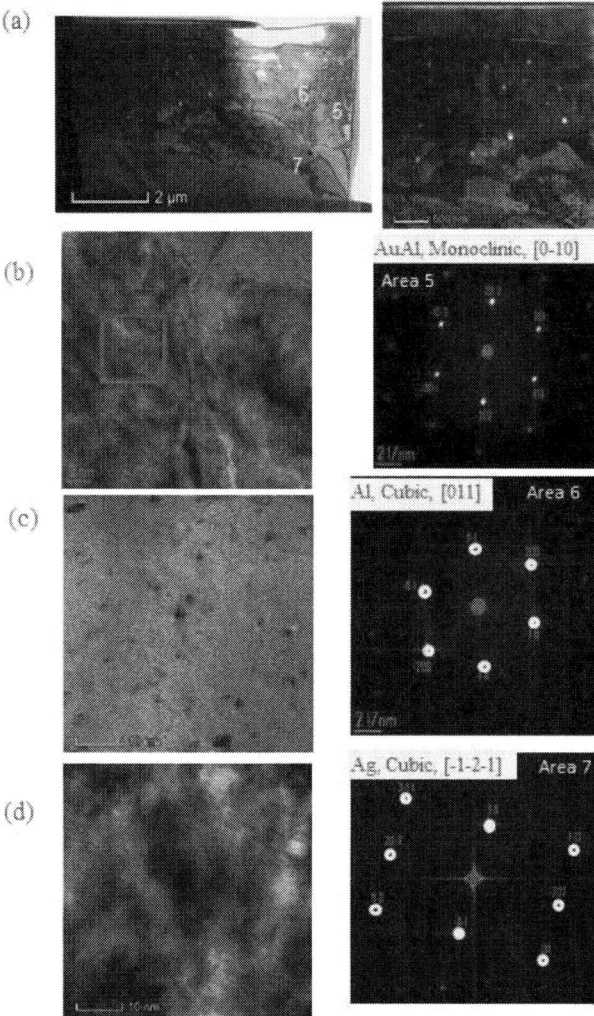

Fig.4a Observations of thermal aged bonded ball at 175°C for 3000h sample for the conditions: (a) TEM image of FIB parted thin sample, (b) spot5 shows formation of AlAu Monoclinic intermetallic phase in Al bond pad region, (c) & (d) confirmed the presence of pure Al and Ag.

Analogous to thermal cycled test samples, bHAST bonded ball interface showed the formation of two layers of intermetallic phases: hexagonal Ag_2Al phase as bottom layer along with a spot of pure Ag, above which cubic $AlAu_4$ co-existed with hexagonal Al_2Au_5. Gold aluminide is identified with [031] zone axis and [-14, 9, 5,-1] zone axis corresponding to cubic and hexagonal atomic structures, respectively. Pure Au region in Ag bonded ball is also seen in the observed regions. Perhaps, the instable, pure Ag in Al bond pad area and pure Au in Ag bonded ball area indicates by further supplying of thermal energy, the three atoms inter-diffuse and transform to stable intermetallic phases. Multi-phase existence signifies instability of the interface and it is likely to be transformed to stable compounds.

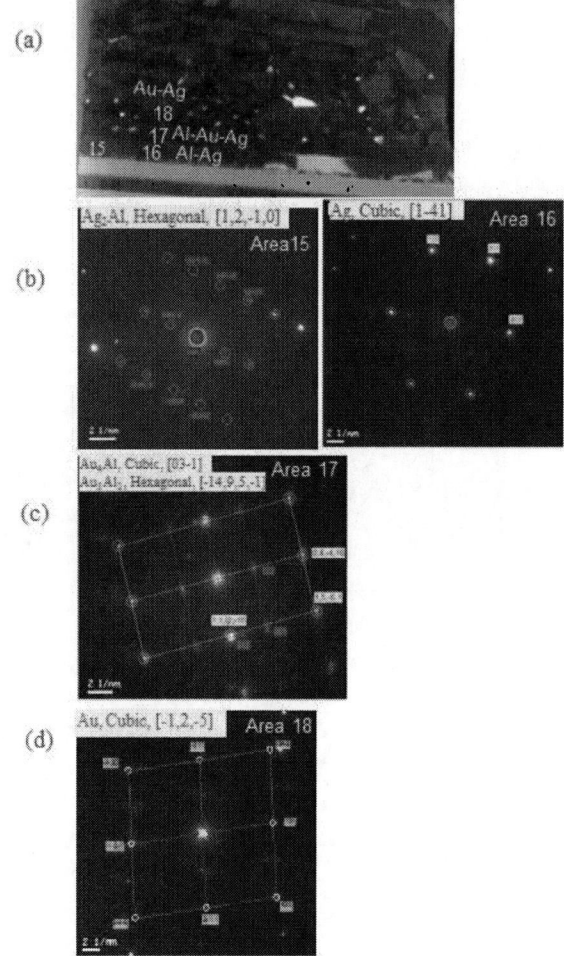

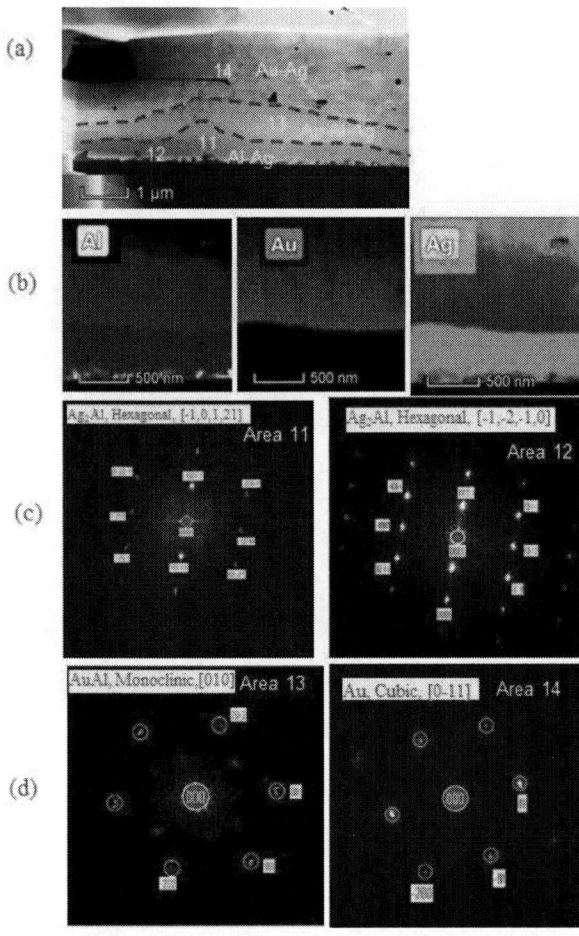

Fig.5 Observations of thermal cycled bonded ball (2000 cycles, -40 to +125°C) epoxy molded device for the conditions: (a) STEM image of FIB parted thin sample, (b) elemental EDX dot mapping revealed the formation of two different layers of silver-aluminide and gold-aluminide, (c) spot 11 & 12 shows formation of hexagonal Ag_2Al structure identified for [-1,0,1,2] zone axis, and (d) spot13 confirmed formation of monoclinic AuAl phase and spot14 revealed Au region in Ag bonded ball area.

Fig.6 Observations of bHAST tested bonded ball (+3.3V, 130°C, 85%RH), epoxy molded device for the conditions: (a) TEM image of FIB parted thin sample, (b) spot15 showed the presence of the formation of hexagonal Ag_2Al structure identified for [0,-1,1,2] zone axis and spot16 revealed pure Ag region (c) spot17 showed two different structures of gold-aluminide Au_4Al and Au_5Al_2 phases co-exists nearby, (d) spot18 revealed Au region in Ag bonded ball area.

Fig.7 shows a typical cross-sectioned image of coated-Ag ball bond after reliability studies and dot mapped for Au element (blue dot). The distribution of Au along the periphery of bonded ball of alloyed Ag core is evident. Thus, chlorine (Cl) or sulfur (S) ion attack from either sides of the bond ball has to react with Au rich Ag core. As shown in Fig.2, AuAg solid solution prevents/reduces diffusion of Cl or S ions.

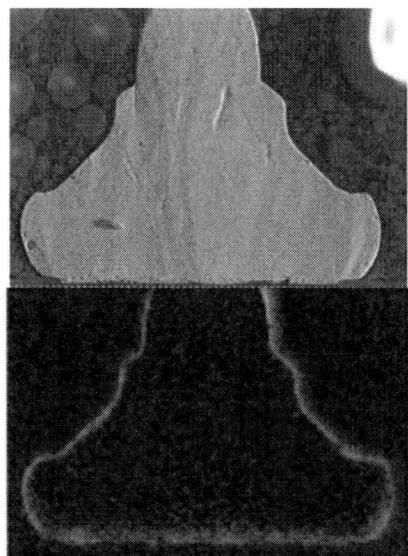

Fig.7 Elemental dot mapping of Au on a cross-sectioned ball bond processed using 0.8mil coated-Ag wire after reliability evaluation

In summary, coated-Ag bond on TC and bHAST tested showed predominant Ag_2Al phase formation along with AuAl, Au_5Al_2 and Au_4Al phases. Moreover, thermal ageing at 175°C for longer period of 3000h revealed Ag_3Al phase formation along with AuAl and Au_2Al phases. Presence of abundant Ag diffuses rapidly as well consuming the Al, transforms to Ag_3Al perhaps a stable phase. Thus, formation of Ag_3Al, Ag_2Al, AuAl, Au_4Al, Au_2Al, and Au_5Al_2 at the bond interface leads to reliable bond.

ACKNOWLEDGEMENT

Authors are grateful to Dr. Liu Bing Hai, WinTechNano, Singapore for his good support on TEM analysis. Authors would also like to thank Mr. Yam Lip Huei and Mr. Balasubramanian Senthil Kumar for experimental support on wire bonding.

REFERENCES

[1] S. Murali, S. S. Eric Tan and C. Y. Jason Wong, "Curriculum of Coated Silver Bonding Wire", IMAPS2018, 51st Symposium on Microelectronics – Wire Bonding Workshop, Pasadena, 8th October, 2018.

[2] S. Murali and C. Y. Jason Wong, "Growth of Intermetallics at Coated Silver Wire Bond Interfaces", IMAPS2018, 51st Symposium on Microelectronics, Pasadena, 11th October, 2018.

[3] S. Murali, B. Senthilkumar, S. S. Eric Tan and C. Y. Jason Wong, "Coated Silver Wire Bond: Reliability of Epoxy Molded Device", EPTC2018, 20th Electronics Packaging Technology Conference, Singapore, December, 2018.

[4] I. T. Kang, Y. D. Tark, M. H. Cho, J. S. Kim, H. S. Jung, T. Y. Kim, Zhang Xi and S. Murali, "Coated Wire", WO2017/091144 A1, June 2017.

[5] B. Senthilkumar, S. Murali, I. T. Kang, L. Y. W. Evonne, T. C. Wei, K. T. Y. James, S. S. Eric Tan and Zhang Xi, "Novel Coated Silver (Ag) Bonding Wire: Bondability and Reliability", EPTC2017, 19th Electronics Packaging Technology Conference Bondability, Singapore, December, 2017.

Research on applied reliability of BGA solder balls in extreme marine environment

Liyuan Liu
Reliability Research and Analysis Centre
China Electronic Product Reliability and Environmental
Testing Research Institute
Guangzhou, China
liuly@ceprei.com

Tao Lu
Reliability Research and Analysis Centre
China Electronic Product Reliability and Environmental
Testing Research Institute
Guangzhou, China
lutao@ceprei.com

Daojun Luo
Reliability Research and Analysis Centre
China Electronic Product Reliability and Environmental
Testing Research Institute
Guangzhou, China
luodj@ceprei.com

Hui Xiao
Reliability Research and Analysis Centre
China Electronic Product Reliability and Environmental
Testing Research Institute
Guangzhou, China
xiaohui@ceprei.com

Abstract— **The paper simulates the pressure of extreme marine environment, carries out pressure test on BGA devices, studies the changes and trends of BGA solder joints before and after test, analyses the reliability changes and the potential risks. The model of BGA solder joints is established by means of finite element modeling, and the influence of the position and size of voids in BGA solder joints after reflow soldering in pressure test is studied. Therefore, the preventive inspection measures for the corresponding reliability risk of BGA solder joints are put forward, which provides reliability assurance advice for BGA solder joints exposed to extreme marine environment.**

Keywords- applied reliability; pressure test; BGA solder joints; extreme marine environment

I. INTRODUCTION

Ocean area accounts for more than 70% of the total surface area of the earth. It is an important resource. In order to explore the deep-sea resource, many submersibles have been developed for exploration. The deepest ocean is now known as Mariana Trench, with a depth about 10994 meters. On January 23, 1960, Jacques Piccard and Lieutenant Don Walsh of the U.S. Navy set a new submarine depth record by descending 10,916 metres (35,814 feet) into the Mariana Trench in the Pacific Ocean using the Trieste [1]. Other 10,000-metre submersibles include: Deep Search，Deep-sea Challenger, Deep Flight challenger etc. [2, 3].

Over the past 60 years, the rapid development of electronic technology has provided important support for the performance improvement and miniaturization of submersibles. However, the reliability of electronic parts has become an important aspect restricting the development of submersibles. For example, the electronic control part of a submersible may need to be exposed to great pressure. The pressure corresponding to the depth of Mariana trench is about 110 MPa, which is a great challenge for electronic components. However, in the existing industrial and commercial electronic components data sheet, there is no clear pressure tolerance value, which brings difficulties to the design of the electronic control part exposed to great pressure and brings corresponding risks. Our team reported earlier on the reliability risks of some components in deep-sea environments due to their own structure and material problems, including capacitors, inductors, discrete devices and hybrid integrated circuits [4].

At the same time, our team focused on the effect of deep-sea high pressure on the reliability of soldering, and studied the changes of BGA (Ball Grid Array) devices before and after pressure test. This paper focuses on BGA solder joints, using experimental analysis and finite element modeling to study the variation characteristics under extreme deep-sea pressure, analysis the application risks and put forward the suggestions for failure prevention.

II. PRESSURE TEST AND PHENOMENON

A. Test Samples

There are several BGA integrated circuits on the electronic control board of a submersible. The local appearance is shown in Fig. 1. The four BGA integrated circuits are numbered #1, #2, #3, #4. This paper pays attention to the solder joints of BGA integrated circuit, and carries out the comparison of the solder joints before and after deep-sea high-pressure simulation test. Take #3 device as an example, which is a CPU, the initial inspection after soldering is performed by 3D X-ray. The typical morphology is shown in Fig. 2.

The 2D X-ray used in this paper is DAGE XD7600NT, with a maximum resolution of 0.95 μm and 3D X-ray used in this paper is YXLON Precision S, with a maximum resolution of 1 μm.

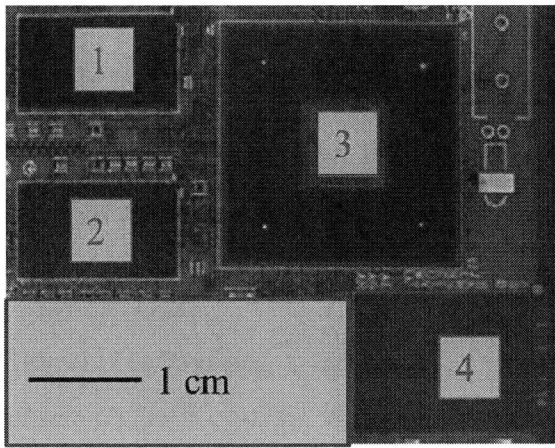

Figure 1. Sample appearance diagram.

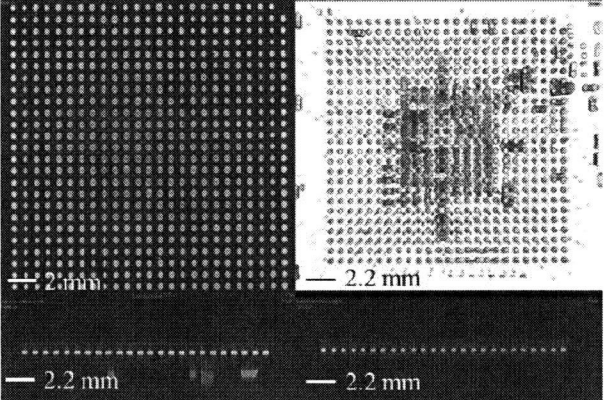

Figure 2. X-ray inspection results of #3 device-CPU soldering condition.

B. Test Conditions

The whole electronic control board is placed in the pressure test cavity and adds pressure according to the following procedures: linear rise pressure from 0 MPa to 114 MPa, maintained 114 MPa for less than 10 hours, and then dropped linearly from 114 MPa to 0 MPa. The purpose of pressure keeping is the process of simulating the operation of equipment under the sea.

A "test" described later contains all the above processes. Unless it is specifically pointed out, the comparison between 'before' and 'after' test in this paper is the result of a single test.

C. Morphology Comparison of BGA solder joints before and after test

Through nondestructive and destructive tracking analysis, we compared the morphology changes of BGA solder joints before and after the test.

1) Nondestructive nondestructive

In the comparative observation of 2D X-ray morphology of solder joints in larger areas, we observed that the voids of solder joints were "obviously compressed" after the test, as shown in Fig. 3. Zoom in on some solder joints, X-ray

pictures of the comparison of the two solder joints (A and B) before and after tests are presented in Fig. 4. It was found that deformation, crack and compressed of the solder joints voids occurred. The void rate of the solder joints can be automatically tested by the software of the X-ray equipment. In Fig. 4 (a) and (c), the void area ratio is 23.4% and 15.5%, respectively.

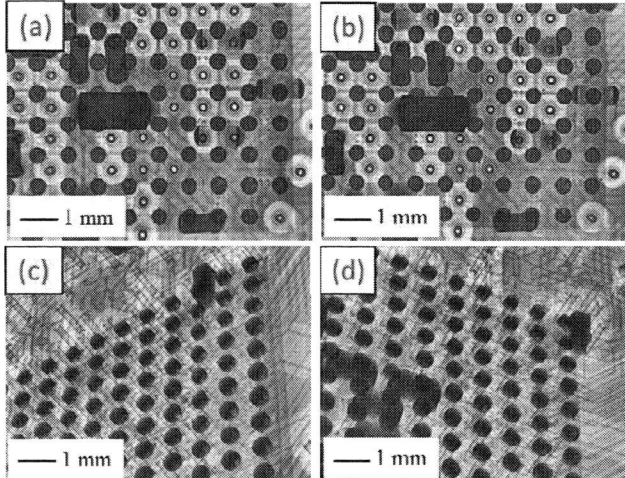

Figure 3. Comparison of 2D X-ray morphology of solder joints before and after test: (a) (c) the pre-test morphology; (b)(d) the post-test morphology.

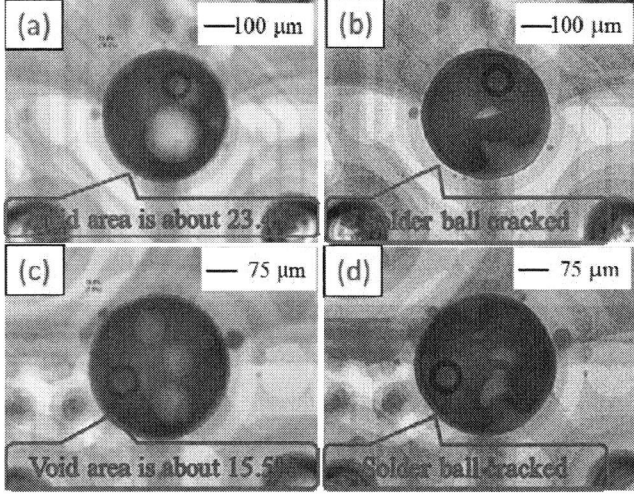

Figure 4. Typical 2D X-ray image comparison of BGA solder joints with voids before and after test: (a) X-ray image of solder joints A before test; (b) X-ray image of solder joints A after test; (c) X-ray image of solder joints B before test; (d) X-ray image of solder joints B after test.

In order to observe the specific location of the void and the specific situation of crack, this paper conducted a 3D X-ray observation of the BGA solder joints of CPU. Before the test, most solder joints have voids of different positions and sizes. Compared with the definition in IPC-7095C, various types of void positions described in table 7-6 of IPC-7095C, such as inside the solder joints and those at the encapsulation interface, are shown in Fig. 5. As the industry knows, at the current level of soldering technology, voids cannot be completely avoided, and they usually appear randomly [5]. At the same time, in the comparison observation of 3D X-ray images before and after the test, the compression deformation of the solder joints can be obviously observed, as shown in Fig. 5 and Fig. 6.

The solder joints in a column of CPU in Fig. 7 are analyzed, and the result is shown in Fig. 8. After the analysis of 25 solder joints, the void rate decreased after the test. The specific changes of void rate are listed in Table 1. The symbol '/' in Table 1 indicates that the 2D X-ray image does not observe the hole or the void rate cannot be measured. At the same time, 3 cases of deformation or crack can be clearly observed.

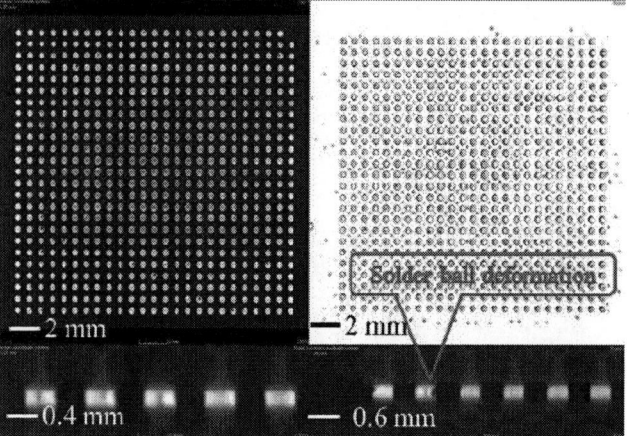

Figure 6. Overview of CPU solder joints morphology after test.

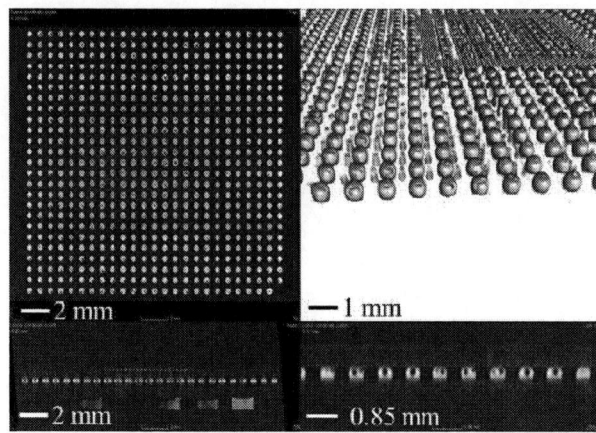

Figure 5. Overview of CPU solder joints morphology before test.

TABLE I. TABLE TYPE STYLES COMPARISON OF SOLDER VOID AREA BEFORE AND AFTER TEST

Void Area(%)	Solder Joints Number				
	1	*2*	*3*	*4*	*5*
Before test	14.9	/	8.2	15	13.9
After test	/	/	/	7.1	13.8
/	*6*	*7*	*8*	*9*	*10*
Before test	16	10.6	/	/	/
After test	15.6	4.8	/	/	/
/	*11*	*12*	*13*	*14*	*15*
Before test	/	/	/	/	/
After test	/	/	/	/	/
/	*16*	*17*	*18*	*19*	*20*
Before test	/	/	/	13.8	/
After test	/	/	/	6.5	/
/	*21*	*22*	*23*	*24*	*25*
Before test	4.9	19.5	19.6	25	1.8
After test	3.1	8.2	7.8	6.3	1.8

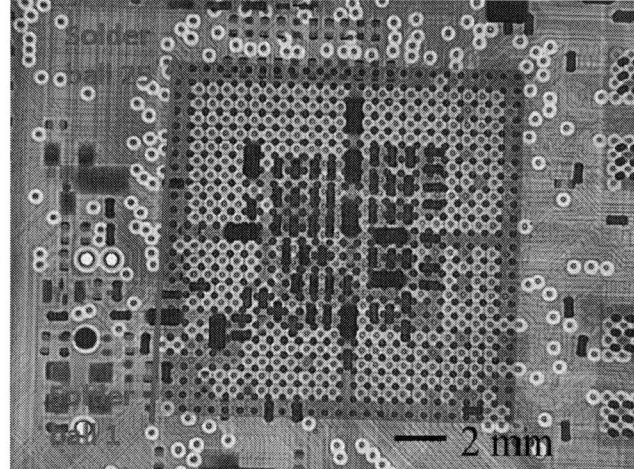

Figure 7. Solder joints analysis location diagram.

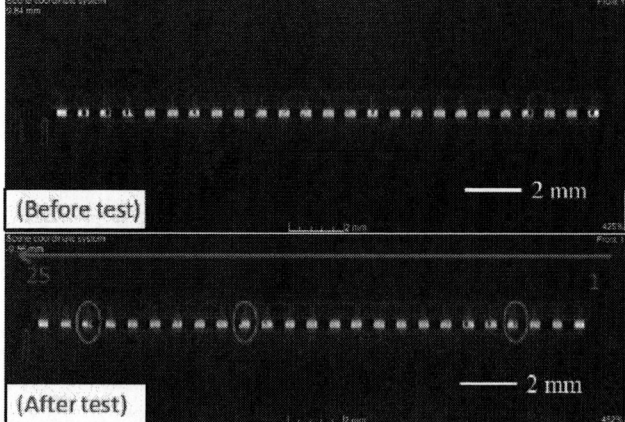

Figure 8. Comparison of BGA solder joints cross-section view before and after test.

2) Destructive analysis

In order to observe the actual change of solder joint cavity, each solder joint in Fig. 8 is analysis by Microsectioning. The typical morphology of solder joints are shown in Fig. 9. Solder joints are numbered corresponds to the number in Fig. 8. It can be seen that the deformation occurs obviously after the test of #1 and #14, the edge of the void is broken in #6, and the cross-sectional area of the solder joint decreases by nearly 30% after the compression of the #23 void. Comparing the morphology of X-ray morphology in Fig. 8 and the morphology of the physical analysis in Fig. 9, it can be found that X-ray morphology can better show the change of solder joints.

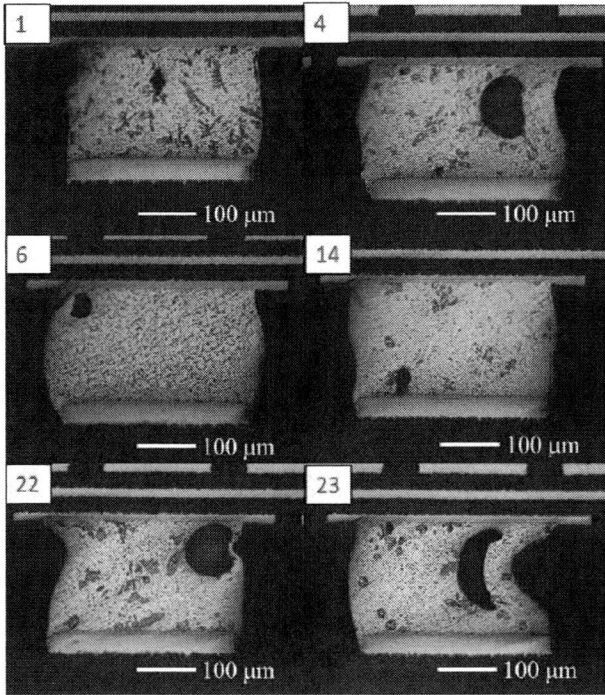

Figure 9. Typical morphology of BGA solder joints analysis by Microsectioning after test.

III. SIMULATION

Under 110 MPa pressure environment, voids cause the solder joints to deform even break. Observing the morphology in previous section, we found that the void in the center area received extrusion deformation, but did not produce crack or break. The void at the edge of the solder joints has the risk of causing to break and the soldering strength to decrease. On this basis, the fact that the void of the solder joint is randomly generated may lead to a great risk in the application of solder joints in extreme marine environment.

The sample shown in Fig. 10 is integrated circuit #4 in Fig. 1. After more than three times high-pressure simulation experiments, there was a serious damage to the soldering joints which remaining connection part was less than 25% of the cross-sectional area. Due to the failure to record the initial state of this solder joint, we only guessed that the position of the solder ball void is special, after several tests, a part of the solder joint ruptured.

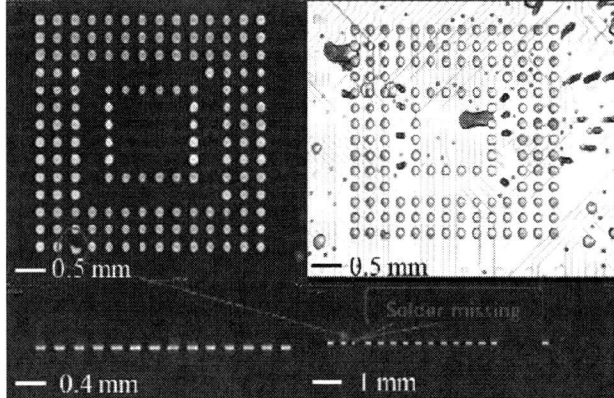

Figure 10. Overview of solder joints morphology after multiple tests of integrated circuits #4 in Fig. 1.

In order to study the influence of void position, this paper analyzes it by finite element modeling.

A. Model

In this paper, the nonlinear finite element modeling was conducted for the 3D board level assembly sample, in which the unified viscoplastic Anand equation was applied to characterize the stress-strain constitutive relationship of the solder material. Considering the symmetry of the sample structure, only the local device structure model is established, as shown in Fig. 11 and Fig. 12. The influence of voids was studied by simulation tests, focusing on the size as well as the location of the voids in the solder joints. The solder joints diameter was set as 0.3 mm and height was set as 0.26 mm. The void diameter was set as 0.05 mm, 0.1 mm and 0.2 mm. The distance of the balls' center for the solder ball and the void in the solder joint was used to characterize the void's position feature. The stress distribution was calculated for the solder joint containing the different types of voids.

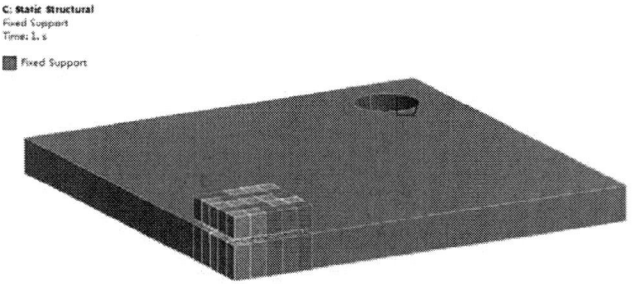

Figure 11. Finite element model.

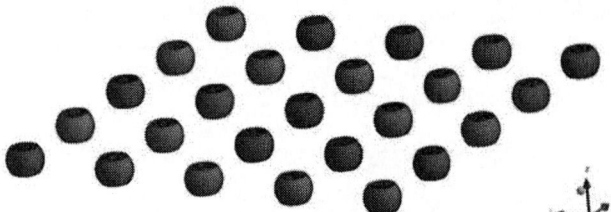

Figure 12. Ball grid array model.

B. Simulation Results

Stress distribution Cloud Map of ball grid array is shown in Fig. 13. Compare the simulation results of the max stress solder ball according to void position and size is shown in Fig. 14. There are differences in the maximum stress of different solder balls in the solder joints array. When the void is located on the edge, the stress extremum appears near the soldering surface. When the void is located internally, the stress extremum appears on the internal surface. When the void deviates from the center, the stress is greater near the side of the edge.

When the voids have the same diameter, the higher stress appears when it is located inside and deviates from the center, and the stress is smaller when the void is at the edge.

When the void in the same position, as the diameter of the hole increases, the stress increases.

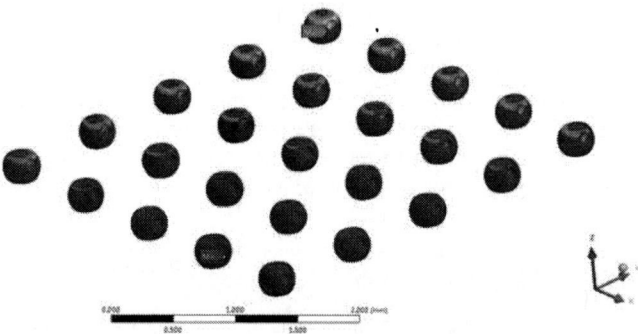

Figure 13. Stress distribution Cloud Map of ball grid array.

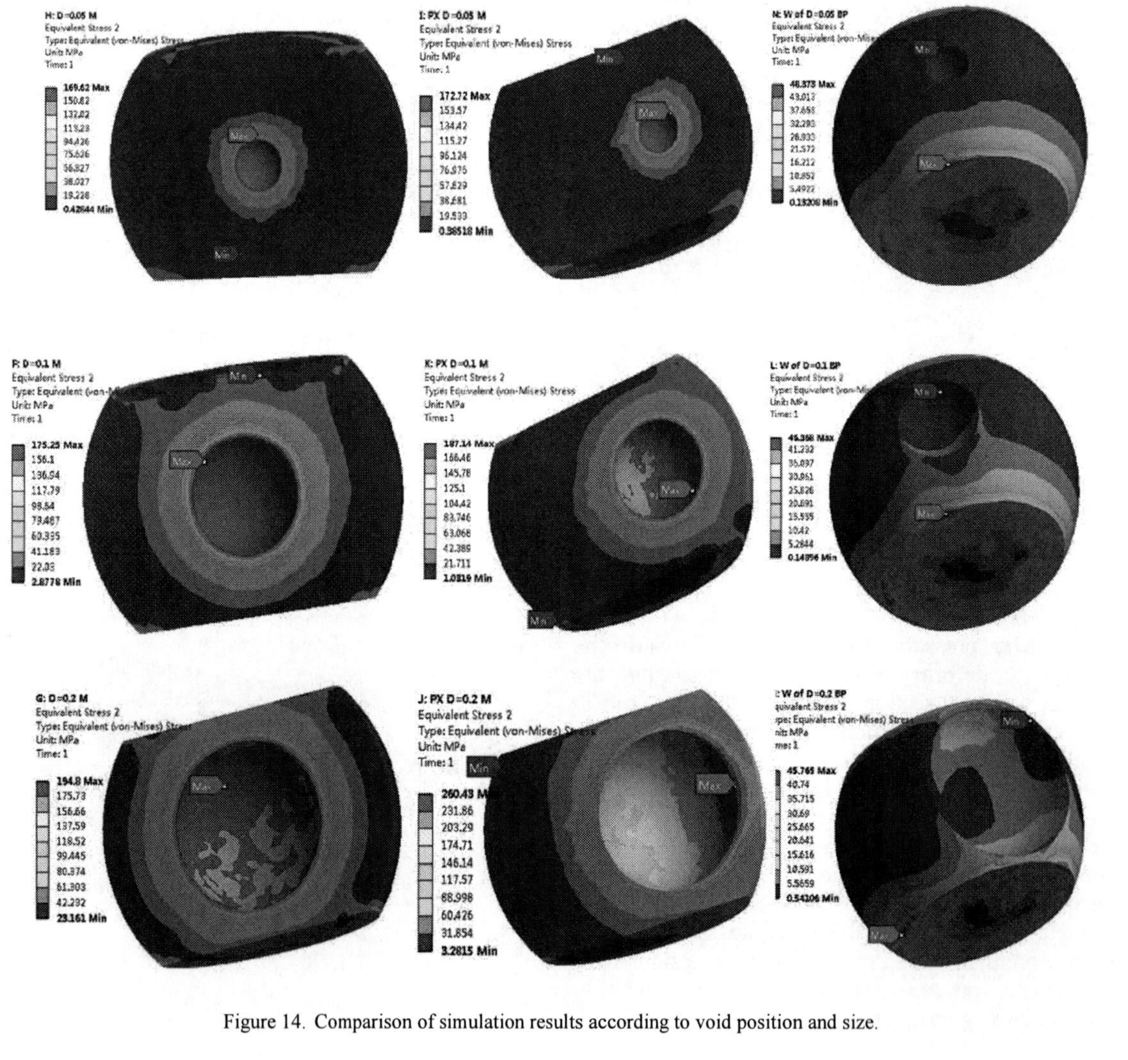

Figure 14. Comparison of simulation results according to void position and size.

IV. POTENTIAL RISK ANALYSIS

A. Universal Interconnect Reliability Risk

In the interconnection process of BGA devices for submersibles, there is also a reliability risk caused by soldering defects, such as pillow effect, solder joint cracks, pad cratering and so on. Of particular concern are the risks of micro-cracks, local cracking in solder joints. Because the crack is small and the function is normal, it is difficult to find the problem in the test. But under the large pressure, the local mechanical distribution is likely to accelerate the crack propagation, thus caused the occurrence of failure. Fig. 15 is an example of local cracking of solder joints [6].

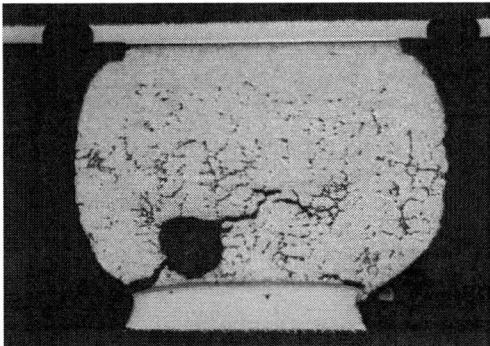

Figure 15. Local cracking diagram of solder joint[6].

B. The risk of BGA solder joints caused by voids

Current industry data suggests that voids in the solder joint are not a reliability concern [5]. However, the random appearance of voids in solder joints poses a risk to BGA used in deep-sea environments. Voids may impact reliability by weakening the solder balls and reducing functionality because the reduced cross-section has lower heat transfer and current carrying capabilities. On the other hand, the deep sea pressure makes the solder joint rupture and deformation in the special position aggravate the above influence, and also affects the soldering strength.

C. The risk of BGA solder joints caused by warp

Material deforms under high pressure, deformation is usually compressed, and difference of compression rate between material and void brings about the problems of interface shear force and void compression. The schematic is shown in Fig. 16. The comprehensive effect of deformation causes the BGA solder joint to have potential risk of cracking.

On the one hand, board-level deformation comes from the soldering process. Due to the high temperature of the soldering process, it is easy to produce warp considering the large difference between the thermal expansion (CTE) of chip package body and the substrate.

The combined effect of each deformation makes the potential risk of cracking or crack enlargement caused by stress in BGA solder joints. Under the deep-sea environment,

if the action of voids and deformation at the same time, the possibility of failure occurring is increased.

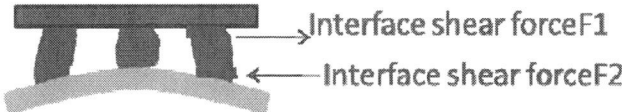

Figure 16. Schematic diagram of interfacial shear force caused by material compression.

V. PREVENTIVE ADVICES

Among the above risks, the random generation of voids is the most uncontrollable risk. Targeted preventive control measures are needed.

In view of the risks posed by solder joints' voids, the existing standards cannot meet the requirements of deep-sea high-pressure environment. Such as now workmanship requirements shall follow the requirements of IPC-A-610. The standard provides that 30% or less voiding of any ball in the X-ray image. This is a 2D projection area for easy implementation. However, voids positions are not specified, and voids located at the edge or near the upper and lower interfaces have a greater impact on the reliability of BGA solder joints under deep-sea environments. IPC-7095C provides for more detailed void position, diameter and area requirements, which can use 2D projection for diameter, area ratio confirmation, and position observation by Micro-sectioning. But this method is destructive and clearly unsuitable for the screening of products. Therefore, it is suggested that in order to prevent the reliability problems caused by the BGA solder joints voids for deep-sea submersible, work should be carried out in the following two aspects.

A. Reduces voids

Reduction of voids through assembly process control, including adjustment of process parameters, solder paste formulation, reflow soldering temperature curve, pad surface treatment [5], etc.

B. Increase 3D X-ray inspection of solder joints

At present, the submersible is not mass-produced. A small number of PCBAs can increase 3D X-ray inspection as screening. By observing the size and location of the voids, screening out the PCBAs with smaller risk. The screening criteria can refer to the requirement of IPC-7095C for voids, and be rigorous according to the specific application. In particular, it is important to note that the use of X-ray requires careful exposure to sensitive materials or components. The implications of radiation dose to radiosensitive components may need to be discussed with component supplier and the dose rates that will be achieved under typical x-ray inspection should be discussed with X-ray equipment supplier [5].

VI. CONCLUSIONS

BGA packaged IC is usually the key integrated circuit of the system, in this paper is CPU, memory, etc. Once it failed, it may cause system failure. However, there is no requirement for static pressure in the device specification. So research on the reliability issues of solder joints under high pressure is very important for the reliability of BGA packaged ICs, also for system applied for extreme marine environment. In this paper, the phenomenon and the apply reliable risk of BGA solder joints in simulation test of extreme marine environment are studied, and the necessary risk prevention measures are suggested, including the reduction of solder joints' voids, the addition of 3D X-ray inspection.

ACKNOWLEDGMENT

At the point of finishing this paper, we'd like to express our sincere thanks to all those who have support us with the experiment and analysis to complete this work.

REFERENCES

[1] Jacques Piccard. https://www.britannica.com/biography/Jacques-Piccard.

[2] L. Taylor, T. Lawso, "Project Deep search:An Innovative Solution for Accessing the Oceans," Marine Technology Society Jounal. vol. 43(5), pp. 169–177, 2009.

[3] Deep flight challenger. http://www.Deepflight.com/project/deepflight-challenger.

[4] L.Y. Liu, T. Tao, W. P. Du, F.Y. Mo, "Typical Components Failure Mode under Compressive Mechanic Stress within All Sea Depth," ICRMS. China: Shangha,2018.

[5] IPC-7095C, IPC Association connecting Electronics Industries, 2013.

[6] Y. B. Zou, H. Xu, Z. H. Wan, G. H. Yuan, "Typical failure cases of component mixing process," Workshop on Failure analysis of aerospace equipment. China: Guizhou, 2012.10 , pp. 569–572.

Influence of Single/Double Sweeping Mode and Sweeping Voltage Increment/Polarity on Measurement of TSV Leakage Current

Qinghua Zeng, Jing Chen*
Institute of Microelectronics
Peking University
Beijing, China
e-mail: zengqinghua@pku.edu.cn, j.chen@pku.edu.cn

Yufeng Jin
Shenzhen Graduate School
Peking University
Shenzhen, China
e-mail: yfjin@pku.edu.cn

Abstract—Leakage current between through-silicon vias (TSVs) and surrounding silicon substrates is a critical electrical reliability problem. Many works have been reported, including factors affecting leakage current, methods to reduce leakage current, leakage current paths and evaluation of TSV yield through leakage current measurement. However, influence of measurement condition or measurement method itself on TSV leakage current was seldom studied in previous works. In this paper, TSV samples with thermal oxidation SiO_2 layer and TSV samples with inductively coupled plasma chemical vapor deposition (ICPCVD) deposited SiO_2 layers were fabricated and their *I-V* characteristics of leakage current were measured and compared. Different measurement methods were applied by changing sweeping voltage increments, voltage polarities and single or double sweeping modes. For TSV samples with thermal oxidation SiO_2 layers, measured TSV leakage current at 20 V was in the order of 10^{-12} A and its detailed numerical value increases with sweeping voltage increments. A hysteresis curve was observed as double sweeping mode was applied due to effect of SiO_2 layers capacitors charge and discharge processes. TSV samples with ICPCVD deposited SiO_2 layers are very leaky and measured leakage current at 20 V was in the order of hundreds of 10^{-9} A. The influence of sweeping voltage increments and sweeping modes can be neglected for such samples. In conclusion, measurement setup and measurement parameters are important factors to evaluate TSV leakage current. It is important to understand the influence of measurement conditions on the leakage current in TSVs.

Keywords-through-silicon via; leakage current; hysteresis; capacitance; sweeping voltage increment; single/double sweeping mode

I. INTRODUCTION

Through-silicon via (TSV) has been studied for years as an interconnection technology with advantages such as shorter interconnection length, lower power consumption and smaller packaging size. A typical TSV structure includes a copper cylinder, adhesion/barrier/seed layers, an isolation layer and a surrounding silicon substrate. Leakage current from TSV copper cylinders to silicon substrates through isolation layers is a significant reliability issue and many works have been reported in this field. In summary, these works mainly focus on the following aspects: factors affecting leakage current, methods to reduce leakage current, leakage current paths and evaluation of TSV yield through leakage current measurement.

Leakage current of TSVs is affected by factors such as roughness of TSV sidewalls, materials selection of isolation layers, deposition processes of isolation layers, temperature treatment and so on. These factors also offer guidance of how to reduce leakage current. It was found that leakage current increases with increasing roughness of TSV scalloping sidewalls due to BOSCH etching process, and possible cracks in isolation layers increase the leakage current because stress is concentrated around the scalloping structures [1][2]. Based on this conclusion, TSV leakage current was decreased by almost three orders of magnitude through reducing the TSV scalloping sidewall roughness [3]. J C Lin *et al* measured the leakage current between adjacent TSVs and found that smooth sidewall topography and bottom cleaning largely improves liner/barrier integrity and reduces leakage current [4].

Leakage current varies with materials of isolation layers and deposition processes. J Charbonnier *et al* measured leakage current as a function of gases ratio O_2/TEOS during SiO_2 PECVD deposition [5]. H Kitada *et al* found that leakage current through silicon-oxy-nitride (SiON) films depends on deposition temperature, thickness of films and chemical properties [6]. D Archard *et al* studied a novel low-temperature (< 200 °C) PECVD TEOS SiO process for TSV isolation and they found that PETEOS SiO:H films are more electrically robust than silane-based films [7]. Electrical leakage current of PETEOS SiO:H films can be reduced further by means of an *in-situ* post deposition plasma treatment [8].

Temperature treatment is another factor that has obvious influence on leakage current. N Ranganathan *et al* found that leakage current increases with an increase of the temperature of different thermal loading conditions during leakage tests [1]. H B Chang *et al* observed slight degradation of leakage current after thermal torture treatment [8]. G Katti *et al* found that leakage current increases with temperature and the oxide isolation property is preserved even at 150 °C maintaining lower leakage current (< 1 μA) [9]. L Zhang *et al* found that annealing in forming gas (N_2/H_2) at 300 °C for 30 min significantly improves leakage current due to reduction of defect states of the Si–dielectric interface [10][11].

In the matter of leakage current paths, Y J Chang *et al* studied leakage current between two TSVs and discharge paths were found to be located at the contact area of titanium layers from redistribution layers (RDLs) and the silicon substrate, which was induced by TSV backside reveal processes [12]. Y Mizushima *et al* applied cross-sectional infrared optical-beam-induced resistance-change (IR-OBIRCH) measurements to localize leakage current paths and successfully measured leakage currents greater than 10^{-8} A, which is the detection threshold for IR-OBIRCH [13]. C L Yu *et al* observed Cu out-diffusion when leakage current was higher than 1 μA while no Cu contamination issue in silicon substrates was observed when leakage current was in the order of 10^{-12} A [14].

In terms of evaluation of TSV yield through leakage current measurements, Y Lou *et al* designed a leakage current monitor circuit to detect pin-holes at TSV sidewalls through measuring the resistance between a TSV and its surrounding substrate [15]. J F Hung *et al* measured the leakage current between two blind TSVs ($\Phi 10$ μm × 60 μm) at a voltage of 10 V and all measurement results are in the order of 10^{-12} A to signify well deposited SiO₂ layers along TSV sidewalls [16].

Measurement itself is included in each work and leakage current was mostly measured at a certain voltage or over a sweeping range of voltage. It means that leakage current is usually evaluated by clarifying the value of leakage current at a certain voltage or by plots of *I-V* characteristics over a sweeping range of voltage. Measurement of *I-V* characteristics usually requires parameters such as start voltage, stop voltage, step size or increment and single or double sweeping mode. It has been reported that leakage current can be affected by the sweeping range of voltage (from start voltage to stop voltage), which determines whether silicon surfaces are accumulation region or depletion region. A Mercha *et al* measured leakage currents both in depletion (2 V) and in accumulation (-6 V) and the absolute leakage current value of each TSV is below 10 pA [17]. M Stucchi *et al* measured leakage current between TSVs and P-type silicon substrates with a voltage sweeping covering both accumulation region and depletion region and found that leakage current could be masked by the high resistance of depletion region [18]. Except the voltage sweeping range, other parameters of TSV leakage current measurement were barely discussed in previous works. To the best of our knowledge, there are few literatures about the influence of increments of sweeping voltage or sweeping modes on measurement of leakage current.

In this paper, we fabricated two kinds of TSV samples with inductively coupled plasma chemical vapor deposition (ICPCVD) SiO₂ layers and thermal oxidation SiO₂ layers. We used multiple sweeping voltage increments and single/double sweeping modes to measure TSV leakage current of these samples. Measured leakage currents of two kinds of TSV samples under different measurement modes were analyzed and compared.

II. Samples Preparation and Measurement

A. TSV Samples for Leakage Current Measurement

In this paper, we designed and fabricated two kinds of TSV samples with different SiO₂ layers for leakage current measurement and comparison, namely sample TO and sample ICP. Sample TO stands for that SiO₂ layers were deposited through thermal oxidation with a thickness of 2 μm on P-type silicon substrates while sample ICP stands for that SiO₂ layers were deposited through ICPCVD with a surface thickness of 1.5 μm on N-type silicon substrates. TSV size of sample TO is $\Phi 100$ μm × 200 μm while TSV size of sample ICP is $\Phi 40$ μm × 180 μm, as shown in Fig. 1 and Fig. 2. The TSV diameter of sample TO is large enough and measurement probes can contact TSV Cu surfaces directly. For sample ICP with smaller TSV diameter, copper pads and copper RDLs were fabricated for more convenient measurement. It is noted that there are many different aspects between sample TO and sample ICP, the purpose of our work is not to compare the leakage current of sample TO and sample ICP directly but to focus on the influence of measurement methods on the leakage current of sample TO and sample ICP.

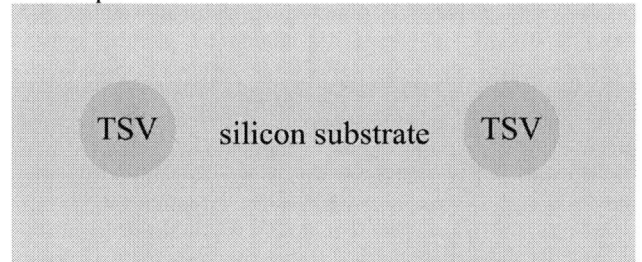

Figure 1. Schematic of sample TO.

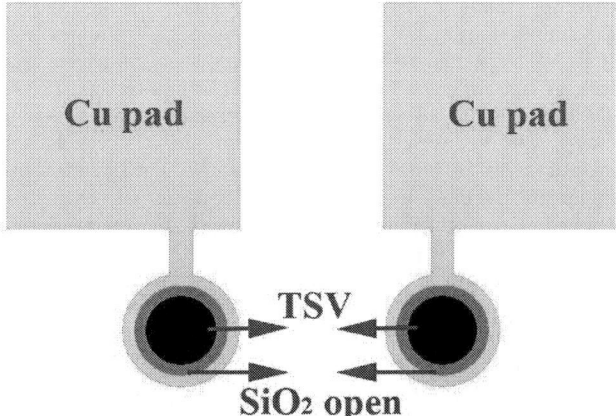

Figure 2. Schematic of sample ICP.

B. TSV Leakage Current Measurement

In this work, we measured leakage current between two adjacent TSVs with two probes by using HP 4156B semiconductor parameter analyzer under room temperature, as shown in Fig. 3. Under "Low Current Measurement" mode, current measurement resolution is 1 fA and offset current is 20 fA.

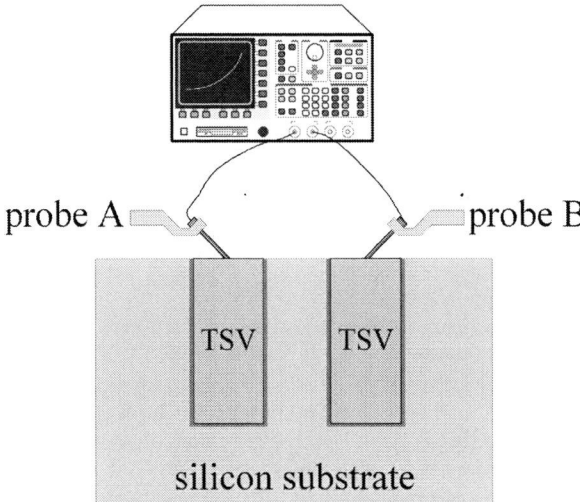

Figure 3. Schematic of TSV leakage current measurement.

III. RESULTS AND DISCUSSION

A. Single Sweeping Mode with Multiple Increments

In this section, single sweeping with diverse voltage increments was carried out. Single sweeping mode stands for that measurement voltage is swept from a start voltage (normally 0 V) to a stop voltage and then one measurement is finished.

The sweeping voltage of TSV leakage current measurement of sample ICP was from 0 V to 20 V with an increment of 100 mV, 200 mV, 500 mV and 1 V, respectively. Measured leakage currents at 20 V were 234 nA, 235 nA, 220 nA and 222 nA, respectively. Although sweeping voltage increments were different, measured leakage currents were nearly the same, as shown in Fig. 4.

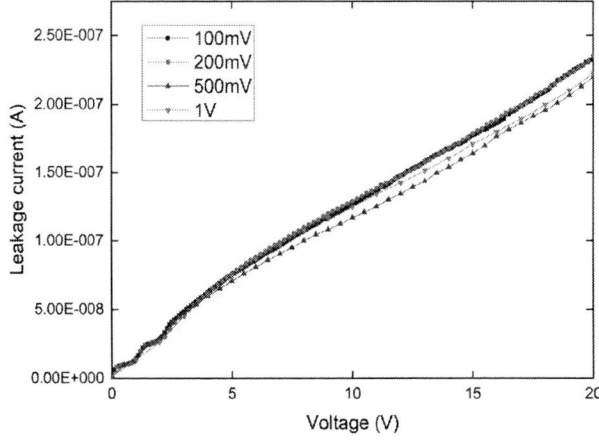

Figure 4. I-V characteristics of TSV leakage current of sample ICP with diverse increments under single sweeping mode.

For sample TO, measurement voltage was also swept from 0 V to 20 V with nine individual increments, namely 50 mV, 100 mV, 200 mV, 400 mV, 800 mV, 1000 mV,

2000 mV and 4000 mV. At the same voltage, measured leakage currents increase as sweeping voltage increments increase, as shown in Fig. 5 in the first quadrant. As measurement voltage is 20 V, measured leakage currents are 50 fA, 40 fA, 140 fA, 290 fA, 490 fA, 570 fA, 910 fA and 1320 fA, respectively. Measurement results showed that measured leakage currents changed with applied sweeping voltage increments. In other words, measured leakage currents were not only determined by applied voltage value, which did not accord with I-V characteristics of an ideal resistor.

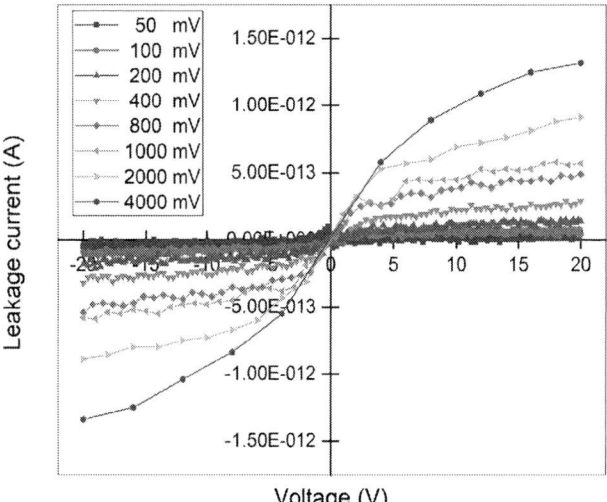

Figure 5. I-V characteristics of TSV leakage current of sample TO with diverse increments under single sweeping mode.

We employed a typical TSV electrical model under low-frequency mode and an equation about current and voltage of capacitors to analyze the difference of I-V characteristics between sample ICP and sample TO.

The electrical model of TSV-silicon substrate-TSV is shown in Fig. 6 [19].

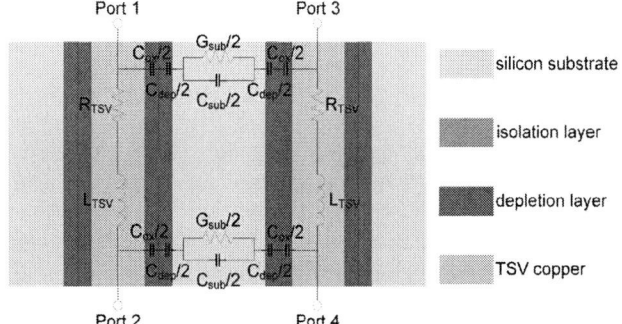

Figure 6. An electrical model of TSV-silicon substrate-TSV under low-frequency mode.

In this model, SiO_2 layers are treated as ideal isolation layers and C_{ox} is the capacitance of SiO_2 layers. C_{dep} is the capacitance of the depletion region at the surface of the silicon substrate. C_{sub} is the capacitance of the silicon substrate. C_{ox}, C_{dep}, C_{sub} are described with (1), (2) and (3), respectively [19].

$$C_{ox} = 2\pi\varepsilon_0\varepsilon_{ox}h_{TSV} / \ln\left(1 + \frac{t_{ox}}{d_{TSV}/2}\right) \qquad (1)$$

where h_{TSV} is the height of TSVs, t_{ox} is the thickness of SiO₂ layers and d_{TSV} is the diameter of TSVs, ε_0 is the electric permittivity of free space, ε_{ox} is the relative permittivity of SiO₂ layers.

$$C_{dep} = 2\pi\varepsilon_0\varepsilon_{Si}h_{TSV} / \ln\left(1 + \frac{w_{dep}}{d_{TSV}/2 + t_{ox}}\right) \qquad (2)$$

where w_{dep} is the width of the depletion region, ε_{ox} is the relative permittivity of the silicon substrate.

$$C_{sub} = \pi\varepsilon_0\varepsilon_{si}h_{TSV} / cosh^{-1}\left(p_{TSV}/d_{TSV}\right) \qquad (3)$$

where p_{TSV} is the pitch of adjacent TSVs.

The equation about current and voltage of capacitors is described with (4).

$$i(t) = C \times \frac{du(t)}{dt} \qquad (4)$$

where $i(t)$ is the current as a function of time, $u(t)$ is the applied voltage, t is time and C is the total capacitance, which includes C_{ox}, C_{dep} and C_{sub} in the TSV-silicon substrate-TSV model.

It was concluded from measurement results of sample TO that TSV leakage current increases with sweeping voltage increments, which can be explained with (4). The interval time between sampling points was same regardless of sweeping voltage increments and therefore dt can be treated as a constant irrespective of sweeping voltage increments. As sweeping voltage increments increase, $du(t)/dt$ increases as well and as a result $i(t)$ increases. If C keeps as a constant, $i(t)$ is supposed to be a constant as well. However, $i(t)$ increases with applied voltage $u(t)$, as shown in Fig. 5 in the first quadrant. On one hand, applied voltage drops across not only capacitors but also resistors, as shown in Fig. 6. As we all know, current flowing through a resistor is proportional to the voltage. On the other hand, C is not a constant because w_{dep} changes with applied voltage, which leads to that C_{dep} changes as well.

On the contrary, TSV leakage currents of sample ICP were nearly the same although sweeping voltage increments increased from 100 mV to 1 V. It inferred that SiO₂ layers of sample ICP played a role of "resistor" more than "capacitor" while SiO₂ of sample TO played a role of "capacitor" more than "resistor". In other words, the capacitance effect of sample ICP was less obvious than that of sample TO. For sample TO and sample ICP, the largest difference of capacitance was C_{ox}, which was mainly influenced by ε_{ox}. The capacitance effect of SiO₂ layers of

sample TO was more obvious than that of sample ICP due to a better quality of SiO₂ layer, which leads to a larger value of ε_{ox}. Further, it was also proved through comparison of the leakage current between sample ICP and sample TO. As measurement voltage was at 20 V and sweeping voltage increment was 1000 mV, the leakage current of sample ICP (222 nA) was almost 6 orders of magnitude larger than that of sample TO (570 fA). Consequently, it was concluded that the isolation effect of thermal oxidation deposited SiO₂ layers was better than ICPCVD deposited SiO₂ layers and the capacitance effect of thermal oxidation deposited SiO₂ layers was more obvious and dominant.

Measurement voltage was swept from 0 V to -20 V under single sweeping mode and measured TSV leakage currents were shown in the Fig. 5 in the third quadrant. It was found out that I-V characteristics in both the first and the third quadrants are symmetric about the origin point, which demonstrates that measured TSV-TSV pairs were perfectly symmetric. The band diagram was determined by applied voltage difference instead of specific voltage value applied on any TSV, as shown in Fig. 7. For example, in the case of perfectly symmetric TSV-TSV pairs, 10 V on TSV-1 and -10 V on TSV-2 is same as -10 V on TSV-1 and 10 V on TSV-2.

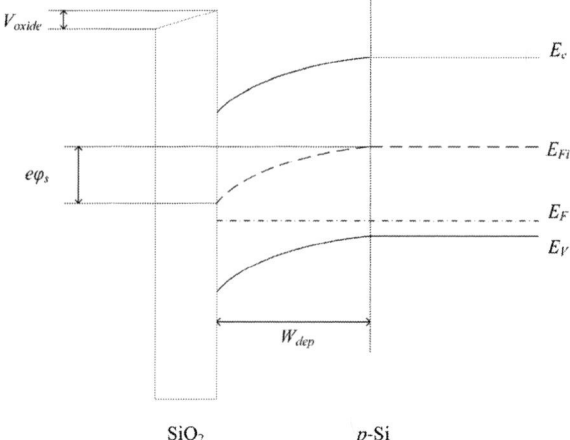

Figure 7. The band diagram of one-side TSV structure.

B. Double Sweeping Mode Measurement

Under double sweeping mode, TSV leakage current of sample ICP was measured with voltage sweeping from 0 V to 20 V (positive sweeping) and then immediately sweeping back to 0 V (negative sweeping), as shown in Fig. 8. The black arrow and the black curve with solid squares is the measured TSV leakage current of positive sweeping while the red arrow and the red curve with solid circles is the measured TSV leakage current of negative sweeping. The black arrow and the red arrow shows the positive and negative sweeping direction, respectively. It was found that I-V characteristics of positive sweeping and negative sweeping are similar.

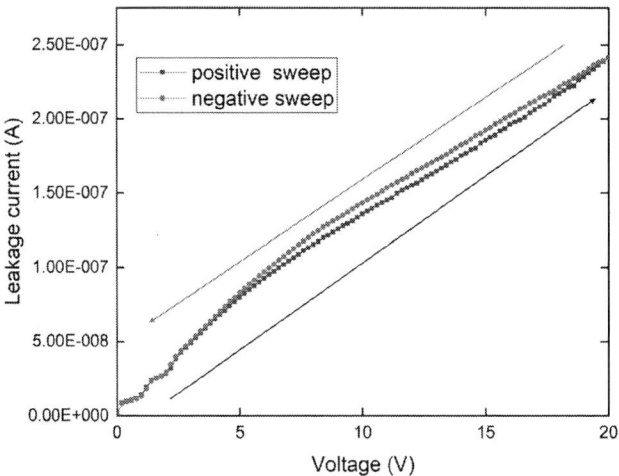

Figure 8. *I-V* characteristics of TSV leakage current of sample ICP under double sweeping mode.

For sample TO, measurement method was similar. We measured TSV leakage current of sample TO as voltage was swept from -20 V to 20 V and then swept back to -20 V. The measurement was lasted continuously for five times to reach a stable state. A hysteresis loop was observed, as shown as the black curve NPD5 (from Negative to Positive Double sweeping for 5 times) with solid squares in Fig. 9. We also measured TSV leakage current of sample TO as voltage was swept from 0 V to 20 V under single sweeping mode, which was plotted as the red curve with solid circles in Fig. 9. The black arrow and the red arrow shows the sweeping direction under double and single sweeping modes, respectively.

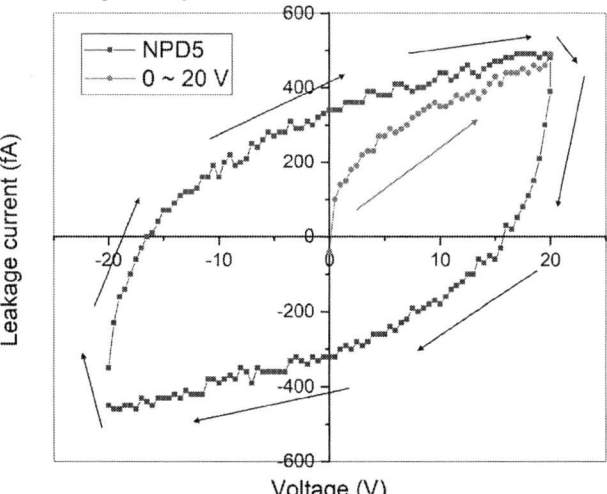

Figure 9. *I-V* characteristics of TSV leakage current of sample TO under single sweeping mode (red curve) and double sweeping mode (black curve).

In Fig. 9, on one hand, the *I-V* characteristic of measured TSV leakage currents under double sweeping mode formed a hysteresis loop. On the other hand, the TSV leakage current that was measured with voltage sweeping from 20 V to -20 V was smaller than the TSV leakage current that was measured with voltage sweeping from -20 V to 20 V. The

measured *I-V* characteristic is similar to the *I-V* characteristic of an ideal capacitor under double sweeping mode, as shown in Fig. 10, which demonstrated again that the capacitance effect of sample TO is obvious. This explanation was further proved by comparison between Fig. 8 and Fig. 9. As it was confirmed in the last section, C_{ox} of sample ICP was less obvious than that of sample TO and SiO$_2$ layers of sample ICP played a role of "resistor" more than a "capacitor". As a result, charges could not be stored in SiO$_2$ layers of sample ICP and therefore measured TSV leakage current decreased to zero along the same path of positive sweeping.

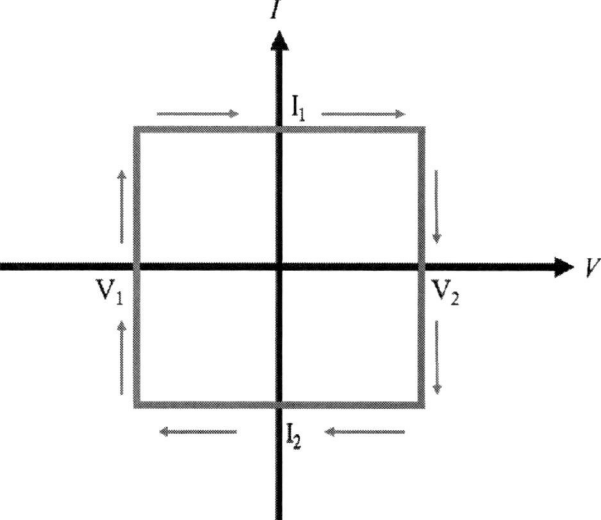

Figure 10. *I-V* characteristic of an ideal capacitor under double sweeping mode.

IV. CONCLUSION

In this work, we studied the influence of single or double sweeping modes and sweeping voltage increments on measurement of TSV leakage current of sample TO and sample ICP. On one hand, measured TSV leakage current of sample TO increased with sweeping voltage increments. However, measured TSV leakage currents of sample ICP were nearly the same although sweeping voltage increments increased. One the other hand, a hysteresis loop was observed when double sweeping mode was applied from -20 V to 20 V for TSV leakage current measurement of sample TO. However, I-V characteristics of measured TSV leakage current of sample ICP during positive sweeping and negative sweeping overlapped. Differences between measured TSV leakage current of sample ICP and sample TO were analyzed with a TSV-silicon substrate-TSV electrical model and I-V relationship of capacitors. It was concluded that capacitance effect of SiO$_2$ layers is a key factor affecting TSV leakage current. For SiO$_2$ layers with higher quality, capacitance effect is more obvious and SiO$_2$ layers played a role of "capacitor" more than "resistor". In other words, when we try to measure leakage current of TSVs with defect-less SiO$_2$ layers, it is important and necessary to set up a uniform measurement method to get confident and comparable results of TSV leakage current. The leakage current between two TSVs is required to be as low as possible and so in such case we should pay attention to the influence of measurement methods because these samples with very low expected leakage current behave like sample TO. If the leakage current between two TSVs is in a similar order of sample ICP, the influence of measurement methods can be neglected.

ACKNOWLEDGMENT

This work is supported by the National Natural Science Foundation of China (Grant No. U1537208).

REFERENCES

[1] Ranganathan N, Prasad K, Balasubramanian N, et al. A study of thermo-mechanical stress and its impact on through-silicon vias[J]. Journal of micromechanics and microengineering, 2008, 18(7): 075018.

[2] Nakamura T, Kitada H, Mizushima Y, et al. Comparative study of side-wall roughness effects on leakage currents in through-silicon via interconnects[C]//3D Systems Integration Conference (3DIC), 2011 IEEE International. IEEE, 2012: 1-4.

[3] Ranganathan N, Lee D Y, Youhe L, et al. Influence of Bosch etch process on electrical isolation of TSV structures[J]. IEEE Transactions on components, packaging and manufacturing technology, 2011, 1(10): 1497-1507.

[4] Lin J C, Chiou W C, Yang K F, et al. High density 3D integration using CMOS foundry technologies for 28 nm node and beyond[C]//Electron Devices Meeting (IEDM), 2010 IEEE International. IEEE, 2010: 2.1. 1-2.1. 4.

[5] Charbonnier J, Hida R, Henry D, et al. Development and characterisation of a 3D technology including TSV and Cu pillars for high frequency applications[C]//Electronic Components and Technology Conference (ECTC), 2010 Proceedings 60th. IEEE, 2010: 1077-1082.

[6] Kitada H, Maeda N, Fujimoto K, et al. Development of low temperature dielectrics down to 150 C for multiple TSVs structure with wafer-on-wafer (WOW) technology[C]//Interconnect Technology Conference (IITC), 2010 International. IEEE, 2010: 1-3.

[7] Archard D, Giles K, Price A, et al. Low temperature PECVD of dielectric films for TSV applications[C]//Electronic Components and Technology Conference (ECTC), 2010 Proceedings 60th. IEEE, 2010: 764-768.

[8] Chang H B, Chen H Y, Kuo P C, et al. High-aspect ratio through silicon via (TSV) technology[C]//VLSI Technology (VLSIT), 2012 Symposium on. IEEE, 2012: 173-174.

[9] Katti G, Mercha A, Stucchi M, et al. Temperature dependent electrical characteristics of through-si-via (TSV) interconnections[C]//Interconnect Technology Conference (IITC), 2010 International. IEEE, 2010: 1-3.

[10] Zhang L, Peng L, Li H Y, et al. Operating TSV in Stable Accumulation Capacitance Region by Utilizing Al2O3-Induced Negative Fixed Charge[J]. IEEE Electron Device Letters, 2012, 33(6): 875-877.

[11] Zhang L, Lim D F, Li H Y, et al. Through silicon via fabrication with low-κ dielectric liner and its implications on parasitic capacitance and leakage current[J]. Japanese Journal of Applied Physics, 2012, 51(4S): 04DB03.

[12] Chang Y J, Ko C T, Yu T H, et al. Backside-process-induced junction leakage and process improvement of Cu TSV based on Cu/Sn and BCB hybrid bonding[J]. IEEE Electron Device Letters, 2013, 34(3): 435-437.

[13] Mizushima Y, Kitada H, Koshikawa K, et al. Novel through silicon vias leakage current evaluation using infrared-optical beam irradiation[J]. Japanese Journal of Applied Physics, 2012, 51(5S): 05EE03.

[14] Yu C L, Chang C H, Wang H Y, et al. TSV process optimization for reduced device impact on 28nm CMOS[C]//VLSI Technology (VLSIT), 2011 Symposium on. IEEE, 2011: 138-139.

[15] Lou Y, Yan Z, Zhang F, et al. Comparing through-silicon-via (TSV) void/pinhole defect self-test methods[J]. Journal of Electronic Testing, 2012, 28(1): 27-38.

[16] Hung J F, Lau J H, Chen P S, et al. Electrical testing of blind Through-Silicon Via (TSV) for 3D IC integration[C]//Electronic Components and Technology Conference (ECTC), 2012 IEEE 62nd. IEEE, 2012: 564-570.

[17] Mercha A, Redolfi A, Stucchi M, et al. Impact of thinning and through silicon via proximity on high-k/metal gate first CMOS performance[C]//VLSI Technology (VLSIT), 2010 Symposium on. IEEE, 2010: 109-110.

[18] Stucchi M, Perry D, Katti G, et al. Test structures for characterization of through-silicon vias[J]. IEEE Transactions on Semiconductor Manufacturing, 2012, 25(3): 355-364.

[19] Xin Sun, "Research on Electrical Characterization of TSV Silicon Interposer," Ph.D. dissertation, Dept. Micro. Electron., Peking Univ., Beijing, China, 2015.

Improving the Solder Wettability via Atmospheric Plasma Technology

[1]Sagung Dewi Kencana
[2]Yu-Lin Kuo
Department of Mechanical Engineering
National Taiwan University of Science and
Technology
Taipei, Taiwan, R.O.C
[1]D10503811@mail.ntust.edu.tw
[2]ylkuo@mail.ntust.edu.tw

[3]Yee-Wen Yen
Department of Materials Science Engineering
National Taiwan University of Science and
Technology
Taipei, Taiwan, R.O.C
[1]ywyen@mail.ntust.edu.tw

[4]Eckart Schellkes
[5]Wallace Chuang
Automotive Electronics Department
Robert Bosch Taiwan Co., Ltd
Taipei, Taiwan, R.O.C
[4]Eckart.Schellkes@tw.bosch.com
[5]Wallace.Chuang@tw.bosch.com

Abstract - **Solder pad oxidation is a well-known issue in the electronic packaging industries where the oxide layer degrades the solder wettability and consequently posts solderability and reliability risks. One common approach on solving this problem is to include an add-on plasma cleaning process. Conventionally, this process has to be operated with a vacuum system, therefore, it is costly and less-uniform. With the recent well development on atmospheric pressure plasma (AP-plasma) technology, the cleaning process is now allowed to be proceeded in the ambient environment. In this work, AP-plasma will be used in the cleaning process. Solder wettability after cleaning and joint reliability after high temperature storage (HTS) will be studied.**

Experimentally, AP-plasma with compressed dry air and H_2O vapour gas sources are used for cleaning PCB substrate with common surface finishes, i.e. OSP, ENIG. The sessile method is then carried out where the SAC305 solder ball is mounted on the substrate. The wetting angle and spreading ratio are then measured to perform the wettability performance. The ball shear test is applied in an atmospheric temperature after HTS at 150 °C up to 1,000 hrs.

Results show that the SAC305 wettability is improved when AP-plasma is applied. The improvement can last for 1 - 4 hours which provides sufficient staging time of the process. It is also worth mentioning that the wettability of the plasma-cleaned-bare Cu without surface finish is better than that of all surface finishes without plasma cleaning, which suggests that it is a dominant factor on the wettability improvement. The ball shear results at post HTS will also be presented and discussed in this paper.

Keywords: surface cleaning, plasma cleaning, AP-plasma, oxidation, wettability

I. INTRODUCTION

The rapid growing of ICs technology influenced semiconductor manufacturers to compete on developing advance technology, not only to produce the ICs packages, but also to develop innovative backend processes in the field of materials research [1-3]. One of the important processes is solder ball mounting, which is required a clean and organic-less-contaminant on the surface of metal substrate, i.e. copper (Cu).

The variety of technology approaches in use today spans a broad range of surface finishes on Cu substrate, e.g. OSP, ENIG [4, 5]. Besides using flux, the vacuum plasma technology is considered for surface cleaning to improve the solder wetting [3, 6-9]. As compared to vacuum plasma, AP-plasma is more reproducible and uniform in terms of plasma reactive formation and surface functionalisation on substrate. Vacuum plasma is also inconvenience of which the vacuum system occupied the space. Therefore, this study has a unique approach to fulfil the purpose, which is using AP-plasma treatment technique for surface functionalisation on substrate surface.

Surface functionalisation in this study is to introduce certain formation of reactive radicals as plasma products to substrate surface, aiming to transform the surface characteristics from hydrophobic to hydrophilic [10, 11].

This study focuses on surface functionalisation, cleaning process, and reliability test i.e. HTS and ball shear test. HTS demonstrates the ability of solder ball on substrate to withstand elevated temperatures for prolonged periods of time and still having a good solder joints as intended. Furthermore, the main objective of this study is to evaluate the feasibility of

978-1-7281-1500-9/19 $31.00 © 2019 IEEE

using AP-plasma as a simple and fast technique to improve the solder wettability.

II. EXPERIMENTAL PROCEDURE

A. Materials

PCB substrates (14 mm x 14 mm x 0.312 mm, with solder pad opening of 600 um diameter) without and with surface finishes, i.e OSP, ENIG are used in this study. BGA SAC305, 760 um in diameter is mounted on the substrate using sessile method. The spreading ratio is then measured quantitatively based on this sessile method.

B. Atmospheric pressure (AP) plasma process

Pre-cleaning process using di-water and ultrasonic is employed before the AP-plasma process. The AP-plasma system used for this work is supplied from Click Sun-Shine Corp. The plasma is generated using two various gases: compressed dry air (CDA) and H_2O vapour (later called air plasma and H_2O vapour plasma, respectively) as shown in Fig. 1. Parameters associated with plasma conditions are given in Table 1. H_2O vapour plasma is produced by di-ionised water that vibrated by piezoelectro chip in the atomizer. This vapour is then flowing to the plasma electrode and acts as main gas, while CDA is still in-used as the working gas. In the plasma electrode, the vapour and air gases are mixed. In a sufficient energy, the electrons and other particles are collided, recombined and produced the plasma.

Thermocouple and thermometer (TES 1307 K / J) are used for plasma gas diagnostic for both air and H_2O vapour plasmas. Optical emission spectroscopy (OES, GIE Optoelectronic, Inc.) is employed to investigate the chemical bonds in the formation of plasma radicals in both air and H_2O vapour plasma processes [12]. This measurement is conducted in the photon-less environment.

Nitrogen concentration analyses is employed to measure the nitrogen level during air and H_2O vapour plasma treatments (Horiba PG250-Portable Gas Analyzer, Ashtead Technology).

TABLE 1. PLASMA PARAMETER

Items	Plasma parameters	
Scanning speed	28 mm/s	
Specimen temperature	22.8 °C	
Plasma torch colour	Air	White + Yellowish
	H_2O vapour	White + Yellowish
Ambient humidity	Air	64 %
	H_2O vapour	69 %

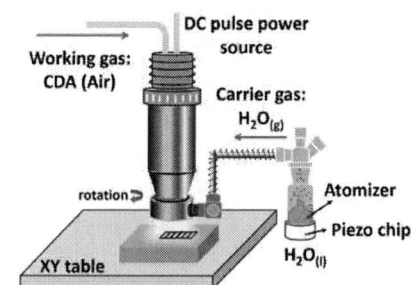

Figure 1. DC-AP system with H_2O vapour introduction and (b) DC-AP gas temperature

C. Characterization techniques

Sessile method is used to define the solder spreading ratio Rs on substrate. Solder spreading ratio Rs is defined as the ratio of two-dimensional projected area of solder ball in the post-and-pre-spreading (Rs = A2/A1) as shown in Fig. 2. The higher the spreading ratio is, the better the wettability will be. In experimental, the substrate is firstly cleaned by the plasma treatment. The mounted BGA solder balls on substrate are placed on hot-plate at a fixed temperature of 250 ± 3 °C. The commercial flux is used in the sessile process. HTS is performed in an oven at 150 °C for 500 and 1,000 hrs as shown in Fig. 2. Shear test is applied on the 24 mounted solder ball on solder pad for each HTS time. The shear test in this study is performed at room temperature using multipurpose bondtester instrument (Dage series 4000, Nordson Dage) in low-speed shear test (100 µm/s, [13]) and 150 µm shear height.

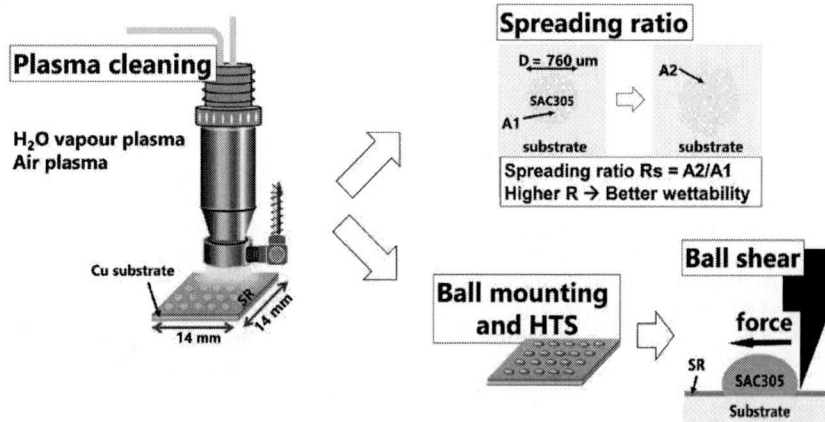

Figure 2. Sequence of event on the solder tests

III. RESULTS

A. Plasma gas temperature diagnostic

The plasma gas temperature measurement in this study shows that both air and H_2O vapour plasmas are giving low initiation temperature of which 150 °C and 170 °C, respectively. Fig. 3 shows that the temperature stability occurred after a short period of initiation time (10 s). AP-plasma is giving an efficient process with only 0.5 s scanning time for each unit.

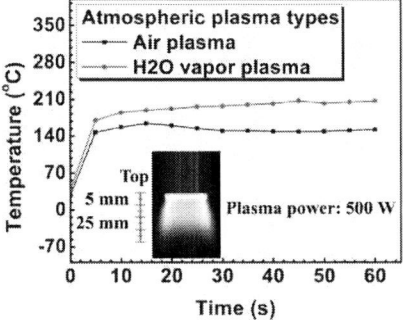

Figure 3. DC-AP gas temperature

B. Spreading ratio in different staging time

Fig. 4 shows that the spreading ratio values, namely the wettability, on bare Cu substrate, OSP and ENIG surface finishes, in an increasing staging time. The staging is defined as the queuing time after plasma treatment to the ball mounting [14].

For bare Cu, the spreading ratio values for the plasma-treated conditions are significantly higher than that without plasma treatment. It is also found that the spreading ratio values after either air or H_2O vapour plasma treatments decreases with increasing staging time. This trend is also observed on OSP and ENIG, as reported in the previous study [15]. These results indicate that the AP-plasma treatment improves the wettability of solder on substrate with or without surface finish. For bare Cu and ENIG, the improvement by air and H_2O vapour plasma is similar. Hence, the results show that the use of H_2O vapour plasma treatment on OSP is the optimum match. H_2O vapour plasma on OSP is enchancing the spreading ratio by around 80% as compared to the no plasma treatment. It is believed that this significant improvement is due to the reactive interaction between the radical formation from H_2O vapour plasma and the OSP layer. For ENIG, although the plasma effect is not as promising as on OSP, but it is more stable and less-sensitive to the staging time. Therefore, ENIG is chosen to perform the HTS test.

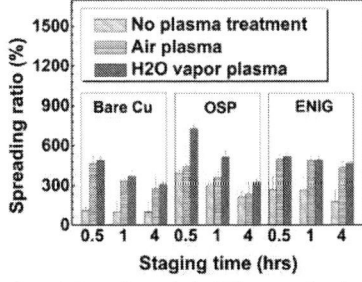

Figure 4. Spreading ratio in different staging time

C. Formation of OH radicals

Surface funtionalisation in this study is affecting the solder wettability of which is necessary that air and H_2O vapour plasma are produce some formation of plasma radicals to recreate the surface properties of substrates. From OES measurement, it is found that O^*, OH^*, and N-related peaks are observed in both air and H_2O vapour plasmas (Figs. 5(a) and (b)). The intensity peaks of O^* and OH^* are slightly decrease from 4,723 to 4,023 arb·units and 4,706 to 4,089 arb·units, respectively. However, N-relative peak is decreased significantly around two times decreases than O^* and OH^* peaks (from 7,608 to 6,207 arb·units). It is assumed that nitrogen has been dissolved in H_2O vapour, therefore, nitrogen concentration during H_2O vapour plasma treatment is decreased. Gas analyzer measurement is conducted to prove the finding assumption. It is found that NOx concentration during air plasma treatment around 164.4 ppm decreased to 146.9 ppm during H_2O vapour plasma introduction.

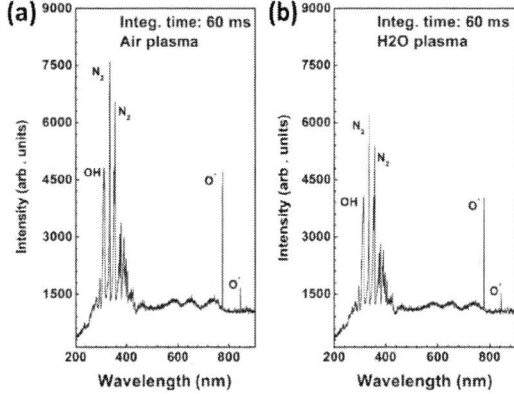

Figure 5. Formation of OH radicals measured by OES in: (a) Air plasma and (b) H_2O vapour plasma

D. Joint reliability

Fig. 6 shows that the ball shear strength degrades from the pre-HTS to the post-HTS (0 to 500 hrs). However, the shear strength become stable after 500 to 1,000 hrs, regardless the plasma sources. It is found that the breaking is taking place in the solder bulk, as shown in Fig. 7(a). Therefore, the cohesion break with a complete solder wetting on solder pad is majorly occurred in SAC305 solder ball as shown in Fig. 7(b). Ductile break mode is the dominant mode. Incomplete solder coverage on pad is also observed, as shown in Fig. 8(a), which is caused by a misplaced solder ball during the ball mounting process. As displayed in Fig. 8(b), the misplaced ball is only wetting partially on the solder pad, and resulted in a lower shear strength (Fig. 6). This decrement is due to a reduced breaking area in the bulk solder. Although solder ball misplacement is inevitably in practical, the solder ball position needs to be better than 20% of the ball diameter even with the plasma treatment.

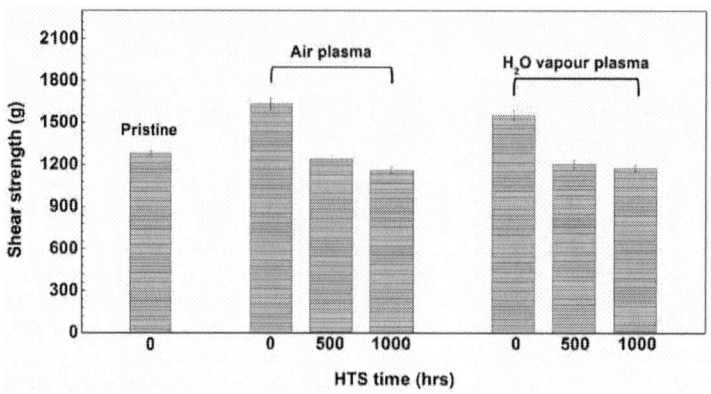

Figure 6. Ball shear strength in pre-and-post-HTS for ENIG

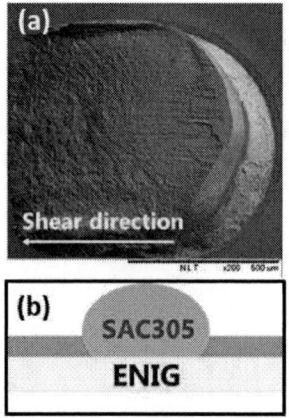

Figure 7. (a) SEM cross-sectional image of cohesion break mode in the solder joint after ball shear test, and (b) Schematic picture of the complete solder wetting on pad

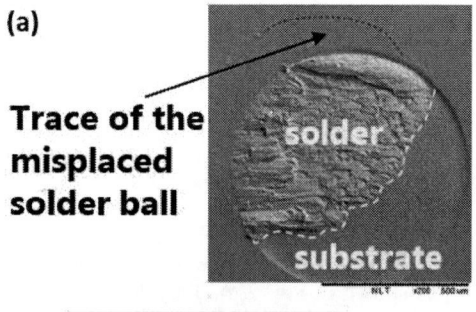

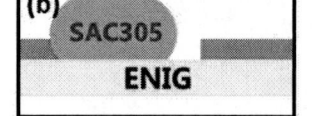

Figure 8. Misplacement of solder ball on pad: (a) SE cross-sectional image, and (b) Schematic picture

SUMMARY

1) AP-plasma treatment with air and H_2O vapour gas sources can improve the solder wettability on the common surface finish efficiently.

2) The best results of the spreading ratio is achieved by the used of H_2O vapour plasma treatment on OSP surface finish.

3) The plasma effect on ENIG is stable and less sensitive to the staging time.

4) The OH radicals are the major roles in regards of plasma functionalisation process. It is also found that NOx concentration decreases during H_2O vapour plasma introduction because of the H_2O dissolved the nitrogen gas during the process.

5) Ball shear strength decreases in between pre HTS and post HTS regardless the plasma gas sources. However, it is stable after 500 hrs.

6) Cohesive break is majorly occurred in solder bulk.

ACKNOWLEDGMENTS

The authors gratefully acknowledge the help and support by many collaborators, especially Robert Bosch Taiwan Co., Ltd, Taipei, Taiwan for financially supporting this research under industrial-academic collaboration project plan number 7806. This work is supported by Click Sun-Shine Corp., New Taipei City, and GIE Optoelectronics Inc., Taipei. The authors appreciate Prof. Albert T. Wu, Mr. Patrick Wen-Chih Lin, Mr. Chia-Yu Liu, Mr. Wei-Shan and all collaborators for the kind assistance during this work.

REFERENCES

[1] G.A. Rinne, "Solder bumping methods for flip chip packaging," in *1997 Proc. 47th Electronic Components and Technology Conference (ECTC)*, 1997.

[2] S. Hou, K.H. Tsai, M.F. Wu, P.H. Tsao, L.H. Chu, "Board Level Reliability Investigation of FO-WLP Package," in *2018 Proc. 68th Electronic Components and Technology Conference (ECTC)*, 2018.

[3] H.-K. Cheng, S.-P. Feng, Y.-J. Lai, K.-C. Liu, Y.-L. Wang, T.-F. Liu, C.-M. Chen, Effect of polyimide baking on bump resistance in flip-chip solder joints. *Microelectronics Reliability*, vol. 54, pp. 629-632, 2014.

[4] S.J. Jo, A.R. Lee, C.Y. Kang, "Improvement of the Solder Joint Strength in a SAC 305 Solder Ball to a ENIG Substrate Using LF Hydrogen Radical Treatment," in *2010 Proc. 60th Electronic*

Components and Technology Conference (ECTC), 2010.

[5] C.C. Lee, H.Y. Chuang, C.K. Chung, R. Kao, "Oxidation behavior of ENIG and ENEPIG surface finish," in *2010 Proc. 5th International Microsystems Pcakaging Assembly and Circuits Technology,* 2010.

[6] M. Godard, D. Drouin, M. Darnon, S. Martel, C. Fortin, "Plasma Treatment for Fluxless Flip-Chip Chip-Joining Process," in *2018 Proc. 68th Electronic Components and Technology Conference (ECTC),* 2018.

[7] R. Deltschew, D. Hirsch, H. Neumann, T. Herzog, K.J. Wolter, M. Nowottnick, K. Wittke, Plasma treatment for fluxless soldering, *Surf. Coat. Tech.*, vol. 142, pp. 803-807, 2001.

[8] A. Paproth, K.J. Wolter, T. Herzog, T. Zerna, "Influence of plasma treatment on the improvement of surface energy," in *Proc. 24th International Spring Seminar on Electronics Technology. Concurrent Engineering in Electronic Packaging. ISSE,* 2001.

[9] K.J. Wolter, T. Zerna, R. Deltschew, H. Neumann, "Plasma treatment process for fluxless reflow soldering," in 2001 *Proc. 51st Electronic Components and Technology Conference,* 2001.

[10] Y.-L.Kuo, K.-H. Chang, C. Chiu, Carbon-free SiOx ultrathin film using atmospheric pressure plasma jet for enhancing the corrosion resistance of magnesium alloys, *Vacuum*, vol. 146, pp. 8-10, 2017.

[11] M. Audronis, S.J. Hinder, P. Mack, V. Bellido-GOnzalez, D. Bussery, A. Matthew, M.A. Baker, A comparison of reactive plasma pre-treatments on PET substrates by Cu and Ti pulsed-DC and HIPIMS discharges, *Thin Solid Films*, vol. 520, pp. 1564-1570, 2011.

[12] G. Takyi, N.N. Ekere, Study of the effects of PCB surface finish on plasma process time for lead-free wave soldering, *Solder. Surf. Mt. Tech.*, vol. 22, pp. 37-42, 2010

[13] JEDEC Standard JESD22-B117B, Solder Ball Shear, 2014. <https://www.jedec.org/standards-documents/docs/jesd-22-b117a>.

[14] H.Y. Chuang, C.H. Lin, J.P. Chu, C.R Kao, Novel Cu–RuNx composite layer with good solderability and very low consumption rate, *J. Alloys Compd.*, vol. 504, pp. L25-L27, 2010.

[15] S.D. Kencana, H.-P.W., A. Laksono, J.-Y. Guo, W. Chuang, E. Schellkes, Y.-W. Yen, Y.-L. Kuo, "Solder wettability improvement in cupper (Cu) substrate using direct current atmospheric pressure plasma," in *Proc. 16th International conference on plasma surface engineering, PSE,* 2018.

Orthogonal Quilt Packaging 3D Integration for High-Energy Particle Detectors

Jason Kulick, Tian Lu, Edit Varga, Gary H.
Bernstein, Carlos Ortega
Indiana Integrated Circuits, LLC,
South Bend, IN USA
e-mail: jason.kulick@indianaic.com

Christopher Kenney, Julie Segal
SLAC National Accelerator Laboratory
Menlo Park, CA USA
e-mail: kenney@slac.stanford.edu

Abstract— **This paper presents Quilt Packaging® (QP), a chip edge-interconnect technology to provide solutions for applications in which multi-chip 3D assemblies are the optimal configuration, but stacking approaches are not desirable. We present an overview of an *orthogonal* multi-chip system enabled by QP chip integration technology, including verification of the chip-to-chip interconnects and QP's compatibility with CMOS, along with fabrication results and testing of the 3D assembly. We then extend this approach to the unique challenge of deep pixel high-energy particle detectors for scientific applications.**

Quilt Packaging, 3D microelectronic packaging, high-energy particle detector, orthogonal assembly, silicon-based sensors, advanced packaging, system-in-package

I. INTRODUCTION

Next-generation microelectronics systems must balance performance increases required by end users against size, weight, power (SWaP) and cost constraints. Over the course of many generations of designs, multi-chip systems for many applications have increasingly advanced towards direct chip-to-chip interconnections in the 3rd dimension such as flip-chip, stacked wire bonding, interposers, and through-silicon-via (TSV) [1, 2]. These all share "stacking approaches" where two or more chips are stacked one on top of the other. For many applications, these methods work well and can provide the desired performance for the cost. However, there are many other applications where multi-chip 3D assemblies are the optimal configuration, but where such stacking approaches are not desirable. Quilt Packaging (QP) chip integration technology offers such an alternative. This paper will present an overview of orthogonal multi-chip system development enabled by QP.

QP is a high-performance, uniquely customizable and versatile edge-interconnect technology that utilizes "nodule" structures that extend out from the vertical facets of microchips to allow for in-package inter-chip communication and mechanical alignment [3-8]. The unique attributes of QP enable a very-low-loss, high-bandwidth interconnection that is also mechanically robust. Shortened chip-to-chip signal paths and minimum transitions enabled through QP allow for an insertion loss of 0.1 dB from DC-100 GHz [9], and less than 1 dB up to 220 GHz [10]. QP I/O connections can be scaled down to a pitch of 10 microns or smaller with a digital bandwidth of 12-100 Gbps per nodule interconnect [6].

QP allows the integration of multiple chips with disparate technologies or substrate materials in both 2D and 3D configurations. Multiple chips are joined (or "quilted") together via nodules to form a monolithic-like "meta-chip" called a "quilt," that performs mechanically and electrically as a single chip. QP differs from conventional packaging methods in that interconnect nodules are fabricated at the wafer level and are lithographically defined, enabling self-aligning nodules that can be assembled easily and precisely at sub-micron scales using commercial microelectronic pick-and-place equipment. This greatly facilitates rapid prototyping, and the process is scalable to commercial production. QP technology can either replace or complement existing 2.5D and 3D SiP approaches such as wirebonds, flip-chip interposers, and TSV chip stacking. The fabrication process for QP nodule interconnects is similar to the via-middle TSV process and may be fabricated concurrently in some instances.

Figure 1 shows the various possible 2D and 3D system chipsets demonstrated using QP. Two elongated Si chips are quilted together in-plane and mounted in a QFN ceramic package (Fig. 1a). Two close-up images show the interconnects on a scanning electron microscope (SEM, Fig. 1b) and an optical microscope (Fig. 1c) images. Figures 1d and 1e show an orthogonal 3D and a planar 2D honeycomb

Figure 1. Various mounting possibilities for silicon based chips. (a) Planar 2D quilt mounted in a QFN ceramic package; (b) and (c) SEM and optical images of the QP nodules, respectively; (d) orthogonal 3D quilt and (e) honeycomb planar 2D quilt; (f) SEM of two-chip quilt bonded onto interposer.

quilt, respectively. Figure 1f is an SEM image of two QP chips quilted together and mounted on an interposer with bump bonding.

II. 3D QUILT PACKAGING STRUCTURES

Current commercial technology commonly uses a motherboard and one or more daughter cards at the board level. The motherboard is the main circuit board of the device while the daughterboard is an expansion card that plugs into and extends the circuitry of a motherboard.

Figure 2 shows schematic drawings of several unique 2D and 3D configurations of quilted chipsets. Beyond standard quilts where various chips are connected at their edges in a planar 2D (Fig. 2a) configuration, QP enables various 3D configurations as well that mimic the motherboard/ daughterboard concept, but here it is at the chip level. The nodules of one chip (daughtercard) can be inserted into sockets on the surface of another chip (motherboard) to form a 90-degree orthogonal connection (Fig. 2b). QP nodules can be designed and fabricated with interdigitated fingers that fit together as a hinged connection to allow for angled edge connections allowing for 3D curved configurations (Fig. 2c). In addition, QP allows for the vertical integration of stacked chiplets using the vertical interconnection of nodules at the side of the chips with the same thickness as the chips themselves (Fig. 2d), or by connecting the nodule to metal tracings on a vertical interposer chip (Fig. 2e).

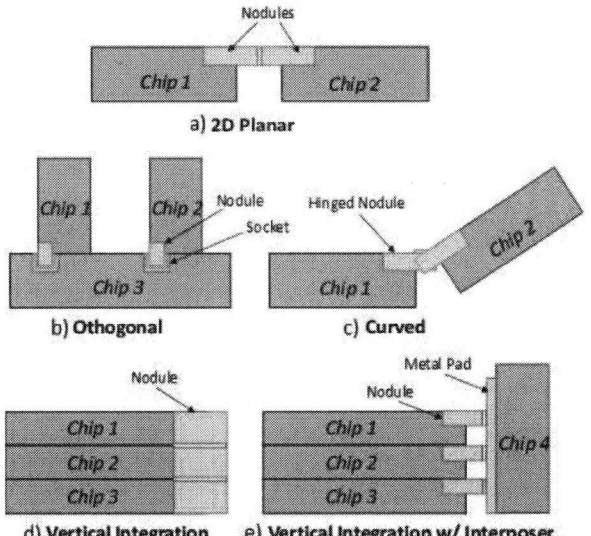

Figure 2. Cross-section schematics of various 2D, 2.5D, and 3D quilt configurations. (a) Standard planar configuration, (b) orthogonal configuration with daughtercard chip vertically inserted into motherboard chip, (c) curved configuration using hinged nodule design, (d) vertically-integrated stacked chips interconnected through edge nodules, and (d) stacked chips integrated through a vertical interposer.

Most microsystem interconnections are formed using wire bonds or bump bonds. Wire bonding is an intrinsically one-dimensional process, while bump bonding is two-dimensional, but is limited to connecting a pair of substrates. In high-energy-particle (HEP) detector applications, partially

3D interconnect technology was developed by Tezzaron, Inc. [11], among others. However, their architecture is effectively limited to connecting 2 to 4 substrates (Fig. 3a). While increasing the amount of circuitry per pixel by a corresponding factor of 2 to 4 is potentially significant, it clearly represents a marginal gain given the increased cost. In contrast, the QP technology provides a more compact design using higher number of chiplets as shown in Figure 3b. This architecture is less expensive, simpler, and can increase the density of devices and circuitry per pixel by a factor between 10 and 100. It can also provide direct, high-bandwidth, low-impedance connections between sensors or circuit chips in spatially separated planes. The two 3D architectures are compared side-by-side in Fig. 3.

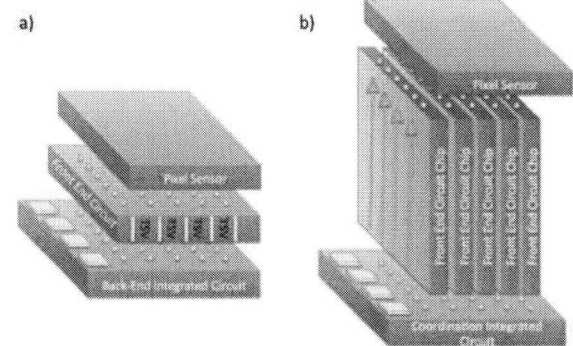

Figure 3. Comparsion of two 3D architectures. (a) State-of-the-art 3D IC stacking technology; (b) example of 3D QP architecture utilizing QP interconnects.

III. QP DEVELOPMENT

In this section we discuss the fabrication of the QP nodules in Si substrates and introduce the fabrication process and challenges for compound semiconductor materials, such as GaAs and InP. We determine the possible effect of our fabrication method on silicon (Si) based CMOS structures, followed by the discussion of the assembly of 3D quilts from Si samples. Our basic testing process is followed by several proposed applications (large-area cameras, high-efficiency silicon-based sensors, high-energy particle detectors) based on the above-mentioned results.

A. Fabrication

The basic fabrication steps of the QP nodules are shown on Figure 4. Nodules are built into the Si wafer (Fig. 4a) by etching (Fig. 4b) and metal sputtering as well as electroplating (Fig. 4c). Then, connections are built with the already existing circuitry on the wafer (Fig. 4d) followed by the singulation etch (Fig. 4e). The result is the singulated chip with nodules built on the sides. These can be used to build the 3D structures shown above.

978-1-7281-1500-9/19 $31.00 © 2019 IEEE

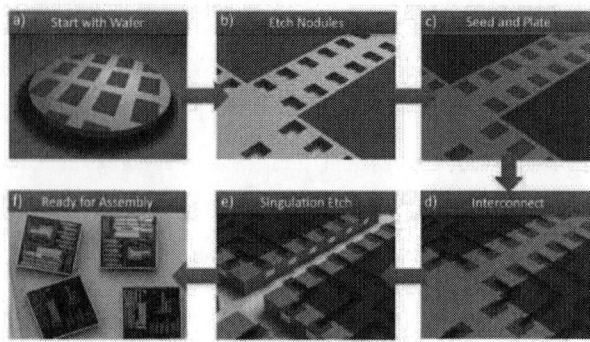

Figure 4. QP fabrication process flow. (a) Finished front end wafer; (b) nodules etched into the Si; (c) etched holes lined with dielectric insulating layer and filled with copper; (d) nodules connected by metallization to existing circuitry; (e) chips singulated by etching; (f) photograph of separated chiplets.

In our specific case, where the front side of the wafer has sensitive structures (Fig. 5a), further development of the QP nodule fabrication is needed. It starts with contact photolithography to define the nodule features (Fig. 5b), which are then etched to a depth of ~30 μm into the wafer using Bosch deep reactive ion etching (DRIE) (Fig. 5c). After nodule shapes are defined, a Ti/Cu seed layer is deposited, and the nodules are filled in a Cu electroplating step (Fig. 5d). Electroplating overburden is removed with a chemical mechanical polishing step. Serpentine patterned Ti/Cu redistribution layer (RDL) daisy-chain lines are then deposited using a liftoff technique (Fig. 5e). A PECVD oxide overglass layer is deposited with etched windows over the nodules and separation streets (Fig. 5f).

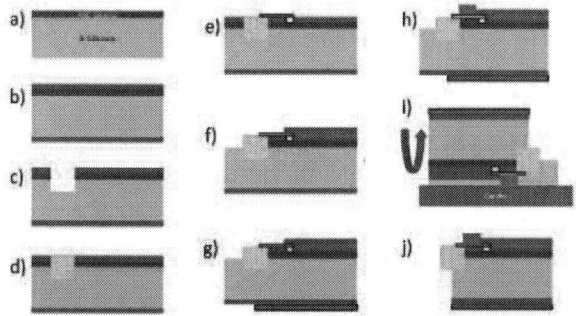

Figure 5. Back-end-of-line (BEOL) fabrication process on ASIC device wafers. (a) Silicon FEOL ASIC wafer from foundry; (b) deposition of SiO2/SiNx passivation layer on both sides; (c) nodule etch; (d) nodule Cu electroplate and CMP planarization; (e) RDL via etch and metal deposition; (f) overglass deposition and window etch; (g) backside separation hard-mask deposition and pattern etch; (h) selective under-bump metalization (UBM) and solder electroplating; (i) mount on carrier for backside separation etch for die singulation; (j) lift-off from carrier wafer for singulated chips.

A chrome hard mask is deposited and patterned on the backside of the wafer for use for backside separation etch through the use of backside alignment (BSA) contact lithography (Fig. 5g). Solder electroplating is performed before the backside separation etch). A Ti/Cu seed layer is deposited on the front side of the wafer to be used for solder plating, which is followed by selective plating of tin (Sn) solder on top of a nickel (Ni) barrier layer. The Ti/Cu seed

layer is then stripped, and the wafer is mounted with the backside facing up on a Si carrier wafer using Crystalbond (Fig. 5i). Separation etch is then performed on the wafers using Bosch DRIE through the thickness of the wafer until the daughtercard chips are singulated (Fig. 5j). The Crystalbond is removed in acetone, and the singulated chips are now fully separated.

The fabrication process for the proof-of-concept motherboard RDL wafers is similar to fabrication process for the daughtercard wafers, with the exception of the socket process steps replacing the nodule steps. Troughs to become "sockets" are fabricated on the motherboard chips using DRIE. The sockets are overlaid with and RDL metal around the openings. The follow-up processing with the oxide overglass window etch and selective solder plating are processed the same as with the daughtercard wafers. Wafer dicing for chip singulation could be used rather than the backside Bosch etch process, although a conventional QP process could be used if additional chips are to be quilted to the motherboard.

B. QP Processing in Compound Semiconductors

Since the QP process is relatively agnostic to wafer material, it can be implemented in materials of interest beyond Si, such as GaAs, SiC, GaN, InP, SiGe, GaSb and other substrate materials. The QP interconnect geometries are lithographically-defined during the fabrication process, allowing for the assembly of a variety of 2D and 3D microelectronic architectures using complex chip shapes and micron-scale interconnects with minimal impact to the fabrication of the components themselves. Utilizing this approach allows for straightforward 2D wafer processing and design and can improve fabrication yields by reducing individual chip area.

One of the most crucial aspects of QP back-end-of-line (BEOL) fabrication is the ability to micromachine the wafer substrate with high precision and process control. This is performed by the DRIE process used during nodule feature etch and chip separation etch steps. A standard substrate material such as Si uses a Bosch DRIE process to create a directional anisotropic etch of deep nodule molds and dicing streets with near vertical sidewalls and can thus be very well-controlled. Unlike Si, there are no standard industry deep anisotropic etch processing for III-V and other compound semiconductor materials that are similar to Bosch DRIE. Anisotropic deep etch recipes for III-V materials such as gallium arsenide (GaAs) and indium phosphide (InP) were developed using the ICP etching tool at the Notre Dame Nanofabrication Facilty [web site]. For deep GaAs etching characterization, a $BCl_3/Cl_2/Ar$-based plasma chemistry was used, with various process parameters such as gas mixture, chamber pressure, flow rate, temperature, time, and RIE/ICP power adjusted to create an anisotropic etch with a near vertical sidewall. Deep InP etching uses a $SiCl_4/Ar$-based plasma at an elevated chamber temperature. Currently, the QP deep GaAs dry etch has been optimized to approach the anisotropy and etch rate of the Si BOSCH process, while the InP etch process requires further investigation and optimization. Figure 6 shows SEM image comparisons of

978-1-7281-1500-9/19 $31.00 © 2019 IEEE

etched Si and GaAs sidewalls. Various scanning angles show the substrate edge along with the nodules.

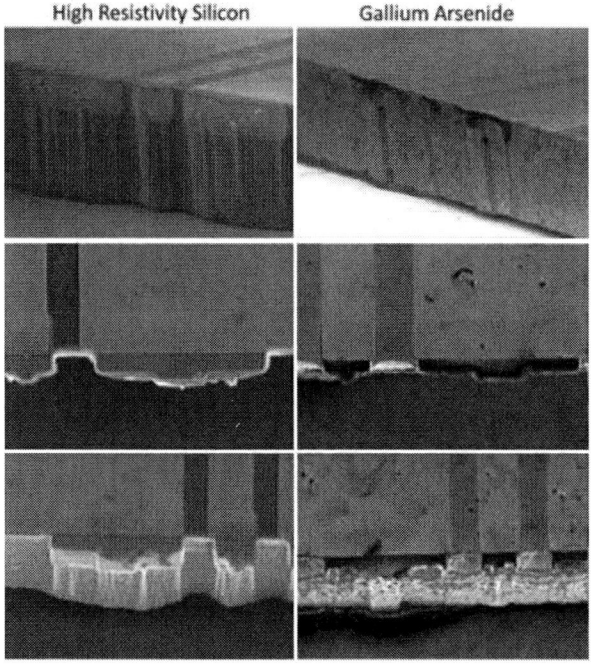

High Resistivity Silicon   **Gallium Arsenide**

Figure 6.   SEM images of post-singulation high-resistivity silicon (left column) and GaAs (right column) chip side-walls.

III-V materials such as GaAs and InP are far more fragile than Si and require more-delicate handling during the QP BEOL process. This is especially crucial for steps such as CMP and backside grinding during separation, which place a lot of mechanical stress on the wafers. These stresses increase the chances of chips and microfractures, which may propagate across the lattice of the wafers, causing breakage. Therefore, special consideration is made for the CMP and backside grinding process parameters as well as general wafer and singulated chip handling to ensure that wafers are not structurally compromised during processing. QP BEOL processing methods on other materials such as silicon carbide (SiC), silicon germanium (SiGe), and gallium nitride (GaN) are currently being developed.

C. CMOS QP Integration

In this section, we demonstrate the compatibility of standard CMOS wafers with the BEOL QP processing and assembly. Implementation of the QP technology and its fabrication is not dependent on the front end of line wafer fabrication, whether that be photonics, CMOS, MMIC, MEMS, power electronics, or other technology. Current adopters of QP technology are customers who have fully vertically integrated foundry capability as well as those who have wafers produced by 3rd parties. In nearly all cases, QP fabrication and assembly occurs after the front-end foundry processes and can be implemented without requiring any alterations to active device or chip design. Implementation of QP typically does not require modification to the front-end-of-line wafer device process, with adjustments to feature location usually the only deviation from existing process flows. This flexibility to "post-process" irrespective of front-end processing can be used to even better leverage integration of many different chip designs and technologies with dense I/Os and 3D architectures.

Our front-end CMOS demonstration was performed using a 2 μm gate process on 4" Si wafers. The QP BEOL process was performed after the CMOS devices and local metal layer were finished, and consisted of nodule plating and CMP, global RDL interconnect metallization, and chip singulation.

Various test structures and devices were fabricated on the chips to test for the effects of the BEOL QP process on on-wafer and on-chip CMOS functionality. These include various devices used for testing ohmic contacts and gate threshold voltages, single CMOS gates for functionality, as well as more-complex IC circuits such as sound drivers and multiple stage ring-oscillators in 77- and 201-stage looped inverter chains. The ring-oscillators were designed to function as on-chip verification circuitry as well as through a series of daisy-chained quilted nodule interconnects. Figure 7 shows the post-singulation CMOS chips (a) with a CMOS inverter device, (b) QP nodules interconnected to CMOS ICs and (c) global RDL interconnect metal layer.

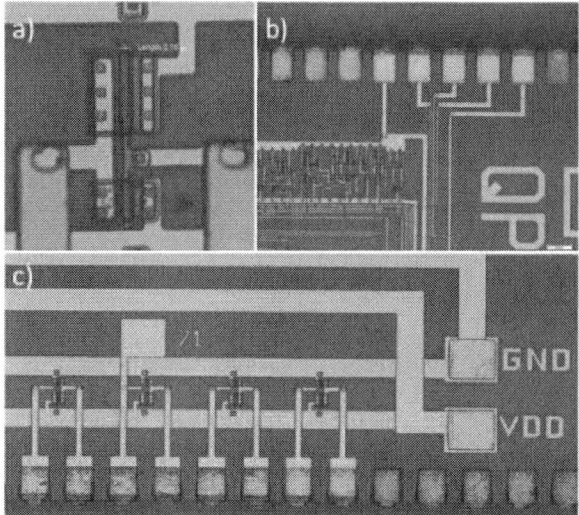

Figure 7.   CMOS/QP integration demonstration. (a) Post-BEOL QP CMOS inverter device; (b) post-BEOL QP CMOS ICs interconnected to nodules; (c) close-up of post-BEOL QP CMOS Ring-Oscillators.

The proper operation of these CMOS device test structures was successfully verified multiple times throughout the QP BEOL process to demonstrate that the process does not degrade CMOS functionality including after front-end-of-line metal fabrication, QP nodule, chip singulation, and QP assembly steps. Figure 8 shows the testing of the ring oscillators after the QP assembly step. The results showed a step-wave output on the RO from a 5V DC power input with varied clock-rate frequency ranging from 10 to 50 MHz. The variation in RO output clock-rates are due to variations in the ohmic contact and gate voltage threshold across the devices on a single wafer as well as different wafers in the fabrication run lot. These variations are the result of NDNF's front-end CMOS process (a multi-process/multi-user research facility

978-1-7281-1500-9/19 $31.00 © 2019 IEEE

instead of a dedicated foundry), and not from the result of the QP BEOL process. Device testing after the fabrication of the CMOS devices but before the start of the QP BEOL displayed the similar variation in clock-rate.

A potential issue that may affect the performance or functionality of the CMOS during the QP BEOL process was discovered during this time. Specifically, this regards the CMP over-planarization during nodule processing steps, which runs the risk of polishing off the underlying metal and poly layers. This problem was prevented with the proper selection of CMP pad and slurry with the appropriate stopping layer underneath the copper with high selectivity to the slurry after the copper has been removed. Deposition of a passivation layer (such as an oxide or nitride) underneath the copper and stopping layers was used to mitigate topological non-uniformity brought on by the running of metal signal lines over poly, which causes the potential for metal line removal during CMP. In addition, a selective copper plating approach is under development to minimize, or even entirely bypass, the CMP process altogether.

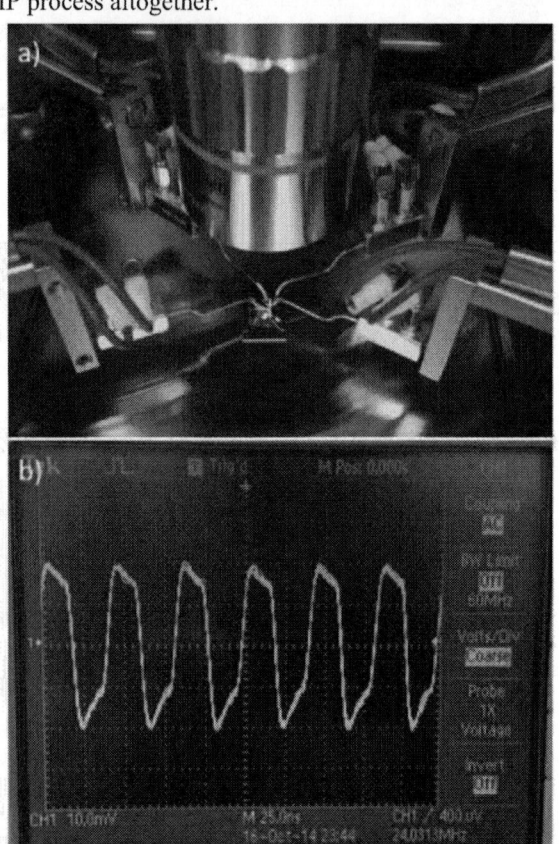

Figure 8. (a) Photograph of our post-BEOL QP fabrication CMOS testing set-up and (b) measurement results for ring-oscilllator output after QP assembly and reflow (24 MHz).

D. 3D Quilt Assembly

The QP interconnect nodule structures can be customized to enable precision sub-micron alignment, dense I/O pitch, extremely low-loss microwave transmission, and high-power current-handling capability. Furthermore, the uniquely customizable nodules facilitate high volume manufacturing by providing a "self-aligned" connection that is directly compatible with commercial pick-and-place tools.

QP technology can enable high-volume alignment and assembly of 2.5D and 3D microsystem architectures using existing tooling (Fig. 9). The 3D QP assembly process can be a highly repeatable, multi-step process that requires a simple 2- or 3- axis tooling system and can be made to be independent of chip substrate thicknesses and any subsequent z-height mismatch issues.

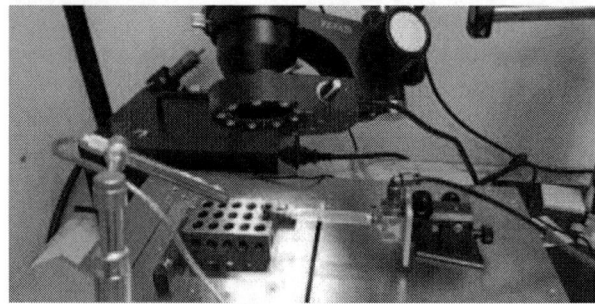

Figure 9. 3D QP Assembly tooling.

Fabrication and assembly of quilts using QP interconnects in an orthogonal 3D architecture is accomplished through the development, fabrication, and verification of 3D connected Si chips. Multiple proof-of-concept test articles have been built consisting of vertically-stacked quilted arrays of mechanical daughtercard chips with daisy chain test structures orthogonally quilted to Si RDL motherboard chips. Figure 10 shows 10×10 mm^2 die assembled perpendicularly. One vertical daughter chip is mounted on the horizontal motherboard chip on Figs. 10a and 10b. Figures 10c and 10d show two and four vertical chips assembled into the horizontal motherboard chip, respectively.

The individual daughtercard chips were plugged orthogonally onto the motherboard chips at the nodule-socket interface. The interconnection of the chips is shown on a schematic image (Fig. 11a), where the daughtercard has the nodules pointing out from the substrate and the motherboard has etched sockets in which to insert the daughtercard nodules. Figure 11b is an SEM image of the connection of the vertical and the horizontal chips.

After the QP nodules are orthogonally inserted onto metallized sockets on the horizontal chip, a solder reflow process is performed that allows the solder to wet between and into the nodules and sockets to complete the interconnection. This reflowed socket also provides the electrical interface between the daughtercard and motherboard chips, while the nodules fitted securely inside the sockets provide additional mechanical stability. A close-up SEM image shows the vertical nodule with the reflow solder metal on top (Fig. 12).

978-1-7281-1500-9/19 $31.00 © 2019 IEEE

Figure 12. SEM image of vertical QP nodules plugged into a horizontal socket interface with reflowed solder joint.

E. Testing of Orthogonal 3D QP quilts

The RDL daisy-chain across the nodule-socket interfaces were tested under a probe station, and electrical continuity across multiple nodule-socket interconnects was verified during electrical probe testing (Fig. 13). Resistance measurements across the daisy-chain, after subtracting the RDL metal and the probe measurement set-up, resulted in a range between 5-12 milli-ohms per reflowed nodule/socket pairing. These results are comparable to soldered QP nodule interconnects reflowed on a flat 2D plane. Further optimization to the solder plating process and placement of solder on the nodule/socket joint for reflow is anticipated to result in even lower measured resistance values.

Figure 10. 10x10mm samples assembled into 3D structures using QP. Isometric back view of assembled and reflowed sample (a); isometric (b) and side view (c) of assembled quilt with four daughtercards.

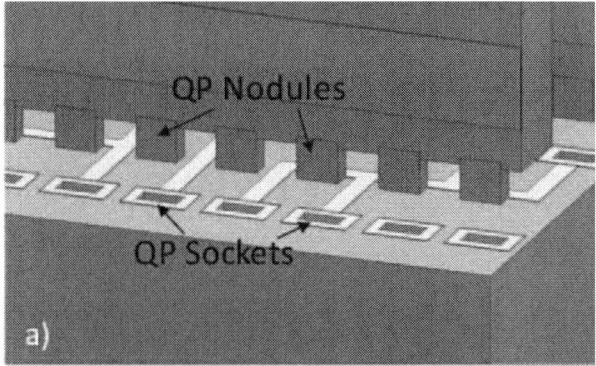

Figure 11. Orthogonal 3D QP assembly. (a) Schematic drawing of the nodules on the vertical chip and the sockets on the horizontal chip; (b) SEM of QP nodules orthogonally inserted onto metallized sockets.

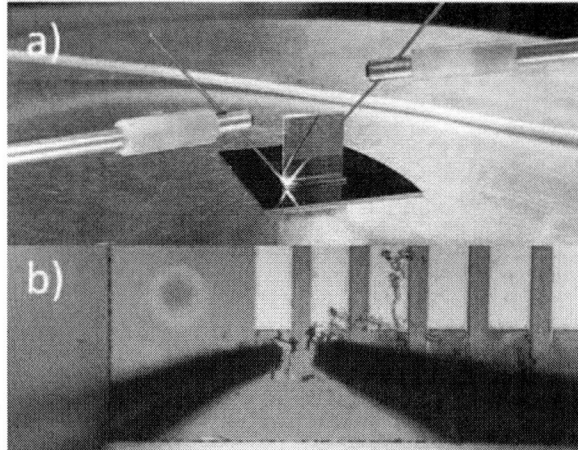

Figure 13. (a) Probe testing of the 3DQP quilt, and (b) de-embedding measurement of a motherboard chip transmission line.

F. Applications

The following applications can greatly benefit from 3D QP technology and in most cases as an enabling technology.

1) High-Energy Particle Detector (HEP)

A typical, general-purpose, detector at a major collider has three core systems that could be improved by the use of QP technology such as the vertexer, tracker, and colorimeters. These have active-Si areas in the ATLAS [12] and SiD [13] detectors on the order of 10 m^2, 100 m^2, and 1000 m^2, respectively. There are several potential new colliders or upgrades to existing ones under discussion such as the Super Large Hadron Collider (SLHC), Circular Electron Positron Collider (CepC), and a linear collider. Any one of these would require multiple large detectors.

Presently, the footprint of modules building up the core systems is limited by lithography. A similar constraint holds for the dimensions of CMOS circuit chips, which are constrained by yield and stepper exposure fields. Larger sensor and circuit units formed via QP interconnects will, in many cases, allow a decrease in channel counts, eliminate insensitive gaps, and simplify assembly. The described technology can break these bottlenecks and enable improvements in hermeticity, decrease mass, and reduce power. By using QP interconnects to form direct bridges between different layers within a tracker or vertex system, novel triggering schemes, such as identifying high PT tracks, become possible. In this way, a trigger processing circuit chip could also serve as a precision mechanical element that is thermally matched to the other main components.

If CMOS image sensors like the Cornell High Energy Synchrotron Source (CHESS) project [14] are incorporated into the pixel or strip subsystems, then the use of our QP process will be valuable as a means to assemble large tiles out of single-stepper field chips. CMOS sensors could potentially also be used in a calorimeter such as that envisioned for SiD. Given the areas involved, the interconnection costs would likely fall in the one to ten million dollar range.

Local trigger processor cores could be incorporated into the thermal-mechanical support structure of the pixel or strip systems. These would perform track finding across multiple sensor planes and generate inputs for global triggers on high-momentum tracks and help distinguish physics jets from pile-up stochastic jets. Quilting would be an enabling technology for this application with potential budgets in the millions of dollars.

Observation of about 10^{16} events is required to reconstruct 100 Higgs bosons, and searches for new physics are looking for even rarer events. So, any improvement in trigger efficiency or flexibility can pay big dividends. Even a 20% improvement in trigger effectiveness could produce the same physics results but save a year of LHC operating costs, which are in the $100M range.

Hadron therapy promises vastly improved outcomes for oncology treatments relative to the present method using photons. By reducing the dose to healthy tissue, there should be fewer and less-severe side effects for patients. More critically, hadron therapy should deliver a higher dose to the tumor for a given patient exposure, and so result in higher survival rates. The main factor limiting the wide adoption of hadron therapy is the cost of the accelerator, but there are promising concepts for reducing the size and cost being actively pursued. Since a hadronic beam deposits its energy in a tight volume, it is crucial that the beam's trajectory and particle energy be monitored and adjusted in real time throughout a treatment session. A tracker system consisting of two or three planes of semiconductor sensors in a static magnetic field would be capable of recording the passage of the treatment beam in a minimally invasive manner. If the sensor planes were bonded to an orthogonal processing circuit chip, this would enable the particle's space points to be collected and analyzed, and particle trajectories, energies, and particle type to be determined. This is obligatory for both efficacy and safety reasons. Several concepts are being pursued to drastically reduce the size and cost of hadron oncology systems. If they are successful, then hadron therapy centers will become ubiquitous with thousands needed worldwide, and each of these would require one or more beam tracker systems using QP technology. Due to radiation damage, the particle trackers would need to be replaced on a regular basis. If a single hadron tracker system costs $50,000, then the global market would be in the $100,000,000 range.

2) High-efficiency Siilicon-based Sensors

The quantum efficiency of Si sensors drops rapidly for photon energies above 12 keV. Non-silicon sensors face many challenges relating to low band gaps or poor material homogeneity. One new approach for high-efficiency Si-based sensors is to employ an edge-on geometry, where the sensor's width serves as the effective absorption depth and can be from mm to cm in depth. The edges of Si sensor chips could be quilted together, arranged into a solid stack of edge detectors, and bonded together using QP to the major surface of a readout chip arranged in an orthogonal manner to the sensor stack. Many areas of X-ray-based science, non-destructive testing, and medicine could benefit from such an innovative and efficient detector.

IV. Future Work

IIC and its partners have an established supply chain for fabrication, assembly, and testing of QP chipsets. For assembly, packaging and testing, IIC is partnered with Plexus Corporation, for applications from prototyping through volume production. Plexus is a leader in providing electronics design, manufacturing, assembly, and packaging services to companies with mid-to-low volume, higher complexity products. Proper documentation for transitioning QP technology from R&D to production volumes (such as FMEA documents, quality assurance documents, design rules, etc.) are being developed.

Throughout the scale-up effort, IIC and its supply chain partners are anticipated to produce larger quantities of QP chipsets to validate the higher volume production and acquire more-statistically-relevant environmental reliability data. More-conclusive experimental data are anticipated for the electrical (analog and digital) and power usage performance. In addition, the QP back-end-of-line (BEOL) production process on other materials such as SiC, SiGe, and GaN are currently being developed.

V. SUMMARY

QP has demonstrated improved analog and digital performance compared with emerging state-of-the-art 2D and 3D microelectronic packaging technologies with comparable metrics in power consumption and environmental robustness. QP offers greater design versatility for 2D, 2.5D and 3D systems with minimum impact to cost and manufacturing complexity. QP technology is designed to be both complimentary or as a replacement for existing integration approaches such as wirebonds, flip-chip interposers, and TSV chip stacking.

Shortened chip-to-chip signal paths and minimum transitions enabled through QP allows for an insertion loss of less than 0.2 dB up to 110 GHz and less than 1 dB up to 220 GHz. QP I/O connections can be scaled down to a pitch of 10 μm or smaller with a digital bandwidth of 12-100 Gbps per nodule interconnect. This allows for a digital performance of beyond terabits/sec per mm of chip edge space, which is a significant improvement in performance over TSV and flip-chip approaches. In terms of power efficiency, QP is superior to standard off-package approaches, and is comparable to emerging 3D TSV-based integration technologies such as Hybrid Memory Cube (HMC) and High Bandwidth Memory (HBM) at less than 10 mW/Gbps. Also, the nodule-socket interfaces were tested and the electrical continuity across multiple nodule-socket interconnects was proven. The initial data from these reliability tests demonstrates the electrical and mechanical robustness of the QP interconnect structures.

The BEOL QP fabrication process in Si is very mature with optimizations in QP BEOL processing of GaAs, InP, SiC, SiGe, and GaN currently being developed or under way. The QP BEOL process has also been successfully demonstrated with NDNF's 2 μm CMOS processed wafers, with a potential processing issue that might otherwise affect the functionality of the on-wafer CMOS devices identified with steps implemented to prevent it. In addition, preliminary tooling and processes were developed for the assembly of 3D QP chipsets. The uniquely self-aligning and customizable QP nodules enable high-volume assembly of 3D microsystem architectures without the need for a complex or expensive tooling procedure.

ACKNOWLEDGMENT

We gratefully acknowledge the U. S. Department of Energy Office of Science (project number: DE-SC0017933) for their support of this work.

REFERENCES

[1] Knickerbocker, J. U. et al., "2.5D and 3D technology challenges and test vehicle demonstrations", 2012 IEEE 62nd Electronic Components and Technology Conference, pp.1068-1076, 2012.

[2] "Assembly and Packaging." The International Technology Roadmap for Semiconductors, 2011.

[3] G. H. Bernstein, Q. Liu, M. Yan, Z. Sun, W. Porod, G. Snider, and P. Fay, "Quilt Packaging: High-Density, High-Speed Interchip Communications," IEEE Trans. Adv. Packaging, vol. 30, no. 4, pp. 731-740, 2007.

[4] J. M. Kulick, and G. H. Bernstein, "Quilt Packaging: A Revolutionary and Flexible Approach to High-Performance System in Package," Advancing Microelectronics, 39(2), March/April 2012.

[5] Q. Liu, P. Fay, and G. H. Bernstein, "A Novel Scheme for Wide Bandwidth Chip-to-Chip Communications," J. Microelectronics and Electronics Packaging, vol. 4, no. 1, pp. 1-7, 2007.

[6] D. Kopp, C. Liang, J. Kulick, M. Khan, G. H. Bernstein, and P. Fay, "Quilt Packaging of RF Systems with Ultrawide Bandwidth," Proc. of the IMAPS - Advanced Technology Workshop on RF and Microwave Packaging, San Diego, 2009.

[7] P. Fay, D. Kopp, T. Lu, D. Neal, G.H. Bernstein and J.M. Kulick, "Ultrawide Bandwidth Chip-to-Chip Interconnects for III-V MMICs," IEEE Microwaves and Wireless Components Letters, Vol. PP., Issue 99, Nov. 2013.

[8] U.S. Patents #7608919, 7612443, 8021965, 8623700

[9] T. Lu, J. Kulick, J. Lannon, G.H. Bernstein, and P. Fay. 2016. "Heterogeneous Microwave and Millimeter-wave System Integration Using Quilt Packaging," Proc. of Microwave Symposium (IMS), 2016 IEEE MTT-S International. May 2016.

[10] T. Lu, J. Kulick, C. Ortega, G.H. Bernstein, S. Ardisson, R. Engelhardt, "Rapid SoC Prototyping Utilizing Quilt Packaging Technology for Modular Functional IC Block Partitioning," Proc. of 27th Internat. Symp. on Rapid System Prototyping, Oct. 2016, pp. 79-85

[11] S. Gupta, M. Hilbert, S. Hong, R. Patti, "Techniquesfor Producing 3D ICs with High-Density Interconnect," Tezzaron Semi. Naperville, IL 2005.

[12] https://atlas.slac.stanford.edu

[13] https://epp.slac.stanford.edu/research/silicon-detector

[14] https://www.chess.cornell.edu/index.php/about/news/particle-physics-detector-makes-way-upgrade

This page intentionally left blank.

Carbonized Electrodes for Electrochemical Sensing

Mohammad Aminul Haque and Nicole McFarlane
Electrical Engineering and Computer Science
The University of Tennessee, Knoxville
Knoxville, TN, USA
mhaque4@vols.utk.edu

Nickolay V. Lavrik and Dale Hensley
The Center for Nanophase Materials Sciences
Oak Ridge National Laboratory
Oak Ridge, TN, USA
lavriknv@ornl.gov

Abstract—**We have fabricated carbonized polymeric 3-D structures on silica substrate and carbonized them within CMOS operating temperature regime towards obtaining an integrated lab-on-CMOS electrochemical sensor. Metal layers of Ti and Au were deposited on silica substrate to provide electrical contact as well as expedite the formation of electrodes on the substrate. Polymeric conical structures were fabricated on metalized silica substrate using 3-D laser writing based on 2-photon polymerization. Desired carbonization of polymeric structures was obtained using a two step annealing process in oxidative and inert environments. Scanning electron microscopy was used to observe structure morphology and Raman spectroscopy verified carbonization. Finally, electrochemical and impedance characterization of the carbonized electrodes was carried out. Experimental results show the potential of these carbonized electrodes to be used in building low-cost and monolithic CMOS electrochemical sensors.**

Index Terms—**lab-on-CMOS, polymer, carbonization, chronoamperometry, sensor**

I. Introduction

Carbon based electrochemical biosensors are known for their low cost, high sensitivity and applicability to a wide range of analytes. An ideal approach is to coat the electrode surface with carbon nanomaterial where a large number of defect sites are present [1]. However, readout electronics for most of these approaches entail circuit boards separate from the electrode interface. Therefore, this work aims to overcome this limitation towards building up an integrated lab-on-CMOS electrochemical sensing system.

Diamond, carbon nanotubes, graphite, and carbon nanofibers are some forms of carbons that has been used as microelectrodes in electrochemical sensing [2]. Carbon nanotubes have been used in sensing biomolecules, such as oligonucleotides, oligopeptides, and proteins [3]–[6]. Modification by CNT has also enabled detection of glucose [7], NADH [8] and hydrogen peroxide [9]. Graphite electrode has been used in detecting hydrogen peroxide [10] and glucose [11]. Vertically oriented graphene flakes on silicon substrate have shown excellent electrochemical sensitivity in determining dopamine, ascorbic acid, and uric acid [12]. Chemically synthesized graphene nanosheets showed higher conductivity, better signal-to-noise ratio, and sensitivity in detecting serotonin and dopamine compared to single walled carbon nanotube [13]. Graphene-based electrodes are

also sensitive to hydrogen peroxide [14] and NADH [15]. Diamond has high conductivity in a degenerate state, and is biocompatible and electrochemically stable in extreme environments [16]. Diamond electrodes have been used in glucose [16], trinitrotoluene (TNT) [17], and DNA sensing [18]. Carbon nanofibers (CNF), structures of which can be described as a stack of graphene cones or cups [19], are useful for gas storage systems [20], scanning probe microscopy [21] and also been utilized in sensing H_2O_2 [22], ethanol [2], and phenol [23] as well as cell impedance sensing [24]. Carbon nanospikes can be grown on gold, silver, and titanium wires and silicon wafer and they are also promising in the electrochemical sensing [25], [26].

These applications show the efficacy of carbon-based electrodes in electrochemical sensing and biosensing. A CMOS lab-on-a-chip solution can process the electric signals from these electrodes in an integrated low-cost and commercially available environment. However, interconnects and parasitic impedances create challenges in designing readout circuits to interface these sensing electrodes within the same chip. To this point, we seek to establish a technological approach to electrochemically active electrodes on a platform suitable for integration with CMOS based readout electronics. Our approach involves 3D printing based on 2-photon polymerization. More specifically, we utilize a commercially available Photonic Professional GT (Nanoscribe GmbH) tool which offers an excellent opportunity to additively create on-chip 3D polymeric structures with the smallest feature sizes down to approximately 0.5 μm. It should be emphasized that carbonization of polymeric precursors entails temperature treatment beyond 700 ^{0}C. However, performance of CMOS circuitry deteriorates if the temperature exceeds 525 ^{0}C. Therefore, this work aims to solve this problem through lower temperature carbonization of polymer structures using CMOS compatible processing sequence. A successful carbonization can be verified by a response in presence of electrochemically active agents.

In the following sections we describe our implemented fabrication sequence (section II), discuss our characterization results (section III), and outline future work with a concluding note (section IV).

A portion of this research was conducted at the Center for Nanophase Materials Sciences, which is a DOE Office of Science User Facility.

978-1-7281-1500-9/19 $31.00 © 2019 IEEE

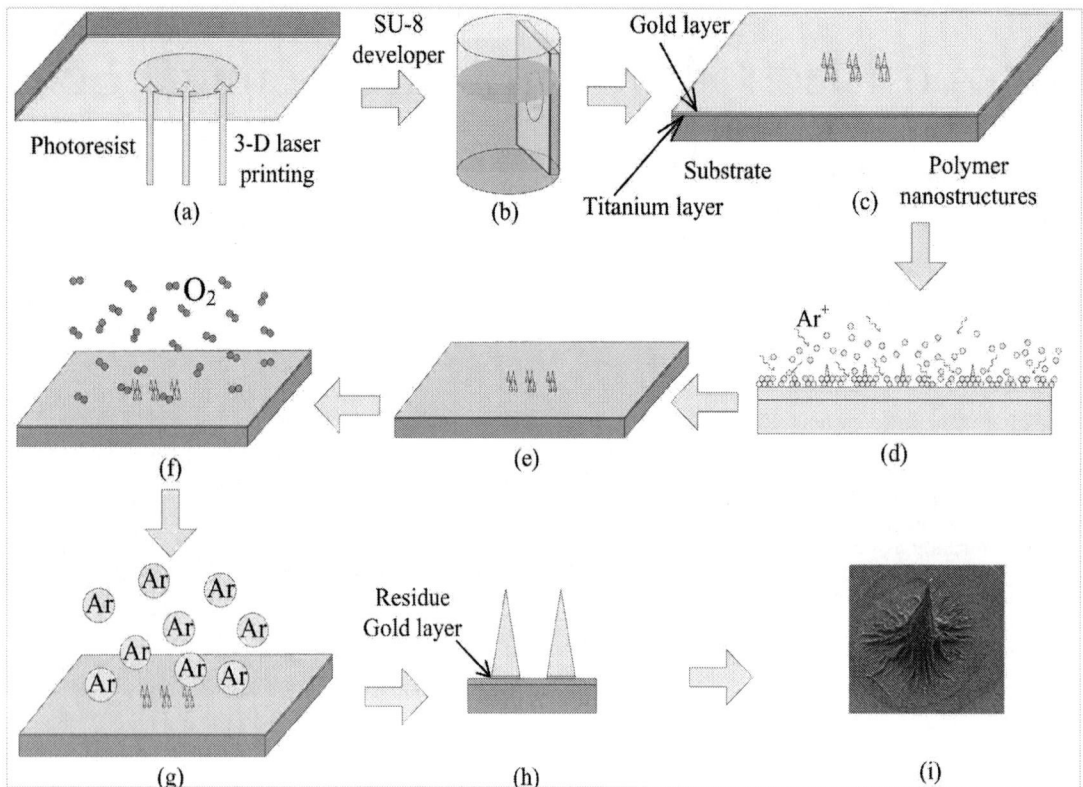

Fig. 1. Sequential steps for fabricating carbonized polymeric conical array on a metalized silica wafer. (a) 3D direct laser writing using on 2-photon polymerization. (b) Developing structures by washing away uncrosslinked IP-S in SU-8 developer. (c) Obtained polymeric cone arrays on metalized silica substrate after developing. (d) Removing gold layer using argon plasma sputter etch to expose titanium. (e) 3-D polymeric cone arrays on silica substrate with exposed titanium layer. (f) Oxidative anneal to partially oxidize titanium. (g) High temperature anneal in argon to achieve carbonization. (h) Schematic to show cross sectional layers after annealing, structures are shrinked due loss of organic material during annealing. (i) SEM image of a single carbonized polyer cone.

II. EXPERIMENTAL STEPS

A. Sequential Steps of Fabrication

Both silicon and fused silica with a metal stack on top were used as substrates for subsequent formation of 3D polymeric structure.Organic solvents were used to clean the substrates ultrasonically before metalization. Two metalization steps were performed to obtain a 50 nm gold layer on top of a 100 nm titanium layer. Metalization was performed in a dual gun electron beam evaporation chamber. After metalization, the wafer was diced using Accretech dicing saw to form 12.5 mm × 12.5 mm chips.

A Photonic Professional GT (Nanoscribe GmbH) 2-photon polymerization (2PP) tool was used to form the structures on the metalized substrate. A drop of IP-S was used on the 12.5 mm × 12.5 mm silica substrate for structure formation followed by laser beam scanning. IP-S is an acrylate based photopolymer supplied by vendor. 2-photon polymerization takes place as the focused laser beam scans across the photopolymer. This process enables spatially selective crosslinking of the precursor. Submicron resolution can be achieved through controlling and optimizing the ultrashort laser pulses using the built in user-friendly software environment [27].

We tried to form polymer structures of different heights and shapes on silica substrate having only one 100 nm thick layer of titanium during the primary experiments. However, we observed cavitation (bubble formation) while the laser beam was scanning in close proximity to the substrate metalized with 100 nm titanium. This phenomenon destabilized the anchoring of polymeric cone structures thereby making the titanium coated substrates unsuitable for fabrication of 3D structures using 2-photon polymerization. The observed cavitation can be explained by the fact that titanium adsorbs laser beam energy in the near-infrared (NIR) producing heat and/or shock waves and ultimately destabilizing the polymer structures. To solve this problem, a gold layer of 50 nm thickness was deposited over the titanium layer to minimize laser absorption by increasing reflectivity of the substrate in NIR. Indeed, polymeric 3D structures were successfully formed on the substrates having two layers of metalization.

After the photopolymer was exposed to the laser, the substrates were soaked in SU-8 developer for 20 minutes and rinsed with isopropanol so that the uncrosslinked IP-S gets washed away. An Oxford Plasma Tech reactive ion etcher was used for plasma sputter etching for 1 minute and 15 seconds

Fig. 2. Scanning electron microscopic image of conical structure fabricated using direct laser writing.

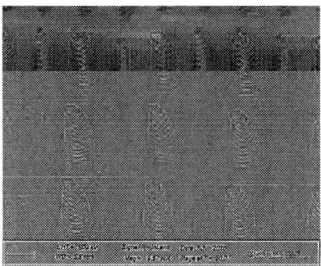

Fig. 3. An array of typical polymeric cones fabricated by 3-D printing using 2-photon polymerization. Distance among the cones and shapes of them can be modified in the CAD.

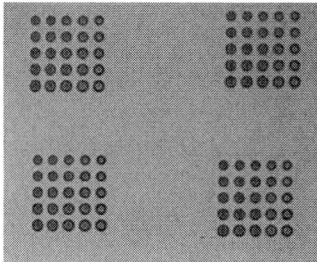

Fig. 4. Optical image of multiple arrays of as-printed conical pillars. This test sample includes polymer cones of varying diameters and heights.

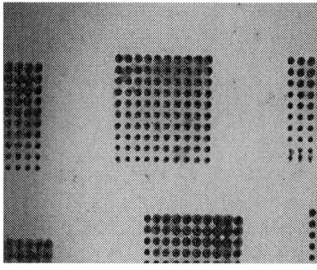

Fig. 5. Optical image of an array of conical pillar structures after processing through two step annealing. Dark mark circling each cone indicates the change of morphology due to polymer spreading during high temperature anneal.

to remove the gold layer so that a layer of insulator can be formed by partially oxidizing the exposed titanium. This leaves only the carbonized structure exposed and sensitive to electrochemically active solution. Following the removal of gold layer, the wafer was processed through oxidation and annealing for pyrolysis.

A two step annealing procedure was followed to carbonize polymeric cone arrays. It started with an oxidative anneal for 10 minutes at 450 ^{0}C to create a titanium oxide insulator around the polymeric structures as well as to expedite the polymer carbonization which ran for 15 minutes at 500 ^{0}C in argon. The overall sequence of the processing steps is shown schematically in Fig. 1.

B. Scanning Electron Microscopy (SEM) to Observe Structure Morphology

We used both a secondary electron detector and an in-lens detector to image the microstructures. Fig. 2 shows an image of a polymeric conical structure obtained using a Zeiss Merlin SEM and an array of such polymers fabricated using Nanoscribe is shown Fig. 3. One single array was first fabricated to verify and troubleshoot the optimized structure. Later on multiple arrays were formed since multiple arrays increase the signal-to-noise ratio and sensitivity of the sensor. Fig. 4 shows an optical microscopy image of such arrays with different heights and diameters.

Fig. 6 and Fig. 7 show SEM images of the cone shaped polymer structure after completing pyrolysis in oxygen and argon. The effect of high temperature annealing is observed as the shrinkage of the structures. Reflow of the polymer induced

by high temperature in the early stage of pyrolysis could also account for the root-like shape at the base of structures.

III. CHARACTERIZATION

A. Verification of Carbonization using Raman Spectroscopy

An InVia Raman microscope (Renishaw, Inc.) was used to verify the carbonization of polymer microstructures through Raman spectroscopy. This tool is equipped with HeNe laser that provides excitation at 632 nm with maximum power at the sample of about 10 mW. Morphological information of the sample, such as structural defects, sp^2 and sp^3 hybridization in C-C bonds, and bond disorder can be studied from Raman spectrum. The Raman spectra of the cone structures before and after annealing is shown in Fig. 8. As can be seen in Fig. 8, thermal annealing caused a significant reduction in the

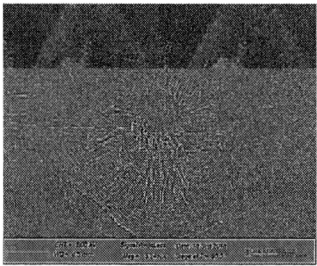

Fig. 6. A single carbonized structure imaged using in-lens detector. The root-like structure with a wide base at the bottom is characteristic of the polymer reflow followed by its shrinking during annealing.

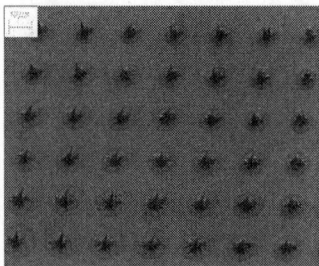

Fig. 7. SEM image of a typical carbonized polymeric array. Pitch length can be optimized depending on user preference and application through CAD programming.

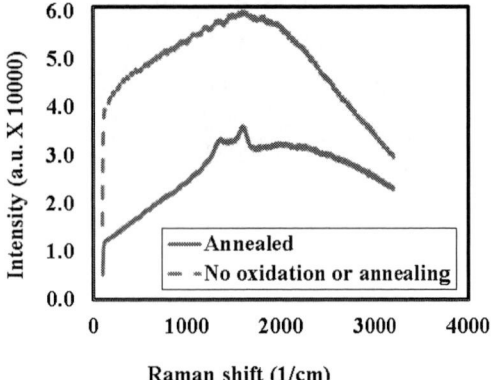

Fig. 8. Raman spectroscopy of carbonized structures on metalized silica substrate compared with substrates at different annealing steps. Laser power was 50% with exposure time of 10 s. Peaks for D-band and G-band verify successful carbonization. Ratio of the peaks of D-band and G-band ($\sim$0.992) confirms plentiful defect sites available for electrochemical reactions. Partially carbonized organic material accounts for the presence of strong broad background in the spectrum.

background signal associated with broadband fluorescence in the photopolymer.

The rise of the D-band and G-band attributable to carbon can be clearly seen in the Raman spectra of the annealed cone structures. However, a substantial broad band signal is still present in the Raman spectrum of the annealed cones. This indicates that our annealing procedures is likely to yield incompletely pyrolyzed material. On the other hand, the presence of both D and G bands confirms the existence of carbon-carbon bonds in multiple arrangements. The G-band comes from planar graphite sheets whereas the D-band (defect band) corresponds to hybridized vibrational modes. A peak ratio of $\sim$0.992 corresponding to the spectrum shown in Fig. 8 is consistent with the presence of numerous defect sites which stem from dangling bonds. These bonds (where acidic and basic groups are available) promote charge transfer on the surface making it electrochemically active and suitable for sensing application [25].

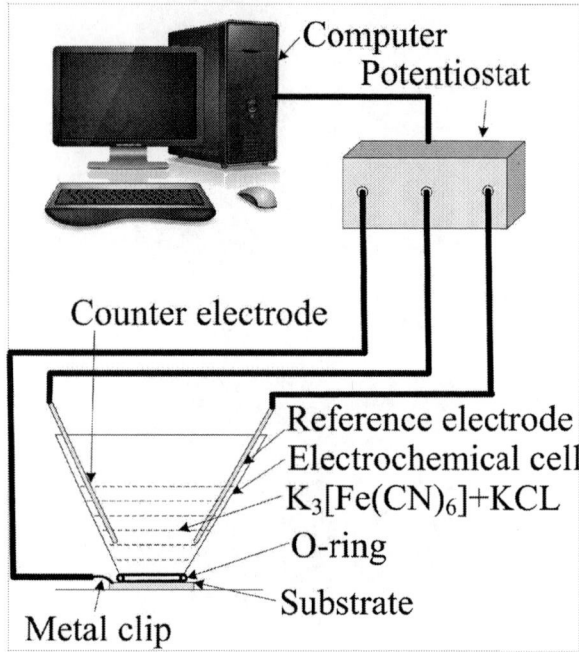

Fig. 9. Standard three electrode electrochemical cell was used to characterize electrochemical activity of carbonized polymer structures. A 3-D printed hollow cell is placed on top of the metalized substrate with carbonized structures. Reference and counter electrodes are placed inside the cell. A metal clip connects the substrate to the working electrode lead of the computer controlled potentiostat.

B. Electrochemical Testing

Electrochemical testing of the carbonized electrodes has been done using a standard 3-electrode electrochemical test setup. The test setup is shown schematically in Fig. 9. The chip containing the carbonized electrodes is placed under a hollow electrochemical cell. The cell was sealed against the chip wafer using an elastomer O-ring. A Ag/AgCl electrode and a platinum wire were used as a reference and a counter electrodes, respectively. A metal clip was used to scratch the insulating oxide and form an electrical contact with the titanium layer. The reference electrode, counter electrode, and the metal clip were connected to the computer controlled potentiostat (Parstate 2273). The counter electrode and working electrode were placed in vicinity of the substrate with carbonized polymer structures to minimize cell impedance.

Chronoamperometric and impedance measurements were carried out to study electrochemical responses caused by presence of carbonized structures in the presence of $K_3[Fe(CN)_6]$ which is used as a model electrochemically active agent added to the background electrolyte. In our preliminary tests, we compared the impedance and chronoamperometric behaviors of the chips with and without carbonized structures (Fig. 10 and Fig. 11). In the chronoamperometric mode, the potential between the working electrode and the reference Ag/AgCl electrode was maintained at 0.8 V. During the impedance measurements the amplitude of the probing AC voltage was 10 mV. The concentrations of $K_3[Fe(CN)_6]$ and KCl were 10

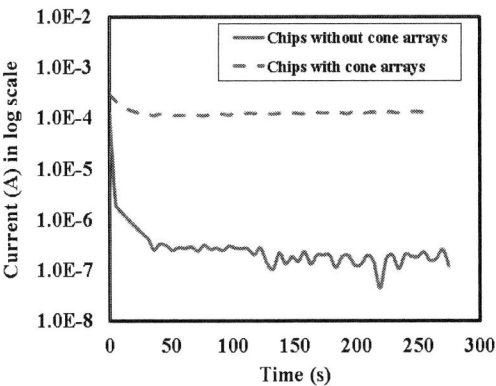

Fig. 10. Chronoamperometric response of a sample with carbonized nanostructure and a sample with no carbonized structure in 10 mM $K_3[Fe(CN)_6]$ and 3.5 M KCl. Potential step was $E_0=0.199$ V to $E_1=0.8$ V

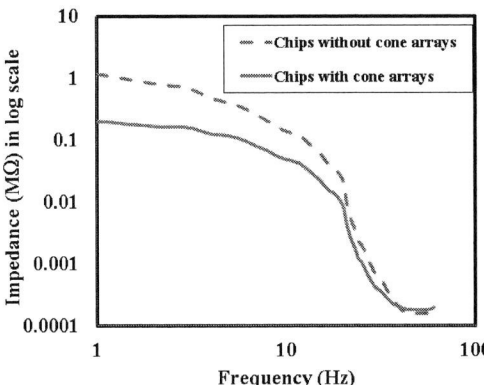

Fig. 11. Impedance spectra of the chips with and without carbonized structures measured in 1 M KCl with addition of 10 mM $K_3[Fe(CN)_6]$. Potential between the working electrode and the reference Ag/AgCl electrode was maintained at 0.8 V.

mM and 1 M respectively.

As can be seen in Fig. 10, the steady state current measured in the chronoamperometric mode is approximately 100 nA when no carbonized structures are present. The current is almost 3 orders of magnitude higher when an array of carbonized structures is present on the chip. This proves that the electrochemical reaction associated with the measured current occurs predominantly on the surface of the carbonized structures while the surrounding area is significantly less electrochemically active due to the presence of an electrically insulating TiO_2 layer. A similar conclusion can be drawn from the impedance spectra shown in Fig. 11. While the high frequency parts of the two spectra converge due to the dominant contribution from the bulk electrolyte, there is nearly an order of magnitude difference at the frequencies below 1 Hz. The latter confirms that the low frequency impedance is dominated by a contribution from active current induced by interfacial electrochemical reactions when carbonized structures are present.

IV. CONCLUSION

This work reports on the successful fabrication and preliminary characterizations of carbon based microelectrodes fabricated by using 2-photon polymerization followed by thermal annealing compatible with CMOS-processing. Raman spectroscopy confirms the presence of carbonized material. Temperature and timing of the annealing process was selected to achieve an acceptable level of carbonization without exceeding the CMOS-compatible temperature window. Our present findings mark a n important step in successful implementation of monolithic lab-on-CMOS low cost and implantable electrochemical platform system for future sensing devices. Our further work in this area will focus on more detailed electrochemical testing of the prepared carbonized structures to verify their satisfactory sensing performance with a goal of developing a complete lab-on-CMOS chip system.

REFERENCES

[1] A. G. Zestos, C. Yang, C. B. Jacobs, D. Hensley, and B. J. Venton, "Carbon nanospikes grown on metal wires as microelectrode sensors for dopamine," *Analyst*, vol. 140, no. 21, pp. 7283-7292, 2015..

[2] J. Huang, Y. Liu, and T. You, "Carbon nanofiber based electrochemical biosensors: A review," *Analytical Methods*, vol. 2, no. 3, pp. 202-211, 2010.

[3] R. J. Chen, S. Bangsaruntip, K. A. Drouvalakis, N. W. S. Kam, M. Shim, Y. Li, , W. Kim, P. J. Utz, and H. Da, "Noncovalent functionalization of carbon nanotubes for highly specific electronic biosensors," *Proceedings of the National Academy of Sciences*, vol. 100, no. 9, pp. 4984-4989, 2003.

[4] K. Besteman, J.-O. Lee, F. G. Wiertz, H. A. Heering, and C. Dekker, "Enzyme-coated carbon nanotubes as single-molecule biosensors," *Nano Letters*, vol. 3, no. 6, pp. 727-730, 2003.

[5] A. Star, J.-C. P. Gabriel, K. Bradley, and G. Grner, "Electronic detection of specific protein binding using nanotube FET devices," *Nano Letters*, vol. 3, no. 4, pp. 459-463, 2003.

[6] S. Boussaad, N. Tao, R. Zhang, T. Hopson, and L. Nagahara, "In situ detection of cytochrome c adsorption with single walled carbon nanotube device," *Chemical Communications*, no. 13, pp. 1502-1503, 2003.

[7] M. Gao, L. Dai, and G. G. Wallace, "Biosensors based on aligned carbon nanotubes coated with inherently conducting polymers," *Electroanalysis: An International Journal Devoted to Fundamental and Practical Aspects of Electroanalysis*, vol. 15, no. 13, pp. 1089-1094, 2003.

[8] M. Musameh, J. Wang, A. Merkoci, and Y. Lin, "Low-potential stable NADH detection at carbon-nanotube-modified glassy carbon electrodes," *Electrochemistry Communications*, vol. 4, no. 10, pp. 743-746, 2002.

[9] J. Wang, M. Musameh, and Y. Lin, "Solubilization of carbon nanotubes by Nafion toward the preparation of amperometric biosensors," *Journal of the American Chemical Society*, vol. 125, no. 9, pp. 2408-2409, 2003.

[10] G. Jnsson and L. Gorton, "An electrochemical sensor for hydrogen peroxide based on peroxidase adsorbed on a spectrographic graphite electrode," *Electroanalysis*, vol. 1, no. 5, pp. 465-468, 1989.

[11] J. Lu, L. T. Drzal, R. M. Worden, and I. Lee, "Simple fabrication of a highly sensitive glucose biosensor using enzymes immobilized in exfoliated graphite nanoplatelets nafion membrane," *Chemistry of Materials*, vol. 19, no. 25, pp. 6240-6246, 2007.

[12] N. G. Shang, P. Papakonstantinou, M. McMullan, M. Chu, A. Stamboulis, A. Potenza, S. S. Dhesi, and H. Marchetto, "Catalystfree efficient growth, orientation and biosensing properties of multilayer graphene nanoflake films with sharp edge planes," *Advanced functional materials*, vol. 18, no. 21, pp. 3506-3514, 2008.

[13] S. Alwarappan, A. Erdem, C. Liu, and C.-Z. Li, "Probing the electrochemical properties of graphene nanosheets for biosensing applications," *The Journal of Physical Chemistry C*, vol. 113, no. 20, pp. 8853-8857, 2009.

[14] M. Zhou, Y. Zhai, and S. Dong, "Electrochemical sensing and biosensing platform based on chemically reduced graphene oxide," *Analytical chemistry*, vol. 81, no. 14, pp. 5603-5613, 2009.

[15] L. Tang, Y. Wang, Y. Li, H. Feng, J. Lu, and J. Li, "Preparation, structure, and electrochemical properties of reduced graphene sheet films," *Advanced Functional Materials,* vol. 19, no. 17, pp. 2782-2789, 2009.

[16] C. E. Troupe, I. C. Drummond, C. Graham, J. Grice, P. John, J. I. B. Wilson, M.G. Jubber, and N.A. Morrison, "Diamond-based glucose sensors," *Diamond and Related Materials,* vol. 7, no. 2-5, pp. 575-580, 1998.

[17] J. De Sanoit, E. Vanhove, P. Mailley, and P. Bergonzo, "Electrochemical diamond sensors for TNT detection in water," *Electrochimica Acta,* vol. 54, no. 24, pp. 5688-5693, 2009.

[18] N. Yang, H. Uetsuka, E. Osawa, and C. E. Nebel, "Vertically aligned diamond nanowires for DNA sensing," *Angewandte Chemie International Edition,* vol. 47, no. 28, pp. 5183-5185, 2008.

[19] V. Vamvakaki, K. Tsagaraki, and N. Chaniotakis, "Carbon nanofiber-based glucose biosensor," *Analytical chemistry,* vol. 78, no. 15, pp. 5538-5542, 2006.

[20] D. J. Browning, M. L. Gerrard, J. B. Lakeman, I. M. Mellor, R. J. Mortimer, and M. C. Turpin, "Studies into the storage of hydrogen in carbon nanofibers: proposal of a possible reaction mechanism," *Nano Letters,* vol. 2, no. 3, pp. 201-205, 2002.

[21] H. Cui, S. Kalinin, X. Yang, and D. Lowndes, "Growth of carbon nanofibers on tipless cantilevers for high resolution topography and magnetic force imaging," *Nano Letters,* vol. 4, no. 11, pp. 2157-2161, 2004.

[22] L. Wu, X. Zhang, and H. Ju, "Highly sensitive flow injection detection of hydrogen peroxide with high throughput using a carbon nanofiber-modified electrode," *Analyst,* vol. 132, no. 5, pp. 406-408, 2007.

[23] J. Zhang, J. Lei, Y. Liu, J. Zhao, and H. Ju, "Highly sensitive amperometric biosensors for phenols based on polyanilineionic liquidcarbon nanofiber composite," *Biosensors and Bioelectronics,* vol. 24, no. 7, pp. 1858-1863, 2009.

[24] Y. Yu, K. A. Al Mamun, A. S. Shanta, S. K. Islam, and N. McFarlane, "Vertically aligned carbon nanofibers as a cell impedance sensor," *IEEE Transactions on Nanotechnology,* vol. 15, no. 6, pp. 856-861, 2016.

[25] S. A. Shanta, S. Shamsir, Y. Song, D. K. Hensley, A. J. Rondinone, S. K. Islam, and N. McFarlane, "Carbon Nanospikes on Silicon Wafer for Amperometric Biosensing Applications," *IEEE Engineering in Medicine and Biology Society,* pp. 4281-4284, 2018.

[26] A. S. Shanta, K. A. Al Mamun, D. Hensley, N. V. Lavrik, S. K. Islam, and N. McFarlane, "Carbon nanospikes for biosensing applications," *IEEE Engineering in Medicine and Biology Society,* pp. 193-196, 2017.

[27] A. Ostendorf and B. N. Chichkov, "Two-photon polymerization: a new approach to micromachining," *Photonics spectra,* vol. 40, no. 10, p. 72, 2006.

Moldability challenges associated with the assembly of thicker IC packages for high voltage and power applications

Sadia Naseem[1], Jack Chiang[2], Megan Chang[2], Bob Lee[2], and Jason Chien[2]

Texas Instruments
[1]Dallas, Texas USA
[2]Taipei, Taiwan
s-naseem@ti.com

Abstract— **Semiconductor technology platforms such as high power and voltage are pushing for greater current carrying capacity and overall functionality improvements by utilizing thicker leadframes, more components per device, and increased wire diameters. Incorporation of these materials and components necessitates thicker packages compared to most IC leaded packages on the market today. The manufacturability however, specifically the moldability of packages thicker than ~3mm continues to be a challenge. Discussed here are the motivations behind designing thicker packages, the associated manufacturability challenges, and methods for optimizing mold tooling and package designs to mitigate IC assembly risks.**

Keywords-high voltage, moldability, voids, wire sweep

I. INTRODUCTION

Semiconductor and electronics industries are generally pushing for smaller form factors, footprints, and overall miniaturization of devices. However, there remain some key performance and functional limitations to how small IC packages can become for high power and high voltage applications. Designing competitive products in these technology areas may require incorporation of more components within an IC package, creative solutions for increasing the overall current carrying capacity of a device, as well as ensuring that the key components and circuitry are sufficiently surrounded by insulative packaging materials. All of these design factors promote the need for larger or thicker packages.

Inclusion of more semiconductor chips or passives in a package naturally requires more space and thus larger final product dimensions. Achieving greater current carrying capacity is typically done through physically increasing the amount of conductive metals within the package. This could include utilization of thicker leadframes that the chips and wirebonds will adhere to, as well as usage of larger diameter wires for interconnection. More metal within a high performance package will physically require more surrounding insulation as well.

Industry safety specifications regulate the amount of required insulation per device with respect to its target operating voltage. Often the technical standards report a minimum required creepage, which is defined as the shortest distance between two opposite electrical potential conductors along the surface of the insulation to which they are fixed [1]. Figure 1 schematically represents how the creepage distance is measured across an IC package. Typically, the required creepage distance of a device increases with higher operating voltage, thus driving the need for larger packages.

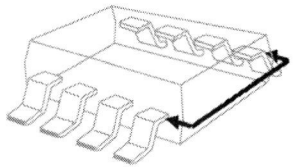

Figure 1. Schematic representation of a creepage measurement on an IC package

II. ASSEMBLY MATERIALS AND PROCESS

In semiconductor packaging, typically an epoxy molding compound is used as the encapsulating and insulating material to surround and protect the delicate circuitry and device components. A transfer molding process continues to be among the most common and cost-effective methods used for wirebonded IC packages. This usually requires the thermosetting mold to start off as a powder that is then compressed under pressure to form pellets or capsules. When leadframe strips arrive to the molding stage, they are clamped in the mold chamber. Mold pellets are preheated and then inserted into the high temperature apparatus that plunges and transfers the melting material across runners, built in mold gates, and ultimately to the package cavity. The mold material is then cured and returns to a solid phase, taking the shape of the desired final package. Figure 2 summarizes the transfer molding process.

978-1-7281-1500-9/19 $31.00 © 2019 IEEE

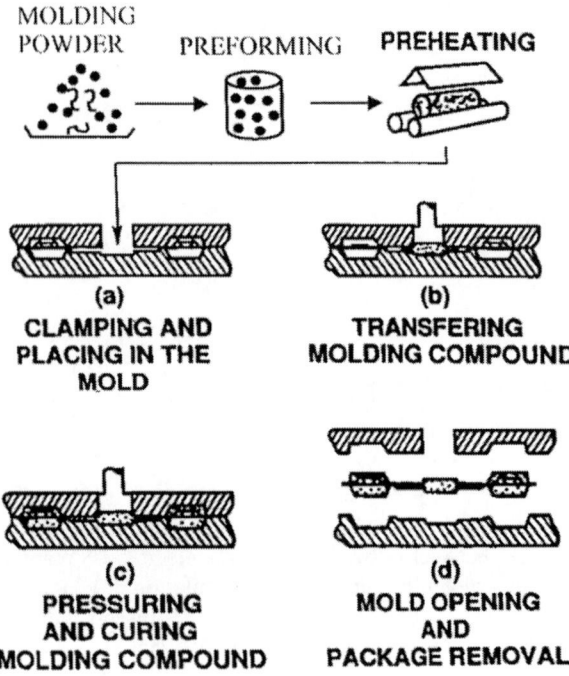

Figure 2. Typical transfer molding process [2]

III. MANUFACTURABILITY CHALLENGES

A. Voiding and Physical Anomalies

Thicker packages naturally have a larger mold cavity that they must fill in approximately the same amount of time that traditional thinner packages are molded, because of the thermal and physical property limitations of the molding material. As described, mold compound must transition from pelletized solid material to viscous liquid that can flow across several rows of a leadframe strip, before solidifying again within a tight time window. Larger packages require more molding material per unit, and therefore units that are further away from the pellet source become more susceptible to incomplete filling, interior voiding, or external physical anomalies.

Topside voiding (Figure 3) was commonly seen on units located in rows furthest from where the pellet is inserted. This is attributed to the fact that the mold material must travel a longer distance from the source to reach the final rows of each leadframe strip. The viscosity of the mold is gradually increasing with time, and thus air is more easily trapped within the cavity in the final rows, resulting in visible air bubbles or voids on the package.

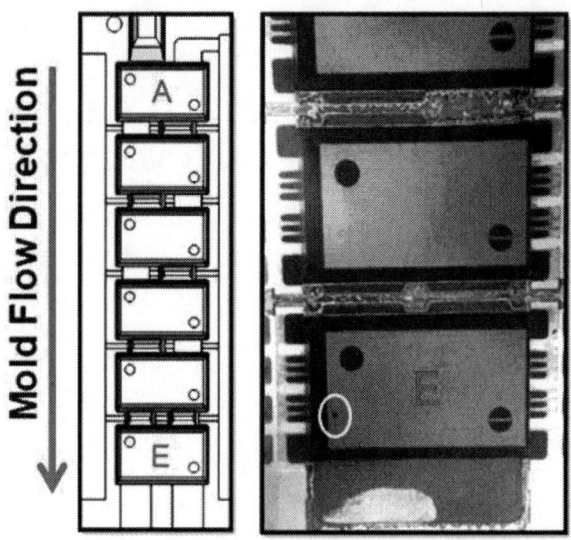

Figure 3. Topside voiding (circled in yellow) seen on units in row E of leadframe strips, furthest from the mold source

Additionally, due to the fact that molding larger packages requires more mold to move across the leadframe quicker, typically the transfer speed and pressure by which pellets are plunged is intentionally increased. However, when molding material is flowing with greater speed and pressure, new challenges arise. Excessive turbulence is induced near the mold gates usually closest to the pellet source, as this is where the greatest volume of material is trying to gush through. This turbulence creates an opportunity for air to get trapped along the side walls of the package, as shown by the visible physical anomalies in Figure 4.

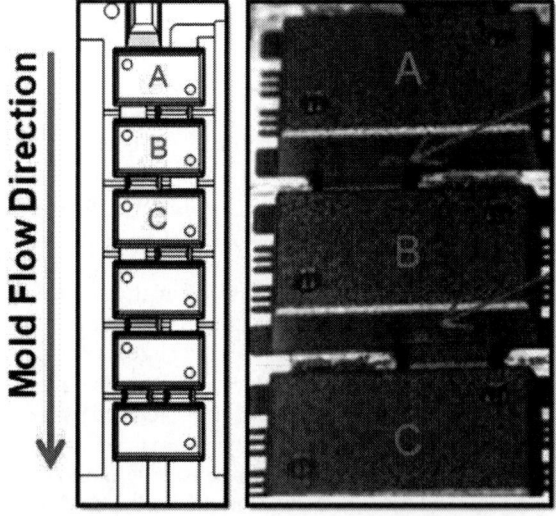

Figure 4. Voiding observed on package sides near the mold gates on units in rows closer to the mold source

These molding defects can pose risks for package reliability, as well as impact the high voltage performance of a device. If air bubbles or voids exist within the final cured package, these locations become susceptible to moisture absorption in humid environments. When these packaged devices are soldered onto a PCB, they can experience temperatures above 200°C during reflow, which is more than enough to cause any trapped moisture within the package to vaporize and induce internal pressures. The vapor pressure can accumulate to the point of creating cracks or crevices, compromising the integrity of the package [3].

Voids within the cured mold can also become a location for charge build up, specifically in high voltage devices. The dielectric constant of air within the void is significantly less than that of the surrounding mold material. This causes the localized electric field in or near the void to be greater than what it would be if the region solely consisted of uniformly molded material. If enough charges accumulate due to voiding in the package, partial discharge can occur across the void, leading to poorer high voltage performance of the respective device.

B. Wire Sweep

Another manufacturability challenge that can arise when molding larger packages is wire sweep. As discussed previously, mold material must flow with increased speed and pressure to swiftly move across the leadframe strip and fill each unit cavity. This can however cause the relatively thin and delicate ~1mil diameter wires within the package to be swept away from their target orientation. Wire sweep, as observed in Figure 5, can directly impact the continuity or functional performance of the device.

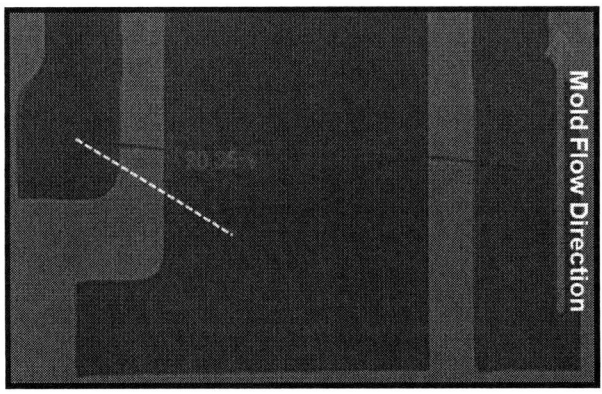

Figure 5. The wire shown has been swept 20.35% when compared to its target orientation shown in yellow

IV. METHODS FOR IMPROVEMENT

A. Mold Tooling Adjustments

For a given leadframe design and target package dimensions, a few key mold tooling adjustments could be made to better control the flow of mold material and mitigate the risk of physical defect formations within or on the surface of IC packages. Firstly, the size and number of mold gates connecting one unit to another directly influences the flowing rate and behavior of the mold across a leadframe strip. As discussed earlier, the leadframe rows closest to the mold source have large volumes of mold trying to gush through the gate. When the cross-sectional area of the mold gate was too small, turbulent flow of material entering a unit cavity resulted in air trapping along small sections of the package side wall, as shown in Figure 6. This effect was mitigated by reducing the transfer speed of the mold, but the tradeoff observed was incomplete filling of units further along the leadframe strip. Therefore, a physical mold gate design change was needed.

The mold gate design was modified to be ~1.5x the depth of the original, increasing the cross-sectional area of each gate region by approximately 1.5x as well. This change allowed a larger volume of mold to be able to flow into a unit per second, and prevented the wave-like turbulent behavior being observed before. Using the narrow gate design, 1-3% of units could be observed to have voiding on the package side walls after inspection of 10 different molded leadframe strips. With the deeper mold gate design schematically represented in Figure 7, 100% of the side wall voiding was mitigated as observed across 30 different molded strips.

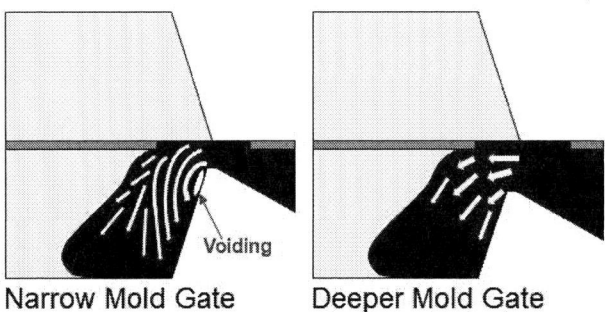

Figure 6. Schematic representation of mold flow as material enters a unit cavity in narrow and deeper mold gate designs

Some package designs experienced mold flow challenges in rows further from the mold source, but here the challenge was very different than in rows close to the mold injection site. As discussed earlier, the mold material becomes more viscous as it travels along the leadframe strip. This could mean that by the time the mold is reaching the final unit in a column, its flow rate has significantly reduced, which can result in incomplete filling or voiding.

To increase the volume of mold entering the final units, one method exercised was increasing the number of mold gates feeding into the last row of each leadframe strip.

For two different packages, A and B, two different mold tooling designs were generated: one tooling design included only single mold gates between all units across a strip, and the second tooling design included single mold gates between all units except for the last row – the last two units in the row were connected with double mold gates. A comparison between these two mold tooling designs for a given package is shown in Figure 7. The final target package volume of design A was ~30% smaller than the final target package volume of design B. The mold flow behavior across the leadframe strip was found to differ between package designs A and B after the inclusion of a second mold gate in the respective final row.

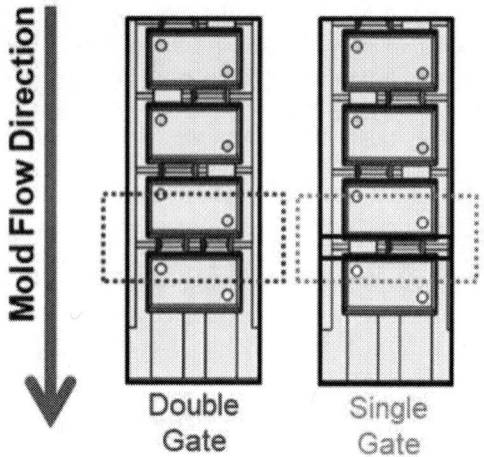

Figure 7. Two different mold tooling designs for a given package. The designs are identical except that the terminal units in each row are connected by two mold gates in one design (red), and one mold gate in the other (green)

In the case of design A, the smaller package design, over 7% of units with a single gate design were observed to have physical defects. After the mold tooling was modified to include a second mold gate to increase mold flow into the final row of units, the rate of mold void occurrence decreased to 0%, where no defects were observed in any row of the package A design.

For the 30% larger package B, inclusion of the second mold gate in the final row of units did not seem to significantly improve the rate of physical defects observed. With a single gate design for package B, voiding was observed on approximately 0.4% of units. After the mold tooling was modified to include a second mold gate in the terminal row, voiding on the package body was still observed on approximately 0.33% of units. The rate of physical defects observed for package designs A and B before and after the mold gate modifications is summarized in Table I.

TABLE I.

Package Design	Mold Void Rate of Occurrence	
	Single Mold Gate Design	*Double Mold Gate Design*
A	~7.6%	0%
B	~0.4%	~0.33%

The final row of units in the larger package B design did not seem to need a higher mold flow-in rate, but rather required a more robust method for air ventilation to prevent air trapping. Air vents are features that can be built into the mold tooling design ultimately to enhance air exhaustion during mold compound injection. Vents can be added into the upper and/or lower mold tooling chase as needed between every unit, or specifically in areas along a given leadframe design that are especially vulnerable to air trapping.

The original mold tooling design for package B already included air vents all along the leadframe strip, but this design evidently was not sufficient for preventing the trapping of air bubbles in the last row of units. The mold chase was redesigned to include a ~3x wider air vent for the final row of units so that any remaining air that had built up could be fully flushed out of the unit cavity. The addition of this larger air vent directly addressed the voiding issue observed in the last row of the package B design, bringing the rate of occurrence to 0% after inspection of 600 molded units.

B. Package Design Modifications

In addition to mold tooling adjustments, there are ways to modify the actual package design that can also improve the overall molding process of larger leaded IC packages. Most IC package bodies are not perfect rectangular prisms but actually have curves and side angles that could influence the moldability or complete filling of the respective cavity design. The corners of larger packages are particularly susceptible to incomplete filling, as mold material is entering each unit from a relatively small gate, and must spread across the large cavity relatively quickly before needing to compress and swiftly exit from another small gate. The central region of each unit cavity fills first and most easily, and the locations along the package fringes are often the most challenging to uniformly mold in a non-optimized design. One method to mitigate this challenge is by changing the draft angle on the target package design. The draft angle of a package is best observed from a side view as seen in Figure 8.

Figure 8. The draft angle, θ, on a given package

Larger draft angles typically result in a reduction of overall package volume, which may create design challenges for high performance applications that require a certain creepage or insulation thickness. However, when kept within the design limits of a given application, a larger draft angle was found to improve the overall moldability of the design. Two different draft angles, C and D, were tested, where angle D was ~2x larger than angle C, and all other package dimensions and design factors remained identical.

The design with draft angle C experienced yield loss due to incomplete filling. 600 units were built with the angle C design, and 59 of them were observed to be incompletely filled. However, the design with the larger draft angle, D, pushed the corners of the package body more toward the center, which helped mold material more easily flow into the package extremities. 1800 units were built using draft angle D, with not a single unit experiencing incomplete mold filling. The rate of incomplete filling for package designs with draft angles C and D is summarized in Table II.

TABLE II.

Draft Angle	Incomplete Mold Fill Rate of Occurrence	Sample Size
C	9.83%	600 units
D	0%	1800 units

Another factor that can cause incomplete filling of mold is the physical presence of several passives and chips in a unit. As discussed previously, high performance applications are pushing the limits in terms of how many components are included in one package. These components typically are attached to the topside of a leadframe, and as the number of components increases, they can begin to impede the flow of mold in the top half region of a given unit cavity. This creates unbalanced mold flow behavior, where mold can move through the bottom half of the cavity at a faster rate than it can travel on the top side, potentially inducing voids or incomplete filling in a unit. Figure 9 schematically shows the difference in component density between the upper and lower halves of a given typical package, and how these components may obstruct mold flow.

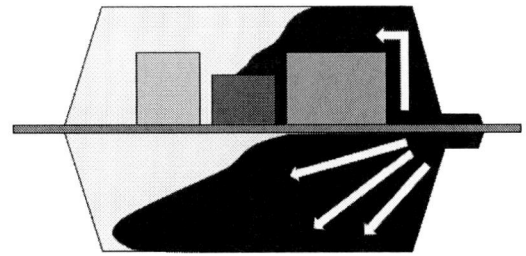

Figure 9. Mold flow comparison above and below the leadframe in a given unit cavity with multiple components

In order to mitigate the risk of unbalanced mold flow, changes can be made to the leadframe design so that the package components sit more centrally in the package instead of entirely in the upper half. The leadframe can be designed to include downset pads below the centerline of a given package for the mounting of Si chips and passives [4].

600 units were built with a typical no-downset leadframe, where all device components sat in the top half of the unit cavity. Approximately 1 % of these units showed signs of unbalanced mold flow and package voiding. 600 units were then built in a mold cavity identical to before, but this time the leadframe design was downset 60% of the total body thickness. In this design (Figure 10) the package components were positioned more centrally in the unit cavity, and no voids or instances of incomplete filling were observed.

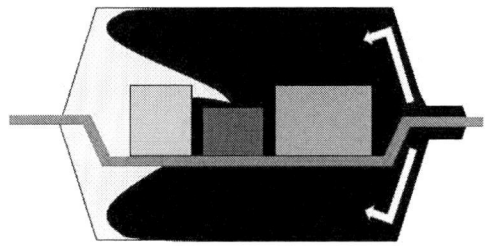

Figure 10. More evenly balanced mold flow on top and bottomside of downset LF in a multi-component package

Not only did downsetting the leadframe mitigate the risk of unbalanced mold flow across the upper and lower regions of the package, it additionally addressed the challenge of wire sweep. By moving the components closer to the centerline of the package, the overall wire length required to bond to the lead fingers can be reduced as the leads are now physically closer to the component bondpads. Shorter wires are typically less susceptible to the drag forces induced by the mold injection process, and thus less likely to be swept away during molding. Figure 11 compares the typical wire loop profiles associated with a downset and traditional leadframe design. Specifically the component to leadframe bonds shown in yellow require a taller loop height in a traditional leadframe design versus a downset leadframe.

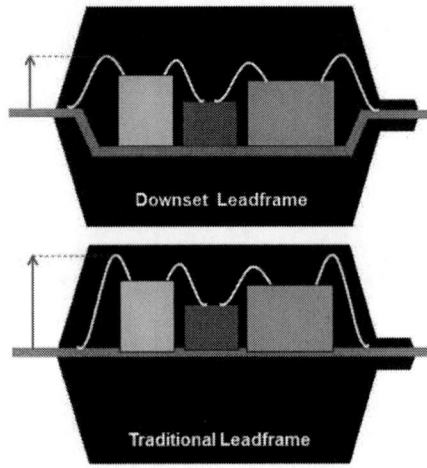

Figure 11. Wire loop height comparison of component to leadframe bonds for downset and traditional leadframe designs

Over 4000 units were built and screened through x-ray inspection for wire sweep. Units built on a traditional leadframe design with no downset were three times more likely to have swept wires than units built with a downset leadframe for the same given package cavity. The average percent wire sweep for the units built on a traditional non-downset leadframe was approximately 4% higher than what would be considered acceptable for most industry specifications. Contrastingly, over 99% of units built on the downset leadframe design showed minimal to no wire sweep.

V. SUMMARY

High voltage and power semiconductor devices require especially robust and reliable packaging, including sufficient insulation, durable interconnections, and minimal to no physical anomalies. Within the package assembly flow, the molding process can be potentially the most disruptive and critical step. The molding epoxy serves as the primary insulative material in a package and thus a non-optimized mold process can result in degradation of the dielectric properties of a device. Further, there can be functional continuity issues if interconnection wires are swept away during mold injection. Finally, the overall package reliability can be compromised if non-uniform molding results in voiding.

In order to mitigate these risks, mold tooling, process parameters, and certain package designs can be tweaked and optimized for a given target package and application. Overall, some combination of larger and more mold gates, increased air ventilation, less components per unit, and a lower set leadframe should help address most thick package moldability challenges, but there is not a universal solution that caters to every package design. These tooling and design modifications each come with non-trivial tradeoffs including turbulent mold flow, incomplete mold filling across the leadframe strip, or higher electrical performance risks. Ultimately, every unique device design will require a balancing of manufacturability and functional performance, and an optimized epoxy molding process is critical. The presented methods for improving the moldability of high performance IC packages can be leveraged in various combinations to specifically meet the needs of a given device application.

REFERENCES

[1] P. v. Schau and W. H. Middendorf, "An International Research Project to Determine New Dimensioning Crules for Creepage Distances," in IEEE Transactions on Electrical Insulation, vol. EI-18, no. 2, pp. 158-162, April 1983. doi: 10.1109/TEI.1983.298661.

[2] C. P. Wong and M. M. Wong, "Recent advances in plastic packaging of flip-chip and multichip modules (MCM) of microelectronics," in IEEE Transactions on Components and Packaging Technologies, vol. 22, no. 1, pp. 21-25, March 1999. doi: 10.1109/6144.759349I.

[3] C.W. Chong, T.F. Guo, and L. Chen, "Popcorn Failure and Unstable Void Growth in Plastic Electronic Packages," in Key Engineering Materials, Vol. 227, pp. 61-66, 2002.

[4] L. D. Kinsman and M. Wolfe, "Downset lead frame for semiconductor packages," U.S. Patent 6246110B1, 12 June, 2001.

Highly compact, multiband composite-right/left-handed(CRLH) transmission line based stub for GPS applications

Hae-In Kim, Seahee Hwangbo, Renuka Bowrothu, and Yong-Kyu Yoon
University of Florida, Gainesville, FL, USA
kimhaein@ufl.edu, ykyoon@ece.ufl.edu

Abstract—A highly compact, multiband composite right/left-handed (CRLH) transmission line based stub for tri-band applications with two GPS (L1, L2) frequencies, and GSM frequency, is demonstrated. The advantages of this work are as follows: 1) The three frequencies are covered by a single $\lambda/4$ stub, providing compactness and reducing system complexity; 2) By using a substrate with a high dielectric constant, additional size reduction is realized. **The lumped element circuit model has been designed using Advanced Design Systems (ADS, Keysight Technologies Inc.) and the parameters are calculated based on the bridged-CRLH (B-CRLH) circuit model.** Distributed components have been designed with **a numerical simulation program, High Frequency Structure Simulator (HFSS, ANSYS Inc.).** The CRLH based quarter wavelength stub is characterized by the frequency-phase dispersion diagram and scattering parameters. **The demonstrated design uses a meander line structure in an M-shape, the so-called M-CRLH, to provide the shunt and series capacitance and inductance, and the bridge inductance.** An open-circuited quarter wavelength ($\lambda/4$) tri-band CRLH transmission line based stub is fabricated and characterized.

Keywords- Composite right/left-hande (CRLH), bridged CRLH (B-CRLH), M shaped CRLH (M-CRLH), metamaterial, Transmission line, Stub, multiband, Global positioning system frequency

I. INTRODUCTION

For portability of a global positioning system (GPS), small form factor and low power consumption are highly desirable. There are various GPS frequency bands for different applications, for example, 1.575 GHz for civilian commercial usage (L1 band), 1.227 GHz for civilian and military usage (L2 band). Often a GPS system equipped with multi-band capability will be advantageous for safety or security purpose which, however, may increase system size. The composite right/left-handed (CRLH) transmission line technology has led to the size reduction of radio frequency (RF)/microwave components [1]. Recent works have been reported on components for the GPS applications with a nominal dimension from 11 mm to 48.67 mm having either dual bands or triple bands including one GPS frequency and the other frequencies [4-8]. Senior et al. reported the bridged composite right/left-handed (B-CRLH) transmission line approach for triple band implementation using a single unit cell [2].

In this work, in the lumped element circuit model of B-CRLH, by adding a bridge inductor between the input and output ports, an additional GSM frequency (865MHz) is achieved. Especially, the modified B-CRLH accommodates a meander line shape to achieve compact RF footprint for triple bands (GSM: 865MHz, GPS L1: 1.575GHz, and GPS L2:1.227GHz) without sacrificing performance. Also, using a

substrate with a high dielectric constant of 10.2 and a thickness of 640 μm helps to decrease the physical dimension. Prototype devices are fabricated using milling machine.

II. COMPOSITE RIGHT/LEFT-HANDED TRANSMISSION LINE THEORY

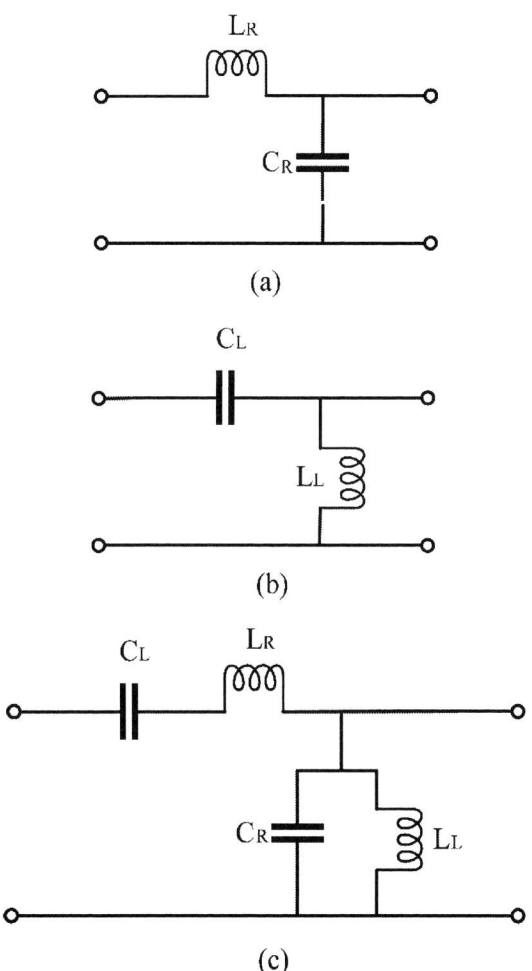

Figure 1. Equivalent lumped element circuit of the transmission line (a) purely right-handed TL, (b) purely left-handed TL, (c) composite right/left-handed TL [1]

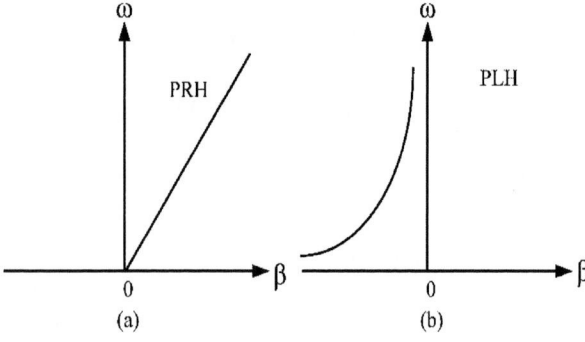

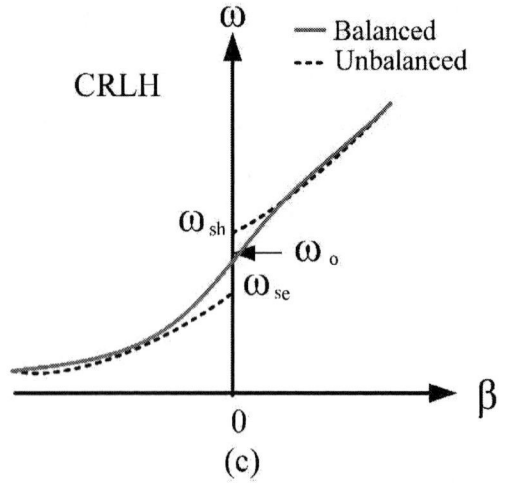

Figure 2. Dispersion diagram of (a) pure right-handed transmission line, (b) pure left-handed transmission line, and (c) composite-right/left-handed transmission line, the red line is a balanced case, and black is an unbalanced case [1].

Metamaterials are defined as artificial effectively homogenous electromagnetic structures with unusual properties not readily available in nature with negative permittivity and permeability. They result in backward electromagnetic (EM) wave propagation, left-handed propagation, which is related to the negative index of refraction having antiparallel phase and group velocities [1]. The CRLH structure is a type of metamaterials, which is based on a transmission line (TL) approach, making the average cell size much smaller than the guided wavelength λ_g [3]. The purely left-handed transmission line is not existing in practice due to parasitic capacitance and inductance. Combining a purely right-handed (PRH) transmission line with forward wave propagation and a purely left-handed (PLH) one with backward wave propagation, a lossless composite-right/left handed (CRLH) transmission line based metamaterial can be realized. As shown in figure 1, both purely right-handed (PRH) and left-handed (PLH) transmission line can be represented by the lumped element circuit components consisting of a per unit length inductance and a per unit length capacitance, to represent CRLH in a more realistic form of a structure.

Figures 2(a), (b), and (c) [1] show the dispersion diagram of a PRH TL with parallel v_g and v_P ($v_g v_P > 0$), PLH TL with antiparallel v_g and v_P ($v_g v_P < 0$), and the CRLH TL

with both LH ($v_g v_P < 0$) and RH ($v_g v_P > 0$) regions where $v_g (= \frac{\partial \omega}{\partial \beta})$ is the group velocity and $v_p (= \frac{\omega}{\beta})$ is the phase velocity.

As shown in Figure 1(a), (b), and (c), the CRLH unit cell lumped element circuit model consists of multiple L, C components: a right-handed (RH) series inductance L_R and a shunt capacitance C_R, and a left handed (LH) series capacitance C_L and a shunt inductance L_L.

Choosing the proper values of inductance and capacitance we are able to design two GPS frequencies: 1.227 GHz, 1.575 GHz. Using a B-CRLH architecture, one can obtain an additional frequency in a lower frequency range. In this work, an additional GSM frequency of 865 MHz, is designed by adding the bridge inductance. For a smooth transition from left-handed to the right-handed operation in the dispersion diagram, the balanced condition design is exploited.

Figures 3 and 4 show the dispersion diagram and the circuit diagram of the bridged CRLH unit cell, respectively [2]. In the dispersion diagram, there are three operation ranges. The top two frequency bands ($\omega > \omega_0$ and $\omega_B < \omega < \omega_0$) are from the conventional CRLH TL approaches while the low frequency band ($0 < \omega < \omega_B$) is from the bridge inductance L_B in Figure 4. A certain phase angle e.g. $\pi/2$ can be realized by three frequency bands (f_1, f_2, f_3) as indicated in Figure 3 enabling a triple band device [2].

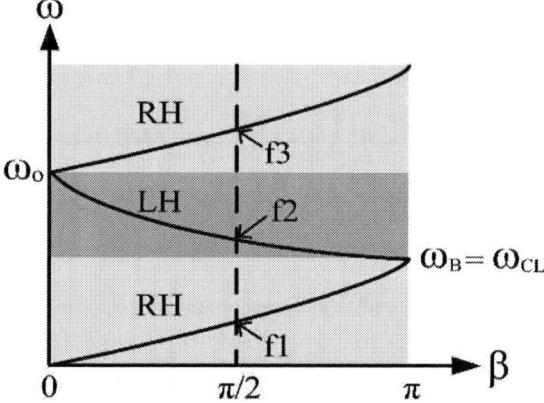

Figure 3. ω-β Dispersion Diagram of Bridged-CRLH [2]

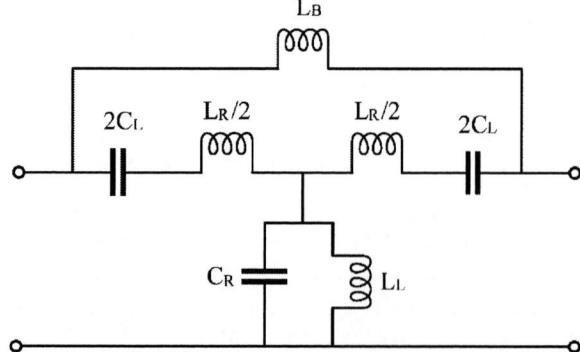

Figure 4. Lumped element circuit of the Bridged Composite Right/Left handed (B-CRLH) structure unit cell [2]

In this work, as a modified B-CRLH, a meander line shape composite-right/left-handed (M-CRLH) transmission line unit cell and its application to a quarter wavelength stub are designed and fabricated. The M-shaped CRLH structure is designed for triple frequencies: two GPS frequencies, and one GSM band frequency. Detail design and simulation is performed using Advanced Design Systems (ADS, Keysight Technologies Inc.) and High Frequency Structure Simulator (HFSS, ANSYS Inc.). The device is fabricated on a Rogers substrate with a dielectric constant ε_r of 10.2 and a thickness of 0.64 mm and characterized using a vector network analyzer (E5071C, Agilent, Inc.). The metal insulator metal (MIM) capacitors and meander line inductors are used for the design.

III. Design Procedure

The three target frequencies are selected (f1=865 MHz, f2=1.227 GHz, and f3=1.575 GHz). The conventional CRLH transmission line approach was used for f2, f3 to calculate L_R, C_R, L_L, C_L [1,7]. The bridged CRLH equalizer network and Bloch analysis is used to investigate f1 along with ABCD matrix [2]. Assuming the balanced condition, $\omega_o = 1/\sqrt[4]{L_R C_R L_L C_L} = \omega_{se} = \omega_{sh}$, and the electrical length of $\lambda/4$ is used for the equation; $\varphi = -N\left(\omega\sqrt{L_R C_R} - 1/\omega\sqrt{L_L C_L}\right)$, where $\omega_{se} = 1/\sqrt{L_R C_L}$, $\omega_{sh} = 1/\sqrt{L_L C_R}$ $\omega_R = 1/\sqrt{L_R C_R}$, $\omega_L = 1/\sqrt{L_L C_L}$ are given where N is the number of unit cells and ω_{se}, ω_{sh} are the series resonance frequency and shunt resonance frequency, respectively. The A parameter of the ABCD matrix is given by

$$ A = 1 - \frac{2\omega^2 L_B C_L \left(1 - \frac{\omega^2}{\omega_{sh}^2}\right)\left(1 - \frac{\omega^2}{\omega_{se}^2}\right)}{\left(1 - \frac{\omega^2}{\omega_{se}^2}\right)^2 \left(1 - \frac{\omega^2}{\omega_{sh}^2}\right) - 4\left(\frac{\omega}{\omega_L}\right)^2 \left(1 - \frac{\omega^2}{\omega_B^2}\right)} $$

where $\omega_B = 1/\sqrt{(L_R + L_B)C_L}$.

The calculated capacitance and inductance parameters are given in Table 1.

TABLE I.   Calculated Capacitance and Inductance Values

$L_R/2$	17.7nH
C_R	14.15pF
L_L	0.9nH
$2C_L$	0.74pF
L_B	12.14nH

The distributed circuit elements are used to design the unit cell based on the calculated capacitance and inductance values. The overall design is shown in Figure 5; the meander line structure is exploited to achieve the series inductance and the shunt capacitance. To get both the large series inductance and small shunt capacitance together, a meander line is used for large Lc. Also, a gap between the meander line, Lc, is used which is four times the linewidth, to minimize the parasitic capacitance. The MIM structure provides high shunt capacitance, C_R, and small shunt

inductance, L_L, is achieved by L_i and Wi. The area under the bridge is removed to minimize the effect of the parasitic capacitance between the bridge inductance and the bottom ground plane. The distributed component dimensions are given in Table 2.

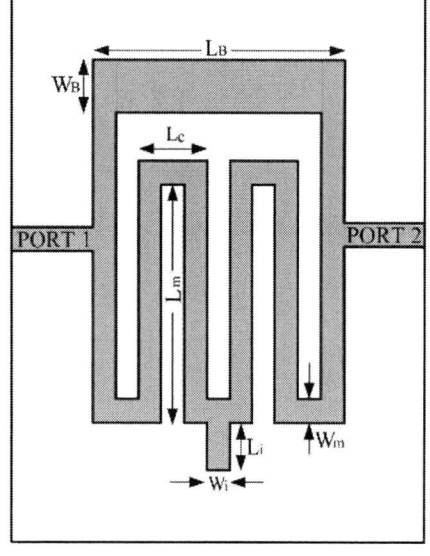

(a)

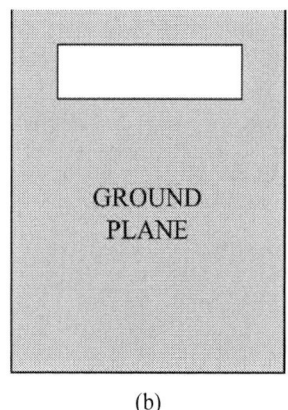

(b)

Figure 5. Design of the M-Shape Composite Right/Left (M-CRLH) handed structure unit cell. (a) top view, (b) bottom view.

TABLE II.   The Used Distributed Component Dimension

L_B	7.8mm
W_B	2.5mm
L_m	32mm
W_m	300um
L_c	1.8mm
L_i	3mm
W_i	0.6mm

978-1-7281-1500-9/19 $31.00 © 2019 IEEE

By using a two-port system with each port located in the side of the unit cell, the phase dispersion diagram is achieved. Figure 6 shows the three frequencies (960 MHz, 1.2 GHz, 1.58 GHz) on 90° within a range between 0 to 180°.

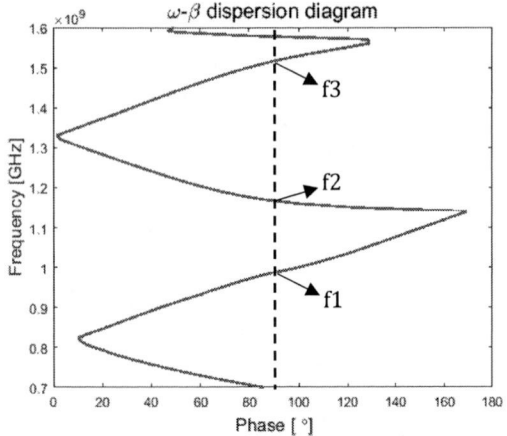

Figure 6. ω-β Dispersion Diagram of M-CRLH unit cell. Three frequencies are obtained, f1=960 MHz, f2=1.2 GHz, f3=1.58 GHz.

Meantime, the insertion loss plot is obtained in M-CRLH stub design. Making the open-circuited M-CRLH λ/4 stub, the same unit cell is connected to the microstrip line port, and the simulation result has been demonstrated in Figure 8.

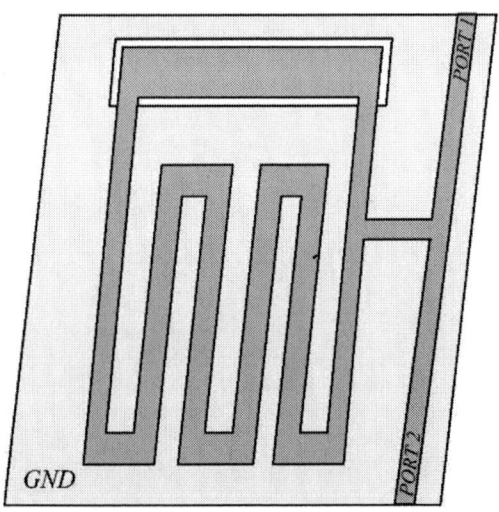

Figure 7. Design of the open-circuited λ/4 M-Shape Composite Right/Left Handed (M-CRLH) structure.

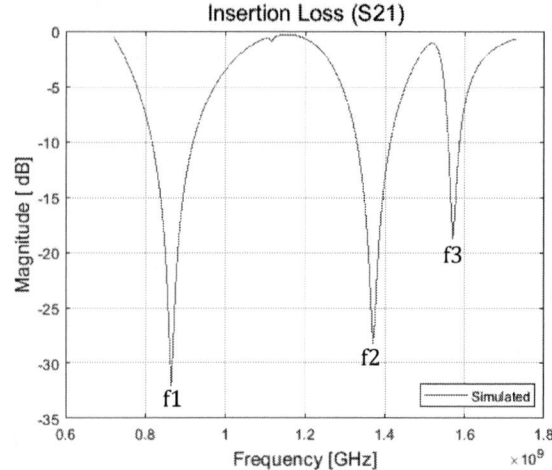

Figure 8. Simulated insertion loss with the open-circuited λ/4 M-CRLH stub. Three frequencies are highlighted f1=865 MHz, f2=1.37 GHz, f3=1.57 GHz.

IV. FABRICATION AND RESULT

The M-CRLH structure is implemented on a printing circuit board (RT/duroid 6010LM, Rogers Inc.) substrate with a thickness of 0.64 mm and a dielectric constant ε_r of 10.2) using LPKF milling machine. SMA connectors are soldered in two ports. Measurement of scattering parameters e.g. the insertion loss (S21) has been carried out using a vector network analyzer (E5071C, Agilent, Inc.) after standard two port calibration, short-open-load-thru (SOLT). Figures 9(a) and (b) show the top and bottom view of the patterned substrate, respectively.

(a)

978-1-7281-1500-9/19 $31.00 © 2019 IEEE   2088

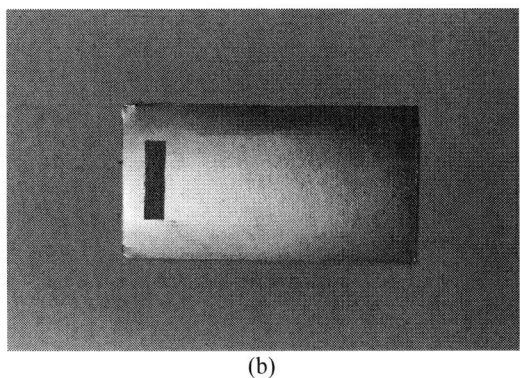

(b)

Figure 9. Fabricated M-CRLH structure (a) top view, (b) bottom view.

TABLE III. COMPARISON BETWEEN THE SIMULATED AND MEASURED RESULTS

	Simulated	Measured
f1	865MHz	747MHz
f2	1.37GHz	1.22GHz
f3	1.57GHz	1.42GHz

The results of both simulated and measured results are shown in Figure 10. The measured result shows f1=747 MHz, f2=1.22 GHz, and f3=1.42 GHz. It shows an approximately 100 MHz overall frequency shift to the lower frequency, which can be attributed to fabrication tolerance. The fabricated device's overall linewidth decides the series inductance value, and it is inversely proportional to the inductance. The comparison between the simulated and fabricated results are given in Table 3.

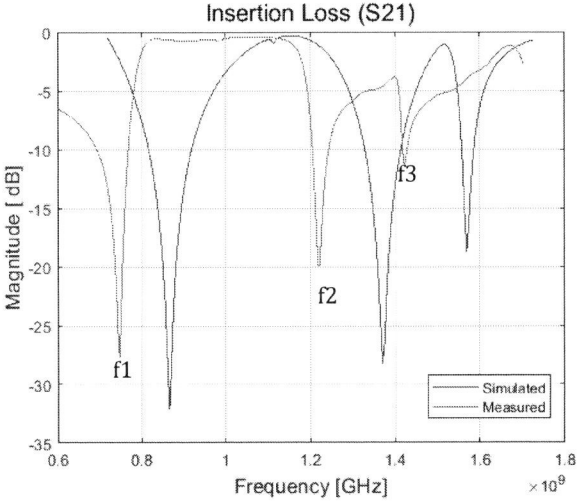

Figure 10. Measured insertion loss with the open-circuited λ/4 M-CRLH stub. Three frequencies are highlighted f1=747 MHz, f2=1.22 GHz, f3=1.42 GHz.

V. CONCLUSION

A triple band quarter wavelength open stub has been designed, fabricated, and characterized using the M shaped CRLH architecture, a modified B-CRLH one. Design is performed using the B-CRLH theory, ADS and HFSS tools. The designed device was fabricated using a Rogers PCB Board, achieving compact RF components. The three frequencies have been covered by a single λ/4 stub, dual GPS frequencies by the conventional CRLH approach, and one additional GSM frequency by adding a bridge structure between the input and output ports, providing compactness and reducing system complexity. The substrate with a high dielectric constant enables to implement more compact RF components. The lumped element circuit model is designed using ADS, and the unit cell is designed and simulated using HFSS. The fabricated quarter wavelength open-circuited stub was characterized by the dispersion diagram and scattering parameters using HFSS. Measurement is performed using a vector network analyzer. Finally, the simulated and measured results were compared with each other. Frequency downshift is observed, which is attributed to fabrication tolerance.

ACKNOWLEDGMENT

Rogers Corporation donates PCB substrates. Fabrication and measurement has been performed in the Interdisciplinary Microsystem Group at the University of Florida.

REFERENCES

[1] A. Lai, T. Itoh and C. Caloz, "Composite right/left-handed transmission line metamaterials," in *IEEE Microwave Magazine*, vol. 5, no. 3, pp. 34-50, Sept. 2004.

[2] D. E. Senior and Y. Yoon, "Bridged Composite Right/Left Handed Unit Cell With All-Pass and Triple Band Response," in *IEEE Microwave and Wireless Components Letters*, vol. 22, no. 11, pp. 568-570, Nov. 2012.)

[3] C. Caloz and T. Itoh, Electromagnetic Metamaterials: Transmission Line Theory and Microwave Applications. New York: Wiley, 2006, ch. 3, pp. 59–131

[4] X. Zhang, H. Deng, L. Zhang, Y. Zhao and L. Qiang, "Compact dual-band BPF for GPS and WLAN with two dual-mode stepped impedance resonators," *2010 Global Mobile Congress*, Shanghai, 2010, pp. 1-3.

[5] H. Liu *et al.*, "Compact and High Selectivity Tri-Band Bandpass Filter Using Multimode Stepped-Impedance Resonator," in *IEEE Microwave and Wireless Components Letters*, vol. 23, no. 10, pp. 536-538, Oct. 2013.

[6] N. Kumar and M. Kumar, "Closely specified wide dual-band microstrip bandpass filter using coupled stepped-impedance resonators," *2015 2nd International Conference on Electronics and Communication Systems (ICECS)*, Coimbatore, 2015, pp. 865-867.

[7] D. S. Elles and Y. Yoon, "Compact dual-band three-way Bagley polygon power divider using composite right/left-handed (CRLH) transmission lines," *2009 IEEE MTT-S International Microwave Symposium Digest*, Boston, MA, 2009, pp. 485-488.

[8] H. -. Liu, Y. -. Lv and W. Zheng, "Compact dual-band bandpass filter using trisection hairpin resonator for GPS and WLAN applications," in *Electronics Letters*, vol. 45, no. 7, pp. 360-362, 26 March 2009.

[9] V. Turgul, Y. Adane and I. Kale, "Compact dual-band bandpass filter for GNSS bands deploying octagonal open-loop resonators," *2013 13th Mediterranean Microwave Symposium (MMS)*, Saida, 2013, pp. 1-4.

Die Thickness Optimization for Preventing Electro-Thermal Fails Induced by Solder Voids in Power Devices

Dario Vitello, Andrea Albertinetti, and Marco Rovitto

STMicroelectronics
Via Camillo Olivetti 2, 20864, Agrate Brianza (MB), Italy
Email: {dario.vitello\andrea.albertinetti\marco.rovitto}@st.com

Abstract— **Vertical power devices require thin dice to reach high electrical performance especially for automotive market. Beside their several advantages, thin power devices reveal issues related to assembly processes. Die bonding process step typically generates solder voids which can lead to thermal-induced fails. The paper deals with die thickness optimization in order to reduce the risk of failures due to the presence of voids by considering manufacturing limitations. For this purpose, electro-thermal modeling is employed to calculate the temperature at which fail occurs. Further, it allows to estimate the impact of die thickness and solder void size on the device temperature distribution during operating life.**

Keywords-thin die; solder voids; power device; finite element method; electrical over stress fail

I. INTRODUCTION

Today's automotive market strongly demands for power devices in order to accommodate high current density. Vertical Intelligent Power® (VIPower) devices are examples of how smart power technologies evolve today in order to better satisfy the market request. VIPower devices employ a fabrication process which allows the integration of complete digital and/or analog control circuits driving a vertical power transistor on the same chip [1]. Because of the transistor vertical operation mode, it is important to shrink silicon die thickness in order to provide a significant reduction of the drain to source on resistance (RDSON) and, subsequently, improve the device electrical performance [2]. On the other hand, when it comes to thin wafers, some weaknesses related to both manufacturing and assembly processes are emphasized. One of the crucial assembly process steps for integrated circuits is die bonding (well known as die attach). It creates a mechanical, thermal, and electrical link between the silicon die and metal die-pad of a leadframe-based package. Further to the mechanical support and electrical connection, it provides a way to remove the heat generated by the silicon die during its operating life reducing the risk of device overheating. For this purpose, mechanical, electrical and thermal requirements make Lead-

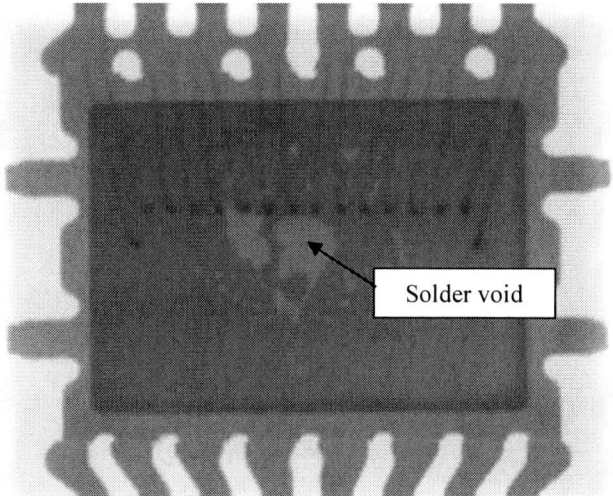

Figure 1: X-ray image of a PSSO package. Solder voids are visible in the die area.

Tin alloys (commonly named solder) suitable for high-temperature applications [3] rather than traditional glues or films. Besides their thermo-electro-mechanical benefits encountered during the manufacturing process, solder materials reveal a number of concerns which could influence the product functionality. One of the most critical issue related to the use of Lead-Tin alloys is the presence of voids between silicon die and metal die-pad. Voids can be formed due to trapped gas during the reaction of materials in the die attachment phase, poor wettability at the joining interfaces [4], or the imperfections of the reflow process [3]. Therefore, in the case of Lead-Tin alloys, the assembly challenge lies in the ability to minimize the amount of solder voids which are generated in the bond line due to the manufacturing process. Figure 1 reports an x-ray image of a Plastic Shirked Small Outline (PSSO) package characterized by a central void generated on purpose in the die attach layer consisting of solder material. It is demonstrated that the presence of voids leads to an increase of the solder joint thermal resistance rising the junction temperature [6] which, especially in power and high-power applications, possibly leads to the device critical overheating. The thermal damage which may occur when an electronic device is subjected to an electric current and/or voltage beyond its specification limit is specifically called electrical over stress (EOS) fail.

978-1-7281-1500-9/19 $31.00 © 2019 IEEE

As previously mentioned, it is well known that thermal damages as EOS fail occurrence increases with the die thickness reduction. For this reason, EOS fail has been recognized as one of the most probable failure mode during first steps of thin die introduction into power packages qualification. Therefore, a set of actions needs to be defined in order to reduce the risk of EOS fail occurrence during package development. For this purpose, thermal simulations are necessary in order to determine the minimum die thickness which prevents EOS fail occurrence by taking into account the intrinsic die attach process limitations. In this work, finite element method (FEM) is employed to perform thermal analysis and allows to investigate the temperature distribution in the silicon die during electrical test. Simulations show promising results in terms of prevention of EOS failures and optimization of die thickness for anticipating electro-thermal issues. Furthermore, FEM analysis could be useful for package qualification and process development.

The paper is divided into three phases. The first one describes a semi-empirical thermal model used in this work in order to determine the temperature at which EOS fail occurs. The second phase investigates the relationship between solder void size, silicon die thickness, and the maximum temperature reached by the device during electrical testing or operating life. In the last phase, the minimum die thickness which allows to prevent EOS fail occurrence is determined by considering die bonding manufacturing limits.

II.   PHASE A: SEMI-EMPIRICAL MODEL

The first section of the paper describes both experimental and numerical model used for the calculation of the temperature at which the EOS fail occurs (T_{EOS}). It should be pointed out that the numerical method is based on empirical evidences. Both experimental process and numerical modelling are described in the following.

A.   Experimental process

As first step, a VIPower test vehicle is assembled into PSSO packages using several die bonding process parameters in order to create different solder void sizes. Then, once the assembly is completed, electrical tests representing operating conditions are performed. The failed samples are subsequently examined by means of X-ray visual characterization so that the solder voids magnitude are measured. The smallest solder void found which brings to EOS fail is taken into account and used as reference for the T_{EOS} calculation via numerical modelling.

B.   Numerical modelling

1)   Geometry and materials

The experimental process described before, consisting of sample configuration and electrical test conditions, is then reproduced by using FEM-based modeling in order to investigate the silicon device temperature distribution

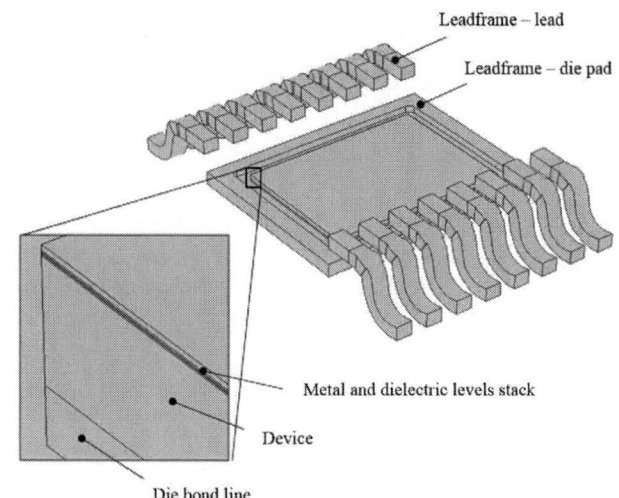

Figure 2: Profile view of the analyzed 3D PSSO package structure with silicon device. All package components are listed. The lower part of the package layout shows the copper die pad. The zoomed-in detail view of the structure depicts device, die bond line, and metal/dielectrics stack-up.

during electrical test.

The three dimensional (3D) model used in the FEM study is sketched in Figure 2, which shows the whole geometry except for the plastic body to make the other domains visible. The test case structure is composed by different domains including plastic body, lead-frame, die bond line, and device made of thermosetting resin, copper, PbSnAg alloy, and silicon, respectively. Moreover, the silicon device presents on top a stack of dielectric and metal levels. Dielectric levels material properties are set by considering both dielectrics and vias, constituted of silicon dioxide (SiO_2) and tungsten, respectively. Metal level domains are composed by aluminium and SiO_2 according to metal circuitry. Die bond line domain is composed by die attach material, which is PbSnAg alloy, and voids, represented by air.

In order to better reproduce metal circuitry and die bond line domains, their material properties are set starting from an image. Data extrapolation is performed on the image in order to extract a proper image function. Its features allow to map material properties data as a function of spatial x-y coordinates and transfer them to scalar output values. Figure 3 shows the procedure of extrapolation of the material properties of die bond line domain starting from a X-ray picture. The void distribution into the die bond line is depicted in Figure 3a. From this picture, it is possible to extract two macro-regions with different electro-thermal properties. The white region in Figure 3b represents air, while the black area defines electrical and thermal properties of PbSnAg alloy. By assuming a two-phase domain (see Figure 3c), it is possible to define a function which represents the distribution of material properties in the x-y coordinates. A similar method is used to model metal level domains using aluminium and SiO2. Both functions are then applied to the FEM model. Following this approach, model complexity and computation time are reduced.

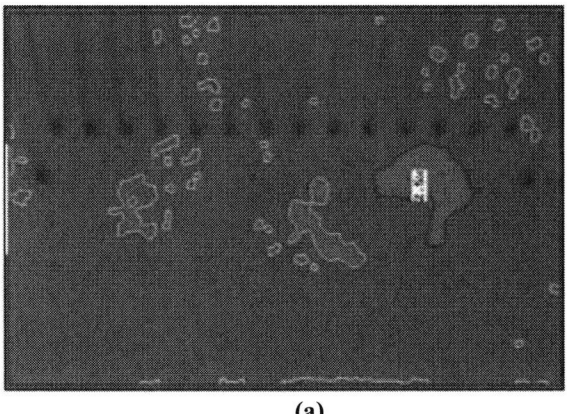

(a)

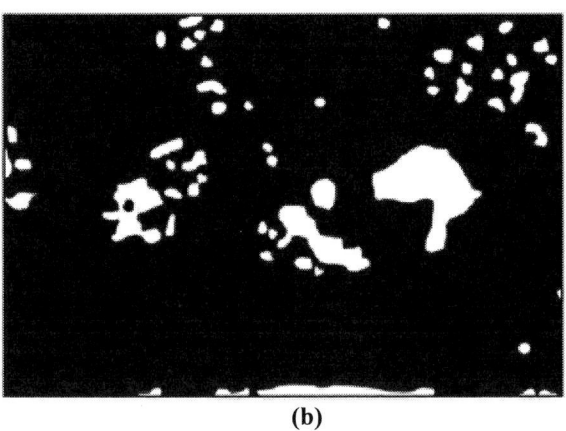

(b)

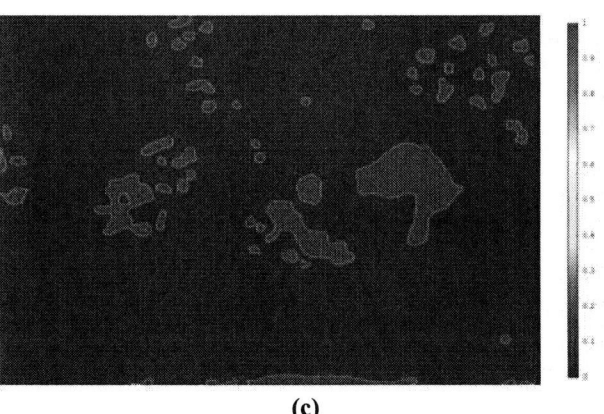

(c)

Figure 3: Schematic procedure employed for the definition of die bond line material properties used in the FEM model. a) X-ray image of an assembled device showing the void distribution into the bond line layer. b) Image tuning in order to highlight void positions (in white) and to remove noise. c) Scalar output map extrapolation useful to define material properties (air in red and PbSnAg alloy in blue).

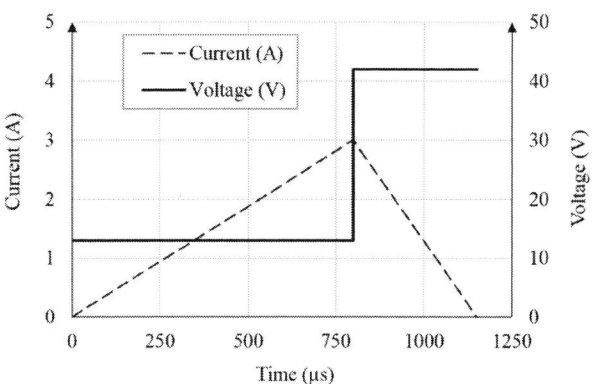

Figure 4: Current and voltage boundary conditions applied on the top of the metal and dielectrics stack.

2) Boundary conditions and calculation

Operating conditions for FEM calculations are set by imposing the following constraints over appropriate domains and boundaries of the case studied. Electrical power is applied on the top surface of the metal and dielectrics stack as shown in Figure 2. Initial temperature is set at 150°C at all domains of the structure as per testing conditions. By imposing the electro-thermal boundary conditions to the test case structure, a power peak occurs at 800µs, as presented in the plot in Figure 4. Current and voltage are imposed by using triangle and step function, respectively. Maximum current value and voltage step positions are set at 800µs. Furthermore, natural convection boundary conditions are applied to the outer surfaces of both plastic body and external leads. Once the boundary conditions are defined, 3D time-dependent numerical simulation is carried out by solving thermal and electrical equations.

3) Simulation results

The electro-thermal analysis results in the maximum temperature reached by the device during the power pulse, as depicted in Figure 5. In general, the device temperature increases with the power applied and it reaches its absolute maximum value, $T_{MAX(ABS)}$, at 800µs. At this pulse time, both input current and voltage reach their maximum values, as shown in Figure 4. In this specific case, by considering the smallest void as the root cause for EOS failure during electrical test, $T_{MAX(ABS)}$ represents the temperature at which the EOS fail occurs ($T_{MAX(ABS)}=T_{EOS}$). The impact of solder void to the device temperature distribution can be better explained by focusing on Figure 6, which shows the EOS fail occurred in the test case mentioned above. The X-ray image in Figure 6a depicts the solder void distribution in VIPower test vehicle assembled into PSSO package. Figure 6b sketches the device temperature map calculated at 800µs as a result of the electro-thermal model previously described. The influence of the presence of solder voids is clearly visible and temperature peaks are located in their correspondence. Moreover, the bigger the solder void, the higher the temperature reached.

978-1-7281-1500-9/19 $31.00 © 2019 IEEE   2093

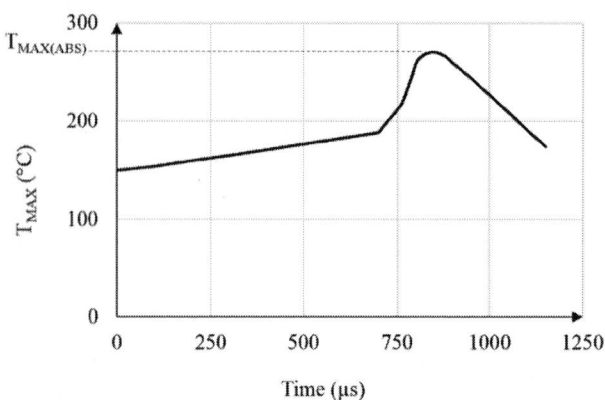

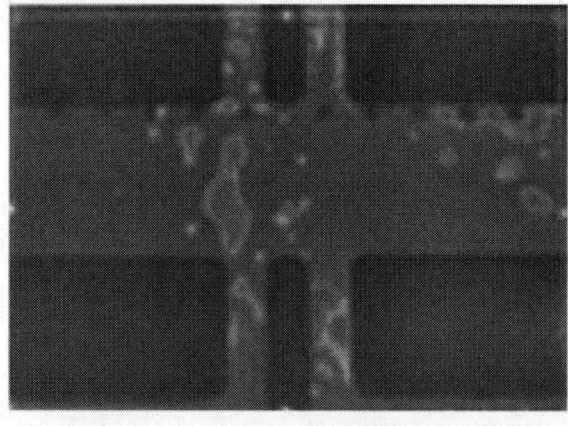

Figure 5: Device maximum temperature as a function of power pulse time. The peak temperature is observed at 800µs.

In the specific case, the highest temperature peak is located in correspondence of the biggest void present in the die bond line domain and represents T_{EOS}. Furthermore, Figure 6c shows the typical EOS fail mark once molding compound has been removed. The correspondence between the biggest die attach defect, the highest temperature peak and EOS fail mark is remarkable and confirms what obtained from the numerical analysis.

Thermal simulations are then carried out to other samples assembled with different solder voids with diverse sizes. Samples with smaller voids do not show any EOS fail and their $T_{MAX(ABS)}$ calculated are lower than T_{EOS} calculated before. On the other hand, samples with wider void size show both electrical fails and EOS marks. Their calculated $T_{MAX(ABS)}$ are higher than T_{EOS}.

III. PHASE B: IMPACT OF VOID SIZE AND DIE THICKNESS

The purpose of this part of the paper is to analyze how die thickness and solder void size affect the device maximum temperature reached during operating life (simulated by electrical test). A new set of thermal simulations is then carried out considering die thickness and solder void size as variables. T_{MAX} is then calculated by applying the same boundary conditions already described in the previous section. Die thicknesses and solder void sizes considered are listed in Table 1. The model is simplified by assuming that solder voids have a round shape. As done in Phase 1, T_{MAX} as a function of power pulse time graphs are built for all the combinations of void size and die thickness and the $T_{MAX(ABS)}$ are then calculated. $T_{MAX(ABS)}$ obtained for each void size/die thickness combination are then summarized in the plot sketched in Figure 7, which represents the $T_{MAX(ABS)}$ as a function of solder void size and die thickness. Each curve refers to a different die thickness. From the results it is possible to observe that the presence of an even small void causes a significant increase of $T_{MAX(ABS)}$. Then, by taking into account wider solder voids, $T_{MAX(ABS)}$ continues to grow but with a smaller slope. This phenomenon is emphasized considering very thin dice (i.e. for thickness A).

(a)

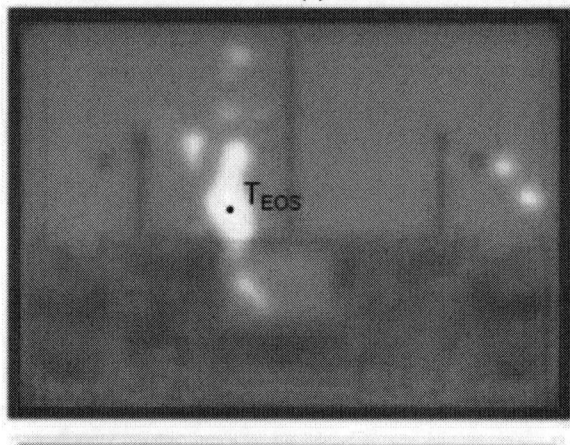

Temperature

(b)

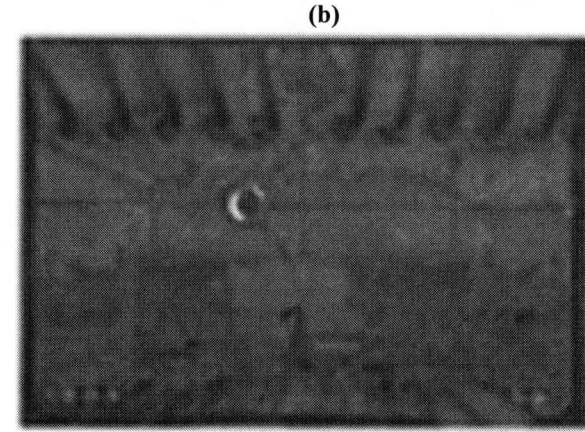

(c)

Figure 6: EOS failed sample in PSSO package. a) X-ray image showing the solder void distribution. b) Device temperature distribution obtained from FEM analysis at 800µs. c) Optical inspection after molding decapsulation showing the typical EOS round fail mark. The correspondence between the biggest solder void, the maximum temperature peak, and EOS mark is visible.

978-1-7281-1500-9/19 $31.00 © 2019 IEEE

By considering different devices having different die thicknesses but with the same level of die attach defects, results show that the thinner the dice, the higher is its sensibility to die attach defects, and the higher the $T_{MAX(ABS)}$ reached. When a solder void free structures ($Ø_{VOID}$ = 0%) is considered, negligible difference in terms of $T_{MAX(ABS)}$ is observed by varying the die thickness. Moreover, in case of very thick dice (i.e. thickness F), solder void presence is not significant for $T_{MAX(ABS)}$ in the range of die attach defect dimensions considered in the study.

IV. PHASE C: IMPACT OF PROCESS LIMITATION

The goal of this final phase consists into the determination of the minimum die thickness which prevents the EOS fail occurrence in presence of solder void by considering the die bonding process limitations. As first step, the critical solder void size ($Ø_C$) for any die thickness can be determined by matching the two results obtained in the previous two analyses. Referring to the plot in Figure 7, the dashed line represents the temperature at which EOS fail occurs (T_{EOS}). The projections on the x-axis of the intersections between T_{EOS} line and $T_{MAX(ABS)}$ curves, represent the critical solder void size for any single die thickness considered. As an example, a solder void as big as the 0.8% of the die area is critical for dice with thickness equal to A. Critical solder void size as a function of the die thickness is then plotted in Figure 8. In general, the thicker the die, the bigger the solder void which can induce an EOS fail. This means that thicker dice can tolerate bigger die attach defects. The dashed line in Figure 8 represents the minimum solder void size guaranteed by the die bonding process considering variability in terms of machines, tools, materials, finishing, and geometries in a large volume production. As an example, in the plot in Figure 8, it is set as big as the 5% of the die area. It means that, despite the proper die bonding recipe is set, a die attach defect as big as the 5% of the die area can be formed because of the intrinsic process limitation mentioned above. The projection on the x-axis of the intersection between the two lines represents the minimum die thickness which prevents the EOS fail considering the die attach process limitation.

In this way, a significant reduction of RDSON is achieved by thinning the device without compromising its thermal and electrical performances.

Table 1: Relative die thicknesses considered in the study. Test case A is considered as reference.

Test case	Die thickness
A	100%
B	115%
C	130%
D	140%
E	155%
F	170%
G	185%
H	215%
I	400%

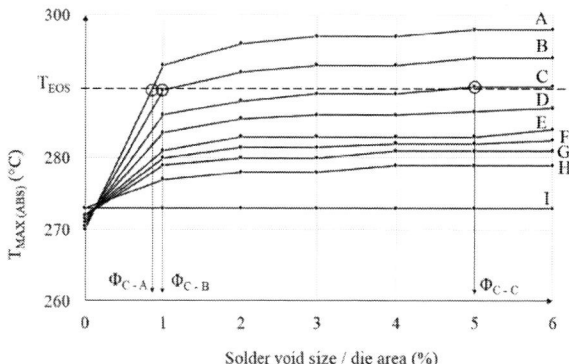

Figure 7: Device maximum temperature as a function of die attach defect size. Nine different curves representing different die thicknesses are plotted. The projections on the x-axis of the intersections between T_{EOS} line and $T_{MAX(ABS)}$ curves, represent the critical solder void size ($Ø_{C-A}$, $Ø_{C-B}$, $Ø_{C-C}$) for any single die thickness considered.

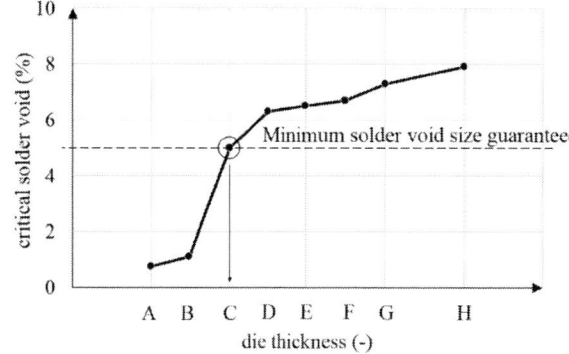

Figure 8: Critical solder void size as a function of silicon die thickness. The dashed line is the reference solder void size guaranteed by die bonding process.

CONCLUSIONS

The paper focuses on the determination of the minimum die thickness which prevents EOS fail due to solder voids in VIP® devices by considering assembly limitations. For this purpose, electro-thermal simulations are carried out by employing FEM. A three step study is then performed. At the beginning, the temperature at which EOS failure occurs is calculated. Then, the second step determines the relationship between die thickness, solder void size, and maximum temperature reached by the device during operating life. In the end, the critical solder void size as a function of the die thickness and the capability of die bonding process is considered in order to achieve the best trade-off between die thickness and maximum solder void allowed. By following this approach, a significant reduction of the RDSON can be reached with the shrinking of die thickness and an improvement of the device electro-thermal performance is accordingly achieved.

ACKNOLWEDGMENT

The authors would like to thank colleagues from STMicroelectronics Agrate and Shenzhen Back End Manufacturing and Technology R&D in particular Xiao Xiang Chen for the support to the whole activity.

REFERENCES

[1] B. Murari, G. Gattavari, G. Ferla, A. Russo "System Level Technologies and High Level Integration Evolution" IEEE The Future of Electronics Power Processing and Conversion, 2001

[2] A. Dhadda, R. Montgomery, Paul Jones, Jason Heirene, Rachel Kuthakis, Florian Bieck "Processing of Ultrathin Wafers for Power Chip Applications" IEEE 14th Electronics Packaging Technology Conference (EPTC), 2012

[3] V.R. Manikam, and K. Y. Cheong, "Die Attach Materials for High Temperature Applications: A Review," IEEE Trans. On Components, Packaging and Manufacturing Technology, vol. 1, no. 4, pp. 457–478, Dec. 2011.

[4] L. Chen, M. Paulasto-Krockel, U. Frohler, D. Schweitzer, H. Pape, "Thermal Impact of Randomly Distributed Solder Voids on Rth-JC of MOSFETs", Proc. 2nd Electron. Syst.-Integr. Technol. Conf. (ESTC), pp. 237-243, Sep. 2008

[5] A. S. Fleischer, L. H. Chang, and B. C. Johnson, "The Effect of Die Attach Voiding on the Thermal Resistance of Chip Level Packages," Microelectron. Reliability, vol. 46, nos. 5–6, pp. 794–804, 2006.

3-T (8-T) Decoupling Capacitors for Improved PDN in LPDDR4/4X/5 System

Sunil Gupta

Qualcomm Technologies, Inc.
5775 Morehouse Drive, San Diego, CA - 92121
sungupta@qti.qualcomm.com

Abstract— The impact of 3-T (3-Terminal) decoupling capacitors on PDN (Power Delivery System) in LPDDR4/4X/5 system is presented. The main advantage of 3-T caps over traditional 2-T caps is their reduced ESL (Equivalent Series Inductance). This reduced ESL leads to lower PDN impedance and hence improved PI (Power Integrity) and eye-apertures of SoC (System-on-chip) – DRAM system. The PDN consists of PMIC, PCB, SoC package and on-die decoupling-capacitor. The 3-T and 2-T caps are used as bulk caps on the PCB. The LPDDR4X system analyzed show PCB inductance in the linear region being ~40% lower in 3-T case compared to 2-T case. The peak system impedance being ~20% lower in 3-T case compared to 2-T case. These effects translate into bigger eye-apertures in 3-T case vs 2-T case. The analysis can be extended to use of 8-T capacitors in PDN to get even lower ESL, lower impedance and larger eyes.

Keywords – 3-Terminal, 8-Terminal, 2-Terminal, PDN, Power Integrity, Signal Integrity, Eye-aperture, SSN, PCB, PoP, External, LPDDR4, LPDDR4X, LPDDR5, ESL.

I. INTRODUCTION

This paper presents LPDDR4/4X/5 (Low Power Double Data Rate 4/4X/5) interface PDN analysis when using 3-T decoupling capacitors. The analysis presented could be applied to other LPDDR/DDR/GDDR parallel interface.

In LPDDR system PDN plays a crucial role in determining the Power Integrity and hence its performance.

Section II covers the background information on LPDDR basics. LPDDR is a parallel interface with single-ended data lanes and differential clock/strobe lanes. The primary PI degradation factors being PSIJ (Power Supply Induced Jitter) and SSO (Simultaneous Switching Output) noise. The primary SI degradation factors being reflections, ISI and single-ended data-lanes crosstalk.

The two main SoC-DRAM configurations of PoP and External DRAM are illustrated along with LPDDR x16 channel description. Electrical representation of 3-T and 2-T decoupling capacitors are presented. Furthermore, extension of the concept of 3-T capacitors to 8-T capacitors is covered for further ESL reduction.

Section III covers the analysis framework. PDN system block diagram along with system schematic showing inductive and capacitive elements is illustrated. Co-SIPI simulation setup is described in detail. It also explains the SI/PI system timing budget and eye-aperture FOM (Figure-of-Merit) based on JEDEC standard.

Section IV presents the results of the analysis of 3-T vs 2-T case in frequency and time domains. Firstly, the PCB impedance profiles are compared, and effective ESL are quantified. Secondly, system peak PDN impedances are compared. Thirdly, the time-domain eye-apertures comparison results are illustrated.

Finally, the key takeaways of 3-T capacitors on PDN improvement summarized in section V.

II. BACKGROUND

A. LPDDR PoP and External DRAM Configurations

Fig. 1a shows the PoP (Package-on-Package) block diagram where the SoC sits on the PCB and the DRAM sits on top of the SoC. PoP configuration is typically found in high-end smartphones. In this paper this configuration is analyzed.

Fig. 1b shows the external SoC-DRAM block diagram where the SoC and DRAM sit side-by-side on the PCB. This type of configuration is typically found in tablets and automotive systems.

Fig. 1a. SoC-DRAM PoP   Fig. 1b. SoC-DRAM External

The PDN is similar in both PoP and External configurations but the SI channel is different. In PoP, the SI channel consists of SoC PKG, interposer and the DRAM PKG. In External, the SI channel consists of SoC PKG, PCB traces and the DRAM PKG.

B. LPDDR x16 channel & DRAM x16 and x8 mode

Fig. 2 shows the schematic of x16 channel with two x8 data buses with CA (Command Address) bus in the middle. LPDDR4/4X/5 interfaces consist of these x16 channels. An SoC might have 4*x16 channels on the periphery connected to single/dual-rank DRAMs. In a high-end system, dual-rank DRAM is employed to get higher capacity. A typical dual-rank 4*x16 system will have DRAM capacity of 8GB (4*2GB).

LPDDR4/4X/5 x16 channel can be connected to DRAM in x16 mode, x8 mode or mixed-mode setup. In x16 mode, x16 channel connects to one x16 DRAM for each rank. In x8 mode, x16 channel connects to two x8 mode DRAMs for each rank. In mixed mode setup, x16 channel connects to combination of x16 mode DRAM for one rank and two x8 mode DRAMs for

978-1-7281-1500-9/19 $31.00 © 2019 IEEE

second rank. When connecting in x8 mode DRAMs, the CA bus is shared for the two x8 DRAMs. Thus, one can obtain various DRAM system densities by utilizing these combinations.

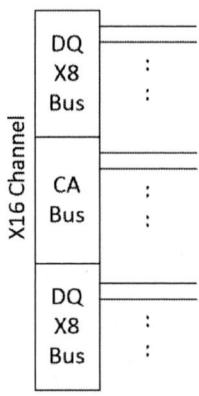

Fig. 2. x16 Channel

C. ESL reduction in 3-T (8-T) capacitors

The ESL reduction in 3-T (8-T) compared to 2-T capacitors happen because of two factors as listed below.

1. Primary factor - Reducing the distance between the terminals to minimize the current loop.

2. Secondary factor – Further reduction in inductance by creating closely coupled adjacent loops with opposing currents.

3-T capacitors are essentially 4-T capacitors. The concept of 3-T/4-T can be extended to 8-T capacitors for further reduction inductance which results in ultra-low ESL.

Fig. 3 shows the terminal connection in 2-T, 3-T and 8-T capacitors. Going from 2-T to 3-T capacitor reduces the distance between the "+" and "-" terminals and furthermore the adjacent currents are in opposite direction. As a decoupling capacitor, the two "+" terminals are combined into one port and two "-" terminals are combined into one port. This gives two reduced ESL capacitors which are in parallel which further gives reduced ESL.

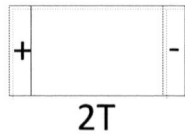

2T

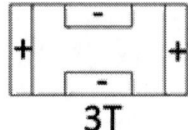

3T

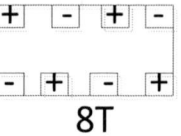

8T

Fig. 3. Terminal locations in 2-T, 3-T and 8-T capacitors.

In this paper, 3-T capacitor analysis is done and compared with 2-T but the analysis using 8-T can be performed along the similar lines and would result in even lower PCB inductance, lower peak system PDN and even better time-domain eye-apertures.

D. Electrical representation of 3-T vs 2-T capacitor

Fig. 4 illustrates electrical representation of 2-T and 3-T capacitor cases. The ESL of 3-T capacitor is lower than that of 2-T capacitor case. In 3-T capacitors case, the partial inductances F1, F2, F3 and F4 are much reduced due to the primary and secondary factors described in subsection C.

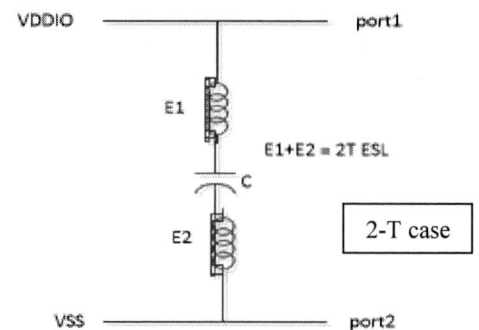

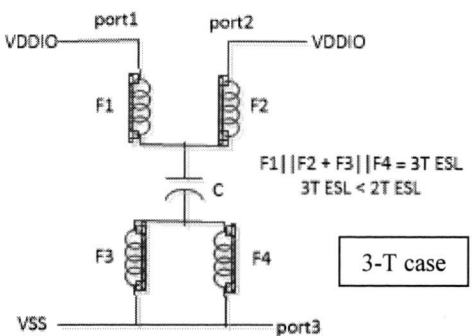

Fig. 4. Electrical Representation of 2-T and 3-T capacitor cases

III. PDN (POWER DELIVERY NETWORK) ANALYSIS FRAMEWORK

To quantify the impact of 3-T capacitors vs 2-T capacitors on the PDN, Co-SIPI analysis is performed using the mobile SoC TX/RX, SoC and DRAM packages and the DRAM RX/TX. Writes from SoC to DRAM were analyzed.

A. LPDDR PDN System Block Diagram

Figure 5 illustrates the LPDDR PHY SoC and DRAM system schematic. The PDN is shown in green text and

978-1-7281-1500-9/19 $31.00 © 2019 IEEE

consists of PMIC, PCB, SoC PKG and on-die decap. This PDN powers the TX (Transmitter) drivers of the SoC during Write traffic when SoC is writing data to DRAM.

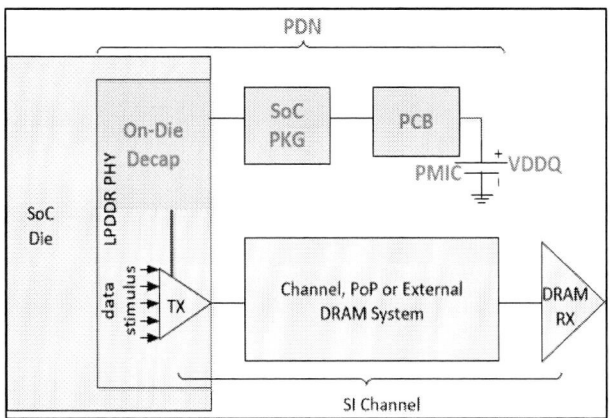

Figure 5. LPDDR System Block Diagram, PDN in green text.

Fig. 6 illustrates the LPDDR system schematic PDN elements. It includes the PMIC, PCB along with 3-T/2-T caps, PKG model and on-die decap.

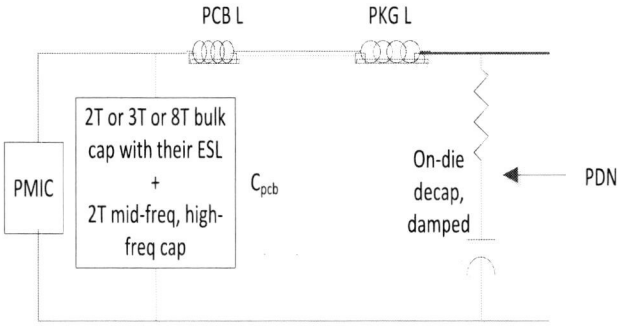

Figure 6. LPDDR PDN system schematic.

B. Co-SIPI Analysis Framework

1. Fig. 5 shows the Co-SIPI schematic of SoC-DRAM system used for analysis.

2. The SI channel consists of SoC TX drivers, PoP channel (SoC PKG and DRAM PKG) and DRAM RX load.

3. The system PDN for the 3-T case consisted 3-T and 2-T capacitors and for the 2-T case, all the capacitors were 2-T capacitors. The total number of capacitors and the total decoupling capacitor value was same for the 3-T and 2-T cases.

4. Co-SIPI simulations were run to obtain the eye-apertures for the 3-T and 2-T capacitor cases in PDN.

5. LPDDR4X interface running at 4266 Mbps was analyzed.

6. Dual-Rank DRAM configuration was analyzed.

7. Write traffic from SoC → DRAM was used.

8. On-die decap values used were 100% damped.

9. The system analyzed was PoP (Package-on-Package) with DRAM connected to SoC package using an interposer.

10. IO transistor FET models were used for SoC TX drivers.

11. Data pattern for TX stimulus consists of PRBS (Pseudo Random Bit Sequence), SSO (Simultaneous Switching Output) and Victim/Aggressor sequences. This pattern represents a worst-case scenario.

12. DRAM RX load consists of ODT in parallel with CIO (Input/Output Capacitance).

13. VDDQ supply rail at 570mV at SS (slow-slow) corner was analyzed. VDDQ = 600mV at TT (Typ.-Typ.) corner.

14. VOH level of the received signal at the DRAM was ~340mV (0.6*VDDQ, SS).

15. Fast HSPICE simulator was used for running simulations.

C. SoC LVSTL TX and DRAM ODT

SoC employs LVSTL (Low Voltage Swing Terminated Logic) TX driver. TX driver sees ground terminated DRAM ODT load in parallel with CIO. In a dual-rank DRAM system, the effective ODT is ODT0 ∥ ODT1. Fig. 7 shows the SoC TX (PU) pull-up and (PD) pull-down segments and the DRAM loads. The channel consists of SoC and DRAM packages.

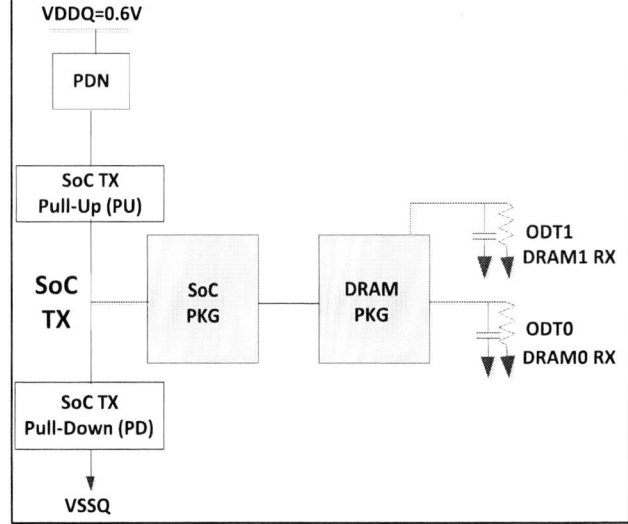

Fig. 7. SoC TX with PDN and dual-rank DRAM loads.

D. SI/PI System Timing Budget and FOM (Figure-of-Merit)

The LPDDR SoC-DRAM link timing budget is divided into three main buckets, namely –

1. SoC/AP timing budget. This consists of items such as clock jitter, DCD, DQS-DQ skew and training errors, among others. Training errors being delay line resolution, step-size

978-1-7281-1500-9/19 $31.00 © 2019 IEEE   2099

non-uniformity, voltage and temperature dependent timing variations.

2. DRAM timing budget. This consists of items such as TdIVW (Time of data Input Valid Window) parameter which is governed by JEDEC standard. LPDDR4X standard allocates 0.25*UI timing budget for TdIVW.

3. SI/PI timing budget. This consists of PI and SI degradation parameters. The main PI parameters being PSIJ and SSO noise and SI degradation factors being reflections, ISI and single-ended data-lanes crosstalk.

Fig. 8 illustrates the system timing budget in one UI (Unit Interval). Any time left over after meeting the timing requirements of the above three main buckets is the extra margin of the data-eye available.

The system PDN and its effects fall under the SI/PI timing budget.

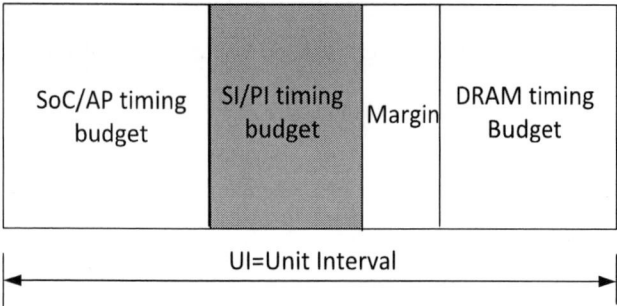

Fig. 8. System timing budget.

The SI/PI FOM (Figure of Merit) is the eye-aperture with eye-mask window VdIVW = +/- 60mV from Vref/Vcent as defined by JEDEC standard. TdIVW is the DRAM RX timing window at VdIVW levels. At 4266 Mbit/s speeds, it's value=0.25*UI. Fig. 9 shows the LPDDR4X one UI data-eye with JEDEC standard parameters of TdIVW and VdIVW.

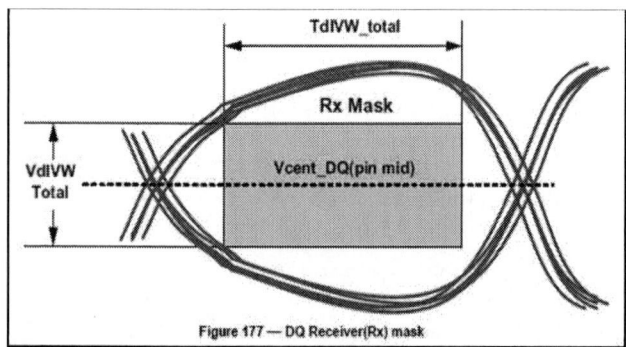

Fig. 9. LPDDR4X JEDEC RX eye-mask.

IV. RESULTS

This section presents the results of frequency-domain and time-domain analysis of 3-T vs 2-T capacitor usage in LPDDR PDN. Frequency-domain analysis investigated PCB and system PDN impedance response.

Time-domain analysis investigated complete LPDDR system consisting of PDN and SI channel. Co-SIPI analysis was performed to obtain comparison results of critical eye-aperture figure-of-merit.

A. Frequency-domain Impact

Frequency-domain PCB impedance and system PDN impedance impact was analyzed for the 3-T and 2-T capacitor cases and differences in the impedance was quantified as below.

a. PCB only comparison when using 3-T vs 2-T capacitor cases

The results show that with 3-T capacitors, PCB has lower impedance and smaller slope compared to 2-T capacitors only case.

In the full system PDN which consists of PCB, SoC PKG and SoC on-die decoupling capacitor, the peak impedance is lower in 3-T capacitor case. This lower peak impedance due to reduced ESL improves the power integrity of the overall system which translates into better eye-apertures.

Figure 10 shows the 3-T vs 2-T cases PCB impedance response (PMIC open), total PCB decoupling capacitance is kept the same. This clearly shows the improved impedance response of the 3-T case which has lower ESL. The system analyzed showed PCB inductance in 3-T case lower by ~40% of the 2-T case inductance.

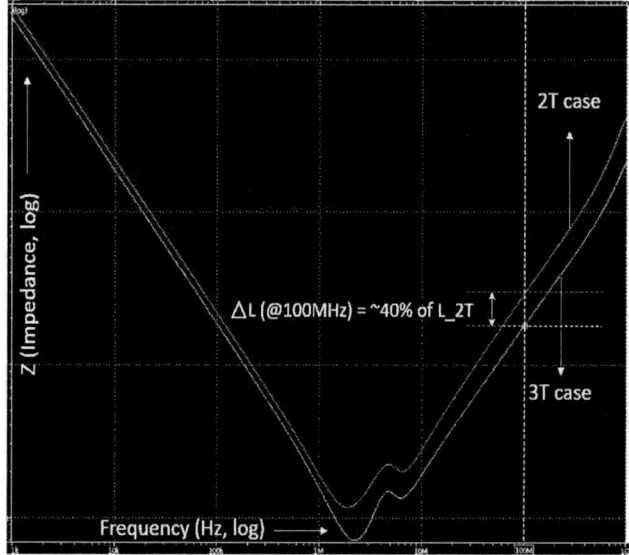

Fig. 10. 3-T vs 2-T PCB impedance response (PMIC open).

b. System PDN comparison when using 3-T vs 2-T

Fig. 11 illustrates the 3-T vs 2-T system impedance response. The system PDN consists of PMIC, PCB, PKG and Die. The total PCB decoupling capacitance is same in both cases.

The 3-T case shows lower peak impedance by about ~20% of 2-T case peak impedance. This improved PDN improves the PI and translates into better time-domain metric of eye-aperture as shown in next subsection.

The peak impedance follows the following relation.

$$Z_{peak} = \sim \frac{L}{R * C_{on-die}}$$

Where, L being the effective PDN inductance which includes the ESL of 3-T and 2-T capacitor cases. C, being the on-die decoupling capacitor.

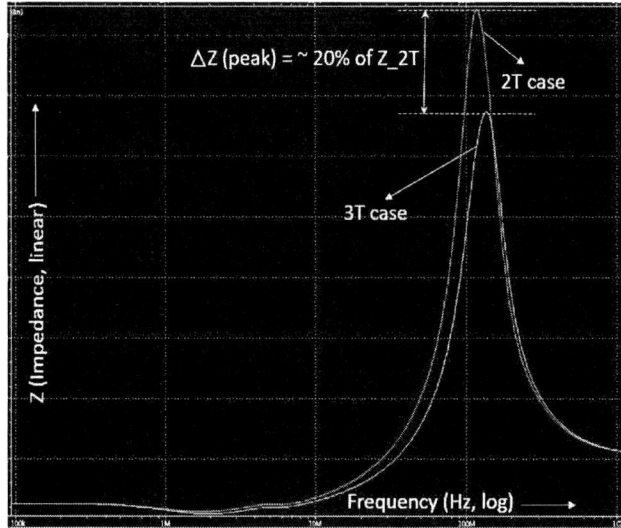

Fig 11. 3-T vs 2-T cases system impedance response.

B. Time-domain Impact

Fig. 12 shows the data-eye apertures comparison of 2-T and 3-T capacitor cases. The eye-aperture of 3-T case "Y" is greater than eye-aperture of 2-T case "X". This shows the impact of improved PDN in time-domain due to lower ESL of 3-T capacitor use. The improvement observed was around ~1.5% of UI.

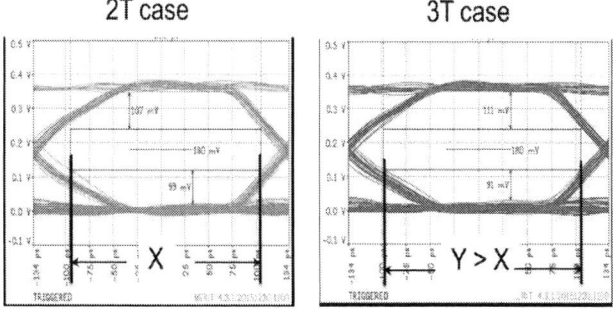

Fig. 12. 2-T and 3-T PCB caps impacting the time-domain eye-apertures. 3-T case shows improved eyes.

V. Conclusions

Summary and key takeaways of this paper are as follows.

1. The impact of 3-T decoupling capacitors on PDN in LPDDR4/4X/5 system was analyzed.

2. 3-T (8-T) multi-terminal capacitor construction and ESL reduction over 2-T capacitor was described.

3. LPDDR4/4X/5 system PDN with 3-T and 2-T capacitors and Co-SIPI setup was described to analyze the impact of 3-T capacitors on system performance.

4. Results of frequency-domain impedance plot of PCB shows inductance in the linear region being ~40% lower when using 3-T capacitors.

5. Results of frequency-domain system impedance plot shows peak impedance being ~20% lower in 3-T case over 2-T case.

6. Results of time-domain analysis shows wider eye-apertures when using 3-T vs 2-T case. The improvement observed was around ~1.5% of UI. This gives extra timing margin when using 3-T capacitors which leads to enhanced performance (higher speeds) or/and reliability.

7. Making use of 8-T decoupling capacitors further reduces the ESL and hence lowers the PCB and system impedance translating into even wider eye-apertures and performance.

8. 3-T capacitors offer improved PDN due to decreased ESL. It could also lead to cost savings over 2-T capacitors due to reduction in components for same PDN response.

9. 3-T capacitors could also reduce expensive on-die decap when keeping same system PDN.

10. The results presented were for a PoP SoC-DRAM system but are equally applicable for an external DRAM system.

11. This technique can be applied to other parallel interfaces such as DDR and GDDR.

Acknowledgment

The author thanks LPDDR, IO, PCB and PKG groups for their assistance.

References

[1] AVX Corporation, "Surface Mount Ceramic Capacitor Products, IDC Low Inductance Capacitors (RoHS), " pp. 82 – 85.

[2] JEDEC STANDARD, LPDDR4/4X, Jan., 2019.

[3] E. Bogatin, Essentail principles of signal integrity, 2014.

[4] S. Gupta, SoC On-die Decap Optimization for PDN in LPDDR, EPEPS, San Jose, CA, October 2018.

[5] S. Gupta, LPDDR4X (3732 Mbps) DBI impact on SI/PI and Power, EPEPS, San Jose, CA, October 2017.

[6] TN_4003_DDR4_network_design_guide, Micron Technical Note, 2014.

[7] A. Ruehli, G. Antonini, L. Jiang, "Circuit Oriented Electromagnetic Modeling Using the PEEC Techniques", Wiley-IEEE Press, 2017.

[8] T. Makharashvili, Y. S. Cao, A. E. Ruehli, J. Drewniak and D. G. Beetner, "Inductance Model of Decoupling Capacitors Including the Local Environment," in Proc. of the IEEE 26th Conference on Electrical Performance of Electronic Packaging and Systems (EPEPS), 2017.

[9] D. G. Figueroa and Y. L. Li, "A technique for the characterization of multi-terminal capacitors for high frequency applications," in Proc. of the Electronic Components and Technology Conference, 2000, pp. 445-448.

978-1-7281-1500-9/19 $31.00 © 2019 IEEE

[10] J. Kim, Y. Takita, K. Araki and J. Fan, "Improved Target Impedance for Power Distribution Network Design With Power Traces Based on Rigorous Transient Analysis in a Handheld Device," in IEEE Transactions on Components, Packaging and Manufacturing Technology, vol. 3, no. 9, pp. 1554-1563, Sep. 2013.

[11] J. W. Zhang, E. K. Chua, K. Y. See, W. J. Koh and W. Y. Chang, "Prelayout multi-layer PDN model for high-speed board," in Proc. of the IEEE Asia-Pacific International Symposium on Electromagnetic Compatibility (APEMC), 2016, pp. 269 – 272.

[12] C. R. Sullivan and Y. Sun, "Physically-Based Distributed Models for Multi-Layer Ceramic Capacitors," in Proc. of the IEEE Electrical Performance of Electrical Packaging, 2003, pp. 185 – 188.

[13] Y. S. Cao, T. Makharashvili, S. Connor, B. Archambeault, L. J. Jiang, A. E. Ruehli, J. Fan and J. L. Drewniak, "Top-layer Inductance Extraction for the Pre-layout Power Integrity Using the Physics-based Model Size Reduction (PMSR) Method," in Proc. of the IEEE International Symposium on Electromagnetic Compatibility (EMC), 2016, pp. 324 – 329.

[14] N. Bondarenko, T. Makharashvili, J. He, P. Berger, J. Drewniak, A. E. Ruehli, D. G. Beetner, "Development of Simple Physics-Based Circuit Macromodel From PEEC," IEEE Transactions on Electromagnetic Compatibility, vol. 58, no. 5, pp. 1485 – 1493, 2016.

[15] Understanding and characterizaing timing jitter, Tektronix, 2012.

Improved Correlation Between Accelerated Board Level Reliability (BLR) Testing and Customer BLR Results Using A Hybrid Closed-Form/Finite Element Methodology

Maxim Serebreni, Natalie Hernandez, Gil Sharon,
Nathan Blattau, Craig Hillman
DfR Solutions
Beltsville, Maryland, USA
e-mail: chillman@dfrsolutions.com

Ken Symonds
Western Digital Corporation
Marlborough, MA, USA
e-mail: ken.symonds@wdc.com

Abstract— **Performing board level reliability (BLR) testing is an industry-standard practice to validate the robustness of semiconductor packaging and provides guidance to the user as to the thermomechanical fatigue lifetime in the field. However, the occurrence of complex loading conditions (triaxiality) during thermal cycling in the actual application can result in significant deviations from expectations, especially when calculating acceleration factors using Coffin-Manson or Darveaux based equations. Attempts to develop more robust acceleration factors have been limited due to a combination of the large scale of electronic systems and the complex loading conditions that can occur during thermal excursions. To address these challenges, a hybrid methodology for reliability assessment of solder joints in complex assemblies was developed. The hybrid methodology is a combination of creep-equivalent finite element modeling and energy-based closed form equations. The finite element modeling (FEM) consists of a coupled linear elastic thermomechanical analysis of the electronic assembly with secant equations to account for creep behavior. Vectorized stress and strain magnitudes, shear and axial, are then extracted from the critical solder joints. Idealized hysteresis loops are used to capture energy dissipation. Time to failure is then determined through closed form energy-based damage models that partition energy dissipation based on the orientation of the stress loading. This approach is computationally efficient and accurately captures system-level effects, such as underfill, mirroring, and housing-board interactions.**

Keywords-Board Level Reliability (BLR); Finite Element, Solder Fatigue, Energy Partitioning, Mirrored BGAs.

I. INTRODUCTION

The adoption of Solid-State Drives (SSD) by industries ranging from mobile devices to data centers has been enabled by the small form factor and low power consumption of NAND memory devices. Implementation of NAND devices in aerospace, military and automotive applications necessitates an appropriate test method to qualify the reliability of NAND Ball Grid Array (BGA) packages. Accelerated thermal cycling is often performed on to ensure failure free operation within the designed life cycle of the product. Electronic packages are exposed to multiple thermal profiles to determine an acceleration factor between various thermal loads. Empirical accelerated life models that are calibrated using experimental data provide information that enables extrapolation of expected useful life at thermal environments that a device has not been validated under with high degree of confidence [1]. A significant concern with such empirical models can occur when a device has been qualified on a test board with different characteristics than the final circuit card that the device will be assembled to in the field [2]. Circuit card properties such as the coefficient of thermal expansion (CTE), board thickness and plating have been shown to influence life of solder joints in BGA packages under thermal cycling environments [3]. Any variation in the test vehicle circuit card properties from the product properties can result in significant deviation in empirical model predictions. Additionally, system level factors such as circuit card configuration have been found to influence the life of surface mount components. Constraining the PCB to a stiff metal housing can increase solder joint strain in surface mount components by locally changing the board warpage around components [4]. Double-sided board configurations have also been found to reduce the life of BGA components under accelerated thermal cycling [5,6]. The relative decrease in fatigue life with these component configurations has been shown to greatly depend on board stiffness [7].

Finite element (FE) simulations have been successfully used for reliability assessment of electronic components as part of a Reliability Physics Analysis (RPA). RPA combines simulation tools and empirical models to predict the reliability of components and advanced packages under complex environmental and operational loads by identifying the susceptibility of components to the dominant failure mode and mechanism [8]. Failure modes can occur in package internal interconnects such as wire bonds, board level solder joints or may occur inside the circuit card itself.

978-1-7281-1500-9/19 $31.00 © 2019 IEEE

Empirical models are necessary to correlate simulation results with wear-out based mechanisms or a failure criterion that is calibrated using wide range experimental data.

A new hybrid approach is developed for reliability assessment of solder joints in BGA devices by combining FE methods with empirical equations. The empirical equations are implemented both to reduce the complexity of the FE analysis and define a more appropriate damage indicator for fatigue life predictions in BGA solder joints.

II. FATIGUE LIFE PREDICTION MODELS

Reliability assessment of BGA components under thermal loading is conducted using a variety of commercially available FEM software packages such as Abaqus, Ansys and NX Nastran. Simulation of a BGA package, solder balls and PCB exposed to a number of thermal cycles is performed to achieve steady state creep deformation in the solder. Once secondary creep deformation is reached, the energy density accumulation in the solder reaches a nearly constant value from cycle to cycle. Only then can a volume of the critical solder joint be taken for averaging dissipated creep energy density. The volume average energy density is then used as the damage indicator in the fatigue equation. Darveaux proposed a model based on the observed time it takes for crack to nucleate and propagate through the solder [9]. Equation 1 provides the term for crack nucleation, equation 2 represents the rate of crack propagation and equation 3 provides the fatigue life time based on the first two terms.

$$N_o = K_1 (\Delta W_{ave})^{K_2} \qquad \text{Eq. 1}$$

$$\frac{da}{dN} = K_3 (\Delta W_{ave})^{K_4} \qquad \text{Eq. 2}$$

$$N_{63.2} = N_o + \frac{a}{\frac{da}{dN}} \qquad \text{Eq. 3}$$

Where K_1 through K_4 are fitting constants, da/dN is the crack propagation rate in the solder and a is the length of solder at the pad interface through which a crack will propagate. The constants provided for the Darveaux model were obtained using a different FE approach and therefore cannot be used for comparative analysis with the approach presented in this study. The energy density used in the Darveaux model is calculated from all the stress and strain components without a particular focus to the loading mode the solder joint experiences. In general, it is known that solder joints are subjected to dominant cyclic shear loading due to CTE mismatch between components and boards. The two dominant loading modes result in different crack behavior as shown in Figure 1. Fatigue experiments on bulk and joint scale Pb-free solder under cyclic shear loading with the addition of a constant axial load have shown decreased fatigue life compared to those without the axial loading component [10, 11].

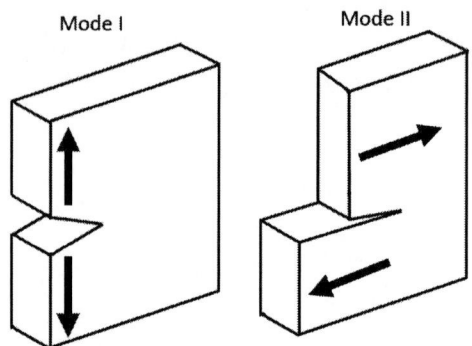

Figure 1. Loading mode of a cracked specimen
Mode I: tensile opening, Mode II: in-plane shear.

This observation indicates that the inclusion of additional loading modes can significantly alter the stress state in solder joints and that a single scalar damage indicator is not suitable for reliability assessment under mixed mode loading. Solder joints in BGA packages are subjected to shear loading due to the in-plane CTE mismatch between the PCB and component as shown in Figure 2. Fatigue cracks tend to propagate either at the package or board interface within the bulk solder region. The location of fatigue cracking will depend on a combination of global and local CTE mismatch as well as the axial Mode I loading components as a result of board or package warpage.

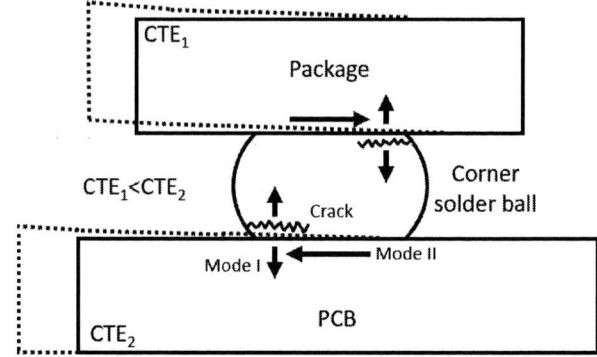

Figure 2. Influence of loading mode on BGA solder joint cracking due to CTE mismatch.

The location in the solder where crack nucleation occurs will experience highest stress concentration that is driven by mixed-mode loading conditions. The rate of crack propagation is then believed to be controlled by the dominant loading mode which is either Mode I or Mode II.

In this analysis, a new approach is introduced that is capable of capturing the contribution of loading mode in BGAs solder joints. Thermal cycling of NAND BGA components is conducted to generate experimental data that will be use to demonstrate the method's capability to predict fatigue life of solder joints in different accelerated thermal environments for double-sided BGA configurations.

978-1-7281-1500-9/19 $31.00 © 2019 IEEE

III. EXPERIMENTAL APPROACH

A. Test Vehicle

The test vehicle consists of a PCB and NAND design based on actual SSD product with modifications that incorporate direct resistance-based measurement across all solder balls in discrete NAND BGAs. Figure 3 shows the daisy-chain layout of BGA solder joints. BGA devices mounted in a mirrored fashion are monitored per PCB. Each test leg is comprised of 2PCBAs and 24 NAND devices per leg.

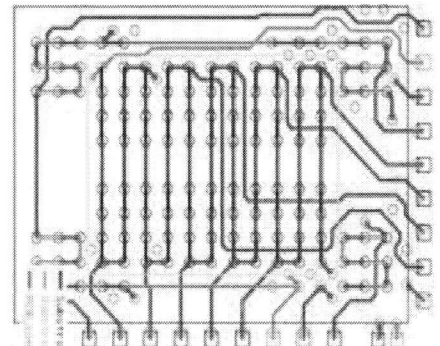

Figure 3. Experimental NAND BGA layout with daisy chained solder balls.

NAND BGA packages consist of multiple stacked die arrangements that are wire bonded. The stacked dies are held together using a die attach film (DAF) and assembled on a thin substrate. A representative cross-section of a stacked die BGA is shown in Figure 4 to illustrate the internal structure of these components. The influence of DAF on board level reliability can be neglected due to the high modulus of the die compared to the DAF.

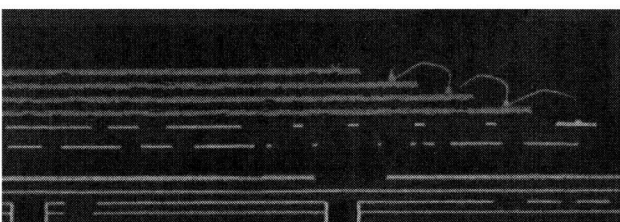

Figure 4. Cross-section representative of stacked-die NAND memory BGA.

Die stacking and DAF thicknesses may vary from package to package. For model simplifications the internal structure can be modeled as a single material block by estimating the effective material properties using the rule of mixtures for composite materials. This will provide an effective CTE and modulus for the package from the substrate, mold and die geometry and material properties.

B. Thermal Cycling

To validate the empirical model for NAND BGAs in SSDs, a controlled experiment was performed which consisted of different hot soak temperatures of 85°C, 105°C and 125°C all with a minimum soak temperature of 0°C. Each thermal profile consisted of a 10-minute dwell at each extreme and a 10°C/minute ramp rate according to JESD22-A104D standard [12]. These profiles are illustrated in Figure 5.

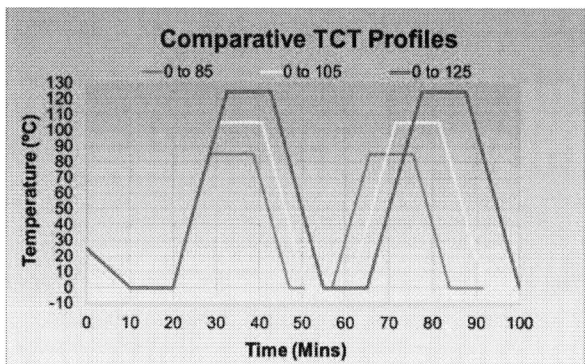

Figure 5. Thermal cycling profiles

For each profile, discrete NAND BGAs assembled on different PCBAs are connected to a data acquisition unit. Both resistance and temperature are captured in 10s increments over repeated thermal cycles. Baseline resistance measurements range from 6 to 8 Ohms; failure criteria is defined as when resistance first hits 10Ω. Testing continues until all BGAs have failed. The chamber configuration is shown in Figure 6.

Figure 6. PCB test coupons inside the thermal chamber.

IV. EXPERIMENTAL RESULTS

Reliability data (cycles to failure) of the NAND BGAs were analyzed using 2-parameter Weibull distributions to determine the characteristic life. An overlay of results from each thermal leg of the DOE are shown in Figure 7 [13]. The unreliability of Weibull distribution is shown in equation 4.

$$F(t) = 1 - exp\left(-\frac{t}{\eta}\right)^{\beta} \qquad \text{Eq. 4}$$

Where F(t) is the number of thermal cycles at which failure occurs. The characteristic life corresponds to the number of cycles at which 63.2 percent of the population is expected to fail.

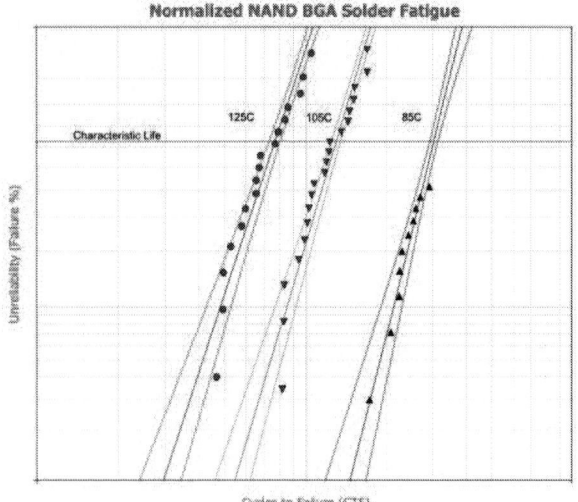

Figure 7. Weibull distributions of recorded failures.

The lowest fatigue life recorded for the temperature profile with highest temperature range. The second lower fatigue life occurred for the second lowest temperature range as expected with a 1.69X increase in characteristic life. The characteristic life with the lowest temperature range of 85°C is found to increase by a factor of 2.32X from 105°C. This trend indicates that the relative increase in fatigue life is not linearly proportional to temperature range but rather depends on other properties such as the creep behavior of the solder alloy. The characteristic lifetime obtained in from these experiments will be used as a metric for validating the accuracy of the proposed hybrid methodology in predicting the characteristic life of the tested components. Figure 8 shows a cross-section taken along the length of a double-sided NAND exposed to 0 to 125°C profile.

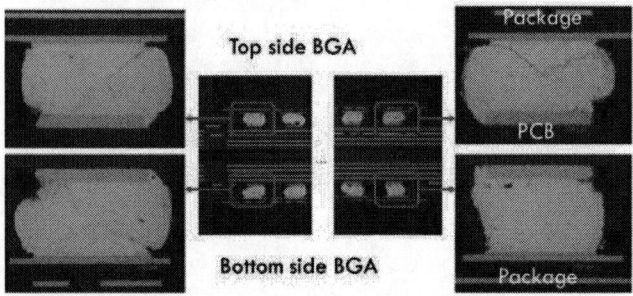

Figure 8. Cross-section of mirrored NAND BGA after 0 to 125°C thermal profiles.

The non-functional corner joints indicate diagonal cracking on one side of the package as well as cracking along the package and board interface of the solder. The shift in cracking mode could be attributed to the placement of the die in the package, solder microstructure and geometry. Cracks under the die region were found to have cracks along the package side of the solder joint.

V. BLR SIMULATIONS

Current methods for reliability assessment of electronic packages using FEM simulations of the package and providing assumptions pertaining to material properties of the package constituents. For the most part, materials such as silicon, copper and FR4 boards are assumed to behave in their linear elastic regime during typical accelerated thermal cycling regimes. Simplification of solder joint behavior can be made using constitutive material models that describe the viscoplastic deformation of the solder alloy.

The actual test vehicle was populated with double-sided components; however, simulations enable modeling both single and double-sided components on the same analysis. The simulated assembly enables virtual assessment of components or test vehicles that have not been tested. Implementing such an approach can drastically reduce qualification efforts of solder joint reliability in real world applications. Therefore, the accuracy of the prediction method is paramount for successful implementation of electronic devices.

The accuracy of the reliability assessment is attributed to assumptions made by modelling approach and quality of empirical model calibration using relevant experimental data to the package under assessment. Knowing which step in the reliability assessment resulted in deviation of fatigue life prediction is often difficult to determine. The culprit that is usually blamed for poor reliability assessment is the applicability of damage model to the analysis performed [15]. Influence of certain simulation methods has been shown to be have minimal effect on fatigue life prediction of solder joints [16]. Deviation of fatigue prediction from experimental results could be attributed to the assumption that the damage model for a certain type of component and loading environment can be used for another type of component in a different loading environment.

One of the most critical approximations in solder joint reliability assessment is the creep behavior of the solder joint itself. A secant modulus approach was previously used to determine the effective creep behavior of the eutectic phase in SnPb solder and for Pb-free solder together with the Mori-Tanaka method [17,18]. The use of the secant modulus allows one to calculate the overall creep strain from the rate-independent plasticity. Figure 9 shows the secant modulus as function of temperature for SAC305 solder determined from rate-independent plasticity corresponding to 2.4 percent strain.

The modulus of elasticity for Pb-free solders decrease linearly with temperature. A similar decrease is observed with the secant modulus. The slight non-linearity of the secant modulus dependence in temperature depends on the hardening behavior derived from the plastic flow. On the other hand, more complex time-dependent viscoplastic constitutive models are used to model solder joint behavior under thermomechanical loading. Such methods can provide more exact estimation of stresses in the solder but also possess limitations of model size and computational cost. Decisions on the exact simulation approach should be made in regard to the desired level of detail in the analysis. The hybrid approach presented in this study is of more practical for assessments that require full scale simulations of PCBs with multiple components and thousands of solder joints, especially when the corner solder joint cannot be assumed as the site of first failure. Figure 9 shows a 3D model of the NAND BGA using in thermal cycling experiments. The 132 solder balls of the BGA are arrange in a non-standard layout.

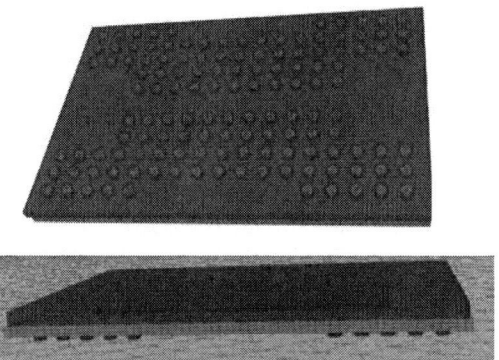

Figure 9. Solder ball layout and model of the NAND BGA used in thermal cycling.

The hybrid methodology process flow is outlines in Figure 10 and consists of four general steps. In the first step design files of the circuit card are loaded into Sherlock software to compute the PCB properties. In the second step material properties are determined from a material library, components are defined from a part database and thermal environment is defined by the user. All this information is then used by Sherlock which automatically generates an input deck. Temperature increments for the analysis are defined by the user. The input deck is then solved by an external FEA engine and returned to the software for post processing of the results. The post processing analysis is used for the reliability assessment of the BGA solder joints.

During thermomechanical loading, solder joints experience non-recoverable deformation. The ductility of the solder alloy is an important parameter in understanding the extent of damage the solder experience.

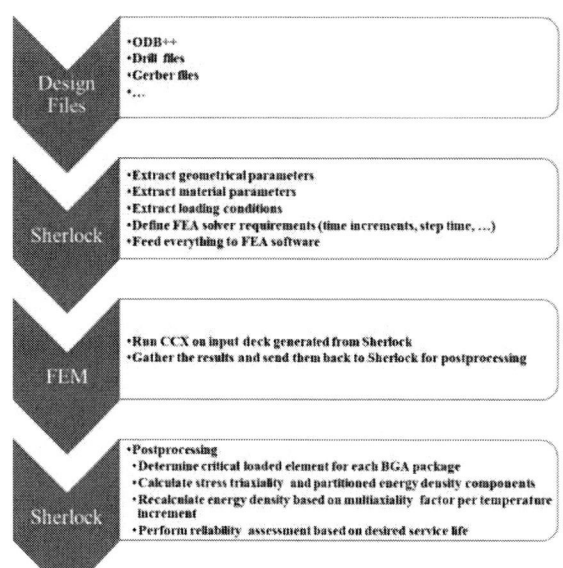

Figure 10. Process flow for the hybrid methodology for reliability assessment of solder joints.

Different thermal excursions will induce a different range of strain. The dominant loading mode solder joint in BGA components experience is shear loading. Solder ductility has been shown to depend on the loading mode whether it is under purely shear or axial loading. Stress triaxiality is an important factor used to describe damage due to plastic deformation in ductile materials as shown in Figure 11. Higher triaxiality factor has been correlated with decrease in ductility. The triaxiality factor is a non-dimensional ratio between the hydrostatic stress and Von-misses stress that is shown in equation 5.

$$TF = \frac{\sigma_{HYD}}{\sigma_{VM}} = \frac{(\sigma_1 + \sigma_2 + \sigma_3)}{\frac{1}{\sqrt{2}}\sqrt{(\sigma_1 - \sigma_2)^2 + (\sigma_2 - \sigma_3)^2 + (\sigma_3 - \sigma_1)^2}} \quad \text{Eq. 5}$$

This TF can be seen to range from -0.33 in region 3 and increase beyond 1 for purely axial stress states. Below a value of -0.33 the TF indicates that the stress state does not contribute to damage accumulation and is often found at compressive or hydrostatic stress states. In region 2 the ductility is government by maximum shear stress and gradually transitions to region 1 that is driven by void nucleation and growth under purely uniaxial loading.

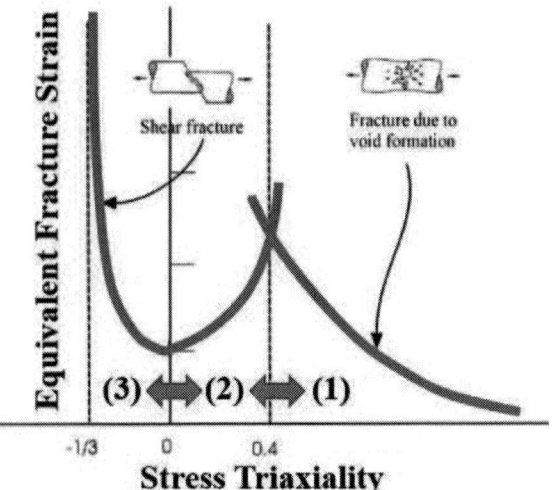

Figure 11. Equivalent fracture strain as function of stress triaxiality factor obtained from [18].

Total accumulated energy density is calculated by the energy under the stress-strain curve under cyclic loading conditions. In this method, the accumulated energy density is treated by partitioning the contribution to energy vectors. The two dominant vectors as previously described are the axial Mode I and shear Mode II components as shown in equation 6.

$$\Delta W_{acc} = \int \varepsilon_{ij}\, \sigma_{ij} = \int (\varepsilon_{ij}\sigma_{ij})_{Mode\,I} + \int (\varepsilon_{ij}\sigma_{ij})_{Mode\,II} \qquad \text{Eq. 6}$$

Mode I energy is taken from the normal stress and strain components to the in-plane direction of the package. Most cracks in solder joints propagate along the solder interface that is perpendicular to the normal direction of the package. The shear energy component is defined as the remaining deviatoric shear and two remaining principal orientations that are parallel to the in-plane configuration of the package and PCB. Manson and Halford [19] proposed the multiaxiality factor (MF) to account for the strain enhancement effect observed for pure torsion loading conditions that result in a TF zero. Equations 7 and 8 show the proposed parameter that retains the TF above a value of 1 and limits TF to 0.5.

$$MF = \frac{1}{2 - TF}; \quad TF \leq 1 \qquad \text{Eq. 7}$$

$$MF = TF; \quad TF \geq 1 \qquad \text{Eq. 8}$$

The MF parameter is multiplied by energy components to provide a stress state correction for accumulated energy density. Figure 12 shows the TF and MF for single and double sided BGAs. Each of the calculated values are performed for the solder element with the highest Von-misses stress. Once the critical element is identified the TF and MF are calculated using stress components that are averaged from 8 integration points. This calculation is performed for the highest loaded element at the ramp up and ramp down step from the same reference temperature. It can be seen that TF is negative at 0°C and positive at 125°C. The

TF at the high temperature is representative of a more shear dominated stress state and for low temperature is outside the ductility driven damage range. During the 0°C it is assumed that the solder is under compressive stress state that does not contribute to damage that is directly aligned with the calculated value in this approach.

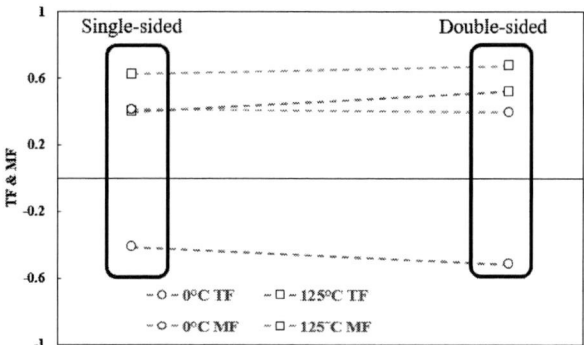

Figure 12. Stress triaxiality and multiaxiality factors for highest loaded solder element in single and double sided BGAs.

The MF in Figure 12 is calculated from the TF and is represented by the blue lines. The MF is found to increase for the double-sided BGA at high temperatures and decrease at low temperatures with respect to the single-sided BGA. The shift in MF proportionally to the temperature and BGA configuration will provide a scaling factor for Mode I and Mode II energy components. An alternative approach using MF to weigh damage indicator has been previously developed although with strain as damage indicator rather than energy density as used in this model [20]. In this method, energy density is first partitioned to mode I and Mode II components and then weighted by the MF.

Figure 13 shows a cross-sectional view of single and double-sided BGAs at a temperature of 125°C. In this image the strain in corner solder joints of the double-sided BGA are higher than for the single-sided one. The difference between decrease in board strain and increase in solder strain is inherent for the BGA packages used in this study. Other BGA package types and board thicknesses might not possess the same behavior and it is important to perform simulations for each package type to determine the increase in strain solder joint can experience between the different configurations.

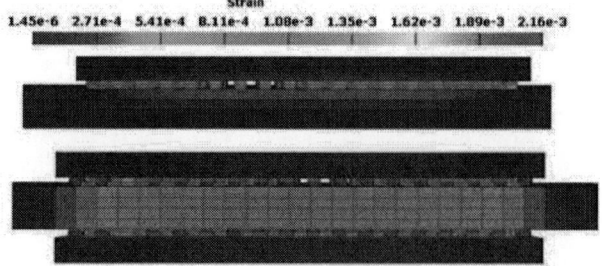

Figure 13. Equivalent strain distribution in single and double-sided BGAs.

Mode I energy density is calculated from the normal stress and strain components to the package surface. Mode II components are calculated from the sum of the deviatoric and hydrostatic stress and strain components. Fatigue life prediction is accomplished using a Morrow type model that correlates the number of cycles to the accumulated energy density. Figure 14 shows energy components for both BGA configurations for the 0 to 125°C thermal cycle. Weighing of the energy density is accomplished using equations 9 and 10. The corrected energy density can be scaled depending on the value of MF that can not be lower than 0.5 and 1 thereby normalizing the energy density with respect to the dominant stress state the solder joint experience.

$$\Delta W_{Mode\,I, MF} = MF x \Delta W_{Mode\,I} \qquad \text{Eq. 9}$$

$$\Delta W_{Mode\,I, MF} = MF x \Delta W_{Mode\,II} \qquad \text{Eq. 10}$$

The increase in Mode II energy density for double-sided BGA is a factor of 2.1X at high temperature and factor of 1.78X for the low temperature range. Single-sided BGAs experience higher variation with a factor of 6X between Mode I and Mode II energies at high temperature and a factor of 4.3X at the low temperature.

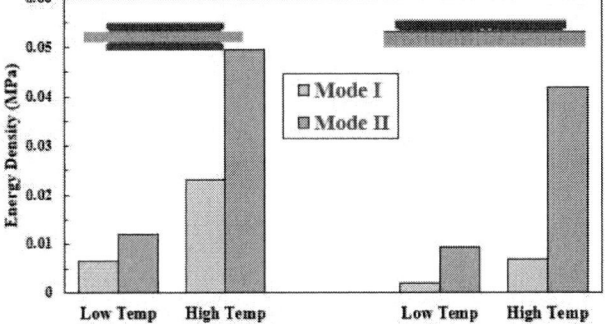

Figure 14. Predicted energy density components for single and double-sided BGAs for 0 to 125°C thermal profile.

These two loading modes can be summed to provide a total energy density as shown in Figure 15. It was found that the relative difference between Mode I and Mode II energy densities decreases with lower temperature range; however, variation in stress state remain consistent. Therefore, the predicted trend in total-energy for double and single-sided BGAs is observed to be consistent between the three thermal profiles.

The final step of the reliability assessment is performed using fatigue equation. In this method the fatigue model is shown in equation 11 is partitioned to two independent damage indicators that have a power law relationship with cycles to failure.

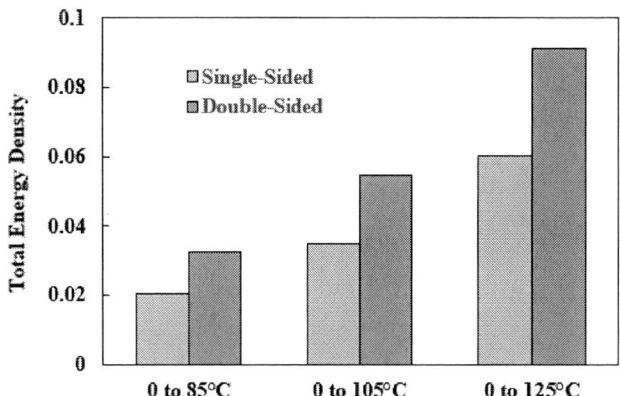

Figure 15. Predicted total energy density for the three temperature profiles.

The Weibull distribution of the predicted cycles to failure for each of the BGA components is plotted against the experimental cycles to failure for components populated on the top side of the BGA in Figure 16. The FEA model shows excellent correlation in predicted the characteristic life and is found to be within 20 percent of the experimental characteristic life. The failure distribution of the predicted time to failure of each NAND device allows for statistically significant result to be obtained. 16.

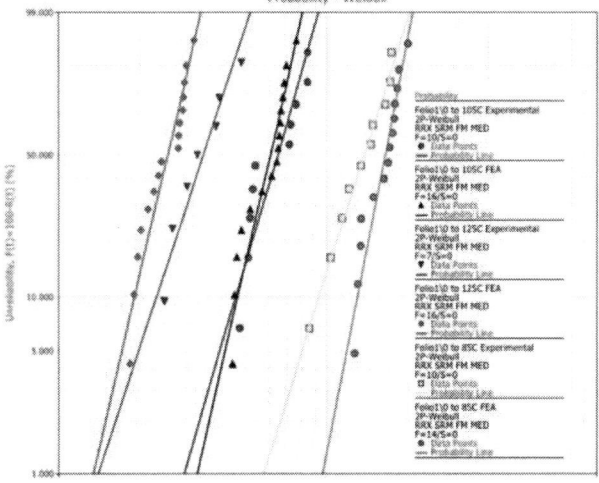

Figure 16. Weibull distribution of experimental and predicted failures of NAND BGAs at three temperature profiles.

The variation in time to failure is captured from location on the board and tolerances in the computational approach itself. Calculated Mode I and Mode II energy densities are then substituted into equation 11 for predicting time to failure for each device.

$$N_f = C_1 \left(\Delta W_{Mode\,I,MF} \right)^{n_1} + C_2 \left(\Delta W_{Mode\,II,MF} \right)^{n_2} \qquad \text{Eq. 11}$$

Fitting constants are C_1, C_2 and n_1 and n_2 are experimentally determined fatigue life exponents. The fitting constants were determined using extensive experimental data that ranged from single-sided BGA components to underfilled components under a variety of thermal cycles and PCBs. In this study, the hybrid methodology presented here is applied for reliability assessment of double-sided BGAs with three thermal profiles. The difference between predicted and experimentally obtained characteristic for the three temperature profiles are shown in Figure 17.

A slightly more conservative prediction was found for the temperature cycle with largest temperature range. The hybrid FEA method presented in this paper provides an accurate virtual qualification approach for board level reliability assessment of BGA components. Fatigue life prediction was shown to correlate well with experimental obtained failure rate of double-sided NAND BGA devices. This method can be extended to provide qualification of alternative configuration of BGA devices and enable cost reduction by minimizing the dependence on costly experimental validation efforts.

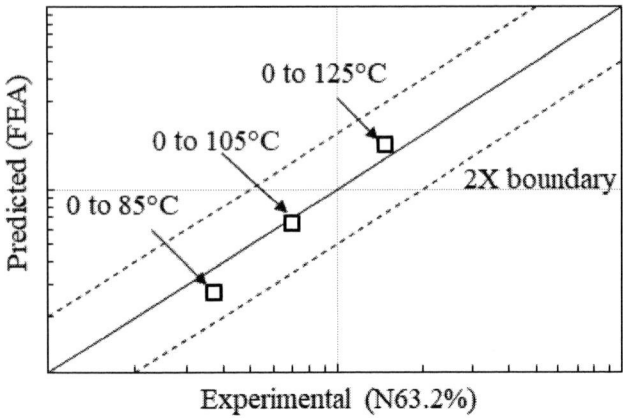

Figure 17. Predicted vs. experimental characteristic lives for mirrored NAND BGAs at three thermal profiles.

VI. CONLUSION

In this study a novel approach for predicting the fatigue life of solder joints in BGA components is presented. The FEA based analysis is coupled with empirical equations to provide a creep equivalent deformation in solder joints under thermomechanical loading using static loading steps. Energy density partitioned into Mode I and Mode II components showed to correlate well with single and double sided BGA configurations. Thermal cycling of double-sided BGAs at various environments conducted to provide experimental results for model validation. The hybrid method used for predicting the fatigue life of double-sided BGAs is found to closely match experimental results between different thermal cycles for double-sided devices. The method presented here can be used to predict board level reliability of BGA devices due to system level effects.

REFERENCES

[1] Van Driel, W. D., G. Q. Zhang, J. W. C. De Vries, M. Jansen, and L. J. Ernst. "Virtual prototyping and qualification of board level assembly." In *Electronics Packaging Technology Conference, 2004. EPTC 2004. Proceedings of 6th*, pp. 772-775. IEEE, 2004.

[2] Maxim Serebrini and Greg Caswell (*2018*) The Impact of Glass Style and Orientation on the Reliability of SMT Components. International Symposium on Microelectronics: Fall 2018, Vol. 2018, No. 1, pp. 000699-000706.

[3] Lall, Pradeep, Naveen Singh, Jeffrey C. Suhling, Mark Strickland, and Jim Blanche. "Thermo-mechanical reliability tradeoffs for deployment of area array packages in harsh environments." *IEEE Transactions on Components and Packaging Technologies* 28, no. 3 (2005): 457-466.

[4] Dudek, R., M. Hildebrand, S. Rzepka, J. Beintner, R. Döring, L. Scheiter, B. Seiler, Th Fries, and R. W. Ortmann. "Board level reliability assessment of consumer components for automotive use by simulation and sophisticated optical deformation analyses."

[5] Ye, Yuming, Sang Liu, Yunhua Tu, Limin Chen, Jian Zhang, and Zhiwei Song. "Assessment on reliability of BGA package double-sided assembled." In *High Density Packaging and Microsystem Integration, 2007. HDP'07. International Symposium on*, pp. 1-4. IEEE, 2007.

[6] Hagberg, Juha, Jussi Putaala, Juha Raumanni, Olli Salmela, and Timo Galkin. "BGA Interconnection Reliability in Mirrored Module Configurations." *IEEE Transactions on Components, Packaging and Manufacturing Technology* 7, no. 10 (2017): 1634-1643.

[7] Chaparala, S., J. M. Pitarresi, S. Parupalli, S. Mandepudi, and M. Meilunas. "Experimental and numerical investigation of the reliability of double-sided area array assemblies." *Journal of electronic packaging* 128, no. 4 (2006): 441-448.

[8] Yang, Decai. "Physics-of-failure-based prognostics and health management for electronic products." In *2014 15th International Conference on Electronic Packaging Technology*, pp. 1215-1218. IEEE, 2014.

[9] Darveaux, Robert. "Effect of simulation methodology on solder joint crack growth correlation." In *Electronic Components & Technology Conference, 2000. 2000 Proceedings. 50th*, pp. 1048-1058. IEEE, 2000.

[10] Liang, J., N. Dariavach, P. Callahan, and D. Shangguan. "Inelastic deformation and fatigue of solder alloys under complicated load conditions." *Journal of electronic packaging* 129, no. 2 (2007): 195-204.

[11] Hsieh, L-Y., H-C. Yang, and T-C. Chiu. "Ratcheting and creep responses of SAC solder joints under cyclic loading." In *Microsystems, Packaging, Assembly and Circuits Technology Conference (IMPACT), 2011 6th International*, pp. 96-99. IEEE, 2011.

[12] Standard, J. E. D. E. C. "Temperature cycling." JESD22-A104D, JEDEC Solid State Technology Association, Arlington, VA (2009): 158-162.

[13] Symonds, K. "Case Study: Solder Fatigue Model Verification using NAND BGAs" 2019 Design for Reliability Conference (2019)

[14] Coyle, Richard, John Osenbach, Maurice N. Collins, Heather McCormick, Peter Read, Debra Fleming, Richard Popowich, Jeff Punch, Michael Reid, and Steven Kummerl. "Phenomenological study of the effect of microstructural evolution on the thermal fatigue resistance of Pb-free solder

joints." *IEEE Transactions on Components, Packaging and Manufacturing Technology* 1, no. 10 (2011): 1583-1593.

[15] Syed, Ahmer, Gil Sharon, and Robert Darveaux. "Factors affecting Pb-free flip chip bump reliability modeling for life prediction." In *2012 IEEE 62nd Electronic Components and Technology Conference*, pp. 1715-1725. IEEE, 2012.

[16] Pei, Min, and Jianmin Qu. "Hierarchal modeling of creep behavior of SnAg solder alloys." *Journal of Electronic Packaging* 130, no. 3 (2008): 031004.

[17] Sharma, Pradeep, Abhijit Dasgupta, Surya Ganti, and James Loman. "Prediction of rate-independent constitutive behavior of Pb-free solders based on first principles." In *ASME 2002 International Mechanical Engineering Congress and Exposition*, pp. 125-134. American Society of Mechanical Engineers, 2002.

[18] Bao, Yingbin, and Tomasz Wierzbicki. "On fracture locus in the equivalent strain and stress triaxiality space." *International Journal of Mechanical Sciences* 46, no. 1 (2004): 81-98.

[19] Manson, S. S., and G. R. Halford. "Discussion:"Multiaxial Low-Cycle Fatigue of Type 304 Stainless Steel"(Blass, JJ, and Zamrik, SY, 1976 Winter Annual Meeting)." *journal of Engineering Materials and Technology* 99, no. 3 (1977): 283-285.

[20] Kuczynska, Marta, Natalja Schafet, U. Becker, B. Métais, A. Kabakchiev, P. Buhl, and S. Weihe. "The role of stress state and stress triaxiality in lifetime prediction of solder joints in different packages utilized in automotive electronics." In *2016 17th International Conference on Thermal, Mechanical and Multi-Physics Simulation and Experiments in Microelectronics and Microsystems (EuroSimE)*, pp. 1-10. IEEE, 2016.

Fabrication and Reliability Demonstration of 3 μm Diameter Photo Vias at 15 μm Pitch in Thin Photosensitive Dielectric Dry Film for 2.5 D Glass Interposer Applications

Daichi Okamoto, Yoko Shibasaki,
Daisuke Shibata and Tadahiko Hanada
TAIYO INK MFG.CO.LTD
900 Hirasawa, Ranzan-machi
Hiki-gun, Saitama, 355-0215, Japan
e-mail: dano@taiyo-america.com

Fuhan Liu, Mohanalingam Kathaperumal
and Rao R. Tummala
Georgia Institute of Technology
3D Systems Packaging Research Center (PRC)
Atlanta, GA, USA

Abstract—In this paper, the authors report on the development of a novel epoxy-based photosensitive dielectric dry film material (PDM). The PDM has two key features: 1) Low CTE value of 30 ppm/°C resulting from the concentration optimization of nano-sized fillers in the material composition, 2) Low Stress of the PDM films, attributed to low-temperature (180 °C) processing of the polymer dielectric which is lower than most of the known advanced dielectric materials. Improvements reported on these two key features will significantly enhance the reliability of high-density packages. To assess the reliability of the PDM, we have fabricated a high-density daisy chain structure on the glass panel consisting of 400 vias of 3 μm diameter at 15 μm pitch. We have also used the PDM as passivation layer on the surface of the test vehicle. Following the fabrication of the test microvia chain, electroless nickel immersion gold (ENIG) process was performed for surface finish.

We have performed reliability measurements of the 3 μm diameter vias after nHAST (Non-Bias Highly Accelerated Stress Tests) at 130 °C, 85 % R.H. for 100 hours followed by thermal cycling test (TCT) with a dwell time of 15 minutes at 125 °C and -55 °C. We have observed no open circuit failure occurred during the TCT. We measured the resistance of the daisy chain circuit by four-point probe method every 100 cycles up to a total of 1500 cycles. The resistance change was less than 5 % even after 1500 cycles which clearly demonstrate the superior reliability of the PDMs. In conclusion, the newly developed PDM is a suitable dielectric material for high-density RDL applications such as 2.5D interposers and fan-out packages.

Keywords- dielectric materials, photosensitive, 3 μm photo vias, reliability, high-density RDL, interposer

I. INTRODUCTION

The demand for higher data bandwidth interconnection among multiple dies is one of the critical driving forces for high-density packaging technologies, such as multi-chip fan-out wafer level packages (FO-WLP) and 2.5D interposers. RDL with fine routing lines of less than 5 μm diameter and small micro-vias less than 5 μm are necessary for the wiring density requirement for such applications. Laser ablation processes are commonly used to form micro-vias, however, it is still very challenging to achieve micro-vias below 5 μm using this technique [1][2]. A recent report has shown that the semi-additive process (SAP) can achieve 2 μm line and space on a smooth surface. Micro-via formation by using photolithography of photosensitive dielectric materials is gaining more momentum among researchers for high performance computing applications.

Photosensitive polyimide (PI) and polybenzoxazole (PBO) are frequently employed to obtain ultra-fine vias. However, their high curing temperature is a major challenge because high temperature curing can cause huge warpage which affects lithographic resolution and reliability. The large shrinkage of photo sensitive PI is a big problem and it has been investigated and reported. [3] High Coefficient thermal expansion (CTE) of the materials (PI and PBO) is another challenge. Larger CTE mismatch between copper and polymer may cause interfacial failure. To avoid reliability failure, dielectric materials with low CTE are important. Adding fillers into matrix resin is one of the common methods to decrease the CTE. However, there is a tradeoff relationship between low CTE and high resolution due to UV light scattering of a filler. The authors found that surface treated nano-sized inorganic fillers can help to achieve an optimum balance of high resolution and low CTE as reported earlier [4]. This paper reports on the fabrication and reliability demonstration of 3 μm diameter photo vias at 15 μm pitch in thin photosensitive dielectric dry film for 2.5 D glass interposer applications. Additionally, to confirm the resistance of one via, a test coupon which has cross-bridge kelvin resistor structure was fabricated. [5]

II. PHOTOSENSITIVE DIELECTRIC MATERIAL (PDM)

1. Material Properties of PDM

The glass transition temperature (T_g) and CTE of the PDM were measured by thermomechanical analysis (TMA). Tensile test was performed to measure the elastic modulus, tensile strength and elongation. Dielectric properties were measured using a network analyzer with split post dielectric resonator (SPDR) fixture. The water absorption rates of PDM films were measured by dipping PDM into boiling water for 1 hour. The material properties of the PDM reported here are listed in Table 1.

978-1-7281-1500-9/19 $31.00 © 2019 IEEE

Table 1. Material Properties of PDM

Properties	Unit	PDM
Tg (@TMA)	(deg.C)	180 - 185
CTE alpha 1	(ppm)	30-35
Elastic Modulus	(GPa)	3.5 - 4.0
Tensile Strength	(MPa)	90 - 95
Elongation	(%)	5.5 - 6.0
Dk (10GHz)		3.3
Df (10GHz)		0.019
Water absorption	(%)	0.84

2. Ultra-small Via formation in PDM and Inspection

To confirm a small via opening performance of the PDM, PDM of 5 μm thickness was laminated on electrolytic plated copper. Then the PDM was exposed by 200 mJ/cm² of UV light (365 nm). After the exposure process, the PDM was developed using 1.0 wt % aqueous sodium carbonate. At the end of the process, a thermal cure at 180 °C for 60 min was performed to cure the material completely. To check the minimum via opening, scanning electron microscope (SEM) was used. As the result, 3 μm diameter via is successfully formed in the 5 μm thickness film. The via shape was characterized by performing cross-section SEM of the vias filled by electrolytic copper. The SEM image is shown in Figure 1.

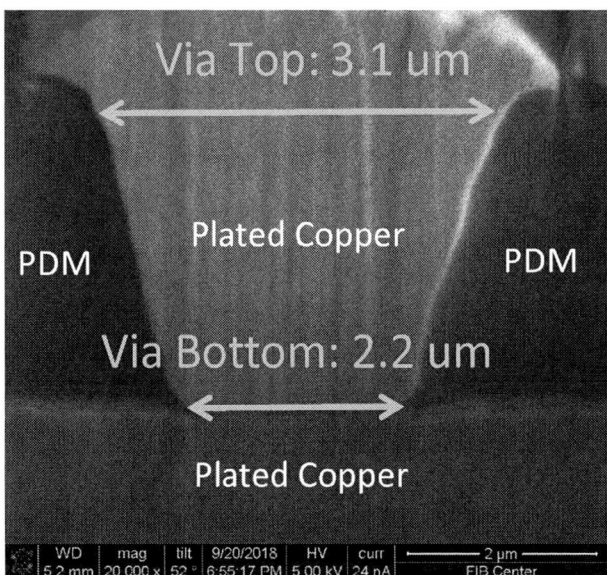

Figure 1. Cross-section image of 3 μm diameter Copper filled Via on electroplated Copper.

3. Adhesion between PDM and Seed Layer

In order to characterize the adhesion between the PDM films and the seed layer, peel strength measurements were performed Titanium (50 nm) and copper (300 nm) were deposited on the PDM by sputtering. Then electrolytic copper plating was employed to add the copper to 25 μm

think, 90 degree peel tests were performed before and after non-bias HAST. The measured peel strengths are shown in Figure 2. As shown in Figure 2, the measured peel strength remained at a high level (5.6 N/cm²) even after 50 hours non-bias HAST compared to the peel strength of the PDM films 7.9 N/cm² prior to subjecting the PDM film to the nHAST test.

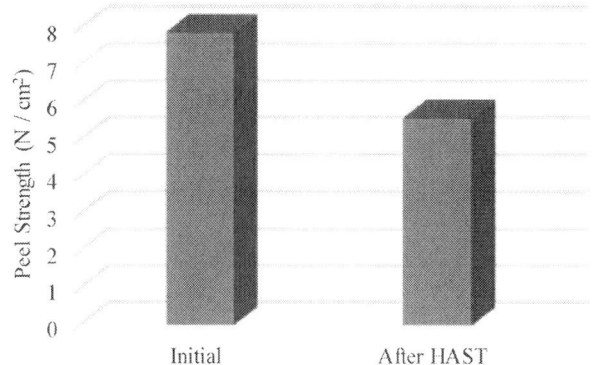

Figure 2. Peel strength between PDM and seed metal. Initial peel strength (Left = 7.9 N/cm²), peel strength after HAST treatment (Right = 5.6 N/cm²).

III. TEST COUPON DESIGN AND FABRICATION PROCESS

1. Four-Terminal Cross-Bridge Kelvin Resistor Structure

Figure 3 shows the four-terminal cross-bridge Kelvin resistor structure for the via resistance measurement of one via (Rv). The Rv can be calculated by measuring the current (I) between PAD_{I1} and PAD_{I2} and voltage drop between PAD_{V1} and PAD_{V2} ($V_{12}=V_1-V_2$). The Rv can be calculated from equation (1)

$$R_m = V_{12} / I \qquad (1)$$

To fabricate the four-terminal cross-bridge kelvin resistor structure, 5μm thick PDM was laminated on a 4 inch silicon wafer then cured. Following this, the laminated bottom layer was patterned with semi-additive process (SAP). Second PDM layer was laminated on the bottom layer and ultra-small vias were formed by photo lithography where 5 different via diameter (3 μm, 4 μm, 5 μm and 6 μm) was formed and then cured. The upper coper layer was then formed. The measured via resistances of one via (Rv) are listed on Table 2. Table 2 also lists the calculated resistance for each via diameter derived from equation (2).

$$R_c = \rho \, h / \, \pi \, r^2 \qquad (2)$$

Where, the via is assumed to have a cylindrical shape, ρ is the resistivity of copper (1.68 x 10⁻⁸ ohm m), r and h are the radius and height of a via respectively.

Table 2. Measured and calculated Via Resistance

Via Diameter	Calculated Value (Rc)	Measured Value (Rm)
3 um	7.1 m ohm	9.8 m ohm
4 um	4.0 m ohm	6.5 m ohm
5 um	2.6 m ohm	4.1 m ohm
6 um	1.8 m ohm	3.3 m ohm

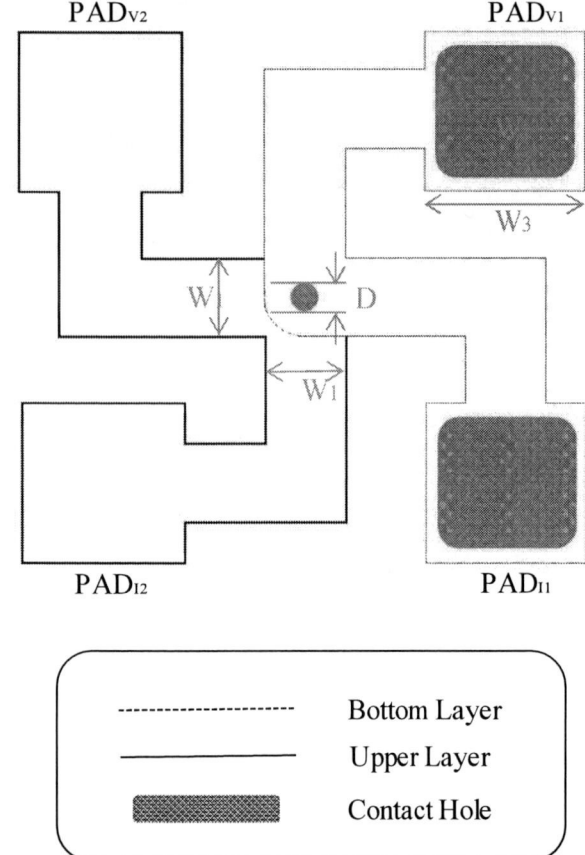

Figure 3. Schematic overview of cross-bridge Kelvin resistor structure; W1=50 μm, W2=1.5mm, W3=2.0mm, D=3.0 μm, 4 μm, 5.0 μm, 6 μm and 8.0 μm.

2. Daisy Chain Structure with Ultra-Small Vias

Figure 4 shows schematic view of the daisy chain structure with ultra-small vias. The test coupon consists of 10 row and 20 column of discontinuous copper lines and each line is connected with a total of 400 ultra-small vias. Figure 5 shows cross-sectional image of the daisy chain. The fabrication process for the test coupon is shown in Figure 6. First, the PDM was laminated on 6 inch glass panel using a vacuum laminator. Then the PDM was exposed by an i-line with a dose of 200 mJ/cm² for patterning, followed by development. Then 2.0 J / cm² of UV light with broad

spectrum was exposed followed by thermal cure at 180 °C for 60 min. To fabricate copper wires on PDM, a standard semi-additive process (SAP) was used. As the first step of SAP, titanium and copper seed layers were deposited on the PDM surface with thicknesses of 50 nm and 300 nm respectively using physical vapor deposition (PVD) or sputtering. Then, a photo resist was coated on the seed layer using a spin coater. The photo resist was exposed through a glass mask which had a Bottom layer pattern followed by development. Then electro copper plating was done until the copper height reached 3 μm. After photoresist stripping, the seed layers were etched. After etching process, annealing was done at 180 °C for 1hour. Next, 3 μm thickness of PDM was formed on bottom copper layer using a vacuum laminator. Then the 2nd layer PDM was exposed by an i-line dose of 200 mJ/cm² for small via patterning, followed by development with 1.0 wt% Na_2CO_3. Then 2.0 J/cm² of UV light with broad spectrum was exposed followed by thermal cure at 180 °C for 60 min. Following the UV exposure and surface cleaning by Ar plasma, titanium and copper seed layers were deposited on the 2nd PDM layer surface as well as vias with thicknesses of 50 nm and 300 nm respectively. An upper layer was formed using the same process for the fabrication of the bottom layer. Overall process flow is shown in Figure 6.

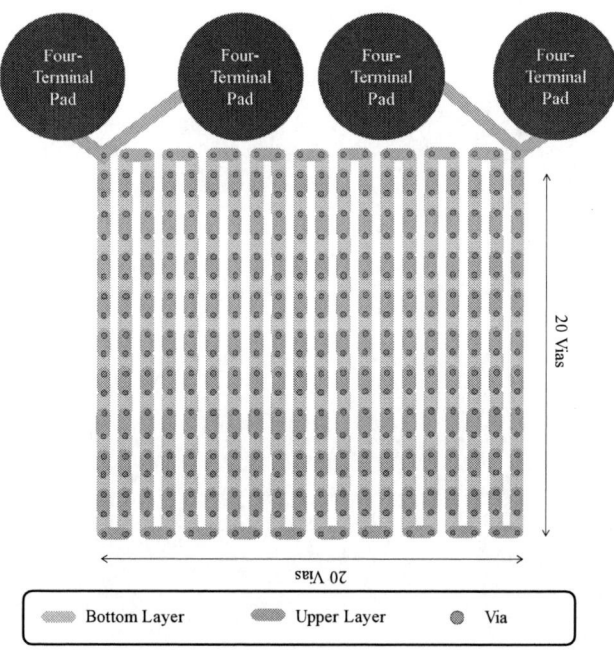

Figure 4. Schematic view of the Daisy chain structure with a total of 400 vias.

978-1-7281-1500-9/19 $31.00 © 2019 IEEE

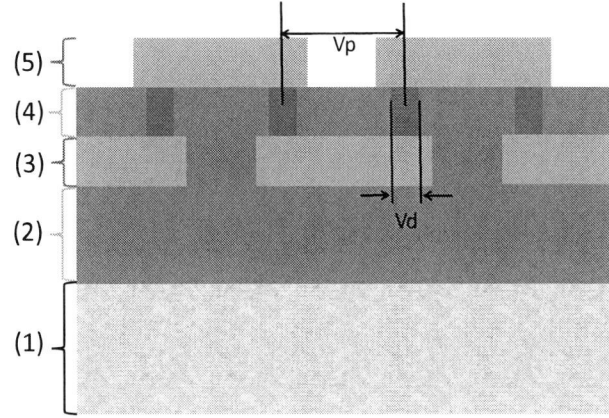

Figure 5 Cross-sectional image of the Daisy chain. (1) Glass panel-300 mm, (2) Bottom PDM layer-5 μm, (3) Bottom Copper layer-3 μm (4) 2nd PDM layer-3 μm, (5) Upper Copper layer-3 μm, Via pitch (Vp)-15 μm, Via diameter (Vd)-3.0 μm.

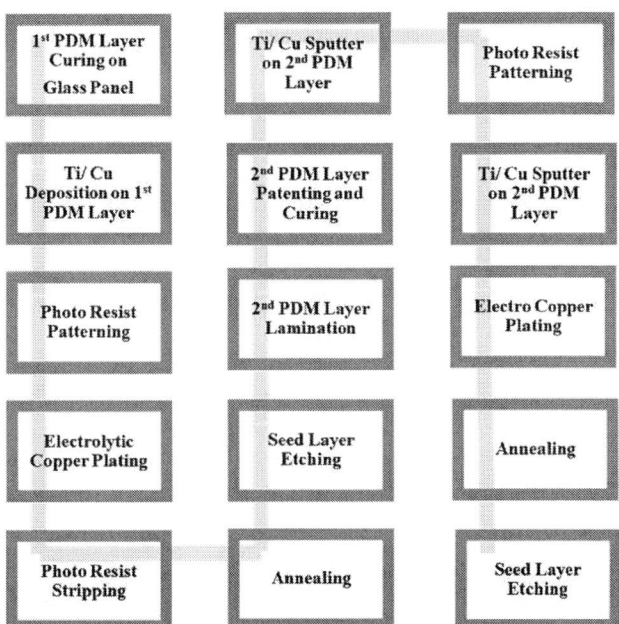

Figure 6. Process flow for Daisy chain structure fabrication.

IV. RESULTS AND DISSCUSSION

1. Reliability Test for One Via

Temperature cycling measurements were conducted to evaluate via connection reliability for one via. Prior to the temperature cycle test, the test coupon was dried at 125 °C for 24 hours. After drying, the sample was put into an environmental chamber that can control and temperature and humidity at 60 °C, 60 % R.H. respectively for 120 hours. After preconditioning, temperature cycle was done with condition B (low temp. -55 °C; high temp.; 125 °C; dwell

time 15 min each). The resistance was measured every 100 cycles until 1,000 cycles using the four-terminal method. Each measured resistance was plotted in Figure 7. Very stable via resistance of each via was confirmed even with 3 μm diameter via.

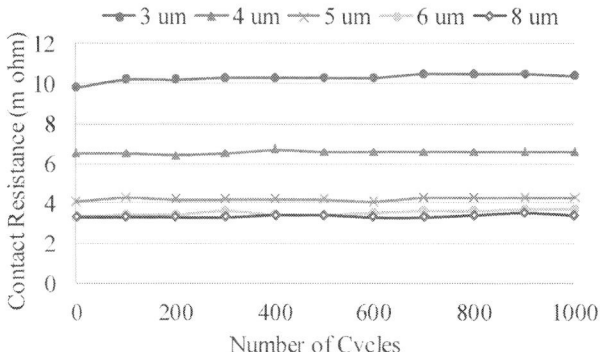

Figure 7. Plots of via resistance of each diameter via

2. Daisy Chain Reliability

As shown in Figure 8, daisy chain structure with ultra-small vias was successfully fabricated in order to perform via connection reliability test. The test coupon was pre-conditioned initially with JEDEC level 2a followed by, temperature cycling with condition B. The resistance was measured every 100 cycles until 1,500 cycles with four-terminal method. The result is shown in Figure 9. As the result, total resistance of the daisy chain which has total of 400 via is stable and the resistance change was less than 5 % even after 1500 cycles.

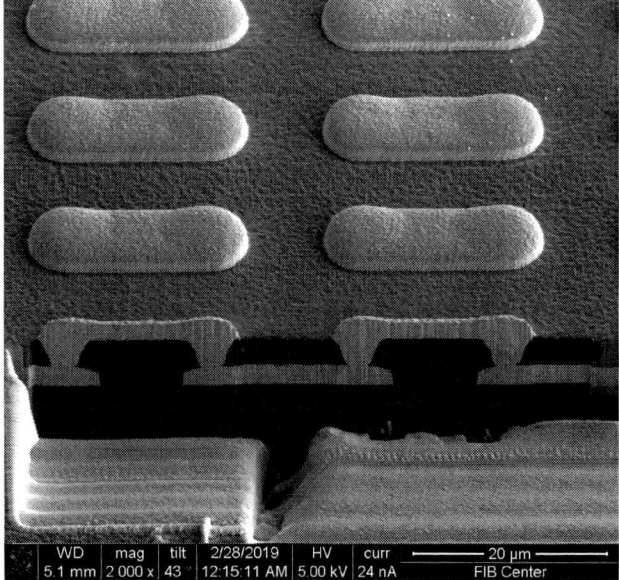

Figure 8. Cross-section image of fabricated daisy chain structure.

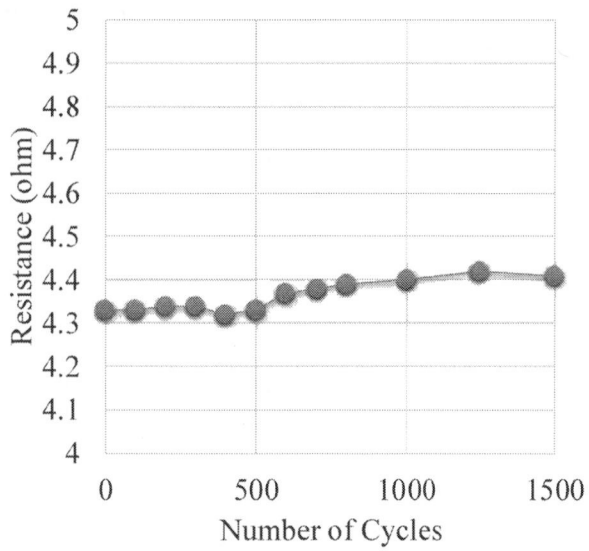

Figure 9. Daisy chain resistance plots over 1,500 cycles of thermal cycle.

V. CONCLUSION

The newly developed PDM reported in this paper can form 3 μm diameter via in 5 μm thickness. Stable via resistance from the PDM films was demonstrated through fabrication of a four-terminal Kelvin bridge structure. A daisy chain test coupon was fabricated to evaluate the reliability of the ultra-small vias. We have demonstrated the total resistance of the daisy chain which has total of 400 via to be very stable and has also shown the resistance change to be less than 5 % even after 1500 cycles. In conclusion, the PDM film reported in this paper is a suitable dielectric material candidate for 2.5D glass interposer applications.

REFERENCES

[1] K. Yamanaka, K. Kobayashi, K. Hayashi, and M. Fukui, "Materials, Processes, and Performance of High-Wiring Density Buildup Substrate With Ultralow-Coefficient of Thermal Expansion," IEEE Transactions on Components and Packaging Technologies, vol. 33, pp. 453461, 2010.

[2] T. Yamada, M. Fukui, K. Terada, M. Harazono, C. Reynolds, J. Audet, et al., "Development of a Low CTE chip scale package," in 63rd Electronic Components and Technology Conference (ECTC), 2013, pp. 944-948.

[3] Daisaku Matsukawa, Nobuyuki Saito, Satoshi Abe, Atsutaro Yoshizawa, Noriyuki Yamazaki, Tetsuya Enomoto, Takeharu Motobe Yuhei Okada and Toshihisa Nonaka "Low Temperature Curable PI/PBO for Wafer Level Packaging" in 14th International Wafer-Level Packaging Conference (IWLPC), 2017.

[4] Xiaozhu Wei, Yoko Shibasaki "A Novel Photosensitive Dry-film Dielectric Material for High Density Package Substrate, Interposer and Wafer Level Package" in 66th IEEE Electronic Components & Technology Conference (ECTC), 2016, pp. 159-164.

[5] Natalie Stavitski, Johan H. Klootwijk, Henk W. van Zeijl, Alexey Y. Kovalgin and Rob A. M. Wolters" Cross-Bridge Kelvin Resistor Structures for Reliable Measurement of Low Contact Resistances and Contact Interface Characterization" IEEE Transactions on Semiconductor Manufacturing Volume: 22, Issue: 1 , Feb. 2009

Pre-cure Modification of Electrically Conductive Adhesive for Low Temperature Interconnection

Jinto George, David Danovitch
University of Sherbrooke,
Bromont, QC, Canada
jinto.george@usherbrooke.ca
david.danovitch@usherbrooke.ca

Alexandre Leblanc, Eric Savage
IBM Canada Ltd,
Bromont, QC, Canada
axleblan@ca.ibm.com
esavage@ca.ibm.com

Michael Ayukawa, Dexter Macaisa
Redlen Technologies,
Saanichton, BC, Canada
michael.ayukawa@redlen.com
dexter.macaisa@redlen.com

Abstract—The temperature sensitivity of CZT medical imaging devices precludes the use of traditional solder attach technologies for package interconnection. Continued advancement in electrically conductive adhesives (ECAs) has resulted in commercially available ultra-low temperature (<60°C) cure candidates that would be compatible with CZT device assembly. However, inherent to their low cure temperature is a rapid onset of room temperature polymerization and associated increase in viscosity. This quickly degrades printability and thereby manufacturing pot life. Conversely, ECA's with a longer pot life typically cure at an unacceptably higher temperature and have lower viscosities that are not compatible with screen printing. To address this dichotomy, we propose an approach that enables low temperature interconnection by initiating the cure process prior to material printing in the assembly process. The approach uses a short time, high temperature pre-heat of a high pot life material to initiate polymerization in a controlled fashion before it is interconnected to the temperature sensitive device then cured at an ultra-low temperature. The results demonstrate that the pre-treatment not only serves to shift the particular material's viscosity to a more acceptable range for screen printing, it also improves low temperature cure resistivity values from 21KΩ-cm to less than 1 mΩ-cm. At the same time, the pre-treatment maintains the long pot life of the material that favors its use in a volume-manufacturing environment. This approach opens the door for exploring a larger portfolio of electrical conductive adhesives to be used in low temperature interconnection applications.

Keywords-electrically conductive adhesive; volume resistivity; viscosity; pot life; screen printing; aging; low temperature assembly; CZT;

I. Introduction

Area array connection, commonly regarded as flip chip, is becoming the prevalent response to furthering both miniaturization and performance in heterogeneously integrated systems. This transition is not always readily compatible with the limitations of individual device or components. For example, a number of photonic components and medical imaging device drift from optimal performance when exposed to the higher temperatures that are inherent to traditional solder attach technologies used in flip chip assembly. This limitation has been frequently addressed by using electrically conductive adhesives (ECAs) to replace the solder interconnections [1–4]. However, when considering additional criteria such as electrical conductivity and manufacturing working span (pot life) for reproducible printing, few commercially available ECA materials make the grade, especially for ultra-low (<60°C) assembly requirements.

Electrical conductivity in an ECA is primarily dependent upon the intimate contact between the silver flakes [2, 5]. This is established through resin shrinkage, often at relatively high temperatures to ensure contraction. Further, high temperatures promote the evaporation of lubricants that coat the flakes and serve to optimize particle distribution and avoid agglomeration during formulation. Any residual lubricant can inhibit contact between the particles and hence produce sub-optimal conductivity. While some studies reported that conductivity only depends on the cure shrinkage [6, 7, 12], other more recent studies [11, 14] revealed that the presence of a small amount of lubricant can increase the resistivity. As such, a number of alternative approaches to remove these lubricants without invoking a high temperature resin cure have been proposed- high temperature pre-annealing, adding short-chain acids, aldehyde, methanol, chloroform, acetone, diluted sulfuric acid, etc. [8-11, 13]. However, most were applied directly to silver flakes prior to ECA formulation or after separation from the epoxy, both with the objective to merely demonstrate the effect of lubricants on conductivity. The challenge therefore lies in effectively eliminating the lubricants in commercially available ECA's without exposing these ECA's to high temperatures in the presence of temperature sensitive components, but only after the lubricants have served their purpose of ensuring processing quality.

Some commercially available ECAs have been developed to cure at lower temperatures while still delivering low resistivity levels, suggesting the absence of lubricants that require high temperature evaporation. However, by their nature, these resins are formulated to polymerize at lower temperatures. Matienzo et al. [15], reported that an ECA that cured at low temperature exhibited changes in viscosity after aging at room temperature before thermal cure, said changes being the result of polymerization, and that the polymerization continues until the gel point is reached, thereby systematically increasing viscosity. In that typical processes to deposit the material for subsequent interconnection, such as screening or dispensing, demand a 'window' of viscosity for effective

978-1-7281-1500-9/19 $31.00 © 2019 IEEE

TABLE I.   PROPERTIES OF ECAs SUPPLIED BY THE MANUFACTURER

Material	Cure Schedule	Recommended Cure Schedule	Viscosity	Volume Resistivity	Pot Life	Shelf Life
ECA-1	45°C@6 hrs -120°C@1 hr	80°C@3 hr	50,000-60,000 cPs	< 5 mΩ-cm	2-3 hrs	1 year
ECA-2	80°C@3 hrs - 175°C@45 sec	150°C@1 hr	3000-4000 cPs	< 0.4 mΩ-cm	3 Days	1 year

manufacturability, any given material has a certain time, or pot life, during which it can be processed at room temperature before becoming unusable. It can be expected that, the lower the gel point and cure temperature, the greater the rate of viscosity increase at room temperature, implying a shorter pot life.

We present herein an approach to breaking this paradigm between cure temperature and pot life through modification of a high temperature ECA immediately prior to assembly. The approach uses a short time, high temperature pre-heat of the material before it is interconnected to the temperature sensitive device and cured at an acceptably low temperature. This initiation not only serves to shift the particular material's viscosity to a more acceptable range for screening, it also produces final resistivity values that were heretofore unattainable during low temperature cure of this material.

II.   EXPERIMENTAL

A. Materials

Two commercially available adhesives were used for the investigation. A first candidate, ECA-1, was selected as a benchmark owing to its excellent volume resistivity performance at a suitably low cure temperature but whose 2-3 hour pot life was not acceptable for production level processing. The other candidate, ECA-2, having a long pot life (3 days) but a correspondingly low viscosity and high optimal cure temperature, was the subject of modification studies. Properties of both ECAs, as provided by the manufacturer, are listed in Table 1.

ECA-1 was supplied as a two-component material and kept at room temperature. Just prior to use, the two components were well mixed at a ratio of 3:1 by weight. ECA-2 was a single component material and was kept in a freezer at -40°C until use, at which time it would be thawed for at least 1 hour prior to use.

The target device for these experiments is a CZT sensor. This device is a crystal that owes its name to its major constituents Cadmium (Cd), Zinc (Zn) and Tellurium (Te). Such a material can absorb X-rays and Gamma rays and generate an electrical charge and thereby serves as an excellent medical imaging and deep space detector. The far side of the crystal is coated with a specific metallurgy and divided into a fine mesh of "pixels". These pixels are the electrodes that connect to an ASIC readout chip as part of a microelectronic module. The ECA material provides such interconnection.

A CZT crystal is particularly sensitive to temperature for two reasons. First, exposing the CZT crystals to elevated temperatures risk increased mobility of impurities and structural defects within the crystal that may degrade detector

performance [16]. In such an instance, the resulting processed X-ray image would be marred by a streak. Second, temperature induced mechanical stresses can warp the X-ray image [17], thereby recommending assembly of the CZT sensor at a temperature close to its imaging operating temperature within the final machine environment.

B. DSC Analysis of ECAs

Cure profiles were studied by Differential Scanning Calorimeter (DSC) using a TA Instruments Q2000 DSC module. Profiles recommended by the manufacturer as well as those required for the target application were analyzed. 20±0.5mg of ECA was placed in a hermetic aluminum pan for non-isothermal study and the sample was heated from 30°C to 250°C at a rate of 10°C/min. Three samples per condition were used to obtain an average percentage of cure.

DSC analysis was also carried out for estimating polymerization of the epoxy during the various aging experiments.

C. Measurement of Volume Resistivity

To measure volume resistivity of the ECA, a 45 x 2 x 0.050 mm strip of material was stencil printed on a glass slide. Three samples per cure condition were prepared. A 4-point probe station was used to measure bulk resistance of the sample while an XYZ microscope was used to measure exact width and thickness of each sample. The measurements were used to calculate volume resistivity using equation (1) [11]

$$\rho = \frac{w \times t}{l} \times R \qquad (1)$$

where ρ is volume resistivity, R is bulk resistance, w is width, t is thickness and l is length of the material. Fig.1 illustrates a sample on the probe station for the measurement.

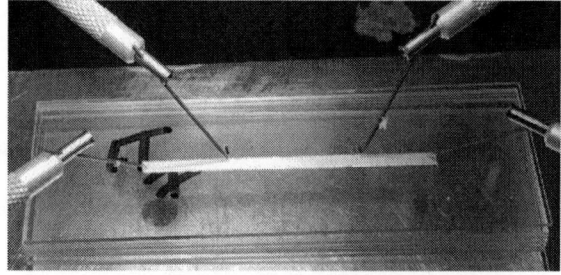

Figure 1.   Bulk resistance measurement of an ECA.

D. Measurement of Viscosity

Viscosity measurements were conducted on the ECAs to identify suitable rheological behavior for the screen printing application. An ARES Rheometer with 25mm diameter parallel plates was used. The material was applied on the bottom plate and the gap between plates was set to 1 mm. Dynamic viscosity was then measured at applied shear rates that progressively increased from $0.0125s^{-1}$ to $125s^{-1}$.

E. Screen Printing on CZT Sensor Device

The ECA was applied on the CZT sensor device using a stencil paste screening tool commonly employed for solder pastes. The sensor was held in place with vacuum in a specialized fixture. A sheet metal stencil with round apertures arranged in a pattern identical to the pixel pattern of the sensor was then aligned with machine vision using fiducials on both the fixture and the stencil. Using a metal squeegee, the ECA was slowly wiped across the top side of the mask to fill the apertures after which the stencil was lifted in a controlled vertical fashion. Specific screening parameters were adjusted in order to optimize possible screening response (aperture fill/release, bridging) for a particular material viscosity. Screening responses at various times following final material preparation were used to determine pot life of the material.

F. Pre-cure Modification of ECAs

Two methods were used for pre-cure modification of ECA. The first was a room temperature aging in order to evaluate in a progressive fashion the correlation between degree of polymerization, viscosity and post cure volume resistivity. The second method was a short, high temperature pre-treatment. Here, the thawed ECA-2 was transferred from the syringe to a glass container. A conventional air-circulating oven was heated to reach the desired temperature after which the glass container was placed inside. Five temperatures between 55°C and 150°C and different time intervals were evaluated.

III. RESULTS

A. Characterization of ECA-1

ECA-1 was characterized to evaluate its suitability at the cure schedule suitable for the target application.

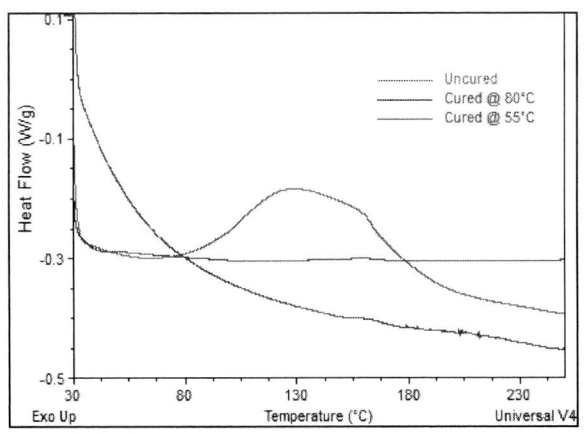

Figure 2. Non-isothermal DSC scan of cured and uncured ECA-1.

Fig. 2 shows the response of the DSC analysis. Two cure schedules were evaluated, comparing curing performance between manufacturer recommended and target cure. As can

be seen in Table II, It was found that both cure schedules exhibited more than 95% cure, suggesting that, by using an extended cure time, ECA-1 could effectively be cured at the target temperature of 55°C instead of the recommended 80°C.

TABLE II. PERCENTAGE OF CURE OF ECA-1 AT DIFFERENT TEMPERATURES

Material	ECA-1	
Cure Temperature (°C)	80	55
Cure Time (hrs)	3	10
Percentage of Cure	98.0	97.0

These same cure schedules were then used for determining volume resistivity, considering a target value of <5 mΩ-cm for this application. As evident from Table III, the 55°C cure schedule produced higher volume resistivity values than its 80°C counterpart, but nonetheless well within the target value.

TABLE III. VOLUME RESISTIVITY OF ECA-1

Cure Schedule	Volume Resistivity
80°C/3 hrs	0.9 mΩ-cm
55°C/10 hrs	2.3 mΩ-cm

The manufacturer recommended pot life of ECA-1 is 2-3 hrs. Therefore, it was decided to measure the viscosity of the ECA-1 immediately after mixing the two components and after 2 hrs of wait time. Fig. 3 compares viscosity as a function of shear rate for the two samples. While the most significant divergence between the two samples appears to have occurred beyond shear rates of 30/s, a notable viscosity increase was still observed in the 0.1-10/s range which is believed to represent typical shear rates exhibited by the screening operation, thus warranting a print screening evaluation of the material as a function of wait time.

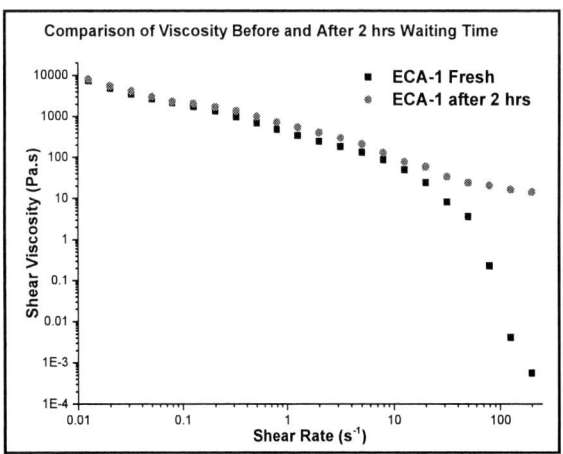

Figure 3. Viscosity of Fresh ECA-1 and aged after 2 hrs.

Fig. 4 shows a top view of ECA-1 printed on a CZT sensor device immediately after mixing while Fig. 5 shows a side view of ECA-1 printed after 2 hrs of working life. It can be seen that there was no bridging of ECA-1 on the CZT sensor device. Beyond 3 hrs from the time of mixing, the material did not uniformly pass through the mask. Considering regular manufacturing production shifts, such a low pot life would incur important downtimes and material waste due to frequent material removal and stencil cleaning.

Figure 4. ECA-1 printed on the CZT sensor device.

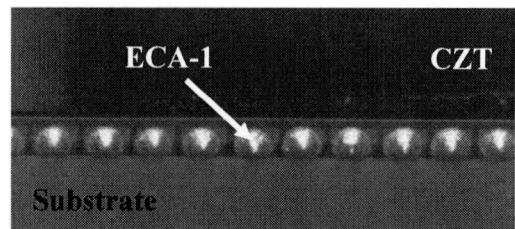

Figure 5. ECA-1 bumps on CZT sensor device during assembly.

B. Aging of ECA-2 at Room Temperature

ECA-2 was identified from the same manufacturer as having a long (3 day) pot life but a higher recommended cure temperature (150°C for ideal resistivity). Further, and believed to be correlated to these properties, viscosity was significantly lower. A baseline printing trial was carried out to confirm the relationship between viscosity and screening performance. The material exhibited very poor results (Fig. 6), as the low viscosity led to significant bleed and even bridging. It was therefore decided to allow the material to age at room temperature in an effort to (1) improve screening performance through an expected increase in viscosity (2) understand viscosity progression with time and (3) correlate other material property responses to this aging.

ECA-2 material samples were aged at room temperature with screening evaluations on a CZT sensor device being performed at two day intervals. It was found that samples aged from 8 days to 12 days produced excellent printing response, suggesting a 4-5 day pot life; a representative image is shown in Fig. 7 after 8 days. Material aged less than 8 days still exhibited bleeding and bridging, although progressively less than the fresh material. Material aged beyond 12 days showed

inconsistent screen deposits indicative of material retention in the stencil.

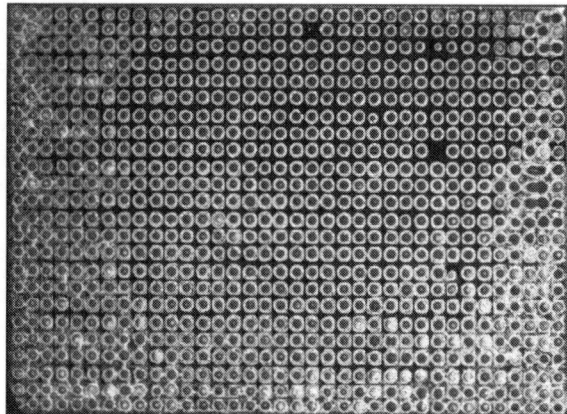

Figure 6. ECA-2 printed on CZT sensor device.

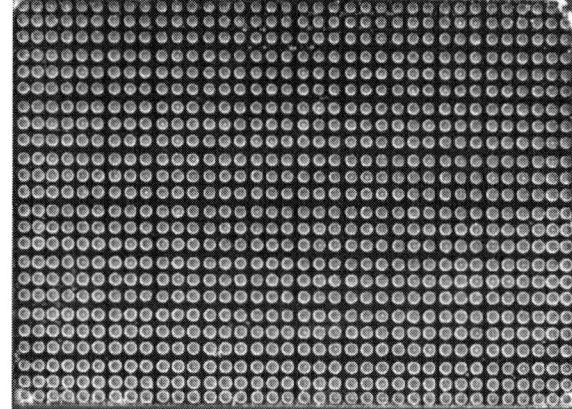

Figure 7. ECA-2 printed on CZT sensor device after 8 days.

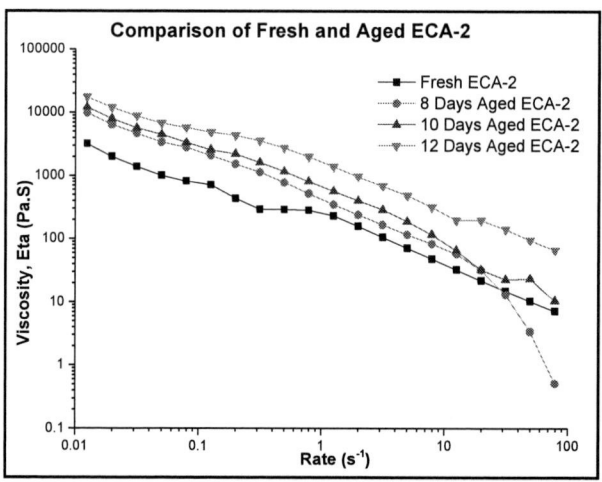

Figure 8. Viscosity plot of fresh and aged ECA-2.

Viscosity measurements were conducted on the various samples. Representative curves after 8, 10 and 12 days of room temperature aging are shown in Fig. 8. Focusing on the previously described shear rate range of 0.1-10 s^{-1}, viscosity is seen to increase by about 2, 3 and 8 times for 8, 10 and 12 days of aging respectively as compared to fresh ECA-2. These values would therefore represent target viscosity increases for alternative means to age the ECA-2 material.

It is also interesting to compare the viscosities of ECA-2 aged 8-12 days to the benchmark ECA-1 in that both produce good printing response. As can be seen in Fig. 9, ECA-2 viscosity at the onset of its printing pot life (8 days aging) is very similar to that of fresh ECA-1, as might be expected. However, the significant viscosity increases seen at 12 days aging suggest that this aged material can sustain a fairly wide viscosity range (4X across its pot life) as opposed to typical recommended viscosity variations of 2X for commercial ECA materials.

In that the viscosity is believe to be caused by some degree of polymerization, DSC analysis was conducted to establish correlation. The cure profiles of the fresh and room temperature aged ECA-2 are shown in Fig. 10. The curves correspond to 36.6%, 45.1% and 53.6% polymerization after 8, 10 and 12 days respectively, suggesting a possible target polymerization range of about 35-55% in order to provide sufficient viscosity for optimum printing performance.

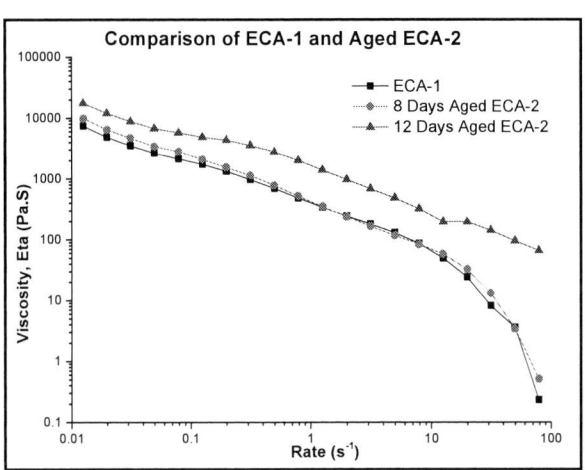

Figure 9. Viscosities of ECA-1 and RT aged ECA-2.

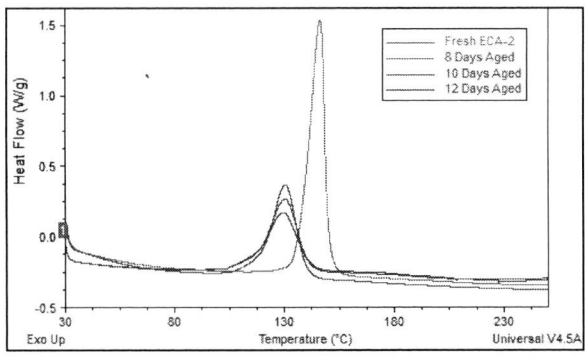

Figure 10. Non-isothermal DSC scan of RT Aged ECA-2.

C. Pre-cure Modification by Accelerated Aging

While room temperature aging enabled the establishment of target ranges for viscosity and polymerization, the 8 days required for acceptable printing would be untenable in a high volume production environment. It was therefore proposed to investigate whether elevated temperatures could effectively obtain similar material behaviors within shorter time spans while still providing an acceptable process window. While degree of polymerization and viscosity values were used as target responses, visual observations were also conducted to assess general material behavior. Candidates approaching target responses were further evaluated through volume resistivity measurements.

Initial accelerated aging experiments were conducted at 55°C for various times, in that this was the intended final cure temperature. Only 21.2% polymerization occurred, even after 6 hrs. Aging was therefore conducted at higher temperatures ranging from 70°C to 150°C

It was found that polymerization of ECA-2 at 70°C/2.5hrs, 100°C/10 min, 120°C/6 min and 150°C/3.5 min reached 34.0%, 4.5%, 15.2% and 29.1% respectively. These times corresponded to the maximum time at a given temperature before the material exhibited prohibitive material behavior such as hardening or granularity. The 70°C/2.5hrs condition came closest to the target polymerization, but preliminary volume resistivity assessments of this material after the target 55°C/10hrs cure schedule demonstrated inadequate performance. The 120°C/6 min. condition provided near target viscosity despite significantly lower polymerization but again could not achieve acceptable volume resistivity at the target cure temperature.

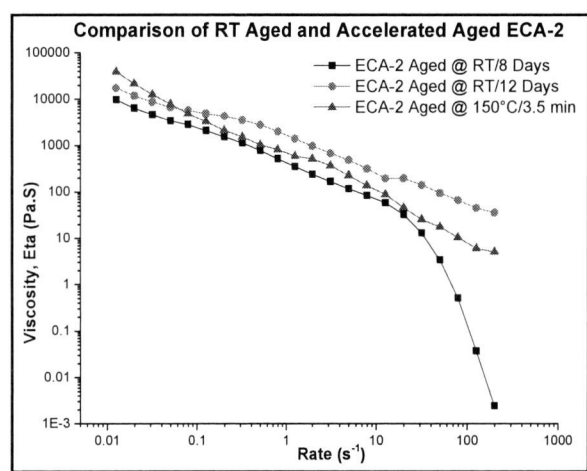

Figure 11. Viscosities of RT aged and accelerated aged ECA-2.

Only the 150°C/3.5 min. condition enabled acceptable volume resistivity at a 55°C cure temperature, as detailed in the subsequent characterization section. Additionally, this condition provided viscosity within the target range despite a polymerization below target. As can be seen in Fig. 11, its viscosity is higher than that of an 8 day room temperature aging yet still below that of a 12 day aging within the shear rate range representative of screen printing (0.1-10/s). These results suggest that solvent evaporation may play a greater role at the higher temperatures than at room temperature, hereby skewing the correlation between polymerization and viscosity.

The accelerated aged ECA-2 using 150°C/3.5 min. was then examined for printing performance, the results of which are shown in Fig. 12. No bridging or bleed out of the material was observed after printing on the device.

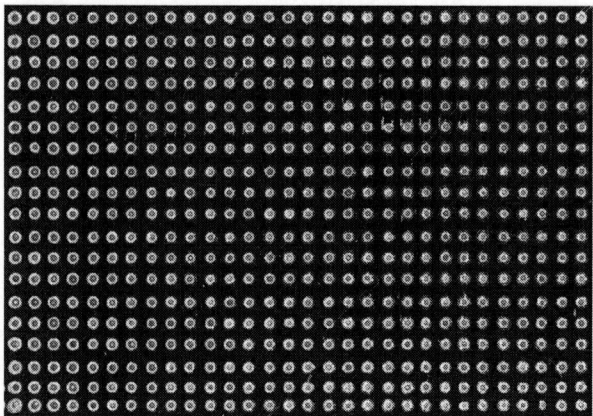

Figure 12. ECA-2 printed on the CZT sensor device after accelerated aging.

D. Characterization of ECA-2

1) DSC Analysis

Percentage of cure as calculated by DSC was determined for the accelerated aged ECA-2 and compared to fresh, un-aged ECA-2, where both materials were cured at the target 55°C cure temperature for 10 hours. The results (96.0% for the modified version vs 95.4% for the fresh material) are considered to be within statistical error, suggesting that the modified ECA-2 achieves adequate and comparable cure at cure temperatures representative of ultra-low temperature assembly.

2) Volume Resistivity

Volume resistivity measurements of ECA-2 were conducted for fresh material (Table IV) as well as under both the room temperature (Table V) and accelerated aging (Table VI) conditions. It can be seen that the fresh material can only achieve target volume resistivity values (<5 mΩ-cm) at the 150°C cure schedule, the highest temperature recommended by the manufacturer, suggesting ideal shrinkage and/or solvent evacuation at this temperature. Room temperature aging within the pot life range provided acceptable resistivity at the 80°C cure schedule, which is not sufficient for the low-temperature assembly requirements of the CZT sensor device. Nonetheless, an improvement in volume resistivity of ECA-2 cured at 55°C/10 hrs was noted after 12 days of aging at RT. This was not achieved with fresh material, suggesting that the reduction of volume resistivity is not due to cure shrinkage alone but depends somewhat on chemical changes inside the material, such as evaporation, induced by the extended time.

For the accelerated (elevated temperature) aged ECA-2, the 55°C/10hrs cure schedule was not used for the first two aging levels since the 80°C/3hrs cure schedule did not meet the target value. It was observed that the ECA-2 aged at 120°C/6 min or 150°C/2.5 min or more provides volume resistivity less than the target value after cure at 80°C/3hrs.

However, only the 150°C/3.5 min schedule met the target of <5 mΩ-cm when cured at the desired 55°C/10hrs, providing values below 1 mΩ-cm.

TABLE IV.   VOLUME RESISTIVITY OF ECA-2

Cure Schedule	Volume Resistivity
55°C/10 hrs	>10000 Ω-cm
80°C/3 hrs	>10 Ω-cm
150°C/1 hr	1.2 mΩ-cm

TABLE V.   VOLUME RESISTIVITY OF ECA-2 AGED AT RT

Age @ RT	55°C/10 hrs	80°C/3 hrs
8 Days	>5000 Ω-cm	2.2 mΩ-cm
10 Days	>1000 Ω-cm	1.5 mΩ-cm
12 Days	>10 Ω-cm	1.9 mΩ-cm

TABLE VI.   VOLUME RESISTIVITY OF ACCELERATED AGED ECA-2

Aged @	55°C/10 hrs	80°C/3 hrs
55°C/6 hrs	--	6.8 mΩ-cm
70°C/2.5 hrs	--	>100 mΩ-cm
100°C/10 min	>100 Ω-cm	>10 mΩ-cm
120°C/6 min	>10 Ω-cm	3.4 mΩ-cm
150°C/2 min	>100 Ω-cm	>10 mΩ-cm
150°C/2.5 min	>10 mΩ-cm	3.1 mΩ-cm
150°C/3 min	7.3 mΩ-cm	3.9 mΩ-cm
150°C/3.5 min	0.7 mΩ-cm	--

3) Pot Life

The ECA-2 aged at 150°C/3.5min was then stored at room temperature for 24 hours in order to characterize its behavior after a suitable manufacturing pot life. The DSC responses of accelerated aged ECA-2 immediately after aging then after 24 additional hours at room temperature are shown in Fig.13.

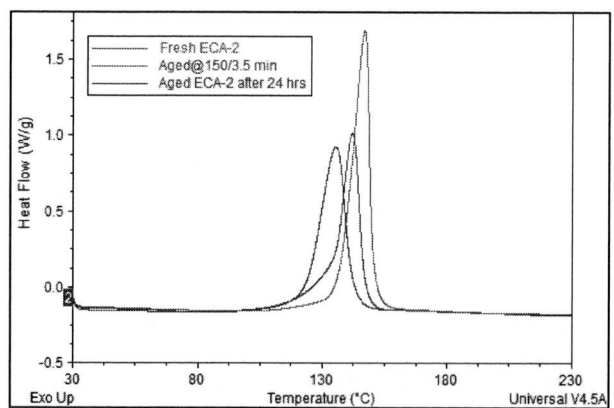

Figure 13. Non-isothermal DSC scan of fresh & accelerated aged ECA-2.

An additional 3.2% of polymerization was measured. Considering that this is well below the 20% polymerization variation that provided good printing performance for the room temperature aged material, it is expected that this modified (accelerated aging) version should provide at least 24 hours of pot life.

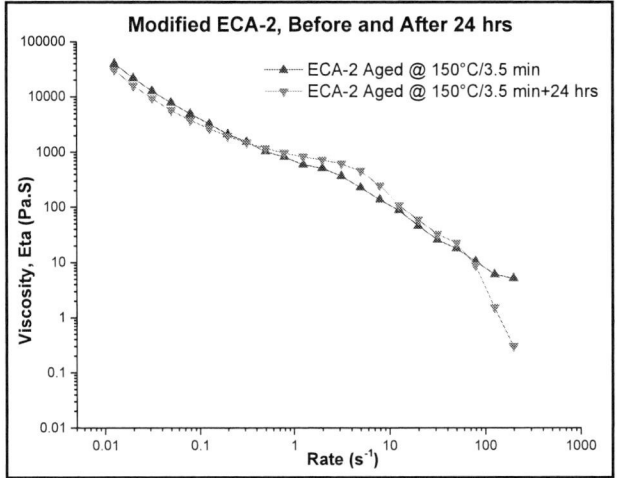

Figure 14. Viscosities of modified ECA-2 before and after 24 hrs.

Viscosity of the modified ECA-2 was also measured after this 24 hour RT storage, as shown in Fig. 14. Increases were calculated to be no more than 1.4 times their original value within the target shear rate range of 0.1-10/s, thus suggesting that the material possesses at least a 24 hour pot life of acceptable printing. This was confirmed on a CZT sensor device as illustrated in Fig. 15 where no bridging or bleed out occurred after printing the material stored for 24 hours.

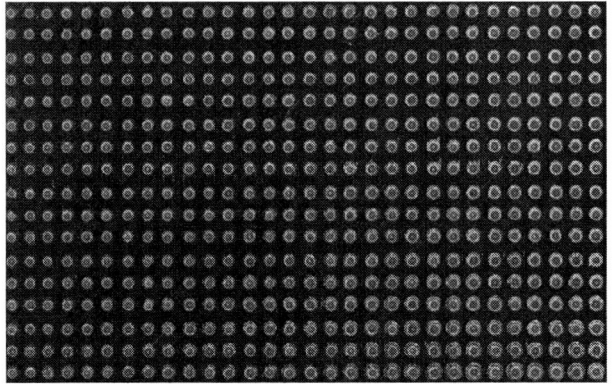

Figure 15. Modified ECA-2 printed on the CZT sensor device after 24 hrs of storage at RT

IV. DISCUSSION

Identification of an off-the-shelf commercial ECA candidate to meet both the requirements of interconnecting a temperature sensitive CZT sensor device and the requirements of volume manufacturing has proven to be daunting. ECA-1 & ECA-2 are merely representative of a larger ECA population that reflect the dilemma between providing long pot life and providing a low temperature cure that produces sufficiently conductive interconnects.

ECA-1 was characterized to establish the benchmark parameters, as this material has demonstrated good initial printing quality and acceptable low-temperature cure electrical properties for the application environment and robustness of the CZT sensor device. Conversely, yet perhaps consequential to these positive qualities, it suffers from insufficient pot life for effective production level assembly.

The objectives of the ECA-2 modification were two-fold. The first was to increase its viscosity, as the poor screen-printing results of the fresh material appears to correlate to its viscosity being significantly lower than ECA-1. The challenge was to do so without reversing its long pot life advantage or introducing other manufacturing issues. The second objective was to improve its resistivity properties when cured at the target interconnect cure temperature of 55°C, since the fresh material cured at this temperature showed extremely high volume resistivity values (>1000 Ω-cm) compared to the <5 mΩ-cm requirement.

To address the first objective, the empirical relationship between polymerization and viscosity was exploited. Room temperature aging was selected premised on previous literature supporting that polymerization does in fact occur under ambient conditions [15]. This approach proved successful in increasing viscosity, and consequently printing performance. Good printing of ECA-2 was achieved after 8 days of aging where viscosity was similar to that of ECA-1, then continued up to 12 days, thereby demonstrating a pot life at least as long as fresh ECA-2 material. Further, DSC analysis supported polymerization as the most likely cause of viscosity increase, as the material showed about 37% polymerization after 8 days and about 54% after 12 days; the results suggest a fairly linear progression of about 4% per day.

From the perspective of the second objective, ie electrical resistivity, the room temperature aging of ECA-2 showed some improvement over its fresh version, especially at 80°C where the < 5 mΩ-cm requirement was achieved. However, improvements at 55°C were much less pronounced, where resistance reductions were observed only after 12 days aging; as such, resistance requirements could not be satisfied prior to the material becoming unusable for printing. Nevertheless, the accelerated difference in resistivity enhancement at the higher cure temperature suggests that temperature plays a more important role than polymerization alone in reducing resistivity. Based on previous literature that recognized the importance of driving off commonly present chemical lubricants from the Ag flake surface in order to promote contact between flakes [9, 11, 14], it is proposed that higher temperatures are progressively more successful in evacuating and/or evaporating such lubricants.

Elevated temperature aging was therefore inspired by this increased probability of driving off the lubricants but also by the opportunity to accelerate the polymerization and therefore viscosity progression, in that the management of 8 day aging cycles in a production environment is logistically cumbersome. Results for various elevated temperatures did indeed support an accelerated increase in viscosity. However, while polymerization also accelerated, it did not correlate as well to viscosity increases. This may be attributed to the fact that lubricant elimination at elevated temperature is an additional contributor to viscosity increase.

While a number of combinations of temperature and time were able to achieve target viscosities for good printing performance, only the 150°C level succeeded in meeting the resistivity objectives, requiring 3.5 minutes of aging to do so. At other temperatures, the material would degrade prior to achieving adequate volume resistivity with a 55°C cure. The 150°C result tracks well with the manufacturer recommendation to use this temperature for optimal volume resistivity; although constituents are proprietary, it is hypothesized that lubricants present in the ECA-2 material are optimally evacuated at this temperature. A preliminary mass spectroscopy analysis revealed that, when comparing fresh ECA-2 cured at both 55°C and 150°C, only the 55°C version exhibited the presence of a carboxylic acid type constituent The intensity of this same constituent was 50% less in pre-treated ECA-2 cured at 55°C compared to untreated ECA-2 cured at the same temperature. This suggests that the high temperature pre-treatment partially evacuates an acid constituent from the material, thereby promoting improved contact between the flakes. While further detailed investigation is required to confirm and understand the hypothesized mechanism of lubricant evacuation, it appears that this short 150°C excursion serves to sufficiently initiate the cure process to allow interconnection then low temperature (55°C) cure.

Successful print screening results for the ECA-2 material after a 150°C/3.5min modification allayed concerns that such an elevated temperature would modify the material in a manner that would render it unprintable. Moreover, continued printability after 24 hour wait times supported a minimum pot life compatible with manufacturing requirements.

V. CONCLUSIONS AND FUTURE WORK

A new approach has been developed that enables low temperature interconnection by initiating the cure process prior to material use in the assembly process. Based on a series of characterizations of a benchmark low temperature material (ECA-1) and a candidate long shelf life material (ECA-2), a high temperature, short time span pre-treatment of ECA-2 has been determined that replicates the screening performance and low temperature cure behavior of ECA-1. At the same time, this pretreatment maintains the long pot life of ECA-2 that favors its use in a volume-manufacturing environment.

The numerous pre-treatment experiments demonstrated that the ECA material increases in viscosity at room temperature, said increase correlating to a linearly progressive polymerization. At elevated temperatures, this correlation seems less evident and is believed to be caused by solvent evacuation, most likely an acid lubricant that often coats the Ag flakes of an ECA. At various temperatures, the lubricant will evacuate at various rates and will induce viscosity increases independent of those induced by polymerization.

Further evidence of lubricant evacuation during the high temperature pre-treatment is supported by ECA volume resistivity behavior. Whereas an untreated ECA-2 cannot achieve low (< 5 mΩ-cm) values, the same material after a

150°C/3.5minute pretreatment readily meets this target value. In that the untreated material is known to produce exceptional volume resistivity after a 150°C/1 hour cure, it is proposed that the short 150°C treatment drives off a sufficient quantity of certain acid lubricants present to promote good contact between the flakes.

Further investigation, including organic characterization, is underway to support the hypothesis of lubricant evacuation. Additionally, mechanical comparisons (visco-elastic properties, adhesion, etc.) of the untreated and pretreated materials will be conducted in conjunction with standard interconnect reliability stress testing.

ACKNOWLEDGMENTS

The authors would like to thank their colleagues for their contribution to this work. In particular, Pierre Beaulieu, Patrice Tremblay, Peng-Thai Chea, Magali M Cote and Valérie Oberson from IBM Bromont are acknowledged for their continuous help throughout the project. This work was funded by the NSERC-IBM Industrial Research Chair in Smarter Microelectronics Packaging for Performance Scaling and Prompt Quebec.

REFERENCES

[1] Li, H., Moon, KS. & Wong, C.P., "A Novel Approach to Stabilize Contact Resistance of Electrically Conductive Adhesives on Lead-Free Alloy Surfaces" Journal of Elec Materi (2004) 33 (2), pp: 106-113.

[2] L. Li and J.E. Morris, "An introduction to electrically conductive adhesives", Int. J. Microelectronic Packaging,1(3), 1998, pp. 159–175

[3] I. Hinz, S. -H. Schulze, F. Barche, M. Schak, "Characterization of Electrically Conductive Adhesives (ECA) for the Photovoltaic-Industry", 27th EU - PVSEC, 2012.

[4] T. Falat and J. Felba. "Conductivity improvement of electrically conductive adhesives by thermal post-curing processes"., International conference of IMAPS Poland Chapter Krakow, 2006, pp. 24-27.

[5] Sancaktar, E., Bai, L., 2011, "Electrically Conductive Epoxy Adhesives," Polymers, 3 (1), pp. 427-466.

[6] J. Haberland, C. Kallmayer, R. Aschenbrenner and H. Reichl, "Fundamental studies of isotropic conductive adhesives focused on the current loadability of ICA for flip chip applications," First International IEEE Conference on Polymers and Adhesives in Microelectronics and Photonics. Potsdam, Germany, 2001, pp. 185-195.

[7] C. P. Wong, D. Lu, L. Meyers, S. A. Vona and Q. K. Tong, "Fundamental study of electrically conductive adhesives (ECAs)," Proceedings. The First IEEE International Symposium on Polymeric Electronics Packaging, Norrkoping, Sweden, 1997, pp. 80-85.

[8] R. Zhang, J. C. Agar and C. P. Wong, "Recent advances on Electrically Conductive Adhesives," 2010 12th Electronics Packaging Technology Conference, Singapore, 2010, pp. 696-704.

[9] Fatang Tan, Xueliang Qiao, Jianguo Chen, "Removal of chemisorbed lubricant on the surface of silver flakes by chemicals", Applied Surface Science, Volume 253, Issue 2, 2006, Pages 703-707.

[10] Daoqiang Lu, Q. K. Tong and C. P. Wong, "A study of lubricants on silver flakes for microelectronics conductive adhesives," in IEEE Transactions on Components and Packaging Technologies, vol. 22, no. 3, pp. 365-371, Sept. 1999.

[11] Chaowei Li, Qiulong Li, Liyao Cheng, Taotao Li, Huifen Lu, Lei Tang, Kai Zhang, Songfeng E, Jun Zhang, Zhuo Li, Yagang Yao, Conductivity enhancement of polymer composites using high-temperature short-time treated silver fillers, Composites Part A: Applied Science and Manufacturing, Volume 100, 2017, Pages 64-70.

978-1-7281-1500-9/19 $31.00 © 2019 IEEE

[12] Daoqiang Lu, Q. K. Tong and C. P. Wong, "Conductivity mechanisms of isotropic conductive adhesives (ICAs)," in IEEE Transactions on Electronics Packaging Manufacturing, vol. 22, no. 3, pp. 223-227, July 1999.

[13] Yi Li, A. Whitman, Kyoung-sik Moon and C. P. Wong, "High performance electrically conductive adhesives (ECAs) modified with novel aldehydes," Proceedings Electronic Components and Technology, 2005. ECTC '05., Lake Buena Vista, FL, 2005, pp. 1648-1652 Vol. 2.

[14] Shigeru Kohinata, Akari Terao, Yosihiko Shiraki, Masahiro Inoue, Keisuke Uenishi, Relationship between the Conductivity of Isotropic Conductive Adhesives (ICAs) and the Lubricant Coated on Silver Filler Particles, Transactions of The Japan Institute of Electronics Packaging, 2013, Volume 6, Issue 1, Pages 104-108,

[15] L. J. Matienzo , R. N. Das & F. D. Egitto (2008) Electrically Conductive Adhesives for Electronic Packaging and Assembly Applications, Journal of Adhesion Science and Technology, 22:8-9, 853-869.

[16] Veale, Matthew C. "CdTe and CdZnTe Small Pixel Imaging Detectors", Chapter 3, Awadalla, S. (Ed.), Iniewski, K. (2015). Solid-State Radiation Detectors. Boca Raton: CRC Press, https://doi.org/10.1201/b18172 (2017) .

[17] S. Taherion, H. Chen, P. Lu, S. Awadalla and G. Bindley, "Optimizing the design parameters of adhesively bonded assemblies to enhance reliability and performance of the CZT detectors," IEEE Nuclear Science Symposuim & Medical Imaging Conference, Knoxville, TN, 2010, pp. 3956-3958.

RDL-1st Fan-Out Panel Level Packaging (FOPLP) for Heterogeneous and Economical Packaging

Vasarla Nagendra Sekhar*, Vempati Srinivasa Rao, F. X. Che, Chong Ser Choong, and Kazunori Yamamoto

Institute of Microelectronics, A*STAR (Agency for Science, Technology and Research),
2 Fusionopolis Way, #08-02 Innovis Tower, Singapore 138634.
*E-mail: vasarla@ime.a-star.edu.sg, Tel: +65-67705383, Fax: +65-67731914

Abstract—Established RDL-1st fan out panel level packaging (FOPLP) processes and modules for Gen 3 panel (550x650 mm) sizes. RDL-1st package test vehicle (TV) has been designed and fabricated with single chip size of 10x10mm and final package size is 15x15mm. IC test chip has been designed and fabricated with Al pad structures and copper pillars with solders for final package assembly. Extensive mechanical modelling and simulation studies have been conducted to provide optimum guide lines for process design and material selection. Multilayer RDL structures up to two layers have been fabricated with copper RDL line width and line spacing (LW/LS) of 8/8um. Number I/Os for this package is around 1000. Prior to final TV fabrication several short loop DOE evaluations have been conducted for material and process feasibility. A special mask set with test structures has been designed for short loop DOE and it includes LW/LS structures and meander structures from 1/1um to 15/15um, via opening structures from 3um to 20um. Established lithography process to resolve LW/LS patterns from 2/2um onwards. Several key modules for panel processing like seed layer deposition, plating, wet etching and final assembly, have been established by closely working with equipment and material suppliers.

Keywords-Fan-Out packaging; Fan-Out Panel Level Packaging (FOPLP); RDL-1st Packaging; Fabrication and Integration; Assembly Processes

I. INTRODUCTION

Semiconductor industry is always looking to develop innovative packaging solutions as there is incessant demand for lower cost, high performance and multifunctional packages. Fan-Out Packaging coupled with heterogeneous integration is one of the best choices to meet these demands from end customers. Fan-out Wafer Level Packaging (FOWLP) is one of the IC packaging technology that can have more number of I/Os when compared to standard wafer level package (WLP). In FOWLP packaging, dies are embedded in low cost epoxy molding compound (EMC) with a predefined space between dies for addition I/O connection routing in addition to silicon surface [1]. Heterogeneous integration focuses on integrating different types of chips at package level other than device scaling, which provides many advantages like better performance, better system integration, coat advantages. Fan-Out packaging is getting attention in semiconductor industry as it can go for high density packaging for advanced applications [2-4]. Other key advantages for fan-out packaging are low thermal resistance, low parasitic effects, better RF performance and its substrate less package [2]. Fan-Out manufacturing is currently being carried out at 300mm or 330mm wafer sizes. However, industry is incessantly looking for Fan-Out manufacturing at lager substrate sizes mainly to gain cost advantages. Industry is more inclined towards panel level packaging mainly for more productivity benefits and leverage on readily available infrastructure from flat panel display/LCD, solar and PCB industries [5]. However main challenge associated with panel level manufacturing is that there is no panel size standardization established yet and it may pose some challenges for equipment manufacturers. Many leading companies in the semiconductor industry have already entered the panel level packaging, while other are closely watching for developments. As per current market scenario it is expected that three big players enter into mass production very soon. Fan-Out panel level packaging (FOPLP) is being discussed thoroughly in semiconductor industry for various packaging platforms like embedded and non-embedded packaging technologies, RDL-1st and Mold-1st fan-out packaging. Many players in the semiconductor packaging are keenly working on high density fan-out packages [6-7].

Present study focuses on RDL-1st packaging development, including module development, process integration and assembly. Basic fabrication flow for fan-out panel processes is same as wafer fabrication except it requires customized tool configurations, materials and process optimization. Typically, RDL-1st fabrication involves several fabrication steps as sacrificial coating on glass panel, dielectric patterning, PVD, photo resist lithography, copper plating, RDL build up, chip to panel (C2P) bonding, panel level molding, carrier debonding and cleaning, solder ball placement and package singulation. In case of materials, different types of release layers, dielectric layers and photo resist materials have been thoroughly evaluated. RDL-1st integration and assembly approach has been established. Slit coating methods have been employed for coating of release layers, dielectric material and photo resist materials. For exposure in lithography, high power laser direct imaging has been employed. In case Cu ECP, plating chemistry and process has been optimized to plate both via opening and RDL structures with minimal dishing. Panel warpage is monitored at each processing step and it is accumulated up to 2mm after completion of integration.

II. TEST VECHILCE DESIGN

A) Package Design

RDL-1st TV is being developed for mobile, baseband, application processors and 2.5D packaging applications. Final package size is 15x15mm and with single chip size of 10x10mm^2. Package schematic is shown in Fig 1 and its specifications are summarized in table 1. Total number of I/Os per package is around 1000. For RDL-1st TV, two Cu re-distribution layer (RDL) structures have been designed with LS/LW of 8/8um and via diameter around 10um and its mask set shown in Fig 2. Target Cu RDL thickness is 5um and photo dielectric material thickness up to 10um. Package and chip under bump metallization is designed with 50um diameter. different dielectric via opening sizes included in the package structures and it is in the range of 10um to 200um. Besides test vehicle design, test structure designs as shown in Fig 2 a) have included and integration is carried out at same time mainly to study the extremes of material and process feasibility at key modules like lithography, electroplating and wet processes. To utilize the maximum space, each reticle consists two test structures and one test vehicle. This test structure includes different LW/LS, meander, via and daisy chain structures as shown in Fig 3. LW/LS structures are designed from 1 to 15um with 1um intervals. Whereas via structures are designed from 3 to 20um. Several daisy chain structures with different via sizes are included in the package to monitor overall layers and individual layers' continuity during fabrication. All module developments and integration activities have been planned and carried out on Gen 3 panel (550x650mm) sizes.

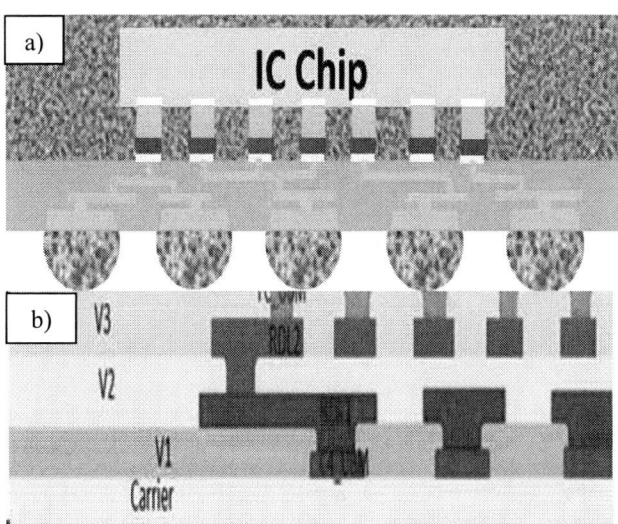

Fig 1. Schematic of RDL-1st test vehicle a) Assembled Package, b) RDL structures

Table 1 RDL-1st package specifications

Package Specifications	
Test Vehicle	RDL-First FOPLP (TV2)
Target Applications	Mobile/Baseband/AP/2.5D
Package Size	15mm x 15mm
# of Chips/Pkg and Size	Single chip of 10 x 10mm^2
RDL LW/LS	8um/8um (up to 2 layers)
# of I/Os	Up to 1000
Panel Size	650 x 550mm (Gen 3)

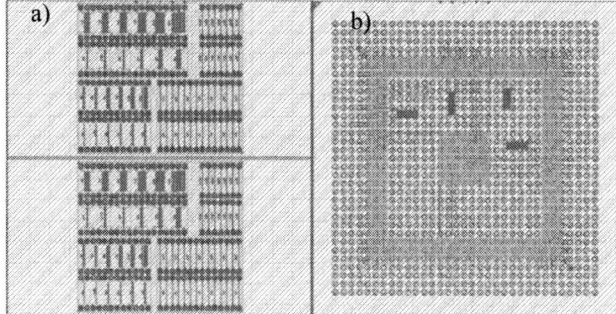

Fig 2. Mask set lay out for a) test structures b) test vehicle

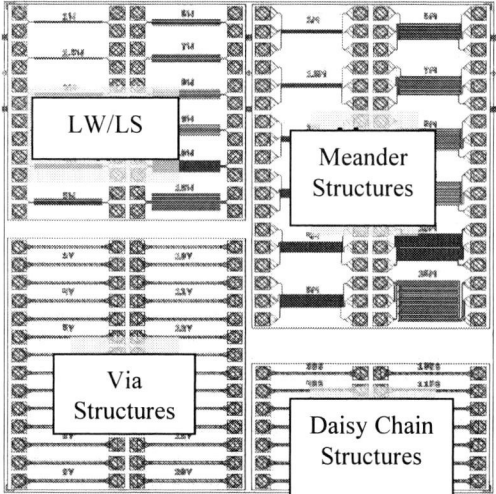

Fig 3 Mask set for different test structures

Various daisy chain structures and meander structures are included in the package as shown in Fig 4. Daisy chain structures used to monitor the electrical continuity between different package Cu RDL layers and chip RDL structures, whereas meander structures have been included to study the parasitics of the Cu RDL.

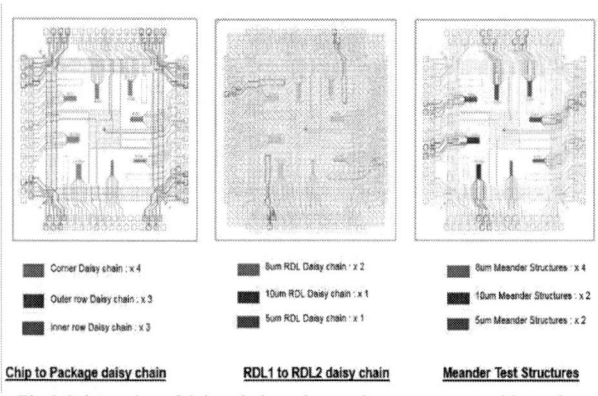

Fig 3 Schematics of daisy chain and meander structures used in package design

B) Test Chip Design and fabrication

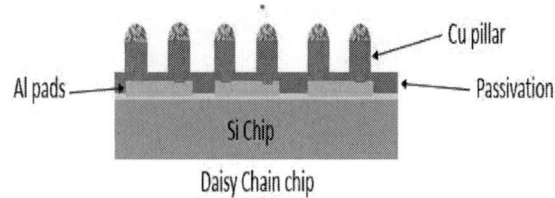

Fig 4. Schematic of test chip

IC test chip of 10x10mm size is designed with Al daisy chain structures as shown in Fig 4 and it is mainly populated at center and edge regions of the chip mainly to study the effect of the respective regions. Middle regions were used for other daisy chain and meander structures at package side. Aluminum pads of size around 70x70um are passivated with oxide layer with an opening of 30um in diameter. Flip-chip UBM structure has been fabricated with 50um diameter copper pillar with solder cap. Fig 5 shows the Cu pillar profile after it has been reflowed using lead free profile.

Fig 5 IC test chip with reflowed Cu pillar structures

III. PROCESS FLOW AND ITS CHALLENGES

A. Process flow for RDL-1st TV fabrication

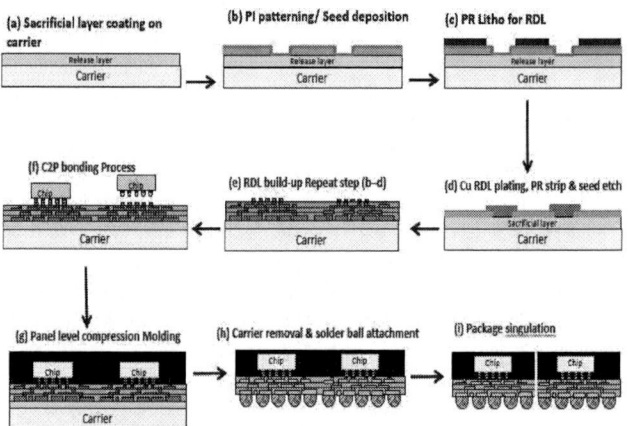

Fig 6. RDL-1st process integration flow

Process integration flow schematic for RDL-1st TV fabrication is shown in Fig 6. All fabrication developments and integration, have been carried out on Gen 3 (550x650mm) glass panels. Before building up multilayer Cu RDL, glass carrier is coated with sacrificial release layers for removing the carrier at end of the fabrication using laser debonding process. It will be followed by photo dielectric lithography process and Ti/Cu seed deposition. Then photo resist lithography and descum process will be carried out for Cu RDL patterning and proceed for Cu electroplating process. Once Cu RDL structures of desired thickness is plated, panels will be subjected to wet processes like PR strip and Ti/Cu seed layer etch back. For multi-layer RDL build-up, PR lithography and wet etch processes need to be repeated. Once multi-layer Cu RDL fabrication is completed and IC chips will be attached to these package structures using flux based chip-to-panel (C2P) flip-chip process. Final permanent bonding will be carried out during reflow process. Then panel level compression molding is carried out using moldable underfill materials which doesn't require any additional underfill process. After completion of all fabrication and assembly processes, glass carrier will be debonded using laser ablation process and debonded panels will be subjected to plasma cleaning process to remove polymer residues. Solder ball attachment is required for all packages for subsequent board assembly and testing. Finally package singulation is carried out to singulate into individual packages.

B. Fabrication Challenges and solutions

FOPLP fabrication involves lot of uncertainties as there are no standards established yet in terms of tool configurations and materials development. In FOPLP integration, stability of the sacrificial release layer will play crucial role in achieve robust multi-layer package structures and in present study few different release layers have been evaluated and selected. Amongst various challenges, panel warpage is a key concern which exist throughout the fabrication and usually it becomes more severe when more processes are added to integration. In multi-layer RDL fabrication, handling warped panels is one of the key concerns during coating and exposure. Special chucks and jigs have been fabricated to overcome warpage issues of the panels. Topography effect arising during multi-layer RDL build-up will cause some CD variation due to out of focus during lithography. Selection of compatible photo resist and dielectric materials is very crucial in achieving better integration results. In Ti/Cu etch back process, control of the etch uniformity over Gen 3 panel is challenging due to its large size. Wet etch process has been optimized to achieve undercut of feature <5% of CD. In Cu electroplating process, tool configurations and processes have been modified to handle warped panels and achieve plating thickness variation below 10%.

IV. KEY PROCESS MODULES

RDL-1st TV fabrication involves various materials, fabrication and assembly processes. Reliability of the RDL-1st packaging greatly depends on established processes and materials. Hence present study focuses on evaluation of key processes and materials like sacrificial release layer coating, lithography of dielectrics and photoresist, seed layer PVD and electroplating.

A. Sacrifical Layer Coating

In RDL-1st TV fabrication, sacrificial layer coating on glass carrier is very first step. Stability of the release layer material is very crucial as it acts as foundation for rest of the processes. In this study two commercially available release layers' materials have been evaluated using slit coating equipment. Slit coater has a wide nozzle with coating material that will be dispensed on glass panel while scanning throughout the panel from one end to other end. Target thickness for sacrificial coating on glass panel is around 0.7um. Coating thickness has been measured in both scanning and nozzle directions and it is in the range of 0.68 to 0.73um as shown in Fig 7.

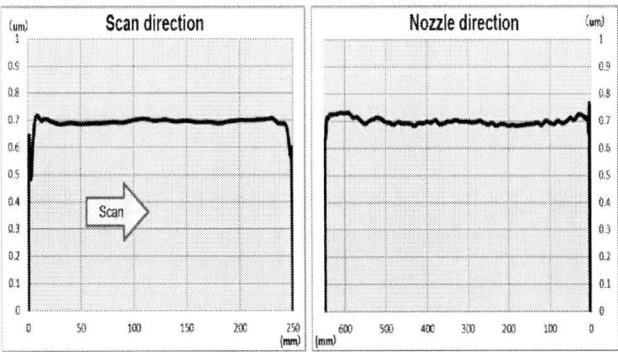
Fig 7. Sacrificial layer coating thickness profile across the panel

B. Lithogrpahy of photo dielectric materials

Two commercially available positive tone dielectric materials have been evaluated. Both of these dielectric materials are TMAH developable. Dielectric materials coating has been optimized using slit coating method and

whereas exposure is carried out using laser direct imaging (LDI) method. Patterning using LDI method is traditionally developed for PCB based technologies and it is slowly being adopted in fan-out packaging technologies due to its ease of implementation and there is match in terms of process requirements and economics. Main advantage of LDI method is it doesn't involve any physical mask as required by traditional lithography process and it helps in reducing the overall cost of the process. Target dielectric thickness is around 10um for both materials. In this development few challenges encountered like achieving coating with less TTV, handling of warped panels and dielectric material cracking. Fig 8 shows the typical dielectric material cracking issue observed in our development. To resolve all developmental issues, extensive DOE has been executed with wide range of process parameters like baking & curing profiles, exposure dose, laser power and depth of focus.

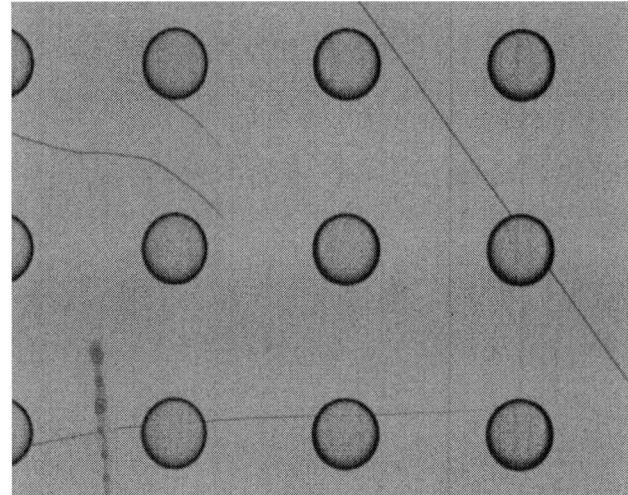
Fig 8. Dielectric material cracking issue during development

In slit coating process dielectric materials thickness has been optimized in the range of 9.7 to 10.47um and achieved TTV is up to 2.7%. Dielectric thickness profile of the panel is measured using contactless probes across scan and nozzle directions as shown in Fig 9. For dielectric material lithography evaluation, a special masking layer has been designed with different via sizes as shown in Fig 10. This layer has via sizes ranging from 3um to 200um and these dielectric openings will be used in final integration for UBM opening, and via connection between metal layers.

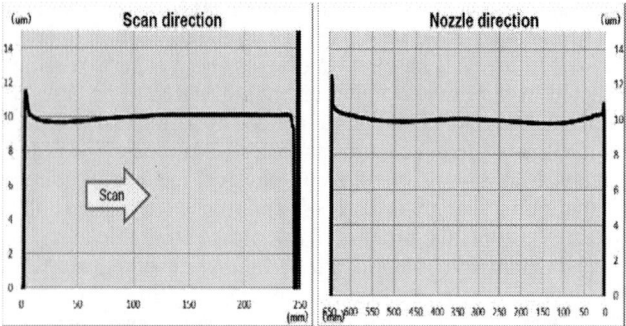

Fig 9 thickness profile of the dielectric material across the panels

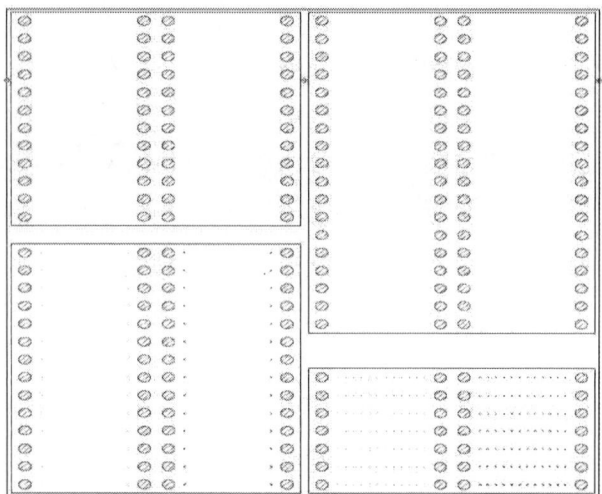

Fig 10. Via opening layout for dielectric material evaluation

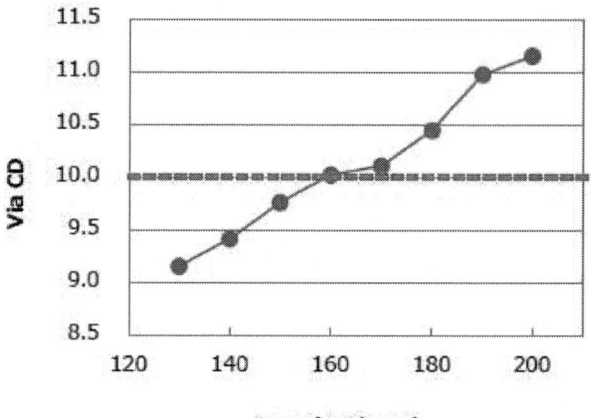

Fig 11. exposure optimization for dielectric via opening

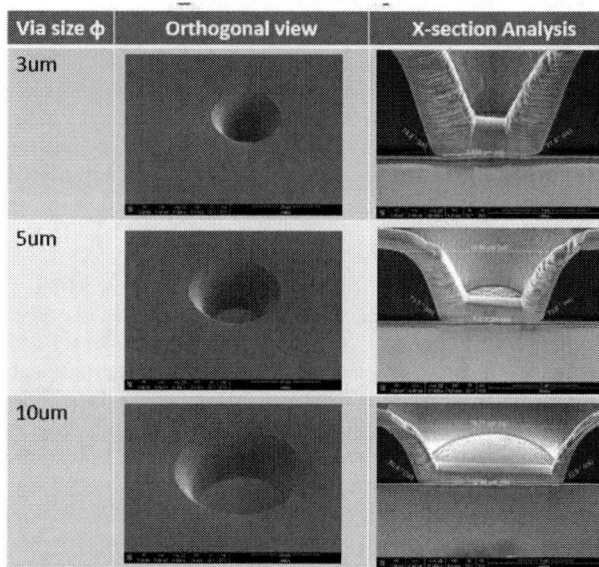

Fig 12 FIB/SEM X-section analysis of different via sizes

For lithography evaluation, a thorough DOE has been planned by considering wide range of process parameters like exposure dose, focus and laser power. All experiments have been carried out on high power LDI tool. Exposure energy used for this evaluation is in the range of 50 mJ/cm2 to 1500 mJ/cm2 and focus is the range of -10um to +10um. Laser power employed is in the range of 0.5 to 1.1W. for a specific CD opening, different exposure energies with increments of 25mJ/cm2 have been used at constant laser power and depth of focus. It has been repeated with different combinations of laser power and depth of focus values. Fig 11 shows a typical example for exposure optimization for 10um via opening. To achieve 10um CD, optimized exposure energy is 160mJ/cm2 and this energy is validated with <10um CD. All via sizes ranging from 3um diameter and above have been opened without any residues. Obtained via CD variations are within the acceptable range. FIB/SEM cross-section analysis has been carried out on different via sizes to understand the CD at bottom and top surface and side wall profile of the via. Fig 12 shows the FIB/SEM X-section analysis on via diameters 3um, 5um and 10um. All via sizes are clearly developed without any residues. Obtained via CD variations are within the acceptable range. Side wall profile of the via is in the range of 70 to 74. This kind of profile very beneficial in seed layer deposition and subsequent plating process. It plays a very important role in obtaining robust multi-level Cu RDL interconnects.

C. Lithography of photo resist material

For this evaluation a commercially available positive tone photo resist (PR) has been selected and it is TMAH developable. Extensive lithography DOE has been planned and executed for optimization of fine line patterns. Slit coating method used for coating process and PR coating thickness has been optimized in the range of 7.08 to 7.11um. High power LDI tool with maximum power up to 2W is used for exposure optimization. Fig 13 shows the coating thickness profile across panels area and obtained TTV is less than 1%. Special masking layer with different LW/LS structures has been designed for thorough evaluation as shown in Fig 14. This layout design includes LW/LS structures ranging from 1/1um to 15/15um with regular intervals. Main intention of this evaluations to evaluate the PR lithography using LDI exposure method for fine patterns. Exposure energy used in this evaluation is in the range of 60 to 380 mJ/cm2 and focus starting from -16um to +16um. During exposure energy and focuses optimization LW/LS,

3/3um structures have been used as reference. Fig 15 shows the exposure energy and focus optimizations DOE for 3/3um structures. These optimized parameters have been used for lithography development for all test structures and able to pattern 2/2um onwards. Fig 16 shows the PR lithography patterns for 2/2um and 5/5um

For all test structures, through CD analysis has been carried out. CD variation for fine LW/LS structures, 2/2um & 3/3um is around 6% and for all remaining it is observed less than the 2%. PR side wall profile is analyzed for few test structures by conducting FIB/SEM cross-section analysis as shown in Fig 17.

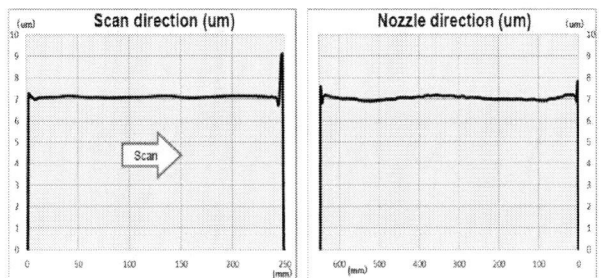

Fig 13 PR coating thickness profile across the panels

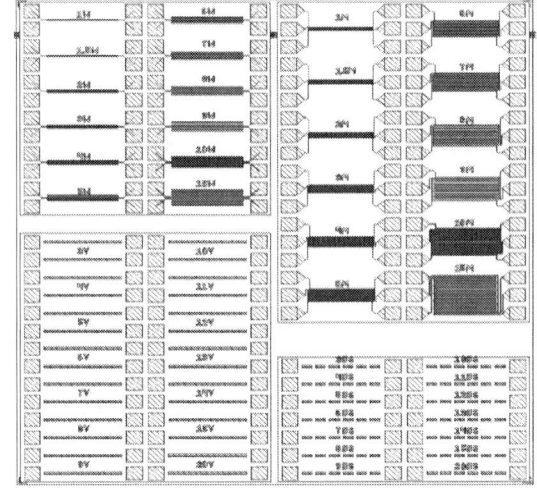

Fig 14 LW/LS test structures for photo resist material evaluation

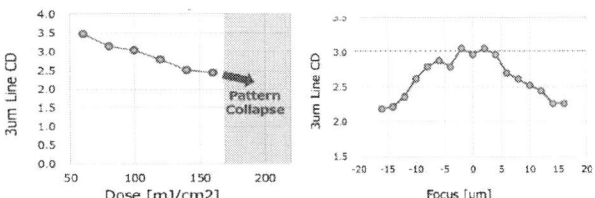

Fig 15 Exposure energy and focus optimization for 3/3um structures

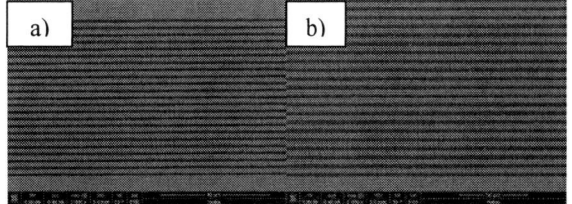

Fig 16 Lithography of fine LW/LS patterns, a) 2/2um and b) 5/5um

Fig 17 FIB/X-SEM analysis for fine LW/LS structures a) 2/2um and b) 5/5um

D. Seed layer deposition and Electroplating

For panel level PVD process a special jig has been fabricated to handle Gen 3 (550X650mm) panels, including warped panels. Before actual seed deposition process, panels are treated with degas process and Argon sputter etching mainly to improve seed layer quality and its adhesion on panels. Target Ti/Cu seed layer thickness is 1KA Ti and 1KA Cu. Obtained Ti/Cu seed layer thickness variation is less than 5%. Fig 18 shows a Ti/Cu seed layer deposited panel.

Fig 18. Ti/Cu seed layer deposition on Gen3 panel

For Cu ECP process, a special jig has been prepared and vertical plating process has been established. Different process methodologies and chemicals have been established for both UBM and fine line RDL processes. Before ECP process, all panels subjected to descum process mainly to remove unwanted residues in patterns and improve the plating uniformity. Target Cu plating thickness is around 7um. Plating thickness is validated by both 3D optical profiling and mechanical cross-sectioning as shown in Fig 19. Obtained Cu ECP thickness is in the range of 7.3 to 7.4

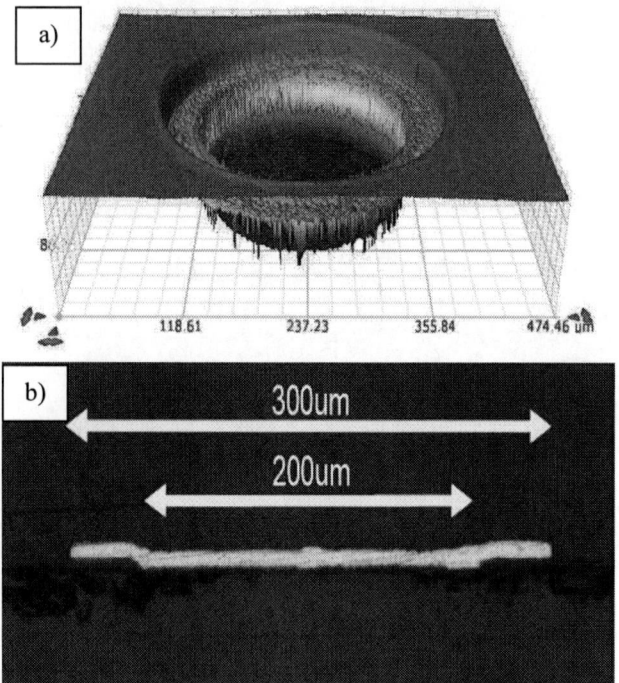

Fig 19. Cu electroplating analysis for UBM pad, a) 3D optical profile and b) cross-sectioning

V. RDL-1ST TV FABRICATION AND ASSEMBLY

A. RDL-1st TV Fabrication

Present TV designed with two layers of Cu routing layers and three layers of photo definable dielectrics. Before starting panel level integration, all panel level fabrication process modules have been optimized and established. First dielectric layer is used as base layer and for UBM fabrication. Dielectric layer 2 used as passivation between two Cu redistribution layers. Dielectric layer 3 is used for final passivation of the package structures. This whole fabrication involves five masking steps, which includes three masking steps for dielectric materials and two for photo resist patterning. For chip assembly, package structures have been fabricated without UBM. This approach will eventually help in reducing minimizing processing steps and reducing cost. Fig 20 shows summary of all key integration processes and results of test vehicle fabrication. Fig 21 shows the RDL-1st TV fabricated panel and integrity of all package structures have been verified by cross sectioning as shown in Fig 21.

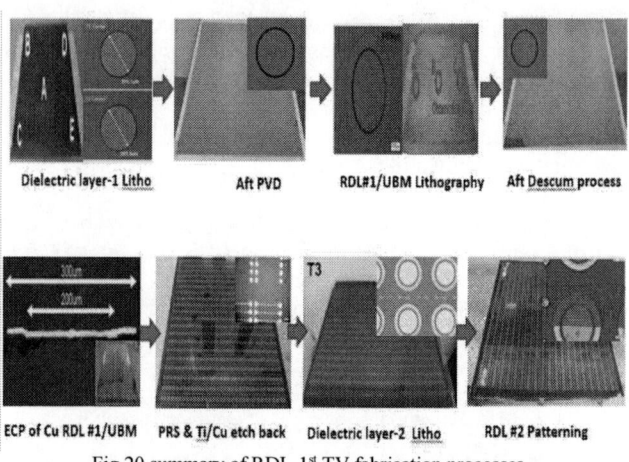

Fig 20 summary of RDL-1st TV fabrication processes

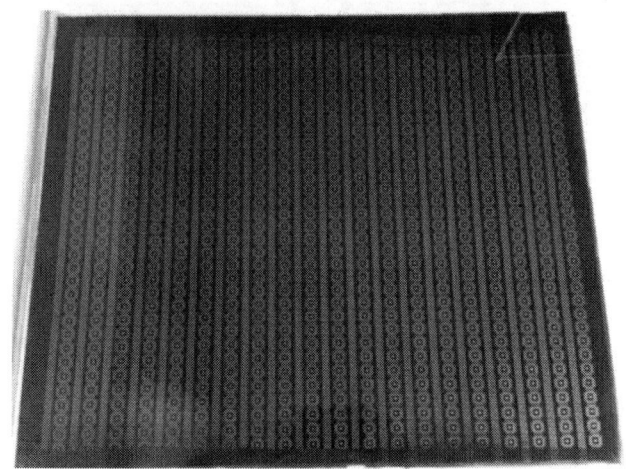

Fig 21 RDl-1st TV fabricated panel with multi-layer RDL

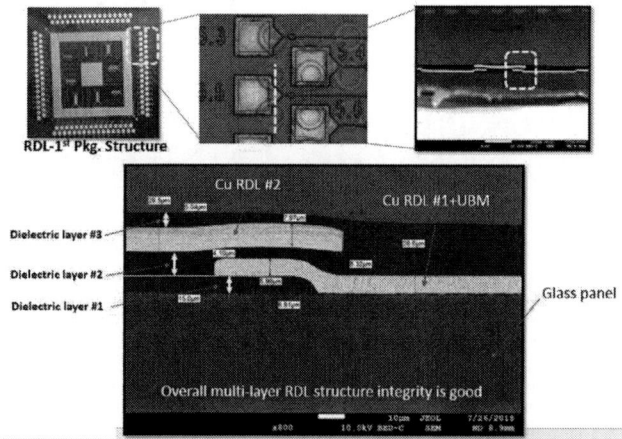

Fig 22 cross-sectioning of fabricated package structres on panel

B. RDL-1st Package Assembly

IC test chip assembly to package structure is carried out by flip-chip method using flux and subsequently permanent bonding by mass reflow process. No clean flux is employed for this process for cleaner assembly process. First IC chip is flipped, then Cu pillars dipped into flux and finally it placed on package substrate. Finally, whole panel substrate populated with chips sent through mass reflow oven for permanent solder joint formation and Fig 23 shows the panel with assembled test chips. Five zone reflow oven used for this purpose and it maintains temperature ranging from 145C to 260C. Solder joint integrity of the package is verified by both X-ray analysis and cross sectioning as shown in Fig 24. As per analysis, assembled packages have exhibited good solder joint integrity with continuous daisy chain readings

Fig 23 RDL-1st fabricated panel with assembled test chips

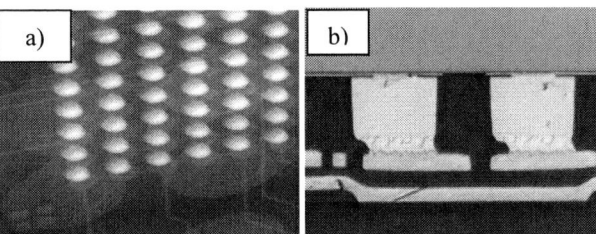

Fig 24 Solder joint analysis of assembled package, a) X-ray and b) cross-sectioning

VI. SUMMARY AND CONCLUSIONS

Present work successfully established RDL-1st approach for fan-out packaging and important results have been summarized below;

1) Laser direct imaging lithography process has been established for two different types of photo sensitive dielectric materials with smallest via opening up to 3um diameter.

2) Positive tone photo resist lithography process has been optimized to achieve fine line LW/LS patterns from 2/2um onwards

3) Panel level ECP process has been established for 7um thickness copper plating with variation <5%.

4) Panel level wet etch processes like photo resist stripping, Ti/Cu etch back have been established.

5) Whole RDL-1st process integration on Gen 3 panel has been demonstrated up to two-layer Cu RDL structures.

6) Final assembled packages on panels have exhibited good solder joint integrity and passed electrical testing

ACKNOWLEDGMENT

This work is the result of a project initiated by Fan-Out Panel Level Packaging (FOPLP) consortium. The Authors greatly appreciate the members' participation in discussion and encouragement throughout the course of the project which makes this research possible

REFERENCES

[1] Mark Lapedus, "Fan-Out Wars Begin" Semiconductor Engineering, 5th February 2018

[2] Vempati Srinivasa Rao et al., 2017 IEEE 67th Electronic Components and Technology Conference, pp. 615-622

[3] Ser Choong Chong et. al., "Development of Package-on-Package Using Embedded Wafer-Level Package Approach" IEEE Transactions on Components, Packaging and Manufacturing Technology, VOL. 3, NO. 10, OCT 2013.

[4] Vempati Srinivasa Rao et. al., "Development of High Density Fan Out Wafer Level Package (HD FOWLP) with Multi-layer Fine Pitch RDL for Mobile Applications" in Proceedings of 66th Electronic Components and Technology Conference, 2016.

[5] Tanja Braun etal., "Panel Level Packaging-A view along the process chain" 2018 IEEE 68th Electronic Components and Technology Conference (ECTC), pp. 70-78

[6] Yole Development " Status of the Panel Level Packaging report" April 2018

[7] Vasarla Nagendra Sekhar et al., "Evaluation of Materials for Fan-Out Panel Level Packaging (FOPLP) Applications" 2018 20th Electronics Packaging Technology Conference.

Epoxy composites with Surface Modified Silicon Carbide Filler for High Temperature Molding Compounds

Fan Wu[1], Nicholas C Mitchell[1], Bo Song[1], Kyoung-Sik Moon[1] and CP Wong[1]

[1]School of Materials Science and Engineering
Georgia Institute of Technology
Atlanta, GA, USA
cp.wong@mse.gatech.edu

Abstract—**With the rapid development of high power, high frequency and high-density electronics, heat management of the package has attracted intensive attention. Thermal conductivity is becoming an increasingly important parameter in the evaluation of packaging materials like EMCs. However, for polymer in encapsulation materials such as epoxy, their low thermal conductivity due to the phonon scattering issue greatly limits their ability in heat dissipation. Therefore, it is worthwhile to study the modification of filler system for the improvement of thermal conductivity of EMCs, since filler occupies 70-90 % of the weight in EMCs. In this work, SiC particles are coated with silane and rubberized layers for molding compounds as a modified filler. The effects of treated SiC filler on thermal conductivity and toughness in epoxy composite are studied at low filler loading.**

Keywords-molding compound; epoxy; thermal conductivity; filler modification

I. INTRODUCTION

Wide bandgap (WBG) semiconductors such as silicon nitride (SiC) and gallium nitride (GaN) have advantages over silicon (Si) in various aspects including larger energy gap, higher electric breakdown field, higher saturated electron velocity and better electron mobility.[1] Therefore, WBG semiconductors have the ability to facilitate the high frequency operation, which is essential in the miniaturization of power devices.[2] One of the major challenges facing the development of WBG semiconductors is thermal management issue. Most of the failures in semiconductor equipment are due to the excessive operation temperature, which becomes increasing concerning, especially for WBG semiconductors with high junction temperature.[2] As a result, electronics encapsulation materials such as EMCs are required to work under higher temperature and have higher heat management efficiency.[3]

With the development of WBG semiconductors, more heat is generated in the package, which imposes more stringent requirements on the heat management. Therefore, the thermal conductivity of EMC, becomes increasingly important. Atomic vibrational waves, which is also referred as phonons, are responsible for the thermal conduction in epoxy resin. When a phonon goes across an interface, it changes from one energy state to another which causes a thermal interfacial resistance. This resistance depends not only on the defects of the surface, but also on the available energy states on both sides of the interface. For most of the polymers, phonons scatter when they transfer from one chain to another, resulting in a low thermal conductivity. It is possible for some of the thermoplastic polymers to exhibit a high thermal conductivity due to their unique chain structure such as π–conjugated polymers (Figure 1).[4] However, for epoxy, as a thermoset polymer, it usually has thermal conductivity as low as 0.2-0.3 W/mK.

Generally, EMCs consist of 10-20 % of epoxy and hardener, which builds up the crosslinking polymer network and gives excellent adhesion strength. A heavy loading of filler, usually more than 70 %, is added into the EMCs to lower the thermal expansion coefficient (CTE), provide higher modulus and lower the moisture absorption of the compounds. In this case, modified filler can be used to increase the thermal conductivity for EMCs. Conventional silica filler for EMC has a thermal conductivity around 1.3 W/mK, which is not quite ideal to make high thermal conductivity composites. In the exploration of high thermal conductivity fillers, various of metal particles are taken into consideration. Zhou *et al* fabricated silica coated self-passivated aluminium fibers and nanoparticles as a filler for polymer composites, which reached a highest thermal conductivity of 15.2 W/mK at 50 vol.% filler loading (Figure 2).[5] Providing high thermal conductivity, metal-based filler also results in high loss factor and high permittivity which is not desirable for high power device packaging. Electrically non–conductive ceramic fillers such as boron nitride (BN) and silicon carbide (SiC) have high thermal conductivity as well as excellent dielectric properties, which make them promising candidate for EMC. Huang *et al* reported a polyhedral oligosilsesquioxane-modified boron nitride nanotube (BNNT) filler for epoxy composites with a thermal conductivity of ~3 W/mK at a 30% filler loading (Figure 3).[6] Filler with high aspect ratio gives more contacts between fillers, which is advantageous for phonon transfer and enables a high thermal conductivity. However, more contacts between fillers also make the interface between filler and resin a critical issue, which affects toughness of the composite.

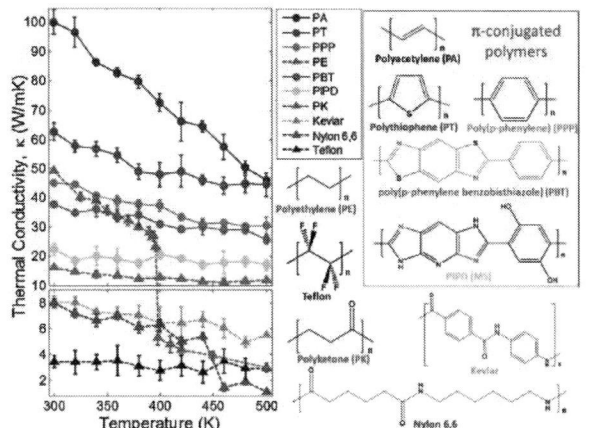

Figure 1. Thermal conductivity and structure of polymer[4] (figure from ref. 4)

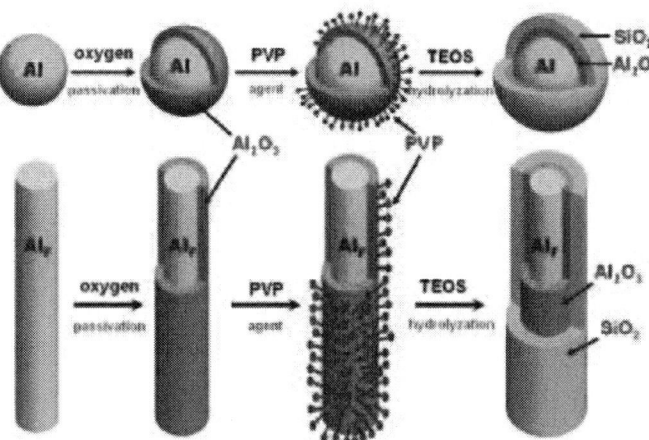

Figure 2. Structure of silica coated self-passivated aluminum fibers[5] (figure from ref. 5)

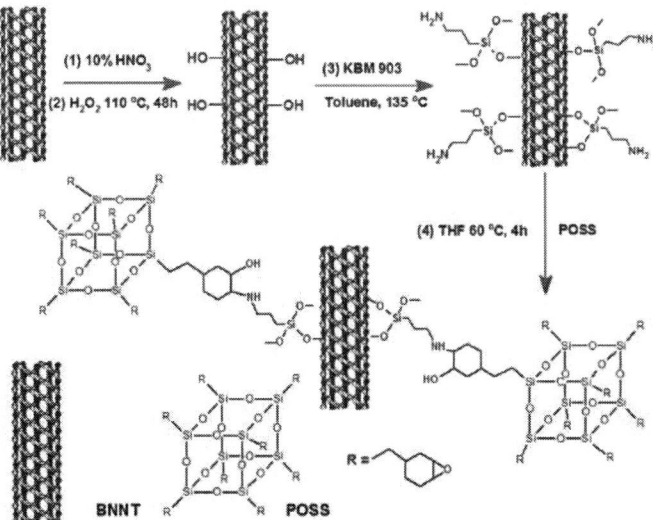

Figure 3. Structure of polyhedral oligosilsesquioxane-modified BNNTs[6] (figure from ref. 6)

In this work, SiC particles are treated with silane and reactive silicone respectively. SiC with different coating layers were filled into epoxy matrix and formed epoxy composites. The effects of different SiC fillers on thermal conductivity and toughness of the epoxy composites were investigated. This study promotes understanding of the effects of filler and filler-matrix interface on thermal conductivity and toughness in EMCs.

II. EXPERIMENTAL

A. Materials

(3-Glycidoxypropyl)trimethoxysilane (GPTMS) and aminopropyl terminated polydimethylsiloxane with molecular weight of 3000 (DMS-A15) were purchased from Gelest, Inc. Formic acid was obtained from Fisher Scientific. Tetra methyl biphenyl epoxy with an epoxide equivalent weight of 197 was purchased from Mitsubishi Chemical Corporation. Aluminum acetylacetonate catalyst was purchased from Aldrich Chemical Company, Inc. Curing agent, hexahydro-4-methylphthalic anhydride (HMPA), was purchased from Lindau Chemicals, Inc. The chemical structures of the major chemicals used are shown in Scheme 1

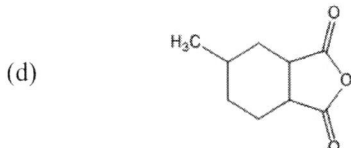

Scheme 1. Chemical structures of (a) aminopropyl terminated polydimethylsiloxane (b) (3-glycidoxypropyl)trimethoxysilane (c) tetra methyl biphenyl epoxy (d) hexahydro-4-methylphthalic anhydride

B. Sample preparation

The surface functionalization of SiC by silanes was conducted according to the method reported by Liang et al.[7] SiC particles were first oxidized at 1200 °C for 3 h in oven. 1 g of SiC with an oxidation layer (SiC-SiO₂) was dispersed in

978-1-7281-1500-9/19 $31.00 © 2019 IEEE   2135

ethanol and sonicated for 30 min. Then 1 g of water and 0.2 ml of GPTMS was added into the solution which had a pH tuned to 4 by formic acid. The solution was refluxed at 80 °C for 24 h to ensure the grafting of the silane on SiC particles. After silane coupling, the silane coated SiC-SiO₂ particles (SiC-silane) were filtered and dried at 120 °C for 6 h. To coat a rubberized layer outside the particles, 1 g of SiC-silane and 0.5 ml of DMS-A15 were mixed in ethanol, refluxed for 24 h, filtered and dried. SiC coated by DMS-A15 is designated as SiC-A15. Scheme 2 shows the preparation steps of SiC-A15.

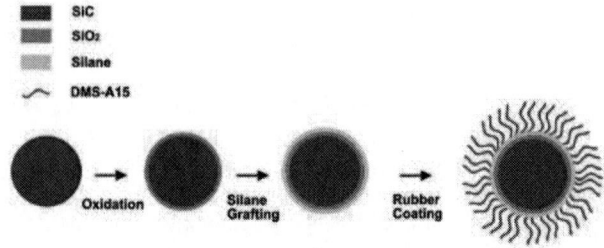

Scheme 2. Schematic representation of sample preparation

In the preparation of epoxy compounds, tetra methyl biphenyl epoxy and hexahydro-4-methylphthalic anhydride were mixed at a ratio of 1:0.85. The mixture was melt and stirred continuously at 100 °C until a homogeneous phase was obtained. Different fillers were loaded into the epoxy melts as shown in Table 1. Filled mixtures were cured at 210 °C for 2 h with 0.3% of catalyst.[3, 8-9]

TABLE I.   EPOXY COMPOUND FORMULAS

Specimen	Epoxy and hardener (%)	SiC-GPTMS (%)	SiC-A15 (%)
Ep	100	-	-
Ep-10SiC-GPTMS	90	10	-
Ep-30SiC-GPTMS	70	30	-
Ep-10SiC-A15	90	-	10
Ep-30SiC-A15	70	-	30

C. Characterization

An XPert PRO Alpha-1 XRD was performed to study the crystal structure of the pristine SiC powder and the oxidized SiC. The rubber coating of particles was chemically characterized by Thermo Scientific iS50 FT-IR Spectrometer to confirm the chemical structures and thermally investigated by TGA Q50 from TA Instruments to obtain the amount of the rubber coated. In order to study the thermal conductivity of the epoxy compounds, an LFA 467 HyperFlash from Netzsch was performed to obtain the thermal diffusivity of the compounds. The heat capacity and density needed in the calculation of thermal conductivity were achieved respectively by DSC Q2000 from TA Instruments and density measurement based on Archimedes' principle.

Thermal conductivity was obtained based on the following equation.

$$\kappa = \alpha \times \rho \times C_p \quad (1)$$

Where κ is the thermal conductivity (W/mK), α is the thermal diffusivity (m2/s), ρ is the density (kg/m3) and C_p is the heat capacity (J/(kg·K)). In the measurement of fracture toughness of the compounds according to ASTM E-399, three-point bending tests were conducted by Instron 5548 MicroTester. The morphology of both fillers and fracture surfaces of the compounds were observed by Hitachi SU8010 SEM. Dielectric properties including the permittivity and the loss factors of the compounds were determined by Agilent E4991A RF Impedance/Material Analyzer at room temperature.

III.   SURFACE TREATMENT OF SiC

In the surface treatment of SiC, it is necessary to form an oxidation layer and thus enhance the reactivity of the particles with silane, since the SiO₂ oxidation layer provides more silanol groups than SiC. XRD scans of SiC and SiC-SiO₂ are conducted as shown in Figure 4, in order to investigate the chemical and crystal structures of the SiC particles before and after oxidation. SiC powder exhibited peaks at 34.04 º, 34.70 º, 35.57 º, 38.08 º, 59.95 º, 65.55 º and 71.70 º, which confirmed its hexagonal crystal structure.[10-11] SiC-SiO₂ showed an extra peak at 21.78 º, which is corresponding to the (011) plane of hexagonal SiO₂[12]

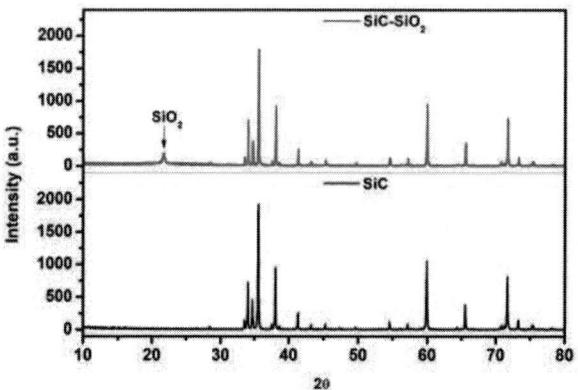

Figure 4. XRD scan of SiC and SiC-SiO₂ particles

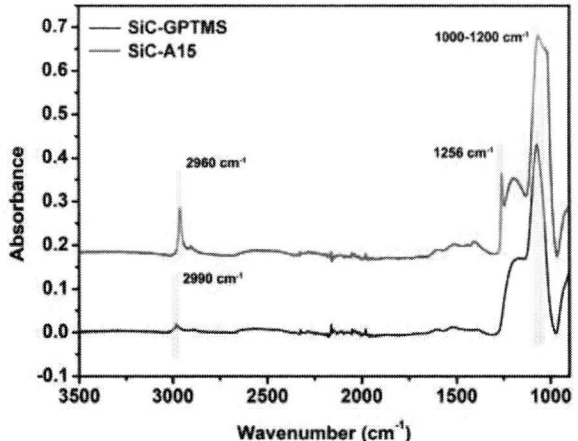

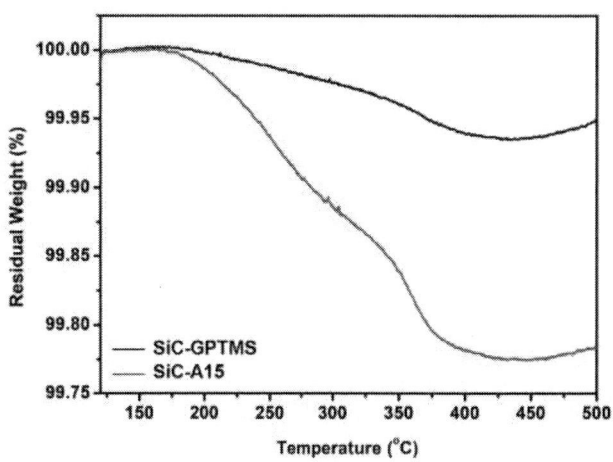

Figure 6. TGA scans of modified SiC

Scheme 3. Chemical reactions of (a) GPTMS grafting (b) DMA-15 coating

With more reactive sites provided by SiO_2 oxidation layer, GPTMS and DMS-A15 were grafted to the particle surface (Scheme 3). GPTMS attached to the silica surface via a hydrolysis reaction, and its epoxide terminated group further reacted with the anime terminated group of DMS-A15, which ensured the rubber layer coating. Chemical structures of the double layer coating were investigated and confirmed by FTIR (Figure 5). For both SiC-GPTMS and SiC-A15, the strong peak at 1000-1200 cm-1 was corresponding to the Si-O-Si bonding. The peak at 2990 cm-1 of SiC-GPTMS can be attributed to –OH groups, while the peak at 2960 cm-1 of SiC-A15 can be attributed to N-H stretching. The sharp peak at 1256 cm-1 represents the C-N stretching of primary amine, confirming the coating of DMS-A15 in SiC-A15.[13] The amount of the coating layer grafted to the SiC surface was investigated by TGA (Figure 6). SiC-GPTMS exhibited a mass loss of 0.06 % when heated from 120 °C to 450 °C, which indicated the amount of silane grafting layer. Compared to SiC-GPTMS, SiC-A15 had a larger mass loss of 0.23 %. Therefore, the weight corresponding to the rubber layer was around 0.17 %.

IV. PROPERTIES OF FILLED EPOXY COMPOUNDS

The effect of treated SiC filler on the thermal conductivity of the compounds is shown in Figure 7. The thermal conductivity increased gradually with the filler loading. The thermal conductivity of Ep-30SiC-A15 reached 0.29 W/mK, which is about three times of the neat epoxy. The thermal conductivity is expected to increase further with more filler loading, when more contacts between fillers are formed and construct a better channel for phonon transfer.[14] Compared with SiC-GPTMS, SiC-A15 offered a larger increase in compound thermal conductivity. To study the difference between SiC-GPTMS and SiC-A15 when applied into epoxy matrix, SEM images of the SiC powder and compound fracture surfaces were taken (Figure 8). As shown in Figure 8a, the irregular polyhedral SiC particle used in this work has a size of 3-5 μm. Silane treated SiC-GPTMS exhibited good dispersion in epoxy matrix, while Ep-30SiC-A15 showed formation of filler regions because of the aggregation trend of rubber layer, which brought together the filler particles. This aggregation of filler created more contacts between these polyhedral particles and more thermally conductive path. Therefore, epoxy with SiC-A15 had a higher thermal conductivity than epoxy with SiC-GPTMS at a same filler loading level.

Figure 5. FTIR spectra of treated SiC

978-1-7281-1500-9/19 $31.00 © 2019 IEEE

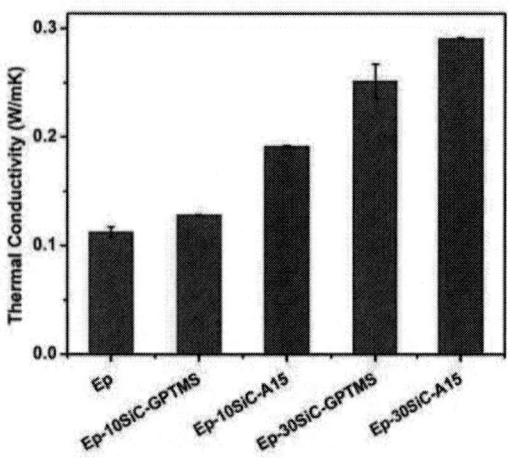

Figure 7. Thermal conductivity of epoxy compounds with modified fillers

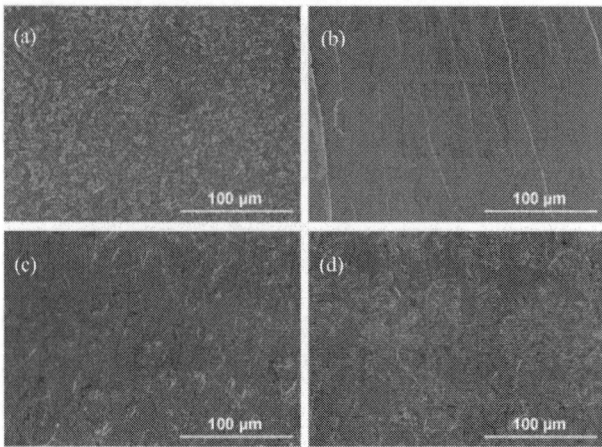

Figure 8. SEM images of (a) SiC particles (b) fracture surface of epoxy resin (c) fracture surface of Ep-30SiC-GPTMS (d) fracture surface of Ep-30SiC-A15

Generally, fracture toughness decreases for high filler loading because of the weak interface between filler and matrix where cracks tend to propagate.[15] The fracture toughness of epoxy with different fillers was shown in Figure 9. K_{IC} decreased from 3 Mpa·m$^{1/2}$ to 1.5 Mpa·m$^{1/2}$ with 30% of filler and a difference not so significant between SiC-GPTMS and SiC-A15. However, SEM images of the fracture surface showed different failure mechanisms for SiC-GPTMS and SiC-A15. Although Ep-30SiC-GPTMS had a better dispersion of filler, the cracks propagated along the filler-matrix interface, leaving a relatively smooth fracture surface (Figure 8c). In Ep-30SiC-A15 the crack broke through the aggregation regions of the rubberized fillers instead of the filler-matrix interface since the rubber layer formed by amine terminated silicone provided a better interface between filler and epoxy (Figure 8d). Figure 8b is the fracture surface of neat epoxy for reference.

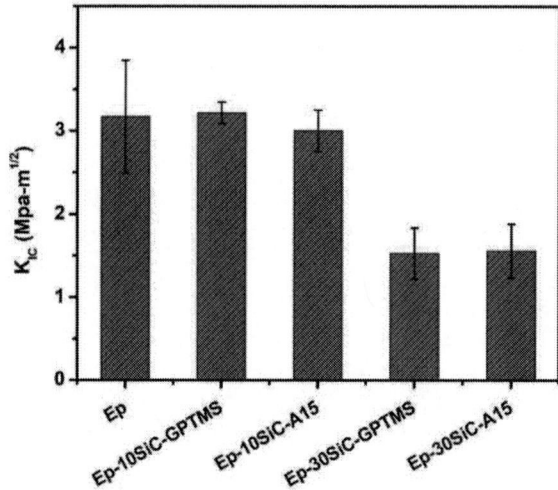

Figure 9. Fracture toughness of epoxy compounds with modified fillers

V. SUMMARY

In this study, SiC particles were coated with silane and silicone rubber layer. SiC treated with silane had a better dispersity in epoxy matrix while SiC with silicone rubber layer tended to aggregate with each other and formed separated filler regions. Despite different filler morphology in composites, both treated fillers increased the thermal conductivity of the epoxy composites. Epoxy with 30% of SiC-A15 had a thermal conductivity of 0.29W/mK. On the other hand, fracture toughness decreased a lot for this formulation after filler loading. Epoxy with SiC-GPTMS and SiC-A15 showed different failure mechanism in crack propagation. For SiC-GPTMS, the filler-matrix interface is the weak point. For SiC-A15, reactive rubber layer provides strengthened interface. Therefore, crack broke though the filler aggregations instead of the interface.

ACKNOWLEDGMENT

The funding support of Georgia Tech Packaging Research Center is gratefully appreciated.

REFERENCES

[1] E. A. Jones, F. F. Wang and D. Costinett. "Review of commercial GaN power devices and GaN-based converter design challenges." *IEEE Trans. Emerg. Sel. Topics Power Electron.* vol. 4, no. 3, 2016, pp. 707-719.

[2] K. Shenai, M. Dudley and R. F. Davis. "Current status and emerging trends in wide bandgap (WBG) semiconductor power switching devices." *ECS J. Solid State Sci. Technol.* vol. 2. no. 8, 2013, pp. 3055-3063.

[3] F. Wu, C. C. Tuan, B. Song, K. S. Moon and C. P. Wong. "Controlled synthesis and evaluation of cyanate ester/epoxy copolymer system for high temperature molding compounds." *J. Polym. Sci. A.* vol. 56, no. 13, 2018, pp. 1337-1345

[4] H. Chen, V. V. Ginzburg, J. Yang, Y. Yang, W. Liu, Y. Huang, L. Du and B. Chen. "Thermal conductivity of polymer-based composites: Fundamentals and applications." *Prog. Polym. Sci.* vol. 59, 2016, pp. 41-85

[5] Y. Zhou, Y. Bai, K. Yu, Y. Kang and H. Wang. "Excellent thermal conductivity and dielectric properties of polyimide

composites filled with silica coated self-passivated aluminum fibers and nanoparticles." *Appl. Phys. Lett.* vol. 102, no. 25, 2013, pp. 252903

[6] X. Huang, C. Zhi, P. Jiang, D. Golberg, Y. Bando and T. Tanaka. "Polyhedral oligosilsesquioxane-modified boron nitride nanotube based epoxy nanocomposites: an ideal dielectric material with high thermal conductivity." *Adv. Funct. Mater.* vol. 23, no. 14, 2013, pp. 1824-1831

[7] Q. Liang, K. S. Moon, H. Jiang and C. P. Wong. "Thermal conductivity enhancement of epoxy composites by interfacial covalent bonding for underfill and thermal interfacial materials in Cu/Low-K application." *IEEE Trans. Compon. Packag. Manuf. Technol.* vol. 2, no. 10, 2013, pp. 1571-1579.

[8] F. Wu, B. Song, J. Hah, C. C. Tuan, K.-S. Moon and C. P. Wong. "Polyimide incorporated cyanate ester/epoxy copolymers for high-temperature molding compounds." *J. Polym. Sci. A.* vol. 56, no. 21, 2018, pp. 2412-2421

[9] C. C. Tuan, F. Wu, K. S. Moon, and C. P. Wong. "Epoxy/Cyanate Ester Copolymer Material for Molding Compounds in High-Temperature Operations." *In Electronic Components and Technology Conference (ECTC),* 2017 IEEE 67th, pp. 1328-1333. IEEE

[10] D. Lundqvist. "On the crystal structure of silicon carbide and its content of impurities." *Acta Chemica Scandinavica.* vol. 2, no. 1, 1948, pp. 177.

[11] G. McIntyre, L. Mélési, M. Guthrie, C. Tulk, J. Xu and J. Parise. "One picture says it all—high-pressure cells for neutron Laue diffraction on VIVALDI." *J. Phys. Condens. Matter* vol. 17, no. 40, 2005, pp. 3017.

[12] J. Martínez, S. Palomares-Sánchez, G. Ortega-Zarzosa, F. Ruiz and Y. Chumakov. "Rietveld refinement of amorphous SiO2 prepared via sol–gel method." *Mater. Lett.* vol. 60, no. 29-30, 2006, pp. 3526-3529.

[13] S. A. Jadhav, R. Bongiovanni, D. L. Marchisio, D. Fontana and C. Egger. "Surface modification of iron oxide (Fe2O3) pigment particles with amino-functional polysiloxane for improved dispersion stability and hydrophobicity." *Pigment & Resin Technology* vol. 43, no. 4, 2014, pp. 219-227.

[14] T. Zhou, X. Wang, X. Liu and D. Xiong "Improved thermal conductivity of epoxy composites using a hybrid multi-walled carbon nanotube/micro-SiC filler." *Carbon.* vol. 48, no. 4, 2010, pp. 1171-1176.

[15] S. Chandrasekaran, N. Sato, F. Tölle, R. Mülhaupt, B. Fiedler and K. Schulte. "Fracture toughness and failure mechanism of graphene based epoxy composites." *Compos. Sci. Technol.* vol. 97, 2014, pp. 90-99.

Ultra Low Resistivity and High Electrical Stability Silo-Ag ECAs Produced from Curing Chemistry Optimization for Flexible Electronics

Xueqiao Wang, Bo Song, Kyoung-Sik Moon, C.P. Wong

School of Materials Science and Engineering
Georgia Institute of Technology
771 Ferst Drive, Atlanta, Georgia, 30332
xwang745@gatech.edu

Abstract— Highly conductive and both mechanically and electrically stable polymer-based adhesive composites are now more extensively used in die attachment, flip-chip interconnections, surface mount interconnections, and are especially important to flexible and wearable electronics. To enhance conductivity, most researches devoted to filler surface engineering and largely neglected network development of polymer matrix. While ultra low resistivity has been achieved, most of these electrically conductive adhesives (ECAs) display poor electrical stability under reversed mechanical deformation. For practical implications to wearable electronic applications, the performance stability of ECA under reversed deformation under relatively small strains of human movement is of critical significance. We investigate the effect of polymer matrix stoichiometry during curing on the conductivity and electrical stability of silver-poly(dimethyl siloxane) (PDMS) ECA composites. Characterization of crosslinking density reveals a competition between crosslinking reaction and surfactant reduction with silver flake sintering during curing, leading to a tradeoff relationship between stretchability and electrical conductivity. By tuning the crosslinking functional group ratio, the best combination of ultra-low resistivity of 1.33×10^{-4} $\Omega \cdot$cm and high stability of 22.2% increase in resistivity after 100 cycles of tensile 0-25% tensile straining is achieved at an intermediate hydride-to-vinyl group ratio of 1:2.

Keywords-ECA; conductivity; electrical stability; PMDS; crosslinking density

I. INTRODUCTION

The rapid development of flexible electronics creates a demand on electrically conductive adhesives (ECAs) for stretchable interconnects [1]. In particular, isotropically conductive adhesives (ICAs) are desired due to their higher flexibility and isotropic conductivity [2]. Common polymers used for ICA matrix include epoxy, polyurethane (PU), silicones, especially poly(dimethyl siloxane) (PDMS),

polyimides, and cyanate esters. PDMS exhibits high performance in terms of flexibility, thermal, chemical and environmental stability, oxidation resistance, low curing temperature, stress dissipation, high elasticity and low dielectric constant [3]. This gives silicone based ICAs great potential in applications such as wearable electronics, stretchable RF devices, microfluidic devices, and biomedical sensors with printable patterns [4-12].

Conductivity development process and filler surface engineering have been the focus of past research on PDMS based ECA composites. Two mechanisms were shown to be responsible for conductivity development during curing of PDMS ECAs with silver fillers: volume shrinkage of PDMS matrix and the reduction of surfactants on silver filler flakes [13-16]. Volume shrinkage is controlled via the degree of crosslinking in PDMS matrix, and surfactant reduction relies on the reaction with hydride functional groups in PDMS precursors. Silver ions are originally stabilized by coordination with a surfactant, and with the addition of PDMS containing hydride groups, the Si-H bonds are oxidized to O-Si-O, thereby reducing the Ag ions to nanoparticles. Contacts of the reduced metal nanoparticles formed on the surface of the micron sized silver flakes can result in solid-state interdiffusion at curing temperature and such a low temperature sintering-like process between the nanoparticles can facilitate electrical transport through the micron sized flakes via conductive bridges formed by the nanoparticles, therefore enhancing conductivity of the whole composite [13-14, 16].

Effect of both mechanisms depends on stoichiometry of PDMS precursors, rending reaction chemistry control an effective means to tune conductivity of silo-Ag ECAs. Further, for applications in wearable electronics, stability in electrical performance under reversed deformation of relatively small strain levels of human movement is critical. As the stretchability of silo-Ag ECAs is directly related to crosslinking density, optimization of reaction stoichiometry proves essential to conductivity preservation of flexible electronics.

We use an exemplary system of hydrosilylation addition cure between vinyl and hydride functional groups with Pt catalyst and peroxide inhibitor, with the addition of micron-sized silver flake conductive fillers. The mechanism of the addition crosslinking reaction is shown in Figure 1 [17]. The ratio of the hydride and vinyl functional groups is varied under the same filler loading, and the conductivity of these ECAs are tested. Crosslinking density is obtained through swelling tests of pure PDMS matrix and the ECAs with different stoichiometric ratios. The variation of conductivity is correlated to that of crosslinking density, which gives insight to reaction kinetics for crosslinking and conductivity development under curing conditions. Electrical stability is characterized by in-situ bulk resistance measurements during tensile fatigue. ECA samples obtained from different reaction stoichiometry are subjected to 100 cycles of reversed straining between 0 and 25% strain. Their resistances are measured before straining, during straining at maximum and minimum strains, and post-straining after complete relaxation to original length. The percent increase in resistance during and post straining is compared for different reaction stoichiometries, and is analyzed as a result of the combined effect of crosslinking density and sintering of Ag nanoparticles. Comparing conductivity and electrical stability across all ECAs, a best performing functional group ratio is determined.

II. Experimental

PDMS matrix is prepared from two precursors, a poly(methylhydrosiloxane) providing hydride groups, and vinyl terminated PMDS with reinforcing additives. The precursors are mixed at hydride: vinyl functional group ratios of 1:4, 1:2, 1:1.5, 1:1, 1.5:1, 2:1, and 4:1. A platinum (Pt) based catalyst is added to the precursors at a catalyst to PDMS resin ratio of 0.5wt%. An inhibitor was used to extend the working time of the ECA before curing. The peroxide based inhibitor was added to the resin at 1.6 wt%. It was shown that the inhibitor disables the catalyst at low temperatures, while volatizes at elevated temperatures above 140°C [17]. Micron sized Ag flakes were added to the PDMS resin at a filler loading of 85wt%. The silo-Ag composite isothermally cured at elevated temperature and under ambient atmosphere.

For conductivity tests, two layers of Kapton polyimide tapes were placed on a glass slide with 6 mm spacings. Silo-Ag composite was applied on the glass slide between polyimide tapes with a doctor's blade. Thickness of the ECA after curing was measured with Heidenhain (thickness measuring equipment, ND 281B, Germany). The width and length of the ECA strip were measured by a digital caliper (VWR). The bulk resistance was measured with the four-point method with Keithley 2000 multimeter (Keithley Instruments Inc.). Conductivity was calculated by Equation 1:

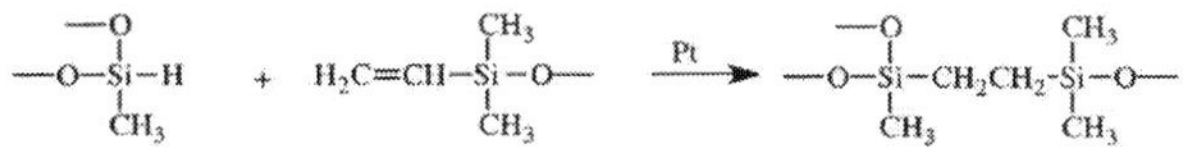

$$\sigma = \frac{l}{Rwt} \tag{1}$$

where R is the bulk resistance, l the length, w the width, and t the thickness of the ECA strip.

For in-situ conductivity measurements during curing, two mechanically polished copper plates were placed on top of the ECA strips at two edges. Electrical contact was secured by clips connecting to a Fluke 116 RMS multimeter. The ECA was placed in the furnace, ramped to 160 °C at 10 °C/min and held at 160 °C for 1 hour. Wires were connected to the clips and extended outside the furnace to the multimeter, which measured the bulk resistance with two-probe method.

For crosslinking density measurement of PDMS matrix, the PDMS precursors were placed on a Teflon mold after mixing and cured under the same conditions. The cured PDMS sheet were peeled off the Teflon mold and cut into 1.5cm x 1.5cm squares with a rubber blade. The square piece was weighed and its thickness measured by a digital caliper. It was then submerged into toluene for 2 hours. After reaching saturation, the swollen piece was weighed. It was then dried for over 4 hours, and weighed again after drying. For silo-Ag composites, three layers of polyimide tapes were placed on the glass slide with one underlayer of polyimide tape spaced in between. ECA was applied onto the single underlayer with doctor's blade and cured under the same conditions. The ECA strips was peeled off with the polyimide tape underlayer and weighed (m_i), then submerged in toluene for 2 hours. After saturation, the ECA strip was peeled off from the underlayer with tweezers and weighed (m_{sat}). The ECA strip and the underlayer were dried and weighed separately ($m_{dried\,ECA}$ and m_{tape}) after drying for 4 hours. Crosslinking density, Xc, was calculated based on the Flory-Rhener equation [18-19]:

$$X_c = -\frac{\ln(1-v_2)+v_2+\chi v_2^2}{V_1(v_2^{\frac{1}{3}}-\frac{v_2}{2})} \tag{2}$$

where χ is the interaction parameter, taken as 0.57 for PDMS in toluene. V_1 is the molar volume of solvent, taken as 106.28 cm³/mol for toluene. v_2 is the volume fraction of crosslinked polymer, calculated from

$$v_2 = \frac{m_{p,dry}\times f/\rho_p}{m_{p,dry}\times\frac{f}{\rho_p}+\frac{m_s}{\rho_s}} \tag{3}$$

where $m_{p,dry}$ is the mass of dried ECA composite. f is the mass fraction of polymer matrix in composite, which is 15% for the ECA and 1 for the pure PDMS matrix. ρ_p is the density of PDMS matrix, taken as 0.965 g/cm³. m_s is the mass of solvent in swelled polymer, calculated as $m_s = m_{sat} - m_{dried\,ECA} - (m_i - m_{tape})$. ρ_s is density of solvent, taken as 0.867 g/cm³ for toluene. It was assumed

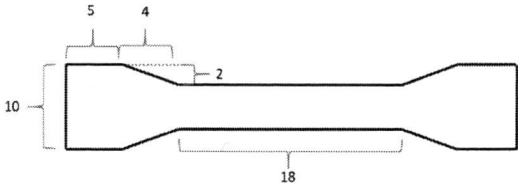

Figure 2. Dimensions of the dogbone substrate in mm.

Figure 1. Hydrosilylation addition crosslinking reaction mechanism [17].

that no swelling of the Ag filler occurred, and the polyimide tape microstructure was unchanged by swelling and drying.

Total attenuated reflectance-Fourier Transform Infrared Spectroscopy (ATR-FTIR) was used on silo-Ag ECAs from 1:1 hydride to vinyl precursor ratio. A Thermo Nicolet iS5-iD7 ATR was used for the characterization, with a resolution of 0.482 cm^{-1} over 16 scans.

Electrical stability was tested under cyclic loading of 25% strain for 100 cycles. Dogbone shaped substrates were produced from PDMS. The dogbone shape, as shown in Figure 2, was cut out by a razor blade. Two layers of polyimide tape were attached along the sides of the gage section with a 3.2 mm spacing. ECA was applied onto the substrate in the spacing with a doctor's blade. Two small pieces of Cu sheet of the same with as the gage section were placed on top of the ECA strip at the two ends. The whole system was cured under the same conditions. Clips were attached to the Cu pieces and connected to a Fluke 116 multimeter. Tensile cyclic testing was performed on an Instron 5548 Microtester at a constant displacement rate of 100 mm/min for 100 cycles. Upper and lower bulk resistances between the Cu electrodes were recorded every 5 cycles.

III. RESULTS AND DISCUSSION

A. Conductivity and conductivity development

The roles of both reaction stoichiometry and inhibitor on conductivity of silo-Ag ECA were investigated. With hydride-to-vinyl group ratio in the range from 1:4 to 4:1, conductivity decreases with increasing hydride content, as shown in Figure 3. Above a ratio of 1:1, conductivity saturates relatively quickly to a minimum value with further

increase in hydride groups. Below the 1:1 ratio, conductivity increases more rapidly without apparent sign of saturation with deviation from perfect stoichiometry.

The inhibitor is shown to enhance conductivity without changing the final composition of the cured ECA. FTIR-ATR spectra of ECAs prepared with and without the inhibitor are compared in Figure 4, and no difference in peak positions are observed, so it can be assumed that no chemical reaction occurs between the inhibitor and the silver flake surfactants. At all stoichiometric ratios tested, ECAs with inhibitor added demonstrate lower resistivities. The mechanism of conductivity enhancement was investigated by in-situ resistance measurement during curing for the exemplary system with 1:1 stoichiometric ratio. Figure 5a shows the initial rapid drop in resistance below 60°C, which can be attributed to volume shrinkage as a result of initial curing [16]. Conductivity development due to volume contraction is more significant without inhibitor, due to the higher rate of hydrosilylation crosslinking. As shown in Figure 5b, when held isothermally at 160°C until the completion of curing, the system without inhibitor experience a slow and limited decrease in resistance, while the system with inhibitor added has a more rapid and greater extent of resistance drop. Previous research has shown that silver flake surfactants can be reduced by hydride groups in PDMS precursors. Possible reaction scheme is shown in Figure 6 [20]. The fatty acid based surfactants are first reduced to aldehydes, which disrupt the chelation between the carboxylic acid and silver with partially positive charges in the flake [21]. In addition, the aldehyde can further react with hydride groups and completely leave the Ag flakes, leaving Ag in its elemental form exposed. Under the relatively high curing temperature of the PDMS matrix, the exposed regions of Ag flake undergo sintering, creating and

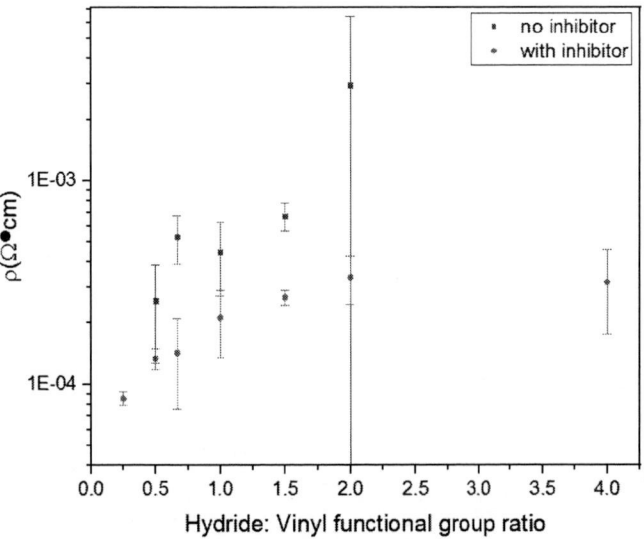

Figure 3. Resistivity of ECAs prepared from hydride: vinyl group ratios of 1:4, 1:2, 1:1.5, 1:1, 1.5:1, 2:1, and 4:1.

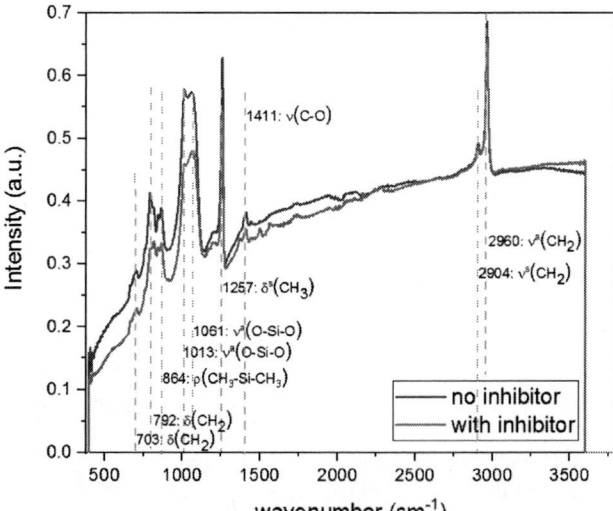

Figure 4. FTIR-ATR spectra of silo-Ag ECA with 1:1 hydride-to-vinyl precursor ratio with and without addition of inhibitor. Chemical composition is not changed by the inhibitor, no reaction occurs between the inhibitor and surfactants on the Ag flake [17].

coalescing Ag nanoparticles on the surface between flakes. These interconnected Ag nanoparticle bridges are responsible for the conductivity development of silo-Ag ECAs to a greater extent than volumetric shrinkage. By delaying hydrosilylation between vinyl and hydride groups in PDMS precursors, more carboxylic acids on Ag flakes are reduced by hydride groups, and thus a higher conductivity is achieved through the higher density of established connections between sintered Ag nanoparticles [16]. This corresponds well to the larger resistance drop of the system with inhibitor added in Figure 5b, enabled by a greater extent of surfactant reduction and higher density of Ag nanoparticle bridges.

However, Figure 3 demonstrates the addition of hydride groups does not facilitate conductivity enhancement by promoting surfactant reduction. It is hypothesized that hydrosilylation has a faster kinetics compared to reduction of surfactants. At low hydride-to-vinyl ratios, due to the high filler loading, surfactants on Ag flakes are more readily available spatially and kinetically to hydride groups compared to vinyl groups. Therefore, hydride groups react preferentially with Ag surfactant to aid in conductivity development while fewer participate in crosslinking reactions to form the PDMS matrix. With increasing hydride-to-vinyl functional group ratio, curing of PDMS matrix occurs much more rapidly than surfactant reduction; viscosity increases sharply due to the development of

crosslinking network, which prevents hydride PDMS precursors to diffuse, come into contact and react with surfactants on silver flakes, thereby suppressing conductivity development. Comparison of crosslinking density of PDMS matrix from different reaction stoichiometry supports this hypothesis. In figure 7a, it is shown that increasing the hydride-to-vinyl functional group ratio increases the crosslinking density of pure PDMS, suggesting that the hydride precursor is the limiting reagent in terms of kinetics and chain mobility. With the addition of inhibitor in pure PDMS, crosslinking density is reduced in all systems due to the delay of crosslinking reaction and disabling of Pt catalyst. However, the decrease in crosslinking density is lower for higher hydride-to-vinyl ratios, implying that the retarding effect of the inhibitor is reduced as a result of the increased concentration of the kinetically limiting species. This trend is also reflected in the matrix crosslinking density in silo-ECA composites, as shown in Figure 7b. There is an overall tendency for increasing crosslinking density with increasing concentration of hydride groups in both systems (with and without inhibitor). While the addition of inhibitor is used to select surfactant reduction reaction over crosslinking reaction, the reason for the unexpected increase in apparent crosslinking density at 1:1.5 and 1.5:1 hydride: vinyl ratios is still unclear. Possible speculations involve the convolution of hydride reduction and hydrosilylation

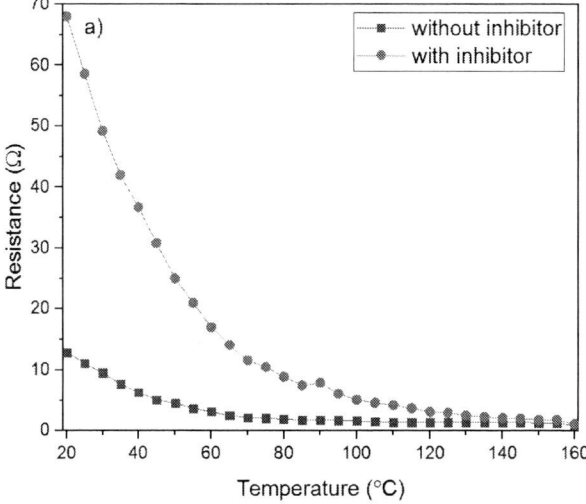

Figure 5. (a) Resistance change with temperature during ramping from room temperature to 160°C at 10°C/min. (b) Resistance change with time during isothermal curing since reaching 160°C.

Physics of Failure Based Simulation and Experimental Testing of Quad Flat No-lead Package

Jia-Shen Lan

Department of Mechanical and Electro-mechanical
Engineering
National Sun Yat-sen University
70 Lien-Hai Rd. Kaohsiung, Taiwan (R.O.C)
E-mail: d023020004@student.nsysu.edu.tw

Mei-Ling Wu*

Department of Mechanical and Electro-mechanical
Engineering
National Sun Yat-sen University
70 Lien-Hai Rd. Kaohsiung, Taiwan (R.O.C)
E-mail: meiling@mail.nsysu.edu.tw

Abstract—In this work, finite element analysis and experimental testing were performed to analyze the delamination failure in quad flat no-lead (QFN). Interfacial delamination during post-mold cure process and precondition test is a critical reliability issue for plastic IC packages. The delamination failure weakens the internal structural strength of the package and affects product reliability. Available evidence indicates that, in the interface delamination, initially generated small voids gradually expand due to thermal stress or hygro-mechanical stress. Furthermore, due to the cracks on the interface, moisture easily penetrates into the package. In plastic IC packages, this can result in not only electrical failure, but also chemical reactions that would lead to corrosion. In the present study, delamination occurred on the interface between the Cu pad and the molding compound, as well as on the interface between the die and the molding compound and the Cu pad and molding during the precondition test. This research can help in improving product reliability by preventing the delamination failure in the QFN package by analyzing physics of failure (PoF).

Keywords-finite element analysis; quad flat no-lead; Interfacial delamination; hygro-mechanical stress; physics of failure

I. INTRODUCTION

Delamination failure is a major concern in QFN and QFP. The QFN package structure is composed of the copper frame, the die, the silver epoxy and the molding compound. However, empirical evidence indicates that the interface between the Cu pad and the molding compound is susceptible to delamination. This issue is ascribed to the growth of cupric oxide (CuO and Cu_2O) on the Cu frame and the Cu pad surfaces during the plasma and baking process. The nonuniformity of the oxide layer causes voids or weakens the bonding strength between the molding compound and the Cu pad. These uncertainty factors make it difficult to precisely define the delamination failure, limiting the accuracy of the predictive model of QFN package delamination failure proposed in this work. Nonetheless, the findings revealed that the results obtained through simulation modeling were consistent with the Scanning Acoustic Tomography (SAT) and Scanning Electron Microscopy (SEM) findings. Shu *et al.* [1] investigated the bonding strength between the Cu pad and the molding compound as a function of the cupric oxide layer thickness. The experimental results reported by the authors indicate that the interface exhibits superior bonding characteristics when the CuO/ Cu_2O ratio exceeds 1. They also noted that the plasma process duration affected the shear strength. Specifically, if it was reduced from 60 to 15 seconds, the probability of delamination declined by about 5%. The authors thus concluded that the slot hole design in the Cu pad could reduce the delamination failure rate. Guojun *et al.* [2] incorporated the moisture factor into their finite element model. Since the polymer material inside the package was characterized by moisture permeability and hygroscopicity, moisture stress was formed between the material interfaces. Consequently, delamination failure occurred at the interface between the Cu pad and the molding compound. In order to simulate the delamination at the interface, the authors adopted the modified virtual crack closure method (MVCCM) to calculate the strain energy release rate. In their work, Kim and Kong [3] calculated thermal stress, hygro-mechanical stress, and vapor pressure in a flip chip package under Moisture Resistance Test (MRT), indicating that all three factors are significant. Ho *et al.* [4] analyzed the surface of the leadframe and developed a Cohesive Zone Model to explore the initial delamination and fracture growth path at high temperatures. On the other hand, Nguyen *et al.* [5] measured the hygro-mechanical stress of the QFN in moisture environment. The authors placed a strain sensor on the silicon chip to measure the hygro-mechanical strain, while the hygro-mechanical stress was verified by finite element modeling and experimental testing. The obtained results showed that the discrepancy between the simulation output and the experimental testing findings was below 15%. In order to further analyze the bonding strength in the interface, Poshtan *et al.* [6] determined the critical (G_c) and sub-critical strain energy release rate between the molding compound and the Cu pad at different temperatures by conducting the bi-material test. The surface roughness and surface composition were established via SEM and Energy Dispersive X-ray (EDX) measurements. Finally, the experimental data obtained by crack growth was applied to the finite element method simulation module. In addition to using the bi-material test to obtain the coefficient of the fracture generation at the interface, Zhang *et al.* [7] proposed the stress ratio analysis to analyze the delamination failure of the QFN by combining the simulation results and those obtained via experimental testing. Hunat *et al.* [8] designed a novel leadframe structure with the aim of improving the thermal stress in QFN. The utility of the new design was verified by performing experimental testing and finite element modeling. The authors concluded that the thermal stress on the Cu pad could be effectively reduced by adopting the groove design. Srikanth, Chan, and Vath [9] conducted both shear and tensile tests to examine the bonding strength between the Cu pad and the molding compound as a function of the oxidation layer thickness. They found that the formation of cupric oxide can increase the surface roughness and improve the shear strength. However, the bonding strength between the molding compound and the Cu gradually decreased once the cupric oxide thickness exceeded 30 nm. Through experimental observations, the authors established that the delamination failure mode occurs between Cu and Cu_2O, as well as between Cu_2O and CuO. In this present research, finite element modeling was utilized to analyze the delamination failure. To accurately determine the delamination site and shear stress via simulation modeling, shear tests for the bonding strength of Cu/molding compound and Ag/molding compound were conducted.

II. METHODOLOGY

In this work, the QFN package was investigated to analyze the delamination failure between the Cu pad and the molding compound. To better understand the delamination failure in the Cu pad, failure analysis, finite element analysis, and material analysis were conducted, as shown in Fig. 1. First, QFN package failure was confirmed via SAT. To investigate the delamination failure modes, SEM measurements were conducted. Through experimental observations, it was established that the interfacial delamination occurred at the Cu pad edge. Next, finite element model was developed to analyze the stress distribution required for establishing the delamination failure index. To identify the delamination cause, experimental testing and simulation modeling must be conducted. During the precondition test, the molding compound and the adhesive, both of which are polymer materials, absorb moisture. In order to incorporate the hygroscopicity of the polymer materials into the simulation model, it is necessary to obtain the moisture diffusivity and the coefficient of moisture expansion of the polymer, which was achieved by using an electronic balance and a thermal mechanical analyzer. Finally, we attempted to reduce the delamination failure rate in the QFN by analyzing the thermal stress and hygro-mechanical stress of the Cu pad.

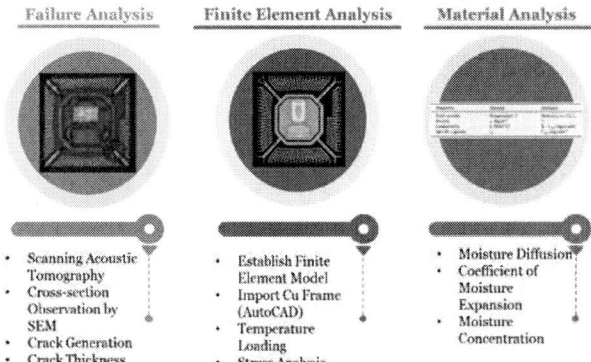

Figure 1. The methods applied in the research

III. FAILURE ANALYSIS

In this study, delamination mostly occurred at the interface between the Cu pad and the molding compound. Specifically, 41.67% of failures were noted at the Cu pad edge, while 8.33% were observed at the corners of the Cu pad, and further 8.33% occurred at the fingers, as shown in Fig. 2. Furthermore, the area covered by Ag plating was identified as the failure site via experimental observations. The bonding strength between the Ag plating and the molding compound was observed to be weaker than that at the interface between the Cu pad and the molding compound. Therefore, the bonding strength between the Ag plating and the molding compound must be improved in order to limit the interfacial delamination in the QFN.

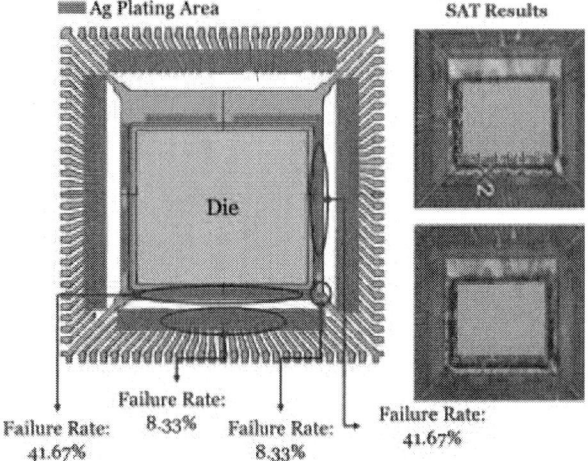

Figure 2. The experimental testing of the QFN

In order to explore the interfacial delamination between the Ag plating and the molding compound, and between the Cu pad and the molding compound as a part of the post-mold cure process and in the moisture sensitivity test, the shear test was conducted, as shown in Fig. 3. The testing specimen for the shear test was of cylindrical shape, with the cross-section diameter of 3.57 mm and the bottom substrate thickness of 10 mm². The shear test results revealed that the bonding strength under the moisture sensitivity test conditions is 20% lower than the bonding strength obtained under the post-mold cure process conditions. Empirical evidence suggests that, during the moisture sensitivity test, moisture slowly penetrates into the interface between the Cu and the molding compound. As moisture penetrates into the package, it weakens the bonding strength at the interface. In addition, interface material type was also found to affect the bonding strength. Specifically, the bonding strength between the molding compound and Ag plating is 50% lower than the bonding strength between the molding compound and the Cu pad.

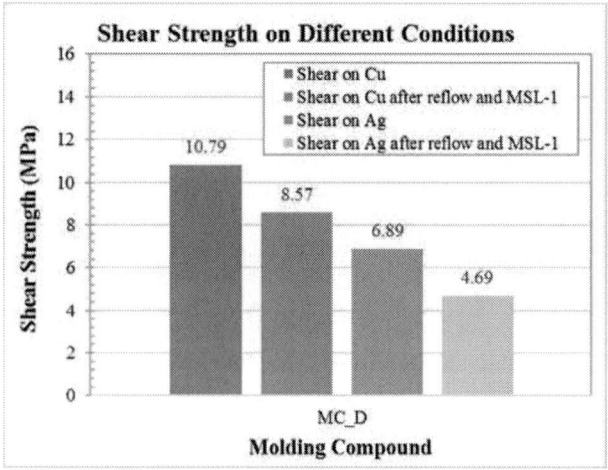

Figure 3. The shear test on Ag plating and Cu after the post mold cure process or the precondition test [10]

IV. FINITE ELEMENT MODELING

The finite element model proposed as a part of this work was incorporated into the ANSYS. The simulation presented in Fig. 4 pertains to a full QFN model. To ensure that the solution is stable and to achieve convergence, the hexahedral elements were applied in the finite element modeling. The thermomechanical loading and hygro-mechanical loading were modeled separately, using temperature changes from 175 °C to 25 °C and 30 °C/60%RH. The simulation results indicated that the shear stress is concentrated on the Ti-bar of the four corners. The extracted path is observed in Fig. 5. As can be seen in this graph, the stress gradually increases from the edge of the die, while it gradually decreases at the end of the Cu pad. The shear stress distribution of the three-dimensional structure does not show significant changes and does not correspond to the experimental testing results. These discrepancies confirm that the Ag/molding compound and Cu/molding compound bonding strengths must be incorporated into the simulation model. To improve the agreement between the experimental testing and simulation modeling results, the shear stress ratio equation was introduced. The shear stress ratio can combine the shear stress obtained via simulations and through experimental testing. The shear stress ratio equation can be defined as:

$$SSR = \frac{\tau_{Cu\_S}}{\tau_{Cu\_E}} + \frac{\tau_{Ag\_S}}{\tau_{Ag\_E}} \quad (1)$$

where SSR is the shear stress ratio, τ is the shear stress, τ_{Cu_S}, τ_{Ag_S} are calculated from the simulation results, and τ_{Cu_E}, τ_{Ag_E} are obtained from the shear test.

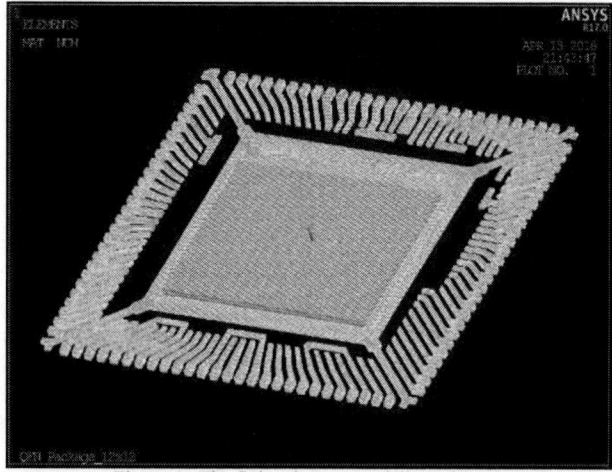

Figure 4. The finite element modeling of QFN

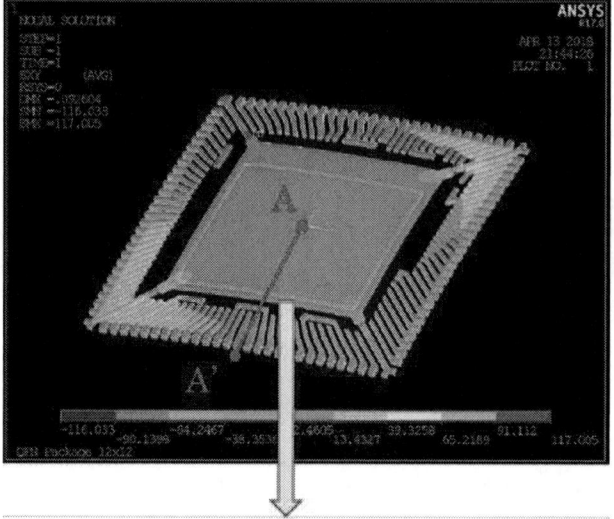

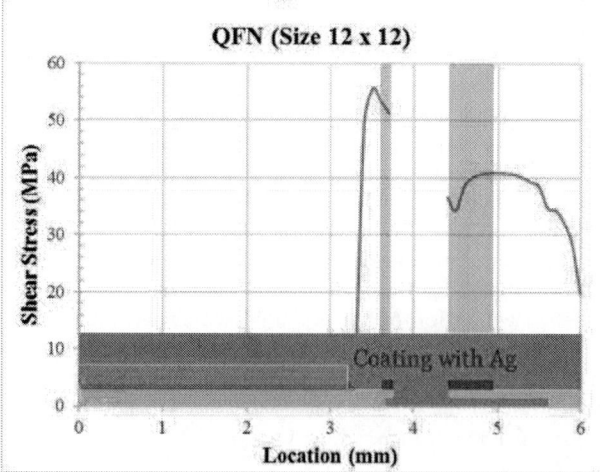

Figure 5. The shear stress distribution and the shear stress in the path

A. Thermo-mechanical Stress Analysis

The post-mold cure process is the first step in QFN manufacturing. In this work, the molding temperature, which is the stress-free temperature, was set at 175 °C. During the post-mold cure process, thermal stress is induced by the coefficient of thermal expansion (CTE) mismatch. The thermal stress is concentrated on the Cu pad (under the die), at the edge of the Cu pad, and in the fingers, as shown in Fig. 6. Since the silver epoxy has good adhesion with the Cu pad, cracks rarely occur under the die. Instead, delamination is predominantly located at the Cu pad edge.

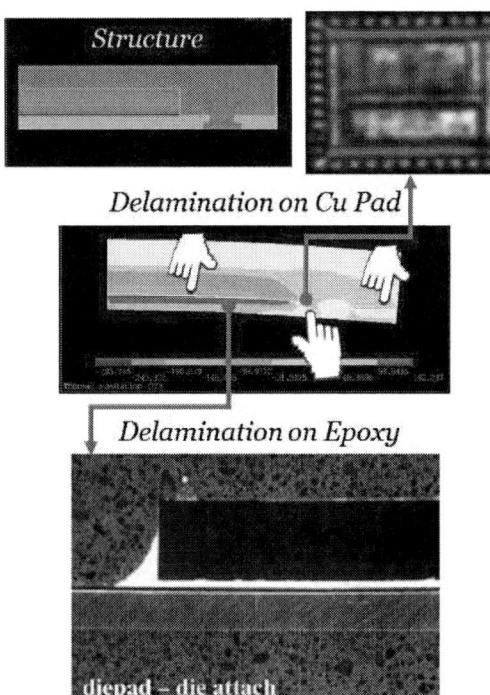

Figure 6. The shear stress distribution and the SAT in the thermo-mechanical stress analysis [11]

B. Hygro-mechanical Stress Analysis

The precondition test is the second step in the QFN manufacture. In the present study, for the precondition test, the moisture sensitivity level 3 was applied at 30 °C/60%RH for a period of 168 hours. The molding compound and the silver epoxy volume expand due to moisture absorption. As metal material cannot absorb moisture, the coefficient of moisture expansion (CME) is not defined for the leadframe and the die. The hygro-mechanical stress is concentrated on the top of the die, at the Cu pad edge, and in the fingers, as shown in Fig. 7. These findings reveal that the simulation results match those obtained experimentally. In this case, during the precondition test, the interfacial delamination is focused on the edge of the Cu pad.

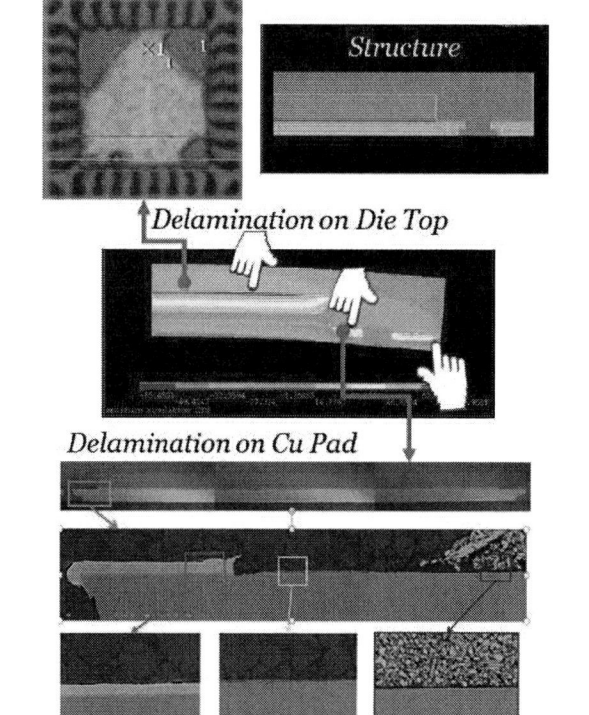

Figure 7. The shear stress distribution and the SAT in the hygro-mechanical analysis

V. RESULTS

A. Die Thickness Analysis

The die thickness is one of the important factors affecting the stress in the Ag plating area on the Cu pad. In this work, die thickness varied from 0.15 mm to 0.38 mm. In the die thickness analysis, the stress ratio equation was combined with the simulation results and shear test results, as shown in Fig. 8 and Fig. 9, respectively. The Ag plating area, which is presented in the light blue color, is the delamination failure site. In the thermo-mechanical analysis, the shear stress is generated on the leadframe due to the thermal mismatch. Maximum shear stress occurred in the middle of the Ag plating area, and it gradually decreased toward the end of the Cu pad. The findings also indicated that 0.12 mm increase in the die thickness would result in about 10% increase in the stress ratio. The hygro-mechanical analysis revealed that the shear stress ratio trend is opposite of that observed in the thermo-mechanical analysis. Greater die thickness represents molding compound volume reduction, and thus reduces its moisture absorption capacity. Therefore, the moisture strain of the molding compound is relatively reduced. In this case, 0.12 mm increase in the die thickness would result in about 10% decrease in the stress ratio.

B. Molding Thickness Analysis

While the investigation included QFN package thicknesses in the 0.8 mm to 1.1 mm range, in this case, the QFN package thickness is 0.85 mm. In the molding thickness analysis, the stress ratio equation is combined with the simulation results and shear test results, as shown in Fig.

10 and Fig. 11, respectively. The results yielded by thermo-mechanical analysis indicate that the package thickness affects the shear stress ratio. In this case, 0.15 mm increase in the molding thickness would result in about 3% increase in the stress ratio. In the hygro-mechanical analysis, 0.15 mm increase in the molding thickness would result about 5% decrease in the stress ratio. When these results are interpreted jointly, it can be surmised the influence of the molding thickness on the shear stress ratio does not exceed 10%. Thus, the effect of the molding thickness on the Cu pad delamination failure pad is limited.

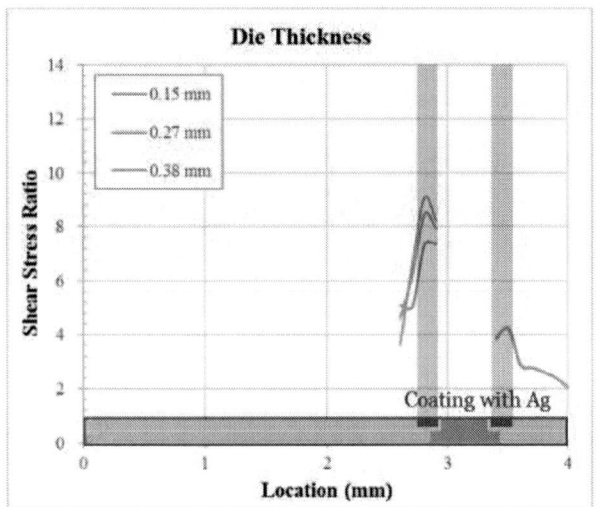

Figure 8. The shear stress ratio Vs. location with different die thickness in the thermo-mechanical loading

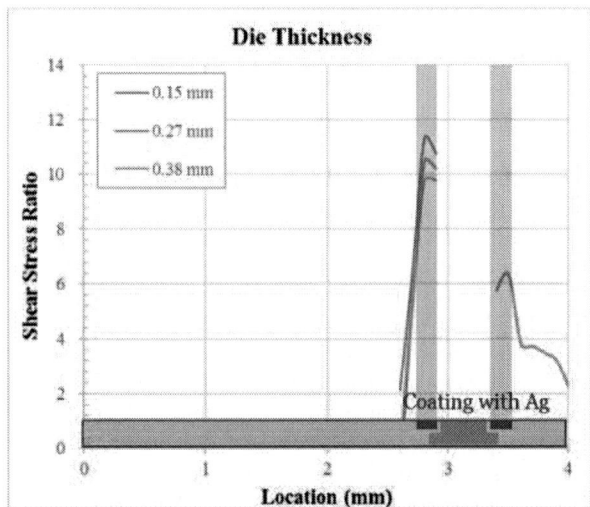

Figure 9. The shear stress ratio Vs. location with different die thickness in the hygro-mechanical loading

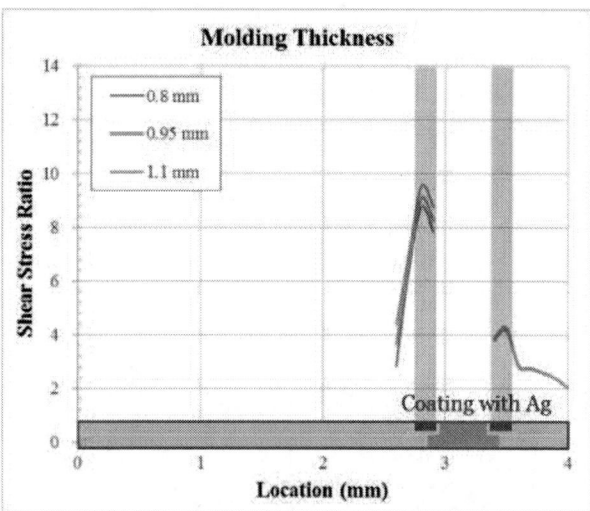

Figure 10. The shear stress ratio Vs. location with different molding thickness in the thermo-mechanical loading

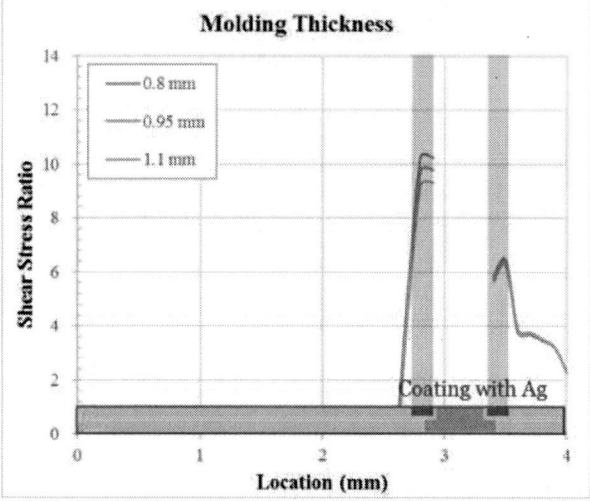

Figure 11. The shear stress ratio Vs. location with different molding thickness in the hygro-mechanical loading

VI. CONCLUSION

In the present research, the package size 12 × 12 (mm × mm) of QFN was investigated during the post-mold cure process and the precondition test. Simulation modeling was performed to predict the delamination failure by combining the finite element method, experimental testing, and the shear stress ratio equation. The findings indicate that delamination failure occurred at the edge of the Cu pad and at the fingers. To identify the critical factors for reducing the delamination failure rate in the Cu pad, parametric analysis was conducted. In the thermo-mechanical analysis, 0.12 mm increase in the die thickness resulted in about 10% increase in the stress ratio. According to the hygro-mechanical analysis, 0.12 mm increase in the die thickness would result in about 10% decrease in the stress ratio. These results indicate that the molding thickness exerts a limited effect on the Cu pad delamination failure. The purpose of this research is to improve the delamination failure in QFN package.

ACKNOWLEDGMENT

The authors would like to gratefully acknowledge financial support from the Advanced Semiconductor Engineering (ASE Group) collaborative research program. Authors would like to thank Director Allen Shen, Manager DavidDW Lo, Supervisor Engineer Ken Chu, and Kuanhui Lin for their advices, technical, and material supports.

REFERENCES

[1] Shu, M. F., Chen, K., Yang, B., Liu, W., & Tseng, Y. H. (2016). Lead-frames copper oxidation effect for a QFN package delamination improvement. In Electronic System-Integration Technology Conference, 13 Sept – 16 Sept. Grenoble, France: World Trade Center.

[2] Guojun, H., Rossi, R., Jing-En, L., & Baraton, X. (2010). Interface Delamination Analysis of TQFP Package during Solder Reflow. Microelectronics Reliability, Volume 50, Issue 7, 1014-1020.

[3] Kim, D. W., & Kong, B. S. (2006). The effect of hygro-mechanical and thermo-mechanical stress on delamination of gold bump. Microelectronics Reliability, Volume 46, Issue 7, 1087-1094.

[4] Ho, S. L., Joshi, S. P., & Tay, A. A. (2013). Experiments and three-dimensional modeling of delamination in an encapsulated microelectronic package under thermal loading. IEEE Transactions on Components, Packaging and Manufacturing Technology, Volume 3, Issue 11, 1859-1867.

[5] Nguyen, Q., Roberts, J. C., Suhling, J. C., & Jaeger, R. C. (2016). Measurement and simulation of moisture induced die stresses in Quad Flat Packages. In Thermal and Thermomechanical Phenomena in Electronic Systems (ITherm), 31 May - 03 Jun. Las Vegas, NV, USA: Cosmopolitan Hotel 3708 South Las Vegas Boulevard.

[6] Poshtan, E. A., Rzepka, S., Silber, C., & Wunderle, B. (2016). An In-situ Numerical–experimental Approach for Fatigue Delamination Characterization in Microelectronic Packages. Microelectronics Reliability, Volume 62, 18-25.

[7] Zhang, W., Luo, W., Hu, A., & Li, M. (2012). Adhesion Improvement of Cu-based Substrate and Epoxy Molding Compound Interface by Hierarchical Structure Preparation. Microelectronics reliability, Volume 52, Issue 6, 1157-1164.

[8] Hunat, C., Lagdameo, C., Galang, R., & Benedicto, E. M. (2006). Design of a More Robust QFN Package through Virtual Prototyping and Advanced Characterization. In Electronic Materials and Packaging, 11 Dec - 14 Dec. Kowloon, China: Hong Kong University of Science and Technology.

[9] Srikanth, N., Chan, L., & Vath, C. J. (2006). Adhesion Improvement of EMC–leadframe Interface Using Brown Oxide Promoters. Thin Solid Films, Volume 504, Issue 1, 397-400.

[10] Zhang, M., Lee, S. R., Zhang, J., Yun, H., Starkey, D., & Chau, H. (2009). Correlation between Material Selection and Moisture Sensitivity Levels of Quad Flat No-lead (QFN) Packages. In Microelectronics and Packaging Conference, 16 Jun - 18 Jun. Rimini, Italy: Palacongressi di Rimini.

[11] Van Driel, W. D., Liu, C. J., Zhang, G. Q., Janssen, J. H. J., Van Silfhout, R. B., van Gils, M. A. J., & Ernst, L. J. (2004). Prediction of Interfacial Delamination in Stacked IC Structures Using Combined Experimental and Simulation Methods. Microelectronics Reliability, Volume 44, Issue 12, 2019-2027.

An Assessment of Electromigration in 2.5D Packaging

Jiefeng Xu[1], Scott McCann[2], Huayan Wang[1], Jing Wang[1], VanLai Pham[1], Stephen R. Cain[1], Gamal Refai-Ahmed[2], S.B. Park[1]

[1] Department of Mechanical Engineering
The State University of New York at Binghamton
Binghamton, NY 13902, USA
[2] Xilinx, Inc.
2100 Logic Drive, San Jose CA 95124

Abstract—In this study, an accelerated Electromigration (EM) test was performed. The test vehicle has four types of common interconnect structure. The first one is a classic Ball Grid Array (BGA), short for BGA; the second one is a solder ball with a copper via on top, short for BGA-Via; the third one is an individual copper via in the substrate, short for Via; the last one is an individual copper Plated Through Hole (PTH), short for PTH in the substrate. The built-in serpentine copper fine lines around each structure were designed to monitor the local temperature in-situ. All test vehicles were stressed at 150ºC temperature with 12A current. The voltage of each test structure and the resistance of the serpentine line were recorded in-situ. The results show that different micro-electrical structures have great effects on EM behavior, especially the time to failure (TTF). In BGA test structure, the failure occurred on the substrate side of solder ball; in BGA-Via, the failure was the depletion of the copper via. No failure was observed in Via and PTH test structures, even after an extremely long testing, although they have higher package temperature. The TTF of BGA-Via is about 2 times shorter than BGA. A finite element simulation based on Atom Flux Divergence (AFD) was performed to understand the failure mechanism and predict the TTF. The results show that via on top of solder ball will cause 10% higher current density than solder ball only. When the void underneath of the via in solder ball was nucleated, the current density will start to redistribute and reduce. In short, Via is the riskiest point for EM when it located near the solder ball.

Keywords: Ball Grid Array (BGA), Via, Plated Through Hole (PTH), Electromigration (EM). Time to Failure (TTF), Atom Flux Divergence (AFD), Finite Element Analysis (FEA), Simulation

I. INTRODUCTION

Nowadays, 2.5D packaging draws great attention in the electronic packaging industry [1-4]. As an advanced packaging solution, 2.5D packaging enables heterogeneous integration with a shorter electrical path between different ICs and achieves more I/Os. However, the need of higher performance in 2.5D packaging has led to a shrinkage of bump size and an increase of power consumption [5]. This trend will not only increase the current density in solder interconnections but also increase the joule heating effect, which makes the Electromigration（EM）become a critical concern in the semiconductor industry [6].

Sn-Ag-Cu (SAC) is a widely used lead-free solder material in semiconductors and the electromigration behavior of SAC solder joint has drawn great attention from researchers [7]. In most of the previous researchers'

studies, the EM acceleration tests were usually performed on SAC solder joints at either ambient temperature or 150C with a current density running from 10^4 to 10^7 A/cm^2 [8]. However, the structural effect of copper via and Plated Through Hole (PTH) were rarely considered in previous studies while they also play important roles in affecting the EM performance of the solder joint.

II. EXPERIMENTAL METHODOLOGY

In this study, four common structures were tested to observe and quantify the Electromigration phenomenon, Fig.1(a)-(d) shows four test structures. The resistance of serpentine fine line can be measured to calculate the local temperature of the test structure. Fig.1. (a) is the traditional FCBGA(SAC305) and a pair of victim solder ball; Fig.1. (b) is a copper via on the top center of solder ball and a pair of victim solder ball with via; Fig.1 (c) is a pair of victim via away from solder ball on the substrate; Fig. (d) is a pair of victim Plated Through Hole (PTH) on the substrate. These pairs of victim solder ball, Via, and PTH were designed to carry all current in the test, which means they are the riskiest points of the test structure.

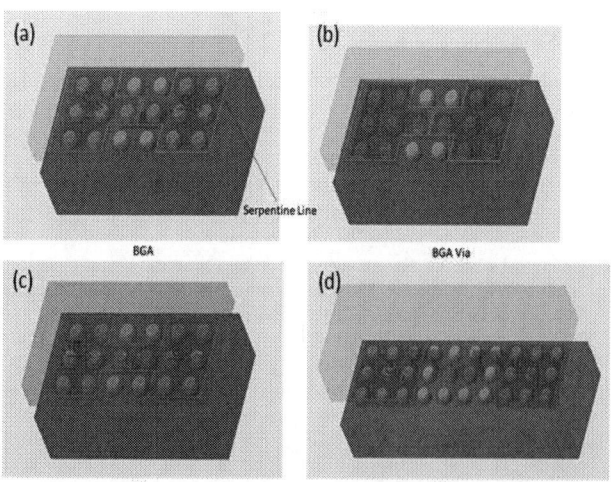

Fig 1 3D schematic of test structures, (a) BGA, (b) BGA-Via(c)Via(d)PTH

Fig.2 shows the setup of the experiment. Each Test Vehicle (TV) have 5 modules and each module have 4 test structures on each corner. A Power Supply Unit (PSU) was used to stress the TV; A Data Acquisition (DAQ) was used to monitor the voltage and resistance in the test vehicle in-situ; An oven chamber was used to maintain the temperature of test vehicle at 150 °C. The failure criteria for experiment is open circuit.

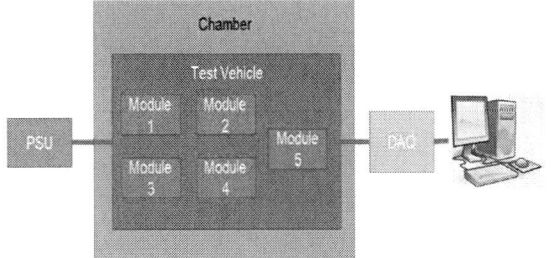

Fig.2 Diagram of Experiment Process

Fig.3 shows the nominal dimensions of the test structures. The pitch of the BGA is 1mm and the solder material was Sn-Ag-Cu (SAC).

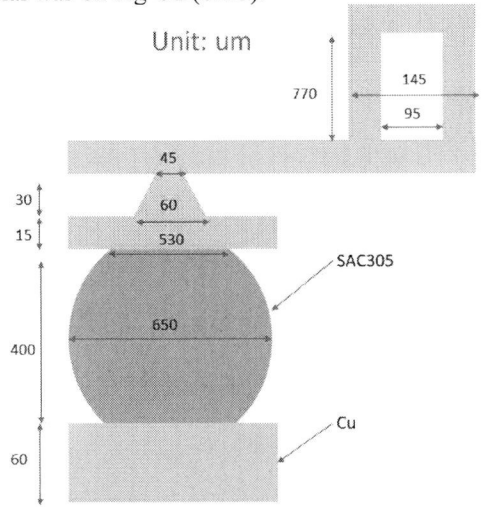

Fig.3 Nominal dimensions of test structures

III. EXPERIMENTAL RESULTS AND DISCUSSION

Fig 4 shows the voltage reading from BGA test structure on the anode side and cathode side, respectively. Only BGA test structure was selected to show the voltage reading is because all four test structures show similar voltage reading. It clearly shows that the EM-induced failure happened on the cathode side of the victims, which well agreed with previous research [9-10]. On the anode side, the slight voltage increase was caused by the thickness increase of Intermetallic compound (IMC). On the cathode side, the voltage increases significantly when it reaches a certain level (0.05V in BGA structure), because, in the late stage of EM failure, the open window will become smaller due to the void propagation. When it reaches a certain size, joule heating effect will generate a large amount of heating to heat up the SAC305 solder to its melting temperature at 217 °C. At this point, the solder joint will fail within a short period of time.

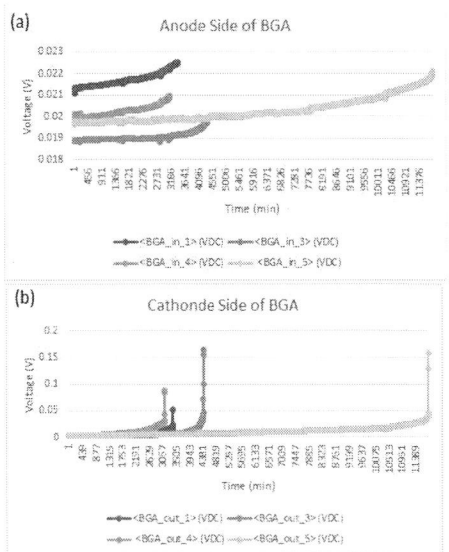

Fig.4 Voltage reading of BGA (a)anode side, (b)cathode side

Fig.5 shows the SEM images of BGA failure. On the anode side, lots of overgrowth CuSn island was accumulated on the substrate side and part of the copper pad was depleted, but the copper trace has no damage, which indicates that the IMC was formed on PCB side and drifted to substrate side due to EM. On Cathode side, the whole copper trace was depleted due to EM and the solder ball was melted because the IMC was uniformly distributed in the solder ball.

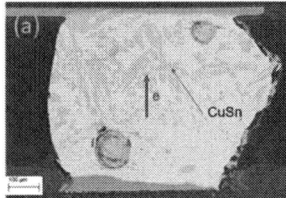 

Fig.5 SEM Images of BGA: (a)anode side, (b)cathode side

In the BGA-Via test structure, the failure was found in the cathode side which is shown in Fig.6. Fig. 6 (a) shows that the cup shape overgrown IMC was accumulated underneath of the via because the current crowding at via is more serious. In Fig.6 (b), the solder ball was distorted because of EM. The whole via was depleted and replaced by solder. In the meantime, the EM enhance joule heating melted the solder ball at late stage of EM.

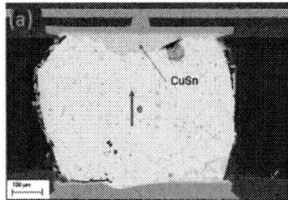 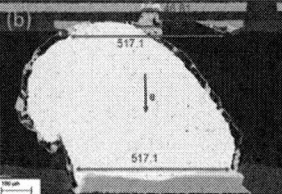

Fig.6 SEM Images of late stage failure in BGA-Via : (a)anode side, (b)cathode side

Fig. 7 and 8 show the SEM images of the Via and PTH test structures. No open circuit failure was found at about 2000h in the Via and PTH test structures, which is totally different from the BGA and BGA-Via test structures. Only a slight depletion of copper trace was observed. Although the package temperature and current are higher in Via and PTH compared to BGA and BGA-Via, the pure copper interconnect have higher EM resistance than Cu-SAC interconnect. It can conclude that Via and PTH structures are more reliable than BGA and BGA-Via structures. BGA and BGA-Via are more sensitive to EM failures.

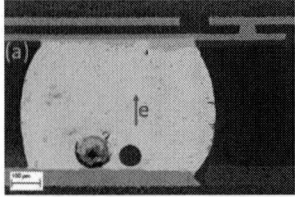

Fig.7 SEM Images of Via: (a)anode side, (b)cathode side

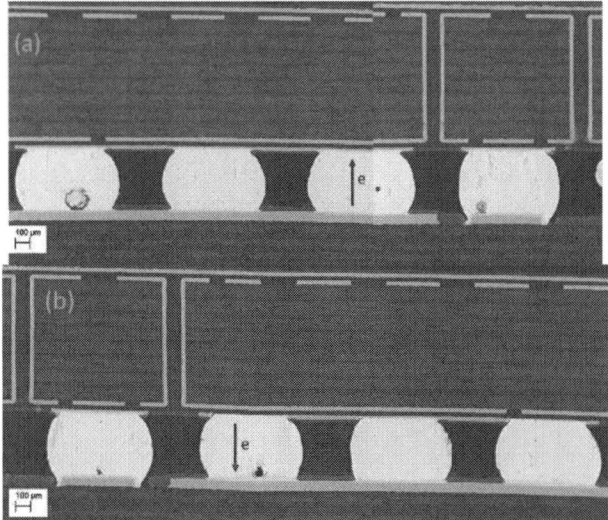

Fig.8 SEM Images of PTH: (a)anode side, (b)cathode side

Table 1 shows the TTF and measured the temperature of BGA and BGA-Via test structures. Because no open circuit failure was found in Via and PTH test structures, the result was ignored. Overall, BGA structure has higher local temperature and smaller TTF than BGA-Via

TABLE 1 TTF AND EXPERIMENTAL TEMPERATURE OF BGA AND BGA-VIA

Loc	BGA		BGA-Via	
	TTF (h)	Temp (°C)	TTF (h)	Temp (°C)
1	61.2	189.3	60.6	185.0
2	57.0	185.1	16.7	184.0
3	202.3	185.5	9.3	186.1
4	465.9	188.6	14.9	183.1
5	86.0	181.9	47.2	186.0

IV. FINITE ELEMENT ANALYSIS

To better understand the EM failure mechanism, the finite element analysis was performed on BGA and BGA-Via structures. Electricity, diffusion, and thermal fields were involved in the analysis. Current density and temperature are the major variables of the EM test. Researches show that the major migration type in solder ball is EM and Thermalmigration (TM) [11]. The conventional numerical model of EM considers Atom Flux Divergence (AFD) [12], which is based on diffusion theory. It is typically expressed as:

$$J_{ew} = [C/(k_B T)]D_0 exp[-E_A/(k_B T)]eZ^* j\rho \qquad (1)$$

$$J_{th} = -C/(k_B T) D_0 exp[-E_A/(k_B T)]Q^* (\nabla T)/T \qquad (2)$$

where J_{ew} is the electronic winds induced vacancy flux; J_{th} is the local thermal gradient induced vacancy flux; C is the vacancy concentration; k_B is the Boltzmann constant; T is the local temperature; D_0 is the pre-factor of the self-diffusion coefficient; E_A is the activation energy; e is the fundamental electronic charge; Z^* is the effective charge number; j is the local current density; ρ is the electrical resistivity, which is given as $\rho = \rho_0[1+\alpha(T-T_0)]$; ρ_0 is the electrical resistivity at temperature T_0, and Q^* is the heat of transport. Equations (1) and (2) show the vacancy flux caused by current and thermal gradient. It can be found that the current density and temperature are the only controllable experiment factors. Equation (3) is the divergence of total flux and Equation (4) is the time-dependent evolution model of EM and TM.

$$J_{total} = \nabla(J_{ew}) + \nabla(J_{th}) = [E_A/(k_B T) - 1/T + \alpha\rho_0/\rho] J_{ew} \nabla T$$
$$+[E_A/(k_B T) - 3/T + \alpha\rho_0/\rho] J_{th} \nabla(T) + CQ^*D/(3k_B^3 T^3)$$
$$j^2\rho^2 e \qquad (3)$$

$$div(J_{total}) + \partial C/\partial t = 0 \qquad (4)$$

Based on Equations (3) and (4), EM and TM can be coupled. In the diffusion simulation, normalization method was used [13-16]. The vacancy concentration of SAC305 at time 0 was normalized to 1. Because of the geometric symmetry, a quarter model was used to perform the analysis.

Fig. 9 (a)-(d) show the detailed mesh and model of four different test structures. The red arrow shows the direction of electron flow. A uniform current was applied at the Cu pad to represent the electrical input. A uniform heat transfer coefficient was applied on the model to represent the convection condition in the oven. Table 2 listed material properties of SAC305 and Copper that were used in the finite element simulation.

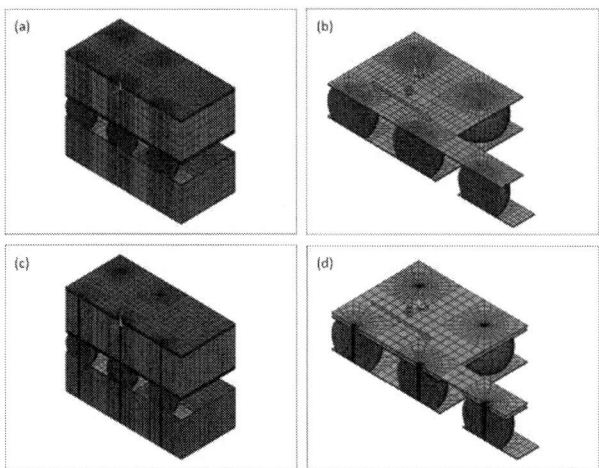

Fig. 9 Mesh and model of test structure: (a) 1/4 model of BGA non-via structure, (b) conductor model of BGA non-via structure, (c) 1/4 model of BGA via structure, and (d) conductor model of BGA via structure.

TABLE 2 EM PROPERTIES OF SAC305 AND COPPER[17]

	Ea (eV)	Q (eV)	Z*	ρ (Ohm*um)	D (um^2/s)
Cu	2	0.3121	4	2.52e-14	7.8e7
SAC 305	0.98	0.0094	23	18.1e-14	4.1e7

To simulate the void growth during EM, it is assumed that the void would form when the vacancy concentration reaches on a specific level. In this simulation, the number 1.4 is used to tune the experiment temperature change during time. The simulation was running in the following step:

1. Steady-state simulation of electric-thermal couple field analysis to obtain the steady state of the temperature distribution at time 0.
2. Apply the time 0 temperature distribution in the EM model
3. Running the electric-thermal-diffusion transient analysis on certain time step (10h).
4. Compare the vacancy concentration with the void criteria (1.4), if the average concentration is larger than 1.4, kill the element.
5. Repeat step 3-4 until the whole connection fails or the Maximum temperature reaches about the melting point of SAC305 (217C).

Because all the major failures happened at the solder ball and no failures were found on Via and PTH structures. The discussion and results only consider the solder ball of BGA and BGA-Via.

Fig 10 shows the TTF and Mean time to failure (MTTF) from experiment. Table 3 compares the simulation results and experiment results. Since the temperature kept increasing during the test, the average value of Time 0 temperature was used for comparison. The results show that the simulation well agreed with the experiment.

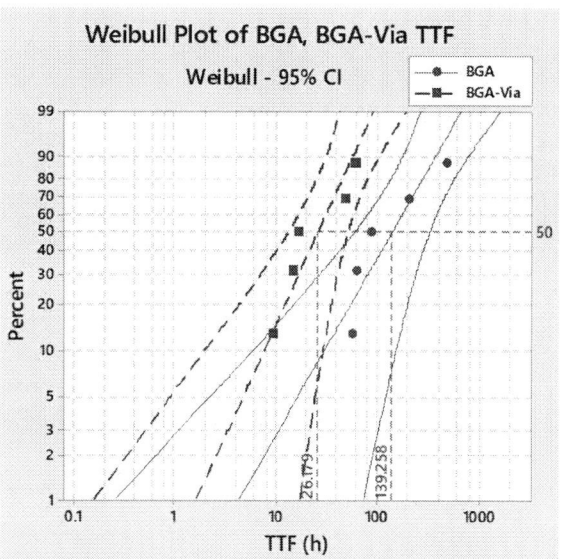

Fig. 10 Schematic Weibull Plot of TTF

TABLE 3 EXPERIMENT AND SIMULATION RESULTS COMPARISON

	BGA			BGA-Via		
	Exp.	Sim.	Err (%).	Exp.	Sim.	Err. (%)
MTTF (h)	139.3	140	0.5	26.2	30	14.5
Temp (C)	186.1	186.7	0.3	184.8	184.6	-0.1

Fig. 11 shows the void propagation during EM. When the resistance of solder increased by 10%, the void almost went across 50% of the open windows in BGA and about 30% in BGA-Via. The void was nucleated on the current entrance site and grew along the Cu-SAC interface. The final shape of the void is a pan-cake like which went across the whole open window. The results well agreed with the experiments.

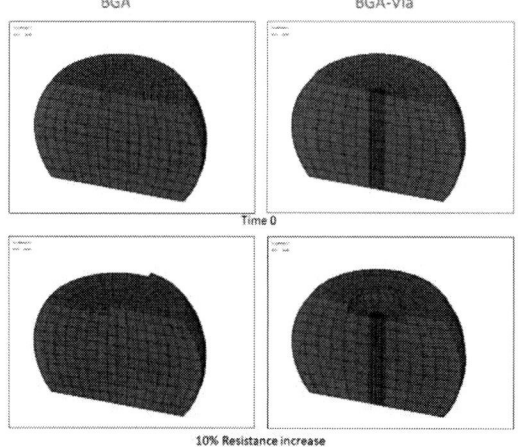

Fig. 11 Void propergation during EM

Fig 12 shows temperature distribution on the solder ball in different stages. It clearly shows that when the resistance increased by 10%, the maximum temperature of solder ball in BGA-Via reached 216.65C. which is almost the melting temperature of SAC305 (217 C) and in the conductor model it was even higher, but, in BGA, it only reached 200C. It was caused by the large amount of joule heating generated in the copper via. In dimension prospected, the Via (60 um) only has 1/10 open window of solder ball (530 um). So, the following discussion takes 10% resistance increase as the criterion of EM failure.

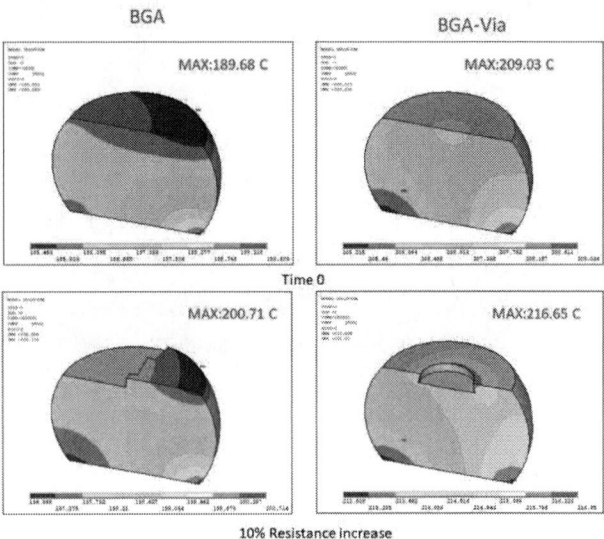

Fig. 12 Temperature distribution during EM

Fig 13 shows the current density distribution in solder ball. At time 0, the maximum current density of solder ball in BGA structure was $5.63*10^4$ A/cm^2. However, it was $6.05*10^4$ A/cm^2 in BGA-Via, which is about 10% larger than BGA. When the resistance of solder ball increased by 10%, the current density was decreased to $5.59*10^4$ A/cm^2. But, in BGA structure, the current density was increased to $5.84*10^4$ A/cm^2. This phenomenon was caused by the void location and the current entrance site.

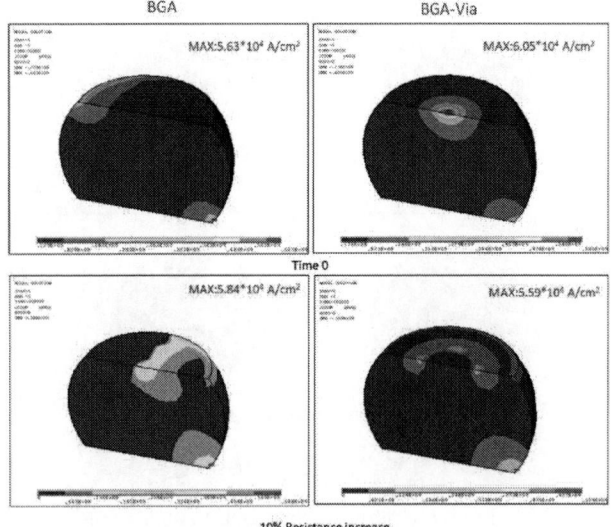

Fig. 13 Current density distribution

Fig. 14 shows the current density vector distribution at different stages. At final stage, there is 10% resistance increase. Since the TTF is different between BGA and BGA-Via, it takes different time step to show the current distribution evolution. In BGA structure, the current crowding effect will become more and more serious as void grows. However, in BGA-Via, the current crowding effect will reduce when the void is larger than the Via. It is caused by the topology of the open windows. In BGA-Via, joule heating caused by via plays the dominant role of the EM failure.

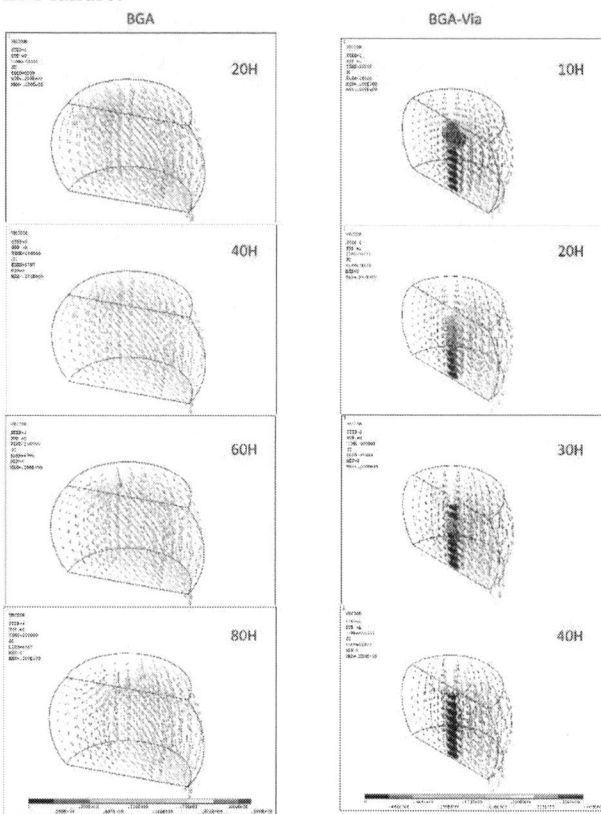

Fig. 14 Current density vector distribution as void evolution

Fig. 15 shows the schematic topology drawing of the open windows. In BGA structure, the open window of solder ball will decrease as the void propagates, which enhances the current crowding and joule heating effects. In BGA-Via structure, the open window shows the same trend. But in topology prospect, these two voids are different in terms of current entrance. In BGA structure, the decreased open windows still maintain the simply connected region compare to time 0. However, in BGA-Via structure, the topology of open window become non-simply connected region (a ring shape) from simply connected when the void generated. This ring shape open window will uniform the current density distribution and reduce the current crowding effect. But the TTF of BGA-Via structure is still smaller than BGA structure because the joule heating generated in via is much larger than solder ball.

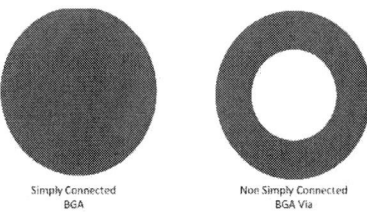

Fig. 15 Schematic topology drawing of the open windows

V. CONCLUSION

In summary, the accelerated EM test of SAC305 solder in different interconnect structures at 150 °C with 12A current were investigated and discussed. Experimental data was collected and used to validate finite element modeling. The AFD numerical simulation matched well with the experimental results. Failure analysis was conducted using cross-sectioning, SEM, and EDX.

EM failure was found in BGA and BGA-Via structures, but no EM failure was observed in Via and PTH structures during the same time frame. In that regard, Via and PTH structures were not a concern in EM failure. Failures show the similar mechanism. Void nucleated on the current entrance site and then grew along the Cu-SAC interface. The copper trace and via would be depleted by EM. In the late stage of EM, the solder ball would melt due to joule heating effect. In BGA-Via structure, the large void formed on the substrate side and deplete the whole copper trace and via. A cup shape overgrowth IMC observed on anode side, which caused by current crowding effect. The TTF of BGA-Via structure is shorter than that of BGA structure.

From the simulation results, it can be known that BGA-Via structure has higher local temperature than BGA structure because the via will induce a large amount of joule heating. In BGA and BGA-Via structures, the current density has the same magnitude on the solder ball. It indicated that joule heating generated by the via is the dominated factor in TTF difference. EM-induced void in BGA-Via structure would uniform the current density distribution because of the ring shape open window is created.

ACKNOWLEDGMENT

Xilinx, Inc. supported this project. The authors are grateful for the assistance from The Integrated Electronics Engineering Center (IEEC) and Opto-Mechanics and Physical Reliability Laboratory, State University of NY at Binghamton.

REFERENCES

[1] S. Shao, Y. Niu, J. Wang, et al, "Comprehensive study on 2.5 D package design for board-level reliability in thermal cycling and power cycling." IEEE 68th Electronic Components and Technology Conference (ECTC), pp. 1668-1675. IEEE, 2018.

[2] J. Wang, Y. Niu, S. Park, et al, "Modeling and design of 2.5 D package with mitigated warpage and enhanced thermo-mechanical reliability." IEEE 68th Electronic Components and Technology Conference (ECTC), pp. 2477-2483, 2018.

[3] Y. Niu, J. Wang, S. Shao, et al. "A comprehensive solution for electronic packages' reliability assessment with digital image correlation (DIC) method." Microelectronics Reliability, 87, pp. 81-88, 2018.

[4] H. Wang, S.Shao, V. Pham, et.al, "Quantification of Underfill Influence to Chip Packaging Interactions of WLCSP",ASME 2018 International Technical Conference and Exhibition on Packaging and Integration of Electronic and Photonic Microsystems, pp. V001T01A004-V001T01A004, 2018

[5] J. Nah, J. O. Suh, K. N. Tu, et al, "Electromigration in flip chip solder joints having a thick Cu column bump and a shallow solder interconnect." Journal of Applied Physics, Vol. 100, 2006.

[6] K. N. Tu, "Recent advances on electromigration in very-large-scale-integration of interconnects." Journal of Applied Physics, pp. 5451-5473, 2003.

[7] E. T. Ogawa, K. Lee, V. A. Blaschk, et al, "Electromigration Reliability Issues in Dual-Damascene Cu Interconnections." Transaction On Reliability, Vol. 51, pp. 403-419, Dec. 2002.

[8] C. Chen, H.M. Tong and K.N. Tu. "Electromigration and Thermomigration in Pb-Free Flip-Chip Solder Joints." Annu. Rev. Mater. Res., Vol. 40, pp. 531-555, 2010.

[9] J. Xu, Y. Niu, S.R.Cain, et.al,"The Expermental and Numerical Study of Electromigration in 2.5D Packaging", IEEE 68th Electronic Components and Technology Conference (ECTC), pp.483-489, 2018

[10] K. Zeng, K.N.Tu, "Six cases of reliability study of Pb-free solder joints in electronic packaging technology." Journal of Applied Physics, pp. 55-105, 2002.

[11] D. Dalleau and K. Weide-Zaage, "Three-Dimensional Voids Simulation in chip Metallization Structures: a Contribution to Reliability Evaluation." Microelectronics Reliability, vol. 41, pp. 1625-1630, 2001.

[12] K. Weide-Zaag, "Simulation of Migration Effects in Solder Bumps." Transctions On Device And Materials Reliability, Vol.8, pp. 442-448, Sep. 2008.

[13] J. Wang, R. Liu, D. Liu, et al, "Advancement in simulating moisture diffusion in electronic packages under dynamic thermal loading conditions." Microelectronics Reliability, 73, pp.42-53, 2017.

[14] J. Wang, Y. Niu, S. Park, "An investigation of moisture-induced interfacial delamination in plastic IC package during solder reflow." ASME InterPack 2017, pp. V001T01A011-V001T01A011, 2017.

[15] D. Liu, J. Wang, R. Liu, et al, "An examination on the direct concentration approach to simulating moisture diffusion in a multi-material system." Microelectronics Reliability, 60, pp. 109-115, 2016.

[16] J. Wang, S. Park, "Non-linear finite element analysis on stacked die package subjected to integrated vapor-hygro-thermal-mechanical stress." IEEE 66th Electronic Components and Technology Conference (ECTC), pp. 1394-1401, 2016.

[17] Y. Liu, L. Liang, S. Irving , et al, "3D Modeling of electromigration combined with thermal–mechanical effect for IC device and package." Microelectronics Reliability, pp. 811-824,2008.

Diffusion enhanced drive sub 100 °C wafer level fine-pitch Cu-Cu thermocompression bonding for 3D IC integration

[1]Asisa Kumar Panigrahy,* [2]Satish Bonam, [2]Tamal Ghosh, [2]Siva Rama Krishna Vanjari and [2]Shiv Govind Singh

[1]Department of ECE, Gokaraju Rangaraju Institute of Engineering & Technology Hyderabad, India
[2]Department of Electrical Engineering, Indian Institute of Technology Hyderabad, India
*E-mail: ee13p1009@iith.ac.in

Abstract— **One of the primary and critical requirements for high quality wafer level thermocompression Copper-Copper (Cu-Cu) bonding is the fast diffusion of Cu atoms across the boundary between two bonding layers. In this paper, we demonstrate low temperature, low pressure and fine pitch Cu-Cu thermocompression bonding by enhancing intrinsic diffusivity at the bonding interface by stress gradient. Stress in the Cu surface was fixed up by simply varying the Ar pressure during physical deposition using the sputtering tool and observed clearly that one of the other deposited wafers had opposite stress and highest differential stress not only led to fine-pitch bonding at sub 100 °C with very low external pressure of 0.25 MPa. Retention of stress even in the smaller patterns was clearly observed using conventional non-destructive wafer bow technique and further corroborated using solid mechanic module of COMSOL Multiphysics simulator. The quality of stress engineered fine-pitch bonded sample was examined using the pull test and Idonus Wafer Bonder IR Inspection (WBI) tool. Furthermore, the absence of voids and defect free bonding interface clearly opens up realistic chances for three dimensional (3D) IC integration. Moreover, this novel stress tailoring Cu surface modification prior to bonding is the primary contestant for future heterogeneous integration.**

Keywords- bumless bonding; Cu-Cu thermocompression bonding; COMSOL Multiphysics; heterogeneous integration; stress engineering; surface modification; 3D IC integration.

I. INTRODUCTION

In order to satisfy, ever-growing demand of customer for integrating more and new functionalities within a single chip is grand challenge to address the semiconductor industries as interconnect delay in planner integration becoming major bottleneck. Furthermore, the limitation of physics and challenge of device scaling has hit back the IC industries to the road block [1]. Hence focus of the IC industries and researchers in the field has completely changed the research from scaling of the device to the development of novel architectures. Development of novel architectures helped a bit to enhance overall circuit performance of IC, but the role of the same was limited [2] as the interconnect performance drag down the overall performance. In order to significantly enhance system performance there is a stringent requirement of paradigm shift from scaling and innovative architectures to novel integration. Three dimensional integrated circuits (3D IC) is one of the premier solution which has potential to reduce the RC delay with better form factor and cost

effective [3] –[5]. Apart from this, 3D IC has ability to achieve heterogeneous integration without performance degradation. Furthermore, 3D IC integration is very much industry adoptable due to its CMOS compatibility. Multiple layers or devices are stacked vertically with electrical interconnects to form 3D IC integration.

Mainly, 3D IC integration can be realized using several stacking options like Wafer-on-Wafer (WoW) bonding, Wafer-on-Chip (WoC) bonding, and Chip-on-Chip (CoC) bonding [6]. From the available techniques, WoW fine-pich bonding is more preferable for 3D IC integration as it not only simple but also CMOS compatible. Cu-Cu WoW thermocompression bonding is the most favorable bonding technique and Cu is most reliable bonding material than Al and Au. Cu - Cu fine-pitch WoW thermocompression not only giving mechanically strength but also provides excellent electrical conductivity without compromising the electromigration resistance [7] - [12].

Wafer level Cu-Cu thermocompression bonding mechanism involves diffusion of Cu atoms across the bonding interface and formation of grain due to the application of thermocompression cycle [13]. Cu is very much reactive to the atmospheric oxygen and quickly forms native oxide even exposure for few seconds which further hinders diffusion of Cu atoms during thermocompression bonding. In order to enhance the diffusion mechanism and bonding at low thermal stress native oxide should be removed prior to Cu-Cu thermocompression bonding as higher thermal requirement may degrade the underneath active devices. Apart from the need of oxidation free Cu surface at the interface, rough surface is also make way for the hindrance for high quality Cu-Cu bonding. Rough Cu surface may require high bonding pressure for quality bonding. But, the application of very high pressure during bonding may damage the delicate Cu interconnects between the stacked layers. Hence for the real multi-layer integration the requirement of low temperature, low pressure and fine-pitch Cu-Cu bonding is very much essential for future 3D IC integration.

In order to reduce the thermal stress and pressure requirement during bonding various researches have proposed several techniques to prevent/protect Cu interconnects from oxidation. Also, some of the researcher's demonstrated low temperature bonding by modifying the Cu surface prior to Cu-Cu interconnect bonding. Surface modification also led to room temperature bonding, the mechanism behind this is increment of intrinsic diffusivity of Cu atoms across the bonding interface. Suga *et. al.* proposed

978-1-7281-1500-9/19 $31.00 © 2019 IEEE

a surface activation bonding technique [14], mechanism behind the surface activation mainly depends on cleanliness of the surface at atomic level which requires insitu cleaning chamber [15], [16]. Mainly, Ar ion bombardment on the Cu surface cleans the native oxide prior to interconnect bonding. But, the usage of highly directive ion bombardment may damage the surface and consequently increases the surface roughness which requires high pressure for quality bonding. Another major drawback for the surface activation bonding is the need of ultra high vacuum (UHV) in the order of 10^{-8} torr. Furthermore, smoothening of rough Cu surface can be efficiently achieved by an extra process step of chemical mechanical polishing (CMP) preceding to interconnect bonding [17], [18]. Therefore the rigorous requirements of atomically clean Cu surface, requirement of UHV during bonding, and an additional process step prior to bonding makes the interconnect bonding very costlier and not manufacturing worthy.

In addition, some of the researchers demonstrated to remove native oxide layer prior to Cu-Cu interconnect thermocompression bonding. Pre-bond chemical treatment on the Cu surface [19] - [24] reduces the need of bonding temperature but, continued dipping of samples inside the strong acids may damage the underneath active devices. Formerly, C.S. Tan *et.al.* demonstrated a non-corrosive method to passivate the Cu surface using self assembled monolayer (SAM) [25] - [27]. SAM of thiol passivate the Cu surface efficiently, but desorption prior to Cu-Cu bonding is more challenging as it require higher temperature of ~250 °C to 300 °C [28]. Alternatively, we have proposed a non-thermal plasma and electrochemical technique to desorb the SAM layer prior to Cu-Cu bonding at 200 °C [29] – [31]. However, SAM treatment on Cu surface and subsequent desorption of the same prior to bonding are ex-situ hence very much unlikely with in-line CMOS process flow.

Recently, Chen *et.al.* proposed a novel in-line process flow compatible WoW Cu-Cu bonding using metal as passivation layer on Cu surface during deposition and subsequent bonding was performed at sub 200 °C and at 1.91 MPa pressure [32], [33]. However, application of very high bonding pressure during Cu-Cu bonding may damage the delicate underneath Cu interconnects and must reduce the system performance. Alternatively, we have reported a low temperature (160 °C) and low pressure (0.25 MPa) WoW Cu - Cu bonding using optimized Ti metal as passivation layer [34]. Apart from this, we have reported efficacy of Cu surface passivation with optimized Ti passivation layer [35]. However, it's clearly observed in the literatures that Ti is getting oxidized at high temperature [36] and Ti is not damascene process compatible.

Recently, we have proposed a CMOS damascene compatible Cu-Cu bonding using systematically optimized metal-alloy as passivation material and bonding at Sub 150 °C with low contact pressure of 4 bar [37], [38]. All the above reported methodologies yielded bonding at reasonably low temperature ranges from 150°C to 300°C. However, in our previous work we have reported a sub 100°C WoW Cu-Cu blanket bonding using stress engineering [39], which not only fulfill the current demand of semiconductor industries but also CMOS process flow compatible. However, blanket WoW Cu-Cu bonding has very restricted application as Cu is conductive and it will short the active devices and finally unreliable. In actual multi-layer integration, Copper fine-pitch WoW bonding is performed with proper isolation between the active layers using inter layer dielectric (ILD). In order ensure the bonding quality both mechanically strong and electrically conductive then all the Cu pads present in the wafer must bond. Hence, the alignment prior to WoW Cu – Cu fine-pitch bonding must accurate and very precise.

In this endeavor, we have proposed a high quality fine-pitch WoW Cu-Cu bonding using stress engineering at Sub 100 °C with 0.25 MPa pressure. Mainly, enhancement of intrinsic diffusivity near bonding interface is the key requirement for low temperature fine-pitch WoW bonding. In this work, we have developed a novel technique to enhance the rate of diffusion at the bonding interface by creating stress difference between both the Cu deposited samples. The concept of creating stress gradient was already utilized in demonstration of out of plane inductor coils [40]. We have also used the same concept to reduce the thermal stress during thermocompression bonding, the schematic of bonding methodology using stress difference is shown in Fig. 1.

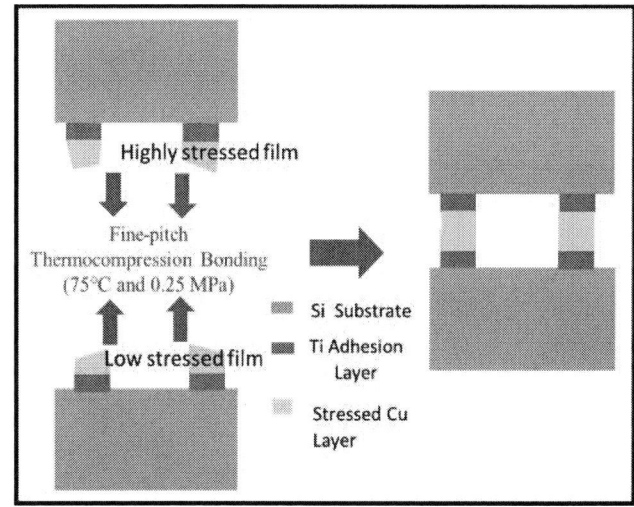

Figure 1. Schematic of stressed Cu pad-to-pad bonding at Sub 100 °C.

EXPERIMENTAL

Using the proposed surface engineering technique, fine pitch (20 μm pitch) thermocompression bonding is demonstrated at sub 100 °C and 0.25 MPa pressure. The process flow comprises of following steps viz. preparation of the mask, cleaning of the wafers, stress gradient study, and finally bonding of wafers.

A. Preparation of the mask for Cu interconnect bonding

Mask design is the most vital and initial requirement for fine-pitch Cu-Cu bonding. Mask design was carried out prudently by keeping in mind that alignment prior to bonding

978-1-7281-1500-9/19 $31.00 © 2019 IEEE

must be simple. Complete 3 inch mask was divided into 8 dies (12 mm × 5 mm). In order to check the mechanical strength and electrical conductivity of the Cu-Cu bonding pads daisy chain structures, SEM line structures, and mechanical stability structures of various pitch size of 20 μm, 35 μm, and 55μm were incorporated in the die. Fig. 2 shows wafer level mask design after complete alignment.

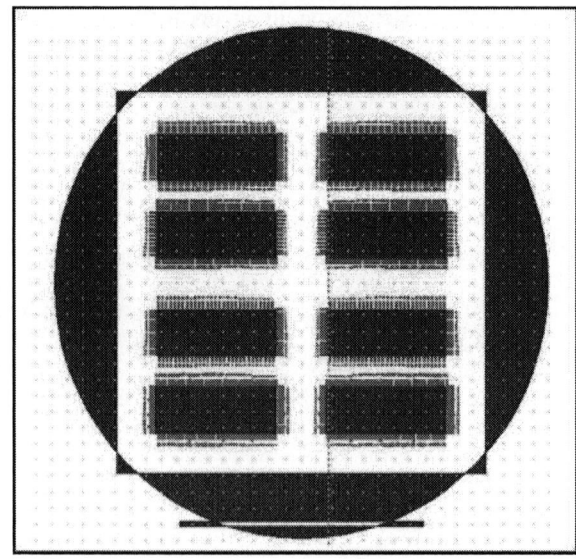

Figure 2.Complete mask design after alignment.

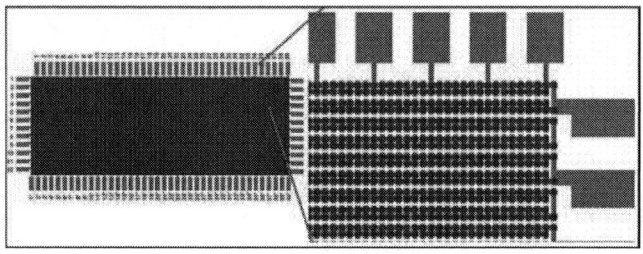

Figure 3. Die level design of daisy chain structure after proper alignment.

B. Cleaning of Wafers

2 inch P-type <100> Double Side Polished (DSP) Silicon wafers were used for the entire experiments. Piranha cleaning followed by standard RCA-1, RCA-2 was used to clean the wafers in order to remove all the foreign particles and contaminants in the substrate. After each step stringent DI water cleaning of wafers also performed to remove any acid residues. Then the wafers get dried by using compressed dry air (CDA) prior to patterning of photoresist.

C. Stress gradient study

Liftoff technique was adapted for patterning of Cu. AZ5214E photoresist was used for patterning of

interconnects in the image reversal mode. Then the photoresist pattern wafers were transferred to the deposition chamber. Metal deposition was performed on the patterned photoresist wafers in order to optimize the stress conditions by sequential deposition 20 nm of Ti film and 185 nm of Cu film was deposited using sputtering tool (AJA International, Phase IIJ). The Ti layer act as adhesion layer between Si and Cu. In order to find out maximum stress conditions the Cu deposition was performed at various Argon pressure by keeping other deposition conditions constant like distance between the wafer and target, speed of the rotation, substrate holder etc. The thin Cu layer was deposited at various inlet Ar pressure starting from 3 mTorr to 21 mTorr with an increment of 3 mTorr. In order to keep the constant thickness of 185 nm of Cu we have just adjusted the time of deposition and thickness was verified by atomic force microscopy (AFM). The whole wafers were then absorbed in acetone to complete the standard lift off process. Stress measurement was performed using wafer bow measurement technique with the help of kSA MOS Ultra (K-Space) ultrascan instruments [39]. The maximum tensile stress was observed for the patterned Cu thin film deposited with Ar pressure of 12 mTorr as observed for the same of blanket bonding and maximum compressive stress was observed for the patterned Cu thin film deposited with 3 mTorr Ar pressure and same as observed for the blanket with 3 mTorr pressure.

Furthermore, the maximum differential stress in the film were cross verified for the patterned Cu samples using solid mechanic module of COMSOL Multiphysics simulator. The simulated result also well corroborated with the wafer bow measurements. Stress profile is retained even for smaller dimension such as 10 μm and 5 μm Cu pads as shown in Fig. 4 (c, d) and Fig. 4(e, f) respectively. This is expected, for, the stress in thin films is due to the change in the lattice constant which is a property getting affected at the atomic level. It is bound to retain it stress based on the conditions on how atoms are arranging rather than the geometry at the micro-level.

As expected, surface displacement reduce as dimension Cu pads reduces. It is worth to mention that surface displacement is still present for pad size even 5 μm x 5 μm (Fig. 4 (a, b)). However just by changing the process conditions we can further create higher stress for smaller dimensions. From Fig. 4 (c, d, e, f) it is clear that surface displacement obtained for 5 μm x 5 μm pad size in different process condition almost same surface displacement for 15 μm x 15 μm size pad.

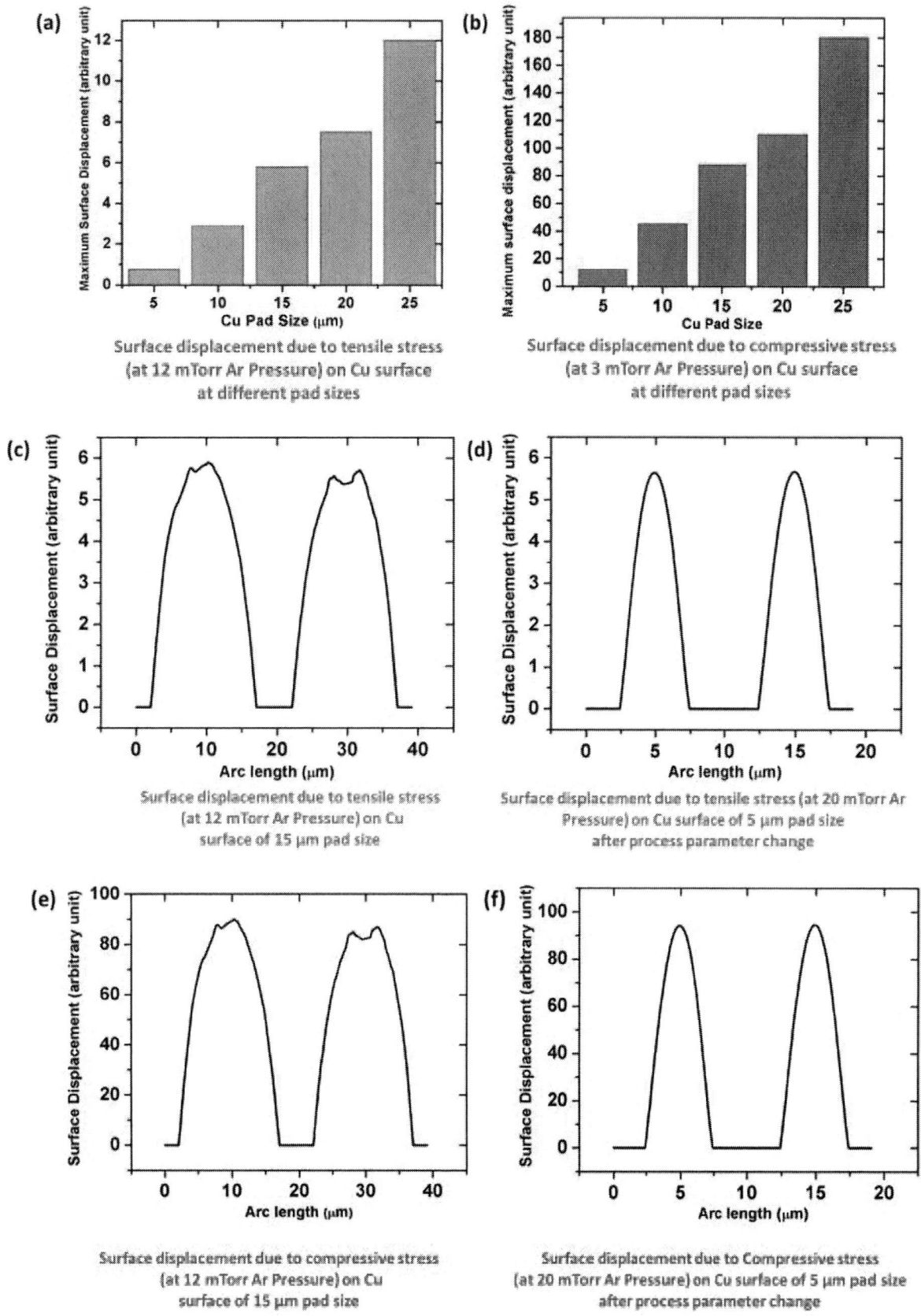

Fig. 4 Stress in the film retained even for the smaller patterns.

D. Fabrication of fine pitch Cu-Cu bonding with maximum differential stress deposited samples

The samples with maximum differential stress (3 mTorr - 12 mTorr sample pair), Cu pads bonded at 75 °C with a nominal contact pressure of 0.25 MPa for 50 mins using AML wafer bonder. Prior to bonding wafers were aligned precisely using IR lamp and microscope. After precise alignment both the wafers were applied the desired thermocompression condition. Bonded sample than removed after cooling down the platens for interface study.

RESULTS AND DISCUSSIONS

A. Microstructure Imaging of bonded sample

Quality of the bonded interface was studied using Infra-red Wafer bonder inspection (IR-WBI) tool. IR-WBI is one of the best qualitative method to evaluate the bond interface, apart from this yield of the bonded Cu pads and misalignment can be observed clearly. IR source act as blocking parameter for the metal pads where the bonding has happened properly. Fig.5 shows the WBI-IR imaging of 2″ bonded Cu pads having highest differential stress, clearly depicts that quality of bonding is excellent without any presence of yield and die break.

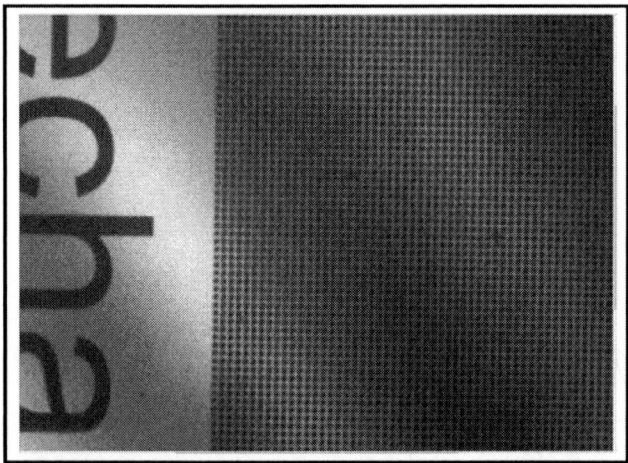

Figure 5. IR-WBI imaging of the Cu-Cu fine-pitch bonded sample.

B. Bond strength analysis of the Cu pad-to-pad fine-pitch bonded samples

Bond strength is the figure of merit of the quality of bonding. In order to measure the bond strength of the Cu-Cu bonded sample with highest differential stress (3 mTorr- 12 mTorr sample pair) using Instron microtester. For this study, we have initially diced the 2″ fine-pitch bonded sample into 1cm^2 pieces using diamond cutter. The intact after such an uneven force during dicing is the initial report of quality and reliable bonding. Then the shear test was performed with the

diced samples, fig 6 shows the bond strength of 144 MPa which is comparably better than the available literatures [33].

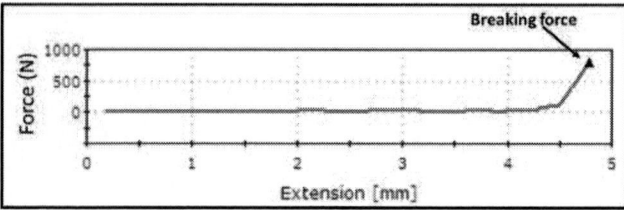

Figure 6. Bond strength qualification.

CONCLUSION

Stress engineering retained even for the Cu film deposited samples of pitch of sub 10 μm. Stress engineering generated on Cu samples enhances the bonding quality. In this work, we achieved high quality fine-pitch (20 μm) Cu-Cu WoW bonding at 75°C with nominal contact pressure of 0.25 MPa. The obtained high bond strength of 120 MPa suggests that surface engineering is the prime contestant for fine-pitch Cu-Cu bonding for 3D IC integration paradigm.

REFERENCES

[1] C.S. Tan, R. J. Gutmann, and L. R. Reif, *Wafer level 3-D ICs process technology*, Springer Science & Business Media, 2009.

[2] P. Garrou, C. Bower, P. Ramm, Handbook of 3D Integration: Technology and Applications of 3D Integrated Circuits, (John Wiley and Sons, Hoboken, New Jersey, 2008).

[3] R. S. Patti, "Three-dimensional integrated circuits and the future of system-on-chip designs," *Proceedings of the IEEE*, vol. 94, no. 6, pp.1214-1224, 2006.

[4] C. C. Liu, I. Ganusov, M. Burtscher, and S. Tiwari, "Bridging the processor-memory performance gap with 3D IC technology," *IEEE Design & Test of Computers*, vol.22, no. 6, pp.556-564, 2005.

[5] Y. Akasaka, "Three-dimensional IC trends," *Proceedings of the IEEE*, vol. 74, no. 12, pp.1703-1714, 1986.

[6] C.S. Tan, R. J. Gutmann, and L.R. Reif, *Overview of Wafer-Level 3D ICs*, Springer, US, 2008, pp. 1-11.

[7] K. Elissa, "Title of paper if known," unpublishe D.F. Lim, J. Wei, K.C. Leong, and C.S. Tan , "Temporary passivation of Cu for low temperature (< 300°C) 3D wafer stacking," In Interconnect Technology Conference and 2011 Materials for Advanced Metallization (IITC/MAM), IEEE International, pp. 1-3. IEEE, 2011.

[8] A. Huffman, J. Lannon, M. Lucck, C. Gregory, and D. Temple, "Fabrication and characterization of metal-to-metal interconnect structures for 3-D integration," In Materials and Technologies for 3-D Integration Symposium, Warrendale, PA, USA, 2009, pp. 107-19, 2008.

[9] K. N. Chen, S. M. Chang, A. Fan, C. S. Tan, L. C. Shen, and R. Reif, "Process development and bonding quality investigations of silicon layer stacking based on copper wafer bonding," Applied Physics Letters, 87(3), p.031909, 2005.

[10] C. Y. Chang and S. M. Sze, ULSI Technology, McGraw-Hill, New York (1996).

[11] D. Save, F. Braud, J. Torres, F. Binder, C. Muller, J. O. Weidner, and W. Hasse, Electromigration resistance of copper interconnects. Microelectronic engineering, 33(1-4), pp.75-84, 1997.

[12] K. N. Tu, "Recent advances on electromigration in very-large-scale-integration of interconnects," Journal of applied physics 94, no. 9, pp. 5451-5473, 2003.

978-1-7281-1500-9/19 $31.00 © 2019 IEEE

[13] A. Fan, A. Rahman, and R. Reif. "Copper wafer bonding," Electrochemical and Solid-State Letters 2, (1999) 534-536.

[14] H. Takagi, K. Kikuchi, R. Maeda, T. Chung, and T. Suga, "Surface activated bonding of silicon wafers at room temperature," Applied physics letters 68, pp. 2222-2224, 1996.

[15] H. Takagi, R. Maeda, T. R. Chung, N. Hosoda, and T. Suga, "Effect of surface roughness on room-temperature wafer bonding by Ar beam surface activation," Japanese journal of applied physics 37, pp.4197, 1998.

[16] T. R. Chung, L. Yang, N. Hosoda, H. Takagi, and T. Suga, "Wafer direct bonding of compound semiconductors and silicon at room temperature by the surface activated bonding method," Applied surface science 117, pp.808-812, 1997.

[17] K. Tsukamoto, E. Higurashi, and T. Suga, "Evaluation of Surface Microroughness for Surface Activated Bonding," In IEEE CPMT Symposium Japan, pp. 1-4, 2010.

[18] T. Wakamatsu, T. Suga, M. Akaike, A. Shigetou, and E. Higurashi, "Effect of SAB process on GaN surfaces for low temperature bonding," In 6th International Conference on Polymers and Adhesives in Microelectronics and Photonics, 2007. Polytronic, pp. 41-44, IEEE, 2007.

[19] J. Fan, D. F. Lim, and C. S. Tan, "Effects of surface treatment on the bonding quality of wafer-level Cu-to-Cu thermo-compression bonding for 3D integration," Journal of Micromechanics and Microengineering, vol. 23, no. 4, p. 045025, 2013.

[20] E.-J. Jang, S. Hyun, H.-J. Lee, and Y.-B. Park, "Effect of wet pretreatment on interfacial adhesion energy of Cu-Cu thermocompression bond for 3D IC packages," Journal of Electronic Materials, vol. 38, pp. 2449–2454, 2009.

[21] B. Swinnen, W. Ruythooren, P. De Moor, L. Bogaerts, L. Carbonell, K. De Munck, B. Eyckens, S. Stoukatch, D. Sabuncuoglu Tezcan, Z. Tokei, J. Vaes, J. Van Aelst, and E. Beyne, "3D integration by Cu-Cu thermo-compression bonding of extremely thinned bulk-Si die containing 10um pitch through-Si vias," in Electron Devices Meeting, 2006. IEDM '06. International, pp. 1–4, 2006.

[22] A. Huffman, J. Lannon, M. Lueck, C. Gregory, and D. Temple, "Fabrication and characterization of metal-to-metal interconnect structures for 3-D integration," Journal of Instrumentation, vol. 4, no. 03, p. P03006, 2009.

[23] K. Chen, C. Tan, A. Fan, and R. Reif, "Copper bonded layers analysis and effects of copper surface conditions on bonding quality for three dimensional integration," Journal of Electronic Materials, vol. 34, pp.1464–1467, 2005.

[24] K. Chen, A. Fan, C. Tan, and R. Reif, "Bonding parameters of blanket copper wafer bonding," Journal of Electronic Materials, vol. 35, pp.230–234, 2006.

[25] D. F. Lim, S. G. Singh, X. F. Ang, J.Wei, C. M. Ng, and C. S. Tan, "Application of self-assembly monolayer (SAM) in lowering the process temperature during Cu-Cu diffusion bonding of 3D IC," In 4th IEEE International Microsystems, Packaging, Assembly and Circuits Technology Conference, IEEE, pp. 68-71, 2009.

[26] C. S. Tan, "Application of self-assembled monolayer (SAM) in low temperature bump-less Cu- Cu bonding for advanced 3D IC," In 5th IEEE International Microsystems Packaging Assembly and Circuits Technology Conference (IMPACT), IEEE, pp. 1-4, 2010.

[27] C. S. Tan, D. F. Lim, S. G. Singh, S. K. Goulet, and M. Bergkvist, "Cu–Cu diffusion bonding enhancement at low temperature by surface passivation using self-assembled monolayer of alkane-thiol," Applied Physics Letters, vol. 95, no. 19, pp. 192108, 2009.

[28] C.S.Tan and D.F.Lim, (Invited) "Cu Surface Passivation with Self-Assembled Monolayer (SAM) and Its Application for Wafer Bonding

at Moderately Low Temperature," ECS Transactions, vol. 50, no. 7, pp.115-123, 2013.

[29] Tamal Ghosh, E. Krushnamurthy, Ch Subrahmanyam, V. SivaRamaKrishna, A. Dutta, and S. G. Singh. "Room temperature desorption of Self Assembled Monolayer from Copper surface for low temperature & low pressure thermocompression bonding." In Electronic Components and Technology Conference (ECTC), 2015 IEEE 65th, pp. 2200-2204, IEEE, 2015.

[30] Tamal Ghosh, Siva Rama Krishna V, Asudeb Dutta and Shiv Govind Singh " Electrochemical self-assembled monolayer desorption assisted low temperature Cu-Cu thermocompression bonding " ICEE 2014, Bangalore, India Dec 3-6,2015.

[31] Tamal Ghosh, K. Krushnamurthy, Asisa Kumar Panigrahi, Asudeb Dutta, Ch Subrahmanyam, Siva Rama Krishna Vanjari, and Shiv Govind Singh. "Facile non thermal plasma based desorption of self assembled monolayers for achieving low temperature and low pressure Cu-Cu thermo-compression bonding." RSC Advances 5, no. 125 (2015): 103643-103648.

[32] Y.P. Huang, Y. S. Chien, R.N.Tzeng, M.S.Shy, T.H.Lin, K.H.Chen, C.T.Chiu, J.C.Chiou, C.T.Chuang, W.Hwang, H.M.Tong, and K.N.Chen, "Novel Cu-to-Cu Bonding With Ti Passivation at 180^0 in 3-D Integration." Electron Device Letters, IEEE 34, vol. 12, 2013, pp. 1551-1553.

[33] Y.P. Huang, Y. S. Chien, R.N. Tzeng, and K.N.Chen, "Demonstration and Electrical Performance of Cu–Cu Bonding at 150 C With Pd Passivation," IEEE Transactions on Electron Devices, vol. 62, no. 8, pp. 2587-2592, 2015.

[34] A.K. Panigrahi, S. Bonam, T. Ghosh, S.G. Singh, and S.R.K. Vanjari, "Ultra-thin Ti passivation mediated breakthrough in high quality Cu-Cu bonding at low temperature and pressure," Materials Letters, vol.169, pp. 269-272, 2016.

[35] A.K. Panigrahi, S. Bonam, T. Ghosh, S.R.K. Vanjari, and S.G. Singh, "Long term efficacy of ultra-thin Ti passivation layer for achieving low temperature, low pressure Cu-Cu Wafer-on-Wafer bonding." In IEEE 3D Systems Integration Conference (3DIC), pp. TS8-13, 2015.

[36] M. C. Burrell, and N.R. Armstrong, "Oxides formed on polycrystalline titanium thin-film surfaces: rates of formation and composition of oxides formed at low and high O2 partial pressures," Langmuir, vol. 2, no. 1, pp. 30-36, 1986.

[37] A.K. Panigrahi, T. Ghosh, S.R.K. Vanjari, and S.G. Singh, "Oxidation Resistive, CMOS Compatible Copper-Based Alloy Ultrathin Films as a Superior Passivation Mechanism for Achieving 150° C Cu–Cu Wafer on Wafer Thermocompression Bonding," IEEE Transactions on Electron Devices, vol. 64, no. 3, pp.1239-1245 (2017).

[38] A.K. Panigrahi, T. Ghosh, S.R.K. Vanjari, and S.G. Singh, "Demonstration of sub 150° C Cu-Cu thermocompression bonding for 3D IC applications, utilizing an ultra-thin layer of Manganin alloy as an effective surface passivation layer," Materials Letters, 194, pp.86-89, 2017.

[39] A.K. Panigrahi, T. Ghosh, C.H. Kumar, S.G. Singh, and S.R.K. Vanjari, "Direct, CMOS In-Line Process Flow Compatible, Sub 100° C Cu–Cu Thermocompression Bonding Using Stress Engineering," Electronic Materials Letters, 14(3), pp.328-335, 2018.

[40] D.-H. Weon, J.-H. Jeong and S. Mohammadi, "High- \$Q\$ micromachined three-dimensional integrated inductors for high-frequency applications, "J. Vac. Sci. Technol. B Microelectron. Process. Phenom, vol. 25, no. 1, pp. 264-270, 2007.

978-1-7281-1500-9/19 $31.00 © 2019 IEEE

DEVELOPMENT OF SHEET TYPE MOLDING COMPOUNDS FOR PANEL LEVEL PACKAGE

Kenichi UENO, Kazuhiro DOHI, Yui Suzuki, Masakazu HIROSE

SANYU REC CO., LTD.

3-5-1 Dou-cho Takatsuki-city Osaka, Japan 569-8558

ueno@sanyu-rec.jp

1. Abstract

Wafer Level Package (WLP) has been widely and commonly used in the electronics market, and getting more popular due to its advantages [1]. Liquid molding compound is a major encapsulation for the package. Panel Level Package (PLP) is considered as one of the next solutions after WLP because of more efficient and effective productivity with bigger panel size than WLP [2] [3]. At the moment, granule encapsulation with Compression Mold (CM) process is a major trend for PLP, however in this paper, we will discuss sheet type molding compound (mold sheet) for PLP as an alternative encapsulation material other than granule types. Same as WLP approach, the warpage is the main topic in PLP approach. Therefore, the warpage is key discussion in this paper, but Tg, Storage Modulus (E'), Co-efficient of Thermal Expansion (CTE) of Mold Sheet and Glass Carrier are also discussed.

Among a lot of parameters, difference of raw materials, solid or liquid, made the biggest impact to warpage. This might be due to difference of capacity for filler content. Making from solid resin can accept more filler than making from liquid resin, which makes CTE of mold sheet much lower. It is expected that Various Mold sheets with Glass carriers combination make more flat products, and we considered about relation between various items.

2. Introduction

The remarkable progress of the semiconductor packaging technology in the field of semiconductor industry today is the coincidence of the public.

The semiconductor packaging technology is a combination of a wide variety of elemental technologies, including mounting methods, various materials, construction methods, processing agents, molding machines, and sub-materials, and is a crystal of advanced technology.

Semiconductor packages are being researched and developed on a daily basis with the aim of being smaller and thinner integrated, composited and modularized by decreasing the size of each package, and further enhanced in functionality.

A method for producing semiconductor packages with higher efficiency while advancing high-performance has been studied. Among them, WLP has been developed, and PLP having a larger area has been developed by the flow of WLP.

The main purpose of PLP is to reduce the production cost per unit area by sealing an area larger than WLP at once time.

In PLP, since only non-defective unit products are lined up on the mold carrier, the chips which have failed in the inspection after the circuit is formed on the wafer are removed, the yield of the package is improved when the final molding is completed.

On the other hand, other than WLP and PLP, molding by tablet type Epoxy mold compound (EMC) with Transfer Molding (TFM) is generally used.

While TFM is a preferred method for molding individual components through runners, but when molding large areas together, it can be subject to very high pressures which give adversely affect to the chip to be molded.

In WLP and PLP, miniaturization of chips and thin packages are progressing, low-pressure molding is required to protect complicated and fragile chips, and high-pressure molding such as TFM by tablet type EMC is not suitable.

Suitable molding materials for low pressure molding include liquids, granules, and sheets, but each molding material has advantages and disadvantages.

For example, in a liquid molding compound, since it is originally liquid resin, its viscosity is low and the filling property is excellent. In addition, since it has a relatively low viscosity without increasing the temperature of the molding equipment, it is easy to cure at a low temperature, and as a result, it is expected that warpage after curing of the wafer and the panel will be reduced.

On the other hand, continuous production of liquid molding compound is difficult, and the price per unit weight becomes relatively high because fine dispersion cannot be achieved unless the batch production is carried out on a small scale.

Fig,1 Granule shape

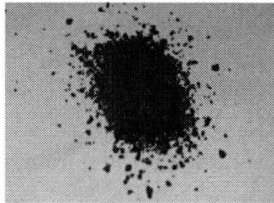

Fig.2 Powder shape

The granule molding compound can be produced at a relatively low cost because the production facility of tablet type EMC can be applied and continuous production is possible.

However, on the other hand, dust is generated in the pulverized powder-like molding material, which affects the clean environment.

In granular form, a process of forming the compound into granules is necessary, and the degree of difficulty in the process increases when the granules become fine particle size. Further, since the viscosity is higher than that of the liquid state, a blocking phenomenon in which the particles solidify before using is occur because of the melting point of the solid resin which is included as the matrix of compound.

Sheet molding compound (Mold sheet) can be made from both liquid and solid materials. It is possible to make products with each 'solid or liquid' characteristic. However, a disadvantage is that the cost per unit weight is higher than that in the liquid molding compound because the process of forming a sheet from the liquid state is performed. A Mold sheet made of a solid material is handled as a large sheet plate and is difficult to handle, and has a disadvantage that it is cracked when stored and handled at a low temperature or a freeze condition.

It can be said that which type of molding compound is best depends on each most important point.

In this report, the research was carried out focusing on the Mold sheet from among the molding materials suitable for the low pressure molding. Two type machines were used for molding: Compression Mold and Vacuum Laminator. In each case, the effect of the combination on the warpage was examined using a glass carrier.

3. Encapsulation Materials and Equipments

Here, the Mold sheet, the glass carrier, used in this experiment and the equipment used in the molding experiment will be described.

3-1. Mold sheet materials

In this report, a sheet type molding compound (Mold sheet) is examined. The raw material of the Mold sheet can be roughly divided into two types, and variations in the cured product characteristics details of each will be described below.

All Mold sheets are manufactured 300mm square size and 500um thickness. They have separation films on both side which are cover film and base film. Firstly, peeling off the cover film and overlap with glass carrier. After the molding and PMC, peeling off the cover film.

3-1-1. Mold sheet which made from liquid material (MSL)

The MSL used in this study which is made from a liquid molding compounds as a raw material. Since the MSL is based on the liquid molding compounds, the cured product characteristics are similar to those of the liquid molding compounds, and it tends to be relatively excellent in liquidity and good in filling up property. The MSL is manufactured by thinly spreading a liquid molding compound and molding it into a sheet-like shape. In this case, a so-called "B stage state" in a semi cured state is produced by mildly heating, and the curing is promoted to such an extent that the thickness does not change, so that the sheet-like shape can be maintained. The sheet can be produced by a roll-to-roll equipment at the time of producing the sheet, and there is an advantage that the sheet can be easily designed in an automated facility even when the sheet is used in the form of a roll-shaped product.

Fig, 3 MSL roll shape

On the other hand, since it is difficult to enlarge batches and continuously produce the original liquid molding compound, the unit price per unit weight is relatively expensive as the molding materials, and the unit price of the MSL using the sealing material as a raw material is further increased. In addition, the liquid resin, which is a liquid component of the liquid molding compound, does not form a liquid state unless otherwise low molecular weight, and the narrow width of the choice of the liquid material is also a weak point.

In this study, 3 types of MSL is considered. Their properties are shown in Table 1.

Table 1. MSL cured product properties

	Tg	CTE	Modulous
MSL-1	190C	20ppm	14GPa
MSL-2	100C	15ppm	17GPa
MSL-3	100C	20ppm	14GPa

3-1-2. Mold sheet which made from Solid material (MSS)

Another sheet type molding compound is MSS which uses a solid molding material as a raw material. Since MSS is based on a solid resin, the cured physical properties are similar to those of a solid 'EMC, powder, and granule' molding compounds, and the fillers can be highly filled and the reliability is excellent.

MSS can be continuously produced and the unit cost per unit weight is relatively low. Since it is impossible to form the sheet by roll-to-roll like a MSL, the sheet is formed in batches by spreading one by one. Therefore, handling of a large-area plate is so difficult. The solid resin used as the material have a high molecular weight, so that various structures can be incorporated into the molecule. Therefore, it is possible to design a compound having a variety of characteristics.

Fig, 4 MSS flat sheet

In this study, 3 types of MSS is considered. Their properties are shown in Table 2.

Table 2. MSS cured product properties

	Tg	CTE	Modulous
MSS-1	190C	9ppm	15GPa
MSS-2	120C	7ppm	25GPa
MSS-3	120C	9ppm	15GPa

3-2. Glass carrier

The compatibility of the glass carrier and the Mold sheet is important because the chips are placed on the glass carrier and molded in the compound. In the case of the glass carrier, it is possible to modify each item such as the coefficient of CTE, thickness, surface roughness, and so on.

In this report, warpage with each Mold sheet was examined by using a 320 mm square glass carrier (mold area is 300mm square) and by giving variations in CTE and its thickness. Total of 6 types of glass carriers, two types (1.1 mm and 0.7 mm) in thickness and three types (10 ppm, 6 ppm, 4 ppm) of different CTE were used. Their properties are shown in Table 3. They were named as 'H' have 10ppm on CTE, for example.

Table 3 Glass carrier properties

	Thickness	CTE
H-0.7	0.7um	10ppm
H-1.1	1.1um	10ppm
M-0.7	0.7um	6ppm
M-1.1	1.1um	6ppm
L-0.7	0.7um	4ppm
L-1.1	1.1um	4ppm

3-3. Molding equipment

In the experiment, a Compression molding machine and a Vacuum laminator machine were used as low-pressure molding equipment. Each will be described below.

3-3-1. Compression Mold

The Compression mold can use molding compounds of various properties such as liquid, granule, and sheet, and is a highly versatile equipment. Since the mold is used, the dimensional stability of the product is excellent. In addition, since the mold is covered with a release film, maintenance against contamination of the mold is unnecessary. Although pressure is applied by pressing, the mold opening area is larger than that of the TFM method, and therefore molding can be performed at a low pressure, and therefore, it has been developed for fine molding and brittle high-end integrated components.

There are two types of Compression molding machines. The first is a face-down system in which the mold is located below and the carrier is set on the upper side. Though there is no large difference in each type, the face-down method places the molding material on the mold, so that the viscosity is quickly reduced by heating and the UPH is better than the other, but the mounting area is slightly reduced because the carrier fixing component is necessary.

Another is a face-up system in which a mold is located on the upper side and a carrier is set on the lower side. In the face-up system, Mold sheet can be attached to the carrier and prepared in advance, and automate of attaching the sheet to the carrier can be prepared separately.

In this study, face-up type is used as compression mold equipment.

3-3-2. Vacuum laminator

Vacuum laminators were made with the aim of holding a large area under uniform pressure with a tough balloon called a diaphragm. Conventionally, it has been widely used for sealing with substrate and film in the field of substrate and flat panel industries. Vacuum laminators can also be used to seal PLP and are easier to accommodate for large areas than compression mold equipment. Since the mold is not used, a jig having a shape surrounding the outer periphery of the carrier is separately required for dimensional stabilization. In addition, the properties of the encapsulant that can be used are limited to sheet shapes for Vacuum laminator.

4. Experiment

In general, it is very important factor to make flat products and keep flat in the process of PLP field. And so, to go through to next following processes, the molding compounds for PLP are much needed to make smaller warpage at after post mold cure (PMC). It was making better to reduce the warpage of Mold sheet used with glass carrier.

Six types of glass carriers with different thickness and CTE and six types of Mold sheets with different Tg and CTE were considered for the researching relationship with the warpage of the product. The warpage was measured after mold and after PMC. And they are compared process with Compression mold and Vacuum laminator, too.

4-1. Combination with Each Mold sheet and Each Glass carrier

Molding experiments were conducted and compared with combination of each Mold sheet and each Glass carrier. Here, all molding conditions were performed under the same conditions. It is investigated that the influence of the combination of the characteristics of the mold sheet and the glass carrier for the warpage.

Molding equipment is Compression mold. In mold condition is 120C / 10min. PMC condition is 150C / 60min. These conditions were fixed for clarity comparison.

4-2. Combination with each Mold sheet and each molding equipment

As the same of above experiment, It is compared each Mold sheet and each molding equipment. Here, all molding conditions were performed under the same conditions. This comparison was carried out only on the MSL-1 and MSS-1. Only M-1.1 Grass carrier was used.

Molding equipment is Compression mold and Vacuum laminator. It is compared combination with each Mold sheet and each molding equipment.

In mold condition is 120C / 10min. PMC condition is 150C / 60min. These conditions were fixed for clarity comparison, too.

4-3. Combination with each Mold sheet and each PMC condition.

As the same of above experiment, It is compared each Mold sheet and each PMC condition. Here, all molding conditions were performed under the same conditions. This comparison was carried out only on the MSL-1 and MSS-1. Only M-1.1 Grass carrier was used. Molding equipment is Compression mold only. In mold condition is 120C / 10min.

In here, it is focused on PMC condition. PMC conditions are '150C / 60min', '140C / 120min', and '100C / 30min + 150C / 60min' step cure schedule. These conditions were fixed for clarity comparison, too.

4-4. Warpage measuring

Warpage was measured with digital caliper or micro scope to check displacement of 4 points which are the four corners of square. Maximum value of warpage of them was reported only. The measurement was done right after mold and just after PMC each. Fig.5 shows how to measure the warpage.

It takes digital caliper or micro scope to measure the how high is warpage. The shape of warpage for upward with the glass carrier facing down is referred to as "smile warpage", and the shape of warpage for downward is defined as "cry warpage". The warpage value is expressed as a positive '+' value when smile warping is performed. And negative '-' value when cry warpage occurs.

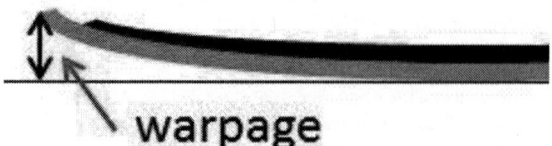

Fig.5 method of warpage measurement

Fig.6 Molded Mold sheet with 320mm square glass carrier

5. Results

5-1. Process study combination with each Mold sheet and each molding equipment.

5-1-1. MSL-1 with each glass carrier

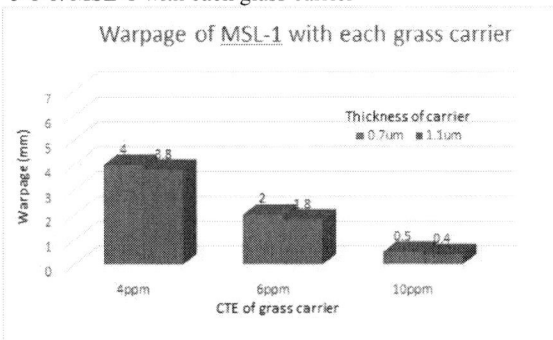

Fig.7 The warpage of MSL-1 with each glass carrier

Fig.7 shows that it was found that there is big difference of the warpage between glass carriers. The thickness differ makes a little variation. They are almost in error range, but surly there is a smaal difference.

5-1-2. MSL-2 with each glass carrier

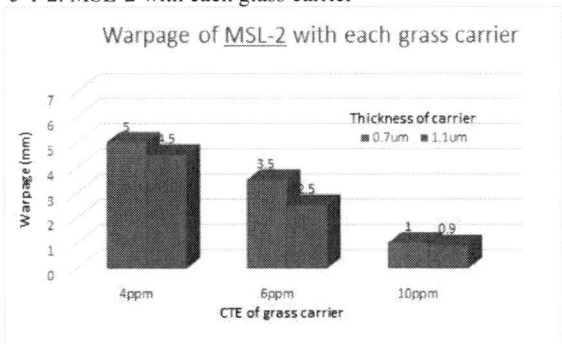

Fig.8 The warpage of MSL-2 with each glass carrier

Fig.8 shows that MSL-2 makes larger warpage than MSL-1. The other trend is the same as MSL-1.

5-1-3. MSL-3 with each glass carrier

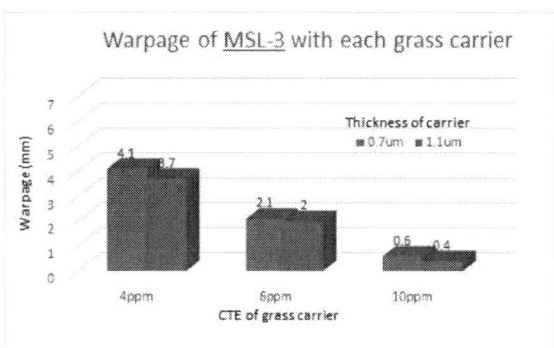

Fig.9 The warpage of MSL-3 with each glass carrier

Fig.9 shows that it was found that MSL-1 and 3 make almost same amount of warpage.

From the results of MSL experiences, lower modulus makes warpage smaller. And, Glass carrier which has smaller CTE makes larger warpage.

5-1-4. MSS-1 with each glass carrier

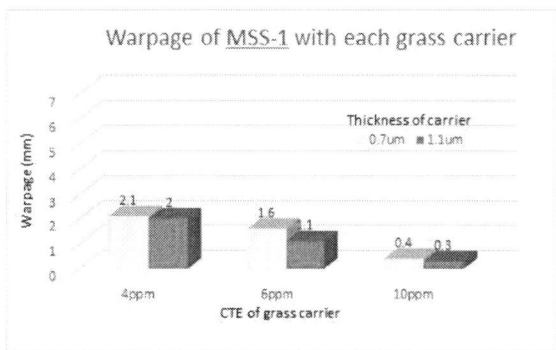

Fig.10 The warpage of MSS-1 with each glass carrier

Fig.10 shows that it was found that MSS-1 makes smaller warpage than MSLs. Even glass carrier which has 4ppm CTE, it makes around 2mm warpage.

5-1-5. MSS-2 with each glass carrier

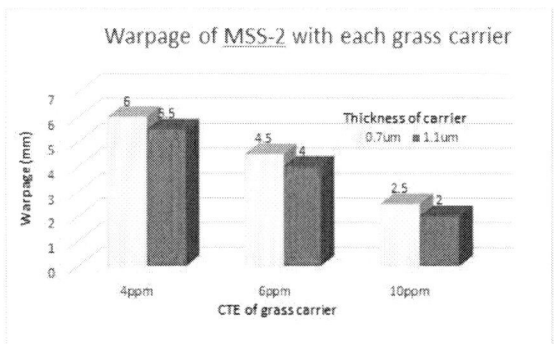

Fig.11 The warpage of MSS-2 with each glass carrier

Fig.11 shows that it was found that MSS-2 makes much larger warpage than MSS-1. The other trend is the same as MSS-1.

5-1-6. MSS-3 with each glass carrier

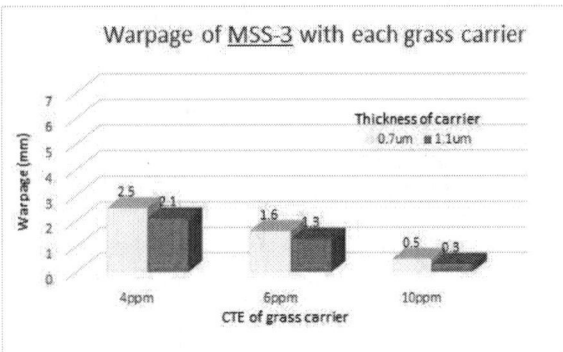

Fig.12 The warpage of MSS-3 with each glass carrier

Fig.12 shows that it was found that MSS-1 and 3 make almost same amount of warpage.

From the results of MSS experiences, a trend similar to MSL experiences was confirmed.

5-2. Process study combination with each Mold sheet and each molding equipment.

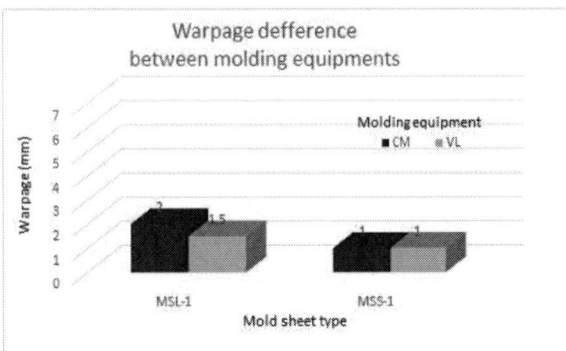

Fig.13 The comparison between Compression mold and Vacuum laminator.

Fig.13 shows that Compression mold and Vacuum laminator make almost same level warpage. As the value was larger, MSL-1 makes difference between Compression mold and Vacuum laminatour, but it seems in error range.

5-3. Process study combination with each Mold sheet and each curing condition.

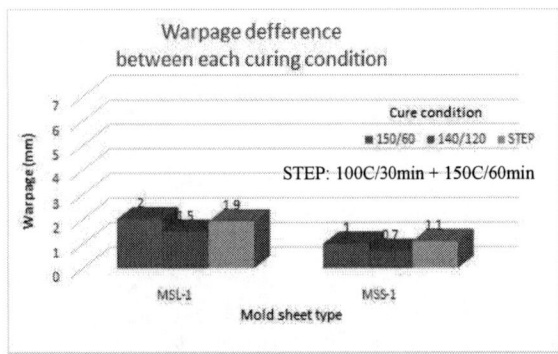

Fig.14 The comparison between each curing condition.

Fig.14 shows that. 150C/60min cure and step cure make almost same level warpage. If 150C cure scedule is included, the results are found to be about the same range of warpage. And, 140C cure

makes smaller warpage, but it is requierd long time curing schedule.

6. Conclusion

According to these results of experiences, following facts are revealed.

Whether it is MSL or MSS, the volume of warpage depends on the properties of the cured products, whether the raw material is liquid or solid. MSL-1 and 3 gave almost same warpage products, and MSL-2 made the largest warpage product in MSL.

From above, Tg of cured product does not matter for warpage. And the low modulus makes smaller warpage clearly. It is found that modulus of cured product has a greater influence on warpage of products than the CTE of them.

And next, in MSS, they have a trend similar to MSL was observed. They are the Tg does not matter for warpage, low modulus makes smaller warpage, and CTE of cured product has no greater influence on the warpage of cured product than their modulus has.

The thickness of glass carrier makes differ to warpage of cured products. The thinner one makes warpage larger. And, the CTE of glass carrier makes differ same as its thickness. The smaller CTE makes warpage larger. This matter seems to be caused by the difference between Mold sheet and glass carrier CTE characteristics.

And, about the molding equipment, researching about the difference between Compression mold and Vacuum lamination. It was observed that there is almost same results among the both equipment. Slightly, Compression mold makes larger warpage than Vacuum laminator. It was thought that the mold of Compression mold has excellent heat retention, it caused the degree of cure to increase more than Vacuum laminator and made larger warpage, in spite of all the same in mold condition and PMC condition.

Last, the difference about cure condition. They have difference in curing schedule. 150C/60min, 140C/120min, and 100C/30min + 150C/60min. According to the results, 140C/120min curing schedule made the smallest warpage in these 3 curing schedule. If 150C cure is included to curing schedule, the warpage results are found to be about the same range.

The knowledge gained from these experiments would give us great hint to develop and progress the new processes and materials which bring us low warpage products for PLP applications.

7. Acknowledgement

The authors would like to acknowledge Towa, Meiki, and AGC for their efforts and contributions to this paper.

8. Reference

[1] Jae Kyu Cho, Shan Gao, Sukeshwar Kannan, Bob Kuo, Miguel Jimarez, and SeokHo Na2 'Chip Package Interaction Analysis for 20-nm Technology with Thermo-Compression Bonding with Non-Conductive Paste' *2015 IEEE 65th Electronic Components and Technology Conference,* : IEEE, 2015, pp. 12-16.

[2] Kihyeok Kwon, Yoonman Lee, Junghwa Kim, Joo Young Chung, Kyunghag Jung, Yong-Yeop Park, Donghwan Lee, Sang Kyun Kim, 'Compression molding encapsulants for wafer-level embedded active devices' *IEEE 67th Electronic Components and Technology Conference, (ECTC)*, IEEE, 2017 Pages: 319 – 324

[3] Toshihisa Nonaka; 'Fan-Out Package that Has Entered the Second Stage Started Leading the System Integration ssembly Technology and Materials to Support This' *2017 Journal of The Japan Institute of Electronics Packaging*, JIEP, vol.20 No.1 2017 Pages: 43 – 47

Defect Detection for the TSV Transmission Channel Using Machine Learning Approach

Huan Liu, Runiu Fang
National Key Laboratory of Science and Technology on
Micro/Nano Fabrication
Peking University
Beijing, China

Min Miao*
Institute of Information Microsystem
Beijing Information Science and Technology University
Beijing, China
*E-mail: miaomin@bistu.edu.cn

Yang Yang, Yufeng Jin
Shenzhen Graduate School
Peking University
Shenzhen, China

Abstract—Through silicon via (TSV) is a key enabler for 3D integration technology to provide high-density and high-speed transmission channels. However, the serious defects including open circuit and short circuit will result in functional failure. In this paper, a method to identify the open and short defect and detect the defect position is developed based on machine learning approach. Taking S parameters as training dataset, the predictive model is trained and evaluated, the performance of models based on bagging method and boosting method are compared. The results demonstrate the predictive model could correctly classify the defect type and predict the defect position accurately.

Keywords- Through Silicon Via; transmission channel; machine learning; bagging method; boosting method

I. INTRODUCTION

Through silicon via (TSV) based 3D integration is a promising technology for the advantages including high density, small form factors and improved performance [1,2]. High-speed transmission channels can be achieved using TSV and redistribution layer (RDL).

As the devices scaling down, the dimension and spacing of transmission channel is decreasing, imposing challenges to manufacture. Significant manufacture defects such as short circuit and open circuit defects may lead to error of logical function and degrade the reliability of the system [3]. It is essential to detect these defects in early stage.

Machine learning may be a potential strategy for defect detection. With the improvement of computing power and resources, it is possible to support complex machine learning algorithm, enabling the application of machine learning in a variety of fields, such as image recognition, data mining and natural language processing. Machine learning has also shown great advantages in failure diagnosis and fault classification [4-5].

In this paper, we aim to propose a defect detection method, combining the electromagnetic characteristics of transmission channel and machine learning approach to identify the open and short defect and detect the defect position. This paper is organized as follows. In section II, the characteristics of transmission channel with defects are obtained by simulation. Section III is devoted to illustrate the machine learning approach adopted and compare the performance of predictive models based on different algorithms. Finally, we conclude this paper in Section IV.

II. MODELING THE TSV BASED SIGNAL TRANSMISSION CHANNEL CONTAINING DEFECTS

TSV is an important component in 3D integration, there have been extensive researches about TSV focusing on the fabrication, modeling and testing. In order to improve the electrical performance and thermo-mechanical reliability or ease the manufacture process, various TSVs with different structure design or material selection have been exploited. For instance, annular TSV [6] with polymer filled inside could alleviate the thermal stress. Tapered TSV [7] is consistent with the DRIE (deep reactive ion etching) process, the taper-shape sidewall is beneficial for bottom-up electroplating. Air-gap TSV [8] could decrease the signal delay by using air as insulator to obtain a lower insular capacitance. TSV with guard ring deployed around it [9] could attenuate the coupling noise between TSV and active devices. However, in dc and low-frequency field, cylindrical copper TSV is widely used due to qualified electrical property and heat management capability, therefore, it is considered as the TSV structure to study in this paper.

TSVs and RDLs could construct the transmission channel in which signal flows between the two sides of individual IC or among different ICs. Fig. 1 shows the transmission channel considered in this paper. It is in signal-ground configuration, and comprised of two channels, each of them is a TSV-RDL chain.

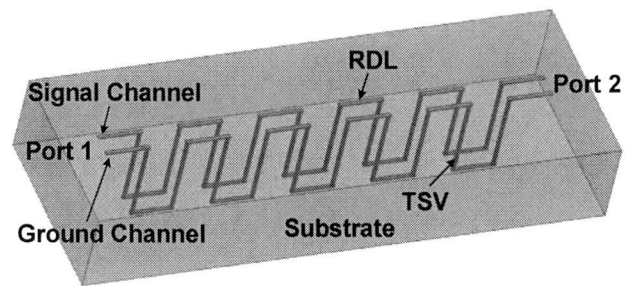

Figure 1. Schematic of the TSV based signal transmission channel which is in signal-ground configuration.

The dimension parameters of TSV and RDL are listed in Table. I. The TSV is 5 μm in radius, 200 μm in height. The length, width and thickness of the RDL are 100 μm, 10 μm and 2 μm respectively. The TSVs and RDLs are isolated from the silicon substrate with 0.1-μm SiO_2 insulation layer. The pitch between adjacent TSVs in signal channel or ground channel is 100 μm and the pitch between these two channels is 50 μm. Because the TSV-RDL chain is made up of 10 TSVs and 11 RDLs, the total length of the channel is 3100 μm.

TABLE I.   DIMENSION PARAMETERS OF THE TSV TRANSMISSION CHANNEL

Dimension parameters	Value
Radius of TSV, rtsv	5 μm
Height of TSV, htsv	200 μm
Length of RDL	100 μm
Width of RDL	10 μm
Thickness of RDL	2 μm
Thickness of SiO_2 insulation layer, tox	0.1 μm
Pitch between adjacent TSVs of individual channel, dtsv	100 μm
Pitch between signal channel and ground channel	50 μm

There may exist various types of defects in an 3D integration system due to imperfect manufacture. Open circuit and short circuit are two types of significant defects, which could cause complete failure of the channel. As shown in Fig. 2, the open defect may occur in TSV section or RDL section, denoted as TSV_open and RDL_open, it is represented by a disconnection of the metal. The short defect usually occurs between RDLs, denoted as RDL_short, it is represented by inserting a metal pad connecting the two adjacent RDLs.

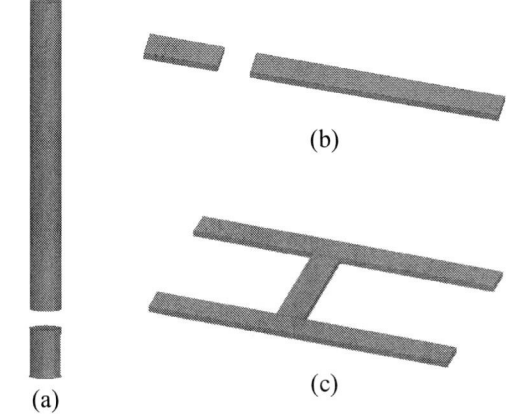

Figure 2.   Open and short defects in the transmission channel. (a) Open defect occurs in TSV section, TSV_open. (b) Open defect occurs in RDL section, RDL_open. (c) Short defect occurs in RDL section, RDL_short.

In frequency domain, the impact of defects on the signal transmission of the channel can be characterized by S parameters. By introducing open defect and short defect at different position in the channel, and modeling the channel with a full-wave simulator HFSS (High Frequency Structure Simulator), the frequency response can be obtained. The excitation and response can be input and monitored through Port 1 and Port 2 in Fig. 1. The sweeping range of frequency is from 0.1 GHz to 4 GHz with a 0.1-GHz-step, 40 frequency points in total. The S parameters of the channel with different defects are depicted in Fig. 3.

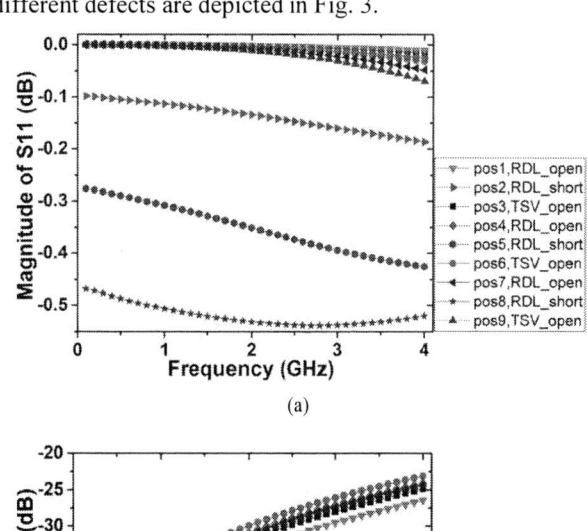

(a)

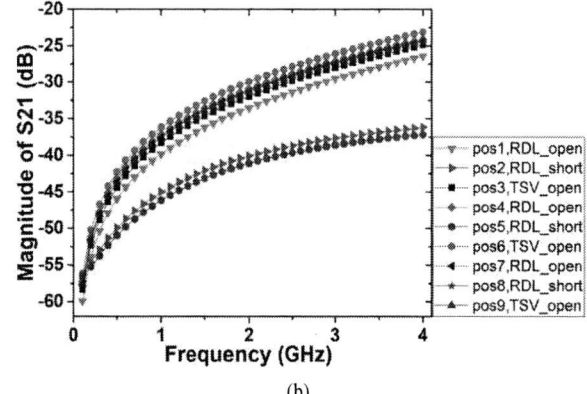

(b)

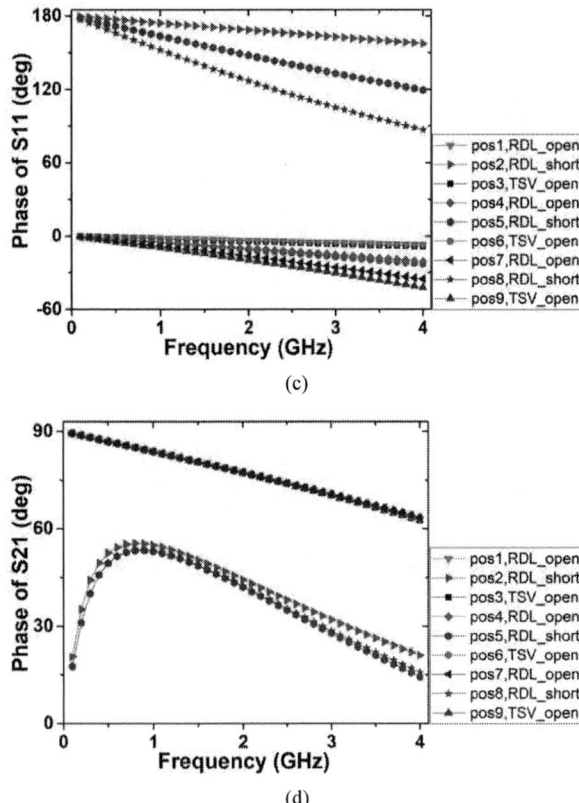

(c)

(d)

Figure 3. S parameters of transmission channel with different types of defects occur at different position. (a) Magnitude of S11 (mag_S11). (b) Magnitude of S21 (mag_S21). (c) Phase of S11 (phase_S11). (d) Phase of S21 (phase_S21).

A coordinate system could be established along the channel by assigning the coordinates of Port 1 and Port 2 are 0 and 3100 respectively. Then the coordinates of pos1 to pos9 in Fig. 3 are 329, 389, 433, 1238, 1244, 1369, 2146, 2155, 2557. It is found that as the coordinates of open defect increases, the magnitude of S11 and phase of S11 is decreases, this is caused by larger conductor loss and substrate loss for the excitation signal in Port 1 if the defect is further away from the Port 1. Similar trend can be observed for short defect. As for the magnitude and phase of S21, there is no explicit trends can be observed, because major source of signal from Port 1 to Port 2 arises from weak electromagnetic coupling, which is affected by many factors.

With a similar simulation procedure, S parameters of transmission channel with different defect types and location can be obtained. The data will be used in next section.

III. DETECTION OF DEFECTS USING ENSEMBLE MACHINE LEARNING APPROACHES

Machine learning could establish the relationship between input and output by learning rules from the data, this property is especially useful when the relationship is not explicit. In this section, a predictive model to identify the defect type and detect its position in the channel is developed based on ensemble machine learning approaches.

Various machine learning algorithms have been proposed in the past decades, such as decision tree, support vector machine, artificial neural network. Besides theses basic individual learner algorithms, stronger learners can be established by combining the weaker learners to obtain better performance. This is known as ensemble method, effective in minimizing the error due to noise, bias and variance. Ensemble method can be divided into two types, bagging and boosting, according to the difference in training process. In the bagging method, each learner is trained on a randomly selected subset of the training dataset, the learners are independent with each other. While in the boosting method, the training is carried out using all the samples of the training dataset in each training step, misclassified data will be assigned a larger weight, subsequent learners will focus on them [10]. Taking decision tree as the base learner algorithm, the concept of bagging and boosting method are depicted in Fig. 4. These two methods will be both applied to develop the predictive model in this paper.

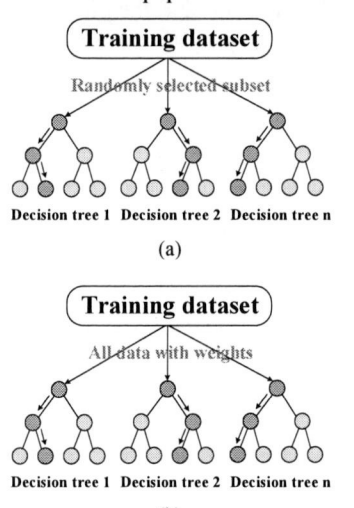

Figure 4. The concepts of bagging and boosting learning method. (a) Bagging method. (b) Boosting method.

The target to identify the defect type is a classification problem, while detect the position of the defect is a regression problem. The modeling procedure is based on the flow shown in Fig. 5. First, the training dataset is prepared, this is the inputs for the machine learning algorithm. In our case, every sample in the training dataset is composed of the S parameters obtained by simulation using HFSS, that is 160 variables is included in every sample. The desired outputs of predictive model are 0 or 1, representing open defect and short defect, in the classification, and a calculated value representing the defect position in the regression. To further validate the accuracy of the model, a testing dataset is prepared. In our case, there are 90 and 9 groups of samples in the training and testing dataset respectively.

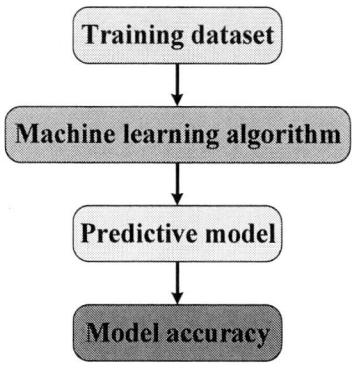

Figure 5.   The procedure to obtain the machine learning model.

To detect the position of the defect, one typical bagging method, random forest (RF) is adopted. In the training procedure of RF, about two thirds of the samples in the training dataset (konwn as in-bag samples) are used to train the trees, and the remaining one third (known as out-of-bag samples) are used for internal cross-validation to evaluate the performance of the model [11].

There are two critical parameters for the RF algorithm. The first one is the number of trees, N_{tree}, usually higher performance can be obtained with a larger N_{tree}, but the consumed computation resource is also larger. The second one is the minimum leaf size, that is the minimum number of samples in the end node of the decision tree. If the minimum leaf size is too large, accuracy of the model will be lowered, while too small minimum leaf size will make the model more prone to be affected by the noise in the training dataset. To determine the optimal value of the two parameters, the out-of-bag mean squared error (MSE) is calculated and shown in Fig. 6.

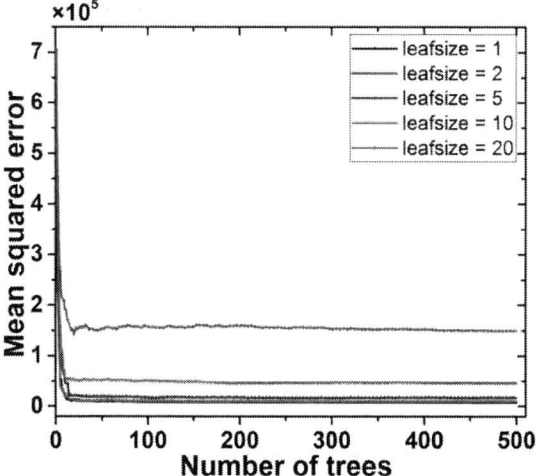

Figure 6.   Mean squared error of the RF model with different number of trees and different minimum leaf size.

While the minimum leaf size decreasing from 20 to 1, the MSE is decreasing, that is the accuracy of the model is increasing. The performance of the model is comparable when the minimum leaf size equals 1 and 2, therefore, the optimal minimum leaf size is set to 2. Similarly, the optimal

number of trees can be determined as 200. The percent error between predicted position and real position of the defect are defined in (1). The model based on RF algorithm shows a maximum percent error on the training dataset of 20%.

$$\text{Err}_{pos} = \left| \frac{\text{pos}_{pred} - \text{pos}_{real}}{\text{pos}_{real}} \right| \times 100\% \qquad (1)$$

A predictive model is also developed with the boosting method. The technique known as k-fold cross-validation is used to estimate the performance of the model on the given training dataset [12]. In the k-fold cross-validation, the training dataset is randomly divided into k folds of subsets, one fold of subset is used as validation data and remaining k-1 folds are for training, this process will be run k times with each subset used exactly once as the validation data. Then the cross-validated mean square error can be calculated, the results are shown in Fig. 7, and the optimal number of trees is determined as 300.

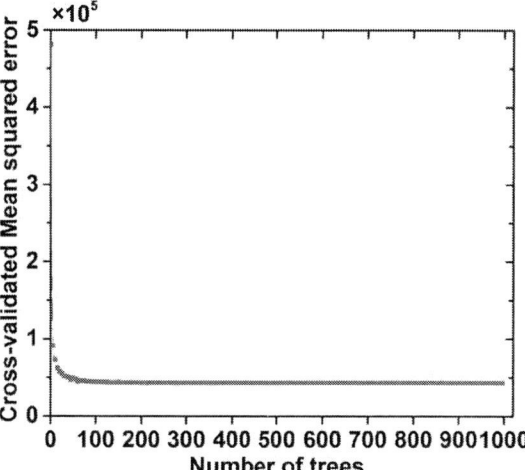

Figure 7.   Mean squared error of the boosting model with different number of trees.

The percent error of defect position on training dataset is calculated, the maximum value is about 8%. The comparison between predicted position and real position on a few samples of the testing dataset is listed in Table. II, the accuracy is well corelated with that on training dataset, demonstrating the model could generalize well to new data. Compared to the model based on RF, the boosting method shows a better performance, therefore it is more suitable for this case. Then a classification model using boosting method is developed to identify the defect type, it shows satisfactory performance on training and testing dataset, classifying the defect type correctly for all samples.

TABLE II. COMPARISON OF THE PREDICTED POSITION USING BOOSTING METHOD AND THE REAL POSITION OF DEFECTS

Defect type	Real position	Predicted position	Percent error
Open	577	596.4	3.4%
Open	1723	1701.3	1.3%
Open	2233	2212.2	0.9%
Short	389	404.1	3.9%
Short	1279	1260.4	1.7%
Short	2792	2775.7	0.6%

IV. CONCLUSION

Open and short defects are major severe defects in the TSV based signal transmission channel. In this paper, we propose a method based on machine learning approach to identify the defect type and detect defect position in the transmission channel. The training process is conducted using S parameters of the channel as dataset. The predictive models based on bagging and boosting learning methods are developed and compared, and it is found that boosting method is more suitable. The model could correctly classify the defect type and accurately predict the defect position with a percent error less than 8%.

ACKNOWLEDGMENT

This work is co-funded by the National Basic Research Program of China (No. 2015CB057201), the National Natural Science Foundation of China (No. 61176102, No. 61674016 and No. U1537208), the Importation and Development of High-Caliber Talents Project of Beijing Municipal Institutions (Great Wall Scholar, No. CIT&TCD20150320), and Beijing Nova Program Interdisciplinary Studies Cooperative Projects (No. Z161100004916036).

REFERENCES

[1] J. U. Knickerbocker, P. S. Andry, B. Dang, et al, "Three-dimensional silicon integration," IBM Journal of Research and Development, vol. 52, no. 6, pp. 553-569, 2008.

[2] Y. Civale, D. S. Tezcan, H. G. G. Philipsen, et al, "3-D wafer-level packaging die stacking using spin-on-dielectric polymer liner through-silicon vias," IEEE Transactions on Components, Packaging and Manufacturing Technology, vol. 1, no. 6, pp. 833-840, 2011.

[3] D. H. Jung, Y. Kim, J. J. Kim, et al, "Through silicon via (TSV) defect modeling, measurement, and analysis," IEEE Transactions on Components, Packaging and Manufacturing Technology, vol. 7, no. 1, pp. 138-152, 2017.

[4] H. Fang, K. Chakrabarty, Z. Wang, et al, "Diagnosis of board-level functional failures under uncertainty using Dempster-Shafer theory," IEEE Transactions on Computer-Aided Design of Integrated Circuits and Systems, vol. 31, no. 10, pp. 1586-1599, 2012.

[5] H. Livani, C. Y. Evrenosoğlu, "A fault classification and localization method for three-terminal circuits using machine learning," IEEE Transactions on Power Delivery, 2013, 28(4): 2282-2290.

[6] F. Wang, Z. Zhu, Y. Yang, et al, "Analytical models for the thermal strain and stress induced by annular through-silicon-via (TSV)," IEICE Electronics Express, vol. 10, no. 20, pp. 20130666, 2013.

[7] X. Chen, J. Tang, G. Xu, et al, "Development of deep reactive ion etching and cu electroplating of tapered via for 3d integration," IEEE Electronic Packaging Technology and High Density Packaging (ICEPT-HDP), 2011, pp. 1-3.

[8] Y. Civale, S. V. Huylenbroeck, A. Redolfi, et al, "Via-middle through-silicon via with integrated airgap to zero TSV-induced stress impact on device performance," IEEE Electronic Components and Technology Conference (ECTC), 2013, pp. 1420-1424.

[9] J. Cho, E. Song, K. Yoon, et al, "Modeling and analysis of through-silicon via (TSV) noise coupling and suppression using a guard ring," IEEE Transactions on Components, Packaging and Manufacturing Technology, vol. 1, no. 2, pp. 220-233, 2011.

[10] Mariana B., Lucian D, "Random forest in remote sensing: A review of applications and future directions," ISPRS Journal of Photogrammetry and Remote Sensing, vol. 114, pp. 24-31, 2016.

[11] Breiman, L., et al. Classification and Regression Trees. Chapman & Hall, Boca Raton, 1993.

[12] Y. Bengio and Y. Grandvalet, "No unbiased estimator of the variance of K-fold cross-validation," Journal of Machine Learning Research, vol. 5, pp. 1089–1105, 2004.

Direct Printing of Heat Sinks, Cases and Power Connectors on Insulated Substrate Using Selective Laser Melting Techniques

Rabih KHAZAKA*, Donatien MARTINEAU, Toni YOUSSEF, Thanh Long LE, Stéphane AZZOPARDI

Safran SA, Safran Tech, Department of Electrical and Electronic Systems,
78114 Magny les Hameaux, France
*rabih.khazaka@safrangroup.com

Abstract—The market introduction of high temperature wide bandgap electronics power semiconductor devices with junction temperature exceeding 200°C significantly accelerates the trend towards high power density and severe ambient temperature electronics applications. Such evolution may have a great impact in aeronautics applications, especially with the development of More Electric Aircraft (MEA), since it can allow to reduce the mass and volume of the power electronics systems. As a consequence, the aircraft operating cost can decrease. However, for electronics used under such harsh conditions, heat evacuation is a very critical issue for the operation and long-term reliability of power modules. Among materials used in the power assembly, Thermal Interface Material (TIM) plays a significant role in improving the thermal contact resistance between the power module package and the heat sink. However, TIM suffers from its high thermal resistance in case of soft material use (thermal grease, phase changing films, elastomers) and from high thermomechanical stresses in case of solder use. This paper focuses on a new approach allowing direct printing of high performance heat sink on the back side of the insulated ceramic substrate leading to the removal of the TIM. In addition, the module case and the electrical connectors can also be built-up on the upper face of the insulated ceramic substrate. Selective Laser Melting (SLM) process is used to achieve complex three dimensional structures with AlSi7Mg0.6 powder alloy on the both sides of direct bonded aluminum substrate. Using this bi-material substrate involves the development of a specific tray to allow ceramic stress relaxation during process. Various heat sink designs including lattice structure, array structures (pin fins, rectangular fins, elliptical fins and water drop shaped fins) as well as channel and cold plate structure were printed. It has been shown that the design has a strong impact on the residual stresses induced during the process, and the latter can induce in some cases, significant substrate warpage and even cracks in the ceramic. Based on the experimental results, design recommendations allowing the reduction of the residual stresses in the structure are briefly introduced. Shear strength measurements have been performed to evaluate the adhesion between the built material and the substrate metallization and shear strength values higher than 20 MPa have been obtained illustrating a good interfacial joint. Finally, the thermal performance of the air cooled direct printed heat sinks was evaluated using thermo-fluidic models and results have been compared to conventional assembly with TIM.

Keywords- Selective laser melting, insulating substrate, heat sink, electronic cooling, power electronics, metal additive manufacturing

I. INTRODUCTION

The aircraft industry has committed to the reduction of gas emissions (CO_2, NO_x), noise and travel costs with electric planes, as shown by the growing investment in lightweight propulsion systems to bring the benefits of electric-cars to air transport. The MEA technology is implemented by replacing the mechanically driven engine accessories (oil pump, fuel pump, hydraulic pump, etc.) as well as pneumatic and hydraulic systems with electrically driven versions. Power electronics control is often based on power electronics converters with power semiconductor devices assembled in power electronics modules. The power module ensures electrical connections, heat management and mechanical robustness. In order to achieve mass and volume gain, an increasing interest to use high density integrated power electronics modules in harsh environment has emerged. As a consequence, higher amount of heat (per square centimeter) needs to be dissipated. The performance and the reliability of the electronics components are directly related to the electronics operating temperature and the device reliability is significantly reduced by increasing the temperature [1]. Therefore, high performance of the cooling systems is requested to insure the operation capability and reliability of the electronic systems [2].

The topology optimization of the heat sink design allows to generate geometry based on a fixed goal such as minimizing of the device temperature and/or the pressure drop in the heat sink. The optimized patterns can be complex to fabricate using traditional machining techniques while metal additive manufacturing techniques can give new opportunities in this field. Accordingly, the use of additive manufacturing to achieve optimized heat sink designs has been shown to be a growing area of research in the close past several years [3-5]. Conventional power module can be thermally coupled to the high performance additive manufacturing heat sinks to improve the thermal management.

In conventional power module, in order to ensure a good thermal dissipation, the metallized substrate is attached to the heat spreader using solder. On one hand, large area joints suffer from high thermomechanical stresses and present limited reliability under harsh thermal cycles. On the other hand, the heat spreader is thermally coupled to the heat sink using TIM. TIMs like thermal grease, phase changing films and elastomers suffer from their high thermal resistance and limited stability at high temperatures [6].

978-1-7281-1500-9/19 $31.00 © 2019 IEEE

In order to bypass the aforementioned limitations of conventional power modules and to take the full advantages of additive manufacturing technologies, our approach detailed in this paper consists in directly printing heat sinks using SLM technique on the back side of the metallized substrate. In addition to the heat sink fabrication, the same technique is used to print the electrical connectors and the case on the upper face of the same metallized substrate. Various conventional and less-conventional extended surfaces have been printed on the substrate and evaluated using numerical methods. The use of extended surface (pin fins, rectangular fins, elliptical fins and water drop shaped fins) can be cooled using air or liquids as heat carrying medium while the direct printed channel are more compatible with liquid cooling. In this paper, particular attention has been paid to the forced air cooled heat sinks, since it is the most widely used cooling system due to its simplicity, reliability and low cost.

II. EXPERIMENTAL TECHNIQUES

A. Selective laser melting

The apparatus used to fabricate the metallic patterns is the SLM 280HL 400W with a 280 x 280 x 365 mm build envelope and a high power laser beam profile. A stainless steel tray has been designed to welcome and bridle 15 ceramic metallized substrates. The Direct Bonded Aluminum (DBA) metallized substrates are commercially available with 1 mm thick AlN ceramic and 400μm thick Al on both sides. The DBA substrate has been chosen for its good thermal properties and thermomechanical reliability [7]. A specific pattern is achieved on the top face in order to achieve the electrical circuit while the Al covers all the surface of the bottom face unless a distance of 1mm from ceramic borders. Once the DBA substrate in inserted in the tray, a planar surface is obtained between the stainless steel tool and the upper metallization of the substrate allowing the spreading of a controlled 50 μm thick layer of metallic AlSi7Mg06 powder on the surface of the substrates. The particles size is in the range of 20 to 60 μm. A laser, with 350W electrical power, scans the pattern according to the 3D model slicer data. The process takes place in argon (Ar) atmosphere to avoid the material oxidation and to prevent fire. The substrate is heated by the tray at a temperature of 150°C during the whole printing process to reduce the residual stresses in the final assembly. The 3D geometry is realized layer by layer. The laser beam locally melts the powder which is quickly solidified to form the slice. Then, the tray is lowered inside the machine and another powder layer is deposited and locally melted. Those steps are repeated until printing the final object. At the end, the tray is cooled down at room temperature and the powder covering the whole printed object is removed, recycled and re-used in the next run. Figure 1 presents the printed objects in the SLM equipment after the partial remove of the powder.

Figure 1. Some printed objects in the SLM equipment partially covered with the AlSi7Mg06 powder.

B. Heat sink designs

The heat sink was directly fabricated on the DBA metallized substrate with 48mm length and 42 mm width as it can be seen in figures 2 and 3. The pin fins array and the rectangular fins array can be realized using traditional manufacturing process. On the other hand, the fabrication of the lattice structure, elliptical and water drop shape arrays and the liquid channel can be complex or impossible, unless additive manufacturing processes are used. Those complex structures can serve as examples to show how the additive manufacturing process can be used to achieve high surface area to volume heat sinks (lattice structure) and complex shapes that allow higher heat transfer with lower pressure drop (water drop shape and elliptical arrays). Moreover, the fabricated structures by SLM technique are advantageously rough and can lead to an enhancement in the heat transfer [8]. As shown in figure 3, electrical connectors and case can also be printed using the SLM technique on the other side of the DBA substrate.

Figure 2. Example of heat sinks printed on the DBA substrates a) lattice structure, b) elliptical fins, c) water drop shape fins, d) and e) rectagular fins and f) pin fin array.

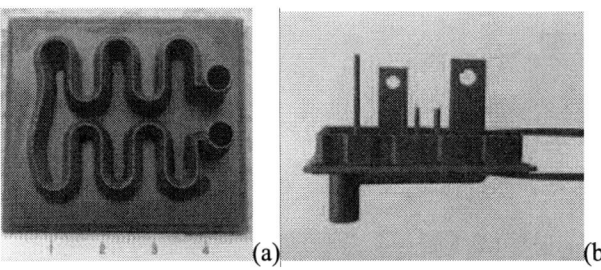

Figure 3. a) Liquid cooling channel with short continuous straigth printing length and b) heat sink, case and connectors printed on both side of a DBA substrate.

III. NUMERICAL METHODS

A 1200V -50A phase leg power module, composed by two switching cells electrically connected in series, has been used in this part with one SiC MOSFET transistor and one SiC Schottky diode per switching cell. The power losses within the MOSFET and the diode were considered to be 40W and 20W respectively. The detailed material properties and thicknesses of the assembly and the cooling fluid (air) are listed in Table I.

TABLE I.   MATERIAL PROPERTIES USED IN THE SIMULATION

	Thickness (mm)	λ (W/mK)	Cp (J/kg.K)	ρ (kg/m³)
SiC	0.2	300	510	4360
Solder	0.1	30	600	9000
Al	0.4	202	871	2719
AlN	1	150	400	3300
AlSiMg	29	120	871	2700
TIM	0.1	2	1000	2000
Air	30	0.025	1006	1

Hereafter, the heat sink geometries and dimensions used in the numerical evaluation are detailed in figure 4 and Table II. The reference heat sink is used in the simulation to compare the thermally coupled substrate to independent heatsink using TIM with the direct printed pattern on the substrate. The base of the reference heat sink is considered to be 0.8 mm thick. The thermal coupling between the reference heat sink and the DBA substrate is ensured by a 100 µm thick thermal grease with a thermal conductivity of 2 W/mK. In order to have a fair comparison between the whole patterns, the heat sink mass slightly varies around 12.1 g (±0.2 g) and the global occupied volume was kept constant.

TABLE II.   HEAT SINK DESING PROPERTIES (FOR EACH HEAT SINK PATTERN a, b AND c ARE INDICATED IN FIGURE 4)

Geometry	Pin fin +base (ref)	Pin fin	Water drop	Rect.	Ellipse
a (mm)	2	2	4	3.7	4
b (mm)			2	2	2
c (mm)	2	2	2	2	2
Number of pins or cell	32	49	32	21	24
Extended heat transfer area (mm²)	5923	9070	8947	7098	7190
Heat sink mass (g)	11.9	12	11.9	12.2	12.3
Heat transfer area/ mass (mm²/g)	498	756	752	582	582

The inlet air coolant temperature has been fixed to 20 °C. The density and heat capacity were 1 kg/m³ and 1006 J/kgK respectively. To reduce the model complexity, some assumptions have been considered: the natural convection, the radiation as well as the roughness of the surface of the heat sink have been neglected. In addition, the variation of the air properties as function of temperature has not been taken into consideration. The 3D steady governing equations for conjugate heat transfer based on Navier-Stokes equations combined with continuity equation, energy equation and momentum equations were solved using ANSYS® Fluent implicit Solver. The turbulent k-ε standard model with enhanced wall treatment was used. The convergence criteria for the residual x, y and z direction velocity and the residual of energy equation were set to 10^{-3} and 10^{-6} respectively. The heat transfer coefficient was calculated using the following equation:

$$h = \frac{C_p m'(Tout - Tin)}{S\Delta T_{lm}} \quad (1)$$

Where Cp is the air heat capacity, m' is the mass flow rate, $Tout$ and Tin are the outlet and inlet temperature respectively, S is the heat transfer surface area and ΔT_{lm} is the log mean temperature difference defined as follow:

$$\Delta T_{lm} = \frac{(Tin - Twall) - (Tout - Twall)}{\ln\left(\frac{Tin - Twall}{Tout - Twall}\right)} \quad (2)$$

Twall is the mean temperature of the surface in contact with the coolant.

The impact of the heat sink geometry on the maximal junction temperature, the heat transfer coefficient and the pressure drop have been evaluated for air flow rate ranking from 10^{-3} up to 0.02 kg/s.

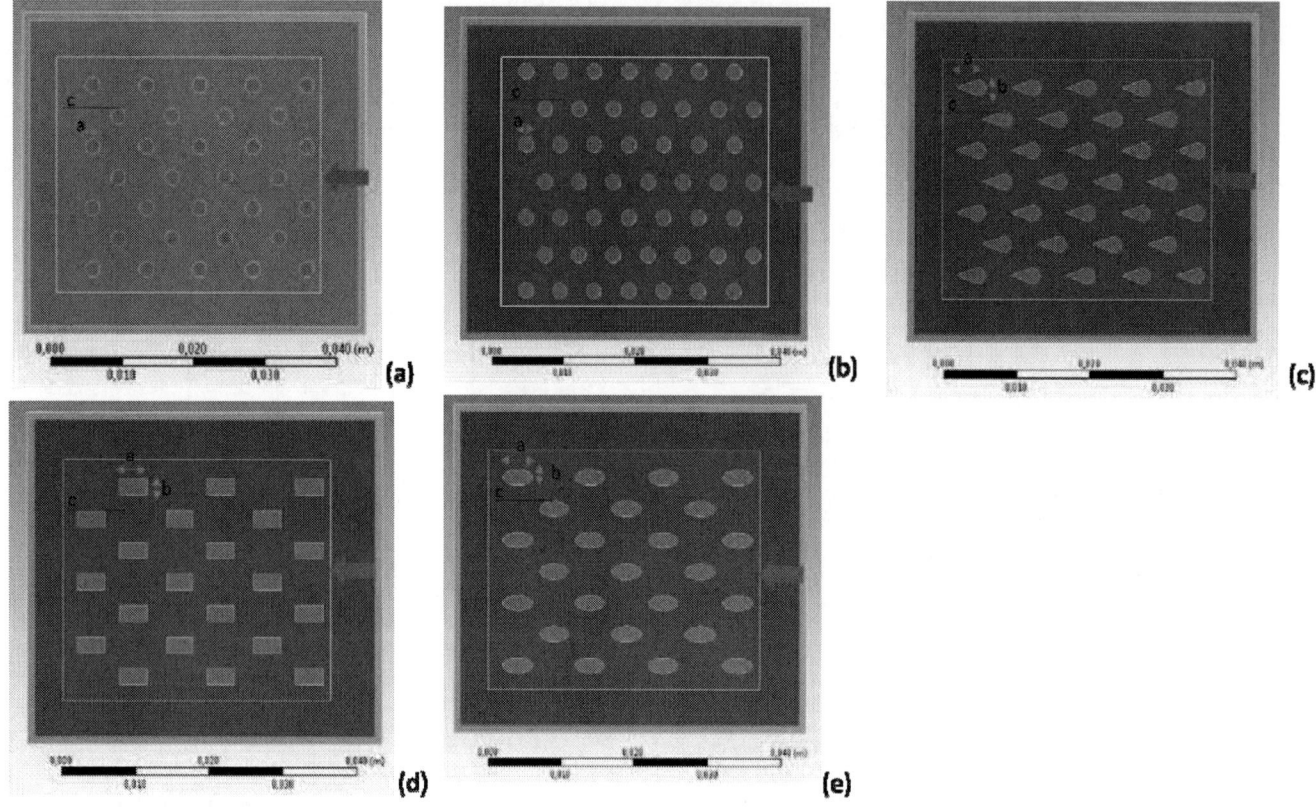

Figure 4. Schematic top view of the heat sinks used in the thermo-fluidic simulation a) the reference used heat sink with 0.8mm thick base and pin fins, b) pin fins, c) water drop shape fins, d) rectangular fins, e) elliptical fins. Blue arrows present the air flow direction.

IV. RESULTS AND DISCUSSION

A. Pattern impact on residual stress

From the beginning of the metal additive manufacturing, residual stresses are considered as a main issue that has to be encountered. In our case, this issue takes an extreme importance because the ceramic of the DBA substrate is brittle and high mechanical stresses can induce the ceramic cracking. The impact of various parameters of the SLM process (laser power, scanning speed, powder layer thickness, preheating and the scanning strategy) on the residual stresses has been detailed in [9]. In our case, we have preheated the tray and the DBA substrates at 150 °C before the beginning of the printing process in order to reduce the residual stresses. The decrease of residual stresses by preheating the substrates is explained by the decrease of the thermal gradient as well as the elastic modulus of the material. Figure 5 shows printed heat sinks with cracks in the ceramic while no crack appears in the heat sinks presented in figure 2. It was clear that when the pattern presents long linear printed vectors, substrates are highly curved and present cracks in the ceramic and even in the Al metallization in the worst cases. In fact, the decrease of the length of scanning vectors plays a crucial role in the reduction of the residual stresses. This fact comes from the

combination of several favorable mechanisms like 1) the local division and redistribution of the stresses in the x and y direction once the melted metal solidifies, and 2) a better distribution of the heat that serves as local preheating and leads to a reduction in the thermal gradient.

Figure 5. Some printed heat sink patterns with cracks apparition in the DBA substrate. From left to right: cold plate, lattice with large surface with the substrate, channels with straight long length and long length rectangular fins.

Accroding to the aforementioned results, the fin array heat sinks patterns and the lattice with short continuous linear length on the substratre are recommended to avoid the substrate bowing and the ceramic cracking. However, in some cases (e.g. to print the case of the module), the use of continuous pattern is necessary. In such cases, continuous crooked line pattern can be successfully used as presented in figure 6.

Figure 6. Cases fabricated with continuous straight lines and crooked lines. The substrate with straight lines case shows cracks in the ceramic (pointed by red arrows) while no cracks appear with the used of crooked lines.

B. Adhesion at the interfaces

Specific samples have been fabricated to evaluate the adhesion between the additive manufacturing materials with the Al metallization of the DBA substrate. Cylindrical pins with a diameter of 350 µm and a thickness of 1 mm have been directly printed on the substrate and tested. The test equipment consists in a load-applying instrument with an accuracy of ±5 % of full scale. A linear motion force is used and the direction of the applied force is parallel with the plane of the substrate and perpendicular to the fins. The shear tool has been set to 50 µm height over the substrate and the shear test velocity has been fixed to 100 µm/s. The values of the applied pressure on the 8 tested samples are presented in figure 7. The type of failure resulting from the application of the force is due to the deformation of the pins. Then, the tool gets over the deformed pins without any deadhesion. Accordingly, the real shear strength values can be considered to be above the mentioned ones (> 20 MPa in the worst case) which remain very satisfying for our applications.

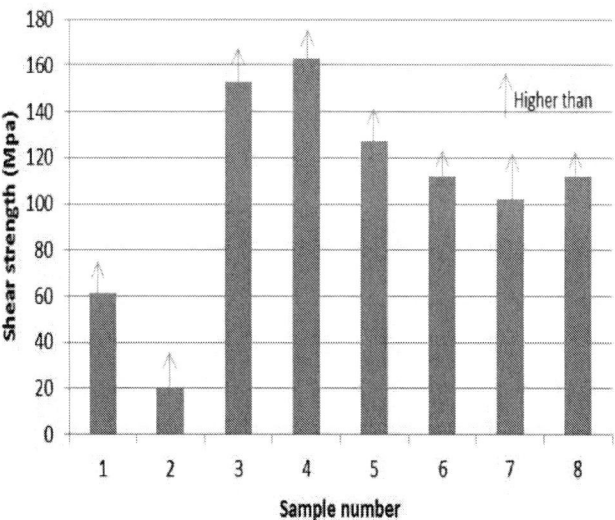

Figure 7. Shear strength applied values tested on 8 samples. None of the sample has been teared off.

C. Thermofluidic simulations

The simulated temperature distribution in the power electronics assembly for the conventional package (TIM and reference heat sink) and the proposed direct printed pin fins on the substrate is illustrated in figure 8. For the two above mentioned cases, the junction temperature can be reduced by 110 °C for an air flow rate of 0.001 kg/s and 30 °C for the flow rate of 0.02 kg/s (see figure 9). The cooling performance improvement when direct heat sink printing is used can be explained by the complete elimination of the low thermal conductivity TIM and the increase of the heat transfer area at isomass heat sinks. The soft TIM elimination is also helpful for the use of the electronic power module at high temperature environments.

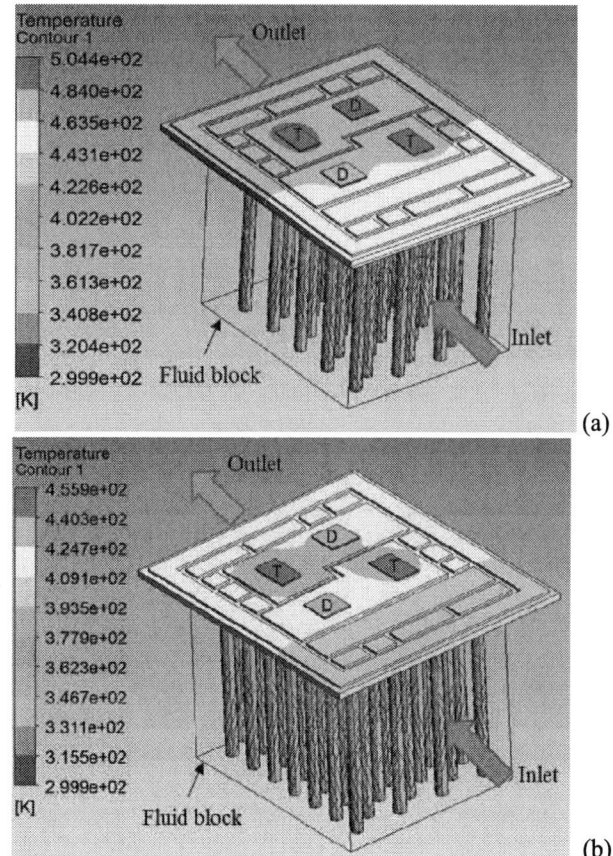

Figure 8. Temperature distribution in the assembly for the reference pin fins heat sink with TIM (a) and the direct printed pin fins (b).

Comparing the direct printed heat sinks presented in figure 4, pin fins presents the lowest junction temperature for a mass flow rate of 0.001 kg/s while for higher flow rate, the water drop shape fins array offers the best results. By considering a maximum junction temperature of 448 K or 175 °C (data sheet of SiC MOSFET), the reference heat sink cannot be used for mass flow rates lower than 0.02 kg/s while for the best case, the water drop shape fins array can be used when the mass flow rate is higher than 0.008 kg/s.

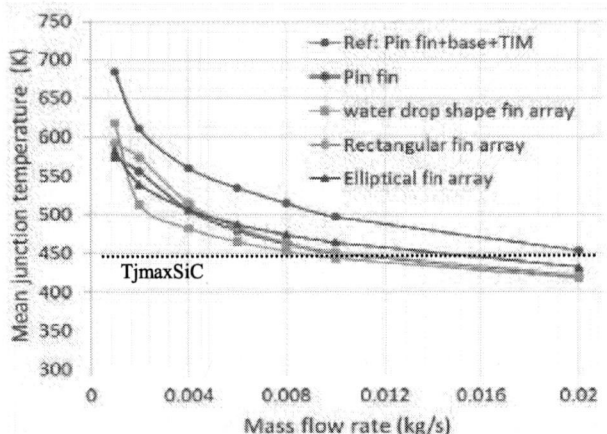

Figure 9. Mean junction temperature of the transistors as function of the mass flow rate for the various simulated patterns.

The heat transfer coefficients (calculated using equation 1) for the various heat sink patterns are presented in figure 10. The heat transfer coefficient for the reference heat sink is higher than the direct printed pin fin in the whole mass flow rate range. This can be explained by the higher wall temperature in contact with the air that promotes the heat convection. The heat transfer area for elliptical and the one for rectangular fins array are very similar. However, the heat transfer coefficient is shown to be lower for the elliptical fin array. The heat transfer performances start diverging at a mass flow rate levels above 0.002 kg/s. This value corresponds to a Reynold number of about 3000 which can be placed above laminar/turbulent transition for strip fins [10, 11]. This suggests that the Vortex shedding obtained for the rectangular fins at around this flow rate level does not take place for the case of streamlined elliptical fins. In reference [4] and [12], the authors have confirmed that the onset of significant vortex shedding coincides with a significant improvement in convection transport.

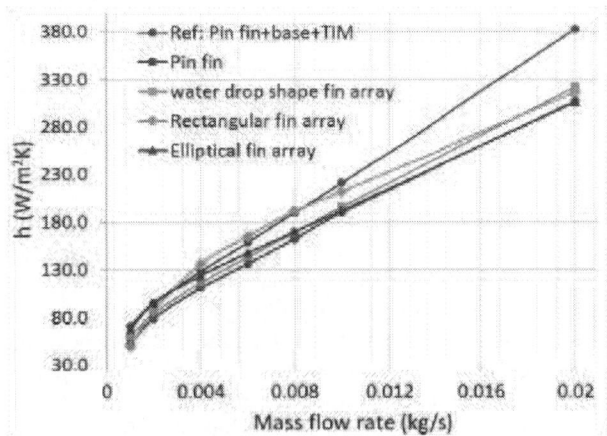

Figure 10. Heat transfer coeficient as function of the mass flow rate for the various simulated heat sink patterns.

The pressure drop versus the flow rate across the six simulated heat sink patterns is presented in figure 11. The superior aerodynamic profile of elliptical and water drop shape array can explain the reduction of the pressure drop across the heat sink using those shapes of offset strip fins even with large contact area between the solid and the fluid domain. In fact, the rounded shape gives a lower form drag that aids the pressure recovery in the wake of each fin. In order to compensate the pressure drop, a higher pumping power should be used which leads to an increase of the mass, the volume and the cost of the cooling systems. Accordingly, for the pin fins and rectangular fins heat sinks that allow to achieve low junction temperature above 0.008 kg/s, the high pressure drop can jeopardize the performance of the cooling systems. The water drop shape consists in the best tradeoff between the junction temperature and the pressure drop while elliptical fins could be penalized by the higher junction temperature. The complex design of the water drop shape geometry clearly illustrates the interest of the additive manufacturing technology to achieve patterns with particular forms that remains complex to achieve using conventional machining technologies. However, further investigations are required to optimize the heat sinks design (by changing the fins dimensions and the spacing between them, etc...), and to validate the simulation results with experimental ones.

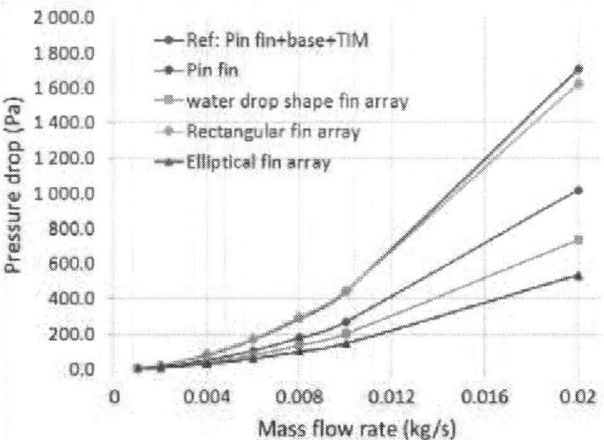

Figure 11. Pressure drop as function of the mass flow rate for the various simulated heat sink patterns.

V. CONCLUSIONS

The rapid metal additive manufacturing technique based on the Selective Laser Melting (SLM) has been used to directly print various shapes of heat sinks on a direct bonded aluminum metallized substrate. The used technology shows the ability to build patterns on both side of the substrate allowing the printing of the heat sink, the electrical connectors and the case on the same substrate. The laser melted AlSi6Mg0.7 alloy presents a good adhesion to the Al metallization of the DBA substrate with shear strength values higher than 20 MPa. The control of residual stress in the assembly has been pointed out to be mandatory. Designs with long straight lines on the substrate lead to high residual stresses in the structure that cause the substrate warpage and the ceramic fracture in most of cases. Hence, it is recommended to use fin array heat sink designs with non-continuous patterns or continuous pattern with crooked lines if necessary. The elimination of thermal interface material by direct printing the heat sink on the substrate allows the improvement of the junction maximum temperature by at least 30 °C in the mass flow rate range from 0.001 to 0.01 kg/s. Thermo-fluidic simulations have shown that the increase of the heat transfer surface is not sufficient to reduce the thermal resistance of package (e.g. pin fins array). The water drop shape heat sink presents the best heat sink performance with low thermal resistance and moderate pressure drop, while the conventionally used pin fins and rectangular fins heat sink suffers from the high pressure drop along them.

REFERENCES

[1] R. Tummala, Fundamentals of Microsystems Packaging, McGraw-Hill, New York, 2001.

[2] A. Sakanova, C. F. Tong, A. Nawawi, R. Simanjorang, K.J. Tseng, A.K. Gupta, Investigation on weight consideration of liquid coolant system for power electronics converter in future aircraft, Applied Thermal Engineering, Vol. 104, pp. 603–615, 2016.

[3] E. M. Dede, S. N. Joshi, F. Zhou, Topology Optimization, Additive Layer Manufacturing, and Experimental Testing of an Air-Cooled Heat Sink, Journal of Mechanical Design, Vol. 137, 2015, DOI: 10.1115/1.4030989

[4] M. Wong, I. Owen, C.J. Sutcliffe, A. Puri, Convective heat transfer and pressure losses across novel heat sinks fabricated by Selective Laser Melting, International Journal of Heat and Mass Transfer, Vol. 52, pp. 281–288, 2009

[5] A. Syed-Khaja, A. Perinan Freire, C. Kaestle, J. Franke, Feasibility Investigations on Selective Laser Melting for the Development of Microchannel Cooling in Power Electronics, IEEE 67th Electronic Components and Technology Conference proceeding, 2017.

[6] R. Skuriat, J.F. Li, P.A. Agyakwa, N. Mattey, P. Evans, C.M. Johnson, Degradation of thermal interface materials for high-temperature power electronics applications, Microelectronics Reliability, Vol. 53, pp. 1933–1942, 2013.

[7] R. Khazaka, L. Mendizabal, D. Henry, R. Hanna, "Survey of high-temperature reliability of power electronics packaging components", IEEE Transactions on Power Electronics, Vol. 30, No. 5, 2015.

[8] L Ventola, E Chiavazzo, F Calignano, D Manfredi, P Asinari, Heat Transfer Enhancement by Finned Heat Sinks with Micro-structured Roughness, Microtechnology and Thermal Problems in Electronics, Journal of Physics: Conference Series 494, 2014. doi:10.1088/1742-6596/494/1/012009.

[9] Bey Vrancken, Study of residual stresses in Selective laser melting, PhD thesis at KU Leuven, Faculty of enginnering science, 2016.

[10] N.C. DeJong, L.W. Zhang, A.M. Jacobi, S. Balachandar, D.K. Tafti, A complementary experimental and numerical study of the flow and heat transfer in offset strip-fin heat exchangers, Transactions of the ASME: Journal of Heat Transfer, Vol. 120, pp. 690–698, 1998.

[11] H.M. Joshi, R.L. Webb, Heat transfer and friction in the offset strip fin heat exchanger, International Journal of Heat and Mass Transfer, Vol. 30, pp. 69–84, 1987.

[12] R.S. Matos, J.V.C. Vargas, T.A. Laursen, F.E.M. Saboya, Optimization study and heat transfer comparison of staggered circular and elliptic tubes in forced convection, International Journal of Heat and Mass Transfer, Vol. 44, pp. 3953–3961, 2001.

Server CPU Package Design Using PoINT Architecture

Arun Chandrasekhar[+], Vijaya Boddu[*], Erich Chuh[*], Krishna Bharath[**], Farzaneh Yahyaei-Moayyed[**], Srikrishnan Venkataraman[+], Sriram Srinivasan[**], Ram Viswanath[**], Ritesh Jain[***], Huthasana Kalyanam[***]

[+] Intel India Technologies Pvt. Ltd., Bangalore, India, [*] Intel Corporation, Santa Clara, CA, United States
[**] Intel Corporation, Chandler, AZ, United States, [***] Intel Corporation, Hillsboro, OR, United States
arun.chandrasekhar@intel.com

Abstract - The package serves as the space transformer between the fine pitch Silicon bumps and the slowly scaling board pins. Over the last few years we have witnessed ~20% CAGR of IO pins driven by performance needs. This leads to the ratio of space transformation to be quite high and introduces multiple challenges. A novel package architecture called PoINT (Patch on Interposer) is used on Intel server CPUs which helps with on-package integration and provide the most optimal performance at the lowest cost. PoINT uses 2 substrates- a HDI substrate that interfaces with the die and an LDI substrate that interfaces with the board. The 2 substrates are interconnected using a Mid-Level Interconnect (MLI). In the paper we illustrate the design requirements and challenges and the interaction of various multi-physics aspects that need co-optimisation to make this architecture successful.

I. INTRODUCTION

Historically, the complexity and cost of server CPU package has been increasing due to an increase in package form factor, layer count and the use of advanced packaging technologies. The standard monolithic package uses High Density Interconnect (HDI) design rules driven by the die to package interface First Level Interconnect (FLI also known as C4) dimensions. The same package also interfaces to the board/PCB through the LGA socket/BGA balls which is known as the Second Level Interconnect (SLI) as illustrated in fig. 1a. The rapid scaling of Silicon features, transistor density and integration means that the FLI pitch also scales at that rate. In fact Intel leads the FLI pitch scaling in the industry. The SLI pitch is dictated by the technology scaling and cost constraints on the board/PCB which scales sub-linearly at a slower rate compared to SLI scaling and the FLI-SLI gap widens. The burden is on the package to bridge this gap. As the CPU feature set and signalling rates increase, package form factor and layer count also grow in tandem. The decreasing silicon to package interconnect pitch results in the use of a finer feature set on the package around the die for First Level Interconnect (FLI) escape.

The goal is to a certain extent decouple the FLI-SLI gap through the use of a packaging technology and hence manage the exponentially growing package complexity. This needs to be done without compromising the product performance [1, 2].

The PoINT packaging technology splits the standard monolithic package into two pieces that are assembled one on

top of another and illustrated in fig. 1b. The substrate that interfaces with the Silicon is called the Patch and uses HDI design rules. The substrate that interfaces with the board is called the Interposer and uses LDI design rules. The high-density substrate provides required density for Silicon fan-out at the lowest possible cost. The low-density substrate with its coarser density acts as a translation layer between the high-density substrate and the socket. The two substrates are connected using a BGA Mid-Level Interconnect (MLI) at a pitch that is roughly mid-way between the FLI and SLI pitch. Such a package construction decouples the FLI and SLI scaling and allows for the packaging technology to scale in a manageable and cost effective manner. PoINT is an acronym for "Patch on Interposer". Fig. 2 illustrates FLI/MLI/SLI.

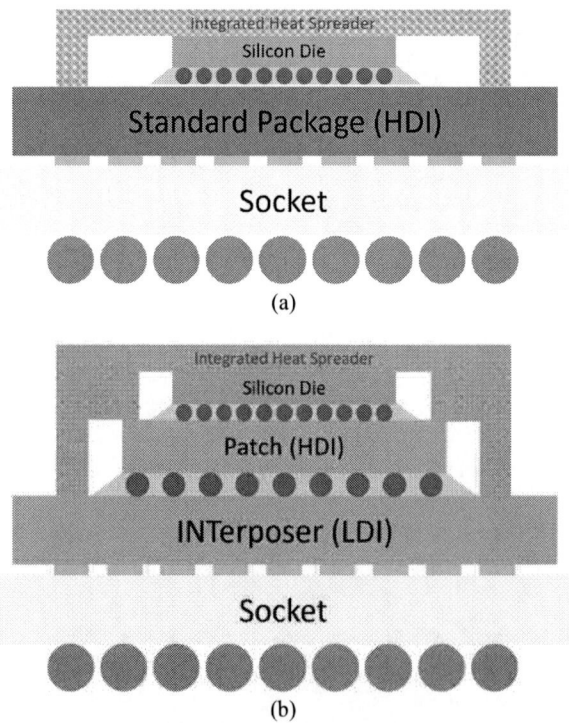

Fig. 1. (a) Standard monolithic package construction
(b) PoINT package construction

The patch is primarily used to escape signals and power supplies from FLI interconnect. Hence, it requires fine features

978-1-7281-1500-9/19 $31.00 © 2019 IEEE

required to route through the densely packed FLI bumps. Thus, most of these fine features are limited to the Patch, leaving the coarser features to the Interposer. The Patch is a standard substrate with HDI build up layers that support IO escape and power distribution under the die. The Interposer has a stack of laminated metal layers with a pre-preg dielectric material between them and can only support coarser design rules. The MLI pitch needs to be sufficiently large to support the coarser design rules on the Interposer needed for IO escape through the MLI BGA.

There are other ways in which this technology can be beneficial. In an ideal scenario, a Patch package resembles client CPU package in both form factor and substrate stack up. This allows feature sharing between client and server substrate technologies like material type, via drill and pad sizes to name a few. Additionally, maintaining the same MLI pin-out between different die designs allows the design teams to either reuse the Patch design or the Interposer design across SKUs. This reduces the time and resource requirement for each programme.

With the selection of optimised material properties and design rules, the PoINT technology enables a path for improved product performance as well as managing packaging complexity. The rest of the paper focusses on the various aspects that need to be addressed to make the technology successful and sustainable.

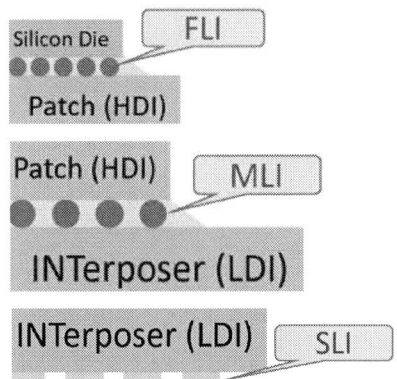

Fig. 2. Hierarchy of Interconnects. FLI and SLI are a part of standard package design. MLI is introduced with PoINT

Fig. 3. Intel Core i9-7920X PoINT package (de-lidded)
Source: www.extremetech.com

One of the Intel Skylake family of server processor packages that shows the PoINT package construction (with the Integrated Heat Spreader removed) can be seen in fig. 3.

II. PHYSICAL DESIGN

The key to successful PoINT package design is taking advantage of the design flexibility the 2 substrates offer as well as co-design of the Silicon floorplan and the board pin-map. The relative sizes of the patch and the Interposer and the location of I/O ports on the die and the pins on the package play a critical role in determining the best use of the architecture. Ideally, the smallest die with the smallest patch size works best from a design flexibility perspective since we will have sufficient real estate on the Interposer which uses coarser design rules to manoeuvre the signals and power planes. However, in most server designs, we will be limited in shrinking the patch size because of the die size and/or the I/O and power pin requirements. Both of these will drive the patch size higher. With the right chip-package co-design process that influences silicon floorplan optimisation, we can get the patch to be of optimal size. Some of the metrics that determine the effective use of the PoINT architecture are:-

a. Die size post co-optimisation
b. Patch size
c. Patch layer count
d. Interposer layer count
e. Product performance trade-offs, if any

An example of how the SLI pin assignments are made on the package-board interface is illustrated in fig. 4. In traditional cases the pin assignment considers board routing capabilities and the location of the I/O and power ports on the die. However, with the PoINT package, we introduce another degree of freedom, namely the MLI and that needs to be considered for optimal design.

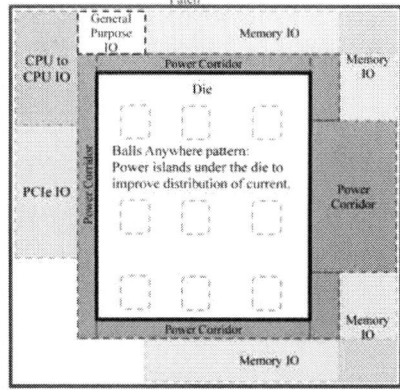

Fig. 4. SLI assignments cognizant of PoINT architecture

The tight MLI BGA ball pitch allows the design to distribute power uniformly under the patch around the MLI cavity to not cause current crowding. The tight MLI pitch also enables more vertical connections [2] in the power islands under the die to reduce resistance further (details in power delivery section). Fig. 4 shows a sample schematic of MLI pattern optimisation for power and I/O domains. This required careful chip-package co-design to optimise I/O floor plans to ensure a wider power entry to the die for lower resistance.

There are advantages with the PoINT architecture which are listed below and upfront design decisions can help take advantage of these.

a. More copper layers available (Patch + Interposer). The additional copper can be used with R_{path} reduction that helps power delivery DC losses (although AC would be comrpomised and discussed in the power delivery section)

b. Ability to decouple ACI (Air Core Inductor) keep out zones on the package base and LSCs (Land Side Capacitors). All Intel server package designs use the Fully Integrated Voltage Regulator (FIVR) [3] and if the ACI design uses the base build-up layers on the package, the base layer cannot be used to place caps or pins. However, with PoINT, the ACI is on the Patch while the caps are on the Interposer.

c. The coarser design rules and thicker copper on the LDI Interposer helps with loss reduction on high speed signals as well as power planes.

d. Additional degree of routing freedom with MLI. The Interposer has a full stack of blind and buried vias and this gives significant routing flexibility. The provides ability to fix the MLI pin-out and re-use the Patch or the Interposer to create multiple package SKUs.

e. PoINT architecture with the appropriate choice of Patch and Interposer sizes offers the best performance for the most optimal cost

The PoINT architecture has some challenges to overcome that include the following:-

a. Additional discontinuities introduced in the form of MLI and longer vertical path that impacts Signal Integrity (SI)

b. Land Side Capacitors (LSCs) are on the Interposer base and have a longer vertical path to reach the die (through the Interposer, MLI and Patch) and hence less effective. This potentially impacts AC noise.

c. Integrated Heat Spreader (IHS) design can be more complex to span two substrates and meet the thermo-mechnaical and reliability requirements.

d. Additional design effort to design two substrates for the same product.

We will look at how some of these challenges can be managed as we discuss each of these areas in the subsequent sections.

Fig. 5a illustrates a small die with an aspectratio ~1:1 on a square patch that is small enough to accommodate the fine pitch FLI escape and the MLI BGA count needed. The Interposer size is determined by the socket size on which the product fits in. In this case, the Interposer has sufficient real estate available outside of the Patch area. Fig. 5b shows the location of two memory channels namely M1 and M2

on the die (adjacent to each other) and it is trasformed to one behind the other on the Patch BGA pin-map and the Interposer LGA pin-map to be in line with the DIMM orientation on the board. Figs. 6a and 6s show the Patch routing zones for these two memory channels which can be accomplsihed on two patch build-up layers, esp. given the ability to route with fine lie width and spacing on the Patch. Figs. 7a and 7b show the Interposer routing zones for the M1 and M2 memory channels which again can be done using two layers as the ports on the Patch and the Interposer are in line-of-sight with respect to each other.

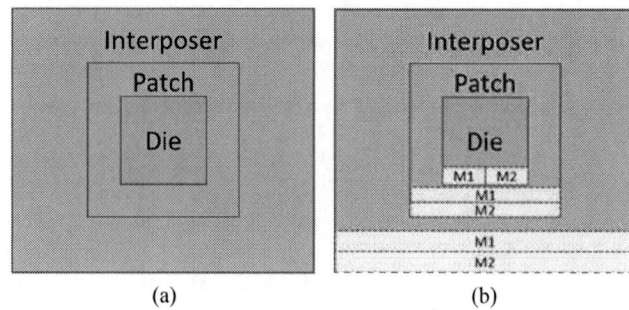

Fig. 5. (a) Small Patch on Interposer (b) Memory channel locations on die, patch and interposer

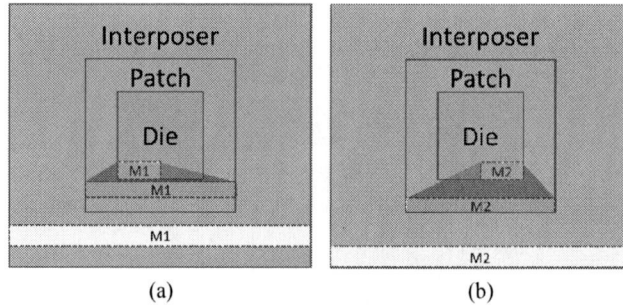

Fig. 6. (a) M1 channel Patch routing (b) M2 channel Patch routing

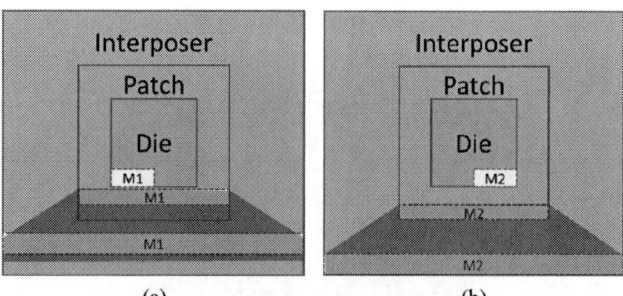

Fig. 7. (a) M1 channel INT routing (b) M2 channel INT routing

Fig. 8a shows another case where the die is large and has an aspect ratio of 1:2 which also grows the patch significantly in the Y direction. Th Interposer area outside the patch in the y dimension is very small. In fig. 8b we can see that the the die has the two memory port locations M1 and M2 on the same south edge of the die and the relative patch BGA and Interposer LGA locations for these memory ports are maintained as in the previos case. However, due to the

978-1-7281-1500-9/19 $31.00 © 2019 IEEE   2182

dimensions, the patch BGA lands have a small overlap with the die port locations while the Interposer LGA land locations have a signifcant overlap with the Patch BGA pin locations. This leads to more complex routing. Figs. 9a and 9b shows that the Patch real estate available is limited and managing the routing in 2 Patch layers may be challenging and we may have to resprt to Patch bask-side build-up layers to complete the routing to the MLI BGA locations. The bigger challenge is on the Interposer routing layers where the escape from the Patch BGA pins have to go out and then jog back in towards the SLI LGA pads which are behind the MLI locations. Given that the Interposer can only accommodate coarse design features, this can consume at least 2 if not more layers per memory channel depending on the overlap.

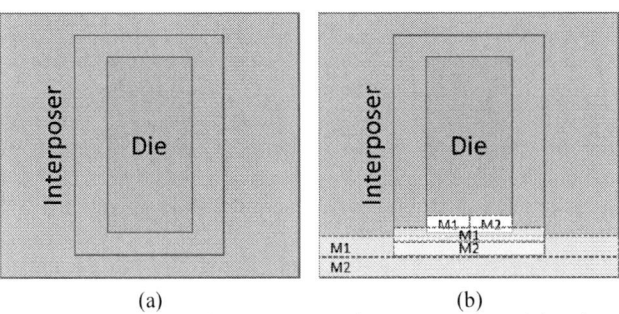

(a)   (b)

Fig. 8. (a) Large Y Patch on Interposer (b) Memory channel locations on die, Patch and Interposer

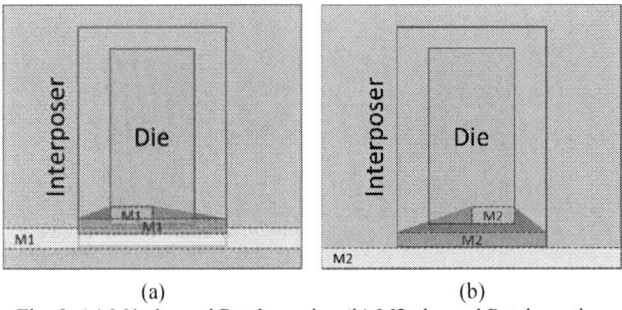

(a)   (b)

Fig. 9. (a) M1 channel Patch routing (b) M2 channel Patch routing

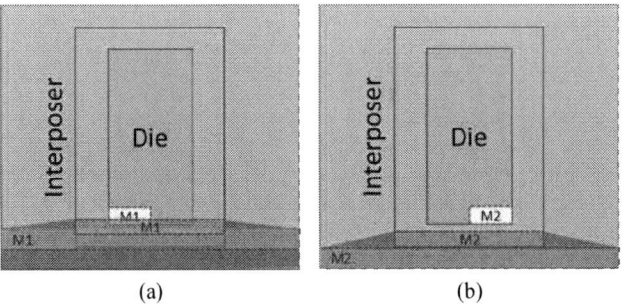

(a)   (b)

Fig. 10. (a) M1 channel INT routing (b) M2 channel INT routing

The Interposer size is usually determined by the socket chosen and the platform constraints. The Patch size is largely a function of the die size and due to the fine BGA pitch it is rarely a size determining factor. Hence it is critical that the die floorplan is co-optimised with the PoINT package architecture

to ensure that the success metrics listed earlier turn out to be positive.

III. POWER DELIVERY

Inherently, a PoINT package has more copper that helps with power delivery DC losses but due to the extra layers the electrical distance from the pins to the die as well as to the Land Side Capacitors (LSCs) is more which hurts the AC noise. However, the perceived impact of the PoINT architecture on power delivery can be resolved using some novel techniques. In order to address the voltage drop issue on the incoming power supply from the motherboard, multiple solutions like adding 'balls anywhere' BGA patterns under the die in the MLI cavity, tighter BGA pitch, increasing copper (Cu) thickness on the Interposer & Patch layers, optimising silicon I/O floor plan and the package physical design have been evaluated. Due to ACI locations, depopulation of MLI BGA balls under the die to avoid metal interference is required. However, there are still limited areas where it is possible to opportunistically populate some BGAs. Populating BGAs in those specific areas amounts to implementing 'balls anywhere' patterns, as illustrated in the schematic in figure 4. These patterns are specific to silicon floor plan and they change from die to die. The design can be optimised to enable BGA patterns in the islands that reduce assembly and (Solder Joint Reliability) SJR risks while providing the intended benefit to package voltage drop issues. Simultaneously, the pin-map has sufficient redundancy for the power and ground BGAs to address any potential SJR failures. The design also takes advantage of the tighter MLI BGA pitch available on this technology to distribute the current uniformly under the Patch by use of a four sided power corridor around the MLI cavity. Tighter BGA pitch has also enabled more vertical connections in the power islands under the die to reduce the resistance further.

Another method to reduce the voltage drop is to increase the Cu thickness of the layers in the Patch and Interposer. Increasing Cu thickness needs interaction with the substrate technology teams and requires analysis of manufacturing feasibility, yield and cost. This needs trade-off with I/O routing since thicker Cu layers impact I/O impedance. Hence, selective layers have their Cu thickness increased (only power and ground) and this is shown to help with the DC power losses (Fig. 11).

In addition to meeting the DC R_{path} specification, it is also a critical requirement to meet the AC noise specification on the incoming power supply. The BGA MLI 'balls anywhere' pattern also help effectively connect the LSCs which are further away from the die due to their location on the base of the interposer. The robustness of the power delivery path is further improved by providing god vertical path using stitching vias (PTHs and µ-vias). The smaller size of the patch is a limiter for the number of Die Side capacitors (DSCs) which can help reduce the transient AC noise. But a smaller patch helps reduce the epoxy Keep Out Zones (KOZs) and also some innovations with the IHS design KOZs and the location of

978-1-7281-1500-9/19 $31.00 © 2019 IEEE   2183

manufacturing marks helps place the optimal number of caps at the right location to help meet the noise specifications.

Other aspects that help mitigate any power delivery losses include co-optimisation of the floorplan with the Silicon design (example is the wide power entry illustrated in fig. 4), physical design optimisation to trade-of vertical and lateral path interconnections, layer assignment optimisation (balance power and ground planes for R_{DC} and loop inductance).

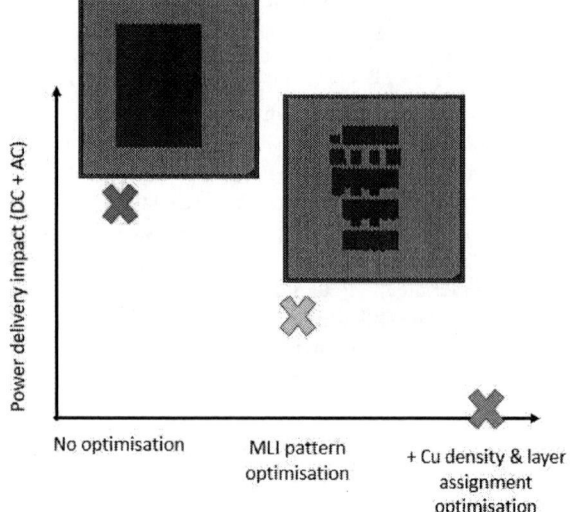

Fig. 11. Power delivery improvements on PoINT architecture

A key benefit with the point technology that design can take advantage of is the ability for ACIs and LSCs to co-exist (ACIs on the Patch and LSCs on the Interposer). This is where it is critical to have the MLI BGAs populated in the zones around the ACI KOZs so that the cap connections are effective. A reduction in the Interposer and Patch layer counts also help with reduction of the vertical parasitics.

IV. SIGNAL INTEGRITY

PoINT architecture has an additional interconnect level (MLI) introduced which adds to the impedance discontinuity. But there are other knobs that can help mitigate this SI bottleneck and achieve performance parity. The two high speed I/O interfaces can be classified under memory I/O (DDR) which is largely single ended and cross-talk dominated and serial I/O (PCIe & UPI) which is largely loss dominated.

Crosstalk is a function of the trace spacing as well as the total routing length. The escape pattern is a function of the FLI bump pitch but any reduction in crosstalk can be managed through looser spacing during fan-out of the traces. By positioning the MLI BGA locations and the SLI pins appropriately the total routing length can be kept within manageable limits. Trade-offs with power corridor width and other GPIO pin positions may be needed.

The primary contributors to the insertion loss that affects serial I/Os are the conductive losses from the copper traces and dielectric loss due to the build-up material loss tangent. Additionally, impedance discontinuities also affect serial I/Os

significantly. Lastly, another critical factor that impacts serial I/Os is the impedance variation between the different routing layers. The more number of layers the signal is routed on the impedance variation gets worse. Moreover the coarser design rules on the interposer also contribute to larger impedance variations. All these aspects can be addressed with design and technology enablers. Although the Interposer traces have wider variation, they are much wider and hence help in keeping insertion loss under control. By tightening the process knobs to keep impedance variations tight and positioning the MLI BGAs and the SLI pins appropriately the number of layers on which the signals are routed can be minimised. The goal has been to keep the Patch routing to 2 layers (on front side and one bottom side) and 1 Interposer layer. Moreover, by keeping the Patch routing length small and having most of the routing on the Interposer, the loss can be kept under check. Appropriate choice of low loss materials for the dielectrics also help with loss but this may add cost. Depending on the bus speed and the margins, length matching within the Patch and Interposer may be needed although that is a bit more difficult to achieve while keeping trace lengths under check. It is easier to length match signals to an overall target across the Patch and the Interposer. Fig. 12 shows how a superposition of the various enablers can help reduce and mitigate the impact on factors contributing to SI degradation.

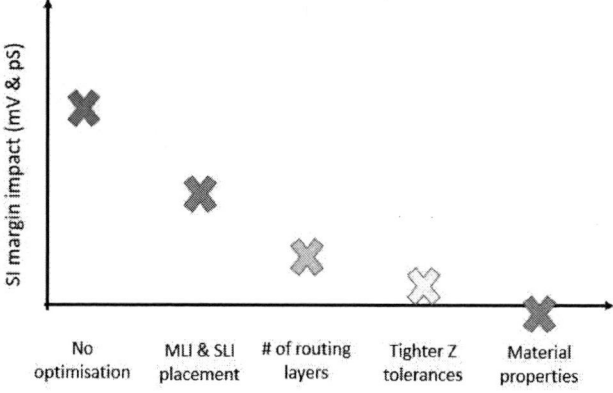

Fig. 12. Signal Integrity improvements on PoINT architecture

V. THERMO-MECHANICAL

Management of the package warpage during socket loading is critical to ensure loading robustness and maintain the range of contact forces at Beginning of Life (BoL) and End of Life (EoL). Novel Integrated Heat Spreader (IHS) designs (Fig. 13) have been developed for the PoINT architecture to mitigate package warpage during enabling and socket loading. This design involves IHS feet landing on the Patch as well as I-shaped feet landing on the interposer to maximise the load coupling all the way to the package edges without compromising on the loading mechanism KOZs. The Z-height of the total stack that includes the socket, Interposer, Patch and the IHS needs to be carefully tuned to enable transparency to end customers and compatibility with non-PoINT package designs. With the IHS optimisation, we are able to keep the

978-1-7281-1500-9/19 $31.00 © 2019 IEEE

interposer (thinner than standard package) warpage within the package warpage specification required for robust socket loading (translated as socket contact force) [4].

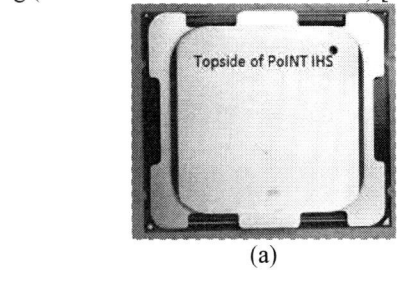

(a)

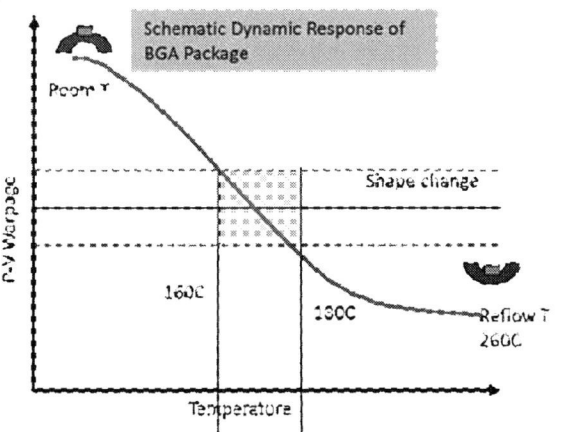

(b)

Fig. 13. Optimised IHS design for PoINT architecture
(a) top view and (b) cross-section

VI. ASSEMBLY

PoINT assembly process flow includes attaching the Silicon die to the Patch followed by attaching this die + Patch complex to the interposer. The most critical aspect in this assembly step is to have a wide enough MLI process window to accommodate the large warpage difference between the die + Patch complex, dominated by the Coefficient of Thermal Expansion (CTE) of Silicon ~3ppm/°C and the CTE of the Interposer ~17-20ppm/°C. Fig. 14 shows the dynamic response of a typical die + Patch complex as a function of temperature [4].

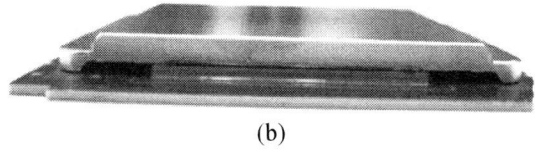

Fig. 14. Dynamic Response of die + Patch BGA complex showing warpage transition region

The key innovation that enables the successful die + Patch complex attach process to the interposer is to tune the attach temperatures to take advantage of the shape cross-over at ~ 160-180°C. This led to the invention and development of a low temperature solder (LTS) chemistry specifically tuned for this MLI attach process. Fig. 15 shows the solder joint

comparison between a standard reflow process and LTS reflow process, highlighting the non-wets in the standard process vs. good solder joints in the LTS process.

(a)   (b)

Fig. 15. Solder joints on the die + Patch complex to the Interposer surface with (a) standard process and (b) Low temp. process

VII. DESIGN TOOLS

PoINT, by the nature of its architecture, has two substrates in place of one which poses a design challenge that requires careful optimisation of netlists, I/O length matching between Patch and Interposer and drives new processes and tools to ensure synchronisation between the Patch and Interposer designs. Customised tools to check the database continuity, interconnect integrity and the handshalke between the Patch and the Interposer make the design process robust and meet flawless design quality requirements.

VIII. SUMMARY

This paper introduces the PoINT package architecture used on several Intel server processor products. The pros and cons of the architecture and the various schemes to address the challenges across a range of disciplines that include physical design, power delivery, signal integrity, thermos-mechanical, assembly and design tools have been outlined.

ACKNOWLEDGEMENTS

The authors wish to acknowledge the contribution of Amar Vuppala, Raja Swaminathan and several members in the Assembly Technology, Data Center and Manufacturing & Validation organisations at Intel who helped in the successful implementation and deployment of the PoINT technology.

REFERENCES

[1] Zhang, Z., Roy M. K., Aygun, K., Prokofiev V., Manusharow M. J., Ravva P. C, "Enabling Package on Interposer Technology for Cost Affordable Server Applications," Intel Assembly & Test Technology Journal, vol. 12, 2009.

[2] Vuppala, A., Chandrasekhar, A., Bharath, K., Swaminathan, R., "Enabling PoINT Packages for Intel CPUs", Intel DTTC 2012.

[3] William J. Lambert, Michael J. Hill, Kaladhar Radhakrishnan, Leigh Wojewoda, Anne E. Augustine, "Package Inductors for Intel Fully Integrated Voltage Regulators", IEEE Transactions on CPMT, Vol. 6, Issue 1, Jan. 2016, pp 3 – 11.

[4] Raja Swaminathan, Ram Viswanath, Sriram Srinivasan, and Arun Chandrasekhar, "Next Generation Xeon Server Package Architecture", International Symposium on Microelectronics: Fall 2017, Vol. 2017, No. 1, pp. 342-345.

Highly Reliable Die-Attach Silver Joint with Pressure-Less Sintering Process

Sihai Chen
Department of R & D
Indium Corporation
Clinton, NY
schen@indium.com

William Shambach
Department of Biomedical Engineering
Rochester Institute of Technology
Rochester, NY

Jordan Plamer
Department of Chemical Engineering
Rochester Institute of Technology
Rochester, NY

Christine LaBarbera
Department of R & D
Indium Corporation
Clinton, NY

Xuanyi Ding
Department of Materials Science and Engineering
Cornell University
Ithaca, NY

Ning-Cheng Lee
Department of R & D
Indium Corporation
Clinton, NY
nclee@indium.com

Abstract—Pressure-less silver sintering pastes with high-reliability judged from temperature cycling test (TCT) under -40°C to 175°C were developed. For 3mm x 3mm sized dies, the silver joints formed between the Ag surface finished Si die and the Ag surface finished Si_3N_4 active metal brazing (AMB) substrate retained the initial shear strength after 1500 cycles, the strength being ~25 Mpa even after 3879 cycles. For 5mm x 5mm dies, the average shear strength was very stable and remained at ~42 Mpa, even after 3879 cycles. For 10mm x 10mm dies, the dies cannot be sheared off (max shear force 200kg) for all the samples before and after 2389 cycles of TCT. Within a standard bondline thickness (BLT) range of 40 to 60μm, the average total voids percentage of the sintered joints for the above different die sizes can be controlled within < 2%. Under ambient atmosphere, the shear strength of sintered joints formed on a bare Cu surface finished AMB substrate reach 62Mpa, slightly stronger than that of Ag- and Au-surface finished ones. TCT shows that these surface finishes have similar decay behaviors, implying that bare Cu surface finished AMB substrate can be used for bonding applications. BLT is critical in determining the bond property and reliability. For a very thin BLT, such as < 10μm, the bond strength decays at about five times the rate that of a joint with ~ 50μm BLT. For a thick BLT, such as > 100 μm with 10mm x 10mm dies, a large amount of voids are generated if using a similar sintering profile, where almost no voids are observed for a ~50 μm BLT joint.

Keywords— *Silver sintering paste, pressure-less, die-attach, surface finish, reliability, temperature cycling.*

I. INTRODUCTION

Development in die-attach materials has been pushed by various semiconductor device needs in applications such as hybrid and electric cars, high speed trains, aircrafts/aviation, and deep-well oil/gas extraction, where power electronic systems are required with increased packing density, integration, and higher reliability. In addition, with the introduction of wide band gap materials, such as silicon carbide or gallium nitride devices, it is theoretically possible to operate the devices at temperatures much higher than present limit (around 150 °C). [1] Therefore, robust and reliable high-temperature (≥ 175 or even 200°C) die-attach materials are in strong demand for the above applications.

Several types of materials are potential candidates, including solder, sintering [2], and transient liquid phase (TLP) materials. [3] Conventional Sn-Ag and Sn-Ag-Cu

alloys are not fit for high-temperature applications due to the low-melting points. High-lead solder is normally not preferred due to environmental and health hazards it poses. Lead-free solders which can operate under ≥ 200°C conditions do exist, such as AuGe, [4] AuSi, [5] AuIn [6] and ZnAl. [7] However, these materials have drawbacks including high price (Au-based alloys) or poor processability (Zn-Al).

In contrast, Ag or Cu sintering materials can withstand a much higher operation temperature due to higher melting points. Also, they can be processed at a lower temperature comparable or even lower than that of solder materials, which will greatly help reduce the thermal stress during processing. They are a promising next generation lead-free die-attach solution due to their capability to work under high-temperature, their high thermal and electrical conductivities, and good thermomechanical properties. [8] Silver sintered joints obtained under pressure conditions have been reported to have much better reliability than that of solder materials through power cycling and temperature cycling tests. [9] However, pressure-sintering has drawbacks such as a) the dies may crack due to the pressure, leading to a low yield; b) large financial investment is required for the pressure-sintering equipment and c) a lower throughput results due to the batch-by-batch process. Therefore, there is a strong desire to have a pressure-less silver sintering paste to overcome the above problems.

In this study, we report the development of pressure-less silver sintering materials with good reliability judged by the temperature cycling test (TCT). The emphasis is focused on finding the best paste materials to achieve high device reliability. We studied different factors, including paste optimization; device structure, such as surface metallization of the substrate; joint BLT; and die size, and explored their effect on joint reliability. The pressure-less silver sintering pastes possess several significant properties. For example, the bonding on a bare Cu surface presents a similar reliability as that on Ag and Au metalized surface, showing promise for the use of a bare Cu surface for bonding. It is also observed that the shear strength of the joints formed using 5mm x 5mm dies is very stable even after 3879 cycles of TCT.

978-1-7281-1500-9/19 $31.00 © 2019 IEEE

II. EXPERIMENTAL

A. Materials and Test

The sizes of the Si die used were 3mm x 3mm, 5mm x 5mm, and 10mm x 10mm. Its surface finish was Ti/Ni/Ag with 75/300/75nm thicknesses, respectively. 2mm x 2mm SiC dies were also used as stated. The surface finish was Ti/Ni/Ag with 100/100/200nm thickness. The substrates used were Ni/Au, Ag, or bare Cu surface finished Si_3N_4 AMB materials; the ceramic thickness was 0.0125", and the Cu thickness was 0.012" on both sides. The samples were assembled by printing Ag sintering paste on the substrate using a 4 mil stencil followed by die placement. The bond line thickness (BLT) was carefully adjusted to a targeted value. The assemblies were pressure-lessly sintered in a box oven under air condition using a sintering profile as below: (1) increasing the oven temperature from 25°C to 250°C with a ramp rate of 3°C/min, (2) keeping the temperature at 250°C for 1h and (3) cooling the sample after switching off the oven.

Reference samples including high-lead solders Indalloy® 151 (92.5Pb5Sn2.5Ag), Indalloy® 171 (95Pb5Sn) and lead-free solder Indalloy® 256 (SAC305, 96.5Sn3Ag0.5Cu) were prepared by using standard reflow profiles. High-lead joints were prepared under an N_2 atmosphere using an infrared reflow oven, while that of lead free samples were prepared under air reflow using a BTU oven.

For the reliability test, a Tenney thermal cycling machine was used. The temperature is set from -40°C to 175°C with a dwell time of 10 minutes at maximum and minimum temperatures, respectively. Each cycle time was 46 minutes.

B. Characterization Method

Shear strength test: Samples before and after temperature cycling were tested at ambient temperature using a XYZTEC Condor 250 shear tester. Shear speed is 100μm/second. Sample shear height was adjusted according to the joint BLT.

A Nanovea Digital Micro Hardness Tester was used in this project to measure the hardness of the samples. The hardness standard test block was first measured, and the measuring lines were zeroed in order to get accurate data. The sample was attached on the mount and the surface was confirmed to be horizontally placed so the sample would not be tilted in the testing. The load was set to 245.2mN, and the dwell time was set to 15s. After each loading, one needs to measure the diagonal distance of indented cracks.

Morphology and Failure analysis: Before and after the shear test, the samples were examined by X-ray to obtain voiding information. For the sheared samples, X-ray, optical microscopy, and scanning electron microscopy (SEM) were used for morphology study; some samples were further cross-sectioned for microstructure analysis via SEM and energy dispersive spectroscopy (EDS). Some of the samples were cross-sectioned without shear testing. The BLT and die

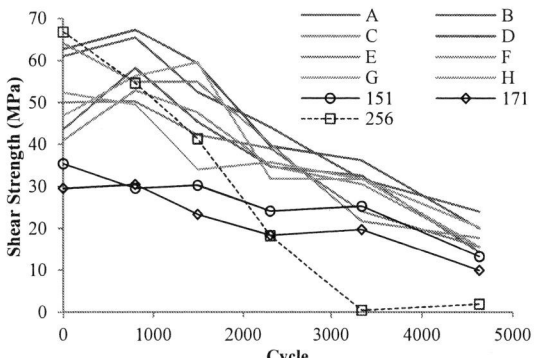

Fig. 1. Shear strength change for the silver sintering joints (from A to H) and solder joints (alloys 151, 171, and 256) before and after at -40 to 175°C temperature cycling test.

tilt angle analysis were conducted using a caliper by measuring the joint thickness at the four die corners.

III. RESULTS AND DISCUSSION

Our experiment was designed to screen a series of silver sintering pastes, with the aim of determining the most stable option through TCT cycles (-40°C to 175°C). At the same time, their stability was also compared to that of references. A total of eight silver pastes (named from A to H) and three reference samples (Indalloys® 151, 171 and 256) were studied. Typically, 3mm x 3mm Si dies with silver finishes are bonded to silver-surface-finished AMB substrates. For each silver paste, more than 30 identical joints are made, and the sintered BLT was controlled at 47 to 57μm. Die tilt angle was normally < 0.3°. The total voids of optimized silver sintering joints can be controlled below the 1% level. [10]

A. Comparison between Silver and Solder Pastes

1) Shear strength before and after TCT

Shear strength is the overall measure to judge the quality of the joint. The average shear strength of at least three specimens for individual samples is shown in Fig. 1. Before thermal cycling, the silver sintering samples have strengths in the range of 41 to 64 Mpa. The high-Pb samples, Indalloys 151 and 171, have strength of 30 to 35 Mpa. The lead-free sample, Indalloy 256 (SAC305) has the strongest strength of 67 Mpa. For samples A, C, D and G, strength after 1500 cycles is as strong as that without TCT; further cycling results in decrease in shear strength. Overall, the pressure-less silver sintered joints have the strength in the range of 20 to 30 Mpa even after 3879 cycles, while high-lead samples (151 and 171) have strength within 15 to 20 Mpa. In contrast, the strength of SAC305 decreased to near zero. In order to compare the properties of different pastes, for the shear strength (y) vs cycle (x) data points shown in Fig. 1, we used linear fit to obtain the slope (a) and intercept (b) of the fitted line,

TABLE 1. ANALYTIC RESULTS OF THE TRENDING LINES FOR DIFFERENT PASTES UNDER -40 TO 175°C TEMPERATURE CYCLING CONDITIONS

	A	B	C	D	E	F	G	H	151	171	256
a	-0.0076	-0.0093	-0.0068	-0.0104	-0.0081	-0.0109	-0.0076	-0.0074	-0.0042	-0.0043	-0.0207
b	54	66	51	69	54	65	57	57	35	31	69
R^2	0.759	0.937	0.7491	0.9102	0.9541	0.9457	0.6776	0.9046	0.8865	0.8958	0.9893
DI	7.6	9.3	6.8	10.4	8.1	10.9	7.6	7.4	4.2	4.3	20.7

978-1-7281-1500-9/19 $31.00 © 2019 IEEE

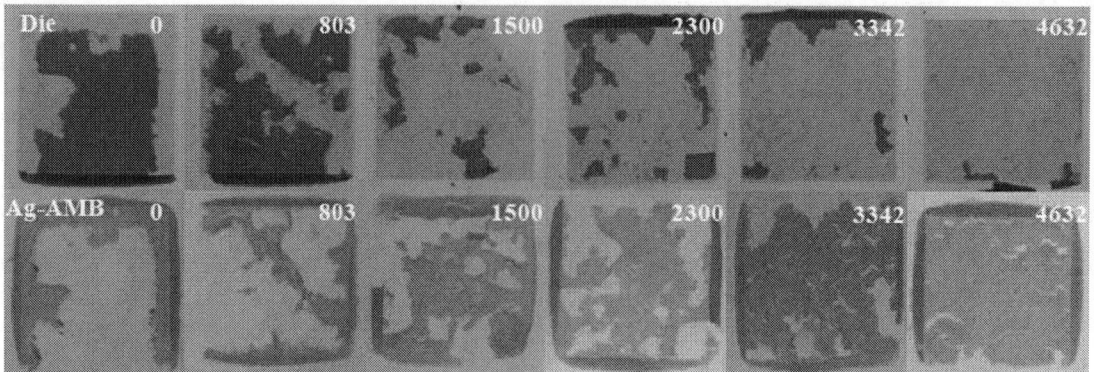

Fig. 2. X-Ray images of sheared 3mm x 3mm Si die and Ag-AMB sides of sample C joint before and after different cycles of TCT. Cycle number is listed at upper-right corner of images.

TABLE 2. AVERAGE PERCENTAGE (%) OF JOINT MATERIALS ON SI AND AMB SIDES BEFORE AND AFTER TCT

Cycles	Side	A	B	C	D	E	F	G	H	151	171	256
0	Die	41.2	77.6	35.6	80.7	59.9	61.7	50.5	52.0	32.9	24.5	24.8
0	AMB	39.4	28.4	39.1	23.7	23.5	29.0	54.6	34.8	88.5	88.8	79.0
800	Die	68.2	39.2	53.2	52.5	47.8	55.9	48.0	53.8	49.2	58.5	n/a
800	AMB	42.2	40.7	30.6	31.3	32.5	48.7	34.9	54.8	54.5	48.7	82.9
1500	Die	15.2	54.1	23.8	24.9	n/a	28.0	33.2	53.3	74.0	70.1	n/a
1500	AMB	69.0	46.2	59.3	54.3	38.7	63.0	41.4	45.5	67.7	44.2	77.9
2300	Die	14.7	11.6	18.3	18.3	23.6	15.2	38.9	37.0	53.9	18.4	23.3
2300	AMB	74.4	73.7	68.4	71.3	64.2	70.9	49.4	46.7	62.3	20.4	44.9
3342	Die	8.9	14.8	13.8	11.9	6.4	27.3	12.9	36.6	52.1	71.1	7.5
3342	AMB	86.7	72.7	74.7	81.8	79.1	59.0	68.6	55.1	53.5	14.0	86.4
4632	Die	10.7	6.6	5.0	18.8	7.8	14.4	15.9	20.7	43.9	50.6	10.9
4632	AMB	81.5	84.4	90.9	69.6	77.2	65.8	72.0	69.1	52.3	42.4	72.9

$$y = ax + b \qquad (1)$$

We also define the degrade index (DI) as

$$DI = -1000 \times a \qquad (2)$$

DI has a unit of kpa/cycle, a smaller value indicating a slower degradation rate. The overall reliability will be judged by both the initial shear strength (intercept b) and DI values. From the data shown in Table 1, silver sintered joints have a higher DI value than that of high-lead solders. However, due to the higher initial shear strength, the sintered joints are able to maintain a high shear strength value even after 3879 cycles of TCT. SAC305 has the worst reliability because of the high DI value of 21kpa/cycle, even though it has a very strong initial shear strength.

2) Sheared sample analysis

After shear testing, X-ray analysis is a useful tool to understand failure mode of the joints. As shown in Fig. 2 for sample C, before TCT, a large percentage of the dense area in the center of die side indicates a strong adhesion of sintered silver on the die, i.e., a strong interfacial reaction between them. At the same time, quite a lot of silver exists at the peripheral area of the substrate. As TCT proceeds, the percentage of the silver on the Si die side decreased gradually while that on the AMB side increased, as shown in Fig. 3a. We observed this similar behavior for all the sintered silver samples (we used Image J to calculate the amount of silver on Si and substrate sides respectively based on the 3mm x 3mm die size; the sum of the two percentages may not amount to 100%, see Table 2). It shows that the interfacial interaction between sintered silver and the AMB substrate is enhanced while that between sintered Ag and the Si dies becomes weaker after TCT. This may be due to the

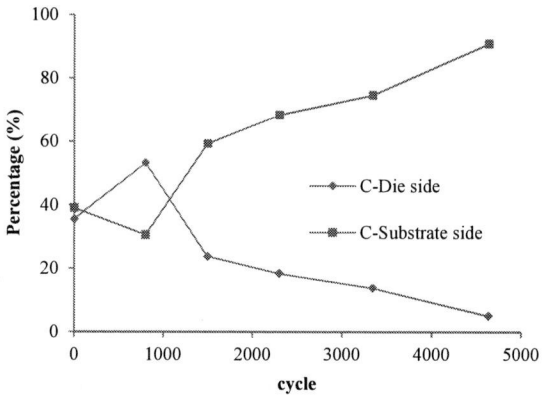

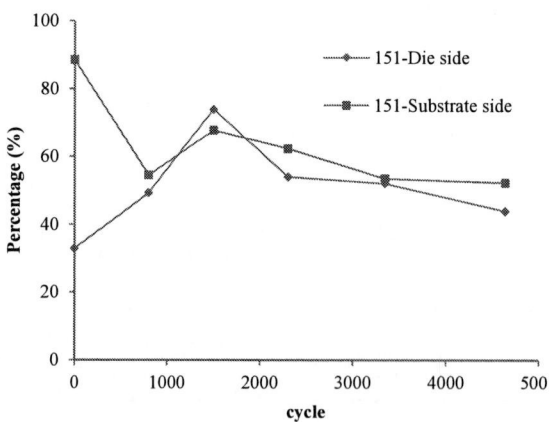

Fig. 3. Percentage of sintered Ag on die and AMB substrate sides before and after -40 to 175°C TCT for a) sample C and b) 151, respectively.

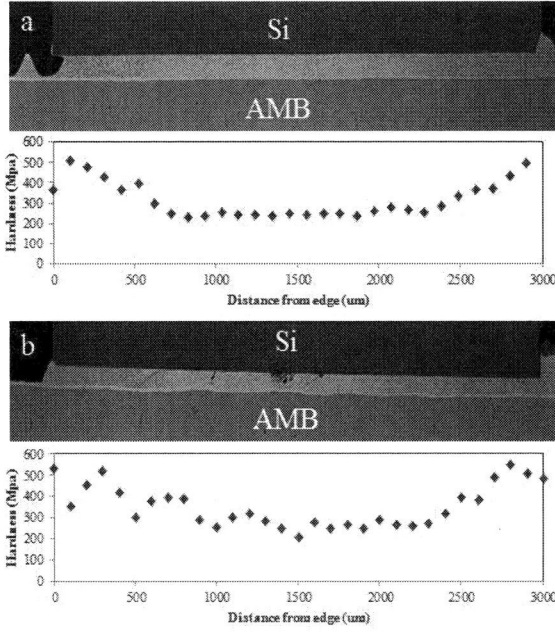

Fig. 4. Cross-section of the silver sintering joint generated by paste A and the hardness value along the joint at a) 0 and b) 1500 cycles of TCT at -40 to 175°C.

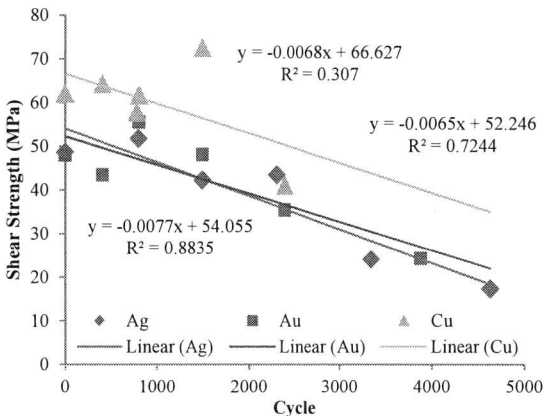

Fig. 5. Shear strength change for the silver sintering joints on different surface finished AMB substrates before and after at -40 to 175°C temperature cycling test. Si die size is 3mm x 3mm.

large CTE mismatch between Si and sintered Ag at the interface. For the reference sample SAC305, the solder on the Si side was reduced to less than 10% after 3342 cycles (Table 2), showing a very weak interfacial interaction between Si and solder. In addition, a lot of cracks are observed in the bulk solder, accounting for the broken of the solder joints. Reference samples 151 and 171 have similar solder amounts on both side of the interface (Figure 3b), implying that the failure is mainly cohesive in nature.

3) Hardness of cross-section of samples

The cross section of the silver sintered joints indicates that the joints have high porosity at the edge and low porosity at the center area. [10] In order to further understand the joint property, Vickers hardness test of the cross-sectioned joints is conducted. For the sample A joint before TCT (Fig. 4a), the hardness test probe along the joint

displays a higher hardness value (~500Mpa) at both of the edge areas and lower hardness (~250Mpa) in the center. Similar results are observed after 1500 cycles of TCT (Fig. 4b), where the edge and center areas have a hardness around 500Mpa and 270Mpa, respectively. This implies that the silver sintering joints are very stable, consistent with the shear strength data (43Mpa vs 45Mpa at cycle 0 and 1500 for the joints, respectively). Noted that the value of the hardness of sintered joint in the low porous center area is very near to that of the pure silver, which is 250Mpa. [11] The reason why the porous edges have a high hardness value needs further study.

B. Effect of Substrate Surface Finish and Sintering Profile on Joint Reliability

The silver pastes developed here show strong shear strength when Si dies are attached on bare copper surface finished substrate, even when sintering in air atmosphere. This is favored by many customers for cost-saving purposes. Previously, oxidation of the Cu surface has been observed and forming gas is needed for sintering joint formation. [12] Despite having high shear strength, the oxidation of copper is often speculated as the cause of poor reliability. To clarify this, the bonding strength of the sintered silver on copper surface-finished substrates was compared with that on Ag

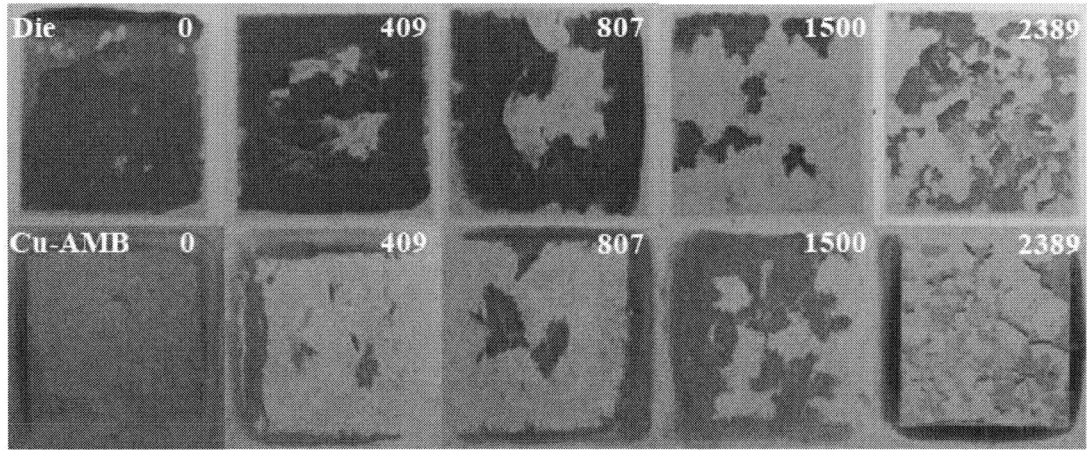

Fig. 6. X-Ray images of sheared 3mm x 3mm Si die and Cu-AMB sides of sample C joint before and after different cycles of TCT. Cycle number is listed at upper right corner of images.

Fig. 7. X-Ray images of sheared 3mm x 3mm Si die and Au-AMB sides of sample C joint before and after different cycles of TCT. Cycle number is listed at upper right corner of images.

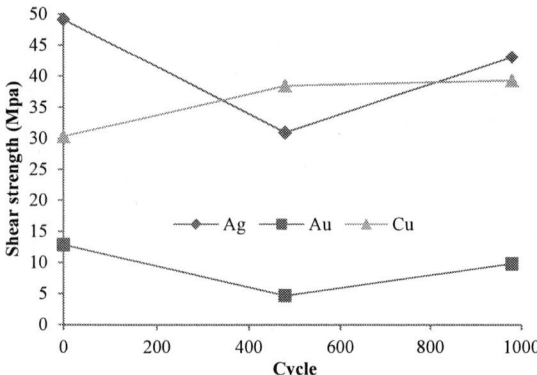

Fig. 8. Shear strength change for the silver sintering joints on different surface finished AMB substrates before and after -40 to 175°C temperature cycling test. The sintered joints are generated with a high Pb solder reflow profile.

and Au surface finished substrates under the same TCT conditions. As shown in Fig. 5, before TCT, the shear strength of the joint on the copper surface was higher than that of Ag or Au surface-finished substrates. Overall, before 1500 cycles, the shear strength was almost unchanged irrespective of different surface finishes, indicating the good stability of silver sintered joints. After 1500 cycles, a decrease in shear strength was observed. The DI values are 6.8, 6.5, and 7.7 for Cu, Ag, and Au finished surfaces, respectively, indicating that they have similar decay behaviors under TCT. This observation shows that sintered joints formed on bare copper surface finished substrates are as good as that on Ag or Au surface finished ones under the TCT test.

The failure mode of the sheared joints on Cu surface finish shows a more cohesive nature; it has the mixed failure modes similar to Ag surface finished substrates before 2879 cycles of TCT (compare Figure 6 to Figure 2). In contrast, the failure location of the Au surface finished one is solely located at the sintered Ag – Au surface finish interfaces as shown in Figure 7; longer cycles of TCT does not change their failure mode.

Sintering profile has a profound effect on the shear strength and failure mode of the sintered joints based on different surface finish on the substrates. Figure 8 shows the shear strength of sintered joints obtained using a commercial BTU reflow oven with a high-Pb profile (320°C peak

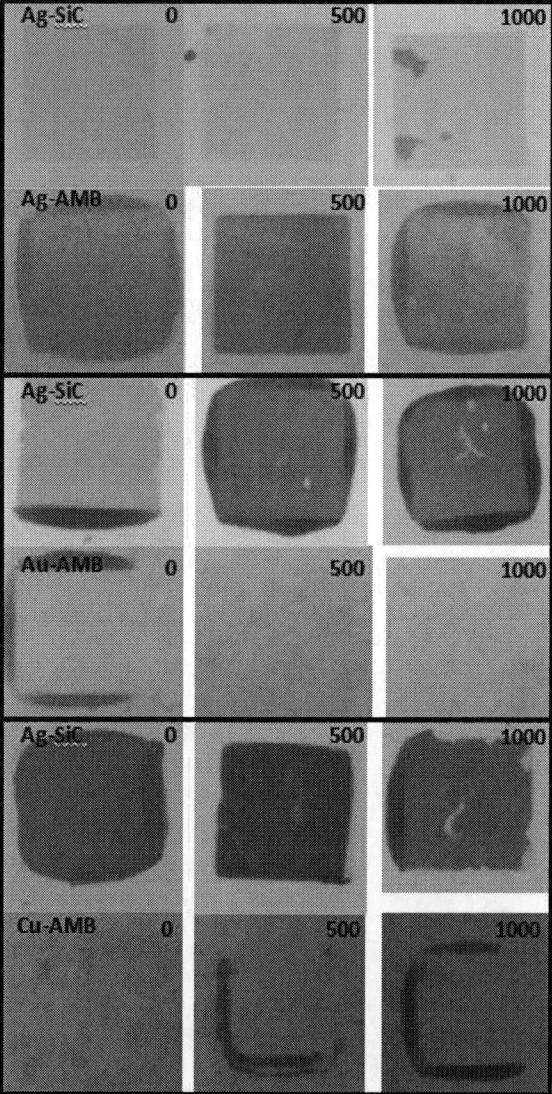

Fig. 9. X-Ray images of sheared 2mm x 2mm Ag-SiC die and Ag-AMB, Au-AMB and Cu-AMB sides of sample C joints before and after different cycles of TCT. Cycle number is listed at up-right corner of images. The sintered joints are generated with a high Pb solder reflow profile.

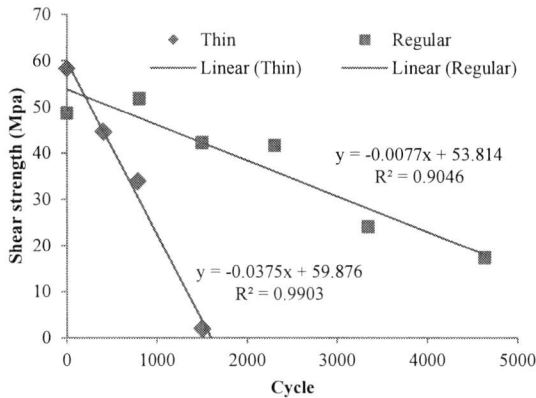

Fig. 10. Shear strength change for silver sintering joints with thin and thick BLTs for the 3mm x 3mm Si die attached on Ag-AMB before and after -40 to 175°C temperature cycling test.

Fig. 11. X-Ray images of sintered Ag joints formed between 10 mm x 10 mm Si die and Ag-AMB substrate before and after different cycles of TCT. Cycle number is listed at up-right corner of images. BLT was at the left-bottom corners, in the range of 44 to 52 μm.

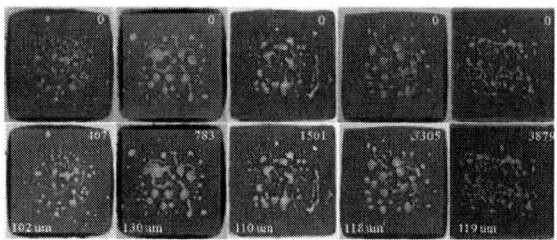

Fig. 12. X-Ray images of sintered Ag joints formed between 10 mm x 10 mm Si die and Ag-AMB substrate before and after different cycles of TCT. Cycle number is listed at up-right corner of images. BLT was at the left-bottom corners, in the range of 102 to 130 μm.

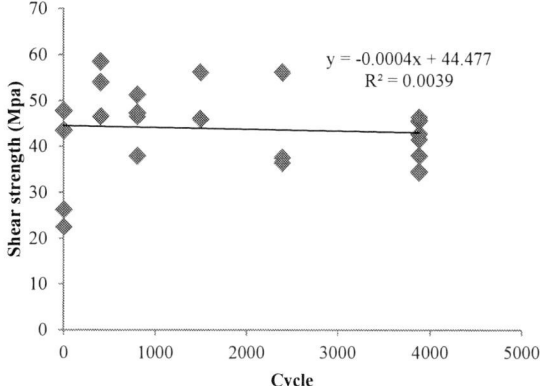

Fig. 13. Shear strength change for silver sintering joints for the 5mm x 5mm Si die attached on Ag-AMB before and after -40 to 175°C temperature cycling test.

temperature and 20 ipm belt speed). Initial shear strength of ~50Mpa was obtained on Ag surface finished substrate, similar to that obtained using box oven (Figure 5). The failure location is at sintered Ag – die interfaces (Figure 9), and TCT does not change the failure mode. This is different from that obtained using a box oven sintering profile (Figure 2).

About half of the shear strength was obtained for Cu-AMB substrate when using high-Pb profiles (Figure 8 vs Figure 5). This strength increased to ~40Mpa after 1000 TCT cycles. The failure location is at the sintered Ag - Cu substrate interfaces, showing an adhesive failure.

The shear strength on the Au-AMB substrate is much weaker when using a high-Pb sintering profile (Figure 8); the failure location is at the sintered Ag – Au surface finish interface, similar to that when using box oven sintering (Figure 7).

C. Effect of BLT on Joint Reliability

BLT has been observed as a critical factor in terms of joint reliability when studying the thermal aging of a silver sintering joint on an Au surface finished DBC. [8] [13] A thin BLT, such as 10 μm, makes the joint easily degrade due to the "depletion layer" formation that is caused by the diffusion of silver atoms into substrate, leaving behind an empty layer. This phenomenon has been observed by several researchers. [14] [15] If a thicker BLT, such as 50 μm, is generated, with a suitable silver sintering paste, the empty

layer can be replenished easily, maintaining the high-reliability of the joints. [8] [13] When considering the observation above, it's still unclear whether the "depletion layer" formation or the thin BLT is the cause of low reliability. Here we studied the sintered joints generated in-between Si die and substrate, both having the same Ag surface finishes, which should eliminate the generation of a "depletion layer". In order to understand the BLT effect under TCT, sintered joints with two BLT thicknesses, 6.5 μm and 50 μm, were prepared with all other conditions the same. The shear strength of the joints using 3mm x 3mm Si dies before and after TCT is shown in Fig. 10. The shear strength of the thin BLT joints decreased much faster than that of the thick BLT joints; the DI value of thin BLT being about five times that of the thick one. It shows BLT is a critical factor to consider when talking about the reliability of the silver sintering joints, regardless of the surface finishes.

The effect of the BLT was also studied with 10mm x 10mm die sizes. The X-ray images of the sintered samples before and after TCT for BLT in the range of 40 to 50 μm was shown in Figure 11. Very low voids are observed, similar to that of the 3mm x 3mm die size case. [10] When the BLT was increased to the range of 102 to 130 μm, large amounts of voids are observed. (Figure 12) The reason for these void formations is the increased amount of solvent, which rapidly evaporated from the center area but could not easily escape from the joint edges.

978-1-7281-1500-9/19 $31.00 © 2019 IEEE

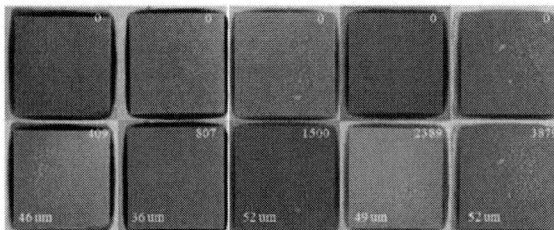

Fig. 14. X-Ray images of sintered Ag joints formed between 5 mm x 5 mm Si die and Ag-AMB substrate before and after different cycles of TCT. Cycle number is listed at up-right corner of images. BLT was at the left-bottom corners, in the range of 36 to 52 μm.

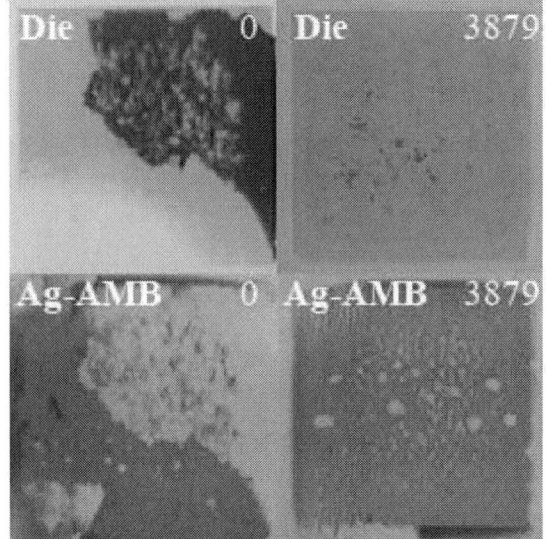

Fig. 15. X-Ray images of sheared 5mm x 5mm Si die and Ag-AMB sides of sample C joint before and after 3879 cycles of TCT. Cycle number is listed at up-right corner of images.

D. Effect of Die Size on Joint Reliability

Die size is a critical factor that affects the bond formation process and bond quality including voids, bond strength, and microstructure. As we have discussed above, the 3mm x 3mm die shows decrease in shear strength after 1500 cycles of TCT (Figures 1 and 5). For the reliability test, normally stresses in the die attach joints increase as the die size increases, and this may pose concern on the reliability. [16] To further understand the die size effect, joints with 5mm x 5mm and 10mm x 10mm dies are prepared and subjected to TCT test. The shear strength of the 10mm x 10mm dies is very strong (>17Mpa) and cannot be sheared off with the available shear tester. The joints normally have very low voids (<1%) for the standard BLT 40 to 50 μm and no delamination was found in the joints even after 2389 cycles of TCT (Figure 11). For 5mm x 5mm dies, Figure 13 displayed the shear strength vs TCT cycle data points for the specimens tested. Interestingly, the shear strength is kept very stable even after 3879 cycles, showing that the silver sintering joints is very reliable. The voids of the joints are normally < 2%, and no delamination was observed for the joints during the TCT (Fig. 14). The failure mode was mixed before TCT, and gradually shifted to that at Si side as the TCT proceeds (Fig. 15); similar to that of the 3 mm x 3mm die cases (Fig. 2).

IV. CONCLUSION

Pressure-less silver sintering pastes with high-reliability judged by TCT under -40°C to 175°C were developed. Several features include:

a). Works for different sized dies with high shear strength. For 3 mm x 3 mm sized dies, the silver joints formed between Si die and Si_3N_4 active metal brazing (AMB) substrate retained the initial shear strength after 1500 cycles, the strength being ~25 Mpa even after 3879 cycles. For 5 mm x 5 mm dies, the average shear strength is very stable and remains at ~42 Mpa even after 3879 cycles. For 10 mm x 10 mm dies, the dies cannot be sheared off (max shear force 200kg) for all the samples before and after 2389 cycles of TCT.

b). The formed joints have very low voids. Within a BLT range of 40 to 60 μm, the average total voids percentage of the sintered joints for the above different die sizes can be controlled within < 2%.

c). Bare Cu surface finish is compatible. The shear strength of sintered joints formed on bare Cu surface finished AMB substrate is slightly stronger than that of Ag- and Au-surface finished ones; TCT shows that these surface finishes have similar decay behaviors, implying that bare Cu surface finished AMB substrate can be used for bonding applications.

d). BLT control is important. BLT is critical in determining the bond property and reliability; for a very thin BLT such as < 10μm, the bond strength decays at about five times that of a joint with ~ 50μm BLT. For a thick BLT, such as > 100 μm with 10mm x 10mm dies, a large amount of voids are generated if using a similar sintering profile, where almost no voids are observed for a ~50 μm BLT joint.

ACKNOWLEDGMENT

The Authors thank Indium Corporation for supporting cooperating programs with universities. William Shambach and Jordan Palmer are co-op students from the Rochester Institute of Technology; Xuanyi Ding is a student from Cornell University.

REFERENCES

[1] C. Buttay, D. Planson, B. Allard, D. Bergogne, P. Bevilacqua, C. Joubert, M. Lazar, C. Martin, H. Morel, D. Tournier and C. Raynaud, "State of the Art of High Temperature Power Electronics," *Mater. Sci. Eng. B: Solid-State Mater. Adv. Technol.*, vol. 176(4), p. 283–288, 2011.

[2] S. Chen and H. Zhang, Die-Attach Materials for High Temperature Applications in Microelectronics Packaging: Materials, Processes, Equipment, and Reliability, K. S. Siow, Ed., Springer, 2019.

[3] K. Guth, D. Siepe, J. Gorlich, H. Torwester, R. Roth, F. Hille and F. Umbach, "New Assembly and Interconnects beyond Sintering Methods," in *Proceedings of PCIM*, 2010.

[4] V. Chidambaram, J. Hald and J. Hattel, "Development of Au–Ge based candidate alloys as an alternative to high-lead content solders," *Journal of Alloys and Compounds*, vol. 490, pp. 170-179, 2010.

[5] V. Chidambaram, H. Yeung and G. Shan, "Reliability of Au-Ge and Au-Si Eutectic Solder Alloys for High-Temperature Electronics," *Journal of Electronic Materials*, vol. 41, no. 8, p. 2107–2117, 2012.

[6] W. So and C. Lee, "Fluxless Process of Fabricating In–Au Joints on Copper Substrates," *IEEE Transition on Components and Packaging Technologies*, vol. 23, no. 2, pp. 377-382, 2000.

[7] Y. Takaku, L. Felicia, I. Ohnuma, R. Kainuma and K. Ishida, "Interfacial reaction between Cu substrates and Zn–Al base high-temperature Pb-free solders," *J Electron Mater*, vol. 37, p. 314–323,

978-1-7281-1500-9/19 $31.00 © 2019 IEEE

2008.

[8] S. Chen, C. LaBarbera and N. C. Lee, "Silver Sintering Paste Rendering Low Porosity Joint for High Power Die Attach Application," in *IMAPS Conference & Exhibition on HiTEN 2016*, Albuquerque, NM, 2016.

[9] M. Knoerr, S. Kraft and A. Schletz, "Riliability Assessment of Sintered Nano-Silver Die Attachment for Power Semiconductors," in *12th Electronics Packaging Technology Conference*, 2010.

[10] S. Chen, W. Shambach, C. LaBarbera and N. C. Lee, "Highly Reliable Pressure-Less Silver Sintering Joints," in *IMAPS Conference & Exhibition on HiTEN 2018*, Albuquerque, NM, 2018.

[11] G. Samsonov, Handbook of the physicochemical properties of the elements, New York: IFI-Plenum, 1968.

[12] S. T. Chua, K. S. Siow and A. Jalar, "Effect of sintering atmosphere on the shear properties of pressureless sintered silver joint," *36th International Electronics Manufacturing Technology Conference*, 2014.

[13] S. Chen, G. Fan, X. Yan, C. LaBarbera, K. Lee and N. C. Lee, "Low-Porosity Pressureless Sintering of a Novel Ag Paste for Die-Attach," in *Proceedings of SMTA International*, Rosemont, IL, 2014.

[14] F. Yu, R. W. Johnson and M. Hamilton, "Pressureless, Low Temperature Sintering of Micro-Scale Silver Paste for Die Attach for 300 C Applications," in *IMAPS Conference & Exhibition on HiTEN 2014*, New Mexico, 2014.

[15] G. Lewis, G. Dumas and S. H. Mannan, "Evaluation of Pressure Free Nanoparticle Sintered Silver Die Attach on Silver and Gold Surface," in *IMAPS Conference & Exhibition on HiTEN 2013*, Oxford, United Kingdom, 2013.

[16] K. S. Siow, "Are Sintered Silver Joints Ready for Use as Interconnect Material in Microelectronic Packaging?," *Journal of Electronic Materials*, vol. 43, pp. 947-961, 2014.

2019 IEEE 69th Electronic Components and Technology Conference (ECTC)

3D Power Packaged Device Thermo-mechanical Modeling and Stress Analysis after Reliability Trials

L. Guarino, L. Zullino, L. Cecchetto, F. Pozzobon and A. Andreini

STMicroelectronics, Via C. Olivetti, 2, 20864 Agrate Brianza (MB), Italy
Email: {lucrezia.guarino\lucia.zullino\luca.cecchetto\fiorella.pozzobon\antonio.andreini}@st.com

Abstract— Power devices technology innovation is moving to new metallization schemes in order to improve electrical performances. The integration of new materials, especially low-k dielectric, requires an increasingly deep understanding of thermo-mechanical stress due to packaging processes and reliability. In this work, a 3D Power Package thermo-mechanical modeling is presented as a powerful tool to predict stress generation not only on fresh die but also after reliability trials. Mechanical properties of two different polymeric materials, available both for as-molded and aged material at different temperatures, are implemented in the model in order to better reproduce strain and deformation at the end of each reliability trial. The above-mentioned package model is then coupled with the silicon chip top surface by the sub-modeling technique to achieve a micro-scale more accurate stress estimate; the focus of stress analysis is on delamination failures mode.

Keywords-package; reliability; thermal cycling; high temperature storage; delamination; molding compound

I. INTRODUCTION

Present challenges for Power technology consist in the selection of suitable materials and in the analysis of the Chip Package Interactions (CPI) to the aim of ensuring and enhancing the ICs reliability performances against mission profiles becoming more severe, in particular in the Automotive market. In High Power applications, device operates in high current and high temperature conditions, so an efficient system of heat dissipation is required. A high thermal conductivity is guaranteed through a family of packages, called Power packages, designed with specific characteristics: a die attach material high conductive and adhesion robust at all temperatures and a metallic lead frame with a large die pad to quickly dissipate heat out of the package ("Figure 1").

Figure 1. Typical Power package

At the same time, specific requirements must be satisfied, as reported in reference quality standards, for example in AEC-Q100 [1]. In particular, it is important the study of Chip Package Interaction, with the monitoring of thermo-mechanical stresses and the prevention of die breakage or lack of adhesion inside the system. Specifically, in Power packages one of the main source of solicitations is the molding compound: its high thickness and strains due to production process causes stress generation on die top surface.

In technology platform qualification, delamination failure mode is addressed with specific trials, for example the Thermal Cycling (effect of 1000 cycles from -55°C to 150°C), as reported in JEDEC standards [2]. Another typical technology qualification test is the High Temperature Storage Life (effect of storage for 1000h @ 175°C). Both treatments enhance stress levels at die / molding compound interface due to thermo-mechanical different response of involved materials. In addition, these reliability trials also induce an ageing effect on the molding compound, so initial stress distribution changes during tests progress. For this reason a Finite Element model, able to study solicitations for as-molded and aged conditions, is needed.

Many works have been presented that analyze delamination, using for example cohesive zone model [3] to describe the interface crack propagation, while molding compound ageing is also analyzed, using for example simplified tests cases, like bi-materials [4].

In this paper a 3D Power Package model is built-up to analyze the electronic device at the end of production process and also after reliability ageing, with the focus on delamination failure mode at die top surface. The model consists of an accurate reproduction of package and die, so simulation results can be linked directly to real structures in order to drive material choice and design layout rules during technology platform qualifications.

II. MODEL

In this section the employed simulation model is described: the geometry and details of the power package employed, the thermo-mechanical model and material properties implemented, and finally the sub-modeling technique approach for a more fine analysis.

A. Geometry

A 3D CAD design is imported in Comsol Multiphysics simulator to reproduce the real geometry of the package under study. The structure is in fact detailed in each part:

978-1-7281-1500-9/19 $31.00 © 2019 IEEE

metallic lead frame, soft solder die attach material, die silicon and molding compound, as shown in "Figure 2".

The metallic lead frame is composed by a pad, on top of which the die is welded with a soft solder die attach material, leads and respective pins. The die, composed by silicon and a passivation top layer, is then covered into a molding compound resin to ensure silicon integrity (no bonding wires are reproduced in this model).

The package symmetry due to its design allows the simulation of only a quarter of the system, simplifying the models in terms of computational time and resources.

B. Physics & Materials

The thermo-mechanical description of the package requires the implementation of a model able to consider material intrinsic properties, stresses due to assembly steps and structure layout. For this last contribution, a detailed geometry is reproduced, while for the other two a specific experimental characterization has been performed.

The physical model used for thermo-mechanical analysis is isotropic linear elastic: Young's Modulus (E), Poisson's ratio (υ), density (ρ) and thermal expansion coefficient (CTE) are the properties required for each material.

Since stress levels on die top surface are mainly due to molding compound contribution, an in-depth study of its mechanical behavior is needed for as-molded and aged materials, as a function of temperature and time.

Young's Modulus values are obtained using Dynamical Mechanical Analysis technique [5] on a molding compound dog-bone of two different materials (namely molding A and B) at three temperatures (-55°C, 25°C and 150°C). The analysis is done on as-molded, after 1000 Thermal Cycles (TC) between -55°C and 150°C and after 1000h of High Temperature Storage Life (HTSL) at 175°C. Experimental results are summarized in "Figure 3".

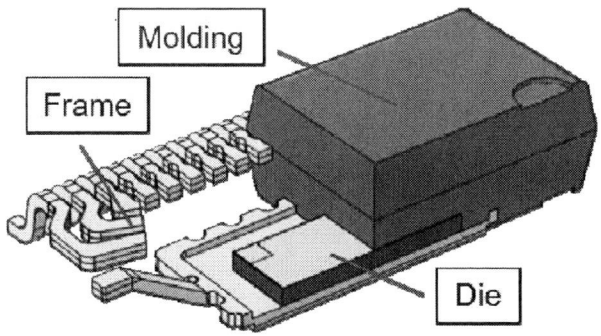

Figure 2.   Power package simulated geometry: metallic frame, die and molding compound.

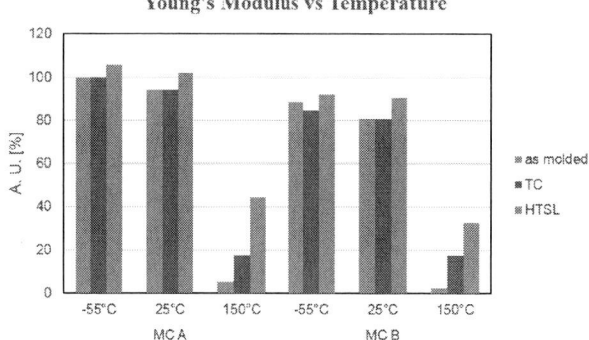

Figure 3.   Ratio of Young's Modulus (reference value set to 100% for MC A at -55°C with as-molded properties) as a function of Temperature for molding compound A (MC A) and B (MC B) as-molded, after TC and after HTSL.

Mechanical characterization highlights a higher stiffness at cold temperatures (which is in fact chosen as a reference value in "Figure 3"), with a decrease of Young's Modulus moving to hot temperatures: molding compounds are in fact polymeric materials with a glass transitions temperature of ≈135°C, so at 150°C they are in viscous state. Regarding as-molded properties, there are no significant differences between molding A and B. Molding compound ageing from reliability trials rises Young's Modulus only at hot temperatures, with HTSL more than TC, and the impact is different for the two materials.

In addition, in order to consider assembly process contribution, intrinsic stresses are introduced for molding compound, die pad and soft solder die attach, as obtained by a warpage package characterization together with FEM modeling, as in [6].

Experimental characterization and geometry are then used to create the package model to study die stress levels in the temperature range of -55°C / 150°C.

C. Sub-modeling

Once the package model is built-up, it is possible to use the sub-modeling technique to explore more in details stresses and related strains on device. In fact the package simulation solutions can be used as input for a smaller model, in which very detailed structures can be reproduced to study package impact with high accuracy. The coupling is made by the implementation of prescribed displacements, variables calculated by the simulator in thermo-mechanical studies, at the external geometry faces of the sub-model.

In this work a small portion of package model is considered: a thin layer of molding compound and passivation at the die corner. Inside this volume metals structures are drawn in a very detailed way, see "Figure 4".

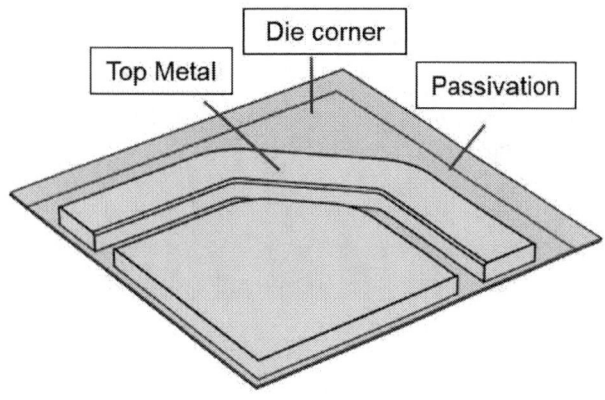

Figure 4. Sub-model geometry: passivation layer and two metal structures (an external ring and a wide plate).

In this study a typical top metal scheme, out from passivation as the last layer of the die, is presented: an external ring and a metal plate are drawn in agreement with device layout rules.

The focus of this work is the evaluation of stresses related to delamination failure mode on top of metal layers, as a function of temperatures and molding (two different materials, as-molded and aged, as mentioned before).

III. RESULTS

In this section the 3D Power Package stress analysis is presented. In particular the focus is on stress levels on die top surface, dealing with delamination failure mechanism. Molding compounds role together with temperature dependence and reliability impact is evaluated.

A. Power Package Stress vs Temperature

The aim of thermo-mechanical analysis is the identification of stress components responsible of delamination at die level. The package model enables the study of solicitations distribution on device passivation due to both package warpage and material properties mismatch. Delamination in fracture mechanics consists of a cohesive fracture generation: an interface crack propagates between the two materials causing the separation between them. The lack of adhesion depends on the nature of applied loads and it is classified as mode I, normal opening, or mode II, in-plane shear sliding of the interface [7].

Therefore the physical parameters under investigation in this work are [8]:

- Peel Stress: the normal stress component (causing opening only if tensile, i.e. positive) perpendicular to materials interface
- Shear Stress: the parameter considering the two in-plane shear components (scalar physical magnitude)

Simulation results for package macro-model are shown in "Figure 5".

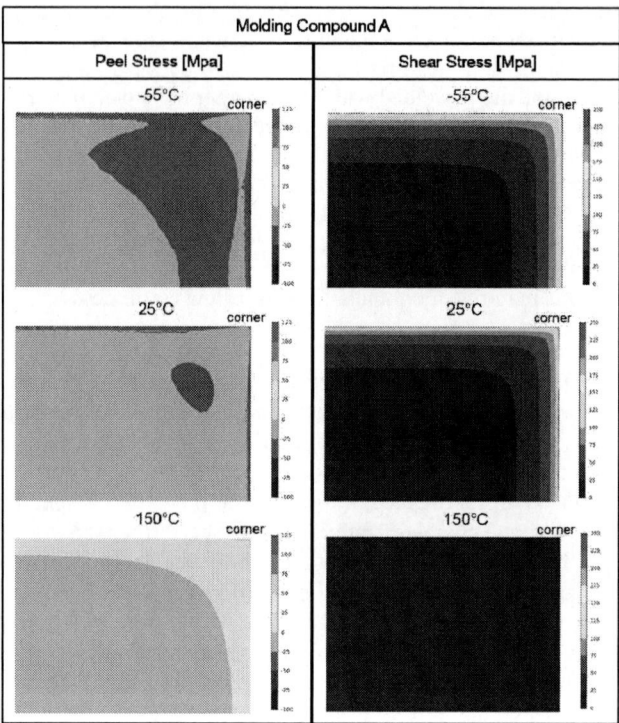

Figure 5. Peel and Shear distributions on die top surface for molding A power package at three temperatures (-55°C, 25°C and 150°C).

Stress distributions at die top surface (passivation level) are reported as a function of temperature for resin A.

Package solicitation are higher at die corner with a more important contribution of Shear type. Temperature role is different for the two evaluated parameters: cold conditions increases Shear stress with a difference in the analyzed range of ≈ 200MPa; on the contrary Peel stress rises from -55°C to 150°C moving from negative (compressive) not significant values, to positive (tensile) values, with a peak at die corner of ≈ 125MPa.

Stress distributions highlight as the most critical area the die corner, suggesting the construction of the detailed sub-model described in section II.C.

B. Molding Compound Comparison

In this section stresses distributions for molding compound A and B are reported. The analyses presented are the results of sub-modeling studies, focusing the attention on metallization structures previously described. In particular, Shear and Peel stresses are evaluated on metals top surface to study delamination failure modes at temperature range limits. Simulation results are shown in "Figure 6" and "Figure 7".

Shear stress behavior observed in package model is confirmed also in sub-model: it is higher at cold temperatures and decreases moving from corner to die center. The distributions are, however, influenced by metal presence: higher stresses are found at external structures edges, with a stress relaxation at the ring internal side. No significant difference is visible between the molding compounds, both at -55°C and 150°C.

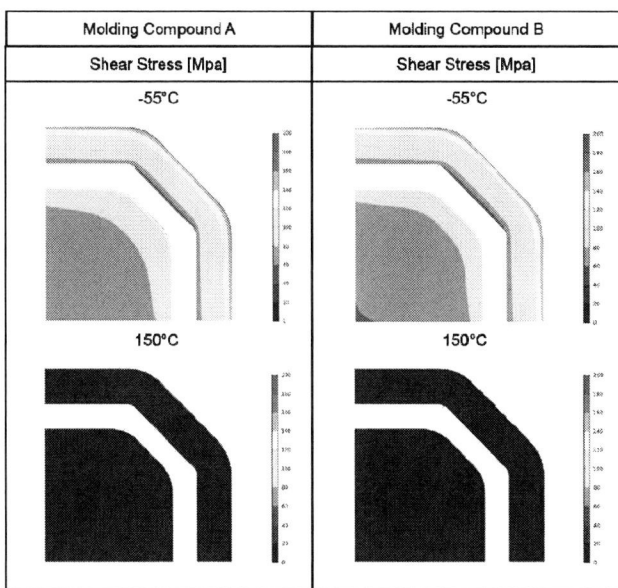 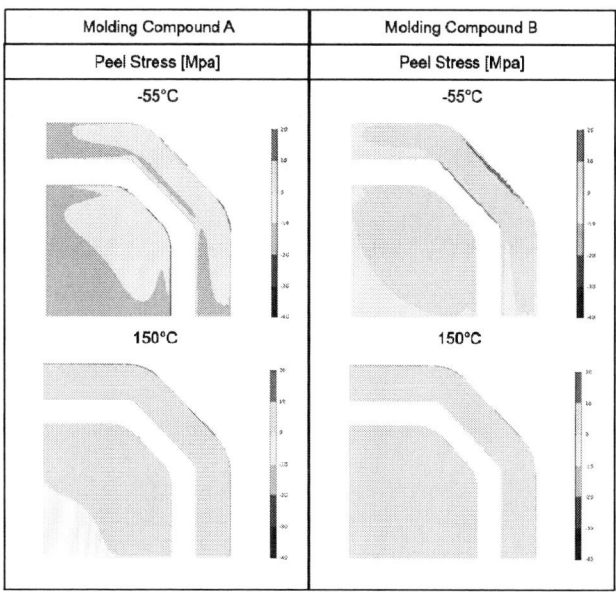

Figure 6. Shear Stress distributions on metal top surface for molding A and B at -55°C and 150°C.

Figure 7. Peel Stress distributions on metal top surface for molding A and B at -55°C and 150°C.

Peel stress also reproduce package model results, shown for molding A in "Figure 5", with negative values at cold temperature and not influenced by metal structures introduction. Stress then moves to positive values rising temperature. A different behavior is found for molding B: at -55°C Peel stress is yet of tensile type and moving to hot conditions it does not change, highlighting a not meaningful temperature dependence. Peel stress values for molding A and B are comparable and very low (quasi-zero), especially if compared with Shear stress levels (one order of magnitude higher).

C. Reliability Impact

Finally, the impact on the stresses at die surface is reported when molding compound properties are aged by reliability trials: this study has been enabled by the previously illustrated experimental characterization of molding compounds. Stress distributions at top metal surface are presented for TC and HTSL aged materials and compared with initial as-molded ones. Simulation results are shown in "Figure 8" (molding A) and in "Figure 9" (molding B) at 150°C, because, as highlighted by experimental characterization, the change of mechanical properties is remarkable only at hot temperatures ("Figure 3").

The molding compound ageing from reliability trials induces a different behavior of stress parameters. Thermal ageing from TC and HTSL lowers Peel stress for molding A, while for molding B values increase at metal ring edges while decrease on top of the metals. Nevertheless, for both molding compounds it can be excluded as a risk factor for delamination failure mode.

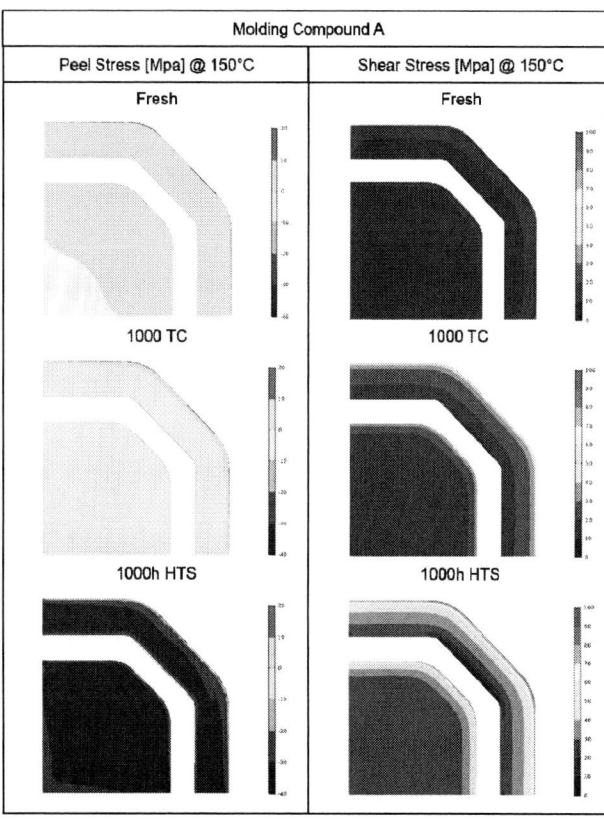

Figure 8. Peel and Shear distributions at 150°C for molding A and B in as-molded, 1000 TC and 1000h HTS aged cases.

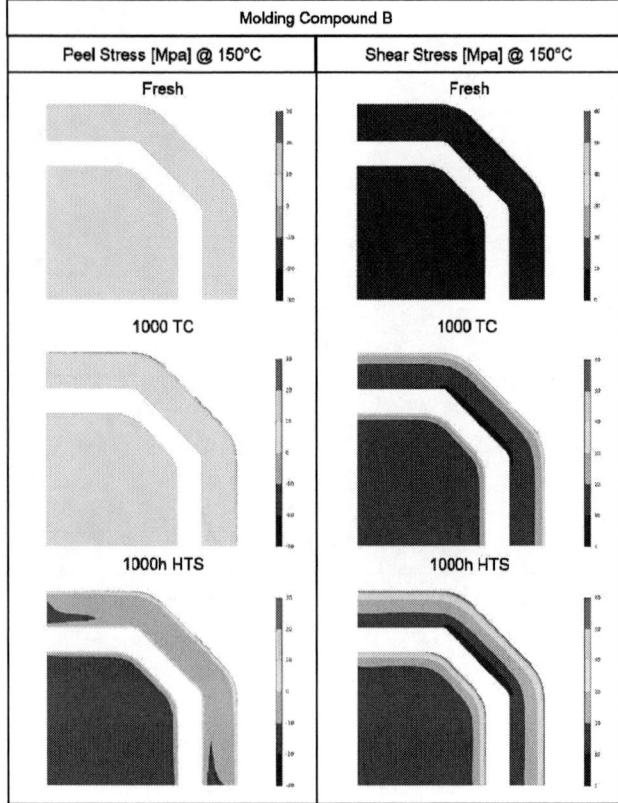

Figure 9. Peel and Shear distributions at 150°C for molding A and B in as-molded, 1000 TC and 1000h HTS aged cases.

On the contrary, Shear stress increases with TC and HTSL aged molding compound, with the second trial showing the highest values. In particular, Shear stress for molding A moves from a maximum value of ≈ 20MPa for as molded material to ≈ 70MPa for TC aged and to ≈ 100MPa for HTSL aged; for molding B maximum values increases from ≈ 10MPa to ≈ 50MPa for TC ageing and to ≈ 60MPa for HTSL ageing.

D. Results Analysis

Stress distributions results highlight that delamination at die surface could be due to solicitation of Shear type, i.e. of mode II sliding of involved materials. Peel contribution can be in fact neglected because of very low observed values, for both molding compounds also after reliability tests. The focus is thus only on Shear and its behavior in analyzed cases is summarized in "Figure 10". Shear stress higher values are found at -55°C, with a reduction of ≈ ten times moving to 150°C for as-molded cases. This behavior points out cold temperatures as the worst conditions for die / molding adhesion. Instead, reliability ageing increases Shear stress at hot temperatures (mechanical property change): the values increases from ≈10% to ≈50% in the worst case, with respect to the cold stress level (Figure 10). So after TC and HTSL, also Shear stress at hot temperatures has to be taken into account.

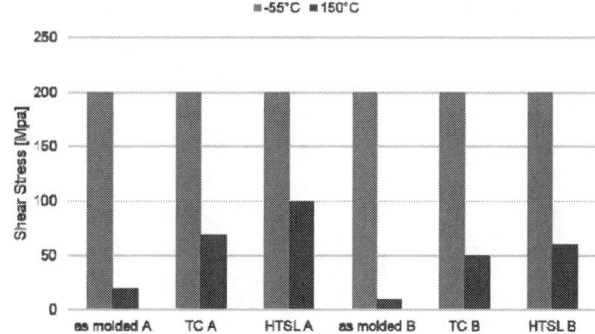

Figure 10. Shear stress maximum levels at different temperatures found for molding A and B as-molded and after ageing from reliability trials. Contributions at -55°C and 150°C are shown for each case.

Comparison between fresh and aged cases is presented in "Figure 11". Results show a Shear stress increase for both molding compounds due to the ageing from reliability trials: the worst increase is found in HTSL model (≈ ×5 for A and ≈ ×6 for B). Instead for Thermal Cycling Shear stress increase is lower: ≈ ×3.5 for molding A and ≈ ×5 for molding B. The stress increase due to materials ageing highlights the importance of the hot temperatures: this contribution can not be neglected after reliability ageing. In fact the molding compounds analyzed are apparently stress equivalent in as-molded models, instead stress distributions are significantly different in TC and HTSL models, with Shear stress values for molding A higher than molding B.

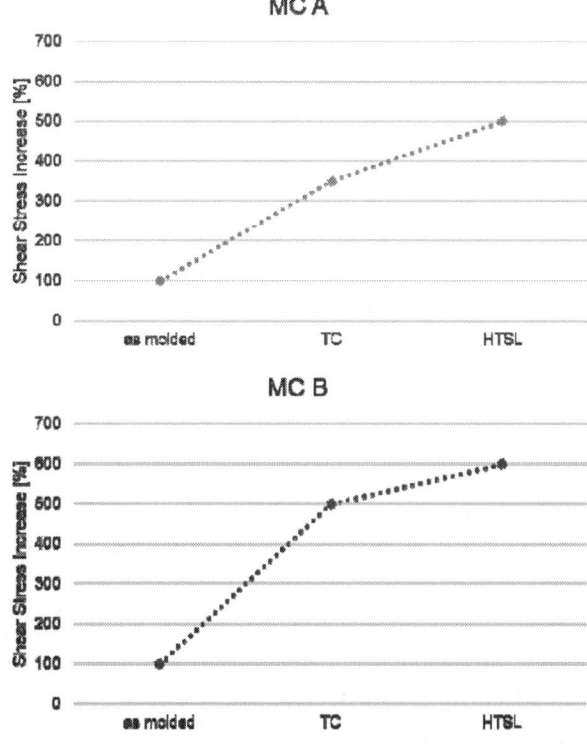

Figure 11. Percentage Shear stress increase at 150°C in aged models for molding A and molding B (reference value set to 100% for as-molded cases).

IV. CONCLUSION

In this work the stress related delamination failure mode on die top surface, for both as-molded and after reliability trials, is evaluated with a 3D Power Package device model. The analysis is done at cold (-55°C) and hot (150°C) temperatures and for two type of molding compounds, characterized by properties measured in the as-molded case and aged by Thermal Cycling and High Temperature Storage Life. The sub-modeling technique implemented allows the evaluation of Peel and Shear stresses due to package thermo-mechanical solicitations on top metals at the die corner. Shear is pointed out as the stress component responsible of delamination and is higher at -55°C. Reliability trials increase stress at 150°C to values not negligible and comparable to cold levels. The package models after ageing allows a benchmark between the two molding compounds that are in fact equivalent in terms of stress if considered only in as-molded case.

ACKNOWLEDGMENT

The authors would like to thank Professors Pasquale Vena and Dario Gastaldi of the Politecnico of Milan University for the molding compound characterization.

REFERENCES

[1] AEC-Q100, Rev. H, "Failure Mechanism based Stress Test Qualification for Integrated Circuits," Sept. 2014

[2] JEDEC, "Standard No. 22-A104C, Temperature Cycling," May 2005

[3] Nadine Pflügler at al, "Advanced Risk Analysis of Interface Delamination in Semiconductor Packages: A Novel Experimental Approach to Calibrating Cohesive Zone Elements for Finite Element Modelling," EuroSimE, 2018

[4] Bingbing Zhang et al, "Thermal aging modeling of molding compound under high-temperature storage and temperature cycling conditions," EuroSimE, 2018

[5] K. P. Menard, "Dynamic Mechanical Analysis: A Practical Introduction", CRC Press, 1999

[6] Marco Rovitto et al, "Dynamic Warpage Analysis of QFP Packages during Soldering Reflow Process and Thermal Cycle", ECTC, 2018

[7] A.V. Pocius and D.A. Dillard, "Adhesion Science and Engineering: Surfaces, Chemistry and Applications," Elsevier, 2002

[8] J.-H. Zhao et al., "Packaging optimization to mitigate stress induced parameter shift in precision devices", 12th IEEE Intersociety Conference on Thermal and Thermomechanical Phenomena in Electronic Systems (Itherm), 2010

Millimeter Wave Dual Polarization Design using Frequency Selective Surface (FSS) for 5G Base-station Applications

Chi-Hau Yang, Chung-Yi Hsu, [1]Lih-Tyng Hwang

Dept. of Electrical Engineering, National Sun Yat-Sen University

Kaohsiung, Taiwan

[1]fiftyohm@mail.nsysu.edu.tw

Abstract— **According to 5G communication bands proposed by various countries, the bands of interest can be divided into 28 GHz/38 GHz and 60 GHz categories, both are millimeter waves. The short wavelengths from the millimeter waves are less likely to propagate (high attenuation), more base stations (i.e., small cells) are required to improve the coverage. Array antennas with beamforming capabilities play another very important role, as a remedy for high attenuation. The antenna proposed in this paper employs a massive dual-polarized MIMO antenna with smart beam switching capability at 38 GHz, in which frequency selective surface was designed to achieve good polarization diversity (decomposition polarization). The frequency selective surface compensates phase delays for electromagnetic waves in space. The phase compensation in our design reached a wider beam angle and had field diversity characteristics. In this article, we first discuss the phase delay of a single FSS and demonstrate the beam diversity using a single antenna with FSS configuration, and finally discuss design improvement using array antenna (that is, replacing the single antenna with an array).**

Keywords— massive MIMO, beamforming, dual-polarization, phase array antenna, frequency selective surface (FSS)

I. Introduction

In [1], an active massive system architecture was proposed, in which power amplifier and low noise amplifier are placed at the front end of each antenna element, and switching between transmission and reception is controlled by a switch. With this architecture, each radiating element can transmit and receive independently. The advantage is that the entire system can still work normally when some antennas fail, and the reliability and stability of the system can be improved.

Phase shift of the antenna front end can be used to synthesize a main beam [2]. This paper also introduces the system specifications, including phase noise, received power, noise index and antenna gain. In [1], it can also be seen that when the beam angle is switched above 20 degrees, the amplitude of the side lobes is also increased, which may cause the receiver to receive non-target signals and cause interference. It is traditionally a very difficult problem on phased antennas.

The beamforming technique and the choice of the best transmit antenna are provided in [3]. In this paper, three different transmission beamforming techniques are compared, namely maximum ratio transmission beamforming (MRTBF), zero forcing beamforming (ZFBF), and minimum mean square error beamforming,

(MMSEBF). In an overall comparison, the number of RF components required for MRTBF needed to achieve the same performance as ZFBF and MMSEBF is the maximum, but E/N ratio of MRTBF is relatively low.

Massive MIMO antennas have the ability of beamforming in space, so it is necessary to effectively and flexibly control the phase of each radiating element in beamforming to avoid destructive interference in space to reduce efficiency. In the case of multiple inputs and multiple outputs, it is also necessary to perform mutual isolation mechanisms for each unit, whether at the ground end of the system, spatial diversity, and polarization diversity, to avoid interference between the antenna elements.

The design methods for improving the isolation of MIMO antennas are as follows: (1) adding a protruded ground plane [4] or a negative group delay between antennas [5]; (2) implanting the slot [6-7] at the ground of the antenna; (3) adding the short-circuit pin [8-9]; (4) controlling the radiation polarization of the antenna [10-12]; (5) using the differential ground path [13] or the bandstop filter [14-15]; (6) using a transmission line to make a wavetrap resonator [16] or join an open ring resonator [17]; (7) adding a decoupling neutralization line [18-19]; and (8) implementing a co-radiation structure design [20-22]. Comparing the above eight methods for increasing the isolation, in which the short-circuit pin, polarization diversity, and decoupling neutralization line can effectively increase the isolation between components by eliminating the need for additional space or components. However, it will cause some disadvantages of field diversity, impedance and frequency band mismatches.

Massive MIMO performs beamforming in space. In order to effectively perform beamforming and the antennas of each other do not interfere with each other, spatial diversity needs to be done to increase spatial multiplexing [23]. The antenna diversity can be further divided into antenna patterns by designing the amplitude of the antenna elements or changing the phase of the antenna so that they do not interfere with each other, and the diversity of the received signals is performed in the space according to the position of the antenna which called spatial diversity. And the design of the polarization characteristics of the antenna, called Polarization diversity [24].

Since large beam switching angles tend to increase the side lobes, conventional zero-pole control [25] or additional antennas, which used to eliminate this side-lobes, will result in system complexity or increase the power consumption. In this paper, a patch antenna integrated with an FSS is designed. The patch antenna unit will have orthogonal polarizations for spatial polarization diversity, and both polarizations can be operated independently, when vertical polarization is received, horizontal polarization can also be used to receive or transmit signals. The phase compensation characteristic of FSS is used to correct the electromagnetic phase delay of the antenna when it radiates in space, and the radiation pattern of high directivity is achieved. The use of FSS isolates the dual-polarized antenna and limits the acceptance of the incident waves to eliminate the radiation lobes while reducing system complexity. Then use of FSS can effectively switch the antenna polarization to achieve good polarization diversity and field diversity in space. The structure of the base-station, patch antenna array with FSS, is shown in Figure 1.

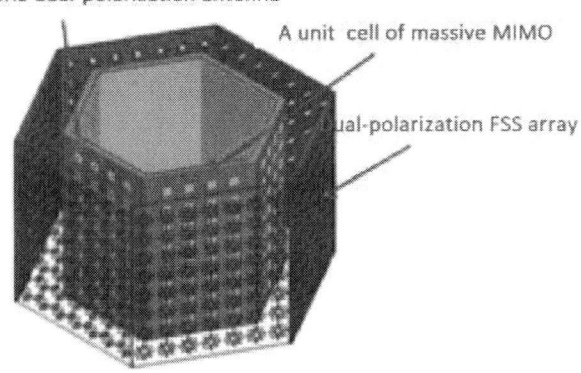

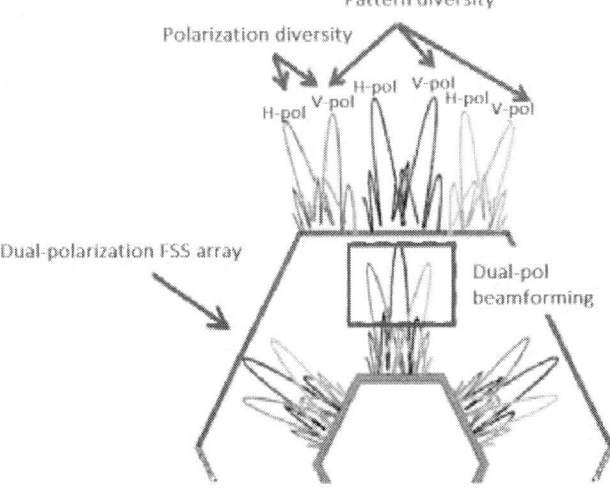

Figure 1. Small cell base station using antenna array integrating with an FSS.

II. FREQUENCY SELECTIVE SURFACE

The frequency selective surface (FSS) is an array metal structure with a two-dimensional periodic structure. The frequency response of the original circuit can be changed by adjusting the shape and size of the metal. The advantage is

that the structure has a low attitude and is easy to design and attach to the object.

The common types of the cell structure are the patch and the aperture as shown in Fig. 2. From the frequency response, the patch type has a band stop and the aperture type has a band pass characteristic, are simple structure and have filter characteristic.

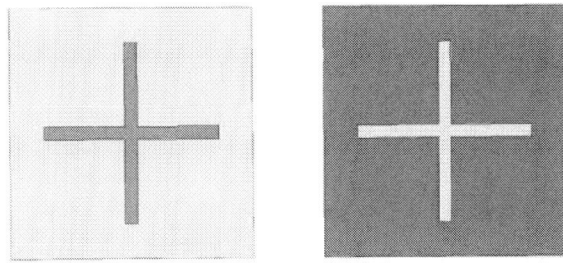

Fig 2. Unit cell structures in an FSS: a patch (left), an aperture (right).

When the electromagnetic wave is incident on the patch type's FSS, the metal on the surface generates an induced current. At this time, the induced current causes inductive effect on the metal, and the coupling between the metal gaps produces a capacitive effect, and the overall circuit can be regarded as an LC series circuit, as shown in the figure. 3, and its frequency response is the characteristics of band stop.

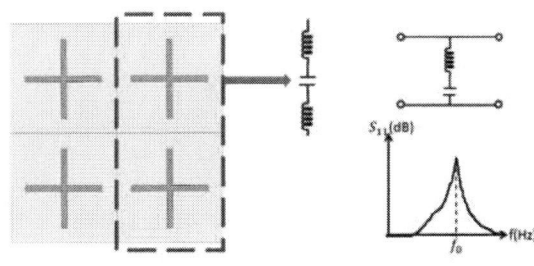

Fig 3. Circuit and frequency response from FSS with patch.

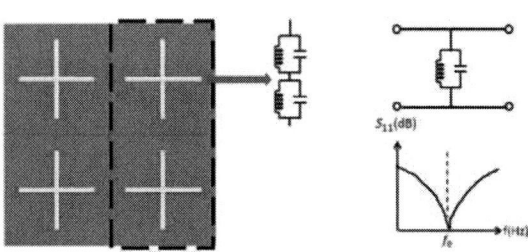

Fig.4 Circuit and frequency response from FSS with aperture.

When electromagnetic waves are incident on the aperture FSS, the induced current around the low-frequency slot is small. As the frequency increases, the slot will induce a large amount of current. At this time, the excited slot can be regarded as an LC parallel circuit, and the frequency rises as the frequency rises. The current is gradually reduced as shown in Figure 4. The frequency response is the characteristic of band stop.

Another application of the polarization selection feature of FSS. When a circularly polarized wave rotated 360 degrees in space incident a dual-polarized FSS, the FSS selects the polarization of the electromagnetic wave passing according to the selectivity of the polarization of the electromagnetic wave. When a circularly polarized wave incident a slot of the same length and orthogonal, its electromagnetic wave polarization is excited by the two orthogonal polarizations. With the rotation of the circularly polarized wave at the polarization angle, the two orthogonal slots are excited at the same time, and the two polarizations are synthesized after passing through the FSS.

It is known that the electromagnetic wave's delay characteristic after passing through FSS, when the slots of the same length are subjected to electromagnetic waves perpendicularly, the two orthogonal slots will have the same electromagnetic wave delay. The electromagnetic waves synthesized through the two orthogonal slots are combined in the phase of the orthogonal polarization by 90 degrees in accordance with the original electromagnetic wave, and passed the circularly polarized wave. The FSS's polarization decomposition characteristics can also be observed after decomposing this FSS.

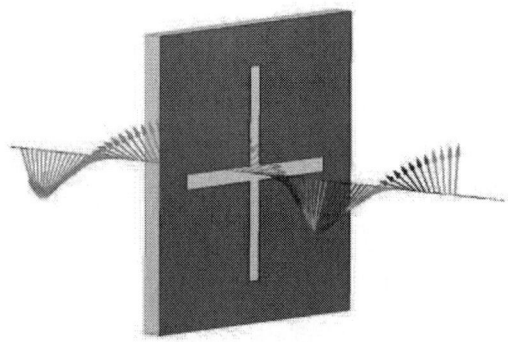

Fig 5. Polarization selectivity of FSS.

As shown in Fig 5, when the circularly polarized wave incident the slot, the FSS only excites electromagnetic waves in the direction of the vertical slot. Using the selectivity of this polarization and the electromagnetic delay characteristics of the FSS, the size in the horizontal and vertical polarization directions can be varied on the FSS. According to the theory of phased arrays, when the spatial electromagnetic waves have different phases, the synthesized field patterns will be offset. If the horizontal

and vertical polarizations are preferably given different phase differences, the two polarizations can be separated from each other in space to form a dual beam radiation pattern, and the two beams are respectively entrained horizontally and Vertically polarized electromagnetic waves, as shown in Figure 4, are driven into the transmitarray with a dual-polarized antenna to obtain two separate beams.

A. Unit Cell Design

Figure 6 shows a single unit cell of the frequency selective surface using a FR-4 (loss tangent = 0.02) substrate with a thickness of 0.4 mm. It can be seen from Fig. 3 that by adjusting the length L of the bent dipole, the phase of the electric field that it penetrates changes accordingly, the single cell can achieve phase delay 50 degrees, within 1 dB loss, as shown in Fig. 7.

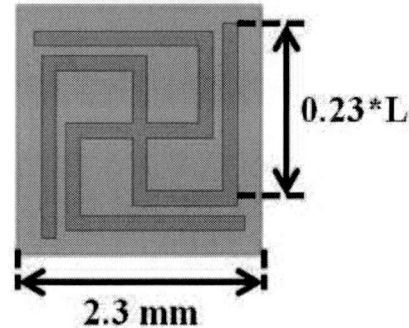

Fig 6. Single (patch) unit cell in our design.

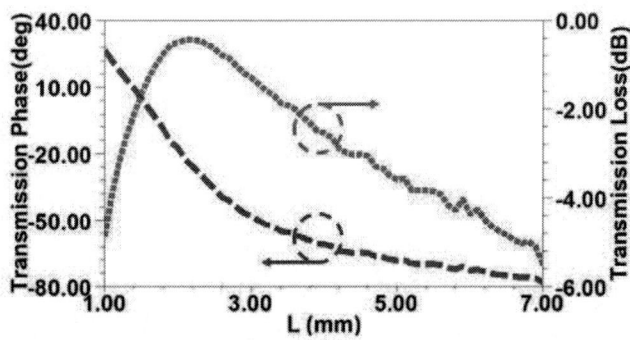

Fig 7. One-layer unit cell trasmisson loss and phase delay.

In order to achieve full-wave (0-360 degrees) phase propagation, the superimposed unit cells are used to accumulate the delay phase Fig 8. The range of the phase can be full wave (0 to 360 degrees) by adjusting the length L, and the phase loss within 5 dB is shown in Fig 9. It can be seen that with the FSS operating at high frequency of 38 GHz, it is still possible to use a high-loss substrate to obtain a conventional transmission line, thereby effectively reducing the cost.

978-1-7281-1500-9/19 $31.00 © 2019 IEEE   2202

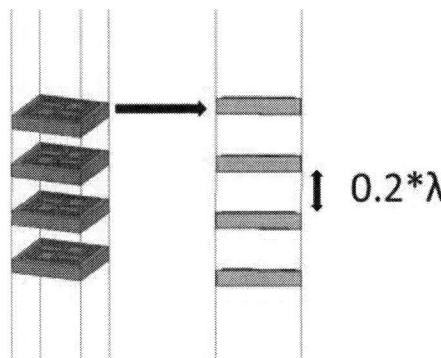

Fig 8. Four-layer unit cell structure.

As can be seen from the above, the penetration phase of the electromagnetic wave can be adjusted by adjusting the length L of the bent dipole.

It can be known from previous research that FSS has the selectivity of polarization, and by using these two characteristics, the two orthogonal characteristic electric fields can be delayed in phase with each other, and then the phase array antenna theory can be used to effectively adjust the beam.

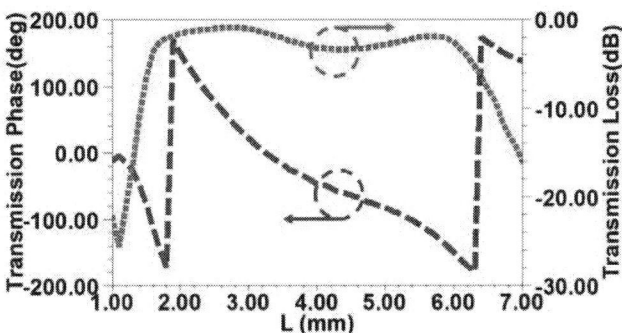

Fig 9. Four-layer unit cell transmission loss and phase delay.

B. Dual-polarization Transmitarray Design

The structure is shown in Figure 10. It can be seen from that the transmitarray of the present study deliberately adjusts the lengths of the bent dipoles in the X and Y directions, so that the phase electric fields of Ex and Ey in the space can reach different phase delays. In order to achieve the expected dual polarization diversity in space, the space beamforming can be used more effectively. More perfect multi-input and multi-output effects in MIMO systems.

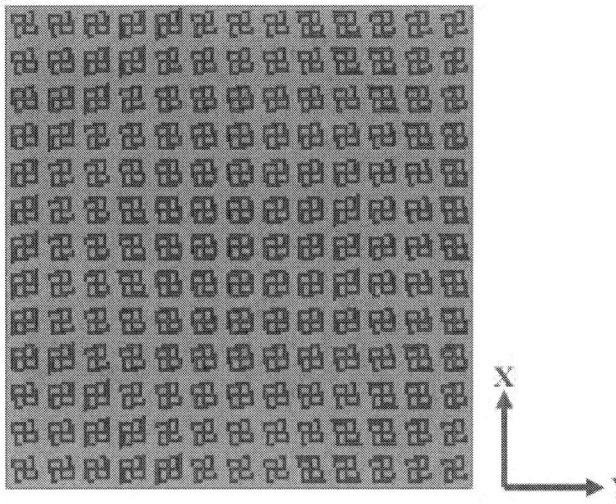

Fig 10. Dual-polarization transmitarray.

C. Antenna Design

In the design of the antenna, we use the rectangular patch antenna to simulate the effect of using a vertical placement to assume a dual-polarized antenna. As Fig 11 and Fig 12, patch width is 1.75mm, using a FR-4 (loss tangent = 0.02) substrate with a thickness of 0.4 mm and Rogers RO4003(tm)(loss tangent = 0.0027) with a thickness of 0.4 mm feeding by 50 ohms strip line.

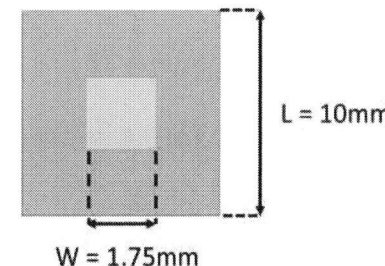

Fig 11. Top view of the patch antenna.

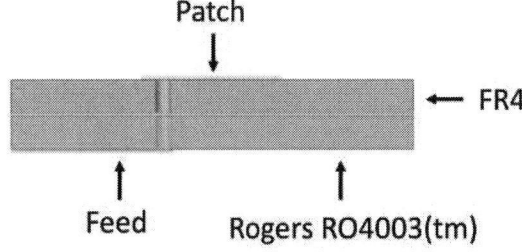

Fig 12. Side view of the patch antenna.

III. SIMULATION AND DISCUSSIONS

Figure 13 shows the configuration of the system in simulation: a single patch antenna at the bottom illuminates a four-layer FSS on top. Figure 14 shows the field patterns after passing through the FSS transmitarray. Presently, the dual polarization was achieved by rotating the bottom patch antenna by 90 degrees. The transmitarray has different phase delays for Ex and Ey, so the beam direction of its antenna changes after passing through the transmitarray. It clearly shows the beam diversity characteristic: the main beams from two different polarization are pointing two different directions, about 30 degrees apart.

Even though the polarization diversity and directivity have been demonstrated, antenna performance, for example, its gain, of the currently used single patch antenna has not be optimized. In the future, the antenna will be improved to secure a better gain. We will also continue to explore the characteristics of space between the transmitarray and the array antenna, such as wide receiving angle, to result in a smaller overall configuration size.

The result demonstrates beam switching capability by using dual polarization using a single patch antenna. In reality, a true dual polarized patch array antenna will be design and used. An array antenna with passive Butler matrix can achieve 2-D beam-switching [26-27], with dual polarization, such an array can deliver even better beam diversification.

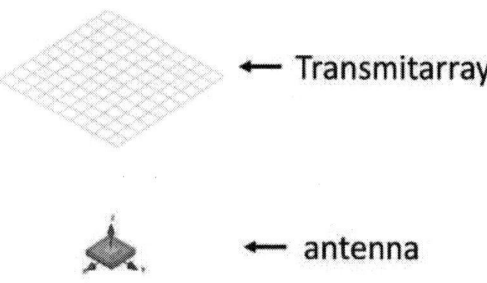

Fig 13. Simulation configuration, a single patch antenna (bottom) with a four-layer FSS (top).

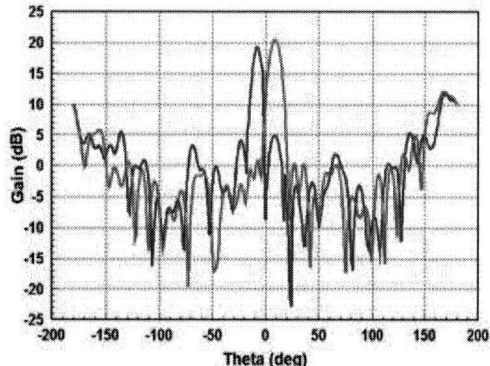

Fig 14. Radiation pattern with FSS transmitarray with dual-polarization from a single patch antenna.

IV. CONCLUSIONS

A dual-polarized patch antenna with FSS transmitarray clearly delivers two separated main beams; that is, it accomplishes main beam spatial diversity. Designed at millimeter wave band of 38 GHz, the dual polarized FSS can be used in 5G Base-station applications. Presently, the dual polarization was achieved by rotating the patch antenna by 90 degrees.

In the future, a MIMO array antenna will be implemented with an FSS. Each antenna in the array will be designed with two ports (two inputs with high isolation) to facilitate dual polarization. An array antenna with Butler matrix can perform 2-D beam forming, such array antenna with dual-polarization can deliver even better beam diversification (twice of that without dual polarization).

ACKNOWLEDGMENT

The authors would like to acknowledge financial support provided by Taiwan Ministry of Science and Technology (Grant MOST 107-2221-E-110-029).

REFERENCES

[1] Pang Xingdong, Hong Wei, Yang Tianyang, and Li Linsheng, "Design and implementation of an active multibeam antenna system with 64 RF channels and 256 antenna elements for massive MIMO application in 5G wireless communications," *China Communications*, vol. 11, issue 11, pp. 16-23, Jan.2015.

[2] Warren L. Stutzman and Gary A Thiele, Antenna Theory and Design, 3rd Ed, John Wiley & Sons, 2012.

[3] Said E. El-khamy, Karim H. Moussa, and Amr A. El-Sherif, "On the performance of massive multiuser MIMO with different transmit beamforming techniques and antenna selection," RadioScience Conference (URSI AT-RASC), 2015 1st URSI Atlantic, Oct. 2015.

[4] T.-Y. Wu, S.-T. Fang, and K.-L. Wong, "Printed diversity monopole antenna for WLAN operation," Electron. Lett., vol.38, Dec. 2002, pp.1625-1626.

[5] J.-Y. Chung, T. Yang, J. Lee, and J. Jeong, "Low correlation MIMO antenna for LTE 700MHz band," IEEE International Symposium on Antennas and Propagation (APS/URSI), 2011, pp. 2202-2204.

[6] H. Minseok and C. Jaehoon, "Dual-band MIMO antenna using a symmetric slotted structure for 4G USB dongle application," IEEE International Symposium on Antennas and Propagation (UPSURSI),2011, pp. 2223-2226.

[7] G.-A. Mavridis, J.-N. Sahalos, and M.-T. Chryssomallis, "Spatial diversity two-branch antenna for wireless devices," Electron. Lett., Vol.42, Mar. 2006, pp. 266-268.

[8] K.-L. Wong and J.-H. Chou, "Integrated 2.4- and 5-GHz WLAN antennas with two isolated feeds for Dual-module applications," Micro. Opt. Technol. Lett., Vol. 47, Nov. 2005, pp. 263-265.

[9] S.-W. Su, J.-H. Chou, and T.-Y. Wu, "Internal broadband diversity dipole antenna," Micro. Opt. Technol. Lett., Vol. 49, Apr. 2007, pp. 810-812.

[10] X. Y. Yao and J. Yu, "Multiband planar monopole antenna for LTE MIMO systems," International Journal of Antennas and Propagation, vol. 2012 (2012), Article ID 890705, 6 pages.

[11] S.-W. Su, "A three-in-one diversity antenna system for 5 GHz WLAN applications," Micro. Opt. Technol. Lett., vol. 51, Oct. 2009, pp. 2477-2481.

[12] S.-W. Su, "A three-in-one diversity antenna system for 5 GHz WLAN applications," Micro. Opt. Technol. Lett., vol. 51, Oct. 2009, pp. 2477-2481.

[13] W.-J. Lee, C. Yoon, S.-U. Kim, and H.-D. Park, "A study on isolation enhancement of MIMO antenna using differential ground path," Micro. Opt. Technol. Lett., Vol. 54, no. 3, Mar. 2012, pp. 802-805.

[14] H. Min-Scok and C. Jachoon, "Multiband MIMO antenna with a band stop filter for high isolation characteristics," IEEE International Symposium in Antennas and Propagation Society (APS/URSI), 2009, pp. 1-4.

[15] Yu-Shin Wang, Jung-Chieh Lu, and Shyh-Jong Chung, "A miniaturized ground edge current choke –design, measurement, and application," IEEE Transactions on Antennas and Propagation, vol.57, issue 5, pp. 1360-1366, May 2009.

[16] T.-W. Kang and K.-L. Wong, "Isolation improvement of 2.4/5.2/5.8 GHz WLAN internal laptop computer antennas using dual-band strip resonator as a wavetrap," Micro. Opt. Technol. Lett., vol. 52, no. 1, Jan. 2010, pp. 2478-2481.

[17] H. Lihao, Z. Huiling, H. Zhang, and C. Quanming, "Reduction of mutual coupling between closely-packed antenna elements with split ring resonator (SRR)," International Conference on Microwave and Millimeter Wave Technology (ICMMT), 2010, pp.1873-1875.

[18] V. Ssorin, A. Artemenko, A. Sevastyanov, and R. Maslennikov, "Compact bandwidth- optimized two element MIMO antenna system for 2.5 – 2.7 GHz band," Proceedings of the 5th European Conference on Antennas and Propagation (EUCAP), 2011, pp. 319-323.

[19] S.-W. Su, C.-T. Tse, and F.-S. Chang, "Printed MIMO-antenna system using neutralization-line technique for wireless USB-dongle applications," IEEE Trans. Antennas Propag., vol. 60, no.2, Jun. 2012, pp. 456-463.

[20] C.-X. Mao and Q.-X. Chu, "Compact co-radiator UWB-MIMO antenna with dual polarization," IEEE Trans. Antennas and Propagation, vol. 62, no. 9, Sep. 2014, pp. 4474-4480.

[21] J. Lu, Z. Kuai, X. Zhu, and N. Zhang, "A high-isolation dual-polarization microstrip patch antenna with quasi-cross-shaped coupling slot," IEEE Trans. Antennas and Propagation, vol. 59, no. 7, July 2011, pp. 2713-2717.

[22] I. Dioum, A. Diallo, S. M. Farssi, and C. Luxey, "A novel compact dual-band LTE antenna-system for MIMO operation," IEEE Trans. Antennas and Propagation, vol. 62, no. 4, Apr. 2014, pp. 2291-2296.

[23] M.A. Jensen and J.W. Wallance, "A review of antennas and propagation for MIMO wireless communications," IEEE Trans. On Antennas and propagation, vol. 52, 2004, pp. 2810-2824.

[24] P. Kyritsi, D. C. Cox, R. A. Valenzuela, and P. W. Wolniansky, "Correlation analysis based on channel measurements in an indoor environment," IEEE Journal on Selected Areas in Communications, vol. 21,no. 5, June 2003, pp.713-720.

[25] Lih-Tyng Hwang, "A nulling technique for microwave imaging with a random thinned array," PhD dissertation, University of Pennsylvania, Philadelphia, PA, USA, 1985.

[26] C.-H. Chen, W.-T. Fang, and Y.-S. Lin "Miniature 2.4-GHz switched beamformer module in IPD and its application to very-low-profile 1D and 2D scanning antenna arrays," ECTC, Orlando, FL, USA, 2017.

[27] Chung-Yi Hsu, Chia-Ling Chiang, Lih-Tyng Hwang, and Fa-Shian Chang, "Design of a Compact Broadband Butler Matrix and its Application in Organic Beam-former at the 5 GHz Band," ECTC, San Diego, CA, USA, 2018.

Direct bonding of low temperature heterogeneous dielectrics

Serena Iacovo, Lan Peng, Alain Phommahaxay, Fumihiro Inoue, Patrick Verdonck, Soon-Wook Kim,
Erik Sleeckx, Andy Miller, Gerald Beyer, Eric Beyne.

IMEC
Leuven
serena.iacovo@imec.be

Abstract—Nowadays, the direct bonding process is embedded in a BEOL manufacturing process where the maximum temperature is $400°C$. For certain applications there is the need to lower such thermal budget. One of the first process steps which will be modified will be the bonding layer deposition step as well as the densification step. It is known that by lowering the deposition temperature the quality of the dielectric will be decreased as well. This change will have a direct consequence on the bonding process which relies on the quality of the dielectric.

It is found that if we use a post bond anneal temperature which exceeds the densification temperature voids originate at the bonding interface. By means of FTIR studies and ERD analysis the origin of the voids is tentatively ascribed to H or H related species. These findings provide a basic understanding on how to tune the deposition condition to select a proper low temperature dielectric which will enable us to obtain a good bonding uniformity and a good bond strength for the described application.

Keywords-Direct wafer to wafer bonding; Low thermal budget; Voids formation; SiCN

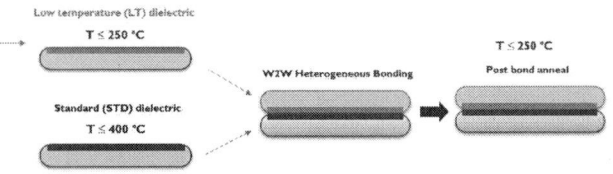

Figure 1. Process flow describing the experiments performed.

I. INTRODUCTION

Three-dimensional integrated circuits (3D-ICs) offer the possibility to solve delay and power problems in the conventional planar 2D-ICs. Different stacking options can be used to enable the use of the third dimension. One attractive option which allows the device stacking at room temperature and the possibility to reach high alignment accuracy is direct wafer to wafer bonding. Typically a post bond anneal step characterized by a temperature lower than or equal to $400°C$ is used to strengthen the adhesion between the bonded stack.

Memory applications are driving towards the reduction of process temperature [1], [2] meaning that an upper limit on the allowable temperature is imposed during deposition, densification and post bond anneal step.

In this work we run a first feasibility test to bond a standard (STD) SiCN dielectric deposited at $370°C$ to a dielectric deposited at temperatures lower than or equal to $200°C$. When we speak about STD dielectric for bonding applications within imec we refer to a SiCN material deposited at $370°C$. The so called "standard process steps" prior to bonding and the bonding process itself have been optimized for this material to obtain outstanding bonding results.

Two different kind of oxides have been selected; two nitrides and two silicon carbon-nitrides. For all the bonding experiments we use as device wafer a wafer where we deposit a low temperature dielectric and as carrier wafer a wafer where we deposit a standard SiCN dielectric. Afterwards the wafers are subjected to a densification step at a temperature lower than or equal to $250°C$. In a following step the wafers are planarized by using the same CMP process. Finally, the wafers are bonded by using the best-known method bonding sequence which includes a cleaning step and a N_2 plasma treatment prior to bonding. In order to complete the bonding process an annealing step is used to enhance the bond strength between the two bonded dielectrics.

Reducing deposition, densification and post bond anneal temperatures tends to go at the expense of the interface quality between the bonded dielectrics [3], [4]. Thus such changes in the process flow must be monitored by means of several characterization techniques enabling the selection of proper materials and integration flow, which can lead to defect-free and high enough bonding energy, ensuring final device reliability.

In this paper a simplified process flow is used to mimic the possible degradation of the interface quality which may occur when process temperatures are reduced to values lower than or equal to $250°C$. Different dielectrics as well as re-tuned and existing deposition, annealing and bonding processes have been fully characterized and explored by means of several characterization techniques being scanning acoustic microscopy (SAM), Fourier transform infrared derivative spectroscopy (FTIR), ellipsometry, electron recoil detection (ERD), transmission electron microscopy (TEM), energy-dispersive X-ray spectroscopy (EDS). Optimized processes and materials which enable a void free bonding with a

978-1-7281-1500-9/19 $31.00 © 2019 IEEE

relative high bond strength have been successfully demonstrated. Void formation mechanisms for the tested materials are proposed.

II. EXPERIMENTAL

For all the experiments a STD SiCN, deposited by PECVD at $375°C$, has been used as a bonding dielectric for the device wafer while as a bonding dielectric for the carrier wafer different dielectrics deposited at temperatures lower than or equal to $200°C$ were used, as described in Fig. 1.

Dielectric films were deposited by PECVD in three tools (A, B and C). We selected two kind of oxides (SiO_2A and SiO_2B), two nitrides ($SiNA$ and $SiNB$) and two silicon carbon-nitrides ($SiCNA$ and $SiCNC$) for 300 mm W2W bonding experiments.

For both SiCN materials NH_3 and $SiH_x(CH)_y$ have been selected as precursors. The remaining SiO_2 and SiN film types were deposited by using different precursors as summarized in Table I. The layers have been characterized by means of ERD analysis revealing atomic composition. In all the cases dielectric thickness was 150 nm with a total thickness variation $< 5\%$ to ensure a good co-planarity for bonding [5].

Table I
MAIN PRECURSORS USED FOR SiO_2 AND SiN FILM TYPES

	SiO_2A	SiO_2B	$SiNA$	$SiNB$
Precursors	SiH_4	TEOS	NH_3	without NH_3

After deposition, the so-called densification step corresponds to a 2-hour annealing step which is used to remove outgas from the dielectrics. Such annealing process is done in pure N_2 atmosphere. In the case of STD SiCN, the temperature used is $350°C$ while for the low temperature (LT) dielectrics two different annealing temperatures $200°C$ and $250°C$ have been tested for this step. Chemical mechanical polishing (CMP) has been used to planarize the surface. AFM, with a scanning area of 4 μm^2, has been used to determine the surface roughness of each dielectric. FTIR has been employed after each process step to monitor the properties of the dielectrics and their evolution. The entire bonding sequence is realized in the EVG GEMINI cluster and it consists of the following steps: plasma activation in N_2 ambient, jet nozzle DI-water rinse, mechanical alignment and bonding at room temperature [6]. The quality of the bonding interface has been evaluated by means of SAM to identify void formation while the bond strength, established between the different combinations of dielectrics, has been evaluated by means of Maszara razor blade test. In some cases, in order to get more information regarding chemical bonding of interfacial layers, TEM together with EDS techniques have been employed.

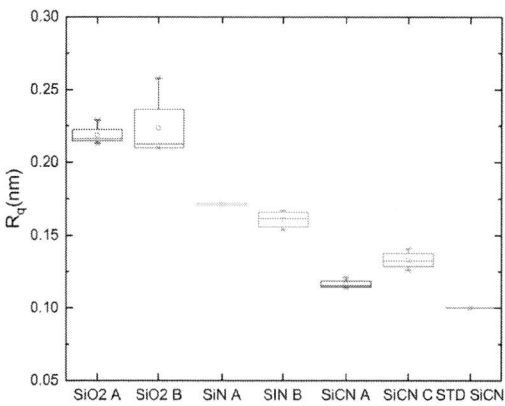

Figure 2. AFM results after CMP. Roughest materials are the SiO_2 ones while the smoothest are the SiCN ones.

III. RESULTS AND DISCUSSION

A critical parameter to be monitored prio to direct bonding is the surface smoothness of the dielectric. R_q values expressed in nm are shown in Fig. 2 for all the dielectrics screened. As it is visible from the graph SiCN films exhibits the lowest roughness while the rougher films are the SiO_2 ones. Nevertheless the maximum R_q value measured for the various films is below 0.5 nm which is identified in many research reports as a maximum required surface roughness enabling optimum hydrophilic low temperature wafer direct bonding [7].

As explained previously the experiments were performed by tuning process temperatures of densification and post bond anneal steps. It has been observed that when deposition, densification and post bond anneal temperatures are kept below or equal to $200°C$, then we obtain good bonding uniformity for all the wafer pairs. However, bond strength results, reported in Fig. 3, point to a weak bonding which might be not fully compatible with subsequent thinning down techniques. On the same graph the bond strength achievable by bonding STD SiCN to STD SiCN, with post bond anneal equal to $200°C$, is reported, highlighting the significant difference in bond strength achievable when we use LT dielectrics.

With the aim of verifying whether we are able to improve these bond strength values, we increase the post bond anneal temperature to $250°C$, which is considered as an optimum post bond anneal temperature value in the standard process [8]. As a result, bonding voids with different shapes and patterns are appearing in the SAM images, presented in Fig. 4, depending on the bonded materials. In particular it has been noticed that when comparing the same kind of materials the ones deposited in tool A were worse compared

978-1-7281-1500-9/19 $31.00 © 2019 IEEE

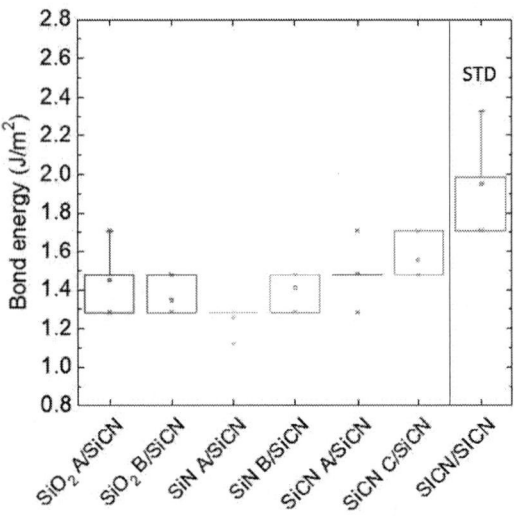

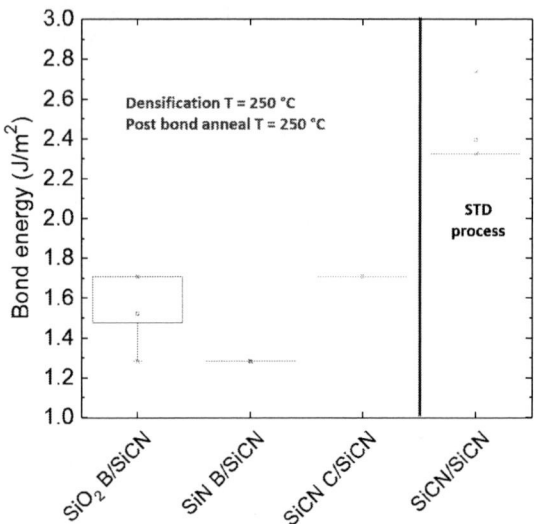

Figure 3. Bond strength results obtained for the different materials tested when densification and post bond anneal temperature are equal to $200°C$

Figure 5. Bond strength results obtained for the different combination of films tested when densification and post bond anneal temperature are equal to $250°C$

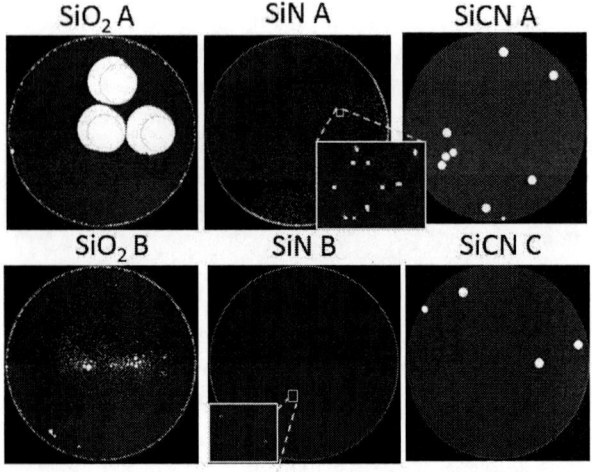

Figure 4. SAM images obtained after bonding and annealing step for the various combinations of films tested. For these experiments densification temperature was $200°C$ and post bond annealing was set to $250°C$. Voids with a variety of shapes are arising after annealing related to out-gassing. If we consider the same kind of material (SiO_2, SiN and SiCN) it appears that films deposited in tool A are worse compared to films deposited in tool B or C.

to the ones deposited in tool B or C. Reasons behind void formation have been analyzed by means of several characterization techniques. The fact that we are not observing any void directly after bonding is excluding particles as possible origin. Whereas one of the most probable root causes can be identified into out-gassing of by-products as it will be discussed later in detail.

If we believe that the possible reason for void formation is out-gassing then a logic choice would be to increase not only the post bond anneal temperature but also the densification temperature to $250°C$ enabling outgassing of the by-products produced at such temperature before the bonding process. By doing so we are able to suppress bonding voids creation and bond strength achieved are shown in Fig. 5. It should be mentioned here that for this test we only used materials which exhibited better void performances meaning SiO_2B , $SiNB$ and $SiCNC$.

Therefore it appears that, when the post bond anneal temperature exceed the densification temperature, bonding voids are generated at the interface. To validate this observation, we subjected the bonded pairs with wafers densified at $250°C$ to a post bond anneal temperature of $300°C$. Once again, voids were created.

Between all the materials tested $SiCNC$ appears to be the most robust material for void formation and the one which allow us to get the highest bond strength meaning 1.7 J/m^2 when bonded to STD SiCN with a process temperature $\leq$ $250°C$. In spite of that, the bond strength we are able to achieve when bonding STD SiCN to STD SiCN by using the same post bond anneal temperature is significantly higher (2.5 J/m^2) and a possible explanation will be given in the following. However, the mechanical stability of STD SiCN bonded to $SiCNC$ was successfully verified by grinding (see Fig. 6) and dicing test. Therefore we can consider $SiCNC$ as a potential material usable for the application described

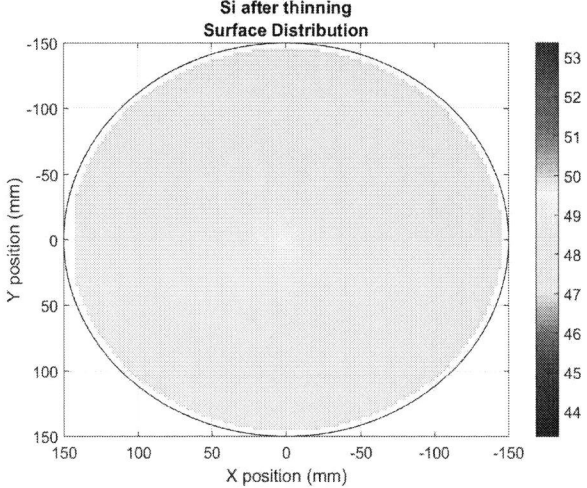

Figure 6. SiCNC/SiCN top wafer surface distribution after grinding to 50 μm

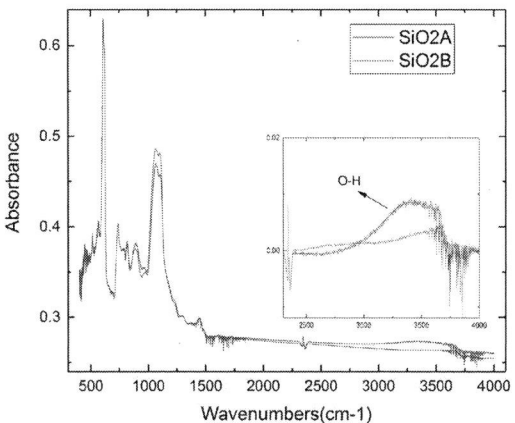

Figure 7. Comparison of FTIR spectra measured for SiO_2A (red line) and SiO_2B (blue line) films. The inset highlight the difference in O-H absorption band (3200-3700 cm $^{-1}$) between the two materials [9].

in the introduction.

A. Void formation mechanisms

In order to understand the failing mechanisms for materials A and in general reasons behind void formation between the various materials, films have been inspected by FTIR directly after deposition.

In Fig. 7 the comparison between FTIR spectra belonging to SiO_2A and SiO_2B is presented. The inset shows that in the case of SiO_2A the O-H stretching absorption band 3200-3700 cm^{-1} is more pronounced than for SiO_2B. Such observation is pointing in the direction of voids generated by physiosorbed water. In order to validate such hypothesis we subjected SiO_2A to an annealing of $250°C$ to simulate

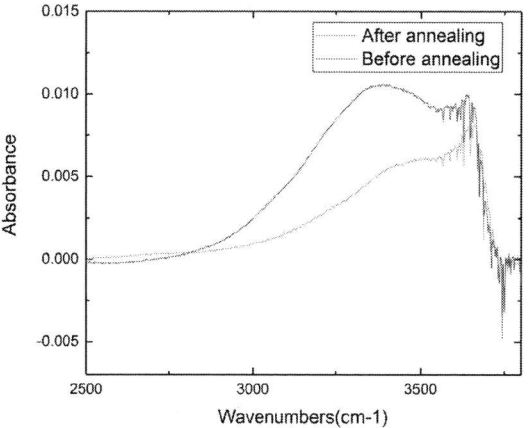

Figure 8. Comparison between spectra belonging to SiO_2 A before and after a 2-hour annealing at $250°C$ centered in the O-H stretching absorption band 3200-3700 cm^{-1}. The broad band can be seen as composed of two regions: molecular water 3225–3400 cm $^{-1}$ and hydroxyl groups 3500–3700 cm $^{-1}$ [9]. It is possible to observe a decrease mainly in the peak corresponding to molecular water.

the post bond anneal step. The comparison between spectra before and after annealing is shown in Fig. 8. It is possible to observe that after annealing the peak corresponding to molecular water (3225-3400 cm $^{-1}$) is decreasing, which make us conclude that one of the reason for void formation is the excessive water in the layer.

Additional characterization analysis has been carried out on SiO_2A. A portion of what it appeared a well bonded area of the wafer pair was submitted to TEM and EDS analysis. Results are shown in Fig. 9. In previous works it was shown that when bonding materials different from SiO_2, the bonding interface was easily located due to the presence of an accumulation of oxygen [8]. For the current case, as it appears from the EDS map, it is difficult to locate what represents the interface region since the difference in oxygen content due to the bonding step cannot be detected.

Interesting features to observe are appearing in the HAADF STEM picture. As highlighted in the yellow circle several darker spots with a diameter of ~5-10 nm appear equally distributed, probably along the region which can be identified as the interface, close to the SiO_2 layer. Such spots can be associated to nanovoids. Similar nanovoids were previously observed by Ventosa et al. [10]. According to Ventosa these are located at the bonded interface and can be ascribed to unbonded zones created to store water during the sealing mechanism.

Remarkably are the small voids at the edge observed whenever we use SiO_2 materials (Fig. 4). Even SiO_2 films deposited at higher temperatures exhibited such edge voids [11], unveiling the inherent weakness of SiO_2 in

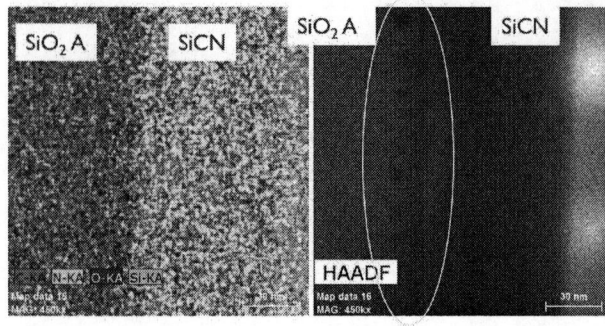

Figure 9. EDS and angular dark field (HAADF) STEM images for SiO₂ A. In the HAADF image are highlighted, in the yellow circle, darker spots assigned to nanovoids and appearing at the interface of SiO₂/SiCN. It is suggested that these nanovoids are filled with water generated during the sealing mechanism of the two wafers.

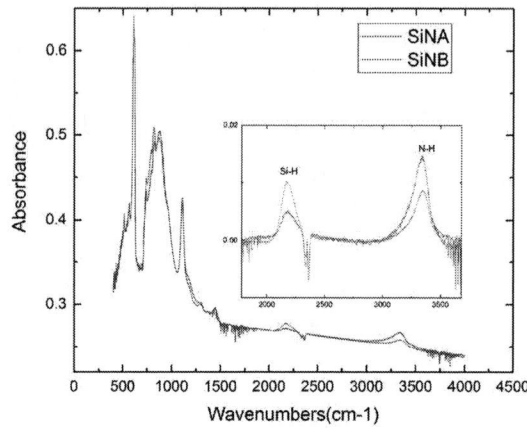

Figure 10. Comparison between SiNA and SiNB FTIR spectra. The inset zooms in the Si-H $\sim$2200 cm^{-1}, and the N-H peak at $\sim$3300 cm^{-1}. For SiNA N-H peak is higher than Si-H.

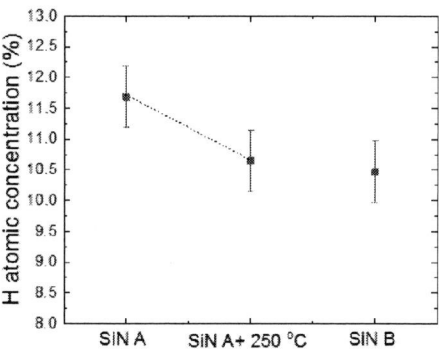

Figure 11. Hydrogen content measured by ERD for as-deposited SiNA, SiNA subjected to $250°C$ anneal and SiNB. It appears that the H% for SiNB is similar to the one of SiNA after annealing.

general for wafer bonding applications. This phenomenon was previously investigated by studying the dynamics of the bonding wave propagation. The voids were ascribed to water droplet nucleation [12]. The fact that in our experiments we are observing this phenomenon exclusively for SiO₂ is confirming our interpretation of water related voids in case of SiO₂.

In the case of nitride type materials we are not able to observe any peak in the region of the O-H absorption band but we can recognize some differences for the peaks around 2100 and 3300 cm^{-1} which represent Si-H and N-H bonds, respectively. In particular, SiNA contains more N-H compared to Si-H bonds, as it appears in Fig. 10. Previous reports are suggesting that this characteristic can potentially identify the layer as N-rich [13], [14]. In the current work we confirmed the N-rich nature of SiN A by means of ERD as shown in Table II.

Table II
ERD COMPOSITION ANALYSIS OF SiNA AND SiNB

	SiNA	SiNB
H	11.7 %	10.5 %
N	50 %	48.1 %
Si	38.3 %	41.4 %

The fact that SiNA can be identified as a N-rich layer is an indication that the hydrogen content is higher compared to SiN B [14]. In Fig. 11 ERD results for the hydrogen concentration are presented. It is possible to notice that H concentration is decreasing after annealing for SiNA and the concentration of H in SiNB is comparable to the one of SiNA after annealing. Taking into account these results we can consider as a possible origin for voids the molecular hydrogen in the bulk or loosely bonded H [15].

B. Lower bond strength of LT materials

As it was stated before, bond strength results achievable when bonding STD SiCN to STD SiCN (STD SiCN/ STD SiCN) are higher compared to the ones for LT materials/ STD SiCN. In order to understand the reason behind the superiority of STD SiCN in terms of W2W bond strength we compare STD SiCN with the SiCN film types.

FTIR spectra of SiCNA and SiCNC did not show any major difference. Contrarily differences can be observed when we compare SiCN STD and SiCNC. As it is shown in Fig. 12, the absorption spectra exhibited peaks at 1260 cm^{-1}, identified as CH₃ symmetric bending in Si-(CH₃), and at 3000 cm^{-1}, attributed to to CH$_x$ stretching vibrations [16]. Such peaks are higher in SiCNC compared to STD

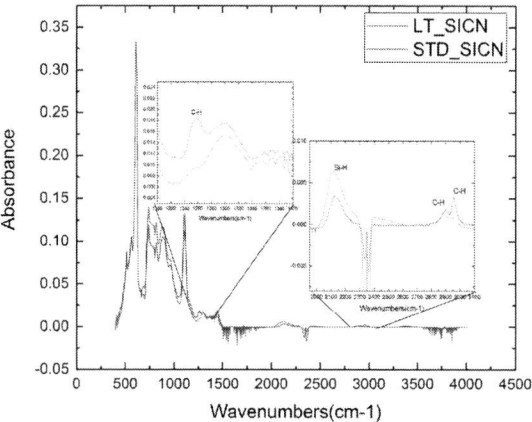

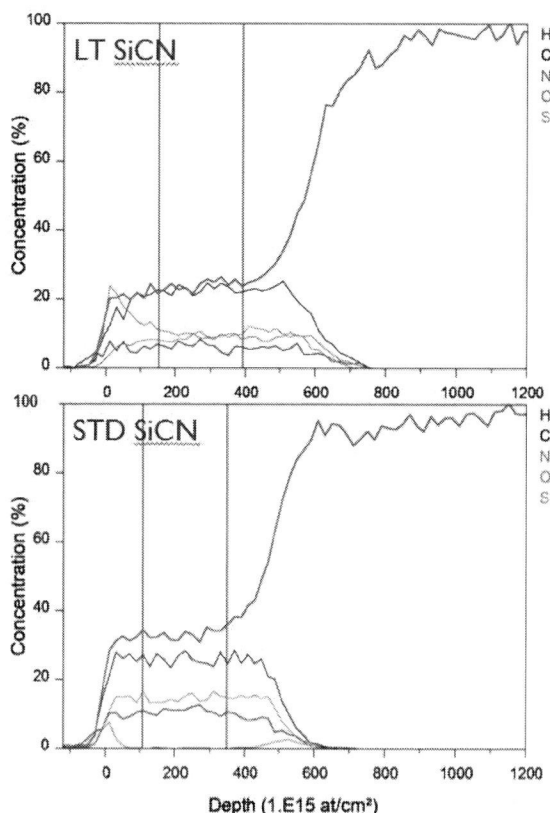

Figure 12. Comparison between SiCNC (LT SiCN) and STD SiCN. The first inset zooms at the C-H stretching band ~ 1260 cm^{-1}, assigned to CH$_3$ symmetric bending mode in Si-$(CH_3)_x$. The second one is zooming at the Si-H ~ 2150 cm^{-1} and CH$_x$ stretching modes. Is it visible that SiCNC is characterized by the presence of an higher C-H concentration compared to STD SiCN. [16]

SiCN suggesting a larger concentration of C-H bonds in SiCNC.

Figure 13 presents ERD results of SiCNC and STD SiCN. In SiCNC a high amount of oxygen is detected even in the bulk identifying the material as SiCNO and not purely SiCN. Density measurement shows that the density of the SiCNC is equal to 1.32 g/cm^3 while the one of STD SiCN is 1.78 g/cm^3. Such low density could point out to a facilitated moisture absorption mechanism which might be the cause of Oxygen presence in the layer.

Also, if we look at the graphs we can see that at the surface we do have a lower Si and C concentration for SiCNC compared to STD SiCN. If considering that SiCNC is characterized by a lower density compared to STD SiCN then we have to expect an even lower amount of Si and C atoms. In the past we have shown some evidence for the importance of Si an C atoms as key elements to increase the W2W bond strength [17]. Then one could ascribe the lower bond strength result that we are able to achieve when bonding SiCNC/STD SiCN compared to STD SiCN/STD SiCN to a decrease amount of of Si and C atoms at the surface.

IV. CONCLUSIONS

A material grown at low temperature ($\leq 200°C$) which guarantees void free interface and a bond strength of 1.7 J/m^2, when bonded to STD SiCN with a process temperature $\leq 250°C$, has been selected. It has been demonstrated that the bond strength is high enough to withstand the wafer thinning and dicing processes.

Figure 13. ERD results for STD SiCN and SiCNC (LT SiCN). Oxygen profile is different from both samples as function of depth. In STD SiCN Oxygen is mainly present at the surface as a result of room temperature oxidation mechanism while for SiCNC oxygen it is distributed in the entire bulk.

The lower bond strength of the best LT SiCN material is tentatively attributed to a lower concentration of Si and C at the bond surface when compared to STD SiCN.

It is found that post bond anneal temperature should not exceed densification temperature in order to avoid bonding voids formation.

A study on void formation mechanism has been carried out. The main source of interface voids can be identified as H$_2$O in the case of SiO$_2$ and H, present in the in the molecular form or loosely bonded, for SiN.

ACKNOWLEDGMENT

The authors would like to thank Prof. Andre Stesmans and Prof. Valeri Afanasiev for the discussions and advice.

This work is carried within the frame of the imec 3D System Integration Industrial Affiliation Program and within a Joint Development Project between imec and EVG and between imec and SPTS.

References

[1] A. K. Panigrahy and K.-N. Chen, "Low Temperature Cu–Cu Bonding Technology in Three-Dimensional Integration: An Extensive Review," *Journal of Electronic Packaging*, vol. 140, no. 1, p. 010801, 2018.

[2] Y. I. Kim, K. H. Yang, and W. S. Lee, "Thermal degradation of DRAM retention time: Characterization and improving techniques," *IEEE International Reliability Physics Symposium Proceedings*, vol. 2004-Janua, no. January, pp. 667–668, 2004.

[3] S. W. Kim, L. Peng, A. Miller, G. Beyer, E. Beyne, and C. S. Lee, "Permanent wafer bonding in the low temperature by using various plasma enhanced chemical vapour deposition dielectrics," *2015 International 3D Systems Integration Conference, 3DIC 2015*, pp. TS7.2.1–TS7.2.4, 2015.

[4] K. Mitani and U. M. G??sele, "Formation of interface bubbles in bonded silicon wafers: A thermodynamic model," *Applied Physics A Solids and Surfaces*, vol. 54, no. 6, pp. 543–552, 1992.

[5] L. Peng, S. W. Kim, M. Soules, M. Gabriel, M. Zoberbier, E. Sleeckx, H. Struyf, A. Miller, and E. Beyne, "W2W permanent stacking for 3D system integration," *Proceedings of the 16th Electronics Packaging Technology Conference, EPTC 2014*, pp. 1–4, 2014.

[6] T. Plach, K. Hingerl, S. Tollabimazraehno, G. Hesser, V. Dragoi, and M. Wimplinger, "Mechanisms for room temperature direct wafer bonding," *Journal of Applied Physics*, vol. 113, no. 9, 2013.

[7] Q.-Y. Tong, G. Cha, R. Gafiteanu, and U. Gosele, "Low temperature wafer direct bonding," *Journal of Microelectromechanical Systems*, vol. 3, no. 1, pp. 29–35, March 1994.

[8] E. Beyne, S. Kim, L. Peng, N. Heylen, J. D. Messemaeker, O. O. Okudur, A. Phommahaxay, T. Kim, M. Stucchi, D. Velenis, A. Miller, and G. Beyer, "Scalable, sub 2μm pitch, cu/sicn to cu/sicn hybrid wafer-to-wafer bonding technology," in *2017 IEEE International Electron Devices Meeting (IEDM)*, Dec 2017, pp. 32.4.1–32.4.4.

[9] C. Sabbione, L. Di Cioccio, L. Vandroux, J. P. Nieto, and F. Rieutord, "Low temperature direct bonding mechanisms of tetraethyl orthosilicate based silicon oxide films deposited by plasma enhanced chemical vapor deposition," *Journal of Applied Physics*, vol. 112, no. 6, 2012.

[10] C. Ventosa, C. Morales, L. Libralesso, F. Fournel, A. M. Papon, D. Lafond, H. Moriceau, J. D. Penot, and F. Rieutord, "Mechanism of Thermal Silicon Oxide Direct Wafer Bonding," *Electrochemical and Solid-State Letters*, vol. 12, no. 10, p. H373, 2009. [Online]. Available: http://esl.ecsdl.org/cgi/doi/10.1149/1.3193533

[11] F. Inoue, L. Peng, A. Phommahaxay, S. . Kim, J. D. Vos, E. Sleeckx, A. Miller, G. Beyer, and E. Beyne, "Characterization of inorganic dielectric layers for low thermal budget wafer-to-wafer bonding," in *2017 5th International Workshop on Low Temperature Bonding for 3D Integration (LTB-3D)*, May 2017, pp. 24–24.

[12] A. Castex, M. Broekaart, F. Rieutord, K. Landry, and C. Lagahe-Blanchard, "Mechanism of Edge Bonding Void Formation in Hydrophilic Direct Wafer Bonding," *ECS Solid State Letters*, vol. 2, no. 6, pp. P47–P50, 2013. [Online]. Available: http://ssl.ecsdl.org/cgi/doi/10.1149/2.006306ssl

[13] Z. Yin and F. W. Smith, "Optical dielectric function and infrared absorption of hydrogenated amorphous silicon nitride films: Experimental results and effective-medium-approximation analysis," *Phys. Rev. B*, vol. 42, pp. 3666–3675, Aug 1990. [Online]. Available: https://link.aps.org/doi/10.1103/PhysRevB.42.3666

[14] T. Domínguez Bucio, A. Z. Khokhar, C. Lacava, S. Stankovic, G. Z. Mashanovich, P. Petropoulos, and F. Y. Gardes, "Material and optical properties of low-temperature NH3-free PECVD SiNxlayers for photonic applications," *Journal of Physics D: Applied Physics*, vol. 50, no. 2, 2017.

[15] V. V. Afanas'ev, P. Ericsson, S. Bengtsson, and M. O. Andersson, "Wafer bonding induced degradation of thermal silicon dioxide layers on silicon," *Applied Physics Letters*, vol. 66, no. 13, pp. 1653–1655, 1995. [Online]. Available: https://doi.org/10.1063/1.113882

[16] E. Ermakova, A. Lis, M. Kosinova, S. Rumyantsev, E. Maximovskii, and V. Rakhlin, "Bis(trimethylsilyl)ethylamine: Synthesis, properties and its use as CVD precursor," *Physics Procedia*, vol. 46, no. Eurocvd 19, pp. 209–218, 2013. [Online]. Available: http://dx.doi.org/10.1016/j.phpro.2013.07.069

[17] L. Peng, S. Kim, S. Iacovo, F. Inoue, A. Phommahaxay, E. Sleeckx, J. D. Vos, A. Miller, G. Beyer, E. Beyne, D. Zinner, T. Wagenleitner, T. Uhrmann, M. Wimplinger, B. Schoenaers, A. Stesmans, and V. V. Afanas'ev, "Advances in sicn-sicn bonding with high accuracy wafer-to-wafer (w2w) stacking technology," in *2018 IEEE International Interconnect Technology Conference (IITC)*, June 2018, pp. 179–181.

Low Temperature Transient Liquid Phase (TLP) bonding using eutectic Sn-In Solder Anisotropic Condctive Films (ACFs) for Flexible Ultrasound Transducer

Jae-Hyeong Park[1], Jongcheol Park[2], and Kyung-Wook Paik[1]

1: KAIST, Materials Science and Engineering Department
291 Daehak-ro, Yuseong-gu, Daejeon, 305-701, South Korea
2: National NanoFab Center
291 Daehak-ro, Yuseong-gu, Daejeon, South Korea
Corresponding author e-mail: kwpaik@kaist.ac.kr

Abstract— A TLP (Transient Liquid Phase) bonding of Ultrasonic transducers using low temperature eutectic Sn-In solder ACFs and Au pads, where the final solder joint consists of Au-In-Sn intermetallics with higher melting point, has been investigated. Sn52In solder ACFs were fabricated and thermo-compression bonded at 130°C for 30 seconds to bond ultrasonic transducer on Flex Printed Circuits (FPCs). Scanning electron microscopy (SEM) / energy dispersive x-ray spectroscopy (EDX) analysis of solder joint cross-sections show that the microstructure of intermetallic compounds (IMCs) were consisted of Au-In and Sn rich-In phases. Based upon EDX analysis, the Au-In-Sn intermetallic compounds are found. The aim of this study is to characterize eutectic Sn-In solder ACFs bonding method suitable for temperature limited bonding environment as well as for flexible ultrasonic transducer applications. Pressure Cooker Test (2 atm, 100% Relative Humidity at 121°C) and Dynamic bending test (R: 7mm) were also carried out to examine the bonding reliability. And these results were compared with conventional metal coated polymer particles ACFs. After a pressure cooker test, the final intermetallic compound was Au-In-Sn ternary phase, which is thermally stable at higher temperatures as high as 380°C. The TLP bonding with Sn-In solder ACFs offers very reliable metallurgical bonding and excellent compatibility for ferroelectric ultrasonic transducers with rough surfaces.

Keywords-solid-liquid interdiffusion bonding; transient liquid phase bonding; anisotropic conductive films; flexible

I. INTRODUCTION

Low-temperature bonding is desirable to avoid the depolarization of piezoelectric ceramics for the production of ultrasound transducers. Non-conductive pastes (NCPs) have been widely used for the commercial applications of ultrasound transducers for several decades. It has the advantages of low curing temperature at 60°C. NCPs are dispensed on the flexible substrates, and the piezoelectric ceramics are bonded with heat and pressure during the NCPs curing. The electrical conduction is established through direct metal to metal contact between metallized piezoelectric ceramics and FPCs. However, several hours are required for NCPs curing which leads to the reduction of production yield. And the electrical properties of NCP interconnection can be also deteriorated during the process, especially in moisture and high temperature environment because of physical metal contact and hygroscopic polymer resin expansion. Therefore, it is desirable for interconnects with reliable electrical conductivity and shorter curing times.

As a result, low-temperature anisotropic conductive films (ACFs), which consist of conductive balls and adhesive polymer films, were introduced. Au-coated polymer balls based ACFs have been previously reported for ultrasound transducer assembly [1]. However, Au-coated polymer balls only make physical contact based joints which are less reliable compared with metallurgical solder joints. Therefore, solder ball based ACFs have been previously investigated as another promising approach in our research group. Solder ACFs have been recently demonstrated as an interconnection material for Flex-On-Board (FOB) applications [2-4]. Solder wetting on the metal electrodes with stable metallurgical joints were successfully demonstrated using solder ACFs. However, the bonding temperature above 200°C for the eutectic SAC305 and Sn58Bi solders melting is too high to maintain the polarization of piezoelectric ceramics. Therefore, eutectic Sn52In solder, one of the low melting temperature lead-free solder, has been selected as an alternative for the low-temperature bonding applications.

In this study, Sn52In eutectic solder ACF was investigated to bond piezoelectric ceramic on FPC. For comparison, various ACFs, including 3 types of conductive balls (Sn52In solder, Au/Ni coated polymer ball and Ni ball) and 3 types of thermosetting polymer resins (acrylic resin, cation epoxy resin and imidazole epoxy resin) were investigated in terms of electrical/mechanical properties before and after the pressure cooker test (PCT) (the test condition: 121°C,

2 atmosphere, and 100% Relative Humidity) and the dynamic bending test with bending radius of 7 mm.

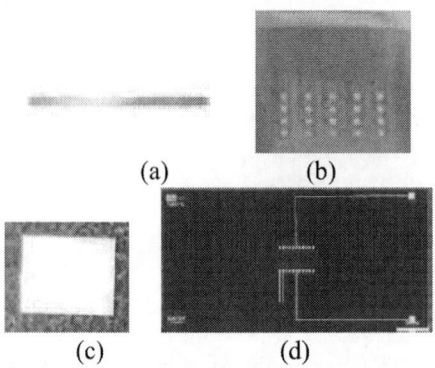

Fig. 1. Test vehicle. (a) Au deposited dummy Si die, (b) FPCs, (c) Au deposited dummy silicon die, (d) FPC for the bending test.

II. EXPERIMENTS

A. Test Vehicle

1) In order to represent piezoelectric ceramic, Si dummy die was designed with the dimensions of 1mm × 20mm × 150 μm and 400 nanometer thickness Au on the silicon dummy die shown in Fig. 1 (a). FPC was made using 55μm thickness of polyimide (PI), 25mm × 25.8mm dimension, and the four-point-kelvin structure was designed with a 500-μm pitch Cu electrodes with ENIG surface finish for measuring contact resistances before/after PCT, as shown in Fig. 1 (b).

2) For Dynamic Bending Test: In order to represent piezoelectric ceramic, another Si dummy die was designed with the dimensions of 4.5mm × 6mm × 150 μm and 400nanomter thickness Au on silicon dummy die shown in Fig. 1 (c). Another FPC was made using a 50μm thickness of polyimide (PI), 20mm × 40mm dimension, and circuit structure was designed for monitoring *in-situ* daisy chain resistances during bending test. The 500-μm pitch Cu electrode of FPC was electroplated and Au/Ni finished. The first, second electrodes were fabricated longer than other electrodes for dicing line shown in Fig. 1 (c) and (d).

B. ACF Material

1) Conductive particles such as 35wt.% 5~20μm diameter Sn52In balls, 15wt.% 10μm diameter Au/Ni coated polymer balls, and 35wt.% 8μm Ni balls are individually added to the each ACF resins to investigate the effect of conductive ball types on ACF joint reliabilities.

The Sn52In solder melting point is 118℃ shown in Fig. 2.

2) Thermosetting resin, acrylic type resin, cationic epoxy resin and imidazole epoxy resin are individually selected as the ACF resin to investigate the effect of polymer resin types on ACF joint reliabilities. Each polymer resin's curing behavior was characterized by Differential Scanning Calorimetry (DSC) as shown in Fig. 3(a). The polymer

resin's viscosity was characterized by rheometer shown in Fig. 3 (b).

Flux activator was added to solder ACF to eliminate solder oxide. Chemical reaction of releasing acid showed about 125°C shown in Fig. 4.

The thickness of Each ACF was 30μm fabricated by comma roll coater.

C. Specimens Preparation

1) Each fabricated ACF was coated on the Au deposited silicon dies and thermo-compressed to FPCs with 2MPa pressure. Si die was placed at the bottom, and FPC was placed on the top of dummy die to guarantee the real temperature of the dummy dies was maintained at 130°C, as shown in Fig. 5. ACF's temperature was monitored by the thermos-couple. Curing degrees were characterized by Fourier Transform Infra-red (FT-IR) spectroscope shown in Fig. 6. Applied bonding time was 30 sec. to ensure cure degree over 80% after thermo-compression bonding. (The degrees of cure were 97.2% for imidazole, 80.7% for cationic, and 100% for acrylic.). However, there is no remaining epoxy polymerization peak from DSC measurement of cationic epoxy after 130°C, 30 seconds iso-thermal bonding which means fully cured shown in Fig. 7.

After bonding, ACFs' electrical resistances were measured by the four-terminal sensing method. And cross-sectional image of the ACFs joint morphology was observed.

*2) B*ending specimens' preparation is shown in Fig. 8. The bonding condition applied was 130°C with a pressure of 2 MPa and for 30 seconds. Si dummies were singulated for building 500μm pitch array. Each of singulated must be electrically and mechanically connected to the corresponding Cu pads of FPC by ACFs' joints. The kerfs were filled using silicone. This process is the same process of the commercial fabrication process.

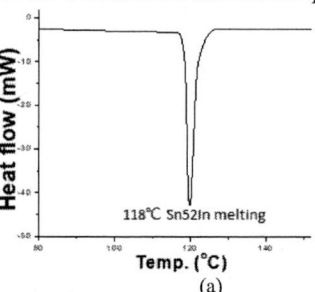

Fig. 2. DSC result of Sn52In solder

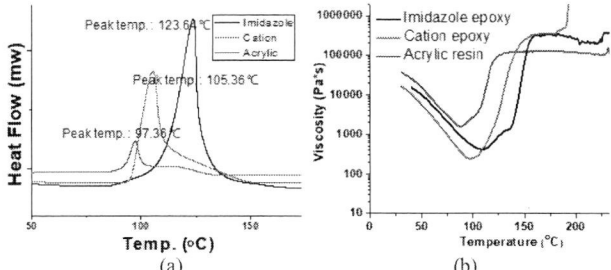

(a)   (b)

Fig. 3. (a) DSC result and (b) viscosities of polymer resins.

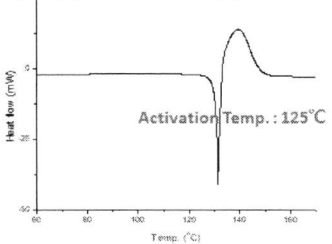

Fig. 4. DSC curve of flux activator.

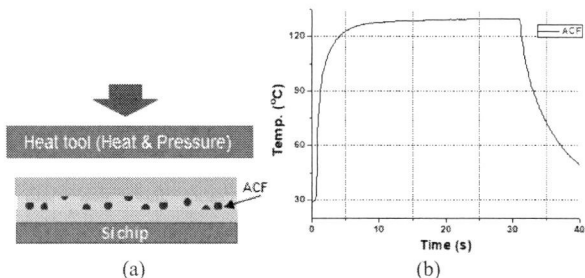

(a)   (b)

Fig. 5. (a) T/C bonding, (b) Temperature profile (130°C, 30 seconds)

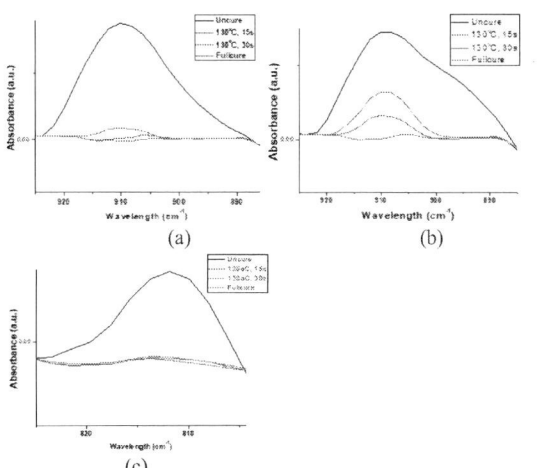

(a)   (b)

(c)

Fig. 6. Cure Degree (a) Imidazole epoxy resin, (b) cationic epoxy resin and (c) acrylic resin after 130°C, 30 seconds

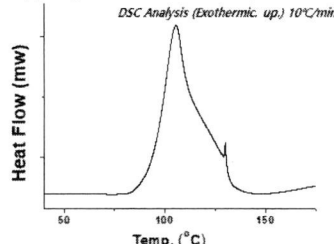

Fig. 7. DSC curve of Cationic epoxy resin after 130°C for 30 seconds iso-thermal bonding

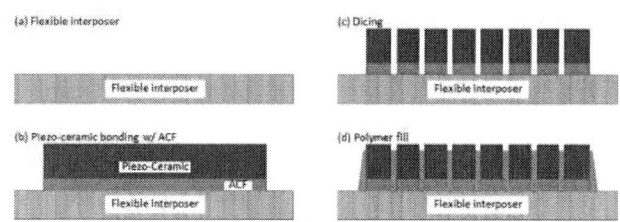

Fig. 8. Bending specimens' preparation.(Schematic images)

Table 1. Polymer resins' viscoelastic and hydroscopic properties

	Coefficient of thermal expansion 1	Glass transient temperature	Coefficient of thermal expansion 2	Hydroscopic rate
Imidazole	643 (ppm/°C)	52.5 °C	1089(ppm/°C)	8.3wt%
Cationic	99(ppm/°C)	82.6 °C	1204(ppm/°C)	2.6wt%
Acrylic	1115(ppm/°C)	53.4 °C	1410(ppm/°C)	4.5wt%

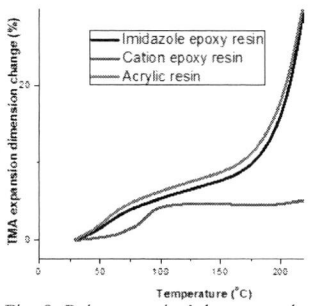

Fig. 9. Polymer resins' thermomechanical property

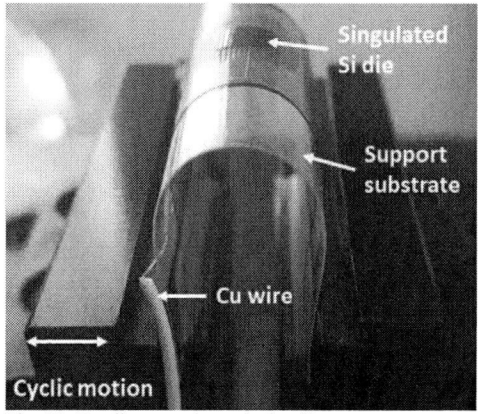

Fig. 10. Bending sample on the bending apparatus.

D. Reliabilities

1) 121°C, 2 atmosphere, and 100% Relative Humidity pressure cooker test was performed. The electrical contact resistance was characterized at every 24 h for 72 h by the method of 4 point probe. Polymer resins' viscoelastic and hydroscopic rate were measured [5] shown in Table 1 and Fig. 9. Cationic polymer shows the best thermal and hygroscopic properties.

2) Bending property was characterized using a bending apparatus. Due to FPC's much compliant for inducing the constant bending, 110 μm thickness of PET plates were selected for the support substrates. The bending specimens were placed on the PET, and side edge was fixed using the scotch tapes shown in Fig. 10. And the specimens were placed on a pair of Al blocks. One block was held, and another ran back and forth. The process ensured the bending specimen for a uniform bended shape of minimum 7mm bending radius at the rate of 1s during the bending test [6]. The changes of *in situ* daisy chain resistances were monitored using a digital multimeter.

III. RESULT

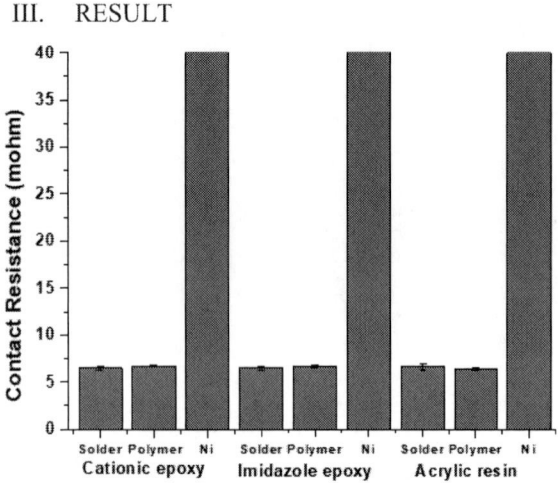

Fig. 11. Results of contact resistance after bonding.

1) Sn52In solder and Au/Ni coated polymer ACFs showed low electrical contact resistances about 6.5 mΩ, whereas Ni ACFs made high electrical contact resistances over 100mΩ shown in Fig. 11.

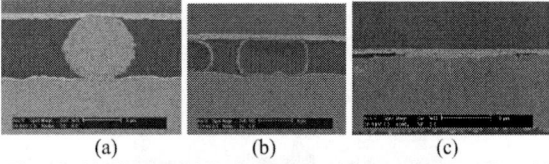

(a)   (b)   (c)
Fig. 12. (a) Ni joint, (b) Au/Ni polymer joint, and (c) Sn52In solder joint.

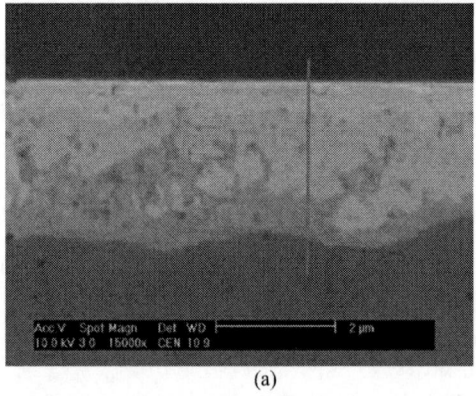

(a)

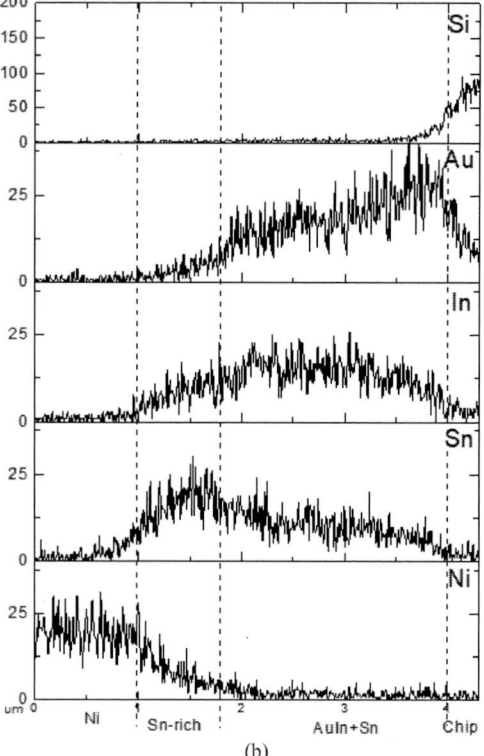

(b)
Fig. 13. (a) Sn52In solder joint and (b) EDS line scan.

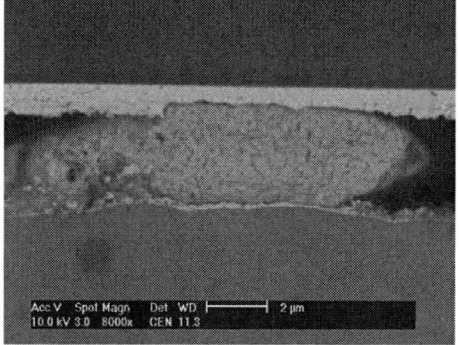

Fig. 14. Acrylic Sn52In ACF joints

Ni ball is not possible to deform due to its hardness during the bonding process. It maintained its original round shape after bonding, resulting in longer electrical path and small contact area shown in Fig. 12(a). While, Ni/Au coated polymer balls and Sn52In Solder balls formed relatively short electrical path and larger contact area, because of polymer ball deformation and solder squeezing shown in Fig. 12 (b) and (c).

During T/C bonding process with Sn52In Solder ACFs, solder oxides of Sn52In are eliminated

$$InO_2 + SnO_2 + 8H^+ + 2e^- = Sn^{2+} + In^{2+} + 4H_2O \quad (1)$$

Therefore, the molten solders provided good wettability on the ENIG surfaces resulting in Au atoms dissolving into the molten solders. The activation energy for the formation of In-Au intermetallic compounds (IMCs) is lower than that of Sn-Au. Therefore, $AuIn_x$ is the major phase formed at the interface shown in Fig. 13. [7]

$$Au + In \rightarrow AuIn \qquad (2)$$

Fig. 13 (b) EDX line scan shows elements present along the substrates, solder joint and interfaces.

However, Acrylic resin based solder ACF was not completely wetted shown in Fig. 14, due to Acrylic polymer's high viscosity which is hard for spreading out shown in Fig. 3 (b). Low contact resistance of Acrylic Sn52In Solder ACFs was due to squeezing of solder balls, however, relatively larger standard deviation was presumably due to not completely wetted solder balls, compared to other resin based solder ACFs. Due to initial contact resistance, every Ni ball ACF was skipped for the Pressure cooker test.

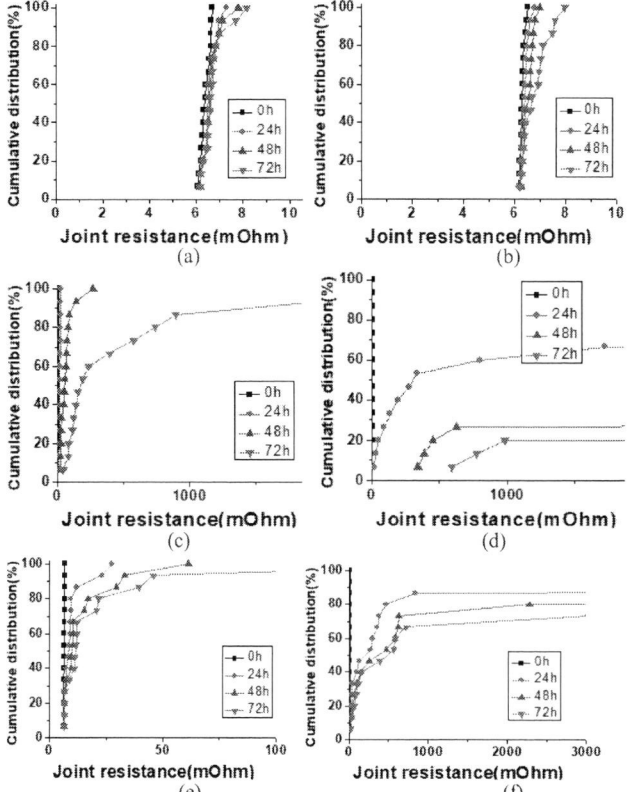

Fig. 15. PCT of Sn52In (a) cationic, (b) imidazole and (c) acrylic ACF and Au/Ni polymer (d) cationic, (e) imidazole, and (f) acrylic ACF

After 72h PCT, contact resistances increased due to the thermal and hygroscopic expansion of polymer resins as shown in Fig. 15. However, Sn52In joints showed far stable contact resistances compared with Au/Ni coated polymer joints. Because metallurgical joint can withstand the expansion of the polymer matrix during the PCT test.

During the PCT, remaining Au atoms diffuse into Sn matrix resulting in further growth of In-Au-Sn IMC [8-9]. The In-Au-Sn IMCs are formed after 72 hours PCT shown in Fig. 16.

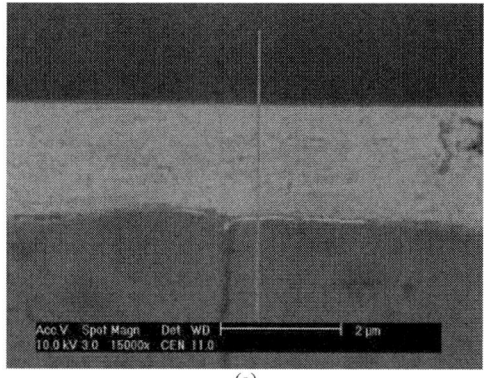

(a)

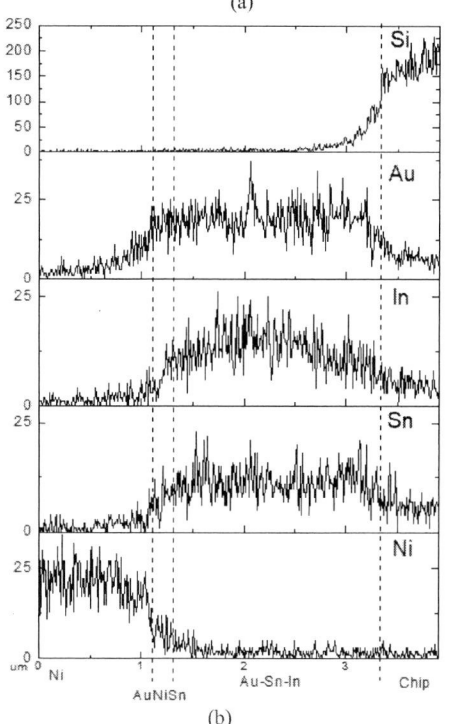

(b)

Fig. 16. (a) Sn52In joints and (b) EDS line scan after 72 h PCT

The dynamic bending test was performed on Au/Ni coated polymer and solder ACFs using two types of ACFs resins, Cationic and Imidazole based Sn52In Solder ACFs.

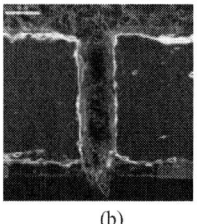

a)   (b)

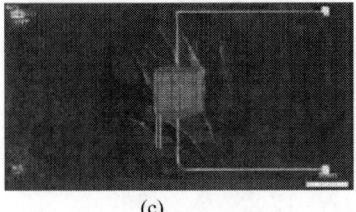

(c)

Fig. 17. (a) After dicing, , (b) kerfs fill, and (c) bending specimens.

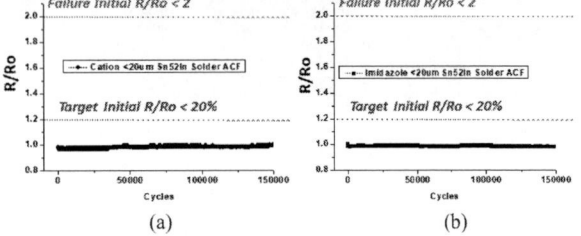

(a)　　　　　　　　　　(b)

Fig. 18. *In-situ* daisy chain resistances change of (a) Cationic and (b) Imidazole Sn52In ACF during bending test.

2) Fig. 17 is the result of each process. The bonded sample was tidily diced with dicing saw. Also Silicone was filled fully. As shown in Fig. 18, the electrical resistances of Cationic and Imidazole Sn52In ACFs were stable even over 150k cycles.

IV. CONCULSIONS

Sn52In solder ACF was selected to bond piezoelectric ceramics and FPCs, and to investigate the effects of metallurgical joints and polymer resin types on the pressure cooker test and the bending test.

Sn52In solder ACF improved the reliabilities because the metallurgical solder joints maintain low electrical contact resistance during the PCT test and the bending test.

By using cationic Sn52In solder ACF, low temperature and the highly reliable solder ACF bonding was demonstrated for ultrasound transducer application.

ACKNOWLEDGMENTS

This work was funded and supported by the Science Ministry, ICT (COI1910M003 and Nano Open Innovation Lab Project through National NanoFab Center

V. REFERENCE

[1] Zhe Wang et al., "A Flexible Ultrasound Transducer Array with Micro-Machined Bulk PZT" *Sensors* 2015, 15(2), 2538-2547.

[2] Julien Rouyer et al., "Conformal Ultrasound Imaging System for Anatomical Breast Inspection" *IEEE Transactions on Ultrasonics, Ferroelectrics, and Frequency Control*, vol. 59, no. 7, July 2012.

[3] Jeanne-Louise Shih et al., "Applications of Flexible Ultrasonic Transducer Array for Defect Detection at 150°C" *Sensors* 2013, 13, 95983

[4] Hoang-Vu Nguyen et al., "Anisotropic Conductive Film Interconnects for Fine-pitch MEMS" presented at the *4th Electronic System-Integration Technology Conference, IEEE*, Amsterdam, Netherlands, Sept. 17-20, 2012

[1] Hoang-Vu Nguyen et al., "Assembly of Transducer Array Using Anisotropic Conductive Film for Medical Imaging Applications" " presented at the *European Microelectronics Packaging Conference (EMPC), IEEE*, Friedrichshafen, Germany, Sept. 14-16, 2015

[2] Seung-Ho Kim et al., "Effect of Flux Activators on the Solder Wettability of Solder Anisotropic Conductive Films" *IEEE Transactions on Components, Packaging and Manufacturing Technology*, Vol. 5, No. 1, Jan. 2015.

[3] Shuye Zhang, Jae-Hyeong Park, and Kyung-Wook Paik, Joint Morphologies and Failure Mechanisms of Anisotropic Conductive Films (ACFs) During a Power Handling Capability Test for Flex-On-Board Applications, *IEEE Transactions on Components, Packaging and Manufacturing Technology*, vol. 6, no.12, pp.1820-1826, Dec. 2016

[4] Shuye Zhang, Tiesong Lin, Peng He, Ning Zhao, Mingliang Huang, and Kyung-Wook Paik, A Study on the Bonding Conditions and Nonconductive Filler Contents on Cationic Epoxy-Based Sn-58Bi Solder ACFs Joints for Reliable Flex-on-Board Applications, *IEEE Transactions on Components, Packaging and Manufacturing Technology*, vol. 7, no. 12, pp.2087-2094, Dec. 2017.

[5] Shuye Zhang et al., "A Study on the Failure Mechanism and Enhanced Reliability of Sn58Bi Solder Anisotropic Conductive Film Joints in a Pressure Cooker Test Due to Polymer Viscoelastic Properties and Hydroswelling" *IEEE Transactions on Components, Packaging and Manufacturing Technology*, Vol. 6, No. 2, Feb. 2016.

[6] S.-I. Park, J.-H. Ahn, X. Feng, S. Wang, Y. Huang, and J. A. Rogers, "Theoretical and experimental studies of bending of inorganic electronic materials on plastic substrates." *Adv. Funct. Mater.*, vol. 18, no. 18, pp. 2673-2648, 2008.

[7] I. Shohji. Et al. "Intermetallic Compound Layer Formation between Au and In-48Sn Solder," Scripta Mater., Vol. 40, No. 7 (1999), pp. 815-820

[8] H.S. Liu, Y. Cui, K. Ishida, and Z.P. Jin, "Thermodynamic Modeling of the Au-In-Sn System", Journal of ELECTRONIC MATERIALS, November 2003, Volume 32, Issue 11, pp 1290–1296

[9] G. Borzone at al, "Phase equilibria in the AuInSn ternary system", CALPHAD: Computer Coupling of Phase Diagrams and Thermochemistry 33 (2009) 1722

Room-Temperature Bonding with Pd Coated Cu Wire on Al Pads: Ball Bond Optimization with 2-Stage Methodology

Nicholas Kam[1], Michael David Hook[1], Celal Con[2], Karim S. Karim[1, 2], Michael Mayer[1]
[1]University of Waterloo, 200 University Ave. W., Waterloo, Ontario, Canada
Ph: 519-888-4567, Ext. 38579
[2]KA Imaging, Kitchener, Ontario, Canada
nkam@edu.uwaterloo.ca

Abstract— **Wirebonding performed at elevated temperatures is the standard interconnect process for integrated circuits, typically with the use of low-cost copper bonding wire. However, for specific applications it is necessary for wire bonds to be reliably joined at room-temperature. This paper details the development of a room-temperature ball bonding process using a 2-stage optimization method. The first stage optimizes ball geometry by applying a 3^2 design of experiment to bonding parameters impact force (IF) and electric flame-off (EFO) current. In the second stage bond shear strength is optimized by stepwise increase in ultrasonic amplitude. Target ball bond values were attained at optimized parameters: IF of 1331 mN, EFO current of 59.9 mA, and an ultrasonic amplitude of 26.46 US%. Pad lift during bonding was observed at excessive ultrasonic amplitudes above 40 US%, as determined by optical images at the bond interface. Bonding parameters at room-temperature (23°C) were increased when compared to a high temperature process (175°C) to account for reduced thermal energy. For the same geometry at room-temperature a 7 % increase to impact force was required. EFO current levels remained relatively constant between the two bonding temperatures. For the same shear strength at room-temperature a 18 % increase in ultrasound amplitude was required. The confirmed average shear strength achieved via the room-temperature process was 116 MPa. Higher values are possible.**

Keywords- Wirebonding; Room-temperature; PCC wire; Al pads; DOE; optimized; 2-stage methodology

I. INTRODUCTION

Amorphous Selenium (a-Se) has been shown to be a feasible photoconductive sensor for use in x-ray detectors [1]. The amorphous structure of selenium is only present at low temperatures and begins to crystallize above temperatures as low as 420 K [2]. To preserve the amorphous structure, wire bonded interconnections to the a-Se layer would be required to be made at low process temperatures, ideally room-temperature. The objective of this study is to develop a methodology to obtain reliable ball bonds for a substrate at room-temperature (23 °C).

We establish optimized ball bonding parameters at room-temperature by modifying a method developed previously at the University of Waterloo's Centre for Advanced Materials Joining [3]. The method relies on optimizing ball geometry and shear strength independently. Standard electronic packaging wire and bond pad materials were used in this study. Copper wire has become an increasingly popular cost-effective alternative to gold wire bonds. Palladium coated copper wire (PCC) was bonded to aluminum pads to allow for comparison to previous optimized parameters obtained at high-temperature (175°C) [4].

II. EXPERIMENTAL PROCEDURE

A. Design of Experiment

Wire bonding was performed using an ESEC 3088 automatic ball bonder. Sample chips used in the study were provided by ON Semiconductor, Phoenix, AZ, an example is shown in Figure 1. The wire and substrate materials used in the experiment are detailed in Table 1. Target values for the bonded ball geometry and shear strength are provided in Table 2. Target geometry values are the same as previously used for a high temperature process, allowing for comparison [4].

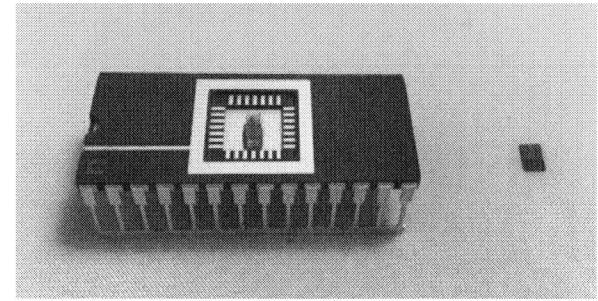

Figure 1 Sample ON Semiconductor chip die-attached to DIP.

TABLE I. BONDING MATERIAL PARAMETERS

Wire Material	PCC
Wire Diameter (μm)	25
Al Pad Thickness (μm)	1
Substrate Temperature (°C)	23
Substrate Metallization	Au

TABLE II. TARGET BONDING SPECIFICATIONS

Target Bond Specification	Value
BDC (μm)	61
BH (μm)	14
Shear Strength (MPa)	120

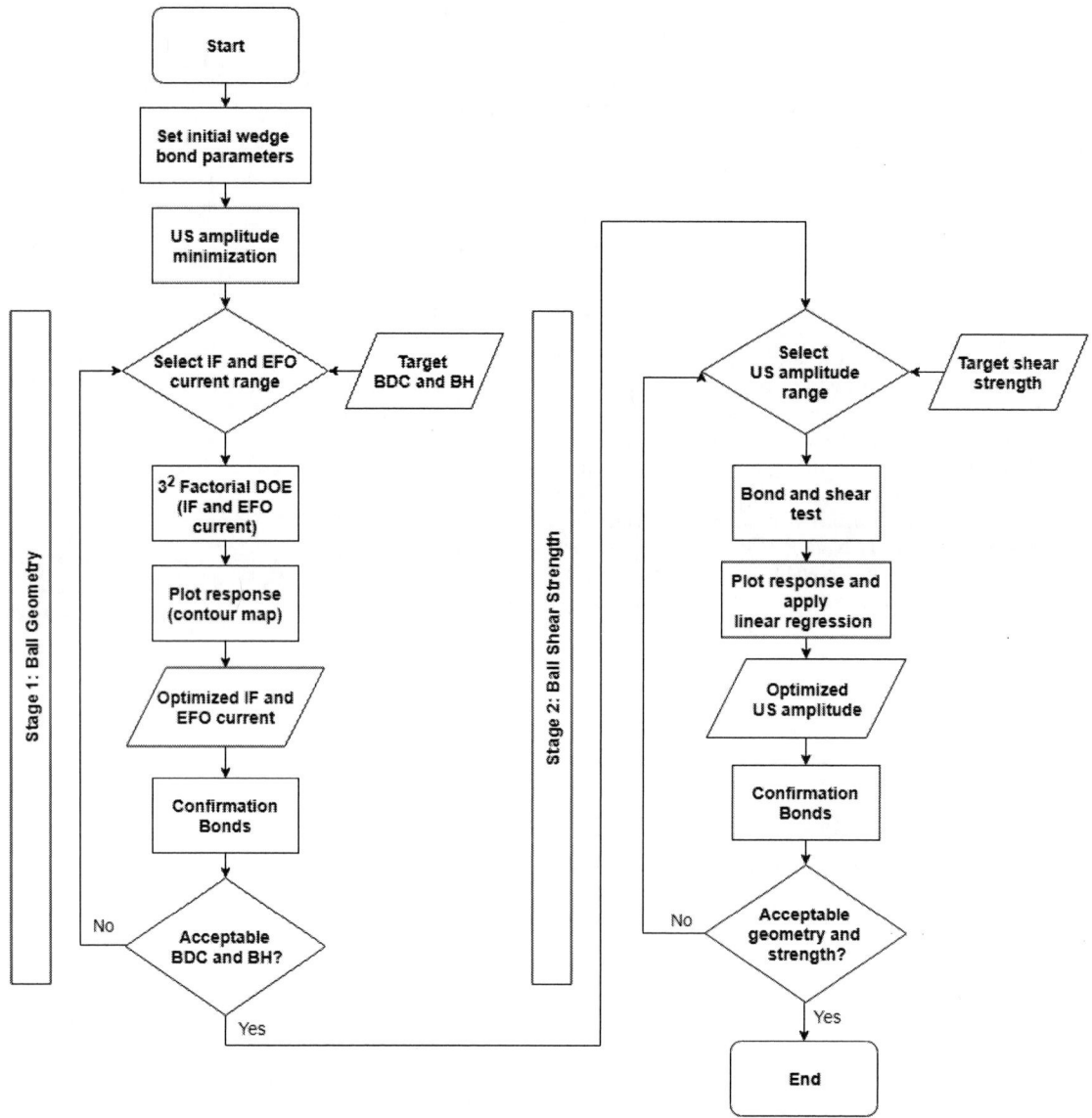

Figure 2 Process flow of 2-stage methodology for optimization of ball bond parameters.

Optimization of the ball bonds occur in two independent stages—ball geometry and ball shear strength—as summarized by the process flow diagram shown in Figure 2. Confirmation bonds are conducted after each stage to verify the optimized bonding parameters.

An initial trial-and-error phase was conducted to setup a consistent un-optimized wedge bond, bonding parameters listed in Table 3. Ultrasonic amplitude for the ball bond was minimized in an additional setup phase to reduce ultrasonic-enhanced deformation, a preparation step for the ball geometry optimization stage. In this pre-phase, wire bonds were made using preliminary parameters and with various ultrasonic amplitude values. Ultrasonic amplitude was iteratively decreased until bonding no longer occurred as experienced by ball non-sticks. Ball non-sticks were

observed at US% values as high as 15 US%. A 5 US% increase to this value was chosen as the initial ultrasound level given in Table 3.

For stage 1 we followed the method described by Gomes et al. [3], for which a 3^2 factorial design of experiment (DOE) was conducted to characterize the effect of impact force (IF) and EFO current on ball diameter at capillary imprint (BDC) and ball height (BH). A total of 10 samples were bonded for each permutation of IF and EFO current. Ball geometries were measured by optical microscope. The contour plot in Figure 3 was then generated from the data relating BDC and BH geometries to applied IF and EFO current. The intersection of target BDC and BH isolines are used to estimate the values of IF and EFO current that yields the target geometry.

In the second stage, ultrasonic amplitude was increased stepwise to optimize ball shear strength. Shear strength was measured using a DAGE4000 shear tester. The data was fitted using a linear regression model with a 95 % confidence interval. The linear fit was used to identify the ultrasonic amplitude that would yield the target ball shear strength.

Table 3 details the initial ball bond parameters based on settings established at high-temperature by Hook et al. [4]. For this study the subject of optimization will be the ball bonds exclusively. A similar methodology can be applied for the wedge bonds.

TABLE III.   INITIAL WIREBONDING MACHINE PARAMETERS

Ball Bond Parameter	Value	Wedge Bond Parameter	Value
Ball IF (mN)	1242	Wedge IF (mN)	500
Ball Bond Force (mN)	169	Wedge Bond Force (mN)	700
Ball Ultrasonic Amplitude (US%)	20	Wedge Bond Time (ms)	20.2
Ball Ultrasonic Time (ms)	27.2	Wedge Ultrasonic Amplitude (US%)	30
Ball EFO Time (ms)	0.7	Wedge Pre Ultrasonic Amplitude (US%)	4.98
Ball EFO Current (mA)	59.84	Wedge Bond Force Dwell Time (ms)	5
Shielding Gas	N$_2$	Wedge Ultrasonic Amplitude On (ms)	5

III.   RESULTS AND DISCUSSIONS

A.   Optimized Geometry

Figure 3 presents the resulting contour plot detailing the effects of EFO current and IF on the target geometry. EFO current was selected for optimization, in contrast to the EFO time used in [3], due to the Cu wire material being used in this study (Au in [3]), for which short EFO times are recommended. The EFO time used for bonding was set close to the minimum possible for the equipment. Constant lines of BDC and BH are determined based on the measured data. The intersection point of target isolines are used to determine the optimized IF and EFO current values given in Table 4.

Compared to bonding at high-temperature, room-temperature bonding required higher impact force to attain the same target geometry (7 % increase). From a materials standpoint the addition of thermal energy to metals reduces the stress necessary for plastic deformation to occur, therefore at low temperatures the yield strength of Cu is typically greater. A room-temperature substrate would supply less thermal energy to the free air ball (FAB) on impact, so a larger impact force would be necessary to induce the stress necessary for plastic deformation of the FAB [5].

EFO current remained expectedly constant (a difference of 0.1%), as any change in the substrate temperature would have no effect on the FAB size created during the separate EFO process.

B.   Optimized Shear Strength

In general, ball bond shear strength at room-temperature was significantly weaker compared to high-temperature. Average shear strength values at all ultrasound levels failed to reach the same 130 MPa target used in the previous high-temperature study [4]. As a consequence, a reduced shear stress target of 120 MPa was selected.

Various ultrasonic levels were initially investigated to find a range near the desired shear strength. The data is split in a lower and higher US% range, shown in Figures 4 and 5, respectively. The plot in Figure 4 compares shear strength against ultrasonic amplitude within the range of 20 – 30 US%. Bonding was generally consistent in this range. A positive correlation can be observed from the linear regression model between ultrasonic amplitude and shear strength. The fitted line was used to estimate the ultrasonic amplitude necessary to achieve the target shear strength of 120 MPa, see Table 4. Compared to optimized parameters at high-temperature [4], a 14 % increase in ultrasonic amplitude was required to achieve the same 120 MPa shear strength.

Figure 5 plots the ultrasonic range above 30 US%. A linear fit of the data in this range shows diminishing shear strength with increasing ultrasound. For the target strength, the optimized ultrasonic amplitude estimated from this plot occurs outside the minimum bounds of the range. Bonding was inconsistent with pad lift, cratering, and loss of FAB common between successive bonds.

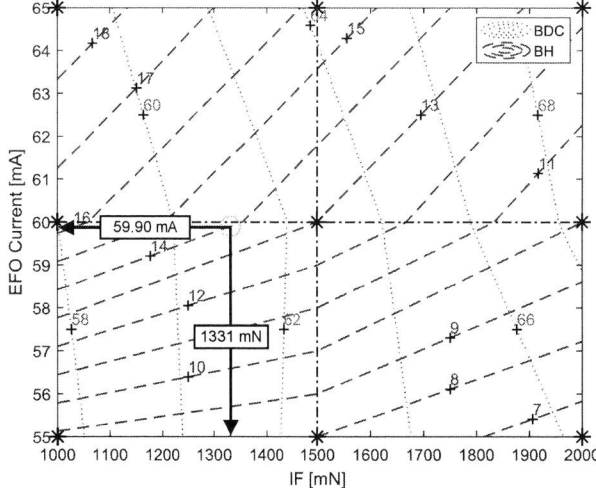

Figure 3 Contour plots for BDC (dashed lines) and BH (dotted lines) [values in μm] used to predict impact force (IF) and EFO current for target geometry BDC = 61 μm and BH = 14 μm. Intersection point of target contour lines marked with a green circle; corresponding IF and EFO current values indicated with arrows. Method similar to that in [3].

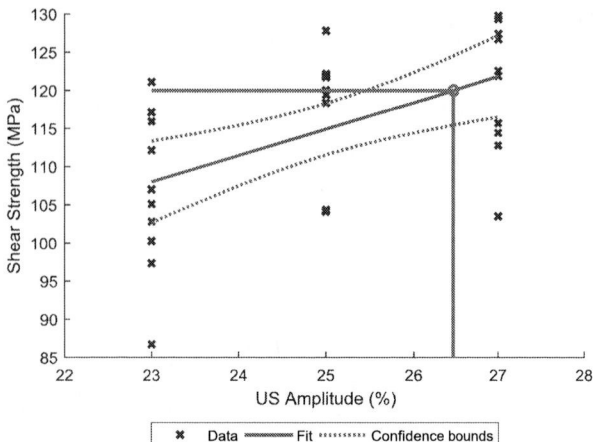

Figure 4 Effect of ultrasonic amplitude on shear strength below 30% US. Linear regression applied to data set to estimate US amplitude required for target shear stress. Dashed lines indicate 95 % confidence intervals.

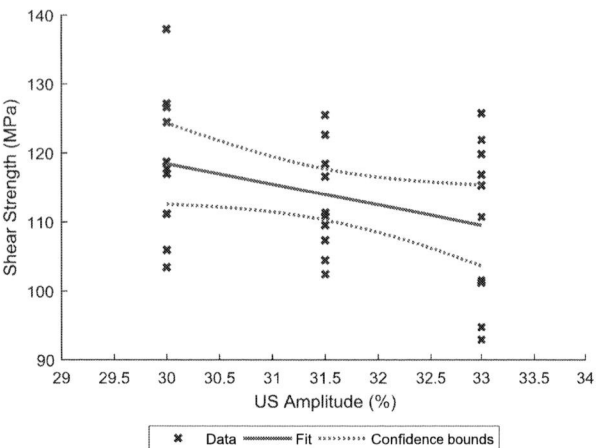

Figure 5 Effect of ultrasonic amplitude on shear strength above 30 US%. Linear regression applied to data set to estimate US amplitude required for target shear stress. Dashed lines indicate 95 % confidence intervals.

Complete bonding failures were observed with ultrasonic amplitudes above 40 US%, as shown in Figures 6 (a) and (b). The primary failure was pad lift-off, indicating bonding failure due to excessive ultrasonic stresses. Damage to the underlying silicon chip (cratering) was also observed on some samples in this US% range.

The purposes of ultrasound in wire bonding is to enable the removal of surface contaminants and the delivery of mechanical friction energy between two bonding interfaces. For the friction energy to result in an ultrasonic weld, asperities at the interface must be broken off (wear) and transported along the interface to form microwelds. A room-temperature bond under reduced thermal energy would therefore require larger stresses (larger ultrasonic amplitudes) to break asperities. This could lead to an increase risk of cracking, pad damage, and bonding failures.

Due to the higher likelihood of bonding failure observed at higher ultrasonic amplitudes, the optimized value obtained at the low range (Figure 4) was accepted for final confirmation. Final optimized bonding parameters are summarized in Table 4.

C. Bond Confirmation

A set of confirmation bonds (n=10) were produced using the optimized parameters listed in Table 4. No issues were observed during bonding. An extended focus image of the as-bonded balls on the chip is shown in Figure 7. Intact wire loops can be observed. After-shear interfaces are present in the adjacent row of pads.

Geometric measurements and shear testing were performed on each confirmation sample. The average response and associated standard deviations are given in Table 5. The bonding response at high-temperature is provided for comparison [4]. A similar variance at both temperatures can be observed in each data set, indicating a high process capability resulting from the applied 2-stage optimization methodology.

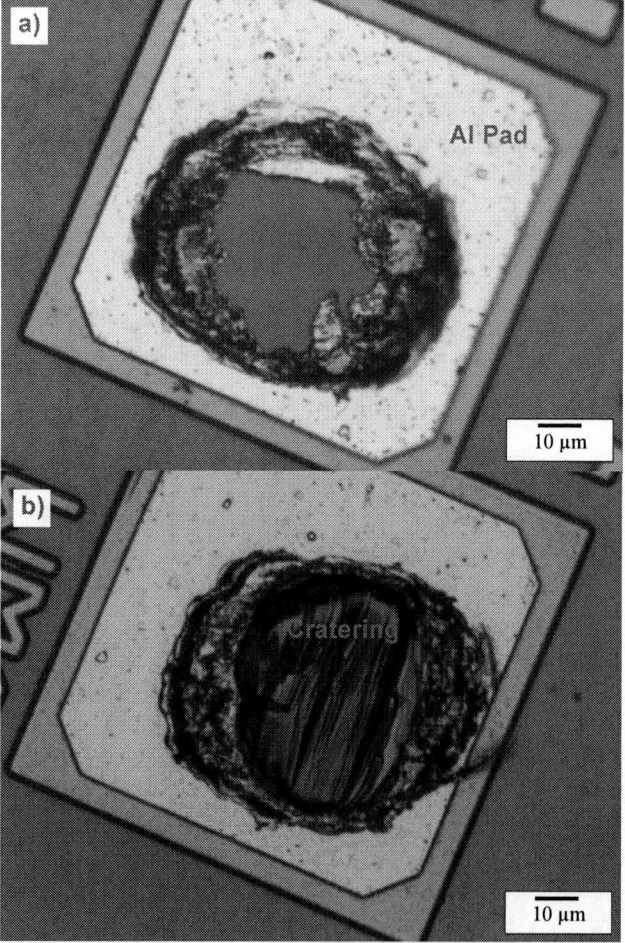

Figure 6 Optical images of damaged pad interfaces after failed bond attempts using 40 US% ultrasonic amplitude. Balls did not stick. a) Pad lift-off observed on Al bond pad. b) Cratering.

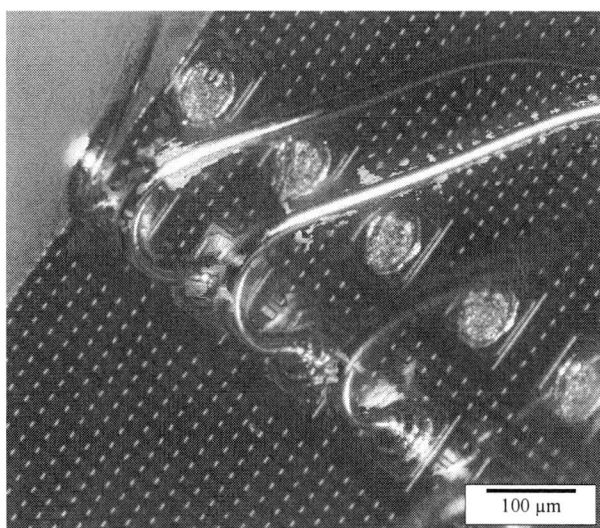

Figure 7 Extended focus image of room-temperature as-bonded balls and shear interfaces after testing.

TABLE IV. BONDING PARAMETERS OPTIMIZED WITH 2-STAGE METHODOLOGY

Ball Bond Parameter	Room-Temperature Values, 23 °C	High-Temperature Values [4], 175 °C
Impact Force (mN)	1331	1242
EFO current (mA)	59.90	59.84
Ultrasonic Amplitude (US%)	26.46	22.50

TABLE V. OPTIMIZED BONDING RESPONSE: AVERAGE AND STANDARD DEVIATION VALUES (SAMPLE SIZE = 10)

Ball Bond Response	Room-Temperature Values, 23 °C	High-Temperature Values [4], 175 °C
Ball Height (µm)	12.02 ± 0.37	13.5 ± 1.0
Ball Diameter (µm)	60.91 ± 0.77	61.97 ± 0.7
Ovality (%)	2 ± 1	NA
Shear Force (mN)	34.34 ± 0.84	36.72 ± 1.95
Shear Strength (MPa)	115.68 ± 4.64	125.6 ± 7.4

Average ball height for the confirmation bonds saw a 14 % reduction from the target ball height, possibly due to a reduction in free-air ball diameter (measurement data missing). Diametric means were also slightly reduced but were acceptable within a standard deviation. Similarly, measured shear strength values were reduced.

Optical images of confirmation bonds were used to characterize bonding quality, see Figure 8 (a) and (b). Overall, ball bonds were observed to have good shape as shown in Figure 8 (a) and as indicated by low average

ovality. As can be seen in the after-shear image, Figure 8 (b), the observed failure mode was either ball lift at the interface or Al shear in the pad. No cratering was observed with any of the confirmation bonds after shearing, a result similar to that from a comparable high temperature process with Au wire [3]; Sufficient ultrasonic energy was thus applied to create a reliable bond while avoiding underlying damage to the chip.

Future research will address ball bond reliability. In general, accelerated testing is achieved by heating the specimen to an elevated temperature. In our case, any a-Se present on the chip would lose its amorphous properties. However, for the limited task of bond reliability it is still possible to preform conventional ball bond reliability testing and to extrapolate the results to room temperature.

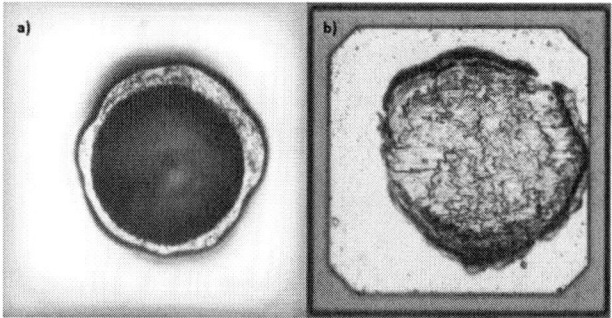

Figure 8 a) Ball bond at 50x magnification as-bonded before shear test. b) Pad interface after shear testing with the bonded ball sheared in direction from left to right.

IV. CONCLUSION

The 2-stage methodology previously used for accelerated optimization of Au ball bonding processes was successfully adapted to optimize ball bonds made with palladium coated copper wire. The methodology is an efficient way to quickly optimize ball bonds with the ability to achieve an exact chosen target geometry for the bonded balls.

While in general room-temperature bond strength was observed to be weaker and more prone to pad damage compared to high-temperature bonding, it was still possible to run a robust process indicating a large enough process window for mass production. For this, higher parameter values were necessary for impact force and ultrasonic amplitude.

It was found that a robust wire bonding process was possible with room-temperature as the process temperature. Bond reliability is expected to be acceptable, but this remains to be confirmed with future experiments. The ability to run a room-temperature process can allow for the efficient interconnection to advanced devices such as those with amorphous selenium layers for high resolution x-ray imaging.

ACKNOWLEDGMENT

We acknowledge the financial support provided by the National Science and Engineering Research Council (NSERC) of Canada. We are thankful for the use of equipment at the University of Waterloo's Centre for Advanced Materials Joining.

REFERENCES

[1] Abbaszadeh, Shiva, Allec, Nicholas, Karim. (2013). Characterization of low dark-current lateral amorphous-selenium metal-semiconductor-metal photodetectors. IEEE Sensors Journal.

[2] Zhang, H. Y., Hu, Z. Q., & Lu, K, "Transformation from the amorphous to the nanocrystalline state of pure selenium," Elsevier Science, 5(1), pp 41-52. 1995.

[3] Gomes, J., Mayer, M., Lin, B. (2015), "Development of a fast method for optimization of Au ball bond process," Microelectronics Reliability, 55 (3-4), pp. 602-607.

[4] M. Hook, S. Hunter, M. Mayer (2018), "Thermosonic ball bonding recipe optimization: comparing Cu and PCC wire on two pad thicknesses," presented at 14th IMAPS Device Packaging Conference, Scottsdale, AZ, 2018..

[5] Frank P. Incropera, David P Dewitt. Fundamentals of Heat and Mass Transfer. Second Edition. 2012

[6] G. Harman. Wire Bonding in Microelectronics, Eight Edition. McGraw-Hill, 2010.

On-Chip ESD Monitor

Kannan K.T., Boris Vaisband, and Subramanian S. Iyer

Centre for Heterogenous Integraton and Performance Scaling (CHIPS)
Department of Electrical and Computer Engineering
University of California, Los Angeles, CA 90025, USA
[kannankt, vaisband, s.s.iyer]@ucla.edu

Abstract—**Electrostatic discharge (ESD) failure results in about 35% of IC field returns, and is the cause of several billion-dollar loss to the semiconductor industry. An on-chip ESD detector can help track the electrostatic history of ICs from manufacturing to end-of-life. Two approaches for on-chip ESD detection are presented: variable dielectric width capacitor, and vertical MOSCAP array. The variable dielectric width capacitor approach employs metal plates terminated with sharp corners to enhance local electric field and facilitate easy breakdown of the thin dielectric between the metal plates. The vertical MOSCAP array consists of a capacitor array connected in series. Both approaches were simulated, fabricated, and experimentally characterized in GlobalFoundries 22 nm fully depleted silicon-on-insulator. Vertical MOSCAP arrays detect ESD events starting from ~6 V with 6V granularity, while the variable dielectric width capacitor is suitable for detection of high ESD voltage from 40 V and above.**

Keywords—*ESD, On-chip ESD Monitor, ESD sensor, ESD voltage detection.*

I. INTRODUCTION

Electrostatic discharge (ESD) in integrated circuits (ICs) occurs due to electrostatic charge transfer between two components in close proximity. As a result of an ESD event, a high transient current (up to few tens of Amps) and large voltage (up to several tens of kV) can develop between the two components. This fast (~150 ns) transient phenomenon can cause serious damage or degrade the performance of affected ICs. Damage from an ESD event can cause local damage to metals, oxides, junctions, and other device components, resulting in either complete or partial device failure [1]. Currently, ESD events significantly contribute to device failure at all stages of IC production, test, assembly, and field usage. About 35% of the IC field returns are reported to be ESD induced, with annual costs estimated to be several billions of dollars [2-3]. ESD protection circuits are included in most modern ICs. Nonetheless, static charge accumulation during transport and handling can exceed the limits of a protection network, causing ESD damage [3].

With the advent of heterogeneous integration, bare dies from a variety of sources need to be integrated using advanced packaging methods such as the silicon interconnect fabric technology [4] or fan-out wafer level packaging such as "FlexTrate" [5]. During die transportation from a variety of sources, ESD events may occur. An accurate ESD sensor allows us to pinpoint the occurrence of an ESD event prior at any point in the supply chain and potentially avoid the assembly of parts that may have been compromised due to ESD events.

In this paper, an on-chip ESD detection circuit, which can track the electrostatic history of the IC from the manufacturing process until the end-of-life, is proposed. This on-chip ESD monitor tracks the unit process module (assuming that testing is possible and performed after each unit process) that suffered an ESD event. The general requirements of an ESD diagnostic structure include: (i) stand alone or easily integrable with heterogeneous dies, (ii) non-volatile, *i.e.*, damage or effects of charge accumulation cannot be reset, modified or erased, (iii) quantitative and correlated to the pad size, and (iv) detect ESD events independent of the packaging unit process, including storage. Two approaches for on-chip ESD evaluation are proposed: (a) variable dielectric width capacitor, and (b) vertical MOS capacitor (MOSCAP) array.

The rest of the paper is composed of the following sections. The variable dielectric width capacitor is presented in Section-II. Vertical MOSCAP array is presented in Section III. Conclusions are offered in Section IV.

II. VARIABLE DIELECTRIC WIDTH CAPACITOR

The basic working principle of the variable dielectric width capacitor is presented in Section-II A. Evaluation of the electric field within the proposed structure, and ESD test simulations are provided in Section-II B. Implementation of the variable dielectric width capacitor in GlobalFoundries (GF) 22 nm fully depleted silicon-on-insulator (FDSOI) technology, along with the experimental results is presented in Section-II C.

A. Working principle

The basic structure of the variable dielectric width capacitor consists of metal plates terminated with sharp corners, as shown in Fig. 1. Sharp corners are incorporated in this structure to increase the local electric field intensity that will ensure that the thin dielectric between the metal plates breaks down easily. The ESD monitor module consists of rows and columns populated with metal plate pairs of varying area and separation. Each plate pair i has a pad area of A_i and a separation of d_i. The structure is subjected to ESD events during the packaging, assembly, or any other unit processing (Fig. 2). Since both metal pads in each pair are of equal area, hence equal capacitance, charge accumulation on the pads will be similar. If charge has accumulated on the metal pads, grounding one of the pads will cause a significant voltage difference between them. Consequently, a large current

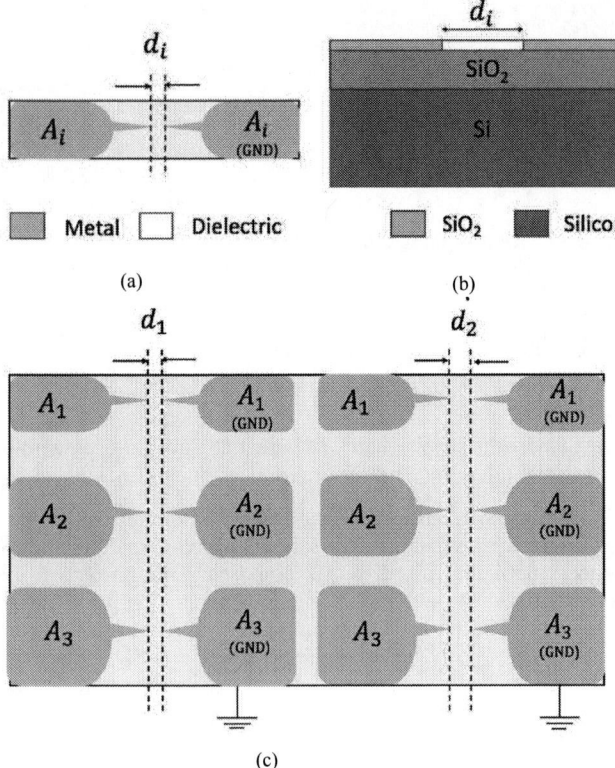

Fig.1. Pairs of metal plates terminated with sharp corners separated by a thin dielectric. (a) Top view, (b) cross-sectional view, and (c) metal plate array.

will flow from the high potential plate to the grounded plate through the thin dielectric causing a dielectric breakdown. The voltage that is developed across a plate pair i, denoted V_i, is directly proportional to the area of the plate A_i. The magnitude of the electric field E_i that is developed between the metal pads, when one of the plates is grounded, depends on the plate separation d_i. The breakdown strength of the dielectric is denoted by $E_{BD,I}$. The criterion for breakdown is $E_i(V_i,d_i) > E_{BD,I}$, i.e., the electric field between the metal pads becomes greater than the breakdown strength of the thin dielectric. In mission mode, one of the metal pads from each single pair in the array is always grounded, resulting in a breakdown of metal plate pairs with corresponding dielectric thickness whenever an ESD event occurs. The I-V curve of each plate pair (all rows and columns) could be measured at any time to observe whether an ESD event has occurred. The induced ESD voltage of the unit process is estimated by identifying the broken plate pairs. A more accurate evaluation of the induced voltage can be performed by increasing the granularity of (A_i and d_i) within the array.

B. Evaluation of electric field and ESD simulations

The electric field $E(\rho,\phi)$, in cylindrical coordinates (ρ,ϕ,z), inside a two-dimension wedge with angle β ($0 \leq \phi \leq \beta$) bounded by a grounded conductor, offers insight into the nature of electric fields near sharp conducting corners (Fig. 3) [6]. When the radial distance ρ is small, the potential near the corner is given by

$$\varphi(\rho,\phi) \approx \rho^{\pi/\beta}\sin\left(\frac{\pi\phi}{\beta}\right).$$

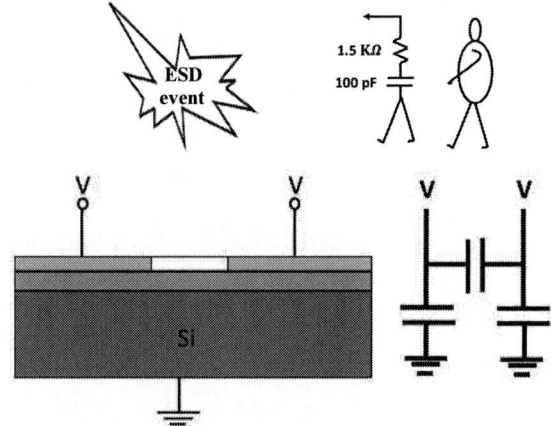

Step A: A Human body touching the ESD sensor. Both the metal plates of the ESD sensor are charged to 'V' volts.

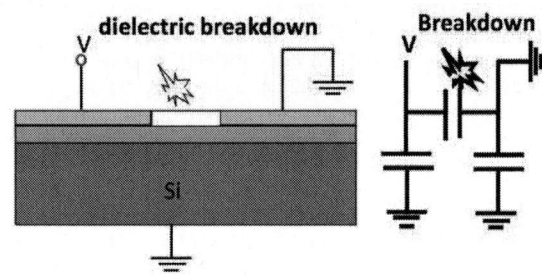

Step B: One of the metal plates of the ESD sensor is grouned. Thin dielectric between the metal plates breakdown due to the potential difference 'V' between the plates.

Fig.2. Working principle of ESD detector unit

The associated electric field is

$$E = -\nabla\varphi = -\frac{\pi}{\beta}\,\rho^{\frac{\pi}{\beta}-1}\{\hat{\boldsymbol{\rho}}\sin\left(\frac{\pi\phi}{\beta}\right) + \hat{\boldsymbol{\phi}}\cos\left(\frac{\pi\phi}{\beta}\right)\}.$$

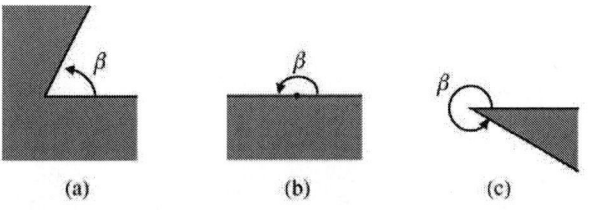

Fig.3. Side view of a two-dimensional wedge ($0 \leq \phi \leq \beta$) cut out of perfectly conducting (shaded) matter. The conductor is held at zero potential and otherwise fills all of the space [6].

Thus, the magnitude of the electric field $|E| \to 0$ as $\rho \to 0$, for $\beta < \pi$, and $|E| \to \infty$ as $\rho \to 0$, for $\beta > \pi$. The intensity of a high local electric fields near sharp corners is exploited in the implementation of the variable dielectric width capacitor. When one of the metal plates in a plate pair is grounded, the charge density on the non-grounded plate near the sharp edge becomes high, resulting in a very high electric field, which can break down the dielectric between the plates.

Electric field simulation of the structure in ANSYS Maxwell as

978-1-7281-1500-9/19 $31.00 © 2019 IEEE

a function of the width of the dielectric between the pads (*i.e.*, separation between sharp corners), is shown in Fig. 4. Break down strength of some dielectric materials that could be employed in the ESD monitor are also indicated in Fig, 4. For example, the maximum electric field between metal plates that are separated by 40 nm of dielectric material, is $1.2 \cdot 10^8$ V/m. If,

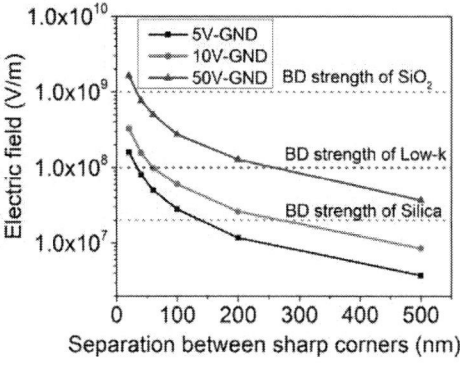

(a)

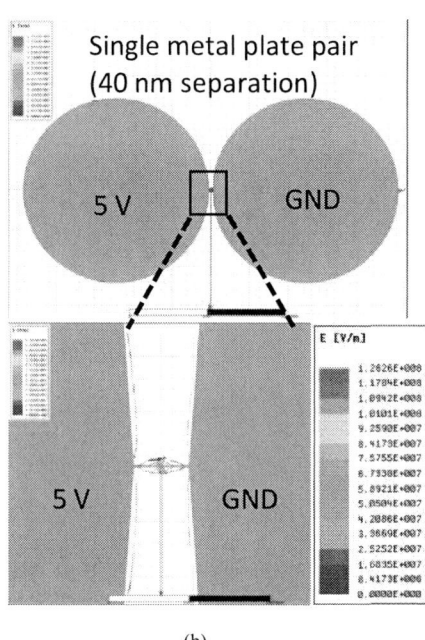

(b)

Fig.4. Electric field profile. (a) Electric field as a function of distance between the sharp corners (BD stands for break down). (b) Electric field lines for a 5 V – GND system at 40 nm separation.

therefore, the dielectric chosen for the 40 nm separation is silica, exhibiting a typical breakdown strength of $\sim 1 \cdot 10^7$ V/m, a voltage of 5 V across the plate pair will suffice to break the silica dielectric.

Classically, damage originating in ESD was considered to be a production problem, and was addressed by ESD safe procedures, like earth mats, operator wrist straps, etc. However, with increased use of plastic and synthetic materials in the modern environment and widespread use of metal oxide

technologies, the effect of ESD on electronic systems has become more significant and is no more considered only as a production problem [7]. ESD induced field returns of ICs during assembly, test, and field usage has led to the development of ESD testing standards to test for robustness and ensure ESD protection of any electronic equipment. For ESD qualification of ICs, test standards have been developed by various organizations, including JEDEC, ESDA, and MIL-STD [8]. There are three main ESD test models based on ESD events that the test is emulating: human body model (HBM), charge device model (CDM), and machine model (MM).

CDM emulates ESD charging followed by a rapid discharge, similar to what is seen in automated handling, manufacturing, and assembly of IC devices. A CDM calibration setup, as shown in Fig. 5, consists of a charge plate, to which a high voltage supply is attached. A thin insulation layer isolates the device under test (DUT) from the charge plate. The DUT is placed on the insulation layer with its pins facing up. This results in the DUT being capacitively coupled through the FR4 dielectric and charged by the charge plate. During the test, a grounded plate approaches the DUT and discharges the current through a 1 Ω resistor. The discharge current through the resistor is monitored by an oscilloscope. JEDEC specifies a standard size metal coin for calibration of the test setup. The test environment from Fig. 5

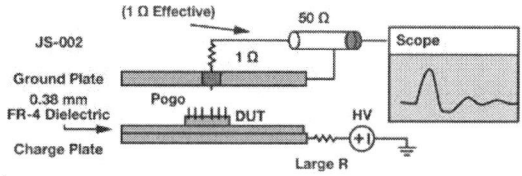

Fig.5. CDM Test setup [9].

was modelled in ANSYS Maxwell using a 25 mm standard size coin as the DUT. All capacitances pertaining to the test setup were extracted. A CDM circuit simulation was performed with the extracted capacitance values, and the discharge current and voltage waveforms were verified with the literature reported values [10-11].

Upon verifying the test setup, a CDM simulation with the ESD detector as the DUT was set up in ANSYS Maxwell and various capacitance values were extracted. CDM circuit schematic and simulation results with the ESD detector as the load are shown in Fig. 6. A test voltage of 5 V is applied to the charge plate. Initially, both metal plates (Met1 and Met2) are equally charged to 4.3 V. At t = 1 ns, metal plate Met1 is grounded. Met1 undergoes a small oscillation and settles to 0 V. Prior to grounding Met1 ($0 \leq t \leq 1$ ns), the voltage difference between the two plates $V_{Met2} - V_{Met1}$ was zero. At t = 1 ns, the difference in voltage shoots up to 7 V and then settles to 4 V. This shows the voltage development across the sharp corners, which results in a high local electric field and a possible dielectric breakdown between the metal plates when one of the metal plates is grounded.

The HBM emulates an electrostatically charged human, touching the pins of an IC, typically generating a discharge current with a rise time of a few nanoseconds, and a decay time

978-1-7281-1500-9/19 $31.00 © 2019 IEEE

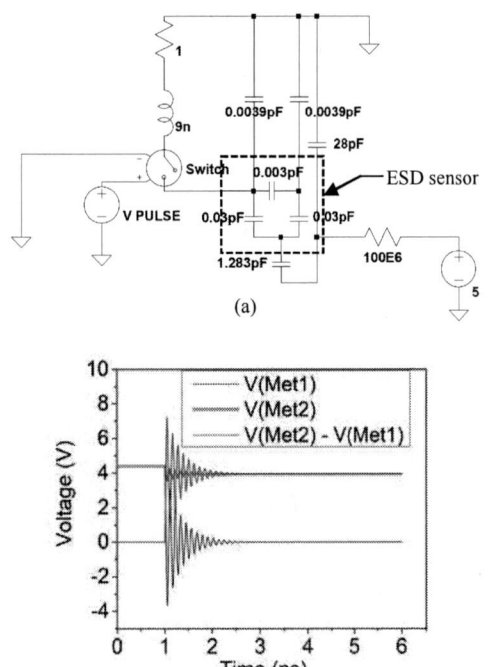

(a)

(b)

Fig.6. CDM simulation. (a) LT-Spice immplementation of CDM with ESD detector as DUT, and (b) corresponding voltage waveforms.

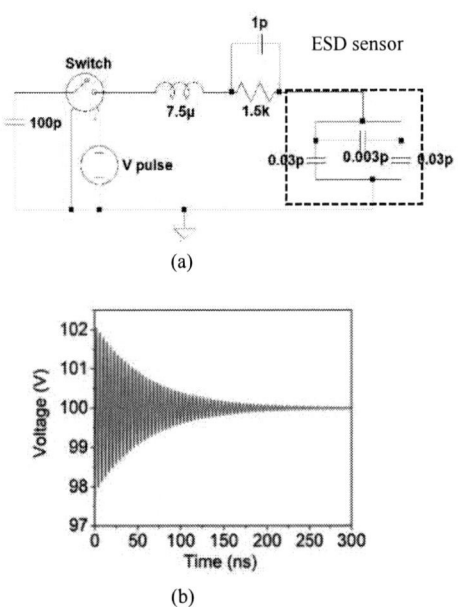

(a)

(b)

Fig.7. HBM simulation. (a) LT-Spice immplementation of HDM with ESD detector as DUT, and (b) corresponding voltage waveforms.

of about 150 ns [12-13]. The small signal impedance of humans standing over a ground plane and holding a metal object varies non-linearly with frequency, from 3 kΩ at low frequency (< 1 MHz), to less than 50 Ω at high frequency (> 1 GHz). This impedance can be modelled with discrete elements and transmission lines with good correlation to human ESD events [14]. In case of an HBM ESD event, the static charge is initially stored in the body of the human, and transferred to the IC when a body part of the human comes in contact with the system. The equivalent circuit diagram of an HBM ESD event along with the ESD detector as the DUT for an input test voltage of 100 V is shown in Fig. 7. Both metal plates of the ESD sensor instantaneously (~150 ns) acquire the applied voltage of 100 V since the capacitance of the ESD sensor is significantly smaller than the typical HBM capacitance. This ensures that a breakdown occurs instantaneously when one of the metal plates is grounded.

The MM ESD standard represents the electrical discharge from a charged conductive source into a component or an object. Unlike the HBM, the MM equivalent capacitor discharges through a small parasitic series resistance, resulting in an oscillatory input pulse. This is comparable to the pulse generated by a charged machine part touching an IC pin. MM simulation of the ESD sensor for an input voltage of 100 V is shown in Fig. 8. Similar to the HBM simulation, both metal plates acquire

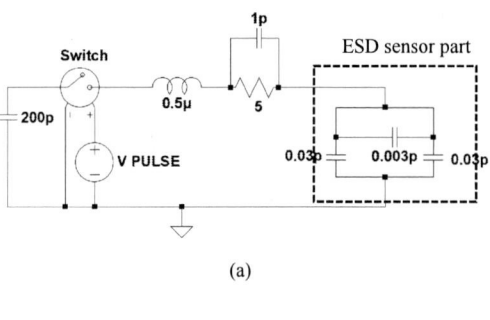

(a)

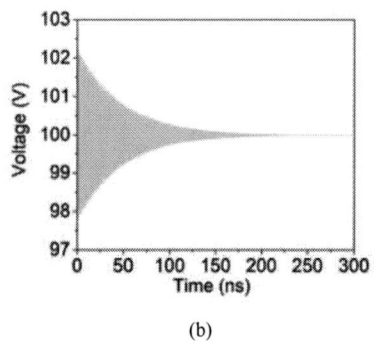

(b)

Fig.8. MM simulation. (a) LT-Spice immplementation of MM with ESD detector as DUT, and (b) corresponding voltage waveforms.

the applied voltage instantaneously, again ensuring that a breakdown would happen when one of the metal plates is grounded. All of the ESD simulations demonstrate the functionality of the ESD detector unit.

978-1-7281-1500-9/19 $31.00 © 2019 IEEE

C. Fabrication in GF 22 nm FDSOI and experimental results

The variable dielectric width capacitor was designed, simulated, and fabricated using GF 22 nm FDSOI technology. Instead of sharp corners, rectangular shaped metal lines (limited by the process technology) were used. The lowest metal line (M1) is used to define the metal structure of the variable dielectric width capacitor. Electrical connection from M1 goes through vias and other higher metallic layers (M1, M2, etc.) and is terminated at the top metal pad of the FDSOI technology, as shown in Fig. 9. The electric field simulation for the metal M1

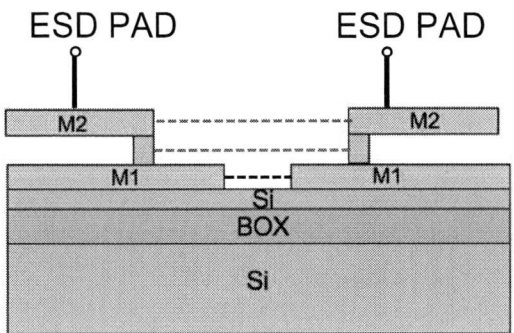

Fig.9. Cross-sectional view of variable dielectric width capacitor in GF 22 nm FDSOI technology.

lines separated by an 80 nm dielectric is shown in Fig. 10. The electric field value is $5 \cdot 10^8$ V/m for an applied voltage of 20 V, which can cause a dielectric breakdown if the chosen thin dielectric is a low-k material [15-17]. Note, the electric field

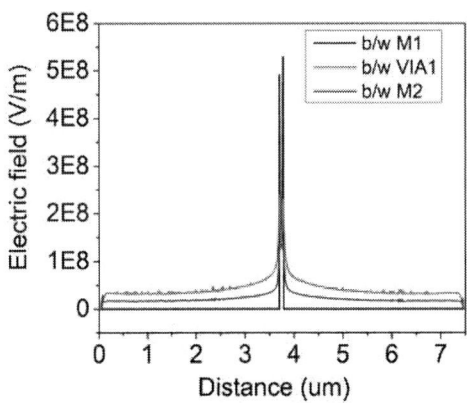

Fig.10. Electric field lines for variable dielectric width capacitor from Fig.9 for a 20 V applied to one ESD pad while the other pad is at zero volts.

between the M1 metal lines is not as high as for the simulated case of sharp corners since the metal lines, limited by the process technology, are rectangular shaped.

Two metal separations, 40 nm and 80 nm, are used in the GF 22 nm FDSOI implementation. For a silicon dioxide inter-layer dielectric exhibiting a breakdown strength of ~1 V/nm, a breakdown voltage of, respectively, 40 V and 80 V is expected. The variable dielectric width capacitor is experimentally characterized in two steps, first, both of the metal pads were

grounded. This ensures that both metal pads are equally charged, mimicking an ESD event, in this case to 0 V. In the second step of the characterization, the applied voltage is varied on one of the pads, while the other metal pad is grounded. The extracted I-V characterization enables to identify the breakdown voltage of the structure. The breakdown characteristics of the two implementations are shown in Fig. 11.

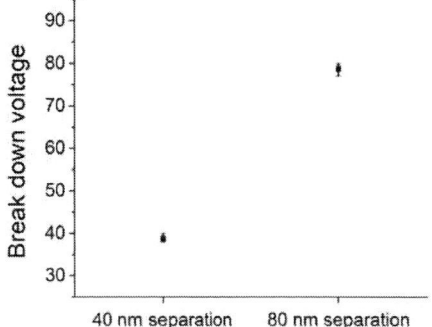

Fig.11. Breakdown voltage characterisitcs of the variable dieletric width capacitor with a dielectric thickness of 40 nm and 80 nm.

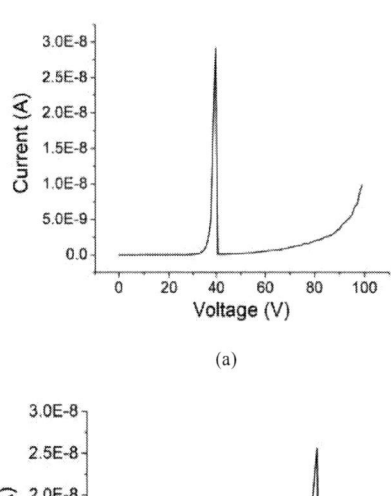

(a)

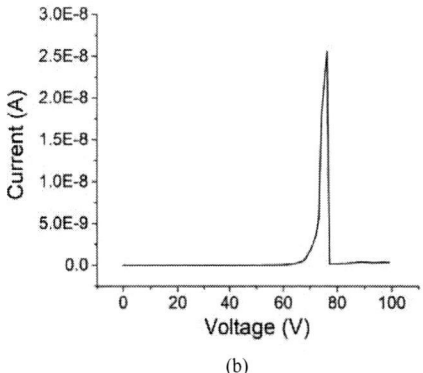

(b)

Fig.12. I-V Characteristics of the variable dielectric width capacitor with (a) 40 nm, and (b) 80 nm metal separation.

The 40 nm dielectric width capacitor breaks down in the range of 38 V to 41 V, while the 80 nm ESD monitor breaks down at a range of 77 V to 81 V. The extracted I-V characteristics of the two structures are shown in Fig. 12. Note, after breakdown, the current drops rapidly due to a possible accelerated electro-migration or a thermally induced void formation in the metal lines caused by the high-power dissipation immediately after the breakdown event. To ensure that the ESD event is stored, the dielectric breakdown needs to be decoupled from generation of thermally induced voids and electro-migration within the metal interconnect, which can happen as a result of high current associated with the breakdown event. For example, the peak current for a 40 V HBM event, assuming 150 ns discharge time would be (100 pF · 40)/150 ns = 26 mA. The current carrying capability of the interconnect has to, therefore, be increased, for example, by adding more vias and increasing interconnect width to support 26 mA of current.

III. VERTICAL MOSCAP ARRAY

The second on-chip ESD monitor approach is the vertical MOSCAP array. The MOSCAPs in the array are realized using MOSFETs in the given technology. The working principle of the vertical MOSCAP array is outlined in III-A. Design and experimental results for vertical MOSCAP arrays in GF 22 nm FDSOI are shown in III-B.

A. Working principle

The basic structure consists of array of capacitors in series. Capacitors in the array are realized using MOSFETs. The gate of the MOSFET is used as one terminal and the source and drain, tied together, serve as the second terminal of the capacitor. The number of capacitors in each array are integer multiples of two (2, 4, 6, 8, etc.). The basic working principle of the vertical MOSCAP array is similar to that of the variable dielectric width capacitor (Fig. 13). Both terminals of the capacitor arrays are exposed to the outside, and are subjected to ESD events. As the array structure is symmetrical, charge accumulation (if any) would be equal on both exposed pads. After any unit process, where a possible ESD event can occur, one of the exposed pads is grounded. A voltage difference is

developed between the ungrounded and grounded pads. The voltage is divided equally across the capacitors within the array,

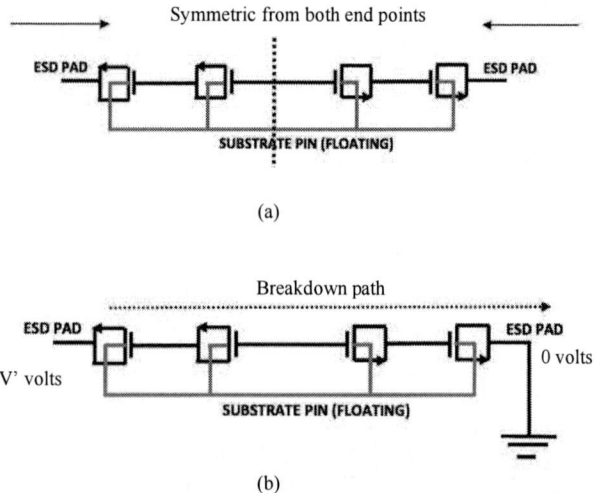

(a)

(b)

Fig.13. Working principle of a vertical MOSCAP array with four transistors. (a) Charge accumulation due to ESD events, results in voltage 'V' being developed on both exposed ESD pads at the end of the array. (b) While one of the ESD pads is grounded, capacitors in the array can break down depending on the magnitude of the voltage 'V' developed between the ESD pads.

as the individual capacitors are of equal size (similar capacitance). Capacitor arrays with a smaller breakdown strength than the developed voltage, will undergo breakdown. In mission mode, similar to the variable dielectric width capacitor array, one of the ESD pads is always tied to the substrate, resulting in a breakdown of the corresponding transistor array, whenever an ESD event occurs. The induced ESD voltage is estimated by measuring the I-V curves of all the transistor arrays and identifying the transistor arrays where breakdown has occurred.

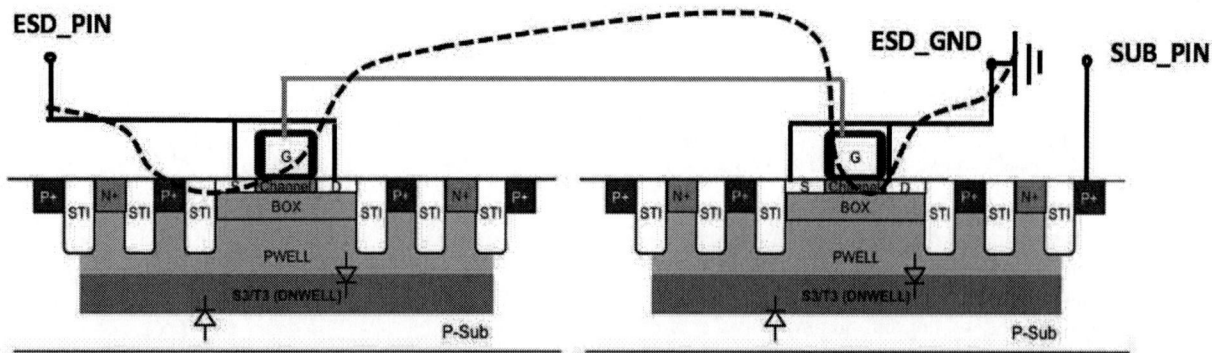

Fig.14. Cross-sectional view of a two transistor vertical MOSCAP array in GF 22 nm FDSOI technology, indicating the breakdown path when one of the ESD pad is grounded.

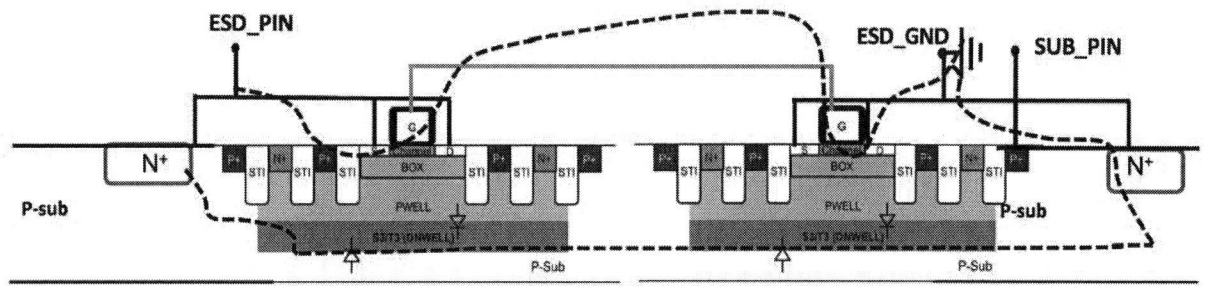

Fig.15. Cross-sectional view of a two transistor vertical MOSCAP array in GF 22 nm FDSOI technology with antenna diodes connected at the input, showing alternate path for breakdown.

B. Fabrication in GF 22 nm FDSOI and experimental results

A schematic of the vertical MOSCAP array fabricated in GF 22 nm FDSOI is shown in Fig. 14. Here the capacitors are realized by using the gate as one terminal, and the source and drain, tied together, as the second terminal. The antenna rules provided with the technology, ensure that the voltage at the internal floating nodes of the circuit do not cause breakdown of the individual capacitors within the array. For example, use of

providing an alternate path for charge leakage (during plasma processing, sputtering, reactive-ion-etching, *etc.*) [18-19]. However, antenna diode protection is not used at the terminals of the ESD monitor since adding the alternate path for breakdown (provided by the antenna diodes) defeats the purpose of the ESD detection circuit, as shown in Fig. 15. Thin and thick gate oxide transistors were used in the design of the MOSCAP arrays. The following arrays were included in the fabricated test chip: two thin oxide transistors (2T thin), two

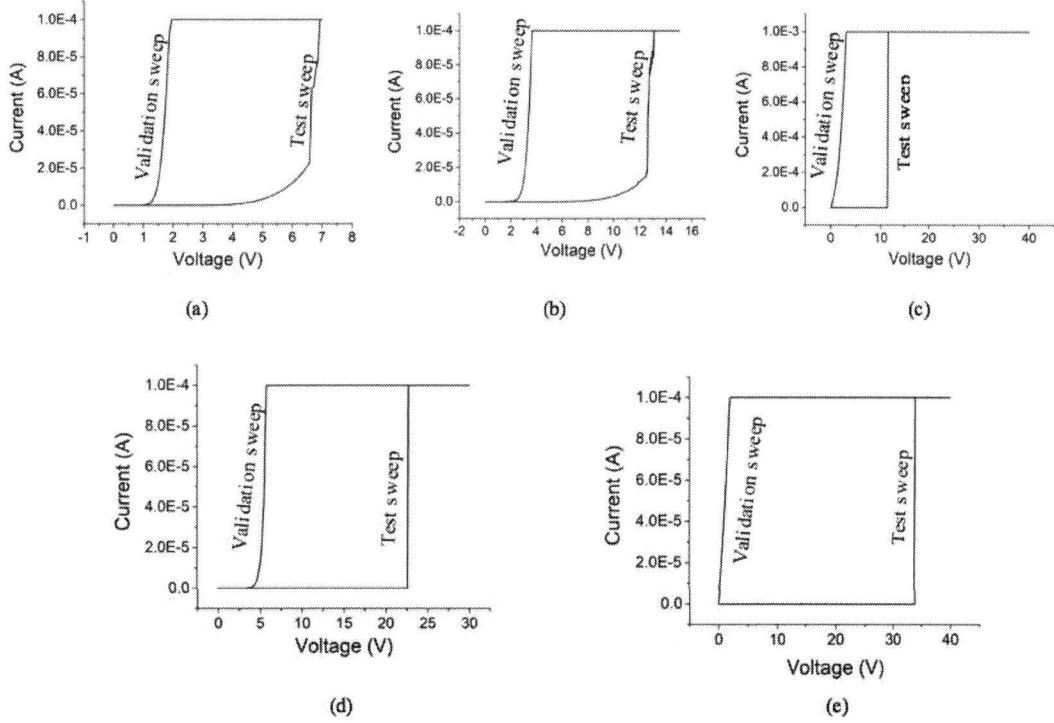

Fig.16. I-V characteristic of vertical MOSCAP array. (a) 2T thin oxide transistor array. (b) 4T thin oxide transistor array, (c) 2T thick oxide transistor array, (d) 4T thick oxide transistor array, and (e) 6T thick oxide transistor array.

higher metal levels (M1-M2-M1) to avoid a long single metal line (M1-M1-M1) in case of antenna violation. Antenna didoes are typically used at the I/O terminal pads and help to protect from breakdown of the gate oxide during fabrication by

thick oxide transistors (2T thick), four thin oxide transistors (4T thin), four thick oxide transistors (4T thick), and six thick oxide transistors (6T thick), with expected breakdown voltages of, respectively, 7 V, 12 V, 14 V, 24 V, and 36 V.

The experimental characterization of the MOSCAP array follows the same procedure as outlined for the variable dielectric width capacitor. In the first step, both terminal pads are grounded to zero volts. This emulates the ESD event which

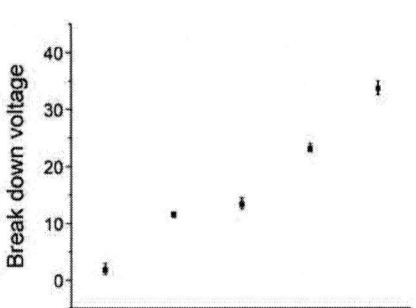

Fig.17. Breakdown characteristics of vertical MOSCAP array.

is common to both pads. In the second step, voltage is varied at one pad, keeping the other pad at ground until a breakdown is observed. The breakdown voltage is noted from the I-V curve. If both ESD pads were initially charged to a voltage greater than the breakdown voltage, as a result of any ESD event, then grounding one of the ESD pads would have resulted in immediate breakdown. Thus, the breakdown voltage of the structure is determined.

In general, an oxide film loses its insulating property in two steps [20]. In the first step, traps are generated in the oxide that increase the leakage current through the oxide. The generation rate of traps depends on the type of dielectric. In case of high-k dielectrics stacked with interfacial SiO_2 (similar to the one used in GF 22 nm FDSOI technology), since high-k dielectrics have higher bulk defect density than SiO_2, the expected generation rate of traps is higher as compared to an SiO_2-only dielectric of the same thickness [21]. These traps could be the growth of an oxygen-deficient filament, facilitated by the grain boundaries of the overlaying high-k film [22]. Eventually, these traps complete a percolation path through the oxide, bridging the two electrodes across the dielectric. Breakdown happens when the number of injected carriers in the dielectric reaches a threshold value [23]. Power dissipation through the percolation path controls the second stage of the breakdown transient, which determines the post-breakdown conduction property of the oxide [20]. A hard breakdown is characterized by a large change in voltage or current during the breakdown transient and a post-breakdown I-V characteristic that is essentially ohmic. The power dissipated during breakdown is high enough to melt silicon near the percolation spot and allow the molten silicon to flow through the oxide. This results in an ohmic short-circuit across the oxide, as the power dissipated exceeded the threshold of irreversible thermal damage. A soft breakdown is detected by a much smaller change of voltage or current after breakdown and by post-breakdown characteristics which can be described by a power law [24]. The thermal damage in soft-breakdown is reversible [25-26].

I-V characteristics of the five vertical MOSCAP array structures are shown in Fig.16. A validation sweep is performed after the test sweep to ensure that breakdown has actually

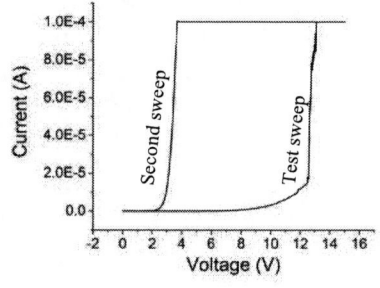

(a)

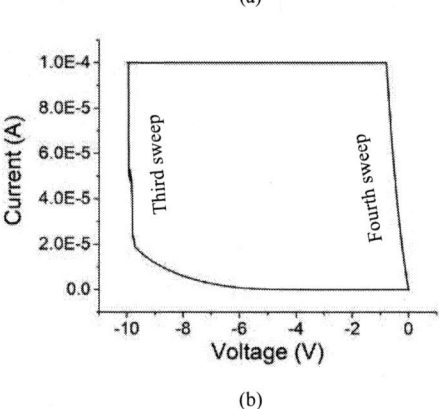

(b)

Fig.18. I-V characteristic showing a negative breakdown when swept in reverse direction. (a) Positve voltage sweep, and (b) reverse voltage sweep showing another breakdown.

happened during the test sweep. Breakdown characteristics of the MOSCAP array are shown in Fig. 17. The 2T thin structure breaks down at 6-7.5 V, 4T thin at 12-14.5 V, 2T thick at 11-12 V, 4T thick at 22-24 V, and 6T thick at 32-35 V. In some cases, with the 2T thin oxide and 4T thin oxide arrays, after a test sweep of voltage in the positive direction, resulting in breakdown (confirmed by 2nd sweep of voltage in positive direction), a 3rd sweep in negative direction, results in another breakdown in the negative direction, as shown in Fig. 18. This possibly indicates that the oxide has undergone a soft breakdown in the positive voltage sweep, confirmed by the second(validation) voltage sweep. A percolation path is created as a result of the positive sweep soft breakdown. When the voltage is swept in the reverse direction (third sweep in the negative direction), the traps which were created in the first soft breakdown reconfigure, resulting in a recovery of the oxide, since the soft breakdown is reversible [27]. The oxide then undergoes another breakdown in the negative voltage sweep direction. However, this behavior is not observed for the high voltage detection MOSACP arrays (4T thick and 6T thick), as the high voltage/current during the breakdown results in higher

978-1-7281-1500-9/19 $31.00 © 2019 IEEE

power dissipation, ensuring a hard breakdown during the first test sweep.

IV. CONCLUSIONS

Two approaches for an on-chip ESD detector are proposed. Variable dielectric width capacitor is suitable for high voltage detection. The lower limit of voltage detection using variable dielectric width capacitor is limited by the minimum metal separation supported by the given technology (40 nm in case of GF 22 nm FDSOI). This method is area efficient since transistors are not used in the implantation of the variable dielectric width capacitor. The second approach, vertical MOSCAP arrays, is ideally suited for low voltage detection. The granularity in ESD voltage level detection can be tuned by using thin and thick gate oxide devices. For high voltage detection that requires a large number of transistors however, this approach becomes area inefficient.

The proposed prototypes, variable dielectric width capacitor and vertical MOSCAP array can be included in any die, manufactured by using any process to help with ESD monitoring during packaging, as well as assembly and transport.

ACKNOWLEDGEMENTS

This work was supported in part by ASCENT, one of six centers in JUMP, a Semiconductor Research Corporation (SRC) program sponsored by DARPA, GlobalFoundries, and the UCLA CHIPS consortium.

REFERENCES

[1] B. A. Unger, "Electrostatic Discharge Failures of Semiconductor Devices," *Proceedings of the IEEE Reliability Physics Symposium*, pp. 193 – 199, 1981.

[2] A. Z. Wang, H. G. Feng, K. Gong, R. Y. Zhan, and J. Stine, "On-Chip ESD Protection Design for Integrated Circuits: An Overview for IC Designers," *Microelectronics Journal*, Vol. 32, No. 9, pp. 733 – 747, September 2001.

[3] J. E. Vinson and J. J. Liou, "Electrostatic Discharge in Semiconductor Devices: Protection Techniques," *Proceedings of the IEEE Conference on Microelectronics*, pp. 311 – 321, May 2000.

[4] S. S. Iyer, "Heterogeneous Integration for Performance and Scaling," *IEEE Transactions on Components, Packaging and Manufacturing Technology*, Vol. 6, No. 7, pp. 973 – 982, 2006.

[5] T. Fukushima, A. Alam, Z. Wan, S. C. Jangam, S. Pal, G. Ezhilarasu, A. Bajwa, and S. S. Iyer, "FlexTrate™—Scaled Heterogeneous Integration on Flexible Biocompatible Substrates Using FOWLP," *Proceedings of the IEEE 67th Electronic Components and Technology Conference (ECTC)*, pp. 649 – 654, May 2017.

[6] A. Zangwill, *Modern Electrodynamics*, Cambridge University Press, 2012.

[7] M. H. Thurlow, "ESD Testing," *Proceedings of the IEE Colloquium on ESD and ESD Counter Measures*, pp. 3/1 – 3/16, March 1995.

[8] M. D. Ker, J. J. Peng, and H. C. Jiang, "ESD Test Methods on Integrated Circuits: An Overview," *Proceedings of the IEEE International Conference on Electronics, Circuits and Systems*, Vol. 2, pp. 1011 – 1014, September 2001.

[9] https://www.analog.com/media/en/analog-dialogue/volume-51/number-4/articles/a-look-at-the-new-ansi-esda-jedec-js-002-cdm-test-standard.pdf

[10] M. Johnson, R. Ashton, and S. Ward, "FCDM Measurements of Small Devices," *Proceedings of the IEEE EOS/ESD Symposium*, pp. 1 – 8, 2009.

[11] R. Ashton, M. Johnson, and S. Ward. "Simulating Small Device CDM Using Spice," *IN Compliance Magazine*, pp. 32 – 36, 2010.

[12] S. H. Voldman, *ESD Testing: From Components to Systems*, John Wiley & Sons Press, 2016.

[13] S. G. Beebe, *Characterization, Modeling, and Design of ESD Protection Circuits*, Ph.D. Thesis, 1998.

[14] W. Rhoades and J. Maas. "New ANSI ESD Standard Overcoming the Deficiencies of Worldwide ESD Standards," *Proceedings of the IEEE International Symposium on Electromagnetic Compatibility*, Vol. 2, pp. 1078 – 1082, August 1998.

[15] S. C. Lee, A. S. Oates, and K. M. Chang, "Fundamental Understanding of Porous Low-k Dielectric Breakdown," *Proceedings of the IEEE International Reliability Physics Symposium*, pp. 481 – 485, April 2009.

[16] D. Shamiryan, T. Abell, F. Iacopi, and K. Maex, "Low-k Dielectric Materials," *Materials Today*, Vol. 7, No. 1, pp. 34 – 39, January 2004.

[17] A. Grill, S. M. Gates, T. E. Ryan, S. V. Nguyen, and D. Priyadarshini, "Progress in the Development and Understanding of Advanced Low k and Ultralow k Dielectrics for Very Large-Scale Integrated Interconnects – State of the Art," *Applied Physics Reviews*, Vol. 1, No. 1, pp. 011306, 2014.

[18] H. C. Shin and C. Hu, "Thin Gate Oxide Damage Due to Plasma Processing," *Semiconductor Science and Technology*, Vol. 11, pp. 463 – 473, 1996.

[19] H. Chen and Y. Chang, "Routing for Manufacturability and Reliability," *Transactions of the IEEE Circuits and Systems Magazine*, Vol. 9, No. 3, pp. 20 – 31, August 2009.

[20] M. A. Alam, B. E. Weir, and P. J. Silverman, "A Study of Soft and Hard Breakdown – Part I: Analysis of Statistical Percolation Conductance," *IEEE Transactions on Electron Devices*, Vol. 49, No. 2, pp. 232 – 238, February 2002.

[21] G. Bersuker, D. Heh, C. D. Young, L. Morassi, A. Padovani, L. Larcher, K. S. Yew, Y. C. Ong, D. S. Ang, K. L. Pey, and W. Taylor, "Mechanism of High-k Dielectric-Induced Breakdown of the Interfacial SiO_2 Layer," *Proceedings of the IEEE International Reliability Physics Symposium*, pp. 373 – 378, May 2010.

[22] G. Bersuker, D. Heh, C. Young, H. Park, P. Khanal, L. Larcher, A. Padovani, P. Lenahan, J. Ryan, B. H. Lee, and H. Tseng, "Breakdown in the Metal/High-k Gate Stack: Identifying the "Weak Link" in the Multilayer Dielectric," *IEEE International Electron Devices Meeting*, pp. 1 – 4, December 2008.

[23] K. Okada, H. Ota, T. Nabatame, and A. Toriumi, "Dielectric Breakdown in High-K Gate Dielectrics-Mechanism and Lifetime Assessment," *Proceedings of the IEEE International Reliability Physics Symposium*, pp. 36 – 43, April 2007.

[24] M. A. Alam, B. Weir, J. Bude, P. Silverman, and D. Monroe, "Explanation of Soft and Hard Breakdown and its Consequences for Area Scaling," *Technical Digest of the IEEE International Electron Devices Meeting*, pp. 449 – 452, December 1999.

[25] A. Ghetti, "Gate Oxide Reliability: Physical and Computational Models," *Predictive Simulation of Semiconductor Processing: Status and Challenges*, J. Dabrowski and E. R. Weber, Springer, pp. 201 – 258, 2004.

[26] T. Pompl, C. Engel, H. Wurzer, and M. Kerber, "Soft Breakdown and Hard Breakdown in Ultra-Thin Oxides," *Microelectronics Reliability*, Vol. 41, No. 4, pp. 543 – 551, April 2001.

[27] M. A. Alam, B. E. Weir, and P. J. Silverman, "A Study of Soft and Hard Breakdown – Part II: Principles of Area, Thickness, and Voltage Scaling," *IEEE Transactions on Electron Devices*, Vol. 49, No. 2, pp. 239 – 246, February 2002.

Preparation and Characterization of Electroplated Cu/Graphene Composite

Xin Wang, Qian Wang*, Jian Cai, Changming
Song, Yang Hu
Department of Microelectronics and Nanoelectronics
Tsinghua University
Beijing, China
wang-qian@tsinghua.edu.cn

Yang Zhao, Yu Pei
Department of Precision Machinery & Precision
Instrumentation
University of Science and Technology of China
Hefei, China
yangz1@ustc.edu.cn

Abstract—Cu and unoxidized graphene composite films were prepared by electroplating under room temperature. Graphene was added to the electrolyte of Cu to form composite materials to improve the performance of Cu for interconnection. Composite electrolyte containing high concentration of graphene (up to 0.5 g/L) was prepared by adding CTAB as surfactant. And ultrasound was also used to increase dispersion degree and reduce graphene agglomeration in electrolyte. Average coefficient of thermal expansion (CTE) of the composite films determined by thermo-mechanical analysis (TMA) shows a decrease from 17.3 ppm/K to 14.2 ppm/K from 260 K to 320 K, which reduce CTE by 18 % compared with Cu. CTE of the composite materials can be reduced to 10 ppm/K at 243 K, which is only 60 % of CTE of Cu. The thermal conductivity of the composite materials measured by phase sensitive transient thermo-reflectance (PSTTR) technique shows an improvement from 385 W/m.K to 468 W/m.K. CTE and thermal conductivity of the composite materials both decrease with the increase of current density. And they also both increase with the increase of graphene concentration. The composite materials also show good mechanical properties. The average hardness is 2.5 GPa, which is about 2 times of Cu. And the average elastic modulus is 145 GPa, which is 38 % higher than that of Cu. The resistivity of the composite materials is basically equivalent to that of pure Cu. Result of the experiment proved that the composite materials have better performance under low temperature conditions. The improvement of material properties makes composite materials have good application prospects for 3D interconnection in the near future.

Keyword- Cu and graphene composite electrodeposition; coefficient of thermal expansion; thermal conductivity; hardness and elastic modulus; graphene concentration and current density; performance optimization.

I. INTRODUCTION

Nowadays, Cu is one of the most applied materials in semiconductor packaging and interconnection, such as electrical traces on substrate, redistribution layer (RDL) for wafer level packaging, wafer-level Cu-Cu bonding, via-filling for through-silicon via (TSV) in 3D integration [1-2], etc. Because Cu has low resistivity and high thermal conductivity. However, CTE of Cu is 7 times that of Si (2.5 ppm/K), which frequently causes critical reliability issues. A typical problem is the heat distribution of TSV. Mismatch in CTE causes large thermal stress between Cu and Si, which may cause local cracking, finally resulting in device failure

[3]. The material properties of Cu can be adjusted by forming composite materials. If other materials can be added into Cu to form composite materials, which has lower CTE and better thermal and mechanical properties, it could benefit the packaging and interconnection of integrated circuits.

Graphene is a good choice because of its extraordinary properties such as CTE, electrical conductivity, thermal conductivity and elastic modulus etc. Comparison of the properties of graphene and Cu is given in Table I . Graphene has a negative CTE below 2700 K, and CTE of graphene reaches (-8.0 ± 0.7) ppm/K at 300 K according to Jiang's model and the Yoon's model [4-5]. Moreover, Graphene has a high thermal conductivity (up to 5000 W/m.K), which is more than 10 times higher than that of Cu. Graphene also has an elastic modulus up to 1 TPa, which is about 9 times higher than that of Cu [6-7]. And the resistivity of graphene is the lowest among the known materials, and the lowest resistivity can reach $1.0 \times 10^{-8} \Omega \cdot m$, which is only about 60 % of the resistivity of Cu.

A simple method to prepare composite materials of Cu and graphene is composite electrodeposition. Because it is a room temperature and low cost process, and also compatible with traditional Cu electroplating process in semiconductor packaging. Composite electrodeposition is to disperse graphene into the Cu electrolyte, and Cu and graphene are co-deposited on the cathode by electroplating. Composite electrodeposited materials can reduce the size of metal grains and improve the mechanical properties of composite materials [8].

Some researches dispersed graphene oxide in electrolyte to form composite materials, and reduce graphene oxide to graphene at high temperature. Chokkakula L. P. Pavithra et al. [9] oxidized graphene and dispersed it into the solution to

TABLE I.   PROPERTIES COMPARISON BETWEEN GRAPHENE AND COPPER

properties comparison between graphene and Cu	Composite materials mechanical properties	
Materials Properties	*Graphene*	*Cu*
CTE (ppm/K)	-8.0 ± 0.7	17.3
Thermal conductivity (W/m.K)	up to 5000	380-400
Elastic modulus (GPa)	1000	105-120
Electrical conductivity ($10^{-8}\ \Omega \cdot m$)	1.0	1.7

form composite materials. Then graphene oxide was reduced to graphene in argon at 300 °C for 30 minutes. The elastic modulus of the composite materials prepared by this method is about 130 GPa, which achieved 25 % higher than the hardness of Cu. K Jagannadham [10] dispersed graphene oxide in the electrolyte and reduced graphene oxide to graphene at 400 °C under hydrogen for 3 hours. The thermal conductivity of composite materials is about 460 W/m.K. However, oxidation of graphene and reduction at high temperatures may not only increase the difficulty of manufacturing, but also destroy the structure of graphene and therefore deteriorate the properties of the composite materials [11]. Furthermore, CTE of the composite materials have not been characterized yet.

The change of electroplating parameters also has a great impact on the properties of the composite materials. The main factors are graphene concentration, current density and pulse current. Generally, the concentration of doped material in the composite materials increases with the increase of the concentration of doped particles in the electrolyte, and finally tends to a stable value [8]. The influence of current on the properties of composite materials is mainly reflected in pulse current and current density. Pulsed current can reduce the agglomeration of graphene and increase dispersion degree [9]. On the other hand, the concentration of doped particles in composite materials generally decreases with the increase of current density.

In this paper, Cu and Graphene Composite films were prepared by dispersing unoxidized graphene in a Cu electrolyte at room temperature. The concentration of graphene in the composite electrolyte was increased up to 0.5 g/L by adding cetyl trimethyl ammonium bromide (CTAB) as surfactants [12]. CTE, thermal conductivity, hardness and elastic modulus of composite materials have been greatly improved compared with Cu, while the electrical conductivity is basically the same as that of Cu. CTE of the composite materials decrease with the increase of graphene concentration and increase with the increase of current density and CTE of the composite materials is lower at low temperature.

II. Experiments

A. Preparation of Cu and graphene composite materials

A serious problem in the composite electrodeposition of Cu and graphene is that the graphene has poor hydrophilicity and it is difficult to increase its dispersion concentration in composite electrolyte. Moreover, graphene is prone to agglomeration. As the number of graphene layers increases, the performance of graphene decreases significantly. CTAB is used as a surfactant to increase the hydrophilicity of graphene. Ultrasound is used to increase dispersion degree and reduce graphene agglomeration. Because the time of composite electrodeposition is relatively short, the time requirement for stable dispersion of the composite electrolyte is also reduced. Therefore, a higher concentration of graphene dispersion can be prepared to enhance the concentration of graphene in the composite materials and the performance of the composite materials. The influence of the concentration of graphene in the electrolyte on the properties of the composite materials will be analyzed in detail later in the experiment. Composite electrolyte containing a maximum concentration of 0.5 g/L of graphene was prepared and no significant sedimentation delamination was observed over a longer period of time.

Since the pulsed current could reduce the agglomeration of graphene, allowing it to be more uniformly dispersed in the composite materials, Cu and graphene composite materials were prepared under pulse current conditions. Fig.1 shows the morphology of a composite materials under a scanning electron microscope. Fig. 1(a) shows the distribution of graphene on the surface of the composite materials by scanning electron microscopy (SEM). It can be seen from the figures that graphene exhibits a relatively uniform distribution on the surface of the composite materials. Fig 1(b) & (c) show the lateral and longitudinal distribution of graphene in the composite materials. The graphene material used in this experiment has a sheet diameter from 2 to 5 μm.

B. Optimization of Cu and graphene composite materials

After successfully preparing the composite materials of Cu and graphene, the properties of the composite materials were optimized. Our main expectation for the Cu-graphene composite materials is to reduce CTE while maintain or further improve the excellent performance of Cu. Therefore, we use CTE of the composite materials to directly characterize the optimization results. In addition, the thermal conductivity of the materials will also be an important reference for its performance optimization.

The concentration of graphene in electrolyte and current density have significant effects on composite electrodeposition. The concentration of graphene in electrolyte was varied among 0.2 g/L, 0.3 g/L, 0.5 g/L, and two different current densities of 1 ASD and 2 ASD were used to prepare the composite samples. The optimum process parameters were determined by the measured CTE and thermal conductivities of the samples.

C. Measurements

CTE of graphene is negative in a wide temperature range. And most of the applications of composite materials such as TSV are at room temperature. Hence, CTE of the composite materials were calibrated at a temperature range from 260 K to 320 K, which basically conforms to the operating temperature range of mobile devices such as mobile phones. CTE of the composite materials at lower temperature (near 243 K) were also measured to confirm whether the composite materials has better performance at lower temperatures. CTE of the composite materials was measured by thermo-mechanical analysis (TMA). By measuring the change in the length of the composite materials with temperature，CTE of composite materials can be calculated.

The measurement of thermal conductivity of composite materials is relatively complicated. Since both Cu and graphene have high thermal conductivities and the film thickness of the composite materials is in micron meter range,

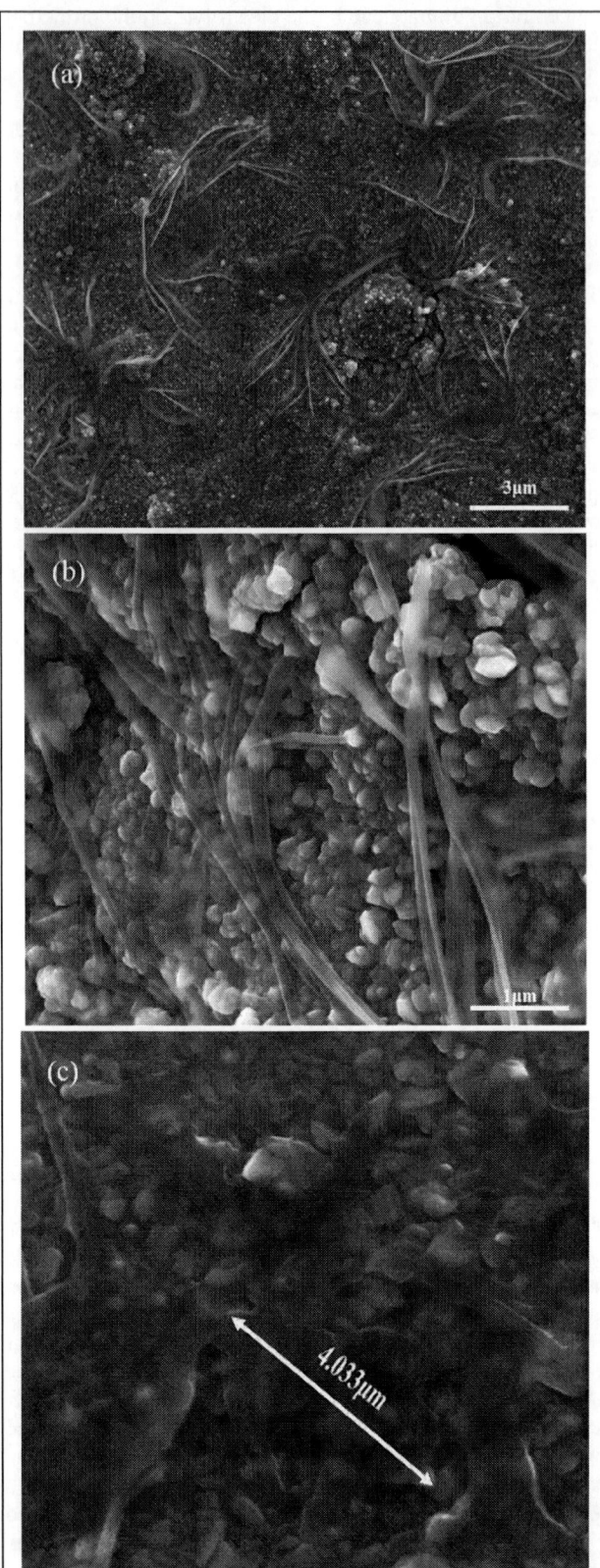

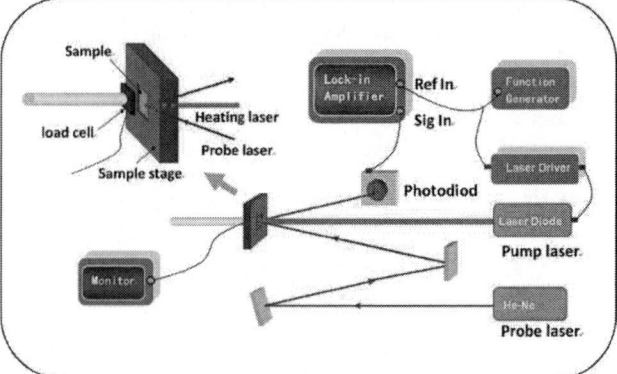

Figure 2. phase sensitive transient thermoreflectance (PSTTR) technique

which makes it difficult to accurately measure the thermal conductivity of the composite materials. We used the phase sensitive transient thermo-reflectance (PSTTR) technique to measure the thermal conductivity of the composite films [13]. The thermal conductivity test system is shown in Fig. 2 [14]. A periodically oscillating heating laser impinges on the surface of the sample, causing periodic oscillations in the temperature of the sample. Owing to the thermal impedance of the sample, the phase of the oscillating surface temperature has a delay compared with the heating laser. And such phase-lag is a function of the thermal conductivity, heat capacity of the materials, and also the modulation frequency of the heating signal. The reflected detection laser carries the information of the temperature of sample surface due to thermo-reflective effect, and is then captured by a photodiode along with a lock-in amplifier. A heat conduction model is developed according to the measured sample, and the thermal conductivity of the film can be extracted from the phase signal of the temperature oscillation at the sample surface. And the electrical conductivity of the composite materials was measured by the four-probe method, and the elastic modulus and hardness were measured using a nanoindentation.

III. RESULTS AND DISCUSSION

A. CTE of composite material and its optimization

The change of CTE of composite materials with temperature under different conditions was measured. Fig. 3(a) shows the CTE of composite materials with different concentrations of graphene in solution at the same current density of 1 ASD. Fig. 3(b) shows the CTE of composite materials with different current density at the same concentrations of graphene in solution of 0.3 g/L. The following four conclusions can be summarized from the experimental results:

- The average CTE of the composite materials is reduced by about 15-20 % compared to the CTE of Cu.
- The CTE of composite materials is more sensitive to temperature than that of Cu.
- The CTE of composite materials decreases more significantly at lower temperature.

Figure 1. (a) Graphene evenly distributed in the composite materials
（b）Longitudinal distribution of graphene combined with Cu formation
（c）Horizontal distribution of graphene in composite materials

- Decrease in current density and the increase in graphene concentration reduce CTE of the composite materials.

CTE of composite materials is more sensitive to temperature than Cu. This phenomenon may be caused by the change of CTE of graphene with temperature. CTE of graphene has many different models due to different measurement methods. The results of this experiment are in good agreement with the Jiang's model and the Yoon's model. The Yoon's model of CTE of the graphene is given in Fig. 4(a). The change of CTE of graphene with temperature of Jiang's model is shown in Fig. 4(b). The " γ " represents the interaction between the substrate and graphene, and $\gamma = 0$ represents the CTE of free graphene. It can be seen from Fig. 4 that CTE of graphene has a significant change in the temperature range from 260 K to 320 K. In this temperature range, CTE of graphene increases significantly with the increase of temperature, and the Tuner model of CTE of composite materials is shown in (2), where α_c, α_p and α_m represent the CTE of the composite materials, graphene and Cu, φ the percentage of graphene, Kc and Kp are the shear modulus of graphene and Cu [15]. Although the proportion of graphene in the composite materials is lower, but according to (2), since the shear modulus of graphene are

about 10 times of the shear modulus of Cu, the increase of CTE of graphene directly leads to a significant increase in CTE of the composite materials with temperature. It explains why CTE of the composite materials is reduced to 11 ppm/K at a temperature of 260 K. A composite material at about 243 K was also measured and CTE is further reduced to 10 ppm/K. This result also indicates that composite materials have good applications in lower temperature environment devices.

$$\alpha_c = \frac{(1 - \varphi) \, K_m \alpha_m + \varphi K_p \alpha_p}{(1 - \varphi) \, K_m + \varphi K_p} \tag{2}$$

The average CTE of the composite materials prepared at different current densities and graphene concentrations from 260 K to 320 K is shown in Fig. 5. It can be seen from Fig. 3(a) and Fig. 5 that in the case of the same current density (1ASD), as the graphene concentration increases, the CTE of the composite materials decrease continuously. In addition, as the concentration of graphene increases, the decrease in CTE of composite materials becomes slower. This indicates that as the concentration of graphene in the solution increases, the concentration of graphene in the composite materials gradually becomes saturated. It can be seen from Fig. 3(b) and Fig. 5 that with the same concentration of graphene, as the current density decreases, CTE of the composite materials decrease continuously.

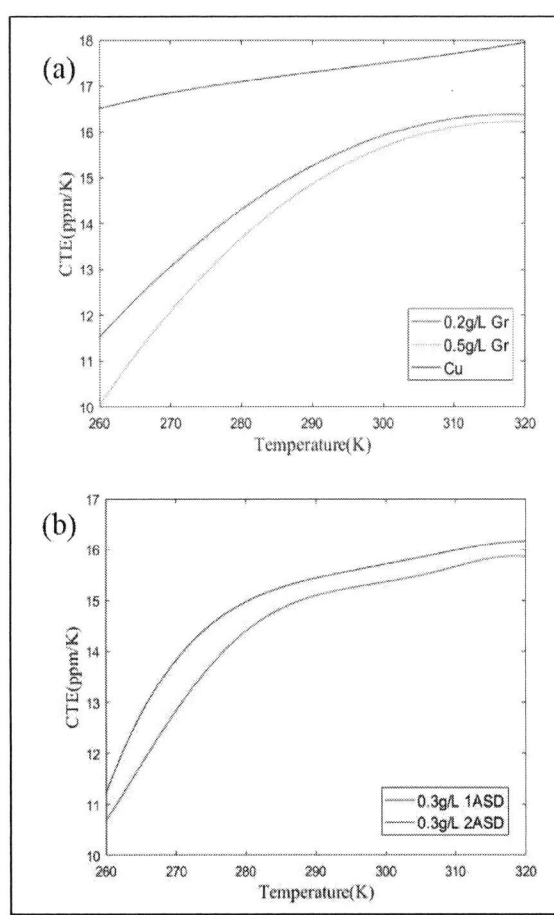

Figure 3. (a) Curve of CTE of composite materials prepared with different graphene concentrations with temperature (b) Curve of CTE of composite materials with different current densities with temperature

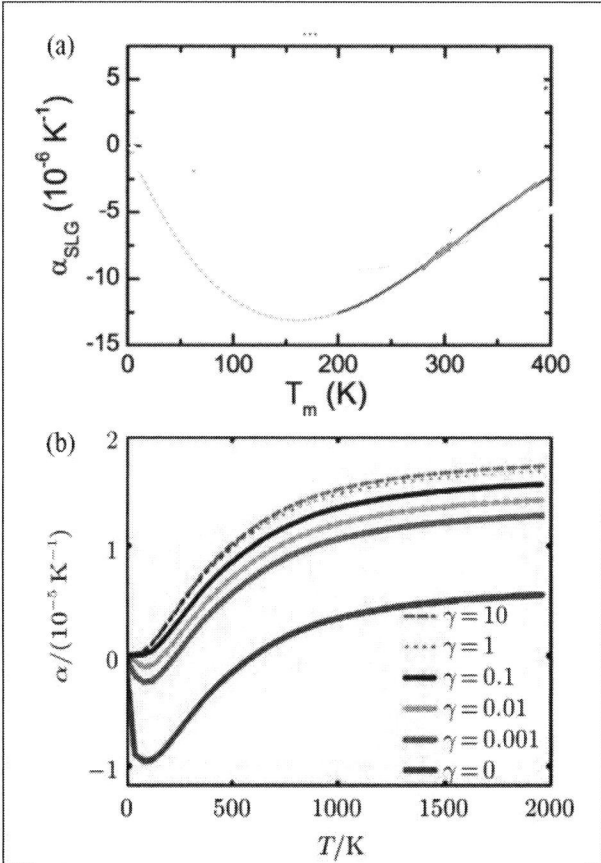

Figure 4. (a) The Yoon's model of CTE of graphene varies with temperature (b) The Jiang's model of CTE of graphene varies with temperature ($\gamma = 0$)

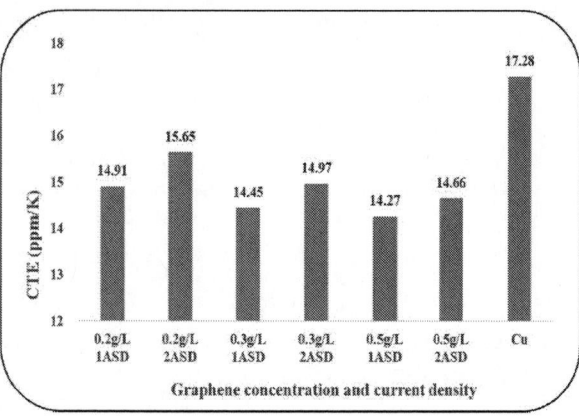

Figure 5. Average CTE of composite materials at different current densities and graphene concentrations

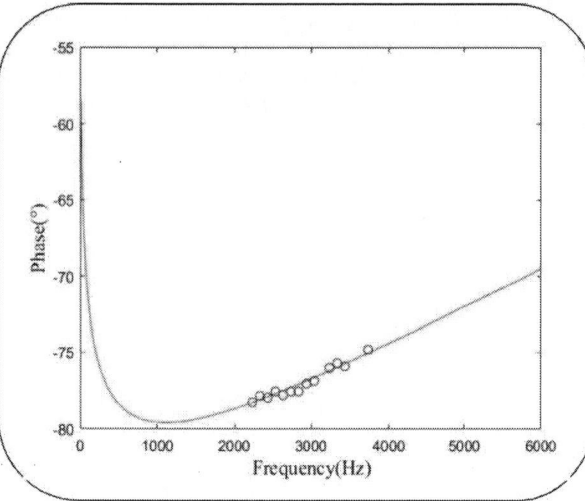

Figure 6. Measurement results and simulation curves of composite thermal conductivity obtained by PSTTR method

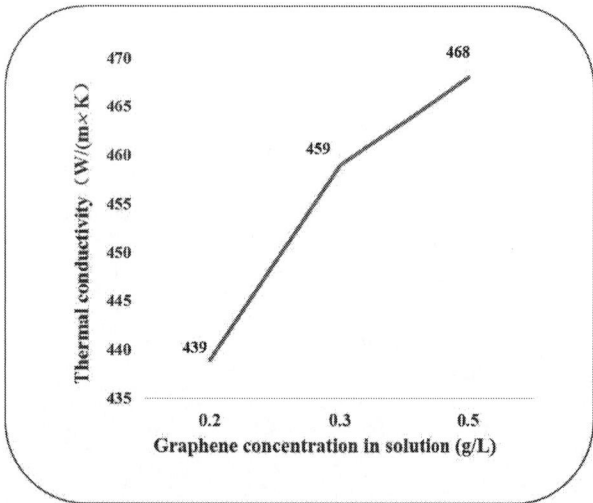

Figure 7. Thermal conductivity of sample composite materials prepared with different graphene concentration solutions

B. Thermal conductivity of composite material and its optimization

The thermal conductivity of composite materials is measured by detecting the phase change of the reflected laser light at different modulation frequencies of the heating laser, and then fitting the experimental results with the thermal model to obtain the thermal conductivity of the composite materials. The experimental results and simulation curves are shown in Fig. 6. The "○" in the figure represents the actual measurement result, and the curve represents the fitting curve to the thermal conductivity when other conditions are determined. It can be seen from the Fig. 6 that the actual measurement results are in good agreement with the theoretical curve. The thermal conductivity of the composite materials under different graphene concentrations obtained by fitting is shown in Fig. 7. It can be seen from Fig. 7 that the thermal conductivity of the composite materials increase with the increase of graphene concentration in the solution, and the thermal conductivity of the composite materials is significantly higher than the thermal conductivity of Cu (385 W/m.K). When the graphene concentration rises from 0.3 g/L to 0.5 g/L, the thermal conductivity of the composite materials increases slower, which is in good agreement with the variation law of CTE of the composite materials.

C. Other properties of composite materials

The four-probe method was used to measure the resistivity of the composite materials. Since the resistivity of Cu can fully meet the requirements of current integrated circuit packaging and interconnection. We only need to ensure that the resistivity of the composite materials is not higher than Cu. The experimental results prove that the average resistivity of the composite materials are 1.67×10^{-8} $\Omega\cdot m$, which is roughly equivalent to the resistivity of Cu (1.7×10^{-8} $\Omega\cdot m$). It indicates that the composite materials maintain the good electrical conductivity of Cu.

Nanoindentation is used to measure the elastic modulus and the hardness of the composite materials. Composite electrodeposition can reduce the grain size and improve the hardness of the composite materials. The measurement results of composite materials hardness and elastic modulus are shown in Table II. The results show that average elastic modulus of the composite materials increases to 145 GPa.

TABLE II. ELASTIC MODULUS AND HARDNESS OF COMPOSITE MATERIALS AND COPPER

Elastic Modulus and hardness of composite materials and Cu	Composite materials mechanical properties	
Group	Elastic Modulus (GPa)	Hardness (GPa)
Group 1	146.1	2.38
Group 2	149.1	2.70
Group 3	141.4	2.37
Group 4	145.5	2.55
Average	145.5	2.50
Pure Cu	105.5	1.19

The elastic modulus of electroplated Cu is 105 GPa. And the average hardness of the composite materials increase to 2.5 GPa, and the elastic modulus of electroplated Cu is about 1.2 GPa. The experimental results prove that the composite materials further enhances the mechanical properties of Cu, which is of great help to improve the reliability of the materials.

IV. CONCLUSIONS

In this paper, Cu and Graphene composite films have been prepared by dispersing unoxidized graphene in a Cu electrolyte at room temperature. The highest concentration of graphene in the composite electrolyte reached 0.5g/L, and the combination of graphene and Cu is well verified and the distribution of graphene in the composite materials is relatively uniform observed by SEM.

In this paper, the CTE, thermal conductivity, hardness, elastic modulus and electrical conductivity of the composite materials were also measured. Average CTE of the composite materials at a range of 260 K to 320 K is about 15-20 % lower than CTE of Cu, and CTE of the composite materials decrease more obviously at lower temperature. CTE of the composite materials reduces to 10 ppm/K at 243 K, which is only 60 % of CTE of Cu. Thermal conductivity of the composite materials rises up to 468 W/m.K, which is about 22 % higher than the thermal conductivity of electroplated Cu. Average elastic modulus of the composite materials is 145 GPa, which is about 38 % higher than the elastic modulus of electroplated Cu. Average hardness of the composite materials is 2.5 GPa, which is more than twice the hardness of the electroplated Cu. While the other properties of the composite materials are improved, the composite materials basically guarantee the same conductivity as Cu.

In this paper, the preparation parameters that affect the performance of the composite materials were also optimized. As the concentration of graphene increases, the thermal conductivity of the composite materials increases gradually, and CTE of the composite materials decreases gradually. With the increasing of the concentration of graphene, the thermal conductivity and CTE change more slowly. As the current density in electrodeposition decreases, CTE of the composite materials decreases gradually, and the thermal conductivity of the composite materials increases gradually, which proves that reducing the current density helps to increase the concentration of graphene in the composite materials. The use of ultrasound pulse current helps to reduce the agglomeration of graphene, and pulse current is also used in composite electrodeposition.

In summary, the properties of the composite materials, especially CTE, are greatly optimized compared with Cu. And the preparation process of composite materials is also compatible with that of Cu. We believe that the composite materials can have a broad applications in integrated circuits, especially in TSV interconnections in the near future.

ACKNOWLEDGMENT

The authors would like to thank Mr. Fengze Hou from the Institute of Microelectronics of Chinese Academy of Sciences for TMA tests. And this work was supported by NSFC under grant 61734004.

REFERENCES

[1] K. Yoon, G. Kim, et al. "Modeling and Analysis of Coupling between TSVs, Metal, and RDL interconnects in TSV-based 3D IC with Silicon Interposer," Electronics Packaging Technology Conference (EPTC 2009), IEEE Press, Dec. 2009, pp. 702-706, doi:10.1109/EPTC.2009.5416458.

[2] J. H. Lau, "Overview and outlook of through‐silicon via (TSV) and 3D integrations," Microelectronics International, Vol. 28, Aug. 2011, pp. 8-22, doi: 10.1108/13565361111127304.

[3] F. Qin, J. Wang, et al. "Review on the Thermal Mechanical Reliability of TSV Structures," Semiconductor Technology, Vol. 37, No. 11, Nov. 2012, pp. 825-831, doi:10.3969/j.issn.1003-353x.2012.11.001.

[4] J. W. Jiang, J. S. Wang, B. W. Li, "Thermal expansion in single-walled carbon nanotubes and graphene:Nonequilibrium Green's function approach," Physical Review B, Vol. 109, Feb. 2011, pp. 1-30, doi:10.1063/1.3531573.

[5] D. Yoon, Y.Son, H. Cheong, "Negative thermal expansion coefficient of graphene measured by Raman spectroscopy," Nano letters, Vol. 11, Aug. 2011, pp. 3227-3231, doi:10.1021/nl201488g.

[6] Y. Zhu, S. Murali, W. Cai, et al, "Graphene and Graphene Oxide: Synthesis, Properties and Applications," Advanced material, Vol. 22, 2010, pp. 3906-3924. doi: 10.1002/adma.201001068.

[7] X. Huang, Z. Yin, S. Wu, et al, "Graphene-Based Materials: Synthesis, Characterization, Properties and Applications." Small, Vol. 14, 2011, pp. 1876-1902, doi: 10.1002/smll.201002009.

[8] C. Low, R. Wills, F. Walsh. "Electrodeposition of composite coatings containing nanoparticles in a metal deposit," Surface & Coatings Technology, Vol. 201, Jan. 2006, pp. 371-383, doi: 10.1016/j.surfcoat.2005.11.123.

[9] C. Pavithra, B. Sarada, et al, "A new electrochemical approach for the synthesis of Cu-graphene nanocomposite foils with high hardness," Scientific reports, Vol.4, Feb. 2014, pp. 1-7, doi: 10.1038/srep04049.

[10] K. Jagannadham, "Thermal conductivity of Cu-graphene composite films synthesized by electrochemical deposition with exfoliated graphene platelets," Metallurgical and Materials Transactions, Vol. 43B, Apr. 2012, pp. 316-324. doi: 10.1007/s11663-011-9597-z.

[11] Z. Wu, J. Cai, J. Wang, Z. Geng, Q. Wang, "Low-Temperature Cu-Cu Bonding Using Silver Nanoparticles Fabricated by Physical Vapor Deposition," Journal of Electronic Materials, Vol. 47, Feb. 2018, pp. 988–993, doi: 10.1007/s11664-017-5831-z.

[12] X. Wang, Q. Wang, Y. Hu, L. Tan, J. Cai, "Copper and graphene composite material prepared by electrodeposition and its potential application for 3D integration," International Conference on Electronic Packaging (ICEPT 2017) , IEEE Press, Sep. 2017, doi:10.1109/ICEPT.2017.8046757.

[13] T. Tong, Y. Zhao, L. Delzeit, A. Kashani, M. Meyyappan, A. Majumdar, "Dense Vertically Aligned Multiwalled Carbon Nanotube Arrays as Thermal Interface Materials," Transactions on Components and Packaging Technologies, Vol. 30, Mar. 2007, pp. 92-100, doi: 10.1109/tcapt.2007.892079.

[14] Y. Pei, H. Zhong, M. Wang, P. Zhang, Y. Zhao, "Effect of Contact Pressure on the Performance of Carbon Nanotube Arrays Thermal Interface Material," Nanomaterials, Vol. 8, Sep. 2018, pp. 732-742, doi:10.3390/nano8090732.

[15] C. Wong, R. Bollampally, "Thermal Conductivity, Elastic Modulus, and Coefficient of Thermal Expansion of Polymer Composites Filled with Ceramic Particles for Electronic Packaging," Journal of Applied Polymer Science, Vol. 74, Apr. 1999, pp. 3396–3403, doi:10.1002/(SICI)1097-4628(19991227)74:14<3396::AID-APP13>3.0.CO;2-3.

Quantifying the Impact of RF Probing Variability on TRL Calibration for LTCC Substrates

Ömer F. Yildiz, David Dahl, Christian Schuster
Institute of Electromagnetic Theory
Hamburg University of Technology (TUHH)
Hamburg, Germany
oemer.yildiz@tuhh.de

Abstract—When conducting radio frequency (RF) measurements using microwave probes, slight misplacements of the probing tips introduce errors and deteriorate measurement results, especially in the case of manual probing. For this reason, short transmission lines in combination with polynomial chaos expansion (PCE) are used to model the imperfect connections between probes and launching structures, and to quantify the impact of probing inaccuracies on the thru-reflect-line (TRL) calibration technique. The analysis is accompanied by a comparison between PCE and Monte Carlo sampling (MCS) in order to assess the accuracy of the proposed method. Additionally, a set of microstrip TRL standards are manufactured on a low temperature cofired ceramic (LTCC) substrate and subsequently measured in the frequency range from 0.05 to 50 GHz for the purpose of studying the correlation with the underlying PCE model.

Keywords-microwave probing; probing variability; polynomial chaos expansion; thru-reflect-line calibration; low temperature cofired ceramics

I. INTRODUCTION

In many areas of engineering, computer simulations are typically accompanied by measurements in order to convey a more accurate picture of the device under test (DUT). As with simulations however, measurements come with their own set of challenges which include systematic errors due to the influence of measurement cables, imperfect connectors, and launching structures between the measurement device itself on the one hand and the actual DUT on the other. In microwave and RF engineering, there is a variety of calibration techniques to choose from among which short-open-load-thru (SOLT) and TRL are the most popular ones [1] that are used to reliably remove the aforementioned errors by effectively moving the plane of reference to the desired location [2]. The TRL calibration in particular has emerged as the favorite for this purpose due to its ease of implementation since only three standards are required. Additionally, it only has non-stringent requirements for the reflect standard which does neither have to be a perfect open nor a perfect short. The main drawback of TRL is the calibration bandwidth which is limited by the length of the line standard. In order to improve the quality of the calibrations, a multiline TRL [3] has been proposed consisting of multiple redundant transmission lines in addition to the single line standard for minimizing random errors. These random errors may stem from nonideal connectors [4], [5] or deviations in the characteristic impedance of the line standard

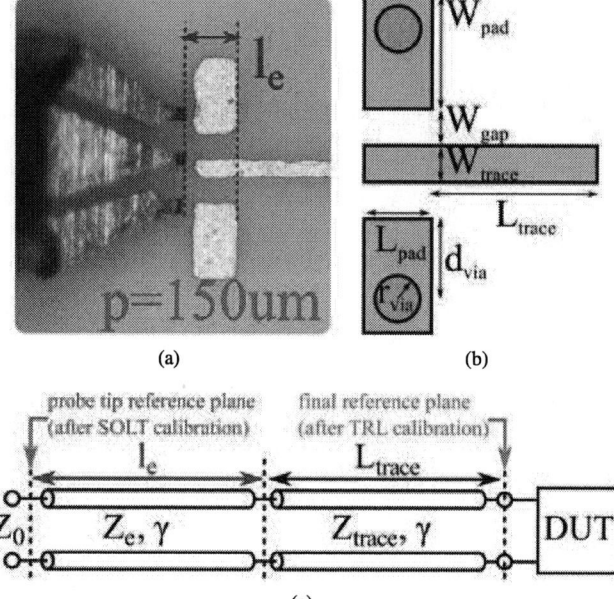

Fig. 1. Microwave probe and microstrip launch structure manufactured on a 10 by 10 cm LTCC substrate. (a) Microwave probe with a tip pitch of 150 μm. (b) Layout of the microstrip launch structure with dimensions referring to Table I. (c) Transmission line model for microwave probe-to-microstrip transition with Z_e as its characteristic impedance, and length l as the distance between probe tips and microstrip junction.

due to manufacturing tolerances [6], and lead to deterioration of measurement results.

In this work, the impact of probing variability caused by the misplacement of microwave probes, as depicted in Fig. 1(a), during the TRL calibration is studied. Even though errors due to imprecise placement of probing tips onto the launching structures are systematic in nature, they can also be treated as random errors, especially in the case of manual probing. The following analysis is essentially comprised of two parts: In the first and theoretical part, a transmission line model [7]–[9] is proposed for representing the transition from probing tips to microstrips as indicated in Fig. 1(c). The characteristic impedances and lengths of these short transmission lines are treated as stochastically independent input random variables. Then, PCE [10]–[12] is introduced to analyze the uncertainty propagation of these input random variables into the calibration as well as into the final measurement results. For a moderate number of input variables PCE is a computationally more efficient alternative to MCS and has been successfully applied in many different areas such as global sensitivity analysis and optimization [13]–[15], model order reduction of equivalent circuits [16], as well as stochastic circuit simulation [17]. In the second and practical part, a set of microstrip TRL standards are manufactured on a 10 by $10\,\text{cm}$ LTCC substrate and repeatedly measured in the frequency range from $0.05\,\text{GHz}$ to $50\,\text{GHz}$ for correlation purposes between the observed variability in measurements and in the proposed theoretical model.

Thus, the remainder of this paper is structured as follows: In Section II, the transmission line model and PCE method are introduced and discussed. In Section III, the impact of the probing variability during TRL calibration is analyzed and quantified. Furthermore, the accuracy of the proposed PCE method is assessed through comparison to MCS. In Section IV, measurements and computer simulations are checked for correlation with each other. Lastly, Section V summarizes the main findings and concludes this work.

II. THEORETICAL BACKGROUND

In general, measuring the DUT before the calibration procedure gives rise to uncalibrated transfer scattering parameters or T-parameters $[T_{mDUT}]$ with

$$[T_{mDUT}] = [T_A][T_{DUT}][T_B], \tag{1}$$

where $[T_A]$ and $[T_B]$ respectively describe the left-hand and right-hand side of the launching structures. Thus, applying TRL [1] results in the extraction of a pair of related error matrices $[X_A]$ and $[X_B]$ in order to obtain the real, calibrated T-parameters,

$$[T_{DUT}] \approx [X_A][T_{mDUT}][X_B]. \tag{2}$$

However, imperfect probing repeatability inevitably introduces errors in the calibration when measuring TRL standards,

$$[T_{mT}] = [T_{e,AT}][T_A][T_T][T_B][T_{e,BT}], \tag{3}$$

$$[T_{mL}] = [T_{e,AL}][T_A][T_L][T_B][T_{e,BL}], \tag{4}$$

$$[T_{mR}] = [T_{e,AR}][T_A][T_R][T_B][T_{e,BR}], \tag{5}$$

with $[T_e]$ as the stochastically independent random error matrices that represent the transtion from probe tips to microstrips. Here, short transmission lines as depicted in Fig. 1(c) are used to model the transition. Using ABCD-parameters [18], the general description of a transmission line becomes

$$[ABCD] = \begin{bmatrix} \cosh(\gamma l) & Z\sinh(\gamma l) \\ \frac{\sinh(\gamma l)}{Z} & \cosh(\gamma l) \end{bmatrix}, \tag{6}$$

and thus may be further transformed into S- or T-parameters. In the case of T-parameters [19] we obtain

$$[T_e] = \frac{1}{2} \begin{bmatrix} -\frac{K_1^2-4}{K_2} & K_1 \\ -K_1 & K_2 \end{bmatrix}, \tag{7}$$

TABLE I
DIMENIONS FOR MICROSTRIP LAUNCH STRUCTURE CORRESPONDING TO FIG. 1(B) WITH SUBSTRATE HEIGHT H_{sub} AND PERMITTIVITY ϵ_r.

parameter	[μm]	[mil]	
W_{trace}	60	2.36	-
L_{trace}	421	16.6	-
W_{pad}	250	9.84	-
L_{pad}	200	7.87	-
W_{gap}	80	3.15	-
r_{via}	50	1.97	-
d_{via}	175	6.9	-
H_{sub}	80	3.1	-
ϵ_r	-	-	7.2

TABLE II
LIST OF BASIS POLYNOMIALS FOR THE UNIVARIATE CASE DEPENDING ON DISTRIBUTION OF THE INPUT VARIABLE (AS FOUND IN [10]).

distribution of ξ	basis polynomials $\phi(\xi)$
Gaussian	Hermite
Gamma	Laguerre
Beta	Jacobi
Uniform	Legendre

with

$$K_1 = \left(\frac{Z_e}{Z_0} - \frac{Z_0}{Z_e}\right)\sinh\gamma l_e, \tag{8}$$

$$K_2 = 2\cosh\gamma l_e + \left(\frac{Z_e}{Z_0} + \frac{Z_0}{Z_e}\right)\sinh\gamma l_e. \tag{9}$$

Z_0 is the fixed reference port impedance at the probing tips and set to $50\,\Omega$, γ the propagation constant of the probe-to-microstrip transition and Z_e its characteristic impedance. Probing variability is modeled by means of uniformly distributed random variables $l_e \sim \mathcal{U}(30\,\mu\text{m}, 170\,\mu\text{m})$ for the distance between probe tips and microstrip junction, as depicted in Fig. 1(a), and $Z_e \sim \mathcal{U}(40\,\Omega, 55\,\Omega)$ for the characteristic impedance.

Using PCE as the method of uncertainty quantification (UQ) [10], [11], any output variable of interest, e.g. the transmission S_{21} of the DUT, can now be described by a polynomial by expanding S_{21} into a finite series,

$$S_{21}(\boldsymbol{\xi}) \approx \sum_{i=0}^{D} \hat{s}_i \phi_i(\boldsymbol{\xi}), \tag{10}$$

with $\boldsymbol{\xi} = (l_{e,AT}, Z_{e,AT}, \ldots, l_{e,RBT}, Z_{e,RBT})$ as the vector of N stochastically independent input random variables, $\{\phi_i\}_0^D$ as the set of known joint orthogonal basis functions, and $\{\hat{s}_i\}_0^D$ as the set of yet unknown PCE coefficients. The joint basis functions $\phi_i(\boldsymbol{\xi})$ are derived from the basis functions $\phi_i(\xi)$ of the univariate case according to Table II. Their orthogonality can be expressed mathematically through the inner product,

$$\langle \phi_i, \phi_j \rangle = \int \phi_i(\boldsymbol{\xi})\phi_j(\boldsymbol{\xi})\rho(\boldsymbol{\xi})d\boldsymbol{\xi} = \gamma_i\delta_{ij}, \tag{11}$$

with $\{\gamma_i\}_0^D$ as the set of multivariate norms, $\rho(\boldsymbol{\xi})$ as the joint probability density function (PDF), and δ_{ij} as the Kronecker delta. The total number of expansion terms $D = \frac{(P+N)!}{P!N!} - 1$

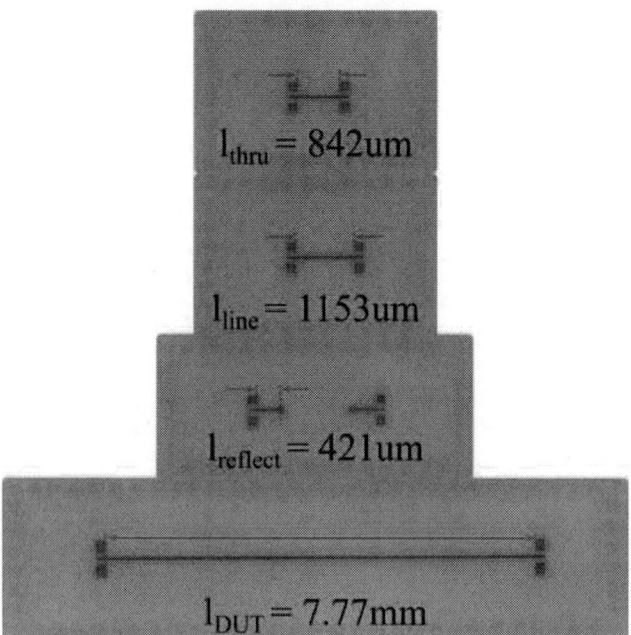

Fig. 2. Set of microstrip TRL standards manufactured on a 10 by 10 cm LTCC substrate with dimensions corresponding to Tabel I. The reflect is a shorted line of length $l_{reflect} = L_{trace}$, and the lengths of the thru and line are $l_{thru} = 2L_{trace}$ and $l_{line} = 1153\,\mu m$, respectively. The DUT is a microstrip line of length $l_{DUT} = 7.77\,mm$. All lines share the same characteristic impedance of approximately $Z_c = 65\,\Omega$.

depends on N and the order of approximation P. The stochastic behavior of $S_{21}(\boldsymbol{\xi})$ is solely described by coefficients $\hat{s}_i$ from which all subsequent measures such as expectation, variance, and stochastic moments can be readily derived. For example, the expectation is given by the very first coefficient:

$$E[S_{21}(\boldsymbol{\xi})] = \int S_{21}(\boldsymbol{\xi})\rho(\boldsymbol{\xi})d\boldsymbol{\xi} = \hat{s}_0 \qquad (12)$$

Exploiting the orthogonality properties of the basis functions, the variance is found be

$$Var(S_{21}(\boldsymbol{\xi})) = E[S_{21}(\boldsymbol{\xi})^2] - E[S_{21}(\boldsymbol{\xi})]^2 \qquad (13)$$

$$= \sum_{i=0}^{D} \hat{s}_i \gamma_i - \hat{s}_0 \qquad (14)$$

$$= \sum_{i=1}^{D} \hat{s}_i \gamma_i. \qquad (15)$$

The use of the PCE in Eq. (10) as a surrogate model allows for fast computation of PDFs as well as conducting sensitivity analysis [15], [16]. The PCE coefficients themselves are obtained through spectral projection onto the orthogonal basis and solving the resulting multidimensional integral using

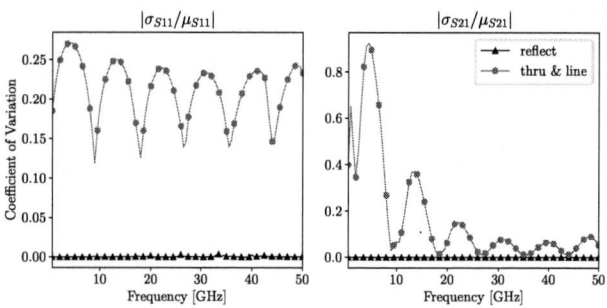

Fig. 3. Coefficient of variation of S_{11} and S_{21} using PCE with $P = 3$ for the case when probing variability is only considered during measurement of the reflect standard, as well as the case when it is only considered during the measurement of the thru and line standards. Uncertainties from the reflect measurements are negligable by comparison. The minima coincide to frequencies where $l_{DUT} = n\lambda/2$.

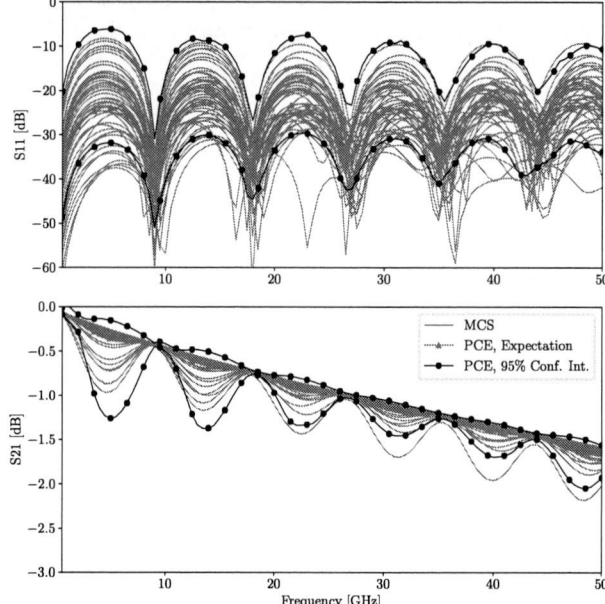

Fig. 4. Comparison of S-parameters between MCS with $N_{MC} = 10^3$ and PCE with $P = 3$ for computed microstrip line of length $l_{DUT} = 7.77\,mm$ after TRL calibration. For the reflection the difference in magnitude differs approximately by $\pm 15\,dB$ range from the expected value. For the transmission, a maximum decrease in magnitude of up to $-1\,dB$ and $-0.5\,dB$ occurs below and above 20 GHz, respectively.

Gaussian quadrature (GQ):

$$\hat{s}_i = \frac{1}{\langle \phi_i^2(\boldsymbol{\xi}) \rangle} \langle S_{21}(\boldsymbol{\xi}), \phi_i(\boldsymbol{\xi}) \rangle \qquad (16)$$

$$= \frac{1}{\gamma_i} \int S_{21}(\boldsymbol{\xi})\phi(\boldsymbol{\xi})\rho(\boldsymbol{\xi})d\boldsymbol{\xi} \qquad (17)$$

$$\approx \frac{1}{\gamma_i} \sum_{k=1}^{Q} S_{21}(\tilde{\boldsymbol{\xi}}_k)\phi_i(\tilde{\boldsymbol{\xi}}_k)w_k, \qquad (18)$$

with $\{\tilde{\boldsymbol{\xi}}_k\}_1^Q$ and $\{w_k\}_1^Q$ as the set of quadrature nodes and weights, respectively, obtained through the full tensor product

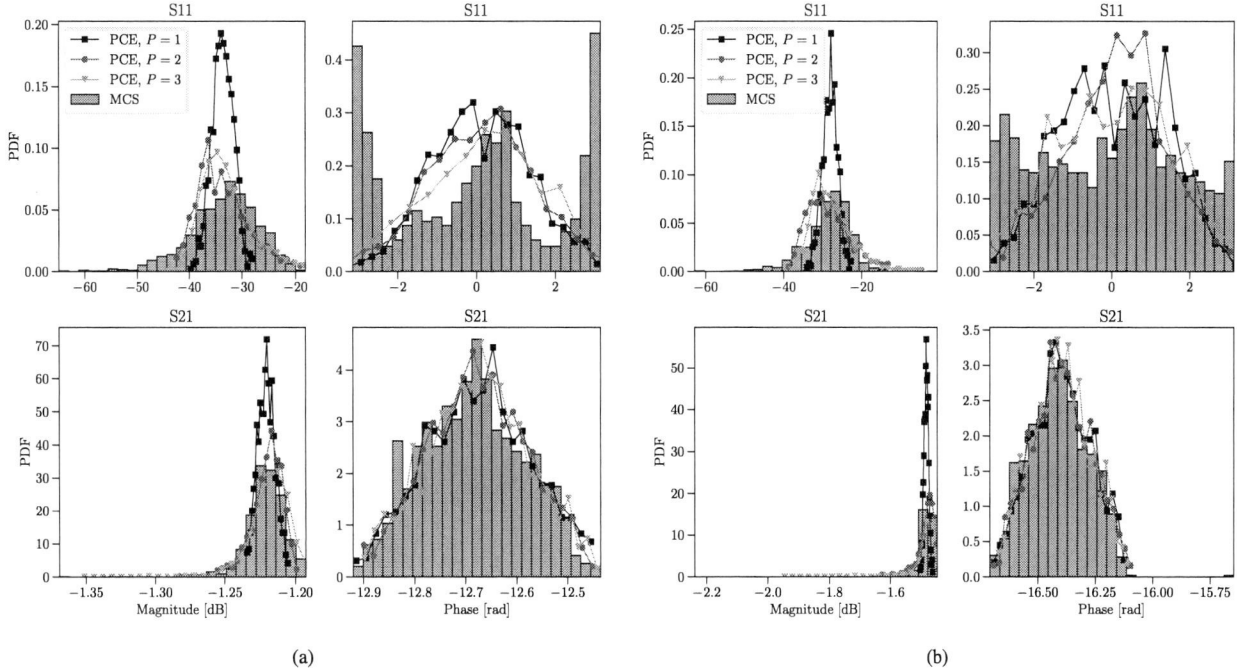

Fig. 5. Probability density function of S-Parameters for computed microstrip line of length $l_{DUT} = 7.77\,\text{mm}$ (a) at frequency $f = 35\,\text{GHz}$ and (b) at frequency $f = 45\,\text{GHz}$. Good agreement bewteen PCE and MCS is achieved with an order of approximiation $P = 3$, despite dicrepencies at the tails of the distribution for the phase of S_{11}.

of GQ nodes for the univariate case. Hence, the total number of evaluations of the forward function $S_{21}(\cdot)$ becomes $Q = (P+1)^N$.

III. IMPACT OF PROBING VARIABILITY

Since the cost of numerical integration through GQ grows exponentially with N, a two-step approach is pursued for the following PCE anlysis: In the first step, only the reflect standard, as described by Eq. (5), is subject to measurement errors, thus resulting in a total number of 4 input random variables. In the second step, both thru and line standards are subjected to probing variability, i.e. Eq. (3) and Eq. (4), leading to 8 input random variables. Setting $P = 3$, the results of both cases are shown in Fig. 3 by comparing the coefficients of variation σ/μ, i.e. the ratio of standard deviation to expectation, of S-parameters of a microstip line of length $l_{DUT} = 7.77\,\text{mm}$ after TRL calibration. The TRL standards used here are depicted in Fig. 2. The dimensions of the microstrip trace are listed in Table I. The reflect standard is a shorted microstrip of length L_{trace}, and the lengths of the thru and line standard are $l_{thru} = 2L_{trace} = 842\,\mu\text{m}$ and $l_{line} = 1153\,\mu\text{m}$ long, respectively. For the analysis in this Section, the microstrips described in Fig. 2 are computed using analytical formulas [7]. From Fig. 3 it is evident that uncertainties during measurements of the reflect standard are negligable by comparison. For the remainder of the analysis, we therefore only consider probing variability for the thru and line standards. The minima and maxima of the coefficient of variation coincide with frequencies where the length of the

DUT is equal to integer multiples of half-wavelengths, i.e. $l_{DUT} = n\lambda/2$, and odd numbers of quarter-wavelengths, i.e. $l_{DUT} = (2n-1)\lambda/4$, respectively.

In Fig. 4, the impact of probing variability for thru and line measurements on S-parameters of the same DUT can be observed. Both MCS with $N_{MC} = 10^3$ samples and the confidence interval obtained by PCE with $P = 3$ show that for the reflection the difference in magnitude can differ by $\pm 15\,\text{dB}$ across the entire frequency range from the expected value. For the transmission, a maximum decrease in magnitude of up to $-1\,\text{dB}$ and $-0.5\,\text{dB}$ occurs below and above $20\,\text{GHz}$, respectively. Similar to the results of Fig. 3, the standard deviation σ_{S21} is once again lowest if the length of the measured microstrip is equal to integer multiples of half-wavelengths, i.e. $l_{DUT} = n\lambda/2$. This behavior can be explained by the impedance transforming characteristic [8] of the measured DUT: The port impedance Z_{Port2} at the end of the transmission line with characteristic impedance Z_c will be transformed into an input impedance at Port 1 as follows:

$$Z_{in} = Z_c \frac{Z_{Port2} + jZ_c \tan\beta l}{Z_c + jZ_{Port2} \tan\beta l} \tag{19}$$

$$= \begin{cases} \frac{Z_c^2}{Z_{Port2}}, & l_{DUT} = \frac{\lambda}{4} \\ Z_{Port2}, & l_{DUT} = \frac{\lambda}{2} \end{cases} . \tag{20}$$

During TRL calibration, the reference impedance at both ports is set to the characteristic impedance of the line standard which is the same as the characteristic impedance Z_c of the DUT. However, in the case of probing variability as described by

Eq. (3) to Eq. (5), the port impedances will differ from Z_c. Consequently, the reflection at Port 1 then becomes

$$S_{11} = \frac{Z_{in} - Z_{Port1}}{Z_{in} + Z_{Port1}}. \tag{21}$$

Since $Z_{Port1} = Z_{Port2}$, the reflection in Eq. (21) will reach its minimum and maximum if $l_{DUT} = \lambda/2$ and $l_{DUT} = \lambda/4$, respectively. Strictly speaking, due to imperfect probing repeatability, the reference port impedances actually become random variables themselves, i.e. $Z_{Port1}(\boldsymbol{\xi})$ and $Z_{Port2}(\boldsymbol{\xi})$, and will therefore differ slightly from each other in practice. For this reason, S_{11} never exactly reaches $-\infty$ in Fig. 4.

In order to assess the accuracy of PCE, the PDFs of both phase and magnitude of the S-parameters are computed and compared to MCS at $f = 35\,\mathrm{GHz}$. As seen in Fig. 5, the results are in good agreement and there is a considerable improvement from $P = 1$ to $P = 3$ for magnitudes $|S_{11}|_{\mathrm{dB}}$ and $|S_{21}|_{\mathrm{dB}}$, with the exception of the phase of S_{11}, where PCE is not able to accurately present the tails of the distribution.

Fig. 6 depicts the Sobol indices [11] of the S-parameters. Sobol indices are a measure of the relative impact of the input random variables onto the output variabilty. Hence, they are representative of a type of global sensitivity analysis [20], [21] wherein the influence of a specific input is projected onto a unitless number in the range $[0, 1]$. When using PCE, Sobol indices do not have to be computed independently. In fact, they are already available once the PCE coefficients in Eq. (18) have been computed and are obtained as follows:

$$\mathrm{Sob}_n = \frac{\sum_{k \in K_n} \hat{s}_k^2 \gamma_k}{\sum_{i=1}^{D} \hat{s}_i^2 \gamma_i} \tag{22}$$

Here, K_n represents the indices corresponding to expansion terms $\hat{s}_k \phi_k(\boldsymbol{\xi})$ that depend on the n-th input variable ξ_n. It can be seen in Fig. 6 that variations in the characteristic impedance Z_e have a slightly lesser impact than variations in l_e. Furthermore, a periodic switch in relative impact of thru and line errors occurs whenever the length of the microstrip is equal to integer multiples of half-wavelengths, i.e. $l_{DUT} = n\lambda/2$.

IV. COMPARISON TO EXPERIMENTS

As depicted in Fig. 2, a set of microstrip TRL standards equivalent to those defined in Section III are manufactured on a 10 by 10 cm LTCC substrate using the dimensions listed in Table I. A vector network analyzer (VNA) is calibrated beforehand up to the tips of the two microwave probes, as illustrated in Fig. 1, using vendor supplied SOLT calibration kits. Then, the TRL calibration kits designed on the LTCC substrate are measured repeatedly in order to obtain a set of $N_X = 20$ different error matrices according to Eq. (2). The DUT, the same microstrip line of length $l_{DUT} = 7.77\,\mathrm{mm}$, is only measured once, and its S-parameters after TRL calibration are plotted in Fig. 7 alongside the computed transmission line model using PCE. It can be observed that the variability is overestimated by PCE, especially at frequencies below $20\,\mathrm{GHz}$. Apart from that, the measurement results correlate sufficiently well with the proposed model: The same

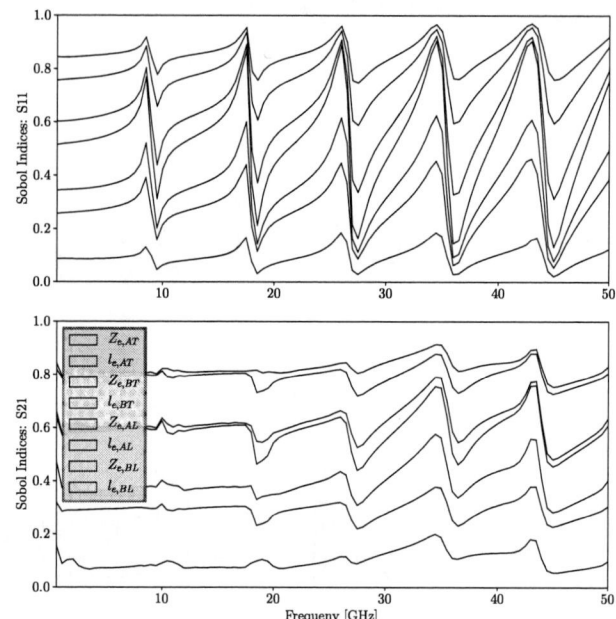

Fig. 6. Sobol indices of S-parameters for computed microstrip line of length $l_{DUT} = 7.77\,\mathrm{mm}$ and $P = 3$. The relative impact of probe misplacements on S-parameter variability is similar for both thru and line measurements. Variations in characteristic impedance Z_e of launching structure have a slightlty lesser overall impact than variations in l_e.

periodicity in the coefficient of variation σ/μ is observed. The difference in measured reflection may be as large as $\Delta|S_{11}|_{\mathrm{dB}} = 30\,\mathrm{dB}$. The maximum decrease in magnitude of the transmission is $\Delta|S_{21}|_{\mathrm{dB}} = 0.25\,\mathrm{dB}$.

There are a few possible explanations for the difference in observed variability of measured and simulated S-parameters: The transmission line model itself may be insufficient. In this case, it needs to be complemented by additional lumped elements in order to account for fringing fields at the pad edges. Another explanation can be found in the presupposed distribution of the input random variables $l_e \sim \mathcal{U}(30\,\mu\mathrm{m}, 170\,\mu\mathrm{m})$ and $Z_e \sim \mathcal{U}(40\,\Omega, 55\,\Omega)$: While the first is an accurate description of possible probe misplacement, the distribution of Z_e may be different in practice. In the future, this problem can be solved using full wave simulations of the probe-to-microstrip transition.

Lastly, a discussion on the usefulness of PCE is appropriate. It is safe to say that the findings of this work could have been produced just as easily by only using MCS which requires much less overhead than PCE and is computationally more efficient if the number of input variables is very large. On the the hand, this argument ignores that global sensitivity analysis as shown in Fig. 6 is generally much more costly when using MCS [20]. In large-scale and practical problems, the authors therefore recommend applying a two-tiered approach: If the number of unknowns or input variables is large, MCS should be used. Once the interesting parameters that warrant further investigation are identified and the number of input variables

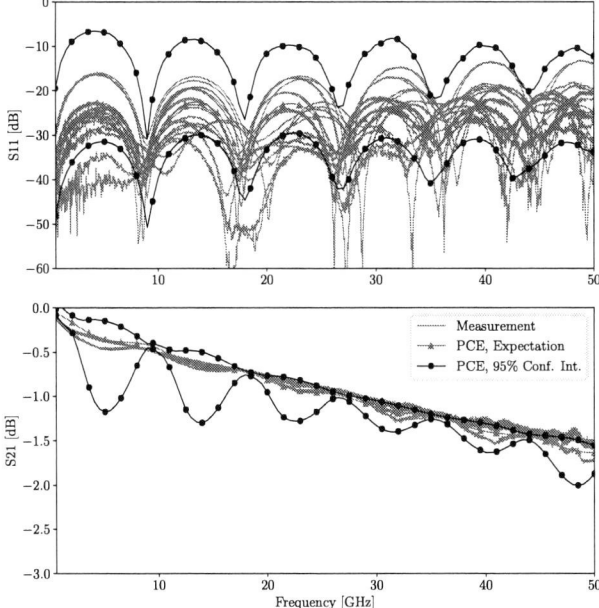

Fig. 7. Comparison of S-parameters between measurements and computed transmission line model using PCE for a microstrip line of length $l_{DUT} = 7.77$ mm after TRL calibration. Impact of probing variability is overestimated by PCE at lower frequencies up 20 GHz. The difference in measured reflection reaches $\Delta|S_{11}|_{dB} = 30$ dB. The maximum decrease in magnitude of the transmission is $\Delta|S_{21}|_{dB} = 0.25$ dB.

greatly reduced, PCE should be applied for the added benefit of efficient and insightful sensitivity analysis.

V. CONCLUSION

Probing variability has a considerable impact on the TRL calibration: Measurements of a microstrip line show that calibrated S-parameters may differ by ± 0.25 dB for the transmission and by ± 15 dB for the reflection from the expected values. From theoretical models, it is found that inaccuracies during the thru and line measurements lead to greater observed variability in S-parameters of the DUT than is the case if only the reflect standard is subject to measuring errors. In the future however, a combined analysis with probing variability for all three TRL standards simultaneously is desired. Measurement results correlate sufficiently well with the proposed method, despite discrepencies at lower frequencies.

ACKNOWLEDGMENT

The authors would like to thank Ole Thomsen, Marc Bochard, and Nico Pathe of KOA Europe for their support during the design and manufacturing of the LTCC substrate.

REFERENCES

[1] V. Teppati, A. Ferrero, and M. Sayed, *Modern RF and Microwave Measurement Techniques*. Cambridge University Press, 2013.

[2] L. Martens, *High-Frequency Characterization of Electronic Packaging*, 2nd ed. Kluwer academic publishers, 2001.

[3] R. B. Marks, "A multiline method of network analyzer calibration," *IEEE Trans. Microw. Theory Tech.*, vol. 39, no. 7, pp. 1205–1215, July 1991.

[4] C. Wu, Y. Xu, J. Li, and S. Gao, "Effects of the length of thru on the measurement precision in TRL technique," *IEEE Microwave and Wireless Components Letters*, vol. 24, no. 12, pp. 905–907, December 2014.

[5] C. Wu, L. Sima, X. Hou, and J. Li, "Periodical measurement error with respect to the length of thru in TRL technique," *IEEE Transactions on Microwave Theory and Techniques*, vol. 65, no. 5, pp. 1605–1614, January 2017.

[6] X. Ye and M. Balogh, "Physics-based fitting to improve PCB loss measurement accuracy," *IEEE International Symposium on Electromagnetic Compatibility (EMC)*, August 2017.

[7] R. E. Collin, *Foundations for Microwave Engineering*, 2nd ed. John Wiley & Sons, Inc., 2001.

[8] D. M. Pozar, *Microwave Engineering*, 4th ed. John Wiley & Sons, Inc., 2012.

[9] T.-M. Winkel, "An accurate and complete frequency dependent transmission line characterization using s-parameter measurements," *IEEE Topical Meeting on Electrical Performance of Electronic Packaging*, October 1999.

[10] D. Xiu, *Numerical Methods for Stochastic Computations: A Spectral Method Approach*. Princeton University Press, 2010.

[11] T. J. Sullivan, *Introduction to Uncertainty Quantification*. Springer, 2015.

[12] J. Feinberg and H. P. Langtangen, "Chaospy: An open source tool for designing methods of uncertainty quantification," *Journal of Computational Science*, vol. 11, pp. 46–57, November 2015.

[13] T. Reuschel, Ö. Yildiz, J. Balachandran, C. Filip, N. Bhagwath, B. Sen, and C. Schuster, "Efficient sensitivity-aware assessment of high-speed links using PCE and implications for COM," *DesignCon*, January 2018.

[14] D. Dahl, Ö. F. Yildiz, E. Frick, C. Seifert, M. Lindner, and C. Schuster, "Feasibility of uncertainty quantification for power distribution network modeling using PCE and a contour integral method," *IEEE International Symposium on Electromagnetic Compatibility and IEEE Asia-Pacific Symposium on Electromagnetic Compatibility (EMC/APEMC)*, May 2018.

[15] Ö. F. Yildiz, J. B. Preibisch, J. Niehof, and C. Schuster, "Sensitivity analysis and empirical optimization of cross-domain coupling on RFICs using polynomial chaos expansion," *IEEE International Symposium on Electromagnetic Compatibility & Signal/Power Integrity (EMCSI)*, August 2017.

[16] Ö. F. Yildiz, H.-D. Brüns, and C. Schuster, "Variance-based iterative model order reduction of equivalent circuits for EMC analysis," *IEEE Transactions on Electromagnetic Compatibility*, vol. 61, no. 1, pp. 128–139, February 2019.

[17] K. Strunz and Q. Su, "Stochastic formulation of SPICE-type electronic circuit simulation with polynomial chaos," *ACM Transactions on Modeling and Computer Simulation*, vol. 18, no. 4, September 2008.

[18] W. R. Eisenstadt and Y. Eo, "S-parameter-based ic interconnect transmission line characterization," *IEEE Trans. Compon., Hybrids, Manuf. Technol.*, vol. 15, no. 4, pp. 483–490, August 1992.

[19] J.-S. Hong, *Microstrip Filters for RF/Microwave Applications*, 2nd ed. John Wiley & Sons, Inc., 2011.

[20] T. Crestaux, O. L. Maître, and J.-M. Marinez, "Polynomial chaos expansion for sensitivity analysis," *Reliability Engineering and System Safety*, vol. 94, no. 7, pp. 1161–1172, July 2009.

[21] B. Sudret and C. V. Mai, "Computing derivative-based global sensitivity mmeasure using polynomial chaos expansions," *Reliability Engineering and System Safety*, vol. 134, pp. 241–250, February 2015.

978-1-7281-1500-9/19 $31.00 © 2019 IEEE

Effects of NCF and UBM Materials on Electromigration Reliabilities of Sn-Ag microbumps for advanced 3D packaging

Kirak Son, Gahui Kim, Hyodong Ryu, Young-Cheon Kim, and Young-Bae Park*

School of Materials Science and Engineering
Andong National University
Andong-si, Korea
*corresponding author's e-mail: ybpark@anu.ac.kr

Gyu-Tae Park
K4 RND project
Amkor Technology Korea Inc.
Gwangju, Korea
e-mail: gytae.park@amkor.co.kr

Ho-Young Son and Nam-Seog Kim
R&D of package development
SK hynix Inc.
Icheon-si, Korea
e-mail: hoyoung.son@sk.com

Cheol-Woong Yang

School of Advanced Materials Science and Engineering
Sungkyunkwan University
Suwon-si, Korea
e-mail: cwyang@skku.edu

Jeong Sam Han
Department of Mechanical Design Engineering
Andong National University
Andong-si, Korea
e-mail: jshan@anu.ac.kr

Abstract— The effect of non-conductive films (NCF) and under-bump metallization (UBM) materials on the electromigration (EM) failure mechanism of Sn-Ag microbumps was investigated under stress conditions at current densities ranging from 0.5~1.3 $\times$ 10^5 A/cm^2 at 150 ℃. In the case of NCF microbumps, the EM failure mechanism were almost similar to that involving Cu/Ni/Sn-Ag microbumps with NCF and non-NCF. However, the NCF applications increased EM lifetime for the Cu/Ni/Sn-Ag microbumps. In the case of Ni and Cu UBM microbumps, the EM failure mechanism varied between Sn-Ag microbumps with Ni UBM and Cu UBM. No EM-induced failure was observed in Cu UBM microbump. However, EM-induced voids were found on the cathode side of the Al trace in the Ni UBM microbump. The three-dimensional (3-D) finite element method was used to simulate the current density and temperature distributions of the microbumps. The results showed that the Joule heating of the Ni UBM microbump was higher than that of the Cu UBM microbump. Thus, the increased Joule heating due to thicker Ni UBM resulted in a negative effect on the EM failure time. Therefore, the Cu UBM microbumps are expected to display enhanced EM resistance.

Keywords- electromigration; non-conductive film; under-bump metallization; Joule heating; finite element simulation

I. INTRODUCTION

Recently, the Internet of things and Big Data environments have intensified the need for high-performance and low-power semiconductor devices [1–4]. High input/output interconnects (I/Os) and fine-pitch were needed to successfully meet the criteria for multi-functional and miniaturization of next-generation electronics [1, 5]. Due to decreased joint size, the evaluation of micro joint reliability is critical during the transition from micro-interconnection to a fine pitch of approximately 10 μm and high I/Os [1,5].

The resulting increase in current density due to the miniaturization of the solder bump and pitch can cause serious electromigration (EM) failure issues [1, 6]. The EM mechanism has been widely investigated in the semiconductor chip interconnect [1, 7–11]. During EM, the metal atoms move in the direction of electron flow, which leads to void formation on the cathode side and extrusions on the anode side [1, 7]. Additionally, significant current crowding is induced by EM as the joint size is reduced [1, 5]. The 3D IC packaging involves a number of candidates representing solder materials and under-bump metallization (UBM) [1]. For instance, Cu pillar reacts with solder Sn and forms well-known Cu_6Sn_5 and Cu_3Sn [1]. On the other hand, Ni is commonly used as the diffusion barrier in Cu pillar to restrain the fast consumption of Cu and is converted into Ni_3Sn_4 during the interfacial reaction with Sn [12]. Selection of UBM materials is crucial for fine-pitch microbump interconnections, because the UBM materials significantly influence the reliability of the solder microbump joints. However, the types of UBM material for fine-pitch solder-capped microbump joint with enhanced electrical reliability has yet to be established. This study systematically investigated the effects of UBM materials and non-conductive film (NCF) structures on the EM performance of microbumps. First, the effect of NCF on EM reliability of Cu/Ni/Sn-Ag microbumps was investigated under stress at a current density of 1.3 x 10^5 A/cm^2 at 130 °C. Second, the effect of UBM materials on EM reliability of Cu/Ni/Sn-Ag microbumps was investigated under stress conditions at a current density of 0.5 x 10^5 A/cm^2 at 150 °C. Finally, finite-element analysis was performed to investigate the temperature and current density distribution in microbumps under current stress.

.

II. EXPERIMENT

In this study, EM samples were prepared various bump structures with daisy-chain structures. The top and bottom chips were bonded using a thermo-compressive bonder after NCF was dispensed over the bottom chip. The peak temperature of the thermo-compressive bonding process was 260 °C.

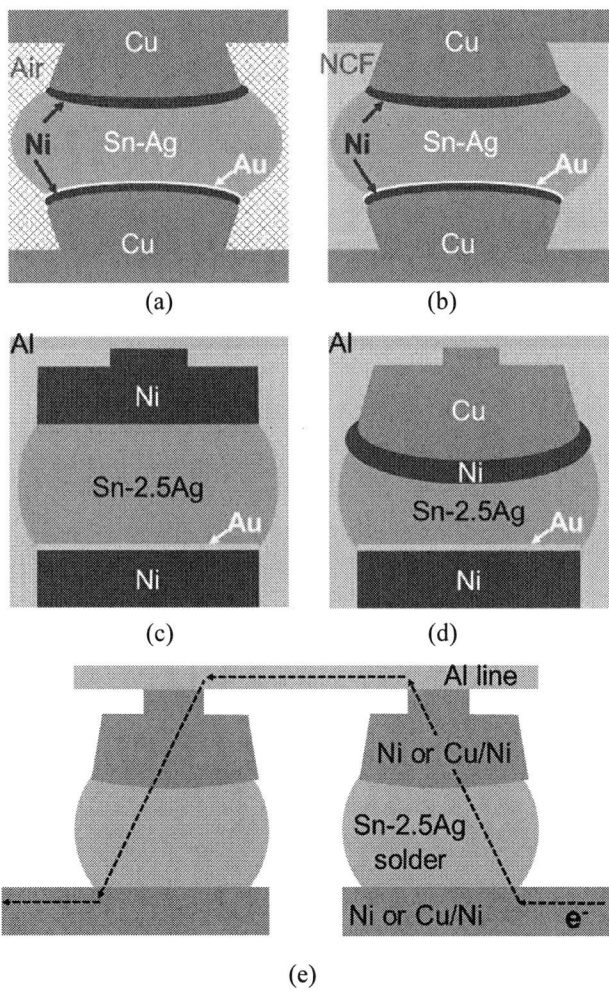

(a)　　　　(b)

(c)　　　　(d)

(e)

Figure 1. Schematics diagrams of EM test structure : (a) non-NCF microbump, (b) NCF microbump, (c) Ni UBM microbump, (d) Cu UBM microbump, and (e) daisy-chain structure for the electromigration tests conducted in this study.

After bonding, the microbump structure, as shown in Fig. 1, consisted of a Cu pillar/Ni barrier/Sn-2.5Ag solder/Ni barrier/Cu pillar or UBM/Sn-2.5Ag solder/Ni trace. NCF and non-NCF microbump samples comprised Cu trace, Cu pillar, Ni barrier, and Sn-2.5Ag on the top chip; and Cu trace, Cu pillar, and Ni barrier on the bottom chip, as shown in Fig. 1(a) and (b).

The Ni UBM microbump sample consisted of Al trace, Ni pillar, and Sn-2.5Ag on the top chip, and a Ni trace on the bottom chip, as shown in Fig. 1(c). The Cu UBM microbump sample consisted of an Al trace, a Cu pillar, a Ni barrier, and Sn-2.5Ag on the top chip, and a Ni trace on the bottom chip, as shown in Fig. 1(d). Also, Au on the bottom chip was used to improve joint wettability. Lead-free solder and UBM chips were electroplated on both sides. The Al trace was deposited using sputtering. Figure 1(e) shows a cross-sectional schematic diagram of the daisy-chain structure. Arrows depict the direction of the electron flow. The resistance changes in the daisy chain under current stress was measured using an in-situ data acquisition system (37970A, Agilent) [13]. The EM test was conducted at 150 °C at a current density of 1.3 x 10^5 A/cm^2 and 0.5 x 10^5 A/cm^2, respectively. Current density was calculated based on the area of the solder/Ni interface [13]. The failure mechanism analysis was conducted using scanning electron microscopy (SEM) and energy dispersive X-ray spectroscopy (EDS), respectively.

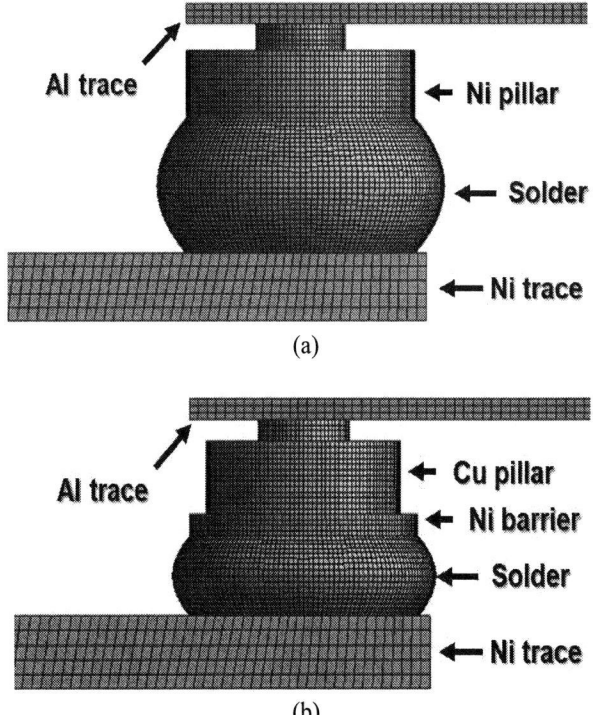

(a)

(b)

Figure 2. Three-dimensional view of the constructed model with meshization : (a) Ni UBM microbump and (b) Cu UBM microbump.

The 3-D finite element analysis was performed to determine the Joule heating behavior within the daisy-chain

978-1-7281-1500-9/19 $31.00 © 2019 IEEE　　　2247

structure under ongoing stress. The Joule heating temperature was predicted via the SOLID 226 3D 20-node coupled-field solid element analysis using ANSYS 18.2 software. Figure 2 shows the mesh geometry used in this study. The electrical resistivities, thermal conductivities, and temperature coefficient resistance of the materials used in this modeling are listed in Table I [14].

TABLE I. MATERIALS PROPERTIES OF THERMAL-ELECTRIC SIMULATION

Materials	Thermal conductivity (W/m · K)	Resistivity (μΩ·cm)	Temperature coefficient of resistivity (TCR) (K-1)	Source
Sn-2.5Ag	33	12.3	4.6×10^{-3}	
Ni	76	28	6.8×10^{-3}	
Cu	403	6.8	4.3×10^{-3}	[14]
NCF	0.55	-	-	
Si	147	-	-	

III. RESULTS AND DISCUSSION

A. NCF effect on Electromigration Reliability of Sn-Ag microbumps

Figure 3 shows the cross-sectional SEM images of NCF microbumps before (as-bonded) and after EM failure for 600 h with NCF microbump at 150 °C and a current density of 1.3×10^5 A/cm^2. Figure 3(a) represent SEM images of NCF microbumps. The phases of Sn, IMCs of (Au,Ni)Sn$_4$ and Ni$_3$Sn$_4$ were observed directly after bonding [13]. The Ni$_3$Sn$_4$ phase following the initial reaction between the Ni barrier and Sn, and the (Au,Ni)Sn$_4$ phase occurring at the interface between the Ni$_3$Sn$_4$ phase and solder, are similar to those reported previously involving similar microbump structures [13,15].

In the case of NCF microbump, as shown in Fig. 3(b), after the EM failure, the Ni barrier was locally consumed at the top chip during the downward electron flow. The (Cu,Ni)$_6$Sn$_5$ phase developed between the upper Ni and the lower Ni barriers. A substantial portion of the Cu pillar on the top chip was transformed into (Cu,Ni)$_6$Sn$_5$ following rapid diffusion across the concentration gradient between the Cu atoms in the Cu pillar and Sn atoms in the solder. Based on the results of the EDS analysis, the IMC interfacing with the Cu pillar was estimated as a Cu$_3$Sn phase due to the diffusion of sufficient Cu atoms from the Cu pillar to the solder across the concentration gradient [13].

Figure 4 shows the cross-sectional SEM images of non-NCF microbumps before (as-bonded) and after EM failure for 180 h with non-NCF microbump at 150 °C and a current

density of 1.3×10^5 A/cm^2. Figure 4(a) represent SEM images of non-NCF microbumps. The phases of Sn, IMCs of (Au,Ni)Sn$_4$ and Ni$_3$Sn$_4$ were observed directly after bonding [13]. The Ni$_3$Sn$_4$ phase following the initial reaction between the Ni barrier and Sn, and the (Au,Ni)Sn$_4$ phase occurring at the interface between the Ni$_3$Sn$_4$ phase and solder, are similar to those reported previously involving similar microbump structures [13,15].

In the case of the non-NCF microbump, as shown in Fig. 4(b), following the EM failure of non-NCF microbump, the Ni barrier was consumed at the top chip in the downward electron flow, and the (Cu,Ni)$_6$Sn$_5$ phase evolved between the upper Ni and the lower Ni barriers.

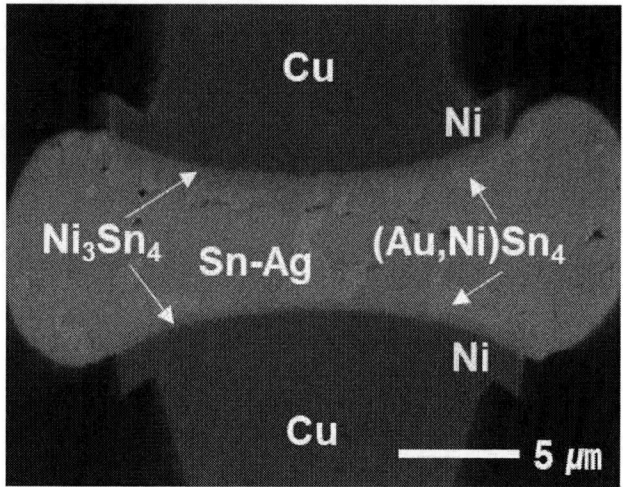

(a)

(b)

Figure 3. Cross-sectional SEM images of NCF and non-NCF microbumps before and after EM failure at 150 °C with a current density of 1.3×10^5 A/cm^2 : (a) before EM test of NCF microbump, (b) after EM test of NCF microbump.

A substantial portion of the Cu pillar on the top chip was transformed into (Cu,Ni)$_6$Sn$_5$ due to rapid diffusion along the concentration gradient between the Cu atoms in the Cu pillar and Sn atoms of the solder. The pure Sn phase at the edge of the bump was also observed. Thus, the EM failure mechanism hardly differed between NCF and non-NCF microbumps. However, obvious effects of prolonged EM failure with NCF were noticed [16].

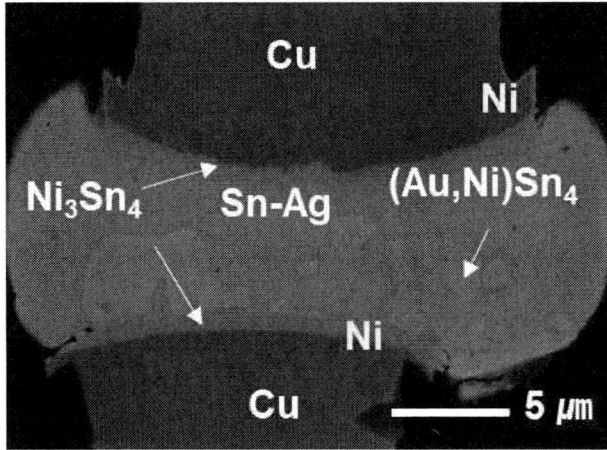

(a)

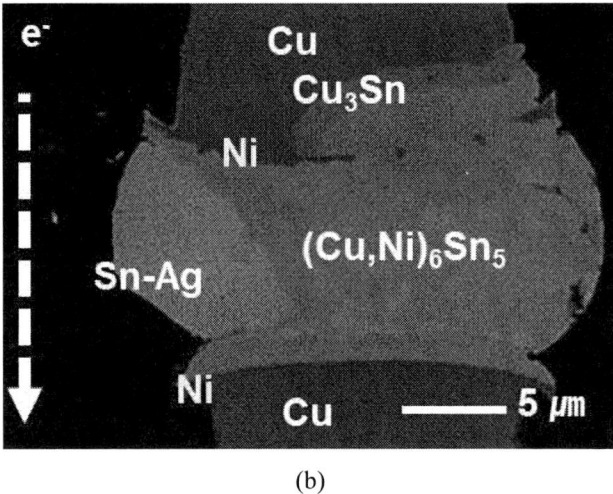

(b)

Figure 4. Cross-sectional SEM images of NCF and non-NCF microbumps before and after EM failure at 150 °C with a current density of 1.3×10^5 A/cm^2 : (a) before EM test of non-NCF microbump, and (b) after EM test of non-NCF microbump.

These results are similar to the underfill effects of the solder bumps reported in previous studies. The back stress was increased by confinement of the underfill, and was relieved by solder deformation. As a result, a longer EM failure was observed [16, 17].

B. UBM materials effect on Electromigration Reliability of Sn-Ag microbumps

Figure 5 shows the cross-sectional SEM images of Ni UBM and Cu UBM microbumps after the EM failure for 600 h with Ni UBM microbump and 900 h with Cu UBM microbump at 150 °C at a current density of 0.5×10^5 A/cm^2.

The Sn inside the whole microbump was totally transformed to $(Au,Ni)Sn_4$ IMCs. The reaction between Ni atoms and Sn atoms to generate Ni_3Sn_4 or $(Ni,Au)_3Sn$ IMC led to a net volume shrinkage of approximately -11.4% for Ni_3Sn_4, resulting in multiple small voids inside the IMC[18, 19]. Void formation at the Al trace was observed with Ni UBM microbump, as highlighted in the dashed line shown in Fig. 5(a). However, void formation at the Al trace was not observed with Cu UBM microbump, as highlighted in the dashed line shown in Fig. 5(b). The failure mode represents the wear-out mechanism involving void formation at the cathode of the Al interconnect above W due to flux divergence, without any void formation at the anode of the Al interconnect [18, 20].

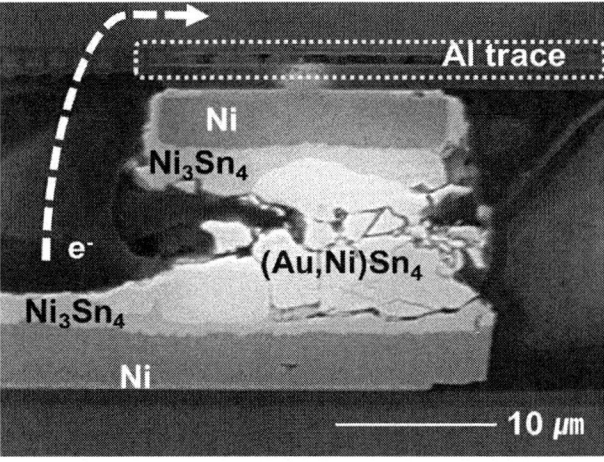

(a)

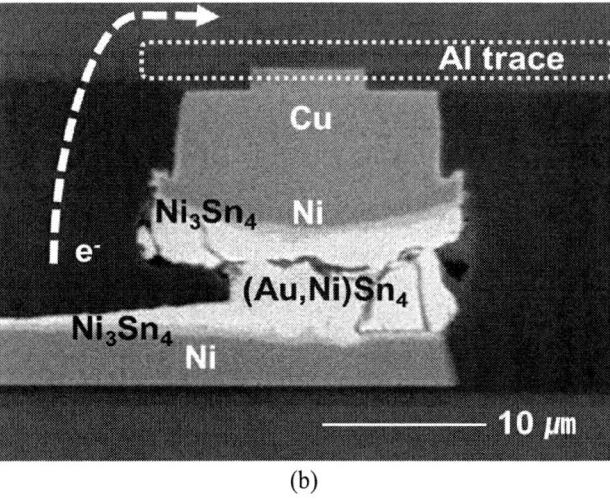

(b)

Figure 5. Cross-sectional SEM images of Sn-Ag microbumps after EM failure at 150 °C with a current density of 0.5×10^5 A/cm^2 : (a) Ni UBM microbump and (b) Cu UBM microbump.

The current density distributions of the microbump with the Ni and Cu UBM under 0.157 amp at 150 °C are presented in Fig. 6. Finite element analysis results for the microbump with the Ni and Cu UBM revealed a maximum current density inside the solder at the point of Al trace entry into the Ni and Cu pillars, without any current crowding effect in the substrate side of the joint.

The temperature distributions of the microbump with the Ni and Cu UBM under 0.157 amp at 150 °C are presented in Fig. 7. Finite element analysis results suggest that the microbump temperature was substantially higher than in serious Joule heating involving the solder. A hot spot exit in the solder near the entrance of the Al trace showed a temperature of 155.4°C in the Ni UBM microbump.

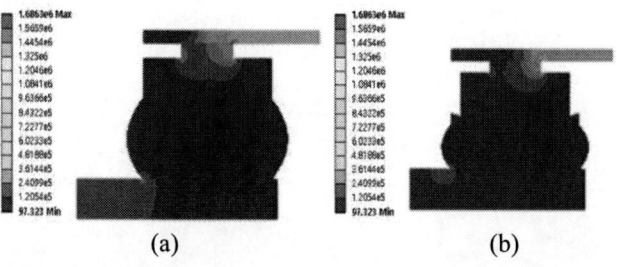

Figure 6. Cross-sectional view of the current density distribution in the microbump with (a) Ni UBM and (b) Cu UBM.

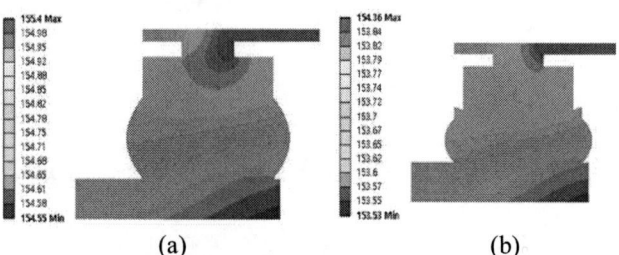

Figure 7. Cross-sectional view of the temperature distribution in the microbump with (a) Ni UBM and (b) Cu UBM

Thus, the Joule heat generation in the current-crowding region apparently accelerated the EM of the Al line on the cathode side. The thermal gradient was also investigated under various electric currents, and the results are displayed in Table 2. The microbump with the Cu UBM showed a low thermal gradient, whereas the microbump with the Ni UBM exhibited a large thermal gradient. The value was about 815.4 °C/cm at 0.301 amp. The thermal gradient might be significant in terms of thermomigration [21, 22].

IV. CONCLUSION

The EM failure mechanisms under the effect of NCF and UBM materials in the Cu/Sn–2.5Ag microbumps were investigated under current stress at 150 °C with current densities ranging from $0.5 \sim 1.3 \times 10^5$ A/ cm^2. Four types of microbumps were fabricated for EM test: microbump encapsulated with NCF; unencapsulated microbump;, Ni UBM microbump; and Cu UBM microbumps.

In the case of NCF microbumps, almost similar EM failure mechanism was observed with Cu/Ni/Sn-Ag microbumps carrying NCF and non-NCF. However, the NCF application prolonged the EM lifetime almost three-fold for the Cu/Ni/Sn-Ag microbumps. It was considered that the increase in back stress due to restricted solder deformation by NCF had a positive effect on prolonging the EM failure time.

In the case of Ni and Cu UBM microbump, the EM failure mechanism differed between Sn-Ag microbump with Ni UBM and Cu UBM. The simulation results showed higher Joule heating with the Ni UBM microbump compared with the Cu UBM microbump. Thus, the increased Joule heating by thicker Ni UBM had a negative effect on EM failure time. Therefore, the Cu UBM microbumps are expected to exhibit better EM resistance.

ACKNOWLEDGMENT

This research was supported by Basic Science Research Program through the National Research Foundation of Korea(NRF) funded by the Ministry of Education(2016R1D1A3B03933937), MOTIE(Ministry of Trade, Industry & Energy (10067804) and KSRC(Korea Semiconductor Research Consortium) support program for the development of the future semiconductor device.

REFERENCES

[1] K. Son, H. Ryu, G. Kim, J. Lee, and Y. B. Park, "Electromigration Polarity Effect of Cu/Ni/Sn-Ag Microbumps for Three-Dimensional Integrated Circuits," Proc. 19th Electronics Packaging Technology Conference, Dec. 2017, doi:10.1109/eptc.2017.8277504.

[2] S. H. Kim, G. T. Park, J. J. Park, and Y. B. Park, "Effects of Annealing, Thermomigration, and Electromigration on the Intermetallic Compounds Growth Kinetics of Cu/Sn-2.5Ag Microbump," J. Nanosci. Nanotechnol., vol. 15, 2015, pp. 8593-8600, doi:10.1166/jnn.2015.11502

[3] H. Y. You, Y. H. Hwang, J. W. Pyun, Y. G. Ryu, and H. S. Kim, "Chip Package Interaction in Micro Bump and TSV Structure," Proc. 62nd Electronic Components and Technology Conference, May. 2012, doi:10.1109/ectc.2012.6248848.

TABLE II. THERMAL GRADIENT IN THE MICROBUMP AS A FUNCTION OF APPLIED CURRENT FOR NI AND CU UBMS

Applied Current[A]	Ni UBM	Cu/Ni UBM
	Temperature gradient (°C/cm)	Temperature gradient (°C/cm)
0.126	140.04	119.16
0.157	219.09	186.40
0.188	316.00	268.81
0.220	430.93	366.51
0.251	564.09	479.64
0.283	715.70	608.42
0.301	815.43	693.08

[4] J. W. Kim, S. J. Jeon, H. J. Lee, S. M. Hyun, and Y. B. Park, "Improvement of Wafer-Level Cu-to-Cu Bonding Quality Using Wet Chemical Pretreatment," J. Nanosci. Nanotechnol., vol. 12, 2012, pp. 3577-3581, doi:10.1166/jnn.2012.5619.

[5] Y. M. Lin, C. J. Zhan, J. Y. Juang, J. H. Lau, T. H. Chen, R. Lo, M. Kao, T. Tian, and K. N. Tu, "Electromigration in Ni/Sn Intermetallic Micro Bump Joint for 3D IC Chip Stacking," Proc. 61st Electronic Components and Technology Conference, May. 2011, doi:10.1109/ectc.2011.5898537.

[6] B. H. Kwak, M. H. Jeong, and Y. B. Park, "Effects of Temperature and Current Stressing on the Intermetallic Compounds Growth Characteristics of Cu Pillar/Sn–3.5Ag Microbump," Jpn. J. Appl. Phys, vol. 51, 2012, pp. 05EE05, doi:10.1143/jjap.51. 05EE05.

[7] K. Murayama, M. Higashi, T. Sakai, and N. Imaizumi, "Electro-migration Behavior in Low Temperature Flip Chip Bonding," Proc. 62nd Electronic Components and Technology Conference, May. 2012, doi:10.1109/ectc.2012.6248893.

[8] B. H. Kwak, M. H. Jeong, and Y. B. Park, "Effect of Intermetallic Compounds Growth Characteristics on the Shear Strength of Cu pillar/Sn-3.5Ag Microbump for a 3-D Stacked IC Package, (in korean)" Korean J. Met. Mater., vol. 50, 2012, pp. 775-783, doi:10.3365/KJMM.2012.50.10.775

[9] H. He, L. Cao, L. Wan, H. Zhao, G. Xu, and F. Guo, "Joule Heating Effect on Oxide Whisker Growth Induced by Current Stressing in Cu/Sn-58Bi/Cu Solder Joint," Electron. Mater. Lett., vol. 4, 2012, pp. 463-466, doi:10.1007/S13391-012-2019-9

[10] S. Kumar, J. Y. Park, and J. P. Jung, "Analysis of high speed shear characteristics of Sn-Ag-Cu solder joints," Electron. Mater. Lett., vol. 7, 2011, pp. 365-373, doi:10.1007/S13391-011-0160-5

[11] W. Tang, Y. Hu, and S. Huang, "Fabrication of Sn-Cu Alloy Solder by Pulse Electroplating on the Metalized Si Wafer," Met. Mater. Int., vol. 18, 2012, pp. 177-183, doi:10.1007/S12540-012-0022-1

[12] K.J. Puttlitz and K.A. Stalter, Handbook of lead-free solder technology for microelectronic assemblies. Marcel Dekker, New York, 2004.

[13] G. T Park, B. R. Lee, K. Son, and Y. B. Park, "Ni Barrier Symmetry Effect on Electromigration Failure Mechanism of Cu/Sn–Ag Microbump," Electron. Mater. Lett., doi:10.1007/S13391-018-00108-5, in press.

[14] C. Y. Hsu, D. J. Yao, S. W. Liang, C. Chen, and C. C. Yeh, "Temperature and Current-Density Distributions in Flip-Chip Solder Joints with Cu Traces," Journal of ELECTRONIC MATERIALS, vol. 35, 2006, pp. 947-953, doi:10.1007/bf02692552

[15] J. W. Yoon, H. S. Chun, J. M. Koo, H. J. Lee, S. B. Jung, "Microstructural evolution of Sn-rich Au–Sn/Ni flip-chip solder joints under high temperature storage testing conditions," Scr. Mater., vol. 56, 2006, pp. 661-664, doi:10.1016/j.scriptamat.2006.12.031

[16] H. Ryu, K. Son, J. S.Han, and Y. B. Park, "NCF Effect on Electrical Reliability of Cu/Ni/Sn-Ag Microbumps," unpublished.

[17] K. Yamanaka, T. Ooyoshi, and T. Nejime, "Effect of underfill on electromigration lifetime in flip chip joints," Journal of Alloys and Compounds, vol. 481, 2009, pp. 659-663, doi:10.1016/j.jallcom.2009.03.063

[18] Fan-Yi Ouyang, Hao Hsu, Yu-Ping Su, and Tao-Chih Chang, "Electromigration induced failure on lead-free micro bumps in three-dimensional integrated circuits packaging," Scr. Mater., vol. 56, 2006, pp. 661-664, doi:10.1016/j.scriptamat.2006.12.031

[19] Y. Liu , Y. T. Chen, S. Gu, D. W. Kim, and K.N. Tu, "Fracture reliability concern of (Au, Ni)Sn$_4$ phase in 3D integrated circuit microbumps using Ni/Au surface finishing", Scripta Mater. vol. 119, 2016, pp. 9-12, doi:10.1016/j.scriptamat.2016.02.025

[20] K. N. Tu, "Recent advances on electromigration in very-large-scale-integration of interconnects", Journal of Applied Physics, vol. 94, 2003, pp. 5451–5473, doi:10.1063/1.1611263

[21] S. W. Liang, Y. W. Chang, and C. Chen, "Three-Dimensional Thermoelectrical Simulation in Flip-Chip Solder Joints with Thick Underbump Metallizations during Accelerated Electromigration Testing," Journal of ELECTRONIC MATERIALS, vol. 36, 2007, pp. 159-167, doi:10.1007/s11664-006-0060-x

[22] 15. Hua Ye, Cemal Basaran, and Douglas Hopkins, Appl. Phys.Lett. 82, 7 (2003).

Ag Diffusion Control Through Sn on a Sequential Plating-Based Bumping Process

A. El Amrani[1,2,*], E. Paradis[1,2], D. Danovitch[1,2], D. Drouin[1,2]

[1]Nanotechnologies & Nanosystems Laboratory (LN2), CNRS UMI-3463, Université de Sherbrooke, J1K 0A5, QC, Canada
[2]Institut Interdisciplinaire d'Innovation Technologique (3IT), Université de Sherbrooke, J1K 0A5, QC, Canada
*email: abderrahim.el.amrani@usherbrooke.ca

Abstract—In this paper, we present our investigations on sequential plating for a low-cost bumping process. Following the demonstration of this process for low Ag concentration (1.4 wt%), we aimed this study on increase Ag concentration to 3 wt% and investigate its impact on Ag diffusion and intermetallics compound formation. Moreover, we created a Ag concentration gradient within the same bump in order to benefit from both favorable effects of low and high Ag concentration on the solder. To this end, we plated different metal sequences to form bumps with 90 µm and 75 µm of diameter and height respectively. The fabricated bumps were characterized, after a cross-sectional preparation, through SEM and EDS analysis to reveal the effects of the reflow profile and the use of a barrier on the Ag diffusion, as well as, the IMC formation.

Keywords-Sequential plating; Ag Concentration gradient; Ni-Barrier.

I. INTRODUCTION

High-density bumping has become one of the most prevalent packaging focus areas over the past decade. Developing SnPb alternatives using lead-free alloys while maintaining comparable bump performance and cost is a challenge to which the electronic packaging community continues to devote considerable effort [1]. The adoption of SnAg and SnAgCu (SAC) alloys for the bumping industry has produced a broader and deeper understanding of the properties and reliability of this alloy group [2]. For example, it has been reported that Ag concentration can represent an important dichotomy. As high Ag concentrations can be beneficial to the ultimate solder joint with respect to electromigration and corrosion resistance [3]. On the other hand, lower Ag concentration (eg 1.3 wt%) are considered in fine pitch bumping applications not only for cost-effectiveness considerations [4] but also to increase the alloy ductility [5] which in turn promotes high fatigue resistance [6]. The optimal bump performance must, therefore, be achieved through a delicate balance, and often times results in compromise. In this paper, we propose that such a compromise may be largely averted by using a sequential plating approach [7] that, when adapted, limits subsequent Ag diffusion and thus

produces a heterogeneous structure that includes temporary regions of more ductile, low Ag solder in the vinicity of the brittle BEOL device materials. The objective of using this approach is to reduce the high stress induced during the chip joint and thus reduce the BEOL materials delamination encountered when using high Ag content SAC alloys. Figure 1 shows the outcome that we aiming to achieve using this approach. Once the solder joints are fully reinforced with underfill, they can be subsequently homogenized, during BGA attach, for example, to impart optimal electromigration resistance.

To accomplish such a structure, we first develop a 3 wt% Ag bumping process using the sequential plating approach. Afterwards, we investigate the viability of various thermal and metallurgical means to exploit sequential plating in a manner that can affect a Ag concentration gradient in flip chip bumps up to and including the critical joint formation stage (chip join) where CTE mismatch can impart maximum stress to the BEOL regions. The first step in this study is establishing a baseline sequential process to fabricate a SnAg bump alloy comprising 3 wt% Ag. Cooling rates of a standard reflow process are then varied to determine their impact on both diffusion activity [8] and microstructure, as previous studies have reported reductions in individual Ag_3Sn IMC formations [9]. The second approach to generate a Ag concentration gradient within the same bump comprises the investigation of an added Ni layer within the plated structure in order to act as a diffusion barrier. Barrier thickness, position within the bump and overall bump shape are explored with respect to their contribution to restricting Ag diffusion and creating a low Ag region near the BEOL structure.

II. EXPERIMENTAL

A. Fabrication methods

1) Process setup: The plating electrolytes and their corresponding current densities are listed in Table I. Cu and Sn electrolytes were warmed up to 35 °C and 45 °C respectively to improve the deposition rate and reduce its roughness.

978-1-7281-1500-9/19 $31.00 © 2019 IEEE

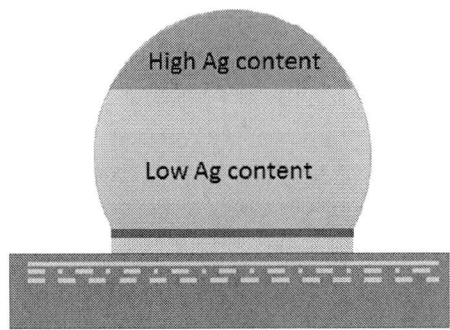

Figure 1: Heterogeneous bump structure

Table I: Plating electrolytes and current densities.

Plating electrolytes		Current density (mA/cm^2)
Copper	Elevate Cu D6370	20
Nickel	Elevate Ni 5910	20
Tin	TECHNI NF JM 6000 WBP	20
Silver	TechniSol Ag	3

All of the electroplating processes were conducted in a lab-scale plating bench adapted for 100 mm wafers. After sputtering the seed layer, a 100 μm thick dry-film photoresist (WBR2100) was deposited using a hot-roll laminator at 100 °C, UV exposed at 150 mJ/cm^2 using a contact mask aligner, and developed by immersion in AZ400 developer 20 % diluted in water. A 1 min pre-exposure, post-exposure and post-development thermal treatments at 100 °C on a hot plate were done in order to strengthen the patterned electroplating mask. The reflow processes were achieved using two differents methods; on one hand the slow-cool reflow was done in an 8 zones industrial furnace. On the other hand, when targeting 15 °C/min cooling rate for an ultra-fast-cool, the reflow was done using a temperature controlled hotplate and a substrate at room temperature. A cleaning step using acetone and water was necessary to remove the flux residue after the reflow.

2) Process flow: Bump fabrication (90 μm diameter) was carried out on a 100 mm wafer using the following electroplating steps: a) TiW-Cu based seed-layer sputtering, b) Dry-film lamination, c) Photolithography patterning, d) Cu-Ni-Cu or Cu-Ni UBM plating, e-i) Sequential plating of solder constituents, j) Dry-film stripping, k) Seed-layer etching, and l) Bumps reflow. For the sequential plating, a number of metallurgical combinations were investigated. For the bump reflow, different parameters were studied (slow vs. fast cool and the variation the dwell (1 min and 4 min)). We used different layers for both UBM and solder constituents. Figure 2 shows a detailed process flow for the

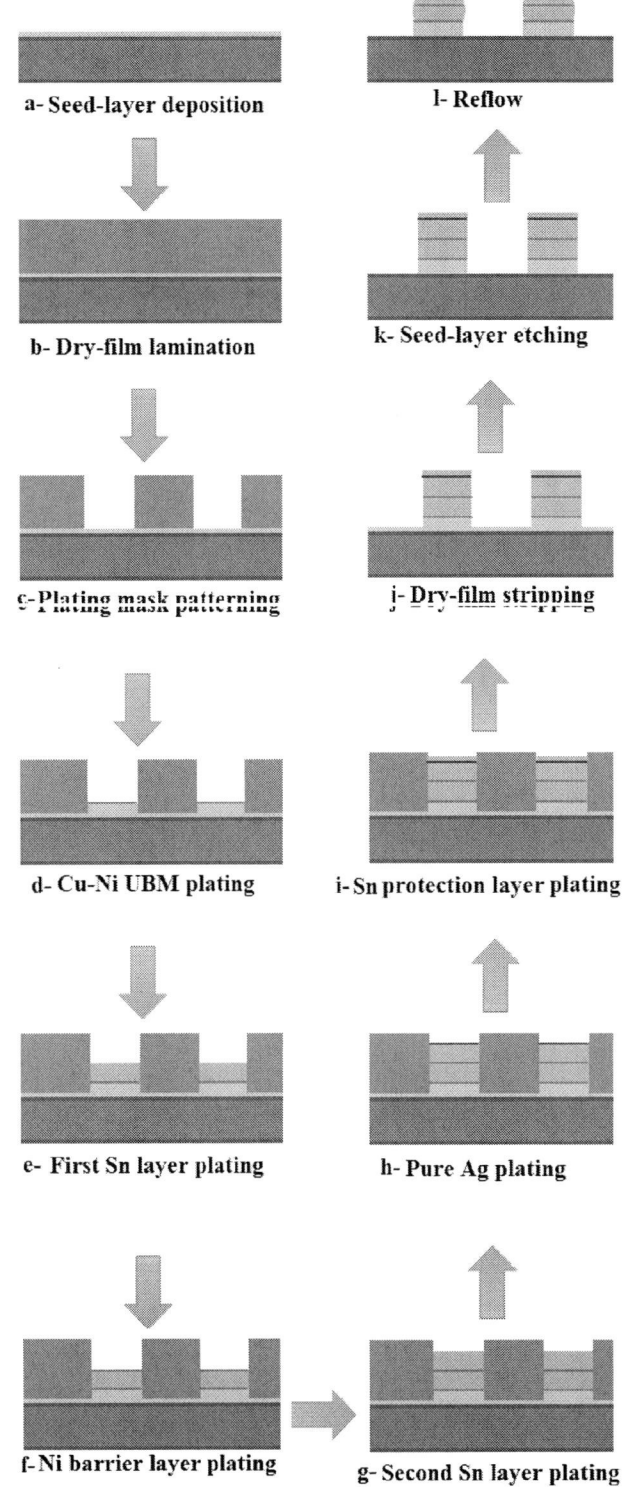

a- Seed-layer deposition

b- Dry-film lamination

c- Plating mask patterning

d- Cu-Ni UBM plating

e- First Sn layer plating

f- Ni barrier layer plating

g- Second Sn layer plating

h- Pure Ag plating

i- Sn protection layer plating

j- Dry-film stripping

k- Seed-layer etching

l- Reflow

Figure 2: Sequential bumping process flow

bump configuration shown in Figure 3-c. Figure 3 shows the main studied sequences. The sequence (a) replicates

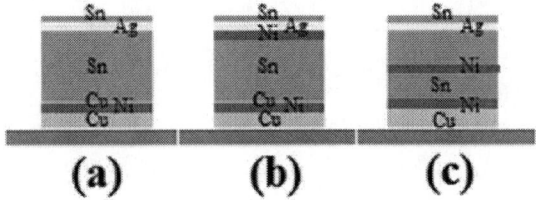

Figure 3: Sequential plating configurations

the same process used in [6] with an increase of Ag concentration from $1.4 \, \text{wt}\%$ to $3 \, \text{wt}\%$. The sequences (b) and (c) included a Ni barrier that was implemented to create a Ag concentration gradient.

B. Characterization methods

1) Samples preparation: For the IMC and the Ag diffusion observation after the reflow, a conventional mechanical cross section was used as an easy access to a lateral view of the bumps, allowing a time-effective characterization. The sample were molded in resin, polished using Silicon Carbide papers and finished using a diamond paste (9 μm to 1 μm). Finally, a thin Au-Pd metal layer is sputtered to avoid saturation during SEM-EDS analysis

2) SEM-EDS mapping: To reveal the IMC formation as well as the Ag diffusion at the same time we used an X-ray energy dispersive spectroscopy (EDS) analysis for an elemental composition mapping throughout the bump. Using this technique, we were able to link specific elements concentrations with a well-known IMC composition, and thus assess the effect of the reflow on the IMC. Moreover, a vertical line-scan with a local composition analysis was used to assess the barrier efficiency.

III. RESULTS

A. Effect of Ag concentration increasing on the bump metallurgy

As reported in [8], in a similar bumping process with $1.4 \, \text{wt}\%$ Ag concentration, the diffusion was found to be fast and uniform for all the studied sequences when using a slow cool reflow process. For this study, we increased the Ag concentration by increasing the associated layer thickness. Figure 4 shows an EDS mapping of the plated elements after a slow cool reflow when aiming a $3 \, \text{wt}\%$ Ag alloy.

Compared to the results reported in [8] for a sequential $1.4 \, \text{wt}\%$ Ag bumping process, the increase of Ag concentration not only altered the diffusion mechanism but also triggered the formation of large Ag_3Sn IMCs, which were nucleated from UBM Cu layer. Instead of getting a uniform

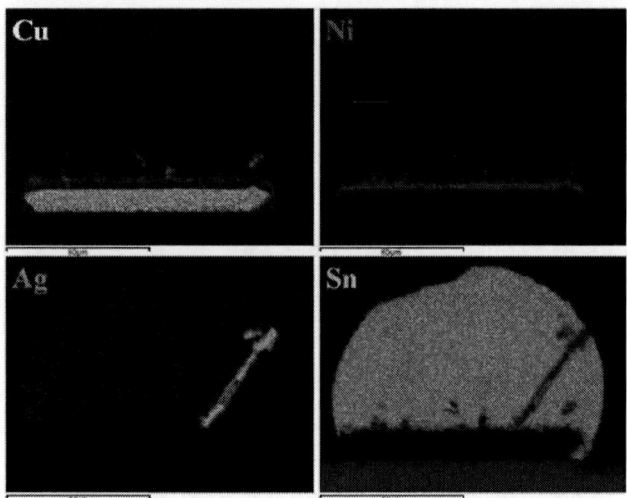

Figure 4: Elements EDS mapping after a slow cool reflow (4 min Dwell and 0.5 °C/min cooling rate)

$3 \, \text{wt}\%$ Ag throughout the bump, the mapping showed that most of Ag was concentrated in some local areas where Ag_3Sn was created (75% Ag) which consequently, exhibited a lower Ag concentration in the solder. This phenomenon was already reported for SnAg alloys plating with $4 \, \text{wt}\%$ Ag concentration when using a slow cool reflow [9]. Using a sequential plating for a bumping process, increases the chances of creating large Ag_3Sn IMCs since Ag is plated in its pure form, that start diffusing in solder during the dwell, which creates a highly concentrated Ag region within the solder in a liquid phase. Because of the temperature gradient, solder is moving and thus the Ag rich region could get near the UBM Cu, which increases the probability of large Ag_3Sn plate-like IMCs formation.

B. Effect of the thermal profile on the bump metallurgy

It was reported that reducing the reflow cooling time prevents the formation of large size Ag_3Sn IMCs [9]. To benefit from this fact and in order to avoid the formation of the Ag_3Sn observed in Figure 4, an ultra-fast cool reflow was achieved. Figure 5 shows an EDS mapping of a cross-sectioned bump.

The EDS mapping showed a uniform Ag distribution and no large Ag_3Sn IMC. As suspected, the ultra-fast cooling prevented the large IMC plates formation and offered a uniform Ag distribution. After the solder is melted and Ag is diffused, Sn grains are quickly formed during the cooling and prevent the creation of Ag rich zone, which prevents the formation of large plate-like Ag_3Sn. The dwell time did not have any effect on the bump metallurgy. Both tested dwell durations, 1 and 4 minutes, offered a fast Ag diffusion with no large IMCs formation. These results implied that

978-1-7281-1500-9/19 $31.00 © 2019 IEEE

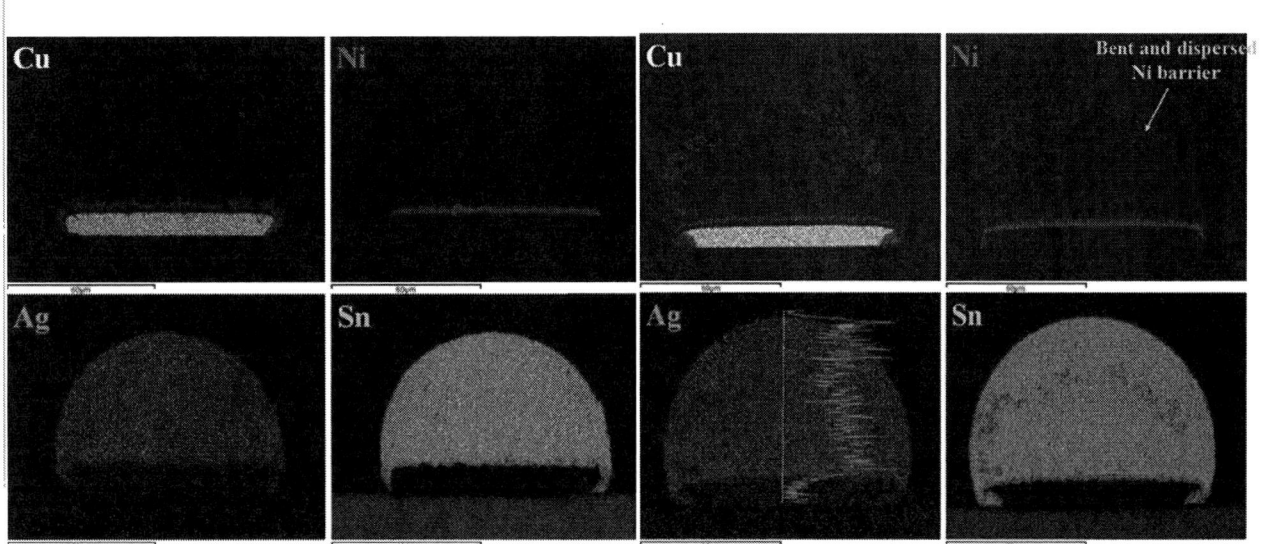

Figure 5: Elements EDS mapping after a fast cool reflow (1 min Dwell and 15 ^{o}C/min cooling rate)

Figure 6: Elements EDS mapping after a slow cool reflow (4 min Dwell and 0.5 ^{o}C/min cooling rate)

the dwell duration could not be used as an effective mean to control Ag diffusion; thereby this outcome needs to be implemented using another metallurgical mean (barrier).

C. Implementing a Ni barrier in the sequential bumping process

In this study, we added a Ni barrier within the sequential plating process (Figure 3-(b-c)).

1) 0.5 μm thick Ni barrier: We started with a test using a 0.5 μm thick Ni barrier located near the top surface of the bump (Figure 3-(b)). Figure 6 shows the elements EDS mapping of the resultant bump cross-section. The elements mapping showed that the Ni barrier has moved from the surface to the mid-height of the bump, it has been dispersed and bent due to the surface tension. Ni has also created $(Ni,Cu)_3Sn_4$ IMCs with the Cu that had diffused from the UBM. However, the barrier was not efficient due to the non-continuity of the Ni layer that invalidated its ability to slow down Ag diffusion. This statement is based on the uniform Ag concentration through the bump, which was confirmed by the line-scan that showed no significant Ag concentration variation between both sides of the Ni barrier.

2) 1 μm thick Ni barrier: In order to improve the barrier efficiency, we increased its thickness to 1 μm and instead of placing it right under the Ag layer, we displaced it to the middle of the bump in order to reduce the preceding plated Sn roughness, which has a direct impact on the barrier thickness. Additionally, this choice had moved the barrier away from the bump surface which reduces the effects of the surface tension. The UBM Cu layer was removed to

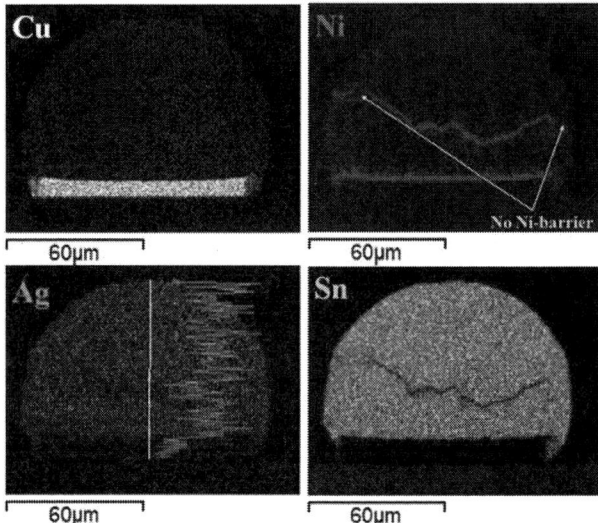

Figure 7: Elements EDS mapping after a slow cool reflow (4 min Dwell and 0.5 ^{o}C/min cooling rate)

prevent its diffusion and the creation of $(Ni,Cu)_3Sn_4$ IMCs and also to avoid the nucleation of Ag_3Sn plates when using a slow cool reflow. Figure 3-(c) shows the plating sequence used for this study. The key element that defines whether the test is successful or not is the Ag concentration difference between solder on each side of the barrier. Figure 7 shows the elemental mapping of a cross-sectioned bump.

There was no large Ag_3Sn plates formation, which con-

978-1-7281-1500-9/19 $31.00 © 2019 IEEE

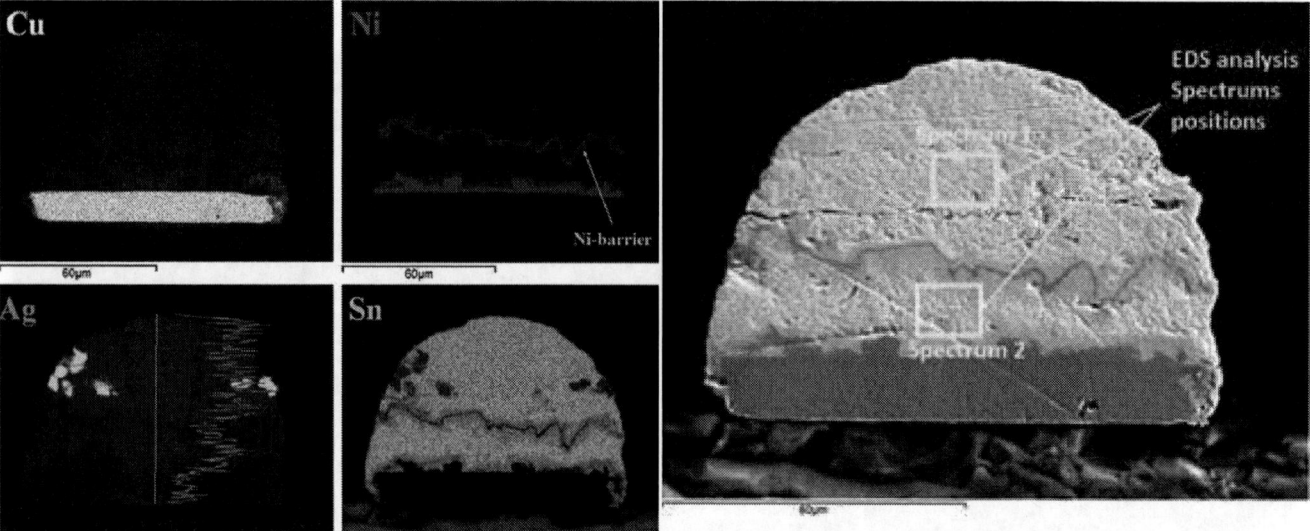

Figure 9: Positions of the EDS local analysis spectrums

Figure 8: Elements EDS mapping after a slow cool reflow
(4 min Dwell and 0.5 °C/min cooling rate)

Table II: Local elements concentrations (in wt%)

Spectrum	Ni	Cu	Ag	Sn
Spectrum 1	0.00	0.00	3.03	96.97
Spectrum 2	1.94	0.80	1.18	96.08

firmed that the presence of a Cu layer enables its nucleation when using a slow cool reflow profile. Moreover, Ag distribution within the bump was uniform and did not show any effect of the used Ni barrier. This was most likely caused by the lack of the Ni-barrier all around the bump, which let a Ag diffusion leakage when the solder is in its liquid state. These openings were created when the bump shape changed from the as-plated cylindrical shape to the reflowed bump shape.

Using the same plating sequence at a lower solder volume, we aimed to get comparable diameters of the bump and the barrier and thus tried to minimize the opening zones that caused the diffusion leakage, which is a factor that can help to get a concentration gradient between the two regions. Figure 8 shows the elemental mapping of a cross-sectioned bump at 60 μm of height (after reflow). The Ag mapping showed the formation of large Ag_3Sn grains on the upper side of the barrier. Moreover, the line-scan showed a slight decrease of Ag concentration. To confirm this observation, we performed a local EDS analysis in order to get the percentage of each element on both sides of the barrier. Figure 9 shows the positions of the EDS analysis for which Table II gives the percentage of each element.

The Ag concentration shown in Table II emphasizes the significant concentration difference between the two spectrums that represent solder in each barrier side. The

sub-barrier side had a 1.18 wt% Ag concentration, should exibith an increase ductility as targeted for the UBM side. On the other side, Ag concentration is 3.03 wt% that is lower than expected since the total plated Ag layer was meant to create an overall concentration of 3 wt%. However, when taking into consideration the average value of all the solder including the IMC grains (75% Ag) this value would increase.

These findings raise the importance of many associated factors that define the barrier efficiency. The first two variables that are linked are the barrier position and the total solder volume. It is crucial to place the barrier according to the solder volume in order to get an efficient Ag diffusion barrier all the way to the edge of the bump, and thus get a concentration gradient. The other variable that is crucial for efficiency is the barrier thickness. This variable is affected by the Sn roughness prior to barrier plating, a high roughness means an increased effective plating surface that may impact the barrier plating thickness. This is an important factor to take into consideration when placing the barrier since the roughness increases with the plated bump height, but also depends on the current density as well as the electrolytes condition. Preliminary work have shown the potential of such technique to control the Ag concentration within a bump. More fine tunings are planned in our future work in order to get a comparable result for 75 μm bumps height. Moreover, further characterizations will be done in order to assess the mechanical characteristics of the resultant metallurgy with respect to both Ag concentration gradient and also the Ni diffused into solder.

IV. CONCLUSION

The effects of using a high Ag concentration sequential plating process for SnAg bumping were studied and compared with a low Ag concentration process. We validated the effects of the cooling rate on the IMCs formation for a sequential plating based-bumping process. We, successfully, induced a Ag concentration gradient within the same bump. The characterization of the resultant bumps, using a slow cool reflow, showed the creation of large Ag_3Sn plates, which do not appear in the case of a fast cool reflow. A Ag concentration gradient was obtained when using the appropriate combination of solder volume, barrier thickness and position. In our future work, we will improve this process for larger solder volume. Moreover, for a better understanding of the mechanical effects, additional characterizations will be achieved.

ACKNOWLEDGMENT

The Authors would like to thank M.Turgeon and C.Fortin for their guidance, expertise, and support in the metallurgical characterization. We are also grateful to S.Gutierrez (CCM-Faculté de génie, Université de Sherbrooke) for his support in EDS analysis.

This work had been made possible by the NSERC/PROMPT/IBM Industrial Research Chair in Smarter Microelectronics Packaging for Performance Scaling.

REFERENCES

[1] C. Melton, "Alternatives of lead bearing solder alloys," in *Electronics and the Environment, 1993., Proceedings of the 1993 IEEE International Symposium on.* IEEE, 1993, pp. 94–97.

[2] J. Woertink, Y. Qin, J. Prange, P. Lopez-Montesinos, I. Lee, Y.-H. Lee, M. Imanari, J. Dong, and J. Calvert, "From c4 to micro-bump: Adapting lead free solder electroplating processes to next-gen advanced packaging applications," in *Electronic Components and Technology Conference (ECTC), 2014 IEEE 64th.* IEEE, 2014, pp. 342–347.

[3] M. Huang, F. Yang, N. Zhao, and Y. Yang, "Synchrotron radiation real-time in situ study on dissolution and precipitation of ag3sn plates in sub-50 μm sn–ag–cu solder bumps," *Journal of Alloys and Compounds*, vol. 602, pp. 281–284, 2014.

[4] R. Pandher and T. Lawlor, "Effect of silver in common lead-free alloys," in *Proceedings of the International Conference on Soldering and Reliabilty*, 2009, pp. 1–14.

[5] F. Che, E. C. Poh, W. Zhu, and B. Xiong, "Ag content effect on mechanical properties of sn-xag-0.5 cu solders," in *Electronics Packaging Technology Conference, 2007. EPTC 2007. 9th.* IEEE, 2007, pp. 713–718.

[6] Y. Kariya, T. Hosoi, S. Terashima, M. Tanaka, and M. Otsuka, "Effect of silver content on the shear fatigue properties of sn-ag-cu flip-chip interconnects," *Journal of electronic materials*, vol. 33, no. 4, pp. 321–328, 2004.

[7] Q. Zhao, Z. Chen, A. Hu, and M. Li, "Formation of snag solder bump by multilayer electroplating," *Microelectronic Engineering*, vol. 106, pp. 33–37, 2013.

[8] A. El Amrani, E. Paradis, D. Danovitch, and D. Drouin, "Investigation of a low-cost sequential plating based process for pb-free bumping," in *2018 7th Electronic System-Integration Technology Conference (ESTC).* IEEE, 2018, pp. 1–5.

[9] C. P. Lin, K. Wan, T. Sun, C.-M. Chen, and R. Lee, "Effects of reflow cooling rate on the growth of ag3sn platelets and deformation of solder balls," in *Electronic Components and Technology Conference (ECTC), 2016 IEEE 66th.* IEEE, 2016, pp. 2075–2081.

Mechanical Reliability Assessment of Cu_6Sn_5 Intermetallic Compound and Multilayer Structures in Cu/Sn Interconnects for 3D IC Applications

J. Y. Wu, C. Robert Kao
Department of Materials Science and Engineering
National Taiwan University
Taipei, Taiwan
e-mail: f03527016@ntu.edu.tw

Jenn-Ming Yang
Department of Materials Science and Engineering
University of California, Los Angeles
Los Angeles, CA, USA
e-mail: jyang@seas.ucla.edu

Abstract—In recent decades, three-dimensional integrated circuit (3D IC) design has been developed for the purpose of moving into a new era of More-than-Moore by vertical stacking and heterogeneous integration of functional chips. However, critical issues arise in the paradigm shift from conventional 2D packaging to hierarchical 3D architecture, and one of them is that solders in micron-sized joints may get totally transformed into intermetallic compounds (IMCs) within a short period of time. With the increased volume fraction of IMCs, the mechanical behaviors was not determined by solders as in conventional ball-grid-array joints anymore, and instead IMCs are anticipated to bear great responsibility of mechanical reliability in 3D IC joints. In light of this, this study put emphasis on mechanical reliability assessment of IMC-occupied micro-joints by means of uniaxial micropillar compression.

At first, mechanical properties of single-crystalline micropillars of Cu_6Sn_5 was characterized to investigate the effect of grain orientation. Besides, Ni element was found to significantly increase Young's modulus of Cu_6Sn_5. Furthermore, micro-compression on multilayer structures, such as $Cu_6Sn_5/Sn/Cu_6Sn_5$ and $Cu_6Sn_5/Cu_3Sn/Cu$, was also conducted to gain insight into overall deformation behaviors of Cu-Sn interconnects. $Cu_6Sn_5/Sn/Cu_6Sn_5$ cylinders exhibited remarkable plasticity in the manner of interface sliding at around 100 MPa, accommodating at least 10% strain and retaining its integrity without any void formation at the interface. On the other hand, multilayer micropillars of $Cu_6Sn_5/Cu_3Sn/Cu$ underwent plastic deformation through slip of Cu substrate, suggesting sufficient strength of intermetallic compounds and the interfaces. These results are beneficial in evaluation of validity and reliability of 3D IC micro-joints.

Keywords: 3D IC Micro Joints; Cu_6Sn_5; Micro-Compression; Multilayered Micropillars; Mechanical Properties

I. INTRODUCTION

As the downsizing of transistors in microchip manufacturing is moving closer to the limits of capabilities in physical dimension, the breakdown of Moore's law has become a matter of serious concern in the foreseeable future [1-2]. To address the scaling limit and to extend the era of Moore's law, three dimensional integrated circuits (3D IC) in electronic packaging has established itself as one of the plausible solutions for continually achieving improved performance of electronic devices [3-6]. By exploring the third dimension in packaging design, 3D IC integration technology vertically stacks multiple heterogeneous chips into a single module by virtue of tunneling through-silicon-vias

(TSVs) and chip-connecting microbumps, offering great benefits of superior performance, advanced multi-functionality, reduced power consumption and a smaller size [4-7].

Nevertheless, various reliability challenges are also present in the practical applications of integrated 3D chip-stacking scheme where the inter-chip spacing and the size of interconnect are considerably diminished [4-5]. On one hand, as silicon dies of reduced thickness are closely-packed, many thorny issues arise from the immense gradient of forces across a limited volume 3D IC architecture, e.g. chip warpage under great stress gradient, thermomigration under large thermal gradient, and electromigration failure due to severe current crowding [4]. On the other hand, the downwards scaling of solder joints from chip-to-module flip chip joints of around 100 μm in diameter to chip-to-chip microbumps of only 10 μm or less in diameter also has substantial impact on interconnects. From the perspective of microstructure, seeing that the volume of microbumps is reduced by three orders of magnitude while the joining conditions of microbump assembly remain the same as those used for conventional flip chip joints, it takes only a short period of time for tin-based solders in microbumps to be consumed during interfacial reaction with UBM (under bump metallization) and get transformed into intermetallic compounds. Due to the excessive IMC formation, the volume ratio of IMCs in microjoints may be as high as almost 100 percent, indicating that IMCs, rather than soft Sn-based solders, are anticipated to take control of mechanical behaviors in 3D IC microbumps [8-10]. On top of that, the shrinkage of joint size down to micron level in diameter also means that merely one or few grains would exist in each microbump, leading to varying mechanical properties of different joints because of the intrinsic anisotropy in crystal structure of common IMCs, such as Cu_6Sn_5 ($P6_3/mmc$), Cu_3Sn (C2/c), Ni_3Sn_4 (C2/m). Unfortunately, we have limited knowledge regarding the mechanical performance of common IMCs in solder joints, and mechanical properties of intermetallic compounds must be better understood. In light of this, this work put emphasis on mechanical reliability assessment of Cu_6Sn_5 intermetallic compounds as well as IMC-occupied joint structures by means of uniaxial micropillar compression.

Since debuted by Uchic et al. in 2004, micro-compression testing has successfully kick-started the small-scale characterization of mechanical behaviors of materials [11-12]. Different from conventional indentation or compression under

978-1-7281-1500-9/19 $31.00 © 2019 IEEE

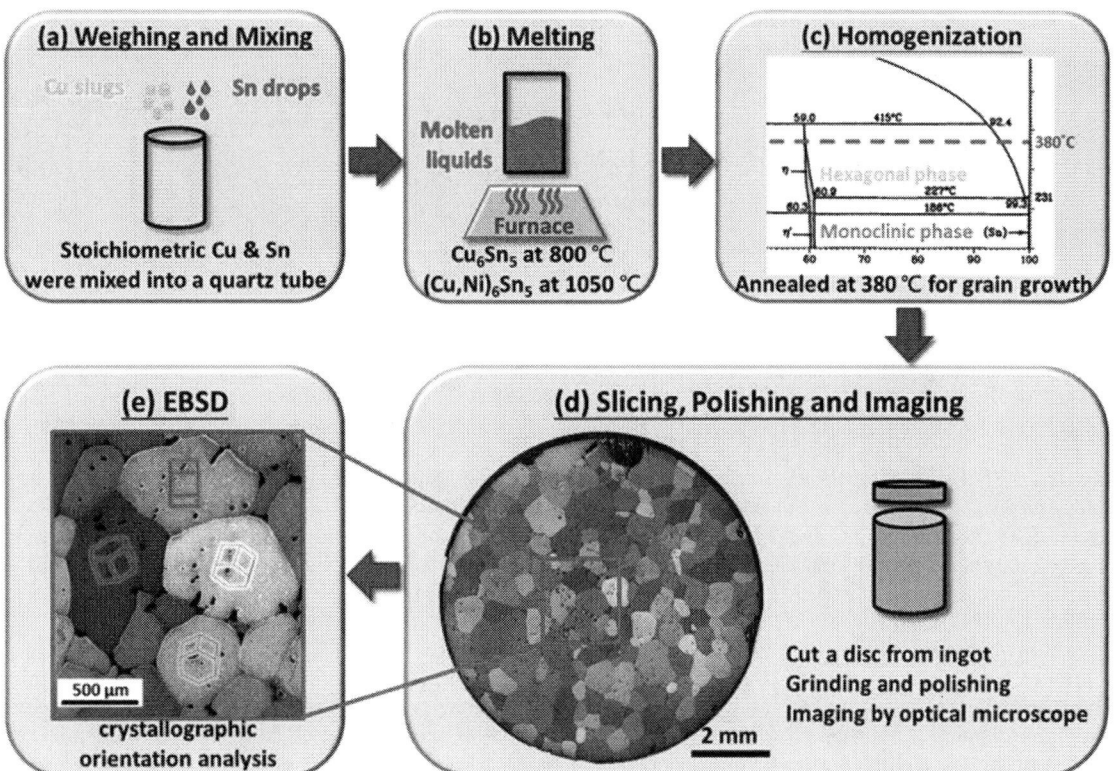

Figure 1. The flowchart of experimental procedures, including casting of intermetallic compounds, thermal annealing, diamond saw cutting, polishing and EBSD crystallographic orientation characterization.

triaxial stresses, this uniaxial compression testing provides a chance to gain valuable insights into fundamental materials science of deformation, plasticity and fractures in a wide range of materials or even multi-layer structures [13-14]. Micro-compression setup paired with focus ion beam (FIB) milling has the capability to probe mechanical properties of pillars with diameters varying from hundreds of nanometers to tens of micrometers, making itself as a perfect tool for studying size effects. Also, the real-time stress-strain responses of testing materials can be directly measured by flat indenter based on controllable and relatively homogeneous stress states across the bounding volume of cylinders, thereby correlating individual slip events in flow curves with microstructural features in in-situ micrographs. More importantly, transmission electron microscope (TEM) examination of deformed pillars can offer opportunities to further elucidate involving deformation mechanisms by visualizing atomic structures change and dislocation activities.

In our previous works, we have characterized mechanical properties of single-crystalline Cu_6Sn_5 and Ni_3Sn_4 IMCs using uniaxial micro-compression, and the preliminary results found that Ni_3Sn_4 showed better fracture stress, Young's modulus and even greater plasticity in particular crystallographic plane than Cu_6Sn_5 [15-16]. However, the Young's moduli measured by micro-compression have much smaller values than those by nanoindentation, which means some other factors during compression should be further calibrated. In addition, many studies in literature reported that the presence of nickel element can have strong influence on the mechanical strength, creep behaviors, crack directions, and phase transformation of Cu_6Sn_5. Moreover, the single phase intermetallic materials may not completely determine the mechanical behaviors of microbumps, given that the microstructure of Cu/Sn/Cu solder joints would gradually evolve into $Cu/Cu_6Sn_5/Sn/Cu_6Sn_5/Cu$ structures and then $Cu/Cu_3Sn/Cu_6Sn_5/Cu_3Sn/Cu$ structures [17]. Therefore, this study conducts micro-compression testing on Cu_6Sn_5 with special focus on calibration technique, and then compares micro-compression result of the Ni-free Cu_6Sn_5 pillars and Ni-doped $(Cu, Ni)_6Sn_5$ pillars, followed by compressive characterization of multi-layered $Cu_6Sn_5/Sn/Cu_6Sn_5$ and $Cu_6Sn_5/Cu_3Sn/Cu$ micropillars for mechanical reliability assessment of solder joints in different time courses.

II. EXPERIMENTS

A. Intermetallic Sample Preparation

As illustrated in Fig. 1, commercially available high-purity Cu slug (4N purity, Alfa Aesar), Sn shot (5N purity, Alfa Aesar), Ni foil (4N purity, Alfa Aesar) were weighed and put into quartz tubes according to the ratio by mass of elements in Cu_6Sn_5, and $(Cu, Ni)_6Sn_5$ with 0.5, 1, 5 at % Ni doping. After sealed in vacuum of 10^{-2} mbar, the Cu_6Sn_5 ampoules were placed in 800 °C box furnace for 5 days, followed by high temperature annealing at 380 °C for 28 days. The Ni-doped samples were cast at higher temperature (1050 °C) for 5 days,

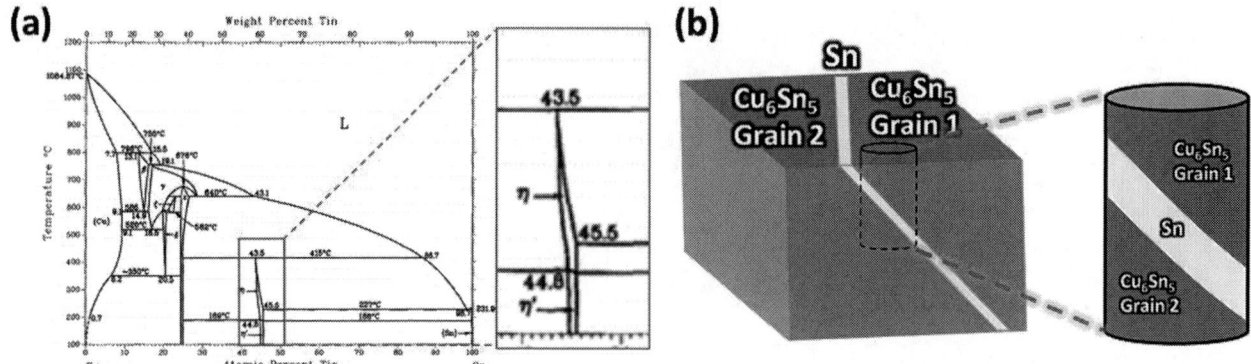

Figure 2. (a) Cu-Sn phase diagram and the enlarged Cu_6Sn_5 phase region, which get skewed toward Cu-rich side at high temperature. (b) Illustration diagrams of direct alloying method for $Cu_6Sn_5/Sn/Cu_6Sn_5$ three-layered micropillars.

and annealed at 380 °C for longer time (56 days) to ensure melting and uniform distribution of nickel element. All samples were quenched in water whenever taken out of furnace so as to retain its crystal structure and microstructure. After intermetallic ingots were cut into slices, specimens were metallographically ground with 800, 1200, 2400 grit SiC paper and carefully polished with 1 μm diamond suspension and 0.025 μm colloidal silica suspension to reveal its cross-sectional microstructure.

Grain orientation maps of polished samples were obtained using a field emission scanning electron microscope (FEI Nova NanoSEM 450) equipped with an electron backscatter diffraction (EBSD) system (EDAX Inc.). After several grains were chosen for further characterization, micropillars were machined by dual beam focus ion beam (FEI Helios 600i) using annular milling technique where multiple circular patterns were applied with descending ion current. For further details of micropillar fabrication, the reader is referred to [18].

The diameter and height of as-fabricated pillars in this work are 2.3 μm and 7.5 μm or so, respectively.

Micropillars compression tests were carried out in Hysitron PI 85 SEM Picoindenter that was installed inside and interfaced with FEI Nova NanoSEM 450. The micro-compression was performed by flat punch at constant displacement rate of 8 nm/s, which in turn means a strain rate of approximately 10^{-3} s^{-1} for the geometry of our micropillars. The analysis of compression result and development of calibration approach will be discussed in the next section.

B. Multi-phase Sample Preparation

As mentioned earlier, the interfacial reaction in micron-sized electronic interconnects may consume the solder and result in microstructural evolution from Sn-remaining into full-IMC joints. Take the common Cu/Solder/Cu system for an example, microstructure changes from $Cu/Cu_6Sn_5/Sn/Cu_6Sn_5/Cu$ at the early stage into $Cu/Cu_3Sn/Cu_6Sn_5/Cu_3Sn/Cu$ after annealing are anticipated

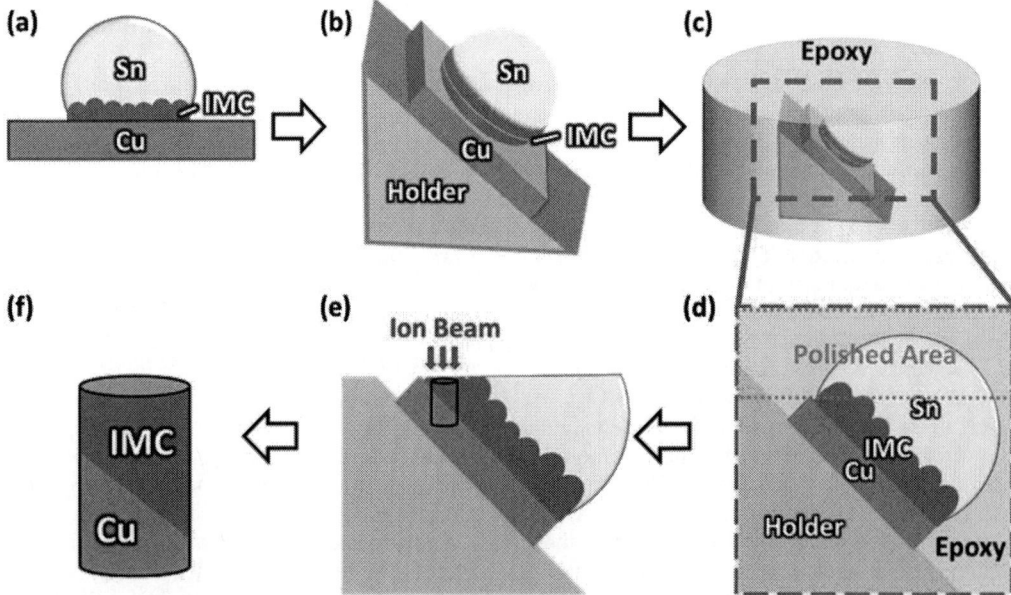

Figure 3. Schematic diagram illustrating $Cu_6Sn_5/Cu_3Sn/Cu$ micropillar preparation by interfacial reaction approach.

[17]. As a result, heterogeneous phase interfaces or homogeneous grain boundaries may be present in real solder joints.

To make the best use of micro-compression and evaluate the mechanical behaviors of solder joints as a whole, two kinds of heterogenous multi-phase structures were also tested using micro-compression. Direct alloying approach, the first way to fabricate multi-layer micropillar, takes advantage of the fact that the Cu_6Sn_5 phase region in Cu-Sn phase diagram get slightly skewed towards Cu-rich side at high temperature, and thus extra Sn would precipitate in grain boundary during the solidification of stoichiometric Cu_6Sn_5 intermetallic samples. As long as we select proper locations near grain boundaries for focus ion beam micro-machining, micropillars with tri-layered $Cu_6Sn_5/Sn/Cu_6Sn_5$ structures can be obtained, as illustrated in Fig. 2 and Fig. 4 (a). Nevertheless, since the shape of Sn phase precipitating in grain boundaries is irregular, the phase interfaces inside micropillars are randomly curved and the inclined angle cannot be experimentally controlled.

The second method, interfacial reaction approach, fixes a Sn/Cu interfacial reaction samples, which are initially reflowed at 260 °C for 5 minutes with use of flux, onto a 45°

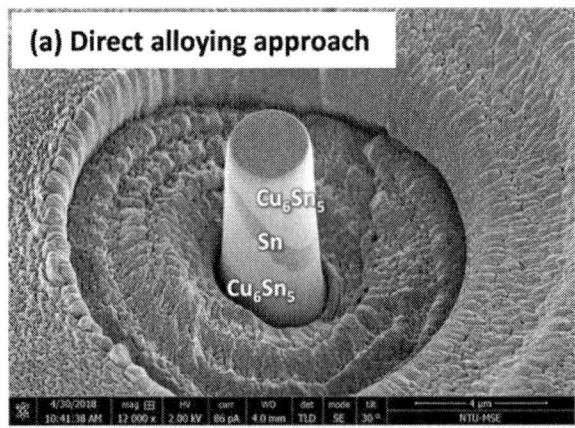

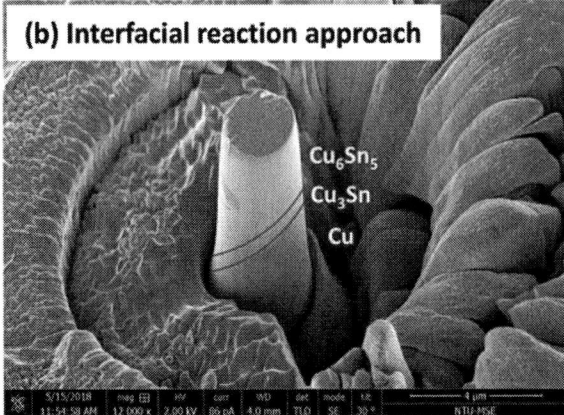

Figure 4. Secondary electron images of $Cu_6Sn_5/Sn/Cu_6Sn_5$ and $Cu_6Sn_5/Cu_3Sn/Cu$ multi-layer micropillars by (a) direct alloying approach and (b) interfacial reaction approach used in this study.

pretilt Cu holder, followed by epoxy mounting. The 45° inclined copper holder is designed to experimentally control the inclination angle between different phases and further to examine the role of interfaces in micropillar compression, otherwise the interfaces will be oriented nearly perpendicular to applied stress. After that, the standard polishing technique was employed to reveal the IMC layer, where micropillars were then fabricated as $Cu_6Sn_5/Cu_3Sn/Cu$ multi-phase structure as seen in Fig. 4 (b). Fig. 3 demonstrates the flowchart of the interfacial reaction approach and an as-fabricated $Cu_6Sn_5/Cu_3Sn/Cu$ cylindrical pillar. Though the interface angle between phases can be manipulated in the second approach, it is much difficult and time-consuming in micropillar fabrication due to differences in milling rates of dissimilar materials surrounding micropillars.

III. RESULTS AND DISCUSSION

A. Calibration approach

Since a large deviation of measured Young's moduli between nanoindentation and micro-compression exists in our previous work [15-16], a calibration method accounting for more factors in micropillar compression experiments is developed in this work. In general, the important factors include gravity effect, thermal drift, load cell compliance, as well as the sink-in depth of indenter and substrate.

At first, as the picoindenter system installed in SEM stage has to be tilted relative to the electron beam for proper vision of in-situ imaging of samples, the test rig is subject to gravitational force under tilt condition. The stage tilting renders the electrostatically-actuated transducer lost in accuracy of loading forces, and hence air indent calibration is indispensable to keep the instrument performing at its maximum efficiency before probing mechanical response of samples. In this study, the air indent calibration is carried out at tilt angle of 30° in vacuum environment of SEM chamber. Also, the drift rate can be deleterious to nanomechanical testing, and hence the average drift rate was measured for 40 seconds prior to each test in order to achieve an equilibrium state of loading machine under the irradiation of electron beam. Besides, although piezo actuators allows for intrinsic displacement controlled mode, the deformation of transducer cannot be neglected on account of the feedback force sensor with a given compliance [19]. The machine compliance as a function of load was measured and developed in use of silicon single crystal as a reference material. Moreover, after the aforementioned factors is calibrated by built-in software package or reference materials, the raw values of measured displacement still consist of three parts, i.e. the displacement from indenter, that from pillar, and that from matrix materials. Assuming elastic deformation in semi-infinite half-spaces, Sneddon's correction was applied for compensating the penetration depth of pillar into both punch and substrate [20-22]. Eventually, the calibrated load-displacement curve is then converted to a engineering stress-strain curve by dividing the load by the cross-section area and dividing the displacement by the initial height of pillar.

978-1-7281-1500-9/19 $31.00 © 2019 IEEE

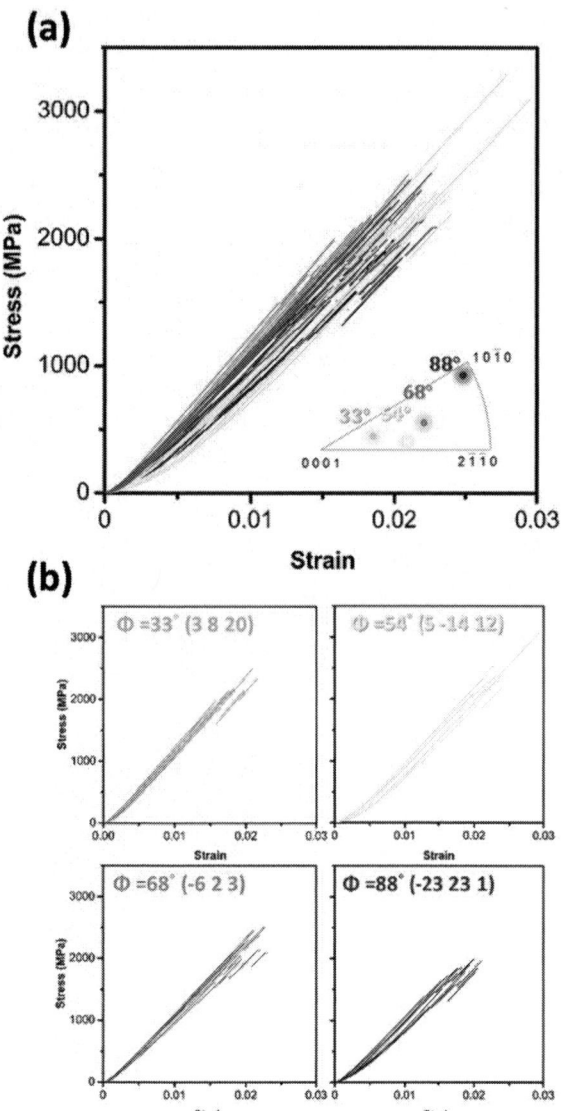

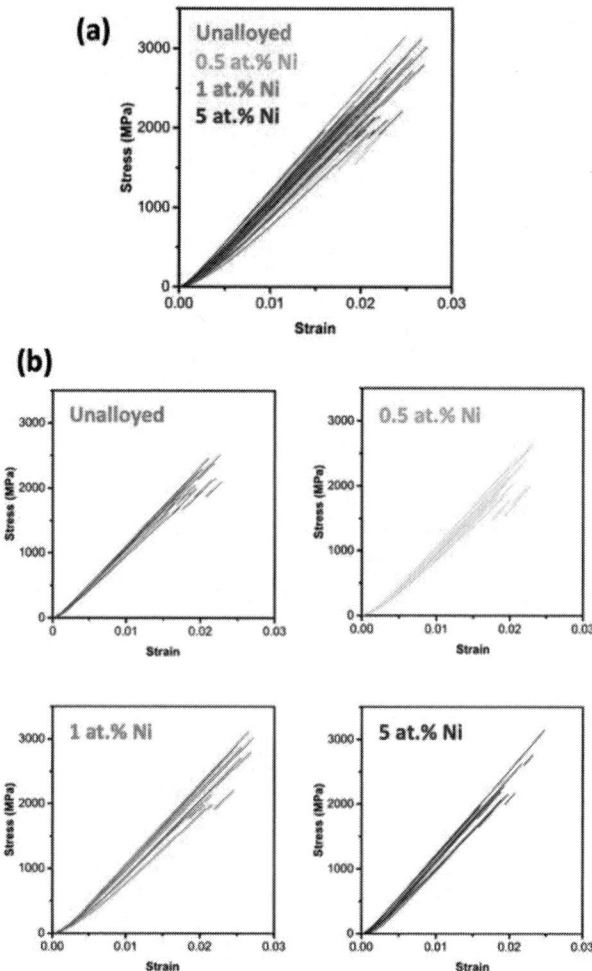

Figure 5. Stress-strain curves of single-crystalline Cu_6Sn_5 micropillar compression with four different crystal orientation.

Figure 6. Stress-strain curves of micropillar compression on $(Cu,Ni)_6Sn_5$ with 0, 0.5, 1, 5 at.% Ni addition.

B. Mechanical Anisotropy in Cu_6Sn_5

According to the phase diagram, Cu_6Sn_5 specimens exist in high-temperature hexagonal crystal structure after annealing at 380 °C. In order to analyze the mechanical anisotropy in hexagonal Cu_6Sn_5, micropillars on four differently-oriented grains were compressed. The angles between plane normals of these four grains and hexagonal c-axis direction, which are denoted by ϕ, range from 34 to 88 degrees. Fig. 5 presents the compression results of micropillars in four grains, and Cu_6Sn_5 micropillars behave in a really brittle fashion with very little plasticity before fracture failure. It is worth noting that these selected grain orientations generate high values (0.4 more or less) of Schmid factor either in basal slip system, prismatic slip system, or pyramidal slip

system, suggesting that the compression stress can theoretically applied shear component on one of the slip directions. However, Cu_6Sn_5 micropillars demonstrate hardly any plasticity but only intermittent strain bursts before fracture failure. It can be seen from the slopes of stress strain curves that the Young's modulus is larger when the grain orientation is close to c-axis, showing a trend of linear decrease from 129 GPa at grain ϕ=33° to 115 GPa at grain ϕ=88°. Additionally, the average fracture strength of grain ϕ=54° and grain ϕ=68° are 2499 and 2340 MPa, respectively, greater than 2202 MPa at grain ϕ=33° and 1896 MPa at grain ϕ=88°. Overall, the anisotropy of Cu6Sn5 results in greater stiffness in grains paralleled to c-axis, and higher fracture strength in grains whose angle with respect to c-axis is close to 45°.

C. The influence of Ni doping

To study the effect if Ni addition, grains with similar ϕ with respect to c-axis are particularly selected for the purpose of downplaying the role of crystal anisotropy. Fig. 6 shows the micro-compression of Ni-added Cu_6Sn_5 micropillars. The slopes of $(Cu, Ni)_6Sn_5$, i.e. Young's moduli, increase with the Ni concentrations, but the fracture strength does not have the

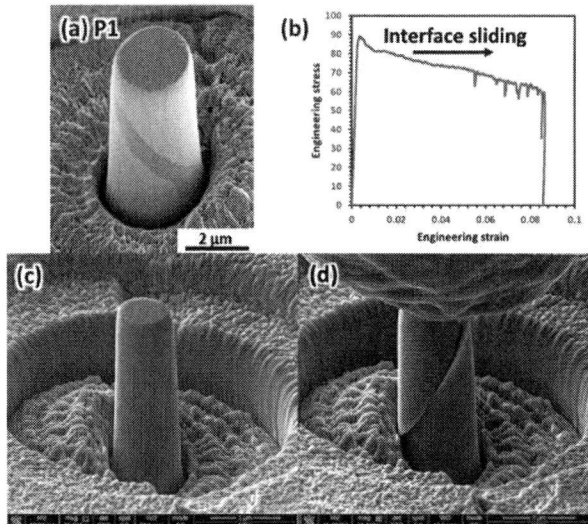

Figure 7. (a) Electron micrograph of as-fabricated P1 micropillar; (b) stress-strain curves of micropillar compression; (c-d) secondary electron image of P1 before and after compression, showing that interface sliding at the upper Cu_6Sn_5/Sn interface.

similar trend. Among them, $(Cu, Ni)_6Sn_5$ with 1 at.% Ni exhibits the greatest fracture strength of 2659 MPa, while the pure Cu_6Sn_5 only has the fracture strength of 2340 MPa. This is likely because fracture strength or strain does not just depend on the ϕ angle. That is to say, the similar ϕ angle in hexagonal structure of different Ni-doped specimens is not enough to guarantee the similar stress state or to activate cleavage fracture along the same crystal planes or directions.

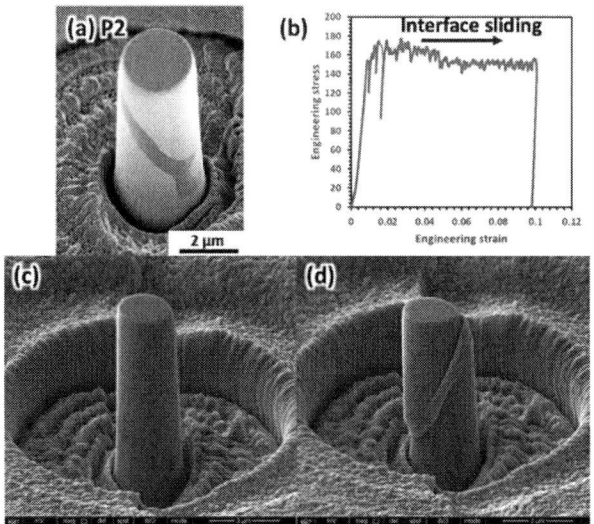

Figure 8. (a) Electron micrograph of as-fabricated P2 micropillar; (b) stress-strain curves of micropillar compression; (c-d) secondary electron image of P2 before and after compression, showing that interface sliding at both the upper and lower Cu_6Sn_5/Sn interfaces.

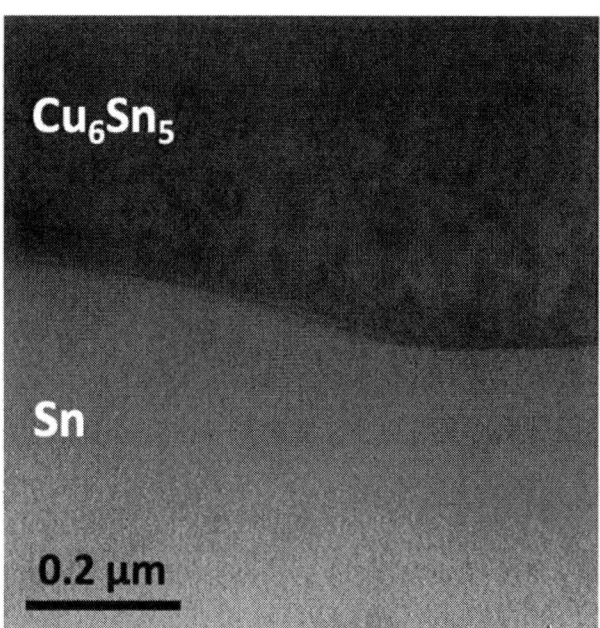

Figure 9. Transmission electron microscopy examination of Cu_6Sn_5/Cu interface of deformed micropillar.

D. Cu_6Sn_5/Sn/Cu_6Sn_5 Micropillar Compression

Two micropillars with three-layered Cu_6Sn_5/Sn/Cu_6Sn_5 structure, marked as P1 and P2, were uniaxially compressed to evaluate the mechanical reliability of Sn-remaining solder joints, and the results are shown in Fig. 7 and 8. It was found that at the beginning of loading, Cu_6Sn_5/Sn/Cu_6Sn_5 micropillar P1 deformed by linear elasticity of Sn layer, as evidenced by the fact that the slope of the initial elastic part is pretty close to elastic modulus of Sn. As the flow stress was increased to around 100 MPa, the Cu_6Sn_5/Sn/Cu_6Sn_5 micropillar P1 underwent plastic deformation by means of interface sliding at the upper Cu_6Sn_5/Sn interface, accommodating up to 10 % strain before the upper part touched the surrounding materials in matrix.

In the case of P2, Cu_6Sn_5/Sn/Cu_6Sn_5 micropillar showed the similar mechanical behaviors by early-stage linear elastic deformation and subsequent interface sliding at higher stress level. However, sliding at both the upper and lower Cu_6Sn_5/Sn interfaces simultaneously took place at 160 MPa during micro-compression of P2. It is believed that the crystallographic orientations and atomic structure at the heterogeneous interface determine the interface strength, causing slightly different phenomena in activated sliding interfaces and sliding stress in P1 and P2.

The deformed Cu_6Sn_5/Sn/Cu_6Sn_5 micropillar was further examined in transmission electron microscope after lamella preparation using focus ion beam. Fig. 9 reveals the bright-field image of the Cu_6Sn_5/Sn interface after sliding, and it is found that the interface retained its integrity with no noticeable void or crack even after experiencing sliding strain of about 10%.

Fig. 10 compares the typical stress-strain response between micro-compression of Cu_6Sn_5 single crystal and

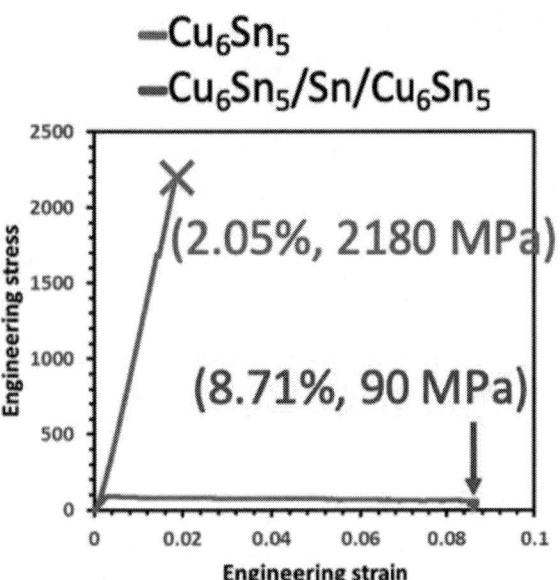

Figure 10. The typical stress-strain curves of micro-compression on Cu_6Sn_5 single crystal and $Cu_6Sn_5/Sn/Cu_6Sn_5$ structure.

$Cu_6Sn_5/Sn/Cu_6Sn_5$ structure. Although the Cu_6Sn_5 intermetallic has fracture strength as high as more than 2000 MPa, Cu_6Sn_5 shows hardly any plastic deformation and is unable to endure large strain greater than 2%. On the other hand, when a Sn layer exists in the middle of Cu_6Sn_5, it turns out that the smooth Cu_6Sn_5/Sn interface is the weakest point in the three-layer structure, exhibiting overall strength of only 100 MPa but great irreversible deformability of more than 10% strain.

E. $Cu_6Sn_5/Cu_3Sn/Cu$ Micropillar Compression

To gain a better understanding of mechanical behaviors of real solder joints after long period service in electronic devices, $Cu_6Sn_5/Cu_3Sn/Cu$ multi-phase structures were fabricated by fixing interfacial reaction samples onto 45° pretilt Cu holder. After reflowed at 260 °C for 5 minutes, both Cu_6Sn_5 and Cu_3Sn intermetallic layers grew in the interface of Sn/Cu. Focus ion beam was used to reveal the cross-sectional microstructure of the 45°-inclined specimen, and Fig. 11 showed that the Cu_3Sn has a layer-type morphology, whereas

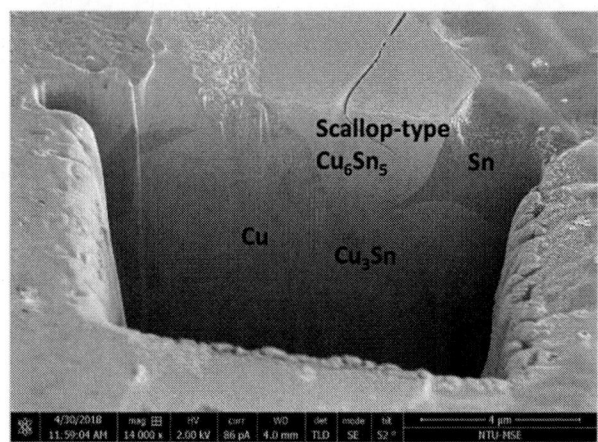

Figure 11. Cross-section microstructure of a Sn/Cu interfacial reaction joint on a 45° pretilt holder after 5-min reflowing at 260 ℃.

the Cu_6Sn_5 forms in a scallop morphology to reduce the interfacial energy in the wetting reaction between molten solder and copper substrate [23]. As a result, the as-fabricated micropillar contain some of Sn solder in the Cu_6Sn_5 layer between two scallops, as demonstrated in Fig. 12 (a).

The stress-strain curve of micro-compression of $Cu_6Sn_5/Cu_3Sn/Cu$ multi-phase structure is shown in Fig. 12 (b). It is found that the residual Sn between Cu_6Sn_5 scallops contribute to the initial elastic deformation, thereby causing bending of cylinder due to the non-uniform distribution of Sn inside the pillar. Afterwards, the pillar underwent plasticity by slip deformation of copper substrate. It is noted that the Cu_6Sn_5/Sn interface is not the weakest point due to the rough interface, and therefore the interface sliding does not occur as in $Cu_6Sn_5/Sn/Cu_6Sn_5$ structures. Besides, our result suggested that the Cu_6Sn_5/Cu_3Sn interface has sufficient strength against compressive stress when void is not present.

IV. SUMMARY

This work firstly evaluates the mechanical behaviors of Cu_6Sn_5 single crystals with different grain orientations via micropillar compression testing. Young's moduli of Cu_6Sn_5 are found to linearly increase with the decreasing angle between plane normal and c-axis, while grains with intermediate angle with respect to c-axis show higher fracture

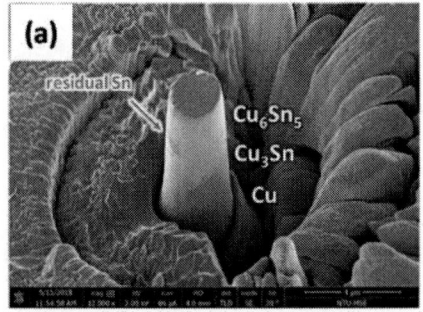
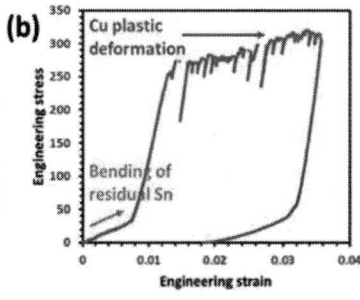
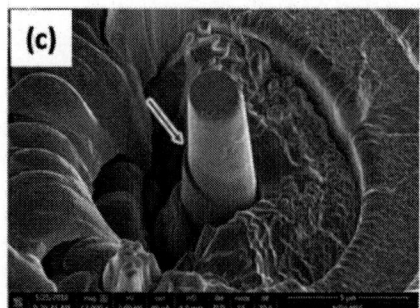

Figure 12. (a) Electron micrograph of as-fabricated $Cu_6Sn_5/Cu_3Sn/Cu$ micropillar; (b) stress-strain curve of micropillar compression; (c) fractographic secondary electron image of the deformed pillar.

978-1-7281-1500-9/19 $31.00 © 2019 IEEE

strength than others. Also, Ni addition can enhance the elastic moduli within the concentration range between 0 to 5 at.%. Furthermore, two different approaches are proposed to fabricate $Cu_6Sn_5/Sn/Cu_6Sn_5$ and $Cu_6Sn_5/Cu_3Sn/Cu$ multiphase micropillars. Under compression, $Cu_6Sn_5/Sn/Cu_6Sn_5$ structures exhibit remarkable plasticity in manner of interface sliding at around 100 MPa, accommodating at least 10% strain and retaining its integrity without any void formation at the interface. In contrast, multilayer $Cu_6Sn_5/Cu_3Sn/Cu$ micropillars undergo plastic deformation through slip in Cu substrate, suggesting sufficient strength of intermetallic compounds and the interfaces.

REFERENCES

[1] M. M. Waldrop, "The chips are down for Moore's law," Nature News, vol. 530, pp. 144–147, 2016.

[2] S. Kumar, "Fundamental limits to Moore's law," arXiv:1511.05956.

[3] W. Arden, M. Brillouët, P. Cogez, M. Graef, B. Huizing, R. Mahnkopf, "Towards a "More-than-Moore" roadmap," Cluster for Application and Technology Research in Europe on NanoElectronics, 2011.

[4] K. N. Tu, "Reliability challenges in 3D IC packaging technology," Microelectronics Reliability, vol. 51, pp. 517-523, 2011.

[5] K. N. Tu, H. Y. Hsiao, C. Chen, "Transition from flip chip solder joint ot 3D IC microbump: Its effect on microstructure anisotropy," Microelectronics Reliability, vol. 53, pp. 2-6, 2013.

[6] K. N. Tu, Y. Liu, "Recent advances on kinetic analysis of solder joint reactions in 3D IC packaging technology," Materuaks Science & Engineering R, vol. 136, pp. 1-12, 2019.

[7] J. H. Lau, "TSV manufacturing yield and hidden costs for 3D IC integration," Electronic Components and Technology Conference, pp. 1031-1042, 2010.

[8] T. L. Yang, J. J. Yu, W. L. Shih, C. H. Hsueh, C. R. Kao, "Effects of silver addition on Cu–Sn microjoints for chip-stacking applications," Journal of Alloys and Compounds, vol. 605, pp. 193-198, 2014.

[9] K. Zeng, K. N. Tu, "Six cases of reliability study of Pb-free solder joints in electronic packaging technology," Materials Science and Engineering: R: Reports, vol. 38, pp. 55-105, 2002.

[10] H. Y. Chuang, W. M. Chen, W. L. Shih, Y. S. Lai, C. R. Kao, "Critical new issues relating to interfacial reactions arising from low solder volume in 3D IC packaging," in Proceedings of the 2011 IEEE 61st Electronic Components and Technology Conference, pp. 1723–1728, 2011.

[11] M. D. Uchic, D. M. Dimiduk, J. N. Florando, W. D. Nix, "Sample dimensions influence strength and crystal plasticity," Science, vol. 305, pp. 986-989, 2004.

[12] D. M. Dimiduk, M. D. Uchic, T. A. Parthasarathy, "Size-affected single-slip behavior of pure nickel microcrystals," Acta Materialia, vol. 53, pp. 4065-4077, 2005.

[13] M. C. Liu, J. C. Huang, Y. T. Fong, S. P. Ju, X. H. Du, H. J. Pei, T. G. Nieh, "Assessing the interfacial strength of an amorphous–crystalline interface," Acta Materialia, vol. 61, pp. 3304-3313, 2013.

[14] D. R. P. Singh, N. Chawla, G. Tang, Y. L. Shen, "Micropillar compression of Al/SiC nanolaminates," Acta Materialia, vol. 58, pp. 6628-6636, 2010.

[15] L. J. Yu, H. W. Yen, J. Y. Wu, J. J. Yu, C. R. Kao, "Micromechanical behavior of single crystalline Ni_3Sn_4 in micro joints for chip-stacking applications", Materials Science and Engineering: A, vol. 685, pp. 123-130, 2017.

[16] J. J. Yu, J. Y. Wu, L. J. Yu, H. W. Yang, C. R. Kao, "Micromechanical behavior of single-crystalline Cu_6Sn_5 by picoindentation", Journal of Materials Science, vol. 52, pp. 7166-7174, 2017.

[17] C. Hang, Y. Tian, R. Zhang, and D. Yang, "Phase transformation and grain orientation of Cu–Sn intermetallic compounds during low temperature bonding process," Journal of Materials Science: Materials in Electronics, vol. 24, pp. 3905-3913, 2013.

[18] J. Y. Wu, Y. S. Chiu, Y. W. Wang, C. R. Kao, "Mechanical characterizations of single-crystalline $(Cu, Ni)_6Sn_5$ through uniaxial micro-compression," unpublished.

[19] G. Dehm, B. N. Jaya, R. Raghavan, C. Kirchlechner, "Overview on micro- and nanomechanical testing: New insights in interface plasticity and fracture at small length scales," Acta Materialia, vol. 142, pp. 248-282, 2018.

[20] I. N. Sneddon, "Boussinesq's problem for a flat-ended cylinder," Mathematical Proceedings of the Cambridge Philosophical Society, vol. 42, pp. 29-39, 1946.

[21] I. N. Sneddon, "The relation between load and penetration in the axisymmetric Boussinesq problem for a punch of arbitrary profile," International Journal of Engineering Science, vol. 3, pp. 47–57, 1965.

[22] H. Fei, A. Abraham, N. Chawla, H. Jiang, "Evaluation of micro-pillar compression tests for accurate determination of elastic-plastic constitutive relations," Journal of Applied Mechanics, vol. 79, pp. 061011, 2012.

[23] K. N. Tu, T. Lee, "Morphological stability of solder reaction products in flip chip technology," Journal of Electronic Materials, vol. 30, pp. 1129-1132, 2001.

A study on the Anchoring Polymer Layer (APL) Anisotropic Conductive Films (ACFs) with self-exposed conductive particles surface for ultra-fine pitch Chip-on-Glass (COG) applications

Dal-Jin Yoon and Kyung-Wook Paik

Department of Materials Science and Engineering
Korea Advanced Institute of Science and Technology
291 Daehak-ro, Yuseong-gu, Daejeon, 305-701, South Korea
Corresponding author: kwpaik@kaist.ac.kr

Abstract— *In this study, APL ACFs combined with self-exposed conductive particles surface are newly introduced. The effects of APL ACFs properties on conductive particle movement and interconnection stability for ultra-fine pitch COG applications were reported earlier. Due to the APL structure with high tensile strength, the movement of conductive particles can be suppressed when the resin flow occurs, thereby achieving 100 % of electrical insulation property in ultra-fine pitch COG interconnection. Due to the high tensile strength property of the thermoplastic polymer as an APL material, APL ACFs have a higher capture rate than conventional ACFs, which can reduce the amount of conductive particles used to make ACFs by one third. However, to achieve a stable electrical interconnection, the APL ACFs require additional process steps such as an oxygen plasma etching process to remove the APL polymer skins coated on the top and bottom surfaces of conductive particles. In this study, the polymer skin can be removed during the APL fabrication process using the self-activate monolayer (SAM) treatment on conductive particles. As a result, stable ACFs joint formation can be achieved without any additional oxygen plasma etching process for APL fabrication processes.*

Keywords-self-exposed surface; anchoring polymer layer; anisotropic conductive films; ultra-fine pitch; electrical properties

I. INTRODUCTION

Display products such as smart phones, smart watches, smart glasses and flat panel televisions need higher quality resolution, increased display panel size, lighter weight and thinner structure. Ultra-fine pitch, which is the center-to-center distance between neighboring bumps of driver ICs or electrodes of glass substrate, interconnection technology becomes major challenges in display electronic products. For 8K UHD (Ultra High Definition) and virtual reality (VR) display products, higher resolution requires ultra-fine pitch assembly. [1] In general, electrical interconnections are formed by interconnecting bumps of driver ICs and electrodes of glass substrates using ACFs [2]. However, as the ultra-fine pitching progresses, the width of the bump or the electrode decreases, but the width of the space between adjacent bumps or electrodes also decreases.

During the assembly process with a narrowed space, when the resin flow occurs, the conductive particles are moved to a narrowed space to form agglomeration and an electrical short circuit is formed between the adjacent bumps, thereby causing a malfunction of the device. If the less amount of conductive particles contents in ACFs are added to avoid the short circuit problem, the number of captured particles can be also decreased between bumps and electrodes resulting in higher contact resistances.

In order to solve the above problems, APL ACFs have been introduced by our research group. The core technology of APL ACFs is to suppress the movement of conductive particles, so that the conductive particles maintain their original position after bonding process. The APL were fabricated using a roll to roll film production equipment capable of mass production and has no adhesion because it's made of a thermoplastic polymer. Therefore, APL is laminated to the non-conductive film(NCF) to give the adhesion and function as the ACFs. By using the APL structure with incorporated conductive particles, the average capture rate of conductive particle was significantly increased up to 90% in 20 μm pitch area. As a summary, APL ACFs can significantly improve the electrical interconnection properties such as contact resistances and electrical insulation, as well as reliability properties. [3-4]

During the ACFs interconnection process, the polymer skins that surround the top and bottom surface of the conductive particles may act as an insulating layer resulting in higher contact resistance between bumps and the electrodes. To achieve a stable electrical interconnection, the APL ACFs require additional process steps such as an oxygen plasma etching process to remove the polymer skins surrounding the conductive particles after the APL fabrication. Alternatively, a vertical ultrasonic bonding process is needed. This is because, the polymer skin can be torn through the vertical directional ultrasonic, and the surface of the conductive particles was exposed to make good electrical interconnection.

In this paper, the method of self-exposed conductive particles surface in APL during a roll-to-roll film coating process without the extra polymer skin etching process. The 20 μm pitch COG test vehicle and average 3.25 μm diameter of conductive particles were used. In addition, high Ultimate Tensile Strength (UTS) Polyacrylonitrile (PAN) and Dimethylmethanamide (DMF) solvent were used. [5] In addition, the Surface Activated Monolayer (SAM) solution was used to expose top and bottom of conductive particles surface in the PAN APL film. In this PAN APL solution, the hydrophilic functional group of the SAM solution selectively binds to the Au surface of the conductive particles, and the hydrophobic functional group is arranged in the PAN. During the APL structure fabrication process, the polymer skins of the conductive particles of APL can be completely removed by the SAM surface treatment. Therefore, stable ACFs joint formation can be achieved without any additional plasma etching process to remove the PAN layer coated on the top and bottom of conductive particles.

II. TEST VEHICLES

Au bump in driver IC (chip) and electrode on glass substrate were prepared, as shown in Table 1. The electrodes on the glass substrate were thin layer (130 nm) of Au and Ti, and the space between adjacent bumps and electrodes was 7 μm. Also, the Kelvin structure and comb structure array in test vehicles was used to measure the contact resistance of single ACFs joint and electrical insulation property, respectively. [6]

III. EXPERIMENTS

A. Materials

In order to maximize the suppression of conductive particles movement, PAN, which is known as a high tensile strength material, was selected as the APL material, and DMF was used as the solvent to dissolve the solid phase PAN. The conductive particles consist of a structure in which Ni and Au layers are covered on a 3 μm diameter polymer core, and the final diameter is 3.25 μm. Since the PAN is thermoplastic polymer, it has no adhesion property, NCF composed of thermosetting epoxy resin and curing agents were prepared and laminated on the APL structure to fabricate ACFs structure.

B. Fabrication of PAN APL structure

The PAN APL structure incorporating conductive particles is fabricated using a roll-to-roll film coater as shown in Fig 1. The cross sectional and surface morphology of PAN APL structure were observed by scanning electron microscope (SEM) analysis.

Table I. Test vehicle information of (a) driver IC (chip) and (b) display glass substrate.

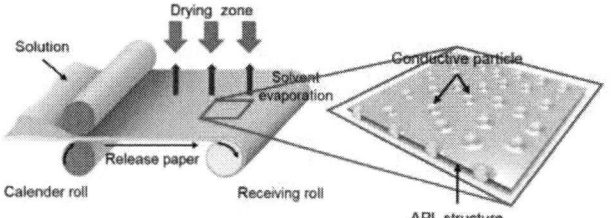

Test vehicle	Size	Bump / Electrode		
		Pitch : 20μm	Size	Height
Si Chip	16mm X 1.5mm		Bump : 12μm X 80 μm	Bump : 12μm (Au)
Glass	26mm X 23mm		Electrode : 12μm X 80 μm	Electrode : 130nm (Ti : 30nm, Au : 100nm)

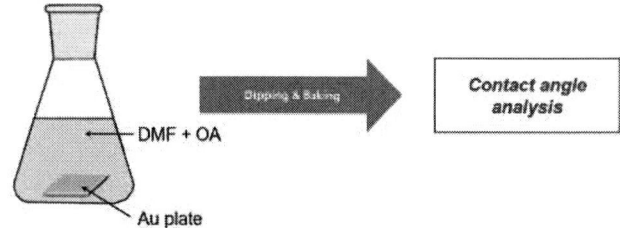

Figure 1. The schematic process image of PAN APLstructure.

C. SAM treatment of conductive particles surface

The conventional oxygen plasma etching process to remove polymer skin is time-consuming process and difficult to be applied on a roll-to-roll system. Therefore, this research has been conducted to expose the conductive particles surface without any additional plasma process.

The contact angles were measured using de-ionized water (D.I. water) to analyze the surface characteristic of PAN and Au. As shown in Fig. 2. Au plate was added in mixture of the DMF and SAM solution to allow the SAM functional group adhere to Au surface. And then contact angle of Au plate was measured as s function of the SAM content.

And the same SAM treatment process were used on Au coated conductive particles by dipping in the SAM solution to coat the conductive particle surface. And then PAN solution was added to the SAM solution containing conductive particles to make PAN APL structure. The SAM treated PAN APL was fabricated in the same process as Fig 1. And the surface and cross section were observed by SEM analysis.

Figure 2. The schematic image of SAM treatment on the Au plate.

D. Characterization of SAM treated PAN APL ACFs

After the SAM treatment process, the PAN APL structure was laminated between two NCFs to make PAN APL ACFs. Before and after the same bonding process (70 MPa, 170 °C, and 10 seconds), an optical microscope was used to observe the movement of the conductive particles in the ACFs. In order to express the movement of the conductive particles, the ratio of the number of particles present on the bumps before and after bonding process was defined as the capture rate of the conductive particles. During the bonding process, as the thermosetting epoxy resin constituting the ACFs passes over a specific temperature range, the viscosity is instantaneously lowered rapidly, and curing immediately follows. When the viscosity is lowered, the epoxy resin moves to space, which is an empty space, and conductive particles move to space together. After using two types of ACFs, the ratio of the circuit that forms an electrical short circuit in the whole circuits was defined as a short circuit rate, and the insulation and contact resistance were measured as shown in Fig. 3. After bonding process, the ACFs joints between bump and electrode were observed by SEM analysis.

IV. RESULTS & DISCUSSION

A. Fabrication of PAN APL structure

The PAN APL structure incorporating conductive particles was fabricated using a roll-to-roll film coating process. After the fabrication of PAN APL structure, top and cross-sectional morphology were observed as shown in Fig. 4. The conductive particles have same phase in the PAN layer, which is PAN APL structure, and the thickness of the PAN layer connecting adjacent conductive particles was about 1.57 μm. However, as shown in Fig. 4 (b), the thin polymer skin remains coated on the top and bottom surfaces of conductive particles, and may affect electrical contact resistance properties.

B. SAM treatment of conductive particles surface

The contact angles of PAN and Au plater were 61.2° and 72.4° respectively, and they have similar hydrophilic characteristic as shown in Fig. 5. In order to change the surface property of the Au plate, the SAM treatment was carried out. As a result, the surface property of the Au plate changed from hydrophilic to hydrophobic at all SAM contents, as shown in Fig. 6. This is because the hydrophilic and hydrophobic functional groups of the SAM material were attached to the Au surface, and the surface of Au is converted to hydrophobic characteristic. To apply the same principle to the conductive particles, as shown in Fig. 7. There was different behavior between the hydrophobic of

PAN and the hydrophilic of Au surface. During the PAN APL fabrication process, the polymer skin cannot be coated on the surface of conductive particles, and the surface of conductive particles is completely exposed after the drying process. And the PAN APL structure was fabricated using a roll-to-roll film coating process, and the cross section was observed by SEM analysis. As the SAM content increased, the degree of exposure of the conductive particle surface increased. As shown in Fig. 8, the surface of the conductive particles is completely exposed at a content of 25 wt. % of SAM. Due to the SAM treatment, self-exposed PAN APL structure can be fabricated without additional plasma etching process. Finally, PAN APL ACFs were successfully fabricated with NCFs using a roll lamination process.

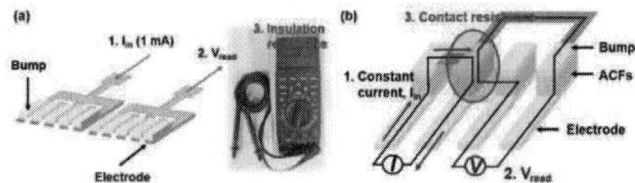

Figure 3. The schematic image of (a) comb structure and (b) Kelvin structure to measure insulation resistance and contact resistance of single ACFs joint.

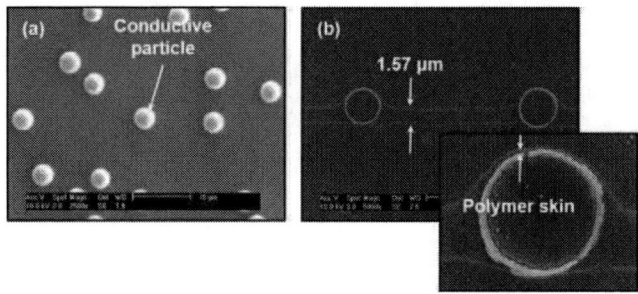

Figure 4. SEM image of (a) top surface and (b) cross sectional PAN APL.

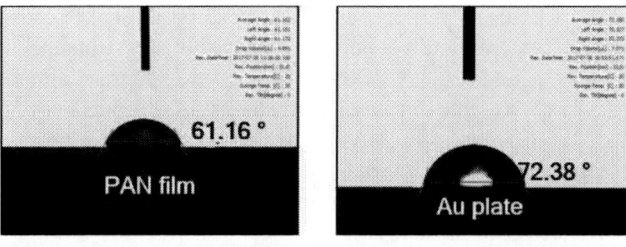

Figure 5. Contact angle results of PAN and Au surface with SAM treatment.

Figure 6. Contact angle results of 10, 20 and 30 wt.% SAM treated Au surface.

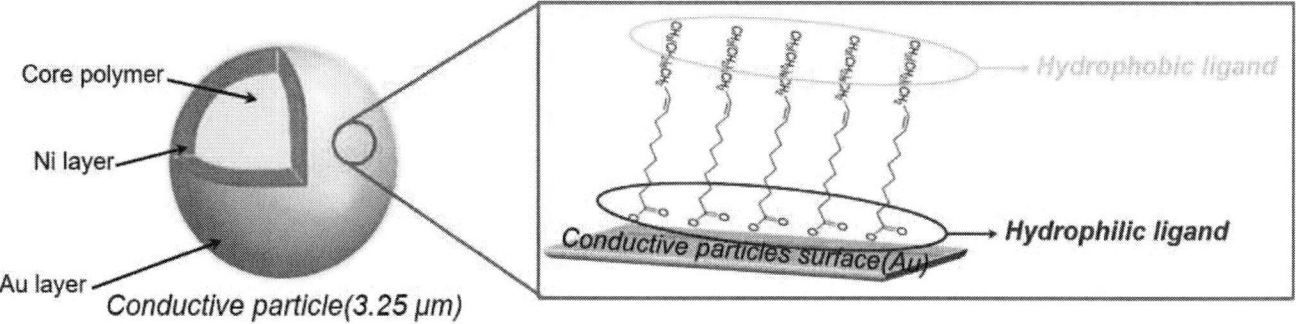

Figure 7. Schematic images of hydrophobic and hydrophilic ligand of SAM materials coated on the conductive particle surface

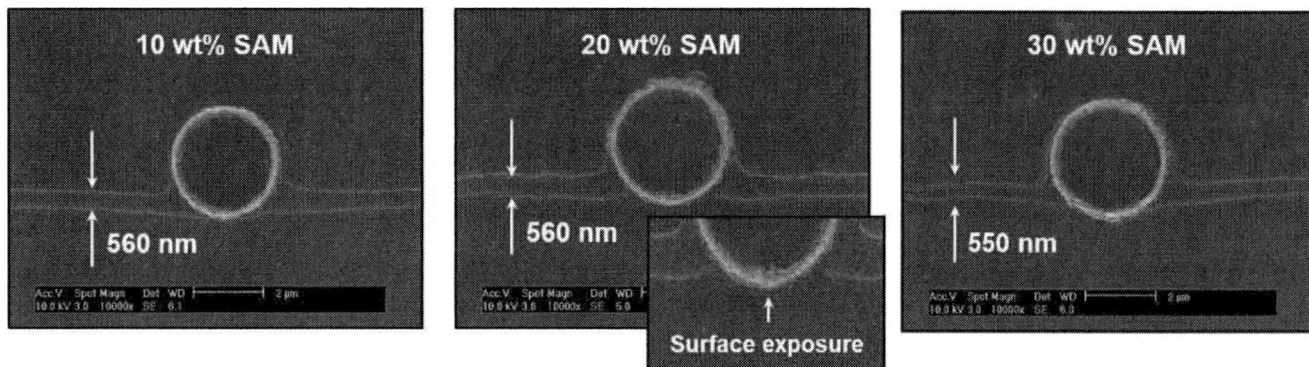

Figure 8. Cross sectional SEM image of SAM treated PAN APL structure at various SAM contents.

C. Characterization of the SAM treated PAN APL ACFs

In conventional ACFs, there were 42 particles on a bump on the average, and most of conductive particles moved to the area outside the bump after bonding process, and the number of captured conductive particles was 13 particles. However, the PAN APL ACFs, the number of conductive particles was initially 16 particles on a bump before ACFs bonding and remained 14 particles after bonding as shown in Fig. 9. In terms of capture rate of conductive particles, Conventional ACFs showed a capture rate of about 30 %, while PAN APL ACFs showed 90 %, which means that about only 1 to 2 conductive particles were lost before and after bonding process. Although the SAM material was added in the PAN APL system, there are no effects on the PAN APL function of suppression of conductive particles movement during resin flow. Finally, the capture rate was significantly increased up to 90 % due to the high tensile strength PAN APL structure.

The average contact resistances of conventional ACFs and PAN APL ACFs were 232 mΩ and 234 mΩ respectively as shown in Fig. 10. Both conventional ACFs and PAN APL ACFs showed the same stable contact resistances, because they have the same number of captured conductive particles per a bump area after ACFs bonding. In addition, the ACFs joints have similar height between the bump and electrode for both conventional and PAN APL ACFs. By the cross sectional SEM images of PAN APL ACFs, there was no polymer skin in ACFs joint morphology,

because the surface of the conductive particles in the PAN APL was completely exposed by the SAM treatment.

As a result of the electrical insulation property, since the capture rate of conventional ACFs was low, an electrical short circuits was formed about 35 % of the whole circuits. However, in the case of PAN APL ACFs, due to the PAN APL structure with high tensile strength, the movement of conductive particles was completely controlled, so no electrical short circuit was formed.

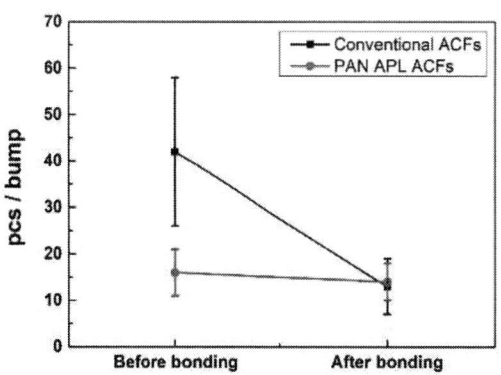

978-1-7281-1500-9/19 $31.00 © 2019 IEEE

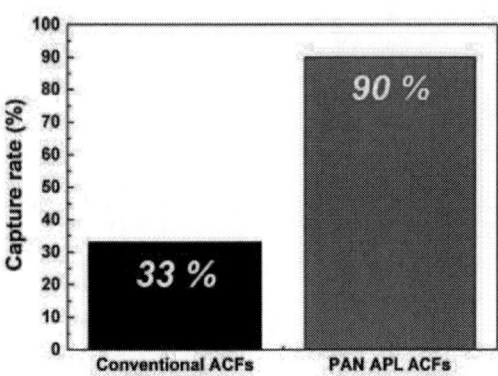

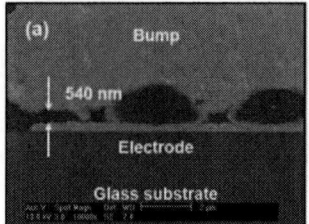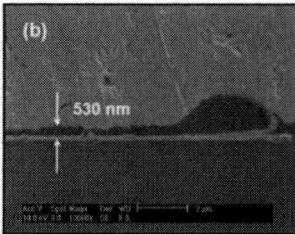

Figure 11. Cross sectional joint SEM images of (a) conventional ACFs and (b) PAN APL ACFs.

Figure 9. The numbers of conductive particles per a bump before and after same bonding process and capture rates of conventional and PAN APL ACFs.

V. CONCLUSION

The SAM treated PAN APL structure were successfully demonstrated, and the exposure effect of the conductive particle surface in the PAN APL structure by the SAM treatment was investigated. Due to the SAM treatment, the surface properties of the Au surface of conductive particles and PAN surface can be made different from each other, and the surface of the conductive particles can be naturally exposed without the polymer skin while the PAN APL structure is being fabricated. Also, since the conductive particles surface was coated by the SAM material, the surface was naturally self-exposed without any additional process to remove polymer skin during a roll-to-roll film coating process.

In addition, the effect of suppressing the movement of the conductive particles occurred during bonding process due to the PAN APL structure was investigated, and stable electrical properties was measured. Since the PAN material has high tensile strength of 65 MPa, the PAN APL ACFs completely suppressed the conductive particle movements resulting in the highest capture rate of 90%. The PAN APL ACFs showed 100 % of insulation property and stable electrical contact resistance in ultra-fine pitch COG applications compared with conventional ACFs. As a result, PAN APL ACFs can completely solve the technical limitation of conventional ACFs interconnection for less than 20 μm ultra-fine pitch assembly.

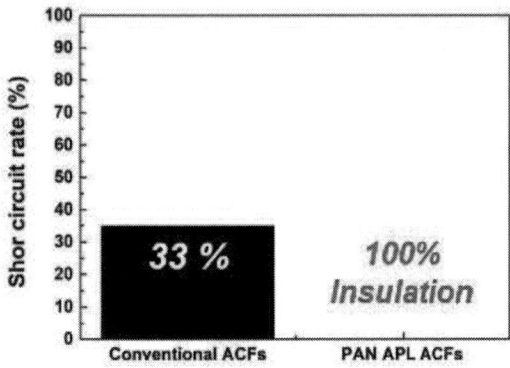

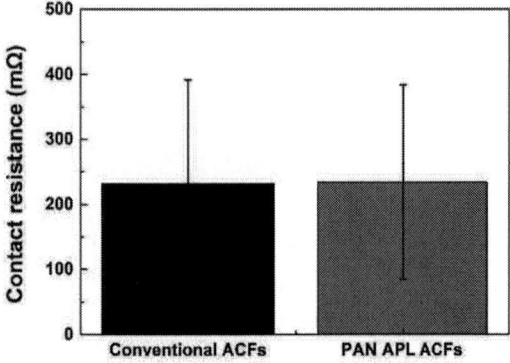

Figure 10. Electrical insulations and contact resistances results of conventional and PAN APL ACFs.

REFERENCES

[1] Y. C. Lin and Jue Zhong., "A Review of the Influencing Factors on Anisotropic Conductive Adhesives Joining Technology in Electrical Applications", *Journal of Materials Science*, Vol. 43, No. 9, 2008, pp. 3072-3093.M. J. Lim and K. W. Paik, "Recent advances on anisotropic conductive adhesion (ACAs) for flat panel displays and semiconductor packaging applications", *Int. J. Adhes. Adhes.*, vol. 26, pp. 304-313, Jul. 2005

[2] M. J. Lim and K. W. Paik, "Recent advances on anisotropic conductive adhesion (ACAs) for flat panel displays and semiconductor packaging applications", *Int. J. Adhes. Adhes.*, vol. 26, pp. 304-313, Jul. 2005

[3] S.H. Lee, D.J. Yoon and K.W. Paik, "A Study on the Conductive Particle Movements in Polyvinylidene fluoride (PVDF) Anchoring

Polymer Layer (APL) Anisotropic Conductive Films (ACFs) for 20 um Fine Pitch Interconnection," *IEEE Trans. Compon. Packag. Manuf. Technol.*, vol. 9, pp. 209-215, Jan. 2019

[4] D.J. Yoon, S.H. Lee, and K.W. Paik, "Effects of the Nylon Anchoring Polymer Layer on the Conductive Particle Movements of Anisotropic Conductive Films for Ultrafine Pitch Chip-on-Glass Applications," *IEEE Trans. Compon. Packag. Manuf. Technol.*, vol. 8, pp. 1723-1728, Oct. 2018

[5] S.Y. Gu, J. Ren and Q.L Wu, "Preparation and structures of electrospun PAN nanofibers as a precursor of carbon nanofibers", Synthetic Metals, vol. 155, Oct. 2005, pp. 157-161, doi:10.1016/synthmet.2005.07.340

Bending Properties of Fine Pitch Flexible CIF (Chip-in-Flex) Packages using APL (Anchoring Polymer Later) ACFs (Anisotropic Conductive Films)

Ji-Hye Kim, Dal-Jin Yoon, and Kyung-Wook Paik*
Department of Materials Science and Engineering
KAIST
Daejeon, Korea
*e-mail) kwpaik@kaist.ac.kr

Abstract— **In this paper, the bending reliability of the CIF package using new APL ACFs was investigated, and compared with that of conventional ACFs.**

APL ACFs are new ACFs structure including conductive particles incorporated polymer film which suppresses the flow of the conductive particles during ACFs bonding process. By using APL ACFs, it is possible to completely solve the short-circuit problem at fine pitch interconnection by suppressing the movement of the conductive particles, and stable electrical resistance can be obtained. APL ACFs can be an excellent packaging solution for future flexible displays requiring very high resolution such as 8K UHD Virtual Reality (VR) and bendable and rollable displays.

Our research group has investigated the Chip-In-Flex (CIF) assembly using ACFs for flexible packaging applications and reported their bending properties before. APL ACFs was fabricated using PAN (Polyacrylonitrile) polymer APL film and Non-conductive Films (NCFs) as adhesive films. And dynamic bending tests of CIF packages using two types of ACFs were also performed up to 150,000 dynamic bending cycles at the bending radius of 6 mm. As a result, it was confirmed that APL ACFs show excellent bending reliability than conventional ACFs, and can be successfully applied for ultra-fine pitch applications such as 8K UHD and VR displays. This is because the PAN APL has very high modulus, which increase the overall modulus of PAN APL ACFs system.

Keywords- chip-in-flex (CIF); anchoring polymer layer (APL); anisotropic conductive films (ACFs); fine pitch assembly; bending properties

I. INTRODUCTION

Wearable devices are still emerging in the semiconductor industry and show rapid technology development over time [1-3]. There are a variety of wearable products such as flexible smartphones, smart watches and rollable displays. Wearable devices work well in repetitive bending environments, but package component-level bending stability is different. Therefore, much research was done for package reliability at bending environment. However, there was a limit that the number of bending was small and the bending radius was large. [4, 5] Our research group has previously introduced the Chip-on-Flex (COF)/Chip-in-Flex (CIF) packages as a flexible package structure. [6] In addition, the material property that has the greatest impact on the bending reliability of ACFs joints was reported as the ACFs resin modulus. [7] Silica fillers can be added to increase the modulus of resin. However, when many silica fillers are added, the ACFs film is easily broken and bendability is reduced. [8] The APL ACFs have been introduced as new ACFs that can suppress conductive particle movements in resin flow. The APL ACFs have a high capture rate of 90% or more of conductive particles, 100% electrical insulation at 20 μm pitch, and provides reliable reliability test results as reported previously. [9] Because of high tensile strength of APL structure, APL ACFs have high modulus, which will affect the bending properties of flexible packaging. In this study, APL ACFs were used to CIF packages and investigated the bending reliability compared with conventional ACFs and APL ACFs. As a result, by using the APL ACFs with higher modulus in CIF packages, higher bending reliability can be obtained for ultra-fine pitch flexible displays.

II. TEST VEHICLES

For test vehicles, 30 μm thick Si chip and 60 μm thick flexible substrates were prepared. Si chip has a dimension with 10 mm x 10 mm size and 298 Au bump. Au bump has a size of 150 μm x 150 μm. Flexible polyimide substrate has size of 40 mm x 20 mm and Cu electrode with electroless nickel immersion gold (ENIG) metal finish. Cu electrode has a dimension of 150 x 150 μm, 12 μm height and 300 μm pitch. Flexible substrates have daisy chain resistance pattern. Figure 1 shows test vehicles design of the silicon chip and the flexible substrate.

III. EXPERIMENTS

A. Fabrication of CIF package structure using PAN APL ACFs

For PAN APL solution, PAN was dissolved in Dimethylformamide (DMF) solvent. And then Au/Ni coated conductive particles (10 μm diameter) were added to the PAN APL solution. After PAN APL coating, the DMF solvent was dried in a drying zone during film coating process, and PAN APL incorporating the conductive particles was fabricated. The PAN APL structure was made by using roll-to-roll film coating process as shown in Fig. 2. [10, 11] To make PAN APL ACFs, Nonconductive films (NCFs) composed of a thermosetting resin and curing agents was also prepared to fabricate ACFs structure. And top and bottom sides of the PAN APL film was laminated by two 10 μm thick NCFs using a roll laminator.

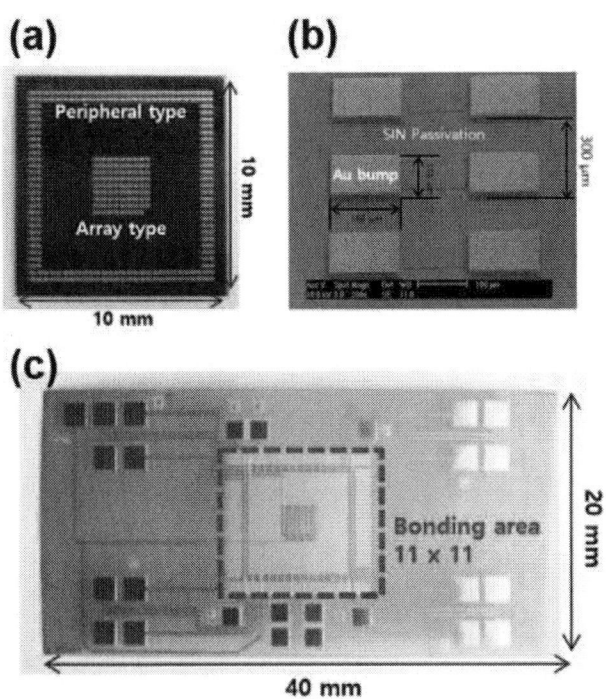

Figure 1. Test vehicles: (a) Chip design, (b) Au bumps on a chip, and (c) flexible substrate.

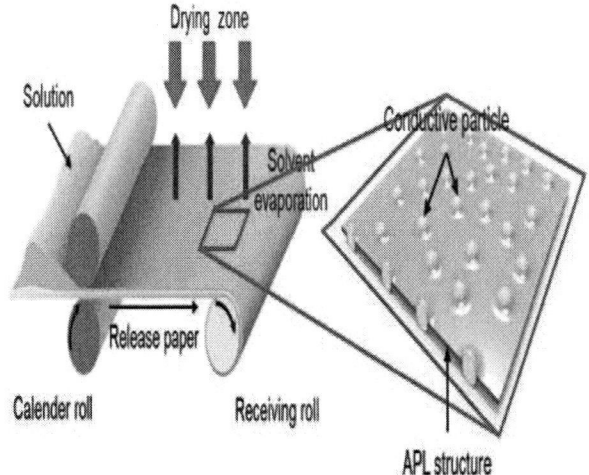

Figure 2. Schematic fabrication images of APL structure using a roll-to-roll film coating process

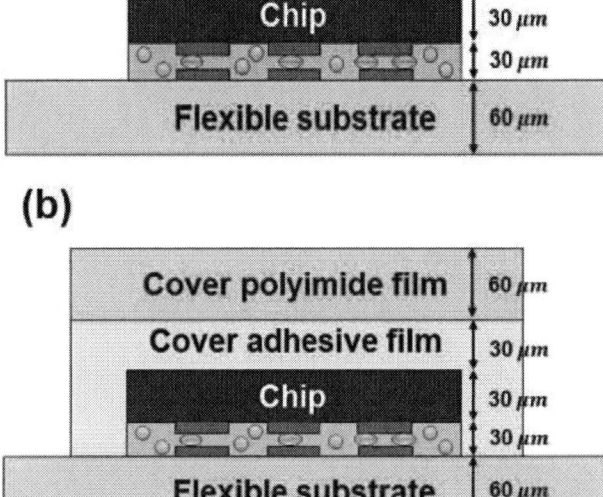

Figure 3. Schematic diagrams of (a) the COF and (b) the optimized CIF packages.

For conventional ACFs, the conductive particles were added to the thermosetting resin and curing agent mixture. 22 μm thick conventional ACFs were made by a roll-to-roll film coating process. Fig. 3 shows the schematic images of COF/CIF packages using APL ACFs. In order to fabricate the CIF packages, COF packages were fabricated first. After lamination of APL ACFs on flexible substrate, Si chip was flip chip bonded on APL ACFs laminated flexible substrate, and thermo-compression bonding was conducted at 170 ℃ for 10 seconds with 1MPa pressure. After fabrication of the COF package, 30 μm thick polymer cover adhesive film and 60 μm thick cover polyimide film was laminated on the COF packages by a roll laminator at 60 ℃. Cover adhesive film is an epoxy based non-conductive film (NCF). For full curing of a cover adhesive film, vacuum lamination was performed at 200 ℃ for 1 minute with 0.5 MPa pressure.

B. Characterization of CIF package using PAN APL ACFs

A tensile test was conducted to clarify the effects of APL structure on the suppression of conductive particles movement. PAN is known as high tensile strength and high modulus material. And the tensile strength of conventional and PAN APL ACFs were also measured. Since conventional ACFs and PAN APL ACFs tensile strength specimen were cured at 170 ℃ for 2 hours in an oven. The tensile test mode was the D638 ASTM condition with the test speed of 10 mm/min at room temperature using universal testing machine. The conductive particles movements were observed by an optical microscope using conventional and PAN APL ACFs at the same bonding condition. The capture rate of conductive particles was defined as the ratio of the number of conductive particles on a bump area before and after bonding process. The applied bonding conditions were 170 ℃, 1 MPa, for 10 seconds. After thermo-compression bonding, final gap height was observed by Scanning Electron Microscope (SEM).

C. Dynamic bending test of the CIF packages using PAN APL ACFs

2-axis dynamic bending test machine was prepared to evaluate the bending reliability of CIF packages as shown in Fig. 4. For uniform bending curvature, the 188 μm thick polyethylene terephthalate (PET) film was attached on the CIF package at the sliding edge sides. The bending direction is a convex bending to apply the cyclic tensile stress on ACFs joints. The cyclic bending test was conducted by repeating the bending radius from 30 mm, which is almost flat, to 6 mm with a frequency of 1 Hz. During the dynamic bending test, in situ daisy chain resistance was recorded by the soldered Cu wires on the flex substrate.

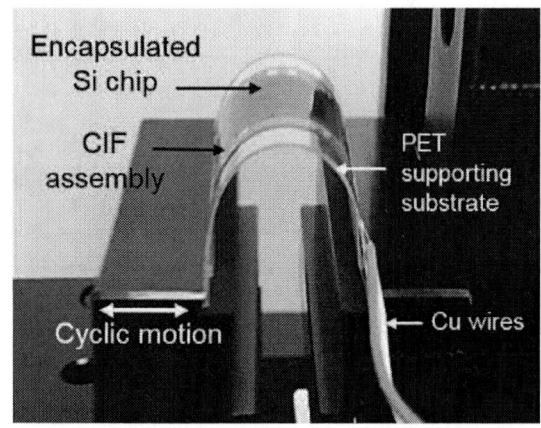

Figure 4. The dynamic bending test apparatus with a loaded CIF specimen

IV. RESULTS & DISCUSSION

A. Fabrication of CIF package structure and PAN APL ACFs

The PAN APL structure incorporating conductive particles was fabricated using a roll-to-roll film coating method. After the fabrication of PAN APL structure, top and cross-sectional morphologies were observed as shown in Fig. 5. The conductive particles have same phase in the PAN layer, which is PAN APL structure, and the thickness of the PAN layer for incorporating the conductive particles is about 1.65 μm. To use PAN APL structure for fabricating ACFs, NCFs were laminated using a roll lamination method. And CIF packages were made by conventional and PAN APL ACFs.

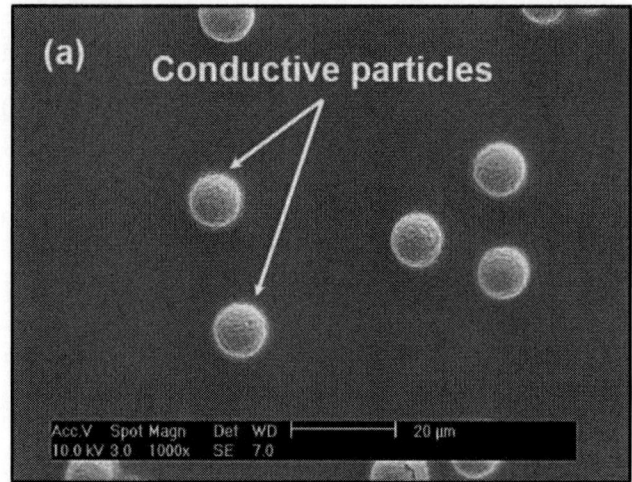

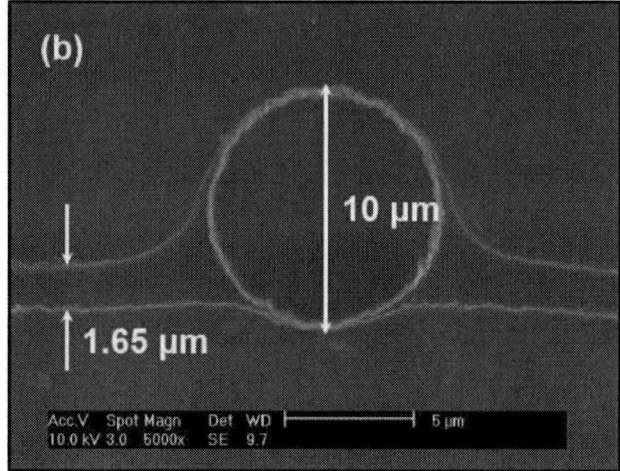

Figure 5. SEM image of the (a) top and (b) cross section of the PAN APL structure

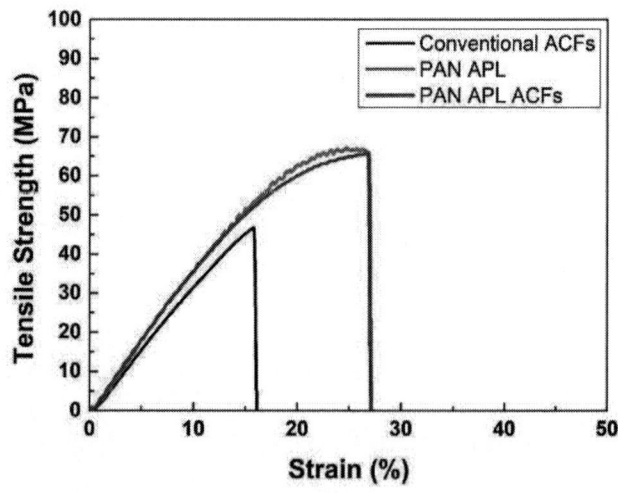

Figure 6. The tensile test results of convention ACFs, PAN APL structure, and PAN APL ACFs

978-1-7281-1500-9/19 $31.00 © 2019 IEEE

B. Characterization of CIF package using PAN APL ACFs

Conventional ACFs showed about 45.3 MPa ultimate tensile strength (UTS), however, PAN APL ACFs have about 67.6 MPa UTS because of high tensile strength of PAN APL structure as shown in Fig. 6. It is expected that higher capture rate of conductive particles and stable bending properties for CIF packages using PAN APL ACFs.

In case of conventional ACFs, there were 37 conductive particles on a bump area, and only 18 particles remained on the bump area after bonding process because of resin flowing. However, for the PAN APL ACFs, the number of conductive particles per a bump area before bonding process were 20, and remained 19 after ACFs bonding as shown in Fig. 7.

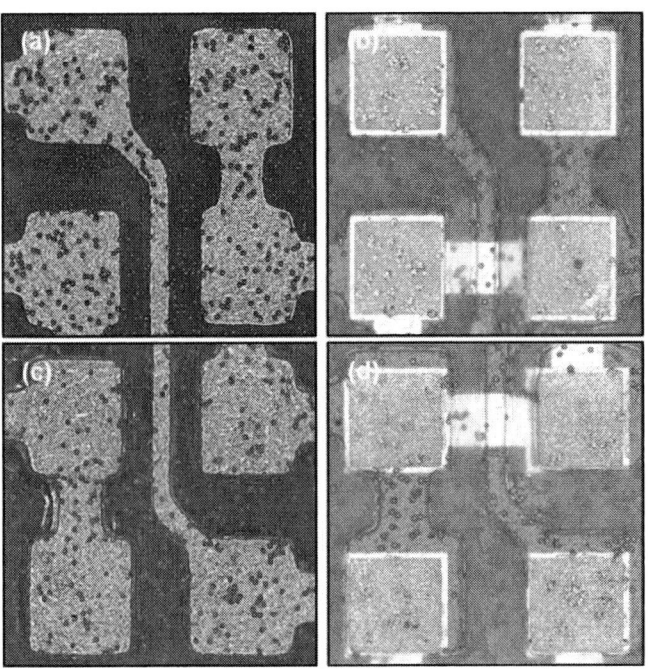

Figure 7. Optical microscope image of (a) before and (b) after bonding process using conventional ACFs and (c) before and (d) after bonding process using PAN APL ACFs.

In terms of capture rate of conductive particles, PAN APL ACFs showed 90 %, about 2 times higher than that of conventional ACFs. And the conventional ACFs and PAN APL ACFs have stable contact joint morphology and similar joint height as 4.6 μm and 4.5 μm respectively as shown in Fig. 8. The average daisy chain resistance of two types of ACFs were similar as 23.5 Ω and 24.9 Ω, because of the similar joint height and the number of captured particles between bump and electrode after ACFs bonding process.

C. Dynamic bending test

In order to evaluate the dynamic bending reliability, the conventional ACFs and APL ACFs bonded CIF packages was prepared. Fig. 9 shows the ACFs joint contact resistance increasing rates versus the dynamic bending cycles at the bending radius of 6 mm dynamic bending test. The failure criterion was defined as the twice of initial resistance. The conventional ACFs and APL ACFs bonded CIF packages failed at 50,000 and 120,000 cycles respectively. APL ACFs are a sandwich structure with the APL structure with 2 NCFs layer, and NCFs are the same resin composition as that of conventional ACFs. Therefore, the adhesion between the polymer resin and the electrode can be considered as same. By adding the APL structure, the ACFs modulus increased from 1.06 GPa to 1.74 GPa, and the dynamic bending reliability was increased about 2.5 times. During the dynamic bending test, The ACF with high modulus kept the conductive ball well without the resin crazing at repetitive bending stress, resulting in high bending reliability.

Fig. 10 shows the cross-sectional SEM images of CIF packages using conventional and APL ACFs after dynamic bending test of 150 k cycles. Because the bending direction is convex bending, tensile stress is applied to ACF joints, and resin delamination usually occurs between ACFs resin and electrode of flexible substrate. It was confirmed that the resin delamination occurred at the interface between ACFs resin and the electrode where the high stress was applied. There was no significant difference in failure mode between two types of ACFs. This is because the same resin composition was used in conventional and APL ACFs as described above, so there is no difference in adhesion.

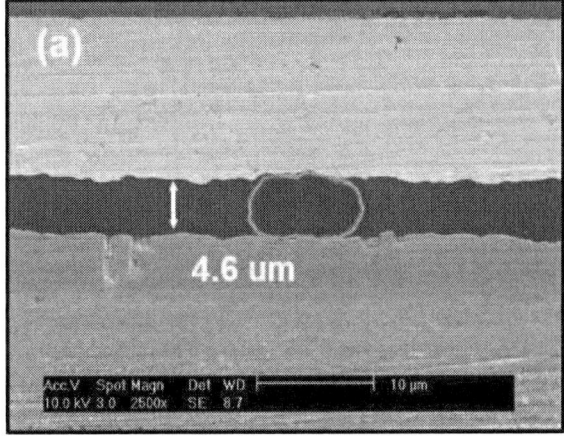

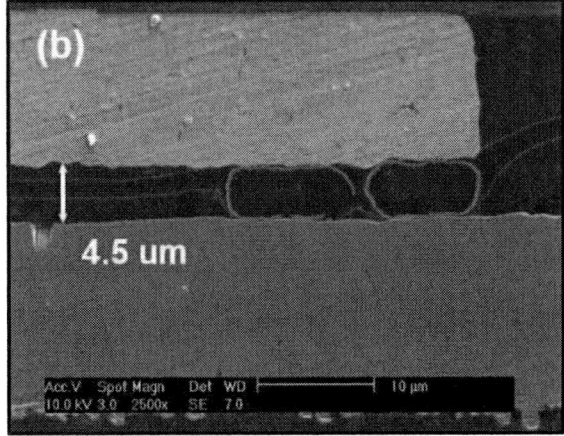

Figure 8. Cross sectional joint SEM images using (a) conventional and (b) PAN APL ACFs.

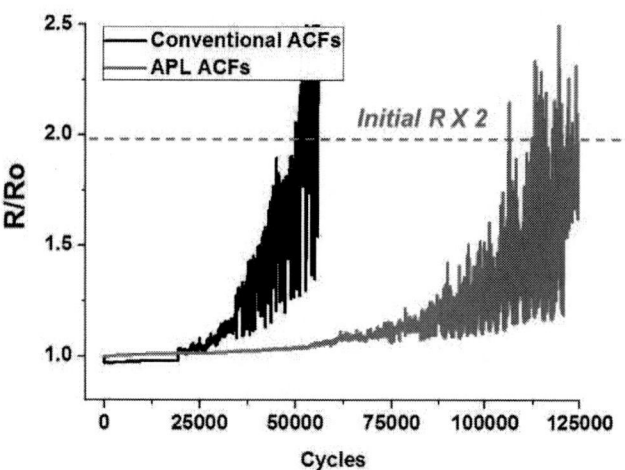

Figure 9. The in-situ daisy chain electrical resistances of CIF packages with the conventional and APL ACFs during a dynamic bending test at 6 mm bending radius.

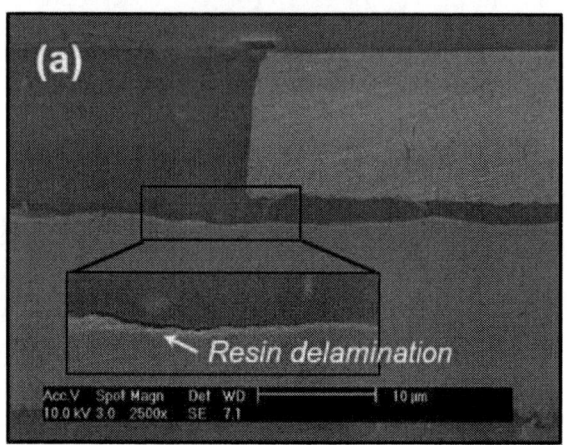

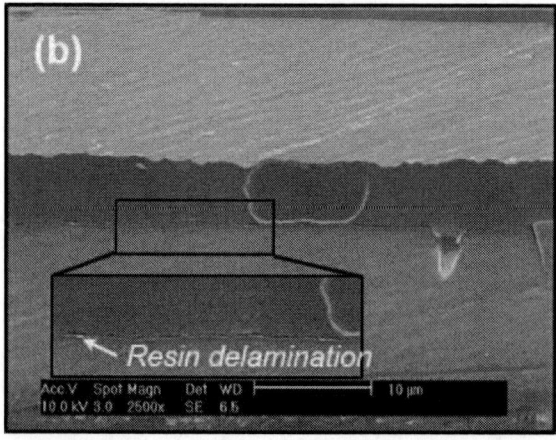

Figure 10. The cross-sectional joint SEM images of CIF packages with (a) conventional ACFs and (b) PAN APL ACFs after dynamic bending test of 150k cycles

V. CONCLUSION

In this study, the PAN APL ACFs were successfully fabricated, and the effects of the PAN APL structure on the dynamic bending properties of CIF packages were investigated. PAN APL ACFs significantly improved the conductive particles capture rate because of the higher tensile strength of PAN APL structure. As a result, PAN APL ACFs can reduce the amount of conductive particles as 1/3 compared with conventional ACFs. The PAN APL ACFs with higher tensile strength and higher modulus showed the stable daisy chain resistance changes compared with that of conventional ACFs in CIF packages. In addition, due to the higher modulus of PAN APL ACFs the bending properties increased more than 2.5 times compared with those of conventional ACFs. Since the PAN APL structure showed high mechanical properties, the highly reliable CIF package with improved dynamic bending performance was demonstrated using PAN APL ACFs. In addition, by combining the effect of suppression of conductive particles movement and CIF packages, 100 % of electrical insulation at 20 μm pitch and stable bending properties were obtained to be successfully used for wearable applications.

ACKNOWLEDGMENT

The authors would like to thank all the anonymous reviewers for their valuable comments and suggestions to improve the quality of this paper. J.-H. Kim and D.-J. Yoon contributed equally to this work. This work was supported by the Wearable Platform Materials Technology Center (WMC) funded by the National Research Foundation of Korea(NRF) Grant by the Korean Government(MSIP) (No. 2016R1A5A1009926).

REFERENCES

[1] Y.-S Rim, S.-H Bae, H. Chen, N.-D Marco, and Y. Yang, "Recent Progress in Materials and Devices toward Printable and Flexible Sensors," in *Advanced Materials*, vol. 28, no. 22, pp. 4415-4440, June. 2016.

[2] N. Palavesam, C. Landesberger, C. Kutter, and K. Bock, "Finite Element Analysis of uniaxial bending of ultra☐thin Silicon dies embedded in flexible foil substrates," in 2015 11th Conference on Ph.D. Research in Microelectronics and Electronics (PRIME), IEEE, Sep, 2015, doi: 10.1109/PRIME.2015.7251353

[3] R. Mitsui, S. Takahashi, and S.-I Nakajima, "Study on an Interconnect Technology toward Flexible Printed Electronics," in 2014 International Conference on Electronics Packaging (ICEP), IEEE, June, 2014, doi: 10.1109/ICEP.2014.6826702

[4] J. Heilmann, I. Nikitin, U. Zschenderlein, D. May, K. Pressel, and B.Wunderle, "Reliability experiments of sintered silver based interconnections by accelerated isothermal bending tests," in *Microelectronics Reliability*, vol. 74, pp. 136-146, July. 2017.

[5] C.-M. Lin, D.-C. Chen, and Y.-C. Liu, "Investigation on fracture and conductivity of flex-on-film flexible bonding using anisotropic conductive film considering repeated bending," in *Microsystem Technologies*, 2018.

[6] Y.-L. Kim, T.-I. Lee, J.-H. Kim, W.-S. Kim, T.-S. Kim and K.-W. Paik, "Effects of the Mechanical Properties of Polymer Resin and Types of the Conductive Ball of Anisotropic Conductive Films (ACFs) on the Bending Properties of Chip in Flex (CIF) Package," *Components, Packaging and Manufacturing Technology, IEEE Transactions on*, vol. 6, no. 2, pp. 200-207, Feb, 2016.

978-1-7281-1500-9/19 $31.00 © 2019 IEEE

[7]   J.-H. Kim, T.-I. Lee, J.-W. Shin, T.-S. Kim and K.-W. Paik, "Bending Properties of Anisotropic Conductive Films Assembled Chip-in-Flex Packages for Wearable Electronics Applications," *Components, Packaging and Manufacturing Technology, IEEE Transactions on*, vol. 6, no. 2, pp. 208-215, Feb, 2016.

[8]   S.-Y. Fu, X.-Q. Feng, B. Lauke, and Y.-W. Mai, "Effects of particle size, particle/matrix interface adhesion and particle loading on mechanical properties of particulate–polymer composites," *Composites Part B: Engineering*, vol. 39, no. 6, pp. 933-961, Sep. 2008.

[9]   D.J. Yoon, S.H. Lee, and K.W. Paik, "Effects of the Nylon Anchoring Polymer Layer on the Conductive Particle Movements of Anisotropic Conductive Films for Ultrafine Pitch Chip-on-Glass Applications," *IEEE Trans. Compon. Packag. Manuf. Technol.*, vol. 8, pp. 1723-1728, Oct. 2018

[10]   Yu, H. Xiang, Y. Long, N. Zhao, X. Zhang, and J. Xu, "Preparation of porous polyacrylonitrile fibers by electrospinning a ternary system of PAN/DMF/H$_2$O," *Mater. Lett.*, vol. 64, pp. 2407-2409, Nov. 2010

[11]   S.Y. Gu, J. Ren and Q.L Wu, "Preparation and structures of electrospun PAN nanofibers as a precursor of carbon nanofibers," *Synth. Met.*, vol. 155, pp. 157-161, Oct. 2005.

Effects of the Curing properties and Viscosities of Non-Conductive Films (NCFs) on the Sn-Ag Solder Bump Joint Morphology and Reliability

HanMin Lee[1], SeYong Lee[1], SangMyung Shin[1] TaeJin Choi[2], SooIn Park[2] and Kyung-Wook Paik[1]

1 Dept. of Materials Science and Engineering, Korea Advanced Institute of Science and Technology (KAIST), Daejeon, Korea
2 Doosan Corporation Electro-Materials BG, Seoul, Korea

E-mail : kwpaik@kaist.ac.kr

Abstract— In this study, solder bump flip chip assembly using NCFs was evaluated for Sn-Ag solder bumps. Flip chip bonding was performed using an isothermal Thermo-Compression (TC) bonding method for 5 seconds. Solder bump joints were evaluated by adjusting the curing properties such as curing onset, peak temperature, and degree of curing and viscosities of NCFs using curing agents and silica contents. And then, the degree of cure and viscosity approximations were conducted to define the precise viscosity of NCFs at the solder melting temperature using measured degree of cures at various bonding temperatures and viscosities. Finally, high temperature and humidity test (85RH%/85°C test) and temperature cycling (T/C) test were performed to evaluate the thermo-mechanical reliability performance depending on solder joint.

Keywords- Solder bump, Cu/solder hybrid bump, Flip-Chip, Thermocompression bonding, Non-conductive films(NCFs), PoP packages

I. Introduction

Recently, small form factor and high performance electronic devices are needed for mobile electronics. For this reason, memory packages are currently stacked on application processor (AP) packages using the Package on Package (PoP) structure. However, the total thickness of the PoP has been continuously increased because of stacking two kind of packages. Therefore, AP package component of the PoP needed to be assembled directly on the main PCB as the Chip on Board (COB) structure, and total thickness of the modules can be significantly reduced. In addition, there will be also an advantage of simplifying the assembly processes due to eliminating PoP assembly process. [1] In this reason, COB solder bump flip chip bonding using NCFs technology is necessary for stable solder joint between AP devices and main PCB using Sn-Ag flip chip solder bumps.

For conventional flip chip bonding processes using Sn-Ag solder bump, there are several processing steps such as flux application, solder reflow, and underfill dispensing processes. However, NCFs, acting as pre-applied type of underfill materials, can be applied to the flip chip wafers with Sn-Ag solder bumps. NCFs assembly processes have several advantages such as reduced processing steps and wafer level process capability. [2, 3] In addition, higher speed isothermal thermo-compression (TC) flip chip bonding method with high heating rates can be also applied to the Sn-Ag solder bump bonding process using NCFs. [2, 4] Therefore, it is necessary to investigate the behavior of Sn-Ag solder and NCFs resin during the NCFs bonding process. Especially, NCFs resin can be trapped at edge of solder joint and solder mask areas during TC bonding process. And then, concave shaped solder bump joint formed at the metallurgical bonding area between solder and PCB metal pads. Eventually, these poor solder bump joint can cause crack initiation and early failure during reliability tests. Figure 1 is solder joint crack from NCFs trap area in solder joint after reliability. To remove the NCFs resin trap at the solder joint clearly, the curing properties and viscosities of NCFs should be investigated and optimized. Especially, the material properties of NCFs at the solder melting temperature which is almost 221°C determines the solder bump joint morphology during TC bonding process. Therefore, it is important to define and optimize the viscosity of NCFs at the solder melting temperature.

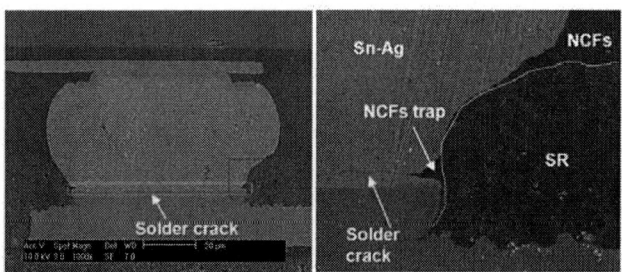

Figure 1. Solder joint crack at the solder joint initiating from NCFs trap area after TC reliability test.

II. Experiment

A. Materials preperation

NCFs were formulated using epoxy resins, curing agents, and additives such as flux, silica. Flux additive was included in NCFs to remove the native solder oxides during the bonding process. The NCFs mixed solution was ball-milled to make homogeneous resin. And then the resin was coated on a releasing film as 30 μm thickness film using a comma roll coater.

B. Test vehicles

Figure 2 and Table1 are the dimension of test chip and PCB. 4 mm × 4 mm size Si chip had 65 µm height and 80 µm diameter Sn-Ag solder bumps, and the pitch between bumps was 140 µm. And 18 mm × 20 mm size PCB had 100 µm width Cu pads and 60 µm width ENEPIG metal surface finishes. As shown in Figure 3, there were daisy chain patterns with 140 µm pitch.

Table 1. Specification of test vehicles

	Item	Size (µm)
Chip	Bump Pitch	140
	Bump Height	65
	Bump Diameter	80
Board	Pad Width	60
	Pad Finish	ENEPIG

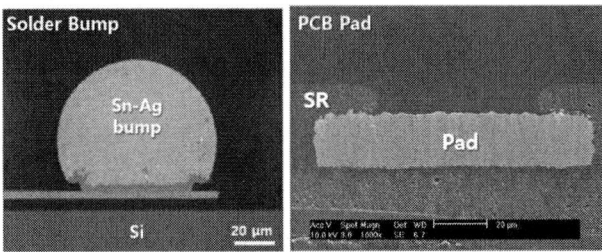

Figure 2. Images of a solder bump and a PCB metal pad

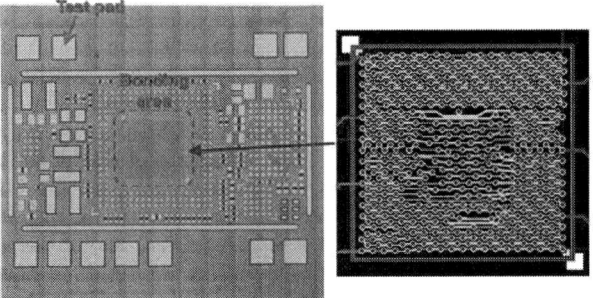

Figure 3. Images of the PCB substrate with the chip on board and daisy chain patterns

C. NCFs chracterization

NCFs were prepared to compare solder joint morphology depending on curing agent and silica contents. As curing properties of NCFs, the peak curing temperature was important, because the NCFs flow properties were rapidly reduced at this temperature because of curing reaction during the TC bonding processes. And the viscosity was important materials property which determines the solder joint formation. Therefore, material characterization was conducted to compare NCFs properties. Viscosity was evaluated using 1 mm thick samples by a rheometer with 5 °C per a minute heating rate during the measurement. And curing properties of NCFs were also measured with a differential scanning calorimeter (DSC).

D. TC flip chip bonding

Thermo-Compression (TC) flip chip bonding was performed with two steps. Alignment was performed at the flip chip bonder with optical system, and pre-bonding was performed at a bonding temperature of 20 N at a temperature of 50°C. After that, the main TC bonding was performed for 5 seconds at a maximum temperature of 190~250°C using a heated hot bar. Figure 4 shows the two step bonding method and measured bonding temperature profile.

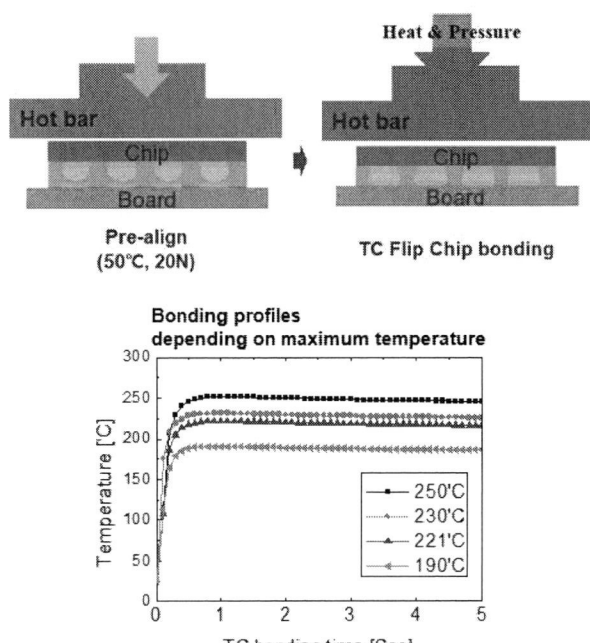

Figure 4. 2 step TC flip chip bonding method and measured bonding temperature profiles

E. Degree of cure and viscosity approximation

It is very difficult to physically measure NCFs degree of cure and viscosity at the solder melting temperature, because the rate of temperature rise in the bonding process is about 140 degrees per second. Therefore, in this study, an approximation was performed to derive NCFs degree of cure and viscosity at the solder melting temperature during the NCFs bonding process. The degree of cure was calculated by using an exponential function, because the changes of degree of cure with temperatures were normally similar as an exponential function. In this case, the exponential fitting was applied through the FT-IR degree of cure measurement value for 1, 2.5, and 5 seconds of bonding times. Figure 5 shows the calculated degree of cures with bonding times from the degree of cure approximation.

Viscosity at solder melting temperature could be obtained by derived degree of cure and viscosity approximation. Viscosity approximation is the process of deriving the viscosity value according to the degree of cure based on the viscosity result measured by the rheometer. At this time, one assumption is that the minimum viscosity point was the point where the degree of cure was 0%, and that the highest viscosity where the viscosity saturates was 100% cured. Figure 6 shows the calculated viscosity with degree of cure from the viscosity approximation.

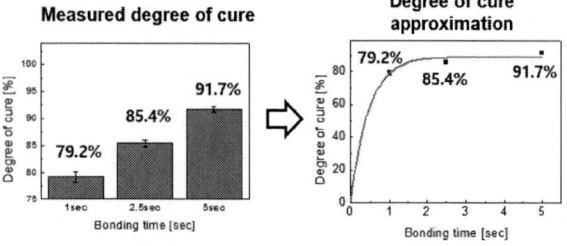

Figure 5. Measured degree of cures using the FT-IR measurement and calculated degree of cures with bonding times from the degree of cure approximation

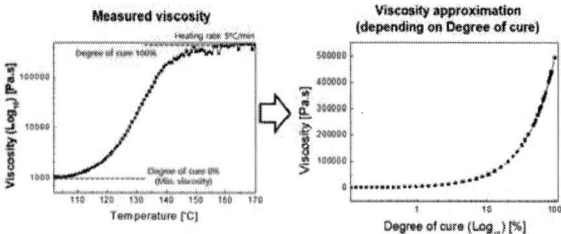

Figure 6. Measured viscosities using a rheometer and calculated viscosities with degrees of cure from the viscosity approximation

III. RESULTS AND DISCUSSION

A. TC bonding and Solder joint depending on curing agent contents

In order to realize a stable solder joint, TC bonding process was conducted depending on curing agent contents. The curing agent was a cationic curing agent having fast curing properties due to the high heating rates of very short isothermal bonding. Figure 7 shows the results of the curing temperature measured by DSC according to the curing agent contents.

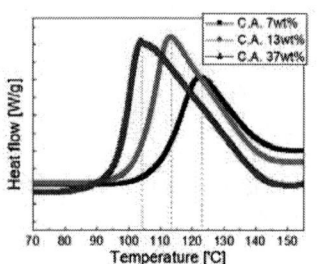

Figure 7. Curing onset and peak temperatures depending on curing agent contents

Using three curing agent contents, TC bonding process was evaluated at the same bonding temperature and time (250°C for 5 seconds), and solder joint were investigated by bonding conditions of 60 N and 180 N. Figure 8 shows the solder joint morphology with various bonding forces and curing agent contents.

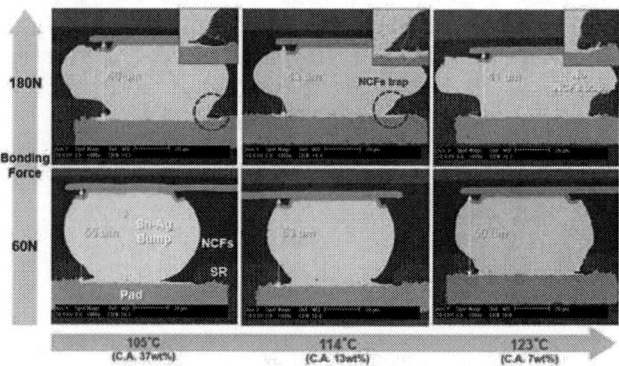

Figure 8. Solder joint morphology depending on bonding forces and curing agent contents

NCFs trap was effectively removed as bonding forces became higher and the curing agent contents were lower. In particular, the NCFs trap was completely removed at a bonding pressure of 180 N under the condition of 123°C curing peak temperature. Therefore, the curing agent content and curing temperature condition could be fixed at 7 wt% of cationic curing agent and 123°C of curing peak temperature.

B. TC bonding and Solder joint morphology depending on silica contents

The effect of NCFs trap formation on the solder joint was investigated according to the minimum viscosity. The properties of the NCFs were evaluated according to the silica content added using the optimized curing agent content. The effect of silica content was evaluated by adding 0, 6, 8, and 10 wt % according to the measured minimum viscosity by adding silica fillers. NCFs trap was not generated under the condition of less than 6 wt%, which was corresponded about 2500 Pa·s. On the other hand, NCFs trap occurred at 8 wt% or more condition. Additional evaluations were made on the influence of the viscosity of the solder melting at 221°C and the minimum viscosity of the NCFs to determine the NCFs trap formation. Figures 9 and 10 show the minimum viscosity, which varies with the silica contents, and the resulting solder joint morphologies.

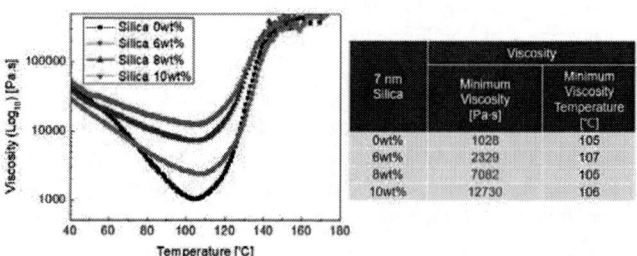

Figure 9. NCFs viscosities depending on various silica contents

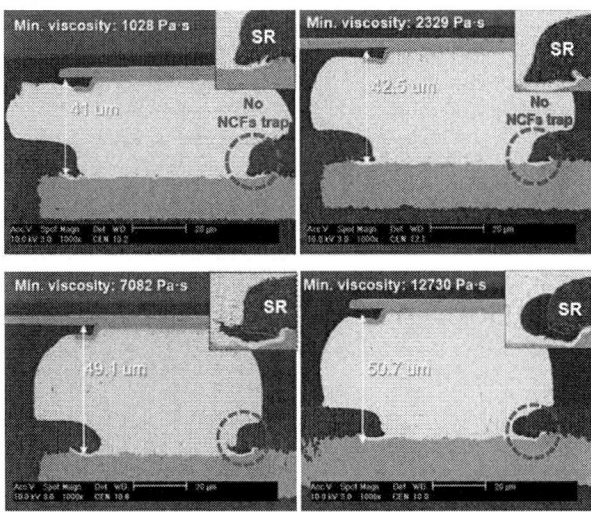

Figure 10. Solder joint morphologies depending on NCFs viscosities

To make NCFs property guidelines for stable solder joints, it is important to define the key factors of NCFs trap generation. Therefore, the solder joint shapes were compared according to the maximum bonding temperatures under the condition of without adding silica and 6 wt% of silica content condition showing no NCFs trap after TC bonding. The maximum bonding temperature was determined before and after solder melting temperature to define effect of minimum viscosity of NCFs on NCFs trap. Figure 11 shows the result of solder joint of without silica and 6 wt% silica content conditions according to the maximum bonding temperatures.

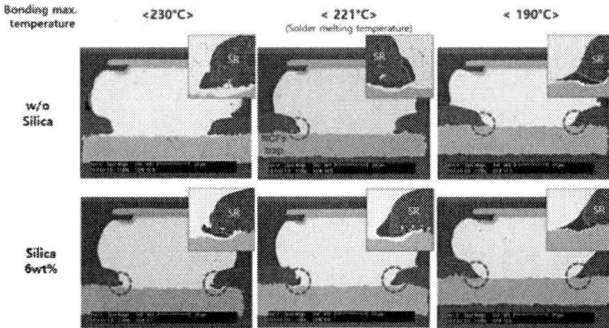

Figure 11. Solder joint morphologies according to the changes of the maximum bonding temperature at the content of without silica and 6wt% silica contents

As a result, the NCFs trap occurred again when the bonding temperature was decreased under the condition of the two silica contents in which the NCFs trap did not occur when the bonding was performed under the 250°C bonding condition. If the NCFs trap was due to the minimum viscosity, then NCFs trap should not occur at all the maximum bonding temperature conditions above the minimum viscosity temperature. It was found out that the occurrence of NCFs trap at the solder joint was not determined by the minimum viscosity. Therefore, the NCFs trap was determined not only by silica contents but also by bonding temperatures. This means that it was determined by the NCFs viscosity which was changing during TC bonding. Therefore, solder joint morphologies were investigated depending on the maximum bonding temperatures and silica contents to define the conditions of NCFs trap generation. As a result, the NCFs trap was determined at over 221°C. And, the silica content at which the trap started occurring at the bonding temperature of 250°C was 7 wt%.

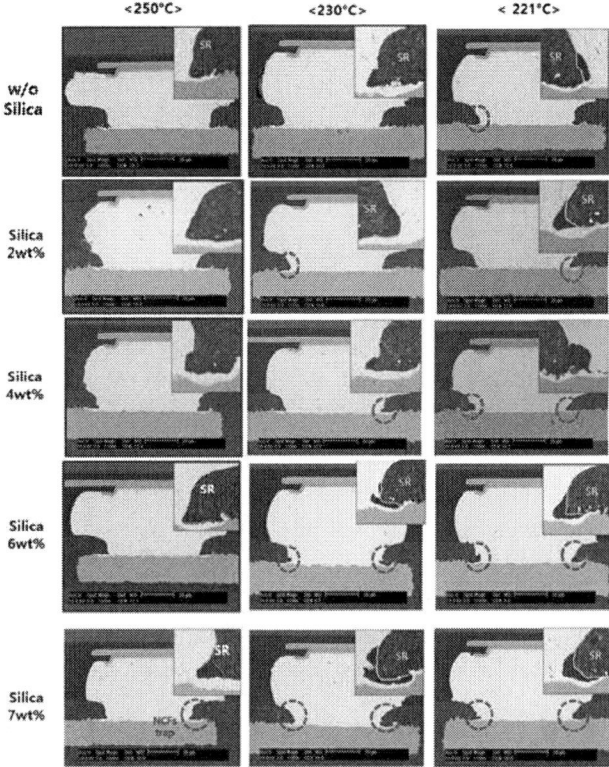

Figure 12. Solder joint morphologies depending on the maximum bonding temperatures and silica contents

C. Degree of cure and viscosity at the solder melting temeprature

The degree of cure was measured at each condition. The results were shown in Figure 13. When silica contents increased, degrees of cure were also increased after bonding. In terms of NCFs trap, it was similar trend in solder joint morphology depending on silica contents. However, the higher the maximum bonding temperature, the less NCFs trap occurred. It can be assumed that the increase of the bonding temperature presumably decreased the degree of curing at the solder melting temperature.

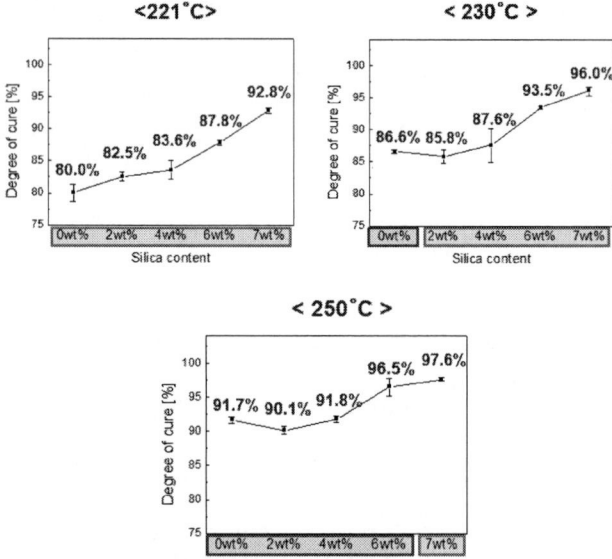

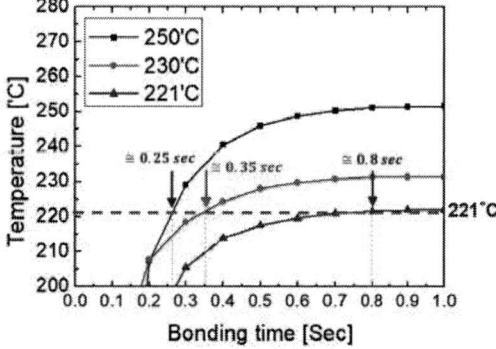

Figure 13. Degrees of cure measured after NCFs bonding depending on silica contents and maximum bonding temperatures

In order to confirm the assumption, the bonding time to reach the solder melting temperature was determined from the results of the bonding temperature profile according to each bonding temperatures. Figure 14 shows the bonding temperature profiles and bonding times for reaching the solder melting temperature. These bonding profiles are measured using thermocouples. Higher the bonding temperatures, shorter the bonding times required to reach the solder melting temperature. In other words, the degrees of cure at the solder melting temperature can be decreased as increasing the maximum bonding temperature. Based on these results, further analysis was conducted to confirm the degrees of cure and viscosities at the solder melting temperature.

Figure 14. Bonding temperature profiles depending on the maximum bonding temperature and times to reach the solder melting temperature

In this study, the degrees of cure at the solder melting temperature were calculated using the exponential approximation. First, for 1, 2.5, and 5 seconds bonding without silica conditions, the bonding points were measured. And the exponential function was obtained from measured

values. And then the degrees of curing according to bonding times were also derived. Figure 15 shows the result of the bonding times according to the bonding time at the without silica condition and the exponential approximation.

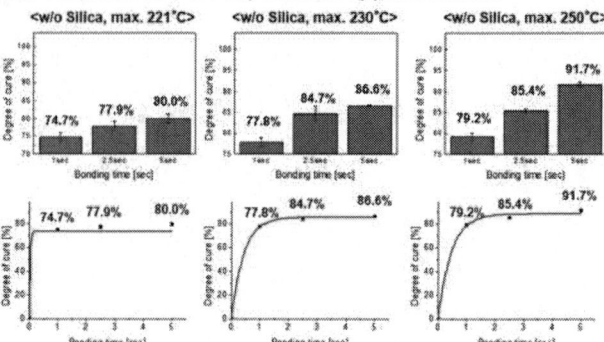

Figure 15. Degrees of cure approximation at without silica content using measured degree of cure

Finally, the degree of cure at the solder melting temperature can be determined with the time to reach the solder melting temperature. From the results, the degree of curing at the solder melting temperature decreased as the bonding temperature became higher as measured from the initial silica contents and bonding temperatures. Especially, at the maximum bonding temperature of 250 ° C, the degree of curing was negligible at 37.8%.

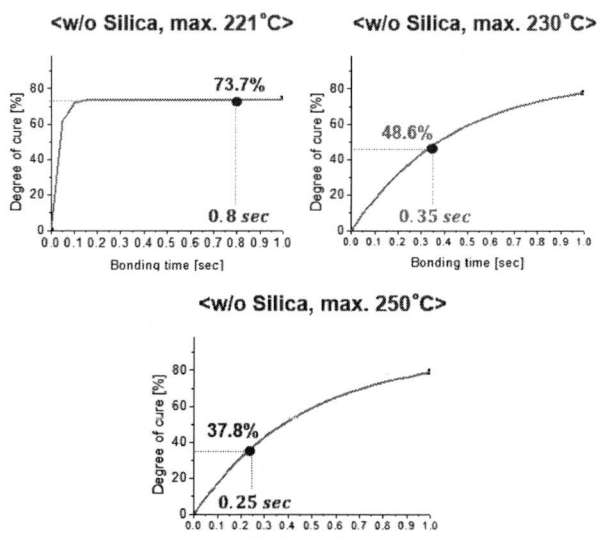

Figure 16. The degrees of cures at the solder melting temperature using the degree of cure approximation

Finally, the viscosity at solder melting temperature could be obtained by derived degree of cure and viscosity approximation. The results were classified according to the NCFs trap, and finally, the range of viscosity at the solder melting temperature with NCFs trap was derived. This result can provide a guide line for forming a stable solder joint. The NCFs trap occurs when the NCFs viscosities at the solder melting temperature were between 171,293 and 188,188 Pa·s, and consequently it can be concluded that the viscosity with stable solder joint without NCFs trap should be less than 171,293 Pa·s. Figure 17 summarizes the results of NCFs viscosity at the solder melting temperature and solder joint shapes for each bonding condition.

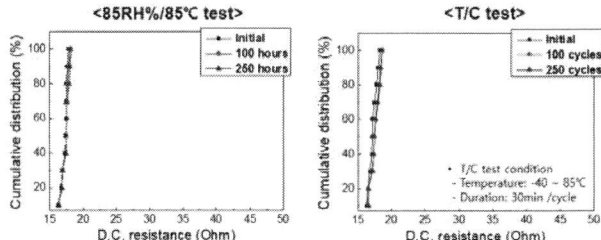

Figure 18. Reliability test results using the optimized NCFs condition

IV. CONCLUSION

In this study, flip chip assembly using NCFs was evaluated using conventional Sn-Ag solder flip chip bump structure. Various NCFs conditions with curing temperatures and viscosities were prepared to find out their effects on solder joint morphology during TC bonding process. And bonding process was also evaluated depending on the maximum bonding temperature. It was found out that stable solder joint without NCFs trap was determined by NCFs viscosity at the solder melting temperature. And then, degree of cures and viscosities at the solder melting temperature were calculated by the approximation based on measured. It was found out that NCFs trap was generated between 171,293 and 188,188 Pa·s of NCFs viscosity at the solder melting temperature. Finally, reliability tests was performed at optimized condition and it successfully passed the 85°C/85RH% and temperature cycling (T/C) reliability tests.

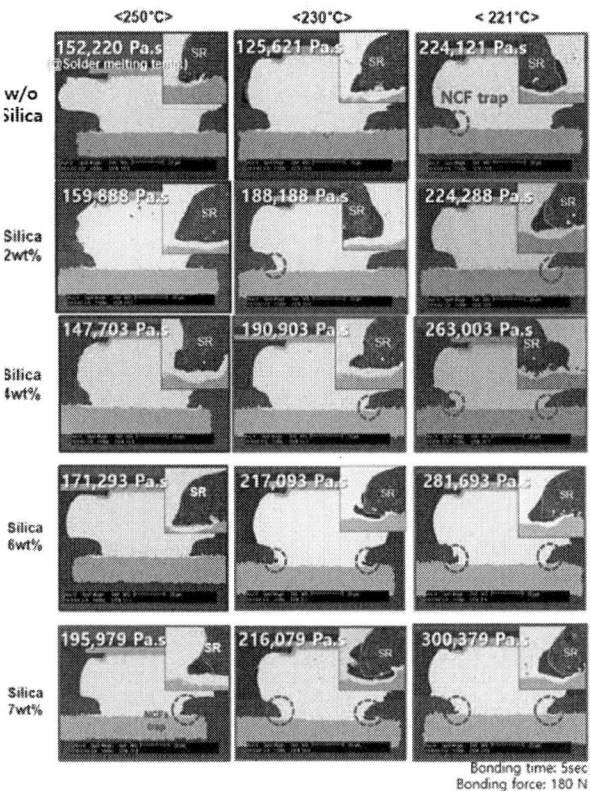

Figure 17. Solder joint shapes with various NCFs viscosities at the solder melting temperature varied by silica contents and bonding temperatures

D. Reliability

Finally, the sample was prepared at a bonding pressure of 180 N with an NCFs composition optimized as the curing peak temperature of 123°C and a silica content of 6 wt% (171,293 Pa · s at the solder melting temperature), and high temperature and high humidity test and thermal cycle test were followed. 10 samples were prepared for each measurement test conditions. As a result, it was confirmed that the resistance changes were stable at an average 10% or less compared to the initial value.

REFERENCES

[1] HanMin Lee, SeYong Lee, JongHo Park, Chan-kyu Chung, Kyung-Woon Jang, Il Kim, SeongWoo Choi, and Kyung-Wook Paik, "A Study on the Curing Properties and Viscosities of Non-Conductive Films (NCFs) for Sn-Ag Solder Bump Flip Chip Assembly", Electronic Components and Technology Conference (ECTC), 2018 IEEE 68th, pp. 2464 - 2469

[2] Hyeong-Gi Lee, Yong-Won Choi, Ji-Won Shin, and Kyung-Wook Paik, "Wafer-Level Packages Using B-Stage Nonconductive Films for Cu Pillar/Sn–Ag Microbump Interconnection", Components, Packaging and Manufacturing Technology, Volume 5 Issue 11, pp. 1567 – 1572

[3] Ji-Won Shin, Yong-Won Choi, Young Soon Kim, Un Byung Kang, Sun Kyung Seo, and Kyung-Wook Paik, "A novel double layer NCF for highly reliable micro-bump interconnection", Electronic Components and Technology Conference (ECTC), 2014 IEEE 64th, pp. 1755 – 1758

[4] Ji-Won Shin, and Kyung-Wook Paik, "A study on Non-conductive Film (NCF) for Fine-pitch Cu-pillar/Sn-Ag Bump Interconnection", Ph.D Dissertation, KAIST, 2014, pp. 1 – 30

Experimental investigations on vertical ultrasonic assisted low temperature sintering process

Henning Seefisch
Institute of Dynamics and Vibration Research
Leibniz Universität Hannover
Appelstraße 11, 30167 Hannover, Germany
Email: seefisch@ids.uni-hannover.de

Jens Twiefel
Institute of Dynamics and Vibration Research
Leibniz Universität Hannover
Appelstraße 11, 30167 Hannover, Germany
Email: twiefel@ids.uni-hannover.de

Abstract—The low temperature joining technology is an important joining technique for high power electronic components. In this process, a substrate and an electronic component are sintered with silver paste by pressure and temperature at approx. 250 °C. The industry often uses a batch process for this technique. Some disadvantages, however, lead to a limited use of this technique. During the process, high pressure and high process temperature have to be applied for a long process time (several minutes). As a consequence, for many components the conventional sintering technique cannot be used. In this work, ultrasonic assisted low temperature joining process is proposed and investigated. The additional ultrasound energy can substitute a part of the thermal energy and the external heating energy can be thus greatly decreased. In this way, temperature sensitive components can be protected from high temperatures. In addition the vertical ultrasound excitation can better compact the sintered material. Finally, the porosity of the connection layer decreases and the strength and the conductivities of the connection increase. The shear strength could even be doubled with very short process times. The joining process is performed as a single placement, which provides for higher positioning accuracy and better control over the process. The ultrasonic assisted low temperature joining process and the connected components will be presented. The results of the investigations and potentials of the introduction of ultrasound will be emphasized. In particular, a comparison is drawn between the ultrasound assisted and the conventional low temperature joining technology. Here the focus is on the mechanical strength of the connection.

Keywords-Ultrasound; Low Temperature Joining Technology (LTJT); Sintering; Die Attachment; Silver Paste

I. INTRODUCTION

Direct chip assembly is one of the most important connection techniques in power electronics. In this connection technique, various methods have been established. These include soldering processes [1], transient liquid-phase sintering (TLPS) [2], bonding processes [3], Adhesive processes [4], as well as the low temperature joining technology (LTJT) [5]. In this process, a silver-coated chip is connected with silver paste with mostly microscaled particles with a silver-coated substrate by a slow speed diffusion process. The paste consists almost exclusively of crushed flake-shaped silver particles which are mixed with organic ad-

ditives to form a suspension [6]. Low-temperature sintering has several advantages over the other processes, including the high mechanical and thermal strength of the compound, as well as due to the almost pure silver compound its very good electrical and thermal conductivity [5]. Furthermore, the pure silver-silver compound significantly increases the temperature change stability compared to solder and adhesion processes. Disadvantages of the low temperature joining technique are very high pressures and temperatures that can affect the sensitive electronic components in the connection process over a long period of time. Temperatures up to 250 °C, pressures up to approx. 40 MPa and process times of several minutes are required [5]. Since these high process parameters often lead to the destruction or damage of the electronic components, there are several investigations and publications to reduce these process parameters. In the diffusion process the particles strive to adopt a state of least free enthalpy, this process is usually very slow [7]. One approach is to reduce the particle size of microscale particles to nanoscale particles [8], [9]. The smaller particles provide a larger specific surface and thus accelerate the diffusion process. A disadvantage of the nanoscale particles is not only the handling and increased safety requirements at work but also the increased proportion of organic binders in the paste. The diffusion process can be further accelerated by a higher temperature and a higher pressure. The higher pressure provides more contact points and the higher temperature ensures greater activity of the particles. Our approach to positively influence these two factors is an additional source of ultrasound. This ensures on the one hand by the excitation of the particles for a greater activity and thus also for more contact points. In addition, a vertical excitation provides additional compaction of the paste. On the other hand, in case of large movement amplitudes, there is a significant energy input into the process in the form of frictional heat. In the next sections we present our results with a vertical excitation. Vertical in this case means that the ultrasonic amplitude is perpendicular to the sintered layer, see Figure 1.

978-1-7281-1500-9/19 $31.00 © 2019 IEEE

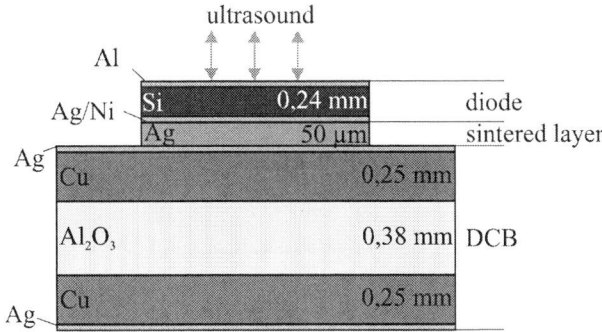

Figure 1. Schematic layers

II. MATERIALS AND EXPERIMENTAL SETUP

A. Materials

For our investigations we have used DCB substrates with a size of 28 mm x 23 mm and a thickness of approx. 0.9 mm from Rogers Corporation, coated on both sides with silver, see Figure 1. The substrate has two contact surfaces so bonding wires can be bonded to the anode side of the diode to investigate electrical properties, see Figure 2a. The sinter paste used is a paste of microscale silver particles from Namics Corporation. The chips are SKCD 09 C 060 I HD diodes of the company Semikron, see Figure 2b. The diodes have a size of about 3 mm x 3 mm and a thickness of about 0.24 mm. We have oriented the paste layer thickness to other publications [5] and have chosen a layer thickness of 50 μm, which we initially keep constant.

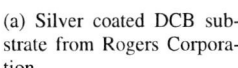
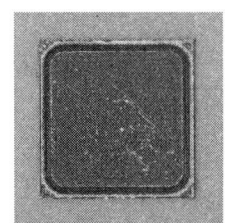

(a) Silver coated DCB substrate from Rogers Corporation

(b) SKCD 09 C 060 I HD diode from Semikron

Figure 2. Used materials

B. Sample Preparation

The experimental preparations and test conditions are very important in order to achieve consistent results that do not scatter too much. The preparation takes place at room temperature and atmosphere conditions. The DCB substrates are manually cleaned with isopropanol prior to coating and the paste is then applied to the substrates by stencil printing. Here, the paste is pulled over a stencil with a squeegee. Subsequently, the diode is placed on the paste before it heats up to sintering temperature. Using a Flir high-speed camera, the heating behaviour was investigated during the experiments. The measured heating profile of the substrate is shown in Figure 3. During these investigations we found that there was a temperature difference between the adjusted and the substrate surface temperature. At 185 °C the difference is 18 °C, so there is only a temperature of 167 °C at the substrate surface. The temperatures mentioned in this paper always refer to the adjusted temperature at the heating device. In order to ensure that a uniform temperature heating and distribution of the substrate takes place, this heat development was also investigated. Figure 4 shows that a very homogeneous temperature distribution in the substrate center is available.

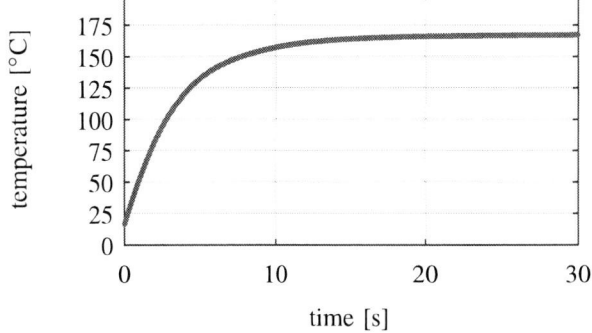

Figure 3. Heating profile of the substrate at 185 °C setpoint

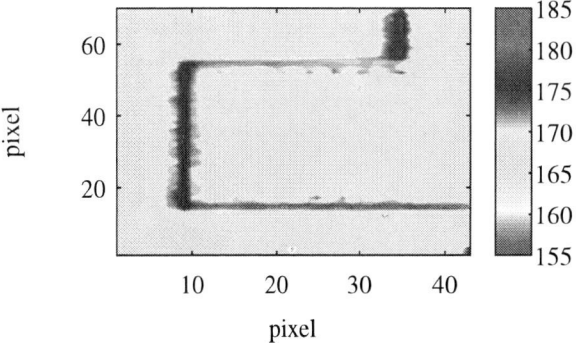

Figure 4. Temperature distribution on the substrate surface

After explaining sample preparation and the experimental preparation, the next section will discuss the experimental setup.

C. Experimental Setup

The test setup was developed entirely at the Institute of Dynamics and Vibration Research (IDS). The Figure 6 shows the linear guide and the drive unit to apply the forces.

With the test setup process forces of up to 1000 N can be applied, this corresponds to a pressure of 10 MPa for a chip size of 100 mm². The force is measured with DMS load cells from HBM and controlled by a suitable LabVIEW program. The substrate is heated from below by a resistance heating system from Hesse GmbH. The test bench control as well as the data acquisition is also carried out via the programming environment LabVIEW. The ultrasonic oscillation system shown in Figure 5b is an in house made $\lambda/2$ oscillator. In the free state, the transducer has a resonant frequency of approx. 20 kHz. In contact and with very high forces, the operating frequency shifts significantly. For ultrasonic signal generation and control, a PLL DPC 500/100k developed at the IDS is used [10], see Figure 5a. The PLL not only enables phase control but also current control. A Brül and Kjaer power amplifier type 2713 is used to amplify the voltage. In order to determine the mechanical vibration, the velocity at the transducer tip was recorded during all experiments by a 3D single-point laser doppler vibrometer CLV 3000 from Polytec.

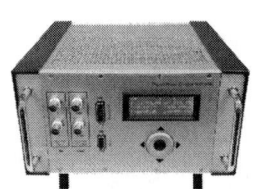 

(a) Phase Locked Loop for controlling   (b) 20 kHz transducer

Figure 5.   Ultrasonic equipment

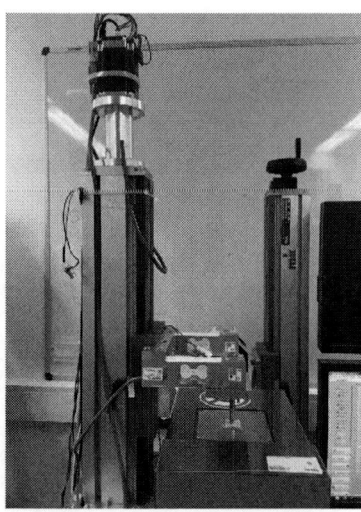

Figure 6.   Experimental test setup

D. Experimental sequence

This section describes the process flow with the different sintering phases. In order to better understand the process, it is divided and described in different phases. These phases are shown for illustration in Figure 7. First, the experiment begins with heating the substrate with the diode already placed. This time is kept as short as possible in the experiments presented here. Directly after the placement on the heater, the force control and thus the linear guidance is started. Only a few seconds pass before the first pressure is applied. During this heating phase, the waxes and other organic components, required for the production of the suspension, evaporate. Furthermore, the diffusion process is slowly accelerated already in this phase by the increasing temperature. The second phase begins with the application of the compression force and is determined by the length of the compression time. In this phase, the fastest diffusion takes place, since now the process temperature is reached and the target process pressure acts on the process. During this phase shortly after reaching the setpoint pressure, the ultrasound is switched on, so that the ultrasonic energy can provide an additional energy input into the connection layer. In case of vertical excitation, the ultrasound also ensures a higher compaction of the particles and a higher particle activity. All three processes accelerate the diffusion process. After the ultrasonic treatment, the pressure was released promptly and the linear guide was moved to its initial position. After a few seconds, the substrate was removed from the heater and temporarily stored on a heating plate at approx. 40 °C for approx. 2 - 3 minutes to prevent rapid cooling. The test bench has been programmed so that the pressure times can be freely adjusted before, during and after the ultrasonic treatment, so that different ultrasound treatments can be examined. In this publication only experiments in which the ultrasonic treatment was carried out during the entire pressure time are presented.

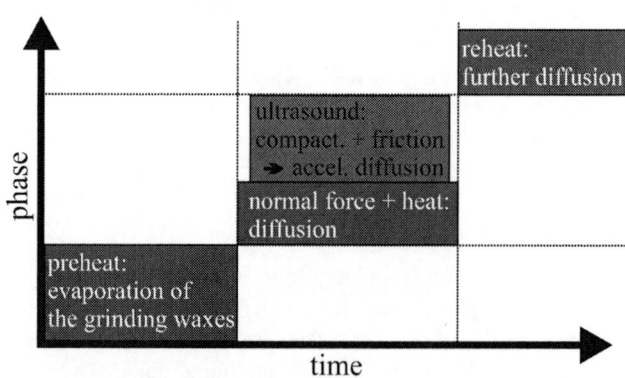

Figure 7.   Process sequence during sintering

E. Characterization techniques

There are many different methods for characterizing the different connection properties. In this publication the focus is on the mechanical strength of the connection, because the mechanical strength is a clear indication of advanced sintering. And thus it can be shown that the ultrasound treatment can provide for a faster and better diffusion. To characterize the mechanical strength, a linear shear tester was developed at the IDS, which is shown in Figure 8. For this purpose, the chips are sheared off the substrate with a shearing tool. The force is determined with DMS load cells from HBM. At present, shear strengths of up to approx. 45 MPa can be measured with the given diode size. This high shear strength has already been achieved in some experiments, as can be seen in the further course. In the shear tests, the diode is sheared from the substrate a few micrometers above the substrate at a constant height. It is possible to control the shear rate or force. In this investigations, the shear tests were carried out using the shear rate control system at 150 μm/s. Initially, the diode is approached very slowly and the shear rate is increased to 150 μm/s at the first increase in force.

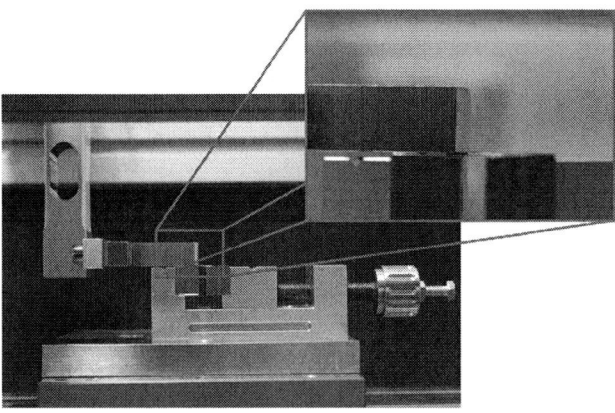

Figure 8.   Manual linear shear tester

III. RESULTS AND DISCUSSION

In the following sections the influence of different process parameters are shown. A comparison is always made between ultrasonic assisted low-temperature sintering technology and conventional LTJT. The experiments without ultrasound were carried out to ensure better comparability with the same process parameters. A comparison to other publications in which LTJT was investigated is not always reasonably possible, as the connection partners, process parameters, process conditions and investigation methods are very different. Since the experiments are subject to a certain variation, at least six experiments were carried out for each parameter set. The other parameters were kept constant during the investigation of one parameter. These constant parameters are located in the middle of the examined parameter

field and are now referred to as standard parameters. The following standard parameters have been defined: process time: $t = 10\,\mathrm{s}$; process temperature: $\vartheta = 195\,°\mathrm{C}$; process pressure: $\sigma = 10\,\mathrm{MPa}$; ultrasound voltage: $U = 10\,\mathrm{V}$. In the following Figures 9 - 11 the blue solid lines are the measurements without ultrasound and the red dotted lines are the measurements with ultrasound.

A. Process time

This section deals with the influence of the process time. The process time refers to the time in which the force and the ultrasound act, see section II-D. The process time was examined from one second to 60 seconds in 10 second steps. It can be clearly seen in Figure 10 that the shear strength increases only slightly during measurements without ultrasound and reaches the maximum of approx. 20.5 MPa after approx. 30 seconds. The shear strength during the experiments with the ultrasonic treatment reaches the same shear strength already after 10 seconds process time, thus the process time can be halved to reach the same shear strength. The shear strength in the ultrasonic experiments also reaches the maximum after approx. 30 seconds with approx. 30 MPa, which means an increase of 75 %. This shows that the movement induced by the ultrasound greatly accelerates the diffusion, thus the time in which the electrical components are exposed to high temperatures can be significantly reduced.

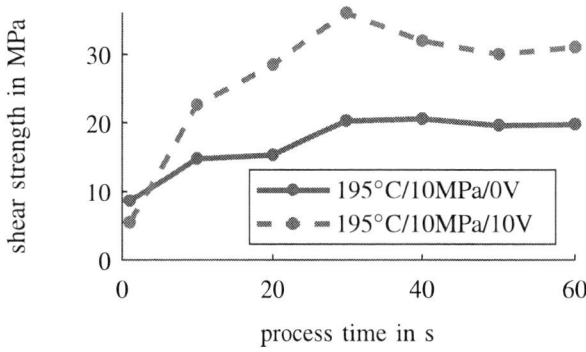

Figure 9.   Comparison of the tests with and without ultrasound variing the processstime

B. Process temperature

Now the parameter temperature is discussed, which is examined from 150 °C to 210 °C in 15 °C steps. As known from many publications, this parameter has a strong influence on the connection quality. This strong influence can also be best illustrated by Figure 10. From about 180 °C a diffusion process begins with assisted ultrasound, below this temperature there is practically no sintering. Therefore, ultrasound has no sufficient influence at low temperatures. However, from 180 °C after the onset of sintering, a positive effect can also be observed through the use of ultrasound. For the experiments without ultrasound the sintering starts only from about 195 °C which corresponds to a temperature reduction of 15 °C. Even at 195 °C higher shear strengths can be achieved with ultrasound than at 15 °C higher experiments at 210 °C without ultrasound.

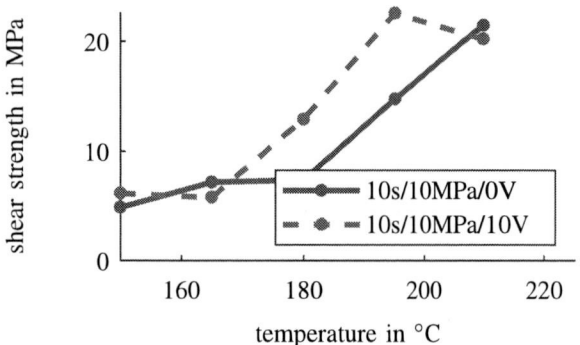

Figure 10.   Comparison of the tests with and without ultrasound variing the temperature

C. Process pressure

Another important parameter which has an influence with the use of microscale particles is the process pressure, which is why we have also examined this parameter more closely. The pressure was investigated from 1 MPa to 25 MPa in 5 MPa steps. Whereby 1 MPa can almost be described as pressureless. Since we carry out ultrasound examinations, it is not possible to carry out pressureless tests. The investigations show that ultrasound has a positive influence on the connection from a process pressure of 1 MPa. At higher pressures up to 25 MPa, this positive effect is even increased, see Figure 11. The measuring points at 25 MPa process pressure with assisted ultrasound were over 45 MPa shear strength and could therefore not be determined accurately with our setup. With these results it can be shown, that

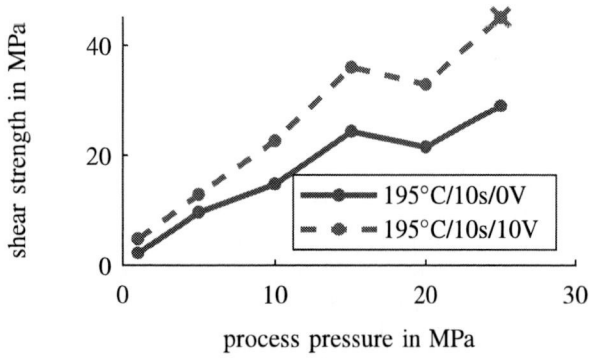

Figure 11.   Comparison of the tests with and without ultrasound variing the process pressure; (value marked with x exceeds the scale)

the transducer causes a compaction of the particles due to its oscillation. This means that the process pressure on the electrical component can also be significantly reduced.

D. Ultrasound voltage

In the previous section, the positive influence of ultrasound on various process parameters was already shown. Now the examination of the ultrasonic amplitude is presented. Here, the ultrasonic peak-to-peak voltage was changed between 0 V to 35 V in 5 V steps. By increasing the ultrasound amplitude, the positive effects of additional ultrasound can be made even clearer. Figure 12 shows, very impressively, that the shear strength was increased from about 15 MPa to about 30 MPa, which corresponds to a doubling of the shear strength. The Figure 12 also shows box plots to illustrate the mentioned scatter of shear strengths. The dispersion of the shear strength results has also decreased slightly towards higher values.

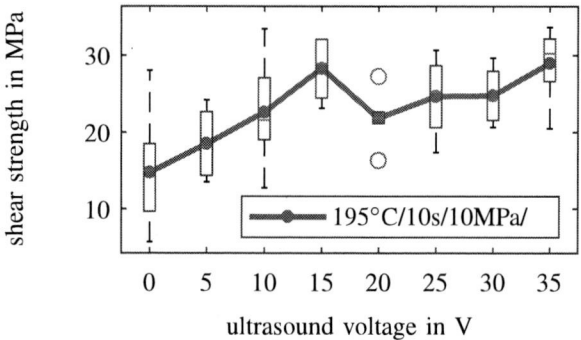

Figure 12.   Influence of the ultrasound on the shear strength

Furthermore, as the results from the previous sections show, it is to be expected that better process parameters than the desired standard parameters exist and that the positive influence of ultrasound can be further increased.

The scattering of the shear strengths corresponds to the scattering that is typical for sintering [8]. The mechanical strength of the diodes was partially achieved with very high shear strengths. And thus the strength of the diode was measured instead of the strength of the compound, as Figure 13 shows. This means that the connection could possibly withstand even higher loads. For this reason, we must either shift the parameter field in the direction of lower shear strengths or switch to another electrical component.

IV. CONCLUSION AND OUTLOOK

The low-temperature connection technology was successfully expanded with an additional vertical ultrasonic excitation. The ultrasound does not only cause an increased compaction of the particles but also an accelerated diffusion process. The mechanical strength of the connection could be doubled with suitable parameters. The process times

Figure 13. Reaching the shear strength of the diode

and process pressure can thus be reduced compared to conventional low temperature joining technology in order to achieve the same high strength. Thus we were able to show that it is possible to reduce the thermal and mechanical loads as well as their exposure time by using ultrasound, with which the application possibility of the LTJT for chip direct assembly can be extended. In addition to mechanical strength, other important parameters such as porosity, electrical conductivity and thermal shock resistance must also be investigated. Especially the temperature cycling load [11] is an important parameter for high performance devices.

ACKNOWLEDGMENT

The authors would like to thank the Hesse GmbH for their support in the investigations.

REFERENCES

[1] D. Feil, T. Herberholz, M. Guyenot, and M. Nowottnick, "Highly variable sn-cu diffusion soldering process for high performance power electronics," *Microelectronics Reliability*, vol. 76-77, pp. 455–459, 2017.

[2] F. Kato, F. Lang, H. Nakagawa, H. Yamaguchi, R. Kimura, K. Okada, H. Shindo, T. Ooi, R. Tamaki, S. Sekine, and H. Sato, "Thermal resistance evaluation of die-attachment made of nano-composite cu/sn tlps paste in sic power module," in *2017 International Conference on Electronics Packaging (ICEP)*. Piscataway, NJ: IEEE, 2017, pp. 125–129.

[3] Y. Long, J. Twiefel, and J. Wallaschek, "A review on the mechanisms of ultrasonic wedge-wedge bonding," *Journal of Materials Processing Technology*, vol. 245, pp. 241–258, 2017.

[4] Mohammed et al., "Fast cure conductive epoxy attach methodology for high speed automated pro-cesses," Patent US 9,566,764 B1, 2017.

[5] T. Herboth, "Gesinterte silber-verbindungsschichten unter thermomechanischer beanspruchung," Ph.D. dissertation, 2015.

[6] C. Mertens, "Die niedertemperatur-verbindungstechnik der leistungselektronik," Zugl.: Braunschweig, Techn. Univ., Diss., 2004, Düsseldorf, 2004.

[7] S. Klaka, "Eine niedertemperatur-verbindungstechnik zum aufbau von leistungshalbleitermodulen," Zugl.: Braunschweig, Techn. Univ., Diss., 1996, Braunschweig, 1996.

[8] G. Bai, "Study of property-microstructure relationships of sintered nanoscale silver for high temperature application," Ph.D. dissertation, Virginia Polytechnic Institute and State University, Blacksburg, Virginia, 2005.

[9] M. Knörr, S. Kraft, and A. Schletz, "Reliability assessment of sintered nano-silver die attachment for power semiconductors: 8 - 10 dec. 2010, singapore," 2010.

[10] S. Mojrzisch and J. Twiefel, "Phase-controlled frequency response measurement of a piezoelectric ring at high vibration amplitude," *Archive of Applied Mechanics*, vol. 86, no. 10, pp. 1763–1769, 2016.

[11] T. Herboth, M. Guenther, A. Fix, and J. Wilde, "Failure mechanisms of sintered silver interconnections for power electronic applications: 28 - 31 may 2013, las vegas, nevada, usa: Ieee 63rd electronic components and technology conference (ectc), 2013," 2013.

Pressureless Transient Liquid Phase Sintering Bonding of Sn-58Bi With Ni Particles for High-Temperature Packaging Applications

Kyung Deuk Min, Kwang-Ho Jung, Choong-Jae Lee, and Seung-Boo Jung*

School of Advanced Materials Science & Engineering
Sungkyunkwan University
Suwon, South Korea
e-mail: sbjung@skku.edu

Abstract—Transient liquid-phase sintering (TLPS) with nickel and Sn-58Bi was investigated for high-temperature packaging applications such as silicon carbide die attachment to direct bonded copper for a power module. Nickel was mixed with various ratios of Sn-58Bi (70, 80, and 90 wt%), which was then mixed with flux by a paste mixer to prevent oxidation and create a printable state. Thermal behaviors of three kinds of TLPS pastes were evaluated by a differential scanning calorimeter. The bonding temperature was 180, 200, and 220 °C for 90 min in air. The mechanical properties were examined by a bond tester. Shear strength increased with the temperature and Sn-58Bi weight ratio. The microstructures were observed by a scanning electron microscope. To investigate the atomic composition of the TLPS joints, an electron probe micro analyzer was used. Our results show that TLPS of nickel and Sn-58Bi is applicable for high-temperature packaging applications with reasonable cost, high re-melting temperature, and robust bonding.

Keywords- Transient liquid-phase sintering (TLPS), Nickel, Sn-58Bi, High heat-endurance bonding, High-temperature packaging, Pressure-less bonding

I. INTRODUCTION

As the importance of electric vehicles increases according to various environmental regulations, it is necessary to develop electronic devices such as power modules [1]. Until now, silicon-based semiconductors have been used in power semiconductors. However, silicon-based semiconductors require additional heat dissipation systems because of unavailability at high temperature. Furthermore, the power conversion efficiency is too low for use in future electric vehicles. A next-generation compound semiconductor has been widely investigated due to a high-power conversion efficiency and high-heat endurance property. Wide-bandgap semiconductors, such as silicon carbide (SiC), operate at about 200 °C, temporarily exceeding 235 °C [1-2]. Therefore, Sn-3.0Ag-0.5Cu (SAC305) solder, which is the most typical bonding material, is difficult to use as a bonding material of SiC because its melting point is 217 °C. As a result, a high heat-endurance bonding technology

is required to attach a SiC die to power module packaging. Recently, transient liquid-phase sintering (TLPS) has been widely investigated due to demands for a low-cost, low-pressure, short-bonding process. The materials of TLPS consist of high- and low-melting point metal particles with a mixed state so that the joint transforms into an intermetallic compound (IMC) faster than conventional transient liquid-phase (TLP) bonding [3-5]. Many studies have investigated TLPS bonding with Sn and high-melting temperature materials. However, the melting temperature and therefore minimum bonding temperature of Sn is 232 °C, which can damage a power module. M.K. Faiz et al. studied the TLPS bonding of Sn-58Bi (T_m = 139 °C) with Cu or Ag particles to reduce the bonding process temperature [4-5]. However, a reduction process is required for TLPS bonding with Cu due to its easy oxidation, and Ag is a high-cost material with limited application. Another unsolved issue is the bonding pressure that causes shortages through the bonding material and damage to the chip [6]. To overcome these issues, we investigated pressure-less TLPS bonding using Sn-58Bi with Ni particles to prevent damage to power module packaging without a reduction process. The TLPS paste was fabricated with a mixture of Sn-58Bi particles (35 μm in diameter), Ni particles (2.6 μm in diameter), and flux. The Cu-Cu bonding processes were performed at 180, 200, and 220 °C for 90 min in air. To investigate the mechanical property of the TLPS joints, a die shear test was performed using a bond tester. The bonding strength of the TLPS joints increased with increasing bonding temperature and Sn-58Bi weight ratio. The thermal behavior of the TLPS paste was analyzed with a differential scanning calorimeter (DSC). The microstructures of the TLPS joints and fracture mode were analyzed using a scanning electron microscope (SEM). The cross-sectional phase compositions were analyzed with an electron probe micro analyzer (EPMA). In our experimental results, a Cu-Cu joint was successfully achieved by the pressure-less TLPS bonding of Sn-58Bi with Ni particles in air. Therefore, this process may be promising for low-cost, heat-endurance bonding of SiC in power module packaging or other high-temperature applications.

978-1-7281-1500-9/19 $31.00 © 2019 IEEE

II. EXPERIMENT

A. Sample preparation

Three kinds of TLPS pastes were fabricated with nickel particles (2.6 μm in diameter), Sn-58Bi particles (35 μm in diameter), and flux. The weight content ratios of nickel and Sn-58Bi were mixed at 10 to 90, 20 to 80, and 30 to 70, respectively, then mixed with flux using a paste mixer (Paste mixer, ARE-310, Thinky, Japan) at 1000 rpm for 5 minutes. After fabrication, the thermal behaviors of the TLPS pastes were evaluated by DSC (DSC7020, SEICO INST., Japan) in a nitrogen atmosphere at a heating rate of 10 °C/min. TLPS pastes were printed on a pickled Cu plate using a stencil mask with a square opening side length of 3 mm x 3 mm. The Cu dummy chips (3 mm x 3 mm) were placed on the printed TLPS paste. The samples were put in the oven at bonding temperatures of 180, 200, and 220 °C for 90 min under a pressure-less condition in air. After the bonding process, the samples were cooled to room temperature. The residual flux was removed by a deflux solution, and the deflux solution was cleaned in ethanol with sonication.

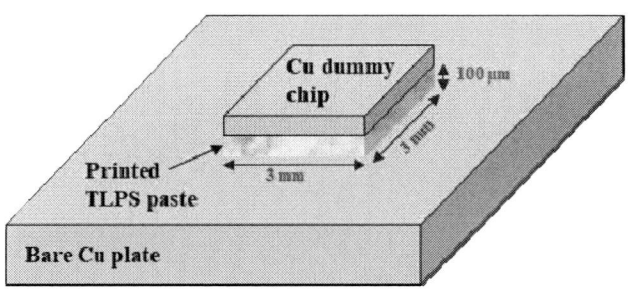

Figure 1. Sample preparation of TLPS bonding (Cu-Cu bonding).

B. Mechanical properties and microstructure

To investigate the mechanical properties of the TLPS bonded joint, a low-speed shear test was conducted by a bond tester (Dage4000, Nordson, UK). The shear speed and shear height were 200 μm/s and 100 μm, respectively. The samples were mounted by epoxy resin and were polished using sandpaper. The cross-sectional micrographs and that of the atomic compositions were observed by a scanning electron microscope (SEM, S-3000H, Hitachi, Japan). The atomic distribution of cross-sectional micrographs was observed using EPMA (JXA-8500F, JEOL, Japan).

III. RESULTS AND DISCUSSION

The representative schematic of our experiment is shown in Fig. 2. Stage 1 is before the bonding state, which is only the Cu chip on the printed TLPS paste. When the temperature was 139 ° C, reaching the bonding temperature [7], the Sn-58Bi melts and surrounds the Ni

particle by surface tension, as shown in Stage 2. The IMC reaction is then activated with Ni, Cu, and Sn-58Bi. After the bonding process (Stage 3), the Sn-58Bi phase reacts with Ni and Cu completely, and a residual Bi phase is formed inside the TLPS joint separately [8].

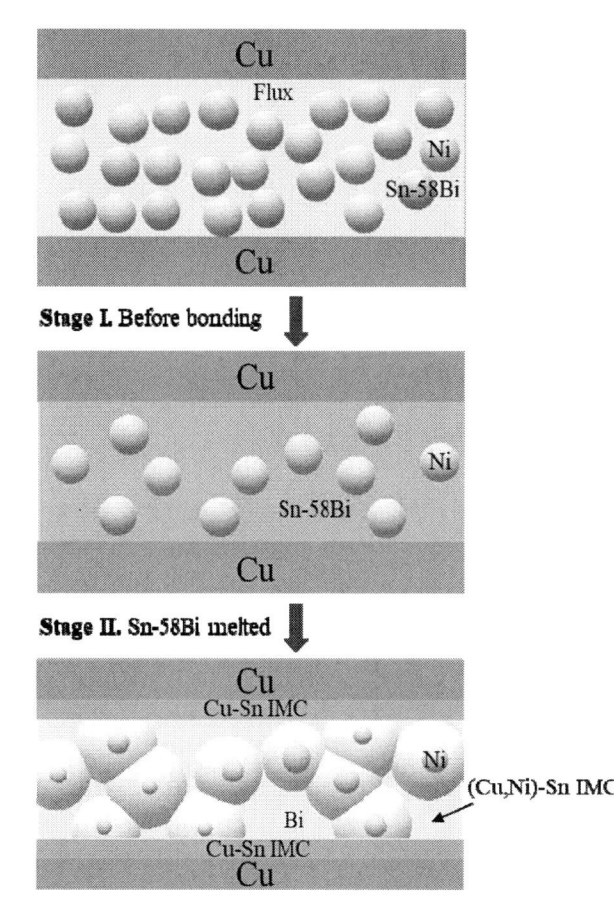

Figure 2. Schematic diagram of the TLPS bonding process with Ni and Sn-58Bi.

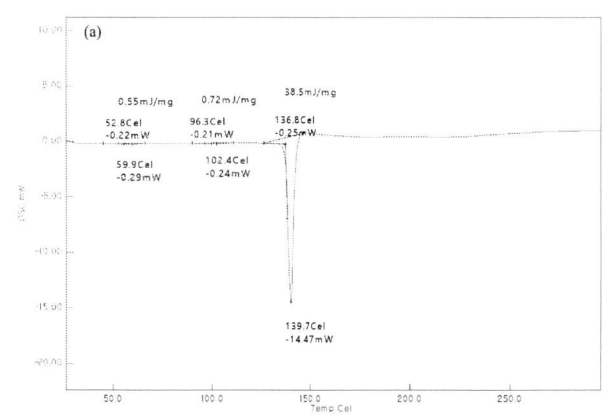

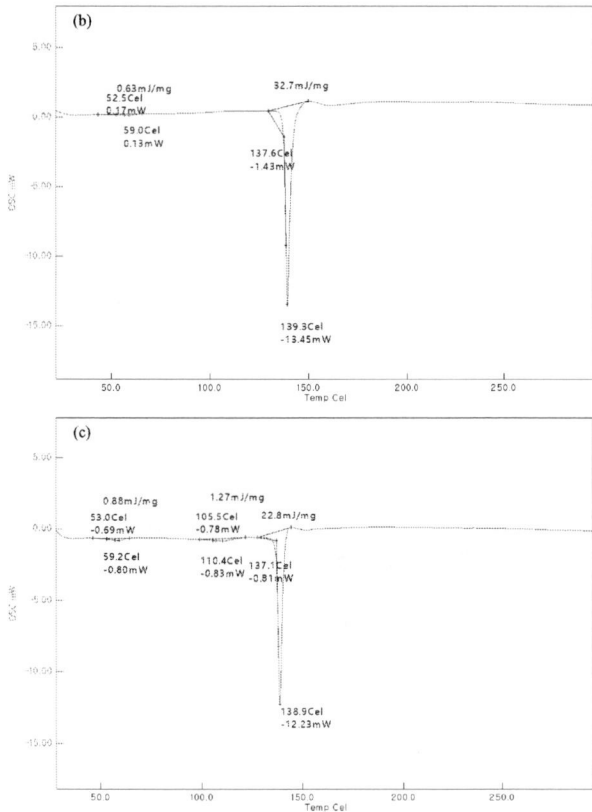

Figure 3. DSC analysis of (a) Ni-90(Sn-58Bi), (b) Ni-80(Sn-58Bi), and (c) Ni-70(Sn-58Bi) TLPS pastes.

Fig. 3 shows the thermal behavior of the fabricated TLPS pastes with various Sn-58Bi weight ratios. The melting temperature of eutectic Sn-58Bi alloy is 139 °C due to the mismatch between the crystal structure of Sn and Bi [9]. Therefore, endothermic peaks occur at 139 °C, which is the melting temperature of Sn-58Bi. The endothermic energy increases with increasing Sn-58Bi content.

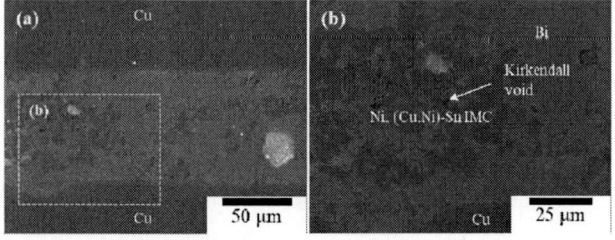

Figure 4. Representative microstructure of Ni and Sn-58Bi TLPS joints (bonding temperature: 220 °C, bonding time: 90 min).

Fig. 4(a)-(b) show the representative microstructure of the Ni and Sn-58Bi TLPS joint at a bonding temperature of 220 °C for 90 min. (Cu,Ni)-Sn IMC was observed inside the TLPS joint with reaction of Cu, Ni, and Sn. A Kirkendall void could be partially induced by forming

(Cu,Ni)-Sn IMC. Since Bi does not form an IMC with Sn, it remains in a single phase [10]. Despite the pressure-less bonding method, voids were rarely formed, except for Kirkendall voids.

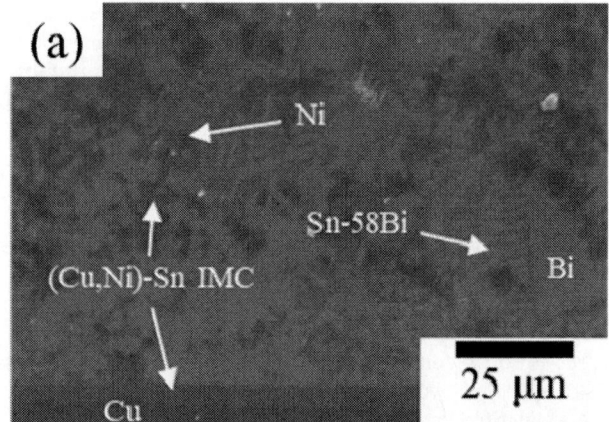

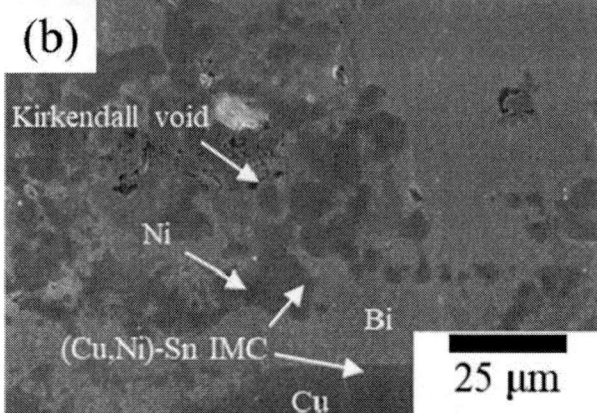

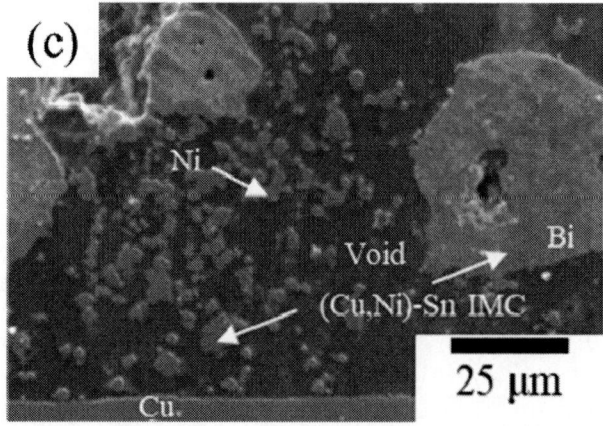

Figure 5. Cross-sectional micrographs of TLPS joints at 220 °C for 90 min (a: Ni-90(Sn-58Bi), b: Ni-80(Sn-58Bi), c: Ni-70(Sn-58Bi)).

TLPS joints with Ni-90(Sn-58Bi), Ni-80(Sn-58Bi), and Ni-70(Sn-58Bi) were produced to investigate the effects of weight ratio of Ni and Sn-58Bi. As shown in Fig. 5(a), the eutectic Sn-Bi lamellar structure remained in the TLPS joints because the Ni and Cu diffusion sources were insufficient in Ni-90(Sn-58Bi) for forming a full IMC with Sn. Some (Cu,Ni)-Sn IMC was observed surrounding Ni. With increasing Ni ratio to 20 wt%, the eutectic Sn-Bi lamellar structure disappeared, as shown in Fig. 5(b). After the bonding process, the re-melting temperature could be increased to the Bi melting temperature. Therefore, the amount of diffusion source (Ni) was enough to react with Sn to form the IMC completely. Additionally, some residual Kirkendall voids were generated with the IMC reaction [11-12]. In Fig. 5(c), voids were widely observed inside the TLPS joints. The melting material, Sn-58Bi, did not fill the bonded area. Therefore, formation of IMC occurred partially inside the joints, and the voids of the TLPS joints would be larger with higher nickel ratio due to Kirkendall voids or lack of low-melting temperature material (Sn-58Bi).

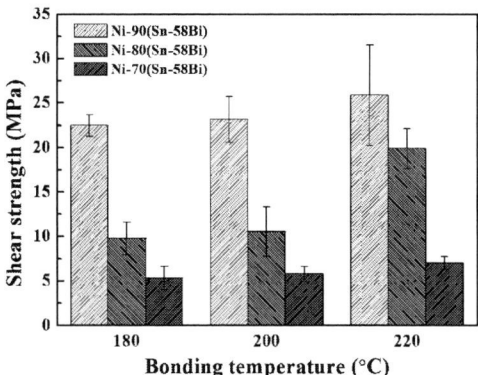

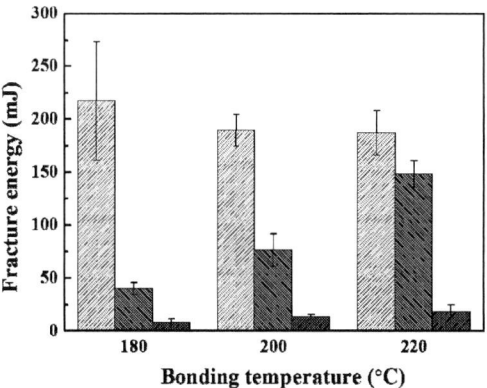

Figure 6. The mechanical properties (shear strength, fracture energy) of Ni and Sn-58Bi TLPS joints.

The shear strength and fracture energy of Ni-90(Sn-58Bi), Ni-80(Sn-58Bi), and Ni-70(Sn-58Bi) TLPS joints at 180, 200, and 220 °C for 90 min were measured by a shear test as shown in Fig. 6. The shear strength of the TLPS joints increased with increasing ratio of low-melting temperature material (Sn-58Bi). These results show that the higher ratio of the low-melting temperature Sn-58Bi in

this experiment could reinforce the mechanical properties of TLPS joints due to fewer voids. Additionally, Ni particles do not sinter well with each other, so there was a larger area of voids at greater Ni ratio. The ductility of the TLPS joints can also be deduced through fracture energy. The tendency of the fracture energy was nearly the same as that of shear strength, except for Ni-90(Sn-58Bi). In the case of Ni-90(Sn-58Bi), the fracture energy decreased with increasing bonding temperature because a larger IMC and coarser Bi grains were formed at the higher temperature [13].

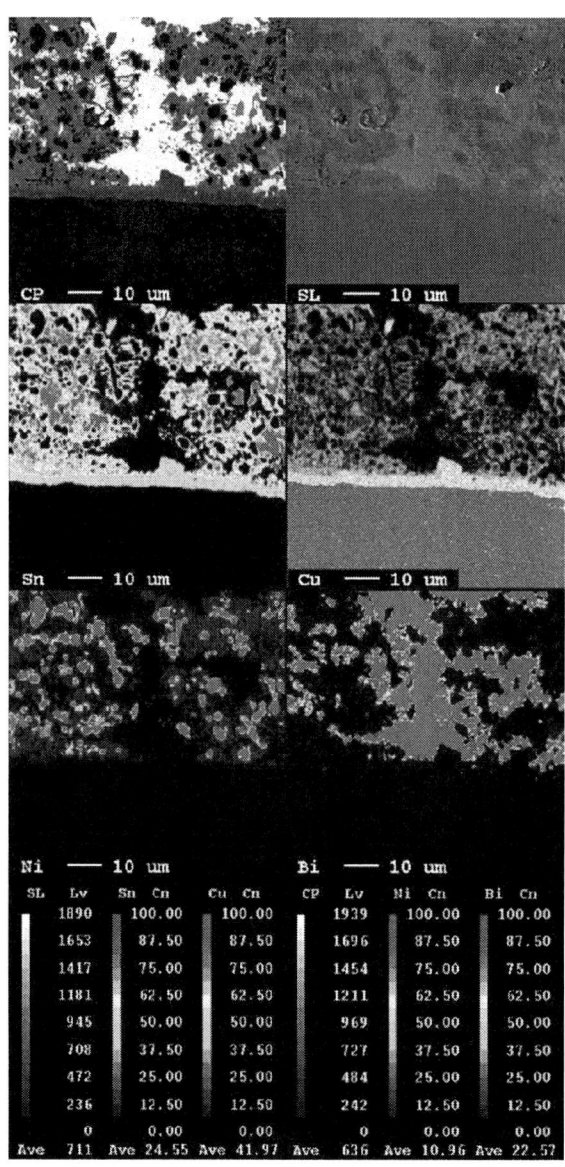

Figure 7. EPMA mapping analysis of cross-sectional micrographs (bonding temperature: 220 °C, bonding time: 90 min).

Fig. 7 shows the representative EPMA mapping analysis of the cross-sectional micrograph of Ni-80(Sn-58Bi) at 220 °C for 90 min, which provides elemental distributions of the TLPS joint. The atoms of Cu and Ni are similar in size. Therefore, they can easily replace with each other by substitution diffusion at high temperature. Through the bonding process, ternary IMC ((Cu,Ni)-Sn) was formed inside the TLPS joints [14]. Therefore, after the bonding process, Sn completely reacted with Ni and Cu such that the single phase of Sn was not observed in the TLPS joints, and (Cu,Ni)-Sn IMC was formed surrounded by Ni particles. Bi remained a single phase since it did not form IMC with Sn. Thus, the re-melting temperature could be increased to the melting temperature of Bi for about 270 °C. In conclusion, Ni-80(Sn-58Bi) is the ideal weight ratio due to the formation of full IMC and such a high bonding strength among the three weight content conditions.

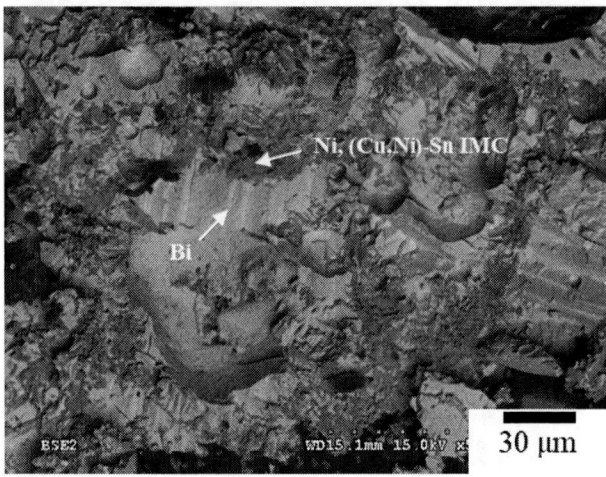

Figure 8. Fracture surface of TLPS joint after the shear test (bonding temperature: 220 °C, bonding time: 90 min).

Fig. 8 shows the fracture surface of TLPS joint at 220 °C for 90 min after the shear test. The fracture surface consisted of Bi, Ni and (Cu,Ni)-Sn IMC. At first, Bi is a brittle material, so that the fracture was propagated with brittle mode [15]. However, brittle fracture modes did not occur in Ni and (Cu,Ni)-Sn IMC parts because their brittle properties were weaker than Bi as shown in Fig. 8.

IV. CONCLUSION

We fabricated TLPS pastes of Ni-x(Sn-58Bi) with various Sn-58Bi contents (10, 20, and 30 wt%). The bonding process was conducted at 180, 200, 220 °C for 90 min, and the samples were cooled to room temperature. We conducted the experiment under a pressure-less condition to prevent chip damage and the paste spreading problem by bonding pressure. Additionally, for simplification, the bonding process was performed in an air atmosphere rather than nitrogen or a hydrogen atmosphere condition. The thermal behavior of the TLPS pastes was tested by DSC. The residual Sn phase remained after bonding of Ni-90(Sn-58Bi), and the eutectic Sn-Bi structure was observed, resulting in a Ni-90(Sn-58Bi) re-melting temperature of 139 °C. Alternatively, in Ni-80(Sn-58Bi) and Ni-70(Sn-58Bi), the Sn phase was not observed, completely transforming to IMC. Therefore, the re-melting temperature could be increased to the melting temperature of Bi in Ni-80(Sn-58Bi) and Ni-70(Sn-58Bi). The shear strength and fracture energy increased with Sn-58Bi due to fewer voids in the TLPS joints due to surface tension. After bonding, Cu-Sn IMC was observed between the Cu plate and TLPS joints. In the TLPS joints, (Cu,Ni)-Sn IMC was observed surrounded with Ni particles, and the Bi phase remained separate.

We investigated pressure-less TLPS bonding in air using Ni and a low-melting temperature material (Sn-58Bi) that melted at 139 °C. Thus, TLPS bonding using Ni and Sn-58Bi may be an alternative bonding method for high-temperature packaging applications.

ACKNOWLEDGMENT

This work was supported by "Human Resources Program in Energy Technology" of the Korea Institute of Energy Technology Evaluation and Planning (KETEP), granted financial resource from the Ministry of Trade, Industry & Energy, Republic of Korea. (No. 20174030201800).

This research was financially supported by the Ministry of Trade, Industry, and Energy (MOTIE), Korea, under the "Regional Specialized Industry Development Program" (reference number P0002867) supervised by the Korea Institute for Advancement of Technology (KIAT).

REFERENCES

[1] H. Zhang, L.M. Tolbert, and B. Ozpineci, "Impact of SiC Devices on Hybrid Electric and Plug-In Hybrid Electric Vehicles," IEEE Transactions on Industry Applications, vol. 47, pp. 912-921, March-April 2011.

[2] J. Son, M. Kim, D.Y. Yu, Y.H. Ko, J.W. Yoon, C.W. Lee, Y.B. Park and J. Bang, "Thermal Aging Characteristics of Sn-xSb Solder for Automotive Power Module", Journal of Welding and Joining, vol. 35, pp. 38-47, September 2017.

[3] H. Feng, J. Huang, J. Yang, S. Zhou, R. Zhang, and S. Chen, "A Transient Liquid Phase Sintering Bonding Process Using Nickel-Tin Mixed Powder for the New Generation of High-Temperature Power Devices", Journal of electronic materials, vol. 46, pp. 4152-4159, July 2017.

[4] M. K. Faiz, K. Bansho, T. Suga, T. Miyashita, and M. Yoshida, "Low temperature Cu-Cu bonding by transient liquid phase sintering of mixed Cu nanoparticles and Sn-Bi eutectic powders", Journal of materials science: matrials in electronics, vol. 28, pp. 16433-16443, November 2017.

[5] M. K. Faiz, T. Yamamoto, and M. Yoshida, "Sn-Bi added Ag-based transient liquid phase sintering for low temperature bonding", 2017 5th International Workshop on Low Temperature Bonding for 3D Integration (LTB-3D), pp. 34, June 2017.

978-1-7281-1500-9/19 $31.00 © 2019 IEEE

[6] H. Tatsumi, A. Lis, H. Yamaguchi, T. Matsuda, T. Sano, Y. Kashiba, and A. Hirose, "Evolution of Transient Liquid-Phase Sintered Cu–Sn Skeleton Microstructure During Thermal Aging", Applied science, vol. 9, pp. 157-168, January 2019.

[7] W. R. Myung, Y. Kim, and S. B. Jung, "Mechanical property of the epoxy-contained Sn–58Bi solder with OSP surface finish", Journal of alloys and compounds, vol. 615, pp. S411-S417, December 2014.

[8] O. Mokhtari, and H. Nishikawa, "Transient liquid phase bonding of Sn–Bi solder with added Cu particles", Journal of materials science: materials in electronics, vol. 27, pp. 4232-4244, January 2016.

[9] J. Wang, L. Wen, J. Zhou, and M. Chung, "Mechanical Properties and Joint Reliability Improvement of Sn-Bi alloy", 2011 IEEE 13th Electronics Packaging Technology Conference, p. 492-496, April 2012.

[10] T. Satoh, T. Ishizaki, and M. Usui, "Nanoparticle/solder hybrid joints for next-generation power semiconductor modules", Materials & Design, vol. 124, pp. 203-210, June 2017.

[11] J. M. Park, S. H. Kim, M. H. Jeong, and Y. B. Park, "Effect of Cu-Sn intermetallic compound reactions on the Kirkendall void growth characteristics in Cu/Sn/Cu microbumps", Japanese Journal of Applied Physics, vol. 53, pp. 05HA06, April 2014.

[12] K. Zeng, R. Strierman, T. C. Chiu, D. Edwards, K. Ano, and K. N. Tu, "Kirkendall void formation in eutectic SnPb solder joints on bare Cu and its effect on joint reliability", Journal of applied physics, vol. 97, pp. 024508, December 2004.

[13] X. Li, F. Li, F. Guo, and Y. Shi, "Effect of Isothermal Aging and Thermal Cycling on Interfacial IMC Growth and Fracture Behavior of SnAgCu/Cu Joints", Journal of electronic materials, vol. 40, pp. 51-61, January 2011.

[14] W. T. Chen, C. E. Ho, and C. R. Kao, "Effect of Cu concentration on the interfacial reactions between Ni and Sn-Cu solders", Journal of materials research, vol. 17, pp. 263-266, February 2002.

[15] Q. Zhang, H. Zou, and Z. F. Zhang, "Improving tensile and fatigue properties of Sn-58Bi/Cu solder joints through alloying substrate", Journal of materials research, vol. 25, pp. 303-314, January 2011.

Epoxy/ triazine copolymer resin system for high temperature encapsulant applications

Jiaxiong Li, Chao Ren, Kyoung-sik Moon and Ching-ping Wong
School of Materials Science and Engineering, Georgia Institute of Technology, Atlanta, GA, USA
cp.wong@mse.gatech.edu

Abstract—The seeking of high temperature stable packaging materials was driven by the progress seen in wide band gap (WBG) semiconductor-based power electronic devices. Epoxy molding compounds (EMCs) are widely used in the packaging industry as the utmost protection of packages. However current EMC chemistry cannot support the high operating temperature in WBG devices. The rigid triazine structure can provide enhanced high temperature performance in epoxy molding compound. In this report, the effect of adding a triazine based epoxy, triglycidyl isocyanurate (TGIC) in cyanate ester/ epoxy copolymer was presented. The curing chemistry, thermo-mechanical properties and high temperature aging performance were evaluated.

Keywords-component; epoxy molding compound; power electronics; high temperature; triazine

I. INTRODUCTION

The next generation power electronic devices will be based on the wide band gap semiconductors including SiC and GaN for their fascinating properties. These materials exhibit significantly higher breakdown voltage, better thermal capabilities, lower on-resistance and switching loss over their Si counterpart. As a result, the wide band gap semiconductors are born suitable for high temperature, high voltage and high frequency applications, offering improved efficiency and shrink chip size [1]. However, these intriguing features also propose higher requirements on the packaging materials. The high operating temperature (200 °C~ 250 °C) is far above the current maximum value for Si (175 °C), and the capability of the polymeric packaging materials [2]. Epoxy-based molding compound (EMC) is usually the outermost encapsulation in electronic packages to protect the electronic system from internal and external forces including short circuiting, dust, moisture as well as impact and pressure. Epoxy provides good adhesion, good processability and low cost and thus is the standard polymer resin used in molding compound. Typical EMCs show glass transition temperature (T_g) below 200 °C and cannot afford the high temperature properties that fits the wide band gap semiconductors application [3].

Approaches on epoxy modifications towards higher temperature application have been found in recent years. Highly aromatic multifunctional groups epoxy and phenolic resin were the standard industrial choice for high temperature application, while researches have been done

in incorporating other high temperature polymers including polyimide, bismaleimide, benzoxazine and cyanate ester (CE) [4-7]. Among these polymer blends, CE shows good compatibility due to its versatile co-reaction with epoxide functional groups. CE is a type of high temperature stable thermosetting materials, making use of the trimerization of cyanate functional groups to form triazine structure and crosslink. The high T_g and high temperature stability were granted from the highly crosslinked structure and the stable triazine rings. Besides, CE is known for its low dielectric constant, low loss and low moisture absorption which can be ascribed to the absence of polar components [8]. The CE and epoxy copolymer chemistry has been studied, which include the triazine formation from trimerization of cyanate ester monomers and the copolymerization of epoxy and CE to form oxazolidinone [9].

The incorporation of triazine structure in epoxy resin was pioneered by Mitsubishi Gas Chemical Company in 1970s. They developed bismaleimide/ triazine (BT) resin for high performance substrates in printed circuit boards (PCBs) application. BT resin is made by copolymerizing bismaleimide, CE and epoxy resin. The mixture takes advantage of high temperature stability of both bismaleimide and CE, and the low dielectric constant and loss from the CE part [10].

The cyanate ester/ epoxy (CE/EP) co-polymer system for molding application has recently been explored by our group [11]. Although higher Tg and thermal decomposition were found when increasing the cyanate ester content, the high temperature aging showed reversed results. The severe hydrolysis degradation in CE part was believed to cause the large weight loss and blistering causing early fail. The trapped unreacted cyanate groups can react with moisture and transform to carbamate that is unstable at over 200 °C. To alleviate this problem while exploit the benefits of the rigid triazine structure, it is worth studying to keep the cyanate ester concentration low in the copolymer system and increase triazine content by other means. Other than the *in-situ* formation of triazine rings, it has been known that epoxy or hardener molecules containing triazine structures can also benefit the high temperature performance. For example, the triglycidyl isocyanurate (TGIC) has been employed by powder coating industry for improving durability of the epoxy coating materials [12]. The triazine structure and high functionality results in high heat resistance and high crosslinking density. Melamine was adopted as hardener for epoxies for the same purpose [13].

In this report, the triglycidyl isocyanurate (TGIC) was selected as a direct incorporation of triazine structure in the CE/EP system consisting bisphenol A type CE and tetramethyl biphenyl epoxy (TMBP). The curing chemistry, thermo-mechanical properties as well as the high temperature stability were evaluated to determine the effects of TGIC in CE/EP system.

II. EXPERIMENTAL

A. Materials

Tetramethyl biphenyl epoxy (TMBP) was provided by Mitsubishi Chemical Corporation. Bisphenol A cyanate ester was purchased from Oakwood Chemical. Tris (2,3-epoxypropyl) isocyanurate (TGIC), Copper (II) acetylacetonate (99.99+%) and nonylphenol were purchased from Aldrich and used as received. The chemical structures of the monomer are shown in Figure 1.

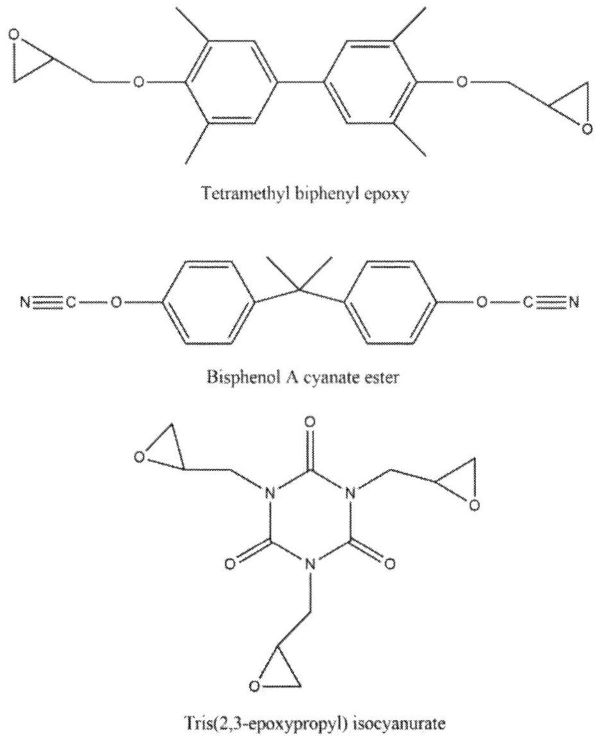

Figure 1. Chemical structure of TMBP, Bis A CE and TGIC.

B. Method

TMBP, CE and TGIC were mixed and melt at 120 °C during stirring. The cyanate functional group and epoxide functional groups were kept at 1:1 molar ratio. Their formulations and designations are shown in Table I. The Cu (II) acetylacetonate catalyst and nonylphenol co-

catalyst for CE were added then at 360 ppm and 3 phr to CE weight, respectively. The mixture was stirred until homogeneous. The samples were then cured at 150 °C for 2 hours, 200 °C for 2 hours and 250 °C for 3 hours.

TABLE I. DESIGNATION AND FORMULATION OF THE CE/EP SYSTEM

Short name	Formulation	
	Cyanate: Epoxide	Epoxide from TGIC
CE	/	/
CE/EP	1:1	0
CE/EP/TGIC1	1:1	10%
CE/EP/TGIC2	1:1	20%
CE/EP/TGIC3	1:1	30%

C. Characteriazation

Differential scanning calorimetry (DSC) was done by DSC Q2000 from TA Instrument to characterize the curing thermo profile of the polymer mix and the T_g of the cure sample. The experiments were ramping at 10 °C/min and were done under nitrogen atmosphere. Thermomechanical analysis (TMA) was done using TMA Q400 from TA Instrument to determine the Tg and CTE. Ramping rate was kept at 20 °C/ min and experiments were under nitrogen atmosphere. The thermal stability of the sample was studied by Thermogravimetric analysis (TGA) both under nitrogen and air atmosphere using TGA Q5000 from TA Instrument. All tests programmed to ramp 20 °C/ min to 800 °C. Fourier-transform infrared spectroscopy (FTIR) was done by Thermo Scientific Nicolet iS5 FT-IR spectrometer using a diamond ATR mode.

III. REUSLTS AND DISCUSSION

A. Curing of CE/EP blend and chemical information

The DSC dynamic scan of the CE and CE/EP were plotted in Figure 2. Two distinct curing peaks were found for the CE/EP system, with one's onset at 150 °C and the other at 250 °C. The identification of these two curing reactions were assisted by FTIR. Typical reaction happen in CE/EP copolymer involves the trimerization of CE to form triazine structure, and the co-reaction of formed triazine rings with epoxy, as shown in Figure 3. Figure 4 shows the revolution of chemical structures of CE/EP during curing and helps the differentiation of reaction happen during two exothermic peaks. At 200 °C, the diminish of cyanate functional groups at 2250 cm^{-1} and the appearance of triazine at 1556 cm^{-1} and 1360 cm^{-1} indicate the trimerization taken place (step *i*) in Figure 3). During ramping the temperature up over the second curing peak, a clear trend of increasing intensity of the oxazolidinone peak at 1747 cm^{-1} can be seen. The reaction of epoxy insertion in triazine rings and formation of oxazolidinone happened in this stage. (step *iv*) in Figure 3). As a result, note that the intensity of triazine peaks decreased along with temperature ramp.

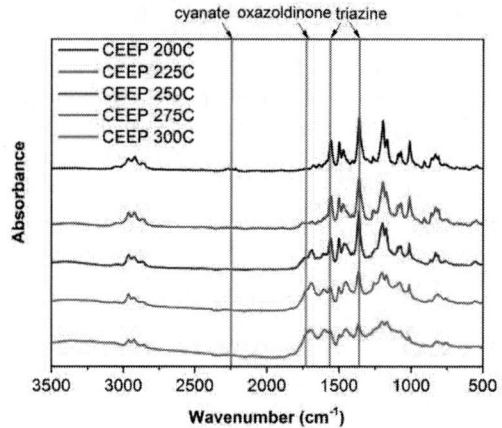

Figure 4. FTIR spectra of CE/EP during curing. The intersects are at 200 °C, 225 °C, 250 °C, 275 °C and 300 °C, respectively.

The effects of TGIC on curing of CE/EP copolymer system was investigated. DSC thermogram (Figure 5) depicted similar profile for the TGIC added polymers showing two distinguished exothermic peaks. With higher concentration of TGIC, the onsets of the two curing peaks were gradually decreasing, which implies a lower kinetic barrier benefitted from the smaller and higher functionalized TGIC molecule compared with TMBP. Also, a lower curing exotherm was found for TGIC added polymers. The FTIR spectra in Figure 6 illustrates the effects of replacing part of the TMBP by TGIC. An increasing intensity can be seen in both the isocyanurate (1692 cm^{-1}) and the triazine functions with increasing concentration of TGIC molecules. The introduction of triazine structure by means of direct insertion of TGIC was confirmed.

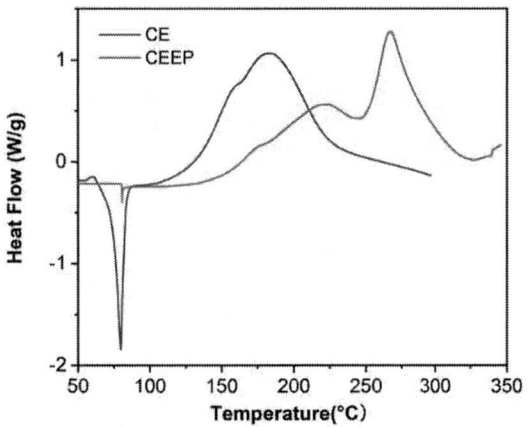

Figure 2. DSC curing profile of CE and CE/EP.

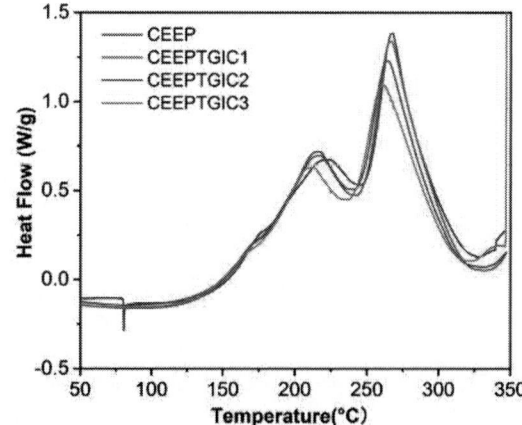

Figure 5. DSC curing profile of CE/EP and TGIC incorporated CE/EP.

Figure 3. Reactions take place in CE/EP copolymer system [8].

978-1-7281-1500-9/19 $31.00 © 2019 IEEE

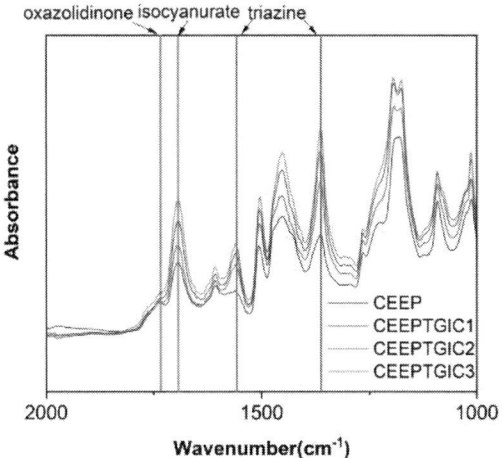

Figure 6. FTIR spectra of cured CE/EP and TGIC incorporated CE/EP.

B. Glass transition temperature

The Tg of the TGIC added CE/EP system were determined by DSC and TMA. The DSC T_g shown in Figure 7 was found to first decrease upon the addition of TGIC epoxy at low concentration. The 10% TGIC in the epoxy lower the DSC Tg from 229.08 °C to 208.97 °C. However, with increasing TGIC concentration, the DSC T_g of the material increased that at 30% concentration a higher Tg (229.72 °C) than CE/EP can be found. The competition between the higher crosslink density and the unreacted TGIC function as plasticizers when replacing part of TMBP might result in this behavior.

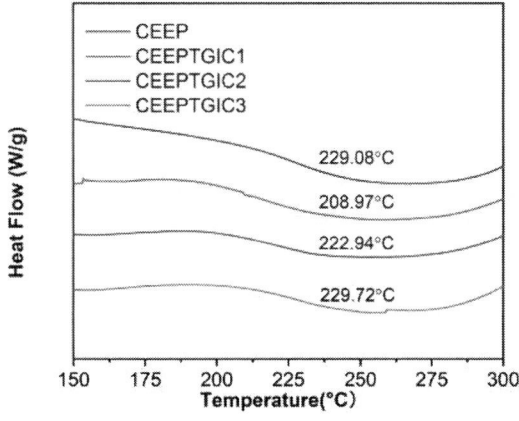

Figure 7. DSC T_g of cured CE/EP and TGIC incorporated CE/EP.

The TMA reuslts of the copolymers are shown in Table II. A clear trend of incresing T_g and the decreasing α_2 implies a higher crosslink density benefitted from the introduction of trifunctional TGIC epoxies and the rigid backbone structure reinforced by triazine rings.

TABLE II. TMA THERMOMECHANICAL PROPERTIES OF CURED CE/EP AND TGIC INCORPORATED CE/EP

	Property		
	$T_g/$ °C	$\alpha_1/$ ppm/ °C	$\alpha_2/$ ppm/ °C
CEEP	211.53	58.94	221.41
CEEPTGIC1	216.64	53.44	213.62
CEEPTGIC2	231.42	60.31	200.05
CEEPTGIC3	237.14	70.71	181.33

C. Thermal decomposition

TGA was used to determine the thermal decomposition profile of the samples both in nitrogen and in air. Figure 8 shows the nitrogen purged thermal decomposition weight loss, with the temperature at 5% weight loss ($T_{5\%}$), temperature at 10% weight loss ($T_{10\%}$) and char yield at 800 °C shown in Table III. It was found that the thermal decomposition onset was increased after blending TGIC in the copolymer where the $T_{5\%}$ raised up to 10 °C and char yield up to 2.7 wt%. This might be the result of an increased triazine functional groups content and the higher crosslink density that provided better integrity of the polymer network against chain scission.

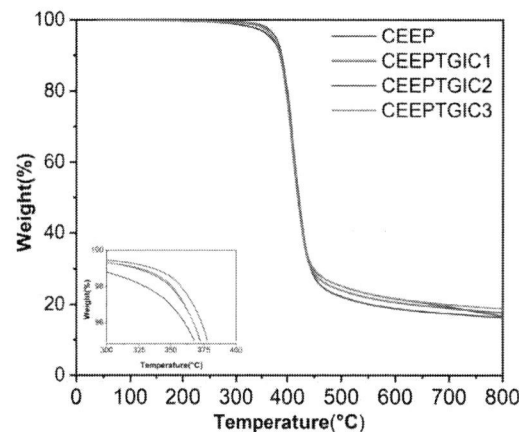

Figure 8. TGA decomposition profile under nitrogen of cured CE/EP and TGIC incorporated CE/EP.

TABLE III. THERMAL DECOMPOSITION PROPERTIES OF CURED CE/EP AND TGIC INCORPORATED CE/EP UNDER NITROGEN

	Property		
	$T_{5\%}/$ °C	$T_{10\%}/$ °C	Char at 800 °C/ %
CEEP	367.11	386.66	16.22
CEEPTGIC1	371.92	386.40	17.71
CEEPTGIC2	377.25	389.28	16.66
CEEPTGIC3	371.91	386.04	18.94

The thermo-oxidative decomposition of the materials was characterized by TGA in air, as shown in Figure 9, and the parameters shown in Table IV. A two stage weight loss was apearent for all copolymers. The first stage of decomposition should relates to the degredation of polymer into carbonaceous residue and relasing small segments as volatiles, and the second corresponds to the oxidation of carbonaceous residues [14]. The effects of introducing TGIC was not obvious during the first stage of weight loss. The decomposition initiated at around 380 °C which is similar to the value in the nitrogen purged case. The oxidation did not dramatically change the kinetic of the polymer degredation. In the second weight loss period, the TGIC incorporated copolymer showed larger tendancy of oxidation weight loss. This may relate to the structure of formed char.

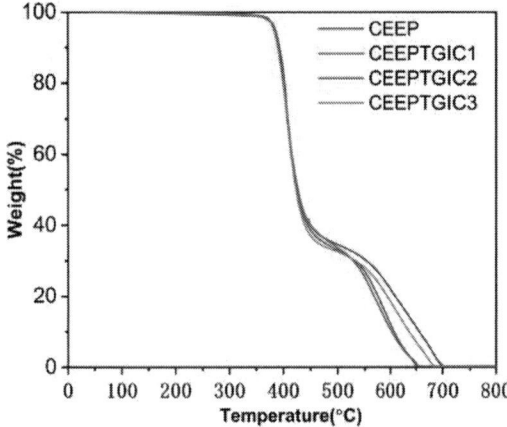

Figure 9. TGA decomposition profile under air of cured CE/EP and TGIC incorporated CE/EP.

TABLE IV. THERMAL DECOMPOSITION PROPERTIES OF CURED CE/EP AND TGIC INCORPORATED CE/EP UNDER AIR

| | Property | |
	$T_{5\%}$/ °C	$T_{10\%}$/ °C
CEEP	384.58	393.12
CEEPTGIC1	385.24	392.63
CEEPTGIC2	385.95	392.86
CEEPTGIC3	382.72	390.30

D. Aging test

The weight loss of the TGIC adopted CE/EP during 250 °C aging test and the comparison with pure CE is shown in Figure 10. A dramatic weight loss of pure CE was found after one week of aging, and the early fail (physical breakdown) due to severe outgas and swelling limited the serving life to 200 hours. This is owned to the well-known severe hydrolysis degradation happening at unreacted cyanate groups. The CE/EP copolymer on the other hand, maintained the integrity over a longer period of time. TGIC incorporated copolymer exhibited larger

weight loss compared to CE/EP despite its better thermal stability from TGA results in nitrogen.

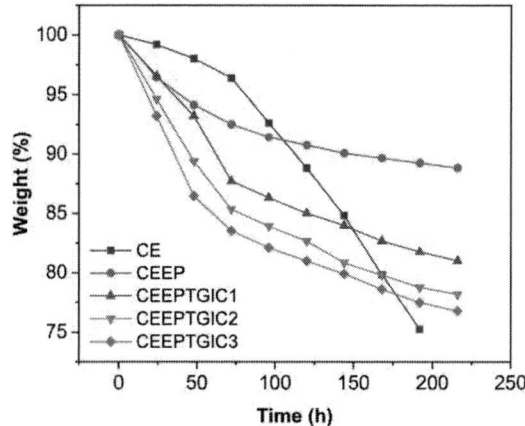

Figure 10. Aging weight loss of CE/EP and TGIC incorporated CE/EP under 250 °C.

The moisture absorption behavior of the samples was evaluated by the 105 °C/ 100 RH test. One day was found to be enough to reach the saturation of the samples and the results are shown in Table V. A higher moisture absorption can be found in the TGIC incorporated copolymer. This might be due to the highly hydrophobic *ortho* substituted TMBP being replaced by TGIC. The poor moisture resistance of the TGIC containing CE/EP should also accout for the results in high temperature aging test as water molecules can act as plasticizers to lower down the Tg of the material and can interact with certain fuctional groups that leads to degredation.

TABLE V. SATURATED MOISTURE ABSORPTION RESULTS OF CE/EP AND TGIC INCORPORATED CE/EP

	Moisture absorption/ %
CE	1.63
CEEP	1.37
CEEPTGIC1	1.41
CEEPTGIC2	1.57
CEEPTGIC3	3.40

IV. SUMMARY

In this report, the TGIC molecule was added in to CE/TMBP copolymer system as a method of increasing triazine content to achieve better thermal stability. Two distinct curing exotherms was found for the CE/EP system. They were differentiated to be CE trimerization and CE/EP co-reaction, respectively. The curing reaction between CE and epoxy was accelerated upon adding of

TGIC, and the triazine content increase upon adding TGIC was confirmed by FTIR. The DSC T_g and TMA T_g after TGIC insertion was found to increase which may arise from the higher crosslink density. TGA under nitrogen and air atmosphere revealed enhanced heat resistance with higher TGIC concentration that is in accordance with the higher triazine content. The high temperature aging behavior was degraded however which might relate to the high moisture absorption. Future work will include hydrophobic materials design for pursuing better aging performance of the CE/EP/TGIC system.

ACKNOWLEDGMENT

The funding from Georgia Tech Packaging Research Center is appreciated.

REFERENCES

[1] J. L. Hudgins, G. S. Simin, E. Santi, and M. A. Khan, "An assessment of wide bandgap semiconductors for power devices," IEEE Trans. Power Electron., vol. 18, pp. 907-914, 2003.

[2] R. Khazaka, L. Mendizabal, D. Henry, and R. Hanna, "Survey of high-temperature reliability of power electronics packaging components," IEEE Trans. Power Electron., vol. 30, pp. 2456-2464, 2015.

[3] Y. Yao, G.-Q. Lu, D. Boroyevich, and K. D. Ngo, "Survey of high-temperature polymeric encapsulants for power electronics packaging," IEEE Trans. Compon. Packag. Manuf. Technol., vol. 5, pp. 168-181, 2015.

[4] C.-H. Chen, K.-W. Lee, C.-H. Lin, M.-J. Ho, M.-F. Hsu, S.-J. Hsiang, et al., "High-Tg, Low-Dielectric Epoxy Thermosets Derived from Methacrylate-Containing Polyimides," Polym., vol. 10, p. 27, 2017.

[5] P. Musto, G. Ragosta, P. Russo, and L. Mascia, "Thermal ‑ Oxidative Degradation of Epoxy and Epoxy ‑ Bismaleimide Networks: Kinetics and Mechanism," Macromol. Chem. Phys., vol. 202, pp. 3445-3458, 2001.

[6] H. T. Lin, C. H. Lin, Y. M. Hu, and W. C. Su, "An approach to develop high-Tg epoxy resins for halogen-free copper clad laminates," Polym., vol. 50, pp. 5685-5692, 2009.

[7] F. Wu, B. Song, J. Hah, C. C. Tuan, K. S. Moon, and C. P. Wong, "Polyimide incorporated cyanate ester/epoxy copolymers for high ‑ temperature molding compounds," J. Polym. Sci., Part A: Polym. Chem., vol. 56, pp. 2412-2421, 2018.

[8] T. Fang and D. A. Shimp, "Polycyanate esters: science and applications," Prog. Polym. Sci., vol. 20, pp. 61-118, 1995.

[9] B. S. Kim, "Effect of cyanate ester on the cure behavior and thermal stability of epoxy resin," J. Appl. Polym. Sci., vol. 65, pp. 85-90, 1997.

[10] L. Ji, A. Gu, G. Liang, and L. Yuan, "Novel modification of bismaleimide–triazine resin by reactive hyperbranched polysiloxane," J. Mater. Sci., vol. 45, pp. 1859-1865, 2010.

[11] F. Wu, C. C. Tuan, B. Song, K. S. Moon, and C. P. Wong, "Controlled synthesis and evaluation of cyanate ester/epoxy copolymer system for high temperature molding compounds," J. Polym. Sci., Part A: Polym. Chem., vol. 56, pp. 1337-1345, 2018.

[12] D. Parra, L. P. Mercuri, J. d. R. Matos, H. F. d. Brito, and R. Romano, "Thermal behavior of the epoxy and polyester powder coatings using thermogravimetry/differential thermal analysis coupled gas chromatography/mass spectrometry (TG/DTA–GC/MS) technique: identification of the degradation products," Thermochim Acta, vol. 386, pp. 143-151, 2002.

[13] I. Hamerton and J. N. Hay, "Recent developments in the chemistry of cyanate esters," Polym. Int., vol. 47, pp. 465-473, 1998.

[14] N. Rose, M. Le Bras, S. Bourbigot, and R. Delobel, "Thermal oxidative degradation of epoxy resins: evaluation of their heat resistance using invariant kinetic parameters," Polym. Degrad. Stab., vol. 45, pp. 387-397, 1994.

Low temperature Ag-Ag direct bonding technology for advanced chip-package interconnection

Jiaqi Wu[a,b,*] and Chin C. Lee[a,b]

[a]Department of Electrical Engineering and Computer Science
[b]Materials and Manufacturing Technology
University of California Irvine
Irvine, CA92697, USA
E-mail: jiaqw10@uci.edu

Abstract—Nowadays, with increasing integration density on very large scale integrated-circuits (VLSI), the pitch between joints and the joint size within the package of advanced device have been scaled down to sub-100 μm. As a result, traditional method which employs solder joint would induce more reliability and yield issues. Direct metal to metal bonding (Cu-Cu and Au-Au) have been proposed as alternative methods. However, high temperature, high bonding pressure, excellent surface finish and high vacuum are required during direct bonding, which restricts their wide acceptance by the industry. In this manuscript, a method to achieve direct Ag-Ag bonding at low temperature (210 °C) with low bonding pressure (1.38 MPa) is proposed by in-situ self-reduction of surface oxides. The temperature is lower than the reflow temperature of lead-free solder (250 °C) and the pressure is much lower than the pressure required in thermal-compression process. The preparation and characterization of surface oxides have been elaborated. During bonding, the in-situ generated Ag "wets" the surfaces and enhanced the surface diffusion, which enables the bonding of materials at atomic scale and merging of grains from two sides. The requirement for surface roughness is much lower that direct Cu-Cu bonding reported in literatures. The quality of the joint is studied by scanning electron microscopy (SEM) and transmission electron microscopy (TEM) and the results show that two Ag films join together to form one piece "bulk-like" polycrystalline materials without any trace of the original interface. We believe that this technique will be promising in flip chip bonding for advanced VLSI packages. In addition, this bonding technique is valuable to die attachment for high power devices as it provides the minimum heat resistance between chip and substrate.

Keywords- silver, silver oxides, in-situ reduction, flip-chip, die attachment

I. INTRODUCTION

Nowadays, with the increase in the integration density on the VLSI chips, the I/O density needed to be addressed during packaging process increases drastically. On the other hand, the technical difficulties and cost of further shrinkage of the critical dimension (CD) on wafer drives the industry to seek for advanced packaging technologies to improve the overall performance of the electronic devices. As a result, the continuously scaling down of

pitch and size of the first level interconnection (FLI) joint is in demand. For traditional flip chip process where solder is the main component, the pitch of the joint is more than one hundred μm. The melting and wetting of solder on other metals (typically Cu or Ni) plays a crucial role in the joint formation and the Sn-Cu or Sn-Ni intermetallic compounds (IMCs) form during reflow and subsequent usage. However, with the advent of 2.5D/3D packaging, the pitches of the micro-bumps will be scaled down to sub-40 μm. Furthermore, the hybrid bonding technologies may enable FLI interconnection with only sub-20 μm pitch. In these fine pitch joints, as the surface to volume ratio decreases with the scaling down of joint, the inter-diffusion rate will increases as the contribution from surface diffusion increases significantly [1], resulting in overgrowth of IMCs, electromigration, thermal migration and voiding. During packaging, the FLI joints may experience multiple reflows during downstream processes and the working temperature is higher than other level interconnection joints, more and more reliability and yield issues are needed to be addresses with solder joints.

In order to avoid IMCs and voiding issues, researchers investigate the direct bonding technology where the materials from both sides are same such as Cu-Cu and Au-Au. The Au-Au direct bonding technology is not widely accepted since it requires high temperature (more than 350 °C) [2] and the materials are highly costive. Cu-Cu direct bonding have attracted lots of research interests in the past few years. However, the hardness of Cu is higher than Au and thus requires higher bonding pressure. Another thing is that the requirement of surface roughness and vacuum to form high quality joint is stringent. It is notable that high bonding pressure (more than 20 MPa) increases the risk of die cracking, resulting in serious yield issue.

The formation of direct metal-to-metal joint without IMCs is illustrated in Fig. 1. In principle, a requirement of forming metal to metal bonding in solid-state is that atoms from both sides are brought within atomic distance [3, 4]. In reality, there are asperities on the surface no matter how the surfaces are prepared, resulting inadequate contact coverage over the bonding area. The root mean square (RMS) of the height of these asperities is one way to represent the surface roughness. During thermal compression bonding, these asperities deform thermo-plastically to increase area where two sides of materials

978-1-7281-1500-9/19 $31.00 © 2019 IEEE

are within atomic distance. With sufficient process time, redistribution of atoms and vacancies by surface/interface diffusion within the interfacial zone help reduce the defects and thus increase the joint strength. As the deformation of asperities and surface diffusion are assisted by pressure and temperature, a reduction in bonding temperature and pressure requires smother surfaces and other factors that can facilitate surface diffusion. In addition, joining materials of low hardness would help.

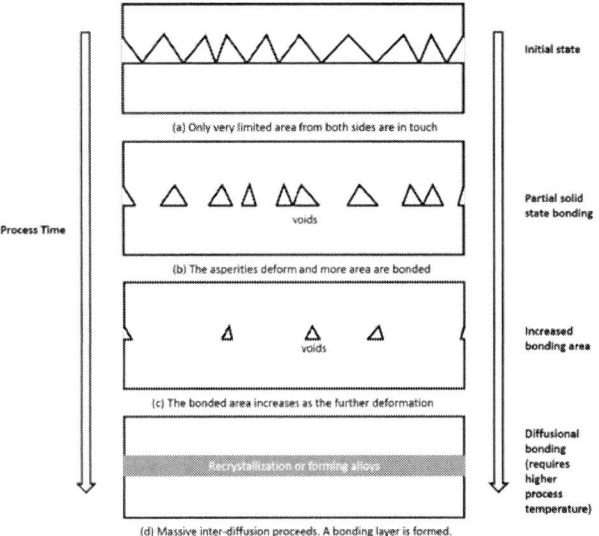

Fig. 1 Evolution of interfacial structure during bonding that only involving solid state materials.

Recent approaches to form Cu-Cu joints under less stringent condition include post-annealing, (111) textured Cu films [5] and surface treatment [6, 7] (i.e. plasma activation, self-assembly monolayer). All of these method focus on the improvement of the surface diffusion rate. However, all the Cu involved solid state joint-forming process requires high vacuum or formic acid atmosphere. All these complexity and cost brought by processes limit their wide acceptance by the industry.

After going over all the metals without oxidation issues, Ag looms large in my mind as silver oxide is not thermally stable. By employing Ag as bonding layer, oxidation is no longer an issue. Moreover, the instability of silver oxide can also be utilized. According to the Ag-O binary phase diagram [8], no silver oxide can exist above 186 °C. Therefore, if the surface of Ag can be pre-oxidized by either reacting with oxygen plasma or anodizing process, the thermal decomposition of silver oxide will generate tiny particles which facilitate the surface diffusion. The bonding temperature doesn't need to be very high, only 10 or 20 °C higher than the temperature of phase transformation (200-210°C) should be sufficient. No reducing atmosphere is needed. Given that the thickness of the oxide layer is very thin, the porous issue should not be serious. The bonding structure is illustrated in Fig. 2.

In the following sections, the experimental design will be described firstly. Then, the microstructure and surface morphology of as-deposited Ag films and surface oxides as a function of plasma treating time will be illustrated.

After reducing silver oxide, the surface morphology changes is discussed. The quality of the joint is evaluated and the optimum condition is determined. The mechanism is discussed in details.

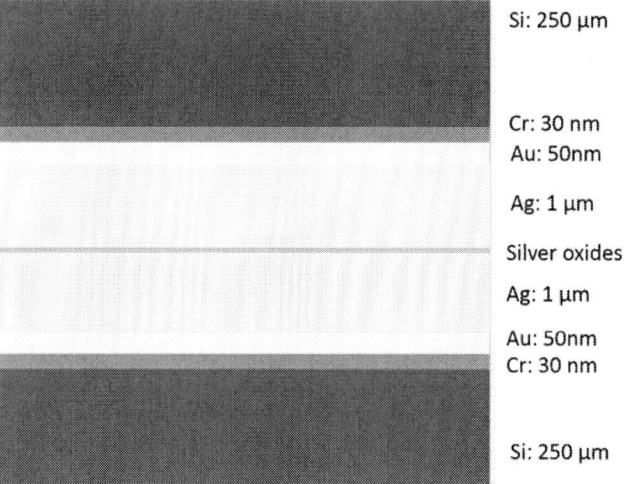

Fig. 2 Pre-bonding Structure of joint formed by in-site reducing silver oxides.

II.   EXPERIMENT DESIGNS AND PROCEDURES

Electron-beam evaporation is selected to deposit Ag films of 1 μm on Si substrate. Prior to the deposition of Ag, 30 nm Cr and 50 nm Au are deposited on Si. Cr serves as adhesion layer and Au is buffer layer to suppress blister issue. The silver oxides is prepared by oxidation reaction in oxygen plasma generated by Gatan plasma cleaner system. The specifications of the normal operation are listed in Table 1. After loading the samples into the chamber, the chamber is pumped down to 70 mTorr in 1 min. Then the oxygen (99.995%) flows into the chamber to reach the operating vacuum (400 mTorr). Then the radio frequency (RF) voltage is applied and the plasma is generated all over the chamber. After reaction of designated duration, the RF is turned off and the chamber is vented by argon. In this study, a few reaction durations including 3s, 5s and 10s are tried. The surface composition and chemical state of silver oxide are analyzed by x-ray photoelectron spectroscopy (XPS). Besides, atomic force microscopy (AFM) is utilized the measure the surface roughness. The joint is formed through the bonding between one piece of film with surface oxides and one piece without surface oxides at 210 °C, which is 40 °C lower than the reflow temperature of lead-free solder. The bonding is conducted under vacuum of 0.1 Torr which prevents the Ag from being corroded by other species in air in a lab environment. After pumping, the temperature is ramped up to 210 °C in 10 min, kept isothermal for 30 min and cooled down to room temperature naturally. Only 200 psi (1.38 MPa) pressure is applied to fix the whole structure. The surface morphology of oxides and quality of the joint is evaluated by scanning electron microscopy (SEM). The joint structure in nano-scale is analyzed by using transmission electron microscopy (TEM).

Parameter	Value
RF frequency	13.56 MHz
RF power	50 W
Vacuum target	70 mTorr
Operating Vacuum	400 mTorr
Oxygen flow rate	33.3 cm^3 per minute

Table. 1 Specifications of Gatan plasma system

III. EXPERIMENTAL RESULTS AND DISCUSSION

A. Characterization of as-deposited Ag film

The top view and cross-sectional view of the as-deposited Ag film are shown in Fig. 3. From Fig. 3a, it can be seen that the surface is compact without voids and cracks. Since e-beam deposited films usually exhibit island-growth, the surface would appear as if there were lots of particles on the surface. In this case, the particle size ranges from 50 nm to 200 nm. Fig. 3b shows that the film is dense without any visible voids inside. The geometries of the grains are random. The one dimension size ranges from a few tens of nanometers to a few hundred nanometers.

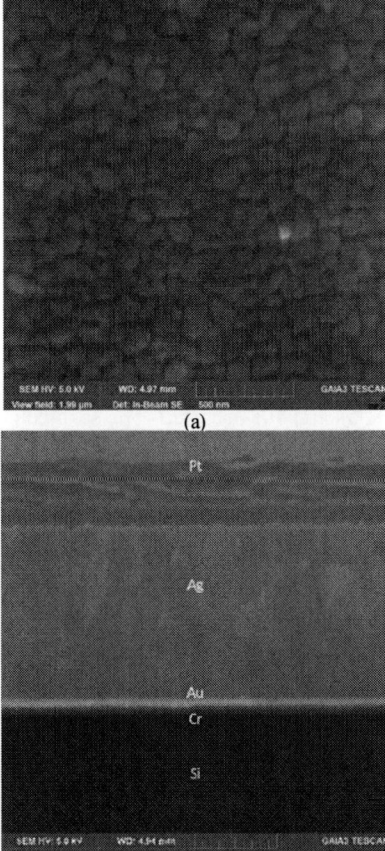

Fig. 3 Microstructure of as-deposited film (a) top view ; (b) cross-sectional view.

B. Surface compostion and chemical state of oxides

The surface composition and chemical state of Ag are studied by XPS and the results are shown in Fig. 4. According to Fig. 4a, only oxygen and silver can be detected from the surface. The quantification is conducted by using the intensity of Ag 3d and O 1s. The algorithm is pre-defined by in the software-CasaXPS. The surface composition determined by this method is Ag 63 at.% and O 37 at.%, which is pretty close the stoichiometry of Ag$_2$O. To further study the chemical state of the Ag and O, regional scans are conducted for Ag 3d and O 1s. For both scan, the resulting curve is smoothened by averaging the results of 20 scans. From Fig. 4b, it can be seen that there is significant right-shift of binding energy of Ag 3d in silver oxide compared to that of Ag 3d in pure Ag. The binding energy of Ag 3d$_{5/2}$ in oxide is measure to be 367.7 eV, which is consistent to the published result of Ag 3d$_{5/2}$ in Ag$_2$O [9]. From Fig. 4c, for O 1s on oxide and Ag, there are broadened peaks centered approximately at 531 eV, which are considered to be originated from a mixture of atomic oxygen and hydroxyl groups [10]. However, for the spectra of O 1s of oxide, there is distinct peak with much higher intensity, symmetry and sharp line width, which is consistent to the published result of O 1s of Ag$_2$O [11]. The regional scan result also explains the deviation of surface composition from 2 to 1. Due to the exposure of sample to air, the oxygen and hydroxyl groups are absorbed on the surface. However, the quantity is small and they not considered to affect the in-situ reduction process.

C. Optimization of oxidation condition

The surface morphologies of as-reacted samples are examined by SEM and shown in Fig. 5. It can be seen that the surface oxides grow in the form of islands. After 3 s plasma treatment, the surface of pure Ag is not fully covered by the oxides. As the increase in reaction time, the islands grow and also the coverage is increased. However, the thickening of the oxide is not uniform. The oxide on lower right corner region of Fig. 5b is thicker than oxides of other regions. Eventually, the islands connect to each other and form large islands. As shown in Fig. 5c, the original Ag grains are invisible and the surface is covered by large islands of silver oxides.

Then these samples are placed on graphite stage, heated up to 210 °C under 0.1 Torr vaccum, kept isothermal for 30 min and cooled down naturally. The surface morphology is examined again by using SEM and shown in Fig. 6. Fig. 6 shows the surface morphologies of samples which are corresponding to the samples shown in Fig. 5. As it can be seen if Fig. 6a, the surface is smoother than that of as-reacted sample. Some grains grow significantly during the reduction. However, there are lots of sub-grains and twins in these large grains, which is due to the low stacking fault energy of Ag. In addition, there are sparse voids on the top of the surface. These voids are very small and most of them are in the range of a few tens of nanometers. Fig. 6b shows that the surface is not as smooth as the surface in Fig. 6a. Although only pure Ag grains can be observed, the size of voids is larger. Lots of voids in the range of a few hundred nanometers can be observed. The surface shown in Fig. 6c is much rougher. It can be seen that Ag grains are covered by lots of particles. A possible reason is that the Ag atoms generated by decomposing silver oxides don't merge with the original grains underneath but forming particles themselves, which may be due to the overgrowth of silver oxides during 10 s plasma treatment.

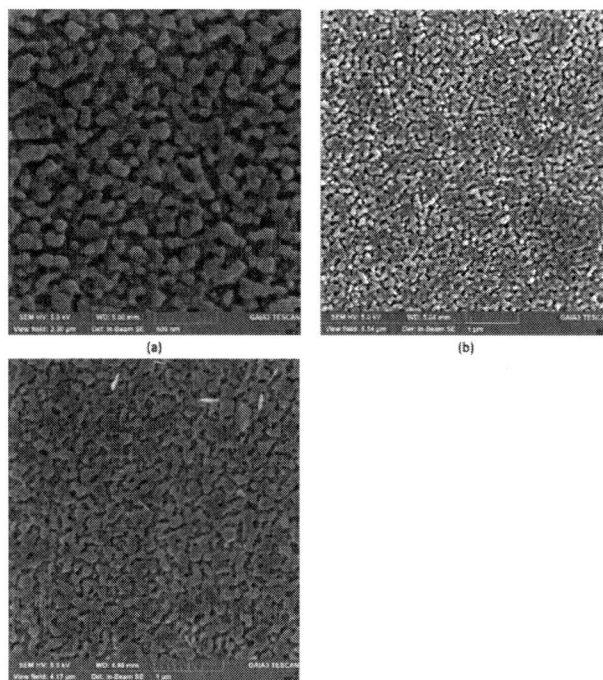

Fig. 5 Surface morphology of oxides formed through different oxidation time (a) 3 s; (b) 5 s; (c) 10s.

Then these samples are placed on graphite stage, heated up to 210 °C under 0.1 Torr vaccum, kept isothermal for 30 min and cooled down naturally. The surface morphology is examined again by using SEM and shown in Fig. 6. Fig. 6 shows the surface morphologies of samples which are corresponding to the samples shown in Fig. 5. As it can be seen if Fig. 6a, the surface is smoother than that of as-reacted sample. Some grains grow significantly during the reduction. However, there are lots of sub-grains and twins in these large grains, which is due to the low stacking fault energy of Ag. In addition, there are sparse voids on the top of the surface. These voids are very small and most of them are in the range of a few tens of nanometers. Fig. 6b shows that the surface is not as smooth as the surface in Fig. 6a. Although only pure Ag grains can be observed, the size of voids is larger. Lots of voids in the range of a few hundred nanometers can be observed. The surface shown in Fig. 6c is much rougher. It can be seen that Ag grains are covered by lots of particles. A possible reason is that the Ag atoms generated by decomposing silver oxides don't merge with the original grains underneath but forming particles themselves, which may be due to the overgrowth of silver oxides during 10 s plasma treatment.

The surface roughness of the as-deposited film, as-oxidized (3s) sample and as-reduced sample are measured by atomic force microscopy (AFM). According to Fig. 7, the root-mean-square (RMS) value of the as-deposited film is around 5 nm. After oxidation, the surface became much rougher, which is consistent with the SEM result in Fig. 5a. However, the film becomes smother after reduction. The RMS value decreases from 49 nm to 19 nm. Same phenomenon is observed in SEM results. The

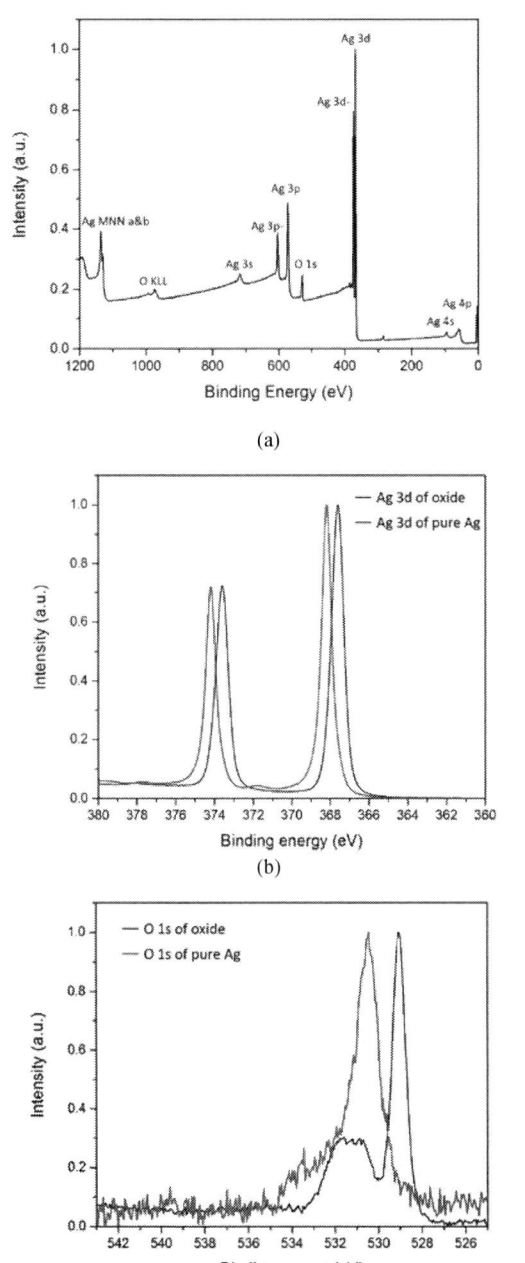

Fig. 4 XPS study of surface oxides (a) survey scan; (b) reginal scan of Ag 3d band; (c) reginal scan of O 1s band.

978-1-7281-1500-9/19 $31.00 © 2019 IEEE

smoothening process is due to the high surface diffusivity of the Ag decomposed from Ag_2O, which facilitates the wetting process of the newly formed Ag on the Ag surface. In other words, the smoothening is a confirmation of the high surface diffusivity of the Ag. Therefore, during bonding, the Ag decomposed from Ag_2O could wet surface of the other piece as well and the bonding could be easier to be formed.

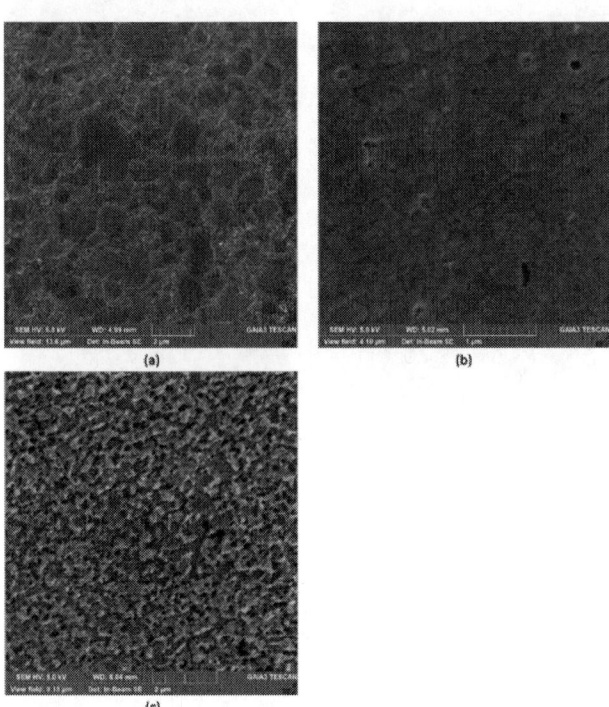

Fig. 6 Surface morphology after reduction with different oxidation time
(a) 3 s; (b) 5 s; (c) 10s.

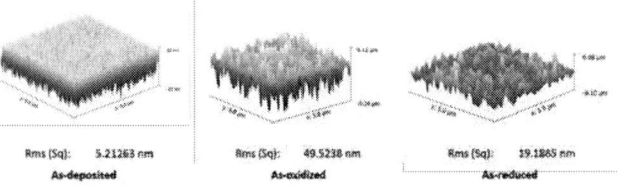

Fig. 7 AFM results of the film of different conditions.

Therefore, given that the surface roughness and size voids are major concerns during bonding, only samples treated by plasma for 3 s and 5 s are used for the following bonding process. After that, the cross-sections of these samples are examined by SEM.

D. Cross-sectional images after FIB cutting

The cross-section cutting sites are randomly chosen from the joints and clean cross-sections are acquired by FIB cutting and polishing. The area of interest is carefully examined by SEM and the results from 3s plasma treatment sample and 5s plasma treatment sample are shown in Fig. 8 and Fig. 9, respectively. Since the surface is cleaned by ion beam, contrast generated by channeling effect is revealed and the outline of grains inside the joint are visible. From Fig. 8a, it can be seen that Ag films from both sides are well bonded together in most regions.

There are sparse voids along the bonding interface. Most voids are very small and the size is within 50 nm. The voids is generated by two reasons. Firstly, the volume of silver oxide is larger than the volume of pure Ag produced by in-situ reduction. Secondly, the surface is not absolute smooth. As indicated in Fig. 8b and 8c, there will be some voids retained in the bonding interface. However, the voids are not always right along the interface. Some voids migrate towards the internal region of the Ag film, which is due to the redistribution of atoms near the interface. There are also some voids in the metallization layer, which is due to the dissolution of Au into Ag and discussed in next section.

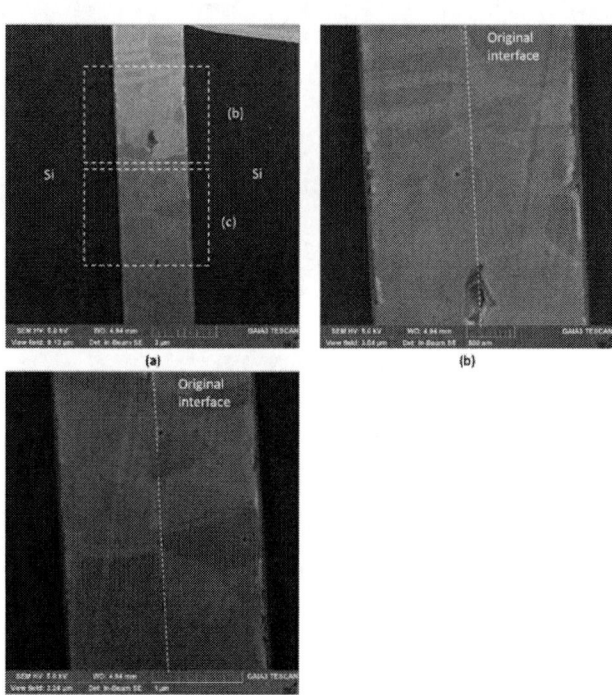

Fig. 8 SEM cross-sectional images of joint with surface oxides formed through 3 s oxidation (a) low mag; (b)&(c) high mag.

From Fig. 9,It can be seen that Ag films from two sides are well bonded. In most cases, the grains grow across the original bonding interface, which is the ideal result of the direct bonding. However, the issue of this joint is associated with the voids. As it can be seen that lots of huge voids (a few μm in one dimension) are inside the joint. Recall that the surface morphology of Ag film after thermal decomposition of Ag oxides is related to the thickness of the original thickness of oxides. There are lots of voids on the surface for the film treated by oxygen plasma for 5 s. In this case, the quantity of oxides is too much so that the volumetric retraction due to the decomposition of oxides is too much to be compensated by surface plastic deformation during bonding. To reduce the size of voids, a higher bonding pressure should be used. However, the die is fragile and any process involving chip requires high pressure is not preferred.

978-1-7281-1500-9/19 $31.00 © 2019 IEEE   2306

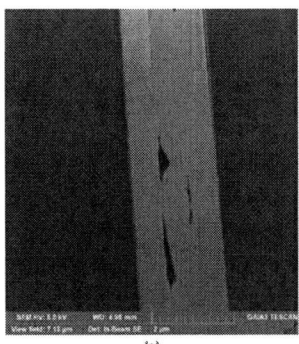 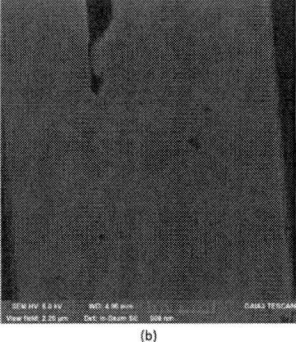

Fig. 9 SEM cross-sectional images of joint with surface oxides formed through 5 s oxidation (a) low mag; (b) high mag.

E. TEM analysis of the bonding interface

In this section, only the sample bonded by Ag films with 3 s plasma treatment is discussed due to the high quality. The sample is randomly cut from the joint and the result is shown in Fig. 10. It can be clearly seen that most grains near the interface grow across the original boundary. Also, only a few grains nucleate at the interface and join the grains from both sides (i.e. the one reside on the right of the red circle in Fig. 10). There are some grains directly bonded at the original bonding interface (i.e. those two grains inside the red circle in Fig. 10). The images with high magnifications of this region is shown in Fig. 11. It shown that the [111] zone axis of one grain is aligned to the direction of incident beam while the [100] zone axis of the other grain is aligned to the direction of incident beam. This is an indication that the grain boundary is large angle grain boundary. On the other hand, this is no voids or oxides within the interface. The black region in Fig. 11b is due to overlapping of the grain boundary along thickness direction. This type of the boundary is same the large angle grain boundary which is commonly observed in polycrystalline materials.

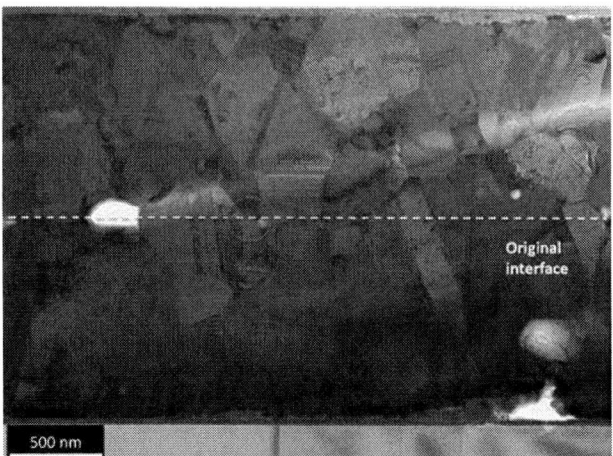

Fig. 10 TEM bright field image of the joint.

It is also worth mentioning that the Au in the metallization layer is almost consumed, leaving some voids between Ag and Cr. This has been reported in other literatures. The effective remedy has been found by others [12], which is to employ Cr and Ni as metallization layer

for Ag. Cr can provide good adhesion between Si and other metals. Pt provides good adhesion between Ag and Cr. The most important thing is that Pt doesn't get dissolved in Ag even at high temperature.

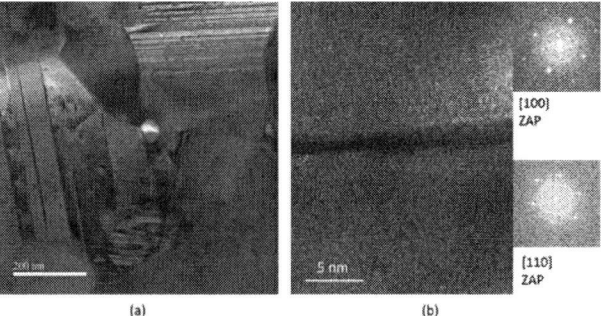

Fig. 11 High mag TEM images (a) circled region in Fig. 10; (b) HRTEM with Fast Fourie Transform (FFT) pattern

Another thing needs to be figured out is whether there is trapped oxygen or residue oxides in the joint. STEM/EDX analysis is conducted and the results are shown in Fig. 12. It can be seen that oxygen can only been detected in the Cr layer. Inside the joint there is no oxygen signal. The oxygen in Cr is quite common since Cr get oxidized easily during the e-beam deposition even though the deposition is conducted under high vacuum. However, the formation of Cr_xO_{1-x} is the reason why Cr can provide good adhesion between Si and other metals.

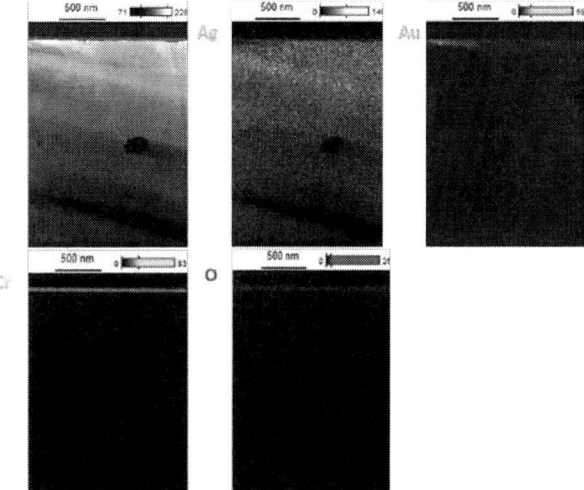

Fig. 12 STEM/EDX mapping of the joint

F. Bonding mechanism

According to previous results and discussions, a high quality joint with high bonded fraction can be formed through solid state bonding of lower pressure and temperature. The mechanism is proposed as follows. As the temperature reaches the decomposition temperature of the silver oxide, the pre-grown silver oxide decompose and Ag atoms with high activity form. This is the nature of the reduction of Ag_2O which has been thoroughly studied and described through the scheme of dissociative evaporation of the reactant with simultaneous condensation of the low-volatile product [13, 14]. During thermal decomposition of Ag_2O, both Ag and O are formed in the form of gas and Ag vapor condenses simultaneously. These Ag atoms wet the both surface of both sides through surface diffusion and the bonding interface becomes dense the grains with low mismatch of alignment merge together. These grains with large mismatch of orientation alignment can't merge and formed large angle grain boundaries. After bonding, two sides of Ag becomes one piece of polycrystalline-like material, which has same properties of polycrystalline Ag.

IV. CONCLUSION

In this paper, an Ag-Ag direct bonding technology is developed through in-situ reduction of surface oxides. The surface oxides is prepared by the reaction between oxygen plasma and the surface of Ag film. To reduce the size and quantity of the interfacial voids, the thickness of surface oxides must be controlled. By using the plasma system mentioned in the context, 3 s is considered to be an optimized reaction time. The bonding temperature is 210 °C, which is much lower than the reflow temperature of lead free solders (240-250 °C). The bonding pressure is only 200 psi (1.38 MPa), which is several orders of magnitude lower than that in traditional thermal compression bonding.

The cross-sectional study shows that the Ag films are bonded with only sparse voids of tens of nanometers along the interface. In most regions, the grains grow across the interface and the original boundary vanishes. This is due to the high surface diffusivity of the Ag decomposed from Ag oxides. In some regions, new grain nucleate at the interface and joins the grains from both sides. It can be barely found that two grains from two sides are directly bonded. HRTEM shows the mismatch of orientation of those two grains are large and high angle grain boundary is formed during bonding. EDX results show that there is no oxides residue within the joint and the content of trapped oxygen is too low to be detected by EDX.

The direct bonding technology reported here is a breakthrough in solid state bonding technology since the pressure and temperature used is too low in the view of traditional thermal compression bonding to form joint where the grains grow across the original boundary. The resulting joint is pure Ag. Unlike soldering or sintering, no IMCs and large fractions of pores formed. Recall that pure Ag has the best thermal and electrical conductivities among metals, high temperature stability and high ductility, this bonding technology exhibits great potentials in the packages high performance ICs and high power/ high temperature electronics and photonics, where high temperature reliability and the reduction of heat resistance of the packages are of great significance.

V. ACKNOWLEDGMENT

SEM/EDX, SEM/FIB, TEM, STEM-EDX work were performed at UC Irvine Materials research Institute (IMRI).

VI. REFERNCES

[1] Y. Liu, Y.C. Chu, K.N. Tu, "Scaling effect of interfacial reaction on intermetallic compound formation in Sn/Cu pillar down to 1 μm diameter," *Acta Mater,* vol. 117, pp. 146-152, 2016.

[2] H.R. Tofteberg, K. Schjølberg-Henriksen, E.J. Fasting, A.S. Moen, M.M. Taklo, E.U. Poppe, C.J. Simensen, "Wafer-level Au–Au bonding in the 350–450° C temperature range," *J Micromech Microeng,* vol. 24, p. 084002, 2014.

[3] G. Chen, Z. Feng, J. Chen, L. Liu, H. Li, Q. Liu, S. Zhang, X. Cao, G. Zhang, Q. Shi, "Analytical approach for describing the collapse of surface asperities under compressive stress during rapid solid state bonding," *Scripta Mater,* vol. 128, pp. 41-44, 2017.

[4] C. C. Lee and L. Cheng, "The quantum theory of solid-state atomic bonding," *2014 IEEE 64th Electronic Components and Technology Conference (ECTC),* Orlando, FL, 2014, pp. 1335-1341.

[5] C.M. Liu, H.W. Lin, Y.S. Huang, Y.C. Chu, C. Chen, D.R. Lyu, K.N. Chen, K.N. Tu, "Low-temperature direct copper-to-copper bonding enabled by creep on (111) surfaces of nanotwinned Cu," *Sci Rep-Uk,* vol. 5, p. 9734, 2015.

[6] T. Kim, M. Howlader, T. Itoh, T. Suga, "Room temperature Cu–Cu direct bonding using surface activated bonding method," *J Vac Sci Technol A,* vol. 21, pp. 449-453, 2003.

[7] C.S. Tan, D.F. Lim, "Cu Surface Passivation with Self-Assembled Monolayer (SAM) and Its Application for Wafer Bonding at Moderately Low Temperature," *ECS Trans,* vol. 50, pp. 115-123, 2012.

[8] I. Karakaya, W. Thompson, "The Ag-O (silver-oxygen) system," *J Phase Equilib,* vol. 13, pp. 137-142, 1992.

[9] X.Y. Gao, S.Y. Wang, J. Li, Y.X. Zheng, R.J. Zhang, P. Zhou, Y.M. Yang, L.Y. Chen, "Study of structure and optical properties of silver oxide films by ellipsometry, XRD and XPS methods," *Thin Solid Films,* vol. 455-456, pp. 438-442, 2004.

[10] J.F. Weaver, G.B. Hoflund, "Surface Characterization Study of the Thermal Decomposition of AgO," *J Phys Chem,* vol. 98, pp. 8519-8524, 1994.

[11] A.I. Boronin, S.V. Koscheev, G.M. Zhidomirov, "XPS and UPS study of oxygen states on silver," *J Electron Spectrosc,* vol. 96, pp. 43-51, 1998.

[12] T. Kunimune, M. Kuramoto, S. Ogawa, T. Sugahara, S. Nagao, K. Suganuma, "Ultra thermal stability of LED die-attach achieved by pressureless Ag stress-migration bonding at low temperature," *Acta Mater,* vol. 89, pp. 133-140, 2015.

[13] B.V. L'Vov, "Kinetics and mechanism of thermal decomposition of silver oxide," *Thermochim Acta,* vol. 333, pp.13-19, 1999.

[14] B.V. L'Vov, "Interpretation of atomization mechanisms in electrothermal atomic absorption spectrometry by analysis of the absolute rates of the processes," *Spectrochim Acta Part B At Spectrosc,* vol. 52, pp. 1-23, 1997.

Reliability of Micro-Alloyed SnAgCu Based Solder Interconnections for Various Harsh Applications

Sinan Su[1], Francy John Akkara[1], Anto Raj[1], Cong Zhao[3], Seth Gordon[1], Sharath Sridhar[1], Sivasubramanian Thirugnanasambandam[1], Sa'd Hamasha[1], Jeffery Suhling[2], John Evans[1]

[1] Department of Industrial and Systems Engineering
[2] Department of Mechanical Engineering
Auburn University, 3301 Shelby Center, Auburn, AL 36849
[3] Apple Inc.
1 Apple Park Way, Cupertino, California, U.S.

Corresponding Author: Sa'd Hamasha
Email: hamasha@auburn.edu

Abstract - **In this study, the reliability of doped solder pastes on 15mm BGA under various accelerated tests was investigated. The primary goal is to compare the solder paste performances under different tests, and ultimately, find a manufacturable solder paste that will demonstrate better reliability under different harsh applications. A total of four tests (thermal cycling, thermal shock, vibration, drop) were conducted on 15mm CABGA208 with three doped solder pastes (Innolot, CycloMax, Ecolloy) and one baseline solder paste SAC305 in order to understand the effect of doped solder paste on conventional packages. Test vehicles were built based on the types of test, and two designs were applied, one for thermal cycling, and another for mechanical cycling. Test vehicles were subjected to the corresponding test conditions with different devices to access the solder paste performances. It was found that Innolot solder paste performed better in thermal cycling, and SAC305 and Ecolloy solder pastes performed better in mechanical cycling.**

Keywords - Solder doping; BGA; Reliability; Harsh applications

I. INTRODUCTION

Reliability of an electronic assembly is limited by the fatigue failure of one of the solder interconnections. Generally, solder joint reliability is questioned in harsh environments, such as automotive and aerospace, because of repetitions (e.g. turn on and off conditions), the result of cyclic working environment changes (e.g. temperature variations from day to night at Alaska), or vibration (e.g. electronics in underhood applications) [1-5]. The electronic module in such environments is subjected to extreme thermo-mechanical cycles throughout its lifetime, where solder joints are exposed to sustained cycles of thermally-induced stress or mechanical-induced stress, and ultimately lead to the accumulated fatigue and failure. Typically, thermal stress is due to the mismatch in the coefficient of thermal expansion (CTE) within the electronic assembly, whereas mechanical cycles such as vibration and drop contribute to the mechanical stress.

Since lead (Pb) was banned under regulations proposed in Japan and the EU given its substantial harmful effects to the human body, the electronic packaging industry has moved to lead-free manufacture. Sn-Ag-Cu (SAC) family has been regarded as one of the potential alternatives; however, solder alloys with just Sn, Ag, and Cu didn't perform well under extreme working conditions [6-10]. Numerous studies have been directed towards the field of SAC-based solder alloys reliability, but the selection of ideal solder alloy that is suitable for all environments is still one of the major challenges in the industry. Currently, several elements such as Bismuth (Bi), Antimony (Sb), and Nickel (Ni) are micro-alloyed or doped to the ternary SAC-based solder alloy in order to improve the thermal and mechanical properties and thus, improve the reliability [11-19]. Current investigations have shown promising results in this field.

In this investigation, four accelerated life testing (ALT) were conducted to study the behaviors of doped solder alloys under different extreme environments and then select solder paste that demonstrates better reliability among all the extreme environments for the industry [20]. The ALT was implemented to accelerate the test process than under normal use conditions, where the product was exposed to conditions (temperature, stress, vibrations, etc.) higher than that of actual service parameters. In this way, enough potential failures could be generated in a shorter period of time without introducing unrealistic failure mechanisms, which would lead the engineers to make reasonable predictions about the reliability of a product, based on the results of ALT. Mainly in this study, mechanical cycling (vibration and drop testing) and thermal cycling (thermal cycling and thermal shock) were conducted to simulate to two major extreme conditions that electronic assemblies may encounter during the realistic applications.

II. EXPERIMENTAL DESIGN & PROCEDURE

Test Vehicle Design

The test vehicles were designed based on two test categories, one for thermal cycling and the other for mechanical cycling. Figure 1 shows the test vehicle for thermal cycling. The test boards have a dimension of 120x125mm with FR-4 substrate and the board surface finish used in this project was Organic Solderability Preservative (OSP) coating. For the scope of this project, only daisy-chained CABGA208 components with SAC305 solder spheres (#1 marked) were selected as test components. The dimension of CABGA208 is 15mm x15mm, with a 0.8mm pitch size. To investigate the effect of doped solder materials on the thermal reliability of conventional BGA, 3 doped solder pastes were applied, along with baseline material SAC305, Table 1 showed the compositions of each solder material. During the assembly, solder pastes are printed onto the board using 6mil stencil, then CABGA components were placed on the solder pastes. After that, test boards are placed in the Vitronics SMR-800 conventional reflow oven with a peak temperature of 245°C and 90s above liquidus as suggested by solder suppliers.

Figure 1. Test vehicle for thermal cycling

Table 1. Solder paste material composition

Solder Materials	Composition
SAC305	96.5Sn - 3Ag - 0.5Cu
Innolot	90.58Sn - 3.5Ag - 0.7Cu - 3Bi - 1.5Sb - 0.125Ni
SAC-Q	92.77Sn - 3.41Ag - 0.52Cu - 3.3Bi
Ecolloy	96.62Sn - 0.92Cu - 2.46Bi

Test vehicle used for the mechanical cycling is shown in Figure 2. The dimension of the board is 280x115mm with FR-4 substrate and OSP surface finish was applied. CABGA208 with SAC305 solder spheres with the same dimension and configuration as thermal cycling test components were selected and compared. Furthermore, the test board was categorized into 4 different zones (#1 - #4),

based on the stress levels that each location was subjected to, during the mechanical cycling. CABGA in zone 1 (#2) was selected as the focus of this study since it was associated with the highest fatigue life degradation during the test. Three doped solder alloys along with SAC305 were applied as solder paste. During the assembly, solder pastes are printed onto the board using 6mil stencil, then CABGA components were placed on the solder pastes. After that, test boards are placed in the Vitronics SMR-800 conventional reflow oven with a peak temperature of 245°C and 90s above liquidus as suggested by solder suppliers

Figure 2. Test vehicle for mechanical cycling

Experimental Design

The goal of thermal cycling (thermal shock) is to compare the thermal behaviors of different solder paste materials when exposed to the thermal stresses induced by alternating high to low temperature extremes. In thermal cycling test, five boards were tested in a thermal chamber until complete failure with a thermal profile ranging from -40°C to 125°C, with a 15mins dwell time at upper peak temperature and 10mins dwell time at lower temperature. Both ramp up and down time are 50 minutes each. The total duration of one cycle is 125mins. Since all BGAs were daisy-chained, a LabView system was used to automatically monitor the failure situation and an increment of component resistance by over 100 Ω for 5 consecutive measurements was defined as failure. Similar to thermal cycling, 5 test boards were subjected to Liquid-to-Liquid Thermal Shock in CSZ TSB chamber. Thermal shock was limited to 3000 cycles and temperature ranges is also from -40°C to 125°C, however, only 5 mins dwell time for each temperature extreme and 2.5 mins of transition time. The components were hand probed every 50 cycles and failure was defined as an increase of electrical resistance over 1000Ω.

Meanwhile, mechanical behaviors of the same solder paste materials were examined by subjecting the test vehicle to repetitive mechanical loadings either through vibration-induced or drop-induced impact environments. Therefore, a comprehensive understanding of the reliability of doped solder alloy under different extreme conditions was achieved. In terms of vibration testing, the natural frequency of the test vehicle was determined to be between 350 to 400Hz. Five test boards (80 components) were mounted on an LDS LV217 electro-dynamic shaker table prior to the testing. Then a random vibration profile with 4.6 G_{rms} was applied and the test was stopped at 30 hours. The components were hand probed every 1 hour to detect failures. A component was determined as a failure when an increase of electrical resistance larger than 1000Ω was detected. For drop testing, a maximum peak acceleration of 1500G and half-sine impact pulse duration of 0.5 milliseconds was maintained throughout the experiment using a Lansmont shock test system. The number of drops was limited to 300 times, and each component was hand probed for every 20 drops. Similarly, the failure was determined when the electrical resistance was larger than 1000Ω.

III. RESULTS AND DISCUSSIONS

Solder Joints Microstructure

Since the mechanical properties of solder joint were largely dependent on the microstructure of the solder and the formation of precipitates, therefore, it is important to the investigate to understand the morphology of precipitates formed after reflowing with different solder alloy compositions. Figure 3 showed the as-reflowed microstructure of SAC305 and the three doped solder alloys. In Figure 3a, the microstructure of SAC305 solder alloy consists of large β-Sn dendrites surrounded by two distinctive Intermetallic Compounds (IMCs) dispersed within the Sn-rich matrix. According to the EDS analysis, the IMCs distributed in the eutectic region are Ag_3Sn and Cu_6Sn_5. After adding 3% Bi into the matrix and removed Ag (Ecolloy), a different morphology was observed as shown in Figure 3b. Adding Bi to SAC305 increased β-Sn dendrites and led to a coarsening of Cu_6Sn_5 IMC. Since Ag was removed from the matrix, no Ag_3Sn precipitates were detected. Moreover, as Bi cannot make any intermetallic compound with other elements [21], no Bi compound was observed in the solder. For SAC-Q (SAC+3.3%Bi), as shown in Figure 3c, Ag_3Sn precipitates degenerated in a chain-like arrangement, while Cu_6Sn_5 IMC further coarsened into a large chunk. Furthermore, with a relatively higher Bi addition (>3%), Bi supersaturated and precipitated in the form of pure Bi phase from the Sn matrix. Lastly, after adding Ni to solder (SAC+Bi+Ni) as shown in Figure 3d, the microstructure has been refined, and the volume fraction of IMCs particle has increased, which would play a critical role in the improvement of mechanical properties of the solder joint. Also, the existence of Ni has influenced the formation of Cu_6Sn_5 IMCs, where the large chunk was replaced by small pieces and distributed into the solder. Most importantly, researches have demonstrated that Bi particle in the solder matrix can 'block' the relatively movement of several grain boundaries, given the 'solid-solution' and precipitate strengthening effect associated of Bi atom in the Sn-rich phase, which would contribute to the mechanical strength of this solder alloy [22, 23].

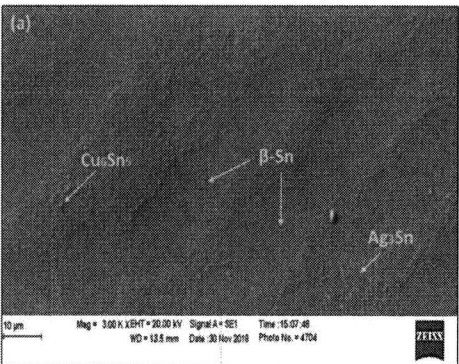
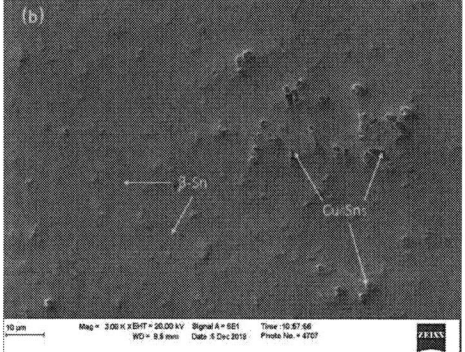
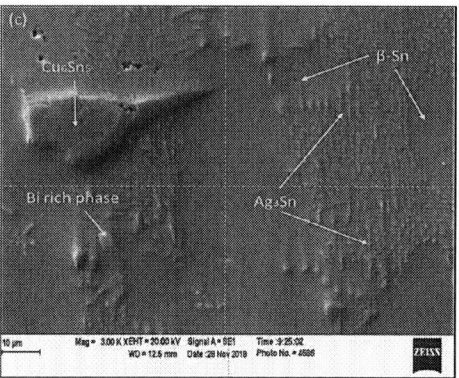
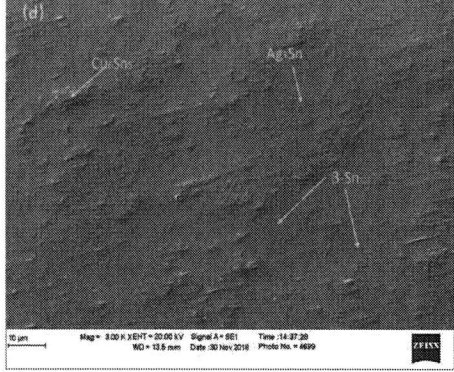

Figure 3. SEM microstructure of: (a) SAC305; (b) Ecolloy; (c) SAC-Q; (d) Innolot

Thermal Cycling & Thermal Shock Testing

Since test boards under thermal cycling test continued until the complete failure, a 2-parameter Weibull distribution was applied to analyze the complete failure data as shown in Equation 1:

$$\lambda(t) = \frac{\beta}{\theta}\left(\frac{t}{\theta}\right)^{\beta-1} \qquad (1)$$

where β is the shape parameter, and θ is the scale parameter. In this study, we take the scale parameter as the characteristic fatigue life, which is the number of cycles at which 63.2% of the overall population is expected to fail, and early failure (B10) is the number of cycles at which 10% of the overall population is expected to fail.

Figure 4 showed the Weibull distribution for thermal cycling testing. Different combination of color and shape represented different solder materials that have been tested (e.g. blue circle represented Innolot solder material). Then characteristic life with each solder material was plotted as a function of solder material name, shown in Figure 5. Innolot was associated with the highest characteristic life, followed by CycloMax. SAC305 showed the least characteristic life among the three doped solder alloys.

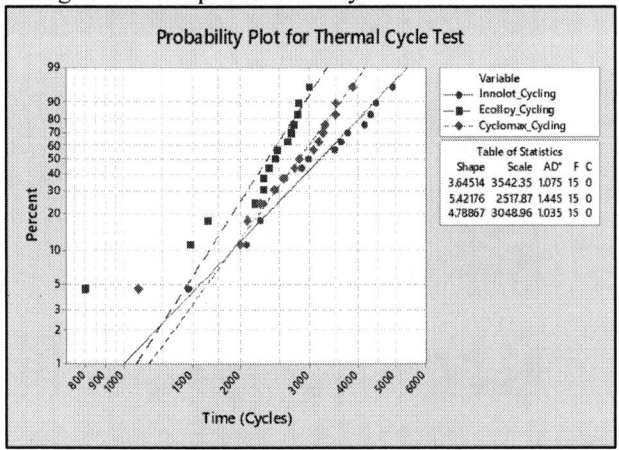

Figure 4. Weibull plot of thermal cycling reliability of CABGA with 4 solder pastes

After characteristic life, early failure (B10) for thermal cycling testing was also examined among different solder materials as shown in Figure 6. The general trend is the same when comparing to characteristic life, as Innolot was associated with the highest B10 life, followed by CycloMax. SAC305 was still associated with the least fatigue life.

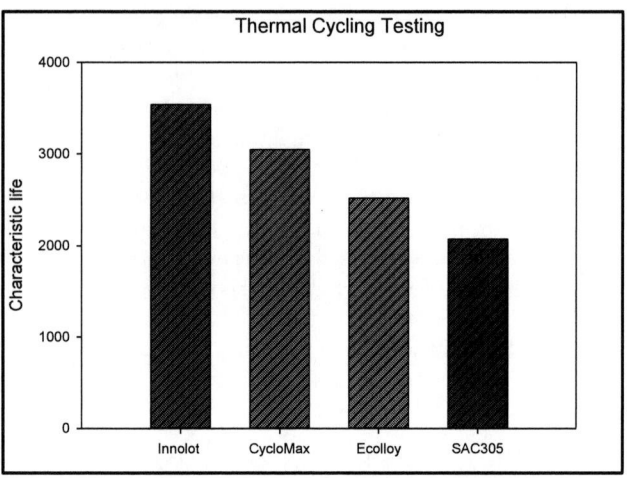

Figure 5. Characteristic life comparisons during thermal cycling

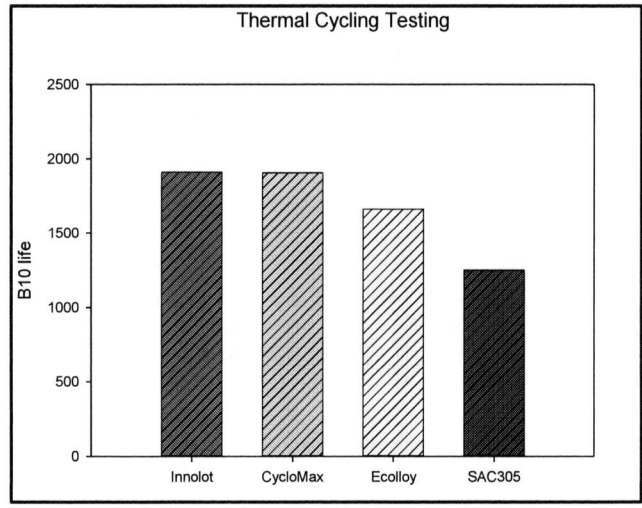

Figure 6. B10 life comparisons during thermal cycling

Failure mode analysis was carried out after all the components failed. Figure 7 represents the most prevalent type of failure during thermal cycling, where the crack was initiated at the corner region on the package side, where the stress concentration was suspected the highest during thermal cycling. Then the crack propagated along the near-intermetallic region since it coarsened continuously during the test. Another type of failure observed in the test was crack initiation within the solder recrystallization area and propagating along the grain boundaries. Instead of propagating along the near-intermetallic region, this time the crack moved away from it and was within the bulk sample where the resistance is the lowest.

978-1-7281-1500-9/19 $31.00 © 2019 IEEE

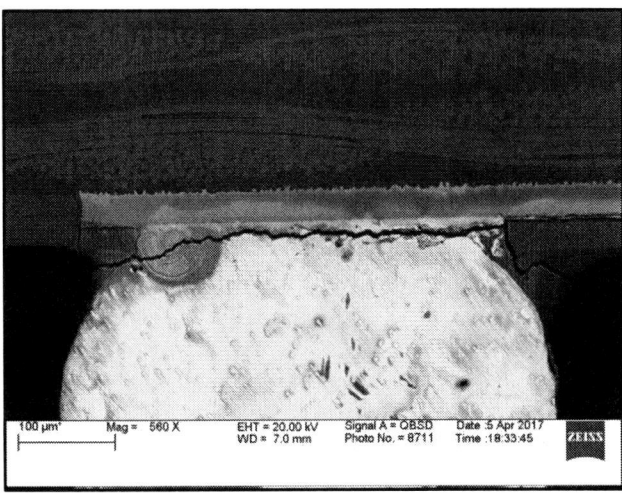

Figure 7. Failure modes analysis for solder joint under thermal cycling testing

In the case of the thermal shock test, since it was limited to only 3000 cycles. By the end of the test, components with CycloMax solder paste showed only 13% failure, and components with Innolot solder paste showed 23% failure. However, Ecolloy solder paste showed 100% failure, which is even higher than the baseline SAC305 (42%). Since the components were probed every 50 cycles, with a considerable number of survivors, a 2-parameter Weibull distribution with right and interval censored method was applied to analyze the thermal shock failure data, as shown in Figure 8. Then the characteristic life associated with each alloy was plotted as a function of solder material name, as shown in Figure 9. In this figure, Innolot solder paste still outperformed the other solder pastes having a characteristic life 5724 cycles followed by CycloMax. Ecolloy solder paste was associated with the least characteristic life and less than the baseline SAC305 solder paste.

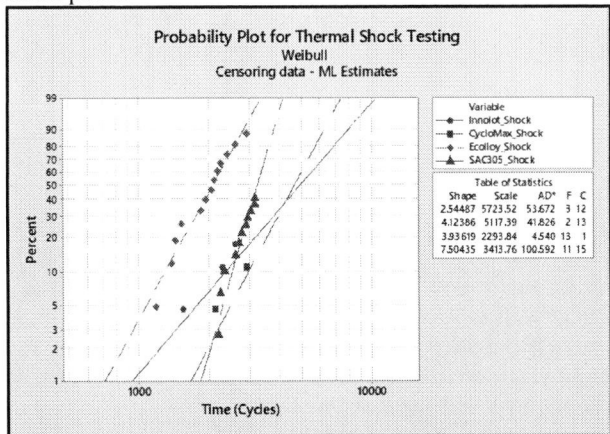

Figure 8. Weibull plot of thermal shock reliability of CABGA with 4 solder pastes

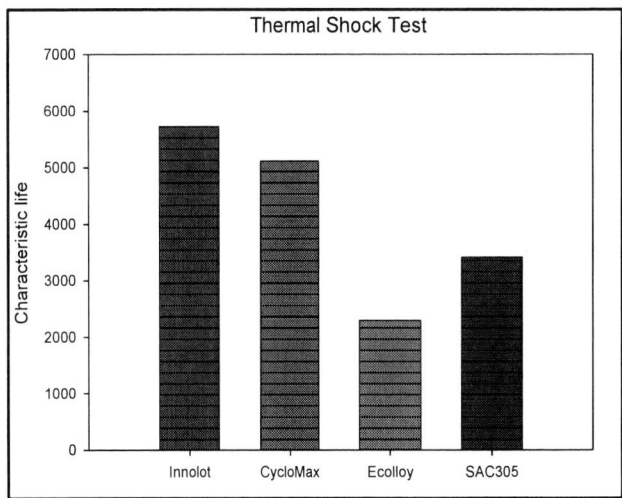

Figure 9. Characteristic life comparisons during thermal shock

B10 life for thermal shock testing among different solder pastes was also compared and the result was shown in Figure 10. CycloMax turned out to be the best one in terms of B10 life comparisons, while Ecolloy was still associated with the least life. Moreover, the degradation of B10 life was observed for Innolot solder alloy.

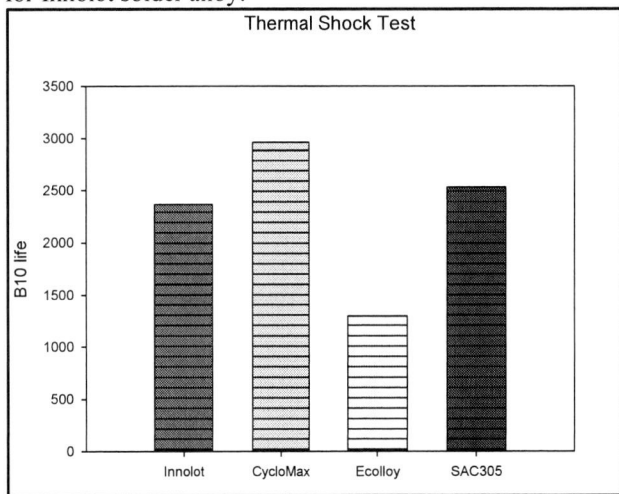

Figure 10. B10 life comparisons during thermal shock

Thermal shock failure modes analysis was carried out after 3000 cycles. Similar to the failure mode for thermal cycling, the most prevalent one for thermal shock was near-intermetallic failure along the package side, as shown in Figure 9. Crack initiated at the upper corner of solder joint where the stress concentration is the highest, and then propagated within the bulk solder along the interfacial region to the other end. Unlike thermal cycling, intermetallic failure was also observed where the crack propagated along the interfacial region between bulk solder and copper pad. Faster transition between temperature extremes (-40°C-125°C) brought higher shear strain rate to the solder joint, given the viscoplastic nature of solder material, the strength of bulk solder against deformation exceeded the strength of intermetallic (IMC), which makes the IMC on the package side the weakest interconnection, therefore, leads to a propagation along the intermetallic region (intermetallic failure).

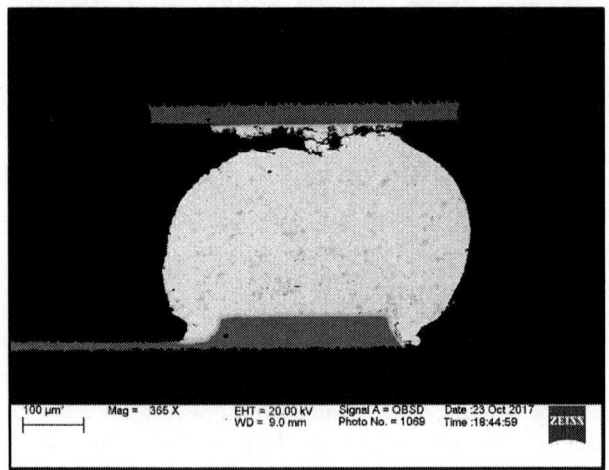

Figure 11. Failure modes analysis for solder joint under thermal shock testing

Overall, Innolot demonstrated the best reliability by having the highest characteristic life and B10 life for thermal cycling test and highest characteristic life for thermal shock test. However, it was also revealed that this material may encounter early failure problem under a higher thermal stress-impact environment (e.g. thermal shock). Solder paste with both higher Ag and Bi demonstrated relatively better thermal reliability since the mechanical strength was reinforced by doping additional elements such as Ag and Bi.

Vibration and Drop Testing

A 2-parameter Weibull distribution with right and interval censored methods was also suitable for vibration test since the test was limited to only 30 hours. Characteristic life of vibration test from Weibull analysis was extracted and plotted as a function of solder paste, as shown in Figure 12. The opposite trend against thermal cycling tests was observed, where SAC305 was observed to outperform the other solder pastes with a characteristic life of 22.3 hours, followed by Ecolloy. Innolot was associated with the lowest characteristic life in terms of vibration testing.

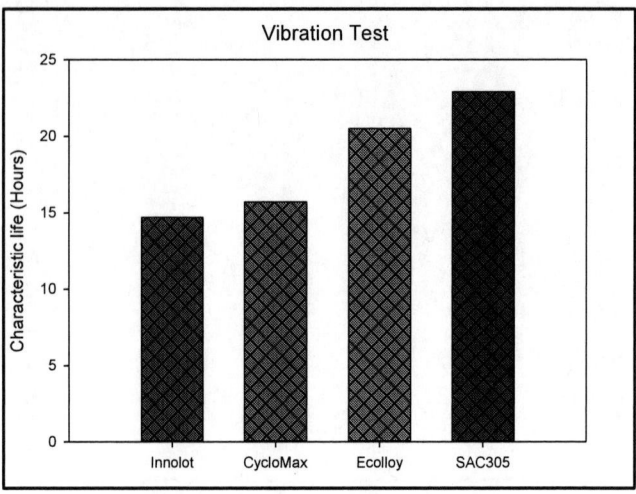

Figure 12. Characteristic life comparisons during the vibration test

Moreover, B10 life associated with vibration test for different solder materials demonstrated the similar trend as shown in Figure 13, where Innolot was observed to have the least B10 life, and Ecolloy outperformed SAC305 and demonstrated the highest B10 life during vibration test.

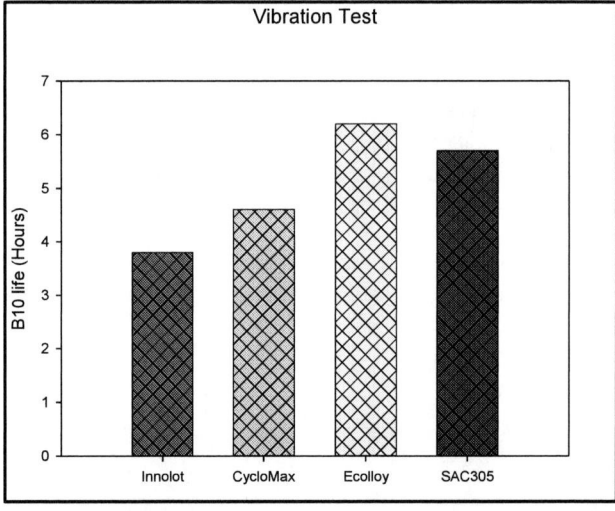

Figure 13. B10 life comparisons during the vibration test

Basic failure analysis for the vibration test was carried out after vibrating the test boards for 30 hours. As for now, the failure mode was solder failure as shown in Figure 14. Typically, crack was observed to initiate at the board side and propagated within the bulk sample near the intermetallic region. Pad cratering could have also occurred; however, we haven't observed any case like that.

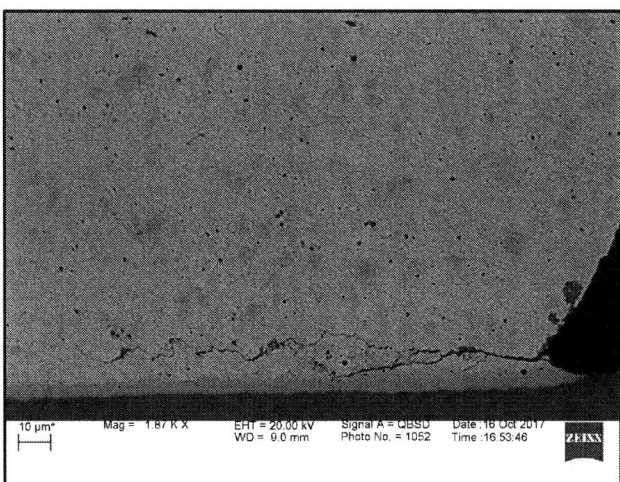

Figure 14. Failure modes analysis for solder joint under vibrational testing

The same data analyzing method was applied to drop test, and characteristic life was compared between each other in a bar chart, as shown in Figure 15. A similar trend was observed in this figure, as compared to the vibrational test result. Ecolloy almost had the same characteristic life as SAC305, while Innolot still demonstrated the lowest life.

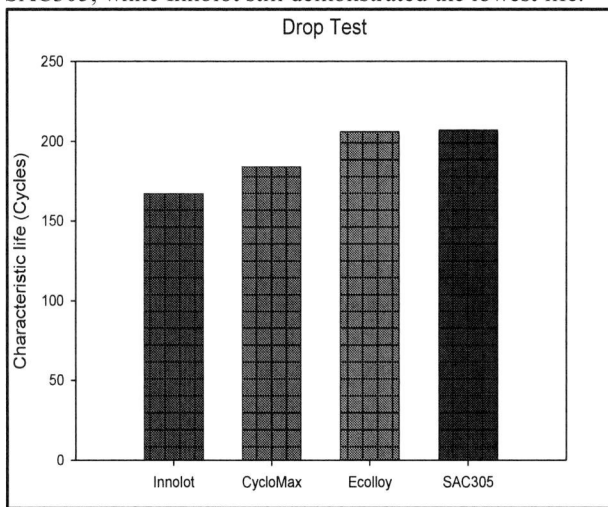

Figure 15. Characteristic life comparisons during drop test

Furthermore, different solder materials associated with drop test were compared in terms of early failures (B10 life), as shown in Figure 16. Ecolloy solder paste turned out to be the best solder pastes in terms of B10 life comparisons for both vibration and drop testing. While CycloMax was associated with the least B10 life for drop test.

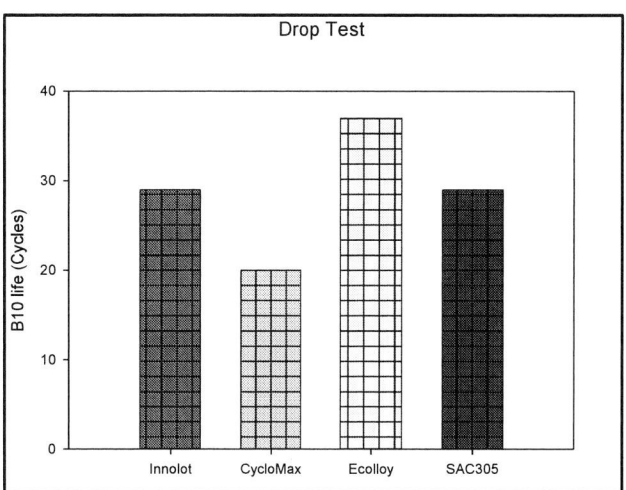

Figure 16. B10 life comparisons during the drop test

Overall, solder pastes that have higher ductility demonstrated better reliability in terms of mechanical cycling (shock) tests, given the higher ductility enables the solder material to absorb more energy by deformation when being subjected to mechanical shock.

IV. CONCLUSIONS

A reliability study for various doped lead-free solder pastes under four different ALT testing has been investigated. The goal is to study the reliability of various doped solder materials under different extreme testing conditions. Mainly, thermal cycling (shock) and mechanical cycling (shock) were introduced to downselect doped solder paste materials with the best thermal and mechanical reliability. The thermal cycling (thermal & shock) performance of 15mm CABGA packages with 4 solder pastes were studied first. Innolot solder paste demonstrated the best thermal reliability, given its outstanding strength and thermal resistance. While Ecolloy demonstrated the worst thermal reliability. After that, mechanical cycling (vibration and drop) performance of 15mm CABGA package with the same 4 solder pastes was compared. In terms of characteristic life, SAC305 demonstrated the best mechanical reliability, while Innolot was associated with the worst mechanical reliability since excessive brittleness affects its reliability under vibration and drop testing, where higher ductility was desired to help mitigate the mechanical shock from mechanical cycling. However, in terms of B10 life, Ecolloy was associated with the best mechanical reliability.

In conclusion, no solder paste was found to demonstrate promising reliability that can be accepted by all the 4 types of accelerated testing. Solder paste with higher Ag and Bi content demonstrated better strength and thermal resistance, at the cost of ductility. An innovational solder material that has both acceptable mechanical strength and ductility is in desperate need so that the reliability under harsh applications that has both thermal and mechanical-induced stress (e.g. electronic modules under-the-hood) can be guaranteed.

V. REFERENCES

[1] A. Raj, S. Thirugnanasambandam, T. Sanders, J. Evans, W. Johnson, M. Carpenter, S. Hamasha, "Comparative Study on Impact of Various Low Creep Doped Lead Free Solder Alloys," in *SMTA International Proceedings*, 2017.

[2] A. Raj, S. Gordon, J. Evans, M. Bozack, W. Johnson, "Long Term Isothermal Aging of BGA Packages Using Doped Lead Free Solder Alloys," in *SMTA International Proceedings*, 2018.

[3] A. Raj, T. Sanders, S. Sridhar, J. Evans, M. Bozack, W. Johnson "Thermal Shock Reliability Test on Multiple Doped Low Creep Lead Free Solder Paste and Solder Ball Grid Array Packages," in *SMTA International Proceedings*, 2015.

[4] S. Thirugnanasambandam, T. Sanders, A. Raj, D. Stone, J. Evans, G. Flowers*, et al.*, "The study of vibrational performance on different doped low creep lead free solder paste and solder ball grid array packages," in *Thermal and Thermomechanical Phenomena in Electronic Systems (ITherm), 2014 IEEE Intersociety Conference on*, 2014, pp. 920-923.

[5] S. Thirugnanasambandam, A. Raj, D. Stone, T. Sanders, S. Sridhar, S. Gordon*, et al.*, "Proportional Hazard Model of doped low creep lead free solder paste under vibration," in *Thermal and Thermomechanical Phenomena in Electronic Systems (ITherm), 2016 15th IEEE Intersociety Conference on*, 2016, pp. 1209-1217.

[6] S. Sridhar, A. Raj, S. Gordon, S. Thirugnanasambandam, J. L. Evans, and W. Johnson, "Drop impact reliability testing of isothermally aged doped low creep lead-free solder paste alloys," in *Thermal and Thermomechanical Phenomena in Electronic Systems (ITherm), 2016 15th IEEE Intersociety Conference on*, 2016, pp. 501-506.

[7] A. Raj, S. Thirugnanasambandam, T. Sanders, S. Sridhar, S. Gordon, J. Evans*, et al.*, "Proportional Hazard Model of doped low creep lead free solder paste under thermal shock," in *Thermal and Thermomechanical Phenomena in Electronic Systems (ITherm), 2016 15th IEEE Intersociety Conference on*, 2016, pp. 1191-1201.

[8] S. Gordon, T. Sanders, A. Raj, S. Sridhar, J. Evans, W. Johnson, S. Hamasha, "Reliability of Doped Ball Grid Array Components in Thermal Cycling after Long-Term Isothermal Aging," in *SMTA International Proceedings*, 2017.

[9] S. Sridhar, J. Evans, M. Bozack, W. Johnson, S. Hamasha, "Reliability Study of Doped Lead Free Solder Paste Alloys by Thermal Cycling Testing," in *SMTA International Proceedings*, 2017.

[10] C. Zhao, "Board Level Reliability of Lead-Free Solder Interconnections with Solder Doping Under Harsh Environment," 2017.

[11] S. Su, N. Fu, F. J. Akkara, and S. Hamasha, "Effect of Long-Term Room Temperature Aging on the Fatigue Properties of SnAgCu Solder Joint," *Journal of Electronic Packaging*, vol. 140, p. 031005, 2018.

[12] S. Su, F. J. Akkara, M. Abueed, M. Jian, J. Suhling, and P. Lall, "Fatigue Properties of Lead-free Doped Solder Joints," in *2018 17th IEEE Intersociety Conference on Thermal and Thermomechanical Phenomena in Electronic Systems (ITherm)*, 2018, pp. 1243-1248.

[13] S. Hamasha, S. Su, F. Akkara, A. Dawahdeh, P. Borgesen, and A. Qasaimeh, "Solder joint reliability in isothermal varying load cycling," in *Thermal and Thermomechanical Phenomena in Electronic Systems (ITherm), 2017 16th IEEE Intersociety Conference on*, 2017, pp. 1331-1336.

[14] F. Akkara, M. Abueed, M. Rababah, C. Zhao, S. Su, J. Suhling*, et al.*, "Effect of Surface Finish and High Bi Solder Alloy on Component Reliability in Thermal Cycling," in *2018 IEEE 68th Electronic Components and Technology Conference (ECTC)*, 2018, pp. 2032-2040.

[15] F. Akkara, S. Su, S. Thirugnanasambandam, A. Dawahdeh, A. Qasaimeh, J. Evans*, et al.*, "Effects of Long-Term Aging on SnAgCu Solder Joints Reliability in Mechanical Cycling Fatigue," in *SMTA International Conference, Rosemont, IL, Sept*, 2017, pp. 17-21.

[16] Sinan Su, Francy John Akkara, Dr. Sa'd Hamasha, "Fatigue and Shear Properties of High Reliable Solder Joints for Harsh Applications," in *SMTA International*, 2018.

[17] Francy John Akkara, Cong Zhao, Sa'd Hamasha, Jeffrey Suhling, "Effects of Mixing Solder Sphere Alloys with Bismuth-Based Pastes on The Component Reliability in Harsh Thermal Cycling," in *SMTA International*, 2018.

[18] C. Zhao, T. Sanders, Z. Hai, C. Shen, and J. L. Evans, "Reliability Analysis of Lead-Free Solder Joints with Solder Doping on Harsh Environment," in *International Symposium on Microelectronics*, 2016, pp. 000117-000122.

[19] C. Zhao, T. Sanders, C. Shen, Z. Hai, J. L. Evans, M. Bozack*, et al.*, "RELIABILITY OF DOPED LEAD-FREE SOLDER JOINTS UNDER ISOTHERMAL AGING AND THERMAL CYCLING."

[20] F. J. Akkara, C. Zhao, R. Athamenh, S. Su, M. Abueed, S. Hamasha*, et al.*, "Effect of Solder Sphere Alloys and Surface Finishes on the Reliability of Lead-Free Solder Joints in Accelerated Thermal Cycling," in *2018 17th IEEE*

Intersociety Conference on Thermal and Thermomechanical Phenomena in Electronic Systems (ITherm), 2018, pp. 1374-1380.

[21] A. Hammad, "Evolution of microstructure, thermal and creep properties of Ni-doped Sn–0.5 Ag–0.7 Cu low-Ag solder alloys for electronic applications," *Materials & Design (1980-2015)*, vol. 52, pp. 663-670, 2013.

[22] M. H. Mahdavifard, M. F. M. Sabri, D. A. Shnawah,

of iron and bismuth addition on the microstructural,

mechanical, and thermal properties of Sn–1Ag–0.5 Cu solder alloy," *Microelectronics Reliability,* vol. 55, pp. 1886-1890, 2015.

[23] A. Fahim, S. Ahmed, J. C. Suhling, and P. Lall, "Mechanical Characterization of Intermetallic Compounds in SAC Solder Joints at Elevated Temperatures," in *2018 17th IEEE Intersociety Conference on Thermal and Thermomechanical Phenomena in Electronic Systems (ITherm)*, 2018, pp. 1081-1090.

Wideband Low-profile Ka-Band Microstrip Antenna with Low cross polarization Using Asymmetry AMC Structure

Mei Xue[1,2,3], Weikang Wan[1,2,3], Qidong Wang[1,2,*], Liqiang Cao[1,2]

[1]National Center for Advanced Packaging (NCAP China), Wuxi, Jiangsu, China 214135
[2]System Packaging and Integration Research Center, Institute of Microelectronics of Chinese Academy of Sciences, Beijing, China 100029
[3]University of Chinese Academy of Sciences, Beijing, China 100049
*e-mail: wangqidong@ime.ac.cn

Abstract—In this work, a 28GHz microstrip antenna based on an asymmetric artificial magnetic conductor structure is proposed, which takes the features of low profile, high bandwidth, high gain, and low cross-polarization level. The principle that the AMC structure servers as a reflection plane of the microstrip antenna to widen bandwidth is deeply investigated. The mechanism of the H-plane cross-polarization level deterioration when the symmetric AMC structure is applied to the microstrip antenna is analyzed for the first time. The asymmetric AMC cell provides a 0° reflection phase for the E-plane horizontal electric field and a 180° reflection phase for the H-plane horizontal electric field. Therefore, it is used as a reflection plane to achieve both large bandwidth and low cross-polarization level. The simulation results demonstrate that the proposed antenna with $0.056\lambda_0$ substrate thickness realizes the high bandwidth of 23% and low H-plane cross-polarization level of less than -20dB in band. The radiation efficiency is up to 97.6% at 27GHz while peak gain reaches to 10.1dBi at 31.2GHz.

Keywords- Antenna-in-Package, mmwave, microstrip antenna, artificial magnetic conductor, low profile, high bandwidth, low cross-polarization level

I. INTRODUCTION

Antenna-in-Package (AiP) is a feasible method for obtaining the millimeter wave transceivers with low profile, miniaturization, high integration and high performance. It possesses wide application prospects in 5G mm-wave mobile communication and satellite communication. Meanwhile, the application of the millimeter wave in the communication field makes AiP possible. For AiP, the microstrip antenna (MSA) comes to be the best choice due to its merits such as low profile, light weight and easy conformal, which make it feasible to integrate in AiP. However, MSA has the characteristics of narrow bandwidth of less than 10%, especially in AiP, where the thickness of the antenna substrate is greatly limited. Various methods have been proposed to improve the bandwidth of MSA, such as parasitic patch [1], optimized feed structure [2] [3], differential feed [4], dual-resonant structure [5], air cavity [6], metasurface antenna [7] [8], and artificial magnetic conductor (AMC) [9]. In [1], a 4×4 antenna array consisting of a radiating element composed of a thin rectangular patch and a surrounding U-shaped parasitic patch with substrate thickness of $0.053\lambda_0$ was proposed. Its bandwidth is 17.7%,

but gain is small. In [2], the L-probe patch antenna array was presented. Although the bandwidth is as high as 29%, the substrate thickness is greater than $0.1\lambda_0$. Another MSA with L-shaped feed was proposed in [3], and the bandwidth is 11.7%. In [4], The TE20-Mode SIW dual-slot-fed patch antenna was put forward, and the bandwidth is 10.2% and the peak gain is 6.48dBi. What's more, for antenna array, the differential feed approach complicates the feed network. In [5], dual-resonant structure including slot and patch performs a high bandwidth of 23%, but antenna substrate thickness is greater than $0.1\lambda_0$, and gain of the antenna element is 3.8dBi. In [6], the antenna substrate contains an air cavity, which achieves 10% bandwidth and is not conducive to integration. In [7] and [8], the radiating element is metasurface antenna with $0.06\lambda_0$ substrate thickness, which realizes 25% bandwidth and 9dBi peak gain, but the radiation pattern has large back lobe. In contrast, the way that AMC structure serves as a reflection plane of the MSA can significantly improve the bandwidth and gain of low profile MSA, however, cross-polarization level is deteriorated compared to reference antenna [9]. The cross-polarization level will directly affect the ability of the antenna to reject noise. In [10] and [11], cross-polarization level is reduced to -20dB by means of differential feed, which causes complicated the feed network.

Therefore, in terms of the above issues, in this paper, we propose a novel and simple asymmetric AMC structure, which is used as the reflection plane of the MSA to obtain low profile, large bandwidth and low cross-polarization level. The paper is organized as follows: Section II describes the theoretical analysis and design of the symmetric AMC structure; Section III demonstrates the MSA based on symmetric AMC structure; MSA based on asymmetric AMC structure is proposed in Section IV, followed by a conclusion in Section V.

II. THEORETICAL ANALYSIS AND DESIGN OF SYMMETRIC AMC STRUCTURE

A. Mechanism of AMC to widen the bandwidth of MSA

For the MSA, the ground plane (GP) is a metal conductor, which serves as the return path of the microstrip line to satisfy the resonance condition, and also serves as a reflection plane to realize the unidirectional pattern. The conductor plane will reflect the incident electromagnetic wave to meet the boundary condition (mirror principle). As

978-1-7281-1500-9/19 $31.00 © 2019 IEEE

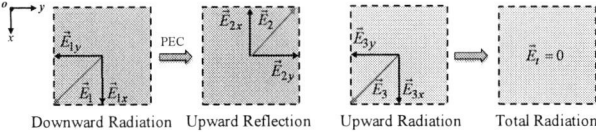

Figure 1. Schematic diagram of electric field distribution when radiation waves are radiated downward to PEC, reflected from PEC, and directly radiated upward.

shown in Figure 1, assuming reflection plane is XOY plane, and downward radiation electromagnetic wave includes electric filed component $\vec{E}_1$ parallel to XOY, $\vec{E}_1$ can be decomposed into two components $\vec{E}_{1x}$ and $\vec{E}_{1y}$ in orthogonal directions. $\vec{E}_{1x}$ and $\vec{E}_{1y}$ are upward reflected by PEC to be $\vec{E}_{2x}$ and $\vec{E}_{2y}$ with $180°$ reflection phase, and $\vec{E}_2 = \vec{E}_{2x} + \vec{E}_{2y}$. When the substrate is very thin, the phase shift of the $\vec{E}_2$ due to the optical path is negligible, direct upward radiation electric filed $\vec{E}_3$ is canceled by $\vec{E}_2$ since they have similar amplitude and conjugate phase. Therefore, the radiation power is reduced, leading to narrow bandwidth. On the contrary, if the magnetic conductor is used as the reflection plane of the MSA, reflection phase of $\vec{E}_1$ is $0°$. $\vec{E}_2$ and $\vec{E}_3$ are in phase and radiation power is increased.

B. Design and analysis of symmetry AMC structure

Figure 2 shows a classic symmetric AMC structure that includes square metal patch array, substrate, and a metal ground plane, and it exhibits same periodicity in both x and y direction. For $\vec{E}_{1x}$ and $\vec{E}_{1y}$, the AMC structure cell can be equivalent to a parallel LC circuit in both x and y direction. The value of equivalent L and C are determined by the width of AMC patch, gap, and thickness and material of substrate [9]. When the frequency of incident electro-magnetic wave is same as the resonant frequency of LC circuit, the impedance of LC circuit is infinite, so that AMC structure serves as a high impedance surface. The reflection phase is $0°$ in both x and y direction shown in Figure 3 leading to that $\vec{E}_t \neq 0$ and consists of both x- and y-direction field components.

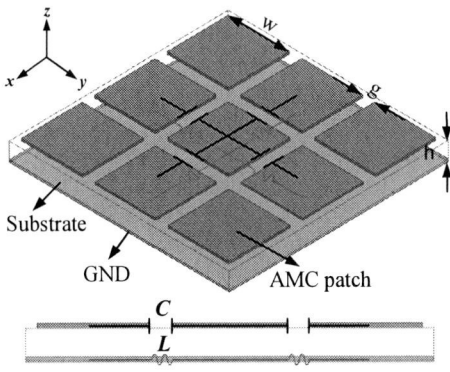

Figure 2. A symmetric AMC structure and equivalent parallel LC circuit.

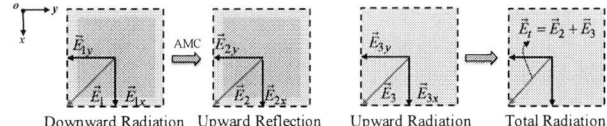

Figure 3. Schematic diagram of electric field distribution when radiation waves are radiated downward to symmetry AMC, reflected from symmetry AMC, and directly radiated upward.

The proposed MSA is proposed to be fabricated on the PCB laminate as shown in the Figure 4, and two layers of PCB substrate Rogers 6006 with thickness of 0.254 mm are laminated with a 0.127mm prepreg FR28, so the total thickness is about $0.056\lambda_0$, while λ_0 is the wavelength of 28GHz in free space. There are four copper cladding layers from M1 to M4 and MSA patch, AMC patch and GND is respectively on M1, M3 and M4, while M2 is intentionally stripped The MSA is excited by a feed pin.

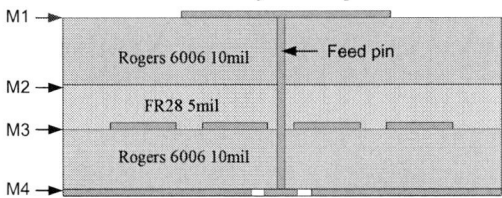

Figure 4. PCB laminate of proposed MSA.

An AMC unit is studied and designed using the 3D electromagnetic simulation software HFSS to characterize infinite-period AMC plane. Detailed simulation methods can refer to [9]. However, it is important to take the dielectric layers on the top of the AMC into account, because they will affect the equivalent L and C of AMC. The Floquet port is de-embedded on the patch of AMC. The material and thickness are fixed, the gap between AMC patch g is is set to 0.1 mm, and binding by the PCB processing tolerance. The width of AMC patch w is adjusted to 1.1mm to make the AMC cell resonate at 28GHz which also is the center frequency of antenna. The AMC cell is symmetrical in the x and y direction, therefore, for a normally incident TE and TM polarized plane wave, their characteristic would be the same.

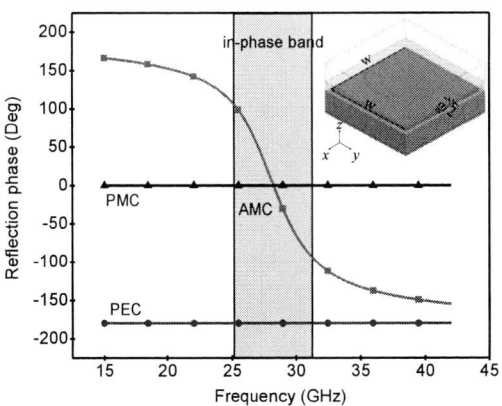

Figure 5. The reflection phase of symmetry AMC, PEC and PMC.

978-1-7281-1500-9/19 $31.00 © 2019 IEEE

As shown in the Figure 5, at resonant frequency of 28GHz, reflection phase is 0°, and the in-phase reflection bandwidth that reflection phase ranges from +90° to -90° is 21.4% at 28GHz (25GHz-31GHz). In contrast, the reflected phase of PEC and PMC are 180° and 0° respectively.

III. MICROSTRIP ANTENNA BASED ON SYMMETRIC AMC STRUCTURE

Firstly, the reference antenna resonanting at 28GHz is designed as shown in Figure 6a, and the key parameters are shown in the Table I. Then, the above symmetric AMC is loaded in the reference antenna (shown in Figure 6b), and patch length and width of MSA, feed pin position, and AMC width are slightly optimized. Under the structural parameters as shown in the Table I, the impedance bandwidth and peak gain versus frequency are shown in Figure 7. The bandwidth is significantly broadened from 7.9% at 28GHz (26.95GHz-29.16GHz) of reference antenna to 25.3% at 28.5GHz (24.9GHz-32.1GHz) of the MSA based on symmetric AMC. However, the H-plane cross-polarization level is increased from -20dB to -11dB shown in Figure 8.

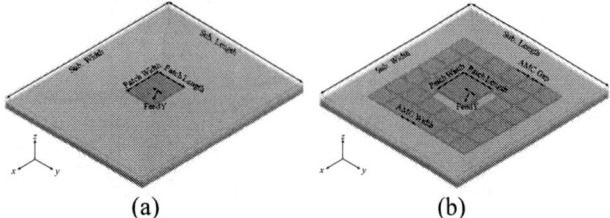

(a)   (b)

Figure 6. The structures of reference antenna (a) and MAS based on symmetry AMC (b).

TABLE I. PARAMATERS OF TWO ANTENNAS IN FIGURE 6

Item	Sub. Width	Sub. Length	Patch Length
Ref. Ant	10mm	11mm	1.94mm
AMC Ant	10mm	11mm	2.1mm
Patch Width	FeedY	AMC Width	AMC Gap
2.4mm	0.5mm	—	—
2.1mm	0.8mm	1.1mm	0.1mm

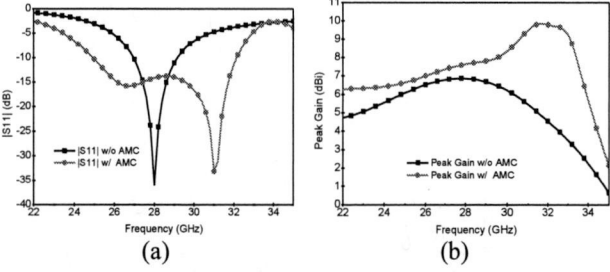

(a)   (b)

Figure 7. Simulated |S11| (a) and peak gain (b) of reference antenna and MSA based on symmetry AMC.

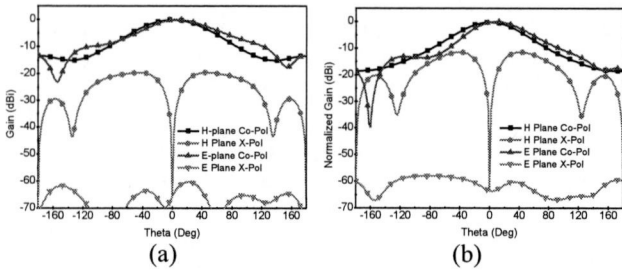

(a)   (b)

Figure 8. Simulated normalized pattern of reference antenna (a) and MAS based on symmetry AMC (b).

From the radiated E-field vector distribution on the plane at a certain height from the MSA surface based on symmetric AMC in Figure 9, we can see that in the red box region, the electric field has a large component in the x direction at t=0 and t=T/2, which partially acts as cross-polarized electric field of the H plane (XOZ plane). Conversely, in the region out of the red box, E-field direction is mainly along the y direction. In essence, although the downward and upward radiated electric field of the reference antenna is mainly along the y direction, a small amount of electric field composed of x-direction electric field component is radiated on the length side and the four corners of the antenna patch. For the PEC reflection plane, x-direction incident electric field has reflection phase of 180° and reflected electric field will cancel the electric field directly radiated upward. Therefore, the H-plane cross-polarization level of the reference antenna is less than -20dB. However, the symmetrical AMC reflects x-direction electric field with 0° reflection phase, which results in the enhancing between reflected electric field and that of direct upward radiation, further lead to increasing of H-plane cross-polarization level.

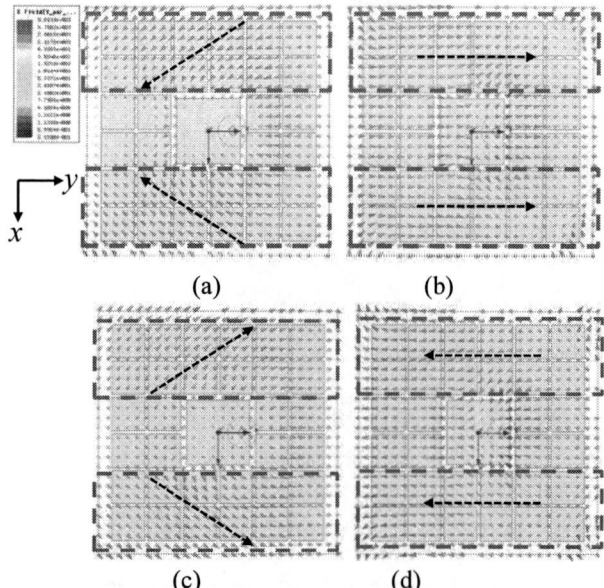

(a)   (b)

(c)   (d)

Figure 9. E-field vector distribution on the plane above the MSA based on symmetric AMC at t=0 (a), T/4 (b), T/2 (c) and T (d).

IV. MICROSTRIP ANTENNA BASED ON ASYMMETRIC AMC STRUCTURE

In order to reduce cross-polarization level, an asymmetric AMC structure is proposed. Figure 10's inset shows an asymmetric AMC cell, there is still 0.1 mm gap in the y direction between adjacent AMC patches but no gap in the x direction. When $AMC\ Width\_x$ = 1.2mm, $AMC\ Width\_y$ =1.1mm, as shown in Figure 10, the asymmetric AMC structure only reflects TM polarization plane wave (y-direction electric field) with 0° reflection phase but for TE polarization, the reflected phase of the plane wave (x-direction electric field) with 180° reflection phase. Correspondingly, as shown in Figure 11, in the MSA, the horizontal component $\vec{E}_{1x}$ in H plane (XOY) is reflected by the asymmetric AMC to be $\vec{E}_{2x}$ with 180° reflection phase, and $\vec{E}_{2x}$ and direct upward radiation $\vec{E}_{3x}$ cancels, which in turn reduces the cross-polarized electric field. At the same time, the asymmetric AMC reflects $\vec{E}_{1y}$ in phase, and finally, $\vec{E}_t = \vec{E}_{2y} + \vec{E}_{3y}$ has no electric field component in the x direction shown in Figure 12. According to the radiation field properties of the MSA, the asymmetric AMC can significantly reduce the H-plane cross-polarization level while ensuring large impedance bandwidth.

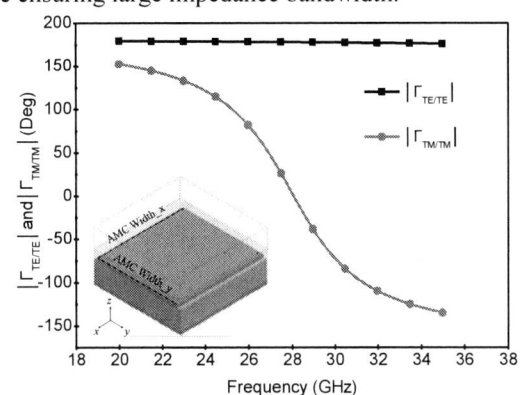

Figure 10. The reflection phase of asymmetric AMC cell for TM and TE polarized plane wave and inset is a AMC cell .

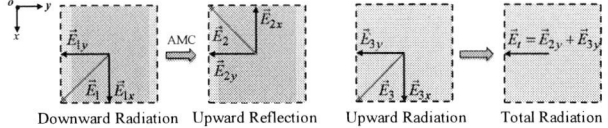

Figure 11. Schematic diagram of electric field distribution when radiation waves are radiated downward to symmetry AMC, reflected from symmetry AMC, and directly radiated upward.

Two schemes of the MSA based on asymmetric AMC are proposed shown in Figure 12. For design A, according to the above analysis, the H-plane cross polarization is mainly derived from radiation field of the red frame region, so only the AMC in this region is set to the asymmetric AMC. In the design B, a complete strip AMC is loaded around the MSA. For both schemes, $Sub.Width$=10mm,

$Sub.Length$=11mm, $AMC\ Gap$=0.1mm, $Patch\ Length$=2.1m, $AMC\ Width$=1.1mm. The other structural parameters of the two schemes are shown in the Table II.

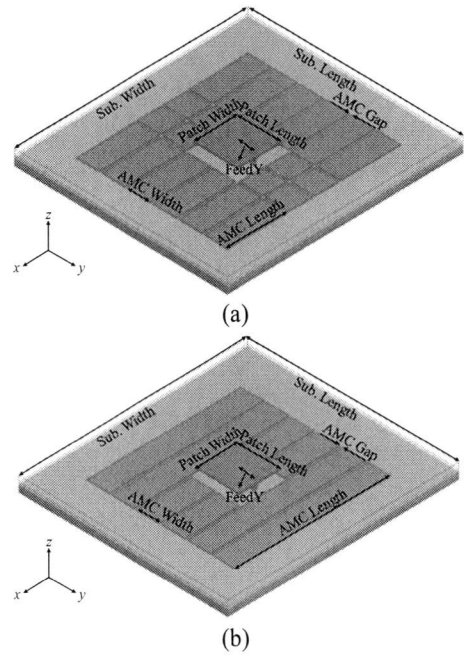

Figure 12. The structures of design A (a) and design B based on asymmetry AMC.

TABLE II. PARAMATERS OF TWO ANTENNAS IN FIGURE 12

Item	Patch Width	FeedY	AMC Length
Design A	2.1mm	0.75mm	2.8mm
Design B	2.0mm	0.8mm	7.4mm

Figure 13. Simulated |S11| (a) and peak gain (b) of design A and design B based on asymmetry AMC.

Figure 13 displays simulated |S11| and peak gain vs. frequency of design A and B. For design A, bandwidth of |S11| < -10dB is 23% at 28.05GHz (24.8GHz to 31.3GHz), while that of design B is 24.6% at 28.5GHz (25GHz to 32GHz). The peak gain of design A is greater than 6.98dBi in the band and the maximum peak gain is up to 10.1dBi at 31.2GHz. Although design B's peak gain from 25GHz to 30.7GHz is basically the same as that of design A, there is a significant dip around 31.5GHz, and the corresponding peak gain drops to 3dBi. Until now, there is no good explanation for this phenomenon.

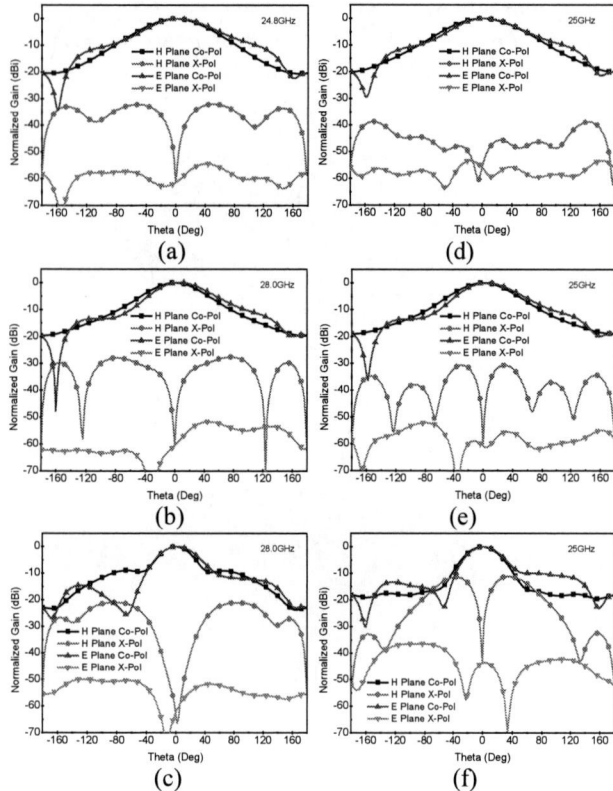

Figure 14. Simulated normalized pattern of design A (a-c) and design B (d-f) at three frequency points.

Figure 14 demonstrates the normalized pattern of both schemes at their upper and lower frequency of the |S11| < -10dB bandwidth and 28GHz. For design A, H-plane cross-polarization levels are -32.1dB, -27.8dB, and -20.9dB respectively at 24.8GHz, 28GHz and 31.3GHz. Design B has H-plane cross-polarization level of less than -30dB at 25GHz and 28GHz, but at 32GHz, it is exacerbated to -10dB, even at some angles greater than the co-polarization. Thus, the overall performance of design A is more stable within |S11| < -10dB band.

The radiation efficiency of design and the front-to-back ratio (FBR) versus frequency are also obtained. As shown in Figure 14, within the band, the radiation efficiency is greater than 76% and the maximum radiation efficiency reaches to 97.6% at 27GHz. The FBR is greater than 19.2dB in the band, and a large FBR is desirable in AiP because it will reduce electromagnetic interference between the antenna and the RF transceiver in a 3D packaging structure.

For design A, to realize an optimized antenna design, parametric analysis will be carried out next. Figure 16 shows quantitative effects of the *AMC Length*, *AMC Width*, *Patch Width*, and *Patch Length* on the impedance bandwidth. There are two deeper resonant points in the band. It can be seen that as the *AMC Length* increases from 2.3 mm to 3.3 mm, the lower frequency of the bandwidth is substantially unchanged, and the upper frequency is significantly reduced, namely, the *AMC Length* mainly affects the second deeper

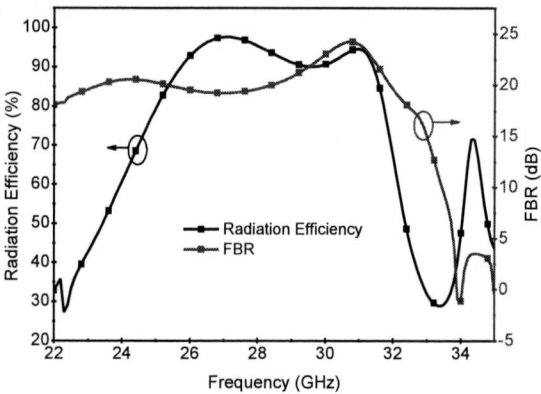

Figure 15. Simulated radiation efficiency and FBR.

resonance point. At *AMC Length* = 3.3mm, the bandwidth attains 28.6% at 28GHz (24GHz to 32GHz) while fluctuation occurs around 25GHz. As the *AMC Width* mainly affects the second deeper resonance point, when it changes from 1.05mm to 1.15mm, the lower frequency does not change, and the upper frequency gradually decreases. *Patch Length* mainly affects the first deep resonance point. When it changes from 2.0mm to 2.2mm, the lower frequency is gradually reduced and the reflection is improved. *Patch Width* has a slight effect on the lower frequency and reflection, which is similar to the effect of *Patch Width* on the reference antenna.

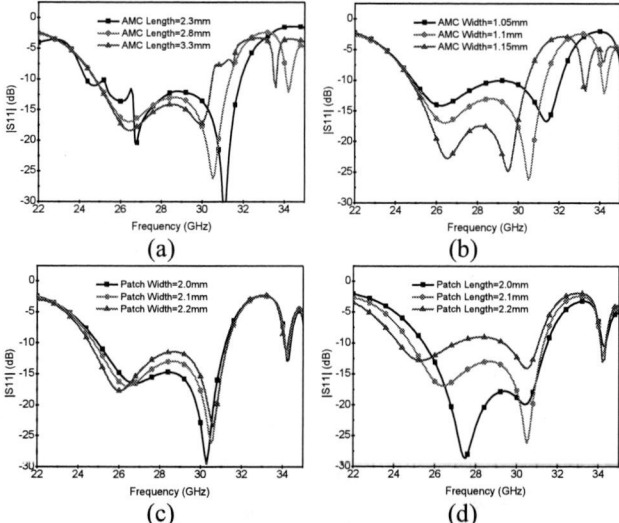

Figure 16. Parametric analysis of the AMC *Length*, AMC *Width*, *Patch Width*, and *Patch Length* for optimized impedance bandwidth.

V. CONCLUSION

In this paper, the mechanism of AMC to widen the bandwidth of MSA is theoretically analyzed from the view of field. Then, the principle of H-plane cross-polarization level deterioration of the symmetric AMC-based MSA is investigated. An asymmetric AMC structure is proposed for this problem. Finally, two schemes about the MSA based on

978-1-7281-1500-9/19 $31.00 © 2019 IEEE

asymmetric AMC structure are proposed and their performance is compared. The simulation results show that the MSA based on asymmetric AMC can not only achieve large bandwidth, high gain, high radiation efficiency and high FBR with low profile, but also significantly reduce the cross-polarization level, furthering enhancing the anti-noise ability of the AiP.

ACKNOWLEDGMENT

The authors acknowledge the support of the National Science and Technology Major Project (Project No. 2014ZX02501).

REFERENCES

[1] Jun Xu, Wei Hong, Zhi Hao Jiang, and Hui Zhang, "Wideband, Low-Profile Patch Array Antenna with Corporate Stacked Microstrip and Substrate Integrated Waveguide Feeding Structure," IEEE Transactions on Antennas and Propagation, vol. 67, Feb. 2019, pp. 1368 – 1373, doi: 10.1109/TAP.2018.2883561.

[2] Lei Wang, Yong-Xin Guo, and Wei-Xing Sheng, "Wideband High-Gain 60-GHz LTCC L-Probe Patch Antenna Array With a Soft Surface," IEEE Transactions on Antennas and Propagation, vol. 61, April 2013, pp. 1802 - 1809, doi: 10.1109/TAP.2012.2220331.

[3] L Huayan Jin, Wenquan Che, Kuo-Sheng Chin, Guangxu Shen, Wanchen Yang, and Quan Xue, "60-GHz LTCC Differential-Fed Patch Antenna Array With High Gain by Using Soft-Surface Structures," IEEE Transactions on Antennas and Propagation, vol. 65, Jan. 2017, pp. 206 - 216, doi: 10.1109/TAP.2016.2631078.

[4] Huayan Jin, Wenquan Che, Kuo-Sheng Chin, Wanchen Yang, and Quan Xue, "Millimeter-Wave TE20-Mode SIW Dual-Slot-Fed Patch Antenna Array With a Compact Differential Feeding Network," IEEE

Transactions on Antennas and Propagation, vol. 66, Jan. 2018, pp. 456 - 461, doi: 10.1109/TAP.2017.2767644.

[5] Kuo-Sheng Chin, Wen Jiang, Wenquan Che, Chih-Chun Chang, and Huayan Jin, "Wideband LTCC 60-GHz Antenna Array With a Dual-Resonant Slot and Patch Structure," IEEE Transactions on Antennas and Propagation, vol. 62, Jan. 2014, pp. 174 - 182, doi: 10.1109/TAP.2013.2287294.

[6] Antti E. I. Lamminen, Jussi Säily, and Antti R. Vimpari, "60-GHz Patch Antennas and Arrays on LTCC With Embedded-Cavity Substrates," IEEE Transactions on Antennas and Propagation, vol. 56, Sept. 2008, pp. 2865 - 2874, doi: 10.1109/TAP.2008.927560.

[7] Wei Liu, Zhi Ning Chen, and Xianming Qing, "Metamaterial-Based Low-Profile Broadband Mushroom Antenna," IEEE Transactions on Antennas and Propagation, vol. 62, March 2014, pp. 1165 - 1172, doi: 10.1109/TAP.2013.2293788.

[8] Wanchen Yang, Si Chen, Wenquan Che, Quan Xue and Qian Meng, "Metamaterial-Based Low-Profile Broadband Mushroom Antenna," IEEE Transactions on Antennas and Propagation, vol. 66, Sept. 2018, pp. 4918 - 4923, doi: 10.1109/TAP.2018.2851659.

[9] W. C. Yang, H. Wang, W. Q. Che, Y. Huang, and J. Wang, "High-Gain and Low-Loss Millimeter-Wave LTCC Antenna Array Using Artificial Magnetic Conductor Structure," IEEE Transactions on Antennas and Propagation, vol. 63, Jan. 2015, pp. 390 - 395, doi: 10.1109/TAP.2014.2364591.

[10] Wanchen Yang, Wenquan Che, and Hao Wang, "High-Gain Design of a Patch Antenna Using Stub-Loaded Artificial Magnetic Conductor," IEEE Antennas and Wireless Propagation Letters, vol. 12, 2013, pp. 1172 - 1175, doi: 10.1109/LAWP.2013.2280576.

[11] Wanchen Yang, Dongxu Chen and Wenquan Che, "High-Efficiency High-Isolation Dual-Orthogonally Polarized Patch Antennas Using Nonperiodic RAMC Structure," IEEE Antennas and Wireless Propagation Letters, vol. 65, Feb. 2017, pp. 887 - 892, doi: 10.1109/TAP.2016.2632700.

Automatic transient thermal impedance tester for quality inspection of soldered and sintered power electronic devices on panel and tile level

Schmid Maximilian, Bhogaraju Sri Krishna, Gordon Elger
Institute of Innovative Mobility (IIMo)
Technische Hochschule Ingolstadt
Ingolstadt, Germany
maximilian.schmid@thi.de

Abstract—Transient thermal analysis (TTA) is an established method to evaluate thermal integrity of interconnects in power semiconductor modules. In this paper the suitability to detect bad interconnects is investigated. In especially for sintered interconnects TTA has potential advantages compared to standard methods like X-Ray where voids but no bad wetted contact can be observed or SAM (scanning acoustic microscopy) where the device has to be immerged in water. An automated TTA equipment prototype was developed to enable fast automatic mass testing for soldered and sintered MOSFETs. The automatic TTA tester was benchmarked to X-Ray, SAM and shear strength tests on sintered MOSFETs. Thereby different packages, surface metallization and sinter profiles were tested. The measurements showed that the evaluation of X-Ray and SAM data is strongly limited for sintered interconnects. With TTA and the destructive shear strength test, small variation in quality were detectable.

Keywords-component; Reliability, Thermal, Silver Sintering, MOSFET, Measurment Equipment

I. INTRODUCTION

Thermal performance is one of the key factors for power electronics of the future. Higher power densities of WBG semiconductors, miniaturization and increased lifetimes raise the requirements. One critical part for the thermal integrity is the quality of the interconnect (soldered or sintered). High thermal stability, reliability and low thermal resistance are necessary to keep the temperatures in the active regions of the semiconductors low.

The most common way to for in-line testing of a solder interconnection of high power devices is X-Ray. Voids and unsoldered areas of solder joints can be detected. For sintered interfaces, X-Ray can't deliver adequate results because the porosity is too small to be resolved and bad contacts due to pad contaminations can't be detected. In especially for sintering under pressure X-Ray is not performed because no useful information is obtained. For pressureless sintering void formation due to binder outgassing can be observed by X-Ray but bad interconnection due to contamination of the pads are not detectable. SAM is a suitable alternative for sintered interconnects. However, the manufacturers want to avoid a contact of their products with water due to the complex drying process. An effective and non-destructive option is the TTA to measure the thermal impedance (Z_{th}) of a semiconductor. By TTA, the complete thermal path from junction of a semiconductor to the heatsink can be observed very detailed. The main drawback of TTA in production is the lack of an automated solution. Measurements are very time consuming and prone to failures. For this reason, a new equipment for automatic Z_{th} measurement was developed and is described in the paper. A current source and electronic module was developed which allows to reduce the required heating current for MOSFETS to 1A by using the gate source voltage (V_{GS}) as temperature sensitive parameter. Lower currents allow higher switching speed and less EMC contamination in the production. The new measurement method was benchmarked to existing equipment where the body-diode is used for heating and as temperature sensitive parameter. To proof feasibility and accuracy, MOSFETS were soldered and sintered to substrates. The automatic TTA measurements were compared to X-Ray, SAM and destructive shear strength tests.

II. TRANSIENT THERMAL ANALYSE ON MOSFETs

The investigated parameter for TTA is Z_{th}. It describes the thermal response of a device according to thermal power step P_{th}. Following equation reveals this relation:

$$\Delta T(t) = P_{th} * Z_{th}(t). \qquad (1)$$

Be analyzing Z_{th}, the properties of the single layers in the setup of a device under test (DUT) can be observed and eventual failures are detected. This includes in a classical semiconductor setup chip, die attach, substrate and connection to the heatsink. The interpretation of the data can be done directly with Z_{th}, its normalized derivation [1] or by a conversion to the structure function [2].

The measurement procedure for semiconductors is defined in [3]. The DUT is fixed to a plate with a regulated constant temperature. Figure 2 shows the principle sequence. First, the semiconductor is heated up for the time t_{Heat} with a heat power of P_{Heat}. The heat is generated by operating the semiconductor with high load currents. In this time, the thermal equilibrium should be reach. Depending on the thermal stack, this can last between a few 100ms and several minutes. Afterwards the semiconductor is switched to a much smaller power loss P_{Sense} and starts to cool down.

978-1-7281-1500-9/19 $31.00 © 2019 IEEE

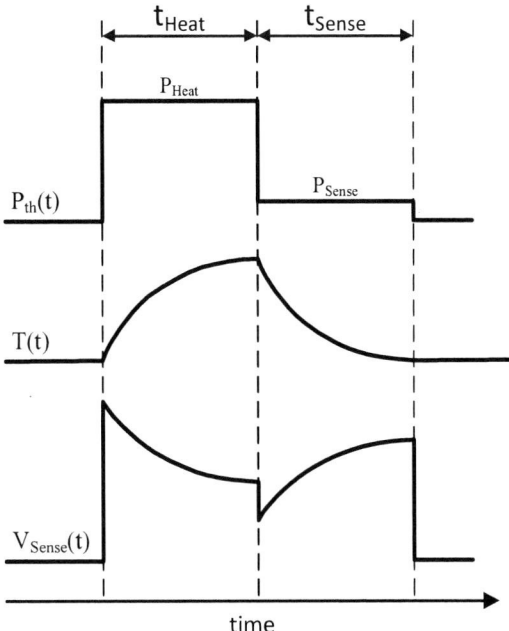

Figure 2: Sequence for TTA on semiconductors

While this time, the temperature of the semiconductor is tract via a voltage measurement of $V_{Sense}(t)$ on a temperature sensitive electrical parameter (TSP). A complete shutdown is not allowed, because the TSP would not be present. Z_{th} is calculated afterwards with $\Delta T(t)$ and the power step between P_{Heat} and P_{Sense} with (1).

This measurement procedure is suited for all kinds of semiconductors. Only the used TSP and the wiring for heating and sensing differs. Independent of chosen TSP the temperature can be calculated with (2).

$$\Delta T(t) = TSP * \Delta V_{Sense}(t) \qquad (2)$$

For MOSFETs, possible TSPs are electrical resistance of the channel R_{DS-on}, the body diode forward voltage V_{f-Body} and the threshold voltage V_{GS-th} [4]. The heating of the MOSFET can be realized over R_{DS-on}, V_{f-Body} and the saturation region (SAT). The principles behind V_{GS-th} and SAT are identical, but a different name is chosen due to definition of V_{GS-th} to low currents. So all TSPs can be used to heat up the MOSFET. Only higher currents must be applied.

Altogether, there are nine possible combinations for TTA on MOSFETs. The recommended setup form the standard is V_{f-Body} as TSP and heating in SAT [5]. More common in commercial equipment is using V_{f-Body} as TSP and for heating [6, 7]. However, the chosen combination for the automated tester is V_{GS-th} as TSP and SAT for heating (V_{GS} method in the following). A schematic electric circuit diagram is shown in Figure 1.

The main reason for this selection is the reduction of the current for heating. SAT heating allows the adjustment of P_{Heat} over I_{DS} and V_{DS}. In contrast to R_{DS-on} and V_{f-Body}, it is

possible to increase the voltage drop over the MOSFET. So smaller currents are sufficient to reach the same P_{Heat}. The current reduction brings three advantages. Through the smaller current step from heating to sensing, the switching time is shortened and electromagnetic interferences (EMI) are reduced. In addition, the use of spring probes for variable automated contacting of the MOSFETs is only sensible with heating currents smaller than 5A. Contact resistances and the size of the spring probe are here the limiting factors. Besides this, the selected combination has additional benefits. Compared to the setup from the standard, only one circuit is necessary for heating and sensing. Further, the current is not inversed between heating and sensing, which is negative for the switching time. In addition, the sensitivity of V_{GS-th} is with $\approx$ -3.5 mV/K higher than the $\approx$ -2.0mV/K of V_{f-Body} [8]. This results a higher measurement signal and therefore a better signal to noise ratio.

III. AUTOMATED TEST EQUIPMENT

All required subsystem for automated TTA were arrange in a setup called automated thermal impedance measurement system (ATIM). All subsystems controlled over a computer. A picture with a list of single parts of the ATIM is shown in Figure 3. For automated testing the DUTs are arranged on panels like in Figure 4. The shown panel consist of three times three separable PCBs with the soldered DUTs. On each PCB are several probing pads to connect the DUT to the measurement equipment via spring probes. The displayed panel is a simple test design for TTA on interconnects without any other functions. Instead, also complete products are testable, if the required test pads are present and accessible. The tester can also be used testing on ceramic tile level before or after wire bonding. Due to the low heating current probing on gate and source is possible. The single subsystems are explained now more detailed:

A. Current source and DAQ unit

The concept of the current source was already explained in II. P_{Heat} can be adjusted up to 50W and switching times less than 20µs from heating to sensing were reached with the prototype. Further advancements shall improve these properties.

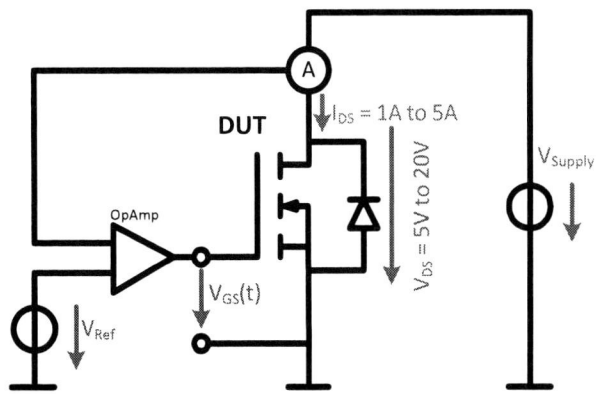

Figure 1: Schematic circuit of measurment principle of prototype

Besides the currents source an analog digital converter (DAC) is needed to measure V_{GS} while the sensing time. A commercial data acquisition unit (DAQ) with 10MHz sampling rate and 16-bit resolution was used. Additional a variable pre-amplifier was connected in front of the input of the DAQ unit. Current source, pre-amplifier and DAQ unit are mounted on top of the system.

B. Temperature stable base plate

For TTA a fixed temperature potential on the bottom of the DUT is required. Four individual regulated peltier elements placed under a 16cm x 16cm copper plate are used to adjust the temperature. The DUT panel is placed on top of the temperature stable base plate with a TIM between. The temperature can be adjusted between 10°C and 85°C.

C. Probing adapter and XYZ table

Key function of the ATIM is the automatic electrical contacting of the DUTs to current source and DAQ unit. The connection is done with a probing adapter. The probing adapter consists of spring probes, which are pushed to the probing pads on the DUT panel and SMA connectors for the connection to current source and the DAQ unit. Spring probes and connectors are soldered to a PCB sandwich for mechanical stability and electrical routing. Depending on the DUT panel design, the layout of the probing adapter has to be adjusted.

To drive the probing adapter to the correct positions a XYZ with a 360° rotation axis is used. The probing adapter is fixed below the rotation and the Z-axis. XY-Axis move the temperature stable base plate with the DUTs on top. This setup was selected to reduce the cable length between current source and probing adapter to a minimum. Inside of the rotation axis, a force sensor measures the pressure on DUT panel and TIM. This serves as a safety feature to avoid failures based on TIM conductivity variations. The force sensor has a measurement range from 0N to 100N.

The driving paths are calculated out of an imported Gerber file. A camera is used for exact position recognition of the DUTs.

D. Automated analyze

The measured V_{GS} signals are saved automatic to a database. A develop automatic analyzing software can afterward evaluate the data. Z_{th}, its deviation and structure function are calculated and the results compared to other components or over aging cycles. Changes and errors in the thermal path of the DUT are tract automatically.

IV. BENCHMARK OF CURRENT SOURCE

To proof feasibility and accuracy of the developed current source using the V_{GS} method, the prototype was benchmarked against two commercial equipment (C-Eq.1 and C-Eq.2), which both use $V_{f\text{-}Body}$ for heating and TSP. Both were operated at their upper current limit of 2A resp. 8A. The prototype was operated with V_{DS} = 10V and I_{DS} = 1A. As reference probe, a MOSFET in TO252 package and a rating of 60V/50A was used. The MOSFET was

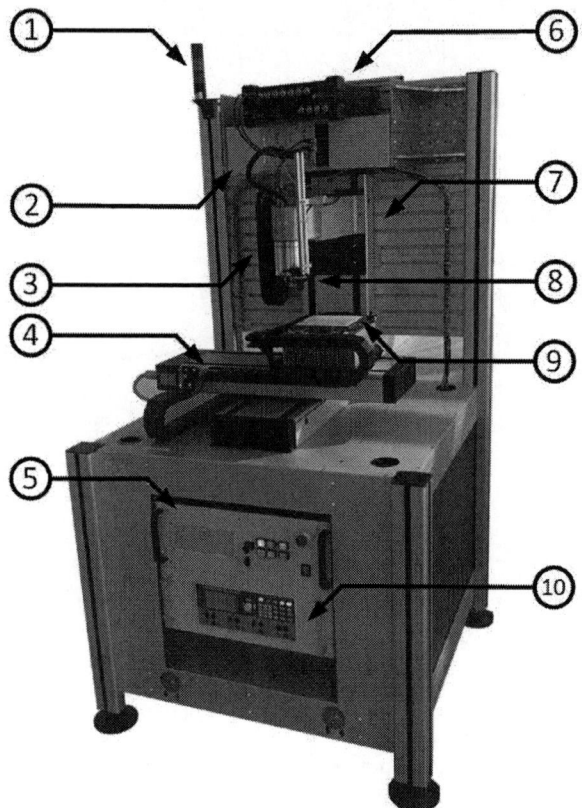

1: Signal column
2: Current source
3: Rotation axis with integrated force sensor and probing adapter
4: XY axis
5: XYZA controller
6: DAQ unit
7: Z axis
8: Camera
9: Temperature stable base plate
10: Power supply

Figure 3: Automated measurement equipment (ATIM)

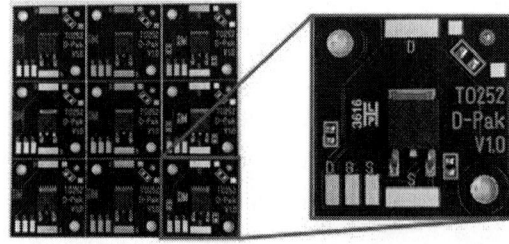

Figure 4: DUTs on panel (left) and focus on single DUT-PCB (right)

soldered with SAC305 on an Al-IMS substrate. The thermal resistant of the TO252 package was specified with 3.0K/W in the datasheet. For the measurement always the same heatsink and the same thermal interface material (TIM) between Al-IMS and heat sink was used.

The results are depicted in Figure 5. All measured Z_{th} (solid lines) were almost identical until the point of separation at approximately 600ms and 4.6K/W. Afterwards, they show a different trend and finally end in different

978-1-7281-1500-9/19 $31.00 © 2019 IEEE

terminal values, representing R_{th} (see legend Figure 5). The reason for the separation is only the TIM and not measurement method or equipment. While the measurements on the three equipment, the pressure to fix the reference probe to the heat sink was not controlled. To low pressure on the TIM downgrades its thermal properties and therefore rises R_{th} and Z_{th} of the measurement. The effect is identical to the Dual-Interface method [9].

To proof that the separation is only influenced by the TIM, transient thermal finite element simulation with a 3D model of the reference probe were run in SolidWorks Flow. The dimensions and thermal parameters of the model are adjusted to the real reference probe. To fit the simulations to the measurements, a variable thermal contact resistant was added between Al-IMS and TIM. All other parameters were kept constant. The simulation results are added to Figure 5 (dotted lines). With a thermal contact resistant of 6.06 $K{*}cm^2/W$ (for C-Eq.1), 1.80 $K{*}cm^2/W$ (for C-Eq.2) and without for Prototype the best matches were reached. This proves that only the TIM causes the separation. The three equipment itself deliver the same Z_{th} results. The feasibility of the prototype is therefore verified.

It should be mentioned, that the behavior of the model differs from the measurements before the point of separation. A better calibration of the Si-die and package parameters will solve this problem. This effect was not in focus for these simulations.

V. ANALYSE OF SILVER SINTERED INTERCONNECTS

The ATIM was used to analyze the thermal integrity of a nanoparticle silver sinter paste. Overall 56 MOSFETs in six categories each nine MOSFETs were analyzed in TTA, X-Ray, SAM and shear strength test. Table I shows the overview of the selected pastes, profiles and surface finishes for all categories. Two different packages were used for the experiments. A standard TO252 package und a bare die MOSFET with dimension of 2.3 x 2.3 mm². This analyze section is spitted in two parts according to the packages.

A. Standard package TO252

The selected MOSFETs was classified for 50A maximal current, 60V breakdown voltage and an R_{th} of 3.0 K/W for the package. The standard surface finish on its Cu-leadframe is a Sn metallization. This finish is not suitable for silver sintering. An alternative surface finish was applied. At first, Sn was mechanical removed and the copper of the leadframe mechanical polished. Nine MOSFET with this Cu-surface were used for panel 1 (TO252_Ag_mech-Cu). For panel 2 (TO252_Ag_ENIG), an electroless nickel immersion gold layer (ENIG) was platted on the Cu-surface. The ENIG finish was applied with four galvanic baths. A galvanic activator (1min), a Pd bath (10min), a Ni bath (60min, 90°C) and an Au bath (5min, 90°C). The thin Pd layer is necessary for the bath chemistry used. An X-ray fluorescence (XRF) analyze on the layer thicknesses showed an average of 10 nm for Pd, 3.68 µm for Ni and 53 nm for Au. The resulting thicknesses are within the specification for ENIG as PCB surface finish [10]. Besides the two sintered panels, a third panel was soldered with a SAC305 paste as reference (TO252_SAC_Ref).

As substrate for all three panels, an Al-IMS-PCB with 1mm aluminum, 100µm dielectric, 70µm copper and ENIG finish was used. Each panel consists of nine MOSFETs. The substrate with MOSFETs was already depicted in Figure 4. The solder/sinter paste was applied over a printing process with a 75µm stencil. For soldering a standard profile (IPC/JEDEC J-STD-020E) was used. The two panels with silver paste were sintered for 60 min at 250°C under N_2 atmosphere (recommended in the datasheet of the paste for pressureless sintering).

The three panels were evaluated with TTA and shear strength test. The results are listed in Table II. Additional X-Ray and SAM analyses were performed. It was observed that the manufactured ENIG finish was not suitable for the Ag paste. Five MOSFETs fell off directly after sintering. One was afterwards destroyed by thermal overload during TTA. The three remaining showed high Z_{th}. While the preparation for shear strength test, they also fell off. The

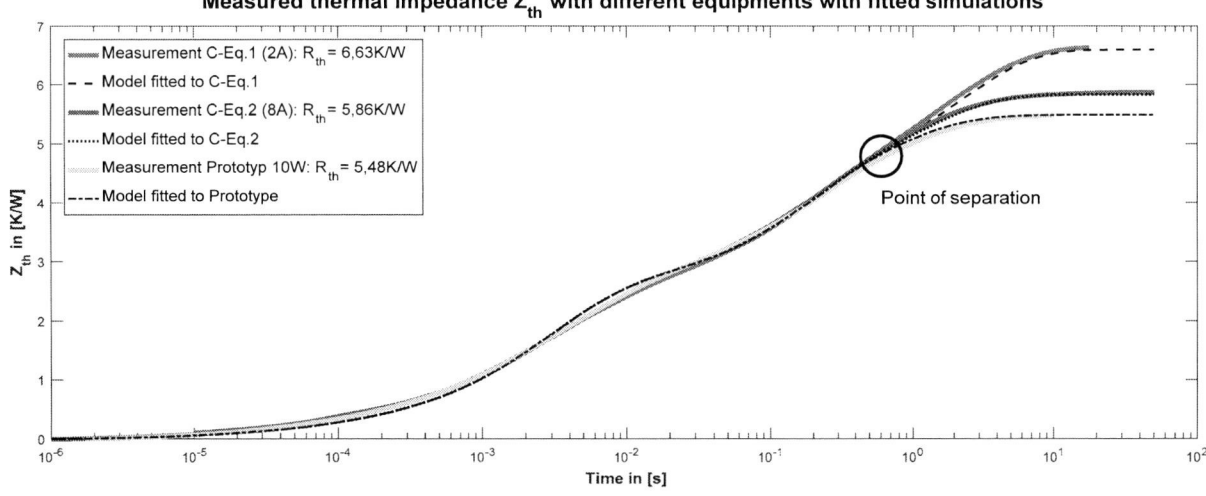

Figure 5: Measured Zth form two commercial equipment and the prototpye with fitted simulations

TABLE I. OVERVIEW MEASURED COMPONENTS

Package	Panel -Nr.	Name	Paste	Surface finish	Profile (Sinter/Solder)
TO252	1	TO252_Ag_mech-Cu	Nanoparticle Ag paste	Mech. polished Cu-leadframe	60 min @ 250°C, N_2
	2	TO252_Ag_ENIG	Nanoparticle Ag paste	ENIG (laboratory process)	60 min @ 250°C, N_2
	3	TO252_SAC_Ref	SAC305	Sn (from manufacturer)	IPC/JEDEC J-STD-020E
Bare Die	4	Die_Ag_pressureless	Nanoparticle Ag paste	Ti/Ni/Ag $(0,1\mu m + 0,2\mu m + 1,0\mu m)$	60 min @ 250°C, N_2
	5	Die_Ag_pressure	Nanoparticle Ag paste	Ti/Ni/Ag $(0,1\mu m + 0,2\mu m + 1,0\mu m)$	Preheat: 15min @ 100°C, air Sinter: 5 min @ 250°C, 10MPa, N_2
	6	Die_SAC_Ref	SAC305	Ti/Ni/Ag $(0,1\mu m + 0,2\mu m + 1,0\mu m)$	IPC/JEDEC J-STD-020E; Vacuum

failure mode was identical for all. They broke between ENIG and silver interconnect. Most probable reason is a lack in quality of the manufactured ENIG despite the right thicknesses. The connection to the ENIG of the substrate was much stronger. Panel 2 was excluded from further investigated.

The sintered MOSFETs with mechanical polished copper of panel 1 reached in average Z_{th} minimal better compared to the SAC references. Only a small improvement can be obtained with Ag sintering for TO252 package, hence the area below the leadframe is huge. Assuming a thickness of 50µm and an area of 25 mm^2 the calculated thermal resistance would be 0.036 K/W for the SAC interconnect (thermal conductivity: 56 W/mK) and 0.005 K/W for the Ag interconnect (thermal conductivity bulk Ag: 400 W/mK). The results are in this range.

Indeed, Z_{th} is varying inside panel 1, which means the quality of the interconnects differs. This variation should also be proved by X-Ray, SAM and shear strength tests. Exemplary the results of the sintered MOSFETs with the lowest (MOSFET 7) and the highest Z_{th} (MOSFET 1) from panel 1 are compared to a SAC reference on the left side of Figure 6. Z_{th} of the SAC reference and the best sintered MOSFET 7 are nearly identical. The separation of sintered MOSFET 1 occurs approximately at 2.5 K/W and 6ms. Time and value suggest to a different quality of interconnect. The X-Ray pictures are the most difficult to evaluate. Even for the SAC sample it is hard to detect the voids, hence the leadframe and the interconnect inside the package overlap the solder joint. For the sintered MOSFETs no difference is visible. The pictures show in fact small formation of gaps due to binder outgassing. This phenomena is known for pressureless sintering. Separation of particles due to binder dynamics creates gaps in vertical direction because of material transport and shrinkage during sintering. The gaps hardly influence initial thermal performance and shear strength but are a concern for long time reliability, e.g. during temperature cycling crack initiation may occur at the gaps. However, quantity and size are hardly estimable. The evaluation of the SAM pictures is easier. A sharp contrast is achieved for the voids in the SAC reference. Also a huge difference is visible between the two sintered MOSFETs. The darker areas represent a better connection. For the MOSFET with the lower Z_{th}, the darker area is lager, but also no entire over the complete lead frame. The evaluation with

the shear strength results was difficult for the tested interconnects. Overall, the values are smaller than 10MPa. The estimated value would be around 40MPa. The most abundant failure mode was a break inside of the sinter interconnect. The Cu surface is therefore suitable for the Ag paste. A possible reason for the low shear strength is the huge surface of the package. However, a correlation with the TTA is detectable (see Table II).

B. Bare die MOSFET

The bare die MOSFETs were specified for 9A maximal current and 200V breakdown voltage. The dimensions are 2.3 * 2.3 mm^2 *300 µm. They come with a 0.1µm Ti, 0.2µm Ni and 1.0µm Ag surface finish suitable for silver sintering. As substrate, the IMS-PCB form the TO252 package and the same stencil for paste printing was used. This design was not optimal for the bare die MOSFET, but should be acceptable for the test. The top side contacts were connected to the substrate over 33µm Al bond wires.

Three panels were build up. Panel 4 for pressureless sintering (Die_Ag_pressureless) with the identical parameters from panel 1 and 2. Panel 5 (Die_Ag_pressure) was sintered under pressure. The paste was first dried for 15 min at 100°C in air. Then the MOSFETs were placed and sintered for 5min at 250°C under 10MPa pressure in N_2 atmosphere. Panel 6 (Die_SAC_Ref) is used as reference and was soldered identical to panel 3, but with high vacuum.

The results for TTA and shear strength test are listed in Table II. An exemplary evaluation of selected MOSFETs is done on the right side of Figure 6. It is visible that for the SAC solder, the voids in the solder are detected by X-Ray and SAM. However, for pressureless sintering X-Ray and SAM deliver hardly any information. Small voids due to binder outgassing are observed for X-Ray. For SAM, using the same settings like for the SAC samples, no significant contrast was reached. For sintering under pressure, X-Ray delivers no information and SAM reveals features, which can be interpreted as non-uniform interconnections (white area bad interconnection).

The TTA measurements reveal that the thermal contact of the sintered devices is better than for the soldered interconnection. In average an improvement 0.37 K/W for the pressureless sintered and 0.49 K/W for the pressure sintered MOSFETs was observed. However, peculiar observation are made. The Z_{th} curve of the soldered device is in the early time domain below the sintered devices (see

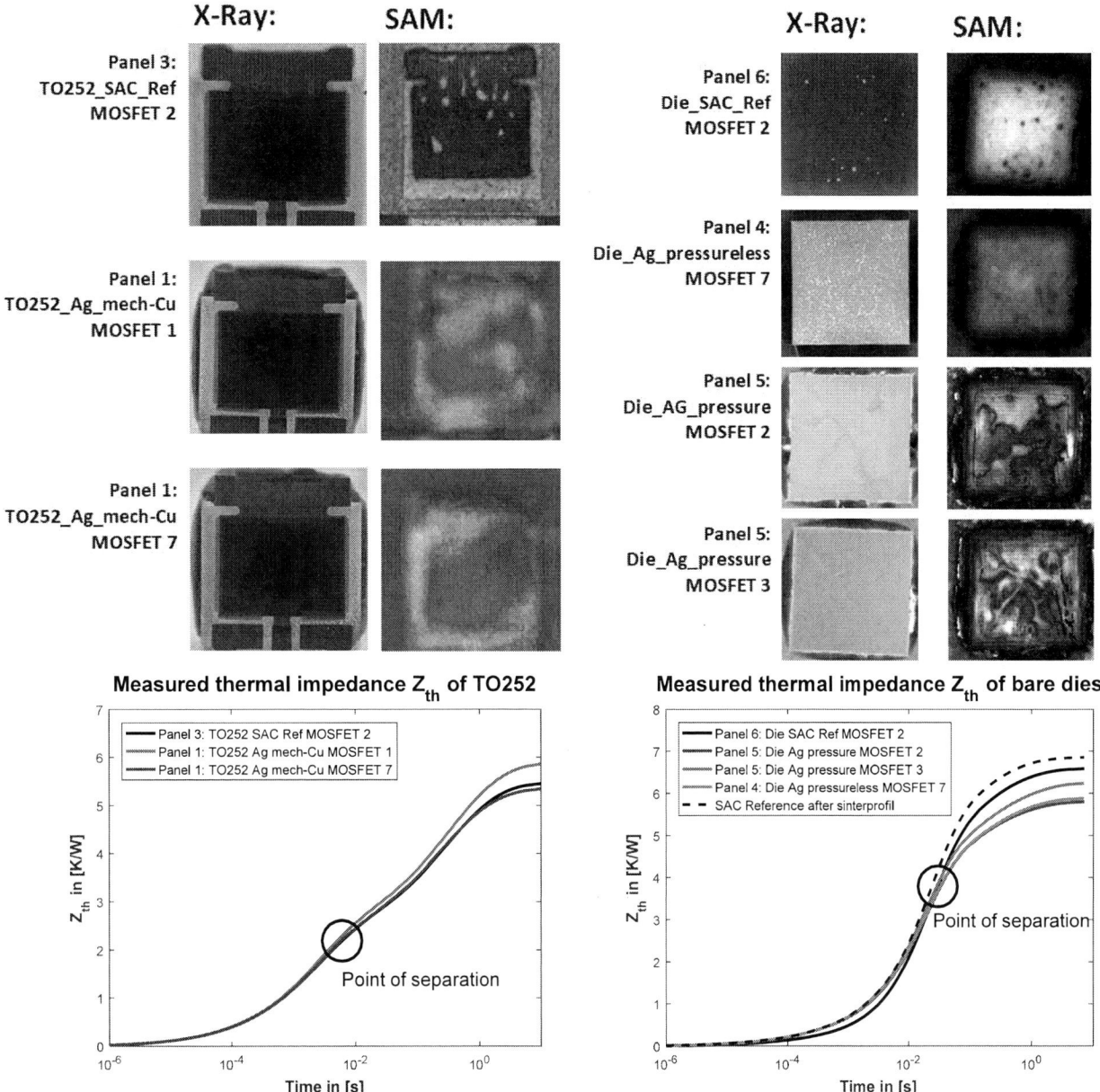

Figure 6: TTA, X-Ray and SAM results for TO252 package (left) and bare die MOSFET (right)

black line Figure 6). This would indicate a worse performance of the sintered interface. However, when applying a sinter temperature profile on the MOSFETS and afterwards soldering the devices the same curve is observed in the early time domain as for the sintered devices (see dashed black line Figure 6). Therefore, the lower thermal performance in the early time domain of the sintered devices is due to temperature induced degradation in the device itself. The Z_{th} curve after longer times of the sintered devices is below the soldered ones. The effect is larger than only obtained by the better interconnection. The large improvement of thermal performance can be caused by better heat spreading due to the additional heat spreading of

the silver layer. The silver paste was printed on the oversized pad, i.e. same stencil as for the packaged MOSFETS.

The shear strength results are much better compared to TO252 package. All pressureless sintered MOSFETs reached values between 10 MPa and 20 MPa. Weakest point was always the interconnect. The MOSFETs sintered under pressure reached 40 MPa in average. This represents the target for silver sintering.

TABLE II. OVERVIEW MEASURED COMPONENTS (DETEILED ANALYSED MOSFETs MARKED GRAY)

		TO252						Bare Die					
		Panel 1		Panel 2		Panel 3		Panel 4		Panel 5		Panel 6	
		Ag_mech-Cu		Ag_ENIG		SAC_Ref		Die_Ag_pressureless		Die_Ag_pressure		Die_SAC_Ref	
		Rth	Shear S.	Rth	Shear S.	Rth	Shear S.	Rth	Shear S.	Rth	Shear S.	Rth	Shear S.
MOSFET	1	5,85 K/W	prep. Err.[a]	9,86 K/W	prep. Err.[b]	5,71 K/W	35,1 MPa[a]	elec. def.	18,1 MPa[a]	5,98 K/W	37,5 MPa[a]	elec. def.	39,0 MPa[a]
	2	5,61 K/W	prep. Err.[a]	fall off	0,0 MPa[b]	5,45 K/W	40,1 MPa[d]	6,11 K/W	17,5 MPa[a]	5,81 K/W	30,4 MPa[c]	6,60 K/W	57,2 MPa[a]
	3	5,58 K/W	prep. Err.[a]	fall off	0,0 MPa[b]	5,66 K/W	42,6 MPa[d]	6,60 K/W	11,3 MPa[a]	6,26 K/W	29,3 MPa[c]	6,87 K/W	63,5 MPa[a]
	4	5,54 K/W	5,0 MPa[a]	elec. def.	prep. Err.[b]	5,69 K/W	44,3 MPa[d]	6,17 K/W	16,2 MPa[a]	6,18 K/W	51,9 MPa[a]	6,72 K/W	28,7 MPa[c]
	5	5,34 K/W	6,0 MPa[a]	fall off	0,0 MPa[b]	5,51 K/W	41,3 MPa[d]	6,36 K/W	15,0 MPa[a]	6,00 K/W	44,9 MPa[a]	elec. def.	56,9 MPa[a]
	6	5,48 K/W	6,5 MPa[a]	7,61 K/W	prep. Err.[b]	5,65 K/W	43,8 MPa[b]	6,17 K/W	10,8 MPa[a]	6,04 K/W	45,7 MPa[a]	6,62 K/W	61,3 MPa[a]
	7	5,34 K/W	7,1 MPa[a]	fall off	0,0 MPa[b]	5,36 K/W	43,7 MPa[d]	5,90 K/W	17,1 MPa[a]	elec. def.	39,6 MPa[a]	elec. def.	61,9 MPa[a]
	8	5,39 K/W	3,5 MPa[a]	7,28 K/W	prep. Err.[b]	5,42 K/W	37,1 MPa[c]	5,96 K/W	16,4 MPa[a]	elec. def.	38,9 MPa[a]	6,19 K/W	65,0 MPa[a]
	9	5,43 K/W	7,2 MPa[a]	fall off	0,0 MPa[b]	5,48 K/W	45,0 MPa[d]	6,03 K/W	Cross Sec.	elec. def.	Cross Sec.	6,24 K/W	63,3 MPa[a]
Aver		5,51 K/W	5,9 MPa	8,25 K/W	0,0 MPa	5,55 K/W	41,4 MPa	6,16 K/W	15,3 MPa	6,04 K/W	39,8 MPa	6,54 K/W	55,2 MPa

a: Break inside the sinter/solder joint
b: Break at interface sinter/solder joint and DUT
c: Break in Si Chip / Package
d: Break of dielectric of Al-IMS

VI. CONCLUSION

An automatic solution (ATIM) for TTA on MOSFETs was introduced. It includes a current source with a new concept for heating and sensing the DUTs over the threshold voltage. Main advantage is the reduction of the heating current. This is necessary to enable automatic probing over spring probes. The current source was benchmarked with commercial equipment. The measured Z_{th} for soldered samples are identical before the separation due to different TIM properties. Reason for the separation was the not controlled pressure on the TIM. Transient thermal finite element simulation proved this influence. The ATIM was afterwards used for evaluation of silver sinter interconnects. Altogether 56 MOSFETs were analyzed by TTA, X-Ray, SAM and shear strength test. The experiments showed that X-Ray is not an adequate inspection for silver sintering. For sintering without pressure only formations of gaps due to outgassing of the paste binder are visible. For sintering under pressure, no information was obtained. With SAM the inspection of sintered interconnects is in general feasible. However, for pressureless sintering of bare die MOSFETs the analyze by SAM was not sensitive. Best non-destructive method to evaluate silver sinter interconnects was the TTA. Final shear strength testing support the TTA results.

ACKNOWLEGEMENT

The authors acknowledge the financial support by the Bavarian Ministry of Science and Art in the project PTTA (H.2-F1116.IN/25/3).

REFERENCES

[1] A. Hanß, M. Schmid, E. Liu and G. Elger, „Transient thermal analysis as measurement method for IC package structural integrity," Chinese Physics B, 2015.J. Clerk Maxwell, A Treatise on Electricity and Magnetism, 3rd ed., vol. 2. Oxford: Clarendon, 1892, pp.68–73.

[2] M. Rencz, E. Kollar, A. Poppe and S. Ress, „EVALUATION ISSUES OF THERMAL MEASUREMENTS BASED ON THE STRUCTURE FUNCTIONS," in Therminic 2013.

[3] JESD51-1, „Integrated Circuits Thermal Measurement Method - Electrical Test Method (Single Semiconductor Device)," DECEMBER 1995.

[4] D. Blackburn and D. Berning, „Power MOSFET temperature measurements," in IEEE Power Electronics Specialists conference, Cambridge, MA, USA, 1982.

[5] JESD24-3, Thermal Impedance Measurements for Vertical Power MOSFETs (Delta Source-Drain Voltage Method), NOVEMBER 1990 (Reaffirmed: OCTOBER 2002).

[6] A. Bhalla, „Thermal resistance characterization of Power MOSFETs," 2003.

[7] P. Panchal, T. v. Essen, M. A. Ras, C. Grosse, B. Wunderle and D. May, „Accurate, versatile and compact transient measurement system for fast thermal package characterization and health monitoring," in 7th Electronic System-Integration Technology Conference (ESTC), 2018.

[8] M. Schmid and G. Elger, „Measurement of the transient thermal impedance of MOSFETs over the sensitivity of the threshold voltage," EPE 2018, accepted, 2018.

[9] JESD51-14, „Transient Dual Interface Test Method for the Measurement of the Thermal Resistance Junction to Case of Semiconductor Devices with Heat Flow Trough a Single Path," JEDEC, 2010.

[10] IPC-4552, Specification for Electroless Nickel/Immersion Gold (ENIG) Plating for Printed Circuit Boards, October 2002.

Time 0 Void Evolution And Effect On Electromigration

Jiefeng Xu[1], Scott McCann[2], Huayan Wang[1], VanLai Pham[1], Stephen R. Cain[1], Gamal Refai-Ahmed[2], S.B. Park[1]

[1] Department of Mechanical Engineering
The State University of New York at Binghamton
Binghamton, NY 13902, USA
[2] Xilinx, Inc.
2100 All Logic Drive, San Jose CA 95124

Abstract—In this study, an Electromigration (EM) acceleration test was conducted at 150 °C ambient temperature with 12A current. The solder joint was made of SAC305. Time 0 void evolution was observed by X-ray. A Finite Element Analysis (FEA) based on Atom Flux Divergence (AFD) and vacancy concentration were introduced. The detail of simulation process is given. The results show that location and size play an essential role in EM failure. On the cathode side of a solder ball, the void position would slightly increase the TTF, temperature and current density, However, a larger void size will significantly increase the TTF, temperature and current density. On the anode side, time 0 void barely affects the TTF, temperature and current density. Avoiding the time 0 void on the substrate side will significantly improve the EM performance because the Cu pad is thicker than Cu trace in general.

Keywords: Ball Grid Array (BGA) Electromigration (EM). Time to Failure (TTF), Atom Flux Divergence (AFD), Finite Element Analysis (FEA), Void, Simulation

I. INTRODUCTION

As the revolution of advanced packaging technics, such as 2.5D and 3D packaging, the electronic package has a shorter electrical path between different ICs and achieves more I/Os, -Furthermore, the scaling down of the package size has led to a shrink of interconnect size and a significant increase in electrical current [1-4]. This trend will result in higher current density and joule heating in the interconnections, which significantly enhance the Electromigration (EM) damage [5]. In this regard, EM becomes a critical issue in the reliability of electronic packaging [6].

Solder and copper are the two most common used conductor materials in an electronic package. Due to the material properties, solder is weaker than copper in EM perspective. In previous research, most of EM acceleration test for solder were performed at 150°C and $10^4 A/cm^2$ or higher, but, for copper, it conducted at 350°C and $10^7 A/cm^2$ at least [7-8]. time 0 void is more natural induced and hard to avoid in solder joint during the manufacturing process. This initial void has a significant impact on the EM because the Time 0 void will accelerate the void nucleation and growing process. There is a lack of research investigating the relationship between time 0 void and EM.

II. EXPERIMENTA METHODOLOGY

In this study, an experiment conducted at 150 °C ambient temperature with 12A current. Temperature and voltage of the test structure monitor in-situ. The initial void is a kind of manufacturing defect and it randomly distributes on either the substrate side or the PCB side, So the experiment picked the sample with void clearly shown in the X-ray At time 0. After the solder joint fails, X-ray was used to examine the time 0 void evolution.

Fig.1 shows the Schematic drawing of the test structure. There are 6 carriers solder ball on each side and a pair of victim solder between carriers. The victims will carry all the current of the test structure. The temperature monitor by a serpentine line around the test structure. The resistance of the serpentine line respect to the local temperature of the test structure. The voltage of test structure and the resistance of serpentine line is monitored in-situ. When the open circuit happens, the structure is considered as failure.

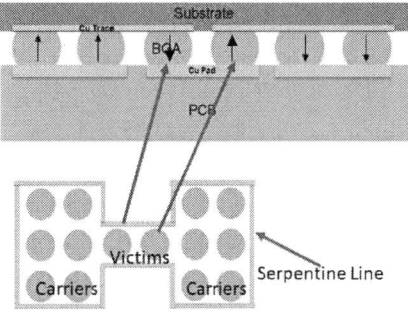

Fig 1   Schematic Drawing of the Test structure

Fig.2 shows the dimensions of the test structure. The BGA pitch is 1mm.

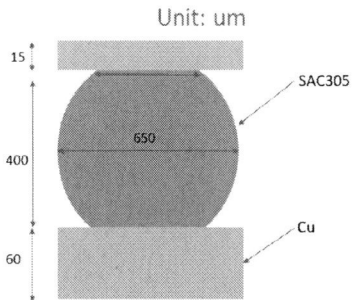

Fig 2   Nominal Dimensions of The Test Structure

III. NUMERICAL MODELING

A conventional numerical model of EM is the Atom Flux Divergence (AFD) [9], which based on diffusion theory. It typically expressed as:

$$J_{ew} = [C/(k_BT)]D_0exp[-E_A/(k_BT)]eZ^*j\rho \qquad (1)$$

$$J_{th} = -C/(k_BT) D_0exp[-E_A/(k_BT)]Q^* (\nabla T)/T \qquad (2)$$

Where J_{ew} is the electronic winds induced vacancy flux; J_{th} is the local thermal gradient induced vacancy flux; C is the vacancy concentration; k_B is the Boltzmann constant; T is the local temperature; D_0 is the pre-factor of the self-diffusion coefficient; E_A is the activation energy; e is the fundamental electronic charge; Z^* is the effective charge number; j is the local current density; ρ is the electrical resistivity, which was given as $\rho = \rho_0[1+\alpha(T-T_0)]$; ρ_0 is the electrical resistivity at temperature T_0, and Q^* is the heat of transport. Equations (1) and (2) are the vacancy flux caused by current and thermal gradient, from which it can be found that the current density and temperature are the only controllable experiment factors. Equation (3) is the divergence of total flux and Equation (4) is the time-dependent evolution model of EM and TM.

$$J_{total}=\nabla(J_{ew})+\nabla(J_{th})=[E_A/(k_BT)-1/T+\alpha\rho_0/\rho]J_{ew}\nabla T+[E_A/(k_BT)-3/T+\alpha\rho_0/\rho]J_{th}\nabla(T) + CQ^*D/(3k_B^3T^3)j^2\rho^2e \qquad (3)$$

$$div(J_{total}) + \partial C/\partial t = 0 \qquad (4)$$

Based on Equations (3) and (4), EM and TM are coupled. The vacancy concentration of SAC305 at time 0 is normalized to 1. Because of the symmetric of model, a quarter model is used to perform analysis.

In the vacancy point of view, EM causes the supersaturation of vacancy on the cathode side. Under supersaturation situation, a void will start to nucleate and grow with existing embryos, such as Kirkendall void, IMC island, etc. The concentration difference between saturation and supersaturation ΔC is the driving force for voiding.

Fig.3 shows the schematic drawing of the vacancy concentration changes during voiding process. Assume that the vacancy saturate at time 0. When current passes through solder joint, vacancies begin to accumulate on the current entrance site. After the concentration difference ΔC reaches a critical level, the voiding process occurs. The voiding process increases the local temperature. In the meantime, the increased temperature will set up a new saturation stage. As the voiding process continues, the vacancy concentration reduces from supersaturation stage until it reaches a new saturation stage. At this point, the voiding process will stop, and the accumulating process will start again. This accumulating and voiding process repeats until the solder joint fail.

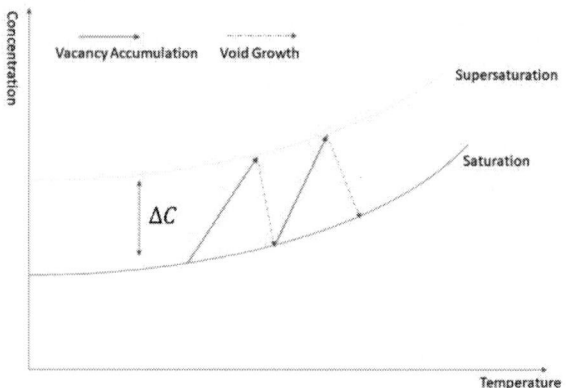

Fig 3. Schematic Drawing of the Vacancy Concentration Change During Voiding

The temperature dependence vacancy concentration can be expressed as Eq. (5). [10]

$$C_v = C e^{\left(-\frac{E_a}{kT}\right)} \qquad (5)$$

Where C_v is the saturation vacancy concentration; C is the atom concentration. Assume the vacancy was saturated at time 0. So, C_v can nomalize to 1.

The saturation vacancy must be adjusted because of the increased temperature after the void formed. The adjustment can be written as

$$C_{new} = C_{Initial} * e^{\left(-\frac{1}{kT_{new}}\right)}/e^{\left(-\frac{1}{kT_{Initial}}\right)} \qquad (6)$$

Where C_{new} is the saturation vacancy concentration after temperature increased; $C_{Initial}$ is the vacancy concentration at time 0. T_{new} is the increased temperature, $T_{Initial}$ is the time 0 temperature.

IV. FINITE ELEMENT MODELING

Based on the mechanism discussed above, the simulation process shows on Fig. 3. ANSYS was used to perform the simulation in the following steps. First, the element type Solid 69 is used to perform static electric-thermal analysis to obtain the steady-state temperature distribution at time 0. Second, apply the steady-state temperature on the EM model at Time 0 with element type Solid 226. Third, solve the EM model uses transient analysis with a specific time step. In this study, 20h time frame is used. step. Fourth, calculate the adjusted vacancy concentration different and then compare it with the simulation results. In this study, the initial voiding criteria $C_{Initial}$ is 1.4. Sixth, use "Ekill" command to dis-active the failure element. Seventh, repeat the simulation process until it reaches the criteria (10% resistance increased). Fig.4 shows the flow chart of the simulation process.

978-1-7281-1500-9/19 $31.00 © 2019 IEEE

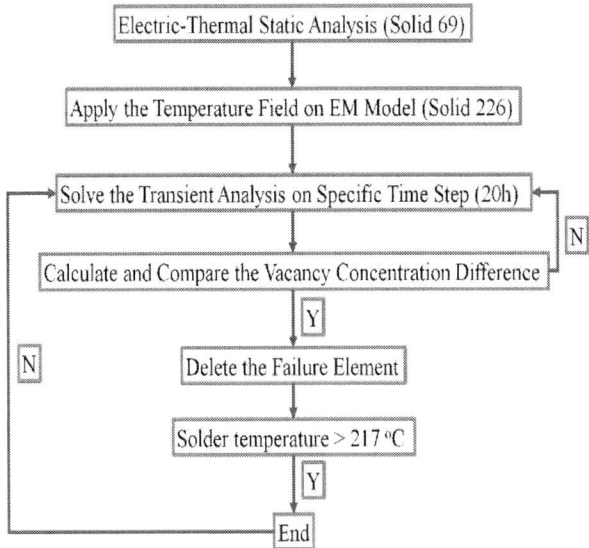

Fig. 4   Flow Chart of Simulation Process

Since the distribution and the pre-existing void size is random. To quantify the effect, there are two assumptions. First, the void projected area on open windows use as the void size and multi-void is considered as a single void. Second, using the geometric center of multi-void as the location of a void.

Based on the discussion above, there 6 cases will be discussed in this study which shows on Fig. 5. In Case 1, there is no void. In Case 2, a void is placed on the cathode side and at the center of the open window with 100um in diameter. In Case 3, a void is placed on the anode side and others are same as case 2. In Case 4, a 100um void is placed near the current entrance with 125um offset. In Case 5, a 100um void is placed away from the current entrance with 125um offset. In Case 6, all conditions are same as Case 2 except the void size is 150um.

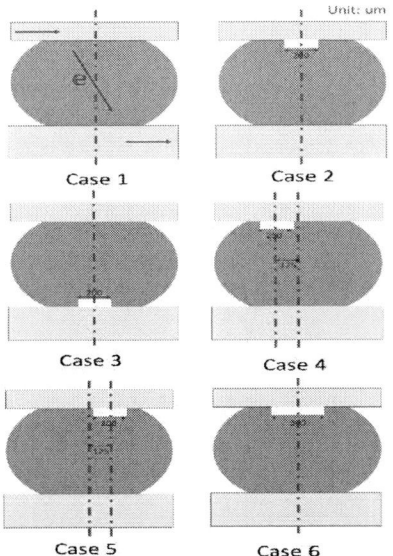

Fig.5   Schematic Drawing of each case

Because of the symmetric of test structure, a quarter model was built. Fig.6 shows the fine detailed mesh and model of test structures. The red arrow shows the direction of electron flow. A uniform current applied at the Cu pad to represent the electrical input. A uniform heat transfer coefficient was applied to the model to represent the convection condition in the oven

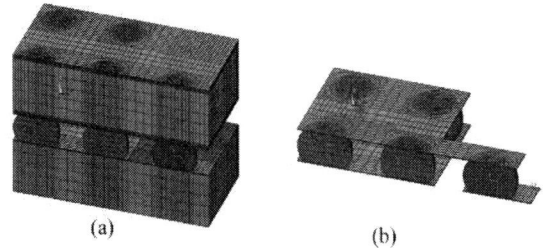

Fig. 6   Mesh and model of test structure: (a) 1/4 model of BGA non-via structure, (b) conductor model of BGA non-via structure

To maintain the mesh quality and density, used "Ekill" command to delete the element on the designed radius near the top or bottom surface in solder ball,which respect to the void projection area. Fig. 7 shows the mesh and model of the solder in each case. The related EM material properties are listed in Table. 1.

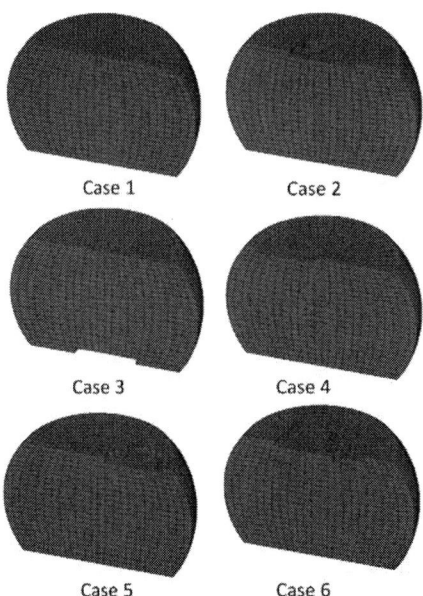

Fig.7   Mesh and Model of Solder Ball in Each Case

TABLE 1   EM PROPERTIES OF SAC305 AND COPPER[11]

	Ea (eV)	Q (eV)	Z*	P (Ohm*um)	D (um^2/s)
Cu	2	0.3121	4	2.52e-14	7.8e7
SAC305	0.98	0.0094	23	18.1e-14	4.1e7

V. RESULTS AND DISCUSSION

In this section, the experimental and simulation results will be discussed. Fig.8 shows the time 0 void evolution before and after the test. It clearly shows that the time 0 void disappeared. A large void form by void merging and propagation. Since the failure happened on the cathode side, the discussion focuses on the cathode side solder ball. Two times 0 voids on the substrate side merged to a large void and propagate to disconnect the solder joint. Meanwhile, the Time To Failure (TTF) of solder joint with voiding is 85h. The TTF of solder joint without voiding is 220h. Thus, the TTF is about 2.5 times shorter in with voiding solder joint.

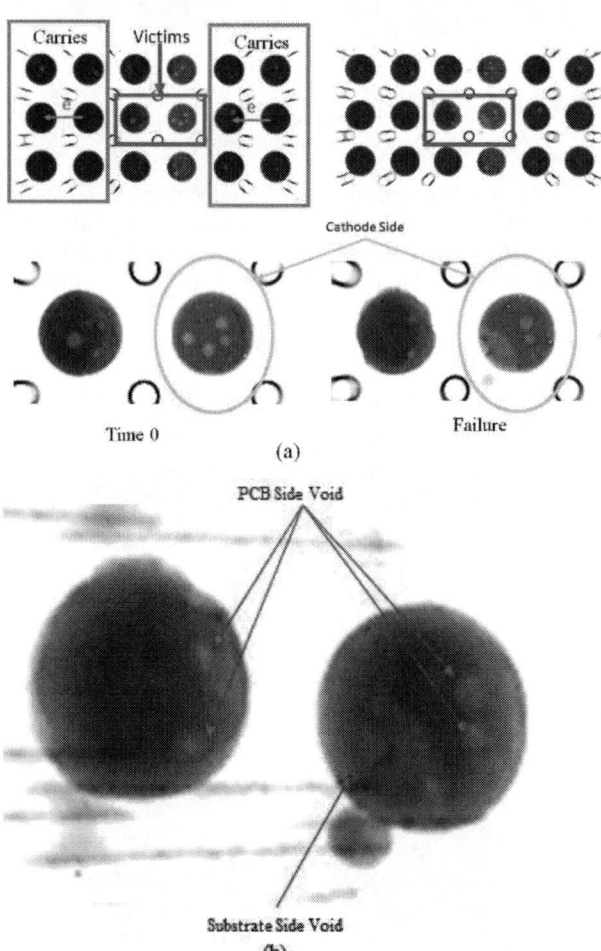

Fig.8   X-ray Image of Time 0 Void Evolution, (a) top view of time 0 void before and after failure, (b) side view of time 0 void after failure

The next goal is to validate the model with the experimental results. Fig.9 Shows the MTTF of BGA in 150 °C 12A. The MTTF is about 139h and the average temperature until the failure occurred in all samples is 186.1 °C. Since the time 0 void randomly distribute, Case 1, the no void case, is used to validate the model. The simulation TTF is 140h and the average temperature is 189.1 °C. The error is 0.7% and 1.6% respectively. It indicates that the simulation well agrees with experiment.

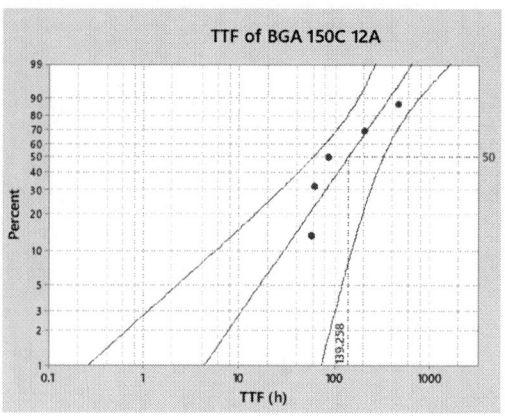

Fig. 9   Weibull plot and MTTF of BGA in 150 °C 12A

In addition to the failure life, the temperature can be validated by comparing experimental data with the model. Due to the temperature of test structure keeps increasing in the simulation and experiment, takes the steady-state at time 0 of the maximum temperature in solder ball for comparison, which is shown on Fig.7. The BGA with no void (Case 1) has the lowest temperature, on the other hand, the case with the largest void (Case 6) has the highest temperature. Case 4 and 5 shows that the void close to the current entrance side would cause a higher temperature. Case 2,3 and 5 shows that the void away from the current entrance has a slight effect on temperature.

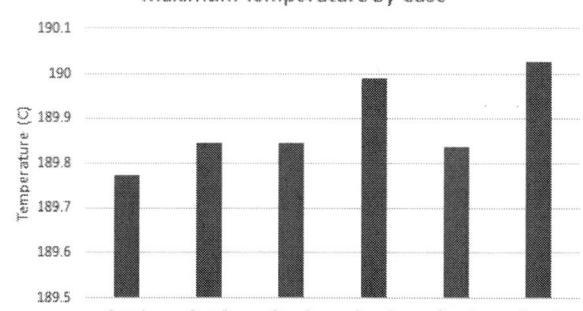

Fig.10   Maximum Temperature at time 0 of solder ball

The current density distribution, which is found to have a similar phenomenon as temperature. They show on Fig.11. It needs to point out that the edge of the void will cause current crowding which enhance EM.

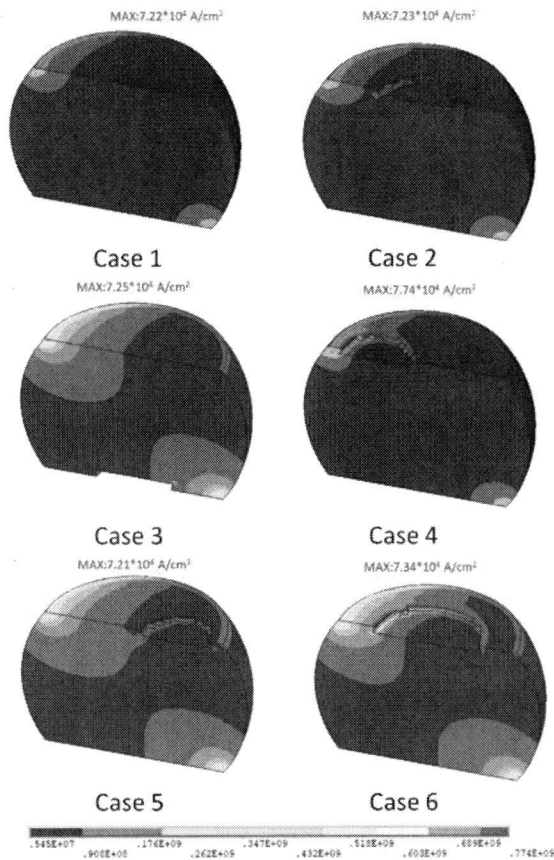

Fig. 11   Current Density Distribution in Each Case

Moving on to the life predictions. Fig. 12 shows the TTF in each case. Case 2,4 and 5 have the same TTF. Those cases have the same size of void and the only difference is location. It means that there no effect in TTF when the void locate on the current entrance side. Case 3 has the same TTF with Case 1, but larger than Case 2. It indicates that the void locates at the current exit site will not affect the TTF. Case 6 has the largest void and the smallest TTF. So the size of void on the current entrance side has a significant impact on TTF.

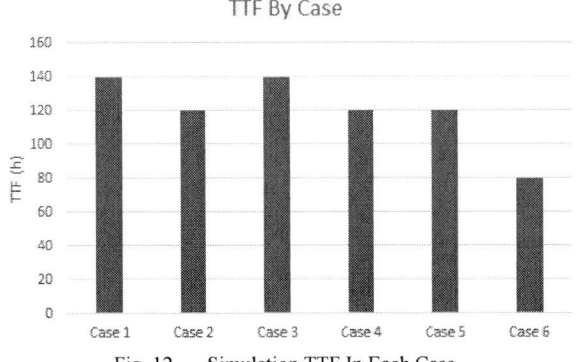

Fig. 12   Simulation TTF In Each Case

We can also compare the final solder shape between cases. Fig. 13 Shows the final shape of void in simulation

and experiment. Case 1,2,4 and 6 are very similar to Case 3, so the results of Case 3 and Case 5 are selected to show. On the cathode side, the solder joint is disconnected by EM. On the anode side, the time 0 void still in the original shape. The results well agree with the experiment and previous research[12].

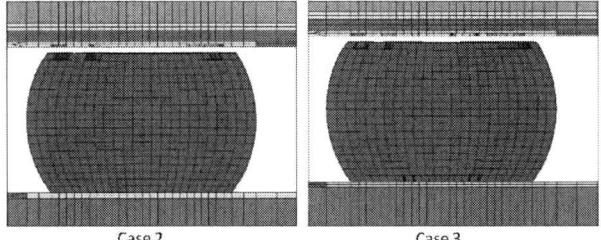

Fig. 13 Final Shape of Void In Simulation.

VI.   Conclusion

In Summary, The EM test performed at 150 °C with 12A current. The experimental results are provided. Multiple time 0 void at the cathode side will merge to a large void and propagate to disconnect the solder joint. A void at the anode side will not change during the EM test.

The detail Finite Element Analysis method and process is given and 6 cases are discussed. The simulation well matches with the experiment. On the cathode side. The void location will not affect the TTF. However, the void size has a significant impact on TTF. On the anode side, the void will not impact the TTF and the void shape will not change by EM. The void on the cathode side slightly increases the temperature and current, but on the anode side, the void barely affects the temperature, current and TTF. In all cases, current crowding effect happens on the void edge.

Avoiding the time 0 void on the substrate side will significantly improve the EM performance because the Cu pad is thicker than Cu trace in general. The thicker Cu connections has higher EM resistance.

ACKNOWLEDGMENT

Xilinx, Inc. supported this project. The authors are grateful for the assistance from The Integrated Electronics Engineering Center (IEEC) and Opto-Mechanics and Physical Reliability Laboratory, the State University of NY at Binghamton.

REFERENCES

[1] J. Wang, Y. Niu, S. Park, et al, "Modeling and design of 2.5 D package with mitigated warpage and enhanced thermo-mechanical reliability." IEEE 68th Electronic Components and Technology Conference (ECTC), pp. 2477-2483, 2018.
[2] J. Hong, K. Choi,D. OH,et.al, "Design Guideline of 2.5D Package with Emphasis on Warpage Control and Thermal Management", IEEE 68th Electronic Components and Technology Conference (ECTC), pp.682-692, 2018
[3] H. Wang, S. Shao, V. Pham, et.al, "Quantification of Underfill Influence to Chip Packaging Interactions of WLCSP",ASME 2018 International Technical Conference and Exhibition on Packaging and Integration of Electronic and Photonic Microsystems, pp. V001T01A004-V001T01A004, 2018

[4] S. Shao, Y. Niu, J. Wang, et al, "Comprehensive study on 2.5 D package design for board-level reliability in thermal cycling and power cycling." IEEE 68th Electronic Components and Technology Conference (ECTC), pp. 1668-1675. IEEE, 2018

[5] K. N. Tu, "Recent advances on electromigration in very-large-scale-integration of interconnects," Journal of Applied Physics, pp. 5451-5473, 2003

[6] E. T. Ogawa, K. Lee, V. A. Blaschke and etc. "Electromigration Reliability Issues in Dual-Damascene Cu Interconnections." Transaction On Reliability, Vol. 51, pp. 403-419, Dec. 2002

[7] C. Chen, H.M. Tong and K.N. Tu. "Electromigration and Thermomigration in Pb-Free Flip-Chip Solder Joints." Annu. Rev. Mater. Res., Vol. 40, pp. 531-555, 2010

[8] K.Zeng, K.N.Tu,"Six cases of reliability study of Pb-free solder joints in electronic packaging technology", Journal of Applied Physics, pp. 55-105, 2002

[9] D. Dalleau and K. Weide-Zaage, "Three-Dimensional Voids Simulation in chip Metallization Structures: a Contribution to Reliability Evaluation," Microelectronics Reliability, vol. 41, pp. 1625-1630, 2001

[10] Siegel, R. W, "Vacancy concentrations in metals", Journal of Nuclear Materials, pp. 117-146,1978

[11] Yong Liu, Lihua Liang, Scott Irving and etc,"3D Modeling of electromigration combined with thermal–mechanical effect for IC device and package", Microelectronics Reliability, PP. 811-824,2008

[12] J. Xu, Y. Niu, S.R.Cain, et.al,"The Expermental and Numerical Study of Electromigration in 2.5D Packaging", IEEE 68th Electronic Components and Technology Conference (ECTC), pp.483-489, 2018

Quintuple Band λ/4 Stub by using Unbalanced Bridged CRLH Transmission Lines

Renuka Bowrothu, Seahee Hwangbo, Haein Kim, and Yong-Kyu Yoon

Department of Electrical and Computer Engineering
University of Florida
Gainesville, FL, USA
rbowrothu93@ufl.edu, ykyoon@ece.ufl.edu

Abstract— In this paper, we demonstrate a quintuple band λ/4 stub using a Bridged Composite Right Left Handed (B-CRLH) transmission line approach. The five bands are achieved using two B-CRLH unit cells. Design is carried out using the B-CRLH transmission line theory and two numerical simulation tools of Advanced Design Systems (ADS, Keysight Technologies, Inc.) and High Frequency Structural Simulator (HFSS, ANSYS Inc.). A prototype device is fabricated on a PCB substrate using milling machine. The demonstrated five stub frequencies are 0.6 GHz, 1.2 GHz, 2.4 GHz, 3.2 GHz and 6.19 GHz. The dimensions of fabricated unit cell-1 and unit cell-2 are 2.52 cm² and 0.25 cm², respectively.

Keywords-Quintuple band; stub; Bridged Composite Right and Left Handed (B-CRLH); HFSS; ADS

I. INTRODUCTION

Circuit theory often assumes that the physical dimensions of the device or network are small compared to the electrical wavelength. In case of the transmission line theory, a distributed parameter network is used, where voltages and currents vary over its length in terms of magnitude and phase [1]. The distributed parameter network of a transmission line is described using the lumped element circuit model as shown in Fig.1, where it consists of series inductors and shunt capacitors and performs like a lossless LC low pass filter. The components are modeled with positive electrical permittivity (ε), and positive magnetic permeability (μ) values. Also, they show a positive group velocity and are termed as a Right-Handed Transmission Line (RHTL) model.

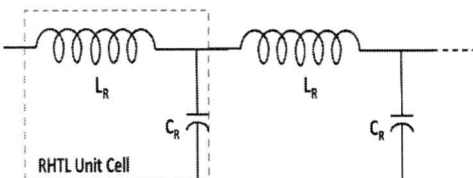

Figure1: Circuit model of lossless Right-Handed Transmission Line

Based on electrical permittivity and magnetic permeability values, all the electromagnetic materials are classified into four categories as shown in Table 1: Double Positive Materials (DPS), ε Negative Materials (ENG), μ Negative Materials (MNG) and Double Negative Materials (DNG).

Table1. Material category

Material	ε (Permittivity)	μ (Permeability)
DPS	>0	>0
ENG	<0	>0
DNG	<0	<0
MNG	>0	<0

In recent years, DNG materials have gotten wide attention due to their unusual material characteristics such as a negative group velocity, the inversion of doppler shift, a negative refractive index [2] and Cerenkov radiation [3] in certain frequency bands, for which they are termed as metamaterials. Early stage metamaterials are implemented using various unit cells such as split ring resonator (SRR) and complementary split ring resonators (CSRR). Also, the wave propagation in such metamaterial media is explained using the conventional transmission line theory, where the inductance and capacitance positions are interchanged. As the electromagnetic wave propagation triode follows a left-handed rule rather than the right-handed rule, the transmission line with swapped inductance and capacitance is called the

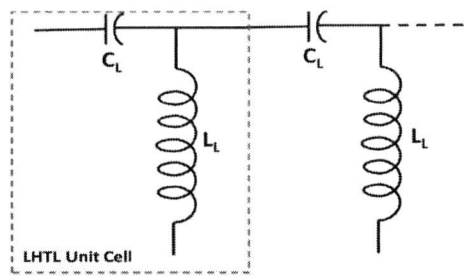

Figure2: Circuit model of lossless Left-Handed Transmission Line

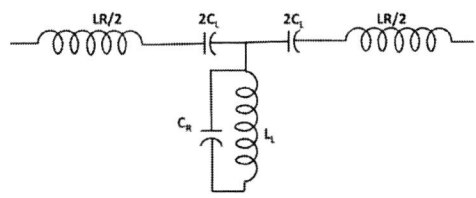

Figure 3: Circuit model of a Composite Right/Left- Handed (CRLH) Transmission Line unit cell

Left- Handed Transmission Line (LHTL) in contrast to the conventional RHTL. From the circuit performance point of view, they possess high pass filter characteristics. Periodic LHTL is shown in Fig.2.

When both RHTL and LHTL are connected, the resultant structure forms a Composite Right/Left-Handed Transmission Line (CRLH TL) [4] and it shows band pass filter characteristics where the lower cut off frequency is set by LHTL and the higher cut off frequency is set by RLTH. A symmetrical lossless CRLH unit cell is shown in Fig.3. Previous literature [5] depicts the CRLH operation with non-linear dispersion characteristics, which are advantageous to realize non-harmonically related multi-band frequencies with the size of the unit cell much smaller than the guided wavelength (λ_g). The multi-band unit cells can be used for an open circuited $\lambda_g/4$ stub with a very compact size.

Meantime, Senior et al. [6] have demonstrated a modified CRLH unit cell, the so called Bridged CRLH (B-CRLH), with an additional low pass band characteristic by adding a bridged inductor between the input and output port of the CRLH circuit as shown in Fig.4. In [6], triple band stub performance is demonstrated with two cascaded unit cells. Such multi-band devices are advantageous for the realization of device and system miniaturization. Especially, the next generation 5G mobile system incorporates diverse functions in a mobile set such as wireless modules for wireless medical sensing, global positioning system (GPS), Wi-Fi, Bluetooth, GSM, and 5G bands to name a few.

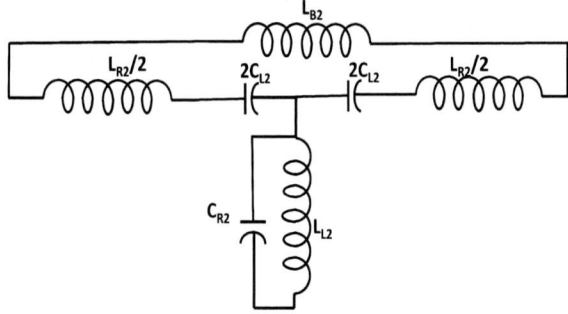

Figure 4: Circuit model of a bridged CRLH Unit Cell

Dispersion diagram of a conventional B-CRLH transmission line is shown in Fig.5. As mention earlier, a lower frequency pass band is achieved by using a Bridged inductor. At ω_L i.e. the cut off frequency of the pure left handed transmission line which is given by $1/\sqrt{L_L C_L}$, the performance of a left-handed transmission line is dominated and the group velocity becomes opposite to the phase velocity. Similarly, above the series (ω_{se}) or shunt resonance frequency (ω_{sh}), the B-CRLH circuit acts like a conventional RHTL and has a positive group velocity. From the phase diagram, it can be observed that a 90^0 phase shift is achievable

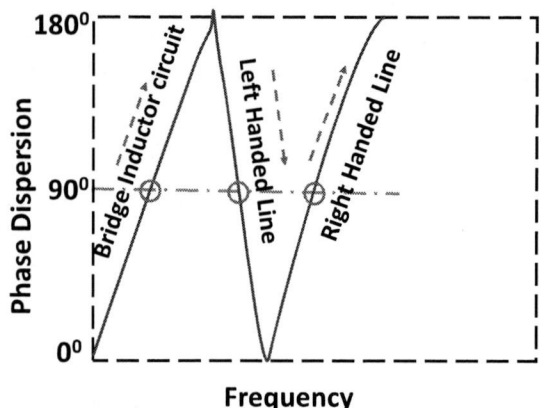

Figure5: Phase Dispersion Diagram of B-CRLH

at three non-harmonically related frequencies and this principle is used to design a 5 band stub

In this paper, the realization of a multi-band quarter wavelength stub with at least five different non-harmonically related frequencies is demonstrated using two B-CRLH unit cells. The target frequencies include 700 MHz for a digital channel TV frequency band, 1.227 GHz for the GPS L2 band, 2.4 GHz for an Wi-Fi and Bluetooth band, 3.5 GHz for a sub 6 GHz 5G band, and 5.8 GHz for an industrial scientific medical (ISM) band. These consist of arbitrarily chosen multi-bands. As the B-CRLH approaches are scalable and expandable, other frequency bands whose frequencies are not necessarily harmonically related could be implementable. Quintuple band design is performed using theoretical analysis and numerical simulation using commercial tools such as High Frequency Structure Simulator (HFSS, ANSYS Inc.) and Advanced Design Systems (ADS, Keysight Technologies Inc.). The designed device is implemented on an Arlon substrate with Metal Insulator Metal (MIM) capacitors and meander line inductors.

II. ANALYSIS AND DESIGN

Bloch analysis [7] and bridged T-network analysis [6] are used to determine the A parameter of the ABCD matrix of the B-CRLH transmission line circuit shown in Fig.4.

$$A = 1 - \frac{2\omega^2 L_B C_L (1 - \frac{\omega^2}{\omega^2_{sh}})(1 - \frac{\omega^2}{\omega^2_{se}})}{\left(1 - \frac{\omega^2}{\omega^2_{se}}\right)^2 \left(1 - \frac{\omega^2}{\omega^2_{sh}}\right) - 4\frac{\omega^2}{\omega^2_L}(1 - \frac{\omega^2}{\omega^2_B})}$$

where ω_{se} is the series resonant frequency $1/\sqrt{L_R C_L}$, ω_{sh} is the shunt resonance i.e. $1/\sqrt{L_L C_R}$, ω_L is the cut off frequency of the pure left handed transmission line given by $1/\sqrt{L_L C_L}$, and ω_B is the resonance frequency $1/\sqrt{(L_R + L_B)C_L}$ due to the bridged inductor LC circuit.

The Bridged CRLH circuit (Fig.4) is tuned using ADS. Values of individual elements for unit cell-1 are shown in Table2.

Table2: Unit Cell-1 ADS parameters

Parameter	Value
L_B	15.873 nH
$L_R/2$	4.7393 nH
$2*C_L$	2.02 pF
C_R	3.78 pF
L_L	2.45 nH

Similarly, circuit parameters and phase diagram for unit cell-1 &2 are shown in Table3 and Fig.5.

Table3: Unit Cell-2 ADS Parameters

Parameter	Value
L_B	4.1976 nH
$L_R/2$	0.82676 nH
$2*C_L$	0.35266 pF
C_R	1 pF
L_L	0.565 nH

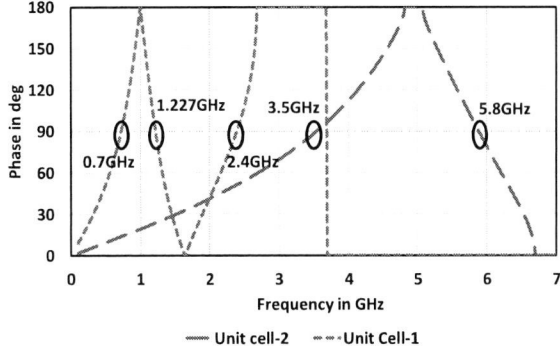

Figure 5: Linear dispersion diagram

It can be observed that unit cell-1 has 90^0 phase shifts at 0.7 GHz, 1.227 GHz and 2.4 GHz. Unit cell-2 gives another 90^0 phase shifts at 3.5 GHz and 5.8 GHz. It is to be noted that 0.7 GHz and 3.5 GHz frequency points are attributed to the bridged inductors for the unit cell-1 and unit cell-2, respectively. The remaining frequencies are from the CRLH transmission line circuits.

Based on parameters obtained via ADS simulation analysis, each unit cell design is fine-tuned and confirmed using HFSS simulation. Fig.6 shows the 3D view of a single unit cell. The patterned top and bottom metal layers and the sandwiched dielectric layer form MIM capacitors (C_L). For the device compactness, a meander line is used for inductor L_B. L_L is achieved by a bottom inductor line on the ground

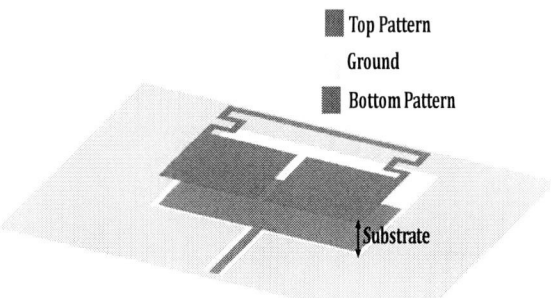

Figure 6: 3D view of each unit cell

pattern. Designs are implemented on Arlon DiClad 880 with a thickness 0.508 mm and a dielectric constant of 2.2.

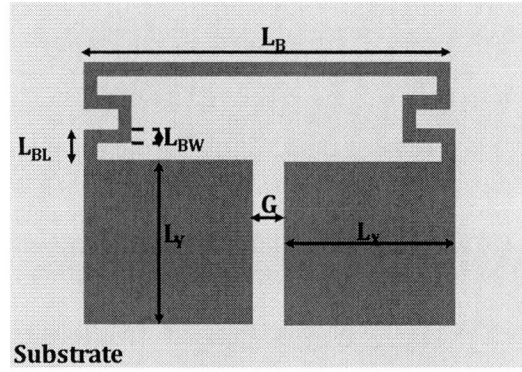

Figure 7: Top View

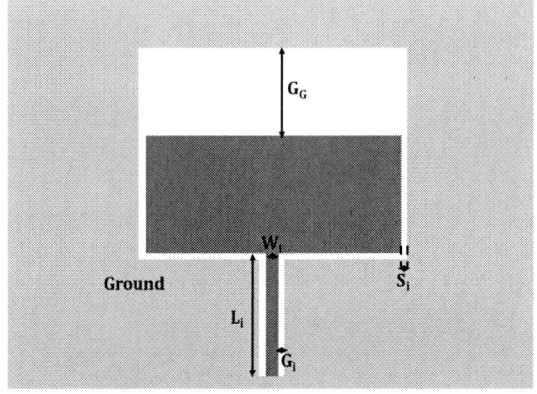

Figure 8: Bottom View

Figure 7 and 8 shoes the top pattern and the bottom pattern of the substrate, respectively. Physical dimensions of each unit cell are given in Table 4. Length of unit cell-1 is 17.6 mm and unit cell-2 is 5.5 mm which is much smaller than a quarter wavelength at respective frequencies.

978-1-7281-1500-9/19 $31.00 © 2019 IEEE

Table4: Dimensions of Unit Cell 1 & 2

Parameter (in mm)	Unit cell-1	Unit Cell-2
L_B	16.372	4.55
L_{BL}	0.372	0.25
L_{BW}	0.435	0.35
L_X	8	2.15
L_Y	8	2
G	0.372	0.275
G_G	3.186	2.5
W_i	0.372	0.246
L_i	4.99	1.8
S_i	0.372	0.3
G_i	0.372	0.25

III. QUINTUPLE BAND DESIGN AND FABRICATION

In order to verify the circuit analysis, a quintuple band design is first simulated using ADS. The schematic of the design is shown in Fig.9. Based on the component values of each unit cell from ADS, fine-tuned design is performed using HFSS. A 3D View of complete design is shown in Fig.10.

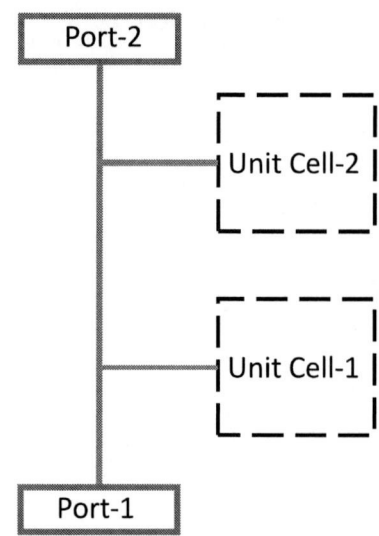

Figure 9: ADS schematic of a quintuple band stub

Unit cells are placed at each side of the transmission line to minimize the coupling effects between the two-unit cells. Using a 0.25 mm diameter milling bit, the quintuple band design is patterned on top and bottom of the Arlon substrate using milling machine. The top and bottom views of the fabricated devices are shown in Fig. 11. The total area of the fabricated device is 2.52 cm^2 and 0.25 cm^2 for the unit cell-1 and 2, respectively. For measurements, the microstrip feed line is soldered using SMA connectors.

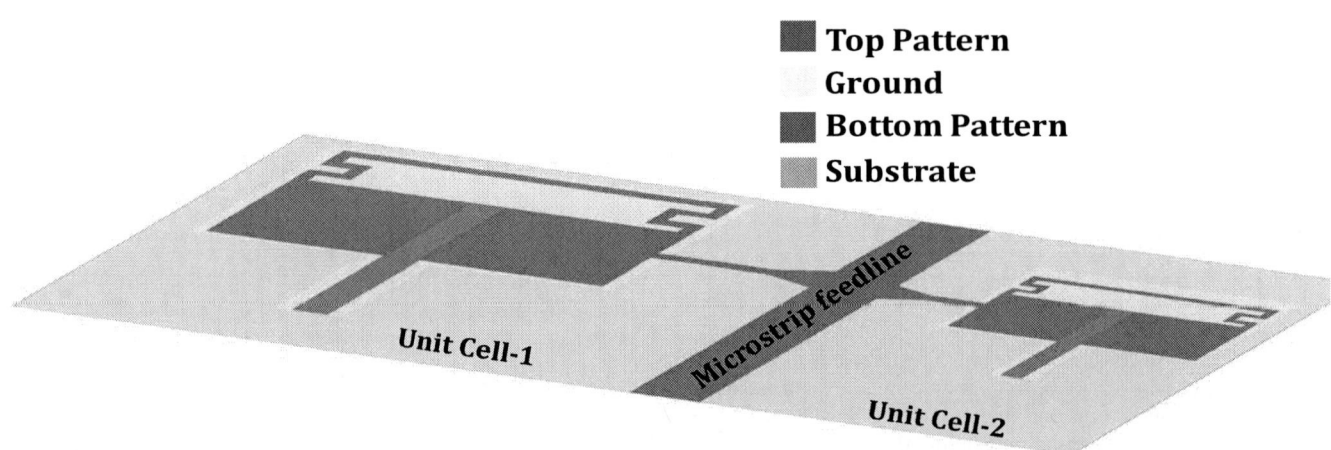

Figure10: 3D View of Quintuple Band Design

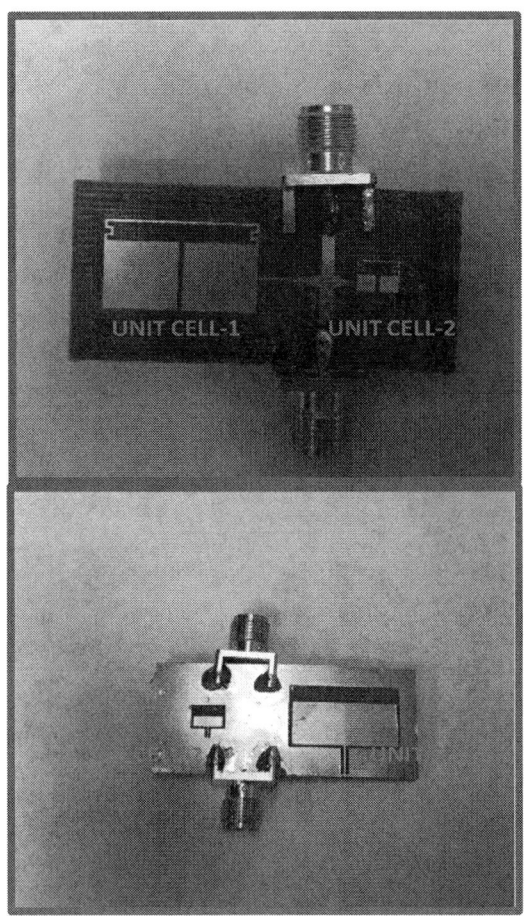

Figure 11: Fabricated Device: Top view (upper) and Bottom view (lower)

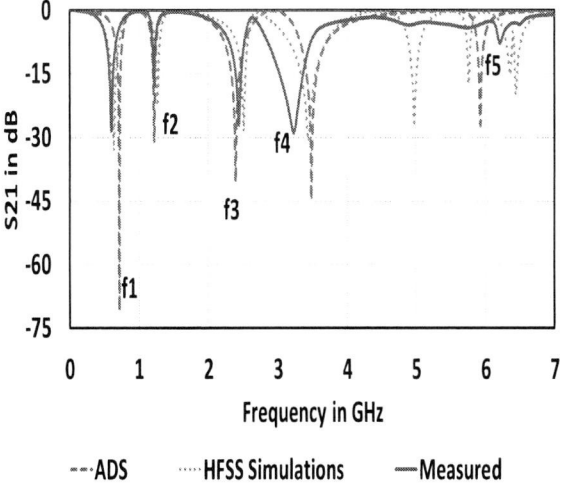

Figure12: Simulation and Measured Results

IV. SIMULATION AND MEASUREMENT RESULTS

Frequency shift in HFSS and ADS shown in Fig 12 are possibly due to the coupling effects between the unit cells and

Table 5. Comparison of Measured and Simulated Data

Frequency (GHz)	ADS (GHz)	HFSS (GHz)	Measured (GHz)
f1	0.7	0.64	0.60
f2	1.227	1.29	1.21
f3	2.4	2.45	2.44
f4	3.5	3.14	3.24
f5	5.8	5.83	6.19

the fabrication tolerance. Note that the smallest bit size used is 250 μm in diameter. The fabrication tolerance is +/- 50 μm. Fabricated devices are measured using vector network analyzer (VNA) and calibrated using a 85052D calibration kit. Table 5 summarizes the comparison of 5 stub frequencies from ADS, HFSS and measurement data.

V. CONCLUSION

Quintuple band stub performance is demonstrated using 2 B-CRLH unit cells. ADS, HFSS and measured results are presented. Frequency bands from the designed and fabricated match well. Some discrepancy in operating frequency is due to coupling and fabrication tolerance. Hence in this design we demonstrate a stub working in five frequencies in 1G-4G communications and 5G communications technologies.

ACKNOWLEDGMENT

Authors would like to thank Rogers Corporation for their generous donation of the substrate boards. Fabrication has been performed in the Interdisciplinary Microsystems Group at the University of Florida.

REFERENCES

[1] D.M.Pozar, Microwave Engineering, 3d ed. New york; Wiley, 2005, ch.2, pp.49-52 and ch.8, pp. 371 - 374

[2] R.A. Shelby, D.R.Smith, and S.Schultz, "Experimented verification of a negative index of refraction", Science, vol.292,pp.77-79, Apr,2001

[3] I. A. Buriak and V. O. Zhurba, "A review of microwave metamaterial structures classifications and applications," *2016 9th International arkiv Symposium on Physics and Engineering of Microwaves, Millimeter and Submillimeter Waves (MSMW)*, Kharkiv, 2016, pp. 1-3.

[4] C. Caloz, "Dual Composite Right/Left-Handed (D-CRLH) Transmission Line Metamaterial," in *IEEE Microwave and Wireless Components Letters*, vol. 16, no. 11, pp. 585-587, Nov. 2006. doi: 10.1109/LMWC.2006.884773 C. Caloz, "Dual Composite

978-1-7281-1500-9/19 $31.00 © 2019 IEEE

Right/Left-Handed (D-CRLH) Transmission Line Metamaterial," in *IEEE Microwave and Wireless Components Letters*, vol. 16, no. 11, pp. 585-587, Nov. 2006. doi: 10.1109/LMWC.2006.884773

[5] F.Bongard, J.Perruisseau-Carrier, and J.R.Mosig, "Enhanced CRLH transmission line performances using lattice network unit cell," *IEEE Microw. Wireless Compon.Lett,* vol19, no.7,pp.431-433, Jul, 2009.

[6] D. E. Senior and Y. Yoon, "Bridged Composite Right/Left Handed Unit Cell With All-Pass and Triple Band Response," in *IEEE Microwave and Wireless Components Letters*, vol. 22, no. 11, pp. 568-570, Nov. 2012

Product Level Design Optimization for 2.5D Package Pad Cratering Reliability during Drop Impact

Huayan Wang[1], Jing Wang[1], Jiefeng Xu[1], Vanlai Pham[1], Ke Pan[1], Seungbae Park[1]
[1]Mechanical Engineering Department
State University of New York at Binghamton
Binghamton, NY, USA

Hohyung Lee[2], Gamal Refai-Ahmed[2]
[2]Xilinx Inc
San Jose, CA, USA

Abstract—**The conversion from using tin-lead solder joint to lead-free solder joint has raised many pad cratering failures for electronics manufacturing. This phenomenon has become more severe to the 2.5D package which has a large heatsink attached on the top. The pad cratering issue has been found during testing, handling or transport due to a single overload. The package qualification process requires that it passes the board level drop test before shipping, however, this doesn't always guarantee that the package will survive from product level drop, especially for a complicated design product. Many types of research have been conducted to improve the package's board-level pad cratering reliability, few emphases were put on the product design. The aim of the present study is to evaluate several product design parameters with respect to PCB stress, using numerical methods. Several different design variables, such as reinforcement structure, heatsink size, pad design, were studied. Data are presented to show the effects of the above factors and to highlight the main factor to cause the pad cratering failure.**

Keywords: Pad cratering, reinforcement structure, heatsink size, pad design

I. INTRODUCTION

2.5D packages have been widely used in the electronics industry for its high integration density and heterogeneous integration ability. It has become one of the best technical solutions for high-end products, such as GPU, computing, and FPGA [1]–[6]. The large heatsink for the 2.5D package due to its high energy density, the usage of lead-free solder and the large package size raise a lot of concerns to the product's drop impact reliability. Thus, it is critical to understand and quantify the drop test reliability of the lead-free solder joints.

There has been a tremendous amount of research done in the last few years on board level drop impact reliability[7]–[10]. It is reported that during a board level drop test, the pad cratering was the most common failure mode and it could lead to trace cracking on the board side [11]. Pad cratering is defined as a separation of the pad from the PCB resin/weave composite or within the composite immediately adjacent to the pad. The pad cratering starts from cracking of the laminate. It may initiate from the intersection of the solder, copper pad and laminate as shown in Figure 1 since this is a stress concentration area for crack initiation [10]. There are several driven factors of the pad cratering issue, such as finer pitch

components, stiffer solders, component size & rigidity, more brittle laminates and presence of a large heat sink.

In this study, the 2.5D package of a PCIe card's BGA failure risk in a drop event was analyzed both numerically and experimentally, and the product design was optimized. In the PCIe card, a 2.5D FPGA package was connected onto a PC board by the BGA. The heat sink was held onto the package with spring screws. The whole assembly was then sandwiched between two metal plates to improve the product structure rigidity. Those parts can add large inertia to the package during the drop, which will pose more reliability risk on the BGA interconnections.

To evaluate the pad cratering strength in the joint level, there are three documented standard test methods, which are pin pull, ball pull and ball shear. The pin pull test method is good for any pad geometry and it is most sensitive to board material and design variables, however it requires pins to be soldered to pads which cost more than the other methods. The ball pull test method is almost as sensitive as pin pull and it doesn't require expensive pins, but it can only measure the BGAs and the result is highly dependent on solder ball, so the process control is critical. For the ball shear test, it is least sensitive to design and material variables and the shear speed will influence results. There is no pass or fail criteria for those method, so the user must define what is acceptable based on design and reliability requirements by themselves.

The initiation of the pad cratering issue usually does not cause any electrical resistant, so it is extremely hard to detect the pad cratering failure, especially at its early stage. There is no standard pad cratering failure detection method yet, but there are some potential warning signs that people should pay attention to, such as excessive BGA repair rate, high percentage of "defective" BGAs and high rate of "retest to pass" at in circuit test. Recently, a research group from Cisco proposed a potential effective method to detect the pad failure at the early stage. Acoustic emission of the elastic waves due to a sudden redistribution of stress in a material is used to report the initiation of the pad cratering failure. Based on this method, they reported that the PCB strains at the onset of pad cratering were found to be significantly lower than prior expectations, which were based on electrical failure of daisy chain circuits [12].

A handful of pad cratering issue mitigation techniques has been introduced by researches. Solder mask define is in favor to reduce pad cratering risk compared to none solder mask define due to its larger pad area to distribute the stress. Intel has defined the solder joints at the corner which have more risk to fail as the non-critical to function solder balls. A group people of Cisco suggested to enlarge the corner most pads by 50% and make them solder mask defined while keep the rest pads as none solder mask defined [13]. Copper pad made of "bullet" geometry was also found to be effective to reduce the pad cratering risk. A more compliant cap layer used between the PCB and copper pad can also block fractures and protect the copper connections(traces) to the pad.

Although there are lots of great board level pad cratering mitigations techniques, there are few means can be used in product level. The package qualification process requires that it passes the board level drop test before shipping, however this doesn't always guarantee that the package will survive from product level drop, especially for a complicated product design. In this paper, our emphasis was put on the product structure design to evaluate various factors' influence on 2.5D package's pad cratering reliability. With numerical method, the parametric study was conducted regarding the product overall rigidity, heatsink size, pad design, package edge bonding.

II. SIMULATION METHODOLOGY

Due to the increasing demand for short time-to-market, drop testing has become a bottleneck for semiconductor industry. Researchers have demonstrated several novel modeling methods to satisfy the requirements in package design analysis [8]. Especially, when it comes to the drop impact simulation, a faster and cheaper solution is most desirable. A proved accurate and reliable model can offer profound insights into the mechanics and physics of failure for design improvement. To setup those advanced drop test modeling techniques, there mainly are three questions needed to be answered, such as analysis type (dynamic vs. static), loading method (free-fall, input-G or direct acceleration input) and solver algorithm (explicit vs. implicit).

A. Analysis type

The nature of a drop test simulation should be a dynamic analysis; however, the static analysis can also be utilized to a certain extent. The main advantage of the static model is that it is much faster than dynamic model. Which makes it possible to utilize more complicated material model and fine mesh within an acceptable solution time, while solution cannot be obtained at all using dynamic model if sub-modeling or mass-scaling modeling technique is not used.

For most design applications, the static model is generally able to deliver the same trend as the dynamic model. It can capture the effect of peak acceleration, however, doesn't not reflect the time-dependent effect or shock wave propagation. Using the static model, large design matrix can be quickly checked to identify critical design variables. Which in the dynamic models can be further analyzed in order to determine whether the design passes the drop test [14].

B. Loading method

There are three widely used loading methods. Each of them has its unique advantages. The applications will determine the choice of loading method.

The most intuitive and conventional method is free-fall. It simulates the actual drop testing process and requires building the whole drop test model including the drop block, felt, contact surface and so on. The PCBA will be mounted to the drop block with screws and drop from a certain height onto a contact surface. The impact pulse generate by the model needs to be finetuned by contact condition to match the experimental requirements. The free-fall method can provide fundamental understanding of different drop test conditions, such as drop block, felt layer on the impact pulse, contact surface, drop height and so on. It is good to be used to predict the impact pulse for the drop test and evaluate the drop tester design variations when no testing result is available. On the other side, due to its full model of the system, the solution time is very high sometimes even unaffordable [8], [14].

To overcome this difficulty, the Input-G method is introduced. In this method, the contact surface, fixture, drop table, and friction of guiding rods in drop test setup are not needed to model. The complex effects of the test setup are considered indirectly by using an impact pulse subjected to the board mounting holes. The problem can be then formulated as (1), with an initial condition of (2) and boundary condition (3):

$$M\ddot{u} = C\dot{u} + K = 0 \tag{1}$$

$$u_{t=0} = 0, \dot{u}_{t=0} = \sqrt{2gH} \tag{2}$$

$$a = \begin{cases} 1500g\sin\dfrac{\pi t}{t_w} & for\ t \leq t_w, \quad t_w = 0.5 \\ 0 & for\ t \geq t_w \end{cases} \tag{3}$$

M: Mass matrix
$\ddot{u}$: Acceleration
C: Damping matrix
$\dot{u}$: Velocity
K: Stiffness matrix
g: Acceleration due to gravity
u: Displacement
t: Time after impact

The boundary condition equation (3) is based on the JEDEC specification for the drop table acceleration profile. It should be noted that, the board response with this method includes a rigid body motion, which is not desired.

In order to counter this effect, the direct acceleration input method was introduced [8]. In which the surfaces of mounting holes are fixed during dynamic responses and the acceleration impulse is applied as a body force to the object. So that it is easier and more straightforward to do input. The direct acceleration input method has been proved analytically equivalent to the input-G method and it eliminate the rigid body motion in the result. So, it will be utilized as the loading method in this study.

978-1-7281-1500-9/19 $31.00 © 2019 IEEE

C. Solver algorithm

The drop test in nature is a transient and dynamic problem. For solving this kind of problem, there are explicit and implicit solvers. The explicit solver applies the central difference approach in the time domain, which is particularly suitable for solving highly nonlinear problems as well as problems concerning wave propagations, but it is conditionally stable and thus prefers a tiny time interval. The implicit solver utilizing the Newmark algorithm, are usually unconditionally stable and feasible to adopt a large time interval, but it is not good to capture the wave propagation effect.

For the current industrial specified drop test, which is a moderate transient process, both explicit and implicit solver give similar results [15]. In order to get results in a relative long-time span, we use the implicit solver in this study.

III. DROP TEST SIMULATION

In the PCIe card ship shock reliability qualification process, the pad cratering failure was found at the FPGA corner solder joints shown as Figure 1. In order to gain more physical understanding of the failure and find out the most influential factors to this kind of failure, a numerical study was carried out to simulate the drop impact process. Since the failure of the resin rich region of the PCB top layer under the copper pad causes the pad cratering issue, the von mises stress of that region was used as the evaluation metric. The full model of the PCIe card was built according to the Figure 2 global level (the actual product was represented by a schematic for confidential purpose), a 2.5D FPGA package is sitting on the PCB. A top and bottom metal plate were used to sandwich the PCBA with screws to increase the overall structure rigidity. Finally, a heatsink was put onto the FPGA die and fixed to the PCB with spring screws.

Since the stress we want to analyze is the one under solder joint's copper pad, a relative dense mesh should be used at the package corner solder joint area. It is clearly that the FEA model will be too large to be handled without using submodeling technique. Figure 2 shows a diagram for the process of submodeling. There are three levels of modeling in this study. In the global level, the solder was modeled as a smeared layer and equivalent material property was used to improve the computation efficiency. At the package level, instead of a smeared layer, the solder was modeled as individual rectangular cubes to better represent the actual situation. From the result of this level, the critical corner can be identified by comparing the max von mises stress of the PCB surface under the 2.5D FPGA package. Then it comes to the package corner level, at which the actual shape of the solder ball, copper pad, and solder mask were modeled. Therefore, the localized structure effect such as SMD and NSMD can be taken account of. The max von mises stress was plotted over the drop impact duration and the peak value was chosen to be compared with across the different cases.

The design of experiment was mainly focused on two categories, board level and product level. In the board level, the board design of solder mask define (SMD), none solder mask define (NSMD), and the usage of the edge bonding were

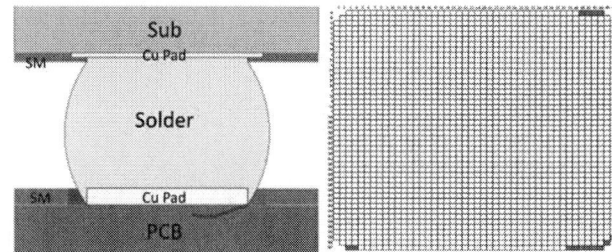

Figure 1. Schematic of failure site observed during drop test.

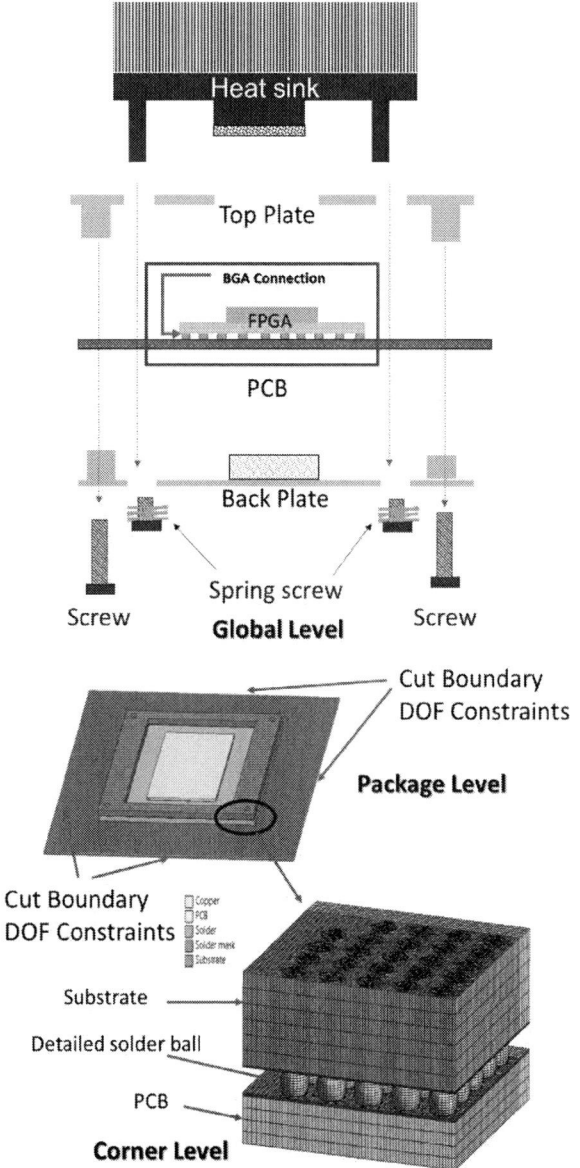

Figure 2. Global and submodeling of the PCIe card

evaluated. In the product level, the top plate material and the heatsink weight will be investigated.

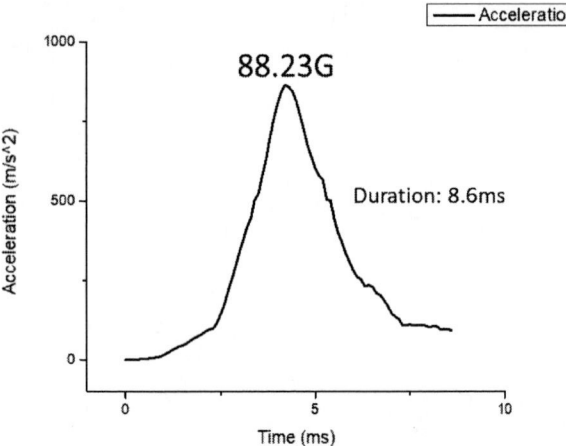

Figure 3. Acceleration during drop

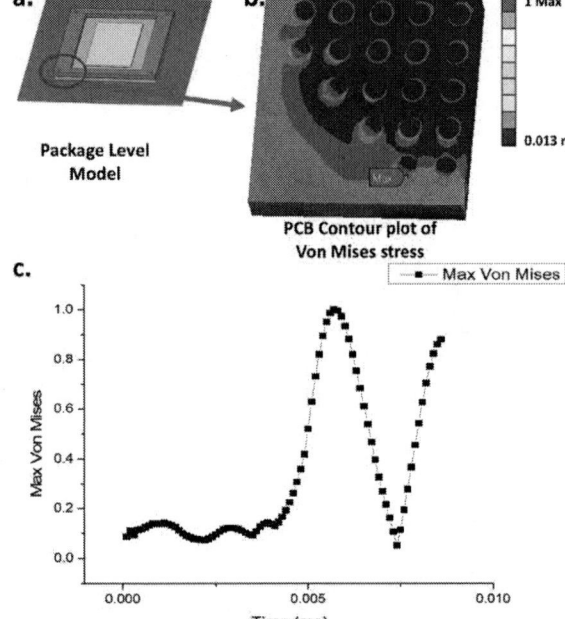

Figure 4. a. Package Level Model, b. Von Mises stress contour plot of the PCB, c. max PCB Von Mises stress over the impact time.

The full DOE cases were listed in the Table 1. The case 1 is the reference case. It is using none solder mask define board design with no edge bonding, aluminum top plate, 11 screws to fasten the top plate and bottom plate to the PCB and has a heatsink weighting 320g. From the case 2 to case 7, each case will change one item to show the difference it makes, the backslash means that it is the same to the reference case. The material properties used in the model are listed in Table 2.

The accelerometers were attached to the product during drop test. The recorded acceleration data Figure 3 was input to the model as a body force according to direct acceleration input method.

IV. RESULT AND DISCUSSION

The reference case result was shown in the Figure 4. The critical corner was identified at the package level modeling. Then, the detailed model of this corner was built to get the final PCB stress response. Figure 4b shows the von mises contour plot of the corner, the stress value was normalized according to its max value over the impact time. The results of the rest cases will also be normalized according to the reference case max von mises stress value. It is observed that the van mises stress of the PCB is higher at the periphery of copper pads and the max value was found at the periphery of a copper pad close to the package edge. During the impact, the link that holds the package and the PCB together is the solder joint, and it is connected to the PCB through the copper pad.

Table 1. Details of DOEs

Case	SMD/NSMD	Edge Bonding	Top Plate Material	Heatsink
1(Ref)	NSMD	No	Al	1
2	SMD	\	\	\
3	\	Yes	\	\
4	SMD	Yes	\	\
5	\	\	Stainless Steel	\
6	\	\	Zinc Alloy	\
7	\	\	\	1.75

Table 2. Material Properties

Material	Mechanical Properties		
	Modulus (GPa)	*Poisson Ratio*	*Density (gm/cm³)*
ABS Plastic	2.3	0.35	1.07
Aluminum	68.9	0.33	2.7
Copper	120	0.34	8.92
Silicon Die	130	0.278	2.5
Solder (SAC)	51	0.36	7.2
Edge Bonding	7.6	0.32	1.2
PCB	16	0.34	1.9
Solder Mask	2.4	0.34	1.9
Stainless Steel	187	0.265	7.87
Zinc Alloy	86	0.25	6.6
Molding Compound	16	0.33	1.9

The holding force will be spread around the copper pad to the PCB, which will show a high stress status at that region. If this stress exceeds the resin strength of the PCB, the pad cratering failure will happen. The rest cases will be compared to the reference case to show each design variation's influence on the pad cratering risk.

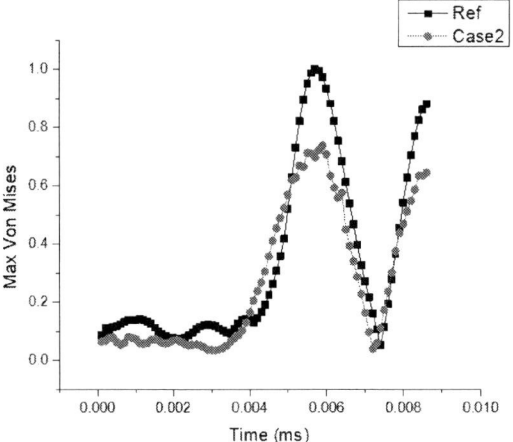

Figure 5. Max Von Mises stress in case2

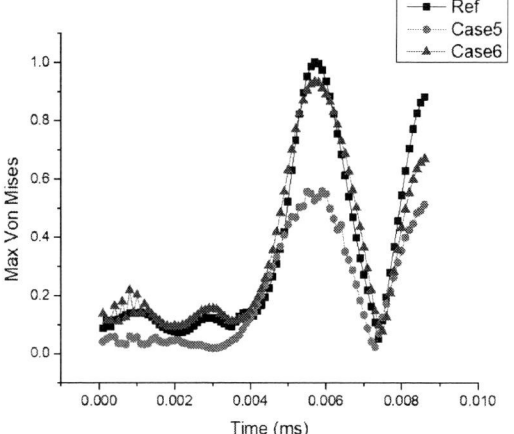

Figure 7.Max Von Mises stress in case5 & case6

A. Board level design parameter

1) SMD vs. NSMD

Figure 5 shows that the peak value of the case2 happens around the same time with the reference, however it brings down 26.2% of the value. Which means that the SMD board design can reduce the pad cratering failure risk compared to the NSMD board design. This confirms with the results reports by other researchers. The reason is attributed to the larger copper in SMD board design could spread out the stress further to reduce the value. However, the drawback of the SMD is that it will shift the failure site to the solder. Therefore, the tradeoff must be made here based on your application.

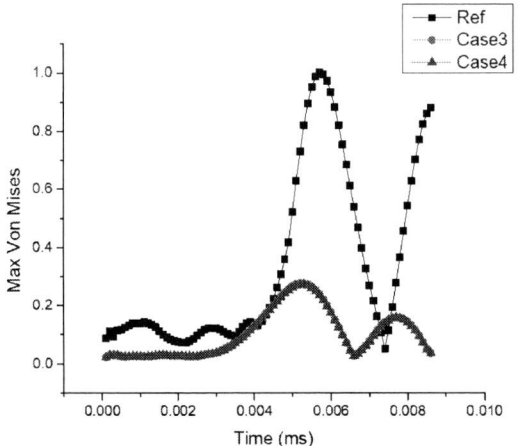

Figure 6. Max Von Mises stress in case3 & case4

2) Edge bonding effect

Figure 6 shows that the peak value drops 72.3% with the help of the edge bonding. Which indicates that the edge

bonding can significantly increase the product pad cratering reliability. Compared with case3, the case4 is using NSMD board design, and both have the edge bonding. With the presence of edge bonding, the improvement of the SMD over NSMD is less than 2%. This can be explained as that the edge bonding will be a dominate factor for the localized stress. Before the break of the edge bonding, the different board design cannot make much difference.

B. Product level design parameter

1) Top plate material

The top plate-PCB-back plate sandwich structure was designed to improve the overall integrity of the product. The back plate was made of stainless steel. Through varying the top plate material, its influence on the product pad cratering failure was investigated. Figure 7 shows that the stainless-steel top plate brings down the peak stress 44.2%, while the zinc alloy brings it down 6.6%. The stainless steel's density is close to the zinc alloy, while its modulus is over two times of the zinc alloy's. Both of their density is much higher than the aluminum. This shows that the modulus instead of the density plays a dominate role on the pad cratering reliability. The larger modulus of the top plate can increase the rigidity of the overall structure thus increase its pad cratering reliability.

2) Heatsink weight effect

A larger heatsink can increase the product's energy dissipation efficiency, but with a larger weight, it also increases the package's pad cratering risk. To what extent the heatsink weight will affect this kind of risk needs to be understood. Figure 8 shows that for an increasement of 1.75 times of the heatsink weight, the stress peak value increases 298.7%. This can be explained as that the heavier the heatsink, the more inertia it has, the inertia will drag the PCB bend more thus introduce more stress on the PCB.

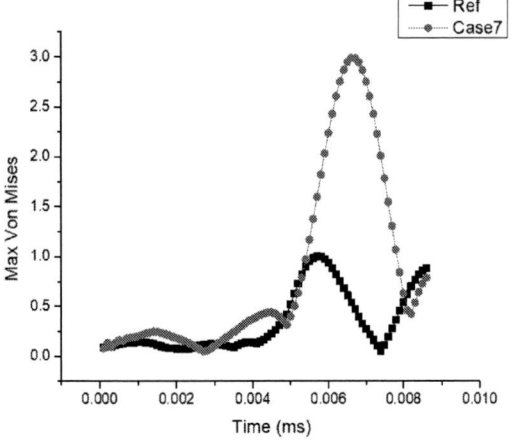

Figure 8. Max Von Mises stress in case7

V. CONCLUSION

Through a numerical method, several key design parameters were evaluated to show theirs influence on the PCIe card pad cratering reliability during drop impact. The results suggest that the pad cratering failure is most sensitive to the heatsink weight. The second dominate factor is the presence of the edge bonding. A rigid top plate can also contribute to the pad cratering reliability by strengthening the rigidity of the overall product. The difference of board design (SMD or NSMD) is neglectable with the presence of the edge bonding.

REFERENCES

[1] J. Hong *et al.*, "Design Guideline of 2.5D Package with Emphasis on Warpage Control and Thermal Management," *Proc. - Electron. Components Technol. Conf.*, vol. 2018–May, pp. 682–692, 2018.

[2] J. Wang, Y. Niu, S. Park, and A. Yatskov, "Modeling and design of 2.5D package with mitigated warpage and enhanced thermo-mechanical reliability," *Proc. - Electron. Components Technol. Conf.*, vol. 2018–May, pp. 2477–2483, 2018.

[3] J. Xu *et al.*, "The Expermental and Numerical Study of

Electromigration in 2.5D Packaging," *Proc. - Electron. Components Technol. Conf.*, vol. 2018–May, pp. 483–489, 2018.

[4] Y. Niu, J. Wang, S. Shao, H. Wang, H. Lee, and S. B. Park, "A comprehensive solution for electronic packages' reliability assessment with digital image correlation (DIC) method," *Microelectron. Reliab.*, vol. 87, no. June, pp. 81–88, 2018.

[5] V. L. Pham *et al.*, "Experimentally minimizing the gap distance between extra tall packages and PCB using the digital image correlation (DIC) method," *Proc. - Electron. Components Technol. Conf.*, vol. 2018–May, no. Dic, pp. 1593–1599, 2018.

[6] H. Wang , S. Shao, P. Shang, C. Zhong, S. Park, and V. Pham, "Quantification of Underfill Influence to Chip Packaging Interactions of WLCSP," p. V001T01A004, 2018.

[7] A. Bansal, C. Guirguis, and K.-C. Liu, "Investigation of pad cratering in large flip-chip BGA using acoustic emission," *Ipc Apex Expo 2012*, vol. 3, pp. 1940–1964, 2012.

[8] H. S. Dhiman, X. Fan, and T. Zhou, "JEDEC board drop test simulation for wafer level packages (WLPs)," *Proc. - Electron. Components Technol. Conf.*, pp. 556–564, 2009.

[9] H. S. Dhiman, F. Xuejun, and Z. Tiao, "Modeling techniques for board level drop test for a wafer-level package," *Proceedings, 2008 Int. Conf. Electron. Packag. Technol. High Density Packag. ICEPT-HDP 2008*, 2008.

[10] D. Xie, D. Shangguan, and H. Kroener, "Pad Cratering Evaluation of PCB," *IPC APEX Expo Conf. Proc.*, 2010.

[11] N. Vickers, K. Rauen, A. Farris, and J. Pan, "Board Level Failure Analysis of Chip Scale Package Drop Test Assemblies," pp. 101–107, 2008.

[12] M. Ahmad, J. Burlingame, C. Guirguis, and C. Systems, "Validated Test Method to Characterize and Quantify Pad Cratering Under Bga Pads on Printed Circuit Boards," *Test.*

[13] A. Bansal, G. Ramakrishna, and K.-C. Liu, "A new approach for early detection of PCB pad cratering failures," *IPC APEX EXPO Proc.*, vol. 3, no. April, pp. 1940– 1964, 2012.

[14] T. Y. Tee, J. Luan, and H. S. Ng, "Development and application of innovational drop impact modeling techniques," *Proc. Electron. Components Technol. 2005. ECTC '05.*, no. 65, pp. 504–512, 2005.

[15] Y. Ma, K.-Y. Goh, and X. Zhang, "Board level drop test simulation using explicit and implicit solvers," *2014 IEEE 16th Electron. Packag. Technol. Conf.*, pp. 405–410, 2014.

Microstructures of Pb-Free Solder Joints by Reflow and Thermo-Compression Bonding (TCB) Processes

Youngja Kim[1], Jinho Hah[2], Patxi Fernandez-Zelaia[3], Sangil Lee[2], Leroy Christie[4], Paul Houston[5], Shreyes Melkote[3], Kyoung-Sik Moon[3*], and Ching-Ping Wong[3**]

[1]Package Engineering Team, Test & Package Center, Samsung Electronics, Asan-si, Chungcheongnam-do, Korea
[2]School of Materials Science and Engineering, Georgia Institute of Technology, Atlanta, GA, USA
[3]George W. Woodruff School of Mechanical Engineering, Georgia Institute of Technology, Atlanta, GA, USA
[4]ASM Pacific Assembly Products, Inc., 7850 S Hardy Drive, Tempe, AZ, USA
[5]Engent Inc., 3140 Northwoods Parkway, Norcross, GA, USA
*Email: jack.moon@mse.gatech.edu
**Email: cp.wong@mse.gatech.edu

Abstract—Differences in microstructures of lead-free solder joints, Sn-Ag-Cu (96.5 wt./3.0 wt./0.5 wt.% SAC-305), made by two different semiconductor packaging processes such as reflow and TCB are discussed. Despite the enormous potential for TCB solder bonding process in microelectronic packaging, there has been few studies regarding the comparative analysis on electro-migration (EM) failure mechanism for the reflow process. We have systematically examined the EM-derived failures of the reflow and TCB-processed solder joints and demonstrated a process-structure-property linkage. This study also includes the analysis performed using generalized spherical harmonics (GSH) representation, a statistical and quantitative measure of material crystallographic informatics, which is novel in this field as to analyzing solder joint microstructures.

Keywords-Thermocompression bonding (TCB); Reflow; Electromigration (EM); Pb-free solder; Intermetallic compound (IMC)

I. INTRODUCTION

In micro-electronic packaging, TCB technology as a bonding process to manufacture flip-chip has been widely employed due to the placement accuracy for fine-pitch flip-chip stacking of thin through-silicon-via (TSV) micro-bumped dies and low board warpage issues. Also, this TCB process is utilized as a micro-bump bonding technology to fabricate 3D electronic packaging devices [1]. In addition, there is an increasing demand for high-performance consumer electronic devices [2]. Therefore, these device modules require finer pitch, which brings an attention to the TCB processed interconnects with copper-posts and copper-pillars. One of the major differences between the two bonding techniques is their process profiles, illustrated in Figure 1. While the conventional reflow process (Figure 1a) achieves the bonding via an in situ thermal process (isothermal environment), the TCB process (Figure 1b) achieves the bonding both thermally and mechanically from the applied force by the preheated bond head, which accounts for a directional cooling. Moreover, the typical temperature ramp-up and cooling rates for the reflow process are only 2.0 °C/s and -2.0 °C/s, respectively, while those for

the TCB process are much faster at 100 °C/s and -50 °C/s [1], respectively. The TCB process operates more rapidly, at least by an order of a magnitude, as its cycle finishes within a few seconds [3, 4]. Therefore, this feature for TCB process highlights its potential for high-throughput assembly yield of the modules with fine-pitch micro-bump joints [5, 6]. Nonetheless, and more importantly, the immense difference in thermal conditions between the two processes affect the quality and reliability performance of TCB-processed modules [1]. Table I summarizes the major differences between the two processes: reflow and TCB.

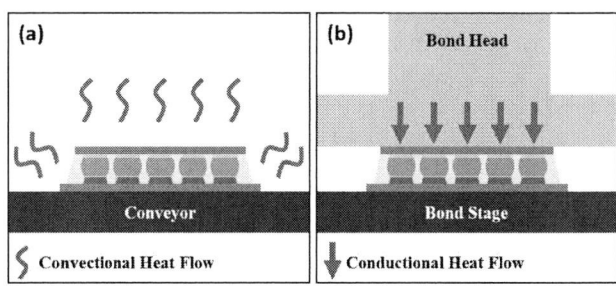

Figure 1. Flip-chip bonding process pertaining to their thermal profiles of (a) traditional reflow, where the multiple packages are placed inside a conveyor oven that isothermally applies heat via convection and (b) TCB, where a package is placed on the bond stage within a TCB tool that directionally applies heat from the bond head via conduction.

TABLE I. SUMMARY OF REFLOW AND TCB PROCESS

	Reflow	TCB
Heat Direction	Isothermal	Top to bottom
Heat Transfer	Convection and Radiation	Conduction
Heating Rate	Slow (1x)	Fast (~50x)
Cooling Rate	Slow (1x)	Fast (~20x)

The solder joints fabricated by TCB process experiences the reflow process because most current packages employing the TCB process also are integrated onto the main substrates together with other packages that need the conventional reflow process. Even though the bonding has been primarily achieved by the TCB process, the entire modules undergo another bonding process again by the reflow process, which means that the solder melting and solidification processes occurs during the reflow process after the actual TCB fabrication process. Hence, these solder joints are believed to be similar to those prepared by a single reflow process. Because of this notion, only few studies have been dedicated into the micro-solder joints fabricated by the TCB processes. However, for some packages, in which their interconnects are formed by a one-step TCB process without an additional reflow process, it would be very important to understand the properties of micro-solder joint formed only by the TCB process. In short, a comparative study on microstructures of these solder joints formed by conventional reflow (multidirectional solder reflow and cooling) and TCB (directional reflow and cooling) processes is lacking, hence solder joint microstructures of reflow and TCB processed packages must be studied to understand their mechanical, electrical, and thermal reliability [2, 4, 7-10].

We observed the major difference in EM reliability between the joints made by TCB and reflow from our previous study. EM-induced failure on the package is primarily due to a combination of formation followed by propagation of voids on the cathode sides as seen in Figure 2 [11] and growth of IMCs that eventually leads to cracking at the solder joint interfaces, e.g. brittle IMC region [12]. Note that the depletion and diffusion of atoms from the surface finishes, Cu bond pad, and from under bump metallurgy (UBM) lead to a formation of voids [13, 14]. Such phenomena eventually leads to an open contact [11]. However, EM-induced failure largely depends on the solder joint composition and its microstructure, in which they provide an important role in understanding the diffusion behavior of Cu and Ni atoms that leads to depletion in these regions upon EM degradation.

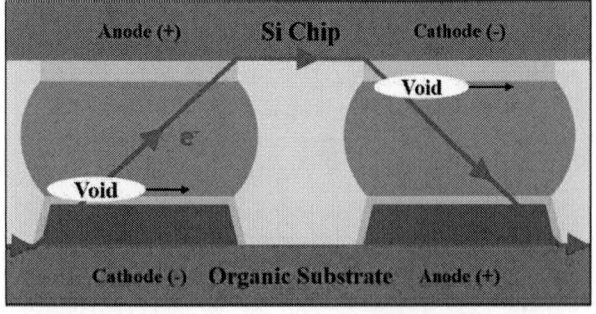

Figure 2. Schematic of EM aging test on flip-chip packages. Regions of void formation on the cathode sides are depicted as white ellipses, and its propagation is illustrated with black arrow. Current direction (electron flow) is illustrated with red arrow.

Since the early 2000s, the ban of using Pb in flip-chip technology and consumer electronics due to the toxic nature of Pb and environmental concerns associated with this element [15] forced the transition into using Pb-free solder alloys. Then, many studies have been focusing on the properties of grain structure and characteristics of the grain boundaries due to recrystallization in Pb-free solder alloys joints [16-19]. Of those studies, Wang et. al. investigated the EM effect on a Sn-based Pb-free solder joint that went through single or multiple reflow processes and derived that having solder joints with a c-axis of a large β-Sn grain aligned to the current flow direction could lead to premature EM failure, and multigrain structured β-Sn, i.e. smaller grains, provides high-angle twin boundaries that increase the EM survival time [18]. In addition to the solder joint structure, formation and growth of IMC, i.e. the kinetics of IMC formation, are another factor that accounts for EM-induced failure in Sn-based solder joints [20]. It is suggested from Bashir and coworkers that a uniform IMC layer could delay EM-induced failure by slowing down the diffusion of Cu atoms into the SAC-305 solder matrix [21]. Furthermore, more extensive studies have been explored, mainly focusing on the influence of β-Sn grain orientation of Sn-based Pb-free solder joints, where the thermomechanical fatigue response and EM behavior strongly depend on the β-Sn grain orientation of the solder joints [14, 22-24]. In addition, the effect of the cooling rate on the microstructure and mechanical properties of solder alloys has been studied as well [25-27].

Due to the anisotropic structure of β-Sn, mechanical, thermal, and electrical anisotropic behaviors are observed in widespread Sn network. The difference in lattice constants of Sn, where a = b = 5.83 Å and c = 3.18 Å, results in much faster interstitial diffusion of Cu and Ni along the c axis than along the a and b axes [28, 29]. Furthermore, Lu et. al. investigated the effect of β-Sn grain orientation on EM degradation mechanism in Sn-based solders and concluded that the failure of the solder joint was due to the highly anisotropic diffusion behaviors of Cu and Ni along the c-axis of the β-Sn matrix that led to a depletion of intermetallic compound (IMC) and under bump metallurgy (UBM) [14].

The crystallographic orientations of solder balls are of great importance because the c-axis orientation in Sn-based solder matrix is a determining factor in the reliability of the flip-chip packages. Based upon many studies discussed in previous paragraphs, this paper explores the effect of different crystallographic textures and microstructures of β-Sn and IMC structures obtained by two different processes (reflow and TCB) on their reliability issues using EM degradation tests. One noticeable difference and the factor that accounts for micro-structural change between reflow and TCB process resides in their thermal profiles, such as cooling rates and directional cooling for TCB. Therefore, we hypothesize that a much faster and directional cooling endured by TCB-processed flip-chip packages would lead to the combination of (1) smaller and more-oriented microstructure of the solder joints, (2) IMC morphologies of high-aspect ratio, and (3) highly-oriented crystallographic orientations of β-Sn, each responsible for their premature failure.

This study analyzes the microstructural results from a comparative study on SAC-305 solder joints fabricated via reflow and TCB process. Typically, in a conventional TCB process, Cu pillar-assisted solder cap is used to bond with the Cu pad on the printed circuit board (PCB) substrate as a test vehicle [2, 4, 8]. However, for the fair comparison, this study utilizes conventional flip-chip architecture without any pillars to effectively explore a comparative study between the two processes. Microstructures of solder joints are investigated and characterized using scanning electron microscopy (SEM). These SEM images are analyzed to determine the morphologies of brittle Cu_6Sn_5 IMCs, which is one of the factors that accounts for premature failure upon the EM aging test. The β-Sn grain size distributions and its crystal orientation are characterized using electron back-scatter diffraction (EBSD), and the established linkages between the process-structure and structure-property are analyzed by using the Principle Component Analysis (PCA). Lastly, EM degraded flip-chip samples are characterized using optical microscopy to verify our hypothesis stated in the previous paragraph. This study will help us elucidate the impact of TCB process on β-Sn microstructures on SAC-305 solder joint reliability.

II. EXPERIMENTAL SECTION

A. Test Coupon and Materials

The flip-chip dies, PB8 (8-mil perimeter pitch) with the SAC 305 solder spheres were provided by Engent Inc. The flip-chips consisted of under bump metallurgy (UBM) layers, and the bump size was 130 μm in diameter. The PCB substrate bond pads were surface finished with a standard electroless nickel immersion gold (ENIG) over the Cu bond pads. During the bonding of the flip-chip to the substrate, flux was used, and after the bonding, underfill material was used to remove flux residues. Epoxy resin, hardeners, and conductive filler were purchased from Buehler, USA. Diamond suspension and colloidal silica suspension for polishing were purchased from Buehler, USA as well.

B. Conventional Reflow Process

During the reflow process, the thermal profile was monitored, and the temperature was measured at each zone in the conveyer from Zone 1 to Zone 9. There were four major steps at specific temperatures as shown in Figure 3a. Between the second and third stage, just before solder reflow, degassing was observed and followed by collapse during solder reflow as the solder balls melted and formed interconnections. Lastly, flux was cleaned as flux residues can initiate corrosion on the package. The cooling rate was determined to be ~2.5 °C/s.

C. TCB Process

Prior to the TCB process, both bond head and substrate pads of the tool were pre-heated to 120 °C. Contact force was applied by the bond head, and the bond head was raised to the peak temperature of 320 °C with a ramp up rate of ~100 °C/s. Before reaching the peak temperature, collapse was observed by the displacement plot from the profile, which triggered to seize the force applied by the bond head. Bond time was determined to be 1.4 s from this step, followed by complete wetting of the solders. The bond head was rapidly quenched to 180 °C with a cooling rate of ~60 °C/s. The total TCB process time was ~5 seconds. This bonding profile is plotted and is illustrated in Figure 3b.

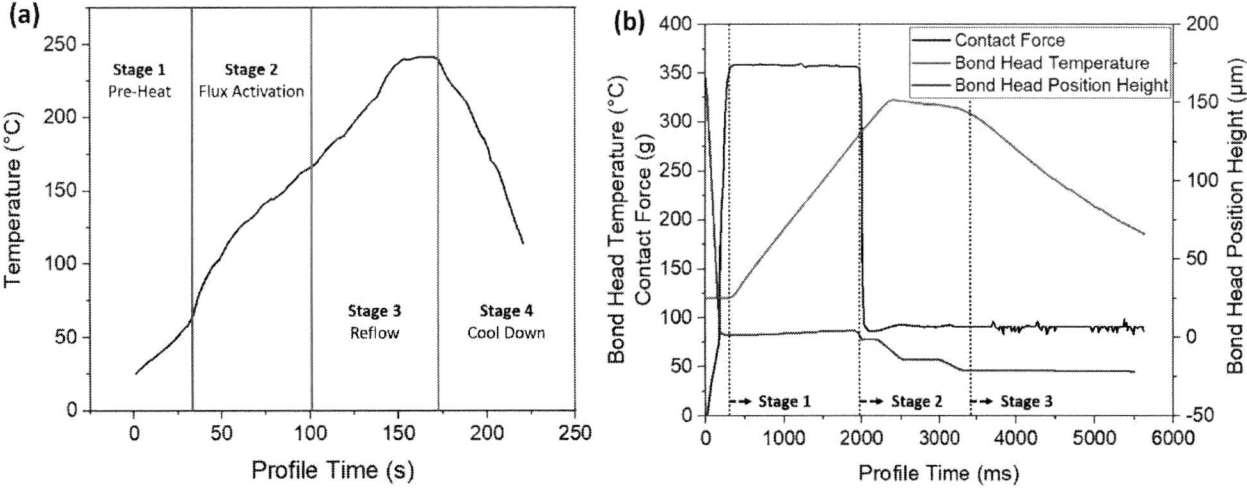

Figure 3. Flip-chip bonding profile of (a) reflow showing thermal profile, where stage 1 is for pre-heating and soaking at ~55 °C to depress any chance in occurrence of thermal shock, stage 2 is for flux activation at ~110 °C to remove any surface oxides, stage 3 represents solder reflow at ~240 °C, and stage 4 is cool down for solidification, and (b) TCB showing thermal, displacement, and force profiles, where stage 1 is when the bond head is in contract with the substrate, stage 2 is during melting of solder alloy, and stage 3 depicts a cooling procedure.

D. Preparation of SEM and EBSD Samples

Both reflow and TCB-bonded flip-chips were as-prepared from the recipe provided in previous sections. Using a drawing map file and QC 7000 PCB Milling Machine, each FR4 PCB substrate were cut into one flip-chip packages in preparation for mounting. Individual flip-chip package was mounted using two-part epoxy resin and hardener, mixed into a ratio provided by manufacturer's specification, and the epoxy mixture and conductive filler were mixed. The mounted samples were grinded and polished, and samples were washed thoroughly with acetone and rinsed with isopropanol (IPA) to remove acetone residue and stains.

E. SEM Imaging for Traditional Reflowed and TCB Samples

As-polished reflowed and TCB samples were characterized using scanning electron microscopy (SEM) for IMC morphology analyses. SEM images were obtained using Tescan Mira FE-SEM with an accelerating voltage of 10 kV and a beam intensity of 20.00 keV using a back-scattered electron (BSE) detector. All images were acquired at a working distance ~10.0 mm.

F. Orientation Imaging Microscopy and Analysis

During the reflow process, the thermal profile was monitored, and the temperature was measured at each zone in the conveyer from Zone 1 to Zone 9. There were four major steps at specific temperatures as shown in Figure 3a. Between the second and third stage, just before solder reflow, degassing was observed and followed by collapse during solder reflow as the solder balls melted and formed interconnections. Lastly, flux was cleaned as flux residues can initiate corrosion on the package. The cooling rate was determined to be ~2.5 °C/s.

A Tescan Mira XMH field emission scanning electron microscope with an EDAX Hikari EBSD detector was utilized for orientation imaging. Indexing of the generated diffraction patterns was performed using TSL OIM analysis. All scans were obtained using a 1 μm raster resolution in both vertical and horizontal directions. Further processing of the results was performed in the open source Matlab toolbox MTEX [30].

Texture and orientation analysis were performed using quantitative and statistical methods. A quantitative description of the crystallographic texture was obtained using a Generalized Spherical Harmonics (GSH) representation [31]. Similar treatments on metallic systems can be found in recent works [32-35]. GSH describes texture using a basis expansion of the orientation distribution function (ODF), f, stated in (1). Here, g is a vector containing the three Euler angles, F_i are the GSH coefficients, and $T_i(g)$ the GSH basis functions.

$$f(g) = \sum_i F_i T_i(g). \qquad (1)$$

The GSH basis is particularly well suited for quantifying crystallographic texture because it preserves crystal symmetries. In this work the basis was truncated at ten terms which was found to be suitable for cubic crystal systems [33, 36].

The GSH representation allows for the crystallographic texture of reflowed and TCB chips to be described numerically simply using the coefficients, F_i, analogous to the Fourier representation of a signal. However, there are ten coefficients and furthermore each is a complex number. Therefore, the EBSD observations require 20 scalar values to describe the observed microstructure ODF. In order to alleviate the analysis and enable the visualization of the data, we utilized principle component analysis (PCA) to perform dimensionality reduction. PCA first numerically identifies efficient basis representations from the observed data. Each observation can then be described compactly simply by retaining a few important basis weights. A reduction to two-dimensional space was found to capture 70% of the variance present in the original data.

In addition to considering mean quantities (mean c-axis orientation, mean crystallographic texture) it is important to consider the spread, or dispersion, of these mean quantities in order to ensure that observations are not the result of chance occurrences (e.g. statistical significance). Consider that there is significant sample-to-sample microstructure variation in the imaged specimens. Since the quantitative description of the microstructure is rather complex (PCA representation of GSH coefficients), we utilized a statistical bootstrapping method to obtain measures of mean dispersion [37]. The procedure is as follows: 1) for a particular flip-chip package class (reflow or TCB), the observed bumps are randomly reselected with replacement, 2) within each of the resampled bumps EBSD pixels are resampled with replacement, 3) from the resampled ensemble the GSH PCA coefficients (which encode the ODF) are computed and this represents a *single bootstrapped sample*, and 4) this procedure is repeated N_b times. The N_b values that constitute the bootstrapped sample quantify the uncertainty associated with microstructure texture. Uncertainties come from step 1, which captures sample-to-sample variation, and step 2 which captures uncertainty within each bump. For visualization, a finite number of suitably spread out points are selected from the bootstrapped sample using a conditional Maximin (cMm) design criteria [38]. Inverse pole figures (IPFs) generated at these points allow for visual interpretation of the mean texture dispersion. A similar procedure was used for computing the mean c-axis misorientation density.

G. EM Aging Test

EM tests were performed for both reflowed and TCB solder joint on the hot plate set to 120 °C, in which ~3×10^4 A/cm^2 of current density was applied to a pair of two solder balls in the flip-chip package at a time as illustrated in Figure 2, in which electrons flowed from cathode (-) to anode (+). These packages were as polished using the same polishing methods described in previous section and were characterized using a Leica digital microscope.

978-1-7281-1500-9/19 $31.00 © 2019 IEEE

III. RESULTS AND DISCUSSION

Electromigration performance of the fine-pitch interconnects under high current density has been increasingly important due to strong demands for small foot print packages and ultra-fine pitch joints requiring high current delivery. Before the EM aging test, both solder joints of reflow and TCB were defect-free. However, after the EM aging test ($3x10^4$ A/cm^2 at 120 °C), only TCB-processed joint was suffered from void defect (Fig. 4b), while reflow-processed joint remained intact. Such defect in the solder joint of the TCB joint results the open circuit after the EM test. Meanwhile, no signs of visible defect within the solder matrix or delamination at the interfaces are observed for reflow-processed solder joints, except for some growth of brittle needle-like Cu_6Sn_5 IMCs. In addition, all solder joints remained intact for reflow-joints upon the same EM test environment that TCB-joint suffered from. Therefore, it is now clearly justified that TCB-joints are relatively unstable and less reliable than those processed by conventional, convection reflow oven.

It was verified from the results of EM aging test that solder joints fabricated via TCB process were heavily degraded with plethora of defects (Figure 4). The packages used in this study were processed differently, one by multidirectional (conventional reflow) and another by directional (TCB) solder reflow and cooling. Cooling rates directly affect the solidification process of the metals, which can control the grain size [22, 26, 39], IMC formations [40-42], and crystallographic orientation of SAC solders [22, 39]. In particular, it is important to evaluate the EM reliability results for the TCB process that has been considered as an excellent bonding technology for next-generation fine-pitch joints. Here, we discuss several factors that account for this phenomenon based upon the difference thermal profiles of two different bonding processes, reflow bonding and TCB.

A. Effects of β-Sn Grain Size on EM Degradation

Grain size is one of the factors that can affect the EM performance.

Figure 5 illustrate the grain size diameter of β-Sn for reflow and TCB joints. These distributions were generated by considering high resolution EBSD scans (raster size 200 nm). In both bonding processes, the mean grain size in the solder joints is ~5 μm. However, it can be noted from

Figure 5 that the higher frequency of grain size (in diameter) is dominated by TCB joints when the grain size in the solder joint is less than ~5 μm, and that is dominated by reflow joints when the grain size in the solder joint is higher than ~5 μm. In short, grain size in the solder joint is smaller when the flip-chip package was TCB-processed. It is, however, important that this difference is negligible when comparing the cooling rates of two bonding processes: ~2.5 °C/s (reflow) vs ~60 °C/s (TCB).

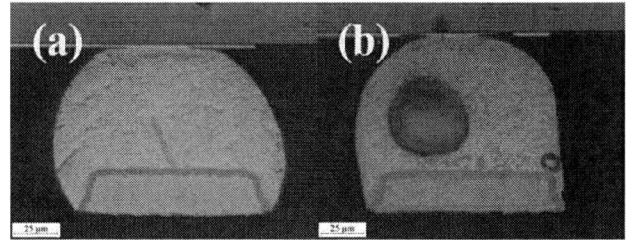

Figure 4. Cross-section digital microscope images of flip-chip packages upon EM aging test: (a) reflow solder joint and (b) TCB solder joint.

Our previous hypothesis states that smaller grain sizes in TCB-prepared solder joints would induce a faster diffusion of Cu and Ni atoms through the SAC 305 solders, one of the influences on a premature EM failure. This hypothesis was based on the experimental study that the grain size of SAC 305 solders decreases with increasing the cooling rate [26].

In addition, a smaller grain size means more grains in the same solder volume, hence a larger fraction of grain boundaries and their paths available for diffusion of Cu and Ni atoms. It has also been known that polycrystalline lattice structures with a low grain-boundary density are potentially more robust to EM failure [43]. Despite the noticeable difference in cooling rates between the two bonding processes (~30-fold difference),
Figure 5 illustrate that the mean grain size and the grain morphology are both similar to one another, which does not correspond with the previously made assumption. This result can be explained from Mueller et. al.'s findings that the influence of cooling rate on the solder microstructure (i.e. grain size and morphology) reduces as decreasing the solder diameter from 1100 μm down to 270 μm and becomes negligible when the solder diameter reaches to 130 μm [26]. In fact, our SAC 305 solder with ~130 μm diameter in size led to a conclusion that the difference in cooling rates between the two process does not yield a significant difference in solder grain size, therefore, a similar grain boundary density between the two samples. No difference in grain size between the two average grain sizes is explained by the degree of undercooling, in which smaller solders (i.e. ~130 μm in diameter) undercools more than the bigger solders [22, 26, 44, 45]. Consequently, grain boundary diffusion kinetics of Cu and Ni atoms would have been similar. Therefore, it was concluded that grain size is not a determining factor for the premature EM failure observed for TCB-joints. Rather, additional mechanism might have been involved that could explain for this phenomenon, such as the effect of IMC and β-Sn crystallographic orientation on susceptibility to EM failure.

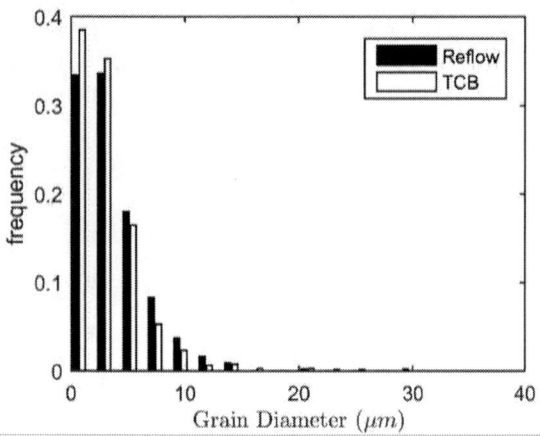

Figure 5. β-Sn grain diameter histogram for reflow and TCB solder joints.

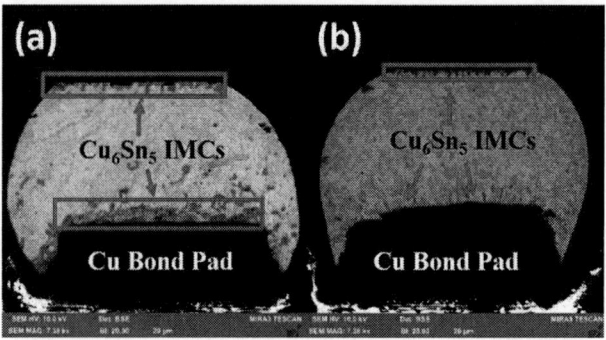

Figure 6. SEM images of solder joint processed via (a) reflow bonding and (b) TCB. Cu_6Sn_5 IMCs are depicted with red arrows in both images.

B. Effects of IMC on EM Degradation

The difference IMC thickness and its morphology affects the kinetics of the Cu and Ni atom diffusion through the SAC 305 solder matrix. As suggested from Bashir's study [21], there is a faster diffusion of Cu and Ni from the bond pad and ENIG surface finishes to the bulk solder if the IMCs do not fully cover those areas or if the IMC scallops have high aspect ratio structures. Fig. 6 illustrates SEM images of the solder joints of reflow and TCB before the EM aging test. On the chip-side, IMC thickness is similar for both samples; however, the low-aspect ratio Cu_6Sn_5 IMC scallops are present in reflow joints. This difference originates from the two bonding process difference, although the IMC morphologies are more obvious at the bond pad interface. When comparing the morphologies of IMCs on the substrate-side, the reflow joint shows are uniformly structured and entirely covers ENIG surface (Fig. 6a) while the TCB joint exhibits a significantly different interface, where only certain regions are discerned with rod-like Cu_6Sn_5 IMCs (Fig. 6b). To better visualize these phenomena, readers are aided with a schematic representation (Fig. 7). In terms of EM effect on these samples, it is likely that TCB joints would suffer from premature failure due to a much faster Cu and Ni diffusion. Eventually, the depletion of those atoms will lead to void formation and propagation, hence delaminating the interface that will short the interconnect.

IMC morphology was shown to have the considerable difference between the solder joints reflowed by two different processes (Fig. 6). The difference in IMC morphology could have been attributed to the cooling rates between the two bonding processes. Because the reflow-processed joints endured a much slower cooling rate, molten solder has more time to wet and form/grow the IMCs before reaching its solidification temperature [26], in which its layer was eventually deposited at the substrate side of the solder joint interface. How the layers of IMCs are structured on the solder-Cu bond pad joint interfaces is essential in understanding both the diffusion of Cu and Ni atoms and Joule Heating that eventually leads to premature failure upon EM-aging test.

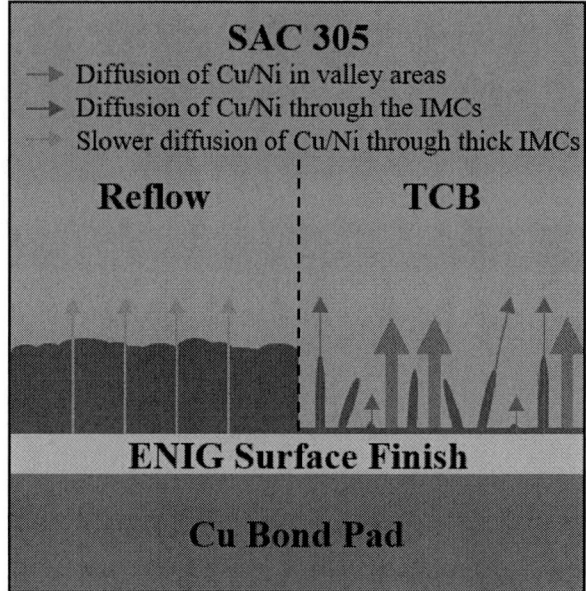

Figure 7. Schematic representation of Cu and Ni atomic diffusion from the substrate side, through the IMC scallops, to the SAC 305 solder matrix. Note that a *thicker red* arrow represents higher amounts of Cu and Ni atoms diffusing through the Sn matrix, and a *shorter blue* arrow represents faster diffusion of Cu and Ni atoms.

The diffusion mechanism of Cu and Ni atoms occurs by a schematic illustrated in Fig. 7, where the IMC layer acts as a diffusion barrier [21]. Therefore, EM-induced failure was not observed for reflow-joints during the EM-aging test period. In contrast, TCB joint failed upon EM aging test as the diffusion of Cu and Ni atoms was highly blocked and impeded by a thick layer of IMC scallops, which left the solder joints intact at both interfaces and integral within the SAC 305 matrix.

A positive feedback loop explains the mechanism of EM-induced damages [43]. Acceleration of the void growth is ceaselessly repeated due to a following situation: void growth increases the local current density that increases the temperature of the flip-chip package due to Joule Heating, and this temperature increase in the package further increases the void growth. This local Joule heating can be expressed as power, P, stated in Ohm's law in (2). Local Joule heating occurs when the current, I, passes through a solder joint producing heat. Therefore, heat accelerates diffusion, which further stimulates the void growth.

$$P = I^2 R. \qquad (2)$$

The temperatures on the top surfaces of silicon dies of the reflow and TCB flip-chip packages are illustrated in Fig. 8. In addition to the temperature of the preheated hot plate at 120 °C, the temperature of both the silicon dies increased with time, but TCB joints exhibited the higher temperature increase than the reflow joints. Consequently, TCB joints failed upon EM aging test as the package was observed with an open circuit, while the temperature of the reflow joints reached a plateau and was maintained. The greater increase in temperature of the TCB joint than that of the reflow joint indicates that the growth of voids within TCB joint takes place more severely during the EM test, which was caused by the positive feedback, i.e. the increased local current density, then the joule heating and the temperature rise, resulting back in the growth of voids. Besides, the positive temperature coefficient of resistance [46] of metal (solder joints) could further increase the joule heating, although this effect is not predominantly governing the temperature increase.

Therefore, a phenomenon known as Joule Heating effect, that increases the temperature of the solder joints, was observed in TCB flip-chip test coupons under EM-aging test. The degree of self-heating between the two processed flip-chips was up to twice as higher for the TCB flip-chip packages (Fig. 8).

Although IMC morphology played a significant role in premature EM failure, a crystallographic orientation of β-Sn should be taken into an account for a holistic diagnosis of a premature failure of solder joints fabricated by the TCB process. Understanding the crystallographic orientation of anisotropic β-Sn is important to fathom the bulk diffusion of Cu and Ni atoms through the Sn crystals. Because of its anisotropy, diffusivity of Cu atoms are 500 times faster along the c-axis than along the other two orthogonal directions (a- and b-axis) at 25 °C [28]. In addition diffusivity of Ni atoms are ~7×10^4 times faster along the c-axis than along the other two orthogonal directions at 120 °C [29]. Note that the temperature of the two-processed flip-chip packages during EM-aging test is higher than both reported temperatures, 25 °C and 120 °C, hence the actual diffusion rate will be much higher, especially for Cu atoms.

C. Effects of Crystallographic Orientation of β-Sn on EM Degradation

Difference in misorientation values (spatial distribution of c-axis) between the two types of joints is important as there is a faster diffusion of Cu and Ni atoms when misorientation value is 90°. However, it is visually difficult to discriminate between reflow and TCB solder joints from Fig. 5. Quantitative measures of the c-axis orientation were measured; however, there appears to be no significant differences (Fig. 9). Although there are some local differences in the angle probability density, there appears to

be no significant differences. In addition, the mean value statistics are nearly identical across the two process conditions. Therefore, it may be concluded from these results that both bonding processes produce similar solder joint microstructures with respect to c-axis orientation.

The c-axis orientation is only one measure of crystallographic orientation. Texture information, represented in the IPF maps (Fig. 10a), also includes additional information about the *secondary* crystal orientations. Therefore, although on average there is no c-axis misorientation difference between the two considered flip-chip configurations, there is a difference in crystallographic texture. Results indicate that there is a clear separation between the texture descriptors (Fig. 10b). This subtle difference may have considerable effect on the physics associated with the process. In this study, we found that the relative orientation between crystal c-axis and the sample geometry was indistinguishable. However, the crystallographic texture, which includes additional information about the crystal orientations, is different across the reflow and TCB joints. The implications of this observation are illustrated schematically in Fig. 11.

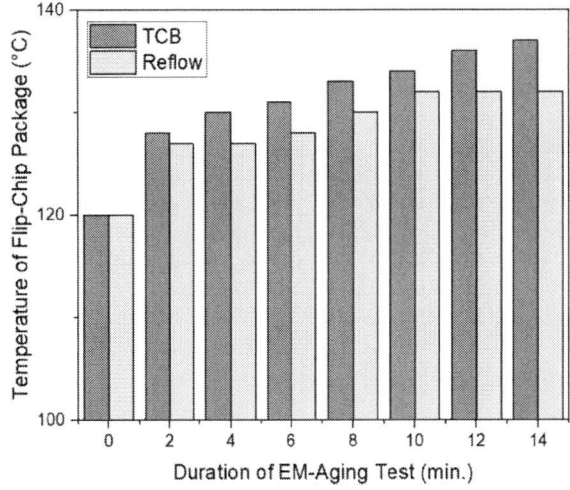

Figure 8. Temperature profiles of Reflow- and TCB-processed flip-chip packages upon EM-aging test at 120 °C under the applied current of 3×10^4 A/cm².

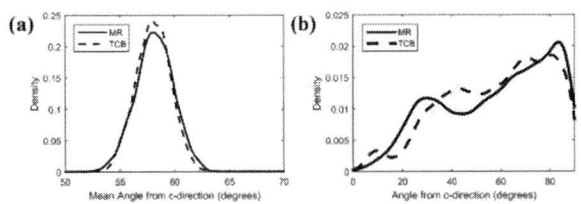

Figure 9. (a) Bootstrapped mean misorientation density and (b) misorientation density for Reflow and TCB samples. Included are both an angle probability density and the distribution of the mean c-axis orientation, obtained using bootstrapping.

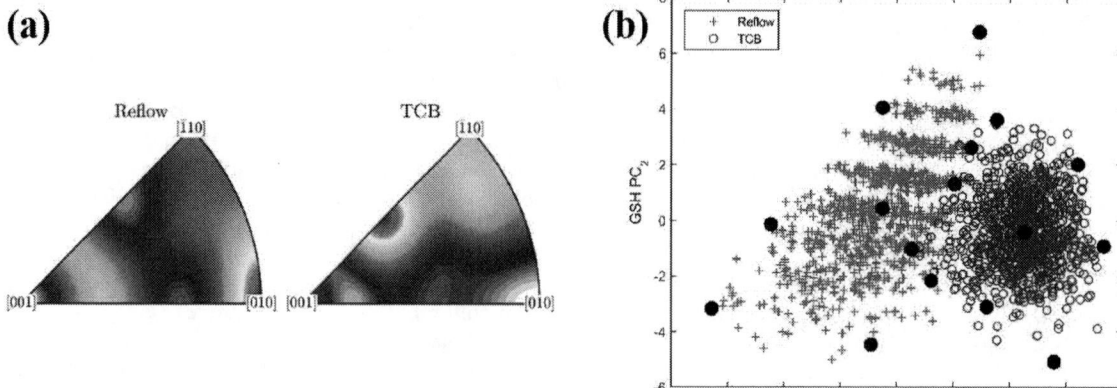

Figure 10. (a) Mean IPF map of reflow and TCB, relative to sample vertical direction (direction of electron flow) and (b) Mean bootstrapped GSH representation samples of Reflow and TCB crystallographic texture in PC space.

Fig. 11 shows two-unit cubes are shown with identical relative orientations (left) and a planar defect is introduced in the form a tilt boundary, but the c-axis alignment remains (right). The tilt boundary restricts interstitial diffusion due to the mismatched lattice structure present at the boundary. In other words, higher lattice coherency was observed for TCB joints that allowed for a much quicker travel of Cu and Ni atoms through the Sn lattice that assisted much faster EM-induced failure for TCB solder joints. Therefore, in this schematic, it is clear that c-axis orientation alone is not sufficient for fully understanding and describing the underlying physics. In addition, grain boundary diffusion is neglected for our analysis, for not only the difference in grain boundary density was similar to one another, but also the diffusion of Cu and Ni atoms through the β-Sn solder matrix is dominated by a bulk diffusion at high temperature [47, 48], as the EM-aging tests were performed at a homologous temperature of 0.8 T_m.

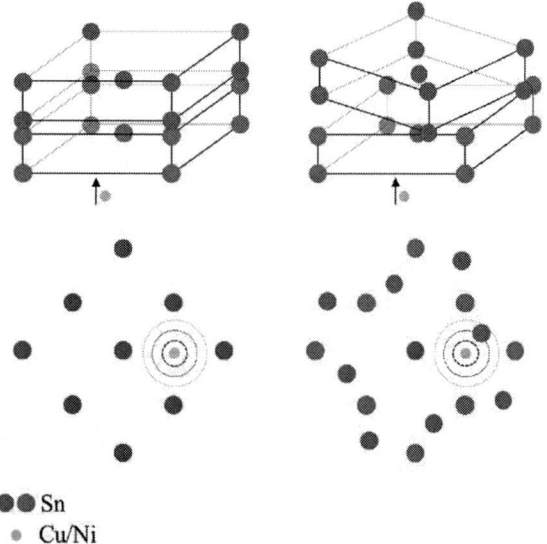

Figure 11. Schematic illustration of interstitial diffusion of Cu and Ni atoms (depicted as gray dot with black arrow) in a β-Sn lattice when a significant tilt boundary is introduced in a lattice structure.

IV. CONCLUSION

The EM performance of TCB and reflow processed solder joints and their microstructure of solder joints has been investigated. The grain morphology, β-Sn orientations, and the IMCs were characterized. This work enhances our understanding on several factors that explain for a premature EM-failure of the solder joints via the sole TCB process. SEM results revealed that the lack of IMC formations in TCB joints. EBSD and its statistically processed results showed that there was no difference in misorientation in respective to c-axis orientation; however, relatively homogenous β-Sn crystallographic textures were observed. Therefore, these factors could have expedited the failure at the interconnection of the TCB processed package upon EM aging test. Further process engineering, i.e. TCB process modification, will be needed to yield the TCB process-3D packages with improved reliability.

ACKNOWLEDGMENT

The authors would like to thank Samsung Electronics for their support for this research.

REFERENCES

[1] S. Lee, Fundamentals of Thermal Compression Bonding Technology and Process Materials for 2.5/3D Packages, pp. 157-203. In: Y. Li and D. Goyal, 3D Microelectronics Packaging, vol. 57, Springer, Cham., 2017.

[2] S. W. Yoon, J. H. Ku, N. Suthiwongsunthorn, P. C. Marimuthu, and F. Carson, "Fabrication and packaging of microbump interconnections for 3D TSV," in 2009 IEEE International Conference on 3D System Integration, 2009, pp. 1-5.

[3] S. W. Lau, "Thermo-Compression Bonding (TCB) for fine-pitch copper pillar flip chip interconnect," in Advanced Packaging Symposium, Taipei, Taiwan, 2014.

[4] A. Eitan and K. Hung, "Thermo-Compression Bonding for fine-pitch copper-pillar flip-chip interconnect - tool features as enablers of unique technology," in 2015 IEEE 65th Electronic Components and Technology Conference (ECTC), 2015, pp. 460-464.

[5] Y. Tomita et al., "Advanced packaging technologies on 3D stacked LSI utilizing the micro interconnections and the layered microthin encapsulation," in 2001 Proceedings. 51st Electronic Components and Technology Conference (Cat. No.01CH37220), 2001, pp. 353-360.

978-1-7281-1500-9/19 $31.00 © 2019 IEEE

[6] J. Jing-Ye et al., "The development of high through-put micro-bump-bonded process with non-conductive paste (NCP)," in 2012 7th International Microsystems, Packaging, Assembly and Circuits Technology Conference (IMPACT), 2012, pp. 114-118.

[7] D. Duffy, C. Gregory, C. Breach, and A. Huffman, "3D and 2.5D packaging assembly with highly silica filled One Step Chip Attach Materials for both thermal compression bonding and mass reflow processes," in 2014 IEEE 64th Electronic Components and Technology Conference (ECTC), 2014, pp. 1803-1809.

[8] C.-L. Liang, K.-L. Lin, and J.-W. Peng, Microstructural Evolution of Intermetallic Compounds in TCNCP Cu Pillar Solder Joints. 2015.

[9] K. Murayama, M. Aizawa, and T. Kurihara, "Low stress bonding for large size die application," in 2015 IEEE 65th Electronic Components and Technology Conference (ECTC), 2015, pp. 1846-1853.

[10] K. Murayama, M. Aizawa, and T. Kurihara, "Study of Crystal Orientation and Microstructure in Sn-Bi and Sn-Ag-Cu Solder with Thermal Compression Bonding and Mass Reflow," in 2016 IEEE 66th Electronic Components and Technology Conference (ECTC), 2016, pp. 909-916.

[11] C. Chen, H. M. Tong, and K. N. Tu, "Electromigration and Thermomigration in Pb-Free Flip-Chip Solder Joints," Annual Review of Materials Research, vol. 40, no. 1, pp. 531-555, 2010.

[12] B. Ebersberger, R. Bauer, and L. Alexa, "Reliability of lead-free SnAg solder bumps: influence of electromigration and temperature," in Proceedings Electronic Components and Technology, 2005. ECTC '05., 2005, pp. 1407-1415 Vol. 2.

[13] S. Härter, R. Dohle, A. Reinhardt, J. Goßler, and J. Franke, Reliability Study of Lead-Free Flip-Chips with Solder Bumps Down to 30 µm Diameter. 2012.

[14] M. Lu, D.-Y. Shih, P. Lauro, C. Goldsmith, and D. W. Henderson, Effect of Sn grain orientation on electromigration degradation mechanism in high Sn-based Pb-free solders. 2008, pp. 211909-211909.

[15] J. Cannis, "Green IC Packaging," Advanced Packaging, vol. 8, p. 33, 2001.

[16] A. U. Telang et al., Grain-boundary character and grain growth in bulk tin and bulk lead-free solder alloys. 2004, pp. 1412-1423.

[17] S. Terashima, K. Takahama, M. Nozaki, and M. Tanaka, Recrystallization of Sn Grains due to Thermal Strain in Sn1.2Ag0.5Cu0.05Ni Solder. 2004, pp. 1383-1390.

[18] Y. Wang et al., Effects of Sn grain structure on the electromigration of Sn-Ag solder joints. 2012, p. 1131.

[19] B. Zhou, T. Bieler, T.-K. Lee, and W. Liu, Characterization of Recrystallization and Microstructure Evolution in Lead-Free Solder Joints Using EBSD and 3D-XRD. 2012.

[20] B. Chao et al., "Electromigration enhanced intermetallic growth and void formation in Pb-free solder joints," Journal of Applied Physics, vol. 100, no. 8, p. 084909, 2006.

[21] Bashir and A. S. M. A. Haseeb, Improving mechanical and electrical properties of Cu/SAC305/Cu solder joints under electromigration by using Ni nanoparticles doped flux. 2017.

[22] T. R. Bieler, H. Jiang, L. P. Lehman, T. Kirkpatrick, and E. J. Cotts, "Influence of Sn grain size and orientation on the thermomechanical response and reliability of Pb-free solder joints," in 56th Electronic Components and Technology Conference 2006, 2006, p. 6 pp.

[23] K. Murayama, M. Higashi, T. Sakai, and N. Imaizumi, "Electro-migration behavior in low temperature flip chip bonding," in 2012 IEEE 62nd Electronic Components and Technology Conference, 2012, pp. 608-614.

[24] N. Zhao, Y. Zhong, W. Dong, M. L. Huang, H. Ma, and C. P. Wong, Formation of highly preferred orientation of β-Sn grains in solidified Cu/SnAgCu/Cu micro interconnects under temperature gradient effect. 2017, p. 093504.

[25] F. Ochoa, X. Deng, and N. Chawla, Effects of cooling rate on creep behavior of a Sn-3.5Ag alloy. 2004, pp. 1596-1607.

[26] M. Mueller, S. Wiese, M. Roellig, and K. Wolter, "Effect of Composition and Cooling Rate on the Microstructure of SnAgCu-Solder Joints," in 2007 Proceedings 57th Electronic Components and Technology Conference, 2007, pp. 1579-1588.

[27] H.-t. Lee and K.-C. Huang, Effects of Cooling Rate on the Microstructure and Morphology of Sn-3.0Ag-0.5Cu Solder. 2015.

[28] B. F. Dyson, T. R. Anthony, and D. Turnbull, "Interstitial Diffusion of Copper in Tin," Journal of Applied Physics, vol. 38, no. 8, pp. 3408-3408, 1967.

[29] D. C. Yeh and H. B. Huntington, "Extreme Fast-Diffusion System: Nickel in Single-Crystal Tin," Physical Review Letters, vol. 53, no. 15, pp. 1469-1472, 1984.

[30] F. Bachmann, R. Hielscher, and H. Schaeben, Texture Analysis with MTEX—Free and Open Source Software Toolbox. 2010.

[31] H. J. Bunge, Texture Analysis in Materials Science: Mathematical Methods. Elsevier, 2013.

[32] Y. Yabansu, D. K. Patel, and S. Kalidindi, Calibrated localization relationships for elastic response of polycrystalline aggregates. 2014, pp. 151–160.

[33] Y. Yabansu and S. Kalidindi, Representation and calibration of elastic localization kernels for a broad class of cubic polycrystals. 2015, pp. 26-35.

[34] N. H. Paulson, M. W. Priddy, D. L. McDowell, and S. Kalidindi, Reduced-order structure-property linkages for polycrystalline microstructures based on 2-point statistics. 2017.

[35] M. W. Priddy, N. H. Paulson, S. Kalidindi, and D. L. McDowell, Strategies for rapid parametric assessment of microstructure-sensitive fatigue for HCP polycrystals. 2017.

[36] R. Liu, Y. C. Yabansu, A. Agrawal, S. R. Kalidindi, and A. N. Choudhary, "Machine learning approaches for elastic localization linkages in high-contrast composite materials," Integrating Materials and Manufacturing Innovation, vol. 4, no. 1, p. 13, 2015.

[37] B. Efron, "Bootstrap Methods: Another Look at the Jackknife," The Annals of Statistics, vol. 7, no. 1, pp. 1-26, 1979.

[38] M. E. Johnson, L. Moore, and D. Ylvisaker, Minimax and Maximin Distance Designs. 1990, pp. 131-148.

[39] P. Darbandi, T. R. Bieler, F. Pourboghrat, and T.-k. Lee, "The Effect of Cooling Rate on Grain Orientation and Misorientation Microstructure of SAC105 Solder Joints Before and After Impact Drop Tests," Journal of Electronic Materials, vol. 43, no. 7, pp. 2521-2529, 2014.

[40] D. W. Henderson et al., "Ag3Sn plate formation in the solidification of near ternary eutectic Sn–Ag–Cu alloys," Journal of Materials Research, vol. 17, no. 11, pp. 2775-2778, 2002.

[41] S. K. Kang et al., "Formation of AgSn plates in Sn-Ag-Cu alloys and optimization of their alloy composition," in 53rd Electronic Components and Technology Conference, 2003. Proceedings., 2003, pp. 64-70.

[42] S. K. Kang et al., "Interfacial reactions of Sn-Ag-Cu solders modified by minor Zn alloying addition," Journal of Electronic Materials, vol. 35, no. 3, pp. 479-485, 2006.

[43] J. L. a. M. Thiele, Fundamentals of Electromigration-Aware Integrated Circuit Design, 1 ed. Springer International Publishing, 2018.

[44] R. Kinyanjui, L. P. Lehman, L. Zavalij, and E. Cotts, Effect of Sample Size on the Solidification Temperature and Microstructure of SnAgCu Near Eutectic Alloys. 2005, pp. 2914-2918.

[45] L. P. Lehman et al., "Microstructure and Damage Evolution in Sn-Ag-Cu Solder Joints," in Proceedings Electronic Components and Technology, 2005. ECTC '05., 2005, pp. 674-681.

[46] S. O. Kasap, "Electrical and Thermal Conduction in Solids," in Principles of Electronic Materials and Devices3 ed.: McGraw-Hill Education, 2006, p. 126.

[47] J. Haimovich and A. Incorporated, Cu-Sn Intermetallic Compound Growth in Hot-Air-Leveled Tin at and below 100°C. 1993.

978-1-7281-1500-9/19 $31.00 © 2019 IEEE   2357

[48] R. Labie, W. Ruythooren, and J. Van Humbeeck, "Solid state diffusion in Cu–Sn and Ni–Sn diffusion couples with flip-chip scale dimensions," Intermetallics, vol. 15, no. 3, pp. 396-403, 2007.

Reduction of Ag Corrosion Rate during Decapsulation of Ag Wire Bond Packages

Yong-Ja Kim[1], Jinho Hah[2], Kyoung-Sik (Jack) Moon[2], and C. P. Wong[2]

[1]Package Engineering Team, Test & Package Center, Samsung Electronics, Asan-si, Chungcheongnam-do, Korea
[2]School of Materials Science and Engineering, Georgia Institute of Technology, Atlanta, GA, USA
*Email: cp.wong@mse.gatech.edu

Abstract—Ag wire-bonded packages have gained a lot of interest as a low-cost substitute material in lieu of Au wire-bonded packages. However, selective decapsulation still remains as challenges due to a corrosion on Ag wires, where Ag has a strong tendency to form a water-soluble Ag salt upon reaction with a conventional nitric acid etchant. This paper serves to present a chemical solution that can reduce corrosion on Ag wires during decapsulation process of the Ag wire-bonded packages. Also, our method is applicable for industry standard needs such as for decapsulation at high temperature.

Keywords-corrosion rate; decapsulation; morphologies; etching; Ag wires;

I. Introduction

In microelectronics package technologies, interconnection fabrication is one of the most important processes in 2.5D and 3D packages. One of the first-level interconnection technologies in microelectronics packages is wire bonding (i.e. sequential bonding), which is a conventional interconnection process for packages that bridges IC chips and substrates using metallic wires. Specifically, these first-level interconnections are formed using Au or Cu wires that bridge between the die and substrate pads via local heat from ultrasonic vibrations. Although wire bonding was first introduced in the late 20th century, it is still a core technology today [1] because it can provide the best solution for stacking dies for high-density and high-throughput memory device packages (flash drives and smart phones) by integrating more functionality in the same footprint [2, 3]. Therefore, wire bonding is still in-practice as an interconnection fabrication method in modern packages on devices requiring a high density interconnection in micro-electronics packaging.

Metallic wires utilized in wire bonding technology are Ag, Au, Cu, and CuPd alloys [4]. Recently, the market share of materials for wires have been replacing Au wires for Ag wires and other metals to meet the requirement of low cost as the price of Au continues to increase [5] and due to comparable properties [6–8] between the two metals. In addition, Ag has the highest electrical conductivity among all metals in the periodic table of elements. To closely match the material properties to those of Au, the percentage of Ag in metal alloy for wires has been adjusted using Au and Pd. Excellent reliability performance of these Ag alloy wire bonding is reported [6–10].

Because wire bond packages are still in their usages, reliability analysis and characterization of those packages are also performed in parallel. Decapsulation of wire bond packages is a crucial process that characterizes for any presence for package failures. Leaving the wires intact in its packages after removal of polymer encapsulants such as epoxy molding compounds (EMCs) is a critical procedure. The decapsulation procedure is practiced by both physical and chemical means, while the standard method [11, 12] utilizes fuming nitric acid via nitration reaction at high temperature to break down strong C–O–C crosslinks [1] present in rigid EMCs.

There have been several studies to selectively etch EMCs from Au and Cu wired packages; however, very few studies have been reported in respective to the decapsulation of Ag wire bond packages. We believe so because Ag wires just recently have been replacing Au and Cu wires, and there is a limited understanding of the chemistry route of this new Ag wire degradation mechanism using conventional acids used for Au and Cu wires. Until now, attempts have been made to successfully decapsulate Ag alloy wire bonded devices via physical means using laser ablation [13]. Decapsulation by physical methods such as laser ablation and plasma are usually practiced and have yielded more successful decapsulation results [13]. Yet, conventional industry standard decapsulation processes by wet chemical etching methods using fuming nitric acid from previous work have failed to selectively etch only EMCs without damaging Ag wires or corroding bond pads [13, 14]. Considerable amount of Ag wire is etched out when using industry standard decapsulation acids, such as nitric acid and sulfuric acid. Etching process of Ag by nitric acid occurs as Ag metal reacts with an acid to precipitate nitrate as shown in (1).

$$Ag + 2\,HNO_3 \rightarrow AgNO_3 + NO_2 \uparrow + H_2O. \qquad (1)$$

A study on reducing the corrosion rate of Ag in nitric acid via using a silver nitrate saturated nitric acid solution has been reported [14]. Suzuki and his co-worker used $AgNO_3$ as a saturation media on fuming nitric acid for their decapsulation at 60 °C, which impeded the Ag corrosion reaction showed (1) [14]. However, this saturation point is elevated at standard decapsulation temperature, 240 °C; therefore, system and process developments are required.

Moreover, when chemical acid mixture between fuming HNO_3 and KI/I_2 (30:1 ratio) was used, no damage was reported on Ag wires and bonding pads [15, 16]; however, this method utilizes one of environmentally unfavorable halogens and was not decapsulated at standard 240 °C.

This paper reports a novel approach using Cu salt-acid mixture that significantly retards Ag corrosion rate during the decapsulation. Also, we systematically studied the chemical reactions responsible for individual decapsulation process by analyzing the microstructure of Ag and performing elemental composition analyses. We present the proposed chemical route by which Ag corrosion rate can significantly be reduced and provide a methodology to leave Ag wires intact in wire bond packages.

II. EXPERIMENTAL SECTION

A. Materials

Ag foil (Takeuchi Metal, 99.99% Japan) with thickness of 0.03 mm was cut into 20 mm x 20 mm in dimension for all silver foil samples utilized in this study. Fuming nitric acid, > 99% purity was obtained from Sigma Aldrich, while diluted nitric acid, 69-70% was purchased from VWR. Copper (II) nitrate hydrate ($CuNO_3$), 99.999% was obtained from Sigma Aldrich, and silver nitrate, S181-25, ($AgNO_3$) in crystalline form was obtained from Fisher Chemical. These metal nitrate salts were used and mixed with nitric acid without further purifications. DI-water was in-house purified using water purification system.

B. Nitric Acid Etchant for Decapsulation and its Process

Three different nitric acid concentrations (50%, 70%, and 99%), each 10 mL, were prepared using fuming nitric acid (> 99% purity) and DI water. Individual Ag foil, as prepared, was fully immersed and reacted inside the 20 mL scintillation glass vial for 10 minutes at room temperature condition. Weight of an individual Ag foil was measured before and after the decapsulation process.

C. Metal Nitrate-based Nitric Acid Etchant for Decapsulation and its Process

The nitrate-based nitric acid solutions were prepared in the following way. 0.10 g of two types of transition metal (II) nitrate (MNO_3) were added to 1 mL of fuming nitric acid (> 99% purity) to formulate metal nitrate nitric acid solutions ($AgNO_3$ + HNO_3 and $CuNO_3$ + HNO_3). After vortex mixing for 30 minutes, the Ag foil, as prepared, was fully immersed and reacted inside the 20 mL scintillation glass vial for 10 minutes at room temperature condition. 1 mL of DI-water was dropped on top of etched Ag foil for cleaning process. In addition, 0.01 g, 0.05 g, and 0.10 g were individually mixed with 1 mL of HNO_3 to produce three different $CuNO_3$ + HNO_3 acid solutions.

D. Chemical Wet Decapsulation at High Temperature using bare Nitric Acid and $CuNO_3$-based Nitric Acid

The procedure for $CuNO_3$-based nitric acid solution is similar to that used for a high temperature chemical wet decapsulation. Two 20 mL scintillation glass vials, each containing bare nitric acid and $CuNO_3$-based nitric acid solution, were placed on the hot plate, pre-heated at 240 °C. This decapsulation process was practiced for 10 minutes. Cleaning process using 1 mL of DI-water was then followed. Image segmentation and meshing using ImageJ were processed to calculate regions of Ag foil affected from the high temperature decapsulation.

E. Ag Foil Surface Characterization

The surfaces of Ag foil were characterized using scanning electron microscopy (SEM) for morphology analyses and energy-dispersive X-ray spectroscopy (EDX) for elemental composition analyses. SEM images were obtained using Hitachi SU8230 FE-SEM with an accelerating voltage of 1 kV and a working distance between 16.0 mm to 16.7 mm using a secondary electron (SE) detector and Tescan Mira FE-SEM with an accelerating voltage of 10 kV and a beam intensity of 15.00 µm using a back-scattered electron (BSE) detector. EDX images were obtained using X-Max^N X-ray detector (Oxford Instruments)-equipped Hitachi SU8230 FE-SEM with an accelerating voltage of 30 kV and a working distance between 14.8 mm to 15.0 mm. AZtec software was used for EDX mapping and analyses.

F. Decapsulation with our Optimized Acid Solution

EMC encapsulated Ag wire bonded package was decapsulated using our optimized acid recipe, $CuNO_3$-based nitric acid solution (0.1 g of $CuNO_3$ with 1 mL of HNO_3) at 240 °C, and its performance was compared with the same test vehicle with conventional decapsulation solution, 1 mL of HNO_3. The visual inspection of the packages was made by using Tescan Mira FE-SEM with an accelerating voltage of 10 kV and a beam intensity of 5.00 um using a back-scattered electron (BSE) detector and optical microscopy using combination of video microscope and Zeiss-equipped camera.

III. RESULTS AND DISCUSSION

A. Chemical Wet Decapsulation with Varying Nitric Concentrations

Temperature and acid concentration are two main factors that affect the decapsulation process and its rate. One of these factors, acid concentration, is investigated to determine the reaction parameter of etching rate on Ag foil using three different nitric acid concentrations, 50% HNO_3, 70% HNO_3, and 99% HNO_3. Table I shows total amounts of an Ag foil that underwent through the decapsulation reaction and the etching rate per three different nitric acid concentrations.

TABLE I. ETCHING RATE OF AN AG FOIL AT THREE DIFFERENT NITRIC ACID CONCENTRATIONS.

Concentration of HNO_3 (%)	50	70	99
Weight of Ag foil before Decapsulation (g)	40.25	40.38	41.28
Weight of Ag foil after Decapsulation (g)	0.00	23.70	40.75
Weight Change of Ag (g)	40.25	16.68	0.53
Etching Rate of Ag (g/min)	4.025	1.668	0.053

The increase in the nitric acid concentration decreased the etching rate of Ag: whereas the etching rate of the Ag foil reacted under 50% HNO_3 was 4.025 g/min, that under 99% HNO_3 was only 0.053 g/min. Therefore, the Ag foil was the least corroded at the highest HNO_3 concentration. The decapsulation reaction that utilizes nitric acid (HNO_3) on Ag metal consumes pure Ag metal to precipitate silver nitrate ($AgNO_3$) salt crystals as previously stated in (1); however, it is still unclear why the etching rate of Ag metal decreases with respect to an increase in nitric acid concentration. Yet, we hypothesized this behavior seen in Table I and compared with (1) that silver nitrate salt crystals ($AgNO_3$) deposit on the Ag foil surface have played a significant role in delaying HNO_3 etching further through the pristine layer of Ag foil.

The surfaces of the Ag foils were compared with and without the cleaning procedure upon an etching procedure to understand the stability and role of $AgNO_3$ and to visualize the surface morphologies of the Ag foil after the etching reaction. As illustrated in Fig. 1a, a granular layer is formed on the surface of the parent Ag foil before the cleaning process. However, after the cleaning process, the granular layer of $AgNO_3$ was rinsed along with DI water as shown in Fig. 1b. This phenomenon suggests that granular layer of $AgNO_3$ is water-soluble. Therefore, a layer of silver nitrate salts formed on the surface of Ag foil at high nitric acid concentration (99%) acts as a barrier layer to prevent additional etching reaction of parent Ag foil, beneath the silver nitrate salts, which accounts for an extremely low corrosion rate of 0.053 g/min. In addition, corrosion of Ag foil is intensified at a lower nitric acid concentration due to a water-soluble characteristic of $AgNO_3$ salts. Hence, it can be assumed that the anhydrous nature of fuming nitric acid at 99% concentration did not affect the dissolution of the $AgNO_3$ barrier layer.

B. Reactivity of Ag Foil to Nitrate-based Nitric Acid

From the previous section, it was observed that corrosion of the parent Ag foil could be impeded by an $AgNO_3$ barrier layer, which occurs when using anhydrous nitric acid. Second part of the results section explores the effect of different types of nitrate-based acid on overall surface morphology and the surface composition of the Ag foil to optimize the best condition that minimizes the corrosion on Ag foil. We have utilized two types of nitrate-salts dissolved nitric acids, $AgNO_3$ + HNO_3 and $CuNO_3$ + HNO_3. Dimensional changes of pure Ag foils were also measured when each was individually reacted with two types of metal nitrate-based nitric acid. Using copper nitrate-based nitric acid ($CuNO_3$ + HNO_3) as the decapsulation solution has shown most resistive to etching. While the Ag foil reacted with $CuNO_3$ has almost maintained its original dimension, those reacted with any others were severely damaged by the solutions.

Similarly, each metal nitrate-based nitric acid was characterized by SEM upon the cleaning process. In all cases, the surface grains (silver nitrate salts) formed on the parent Ag foil surface, and those grains were completely rinsed during cleaning process. The morphology of the salt grains, when used HNO_3 and $AgNO_3$ + HNO_3 as the etchants, exhibits blunt 3D grain textures and is distributed similar in size (Fig. 1a and 1b). However, those of the salt grains are much larger in size, and these salt grains have increased aspect ratios (Fig. 1c) when $CuNO_3$ + HNO_3 was used as an etchant. When the average grain sizes were compared between Fig. 1a and Fig. 1c, the former (~5,000 μm^2) was five times smaller than the latter (~25,000 μm^2). In addition, a new material phase was observed at the interface between the grains as well as on top of the parent Ag surface in Fig. 1c. Specifically, surface morphology in Fig. 1c rather possesses a much smoother texture on the area where it is not filled with $AgNO_3$ salts, unlike the porous and rough surface structure of the parent Ag film. To further understand this phase, EDX maps were analyzed to confirm the hypothesis that $CuNO_3$ + HNO_3 as an etchant for decapsulation generates a different surface morphology and introduces a new material phase.

From Fig. 2, EDX results also confirm that grains are composed of elements consisting Ag, N, and O, which is the $AgNO_3$ particles precipitated by the reaction from (1). The grains precipitated by reaction using any metal nitrate-based nitric acid are determined to be $AgNO_3$. In addition, the parent Ag foil is not covered with any other new elements for the sample that is reacted only with pure HNO_3, while Ag foil that reacts with the solution (mixture of copper nitrate and nitric acid) precipitated copper (II) nitrate, $(CuNO_3)_2$, between the grain boundaries, at the interface, and on the regions not covered with $AgNO_3$ grains (Fig. 2). Therefore, the smooth textured layer from Fig. 1(c) was determined to be a water-insoluble $Cu(NO_3)_2$, instead of pure silver from the parent Ag foil.

The least amount of corrosion on Ag was confirmed when copper nitrate nitric acid was used as an etchant due to the formation of a new layer at regions not covered with silver nitrate salts. Due to these phenomena, a supplementary set of experiment was prepared to investigate the effect of $CuNO_3$-based nitric acid concentration on the corrosion behavior of Ag. Qualitative results indicated that after submerging the Ag foil into the $CuNO_3$ acid, the color of the Ag foil became the darkest at highest concentration, and the Ag foil dipped into the acid with the lowest concentration dissolved the fastest upon cleaning process.

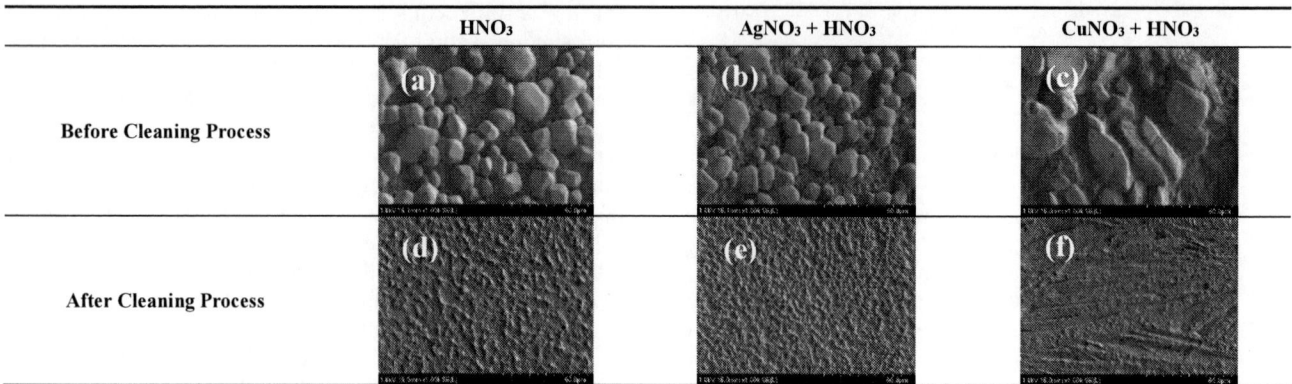

Figure 1. SEM images of Ag foil surface, where (a) – (c) are images taken after the reaction with different types of metal nitrate-based solution, followed by a cleaning process (d) – (f) for 10 minutes.

	Electron Image	Oxygen (O)	Nitrogen (N)	Silver (Ag)	Copper (Cu)
HNO₃					
CuNO₃ + HNO₃					

Figure 2. EDX elemental composition map of Ag foil surface showing different surface morphologies after decapsulation using (top) HNO_3 and (bottom) $CuNO_3$ + HNO_3.

Fig. 3 illustrates several trends based on SEM image analysis per three different $CuNO_3$-based nitric acid concentrations. Surface area of individual silver nitrate salt particle increases with $CuNO_3$-based nitric acid. Therefore, the increase in area fraction covered by $AgNO_3$ grains was observed at the highest solution concentration. The change in grain size was most noticeable between Fig. 3a and 3b. Once the $CuNO_3$ was added up to 0.05 g to increase the solution concentration, saturation point in grain size was reached (Fig. 3b and 3c). However, a new $Cu(NO_3)_2$ layer was observed, surrounding the interface between the grains and the grain boundaries as signified by red arrows in Fig. 3b and Fig. 3c. This $Cu(NO_3)_2$ layer becomes thicker, more compact, and more evident when the highest amount of $CuNO_3$ is added to the etchant, therefore covering the parent Ag foil, which could have protected the pure Ag surface from further etching induced by acid. These analyses are also supported from chemical reaction equations stated from (2) to (4). The reaction of Ag and nitric acid in (1) is exothermic, and with this generated heat will react with $Cu(NO_3)_2$ to form CuO as shown in (2).

$$2\ Cu(NO_3)_2 \rightarrow 2\ CuO + 4\ NO_2 \uparrow + O_2. \qquad (2)$$

This CuO phase is further reacted with the already present nitric acid to produce copper nitrate back as stated in (3a); however, it also reacts with $AgNO_3$ from (1) and yields Ag_2O as stated in (3b).

$$CuO + 2\ HNO_3 \rightarrow Cu(NO_3)_2 + H_2O. \qquad (3a)$$
$$CuO + 2\ AgNO_3 \rightarrow Cu(NO_3)_2 + Ag_2O. \qquad (3b)$$

Finally, precipitated Ag_2O in (3b) is further reacted with the nitric acid solution to generate more $AgNO_3$ as stated in (4). The consumption of CuO to generate $AgNO_3$ explains the growth of $AgNO_3$ particles as seen in Fig. 3a to Fig. 3c because the higher concentration of $Cu(NO_3)_2$ fuels the reaction pathways from (2) to (4).

$$Ag_2O + 2\ HNO_3 \rightleftharpoons 2\ AgNO_3 + H_2O. \qquad (4)$$

Therefore, the Ag foil is eventually covered by $AgNO_3$ particles and the interfaces between these particles are covered by $Cu(NO_3)_2$, therefore protecting the Ag foil from direct etching at the highest concentration of $CuNO_3$-based nitric acid solution. Fig. 4 illustrates the reaction pathway expressed in above equations.

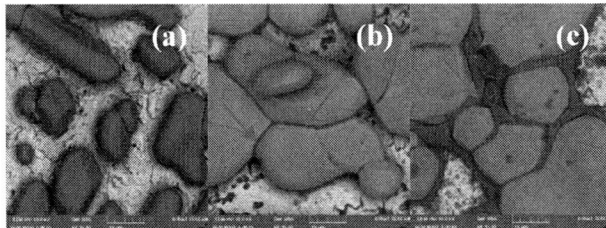

Figure 3. SEM images of surface of Ag foil showing different surface morphologies at 5000X magnification per varying CuNO₃-based acid solution concentration for chemical wet decapsulation: (a) 0.01 g of CuNO₃ + 1 mL HNO₃, (b) 0.05 g of CuNO₃ + 1 mL HNO₃, and (c) 0.10 g of CuNO₃ + 1 mL HNO₃. New layers of Cu(NO₃)₂ on the grain boundaries are denoted by red arrows.

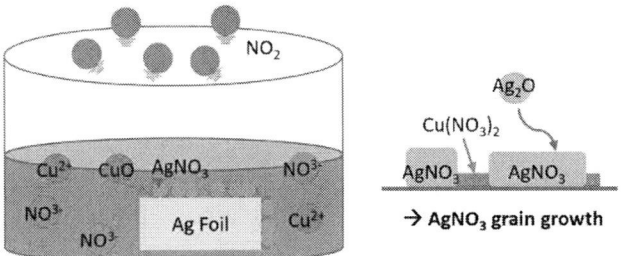

Figure 4. Schematic showing grain growth of AgNO3 in the presence of Cu(NO₃)₂ from the reaction mechanism between Ag foil and CuNO₃-based nitric acid solution.

C. Chemical Wet Decapsulation at High Temperature using bare Nitric Acid and CuNO₃-based Nitric Acid

Temperature is another main factor that affects the corrosion (decapsulation) rate, and decapsulation process is performed at high temperatures (e.g. 240 °C or higher) in many semiconductor industries to expedite its process. This section investigates the effect of temperature on corrosion rate. Decapsulation processes were performed with Ag foils at 240 °C using a CuNO₃-based nitric acid decapsulation etchant. Two Ag foils were submerged in each solution bath (HNO₃ and CuNO₃ + HNO₃).

When the corrosion reaction was occurred at an elevated temperature of 240 °C, both Ag foils released more NO₂ gas than those did at room temperature; however, both Ag foils did not change significantly in their sizes. It was observed that a pure Ag foil began to dissolve from the outer edges to inwards and decreased in size within 3 minutes in HNO₃ upon a cleaning process; therefore, an Ag foil was able to maintain its original shape but suffered from its size change. As a result, the remaining area of the Ag foil (61.95 mm²) was calculated to be 33.9% of the original dimension (182.69 mm²) using an ImageJ processing tool.

D. Decapsulation with our Optimized Solution

Throughout the results, we have optimized the solution for decapsulation and chose the solution chemistry that has the lowest corrosion rate on Ag. Using this optimized CuNO₃ + HNO₃ solution, the test vehicle was decapsulated at normal decapsulation temperature (240 °C).

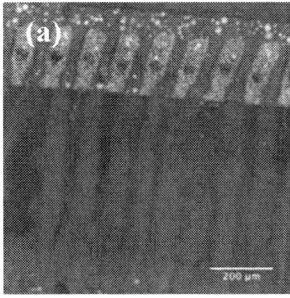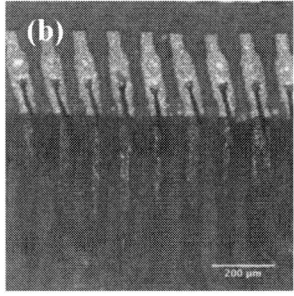

Figure 5. Optical microscope images of decapsulated Ag-wire bond package vehicle using (a) conventional and (b) optimized solution etchant.

The optimized acid solution demonstrated its performance, as can be seen in Fig. 5. When conventional HNO₃ acid was used for decapsulation, no Ag wires were left on the bond pads (Fig. 5a), but CuNO₃ + HNO₃ solution was able to selectively decapsulate the package and leave the Ag wires intact (Fig. 5b).

IV. CONCLUSION

A systematic study was conducted to retard a corrosion speed on Ag metal during a decapsulation process of Ag wire-bonded packages. We investigated key factors of the selective decapsulation for Ag-wire packages and reported a chemical etchant using a CuNO₃–nitric acid solution. This new approach demonstrated an improvement to the order of three of the Ag damages compared to the conventional HNO₃ solution even at an elevated temperature. We have confirmed our proposed chemistry route of this new method that delays the Ag damage in CuNO₃-based nitric acid due to a formation of a unique barrier layer consisting of enlarged AgNO₃ particles and Cu(NO₃)₂ that fills between the salt particles throughout the entire surface. In summary, our new decapsulation method using CuNO₃-based nitric acid serves to represent and meet the industry standard needs of cost-effectiveness, environmentally-friendliness, and the high temperature decapsulation application.

ACKNOWLEDGMENT

The authors would like to thank Samsung Electronics for the financial support on this project.

REFERENCES

[1] B. L. Gehman. Bonding Wire Microelectronic Interconnections. *IEEE Transactions on Components, Hybrids, and Manufacturing Technology (CHMT).* 1980, 3, 375-383.

[2] H. Nakanishi, T. Maruyama, K. Miyata, T. Ishio, Y. Sota, A. Narai, S. Fukunaga, K. Toyosawa, K. Fujita, M. Kada. Development of High Density Memory IC Package by Stacking IC Chips. *IEEE Transactions on Electronic Components and Technology Conference (ECTC).* 1995, 635-640.

[3] I. Qin, O. Yauw, G. Schulze, A. Shah, B. Chylak, N. Wong. Advances in Wire Bonding Technology for 3D Die Stacking and Fan Out Wafer Level Package. *IEEE 67th Electronic Components and Technology Conference (ECTC).* 2017.

[4] S. Fan, L. Huang, M. Ho, and K. Hsieh, "The intermetallic compound (IMC) growth and phase identification of different kinds of copper wire and Al pad thickness," in 2015 IEEE 65th Electronic

Components and Technology Conference (ECTC), 2015, pp. 1412-1416.

[5] "Gold Spot Price & Charts." J. M. Bullion, 2018.

[6] T. B. Wei, N. Y. Hua, W. K. Sheng. Silver Alloy Wire for IC Packaging Solution. *IEEE 36th International Electronics Manufacturing Technology Conference (IEMT)*. 2015.

[7] C. H. Cheng, H. L. Hsiao, S. I. Chu, Y. Y. Shieh, C. Y. Sun, C. Peng. Low Cost Silver Alloy Wire Bonding with Excellent Reiliability Performance. *IEEE 63rd Electronic Components and Technology Conference (ECTC)*. 2013.

[8] T. Oyamada, T. Uno, T. Yamada, D. Oda. High-Performance Silver Alloy Bonding Wire for Memory Devices. *IEEE 67th Electronic Components and Technology Conference (ECTC)*. 2017.

[9] A. Lan, J. Tsai, J. Huang, O. Hung. Interconnection Challenge of Wire Bonding – Ag Alloy Wire. *IEEE 15th Electronic Packaging and Technology Conference (EPTC)*. 2013, 504-509.

[10] K. A. Yoo, C. Uhm, T. J. Kwon, J. S. Cho, J. T. Moon. Reliability Study of Low Cost, Alternative Ag Bonding Wire with Various Bond Pad, Materials. *IEEE 11th Electronic Packaging and Technology Conference (EPTC)*. 2009, 851-857.

[11] V. S. Bhide. Decapsulation of Silicon-Epoxy Copolymer Packages. *IEEE 20th International Reliability Physics Symposium (IRPS)*. 1982.

[12] S. W. Ng, H. B. Zhang, K. N. Liew, W. Lee, R. D. Lin. Copper Wire Bond Package Decapsulation Technique using Mixed Acid Chemistry. *IEEE 19th International Symposium on the Physical and Failure Analysis of Integrated Circuits (IPFA)* 2012.

[13] F. Kerisit, M. J. Lefevre, B. Domenges, W. Prellier, M. Obein. Decapsulation of Silver-Alloy Wire-Bonded Devices. *IEEE 21st International Symposium on the Physical and Failure Analysis of Integrated Circuits (IPFA)* 2014.

[14] S. Suzuki, M. Yamaguchi. Acid Decapsulation for Silver Wire Bonded Package. *IEEE 22nd International Symposium on the Physical and Failure Analysis of Integrated Circuits (IPFA)* 2015.

[15] M. Obein, F. Berger, P. Poirier. New Decapsulation Methods for ICs with Cu and Ag Wires. *IEEE 17th Electronic Packaging and Technology Conference (EPTC)*. 2015.

[16] H. Wu, Y. Liang, W. Du, S. He, Y. Wang, D. Lei. Study of Silver Alloy Wire Decapsulation Methods for Failure Analysis. *IEEE 24th International Symposium on the Physical and Failure Analysis of Integrated Circuits (IPFA)* 2017.

Author Index

Aasmundtveit, Knut E 141
Abe, Takatoshi 1272
Abhijit, Dasgupta 498
Abrami, Avner 270
Abrol, Amrit 370, 1347
Agarwal, Amit 718
Ahari, Arman 1099
Ahasan, Kawkab 1860
Ahmed, Omar 1106
Ahmed, Sudan 1347
Ahn, Geun-Sik 197
Akahoshi, Tomoyuki 1294
Akashi, Takahiro 1022
Akazawa, Miyuki 94
Akejima, Shuzo 1140
Akiyama, Kentaro 1641
Alam, Arsalan 277
Alam, Mohammad S. 1815, 1958
Albertinetti, Andrea 2091
Albrecht, John 948
Alexandre, Giry 1279
Alhendi, M. 1946
Alhendi, Mohammed 768, 1581
Ali, Muhammad 960
Alizadeh, Azar 1581
Allain, Fabienne 1622
Allouti, Nacima 569
Alvarez, Claudio 1672
Amandine, Jouve 225
Amano, Takeru 2042
Ambat, Rajan 515
Ambhore, Pranav 620, 800, 1605
Amiran, Johnny 995
An, Sang-Ho 204
Anai, Kei 76
Andersson, Rickard 1870
Andreini, Antonio 2194
Andriani, Yosephine 1543

Antoniou, Antonia 655
Antretter, Thomas 1509, 2029
Apriyana, Anak Agung Alit 1735
Araki, Hitoshi 346
Araki, Naoko 1002
Araki, Noritoshi 175
Arayama, Chika 1022
Argoud, Maxime 569
Arnaud, Lucile 569, 1926
Arnold, Kim 340
Aschenbrenner, R. 861
Aslani Amoli, Nahid 249, 1939
Assous, Myriam 1622
Audet, Jean 1179
Aue, Maximilian 1883
Auer, Benedikt 1789
Aumont, Christophe 569
Aygun, Kemal 667
Ayssar, Serhan 1279
Ayukawa, Michael 2117
Azizi, Arad 1970
Azzopardi, Stephane 2173
Baelmans, Martine 126
Bajwa, Adeel Ahmed 620
Bakir, Muhannad S. 1803
Balaraman, Devarajan 163
Banerjee, Deepayan 1770
Barth, Maximilian 868
Bartl, Ulf 855
Barut, Atila 825
Barwicz, Tymon 528, 1060
Baty, Greg 1099
Bauwelinck, Johan 1052, 1757
Bea, J.C 1047
Becker, Karl-Friedrich 861
Bécu, Stéphane 168
Bedjaoui, Messaoud 995
Bedoin, Alexis 535

Behr, Andy 1272

Beica, Rozalia 14, 903

Beigné, Edith 1926

Bejugam, Vinith 47

Bellaredj, Mohamed 1672, 1939

Bellaredj, Mohamed L. F. 249

Beltritti, Jérôme 569

Bengsch, Sebastian 1883

Berger, Frédéric 569

Bernstein, Gary H. 2072

Bertheau, Julien 340

Bex, Pieter 340, 607, 674

Beyer, Gerald 340, 607, 1035, 2206

Beyne, Eric 126, 340, 437, 607, 674, 1035, 1215, 2206

Bharath, Krishna 2180

Bhattacharya, Surya 587, 917

Bhuvanendran Nair Gourikutty, Sajay 1227

Bicer, Mehmet 1622

Bilgen, Halim 1926

Billard, Christophe 1680

Biscarrat, Jérôme 168

Bito, Jo 896

Blachier, Denis 168

Blackshear, Edmund 1179

Blattau, Nathan 2103

Blecker, Ken 505

Boddu, Vijaya 2180

Bolkhovsky, Vladimir 1611

Bonam, Satish 2156

Boo, Hyunpil 277

Borgesen, P. 1946

Borgesen, Peter 768, 1581

Bouillard, Boris 428

Bowrothu, Renuka 695, 983, 2085, 2337

Bozano, Luisa 417

Brandl, Elisabeth 1789

Brandstätter, Birgit 1789

Braun, Tanja 363, 861, 861

Bravin, Julian 1789

B. Reddy, Vishnu V. 1333

Bruckner, Gudrun 878

Bruderer, Alex 726

Brun, Jean 995

Bu, Lin 1152, 1419, 1735

Butylkov, Sergey 1312

Bylund, Maria 1870

Byong Jin, Kim 1457

Cahu, M. 1339

Cai, Biao 1200

Cai, Jian 69, 1306, 2234

Cain, Stephen R. 2150, 2331

Campos, Didier 569

Cao, Liqiang 2318

Cao, Xinpei 753

Capecchi, Simone 910

Caplet, Stéphane 535

Carlsson, Mats 968

Cartier, Mathilde 1535

Castagné, Laetitia 1575

Castan, Clément 225

Cecchetto, Luca 2194

Chacon, Oswaldo 1396

Chahal, Premjeet 113, 948, 1240, 1687

Chai, Tai Chong 21, 1419

Chambion, Bertrand 535

Chan, Alex 1272, 1902

Chan, Chin-Wei 1751

Chandrasekhar, Arun 2180

Chang, Chia-Cheng 758, 1328

Chang, Chieh-Lin 14

Chang, Grace 1595

Chang, K.C. 763

Chang, Keng Tuan 41

Chang, Kuo-Chin 397

Chang, Megan 2079

Chang, Tao-Chih 1463

Chang, Victor C. Y.	688
Chang, Yu-Chen	1359
Chang Chien, Chien-Lin	461
Chao, Tz-Yuan	1170
Charbonnier, Jean	569, 1622
Charles, Matthew	168
Chausse, Pascal	569
Che, Fa Xing	842
Che, F. X.	1152, 2126
Chen, Allenyl	1426
Chen, Chen	661
Chen, Cheng-Chih	1826
Chen, Chih	642, 758, 1328
Chen, Chi-Jen	7
Chen, Chi-Yuan	289
Chen, Chuantong	474
Chen, C.S.	931, 1595
Chen, Dao-Long	1710, 1902
Chen, Fang-Cheng	594
Chen, J.H.	1693
Chen, Jie	1221, 1653
Chen, Jing	2061
Chen, J. Y.	700
Chen, Kang	1165
Chen, KarenYU	1710
Chen, Kuan-Neng	1463
Chen, Kuan-Ta	1704
Chen, Liangbiao	1521
Chen, Louis	1426
Chen, Ming-Fa	594
Chen, Qianwen	1246
Chen, Qiaoli	1200
Chen, Rui	249, 1939
Chen, S.	1629
Chen, Shujing	1564
Chen, Si	1324
Chen, Sihai	2186
Chen, Simon	700

Chen, S.M.	931
Chen, Tangsheng	1842
Chen, Tianfang	1983
Chen, Tony	14, 325, 903
Chen, Wei-Han	1877
Chen, Xi-Hong	1359
Chen, Xu	1889
Chen, Yan-Hao	1729
Chen, Y. F.	1175
Chen, YH	903
Chen, Y. H.	14
Chen, Yih-Sin	1170
Chen, Y.N.	1595
Chen, Yu-Hua	1413, 1463
chen, zhaohui	1543
Chen, Zhaoqing	1989
Chen, Zhiwen	63, 850, 1377
Chen, Zihao	917
Cheng, Po Wen	1246
Cheng, Shau-Fei	1877, 1877
Cheng, Ta-Chien	910
Cheng, Wei-Yuan	1877, 2009
Chenhsiu, Sung	1933
Chéramy, Séverine	569
Cheramy, Severine	225, 1926
Cherman, Vladimir	126
Cheung, Y. M.	14, 903
Chew, Ly May	87
Chew, Nam Piau	1735
Chi, Yenyao	1165
Chiang, Jack	2079
Chiang, K.N.	1515
Chiang, W.S.	493
Chien, Feng-Lung	1170
Chien, Feng Lung	1635
Chien, Han	800
Chien, Jason	1902, 2079
Chim, Weng Tuck	1457

Ching, Eva Wai Leong	917
Chiou, Wen-Chih	594
Chiu, Chi-Tsung	1426
Chiu, Jason	763
Chiu, J. M.	1175
Chiu, Julia	453
Chiu, Ryan	700
Chiu, Steve	2009
Chiu, Tz-Cheng	1359
Chiu, Yihsiang	1503
Chiu, Yung-Da	1426
Chiu, Yu-Shan	235
Cho, Cheng-Lin	258
Cho, Jae Kyu	910
Cho, NamJu	289
Cho, Sung-Il	204
Cho, Youngsang	300
Choi, Heejung	300
Choi, Hyun-Seok	204
Choi, Jinu	753
Choi, Kwang-Seong	197
Choi, Kwang Won	1179
Choi, TaeJin	2278
Choi, Won Kyung	968, 1165
Chong, Ser Choong	21, 191
Choong, Chong Ser	2126
Chou, P. H.	1515
Chow, Eugene M.	1312
Chow, Justin H.	785
Chow, Seng Guan	1165
Chowdhury, Md Mahmudur	792
Christie, Leroy	2349
Chu, Fu-Cheng	977
Chuang, Chun-Hsiang	258
Chuang, Oscar	2009
Chuang, Wallace	2067
Chuh, Erich	2180
Chung, C. Key	7, 1287

Chylak, Bob	620
Cibié, Anthony	168
Colonna, Jean-Philippe	168, 1535
Colosimo, Tom	620
Con, Celal	2219
Connolly, Brian	1200
Coquand, Rémi	1622
Coudrain, Perceval	168, 569
Crawford, Lara S.	1312
Cromwell, Kevin	1883
Crump, Cameron	948
Cui, Xiaole	1983
Cyr, Élaine	1074
Daeumer, Matthias A.	1970
Dahl, David	2240
Dahlbäck, Robin	968
Dalmia, Sidharth	294, 954, 1666
Dang, Bing	1246
Danovitch, David	467, 2117, 2252
Das, Rabindra	1611
David, Leslie	498
Day, Doug	1933
Decker, Michael	1509
Dede, Ercan M.	1437
Deep, John	1366
Dehe, Alfons	855
De Heyn, Peter	1757
Del Nero, Daniel	1258, 1848
Deloffre, Emilie	1926
DeProspo, Bartlet	334, 1588
DeProspo, Bartlet H.	924
Deschamps, Jerôme	535
Desmaris, Vincent	1870
Desmet, Andres	1757
De Vos, Joeri	1035
De Wolf, Ingrid	126
Dhandapani, Karthik	819
Dias, Rajen	163

Di Cioccio, Léa 168

Dincau, Brian 1860

Ding, Guifu 707

Ding, Kunpeng 1306

Ding, Qian 1897

Ding, Xuanyi 2186

Docanto, Manuel 1611

Dohi, Kazuhiro 2162

Dolores-Calzadilla, Victor 1060

Dorduncu, Mehmet 825

Dornala, Kalyan 1366

Dreissigacker, Marc 861

Dreps, Daniel 1200

Dressler, Marc 1113

Drouin, Dominique 1396, 2252

Du, Ke 69

Dubis, Monique 87

Dunkel, Christian 1498

Duran-Martinez, Adriana Carolina 1258

Durgun, Ahmet C. 667

Dutoit, Denis 569

Dwarakanath, Shreya 718

Ecker, Melanie 1258

Economou, Manthos 486

Edouard, Deschaseaux 1279

Eichhammer, Yann 1052

Eid, Aline 896

Ekstrom, Noah 753

EL Amrani, Abderrahim 2252

Eleouet, Raphaël 569

Elger, Gordon 2324

El-Mekki, Zaid 1035

Elmogi, Ahmed 1757

En, Yunfei 1324

England, Luke 600

Enomoto, Tetsuya 352

Eom, Yong-Sung 197

Erhart, Andreas 1833

Escoffier, René 168

Eto, Motoki 175

Evans, John 2309

Evans, John L. 792

Evertsen, Rogier 423

Exbrayat, Yorrick 569

Ezawa, Hirokazu 1140

Ezhilarasu, Goutham 277, 1470

Fager, Christian 1405

Fan, Nelson 14, 903

Fan, Xuejun 806

Fan, Zhineng 1200

Fana, Jilei 81

Fang, Bo-Siang 1432, 1704

Fang, Runiu 2168

Fang, Sheng-Po 1647, 1809

Fang, T.J. 931

Fang, Y.H. 493

Farcy, Alexis 569, 1926

Farrugia, M-L 479, 777

Fasoli, Andrea 417

Feng, Qingming 878

Fernandez, Hector 1106

Fernandez, Maïlys 535

Fernandez-Zelaia, Patxi 2349

Fettke, Matthias 47, 210

Feuchter, Michael 2029

Filipp Fuchs, Peter 2029

Finn, Daragh 453

Fischer, Thomas 855

Fischer, Thorsten 1475

Fisher, Daniel 600

Fisher, Timothy 1605

Fisher, Timothy S 277

Fitzgerald, Padraig 1660

Flaim, Tony 1722

Fleischman, Martin 811

Fortier, Paul 1074

Fortin, Clément	306
Fountain, Gill	628, 1041
Fournel, Frank	225
Fowler, Michelle	363
Franiatte, Rémi	225, 1622
Franieck, Erick	811
Fraschke, Mirko	218
Friedmann, T. A.	648
Friedrich, Georg	47, 210
Fritsche, Carola	1475
Fu, Haley	318
Fu, Xing	1324
Fuchs, Peter Filipp	1509
Fuguet Tortolero, César	569
Fujimagari, Junichiro	1641
Fujinaga, Tetsushi	358, 1865
Fujisaki, Hidehiko	1294, 1952
Fujiwara, Atsushi	1641
Fukuda, Takafumi	1140
Fukui, Kei	1294
Fukuomori, Minoru	1451
Fukushima, Takafumi	264, 1047
Furuya, Akira	1067
Gagnon, Pascale	306, 1744
Galbraith, Christopher	1611
Gao, Guilian	628, 1041
Garnier, Arnaud	569
Gaschet, Christophe	535
Geissler, Christian	855
George, Jinto	2117
Gerber, Mark	1902
Gernhardt, Robert	363
Ghannam, Ayad	1789
Ghosh, Tamal	2156
Giesen, Kyle	1200
Gillot, Charlotte	168
Gjokaj, Vincens	948
Glodde, Martin	1060

Goemare, Charlotte	1870
Goggin, Ray	1660
Goller, Bernd	855
Gong, Dan	1503
Goodelle, Jason	28
Goorsky, Mark	1605
Gordon, Seth	2309
Gore, Aaron	453
Gore, Brandon T.	726
Gorrell, Robin	330
Goto, Yoshio	101
Gottardi, Mathilde	569
Gottwald, Thomas	726
Goumans, L.	777
Gourvest, Emmanuel	428
Graap, Pascal	861
Graham, Samuel	1977
Green, Ryan B.	1782
Green, William M.J.	1060
Gromala, Przemyslaw	811, 1529
Gschwandl, Mario	1509, 2029
Gu, Han	243, 1916
Guarino, Lucrezia	2194
Guerrero, Alice	340
Gueugnot, Alain	569
Guevara, Gabe	628
Guidoni, Luca	1735
Guo, Huaixin	1842
Gupta, Sunil	1194, 2097
Gupte, Omkar	1028
Guthmuller, Eric	569
Guthrie, Bill	600
Hagn, Josef	954
Hah, Jinho	1977, 2349, 2359
Hai, Joe	486
Hajjar, Jean-Jacques	1660
Hama, Hiroki	550
Hamasha, Sa'd	792

Hamasha, Sa'd	2309
Han, Bongtae	811, 1382, 1529
Han, Jeong Sam	2246
Han, Kwangwoo	707
Han, Yong	21, 1543
Hanada, Tadahiko	2112
Hanisch, Anke	628
Hanna, Amir	277, 579, 800, 1470
Haque, Mohammad Aminul	2073
Harrison, Todd	1653
Hasegawa, Yasuo	101
Hashimoto, Keika	346
Hashmi, Mohammad	1770
Hassan, KM Rafidh	1815, 1958
Haumesser, Paul-Henri	168
Hayashi, Kazutaka	712
Hayashi, Toshihiko	1641
Hazellah, Muhammad Hadhari	1457
He, Eric	700
He, Jiangling	135
He, Peng	2022
He, Quanfeng	1716
He, Xuanke	119
Heinig, Andy	314
Heisig, Stephen J	270
Hejase, Jose	1200
Helbig, Stephan	855
Helou, Assaad	405
Henrion, Yann	1926
Henry, David	535
Henry, M. David	648
Hensley, Dale	2073
Herbert, Robert	1233
Hernandez, Natalie	2103
Hernandez, Selene	753
Herrmann, Matthias	855
Hester, Jimmy	896
Hikita, Masayuki	1451

Hillman, Craig	2103
Hirabayashi, Keiichi	1179
Hirano, Mitsuharu	1067
Hirose, Masakazu	2162
Hirt, Etienne	868
Ho, Bin-En	1693
Ho, Cheng-Yu	977
Ho, David	1432
Ho, David Soon Wee	917
Ho, Soon Wee	21
Hoang, Tim Tri	667
Hoelck, Ole	861
Hokari, Ryohei	1764
Honda, Kazutaka	446, 740
Hong, Xuan	753
Hook, Michael David	2219
Hooshmand, Nasrin	1588
Hopsch, Fabian	314
Hoque, Mohd Aminul	792
Horibe, Akihiro	1921
Hoshino, Hitoshi	437
Hosseini, Seyedmahmoud	1258
Hou, Xinnan	69
Hou, Zhuangzhuang	1716
Houston, Paul	2349
Hsiao, Andy	1099
Hsiao, Hsiang-Yao	21
Hsiao, H. Y.	1515
Hsiao, Yu-Hsiang	461
Hsieh, Chia-Ping	2009
Hsieh, Jeng-Shien	688
Hsieh, Ming-Che	289
Hsieh, Ricky	977
Hsieh, Tsun-Lung	41
Hsieh, Yi-Chen	1751
Hsu, C.C.	1595
Hsu, Che-Ming	461, 600
Hsu, Chieh-Hao	397

Hsu, Chih Chung	318
Hsu, Chih-Hsun	7
Hsu, Chung-Yi	2200
Hsu, C.K.	931
Hsu, C. K.	1550
Hsu, F.C.	931
Hsu, Fussen	1413
Hsu, Hsiang-Han	1074
Hsu, Ian	289, 493
Hsu, Po-Ning	642
Hsu, Steven	397, 763, 1175
Hsu, Y.N.	1693
Hu, Ian	1710
Hu, Yang	2234
Hu, Yougen	243, 1916
Hu, Yuan	277
Huang, Baron	1722
Huang, Chen-Yu	1287
Huang, Chih-Yi	41
Huang, Dick	700
Huang, Dinos	1710
Huang, G.C.	1595
Huang, H.L.	1595
Huang, Mian	1306
Huang, Mingliang	2022
Huang, Mingliang L.	1774, 2036
Huang, Pei-Chen	1413
Huang, P.S.	493
Huang, Rocky	1200
Huang, Shih-Ya	688
Huang, Shin-Yi	1463
Huang, Yifan	1200
Huesgen, Till	1443
Hung, Han-Tang	1729
Hung, Mi-Chun	41
Huo, Jia Ren	1485
Huo, Yongjun	150
Huynh, Michael	628

Hwang, Jisoo	300, 682
Hwang, Jung Woo	733
Hwang, Kihyun	636
Hwang, Kyo-sung	330
Hwang, Lih-Tyng	1751, 2200
Hwang, Taejoo	614
Hwangbo, Seahee	695, 983, 2085, 2337
Hyun, Sangjin	636
Iacovo, Serena	607, 2206
Ihori, Atsuhito	1865
Iida, Kenji	1952
Iizuka, Tomonori	1451
Im, Yunhyeok	300
Inaba, Takayuki	1952
Inamdar, Adwait	811
Ingelhag, Per	1405
Inoue, Fumihiro	437, 607, 2206
Inoue, Junishi	556
Irwin, Randall	277
Ishigure, Takaaki	550
Islam, Nokibul	325
Itawi, Ahmad	1575
Iwai, Toshiki	1952
Iwamoto, Hayato	1641
Iyer, S. S.	277
Iyer, Subramanian	620, 1470
Iyer, Subramanian S.	543, 579, 800, 1605, 2225
Jacquemond, Achille	264
Jacques, Patrick	1074
Jain, Ritesh	2180
Jalilvand, Golareh	1106, 1909
Jalink, J.	1339
Jamieson, Geraldine	1035
Janek, Florian	868
Jang, Joohee	636
Jangam, SivaChandra	543, 620
Jangam, Siva Chandra	800
Jani, Imed	1926

Janta-Polczynski, Alexander	1074
Jarecki, Robert	648
Jayabalan, Jayasanker	587
Jean-Philippe, Michel	1279
Jemaa, Salwa Ben	1744
Jeng, Shin-Puu	931, 1550
Jeon, HyeongIl	1457
Jeong, James	300
Jeong, leeseul	197
Jeong, Minsu	1146
Jeong, Se Young	733
Jhong, Ming-Fong	41
Ji, Hongjun	183
Jiang, Don-Son	1704
Jiang, Han	1569
Jiang, Jing	1485
Jiang, Tengfei	1106, 1909
Jiang, Yih-Jenn	7
Jin, Yufeng	1503, 1983, 2016, 2061, 2168
Jo, Chanmin	1188
Jo, Jung-Lae	76
Jo, Paul K.	1803
John Akkara, Francy	2309
Johnson, Leonard	1611
Joly, Pierre	535
Jong, Ming Chinq	1543
Joo, Jiho	197
Joo, Kisu	733
Jordan, Matthew B.	648
Joshi, Rahul	806
Joshi, Shailesh N.	1437
Joshi-Imre, Alexandra	1258, 1848
Jourdon, Joris	1926
Juang, Jing Ye	642
June Rebibis, Kenneth	437
Jung, Jin-San	204
Jung, Kwang-Ho	2290
Jung, Seung-Boo	2290

Jung, Seung-Yoon	283
Kabir, Mohammed	1870
Kahle, Ruben	861
Kalappurakal Thankappan, Kannan	2225
Kalb, Jamie	1312
Kalnitsky, Alex	1595
Kalyanam, Huthasana	2180
Kam, Nicholas	2219
Kamimura, Rikiya	1451
Kamlapurkar, Swetha	528, 1060
Kanagawa, Naoki	1022
Kandanur, Sashi	1588
Kaneko, Junichi	1933
Kang, Kuo-Chang	600
Kang, Minsoo	1977
Kang, Pilkyu	636
Kang, Qiushi	1266
Kannan, Jenefa	334
Kao, C. Robert	235, 1729, 2258
Kao, Feng	700
Kao, Hsuan-Ling	258
Karim, Karim S.	2219
Karlheinz, Bock	498
Karsten, Meier	498
Karuppuswami, Saranraj	113, 1240
Katagiri, Shunsuke	1009
Kathaperumal, Mohan	718
Kathaperumal, Mohanalingam	1796, 2112
Kathaperumal, Mohananlingam	334
Katkar, Rajesh	1041
Kavle, Pravin	931
Kawanabe, Naoki	1451
Kawano, Masaya	1996
Kaynak, Mehmet	218, 942
Ke, Chang-Bo	410
Ke, C.N.	1595
Keith Newman, Keith	806
Kelly, Mike	163

Kencana, Sagung Dewi 2067
Kennes, Koen 607
Kenney, Christopher 2072
Kerepesi, Peter 218, 942
Kersjes, Sebastiaan 1789
Keser, Beth 1159
Khazaka, Rabih 2173
Khim, Jin Young 1457
Khinda, G.S. 1946
Khurana, Gaurav 924
Kida, Tsuyoshi 1009
Kidera, Nobutaka 712
Kilger, Thomas 855
Kim, Changsu 1382
Kim, Choong-Un 1316
Kim, Dogeun 937
Kim, Dongsu 1647
Kim, Dong wook 204
Kim, Gahui 937, 2246
Kim, Hae-In 983, 2085
Kim, Haein 2337
Kim, Jaechoon 614
Kim, Ji-Hye 2272
Kim, Ji-Min 204
Kim, Jong Heon 35, 563
Kim, Jong-Hoon 1860
Kim, Ju hyeon 197
Kim, Jung Hak 197
Kim, Junghwa 300
Kim, JunMo 1146
Kim, Kilsoo 614
Kim, Kyoung-Tae 1809
Kim, KyuHyoun 1200
Kim, Nam Chul 35, 563
Kim, Nam-Seog 2246
Kim, Seokho 636
Kim, Soon-Wook 2206
Kim, Taehun 614

Kim, Taehwan 614
Kim, Taek-Soo 1146
Kim, Taeyeong 636
Kim, Yi-Ram 1316
Kim, Yoon-Hyun 733
Kim, Young-Cheon 2246
Kim, Young Ho 35
Kim, Youngja 2349
Kim, Young-Ja 2359
Kino, Hisashi 264
Kintaka, Kenji 556
Kirchner, Lisa 1722
Klengel, Robert 175
Klengel, Sandy 175
Klingler, Hannes 1789
Knickerbocker, John 270, 1246
Ko, Cheng-Ta 14, 903, 1413, 1463
Ko, T. 688
Kobayashi, Naoki 1599
Kodama, Shoichi 1002
Kohl, Paul A. 249
Koide, Masateru 1294
Kojima, Ryoji 1933
Kokash, M.Z. 1946
Kolbasow, Andrej 47, 210
Kong, Yuechan 1842
Kothari, Nakul 1347
Koyama, Koichi 1067
Koyama, Toshinori 1599
Koyama, Yutaro 346
Koyanagi, Mitsumasa 1047
Kozlovsky, William J. 726
Kraetschmer, Daniel 1113
Kraft, Jochen 1052
Krishna, Bhogaraju Sri 2324
Krivec, Thomas 1509
Kröhnert, Kevin 1475
Krumbein, Ulrich 855

Ku, Harry 1175, 1595, 1693

Ku, Terry 1

Kuah, Eric 14, 903

Kuang, Jiameng M. 2036

Kubo, Atsushi 334, 718, 924

Kubsch, Timo 210

Kudo, Hiroshi 94

Kudo, Tomoya 1015

Kuechenmeister, Frank 910

Kulick, Jason 2072

Kulterman, Ron W. 318

Kumar, Deepak 1687

Kumazawa, Yune 1009

Kuo, C.C. 1595

Kuo, C.H. 1595

Kuo, Hung-Chun 41

Kuo, Kuei Hsiao (Frank) 1635

Kuo, Ping-Jui 600

Kuo, Yu-Lin 2067

Kurihara, Kazuma 1764

Kurosaka, Seigo 474

Kurz, Helmut 218, 942

Kwon, Odal 55

Kwon, Yong Tae 35, 563

Kyung, Youjin 1146

Labarbera, Christine 2186

Lai, Chia-Chu 1704

Lai, Chieh-Lung 1170

Lai, Hsin-Cheng 1877

Lai, P. C. 1550

Lai, T.M. 931

Lai, Yen-Kun 397

Lall, Pradeep 370, 505, 792, 1087, 1347, 1366, 1815, 1958

Lambert, Renée 1611

Lambrecht, Joris 1757

Lan, Jia-Shen 2144

Lang, Klaus-Dieter 861, 1475, 1853

Langlois, Richard 1074

Larsson, Andreas 141

Lasfargues, Gilles 535

Lattard, Didier 569, 1926

Lau, Boon Long 1419, 1543

Lau, Chun Sean 1387

Lau, John 903

Lau, John H. 14

Laugier, Maxence 225

Lauser, Simone 515

Lavrik, Nickolay V. 2073

Le, Thanh Long 2173

Le, Wen-Kai 410

Leblanc, Alexandre 2117

Lee, Bob 2079

Lee, Bongsub 628, 1041

Lee, Chang-Chi 600

Lee, Chang-Chun 1413, 2009

Lee, Chang Woo 35, 563

Lee, Chia-Hsin 1463

Lee, Chin C. 150, 2302

Lee, Choong-Jae 2290

Lee, Chul-Hee 197

Lee, Chul Hyo 35

Lee, HanMin 1146, 2278

Lee, Heeseok 300

Lee, Heesok 682

Lee, Hohyung 2343

Lee, Hoi-jin 682

Lee, Hung-Ho 1287

Lee, Hungping 1503

Lee, Hyeong Gi 204

Lee, Hyun-Seop 1382

Lee, Hyun Seop 1529

Lee, Jae Cheon 35

Lee, Jeffrey ChangBing 1826

Lee, Joungphil 614

Lee, Jun Kyu 35, 563

Lee, Kang Hai 1165

Lee, K.C.	931
Lee, K. C.	1550
Lee, Kwang-Hee	197
Lee, Kwangjoo	197, 1146
Lee, Kyuha	636
Lee, Kyu Jae	733
Lee, NC	903
Lee, N. C.	14
Lee, Ning-Cheng	2186
Lee, Rick	1635
Lee, Ricky	14, 903
Lee, Sangil	2349
Lee, Seok-hyun	937
Lee, Seung Jae	733
Lee, SeYong	2278
Lee, Sung Hyuk	35
Lee, Tae-Ik	1146
Lee, Tae-Kyu	1099, 1106
Lee, Yisang	1047
Lee, Yuh-Zheng	1877
Leever, Ben	370, 1347
Legalland, Corinne	569
Li, Gang	81, 746
Li, Guanglin	243
Li, Hong Yu	1735
Li, Ji	135
Li, Jiahui	2003
Li, Jiaxiong	2296
Li, Junjie	661
Li, Kunkun	1983
Li, Ming	14, 903
Li, Mingyu	183
Li, Na	1983
Li, Ping	28
Li, Wen-Yang	7
Li, Yu-Jin	758, 1328
Li, Yu Jin	642
Li, Zhang	14, 903

Liang, Qi	661
Liang, Shui-Bao	410
Liang, Xianwen	746
Liao, Guanglan	661
Liao, Kuo-Hsien	1902
Liao, Marvin	1175, 1595, 1693
Liao, Siyuan	81
Lii, Mirng-Ji	397
Lii, M.J.	763
Lim, Francis Chee Peng	968
Lim, Jun Su	204
Lim, Ruiqi	1227
Lim, Sharon Pei Siang	1543
Lim, Simon Siak Boon	21, 1543
Lim, Sze Pei	14, 903
Lim, Teck Guan	917, 1419
Lim, Yeow Kheng	1165
Lim, Yew Kheng	968
Lim, Yu Dian	1735
Lim Sharon, Pei Siang	191
Lin, Ang-Ying	1463
Lin, Benson	493, 758, 1165, 1328
Lin, C.H.	931
Lin, Chang-Fu	7, 1287
Lin, Cheng Ping	924
Lin, Curry	14
Lin, Gu-Yan	1170
Lin, Marc	14
Lin, M.J.	493
Lin, M.Z.	493
Lin, Puru Bruce	1413, 1463
Lin, P. Y.	1550
Lin, P.Y.	931
Lin, Stanley	289
Lin, Tiesong	2022
Lin, Tong-Hong	896, 960
Lin, Vito	1287
Lin, W. Y.	1550

Lin, Yi-Hang 931
Lin, Yi-Sheng 461
Lin, Yu-Min 1463
Lin, Zhibin 453
Liou, Yan-Yu 1413
Litzenberger, Lorenz 1443
Liu, Canyu 63, 850
Liu, Changqing 63, 850, 1569
Liu, Chan-Yuan 1902
Liu, Chun-Chen 1693
Liu, C. S. 1175
Liu, Fuhan 334, 718, 924, 1796, 2112
Liu, Handa 135
Liu, Hao-Chun 397
Liu, Huan 1983, 2168
Liu, Hui 1897
Liu, Johan 1564
Liu, Kai 1246
Liu, K.C. 1595, 1693
Liu, K. C. 1175
Liu, Li 63, 850, 1377
Liu, Liyuan 2054
Liu, Meng-Hsiang 579
Liu, M.S. 931
Liu, N.W. 493
Liu, NW 1165
Liu, Penglin 1897
Liu, Ping 628
Liu, Sheng 850, 1377
Liu, Shengfa 63
Liu, Songlin 1543
Liu, Weidong 28
Liu, Weifeng 1272, 1826
Liu, Xiao 1463, 1722
Liu, Ya 1564
Liu, Yanghe 1437
Liu, Yingia 1716
Liu, Yong 1521

Lo, ChangHo 1826
Lo, I-Fang 1751
Lo, Jeffery 14, 903
Lo, Penny 14, 903
Lodermeyer, Johannes 855
Loerke, Friederike 1113
Loh, Wei Keat 318
Lombard, Marc 535
Lombardi, Jack 1581
Lombardi, J.P. 1946
Lord, David 453
Lowe, Ryan 1366
Lu, Calvin 1175, 1693
Lu, Chun-Lin 1697
Lu, JengPing 1312
Lu, Tan 1916
Lu, Tao 2054
Lu, Tian 2072
Lu, Ying-Wei 1704
Luo, Bin 890
Luo, Daojun 2054
Luo, Jiangbo 707
Luo, Yandong 579
Luo, Yu 1246
Luu, Thi-Thuy 141
Luu Trung Duong, Pham 834
Ma, B.H. 1432
Ma, H.T. 1629
Ma, Kun 1377
Ma, Li 28
Ma, Lulu 806
Ma, Shenglin 1503
Ma, Shuying 28
Ma, Xiao 410
Macaisa, Dexter 2117
Machida, Hideki 1067
Mackowiak, Piotr 1475
Madanipour, Hossein 1316

Madenci, Erdogan	825
Maeda, Toru	1933
Maehara, Masataka	1641
Maetani, Shinji	1002
Mahajan, Ravi	667
Maier, Dominic	855
Makita, Toshiyuki	924
Mandal, Rathin	834
Manepalli, Rahul	1588
Mansoor, Bilal	1081
Mantysalo, Matti	1252
Marchack, Nathan	528
Maria, Winkler	498
Marnat, Loic	1535
Martin, Letz	726
Martin, Yves	528, 1060
Martina, Manuel	726
Martineau, Donatien	2173
Maslyk, Dan	753
Massey, John P.	363
Masuda, Koji	1074
Masuda, Yuki	346
Matsukawa, Daisaku	352
Matsumoto, Keiji	417
Matthias, Jost	726
Maune, Holger	726
Mavinkurve, A.	479
Mavinkurve, Amar	777
Mayer, Michael	2219
Mayr, Andreas	1492
McCann, Scott	2150, 2331
McFarlane, Nicole	2073
Mei, Ping	1312
Meiler, Josef	942
Melin, Peter	1405
Melkote, Shreyes	2349
Mellen, Jon	453
Mercier, Denis	1680

Meth, Jeffrey	785
Meunier, Philippe	1789
Miao, Min	1983, 2168
Michailos, Jean	569
Michel, Jean-Philippe	1680
Michihiro, Toshiaki	2042
Miki, Shota	1599
Miller, Andy	340, 437, 1035, 2206
Miller, Scott	370, 1347
Milton, Basil	55
Min, Fan-Yu	600
Min, Kyung Deuk	2290
Ming Chinq, Jong	587
Minoret, Stéphane	569, 1622
Mirkarimi, Laura	628, 1041
Mishra, Dibyajat	1316
Missinne, Jeroen	1757
Mitchell, Nicholas C	2134
Mitev, Ivaylo	1509
Miura, Seiya	101
Miyazawa, Risa	1921
Miyazawa, Yoshinori	1952
Mizutani, Daisuke	1294, 1952
Moehrle, Martin	1060
Moeller, Berthold	437
Mogera, Umesh	620
Mogera, Umesha	1605
Mohan, Kashyap	655
Mohd. Ghazali, Mohd. Ifwat	1687
Mohd Ghazali, Mohd Ifwat	113
Momozawa, Aya	334
Mondal, Saikat	113, 1240, 1687
Montmayeul, Brigitte	225
Moon, Kwangjin	636
Moon, Kyoung-Sik	157, 1977, 2134, 2140, 2296, 2349
Moon, Kyoung-Sik (Jack)	2359
Moon, Seok Hwan	197
Moon, Sungwook	1188

Mori, Daichi	1933
Mori, Hiroyuki	417, 1921
Mori, Ken-Ichiro	101
Mori, Kentaro	1140
Mori, Kiyoharu	1047
Mori, Takahiro	1921
Morikawa, Yasuhiro	1865
Morisako, Isamu	1451
Motobe, Takeharu	352
Motoyoshi, Makoto	1047
Mourier, Therry	569
Mourier, Thierry	1622
Moussodji Moussodji, Jeff	1396
Mrozek, Pawel	628, 1041
Mu, Fengwen	989
Mudrick, John P.	648
Muehlbauer, Franz-Xaver	855
Muga, Karthik	1035
Müller, Ernst	868
Murakami, Yasunori	1067
Murayama, Takahide	1865
Murray, Bruce T.	1970
Murtagian, Gregorio	1028
Murugan, Rajen	1221, 1653
Murugesan, Murugesan	1047
Mydlak, Mathias	726
Na, Hoonjoo	636
Na, Nanju	1208
Nachiappan, Vivek Chidambaram	587
Nagai, Koji	1599
Nagamatsu, Tatsuo	1933
Nah, Jae-woong	528, 1246
Nahalingam, Kirthika	1666
Nair, Chandrasekharan	334, 924
Nakamura, Ai	1047
Nakamura, Eiji	417
Nakamura, Takuya	1641
Nakamura, Tomonori	1933

Nakayama, Tomoki	550
Nakazaki, Fukino	550
Nam, Ju Hyun	563
Nam, Seungki	1188
Narayanan, Rajeev	270
Naseem, Sadia	2079
Nauroze, Syed Abdullah	119
Ndip, Ivan	1475
Nedumthakady, Nithin	1588
Nemeth, Csaba	811
Neumeyr, Christian	1052
Ng, Daniel	1287
Ng, Eric	14
Ngo, Ha-Duong	1475
Nguyen, Hoang-Vu	141
Nguyen, Luu	655, 1316, 1333
Nguyen, Thong	1889
Nieh, Simon	1246
Nilsson, Torbjörn M. J.	1405
Nishikawa, Hiroshi	1081, 2003
Niu, Mengnian	1246
Niu, Yuling	819
Noguet, Dominique	1535
Nolmans, Philip	674
Nonaka, Toshihisa	446, 740
Noriki, Akihiro	2042
Oates, Daniel	1611
Oberndorff, P.	479
Ogawa, Tsuyoshi	446, 740
Ogura, Nobuo	972
Oh, Dan(Kyung Suk)	614
Oh, KwangSeok	163
O'Halloran, G.M.	479
O'Halloran, Orla	777
Ohba, Takayuki	1002
Ohde, Christian	106
Ohkubo, Tomohiro	1641
Oi, Kiyoshi	1599

Öjefors, Erik 968
Okada, Kazuya 1015
Okamoto, Daichi 718, 2112
On, JY 1710
Onishi, Tetsuya 726
Onitake, Shigeo 726
Oo, Aung Kyaw 968
Oppermann, Hermann 1052
Oprins, Herman 126
Orcutt, Jason S. 1060
Ortega, Carlos 2072
Osborn, Tyler 453
Ötzlinger, Herbert 1498
Owens, N. 479
Pacot, Guilhem 1870
Paeck, Marcus 1853
Paik, Kyung-Wook 283, 1146, 2022, 2213, 2266, 2272, 2278
Palmer, Jordan 2186
Palys, Anna 47
Pan, Jhih-Yuan 14
Pan, Ke 2343
Pan, Ponder 1693
Pancoast, Leanna 1246
Pang, Ponder 1175
Panigrahy, Asisa Kumar 2156
Panikkanvalappil, Sajanlal 1588
Pantano, Nicolas 674, 1215
Pantouvaki, Marianna 1757
Papapolymerou, John 948
Paradis, Etienne 2252
Paranjpe, Ajit 1470
Pares, G. 1279
Pares, Gabriel 1535, 1622
Parikh, Bakul 1121
Park, Gyu-Tae 2246
Park, Hyun Ho 733
Park, Jae-Hyeong 2213
Park, Jongcheol 2213

Park, JoonYoung 163
Park, Sang Yong 35, 563
Park, S.B. 2150, 2331
Park, Seungbae 1130, 2343
Park, SooIn 2278
Park, Yong-Jin 204
Park, Yong Sung 204
Park, Young-Bae 937, 2246
Parker, David 428
Parthasarathy, Srivatsan 1660
Paul, Jens 910
Pei, Yu 2234
Pei Siang, Sharon Lim 587
Peng, lan 607, 2206
Peray, Patrick 535
Petzold, Matthias 175
Pfost, Martin 1509
Pham, Van-Lai 1130
Pham, Vanlai 2150, 2331, 2343
Phansalkar, Sukrut 1382
Philip, Pierre-Emile 569
Phommahaxay, Alain 340, 437, 607, 2206
Pierre, Ferris 1279
Pietryga, Christoph 1159
Plant, Jason 1611
Plochowietz, Anne 1312
Podpod, Arnita 340, 437
Polezhaev, Vladimir 1443
Poliks, Mark D. 768, 1581, 1946
Ponthenier, Fabienne 569
Posthill, John 1041
Pozzobon, Fiorella 2194
Premerlani, Romeo 726
Prenger, Luke 1463
Prisacaru, Alexandru 811
Pristauz, Hugo 1492
Proschwitz, Jan 1159
Pu, Han-Ping 688

Pu, Li 1716

Puligadda, Rama 1722

Pulugurtha, Markondeya Raj 960

Pulugurtha, P. Raj 1300

Qi, Tao 1509

Qian, Zhiguo 667

Qiang, Song Guan 1485

Qiao, Y.Y. 1629

Qin, Ivy 55

Raad, Peter 405

Raghavan, Nagarajan 834

Rahim, Kaysar 800

Raj, Anto 2309

Raj, P. Markondeya 718, 972

Rajagoapal, Varun 334

Ramon, Hannes 1757

Rao, Vempati Srinivasa 842, 1152, 2126

Rasilainen, Kimmo 1405

Rastogi, Ravi 1611

Ravichandran, Siddharth 726, 1796

Ravinder, Pal Singh 1419

Raychaudhuri, Sourobh 1312

Raynaud, Christine 1680

Rebhan, Bernhard 218, 942

Refai-Ahmed, Gamal 2150, 2331, 2343

Ren, Chao 1977, 2296

Ren, Linlin 1556

Ren, Qin 1996

Ribière, Céline 569, 1622

Richter, Theresia 515

Rivera, Katie 600

Robertson, Stuart 63, 1569

Rogers, Jeff 270

Rolland, Emmanuel 225

Romano, Giovanni 569

Romero, Gilles 569

Rongen, Rene T.H. 479

Rongen, R.T.H. 777, 1339

Ross, Joseph 1121

Roucou, R. 1339

Rovitto, Marco 2091

R. Tummala, Rao 718

Rudolph, Catharina 628

Rupp, Bradley B. 1312

Ruzicka, Klaus 868

Ryu, Dong Su 1457

Ryu, Hyodong 2246

Sakai, Taiji 1952

Sakakibara, Shiori 352

Sakaue, Takahiko 76

Sakuma, Katsuyuki 270

Sakuyama, Seiki 1952

Salahoueldhadj, Abdellah 340

Salcedo, Javier 1660

Saleem, Amin 1870

Saleh, Rafat 868

Sammakia, Bahgat G. 1970

Sanchez, Juliet 753

Sanchez, Loïc 225

Santerre, Francis 1396

Sarangapani, Murali 2048

Sato, Muneyuki 1865

Sato, Nobuaki 1451

Sato, Yoichiro 712

Savage, Eric 2117

Saxena, Antra 1770

Scevola, Daniel 569

Schares, Laurent 1060

Scheller, Britta 106

Schellkes, Eckart 2067

Schempp, Fabian 1113

Schiffmacher, Alexander 1443

Schiffres, Scott N. 1970

Schingale, Angelika 1509

Schischka, Jan 175

Schmid, Maximilian 2324

Schmitt, Wolfgang	87
Schneider-Ramelow, Martin	861
Schroeder, Raoul	1498
Schulze, Gary	55
Schulze, Sebastian	218, 942
Schumann, Todd	695, 1647, 1809
Schuster, Christian	2240
Schutt-Aine, Jose	1889
Schwarz, Mark	392, 819
Schwenk, Erika	87
Seefisch, Henning	2284
Segal, Julie	2072
Segaud, Roselyne	569
Sekhar, Vasarla Nagendra	842
Sekiguchi, Masahiro	1140
Selhofer, Hubert	1492
Selvanayagam, Cheryl	834
Serebreni, Maxim	2103
Shaddock, David M.	768
Shah, Aashish	55
Shah, Ujash	1605
Shahane, Ninad	1316
Shakoorzadeh, Niloofar	800
Shambach, William	2186
Shang, Jintang	522, 890
Shangguan, Dongkai	1272
Shapiro, Dmitri	1611
Sharma, Himani	1300, 1588
Sharon, Gil	2103
Sharon Lim, Pei Siang	21
Sheikhi, Roozbeh	150
Sheikhnejad, Ommeaymen	243
Shelton, Douglas	101
Shen, Yu-An	1081, 2003
Shi, Aihua	884
Shi, Tielin	661
Shibahara, Hiromi	1933
Shibasaki, Yoko	2112

Shibata, Daisuke	2112
Shie, Kai Cheng	642
Shigetou, Akitsu	235
Shih, Andy	1246
Shih, Meng-Kai	1710, 1902
Shika, Seiji	1009
Shim, Ji Ni	563
Shim, Moo-Sup	197
Shimada, Sawako	1015
Shimatsu, Takehito	989
Shimazu, Takayuki	1764
Shimizu, Kan	1641
Shimizu, Koji	1451
Shin, SangMyung	1146, 2278
Shin, Youngmin	300, 682
Shiraiwa, Tomio	163
Shirley, Tim	543
Shoji, Hideaki	163
Shoji, Yu	346
Shreve, Matthew	1312
Shumarayev, Sergey Yuryevich	667
Shunmugasamy, Vasanth	1081
Sidorov, Victor	1052
Sierra-Suarez, Jonatan A.	648
Sigl, Alfred	855
Sigmund, Ariane	1060
Sikka, Kamal	1121
Silberer, Gerald	942
Simmons, Jacob C.	1970
Simon, Gilles	569
Singh, Chrandeep	1130
Singh, Shiv Govind	2156
Sirbu, Bogdan	1052
Sitaraman, Suresh	1521
Sitaraman, Suresh K.	249, 785, 1939
Sitaraman, Suresh K	382
Sivapurapu, Sridhar	249, 1939
Sivasubramony, Rajesh S.	1581

Sivasubramony, Rajesh Sharma ... 768
Slabbekoorn, John ... 340, 607
Sleeckx, Erik ... 340, 437, 607, 2206
Smet, Vanessa ... 655, 972, 1028, 1796
Smith, Stephen ... 1200
So, R. ... 14
So, Raymond ... 903
Soares, Francisco ... 1052
Son, Ho-Young ... 2246
Son, Kirak ... 937, 2246
Song, Bo ... 157, 2134, 2140
Song, Changming ... 2234
Song, Euseok ... 614
Soon-Wook, Kim ... 1215
Sorensen, Eric ... 543
Sosa, Ramón A. ... 655
Souriau, Jean-Charles ... 1575
Southard, Arthur ... 1722
Sover, Raanan ... 294
Spurney, Robert G. ... 1300
Sridhar, Sharath ... 2309
Srinivasan, Sriram ... 2180
Stegmann, Tamira ... 87
Steinhorst, Rachel ... 1240
Stephan, Tino ... 175
Stewart, Benjamin G ... 382
Stoffel, Nancy C. ... 768, 1946
Stone, Bill ... 392
Stone, David ... 1179
Stucchi, Michele ... 1035
Su, An-Jhih ... 1
Su, C.H. ... 1175
Su, Jay ... 1463
Su, Ming-Sin ... 1175
Su, Peng ... 1099, 1106
Su, Sinan ... 792, 2309
Su, Zhaoxi ... 890
Subbiah, Nilavazhagan ... 878

Sueoka, Kuniaki ... 1921
Suga, Tadatomo ... 989
Suganuma, Katsuaki ... 474
Sugiura, Kazuhiko ... 1451
Suhard, Samuel ... 437, 607
Suhling, Jeff ... 505, 1347
Suhling, Jeffery ... 2309
Suhling, Jeffrey C. ... 792, 1087, 1815, 1958
Sulkis, Michael ... 1977
Sun, Hongyu Y. ... 2036
Sun, Rong ... 81, 243, 746, 1556, 1916
Sun, Teng ... 1300
Sun, Xiao ... 1215
Sun, Yangyang ... 392
Sun, Yunna ... 707
Sun, Yunting ... 707
Sung, Yun Hyun ... 35
Surillo, Emanuel ... 334
Suryoatmojo, Heri ... 1751
Susumago, Yuki ... 264
Suzuki, Akiyoshi ... 1865
Suzuki, Takuya ... 1009
Suzuki, Yui ... 2162
Suzuki, Yuya ... 1015
Swaminathan, Madhavan ... 249, 1672, 1939
Syed, Ahmer ... 392, 819
Sylvestre, Julien ... 1744
Symonds, Ken ... 2103
Tai, Jui-Feng ... 7
Tak, Coen ... 2029
Takahashi, Noriyuki ... 264
Takano, Takamasa ... 94
Takiar, Hem ... 1387
Tal, Sharon ... 294
Talebbeydokhti, Pouya ... 294, 954, 1666
Tamura, Akira ... 1952
Tan, Chuan Seng ... 1735
Tan, KH ... 325

Tan, Kim Hwee	14, 903
Tanaka, Kazunori	1067
Tanaka, Masaya	94
Tanaka, Tetsu	264
Taneda, Hiroshi	1599
Tang, Gongyue	1419
Tang, Junyan	1200
Tang, Tom	1432
Tang, Zirong	661
Tani, Daisuke	1933
Tan Swee Seng, Eric	2048
Tao, Jing	1735
Tao, Mian	14, 903
Tao, Qi	2029
Tao, Wang Jun	1485
Tarng, David	1426, 1710
Tasi, Mike	1704
Tatsumi, Kohei	1451
Tehrani, Bijan	896
Tekin, Tolga	1052
Tentzeris, Manos M.	119, 896, 960, 972
Teoh, Kristie	1028
Tetsuya, Onishi	1498
Teutsch, Thorsten	47, 210
Thai, Trang	294, 954
Theil, Jeremy	628, 1041
Theng, Chih-Han	758
Theuss, Horst	855
Thirugnanasambandam, Sivasubramanian	2309
Thomas, Tony	505
Thorsell, Mattias	1405
Thukral, Varun	1339
Tian, Yanhong	1266
Tissier, Pierre	1622
To, Hing "Thomas"	1208
Toda, Keiji	1451
Toepper, Michael	1853

Tokunari, Masao	1074
Tollefsen, Torleif A	141
Tolunay Wipf, Selin	942
Tomikawa, Masao	346
Tomita, Yasunari	1022
Tomohiro, Fukao	1272
Topsakal, Erdem	1782
Tremble, Eric	1179
Trombley, Django	1221
Tsai, Chung-Hao	1
Tsai, Clair	1175
Tsai, Jensen	700, 1432
Tsai, Mike	700
Tsai, Sheng-Han	397
Tsao, Pei-Haw	763
Tschoban, Christian	1475
Tseng, I-Hsin	758, 1328
Tseng, Yi-Hsiu	1359
Tseng, Yu-Chou	1902
Tsfati, Yossi	954
Tsunoda, Masatoshi	2042
Tsuruta, Kazuhiro	1451
Tu, K N	642
Tu, K. N.	1716, 2003
Tummala, Rao	334, 655, 1028, 1300, 1588, 1796
Tummala, Rao R.	924, 960, 972, 2112
Tunga, Krishna	1121
Tuominen, Samuli	1252
Turcotte, Eric	467
Tutunjyan, Nina	1035
Twiefel, Jens	2284
Ueda, Kazutoshi	1451
Ueno, Keiko	446, 740
Ueno, Kenichi	2162
Ueta, Chiho	1015
Ume, I. Charles	1333
Uomoto, Miyuki	989
Ura, Shogo	556

Uresti, Tiffani 1081

Utano, Tetsuya 1933

Uy, William 1272

Vadimas, Verdingovas 515

Vaisband, Boris 543, 579, 1605, 2225

van Borkulo, Jeroen 423

Van Campenhout, Joris 1757

Vandendaele, William 168

Vandeneynde, Aurélie 535

Van der Plas, Geert 674, 1215

van der Stam, Richard 423

van Haare, Niek 1789

Van Huylenbroeck, Stefaan 1035

Vanjari, Siva Rama Krishna 2156

van Olst, E. 479, 777

van Soestbergen, M. 479, 777

Van Steenberge, Geert 1757

Varga, Edit 2072

Vasarla, Nagendra Sekhar 2126

Vélard, Rémi 569

Velenis, Dimitrios 674

Venkataraman, Srikrishnan 2180

Venugopal, Archana 405

Verdonck, Patrick 2206

Veres, Agnes 811

Verhelst, Marian 674

Verrun, Sophie 1622

Viehweger, Kay 1833

Vijayakumar, Swathi 1666

Vinci, Andrea 569

Viswanath, Ram 2180

Vitello, Dario 2091

Vivet, Pascal 569, 1926

Vladimirova, Kremena 168

Vobl, Matthias 855

Voges, Steve 363, 861

Voit, Walter 1848

Voit, Walter E. 1258

von Waechter, Claus 855

Wada, Keiko 1451

Wagner, Juergen 855

Wagner, Thomas 1159

Wai, Leong Ching 21, 1419

Waidhas, Bernd 1159

Wan, Weikang 2318

Wan, Zhe 579

Wang, Chang-Ning 1175

Wang, Chen-Chao 41, 977

Wang, Chengqian 28

Wang, Chenxi 1266

Wang, Chuei-Tang 688

Wang, Chun-Min 1463

Wang, Daixing 2016

Wang, Huayan 1130, 2150, 2331, 2343

Wang, Hui 243

Wang, Huiying 707

Wang, J. H. 1550

Wang, Jin 69

Wang, Jing 1130, 2150, 2343

Wang, Jiunn Jie 1635

Wang, Kirin 1175, 1595, 1693

Wang, Lejun 392

Wang, Liyuan 1983

Wang, Nan 1564

Wang, Peiren 135

Wang, Qian 69, 1312, 2234

Wang, Qidong 2318

Wang, Rung-De 1693

Wang, Shiu-Chih 1426

Wang, Tai-Jui 1877

Wang, Wei 392, 819, 2016

Wang, Xiangy-Yu 1996

Wang, Xiaobai 1543

Wang, Xin 2234

Wang, Xinying 1889

Wang, Xueqiao 2140

Wang, Yan 707
Wang, Yang 135
Wang, Yen Neng 1635
Wang, Yiteng 972
Wang, Y.P. 1629
Wang, Yu 1312
Wang, Yu-Cheng 1693
Wang, Yunda 1312
Wang, Yu-Po 700, 1432
Wang, Zhijie 819
Watanabe, Atom 924, 960, 1300
Watanabe, Atom O. 972
Watanabe, Manabu 1294
Watanabe, Naoki 924
Watanabe, Osamu 1933
Watanabe, Takuro 1764
Watariguchi, Shigeru 1047
Webb, Bucknell 270
Weber, Y. 479
Weerawarne, Darshana L. 1581
Weerawarne, D.L. 1946
Wei, Cheng 410
Wei, Tiwei 126
Weichart, Johannes 1833
Weichart, Jüergen 1833
Weir, Terence 1611
Weiss, Thomas 306
Weng, Chen-Yuan 600
Weng, I-An 1729
Wen Kong, Ling 1485
Werner, Thomas 628
Widiez, Julie 168
Wietstruck, Matthias 218, 942
Wigger, Benedikt 868
Wijewardena, Kanishka 1687
Wilde, Juergen 878, 1113, 1443
Williamson, Jaimal 1333
Wipf, Christian 942

Wohrmann, Markus 363
Wöhrmann, Markus 861, 1853
Wolfberger, Archim 2029
Wong, Chee Wei 277
Wong, Ching-Ping 81, 243, 746, 1556, 1916, 2296, 2349
Wong, CP 157, 2134
Wong, C. P. 1977, 2140, 2359
Wong, Nelson 55
Wong Chin Yeung, Jason 2048
Workman, Thomas 628
Wu, An-Tai 1426
Wu, C.M.L. 1629
Wu, Dapeng 968
Wu, Fan 157, 2134
Wu, Hsing-Hui 14
Wu, Jiaqi 2302
Wu, Jing 1087
Wu, Jui-Yang 2258
Wu, Mei-Ling 2144
Wu, Sheng-Tsai 1463
Wu, W. C. 1175
Wu, Yang 2022
Wu, Y.H. 1693
Wu, Zhichao 1306
Wu, Zhongming 486
Wurz, Marc Christopher 1883
Xi, Cao 14, 903
Xiang, Gengzhao 135
Xiang, Hui 63
Xiangyu, Wang 587
Xiao, Hui 2054
Xiao, Zhiyi 884
Xie, Dongji 486
Xie, Hong 28
Xie, Yong 1221
Xiong, Chi 1060
Xiong, Yaoxu 243
Xu, Hui 55

Xu, Iris	14, 903
Xu, Jianbin	1556
Xu, Jiefeng	1130, 2150, 2331, 2343
Xu, Jikai	1266
Xu, Xirui	884
Xue, Mei	2318
Yadav, M.	1946
Yadav, Manu	768
Yagi, Hidekazu	1933
Yahyaei-Moayyed, Farzaneh	2180
Yakabe, Sho	1764
Yalagach, Mahesh	2029
Yamada, Shuhei	1796
Yamada, Takashi	175
Yamamoto, Kazunori	842, 1419, 2126
Yamashita, Soichi	1140
Yamauchi, Sinichi	76
Yamawaki, Seigo	1294
Yamazaki, Noriyuki	352
Yang, C. C.	1550
Yang, Chen-Tsai	1877
Yang, Cheol-Woong	2246
Yang, Chi-Hau	2200
Yang, Fan	850
Yang, Henry	14, 903
Yang, Jenn-Ming	2258
Yang, Ming	2022
Yang, Rolance	1175
Yang, S.B.	1595
Yang, Sean	1729
Yang, Tilo H.	235
Yang, T. L.	1175
Yang, Xiaobing	28, 884
Yang, Yang	1983, 2168
Yang, Yong	1716
Yang, Yong-suk	330
Yang, Yu-Hsiang	811
Yang, Yu-Ting	1359

Yang, Zhuoqing	707
Yao, Ruohe	1324
Ye, Chen	522
Ye, Lilei	1564
Ye, Ning	1387
Ye, Yong Liang	1419
Yee, Kuo-Chung	1
Yeh, Meng-Kao	1697
Yeh, Shu-Shen	1550
Yen, Yee-Wen	2067
Yeo, Woon-Hong	1233
Yess, Kim	340
Yew, M.C.	931
Yew, M. C.	1550
Yi, Luyun	1200
Yildiz, Ömer Faruk	2240
Yin, Liang	768
Yin, Xin	1052
Yook, Jongmin	1647
Yoon, Dal-Jin	2266, 2272
Yoon, Gil-Sang	197
Yoon, Seung Wook	325, 968, 1165
Yoon, Yong-Kyu	695, 983, 1647, 1809, 2085, 2337
Yoshida, Shu	1009
Yoshihiro, Furukawa	1300
Yotsuyanagi, Hiroko	352
Young Suk, Kim	1002
Youssef, Toni	2173
Yu, C.K.	493
Yu, C.T.	931
Yu, Daquan	28, 884
Yu, Doug C.H.	594
Yu, Douglas	1, 688
Yu, Hai-Yang	235
Yu, Jambo	28
Yu, Ji-In	204
Yu, Shiang-Hwua	1751
Yu, Ta-Jen	289

Yuan, K.S.	1595	Zhou, Han	1569
Yue, Xiang	522	Zhou, Jianwen	746
Zaal, J.J.M.	1339	Zhou, Jie-Ying	410
Zachariah, Ashwin Varkey	768	Zhou, Min-Bo	410
Zarr, Scott	1611	Zhou, Shicheng	1266
Zeb, Gul	467	Zhou, Shiqi	1081, 2003
Zeng, Qinghua	2061	Zhou, Yi	249, 1521, 1939
Zeng, Xiaoliang	1556	Zhou, Zhaoxia	63, 850, 1569
Zhang, Baotan	81	Zhu, Pengli	243, 746, 1916
Zhang, Bowei	1306	Zhua, Pengli	81
Zhang, Dongxiao	63	Zhuo, Qizhuo Zhuo	753
Zhang, Eric J.	1060	Zoberbier, Ralph	106
Zhang, Hong	1722	Zoschke, Kai	1475
Zhang, Hongqing	1246	Zou, Lin	1774
Zhang, Jianfeng	890	Zou, Yichao	884
Zhang, Jincan	1983	Zuber, Fabien	535
Zhang, Shuye	2022	Zullino, Lucia	2194
Zhang, Wenwu	183	Zussy, Marc	535
Zhang, Xiaowu	1152, 1419, 1543		
Zhang, Xin-Ping	410		
Zhang, Xuefeng	392		
Zhang, Yao	1377		
Zhang, Zheng	474		
Zhao, Cong	2309		
Zhao, Liguo	1569		
Zhao, Lily	392		
Zhao, Lixin	69		
Zhao, N.	1629		
Zhao, Peng	1735		
Zhao, Tao	81, 746, 1916		
Zhao, Weiwei	183		
Zhao, Xiuchen	1716		
Zhao, Yang	2234		
Zheng, Fengxia	28		
Zheng, Hao	1377		
Zheng, Ting	1803		
Zhengyang, Qian	264		
Zhou, Bin	1324		